Comparison of *Bacteria*, *Archaea*, and *Eucarya*

Property	Bacteria	Archaea	Eucarya
Membrane-Enclosed Nucleus with Nucleolus	Absent	Absent	Present
Complex Internal Membranous Organelles	Absent	Absent	Present
Cell Wall	Almost always have peptidoglycan containing muramic acid	Variety of types, no muramic acid	No muramic acid
Membrane Lipid	Have ester-linked, straight-chained fatty acids	Have ether-linked, branched aliphatic chains	Have ester-linked, straight-chained fatty acids
Gas Vesicles	Present	Present	Absent
Transfer RNA	Thymine present in most tRNAs	No thymine in T or TψC arm of tRNA	Thymine present
	N-formylmethionine carried by initiator tRNA	Methionine carried by initiator tRNA	Methionine carried by initiator tRNA
Polycistronic mRNA	Present	Present	Absent
mRNA Introns	Absent	Absent	Present
mRNA Splicing, Capping, and Poly A Tailing	Absent	Absent	Present
Ribosomes			
Size	70S	70S	80S (cytoplasmic ribosomes)
Elongation factor 2 with diphtheria toxin	Does not react	Reacts	Reacts
Sensitivity to chloramphenicol and kanamycin	Sensitive	Insensitive	Insensitive
Sensitivity to anisomycin	Insensitive	Sensitive	Sensitive
DNA-Dependent RNA Polymerase			
Number of enzymes	One	One	Three
Structure	Simple subunit pattern (4 subunits)	Complex subunit pattern similar to eucaryotic enzymes (8–12 subunits)	Complex subunit pattern (12–14 subunits)
Rifampicin sensitivity	Sensitive	Insensitive	Insensitive
RNA Polymerase II Type Promoters	Absent	Present	Present
Metabolism			
Similar ATPase	No	Yes	Yes
Methanogenesis	Absent	Present	Absent
Nitrogen fixation	Present	Present	Absent
Chlorophyll-based photosynthesis	Present	Absent	Present[a]
Chemolithotrophy	Present	Present	Absent

[a]Present in chloroplasts (of bacterial origin).

(repeated as Table 17.1)

Prescott's Principles

of MICROBIOLOGY

Joanne M. Willey
Hofstra University

Linda M. Sherwood
Montana State University

Christopher J. Woolverton
Kent State University

McGraw-Hill
Higher Education

Boston   Burr Ridge, IL   Dubuque, IA   New York   San Francisco   St. Louis
Bangkok   Bogotá   Caracas   Kuala Lumpur   Lisbon   London   Madrid   Mexico City
Milan   Montreal   New Delhi   Santiago   Seoul   Singapore   Sydney   Taipei   Toronto

The McGraw-Hill Companies

McGraw-Hill Higher Education

PRESCOTT'S PRINCIPLES OF MICROBIOLOGY

Published by McGraw-Hill, a business unit of The McGraw-Hill Companies, Inc., 1221 Avenue of the Americas, New York, NY 10020. Copyright © 2009 by The McGraw-Hill Companies, Inc. All rights reserved. No part of this publication may be reproduced or distributed in any form or by any means, or stored in a database or retrieval system, without the prior written consent of The McGraw-Hill Companies, Inc., including, but not limited to, in any network or other electronic storage or transmission, or broadcast for distance learning.

Some ancillaries, including electronic and print components, may not be available to customers outside the United States.

This book is printed on acid-free paper.

1 2 3 4 5 6 7 8 9 0 DOW/DOW 0 9 8

ISBN 978–0–07–337523–6
MHID 0–07–337523–3

Publisher: *Michelle Watnick*
Senior Sponsoring Editor: *James F. Connely*
Senior Developmental Editor: *Lisa A. Bruflodt*
Senior Marketing Manager: *Tami Petsche*
Project Coordinator: *Mary Jane Lampe*
Lead Production Supervisor: *Sandy Ludovissy*
Lead Media Project Manager: *Stacy A. Patch*
Designer: *John Joran*
Lead Photo Research Coordinator: *Carrie Burger*
Photo Research: *Mary Reeg*
Compositor: *Aptara*
Typeface: *10/12 Times Roman*
Printer: *R. R. Donnelley Willard, OH*
(USE) Cover Image (Front and Back): *©Dennis Kunkel Microscopy, Inc.*

The credits section for this book begins on page C-1 and is considered an extension of the copyright page.

Library of Congress Cataloging-in-Publication Data

Willey, Joanne M.
  Prescott's principles of microbiology / Joanne M. Willey, Linda M. Sherwood, Christopher J. Woolverton.—1st ed.
      p. cm.
  "In using the seventh edition of PHK's microbiology as the foundation for the development of principles, we identified two overarching goals."
  Includes index.
  ISBN 978–0–07–337523–6 — ISBN 0–07–337523–3 (hard copy : alk. paper) 1. Microbiology.
  I. Sherwood, Linda. II. Woolverton, Christopher J. III. Prescott, Lansing M. Microbiology. IV. Title.
V. Title: Microbiology.

QR41.2.W545   2009

616.9'041—dc22

                                                                2007040352

www.mhhe.com

Brief Table of Contents

Table of Contents

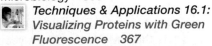

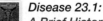

Part Seven   HOST DEFENSES

Part Eight   MICROBIAL DISEASES AND THEIR CONTROL

Part Nine   APPLIED MICROBIOLOGY

About the Authors

Joanne M. Willey is Professor of Biology at Hofstra University on Long Island, N.Y. Dr. Willey received her BA in biology from the University of Pennsylvania, where her interest in microbiology began with work on cyanbacterial growth in eutrophic streams. She earned her PhD in biological oceanography (specializing in marine microbiology) from the Massachusetts Institute of Technology–Woods Hole Oceanographic Institution Joint Program in 1987. She then went to Harvard University, where she spent four years as a postdoctoral fellow studying the filamentous soil bacterium *Streptomyces coelicolor.* Dr. Willey continues to actively investigate this fascinating microbe through funding provided by the National Institutes of Health and the National Science Foundation. She has coauthored a number of publications that focus on the complex developmental cycle of the streptomycetes. She is an active member of the American Society for Microbiology (ASM) and has served on the editorial board of the journal *Applied and Environmental Microbiology* since 2000. Dr. Willey regularly teaches microbiology to biology majors as well as allied health students. She also teaches courses in cell biology, marine microbiology, and laboratory techniques in molecular genetics. Dr. Willey lives on the north shore of Long Island with her husband and two sons. She is an avid runner and enjoys skiing, hiking, sailing, and reading. She can be reached at biojmw@hofstra.edu.

Linda M. Sherwood is a member of the Department of Microbiology at Montana State University. Her interest in microbiology was sparked by the last course she took to complete a BS degree in psychology at Western Illinois University. She went on to complete an MS degree in microbiology at the University of Alabama, where she studied *Pseudomonas acidovorans* physiology. She subsequently earned a PhD in genetics at Michigan State University, where she studied sporulation in *Saccharomyces cerevisiae.* Dr. Sherwood has always had a keen interest in teaching, and her psychology training has helped her to understand current models of cognition and learning and their implications for teaching. Over the years, she has taught courses in general microbiology, genetics, biology, microbial genetics, and microbial physiology. She has served as the editor for ASM's *Focus on Microbiology Education* and has participated in and contributed to numerous ASM Conferences for Undergraduate Educators. She also has worked with K-12 teachers to develop a kit-based unit to introduce microbiology into the elementary school curriculum and has coauthored with Barbara Hudson a general microbiology laboratory manual, *Explorations in Microbiology: A Discovery Approach,* published by Prentice-Hall. Her nonacademic interests focus primarily on her family. She also enjoys reading, hiking, gardening, and traveling. She can be reached at lsherwood@montana.edu.

Christopher J. Woolverton is Professor of Biological Sciences and a member of the graduate faculty in Biological Sciences and the School of Biomedical Sciences at Kent State University in Kent, Ohio. Dr. Woolverton also serves as the director of the KSU Center for Public Health Preparedness, overseeing its BSL-3 Training Facility. He earned his BS from Wilkes College, Wilkes-Barre, Pa., and a MS and a PhD in medical microbiology from West Virginia University, College of Medicine. He spent two years as a postdoctoral fellow at the University of North Carolina at Chapel Hill, studying cellular immunology. Dr. Woolverton's research interests are focused on the detection and control of bacterial pathogens. Dr. Woolverton and his colleagues have developed the first liquid crystal biosensor for the immediate detection and identification of microorganisms and a natural polymer system for controlled antibiotic delivery. He publishes and frequently lectures on these two technologies. Dr. Woolverton has taught microbiology to science majors and allied health students, as well as graduate courses in immunology and microbial physiology. He is an active member of ASM, serving as the editor of ASM's *Microbiology Education.* He has participated in and contributed to numerous ASM Conferences for Undergraduate Educators, serving as cochair of the 2001 conference. Dr. Woolverton resides in Kent with his wife and three daughters. When not in the lab or classroom, he enjoys hiking, biking, tinkering with technology, and just spending time with his family. His email address is cwoolvcr@kent.edu.

Prescott's Principles of Microbiology continues in the tradition of *Prescott, Harley, and Klein's Microbiology* by covering the broad discipline of microbiology at a depth not found in any other textbook. In using the 7th edition of *PHK's Microbiology* as the foundation for the development of *Principles*, we identified two overarching goals. First, we sought to present material likely to be covered in a single semester microbiology course, with the knowledge that not all introductory microbiology courses cover the same topics. Therefore, each chapter in Prescott's Principles of Microbiology was revised from the 7th edition of *PHK's Microbiology* to provide a streamlined, briefer discussion of key concepts that include only the most relevant, up-to-date examples. Secondly, we strove to further extend the student-friendly approach used in the 7th edition by enhancing readability and adding tools designed to promote learning.

OUR STRENGTHS

Connecting with Students

We have retained the relatively simple and direct writing style used in *PHK's Microbiology*, but have added style elements designed to further engage students. For example, we frequently use the first person voice to describe important concepts—especially those that our students find most difficult. Each chapter is divided into numbered section headings and organized in an outline format—the same outline format that is presented in the end-of-chapter summaries. Key terminology is boldfaced and clearly defined. We have introduced a glossary of essential terms at the beginning of each chapter to serve as an easy reference for students, while retaining the full glossary in the back of the book. Our belief that concepts are just as important as facts, if not more, is also reflected in the questions for review and reflection that appear throughout each chapter. These questions are of two types: those that quiz student retention of key facts and vocabulary and those designed to foster critical thinking.

Instructive Artwork

To truly engage students, a textbook must do more than offer words and images that just adequately describe the topic at hand. We view the artwork of a text as a critical tool in enticing students to read the text. *Principles* features the art program introduced in the 7th edition of *PHK's Microbiology*. The three-dimensional renderings help students appreciate the beauty and elegance of the cell, while at the same time make the material more comprehendible. Of course we also believe that figures should be content-rich, not just pretty to look at. Therefore, the art program also includes pedagogical features such as concept maps (e.g., see figures 9.1 and 13.1) and annotation of key pathways and processes (e.g., see figures 10.8 and 12.12).

x

Unique Organization Around Key Themes

With the advent of genomics, proteomics, metabolomics and the increased reach of cell biology, the divisions among microbiology subdisciplines have become blurred. This is reflected in the emergence of fields like disease ecology and metagenomics. In addition, today's microbiologist must be acquainted with all members of the microbial world: viruses, bacteria, archaea, protists, and fungi. It follows that students new to microbiology are asked to assimilate vocabulary, facts, and most importantly, concepts, from a seemingly vast array of subjects. The challenge to the professor of microbiology is to effectively communicate essential concepts while conveying the ingenuity of microbes and excitement of this dynamic field.

Microbial Evolution and Ecology

Because microbial evolution and ecology are no longer subdisciplines to be ignored by those interested in microbial genetics, physiology, or pathogenesis, *Principles* strives to integrate these themes throughout the text. We begin in chapter 1 with a discussion of the universal tree of life and whenever possible, discuss diverse microbial species so that students can begin to appreciate the tremendous variation in the microbial world. In addition, *Principles* uses the topics of intercellular communication (chapters 6 and 13), biofilms (throughout the text, but specifically in chapters 6, 13, and 29), microbial evolution (chapter 17), and polymicrobial diseases (chapter 33) to emphasize that evolution must be linked to genetics, physiology to diversity, and ecology to pathogenesis.

Microbial Pathogenicity and Diversity

Unique to *Principles* is the inclusion of microbial pathogens into the diversity chapters (chapters 19–24). Thus when students read about the metabolic and genetic diversity of each bacterial, protist, and viral taxon, they are also presented with the important pathogens. In this way, the physiological adaptations that make a given organism successful can be immediately related to its role as a pathogen and pathogens can be readily compared to phylogenetically related nonpathogenic microbes.

In addition, *Principles* introduces viruses and other acellular agents in chapter 5, following the chapters of Procaryotic and Eucaryotic Cell Structure and Function (chapters 3 and 4, respectively). By placing a similarly themed chapter on viruses here, professors can introduce all divisions of the microbial world to their students early in the term. For those professors who include more in-depth coverage of viruses, chapter 24 explores the molecular genetics of bacteriophages and other viruses as well as the pathogenicity of important animal and plant viruses. As in *PHK's Microbiology*, we use the classification schemes set forth in the second edition of *Bergey's Manual of Systematic Bacteriology*, the Baltimore System of virus classification (chapters 5 and 24), and the International Society of Protistologists' new classification scheme for eucaryotes (chapter 23).

Visual Tour

STUDENT RESOURCES

Laboratory Exercises in Microbiology

The seventh edition of *Laboratory Exercises in Microbiology* by John P. Harley has been prepared to accompany the text. Like the text, the laboratory manual provides a balanced introduction in each area of microbiology. The class-tested exercises are modular and short so that instructors can easily choose those exercises that fit their course.

ARIS™
(www.mhhe.com/prescottprinciples)

McGraw-Hill's ARIS—Assessment, Review, and Instruction System—for *Prescott's Principles of Microbiology* provides helpful online study materials and resources that support each chapter in the book. Features include:

- Self-quizzes
- Animations (with quizzing)
- Flash cards
- Clinical case studies
- Recommended readings
  and more!

INSTRUCTOR RESOURCES

ARIS

McGraw-Hill's ARIS (Assessment, Review, and Instruction System) for *Prescott's Principles of Microbiology* is a complete, online tutorial, electronic homework, and course management system, designed for greater ease of use than any other system available. For students, ARIS contains self-study tools such as animations, interactive quizzes, and more. This program enables students to complete their homework online, as assigned by their instructors. ARIS provides all instructor resources online, as well provides the ability to create or edit questions from the question bank, import your own content, and automatically grade and report easy-to-assign homework, quizzing, and testing.

Go to www.aris.mhhe.com to learn more.

Presentation Center

Build instructional materials wherever, whenever, and however you want!

Presentation Center is an online digital library containing assets such as photos, artwork, animations, PowerPoints, and other types of media that can be used to create customized lectures, visually enhanced tests and quizzes, compelling course websites, or attractive printed support materials.

Access to your book, access to all books! This ever-growing resource gives instructors the power to utilize assets specific to their adopted textbook as well as content from other McGraw-Hill books in the library. Presentation Center's dynamic search engine allows you to explore by discipline, course, textbook chapter, asset type, or keyword. Simply browse, select, and download the files you need to build engaging course materials. All assets are copyrighted by McGraw-Hill Higher Education but can be used by instructors for classroom purposes.

Instructor's Manual

The *Instructor's Manual* is available in both Word and PDF formats and contains chapter overviews, objectives, and answer guidelines for Critical Thinking Questions.

Test Bank

The Test Bank provides questions that can be used for homework assignments or the preparation of exams. The computerized test bank allows the user to quickly create customized exams. This user-friendly program allows instructors to search for questions by topic, format, or difficulty level; edit existing questions or add new ones; and scramble questions and answer keys for multiple versions of the same test.

Transparencies

A set of transparency masters can be customized for your course. Please contact your McGraw-Hill sales representative for details.

Electronic Books

If you or your students are ready for an alternative version of the traditional textbook, McGraw-Hill and VitalSource have partnered to bring you innovative and inexpensive electronic textbooks. By purchasing E-books from McGraw-Hill & VitalSource, students can save as much as 50% on selected titles delivered on the most advanced E-book platform available, VitalSource Bookshelf.

E-books from McGraw-Hill & VitalSource are smart, interactive, searchable, and portable. VitalSource Bookshelf comes with a powerful suite of built-in tools that allow detailed searching, highlighting, note taking, and student-to-student or instructor-to-student note sharing. E-books from McGraw-Hill & VitalSource will help students study smarter and quickly find the information they need. And they will save money. Contact your McGraw-Hill sales representative to discuss E-book packaging options.

TOOLS FOR LEARNING

The History and Scope of Microbiology

1

Louis Pasteur, one of the greatest scientists of the nineteenth century, maintained that "Science knows no country, because knowledge belongs to humanity, and is a torch which illuminates the world."

Archaea The domain of life that contains procaryotic cells with cell walls that lack peptidoglycan; they have unique lipids in their membranes and archaeal rRNA (among many differences).

Bacteria The domain of life that contains procaryotic cells with cell walls that contain the structural molecule peptidoglycan; they have bacterial rRNA.

Eucarya The domain of life that features organisms made of cells that have a membrane-delimited nucleus and differ in many other ways from procaryotic cells; includes protists, fungi, plants, and animals.

fungi A diverse group of microorganisms that range from unicellular forms (yeasts) to multicellular molds and mushrooms.

Chapter Glossary

Koch's postulates A set of rules for proving that a specific microorganism causes a particular disease.

microbiology The study of organisms that are usually too small to be seen with the naked eye; special techniques are required to isolate and grow them.

microorganism An organism that is too small to be seen clearly with the naked eye and lacks highly differentiated cells and distinct tissues.

prions Infectious agents that cause spongiform encephalopathies such as scrapie in sheep; they are composed only of protein.

procaryotic cells Cells that lack a true, membrane-enclosed nucleus; Bacteria and Archaea are procaryotic and have their genetic material located in a nucleoid.

protists Mostly unicellular eucaryotic organisms that lack cellular differentiation into tissues; cell differentiation is limited to cells involved in sexual reproduction, alternate vegetative morphology, or resting states such as cysts; includes organisms often referred to as algae and protozoa.

spontaneous generation An early belief, now discredited, that living organisms could develop from nonliving matter.

viroids Infectious agents composed only of single-stranded, circular RNA; they cause numerous plant diseases.

viruses Infectious agents having a simple acellular organization with a protein coat and a nucleic acid genome, lacking independent metabolism, and reproducing only within living host cells.

virusoids Infectious agents composed only of single-stranded RNA; they are unable to replicate without the aid of specific viruses that coinfect the host cell.

Dans les champs de l'observation, le hasard ne favorise que les esprits préparés.
(In the field of observation, chance favors only prepared minds.)

—Louis Pasteur

The importance of microorganisms cannot be overemphasized. In terms of sheer number and mass—microbes contain an estimated 50% of the biological carbon and 90% of the biological nitrogen on Earth—they greatly exceed every other group of organisms on the planet. Furthermore, they are found everywhere: from geothermal vents in the ocean depths to the coldest Arctic ice. They are major contributors to the functioning of the biosphere, being indispensable for the cycling of the elements essential for life. They also are a source of nutrients at the base of all ecological food webs. Most important, certain microorganisms carry out photosynthesis, rivaling plants in their role of capturing carbon dioxide and releasing oxygen into the atmosphere. Those microbes that inhabit humans are also important, helping the body digest food and producing vitamins B and K. In addition, society in general benefits from microorganisms. Indeed, modern biotechnology rests upon a microbiological foundation, as microbes are necessary for the production of bread, cheese,

Chapter Glossary

Each chapter begins with a glossary—a list of key terms discussed in the chapter. Each term is succinctly defined.

10.12 Phototrophy   **215**

Figure 10.30 The Mechanism of Photosynthesis. An illustration of the chloroplast thylakoid membrane showing photosynthetic ETC function and noncyclic photophosphorylation. The chain is composed of three complexes: PS I, the cytochrome bf complex, and PS II. Two diffusible electron carriers connect the three complexes. Plastoquinone (PQ) connects PS I with the cytochrome bf complex, and plastocyanin (PC) connects the cytochrome bf complex with PS II. The light-driven electron flow pumps protons across the thylakoid membrane and generates an electrochemical gradient, which can then be used to make ATP. Water is the source of electrons and the oxygen-evolving complex (OEC) produces oxygen.

Anoxygenic phototrophs have photosynthetic pigments called **bacteriochlorophylls** (figure 10.27). In some bacteria, these are located in membranous vesicles called chlorosomes. The absorption maxima of bacteriochlorophylls (Bchl) are at longer wavelengths than those of chlorophylls. Bacteriochlorophylls a and b have maxima in either at 775 and 790 nm, respectively. In vivo maxima are about 830 to 890 nm (Bchl a) and 1,020 to 1,040 nm (Bchl b). This shift of absorption maxima into the infrared region better adapts these bacteria to their ecological niches. >> *Photosynthetic bacteria (section 19.3)*

Many differences found in anoxygenic phototrophs are because they have a single photosystem. Because of this, they are restricted to cyclic photophosphorylation and are unable to produce O_2 from H_2O. Indeed, almost all anoxygenic phototrophs are strict anaerobes. A tentative scheme for the photosynthetic ETC of a purple nonsulfur bacterium is given in **figure 10.31**. When the reaction-center bacteriochlorophyll P870 is excited, it donates an electron to bacteriopheophytin. Electrons then flow to quinones and through an ETC back to P870 while generating sufficient PMF to drive ATP synthesis by ATP synthase. Note that although both green and purple bacteria lack two photosystems, the purple bacteria have a photosynthetic apparatus similar to photosystem II of oxygenic phototrophs, whereas the green sulfur bacteria have a system similar to photosystem I. >> *Class Alphaproteobacteria: Purple nonsulfur bacteria (section 20.1)*

Anoxygenic photoautotrophs face a further problem because they also require reducing power (NAD[P]H or reduced ferredoxin) for CO_2 fixation and other biosynthetic processes. They are able to generate reducing power in at least three ways, depending on the bacterium. Some have hydrogenases that are used to produce NAD(P)H directly from the oxidation of hydrogen gas. This is possible because hydrogen gas has a more negative reduction potential than NAD$^+$ (*see table 9.1*). Others, such as the photosynthetic purple bacteria, use reverse electron flow to generate NAD(P)H (figure 10.31). In this mechanism, electrons are drawn off the photosynthetic ETC and "pushed" to NAD(P)$^+$ using PMF. Electrons from electron donors such as

Cross-Referenced Notes

In-text references with icons refer students to other parts of the book to review.

ACKNOWLEDGMENTS

We would like to thank the Board of Advisors, who provided constructive reviews of every chapter, including the line art and photos in the book. Their specialized knowledge helped assimilate more reliable sources of informaton, and find more effective ways of expressing an idea for the student reader.

Board of Advisors

Morad Abou-Sabe, *Rutgers University*
Shivanthi Anandan, *Drexel University*
Penny Antley, *University of Louisiana–Lafayette*
Phil Cunningham, *Wayne State University*
Bernard Frye, *University of Texas at Arlington*
Mike Henson, *Clemson University*

Mike Hyman, *North Carolina State University–Raleigh*
Michael Ibba, *The Ohio State University*
Jeffrey Leblond, *Middle Tennessee State University*
S. N. Rajagopal, *University of Wisconsin–Lacrosse*
William Safranek, *University of Central Florida*
Lisa Stein, *University of California–Riverside*
Herman Witmer, *University of Illinois at Chicago*

We owe our collective thanks to Lisa Bruflodt, Jim Connely, Tami Petsche, Mary Powers, Mary Jane Lampc and Sandy Ludovissy. We would also like to thank our design editor John Joran and photo editor Mary Reeg.

This text is dedicated to our families for their patience and to our students for teaching us how to teach better.

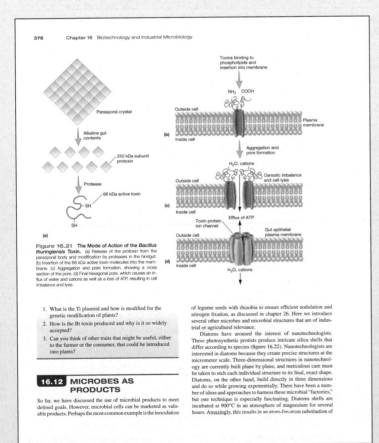

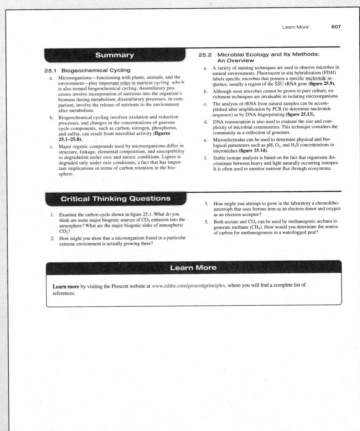

Review and Reflection Questions Within Narrative
Review questions throughout each chapter assist students in mastering section concepts before moving on to other topics.

End-of-Chapter Material
• End-of-chapter summaries are organized by numbered headings and provide a snapshot of important chapter concepts.
• Critical Thinking Questions supplement the questions for review and reflection found throughout each chapter; they are designed to stimulate analytical problem-solving skills.

Microbial Diversity & Ecology

35.2 A Fungus with a Voracious Appetite

The basidiomycete *Phanerochaete chrysosporium* (the scientific name means "visible hair, golden spore") is a fungus with unusual degradative capabilities. This organism is termed a "white rot fungus" because of its ability to degrade lignin, a randomly linked, phenylpropene-based polymeric component of wood. The cellulosic portion of wood is attacked to a lesser extent, resulting in the characteristic white color of the degraded wood. This organism also degrades a truly amazing range of xenobiotic compounds (nonbiological foreign chemicals) using both intracellular and extracellular enzymes.

As examples, the fungus degrades benzene, toluene, ethylbenzene, and xylenes (the so-called BTEX compounds), chlorinated compounds such as 2,4,5-trichloroethylene (TCE), and trichlorophenols (figure 35.12). The latter are present as contaminants in wood preservatives and are used as pesticides. In addition, other chlorinated benzenes can be degraded with or without toluenes being present. Even the insecticide Hydramethylnon is degraded.

How does this microorganism carry out such feats? Apparently most xenobiotic degradation occurs after active growth, during secondary metabolic lignin degradation. Degradation of some compounds involves important extracellular enzymes including lignin peroxidase, manganese-dependent peroxidase, and glyoxal oxidase. A critical enzyme is pyranose oxidase, which releases H_2O_2 for use by the manga[…] peroxidase enzyme. The H_2O_2 also is a precursor of the […] hydroxyl radical, which participates in wood degradati[…] the pyranose oxidase enzyme is located in the interperip[…] the fungal cell wall, where it can function either as a pa[…] or be released and penetrate into the wood substrate. It […] nonspecific enzymatic system that releases these oxid[…] degrades many cyclic, aromatic, and chlorinated compo[…] lignins.

We can expect to continue hearing of many new adva[…] this organism. Potentially valuable applications being […] growth in bioreactors, where intracellular and extracellul[…] be maintained in the bioreactor while liquid wastes flow […] bilized fungi.

organism physical protection, as well as possibly supplying nutrients. This makes it possible for the microorganism to survive in spite of the intense competitive pressures that exist in the natural environment, including pressure from protozoan predators. Microhabitats may be either living or inert. Specialized living microhabitats include the surface of a seed, a root, or a leaf. Here, higher nutrient fluxes and rates of initial colonization by the added microorganisms protect against the fierce competitive conditions in the natural environment. For example, to ensure that the nitrogen-fixing microbe *Rhizobium* is in close association with the legume, seeds are coated with the microbe using an oil-organism mixture or the bacteria are placed in a band under the seed where the newly developing primary root will penetrate. << *Microorganisms in terrestrial environments: The Rhizobia (section 26.2)*

Recently it has been found that microorganisms can be added to natural communities together with protective inert microhabi-tats. As an example, if microbes are added t[…] microporous glass, the survival of added microorg[…] markedly enhanced. Other microbes have been ob[…] ate their own microhabitats. Microorganisms in the […] overlying PCB-contaminated sand-clay soils have […] to create their own "clay hutches" by binding clays […] surfaces with exopolysaccharides. Thus the application of principles of microbial ecology can facilitate the successful management of microbial communities in nature.

1. What factors might limit the ability of microorganisms, after addition to a soil or water, to persist and carry out desired functions?
2. What types of microhabitats can be used with microorganisms when they are added to a complex […] environment?

Techniques and Applications

30.1 Detection and Removal of Endotoxins

Bacterial endotoxins plagued the pharmaceutical industry and medical device producers for years. For example, administration of drugs contaminated with endotoxins resulted in complications—even death—to patients. In addition, endotoxins can be problematic for individuals and firms working with cell cultures and genetic engineering. The result has been the development of sensitive tests and methods to identify and remove these endotoxins. The procedures must be very sensitive to trace amounts of endotoxins. Most firms have set a limit of 0.25 endotoxin units (E.U.), 0.025 ng/ml, or less as a release standard for their drugs, media, or products.

One of the most accurate tests for endotoxins is the in vitro *Limulus* amoebocyte lysate (LAL) assay. The assay is based on the observation that when an endotoxin contacts the clot protein from circulating amoebocytes of the horseshoe crab (*Limulus*), a gel-clot forms. The assay kits contain calcium, proclotting enzyme, and procoagulogen. The proclotting enzyme is activated by bacterial endotoxin (lipopolysaccharide) and calcium to form active clotting enzyme (box figure). Active clotting enzyme then catalyzes the cleavage of procoagulogen into polypeptide subunits (coagulogen). The subunits join by disulfide bonds to form a gel-clot. Spectrophotometry is then used to measure the protein precipitated by the lysate. The LAL test is sensitive at the nanogram level but must be standardized against U.S. Food and Drug Administration Bureau of Biologics endotoxin reference standards. Results are reported in endotoxin units per milliliter and reference made to the particular reference standards used.

Removal of endotoxins presents more of a problem than their detection. Those present on glassware or medical devices can be inactivated if the equipment is heated at 250°C for 30 minutes. Soluble endotoxins range in size from 20 kDa to large aggregates with diameters up to 0.1 μm. Thus they cannot be removed by conventional filtration systems. Manufacturers have developed special filtration systems and filtration cartridges that retain these endotoxins and help alleviate contamination problems.

Historical Highlights

33.1 John Snow—The First Epidemiologist

Much of what we know today about the epidemiology of cholera is based on the classic studies conducted by the British physician John Snow between 1849 and 1854. During this period, a series of cholera outbreaks occurred in London, England, and Snow set out to find the source of the disease. Some years earlier when he was still a medical apprentice, Snow had been sent to help during an outbreak of cholera among coal miners. His observations convinced him that the disease was usually spread by unwashed hands and shared food, not by "bad" air or casual direct contact.

T[…] when the outbreak of 1849 occurred, Snow believed that […]as spread among the poor in the same way as among the […]rs. He suspected that water, and not unwashed hands and […]d, was the source of the cholera infection among the […]residents. Snow examined official death records and dis[…]at most of the victims in the Broad Street area had lived […]e Broad Street water pump or had been in the habit of […]om it. He concluded that cholera was spread by drinking […] the Broad Street pump, which was contaminated with raw […]ntaining the disease agent. When the pump handle was […]he number of cholera cases dropped dramatically.

In 1854 another cholera outbreak struck London. Part of the city's water supply came from two different suppliers: the Southwark and Vauxhall Company, and the Lambeth Company. Snow interviewed cholera patients and found that most of them purchased their drinking water from the Southwark and Vauxhall Company. He also discovered that this company obtained its water from the Thames River below locations where Londoners had discharged their sewage. In contrast, the Lambeth Company took its water from the Thames before the river reached the city. The death rate from cholera was over eightfold lower in households supplied with Lambeth Company water. Water contaminated by sewage was transmitting the disease. Finally, Snow concluded that the cause of the disease must be able to multiply in water. Thus he nearly recognized that cholera was caused by a microorganism, though Robert Koch did not discover the causative bacterium (*Vibrio cholerae*) until 1883.

To commemorate these achievements, the John Snow Pub now stands at the site of the old Broad Street pump. Those who complete the Epidemiologic Intelligence Program at the Centers for Disease Control and Prevention receive an emblem bearing a replica of a barrel of Whatney's Ale—the brew dispensed at the John Snow Pub.

Disease

31.1 Antibiotic Misuse and Drug Resistance

The sale of antimicrobial drugs is big business. In the United States, millions of pounds of antibiotics valued at billions of dollars are produced annually. As much as 70% of these antibiotics are added to livestock feed.

Because of the massive quantities of antibiotics being prepared and used, an increasing number of diseases are resisting treatment due to the spread of drug resistance. A good example is *Neisseria gonorrhoeae*, the causative agent of gonorrhea. Gonorrhea was first treated successfully with sulfonamides in 1936, but by 1942 most strains were resistant and physicians turned to penicillin. Within 16 years, a penicillin-resistant strain emerged in Asia. A penicillinase-producing gonococcus reached the United States in 1976 and is still spreading in this country. Thus penicillin is no longer used to treat gonorrhea.

In late 1968 an epidemic of dysentery caused by *Shigella* broke out in Guatemala and affected at least 112,000 persons; 12,500 deaths resulted. The strains responsible for this devastation carried an R plasmid conferring resistance to chloramphenicol, tetracycline, streptomycin, and sulfonamide. In 1972 a typhoid epidemic swept through Mexico producing 100,000 infections and 14,000 deaths. It was due to a *Salmonella* strain with the same multiple-drug-resistance pattern seen in the previous *Shigella* outbreak.

Haemophilus influenzae type b is responsible for many cases of childhood pneumonia and middle ear infections, as well as respiratory infections and meningitis. It is now becoming increasingly resistant to tetracyclines, ampicillin, and chloramphenicol. Similarly, the worldwide rate of penicillin-nonsusceptible (i.e., resistant) *Streptococcus pneumoniae* (PNSP) continues to increase. There is a direct correlation between the daily use of antibiotics (expressed as defined daily dose [DDD] per day) and the percent of PNSP isolates cultured (box figure). This dramatic correlation is alarming. More alarming is the continued indiscriminant use of antibiotics in light of these data.

In 1946 almost all strains of *Staphylococcus* were penicillin sensitive. Today most hospital strains are resistant to penicillin G, and some are now also resistant to methicillin and gentamicin and only can be treated with vancomycin. Strains of *Enterococcus* have become resistant to most antibiotics, including vancomycin, and a few cases of vancomycin-resistant *S. aureus* have been reported in the United States and Japan.

It is clear from these and other examples (e.g., […] *Mycobacterium tuberculosis*) that drug resistance is […] serious public health problem. Much of the difficulty […]

Microbial Tidbits

21.1 Spores in Space

During the nineteenth-century argument over the question of the evolution of life, the panspermia hypothesis became popular. According to this hypothesis, life did not evolve from inorganic matter on Earth but arrived as viable bacterial spores that escaped from another planet. More recently the British astronomer Fred Hoyle has revived the hypothesis based on his study of the absorption of radiation by interstellar dust. Hoyle maintains that dust grains were initially viable cells that have been degraded and that the beginning of life on Earth was due to the arrival of bacterial spores that had survived their trip through space.

Even more recently Peter Weber and J. Mayo Greenberg from the University of Leiden in the Netherlands have studied the effect of very high vacuum, low temperature, and UV radiation on the survival of *Bacillus subtilis* spores. Their data suggest that spores within an interstellar molecular cloud might be able to survive between 4.5 to 45 million years. Molecular clouds move through space at speeds sufficient to transport spores between solar systems in this length of time. Although these results do not prove the panspermia hypothesis, they are consistent with the possibility that bacteria might be able to travel between planets capable of supporting life.

Special Interest Essays

Interesting essays on relevant topics are included in most chapters. Readings are organized into these topics: Historical Highlights, Techniques & Applications, Microbial Diversity & Ecology, Disease, and Microbial Tidbits.

The History and Scope of Microbiology

1

Louis Pasteur, one of the greatest scientists of the nineteenth century, maintained that "Science knows no country, because knowledge belongs to humanity, and is a torch which illuminates the world."

Archaea The domain of life that contains procaryotic cells with cell walls that lack peptidoglycan; they have unique lipids in their membranes and archaeal rRNA (among many differences).

Bacteria The domain of life that contains procaryotic cells with cell walls that contain the structural molecule peptidoglycan; they have bacterial rRNA.

Eucarya The domain of life that features organisms made of cells that have a membrane-delimited nucleus and differ in many other ways from procaryotic cells; includes protists, fungi, plants, and animals.

fungi A diverse group of microorganisms that range from unicellular forms (yeasts) to multicellular molds and mushrooms.

Chapter Glossary

Koch's postulates A set of rules for proving that a specific microorganism causes a particular disease.

microbiology The study of organisms that are usually too small to be seen with the naked eye; special techniques are required to isolate and grow them.

microorganism An organism that is too small to be seen clearly with the naked eye and lacks highly differentiated cells and distinct tissues.

prions Infectious agents that cause spongiform encephalopathies such as scrapie in sheep; they are composed only of protein.

procaryotic cells Cells that lack a true, membrane-enclosed nucleus; *Bacteria* and *Archaea* are procaryotic and have their genetic material located in a nucleoid.

protists Mostly unicellular eucaryotic organisms that lack cellular differentiation into tissues; cell differentiation is limited to cells involved in sexual reproduction, alternate vegetative morphology, or resting states such as cysts; includes organisms often referred to as algae and protozoa.

spontaneous generation An early belief, now discredited, that living organisms could develop from nonliving matter.

viroids Infectious agents composed only of single-stranded, circular RNA; they cause numerous plant diseases.

viruses Infectious agents having a simple acellular organization with a protein coat and a nucleic acid genome, lacking independent metabolism, and reproducing only within living host cells.

virusoids Infectious agents composed only of single-stranded RNA; they are unable to replicate without the aid of specific viruses that coinfect the host cell.

Dans les champs de l'observation, le hasard ne favorise que les esprits préparés.
(In the field of observation, chance favors only prepared minds.)

—*Louis Pasteur*

The importance of microorganisms cannot be overemphasized. In terms of sheer number and mass—microbes contain an estimated 50% of the biological carbon and 90% of the biological nitrogen on Earth—they greatly exceed every other group of organisms on the planet. Furthermore, they are found everywhere: from geothermal vents in the ocean depths to the coldest Arctic ice. They are major contributors to the functioning of the biosphere, being indispensable for the cycling of the elements essential for life. They also are a source of nutrients at the base of all ecological food webs. Most important, certain microorganisms carry out photosynthesis, rivaling plants in their role of capturing carbon dioxide and releasing oxygen into the atmosphere. Those microbes that inhabit humans are also important, helping the body digest food and producing vitamins B and K. In addition, society in general benefits from microorganisms. Indeed, modern biotechnology rests upon a microbiological foundation, as microbes are necessary for the production of bread, cheese, beer, antibiotics, vaccines, vitamins, enzymes, and many other products. Their ability to produce biofuels such as ethanol is also being intensively explored. These alternative fuels are both renewable and can help decrease pollution associated with burning fossil fuels.

Although most microorganisms play beneficial or benign roles, some harm humans and have disrupted society over the millennia. Microbial diseases undoubtedly played a major role in historical events such as the decline of the Roman Empire and the conquest of the New World. In 1347, plague (Black Death), an arthropod-borne disease, struck Europe with brutal force, killing one-third of the population (about 25 million people) within four years. Over the next 80 years, the disease struck repeatedly, eventually wiping out 75% of the European population. The plague's effect was so great that some historians believe it changed European culture and prepared the way for the Renaissance. Today the struggle by microbiologists and others against killers such as AIDS and malaria continues.

In this chapter, we introduce the microbial world to provide a general idea of the organisms and agents that microbiologists study. We next discuss the scope and relevance of modern microbiology. Finally, we describe the historical development of the science of microbiology and its relationship to medicine and other areas of biology.

1.1   MEMBERS OF THE MICROBIAL WORLD

Microbiology often has been defined as the study of organisms and agents too small to be seen clearly by the unaided eye—that is, the study of **microorganisms.** Because objects less than about 1 millimeter in diameter cannot be seen clearly and must be examined with a microscope, microbiology is concerned primarily with organisms and agents this small and smaller. However, some microorganisms, particularly some eucaryotic microbes, are visible without microscopes. For example, bread molds and filamentous algae are studied by microbiologists yet are visible to the naked eye, as are the two bacteria *Thiomargarita* and *Epulopiscium.*
>> *Microbial Diversity & Ecology 3.1: Monstrous Microbes*

The difficulty in setting the boundaries of microbiology has led to the suggestion of other criteria for defining the field. For instance, an important characteristic of microorganisms, even those that are large and multicellular, is that they are relatively simple in their construction, lacking highly differentiated cells and distinct tissues. Another suggestion, made by Roger Stanier, is that the field also be defined in terms of its techniques. Microbiologists usually first isolate a specific microorganism from a population and then culture it. Thus microbiology employs techniques—such as sterilization and the use of culture media—that are necessary for successful isolation and growth of microorganisms.

Microorganisms are diverse, and their classification has always been a challenge for microbial taxonomists. Their early descriptions as either plants or animals were too simple. For instance, some microbes are motile like animals but also have cell walls and

are photosynthetic like plants. Such microbes cannot be placed easily into one kingdom or another. Another important factor in classifying microorganisms is that some are composed of procaryotic cells and others of eucaryotic cells. **Procaryotic cells** (Greek *pro,* before, and *karyon,* nut or kernel; organisms with a primordial nucleus) have a much simpler morphology than eucaryotic cells and lack a true membrane-delimited nucleus. In contrast, **eucaryotic cells** (Greek *eu,* true, and *karyon,* nut or kernel) have a membrane-enclosed nucleus; they are more complex morphologically and are usually larger than procaryotes. These observations eventually led to the development of a classification scheme that divided organisms into five kingdoms: the *Monera, Protista, Fungi, Animalia,* and *Plantae.* Microorganisms (except for viruses and other acellular infectious agents, which have their own classification system) were placed in the first three kingdoms.

In the last few decades, great progress has been made in three areas that profoundly affect microbial classification. First, much has been learned about the detailed structure of microbial cells from the use of electron microscopy. Second, microbiologists have determined the biochemical and physiological characteristics of many different microorganisms. Third, the sequences of nucleic acids and proteins from a wide variety of organisms have been compared. The comparison of ribosomal RNA (rRNA), begun by Carl Woese in the 1970s, was instrumental in demonstrating that there are two very different groups of procaryotic organisms: *Bacteria* and *Archaea,* which had been classified together as *Monera* in the five-kingdom system. Later studies based on rRNA comparisons showed that *Protista* is not a cohesive taxonomic unit and that it should be divided into three or more kingdoms. These studies and others have led many taxonomists to conclude that the five-kingdom system is too simple. A number of alternatives have been suggested, but currently most microbiologists believe that organisms should be divided among three domains: *Bacteria* (the true bacteria or eubacteria), *Archaea,*[1] and *Eucarya* (all eucaryotic organisms) (**figure 1.1**). We use this system throughout the text, and it is discussed in detail in chapter 17. However, a brief description of the three domains and of the microorganisms placed in them follows.

Bacteria[2] are procaryotes that are usually single-celled organisms. Most have cell walls that contain the structural molecule peptidoglycan. They are abundant in soil, water, and air, and are major inhabitants of our skin, mouth, and intestines. Some bacteria live in environments that have extreme temperatures, pH, or salinity. Although some bacteria cause disease, many more play beneficial roles such as cycling elements in the biosphere, breaking down dead plant and animal material, and producing vitamins. Cyanobacteria (once called blue-green algae) produce significant amounts of oxygen through the process of photosynthesis.

Archaea are procaryotes that are distinguished from *Bacteria* by many features, most notably their unique ribosomal RNA sequences. They lack peptidoglycan in their cell walls and have unique membrane lipids. Some have unusual metabolic

---

[1] Although this is discussed further in chapter 17, it should be noted here that several names have been used for the *Archaea.* The two most important are archaeobacteria and archaebacteria. In this text, we use only the name *Archaea.*

[2] In this text, the term bacteria (s., bacterium) is used to refer to procaryotes that belong to domain *Bacteria,* and the term archaea (s., archaeon) is used to refer to procaryotes that belong to domain *Archaea.* In some publications, the term bacteria is used to refer to all procaryotes. That is not the case in this text.

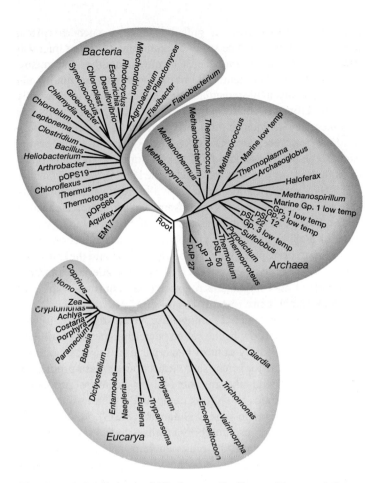

Figure 1.1 Universal Phylogenetic Tree. These evolutionary relationships are based on rRNA sequence comparisons. The human genus (*Homo*) is highlighted in red.

characteristics, such as the methanogens, which generate methane gas. Many archaea are found in extreme environments, including those with high temperatures (thermophiles) and high concentrations of salt (extreme halophiles). Pathogenic archaea have not yet been identified.

Domain *Eucarya* includes microorganisms classified as protists or fungi. Animals and plants are also placed in this domain. **Protists** are generally unicellular but larger than procaryotes. Photosynthetic protists, together with the cyanobacteria, produce about 75% of the planet's oxygen. These phytoplankton are the foundation of aquatic food chains. **Protozoa** are unicellular, animal-like protists that are usually motile. Many free-living protozoa function as the principal hunters and grazers of the microbial world. They obtain nutrients by ingesting organic matter and other microbes. They can be found in many different environments, and some are normal inhabitants of the intestinal tracts of animals, where they aid in digestion of complex materials such as cellulose. A few cause disease in humans and other animals. **Slime molds** are protists that are like protozoa in one stage of their life cycle but like fungi in another. In the protozoan phase, they hunt for and engulf food particles, consuming decaying vegetation and other microbes. **Water molds** are protists that grow on the surface of freshwater and moist soil. They feed on decaying vegetation such as logs and mulch. Some water molds have produced devastating plant infec-

tions, including the Great Potato Famine of 1846–1847 in Ireland. **Fungi** are a diverse group of microorganisms that range from unicellular forms (yeasts) to molds and mushrooms. Molds and mushrooms are multicellular fungi that form thin, threadlike structures called hyphae. They absorb nutrients from their environment, including the organic molecules that they use as a source of carbon and energy. Because of their metabolic capabilities, many fungi play beneficial roles, including making bread rise, producing antibiotics, and decomposing dead organisms. Some fungi associate with plant roots to form mycorrhizae. Mycorrhizal fungi transfer nutrients to the roots, improving the growth of the plants, especially in poor soils. Other fungi cause plant diseases (e.g., rusts, powdery mildews, and smuts) and diseases in humans and other animals.

The microbial world also includes numerous acellular infectious agents. **Viruses** are acellular entities that must invade a host cell to replicate. The simplest viruses are composed only of proteins and a nucleic acid, and can be extremely small (the smallest is 10,000 times smaller than a typical bacterium). However, their small size belies their power—they cause many animal and plant diseases and have caused epidemics that have shaped human history. The diseases they cause include smallpox, rabies, influenza, AIDS, the common cold, and some cancers. **Viroids** and **virusoids** are infectious agents composed only of ribonucleic acid (RNA). Viroids cause numerous plant diseases, whereas virusoids cause some important animal diseases such as hepatitis. Finally, **prions,** infectious agents composed only of protein, are responsible for causing a variety of spongiform encephalopathies such as scrapie and "mad cow disease."

1. Describe the field of microbiology in terms of the size of its subject material and the nature of its techniques.
2. Describe and contrast procaryotic and eucaryotic cells.

1.2 SCOPE AND RELEVANCE OF MICROBIOLOGY

As the scientist-writer Steven Jay Gould (1941–2002) emphasized, we live in the age of bacteria. They were the first living organisms on our planet, likely created the atmosphere that allowed the evolution of oxygen-consuming life-forms, and now live virtually everywhere life is possible. Furthermore, the biosphere depends on their activities, and they influence human society in countless ways. Because microorganisms play such diverse roles, modern microbiology is a large discipline with many different specialties; it has a great impact on fields such as medicine, agricultural and food sciences, ecology, genetics, biochemistry, and molecular biology. One indication of the importance of microbiology is the Nobel Prize given for work in physiology or medicine. About one-third of these prizes have been awarded to scientists working on microbiological problems (*see inside front cover*).

Microbiology has both basic and applied aspects. The basic aspects are concerned with the biology of microorganisms themselves. The applied aspects are concerned with practical problems such as disease, water and wastewater treatment, food spoilage and food production, and industrial uses of microbes. It is important to note that the basic and applied aspects of microbiology are intertwined. Basic research is often conducted in applied fields, and applications often arise out of basic research. A discussion of some of the major fields of microbiology and the occupations within them follows.

Although pathogenic microbes are the minority, they garner considerable interest. Thus, one of the most active and important fields in microbiology is medical microbiology, which deals with diseases of humans and animals. Medical microbiologists identify the agents causing infectious diseases and plan measures for their control and elimination. Frequently they are involved in tracking down new, unidentified pathogens such as the agent that causes variant Creutzfeldt-Jakob disease (the human version of "mad cow disease"), hantavirus, West Nile virus, and the virus responsible for SARS. These microbiologists also study the ways in which microorganisms cause disease. >> *Microbial Diversity & Ecology 24.1: SARS: Evolution of a virus*

As noted earlier, major epidemics have regularly affected human history. The 1918 influenza pandemic is of particular note; it killed more than 20 million people in about one year. Public health microbiology is concerned with the control and spread of such communicable diseases. Public health microbiologists and epidemiologists monitor the amount of disease in populations. Based on their observations, they can detect outbreaks and developing epidemics, and implement appropriate control measures in response. They also conduct surveillance for new diseases as well as bioterrorism events. Those public health microbiologists working for local governments monitor community food establishments and water supplies in an attempt to keep them safe and free from infectious disease agents.

Immunology is concerned with how the immune system protects the body from pathogens and the response of infectious agents. It is one of the fastest growing areas in science. Much of the growth began with the discovery of HIV, which specifically targets cells of the immune system. Immunology also deals with health problems such as the nature and treatment of allergies and autoimmune diseases such as rheumatoid arthritis. >> *Techniques & Applications 29.1: Monoclonal antibody technology*

Agricultural microbiology is concerned with the impact of microorganisms on agriculture. Microbes such as nitrogen-fixing bacteria play critical roles in the nitrogen cycle and affect soil fertility. Other microbes live in the digestive tracts of ruminants such as cattle and break down the plant materials these animals ingest. There are also plant and animal pathogens that can have significant economic impacts if not controlled. Agricultural microbiologists work on methods to increase soil fertility and crop yields, study rumen microorganisms in order to increase meat and milk production, and try to combat plant and animal diseases. Currently many agricultural microbiologists are studying the use of bacterial and viral insect pathogens as substitutes for chemical pesticides.

Microbial ecology is concerned with the relationships between microorganisms and the components of their living and nonliving habitats. Microbial ecologists study the global and local contributions of microorganisms to the carbon, nitrogen, and sulfur cycles, including the role of microbes in both the production and removal of greenhouse gases such as carbon dioxide and methane. The study of pollution effects on microorganisms also is important because of the impact these organisms have on the environment. Microbial ecologists are employing microorganisms in bioremediation to reduce pollution. The study of the microbes normally associated with the human body has become a new frontier in microbial ecology.

Numerous foods are made using microorganisms. On the other hand, some microbes cause food spoilage or are pathogens spread through food. An excellent example of the latter is *Escherichia coli* O157:H7, which in 2006 caused a widespread outbreak of disease when it contaminated a major source of spinach in the United States. Scientists working in food and dairy microbiology continue to explore the use of microbes in food production. They also work to prevent microbial spoilage of food and the transmission of food-borne diseases. There is also considerable research on the use of microorganisms themselves as a nutrient source for livestock and humans. >> *Microbiology of food (chapter 34)*

In 1929 Alexander Fleming discovered that the fungus *Penicillium* produced what he called penicillin, the first antibiotic that could successfully control bacterial infections. Although it took World War II for scientists to learn how to mass-produce it, scientists soon found other microorganisms capable of producing additional antibiotics as well as compounds such as citric acid, vitamin B_{12}, and monosodium glutamate (MSG). Today, industrial microbiologists use microorganisms to make products such as antibiotics, vaccines, steroids, alcohols and other solvents, vitamins, amino acids, and enzymes. Industrial microbiologists identify microbes of use to industry. They also utilize techniques to improve production by microbes and devise systems for culturing them and isolating the products they make.

Microbes are metabolically diverse and can employ a wide variety of energy sources, including organic matter, inorganic molecules (e.g., H_2 and NH_3), and sunlight. Microbiologists working in microbial physiology and biochemistry study many aspects of the biology of microorganisms, including their metabolic capabilities. They may also study the synthesis of antibiotics and toxins, the ways in which microorganisms survive harsh environmental conditions, and the effects of chemical and physical agents on microbial growth and survival.

Microbial genetics and molecular biology focus on the nature of genetic information and how it regulates the development and function of cells and organisms. The bacteria *E. coli* and *Bacillus subtilis,* the yeast *Saccharomyces cerevisiae* (baker's yeast), and bacterial viruses such as T4 and lambda continue to be important model organisms used to understand biological phenomena. Microbial geneticists also play a significant role in applied microbiology because they develop techniques that are useful in agricultural microbiology, industrial microbiology, food and dairy microbiology, and medicine.

Because of the practical importance of microbes and their use as model organisms, the future of microbiology is bright. However, it is important to remember that future advances in microbiology will build on the foundations laid by earlier scientists. The development of microbiology as a science is described in sections 1.3 to 1.5. **Figure 1.2** presents a summary of some of

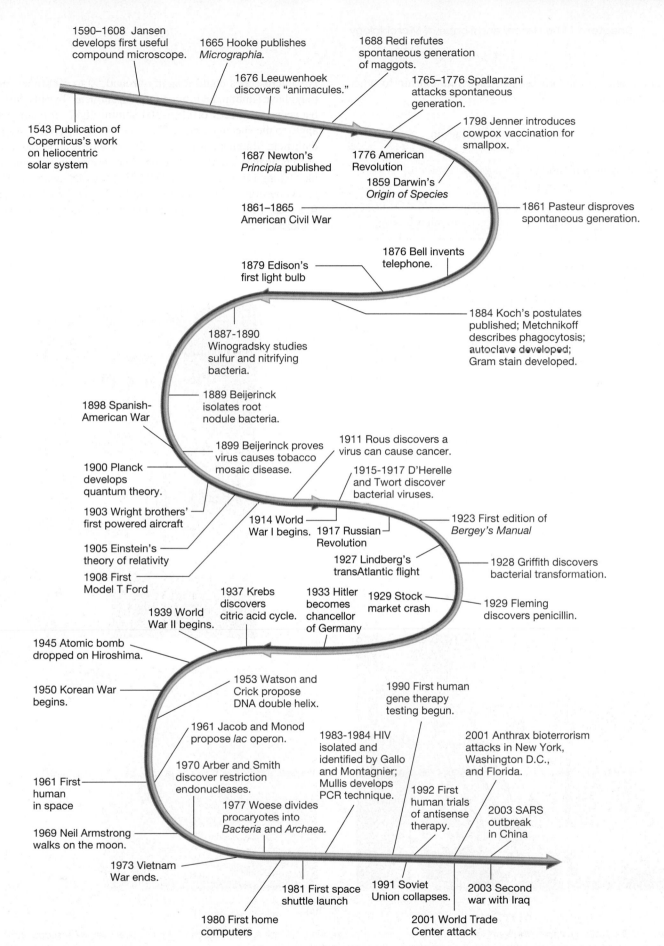

Figure 1.2 **Some Important Events in the Development of Microbiology.** Milestones in microbiology are marked in red; other historical events are in black.

the major events in this process and their relationship to other historical landmarks.

1. Briefly describe the major subdisciplines in microbiology.
2. Why do you think microorganisms are useful to biologists as experimental models?
3. List all the activities or businesses you can think of in your community that directly depend on microbiology.

1.3 DISCOVERY OF MICROORGANISMS

Even before microorganisms were seen, some investigators suspected their existence and responsibility for disease. Among others, the Roman philosopher Lucretius (about 98–55 BCE) and the physician Girolamo Fracastoro (1478–1553) suggested that disease was caused by invisible living creatures. The earliest microscopic observations appear to have been made between 1625 and 1630 on bees and weevils by the Italian Francesco Stelluti, using a microscope probably supplied by Galileo. In 1665 the first drawing of a microorganism was published in Robert Hooke's *Micrographia*. However, the first person to publish extensive, accurate observations of microorganisms was the amateur microscopist Antony van Leeuwenhoek (1632–1723) of Delft, the Netherlands (**figure 1.3a**). Leeuwenhoek earned his living as a draper and haberdasher (a dealer in men's clothing and accessories) but spent much of his spare time constructing simple microscopes composed of double convex glass lenses held between two silver plates (figure 1.3b).

His microscopes could magnify around 50 to 300 times, and he may have illuminated his liquid specimens by placing them between two pieces of glass and shining light on them at a 45° angle to the specimen plane. This would have provided a form of dark-field illumination in which the organisms appeared as bright objects against a dark background and made bacteria clearly visible (figure 1.3c). Beginning in 1673, Leeuwenhoek sent detailed letters describing his discoveries to the Royal Society of London. It is clear from his descriptions that he saw both procaryotes and protozoa.

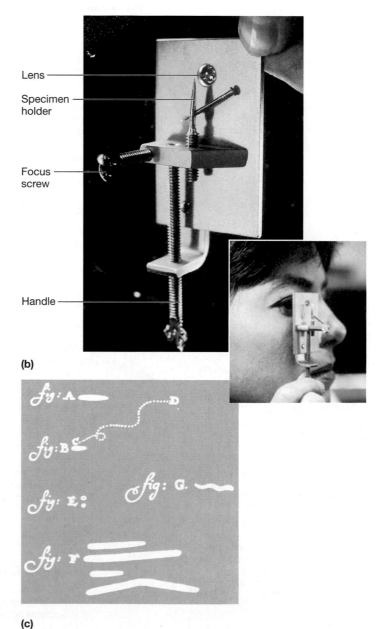

Lens

Specimen holder

Focus screw

Handle

(b)

(a)

(c)

Figure 1.3 Antony van Leeuwenhoek. (a) An oil painting of Leeuwenhoek. (b) A brass replica of the Leeuwenhoek microscope. Inset photo shows how it is held. (c) Leeuwenhoek's drawings of bacteria from the human mouth.

As important as Leeuwenhoek's observations were, the development of microbiology essentially languished for the next 200 years. Little progress was made primarily because microscopic observations of microorganisms do not provide sufficient information to understand their biology. For the discipline to develop, techniques for isolating and culturing microbes in the laboratory were needed. Many of these techniques began to be developed as scientists grappled with the conflict over the theory of spontaneous generation. This conflict and the subsequent studies on the role played by microorganisms in causing disease ultimately led to what is now called the golden age of microbiology.

1. Give some examples of the kind of information you think can be provided by microscopic observations of microorganisms.

2. Give some examples of the kind of information you think can be provided by isolating microorganisms from their natural environment and culturing them in the laboratory.

1.4 CONFLICT OVER SPONTANEOUS GENERATION

From earliest times, people had believed in **spontaneous generation**—that living organisms could develop from nonliving matter. Even Aristotle (384–322 BCE) thought some of the simpler invertebrates could arise by spontaneous generation. This view finally was challenged by the Italian physician Francesco Redi (1626–1697), who carried out a series of experiments on decaying meat and its ability to produce maggots spontaneously. Redi placed meat in three containers. One was uncovered, a second was covered with paper, and the third was covered with fine gauze that would exclude flies. Flies laid their eggs on the uncovered meat and maggots developed. The other two pieces of meat did not produce maggots spontaneously. However, flies were attracted to the gauze-covered container and laid their eggs on the gauze; these eggs produced maggots. Thus the generation of maggots by decaying meat resulted from the presence of fly eggs, and meat did not spontaneously generate maggots as previously believed. Similar experiments by others helped discredit the theory for larger organisms.

Leeuwenhoek's discovery of microorganisms renewed the controversy. Some proposed that microorganisms arose by spontaneous generation even though larger organisms did not. They pointed out that boiled extracts of hay or meat gave rise to microorganisms after sitting for a while. In 1748 the English priest John Needham (1713–1781) reported the results of his experiments on spontaneous generation. Needham boiled mutton broth in flasks that he then tightly stoppered. Eventually many of the flasks became cloudy and contained microorganisms. He thought organic matter contained a vital force that could confer the properties of life on nonliving matter. A few years later, the Italian priest and naturalist Lazzaro Spallanzani (1729–1799) improved on Needham's experimental design by first sealing glass flasks that contained water and seeds. If the sealed flasks were placed in boiling water for three-quarters of an hour, no growth took place as long as the flasks remained sealed. He proposed that air carried germs to the culture medium but also commented that the external air might be required for growth of animals already in the medium. The supporters of spontaneous generation maintained that heating the air in sealed flasks destroyed its ability to support life.

Several investigators attempted to counter such arguments. Theodore Schwann (1810–1882) allowed air to enter a flask containing a sterile nutrient solution after the air had passed through a red-hot tube. The flask remained sterile. Subsequently Georg Friedrich Schroder (1810–1885) and Theodor von Dusch (1824–1890) allowed air to enter a flask of heat-sterilized medium after it had passed through sterile cotton wool. No growth occurred in the medium even though the air had not been heated. Despite these experiments, the French naturalist Felix Pouchet (1800–1872) claimed in 1859 to have carried out experiments conclusively proving that microbial growth could occur without air contamination. This claim provoked Louis Pasteur (1822–1895) to settle the matter. Pasteur (**figure 1.4**)

Figure 1.4 Louis Pasteur. Pasteur working in his laboratory.

first filtered air through cotton and found that objects resembling plant spores had been trapped. If a piece of the cotton was placed in sterile medium after air had been filtered through it, microbial growth occurred. Next he placed nutrient solutions in flasks, heated their necks in a flame, and drew them out into a variety of curves. The swan neck flasks that he produced in this way had necks open to the atmosphere (**figure 1.5**). Pasteur then boiled the solutions for a few minutes and allowed them to cool. No growth took place even though the contents of the flasks were exposed to the air. Pasteur pointed out that no growth occurred because dust and germs had been trapped on the walls of the curved necks. If the necks were broken, growth commenced immediately. Pasteur had not only resolved the controversy by 1861 but also had shown how to keep solutions sterile.

The English physicist John Tyndall (1820–1893) and the German botanist Ferdinand Cohn (1828–1898) dealt a final blow to spontaneous generation. In 1877 Tyndall demonstrated that dust did indeed carry germs and that if dust was absent, broth remained sterile even if directly exposed to air. During the course of his studies, Tyndall provided evidence for the existence of exceptionally heat-resistant forms of bacteria. Working independently, Cohn discovered that the heat-resistant bacteria recognized by Tyndall were species capable of producing bacterial endospores. Cohn later played an instrumental role in establishing a classification system for procaryotes based on their morphology and physiology. **>>** *Bacterial endospores (section 3.8)*

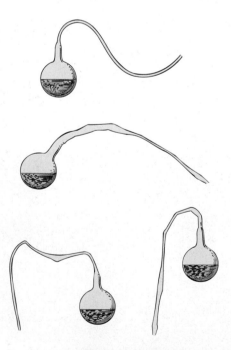

Figure 1.5 **Pasteur's Swan Neck Flasks.** These flasks were used in his experiments on the spontaneous generation of microorganisms. *Source: Annales Sciences Naturelle. 4th series, vol. 16, pp. 1–98, Pasteur, L., 1861, "Mémoire sur les Corpuscules Organisés Qui Existent Dans L'Apmosphére: Examen de la Doctrine des Générations Spontanées."*

1. How did Pasteur, Tyndall, and Cohn finally settle the spontaneous generation controversy?
2. Why was the belief in spontaneous generation an obstacle to the development of microbiology as a scientific discipline?

1.5 GOLDEN AGE OF MICROBIOLOGY

Pasteur's work with swan neck flasks ushered in the golden age of microbiology. Within 60 years (1857–1914), a number of disease-causing microbes were discovered, great strides in understanding microbial metabolism were made, and techniques for isolating and characterizing microbes were improved. Scientists also identified the role of immunity in preventing disease and controlling microbes, developed vaccines, and introduced techniques used to prevent infection during surgery.

Microorganisms and Disease

Although Fracastoro and a few others had suggested that invisible organisms produced disease, most people believed that disease was due to causes such as supernatural forces, poisonous vapors called miasmas, and imbalances among the four humors thought to be present in the body. The role of the four humors (blood, phlegm, yellow bile [choler], and black bile [melancholy]) in disease had been widely accepted since the time of the Greek physician Galen (129–199). Support for the idea that microorganisms cause disease—that is, the germ theory of disease—began to accumulate in the early nineteenth century from diverse fields. Agostino Bassi (1773–1856) first showed a microorganism could cause disease when he demonstrated in 1835 that a silkworm disease was due to a fungal infection. He also suggested that many diseases were due to microbial infections. In 1845 M. J. Berkeley (1803–1889) proved that the great potato blight of Ireland was caused by a water mold, and in 1853 Heinrich de Bary (1831–1888) showed that smut and rust fungi caused cereal crop diseases. Following his successes with the study of fermentation, Pasteur was asked by the French government to investigate the *pèbrine* disease of silkworms that was disrupting the silk industry. After several years of work, he showed that the disease was due to a protozoan parasite.

Indirect evidence for the germ theory of disease came from the work of the English surgeon Joseph Lister (1827–1912) on the prevention of wound infections. Lister, impressed with Pasteur's studies on the involvement of microorganisms in fermentation and putrefaction, developed a system of antiseptic surgery designed to prevent microorganisms from entering wounds. Instruments were heat sterilized, and phenol was used on surgical dressings and at times sprayed over the surgical area. The approach was remarkably successful and transformed surgery. It also provided strong indirect evidence for the role of microorganisms in disease because phenol, which kills bacteria, also prevented wound infections.

Koch's Postulates

The first direct demonstration of the role of bacteria in causing disease came from the study of anthrax by the German physician Robert Koch (1843–1910). Koch (**figure 1.6**) used the criteria proposed by his former teacher Jacob Henle (1809–1885) to establish the relationship between *Bacillus anthracis* and anthrax; he published his findings in 1876. Koch injected healthy mice with material from diseased animals, and the mice became ill. After transferring anthrax by inoculation through a series of 20 mice, he incubated a piece of spleen containing the anthrax bacillus in beef serum. The bacilli grew, reproduced, and produced

Figure 1.6   Robert Koch.   Koch examining a specimen in his laboratory.

endospores. When the isolated bacilli or their spores were injected into healthy mice, anthrax developed. His criteria for proving the causal relationship between a microorganism and a specific disease are known as **Koch's postulates.** Koch's proof that *B. anthracis* caused anthrax was independently confirmed by Pasteur and his coworkers. They discovered that after burial of dead animals, anthrax spores survived and were brought to the surface by earthworms. Healthy animals then ingested the spores and became ill.

After completing his anthrax studies, Koch fully outlined his postulates in his work on the cause of tuberculosis (**table 1.1**). In 1884 he reported that this disease was caused by the rod-shaped bacterium *Mycobacterium tuberculosis,* and he was awarded the Nobel Prize in Physiology or Medicine in 1905 for his work. Koch's postulates were quickly adopted by others and used to connect many diseases to their causative agent. However, their use is at times not feasible (**Disease 1.1**). For instance, organisms such as *Mycobacterium leprae,* the causative agent of leprosy, cannot be isolated in pure culture.

Development of Techniques for Studying Microbial Pathogens

During Koch's studies on bacterial diseases, it became necessary to isolate suspected bacterial pathogens in pure culture—a culture containing only one type of microorganism. At first, Koch cultured bacteria on the sterile surfaces of cut, boiled potatoes, but the bacteria would not always grow well. Eventually he developed culture media using meat extracts and protein digests, reasoning these were similar to body fluids. Initially he tried to solidify the media by adding gelatin. Separate bacterial colonies developed after the surface of the solidified medium had been streaked with a bacterial sample. The sample could also be mixed with liquefied gelatin medium. When the gelatin medium hardened, individual bacteria produced separate colonies. Despite its advantages, gelatin was not an ideal solidifying agent because it can be digested by many microbes and melts at temperatures

Table 1.1	Koch's Application of His Postulates Demonstrated That *Mycobacterium tuberculosis* Is the Causative Agent of Tuberculosis	
Postulate	**Experimentation**
1. The microorganism must be present in every case of the disease but absent from healthy organisms.	Koch developed a staining technique to examine human tissue. *M. tuberculosis* cells could be identified in diseased tissue.
2. The suspected microorganisms must be isolated and grown in a pure culture.	Koch grew *M. tuberculosis* in pure culture on coagulated blood serum.
3. The same disease must result when the isolated microorganism is inoculated into a healthy host.	Koch injected cells from the pure culture of *M. tuberculosis* into guinea pigs. The guinea pigs subsequently died of tuberculosis.
4. The same microorganism must be isolated again from the diseased host.	Koch isolated *M. tuberculosis* from the dead guinea pigs and was able to again culture the microbe in pure culture on coagulated blood serum.

Disease

1.1   A Molecular Approach to Koch's Postulates

Although the criteria that Koch developed for proving a causal relationship between a microorganism and a specific disease have been of great importance in medical microbiology, it is not always possible to apply them in studying human diseases. For example, some pathogens cannot be grown in pure culture outside the host; because other pathogens grow only in humans, their study would require experimentation on people. The identification, isolation, and cloning of genes responsible for virulence have made possible a new molecular form of Koch's postulates that resolves some of these difficulties. The emphasis is on the virulence genes present in the infectious agent rather than on the agent itself. The molecular postulates can be briefly summarized as follows:

1. The virulence trait under study should be associated much more with pathogenic strains of the species than with non-pathogenic strains.

2. Inactivation of the gene or genes associated with the suspected virulence trait should substantially decrease pathogenicity.

3. Replacement of the mutated gene with the normal wild-type gene should fully restore pathogenicity.

4. The gene should be expressed at some point during the infection and disease process.

5. Antibodies or immune system cells directed against the gene products should protect the host.

The molecular approach cannot always be applied because of problems such as the lack of an appropriate animal system. It also is difficult to employ the molecular postulates when the pathogen is not well characterized genetically.

---

above 28°C. A better alternative was provided by Fanny Eilshemius Hesse (1850–1934), the wife of Walther Hesse (1846–1911), one of Koch's assistants. She suggested the use of agar as a solidifying agent, which she used to make jellies. Agar was not attacked by most bacteria. Furthermore, it did not melt until reaching a temperature of 100°C and, once melted, did not solidify until reaching a temperature of 50°C; this eliminated the need to handle boiling liquid. Some of the media developed by Koch and his associates, such as nutrient broth and nutrient agar, are still widely used. Another important tool developed in Koch's laboratory was a container for holding solidified media—the petri dish (plate), named after Richard Petri (1852–1921), who devised it. These developments directly stimulated progress in all areas of microbiology. >> *Culture media (section 6.7); Isolation of pure cultures (section 6.8)*

Viral pathogens were also studied during this time. The discovery of viruses and their role in disease was made possible when Charles Chamberland (1851–1908), one of Pasteur's associates, constructed a porcelain bacterial filter in 1884. Dimitri Ivanowski (1864–1920) and Martinus Beijerinck (pronounced "by-a-rink;" 1851–1931) used the filter to study tobacco mosaic disease. They found that plant extracts and sap from diseased plants were infectious, even after being filtered with Chamberland's filter. Because the infectious agent passed through a filter that was designed to trap bacterial cells, they reasoned that the agent must be something smaller than a bacterium. Beijerinck proposed that the agent was a "filterable virus." Eventually viruses were shown to be tiny, acellular infectious agents.

Immunological Studies

In this period, progress also was made in determining how animals resisted disease and in developing techniques for protecting humans and livestock against pathogens. During studies on chicken cholera, Pasteur and Pierre Roux (1853–1933) discovered that

incubating the cultures for long intervals between transfers would attenuate the bacteria, which meant they had lost their ability to cause the disease. If the chickens were injected with these attenuated cultures, they remained healthy and developed the ability to resist the disease when exposed to virulent cultures. Pasteur called the attenuated culture a vaccine (Latin *vacca*, cow) in honor of Edward Jenner (1749–1823) because, many years earlier, Jenner had used material from cowpox lesions to protect people against smallpox. Shortly after this, Pasteur and Chamberland developed an attenuated anthrax vaccine. >> *Control of epidemics: Vaccines and immunizations (section 33.8)*

Pasteur also prepared a rabies vaccine using an attenuated strain of *Rabies virus*. During the course of these studies, Joseph Meister, a nine-year-old boy who had been bitten by a rabid dog, was brought to Pasteur. Since the boy's death was certain in the absence of treatment, Pasteur agreed to try vaccination. Joseph was injected 13 times over the next 10 days with increasingly virulent preparations of the attenuated virus. He survived. In gratitude for Pasteur's development of vaccines, people from around the world contributed to the construction of the Pasteur Institute in Paris, France. One of the initial tasks of the institute was vaccine production.

After the discovery that the diphtheria bacillus produced a toxin, Emil von Behring (1854–1917) and Shibasaburo Kitasato (1852–1931) injected inactivated toxin into rabbits, inducing them to produce an antitoxin, a substance in the blood that would inactivate the diphtheria toxin and protect against the disease. A tetanus antitoxin was then prepared, and both antitoxins were used in the treatment of people.

The antitoxin work provided evidence that immunity could result from soluble substances in the blood, now known to be antibodies (humoral immunity). It became clear that blood cells were also important in immunity (cellular immunity) when Elie Metchnikoff (1845–1916) discovered that some white blood cells could engulf disease-causing bacteria (**figure 1.7**). He called

Figure 1.7 Elie Metchnikoff. Metchnikoff at work in his laboratory.

these cells phagocytes and the process phagocytosis (Greek *phagein,* eating).

1. Discuss the contributions of Lister, Pasteur, and Koch to the germ theory of disease and the treatment or prevention of diseases.

2. What other contributions did Koch make to microbiology?

3. Describe Koch's postulates. What is a pure culture? Why are pure cultures important to Koch's postulates?

4. Would microbiology have developed more slowly if Fanny Hesse had not suggested the use of agar? Give your reasoning.

5. Some individuals can be infected by a pathogen yet not develop disease. In fact, some become chronic carriers of the pathogen. How does this observation affect Koch's postulates? How might the postulates be modified to account for the existence of chronic carriers?

6. How did Jenner, von Behring, Kitasato, and Metchnikoff contribute to the development of immunology?

1.6 DEVELOPMENT OF INDUSTRIAL MICROBIOLOGY AND MICROBIAL ECOLOGY

Although humans had unknowingly exploited microbes for thousands of years, industrial microbiology developed in large part from the work of Louis Pasteur and others on the alcoholic fermentations that yielded wine and other alcoholic beverages. In 1837, when Theodore Schwann and others proposed that yeast cells were responsible for the conversion of sugars to alcohol, the leading chemists of the time believed microorganisms were not involved. They were convinced that fermentation was due to a chemical instability that degraded the sugars to alcohol. Pasteur did not agree; he believed that fermentations were carried out by living organisms. In 1856 M. Bigo, an industrialist in Lille, France, where Pasteur worked, requested Pasteur's assistance. His business produced ethanol from the fermentation of beet sugars, and the alcohol yields had recently declined and the product had become sour. Pasteur discovered that the fermentation was failing because the yeast normally responsible for alcohol formation had been replaced by bacteria that produced acid rather than ethanol. In solving this practical problem, Pasteur demonstrated that all fermentations were due to the activities of specific yeasts and bacteria, and he published several papers on fermentation between 1857 and 1860. His success led to a study of wine diseases and the development of pasteurization to preserve wine during storage. Pasteur's studies on fermentation continued for almost 20 years. One of his most important discoveries was that some fermentative microorganisms were anaerobic and could live only in the absence of oxygen, whereas others were able to live either aerobically or anaerobically. >> *Controlling food spoilage (section 34.3)*

Microbial ecology developed when a few of the early microbiologists chose to investigate the ecological role of microorganisms. In particular, they studied microbial involvement in the carbon, nitrogen, and sulfur cycles taking place in soil and aquatic habitats. The Russian microbiologist Sergei Winogradsky (1856–1953) made many contributions to soil microbiology. He discovered that soil bacteria could oxidize iron, sulfur, and ammonia to obtain energy and that many of these bacteria could incorporate CO_2 into organic matter much as photosynthetic organisms do. Winogradsky also isolated anaerobic nitrogen-fixing soil bacteria and studied the decomposition of cellulose. Martinus Beijerinck was one of the great general microbiologists who made fundamental contributions to microbial ecology and many other fields. He isolated the aerobic nitrogen-fixing bacterium *Azotobacter,* a root nodule bacterium also capable of fixing nitrogen (later named *Rhizobium*), and sulfate-reducing bacteria. Beijerinck and Winogradsky also developed the enrichment-culture technique and the use of selective media, which have been of great importance in microbiology. >> *Biogeochemical cycling (section 25.1); Culture media (section 6.7)*

1. Briefly describe Pasteur's work on microbial fermentations.

2. How did Winogradsky and Beijerinck contribute to the study of microbial ecology?

3. Leeuwenhoek is often referred to as the father of microbiology. However, many historians feel that Louis Pasteur, Robert Koch, or perhaps both, deserve that honor. Who do you think is the father of microbiology? Why?

4. Consider the discoveries described in sections 1.3 to 1.6. Which do you think were the most important to the development of microbiology? Why?

Summary

1.1 Members of the Microbial World

a. Microbiology studies microscopic organisms that are often unicellular or, if multicellular, do not have highly differentiated tissues. The discipline is also defined by the techniques it uses—in particular, those used to isolate and culture microorganisms.

b. Procaryotic cells differ from eucaryotic cells in their lack of a membrane-delimited nucleus and other ways as well.

c. Microbiologists divide organisms into three domains: *Bacteria, Archaea,* and *Eucarya.*

d. Domains *Bacteria* and *Archaea* consist of procaryotic microorganisms. The eucaryotic microbes (protists and fungi) are placed in *Eucarya.* Viruses, viroids, and virusoids are acellular entities that are not placed in any of the domains but are classified by a separate system.

1.2 Scope and Relevance of Microbiology

a. Microbiology has contributed greatly to the fields of medicine, genetics, agriculture, food science, biochemistry, and molecular biology.

b. There are many fields in microbiology. These include the more applied disciplines such as medical, public health, industrial, and food and dairy microbiology. Microbial ecology, physiology, biochemistry, and genetics are examples of basic microbiological research fields.

1.3 Discovery of Microorganisms

a. Antony van Leeuwenhoek was the first person to extensively describe microorganisms.

1.4 Conflict over Spontaneous Generation

a. Experiments by Redi and others disproved the theory of spontaneous generation of larger organisms.

b. The spontaneous generation of microorganisms was disproved by Spallanzani, Pasteur, Tyndall, Cohn, and others.

1.5 Golden Age of Microbiology

a. Support for the germ theory of disease came from the work of Bassi, Pasteur, Koch, and others. Lister provided indirect evidence with his development of antiseptic surgery.

b. Koch's postulates are used to prove a direct relationship between a suspected pathogen and a disease.

c. Koch and his coworkers developed the techniques required to grow bacteria on solid media and to isolate pure cultures of pathogens.

d. Vaccines against anthrax and rabies were made by Pasteur; von Behring and Kitasato prepared antitoxins for diphtheria and tetanus.

e. Metchnikoff discovered some white blood cells could phagocytose and destroy bacterial pathogens.

1.6 Development of Industrial Microbiology and Microbial Ecology

a. Pasteur showed that fermentations were caused by microorganisms and that some microorganisms could live in the absence of oxygen.

b. The role of microorganisms in carbon, nitrogen, and sulfur cycles was first studied by Winogradsky and Beijerinck.

Critical Thinking Questions

1. Consider the impact of microbes on the course of world history. History is full of examples of instances or circumstances under which one group of people lost a struggle against another. In fact, when examined more closely, the "losers" often had the misfortune of being exposed to, more susceptible to, or unable to cope with an infectious agent. Thus weakened in physical strength or demoralized by the course of a devastating disease, they were easily overcome by human "conquerors."

   a. Choose an example of a battle or other human activity such as exploration of new territory and determine the impact of microorganisms, either indigenous or transported to the region, on that activity.

   b. Discuss the effect that the microbe(s) had on the outcome in your example.

   c. Suggest whether the advent of antibiotics, food storage and preparation technology, or sterilization technology would have made a difference in the outcome.

2. Vaccinations against various childhood diseases have contributed to the entry of women, particularly mothers, into the full-time workplace.

   a. Is this statement supported by data comparing availability and extent of vaccination with employment statistics in different places or at different times?

   b. Before vaccinations for measles, mumps, and chickenpox, what was the incubation time and duration of these childhood diseases? What impact would such diseases have on mothers with several elementary schoolchildren at home if they had full-time jobs and lacked substantial child care support?

   c. What would be the consequence if an entire generation of children (or a group of children in one country) was not vaccinated against any diseases? What do you predict would happen if these children went to college and lived in a dormitory in close proximity with others who had received all of the recommended childhood vaccines?

Learn More

Learn More by visiting the Prescott website at www.mhhe.com/prescottprinciples, where you will find a complete list of references.

Microscopes and the Study of Microbial Structure

2

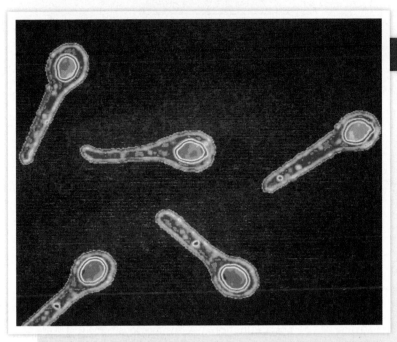

Clostridium tenani is a rod-shaped bacterium that forms endospores and releases tetanus toxin, the cause of tetanus. In this colorized phase-contrast micrograph, the endospores are the bright, oval objects located at the ends of the rods.

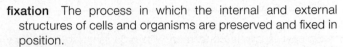
Chapter Glossary

bright-field microscope A microscope that illuminates the specimen directly with bright light and forms a dark image on a brighter background.

confocal scanning laser microscope (CSLM) A light microscope in which laser-derived light scans across the specimen at a specific level and illuminates one area at a time; stray light from other parts of the specimen is blocked out to give an image with excellent contrast and resolution.

dark-field microscope A microscope that brightly illuminates the specimen while leaving the background dark.

differential interference contrast (DIC) microscopy A type of microscopy that combines two beams of plane polarized light after passing through a specimen; their interference is used to create the image.

differential staining Staining procedures that divide microbes into separate groups based on staining properties.

fixation The process in which the internal and external structures of cells and organisms are preserved and fixed in position.

fluorescence microscopy A type of microscopy that exposes a specimen to light of a specific wavelength and then forms an image from the fluorescent light produced; usually the specimen is stained with a fluorescent dye (fluorochrome).

Gram stain A differential staining procedure that divides bacteria into gram-positive and gram-negative groups based on their ability to retain crystal violet when decolorized with an organic solvent such as ethanol.

negative staining A staining procedure in which a dye is used to make the background dark while the specimen is unstained.

parfocal A microscope that retains proper focus when the objectives are changed.

phase-contrast microscope A microscope that converts slight differences in refractive index and cell density into easily observed differences in light intensity.

refractive index A measure of how much a substance deflects a light ray from a straight path as it passes from one medium (e.g., glass) to another (e.g., air).

resolution The ability of a microscope to separate or distinguish between small objects that are close together.

scanning electron microscope (SEM) An electron microscope that scans a beam of electrons over the surface of a specimen and forms an image of the surface from the electrons that are emitted by it.

scanning probe microscope A microscope used to study surface features by moving a sharp probe over the object's surface (e.g., the scanning tunneling microscope).

simple staining Staining a specimen with a single dye.

transmission electron microscope (TEM) A microscope in which an image is formed by passing an electron beam through a specimen and focusing the scattered electrons with magnetic lenses.

There are more animals living in the scum on the teeth in a man's mouth than there are men in a whole kingdom.

—*Antony van Leeuwenhoek*

Table 2.1	Common Units of Measurement	
Unit	Abbreviation	Value
1 centimeter	cm	10^{-2} meter or 0.394 inches
1 millimeter	mm	10^{-3} meter
1 micrometer	μm	10^{-6} meter
1 nanometer	nm	10^{-9} meter
1 Angstrom	Å	10^{-10} meter

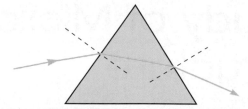

Figure 2.1   The Bending of Light by a Prism.   Normals (lines perpendicular to the surface of the prism) are indicated by dashed lines. As light enters the glass, it is bent toward the first normal. When light leaves the glass and returns to air, it is bent away from the second normal. As a result the prism bends light passing through it.

Microbiology usually is concerned with organisms so small they cannot be seen distinctly with the unaided eye. The organisms and entities studied by microbiologists typically range in size from viruses, which are measured in nanometers (nm), to protists, the largest of which are about 200 micrometers (μm) in diameter (**table 2.1**). Thus, the microscope is of crucial importance, and it is important to understand how microscopes work and how specimens are prepared for examination.

In this chapter we begin with a detailed treatment of the standard bright-field microscope and then describe other common types of light microscopes. Next we discuss preparation and staining of specimens for examination with the light microscope. This is followed by a description of transmission and scanning electron microscopes, both of which are used extensively in current microbiological research. We close the chapter with a brief introduction to two newer forms of microscopy: confocal microscopy and scanning probe microscopy.

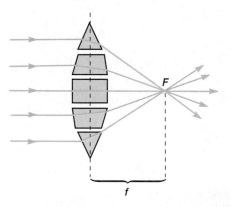

Figure 2.2   Lens Function.   A lens functions somewhat like a collection of prisms. Light rays from a distant source are focused at the focal point F. The focal point lies a distance f, the focal length, from the lens center.

2.1   LENSES AND THE BENDING OF LIGHT

To understand how a light microscope operates, one must know something about the way lenses bend and focus light to form images. When a ray of light passes from one medium to another, refraction occurs—that is, the ray is bent at the interface. The **refractive index** is a measure of how greatly a substance slows the velocity of light; the direction and magnitude of bending are determined by the refractive indices of the two media forming the interface. For example, when light passes from air into glass, a medium with a greater refractive index, it is slowed and bent toward the normal, a line perpendicular to the surface (**figure 2.1**). As light leaves glass and returns to air, a medium with a lower refractive index, it accelerates and is bent away from the normal. Thus a prism bends light because glass has a different refractive index from air, and the light strikes its surface at an angle.

Lenses act like a collection of prisms operating as a unit. When the light source is distant so that parallel rays of light strike the lens, a convex lens focuses these rays at a specific point, the focal point (F in **figure 2.2**). The distance between the center of the lens and the focal point is called the focal length (f in figure 2.2).

Our eyes cannot focus on objects nearer than about 25 cm or 10 inches. This limitation may be overcome by using a convex lens as a simple magnifier (or microscope) and holding it close to an object. A magnifying glass provides a clear image at much closer range, and the object appears larger. Lens strength is related to focal length; a lens with a short focal length magnifies an object more than a weaker lens having a longer focal length.

1. Define refraction, refractive index, focal point, and focal length.
2. Describe the path of a light ray through a prism or lens.
3. How is lens strength related to focal length? How do you think this principle is applied to corrective eyeglasses?

2.2   LIGHT MICROSCOPES

Microbiologists currently employ a variety of light microscopes in their work; bright-field, dark-field, phase-contrast, and fluorescence microscopes are most commonly used. Each is useful for certain applications. Modern microscopes are all

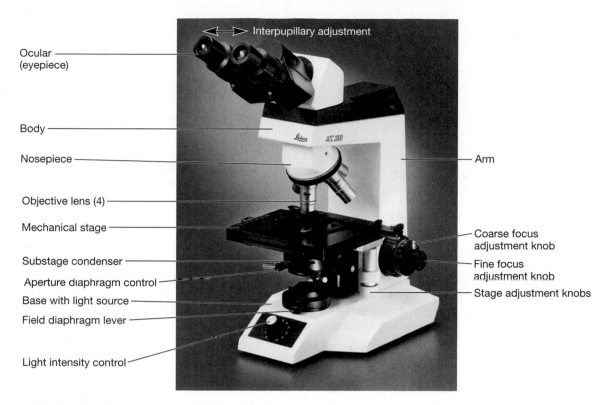

Ocular (eyepiece)

Body

Nosepiece

Objective lens (4)

Mechanical stage

Substage condenser

Aperture diaphragm control

Base with light source

Field diaphragm lever

Light intensity control

Interpupillary adjustment

Arm

Coarse focus adjustment knob

Fine focus adjustment knob

Stage adjustment knobs

Figure 2.3 A Bright-Field Microscope. The parts of a modern bright-field microscope. The microscope pictured is somewhat more sophisticated than those found in many student laboratories. For example, it is binocular (has two eyepieces) and has a mechanical stage, an adjustable substage condenser, and a built-in illuminator.

compound microscopes. That is, the magnified image formed by the **objective lens** (the lens closest to the object being examined) is further enlarged by one or more additional lenses.

Bright-Field Microscope

The **bright-field microscope** is routinely used in microbiology labs, where it can be employed to examine both stained and unstained specimens. It is called a bright-field microscope because it forms a dark image against a brighter background. It consists of a sturdy metal body or stand composed of a base and an arm to which the remaining parts are attached (**figure 2.3**). A light source, either a mirror or an electric illuminator, is located in the base. Two focusing knobs, the fine and coarse adjustment knobs, are located on the arm and can move either the stage or the nosepiece to focus the image.

The stage is positioned about halfway up the arm. It holds microscope slides either by simple slide clips or by a mechanical stage clip. A mechanical stage allows the operator to move a slide smoothly during viewing by use of stage control knobs. The **substage condenser** (or simply condenser) is mounted within or beneath the stage and focuses a cone of light on the slide. Its position often is fixed in simpler microscopes but can be adjusted vertically in more advanced models.

The curved upper part of the arm holds the body assembly, to which a nosepiece and one or more **eyepieces** or **ocular lenses** are attached. More advanced microscopes have eyepieces for both eyes and are called binocular microscopes. The body assembly itself contains a series of mirrors and prisms so that the barrel holding the eyepiece may be tilted for ease in viewing. The nosepiece holds three to five objective lenses of differing magnifying power and can be rotated to position any objective beneath the body assembly. Ideally a microscope should be **parfocal**—that is, the image should remain in focus when objectives are changed.

The image one sees when viewing a specimen with a compound microscope is created by the objective and ocular lenses working together. Light from the illuminated specimen is focused by the objective lens, creating an enlarged image within the microscope. The ocular lens further magnifies this primary image. The total magnification is calculated by multiplying the objective and eyepiece magnifications together. For example, if a 45× objective is used with a 10× eyepiece, the overall magnification of the specimen is 450×.

Microscope Resolution

The most important part of the microscope is the objective lens, which must produce a clear image, not just a magnified one. Thus resolution is extremely important. **Resolution** is the ability of a

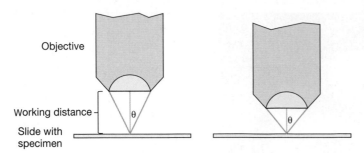

Figure 2.4   Numerical Aperture in Microscopy.   The angular aperture θ is 1/2 the angle of the cone of light that enters a lens from a speci-men, and the numerical aperture is *n* sin θ. In the right-hand illustration the lens has larger angular and numerical apertures; its resolution is greater and its working distance smaller.

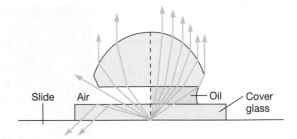

Figure 2.5   The Oil Immersion Objective.   An oil immersion objective operating in air and with immersion oil.

lens to separate or distinguish between small objects that are close together.

Resolution is described mathematically by an equation devel-oped in the 1870s by Ernst Abbé, a German physicist responsible for much of the optical theory underlying microscope design. The Abbé equation states that the minimal distance (*d*) between two objects that reveals them as separate entities depends on the wavelength of light (λ) used to illuminate the specimen and on the **numerical aperture** of the lens (*n* sin θ), which is the ability of the lens to gather light.

$$d = \frac{0.5\lambda}{n \sin \theta}$$

As *d* becomes smaller, the resolution increases, and finer detail can be discerned in a specimen; *d* becomes smaller as the wave-length of light used decreases and as the numerical aperture (NA) increases. Thus the greatest resolution is obtained using a lens with the largest possible NA and light of the shortest wavelength, light at the blue end of the visible spectrum (in the range of 450 to 500 nm; *see figure 7.24*).

The numerical aperture (*n* sin θ) of a lens is defined by two components: *n* is the refractive index of the medium in which the lens works (e.g., air) and θ is 1/2 the angle of the cone of light entering an objective (**figure 2.4**). When this cone has a

narrow angle and tapers to a sharp point, it does not spread out much after leaving the slide and therefore does not adequately separate images of closely packed objects. If the cone of light has a very wide angle and spreads out rapidly after passing through a specimen, closely packed objects appear widely sep-arated and are resolved. The angle of the cone of light that can enter a lens depends on the refractive index (*n*) of the medium in which the lens works, as well as on the objective itself. The refractive index for air is 1.00 and sin θ cannot be greater than 1 (the maximum θ is 90° and sin 90° is 1.00). Therefore no lens working in air can have a numerical aperture greater than 1.00. The only practical way to raise the numerical aperture above 1.00, and therefore achieve higher resolution, is to increase the refractive index with immersion oil, a colorless liquid with the same refractive index as glass (**table 2.2**). If air is replaced with immersion oil, many light rays that did not enter the objective due to reflection and refraction at the surfaces of the objective lens and slide will now do so (**figure 2.5**). This results in an increase in numerical aperture and resolution.

Numerical aperture is related to another characteristic of an objective lens, the working distance. The working distance of an objective is the distance between the front surface of the lens and the surface of the cover glass (if one is used) or the specimen when it is in sharp focus (figure 2.4). Objectives with large numerical apertures and great resolving power have short work-ing distances (table 2.2).

Table 2.2	The Properties of Microscope Objectives			
	Objective			
Property	**Scanning**	**Low Power**	**High Power**	**Oil Immersion**
Magnification	4×	10×	40–45×	90–100×
Numerical aperture	0.10	0.25	0.55–0.65	1.25–1.4
Approximate focal length (*f*)	40 mm	16 mm	4 mm	1.8–2.0 mm
Working distance	17–20 mm	4–8 mm	0.5–0.7 mm	0.1 mm
Approximate resolving power with light of 450 nm (blue light)	2.3 μm	0.9 μm	0.35 μm	0.18 μm

The preceding discussion has focused on the resolving power of the objective lens. The resolution of an entire microscope must take into account the numerical aperture of its condenser, as is evident from the following equation.

$$d_{\text{microscope}} = \frac{\lambda}{(\text{NA}_{\text{objective}} + \text{NA}_{\text{condenser}})}$$

The condenser is a large, light-gathering lens used to project a wide cone of light through the slide and into the objective lens. Most microscopes have a condenser with a numerical aperture between 1.2 and 1.4. However, the condenser's numerical aperture will not be much above about 0.9 unless the top of the condenser is oiled to the bottom of the slide. During routine microscope operation, the condenser usually is not oiled and this limits the overall resolution, even with an oil immersion objective.

Although the resolution of the microscope must consider both the condenser and the objective lens, in most cases the limit of resolution of a light microscope is calculated using the Abbé equation, which considers the objective lens only. The maximum theoretical resolving power of a microscope with an oil immersion objective (numerical aperture of 1.25) and blue-green light is approximately 0.2 μm.

$$d = \frac{(0.5)(530 \text{ nm})}{1.25} = 212 \text{ nm or } 0.2 \text{ μm}$$

At best, a bright-field microscope can distinguish between two dots about 0.2 μm apart (the same size as a very small bacterium). Thus, the vast majority of viruses cannot be examined with a light microscope.

Given the limit of resolution of a light microscope, the largest useful magnification—the level of magnification needed to increase the size of the smallest resolvable object to be visible with the light microscope—can be determined. Our eye can just detect a speck 0.2 mm in diameter, and consequently the useful limit of magnification is about 1,000 times the numerical aperture of the objective lens. Most standard microscopes come with 10× eyepieces and have an upper limit of about 1,000× with oil immersion. A 15× eyepiece may be used with good objectives to achieve a useful magnification of 1,500×. Any further magnification does not enable a person to see more detail. Indeed, a light microscope can be built to yield a final magnification of 10,000×, but it would simply be magnifying a blur. Only the electron microscope provides sufficient resolution to make higher magnifications useful.

Dark-Field Microscope

Although unpigmented living cells can be viewed in the bright-field microscope, they are not clearly visible because there is little difference in contrast between the cells and water. As discussed in section 2.3, one solution to this problem is to kill and stain cells before observation to increase contrast and create variations in color between cell structures. But what if an investigator must view living cells to observe a dynamic process such as movement

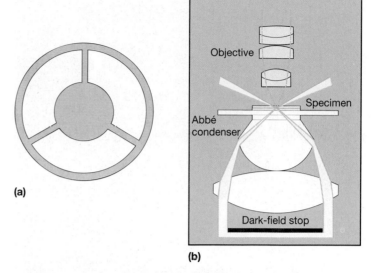

(a)

(b)

Figure 2.6 Dark-Field Microscopy. The simplest way to convert a microscope to dark-field microscopy is to place (a) a dark-field stop underneath (b) the condenser lens system. The condenser then produces a hollow cone of light so that the only light entering the objective comes from the specimen.

or phagocytosis? Three types of light microscopes create detailed, clear images of living specimens: the dark-field microscope, the phase-contrast microscope, and the differential interference contrast microscope.

The **dark-field microscope** produces much more detailed images of living, unstained cells and organisms by simply changing the way in which they are illuminated. A hollow cone of light is focused on the specimen in such a way that unreflected and unrefracted rays do not enter the objective. Only light that has been reflected or refracted by the specimen forms an image (**figure 2.6**). The field surrounding a specimen appears black, while the object itself is brightly illuminated (**figure 2.7a,b**). The dark-field microscope can reveal considerable internal structure in larger eucaryotic microorganisms (figure 2.7b). It also is used to identify certain bacteria such as the thin and distinctively shaped *Treponema pallidum* (figure 2.7a), the causative agent of syphilis.

Phase-Contrast Microscope

Dark-field microscopy is one solution to viewing unpigmented living cells. Phase-contrast microscopy is another. A **phase-contrast microscope** converts slight differences in refractive index and cell density into easily detected variations in light intensity. It is an excellent way to observe living cells (figure 2.7c–e).

The condenser of a phase-contrast microscope has an annular stop, an opaque disk with a thin transparent ring, which produces a hollow cone of light (**figure 2.8**). As this cone passes through a cell, some light rays are bent due to variations in density and refractive index within the specimen and are retarded by about 1/4 wavelength. The deviated light is focused to form an image of the

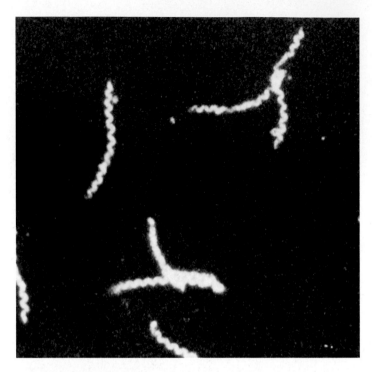

(a) *T. pallidum:* dark-field microscopy

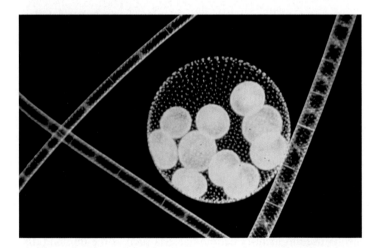

(b) *Volvox* and *Spirogyra:* dark-field microscopy

Figure 2.7 Examples of Dark-Field and Phase-Contrast Microscopy.   (a) *Treponema pallidum,* the spirochete that causes syphilis; dark-field microscopy. (b) *Volvox* and *Spirogyra;* dark-field microscopy (×175). Note daughter colonies within the mature *Volvox* colony (center) and the spiral chloroplasts of *Spirogyra* (left and right). (c) A phase-contrast micrograph of *Pseudomonas* cells, which range from 1–3 μm in length. (d) *Desulfotomaculum acetoxidans* with spherical endospores and oblong gas vacuoles; phase contrast (×2,000). (e) *Paramecium* stained to show a large central macronucleus with a small spherical micronucleus at its side; phase-contrast microscopy (×100).

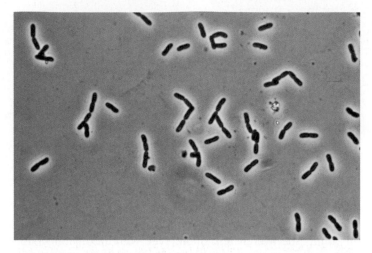

(c) *Pseudomonas:* phase-contrast microscopy

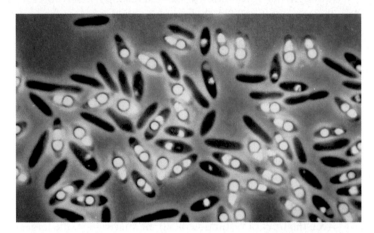

(d) *Desulfotomaculum:* phase-contrast microscopy using a color filter

Micronucleus   Macronucleus

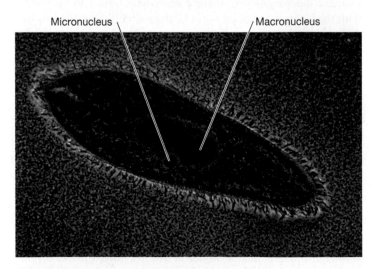

(e) *Paramecium:* phase-contrast microscopy with stained specimen

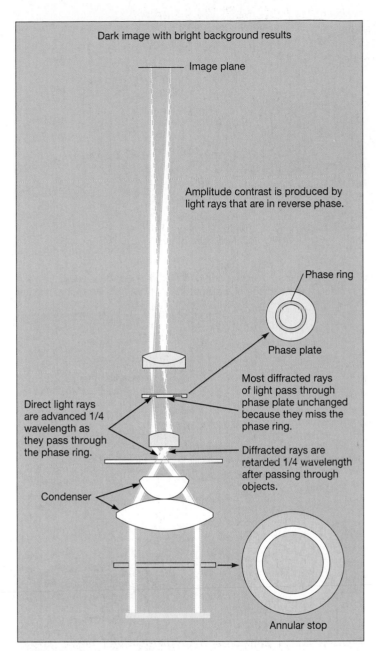

Figure 2.8 Phase-Contrast Microscopy. The optics of a dark-phase-contrast microscope.

Labels within figure:
Dark image with bright background results
Image plane
Amplitude contrast is produced by light rays that are in reverse phase.
Phase ring
Phase plate
Most diffracted rays of light pass through phase plate unchanged because they miss the phase ring.
Direct light rays are advanced 1/4 wavelength as they pass through the phase ring.
Diffracted rays are retarded 1/4 wavelength after passing through objects.
Condenser
Annular stop

and well defined. This type of microscopy is called dark-phase-contrast microscopy. Color filters often are used to improve the image (figure 2.7*d*).

Phase-contrast microscopy is especially useful for studying microbial motility, determining the shape of living cells, and detecting bacterial components such as endospores and inclusion bodies that contain poly-β-hydroxyalkanoates (e.g., poly-β-hydroxybutyrate), polymetaphosphate, sulfur, or other substances. These are clearly visible (figure 2.7*d*) because they have refractive indices markedly different from that of water. Phase-contrast microscopes also are widely used in studying eucaryotic cells. **>>** *Procaryotic cytoplasm: Inclusion bodies (section 3.3)*

Differential Interference Contrast Microscope

The **differential interference contrast (DIC) microscope** is similar to the phase-contrast microscope in that it creates an image by detecting differences in refractive indices and thickness. Two beams of plane-polarized light at right angles to each other are generated by prisms. In one design, the object beam passes through the specimen, while the reference beam passes through a clear area of the slide. After passing through the specimen, the two beams are combined and interfere with each other to form an image. A live, unstained specimen appears brightly colored and three-dimensional (**figure 2.10**). Structures such as cell walls, endospores, granules, vacuoles, and eucaryotic nuclei are clearly visible.

Fluorescence Microscope

The microscopes thus far considered produce an image from light that passes through a specimen. An object also can be seen because it actually emits light: this is the basis of fluorescence microscopy. When some molecules absorb radiant energy, they become excited and release much of their trapped energy as light. Any light emitted by an excited molecule will have a longer wavelength (or be of lower energy) than the radiation originally absorbed. **Fluorescent light** is emitted very quickly by the excited molecule as it gives up its trapped energy and returns to a more stable state.

The **fluorescence microscope** exposes a specimen to ultraviolet, violet, or blue light and forms an image of the object with the resulting fluorescent light. The most commonly used fluorescence microscopy is epifluorescence microscopy, also called incident light or reflected light fluorescence microscopy. Epifluorescence microscopes employ an objective lens that also acts as a condenser (**figure 2.11**). A mercury vapor arc lamp or other source produces an intense beam of light that passes through an exciter filter. The exciter filter transmits only the desired wavelength of excitation light. The excitation light is directed down the microscope by a special mirror called the dichromatic mirror. This mirror reflects light of shorter wavelengths (i.e., the excitation light) but allows light of longer

object. Undeviated light rays strike a phase ring in the phase plate, a special optical disk located in the objective, while the deviated rays miss the ring and pass through the rest of the plate. If the phase ring is constructed in such a way that the undeviated light passing through it is advanced by 1/4 wavelength, the deviated and undeviated waves will be about 1/2 wavelength out of phase and will cancel each other when they come together to form an image (**figure 2.9**). The background, formed by undeviated light, is bright, while the unstained object appears dark

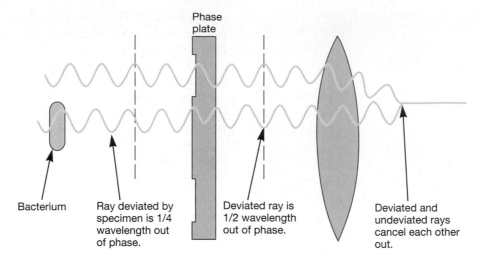

Figure 2.9   The Production of Contrast in Phase Microscopy.   The behavior of deviated and undeviated or undiffracted light rays in the dark-phase-contrast microscope. Because the light rays tend to cancel each other out, the image of the specimen will be dark against a brighter background.

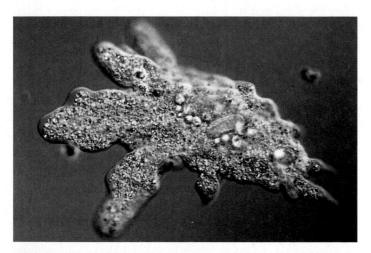

Figure 2.10   Differential Interference Contrast Microscopy. A micrograph of the protozoan *Amoeba proteus*. The three-dimensional image contains considerable detail and is artificially colored (×160).

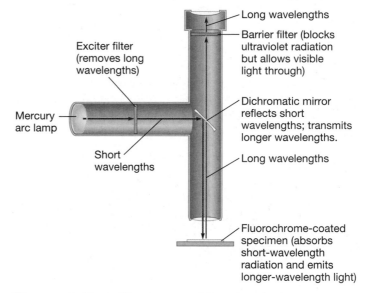

Figure 2.11   Epifluorescence Microscopy.   The principles of operation of an epifluorescence microscope.

wavelengths to pass through. The excitation light continues down, passing through the objective lens to the specimen, which is usually stained with molecules called **fluorochromes** (**table 2.3**). The fluorochrome absorbs light energy from the excitation light and fluoresces brightly. The emitted fluorescent light travels up through the objective lens into the microscope. Because the emitted fluorescent light has a longer wavelength, it passes through the dichromatic mirror to a barrier filter, which blocks out any residual excitation light. Finally, the emitted light passes through the barrier filter to the eyepieces.

The fluorescence microscope has become an essential tool in microbiology. Bacterial pathogens (e.g., *Mycobacterium tuberculosis,* the cause of tuberculosis) can be identified after staining with fluorochromes or specifically tagging them with

fluorescently labeled antibodies using immunofluorescence procedures. In ecological studies, the fluorescence microscope is used to observe microorganisms stained with fluorochrome-labeled probes or fluorochromes that bind specific cell constituents (table 2.3). In addition, microbial ecologists use epifluorescence microscopy to visualize photosynthetic microbes, as their pigments naturally fluoresce when excited by light of specific wavelengths. It is even possible to distinguish live bacteria from dead bacteria by the color they fluoresce after treatment with a specific mixture of stains (**figure 2.12a**). Thus the microorganisms can be viewed and directly counted in a relatively undisturbed ecological niche. >> *Identification of microorganisms from specimens: Microscopy (section 32.2)*

Table 2.3	Commonly Used Fluorochromes
Fluorochrome	Uses
Acridine orange	Stains DNA; fluoresces orange
Diamidino-2-phenyl indole (DAPI)	Stains DNA; fluoresces green
Fluorescein isothiocyanate (FITC)	Often attached to antibodies that bind specific cellular components or to DNA probes; fluoresces green
Tetramethyl rhodamine isothiocyanate (TRITC or rhodamine)	Often attached to antibodies that bind specific cellular components; fluoresces red

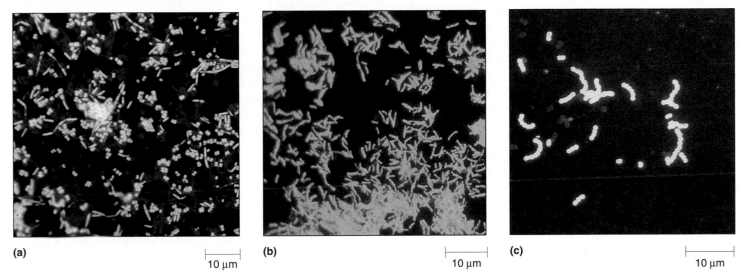

(a)   10 μm   (b)   10 μm   (c)   10 μm

Figure 2.12 Fluorescent Dyes and Tags.   (a) Dyes that cause live cells to fluoresce green and dead ones red; (b) Auramine is used to stain *Myco-bacterium* species in a modification of the acid-fast technique; (c) Fluorescent antibodies tag specific molecules. In this case, the antibody binds to a molecule that is unique to *Streptococcus pyogenes*.

1. List the parts of a light microscope and describe their functions.

2. Define resolution, numerical aperture, working distance, and fluorochrome.

3. If a specimen is viewed using a 5× objective in a microscope with a 15× eyepiece, how many times has the image been magnified?

4. How does resolution depend on the wavelength of light, refractive index, and numerical aperture? How are resolution and magnification related?

5. What is the function of immersion oil?

6. Why don't most light microscopes use 30× ocular lenses for greater magnification?

7. Briefly describe how dark-field, phase-contrast, differential interference contrast, and epifluorescence microscopes work and the kind of image provided by each. Give a specific use for each type.

2.3 PREPARATION AND STAINING OF SPECIMENS

Although living microorganisms can be directly examined with the light microscope, they often are fixed and stained. Such preparation serves to increase the visibility of the microorganisms, accentuate specific morphological features, and preserve them for future study.

Fixation

The stained cells seen in a microscope should resemble living cells as closely as possible. **Fixation** is the process by which the internal and external structures of cells and microorganisms are preserved and fixed in position. It inactivates enzymes that might disrupt cell morphology and toughens cell structures so that they do not change during staining and observation. A microorganism usually is killed and attached firmly to the microscope slide during fixation.

There are two fundamentally different types of fixation. **Heat fixation** is routinely used to observe procaryotes. Typically, a film of cells (a smear) is gently heated as a slide is passed through a flame. Heat fixation preserves overall morphology but not structures within cells. **Chemical fixation** is used to protect fine cellular substructure and the morphology of larger, more delicate microorganisms. Chemical fixatives penetrate cells and react with cellular components, usually proteins and lipids, to render them inactive, insoluble, and immobile. Common fixative mixtures contain such components as ethanol, acetic acid, mercuric chloride, formaldehyde, and glutaraldehyde.

Dyes and Simple Staining

The many types of dyes used to stain microorganisms have two features in common: they have **chromophore groups,** groups with conjugated double bonds that give the dye its color, and they can bind with cells by ionic, covalent, or hydrophobic bonding. Most dyes are used to directly stain the cell or object of interest, but some dyes (e.g., India ink and nigrosin) are used in **negative staining,** where the background but not the cell is stained; the unstained cells appear as bright objects against a dark background.

Dyes that bind cells by ionic interactions are probably the most commonly used dyes. These ionizable dyes may be divided into two general classes based on the nature of their charged group.

1. **Basic dyes**—methylene blue, basic fuchsin, crystal violet, safranin, malachite green—have positively charged groups (usually some form of pentavalent nitrogen) and are generally sold as chloride salts. Basic dyes bind to negatively charged molecules such as nucleic acids, many proteins, and the surfaces of procaryotic cells.

2. **Acidic dyes**—eosin, rose bengal, and acid fuchsin—possess negatively charged groups such as carboxyls (—COOH) and phenolic hydroxyls (—OH). Acidic dyes, because of their negative charge, bind to positively charged cell structures.

The staining effectiveness of ionizable dyes may be altered by pH, since the nature and degree of the charge on cell components change with pH. Thus acidic dyes stain best under acidic conditions when proteins and many other molecules carry a positive charge; basic dyes are most effective at higher pHs.

Dyes that bind through covalent bonds or because of their solubility characteristics are also useful. For instance, DNA can be stained by the Feulgen procedure in which the staining compound (Schiff's reagent) is covalently attached to its deoxyribose sugars. Sudan III (Sudan Black) selectively stains lipids because it is lipid soluble but does not dissolve in aqueous portions of the cell.

Microorganisms often can be stained very satisfactorily by **simple staining,** in which a single dye is used (**figure 2.13a,b**).

Simple staining's value lies in its simplicity and ease of use. One covers the fixed smear with stain for a short period, washes the excess stain off with water, and blots the slide dry. Basic dyes such as crystal violet, methylene blue, and carbolfuchsin are frequently used in simple staining to determine the size, shape, and arrangement of procaryotic cells.

Differential Staining

The **Gram stain,** developed in 1884 by the Danish physician Christian Gram, is the most widely employed staining method in bacteriology. It is an example of **differential staining**—procedures that are used to distinguish organisms based on their staining properties. Use of the Gram stain divides *Bacteria* into two groups—gram negative and gram positive.

The Gram-staining procedure is illustrated in **figure 2.14**. In the first step, the smear is stained with the basic dye crystal violet, the primary stain. This is followed by treatment with an iodine solution functioning as a **mordant,** a substance that helps bind the dye to a cell. Iodine increases the interaction between the cell and the dye so that the cell is stained more strongly. The smear is next decolorized by washing with ethanol or acetone. This step generates the differential aspect of the Gram stain; gram-positive bacteria retain the crystal violet, whereas gram-negative bacteria lose the crystal violet and become colorless. Finally, the smear is counterstained with a simple, basic dye different in color from crystal violet. Safranin, the most common counterstain, colors gram-negative bacteria pink to red and leaves gram-positive bacteria dark purple (figures 2.13c and 2.14b). >> *Bacterial cell walls (section 3.4)*

Acid-fast staining is another important differential staining procedure. It is most commonly used to identify *Mycobacterium tuberculosis* and *M. leprae* (figure 2.13d), the pathogens responsible for tuberculosis and leprosy, respectively. These bacteria have cell walls with high lipid content, in particular mycolic acids—a group of branched-chain hydroxy lipids, which prevent dyes from readily binding to the cells. However, *M. tuberculosis* and *M. leprae* can be stained by harsh procedures such as the Ziehl-Neelsen method, which uses heat and phenol to drive basic fuchsin into cells. Once basic fuchsin has penetrated, *M. tuberculosis* and *M. leprae* are not easily decolorized by acidified alcohol (acid-alcohol) and thus are said to be acid-fast. Non-acid-fast bacteria are decolorized by acid-alcohol and thus are stained blue by methylene blue counterstain.

Staining Specific Structures

Many special staining procedures have been developed to study specific structures with the light microscope. **Endospore staining,** like acid-fast staining, also requires harsh treatment to drive dye into a target, in this case an endospore. An endospore is an exceptionally resistant structure produced by some bacterial genera (e.g., *Bacillus* and *Clostridium*). It can survive for

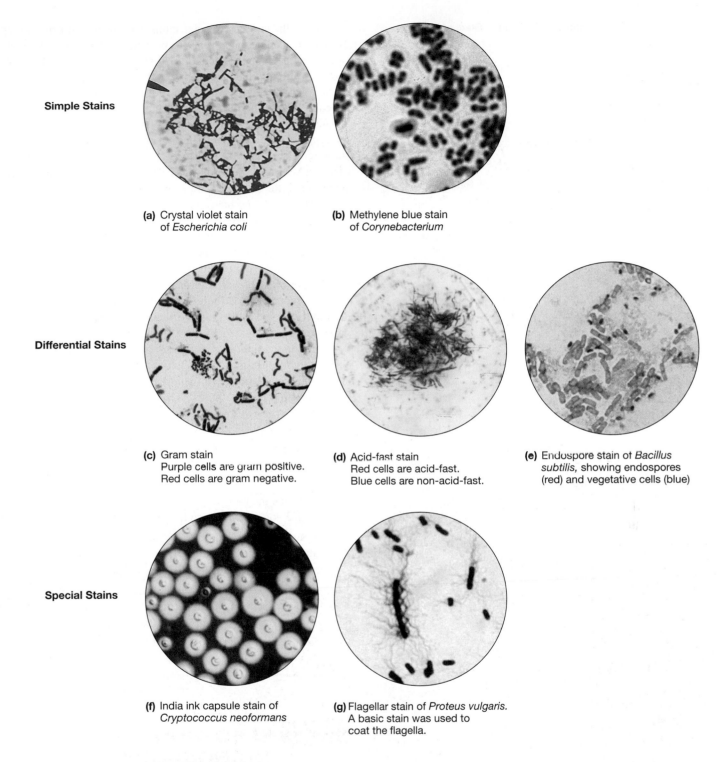

Simple Stains

(a) Crystal violet stain of *Escherichia coli*

(b) Methylene blue stain of *Corynebacterium*

Differential Stains

(c) Gram stain
Purple cells are gram positive.
Red cells are gram negative.

(d) Acid-fast stain
Red cells are acid-fast.
Blue cells are non-acid-fast.

(e) Endospore stain of *Bacillus subtilis,* showing endospores (red) and vegetative cells (blue)

Special Stains

(f) India ink capsule stain of *Cryptococcus neoformans*

(g) Flagellar stain of *Proteus vulgaris.*
A basic stain was used to coat the flagella.

Figure 2.13 Types of Microbiological Stains.

long periods in an unfavorable environment, and it is called an endospore because it develops within the parent bacterial cell. Endospore morphology and location vary with species and often are valuable in identification; endospores may be spherical to elliptical and either smaller or larger than the diameter of the parent bacterium. Endospores are not stained well by most dyes, but once stained they strongly resist decolorization. This property is the basis of most endospore staining methods (figure 2.13e). In the Schaeffer-Fulton procedure, endospores are first stained by heating bacteria with malachite green, which is a very strong stain

Steps in Staining	State of Bacteria
Step 1: Crystal violet (primary stain)	Cells stain purple.
Step 2: Iodine (mordant)	Cells remain purple.
Step 3: Alcohol (decolorizer)	Gram-positive cells remain purple; Gram-negative cells become colorless.
Step 4: Safranin (counterstain)	Gram-positive cells remain purple; Gram-negative cells appear red.

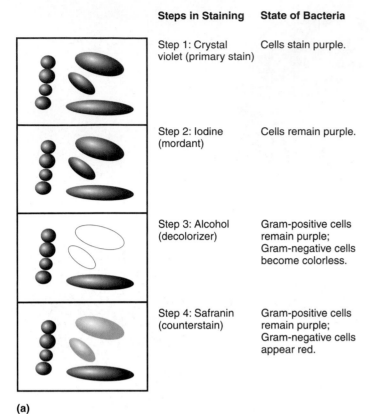

(a)

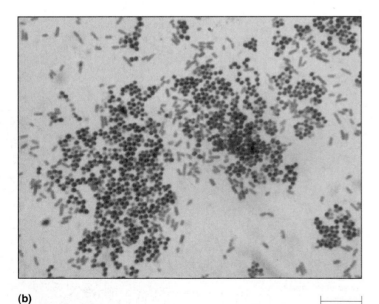

(b)

|———————|
10 µm

Figure 2.14   Gram Stain.   (a) Steps in the Gram stain procedure. (b) Results of a Gram stain. The gram-positive cells (purple) are *Staphylococcus aureus;* the gram-negative cells (reddish-pink) are *Escherichia coli.*

that can penetrate endospores. After malachite green treatment, the rest of the cell is washed free of dye with water and is counterstained with safranin. This technique yields a green endospore resting in a pink to red cell.   >>   *Bacterial endospores (section 3.8); Class* Clostridia *(section 21.3); and Class* Bacilli *(section 21.4)*

One of the simplest special stains is **capsule staining** (figure 2.13*f*), a technique that reveals the presence of capsules, a network usually made of polysaccharides that surrounds many bacteria and some fungi. Cells are mixed with India ink or nigrosin dye and spread out in a thin film on a slide. After air-drying, the cells appear as lighter bodies in the midst of a blue-black background because ink and dye particles cannot penetrate either the cell or its capsule. Thus capsule staining is an example of negative staining. The extent of the light region is determined by the size of the capsule and of the cell itself. There is little distortion of cell shape, and the cell can be counterstained for even greater visibility.   >>   *Components external to the cell wall: Capsules and slime layers (section 3.6)*

Flagella staining provides taxonomically valuable information about the presence and distribution pattern of flagella on procaryotic cells (figure 2.13*g; see also figure 3.35*). Procaryotic flagella are fine, threadlike organelles of locomotion that are so slender (about 10 to 30 nm in diameter) they can only be seen directly using the electron microscope (although bundles of flagella can be visualized by dark-field microscopy). To observe flagella with the light microscope, their thickness is increased by coating them with mordants such as tannic acid and potassium alum, and then staining with pararosaniline (Leifson method) or basic fuchsin (Gray method).   >>   *Components external to the cell wall: Flagella (section 3.6)*

1. Define fixation, dye, chromophore, basic dye, acidic dye, simple staining, differential staining, mordant, negative staining, and acid-fast staining.

2. Describe the two general types of fixation. Which would you normally use for procaryotes? For protozoa?

3. Why would one expect basic dyes to be more effective under alkaline conditions?

4. Describe the Gram stain procedure and explain how it works. What step in the procedure could be omitted without losing the ability to distinguish between gram-positive and gram-negative bacteria? Why?

5. How would you visualize capsules, endospores, and flagella?

2.4   ELECTRON MICROSCOPY

For centuries the light microscope has been the most important instrument for studying microorganisms. However, even the best light microscopes have a resolution limit of about 0.2 µm, which greatly compromises their usefulness for detailed studies of many microorganisms. Viruses, for example, are too small to be seen with light microscopes. Procaryotes can be observed, but because they are usually only 1 to 2 µm in diameter, only their general shape and major morphological features are visible. The detailed

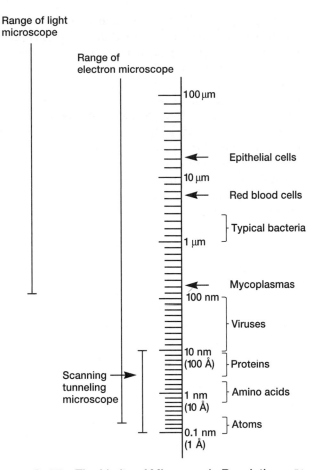

Figure 2.15 **The Limits of Microscopic Resolution.** Dimensions are indicated with a logarithmic scale (each major division represents a tenfold change in size). To the right side of the scale are the approximate sizes of cells, bacteria, viruses, molecules, and atoms.

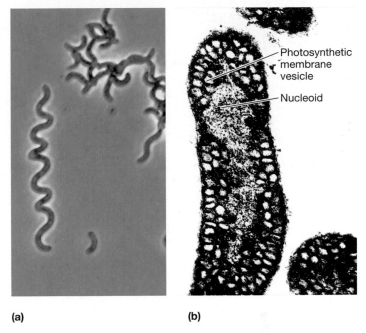

(a)   **(b)**

Figure 2.16 **Light and Electron Microscopy.** A comparison of light and electron microscopic resolution. (a) *Rhodospirillum rubrum* in phase-contrast light microscope (×600). (b) A thin section of *R. rubrum* in transmission electron microscope (×100,000).

internal structure of larger microorganisms also cannot be effectively studied by light microscopy. These limitations arise from the nature of visible light waves, not from any inadequacy of the light microscope itself. Electron microscopes have much greater resolution. They have transformed microbiology and added immeasurably to our knowledge. The nature of the electron microscope and the ways in which specimens are prepared for observation are reviewed briefly in this section.

Transmission Electron Microscope

Electron microscopes use a beam of electrons to illuminate and create magnified images of specimens. Recall that the resolution of a light microscope increases with a decrease in the wavelength of the light it uses for illumination. Electrons replace light as the illuminating beam. They can be focused, much as light is in a light microscope, but their wavelength is about 100,000 times shorter than that of visible light. Therefore electron microscopes have a practical resolution roughly 1,000 times better than the light microscope; with many electron microscopes, points closer than 0.5 nm can be distinguished, and the useful magnification is well over 100,000× (**figure 2.15**). The value of the

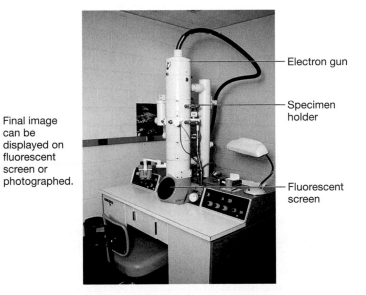

Figure 2.17 **A Transmission Electron Microscope.** The electron gun is at the top of the central column, and the magnetic lenses are within the column. The image on the fluorescent screen may be viewed through a magnifier positioned over the viewing window. The camera is in a compartment below the screen.

electron microscope is evident on comparison of the photographs in **figure 2.16**—microbial morphology can now be studied in great detail.

A modern **transmission electron microscope (TEM)** is complex, but the basic principles behind its operation can be readily understood (**figure 2.17**). A heated tungsten filament in

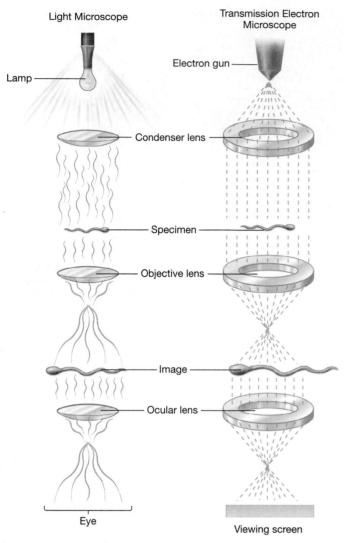

Figure 2.18 A Comparison of the Operation of Light and Transmission Electron Microscopes.

the electron gun generates a beam of electrons that is focused on the specimen by the condenser (**figure 2.18**). Since electrons cannot pass through a glass lens, doughnut-shaped electromagnets called magnetic lenses are used to focus the beam. The column containing the lenses and specimen must be under high vacuum to obtain a clear image because electrons are deflected by collisions with air molecules. The specimen scatters some electrons, but those that pass through are used to form an enlarged image of the specimen on a fluorescent screen. A denser region in the specimen scatters more electrons and therefore appears darker in the image since fewer electrons strike that area of the screen; these regions are said to be "electron dense." In contrast, electron-transparent regions are brighter. The image can be captured on photographic film as a permanent record.

Table 2.4 compares some of the important features of light and transmission electron microscopes. The TEM has distinctive features that place harsh restrictions on the nature of samples that can be viewed and the means by which those samples must be prepared. Since electrons are deflected by air molecules and are easily absorbed and scattered by solid matter, only extremely thin slices (20 to 100 nm) of a specimen can be viewed in the average TEM. Such a thin slice cannot be cut unless the specimen has support of some kind; the necessary support is provided by plastic. After fixation with chemicals such as glutaraldehyde or osmium tetroxide to stabilize cell structure, the specimen is dehydrated with organic solvents (e.g., acetone or ethanol). Complete dehydration is essential because most plastics used for embedding are not water soluble. Next the specimen is soaked in unpolymerized, liquid epoxy plastic until it is completely permeated, and then the plastic is hardened to form a solid block. Thin sections are cut from this block with a glass or diamond knife using a special instrument called an ultramicrotome.

As with bright-field microscopy, cells usually must be stained before they can be seen clearly. The probability of electron scattering is determined by the density (atomic number) of atoms in the specimen. Biological molecules are composed primarily of

Table 2.4	Characteristics of Light and Transmission Electron Microscopes	
Feature	**Light Microscope**	**Transmission Electron Microscope**
Highest practical magnification	About 1,000–1,500	Over 100,000
Best resolution[a]	0.2 μm	0.5 nm
Radiation source	Visible light	Electron beam
Medium of travel	Air	High vacuum
Type of lens	Glass	Electromagnet
Source of contrast	Differential light absorption	Scattering of electrons
Focusing mechanism	Adjust lens position mechanically	Adjust current to the magnetic lens
Method of changing magnification	Switch the objective lens or eyepiece	Adjust current to the magnetic lens
Specimen mount	Glass slide	Metal grid (usually copper)

[a]The resolution limit of a human eye is about 0.2 mm.

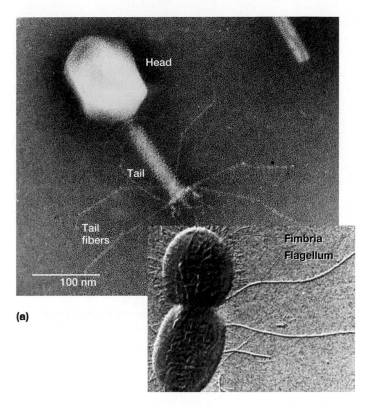

(a)

(b) *P. mirabilis*

Figure 2.19 Staining Microorganisms for the TEM. Examples of organisms viewed in the TEM after negative staining and shadowing. (a) Negatively stained T4, a virus that infects *E. coli*. (b) *Proteus mirabilis* after shadowing (×42,750); note flagella and fimbriae.

atoms with low atomic numbers (H, C, N, and O), and electron scattering is fairly constant throughout an unstained cell. Therefore specimens are prepared for observation by soaking thin sections with solutions of heavy metal salts such as lead citrate and uranyl acetate. The lead and uranium ions bind to cell structures and make them more electron opaque, thus increasing contrast in the material. Heavy osmium atoms from the osmium tetroxide fixative also stain cells and increase their contrast. The stained thin sections are then mounted on tiny copper grids and viewed.

Two other important techniques for preparing specimens are negative staining and shadowing. In negative staining, the specimen is spread out in a thin film with either phosphotungstic acid or uranyl acetate. Just as in negative staining for light microscopy, heavy metals do not penetrate the specimen but render the background dark, whereas the specimen appears bright in photographs. Negative staining is an excellent way to study the structure of viruses, bacterial gas vacuoles, and other similar objects (**figure 2.19a**). In shadowing, a specimen is coated with a thin film of platinum or other heavy metal by evaporation at an angle of about 45° from horizontal so that the metal strikes the microorganism on only one side. In one commonly used imaging method, the area coated with metal appears dark in photographs, whereas the uncoated side and the shadow region created by the object are light (figure 2.19b). This technique is particularly useful in studying virus morphology, procaryotic flagella, and DNA.

The shapes of organelles within cells can be observed by TEM if specimens are prepared by the freeze-etching procedure. When cells are rapidly frozen in liquid nitrogen, they become very brittle and can be broken along lines of greatest weakness, usually down the middle of internal membranes (**figure 2.20**). The exposed surfaces are then shadowed and coated with layers of platinum and carbon to form a replica of the surface. After the specimen has been removed chemically, this replica is studied in the TEM, providing a detailed, three-dimensional view of intracellular structure. An advantage of freeze-etching is that it minimizes the danger of artifacts because the cells are frozen quickly rather than being subjected to chemical fixation, dehydration, and plastic embedding.

Scanning Electron Microscope

Transmission electron microscopes form an image from radiation that has passed through a specimen. The **scanning electron microscope (SEM)** works in a different manner. It produces an image from electrons released from atoms on an object's surface. The SEM has been used to examine the surfaces of microorganisms in great detail; many SEMs have a resolution of 7 nm or less.

Specimen preparation for SEM is relatively easy, and in some cases, air-dried material can be examined directly. However, microorganisms usually must first be fixed, dehydrated, and dried to preserve surface structure and prevent collapse of the cells when they are exposed to the SEM's high vacuum. Before viewing, dried samples are mounted and coated with a

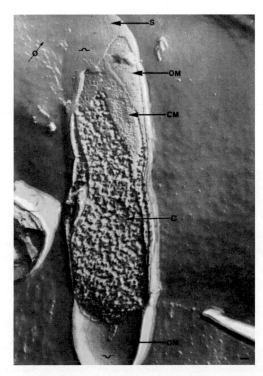

Figure 2.20 Example of Freeze-Etching. A freeze-etched preparation of the bacterium *Thiobacillus kabobis*. Note the differences in structure between the outer surface, S; the outer membrane of the cell wall, OM; the cytoplasmic membrane, CM; and the cytoplasm, C. Bar = 0.1 μm.

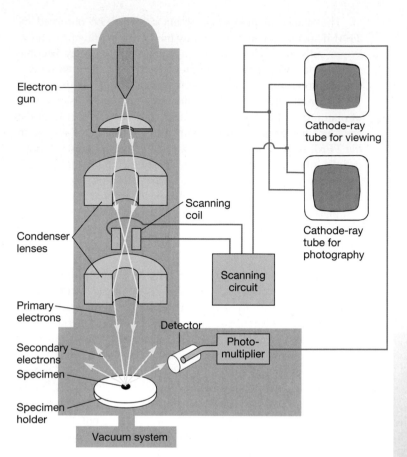

Figure 2.21   The Scanning Electron Microscope.

thin layer of metal to prevent the buildup of an electrical charge on the surface and to give a better image.

To create an image, the SEM scans a narrow, tapered electron beam back and forth over the specimen (**figure 2.21**). When the beam strikes a particular area, surface atoms discharge a tiny shower

of electrons called secondary electrons, and these are trapped by a detector. Secondary electrons entering the detector strike a scintillator, causing it to emit light flashes that a photomultiplier converts to an electrical current and amplifies. The signal is sent to a cathode-ray tube and produces an image like a television picture, which can be viewed or photographed.

The number of secondary electrons reaching the detector depends on the nature of the specimen's surface. When the electron beam strikes a raised area, a large number of secondary electrons enter the detector; in contrast, fewer electrons escape a depression in the surface and reach the detector. Thus raised areas appear lighter on the screen and depressions are darker. A realistic three-dimensional image of the microorganism's surface results (**figure 2.22**). The actual in situ location of microorganisms in ecological niches such as the human skin and the lining of the gut also can be examined.

1. Why does the transmission electron microscope have much greater resolution than the light microscope?

2. Describe in general terms how the TEM functions. Why must the TEM use high vacuum and very thin sections?

3. Material is often embedded in paraffin before sectioning for light microscopy. Why can't this approach be used when preparing a specimen for the TEM?

4. Under what circumstances would it be desirable to prepare specimens for the TEM by use of negative staining? Shadowing? Freeze-etching?

5. How does the scanning electron microscope operate and in what way does its function differ from that of the TEM? The SEM is used to study which aspects of morphology?

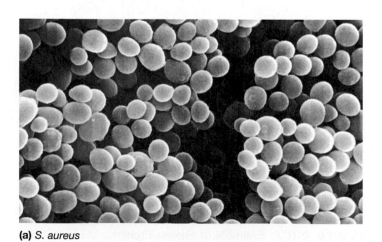

(a) *S. aureus*

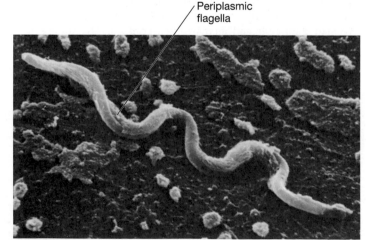

(b) *Cristispira*

Figure 2.22   Scanning Electron Micrographs of Bacteria.   (a) *Staphylococcus aureus* (×32,000). (b) *Cristispira*, a spirochete from the crystalline style of the oyster, *Ostrea virginica*. The axial fibrils or periplasmic flagella are visible around the protoplasmic cylinder (×6,000).

2.5 NEWER TECHNIQUES IN MICROSCOPY

Although the first microscopes were developed about 400 years ago, they continue to be improved and new types of microscopy continue to be developed. Two of the most important recent advances in microscopy are confocal microscopy and scanning probe microscopy. They are the focus of this section.

Confocal Microscopy

Like the large and small beads illustrated in **figure 2.23a**, biological specimens are three-dimensional. When three-dimensional objects are viewed with traditional light microscopes, light from all areas of the object, not just the plane of focus, enters the microscope and is used to create an image. The resulting image is murky and fuzzy (figure 2.23b). This problem has been solved by the development of the **confocal scanning laser microscope (CSLM),** or simply, confocal microscope. The confocal microscope uses a laser beam to illuminate a specimen, usually one that has been fluorescently stained. A major component of the confocal microscope is an aperture placed above the objective lens, which eliminates stray light from parts of the specimen that lie above and below the plane of focus (**figure 2.24**). Thus the only light used to create the image is from the plane of focus, and a much sharper image is formed.

Computers are integral to the process of creating confocal images. A computer interfaced with the confocal microscope receives digitized information from each plane in the specimen that is examined. This information can be used to create a composite image that is very clear and detailed (figure 2.23c) or to create a three-dimensional reconstruction of the specimen (figure 2.23d). Images of x-z plane cross sections of the specimen can also be generated, giving the observer views of the specimen from three perspectives (figure 2.23e). Confocal microscopy has numerous applications, including the study of biofilms, which can form on many different types of surfaces including indwelling medical devices such as hip joint replacements. As shown in figure 2.23f, it is difficult to kill all cells in a biofilm. This makes them a particular concern to the medical field because formation of biofilms on medical devices can result in infections that are difficult to treat. **>>** *Microbial growth in natural environments: Biofilms (section 7.6)*

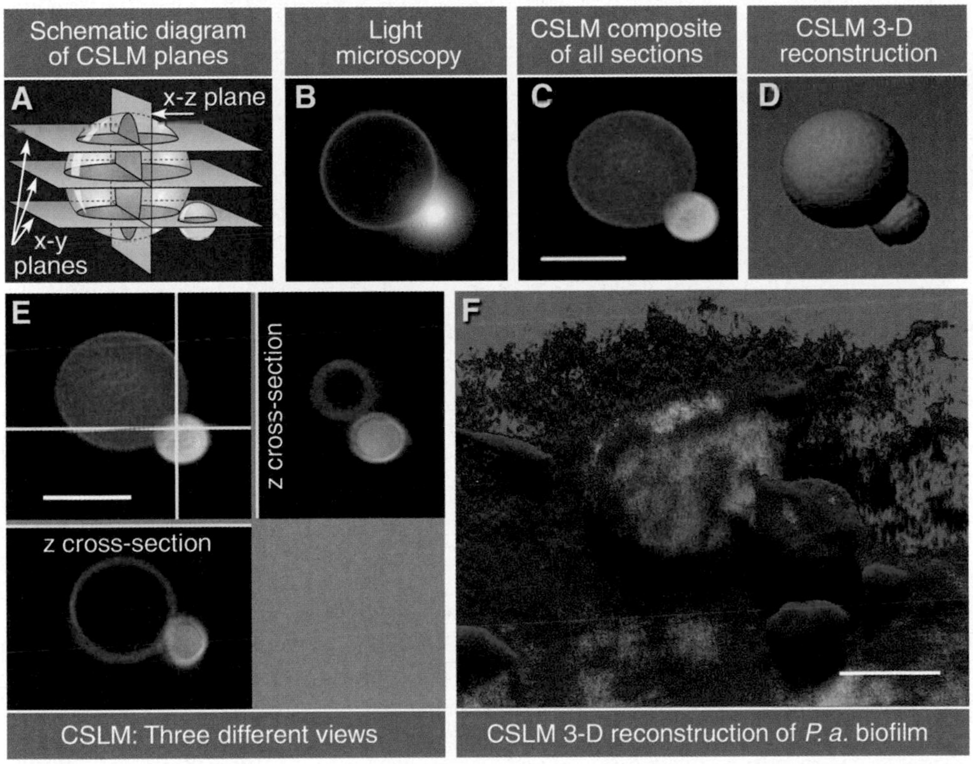

| Schematic diagram of CSLM planes | Light microscopy | CSLM composite of all sections | CSLM 3-D reconstruction |

A x-z plane / x-y planes
B
C
D
E z cross-section / z cross-section
F

CSLM: Three different views | CSLM 3-D reconstruction of *P. a.* biofilm

Figure 2.23 **Light and Confocal Microscopy.** Two beads examined by light and confocal microscopy. Light microscope images are generated from light emanating from many areas of a three-dimensional object. Confocal images are created from light emanating from only a single plane of focus. Multiple planes within the object can be examined and used to construct clear, finely detailed images. (a) The planes observable by confocal microscopy. (b) The light microscope image of the two beads shown in (a). Note that neither bead is clear and that the smaller bead is difficult to recognize as a bead. (c) A computer connected to a confocal microscope can make a composite image of the two beads using digitized information collected from multiple planes within the beads. The result is a much clearer and more detailed image. (d) The computer can also use digitized information collected from multiple planes within the beads to generate a three-dimensional reconstruction of the beads. (e) Computer generated views of a specimen: the top left panel is the image of a single x-y plane (i.e., looking down from the top of the specimen). The two lines represent the two x-z planes imaged in the other two panels. The vertical line indicates the x-z plane shown in the top right panel (i.e., a view from the right side of the specimen) and the horizontal line indicates the x-z plane shown in the bottom panel (i.e., a view from the front face of the specimen). (f) A three-dimensional reconstruction of a *Pseudomonas aeruginosa* biofilm. The biofilm was exposed to an antibacterial agent and then stained with dyes that distinguish living (green) from dead (red) cells. The cells on the surface of the biofilm have been killed, but those in the lower layers of the biofilm are still alive.

Scanning Probe Microscopy

Among the most powerful new microscopes are **scanning probe microscopes.** These microscopes measure surface features of an object by moving a sharp probe over the object's surface. The **scanning tunneling microscope** was invented in 1980. It can achieve magnifications of 100 million times, and it allows scientists to view atoms on the surface of a solid. The electrons surrounding surface

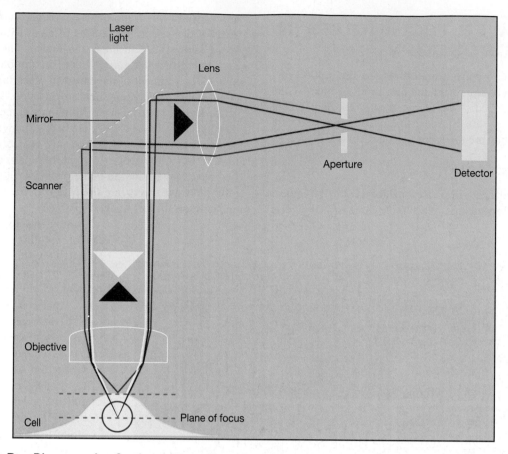

Figure 2.24 A Ray Diagram of a Confocal Laser Scanning Microscope. The yellow lines represent laser light used for illumination. Red lines symbolize the light arising from the plane of focus, and the blue lines stand for light from parts of the specimen above and below the focal plane.

atoms tunnel or project a very short distance out from the surface boundary. The scanning tunneling microscope has a needle-like probe with a point so sharp that often there is only one atom at its tip. The probe is lowered toward the specimen surface until its electron cloud just touches that of the surface atoms. If a small voltage is applied between the tip and specimen, electrons flow through a narrow channel in the electron clouds. This tunneling current, as it is called, is extraordinarily sensitive to distance and will decrease about a thousandfold if the probe is moved away from the surface by a distance equivalent to the diameter of an atom.

The arrangement of atoms on the specimen surface is determined by moving the probe tip back and forth over the surface while keeping the probe at a constant height above the specimen. As the tip moves up and down while following the surface contours, its motion is recorded and analyzed by a computer to create an accurate three-dimensional image of the surface atoms. The surface map can be displayed on a computer screen or plotted on paper. The resolution is so great that individual atoms are observed easily. Also because the microscope can examine objects when

they are immersed in water, it can be used to study biological molecules such as DNA (**figure 2.25**). The microscope's inventors, Gerd Binnig and Heinrich Rohrer, shared the 1986 Nobel Prize in Physics for their work, together with Ernst Ruska, the designer of the first transmission electron microscope.

More recently a second type of scanning probe microscope has been developed. The **atomic force microscope** moves a sharp probe over the specimen surface while keeping the distance between the probe tip and the surface constant. It does this by exerting a very small amount of force on the tip, just enough to maintain a constant distance but not enough force to damage the surface. The vertical motion of the tip usually is followed by measuring the deflection of a laser beam that strikes the lever holding the probe (**figure 2.26**). Unlike the scanning tunneling microscope, the atomic force microscope can be used to study surfaces that do not conduct electricity well. The atomic force microscope has been used to study the interactions of proteins, to follow the behavior of living bacteria and other cells, and to visualize membrane proteins such as aquaporins (**figure 2.27**).

1. How does a confocal microscope operate? Why does it provide better images of thick specimens than does the standard compound light microscope?

2. Briefly describe the scanning probe microscope and compare and contrast its most popular versions—the scanning tunneling microscope and the atomic force microscope. What are these microscopes used for?

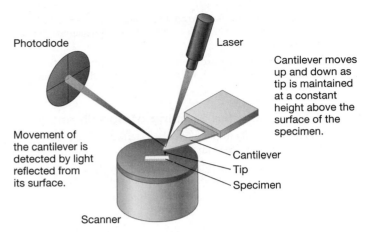

Figure 2.26 **The Basic Elements of an Atomic Force Microscope.** The tip used to probe the specimen is attached to a cantilever. As the probe passes over the "hills and valleys" of the specimen's surface, the cantilever is deflected vertically. A laser beam directed at the cantilever is used to monitor these vertical movements. Light reflected from the cantilever is detected by the photodiode and used to generate an image of the specimen.

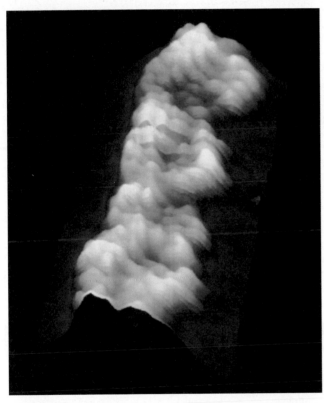

Figure 2.25 **Scanning Tunneling Microscopy of DNA.** The DNA double helix with approximately three turns shown (false color; ×2,000,000).

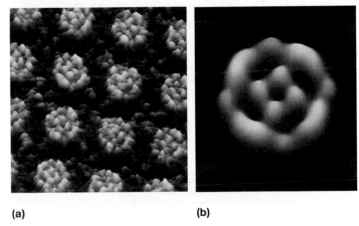

(a)　　　　　　　　　　　　　　**(b)**

Figure 2.27 **The Membrane Protein Aquaporin Visualized by Atomic Force Microscopy.** Aquaporin is a membrane-spanning protein that allows water to move across the membrane. (a) Each circular structure represents the surface view of a single aquaporin protein. (b) A single aquaporin molecule observed in more detail and at higher magnification.

Summary

2.1 Lenses and the Bending of Light

a. A light ray moving from air to glass or vice versa is bent in a process known as refraction.

b. Lenses focus light rays at a focal point and magnify images **(figure 2.2).**

2.2 Light Microscopes

a. In a compound microscope such as the bright-field microscope, the primary image is formed by an objective lens and enlarged by the eyepiece or ocular lens to yield the final image **(figure 2.3).** A substage condenser focuses a cone of light on the specimen.

b. Microscope resolution increases as the wavelength of radiation used to illuminate the specimen decreases. The maximum resolution of a light microscope is about 0.2 μm.

c. The dark-field microscope uses only refracted light to form an image, and objects glow against a black background **(figure 2.6).**

d. The phase-contrast microscope converts variations in the refractive index and density of cells into changes in light intensity and thus makes colorless, unstained cells visible **(figure 2.8).**

e. The differential interference contrast microscope uses two beams of light to create high-contrast, three-dimensional images of live specimens.

f. The fluorescence microscope illuminates a fluorochrome-labeled specimen and forms an image from its fluorescence **(figure 2.11)**.

2.3 Preparation and Staining of Specimens

a. Specimens usually must be fixed and stained before viewing them in the bright-field microscope.

b. Most dyes are either positively charged basic dyes or negatively charged acidic dyes that bind to ionized parts of cells.

c. In simple staining, a single dye is used to stain microorganisms.

d. Differential staining procedures such as the Gram stain and acid-fast stain distinguish between microbial groups by staining them differently **(figures 2.13c, d and 2.14)**.

e. Some staining techniques are specific for particular structures such as endospores, bacterial capsules, and flagella **(figure 2.13f, g)**.

2.4 Electron Microscopy

a. The transmission electron microscope uses magnetic lenses to form an image from electrons that have passed through a very thin section of a specimen **(figure 2.18)**. Resolution is high because the wavelength of electrons is very short.

b. Thin section contrast can be increased by treatment with solutions of heavy metals such as osmium tetroxide, uranium, and lead.

c. Specimens are also prepared for the TEM by negative staining, shadowing with metal, or freeze-etching.

d. The scanning electron microscope is used to study external surface features of microorganisms **(figure 2.21)**.

2.5 Newer Techniques in Microscopy

a. The confocal scanning laser microscope is used to study thick, complex specimens **(figures 2.23 and 2.24)**.

b. Scanning probe microscopes reach very high magnifications that allow scientists to observe biological molecules **(figures 2.25 and 2.27)**.

Critical Thinking Questions

1. If you prepared a sample of a specimen for light microscopy, stained it with the Gram stain, and failed to see anything when you looked through your light microscope, list the things that you may have done incorrectly.

2. In a journal article, find an example of a light micrograph, a scanning or transmission electron micrograph, or a confocal image. Discuss why the figure was included in the article and why that particular type of microscopy was the method of choice for the research. What other figures would you like to see used in this study? Outline the steps that the investigators would take to obtain such photographs or figures.

Learn More

Learn more by visiting the Prescott website at www.mhhe.com/prescottprinciples, where you will find a complete list of references.

Procaryotic Cell Structure and Function

Chapter Glossary

integral proteins buried in the lipid and peripheral proteins more loosely attached to the membrane surface.

gas vacuole A proteinaceous cytoplasmic organelle composed of clusters of gas-filled vesicles; used by aquatic procaryotes to change location in a water column.

glycocalyx A network of polysaccharides extending from the surface of procaryotes.

inclusion bodies Granules of organic or inorganic material in the cytoplasm of bacteria.

lipopolysaccharides (LPSs) Molecules containing both lipid and polysaccharide, which are located in the outer membrane of the cell wall of gram-negative bacteria.

nucleoid An irregularly shaped region in the procaryotic cell that consists of its genetic material.

peptidoglycan An essential component of the bacterial cell wall composed of long chains of alternating *N*-acetylglucosamine and *N*-acetylmuramic acid residues; chains are linked to each other by short chains of amino acids.

periplasmic space The space between the plasma membrane and outermost layers of the cell wall in procaryotes.

plasmid A double-stranded DNA molecule that can exist and replicate independently of the chromosome or may be integrated with it; it is not required for the cell's growth and reproduction.

porin proteins Proteins that form channels for the transport of small molecules across the outer membrane of gram-negative bacterial cell walls.

sex pili Thin protein appendages required for bacterial conjugation; cells with sex pili donate DNA to recipient cells.

S-layer A regularly structured layer composed of protein or glycoprotein that lies on the surface of many procaryotes.

slime layer A layer of diffuse, unorganized, easily removed material lying outside the cell wall.

spirillum A rigid, spiral-shaped bacterium.

spirochete A flexible, spiral-shaped bacterium with periplasmic flagella.

Although usually unicellular and structurally simple, some procaryotes can have complex life cycles. The myxobacteria are rod-shaped, gram-negative bacteria that exhibit gliding motility. In response to nutrient deprivation, many cells swarm together and form elaborate structures. The fruiting body of *Myxococcus stipitatus* is shown here.

bacillus A rod-shaped procaryote.

capsule A layer of well-organized material, not easily washed off, lying outside the bacterial cell wall.

cell envelope All the structures of a procaryote from the plasma membrane outward.

chemotaxis The movement of a microorganism toward chemical attractants and away from chemical repellents.

coccus A spherical procaryotic cell.

endospore An extremely heat- and chemical-resistant, dormant, thick-walled spore that develops within some gram-positive bacteria.

fimbriae Fine, hairlike protein appendages on some procaryotes; some help attach cells to surfaces and others are involved in a type of twitching motility.

fluid mosaic model The currently accepted model of cell membranes in which the membrane is a lipid bilayer with

The era in which workers tended to look at bacteria as very small bags of enzymes has long passed.

—*Howard J. Rogers*

Even a superficial examination of life on Earth shows that procaryotes are one of the most important groups by any criterion: numbers of organisms, general ecological importance, or practical importance for humans. Indeed, much of our understanding of phenomena in biochemistry and molecular biology comes from research on procaryotes. As mentioned in chapter 1, there are two quite different groups of procaryotes: *Bacteria* and *Archaea*. Although considerably less is known about archaeal cell structure and biochemistry, certain features distinguish the two domains. Whenever possible, these distinctions are noted. A more detailed discussion of the *Archaea* is provided in chapter 18. A comment about nomenclature is necessary to avoid confusion. The word procaryote is used in a general sense to include both *Bacteria* and *Archaea;* the term bacterium refers specifically to a member of the *Bacteria* and archaeon to a member of the *Archaea.* ◄◄ *Members of the microbial world (section 1.1)*

3.1 OVERVIEW OF PROCARYOTIC CELL STRUCTURE

Much of this chapter is devoted to a discussion of individual cell components. Therefore a preliminary overview of the procaryotic cell as a whole is in order.

Shape, Arrangement, and Size

One might expect that small, relatively simple organisms like procaryotes would be uniform in shape and size. This is not the case, as the microbial world offers almost endless variety in terms of morphology. However, the two most common shapes are cocci and rods (**figure 3.1**). **Cocci** (s., **coccus**) are roughly spherical cells. They can exist singly or can be associated in characteristic arrangements that are frequently useful in their identification. **Diplococci** (s., **diplococcus**) arise when cocci divide and remain together to form pairs. Long chains of cocci result when cells adhere after repeated divisions in one plane; this pattern is seen in the genera *Streptococcus, Enterococcus,* and *Lactococcus* (figure 3.1*a*). *Staphylococcus* divides in random planes to generate irregular, grapelike clumps (figure 3.1*b*). Divisions in two or three planes can produce symmetrical clusters of cocci. Members of the genus *Micrococcus* often divide in two planes to form square groups of four cells called tetrads. In the genus *Sarcina*, cocci divide in three planes, producing cubical packets of eight cells.

Bacillus megaterium is an example of a bacterium with a **rod** shape (figure 3.1*c*). Rods, sometimes called **bacilli** (s., **bacillus**), differ considerably in their length-to-width ratio, the coccobacilli being so short and wide that they resemble cocci. The shape of the rod's end often varies between species and may be flat, rounded, cigar-shaped, or bifurcated. Although many rods occur singly, some remain together after division to form pairs or chains (e.g., *Bacillus megaterium* is found in long chains).

Although procaryotes are most often simple spheres or rods, other cell shapes and arrangements are observed. **Vibrios** most closely resemble rods, as they are comma-shaped (**figure 3.2*a***). **Spirilla** are rigid, spiral-shaped procaryotes that usually have tufts of flagella at one or both ends of the cell (figure 3.2*b*). **Spirochetes** are flexible, spiral-shaped bacteria that have a unique, internal flagellar arrangement (figure 3.2*c*). Actinomycetes typically form long filaments called hyphae that may branch to produce a network called a **mycelium** (figure 3.2*d*). In this sense, they are similar to filamentous fungi, a group of eucaryotic microbes. The oval- to pear-shaped *Hyphomicrobium* (figure 3.2*e*) produces a bud at the end of a long hypha. A few procaryotes actually are flat. For example, square archaea living in salt ponds are shaped like flat, square-to-rectangular boxes about 2 μm by 2 to 4 μm and only 0.25 μm thick (figure 3.2*f*). The myxobacteria are of particular note. These bacteria sometimes aggregate to form complex structures called fruiting bodies (**chapter opener figure**). Finally,

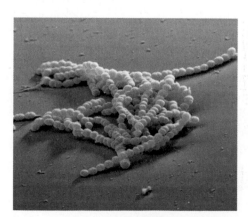

(a) *S. agalactiae*—cocci in chains

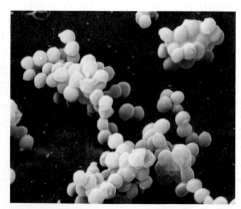

(b) *S. aureus*—cocci in clusters

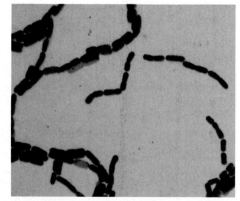

(c) *B. megaterium*—rods in chains

Figure 3.1   Cocci and Rods.   (a) *Streptococcus agalactiae,* the cause of Group B streptococcal infections; cocci arranged in chains; color-enhanced scanning electron micrograph (×4,800). (b) *Staphylococcus aureus* cocci arranged in clusters; color-enhanced scanning electron micrograph; average cell diameter is about 1 μm. (c) *Bacillus megaterium,* a rod-shaped bacterium arranged in chains, Gram stain (×600).

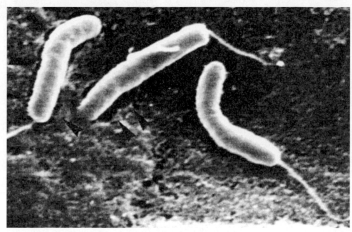

(a) *V. cholerae*—comma-shaped vibrios

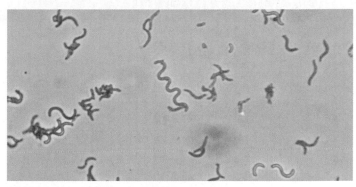

(b) *R. rubrum*—spiral-shaped spirilla

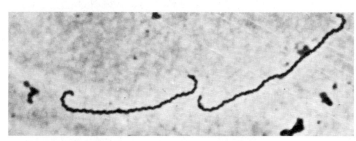

(c) *Leptospira interrogans*—a spirochete

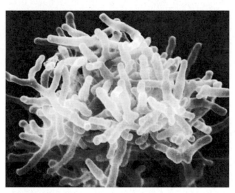

(d) *Actinomyces*—a filamentous bacterium

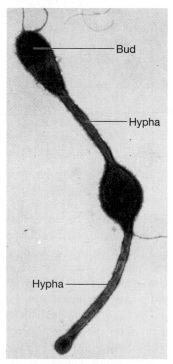

(e) *Hyphomicrobium*

(f) *Haloquadratum walsbyi,* a square archaeon

Figure 3.2 Other Procaryotic Cell Shapes. (a) *Vibrio cholerae,* curved rods with polar flagella; scanning electron micrograph. (b) *Rhodospirillum rubrum,* phase contrast (×500). (c) *Leptospira interrogans,* the spirochete that causes the waterborne disease leptospirosis (×2,200). (d) *Actinomyces,* SEM (×21,000). (e) *Hyphomicrobium* with hyphae and bud, electron micrograph with negative staining. (f) A square archaeon.

Microbial Diversity & Ecology

3.1 Monstrous Microbes

Biologists often have distinguished between procaryotes and eucaryotes based in part on cell size, usually procaryotic cells being the smaller of the two. Procaryotes grow extremely rapidly compared to most eucaryotes and lack the complex vesicular transport systems of eucaryotic cells described in chapter 4. Therefore it has been thought that procaryotes must be small so that the diffusion of nutrients into the cell can occur rapidly enough to support their metabolic activities. Thus in the 1980s, when Fishelson, Montgomery, and Myrberg discovered a large, cigar-shaped microorganism in the intestinal tract of the Red Sea brown surgeonfish, *Acanthurus nigrofuscus,* they suggested that it was a protist. It seemed too large to be anything else. However, in 1993, their suggestion was shown to be wrong, when Angert, Clemens, and Pace used rRNA sequence comparisons to show that the microorganism, now called *Epulopiscium fishelsoni,* is a procaryote related to the gram-positive bacterial genus *Clostridium.*

E. fishelsoni (Latin, *epulum,* a feast or banquet, and *piscium,* fish) can reach a size of 80 μm by 600 μm and normally ranges from 200 to 500 μm in length (**box figure**). It is about a million times larger in volume than *Escherichia coli.* Despite its huge size, the organism has a procaryotic cell structure. It is motile and swims about 2 body lengths a second (approximately 2.4 cm/min) using peritrichous flagella. The cytoplasm contains large nucleoids and many ribosomes, as would be required for such a large cell. *E. fishelsoni* appears to overcome the size limits set by diffusion by having a highly convoluted plasma membrane. This increases the cell's surface area and aids in nutrient transport.

In 1997, Schulz discovered an even larger procaryote in the ocean sediment off the coast of Namibia. *Thiomargarita namibiensis* is a spherical bacterium, between 100 and 750 μm in diameter, that often forms chains of cells enclosed in slime sheaths. It is over 100 times larger in volume than *E. fishelsoni.* A vacuole occupies about 98% of the cell and contains fluid rich in nitrate; it is surrounded by a 0.5 to 2.0 μm layer of cytoplasm filled with sulfur granules. The cytoplasmic layer is the same thickness as most bacteria and sufficiently thin for adequate diffusion rates. It uses sulfur as an energy source and nitrate as the electron acceptor for the electrons released when sulfur is oxidized in energy-conserving processes.

The discovery of these procaryotes greatly weakens the distinction between procaryotes and eucaryotes based on cell size. They are certainly larger than a normal eucaryotic cell. The size distinction between procaryotes and eucaryotes has been further weakened by the discovery of eucaryotic cells that are smaller than previously thought possible. The best example is *Nanochlorum eukaryotum.* It is only about 1 to 2 μm in diameter yet is truly eucaryotic and has a nucleus, a chloroplast, and a mitochondrion. Our understanding of the factors limiting procaryotic cell size must be reevaluated. It is no longer safe to assume that large cells are eucaryotic and small cells are procaryotic.

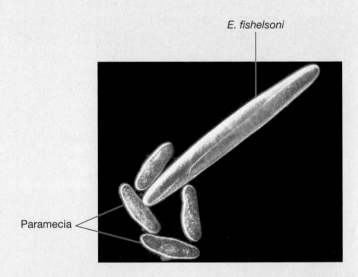

E. fishelsoni

Paramecia

Giant Bacteria. This photograph, taken with pseudo dark-field illumination, shows *Epulopiscium fishelsoni* at the top of the figure dwarfing the paramecia at the bottom (×200).

---

some procaryotes are variable in shape and lack a single, characteristic form. These are called **pleomorphic.** >> *Phylum* Spirochaetes *(section 19.6); Class* Alphaproteobacteria: *The* Caulobacteraceae *and* Hyphomicrobiaceae *(section 20.1); Class* Deltaproteobacteria: *Order* Mxyococcales *(section 20.4)*

Procaryotes vary in size as much as in shape (**figure 3.3**). *Escherichia coli* is a rod of about average size, 1.1 to 1.5 μm wide by 2.0 to 6.0 μm long. Near the small end of the size continuum are members of the genus *Mycoplasma,* an interesting group of bacteria that lack cell walls. For many years, it was thought that they were the smallest procaryotes at about 0.3 μm in diameter, approximately the size of the poxviruses. However, even smaller procaryotes have been discovered. Nanobacteria and nanoarchaea range from around 0.2 μm to less than 0.05 μm in diameter. Their discovery was quite surprising because theoretical calculations predicted that the smallest cells would be about 0.14 to 0.2 μm in diameter. At the other end of the continuum are bacteria such as the spirochetes, which can reach 500 μm in length, and the photosynthetic bacterium *Oscillatoria,* which is about 7 μm in diameter (the same diameter as a red blood cell). The huge bacterium *Epulopiscium fishelsoni* lives in the intestine of the brown surgeonfish, *Acanthurus nigrofuscus. E. fishelsoni* grows as large as 600 by 80 μm, a little smaller than a printed hyphen. An even larger bacterium, *Thiomargarita namibiensis,* has been discovered in ocean sediment (**Microbial Diversity & Ecology 3.1**). Thus a few bacteria are much larger than the average eucaryotic cell (typical plant and animal cells are around 10 to 50 μm in diameter).

Specimen	Approximate diameter or width × length in nm
Oscillatoria Red blood cell	7,000
E. coli	1,300 × 4,000
Streptococcus	800–1,000
Poxvirus	230 × 320
Influenza virus	85
T2 *E.coli* bacteriophage	65 × 95
Tobacco mosaic virus	15 × 300
Poliomyelitis virus	27

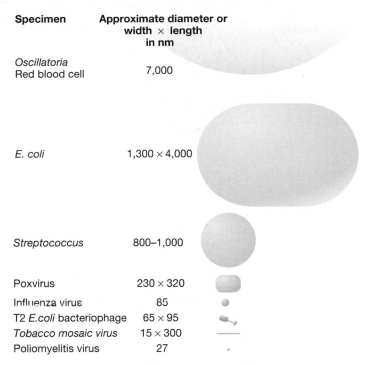

Figure 3.3 Sizes of Procaryotes and Viruses. The sizes of selected bacteria relative to a red blood cell and viruses. Recall that 1,000 nm = 1 μm. Thus *E. coli* is 1.3 × 4 μm.

Procaryotic Cell Organization

Procaryotic cells are morphologically simpler than eucaryotic cells, but they are not just simpler versions of eucaryotes. Although many structures are common to both cell types, some are unique to procaryotes. The major procaryotic structures and their functions are summarized and illustrated in **table 3.1** and **figure 3.4,** respectively. Note that no single procaryote possesses all of these structures at all times. Some are found only in certain cells in certain conditions or in certain phases of the life cycle. Despite these variations, procaryotes are consistent in their fundamental structure and most important components.

Procaryotic cells usually are bounded by a chemically complex cell wall, which covers the plasma membrane. The plasma membrane in turn surrounds the cytoplasm and its contents. Because most procaryotic cells do not contain internal, membrane-bound organelles, their interior appears morphologically simple. The genetic material is localized in a discrete region, the nucleoid, and usually is not separated from the surrounding cytoplasm by membranes. Ribosomes and larger masses called inclusion bodies are scattered about the cytoplasm. Many procaryotes use flagella for locomotion. In addition, many are surrounded by a capsule or slime layer external to the cell wall.

In the remaining sections of this chapter, we describe the major procaryotic structures in more detail. We begin with the plasma membrane, a structure that defines all cells. We then proceed inward to consider structures located within the cytoplasm. Then the discussion moves outward, first to the cell wall and then to structures outside the cell wall. Finally, we consider a structure unique to bacteria, the bacterial endospore.

1. What characteristic shapes can bacteria assume? Describe the ways in which bacterial cells cluster together.

2. What advantages might a microbial species that forms multicellular arrangements (e.g., clusters or chains) have that are not afforded unicellular microbes?

3. Draw a bacterial cell and label all important structures.

Table 3.1	Procaryotic Structures and Their Functions
Plasma membrane	Selectively permeable barrier, mechanical boundary of cell, nutrient and waste transport, location of many metabolic processes (respiration, photosynthesis), detection of environmental cues for chemotaxis
Gas vacuole	Buoyancy for floating in aquatic environments
Ribosomes	Protein synthesis
Inclusion bodies	Storage of carbon, phosphate, and other substances
Nucleoid	Localization of genetic material (DNA)
Periplasmic space	Contains hydrolytic enzymes and binding proteins for nutrient processing and uptake
Cell wall	Provides shape and protection from osmotic stress
Capsules and slime layers	Resistance to phagocytosis, adherence to surfaces
Fimbriae and pili	Attachment to surfaces, bacterial mating
Flagella	Swimming motility
Endospore	Survival under harsh environmental conditions

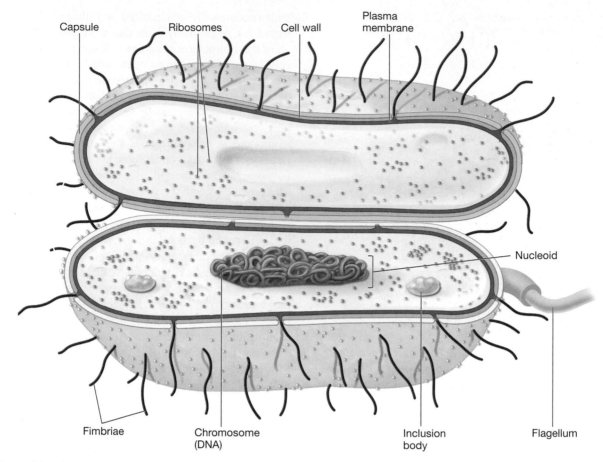

Figure 3.4   Morphology of a Procaryotic Cell.

3.2   PROCARYOTIC CELL MEMBRANES

Membranes are an absolute requirement for all living organisms. Cells must interact in a selective fashion with their environment, acquire nutrients, and eliminate waste. They also have to maintain their interior in a constant, highly organized state in the face of external changes.

The **plasma membrane** encompasses the cytoplasm of both procaryotic and eucaryotic cells. It is the chief point of contact with the cell's environment and thus is responsible for much of its relationship with the outside world. The plasma membranes of procaryotic cells are particularly important because they must fill an incredible variety of roles. In addition to retaining the cytoplasm, the plasma membrane also serves as a selectively permeable barrier: it allows particular ions and molecules to pass, either into or out of the cell, while preventing the movement of others. Thus the membrane prevents the loss of essential components through leakage while allowing the movement of other molecules. Because many substances cannot cross the plasma membrane without assistance, it must aid such movement when necessary. Transport systems are used for such tasks as nutrient uptake, waste excretion, and

protein secretion. The procaryotic plasma membrane also is the location of a variety of crucial metabolic processes: respiration, photosynthesis, and the synthesis of lipids and cell wall constituents. Finally, the membrane contains special receptor molecules that help procaryotes detect and respond to chemicals in their surroundings. Clearly the plasma membrane is essential to the survival of microorganisms. **>>** *Uptake of nutrients (section 6.6); Protein maturation and secretion (section 12.8)*

All membranes have a common, basic design. However, procaryotic membranes can differ dramatically in terms of the lipids they contain. Indeed, membrane chemistry can be used to identify particular procaryotic species. To understand these chemical differences and the many functions of the plasma membrane, it is necessary to become familiar with membrane structure. In this section, the fundamental design of all membranes is discussed. This is followed by a consideration of the significant differences between bacterial and archaeal membranes.

Fluid Mosaic Model of Membrane Structure

The most widely accepted model for membrane structure is the **fluid mosaic model** of Singer and Nicholson, which proposes that membranes are lipid bilayers within which proteins float

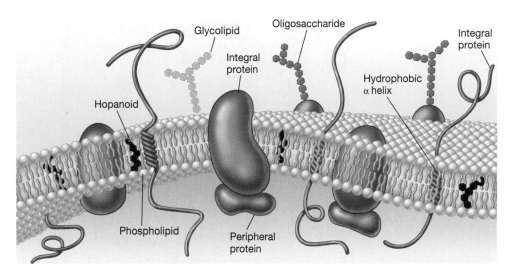

Figure 3.5   The Fluid Mosaic Model of Bacterial Membrane Structure.   This diagram shows the integral proteins (blue) floating in a lipid bilayer. Peripheral proteins (purple) are associated loosely with the inner membrane surface. Small spheres represent the hydrophilic ends of membrane phospholipids, and wiggly tails are the hydrophobic fatty acid chains. Other membrane lipids such as hopanoids (red) may be present. For the sake of clarity, phospholipids are shown in proportionately much larger size than in real membranes.

(figure 3.5). The model is based on studies of eucaryotic and bacterial membranes, and a variety of experimental approaches was used to establish it. Transmission electron microscopy (TEM) studies were particularly important. When cell membranes are stained and examined by TEM, they are revealed as very thin structures, about 5 to 10 nm thick, that appear as two dark lines on either side of a light interior. This characteristic appearance has been interpreted to mean that the membrane is organized in two sheets of molecules arranged end-to-end (figure 3.5). When membranes are cleaved by the freeze-etching technique, they can be split down the center of the lipid bilayer, exposing the complex internal structure. Within the lipid bilayer, small globular particles are visible; these have been suggested to be membrane proteins lying within the membrane lipid bilayer. The use of atomic force microscopy has provided powerful images to support this interpretation. << *Electron microscopy (section 2.4); Newer techniques in microscopy: Scanning probe microscopy (section 2.5)*

The chemical nature of membrane lipids is critical to their ability to form bilayers. Most membrane-associated lipids (e.g., the phospholipids shown in figure 3.5) are structurally asymmetric, with polar and nonpolar ends, and are called **amphipathic** (figure 3.6). The polar ends interact with water and are **hydrophilic;** the nonpolar **hydrophobic** ends are insoluble in water and tend to associate with one another. In aqueous environments, amphipathic lipids can interact to form a bilayer. The outer surfaces of the bilayer membrane are hydrophilic, whereas hydrophobic ends are buried in the interior away from the surrounding water (figure 3.5). >> *Lipids (appendix I)*

Two types of membrane proteins have been identified based on their ability to be separated from the membrane. **Peripheral proteins** are loosely connected to the membrane

and can be easily removed (figure 3.5). They are soluble in aqueous solutions and make up about 20 to 30% of total membrane protein. About 70 to 80% of membrane proteins are **integral proteins.** These are not easily extracted from membranes and are insoluble in aqueous solutions when freed of lipids. Integral proteins, like membrane lipids, are amphipathic; their hydrophobic regions are buried in the lipid while the hydrophilic portions project from the membrane surface (figure 3.5). Integral proteins can diffuse laterally in the membrane to new locations but do not flip-flop or rotate through the lipid layer. Carbohydrates often are attached to the outer surface of plasma membrane proteins, where they have important functions. >> *Proteins (appendix I)*

Bacterial Membranes

Bacterial membranes are similar to eucaryotic membranes in that many of their amphipathic lipids are phospholipids (figure 3.6), but they usually differ from eucaryotic membranes in lacking sterols (steroid-containing lipids) such as cholesterol (**figure 3.7a**). However, many bacterial membranes contain sterol-like molecules called hopanoids (figure 3.7b). **Hopanoids** are synthesized from the same precursors as steroids, and like the sterols in eucaryotic membranes, they probably stabilize the membrane. Hopanoids are also of interest to ecologists and geologists: the total mass of

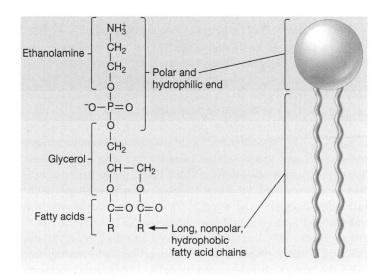

Figure 3.6   The Structure of a Polar Membrane Lipid.   Phosphatidylethanolamine, an amphipathic phospholipid often found in bacterial membranes. The R groups are long, nonpolar fatty acid chains.

(a) Cholesterol (a steroid) is found in eucaryotes

(b) A bacteriohopanetetrol (a hopanoid) is found in bacteria

Figure 3.7 **Membrane Steroids and Hopanoids.** Common examples.

hopanoids in sediments is estimated to be around $10^{11\text{-}12}$ tons—about as much as the total mass of organic carbon in all living organisms (10^{12} tons)—and evidence exists that hopanoids have contributed significantly to the formation of petroleum.

The emerging picture of bacterial plasma membranes is one of a highly organized and asymmetric system that also is flexible and dynamic. Numerous studies have demonstrated that lipids are not homogeneously distributed in the plasma membrane. Rather, there are domains in which particular lipids are concentrated. It has also been demonstrated that the lipid composition of bacterial membranes varies with environmental temperature in such a way that the membrane remains fluid during growth. For example, bacteria growing at lower temperatures have more unsaturated fatty acids in their membrane phospholipids—that is, there are one or more double covalent bonds in the long hydrocarbon chain. At higher temperatures, their phospholipids have more saturated fatty acids—those in which the carbon atoms are connected only with single covalent bonds. >> *Influences of environmental factors on growth: Temperature (section 7.5)*

Although procaryotes do not contain complex membranous organelles like mitochondria or chloroplasts, internal membranous structures are observed in some bacteria (**figure 3.8**). These can be extensive and complex in photosynthetic bacteria and in bacteria with very high respiratory activity, such as the nitrifying bacteria. The internal membranes of cyanobacteria are called thylakoids and are analogous to the thylakoids of chloroplasts. They contain chlorophyll and the photosynthetic reaction centers responsible for converting light energy into ATP, the energy currency used by cells. The internal membranous structures observed in bacteria may be aggregates of spherical vesicles, flattened vesicles, or tubular membranes. Their function may be to provide a larger membrane surface for greater metabolic activity. >> *Photosynthetic bacteria*

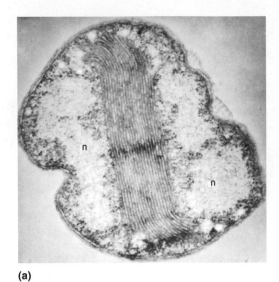

(a)

(b)

Figure 3.8 **Internal Bacterial Membranes.** Membranes of nitrifying and photosynthetic bacteria. (a) *Nitrocystis oceanus* with parallel membranes traversing the whole cell. Note nucleoid (n) with fibrillar structure. (b) *Ectothiorhodospira mobilis* with an extensive intracytoplasmic membrane system (×60,000).

(section 19.3); Class Alphaproteobacteria: *Nitrifiying bacteria (section 20.1); Organelles involved in energy conservation (section 4.6)*

Archaeal Membranes

One of the most distinctive features of the *Archaea* is the nature of their membrane lipids. They differ from both *Bacteria* and *Eucarya* in two ways. First, they contain hydrocarbons derived from isoprene units—five-carbon, branched molecules. Thus the hydrocarbons are branched as shown in **figure 3.9**. Second, the hydrocarbons are attached to glycerol by ether links rather than ester links. When two hydrocarbons are attached to glycerol, the lipids are called diether lipids. Usually the diether hydrocarbon chains are 20 carbons in length. Sometimes tetraether lipids are formed when two glycerol residues are linked by two long hydrocarbons that are 40 carbons in

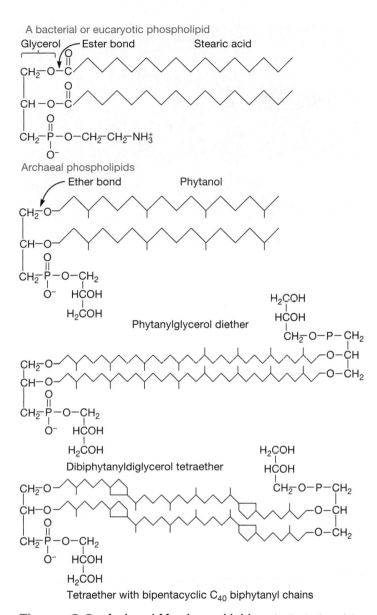

A bacterial or eucaryotic phospholipid

Archaeal phospholipids

Phytanylglycerol diether

Dibiphytanyldiglycerol tetraether

Tetraether with bipentacyclic C_{40} biphytanyl chains

Figure 3.9 Archaeal Membrane Lipids. An illustration of the difference between archaeal lipids and those of *Bacteria*. Archaeal lipids are derivatives of isoprenyl glycerol ethers rather than the glycerol fatty acid esters in *Bacteria*.

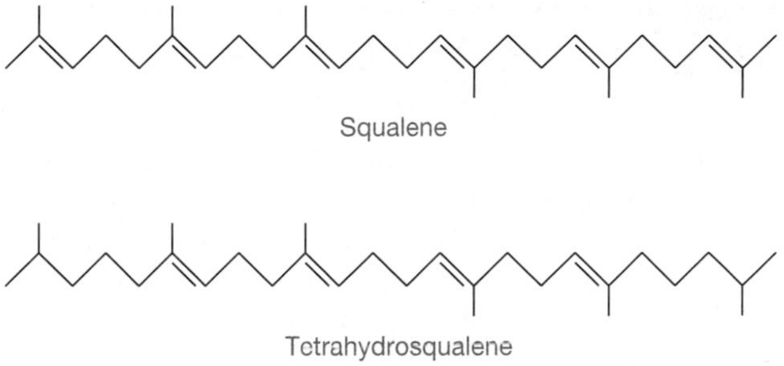

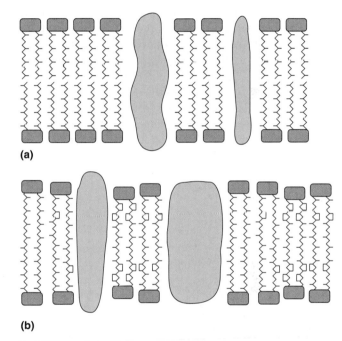

Figure 3.10 Nonpolar Lipids of *Archaea.* Two examples of the most predominant nonpolar lipids are the C_{30} isoprenoid squalene and one of its hydroisoprenoid derivatives, tetrahydrosqualene.

(a)

(b)

Figure 3.11 Examples of Archaeal Membranes. (a) A membrane composed of integral proteins and a bilayer of C_{20} diethers. (b) A rigid monolayer composed of integral proteins and C_{40} tetraethers.

length. Cells can adjust the overall length of the tetraethers by cyclizing the chains to form pentacyclic rings (figure 3.9). Phosphate-, sulfur-, and sugar-containing groups can be attached to the third carbons of the glycerol moieties in the diethers and tetraethers, making them polar lipids. Seventy to 93% of archaeal membrane lipids are polar. The remaining lipids are nonpolar and are usually derivatives of squalene (**figure 3.10**).

Despite these significant differences in membrane lipids, the basic design of archaeal membranes is similar to that of *Bacteria* and eucaryotes—there are two hydrophilic surfaces and a hydrophobic core. When C_{20} diethers are used, a regular bilayer membrane is formed (**figure 3.11a**). When the membrane is constructed of C_{40} tetraethers, a monolayer membrane with much more rigidity is formed (figure 3.11b). As might be expected

from their need for stability, the membranes of extreme thermophiles such as *Thermoplasma* and *Sulfolobus,* which grow best at temperatures over 85°C, are almost completely tetraether monolayers. Archaea that live in moderately hot environments have membranes containing some regions with monolayers and some with bilayers. ▶▶ *Phylum* Euryarchaeota: Thermoplasms *(section 18.3); Phylum* Crenarchaeota: Sulfolobus *(section 18.2)*

1. List the functions of the procaryotic plasma membrane.

2. Describe in words and with a labeled diagram the fluid mosaic model for cell membranes.

3. Compare and contrast bacterial and archaeal membranes.

4. Discuss the ways bacteria and archaea adjust the lipid content of their membranes in response to environmental conditions.

3.3 PROCARYOTIC CYTOPLASM

The **cytoplasm** is bounded by the plasma membrane and contains inclusion bodies, ribosomes, the nucleoid, and plasmids. It usually lacks membrane-delimited organelles and is largely water (about 70% of procaryotic mass is water). Until recently, it was thought to lack a cytoskeleton. The plasma membrane and everything within is called the **protoplast;** thus the cytoplasm is a major part of the protoplast.

Procaryotic Cytoskeleton

For many years it was thought that procaryotes lacked the high level of cytoplasmic organization present in eucaryotic cells because they lacked a cytoskeleton. Recently homologs of all three eucaryotic cytoskeletal elements (microfilaments, intermediate filaments, and microtubules) have been identified in bacteria, and two have been identified in archaea (**table 3.2**). The cytoskeletal filaments of procaryotes are structurally similar to their eucaryotic counterparts and carry out similar functions: they participate in cell division, localize proteins to certain sites in the cell, and determine cell shape (table 3.2 and **figure 3.12**). **>>** *Eucaryotic cytoplasm (section 4.3); Bacterial cell cycle: Cytokinesis (section 7.1)*

Inclusion Bodies

Inclusion bodies, granules of organic or inorganic material that often are clearly visible in a light microscope, are present in the cytoplasm. These bodies usually are used for storage (e.g., of carbon compounds, inorganic substances, and energy) or to reduce osmotic pressure by tying up molecules in particulate form. Some inclusion bodies lie free in the cytoplasm—for example, polyphosphate granules, cyanophycin granules, and some glycogen granules. Other inclusion bodies are enclosed by a shell about 2 to 4 nm thick, which is single-layered and may consist of proteins or a membranous structure composed of proteins and phospholipids. Examples of enclosed inclusion bodies are poly-β-hydroxybutyrate granules, some glycogen and sulfur granules, carboxysomes, and gas vacuoles. The quantity of inclusion bodies used for storage varies with the nutritional status of the cell. For example, polyphosphate granules are

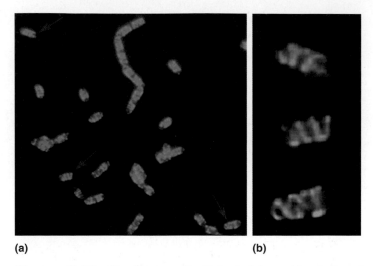

(a)   **(b)**

Figure 3.12   The Procaryotic Cytoskeleton. Visualization of the MreB-like cytoskeletal protein (Mbl) of *Bacillus subtilis*. The Mbl protein has been fused with green fluorescent protein, and live cells have been examined by fluorescence microscopy. (a) Arrows point to the helical cytoskeletal cables that extend the length of the cells. (b) Three of the cells from (a) are shown at a higher magnification.

depleted in freshwater habitats that are phosphate limited. A brief description of several important inclusion bodies follows.

Organic inclusion bodies usually contain either glycogen or poly-β-hydroxyalkanoates (e.g., poly-β-hydroxybutyrate). Glycogen is a polymer of glucose units composed of long chains formed by $\alpha(1\rightarrow4)$ glycosidic bonds and branching chains connected to them by $\alpha(1\rightarrow6)$ glycosidic bonds. **Poly-β-hydroxybutyrate (PHB)** contains β-hydroxybutyrate molecules joined by ester bonds between the carboxyl and hydroxyl groups of adjacent molecules. Usually only one of these polymers is found in a species, but some photosynthetic bacteria have both glycogen and PHB. PHB accumulates in distinct bodies, around 0.2 to 0.7 μm in diameter, that are readily stained with Sudan black for light microscopy and are seen as empty "holes" in the electron microscope (**figure 3.13a**). This is because the solvents used to prepare specimens for electron microscopy dissolve these hydrophobic inclusion bodies. Glycogen is dispersed more evenly throughout the cytoplasm as small granules (about 20 to 100 nm in diameter) and often can be seen only with the electron microscope. If cells contain a large amount of glycogen, staining with an iodine solution will

Table 3.2 Procaryotic Cytoskeletal Proteins		
Procaryotic Protein (Eucaryotic Counterpart)	**Function**	**Comments**
FtsZ (tubulin)	Cell division	Widely observed in *Bacteria* and *Archaea*
MreB (actin)	Cell shape	Observed in many rod-shaped bacteria and some archaea; in *Bacillus subtilis* is called Mbl
Crescentin (intermediate filament proteins)	Cell shape	Discovered in *Caulobacter crescentus*

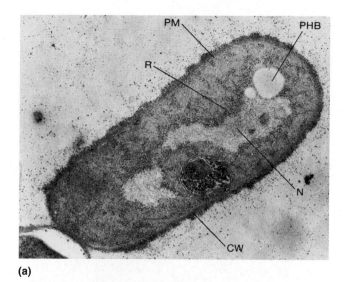

(a)

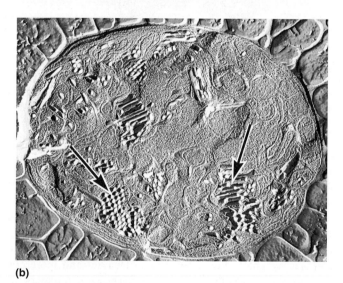

(b)

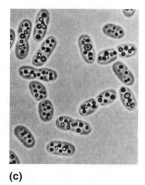

(c)

Figure 3.13 Inclusion Bodies in Bacteria. (a) Electron micrograph of *Bacillus megaterium* (×30,500). Poly-β-hydroxybutyrate inclusion body, PHB; cell wall, CW; nucleoid, N; plasma membrane, PM; and ribosomes, R. (b) A freeze-fracture preparation of *Anabaena flosaquae* (×89,000) showing gas vesicles and gas vacuoles. Clusters of the cigar-shaped vesicles form gas vacuoles. Both longitudinal and cross-sectional views of gas vesicles are indicated by arrows. (c) *Chromatium vinosum,* a purple sulfur bacterium, with intracellular sulfur granules, bright-field microscopy (×2,000).

turn them reddish-brown. Glycogen and PHB inclusion bodies are carbon storage reservoirs providing material for energy and biosynthesis. Many bacteria also store carbon as lipid droplets. >> *Carbohydrates (appendix I)*

Cyanobacteria, a group of photosynthetic bacteria, have two distinctive organic inclusion bodies. **Cyanophycin granules** are composed of large polypeptides containing approximately equal amounts of the amino acids arginine and aspartic acid. The granules often are large enough to be visible in the light microscope and store extra nitrogen for the bacteria. **Carboxysomes** are present in many cyanobacteria and other CO_2-fixing bacteria. They are polyhedral, about 100 nm in diameter, and contain the enzyme ribulose-1,5-bisphosphate carboxylase (Rubisco). Rubisco is the critical enzyme for CO_2 fixation, the process of converting CO_2 into sugar. The enzyme assumes a paracrystalline arrangement in the carboxysome, which serves as a reserve of the enzyme. Carboxysomes also may be a site of CO_2 fixation. >> *CO_2 fixation (section 11.3)*

A most remarkable organic inclusion body is the **gas vacuole,** a structure that provides buoyancy to some aquatic procaryotes. Gas vacuoles are present in many photosynthetic bacteria and a few other aquatic procaryotes such as *Halobacterium* (a salt-loving archaeon) and *Thiothrix* (a filamentous bacterium). Gas vacuoles are aggregates of enormous numbers of small, hollow, cylindrical structures called **gas vesicles** (figure 3.13b). Gas vesicle walls are composed entirely of a single small protein. These protein subunits assemble to form a rigid, enclosed cylinder that is hollow and impermeable to water but freely permeable to atmospheric gases. Procaryotes with gas vacuoles can regulate their buoyancy to float at the depth necessary for proper light intensity, oxygen concentration, and nutrient levels. They descend by simply collapsing vesicles and float upward when new ones are constructed.

Two major types of inorganic inclusion bodies are seen in procaryotes: polyphosphate granules and sulfur granules. Many bacteria store phosphate as **polyphosphate granules,** also called **volutin granules** or **metachromatic granules.** Polyphosphate is a linear polymer of orthophosphates joined by ester bonds. Thus polyphosphate granules store the phosphate needed for synthesis of important cell constituents such as nucleic acids. In some cells they act as an energy reserve, and polyphosphate can serve as an energy source in some reactions. Polyphosphate granules are sometimes called metachromatic granules because they show the metachromatic effect; that is, they appear red or a different shade of blue when stained with the blue dyes methylene blue or toluidine blue. Sulfur granules are used by some procaryotes to store sulfur temporarily (figure 3.13c). For example, photosynthetic bacteria can use hydrogen sulfide as a photosynthetic electron donor and accumulate the resulting sulfur either in the periplasmic space or in special cytoplasmic globules. >> *Phototrophy: Light reactions in anoxygenic photosynthesis (section 10.12)*

Inorganic inclusion bodies can be used for purposes other than storage. An excellent example is the **magnetosome;** aquatic magnetotactic bacteria use these inclusions to orient themselves in Earth's magnetic field. Magnetosomes are intracellular chains of magnetite (Fe_3O_4) particles (**figure 3.14**). They are around 35 to 125 nm in diameter, and they have the unusual characteristic of

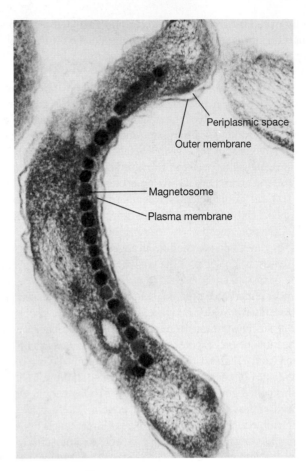

Figure 3.14 Magnetosomes. Transmission electron micrograph of the magnetotactic bacterium *Aquaspirillum magnetotacticum* (×123,000).

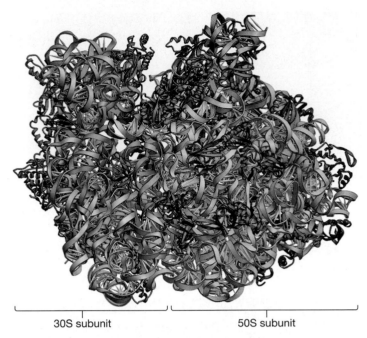

Figure 3.15 Procaryotic Ribosome. The two subunits of a bacterial ribosome are shown. The 50S subunit includes 23S rRNA (gray) and 5S rRNA (lavender), while 16S rRNA (turquoise) is found in the 30S subunit. A molecule of tRNA (gold) is shown in the A site. To generate this ribbon diagram, crystals of purified bacterial ribosomes were grown, exposed to X rays, and the resulting diffraction pattern analyzed.

being bounded by a lipid bilayer. Since each iron particle is a tiny magnet, the Northern Hemisphere bacteria use their magnetosome chain to determine northward and downward directions, and swim down to nutrient-rich sediments or locate the optimum depth in freshwater and marine habitats. Magnetotactic bacteria in the Southern Hemisphere generally orient southward and downward, with the same result.

Ribosomes

When examined with the electron microscope, the cytoplasm of procaryotes is often packed with **ribosomes,** and others may be loosely attached to the plasma membrane. Ribosomes are very complex structures made of both protein and ribonucleic acid (RNA). They are the site of protein synthesis; cytoplasmic ribosomes synthesize proteins destined to remain within the cell, whereas plasma membrane-associated ribosomes make proteins for transport to the outside. Protein synthesis is discussed in detail in chapter 12.

Procaryotic ribosomes are smaller than the ribosomes of eucaryotic cells. Procaryotic ribosomes are called 70S ribosomes (as opposed to 80S in eucaryotes), have dimensions of about 14 to 15 nm by 20 nm, a molecular weight of approximately 2.7 million,

and are constructed of a 50S and a 30S subunit (**figure 3.15**). The S in 70S and similar values stands for **Svedberg unit.** This is the unit of the sedimentation coefficient, a measure of the sedimentation velocity in a centrifuge; the faster a particle travels when centrifuged, the greater its Svedberg value or sedimentation coefficient. The sedimentation coefficient is a function of a particle's molecular weight, volume, and shape. Heavier and more compact particles normally have larger Svedberg numbers and sediment faster.

Nucleoid

Probably the most striking difference between procaryotes and eucaryotes is the way their genetic material is packaged. Eucaryotic cells have two or more chromosomes contained within a membrane-bound organelle, the nucleus. In contrast, procaryotes lack a membrane-delimited nucleus. Instead, the procaryotic chromosome is located in an irregularly shaped region called the **nucleoid** (other names are also used: the nuclear body, chromatin body, and nuclear region) (**figure 3.16**). Most procaryotes contain a single circle of double-stranded **deoxyribonucleic acid (DNA),** but some have a linear chromosome and some, such as *Vibrio cholerae* and *Borrelia burgdorferi* (the causative agents of cholera and Lyme disease, respectively), have more than one chromosome.

It is possible to isolate pure nucleoids. Chemical analysis of purified nucleoids reveals that they are composed of about 60% DNA, 30% RNA, and 10% protein by weight. In *Escherichia*

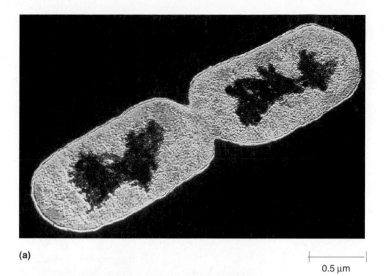

(a)

|———————|———————|
0.5 μm

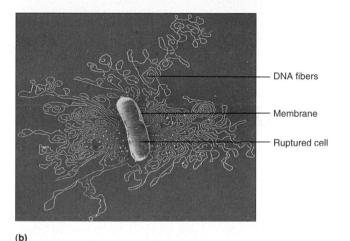

———— DNA fibers

———— Membrane

———— Ruptured cell

(b)

Figure 3.16 Procaryotic Nucleoids and Chromosomes. Procaryotic chromosomes are located in the nucleoid, an area in the cytoplasm. (a) A color-enhanced transmission electron micrograph of a thin section of a dividing *E. coli* cell. The red areas are the nucleoids present in the two daughter cells. (b) Chromosome released from a gently lysed *E. coli* cell. Note how tightly packaged the DNA must be inside the cell.

coli, the closed DNA circle measures approximately 1,400 μm or about 230–700 times longer than the cell (figure 3.16*b*). Obviously it must be very efficiently packaged to fit within the nucleoid. The DNA is looped and coiled extensively (*see figure 12.5*), probably with the aid of RNA and a variety of nucleoid proteins. These include proteins called condensins, which are conserved in both *Bacteria* and *Archaea.* Unlike eucaryotes and some archaea, *Bacteria* do not use histone proteins to package their DNA.

There are a few exceptions to the preceding picture. Membrane-bound DNA-containing regions are present in two genera of the unusual bacterial phylum *Planctomycetes* (*see figure 19.11*). *Pirellula* has a single membrane that surrounds a region called the pirellulosome, which contains a fibrillar nucleoid and ribosome-like particles. The nuclear body of *Gemmata obscuriglobus* is bounded by two membranes. More work will be required to determine the functions of these membranes and how widespread this phenomenon is. ⟩⟩ *Phylum* Planctomycetes *(section 19.4)*

Plasmids

In addition to the genetic material present in the nucleoid, many procaryotes (and some yeasts and other fungi) contain extrachromosomal DNA molecules called plasmids. Indeed, most of the bacterial and archacal genomes sequenced thus far include plasmids. In some cases, numerous different plasmids within a single species have been identified. For instance, *Borrelia burgdorferi,* which causes Lyme disease, carries 12 linear and 9 circular plasmids. Plasmids play many important roles in the lives of the organisms that have them. They also have proved invaluable to microbiologists and molecular geneticists in constructing and transferring new genetic combinations and in cloning genes, as described in chapter 16. This section discusses the different types of procaryotic plasmids.

Plasmids are small, double-stranded DNA molecules that can exist independently of the chromosome. Both circular and linear plasmids have been documented, but most known plasmids are circular. Plasmids have relatively few genes, generally less than 30. Their genetic information is not essential to the host, and cells that lack them usually function normally. However, many plasmids carry genes that confer a selective advantage to their hosts in certain environments.

Plasmids are able to replicate autonomously. That is, plasmid and chromosomal replication are independent. Single-copy plasmids produce only one copy per host cell. Multicopy plasmids may be present at concentrations of 40 or more per cell. Some plasmids are able to integrate into the chromosome and thus are replicated with the chromosome. Such plasmids are called **episomes.** Plasmids are inherited stably during cell division, but they are not always equally apportioned into daughter cells and sometimes are lost. The loss of a plasmid is called **curing.** It can occur spontaneously or be induced by treatments that inhibit plasmid replication but not host cell reproduction. Some commonly used curing treatments are acridine mutagens, ultraviolet and ionizing radiation, thymine starvation, antibiotics, and growth above optimal temperatures.

Plasmids may be classified in terms of their mode of existence, spread, and function. A brief summary of the types of bacterial plasmids and their properties is given in **table 3.3.** **Conjugative plasmids** are of particular note. They have genes for the construction of hairlike structures called pili and can transfer copies of themselves to other bacteria during conjugation. Perhaps the best-studied conjugative plasmid is the **F factor** (fertility factor or F plasmid) of *E. coli,* which is discussed in detail in chapter 14. Some conjugative plasmids are also **R plasmids (resistance factors, R factors).** R plasmids confer antibiotic resistance to the cells that contain them. Conjugative R factors are therefore important in the spread of antibiotic resistance among bacteria. ⟩⟩ *Bacterial conjugation (section 14.7)*

Several other important types of plasmids have been discovered. These include bacteriocin-encoding plasmids, virulence plasmids, and metabolic plasmids. Bacteriocin-encoding plasmids

Table 3.3　Major Types of Bacterial Plasmids

Type	Representatives	Approximate Size (kbp)	Copy Number (Copies/Chromosome)	Hosts	Phenotypic Features[a]
Conjugative Plasmids[b]	F factor	95–100	1–3	*E. coli, Salmonella, Citrobacter*	Sex pilus, conjugation
R Plasmids	RP4	54	1–3	*Pseudomonas* and many other gram-negative bacteria	Sex pilus, conjugation, resistance to Amp, Km, Nm, Tet
	pSH6	21		*Staphylococcus aureus*	Resistance to Gm, Tet, Km
Col Plasmids	ColE1	9	10–30	*E. coli*	Colicin E1 production
	CloDF13	10	50–70	*E. coli*	Cloacin DF13
Virulence Plasmids	Ent (P307)	83		*E. coli*	Enterotoxin production
	Ti	200		*Agrobacterium tumefaciens*	Tumor induction in plants
Metabolic Plasmids	CAM	230		*Pseudomonas*	Camphor degradation
	TOL	75		*Pseudomonas putida*	Toluene degradation

[a]Abbreviations used for resistance to antibiotics: Amp, ampicillin; Gm, gentamycin; Km, kanamycin; Nm, neomycin; Tet, tetracycline.
[b]Many R plasmids, metabolic plasmids and others are also conjugative.

may give the bacteria that harbor them a competitive advantage in the microbial world. Bacteriocins are bacterial proteins that destroy other, closely related bacteria. **Col plasmids** contain genes for the synthesis of bacteriocins known as colicins, which are produced by and directed against strains of *E. coli*. Plasmids in other bacteria carry genes for bacteriocins against other species. **Virulence plasmids** encode factors that make their hosts more pathogenic. For example, enterotoxigenic strains of *E. coli* cause traveler's diarrhea because they contain a plasmid that codes for an enterotoxin. **Metabolic plasmids** carry genes for enzymes that degrade substances such as aromatic compounds (toluene), pesticides (2,4-dichlorophenoxyacetic acid), and sugars (lactose). Metabolic plasmids even carry the genes required for some strains of *Rhizobium* to induce legume nodulation and carry out nitrogen fixation.

1. Briefly describe the nature and function of the cytoplasm, and the regions and structures within it.
2. List the most common kinds of inclusion bodies.
3. Relate the structure of a gas vacuole to its function.
4. List three genera that are exceptional in terms of their chromosome or nucleoid structure. Suggest how the differences observed in these genera might impact how they function.
5. Give the major features of plasmids. How do they differ from chromosomes? What is an episome?
6. Describe each of the following plasmids and explain their importance: conjugative plasmid, F factor, R factor, Col plasmid, virulence plasmid, and metabolic plasmid.

3.4　BACTERIAL CELL WALLS

The cell wall is the layer, usually fairly rigid, that lies just outside the plasma membrane. It is one of the most important procaryotic structures for several reasons: it helps determine the shape of the cell; it helps protect the cell from osmotic lysis; it can protect the cell from toxic substances; and in pathogens, it can contribute to pathogenicity. Cell walls are so important that relatively few procaryotes lack them. Those that do have other features that fulfill cell wall function. The bacterial cell wall also is the site of action of several antibiotics. Therefore, it is important to understand its structure.

The cell walls of *Bacteria* and *Archaea* are distinctive and serve as another example of the fundamental difference between these organisms. In this section, we focus on bacterial cell walls. An overview of bacterial cell wall structure is provided first. This is followed by more detailed discussions of particular aspects of cell wall structure and function. Archaeal cell walls are discussed in section 3.5.

Overview of Bacterial Cell Wall Structure

After Christian Gram developed the Gram stain in 1884, it soon became evident that most bacteria could be divided into two major groups based on their response to the Gram-stain procedure (*see table 17.7*). Gram-positive bacteria stained purple, whereas gram-negative bacteria were colored pink or red by the technique. The true structural difference between these two groups did not become clear until the advent of the transmission electron microscope. The gram-positive cell wall consists of a single, 20 to 80 nm thick homogeneous layer of **peptidoglycan (murein)** lying outside the plasma membrane (**figure 3.17**). In contrast, the gram-negative cell wall is quite complex. It has a

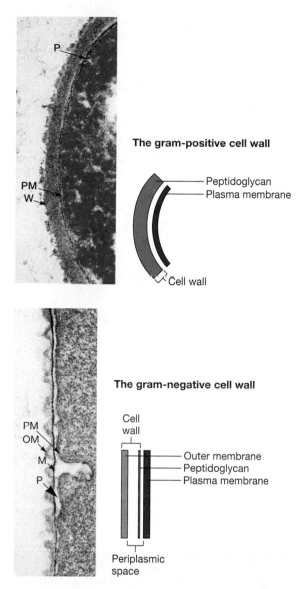

The gram-positive cell wall

- Peptidoglycan
- Plasma membrane

Cell wall

The gram-negative cell wall

Cell wall

- Outer membrane
- Peptidoglycan
- Plasma membrane

Periplasmic space

Figure 3.17 Gram-Positive and Gram-Negative Cell Walls. The gram-positive envelope is from *Bacillus licheniformis* (top), and the gram-negative micrograph is of *Aquaspirillum serpens* (bottom). M, peptidoglycan or murein layer; OM, outer membrane; PM, plasma membrane; P, periplasmic space; W, gram-positive peptidoglycan wall.

2 to 7 nm peptidoglycan layer covered by a 7 to 8 nm thick **outer membrane.** Because of the thicker peptidoglycan layer, the walls of gram-positive cells are more resistant to osmotic pressure than those of gram-negative bacteria. Microbiologists often call all the structures from the plasma membrane outward the **cell envelope.** Therefore this includes the plasma membrane, cell wall, and structures such as capsules (p. 53) when present. ◄◄ *Preparation and staining of specimens: Differential staining (section 2.3)*

One important feature of the cell envelope is a space that is frequently seen between the plasma membrane and the outer membrane in electron micrographs of gram-negative bacteria. It also is sometimes observed between the plasma membrane and the wall in gram-positive bacteria. This space is called the **periplasmic space.** The substance that occupies the periplasmic space is the periplasm. The nature of the periplasmic space and periplasm differs in gram-positive and gram-negative bacteria. These differences are pointed out in the more detailed discussions of gram-positive and gram-negative cell walls that follow.

Peptidoglycan Structure

Peptidoglycan is an enormous, meshlike polymer composed of many identical subunits. The polymer contains two sugar derivatives, *N*-acetylglucosamine and *N*-acetylmuramic acid (the lactyl ether of *N*-acetylglucosamine), and several different amino acids. Three of these amino acids are not found in proteins: D-glutamic acid, D-alanine, and *meso*-diaminopimelic acid. The presence of D-amino acids protects against degradation by most peptidases, which recognize only the L-isomers of amino acid residues. The peptidoglycan subunit present in most gram-negative and many gram-positive bacteria is shown in **figure 3.18.** The backbone of this polymer is composed of alternating

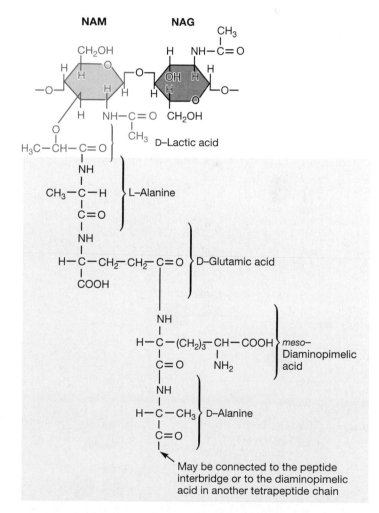

Figure 3.18 Peptidoglycan Subunit Composition. The peptidoglycan subunit of *E. coli*, most other gram-negative bacteria, and many gram-positive bacteria. NAG is *N*-acetylglucosamine. NAM is *N*-acetylmuramic acid (NAG with lactic acid attached by an ether linkage). The tetrapeptide side chain is composed of alternating D- and L-amino acids since *meso*-diaminopimelic acid is connected through its L-carbon. NAM and the tetrapeptide chain attached to it are shown in different shades of color for clarity.

N-acetylglucosamine and *N*-acetylmuramic acid residues. A peptide chain of four alternating D- and L-amino acids is connected to the carboxyl group of *N*-acetylmuramic acid. Many bacteria replace *meso*-diaminopimelic acid with another diaminoacid, usually L-lysine (**figure 3.19**). >> *Carbohydrates (appendix I); Peptidoglycan and endospore structure (section 21.2); Proteins (appendix I)*

To make a strong, meshlike polymer, peptidoglycan chains must be joined by cross-links between the peptides. Often the carboxyl group of the terminal D-alanine is connected directly to the amino group of diaminopimelic acid, but a peptide interbridge may be used instead (**figure 3.20**). Most gram-negative cell wall peptidoglycan lacks the **peptide interbridge.** With or without a peptide interbridge, cross-linking results in an enormous peptidoglycan sac that is actually one dense, interconnected network

(**figure 3.21**). These sacs have been isolated from gram-positive bacteria and are strong enough to retain their shape and integrity, yet they are relatively porous and elastic (**figure 3.22**).

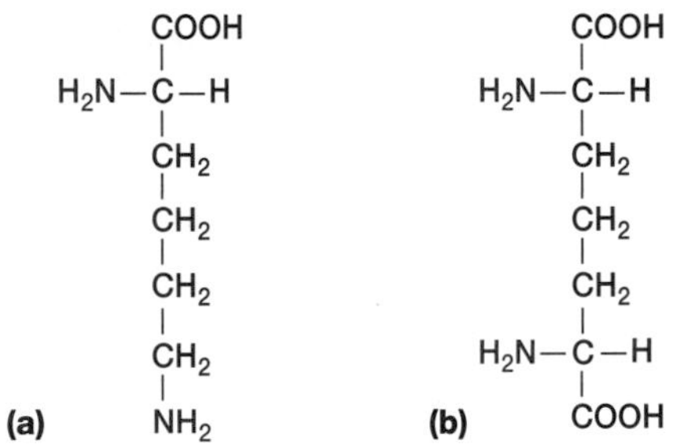

Gram-Positive Cell Walls

Gram-positive bacteria normally have cell walls that are thick and composed primarily of peptidoglycan. Peptidoglycan in gram-positive bacteria often contains a peptide interbridge (figures 3.20 and 3.21). In addition, gram-positive cell walls usually contain large amounts of **teichoic acids,** polymers of

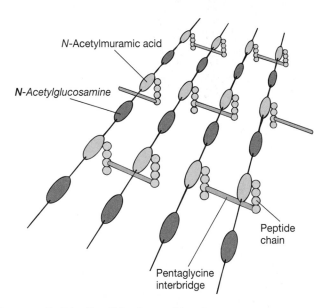

Figure 3.21   Peptidoglycan Structure.   A schematic diagram of one model of peptidoglycan. Shown are the polysaccharide chains, tetrapeptide side chains, and peptide interbridges.

Figure 3.19 Diaminoacids Present in Peptidoglycan.
(a) L-lysine. (b) *meso*-diaminopimelic acid.

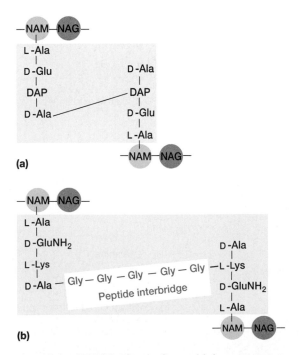

Figure 3.20   Peptidoglycan Cross-Links.   (a) *E. coli* peptidoglycan with direct cross-linking, typical of many gram-negative bacteria. (b) *Staphylococcus aureus* peptidoglycan. *S. aureus* is a gram-positive bacterium. NAM is *N*-acetylmuramic acid. NAG is *N*-acetylglucosamine. Gly is glycine. D-GluNH$_2$ is D-glutamic acid amidated on the α carbon, as found in some gram-positives.

Figure 3.22   Isolated Gram-Positive Cell Wall.   The peptidoglycan wall from *Bacillus megaterium,* a gram-positive bacterium. The latex spheres have a diameter of 0.25 μm.

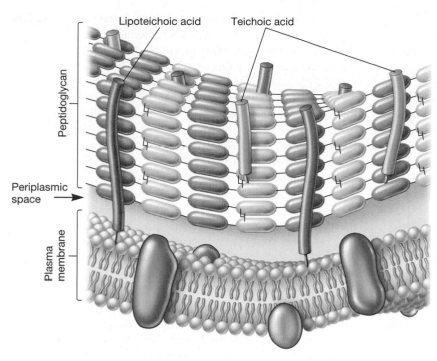

Figure 3.23 The Gram-Positive Envelope.

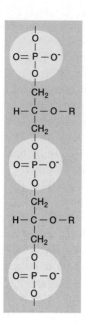

Figure 3.24 Teichoic Acid Structure. The segment of a teichoic acid made of phosphate, glycerol, and a side chain, R. R may represent D-alanine, glucose, or other molecules.

glycerol or ribitol joined by phosphate groups (**figure 3.23** and **figure 3.24**). Amino acids such as D-alanine or sugars such as glucose are attached to the glycerol and ribitol groups. The teichoic acids are covalently connected to the peptidoglycan itself or to plasma membrane lipids; in the latter case, they are called lipoteichoic acids. Teichoic acids appear to extend to the surface of the peptidoglycan. Because they are negatively charged, they help give the gram-positive cell wall its negative charge. The functions of techoic acids are still unclear, but they may be important in maintaining the structure of the wall. Teichoic acids are not present in gram-negative bacteria.

The periplasmic space of gram-positive bacteria lies between the plasma membrane and the cell wall, and is smaller than that of gram-negative bacteria. The periplasm has relatively few proteins; this is probably because the peptidoglycan sac is porous and any proteins secreted by the cell usually pass through it. Enzymes secreted by gram-positive bacteria are called **exoenzymes.** They often serve to degrade polymeric nutrients that would otherwise be too large for transport across the plasma membrane. Those proteins that remain in the periplasmic space are usually attached to the plasma membrane.

Staphylococci and most other gram-positive bacteria have a layer of proteins on the surface of the peptidoglycan. These proteins are involved in interactions of the cell with its environment. Some are noncovalently attached by binding to the peptidoglycan, teichoic acids, or other receptors. For example, the S-layer proteins (p. 54) bind noncovalently to polymers scattered throughout the cell wall. Enzymes involved in peptidoglycan synthesis and turnover also seem to interact noncovalently with the cell wall. Other surface proteins are covalently attached to the peptidoglycan. Many covalently attached proteins, such as the M protein of pathogenic streptococci,

have roles in virulence, such as aiding in adhesion to host tissues and interfering with host defenses. In staphylococci, these surface proteins are covalently joined to the pentaglycine interbridge of the peptidoglycan (figure 3.20b). An enzyme called sortase catalyzes the attachment of these surface proteins to the peptidoglycan. Sortases are attached to the plasma membrane of the cell.

Gram-Negative Cell Walls

Even a brief inspection of figure 3.17 shows that gram-negative cell walls are much more complex than gram-positive walls. The thin peptidoglycan layer next to the plasma membrane and bounded on either side by the periplasmic space usually constitutes only 5 to 10% of the wall weight. In *E. coli*, it is about 2 nm thick and contains only one or two sheets of peptidoglycan.

The periplasmic space of gram-negative bacteria is also strikingly different from that of gram-positive bacteria. It ranges in width from 1 nm to as great as 71 nm. Some recent studies indicate that it may constitute about 20 to 40% of the total cell volume, and it is usually 30 to 70 nm wide. When cell walls are disrupted carefully or removed without disturbing the underlying plasma membrane, periplasmic enzymes and other proteins are released and may be easily studied. Some periplasmic proteins participate in nutrient acquisition—for example, hydrolytic enzymes and transport proteins. Some periplasmic proteins are involved in energy conservation. For example, the denitrifying bacteria, which convert nitrate to nitrogen gas, and bacteria that use inorganic molecules as energy sources (chemolithotrophs) have electron transport proteins in their periplasm. Other periplasmic proteins are involved in peptidoglycan synthesis and the modification of toxic compounds that could harm the cell. **>>** *Chemolithotrophy (section 10.11); Biogeochemical cycling: The nitrogen cycle (section 25.1)*

The outer membrane lies outside the thin peptidoglycan layer and is linked to the cell in two ways (**figure 3.25**). The first is by Braun's lipoprotein, the most abundant protein in the outer membrane. This small lipoprotein is covalently joined to the underlying peptidoglycan and is embedded in the outer membrane by its hydrophobic end. The second linking mechanism involves the many adhesion sites joining the outer membrane and the plasma membrane. The two membranes appear to be in direct contact at these sites. In *E. coli,* 20 to 100 nm areas of contact between the two membranes can be seen. Adhesion sites may be regions of direct contact or possibly true membrane fusions.

Possibly the most unusual constituents of the outer membrane are its **lipopolysaccharides (LPSs).** These large, complex molecules contain both lipid and carbohydrate, and consist of three parts: (1) lipid A, (2) the core polysaccharide, and (3) the O side chain. The LPS from *Salmonella* has been studied most, and its general structure is described here (**figure 3.26**). The **lipid A** region contains two glucosamine sugar derivatives, each with three fatty acids and phosphate or pyrophosphate attached. The fatty acids of lipid A are embedded in the outer membrane, while the remainder of the LPS molecule projects from the surface. The **core polysaccharide** is joined to lipid A. In *Salmonella,* it is constructed of 10 sugars, many of them unusual in structure. The **O side chain** or **O antigen** is a polysaccharide chain extending outward from the core. It has several peculiar sugars and varies in composition between bacterial strains.

LPS has many important functions. Because the core polysaccharide usually contains charged sugars and phosphate (figure 3.26), LPS contributes to the negative charge on the bacterial surface. LPS helps stabilize outer membrane structure because lipid A is a major constituent of the exterior leaflet of the outer membrane. LPS may contribute to bacterial attachment to surfaces and biofilm formation. A major function of LPS is that it helps create a permeability barrier. The geometry of LPS (figure 3.26b) and interactions between neighboring LPS molecules are thought to restrict the entry of bile salts, antibiotics, and other toxic substances that might kill or injure the bacterium. LPS also plays a role in protecting pathogenic gram-negative bacteria from host defenses. The O side chain of LPS is also called the O antigen because it elicits an immune response by an infected host. This response involves the production of antibodies that bind the strain-specific form of LPS that elicited the response. However, many gram-negative bacteria can rapidly change the antigenic nature of their O side chains, thus thwarting host defenses. Importantly, the lipid A portion of LPS is toxic; as a result, LPS can act as an endotoxin and cause some of the symptoms that arise in gram-negative bacterial infections. If LPS or lipid A enters the bloodstream, a form of septic shock develops, for which there is no direct treatment. **>>** *Antibodies (section 29.7); Overview of bacterial pathogenesis (section 30.3)*

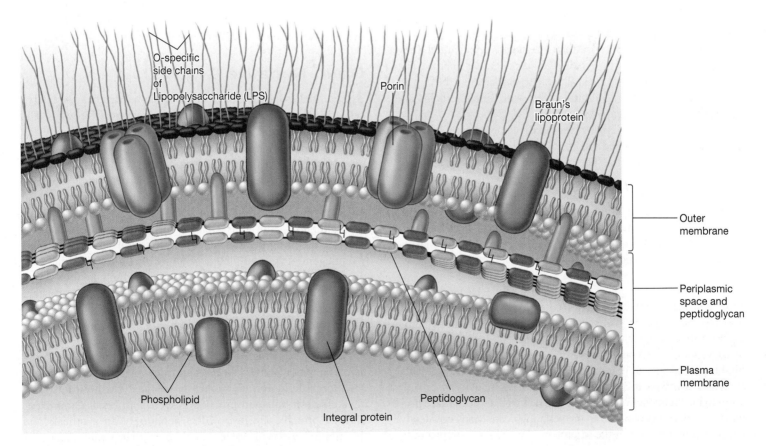

O-specific side chains of Lipopolysaccharide (LPS)

Porin

Braun's lipoprotein

Outer membrane

Periplasmic space and peptidoglycan

Plasma membrane

Phospholipid

Integral protein

Peptidoglycan

Figure 3.25   The Gram-Negative Envelope.

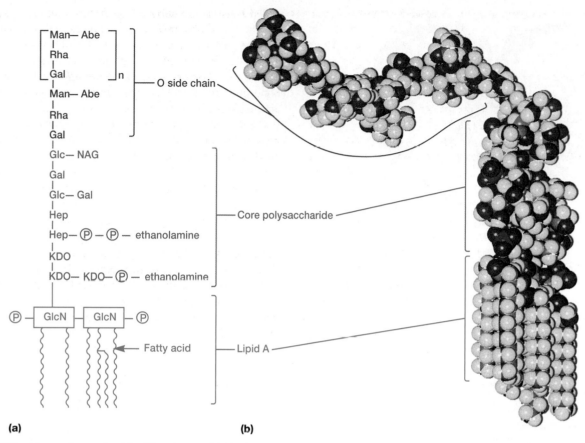

Figure 3.26 Lipopolysaccharide Structure. (a) The lipopolysaccharide from *Salmonella*. This slightly simplified diagram illustrates one form of the LPS. Abbreviations: Abe, abequose; Gal, galactose; Glc, glucose; GlcN, glucosamine; Hep, heptulose; KDO, 2-keto-3-deoxyoctonate; Man, mannose; NAG, *N*-acetylglucosamine; P, phosphate; Rha, L-rhamnose. Lipid A is buried in the outer membrane. (b) Molecular model of an *Escherichia coli* lipopolysaccharide. The lipid A and core polysaccharide are straight; the O side chain is bent at an angle in this model.

Despite the role of LPS in creating a permeability barrier, the outer membrane is more permeable than the plasma membrane and permits the passage of small molecules such as glucose and other monosaccharides. This is due to the presence of **porin proteins.** Most porin proteins cluster together to form a trimer in the outer membrane (figure 3.25 and **figure 3.27**). Each porin protein spans the outer membrane and is more or less tube-shaped; its narrow channel allows passage of molecules smaller than about 600 to 700 daltons. However, larger molecules such as vitamin B_{12} also cross the outer membrane. Such large molecules do not pass through porins; instead, specific carriers transport them across the outer membrane. >> *Uptake of nutrients (section 6.6)*

Mechanism of Gram Staining

The difference between gram-positive and gram-negative bacteria is thought to be due to the physical nature of their cell walls. If the cell wall is removed from gram-positive bacteria, they stain gram negative. Furthermore, genetically wall-less bacteria such as the mycoplasmas also stain gram negative. During the procedure, bacteria are first stained with crystal violet and next treated with iodine to promote dye retention. When bacteria are treated with ethanol in the decolorization step, the alcohol is thought to shrink the pores of the thick peptidoglycan found in gram-positive

bacteria, causing the peptidoglycan to act as a permeability barrier that prevents loss of crystal violet. Thus the dye-iodine complex is retained during the decolorization step and the bacteria remain purple. In contrast, gram-negative peptidoglycan is very thin, not as highly cross-linked, and has larger pores. Alcohol treatment also may extract enough lipid from the outer membrane to increase the cell wall's porosity further. For these reasons, alcohol more readily removes the crystal violet-iodine complex from gram-negative bacteria. Thus gram-negative bacteria are easily stained red or pink by the counterstain safranin.

Cell Walls and Osmotic Protection

Microbes have several mechanisms for responding to changes in osmotic pressure. This pressure arises when the concentration of solutes inside the cell differs from that outside, and the adaptive responses work to equalize the solute concentrations. However, in certain situations, the osmotic pressure can exceed the cell's ability to adapt. In these cases, additional protection is provided by the cell wall. When cells are in hypotonic solutions—ones in which the solute concentration is less than that in the cytoplasm—water moves into the cell, causing it to swell. Without the cell wall, the pressure on the plasma membrane would become so great that the membrane would be disrupted and the cell would burst—a process

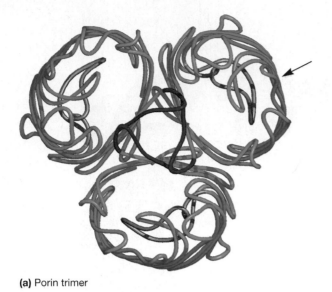

(a) Porin trimer

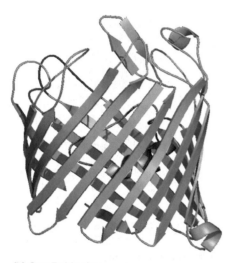

(b) OmpF side view

Figure 3.27 Porin Proteins. Two views of the OmpF porin of *E. coli.* (a) Porin structure observed when looking down at the outer surface of the outer membrane (i.e., top view). The three porin proteins forming the protein each form a channel. Each OmpF porin can be divided into three loops: the green loop forms the channel, the blue loop interacts with other porin proteins to help form the trimer, and the orange loop narrows the channel. The arrow indicates the area of a porin molecule viewed from the side in panel (b). Side view of a porin monomer showing the β-barrel structure characteristic of porin proteins.

called **lysis.** Conversely, in hypertonic solutions, water flows out and the cytoplasm shrivels up—a process called **plasmolysis.**

The protective nature of the cell wall is most clearly demonstrated when bacterial cells are treated with lysozyme or penicillin. The enzyme **lysozyme** attacks peptidoglycan by hydrolyzing the bond that connects *N*-acetylmuramic acid with *N*-acetylglucosamine (figure 3.18). Penicillin works by a different mechanism. It inhibits the enzyme transpeptidase, which is responsible for making the cross-links between peptidoglycan chains. If bacteria are treated with either of these substances while in a hypotonic solution, they lyse. However, if they are in an isotonic solution, they can survive and grow normally. If they are gram positive, treatment with lysozyme or penicillin results in the complete loss of the cell wall, and the cell becomes a protoplast. When gram-negative bacteria are exposed to lysozyme or penicillin, the peptidoglycan layer is lost, but the outer membrane remains. These cells are called **spheroplasts.** Because they lack a complete cell wall, both protoplasts and spheroplasts are osmotically sensitive. If they are transferred to a hypotonic solution, they lyse due to uncontrolled water influx (**figure 3.28**). **>>** *Antibacterial drugs (section 31.4)*

Although most bacteria require an intact cell wall for survival, some have none at all. For example, the mycoplasmas lack a cell wall and are osmotically sensitive, yet often can grow in dilute media or terrestrial environments because their plasma membranes are more resistant to osmotic pressure than those of bacteria having walls. The precise reason for this is not clear, although the presence of sterols in the membranes of many species may provide added strength. Without a rigid cell wall, mycoplasmas tend to be pleomorphic or variable in shape (*see figure 21.3*).

1. List the functions of the cell wall.

2. Describe in detail the composition and structure of peptidoglycan. Why does peptidoglycan contain the unusual D-isomers of alanine and glutamic acid rather than the L-isomers observed in proteins?

3. Compare and contrast the cell walls of gram-positive bacteria and gram-negative bacteria. Include labeled drawings in your discussion.

4. When protoplasts and spheroplasts are made, the shape of the cell becomes spherical regardless of the original cell shape. Why does this occur?

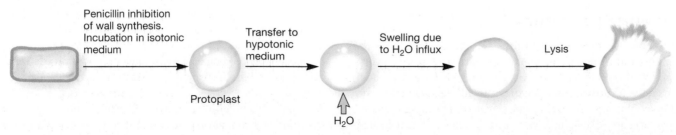

Penicillin inhibition of wall synthesis. Incubation in isotonic medium

Protoplast

Transfer to hypotonic medium

H₂O

Swelling due to H₂O influx

Lysis

Figure 3.28 Protoplast Formation and Lysis. Protoplast formation induced by incubation with penicillin in an isotonic medium. Transfer to hypotonic medium will result in lysis.

5. Design an experiment that illustrates the cell wall's role in protecting against lysis.

6. With a few exceptions, the cell walls of gram-positive bacteria lack porins. Why is this the case?

3.5 ARCHAEAL CELL WALLS

Before they were distinguished as a unique domain of life, the *Archaea* were characterized as being either gram positive or gram negative. However, their staining reaction does not correlate as reliably with a particular cell wall structure. Archaeal wall structure and chemistry differ from those of the *Bacteria*. Archaeal cell walls lack peptidoglycan and exhibit considerable variety in terms of their chemical makeup. Some of the major features of archaeal cell walls are described in this section.

Many archaea have a wall with a single, thick homogeneous layer resembling that in gram-positive bacteria (**figure 3.29a**). These archaea often stain gram positive. Their wall chemistry varies from species to species but usually consists of complex heteropolysaccharides. For example, *Methanobacterium* and some other methane-generating archaea (methanogens) have walls containing **pseudomurein,** a peptidoglycan-like polymer that has L-amino acids instead of D-amino acids in its cross-links, *N*-acetyltalosaminuronic acid instead of *N*-acetylmuramic acid, and β(1→3) glycosidic bonds instead of β(1→4) glycosidic bonds (**figure 3.30**). Other archaea, such as *Methanosarcina* and the salt-loving *Halococcus,* contain complex polysaccharides similar to the chondroitin sulfate of animal connective tissue. >> *Phylum* Euryarchaeota *(section 18.3)*

Many archaea that stain gram negative have either a layer of glycoprotein or protein outside their plasma membrane (figure 3.29b). The layer may be as thick as 20 to 40 nm. Sometimes there are two layers—an electron-dense layer and a sheath surrounding it. Some methanogens (*Methanolobus*), salt-loving archaea (*Halobacterium*), and extreme thermophiles (*Sulfolobus,*

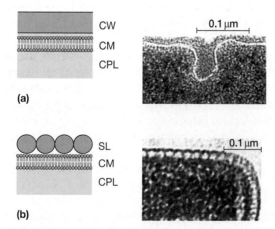

(a)

(b)

Figure 3.29 **Cell Envelopes of *Archaea.*** Schematic representations and electron micrographs of (a) *Methanobacterium formicicum* and (b) *Thermoproteus tenax.* CW, cell wall; SL, surface layer; CM, cell membrane or plasma membrane; CPL, cytoplasm.

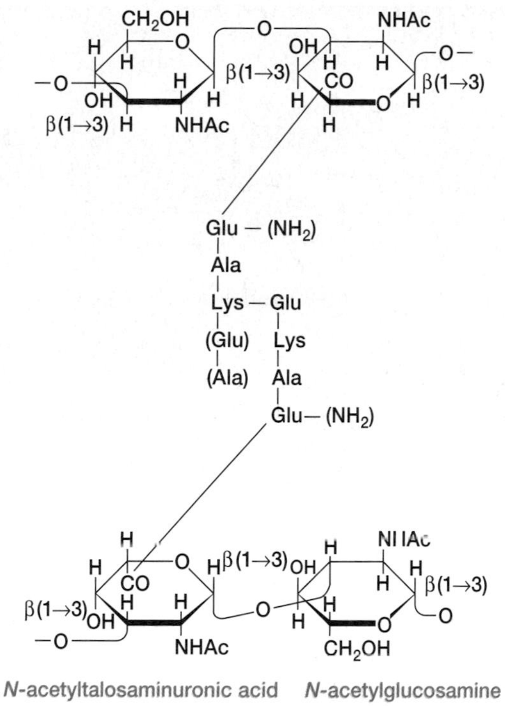

N-acetyltalosaminuronic acid   *N*-acetylglucosamine

Figure 3.30 **The Structure of Pseudomurein.** The amino acids and amino groups in parentheses are not always present. Ac represents the acetyl group.

Thermoproteus, and *Pyrodictium*) have glycoproteins in their walls. In contrast, other methanogens (*Methanococcus, Methanomicrobium,* and *Methanogenium*) and the extreme thermophile *Desulfurococcus* have protein walls. >> *Phylum* Crenarchaeota *(section 18.2); Phylum* Euryarchaeota *(section 18.3)*

1. How do the cells walls of *Archaea* differ from those of *Bacteria*?

2. What is pseudomurein? How is it similar to peptidoglycan? How is it different?

3. Archaea with cell walls consisting of a thick, homogeneous layer of complex polysaccharides often retain the crystal violet dye when stained using the Gram-staining procedure. Why do you think this is so?

3.6 COMPONENTS EXTERNAL TO THE CELL WALL

Procaryotes have a variety of structures outside the cell wall that can function in protection, attachment to objects, and cell movement. Several of these are discussed in this section.

Capsules and Slime Layers

Some procaryotes have a layer of material lying outside the cell wall. This layer has different names depending on its characteristics. When the layer is well organized and not easily washed off, it is

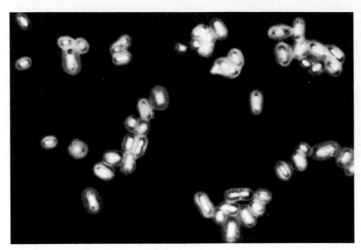

(a) *K. pneumoniae*

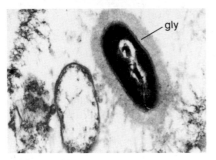

(b) *Bacteroides*

Figure 3.31 Bacterial Capsules. (a) *Klebsiella pneumoniae* with its capsule stained for observation in the light microscope (×1,500). (b) *Bacteroides* glycocalyx (gly), TEM (×71,250).

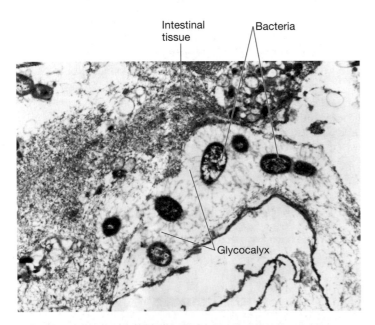

Figure 3.32 Bacterial Glycocalyx. Bacteria connected to each other and to the intestinal wall by their glycocalyxes, the extensive networks of fibers extending from the cells (×17,500).

called a **capsule** (**figure 3.31***a*). It is called a **slime layer** when it is a zone of diffuse, unorganized material that is removed easily. When the layer consists of a network of polysaccharides extending from the surface of the cell, it is referred to as the **glycocalyx** (figure 3.31*b*), a term that can encompass both capsules and slime layers because they usually are composed of polysaccharides. However, some slime layers and capsules are constructed of other materials. For example, *Bacillus anthracis* has a proteinaceous capsule composed of poly-D-glutamic acid. Capsules are clearly visible in the light microscope when negative stains or special capsule stains are employed (figure 3.31*a*); they also can be studied with the electron microscope (figure 3.31*b*).

Although capsules are not required for growth and reproduction in laboratory cultures, they confer several advantages when procaryotes grow in their normal habitats. They help pathogenic bacteria resist phagocytosis by host phagocytes. *Streptococcus pneumoniae* provides a dramatic example. When it lacks a capsule, it is destroyed easily and does not cause disease. On the other hand, the capsulated variant quickly kills mice. Capsules contain a great deal of water and can protect against desiccation. They exclude viruses and most hydrophobic toxic materials such as detergents. The glycocalyx also aids in attachment to solid surfaces, including tissue surfaces in plant and animal hosts (**figure 3.32**). Gliding bacteria often produce slime, which in some cases has been shown to facilitate motility (section 3.7). **>>** *Overview of bacterial pathogenesis (section 30.3)*

S-Layers

Many procaryotes have a regularly structured layer called an **S-layer** on their surface. In bacteria, the S-layer is external to the cell wall. In some archaea, the S-layer is the only structure outside the plasma membrane where it serves as the cell wall. The S-layer has a pattern something like floor tiles and is composed of protein or glycoprotein (**figure 3.33**). In gram-negative bacteria, the S-layer adheres directly to the outer membrane; it is associated with the peptidoglycan surface in gram-positive bacteria.

Currently S-layers are of considerable interest not only for their biological roles but also in the growing field of nanotechnology. Their biological roles include protecting the cell against ion and pH fluctuations, osmotic stress, enzymes, or predacious bacteria. The S-layer also helps maintain the shape and envelope rigidity of some cells, and it can promote cell adhesion to surfaces. Finally, the S-layer seems to protect some bacterial pathogens against host defenses, thus contributing to their virulence. The potential use of S-layers in nanotechnology is due to the ability of S-layer proteins to self-assemble. That is, the S-layer proteins contain the information required to associate and form the S-layer without the aid of any special enzymes or other factors. Thus S-layer proteins could be used as building blocks for the creation of technologies such as drug-delivery systems and novel detection systems for toxic chemicals or bioterrorism agents.

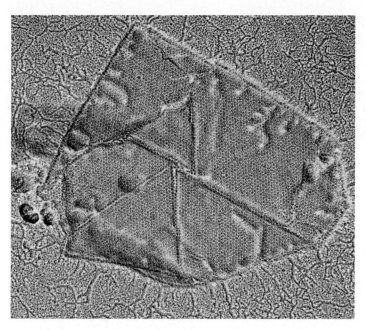

Figure 3.33 The S-Layer. An electron micrograph of the S-layer of the bacterium *Deinococcus radiodurans* after shadowing.

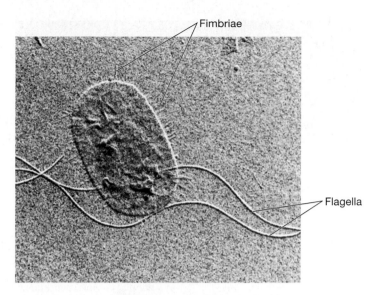

Figure 3.34 Flagella and Fimbriae. The long flagella and the numerous shorter fimbriae are evident in this electron micrograph of the bacterium *Proteus vulgaris* (×39,000).

Pili and Fimbriae

Many procaryotes have short, fine, hairlike appendages that are thinner than flagella. These are usually called **fimbriae** (s., **fimbria**) or pili (s., pilus). Although many people use the terms fimbriae and pili interchangeably, we distinguish between fimbriae and sex pili. A cell may be covered with up to 1,000 fimbriae, but they are only visible in an electron microscope due to their small size (**figure 3.34**). They are slender tubes composed of helically arranged protein subunits and are about 3 to 10 nm in diameter and up to several micrometers long. Some types of fimbriae attach bacteria to solid surfaces such as rocks in streams and host tissues, and some are involved in motility (section 3.7).

Many bacteria have about one to 10 **sex pili** (s., **sex pilus**) per cell. These hairlike structures differ from fimbriae in the following ways. Pili often are larger than fimbriae (around 9 to 10 nm in diameter). They are genetically determined by conjugative plasmids and are required for conjugation. Some bacterial viruses attach specifically to receptors on sex pili at the start of their reproductive cycle. >> *Bacterial conjugation (section 14.7)*

Flagella

Most motile procaryotes move by use of **flagella** (s., **flagellum**), threadlike locomotor appendages extending outward from the plasma membrane and cell wall. Bacterial flagella are the best studied and are the focus of this discussion.

Bacterial flagella are slender, rigid structures, about 20 nm across and up to 20 μm long. Flagella are so thin they cannot be observed directly with a bright-field microscope but must be stained with special techniques designed to increase their thickness. The detailed structure of a flagellum can only be seen in the electron microscope.

Bacterial species often differ distinctively in their patterns of flagella distribution, and these patterns are useful in identifying bacteria. **Monotrichous** bacteria (*trichous* means hair) have one flagellum; if it is located at an end, it is said to be a **polar flagellum** (**figure 3.35a**). **Amphitrichous** bacteria (*amphi* means on both sides) have a single flagellum at each pole. In contrast, **lophotrichous** bacteria (*lopho* means tuft) have a cluster of flagella at one or both ends (figure 3.35b). Flagella are spread evenly over the whole surface of **peritrichous** (*peri* means around) bacteria (figure 3.35c).

Flagellar Ultrastructure

Transmission electron microscope studies have shown that the bacterial flagellum is composed of three parts. (1) The longest and most obvious portion is the **flagellar filament,** which extends from the cell surface to the tip. (2) The **basal body** is embedded in the cell; and (3) a short, curved segment, the **flagellar hook,** links the filament to its basal body and acts as a flexible coupling. The filament is a hollow, rigid cylinder constructed of subunits of the protein **flagellin,** which ranges in molecular weight from 30,000 to 60,000 daltons, depending on the bacterial species. The filament ends with a capping protein. Some bacteria have sheaths surrounding their flagella. For example, *Vibrio cholerae* has a lipopolysaccharide sheath.

The hook and basal body are quite different from the filament (**figure 3.36**). Slightly wider than the filament, the hook is made of different protein subunits. The basal body is the most complex part of a flagellum. In transmission electron micrographs of the basal bodies of *E. coli* and most other gram-negative bacteria, the basal body appears to have four rings, L ring, P ring, S ring, and M ring, connected to a central rod (figure 3.36a). It is now known that the S ring and M ring are different portions of the same protein, and they are now referred to as the MS ring. On the cytoplasmic side of the MS ring is the C ring, which was discovered

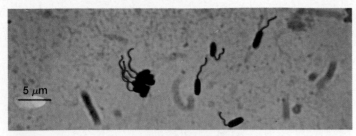

(a) *Pseudomonas*—monotrichous polar flagellation

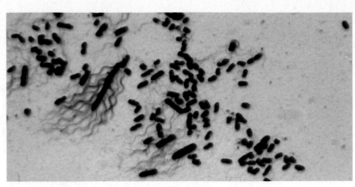

(c) *P. vulgaris*—peritrichous flagellation

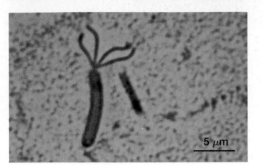

(b) *Spirillum*—lophotrichous flagellation

Figure 3.35   Flagellar Distribution.   Examples of various patterns of flagellation as seen in the light microscope. (a) Monotrichous polar (*Pseudomonas*). (b) Lophotrichous (*Spirillum*). (c) Peritrichous (*Proteus vulgaris*, ×600).

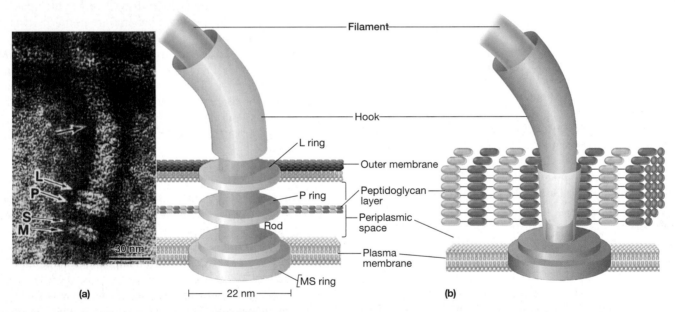

Figure 3.36   The Ultrastructure of Bacterial Flagella.   Flagellar basal bodies and hooks in (a) gram-negative and (b) gram-positive bacteria. The inset photo shows an enlarged view of the basal body of an *E. coli* flagellum. All three rings (L, P, and MS) can be clearly seen. The uppermost arrow is at the junction of the hook and filament.

later. Gram-positive bacteria have only two rings—an inner ring connected to the plasma membrane and an outer one probably attached to the peptidoglycan (figure 3.36*b*).

Flagellar Synthesis

The synthesis of bacterial flagella is a complex process involving at least 20 to 30 genes. Besides the gene for flagellin, 10 or more genes code for hook and basal body proteins; other genes are concerned with the control of flagellar construction or function. How the cell regulates or determines the exact location of flagella is not known.

Because many components of the flagellum lie outside the cell wall, they must be transported across the plasma membrane and cell wall. Transport of many flagellar components is carried out by an apparatus in the basal body. It is thought that flagellin subunits are transported through the filament's hollow internal core. When they reach the tip, the subunits spontaneously aggregate under the direction of a special filament cap so that the filament grows at its tip rather than at the base (**figure 3.37**). Thus filament synthesis, like S-layer formation, is an example of **self-assembly.**

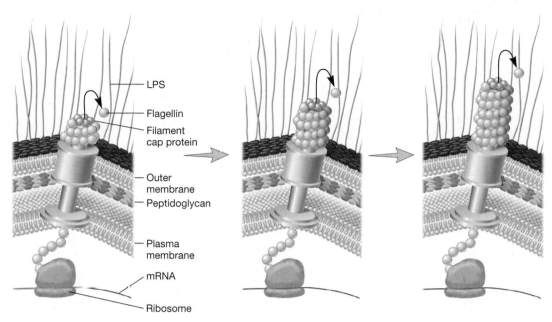

Figure 3.37   Growth of Flagellar Filaments.   Flagellin subunits travel through the flagellar core and attach to the growing tip. Their attachment is directed by the filament cap protein.

1. Briefly describe capsules, slime layers, glycocalyxes, and S-layers. What are their functions?

2. Distinguish between fimbriae and sex pili, and give the function of each.

3. Discuss flagella distribution patterns and structure and synthesis of flagella.

4. What is self-assembly? Why does it make sense that the flagellar filament is assembled in this way?

3.7   BACTERIAL MOTILITY AND CHEMOTAXIS

As we note in section 3.6, several structures outside the cell wall contribute to the motility of procaryotes. Four major methods of movement have been observed in *Bacteria:* the swimming movement conferred by flagella; the corkscrew movement of spirochetes; the twitching motility associated with fimbriae; and gliding motility. Bacteria have not evolved motility to move aimlessly. Rather, motility is used to move toward nutrients such as sugars and amino acids and away from many harmful substances and bacterial waste products. Bacteria also can respond to environmental cues such as temperature (thermotaxis), light (phototaxis), oxygen (aerotaxis), osmotic pressure (osmotaxis), and gravity. Movement toward chemical attractants and away from repellents is known as chemotaxis.

Flagellar Movement

Procaryotic flagella operate differently from eucaryotic flagella. Eucaryotic flagella flex and bend, resulting in a whiplash that moves the cell. The filament of a procaryotic flagellum is in the shape of a rigid helix, and the cell moves when this helix rotates like a propeller on a boat. The flagellar motor can rotate very rapidly. The *E. coli* motor rotates 270 revolutions per second (rps); *Vibrio alginolyticus* averages 1,100 rps. **>>** *Structures external to the plasma membrane: Cilia and flagella (section 4.7)*

The direction of flagellar rotation determines the nature of bacterial movement. Monotrichous, polar flagella rotate counterclockwise (when viewed from outside the cell) during normal forward movement, whereas the cell itself rotates slowly clockwise. The rotating helical flagellar filament thrusts the cell forward with the flagellum trailing behind (**figure 3.38**). Monotrichous bacteria stop and tumble randomly by reversing the direction of flagellar rotation. Peritrichously flagellated bacteria operate in a somewhat similar way. To move forward, the flagella rotate counterclockwise. As they do so, they bend at their hooks to form a rotating bundle that propels the cell forward. Clockwise rotation of the flagella disrupts the bundle and the cell tumbles.

The motor that drives flagellar rotation is located at the base of the flagellum, where it is associated with the basal body. Torque generated by the motor is transmitted by the basal body to the hook and filament. The motor is composed of two components: the rotor and the stator. It is thought to function like an electrical motor, where the rotor turns in the center of a ring of electromagnets, the stator. In gram-negative bacteria, the rotor is composed of the MS ring and the C ring (**figure 3.39**). The flagellar protein FliG is a particularly important component of the rotor as it is thought to interact with the stator. The stator is composed of the proteins MotA and MotB. Both form a channel through the plasma membrane, and MotB also anchors MotA to cell wall peptidoglycan.

As with all motors, the flagellar motor must have a power source that allows it to generate torque and cause flagellar rotation. The power used by most flagellar motors is a difference in charge

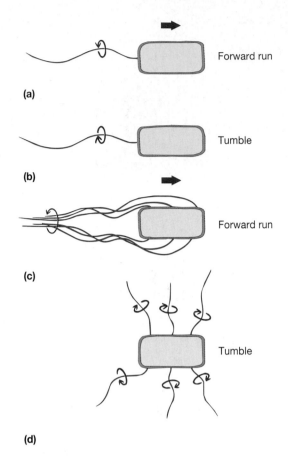

(a) Forward run

(b) Tumble

(c) Forward run

(d) Tumble

Figure 3.38 **Flagellar Motility.** The relationship of flagellar rotation to bacterial movement. Parts (a) and (b) describe the motion of monotrichous, polar bacteria. Parts (c) and (d) illustrate the movements of peritrichous organisms.

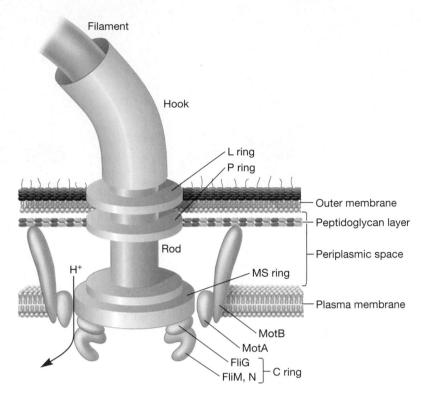

Figure 3.39 **Mechanism of Flagellar Movement.** This diagram of a gram-negative flagellum shows some of the more important components and the flow of protons that drives rotation. Five of the many flagellar proteins are labeled (MotA, MotB, FliG, FliM, FliN).

and pH across the plasma membrane. This difference is called the proton motive force (PMF). PMF is largely created by the metabolic activities of organisms as described in chapter 10. One important metabolic process carried out by cells is the transfer of electrons from an electron donor to a terminal electron acceptor via a chain of electron carriers called the electron transport chain (ETC). In procaryotes, the ETC is located in the plasma membrane. As electrons are transported down the ETC, protons are transported from the cytoplasm to the outside of the cell. Because there are more protons outside the cell than inside, the outside has more positively charged ions (the protons) and has a lower pH. PMF is a type of potential energy that can be used to do work: mechanical work, as in the case of flagellar rotation; transport work, the movement of materials into or out of the cell; or chemical work such as the synthesis of ATP, the cell's energy currency. So how can PMF be used to power the flagellar motor? The channels created by the MotA and MotB proteins allow protons to move across the plasma membrane from the outside to the inside. Thus they move down the charge and pH gradient. This movement releases energy that is used to rotate the flagellum. In essence, the entry of a proton into the channel is like the entry of a person into a revolving door. The "power" of the proton generates torque, rather like a person pushing the revolving door.

Indeed, the speed of flagellar rotation is proportional to the magnitude of the PMF. **>>** *Electron transport and oxidative phosphorylation (section 10.5); Uptake of nutrients (section 6.6)*

The flagellum is a very effective swimming device. From the bacterium's point of view, swimming is quite a difficult task because the surrounding water seems as viscous as molasses. The cell must bore through the water with its corkscrew-shaped flagella, and if flagellar activity ceases, it stops almost instantly. Despite such environmental resistance to movement, bacteria can swim from 20 to almost 90 μm/second. This is equivalent to traveling from 2 to over 100 cell lengths per second. In contrast, an exceptionally fast human might be able to run around 5 to 6 body lengths per second.

Spirochete Motility

Although spirochetes have flagella, they work in a different manner. In many spirochetes, multiple flagella arise from each end of the cell and associate to form an axial fibril, which winds around the cell (**figure 3.40**). The flagella do not extend outside the cell wall but rather remain in the periplasmic space and are covered by an outer sheath. The way in which axial fibrils propel the cell has not been fully established. They are thought to rotate like the external flagella of other bacteria, causing the corkscrew-shaped outer sheath to rotate and move the cell through the surrounding liquid, even very viscous liquids. Flagellar rotation may also flex

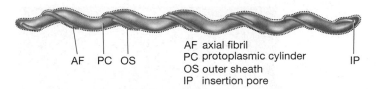

Figure 3.40 Spirochete Flagella. Generally, numerous flagella arise from each end of the spirochete. These intertwine to form an axial fibril. The axial fibril winds around the cell, usually overlapping in the middle.

AF axial fibril
PC protoplasmic cylinder
OS outer sheath
IP insertion pore

or bend the cell and account for the creeping or crawling movement observed when spirochetes are in contact with a solid surface. >> *Phylum* Spirochaetes *(section 19.6)*

Twitching and Gliding Motility

Twitching and gliding motility occur when cells are on a solid surface. Both types of motility can involve fimbriae, the production of slime, or both. Thus they are considered together.

Several types of fimbriae have been identified on procaryotic cells. Type IV fimbriae are present at one or both poles of some bacteria and are involved in twitching motility and in the gliding motility of some bacteria. Twitching motility is characterized by short, intermittent, jerky motions of up to several micrometers in length and is normally seen on very moist surfaces. It occurs only when cells are in contact with each other; isolated cells rarely move by this mechanism. Considerable evidence exists that the fimbriae alternately extend and retract to move bacteria during twitching motility.

Gliding motility is smooth and varies greatly in rate (from 2 to over 600 μm per minute) and in the nature of the motion. Although first observed over 100 years ago, the mechanism by which many bacteria glide remains a mystery. Some glide along in a direction parallel to the longitudinal axis of their cells. Others travel with a screwlike motion or even move in a direction perpendicular to the long axis of the cells. Still others rotate around their longitudinal axis while gliding. Such diversity in gliding movement correlates with the observation that more than one mechanism for gliding motility exists. Some types involve type IV fimbriae, some involve slime, and some involve mechanisms that have not yet been elucidated.

Gliding motility is best understood in the bacterium *Myxococcus xanthus.* This rod-shaped microbe has a complex life cycle that includes the aggregation of cells to form a complex fruiting body in response to nutrient starvation. *M. xanthus* exhibits two types of motility. The first is called social (S) motility because it occurs when large groups of cells move together in a coordinated fashion. The second is called adventurous (A) motility, and it is observed when single cells move independently. Both S and A motility can occur during aggregation and fruiting body formation. S motility is mediated by the extension and retraction of type IV fimbriae at the front pole of the cell. To reverse its direction, the bacterium disassembles fimbriae at one pole and moves them to the opposite pole. The mechanism of A motility is not as well understood. One hypothesis is that the cells contain pores through which slime is secreted

and that this propels the cell forward. A more recent hypothesis is that adhesion complexes are located along the length of the cell and that these attach the cell to the surface. The adhesion complexes are thought to span all the layers of the cell envelope, such that some portions are external and in contact with the surface and other portions are in the cytoplasm. The adhesion complexes remain stationary relative to the surface on which the cell is gliding but move along a "track" within the cell. >> *Class* Deltaproteobacteria: *Order* Myxococcales *(section 20.4)*

Chemotaxis

The movement of cells toward chemical attractants or away from chemical repellents is called **chemotaxis.** Chemotaxis is readily observed in petri dish cultures. If bacteria are placed in the center of a dish of semisolid agar containing an attractant, the bacteria will exhaust the local supply of the nutrient and swim outward following the attractant gradient they have created. The result is an expanding ring of bacteria (**figure 3.41a**). When a disk of repellent is placed in a petri dish of semisolid agar and bacteria, the bacteria will swim away from the repellent, creating a clear zone around the disk (figure 3.41b).

Attractants and repellents are detected by **chemoreceptors,** proteins that bind chemicals and transmit signals to other components of the chemosensing system. The chemosensing systems are very sensitive and allow the cell to respond to very low levels of attractants (about 10^{-8} M for some sugars). In gram-negative bacteria, the chemoreceptor proteins are located in the periplasmic space or in the plasma membrane. Some receptors also participate in the initial stages of sugar transport into the cell.

The chemotactic behavior of bacteria has been studied using the tracking microscope, a microscope with a moving stage that automatically keeps an individual bacterium in view. In the absence of a chemical gradient, bacteria move randomly, switching back and forth between a phase called a run and a phase called a tumble. For a bacterium with peritrichous flagella, a **run** occurs when its flagella are organized into a coordinated, corkscrew-shaped bundle (figure 3.38c). During a run, the bacterium travels in a straight or slightly curved line. After a few seconds, the flagella "fly apart" and the bacterium will stop and **tumble.** The tumble randomly reorients the bacterium so that it often is facing in a different direction. Therefore when it begins the next run, it usually goes in a different direction (**figure 3.42a**). In contrast, when the bacterium is exposed to an attractant, it tumbles less frequently (or has longer runs) when traveling toward the attractant. Although the tumbles can still orient the bacterium away from the attractant, over time, the bacterium gets closer and closer to the attractant (figure 3.42b). The opposite response occurs with a repellent. Tumbling frequency decreases (the run time lengthens) when the bacterium moves away from the repellent.

Clearly, the bacterium must have some mechanism for sensing that it is getting closer to the attractant (or moving away from the repellent). The behavior of the bacterium is shaped by temporal changes in chemical concentration. The bacterium moves toward the attractant because it senses that the concentration of

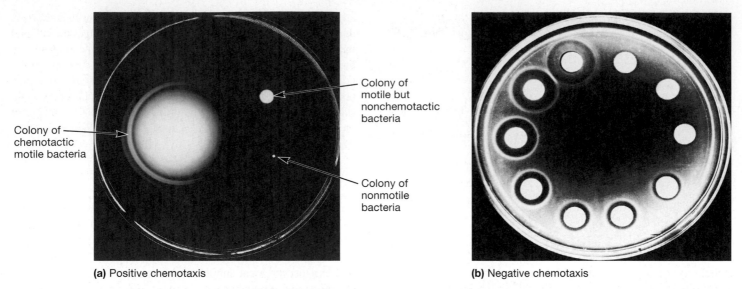

(a) Positive chemotaxis

(b) Negative chemotaxis

Colony of chemotactic motile bacteria

Colony of motile but nonchemotactic bacteria

Colony of nonmotile bacteria

Figure 3.41   Positive and Negative Bacterial Chemotaxis.   (a) Positive chemotaxis can be demonstrated on an agar plate that contains various nutrients; positive chemotaxis by *E. coli* is shown on the left. The outer ring is composed of bacteria consuming serine. The second ring was formed by *E. coli* consuming aspartate, a less powerful attractant. The upper right colony is composed of motile but nonchemotactic mutants. The bottom right colony is formed by nonmotile bacteria. (b) Negative chemotaxis by *E. coli* in response to the repellent acetate. The bright disks are plugs of concentrated agar containing acetate that have been placed in dilute agar inoculated with *E. coli*. Acetate concentration increases from zero at the top right to 3 M at top left. Note the increasing size of bacteria-free zones with increasing acetate. The bacteria have migrated for 30 minutes.

Tumble   Run

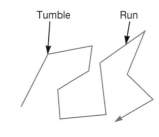

(a)

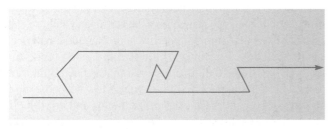

(b)

Figure 3.42   Directed Movement in Bacteria.   (a) Random movement of a bacterium in the absence of a concentration gradient. Tumbling frequency is fairly constant. (b) Movement in an attractant gradient. Tumbling frequency is reduced when the bacterium is moving up the gradient. Therefore runs in the direction of increasing attractant are longer.

1. Describe the way flagella operate to move a bacterium. Define chemotaxis, run, and tumble.

2. Explain in a general way how bacteria move toward substances such as nutrients and away from toxic materials.

3. Why do you think chemotaxis is sometimes called a "biased random walk"?

the attractant is increasing. Likewise, it moves away from a repellent because it senses that the concentration of the repellent is decreasing. The bacterium's chemoreceptors play a critical role in this process. The molecular events that enable bacterial cells to sense a chemical gradient and respond appropriately are presented in chapter 9.

3.8   BACTERIAL ENDOSPORES

Several genera of gram-positive bacteria, including *Bacillus* and *Clostridium* (rods), and *Sporosarcina* (cocci), can form a resistant, dormant structure called an **endospore.** Endospores develop within vegetative bacterial cells and are extraordinarily resistant to environmental stresses such as heat, ultraviolet radiation, gamma radiation, chemical disinfectants, and desiccation. In fact, some endospores have remained viable for around 100,000 years. Because of their resistance and the fact that several species of endospore-forming bacteria are dangerous pathogens, endospores are of great practical importance in food, industrial, and medical microbiology. This is because it is essential to be able to sterilize solutions and solid objects. Endospores often survive boiling for an hour or more; therefore autoclaves must be used to sterilize many materials. In the environment, endospores aid in survival when moisture or nutrients are scarce. Endospores are also of considerable theoretical interest. Because bacteria manufacture these intricate structures in a very organized fashion over a period of a few hours, spore formation is well suited for research on the construction of complex biological structures. This has made the endospore-forming *Bacillus subtilis* an important model organism. >> *The use of physical methods in control: Heat (section 8.4)*

Endospores can be examined with both light and electron microscopes. Because endospores are impermeable to most stains, they often are seen as colorless areas in bacteria treated with methylene blue and other simple stains; special endospore stains are used to make them clearly visible. Endospore position in the mother cell (**sporangium**) frequently differs among species, making it of value in identification. Endospores may be centrally located, close to one end (subterminal), or terminal (**figure 3.43**). Sometimes an endospore is so large that it swells the sporangium. << *Preparation and staining of specimens (section 2.3)*

Electron micrographs show that endospore structure is complex (**figure 3.44**). The spore often is surrounded by a thin, delicate covering called the exosporium. A spore coat lies beneath the exosporium, is composed of several protein layers, and may be fairly thick. It is impermeable to many toxic

molecules and is responsible for the spore's resistance to chemicals. The coat also is thought to contain enzymes involved in germination. The cortex, which may occupy as much as half the spore volume, rests beneath the spore coat. It is made of a peptidoglycan that is less cross-linked than that in vegetative cells. The spore cell wall (or core wall) is inside the cortex and surrounds the protoplast or spore core. The core has normal cell structures such as ribosomes and a nucleoid but is metabolically inactive.

It is still not known precisely why the endospore is so resistant to heat and other lethal agents. As much as 15% of the spore's dry weight consists of dipicolinic acid complexed with calcium ions, which is located in the core (**figure 3.45**). It has long been thought that dipicolinic acid was directly involved in heat resistance, but heat-resistant mutants lacking dipicolinic acid have been isolated. Calcium aids in resistance to wet heat, oxidizing agents, and sometimes dry heat. It may be that calcium-dipicolinate stabilizes the spore's nucleic acids. In addition, small, acid-soluble DNA-binding proteins (SASPs) are found in the endospore. They saturate spore DNA and protect it from heat, radiation, desiccation, and chemicals. Dehydration of the protoplast appears to be very important in heat resistance. The cortex may osmotically remove water from the protoplast, thereby protecting it from both heat and radiation damage. The spore coat also seems to protect against enzymes and chemicals such as hydrogen peroxide. Finally, spores contain some DNA repair enzymes. DNA is repaired once the spore germinates and the cell becomes active again. In summary, endospore heat resistance probably is due to several factors: calcium-dipicolinate and acid-soluble protein stabilization of DNA, protoplast dehydration, the spore coat, DNA repair, the greater stability of cell proteins in bacteria adapted to growth at high temperatures, and others.

Endospore formation (**sporulation**) normally commences when growth ceases due to lack of nutrients. It is a complex process and may be divided into seven stages (**figure 3.46**). An axial filament of nuclear material forms (stage I), followed by an inward folding of the cell membrane to enclose part of the DNA and produce the forespore septum (stage II). The membrane continues to grow and engulfs the immature endospore in a second membrane (stage III). Next, cortex is laid down in the space between the two membranes, and both calcium and dipicolinic acid are accumulated (stage IV). Protein coats then are formed around the cortex (stage V), and maturation of the endospore occurs (stage VI). Finally, lytic enzymes destroy the sporangium, releasing the spore (stage VII). Sporulation requires about 10 hours in *Bacillus megaterium*. >> *Global regulatory systems: Sporulation in* Bacillus subtilis *(section 13.5)*

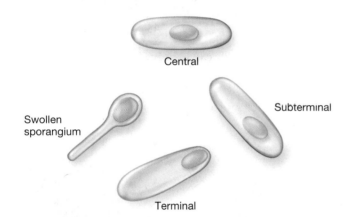

Central

Subterminal

Swollen sporangium

Terminal

Figure 3.43 **Examples of Endospore Location and Size.**

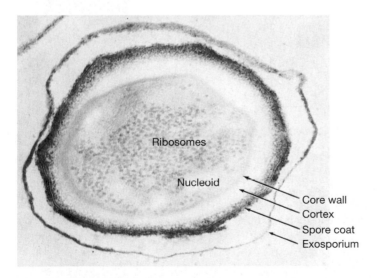

Ribosomes

Nucleoid

Core wall
Cortex
Spore coat
Exosporium

Figure 3.44 **Endospore Structure.** *Bacillus anthracis* endospore (×151,000).

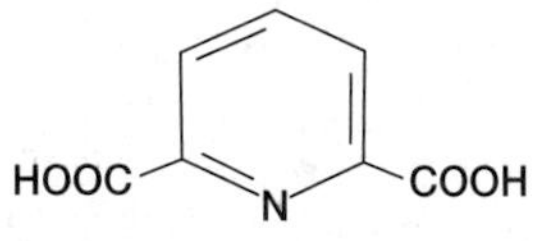

$$HOOC \quad \underset{N}{\bigcirc} \quad COOH$$

Figure 3.45 **Dipicolinic Acid.**

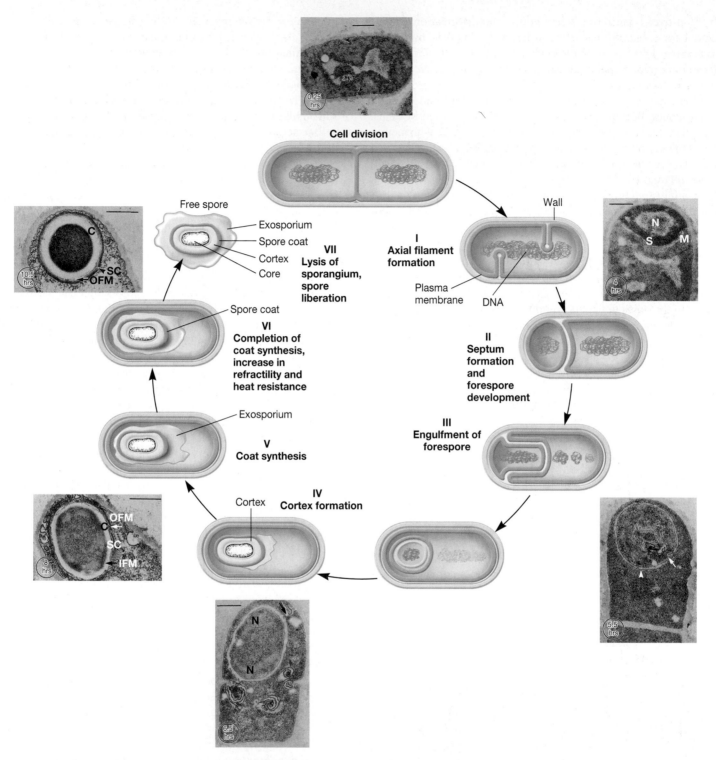

Figure 3.46 **Endospore Formation: Life Cycle of *Bacillus megaterium.*** The stages are indicated by Roman numerals. The circled numbers in the photographs refer to the hours from the end of the logarithmic phase of growth: 0.25 h—a typical vegetative cell; 4 h—stage II cell, septation; 5.5 h—stage III cell, engulfment; 6.5 h—stage IV cell, cortex formation; 8 h—stage V cell, coat formation; 10.5 h—stage VI cell, mature spore in sporangium. Abbreviations used: C, cortex; IFM and OFM, inner and outer forespore membranes; M, mesosome; N, nucleoid; S, septum; SC, spore coats. Bars = 0.5 μm.

The transformation of dormant spores into active vegetative cells seems almost as complex a process as sporulation. It occurs in three stages: (1) activation, (2) germination, and (3) outgrowth (**figure 3.47**). Activation is a process that prepares spores for ger-

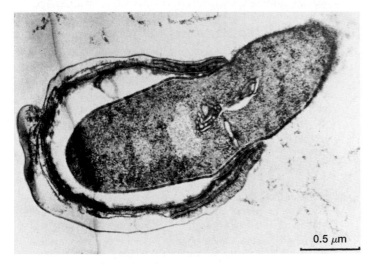

Figure 3.47   Endospore Germination. *Clostridium pectinovorum* emerging from the spore during germination.

mination and usually results from treatments such as heating. This is followed by **germination,** the breaking of the spore's dormant state. This process is characterized by spore swelling, rupture, or absorption of the spore coat, loss of resistance to heat and other stresses, loss of refractility, release of spore components, and increase in metabolic activity. Many normal metabolites or nutrients (e.g., amino acids and sugars) can trigger germination after activation. Germination is followed by the third stage, outgrowth. The spore protoplast makes new components, emerges from the remains of the spore coat, and develops again into an active bacterium.

1. Describe the structure of the bacterial endospore using a labeled diagram.

2. Briefly describe endospore formation and germination. What is the importance of the endospore? What might account for its heat resistance?

3. How might one go about showing that a bacterium forms true endospores?

4. Why do you think dehydration of the protoplast is an important factor in the ability of endospores to resist environmental stress?

Summary

3.1   Overview of Procaryotic Cell Structure

a. Procaryotes may be spherical (cocci), rod-shaped (bacilli), spiral, or filamentous; they may form buds and stalks; or they may have no characteristic shape (pleomorphic) (**figures 3.1** and **3.2**).

b. Procaryotic cells can remain together after division to form pairs, chains, and clusters of various sizes and shapes.

c. Procaryotes are much simpler structurally than eucaryotes, but they do have unique structures (**figure 3.4**). **Table 3.1** summarizes the major functions of procaryotic cell structures.

3.2   Procaryotic Cell Membranes

a. The plasma membrane fulfills many roles, including acting as a semipermeable barrier, carrying out respiration and photosynthesis, and detecting and responding to chemicals in the environment.

b. The fluid mosaic model proposes that cell membranes are lipid bilayers in which integral proteins are buried. Peripheral proteins are loosely associated with the membrane (**figure 3.5**).

c. Bacterial membranes are bilayers composed of phospholipids constructed of fatty acids connected to glycerol by ester linkages (**figure 3.6**). Bacterial membranes usually lack sterols but often contain hopanoids (**figure 3.7**).

d. Some bacteria have simple internal membrane systems containing photosynthetic and respiratory machinery (**figure 3.8**).

e. Archaeal membranes are composed of glycerol diether and diglycerol tetraether lipids (**figure 3.9**). Membranes composed of glycerol diether are lipid bilayers. Membranes composed of diglycerol tetraethers are lipid monolayers (**figure 3.11**). The overall structure of a monolayer membrane is similar to that of the bilayer membrane in that the membrane has a hydrophobic core and its surfaces are hydrophilic.

3.3   Procaryotic Cytoplasm

a. The cytoplasm of procaryotes contains proteins that are similar in structure and function to the cytoskeletal proteins observed in eucaryotes (**figure 3.12**).

b. The cytoplasm of procaryotes contains inclusion bodies. Most are used for storage (glycogen inclusions, PHB inclusions, cyanophycin granules, carboxysomes, and polyphosphate granules) (**figure 3.13**), but some are used for other purposes (magnetosomes and gas vacuoles) (**figure 3.14**).

c. The cytoplasm of procaryotes is packed with 70S ribosomes (**figure 3.15**).

d. Procaryotic genetic material is located in an area within the cytoplasm called the nucleoid. The nucleoid is not usually enclosed by a membrane (**figure 3.16**).

e. In most procaryotes, the nucleoid contains a single chromosome. The chromsosome usually consists of a double-stranded, covalently closed, circular DNA molecule.

f. Plasmids are extrachromosomal DNA molecules found in many procaryotes. Some are episomes—plasmids that are able to exist freely in the cytoplasm or can be integrated into the chromosome.

g. Although plasmids are not required for survival in most conditions, they can encode traits that confer selective advantage in some environments.

h. Many types of plasmids have been identified. Conjugative plasmids encode genes that promote their transfer from one cell to another. Resistance factors have genes conferring resistance to antibiotics. Col plasmids contain genes for the synthesis of

colicins, proteins that kill *E. coli.* Other plasmids encode virulence factors or metabolic capabilities **(table 3.3).**

3.4 Bacterial Cell Walls

a. The vast majority of procaryotes have a cell wall outside the
plasma membrane to give them shape and protect them from
osmotic stress.

b. Bacterial walls are chemically complex and usually contain
peptidoglycan **(figures 3.17–3.21).**

c. Bacteria often are classified as either gram positive or gram
negative based on differences in cell wall structure and their
response to Gram staining.

d. Gram-positive walls have thick, homogeneous layers of peptidoglycan and teichoic acids **(figure 3.23).** Gram-negative bacteria
have a thin peptidoglycan layer surrounded by a complex outer
membrane containing lipopolysaccharides (LPSs) and other
components **(figure 3.25).**

e. The mechanism of the Gram stain is thought to depend on the
peptidoglycan, which binds crystal violet tightly, preventing the
loss of crystal violet during the ethanol wash.

3.5 Archaeal Cell Walls

a. Archaeal cell walls do not contain peptidoglycan **(figure 3.29).**

b. Archaea exhibit great diversity in their cell wall makeup. Some
archaeal cell walls are composed of heteropolysaccharides, some
are composed of glycoprotein, and some are composed of protein.

3.6 Components External to the Cell Wall

a. Capsules, slime layers, and glycocalyxes are layers of material
lying outside the cell wall. They can protect procaryotes from
certain environmental conditions, allow procaryotes to attach
to surfaces, and protect pathogenic bacteria from host defenses
(figures 3.31 and 3.32).

b. S-layers are observed in some bacteria and many archaea. They
are composed of proteins or glycoprotein and have a characteristic geometric shape. In many archaea, the S-layer serves as the
cell wall **(figure 3.33).**

c. Pili and fimbriae are hairlike appendages. Fimbriae function
primarily in attachment to surfaces, but some are involved in
motility. Sex pili participate in the transfer of DNA from one
bacterium to another **(figure 3.34).**

d. Many procaryotes are motile, usually by means of threadlike,
locomotory organelles called flagella **(figure 3.35).** Bacterial
species differ in the number and distribution of their flagella.

3.7 Bacterial Motility and Chemotaxis

a. Several types of bacterial motility have been observed. These
include movement by flagella, spirochete motility, twitching
motility, and gliding motility.

b. The bacterial flagellar filament is a rigid helix that rotates like a
propeller to push the bacterium through water **(figure 3.38).**

c. Spirochete motility is brought about by flagella that are wound
around the cell and remain within the periplasmic space. When
they rotate, the outer sheath of the spirochete is thought to rotate,
thus moving the cell **(figure 3.40).**

d. Twitching and gliding motility are similar in that both occur on
moist surfaces and can involve type IV fimbriae and the secretion of slime. Twitching motility is a jerky movement, whereas
gliding motility is smooth.

e. Motile procaryotes can respond to gradients of attractants and
repellents, a phenomenon known as chemotaxis.

f. A peritrichously flagellated bacterium accomplishes movement
toward an attractant by increasing the length of time it spends
moving toward the attractant and shortening the time it spends
tumbling **(figure 3.42).** Conversely, a bacterium increases its run
time when it moves away from a repellent.

3.8 Bacterial Endospores

a. Some bacteria survive adverse environmental conditions by
forming endospores, dormant structures resistant to heat, desiccation, and many chemicals **(figure 3.44).**

b. Both endospore formation and germination are complex processes that begin in response to certain environmental signals and
involve numerous stages **(figures 3.46 and 3.47).**

Critical Thinking Questions

1. Propose a model for the assembly of a flagellum in a gram-
positive cell envelope. How would that model need to be modified
for the assembly of a flagellum in a gram-negative cell envelope?

2. If you could not use a microscope, how would you determine
whether a cell is procaryotic or eucaryotic? Assume the organism can be cultured easily in the laboratory.

3. The peptidoglycan of bacteria has been compared with the
chain mail worn beneath a medieval knight's suit of armor. It
provides both protection and flexibility. Can you describe other
structures in biology that have an analogous function? How are
they replaced or modified to accommodate the growth of the
inhabitant?

Learn More

Learn More by visiting the Prescott website at www.mhhe.com/prescottprinciples, where you will find a complete list of references.

Eucaryotic Cell Structure and Function

4

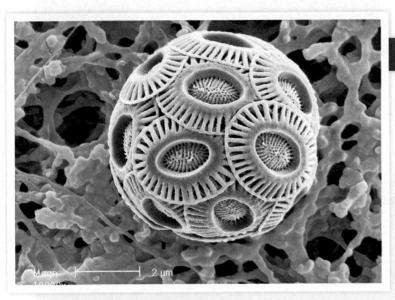

Emiliana huxleyi: a protist covered with calcite scales.

The key to every biological problem must finally be sought in the cell.

—E. B. Wilson

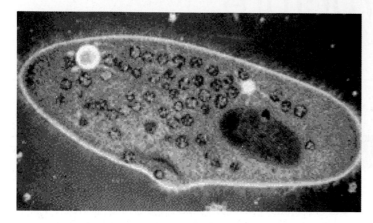

(a) *Paramecium*

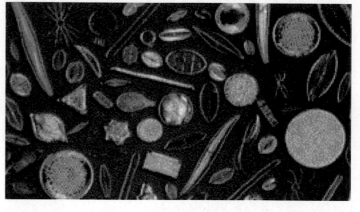

(b) Diatom frustules

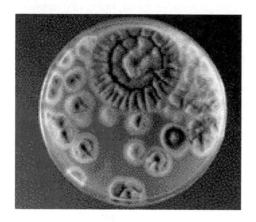

(c) *Penicillium*

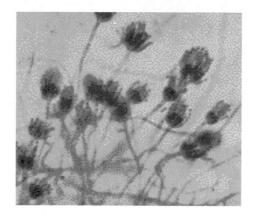

(d) *Penicillium*

(e) *Stentor*

Figure 4.1 Representative Examples of Eucaryotic Microorganisms. (a) *Paramecium* as seen with interference-contrast microscopy (×115). (b) Mixed diatom frustules (×100). (c) *Penicillium* colonies, and (d) a microscopic view of the mold's hyphae and conidia (×220). (e) *Stentor.* The ciliated protozoa are extended and actively feeding, dark-field microscopy (×100).

Although procaryotes are immensely important in microbiology and have occupied a large portion of microbiologists' attention in the past, protists and fungi also are microorganisms and have been extensively studied. These eucaryotes often are extraordinarily complex and prominent members of ecosystems (**figure 4.1**). In addition, many protists and fungi are important model organisms, as well as being exceptionally useful in industrial microbiology. A number of protists and fungi are also major human pathogens; one only need think of candidiasis, malaria, or African sleeping sickness to appreciate the significance of eucaryotes in medical microbiology. So, although this text emphasizes procaryotes, eucaryotic microorganisms also demand attention and are briefly discussed in this chapter.

Chapter 4 focuses on eucaryotic cell structure and its relationship to cell function. Because many valuable studies on eucaryotic cell ultrastructure have used organisms other than microorganisms, some work on nonmicrobial cells is presented. Then procaryotic and eucaryotic cells are compared in some depth. At the end of the chapter, an overview of protists and fungi is provided.

4.1 OVERVIEW OF EUCARYOTIC CELL STRUCTURE

The most obvious difference between eucaryotic and procaryotic cells is in their use of membranes. Eucaryotic cells have membrane-delimited nuclei, and membranes play a prominent part in the structure of many other organelles (**figures 4.2** and **4.3**). **Organelles** are intracellular structures that perform specific functions in cells analogous to the functions of organs in the body. A comparison of figures 4.2 and 4.3 with figures 3.4 and 3.13*a* shows how structurally complex the eucaryotic cell is. This complexity is due chiefly to the use of internal membranes for several purposes. The partitioning of the eucaryotic cell interior by membranes makes possible the placement of different biochemical and physiological functions in separate compartments so that they can more easily take place simultaneously under

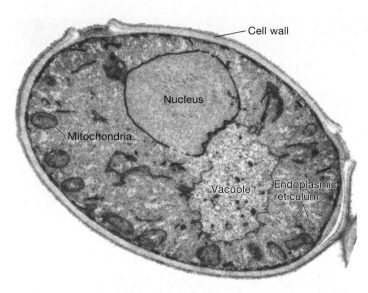

Figure 4.2 Eucaryotic Cell Ultrastructure. The yeast *Saccharomyces* (×7,200). Note the nucleus, mitochondrion, vacuole, endoplasmic reticulum, and cell wall.

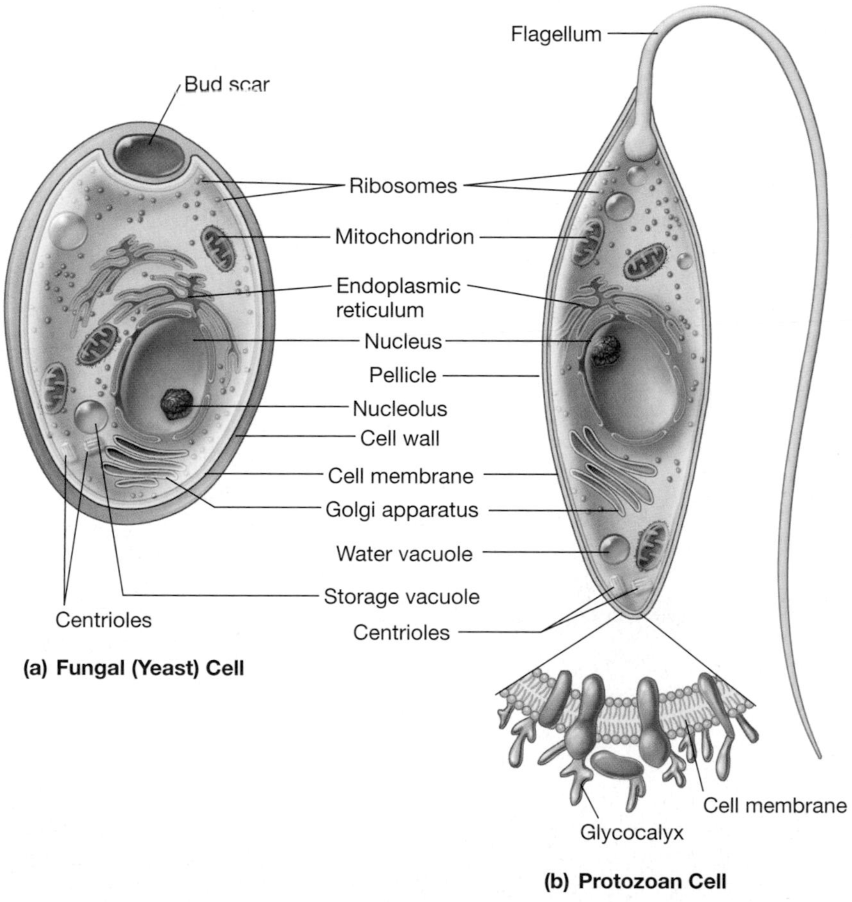

independent control and proper coordination. Large membrane surfaces make possible greater respiratory and photosynthetic activity because these processes are located exclusively in membranes. The intracytoplasmic membrane complex also serves as a transport system to move materials between different cell locations. Thus abundant membrane systems probably are necessary in eucaryotic cells because of their large volume and the need for adequate regulation, metabolic activity, and transport.

Figures 4.2 and 4.3 illustrate most of the organelles to be discussed here. **Table 4.1** briefly summarizes the functions of the major eucaryotic organelles. Our detailed discussion of eucaryotic cell structure begins with the eucaryotic membrane. We then proceed to organelles within the cytoplasm and finally to components outside the membrane.

1. What is an organelle?
2. Why is the compartmentalization of the cell interior advantageous to eucaryotic cells?

4.2 EUCARYOTIC MEMBRANES

As discussed in chapter 3, the fluid mosaic model of membrane structure is based largely on studies of eucaryotic membranes. In eucaryotes, the major membrane lipids are phosphoglycerides, sphingolipids, and cholesterol (**figure 4.4**). The distribution of these lipids is asymmetric. Lipids in the outer monolayer differ from those of the inner monolayer. Although most lipids in individual monolayers mix freely with each other, there are microdomains that differ in lipid and protein composition. One such microdomain is the **lipid raft,** which is enriched in cholesterol and lipids with many saturated fatty acids, including some sphingolipids. The lipid raft spans the membrane bilayer, and lipids in the adjacent monolayers interact. Lipid rafts appear to participate in a variety of cellular processes (e.g., cell movement and signal transduction). They also may be involved in the entrance of some viruses into their host cells and the assembly of some viruses before they are released from their host cells.

4.3 EUCARYOTIC CYTOPLASM

The **cytoplasm** is one of the most important and complex parts of a cell. It consists of a liquid component, the cytosol, in which many organelles are located. It is the location of many important biochemical processes

Flagellum

Bud scar

Ribosomes

Mitochondrion

Endoplasmic reticulum

Nucleus

Pellicle

Nucleolus

Cell wall

Cell membrane

Golgi apparatus

Water vacuole

Storage vacuole

Centrioles

Centrioles

(a) Fungal (Yeast) Cell

Cell membrane

Glycocalyx

(b) Protozoan Cell

Figure 4.3 The Structure of Two Representative Eucaryotic Microbes. Illustrations of a yeast cell (fungus) (a) and the flagellated protozoan *Peranema* (b).

Table 4.1	Functions of Eucaryotic Organelles
Plasma membrane	Mechanical cell boundary; selectively permeable barrier with transport systems; mediates cell-cell interactions and adhesion to surfaces; secretion
Cytoplasm	Environment for other organelles; location of many metabolic processes
Microfilaments, intermediate filaments, and microtubules	Cell structure and movements; form the cytoskeleton
Endoplasmic reticulum	Transport of materials; protein and lipid synthesis
Ribosomes	Protein synthesis
Golgi apparatus	Packaging and secretion of materials for various purposes; lysosome formation
Lysosomes	Intracellular digestion
Mitochondria	Energy production through use of the tricarboxylic acid cycle, electron transport, oxidative phosphorylation, and other pathways
Chloroplasts	Photosynthesis—trapping light energy and formation of carbohydrate from CO_2 and water
Nucleus	Repository for genetic information; control center for cell
Nucleolus	Ribosomal RNA synthesis; ribosome construction
Cell wall and pellicle	Strengthen and give shape to the cell
Cilia and flagella	Cell movement
Vacuole	Temporary storage and transport; digestion (food vacuoles); water balance (contractile vacuole)

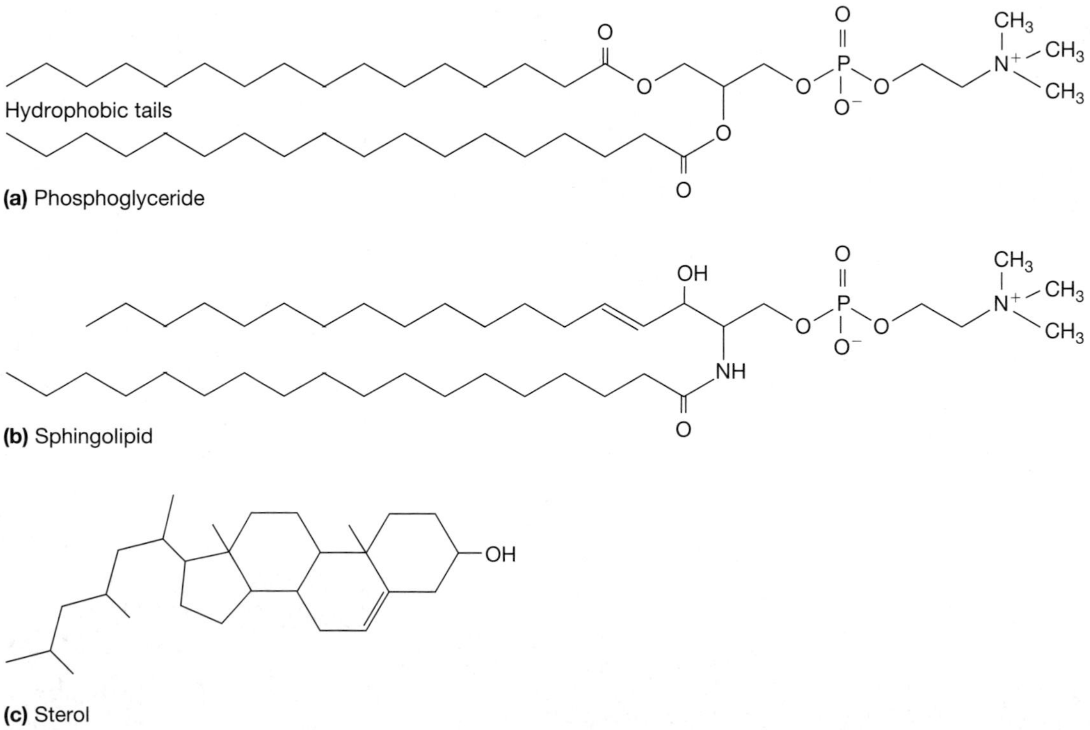

(a) Phosphoglyceride

(b) Sphingolipid

(c) Sterol

Figure 4.4 Examples of Eucaryotic Membrane Lipids. (a) Phosphatidylcholine, a phosphoglyceride. (b) Sphingomyelin, a sphingolipid. (c) Cholesterol, a sterol.

and several physical changes seen in cells—viscosity changes, cytoplasmic streaming, and others—also are due to cytoplasmic activity.

A major component of the cytoplasm is a vast network of interconnected filaments called the **cytoskeleton.** The cytoskeleton plays a role in both cell shape and movement. Three types of filaments form the cytoskeleton: microfilaments, microtubules, and intermediate filaments (**figure 4.5**). **Microfilaments** are minute protein filaments, 4 to 7 nm in diameter, that are either scattered within the cytoplasm or organized into networks and parallel arrays. Microfilaments are composed of an actin protein that is similar to the actin contractile protein of muscle tissue. Microfilaments are involved in cell motion and shape changes such as the motion of pigment granules, amoeboid movement, and protoplasmic streaming in slime molds. Interestingly, some pathogens use the actin proteins of their eucaryotic hosts to move rapidly through the host cell and to propel themselves into new host cells (*see figure 30.4*). ▶▶ *Protist classification:* Eumycetozoa *and* Stramenopiles *(section 23.2).*

Microtubules are shaped like thin cylinders about 25 nm in diameter. They are complex structures constructed of two spherical protein subunits—α-tubulin and β-tubulin. The two proteins are the same molecular weight and differ only slightly in terms of their amino acid sequence and tertiary structure. Each tubulin is approximately 4 to 5 nm in diameter. These subunits are assembled in a helical arrangement to form a cylinder with an average of 13 subunits in one turn (figure 4.5).

Microtubules serve at least three purposes: (1) they help maintain cell shape, (2) they are involved with microfilaments in cell movements, and (3) they participate in intracellular transport processes. Microtubules are found in long, thin cell structures requiring support such as the axopodia (long, slender, rigid pseudopodia) of protists (**figure 4.6**). Microtubules also are present in structures that participate in cell or organelle movements—the mitotic spindle, cilia, and flagella.

Intermediate filaments are heterogeneous elements of the cytoskeleton. They are about 10 nm in diameter and are assembled from a group of proteins that can be divided into several classes. Intermediate filaments having different functions are assembled from one or more of these classes of proteins. The role of intermediate filaments in eucaryotic microorganisms is unclear. Thus far, they have been identified and studied only in animals: some intermediate filaments have been shown to form the nuclear lamina, a structure that provides support for the nuclear envelope (p. 73); and other intermediate filaments help link cells together to form tissues.

1. Compare the membranes of *Eucarya, Bacteria,* and *Archaea.* How are they similar? How do they differ?

2. What are lipid rafts? What roles do they play in eucaryotic cells?

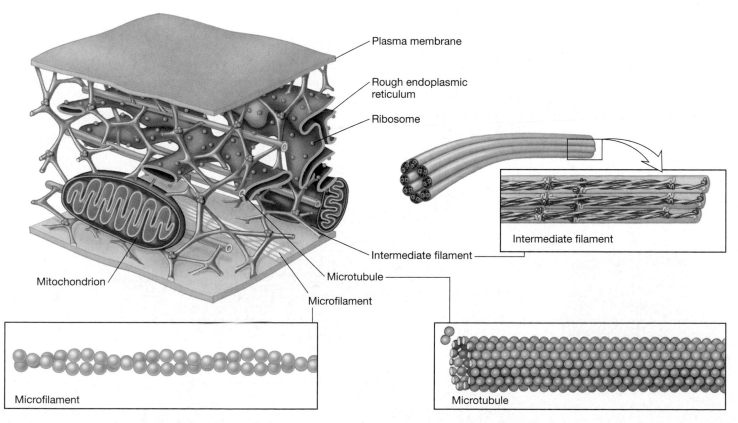

Figure 4.5   The Eucaryotic Cytoplasm and Cytoskeleton.   The cytoplasm of eucaryotic cells contains many important organelles. The cytoskeleton helps form a framework within which the organelles lie. The cytoskeleton is composed of three elements: microfilaments, microtubules, and intermediate filaments.

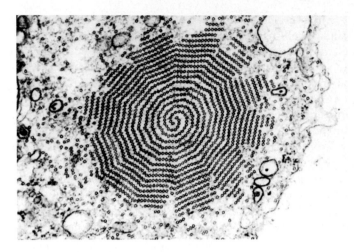

Figure 4.6   Cytoplasmic Microtubules. Electron micrograph of a transverse section through the axopodium of a protist known as a heliozoan (×48,000). Note the parallel array of microtubules organized in a spiral pattern.

3. Define cytoplasm, microfilament, microtubule, and tubulin. Discuss the roles of microfilaments, intermediate filaments, and microtubules.

4. What are the functions of the cytoskeleton?

4.4   ORGANELLES OF THE BIOSYNTHETIC-SECRETORY AND ENDOCYTIC PATHWAYS

In addition to the cytoskeleton, the cytoplasm is permeated with an intricate complex of membranous organelles and vesicles that move materials into the cell from the outside (endocytic pathway) and from the inside of the cell out, as well as from location to location within the cell (biosynthetic-secretory pathway). In this section, some of these organelles are described. This is followed by a summary of how the organelles function in the biosynthetic-secretory and endocytic pathways.

Endoplasmic Reticulum

The **endoplasmic reticulum (ER)** (figure 4.3) is an irregular network of branching and fusing membranous tubules, around 40 to 70 nm in diameter, and many flattened sacs called cisternae (s., cisterna). The nature of the ER varies with the functional and physiological status of the cell. In cells synthesizing a great deal of protein for purposes such as secretion, a large part of the ER is studded on its outer surface with ribosomes and is called **rough endoplasmic reticulum (RER)**. Other cells, such as those producing large quantities of lipids, have ER that lacks ribosomes. This is **smooth endoplasmic reticulum (SER)**.

The endoplasmic reticulum has many important functions. Not only does it transport proteins, lipids, and other materials through the cell, it is also involved in the synthesis of many of the materials it transports. Lipids and proteins are synthesized by ER-associated enzymes and ribosomes. Polypeptide chains synthesized on RER-bound ribosomes may be inserted either into the ER membrane or into its lumen for transport elsewhere. The ER is also a major site of cell membrane synthesis.

Golgi Apparatus

The **Golgi apparatus** is composed of flattened, saclike cisternae stacked on each other (**figure 4.7**). These membranes, like the smooth ER, lack bound ribosomes. Usually around four to eight cisternae are in a stack, although there may be

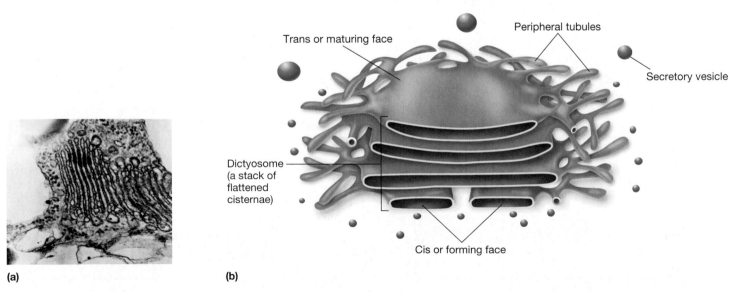

(a)   (b)

Figure 4.7   Golgi Apparatus Structure. Golgi apparatus of *Euglena gracilis*. Cisternal stacks are shown in the electron micrograph (×165,000) in (a) and diagrammatically in (b).

many more. Each is 15 to 20 nm thick and separated from other cisternae by 20 to 30 nm. A complex network of tubules and vesicles (20 to 100 nm in diameter) is located at the edges of the cisternae. The stack of cisternae has two faces that are quite different from one another. The sacs on the cis or forming face often are associated with the ER and differ from the sacs on the trans or maturing face in thickness, enzyme content, and degree of vesicle formation.

The Golgi apparatus is present in most eucaryotic cells, but many fungi and ciliate protozoa lack a well-formed structure. Sometimes the Golgi consists of a single stack of cisternae; however, many cells may contain 20 or more separate stacks. These stacks of cisternae, often called dictyosomes, can be clustered in one region or scattered about the cell.

The Golgi apparatus packages materials and prepares them for secretion, the exact nature of its role varying with the organism. For instance, the surface scales of some flagellated photosynthetic and radiolarian protists appear to be constructed within the Golgi apparatus and then transported to the surface in vesicles. The Golgi often participates in the development of cell membranes and the packaging of cell products. The growth of some fungal hyphae occurs when Golgi vesicles contribute their contents to the wall at the hyphal tip. **>>** *Eucaryotic microbes (chapter 23)*

Lysosomes

Lysosomes are found in most eucaryotic organisms, including protists, fungi, plants, and animals. Lysosomes are roughly spherical and enclosed in a single membrane; they average about 500 nm in diameter but range from 50 nm to several μm in size. They are involved in intracellular digestion and contain the enzymes needed to digest all types of macromolecules. These enzymes, called hydrolases, catalyze the hydrolysis of molecules and function best under slightly acidic conditions (usually around pH 3.5 to 5.0). Lysosomes maintain an acidic environment by pumping protons into their interior.

Biosynthetic-Secretory Pathway

The **biosynthetic-secretory pathway** is used to move materials to lysosomes as well as from the inside of the cell to either the plasma membrane or cell exterior. The process is complex and not fully understood. The movement of proteins is of particular importance and is the focus of this discussion.

Proteins destined for the cell membrane, lysosomes, or secretion are synthesized by ribosomes attached to the rough endoplasmic reticulum (RER). These proteins have sequences of amino acids that target them to the lumen of the RER through which they move until released in small vesicles that bud from the ER. As the proteins pass through the ER, they are often modified by the addition of sugars—a process known as glycosylation.

The vesicles released from the ER travel to the cis face of the Golgi apparatus (figure 4.7). A popular model of the biosynthetic-

secretory pathway posits that these vesicles fuse to form the cis face of the Golgi. The proteins then proceed to the trans face of the Golgi by a process called cisternal maturation. As the proteins proceed from the cis to the trans side, they are further modified. Some of these modifications target the proteins for their final location.

Transport vesicles are released from the trans face of the Golgi. Some deliver their contents to lysosomes. Others deliver proteins and other materials to the cell membrane. Two types of vesicles transport materials to the cell membrane. One type constitutively delivers proteins in an unregulated manner, releasing them to the outside of the cell as the transport vesicle fuses with the plasma membrane. Other vesicles, called secretory vesicles, are found only in multicellular eucaryotes, where they are observed in secretory cells such as mast cells and other cells of the immune system. Secretory vesicles store the proteins to be released until the cell receives an appropriate signal. Once received, the secretory vesicles move to the plasma membrane, fuse with it, and release their contents to the cell exterior. **>>** *Cells, tissues, and organs of the immune system (section 28.2)*

One interesting and important feature of the biosynthetic-secretory pathway is its quality-assurance mechanism. Proteins that fail to fold or have misfolded are not transported to their intended destination. Instead they are secreted into the cytosol, where they are targeted for destruction by the attachment of several small ubiquitin polypeptides, as detailed in **figure 4.8.** Ubiquitin marks the protein for degradation, which is accomplished by a huge, cylindrical complex called a 26S **proteasome.** The protein is broken down to smaller peptides using energy supplied by adenosine triphosphate (ATP) when the terminal phosphate is removed. As the protein is broken down, ubiquitins are released. The proteasome also is involved in producing peptides for antigen presentation during many immunological responses described in chapter 29. **>>** *ATP (section 9.4)*

Endocytic Pathway

Endocytosis is used to bring materials into the cell from the outside. During endocytosis, a cell takes up solutes or particles by enclosing them in vesicles pinched off from the plasma membrane. In most cases, these materials are delivered to a lysosome where they are digested. Endocytosis occurs regularly in all cells as a mechanism for recycling molecules in the membrane. In addition, some cells have specialized **endocytic pathways** that allow them to concentrate materials outside the cell before bringing them in. Others use endocytic pathways as a feeding mechanism. Many viruses and other intracellular pathogens use endocytic pathways to enter host cells.

Numerous types of endocytosis have been described. **Phagocytosis** involves the use of protrusions from the cell surface to surround and engulf particulates. It is carried out by certain immune system cells and many eucaryotic microbes. The endocytic vesicles formed by phagocytosis are called **phagosomes**

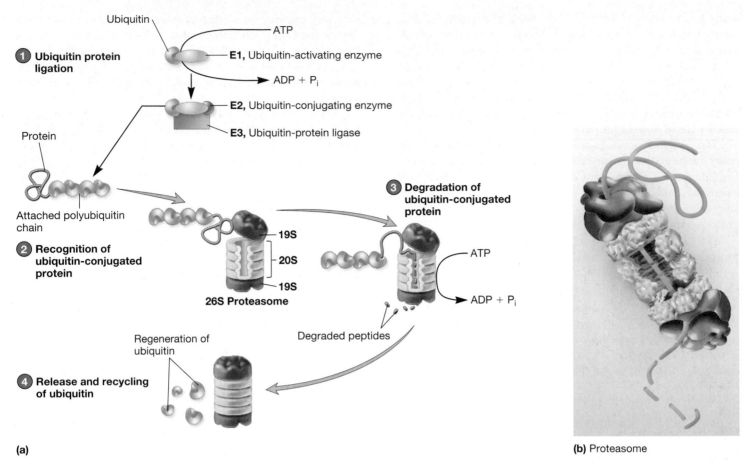

(a)

(b) Proteasome

Figure 4.8 **Proteasome Degradation of Proteins.** (a) The first step in this protein degradation pathway is to tag the target protein with a small polypeptide called ubiquitin. This requires the action of three enzymes and energy is consumed. Once tagged, the protein is recognized by the 26S proteasome. It passes into the large, cylindrical proteasome and is cleaved into smaller peptides, which are released into the cytoplasm. The amino acids in the small peptides can be recycled and used in the synthesis of new proteins. The ubiquitin polypeptides are regenerated and can participate in the degradation of other proteins. (b) A model of the 26S proteasome, showing its cylindrical structure and the location of the tagged protein within the cylinder.

(figure 4.9). Other types of endocytosis involve invagination of the plasma membrane. As the membrane invaginates, it encloses liquid, soluble matter and sometimes particulates in the resulting endocytic vesicle. One example of endocytosis by invagination is clathrin-dependent endocytosis. Clathrin-dependent endocytosis begins with coated pits, which are specialized membrane regions coated with the protein **clathrin** on the cytoplasmic side. The endocytic vesicles formed when these regions invaginate are called **coated vesicles.** Coated pits have receptors on their extracellular side that specifically bind macromolecules, concentrating them before they are endocytosed. Therefore this endocytic mechanism is referred to as **receptor-mediated endocytosis.** Clathrin-dependent endocytosis is used to internalize such things as hormones, growth factors, iron, and cholesterol. Another example of endocytosis by invagination is caveolae-dependent endocytosis. **Caveolae** ("little caves") are tiny, flask-shaped invaginations of the plasma membrane (about 50 to 80 nm in diameter) that are enriched in cholesterol and the membrane protein caveolin. The vesicles formed when caveolae pinch off are called caveolar vesicles. Caveolae-dependent endocytosis has

been implicated in signal transduction, transport of small molecules such as folic acid, as well as transport of macromolecules. Evidence exists that toxins such as cholera toxin enter their target cells via caveolae. Caveolae also appear to be used by many viruses, bacteria, and protozoa to enter host cells.

With the exception of caveolar vesicles, all other endocytic vesicles eventually deliver their contents to lysosomes. However, the route used varies. Coated vesicles fuse with small organelles containing lysosomal enzymes. These organelles are called early endosomes (figure 4.9). Early endosomes mature into late endosomes, which fuse with transport vesicles from the Golgi delivering additional lysosomal enzymes. Late endosomes eventually become lysosomes. The development of endosomes into lysosomes is not well understood. It appears that maturation involves the movement of the organelles to a more central location in the cell and the selective retrieval of membrane proteins. Phagosomes take a slightly different route to lysosomes; they fuse with late endosomes rather than early endosomes (figure 4.9).

Materials for digestion also can be delivered to lysosomes by a route that does not involve endocytosis. Cells selectively digest and

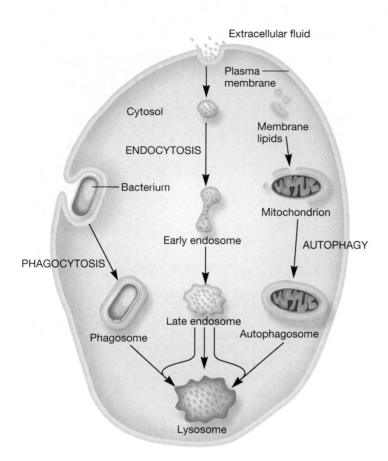

Figure 4.9   The Endocytic Pathway.   Materials ingested by endocytic processes (except caveolae-dependent endocytosis) are delivered to lysosomes. The pathway to lysosomes differs, depending on the type of endocytosis. In addition, cell components are recycled when autophagosomes deliver them to lysosomes for digestion. This process is called autophagy.

recycle cytoplasmic components (including organelles such as mitochondria) by a process called **autophagy.** It is believed that the cell components to be digested are surrounded by a double membrane, as shown in figure 4.9. The source of the membrane is unknown, but it has been suggested that a portion of the ER is used. The resulting **autophagosome** fuses with a late endosome in a manner similar to that seen for phagosomes.

No matter the route taken, digestion occurs once the lysosome is formed. Amazingly, the lysosome accomplishes this without releasing its digestive enzymes into the cytoplasm. As the contents of the lysosome are digested, small products of digestion leave the lysosome and are used as nutrients or for other purposes. The resulting lysosome containing undigested material is often called a **residual body.** In some cases, the residual body can release its contents to the cell exterior by a process called lysosome secretion.

1. How do RER and SER differ from one another in terms of structure and function? List the processes in which the ER is involved.

2. Describe the structure of a Golgi apparatus in words and with a diagram. How do the cis and trans faces of the Golgi apparatus differ? List the major Golgi apparatus functions.

3. What is a proteasome? Why is it important to the proper functioning of the endoplasmic reticulum?

4. What are lysosomes? How do they participate in intracellular digestion?

5. Describe the biosynthetic-secretory pathway. To what destinations does this pathway deliver proteins and other materials?

6. Define endocytosis. Describe the endocytic pathway and the three routes that deliver materials to lysosomes for digestion. Which type of endocytosis does not deliver ingested material to lysosomes?

7. Define autophagy, autophagosome, phagosome, phagocytosis, and residual body.

8. Caveolae-mediated endocytosis is used by a number of pathogens to enter their host cells. Why might this route of entry be advantageous to the pathogens that use it?

4.5   ORGANELLES INVOLVED IN GENETIC CONTROL OF THE CELL

DNA is the molecule that houses the genetic blueprint of the cell. Eucaryotic cells differ dramatically from procaryotic cells in the way DNA is stored and used. In this section, the organelles involved with these important cellular functions are introduced.

Nucleus

The nucleus is by far the most visually prominent organelle in eucaryotic cells. It was discovered early in the study of cell structure and was shown by Robert Brown in 1831 to be a constant feature of eucaryotic cells. The **nucleus** is the repository for the cell's genetic information and its control center.

Nuclei are membrane-delimited spherical bodies about 5 to 7 μm in diameter (figures 4.2 and 4.3). Dense fibrous material called **chromatin** can be seen within the nucleoplasm of the nucleus of a stained cell. This is the DNA-containing part of the nucleus. In nondividing cells, chromatin is dispersed, but it condenses during cell division to become visible as **chromosomes.** Some chromatin, the euchromatin, is loosely organized, and it contains those genes that are actively expressed. In contrast, heterochromatin is coiled more tightly, appears darker in the electron microscope, and is genetically inactive most of the time.

The nucleus is bounded by the **nuclear envelope** (figures 4.2 and 4.3), a complex structure consisting of inner and outer membranes separated by a 15 to 75 nm perinuclear space. The

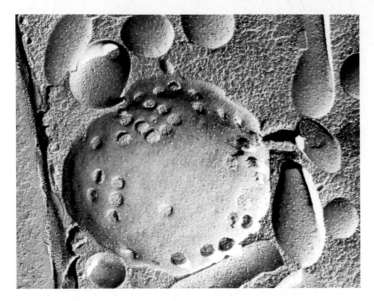

Figure 4.10 The Nucleus. A freeze-etch preparation of the conidium of the fungus *Geotrichum candidum* (×44,600). Note the large, convex nuclear surface with nuclear pores scattered over it.

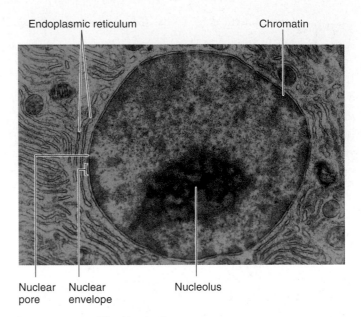

Endoplasmic reticulum   Chromatin

Nuclear pore   Nuclear envelope   Nucleolus

Figure 4.11 The Nucleolus. The nucleolus is a prominent feature of the nucleus. It functions in rRNA synthesis and the assembly of ribosomal subunits. Chromatin, nuclear pores, and the nuclear envelope are also visible in this electron micrograph of an interphase nucleus.

envelope is continuous with the ER at several points, and its outer membrane is covered with ribosomes. A network of inter-mediate filaments, called the nuclear lamina, is observed in animal cells. It lies against the inner surface of the nuclear envelope and supports it. Chromatin usually is associated with the inner membrane. Many **nuclear pores** penetrate the enve-lope, and each pore is formed by a fusion of the outer and inner membranes (**figure 4.10**). Pores are about 70 nm in diameter and collectively occupy about 10 to 25% of the nuclear surface. A complex, ringlike arrangement of granular and fibrous material called the annulus is located at the edge of each pore. The nuclear pores serve as transport routes between the nucleus and sur-rounding cytoplasm. Although the function of the annulus is not understood, it may either regulate or aid the movement of mate-rial through the pores. Substances also move directly through the nuclear envelope by unknown mechanisms.

Often the most noticeable structure within the nucleus is the **nucleolus** (**figure 4.11**). A nucleus may contain from one to many nucleoli. Although the nucleolus is not membrane-enclosed, it is a complex organelle with separate granular and fibrillar regions. It is present in nondividing cells but frequently disappears during mitosis. After mitosis, the nucleolus reforms around the nucleolar organizer, a particular part of a specific chromosome.

The nucleolus plays a major role in ribosome synthesis. The DNA of the nucleolar organizer directs the production of ribo-somal RNA (rRNA). This RNA is synthesized in a single long piece that is cut to form the final rRNA molecules. The processed rRNAs combine with ribosomal proteins (which have been syn-thesized in the cytoplasm) to form partially completed ribosomal subunits. The granules seen in the nucleolus are probably these

subunits. Immature ribosomal subunits then leave the nucleus, presumably by way of the nuclear envelope pores, and mature in the cytoplasm.

Eucaryotic Ribosomes

The eucaryotic ribosome (i.e., one not found in mitochondria and chloroplasts) is larger than the procaryotic 70S ribosome. It is a dimer of a 60S and a 40S subunit, is about 22 nm in diameter, and has a sedimentation coefficient of 80S and a molecular weight of 4 million. Eucaryotic ribosomes are either associated with the endoplasmic reticulum or free in the cytoplasm. When bound to the endoplasmic reticulum to form rough ER, they are attached through their 60S subunits.

Both free and ER-bound ribosomes synthesize proteins. Proteins made on the ribosomes of the RER are often secreted or are inserted into the ER membrane as integral membrane pro-teins. Free ribosomes are the sites of synthesis for nonsecretory and nonmembrane proteins. Some proteins synthesized by free ribosomes are inserted into organelles such as the nucleus, mito-chondrion, and chloroplast. As discussed in chapter 12, proteins called molecular chaperones aid the proper folding of proteins after synthesis. They also assist the transport of proteins into eucaryotic organelles such as mitochondria.

1. Describe the structure of the eucaryotic 80S ribosome and contrast it with the procaryotic ribosome.

2. How do free ribosomes and those bound to the ER differ in function?

4.6 ORGANELLES INVOLVED IN ENERGY CONSERVATION

The two most important organelles involved in energy conservation are mitochondria and chloroplasts. They are of scientific interest not only for this role but also because of their evolutionary history. Both are thought to be derived from bacterial cells that invaded or were ingested by early ancestors of eucaryotic cells. This hypothesis, called the endosymbiotic hypothesis, is supported by several lines of evidence, including the fact that both organelles contain DNA and ribosomes. These ribosomes are the same size as bacterial ribosomes, and the 16S ribosomal RNA sequences are most similar to those of *Bacteria (see figure 1.1)*. Other bacteria-like features are noted in the discussions that follow. << *Procaryotic cytoplasm: Ribosomes (section 3.3);* >> *Microbial evolution: The endosymbiotic origin of mitochondria and chloroplasts (section 17.1).*

Mitochondria

Found in most eucaryotic cells, **mitochondria** (s., **mitochondrion**) frequently are called the "powerhouses" of the cell (**figure 4.12**). Metabolic processes such as the tricarboxylic acid cycle and the generation of ATP, the energy currency of all life-forms, take place here. In the transmission electron microscope, mitochondria usually are cylindrical structures and measure approximately 0.3 to 1.0 μm by 5 to 10 μm. (In other words, they are about the same size as procaryotic cells.) Although some cells possess 1,000 or more mitochondria, others (some yeasts, unicellular algae, and trypanosome protozoa) have a single, giant, tubular mitochondrion twisted into a continuous network permeating the cytoplasm. >> *Tricarboxylic acid cycle (section 10.4)*; *Electron transport and oxidative phosphorylation (section 10.5)*

The mitochondrion is bounded by two membranes, an outer mitochondrial membrane separated from an inner mitochondrial membrane by a 6 to 8 nm intermembrane space (figure 4.12). The outer mitochondrial membrane contains porins and thus is similar to the outer membrane of gram-negative bacteria. The inner membrane has special infoldings called **cristae** (s., **crista**), which greatly increase its surface area. The shape of cristae differs in mitochondria from various species. Fungi have platelike (laminar) cristae, whereas euglenoid flagellates may have cristae shaped like disks. Tubular cristae are found in a variety of eucaryotes; however, amocbae can possess mitochondria with cristae in the shape of vesicles (**figure 4.13**). The inner membrane encloses the mitochondrial matrix, a dense matrix containing ribosomes, DNA, and often large calcium phosphate granules. In many organisms, mitochondrial DNA is a closed circle, like bacterial DNA. However, in some protists, mitochondrial DNA is linear.

Each mitochondrial compartment is different from the others in chemical and enzymatic composition. The outer and inner mitochondrial membranes, for example, possess different lipids. Enzymes and electron carriers involved in electron transport and oxidative phosphorylation (the formation of ATP during respiration) are located only in the inner membrane. Enzymes of the tricarboxylic acid cycle and those involved with the catabolism

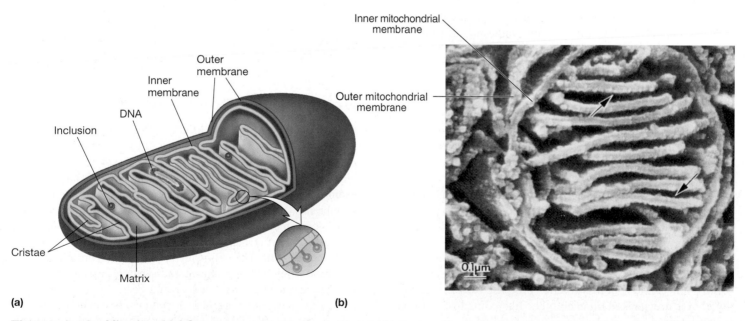

(a)   **(b)**

Figure 4.12   Mitochondrial Structure.   (a) A diagram of mitochondrial structure. The insert shows the ATP-synthesizing enzyme ATP synthase lining the inner surface of the cristae. (b) Scanning electron micrograph (×70,000) of a freeze-fractured mitochondrion showing the cristae (arrows). The outer and inner mitochondrial membranes also are evident.

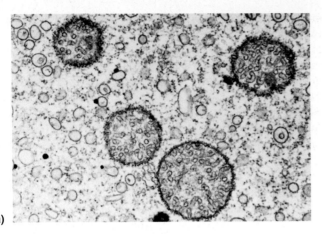

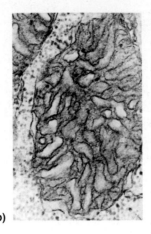

(a)   (b)

Figure 4.13   Mitochondrial Cristae.   Mitochondria with a variety of cristae shapes. (a) Mitochondria from the slime mold *Schizoplasmodiopsis micropunctata*. Note the tubular cristae (×49,500). (b) The protist *Actinosphaerium* with vesicular cristae (×75,000).

(breaking down) of fatty acids are located in the matrix. >> *Lipid catabolism (section 10.9)*

The mitochondrion uses its DNA and ribosomes to synthesize some of its own proteins. In fact, mutations in mitochondrial DNA often lead to serious diseases in humans. Most mitochondrial proteins, however, are manufactured under the direction of the nucleus. Mitochondria reproduce by binary fission, a reproductive process used by most bacteria. >> *Bacterial cell cycle (section 7.1)*

Chloroplasts

Plastids are cytoplasmic organelles of photosynthetic protists and plants. They often possess pigments such as chlorophylls and carotenoids, and are the sites of synthesis and storage of food reserves. The most important type of plastid is the chloroplast. **Chloroplasts** contain chlorophyll and use light energy to convert CO_2 and water to carbohydrates and O_2. That is, they are the site of photosynthesis.

Although chloroplasts are quite variable in size and shape, they share many structural features. Most are oval with dimensions of 2 to 4 μm by 5 to 10 μm, but some algae possess one huge chloroplast that fills much of the cell. Like mitochondria, chloroplasts are encompassed by two membranes (**figure 4.14**). A matrix called the stroma is enclosed by the inner membrane. It contains DNA, ribosomes, lipid droplets, starch granules, and a complex internal membrane system whose most prominent components are flattened, membrane delimited sacs, called **thylakoids.** Clusters of two or more thylakoids are dispersed within the stroma of most algal chloroplasts (figure 4.14*b*). In some photosynthetic protists, several disklike thylakoids are stacked on each other like coins to form grana (s., granum).

Photosynthetic reactions are separated structurally in the chloroplast just as electron transport and the tricarboxylic acid cycle are in the mitochondrion. The trapping of light energy to generate ATP, NADPH, and O_2 is referred to as the light reactions. These reactions are located in the thylakoid membranes, where chlorophyll and electron transport components are also found. The ATP and NADPH formed by the light reactions are used to form carbohydrates from CO_2 and water in the dark reactions. The dark reactions take place in the stroma. >> *Phototrophy (section 10.12)*

1. Describe in detail the structure of mitochondria and chloroplasts. Where are the different components of these organelles' energy-trapping systems located?

2. Define plastid, dark reactions, and light reactions.

3. What is the role of mitochondrial DNA?

4. What features of chloroplasts and mitochondria support the endosymbiotic hypothesis of their evolution?

4.7   STRUCTURES EXTERNAL TO THE PLASMA MEMBRANE

Eucaryotic cells have a variety of structures external to the plasma membrane. These include external coverings such as cell walls, cilia, and flagella. Although the functions of these structures are analogous to those of procaryotes, they are structurally distinct and function by different mechanisms.

External Cell Coverings

Eucaryotic microorganisms differ greatly from procaryotes in the supporting and protective structures they have external to the plasma membrane. In contrast to most procaryotes, many eucaryotes lack an external cell wall. The amoeba is an excellent example. Eucaryotic cell membranes, unlike most procaryotic membranes, contain sterols such as cholesterol, and this may make them mechanically stronger, thus reducing the need for external support. Of course many eucaryotes do have a rigid external **cell wall.** The cell walls of photosynthetic protists usually have a layered appearance and contain large quantities of polysaccharides such as cellulose and pectin. In addition, inorganic substances such as silica (in diatoms) or calcium carbonate may be present. Fungal cell walls normally are rigid. Their exact composition varies with the organism; usually cellulose, chitin, or glucan (a glucose polymer different from cellulose) are present. Despite their nature, the rigid materials in eucaryotic cell walls are

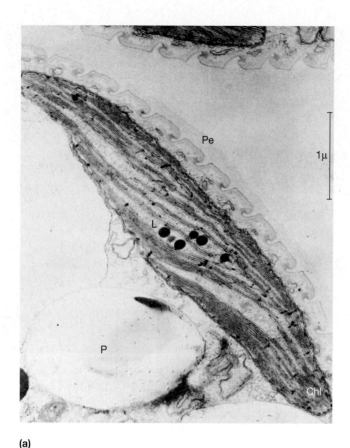

(a)

Chloroplast envelope (double membrane)   Stroma lamella   Stroma matrix

Granum

Thylakoids

(b)

Figure 4.14   Chloroplast Structure.   (a) The chloroplast (Chl) of the euglenoid flagellate *Colacium cyclopicolum*. The chloroplast is bounded by a double membrane and has its thylakoids in groups of three or more. A paramylon granule (P), lipid droplets (L), and the pellicular strips (Pe) can be seen (×40,000). (b) A diagram of chloroplast structure.

chemically simpler than bacterial peptidoglycan. ≪ *Bacterial cell walls (section 3.4)*

Many protists have a supportive mechanism called the **pellicle** (figure 4.14*a*). This is a relatively rigid layer of components just beneath the plasma membrane (sometimes the plasma membrane is also considered part of the pellicle). The pellicle may be

simple in structure. For example, *Euglena* has a series of overlapping strips with a ridge at the edge of each strip fitting into a groove on the adjacent one. In contrast, the pellicles of ciliate protozoa are exceptionally complex with two membranes and a variety of associated structures. Although pellicles are not as strong and rigid as cell walls, they give their possessors a characteristic shape.

Cilia and Flagella

Cilia (s., **cilium**) and **flagella** (s., **flagellum**) are the most prominent organelles associated with motility in eucaryotes. Although both are whiplike and beat to move the microorganism along, they differ from one another in two ways. First, cilia are typically only 5 to 20 µm in length, whereas flagella are 100 to 200 µm long. Second, their patterns of movement are usually distinctive (**figure 4.15**). Flagella move in an undulating fashion and generate planar or helical waves originating at either the base or the tip. If the wave moves from base to tip, the cell is pushed along; a beat traveling from the tip toward the base pulls the cell through the water. Sometimes the flagellum has lateral hairs called flimmer filaments (thicker, stiffer hairs are called mastigonemes). These filaments change flagellar action so that a wave moving down the filament toward the tip pulls the cell along instead of pushing it. Such a flagellum often is called a tinsel flagellum,

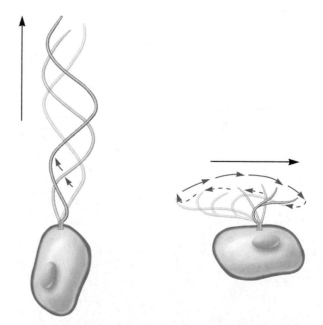

Figure 4.15   Patterns of Flagellar and Ciliary Movement. Flagellar and ciliary movement often takes the form of waves. Flagella (left illustration) move either from the base of the flagellum to its tip or in the opposite direction. The motion of these waves propels the organism along. The beat of a cilium (right illustration) may be divided into two phases. In the effective stroke, the cilium remains fairly stiff as it swings through the water. This is followed by a recovery stroke in which the cilium bends and returns to its initial position. The black arrows indicate the direction of water movement in these examples.

whereas the naked flagellum is referred to as a whiplash flagellum (**figure 4.16**). Cilia, on the other hand, normally have a beat with two distinctive phases. In the effective stroke, the cilium strokes through the surrounding fluid like an oar, thereby propelling the organism. The cilium next bends along its length while it is pulled forward during the recovery stroke in preparation for another effective stroke (figure 4.15). A ciliated microorganism actually coordinates the beats so that some of its cilia are in the recovery phase while others are carrying out their effective stroke (**figure 4.17**). This coordination allows the organism to move smoothly through the water.

Despite their differences, cilia and flagella are very similar in ultrastructure. They are membrane-bound cylinders about 0.2 µm in diameter. Located in the matrix of the organelle is the axoneme, which consists of nine pairs of microtubule doublets arranged in a circle around two central tubules (**figure 4.18**). This is called the 9 + 2 pattern of microtubules. Each doublet has pairs of arms projecting from subtubule A (the complete microtubule) toward a neighboring doublet. A radial spoke extends from subtubule A toward the internal pair of microtubules with their central sheath. These microtubules are similar to those found in the cytoplasm. << *Components external to the cell wall: Flagella (section 3.6)*

A basal body lies in the cytoplasm at the base of each cilium or flagellum. It is a short cylinder with nine microtubule triplets around its periphery (a 9 + 0 pattern) and is separated from the rest of the organelle by a basal plate. The basal body directs the construction of these organelles. Cilia and flagella appear to grow through the addition of preformed microtubule subunits at their tips.

Cilia and flagella bend because adjacent microtubule doublets slide along one another while maintaining their individual

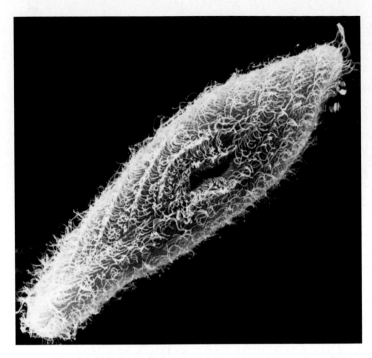

Figure 4.17 Coordination of Ciliary Activity. A scanning electron micrograph of *Paramecium* showing cilia (×1,500). The ciliary beat is coordinated and moves in waves across the protozoan's surface, as can be seen in the photograph.

lengths. The doublet arms, about 15 nm long, are made of the protein dynein (figure 4.18). It appears that dynein arms interact with the B subtubules of adjacent doublets to cause the sliding. The radial spokes also participate in this sliding motion. ATP powers the movement of cilia and flagella.

Cilia and flagella beat at a rate of about 10 to 40 strokes or waves per second and propel microorganisms rapidly. The record holder is the flagellate protist *Monas stigmatica*, which swims at a rate of 260 µm/second (approximately 40 cell lengths per second); the common euglenoid flagellate *Euglena gracilis* travels at around 170 µm or 3 cell lengths per second. The ciliate protozoan *Paramecium caudatum* swims about 2,700 µm/second (12 lengths per second). Such speeds are equivalent to or much faster than those seen in higher animals but not as fast as those in procaryotes.

1. How do eucaryotic microorganisms differ from procaryotes with respect to supporting or protective structures external to the plasma membrane? Describe the pellicle and indicate which microorganisms have one.

2. Prepare and label a diagram showing the detailed structure of a cilium or flagellum. How do cilia and flagella move, and what is dynein's role in the process? Contrast the ways in which flagella and cilia propel eucaryotic microorganisms through water.

3. Compare the structure and mechanism of action of procaryotic and eucaryotic flagella.

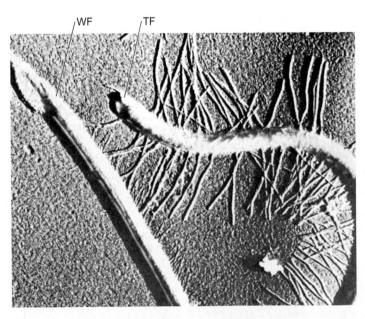

Figure 4.16 Whiplash and Tinsel Flagella. Transmission electron micrograph of a shadowed whiplash flagellum, WF, and a tinsel flagellum, TF, with mastigonemes.

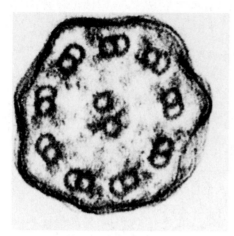

(a)

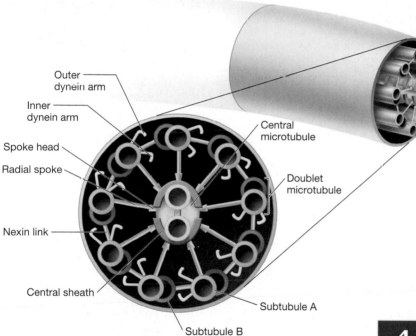

Outer
dynein arm

Inner
dynein arm

Spoke head

Radial spoke

Nexin link

Central sheath

Central
microtubule

Doublet
microtubule

Subtubule A

Subtubule B

(b)

Figure 4.18   Cilia and Flagella Structure.   (a) An electron micrograph of a cilium cross section. Note the two central microtubules surrounded by nine microtubule doublets (×160,000). (b) A diagram of cilia and flagella structure.

4.8   COMPARISON OF PROCARYOTIC EUCARYOTIC CELLS

A comparison of the cells in **figure 4.19** demonstrates that there are many fundamental differences between eucaryotic and procaryotic cells. **Eucaryotic cells** have a membrane-enclosed nucleus. In contrast, **procaryotic cells** lack a true, membrane-delimited nucleus. *Bacteria* and *Archaea* are procaryotes; all other organisms—fungi,

protists, plants, and animals—are eucaryotic. Most procaryotes are smaller than eucaryotic cells, often about the size of eucaryotic mitochondria and chloroplasts.

The presence of the eucaryotic nucleus is the most obvious difference between these two cell types, but many other major distinctions exist. It is clear from **table 4.2** that procaryotic cells are much simpler structurally. In particular, an extensive and diverse collection of membrane-delimited organelles is missing. Furthermore, procaryotes are functionally simpler in several ways. They lack mitosis and meiosis, and have a simpler genetic organization. Many complex eucaryotic processes are absent in procaryotes: endocytosis, intracellular digestion, directed cytoplasmic streaming, and ameboid movement, are just a few.

Despite the many significant differences between these two basic cell forms, they are remarkably similar on the biochemical level, as we discuss in succeeding chapters. With a few exceptions, the genetic code is the same in both, as is the way in which the genetic information in DNA is expressed. The principles underlying metabolic processes and many important metabolic pathways are identical. Thus beneath the profound structural and functional differences between procaryotes and eucaryotes, there is an even more fundamental unity: a molecular unity that is basic to all known life processes.

1. Outline the major differences between procaryotes and eucaryotes. How are they similar?

2. What characteristics make *Archaea* more like eucaryotes? What features make them more like *Bacteria?*

4.9   OVERVIEW OF PROTIST STRUCTURE AND FUNCTION

Because protists are eucaryotic cells, in many respects their morphology and physiology are the same as the cells of multicellular plants and animals. However, because all of life's various functions must be performed within a single cell, complexity arises at the level of specialized organelles rather than at the tissue level. Protists must remain relatively small because without multicellularity to facilitate the exchange of nutrients and metabolites, they need a high ratio of cell surface to intracellular volume. This is because the distance from the cell membrane to the center of the cell cannot exceed the distance a molecule can diffuse. Even the largest algae (the seaweeds) are long, thin, and often flattened—a shape that maximizes the surface-to-volume ratio. We now discuss some of the fundamental features of protist morphology and reproduction.

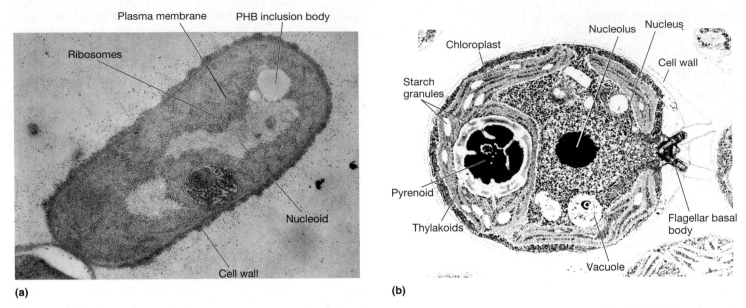

Figure 4.19   Comparison of Procaryotic and Eucaryotic Cell Structure. (a) The procaryote *Bacillus megaterium* (×30,500). (b) The eucaryotic alga *Chlamydomonas reinhardtii.* Note the large chloroplast with its pyrenoid body (×30,000).

Table 4.2	**Comparison of Procaryotic and Eucaryotic Cells**		
	Procaryotes		**Eucaryotes**
Property	*Bacteria*	*Archaea*	*Eucarya*
Organization of Genetic Material			
True membrane-bound nucleus	No	No	Yes
DNA complexed with histones	No	Some	Yes
Chromosomes	Usually one circular chromosome	Usually one circular chromosome	More than one; chromosomes are linear
Plasmids	Very common	Very common	Rare
Introns in genes	Rare	Rare	Yes
Nucleolus	No	No	Yes
Mitochondria	No	No	Yes
Chloroplasts	No	No	Yes
Plasma Membrane Lipids	Ester-linked phospholipids and hopanoids; some have sterols	Glycerol diethers and diglycerol tetraethers; some have sterols	Ester-linked phospholipids and sterols
Flagella	Submicroscopic in size; composed of one protein fiber	Submicroscopic in size; composed of one protein fiber	Microscopic in size; membrane bound; usually 20 microtubules in 9 + 2 pattern
Endoplasmic Reticulum	No	No	Yes
Golgi Apparatus	No	No	Yes
Peptidoglycan in Cell Walls	Yes	No	No
Ribosome Size	70S	70S	80S
Lysosomes	No	No	Yes
Cytoskeleton	Rudimentary	Rudimentary	Yes
Gas Vesicles	Yes	Yes	No

Protist Morphology

The protist cell membrane is called the **plasmalemma** and is identical to that of multicellular organisms. In some protists, the cytoplasm immediately under the plasmalemma is divided into an outer gelatinous region called the **ectoplasm** and an inner fluid region, the **endoplasm.** The ectoplasm imparts rigidity to the cell body. The plasmalemma and structures immediately beneath it make up the pellicle.

One or more vacuoles are usually present in the cytoplasm of protozoa. These are differentiated into contractile, secretory, and food vacuoles. **Contractile vacuoles** function as osmoregulatory organelles in those protists that live in hypotonic environments, such as freshwater lakes. Osmotic balance is maintained by continuous water expulsion. **Phagocytic vacuoles** are conspicuous in protists that ingest food by phagocytosis (holozoic protists) and in parasitic species. Phagocytic vacuoles are the sites of food digestion. In some organisms, they may occur anywhere on the cell surface, while others have a specialized structure for phagocytosis called the **cytostome** (cell mouth). When digestion commences, the phagocytic vacuole is acidic, but as it proceeds, the vacuolar pH increases and the membrane forms small blebs. These pinch off and carry nutrients throughout the cytoplasm. The undigested contents of the original phagocytic vacuole are expelled from the cell either at a random site on the cell membrane or at a designated position called the **cytoproct.**

Several energy-conserving organelles are observed in protists. Most aerobic protists that use organic matter as their source of energy (chemoorganotrophs) have mitochondria. The majority of anaerobic chemoorganotrophic protists (such as

Trichonympha, which lives in the gut of termites) lack mitochondria. However, some have small, membrane-bound organelles termed **hydrogenosomes.** Hydrogenosomes, like mitochondria and chloroplasts, are thought to have evolved by the endosymbiotic hypothesis. Within hydrogenosomes, energy released by the catabolism of pyruvate is used to make ATP. In the process, CO_2, H_2, and acetate are formed (**figure 4.20**). These metabolic products may be consumed by symbiotic procaryotes living within the protist. In some cases, methanogenic archaea consume the CO_2 and H_2 and generate methane (CH_4). Photosynthetic protists have chloroplasts. A dense proteinaceous area, the **pyrenoid,** which is associated with the synthesis and storage of starch, may be present in the chloroplasts (figure 4.19*b*). >> *Microbial evolution (section 17.1)*

Many protists feature cilia or flagella at some point in their life cycle. Their formation is associated with a basal bodylike organelle called the kinetosome, which is similar in structure to the centriole. In addition to aiding in motility, these organelles may be used to generate water currents for feeding and respiration.

Encystment and Excystment

Many protists are capable of **encystment.** During encystment, the organism becomes simpler in morphology and develops into a resting stage called a cyst. The cyst is a dormant form marked by the presence of a cell wall and very low metabolic activity. Cyst formation is particularly common among aquatic, free-living protists and parasitic forms. Cysts serve three major functions: (1) they protect against adverse changes in the

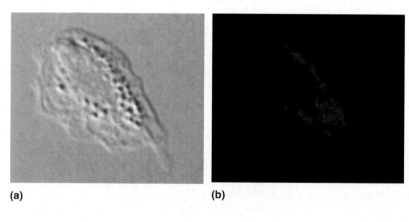
(a)   (b)

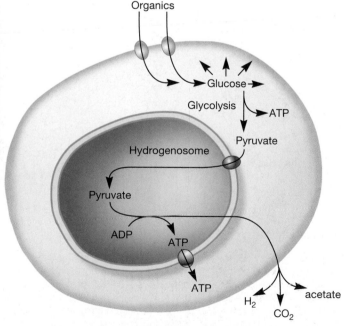
(c)

Figure 4.20 Hydrogenosomes of *Trichomonas vaginalis*. (a) Differential interference contrast image of the protist *Trichomonas vaginalis*. (b) Hydrogenosomes of the *T. vaginalis* cell shown in (a) labeled with red fluorescent antibody that recognizes the hydrogenosome-specific malic enzyme. (c) Schematic diagram of a hydrogenosome and its associated metabolic activities.

environment, such as nutrient deficiency, desiccation, adverse pH, and low levels of O_2; (2) they are sites for nuclear reorganization and cell division (reproductive cysts); and (3) they serve as a means of transfer between hosts in parasitic species. Protists escape from cysts by a process called **excystment.** Although the exact stimulus for excystment is unknown, it is generally triggered by a return to favorable environmental conditions. For example, cysts of parasitic species excyst after ingestion by the host.

Protist Reproduction

Most protists have both asexual and sexual reproductive phases in their life cycles. The most common method of asexual reproduction is **binary fission.** During this process, the nucleus first undergoes mitosis, and then the cytoplasm divides by cytokinesis to form two identical individuals (**figure 4.21**). Multiple fission is also common, as is budding. Some filamentous, photosynthetic protists undergo fragmentation so that each piece of the broken filament grows independently.

Sexual reproduction involves the formation of gametes. Protist cells that produce gametes are termed **gamonts.** The fusion of haploid gametes is called **syngamy.** Among protists, syngamy can involve the fusion of two morphologically similar gametes (isogamy) or two morphologically different types (anisogamy). Meiosis may occur before the formation and union of gametes, as in most animals, or just after fertilization, as is the case with lower plants. Furthermore, the exchange of nuclear material may occur in the familiar fashion—between two different individuals (**conjugation**)—or by the development of a genetically distinct nucleus within a single individual (autogamy).

With this level of reproductive complexity, perhaps it is not surprising that the nuclei among protists show considerable diversity. Most commonly, a **vesicular nucleus** is present. This is 1 to 10 μm in diameter, spherical, and has a distinct nucleolus and uncondensed chromosomes. Ovular nuclei are up to 10 times this size and possess many peripheral nucleoli. Still others have chromosomal nuclei, in which the chromosomes remain condensed throughout the cell cycle. Finally, the *Ciliophora* have two types of nuclei: a large **macronucleus** with distinct nucleoli and condensed chromatin, and a smaller, diploid **micronucleus** with dispersed chromatin but lacking nucleoli. Macronuclei are engaged in trophic activities and regeneration processes, whereas micronuclei are involved only in genetic recombination during sexual reproduction and the regeneration of the macronucleus. >> *Protist classification (section 23.2)*

1. What is a hydrogenosome? How does it differ from a mitochondrion?
2. Trace the path of a food item from the phagocytic vacuole to the cytoproct.
3. What functions do cysts serve for a typical protist? What causes excystment to occur?
4. What is a gamont? What is the difference between anisogamy and syngamy?
5. How does conjugation differ from autogamy?
6. Describe vesicular, ovular, and chromosomal nuclei. Which is most like the nuclei found in higher eucaryotes?

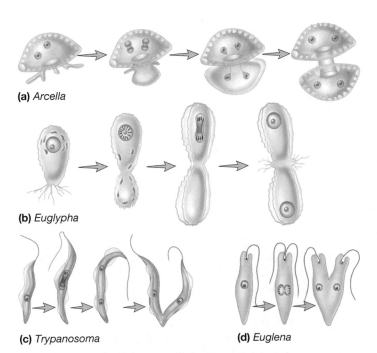

(a) *Arcella*

(b) *Euglypha*

(c) *Trypanosoma*

(d) *Euglena*

Figure 4.21   Binary Fission in Protists.   (a) The two nuclei of the testate (shelled) amoeba *Arcella* divide as some of its cytoplasm is extruded and a new test for the daughter cell is secreted. (b) In another testate amoeba, *Euglypha,* secretion of new platelets is begun before cytoplasm begins to move out of the aperture. The nucleus divides while the platelets are used to construct the test for the daughter cell. For many protists, all organelles must be replicated before the cell divides. This is also the case with (c) *Trypanosoma* and (d) *Euglena.*

4.10   OVERVIEW OF FUNGAL STRUCTURE AND FUNCTION

The term *fungus* (pl., fungi; Latin *fungus,* mushroom) describes eucaryotic organisms that are spore-bearing, have absorptive nutrition, lack chlorophyll, and reproduce sexually and asexually.

Fungal Structure

The body or vegetative structure of a fungus is called a **thallus** (pl., **thalli**). It varies in complexity and size, ranging from the single-cell microscopic yeasts to multicellular molds, macroscopic puffballs, and mushrooms. The fungal cell usually is encased in a cell wall of **chitin.** Chitin is a strong but flexible nitrogen-containing polysaccharide consisting of *N*-acetylglucosamine residues.

A **yeast** is a unicellular fungus that has a single nucleus and reproduces either asexually by budding and transverse division or sexually through spore formation. Each bud that separates can grow into a new cell, and some group together to form colonies. Generally yeast cells are larger than bacteria, vary considerably in size, and are commonly spherical to egg shaped. They lack flagella and cilia but possess most other eucaryotic organelles (figure 4.3*a*).

The thallus of a **mold** consists of long, branched, threadlike filaments of cells called **hyphae** (s., hypha; Greek *hyphe,* web) that form a **mycelium** (pl., **mycelia**), a tangled mass or tissue-like aggregation of hyphae. In some fungi, protoplasm streams through hyphae, uninterrupted by cross walls. These hyphae are called **coenocytic** or aseptate (**figure 4.22*a***). The hyphae of other fungi (figure 4.22*b*) have cross walls called **septa** (s., **septum**) with either a single pore (figure 4.22*c*) or multiple pores (figure 4.22*d*) that enable cytoplasmic streaming. These hyphae are termed **septate.**

Hyphae are composed of an outer cell wall and an inner lumen, which contains the cytosol and organelles. A plasma membrane surrounds the cytoplasm and lies next to the cell wall. The filamentous nature of hyphae results in a large surface area relative to the volume of cytoplasm. This makes adequate nutrient absorption possible.

Fungal Reproduction

Reproduction in fungi can be either asexual or sexual. Asexual reproduction is accomplished in several ways: (1) a parent cell can undergo mitosis and divide into two daughter cells by a central constriction and formation of a new cell wall (**figure 4.23*a***), and (2) mitosis in vegetative cells may be concurrent with budding to produce a daughter cell. This is very common in the yeasts. Often accompanying asexual reproduction is the formation of asexual spores. Many asexual spores are formed as a means of dispersal. These spores are generally small and easily released from the fungus by air currents. There are many types of asexual spores, each with its own name. **Arthroconidia (arthrospores)** are formed when hyphae fragment through splitting of the cell wall or septum (figure 4.23*b*). **Sporangiospores** develop within a sac (sporangium; pl., sporangia) at a hyphal tip (figure 4.23*c*). **Conidiospores** are spores that are not enclosed in a sac but produced at the tips or sides of the hypha (figure 4.23*d*). **Blastospores** are produced from a vegetative mother cell by budding (figure 4.23*e*).

Sexual reproduction in fungi involves the fusion of compatible nuclei. Homothallic fungal species are self-fertilizing and produce sexually compatible gametes on the same mycelium. Heterothallic species require outcrossing between different but sexually compatible mycelia. It has long been held that sexual

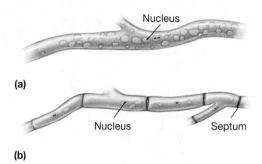

(a)

(b)

Nucleus

Nucleus   Septum

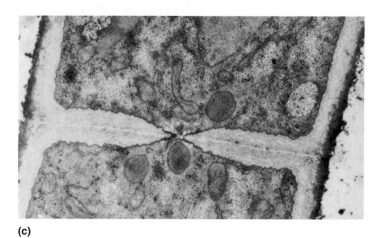

(c)

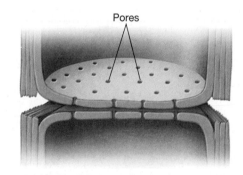

Pores

(d)

Figure 4.22 Hyphae. Drawings of (a) coenocytic hyphae (aseptate) and (b) hyphae divided into cells by septa. (c) Electron micrograph (×40,000) of a section of *Drechslera sorokiniana* showing wall differentiation and a single pore. (d) Drawing of a multiperforate septal structure.

reproduction must occur between mycelia of opposite **mating types (MAT).** However, one instance of same-sex mating was discovered following an outbreak of the pathogenic yeast *Cryptococcus gatti* in Canada. Depending on the species, sexual fusion may occur between haploid gametes, gamete-producing bodies called **gametangia,** or hyphae. Sometimes both the cytoplasm and haploid nuclei fuse immediately to produce the

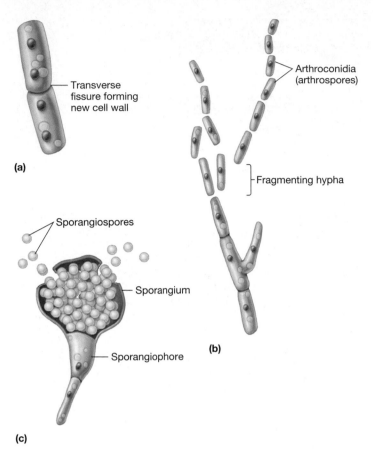

(a)

(b)

(c)

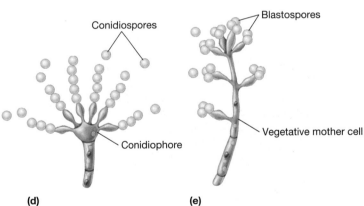

(d)   (e)

Figure 4.23 Asexual Reproduction in Fungi and Some Representative Asexual Spores. (a) Transverse fission. (b) Hyphal fragmentation resulting in arthroconidia (arthrospores). (c) Sporangiospores in a sporangium. (d) Conidiospores arranged in chains at the end of a conidiophore. (e) Blastospores are formed from buds off of the parent cell.

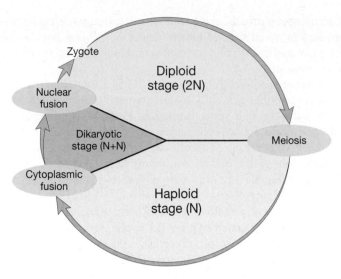

Figure 4.24   Reproduction in Fungi.   A drawing of the generalized life cycle for fungi showing the alternation of haploid and diploid stages. Some fungal species do not pass through the dikaryotic stage indicated in this drawing. The asexual (haploid) stage is used to produce spores that aid in the dissemination of the species. The sexual (diploid) stage involves the formation of spores that survive adverse environmental conditions (e.g., cold, dryness, heat).

Fungal spores, both asexual and sexual, are important for several reasons. They enable fungi to survive environmental stresses such as desiccation, nutrient limitation, and extreme temperatures, although they are not as stress resistant as bacterial endospores. They aid in fungal dissemination, which helps explain their wide distribution. Because spores are often small and light, they can remain suspended in air for long periods. Thus fungal spores often spread by adhering to the bodies of insects and other animals. The bright colors and fluffy textures of many molds often are due to their aerial hyphae and spores. Finally, the size, shape, color, and number of spores are useful in the identification of fungal species.

1. How can a fungus be defined? What is the structural difference between a yeast and a mold?

2. What organelles would you expect to find in the cytoplasm of a typical fungus?

3. How is asexual reproduction accomplished in the fungi? Sexual reproduction?

4. Describe each of the following types of asexual fungal spores: sporangiospore, arthrospore, conidiospore, and blastospore.

5. How are fungi dispersed in the environment?

diploid zygote. Usually, however, there is a delay between cytoplasmic and nuclear fusion. This produces a **dikaryotic stage** in which cells contain two separate haploid nuclei (N + N), one from each parent (**figure 4.24**). After a period of dikaryotic existence, the two nuclei fuse and undergo meiosis to yield haploid spores.

Summary

4.1 Overview of Eucaryotic Cell Structure

a. The eucaryotic cell has a true, membrane-delimited nucleus and many membranous organelles (**table 4.1; figures 4.2** and **4.3**).

b. The membranous organelles compartmentalize the cytoplasm of the cell. This allows the cell to carry out a variety of biochemical reactions simultaneously. It also provides more surface area for membrane-associated activities such as respiration.

4.2 Eucaryotic Membranes

a. Eucaryotic membranes are similar in structure and function to those of *Bacteria*. The two differ in terms of their lipid composition.

b. Eucaryotic membranes contain microdomains called lipid rafts. They are enriched for certain lipids and proteins, and participate in a variety of cellular processes.

4.3 Eucaryotic Cytoplasm

a. The cytoplasm consists of a cytosol in which is found the organelles and cytoskeleton.

b. The cytoskeleton is organized from microfilaments, intermediate filaments, and microtubules. These cytoskeletal elements are partly responsible for cell structure and movement (**figure 4.5**).

c. Microfilaments and microtubules have been observed in eucaryotic microbes. Microfilaments are composed of actin proteins; microtubules are composed of α-tubulin and β-tubulin.

d. Intermediate filaments are assembled from a heterogeneous family of proteins. They have not been identified or studied in eucaryotic microbes.

4.4 Organelles of the Biosynthetic-Secretory and Endocytic Pathways

a. The cytoplasm is permeated with a complex of membranous organelles and vesicles. Some are involved in the synthesis and secretion of materials (biosynthetic-secretory pathway). Some are involved in the uptake of materials from the extracellular milieu (endocytic pathway).

b. The endoplasmic reticulum (ER) is an irregular network of tubules and flattened sacs (cisternae). The ER may have attached ribosomes and may be active in protein synthesis (rough endoplasmic reticulum), or it may lack ribosomes (smooth ER).

c. The ER can donate materials to the Golgi apparatus, an organelle composed of one or more stacks of cisternae (**figure 4.7**). This organelle prepares and packages cell products for secretion.

d. The Golgi apparatus also buds off vesicles that deliver hydrolytic enzymes and other proteins to lysosomes. Lysosomes are organelles that contain digestive enzymes and aid in intracellular digestion of extracellular materials delivered to them by endocytosis (**figure 4.9**).

e. Eucaryotes ingest materials using several kinds of endocytosis. These include phagocytosis, clathrin-dependent endocytosis, and caveolae-dependent endocytosis. Some macromolecules are bound to receptors prior to endocytosis in a process called receptor-mediated endocytosis.

4.5 Organelles Involved in Genetic Control of the Cell

a. The nucleus is a large organelle containing the cell's chromosomes. It is bounded by a complex, double-membrane envelope perforated by pores through which materials can move (**figure 4.10**).

b. The nucleolus lies within the nucleus and participates in the synthesis of ribosomal RNA and ribosomal subunits (**figure 4.11**).

c. Eucaryotic ribosomes are either found free in the cytoplasm or bound to the ER. They are 80S in size.

4.6 Organelles Involved in Energy Conservation

a. Mitochondria are organelles bounded by two membranes, with the inner membrane folded into cristae (**figure 4.12**). They are responsible for energy conservation by cell respiration, which involves the tricarboxylic acid cycle, electron transport, and oxidative phosphorylation.

b. Chloroplasts are pigment-containing organelles that serve as the site of photosynthesis (**figure 4.14**). The trapping of light energy takes place in the thylakoid membranes of the chloroplast, whereas CO_2 fixation occurs in the stroma.

4.7 Structures External to the Plasma Membrane

a. When a cell wall is present, it is constructed from polysaccharides (e.g., cellulose) that are chemically simpler than peptidoglycan, the molecule found in bacterial cell walls. Many protozoa have a pellicle rather than a cell wall.

b. Many eucaryotic cells are motile because of cilia and flagella, membrane-enclosed organelles with nine microtubule doublets surrounding two central microtubules (**figure 4.18**). The cell moves when the microtubule doublets slide along each other, causing the cilium or flagellum to bend.

4.8 Comparison of Procaryotic and Eucaryotic Cells

a. Despite the fact that eucaryotes and procaryotes differ structurally in many ways (**table 4.2**), they are quite similar biochemically.

4.9 Overview of Protist Structure and Function

a. Because protists are eucaryotic cells, in many respects their morphology and physiology resemble those of multicellular plants and animals. However, because all their functions must be performed within the individual protist, many morphological and physiological features are unique.

b. The protistan cell membrane is called the plasmalemma, and the cytoplasm can be divided into the ectoplasm and the endoplasm. The cytoplasm contains contractile, secretory, and food (phagocytic) vacuoles.

c. Energy metabolism occurs within mitochondria, hydrogenosomes, or chloroplasts.

d. Some protists can secrete a resistant covering and go into a resting stage (encystment) called a cyst. Cysts protect the organism against adverse environments, function as a site for nuclear reorganization, and serve as a means of transmission in parasitic species.

e. Most protists reproduce asexually by binary or multiple fission or budding (**figure 4.21**).

f. Some protists also use sexual reproduction via a variety of sexual reproductive strategies including syngamy and autogamy.

4.10 Overview of Fungal Structure and Function

a. A fungus is a eucaryotic, spore-bearing organism that has absorptive nutrition and lacks chlorophyll; reproduces asexually, sexually, or by both methods; and normally has a cell wall containing chitin.

b. The body or vegetative structure of a fungus is called a thallus.

c. Yeasts are unicellular fungi that have a single nucleus and reproduce either asexually by budding and transverse division or sexually through spore formation **(figure 4.3a).**

d. A mold consists of long, branched, threadlike filaments of cells, the hyphae, that form a tangled mass called a mycelium. Hyphae may be either septate or coenocytic (nonseptate). The mycelium can produce reproductive structures.

e. Asexual reproduction in the fungi often leads to the production of specific types of spores that are easily dispersed **(figure 4.23).**

f. Sexual reproduction is initiated in fungi by the fusion of hyphae (or cells) of different mating strains. In some fungi, the nuclei in the fused hyphae immediately combine to form a zygote. In others, the two genetically distinct nuclei remain separate, forming pairs that divide synchronously. Eventually some nuclei fuse **(figure 4.24).**

Critical Thinking Questions

1. Discuss the statement: "The most obvious difference between eucaryotic and procaryotic cells is in their use of membranes." Specifically, compare the roles of the procaryotic plasma membrane with those of the organelle membranes in eucaryotes.

2. Protist encystment is usually triggered by changes in the environment. How might this be similar to or different from endospore formation in gram-positive bacteria?

3. Yeasts and some protists tend to reproduce asexually when nutrients are plentiful and conditions are favorable for growth but reproduce sexually when environmental or nutrient conditions are not favorable. Why is this an evolutionarily important and successful strategy?

Learn More

Learn More by visiting the Prescott website at www.mhhe.com/prescottprinciples, where you will find a complete list of references.

Viruses and Other Acellular Agents

5

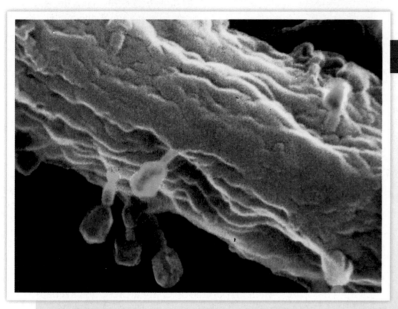

A scanning electron micrograph of T-even bacteriophages infecting *E. coli*. The phages are colored blue.

bacteriophage A virus that uses bacteria as its host; often called a phage.

capsid The protein coat or shell that surrounds a virion's nucleic acid.

enveloped virus A virus with an outer membranous layer that surrounds the nucleocapsid.

induction An event in the life cycle of some viruses (e.g., temperate bacteriophage) that results in the provirus initiating synthesis of mature virions and entering the lytic cycle.

lysogeny The state in which a viral genome remains within a bacterial or archaeal cell after infection and reproduces along with it rather than taking control of the host cell and destroying it.

lytic cycle The infection of a host cell by a virus that induces cell death through lysis and release of newly replicated virus.

minus strand or negative strand A DNA or RNA viral genome that is complementary to the virus's mRNA.

naked virus A virus that lacks an envelope.

nucleocapsid That portion of a virus that consists of viral nucleic acid surrounded by a protein coat called the capsid.

Chapter Glossary

peplomers or **spikes** Proteins that project from a viral envelope.

plaque A clear area in a lawn or layer of host cells that results from the lysis or killing of the host cells by viruses.

plus strand or positive strand A DNA or RNA viral genome that is identical in nucleotide sequence to the virus's mRNA; in some RNA viruses, the genome also serves as mRNA.

prion An infectious agent composed only of protein that is responsible for causing a variety of spongiform encephalopathies (e.g., scrapie).

provirus (prophage) The latent form of a virus that remains within a host cell, usually integrated into the host chromosome; in archaeal and bacterial hosts, this occurs during the lysogenic cycle; when the virus is a bacteriophage, the provirus is called a prophage.

retrovirus A member of a group of viruses with RNA genomes that carry the enzyme reverse transcriptase and form a DNA copy of their genome during their reproductive cycle.

segmented genome A viral genome that is divided into several fragments, each usually coding for a single polypeptide.

temperate phage A bacteriophage that infects bacteria and establishes a lysogenic relationship with its host rather than causing immediate lysis.

virion A complete virus particle.

viroid An infectious agent of plants that consists only of RNA.

virulent viruses Viruses that lyse their host cells during the reproductive cycle.

virus An infectious agent having a simple acellular organization, often just a protein coat and a nucleic acid genome; lacking independent metabolism; and reproducing only within living host cells.

virusoid An infectious agent composed only of RNA; its RNA encodes one or more proteins, but it can only replicate in cells also infected by a virus.

The Virus. Observe this virus: think how small Its arsenal, and yet how loud its call; It took my cell, now takes your cell, And when it leaves will take our genes as well. Genes that are master keys to growth That turn it on, or turn it off, or both; Should it return to me or you It will own the skeleton keys to do a number on our tumblers; stage a coup.

—Michael Newman

The microbial world consists not only of cellular organisms but also of acellular infectious agents. In this chapter, we turn our attention to these infectious agents: viruses, viroids, virusoids, and prions. These entities are composed simply of protein and nucleic acid (viruses), RNA only (viroids and virusoids), or protein only (prions). Yet they are major causes of disease. For instance, many human diseases are caused by viruses, and more are discovered every year, as demonstrated by the appearance of SARS and new avian influenza viruses.

Although viruses are most often discussed in terms of their ability to cause disease, it is important to remember that viruses are significant for other reasons. Recent ecological studies have shown that viruses are important members of aquatic ecosystems. There they interact with cellular microbes and contribute to the movement of organic matter from particulate forms to dissolved forms. They also affect population sizes of cellular microbes in these habitats. Finally, bacterial viruses transfer genes from bacterium to bacterium at a high rate, thus contributing to the evolution of bacteria. Bacterial viruses are also receiving renewed interest as therapies for bacterial infections due to the increasing number of drug-resistant bacterial pathogens. Viruses also serve as models for understanding important processes such as DNA replication, RNA synthesis, and protein synthesis. Therefore the study of viruses has contributed significantly to the discipline of molecular biology. In fact, the field of genetic engineering is based in large part on discoveries of viruses. >> *Marine and freshwater microbiology (section 26.2); Drug resistance (section 31.6); Biotechnology and industrial microbiology (chapter 16)*

We begin this chapter by reviewing the general properties and structure of viruses and the ways in which viruses are cultured and studied. The chapter ends with a brief introduction to viroids, virusoids, and prions.

5.1   INTRODUCTION TO VIRUSES

The discipline of **virology** studies **viruses,** a unique group of infectious agents whose distinctiveness resides in their simple, acellular organization and pattern of reproduction. A complete virus particle, called a **virion,** consists of one or more molecules of DNA or RNA enclosed in a coat of protein. Some viruses have additional layers that can be very complex and contain carbohydrates, lipids, and additional proteins (**figure 5.1**). Viruses can exist in two phases: extracellular and intracellular. They possess few, if any, enzymes and cannot reproduce outside of living cells. In the intracellular phase, viruses exist primarily as replicating nucleic acids that induce host metabolism to synthesize viral components, from which virions are assembled and eventually released.

Viruses can infect either eucaryotic or procaryotic cells. Viruses that infect bacteria are called **bacteriophages,** or **phages** for short. Relatively few viruses are known to use archaea as their hosts and most have not yet been assigned to viral taxa (**Microbial Diversity & Ecology 5.1**). Despite the abundance of phages, most known viruses infect eucaryotic organisms, including

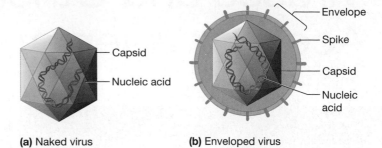

(a) Naked virus   **(b)** Enveloped virus

Figure 5.1   Generalized Structure of Viruses.   (a) The simplest virus is a naked virus (nucleocapsid) consisting of a geometric capsid assembled around its nucleic acid. (b) An enveloped virus is composed of a nucleocapsid surrounded by a flexible membrane called an envelope. The envelope usually has viral proteins called spikes inserted into it.

plants, animals, protists, and fungi. All viruses have been classified into numerous families, based primarily on genome structure, life cycle, morphology, and genetic relatedness. These families have been designated by the International Committee for the Taxonomy of Viruses (ICTV), the agency responsible for standardizing the classification of all viruses. >> *Principles of virus taxonomy (section 24.1)*

In this chapter, we introduce the structure and multiplication strategies of viruses. Their taxonomic diversity is presented in chapter 24, and their ecological relevance is discussed in chapter 26.

5.2   STRUCTURE OF VIRUSES

Viral morphology has been intensely studied over the past decades because of the importance of viruses and the realization that their structure was simple enough to be understood in detail. Progress has come from the use of several different techniques: electron microscopy, X-ray diffraction, biochemical analysis, and immunology. Although our knowledge is incomplete due to the large number of different viruses, we can discuss the general nature of viral structure.

Virion Size

Virions range in size from about 10 to 400 nm in diameter (**figure 5.2**). The smallest viruses are a little larger than ribosomes, whereas poxviruses (e.g., *Variola virus,* the causative agent of small pox) are about the same size as the smallest bacteria and can be seen in the light microscope. Most viruses, however, are too small to be visible in the light microscope and must be viewed with scanning and transmission electron microscopes. >> *Viruses with double-stranded DNA genomes (Group I):* Variola virus *(section 24.2);* << *Electron microscopy (section 2.4)*

General Structural Properties

The simplest viruses are constructed of a **nucleocapsid.** The nucleocapsid is composed of a nucleic acid, either DNA or RNA, held within a protein coat called the **capsid,** which protects viral

Microbial Diversity & Ecology

5.1 Host-Independent Growth of an Archaeal Virus

The fact that viruses cannot multiply without first infecting a host cell has resulted in their classification as "acellular entities" or "forms"—they are not cells. So it was quite a surprise when an archaeal virus that develops long tails only when outside its host was discovered (**box figure**). This archaeal virus was found in acidic hot springs (pH 1.5, 85–93°C) in Italy, where it infects the hyperthermophilic archaeon *Acidianus convivator*. When the virus infects its host, lemon-shaped virions are assembled. Following host lysis, two "tails" begin to form at either end of the virion. These projections continue to assemble until they reach a length at least that of the viral capsid. Curiously, tails are only produced if virions are incubated at high temperature, leading to the hypothesis that they are part of a survival strategy. The virus is thus called ATV for *Acidianus* two-tailed virus.

To find out more about the structure of the tails, the ATV genome was sequenced. ATV is a double-stranded DNA virus that encodes only nine structural proteins. The tail protein is an 800 amino acid protein that bears homology to eucaryotic intermediate filament proteins. Both intermediate filaments and purified ATV tail protein assemble into filamentous structures without additional energy or cofactors. **<<** *Eucaryotic cytoplasm (section 4.3)*

It is suspected that the development of tails only at high temperatures may be a survival strategy for the virus when host cell density is low. So far, ATV is the only virus that infects procaryotes living in acidic hot springs to induce lysis rather than lysogeny. It would thus seem that all other such viruses have evolved lysogeny

as a means to survive these harsh conditions. Why this virus has evolved lysis and tail development is unknown, but it suggests that viruses may be more complicated than simple "entities."

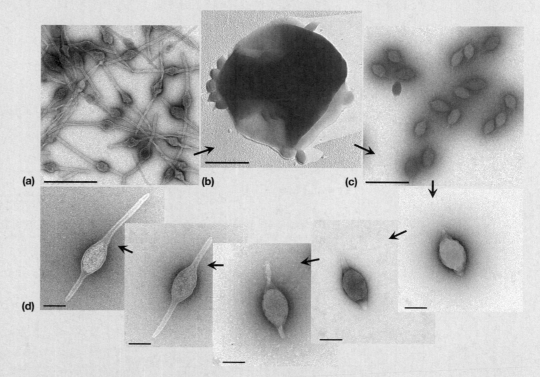

(a) **(b)** **(c)** **(d)**

***Acidianus* Two-Tailed Virus, ATV.** **(a)** Virions collected from an acidic hot spring bear two long projections, or tails, at each end. **(b, c)** The hyperthermophilic archaeon *A. convivator* extrudes lemon-shaped virions. **(d)** These subsequently develop tail-like structures independent of the host and only when at high temperature. Scale bars: *a–c*, 0.5 μm; *d*, 0.1 μm.

Source: Häring, M.; Vestergaard, G.; Rachel, R.; Chen, L.; Garret, R. A.; and Prangishvili, D. 2005. Independent virus development outside a host. Nature *436:1101–02.*

genetic material and aids in its transfer between host cells. Capsids are large macromolecular structures made of proteins called **protomers.** Capsids self-assemble by a process that is not fully understood. Some viruses use noncapsid proteins as scaffolding upon which the capsids are assembled.

Probably the most important advantage of this design strategy is that viral genetic material is used with maximum efficiency. For example, the *Tobacco mosaic virus* (TMV) capsid is constructed using a single type of protomer that is 158 amino acids in length (**figure 5.3**). Therefore, of the 6,400 nucleotides in the TMV genome, only about 474 nucleotides are required to code for the coat protein. Suppose, however, that the TMV capsid was composed of six different

protomers all about 150 amino acids in length. If this were the case, about 2,900 of the 6,400 nucleotides in the TMV genome would be required just for capsid construction, and much less genetic material would be available for other purposes. **>>** *Viruses with plus-strand RNA genomes (Group IV): Tobacco mosaic virus (section 24.5)*

The various morphological types of viruses primarily result from the combination of a particular type of capsid symmetry with the presence or absence of an envelope—a lipid layer external to the nucleocapsid. There are three types of capsid symmetry: helical, icosahedral, and complex. Virions having an envelope are called **enveloped viruses,** whereas those lacking an envelope are called **naked viruses** (figure 5.1).

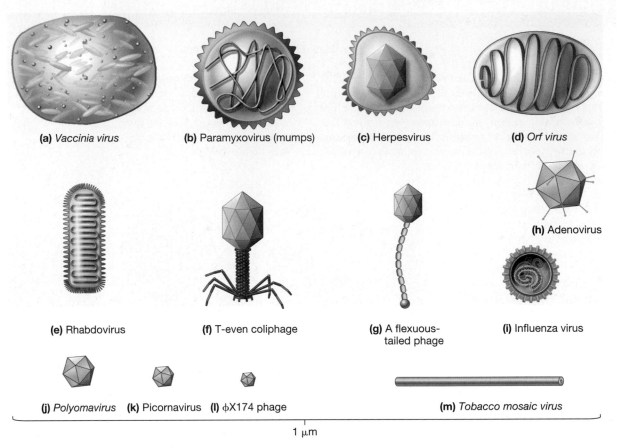

(a) *Vaccinia virus*

(b) Paramyxovirus (mumps)

(c) Herpesvirus

(d) *Orf virus*

(e) Rhabdovirus

(f) T-even coliphage

(g) A flexuous-tailed phage

(h) Adenovirus

(i) Influenza virus

(j) *Polyomavirus*   **(k)** Picornavirus   **(l)** φX174 phage

(m) *Tobacco mosaic virus*

1 μm

Figure 5.2   The Size and Morphology of Selected Viruses.   The viruses are drawn to scale. A 1 μm line is provided at the bottom of the figure.

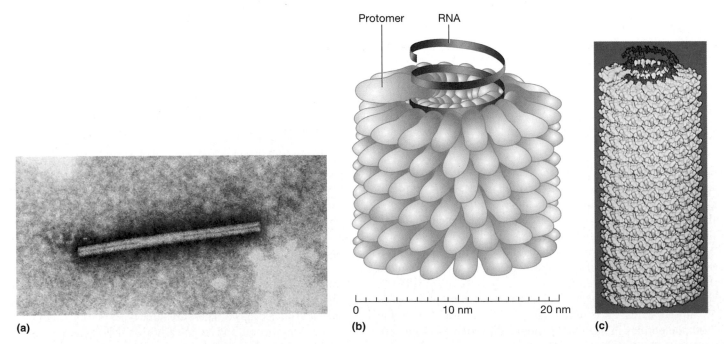

Protomer   RNA

0   10 nm   20 nm

(a)

(b)

(c)

Figure 5.3   *Tobacco Mosaic Virus* Structure.   (a) An electron micrograph of the negatively stained helical capsid (×400,000). (b) Illustration of TMV structure. Note that the nucleocapsid is composed of a helical array of protomers with the RNA spiraling on the inside. (c) A model of TMV.

Helical Capsids

Helical capsids are shaped like hollow tubes with protein walls. The *Tobacco mosaic virus* is a well-studied example of helical capsid structure (figure 5.3). In this virus, the self-assembly of protomers in a helical or spiral arrangement produces a long, rigid tube, 15 to 18 nm in diameter by 300 nm long. The capsid encloses an RNA genome, which is wound in a spiral and lies within a groove formed by the protein subunits. Not all helical capsids are as rigid as the TMV capsid. The influenza virus genome is enclosed in thin, flexible helical capsids that are folded within an envelope (**figure 5.4**). **>>** *Viruses with minus-strand RNA genomes (Group V): Influenza (flu) (section 24.6)*

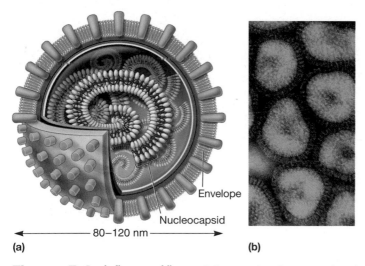

Envelope

Nucleocapsid

(a) ◀——— 80–120 nm ———▶

(b)

Figure 5.4 Influenza Virus. Influenza virus is an enveloped virus with a helical nucleocapsid. (a) Schematic view. Influenza viruses have segmented genomes consisting of 7 to 8 different RNA molecules. Each is coated by capsid proteins. (b) Because there are 7 to 8 flexible nucleocapsids enclosed by an envelope, the virions are pleomorphic. Electron micrograph (×350,000).

The size of a helical capsid is influenced by both its protomers and the nucleic acid enclosed within the capsid. The diameter of the capsid is a function of the size, shape, and interactions of the protomers. The nucleic acid appears to determine its length because a helical capsid does not extend much beyond the end of the viral genome.

Icosahedral Capsids

The icosahedron is a regular polyhedron with 20 equilateral triangular faces and 12 vertices (figure 5.2*h, j–l*). The **icosahedral capsid** is the most efficient way to enclose a space. A few genes, sometimes only one, can code for proteins that self-assemble to form the capsid. In this way, a small number of genes can specify a large three-dimensional structure.

Icosahedral capsids are constructed from ring- or knob-shaped units called **capsomers,** each usually made of five or six protomers (**figure 5.5**). Pentamers (pentons) have five protomers; hexamers (hexons) possess six. Pentamers are usually at the vertices of the icosahedron, whereas hexamers generally form its edges and triangular faces. The icosahedron in **figure 5.6** is constructed of 42 capsomers; larger icosahedra are made if more hexamers are used to form the edges and faces (e.g., adenoviruses have a capsid with 252 capsomers, as shown in figure 5.5*b*). In some RNA viruses, both the pentamers and hexamers of a capsid are constructed with only one type of subunit. In other viruses, pentamers are composed of different proteins than are the hexamers. Although most icosahedral capsids appear to contain both pentamers and hexamers, some have only pentamers.

Capsids of Complex Symmetry

Although most viruses have either icosahedral or helical capsids, some viruses do not fit into either category. Poxviruses and large bacteriophages are two important examples.

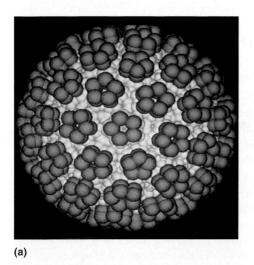

(a)

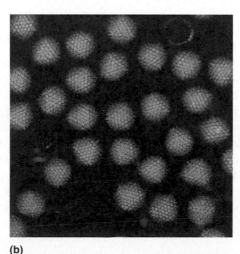

(b)

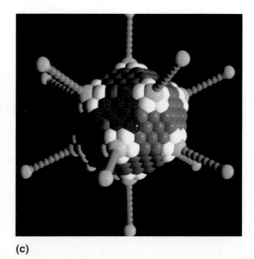

(c)

Figure 5.5 Examples of Icosahedral Capsids. (a) Computer-simulated image of the polyomavirus (72 capsomers) that causes a rare demyelinating disease of the central nervous system. (b) Adenovirus, 252 capsomers (×171,000). (c) Computer-simulated model of adenovirus.

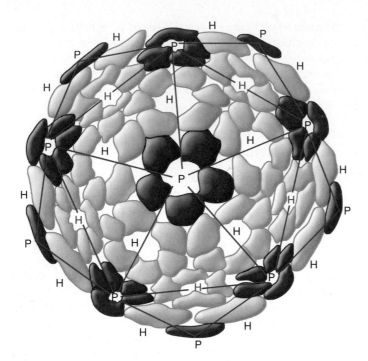

Figure 5.6   The Structure of an Icosahedral Capsid Formed from a Single Type of Protomer. The protomers associate to form either pentons (P), shown in red, or hexons (H), shown in gold. The blue lines define the triangular faces of the icosahedron. Notice that pentons are located at the vertices and that the hexons form the edges and faces of the icosahedron. This capsid contains 42 capsomers.

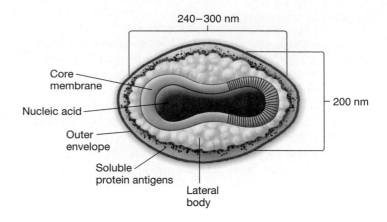

Figure 5.7   *Vaccinia Virus* Morphology. (a) Diagram of vaccinia structure. (b) Micrograph of the virion clearly showing the nucleoid (×200,000).

Poxviruses are the largest of the animal viruses (about 400 by 240 by 200 nm in size) and can even be seen with a light microscope. They possess an exceptionally complex internal structure with an ovoid- to brick-shaped exterior. **Figure 5.7** shows the morphology of *Vaccinia virus,* a poxvirus. Its double-stranded DNA genome is associated with proteins and contained in the nucleoid, a central structure shaped like a biconcave disk and surrounded by a membrane. Two lateral bodies lie between the nucleoid and the virus's outer envelope.

Some large bacteriophages are even more elaborate than the poxviruses. The T2, T4, and T6 phages (T-even phages) that infect *Escherichia coli* are said to have **binal symmetry** because they have a head that resembles an icosahedron and a tail that is helical. The icosahedral head is elongated by one or two rows of hexamers in the middle and contains the DNA genome (**figure 5.8**). The tail is composed of a collar joining it to the head, a central hollow tube, a sheath surrounding the tube, and a complex baseplate. The sheath is made of 144 copies of a single protein (gp18) arranged in 24 rings, each containing six copies. In T-even phages, the baseplate is hexagonal and has a pin and a jointed tail fiber at each corner. >> *Viruses with double-stranded DNA genomes (Group I): Bacteriophage T4: a virulent bacteriophage (section 24.2)*

Considerable variation in structure exists among the large bacteriophages, even those infecting a single host. In contrast with the T-even phages, many other coliphages (phages that infect

E. coli) have true icosahedral heads. T1, T5, and lambda phages have sheathless tails that lack a baseplate and terminate in rudimentary tail fibers. Coliphages T3 and T7 have short, noncontractile tails without tail fibers. >> *Viruses with double-stranded DNA genomes (Group I): Bacteriophage Lambda: A temperate bacteriophage (section 24.2)*

Viral Envelopes and Enzymes

Many animal viruses, some plant viruses, and at least one bacterial virus are bounded by an outer membranous layer called an **envelope** (**figure 5.9**). Animal virus envelopes usually arise from host cell plasma or nuclear membranes. Envelope lipids and carbohydrates are therefore acquired from the host. In contrast, envelope proteins are coded for by viral genes and may even project from the envelope surface as **spikes,** which are also called **peplomers** (figure 5.9). In many cases, these spikes are involved in virus attachment to the host cell surface. Because they differ among viruses, they also can be used to identify some viruses. Many enveloped viruses have a somewhat variable shape and are called pleomorphic. However, the envelopes of viruses such as the bullet-shaped *Rabies virus* are firmly attached to the underlying nucleocapsid and endow the virion with a constant, characteristic shape (figure 5.9*a*).

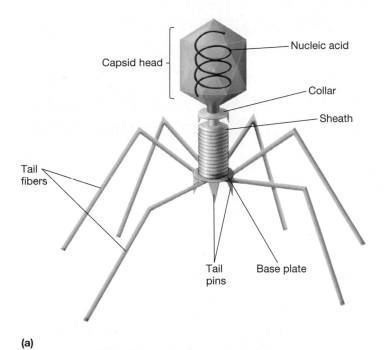

(a)

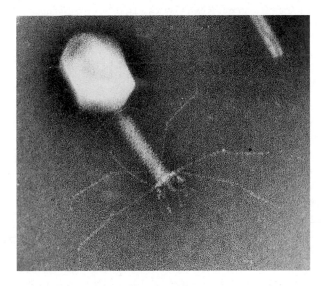

(b)

Figure 5.8   T-Even Coliphages.   (a) The structure of T4 bacteriophage. (b) The micrograph shows the phage before injection of its DNA.

Influenza virus (figure 5.4) is a well-studied example of an enveloped virus. Spikes project about 10 nm from the surface at 7 to 8 nm intervals. Some spikes consist of the enzyme neuraminidase, which functions in the release of mature virions from the host cell. Other spikes are hemagglutinin proteins, so named because they bind virions to red blood cells and cause the cells to clump together—a process called hemagglutination. Hemagglutinins participate in virion attachment to host cells. Most of the influenza virus's envelope proteins are glycoproteins—proteins that have carbohydrate attached to them. A nonglycosylated protein, the M

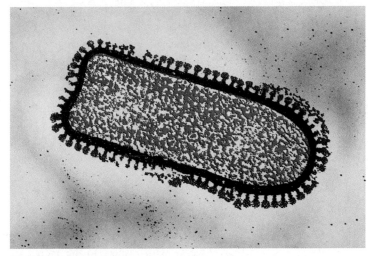

(a) *Rabies virus*

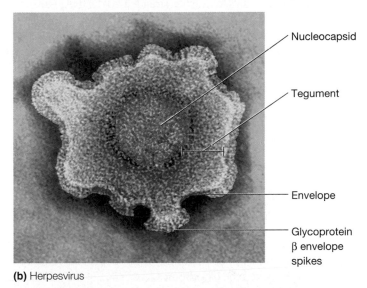

(b) Herpesvirus

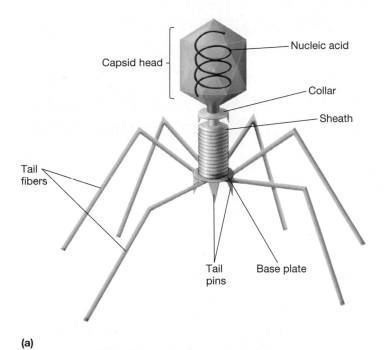

(c) *Semliki Forest virus*

Figure 5.9   Examples of Enveloped Viruses.   (a) Negatively stained *Rabies virus*. (b) Herpesvirus. (c) Computer image of the *Semliki Forest virus,* a virus that occasionally causes encephalitis in humans. Images (a) and (b) are artificially colorized.

or matrix protein, is found on the inner surface of the envelope and helps stabilize it.

It was originally thought that all virions lacked enzymes. However, this is not the case. In some instances, enzymes are associated with the envelope or capsid (e.g., influenza neuraminidase). Enzymes within the capsid are usually involved in nucleic acid replication. For example, the influenza virus has an RNA genome and carries an enzyme that synthesizes RNA using an RNA template. Thus although viruses lack true metabolism and cannot reproduce independently of living cells, they may carry one or more enzymes essential to the completion of their life cycles.

Viral Genomes

Viruses are exceptionally diverse with respect to the nature of their genomes. They employ all four possible nucleic acid types: single-stranded (ss) DNA, double-stranded (ds) DNA, ssRNA, and dsRNA. All four types are found in animal viruses. Most plant viruses have ssRNA genomes, and most bacterial viruses contain dsDNA. The size of viral genetic material also varies greatly. Very small genomes (e.g., those of the MS2 and Qβ viruses) are around 4,000 nucleotides; nucleotides are the building blocks from which nucleic acids are made. These genomes are just large enough to code for three or four proteins. MS2, Qβ, and some other viruses even save space by using overlapping genes. At the other extreme, T-even bacteriophages, herpesviruses, and *Vaccinia virus* have genomes of 1.0 to 2.0×10^5 nucleotides and may be able to direct the synthesis of over 100 proteins. The nature of each nucleic acid type is briefly summarized next. >> *Viruses with plus-strand RNA genomes (Group IV): Bacteriophages MS2 and Qβ (section 24.5); Gene structure (section 12.4); Viruses with double-stranded DNA genomes (Group I): Herpesviruses (section 24.2)*

Most DNA viruses use dsDNA as their genetic material. However, some have ssDNA genomes (e.g., φX174 and M13). In both cases, the genomes may be either linear or circular. Some DNA genomes can switch from one form to the other. For instance, the *E. coli* phage lambda has a genome that is linear in the capsid but is converted into a circular form once the genome enters the host cell. Another important characteristic of DNA bacteriophages is that their genomes often contain unusual nitrogenous bases. >> *Viruses with single-stranded DNA genomes (Group II): φX174 and fd (section 24.3)*

RNA viruses also can be either dsRNA or ssRNA. Although relatively few RNA viruses have dsRNA genomes, dsRNA viruses are known to infect animals, plants, fungi, and at least one bacterial species (e.g., φ6 and rotaviruses). More common are viruses with ssRNA genomes. Some ssRNA genomes have a base sequence that is identical to that of viral mRNA, in which case the genomic RNA is called **plus-strand** or **positive-strand RNA** (**figure 5.10**). Polio, tobacco mosaic, SARS, and brome mosaic viruses are all positive-strand RNA viruses. The genomes of plus-strand RNA viruses can direct protein synthesis immediately after entering the cell. Some plus-strand viruses of eucaryotes

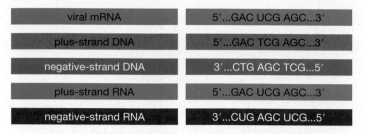

viral mRNA	5'...GAC UCG AGC...3'
plus-strand DNA	5'...GAC TCG AGC...3'
negative-strand DNA	3'...CTG AGC TCG...5'
plus-strand RNA	5'...GAC UCG AGC...3'
negative-strand RNA	3'...CUG AGC UCG...5'

Figure 5.10 Plus-Strand and Negative-Strand Viral Genomes. The genomes and replication intermediates of viral genomes can be either plus strand or negative strand. This designation is relative to the sequence of nucleotides in the mRNA of the virus. Plus-strand genomes have the same sequence as the mRNA, either using DNA nucleotides if a DNA genome or RNA nucleotides if an RNA genome. Negative-strand genomes are complementary to the viral mRNA.

have characteristic structures at each end of their RNA genomes that trick the host into behaving as if the virus's genome is the host's own mRNA. Viral RNA genomes that are complementary rather than identical to viral mRNA are called **minus-** or **negative-strand RNA.** Rabies, mumps, measles, and influenza viruses are examples of negative-strand RNA viruses. >> *Viruses with double-stranded RNA genomes (Group III) (section 24.4); Viruses with plus-strand RNA genomes (Group IV) (section 24.5); Viruses with minus-strand RNA genomes (Group V) (section 24.6)*

Many RNA viruses have **segmented genomes**—genomes that consist of more than one piece (segment) of RNA. In many cases, each segment codes for one protein and there may be as many as 10 to 12 segments. Usually all segments are enclosed in the same capsid; however, this is not always the case. For example, the genome of *Brome mosaic virus,* a virus that infects certain grass species, is composed of three segments distributed among three different virus particles. Despite this complex and seemingly inefficient arrangement, the different brome mosaic virions successfully manage to infect the same host.

Finally, **retroviruses,** such as *h*uman *i*mmunodeficiency *v*irus (HIV), differ from other ssRNA viruses in that their plus-strand RNA genomes are converted to DNA rather than serving as mRNA. The result is the synthesis of a dsDNA copy of the ssRNA genome. This DNA, called a **provirus,** is integrated into the host genome. >> *Viruses with single-stranded RNA genomes (Group VI—Retroviruses) (section 24.7); Transcription (section 12.5); Translation (section 12.7)*

1. How are viruses similar to cellular organisms? How do they differ?
2. What is the difference between a nucleocapsid and a capsid?
3. Compare the structure of an icosahedral capsid with that of a helical capsid. How do pentamers and hexamers associate to form a complete icosahedron; what determines helical capsid length and diameter?

4. What is an envelope? What are spikes (peplomers)? Why are some enveloped viruses pleomorphic? Give two functions spikes might serve in the viral life cycle and the proteins that the influenza virus uses in these processes.

5. All four nucleic acid forms can serve as viral genomes. Describe each. What are the following: plus-strand, minus-strand, and segmented RNA genome?

4. What advantage would an RNA virus gain by having its genome resemble eucaryotic mRNA?

5.3 VIRAL MULTIPLICATION

The differences in viral structure and genomes have important implications for the mechanism a virus uses to reproduce within its host cell. Indeed, even among viruses with similar structures and genomes, each can exhibit unique life cycles. Despite these differences, a general pattern of viral life cycles can be discerned. Because viruses need a host cell in which to reproduce, the first step in the life cycle of a virus is attachment (often called adsorption) to a host (figure 5.11). This is followed by entry of either the nucleocapsid or the viral nucleic acid into the host. If the nucleocapsid enters, uncoating of the genome usually occurs before the life cycle continues. Once inside the host cell, the synthesis stage begins. During this stage, genes encoded by the viral genome are expressed. That is, the viral genes are transcribed and translated. This allows the virus to take control of the host cell's biosynthetic machinery so that the viral genome can be replicated and viral proteins synthesized. This is followed by the assembly stage, during which new nucleocapsids are constructed by self-assembly of coat proteins with the nucleic acids. Finally, during the release step, mature viruses escape the host.

Attachment (Adsorption)

Encounters of the virus and the host cell surface are thought to occur through a random collision of the virion with a potential host. However, viruses do not randomly attach to the surface of a host cell; rather, adsorption to the host is mediated by an interaction between **receptors** on the surface of the host cell and molecules (ligands) on the surface of the virion. The nature of cellular receptors varies. For instance, some bacteriophages use cell wall lipopolysaccharides and proteins as receptors, while others use teichoic acids, flagella, or pili.

Variation in receptor properties is at least partly responsible for host preferences. Bacteriophages not only infect a particular species but often infect only certain strains within a given species. Receptor specificity also accounts for the observation that viruses of eucaryotes (i.e., eucaryotic viruses) infect specific

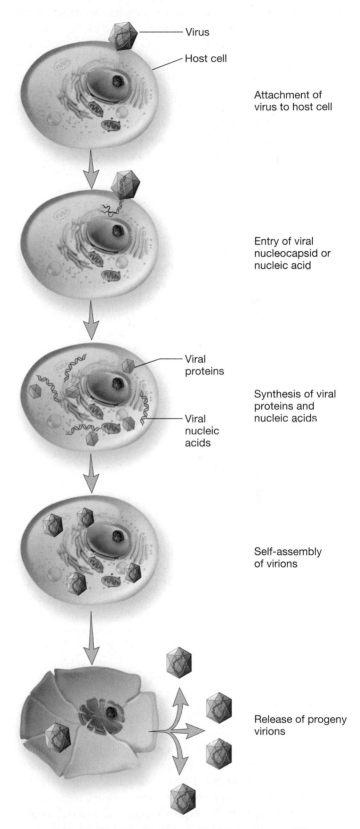

Figure 5.11 Generalized Illustration of Virus Reproduction. There is great variation in the details of virus reproduction for individual virus species.

organisms and, in some cases, only particular tissues within that host. It is important to note that the receptors on the host cell have specific cellular functions. For instance, they may normally bind hormones or other molecules essential to the cell's function. In some cases, it appears that two or more host cell receptors are involved in attachment. HIV uses CD4 and CXCR-4 (fusin) or the CCR5 (CC-CKR-5) receptor. Both of these host molecules normally bind chemokines—signaling molecules produced by the immune system. >> *Chemical mediators in nonspecific (innate) resistance: Cytokines (section 28.6)*

The distribution of the host cell receptors to which animal viruses attach also varies at the cellular level. Eucaryotic cell membranes have microdomains called lipid rafts that seem to be involved in both virion entrance and assembly. For example, the receptors for enveloped viruses such as HIV and Ebola are concentrated in lipid rafts. Distribution at the tissue level plays a crucial role in determining the tropism of the virus and the outcome of infection. For example, *Poliovirus* receptors are found only in the human nasopharynx, gut, and anterior horn cells of the spinal cord. Therefore *Poliovirus* infects these tissues, causing gastrointestinal disease in its milder forms and paralytic disease in its more serious forms. In contrast, *Measles virus* receptors are present in most tissues and disease is disseminated throughout the body, resulting in the widespread rash characteristic of measles. << *Eucaryotic membranes (section 4.2)*

Entry into the Host

After attachment to the host cell, the virus's genome or the entire nucleocapsid enters the cytoplasm. Many bacteriophages inject their nucleic acid into the cytoplasm of their host, leaving the capsid outside and attached to the cell wall. In contrast to phages, the nucleocapsid of eucaryotic viruses penetrates the cell membrane and enters the cytoplasm with the genome still enclosed. Once inside the cytoplasm, some viruses shed some or all of their capsid proteins, in a process called uncoating, whereas other viruses remain encapsidated. Because penetration and uncoating are often coupled, we consider them together.

The mechanisms of penetration and uncoating vary with the type of virus, and for many eucaryotic viruses, detailed mechanisms of penetration are unclear; however, it appears that one of two different modes of entry is usually employed (**figure 5.12**).

1. Fusion of the viral envelope with the host cell membrane—The envelopes of some viruses (e.g., HIV) fuse directly with the host cell plasma membrane (figure 5.12*a*). Fusion may involve envelope glycoproteins that bind to plasma membrane proteins. For example, after attachment of paramyxoviruses (negative-strand RNA viruses), membrane lipids rearrange, the adjacent halves of the contacting membranes merge, and a proteinaceous fusion pore forms. The nucleocapsid then enters the host cell cytoplasm, where a viral enzyme carried within the nucleocapsid begins synthesizing viral mRNA while it is still within the capsid.

2. Entry by endocytosis—Naked viruses and some enveloped viruses enter cells by endocytosis. They may be engulfed by receptor-mediated endocytosis to form coated vesicles (figure 5.12*b*). The virions attach to clathrin-coated pits, and the pits then pinch off to form coated vesicles filled with viruses. These vesicles fuse with endosomes after the clathrin has been removed; depending on the virus, escape of the nucleocapsid or its genome may occur either before or after vesicle fusion. Endosomal enzymes can aid in virus uncoating and low pH often triggers the uncoating process. In at least some instances, the viral envelope fuses with the endosomal membrane, and the nucleocapsid is released into the cytoplasm (the capsid proteins may have been partially removed by endosomal enzymes). Once in the cytoplasm, viral nucleic acid may be released from the capsid upon completion of uncoating or may function while still attached to capsid components. << *Organelles of the biosynthetic-secretory and endocytic pathways (section 4.4)*

Naked eucaryotic viruses lack an envelope and thus cannot employ the membrane fusion mechanism for release from the endosome (figure 5.12*c*). In this case, it appears that vesicle acidification causes a conformational change in the capsid. The altered capsid contacts the vesicle membrane and either releases the viral nucleic acid into the cytoplasm through a membrane pore (picornaviruses) or ruptures the membrane to release the virion (adenovirus). Viruses may also enter the host cell by way of caveolae formation.

Synthesis Stage

This stage of the viral life cycle exhibits dramatic diversity because the genome of each virus dictates the events that occur. For dsDNA viruses, the events are very similar to the typical flow of information in cells. That is, the genetic information is stored in DNA and replicated by enzymes called DNA polymerases, recoded as mRNA (transcription), and decoded during protein synthesis (translation). Because of this similarity, some dsDNA viruses have the luxury of depending solely on their host cells' biosynthetic machinery to replicate their genomes and synthesize their proteins.

The same is not true for RNA viruses. Cellular organisms (except for plants) lack the enzymes needed to replicate RNA or to synthesize mRNA from an RNA genome. Therefore negative-strand RNA viruses, retroviruses, and dsRNA viruses must carry in their nucleocapsids the enzymes needed to complete the synthesis stage. Likewise, positive-strand viruses must encode the enzymes they need.

No matter what the genome, synthesis of viral proteins is tightly regulated. Some proteins, often called early proteins, are synthesized early in the infection, whereas other proteins are synthesized later. Early proteins are most often involved in taking over the host cell. Late proteins usually include capsid proteins and other proteins involved in self-assembly and release.

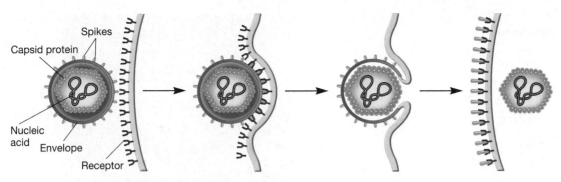

(a) Entry of enveloped virus by fusing with plasma membrane

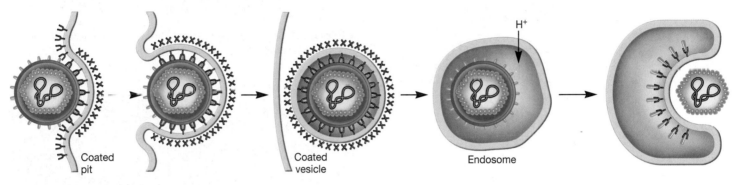

(b) Entry of enveloped virus by endocytosis

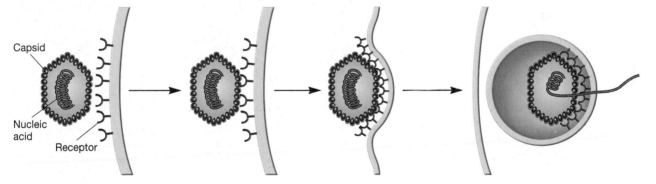

(c) Entry of naked virus by endocytosis

Figure 5.12   Animal Virus Entry.   Examples of animal virus attachment and entry into host cells. Enveloped viruses can (a) enter after fusion of the envelope with the plasma membrane or (b) escape from the vesicle after endocytosis. (c) Naked viruses such as *Poliovirus*, a picornavirus, may be taken up by endocytosis and then insert their nucleic acid into the cytoplasm through the vesicle membrane. It also is possible that they insert the nucleic acid directly through the plasma membrane within a coated pit.

Assembly

Several kinds of late proteins are involved in the assembly of mature viruses. Some are nucleocapsid proteins, some are not incorporated into the nucleocapsid but participate in its assembly, and still other late proteins are involved in cell lysis and virus release. In addition, proteins and other factors synthesized by the host may be involved in assembling mature viruses, as is the case in bacteriophage T4.

The assembly process can be quite complex with multiple subassembly lines functioning independently and converging in later steps to complete nucleocapsid construction. As shown in **figure 5.13**, bacteriophage T4 assembles the base plate, tail fibers, and viral head components separately. Once the baseplate is finished, the tail tube is built on it and the sheath is assembled around the tube. The phage prohead (procapsid) is assembled with the aid of scaffolding proteins that are degraded or removed after construction is completed. DNA is incorporated into the prohead by a complex of proteins sometimes called the "packasome." The packasome consists of a protein called the portal protein, which is located at the base of the prohead, and a set of

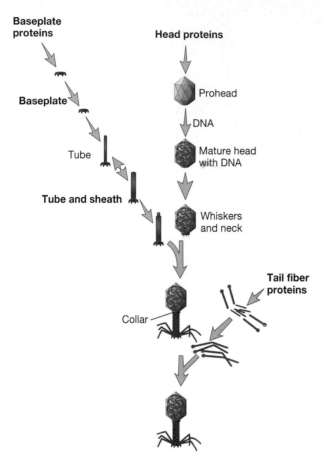

Figure 5.13 The Assembly of T4 Bacteriophage. Note the subassembly lines for the baseplate, tail tube and sheath, tail fibers, and head.

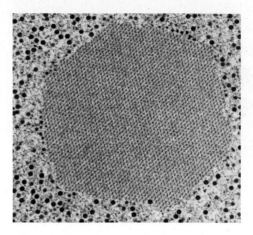

Figure 5.14 Paracrystalline Clusters. A crystalline array of adenoviruses within the cell nucleus (×35,000).

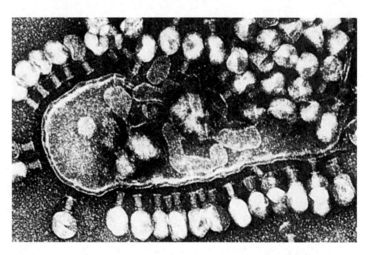

Figure 5.15   Release of T4 Bacteriophages by Lysis of the Host Cell. The host cell has been lysed (upper right portion of the cell) and virions have been released into the surroundings. Progeny virions also can be seen in the cytoplasm. In addition, empty capsids of the infecting phages coat the outside of the cell (×36,500).

proteins called the terminase complex, which moves DNA into the prohead. The movement of DNA consumes energy in the form of ATP, which is supplied by the metabolic activity of the host bacterium. After the mature head is completed, it spontaneously combines with the tail assembly.

While bacteriophages are assembled in the host cytoplasm, the site of morphogenesis varies with plant and animal viruses. Large paracrystalline clusters of either complete virions or procapsids are often seen at the site of virus maturation (**figure 5.14**). Some plant and animal viruses are assembled in the nucleus; others such as poxviruses are assembled in the cytoplasm.

Virion Release

Many viruses, especially naked viruses, lyse their host cells at the end of the intracellular phase. This process involves the activity of additional viral proteins. For instance, the lysis of *E. coli* by T4 requires two specific proteins (**figure 5.15**). One is lysozyme, an enzyme that attacks peptidoglycan in the host's cell wall. Another T4 protein creates holes in the *E. coli* plasma membrane, enabling T4 lysozyme to move from the cytoplasm to the peptidoglycan.

Another common release mechanism is budding. This is frequently observed in enveloped viruses—in fact, envelope formation and virus release are usually concurrent processes. Because this is a more insidious process than lysis, the host cell may continue releasing virions for some time. All envelopes of eucaryotic viruses are derived from host cell membranes by a multistep process. First, virus-encoded proteins are incorporated into the membrane. Then the nucleocapsid is simultaneously released and the envelope formed by membrane budding (**figure 5.16**). In several virus families, a matrix (M) protein attaches to the plasma membrane and aids in budding. Most envelopes arise from the plasma membrane. However, in herpesviruses, budding and envelope formation usually involve the nuclear membrane. The endoplasmic reticulum, Golgi apparatus, and other internal membranes also can be used to form envelopes.

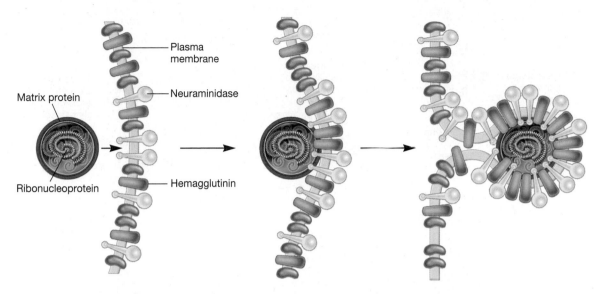

Figure 5.16   Release of Influenza Virus by Plasma Membrane Budding.   First, viral envelope proteins (hemagglutinin and neuraminidase) are inserted into the host plasma membrane. Then the nucleocapsid approaches the inner surface of the membrane and binds to it. At the same time, viral proteins collect at the site and host membrane proteins are excluded. Finally, the plasma membrane buds to simultaneously form the viral envelope and release the mature virion.

Interestingly, it has been discovered that actin filaments can aid in the release of many eucaryotic viruses. These viruses alter the actin microfilaments of the host cell cytoskeleton. For example, *Vaccinia virus* appears to form long actin tails and use them to move intracellularly at speeds up to 2.8 μm per minute. The actin filaments also propel vaccinia through the plasma membrane. In this way, the virion escapes without destroying the host cell and infects adjacent cells. This is similar to the mechanism of pathogenesis used by some intracellular pathogenic bacteria such as *Listeria monocytogenes*.   << *Eucaryotic cytoplasm (section 4.3)*

1. What probably plays the most important role in determining the tissue and host specificity of viruses? Give some specific examples.

2. In general, DNA viruses can be much more dependent on their host cells than can RNA viruses. Why is this so?

3. Given the origin of viral envelopes, why do you think enveloped viruses that infect plants and procaryotes are rare?

5.4   TYPES OF VIRAL INFECTIONS

So far, our discussion has focused on viral structures and modes of multiplication with little mention of the cost to host cells. The dependence of viruses on their host cells has many consequences—as anyone who has ever had a cold or the flu well understands. Next, we discuss the interaction between virus and host cell.

Infections of Procaryotic Cells: Lysis and Lysogeny

Figure 5.11 illustrates the life cycle of a **virulent phage**—one that has only one reproductive option: to begin multiplying immediately upon entering its host, followed by release from the host by lysis. However, many phages are **temperate phages** that have two reproductive options: upon entry into the host, they can reproduce like the virulent phages and lyse the host cell, or they can remain within the host without destroying it (**figure 5.17**). Many temperate phages accomplish this by integrating their genome into the host cell's chromosome.

The relationship between a temperate phage and its host is called **lysogeny.** The form of the virus that remains within its host is called a **prophage,** and the infected bacteria are called **lysogens** or **lysogenic bacteria.** Lysogenic bacteria reproduce and in most other ways appear to be perfectly normal. However, they have two distinct characteristics. The first is that they cannot be reinfected by the same virus—that is, they have immunity to superinfection. The second is that they can switch from the lysogenic cycle to the **lytic cycle.** This results in host cell lysis and release of phage particles. This occurs when conditions within the cell cause the prophage to initiate synthesis of phage proteins and to assemble new virions, a process called **induction.** Induction is commonly caused by changes in growth conditions or ultraviolet irradiation of the host cell. Why and how these phenomena occur are discussed in chapter 24.

Another important outcome of lysogeny is **lysogenic conversion.** This occurs when a temperate phage changes the phenotype of its host. Lysogenic conversion often involves alteration in surface characteristics of the host. For example, when *Salmonella* is infected by epsilon phage, the phage

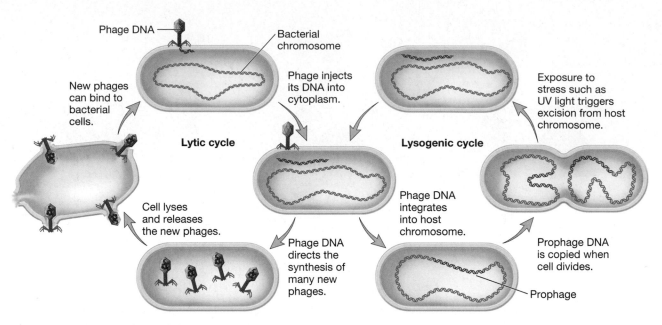

Figure 5.17   Lytic and Lysogenic Cycles of Temperate Phages.   Temperate phages have two phases to their life cycles. The lysogenic cycle allows the genome of the virus to be replicated passively as the host cell's genome is replicated. Certain environmental factors such as UV light can cause a switch from the lysogenic cycle to the lytic cycle. In the lytic cycle, new virus particles are made and released when the host cell lyses. Virulent phages are limited to just the lytic cycle.

changes the activities of several enzymes involved in construction of the carbohydrate component of the bacterium's lipopolysaccharide. This alters the antigenic properties of the bacterium as well as eliminates the receptor for epsilon phage, so the bacterium becomes immune to infection by another epsilon phage. Many other lysogenic conversions give the host pathogenic properties. This is the case when *Corynebacterium diphtheriae,* the cause of diphtheria, is infected with phage β. The phage genome encodes diphtheria toxin, which is responsible for the disease. Thus only those strains of *C. diphtheriae* that are infected by the phage (i.e., lysogens) cause disease.

Clearly, the infection of a bacterium by a temperate phage has significant impact on the host, but why would viruses evolve this alternate life cycle? Two advantages of lysogeny have been recognized. The first is that lysogeny allows a virus to remain viable within a dormant host. Bacteria often become dormant due to nutrient deprivation, and while in this state, they do not synthesize nucleic acids or proteins. In such situations, a prophage would survive but most virulent bacteriophages would not be replicated, as they require active cellular biosynthetic machinery. The second advantage arises when there are many more phages in an environment than there are host cells, a situation virologists refer to as a high multiplicity of infection (MOI). In these conditions, lysogeny enables the survival of host cells so that the virus can continue to reproduce.

Archaeal viruses can also be virulent or temperate. Most archaeal viruses discovered thus far are temperate. Unfortunately, little is known about the mechanisms they use to establish lysogeny.

1. Define the terms lysogeny, temperate phage, lysogen, prophage, immunity, and induction.
2. What advantages might a phage gain by being capable of lysogeny?
3. Describe lysogenic conversion and its significance.

Infection of Eucaryotic Cells

Viruses can harm their eucaryotic host cells in many ways. An infection that results in cell death is a cytocidal infection. As with procaryotic viruses, this can occur by lysis of the host (**figure 5.18a**). Viral growth does not always result in the lysis of host cells. Some viruses (e.g., herpesviruses) can establish persistent infections lasting many years (figure 5.18b,c). Animal viruses, in particular, can cause microscopic or macroscopic degenerative changes or abnormalities in host cells and in tissues that are distinct from lysis. These are called **cytopathic effects (CPEs).** Seven possible mechanisms of host cell damage are briefly described here. However, it should be emphasized that more than one of these mechanisms may be involved in any given cytopathic or cytocidal effect.

1. Many cytocidal viruses inhibit host DNA, RNA, and protein synthesis. The mechanisms of inhibition are not yet clear.
2. Cell endosomes may be damaged, resulting in the release of hydrolytic enzymes and cell destruction.

Figure 5.18 Types of Infections and Their Effects on Host Cells. (a) Lytic infections of host cells can lead to disease states in animal hosts called acute infections. (b) Latent infections and (c) chronic infections are two different types of persistant infections. (d) Some infections of animal cells cause the cell to be transformed into a malignant cell, which can cause cancer in the animal host.

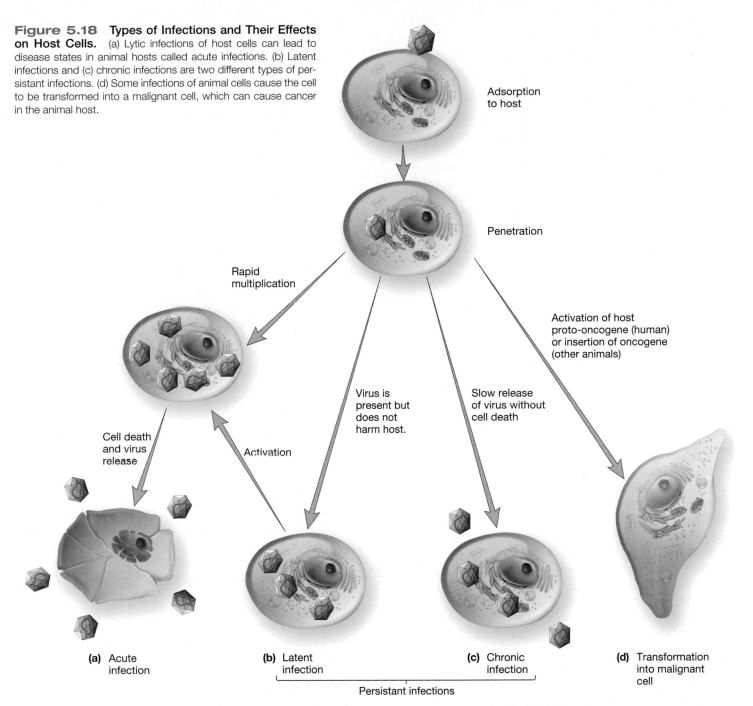

Adsorption to host

Penetration

Rapid multiplication

Activation of host proto-oncogene (human) or insertion of oncogene (other animals)

Virus is present but does not harm host.

Slow release of virus without cell death

Cell death and virus release

Activation

(a) Acute infection

(b) Latent infection

(c) Chronic infection

(d) Transformation into malignant cell

Persistant infections

3. Viral infection can drastically alter plasma membranes through the insertion of virus-specific proteins so that the infected cells are attacked by the immune system. When infected by viruses such as herpesviruses and *Measles virus*, as many as 50 to 100 cells may fuse into one abnormal, giant, multinucleated cell called a syncytium. HIV appears to destroy cells of the immune system called CD4$^+$ T-helper cells at least partly through its effects on their plasma membranes.

4. High concentrations of proteins from several viruses (e.g., *Mumps virus* and influenza virus) can have a direct toxic effect on cells and organisms.

5. Inclusion bodies that directly disrupt cell structure are formed during infections by many viruses. These intracellular structures may result from the clustering of viral components, virions, or even cell structures (e.g., ribosomes or chromatin).

6. Chromosomal disruptions result from infections by herpesviruses and others.

7. Finally, the host cell may not be directly destroyed but transformed into a malignant cell (figure 5.18*d*). This is discussed next.

Viruses and Cancer

Cancer is one of the most serious medical problems in developed nations, and it is the focus of an immense amount of research. A tumor is a growth or lump of tissue resulting from **neoplasia**—abnormal new cell growth and reproduction due to loss of regulation. Tumor cells have aberrant shapes and altered plasma membranes that may contain distinctive tumor antigens. Their unregulated proliferation and loss of differentiation result in invasive growth that forms unorganized cell masses. This reversion to a more primitive or less differentiated state is called **anaplasia.**

Two major types of tumor growth patterns exist. If the tumor cells remain in place to form a compact mass, the tumor is benign. In contrast, cells from malignant or cancerous tumors actively spread throughout the body in a process known as metastasis. Some cancers are not solid but cell suspensions. For example, leukemias are composed of undifferentiated malignant white blood cells that circulate throughout the body. Indeed, dozens of kinds of cancers arise from a variety of cell types and afflict all kinds of organisms.

As one might expect from the wide diversity of cancers, cancer has many causes, only a few of which are directly related to viruses. Carcinogenesis is a complex, multistep process that involves the mutation of multiple genes. Genes involved in carcinogenesis are called **oncogenes.** Some oncogenes are contributed to a cell by viruses; others arise from normal genes within the cell called **proto-oncogenes.** Proto-oncogenes are cellular genes required for normal growth, but when mutated or overexpressed, they become oncogenes. That is, their products contribute to the malignant transformation of the cell. Many oncogenes are involved in the regulation of cell growth and signal transduction; for example, some code for growth factors that regulate cell reproduction. Proto-oncogenes can be transformed into oncogenes by spontaneous mutation or through the activity of a mutation-causing agent, called a mutagen. >> *Mutations and their chemical basis (section 14.1)*

Although viruses are known to cause many animal cancers, currently only seven human cancers are known to be of viral origin. Most of these viruses have dsDNA genomes.

1. Two herpesviruses, *Human herpesvirus 8* and Epstein-Barr virus (EBV), are linked to cancer. *Human herpesvirus 8* is associated with the development of Kaposi's sarcoma, which has a high incidence in AIDS patients. EBV is the cause of two cancers. Burkitt's lymphoma is a malignant tumor of the jaw and abdomen found in children of central and western Africa. EBV also causes nasopharyngeal carcinoma. Interestingly, some evidence exists that a person also must have had malaria to develop Burkitt's lymphoma. Environmental factors must play a role, because EBV-associated cancer is rare in the United States despite the prevalence of the virus. This may be due to a low incidence of malaria in the United States. >> *Viruses with double-stranded DNA genomes (Group I): Herpesviruses (section 24.2)*

2. Two viruses that cause hepatitis are associated with human cancers. *Hepatitis B virus* is linked with one form of liver cancer (hepatocellular carcinoma). *Hepatitis C virus* causes cirrhosis of the liver, which can lead to liver cancer.

3. Some strains of human papillomaviruses (HPV) cause cervical cancer.

4. The retrovirus human T-cell lymphotropic virus I (HTLV-1) is associated with adult T-cell leukemia.

Viruses known to cause cancer are called **oncoviruses.** All known human dsDNA oncoviruses trigger cancerous transformation of cells by a similar mechanism. They encode proteins that bind to and thereby inactivate host proteins known as **tumor suppressors.** Tumor-suppressor proteins regulate cell cycling, or monitor or repair DNA damage. Two tumor suppressors known to be targets of human oncovirus proteins are called Rb and p53. Rb has multiple functions in the nucleus, all of which are critical to normal cell cycling. When Rb molecules are rendered inactive by an oncoviral protein, cells undergo uncontrolled reproduction and are said to be hyperproliferative. The protein p53 is often referred to as "the guardian of the genome." This is because p53 normally initiates either cell cycle arrest or programmed cell death in response to DNA damage. However, when p53 is inactivated by the binding of an oncoviral protein, it cannot do so and genetic damage persists. From the point of view of the virus, hyperproliferation and the lack of programmed cell death are beneficial. For the cell, however, it can be catastrophic. Cells can rapidly accumulate the additional mutations needed for oncogenic transformation.

Retroviruses exert their oncogenic powers in a different manner. Some carry oncogenes captured from host cells many, many generations ago. Thus they transform the host cell by bringing the oncogene into the cell. For example, the retrovirus HTLV-1 transforms a group of immune system cells called T cells by producing a regulatory protein that sometimes activates genes involved in cell division and stimulates viral multiplication. The second transformation mechanism used by retroviruses such as *Avian leukosis virus* involves the integration of a viral genome into the host chromosome such that strong, viral regulatory elements are near a cellular proto-oncogene. These elements cause the nearby proto-oncogene to be transcribed at a high level, causing the gene to be considered an oncogene.

1. What is a cytocidal infection? What is a cytopathic effect? Outline the ways in which viruses can damage host cells during cytocidal infections.

2. Define the following terms: tumor, neoplasia, anaplasia, metastasis, and oncogene.

3. Distinguish the mechanism by which DNA viruses cause cancer from that of retroviruses. Why do you think there is an environmental component to many kinds of cancer?

5.5 CULTIVATION AND ENUMERATION OF VIRUSES

Because they are unable to reproduce outside of living cells, viruses cannot be cultured in the same way as cellular microorganisms. It is relatively simple to grow bacterial and archaeal viruses as long as the host cell is easily grown in culture. The cells and viruses are simply mixed such that the concentration of the host microbe far exceeds the concentration of viruses. Those cells that are not infected will, when plated on suitable medium, form a confluent lawn of cells that appears turbid. However, those cells that are infected will lyse (as long as the viruses do not enter the lysogenic cycle), resulting in clear areas in the lawn called **plaques** (figures 5.19 and 5.20*a*). Each plaque contains many thousands of virions, which spread ever outward as they infect new host cells.

For many years, researchers have cultivated animal viruses by inoculating suitable host animals or embryonated eggs—fertilized chicken eggs incubated about six to eight days after laying. Animal viruses also are grown in tissue (cell) culture on monolayers of animal cells. If the virus causes its host cell to lyse, plaques often are formed (figure 5.20*b*) and may be detected if stained with dyes, such as neutral red or trypan blue, that can distinguish living from dead cells. Viruses that cause cytopathic effects are also grown and detected in tissue culture (**figure 5.21**).

Plant viruses are cultivated in a variety of ways. Plant tissue cultures, cultures of separated cells, or cultures of protoplasts (cells lacking cell walls) may be used. Viruses also can be grown in whole plants, usually after the leaves are mechanically inoculated by rubbing them with a mixture of viruses and an abrasive. When the cell walls are broken by the abrasive, the viruses directly contact the plasma membrane and infect the exposed host cells. In nature, the role of the abrasive is frequently filled by insects that suck or crush plant leaves and thus transmit viruses. Some plant viruses can be transmitted only if a diseased part is grafted onto a healthy plant. A localized necrotic lesion often develops in infected plants due to the rapid death of cells in the infected area (**figure 5.22**). Even when lesions do not occur, the infected plant may show symptoms such as changes in pigmentation or leaf shape.

It is often necessary to determine the number of virions in a preparation of viruses. The most direct approach is to count virions with the electron microscope. In one procedure, the virus-containing sample is mixed with a known concentration of small latex beads and sprayed on a coated specimen grid. The beads and virions are counted; the virus concentration is calculated from these counts and from the bead concentration. This technique often works well with concentrated preparations of viruses of known morphology. Viruses can be concentrated by centrifugation before counting if the preparation is too dilute. However, if the beads and viruses are not evenly distributed (as sometimes happens), the final count will be inaccurate.

An indirect method of counting viruses is the **hemagglutination assay.** Many viruses bind to the surface of red blood cells (*see figure 32.7*). If the ratio of viruses to cells is large enough, virus

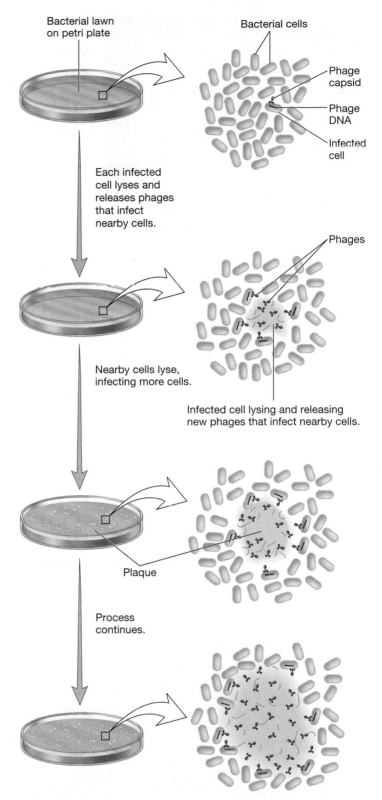

Figure 5.19 The Formation of Phage Plaques. When phages and host bacterial cells are mixed at an appropriate ratio, only a portion of the cells are initially infected. When this mixture is plated, the infected cells will be separated from each other. The infected cells eventually lyse, releasing progeny phages. They infect nearby cells, which eventually lyse, releasing more phages. This continues and ultimately gives rise to a clear area within a lawn of bacteria. The clear area is a plaque.

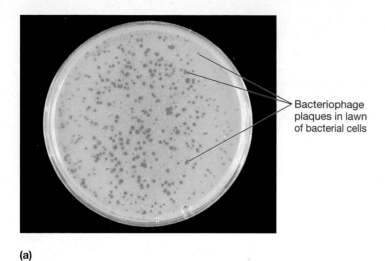

(a)

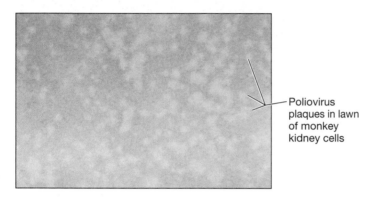

(b)

Figure 5.20 Viral Plaques. (a) Plaques formed by bacteriophages growing on a lawn of bacterial cells. (b) *Poliovirus* plaques in a monkey kidney cell culture.

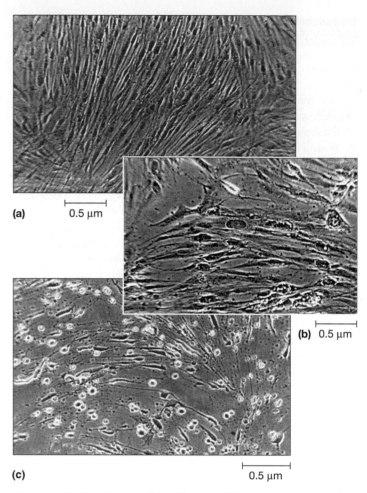

(a)   0.5 μm

(b) 0.5 μm

(c)   0.5 μm

Figure 5.21 Cytopathic Effects of Viruses. (a) A monolayer of normal fibroblast cells from fetal tonsils. (b) Cytopathic effects caused by infection of fetal tonsil fibroblasts with adenovirus. (c) Cytopathic effects caused by infection of fetal tonsil fibroblasts with herpes simplex virus.

Figure 5.22 Necrotic Lesions on Plant Leaves. *Tobacco mosaic virus* infection of an orchid showing leaf color changes.

particles join the red blood cells together—that is, they agglutinate, forming a network that settles out of suspension. In practice, red blood cells are mixed with diluted samples of the virus and each mixture is examined. The hemagglutination titer is the highest dilution of virus (or the reciprocal of the dilution) that still causes hemagglutination. This assay is an accurate, rapid method for determining the relative quantity of viruses such as the influenza virus. If the actual number of viruses needed to cause hemagglutination is determined by another technique, the assay can be used to ascertain the number of virions present in a sample.

Other indirect assays determine virus numbers based on their infectivity, and many of these are based on the same techniques used for virus cultivation. For example, in the **plaque assay,** several dilutions of viruses are plated with appropriate host cells. When the number of viruses plated is much lower than the number of host cells available for infection and when the viruses are distributed evenly, each plaque in a layer of host cells is assumed to have arisen from the reproduction of a single virion. Therefore a count of the plaques produced at a particular dilution gives the number of infectious virions, called **plaque-forming units (PFU),** and the concentration of infectious units in the original

sample can be easily calculated. For instance, suppose that 0.1 mL of a 10^{-6} dilution of the virus preparation yields 75 plaques. The original concentration of plaque-forming units is

$$\text{PFU/mL} = (75 \text{ PFU}/0.10 \text{ mL})(10^6) = 7.5 \times 10^8.$$

The number of PFU does not equal the number of virions as not all virions may be infective. Furthermore, even though there are far fewer viruses than host cells, it is still theoretically possible for more than one virus to infect the same cell. However, the number of PFU is proportional to the number of viruses: a preparation with twice as many viruses will have twice the plaque-forming units.

The same approach employed in the plaque assay may be used with embryos and plants. Chicken embryos can be inoculated with a diluted preparation or plant leaves rubbed with a mixture of diluted virus and abrasive. The number of pocks on embryonic membranes or necrotic lesions on leaves is multiplied by the dilution factor and divided by the inoculum volume to obtain the concentration of infectious units.

When biological effects are not readily quantified in these ways, the amount of virus required to cause disease or death can be determined by the endpoint method. Organisms or cell cultures are inoculated with serial dilutions of a virus suspension. The results are used to find the endpoint dilution at which 50% of the host cells or organisms are killed (**figure 5.23**). The **lethal dose (LD_{50})** is the dilution that contains a dose large enough to destroy 50% of the host cells or organisms. In a similar sense, the **infectious dose (ID_{50})** is the dose that, when given to a number of hosts, causes an infection of 50% of the hosts under the conditions employed.

1. Discuss the ways that viruses can be cultivated. Define the terms plaque, cytopathic effect, and necrotic lesion.

2. Given that viruses must be cultivated to make vaccines against viral diseases, discuss the advantages and disadvantages that might occur with each approach (embryonated egg versus cell culture) to growing animal viruses.

3. What is the original concentration of PFU in a phage suspension that yields 125 plaques when 0.20 mL of a 10^{-8} dilution of the virus preparation is plated in a plaque assay?

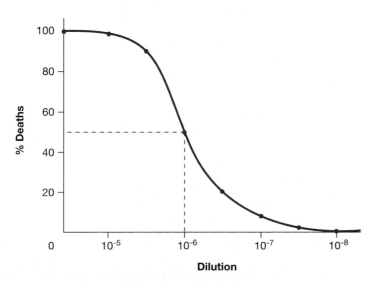

Figure 5.23 **A Hypothetical Dose-Response Curve.** The LD_{50} is indicated by the dashed line.

5.6 VIROIDS AND VIRUSOIDS

Although some viruses are exceedingly small and simple, even simpler infectious agents exist. **Viroids** are infectious agents that consist only of RNA. They cause over 20 different plant diseases, including potato spindle-tuber disease, exocortis disease of citrus trees, and chrysanthemum stunt disease. Viroids are covalently closed, circular ssRNAs, about 250 to 370 nucleotides long (**figure 5.24**). The circular RNA normally exists as a rodlike shape due to intrastrand base pairing, which forms double-stranded regions with single-stranded loops (**figure 5.25**). Some viroids are found in the nucleolus of infected host cells where between 200 and 10,000 copies may be present. Others are located within chloroplasts. Interestingly, the RNA of viroids does not encode any gene products, so they cannot replicate themselves. Rather, it is thought that the viroid is replicated by a host cell enzyme called a DNA-dependent RNA polymerase. This enzyme is normally used by the host to synthesize RNA using DNA as the template. However, when infected by a viroid, the host

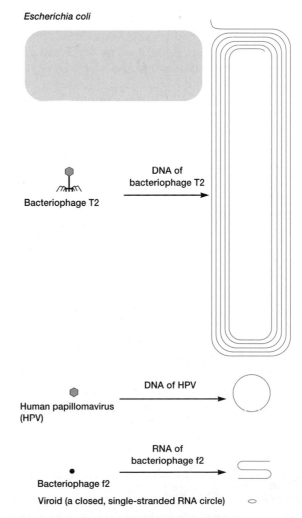

Figure 5.24 **Viroids, Viruses, and Bacteria.** A comparison of *Escherichia coli*, several viruses, and the potato spindle-tuber viroid with respect to size and the amount of nucleic acid possessed. (All dimensions are enlarged approximately ×40,000.)

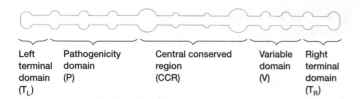

Figure 5.25 Viroid Structure. This schematic diagram shows the general organization of a viroid. The closed single-stranded RNA circle has extensive intrastrand base pairing and interspersed unpaired loops. Viroids have five domains. Most changes in viroid pathogenicity seem to arise from variations in the P and T_L domains.

polymerase evidently uses the viroid RNA as a template for RNA synthesis, rather than host DNA. The host polymerase synthesizes a complementary RNA molecule (a negative-strand RNA), which then serves as the template for synthesis of new viroid RNAs.

A plant may be infected with a viroid without showing symptoms—that is, it may have a latent infection. However, the same viroid in another host species may cause severe disease. The pathogenicity of viroids is not well understood, but it is known that particular regions of the RNA are required; studies have shown that removing these regions blocks the development of disease (figure 5.25). Some data suggest that viroids cause disease by triggering a eucaryotic response called **RNA silencing,** which normally functions to protect against infection by dsRNA viruses. During RNA silencing, the cell detects the presence of dsRNA and selectively degrades it. Viroids may usurp this response by hybridizing to specific host mRNA molecules to which they have a complementary nucleotide sequence. Formation of the viroid:host hybrid dsRNA molecule is thought to elicit RNA silencing. This results in destruction of the host mRNA and therefore silencing of the host gene. Failure to express a required host gene leads to disease in the host plant.

Virusoids, formerly called satellite RNAs, are similar to viroids in that they also consist only of covalently closed, circular ssRNA molecules with regions capable of intrastrand base pairing. In contrast to viroids, they encode one or more gene products, and they typically need a helper virus to infect host cells. The helper virus supplies gene products and other materials needed by the virusoid for completion of its replication cycle. The best-studied virusoid is the human hepatitis D virusoid, which is 1,700 nucleotides long. It uses the hepatitis B virus as its helper virus (a virus with an interesting gapped dsDNA genome). If a host cell contains both the hepatitis B virus and the hepatitis D virusoid, the virusoid RNA and its gene product, called delta antigen, can be packaged within the envelope of the virus. These enveloped virusoids and delta antigens are capable of entering other host cells and initiating infection.

>> *Viruses with gapped DNA genomes (Group VII) (section 24.8)*

5.7 PRIONS

Prions (for *proteinaceous infectious particle*) cause a variety of neurodegenerative diseases in humans and animals. The best-studied prion is the scrapie prion, which causes the disease scrapie

in sheep. Afflicted animals lose coordination of their movements, tend to scrape or rub their skin, and eventually cannot walk.

Researchers have shown that scrapie is caused by an abnormal form of a cellular protein. The abnormal form is called PrP^{Sc} (for *sc*rapie-associated *prion protein*), and the normal cellular form is called PrP^C. Evidence supports a model in which entry of PrP^{Sc} into the brain of an animal causes the PrP^C protein to change from its normal conformation to the abnormal form (**figure 5.26**). The newly produced PrP^{Sc} molecules then convert more PrP^C molecules into the abnormal PrP^{Sc} form. How the PrP^{Sc} causes this conformational change is unclear. However, the best-supported

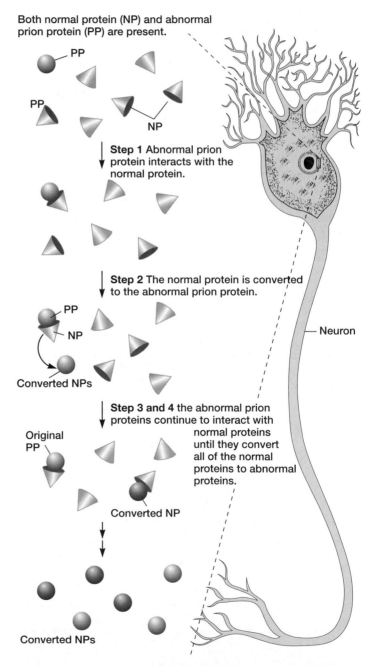

Figure 5.26 Proposed Mechanism by Which Prions Replicate. The normal and prion proteins differ in their tertiary structures.

model is that the PrP^Sc directly interacts with PrP^C, causing the change. It is noteworthy that mice lacking the PrP gene cannot be infected with PrP^Sc. Although the evidence is strong that PrP^Sc causes PrP^C to fold abnormally, how this triggers neuron loss is poorly understood. Recent evidence suggests that the interaction of PrP^Sc with PrP^C serves to cross-link PrP^C molecules. The cross-linked PrP^C molecules trigger a series of events called apoptosis or programmed cell death. Thus the normal but cross-linked protein causes neuron loss, whereas the abnormal protein acts as the infectious agent.

In addition to scrapie, prions are responsible for bovine spongiform encephalopathy (BSE or "mad cow disease") and the human diseases kuru, fatal familial insomnia, Creutzfeldt-Jakob disease (CJD), and Gerstmann-Strässler-Scheinker syndrome (GSS). All result in progressive degeneration of the brain and eventual death. At present, no effective treatment exists. Mad cow disease reached epidemic proportions in Great Britain in the 1990s and initially spread because cattle were fed meal made from all parts of cattle, including brain tissue. It has now been shown that eating meat from cattle with BSE can cause a variant of Creutzfeldt-Jakob disease in humans (vCJD). More than 90 people have died in the United Kingdom and France from this source. Variant CJD differs from CJD in origin only: people acquire vCJD by eating contaminated meat, while CJD is an extremely rare condition caused by spontaneous mutation of the gene that encodes the prion protein. CJD and GSS are rare and cosmopolitan in distribution among middle-aged people, while kuru has been found only in the Fore, an eastern New Guinea tribe. This tribe had a custom of consuming dead kinsmen. Women and children were given the less desirable body parts to eat, including the brain, and were thus infected. Cannibalism was stopped many years ago, and kuru has been eliminated.

1. What are viroids and why are they of great interest?

2. How does a viroid differ from a virus?

3. What is a prion? In what way does a prion appear to differ fundamentally from viruses and viroids?

4. Prions are difficult to detect in host tissues. Why do you think this is so? Why do you think we have not been able to develop effective treatments for these diseases?

Summary

5.1 Introduction to Viruses

a. A virion is composed of either DNA or RNA enclosed in a coat of protein (and sometimes other substances as well). It cannot reproduce independently of living cells.

5.2 Structure of Viruses

a. All virions have a nucleocapsid composed of a nucleic acid, either DNA or RNA, held within a protein capsid made of one or more types of protein subunits called protomers (**figure 5.1**).

b. There are four types of viral morphology: naked icosahedral, naked helical, enveloped icosahedral and helical, and complex.

c. Helical capsids resemble long hollow protein tubes and may be either rigid or quite flexible. The nucleic acid is coiled in a spiral on the inside of the cylinder (**figure 5.3**).

d. Icosahedral capsids are usually constructed from two types of capsomers: pentamers (pentons) at the vertices and hexamers (hexons) on the edges and faces of the icosahedron (**figure 5.5**).

e. Complex viruses (e.g., poxviruses and large phages) have complicated morphology not characterized by icosahedral and helical symmetry (**figure 5.7**). Large phages often have binal symmetry: their heads are icosahedral and their tails, helical (**figure 5.8**).

f. Viruses can have a membranous envelope surrounding their nucleocapsid. The envelope lipids usually come from the host cell; in contrast, many envelope proteins are viral and may project from the envelope surface as spikes or peplomers (**figure 5.9**).

g. Viral nucleic acids can be either single stranded (ss) or double stranded (ds), DNA or RNA.

h. Most DNA viruses have dsDNA genomes that may be linear or closed circles.

i. RNA viruses usually have ssRNA that may be either plus (positive) or minus (negative) when compared with mRNA (positive) (**figure 5.10**). Some RNA genomes are segmented.

j. Although viruses lack true metabolism, some contain a few enzymes necessary for their reproduction.

5.3 Viral Multiplication

a. Viral life cycles can be divided into five steps: (1) attachment to host; (2) entry into host; (3) synthesis of viral nucleic acids and proteins; (4) self-assembly of virions and; (5) release from host (**figures 5.11** to **5.16**).

b. The synthesis stage of virus's life cycle depends on the nature of its genome. DNA viruses use enzymes that are similar to host enzymes. In some cases, they can rely solely on their hosts for synthesis of their nucleic acids and proteins.

c. RNA viruses must either carry in their capsids or encode the enzymes they need to make mRNA and replicate their genomes

5.4 Types of Viral Infections

a. Virulent phages and archaeal viruses lyse their host. In addition to host cell lysis, temperate viruses of procaryotes can enter the lysogenic cycle in which they remain dormant in the host cell. This is often accomplished by integrating the viral genome into that of the host (**figure 5.17**).

b. Viruses that infect eucaryotic cells can cause either host cell lysis or a more insidious cellular death; such viruses are often said to cause cytocidal infection.

c. One relatively rare outcome of animal virus infection is the transformation of normal host cells into malignant or cancerous cells (**figure 5.18**).

5.5 Cultivation and Enumeration of Viruses

a. Viruses are cultivated using tissue cultures, embryonated eggs, bacterial cultures, and other living hosts.

b. Phages produce plaques in bacterial lawns. Sites of animal viral infection may be characterized by cytopathic effects such as pocks and plaques. Plant viruses can cause localized necrotic lesions in plant tissues (**figures 5.20** to **5.22**).

c. Virions can be counted directly with the transmission electron microscope or indirectly by the hemagglutination assay.

d. Infectivity assays can be used to estimate virus numbers in terms of plaque-forming units, lethal dose (LD_{50}), or infectious dose (ID_{50}) (**figure 5.23**).

5.6 Viroids and Virusoids

a. Infectious agents simpler than viruses exist. For example, several plant diseases are caused by small, circular ssRNA molecules called viroids (**figures 5.24** and **5.25**).

b. Virusoids are infectious RNAs that encode one or more gene products. They require a helper virus for replication.

5.7 Prions

a. Prions are small proteinaceous agents associated with at least six degenerative nervous system disorders: scrapie, bovine spongiform encephalopathy, kuru, fatal familial insomnia, Gerstmann-Strässler-Scheinker syndrome, and Creutzfeldt-Jakob disease.

b. Most evidence supports the hypothesis that prion proteins exist in two forms: the infectious, abnormally folded form and a normal cellular form. The interaction between the abnormal form and the cellular form converts the cellular form into the abnormal form (**figure 5.26**).

Critical Thinking Questions

1. Many classification schemes are used to identify bacteria. These start with Gram staining, progress to morphology and arrangement characteristics, and include a battery of metabolic tests. Build an analogous scheme that could be used to identify viruses. You might start by considering the host, or you might start with viruses found in a particular environment, such as a marine filtrate.

2. The origin and evolution of viruses is controversial. Discuss whether you think viruses evolved before the first procaryote or whether they have coevolved and are perhaps still coevolving with their hosts.

3. Viruses with ssRNA genomes tend to mutate faster than dsDNA viruses. Why do you think this is the case? In addition, why do you think segmented viruses (e.g., influenza viruses) are able to change host species faster than a virus with a dsDNA genome?

4. Consider the separate stages of an animal virus life cycle. Assemble a short list of structures and processes that are unique to the virus and would make good drug targets for an antiviral agent. Explain your rationale for each choice.

Learn More

Learn more by visiting the Prescott website at www.mhhe.com/prescottprinciples, where you will find a complete list of references.

Microbial Nutrition

6

Beautiful snowflakelike colonies produced by *Bacillus subtilis* when grown on nutrient-poor agar.

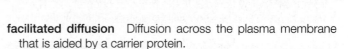

active transport The transport of solute molecules across a membrane against an electrochemical gradient; it requires a carrier protein and the input of energy.

agar A complex sulfated polysaccharide, usually from red algae, that is used as a solidifying agent in the preparation of culture media.

autotroph An organism that uses CO_2 as its sole or principal source of carbon.

chemolithoautotroph A microorganism that oxidizes reduced inorganic compounds to derive both energy and electrons; CO_2 is the carbon source.

chemolithoheterotroph A microorganism that uses reduced inorganic compounds to derive both energy and electrons; organic molecules are used as the carbon source.

chemoorganoheterotroph An organism that uses organic compounds as sources of energy, electrons, and carbon.

chemotroph An organism that uses chemicals, either organic or inorganic, as its source of energy.

colony An assemblage of microorganisms growing on a solid surface.

facilitated diffusion Diffusion across the plasma membrane that is aided by a carrier protein.

group translocation A transport process in which a molecule is moved across a membrane by carrier proteins while being chemically altered at the same time (e.g., phosphoenolpyruvate: sugar phosphotransferase system).

heterotroph An organism that uses reduced, preformed organic molecules as its principal carbon source.

lithotroph An organism that uses reduced inorganic compounds as its electron source.

macroelements Nutrients such as carbon, hydrogen, oxygen, and nitrogen that are required in relatively large amounts; also called macronutrients.

micronutrients Nutrients such as manganese, zinc, and copper that are required in very small amounts; also called trace elements.

organotroph An organism that uses reduced organic compounds as its electron source.

permease A membrane-bound carrier protein or a system of two or more proteins that transports a substance across the plasma membrane.

photoautotroph An organism that uses light for energy and CO_2 as its carbon source.

photolithoautotroph An organism that uses light for energy, an inorganic electron source (e.g., H_2O, H_2, H_2S), and CO_2 as its carbon source.

photoorganoheterotroph A microorganism that uses light energy, organic electron sources, and organic molecules as a carbon source.

phototroph An organism that uses light as its source of energy.

pure culture A population of cells that are identical because they arose from a single cell.

siderophore A small molecule that complexes with ferric iron and supplies it to a cell by aiding in its transport across the plasma membrane.

The whole of nature, as has been said, is a conjugation of the verb to eat, in the active and passive.

—*William Ralph Inge*

As discussed in chapters 3 and 4, microbial cells are structurally complex, and they carry out numerous functions. These functions are geared toward the primary goal of all life forms: to reproduce. To thrive and reproduce, organisms must have a source of energy. The energy source is used to generate the cell's energy currency—the high-energy molecule ATP. ATP is used to fuel the various types of work done by the cell, including biosynthesis. Biosynthesis not only requires energy but also a supply of elements that are used to construct the molecules of life. In this chapter, we describe the nutritional requirements of microorganisms, how nutrients are acquired, and how microbes are cultivated. How microbes make ATP from their energy sources is the focus of chapter 10. The use of nutrients and ATP in biosynthesis is discussed in chapter 11.

6.1   ELEMENTS OF LIFE

Chemical analysis of cells shows that over 95% of cell dry weight is made up of a few major elements: carbon, oxygen, hydrogen, nitrogen, sulfur, phosphorus, potassium, calcium, magnesium, and iron. These are called **macroelements** or mac-ronutrients because they are required in relatively large amounts. The first six (C, O, H, N, S, and P) are components of carbohydrates, lipids, proteins, and nucleic acids. The remaining four macroelements exist in the cell as cations and play a variety of roles. For example, potassium (K^+) is required for activity by a number of enzymes, including some involved in protein synthesis. Calcium (Ca^{2+}), among other functions, contributes to the heat resistance of bacterial endospores. Magnesium (Mg^{2+}) serves as a cofactor for many enzymes, complexes with ATP, and stabilizes ribosomes and cell membranes. Iron (Fe^{2+} and Fe^{3+}) is a part of some molecules involved in the synthesis of ATP by electron transport-related processes. >> *Enzymes (section 9.7); Electron transport chains (section 9.6)*

In addition to macroelements, all microorganisms require several nutrients in small amounts—amounts so small that in the lab they are often obtained as contaminants in water, glassware, and growth media. Likewise in nature, they are ubiquitous and usually present in adequate amounts to support the growth of microbes. These nutrients are called **micronutrients** or **trace elements.** The micronutrients—manganese, zinc, cobalt, molybdenum, nickel, and copper—are needed by most cells. Micronutrients are normally a part of enzymes and cofactors, and they aid in the catalysis of reactions and maintenance of protein structure. For example, zinc (Zn^{2+}) is present at the active site of some enzymes but can also be involved in the association of different subunits of a multimeric protein. Manganese (Mn^{2+}) aids many enzymes that catalyze the transfer of phosphate. Molybdenum (Mo^{2+}) is required for nitrogen fixation, and cobalt (Co^{2+}) is a component of vitamin B_{12}. >> *Proteins (appendix I)*

Besides the common macroelements and trace elements, some microorganisms have particular requirements that reflect their specific morphology or metabolic capabilities. For instance, diatoms, members of the eucaryotic taxon *Stramenopiles,* need silicic acid (H_4SiO_4) to construct their beautiful cell walls of silica [$(SiO_2)_n$]. But no matter what their nutritional requirements, microbes require a balanced mixture of nutrients. If an essential nutrient is in short supply, microbial growth will be limited regardless of the concentrations of other nutrients. The importance of the most critical macroelements is discussed next. >> *Protist classification:* Stramenopiles *(section 23.2)*

6.2   REQUIREMENTS FOR CARBON, HYDROGEN, OXYGEN, AND ELECTRONS

All organisms need carbon, hydrogen, oxygen, and a source of electrons. Carbon is needed to synthesize the organic molecules from which organisms are built. Hydrogen and oxygen are also important elements found in many organic molecules. Electrons are needed for two reasons. As described more completely in chapter 10, the movement of electrons through electron transport chains and during other oxidation-reduction reactions can provide energy for use in cellular work. Electrons also are needed to reduce molecules during biosynthesis (e.g., the reduction of CO_2 to form organic molecules). >> *Oxidation-reduction reactions (section 9.5); CO_2 fixation (section 11.3)*

The requirements for carbon, hydrogen, and oxygen usually are satisfied together because molecules serving as carbon sources often contribute hydrogen and oxygen as well. For instance, many **heterotrophs**—organisms that use reduced, preformed organic molecules as their carbon source—can also obtain hydrogen, oxygen, and electrons from the same molecules (**table 6.1**). Because the electrons provided by these organic carbon sources can be used in electron transport as well as in other oxidation-reduction reactions, many heterotrophs also use their carbon

Table 6.1	Sources of Carbon, Energy, and Electrons
Carbon Sources	
Autotrophs	CO_2 sole or principal biosynthetic carbon source (*section 11.3*)
Heterotrophs	Reduced, preformed, organic molecules from other organisms (*chapters 10 and 11*)
Energy Sources	
Phototrophs	Light (*section 10.12*)
Chemotrophs	Oxidation of organic or inorganic compounds (*chapter 10*)
Electron Sources	
Lithotrophs	Reduced inorganic molecules (*section 10.11*)
Organotrophs	Organic molecules (*chapters 9 and 10*)

source as an energy source. Indeed, the more reduced the organic carbon source (i.e., the more electrons it carries), the higher its energy content. Thus lipids have a higher energy content than carbohydrates. >> *Carbohydrates (appendix I); Lipids (appendix I)*

A most remarkable characteristic of heterotrophic microorganisms is their extraordinary flexibility with respect to carbon sources. Laboratory experiments indicate that all naturally occurring organic molecules can be used as a source of carbon or energy or both by at least some microorganisms. Actinomycetes, common soil bacteria, degrade amyl alcohol, paraffin, and even rubber. The bacterium *Burkholderia cepacia* can use over 100 different carbon compounds. Microbes can degrade even relatively indigestible human-made substances such as pesticides. This is usually accomplished in complex microbial communities. These molecules sometimes are degraded in the presence of a growth-promoting nutrient that is metabolized at the same time—a process called cometabolism. Other microorganisms can use the products of this breakdown process as nutrients. In contrast to these bacterial omnivores, some microbes are exceedingly fastidious and catabolize only a few carbon compounds. Cultures of methylotrophic bacteria metabolize methane, methanol, carbon monoxide, formic acid, and related one-carbon molecules. Parasitic members of the genus *Leptospira* use only long-chain fatty acids as their major source of carbon and energy. >> *Biodegradation and bioremediation by natural communities (section 35.3)*

Other microbes are **autotrophs**—organisms that use carbon dioxide (CO_2) as their sole or principal source of carbon (table 6.1). Although CO_2 is plentiful, its use as a carbon source presents a problem to autotrophs. CO_2 is the most oxidized form of carbon, lacks hydrogen, and is unable to donate electrons during oxidation-reduction reactions. Therefore CO_2 cannot be used as a source of hydrogen, electrons, or energy. Because CO_2 cannot supply their energy needs, autotrophs must obtain energy from other sources, such as light or reduced inorganic molecules.

1. What are nutrients? On what basis are they divided into macroelements and trace elements?
2. What are the six most important macroelements? How do cells use them?
3. List two trace elements. How do cells use them?
4. Define heterotroph and autotroph.

6.3 NUTRITIONAL TYPES OF MICROORGANISMS

Because the need for carbon, energy, and electrons is so important, biologists use specific terms to define how these requirements are fulfilled. We have already seen that microorganisms can be classified as either heterotrophs or autotrophs with respect to their preferred source of carbon (table 6.1). Only two sources of energy are available to organisms: (1) light energy, and (2) the energy derived from oxidizing organic or inorganic molecules.

Phototrophs use light as their energy source; **chemotrophs** obtain energy from the oxidation of chemical compounds (either organic or inorganic). Microorganisms also have only two sources for electrons. **Lithotrophs** (i.e., "rock-eaters") use reduced inorganic substances as their electron source, whereas **organotrophs** extract electrons from reduced organic compounds.

Despite the great metabolic diversity seen in microorganisms, most may be placed in one of five nutritional classes based on their primary sources of carbon, energy, and electrons (**table 6.2**). The majority of microorganisms thus far studied are either photolithoautotrophic or chemoorganoheterotrophic.

Photolithoautotrophs (often called simply **photoautotrophs**) use light energy and have CO_2 as their carbon source. Photosynthetic protists and cyanobacteria employ water as the electron donor and

Table 6.2	Major Nutritional Types of Microorganisms			
Nutritional Type	**Carbon Source**	**Energy Source**	**Electron Source**	**Representative Microorganisms**
Photolithoautotroph	CO_2	Light	Inorganic e^- donor	Purple and green sulfur bacteria, cyanobacteria
Photoorganoheterotroph	Organic carbon	Light	Organic e^- donor	Purple nonsulfur bacteria, green nonsulfur bacteria
Chemolithoautotroph	CO_2	Inorganic chemicals	Inorganic e^- donor	Sulfur-oxidizing bacteria, hydrogen-oxidizing bacteria, methanogens, nitrifying bacteria, iron-oxidizing bacteria
Chemolithoheterotroph	Organic carbon	Inorganic chemicals	Inorganic e^- donor	Some sulfur-oxidizing bacteria (e.g., *Beggiatoa*)
Chemoorganoheterotroph	Organic carbon	Organic chemicals often same as C source	Organic e^- donor, often same as C source	Most nonphotosynthetic microbes, including most pathogens, fungi, and many protists and archaea

release oxygen (**figure 6.1a**). Other photolithoautotrophs, such as the purple and green sulfur bacteria (figure 6.1b), cannot oxidize water but extract electrons from inorganic donors such as hydrogen, hydrogen sulfide, and elemental sulfur. **Chemoorganoheterotrophs** (sometimes called **chemoheterotrophs** or chemoorganotrophs) use organic compounds as sources of energy, hydrogen, electrons, and carbon. Frequently the same organic nutrient will satisfy all these requirements. Nearly all pathogenic microorganisms are chemoorganoheterotrophs.

The other nutritional types have fewer known microorganisms but are very important ecologically. Some photosynthetic bacteria (purple and green bacteria) use organic matter as their electron donor and carbon source. These **photoorganoheterotrophs** are common inhabitants of polluted lakes and streams. Some of these bacteria also can grow as photolithoautotrophs with molecular hydrogen as an electron donor. **Chemolithoautotrophs** oxidize

reduced inorganic compounds such as iron, nitrogen, or sulfur molecules to derive both energy and electrons for biosynthesis (**figure 6.2a**). Carbon dioxide is the carbon source. **Chemolithoheterotrophs** use reduced inorganic molecules as their energy and electron source but derive their carbon from organic sources (figure 6.2b). Chemolithotrophs contribute greatly to the chemical transformations of elements (e.g., the conversion of ammonia to nitrate or sulfur to sulfate) that continually occur in ecosystems. >> *Photosynthetic bacteria (section 19.3); Class* Alphaproteobacteria: *Nitrifying bacteria (section 20.1); Biogeochemical cycling (section 25.1)*

Although a particular species usually belongs in only one of the major nutritional classes, some show great metabolic flexibility and alter their metabolism in response to environmental

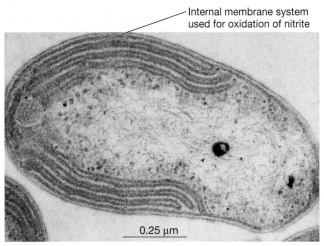

Internal membrane system used for oxidation of nitrite

0.25 µm

(a) *Nitrobacter winogradskyi,* a chemolithoautotroph

(a) Bloom of cyanobacteria (photolithoautotrophic bacteria)

Sulfur granule within filaments

10 microns

(b) *Beggiatoa alba,* a chemolithoheterotroph

(b) Purple sulfur bacteria (photoheterotrophs)

Figure 6.1 Phototrophic Bacteria. Phototrophic microbes play important roles in aquatic ecosystems, where they can cause blooms. (a) A cyanobacterial bloom in a eutrophic pond. (b) Purple sulfur bacteria growing in a bog.

Figure 6.2 Chemolithotrophic Bacteria. (a) Transmission electron micrograph of *Nitrobacter winogradskyi,* an organism that uses nitrite as its source of energy (×213,000). (b) Light micrograph of *Beggiatoa alba,* an organism that uses hydrogen sulfide as its energy source and organic molecules as carbon sources. The dark spots within the filaments are granules of elemental sulfur produced when hydrogen sulfide is oxidized.

changes. For example, many purple nonsulfur bacteria act as photoorganoheterotrophs in the absence of oxygen but oxidize organic molecules and function chemoorganotrophically at normal oxygen levels. When oxygen is low, photosynthesis and chemoorganotrophic metabolism may function simultaneously. This sort of flexibility gives these microbes a distinct advantage if environmental conditions frequently change.

1. Discuss the ways in which microorganisms are classified based on their requirements for energy, carbon, and electrons.
2. Describe the nutritional requirements of the major nutritional groups and give some microbial examples of each.

6.4 REQUIREMENTS FOR NITROGEN, PHOSPHORUS, AND SULFUR

To grow and reproduce, a microorganism must be able to incorporate large quantities of nitrogen, phosphorus, and sulfur. Although these elements may be acquired from the same nutrients that supply carbon, microorganisms usually employ inorganic sources as well.

Nitrogen is needed for the synthesis of amino acids, purines, pyrimidines, some carbohydrates and lipids, enzyme cofactors, and other substances. Many microorganisms can use the nitrogen in amino acids. Others can incorporate ammonia directly through the action of enzymes such as glutamate dehydrogenase or glutamine synthetase and glutamate synthase (*see figures 11.14 to 11.16*). Most phototrophs and many chemotrophic microorganisms reduce nitrate to ammonia and incorporate the ammonia in a process known as assimilatory nitrate reduction. A variety of bacteria (e.g., many cyanobacteria and the symbiotic bacterium *Rhizobium)* can assimilate atmospheric nitrogen (N_2) by reducing it to ammonia (NH_3). This is called nitrogen fixation. >> *Synthesis of amino acids (section 11.5); Biogeochemical cycling: Nitrogen cycle (section 25.1)*

Phosphorus is present in nucleic acids, phospholipids, nucleotides such as ATP, several cofactors, some proteins, and other cell components. Almost all microorganisms use inorganic phosphate as their phosphorus source and incorporate it directly. Low phosphate levels can limit microbial growth in aquatic environments. Some microbes, such as *Escherichia coli,* use both organic and inorganic phosphate. Some organophosphates such as hexose 6-phosphates are taken up directly by the cell. Other organophosphates are hydrolyzed in the periplasm by the enzyme alkaline phosphatase to produce inorganic phosphate, which then is transported across the plasma membrane. >> *Synthesis of purines, pyrimidines, and nucleotides (section 11.6)*

Sulfur is needed for the synthesis of substances such as the amino acids cysteine and methionine, some carbohydrates, biotin, and thiamine. Most microorganisms use sulfate as a source of sulfur after reducing it; a few microorganisms require a prereduced form of sulfur such as cysteine.

1. Briefly describe how microorganisms use the various forms of nitrogen, phosphorus, and sulfur.
2. Why do you think ammonia (NH_3) can be directly incorporated into amino acids while other forms of combined nitrogen (e.g., NO_2^- and NO_3^-) are not?

6.5 GROWTH FACTORS

Some microorganisms lack the enzymes or biochemical pathways needed to synthesize all cell components directly from macroelements and trace elements. Therefore they must obtain these constituents or their precursors from the environment. Organic compounds that are essential cell components or precursors of such components but cannot be synthesized by the organism are called **growth factors.**

There are three major classes of growth factors: (1) amino acids, (2) purines and pyrimidines, and (3) vitamins. Amino acids are needed for protein synthesis, purines and pyrimidines for nucleic acid synthesis. **Vitamins** are small organic molecules that usually make up all or part of enzyme cofactors. They are needed in only very small amounts to sustain growth. The functions of selected vitamins and examples of microorganisms requiring them are given in **table 6.3.** Some microorganisms require many vitamins; for example, *Enterococcus faecalis* needs eight different vitamins for growth. Other growth factors include heme (for the synthesis of cytochromes), which is required by *Haemophilus influenzae,* and cholesterol, which is needed by some mycoplasmas. >> *Enzymes (section 9.7)*

Some microorganisms are able to synthesize large quantities of vitamins needed by humans. These microbes can be used to manufacture these vitamins for human use. Several water-soluble and fat-soluble vitamins are produced partly or completely using industrial fermentations. Examples of such vitamins and the microorganisms that synthesize them are riboflavin (*Clostridium, Candida*), coenzyme A (*Brevibacterium*), vitamin B_{12} (*Streptomyces, Propionibacterium, Pseudomonas*), vitamin C (*Gluconobacter, Erwinia, Corynebacterium*), β-carotene (*Dunaliella*), and vitamin D (*Saccharomyces*). Current research focuses on improving yields and finding microorganisms that can produce large quantities of other vitamins. >> *Biotechnology and industrial microbiology (chapter 16)*

1. What are growth factors? What are vitamins?
2. List the growth factors that microorganisms produce industrially.
3. Why do you think amino acids, purines, and pyrimidines are often growth factors, whereas glucose is not?

Table 6.3	Functions of Some Common Vitamins in Microorganisms	
Vitamin	**Functions**	**Examples of Microorganisms Requiring Vitamin[a]**
Biotin	Carboxylation (CO_2 fixation) One-carbon metabolism	*Leuconostoc mesenteroides* (B) *Saccharomyces cerevisiae* (F) *Ochromonas malhamensis* (P)
Cyanocobalamin (B_{12})	Molecular rearrangements One-carbon metabolism—carries methyl groups	*Lactobacillus* spp. (B) *Euglena gracilis* (P)
Folic acid	One-carbon metabolism	*Enterococcus faecalis* (B) *Tetrahymena pyriformis* (P)
Lipoic acid	Transfer of acyl groups	*Lactobacillus casei* (B) *Tetrahymena* spp. (P)
Pantothenic acid	Precursor of coenzyme A—carries acyl groups (pyruvate oxidation, fatty acid metabolism)	*Proteus morganii* (B) *Hanseniaspora* spp. (F) *Paramecium* spp. (P)
Pyridoxine (B_6)	Amino acid metabolism (e.g., transamination)	*Lactobacillus* spp. (B) *Tetrahymena pyriformis* (P)
Niacin (nicotinic acid)	Precursor of NAD and NADP—carry electrons and hydrogen atoms	*Haemophilus influenzae* (B) *Blastocladia pringsheimii* (F) *Crithidia fasciculata* (P)
Riboflavin (B_2)	Precursor of FAD and FMN—carry electrons or hydrogen atoms	*Caulobacter vibrioides* (B) *Dictyostelium* spp. (P)
Thiamine (B_1)	Aldehyde group transfer (pyruvate decarboxylation, α-keto acid oxidation)	*Bacillus anthracis* (B) *Phycomyces blakesleeanus* (F) *Colpidium campylum* (P)

[a]The representative microorganisms are members of the following groups: *Bacteria* (B), fungi (F), and protists (P).

6.6 UPTAKE OF NUTRIENTS

The first step in nutrient use is their uptake by the microbial cell. Uptake mechanisms must be specific—that is, the necessary substances, and not others, must be acquired. It does a cell no good to take in a substance that it cannot use. Because microorganisms often live in nutrient-poor habitats, they must be able to transport nutrients from dilute solutions into the cell against a concentration gradient. Finally, nutrient molecules must pass through a selectively permeable plasma membrane that prevents the free passage of most substances. In view of the enormous variety of nutrients and the complexity of the task, it is not surprising that microorganisms make use of several different transport mechanisms. The most important of these are facilitated diffusion, active transport, and group translocation. Eucaryotic microorganisms do not appear to employ group translocation but take up nutrients by the process of endocytosis. << *Organelles of the biosynthetic-secretory and endocytic pathways (section 4.4)*

Passive Diffusion

A few substances, such as glycerol, can cross the plasma membrane by **passive diffusion.** Passive diffusion, often called diffusion or simple diffusion, is the process by which molecules move from a region of higher concentration to one of lower concentration. The rate of passive diffusion depends on the size of the concentration gradient between a cell's exterior and its interior (**figure 6.3**). A large concentration gradient is required for adequate nutrient uptake by passive diffusion (i.e., the external nutrient concentration must be high while the internal concentration is low). Unless the nutrient is used immediately upon entry, the rate of diffusion decreases as more nutrient accumulates in the cell. Very small molecules such as H_2O, O_2, and CO_2 often move across membranes by passive diffusion. Larger molecules, ions, and polar substances must enter the cell by other mechanisms.

Facilitated Diffusion

The rate of diffusion across selectively permeable membranes is greatly increased by using carrier proteins, sometimes called **permeases,** which are embedded in the plasma membrane. Diffusion involving carrier proteins is called **facilitated diffusion.** The rate of facilitated diffusion increases with the concentration gradient much more rapidly and at lower concentrations of the diffusing molecule than that of passive diffusion (figure 6.3). Note that the diffusion rate reaches a plateau above a specific gradient value because the carrier is saturated—that is, the carrier protein is binding and transporting as many solute molecules as possible.

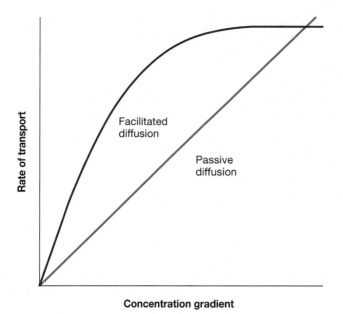

Concentration gradient

Figure 6.3 Passive and Facilitated Diffusion. The dependence of diffusion rate on the size of the solute's concentration gradient (the ratio of the extracellular concentration to the intracellular concentration). Note the saturation effect or plateau above a specific gradient value when a facilitated diffusion carrier is operating. This saturation effect is seen whenever a carrier protein is involved in transport.

The resulting curve resembles an enzyme-substrate curve (*see figure 9.15*) and is different from the linear response seen with passive diffusion. Carrier proteins also resemble enzymes in their specificity for the substance to be transported; each carrier is selective and transports only closely related solutes. Although a carrier protein is involved, facilitated diffusion is truly diffusion. A concentration gradient spanning the membrane drives the movement of molecules, and no metabolic energy input is required. If the concentration gradient disappears, net inward movement ceases. The gradient can be maintained by transforming the transported nutrient to another compound. Eucaryotic cells can maintain a gradient by moving the nutrient to another membranous compartment. Some permeases are related to the major intrinsic protein (MIP) family of proteins. MIPs facilitate diffusion of small polar molecules. They are observed in virtually all organisms. The two most widespread MIP channels in bacteria are aquaporins (*see figure 2.27*). These transport water and glycerol facilitators, which aid glycerol diffusion.

Although much work has been done on the mechanism of facilitated diffusion, the process is not understood completely. It appears that the carrier protein complex spans the membrane (**figure 6.4**). After the solute molecule binds to the outside, the carrier may change conformation and release the molecule on the cell interior. The carrier subsequently changes back to its original shape and is ready to pick up another molecule. The net effect is that a hydrophilic molecule can enter the cell in response to its concentration gradient. Remember that the mechanism is driven by concentration gradients and therefore is reversible. If the solute's concentration is greater inside the cell, it will move outward. However, because the cell metabolizes nutrients upon entry, influx is favored.

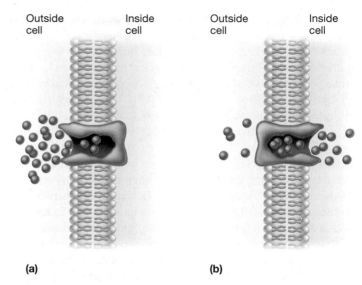

(a)   **(b)**

Figure 6.4 A Model of Facilitated Diffusion. The membrane carrier (a) changes conformation after binding an external molecule and subsequently releases the molecule on the cell interior. (b) It then returns to the outward oriented position and is ready to bind another solute molecule. Because there is no energy input, molecules continue to enter only as long as their concentration is greater on the outside.

Although glycerol is transported by facilitated diffusion in many procaryotes, facilitated diffusion does not seem to be the major uptake mechanism. This is because nutrient concentrations often are lower outside the cell. Facilitated diffusion is much more prominent in eucaryotic cells, where it is used to transport a variety of sugars and amino acids.

Active Transport

Because facilitated diffusion can efficiently move molecules to the interior only when the solute concentration is higher on the outside of the cell, microbes must have transport mechanisms that can move solutes against a concentration gradient. This is important because microorganisms often live in habitats with very low nutrient concentrations. Microbes use two important transport processes in such situations: active transport and group translocation. Both are energy-dependent processes.

Active transport is the transport of solute molecules to higher concentrations, or against a concentration gradient, with the input of metabolic energy. Because active transport involves permeases, it resembles facilitated diffusion in some ways. The permeases bind particular solutes with great specificity. Similar solute molecules can compete for the same carrier protein in both facilitated diffusion and active transport. Active transport is also characterized by the carrier saturation effect at high solute concentrations (figure 6.3). Nevertheless, active transport differs from facilitated diffusion in its use of metabolic energy and its ability to concentrate substances. Metabolic inhibitors that block energy production inhibit active transport but do not immediately affect facilitated diffusion.

ATP-binding cassette transporters (ABC transporters) are important examples of active transport systems. They are observed in *Bacteria, Archaea,* and eucaryotes. Usually ABC transporters consist of two hydrophobic membrane-spanning domains associated on their cytoplasmic surfaces with two ATP-binding domains (**figure 6.5**). The membrane-spanning domains form a pore in the membrane, and the ATP-binding domains bind and hydrolyze ATP to drive uptake. ABC transporters employ substrate-binding proteins, which are located in the periplasmic space of gram-negative bacteria (*see figure 3.25*) or are attached to membrane lipids on the external face of the plasma membrane of gram-positive bacteria. These proteins bind the molecule to be transported and then interact with the membrane transport proteins to move the molecule into the cell. Because a single molecule is transported, this is termed uniport transport. *E. coli* transports a variety of sugars (arabinose, maltose, galactose, ribose) and amino acids (glutamate, histidine, leucine) by this mechanism.

Substances entering gram-negative bacteria must pass through the outer membrane before ABC transporters and other active transport systems can take action. This is accomplished in several ways. When the substance is small, a generalized porin protein such as OmpF (outer membrane protein) can be used (*see figure 3.27*). An example of the movement of small molecules across the outer membrane is provided by the phosphate uptake systems of *E. coli*. Inorganic phosphate crosses the outer membrane by the use of a porin protein channel. Then one of two transport systems moves the phosphate across the plasma membrane. Which system is used depends on the concentration of phosphate. The PiT (inorganic phosphate transporter) system functions at high phosphate concentrations. When phosphate concentrations are low, an ABC transporter system called PST (phosphate-specific transport) brings phosphate into the cell, using a periplasmic binding protein. In contrast to small molecules such as phosphate, the transport of larger molecules, such as vitamin B_{12}, requires the use of specialized, high-affinity outer-membrane receptors that function in association with specific transporters in the plasma membrane.

As discussed in chapter 10, electron transport during energy-conserving processes generates a proton gradient (in procaryotes, the protons are at a higher concentration outside the cell than inside). The proton gradient can be used to do cellular work, including active transport. The uptake of lactose by the lactose permease of *E. coli* is a well-studied example. The permease is a single protein that transports a lactose molecule inward as a proton simultaneously enters the cell. Such linked transport of two substances in the same direction is called **symport.** Here, energy in the form of a proton gradient drives solute transport. Although the mechanism of lactose symport is not completely understood, X-ray diffraction studies show that the transport protein exists in outward- and inward-facing conformations. When lactose and a proton bind to separate sites on the outward-facing conformation, the protein changes to its inward-facing conformation. Then the sugar and proton are released into the cytoplasm. *E. coli* also uses proton symport to take up amino acids and organic acids such as succinate and malate; the phosphate transporter PiT is another example of proton symport. >> *Electron transport and oxidative phosphorylation (section 10.5)*

A proton gradient also can power active transport indirectly, often through the formation of a sodium ion gradient. For example, an *E. coli* sodium transport system pumps sodium outward in response to the inward movement of protons (**figure 6.6**). Such

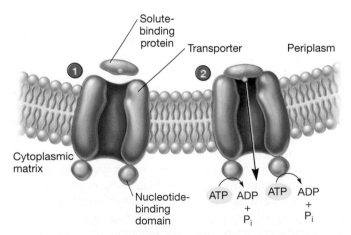

Figure 6.5 **ABC Transporter Function.** (1) The solute binding protein binds the substrate to be transported and approaches the ABC transporter complex. (2) The solute binding protein attaches to the transporter and releases the substrate, which is moved across the membrane with the aid of ATP hydrolysis.

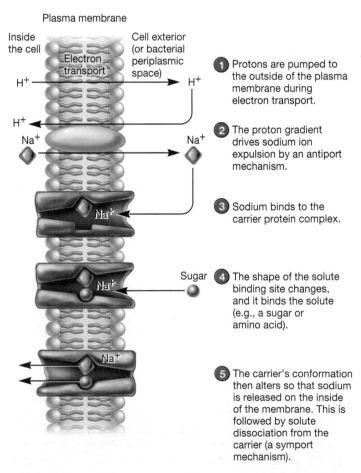

Plasma membrane

Inside the cell | Cell exterior (or bacterial periplasmic space)

Electron transport

1 Protons are pumped to the outside of the plasma membrane during electron transport.

2 The proton gradient drives sodium ion expulsion by an antiport mechanism.

3 Sodium binds to the carrier protein complex.

4 The shape of the solute binding site changes, and it binds the solute (e.g., a sugar or amino acid).

5 The carrier's conformation then alters so that sodium is released on the inside of the membrane. This is followed by solute dissociation from the carrier (a symport mechanism).

Figure 6.6 Active Transport Using Proton and Sodium Gradients.

linked transport in which the transported substances move in opposite directions is termed **antiport.** The sodium gradient generated by this proton antiport system drives the uptake of sugars and amino acids. It is thought that a sodium ion attaches to a carrier protein, causing it to change shape. The carrier then binds the sugar or amino acid tightly and orients its binding sites toward the cell interior. Because of the low intracellular sodium concentration, the sodium ion dissociates from the carrier, and the other molecule follows. *E. coli* transport proteins carry the sugar melibiose and the amino acid glutamate when sodium simultaneously moves inward. Sodium symport or cotransport also is an important process in eucaryotic cells where it is used in sugar and amino acid uptake. However, ATP, rather than proton motive force, usually drives sodium transport in eucaryotic cells.

Often a microorganism has more than one transport system for a nutrient, as can be seen with *E. coli.* This bacterium has at least five transport systems for the sugar galactose, three systems each for the amino acids glutamate and leucine, and two potassium transport complexes. When several transport systems exist for the same substance, the systems differ in such properties as their energy source, their affinity for the solute transported, and the nature of their regulation. This diversity gives the microbe an added competitive advantage in a variable environment.

Group Translocation

In active transport, solute molecules move across a membrane without modification. Another type of transport, called **group translocation,** chemically modifies the molecule as it is brought into the cell. Group translocation is a type of active transport because metabolic energy is used during uptake of the molecule. This is clearly demonstrated by the best-known group translocation system, the **phosphoenolpyruvate: sugar *phosphotransferase* system (PTS),** which is observed in many bacteria. The PTS transports a variety of sugars while phosphorylating them, using phosphoenolpyruvate (PEP) as the phosphate donor.

PEP + sugar (outside) → pyruvate + sugar-phosphate (inside)

PEP is an important intermediate of a biochemical pathway used by many chemoorganoheterotrophs to extract energy from organic energy sources. PEP is a high-energy molecule that can be used to synthesize ATP, the cell's energy currency. However, when it is used in PTS reactions, the energy present in PEP is used to energize uptake rather than ATP synthesis. >> *ATP (section 9.4); Breakdown of glucose to pyruvate (section 10.3)*

The transfer of phosphate from PEP to the incoming molecule involves several proteins and is an example of a **phosphorelay system.** In *E. coli* and *Salmonella,* the PTS consists of two enzymes and a low molecular weight heat-stable protein (HPr). HPr and enzyme I (EI) are cytoplasmic. Enzyme II (EII) is more variable in structure and often composed of three subunits or domains. EIIA is cytoplasmic and soluble. EIIB also is hydrophilic and frequently is attached to EIIC, a hydrophobic protein that is embedded in the membrane. A phosphate is transferred from PEP to enzyme II with the aid of enzyme I and HPr (**figure 6.7**). Then a sugar molecule is phosphorylated as it is carried across the membrane by enzyme II. Enzyme II transports only specific sugars and varies with the PTS, whereas enzyme I and HPr are common to all PTSs. >> *Enzymes (section 9.7)*

PTSs are widely distributed in bacteria. Most members of the genera *Escherichia, Salmonella, Staphylococcus,* as well as many other facultatively anaerobic bacteria (bacteria that grow in either the presence or absence of O_2), have phosphotransferase systems; some obligately anaerobic bacteria (e.g., *Clostridium*) also have PTSs. However, most aerobic bacteria, with the exception of some species of *Bacillus,* seem to lack PTSs. Many carbohydrates are transported by PTSs. *E. coli* takes up glucose, fructose, mannitol, sucrose, N-acetylglucosamine, cellobiose, and other carbohydrates by group translocation. Besides their role in transport, PTS proteins can bind chemical attractants, toward which bacteria move by the process of chemotaxis. >> *Influences of environmental factors on growth: Oxygen concentration (section 7.5)*

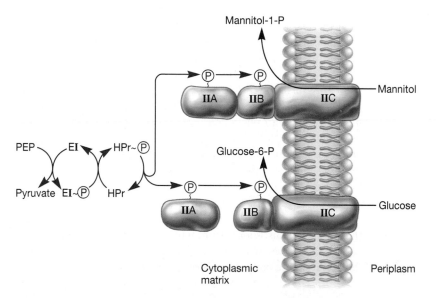

Figure 6.7 Group Translocation: Bacterial PTS Transport. Two examples of the phosphoenolpyruvate: sugar phosphotransferase system (PTS) are illustrated. The following components are involved in the system: phosphoenolpyruvate (PEP), enzyme I (EI), the low molecular weight heat-stable protein (HPr), and enzyme II (EII). The high-energy phosphate is transferred from HPr to the soluble EIIA. EIIA is attached to EIIB in the mannitol transport system and is separate from EIIB in the glucose system. In either case the phosphate moves from EIIA to EIIB and then is transferred to the sugar during transport through the membrane. Other relationships between the EII components are possible. For example, IIA and IIB may form a soluble protein separate from the membrane complex; the phosphate still moves from IIA to IIB and then to the membrane domain(s).

Iron Uptake

Almost all microorganisms require iron for use in cytochromes and many enzymes. Iron uptake is made difficult by the extreme insolubility of ferric iron (Fe^{3+}) and its derivatives, which leaves little free iron available for transport. Many bacteria and fungi have overcome this difficulty by secreting **siderophores** (Greek for iron bearers). Siderophores are low molecular weight organic molecules that bind ferric iron and supply it to the cell. Two examples of siderophores are ferrichrome, which is produced by many fungi, and enterobactin, which is formed by *E. coli* (**figure 6.8**).

Microorganisms secrete siderophores when iron is scarce in the medium. Once the iron-siderophore complex has reached the cell surface, it binds to a siderophore-receptor protein. Then either the iron is released to enter the cell directly or the whole iron-siderophore complex is transported inside by an ABC transporter. In *E. coli,* the siderophore receptor is in the outer membrane of the cell envelope; when the iron reaches the periplasmic space, it moves through the plasma membrane with the aid of the transporter. After the iron has entered the cell, it is reduced to the ferrous form (Fe^{2+}). Iron is so crucial to microorganisms that they may use more than one route of iron uptake to ensure an adequate supply.

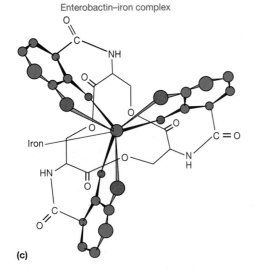

Ferrichrome

Enterobactin

(a)

(b)

Enterobactin–iron complex

Iron—

(c)

Figure 6.8 Siderophore Ferric Iron Complexes. (a) Ferrichrome is a cyclic hydroxamate [—CO—N(O⁻)—] molecule formed by many fungi. (b) *E. coli* produces the cyclic catecholate derivative, enterobactin. (c) Ferric iron probably complexes with three siderophore groups to form a six-coordinate, octahedral complex as shown in this illustration of the enterobactin-iron complex.

1. Describe facilitated diffusion, active transport, and group translocation in terms of their distinctive characteristics and mechanisms. What advantage does a microbe gain by using active transport rather than facilitated diffusion?
2. What are uniport, symport, and antiport?
3. What two mechanisms allow the passage of nutrients across the outer membrane of gram-negative bacteria before they are actively transported across the plasma membrane?
4. What is the difference between an ABC transporter and a porin protein in terms of function and cellular location?
5. What are siderophores? Why are they important?

6.7 CULTURE MEDIA

Microbiology research depends largely on the ability to grow and maintain microorganisms in the laboratory, and this is possible only if suitable culture media are available. A culture medium is a solid or liquid preparation used to grow, transport, and store microorganisms. To be effective, the medium must contain all the nutrients the microorganism requires for growth. Specialized media are essential in the isolation and identification of microorganisms, the testing of antibiotic sensitivities, water and food analysis, industrial microbiology, and other activities. Although all microorganisms need sources of energy, carbon, nitrogen, phosphorus, sulfur, and various minerals, the precise composition of a satisfactory medium depends on the species one is trying to cultivate because nutritional requirements vary so greatly. Knowledge of a microorganism's normal habitat often is useful in selecting an appropriate culture medium because its nutrient requirements reflect its natural surroundings. Frequently a medium is used to select and grow specific microorganisms or to help identify a particular species. Media can also be specifically designed to enrich for microbes in a sample from nature. The resulting culture is called an **enrichment culture,** and it can be used to isolate a species of interest for study in the lab. ▶▶ *Microbial ecology and its methods: Culturing techniques (section 25.2)*

Culture media can be classified based on several parameters: the chemical constituents from which they are made, their physical nature, and their function (**table 6.4**). The types of media defined by these parameters are described here.

Table 6.4	Types of Media
Basis for Classification	**Types**
Chemical composition	Defined (synthetic), complex
Physical nature	Liquid, semisolid, solid
Function	Supportive (general purpose), enriched, selective, differential

Chemical and Physical Types of Culture Media

A medium in which all chemical components are known is a **defined** or **synthetic medium.** It can be in a liquid form (broth) or solidified by an agent such as agar, as described in the following sections. Defined media are often used to culture photolithoautotrophs such as cyanobacteria and photosynthetic protists. They can be grown on media containing CO_2 as a carbon source (often added as sodium carbonate or bicarbonate), nitrate or ammonia as a nitrogen source, sulfate, phosphate, and other minerals (**table 6.5**). Many chemoorganoheterotrophs also can be grown in defined media with glucose as a carbon source and an ammonium salt as a nitrogen source. Not all defined media are as simple as the examples in table 6.5 but may be constructed from dozens of components. Defined media are used widely in research, as it is often desirable to know what the microorganism is metabolizing.

Media that contain some ingredients of unknown chemical composition are **complex media.** Such media are very useful, as a single complex medium may be sufficiently rich to meet all the nutritional requirements of many different microorganisms. In addition, complex media often are needed because the nutritional requirements of a particular microorganism are unknown, and thus a defined medium cannot be constructed. Complex media are also used to culture fastidious microbes, microbes with complex nutritional or cultural requirements. Some fastidious microbes may even require a medium containing blood or serum.

Most complex media contain undefined components such as peptones, meat extract, and yeast extract. Peptones are protein hydrolysates prepared by partial proteolytic digestion of meat, casein, soya meal, gelatin, and other protein sources. They serve as sources of carbon, energy, and nitrogen. Beef extract and yeast extract are aqueous extracts of lean beef and brewer's yeast, respectively. Beef extract contains amino acids, peptides, nucleotides, organic acids, vitamins, and minerals. Yeast extract is an excellent source of B vitamins as well as nitrogen and carbon compounds. Three commonly used complex media are (1) nutrient broth, (2) tryptic soy broth, and (3) MacConkey agar (**table 6.6**).

Although both liquid and solidified media are routinely used, solidified media are particularly important because they can be

Table 6.5	Examples of Defined Media
BG–11 Medium for Cyanobacteria	**Amount (g/liter)**
$NaNO_3$	1.5
$K_2HPO_4 \cdot 3H_2O$	0.04
$MgSO_4 \cdot 7H_2O$	0.075
$CaCl_2 \cdot 2H_2O$	0.036
Citric acid	0.006
Ferric ammonium citrate	0.006
EDTA (Na_2Mg salt)	0.001
Na_2CO_3	0.02
Trace metal solution[a]	1.0 ml/liter
Final pH 7.4	
Medium for *Escherichia coli*	**Amount (g/liter)**
Glucose	1.0
Na_2HPO_4	16.4
KH_2PO_4	1.5
$(NH_4)_2SO_4$	2.0
$MgSO_4 \cdot 7H_2O$	200.0 mg
$CaCl_2$	10.0 mg
$FeSO_4 \cdot 7H_2O$	0.5 mg
Final pH 6.8–7.0	

Sources: Data from Rippka, et al. 1979. *Journal of General Microbiology,* 111:1–61; and Cohen, S. S., and Arbogast, R. 1950. *Journal of Experimental Medicine,* 91:619.
[a]The trace metal solution contains H_3BO_3, $MnCl_2 \cdot 4H_2O$, $ZnSO_4 \cdot 7H_2O$, $Na_2Mo_4 \cdot 2H_2O$, $CuSO_4 \cdot 5H_2O$, and $Co(NO_3)_2 \cdot 6H_2O$.

Table 6.6	Some Common Complex Media
Nutrient Broth	**Amount (g/liter)**
Peptone (gelatin hydrolysate)	5
Beef extract	3
Tryptic Soy Broth	
Tryptone (pancreatic digest of casein)	17
Peptone (soybean digest)	3
Glucose	2.5
Sodium chloride	5
Dipotassium phosphate	2.5
MacConkey Agar	
Pancreatic digest of gelatin	17.0
Pancreatic digest of casein	1.5
Peptic digest of animal tissue	1.5
Lactose	10.0
Bile salts	1.5
Sodium chloride	5.0
Neutral red	0.03
Crystal violet	0.001
Agar	13.5

used to isolate different microbes from each other to establish pure cultures. As discussed in chapter 1, this is a critical step in demonstrating the relationship between a microbe and a disease using Koch's postulates. **Agar** is the most commonly used solidifying agent. It is a sulfated polymer composed mainly of D-galactose, 3,6-anhydro-L-galactose, and D-glucuronic acid. It usually is extracted from red algae. Agar is well suited as a solidifying agent for several reasons. One is that it melts at about 90°C but once melted does not harden until it reaches about 45°C. Thus after being melted in boiling water, it can be cooled to a temperature that is tolerated by human hands as well as microbes. Furthermore, microbes growing on agar medium can be incubated at a wide range of temperatures. Finally, agar is an excellent hardening agent because most microorganisms cannot degrade it.

Functional Types of Media

Media such as tryptic soy broth and tryptic soy agar are called general purpose or **supportive media** because they sustain the growth of many microorganisms. Blood and other special nutrients may be added to supportive media to encourage the growth of fastidious microbes. These specially fortified media (e.g., blood agar) are called **enriched media** (**figure 6.9**).

Selective media favor the growth of particular microorganisms (**table 6.7**). Bile salts or dyes such as basic fuchsin and crystal violet favor the growth of gram-negative bacteria by inhibiting the growth of gram-positive bacteria; the dyes have no effect on gram-negative organisms. Endo agar, eosin methylene blue agar, and MacConkey agar (tables 6.6 and 6.7) are three media widely used for the detection of *E. coli* and related bacteria in water supplies and elsewhere. These media contain dyes that suppress the growth of gram-positive bacteria. MacConkey agar also contains bile salts. Bacteria also may be selected by incubation with nutrients that they specifically can use. A medium containing only cellulose as a carbon and energy source is quite effective in the isolation of cellulose-digesting bacteria from samples such as soil. Thus selective media are useful as media for enrichment cultures. The possibilities for selection are endless, and dozens of special selective media are in use (**Techniques & Applications 6.1**).

Differential media are media that distinguish among different groups of microbes and even permit tentative identification of microorganisms based on their biological characteristics. Blood agar is both a differential medium and an enriched one. It distinguishes between hemolytic and nonhemolytic bacteria. Hemolytic bacteria (e.g., many streptococci and staphylococci isolated from throats) produce clear zones around their colonies because of red blood cell destruction (figure 6.9*a*). MacConkey agar is both differential and selective. Since it contains lactose and neutral red dye, bacteria that catabolize lactose by fermenting it release acidic waste products that make colonies appear pink to red in color. These are easily distinguished from colonies of bacteria that do not ferment lactose.

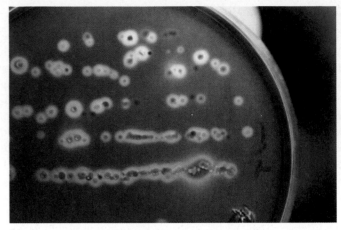

(a)

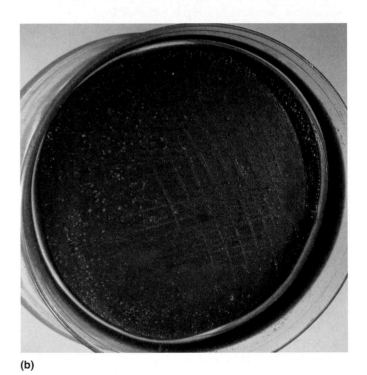

(b)

Figure 6.9   Enriched Media.   (a) Blood agar culture of bacteria from the human throat. (b) Chocolate agar, an enriched medium used to grow fastidious organisms such as *Neisseria gonorrhoeae.* The brown color is the result of heating red blood cells and lysing them before adding them to the medium. It is called chocolate agar because of its chocolate brown color.

1. Describe the following kinds of media and their uses: defined media, complex media, supportive media, enriched media, selective media, and differential media. Give an example of each kind.

2. What are peptones, yeast extract, beef extract, and agar? Why are they used in media?

Table 6.7	Mechanisms of Action of Selective and Differential Media	
Medium	**Functional Type**	**Mechanism of Action**
Blood agar	Enriched and differential	Blood agar supports the growth of many fastidious bacteria. These can be differentiated based on their ability to produce hemolysins—proteins that lyse red blood cells. Hemolysis appears as a clear zone (β-hemolysis) or greenish halo around the colony (α-hemolysis) (e.g., *Streptococcus pyogenes,* a β-hemolytic streptococcus).
Eosin methylene blue (EMB) agar	Selective and differential	Two dyes, eosin Y and methylene blue, inhibit the growth of gram-positive bacteria. They also react with acidic products released by certain gram-negative bacteria when they use lactose or sucrose as carbon and energy sources. Colonies of gram-negative bacteria that produce large amounts of acidic products have a green, metallic sheen (e.g., fecal bacteria such as *E. coli*).
MacConkey (MAC) agar	Selective and differential	The selective components in MAC are bile salts and crystal violet, which inhibit the growth of gram-positive bacteria. The presence of lactose and neutral red, a pH indicator, allows the differentiation of gram-negative bacteria based on the products released when they use lactose as a carbon and energy source. The colonies of those that release acidic products are red (e.g., *E. coli*).
Mannitol salt agar	Selective and differential	A concentration of 7.5% NaCl selects for the growth of staphylococci. Pathogenic staphylococci can be differentiated based on the release of acidic products when they use mannitol as a carbon and energy source. The acidic products cause a pH indicator (phenol red) in the medium to turn yellow (e.g., *Staphylococcus aureus*).

Techniques and Applications

6.1   Enrichment Cultures

A major practical problem is the preparation of pure cultures when microorganisms are present in very low numbers in a sample. Plating methods can be combined with the use of selective or differential media to enrich and isolate rare microorganisms. A good example is the isolation of bacteria that degrade the herbicide 2,4-dichlorophenoxyacetic acid (2,4-D). Bacteria able to metabolize 2,4-D can be obtained with a liquid medium containing 2,4-D as its sole carbon source and the required nitrogen, phosphorus, sulfur, and mineral components. When this medium is inoculated with soil, only bacteria able to use 2,4-D will grow. After incubation, a sample of the culture is transferred to a fresh flask of selective medium for further enrichment of 2,4-D metabolizing bacteria. A mixed population of 2,4-D degrading bacteria will arise after several such transfers. Pure cultures can be obtained by plating this mixture on agar containing 2,4-D as the sole carbon source. Only bacteria able to grow on 2,4-D form visible colonies, and these can be subcultured. This same general approach is used to isolate and purify a variety of bacteria by selecting for specific physiological characteristics.

The preceding techniques require the use of special culture dishes named petri dishes after their inventor Julius Richard Petri, a member of Robert Koch's laboratory; Petri developed these dishes around 1887, and they immediately replaced agar-coated glass plates. They consist of two round halves, the top half overlapping the bottom. Petri dishes are very easy to use, may be stacked on each other to save space, and are one of the most common items in microbiology laboratories.

6.8   ISOLATION OF PURE CULTURES

In natural habitats, microorganisms usually grow in complex, mixed populations with many species. This presents a problem for microbiologists because a single type of microorganism cannot be studied adequately in a mixed culture. One needs a **pure culture,** a population of cells arising from a single cell, to characterize an individual species. Pure cultures are so important that the development of pure culture techniques by the German bacteriologist Robert Koch transformed microbiology. Within about 20 years after the development of pure culture techniques, most pathogens responsible for the major human bacterial diseases had been isolated. Pure cultures can be prepared in several ways; a few of the more common approaches are reviewed here.

Streak Plate

If cells from a mixture of microbes can be spatially isolated from each other, each cell will give rise to a completely separate **colony**—a macroscopically visible growth or cluster of microorganisms in or on a solid medium. Because each colony arises from a single cell, each colony represents a pure culture. One method for separating cells is the **streak plate.** In this technique, cells are transferred to the edge of an agar plate with an inoculating loop or swab and then streaked out over the surface in one of several patterns (**figure 6.10**). After the first sector is streaked, the inoculating loop is sterilized and an inoculum for the second sector is obtained from the first sector. A similar process is followed for streaking the third sector, except that the inoculum is from the second sector. Thus this is essentially a dilution process. Eventually very few cells will be on the loop, and single cells will drop from it as it is rubbed along the agar surface. These develop into separate colonies.

Spread Plate and Pour Plate

Spread-plate and pour-plate techniques are similar in that they both dilute a sample of cells before separating them spatially. They differ in that the spread plate spreads the cells on the surface of the agar, whereas the pour plate embeds the cells within the agar.

For the **spread plate,** a small volume of a diluted mixture containing around 30 to 300 cells (25 to 250 cells if examining microbes in food or water samples) is transferred to the center of an agar plate and spread evenly over the surface with a sterile bent rod (**figure 6.11**). The dispersed cells develop into isolated colonies.

The **pour plate** is extensively used with procaryotes and fungi. The original sample is diluted several times to reduce the microbial population sufficiently to obtain separate colonies when plating

(**figure 6.12**). Then small volumes of several diluted samples are mixed with liquid agar that has been cooled to about 45°C, and the mixtures are poured immediately into sterile culture dishes. Most bacteria and fungi survive a brief exposure to the warm agar. Each cell becomes fixed in place to form an individual colony after the agar hardens. Like the spread plate, the pour plate can be used to determine the number of cells in a population. For both methods, the total number of colonies equals the number of viable microorganisms in the sample that are capable of growing in the medium used. Colonies growing on the surface also can be used to inoculate fresh medium and prepare pure cultures (Techniques & Applications 6.1).

Although the preparation of serial dilutions is the bane of many microbiology students, it has many applications other than the spread-plate and pour-plate methods. The numbers of cells in a solution can be diluted to the point where no cells are present in the dilution tube. If replicate dilutions are prepared and the number of tubes yielding growth determined, then the number of viable cells in the sample can be estimated by the most probable number method (MPN). Similarly, the number of viruses in a solution or the number of antibodies in a blood sample can be determined by first diluting the sample and then testing for the presence of viruses or antibodies. << *Cultivation and enumeration of viruses (section 5.5)* >> *Measurement of microbial growth (section 7.3); Microbial ecology and its methods (section 25.2); Antibodies (section 29.7)*

Microbial Growth on Agar Surfaces

Colony development on agar surfaces aids microbiologists in identifying microorganisms because individual species often form colonies of characteristic size and appearance (**figure 6.13**). When a mixed population has been plated properly, it sometimes is possible to identify the desired colony based on its overall appearance and use it to obtain a pure culture.

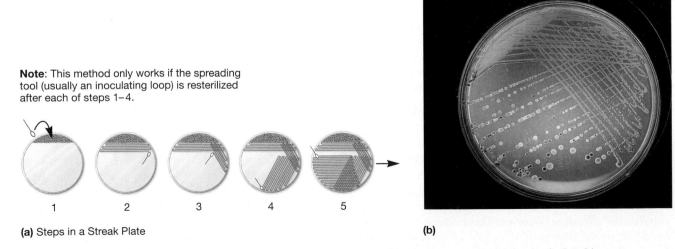

Note: This method only works if the spreading tool (usually an inoculating loop) is resterilized after each of steps 1–4.

1   2   3   4   5

(a) Steps in a Streak Plate

(b)

Figure 6.10   Streak-Plate Technique.   A typical streaking pattern is shown (a) as well as an example of a streak plate (b).

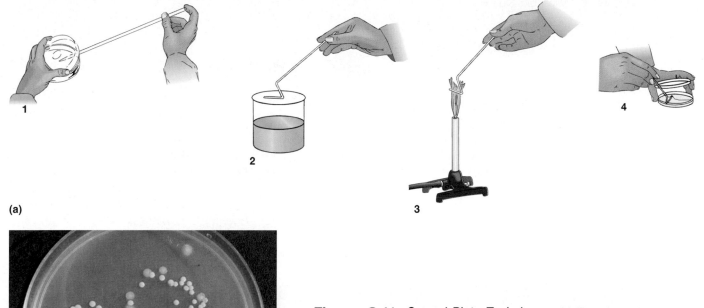

(a)

(b)

Figure 6.11 Spread-Plate Technique. (a) The preparation of a spread plate. (1) Pipette a small sample onto the center of an agar medium. (2) Dip a glass spreader into a beaker of ethanol. (3) Briefly flame the ethanol-soaked spreader and allow it to cool. (4) Spread the sample evenly over the agar surface with the sterilized spreader. Incubate. (b) Typical result of spread-plate technique.

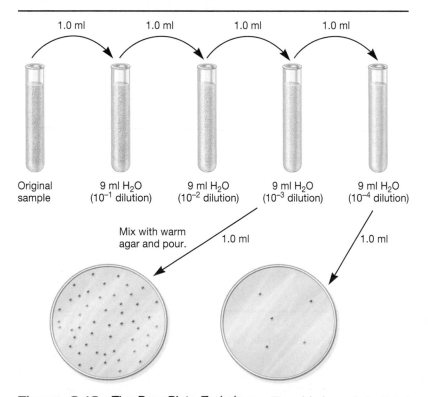

Figure 6.12 The Pour-Plate Technique. The original sample is diluted several times. The most diluted samples are then mixed with warm agar and poured into petri dishes. Isolated cells grow into colonies and can be used to establish pure cultures. The surface colonies are circular; subsurface colonies are lenticular (lens shaped).

In nature, microorganisms often grow on surfaces in biofilms—slime-encased aggregations of microbes. However, sometimes they form discrete colonies. Therefore an understanding of colony growth is important, and the growth of colonies on agar has been well studied. Generally the most rapid cell growth occurs at the colony edge. Growth is much slower in the center, and cell autolysis takes place in the older central portions of some colonies. These differences in growth are due to gradients of oxygen, nutrients, and toxic products within the colony. At the colony edge, oxygen and nutrients are plentiful. The colony center is much thicker than the edge. Consequently oxygen and nutrients do not diffuse readily into the center, toxic metabolic products cannot be quickly eliminated, and growth in the colony center is slowed or stopped. Because of these environmental variations within a colony, cells on the periphery can be growing at maximum rates while cells in the center are dying. ≫ *Microbial growth in natural environments: Biofilms (section 7.6)*

It is obvious from the colonies pictured in figure 6.13 that bacteria growing on solid surfaces such as agar can form quite complex and intricate colony shapes. These patterns vary with nutrient availability and the hardness of the agar surface. It is not yet clear how characteristic colony patterns develop. Nutrient diffusion and availability, bacterial chemotaxis, and the presence of liquid on the surface all appear to play a role in pattern formation. Cell-cell communication is important as well. Much research is currently focused on understanding the formation of bacterial colonies and biofilms.

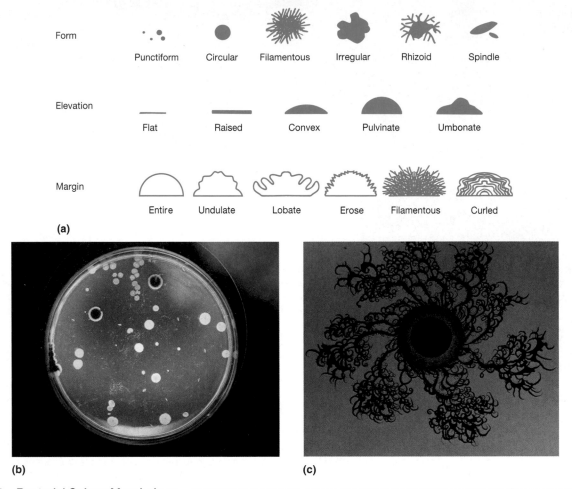

(a)

(b)   (c)

Figure 6.13 **Bacterial Colony Morphology.** (a) Variations in bacterial colony morphology seen with the naked eye. The general form of the colony and the shape of the edge or margin can be determined by looking down at the top of the colony. The nature of colony elevation is apparent when viewed from the side as the plate is held at eye level. (b) Examples of commonly observed colony morphologies. (c) Colony morphology can vary dramatically with the medium on which the bacteria are growing. These beautiful snowflakelike colonies were formed by *Bacillus subtilis* growing on nutrient-poor agar. The bacteria apparently behave cooperatively when confronted with poor growth conditions.

1. What are pure cultures, and why are they important? How are spread plates, streak plates, and pour plates prepared?

2. In what way does microbial growth vary within a colony? What factors might cause these variations in growth?

3. How might an enrichment culture be used to isolate bacteria capable of growing photoautotrophically from a mixed microbial assemblage?

Summary

6.1 Elements of Life

a. Microorganisms require nutrients, materials that are used in energy conservation and biosynthesis.

b. Macronutrients (macroelements: C, O, H, N, S, P, K, Ca, Mg, and Fe) are needed in relatively large quantities.

c. Micronutrients (trace elements: e.g., Mn, Zn, Co, Mo, Ni, and Cu) are used in very small amounts.

6.2 Requirements for Carbon, Hydrogen, Oxygen, and Electrons

a. All organisms require a source of carbon, hydrogen, oxygen, and electrons.

b. Heterotrophs use reduced organic molecules as their source of carbon. These molecules often supply hydrogen, oxygen, and electrons as well. Some heterotrophs also derive energy from their organic carbon source.

c. Autotrophs use CO_2 as their primary or sole carbon source; they must obtain hydrogen, energy, and electrons from other sources.

6.3 Nutritional Types of Microorganisms

a. Microorganisms can be classified based on their energy and electron sources (**table 6.1**). Phototrophs use light energy, and chemotrophs obtain energy from the oxidation of chemical compounds.

b. Electrons are extracted from reduced inorganic substances by lithotrophs and from organic compounds by organotrophs (**table 6.2**).

6.4 Requirements for Nitrogen, Phosphorus, and Sulfur

a. Nitrogen, phosphorus, and sulfur may be obtained from the same organic molecules that supply carbon, from the direct incorporation of ammonia and phosphate, and by the reduction and assimilation of oxidized inorganic molecules.

6.5 Growth Factors

a. Many microorganisms need growth factors.

b. The three major classes of growth factors are amino acids, purines and pyrimidines, and vitamins. Vitamins are small organic molecules that usually are components of enzyme cofactors.

6.6 Uptake of Nutrients

a. Although some nutrients can enter cells by passive diffusion, a membrane carrier protein is usually required.

b. In facilitated diffusion, the transport protein simply carries a molecule across the membrane in the direction of decreasing concentration, and no metabolic energy is required (**figure 6.4**).

c. Active transport systems use metabolic energy and membrane carrier proteins to concentrate substances by transporting them against a gradient. ATP is used as an energy source by ABC transporters (**figure 6.5**). Gradients of protons and sodium ions also drive solute uptake across membranes (**figure 6.6**).

d. Bacteria also transport organic molecules while modifying them, a process known as group translocation. For example,

many sugars are transported and phosphorylated simultaneously (**figure 6.7**).

e. Iron is accumulated by the secretion of siderophores, small molecules able to complex with ferric iron. When the iron-siderophore complex reaches the cell surface, it is taken inside and the iron is reduced to the ferrous form (**figure 6.8**).

6.7 Culture Media

a. Culture media can be constructed completely from chemically defined components (defined media or synthetic media) or constituents, such as peptones and yeast extract, whose precise composition is unknown (complex media).

b. Culture media can be solidified by the addition of agar, a complex polysaccharide from red algae.

c. Culture media are classified based on function and composition as supportive media, enriched media, selective media, and differential media. Supportive media are used to culture a wide variety of microbes. Enriched media are supportive media that contain additional nutrients needed by fastidious microbes. Selective media contain components that select for the growth of some microbes. Differential media contain components that allow microbes to be differentiated from each other, usually based on some metabolic capability.

6.8 Isolation of Pure Cultures

a. Pure cultures usually are obtained by isolating individual cells with any of three plating techniques: the streak-plate, spread-plate, and pour-plate methods.

b. The streak-plate technique uses an inoculating loop to spread cells across an agar surface (**figure 6.10**).

c. The spread-plate (**figure 6.11**) and pour-plate (**figure 6.12**) methods usually involve diluting a culture or sample and then plating the dilutions. In the spread-plate technique, a specially shaped rod is used to spread the cells on the agar surface; in the pour-plate technique, the cells are first mixed with cooled agar-containing media before being poured into a petri dish.

d. Microorganisms growing on solid surfaces tend to form colonies with distinctive morphology (**figure 6.13**). Colonies usually grow most rapidly at the edge where larger amounts of required resources are available.

Critical Thinking Questions

1. Discuss the advantages and disadvantages of group translocation versus endocytosis.

2. If you wished to obtain a pure culture of bacteria that could degrade benzene and use it as a carbon and energy source, how would you proceed?

Learn More

Learn more by visiting the Prescott website at www.mhhe.com/prescottprinciples, where you will find a complete list of references.

7

Microbial Growth

A black smoker: one extreme habitat where microbes can be found.

acidophile A microorganism that has its growth optimum between pH 0 and about 5.5.

aerobe An organism that grows in the presence of atmospheric oxygen.

aerotolerant anaerobe A microbe that grows equally well whether or not oxygen is present.

alkalophile (alkaliphile) A microorganism that grows best at pH values from about 8.5 to 11.5.

anaerobe An organism that can grow in the absence of free oxygen.

batch culture A culture of microorganisms produced by inoculating a closed culture vessel containing a single batch of medium.

biofilm Organized microbial communities consisting of layers of microbial cells associated with surfaces.

chemostat A continuous culture apparatus that feeds medium into the culture vessel at the same rate as medium containing microorganisms is removed; the medium contains a limiting quantity of one essential nutrient.

colony forming units (CFU) The number of microorganisms that form colonies when cultured using spread plates or pour plates; used as a measure of the number of viable microorganisms in a sample.

cytokinesis Processes that apportion the cytoplasm and organelles, synthesize a septum, and divide a cell into two daughter cells during cell division.

exponential (log) phase The phase of a growth curve during which the microbial population is growing at a constant and maximum rate, dividing and doubling at regular intervals.

extremophiles Microorganisms that grow under harsh or extreme environmental conditions.

facultative anaerobe A microorganism that does not require oxygen for growth but grows better in its presence.

generation (doubling) time The time required for a microbial population to double in number.

halophile A microorganism that requires high levels of sodium chloride for growth.

hyperthermophile A procaryote with a growth optimum above 85°C.

lag phase A period following the introduction of microorganisms into fresh batch culture medium when there is no increase in cell numbers or mass.

mesophile A microorganism with a growth optimum around 20 to 45°C, a minimum of 15 to 20°C, and a maximum of less than 45°C.

microaerophile A microorganism that requires low levels of oxygen for growth (2 to 10%) but is damaged by normal atmospheric oxygen levels.

neutrophile A microorganism that grows best at a neutral pH range between pH 5.5 and 8.0.

obligate aerobe An organism that grows only when oxygen is present.

obligate anaerobe An organism that grows only when oxygen is absent.

osmotolerant Organisms that grow over a wide range of water activity or solute concentration.

psychrophile A microorganism that grows well at 0°C, has an optimum growth temperature of 15°C or lower, and a temperature maximum of around 20°C.

quorum sensing The exchange of extracellular molecules that allows microbial cells to sense cell density.

stationary phase The phase of microbial growth in a batch culture when population growth ceases and the growth curve levels off.

thermophile A microorganism that can grow at temperatures of 55°C or higher, with a minimum of around 45°C.

water activity (a_w) A quantitative measure of water availability in a habitat.

The paramount evolutionary accomplishment of bacteria as a group is rapid, efficient cell growth in many environments.

—J. L. Ingraham, O. Maaløe, and F. C. Neidhardt

In chapter 6, we emphasize that microorganisms need access to a source of energy and the raw materials essential for the construction of cellular components. Chapter 7 concentrates more directly on microbial reproduction and growth. First we describe binary fission, the type of cell division most frequently observed among procaryotes, and the bacterial cell cycle. Cell reproduction leads to an increase in population size, so next we consider growth and the ways in which it can be measured. Then we discuss continuous culture techniques. An account of the influence of environmental factors on microbial growth and microbial growth in natural environments completes the chapter.

Growth may be defined as an increase in cellular constituents. It leads to a rise in cell number when microorganisms reproduce. Growth also results when cells simply become longer or larger. If the microorganism is **coenocytic**—that is, a multinucleate organism in which chromosomal replication is not accompanied by cell division—growth results in an increase in cell size but not cell number. It is usually not convenient to investigate the growth and reproduction of individual microorganisms because of their small size. Therefore, when studying growth, microbiologists normally follow changes in the total population number.

7.1   BACTERIAL CELL CYCLE

The **cell cycle** is the complete sequence of events extending from the formation of a new cell through the next division. It is of intrinsic interest to microbiologists as a fundamental biological process. However, understanding the cell cycle has practical importance as well. For instance in bacteria, the synthesis of peptidoglycan is the target of numerous antibiotics. **>>** *Antibacterial drugs: Inhibitors of cell wall synthesis (section 31.4)*

Although some procaryotes reproduce by budding, fragmentation, and other means, most procaryotes reproduce by **binary fission** (**figure 7.1**). Binary fission is a relatively simple type of cell division: the cell elongates, replicates its chromosome, and separates the newly formed DNA molecules so there is one chromosome in each half of the cell. Finally, a septum (cross wall) is formed at midcell, dividing the parent cell into two progeny cells, each having its own chromosome and a complement of other cellular constituents.

Despite the apparent simplicity of the procaryotic cell cycle, it is poorly understood. The cell cycles of several bacteria—*Escherichia coli, Bacillus subtilis,* and the aquatic bacterium *Caulobacter crescentus*—have been examined extensively, and our understanding of the bacterial cell cycle is based largely on these studies. Two pathways function during the bacterial cell cycle: one pathway replicates and partitions the DNA into the progeny cells, the other carries out cytokinesis—formation of the septum and progeny cells. Although these pathways overlap, it is easiest to consider them separately.

Chromosome Replication and Partitioning

Recall that most bacterial chromosomes are circular. Each circular chromosome has a single site at which replication starts called the **origin of replication,** or simply the origin (**figure 7.2**).

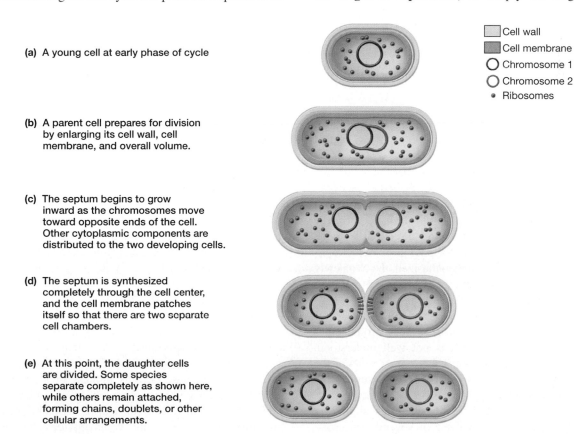

(a) A young cell at early phase of cycle

(b) A parent cell prepares for division by enlarging its cell wall, cell membrane, and overall volume.

(c) The septum begins to grow inward as the chromosomes move toward opposite ends of the cell. Other cytoplasmic components are distributed to the two developing cells.

(d) The septum is synthesized completely through the cell center, and the cell membrane patches itself so that there are two separate cell chambers.

(e) At this point, the daughter cells are divided. Some species separate completely as shown here, while others remain attached, forming chains, doublets, or other cellular arrangements.

- Cell wall
- Cell membrane
- Chromosome 1
- Chromosome 2
- Ribosomes

Figure 7.1   Binary Fission.

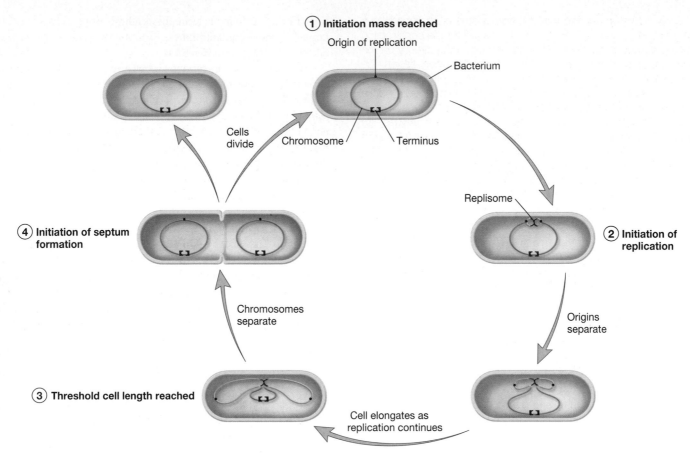

Figure 7.2   Cell Cycle of *E. coli.*   As the cell readies for replication, the origin migrates to the center of the cell and proteins that make up the replisome assemble. As replication proceeds, newly synthesized chromosomes move toward poles so that upon cytokinesis, each daughter cell inherits only one chromosome. In this illustration, one round of DNA replication is completed before the cell divides. In rapidly growing cultures (generation times of about 20 minutes), second and third rounds of replication are initiated before division of the original cell is completed. Thus the daughter cells inherit partially replicated DNA.

Replication is completed at the terminus, which is located directly opposite the origin. In a newly formed *E. coli* cell, the chromosome is compacted and organized so that the origin and terminus are in opposite halves of the cell. Early in the cell cycle, the origin and terminus move to midcell and a group of proteins needed for DNA synthesis assemble at the origin to form the **replisome.** DNA replication proceeds in both directions from the origin, and the parent DNA is thought to spool through the replisomes. As progeny chromosomes are synthesized, the two newly formed origins move toward opposite ends of the cell, and the rest of the chromosome follows in an orderly fashion.

Although the process of DNA synthesis and movement seems rather straightforward, the mechanism by which chromosomes are partitioned to each daughter cell is not well understood. Surprisingly, a picture is emerging in which components of the cytoskeleton are involved. For many years, it was assumed that procaryotes were too small for eucaryotic-like cytoskeletal structures. However, a protein called MreB, which is similar to eucaryotic actin, seems to be involved in several processes, including determining cell shape and chromosome movement.

MreB polymerizes to form a spiral around the inside periphery of the cell (**figure 7.3*a***). One model suggests that the origin of each newly replicated chromosome associates with MreB, which then moves them to opposite poles of the cell. The notion that procaryotic chromosomes may be actively moved to the poles is further suggested by the fact that if MreB is mutated so that it can no longer hydrolyze ATP, its source of energy, chromosomes fail to segregate properly. << *Procaryotic cytoplasm: Procaryotic cytoskeleton (section 3.3)*

Cytokinesis

Septation is the process of forming a cross wall between two daughter cells. **Cytokinesis,** a term that has traditionally been used to describe the formation of two eucaryotic daughter cells, is now used to describe this process in procaryotes as well. Septation is divided into several steps: (1) selection of the site where the septum will be formed; (2) assembly of a specialized structure called the Z ring, which divides the cell in two by constriction; (3) linkage of the Z ring to the plasma membrane

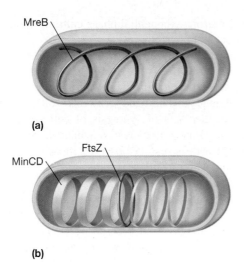

(a)

(b)

Figure 7.3   Cytoskeletal Proteins Involved in Cytokinesis in Rod-Shaped Bacteria.   (a) The actin homolog MreB forms spiral filaments around the inside of the cell that help determine cell shape and may serve to move chromosomes to opposite cell poles. (b) The tubulin-like protein FtsZ assembles in the center of the cell to form a Z ring, which is essential for septation. MinCD, together with other Min proteins, oscillates from pole to pole, thereby preventing the formation of an off-center Z ring.

and perhaps components of the cell wall; (4) assembly of the cell wall-synthesizing machinery (i.e., for synthesis of peptidoglycan and other cell wall constituents); and (5) constriction of the cell and septum formation. >> *Synthesis of sugars and polysaccharides: Synthesis of peptidoglycan (section 11.4)*

The assembly of the Z ring is a critical step in septation, as it must be formed if subsequent steps are to occur. The FtsZ protein, a tubulin homologue found in most bacteria and many archaea, forms the Z ring. FtsZ, like tubulin, polymerizes to form filaments, which are thought to create the meshwork that constitutes the Z ring. Numerous studies show that the Z ring is very

dynamic, with portions being exchanged constantly with newly formed, short FtsZ polymers from the cytosol. Another protein, called MinCD, is an inhibitor of Z-ring assembly. Like FtsZ, it is very dynamic, oscillating its position from one end of the cell to the other, forcing Z-ring formation only at the center of the cell (figure 7.3*b*). Once the Z-ring forms, the rest of the division machinery is constructed, as illustrated in **figure 7.4**. First one or more anchoring proteins link the Z ring to the cell membrane. Then the cell wall-synthesizing machinery is assembled.

The final steps in division involve constriction of the cell by the Z ring, accompanied by invagination of the cell membrane and synthesis of the septal wall. Several models for Z-ring function have been proposed. One model holds that the FtsZ filaments are shortened by losing FtsZ subunits (i.e., depolymerization) at sites where the Z ring is anchored to the plasma membrane. This model is supported by the observation that Z rings of cells producing an excessive amount of FtsZ subunits fail to constrict the cell.

The preceding discussion of the cell cycle describes what occurs in slowly growing *E. coli* cells. In these cells, the cell cycle takes approximately 60 minutes to complete. However, *E. coli* can reproduce at a much more rapid rate, completing the entire cell cycle in about 20 minutes, despite the fact that DNA replication always requires at least 40 minutes. *E. coli* accomplishes this by beginning a second round of DNA replication (and sometimes even a third or fourth round) before the first round of replication is completed. Thus the progeny cells receive two or more replication forks, and replication is continuous because the cells are always copying their DNA.

1. What two pathways function during the procaryotic cell cycle?

2. How does the bacterial cell cycle compare with the eucaryotic cell cycle? List two ways they are similar and two ways they differ.

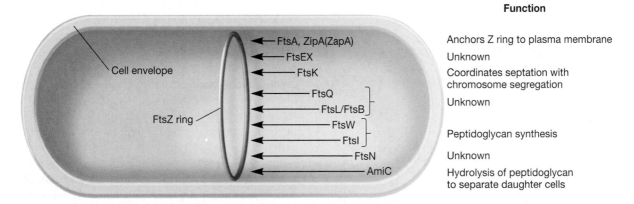

Figure 7.4   Formation of the Cell Division Apparatus in *E. coli*.   The cell division apparatus is composed of numerous proteins that are thought to assemble in the order shown. The process begins with the polymerization of FtsZ to form the Z ring. Then FtsA and ZipA (possibly ZapA in *Bacillus subtilis*) proteins anchor the Z ring to the plasma membrane. Although numerous proteins are known to be part of the cell division apparatus, the functions of relatively few are known.

7.2 GROWTH CURVE

Binary fission and other cell division processes bring about an increase in the number of cells in a population. Population growth is studied by analyzing the growth curve of a microbial culture. When microorganisms are cultivated in liquid medium, they usually are grown in a **batch culture**—that is, they are incubated in a closed culture vessel with a single batch of medium. Because no fresh medium is provided during incubation, nutrient concentrations decline and concentrations of wastes increase. The growth of microorganisms reproducing by binary fission can be plotted as the logarithm of the number of viable cells versus the incubation time. The resulting curve has four distinct phases (**figure 7.5**).

Lag Phase

When microorganisms are introduced into fresh culture medium, usually no immediate increase in cell number occurs. This period is called the **lag phase.** However, cells in the culture are synthesizing new components. A lag phase can be necessary for a variety of reasons. The cells may be old and depleted of ATP, essential cofactors, and ribosomes; these must be synthesized before growth can begin. The medium may be different from the one the microorganism was growing in previously. Here new enzymes would be needed to use different nutrients. Possibly the microorganisms have been injured and require time to recover. Whatever the causes, eventually the cells begin to replicate their DNA, increase in mass, and finally divide.

Exponential Phase

During the **exponential (log) phase,** microorganisms are growing and dividing at the maximal rate possible given their genetic potential, the nature of the medium, and the environmental conditions. Their rate of growth is constant during the exponential phase; that is, they are completing the cell cycle and doubling in number at regular intervals (figure 7.5). The population is most uniform in terms of chemical and physiological properties during

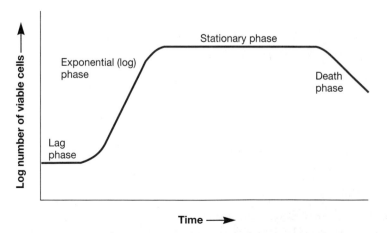

Figure 7.5 Microbial Growth Curve in a Closed System. The four phases of the growth curve are identified on the curve.

this phase; therefore exponential phase cultures are usually used in biochemical and physiological studies.

Exponential (logarithmic) growth is **balanced growth.** That is, all cellular constituents are manufactured at constant rates relative to each other. If nutrient levels or other environmental conditions change, **unbalanced growth** results. During unbalanced growth, the rates of synthesis of cell components vary relative to one another until a new balanced state is reached. Unbalanced growth is readily observed in two types of experiments: shift-up, where a culture is transferred from a nutritionally poor medium to a richer one; and shift-down, where a culture is transferred from a rich medium to a poor one. In a shift-up experiment, there is a lag while the cells first construct new ribosomes to enhance their capacity for protein synthesis. In a shift-down experiment, there is a lag in growth because cells need time to make the enzymes required for the biosynthesis of unavailable nutrients. Once the cells are able to grow again, balanced growth is resumed and the culture enters the exponential phase. These shift-up and shift-down experiments demonstrate that microbial growth is under precise, coordinated control and responds quickly to changes in environmental conditions.

When microbial growth is limited by the low concentration of a required nutrient, the final net growth or yield of cells increases with the initial amount of the limiting nutrient present (**figure 7.6a**). The rate of growth also increases with nutrient concentration (figure 7.6b) but in a hyperbolic manner much like that seen with many enzymes (see figure 9.17). The shape of the curve seems to reflect the rate of nutrient uptake by microbial transport proteins. At sufficiently high nutrient levels, the transport systems are saturated, and the growth rate does not rise further with increasing nutrient concentration. << *Uptake of nutrients (section 6.6)*

Stationary Phase

In a closed system such as a batch culture, population growth eventually ceases and the growth curve becomes horizontal (figure 7.5). This **stationary phase** usually is attained by bacteria at a population level of around 10^9 cells per ml. Other microorganisms normally do not reach such high population densities. For instance, protist cultures often have maximum concentrations of about 10^6 cells per ml. Final population size depends on nutrient availability and other factors, as well as the type of microorganism being cultured. In the stationary phase, the total number of viable microorganisms remains constant. This may result from a balance between cell division and cell death, or the population may simply cease to divide but remain metabolically active.

Microbial populations enter the stationary phase for several reasons. One obvious factor is nutrient limitation; if an essential nutrient is severely depleted, population growth will slow. Aerobic organisms often are limited by O_2 availability. Oxygen is not very soluble and may be depleted so quickly that only the surface of a culture will have an O_2 concentration adequate for growth. The cells beneath the surface will not be able to grow unless the culture

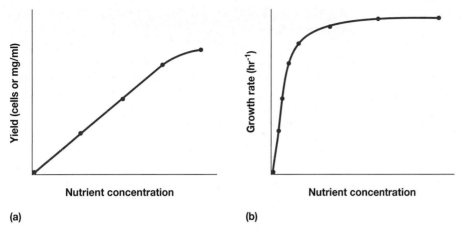

Figure 7.6 Nutrient Concentration and Growth. (a) The effect of changes in limiting nutrient concentration on total microbial yield. At sufficiently high concentrations, total growth will plateau. (b) The effect on growth rate.

is shaken or aerated in another way. Population growth also may cease due to the accumulation of toxic waste products. This factor seems to limit the growth of many anaerobic cultures (cultures growing in the absence of O_2). For example, streptococci can produce so much lactic acid and other organic acids from sugar fermentation that their medium becomes acidic and growth is inhibited. Finally, some evidence exists that growth may cease when a critical population level is reached. Thus entrance into the stationary phase may result from several factors operating in concert.

As we have seen, bacteria in a batch culture may enter stationary phase in response to starvation. This probably occurs often in nature because many environments have low nutrient levels. Procaryotes have evolved a number of strategies to survive starvation. Some bacteria respond with obvious morphological changes such as endospore formation, but many only decrease somewhat in overall size. This is often accompanied by protoplast shrinkage and nucleoid condensation. The more important changes during starvation are in gene expression and physiology. Starving bacteria frequently produce a variety of **starvation proteins,** which make the cell much more resistant to damage. Some increase peptidoglycan crosslinking and cell wall strength. The Dps (*D*NA-binding *p*rotein from *s*tarved cells) protein protects DNA. Proteins called chaperone proteins prevent protein denaturation and renature damaged proteins. Because of these and many other mechanisms, starved cells become harder to kill and more resistant to starvation, damaging temperature changes, oxidative and osmotic damage, and toxic chemicals such as chlorine. These changes are so effective that some bacteria can survive starvation for years. There is even evidence that

Salmonella enterica serovar Typhimurium *(S. typhimurium)* and some other bacterial pathogens become more virulent when starved. Clearly, these considerations are of great practical importance in medical and industrial microbiology.

Senescence and Death

For many years, the decline in viable cells following the stationary phase was described simply as the "death phase." It was assumed that detrimental environmental changes such as nutrient deprivation and the buildup of toxic wastes caused irreparable harm and loss of viability. That is, even when bacterial cells were transferred to fresh medium, no cellular growth was observed. Because loss of viability was often not accompanied by a loss in total cell number, it was assumed that cells died but did not lyse.

This view is currently under debate. There are two alternative hypotheses (**figure 7.7**). Some microbiologists think starving cells that show an exponential decline in density have not irreversibly lost their ability to reproduce. Rather, they suggest that microbes are temporarily unable to grow, at least under the laboratory conditions used. This phenomenon, in which the cells are called

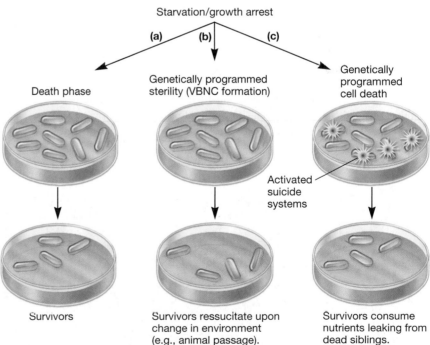

Figure 7.7 Loss of Viability. (a) It has long been assumed that as cells leave stationary phase due to starvation or toxic waste accumulation, the exponential decline in culturability is due to cellular death. (b) The viable but nonculturable (VBNC) hypothesis posits that when cells are starved, they become temporarily nonculturable under laboratory conditions. When exposed to appropriate conditions, some cells will regain the capacity to reproduce. (c) Some believe that a fraction of a microbial population dies due to activation of programmed cell death genes. The nutrients that are released by dying cells support the growth of other cells.

viable but nonculturable (VBNC), is thought to be the result of a genetic response triggered in starving, stationary phase cells. Just as some bacteria form endospores as a survival mechanism, it is argued that others are able to become dormant without changes in morphology (figure 7.7*b*). Once the appropriate conditions are available (for instance, a change in temperature or passage through an animal), VBNC microbes resume growth. VBNC microorganisms could pose a public health threat, as many assays that test for food and drinking water safety are culture-based.

The second alternative to a simple death phase is **programmed cell death** (figure 7.7*c*). In contrast to the VBNC hypothesis whereby cells are genetically programmed to survive, programmed cell death predicts that a fraction of the microbial population is genetically programmed to die after growth ceases. In this case, some cells die and the nutrients they leak enable the eventual growth of those cells in the population that did not initiate cell death. The dying cells are thus "altruistic"—they sacrifice themselves for the benefit of the larger population.

Long-term growth experiments reveal that an exponential decline in viability is sometimes replaced by a gradual decline in the number of culturable cells. This decline can last months to years (**figure 7.8**). During this time, the bacterial population continually evolves so that actively reproducing cells are those best able to use the nutrients released by their dying brethren and best able to tolerate the accumulated toxins. This dynamic process is marked by successive waves of genetically distinct variants. Thus natural selection can be witnessed within a single culture vessel.

Mathematics of Growth

Knowledge of microbial growth rates during the exponential phase is indispensable to microbiologists. Growth rate studies contribute to basic physiological and ecological research and are applied in

Table 7.1	An Example of Exponential Growth			
Time[a]	Division Number	2^n	Population[b] ($N_0 \times 2^n$)	$\log_{10}N_t$
0	0	$2^0 = 1$	1	0.000
20	1	$2^1 = 2$	2	0.301
40	2	$2^2 = 4$	4	0.602
60	3	$2^3 = 8$	8	0.903
80	4	$2^4 = 16$	16	1.204

[a]The hypothetical culture begins with one cell having a 20-minute generation time.
[b]Number of cells in the culture.

industry. The quantitative aspects of exponential phase growth discussed here apply to microorganisms that divide by binary fission.

During the exponential phase, each microorganism is dividing at constant intervals. Thus the population doubles in number during a specific length of time called the **generation (doubling) time.** This can be illustrated with a simple example. Suppose that a culture tube is inoculated with one cell that divides every 20 minutes (**table 7.1**). The population will be 2 cells after 20 minutes, 4 cells after 40 minutes, and so forth. Because the population is doubling every generation, the increase in population is always 2^n where *n* is the number of generations. The resulting population increase is exponential—that is, logarithmic (**figure 7.9**).

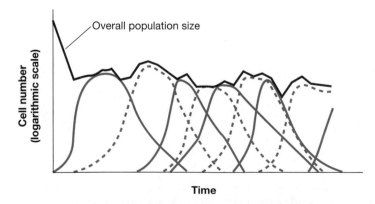

Figure 7.8 Prolonged Decline in Cell Numbers. Instead of a distinct death phase, successive waves of genetically distinct subpopulations of microbes better able to use the released nutrients and accumulated toxins survive. Each successive solid or dashed blue curve represents the growth of a new subpopulation.

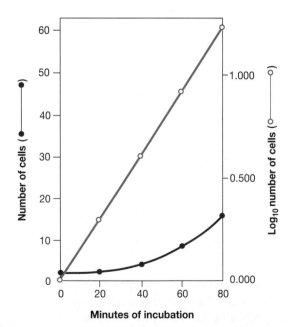

Figure 7.9 Exponential Microbial Growth. Four generations of growth are plotted directly (●——●) and in the logarithmic form (○——○). The growth curve is exponential, as shown by the linearity of the log plot.

These observations can be expressed as equations for the generation time.

Let N_0 = the initial population number

N_t = the population at time t

n = the number of generations in time t

Then inspection of the results in table 7.1 shows that

$$N_t = N_0 \times 2^n$$

Solving for n, the number of generations, where all logarithms are to the base 10,

$$\log N_t = \log N_0 + n \cdot \log 2, \text{ and}$$

$$n = \frac{\log N_t - \log N_0}{\log 2} = \frac{\log N_t - \log N_0}{0.301}$$

The rate of growth during the exponential phase in a batch culture can be expressed in terms of the **mean growth rate constant** (**k**). This is the number of generations per unit time, often expressed as the generations per hour.

$$k = \frac{n}{t} = \frac{\log N_t - \log N_0}{0.301t}$$

The time it takes a population to double in size—that is, the **mean generation (doubling) time** (g)—can now be calculated. If the population doubles ($t = g$), then

$$N_t = 2N_0$$

Substitute $2N_0$ into the mean growth rate equation and solve for k.

$$k = \frac{\log (2N_0) - \log N_0}{0.301g} = \frac{\log 2 + \log N_0 - \log N_0}{0.301g}$$

$$k = \frac{1}{g}$$

The mean generation time is the reciprocal of the mean growth rate constant.

$$g = \frac{1}{k}$$

The mean generation time (g) can be determined directly from a semilogarithmic plot of the growth data (**figure 7.10**) and the growth rate constant calculated from the g value. The generation time also may be calculated directly from the previous equations. For example, suppose that a bacterial population increases from 10^3 cells to 10^9 cells in 10 hours.

$$k = \frac{\log 10^9 - \log 10^3}{(0.301)(10 \text{ hr})} = \frac{9 - 3}{3.01 \text{ hr}} = 2.0 \text{ generations/hr}$$

$$g = \frac{1}{2.0 \text{ gen. hr}} = 0.5 \text{ hr/gen. or } 30 \text{ min/gen.}$$

Generation times vary markedly with the species of microorganism and environmental conditions. They range from less than

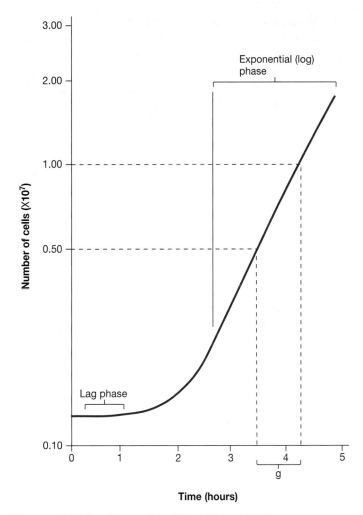

Figure 7.10 Generation Time Determination. The generation time can be determined from a microbial growth curve. The population data are plotted with the logarithmic axis used for the number of cells. The time to double the population number is then read directly from the plot. The log of the population number can also be plotted against time on regular axes.

10 minutes (0.17 hours) to several days (**table 7.2**). Generation times in nature are usually much longer than in culture.

1. Define growth. Describe the four phases of the growth curve and discuss the causes of each.

2. Why would cells that are vigorously growing when inoculated into fresh culture medium have a shorter lag phase than those that have been stored in a refrigerator?

3. List two physiological changes that are observed in stationary cells. How do these changes impact the organism's ability to survive?

4. Define balanced growth and unbalanced growth. Why do shift-up and shift-down experiments cause cells to enter unbalanced growth?

5. Define generation (doubling) time and mean growth rate constant. Calculate the mean growth rate and generation time of a culture that increases in the exponential phase from 5×10^2 to 1×10^8 in 12 hours.

6. Suppose the generation time of a bacterium is 90 minutes and the initial number of cells in a culture is 10^3 cells at the start of the log phase. How many bacteria will there be after 8 hours of exponential growth?

7. What effect does increasing a limiting nutrient have on the yield of cells and the growth rate?

8. Contrast and compare the viable but nonculturable status of microbes with that of programmed cell death as a means of responding to starvation.

Table 7.2	Examples of Generation Times[a]	
Microorganism	Incubation Temperature (°C)	Generation Time (Hours)
Bacteria		
Escherichia coli	40	0.35
Bacillus subtilis	40	0.43
Staphylococcus aureus	37	0.47
Pseudomonas aeruginosa	37	0.58
Clostridium botulinum	37	0.58
Mycobacterium tuberculosis	37	≈ 12
Treponema pallidum	37	33
Protists		
Tetrahymena geleii	24	2.2–4.2
Chlorella pyrenoidosa	25	7.75
Paramecium caudatum	26	10.4
Euglena gracilis	25	10.9
Giardia lamblia	37	18
Ceratium tripos	20	82.8
Fungi		
Saccharomyces cerevisiae	30	2
Monilinia fructicola	25	30

[a]Generation times differ depending on the growth medium and environmental conditions used.

7.3 MEASUREMENT OF MICROBIAL GROWTH

There are many ways to measure microbial growth to determine growth rates and generation times. Either population number or mass may be followed because growth leads to increases in both. Here the most commonly employed techniques for determining population size are examined briefly and the advantages and disadvantages of each noted. No single technique is always best; the most appropriate approach depends on the experimental situation.

Measurement of Cell Numbers

The most obvious way to determine microbial numbers is by **direct counts.** Using a counting chamber is easy, inexpensive, and relatively quick; it also gives information about the size and morphology of microorganisms. Petroff-Hausser counting chambers can be used for counting procaryotes; hemocytometers can be used for both procaryotes and eucaryotes. Both of these specially designed slides have chambers of known depth with an etched grid on the chamber bottom (**figure 7.11**). Procaryotes are more easily counted if they are stained or when a phase-contrast or a fluorescence microscope is employed. The number of microorganisms in a sample can be calculated by taking into account the chamber's volume and any sample dilutions required. One disadvantage is that to determine population size accurately, the microbial population must be relatively large because only a small volume of the population is sampled.

Larger microorganisms such as protists and yeasts can be directly counted with electronic counters such as the Coulter Counter, although the flow cytometer is increasingly used. In the Coulter Counter, the microbial suspension is forced through a small hole. Electrical current flows through the hole, and electrodes placed on both sides of the hole measure electrical resistance. Every time a microbial cell passes through the hole, electrical resistance increases (i.e., the conductivity drops), and the cell is counted. Flow cytometry gives accurate results with larger cells, and is extensively used in hospital laboratories to count red and white blood cells. It is not as useful in counting bacteria because of interference by small debris particles, the formation of filaments, and other problems. >> *Identification of microorganisms from specimens (section 32.2)*

The number of bacteria in aquatic samples is frequently determined from direct counts after the bacteria have been trapped on membrane filters. In the membrane filter technique, the sample is first filtered through a black polycarbonate membrane filter. Then the bacteria are stained with a fluorescent dye such as acridine orange or the DNA stain DAPI and observed microscopically. The stained cells are easily observed against the black background of the membrane filter and can be counted when viewed with an epifluorescence microscope. << *Light microscopes: Fluorescence microscope (section 2.2)*

Traditional methods for directly counting microbes in a sample usually yield cell densities that are much higher than the plating methods described next because direct counting procedures do not distinguish dead cells from culturable cells. Newer methods for direct counts avoid this problem. Commercial kits that use fluorescent reagents to stain live and dead cells differently are now available, making it possible to count directly the number of live and dead microorganisms in a sample *(see figures 2.12a and 25.10).*

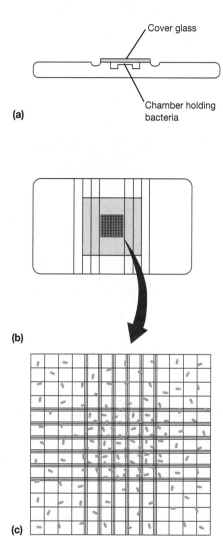

Figure 7.11   The Petroff-Ha[u]sser Counting Chamber. (a) Side view of the chamber showing the cover glass and the space beneath it that holds a bacterial suspension. (b) A top view of the chamber. The grid is located in the center of the slide. (c) An enlarged view of the grid. The bacteria in several of the central squares are counted, usually at ×400 to ×500 magnification. The average number of bacteria in these squares is used to calculate the concentration of cells in the original sample. Since there are 25 squares covering an area of 1 mm², the total number of bacteria in 1 mm² of the chamber is (number/square)(25 squares). The chamber is 0.02 mm deep and therefore,

$$\text{bacteria/mm}^3 = (\text{bacteria/square})(25 \text{ squares})(50).$$

The number of bacteria per cm³ is 10^3 times this value. For example, suppose the average count per square is 28 bacteria:

$$\text{bacteria/cm}^3 = (28 \text{ bacteria})(25 \text{ squares})(50)(10^3) = 3.5 \times 10^7.$$

Several plating methods can be used to determine the number of viable microbes in a sample. These are referred to as **viable counting methods** (plate counts) because they count only those cells that are able to reproduce when cultured. Two commonly used procedures are the spread-plate and the pour-plate techniques. In both of these methods, a diluted sample of microorganisms is dispersed over or within agar. If each cell is far enough away from other cells, then each cell will develop into a distinct colony. The original number of viable microorganisms in the sample can be calculated from the number of colonies formed and the sample dilution. For example, if 1.0 ml of a 1×10^6 dilution yielded 150 colonies, the original sample contained around 1.5×10^8 cells per ml. Usually the count is made more accurate by use of a colony counter. In this way the spread-plate and pour-plate techniques may be used to find the number of microorganisms in a sample. << *Isolation of pure cultures: Spread plate and pour plate (section 6.8)*

Another commonly used plating method first traps bacteria in aquatic samples on a membrane filter. The filter is then placed on an agar medium or on a pad soaked with liquid media (**figure 7.12**) and incubated until each cell forms a separate colony. A colony count gives the number of microorganisms in the filtered sample, and selective media can be used to select for specific microorganisms (**figure 7.13**). This technique is especially useful in analyzing water purity. << *Culture media (section 6.7);* >> *Water purification and sanitary analysis (section 35.1).*

Plating techniques are simple, sensitive, and widely used for viable counts of bacteria and other microorganisms in samples of food, water, and soil. Several problems, however, can lead to inaccurate counts. Low counts will result if clumps of cells are not broken up and the microorganisms well dispersed. Because it is not possible to be certain that each colony arose from an individual cell, the results are often expressed in terms of **colony forming units (CFU)** rather than the number of microorganisms. The samples should yield between 30 and 300 colonies for most accurate counting (counts of 25 to 250 are used in some applications). Of course the counts will also be low if the medium employed cannot support growth of all the viable microorganisms present. The hot agar used in the pour-plate technique may injure or kill sensitive cells; thus spread plates sometimes give higher counts than pour plates.

Measurement of Cell Mass

Techniques for measuring changes in cell mass also can be used to follow growth. The most direct approach is the determination of microbial dry weight. Cells growing in liquid medium are collected by centrifugation, washed, dried in an oven, and weighed. This is an especially useful technique for measuring the growth of filamentous fungi. It is time-consuming, however, and not very sensitive. Because bacteria weigh so little, it may be necessary to centrifuge several hundred milliliters of culture to collect a sufficient quantity.

A more rapid and sensitive method for measuring cell mass is spectrophotometry. Spectrophotometry depends on the fact that microbial cells scatter light that strikes them. Because microbial

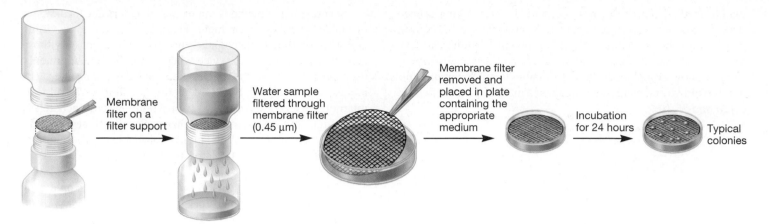

Figure 7.12 The Membrane Filtration Procedure. Membranes with different pore sizes are used to trap different microorganisms. Incubation times for membranes also vary with the medium and microorganism.

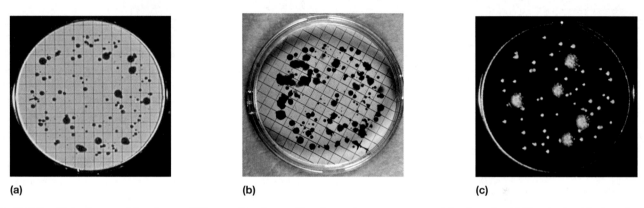

(a)   (b)   (c)

Figure 7.13 Colonies on Membrane Filters. Membrane-filtered samples grown on a variety of media. (a) Standard nutrient media for a total bacterial count. An indicator colors colonies dark red for easy counting. (b) Fecal coliform medium for detecting fecal coliforms that form blue colonies. (c) Wort agar for the culture of yeasts and molds.

cells in a population are of roughly constant size, the amount of scattering is directly proportional to the biomass of cells present and indirectly related to cell number. When the concentration of bacteria reaches about a million (10^6) cells per ml, the medium appears slightly cloudy or turbid. Further increases in concentration result in greater turbidity, and less light is transmitted through the medium. The extent of light scattering (i.e., decrease in transmitted light) can be measured by a spectrophotometer and is called the absorbance (optical density) of the medium. Absorbance is almost linearly related to cell concentration at absorbance levels less than about 0.5 (**figure 7.14**). If the sample exceeds this value, it must first be diluted and then absorbance measured. Thus population growth can be easily measured as long as the population is high enough to give detectable turbidity.

Cell mass can also be estimated by measuring the concentration of some cellular substance, as long as its concentration is constant in each cell. For example, a sample of cells can be analyzed for total protein or nitrogen. An increase in the microbial population will be reflected in higher total protein levels. Similarly, chlorophyll determinations can be used to measure phototrophic protist and cyanobacterial populations, and the quantity of ATP can be used to estimate the amount of living microbial mass.

1. Briefly describe each technique by which microbial population numbers may be determined and give its advantages and disadvantages.

2. When using direct cell counts to follow the growth of a culture, it may be difficult to tell when the culture enters the phase of senescence and death. Why?

3. Why are plate count results expressed as colony forming units?

7.4 CONTINUOUS CULTURE OF MICROORGANISMS

Thus far, our focus has been on closed systems called batch cultures in which nutrients are not renewed nor wastes removed. Exponential (logarithmic) growth lasts for only a few generations and soon stationary phase is reached. However, it is possible to grow microorganisms in a system with constant environmental conditions maintained through continual provision of nutrients and removal of wastes. Such a system is called

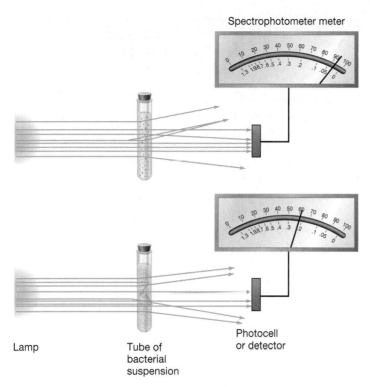

Figure 7.14 **Turbidity and Microbial Mass Measurement.** Determination of microbial mass by measurement of light absorption. As the population and turbidity increase, more light is scattered and the absorbance reading given by the spectrophotometer increases. The spectrophotometer meter has two scales. The bottom scale displays absorbance and the top scale, percent transmittance. Absorbance increases as percent transmittance decreases.

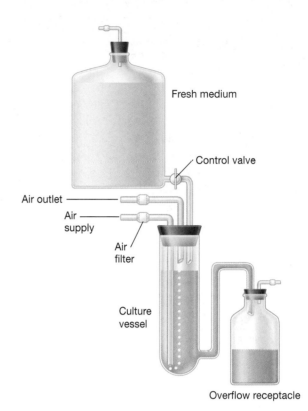

Figure 7.15 **A Continuous Culture System: The Chemostat.** Schematic diagram of the system. The fresh medium contains a limiting amount of an essential nutrient. Although not shown, usually the incoming medium is mechanically mixed with that already in the culture vessel. Growth rate is determined by the rate of flow of medium through the culture vessel.

a **continuous culture system.** These systems can maintain a microbial population in exponential growth, growing at a known rate and at a constant biomass concentration for extended periods. Continuous culture systems make possible the study of microbial growth at very low nutrient levels, concentrations close to those present in natural environments. These systems are essential for research in many areas, including ecology. For example, interactions between microbial species in environmental conditions resembling those in a freshwater lake or pond can be modeled. Continuous culture systems also are used in food and industrial microbiology. Two major types of continuous culture systems commonly are used: chemostats and turbidostats. >> *Microbiology of food (chapter 34); Applied environmental microbiology (chapter 35)*

Chemostats

A **chemostat** is constructed so that the rate at which sterile medium is fed into the culture vessel is the same as the rate at which the media containing microorganisms is removed (**figure 7.15**). The culture medium for a chemostat possesses an essential nutrient (e.g., a vitamin) in limiting quantities. Because one nutrient is limiting, growth rate is determined by the rate at which new medium is fed into the growth chamber; the final cell density

depends on the concentration of the limiting nutrient. The rate of nutrient exchange is expressed as the dilution rate (D), the rate at which medium flows through the culture vessel relative to the vessel volume, where f is the flow rate (ml/hr) and V is the vessel volume (ml).

$$D = f/V$$

For example, if f is 30 ml/hr and V is 100 ml, the dilution rate is 0.30 hr^{-1}.

Both population size and generation time are related to the dilution rate, and population density remains unchanged over a wide range of dilution rates (**figure 7.16**). The generation time decreases (i.e., the rate of growth increases) as the dilution rate increases. The limiting nutrient will be almost completely depleted under these balanced conditions. If the dilution rate rises too high, microorganisms can actually be washed out of the culture vessel before reproducing because the dilution rate is greater than the maximum growth rate. This occurs because fewer microorganisms are present to consume the limiting nutrient.

At very low dilution rates, an increase in D causes a rise in both cell density and the growth rate. This is because of the effect of nutrient concentration on the growth rate, sometimes called the Monod relationship (figure 7.6b). When dilution rates

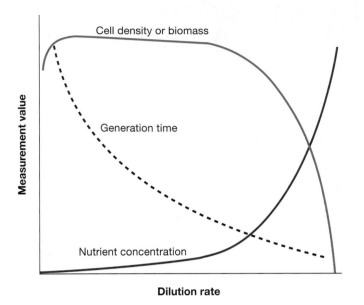

Figure 7.16 Chemostat Dilution Rate and Microbial Growth. The effects of changing the dilution rate in a chemostat.

are low, only a limited supply of nutrient is available and the microbes can conserve only a limited amount of energy. Much of that energy must be used for cell maintenance, not for growth and reproduction. As the dilution rate increases, the amount of nutrients and the resulting cell density rise because energy is available for both maintenance and reproduction. The growth rate increases when the total available energy exceeds the **maintenance energy.**

Turbidostats

The second type of continuous culture system, the **turbidostat,** has a photocell that measures the turbidity (absorbance) of the culture in the growth vessel. The flow rate of media through the vessel is automatically regulated to maintain a predetermined turbidity. Because turbidity is related to cell density, the turbidostat maintains a desired cell density. The turbidostat differs from the chemostat in several ways. The dilution rate in a turbidostat varies rather than remaining constant, and a turbidostat's culture medium contains all nutrients in excess. That is, none of the nutrients is limiting. The turbidostat operates best at high dilution rates; the chemostat is most stable and effective at lower dilution rates.

1. How does a continuous culture system differ from a closed culture system or batch culture?
2. Describe how chemostats and turbidostats operate. How do they differ?
3. What is the dilution rate? What is maintenance energy? How are they related?

7.5 INFLUENCES OF ENVIRONMENTAL FACTORS ON GROWTH

As we have seen, microorganisms must be able to respond to variations in nutrient levels. Microorganisms also are greatly affected by the chemical and physical nature of their surroundings. An understanding of environmental influences aids in the control of microbial growth and the study of the ecological distribution of microorganisms.

The adaptations of some microorganisms to extreme and inhospitable environments are truly remarkable. Microbes are present virtually everywhere on Earth. Many habitats in which microbes thrive would kill most other organisms. Bacteria such as *Bacillus infernus* are able to live over 2.4 kilometers below Earth's surface, without oxygen and at temperatures above 60°C. Other microbes live in acidic hot springs, at great ocean depths, or in lakes such as the Great Salt Lake in Utah (USA) that have high sodium chloride concentrations. Microorganisms that grow in such harsh conditions are called **extremophiles.** >> *Microorganisms in natural environments (Chapter 26)*

In this section, we briefly review the effects of the most important environmental factors on microbial growth. Major emphasis is given to solutes and water activity, pH, temperature, oxygen level, pressure, and radiation. **Table 7.3** summarizes how microorganisms are categorized in terms of their response to these factors. It is important to note that for most environmental factors, a range of levels supports growth of a microbe. For example, a microbe might exhibit optimum growth at pH 7 but grows, though not optimally, at pH values down to pH 6 (its pH minimum) and up to pH 8 (its pH maximum). Furthermore, outside this range, the microbe might cease reproducing but remain viable for some time. Clearly, each microbe must have evolved adaptations that allow it to adjust its physiology within its preferred range, and it may also have adaptations that protect it in environments outside this range. These adaptations also are discussed in this section.

Solutes and Water Activity

Because a selectively permeable plasma membrane separates microorganisms from their environment, they can be affected by changes in the osmotic concentration of their surroundings. If a microorganism is placed in a hypotonic solution (one with a lower osmotic concentration), water will enter the cell and cause it to burst unless something is done to prevent the influx or inhibit plasma membrane expansion. Conversely if it is placed in a hypertonic solution (one with a higher osmotic concentration), water will flow out of the cell. In microbes that have cell walls, the membrane shrinks away from the cell wall—a process called plasmolysis. Dehydration of the cell in hypertonic environments may damage the cell membrane and cause the cell to become metabolically inactive.

Clearly it is important that microbes be able to respond to changes in the osmotic concentrations of their environment.

Table 7.3	**Microbial Responses to Environmental Factors**	
Descriptive Term	**Definition**	**Representative Microorganisms**
Solute and Water Activity		
Osmotolerant	Able to grow over wide ranges of water activity or osmotic concentration	*Staphylococcus aureus, Saccharomyces rouxii*
Halophile	Requires high levels of sodium chloride, usually above about 0.2 M, to grow	*Halobacterium, Dunaliella, Ectothiorhodospira*
pH		
Acidophile	Growth optimum between pH 0 and 5.5	*Sulfolobus, Picrophilus, Ferroplasma, Acontium*
Neutrophile	Growth optimum between pH 5.5 and 8.0	*Escherichia, Euglena, Paramecium*
Alkalophile	Growth optimum between pH 8.0 and 11.5	*Bacillus alcalophilus, Natronobacterium*
Temperature		
Psychrophile	Grows at 0°C and has an optimum growth temperature of 15°C or lower	*Bacillus psychrophilus, Chlamydomonas nivalis*
Psychrotroph	Can grow at 0–7°C; has an optimum between 20 and 30°C and a maximum around 35°C	*Listeria monocytogenes, Pseudomonas fluorescens*
Mesophile	Has growth optimum around 20–45°C	*Escherichia coli, Trichomonas vaginalis*
Thermophile	Can grow at 55°C or higher; optimum often between 55 and 65°C	*Geobacillus stearothermophilus, Thermus aquaticus, Cyanidium caldarium, Chaetomium thermophile*
Hyperthermophile	Has an optimum between 80 and about 113°C	*Sulfolobus, Pyrococcus, Pyrodictium*
Oxygen Concentration		
Obligate aerobe	Completely dependent on atmospheric O_2 for growth	*Micrococcus luteus,* most protists and fungi
Facultative anaerobe	Does not require O_2 for growth but grows better in its presence	*Escherichia, Enterococcus, Saccharomyces cerevisiae*
Aerotolerant anaerobe	Grows equally well in presence or absence of O_2	*Streptococcus pyogenes*
Obligate anaerobe	Does not tolerate O_2 and dies in its presence	*Clostridium, Bacteroides, Methanobacterium*
Microaerophile	Requires O_2 levels between 2–10% for growth and is damaged by atmospheric O_2 levels (20%)	*Campylobacter, Spirillum volutans, Treponema pallidum*
Pressure		
Barophile	Growth more rapid at high hydrostatic pressures	*Photobacterium profundum, Shewanella benthica*

Microbes in hypotonic environments can reduce the osmotic concentration of their cytoplasm. This can be achieved using inclusion bodies or other mechanisms. For example, some procaryotes have mechanosensitive (MS) channels in their plasma membrane. In a hypotonic environment, the membrane stretches due to an increase in hydrostatic pressure and cellular swelling. MS channels then open and allow solutes to leave. Thus, MS channels act as escape valves to protect cells from bursting. Because many protists do not have a cell wall, they must use contractile vacuoles to expel excess water. Many microorganisms, whether in hypotonic or hypertonic environments, keep the osmotic concentration of their cytoplasm somewhat above that of the habitat by the use of compatible solutes, so that the plasma membrane is always pressed firmly against their cell wall. **Compatible solutes** are solutes that do not interfere with metabolism and growth when at high intracellular concentrations. Most procaryotes increase their internal osmotic concentration in a hypertonic environment through the synthesis or uptake of choline, betaine, proline, glutamic acid, and other amino acids; elevated levels of potassium ions may also be used. Photosynthetic protists and fungi employ sucrose and polyols—for example, arabitol, glycerol, and mannitol—for the same purpose. Polyols and amino acids are ideal compatible solutes because they normally do not disrupt enzyme structure and function. << *Procaryotic cytoplasm: Inclusion bodies (section 3.3)*

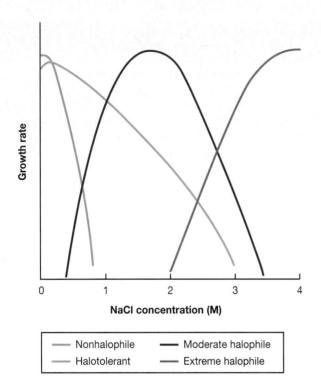

Figure 7.17 **The Effects of Sodium Chloride on Microbial Growth.** Four different patterns of microbial dependence on NaCl concentration are depicted. The curves are only illustrative and are not meant to provide precise shapes or salt concentrations required for growth.

Some microbes are adapted to extreme hypertonic environments. **Halophiles** grow optimally in the presence of NaCl or other salts at a concentration above about 0.2 M (**figure 7.17**). Extreme halophiles have adapted so completely to hypertonic, saline conditions that they require high levels of sodium chloride to grow—concentrations between about 2 M and saturation (about 6.2 M). The archaeon *Halobacterium* can be isolated from the Dead Sea (a salt lake between Israel and Jordan), the Great Salt Lake in Utah, and other aquatic habitats with salt concentrations approaching saturation. *Halobacterium* and other extremely halophilic procaryotes accumulate enormous quantities of potassium in order to remain hypertonic to their environment; the internal potassium concentration may reach 4 to 7 M. Furthermore, their enzymes, ribosomes, and transport proteins require high potassium levels for stability and activity. In addition, the plasma membrane and cell wall of *Halobacterium* are stabilized by high concentrations of sodium ion. If the sodium concentration decreases too much, the wall and plasma membrane disintegrate. Extreme halophiles have successfully adapted to environmental conditions that would destroy most organisms. In the process, they have become so specialized that they have lost ecological flexibility and can prosper only in a few extreme habitats. ▸▸ *Phylum* Euryarchaeota: Halobacteria *(section 18.3)*

Because the osmotic concentration of a habitat has such profound effects on microorganisms, it is useful to express quantitatively the degree of water availability. Microbiologists generally use **water activity (a_w)** for this purpose (water avail-

ability also may be expressed as water potential, which is related to a_w). The water activity of a solution is 1/100 the relative humidity of the solution (when expressed as a percent). It is also equivalent to the ratio of the solution's vapor pressure (P_{soln}) to that of pure water (P_{water}).

$$a_w = \frac{P_{soln}}{P_{water}}$$

The water activity of a solution or solid can be determined by sealing it in a chamber and measuring the relative humidity after the system has come to equilibrium. Suppose after a sample is treated in this way, the air above it is 95% saturated—that is, the air contains 95% of the moisture it would have when equilibrated at the same temperature with a sample of pure water. The relative humidity would be 95% and the sample's water activity, 0.95. Water activity is inversely related to osmotic pressure; if a solution has high osmotic pressure, its a_w is low.

Microorganisms differ greatly in their ability to adapt to habitats with low water activity. A microorganism must expend extra effort to grow in a habitat with a low a_w value because it must maintain a high internal solute concentration to retain water. Some microorganisms can do this and are **osmotolerant;** they grow over wide ranges of water activity. For example, *Staphylococcus aureus* is halotolerant (figure 7.17), can be cultured in media containing sodium chloride concentration up to about 3 M, and is well adapted for growth on the skin. The yeast *Saccharomyces rouxii* grows in sugar solutions with a_w values as low as 0.6. The photosynthetic protist *Dunaliella viridis* tolerates sodium chloride concentrations from 1.7 M to a saturated solution. Some microbes (e.g., *Halobacterium*) are true xerophiles. That is, they grow best at low a_w. However, most microorganisms only grow well at water activities around 0.98 (the approximate a_w for seawater) or higher. This is why drying food or adding large quantities of salt and sugar effectively prevents food spoilage. ▸▸ *Controlling food spoilage (section 34.3)*

1. How do microorganisms adapt to hypotonic and hypertonic environments? What is plasmolysis?

2. Define water activity and briefly describe how it can be determined. Why is it difficult for microorganisms to grow at low a_w values?

3. What are halophiles and why does *Halobacterium* require sodium and potassium ions?

pH

pH is a measure of the relative acidity of a solution and is defined as the negative logarithm of the hydrogen ion concentration (expressed in terms of molarity).

$$pH = -\log[H^+] = \log(1/[H^+])$$

The pH scale extends from pH 0.0 (1.0 M H^+) to pH 14.0 (1.0 × 10^{-14} M H^+), and each pH unit represents a tenfold change in hydrogen ion concentration. **Figure 7.18** shows that microbial habitats vary widely in pH—from pH 0 to 2 at the acidic end to alkaline lakes and soil with pH values between 9 and 10.

Each species has a definite pH growth range and pH growth optimum. **Acidophiles** have their growth optimum between pH 0 and 5.5; **neutrophiles,** between pH 5.5 and 8.0; and **alkalophiles (alkaliphiles),** between pH 8.0 and 11.5. Extreme alkalophiles have growth optima at pH 10 or higher. In general, different microbial groups have characteristic pH preferences. Most bacteria and protists are neutrophiles. Most fungi prefer more acidic surroundings, about pH 4 to 6; photosynthetic protists also seem to favor slight acidity. Many archaea are acidophiles. For example, the archaeon *Sulfolobus acidocaldarius* is a common inhabitant of acidic hot springs; it grows well from pH 1 to 3 and at high temperatures. The archaea *Ferroplasma acidarmanus* and *Picrophilus oshimae* can actually grow very close to pH 0. Alkalophiles are distributed among all three domains of life. They include bacteria belonging to the genera *Bacillus,* *Micrococcus, Pseudomonas,* and *Streptomyces;* yeasts and filamentous fungi; and numerous archaea.

Although microorganisms often grow over wide ranges of pH and far from their optima, there are limits to their tolerance. When the external pH is low, the concentration of H^+ is greater outside than inside, and H^+ will move into the cytoplasm and lower the cytoplasmic pH. Drastic variations in cytoplasmic pH can harm microorganisms by disrupting the plasma membrane or inhibiting the activity of enzymes and membrane transport proteins. Most procaryotes die if the internal pH drops much below 5.0 to 5.5. Changes in the external pH also might alter the ionization of nutrient molecules and thus reduce their availability to the organism.

Microorganisms respond to external pH changes using mechanisms that maintain a neutral cytoplasmic pH. Several mechanisms for adjusting to small changes in external pH have been proposed. Neutrophiles appear to exchange potassium for protons using an antiport transport system. Internal buffering also may contribute to pH homeostasis. However, if the external pH becomes too acidic, other mechanisms come into play. When the pH drops below about 5.5 to 6.0, *Salmonella enterica* serovar Typhimurium and *E. coli* synthesize an array of new proteins as part of what has been called their acidic tolerance response. A proton-translocating ATPase enzyme contributes to this protective response, either by making more ATP or by pumping protons out of the cell. If the external pH decreases to 4.5 or lower, acid shock proteins and heat shock proteins are synthesized. These prevent the denaturation of proteins and aid in the refolding of denatured proteins in acidic conditions. << *Uptake of nutrients (section 6.6);* >> *Protein maturation and secretion (section 12.8)*

What about microbes that live at pH extremes? Extreme alkalophiles such as *Bacillus alcalophilus* maintain their internal pH close to neutrality by exchanging internal sodium ions for external protons. Acidophiles use a variety of measures to maintain a neutral internal pH. These include the transport of cations (e.g., potassium ions) into the cell, thus decreasing the movement of H^+ into the cell; proton transporters that pump H^+ out if they get in; and highly impermeable cell membranes.

pH	[H⁺] (Molarity)		Environmental examples	Microbial examples
0	$10^{-0}(1.0)$	Increasing acidity	Concentrated nitric acid	*Ferroplasma* *Picrophilus oshimae*
1	10^{-1}		Gastric contents, acid thermal springs	*Dunaliella acidophila*
2	10^{-2}		Lemon juice Acid mine drainage	*Cyanidium caldarium* *Thiobacillus thiooxidans* *Sulfolobus acidocaldarius*
3	10^{-3}		Vinegar, ginger ale Pineapple	
4	10^{-4}		Tomatoes, orange juice Very acid soil	
5	10^{-5}		Cheese, cabbage Bread	*Physarum polycephalum* *Acanthamoeba castellanii*
6	10^{-6}		Beef, chicken Rain water Milk Saliva	*Lactobacillus acidophilus* *E. coli, Pseudomonas aeruginosa, Euglena gracilis, Paramecium bursaria*
7	10^{-7}	Neutrality	Pure water Blood	*Staphylococcus aureus*
8	10^{-8}		Seawater	*Nitrosomonas* spp.
9	10^{-9}		Strongly alkaline soil Alkaline lakes	
10	10^{-10}		Soap	*Microcystis aeruginosa* *Bacillus alcalophilus*
11	10^{-11}		Household ammonia	
12	10^{-12}		Saturated calcium hydroxide solution	
13	10^{-13}		Bleach Drain opener	
14	10^{-14}	Increasing alkalinity		

Figure 7.18   The pH Scale. The pH scale and examples of substances with different pH values. The microorganisms are placed at their growth optima.

1. Define pH, acidophile, neutrophile, and alkalophile.

2. Classify each of the following organisms as an alkalophile, a neutrophile, or an acidophile: *Staphylococcus aureus, Microcystis aeruginosa, Sulfolobus acidocaldarius,* and *Pseudomonas aeruginosa.* Which might be pathogens? Explain your choices.

3. Describe the mechanisms microbes use to maintain a neutral pH. Explain how extreme pH values might harm microbes.

Temperature

Microorganisms are particularly susceptible to external temperatures because they cannot regulate their internal temperature. An important factor influencing the effect of temperature on growth is the temperature sensitivity of enzyme-catalyzed reactions. Each enzyme has a temperature at which it functions optimally. At some temperature below the optimum, it ceases to be catalytic. As the temperature rises from this low point, the rate of catalysis increases to that observed for the optimal temperature. The velocity of the reaction roughly doubles for every 10°C rise in temperature. When all enzymes in a microbe are considered together, as the rate of each reaction increases, metabolism as a whole becomes more active, and the microorganism grows faster. However, beyond a certain point, further increases actually slow growth, and sufficiently high temperatures are lethal. High temperatures denature enzymes, transport carriers, and other proteins. Temperature also has a significant effect on microbial membranes. At very low temperatures, membranes solidify. At high temperatures, the lipid bilayer simply melts and disintegrates. Thus when organisms are above their optimum temperature, both function and cell structure are affected. If temperatures are very low, function is affected but not necessarily cell chemical composition and structure.

Because of these opposing temperature influences, microbial growth has a characteristic temperature dependence with distinct **cardinal temperatures**—minimum, optimum, and maximum growth temperatures (**figure 7.19**). Although the shape of temperature dependence curves varies, the temperature optimum is always closer to the maximum than to the minimum. The cardinal temperatures are not rigidly fixed. Instead they depend to some extent on other environmental factors such as pH and available nutrients. For example, *Crithidia fasciculate,* a flagellated protist living in the gut of mosquitoes, grows in a simple medium at 22 to 27°C. However, it cannot be cultured at 33 to 34°C without the addition of extra metals, amino acids, vitamins, and lipids.

The cardinal temperatures vary greatly among microorganisms (**table 7.4**). Optima usually range from 0°C to 75°C, whereas microbial growth occurs at temperatures extending from less than −20°C to over 120°C. Some archaea even grow at 121°C (250°F), the temperature normally used in autoclaves (**Microbial Diversity & Ecology 7.1**). A major factor determining growth range seems to be water. Even at the most extreme temperatures, microorgan-

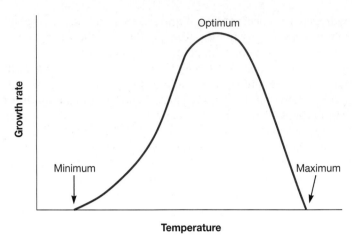

Figure 7.19 Temperature and Growth. The effect of temperature on growth rate.

isms need liquid water to grow. The growth temperature range for a particular microorganism usually spans about 30 degrees. Some species (e.g., *Neisseria gonorrhoeae*) have a small range; others, such as *Enterococcus faecalis,* grow over a wide range of temperatures. The major microbial groups differ from one another regarding their maximum growth temperatures. The upper limit for protists is around 50°C. Some fungi grow at temperatures as high as 55 to 60°C. Procaryotes can grow at much higher temperatures than eucaryotes. It has been suggested that eucaryotes are not able to manufacture stable and functional organellar membranes at temperatures above 60°C. The photosynthetic apparatus also appears to be relatively unstable because photosynthetic organisms are not found growing at very high temperatures.

Microorganisms such as those listed in table 7.4 can be placed in one of five classes based on their temperature ranges for growth (**figure 7.20**).

1. **Psychrophiles** grow well at 0°C and have an optimum growth temperature of 10°C or lower; the maximum is around 15°C. They are readily isolated from Arctic and Antarctic habitats. Oceans constitute an enormous habitat for psychrophiles because 90% of ocean water is 5°C or colder. The psychrophilic protist *Chlamydomonas nivalis* can actually turn a snowfield or glacier pink with its bright red spores. Psychrophiles are widespread among bacterial taxa and are found in such genera as *Pseudomonas, Vibrio, Alcaligenes, Bacillus, Photobacterium,* and *Shewanella.* A psychrophilic archaeon, *Methanogenium,* has been isolated from Ace Lake in Antarctica. Psychrophilic microorganisms have adapted to their environment in several ways. Their enzymes, transport systems, and protein synthetic machinery function well at low temperatures. The cell membranes of psychrophilic microorganisms have high levels of unsaturated fatty acids and remain semifluid when cold. Indeed, many psychrophiles begin to leak

Table 7.4	Temperature Ranges for Microbial Growth		
	Cardinal Temperatures (°C)		
Microorganism	**Minimum**	**Optimum**	**Maximum**
Nonphotosynthetic Procaryotes			
Bacillus psychrophilus	−10	23–24	28–30
Pseudomonas fluorescens	4	25–30	40
Enterococcus faecalis	0	37	44
Escherichia coli	10	37	45
Neisseria gonorrhoeae	30	35–36	38
Thermoplasma acidophilum	45	59	62
Thermus aquaticus	40	70–72	79
Pyrococcus abyssi	67	96	102
Pyrodictium occultum	82	105	110
Pyrolobus fumarii	90	106	113
Photosynthetic Bacteria			
Anabaena variabilis	ND[a]	35	ND
Synechococcus eximius	70	79	84
Protists			
Chlamydomonas nivalis	−36	0	4
Amoeba proteus	4–6	22	35
Skeletonema costatum	6	16–26	>28
Trichomonas vaginalis	25	32–39	42
Tetrahymena pyriformis	6–7	20–25	33
Cyclidium citrullus	18	43	47
Fungi			
Candida scotti	0	4–15	15
Saccharomyces cerevisiae	1–3	28	40
Mucor pusillus	21–23	45–50	50–58

[a]ND, not determined

cellular constituents at temperatures higher than 20°C because of cell membrane disruption.

2. **Psychrotrophs (facultative psychrophiles)** grow at 0 to 7°C even though they have optima between 20 and 30°C, and maxima at about 35°C. Psychrotrophic bacteria and fungi are major causes of refrigerated food spoilage, as described in chapter 34.

3. **Mesophiles** are microorganisms with growth optima around 20 to 45°C. They often have a temperature minimum of 15 to 20°C, and their maximum is about 45°C or lower. Most microorganisms probably fall within this category. Almost all human pathogens are mesophiles, as might be expected because the human body is a fairly constant 37°C.

4. **Thermophiles** grow at temperatures between 55 and 85°C. Their growth minimum is usually around 45°C, and they often have optima between 55 and 65°C. The vast majority are procaryotes, although a few photosynthetic protists and fungi are thermophilic (table 7.4). These organisms flourish in many habitats including composts, self-heating hay stacks, hot water lines, and hot springs.

5. **Hyperthermophiles** have growth optima between 85°C and about 113°C. They usually do not grow well below 55°C. *Pyrococcus abyssi* and *Pyrodictium occultum* are examples of marine hyperthermophiles found in hot areas of the seafloor.

Thermophiles and hyperthermophiles differ from mesophiles in many ways. They have heat-stable enzymes and protein synthesis systems that function properly at high temperatures. These proteins are stable for a variety of reasons. Heat-stable proteins have highly organized hydrophobic interiors and more hydrogen and other noncovalent bonds. Larger quantities of amino acids such as proline also make polypeptide chains less flexible and more heat stable. In addition, the proteins are stabilized and aided in folding by proteins called chaperone proteins. Evidence exists that histonelike proteins stabilize the DNA of thermophilic bacteria. The membrane lipids of thermophiles and hyperthermophiles are also quite temperature stable. They tend to be more saturated, more branched, and of higher molecular weight. This increases the melting points of membrane lipids. Archaeal thermophiles have membrane lipids with ether linkages, which protect the lipids from hydrolysis at high temperatures. Sometimes archaeal lipids actually span the membrane to form a rigid, stable monolayer. >> *Proteins (appendix I);* << *Procaryotic cell membranes (section 3.2)*

1. What are cardinal temperatures?
2. Why does the growth rate rise with increasing temperature and then fall again at higher temperatures?
3. Define psychrophile, psychrotroph, mesophile, thermophile, and hyperthermophile.
4. What metabolic and structural adaptations for extreme temperatures do psychrophiles and thermophiles have?

Microbial Diversity & Ecology

7.1 Life Above 100°C

Until recently the highest reported temperature for procaryotic growth was 105°C. It seemed that the upper temperature limit for life was about 100°C, the boiling point of water. Now thermophilic procaryotes have been reported growing in sulfide chimneys or "black smokers," located along rifts and ridges on the ocean floor, that spew sulfide-rich, super-heated vent water with temperatures above 350°C (**chapter opener**). Evidence suggests that these microbes can grow and reproduce at 121°C and can survive temperatures to 130°C for up to 2 hours. The pressure present in their habitat is sufficient to keep water liquid (at 265 atm, seawater doesn't boil until 460°C).

The implications of this discovery are many. The proteins, membranes, and nucleic acids of these procaryotes are remarkably temperature stable and provide ideal subjects for studying the ways in which macromolecules and membranes are stabilized. In the future it may be possible to design enzymes that operate at very high temperatures. Some thermostable enzymes from these organisms have important industrial and scientific uses. For example, the Taq polymerase from the thermophile *Thermus aquaticus* is used extensively in the polymerase chain reaction. >> *Polymerase chain reaction (section 16.2)*

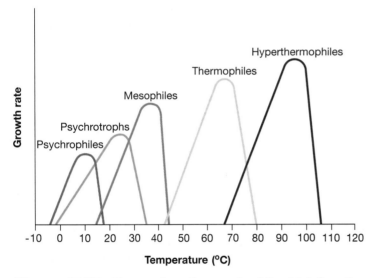

Figure 7.20 Temperature Ranges for Microbial Growth. Microorganisms are placed in different classes based on their temperature ranges for growth. They are ranked in order of increasing growth temperature range as psychrophiles, psychrotrophs, mesophiles, thermophiles, and hyperthermophiles. Representative ranges and optima for these five types are illustrated.

Oxygen Concentration

The importance of oxygen to the growth of an organism correlates with its metabolism—in particular, with the processes it uses to conserve the energy supplied by its energy source. Almost all energy-conserving metabolic processes involve the movement of electrons through a series of membrane-bound electron carriers called the electron transport chain (ETC). For chemotrophs, an externally supplied terminal electron acceptor is critical to the functioning of the ETC. The nature of the terminal electron accep-

tor is related to an organism's oxygen requirement. >> *Electron transport chains (section 9.6)*

An organism able to grow in the presence of atmospheric O_2 is an **aerobe,** whereas one that can grow in its absence is an **anaerobe.** Almost all multicellular organisms are completely dependent on atmospheric O_2 for growth—that is, they are **obligate aerobes** (table 7.3). Oxygen serves as the terminal electron acceptor for the ETC in the metabolic process called aerobic respiration. In addition, aerobic eucaryotes employ O_2 in the synthesis of sterols and unsaturated fatty acids. **Microaerophiles** such as *Campylobacter* are damaged by the normal atmospheric level of O_2 (20%) and require O_2 levels in the range of 2 to 10% for growth. **Facultative anaerobes** do not require O_2 for growth but grow better in its presence. In the presence of oxygen, they use O_2 as the terminal electron acceptor during aerobic respiration. **Aerotolerant anaerobes** such as *Enterococcus faecalis* simply ignore O_2 and grow equally well whether it is present or not; chemotrophic aerotolerant anaerobes are often described as having strictly fermentative metabolism. In contrast, strict or **obligate anaerobes** (e.g., *Bacteroides, Clostridium pasteurianum, Methanococcus*) are usually killed in the presence of O_2. Strict anaerobes cannot generate energy through aerobic respiration and employ other metabolic strategies such as fermentation or anaerobic respiration, neither of which requires O_2. The nature of bacterial O_2 responses can be readily determined by growing the bacteria in culture tubes filled with a solid culture medium or a medium such as thioglycollate broth, which contains a reducing agent to lower O_2 levels (**figure 7.21**). >> *Aerobic respiration (section 10.2); Anaerobic respiration (section 10.6); Fermentation (section 10.7)*

A microbial group may show more than one type of relationship to O_2. All five types are found among the procaryotes and protists. Fungi are normally aerobic, but a number of species—particularly among the yeasts—are facultative anaerobes. Photosynthetic protists are usually obligate aerobes. Although obligate anaerobes are killed by O_2, they may be recovered from

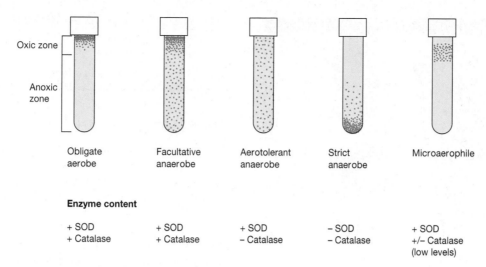

Figure 7.21 Oxygen and Bacterial Growth. Each dot represents an individual bacterial colony within the agar or on its surface. The surface, which is directly exposed to atmospheric oxygen, is oxic. The oxygen content of the medium decreases with depth until the medium becomes anoxic toward the bottom of the tube. The presence and absence of the enzymes superoxide dismutase (SOD) and catalase for each type are shown.

habitats that appear to be oxic. In such cases they associate with facultative anaerobes that use up the available O_2 and thus make the growth of strict anaerobes possible. For example, the strict anaerobe *Bacteroides gingivalis* lives in the mouth where it grows in the anoxic crevices around the teeth. Clearly the ability to grow in both oxic and anoxic environments provides considerable flexibility and is an ecological advantage.

The different relationships with O_2 are due to several factors, including the inactivation of proteins and the effect of toxic O_2 derivatives. Enzymes can be inactivated when sensitive groups such as sulfhydryls are oxidized. A notable example is the nitrogen-fixation enzyme nitrogenase, which is very oxygen sensitive. Toxic O_2 derivatives are formed when proteins such as flavoproteins promote oxygen reduction. The reduction products are superoxide radical, hydrogen peroxide, and hydroxyl radical.

$$O_2 + e^- \rightarrow O_2^{\cdot -} \text{ (superoxide radical)}$$

$$O_2^{\cdot -} + e^- + 2H^+ \rightarrow H_2O_2 \text{ (hydrogen peroxide)}$$

$$H_2O_2 + e^- + H^+ \rightarrow H_2O + OH\cdot \text{ (hydroxyl radical)}$$

These products are extremely toxic because they oxidize and rapidly destroy cellular constituents. A microorganism must be able to protect itself against such oxygen products or it will be killed. Indeed, neutrophils and macrophages, two important immune system cells, use these toxic oxygen products to destroy invading pathogens. >> *Synthesis of amino acids: Nitrogen assimilation (section 11.5); Phagocytosis (section 28.3)*

Many microorganisms possess enzymes that protect against toxic O_2 products (figure 7.21). Obligate aerobes and facultative anaerobes usually contain the enzymes **superoxide dismutase (SOD)** and **catalase,** which catalyze the destruction of superoxide radical and hydrogen peroxide, respectively. Peroxidase also can be used to destroy hydrogen peroxide.

$$2O_2^{\cdot -} + 2H^+ \xrightarrow{\text{superoxide dismutase}} O_2 + H_2O_2$$

$$2H_2O_2 \xrightarrow{\text{catalase}} 2H_2O + O_2$$

$$H_2O_2 + NADH + H^+ \xrightarrow{\text{peroxidase}} 2H_2O + NAD^+$$

Aerotolerant microorganisms may lack catalase but usually have superoxide dismutase. The aerotolerant bacterium *Lactobacillus plantarum* uses manganous ions instead of superoxide dismutase to destroy the superoxide radical. All strict anaerobes lack both enzymes or have them in very low concentrations and therefore cannot tolerate O_2. However, some microaerophilic bacteria and anaerobic archaea protect themselves from the toxic effects of O_2 with the enzymes superoxide reductase and peroxidase. Superoxide reductase reduces superoxide to H_2O_2 without producing O_2. The H_2O_2 is then converted to water by peroxidase. >> *Oxidation-reduction reactions (section 9.5)*

Because aerobes need O_2 and anaerobes are killed by it, radically different approaches must be used when they are cultured. When large volumes of aerobic microorganisms are cultured, either they must be shaken to aerate the culture medium or sterile air must be pumped through the culture vessel. Precisely the opposite problem arises with anaerobes—all O_2 must be excluded. This is accomplished in several ways. (1) Special anaerobic media containing reducing agents such as thioglycollate or cysteine may be used. The medium is boiled during preparation to dissolve its components and drive off oxygen. The reducing agents eliminate any residual dissolved O_2 in the medium so that anaerobes can grow beneath its surface. (2) Oxygen also may be eliminated by removing air with a vacuum pump and flushing out residual O_2 with nitrogen gas (**figure 7.22**). Often CO_2 as well as nitrogen is added to the chamber since many anaerobes require a small amount of CO_2 for best growth. (3) One of the most popular ways of culturing small numbers of anaerobes is by use of a

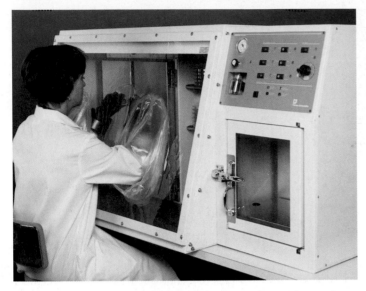

Figure 7.22 An Anaerobic Work Chamber and Incubator.
This anaerobic system contains an oxygen-free work area and an incubator. The interchange compartment on the right of the work area allows materials to be transferred inside without exposing the interior to oxygen. The anoxic atmosphere is maintained largely with a vacuum pump and nitrogen purges. The remaining oxygen is removed by a palladium catalyst and hydrogen. The oxygen reacts with hydrogen to form water, which is absorbed by desiccant.

GasPak jar, which uses hydrogen and a palladium catalyst to remove O_2 (**figure 7.23**). (4) A similar approach uses plastic bags or pouches containing calcium carbonate and a catalyst, which produce an anoxic, carbon dioxide–rich atmosphere.

1. Describe the five types of O_2 relationships seen in microorganisms.

2. What are the toxic effects of O_2? How do aerobes and other oxygen-tolerant microbes protect themselves from these effects?

3. Describe four ways in which anaerobes may be cultured.

Pressure

Organisms that spend their lives on land or the surface of water are always subjected to a pressure of 1 atmosphere (atm) and are never affected significantly by pressure. It is thought that high hydrostatic pressure affects membrane fluidity and membrane-associated function. Yet many procaryotes live in the deep sea (ocean depths of 1,000 m or more) where the hydrostatic pressure can reach 600 to 1,100 atm and the temperature is about 2 to 3°C. Many of these procaryotes are **barotolerant:** increased pressure adversely affects them but not as much as it does nontolerant microbes. Some procaryotes are truly **barophilic**—they grow more rapidly at high pressures. A barophile recovered from the Mariana trench near the Philippines (depth about 10,500 m) grows only at pressures between about 400 to 500 atm when incubated at 2°C. Barophiles may play an important role in nutrient recycling in the deep sea. Thus far, they have been found among several bacterial genera (e.g., *Photobacterium, Shewanella, Colwellia*). Some archaea are thermobarophiles (e.g., *Pyrococcus* spp., *Methanocaldococcus jannaschii*). >> *Microorganisms in marine environments (section 26.1)*

Radiation

Our world is bombarded with electromagnetic radiation of various types (**figure 7.24**). Radiation behaves as if it were composed of waves moving through space like waves traveling on the surface of water. The distance between two wave crests or troughs is the wavelength. As the wavelength of electromagnetic radiation decreases, the energy of the radiation increases; gamma rays and X rays are much more energetic than visible light or infrared waves. Electromagnetic radiation also acts like a stream of energy packets called photons, each photon having a quantum of energy whose value depends on the wavelength of the radiation.

Sunlight is the major source of radiation on Earth. It includes visible light, ultraviolet (UV) radiation, infrared rays, and radio waves. Visible light is a most conspicuous and important aspect of our environment: most life depends on the ability of photosynthetic

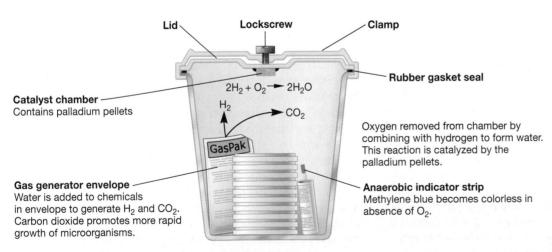

Lid   Lockscrew   Clamp

Rubber gasket seal

Catalyst chamber
Contains palladium pellets

$2H_2 + O_2 \rightarrow 2H_2O$

H_2

CO_2

GasPak

Oxygen removed from chamber by combining with hydrogen to form water. This reaction is catalyzed by the palladium pellets.

Gas generator envelope
Water is added to chemicals in envelope to generate H_2 and CO_2. Carbon dioxide promotes more rapid growth of microorganisms.

Anaerobic indicator strip
Methylene blue becomes colorless in absence of O_2.

Figure 7.23 The GasPak Anaerobic System. Hydrogen and carbon dioxide are generated by a GasPak envelope. The palladium catalyst in the chamber lid catalyzes the formation of water from hydrogen and oxygen, thereby removing oxygen from the sealed chamber.

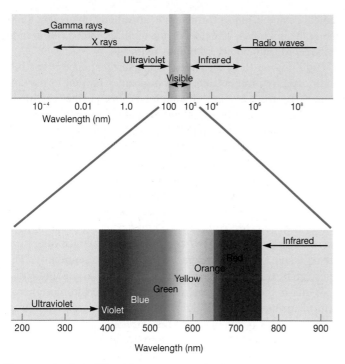

Figure 7.24 The Electromagnetic Spectrum. A portion of the spectrum is expanded at the bottom of the figure.

organisms to trap the light energy of the sun. Almost 60% of the sun's radiation is in the infrared region rather than the visible portion of the spectrum. Infrared is the major source of Earth's heat. At sea level, one finds very little ultraviolet radiation below about 290 to 300 nm. UV radiation of wavelengths shorter than 287 nm is absorbed by O_2 in Earth's atmosphere; this process forms a layer of ozone between 40 and 48 kilometers above Earth's surface. The ozone layer absorbs somewhat longer UV rays and reforms O_2. The even distribution of sunlight throughout the visible spectrum accounts for the fact that sunlight is generally "white." >> *Phototrophy (section 10.12)*

Many forms of electromagnetic radiation are very harmful to microorganisms. This is particularly true of **ionizing radiation,** radiation of very short wavelength and high energy, which can cause atoms to lose electrons (ionize). Two major forms of ionizing radiation are (1) X rays, which are artificially produced, and (2) gamma rays, which are emitted during radioisotope decay. Low levels of ionizing radiation may produce mutations and may indirectly result in death, whereas higher levels are directly lethal. Although microorganisms are more resistant to ionizing radiation than larger organisms, they are still destroyed by a sufficiently large dose. Ionizing radiation can be used to sterilize items. Some procaryotes (e.g., *Deinococcus radiodurans*) and bacterial endospores can survive large doses of ionizing radiation. >> *The use of physical methods in control: Radiation (section 8.4); Deinococcus-Thermus (section 19.2)*

Ionizing radiation causes a variety of changes in cells. It breaks hydrogen bonds, oxidizes double bonds, destroys ring

structures, and polymerizes some molecules. Oxygen enhances these destructive effects, probably through the generation of hydroxyl radicals (OH•). Although many types of constituents can be affected, the destruction of DNA is probably the most important cause of death.

Ultraviolet (UV) radiation can kill microorganisms due to its short wavelength (approximately from 10 to 400 nm) and high energy. The most lethal UV radiation has a wavelength of 260 nm, the wavelength most effectively absorbed by DNA. The primary mechanism of UV damage is the formation of thymine dimers in DNA. Two adjacent thymines in a DNA strand are covalently joined to inhibit DNA replication and function. The damage caused by UV light can be repaired by several DNA repair mechanisms, as discussed in chapter 14. Excessive exposure to UV light outstrips the organism's ability to repair the damage and death results. Longer wavelengths of UV light (near-UV radiation; 325 to 400 nm) can also harm microorganisms because they induce the breakdown of tryptophan to toxic photoproducts. It appears that these toxic tryptophan photoproducts plus the near-UV radiation itself produce breaks in DNA strands. The precise mechanism is not known, although it is different from that seen with 260 nm UV. >> *Mutations and their chemical basis (section 14.1)*

Even visible light, when present in sufficient intensity, can damage or kill microbial cells. Usually pigments called photosensitizers and O_2 are involved. Photosensitizers include pigments such as chlorophyll, bacteriochlorophyll, cytochromes, and flavins, which can absorb light energy and become excited or activated. The excited photosensitizer (P) transfers its energy to O_2, generating singlet oxygen ($^1O_2$).

$$P \xrightarrow{\text{light}} P \text{ (activated)}$$

$$P \text{ (activated)} + O_2 \rightarrow P + {}^1O_2$$

Singlet oxygen is a very reactive, powerful oxidizing agent that quickly destroys a cell.

Many microorganisms that are airborne or live on exposed surfaces use carotenoid pigments for protection against photooxidation. Carotenoids effectively quench singlet oxygen—that is, they absorb energy from singlet oxygen and convert it back into the unexcited ground state. Both photosynthetic and nonphotosynthetic microorganisms employ pigments in this way.

1. What are barotolerant and barophilic bacteria? Where would you expect to find them?

2. List the types of electromagnetic radiation in the order of decreasing energy or increasing wavelength.

3. What is the importance of ozone formation?

4. How do ionizing radiation, ultraviolet radiation, and visible light harm microorganisms? How do microorganisms protect themselves against damage from UV and visible light?

7.6 MICROBIAL GROWTH IN NATURAL ENVIRONMENTS

The microbial environment is complex and constantly changing. It often contains low nutrient concentrations (**oligotrophic environment**) and exposes microbes to many overlapping gradients of nutrients and other environmental factors. The growth of microorganisms depends on both the nutrient supply and their tolerance of the environmental conditions present in their habitat at any particular time. Inhibitory substances in the environment can also limit microbial growth. For instance, rapid, unlimited growth ensues if a microorganism is exposed to excess nutrients. Such growth quickly depletes nutrients and often results in the release of toxic products. Both nutrient depletion and the toxic products limit further growth.

Biofilms

Although ecologists observed as early as the 1940s that more microbes in aquatic environments were found attached to surfaces (sessile) than were free-floating (planktonic), only relatively recently has this fact gained the attention of microbiologists. These attached microbes are members of complex, slime-encased communities called **biofilms.** Biofilms are ubiquitous in nature, where they are most often seen as layers of slime on rocks or other objects in water (**figure 7.25a**). When they form on the hulls of boats and ships, they cause corrosion, which limits the life of the ships and results in economic losses. Of major concern is the formation of biofilms on medical devices such as hip and knee implants (figure 7.25b). These biofilms often cause serious illness and failure of the medical device. Biofilm formation is apparently an ancient ability among the microbes, as evidence for biofilms can be found in the fossil record from about 3.4 billion years ago.

Biofilms can form on virtually any surface, once it has been conditioned by proteins and other molecules present in the environment (**figure 7.26**). Initially microbes attach to the conditioned surface but can readily detach. Eventually they begin releasing polysaccharides, proteins, and DNA and these polymers allow the microbes to stick more stably to the surface. As the biofilm thickens and matures, the microbes reproduce and secrete additional polymers. The result is a complex, dynamic community of microorganisms. The microbes interact in a variety of ways. For instance, the waste products of one microbe may be the energy source for another microbe. The cells also communicate with each other, as described next. Finally, DNA present in the extracellular slime can be taken up by members of the biofilm community. Thus genes can be transferred from one cell (or species) to another.

While in the biofilm, microbes are protected from numerous harmful agents such as UV light, antibiotics, and other antimicrobial agents. This is due in part to the extracellular matrix in which they are embedded, but it also is due to physiological changes. Indeed, numerous proteins synthesized or activated in biofilm cells are not observed in planktonic cells and vice versa. The resistance of biofilm cells to antimicrobial agents has serious consequences. When biofilms form on a medical device such as a hip implant (figure 7.25b), they are difficult to kill and can cause serious illness. Often the only way to treat patients in this situation is by removing the implant. Another problem with biofilms is that cells are regularly sloughed off (figure 7.26). This can have many consequences. For instance, biofilms in a city's water distribution pipes can serve as a source of contamination after the water leaves a water treatment facility.

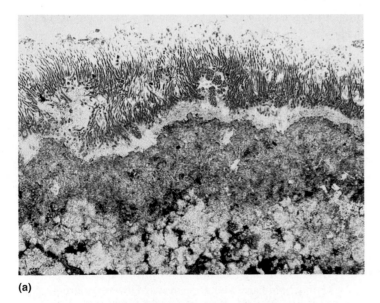

(a)

(b)

Figure 7.25 Examples of Biofilms. Biofilms form on almost any surface exposed to microorganisms. (a) Biofilm on the surface of a stromatolite in Walker Lake (Nevada, USA), an alkaline lake. The biofilm consists primarily of the cyanobacterium *Calothrix.* (b) Photograph taken during surgery to remove a biofilm-coated artificial joint. The white material is composed of pus, bacterial and fungal cells, and the patient's white blood cells.

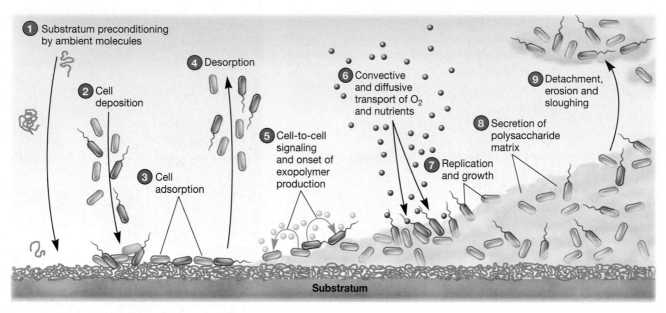

Figure 7.26 Biofilm Formation.

Cell-Cell Communication Within Microbial Populations

For decades, microbiologists tended to think of bacterial populations as collections of individual cells growing and behaving independently. But about 30 years ago, it was discovered that the marine luminescent bacterium *Vibrio fischeri* controls its ability to glow by producing a small, diffusible substance called autoinducer. The autoinducer molecule was later identified as an **N-acylhomoserine lactone (AHL).** It is now known that many gram-negative bacteria make AHL molecular signals that vary in length and substitution at the third position of the acyl side chain (**figure 7.27**). In many of these species, AHL is freely diffusible across the plasma membrane. Thus at a low cell density, it diffuses out of the cell. However, when the cell population increases and AHL accumulates outside the cell, the diffusion gradient is reversed so that the AHL enters the cell. Because the influx of AHL is cell density–dependent, it enables individual cells to assess population density. This is referred to as **quorum sensing**—a quorum usually refers to the minimum number of members in an organization, such as a legislative body, needed to conduct business. When AHL reaches a threshold level inside the cell, it induces the expression of target genes that regulate a number of functions, depending on the microbe. These functions are most effective only if a large number of microbes are present. For instance, the light produced by one cell is not visible, but cell

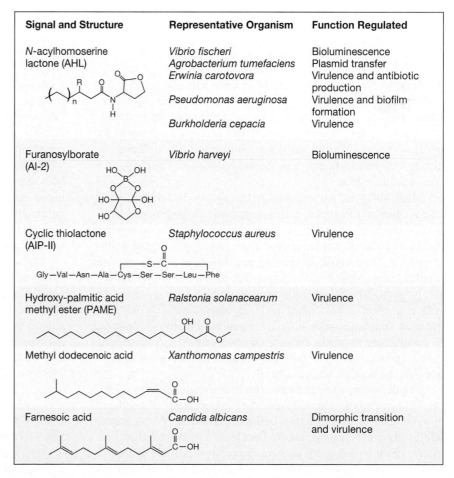

Signal and Structure	Representative Organism	Function Regulated
N-acylhomoserine lactone (AHL)	*Vibrio fischeri* *Agrobacterium tumefaciens* *Erwinia carotovora* *Pseudomonas aeruginosa* *Burkholderia cepacia*	Bioluminescence Plasmid transfer Virulence and antibiotic production Virulence and biofilm formation Virulence
Furanosylborate (AI-2)	*Vibrio harveyi*	Bioluminescence
Cyclic thiolactone (AIP-II)	*Staphylococcus aureus*	Virulence
Hydroxy-palmitic acid methyl ester (PAME)	*Ralstonia solanacearum*	Virulence
Methyl dodecenoic acid	*Xanthomonas campestris*	Virulence
Farnesoic acid	*Candida albicans*	Dimorphic transition and virulence

Figure 7.27 Representative Cell-Cell Communication Molecules.

(a) *E. scolopes*, the bobtail squid

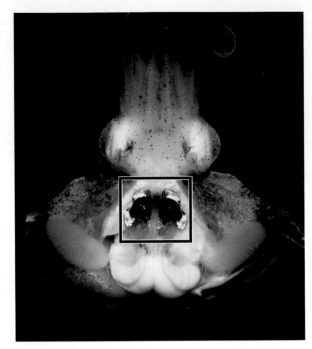

(b) Light organ

Figure 7.28   *Euprymna scolopes*.   (a) *E. scolopes* is a warm-water squid that remains buried in sand during the day and feeds at night. (b) When feeding it uses its light organ (boxed, located on its ventral surface) to provide camouflage by projecting light downward. Thus the outline of the squid appears as bright as the water's surface to potential predators looking up through the water column. The light organ is colonized by a large number of *Vibrio fischeri* so autoinducer accumulates to a threshold concentration, triggering light production.

densities within the light organ of marine fish and squid reach 10^{10} cells per milliliter. This provides the animal with a flashlight effect while the microbes have a safe and nutrient-enriched habitat (**figure 7.28**). In fact, many of the processes regulated by quorum sensing involve host-microbe interactions such as symbioses and pathogenicity. ▶▶ *Global regulatory systems: Quorum sensing (section 13.5)*

Many different bacteria use AHL signals. In addition to *V. fischeri* bioluminescence, the opportunistic pathogens *Burkholderia cepacia* and *Pseudomonas aeruginosa* use AHLs to regulate the expression of virulence factors (figure 7.27). These gram-negative bacteria cause debilitating pneumonia in people who are immunocompromised and are important pathogens in cystic fibrosis patients. The plant pathogens *Agrobacterium tumefaciens* will not infect a host plant and *Erwinia carotovora* will not produce antibiotics without AHL signaling. Finally, *B. cepacia*, *P. aeruginosa*, as well as *Vibrio cholerae* use AHL intercellular communication to control biofilm formation, an important strategy to evade the host's immune system.

The discovery of additional molecular signals made by a variety of microbes underscores the importance of cell-cell communication in regulating cellular processes. For instance, while only gram-negative bacteria are known to make AHLs, both gram-negative and gram-positive bacteria make autoinducer-2 (AI-2). Gram-positive bacteria also exchange short peptides called oligopeptides instead of autoinducer-like molecules.

Examples include *Enterococcus faecalis,* whose oligopeptide signal is used to determine the best time to conjugate (transfer genes). Oligopeptide communication by *Staphylococcus aureus* and *Bacillus subtilis* is used to trigger the uptake of DNA from the environment. The soil microbe *Streptomyces griseus* produces a γ-butyrolactone known as A-factor. This small molecule regulates both morphological differentiation and the production of the antibiotic streptomycin. Eucaryotic microbes also rely on cell-cell communication to coordinate key activities within a population. For example, the pathogenic fungus *Candida albicans* secretes farnesoic acid to govern morphology and virulence.

These examples of cell-cell communication demonstrate what might be called multicellular behavior in that many individual cells communicate and coordinate their activities to act as a unit. Other examples of such complex behavior are pattern formation in colonies and fruiting body formation in the myxobacteria. ◀◀ *Isolation of pure cultures: Microbial growth on agar surfaces (section 6.8);* ▶▶ *Class* Deltaproteobacteria: *Order* Myxococcales *(section 20.4)*

---

1. What is a biofilm? Why might life in a biofilm be advantageous for microbes?

2. What is quorum sensing? Describe how it occurs and briefly discuss its importance to microorganisms.

Summary

7.1 Bacterial Cell Cycle

a. Growth is an increase in cellular constituents and results in an increase in cell size, cell number, or both.

b. Most procaryotes reproduce by binary fission. During binary fission, the cell elongates, the chromosome is replicated, and then it segregates to opposite poles of the cell prior to the formation of a septum, which divides the cell into two progeny cells (**figures 7.1** and **7.2**).

c. Two overlapping pathways function during the procaryotic cell cycle: the pathway for chromosome replication and segregation and the pathway for septum formation (**figure 7.2**). Both are complex and poorly understood. The partitioning of the progeny chromosomes may involve homologues of eucaryotic cytoskeletal proteins (**figure 7.3**).

d. In rapidly dividing cells, initiation of DNA synthesis may occur before the previous round of synthesis is completed. This allows the cells to shorten the time needed for completing the cell cycle.

7.2 Growth Curve

a. When microorganisms are grown in a batch culture, the resulting growth curve usually has four phases: lag, exponential (log), stationary, and death (**figure 7.5**).

b. In the exponential phase, the population number of cells undergoing binary fission doubles at a constant interval called the generation or doubling time (**figure 7.9**). The mean growth rate constant (k) is the reciprocal of the generation time.

c. Exponential growth is balanced growth; that is, cell components are synthesized at constant rates relative to one another. Changes in culture conditions (e.g., in shift-up and shift-down experiments) lead to unbalanced growth. A portion of the available nutrients is used to supply maintenance energy.

7.3 Measurement of Microbial Growth

a. Microbial populations can be counted directly with counting chambers, electronic counters, or fluorescence microscopy. Viable counting techniques such as the spread plate, the pour plate, or the membrane filter can be employed (**figures 7.11, 7.12,** and **7.13**).

b. Population changes also can be followed by determining variations in microbial mass through the measurement of dry weight, turbidity, or the amount of a cell component (**figure 7.14**).

7.4 Continuous Culture of Microorganisms

a. Microorganisms can be grown in an open system in which nutrients are constantly provided and wastes removed.

b. A continuous culture system can maintain a microbial population in log phase. There are two types of these systems: chemostats and turbidostats (**figure 7.15**).

7.5 Influences of Environmental Factors on Growth

a. Most bacteria, photosynthetic protists, and fungi have rigid cell walls and are hypertonic to the habitat because of solutes such as amino acids, polyols, and potassium ions. The amount of water actually available to microorganisms is expressed in terms of the water activity (a_w).

b. Although most microorganisms do not grow well at water activities below 0.98 due to plasmolysis and associated effects, osmotolerant organisms survive and even flourish at low a_w values. Halophiles actually require high sodium chloride concentrations for growth (**figure 7.17** and **table 7.3**).

c. Each species of microorganism has an optimum pH for growth, and it can be classified as an acidophile, neutrophile, or alkalophile (**figure 7.18**).

d. Microorganisms have distinct temperature ranges for growth with minima, maxima, and optima—the cardinal temperatures. These ranges are determined by the effects of temperature on the rates of catalysis, protein denaturation, and membrane disruption (**figure 7.19**).

e. There are five major classes of microorganisms with respect to temperature preferences: (1) psychrophiles, (2) psychrotrophs (facultative psychrophiles), (3) mesophiles, (4) thermophiles, and (5) hyperthermophiles (**figure 7.20** and **table 7.3**).

f. Microorganisms can be placed into at least five different categories based on their response to the presence of O_2: obligate aerobes, microaerophiles, facultative anaerobes, aerotolerant anaerobes, and strict or obligate anaerobes (**figure 7.21** and **table 7.3**).

g. Oxygen can become toxic because of the production of hydrogen peroxide, superoxide radical, and hydroxyl radical. These are destroyed by the enzymes superoxide dismutase, catalase, and peroxidase. In some organisms, superoxide reductase and peroxidase are used instead.

h. Most deep-sea microorganisms are barotolerant, but some are barophilic and require high pressure for optimal growth.

i. High-energy or short-wavelength radiation harms organisms in several ways. Ionizing radiation—X rays and gamma rays—ionizes molecules and destroys DNA and other cell components. Ultraviolet (UV) radiation induces the formation of thymine dimers and strand breaks in DNA.

j. Visible light can provide energy for the formation of reactive singlet oxygen, which will destroy cells.

7.6 Microbial Growth in Natural Environments

a. Microbial growth in natural environments is profoundly affected by nutrient limitations and other adverse factors.

b. Many microbes form biofilms, aggregations of microbes growing on surfaces and held together by extracellular polysaccharides (**figure 7.26**). Life in a biofilm has several advantages, including protection from harmful agents.

c. Bacteria often communicate with one another in a density-dependent way and carry out a particular activity only when a certain population density is reached. This phenomenon is called quorum sensing (**figure 7.27**).

Critical Thinking Questions

1. As an alternative to diffusable signals, suggest another mechanism by which bacteria can quorum sense.

2. Design an enrichment culture medium and a protocol for the isolation and purification of a soil bacterium (e.g., *Bacillus subtilis*) from a sample of soil. Note possible contaminants and competitors. How will you adjust conditions of growth and what conditions will be adjusted to enhance preferentially the growth of the *Bacillus*?

3. Design an experiment to determine if a slow-growing microbial culture is exiting lag phase or is in exponential phase.

4. Why do you think the cardinal temperatures of some microbes change depending on other environmental conditions (e.g., pH)? Suggest one specific mechanism underlying such change.

5. Consider cell-cell communication: bacteria that "subvert" and "cheat" have been described. Describe a situation in which it would be advantageous for one species to subvert another, that is, degrade an intercellular signal made by another species. Also, describe a scenario whereby bacterial cheaters—defined as bacteria that do not make a molecular signal but profit by the uptake and processing of signal made by another microbe—might have a growth advantage.

Learn More

Learn more by visiting the Prescott website at www.mhhe.com/prescottprinciples, where you will find a complete list of references.

Control of Microorganisms

8

Chapter Glossary

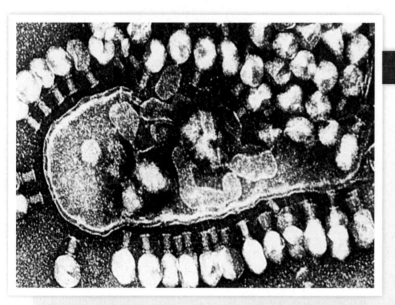

Release of T4 Bacteriophages by Lysis of Host Cell. The host cell has been lysed (upper right portion of the cell) and virions have been released into the surrounding area. Progeny virions also can be seen in the cytoplasm. In addition, empty capsids of the infecting phages coat the outside of the host (×36,500).

antisepsis   The prevention of infection or sepsis.

autoclave   An apparatus for sterilizing objects by the use of steam under pressure.

bactericide   An agent that kills bacteria.

bacteriostatic   Inhibiting the growth and reproduction of bacteria.

chemotherapy   The use of chemical agents to kill or inhibit microbial growth in tissues.

decimal reduction time (D value)   The time required to kill 90% of microorganisms or spores at a specific temperature.

depth filter   Fibrous or granular materials bound as a thick layer and used to retain microbial cells when contaminated liquids are passed through it.

detergent   An organic molecule, other than a soap, that serves as a wetting agent and emulsifier; it is normally used as cleanser, but some may be used as antimicrobial agents.

disinfection   The killing, inhibition, or removal of microorganisms that may cause disease. It usually refers to the treatment of inanimate objects with chemicals.

fungicide   An agent that kills fungi.

fungistatic   Inhibiting the growth and reproduction of fungi.

germicide   An agent that kills pathogens and many nonpathogens but not necessarily bacterial endospores.

high-efficiency particulate air (HEPA) filter   Depth filter constructed to remove 99.97% of particles that are 0.3 μm or larger.

iodophor   Antiseptic formed as a complex of iodine and an organic carrier.

ionizing radiation   Radiation of very short wavelength and high energy that causes atoms to lose electrons or ionize.

laminar flow biological safety cabinets   Cabinets that use HEPA filters to project a curtain of sterile air across its opening, preventing microbes from entering or exiting the cabinet.

pasteurization   The process of heating milk and other liquids to destroy microorganisms that can cause spoilage or disease.

sanitization   Reduction of the microbial population on an inanimate object to levels judged safe by public health standards.

sterilization   The process by which all living cells, viable spores, viruses, viroids, virisoids, and prions are either destroyed or removed from an object or habitat.

tyndallization   The process of repeated heating and incubation of liquids to destroy bacterial spores.

viricide   An agent that inactivates viruses so that they cannot reproduce within host cells.

We all labour against our own cure, for death is the cure of all diseases.

—Sir Thomas Browne

In this chapter, we address the subject of the control and destruction of microorganisms, a topic of immense practical importance. Although most microorganisms are beneficial, some microbial activities have undesirable consequences, such as food spoilage and disease. Therefore it is essential to be able to kill a wide variety of microorganisms or inhibit their growth to minimize their destructive effects. The goal is twofold: (1) to destroy pathogens and prevent their transmission, and (2) to reduce or eliminate microorganisms responsible for the contamination of water, food, and other substances. Thus this chapter focuses on the control of microorganisms by physical, chemical, and biological agents. Chemotherapeutic agents are discussed in chapter 31.

8.1 DEFINITIONS OF FREQUENTLY USED TERMS

Terminology is especially important when the control of microorganisms is discussed because words such as disinfectant and antiseptic often are used loosely. The situation is even more confusing because a particular treatment can either inhibit growth or kill, depending on the conditions. The types of control agents and their uses are outlined in **figure 8.1.**

Sterilization (Latin *sterilis,* unable to produce offspring or barren) is the process by which all living cells, spores, and acellular entities (e.g., viruses, viroids, and prions) are either destroyed or removed from an object or habitat. A sterile object is totally free of viable microorganisms, spores, and other infectious agents. When sterilization is achieved by a chemical agent, the chemical is called a sterilant. In contrast, **disinfection** is the killing, inhibition, or removal of microorganisms that may cause disease; disinfection is the substantial reduction of the total microbial population and the destruction of potential pathogens. **Disinfectants** are agents, usually chemical, used to carry out disinfection and normally used only on inanimate objects. A disinfectant does not necessarily sterilize an object because viable spores and a few microorganisms may remain. **Sanitization** is closely related to disinfection. In sanitization, the microbial population is reduced to levels that are considered safe by public health standards. The inanimate object is usually cleaned as well as partially disinfected. For example, sanitizers are used to clean eating utensils in restaurants. ◄◄ *Prions (section 5.7); Viroids and virusoids (section 5.6)*

It also is frequently necessary to control microorganisms on or in living tissue with chemical agents. **Antisepsis** (Greek *anti,* against, and *sepsis,* putrefaction) is the prevention of infection or sepsis and is accomplished with **antiseptics.** These are chemical agents applied to tissue to prevent infection by killing or inhibiting pathogen growth; they also reduce the total microbial population. Because they must not destroy too much host tissue, antiseptics are generally not as toxic as disinfectants. **Chemotherapy** is the use of chemical agents to kill or inhibit the growth of microorganisms within host tissue.

A suffix can be employed to denote the type of antimicrobial agent. Substances that kill organisms often have the suffix *–cide* (Latin *cida,* to kill); a **germicide** kills pathogens (and many nonpathogens) but not necessarily endospores. A disinfectant or antiseptic can be particularly effective against a specific group, in which case it may be called a **bactericide, fungicide,** or **viricide.** Other chemicals do not kill but rather prevent growth. If these agents are removed, growth will resume. Their names end in *–static* (Greek *statikos,* causing to stand or stopping)—for example, **bacteriostatic** and **fungistatic.**

1. Define the following terms: sterilization, sterilant, disinfection, disinfectant, sanitization, antisepsis, antiseptic, chemotherapy, germicide.
2. What is the difference between bactericidal and bacteriostatic? To which category do you think most household cleaners belong?

8.2 THE PATTERN OF MICROBIAL DEATH

A microbial population is not killed instantly when exposed to a lethal agent. Population death is generally exponential (logarithmic)—that is, the population will be reduced by the same fraction at constant intervals (**table 8.1**). If the logarithm of the population number remaining is plotted against the time of exposure of the microorganism to the agent, a straight-line plot will result (**figure 8.2**). When the population has been greatly reduced, the rate of killing may slow due to the survival of a more resistant strain of the microorganism. It is essential to have a precise measure of an agent's killing efficiency. One such measure is the **decimal reduction time** (D) or **D value.** The decimal reduction time is the time required to kill 90% of the microorganisms or spores in a sample under specified conditions. For example, in a semilogarithmic plot of the population remaining versus the time of heating, the D value is the time required for the line to drop by one log cycle or tenfold. The D value is usually written with a subscript to indicate the temperature for which it applies (figure 8.2).

To study the effectiveness of a lethal agent, one must be able to decide when microorganisms are dead, which may present some challenges. A microbial cell is often defined as dead if it does not grow and reproduce when inoculated into culture medium that would normally support its growth. In like manner, an inactive virus cannot infect a suitable host. This definition has flaws, however. It has been demonstrated that when bacteria are exposed to certain conditions, they can remain alive but are temporarily unable to reproduce. When in this state, they are referred to as viable but nonculturable (VBNC) *(see figure 7.7).* In conventional tests to demonstrate killing by an antimicrobial agent, VBNC bacteria would be thought to be dead. This is a serious problem because after a period of recovery, the bacteria may regain their ability to reproduce and cause infection. ◄◄ *The growth curve: Senescence and death (section 7.2)*

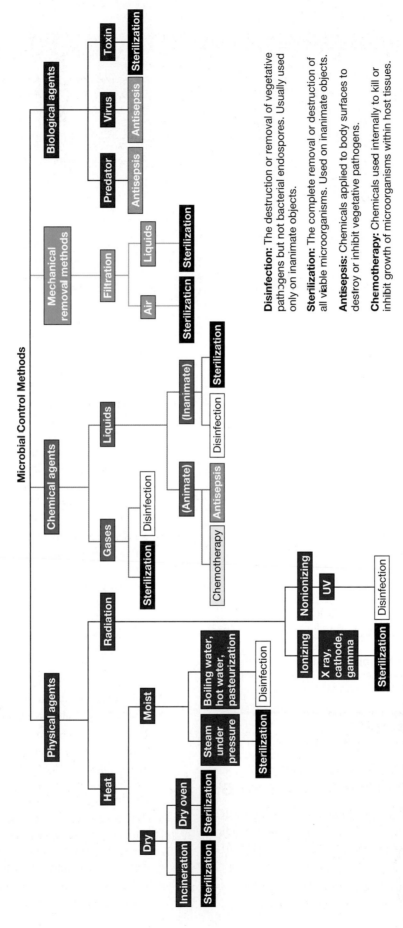

Figure 8.1 Microbial Control Methods.

Disinfection: The destruction or removal of vegetative pathogens but not bacterial endospores. Usually used only on inanimate objects.

Sterilization: The complete removal or destruction of all viable microorganisms. Used on inanimate objects.

Antisepsis: Chemicals applied to body surfaces to destroy or inhibit vegetative pathogens.

Chemotherapy: Chemicals used internally to kill or inhibit growth of microorganisms within host tissues.

Table 8.1		A Theoretical Microbial Heat-Killing Experiment		
Minute	Microbial Number at Start of Minute[a]	Microorganisms Killed in 1 Minute (90% of total)[a]	Microorganisms at End of 1 Minute	Log_{10} of Survivors
1	10^6	9×10^5	10^5	5
2	10^5	9×10^4	10^4	4
3	10^4	9×10^3	10^3	3
4	10^3	9×10^2	10^2	2
5	10^2	9×10^1	10	1
6	10^1	9	1	0
7	1	0.9	0.1	−1

[a]Assume that the initial sample contains 10^6 vegetative microorganisms per ml and that 90% of the organisms are killed during each minute of exposure. The temperature is 121°C.

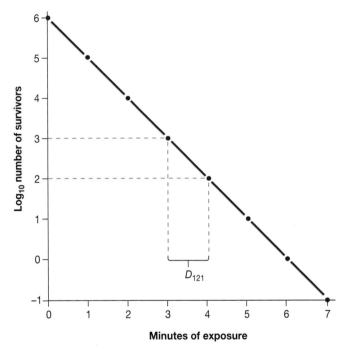

Figure 8.2   The Pattern of Microbial Death.   An exponential plot of the survivors versus the minutes of exposure to heating at 121°C. In this example, the D_{121} value is 1 minute. The data are from table 8.1.

8.3   CONDITIONS INFLUENCING THE EFFECTIVENESS OF ANTIMICROBIAL AGENTS

Destruction of microorganisms and inhibition of microbial growth are not simple matters because the efficiency of an **antimicrobial agent** (an agent that kills microorganisms or inhibits their growth) is affected by at least six factors.

1. **Population size.** Because an equal fraction of a microbial population is killed during each interval, a larger

population requires a longer time to die than a smaller one (table 8.1 and figure 8.2).

2. **Population composition.** The effectiveness of an agent varies greatly with the nature of the organisms being treated because microorganisms differ markedly in susceptibility. Bacterial spores are much more resistant to most antimicrobial agents than are vegetative forms, and younger cells are usually more readily destroyed than mature organisms. Some species are able to withstand adverse conditions better than others. For instance, *Mycobacterium tuberculosis,* which causes tuberculosis, is much more resistant to antimicrobial agents than most other bacteria.

3. **Concentration or intensity of an antimicrobial agent.** Often, but not always, the more concentrated a chemical agent or intense a physical agent, the more rapidly microorganisms are destroyed. However, agent effectiveness usually is not directly related to concentration or intensity. Over a short range, a small increase in concentration leads to an exponential rise in effectiveness; beyond a certain point, increases may not raise the killing rate much at all. Sometimes an agent is more effective at lower concentrations. For example, 70% ethanol is more bacteriocidal than 95% ethanol because the activity of ethanol is enhanced by the presence of water.

4. **Duration of exposure.** The longer a population is exposed to a microbicidal agent, the more organisms are killed (figure 8.2). To achieve sterilization, exposure should be long enough to reduce the probability of survival to 10^{-6} or less.

5. **Temperature.** An increase in the temperature at which a chemical acts often enhances its activity. Frequently a lower concentration of disinfectant or sterilizing agent can be used at a higher temperature.

6. **Local environment.** The population to be controlled is not isolated but surrounded by environmental factors that

may either offer protection or aid in its destruction. For example, because heat kills more readily at an acidic pH, acidic foods and beverages such as fruits and tomatoes are easier to pasteurize than more alkaline foods such as milk. A second important environmental factor is organic matter, which can protect microorganisms against physical and chemical disinfecting agents. Biofilms are a good example. The organic matter in a biofilm protects the biofilm's microorganisms. Furthermore, it has been clearly documented that bacteria in biofilms are altered physiologically, and this makes them less susceptible to many antimicrobial agents. Because of the impact of organic matter, it may be necessary to clean objects, especially medical and dental equipment, before they are disinfected or sterilized. << *Microbial growth in natural environments: Biofilms (section 7.6)*

1. Briefly explain how the effectiveness of antimicrobial agents varies with population size, population composition, concentration or intensity of the agent, treatment duration, temperature, and local environmental conditions.

2. How does being in a biofilm affect an organism's susceptibility to antimicrobial agents?

3. Suppose hospital custodians have been assigned the task of cleaning all showerheads in patient rooms to prevent the spread of infectious disease. What two factors would have the greatest impact on the effectiveness of the disinfectant the custodians use? Explain what that impact would be.

8.4 THE USE OF PHYSICAL METHODS IN CONTROL

Heat and other physical agents are normally used to control microbial growth and sterilize objects, as can be seen from the continual operation of the autoclave in every microbiology laboratory. The most frequently employed physical agents are heat, filtration, and radiation.

Heat

Moist heat readily destroys viruses, bacteria, and fungi (**table 8.2**). Moist heat kills by degrading nucleic acids and denaturing enzymes and other essential proteins. It also disrupts cell membranes. Exposure to boiling water for 10 minutes is sufficient to destroy vegetative cells and eucaryotic spores. Unfortunately the temperature of boiling water (100°C or 212°F at sea level) is not sufficient to destroy bacterial spores, which may survive hours of boiling. Therefore boiling can be used for disinfection of drinking water and objects not harmed by water, but boiling does not sterilize.

Table 8.2	Approximate Conditions for Moist Heat Killing	
Organism	**Vegetative Cells**	**Spores**
Yeasts	5 minutes at 50–60°C	5 minutes at 70–80°C
Molds	30 minutes at 62°C	30 minutes at 80°C
Bacteria[a]	10 minutes at 60–70°C	2 to over 800 minutes at 100°C 0.5–12 minutes at 121°C
Viruses	30 minutes at 60°C	

[a]Conditions for mesophilic bacteria.

To destroy bacterial spores, moist heat sterilization must be carried out at temperatures above 100°C, and this requires the use of saturated steam under pressure. Steam sterilization is carried out with an **autoclave** (**figure 8.3**), a device somewhat like a fancy pressure cooker. The development of the autoclave by Chamberland in 1884 tremendously stimulated the growth of microbiology. Water is boiled to produce steam, which is released into the autoclave's chamber (figure 8.3b). The air initially present in the chamber is forced out until the chamber is filled with saturated steam and the outlets are closed. Hot, saturated steam continues to enter until the chamber reaches the desired temperature and pressure, usually 121°C and 15 pounds of pressure. At this temperature saturated steam destroys all vegetative cells and spores in a small volume of liquid within 10 to 12 minutes. Treatment is continued for at least 15 minutes to provide a margin of safety. Of course larger containers of liquid such as flasks and carboys require much longer treatment times.

Autoclaving must be carried out properly or the processed materials will not be sterile. If all air has not been flushed out of the chamber, it will not reach 121°C, even though it may reach a pressure of 15 pounds. The chamber should not be packed too tightly because the steam needs to circulate freely and contact everything in the autoclave. Bacterial spores will be killed only if they are kept at 121°C for 10 to 12 minutes. When a large volume of liquid must be sterilized, an extended sterilization time is needed because it takes longer for the center of the liquid to reach 121°C; 5 liters of liquid may require about 70 minutes. In view of these potential difficulties, a biological indicator is often autoclaved along with other material. This indicator commonly consists of a culture tube containing a sterile ampule of medium and a paper strip covered with spores of *Geobacillus stearothermophilus*. After autoclaving, the ampule is aseptically broken and the culture incubated for several days. If the test bacterium does not grow in the medium, the sterilization run has been successful. Sometimes either special indicator tape or paper that changes color upon sufficient heating is autoclaved with a load of material. These approaches are convenient and save time but are not as reliable as the killing of bacterial spores.

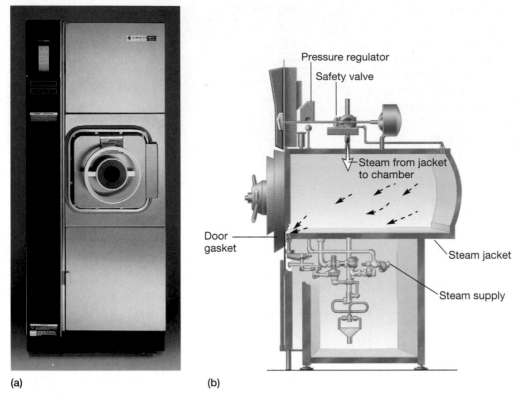

(a)   **(b)**

Pressure regulator

Safety valve

Steam from jacket to chamber

Door gasket

Steam jacket

Steam supply

Figure 8.3 **The Autoclave or Steam Sterilizer.** (a) A modern, automatically controlled autoclave or sterilizer. (b) Longitudinal cross section of a typical autoclave showing some of its parts and the pathway of steam. *From John J. Perkins,* Principles and Methods of Sterilization in Health Science, *2d ed., 1969. Courtesy of Charles C. Thomas, Publisher, Springfield, Illinois.*

However, steam exposure is repeated for a total of three times with 23 to 24 hour incubations between steam exposures. The incubations permit remaining spores to germinate into heat-sensitive vegetative cells that are then destroyed upon subsequent steam exposures.

Many objects are best sterilized in the absence of water by dry heat sterilization. Some items are sterilized by incineration. For instance, inoculating loops, which are used routinely in the laboratory, can be sterilized in a small, bench-top incinerator (**figure 8.4**). Other items are sterilized in an oven at 160 to 170°C for 2 to 3 hours. Microbial death results from the oxidation of cell constituents and denaturation of proteins. Dry air heat is less effective than moist heat. The spores of *Clostridium botulinum,* the cause of botulism, are killed in 5 minutes at 121°C by moist heat but only after 2 hours at 160°C with dry heat. However, dry heat has some definite advantages. It does not corrode glassware and metal instruments as moist heat does, and it can be used to sterilize powders, oils, and similar items. Despite these advantages, dry heat sterilization is slow and not suitable for heat-sensitive materials such as many plastic and rubber items.

Many heat-sensitive substances, such as milk, are treated with controlled heating at temperatures well below boiling, a process known as **pasteurization** in honor of its developer, Louis Pasteur. In the 1860s the French wine industry was plagued by the problem of wine spoilage, which made wine storage and shipping difficult. Pasteur examined spoiled wine under the microscope and detected microorganisms that looked like the bacteria responsible for lactic acid and acetic acid fermentations. He then discovered that a brief heating at 55 to 60°C would destroy these microorganisms and preserve wine for long periods. In 1886 the German chemists V. H. and F. Soxhlet adapted the technique for preserving milk and reducing milk-transmissible diseases. Milk pasteurization was introduced into the United States in 1889. Milk, beer, and many other beverages are now pasteurized. Pasteurization does not sterilize a beverage, but it does kill any pathogens present and drastically slows spoilage by reducing the level of nonpathogenic spoilage microorganisms.

Some materials cannot withstand the high temperature of the autoclave, and spore contamination precludes the use of other methods to sterilize them. For these materials, a process of intermittent sterilization, also known as **tyndallization** (for John Tyndall, the British physicist who used the technique to destroy heat-resistant microorganisms in dust) is used. The process also uses steam (30–60 minutes) to destroy vegetative bacteria.

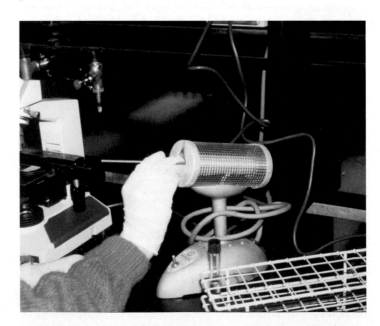

Figure 8.4 **Dry Heat Incineration.** Bench-top incinerators are routinely used to sterilize inoculating loops used in microbiology laboratories.

1. Describe how an autoclave works. What conditions are required for sterilization by moist heat? What three things must one do when operating an autoclave to help ensure success?

2. In the past, spoiled milk was responsible for a significant proportion of infant deaths. Why is untreated milk easily spoiled?

Filtration

Filtration is an excellent way to reduce the microbial population in solutions of heat-sensitive material, and sometimes it can be used to sterilize solutions. Rather than directly destroying contaminating microorganisms, the filter simply removes them. There are two types of filters. **Depth filters** consist of fibrous or granular materials that have been bonded into a thick layer filled with twisting channels of small diameter. The solution containing microorganisms is sucked through this layer under vacuum, and microbial cells are removed by physical screening or entrapment and by adsorption to the surface of the filter material. Depth filters are made of diatomaceous earth (Berkefeld filters), unglazed porcelain (Chamberlain filters), asbestos, or other similar materials.

Membrane filters have replaced depth filters for many purposes. These circular filters are porous membranes, a little over 0.1 mm thick, made of cellulose acetate, cellulose nitrate, polycarbonate, polyvinylidene fluoride, or other synthetic materials. Although a wide variety of pore sizes are available, membranes with pores about 0.2 μm in diameter are used to remove most vegetative cells, but not viruses, from solutions ranging in volume from less than 1 ml to many liters. The membranes are held in special holders (**figure 8.5**) and are often preceded by depth filters made of glass fibers to remove larger particles that might clog the membrane filter. The solution is pulled or forced through the filter with a vacuum or with pressure from a syringe, peristaltic pump, or nitrogen gas, and collected in previously sterilized containers. Membrane filters remove microorganisms by screening them out much as a sieve separates large sand particles from small ones (**figure 8.6**). These filters are used to sterilize pharmaceuticals, ophthalmic solutions, culture media, oils, antibiotics, and other heat-sensitive solutions.

Air also can be sterilized by filtration. Two common examples are N-95 disposable masks used in hospitals and labs, and cotton plugs on culture vessels that let air in but keep microorganisms out. N-95 masks exclude 95% of particles that are larger than 0.3 μm. Other important examples are **laminar flow biological safety cabinets,** which employ **high-efficiency particulate air (HEPA) filters** (a type of depth filter) to remove 99.97% of particles 0.3 μm or larger. Laminar flow biological safety cabinets or hoods force air through HEPA filters, then project a vertical curtain of sterile air across the cabinet opening. This protects a worker from microorganisms being handled within the cabinet and prevents contamination of the room (**figure 8.7**). A person uses these cabinets when working with dangerous agents such as *M. tuberculosis* and tumor viruses. They are also employed in

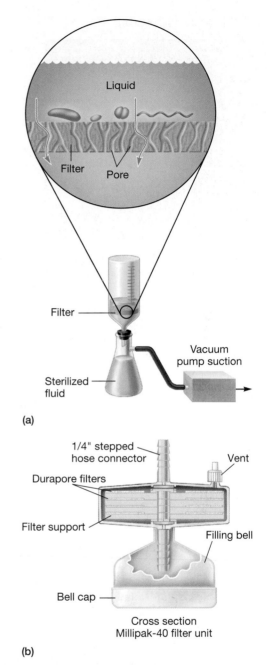

(a)

(b)

Figure 8.5 Membrane Filter Sterilization. The liquid to be sterilized is pumped through a membrane filter and into a sterile container. (a) Schematic representation of a membrane filtration setup that uses a vacuum pump to force liquid through the filter. The inset shows a cross section of the filter and its pores, which are too small for microbes to pass through. (b) Cross section of a membrane filtration unit. Several membranes are used to increase its capacity.

research labs and industries, such as the pharmaceutical industry, when a sterile working environment is needed.

Radiation

Ultraviolet (UV) radiation around 260 nm (*see figure 7.24*) is quite lethal. UV radiation causes thiamine-thiamine dimerization

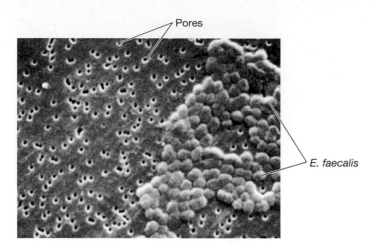

Figure 8.6   Membrane Filter. *Enterococcus faecalis* resting on a polycarbonate membrane filter with 0.4 μm pores (×5,900).

of the microbial DNA, preventing polymerase-mediated replication and transcription. However, UV does not penetrate glass, dirt films, water, and other substances very effectively. Because of this disadvantage, UV radiation is used as a sterilizing agent only in a few specific situations. UV lamps are sometimes placed on the ceilings of rooms or in biological safety cabinets to sterilize the air and any exposed surfaces. Because UV radiation burns the skin and damages eyes, people working in such areas must be certain the UV lamps are off when the areas are in use. Commercial UV units are available for water treatment. Pathogens and other microorganisms are destroyed when a thin layer of water is passed under the lamps.

Ionizing radiation is an excellent sterilizing agent and penetrates deep into objects. It will destroy bacterial spores and vegetative cells, both procaryotic and eucaryotic; however, ionizing radiation is not always effective against viruses. Gamma radiation from a cobalt 60 source and accelerated electrons from high-voltage electricity are used in the cold sterilization of antibiotics, hormones, sutures, and plastic disposable supplies such as syringes. Gamma radiation and electron beams have also been used to sterilize and "pasteurize" meat and other foods (**figure 8.8**). Irradiation can eliminate the threat of such pathogens as *Escherichia coli* O157:H7, *Staphylococcus aureus,* and *Campylobacter jejuni.* Based on the results of numerous studies, both the U.S. Food and Drug Administration and the World Health Organization have approved irradiated food and declared it safe for human consumption. Currently irradiation is being used to treat poultry, beef, pork, veal, lamb, fruits, vegetables, and spices. >> *Controlling food spoilage (section 34.3)*

---

1. What are depth filters and membrane filters, and how are they used to sterilize liquids? Describe the operation of a biological safety cabinet.
2. Give the advantages and disadvantages of ultraviolet light and ionizing radiation as sterilizing agents. Provide a few examples of how each is used for this purpose.

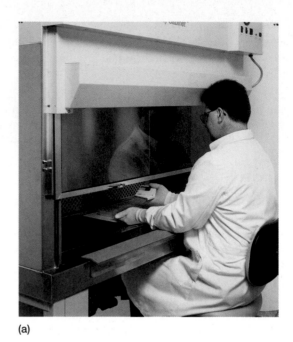

(a)

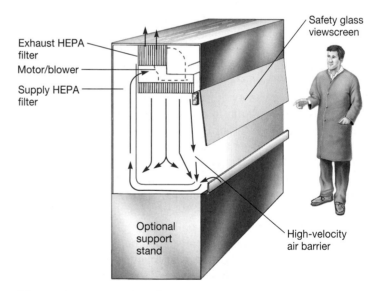

(b)

Figure 8.7   A Laminar Flow Biological Safety Cabinet.
(a) A technician pipetting potentially hazardous material in a safety cabinet.
(b) A schematic diagram showing the airflow pattern.

8.5   THE USE OF CHEMICAL AGENTS IN CONTROL

Physical agents are generally used to sterilize objects. Chemicals, on the other hand, are more often employed in disinfection and antisepsis. The proper use of chemical agents is essential to laboratory and hospital safety (**Techniques & Applications 8.1**). Chemicals also are employed to prevent microbial growth in

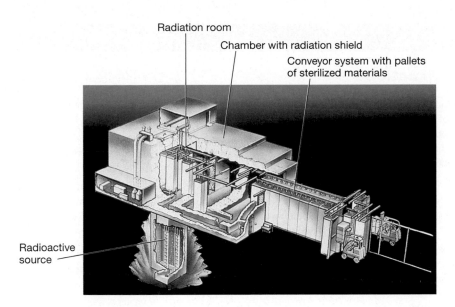

Radiation room

Chamber with radiation shield

Conveyor system with pallets of sterilized materials

Radioactive source

Figure 8.8 Sterilization with Ionizing Radiation. An irradiation machine that uses radioactive cobalt 60 as a gamma radiation source to sterilize fruits, vegetables, meats, fish, and spices.

food, and certain chemicals are used to treat infectious disease. The use of chemical agents for chemotherapy in humans is covered in chapter 31. Next we discuss chemicals used outside the body.

Many different chemicals are available for use as disinfectants, and each has its own advantages and disadvantages. Ideally the disinfectant must be effective against a wide variety of infectious agents (gram-positive and gram-negative bacteria, acid-fast bacteria, bacterial spores, fungi, and viruses) at low concentrations and in the presence of organic matter. Although the chemical must be toxic for infectious agents, it should not be toxic to people or corrosive for common materials. In practice, this balance between effectiveness and low toxicity for animals is hard to achieve. Some chemicals are used despite their low effectiveness because they are relatively nontoxic. The ideal disinfectant should be stable upon storage, odorless or with a pleasant odor, soluble in water and lipids for penetration into microorganisms, have a low surface tension so that it can enter cracks in surfaces, and be relatively inexpensive.

One potentially serious problem is the overuse of antiseptics. For instance, the antibacterial agent triclosan is found in products such as deodorants, mouthwashes, soaps, cutting boards, and baby toys. Unfortunately, the emergence of triclosan-resistant bacteria has become a problem. For example, *Pseudomonas aeruginosa* actively pumps the antiseptic out of the cell. There is now evidence that extensive use of triclosan also increases the frequency of bacterial resistance to antibiotics. Thus overuse of antiseptics can have unintended harmful consequences. >> *Drug resistance (section 31.6)*

The properties and uses of several groups of common disinfectants and antiseptics are surveyed next. Many of the characteristics of disinfectants and antiseptics are summarized

in **tables 8.3** and **8.4**. Structures of some common agents are shown in **figure 8.9**.

Phenolics

Phenol was the first widely used antiseptic and disinfectant. In 1867 Joseph Lister employed it to reduce the risk of infection during surgery. Today phenol and phenolics (phenol derivatives) such as cresols, xylenols, and orthophenylphenol are used as disinfectants in laboratories and hospitals. The commercial disinfectant Lysol is made of a mixture of phenolics. Phenolics act by denaturing proteins and disrupting cell membranes. They have some real advantages as disinfectants: phenolics are tuberculocidal, effective in the presence of organic material, and remain active on surfaces long after application. However, they have a disagreeable odor and can cause skin irritation.

Alcohols

Alcohols are among the most widely used disinfectants and antiseptics. They are bactericidal and fungicidal but not sporicidal; some lipid-containing viruses are also destroyed. The two most popular alcohol germicides are ethanol and isopropanol, usually used in about 70 to 80% concentration. They act by denaturing proteins and possibly by dissolving membrane lipids. A 10 to 15 minute soaking is sufficient to disinfect small instruments.

Halogens

A halogen is any of the five elements (fluorine, chlorine, bromine, iodine, and astatine) in group VIIA of the periodic table. They exist as diatomic molecules in the free state and form saltlike compounds with sodium and most other metals. The halogens iodine and chlorine are important antimicrobial agents. Iodine is used as a skin antiseptic and kills by oxidizing cell constituents and iodinating cell proteins. At higher concentrations, it may even kill some spores. Iodine often has been applied as tincture of iodine, 2% or more iodine in a water-ethanol solution of potassium iodide. Although it is an effective antiseptic, the skin may be damaged, a stain is left, and iodine allergies can result. Iodine has been complexed with an organic carrier to form an **iodophor.** Iodophors are water soluble, stable, and nonstaining, and release iodine slowly to minimize skin burns and irritation. They are used in hospitals for cleansing preoperative skin and in hospitals and laboratories for disinfecting. Some popular brands are Wescodyne for skin and laboratory disinfection and Betadine for wounds.

Chlorine is the usual disinfectant for municipal water supplies and swimming pools and is also employed in the dairy and food industries. It may be applied as chlorine gas (Cl_2), sodium hypochlorite (bleach, NaClO), or calcium hypochlorite [$Ca(OCl)_2$], all of which yield hypochlorous acid (HClO)

Techniques and Applications

8.1   Standard Microbiological Practices

The identification of potentially fatal, blood-borne infectious agents (HIV, hepatitis B virus, and others) spurred the codification of standard microbiological practices to limit exposure to such agents. These standard microbiological practices are *minimum* guidelines that should be supplemented with other precautions based on the potential exposure risks and biosafety level regulations for the lab. Briefly:

1. Eating, drinking, manipulation of contact lenses, and the use of cosmetics, gum, and tobacco products are strictly prohibited in the lab.

2. Hair longer than shoulder length should be tied back. Hands should be kept away from face at all times. Items (e.g., pencils) should not be placed in the mouth while in the lab. Protective clothing (lab coat, smock, etc.) is recommended while in the lab. Exposed wounds should be covered and protected.

3. Lab personnel should know how to use the emergency eyewash and/or shower stations.

4. Work space should be disinfected at the beginning and completion of lab time. Hands should be washed thoroughly after any exposure and before leaving the lab.

5. Precautions should be taken to prevent injuries caused by sharp objects (needles, scalpels, etc.). Sharp instruments should be discarded for disposal in specially marked containers.

Recommended guidelines for additional precautions should reflect the laboratory's biosafety level (BSL). The following table defines the BSL for the four categories of biological agents and suggested practices.

BSL	Agents	Practices
1	Not known to consistently cause disease in healthy adults (e.g., *Lactobacillus casei, Vibrio fischeri*)	Standard Microbiological Practices
2	Associated with human disease, potential hazard if percutaneous injury, ingestion, mucous membrane exposure occurs (e.g., *Salmonella typhi, E. coli* O157:H7, *Staphylococcus aureus*)	BSL-1 practice plus: • Limited access • Biohazard warning signs • "Sharps" precautions • Biosafety manual defining any needed waste decontamination or medical surveillance policies
3	Indigenous or exotic agents with potential for aerosol transmission; disease may have serious or lethal consequences (e.g., *Coxiella burnetti, Yersinia pestis,* herpes viruses)	BSL-2 practice plus: • Controlled access • Decontamination of all waste • Decontamination of lab clothing before laundering • Baseline serum values determined in workers using BSL-3 agents
4	Dangerous/exotic agents that pose high risk of life-threatening disease; aerosol-transmitted lab infections; or related agents with unknown risk of transmission (e.g., *Variola major* (smallpox virus), hemorrhagic fever viruses)	BSL-3 practices plus: • Clothing change before entering • Shower on exit • All material decontaminated on exit from facility

(see chemical reactions below). The result is oxidation of cellular materials and destruction of vegetative bacteria and fungi.

$$Cl_2 + H_2O \rightarrow HCl + HClO$$

$$NaOCl + H_2O \rightarrow NaOH + HClO$$

$$Ca(OCl)_2 + 2H_2O \rightarrow Ca(OH)_2 + 2HClO$$

Death of almost all microorganisms usually occurs within 30 minutes. One potential problem is that chlorine reacts with organic compounds to form carcinogenic trihalomethanes, which must be monitored in drinking water. Ozone has been successfully as an alternative to chlorination in Europe and Canada.

Chlorine is also an excellent disinfectant for individual use because it is effective, inexpensive, and easy to employ. Small quantities of drinking water can be disinfected with halazone tablets. Halazone (parasulfone dichloramidobenzoic acid) slowly releases chloride when added to water and disinfects it in about a half hour. It is frequently used by campers lacking access to uncontaminated drinking water.

Heavy Metals

For many years the ions of heavy metals such as mercury, silver, arsenic, zinc, and copper were used as germicides. These have now been superseded by other less toxic and more effective

Table 8.3	Activity Levels of Selected Germicides	
Class	Use Concentration of Active Ingredient	Activity Level[a]
Gas		
Ethylene oxide	450–500 mg/l[b]	High
Liquid		
Glutaraldehyde, aqueous	2%	High to intermediate
Formaldehyde + alcohol	8 + 70%	High
Stabilized hydrogen peroxide	6–30%	High to intermediate
Formaldehyde, aqueous	6–8%	High to intermediate
Iodophors	750–5,000 mg/l[c]	High to intermediate
Iodophors	75–150 mg/l[c]	Intermediate to low
Iodine + alcohol	0.5 + 70%	Intermediate
Chlorine compounds	0.1–0.5%[d]	Intermediate
Phenolic compounds, aqueous	0.5–3%	Intermediate to low
Iodine, aqueous	1%	Intermediate
Alcohols (ethyl, isopropyl)	70%	Intermediate
Quaternary ammonium compounds	0.1–0.2% aqueous	Low
Chlorhexidine	0.75–4%	Low
Hexachlorophene	1–3%	Low
Mercurial compounds	0.1–0.2%	Low

Source: From Seymour S. Block, *Disinfection, Sterilization and Preservation.* Copyright © 1983 Lea & Febiger, Malvern, Pa. Reprinted by permission.
[a]High-level disinfectants destroy vegetative bacterial cells including *M. tuberculosis*, bacterial endospores, fungi, and viruses. Intermediate-level disinfectants destroy all of the above except spores. Low-level agents kill bacterial vegetative cells except for *M. tuberculosis*, fungi, and medium-sized lipid-containing viruses (but not bacterial endospores or small, nonlipid viruses).
[b]In autoclave-type equipment at 55 to 60°C.
[c]Available iodine.
[d]Free chlorine.

germicides (many heavy metals are more bacteriostatic than bactericidal). There are a few exceptions. In some hospitals, a 1% solution of silver nitrate is added to the eyes of infants to prevent ophthalmic gonorrhea. Silver sulfadiazine is used on burns. Copper sulfate is an effective algicide in lakes and swimming pools. Heavy metals combine with proteins, often with their sulfhydryl groups, and inactivate them. They may also precipitate cell proteins.

Quaternary Ammonium Compounds

Quaternary ammonium compounds are detergents that have antimicrobial activity and are effective disinfectants. **Detergents** (Latin *detergere,* to wipe away) are organic cleansing agents that are amphipathic, having both polar hydrophilic and nonpolar hydrophobic components. The hydrophilic portion of a quaternary ammonium compound is a positively charged quaternary nitrogen; thus quaternary ammonium compounds are cationic detergents. Their antimicrobial activity is the result of their ability to disrupt microbial membranes; they may also denature proteins.

Cationic detergents such as benzalkonium chloride and cetylpyridinium chloride kill most bacteria but not *M. tuberculosis* or spores. They have the advantages of being stable and nontoxic, but they are inactivated by hard water and soap. Cationic detergents are often used as disinfectants for food utensils and small instruments and as skin antiseptics.

Aldehydes

Both of the commonly used aldehydes, formaldehyde and glutaraldehyde (figure 8.9), are highly reactive molecules that combine with nucleic acids and proteins, and inactivate them, probably by cross-linking and alkylating molecules (**figure 8.10**). They are sporicidal and can be used as chemical sterilants. Formaldehyde is usually dissolved in water or alcohol before use. A 2% buffered

Table 8.4	Relative Efficacy of Commonly Used Disinfectants and Antiseptics		
Class	Disinfectant	Antiseptic	Comment
Gas			
Ethylene oxide	3–4[a]	0[a]	Sporicidal; toxic; good penetration; requires relative humidity of 30% or more; microbicidal activity varies with apparatus used; absorbed by porous material; dry spores highly resistant; moisture must be present, and presoaking is most desirable
Liquid			
Glutaraldehyde, aqueous	3	0	Sporicidal; active solution unstable; toxic
Stabilized hydrogen peroxide	3	0	Sporicidal; solution stable up to 6 weeks; toxic orally and to eyes; mildly skin toxic; little inactivation by organic matter
Formaldehyde + alcohol	3	0	Sporicidal; noxious fumes; toxic; volatile
Formaldehyde, aqueous	1–2	0	Sporicidal; noxious fumes; toxic
Phenolic compounds	3	0	Stable; corrosive; little inactivation by organic matter; irritates skin
Chlorine compounds	1–2	0	Fast action; inactivation by organic matter; corrosive; irritates skin
Alcohol	1	3	Rapidly microbicidal except for bacterial spores and some viruses; volatile; flammable; dries and irritates skin
Iodine + alcohol	0	4	Corrosive; very rapidly microbicidal; causes staining; irritates skin; flammable
Iodophors	1–2	3	Somewhat unstable; relatively bland; staining temporary; corrosive
Iodine, aqueous	0	2	Rapidly microbicidal; corrosive; stains fabrics; stains and irritates skin
Quaternary ammonium compounds	1	0	Inactivated by soap and anionics; compounds absorbed by fabrics; old or dilute solution can support growth of gram-negative bacteria
Hexachlorophene	0	2	Insoluble in water, soluble in alcohol; not inactivated by soap; weakly bactericidal
Chlorhexidine	0	3	Soluble in water and alcohol; weakly bactericidal
Mercurial compounds	0	±	Greatly inactivated by organic matter; weakly bactericidal

Source: From Seymour S. Block, *Disinfection, Sterilization and Preservation*. Copyright © 1983 Lea & Febiger, Malvern, Pa. Reprinted by permission.
[a]Subjective ratings of practical usefulness in a hospital environment—4 is maximal usefulness; 0 is little or no usefulness; ± signifies that the substance is sometimes useful but not always.

solution of glutaraldehyde is an effective disinfectant. It is less irritating than formaldehyde and is used to disinfect hospital and laboratory equipment. Glutaraldehyde usually disinfects objects within about 10 minutes but may require as long as 12 hours to destroy all spores.

Sterilizing Gases

Many heat-sensitive items such as disposable plastic petri dishes and syringes, heart-lung machine components, sutures, and catheters are sterilized with ethylene oxide gas (figure 8.9). Ethylene oxide (EtO) is both microbicidal and sporicidal. It is a very strong alkylating agent that kills by reacting with functional groups of DNA and proteins to block replication and enzymatic activity. It

is a particularly effective sterilizing agent because it rapidly penetrates packing materials, even plastic wraps.

Sterilization is carried out in a special ethylene oxide sterilizer, very much resembling an autoclave in appearance, that controls the EtO concentration, temperature, and humidity (**figure 8.11**). Because pure EtO is explosive, it is usually supplied in a 10 to 20% concentration mixed with either CO_2 or dichlorodifluoromethane. The ethylene oxide concentration, humidity, and temperature influence the rate of sterilization. A clean object can be sterilized if treated for 5 to 8 hours at 38°C or 3 to 4 hours at 54°C when the relative humidity is maintained at 40 to 50% and the EtO concentration at 700 mg/l. Because it is so toxic to humans, extensive aeration of the sterilized materials is necessary to remove residual EtO.

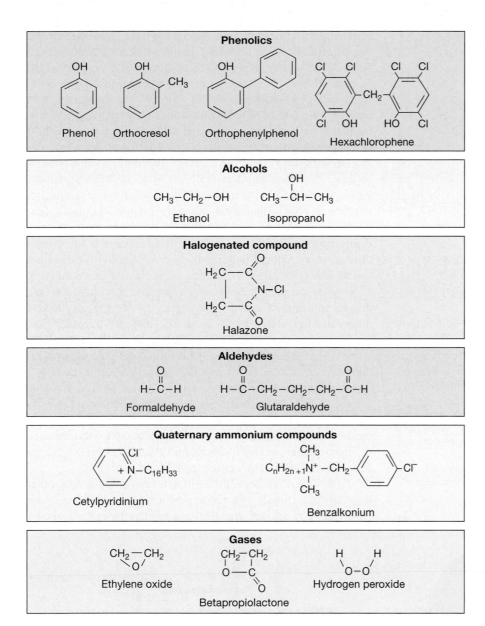

Figure 8.9 Disinfectants and Antiseptics. The structures of some frequently used disinfectants and antiseptics.

variety of microorganisms. During the course of the decontamination process, it breaks down to water and oxygen, both of which are harmless. Other advantages of these systems are that they can be used at a wide range of temperatures (4 to 80°C) and they do not damage most materials.

1. Why are most antimicrobial chemical agents disinfectants rather than sterilants? What general characteristics should one look for in a disinfectant?

2. Describe each of the following agents in terms of its chemical nature, mechanism of action, mode of application, common uses and effectiveness, and advantages and disadvantages: phenolics, alcohols, halogens, heavy metals, quaternary ammonium compounds, aldehydes, and ethylene oxide.

3. Which disinfectants or antiseptics would be used to treat the following: laboratory bench top, drinking water, patch of skin before surgery, small medical instruments (probes, forceps, etc.)? Explain your choices.

4. How do phendic agents differ from the other chemical control agents described in this chapter?

5. Which physical or chemical agent would be the best choice for sterilizing the following items: glass pipettes, tryptic soy broth tubes, nutrient agar, antibiotic solution, interior of a biological safety cabinet, wrapped package of plastic petri plates? Explain your choices.

Betapropiolactone (BPL) is occasionally employed as a sterilizing gas. In the liquid form it has been used to sterilize vaccines and sera. BPL decomposes to an inactive form after several hours and is therefore not as difficult to eliminate as EtO. It also destroys microorganisms more readily than ethylene oxide but does not penetrate materials well and may be carcinogenic. For these reasons, BPL has not been used as extensively as EtO.

Vaporized hydrogen peroxide can be used to decontaminate biological safety cabinets, operating rooms, and other large facilities. These systems introduce vaporized hydrogen peroxide into the enclosure for some time, depending on the size of the enclosure and the materials within. Hydrogen peroxide and its oxy-radical byproducts are toxic and kill a wide

8.6 EVALUATION OF ANTIMICROBIAL AGENT EFFECTIVENESS

Testing of antimicrobial agents is a complex process regulated by two different federal agencies. The U.S. Environmental Protection Agency regulates disinfectants, whereas agents used on humans and animals are under the control of the Food and Drug Administration. Testing of antimicrobial agents often begins with an initial screening to see if they are effective and at what concentrations. This may be followed by more realistic in-use testing.

The best-known disinfectant screening test is the **phenol coefficient test** in which the potency of a disinfectant is

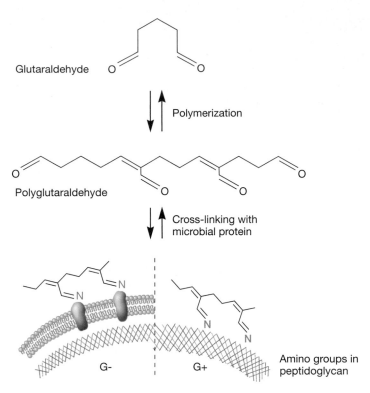

Glutaraldehyde

Polymerization

Polyglutaraldehyde

Cross-linking with
microbial protein

G-   G+

Amino groups in
peptidoglycan

Figure 8.10 Effects of Glutaraldehyde. Glutaraldehyde polymerizes and then interacts with amino acids in proteins (left) or in peptidoglycan (right). As a result, the proteins are alkylated and cross-linked to other proteins, which inactivates them. The amino groups in peptidoglycan are also alkylated and cross-linked, which prevents them from participating in other chemical reactions such as those involved in peptidoglycan synthesis.

compared with that of phenol. A series of dilutions of phenol and the disinfectant being tested are prepared. Standard amounts of *Salmonella enterica* Typhi and *Staphylococcus aureus* are added to each dilution; the dilutions are then placed in a 20 or 37°C water bath. At 5-minute intervals, samples are withdrawn from each dilution and used to inoculate a growth medium, which is incubated for two or more days and then examined for growth. If there is no growth in the medium, the dilution at that particular time of sampling killed the bacteria. The highest dilution (i.e., the lowest concentration) that kills the bacteria after a 10-minute exposure but not after 5 minutes is used to calculate the phenol coefficient. This is done by dividing the reciprocal of the appropriate dilution for the disinfectant being tested by the reciprocal of the appropriate phenol dilution. For instance, if the phenol dilution was 1/90 and maximum effective dilution for disinfectant X was 1/450, then the phenol coefficient of X would be 5. The higher the phenol coefficient value, the more effective the disinfectant under these test conditions. A value greater than 1 means that the disinfectant is more effective than phenol. A few representative phenol coefficient values are given in **table 8.5**.

The phenol coefficient test is a useful initial screening procedure, but the phenol coefficient can be misleading if taken as a direct indication of disinfectant potency during normal use. This is because the phenol coefficient is determined under carefully controlled conditions with pure bacterial cultures, whereas disinfectants are normally used on complex populations in the presence of organic matter and with significant variations in environmental factors such as pH, temperature, and presence of salts.

To more realistically estimate disinfectant effectiveness, other tests are often used. The rates at which selected bacteria are destroyed with various chemical agents may be experimentally

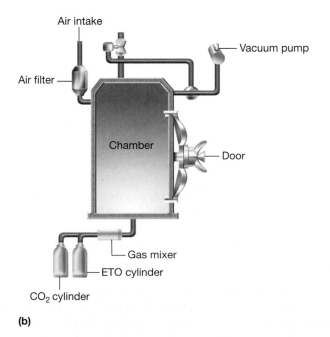

Air intake

Vacuum pump

Air filter

Chamber

Door

Gas mixer

ETO cylinder

CO$_2$ cylinder

(a)   (b)

Figure 8.11   An Ethylene Oxide Sterilizer.   (a) An automatic ethylene oxide (EtO) sterilizer. (b) Schematic of an EtO sterilizer. Items to be sterilized are placed in the chamber, and EtO and carbon dioxide are introduced. After the sterilization procedure is completed, the EtO and carbon dioxide are pumped out of the chamber and air enters.

Table 8.5	Phenol Coefficients for Some Disinfectants	
	Phenol Coefficients[a]	
Disinfectant	*Salmonella enterica* Typhi	*Staphylococcus aureus*
Phenol	1	1
Cetylpyridinium chloride	228	337
O-phenylphenol	5.6 (20°C)	4.0
p-cresol	2.0–2.3	2.3
Hexachlorophene	5–15	15–40
Merthiolate	600	62.5
Mercurochrome	2.7	5.3
Lysol	1.9	3.5
Isopropyl alcohol	0.6	0.5
Ethanol	0.04	0.04
2% I_2 solution in EtOH	4.1–5.2 (20°C)	4.1–5.2 (20°C)

[a]All values were determined at 37°C except where indicated.

determined and compared. A use dilution test can also be carried out. Stainless steel cylinders are contaminated with specific bacterial species under carefully controlled conditions. The cylinders are dried briefly, immersed in the test disinfectants for 10 minutes, transferred to culture media, and incubated for two days. The disinfectant concentration that kills the organisms in the sample with a 95% level of confidence under these conditions is determined. Disinfectants also can be tested under conditions designed to simulate normal in-use situations. In-use testing techniques allow a more accurate determination of the proper disinfectant concentration for a particular situation.

1. Briefly describe the phenol coefficient test.
2. Why might it be necessary to employ procedures such as the use dilution and in-use tests?

8.7 BIOLOGICAL CONTROL OF MICROORGANISMS

The emerging field of biological control of microorganisms demonstrates great promise. Scientists are learning to exploit natural control processes such as predation of one microorganism on another, viral-mediated lysis, and toxin-mediated killing. While these control mechanisms occur in nature, their approval and use by humans is relatively new. Studies evaluating control of *Salmonella, Shigella,* and *E. coli* by gram-negative predators such as *Bdellovibrio* suggest that poultry farms may be sprayed with the predator to reduce potential contamination. The control of human pathogens using bacteriophage is gaining wide support and appears to be effective in the eradication of a number of bacterial species by lysing the pathogenic host **(opening figure)**. This seems intuitive, knowing that the virus lyses its specific bacterial host, yet unnerving when one thinks about maybe swallowing, injecting, or applying a virus to the human body. The use of microbial toxins (such as bacteriocins) to control susceptible populations suggests yet another method for potential control of other microorganisms. << *Introduction to viruses (chapter 5)*

Summary

8.1 Definitions of Frequently Used Terms

a. Sterilization is the process by which all living cells, viable spores, viruses, and viroids are either destroyed or removed from an object or habitat. Disinfection is the killing, inhibition, or removal of microorganisms (but not necessarily endospores) that can cause disease.

b. The main goal of disinfection and antisepsis is the removal, inhibition, or killing of pathogenic microbes. Both processes also reduce the total number of microbes. Disinfectants are chemicals used to disinfect inanimate objects; antiseptics are used on living tissue.

c. Antimicrobial agents that kill organisms often have the suffix *-cide,* whereas agents that prevent growth and reproduction have the suffix *-static.*

8.2 The Pattern of Microbial Death

a. Microbial death is usually exponential or logarithmic **(figure 8.2).**

8.3 Conditions Influencing the Effectiveness of Antimicrobial Agents

a. The effectiveness of a disinfectant or sterilizing agent is influenced by population size, population composition, concentration or intensity of the agent, exposure duration, temperature, and nature of the local environment.

8.4 The Use of Physical Methods in Control

a. Moist heat kills by degrading nucleic acids, denaturing enzymes and other proteins, and disrupting cell membranes.

b. Although treatment with boiling water for 10 minutes kills vegetative forms, an autoclave must be used to destroy endospores by heating at 121°C and 15 pounds of pressure **(figure 8.3).**

c. Glassware and other heat-stable items may be sterilized by dry heat at 160 to 170°C for 2 to 3 hours.

d. Microorganisms can be efficiently removed by filtration with either depth filters or membrane filters **(figure 8.5).**

e. Biological safety cabinets with high-efficiency particulate filters sterilize air by filtration **(figure 8.7).**

f. Radiation of short wavelength or high-energy ultraviolet and ionizing radiation can be used to sterilize objects (**figure 8.8**).

8.5 The Use of Chemical Agents in Control

a. Chemical agents usually act as disinfectants because they cannot readily destroy bacterial spores. Disinfectant effectiveness depends on concentration, treatment duration, temperature, and presence of organic material (**tables 8.3 and 8.4**).

b. Phenolics and alcohols are popular disinfectants that act by denaturing proteins and disrupting cell membranes (**figure 8.9**).

c. Halogens (iodine and chlorine) kill by oxidizing cellular constituents; cell proteins may also be iodinated. Iodine is applied as a tincture or iodophor. Chlorine may be added to water as a gas, hypochlorite, or an organic chlorine derivative.

d. Heavy metals tend to be bacteriostatic agents. They are employed in specialized situations such as the use of silver nitrate in the eyes of newborn infants and copper sulfate in lakes and pools.

e. Cationic detergents are often used as disinfectants and antiseptics; they disrupt membranes and denature proteins.

f. Aldehydes such as formaldehyde and glutaraldehyde can sterilize as well as disinfect because they kill spores.

g. Ethylene oxide gas penetrates plastic wrapping material and destroys all life forms by reacting with proteins. It is used to sterilize packaged, heat-sensitive materials.

h. Vaporized hydrogen peroxide is used to decontaminate enclosed spaces (e.g., safety cabinets and small rooms). The vaporized hydrogen peroxide is a mist that can be circulated throughout the space. The peroxide and its oxy-radical byproducts are toxic to most microorganisms.

8.6 Evaluation of Antimicrobial Agent Effectiveness

a. A variety of procedures can be used to determine the effectiveness of disinfectants, among them: phenol coefficient test, measurement of killing rates with germicides, use dilution testing, and in-use testing.

8.7 Biological Control of Microorganisms

a. Control of microorganisms by natural means such as through predation, viral lysis, and toxins is emerging as a promising field.

Critical Thinking Questions

1. Throughout history, spices have been used as preservatives and to cover up the smell or taste of food that is slightly spoiled. The success of some spices led to a magical, ritualized use of many of them, and possession of spices was often limited to priests or other powerful members of the community.

   a. Choose a spice and trace its use geographically and historically. What is its common use today?

   b. Spices grow and tend to be used predominantly in warmer climates. Explain.

2. Design an experiment to determine whether an antimicrobial agent is acting as a cidal or static agent. How would you determine whether an agent is suitable for use as an antiseptic rather than as a disinfectant?

3. Suppose that you are testing the effectiveness of disinfectants with the phenol coefficient test and obtained the following results.

What disinfectant can you safely say is the most effective? Can you determine its phenol coefficient from these results?

	Bacterial Growth after Treatment		
Dilution	Disinfectant A	Disinfectant B	Disinfectant C
1/20	−	−	−
1/40	+	−	−
1/80	+	−	+
1/160	+	+	+
1/320	+	−	+

Learn More

Learn more by visiting the Prescott website at www.mhhe.com/prescottprinciples, where you will find a complete list of references.

Introduction to Metabolism

Side view of the enzyme glutamine synthetase, an important enzyme in nitrogen assimilation.

Chapter Glossary
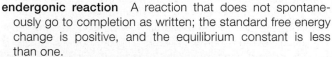

activation energy The energy required to bring reacting molecules together to reach the transition state in a chemical reaction.

active site The part of an enzyme that binds the substrate to form an enzyme-substrate complex and catalyze the reaction; also called the catalytic site.

allosteric enzyme An enzyme whose activity is altered by noncovalent binding of a small effector molecule at a regulatory site separate from the catalytic site.

anabolism The synthesis of complex molecules from simpler molecules with the input of energy.

apoenzyme The protein part of an enzyme that also has a nonprotein component.

catabolism The breakdown of larger, more complex molecules into smaller, simpler molecules with the release of energy.

catalyst A substance that accelerates a reaction without being permanently changed itself.

coenzyme A loosely bound cofactor that often dissociates from the enzyme active site after product has been formed.

denaturation A change in protein shape that destroys the protein's activity; also can refer to changes in nucleic-acid structure.

electron transport chain (ETC) A series of electron carriers that operate together to transfer electrons from donors such as the energy source of a chemotroph to acceptors such as oxygen; also called an electron transport system (ETS).

endergonic reaction A reaction that does not spontaneously go to completion as written; the standard free energy change is positive, and the equilibrium constant is less than one.

entropy A measure of the randomness or disorder of a system; a measure of that part of the total energy in a system that is unavailable for useful work.

enzyme A protein catalyst with specificity for both the reaction catalyzed and its substrates.

equilibrium The state of a system in which no net change is occurring and free energy is at a minimum; in a chemical reaction at equilibrium, the rates in the forward and reverse directions are equal.

exergonic reaction A reaction that spontaneously goes to completion as written; the standard free energy change is negative, and the equilibrium constant is greater than one.

feedback inhibition A regulatory mechanism in which the end product of a biochemical pathway inhibits the activity of one of the pathway's enzymes.

free energy change The total energy change in a system that is available to do useful work as the system goes from its initial state to its final state at constant temperature and pressure.

metabolism The total of all chemical reactions carried out by a cell.

Michaelis constant (K_m) A kinetic constant for an enzyme reaction that equals the substrate concentration required for the enzyme to operate at half-maximal velocity.

phosphorelay system A mechanism for altering enzyme (or other protein) activity that involves the transfer of phosphate from one molecule to another.

reducing power Molecules such as NADH and NADPH that store electrons until they are used in anabolic reactions.

reversible covalent modification A mechanism of enzyme regulation in which the enzyme's activity is altered by the reversible covalent addition of a group such as phosphate.

standard reduction potential A measure of the tendency of a chemical to lose electrons in an oxidation-reduction (redox) reaction.

Fresh oxygen flows from the open stomata the whole world inhales.

—Crystal Cunningham

In the early chapters of this text, we focus on a series of "what" questions about microorganisms: What are they? What do they look like? What are they made of? We now begin to consider a number of "how" questions: How do microbes extract energy from their energy source? How do they use the nutrients obtained from their environment? How do they build themselves? To begin to answer these "how" questions, we must turn our attention more fully to the chemistry of cells; that is, their metabolism. Chapters 9 through 11 consider metabolism, focusing on those processes that conserve the energy supplied by an organism's energy source and how that energy is used to synthesize the building blocks from which an organism is constructed.

Metabolism is the total of all chemical reactions occurring in the cell. Some metabolic reactions are **fueling reactions.** The fueling reactions are part of **catabolism.** They conserve energy from the organisms' energy source, generate a ready supply of electrons **(reducing power),** and generate precursors for biosynthesis. The products of the fueling reactions are used in another set of metabolic reactions that build new organic molecules from smaller inorganic and organic compounds. These biosynthetic reactions are called **anabolism.**

To understand metabolism, the nature of energy and the laws of thermodynamics must be considered, so we start this chapter with these topics. Microorganisms display an amazing array of metabolic diversity, especially in terms of the energy sources and energy-conserving processes they employ. Despite this diversity, several basic principles and processes are common to the metabolism of all microbes. These are the focus of most of the chapter. The chapter ends with a discussion of metabolic regulation.

9.1   ENERGY AND WORK

Energy may be most simply defined as the capacity to do work. This is because all physical and chemical processes are the result of the application or movement of energy. Living cells carry out three major types of work. **Chemical work** involves the synthesis of complex biological molecules from much simpler precursors (i.e., anabolism); energy is needed to increase the molecular complexity of a cell. **Transport work** requires energy to take up nutrients, eliminate wastes, and maintain ion balances. Energy input is needed because molecules and ions often must be transported across cell membranes against an electrochemical gradient. The third type of work is **mechanical work,** perhaps the most familiar of the three. Energy is required for cell motility and the movement of structures within cells.

1. Define metabolism, energy, catabolism, and anabolism.
2. What kinds of work are carried out in a cell? Suppose a bacterium was doing the following: synthesizing peptidoglycan, rotating its flagellum and swimming, and secreting siderophores. What type of work is the bacterium doing in each case?

9.2   LAWS OF THERMODYNAMICS

To understand how energy is conserved in ATP and how ATP is used to do cellular work, some knowledge of the basic principles of thermodynamics is required. The science of **thermodynamics** analyzes energy changes in a collection of matter (e.g., a cell or a plant) called a system. All other matter in the universe is called the surroundings. Thermodynamics focuses on the energy differences between the initial state and the final state of a system. It is not concerned with the rate of the process. For instance, if a pan of water is heated to boiling, only the condition of the water at the start and at boiling is important in thermodynamics, not how fast it is heated or on what kind of stove.

Two important laws of thermodynamics must be understood. The **first law of thermodynamics** says that energy can be neither created nor destroyed. The total energy in the universe remains constant, although it can be redistributed, as it is during the many energy exchanges that occur during chemical reactions. For example, heat is given off by exothermic reactions and absorbed during endothermic reactions. However, the first law alone cannot explain why heat is released by one chemical reaction and absorbed by another. Explanations for this require the **second law of thermodynamics** and a condition of matter called entropy. **Entropy** is a measure of the randomness or disorder of a system. The greater the disorder of a system, the greater is its entropy. The second law states that physical and chemical processes proceed in such a way that the randomness or disorder of the universe (the system and its surroundings) increases. However, even though the entropy of the universe increases, the entropy of any given system within the universe can increase, decrease, or remain unchanged.

It is necessary to specify the amount of energy used in or evolving from a particular process, and two types of energy units are employed. A **calorie** (cal) is the amount of heat energy needed to raise one gram of water from 14.5 to 15.5°C. The amount of energy also may be expressed in terms of **joules** (J), the units of work capable of being done. One cal of heat is equivalent to 4.1840 J of work. One thousand calories or a kilocalorie (kcal) is enough energy to boil 1.9 ml of water. A kilojoule is enough energy to boil about 0.44 ml of water or enable a person weighing 70 kg to climb 35 steps. The joule is normally used by chemists and physicists. Because biologists most often speak of energy in terms of calories, this text employs calories when discussing energy changes.

9.3   FREE ENERGY AND REACTIONS

The first and second laws of thermodynamics can be combined in a useful equation, relating the changes in energy that can occur in chemical reactions and other processes.

$$\Delta G = \Delta H - T \Delta S$$

ΔG is the change in free energy, ΔH is the change in enthalpy, T is the temperature in Kelvin (°C + 273), and ΔS is the change in entropy occurring during the reaction. The change in **enthalpy** is the change in heat content. Cellular reactions occur under conditions of constant pressure and volume. Thus the change in enthalpy is about the same as the change in total energy during the reaction. The **free energy change** is the amount of energy in a system (or cell) available to do useful work at constant temperature and pressure. Therefore the change in entropy (ΔS) is a measure of the proportion of the total energy change that the system cannot use in performing work. Free energy and entropy changes do not depend on how the system gets from start to finish. A reaction will occur spontaneously—that is, without any external cause—if the free energy of the system decreases during the reaction or, in other words, if ΔG is negative. It follows from the equation that a reaction with a large positive change in entropy will normally tend to have a negative ΔG value and therefore occur spontaneously. A decrease in entropy will tend to make ΔG more positive and the reaction less favorable.

It is helpful to think of the relationship between entropy (ΔS) and change in free energy (ΔG) in terms that are more concrete. Consider the Greek myth of Sisyphus, king of Corinth. For his assorted crimes against the gods, he was condemned to roll a large boulder to the top of a steep hill for all eternity. This represents a very negative change in entropy—a boulder poised at the top of a hill is neither random nor disordered—and this activity (reaction) has a very positive ΔG. That is to say, Sisyphus had to put a lot of energy into the system. Unfortunately for Sisyphus, as soon as the boulder was at the top of the hill, it spontaneously rolled back down the hill. This represents a positive change in entropy and a negative ΔG. Sisyphus did not need to put energy into the system. He probably just stood at the top of the hill and watched the reaction proceed.

The change in free energy also has a definite, concrete relationship to the direction of chemical reactions. Consider this simple reaction.

$$A + B \rightleftharpoons C + D$$

If molecules A and B are mixed, they will combine to form the products C and D. Eventually C and D will become concentrated enough to combine and produce A and B at the same rate as C and D are formed from A and B. The reaction is now at **equilibrium:** the rates in both directions are equal and no further net change occurs in the concentrations of reactants and products. This situation is described by the **equilibrium constant (K_{eq}),** relating the equilibrium concentrations of products and substrates to one another.

$$K_{eq} = \frac{[C][D]}{[A][B]}$$

If the equilibrium constant is greater than one, the products are in greater concentration than the reactants at equilibrium—that is, the reaction tends to go to completion as written.

The equilibrium constant of a reaction is directly related to its change in free energy. When the free energy change for a process

Exergonic reactions	Endergonic reactions
A + B $\rightleftharpoons$ C + D	A + B $\rightleftharpoons$ C + D
$K_{eq} = \dfrac{[C][D]}{[A][B]} > 1.0$	$K_{eq} = \dfrac{[C][D]}{[A][B]} < 1.0$
$\Delta G^{\circ\prime}$ is negative.	$\Delta G^{\circ\prime}$ is positive.

Figure 9.1 $\Delta G^{\circ\prime}$ **and Equilibrium.** The relationship of $\Delta G^{\circ\prime}$ to the equilibrium of reactions. Note the differences between exergonic and endergonic reactions.

is determined at carefully defined standard conditions of concentration, pressure, pH, and temperature, it is called the **standard free energy change** (ΔG°). If the pH is set at 7.0 (which is close to the pH of living cells), the standard free energy change is indicated by the symbol $\Delta G^{\circ\prime}$. The change in standard free energy may be thought of as the maximum amount of energy available from the system for useful work under standard conditions. Using $\Delta G^{\circ\prime}$ values allows one to compare reactions without worrying about variations in ΔG due to differences in environmental conditions. The relationship between $\Delta G^{\circ\prime}$ and K_{eq} is given by this equation.

$$\Delta G^{\circ\prime} = -2.303RT \cdot \log K_{eq}$$

R is the gas constant (1.9872 cal/mole-degree or 8.3145 J/mole-degree), and T is the absolute temperature. Inspection of this equation shows that when $\Delta G^{\circ\prime}$ is negative, the equilibrium constant is greater than one and the reaction goes to completion as written. It is said to be an **exergonic reaction** (figure 9.1). In an **endergonic reaction,** $\Delta G^{\circ\prime}$ is positive and the equilibrium constant is less than one. That is, the reaction is not favorable, and little product will be formed at equilibrium under standard conditions. Keep in mind that the $\Delta G^{\circ\prime}$ value shows only where the reaction lies at equilibrium, not how fast the reaction reaches equilibrium.

9.4 ATP

Considerable metabolic diversity exists in the microbial world. However, several biochemical principles are common to all types of metabolism. These are (1) the use of ATP to conserve energy released during exergonic reactions, so it can be used to drive endergonic reactions; (2) the organization of metabolic reactions into pathways and cycles; (3) the catalysis of metabolic reactions by enzymes; and (4) the importance of oxidation-reduction reactions in energy conservation. This section considers the role of ATP in metabolism.

Energy is released from a cell's energy source in exergonic reactions (i.e., those reactions with a negative ΔG). Rather than wasting this energy, much of it is trapped in a practical form that allows its transfer to the cellular systems doing work. These systems carry out endergonic reactions (e.g., anabolism), and the energy captured by the cell is used to drive these reactions to completion. In living organisms, this practical form of energy is **adenosine 5'-triphosphate (ATP; figure 9.2).** In a sense, cells carry out certain processes so that they can "earn" ATP and carry

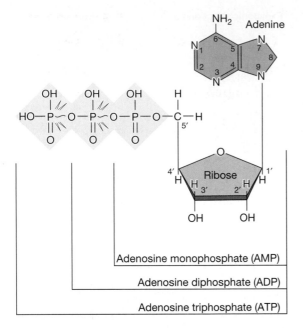

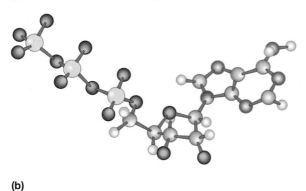

(a) ⌒ Bond that releases energy when broken

(b)

Figure 9.2   Adenosine Triphosphate. (a) Structure of ATP, ADP, and AMP. The two red bonds (~) are more easily broken and have a high phosphate group transfer potential. The pyrimidine ring atoms have been numbered as have the carbon atoms in ribose. (b) A model of ATP. Carbon is in green; hydrogen in light blue; nitrogen in dark blue; oxygen in red; and phosphorus in yellow.

out other processes in which they "spend" their ATP. Thus ATP is often referred to as the cell's energy currency. In the cell's economy, ATP serves as the link between exergonic reactions and endergonic reactions (**figure 9.3**).

What makes ATP suited for this role as energy currency? ATP is a high-energy molecule. That is, it breaks down or hydrolyzes almost completely to the products **adenosine diphosphate (ADP)** and orthophosphate (P_i) and is strongly exergonic, having a $\Delta G^{\circ\prime}$ of −7.3 kcal/mole.

$$ATP + H_2O \longrightarrow ADP + P_i + H^+$$

Because ATP readily transfers its phosphate to water, it is said to have a high phosphate group transfer potential, defined as the negative of $\Delta G^{\circ\prime}$ for the hydrolytic removal of phosphate. A molecule with a higher group transfer potential donates phosphate

Figure 9.3   ATP as a Coupling Agent.   The use of ATP to make endergonic reactions more favorable. It is formed by exergonic reactions and then used to drive endergonic reactions.

to one with a lower potential. Thus ATP readily donates a phosphate to molecules such as glucose and glucose 6-phosphate in reactions such as those found in some catabolic pathways. >> *Breakdown of glucose to pyruvate (section 10.3)*

Although the free energy change for hydrolysis of ATP is quite large, metabolic reactions exist that release even greater amounts of free energy. This energy is used to resynthesize ATP from ADP and P_i during fueling reactions. Likewise, catabolism can generate molecules with a phosphate group transfer potential that is even higher than that of ATP. Phosphoenolpyruvate (PEP) and guanosine 5′-triphosphate (GTP) are two important examples. Cells use these molecules to regenerate ATP from ADP by a mechanism called substrate-level phosphorylation. Thus ATP, ADP, and P_i form an energy cycle (**figure 9.4**). The fueling reactions conserve energy released from an energy source by using it to synthesize ATP from ADP and P_i. When ATP is hydrolyzed, the energy released drives endergonic processes such as anabolism, transport, and mechanical work. The mechanisms for synthesizing ATP are described in more detail in chapter 10.

1. What is thermodynamics? Summarize the first and second laws of thermodynamics.

2. Define entropy and enthalpy. Do living cells increase entropy within themselves? Do they increase entropy in the environment?

3. Define free energy. What are exergonic and endergonic reactions?

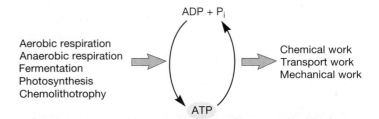

Figure 9.4   The Cell's Energy Cycle.   ATP is formed from energy made available during aerobic respiration, anaerobic respiration, fermentation, chemolithotrophy, and photosynthesis. Its breakdown to ADP and phosphate (P_i) makes chemical, transport, and mechanical work possible.

4. Suppose that a chemical reaction had a large negative $\Delta G^{\circ\prime}$ value. Is the reaction endergonic or exergonic? What would this indicate about its equilibrium constant?

5. Describe the energy cycle and ATP's role in it. What characteristics of ATP make it suitable for this role? Why is ATP called a high-energy molecule?

Table 9.1	Selected Biologically Important Redox Couples	
Redox Couple[a]		**E'_0 (Volts)[b]**
$2H^+ + 2e^- \rightarrow H_2$		−0.42
Ferredoxin (Fe^{3+}) + e^- → ferredoxin (Fe^{2+})		−0.42
$NAD(P)^+ + H^+ + 2e^- \rightarrow NAD(P)H$		−0.32
$S + 2H^+ + 2e^- \rightarrow H_2S$		−0.27
Acetaldehyde + $2H^+ + 2e^-$ → ethanol		−0.20
$Pyruvate^- + 2H^+ + 2e^-$ → $lactate^{2-}$		−0.19
$FAD + 2H^+ + 2e^- \rightarrow FADH_2$		−0.18[c]
$Oxaloacetate^{2-} + 2H^+ + 2e^-$ → $malate^{2-}$		−0.17
$Fumarate^{2-} + 2H^+ + 2e^-$ → $succinate^{2-}$		0.03
Cytochrome b (Fe^{3+}) + e^- → cytochrome b (Fe^{2+})		0.08
Ubiquinone + $2H^+ + 2e^-$ → ubiquinone H_2		0.10
Cytochrome c (Fe^{3+}) + e^- → cytochrome c (Fe^{2+})		0.25
Cytochrome a (Fe^{3+}) + e^- → cytochrome a (Fe^{2+})		0.29
Cytochrome a_3 (Fe^{3+}) + e^- → cytochrome a_3 (Fe^{2+})		0.35
$NO_3^- + 2H^+ + 2e^- \rightarrow NO_2^- + H_2O$		0.42
$NO_2^- + 8H^+ + 6e^- \rightarrow NH_4^+ + 2H_2O$		0.44
$Fe^{3+} + e^- \rightarrow Fe^{2+}$		0.77[d]
$O_2 + 4H^+ + 4e^- \rightarrow 2H_2O$		0.82

[a]Redox couples are written as half-reactions. They can also be written as acceptor/donor. Any two half reactions (i.e., redox couples) can be combined in a redox reaction. The E'_0 of each couple determines which half reaction will function as the electron-donating half reaction and which will function as the electron-accepting half reaction.
[b]E'_0 is the standard reduction potential at pH 7.0
[c]The value for FAD/FADH$_2$ applies to the free cofactor because it can vary considerably when bound to an apoenzyme.
[d]The value for free Fe, not Fe complexed with proteins (e.g., cytochromes).

9.5   OXIDATION-REDUCTION REACTIONS

Free energy changes are related to the equilibria of all chemical reactions including the equilibria of oxidation-reduction reactions. The release of energy from an energy source normally involves oxidation-reduction reactions. **Oxidation-reduction (redox) reactions** are those in which electrons move from an **electron donor** to an **electron acceptor.**[1] By convention, such a reaction is written with the donor to the right of the acceptor and the number (n) of electrons (e^-) transferred.

$$Acceptor + ne^- \rightleftharpoons donor$$

The acceptor and donor pair is referred to as a redox couple (**table 9.1**). When an acceptor accepts electrons, it then becomes the donor of the couple. The equilibrium constant for the reaction is called the **standard reduction potential** (E_0) and is a measure of the tendency of the donor to lose electrons. By convention, the standard reduction potentials for redox couples such as those in table 9.1 are determined at pH 7 and are represented by E'_0. Standard reduction potentials are measured in volts, a unit of electrical potential or electromotive force. Therefore redox couples are a potential source of energy.

The reduction potential has a concrete meaning. Redox couples with more negative reduction potentials will donate electrons to couples with more positive potentials and greater affinity for electrons. Thus electrons tend to move from donors at the top of the list in table 9.1 to acceptors at the bottom because the latter have more positive potentials. This may be expressed visually in the form of an electron tower in which the most negative reduction potentials are at the top (**figure 9.5**). Electrons move from donors to acceptors down the potential gradient or fall down the tower to more positive potentials. Consider the case of the electron acceptor **nicotinamide adenine dinucleotide (NAD$^+$)**. The NAD$^+$/NADH couple has a very negative E'_0 and can therefore give electrons to many acceptors, including O_2.

$$NAD^+ + 2H^+ + 2e^- \rightleftharpoons NADH + H^+ \quad E'_0 = -0.32 \text{ volts}$$

$$1/2 O_2 + 2H^+ + 2e^- \rightleftharpoons H_2O \quad\quad E'_0 = +0.82 \text{ volts}$$

Because the reduction potential of NAD$^+$/NADH is more negative than that of $1/2 O_2/H_2O$, electrons flow from NADH (the donor) to O_2 (the acceptor), as shown in figure 9.5.

$$NADH + H^+ + 1/2 O_2 \rightarrow H_2O + NAD^+$$

Because the NAD$^+$/NADH couple has a relatively negative E'_0, it stores more potential energy than redox couples with less negative (or more positive) E'_0 values. It follows that when electrons move from a donor to an acceptor with a more positive redox potential, free energy is released. The $\Delta G^{\circ\prime}$ of the reaction is directly related to the magnitude of the difference between the reduction potentials of the two couples ($\Delta E'_0$). The larger the $\Delta E'_0$, the greater the amount of free energy made available, as is evident from the equation

$$\Delta G^{\circ\prime} = -nF \cdot \Delta E'_0$$

---

[1] In redox reactions, the electron donor is often called the reducing agent or reductant because it is donating electrons to the acceptor and thus reducing it. The electron acceptor is called the oxidizing agent or oxidant because it is removing electrons from the donor and oxidizing it.

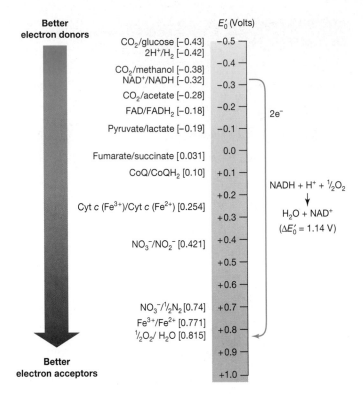

Better electron donors

	E'_0 (Volts)
CO_2/glucose [−0.43]	−0.5
$2H^+/H_2$ [−0.42]	
	−0.4
CO_2/methanol [−0.38]	
$NAD^+/NADH$ [−0.32]	−0.3
CO_2/acetate [−0.28]	
$FAD/FADH_2$ [−0.18]	−0.2
Pyruvate/lactate [−0.19]	−0.1
	0.0
Fumarate/succinate [0.031]	
$CoQ/CoQH_2$ [0.10]	+0.1
	+0.2
Cyt c (Fe^{3+})/Cyt c (Fe^{2+}) [0.254]	+0.3
	+0.4
NO_3^-/NO_2^- [0.421]	+0.5
	+0.6
$NO_3^-/\frac{1}{2}N_2$ [0.74]	+0.7
Fe^{3+}/Fe^{2+} [0.771]	+0.8
$\frac{1}{2}O_2/H_2O$ [0.815]	+0.9
	+1.0

Better electron acceptors

$2e^-$

$NADH + H^+ + \frac{1}{2}O_2$
↓
$H_2O + NAD^+$
$(\Delta E'_0 = 1.14\ V)$

Figure 9.5   Electron Movement and Reduction Potentials. Electrons spontaneously move from donors higher on the tower (more negative potentials) to acceptors lower on the tower (more positive potentials). That is, the donor is always higher on the tower than the acceptor. For example, NADH will donate electrons to oxygen and form water in the process. Some typical donors and acceptors are shown on the left, and their reduction potentials are given in brackets.

in which n is the number of electrons transferred and F is the Faraday constant (23,062 cal/mole-volt). For every 0.1 volt change in $\Delta E'_0$, there is a corresponding 4.6 kcal change in $\Delta G^{\circ\prime}$ when a two-electron transfer takes place. This is similar to the relationship of $\Delta G^{\circ\prime}$ and K_{eq} in other chemical reactions—the larger the equilibrium constant, the greater the $\Delta G^{\circ\prime}$. The difference in reduction potentials between $NAD^+/NADH$ and $1/2O_2/H_2O$ is 1.14 volts, a large $\Delta E'_0$ value. When electrons move from NADH to O_2, a large amount of free energy is made available and can be used to synthesize ATP and do other work.

9.6   ELECTRON TRANSPORT CHAINS

We have focused our attention on the reduction of O_2 by NADH because NADH plays a central role in the metabolism of many organisms, especially chemoorganotrophs. Many chemoorganotrophs use glucose as a source of energy. As glucose is catabolized, it is oxidized. Many of the electrons released from glucose are accepted by NAD^+, which is then reduced to NADH. NADH next transfers the electrons to O_2. However, it does not do so directly. Instead, the electrons are transferred to O_2 via a series

of electron carriers that are organized into a system called an **electron transport chain (ETC).** The first electron carrier in the ETC has the most negative E'_0, and each successive carrier is slightly less negative (**figure 9.6**). In this way, the potential energy stored in the redox couple whose electrons initiate electron flow is released and used to form ATP.

The ETCs of chemoorganotrophs are located in the plasma membrane in procaryotes and the internal mitochondrial membranes in eucaryotes. Electron transport chains also play a pivotal role in the metabolism of chemolithotrophs and phototrophs, where they are used to conserve energy from inorganic chemicals and light, respectively. These ETCs are located in the plasma membrane or internal membrane systems of chemolithotrophs, which are all procaryotes. They are located in the plasma membrane and internal membrane systems of procaryotic phototrophs and in the thylakoid membranes of chloroplasts in eucaryotic phototrophs (figure 9.6).   ◄◄ *Nutritional types of microorganisms (section 6.3)*

The carriers that make up ETCs differ in terms of their chemical nature and the way they carry electrons. NAD^+ and its chemical relative **nicotinamide adenine dinucleotide phosphate ($NADP^+$)** contain a nicotinamide ring (**figure 9.7**). This ring accepts two electrons and one proton from a donor (e.g., an intermediate formed during the catabolism of glucose), and a second proton is released. **Flavin adenine dinucleotide (FAD)** and **flavin mononucleotide (FMN)** bear two electrons and two protons on the complex ring system shown in **figure 9.8**. Proteins bearing FAD and FMN are often called flavoproteins. **Coenzyme Q (CoQ)** or **ubiquinone** is a quinone that transports two electrons and two protons (**figure 9.9**). **Cytochromes** and several other carriers use iron atoms to transport one electron at a time. In cytochromes, the iron atoms are part of a heme group or other similar iron-porphyrin rings (**figure 9.10**). There are several different cytochromes, each of which consists of a protein and an iron-porphyrin ring. Some iron-containing electron-carrying proteins lack a heme group and are called **nonheme iron proteins.** They are also commonly called **iron-sulfur (Fe-S) proteins** because the iron is associated with sulfur atoms. The sulfur atoms are often present in the cysteine residues of the protein. **Ferredoxin** is an Fe-S protein active in photosynthetic electron transport and several other electron transport processes. Like cytochromes, Fe-S proteins carry only one electron at a time. This difference in the number of electrons and protons transported by carriers in ETCs is of great importance in their operation and is discussed further in chapter 10.

1. Write a generalized equation for a redox reaction. Define standard reduction potential.

2. How is the direction of electron flow between redox couples related to the standard reduction potential and the release of free energy?

3. When electrons flow from the $NAD^+/NADH$ redox couple to the $1/2O_2/H_2O$ redox couple, does the reaction begin with NAD^+ or with NADH? What is produced—O_2 or H_2O?

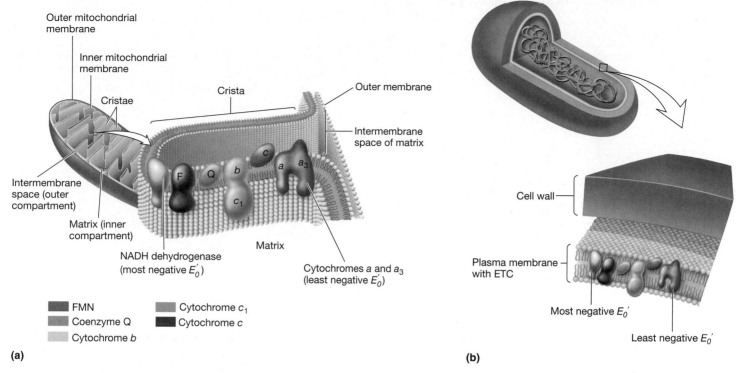

Figure 9.6 Electron Transport Chains. Electron transport chains (ETCs) are located in membranes. Electrons flow from the electron carrier having the most negative reduction potential to the carrier having the most positive reduction potential. During respiratory processes (aerobic respiration, anaerobic respiration, and chemolithotrophy), an exogenous molecule such as oxygen serves as the terminal electron acceptor. (a) The mitochondrial ETC (b) A typical bacterial ETC.

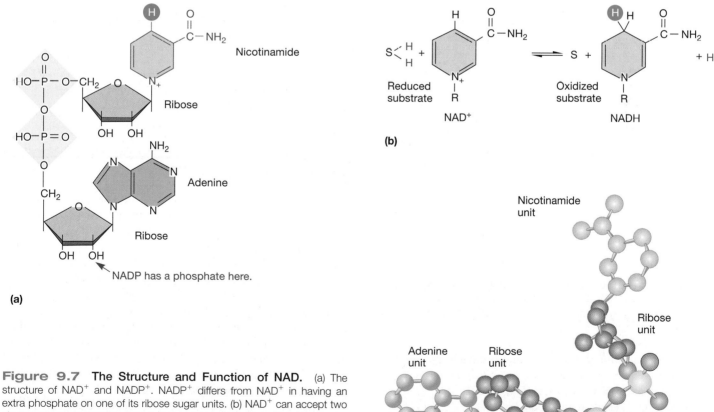

Figure 9.7 The Structure and Function of NAD. (a) The structure of NAD^+ and $NADP^+$. $NADP^+$ differs from NAD^+ in having an extra phosphate on one of its ribose sugar units. (b) NAD^+ can accept two electrons and one proton from a reduced substrate (e.g., SH_2 of a cysteine residue). In this example, the electrons and proton are supplied by the hydrogen atoms of the reduced substrate; recall that each hydrogen atom consists of one proton and one electron. (c) Model of NAD^+ when bound to the enzyme lactate dehydrogenase.

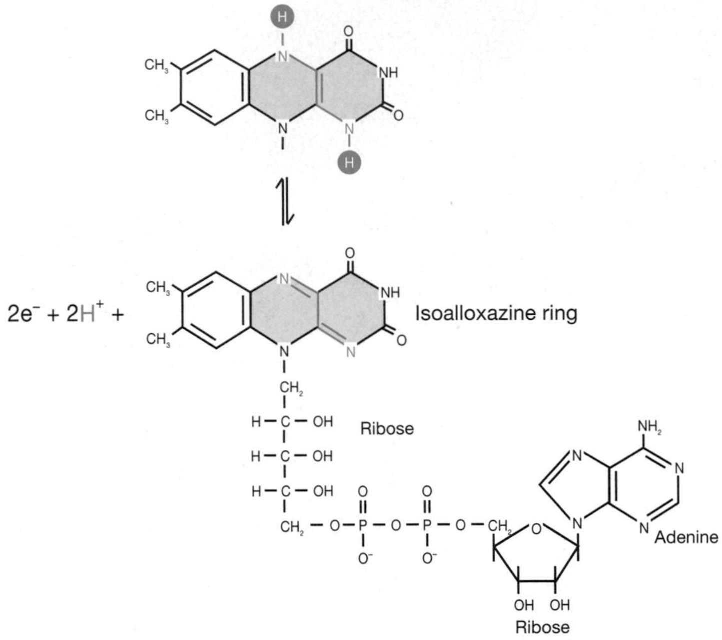

Figure 9.8 **The Structure and Function of FAD.** The vitamin riboflavin is composed of the isoalloxazine ring and its attached ribose sugar. FMN is riboflavin phosphate. The portion of the ring directly involved in oxidation-reduction reactions is in color.

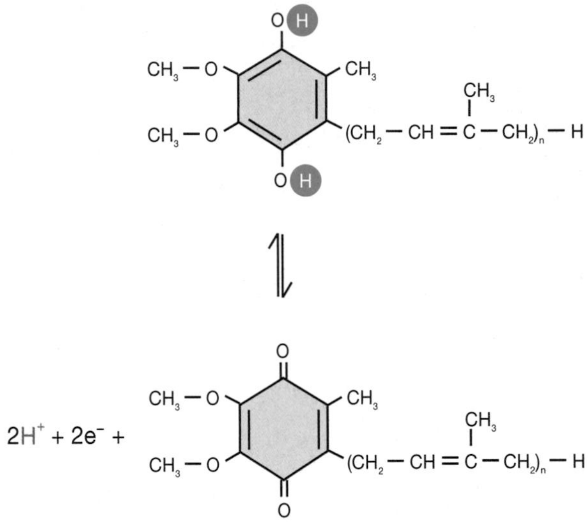

Figure 9.9 **The Structure and Function of Coenzyme Q or Ubiquinone.** The length of the side chain varies among organisms from n = 6 to n = 10.

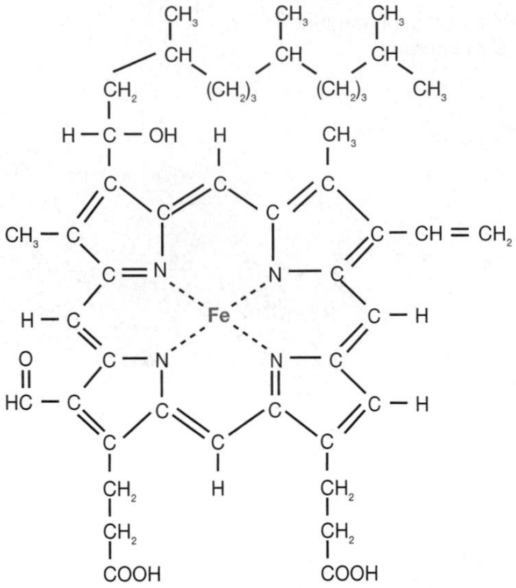

4. Which among the following would be the best electron donor? Which would be the worst? Ubiquinone/ubiquinoneH$_2$, NAD$^+$/NADH, FAD/FADH$_2$, NO$_3^-$/NO$_2^-$. Explain your answers.

5. In general terms, how is $\Delta G^{\circ\prime}$ related to $\Delta E'_0$? What is the $\Delta E'_0$ when electrons flow from the NAD$^+$/NADH redox couple to the Fe^{3+}/Fe^{2+} redox couple? How

Figure 9.10 **The Structure of Heme.** Heme is composed of a porphyrin ring and an attached iron atom. It is the nonprotein component of many cytochromes. The iron atom alternatively accepts and releases an electron.

does this compare to the $\Delta E'_0$ when electrons flow from the Fe$_3^+$/Fe$_2$ redox couple to the 1/2O$_2$/H$_2$O couple? Which will yield the largest amount of free energy to the cell?

6. Name and briefly describe the major electron carriers found in cells. Why is NADH a good electron donor? Why is ferredoxin an even better electron donor?

9.7 ENZYMES

Recall that an exergonic reaction is one with a negative $\Delta G^{\circ\prime}$ and an equilibrium constant greater than one. An exergonic reaction proceeds to completion in the direction written (i.e., toward the right of the equation). Nevertheless, one often can combine the reactants for an exergonic reaction with no obvious result. For instance, if a polysaccharide such as starch is mixed in water, the hydrolysis of the starch into its component monosaccharides (glucose) is exergonic and will occur spontaneously—that is, it will occur on its own, given enough time. However, the time needed is very long. Thus even if an organic chemist carried out this reaction in 6 moles/liter (M) HCl and at 100°C, it would still take several hours to go to completion. A cell, on the other hand, can accomplish the same reaction at neutral pH, at a much lower temperature, and in just fractions of a second. Cells can do this because they manufacture proteins called enzymes that speed up chemical reactions. Enzymes are critically important to cells, since most biological reactions occur very slowly without them. Indeed, enzymes make life possible.

Structure and Classification of Enzymes

Enzymes are protein catalysts that have great specificity for the reaction catalyzed and the molecules acted on. A **catalyst** is a substance that increases the rate of a chemical reaction without being permanently altered itself. Thus enzymes speed up cellular reactions. The reacting molecules are called **substrates,** and the substances formed are the **products. >>** *Proteins (appendix I)*

Many enzymes are composed only of proteins. However, some enzymes are composed of two parts: a protein component called the **apoenzyme** and a nonprotein component called a **cofactor.** The complete enzyme consisting of the apoenzyme and its cofactor is called the **holoenzyme.** If the cofactor is firmly attached to the apoenzyme, it is a **prosthetic group.** If the cofactor is loosely attached and can dissociate from the apoenzyme after products have been formed, it is called a **coenzyme.** Many coenzymes carry one of the products to another enzyme or transfer chemical groups from one substrate to another (**figure 9.11**). For example, NAD^+ is a coenzyme that carries electrons within the cell. Many vitamins that humans require serve as coenzymes or as their precursors. Niacin is incorporated into NAD^+ and riboflavin into FAD. Metal ions may also be bound to apoenzymes and act as cofactors.

Enzymes may be placed in one of six general classes and usually are named in terms of the substrates they act on and the type of reaction catalyzed (**table 9.2**). For example, lactate dehydrogenase (LDH) removes hydrogens from lactate. Lactate dehydrogenase can also be given a more complete and detailed name, L-lactate:NAD^+ oxidoreductase. This name describes the substrates and reaction type with even more precision.

Mechanism of Enzyme Reactions

It is important to keep in mind that enzymes increase the rates of reactions but do not alter their equilibrium constants. If a reaction is endergonic, the presence of an enzyme will not shift its

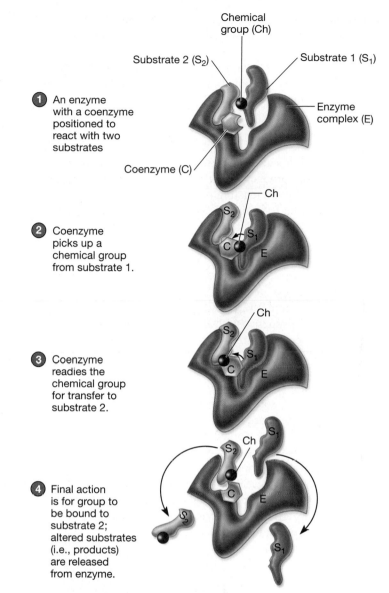

1. An enzyme with a coenzyme positioned to react with two substrates

2. Coenzyme picks up a chemical group from substrate 1.

3. Coenzyme readies the chemical group for transfer to substrate 2.

4. Final action is for group to be bound to substrate 2; altered substrates (i.e., products) are released from enzyme.

Figure 9.11 Coenzymes as Carriers.

Table 9.2	Enzyme Classification	
Type of Enzyme	**Reaction Catalyzed by Enzyme**	**Example of Reaction**
Oxidoreductase	Oxidation-reduction reactions	Lactate dehydrogenase: Pyruvate + NADH + H^+ $\rightleftharpoons$ lactate + NAD^+
Transferase	Reactions involving the transfer of groups between molecules	Aspartate carbamoyltransferase: Aspartate + carbamoylphosphate $\rightleftharpoons$ carbamoylaspartate + phosphate
Hydrolase	Hydrolysis of molecules	Glucose-6-phoshatase: Glucose 6-phosphate + H_2O → glucose + P_i[a]
Lyase	Cleaving of C-C, C-O, C-N and other bonds by a means other than hydrolysis	Fumarate hydratase: L-malate $\rightleftharpoons$ fumarate + H_2O
Isomerase	Reactions involving isomerizations	Alanine racemase: L-alanine $\rightleftharpoons$ D-alanine
Ligase	Joining of two molecules using ATP (or the energy of other nucleoside triphosphates)	Glutamine synthetase: Glutamate + NH_3 + ATP → glutamine + ADP + P_i

[a]P_i is inorganic phosphate

equilibrium so that more products can be formed. Enzymes simply speed up the rate at which a reaction proceeds toward its final equilibrium.

How do enzymes catalyze reactions? Some understanding of the mechanism can be gained by considering the course of a simple exergonic chemical reaction.

$$A + B \rightleftharpoons C + D$$

When molecules A and B approach each other to react, they form a transition-state complex, which resembles both the substrates and the products (**figure 9.12**). **Activation energy** is required to bring the reacting molecules together in the correct way to reach the transition state. The transition-state complex can then resolve to yield the products C and D. The difference in free energy level between reactants and products is $\Delta G^{\circ\prime}$. Thus the equilibrium in our example lies toward the products because $\Delta G^{\circ\prime}$ is negative (i.e., the products are at a lower energy level than the substrates).

As seen in figure 9.12, A and B will not be converted to C and D if they are not supplied with an amount of energy equivalent to the activation energy. Enzymes accelerate reactions by lowering the activation energy; therefore more substrate molecules will have sufficient energy to come together and form products. Even though the equilibrium constant (or $\Delta G^{\circ\prime}$) is unchanged, equilibrium is reached more rapidly in the presence of an enzyme because of this decrease in activation energy.

Researchers have worked hard to discover how enzymes lower the activation energy of reactions, and the process is becoming clearer. Enzymes bring substrates together at a specific location in the enzyme called the **active** or **catalytic site** to form an enzyme-substrate complex (**figures 9.13** and **9.14**; *see also appendix figure AI.19*). An enzyme can interact with its substrate in two general ways. In the lock-and-key model, the active site is

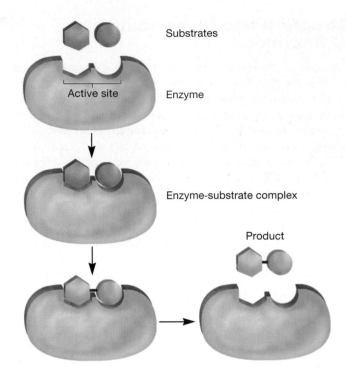

Figure 9.13 Lock-and-Key Model of Enzyme Function. In this model, the active site is a relatively rigid structure that accommodates only those molecules with the correct corresponding shape. The formation of the enzyme-substrate complex and its conversion to product are shown.

rigid and precisely shaped to fit the substrate, so that a specific substrate binds and is positioned properly for the reaction (figure 9.13). An enzyme also may change shape when it binds the substrate so that the active site surrounds and precisely fits the substrate. This has been called the induced fit model and is used by hexokinase and many other enzymes (figure 9.14). The formation of an enzyme-substrate complex can lower the activation energy in many ways. For example, by bringing the substrates together at the active site, the enzyme is, in effect, concentrating them and speeding up the reaction. An enzyme does not simply concentrate its substrates, however. It also binds them so that they are correctly oriented with respect to each other. Such an orientation lowers the amount of energy that the substrates require to reach the transition state. These and other catalytic site activities speed up a reaction by hundreds of thousands of times.

Environmental Effects on Enzyme Activity

Enzyme activity varies greatly with changes in environmental factors, one of the most important being substrate concentration. Substrate concentrations are usually low within cells. At very low substrate concentrations, an enzyme makes product slowly because it seldom contacts a substrate molecule. If more substrate molecules are present, an enzyme binds substrate more often, and the velocity of the reaction (usually expressed in terms of the rate of product formation) is greater than at a lower substrate

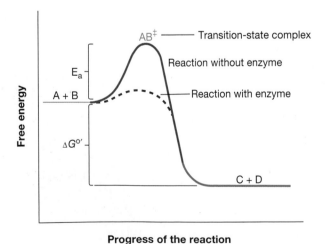

Progress of the reaction

Figure 9.12 Enzymes Lower the Energy of Activation. This figure traces the course of a chemical reaction in which A and B are converted to C and D. The transition-state complex is represented by AB‡, and the activation energy required to reach it, by E_a. The red line represents the course of the reaction in the presence of an enzyme. Note that the activation energy is much lower in the enzyme-catalyzed reaction.

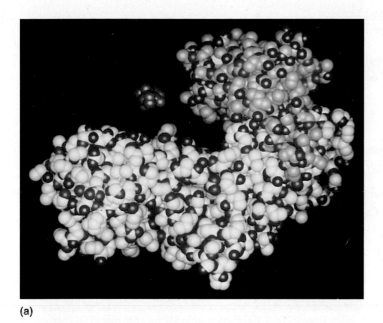
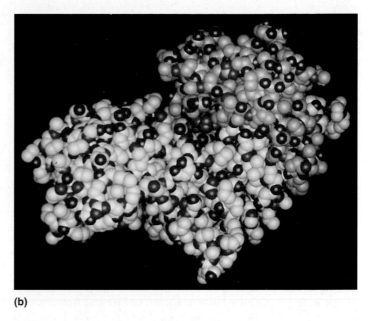

(a)

(b)

Figure 9.14 The Induced Fit Model of Enzyme Function. (a) A space-filling model of yeast hexokinase and its substrate glucose (purple). The active site is in the cleft formed by the enzyme's small lobe (green) and large lobe (blue). (b) When glucose binds to form the enzyme-substrate complex, hexokinase changes shape and surrounds the substrate.

concentration. Thus the rate of an enzyme-catalyzed reaction increases with substrate concentration (**figure 9.15**). Eventually further increases in substrate concentration do not result in a greater reaction velocity because the available enzyme molecules are binding substrate and converting it to product as rapidly as possible. That is, the enzyme is saturated with substrate and operating at maximal velocity (V_{max}). The resulting substrate concentration curve is a hyperbola (figure 9.15). It is useful to

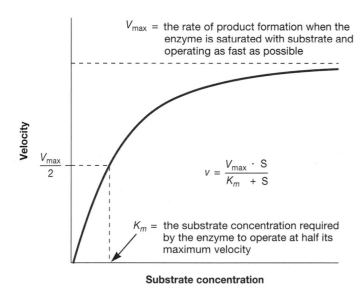

V_{max} = the rate of product formation when the enzyme is saturated with substrate and operating as fast as possible

$$v = \frac{V_{max} \cdot S}{K_m + S}$$

K_m = the substrate concentration required by the enzyme to operate at half its maximum velocity

Substrate concentration

Figure 9.15 Michaelis-Menten Kinetics. The dependence of enzyme activity upon substrate concentration. This substrate curve fits the Michaelis-Menten equation given in the figure, which relates reaction velocity (v) to the substrate concentration (S) using the maximum velocity and the Michaelis constant (K_m).

know the substrate concentration an enzyme needs to function adequately. Usually the **Michaelis constant (K_m),** the substrate concentration required for the enzyme to achieve half-maximal velocity, is used as a measure of the apparent affinity of an enzyme for its substrate. The lower the K_m value, the lower the substrate concentration at which an enzyme catalyzes its reaction. Enzymes with a low K_m value are said to have a high affinity for their substrates. Since the concentrations of substrates in cells are often low, enzymes with lower K_m values are able to function better.

Enzyme activity is also changed with alterations in pH and temperature. Each enzyme functions most rapidly at a specific pH optimum. When the pH deviates too greatly from an enzyme's optimum, activity slows and the enzyme may be damaged. Enzymes likewise have temperature optima for maximum activity. If the temperature rises too much above the optimum, an enzyme's structure will be disrupted and its activity lost. This phenomenon, known as **denaturation,** may be caused by extremes of pH and temperature or other factors. The pH and temperature optima of a microorganism's enzymes often reflect the pH and temperature of its habitat. Not surprisingly, bacteria that grow best at high temperatures often have enzymes with high temperature optima and great heat stability. ◄◄ *Influences of environmental factors on growth (section 7.5)*

Enzyme Inhibition

Microorganisms can be poisoned by a variety of chemicals, and many of the most potent poisons are enzyme inhibitors. A **competitive inhibitor** directly competes with the substrate at an enzyme's catalytic site and prevents the enzyme from forming product (**figure 9.16**). Competitive inhibitors usually resemble normal substrates, but they cannot be converted to products.

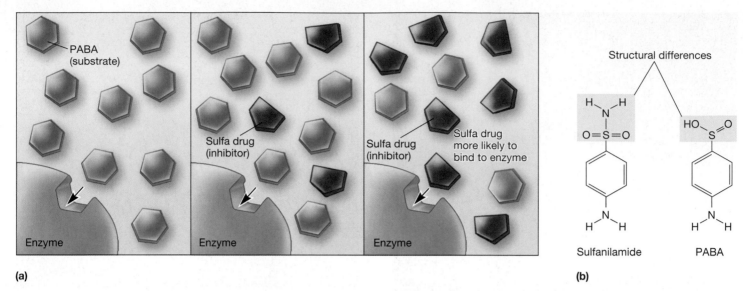

Figure 9.16 Competitive Inhibition of Enzyme Activity.   (a) A competitive inhibitor is usually similar in shape to the normal substrate of the enzyme, and therefore can bind the active site of the enzyme. This prevents the substrate from binding, and the reaction is blocked. (b) Structure of sulfanilamide, a structural analog of PABA. PABA is the substrate of an enzyme involved in folic acid biosynthesis. When sulfanilamide binds the enzyme, activity of the enzyme is inhibited and synthesis of folic acid is stopped.

Competitive inhibitors are important in the treatment of many microbial diseases. Sulfa drugs such as sulfanilamide (figure 9.16*b*) resemble *p*-aminobenzoate (PABA), a molecule used in the formation of the coenzyme folic acid. The drugs compete with PABA for the catalytic site of an enzyme involved in folic acid synthesis. This blocks the production of folic acid and inhibits bacterial growth. Humans are not harmed because they do not synthesize folic acid but rather obtain it in their diet. >>
Antibacterial drugs: Metabolic antagonists (section 31.4)

Noncompetitive inhibitors affect enzyme activity by binding to the enzyme at some location other than the active site. This alters the enzyme's shape, rendering it inactive or less active. These inhibitors are called noncompetitive because they do not directly compete with the substrate. Heavy metal poisons such as mercury frequently are noncompetitive inhibitors of enzymes.

1. What is an enzyme? How does it speed up reactions? How are enzymes named? Define apoenzyme, holoenzyme, cofactor, coenzyme, prosthetic group, active or catalytic site, and activation energy.

2. Draw a diagram showing how enzymes catalyze reactions by altering the activation energy. What is a transition state complex? Use the diagram to explain why enzymes do not change the equilibria of the reactions they catalyze.

3. What is the difference between the lock-and-key and the induced-fit models of enzyme-substrate complex formation?

4. Define the terms Michaelis constant and maximum velocity. How does enzyme activity change with substrate concentration, pH, and temperature?

5. What special properties might an enzyme isolated from a psychrophilic bacterium have? Will enzymes need to lower the activation energy more or less in thermophiles than in psychrophiles?

6. What are competitive and noncompetitive inhibitors, and how do they inhibit enzymes?

9.8 RIBOZYMES

Biologists once thought that all cellular reactions were catalyzed by enzymes. However, in the early 1980s Thomas Cech and Sidney Altman discovered that some RNA molecules also can catalyze reactions. These RNA molecules are called **ribozymes.** One important ribozyme is located in ribosomes and is responsible for catalyzing peptide bond formation during protein synthesis. However, the best-studied ribozymes catalyze self-splicing of RNA. This process is widespread and occurs in *Tetrahymena* (a protist) pre-rRNA; in the mitochondrial rRNA and mRNA of yeast and other fungi; in chloroplast tRNA, rRNA, and mRNA; in mRNA from some bacterial viruses (e.g., the T4 phage of *E. coli*); and in the hepatitis delta virusoid. Just as with enzymes, the shape of a ribozyme is essential to catalytic efficiency. Ribozymes even have Michaelis-Menten kinetics (figure 9.15). The ribozyme from the hepatitis delta virusoid catalyzes RNA cleavage that is involved in its replication. It is unusual in that the same RNA can fold into two shapes with quite different catalytic activities: the regular RNA cleavage activity and an RNA ligation reaction. >> *Translation (section 12.7); Microbial evolution (section 17.1)*

9.9 REGULATION OF METABOLISM

Microorganisms must regulate their metabolism to conserve raw materials and energy and to maintain a balance among various cell components. Because they live in environments where the nutrients, energy sources, and physical conditions often change rapidly, they must continuously monitor internal and external conditions and respond accordingly. This involves activating or inactivating pathways as needed. For instance, if a particular energy source is unavailable, the enzymes required for its use are not needed and their further synthesis is a waste of carbon, nitrogen, and energy. Similarly it would be extremely wasteful for a microorganism to synthesize the enzymes required to manufacture a certain end product if that end product were already present in adequate amounts.

The drive to maintain balance and conserve energy is evident in the regulatory responses of a bacterium such as *E. coli*. If the bacterium is grown in a very simple medium containing only glucose as a carbon and energy source, it will synthesize all needed cell components in balanced amounts. However, if the amino acid tryptophan is added to the medium, the pathway synthesizing tryptophan will be immediately inhibited and synthesis of the pathway's enzymes will slow or cease. Likewise, if *E. coli* is transferred to a medium containing only the sugar lactose, it will synthesize the enzymes required for catabolism of this nutrient. In contrast, when *E. coli* grows in a medium possessing both glucose and lactose, glucose (the sugar supporting most rapid growth) is catabolized first. The culture will use lactose only after the glucose supply has been exhausted. Metabolic pathways can be regulated in three major ways: (1) metabolic channeling, (2) regulation of the amount of synthesis of a particular enzyme, and (3) direct stimulation or inhibition of the activity of critical enzymes.

Metabolic channeling influences pathway activity by localizing metabolites and enzymes into different parts of a cell. One of the most common metabolic channeling mechanisms is **compartmentation,** the differential distribution of enzymes and metabolites among separate cell structures or organelles. Compartmentation is particularly important in eucaryotic microorganisms with their many membrane-bound organelles. For example, fatty acid catabolism is located within the mitochondrion, whereas fatty acid synthesis occurs in the cytoplasm. The periplasm in procaryotes can also be considered an example of compartmentation. Compartmentation makes possible the simultaneous but separate operation and regulation of similar pathways. Furthermore, pathway activities can be coordinated through regulation of the transport of metabolites and coenzymes between cell compartments.

Metabolic channeling can generate marked variations in metabolite concentrations and therefore directly affect enzyme activity. Substrate levels are generally around 10^{-3} M to 10^{-6} M or even lower. Thus they may be in the same range as enzyme concentrations and equal to or less than the Michaelis constants (K_m) of many enzymes (figure 9.15). Under these conditions, the concentration of an enzyme's substrate may control its activity because the substrate concentration is in the rising portion of the hyperbolic substrate saturation curve (**figure 9.17**). As the substrate level increases, it is converted to product more rapidly; a decline in substrate concentration automatically leads to lower enzyme activity. If two enzymes in different pathways use the same metabolite, they may directly compete for it. The pathway winning this competition—the one with the enzyme having the lowest K_m value for the metabolite—will operate closer to full capacity. Thus channeling within a cell compartment can regulate and coordinate metabolism through variations in metabolite and coenzyme levels.

In the second regulatory mechanism—regulation of the amount of synthesis of a particular enzyme—transcription and translation can be regulated to control the amount of an enzyme present in the cell. Regulation at this level is relatively slow, but it saves the cell considerable energy and raw material. In contrast, the direct stimulation or inhibition of the activity of critical enzymes rapidly alters pathway activity. It is often called posttranslational regulation because it occurs after the enzyme has been synthesized. This type of regulation is discussed next. **>>** *Transcription (section 12.5); Translation (section 12.7); Regulation of gene expression (chapter 13).*

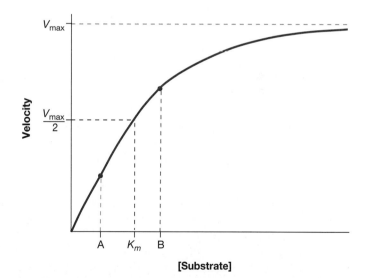

Figure 9.17 Control of Enzyme Activity by Substrate Concentration. An enzyme-substrate saturation curve with the Michaelis constant (K_m) and the velocity equivalent to half the maximum velocity (V_{max}) indicated. The initial velocity of the reaction (v) is plotted against the substrate concentration [Substrate]. The maximum velocity is the greatest velocity attainable with a fixed amount of enzyme under defined conditions. When the substrate concentration is equal to or less than the K_m, the enzyme's activity will vary almost linearly with the substrate concentration. Suppose the substrate increases in concentration from level A to B. Because these concentrations are in the range of the K_m, a significant increase in enzyme activity results. A drop in concentration from B to A will lower the rate of product formation.

1. Briefly describe the three ways a metabolic pathway may be regulated.

2. Define the terms metabolic channeling and compartmentation. How are they involved in the regulation of metabolism?

9.10 POSTTRANSLATIONAL REGULATION OF ENZYME ACTIVITY

Adjustment of the activity of regulatory enzymes and other proteins controls the functioning of many metabolic pathways and cellular processes. A number of posttranslational regulatory mechanisms are known. Some are irreversible—for instance, cleavage of a protein can either activate or inhibit its activity. Others, such as allosteric regulation and covalent modification, are reversible. Although our focus is on the regulation of metabolic pathways by reversible mechanisms, it is important to remember that not all proteins or enzymes function in metabolic pathways. Instead, some are involved in cellular behaviors. At the end of this section, we consider the regulation of one of these behaviors—chemotaxis.

Allosteric Regulation

Most regulatory enzymes are **allosteric enzymes.** The activity of an allosteric enzyme is altered by a small molecule known as an **allosteric effector.** The effector binds reversibly by noncovalent forces to a **regulatory site** separate from the catalytic site and causes a change in the shape (conformation) of the enzyme (**figure 9.18**). The activity of the catalytic site is altered as a result. A positive effector increases enzyme activity, whereas a negative effector decreases activity (i.e., inhibits the enzyme). These changes in activity often result from alterations in the apparent affinity of the enzyme for its substrate, but changes in maximum velocity also can occur.

Covalent Modification of Enzymes

Regulatory enzymes also can be switched on and off by **reversible covalent modification.** Usually this occurs through the addition and removal of a particular chemical group, typically a phosphoryl, methyl, or adenylyl group.

One of the most intensively studied regulatory enzymes is *E. coli* glutamine synthetase, an enzyme involved in nitrogen assimilation. It is a large, complex enzyme consisting of 12 subunits, each of which can be covalently modified by an adenylic acid residue (**figure 9.19**). When an adenylic acid residue is attached to all of its 12 subunits, glutamine synthetase is not very active. Removal of AMP groups produces more active deadenylylated glutamine synthetase, and glutamine is formed.
>> *Synthesis of amino acids: Nitrogen assimilation (section 11.5)*

Using covalent modification for the regulation of enzyme activity has some advantages. These interconvertible enzymes often are also allosteric. For instance, glutamine synthetase also is regulated allosterically. Because each form can respond differently to allosteric effectors, systems of covalently modified enzymes are able to

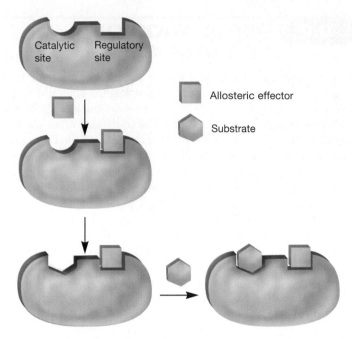

Figure 9.18 Allosteric Regulation. The structure and function of an allosteric enzyme. In this example, the effector first binds to a separate regulatory site and causes a change in enzyme conformation that results in an alteration in the shape of the active site. The active site can now more effectively bind the substrate. This effector is a positive effector because it stimulates substrate binding and catalytic activity.

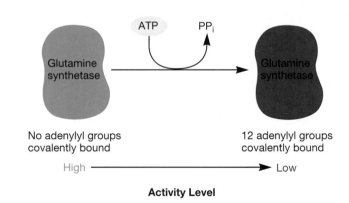

Figure 9.19 Regulation of Glutamine Synthetase Activity. Glutamine synthetase consists of 12 subunits, each of which can be adenylylated. As the number of adenylyl groups increases, the activity of the enzyme decreases.

respond to more stimuli in varied and sophisticated ways. Regulation can also be exerted on the enzymes that catalyze the covalent modifications, which adds a second level of regulation to the system.

Feedback Inhibition

The rate of many metabolic pathways is adjusted through control of the activity of the regulatory enzymes described in the preceding sections. Every pathway has at least one pacemaker enzyme that catalyzes the slowest or rate-limiting reaction in the pathway. Because other reactions proceed more rapidly than the pacemaker reaction, changes in the activity of this enzyme directly alter the speed with which a pathway operates. Usually the first step in a pathway is a pacemaker reaction catalyzed by a regulatory enzyme. The end product of the pathway often inhibits this regulatory enzyme, a process known as **feedback inhibition** or **end product inhibition.** Feedback inhibition ensures balanced production of a pathway end product. If the end product becomes too concentrated, it inhibits the regulatory enzyme and slows its own synthesis. As the concentration of the end product decreases, pathway activity again increases and more product is formed. In this way, feedback inhibition automatically matches end product supply with the demand.

Frequently a biosynthetic pathway branches to form more than one end product. In such a situation, the synthesis of all end products must be coordinated precisely. It would not do to have one end product present in excess while another is lacking. Branching biosynthetic pathways usually achieve a balance among end products by using regulatory enzymes at branch points (**figure 9.20**). If an end product is present in excess, it often inhibits the branch-point enzyme on the sequence of reactions leading to its formation, in this way regulating its own formation without affecting the synthesis of other products. In figure 9.20, notice that both products also inhibit the initial enzyme in the pathway. An excess of one product slows the flow of carbon into the whole pathway while inhibiting the appropriate branch-point enzyme. Because less carbon is required when a branch is not functioning, feedback inhibition of the initial pacemaker enzyme helps match the supply with the demand in branching pathways. The regulation of multiple branched pathways is often made even more sophisticated by the presence of **isoenzymes,** different forms of

an enzyme that catalyze the same reaction. The initial pacemaker step may be catalyzed by several isoenzymes, each under separate control. In this case, an excess of a single end product reduces pathway activity but does not completely block pathway function because some isoenzymes are still active.

1. Define allosteric enzyme and allosteric effector.
2. How can regulatory enzymes be influenced by reversible covalent modification? What group is used for this purpose with glutamine synthetase, and which form of this enzyme is active?
3. What is a pacemaker enzyme? Feedback inhibition? How does feedback inhibition automatically adjust the concentration of a pathway end product?
4. What is the significance of the fact that regulatory enzymes often are located at pathway branch points? What are isoenzymes, and why are they important in pathway regulation?

Chemotaxis

Thus far in our discussion of the regulation of enzyme activity, we have focused on the control of metabolic pathways brought about by modulating the activity of certain regulatory enzymes that function in the pathway. However, not all enzymes function in metabolic pathways. Rather, some enzymes are involved in processes that are more complex. These include behavioral changes made by microbes in response to their environment. Chemotaxis illustrates the roles enzymes play in microbial behavior and how controlling enzyme activity changes that behavior.

In chapter 3, chemotaxis is briefly introduced. Recall that microorganisms are able to sense chemicals in their environment and move either toward them or away from them, depending on whether the chemical is an attractant or a repellant. For simplicity, we only concern ourselves with movement toward an attractant. The best-studied chemotactic system is that of *E. coli,* which, like many other bacteria, exhibits two movement modalities: a smooth swimming motion called a run, interrupted by tumbles. A run occurs when the flagellum rotates in a counterclockwise direction (CCW), and a tumble occurs when the flagellum rotates clockwise (CW) *(see figures 3.38 and 3.42).* The cell alternates between these two types of movements, with the tumble establishing the direction of movement in the run that follows. When *E. coli* is in an environment that is homogenous—that is, the concentration of all chemicals in the environment is the same throughout its habitat—the cell moves about randomly, with no apparent direction or purpose; this is called a random walk. However, if a chemical gradient exists in its environment, the frequency of tumbles decreases as long as the cell is moving toward the attractant. In other words, the length of time spent moving toward the attractant is increased and eventually the cell gets closer to the attractant. The process is not perfect, and the cell must continually readjust its direction through a trial-and-error process that is mediated by tumbling. When one examines

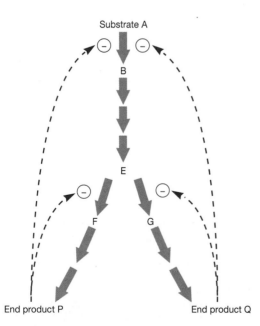

Figure 9.20 Feedback Inhibition. Feedback inhibition in a branching pathway with two end products. The branch-point enzymes, those catalyzing the conversion of intermediate E to F and G, are regulated by feedback inhibition. Products P and Q also inhibit the initial reaction in the pathway. A colored line with a minus sign at one end indicates that an end product, P or Q, is inhibiting the enzyme catalyzing the step next to the minus.

the path taken by the cell, it is similar to a random walk but is biased toward the attractant.

For over three decades, scientists have been dissecting this complex behavior in order to understand how *E. coli* senses the presence of an attractant, how it switches from a run to a tumble and back again, and how it "knows" it is heading in the correct direction. These studies reveal that the chemotactic response of *E. coli* involves a number of enzymes and other proteins that are regulated by covalent modification. One important component is a phosphorelay system. **Phosphorelay systems** consist of at least

two proteins: a **sensor kinase** and a **response regulator.** As described here, a phosphorelay system is used to regulate enzyme activity. Other phosphorelay systems are used to regulate the amount of protein synthesized and generally use only these two components. These systems are described in chapter 13.

For chemotaxis to occur, *E. coli* must determine if an attractant is present and then modulate the activity of the phosphorelay system that dictates the rotational direction of the flagellum (i.e., either run or tumble). *E. coli* senses chemicals in its environment when they bind to chemoreceptors (**figure 9.21**). Numerous chemoreceptors have been identified. We will focus on one class of receptors called methyl-accepting chemotaxis proteins (MCPs). The phosphorelay system that controls direction of flagellar rotation consists of the sensor kinase CheA and the response regulator CheY. When activated, CheA phosphorylates itself using ATP (figure 9.21*c*). The phosphoryl group is then quickly transferred to CheY. Phosphorylated CheY diffuses through the cytoplasm to the flagellar motor. Upon interacting with the motor, the direction of rotation is switched from CCW to CW, and a tumble ensues. When CheA is inactive, the flagellum rotates in its default mode (CCW), and the cell moves forward in a smooth run.

As implied by the preceding discussion, the state of the MCPs must be communicated to the CheA/CheY phosphorelay system. How is this accomplished? The MCPs are buried in the plasma membrane with different parts exposed on each side of the membrane (figure 9.21*c*). The periplasmic side of each MCP has a binding site for

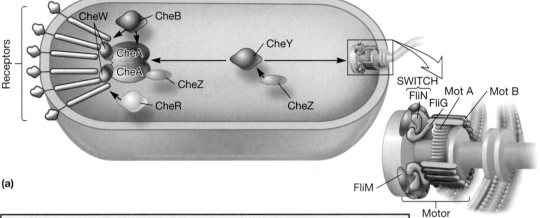

(a)

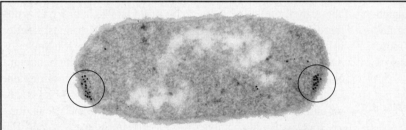

(b)

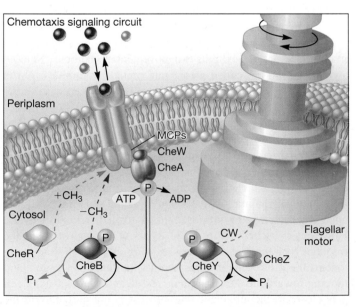

(c)

Figure 9.21 Proteins and Signaling Pathways of the Chemotaxis Response in *E. coli.* (a) The methyl-accepting chemotaxis proteins (MCPs) form clusters associated with the CheA and CheW proteins. CheA is a sensor kinase that when activated phosphorylates CheB, a methylesterase, or CheY. Phosphorylated CheY interacts with the FliM protein of the flagellar motor, causing rotation of the flagellum to switch from counterclockwise (CCW) to clockwise (CW). This results in a switch from a run (CCW rotation) to a tumble (CW rotation). (b) MCPs, CheW, CheA complexes form large clusters of receptors at either end of the cell, as shown in this electron micrograph of *E. coli.* Gold-tagged antibodies were used to label the receptor clusters, which appear as black dots (encircled). (c) The chemotactic signaling pathways of *E. coli.* The pathways that increase the probability of CCW rotation are shown in red. CCW rotation is the default rotation. It is periodically interrupted by CW rotation, which causes tumbling. The pathways that lead to CW rotation are shown in green. Molecules shown in gray are unphosphorylated and inactive. Note that MCP, CheA, and CheZ are homodimers. CheW, CheB, CheY, and CheR are monomers.

one or more attractant molecules. The cytoplasmic side of an MCP interacts with two proteins, CheW and CheA. The CheW protein binds to the MCP and helps attach the CheA protein. Together with CheW and CheA, the MCP receptors form large receptor clusters at one or both poles of the cell (figure 9.21*b*). It is thought that smaller aggregations of the MCPs, CheA, and CheW function as signaling teams and are the building blocks of the receptor clusters. The number of each of these molecules in the signaling team is not clear, but it has been suggested that each team includes three receptors, often of different types, two CheW molecules, and one CheA dimer (**figure 9.22**). It has also been suggested that the signaling teams aggregate with each other to form "signaling leagues." Finally, the signaling teams (and perhaps signaling leagues) become interconnected by an unknown mechanism to form the receptor clusters visible at the poles of the cell.

No matter what the precise stoichiometry or architecture of the receptor clusters, evidence exists that the MCPs in each signaling team work cooperatively to modulate CheA activity. When any one of the MCPs in the signaling team is bound to an attractant, CheA autophosphorylation is inhibited, the flagellum continues rotating CCW, and the cell continues in its run. Because of this cooperation, the cell can respond to very low concentrations of attractant. Furthermore, it can integrate signals from all receptors in the team (figure 9.22). On the other hand, if attractant levels decrease, so that the level of attractant bound to the

MCPs in a signaling team decreases, CheA is stimulated to autophosphorylate, the phosphorelay is set into motion, and the cell begins to tumble. However, tumbling does not continue indefinitely. About 10 seconds after the switch to CW rotation occurs, the phosphoryl group is removed from CheY by the CheZ protein, and CCW rotation is resumed.

But how does *E. coli* measure the concentration of attractant in its environment, and how does it know when it is moving toward the attractant? *E. coli* measures the concentration of an attractant every few seconds and determines if the concentration is increasing or decreasing over time. As long as the concentration increases, the cell continues a run. If the concentration decreases, a tumble is triggered. To compare concentrations of the attractant over time, *E. coli* must have a mechanism for "remembering" the previous concentration. *E. coli* accomplishes this by comparing the overall methylation level of the MCPs (on the cytoplasmic side) with the overall amount of attractant bound (on the periplasmic face). The cytoplasmic portion of each MCP has four to six glutamic acid residues that can be methylated. Addition and removal of methyl groups is catalyzed by two different enzymes. Methylation is catalyzed by the MCP-specific methyltransferase CheR. Demethylation is catalyzed by the MCP-specific methylesterase CheB. Methylation occurs at a steady rate regardless of the attractant level. However, an MCP-attractant complex is a better substrate for CheR than is an MCP that is not bound to attractant. Thus when attractant is bound, methylation of the MCP is favored. The methylesterase activity of CheB is also modified by the CheA protein. As long as the concentration of the attractant keeps increasing, the number of MCPs bound to attractant remains high, and the MCP methylation level remains high. However, if the attractant concentration decreases, the level of methylation will exceed the level of attractant bound. This disparity in methylation level and MCP-bound attractant stimulates CheA to autophosphorylate. As a result, the phosphorelay signal for CW flagellar rotation is initiated and the cell tumbles in an attempt to reorient itself in the gradient so that it is moving up the gradient (toward the attractant) rather than down the gradient (away from the attractant). At the same time, some of the phosphoryl groups on CheA are transferred to CheB. This activates CheB, and it removes methyl groups from the MCPs. This lowers the methylation level so that it is commensurate with the number of MCPs bound to attractant. A few seconds later, the number of MCPs bound to attractant will be compared to this new methylation level. Based on the correspondence of the two, the cell will determine if it is again moving up the gradient. If it is, tumbling will be suppressed (as will methylesterase activity) and the run will continue.

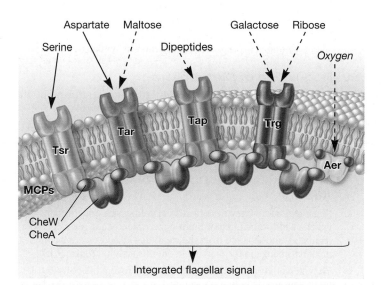

Figure 9.22 The Methyl-Accepting Chemotaxis Proteins of *E. coli*. The attractants sensed by each methyl-accepting chemotaxis protein (MCP) are shown. Some are sensed directly, when the attractant binds the MCP (solid lines). Others are sensed indirectly (dashed lines). The attractants maltose, dipeptides, galactose, and ribose are detected by their interaction with periplasmic binding proteins. Oxygen is detected indirectly by the Aer chemoreceptor, which differs from other MCPs in that it lacks a periplasmic sensing domain. Instead, the cytoplasmic domain has a binding site for FAD. FAD is an important electron carrier found in many electron transport systems. The redox state of the MCP-bound FAD molecule is used to monitor the functioning of the electron transport system. This in turn mediates a tactic response to oxygen.

1. What is a phosphorelay system?
2. Describe the MCP-CheW-CheA receptor complex. What two proteins are phosphorylated by CheA? What is the role of each?
3. How does the MCP regulate the rate of CheA autophosphorylation? How does this mediate chemotaxis?

Summary

9.1 Energy and Work

a. Metabolism is the total of all chemical reactions that occur in cells. It can be divided into two parts: the energy-conserving fueling reactions (called catabolism) and anabolism.

b. Energy is the capacity to do work. Living cells carry out three major kinds of work: chemical work of biosynthesis, transport work, and mechanical work.

9.2 Laws of Thermodynamics

a. The first law of thermodynamics states that energy is neither created nor destroyed.

b. The second law of thermodynamics states that changes occur in such a way that the randomness or disorder of the universe increases to the maximum possible. That is, entropy always increases during spontaneous processes.

9.3 Free Energy and Reactions

a. The first and second laws can be combined to determine the amount of energy made available for useful work.

$$\Delta G = \Delta H - T \cdot \Delta S$$

In this equation the change in free energy (ΔG) is the energy made available for useful work, the change in enthalpy (ΔH) is the change in heat content, and the change in entropy is ΔS.

b. The standard free energy change ($\Delta G^{\circ\prime}$) for a chemical reaction is directly related to the equilibrium constant.

c. In exergonic reactions, $\Delta G^{\circ\prime}$ is negative and the equilibrium constant is greater than one; the reaction goes to completion as written. Endergonic reactions have a positive $\Delta G^{\circ\prime}$ and an equilibrium constant less than one (**figure 9.1**).

9.4 ATP

a. ATP is a high-energy molecule that serves as energy currency; it transports energy in a useful form from one reaction or location in a cell to another (**figure 9.2**).

b. ATP is readily synthesized from ADP and P_i using energy released from exergonic reactions; when hydrolyzed back to ADP and P_i, it releases the energy, which is used to drive endergonic reactions. This cycling of ATP with ADP and P_i is called the cell's energy cycle (**figure 9.4**).

9.5 Oxidation-Reduction Reactions

a. In oxidation-reduction (redox) reactions, electrons move from an electron donor to an electron acceptor. The standard reduction potential measures the tendency of the donor to give up electrons.

b. Redox couples with more negative reduction potentials donate electrons to those with more positive potentials, and energy is made available during the transfer (**figure 9.5** and **table 9.1**).

9.6 Electron Transport Chains

a. Some of the most important electron carriers in cells are NAD^+, $NADP^+$, FAD, FMN, coenzyme Q, cytochromes, and the non-heme iron (FeS) proteins.

b. Electron carriers are often organized into electron transport chains that are located in membranes. These chains are critical to the energy-conserving processes observed during aerobic respiration, anaerobic respiration, chemolithotrophy, and photosynthesis (**figure 9.6**).

9.7 Enzymes

a. Enzymes are protein catalysts that catalyze specific reactions.

b. Many enzymes consist of a protein component, the apoenzyme, and a nonprotein cofactor that may be a prosthetic group, a coenzyme, or a metal activator.

c. Enzymes speed reactions by binding substrates at their active sites and lowering the activation energy (**figure 9.12**).

d. The rate of an enzyme-catalyzed reaction increases with substrate concentration at low substrate levels and reaches a plateau (the maximum velocity) at saturating substrate concentrations. The Michaelis constant is the substrate concentration that the enzyme requires to achieve half maximal velocity (**figure 9.15**).

e. Enzymes have pH and temperature optima for activity.

f. Enzyme activity can be slowed by competitive and noncompetitive inhibitors (**figure 9.16**).

9.8 Ribozymes

a. Some RNA molecules have catalytic activity.

b. Some ribozymes alter either their own structure or that of other RNAs. An important ribozyme is located in the ribosome, where it links amino acids together during protein synthesis.

9.9 Regulation of Metabolism

a. The regulation of metabolism keeps cell components in proper balance and conserves metabolic energy and material.

b. Metabolic channeling localizes metabolites and enzymes in different parts of the cell and influences pathway activity. A common channeling mechanism is compartmentation.

9.10 Posttranslational Regulation of Enzyme Activity

a. Many regulatory enzymes are allosteric enzymes, enzymes in which an allosteric effector binds noncovalently and reversibly to a regulatory site separate from the catalytic site and causes a conformational change in the enzyme to alter its activity (**figure 9.18**).

b. Enzyme activity also can be regulated by reversible covalent modification. Usually a phosphoryl, methyl, or adenylyl group is attached to the enzyme.

c. The first enzyme in a pathway and enzymes at branch points often are subject to feedback inhibition by one or more end products. Excess end product slows its own synthesis (**figure 9.20**).

d. Complex behaviors such as chemotaxis can also be regulated by altering enzyme activity (**figures 9.21** and **9.22**).

Critical Thinking Questions

1. How could electron transport be driven in the opposite direction? Why would it be desirable to do this?

2. Suppose that a chemical reaction had a large negative $\Delta G°'$ value. What would this indicate about its equilibrium constant? If displaced from equilibrium, would it proceed rapidly to completion? Would much or little free energy be made available?

3. Examine the structures of macromolecules in appendix I. Which type has the most electrons to donate? Why are carbohydrates usually the primary source of electrons for chemoorganotrophic bacteria?

4. Most enzymes do not operate at their biochemical optima inside cells. Why not?

5. Examine the branched pathway shown here for the synthesis of the amino acids aspartate, methionine, lysine, threonine, and isoleucine. For each of these two scenarios, answer the following questions:

   a. Which portion(s) of the pathway would need to be shut down in this situation?

   b. How might allosteric control be used to accomplish this?

Scenario 1: The microbe is cultured in a medium containing aspartate and lysine but lacking methionine, threonine, and isoleucine.

Scenario 2: The microbe is cultured in a medium containing a rich supply of all five amino acids.

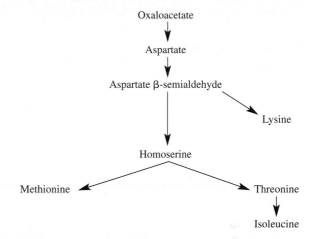

Learn More

Learn more by visiting the Prescott website at www.mhhe.com/prescottprinciples, where you will find a complete list of references.

Catabolism: Energy Release and Conservation

10

aerobic respiration An energy-yielding process in which molecules, often organic, are oxidized with oxygen as the final electron acceptor.

anaerobic respiration An energy-yielding process in which the terminal acceptor for an electron transport chain is a molecule other than oxygen.

anoxygenic photosynthesis Photosynthesis that does not oxidize water to produce oxygen.

ATP synthase A membrane-bound enzyme that synthesizes ATP from ADP + P_i using the energy derived from the proton motive force.

bacteriochlorophyll A modified chlorophyll that is the primary light-trapping pigment in anoxygenic photosynthetic bacteria.

bacteriorhodopsin A transmembranous protein to which retinal is bound; it functions as a light-driven proton pump.

chemiosmotic hypothesis The hypothesis that a proton and electrochemical gradient are generated by electron transport and then used to perform work (e.g., drive ATP synthesis).

chemolithotroph A microorganism that oxidizes reduced inorganic compounds to derive both energy and electrons.

chlorophyll The green photosynthetic pigment that consists of a tetrapyrrole ring with a central magnesium atom.

cyclic photophosphorylation The formation of ATP when light energy is used to move electrons cyclically through an electron transport chain during photosynthesis.

dark reactions Pathways in which photosynthetically derived energy is used to drive CO_2 fixation.

Embden-Meyerhof pathway A glycolytic pathway that degrades glucose to pyruvate and also generates several precursor metabolites.

Entner-Doudoroff pathway A glycolytic pathway that converts glucose to pyruvate and glyceraldehyde 3-phosphate.

fermentation An energy-yielding process in which an organic molecule is oxidized without an exogenous electron acceptor or participation of an electron transport chain.

glycolysis The conversion of glucose to pyruvic acid by use of the Embden-Meyerhof pathway, pentose phosphate pathway, or Entner-Doudoroff pathway.

light reactions Photochemical events that lead to the synthesis of ATP and, in some cases, the reduction of $NAD(P)^+$ to $NAD(P)H$.

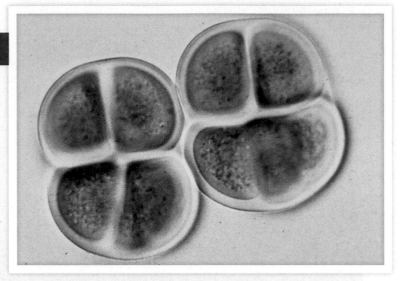

The cyanobacterium *Chroococcus turgidus,* an oxygenic phototroph.

noncyclic photophosphorylation The process in which light energy is used to make ATP and reducing power when electrons are moved from water to $NADP^+$ during oxygenic photosynthesis.

oxidative phosphorylation The synthesis of ATP from ADP + P_i using energy made available during electron transport initiated by the oxidation of a chemical energy source.

oxygenic photosynthesis Photosynthesis that oxidizes water to form oxygen; the form of photosynthesis characteristic of plants, protists, and cyanobacteria.

pentose phosphate pathway A glycolytic pathway that forms reducing power (NADPH) for biosynthesis and several precursor metabolites.

photophosphorylation The synthesis of ATP from ADP + P_i using energy made available during electron transport initiated by the absorption of light energy.

photosynthesis The trapping of light energy and its conversion to chemical energy, which is then used to reduce CO_2 and incorporate it into organic molecules.

proton motive force (PMF) The potential energy arising from a chemical and charge gradient of protons across a membrane.

substrate-level phosphorylation The synthesis of ATP from ADP by phosphorylation coupled with the exergonic breakdown of a high-energy organic molecule.

tricarboxylic acid (TCA) cycle The cycle that oxidizes acetyl coenzyme A to CO_2 and generates NADH and $FADH_2$ for oxidation in the electron transport chain; the cycle also supplies precursor metabolites; also called citric acid cycle and Kreb's cycle.

It is in the fueling reactions that bacteria display their extraordinary metabolic diversity and versatility. Bacteria have evolved to thrive in almost all natural environments, regardless of the nature of available sources of carbon, energy, and reducing power. . . . The collective metabolic capacities of bacteria allow them to metabolize virtually every organic compound on this planet. . . .

—*F. C. Neidhardt, J. L. Ingraham, and M. Schaechter*

Microbes are the most successful organisms on Earth as witnessed by their growth under almost every conceivable condition. In large part, their success results from the diversity of their fueling reactions—those reactions that convert energy from an organism's energy source into ATP. The fueling reactions of the major nutritional types of organisms are the focus of this chapter (**figure 10.1**).

Animals and many microbes are chemoorganoheterotrophs. These organisms use organic molecules as their source of energy, carbon, and electrons. In other words, the same molecule that supplies them with energy also supplies them with carbon and electrons. Chemoorganoheterotrophs (often simply referred to as chemoorganotrophs or chemoheterotrophs) can use one or more of the following catabolic processes: aerobic respiration, anaerobic respiration, or fermentation. Chemolithoautotrophs use CO_2 as a carbon source and reduced inorganic molecules as sources of both energy and electrons. Their energy-conserving processes are sometimes referred to as respiration because they are similar to the respiratory processes carried out by chemoorganoheterotrophs. Photolithotrophic microbes use light as their source of energy and inorganic molecules as a source of electrons. Some use water as their electron source, as do plants, and release oxygen into the atmosphere in a process called oxygenic photosynthesis. Some photosynthetic bacteria use molecules such as hydrogen sulfide (H_2S) rather than water as an electron source, so do not release oxygen into the atmosphere; they are called anoxygenic phototrophs. Photolithotrophs are usually autotrophic, using CO_2 as a carbon source. However, some phototrophic microbes are heterotrophic.

We begin our consideration of fueling reactions with an overview of the metabolism of chemoorganotrophs. This is followed by an introduction to the oxidation of carbohydrates, especially glucose, and a discussion of ATP synthesis during aerobic and anaerobic respiration. Fermentation is then described, followed by a survey of the breakdown of other carbohydrates and organic

substances. We end the chapter with sections on chemolithotrophy and phototrophy.

10.1 CHEMOORGANO-TROPHIC FUELING PROCESSES

When chemoorganotrophs oxidize an organic energy source, the electrons released are accepted by a variety of electron acceptors. When the electron acceptor is exogenous (i.e., externally supplied), the metabolic process is called **respiration** and may be divided into two different types (**figure 10.2**). In aerobic respiration, the final electron acceptor is oxygen, whereas the terminal acceptor in anaerobic respiration is a different exogenous molecule such as NO_3^-, SO_4^{2-}, CO_2, Fe^{3+}, and SeO_4^{2-}. Organic acceptors such as fumarate and humic acids also may be used. Respiration involves the activity of an electron transport chain. As electrons pass through the chain to the final electron acceptor, a type of potential energy called the proton motive force (PMF) is generated and used to synthesize ATP from ADP and P_i. In contrast, **fermentation** (Latin *fermentare*, to cause to rise) uses an electron acceptor that is endogenous (from within the cell) and does not involve an electron transport chain or the generation of PMF. The endogenous electron acceptor is usually an intermediate (e.g., pyruvate) of the catabolic pathway used to degrade and oxidize the organic energy source. During fermentation, ATP is synthesized only by substrate-level phosphorylation, a process in which a phosphate group is transferred to ADP from a high-energy molecule (e.g., phosphoenolpyruvate) generated by catabolism of the energy source.

By convention, aerobic respiration, anaerobic respiration, and fermentation are usually described with glucose as the energy source. This is done for several reasons. One is that glucose is used by many chemoorganotrophs as an energy source. But perhaps more important is the way catabolic pathways are

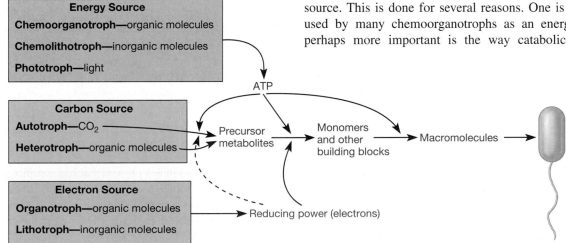

Figure 10.1 Overview of Metabolism. The cell structures of organisms are assembled from various macromolecules (e.g., nucleic acids and proteins). Macromolecules are synthesized from monomers and other building blocks (e.g., nucleotides and amino acids), which are the products of biochemical pathways that begin with precursor metabolites (e.g., pyruvate and α-ketoglutarate). In autotrophs, the precursor metabolites arise from CO_2-fixation and related pathways; in heterotrophs, they arise from reactions of the central metabolic pathways. Reducing power and ATP are consumed in many metabolic pathways. As indicated in colored boxes, all organisms can be defined metabolically in terms of their energy source, carbon source, and electron source.

Chemoorganotrophic Fueling Processes

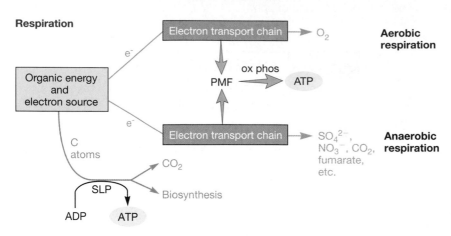

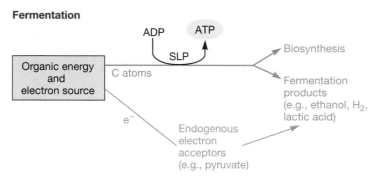

Figure 10.2 Chemoorganotrophic Fueling Processes. Organic molecules serve as energy and electron sources for all three fueling processes used by chemoorganotrophs. In aerobic respiration and anaerobic respiration, the electrons pass through an electron transport chain. This generates a proton motive force (PMF), which is used to synthesize most of the cellular ATP by a mechanism called oxidative phosphorylation (ox phos); a small amount of ATP is made by a process called substrate-level phosphorylation (SLP). In aerobic respiration, O_2 is the terminal electron acceptor, whereas in anaerobic respiration exogenous molecules other than O_2 serve as electron acceptors. During fermentation, endogenous organic molecules act as electron acceptors, electron flow is not coupled with ATP synthesis, and ATP is synthesized only by substrate-level phosphorylation.

organized. Most organisms can use a wide variety of organic molecules as energy sources, including macromolecules such as proteins, polysaccharides, and lipids (**figure 10.3**). These molecules are broken down into their constituent parts: amino acids, monosaccharides, and lipids. This is often referred to as the first stage of catabolism; it releases relatively little energy. The monomers are subsequently converted to glucose or to intermediates of the pathways used in its catabolism. Thus nutrient molecules are funneled into ever fewer metabolic intermediates. Indeed a common pathway often degrades many similar molecules (e.g., several different sugars). The existence of a few metabolic pathways, each degrading many nutrients, greatly increases metabolic efficiency by avoiding the need for a large number of less metabolically flexible pathways.

The catabolic pathways of greatest importance to chemoorganotrophs are the glycolytic pathways (section 10.3) and the tricarboxylic acid (TCA) cycle (section 10.4). Each pathway

consists of a set of enzyme-catalyzed reactions arranged so that the product of one reaction serves as a substrate for the next. They are important not only for their role in catabolism but also for their roles in anabolism. They supply material needed for biosynthesis, including precursor metabolites and reducing power. Precursor metabolites serve as the starting molecules for biosynthetic pathways. Reducing power is used in redox reactions that reduce the precursor metabolites as they are transformed into amino acids, nucleotides, and other small molecules needed for synthesis of macromolecules (figure 10.1).

Pathways that function both catabolically and anabolically are called **amphibolic pathways** (Greek *amphi*, on both sides). Many of the reactions of amphibolic pathways are freely reversible and can be used to synthesize or degrade molecules, depending on the nutrients available and the needs of the microbe. The few irreversible catabolic steps are bypassed in biosynthesis with alternate enzymes that catalyze the reverse reaction (**figure 10.4**). The presence of two separate enzymes, one catalyzing the reversal of the other's reaction, permits independent regulation of the catabolic and anabolic functions of these amphibolic pathways. ⟫ *Precursor metabolites (section 11.2)*

In the following sections, we explore the metabolism of chemoorganotrophs in more detail. The discussion begins with aerobic respiration and introduces the glycolytic pathways, TCA cycle, and other important processes. Many of these also occur during anaerobic respiration. Thus in many cases, anaerobic respiration differs from aerobic respiration mainly in terms of the terminal electron acceptor. Fermentation is quite distinct from respiration. It involves only a subset of the pathways that function during respiration and only partially catabolizes the energy source. However, it is widely used by microbes and has important practical applications. ⟫ *Microbiology of food (chapter 34)*

10.2 AEROBIC RESPIRATION

Aerobic respiration is a process that can completely catabolize an organic energy source to CO_2 using the glycolytic pathways and TCA cycle with O_2 as the terminal electron acceptor (figure 10.2). This is accomplished in three stages (figure 10.3). The first has already been noted: the breakdown of macromolecules into monomers. The monomers are further degraded in the second stage, which includes the glycolytic and other pathways that yield pyruvate or acetyl coenzyme A. This stage produces some ATP as well as NADH, FADH$_2$, or both. Finally, during the third stage, partially oxidized carbon is fed into the TCA cycle and oxidized completely to CO_2 with the production of some ATP, NADH, and FADH$_2$. Also

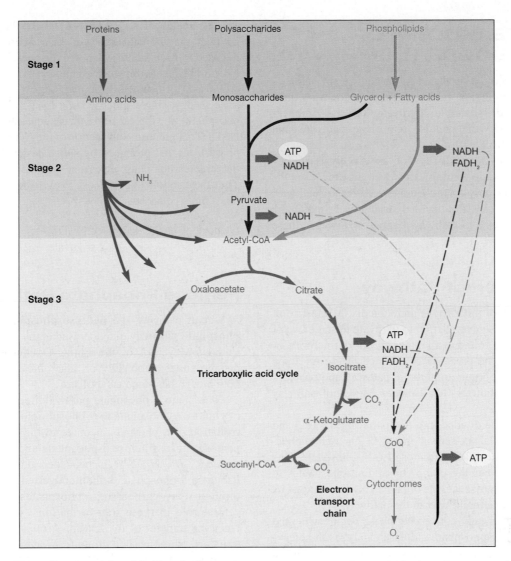

Figure 10.3   The Three Stages of Aerobic Respiration.   A general diagram of aerobic respiration in a chemoorganoheterotroph showing the three stages in this process and the central position of the tricarboxylic acid cycle. Although there are many different proteins, polysaccharides, and lipids, they are degraded through the activity of a few common metabolic pathways. The dashed lines show the flow of electrons, carried by NADH and FADH$_2$, to the electron transport chain.

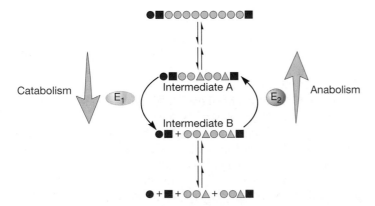

Figure 10.4   Amphibolic Pathway.   A simplified diagram of an amphibolic pathway, such as glycolysis. Note that the interconversion of intermediates A and B is catalyzed by two separate enzymes, E$_1$ operating in the catabolic direction and E$_2$ in the anabolic.

in stage three the NADH and FADH$_2$ that have been formed are oxidized by an electron transport chain, using O$_2$ as the terminal electron acceptor. It is the functioning of the electron transport chain that yields the most ATP for the cell during aerobic respiration.

1. Compare and contrast fermentation and respiration. Give examples of the types of electron acceptors used by each process. What is the difference between aerobic respiration and anaerobic respiration?

2. Why is it to the cell's advantage to catabolize diverse organic energy sources by funneling them into a few common pathways?

3. What is an amphibolic pathway? Why are amphibolic pathways important?

10.3 BREAKDOWN OF GLUCOSE TO PYRUVATE

Microorganisms employ several metabolic pathways to catabolize glucose and other sugars. In this section, we focus on three pathways that degrade sugars to pyruvate: (1) the Embden-Meyerhof pathway, (2) the pentose phosphate pathway, and (3) the Entner-Doudoroff pathway. We refer to these pathways collectively as **glycolytic pathways** or as **glycolysis** (Greek *glyco,* sweet, and *lysis,* a loosening). However, in some texts, the term glycolysis refers only to the Embden-Meyerhof pathway. For the sake of simplicity, the detailed structures of metabolic intermediates are not used in pathway diagrams. However, these can be found in appendix II.

Embden-Meyerhof Pathway

The **Embden-Meyerhof pathway** is undoubtedly the most common pathway for glucose degradation to pyruvate in stage two of aerobic respiration. It is found in all major groups of microorganisms and functions in the presence or absence of O_2. As noted earlier, it is also an important amphibolic pathway and provides several precursor metabolites. The Embden-Meyerhof pathway occurs in the cytoplasm.

The pathway may be divided into two parts (**figure 10.5** and *appendix II*). In the initial six-carbon phase, ATP is used to phosphorylate glucose twice, yielding fructose 1,6-bisphosphate. This preliminary phase "primes the pump" by adding phosphates to each end of the sugar. In essence, the organism invests some of its ATP so that more can be made later in the pathway.

The three-carbon, energy-conserving phase begins when the enzyme fructose 1,6-bisphosphate aldolase catalyzes the cleavage of fructose 1,6-bisphosphate into two halves, each with a phosphate. One of the products, dihydroxyacetone phosphate, is immediately converted to glyceraldehyde 3-phosphate. This yields two molecules of glyceraldehyde 3-phosphate, which are then catobolized to pyruvate in a five-step process. Because dihydroxyacetone phosphate can be easily converted to glyceraldehyde 3-phosphate, both halves of fructose 1,6-bisphosphate are used in the three-carbon phase. First, glyceraldehyde 3-phosphate is oxidized with NAD^+ as the electron acceptor (to form NADH), and a phosphate (P_i) is simultaneously incorporated to give a high-energy molecule called 1,3-bisphosphoglycerate. The high-energy phosphate on the first carbon is subsequently donated to ADP to produce ATP. This is an example of **substrate-level phosphorylation** because ADP phosphorylation is coupled with the exergonic breakdown of a high-energy bond. ≪ *ATP (section 9.4)*

A somewhat similar process generates a second ATP by substrate-level phosphorylation. The phosphate on 3-phosphoglycerate shifts to the second carbon, and 2-phosphoglycerate is dehydrated to form the high-energy molecule phosphoenolpyruvate. This molecule donates its phosphate to ADP forming a second ATP and pyruvate, the final product of the pathway.

The yields of ATP and NADH by the Embden-Meyerhof pathway may be calculated. In the six-carbon phase, two ATPs are used to form fructose 1,6-bisphosphate. For each glyceraldehyde 3-phosphate transformed into pyruvate, one NADH and two ATPs are formed. Because two glyceraldehyde 3-phosphates arise from a single glucose (one by way of dihydroxyacetone phosphate), the three-carbon phase generates four ATPs and two NADHs per glucose. Subtraction of the ATP used in the six-carbon phase from that produced by substrate-level phosphorylation in the three-carbon phase gives a net yield of two ATPs per glucose. Thus the catabolism of glucose to pyruvate can be represented by this simple equation.

$$\text{Glucose} + 2\text{ADP} + 2\text{P}_i + 2\text{NAD}^+ \rightarrow$$
$$2 \text{ pyruvate} + 2\text{ATP} + 2\text{NADH} + 2\text{H}^+$$

Pentose Phosphate Pathway

A second pathway, the **pentose phosphate** or **hexose monophosphate pathway,** may be used at the same time as either the Embden-Meyerhof or the Entner-Doudoroff pathways. It can operate either aerobically or anaerobically and is important in both biosynthesis and catabolism.

The pentose phosphate pathway begins with the oxidation of glucose 6-phosphate to 6-phosphogluconate followed by the oxidation of 6-phosphogluconate to the five-carbon sugar (i.e., pentose sugar) ribulose 5-phosphate and CO_2 (**figure 10.6** and *appendix II*). NADPH is produced during these oxidations. Ribulose 5-phosphate is then converted to a mixture of three-through seven-carbon sugar phosphates. Two enzymes play a central role in these transformations: (1) transketolase catalyzes the transfer of two-carbon groups, and (2) transaldolase transfers a three-carbon group from sedoheptulose 7-phosphate (a seven-carbon molecule) to glyceraldehyde 3-phosphate (a three-carbon molecule). The overall result is that three glucose 6-phosphates are converted to two fructose 6-phosphates, glyceraldehyde 3-phosphate, and three CO_2 molecules, as shown in this equation.

$$3 \text{ glucose 6-phosphate} + 6\text{NADP}^+ + 3\text{H}_2\text{O} \rightarrow$$
$$2 \text{ fructose 6-phosphate} + \text{glyceraldehyde 3-phosphate} +$$
$$3\text{CO}_2 + 6\text{NADPH} + 6\text{H}^+$$

These intermediates are used in two ways. The fructose 6-phosphate can be changed back to glucose 6-phosphate while glyceraldehyde 3-phosphate is converted to pyruvate by enzymes of the Embden-Meyerhof pathway. Alternatively two glyceraldehyde 3-phosphates may combine to form fructose 1,6-bisphosphate, which is eventually converted back into glucose 6-phosphate. This results in the complete degradation of glucose 6-phosphate to CO_2 and the production of a great deal of NADPH.

$$\text{Glucose 6-phosphate} + 12\text{NADP}^+ + 7\text{H}_2\text{O} \rightarrow$$
$$6\text{CO}_2 + 12\text{NADPH} + 12\text{H}^+ + \text{P}_i$$

Glucose is phosphorylated at the expense of one ATP, creating glucose 6-phosphate, a precursor metabolite and the starting molecule for the pentose phosphate pathway.

Isomerization of glucose 6-phosphate (an aldehyde) to fructose 6-phosphate (a ketone and a precursor metabolite)

ATP is consumed to phosphorylate C1 of fructose. The cell is spending some of its energy currency in order to earn more in the next part of the pathway.

Fructose 1, 6-bisphosphate is split into two 3-carbon molecules, one of which is a precursor metabolite.

Glyceraldehyde 3-phosphate is oxidized and simultaneously phosphorylated, creating a high-energy molecule. The electrons released reduce NAD^+ to NADH.

ATP is made by substrate-level phosphorylation. Another precursor metabolite is made.

Another precursor metabolite is made.

The oxidative breakdown of one glucose results in the formation of two pyruvate molecules. Pyruvate is one of the most important precursor metabolites.

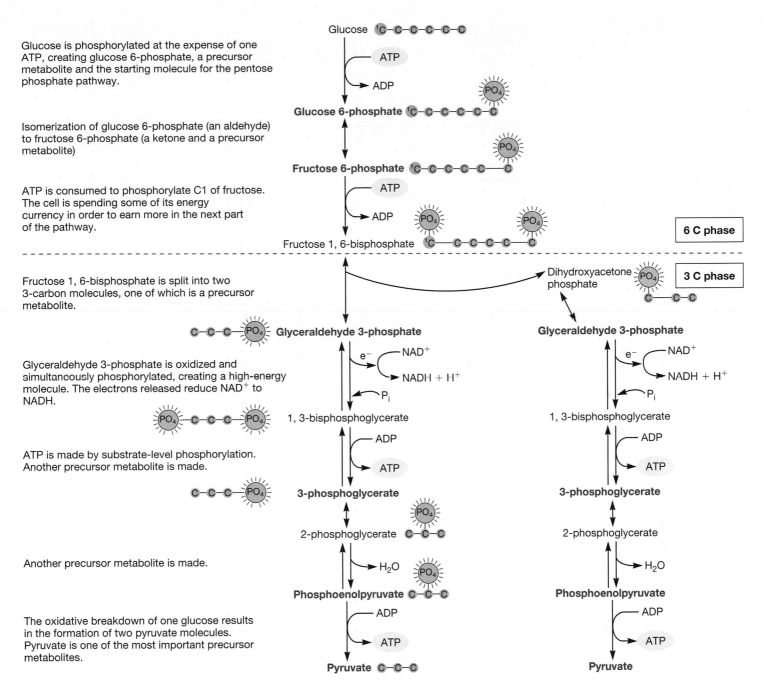

Figure 10.5 Embden-Meyerhof Pathway. This is one of three glycolytic pathways used to catabolize glucose to pyruvate, and it can function during aerobic respiration, anaerobic respiration, and fermentation. When used during respiration, the electrons accepted by NAD^+ are transferred to an electron transport chain and are ultimately accepted by an exogenous electron acceptor. When used during fermentation, the electrons accepted by NAD^+ are donated to an endogenous electron acceptor (e.g., pyruvate). The Embden-Meyerhof pathway is also an important amphibolic pathway, as it generates several precursor metabolites (shown in blue).

The pentose phosphate pathway is also an amphibolic pathway because: (1) NADPH produced by the pentose phosphate pathway serves as a source of electrons for the reduction of molecules during biosynthesis. Indeed, the pentose phosphate pathway is the major source of reducing power for cells, yielding 2 NADPH molecules for each glucose metabolized to pyruvate in this way. (2) The pathway produces two important precursor metabolites: erythrose 4-phosphate, which is used to synthesize aromatic amino acids and vitamin B_6 (pyridoxal); and ribose 5-phosphate, which is a major component of nucleic acids. Furthermore, when a microorganism is growing on a five-carbon sugar, the pathway can function biosynthetically to supply hexose sugars (e.g., glucose needed for peptidoglycan synthesis). (3) Intermediates in the pathway may be used to produce ATP. For instance, glyceraldehyde 3-phosphate from the pathway can enter the three-carbon phase of the Embden-Meyerhof pathway. As it

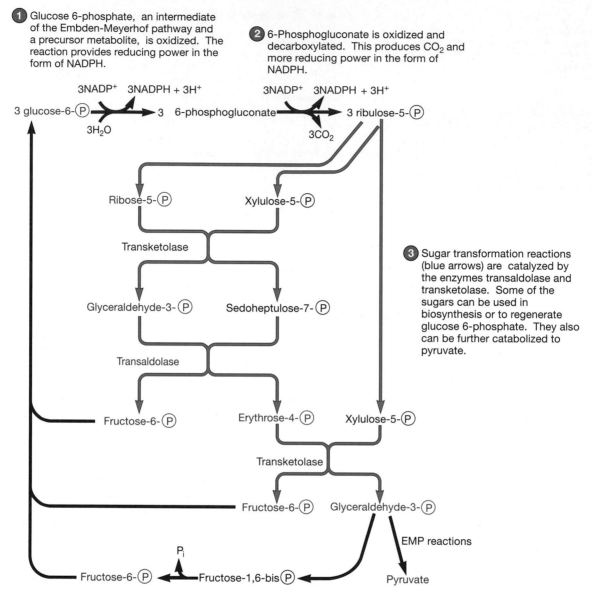

Figure 10.6 **The Pentose Phosphate Pathway.** The catabolism of three glucose 6-phosphate molecules to two fructose 6-phosphates, a glyceraldehyde 3-phosphate, and three CO_2 molecules is traced. Note that the pentose phosphate pathway generates several intermediates that are also intermediates of the Embden-Meyerhof pathway (EMP). These intermediates can be fed into the EMP with two results: (1) continued degradation to pyruvate or (2) regeneration of glucose 6-phosphate. The pentose phosphate pathway also plays a major role in producing reducing power (NADPH) and several precursor metabolites (shown in blue). The sugar transformations are indicated with blue arrows. These reactions are catalyzed by the enzymes transketolase and transaldolase.

is degraded to pyruvate, two ATPs are formed by substrate-level phosphorylation.

Entner-Doudoroff Pathway

Although the Embden-Meyerhof pathway is the most common route for the conversion of hexoses to pyruvate, the **Entner-Doudoroff pathway** is used by some soil microbes (e.g., *Pseudomonas, Rhizobium, Azotobacter,* and *Agrobacterium*) and a few other gram-negative bacteria. To date, very few gram-positive bacteria have been found to use this pathway,

with the intestinal bacterium *Enterococcus faecalis* being a rare exception.

The Entner-Doudoroff pathway begins with the same reactions as the pentose phosphate pathway: the formation of glucose 6-phosphate, which is then converted to 6-phosphogluconate (**figure 10.7** and *appendix II*). Instead of being further oxidized, 6-phosphogluconate is dehydrated to form 2-keto-3-deoxy-6-phosphogluconate (KDPG), the key intermediate of the pathway. KDPG is then cleaved by KDPG aldolase to pyruvate and glyceraldehyde 3-phosphate. The glyceraldehyde 3-phosphate is converted to pyruvate in the Embden-Meyerhof pathway. If the

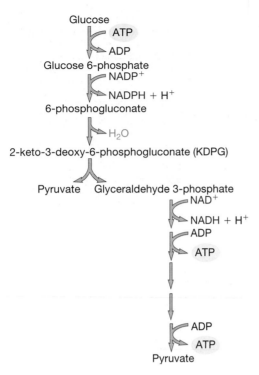

Figure 10.7   The Entner-Doudoroff Pathway. The sequence leading from glyceraldehyde 3-phosphate to pyruvate is catalyzed by enzymes common to the Embden-Meyerhof pathway.

Entner-Doudoroff pathway degrades glucose to pyruvate in this way, it yields one ATP, one NADPH, and one NADH per glucose metabolized.

1. Summarize the major features of the Embden-Meyerhof, pentose phosphate, and Entner-Doudoroff pathways. Include the starting points, the products of the pathways, any critical or unique enzymes, the ATP yields, and the metabolic roles each pathway has.

2. What is substrate-level phosphorylation?

10.4 TRICARBOXYLIC ACID CYCLE

In the glycolytic pathways, glucose is oxidized to pyruvate. During aerobic respiration, the catabolic process continues by oxidizing pyruvate to three CO_2. The first step of this process employs a multienzyme system called the pyruvate dehydrogenase complex. It oxidizes and cleaves pyruvate to form one CO_2 and the two-carbon molecule **acetyl-coenzyme A (acetyl-CoA) (figure 10.8)**. Acetyl-CoA is energy-rich because a high-energy thiol links acetic acid to coenzyme A. As shown in figure 10.3, during stage three of aerobic respiration, carbohy-

drates as well as fatty acids and amino acids can be converted to acetyl-CoA.

Acetyl-CoA then enters the **tricarboxylic acid (TCA) cycle,** which is also called the **citric acid cycle** or the **Krebs cycle** (figure 10.8 and *appendix II*). In the first reaction, acetyl-CoA is condensed with (i.e., added to) the four-carbon intermediate oxaloacetate to form citrate, a molecule with six carbons. Citrate (a tertiary alcohol) is rearranged to give isocitrate, a more readily oxidized secondary alcohol. Isocitrate is subsequently oxidized and decarboxylated twice to yield α-ketoglutarate (five carbons) and then succinyl-CoA (four carbons), a molecule with a high-energy bond. At this point, two NADH molecules have been formed and two carbons lost from the cycle as CO_2. The cycle continues when succinyl-CoA is converted to succinate. This involves breaking the high-energy bond in succinyl-CoA and using the energy released to form one GTP by substrate-level phosphorylation. GTP is also a high-energy molecule, and it is functionally equivalent to ATP. Two oxidation steps follow, yielding one FADH$_2$ and one NADH. The last oxidation step regenerates oxaloacetate, and as long as there is a supply of acetyl-CoA, the cycle can repeat itself. Inspection of figure 10.8 shows that the TCA cycle generates two CO_2 molecules, three NADH molecules, one FADH$_2$, and one GTP for each acetyl-CoA molecule oxidized.

TCA cycle enzymes are widely distributed among microorganisms. In procaryotes, they are located in the cytoplasm. In eucaryotes, they are found in the mitochondrial matrix. The complete cycle appears to be functional in many aerobic bacteria, free-living protists, and fungi. This is not surprising because the cycle plays an important role in energy conservation by producing numerous NADH and FADH$_2$ (section 10.5). Even those microorganisms that lack the complete TCA cycle usually have most of the cycle enzymes, because the TCA cycle is also a key source of precursor metabolites for use in biosynthesis. >> *Synthesis of amino acids: Anaplerotic reactions and amino acid biosynthesis (section 11.5)*

1. Give the substrate and products of the TCA cycle. Describe its organization in general terms. What are its two major functions?

2. What chemical intermediate links pyruvate to the TCA cycle?

3. How many times must the TCA cycle be performed to oxidize one molecule of glucose completely to six molecules of CO_2? Why?

4. In what eucaryotic organelle is the TCA cycle found? Where is the cycle located in procaryotes?

5. Why might it be desirable for a microbe with the Embden-Meyerhof pathway and the TCA cycle also to have the pentose phosphate pathway?

6. Why is GTP functionally equivalent to ATP?

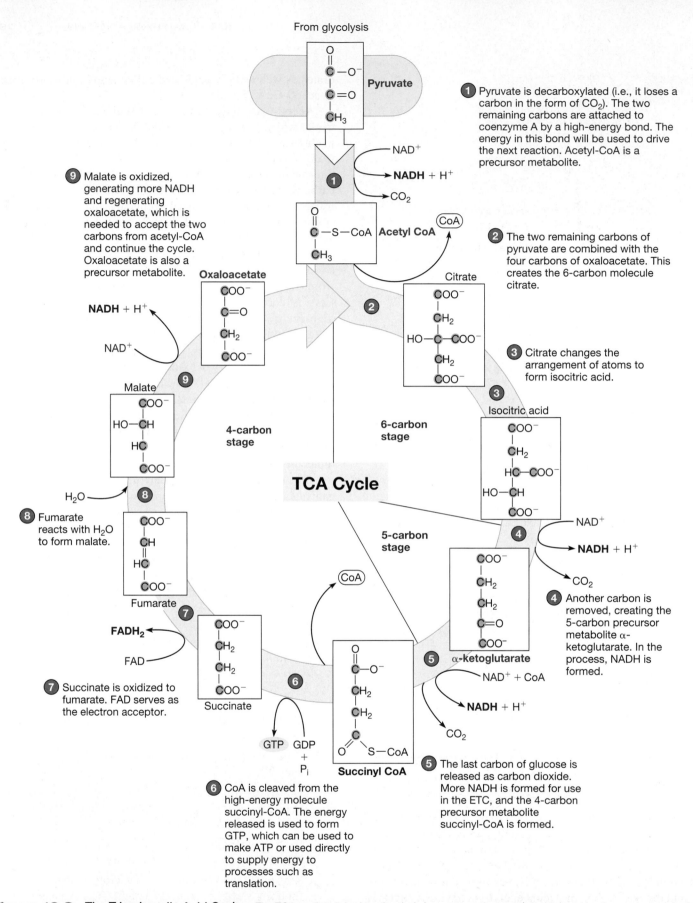

Figure 10.8   The Tricarboxylic Acid Cycle.   The TCA cycle is linked to glycolysis by a connecting reaction catalyzed by the pyruvate dehydrogenase complex. The reaction decarboxylates pyruvate (removes a carboxyl group as CO_2) and generates acetyl-CoA. The cycle may be divided into three stages based on the size of its intermediates. The three stages are separated from one another by two decarboxylation reactions. Precursor metabolites, carbon skeletons used in biosynthesis, are shown in blue. NADH and $FADH_2$ are shown in purple; they can transfer electrons to the electron transport chain (ETC).

ELECTRON TRANSPORT AND OXIDATIVE PHOSPHORYLATION

During the oxidation of glucose to six CO_2 molecules by glycolysis and the TCA cycle, as many as four ATP molecules are generated by substrate-level phosphorylation. Thus at this point, the work done by the cell has yielded relatively little ATP. However, in oxidizing glucose, the cell has also generated numerous molecules of NADH and $FADH_2$. Both of these molecules have a relatively negative E'_0 and can be used to conserve energy (*see table 9.1*). In fact, most of the ATP generated during respiration comes from the oxidation of these electron carriers in the electron transport chain. The mitochondrial electron transport chain is examined first because it has been so well studied. Then we turn to bacterial chains and finish with a discussion of ATP synthesis.

Electron Transport Chains

The mitochondrial **electron transport chain (ETC)** is composed of a series of electron carriers that operate together to transfer electrons from donors, such as NADH and $FADH_2$, to O_2 (**figure 10.9**). The electrons flow from carriers with more negative

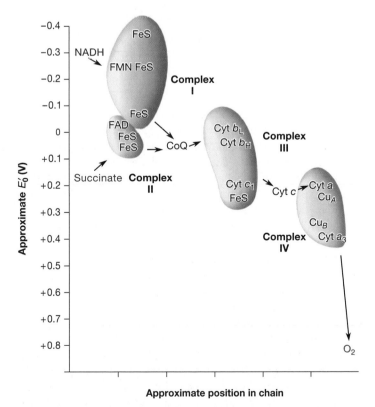

reduction potentials to those with more positive potentials and eventually combine with O_2 and H^+ to form water. This pattern of electron flow is exactly the same as seen in the electron tower that is described in chapter 9 (*see figure 9.5*). The electrons move down this potential gradient much like water flowing down a series of rapids. The difference in reduction potentials between O_2 and NADH is large, about 1.14 volts, which makes possible the release of a great deal of energy. The differences in reduction potential at several points in the chain are large enough to provide sufficient energy for ATP production, much as the energy from waterfalls can be harnessed by waterwheels and used to generate electricity. Thus the ETC breaks up the large overall energy release into small steps. As will be seen shortly, electron transport generates proton and electrical gradients. These gradients can drive ATP synthesis and perform other work.

In eucaryotes, the ETC carriers reside within the inner membrane of the mitochondrion where they are arranged into four complexes, each capable of transporting electrons part of the way to O_2 (**figure 10.10**). Coenzyme Q and cytochrome *c* connect the complexes with each other. Although some procaryotic ETCs resemble the mitochondrial chain, they are frequently very different. First, their ETCs are located within the plasma membrane. They also can be composed of different electron carriers (e.g., their cytochromes) and may be extensively branched. Electrons often can enter the chain at several points and leave through several terminal oxidases. Procaryotic ETCs also may be shorter, resulting in the release of less energy. Although procaryotic and eucaryotic ETCs differ in details of construction, they operate using the same fundamental principles.

The ETC of *E. coli* will serve as an example of these differences. A simplified view of the *E. coli* chain is shown in **figure 10.11**. NADH generated by the oxidation of organic substrates (during glycolysis and the TCA cycle) is oxidized to NAD^+ by the first component in the ETC, membrane-bound NADH dehydrogenase. The electrons from NADH are transferred to carriers with progressively more positive reduction potentials. As electrons move through the carriers, protons are moved across the plasma membrane to the periplasmic space (i.e., outside the cell) rather than to an intermembrane space, as seen in the mitochondria (compare figures 10.10 and 10.11). Another significant difference between the *E. coli* chain and the mitochondrial chain is that the bacterial ETC contains a different array of cytochromes. Furthermore, *E. coli* has evolved two branches of the ETC that operate under different aeration conditions. When oxygen is readily available, the cytochrome *bo* branch is used. When oxygen levels are reduced, the cytochrome *bd* branch is used because it has a higher affinity for oxygen. However, it is less efficient than the *bo* branch because the *bd* branch moves fewer protons into the periplasmic space (figure 10.11).

Oxidative Phosphorylation

Oxidative phosphorylation, sometimes called respiratory phosphorylation, is the process by which ATP is synthesized as the result of electron transport driven by the oxidation of a chemical

Figure 10.9 The Mitochondrial Electron Transport Chain. Many of the more important carriers are arranged at approximately the correct reduction potential and sequence. In mitochondria, they are organized into four complexes that are linked by coenzyme Q (CoQ) and cytochrome *c* (Cyt *c*). Electrons flow from NADH and succinate down the reduction potential gradient to oxygen.

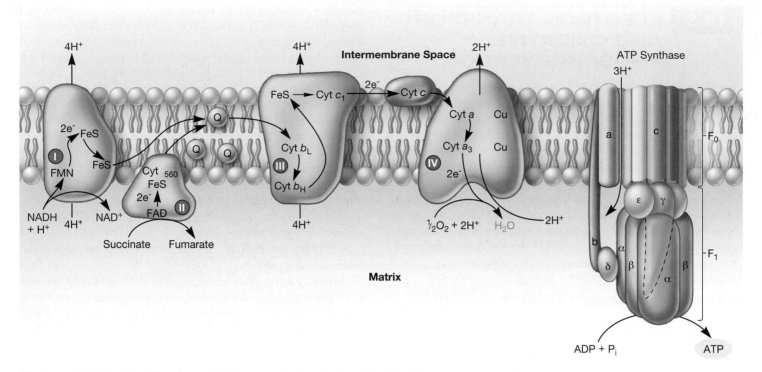

Figure 10.10 **The Chemiosmotic Hypothesis Applied to Mitochondria.** In this scheme, the carriers are organized asymmetrically within the inner membrane so that protons are transported across as electrons move along the chain. Proton release into the intermembrane space occurs when electrons are transferred from carriers, such as FMN and coenzyme Q (Q), that carry both electrons and protons to components such as nonheme iron proteins (FeS proteins) and cytochromes (Cyt) that transport only electrons. Complex IV pumps protons across the membrane as electrons pass from cytochrome a to oxygen. Coenzyme Q transports electrons from complexes I and II to complex III. Cytochrome c moves electrons between complexes III and IV. The number of protons moved across the membrane at each site per pair of electrons transported is still somewhat uncertain; the current consensus is that at least 10 protons must move outward during NADH oxidation. One molecule of ATP is synthesized and released from the enzyme ATP synthase for every three protons that cross the membrane by passing through it.

energy source. The mechanism by which oxidative phosphorylation takes place has been studied intensively for years. The most widely accepted hypothesis is the **chemiosmotic hypothesis,** which was formulated by British biochemist Peter Mitchell. According to the chemiosmotic hypothesis, the ETC is organized so that protons move outward from the mitochondrial matrix as electrons are transported down the chain (figure 10.10). In procaryotes, the protons are moved across the plasma membrane from the cytoplasm to the periplasmic space.

The movement of protons across the membrane is not completely understood. However, in some cases, the protons are actively pumped across the membrane (e.g., by complex IV of the mitochondrial chain; figure 10.10). In other cases, translocation of protons results from the juxtaposition of carriers that accept both electrons and protons with carriers that accept only electrons. For instance, coenzyme Q carries two electrons and two protons to cytochrome b in complex III of the mitochondrial chain. Cytochrome b accepts only one electron at a time and does not accept protons. Thus for each electron transferred by coenzyme Q, one proton is released to the intermembrane space.

The result of proton expulsion during electron transport is the formation of a concentration gradient of protons (Δ pH; chemical potential energy) and a charge gradient ($\Delta\psi$; electrical potential

energy). Thus the mitochondrial matrix is more alkaline and more negative than the intermembrane space. Likewise with procaryotes, the cytoplasm is more alkaline and more negative than the periplasmic space. The combined chemical and electrical potential differences make up the **proton motive force (PMF).** The PMF is used to perform work when protons flow back across the membrane, down the concentration and charge gradients, and into the mitochondrial matrix (or procaryotic cytoplasm). This flow is exergonic and is often used to phosphorylate ADP to ATP. The PMF is also used to transport molecules into the cell directly (i.e., without the hydrolysis of ATP) and to rotate the bacterial flagellar motor. Thus the PMF plays a central role in a cell's physiology (**figure 10.12**). << *Uptake of nutrients (section 6.6); Components external to the cell wall: Flagella (section 3.6)*

The use of PMF for ATP synthesis is catalyzed by **ATP synthase,** a multisubunit enzyme also known as F_1F_0 ATPase because it consists of two components and can catalyze ATP hydrolysis (**figure 10.13**). The mitochondrial F_1 component appears as a spherical structure attached to the mitochondrial inner membrane surface by a stalk. The F_0 component is embedded in the membrane. ATP synthase is on the inner surface of the plasma membrane in procaryotes. F_0 participates in proton movement across the membrane. F_1 is a large complex in which three a subunits alternate with three b subunits. The catalytic

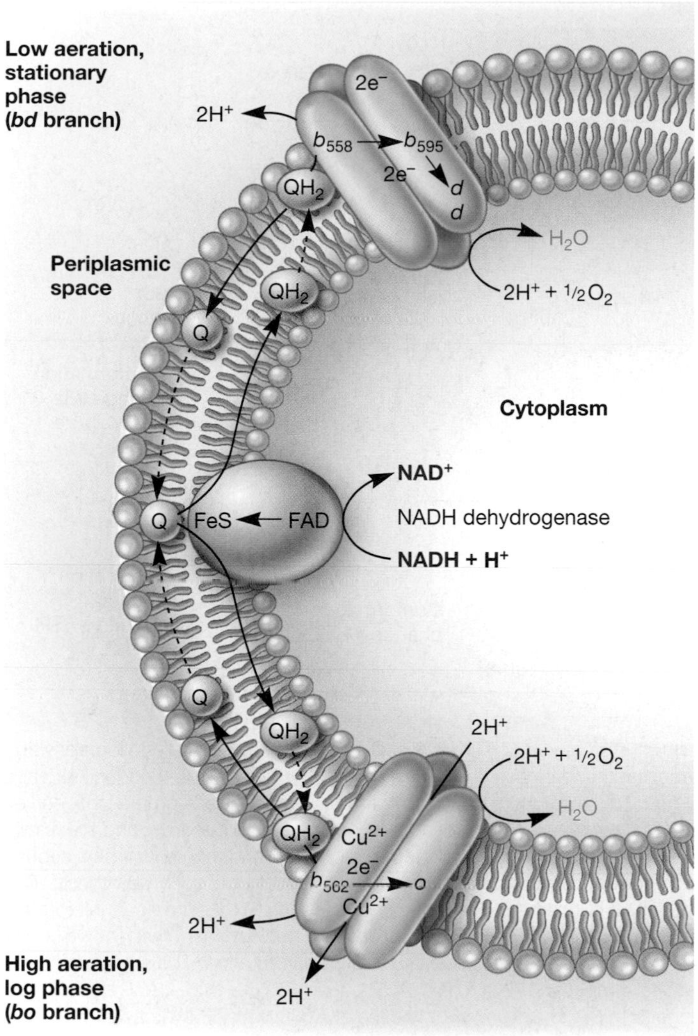

Low aeration, stationary phase (*bd* branch)

$2H^+$

$2e^-$

b_{558} → b_{595}

QH_2 $2e^-$ d d

H_2O

Periplasmic space QH_2 $2H^+ + \frac{1}{2}O_2$

Q

Q FeS ← FAD

Cytoplasm

NAD$^+$

NADH dehydrogenase

NADH + H$^+$

Q

QH_2 $2H^+$

$2H^+ + \frac{1}{2}O_2$

QH_2 Cu^{2+} H_2O

b_{562} $2e^-$ o

Cu^{2+}

$2H^+$

High aeration, log phase (*bo* branch)

$2H^+$

Figure 10.11 The Electron Transport Chain of *E. coli.* NADH transfers electrons from the organic energy and electron source to the electron transport chain. Ubiquinone-8 (Q) connects the NADH dehydrogenase with two terminal oxidase systems. The upper branch operates when the bacterium is in stationary phase and there is little oxygen. At least five cytochromes are involved: b_{558}, b_{595}, b_{562}, d, and o. The lower branch functions when *E. coli* is growing rapidly with good aeration.

sites for ATP synthesis are located on the b subunits. At the center of F_1 is the γ subunit. The γ subunit extends through F_1 and interacts with F_0.

It is now known that ATP synthase functions like a rotary engine. It is thought the flow of protons down the proton gradient through the F_0 subunit causes F_0 and the γ subunit to rotate. As the γ subunit rotates rapidly within the F_1 (much like a car's crankshaft), conformation changes occur in the b subunits (figure 10.13*b*). One conformation change (β_E to β_{HC}) allows entry of ADP and P_i into the catalytic site. Another conformation change (β_{HC} to β_{DP}) loosely binds ADP and P_i in the catalytic site. ATP is synthesized when the β_{DP} conformation is changed to the β_{TP} conformation, and ATP is released when β_{TP} changes to the β_E conformation, to start the synthesis cycle anew.

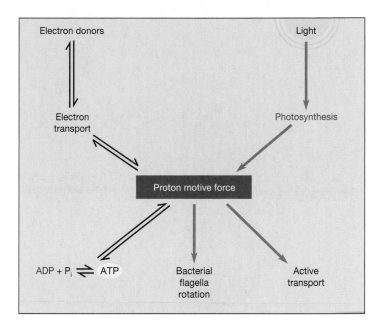

Figure 10.12 The Central Role of Proton Motive Force. Active transport is not always driven by PMF.

ATP Yield During Aerobic Respiration

It is possible to estimate the number of ATP molecules synthesized per NADH or FADH$_2$ oxidized by an ETC. During aerobic respiration, a pair of electrons from NADH is donated to the electron transport chain and ultimately used to reduce an atom of oxygen to H$_2$O. This releases enough energy to drive the synthesis of three ATP. This is referred to as the phosphorus to oxygen (P/O) ratio because it measures the number of ATP molecules (phosphorus) generated per oxygen (O) reduced. Because FADH$_2$ has a more positive reduction potential than NADH (*see figure 9.5*), electrons arising from its oxidation flow down a shorter chain, releasing less energy. Thus while the P/O ratio for NADH is 3, only two ATP can be made from the oxidation of a single FADH$_2$.

Given this information, we can calculate the maximum ATP yield of aerobic respiration (**figure 10.14**). Substrate-level phosphorylation during glycolysis yields at most two ATP molecules per glucose converted to pyruvate (figures 10.5, 10.6, and 10.7). Two additional GTP (ATP equivalents) are generated by substrate-level phosphorylation during the two turns of the TCA cycle needed to oxidize two acetyl-CoA molecules (figure 10.8). However, most of the ATP made during aerobic respiration is generated by oxidative phosphorylation. Up to 10 NADH (2 from glycolysis, 2 from pyruvate conversion to acetyl-CoA, and 6 from the TCA cycle) and 2 FADH$_2$ (from the TCA cycle) are generated when glucose is oxidized completely to 6 CO$_2$. Assuming a P/O ratio of 3 for NADH oxidation and 2 for FADH$_2$ oxidation, the 10 NADH could theoretically drive the synthesis of 30 ATP, while oxidation of the 2 FADH$_2$ molecules would add another 4 ATP for a maximum of 34 ATP generated via oxidative phosphorylation. Thus oxidative phosphorylation accounts for at least eight times more ATP than does substrate-level phosphorylation. The maximum total yield of

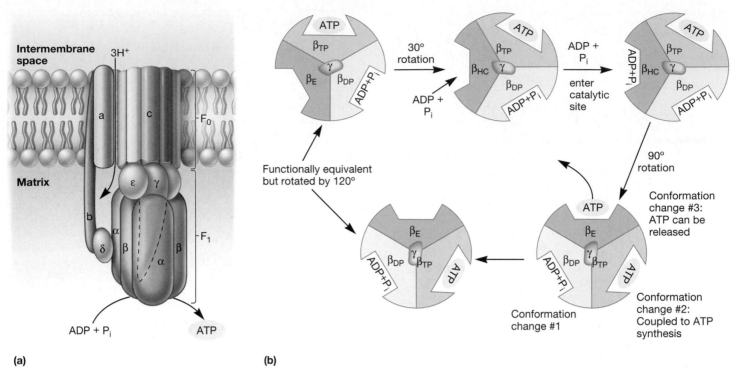

(a)   **(b)**

Figure 10.13   ATP Synthase Structure and Function.   (a) The major structural features of ATP synthase deduced from X-ray crystallography and other studies. F_1 is a spherical structure composed largely of alternating α and β subunits; the three active sites are on the β subunits. The γ subunit extends upward through the center of the sphere and can rotate. The stalk (γ and ϵ subunits) connects the sphere to F_0, the membrane embedded complex that serves as a proton channel. F_0 contains one a subunit, two b subunits, and 9–12 c subunits. The stator arm is composed of subunit a, two b subunits, and the δ subunit; it is embedded in the membrane and attached to F_1. A ring of c subunits in F_0 is connected to the stalk and may act as a rotor and move past the a subunit of the stator. As the c subunit ring turns, it rotates the shaft ($\gamma\epsilon$ subunits). (b) The binding change mechanism is a widely accepted model of ATP synthesis. This simplified drawing of the model shows the three catalytic β subunits and the γ subunit, which is located at the center of the F_1 complex. As the γ subunit rotates, it causes conformational changes in each subunit. The β_E (empty) conformation is an open conformation, which does not bind nucleotides. When the γ subunit rotates 30°, β_E is converted to the β_{HC} (half closed) conformation. P_i and ADP can enter the catalytic site when it is in this conformation. The subsequent 90° rotation by the γ subunit is critical because it brings about three significant conformational changes: (1) β_{HC} to β_{DP} (ADP bound), (2) β_{DP} to β_{TP} (ATP bound), and (3) β_{TP} to β_E. Change from β_{DP} to β_{TP} is accompanied by the formation of ATP; change from β_{TP} to β_E allows for release of ATP from ATP synthase.

ATP during aerobic respiration is 38 ATPs. However, these calculations and those presented in figure 10.14 are theoretical. In fact, the P/O ratios are more likely about 2.5 for NADH and 1.5 for $FADH_2$. Thus the total ATP yield from glucose may be closer to 30 ATPs rather than 38.

Because procaryotic ETCs often have lower P/O ratios than eucaryotic chains, procaryotic ATP yields are less. For example, *E. coli*, with its truncated ETC, has a P/O ratio around 1.3 when using the cytochrome *bo* path at high oxygen levels and only a ratio of about 0.67 when employing the cytochrome *bd* branch (figure 10.11) at low oxygen concentrations. In this case, ATP production varies with environmental conditions. Perhaps because *E. coli* normally grows in habitats such as the intestinal tract that are very rich in nutrients, it does not have to be particularly efficient in ATP synthesis. Presumably the ETC functions when *E. coli* is in an oxic freshwater environment between hosts.

1. Briefly describe the structure of ETCs and their role in ATP formation. How do mitochondrial and procaryotic chains differ?

2. Describe the current model of oxidative phosphorylation. Briefly describe the structure of ATP synthase and explain how it is thought to function.

3. How do substrate-level phosphorylation and oxidative phosphorylation differ?

4. Calculate the ATP yield when glucose is catabolized completely to six CO_2 by a eucaryotic microbe. How does this value compare to the ATP yield observed for a bacterium? Suppose a bacterium used the Entner-Doudoroff pathway to degrade glucose to pyruvate. How would this impact the total ATP yield? Explain your reasoning.

10.6   ANAEROBIC RESPIRATION

As we have seen, during aerobic respiration sugars and other organic molecules are oxidized and their electrons transferred to NAD^+ and FAD to generate NADH and $FADH_2$, respectively. These carriers then donate the electrons to an ETC that

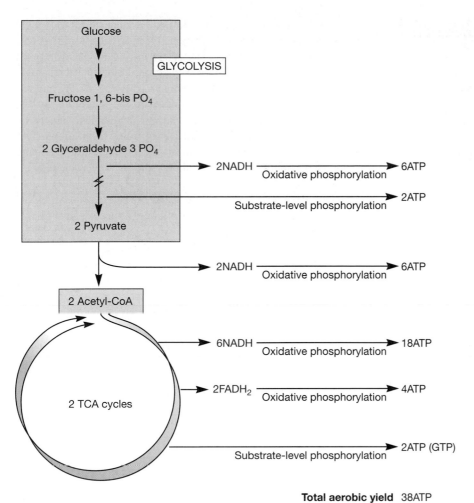

Figure 10.14 Maximum Theoretic ATP Yield from Aerobic Respiration. To attain the theoretic maximum yield of ATP, one must assume a P/O ratio of 3 for the oxidation of NADH and 2 for FADH$_2$. The actual yield is probably significantly less and varies between eucaryotes and procaryotes and among procaryotic species.

uses O$_2$ as the terminal electron acceptor. However, it is also possible for other terminal electron acceptors to be used for electron transport. **Anaerobic respiration,** a process whereby an exogenous terminal electron acceptor other than O$_2$ is used for electron transport, is carried out by many bacteria and archaea. The most common terminal electron acceptors used during anaerobic respiration are nitrate, sulfate, and CO$_2$, but metals and a few organic molecules can also be reduced (**table 10.1**). Some procaryotes oxidize sugars and other organic molecules to generate NADH and FADH$_2$ using the exact pathways used for aerobic respiration. Thus, depending on the organism, glycolysis and the TCA cycle function in the same way during both aerobic respiration and anaerobic respiration. Furthermore, most of the ATP generated during anaerobic respiration is made by oxidative phosphorylation, as is the case during aerobic respiration.

An example of a bacterium that can carry out both anaerobic respiration and aerobic respiration is *Paracoccus denitrificans*. It is a gram-negative, facultative, anaerobic soil bacterium that is extremely versatile metabolically. Under anoxic conditions, *P. denitrificans* uses NO$_3^-$ as its electron acceptor. As shown in **figure 10.15**, it uses two different ETCs, one for aerobic respiration and one for anaerobic respiration. NADH transfers electrons from the microbe's electron source to the chains. The aerobic chain has four complexes that correspond to the mitochondrial chain (figure 10.15*a*). When *P. denitrificans* grows without oxygen using NO$_3^-$ as the terminal electron acceptor, the ETC is more complex (figure 10.15*b*). The chain is highly branched and the cytochrome *aa* complex is replaced. Electrons are passed from coenzyme Q to cytochrome *b* for the reduction of nitrate to nitrite (catalyzed by nitrate reductase). Electrons then flow through cytochrome *c* for the sequential oxidation of nitrite to gaseous dinitrogen (N$_2$). Not as many protons are pumped across the membrane during anaerobic growth, but nonetheless a PMF is established.

The anaerobic reduction of nitrate makes it unavailable for assimilation into the cell. Therefore this process is called **dissimilatory nitrate reduction.** Nitrate reductase replaces cytochrome oxidase to catalyze the reaction:

$$NO_3^- + 2e^- + 2H^+ \rightarrow NO_2^- + H_2O$$

However, reduction of nitrate to nitrite is not a particularly efficient way of making ATP because only two electrons are accepted by nitrate when it is reduced to nitrite. Furthermore, nitrite is quite toxic. Bacteria such as *P. denitrificans* avoid the toxic effects of nitrite by reducing it to nitrogen gas, a process known as **denitrification.** By donating five electrons to a nitrate molecule, NO$_3^-$ is converted into a nontoxic product.

$$2NO_3^- + 10e^- + 12H^+ \rightarrow N_2 + 6H_2O$$

As illustrated in figure 10.15, denitrification is a multistep process with four enzymes participating: nitrate reductase, nitrite reductase, nitric oxide reductase, and nitrous oxide reductase.

$$NO_3^- \rightarrow NO_2^- \rightarrow NO \rightarrow N_2O \rightarrow N_2$$

Two types of bacterial nitrite reductases catalyze the formation of NO in bacteria. One contains cytochromes *c* and *d$_1$* (e.g., *Paracoccus* and *Pseudomonas aeruginosa*), and the other is a copper-containing protein (e.g., *Alcaligenes*). Nitrite reductase seems to be periplasmic in gram-negative bacteria. Nitric oxide reductase catalyzes the formation of nitrous oxide from NO and is a membrane-bound cytochrome *bc* complex. In *P. denitrificans,* the nitrate reductase and nitric oxide reductase are membrane-bound, whereas nitrite reductase and nitrous oxide reductase are periplasmic (figure 10.15*b*).

In addition to *P. denitrificans,* some members of the genera *Pseudomonas* and *Bacillus* carry out denitrification. All three

Table 10.1	Some Electron Acceptors Used in Respiration		
	Electron Acceptor	Reduced Products	Examples of Microorganisms
Aerobic	O_2	H_2O	All aerobic bacteria, fungi, and protists
Anaerobic	NO_3^-	NO_2^-	Enteric bacteria
	NO_3^-	NO_2^-, N_2O, N_2	*Pseudomonas, Bacillus,* and *Paracoccus*
	SO_4^{2-}	H_2S	*Desulfovibrio* and *Desulfotomaculum*
	CO_2	CH_4	All methanogens and acetogens
	S^0	H_2S	*Desulfuromonas* and *Thermoproteus*
	Fe^{3+}	Fe^{2+}	*Pseudomonas, Bacillus,* and *Geobacter*
	$HAsO_4^{2-}$	$HAsO_2$	*Bacillus, Desulfotomaculum, Sulfurospirillum*
	SeO_4^{2-}	Se, $HSeO_3^-$	*Aeromonas, Bacillus, Thauera*
	Fumarate	Succinate	*Wolinella*

genera use denitrification as an alternative to aerobic respiration and are considered facultative anaerobes. Indeed, if O_2 is present, these bacteria use aerobic respiration, which is far more efficient in capturing energy. In fact, the synthesis of nitrate reductase is repressed by O_2. Denitrification in anoxic soil results in the loss of soil nitrogen and adversely affects soil fertility. >> *Biogeochemical cycling: Nitrogen cycle (section 25.1)*

Not all microbes employ anaerobic respiration facultatively. Some are obligate anaerobes that carry out only anaerobic respiration. The methanogens are an example. These archaea use CO_2 or carbonate as a terminal electron acceptor. They are called methanogens because the electron acceptor is reduced to methane. Bacteria such as *Desulfovibrio* are another example. They donate eight electrons to sulfate, reducing it to sulfide (S^{2-} or H_2S).

$$SO_4^{2-} + 8e^- + 8H^+ \rightarrow S^{2-} + 4H_2O$$

It should be noted that both methanogens and *Desulfovibrio* are able to function as chemolithotrophs, using H_2 as an energy source (section 10.11).

As we saw for denitrification, anaerobic respiration using sulfate or CO_2 as the terminal electron acceptors is not as efficient in

ATP synthesis as is aerobic respiration. Reduction in ATP yield arises from the fact that these alternate electron acceptors have less positive reduction potentials than O_2 (*see table 9.1*). The difference in standard reduction potential between a donor such as NADH and nitrate is smaller than the difference between NADH and O_2. Because energy yield is directly related to the magnitude of the reduction potential difference, less energy is available to make ATP in anaerobic respiration. Nevertheless, anaerobic respiration is useful because it allows ATP synthesis by electron transport and oxidative phosphorylation in the absence of O_2. Anaerobic respiration is prevalent in oxygen-depleted soils and sediments.

The ability of microbes to use a variety of electron acceptors has ecological consequences. Often one sees a succession of microorganisms in an environment when several electron acceptors are present. For example, if O_2, nitrate, manganese ion, ferric ion, sulfate, and CO_2 are available in a particular environment, a predictable sequence of electron acceptor use takes place when an oxidizable substrate is available to the microbial population. Oxygen is employed as an electron acceptor first because it inhibits nitrate use by microorganisms capable of respiration with either O_2 or nitrate. While O_2 is available, sulfate reducers and methanogens are inhibited because these groups are obligate anaerobes. Once the O_2 and nitrate are exhausted and fermentation products (section 10.7), including hydrogen, have accumulated, competition for use of other electron acceptors begins. Manganese and iron are used first, followed by competition between sulfate reducers and methanogens. >> *Microorganisms in natural environments (chapter 26)*

1. Describe the process of anaerobic respiration. Does anaerobic respiration yield as much ATP as aerobic respiration? Why or why not?

2. What is denitrification? Why do farmers dislike this process?

3. *E. coli* can use O_2, fumarate, or nitrate as a terminal electron acceptor under different conditions. What is the order of energy yield from highest to lowest for these electron acceptors? Explain your answer in thermodynamic terms.

10.7 FERMENTATION

Despite the tremendous ATP yield obtained by oxidative phosphorylation, some chemoorganotrophic microbes do not respire because either they lack ETCs or they repress the synthesis of ETC components under anoxic conditions, making anaerobic respiration impossible. Yet NADH produced by the Embden-Meyerhof pathway reactions during glycolysis (figure 10.5) must still be oxidized back to NAD^+. If NAD^+ is not regenerated, the oxidation of glyceraldehyde 3-phosphate will cease and glycolysis will stop. Many microorganisms solve this problem by slowing or stopping pyruvate dehydrogenase activity and using pyruvate

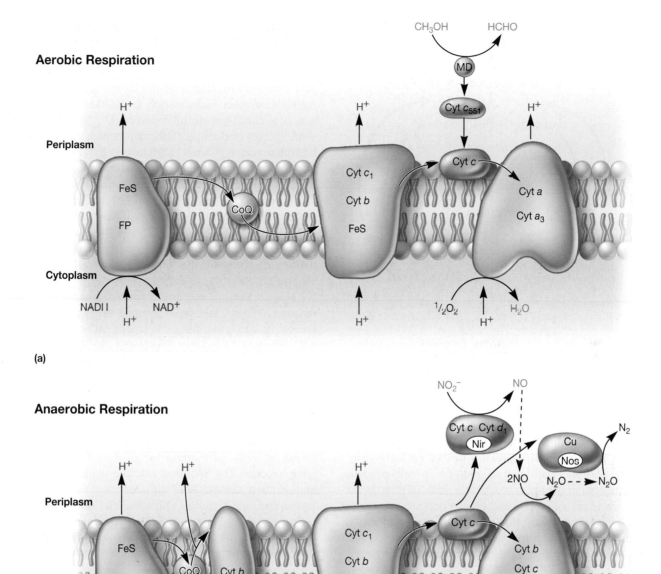

Figure 10.15 *Paracoccus denitrificans* **Electron Transport Chains.** (a) The aerobic ETC resembles a mitochondrial ETC and uses oxygen as its acceptor. Methanol and methylamine can contribute electrons at the cytochrome *c* level. (b) The highly branched anaerobic chain is made of both membrane and periplasmic proteins. Nitrate is reduced to diatomic nitrogen by the collective action of four different reductases that receive electrons from CoQ and cytochrome *c*. Locations of proton movement are shown, but the number of protons involved has not been indicated. Abbreviations used: flavoprotein (FP), methanol dehydrogenase (MD), nitrate reductase (Nar), nitrite reductase (Nir), nitric oxide reductase (Nor), and nitrous oxide reductase (Nos).

or one of its derivatives as an electron acceptor for the reoxidation of NADH in a fermentation process (**figure 10.16**). There are many kinds of fermentations, and they often are characteristic of particular microbial groups (**figure 10.17**). A few of the more

common fermentations are introduced here. Three unifying themes should be kept in mind when microbial fermentations are examined: (1) NADH is oxidized to NAD⁺, (2) the electron acceptor is often either pyruvate or a pyruvate derivative, and

Glycolysis

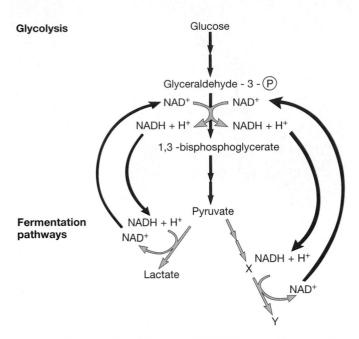

Figure 10.16   Reoxidation of NADH During Fermentation.
NADH from glycolysis is reoxidized by being used to reduce pyruvate or a pyruvate derivative (X). Either lactate or reduced product Y result.

(3) oxidative phosphorylation cannot operate, reducing the ATP yield per glucose significantly. In fermentation, the substrate (e.g., glucose) is only partially oxidized, ATP is formed exclusively by substrate-level phosphorylation, and oxygen is not needed. Although fermentation yields considerably less ATP, it is an important component of the metabolic repertoire of many microbes, allowing them to adjust to changes in their habitats.

Many fungi, protists, and some bacteria ferment sugars to ethanol and CO_2 in a process called **alcoholic fermentation.** Pyruvate is decarboxylated to acetaldehyde, which is then reduced to ethanol by alcohol dehydrogenase with NADH as the electron donor (figure 10.17, number 2). Lactic acid fermentation, the reduction of pyruvate to lactate (figure 10.17, number 1), is even more common. It is present in bacteria (lactic acid bacteria, *Bacillus*), protists (*Chlorella* and some water molds), and even in animal skeletal muscle. Lactic acid fermenters can be separated into two groups. **Homolactic fermenters** use the Embden-Meyerhof pathway and directly reduce almost all their pyruvate to lactate with the enzyme lactate dehydrogenase. **Heterolactic fermenters** form substantial amounts of products other than lactate; many also produce ethanol and CO_2. Lactic acid bacteria are important to the food industry, where they are used to make a variety of fermented foods (e.g., yogurt). >> *Class* Gammaproteobacteria: *Order* Enterobacteriales *(section 20.3); Microbiology of food (chapter 34)*

Many bacteria, especially members of the family *Enterobacteriaceae,* can metabolize pyruvate to numerous products using several pathways simultaneously. One such complex fermentation is the **mixed acid fermentation,** which results in the excretion of ethanol and a mixture of acids, particularly acetic, lactic, succinic, and formic acids **(table 10.2)**. Members

of the genera *Escherichia, Salmonella,* and *Proteus* carry out mixed acid fermentation. They use pathways numbered 1, 5, 8, and 9 in figure 10.17 to make all the fermentation products, except succinic acid. If formic hydrogenlyase is present (figure 10.17, number 6), formic acid will be degraded to H_2 and CO_2. Another complex fermentation is the butanediol fermentation, which is characteristic of *Enterobacter, Serratia, Erwinia* and some species of *Bacillus.* The predominant pathway used during this fermentation process yields butanediol (figure 10.17, number 4). However, large amounts of ethanol also are produced (figure 10.17, number 9), as are smaller amounts of lactic acid (figure 10.17, number 1), and formic acid (figure 10.17, number 5); in some bacteria, the formic acid is further catabolized to H_2 and CO_2 (figure 10.17, number 6). The use of either mixed acid fermentation or butanediol fermentation is important in differentiating members of the *Enterobacteriaceae* as described in chapter 32.   >>   *Class* Bacilli *(section 21.4)*

Microorganisms carry out a vast array of fermentations using numerous sugars and other organic substrates as their energy source. Protozoa and fungi often ferment sugars to lactate, ethanol, glycerol, succinate, formate, acetate, butanediol, and additional products. Some members of the genus *Clostridium* ferment mixtures of amino acids. Proteolytic clostridia such as the pathogens *C. sporogenes* and *C. botulinum* carry out the **Stickland reaction** in which one amino acid is oxidized and a second amino acid acts as the electron acceptor (**figure 10.18**). Some ATP is formed by substrate-level phosphorylation, and the fermentation is quite useful for growing in anoxic, protein-rich environments. The Stickland reaction is used to oxidize several amino acids: alanine, leucine, isoleucine, valine, phenylalanine, tryptophan, and histidine. Other bacteria ferment amino acids by different mechanisms. In addition to sugars and amino acids, organic acids such as acetate, lactate, propionate, and citrate can be fermented. Some of these fermentations are of great practical importance. For example, citrate can be converted to diacetyl and give flavor to fermented milk.   >>   *Class* Clostridia *(section 21.3); Microbiology of fermented foods (section 34.6)*

1. What are fermentations and why are they useful to many microorganisms?

2. How do the electron acceptors used in fermentation differ from the terminal electron acceptors used during either aerobic respiration or anaerobic respiration?

3. Briefly describe alcoholic, lactic acid, mixed acid and butanediol fermentations. How do homolactic fermenters and heterolactic fermenters differ? How do mixed acid fermenters and butanediol fermenters differ?

4. What is the net yield of ATP during homolactic, acetate, and butyrate fermentations? How do these yields compare to aerobic respiration in terms of both quantity and mechanism of phosphorylation?

5. Some bacteria carry out fermentation only because they lack ETCs. Yet these bacteria still

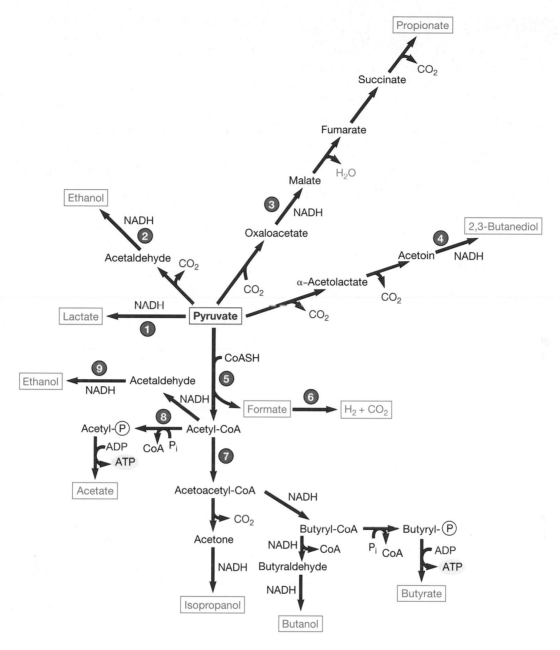

Figure 10.17 Some Common Microbial Fermentations. Only pyruvate fermentations are shown for the sake of simplicity; many other organic molecules can be fermented. Most of these pathways have been simplified by deletion of one or more steps and intermediates. Pyruvate and major end products are shown in color.

1. Lactic acid bacteria (*Streptococcus, Lactobacillus*), *Bacillus,* enteric bacteria (*Escherichia, Enterobacter, Salmonella, Proteus*)

2. Yeast, *Zymomonas*

3. Propionic acid bacteria (*Propionibacterium*)

4. *Enterobacter, Serratia, Bacillus*

5. Enteric bacteria

6. Enteric bacteria

7. *Clostridium*

8. Enteric bacteria

9. Enteric bacteria

have a membrane-bound ATPase. Why do you think this is the case? How do you think these bacteria use the ATPase? (*Hint: Examine figure 10.12 closely.*)

6. When bacteria carry out fermentation, only a few reactions of the TCA cycle operate. What purpose do you think these reactions might serve? Why do you think some parts of the cycle are shut down?

Table 10.2	**Mixed Acid Fermentation Products of *Escherichia coli***	
	Fermentation Balance (µM Product/100 µM Glucose)	
	Acid Growth (pH 6.0)	**Alkaline Growth (pH 8.0)**
Ethanol	50	50
Formic acid	2	86
Acetic acid	36	39
Lactic acid	80	70
Succinic acid	11	15
Carbon dioxide	88	2
Hydrogen gas	75	0.5

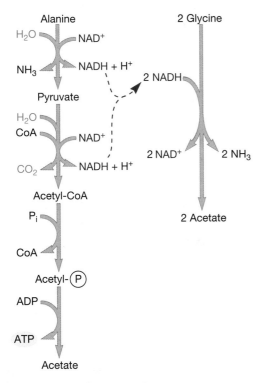

Figure 10.18 The Stickland Reaction. Alanine is oxidized to acetate and glycine is used to reoxidize the NADH generated during alanine degradation. The fermentation also produces some ATP.

10.8 CATABOLISM OF CARBOHYDRATES AND INTRACELLULAR RESERVE POLYMERS

Thus far, our main focus has been on the catabolism of glucose. However, microorganisms can catabolize many other carbohydrates. These carbohydrates may come either from outside the cell or from internal sources generated during normal metabolism. Often the initial steps in the degradation of external carbohydrate polymers differ from those employed with internal reserves.

Carbohydrates

Figure 10.19 outlines some catabolic pathways for the monosaccharides (single sugars) glucose, fructose, mannose, and galactose. The first three are phosphorylated using ATP and easily enter the Embden-Meyerhof pathway. In contrast, galactose must be converted to uridine diphosphate galactose (*see figure 11.10*) after initial phosphorylation, then changed into glucose 6-phosphate in a three-step process.

The common disaccharides are cleaved to monosaccharides by at least two mechanisms (figure 10.19). Maltose, sucrose, and lactose can be directly hydrolyzed to their constituent sugars. Many disaccharides (e.g., maltose, cellobiose, and sucrose) are also split by a phosphate attack on the bond joining the two sugars, a process called phosphorolysis.

Polysaccharides, like disaccharides, are cleaved by both hydrolysis and phosphorolysis. Procaryotes and fungi degrade

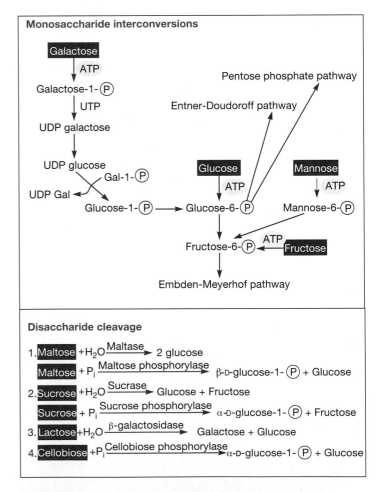

Figure 10.19 Carbohydrate Catabolism. Examples of enzymes and pathways used in disaccharide and monosaccharide catabolism. UDP is an abbreviation for uridine diphosphate.

external polysaccharides by secreting hydrolytic enzymes. These exoenzymes cleave polysaccharides that are too large to cross the plasma membrane into smaller molecules that can then be assimilated. Starch and glycogen are hydrolyzed by amylases to glucose, maltose, and other products. Cellulose is more difficult to digest; many fungi and a few bacteria (some gliding bacteria, clostridia, and actinomycetes) produce extracellular cellulases that hydrolyze cellulose to cellobiose and glucose. Some actinomycetes and members of the bacterial genus *Cytophaga,* isolated from marine habitats, excrete an agarase that degrades agar. Many soil bacteria and bacterial plant pathogens degrade pectin, a polymer of galacturonic acid (a galactose derivative) that is an important constituent of plant cell walls and tissues. Lignin, another important component of plant cell walls, is usually degraded only by certain fungi that release peroxide-generating enzymes. >> *Microorganisms in the terrestrial environments (section 26.2)*

Reserve Polymers

Microorganisms often survive for long periods in the absence of exogenous nutrients. Under such circumstances, they catabolize intracellular stores of glycogen, starch, poly-β-hydroxybutyrate, and other carbon and energy reserves. Glycogen and starch are degraded by phosphorylases. Phosphorylases catalyze a phosphorolysis reaction that shortens the polysaccharide chain by one glucose and yields glucose 1-phosphate.

$$(\text{Glucose})_n + P_i \rightarrow (\text{glucose})_{n-1} + \text{glucose-1-P}$$

Glucose 1-phosphate can enter glycolytic pathways by way of glucose 6-phosphate (figure 10.19).

Poly-β-hydroxybutyrate (PHB) is an important, widespread reserve material. Its catabolism has been studied most thoroughly in the soil bacterium *Azotobacter.* This bacterium hydrolyzes PHB to 3-hydroxybutyrate, then oxidizes the hydroxybutyrate to acetoacetate. Acetoacetate is converted to acetyl-CoA, which can be oxidized in the TCA cycle.

10.9 LIPID CATABOLISM

Chemoorganotrophic microorganisms frequently use lipids as energy sources. Triglycerides or triacylglycerols, esters of glycerol and fatty acids, are common energy sources and serve as our examples (figure 10.20). They can be hydrolyzed to glycerol and

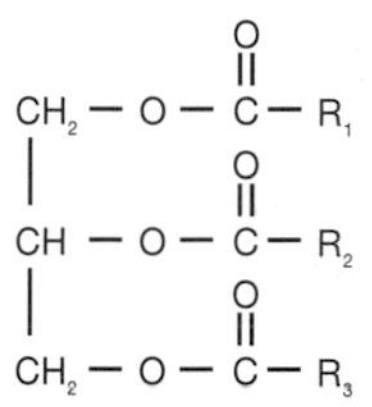

Figure 10.20 A Triacylglycerol or Triglyceride. The R groups represent the fatty acid side chains.

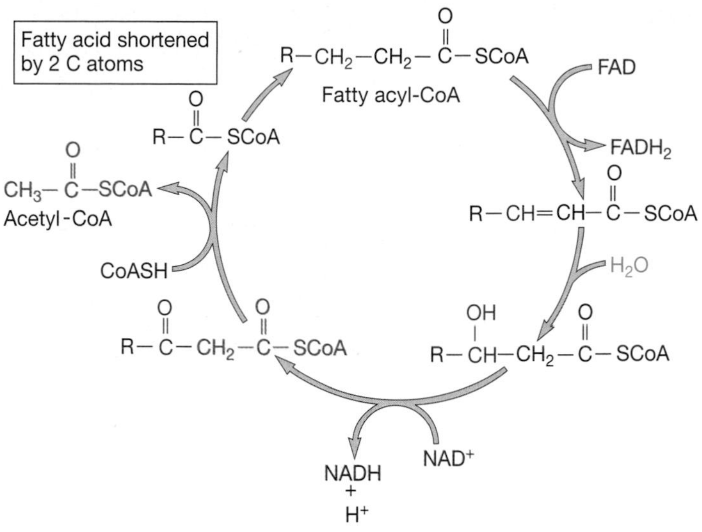

Figure 10.21 Fatty Acid β-Oxidation. The portions of the fatty acid being modified are shown in red.

fatty acids by microbial lipases. The glycerol is then phosphorylated, oxidized to dihydroxyacetone phosphate, and catabolized in the Embden-Meyerhof pathway (figure 10.5).

Fatty acids from triacylglycerols and other lipids are often oxidized in the **β-oxidation pathway** after conversion to coenzyme A esters (**figure 10.21**). In this pathway, fatty acids are shortened by two carbons with each turn of the cycle. The two carbon units are released as acetyl-CoA, which can be fed into the TCA cycle or used in biosynthesis. One turn of the cycle produces acetyl-CoA, NADH, and FADH$_2$; NADH and FADH$_2$ can be oxidized by an ETC to provide more ATP. Fatty acids are a rich source of energy for microbial growth. In a similar fashion, some microorganisms grow well on petroleum hydrocarbons under oxic conditions.

10.10 PROTEIN AND AMINO ACID CATABOLISM

Some bacteria and fungi—particularly pathogenic, food spoilage, and soil microorganisms—use proteins as their source of carbon and energy. They secrete enzymes called **proteases** that hydrolyze proteins to amino acids, which are transported into the cell and catabolized.

The first step in amino acid catabolism is **deamination,** the removal of the amino group from an amino acid. This is often accomplished by **transamination.** The amino group is transferred from an amino acid to an α-keto acid acceptor (**figure 10.22**). The organic acid resulting from deamination can be converted to pyruvate, acetyl-CoA, or a TCA cycle intermediate and eventually oxidized in the TCA cycle to release energy. It also can be used as a source of carbon for the synthesis of cell constituents. Excess nitrogen from deamination may be excreted as ammonium ion, thus making the medium alkaline.

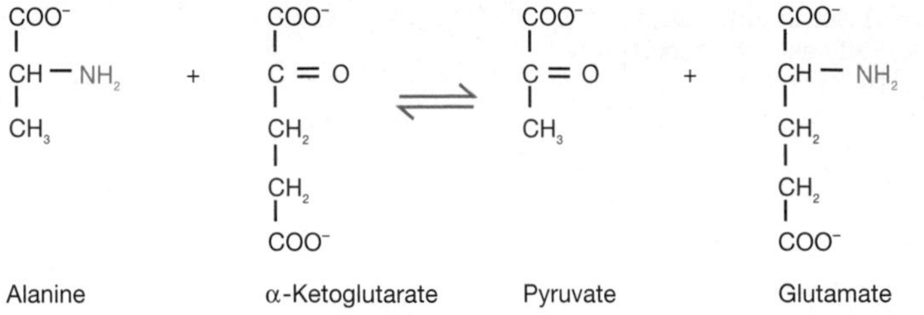

| COO⁻ | | COO⁻ | | COO⁻ | | COO⁻ |

$$CH-NH_2 + C=O \rightleftharpoons C=O + CH-NH_2$$

Alanine   α-Ketoglutarate   Pyruvate   Glutamate

Figure 10.22 Transamination. A common example of this process. The α-amino group (blue) of alanine is transferred to the acceptor α-ketoglutarate forming pyruvate and glutamate. The pyruvate can be catabolized in the tricarboxylic acid cycle or used in biosynthesis.

1. Briefly discuss the ways in which microorganisms degrade and use common monosaccharides, disaccharides, and polysaccharides from both external and internal sources.
2. Can members of the genus *Cytophaga* be cultured on standard solidified media? Why or why not?
3. Describe how a microorganism might derive carbon and energy from the lipids and proteins in its diet. What is β-oxidation? Deamination? Transamination?

10.11 CHEMOLITHOTROPHY

So far, we have considered microbes that synthesize ATP with the energy liberated by the oxidation of organic substrates such as carbohydrates, lipids, and proteins. Certain bacteria and archaea are **chemolithotrophs.** These microbes obtain electrons for their ETCs from the oxidation of inorganic molecules rather than from the NADH generated by the oxidation of organic nutrients (**figure 10.23**). Each species is specific in its preferences for electron donors and acceptors (**table 10.3**). The most common energy sources (electron donors) are hydrogen, reduced nitrogen compounds, reduced sulfur compounds, and ferrous iron (Fe^{2+}). The acceptor is usually O_2, but sulfate and nitrate may also be used.

Much less energy is available from the oxidation of inorganic molecules than from the complete oxidation of glucose to

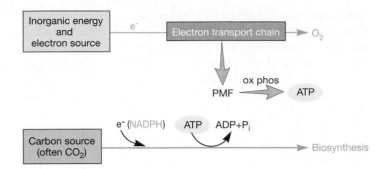

Figure 10.23 Chemolithotrophic Fueling Processes. Chemolithotrophic bacteria and archaea oxidize inorganic molecules (e.g., H_2S and NH_3), which serve as energy and electron sources. The electrons released pass through an electron transport chain, generating a proton motive force (PMF). ATP is synthesized by oxidative phosphorylation (ox phos). Most chemolithotrophs use O_2 as the terminal electron acceptor. However, some can use other exogenous molecules as terminal electron acceptors. Note that a molecule other than the energy source provides carbon for biosynthesis. Many chemolithotrophs are autotrophs.

CO_2 (**table 10.4**). This is because the NADH that donates electrons to the chain following the oxidation of an organic substrate has a more negative reduction potential than most of the inorganic substrates that chemolithotrophs use as direct electron donors to their ETC. Thus the P/O ratios for oxidative phosphorylation for many chemolithotrophs are probably around 1.0. Because the yield of ATP is so low, chemolithotrophs must oxidize a large quantity of inorganic material to grow and reproduce. This is particularly true of chemolithoautotrophs, which fix CO_2 into carbohydrates. CO_2-fixation pathways consume considerable amounts of ATP and reducing power (usually NADPH). Because they must consume a large amount of inorganic material, chemolithotrophs have significant ecological impact. ▶▶ *CO_2 fixation (section 11.3)*

Several procaryotic genera can oxidize hydrogen gas using a hydrogenase enzyme (table 10.3).

$$H_2 \rightarrow 2H^+ + 2e^-$$

Because the $H_2/2H^+$ redox couple has a very negative standard reduction potential, the electrons are donated either to an ETC or to NAD^+, depending on the hydrogenase. If NADH is produced,

Table 10.3	Representative Chemolithotrophs and Their Energy Sources			
Bacteria	**Electron Donor**	**Electron Acceptor**	**Products**
Alcaligenes, Hydrogenophaga, and *Pseudomonas* spp.	H_2	O_2	H_2O
Nitrobacter	NO_2^-	O_2	NO_3^-, H_2O
Nitrosomonas	NH_4^+	O_2	NO_2^-, H_2O
Thiobacillus denitrificans	S^0, H_2S	NO_3^-	SO_4^{2-}, N_2
Thiobacillus ferrooxidans	Fe^{2+}, S^0, H_2S	O_2	Fe^{3+}, H_2O, H_2SO_4

Table 10.4	Energy Yields from Oxidations Used by Chemolithotrophs	
Reaction		**$\Delta G^{\circ\prime}$ (kcal/mole)**[a]
$H_2 + 1/2\,O_2 \rightarrow H_2O$		-56.6
$NO_2^- + 1/2\,O_2 \rightarrow NO_3^-$		-17.4
$NH_4^+ + 1\,1/2\,O_2 \rightarrow NO_2^- + H_2O + 2H^+$		-65.0
$S^0 + 1\,1/2\,O_2 + H_2O \rightarrow H_2SO_4$		-118.5
$S_2O_3^{2-} + 2O_2 + H_2O \rightarrow 2SO_4^{2-} + 2H^+$		-223.7
$2Fe^{2+} + 2H^+ + 1/2\,O_2 \rightarrow 2Fe^{3+} + H_2O$		-11.2

[a]The $\Delta G^{\circ\prime}$ for complete oxidation of glucose to CO_2 is -686 kcal/mole. A kcal is equivalent to 4.184 kJ.

it can be used in ATP synthesis by electron transport and oxidative phosphorylation, with O_2, Fe^{3+}, S^0, and even carbon monoxide (CO) as the terminal electron acceptors. Often these hydrogen-oxidizing microorganisms use organic compounds as energy sources when such nutrients are available.

Some bacteria use the oxidation of nitrogenous compounds as a source of electrons. Among these chemolithotrophs, the **nitrifying bacteria** are best understood. They are soil and aquatic bacteria of considerable ecological significance that carry out **nitrification**—the oxidation of ammonia to nitrate. Nitrification is a two-step process that depends on the activity of at least two different genera. In the first step, ammonia is oxidized to nitrite by a number of genera, including *Nitrosomonas*:

$$NH_4^+ + 1\,1/2\,O_2 \rightarrow NO_2^- + H_2O + 2H^+$$

In the second step, the nitrite is oxidized to nitrate by genera such as *Nitrobacter*:

$$NO_2^- + 1/2\,O_2 \rightarrow NO_3^-$$

Nitrification differs from denitrification in that nitrification involves the oxidation of inorganic nitrogen compounds to yield nitrate, whereas denitrification is the reduction of oxidized nitrogenous compounds to nitrogen gas (p. 201). In nitrification, electrons are donated to an ETC, while in denitrification, nitrogen species are used as electron acceptors and nitrogen is lost to the atmosphere. >> *Biogeochemical cycling: Nitrogen cycle (section 25.1)*

Energy released upon the oxidation of both ammonia and nitrite is used to make ATP by oxidative phosphorylation. However, autotrophic microorganisms also need NAD(P)H (reducing power) as well as ATP to reduce CO_2 and other molecules (figure 10.23). Because molecules such as ammonia and nitrite have more positive reduction potentials than NAD^+, they cannot directly donate their electrons to form the required NADH and NADPH. Recall that electrons spontaneously move only from donors with more negative reduction potentials to acceptors with more positive potentials (*see figure 9.5*). Sulfur-oxidizing bacteria (described next) face the same difficulty. Both types of chemolithotrophs solve this problem by moving the electrons derived from the oxidation of their inorganic substrate (reduced nitrogen or sulfur compounds) up their ETCs to reduce $NAD(P)^+$ to NAD(P)H (**figure 10.24**).

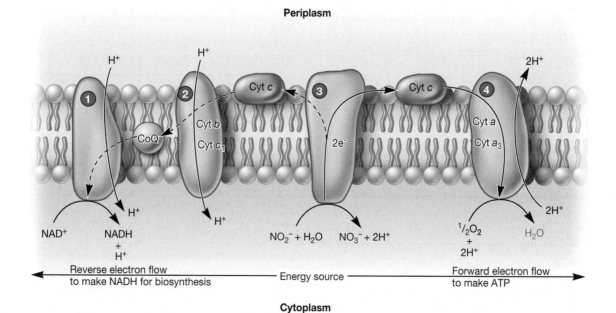

Figure 10.24 Electron Flow in *Nitrobacter* Electron Transport Chain. *Nitrobacter* oxidizes nitrite and carries out normal electron transport to generate proton motive force for ATP synthesis. This is the right-hand branch of the diagram. Some of the proton motive force also is used to force electrons to flow up the reduction potential gradient from nitrite to NAD^+ (left-hand branch). Cytochrome *c* and four complexes are involved: NADH-ubiquinone oxidoreductase (1), ubiquinol-cytochrome *c* oxidoreductase (2), nitrite oxidase (3), and cytochrome aa_3 oxidase (4).

Microbial Diversity & Ecology

10.1 Acid Mine Drainage

Each year millions of tons of sulfuric acid flow to the Ohio River from the coal mines of the Appalachian Mountains. This sulfuric acid is of microbial origin and leaches enough metals from the mines to make the river reddish and acidic. The primary culprit is *Thiobacillus ferrooxidans,* a chemolithotrophic bacterium that derives its energy from oxidizing ferrous ion to ferric ion and sulfide ion to sulfate ion. The combination of these two energy sources is important because of the solubility properties of iron. Ferrous ion is somewhat soluble and can be formed at pH values of 3.0 or less in moderately reducing environments. However, when the pH is greater than 4.0 to 5.0, ferrous ion is spontaneously oxidized to ferric ion by O_2 in the water and precipitates as a hydroxide. If the pH drops below 2.0 to 3.0 because of sulfuric acid production by spontaneous oxidation of sulfur or sulfur oxidation by thiobacilli and other bacteria, the ferrous ion remains reduced, soluble, and available as an energy source. Remarkably, *T. ferrooxidans* grows well at such acidic pHs and actively oxidizes ferrous ion to an insoluble ferric precipitate. The water is rendered toxic for most aquatic life and unfit for human consumption.

The ecological consequences of this metabolic life-style arise from the common presence of pyrite (FeS_2) in coal mines. The bacteria oxidize both elemental components of pyrite for their growth and in the process form sulfuric acid, which leaches the remaining minerals.

Autoxidation or bacterial action

$$2FeS_2 + 7O_2 + 2H_2O \rightarrow 2Fe^{2+} + 4SO_4^{2-} + 4H^+$$

T. ferrooxidans

$$2Fe^{2+} + 1/2O_2 + 2H^+ \rightarrow 2Fe^{3+} + H_2O$$

Pyrite oxidation is further accelerated because the ferric ion generated by bacterial activity readily oxidizes more pyrite to sulfuric acid and ferrous ion. In turn the ferrous ion supports further bacterial growth. It is difficult to prevent *T. ferrooxidans* growth as it requires only pyrite and common inorganic salts. Because *T. ferrooxidans* gets its O_2 and CO_2 from the air, the only feasible method of preventing its damaging growth is to seal the mines to render the habitat anoxic.

This is called **reverse electron flow.** Of course, this is not thermodynamically favorable, so energy in the form of the proton motive force must be diverted from performing other cellular work (e.g., ATP synthesis, transport, motility) to "push" the electrons from molecules of relatively positive reduction potentials to those that are more negative. Chemolithotrophs can afford this inefficiency, as they have no serious competitors for their unique energy sources. << *Electron transport chains (section 9.6)*

Sulfur-oxidizing microbes are the third major group of chemolithotrophs. The metabolism of *Thiobacillus* has been best studied. These bacteria oxidize sulfur (S^0), hydrogen sulfide (H_2S), thiosulfate ($S_2O_3^{2-}$), and other reduced sulfur compounds to sulfuric acid; therefore they have a significant ecological impact (**Microbial Diversity & Ecology 10.1**). Interestingly, they generate ATP by both oxidative phosphorylation and substrate-level phosphorylation involving adenosine 5′-phosphosulfate (APS). APS is a high-energy molecule formed from sulfite and adenosine monophosphate (**figure 10.25**).

Some sulfur-oxidizing procaryotes are extraordinarily flexible metabolically. For example, *Sulfolobus brierleyi,* an archaeon, and some bacteria can grow aerobically by oxidizing sulfur with oxygen as the electron acceptor. However, in the absence of O_2, they carry out anaerobic respiration and oxidize organic material with sulfur as the electron acceptor. Many sulfur-oxidizing chemolithotrophs use CO_2 as their carbon source but will grow heterotrophically if they are

(a) Direct oxidation of sulfite

$$SO_3^{2-} \xrightarrow{\text{sulfite oxidase}} SO_4^{2-} + 2e^-$$

(b) Formation of adenosine 5′-phosphosulfate

$$2SO_3^{2-} + 2AMP \longrightarrow 2APS + 4e^-$$

$$2APS + 2P_i \longrightarrow 2ADP + 2SO_4^{2-}$$

$$2ADP \longrightarrow AMP + ATP$$

$$2SO_3^{2-} + AMP + 2P_i \longrightarrow 2SO_4^{2-} + ATP + 4e^-$$

Adenosine 5′-phosphosulfate

(c)

Figure 10.25 Energy Generation by Sulfur Oxidation. (a) Sulfite can be directly oxidized to provide electrons for electron transport and oxidative phosphorylation. (b) Sulfite can also be oxidized and converted to adenosine 5′-phosphosulfate (APS). This route produces electrons for use in electron transport and ATP by substrate-level phosphorylation with APS. (c) The structure of APS.

supplied with reduced organic carbon sources such as glucose or amino acids.

1. How do chemolithotrophs obtain their ATP and NADH? What is their most common source of carbon?
2. Describe energy production by hydrogen-oxidizing bacteria, nitrifying bacteria, and sulfur-oxidizing bacteria.
3. Why can hydrogen-oxidizing bacteria and archaea donate electrons to NAD^+ while sulfur- and ammonia-oxidizing bacteria and archaea cannot?
4. What is reverse electron flow and why do many chemolithotrophs perform it?
5. Arsenate is a compound that inhibits substrate-level phosphorylation. Compare the effect of this compound on an H_2-oxidizing chemolithotroph, on a sulfite-oxidizing chemolithotroph, and on a chemoorgano-troph carrying out fermentation.

10.12 PHOTOTROPHY

Many microorganisms derive energy from the oxidation of inorganic and organic compounds, but many others capture the energy in light and use it to synthesize ATP and reducing power (e.g., NADPH) (figure 10.1). When the ATP and reducing power are used to reduce and incorporate CO_2 (CO_2 fixation), the process is called **photosynthesis.** Photosynthesis is one of the most significant metabolic processes on Earth because almost all our energy is ultimately derived from solar energy. Photosynthetic organisms serve as the base of most food chains in the biosphere. One type of photosynthesis is also responsible for replenishing our supply of O_2. Although most people associate photosynthesis with plants, over half the photosynthesis on Earth is carried out by microorganisms (table 10.5).

Photosynthesis is divided into two parts. In the **light reactions,** light energy is trapped and converted to chemical energy

Table 10.5	Diversity of Phototrophic Organisms	
Eucaryotic Organisms	**Procaryotic Organisms**	
Plants	Cyanobacteria	
Multicellular green, brown, and red algae	Green sulfur bacteria Green nonsulfur bacteria	
Unicellular protists (e.g., euglenoids, dinoflagellates, diatoms)	*Halobacterium* (archaeon) Purple sulfur bacteria Purple nonsulfur bacteria	

Chlorophyll-based phototrophy

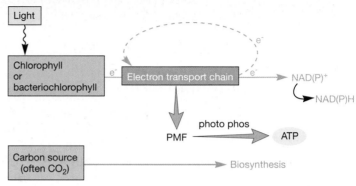

Rhodopsin-based phototrophy

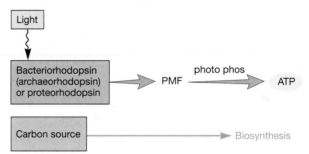

Figure 10.26 Phototrophic Fueling Reactions. Phototrophs use light to generate a proton motive force (PMF), which is then used to synthesize ATP by a process called photophosphorylation (photo phos). The process requires light-absorbing pigments. When the pigments are chlorophyll or bacteriochlorophyll, the absorption of light triggers electron flow through an electron transport chain, accompanied by the pumping of protons across a membrane. The electron flow can be either cyclic (dashed line) or noncyclic (solid line), depending on the organism and its needs. Rhodopsin-based phototrophy differs in that the PMF is formed directly by the light-absorbing pigment, which is a light-driven proton pump. Many phototrophs are autotrophs and must use much of the ATP and reducing power they make to fix CO_2.

and reducing power. These are then used to fix CO_2 and synthesize cell constituents in the **dark reactions.** In this section, three types of phototrophy are discussed: oxygenic photosynthesis, anoxygenic photosynthesis, and rhodopsin-based phototrophy (**figure 10.26**). The dark reactions of photosynthesis are reviewed in chapter 11. >> *CO_2 fixation (section 11.3)*

Light Reactions in Oxygenic Photosynthesis

Phototrophic eucaryotes and cyanobacteria carry out **oxygenic photosynthesis,** so named because oxygen is generated and released into the environment when light energy is converted to chemical energy. Central to this process, and to all other phototrophic processes, are light-absorbing pigments (**table 10.6**). In oxygenic phototrophs, the most important pigments are the **chlorophylls.** Chlorophylls are large planar molecules composed of four substituted pyrrole rings with a magnesium atom coordinated

Table 10.6	Properties of Chlorophyll-Based Photosynthetic Systems		
Property	Eucaryotes	Cyanobacteria	Green Bacteria, Purple Bacteria, and Heliobacteria
Photosynthetic pigment	Chlorophyll *a*	Chlorophyll *a*	Bacteriochlorophyll
Number of photosystems	2	2	1
Photosynthetic electron donors	H_2O	H_2O	H_2, H_2S, S, organic matter
O_2 production pattern	Oxygenic	Oxygenic[a]	Anoxygenic
Primary products of energy conversion	ATP + NADPH	ATP + NADPH	ATP
Carbon source	CO_2	CO_2	Organic or CO_2

[a]Some cyanobacteria can function anoxygenically under certain conditions. For example, *Oscillatoria* can use H_2S as an electron donor instead of H_2O.

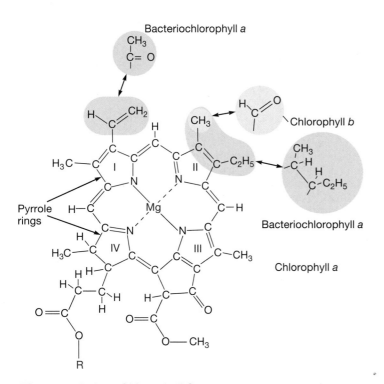

Figure 10.27   Chlorophyll Structure.   The structures of chlorophyll *a*, chlorophyll *b*, and bacteriochlorophyll *a*. The complete structure of chlorophyll *a* is given. Only one group is altered to produce chlorophyll *b*, and two modifications in the ring system are required to change chlorophyll *a* to bacteriochlorophyll *a*. The side chain (R) of bacteriochlorophyll *a* may be either phytyl (a 20-carbon chain also found in chlorophylls *a* and *b*) or geranylgeranyl (a 20-carbon side chain similar to phytyl but with three more double bonds).

to the four central nitrogen atoms (**figure 10.27**). Several chlorophylls are found in eucaryotes; the two most important are chlorophyll *a* and chlorophyll *b*. These two molecules differ slightly in their structure and spectral properties. When dissolved in acetone, chlorophyll *a* has a light absorption peak at 665 nm; the corresponding peak for chlorophyll *b* is at 645 nm. In addition to absorbing red light, chlorophylls also absorb blue light strongly

(the second absorption peak for chlorophyll *a* is at 430 nm). Because chlorophylls absorb primarily in the red and blue ranges, green light is transmitted, and these organisms appear green. A long hydrophobic tail attached to the chlorophyll ring aids in its attachment to membranes, the site of the light reactions.

Other photosynthetic pigments also trap light energy. The most widespread of these are the carotenoids, long molecules, usually yellowish in color, that possess an extensive conjugated double bond system (**figure 10.28**). β-Carotene is present in cyanobacteria belonging to the genus *Prochloron* and most photosynthetic protists; fucoxanthin is found in protists such as diatoms and dinoflagellates. Red algae and cyanobacteria have photosynthetic pigments called **phycobiliproteins,** consisting of a protein with a linear tetrapyrrole attached (figure 10.28). Phycoerythrin is a red pigment with a maximum absorption around 550 nm, and phycocyanin is blue (maximum absorption at 620 to 640 nm).

Carotenoids and phycobiliproteins are often called **accessory pigments** because of their role in photosynthesis. They are important because they absorb light in the range not absorbed by chlorophylls (the blue-green through yellow range; about 470–630 nm) (*see figure 19.4*). This light is transferred to chlorophyll. In this way accessory pigments make photosynthesis more efficient over a broader range of wavelengths. In addition, this allows organisms to use light not used by other phototrophs in their habitat. Accessory pigments also protect microorganisms from intense sunlight, which could oxidize and damage the photosynthetic apparatus.

Chlorophylls and accessory pigments are assembled in highly organized arrays called antennas, whose purpose is to create a large surface area to trap as many photons as possible. An antenna has about 300 chlorophyll molecules. Light energy is captured in an antenna and transferred from chlorophyll to chlorophyll until it reaches a **reaction-center chlorophyll pair** that is directly involved in photosynthetic electron transport. In oxygenic phototrophs, there are two kinds of antennas associated with two different photosystems (**figure 10.29**). **Photosystem I** absorbs longer wavelength light (≥680 nm) and funnels the energy to a reaction center chlorophyll *a* pair called P700. The term P700

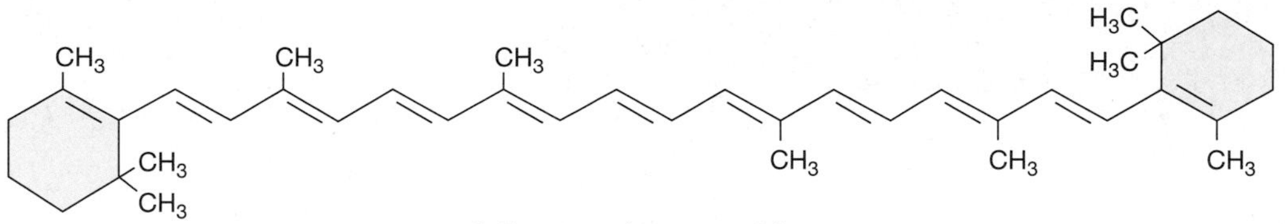

Figure 10.28 Representative Accessory Pigments. Beta-carotene is a carotenoid found in photosynthetic protists and plants. Note that it has a long chain of alternating double and single bonds called conjugated double bonds. Fucoxanthin is a carotenoid accessory pigment in several divisions of algae (the dot in the structure represents a carbon atom). Phycocyanobilin is an example of a linear tetrapyrrole that is attached to a protein to form a phycobiliprotein.

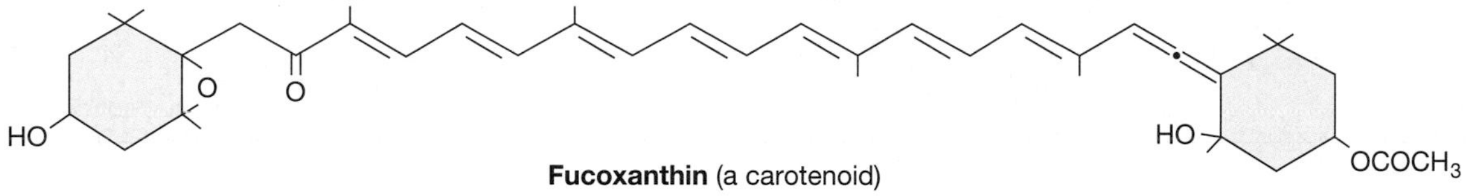

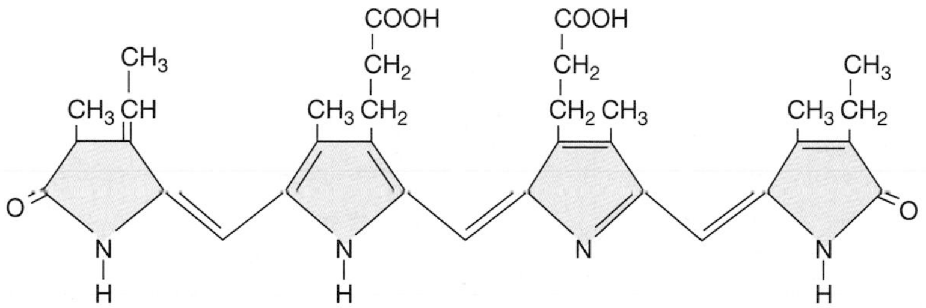

signifies that this molecule most effectively absorbs light at a wavelength of 700 nm. **Photosystem II** traps light at shorter wavelengths ($\leq$680 nm) and transfers its energy to the reaction center chlorophyll pair P680.

When the photosystem I antenna transfers light energy to P700, P700 absorbs the energy and is excited, and its reduction potential becomes very negative. This allows P700 to donate its excited, high-energy electron to a specific acceptor, probably a special chlorophyll a molecule or an iron-sulfur protein. The electron is eventually transferred to ferredoxin and then travels in either of two directions. In the cyclic pathway (the dashed lines in figure 10.29), the electron moves in a cyclic route through a series of electron carriers and back to the oxidized P700. The pathway is termed cyclic because the electron from P700 returns to P700 after traveling through the photosynthetic ETC. PMF is formed during cyclic electron transport and used to synthesize ATP. This process is called **cyclic photophosphorylation** because electrons travel in a circle and ATP is formed. Only photosystem I participates.

Electrons also can travel in a noncyclic pathway involving both photosystems. P700 is excited and donates electrons to ferredoxin as before. In the noncyclic route, however, reduced ferredoxin reduces NADP$^+$ to NADPH (figure 10.29). Because the electrons contributed to NADP$^+$ cannot be used to reduce oxidized P700, photosystem II participation is required. It donates electrons to oxidized P700 and generates ATP in the process. The

photosystem II antenna absorbs light energy and excites P680, which then reduces pheophytin a. Pheophytin a is chlorophyll a in which two hydrogen atoms have replaced the central magnesium. Electrons subsequently travel to the plastoquinone pool and down an ETC to reduce P700. Now P680 must also be reduced if it is to accept more light energy. Figure 10.29 indicates that the standard reduction potential of P680 is more positive than that of the $1/2O_2/H_2O$ redox couple. Thus H_2O can be used to donate electrons to P680 resulting in the release of oxygen. Because electrons flow from water to NADP$^+$ with the aid of energy from two photosystems, ATP is synthesized by **noncyclic photophosphorylation.** It appears that one ATP and one NADPH are formed when two electrons travel through the noncyclic pathway.

It is worth reemphasizing that although light is the source of energy for chlorophyll-based phototrophy, the process used to make ATP is virtually the same as seen for chemotrophs: oxidation-reduction reactions occurring in ETCs generate a PMF that is used by ATP synthase to make ATP. Furthermore, just as is true of mitochondrial electron transport, photosynthetic electron transport takes place within a membrane. Chloroplast granal membranes contain both photosystems and their antennas. **Figure 10.30** shows a thylakoid membrane carrying out noncyclic photophosphorylation by the chemiosmotic mechanism. Protons move to the thylakoid interior during photosynthetic electron transport and return to the stroma when ATP is formed.

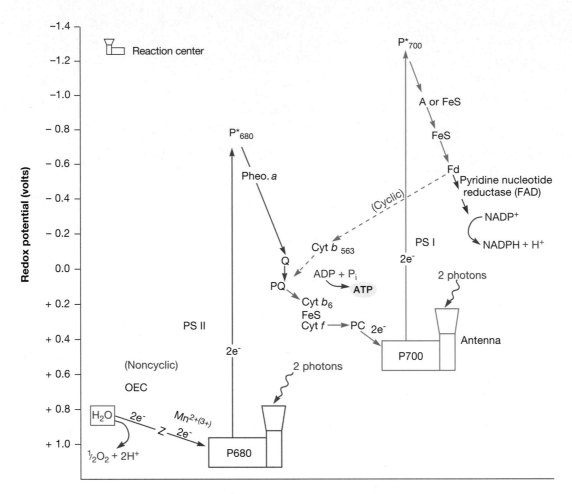

Figure 10.29   Green Plant Photosynthesis.   Electron flow during photosynthesis in higher plants. Cyanobacteria and eucaryotic algae are similar in having two photosystems, although they may differ in some details. The carriers involved in electron transport are ferredoxin (Fd) and other FeS proteins; cytochromes b_6, b_{563}, and f; plastoquinone (PQ); copper containing plastocyanin (PC); pheophytin a (Pheo. a); possibly chlorophyll a (A); and the unknown quinone Q, which is probably a plastoquinone. Both photosystem I (PS I) and photosystem II (PS II) are involved in noncyclic photophosphorylation; only PS I participates in cyclic photophosphorylation. The oxygen evolving complex (OEC) that extracts electrons from water contains manganese ions and the substance Z, which transfers electrons to the PS II reaction center.

It is believed that stromal lamellae possess only photosystem I and are involved in cyclic photophosphorylation alone. In cyanobacteria, photosynthetic light reactions are located in thylakoid membranes within the cell.

The dark reactions of oxygenic phototrophs use three ATPs and two NADPHs to reduce one CO_2 to carbohydrate (CH_2O).

$$CO_2 + 3ATP + 2NADPH + 2H^+ + H_2O \rightarrow$$
$$(CH_2O) + 3ADP + 3P_i + 2NADP^+$$

The noncyclic system generates one NADPH and one ATP per pair of electrons; therefore four electrons passing through the system produce two NADPHs and two ATPs. A total of 8 quanta of light energy (4 quanta for each photosystem) is needed to propel the four electrons from water to $NADP^+$. Because the ratio of ATP to NADPH required for CO_2 fixation is 3:2, at least one more ATP must be supplied. Cyclic photophosphorylation probably operates independently to generate the extra ATP. This requires absorption of another 2 to 4 quanta. It follows that around

10 to 12 quanta of light energy are needed to reduce and incorporate one molecule of CO_2 during photosynthesis.

Light Reactions in Anoxygenic Photosynthesis

Certain bacteria carry out a second type of photosynthesis called **anoxygenic photosynthesis.** This phototrophic process derives its name from the fact that molecules other than water are used as an electron source and therefore O_2 is not produced. The process also differs in terms of the photosynthetic pigments used, the participation of just one photosystem, and the mechanisms used to generate reducing power. Three groups of bacteria carry out anoxygenic photosynthesis: phototrophic green bacteria, phototrophic purple bacteria, and heliobacteria. The biology and ecology of these organisms are described in more detail in chapters 18, 19, and 20.

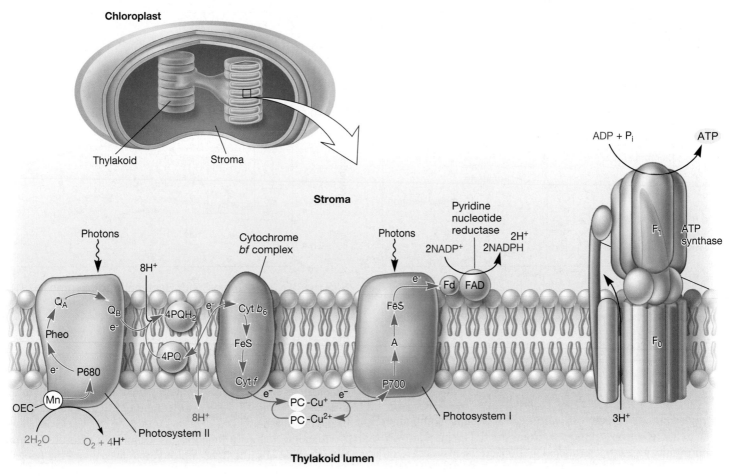

Figure 10.30   The Mechanism of Photosynthesis.   An illustration of the chloroplast thylakoid membrane showing photosynthetic ETC function and noncyclic photophosphorylation. The chain is composed of three complexes: PS I, the cytochrome *bf* complex, and PS II. Two diffusible electron carriers connect the three complexes. Plastoquinone (PQ) connects PS I with the cytochrome *bf* complex, and plastocyanin (PC) connects the cytochrome *bf* complex with PS II. The light-driven electron flow pumps protons across the thylakoid membrane and generates a PMF, which can then be used to make ATP. Water is the source of electrons and the oxygen-evolving complex (OEC) produces oxygen.

Anoxygenic phototrophs have photosynthetic pigments called **bacteriochlorophylls** (figure 10.27). In some bacteria, these are located in membranous vesicles called chlorosomes. The absorption maxima of bacteriochlorophylls (Bchl) are at longer wavelengths than those of chlorophylls. Bacteriochlorophylls *a* and *b* have maxima in ether at 775 and 790 nm, respectively. In vivo maxima are about 830 to 890 nm (Bchl *a*) and 1,020 to 1,040 nm (Bchl *b*). This shift of absorption maxima into the infrared region better adapts these bacteria to their ecological niches. ▸▸ *Photosynthetic bacteria (section 19.3)*

Many differences found in anoxygenic phototrophs are because they have a single photosystem. Because of this, they are restricted to cyclic photophosphorylation and are unable to produce O_2 from H_2O. Indeed, almost all anoxygenic phototrophs are strict anaerobes. A tentative scheme for the photosynthetic ETC of a purple nonsulfur bacterium is given in **figure 10.31**. When the reaction-center bacteriochlorophyll P870 is excited, it donates an electron to bacteriopheophytin. Electrons then flow to quinones and through an ETC back to

P870 while generating sufficient PMF to drive ATP synthesis by ATP synthase. Note that although both green and purple bacteria lack two photosystems, the purple bacteria have a photosynthetic apparatus similar to photosystem II of oxygenic phototrophs, whereas the green sulfur bacteria have a system similar to photosystem I. ▸▸ *Class* Alphaproteobacteria: *Purple nonsulfur bacteria (section 20.1)*

Anoxygenic photoautotrophs face a further problem because they also require reducing power (NAD[P]H or reduced ferredoxin) for CO_2 fixation and other biosynthetic processes. They are able to generate reducing power in at least three ways, depending on the bacterium. Some have hydrogenases that are used to produce NAD(P)H directly from the oxidation of hydrogen gas. This is possible because hydrogen gas has a more negative reduction potential than NAD^+ (*see table 9.1*). Others, such as the photosynthetic purple bacteria, use reverse electron flow to generate NAD(P)H (figure 10.31). In this mechanism, electrons are drawn off the photosynthetic ETC and "pushed" to $NAD(P)^+$ using PMF. Electrons from electron donors such as

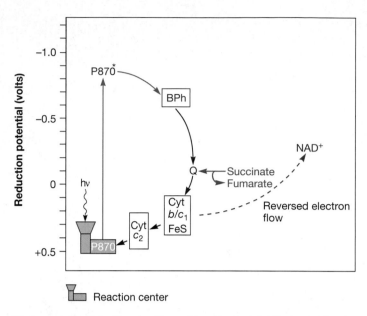

Figure 10.31   Purple Nonsulfur Bacterial Photosynthesis. The photosynthetic ETC in the purple nonsulfur bacterium *Rhodobacter sphaeroides*. This scheme is incomplete and tentative. Ubiquinone (Q) is very similar to coenzyme Q. BPh stands for bacteriopheophytin. The electron source succinate is in blue.

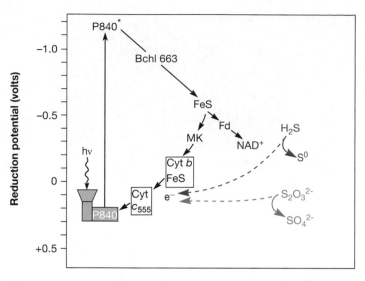

Figure 10.32   Green Sulfur Bacterial Photosynthesis. The photosynthetic ETC in the green sulfur bacterium *Chlorobium limicola*. Light energy is used to make ATP by cyclic photophosphorylation and to move electrons from thiosulfate ($S_2O_3^{2-}$) and H_2S (green and blue) to NAD^+. The ETC has a quinone called menaquinone (MK).

hydrogen sulfide, elemental sulfur, and organic compounds replace the electrons removed from the ETC in this way. Phototrophic green bacteria and heliobacteria also draw off electrons from their ETCs. However, because the reduction potential of the component of the chain where this occurs is more negative than NAD^+ and oxidized ferredoxin, the electrons flow spontaneously to these electron acceptors. Thus these bacteria exhibit a simple form of noncyclic photosynthetic electron flow (**figure 10.32**).

Rhodopsin-Based Phototrophy

Oxygenic and anoxygenic photosynthesis are chlorophyll-based types of phototrophy—that is, chlorophyll or bacteriochlorophyll is the major pigment used to absorb light and initiate the conversion of light energy to chemical energy. This type of phototrophy is observed only in eucaryotes and bacteria; it has not been observed in any archaea to date. However, some archaea are able to use light as a source of energy. Instead of using chlorophyll, these microbes use a membrane protein called **bacteriorhodopsin** (more correctly called archaeorhodopsin). One such archaeon is the halophile *Halobacterium salinarum.*

H. salinarum normally depends on aerobic respiration for the release of energy from an organic energy source. It cannot grow anaerobically by anaerobic respiration or fermentation. However, under conditions of low oxygen and high light intensity, it synthesizes bacteriorhodopsin, a deep-purple pigment that closely resembles the rhodopsin found in the rods and cones of vertebrate eyes. Bacteriorhodopsin's chromophore is retinal, a type of carotenoid. The chromophore is covalently

attached to the pigment protein, which is embedded in the plasma membrane in such a way that the retinal is in the center of the membrane.

Bacteriorhodopsin functions as a light-driven proton pump. When retinal absorbs light, a proton is released and the bacteriorhodopsin undergoes a sequence of conformation changes that translocate the proton into the periplasmic space (*see figure 18.14*). The light-driven proton pumping generates a pH gradient that can be used to power the synthesis of ATP by chemiosmosis. This phototrophic capacity is particularly useful to *Halobacterium* because oxygen is not very soluble in concentrated salt solutions and may decrease to an extremely low level in *Halobacterium*'s habitat. When the surroundings become temporarily anoxic, the archaeon uses light energy to synthesize sufficient ATP to survive until oxygen levels rise again. Note that this type of phototrophy does not involve electron transport. It had been thought that rhodopsin-based phototrophy is unique to *Archaea*. However, proton-pumping rhodopsins have recently been discovered in some proteobacteria (proteorhodopsin) and a fungus. ▶▶ *Environmental genomics (section 15.8); Marine and freshwater microbiology (section 26.1)*

1. Define the following terms: light reactions, dark reactions, chlorophyll, carotenoid, phycobiliprotein, antenna, and photosystems I and II.

2. What happens to a reaction center chlorophyll pair, such as P700, when it absorbs light?

3. What is the function of accessory pigments?

4. What is photophosphorylation? What is the difference between cyclic and noncyclic photophosphorylation?

5. Compare and contrast anoxygenic photosynthesis and oxygenic photosynthesis. How do these two types of phototrophy differ from rhodopsin-based phototrophy?

6. Suppose you isolated a bacterial strain that carried out oxygenic photosynthesis. What photosystems would it possess and what group of bacteria would it most likely belong to?

Summary

10.1 Chemoorganotrophic Fueling Processes

a. Chemotrophic organisms use chemical sources of electrons and energy.

b. Chemoorganotrophic microorganisms can use three kinds of electron acceptors during energy metabolism (**figure 10.2**). The nutrient may be oxidized with an endogenous electron acceptor (fermentation), with oxygen as an exogenous electron acceptor (aerobic respiration), or with another external electron acceptor (anaerobic respiration).

10.2 Aerobic Respiration

a. Aerobic respiration can be divided into three stages: (1) breakdown of macromolecules into their constituent parts, (2) catabolism to pyruvate, acetyl-CoA, and other molecules by pathways that converge on glycolytic pathways and the TCA cycle, and (3) completion of catabolism by the TCA cycle. Most energy is produced at this stage and results from oxidation of NADH and $FADH_2$ by the electron transport chain and oxidative phosphorylation (**figure 10.3**).

b. The pathways used during aerobic respiration are amphibolic, having both catabolic and anabolic functions (**figure 10.4**).

10.3 Breakdown of Glucose to Pyruvate

a. Glycolysis, used in its broadest sense, refers to all pathways used to break down glucose to pyruvate.

b. The Embden-Meyerhof pathway has a net production of two NADHs and two ATPs, the latter being produced by substrate-level phosphorylation. It also produces several precursor metabolites (**figure 10.5**).

c. In the pentose phosphate pathway, glucose 6-phosphate is oxidized twice and converted to pentoses and other sugars. It is a source of NADPH, ATP, and several precursor metabolites (**figure 10.6**).

d. In the Entner-Doudoroff pathway, glucose is oxidized to 6-phosphogluconate, which is then dehydrated and cleaved to pyruvate and glyceraldehyde 3-phosphate (**figure 10.7**). The latter product can be oxidized by enzymes of the Embden-Meyerhof pathway to provide ATP, NADH, and another molecule of pyruvate.

10.4 Tricarboxylic Acid Cycle

a. The tricarboxylic acid cycle is the final stage of catabolism during aerobic respiration (**figure 10.8**).

b. It oxidizes acetyl-CoA to CO_2 and forms one GTP, three NADHs, and one $FADH_2$ per acetyl-CoA. It also generates several precursor metabolites.

10.5 Electron Transport and Oxidative Phosphorylation

a. The NADH and $FADH_2$ produced from the oxidation of carbohydrates, fatty acids, and other nutrients can be oxidized in an electron transport chain (ETC). Electrons flow from carriers with more negative reduction potentials to those with more positive potentials (**figure 10.9**; *see also figure 9.5*), and free energy is released for ATP synthesis by oxidative phosphorylation.

b. Procaryotic ETCs are often different from eucaryotic chains with respect to such aspects as carriers and branching (**figure 10.11**). In eucaryotes, the P/O ratio for NADH is about 3 and that for $FADH_2$ is around 2; P/O ratios are usually much lower in bacteria.

c. ATP synthase catalyzes the synthesis of ATP (**figure 10.13**). In eucaryotes, it is located on the inner surface of the inner mitochondrial membrane. Procaryotic ATP synthase is on the inner surface of the plasma membrane.

d. The most widely accepted mechanism of oxidative phosphorylation is the chemiosmotic hypothesis in which proton motive force (PMF) drives ATP synthesis (**figure 10.10**).

e. Aerobic respiration in eucaryotes can theoretically yield a maximum of 38 ATPs (**figure 10.14**).

10.6 Anaerobic Respiration

a. Anaerobic respiration is the process of ATP production by electron transport in which the terminal electron acceptor is an exogenous molecule other than O_2. The most common acceptors are nitrate, sulfate, and CO_2 (**figure 10.15**).

b. For some microorganisms, the same pathways used to aerobically respire an organic energy source are also used for anaerobic respiration.

10.7 Fermentation

a. During fermentation, an endogenous electron acceptor is used to reoxidize any NADH generated by the catabolism of glucose to pyruvate (**figure 10.16**).

b. Flow of electrons from the electron donor to the electron acceptor does not involve an ETC, and ATP is synthesized only by substrate-level phosphorylation.

c. There are many different fermentation pathways. These are of practical importance in clinical and industrial settings (**figure 10.17**).

10.8 Catabolism of Carbohydrates and Intracellular Reserve Polymers

a. Microorganisms catabolize many extracellular carbohydrates. Monosaccharides are taken in and phosphorylated; disaccharides may be cleaved to monosaccharides by either hydrolysis or phosphorolysis.

b. External polysaccharides are degraded by hydrolysis and the products are absorbed. Intracellular glycogen and starch are converted to glucose 1-phosphate by phosphorolysis (**figure 10.19**).

10.9 Lipid Catabolism

a. Triglycerides are hydrolyzed to glycerol and fatty acids by enzymes called lipases.

b. Fatty acids are usually oxidized to acetyl-CoA in the β-oxidation pathway (**figure 10.21**).

10.10 Protein and Amino Acid Catabolism

a. Proteins are hydrolyzed to amino acids that are then deaminated (**figure 10.22**).

b. The carbon skeletons produced by deamination are fed into the TCA cycle (**figure 10.3**).

10.11 Chemolithotrophy

a. Chemolithotrophs synthesize ATP by oxidizing inorganic compounds—usually hydrogen, reduced nitrogen and sulfur compounds, or ferrous iron—with an ETC and O_2 as the electron acceptor (**figure 10.23** and **table 10.3**). The PMF produced is used by ATP synthase to make ATP.

b. Many of the energy sources used by chemolithotrophs have a more positive standard reduction potential than the $NAD^+/$ NADH redox pair. These chemolithotrophs must expend energy (PMF) to drive reverse electron flow and produce the NADH they need for CO_2 fixation and other processes (**figure 10.24**).

10.12 Phototrophy

a. In oxygenic photosynthesis, eucaryotes and cyanobacteria trap light energy with chlorophyll and accessory pigments and move electrons through photosystems I and II to make ATP and NADPH (the light reactions). The ATP and NADPH are used in the dark reactions to fix CO_2.

b. Cyclic photophosphorylation involves the activity of photosystem I alone and generates ATP only. In noncyclic photophosphorylation, photosystems I and II operate together to move electrons from water to $NADP^+$ producing ATP, NADPH, and O_2 (**figure 10.29**). In both cases, electron flow generates PMF, which is used by ATP synthase to make ATP.

c. Anoxygenic phototrophs differ from oxygenic phototrophs in possessing bacteriochlorophyll and having only one photosystem (**figures 10.31** and **10.32**). Cyclic electron flow generates a PMF, which is used by ATP synthase to make ATP (i.e., cyclic photophosphorylation). They are anoxygenic because they use molecules other than water as an electron donor for electron flow and the production of reducing power.

d. Some archaea use a type of phototrophy that involves the proton-pumping pigment bacteriorhodopsin. This type of phototrophy generates PMF but does not involve an ETC.

Critical Thinking Questions

1. Without looking in chapters 19 and 20, predict some characteristics that would describe niches occupied by green and purple photosynthetic bacteria.

2. From an evolutionary perspective, discuss why most microorganisms use aerobic respiration to generate ATP.

3. How would you isolate a thermophilic chemolithotroph that uses sulfur compounds as a source of energy and electrons? What changes in the incubation system would be needed to isolate bacteria using sulfur compounds in anaerobic respiration? How would you tell which process is taking place through an analysis of the sulfur molecules present in the medium?

4. Certain chemicals block ATP synthesis by allowing protons and other ions to "leak across membranes," disrupting the charge and proton gradients established by electron flow through an ETC. Does this observation support the chemiosmosis hypothesis? Explain your reasoning.

5. Two flasks of *E. coli* are grown in batch culture in the same medium (2% glucose and amino acids; no nitrate) and at the same temperature (37°C). Culture #1 is well aerated. Culture #2 is anoxic. After 16 hours the following observations are made:

- Culture #1 has a high cell density; the cells appear to be in stationary phase, and the glucose level in the medium is reduced to 1.2%.

- Culture #2 has a low cell density; the cells appear to be in logarithmic phase, although their doubling time is prolonged (over one hour). The glucose level is reduced to 0.2%.

Why does culture #2 have so little glucose remaining relative to culture #1, even though culture #2 displayed slower growth and has less biomass?

6. Although rhodopsin has been found among many marine proteobacteria, it is not known if all these microbes actually use their proteorhodopsin to generate a proton motive force. Assuming you could grow these organisms in pure culture, how would you go about determining whether or not proteorhodopsin-based phototrophy is occurring in these otherwise chemoorganoheterotrophic bacteria?

Learn More

Learn more by visiting the Prescott website at www.mhhe.com/prescottprinciples, where you will find a complete list of references.

Anabolism: The Use of Energy in Biosynthesis

11

Root nodules formed by a nitrogen-fixing bacterium. This bacterium can use N_2 gas as a source of nitrogen for amino acid and nucleotide synthesis.

anaplerotic reactions Reactions that replenish depleted tricarboxylic acid cycle intermediates.

assimilatory nitrate reduction The reduction of nitrate and its incorporation into organic material.

assimilatory sulfate reduction The reduction of sulfate and its incorporation into organic material.

Calvin cycle The main pathway for the fixation (i.e., reduction and incorporation) of CO_2 into organic material by photoautotrophs and chemolithoautotrophs.

gluconeogenesis The synthesis of glucose from noncarbohydrate precursors such as lactate and amino acids.

Chapter Glossary

glyoxylate cycle A modified tricarboxylic acid cycle used to replenish precursor metabolites normally provided by the TCA cycle.

macromolecule A large molecule that is a polymer of smaller units joined together.

nitrogen fixation The metabolic process in which atmospheric molecular nitrogen (N_2) is reduced to ammonia.

nucleoside A combination of ribose or deoxyribose with a purine or pyrimidine nitrogenous base.

nucleotide A nucleoside plus one or more phosphates.

purine A basic, heterocyclic (two joined rings), nitrogen-containing molecule found in nucleic acids and other cell constituents; includes adenine and guanine.

pyrimidine A basic, heterocyclic (one ring), nitrogen-containing molecule found in nucleic acids and other cell constituents; includes cytosine, thymine, and uracil.

ribulose-1,5-bisphosphate carboxylase The enzyme that catalyzes the incorporation of CO_2 in the Calvin cycle.

transaminases Enzymes that catalyze the transfer of an amino group from an amino acid to an α-keto acid.

transpeptidation The reaction that forms the peptide cross-links during peptidoglycan synthesis.

Biological structures are almost always constructed in a hierarchical manner, with subassemblies acting as important intermediates en route from simple starting molecules to the end products of organelles, cells, and organisms.

—W. M. Becker and D. W. Deamer

As chapter 10 makes clear, microorganisms can obtain energy in many ways. Much of this energy is used in anabolism. During anabolism, an organism begins with simple inorganic molecules and a carbon source and constructs ever more complex molecules until new organelles and cells arise (**figure 11.1**). Although there is considerably less diversity in anabolic processes as compared to catabolic processes, anabolism is amazing in its own right. From just 12 precursor metabolites, the cell is able to manufacture the myriad of molecules from which it is constructed. Furthermore, numerous antibiotics exert their control over microbial growth by inhibiting anabolic pathways.

In this chapter, we discuss the synthesis of some of the most important types of cell constituents. We begin with a general introduction to anabolism and the role played by the precursor metabolites in biosynthetic pathways. We then focus on CO_2 fixation

and the synthesis of carbohydrates, amino acids, purines and pyrimidines, and lipids. Because protein and nucleic acid synthesis are so significant and complex, the polymerization reactions that yield these macromolecules are described separately in chapter 12.

Anabolism is the creation of order. Because a cell is highly ordered and immensely complex, much energy is required for biosynthesis. This is readily apparent from estimates of the biosynthetic capacity of rapidly growing *Escherichia coli* (**table 11.1**). Although most ATP dedicated to biosynthesis is employed in protein synthesis, ATP is also used to make other cell material.

It is intuitively obvious why rapidly growing cells need a large supply of ATP. But even nongrowing cells need energy for the biosynthetic processes they carry out. This is because nongrowing cells continuously degrade and resynthesize cellular molecules during a process known as turnover. Thus cells are never the same from one instant to the next. Clearly metabolism must be carefully regulated if the rate of turnover is to be balanced by the rate of biosynthesis. It must also be regulated in response to a microbe's environment. Some of the mechanisms of metabolic regulation have already been introduced in chapter 9; others are discussed in chapter 13.

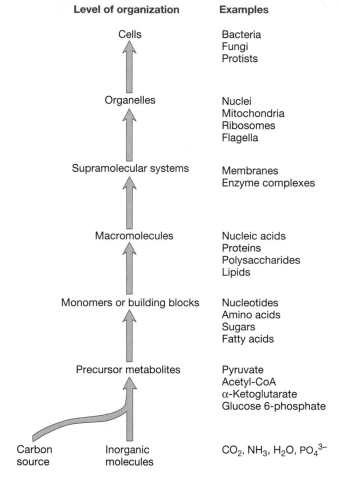

Figure 11.1 The Construction of Cells. The biosynthesis of cells and their constituents is organized in levels of ever greater complexity.

11.1 PRINCIPLES GOVERNING BIOSYNTHESIS

The problem faced by all cells is how to make the many molecules they need as efficiently as possible. Cells have solved this problem by carrying out biosynthesis using a few basic principles. Six are now briefly discussed.

1. **Large molecules are made from small molecules.** The construction of large **macromolecules** (complex molecules) from a few simple structural units (monomers) saves much genetic storage capacity, biosynthetic raw material, and energy. A consideration of protein synthesis clarifies this. Proteins—whatever size, shape, or function—are made of only 20 common amino acids joined by peptide bonds. Different proteins simply have different amino acid sequences but not new and dissimilar amino acids. Suppose that proteins were composed of 40 different amino acids instead of 20. The cell would then need the enzymes to manufacture twice as many amino acids (or would have to obtain the extra amino acids in its diet). Genes would be required for the extra enzymes, and the cell would have to invest raw materials and energy in the synthesis of these additional genes, enzymes, and amino acids. Clearly the use of a few monomers linked together by a single type of covalent bond makes the synthesis of macromolecules a highly efficient process. >> *Proteins and amino acids (appendix I)*

2. **Many enzymes do double duty.** Many enzymes are used for both catabolic and anabolic processes, saving additional materials and energy. For example, most glycolytic enzymes are involved in both the synthesis and the degradation of glucose.

3. **Some enzymes function in one direction only.** Although many steps of amphibolic pathways are catalyzed by enzymes that act reversibly, some are not. These steps require the use of separate enzymes: one to catalyze the catabolic reaction, the other to catalyze the anabolic reaction. The use of two enzymes allows independent

Table 11.1	Biosynthesis in *Escherichia coli*		
Cell Constituent	Number of Molecules per Cell[a]	Molecules Synthesized per Second	Molecules of ATP Required per Second for Synthesis
DNA	1[b]	0.00083	60,000
RNA	15,000	12.5	75,000
Polysaccharides	39,000	32.5	65,000
Lipids	15,000,000	12,500.0	87,000
Proteins	1,700,000	1,400.0	2,120,000

From *Bioenergetics* by Albert Lehninger. Copyright © 1971 by the Benjamin/Cummings Publishing Company. Reprinted by permission.
[a]Estimates for a cell with a volume of 2.25 μm^3, a total weight of 1×10^{-12} g, a dry weight of 2.5×10^{-13} g, and a 20 minute cell division cycle.
[b]It should be noted that bacteria can contain multiple copies of their genomic DNA.

regulation of catabolism and anabolism (**figure 11.2**). Thus catabolic and anabolic pathways are never identical, although many enzymes are shared. Although both types of pathways can be regulated by their end products as well as by the concentrations of ATP, ADP, AMP, and NAD^+, end-product regulation generally assumes more importance in anabolic pathways.

4. **Anabolic pathways are irreversible.** To synthesize molecules efficiently, anabolic pathways must operate irreversibly in the direction of biosynthesis. Cells achieve this by connecting some biosynthetic reactions to the breakdown of ATP and other nucleoside triphosphates. When these two processes are coupled, the free energy made available during nucleoside triphosphate breakdown drives the biosynthetic reaction to completion. ◄◄ *ATP (section 9.4)*

5. **Catabolism and anabolism are physically separated.** In eucaryotic cells, catabolic and anabolic pathways can be localized into distinct cellular compartments—a process called compartmentation. Compartmentation makes it easier for catabolic and anabolic pathways to operate simultaneously yet independently.

6. **Catabolism and anabolism use different cofactors.** Usually catabolic oxidations produce NADH, a substrate for electron transport. In contrast, when an electron donor is needed during biosynthesis, NADPH serves as the donor.

After macromolecules have been constructed from simpler precursors, they are assembled into larger, more complex structures such as supramolecular systems and organelles (figure 11.1). Macromolecules normally contain the necessary information to form supramolecular systems spontaneously in a process known as **self-assembly.** For example, ribosomes are large assemblages of many proteins and ribonucleic acid molecules, yet they arise by the self-assembly of their components without the involvement of extra factors.

1. Define anabolism, turnover, and self-assembly.
2. Summarize the six principles by which biosynthetic pathways are organized.

11.2 PRECURSOR METABOLITES

The generation of the **precursor metabolites** is a critical step in anabolism, because they give rise to all other molecules made by the cell. Precursor metabolites are carbon skeletons (i.e., carbon chains) used as the starting substrates for the synthesis of monomers and other building blocks needed for the synthesis of macromolecules. Precursor metabolites are referred to as carbon skeletons because they are molecules that lack functional moieties such as amino and sulfhydryl groups; these are added during the biosynthetic process. The precursor metabolites and their use in biosynthesis are shown in **figures 11.3** and **11.4.** Several things should be noted in figure 11.3. First, all the precursor metabolites are intermediates of the glycolytic pathways (Embden-Meyerhof, Entner-Doudoroff, and the pentose phosphate pathways) and the tricarboxylic acid (TCA) cycle. Therefore these pathways play a central role in metabolism and are often referred to as the **central metabolic pathways.** Note, too, that most of the precursor metabolites are used for synthesis of amino acids and nucleotides. ◄◄ *Breakdown of glucose to pyruvate (section 10.3); Tricarboxylic acid cycle (section 10.4)* ►► *Common metabolic pathways (appendix II)*

If an organism is a chemoorganotroph using glucose as its energy, electron, and carbon source (either aerobically or anaerobically), it generates the precursor metabolites as it generates ATP and reducing power. But what if the chemoorganotroph is using an amino acid as its sole source of carbon, electrons, and energy? And what about autotrophs? How do they generate precursor metabolites from CO_2, their carbon source? Heterotrophs growing on something other than glucose degrade that carbon and energy source into one or more intermediates of the central metabolic pathways. From there, they can generate the remaining precursor metabolites. Autotrophs must first convert CO_2 into organic carbon from which they can generate the precursor metabolites. Many of the reactions that autotrophs use to generate the precursor metabolites are reactions of the central metabolic

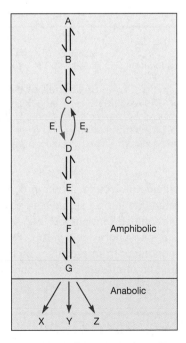

Figure 11.2 A Hypothetical Biosynthetic Pathway. The routes connecting G with X, Y, and Z are purely anabolic because they are used only for synthesis of the end products. The pathway from A to G is amphibolic—that is, it has both catabolic and anabolic functions. Most reactions are used in both roles; however, the interconversion of C and D is catalyzed by two separate enzymes, E_1 and E_2.

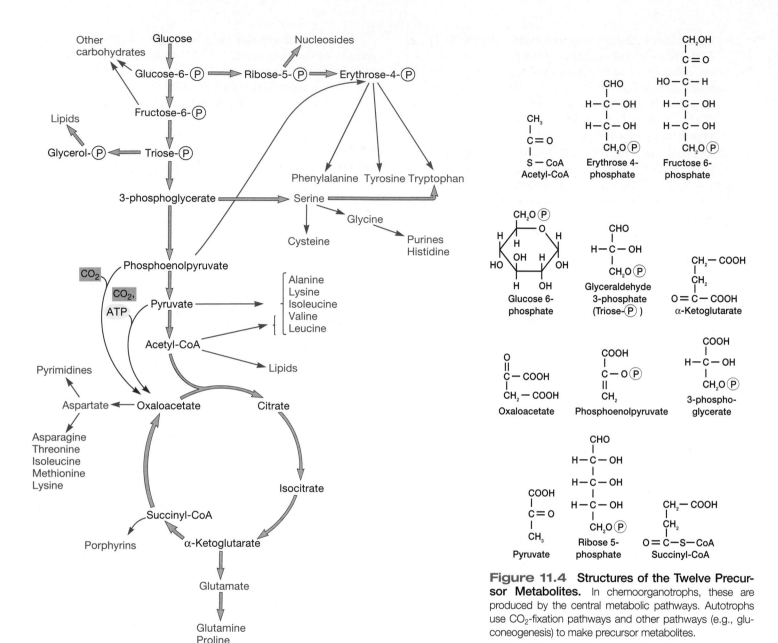

Figure 11.3 The Organization of Anabolism. Biosynthetic products (in blue) are derived from precursor metabolites, which are intermediates of amphibolic pathways. Two major anaplerotic reactions are shown in red. These reactions ensure an adequate supply of TCA cycle–derived precursor metabolites and are especially important to fermentative organisms, in which only certain TCA cycle reactions operate.

Figure 11.4 Structures of the Twelve Precursor Metabolites. In chemoorganotrophs, these are produced by the central metabolic pathways. Autotrophs use CO_2-fixation pathways and other pathways (e.g., gluconeogenesis) to make precursor metabolites.

pathways, operating either in the catabolic direction or in the anabolic direction. Thus the central metabolic pathways are important to the anabolism of both heterotrophs and autotrophs.

We begin our discussion of anabolism by first considering CO_2 fixation by autotrophs. Once CO_2 is converted to organic carbon, the synthesis of other precursor metabolites, amino acids, nucleotides, and additional building blocks is essentially the same in both autotrophs and heterotrophs. Recall that the precursor metabolites provide the carbon skeletons for the synthesis of

other important organic molecules. In the process of transforming a precursor metabolite into an amino acid or a nucleotide, the carbon skeleton is modified in a number of ways, including the addition of nitrogen, phosphorus, and sulfur. Thus as we discuss the synthesis of monomers from precursor metabolites, we also address the assimilation of nitrogen, sulfur, and phosphorus.

11.3 CO_2 FIXATION

Autotrophs use CO_2 as their sole or principal carbon source, and the reduction and incorporation of CO_2 requires much energy. Many autotrophs obtain energy by trapping light during the light reactions of photosynthesis, but some derive energy from the oxidation of inorganic electron donors. Autotrophic CO_2 fixation is crucial to life on Earth because it provides the organic matter on which heterotrophs depend. << *Chemolithotrophy (section 10.11); Phototrophy (section 10.12);* >> *Biogeochemical cycling: Carbon cycle (section 25.1)*

Four different CO_2-fixation pathways have been identified in microorganisms. Most autotrophs use the **Calvin cycle,** which is also called the Calvin-Benson cycle or the reductive pentose phosphate cycle. The Calvin cycle is found in photosynthetic eucaryotes and most photosynthetic bacteria. It is absent in some obligatory anaerobic and microaerophilic bacteria. Autotrophic archaea also use an alternative pathway for CO_2 fixation. We consider the Calvin cycle first and then briefly introduce the three other CO_2-fixation pathways.

Calvin Cycle

The Calvin cycle is also called the reductive pentose phosphate cycle because it is essentially the reverse of the pentose phosphate pathway. Thus many of the reactions are similar, in particular the sugar transformations. The reactions of the Calvin cycle occur in the chloroplast stroma of eucaryotic autotrophs. In cyanobacteria, some nitrifying bacteria, and thiobacilli (sulfur-oxidizing chemolithotrophs), the Calvin cycle is associated with inclusion bodies called **carboxysomes.** These polyhedral structures contain the enzyme critical to the Calvin cycle and may be the site of CO_2 fixation. << *Breakdown of glucose to pyruvate: Pentose phosphate pathway (section 10.3)*

The Calvin cycle is divided into three phases: carboxylation phase, reduction phase, and regeneration phase (**figure 11.5** and *appendix II*). During the carboxylation phase, the enzyme **ribulose 1,5-bisphosphate carboxylase,** also called ribulose bisphosphate carboxylase/oxygenase (Rubisco), catalyzes the addition of CO_2 to the five-carbon molecule ribulose 1,5-bisphosphate (RuBP), forming a six-carbon intermediate that rapidly and spontaneously splits into two molecules of 3-phosphoglycerate (PGA). Note that PGA is an intermediate of the Embden-Meyerhof pathway (EMP), and in the reduction phase, PGA is reduced to glyceraldehyde 3-phosphate by two reactions that are essentially the reverse of two EMP reactions. The difference is that the Calvin cycle enzyme glyceraldehyde 3-phosphate dehydrogenase uses $NADP^+$ rather than NAD^+ (compare figures 11.5 and 10.5). Finally, in the regeneration phase, RuBP is regenerated, so that the cycle can repeat. In addition, this phase produces carbohydrates such as glyceraldehyde 3-phosphate, fructose 6-phosphate, and glucose 6-phosphate, all of which are precursor metabolites (figures 11.3 and 11.4). This portion of the cycle is similar to the pentose phosphate pathway and involves the transketolase and transaldolase reactions.

To synthesize fructose 6-phosphate or glucose 6-phosphate from CO_2, the cycle must operate six times to yield the desired hexose and reform the six RuBP molecules.

$$6RuBP + 6CO_2 \rightarrow 12PGA \rightarrow 6RuBP + \text{fructose-6-P}$$

The incorporation of one CO_2 into organic material requires three ATPs and two NADPHs. The formation of glucose from CO_2 may be summarized by the following equation.

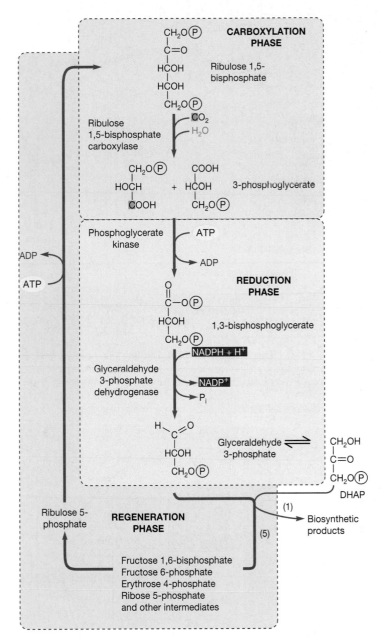

Figure 11.5 The Calvin Cycle. This overview of the cycle shows only the carboxylation and reduction phases in detail. Three ribulose 1,5-bisphosphates are carboxylated to give six 3-phosphoglycerates in the carboxylation phase. These are converted to six glyceraldehyde 3-phosphates, which can be converted to dihydroxyacetone phosphate (DHAP). Five of the six trioses (glyceraldehyde phosphate and dihydroxyacetone phosphate) are used to reform three ribulose 1,5-bisphosphates in the regeneration phase. The remaining triose is used in biosynthesis. The numbers in parentheses at the lower right indicate this carbon flow.

$$6CO_2 + 18ATP + 12NADPH + 12H^+ + 12H_2O \rightarrow$$
$$\text{glucose} + 18ADP + 18P_i + 12NADP^+$$

Other CO₂-Fixation Pathways

Certain bacteria and archaea fix CO_2 using the reductive TCA cycle, the 3-hydroxypropionate cycle, or the acetyl-CoA pathway. The **reductive TCA cycle (figure 11.6)** is used by some

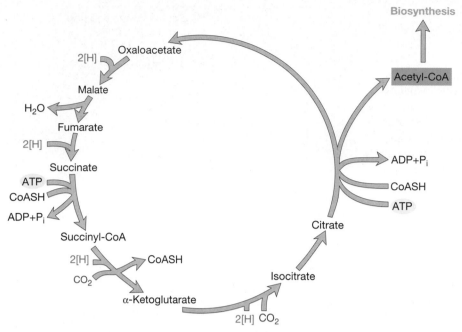

Figure 11.6   The Reductive TCA Cycle.   This cycle is used by green sulfur bacteria and some chemolithotrophic archaea to fix CO_2. The cycle runs in the opposite direction as the TCA cycle. ATP and reducing equivalents [H] power the reversal. In green sulfur bacteria, the reducing equivalents are provided by reduced ferredoxin. The product of this process is acetyl-CoA, which can be used to synthesize other organic molecules and precursor metabolites.

bacteria (another group of anoxygenic phototrophs) use the **3-hydroxypropionate cycle** to fix CO_2. **Figure 11.7** shows the cycle as it is thought to function in the green nonsulfur bacterium *Chloroflexus aurantiacus*. How its product, glyoxylate, is assimilated is unclear. Methanogens use portions of the **acetyl-CoA pathway** for carbon fixation; the pathway as it is used by *Methanobacterium thermoautotrophicum* is illustrated in **figure 11.8**. Acetogens use the pathway in its entirety. Both the acetyl-CoA pathway and methanogenesis involve the activity of a number of unusual enzymes and coenzymes. These are described in more detail in chapter 18. >> *Phylum* Crenarchaeota *(section 18.2);* Aquificae *and* Thermotogae *(section 19.1); Photosynthetic bacteria (section 19.3)*

1. Briefly describe the three phases of the Calvin cycle. What other pathways are used to fix CO_2?

2. Which two enzymes are specific to the Calvin cycle?

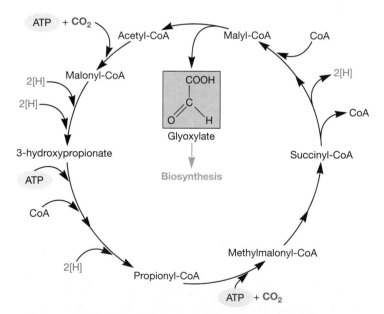

Figure 11.7   The 3-Hydroxypropionate Pathway.   This pathway functions in green nonsulfur bacteria, a group of anoxygenic phototrophs. The product of the cycle is glyoxylate, which is used in biosynthesis by mechanisms that have not been definitively elucidated.

chemolithoautotrophs (e.g., *Thermoproteus* and *Sulfolobus,* two archaeal genera, and the bacterial genus *Aquifex*) and anoxygenic phototrophs such as *Chlorobium,* a green sulfur bacterium. The reductive TCA cycle is so named because it runs in the reverse direction of the normal, oxidative TCA cycle (compare figures 11.6 and 10.8). A few archaeal genera and the green nonsulfur

11.4   SYNTHESIS OF SUGARS AND POLYSACCHARIDES

Autotrophs using CO_2-fixation processes other than the Calvin cycle and heterotrophs growing on carbon sources other than sugars must be able to synthesize glucose. The synthesis of glucose from noncarbohydrate precursors is called **gluconeogenesis.** The gluconeogenic pathway shares seven enzymes with the Embden-Meyerhof pathway. However, the two pathways are not identical (**figure 11.9**). Three glycolytic steps are irreversible: (1) the conversion of phosphoenolpyruvate to pyruvate, (2) the formation of fructose 1,6-bisphosphate from fructose 6-phosphate, and (3) the phosphorylation of glucose. These must be bypassed when the pathway is operating biosynthetically. For example, the formation of fructose 1,6-bisphosphate by phosphofructokinase is reversed by the enzyme fructose bisphosphatase, which hydrolytically removes a phosphate from fructose bisphosphate. Usually at least two enzymes are involved in the conversion of pyruvate to phosphoenolpyruvate (the reversal of the pyruvate kinase step).

Synthesis of Monosaccharides

As can be seen in figure 11.9, gluconeogenesis synthesizes fructose 6-phosphate and glucose 6-phosphate. Once these two precursor metabolites have been formed, other common sugars can be manufactured. For example, mannose comes directly from

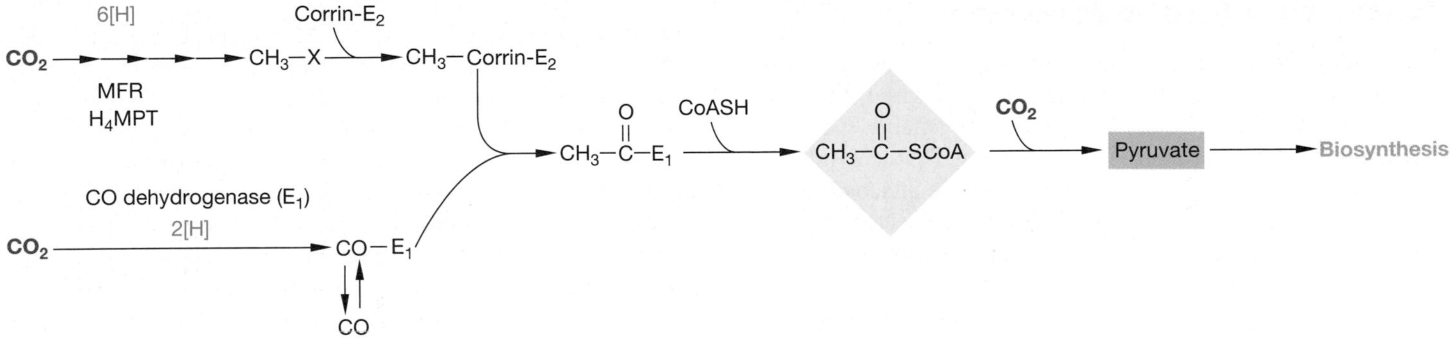

Figure 11.8 The Acetyl-CoA Pathway. Methanogens reduce two molecules of CO_2, each by a different mechanism, and combine them to form acetyl and then acetyl-CoA. Acetogenic bacteria use a slightly different version of the pathway.

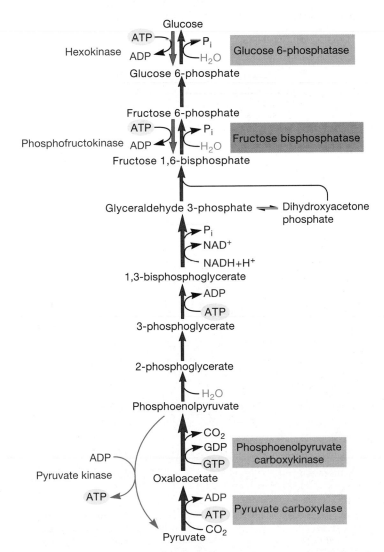

Figure 11.9 Gluconeogenesis. The gluconeogenic pathway used in many microorganisms. The names of the four enzymes catalyzing irreversible reactions different from those found in the Embden-Meyerhof pathway (EMP) are in shaded boxes. EMP steps are shown in blue for comparison.

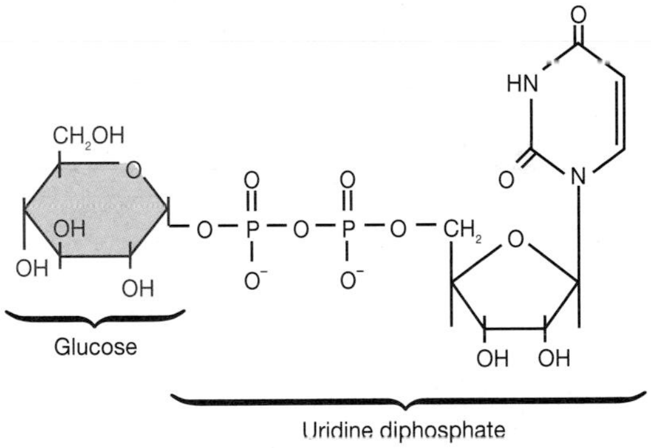

Figure 11.10 Uridine Diphosphate Glucose.

fructose 6-phosphate by a simple rearrangement of a hydroxyl group *(see figure AI.9)*. Several sugars are synthesized while attached to a nucleoside diphosphate. The most important nucleoside diphosphate sugar is **uridine diphosphate glucose (UDPG),** which is formed when glucose reacts with uridine triphosphate (**figure 11.10**). UDP carries glucose around the cell for participation in enzyme reactions much like ADP bears phosphate in the form of ATP. Other important uridine diphosphate sugars are UDP-galactose and UDP-glucuronic acid.

Synthesis of Polysaccharides

Nucleoside diphosphate sugars also play a central role in the synthesis of polysaccharides such as starch and glycogen, both of which are long chains of glucose. Again, biosynthesis is not simply a direct reversal of catabolism. For instance, during the synthesis of glycogen and starch in bacteria and protists, adenosine diphosphate glucose (ADP-glucose) is formed from glucose 1-phosphate and ATP. It then donates glucose to the end of growing glycogen and starch chains.

$$\text{ATP} + \text{glucose 1-phosphate} \rightarrow \text{ADP-glucose} + \text{PP}_i$$

$$(\text{Glucose})_n + \text{ADP-glucose} \rightarrow (\text{glucose})_{n+1} + \text{ADP}$$

Synthesis of Peptidoglycan

Nucleoside diphosphate sugars also participate in the synthesis of peptidoglycan. Recall that peptidoglycan is a large, complex molecule consisting of long polysaccharide chains made of alternating N-acetylmuramic acid (NAM) and N-acetylglucosamine (NAG) residues. Pentapeptide chains are attached to the NAM groups. The polysaccharide chains are connected through their pentapeptides or by interbridges (*see figures 3.20 and 3.21*). ◄◄ *Bacterial cell walls (section 3.4)*

Not surprisingly, such an intricate structure requires an equally intricate biosynthetic process, especially because some reactions occur in the cytoplasm, others in the membrane, and others in the periplasmic space. Peptidoglycan synthesis involves two carriers (**figure 11.11**). The first, uridine diphosphate (UDP), functions in the cytoplasmic reactions. In the first step of peptidoglycan synthesis, UDP derivatives of NAM and NAG are formed. Amino acids are then added sequentially to UDP-NAM

to form the pentapeptide chain. NAM-pentapeptide is then transferred to the second carrier, bactoprenol phosphate, which is located at the cytoplasmic side of the plasma membrane. The resulting intermediate is often called Lipid I. Bactoprenol is a 55-carbon alcohol and is linked to NAM by a pyrophosphate group (**figure 11.12**). Next, UDP transfers NAG to the bactoprenol-NAM-pentapeptide complex (Lipid I) to generate Lipid II. This creates the peptidoglycan repeat unit. The repeat unit is transferred across the membrane by bactoprenol. If the peptidoglycan unit requires an interbridge, it is added while the repeat unit is within the membrane. Bactoprenol stays within the membrane and does not enter the periplasmic space. After releasing the peptidoglycan repeat unit into the periplasmic space, bactoprenol-pyrophosphate is dephosphorylated and returns to the cytoplasmic side of the plasma membrane, where it can function in the next round of synthesis. Meanwhile, the peptidoglycan repeat unit is added to the growing end of a peptidoglycan chain.

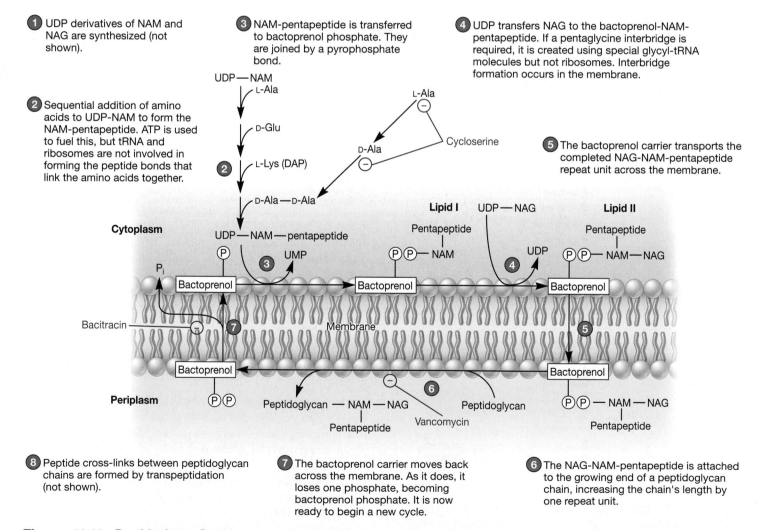

Figure 11.11 Peptidoglycan Synthesis. NAM is *N*-acetylmuramic acid and NAG is *N*-acetylglucosamine. The pentapeptide contains L-lysine in *Staphylococcus aureus* peptidoglycan, and diaminopimelic acid (DAP) in *E. coli*. Inhibition by bacitracin, cycloserine, and vancomycin also is shown. Stage eight is depicted in figure 11.13.

The final step in peptidoglycan synthesis is **transpeptidation** (**figure 11.13**), which creates the peptide cross-links between the peptidoglycan chains. The enzyme that catalyzes the reaction removes the terminal D-alanine as the cross-link is formed.

To grow and divide efficiently, a bacterial cell must add new peptidoglycan to its cell wall in a precise and well-regulated way while maintaining wall shape and integrity in the presence of high osmotic pressure. Because the cell wall peptidoglycan is essentially a single, enormous network, the growing bacterium must be able to degrade it just enough to provide acceptor ends for the incorporation of new peptidoglycan units. It must also reorganize peptidoglycan structure when necessary. This limited peptidoglycan digestion is accomplished by enzymes known as **autolysins,** some of which attack the polysaccharide chains, while others hydrolyze the peptide cross-links. Autolysin inhibitors are produced to keep the activity of these enzymes under tight control.

Because of the importance of peptidoglycan to bacterial cell wall structure and function, its synthesis is a particularly effective target for antimicrobial agents. Inhibition of any stage of synthesis weakens the cell wall and can lead to lysis. Many commonly used antibiotics interfere with peptidoglycan synthesis. For example, penicillin inhibits the transpeptidation reaction (figure 11.13), and bacitracin blocks the dephosphorylation of bactoprenol pyrophosphate (figure 11.12). **>>** *Antibacterial drugs: Inhibitors of cell wall synthesis (section 31.4)*

1. What is gluconeogenesis? Why is it important?
2. Describe the formation of mannose, galactose, starch, and glycogen. What are nucleoside diphosphate sugars? How do microorganisms use them?
3. Suppose that a microorganism is growing on a medium that contains amino acids but no sugars. In general terms, how would it synthesize the pentoses and hexoses it needs?
4. Diagram the steps involved in the synthesis of peptidoglycan and show where they occur in the cell. What are the roles of bactoprenol and UDP? What is unusual about the synthesis of the pentapeptide chain?

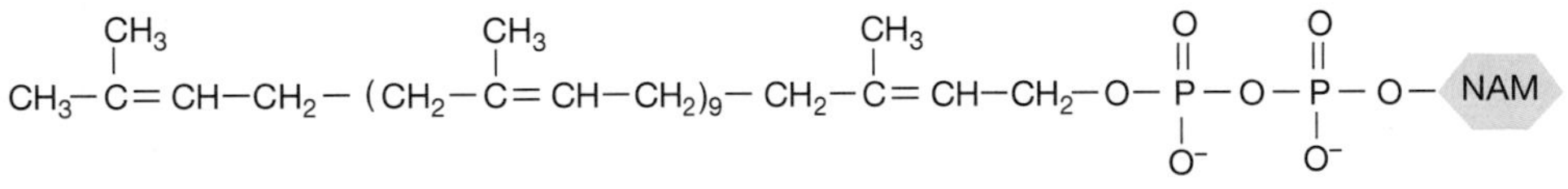

$$CH_3-\underset{\underset{CH_3}{|}}{C}=CH-CH_2-(CH_2-\underset{\underset{CH_3}{|}}{C}=CH-CH_2)_9-CH_2-\underset{\underset{CH_3}{|}}{C}=CH-CH_2-O-\underset{\underset{O^-}{|}}{\overset{\overset{O}{\|}}{P}}-O-\underset{\underset{O^-}{|}}{\overset{\overset{O}{\|}}{P}}-O-\text{NAM}$$

Figure 11.12   Bactoprenol NAM.   Bactoprenol is connected to *N*-acetylmuramic acid (NAM) by pyrophosphate.

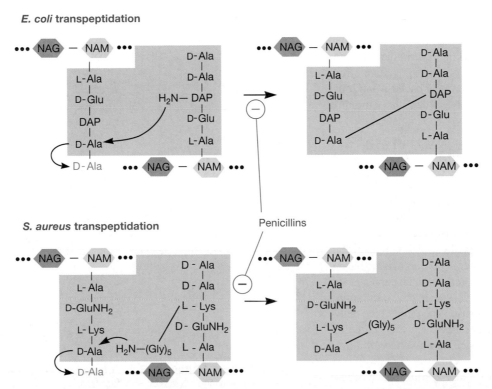

Figure 11.13   Transpeptidation.   The transpeptidation reactions in the formation of the peptidoglycans of *Escherichia coli* and *Staphylococcus aureus.*

11.5 SYNTHESIS OF AMINO ACIDS

Many of the precursor metabolites serve as starting substrates for the synthesis of amino acids (figure 11.3). In the amino acid biosynthetic pathways, the carbon skeleton is remodeled and an amino group, and sometimes sulfur, is added. In this section, we first examine the mechanisms by which nitrogen and sulfur are assimilated and incorporated into amino acids. This is followed by a brief consideration of the organization of amino acid biosynthetic pathways.

Nitrogen Assimilation

Nitrogen is a major component not only of proteins but also of nucleic acids, coenzymes, and many other cell constituents. Thus the cell's ability to assimilate inorganic nitrogen is exceptionally important. Although nitrogen gas is abundant in the atmosphere, only a few procaryotes can reduce the gas and use it as a nitrogen source. Most must incorporate either ammonia or nitrate. We examine ammonia and nitrate assimilation first and then briefly discuss nitrogen assimilation in microbes that fix N_2.

Ammonia Incorporation

Ammonia can be incorporated into organic material relatively easily and directly because it is more reduced than other forms of inorganic nitrogen. Ammonia is initially incorporated into carbon skeletons by one of two mechanisms: reductive amination or the glutamine synthetase–glutamate synthase system. Once incorporated, the nitrogen can be transferred to other carbon skeletons by enzymes called transaminases. The major reductive amination pathway involves the formation of glutamate from α-ketoglutarate, catalyzed in many bacteria and fungi by glutamate dehydrogenase when the ammonia concentration is high.

$$\alpha\text{-ketoglutarate} + NH_4^+ + NAD(P)H + H^+ \rightleftharpoons$$
$$\text{glutamate} + NAD(P)^+ + H_2O$$

Once glutamate has been synthesized, the newly formed α-amino group can be transferred to other carbon skeletons by transamination reactions to form different amino acids. **Transaminases** possess the coenzyme pyridoxal phosphate, which is responsible for the amino group transfer. Microorganisms have a number of transaminases, each of which catalyzes the formation of several amino acids using the same amino acid as an amino group donor. When glutamate dehydrogenase works in cooperation with transaminases, ammonia can be incorporated into a variety of amino acids (**figure 11.14**).

The **glutamine synthetase–glutamate synthase (GS-GOGAT)** system is observed in *E. coli, Bacillus megaterium,* and other bacteria (**figure 11.15**). It

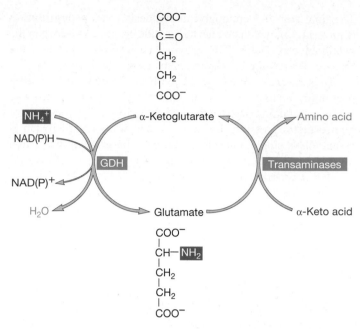

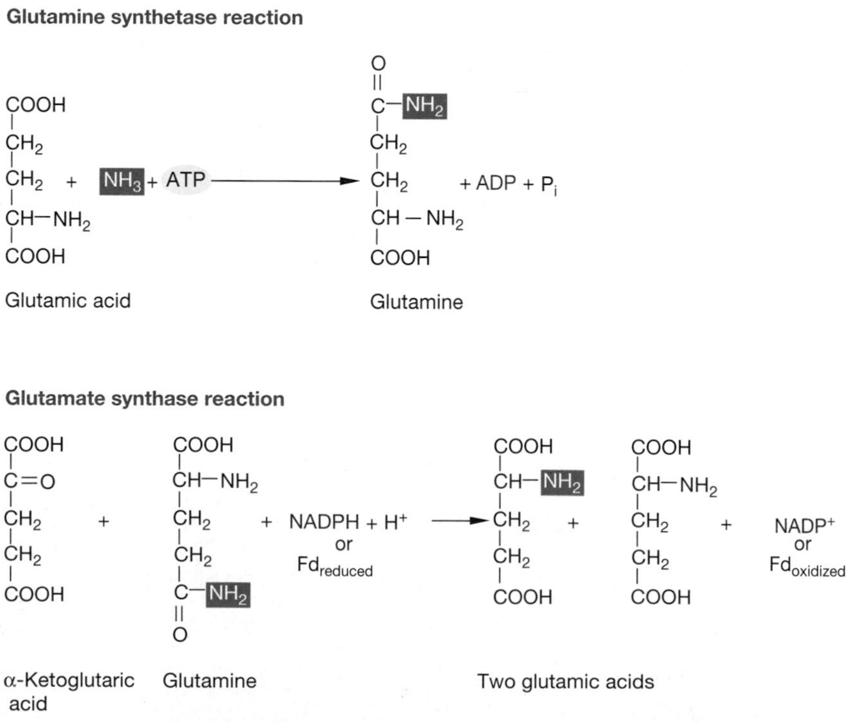

Figure 11.14   The Ammonia Assimilation Pathway. Ammonia assimilation by use of glutamate dehydrogenase (GDH) and transaminases. Either NADPH- or NADH-dependent glutamate dehydrogenases may be involved. This route is most active at high ammonia concentrations.

Glutamine synthetase reaction

Glutamic acid + NH_3 + ATP → Glutamine + ADP + P_i

Glutamic acid   Glutamine

Glutamate synthase reaction

α-Ketoglutaric acid + Glutamine + NADPH + H$^+$ or Fd$_{reduced}$ → Two glutamic acids + NADP$^+$ or Fd$_{oxidized}$

α-Ketoglutaric   Glutamine   Two glutamic acids
acid

Figure 11.15   Glutamine Synthetase and Glutamate Synthase. The glutamine synthetase and glutamate synthase reactions involved in ammonia assimilation. Some glutamine synthases use NADPH as an electron source; others use reduced ferredoxin (Fd).

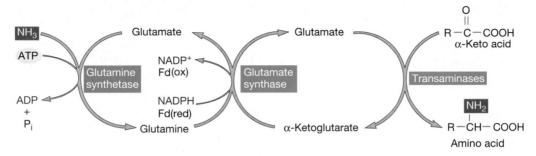

Figure 11.16   Ammonia Incorporation Using Glutamine Synthetase and Glutamate Synthase.   This route is effective at low ammonia concentrations.

functions when ammonia levels are low. Incorporation of ammonia by this system begins when ammonia is used to synthesize glutamine from glutamate in a reaction catalyzed by **glutamine synthetase** (**figure 11.16**). Then the amide nitrogen of glutamine is transferred to α-ketoglutarate to generate a new glutamate molecule. This reaction is catalyzed by **glutamate synthase.** Because glutamate acts as an amino donor in transaminase reactions, ammonia may be used to synthesize all common amino acids when suitable transaminases are present.

Assimilatory Nitrate Reduction

The nitrogen in nitrate (NO_3^-) is much more oxidized than that in ammonia. Therefore nitrate must first be reduced to ammonia before the nitrogen can be converted to an organic form. This reduction of nitrate is called **assimilatory nitrate reduction,** which is not the same as that which occurs during anaerobic respiration (dissimilatory nitrate reduction). In assimilatory nitrate reduction, nitrate is incorporated into organic material and does not participate in energy conservation. The process is widespread among bacteria, fungi, and photosynthetic protists, and it is an important step in the nitrogen cycle. **<<** *Anaerobic respiration (section 10.6);* **>>** *Biogeochemical cycling: Nitrogen cycle (section 25.1)*

Assimilatory nitrate reduction takes place in the cytoplasm in bacteria. The first step in nitrate assimilation is its reduction to nitrite by **nitrate reductase,** an enzyme that contains both FAD and molybdenum (**figure 11.17**). NADPH is the electron source.

$$NO_3^- + NADPH + H^+ \rightarrow NO_2^- + NADP^+ + H_2O$$

Nitrite is next reduced to ammonia with a series of two electron additions catalyzed by nitrite reductase and possibly other enzymes. The ammonia is then incorporated into amino acids by the routes already described.

Nitrogen Fixation

The reduction of atmospheric gaseous nitrogen to ammonia is called **nitrogen fixation.** Because ammonia and nitrate levels often are low and only a few bacteria and archaea can carry

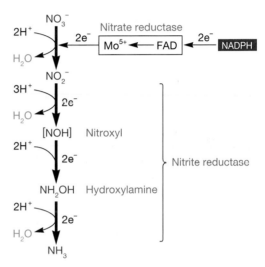

Figure 11.17   Assimilatory Nitrate Reduction.   This sequence is thought to operate in bacteria that can reduce and assimilate nitrate nitrogen.

out nitrogen fixation (eucaryotic cells completely lack this ability), the rate of this process limits plant growth in many situations. Nitrogen fixation occurs in (1) free-living chemotrophic bacteria and archaea (e.g., *Azotobacter, Klebsiella, Clostridium,* and *Methanococcus*), (2) bacteria living in symbiotic associations with plants such as legumes (e.g., *Rhizobium*), and (3) cyanobacteria (e.g., *Nostoc, Anabaena,* and *Trichodesmium*). These microbes play a critical role in the nitrogen cycle. They complete the cycle from NO_3^- (the most oxidized form of nitrogen) to ammonia (the most reduced form of nitrogen) via N_2, which has an intermediate oxidation state. The biological aspects of nitrogen fixation are discussed in chapters 25 and 26. The biochemistry of nitrogen fixation is the focus of this section.

The reduction of nitrogen to ammonia is catalyzed by the enzyme **nitrogenase.** Although the enzyme-bound intermediates in this process are still unknown, it is believed that nitrogen is reduced by two-electron additions in a way similar to that illustrated in **figure 11.18**. The reduction of molecular nitrogen to ammonia is quite exergonic, but the reaction has a high activation energy because molecular nitrogen is an unreactive

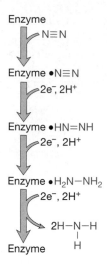

Figure 11.18 **Nitrogen Reduction.** A hypothetical sequence of nitrogen reduction by nitrogenase.

gas with a triple bond connecting the two nitrogen atoms. Therefore nitrogen reduction is expensive, requiring a large ATP expenditure—at least 8 electrons and 16 ATP molecules (4 ATPs per pair of electrons). Ferridoxin is used as the electron donor.

$$N_2 + 8H^+ + 8e^- + 16ATP \rightarrow 2NH_3 + H_2 + 16ADP + 16P_i$$

Nitrogenase is a complex enzyme consisting of two major protein components, a MoFe protein (MW 220,000) joined with one or two Fe proteins (MW 64,000). The MoFe protein contains 2 atoms of molybdenum and 28 to 32 atoms of iron; the Fe protein has 4 iron atoms. Fe protein is first reduced by ferredoxin and then binds ATP (**figure 11.19**). ATP binding changes the conformation of the Fe protein and lowers its reduction potential, enabling it to reduce the MoFe protein. ATP is hydrolyzed when this electron transfer occurs. Finally, reduced MoFe protein donates electrons to atomic nitrogen. Nitrogenase is quite sensitive to O_2 and must be protected from O_2 inactivation within the cell. Microbes use a variety of strategies to protect nitrogenase, as discussed more fully in chapters 21 and 27.

The reduction of N_2 to NH_3 occurs in three steps, each of which requires an electron pair (figures 11.18 and 11.19). Six electron transfers take place, and this uses a total of 12 ATPs per N_2 reduced. The overall process actually requires at least 8 electrons and 16 ATPs because nitrogenase also reduces protons to H_2. The H_2 reacts with diimine (HN=NH) to form N_2 and H_2. This futile cycle produces some N_2 even under favorable conditions and makes nitrogen fixation even more expensive. Symbiotic nitrogen-fixing bacteria can consume almost 20% of the ATP produced by the host plant. Once molecular nitrogen has been reduced to ammonia, the ammonia can be incorporated into organic compounds.

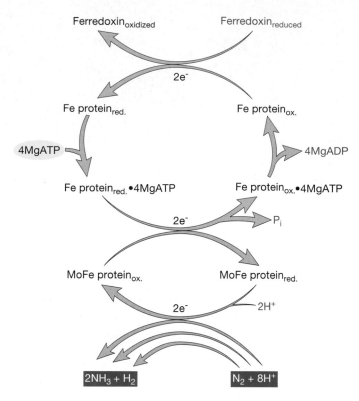

Figure 11.19 **Mechanism of Nitrogenase Action.** The flow of two electrons from ferredoxin to nitrogen is outlined. This process is repeated three times in order to reduce N_2 to two molecules of ammonia. The stoichiometry at the bottom includes proton reduction to H_2.

Sulfur Assimilation

Sulfur is needed for the synthesis of the amino acids cysteine and methionine. It is also needed for the synthesis of several coenzymes (e.g., coenzyme A and biotin). Sulfur is obtained from two sources. Many microorganisms use cysteine and methionine, obtained from either external sources or intracellular amino acid reserves. In addition, sulfate can provide sulfur for biosynthesis. The sulfur atom in sulfate is more oxidized than it is in cysteine and other organic molecules; thus sulfate must be reduced before it can be assimilated. This process is known as **assimilatory sulfate reduction** to distinguish it from the dissimilatory sulfate reduction that takes place when sulfate acts as an electron acceptor during anaerobic respiration. << *Anaerobic respiration (section 10.6);* >> *Biogeochemical cycling: Sulfur cycle (section 25.1)*

Assimilatory sulfate reduction involves sulfate activation through the formation of phosphoadenosine 5'-phosphosulfate (**figure 11.20**), followed by reduction of the sulfate. The process is complex (**figure 11.21**). Sulfate is first reduced to sulfite (SO_3^{2-}), then to hydrogen sulfide. Cysteine can be synthesized from hydrogen sulfide in two ways. Fungi appear to combine hydrogen sulfide with serine to form cysteine (process 1), whereas

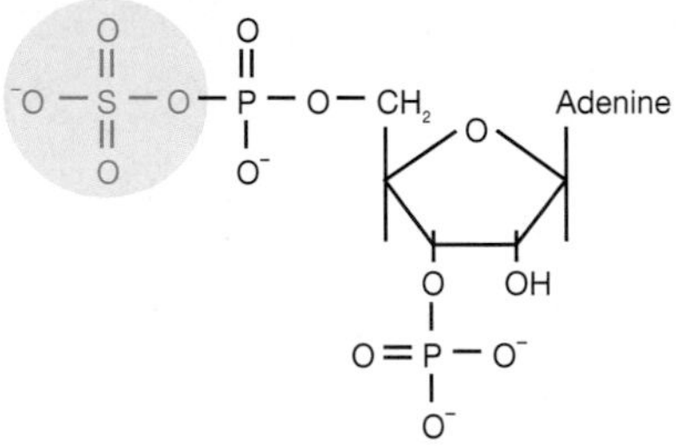

Figure 11.20 Phosphoadenosine 5′-phosphosulfate (PAPS).
The sulfate group is in color.

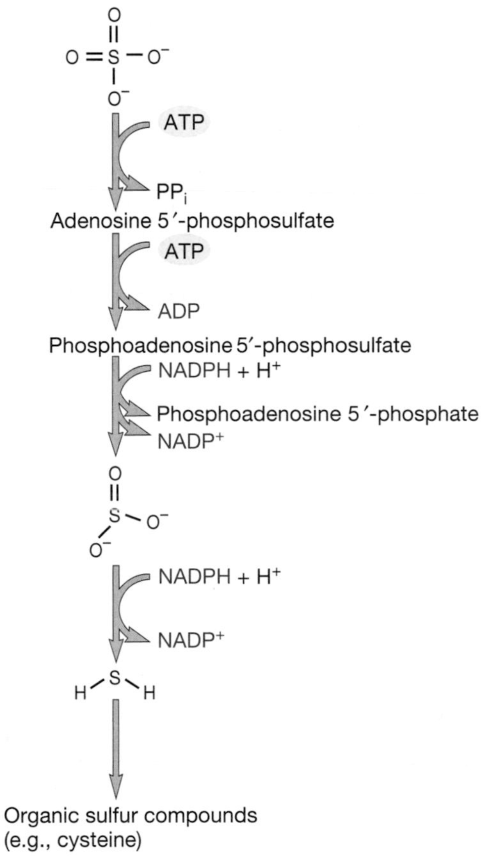

Figure 11.21 The Sulfate Reduction Pathway.

many bacteria join hydrogen sulfide with O-acetylserine instead (process 2).

(1) H_2S + serine → cysteine + H_2O

Once formed, cysteine can be used in the synthesis of other sulfur-containing organic compounds, including the amino acid methionine.

Amino Acid Biosynthetic Pathways

Some amino acids are made directly by transamination of a precursor metabolite. For example, alanine and aspartate are made directly from pyruvate and oxaloacetate, respectively, using glutamate as the amino group donor. However, most precursor metabolites must be altered by more than just the addition of an amino group. In many cases, the carbon skeleton must be reconfigured, and for cysteine and methionine, the carbon skeleton must be amended by the addition of sulfur. These biosynthetic pathways are more complex. They often involve many steps and are branched. By using branched pathways, a single precursor metabolite can be used for the synthesis of a family of related amino acids. For example, the amino acids lysine, threonine, isoleucine, and methionine are synthesized from oxaloacetate by a branching route (**figure 11.22**). The biosynthetic pathway for the aromatic amino acids phenylalanine, tyrosine, and tryptophan is also branched; it begins with two precursor metabolites, phosphoenolpyruvate and erythrose 4-phosphate (**figure 11.23**). Because of the need to conserve nitrogen, carbon, and energy, amino acid synthetic pathways are usually tightly regulated by allosteric and feedback mechanisms. << *Posttranslational regulation of enzyme activity (section 9.10)*

Anaplerotic Reactions and Amino Acid Biosynthesis

When an organism is actively synthesizing amino acids, a heavy demand for precursor metabolites is placed on the central metabolic pathways, especially the TCA cycle. Therefore it is critical that TCA cycle intermediates be readily available. This is especially true for organisms carrying out fermentation, where the TCA cycle does not function in the catabolism of glucose. To ensure an adequate supply of TCA cycle–generated precursor metabolites, microorganisms use reactions that replenish TCA cycle intermediates—these reactions are called **anaplerotic reactions** (Greek *anaplerotic,* filling up).

Most microorganisms can replace TCA cycle intermediates using two reactions that generate oxaloacetate from either phosphoenolpyruvate or pyruvate, both of which are intermediates of the Embden-Meyerhof pathway (figure 11.3). These 3-carbon molecules are converted to oxaloacetate by a carboxylation reaction (i.e., CO_2 is added to the molecule, forming a carboxyl group).

The conversion of pyruvate to oxaloacetate is catalyzed by the enzyme pyruvate carboxylase, which requires the cofactor biotin.

$$\text{Pyruvate} + CO_2 + ATP + H_2O \xrightarrow{\text{biotin}} \text{oxaloacetate} + ADP + P_i$$

Biotin is often the cofactor for enzymes catalyzing carboxylation reactions. Because of its importance, biotin is a required growth factor for many species. The pyruvate carboxylase reaction is observed in yeasts and some bacteria. Other microorganisms, such as the bacteria *E. coli* and *Salmonella* spp., have the enzyme

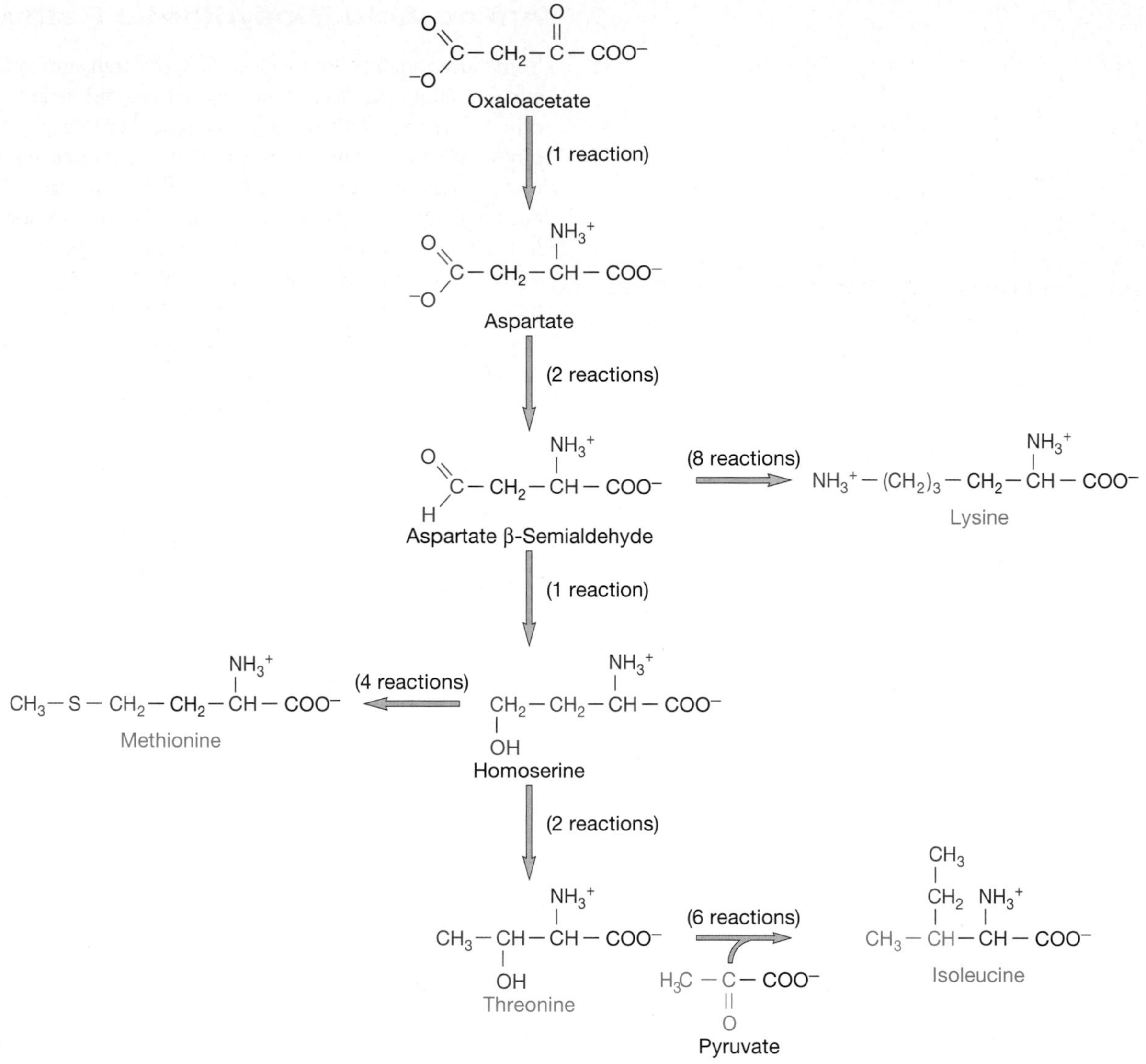

Figure 11.22 **The Branching Synthetic Pathway to Methionine, Threonine, Isoleucine, and Lysine.** Notice that most arrows represent numerous enzyme-catalyzed reactions. Also not shown is the consumption of reducing power and ATP. For instance, the synthesis of isoleucine consumes two ATPs and three NADPHs.

phosphoenolpyruvate carboxylase, which catalyzes the carboxylation of phosphoenolpyruvate.

$$\text{Phosphoenolpyruvate} + CO_2 \rightarrow \text{oxaloacetate} + P_i$$

Other anaplerotic reactions are part of the **glyoxylate cycle,** which functions in some bacteria, fungi, and protists (**figure 11.24**). This cycle is made possible by two unique enzymes, isocitrate lyase and malate synthase. The glyoxylate cycle is actually a modified TCA cycle. The two decarboxylations of the TCA cycle (the isocitrate dehydrogenase and α-ketoglutarate dehydrogenase steps) are bypassed, making possible the conversion of acetyl-CoA to form oxaloacetate without loss of acetyl-CoA carbon as CO_2. In this fashion, acetate and any molecules that give rise to it can contribute carbon to the cycle

and support microbial growth. ◄◄ *Tricarboxylic acid cycle (section 10.4)*

1. Describe the roles of glutamate dehydrogenase, glutamine synthetase, glutamate synthase, and transaminases in ammonia assimilation.

2. How is nitrate assimilated? How does assimilatory nitrate reduction differ from dissimilatory nitrate reduction? What is the fate of nitrate following assimilatory nitrate reduction versus its fate following denitrification?

3. What is nitrogen fixation? Briefly describe the structure and mechanism of action of nitrogenase.

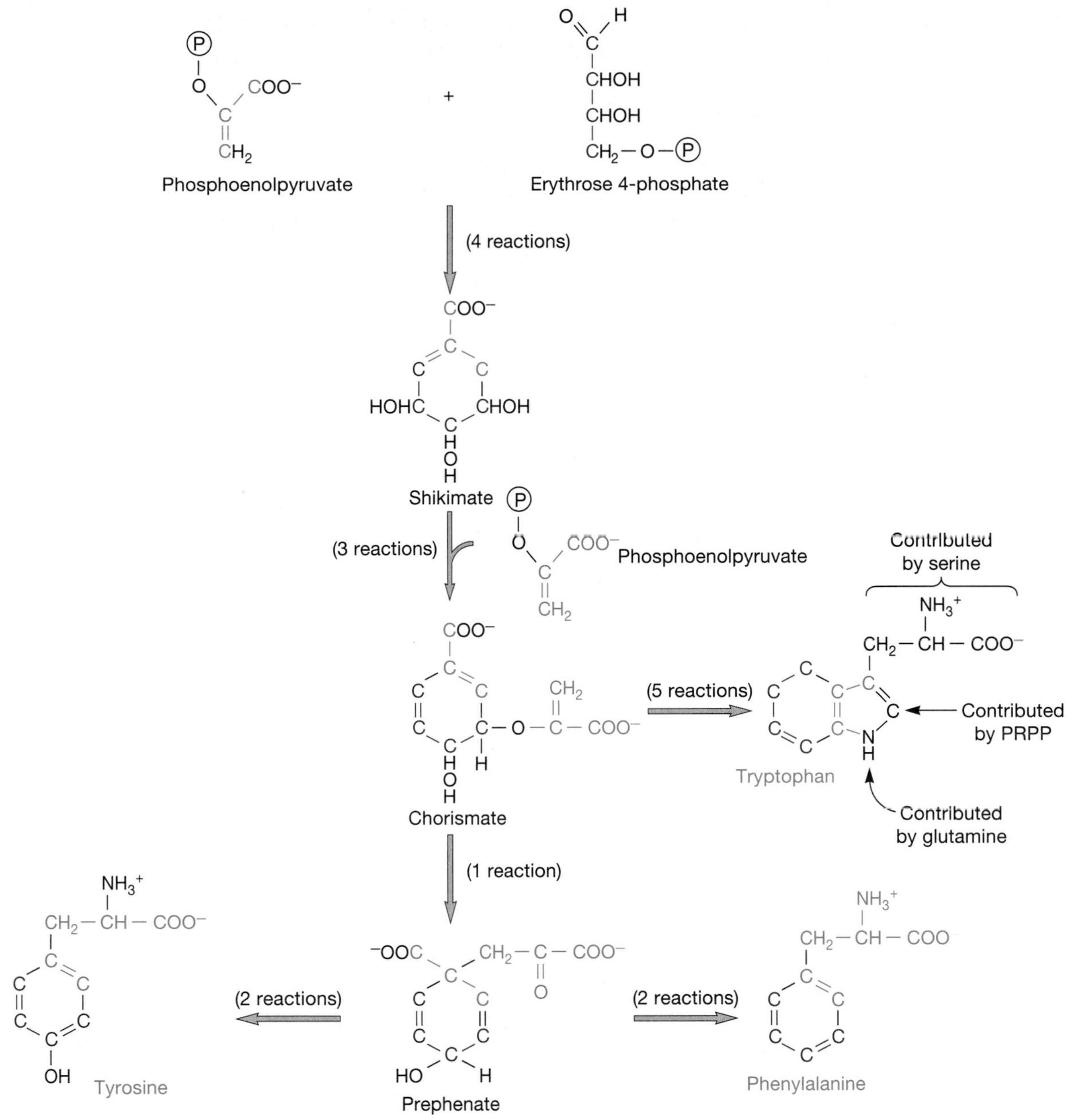

Figure 11.23 Aromatic Amino Acid Synthesis. The carbons arising from phosphoenolpyruvate (green) and erythrose 4-phosphate (red) are shown. The remaining carbons present in trytophan are provided by 5-phosphoribosyl-1-pyrophosphate (PRPP) and the amino acid serine. PRPP also is important in purine biosynthesis (*see figure AII.9*).

4. How do organisms assimilate sulfur? How does assimilatory sulfate reduction differ from dissimilatory sulfate reduction?

5. Why is using branched pathways an efficient mechanism for synthesizing amino acids?

6. Define an anaplerotic reaction. Give three examples of anaplerotic reactions.

7. Describe the glyoxylate cycle. How is it similar to the TCA cycle? How does it differ?

11.6 SYNTHESIS OF PURINES, PYRIMIDINES, AND NUCLEOTIDES

Purine and pyrimidine biosynthesis is critical for all cells because these molecules are used to synthesize ATP, several cofactors, ribonucleic acid (RNA), deoxyribonucleic acid (DNA), and other important cell components. Nearly all microorganisms can synthesize their own purines and pyrimidines as these are so crucial to cell function. ▶▶ *DNA replication (section 12.3); Transcription (section 12.5)*

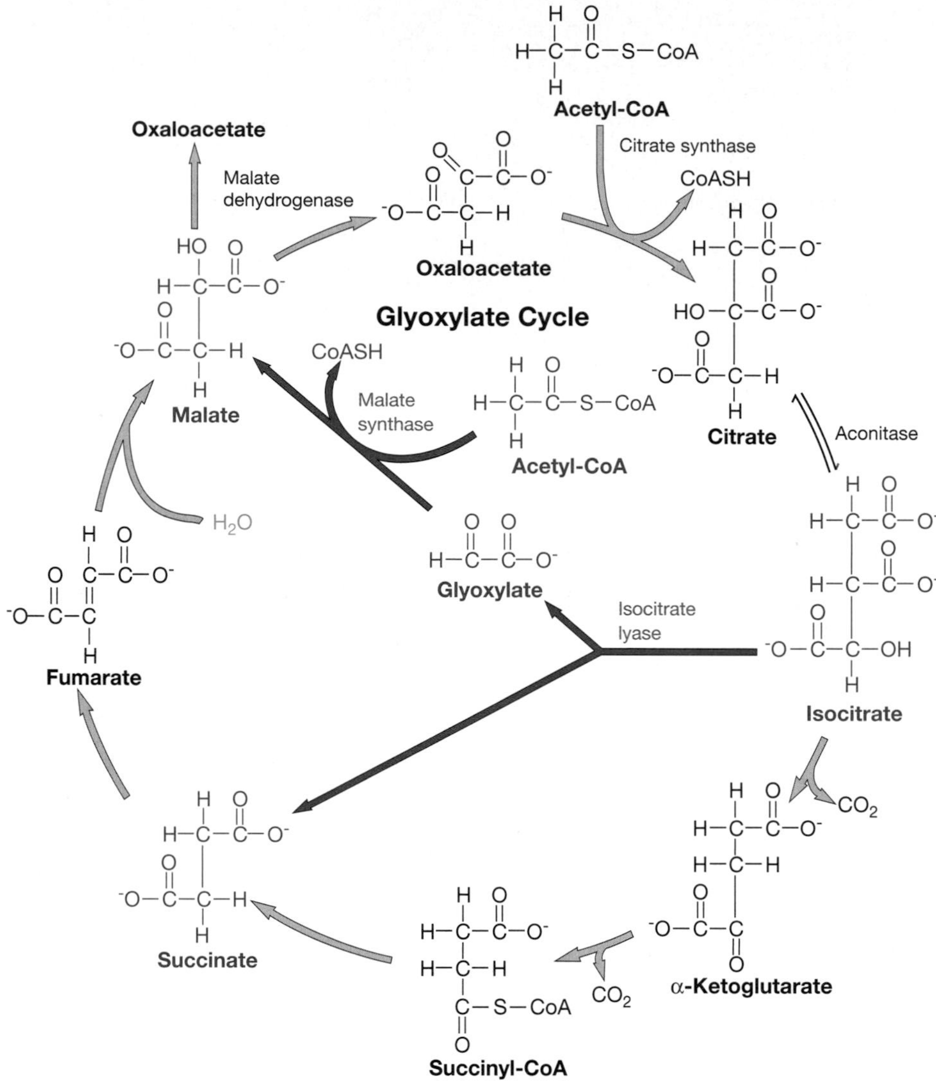

Overall equation:

$$2 \text{ Acetyl-CoA} + \text{FAD} + 2\text{NAD}^+ + 3\text{H}_2\text{O} \longrightarrow \text{Oxaloacetate} + 2\text{CoA} + \text{FADH}_2 + 2\text{NADH} + 2\text{H}^+$$

Figure 11.24   The Glyoxylate Cycle.   The reactions and enzymes unique to the cycle are shown in red and blue, respectively.

Purines and **pyrimidines** are cyclic nitrogenous bases with several double bonds. **Adenine and guanine** are purines, consisting of two joined rings, whereas pyrimidines (**uracil, cytosine, and thymine**) have only one ring. A purine or pyrimidine base joined with a pentose sugar, either ribose or deoxyribose, is a **nucleoside.** A **nucleotide** is a nucleoside with one or more phosphate groups attached to the sugar.

Amino acids participate in the synthesis of nitrogenous bases and nucleotides in a number of ways, including providing the nitrogen that is part of all purines and pyrimidines. The phosphorus present in nucleotides is provided by other mechanisms. We begin this section by examining phosphorus assimilation.

We then examine the pathways for synthesis of nitrogenous bases and nucleotides.

Phosphorus Assimilation

In addition to nucleic acids, phosphorus is found in proteins (i.e., phosphorylated proteins), phospholipids, and coenzymes such as NADP^+. The most common phosphorus sources are inorganic phosphate and organic phosphate esters. Inorganic phosphate is incorporated through the formation of ATP in one of three ways: (1) photophosphorylation, (2) oxidative phosphorylation, and (3) substrate-level phosphorylation.

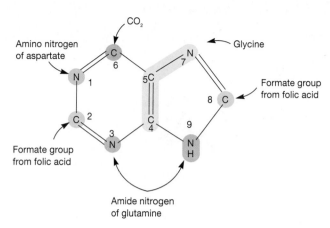

Figure 11.25 Purine Biosynthesis. The sources of nitrogen and carbon are indicated.

<< *Breakdown of glucose to pyruvate (section 10.3); Electron transport and oxidative phosphorylation (section 10.5); Phototrophy (section 10.12)*

Microorganisms may obtain organic phosphates from their surroundings in dissolved or particulate form. **Phosphatases** very often hydrolyze organic phosphate esters to release inorganic phosphate. Gram-negative bacteria have phosphatases in the periplasmic space, which allows phosphate to be taken up immediately after release. On the other hand, protists can directly use organic phosphates after ingestion or hydrolyze them in lysosomes and incorporate the phosphate.

Purine Biosynthesis

The biosynthetic pathway for purines is a complex, 11-step sequence *(see appendix II)* in which seven different molecules contribute parts to the final purine skeleton (**figure 11.25**). The pathway begins with ribose 5-phosphate and the purine skeleton is constructed on this sugar. Therefore the first purine product of the pathway is the nucleotide inosinic acid, not a free purine base. The cofactor folic acid is very important in purine biosynthesis. Folic acid derivatives contribute carbons two and eight to the purine skeleton.

Once inosinic acid has been formed, relatively short pathways synthesize adenosine monophosphate and guanosine monophosphate (**figure 11.26**), and produce nucleoside diphosphates and triphosphates by phosphate transfers from ATP. DNA contains deoxyribonucleotides (the ribose lacks a hydroxyl group on carbon two) instead of the ribonucleotides found in RNA. Deoxyribonucleotides arise from the reduction of nucleoside diphosphates or nucleoside triphosphates by two different routes. Some microorganisms reduce the triphosphates with a system requiring vitamin B_{12} as a cofactor. Others, such as *E. coli*, reduce the ribose in nucleoside diphosphates. Both systems employ a small, sulfur-containing protein called thioredoxin as their reducing agent.

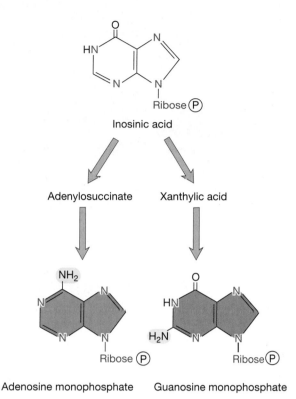

Figure 11.26 Synthesis of Adenosine Monophosphate and Guanosine Monophosphate. The highlighted groups differ from those in inosinic acid.

Pyrimidine Biosynthesis

Pyrimidine biosynthesis begins with aspartic acid and carbamoyl phosphate, a high-energy molecule synthesized from bicarbonate and ammonia provided by the amino acid glutamine (**figure 11.27**). Aspartate carbamoyltransferase catalyzes the condensation of these two substrates to form carbamoylaspartate, which is then converted to the initial pyrimidine product, orotic acid. A nucleotide is produced by the addition of ribose 5-phosphate, using the high-energy intermediate 5-phosphoribosyl 1-pyrophosphate. Thus construction of the pyrimidine ring is completed before ribose is added, in contrast with purine ring synthesis, which begins with ribose 5-phosphate.

Orotidine monophosphate is decarboxylated, yielding uridine monophosphate. This is followed by formation of uridine triphosphate and cytidine triphosphate. These two nucleotides are reduced in the same way that purine nucleotides are to make the deoxy-forms needed for DNA synthesis. Deoxythymidine monophosphate is made when deoxyuridine monophosphate is methylated with a folic acid derivative (**figure 11.28**).

1. How is phosphorus assimilated? What roles do phosphatases play in phosphorus assimilation? Why can phosphate be directly incorporated into cell constituents, whereas nitrate, nitrogen gas, and sulfate cannot?

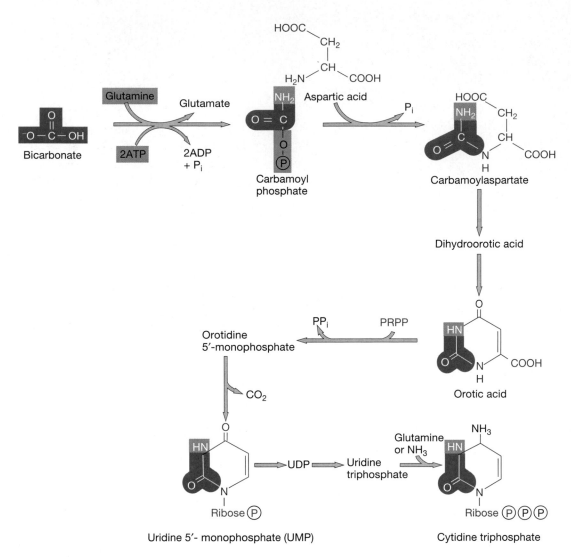

Figure 11.27 Pyrimidine Synthesis. PRPP stands for 5-phosphoribosyl 1-pyrophosphate, which provides the ribose 5-phosphate. The parts derived from bicarbonate and glutamine are shaded.

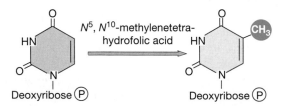

Deoxyuridine monophosphate   Deoxythymidine monophosphate

Figure 11.28 Deoxythymidine Monophosphate Synthesis. Deoxythymidine differs from deoxyuridine in having the shaded methyl group.

2. Explain the difference between a purine and a pyrimidine, and between a nucleoside and a nucleotide.

3. Outline the way in which purines and pyrimidines are synthesized. How is the deoxyribose component of deoxyribonucleotides made?

11.7 LIPID SYNTHESIS

Lipids are absolutely required by cells, as they are the major components of cell membranes. Most bacterial and eucaryal lipids contain fatty acids or their derivatives. Fatty acids are monocarboxylic acids with long alkyl chains that usually have an even number of carbons (the average length is 18 carbons). Some may be unsaturated—that is, have one or more double bonds. Most microbial fatty acids are straight chained, but some are branched. Gram-negative bacteria often have cyclopropane fatty acids (fatty acids with one or more cyclopropane rings in their chains). The synthesis of lipids is complex and is the target of some antimicrobial agents used to treat infectious disease caused by eucaryotic pathogens. >> *Lipids (appendix I); Antifungal drugs (section 31.7)*

Fatty acid synthesis is catalyzed by the **fatty acid synthase** complex with acetyl-CoA and malonyl-CoA as the substrates and NADPH as the electron donor. Malonyl-CoA arises from the ATP-driven carboxylation of acetyl-CoA (**figure 11.29**). Synthesis takes place after acetate and malonate have been transferred from coenzyme A (CoA) to the sulfhydryl group of the acyl carrier protein (ACP), a small protein that carries the growing fatty acid chain during synthesis. Fatty acid synthase adds two carbons at a time to the carboxyl end of the growing fatty acid chain in a two-stage process (figure 11.29). First, malonyl-ACP reacts with the fatty acyl-ACP to yield CO_2 and a fatty acyl-ACP two carbons longer. The loss of CO_2 drives this reaction to completion. Notice that ATP is used to add CO_2 to acetyl-CoA, forming malonyl-CoA. The same CO_2 is lost when malonyl-ACP donates carbons to the chain. Thus carbon dioxide is essential to fatty acid synthesis but is not permanently incorporated. Indeed, some microorganisms require CO_2 for optimum growth, but they can do without it in the presence of a fatty acid such as oleic acid (an 18-carbon unsaturated fatty acid). In the second stage of synthesis, the α-keto group arising from the initial condensation reaction is removed in a three-step process involving two reductions and a dehydration. The fatty acid is then ready for the addition of two more carbon atoms.

Unsaturated fatty acids are synthesized in two ways. Eucaryotes and aerobic bacteria such as *B. megaterium* employ an aerobic pathway using both NADPH and O_2.

$$R-(CH_2)_9-\overset{\overset{\text{O}}{\|}}{C}-SCoA + NADPH + H^+ + O_2 \rightarrow$$

$$R-CH=CH(CH_2)_7-\overset{\overset{\text{O}}{\|}}{C}-SCoA + NADP^+ + 2H_2O$$

A double bond is formed between carbons nine and ten, and O_2 is reduced to water with electrons supplied by both the fatty acid and NADPH. Anaerobic bacteria and some aerobes create double bonds during fatty acid synthesis by dehydrating hydroxy fatty acids. Oxygen is not required for double bond synthesis by this pathway. The anaerobic pathway is present in a number of common gram-negative bacteria (e.g., *E. coli* and *Salmonella* spp.), gram-positive bacteria (e.g., *Lactobacillus plantarum* and *Clostridium pasteurianum*), and cyanobacteria.

Eucaryotic microorganisms frequently store carbon and energy as **triacylglycerol,** glycerol esterified to three fatty acids. Glycerol arises from the reduction of the precursor metabolites dihydroxyacetone phosphate to glycerol 3-phosphate, which is then esterified with two fatty acids to give phosphatidic acid (**figure 11.30**). Phosphate is hydrolyzed from phosphatidic acid, giving a diacylglycerol, and the third fatty acid is attached to yield a triacylglycerol.

Phospholipids are major components of eucaryotic and bacterial cell membranes. Their synthesis usually proceeds by way of phosphatidic acid and a cytidine diphosphate (CDP) carrier that plays a role similar to that of uridine and adenosine diphosphate carriers in carbohydrate biosynthesis. For example, bacteria synthesize phosphatidylethanolamine, a major cell membrane component, through the initial formation of CDP-diacylglycerol (figure 11.30). This CDP derivative then reacts with serine to form the phospholipid phosphatidylserine, and decarboxylation yields phosphatidylethanolamine. In this way, a complex membrane lipid is constructed from the products of glycolysis, fatty acid biosynthesis, and amino acid biosynthesis.

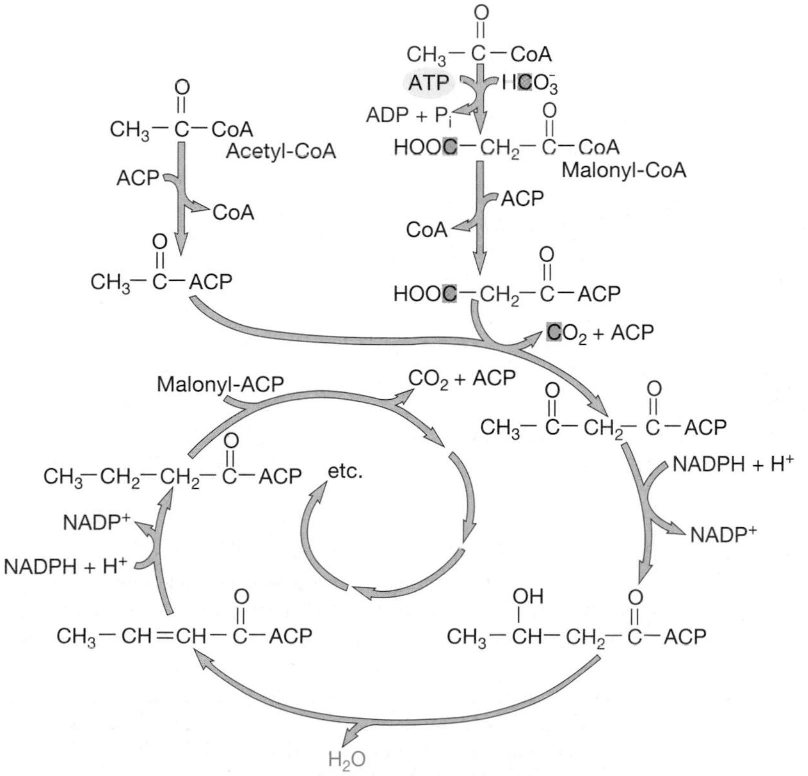

Figure 11.29 Fatty Acid Synthesis. The cycle is repeated until the proper chain length has been reached. Carbon dioxide carbon and the remainder of malonyl-CoA are shown in red. ACP stands for acyl carrier protein.

1. What is a fatty acid? Describe in general terms how fatty acid synthase manufactures a fatty acid.

2. How are unsaturated fatty acids made?

3. Briefly describe the pathways for triacylglycerol and phospholipid synthesis. Of what importance are phosphatidic acid and CDP-diacylglycerol?

4. Activated carriers participate in carbohydrate, peptidoglycan, and lipid synthesis. Briefly describe these carriers and their roles. Are there any features common to all the carriers? Explain your answer.

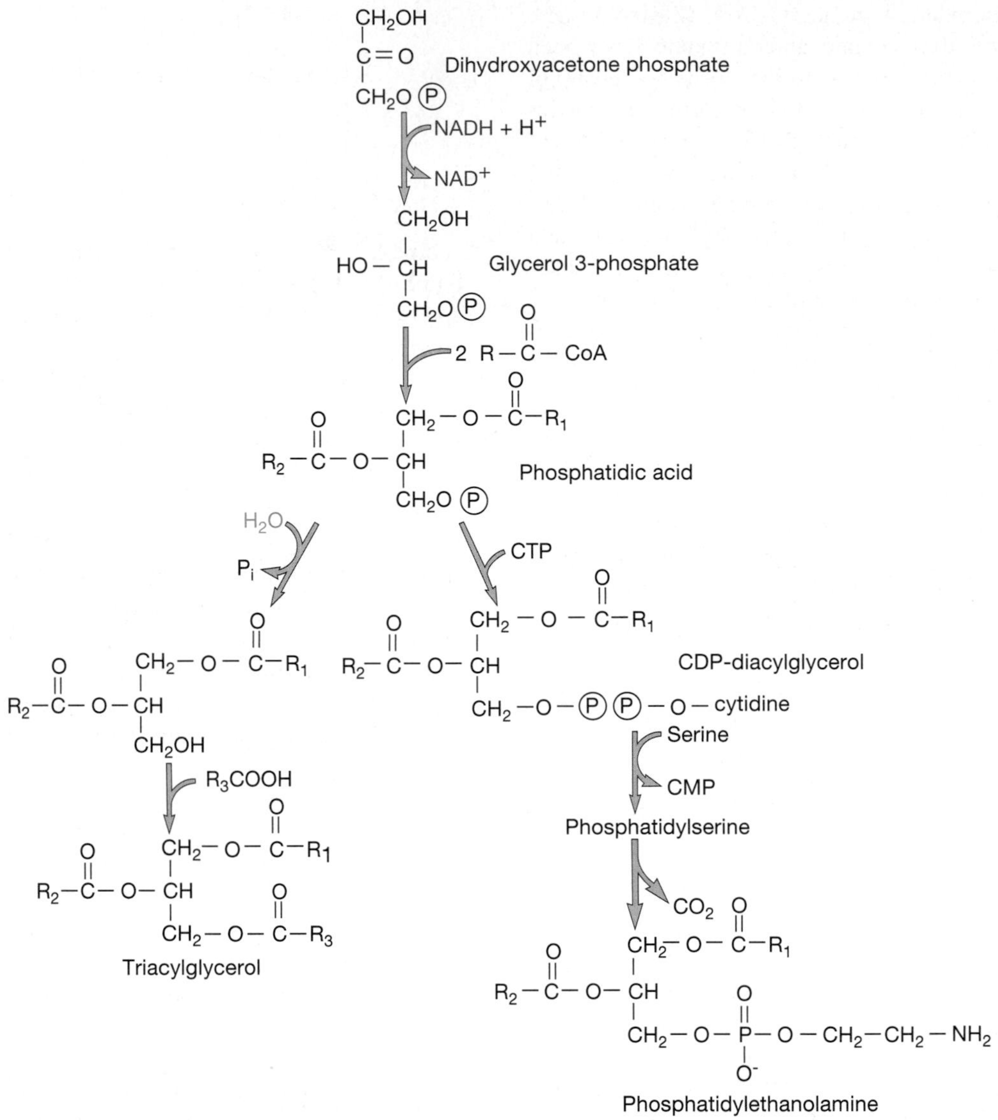

Figure 11.30   Triacylglycerol and Phospholipid Synthesis.

Summary

11.1 Principles Governing Biosynthesis

a. Many important cell constituents are macromolecules, large polymers constructed of simple monomers.

b. Although many catabolic and anabolic pathways share enzymes for the sake of efficiency, some of their enzymes are separate and independently regulated.

c. Macromolecular components often undergo self-assembly to form the final molecule or complex.

11.2 Precursor Metabolites

a. Precursor metabolites are carbon skeletons used as the starting substrates for biosynthetic pathways. They are intermediates of glycolytic pathways and the TCA cycle (i.e., the central metabolic pathways) (**figures 11.3** and **11.4**).

b. Most precursor metabolites are used for amino acid biosynthesis; others are used for synthesis of purines, pyrimidines, and lipids.

11.3 CO₂ Fixation

a. Four different CO_2-fixation pathways have been identified in autotrophic microorganisms: the Calvin cycle, the reductive TCA cycle, the acetyl-CoA pathway, and the hydroxypropionate cycle.

b. The Calvin cycle is used by most autotrophs to fix CO_2. It can be divided into three phases: the carboxylation phase, the reduction phase, and the regeneration phase (**figure 11.5**). Three ATPs and two NADPHs are used during the incorporation of one CO_2.

c. The reductive TCA cycle, acetyl-CoA pathway, and hydroxypropionate cycle are used by certain other bacteria to fix CO_2 (**figures 11.6–11.8**).

11.4 Synthesis of Sugars and Polysaccharides

a. Gluconeogenesis is the synthesis of glucose and related sugars from nonglucose precursors.

b. Glucose, fructose, and mannose are gluconeogenic intermediates or are made directly from them (**figure 11.9**); galactose is synthesized with nucleoside diphosphate derivatives (**figure 11.10**). Bacteria and protists synthesize glycogen and starch from adenosine diphosphate glucose.

c. Peptidoglycan synthesis is a complex process involving both UDP derivatives and the lipid carrier bactoprenol, which transports NAG-NAM-pentapeptide units across the cell membrane. Cross-links are formed by transpeptidation (**figures 11.11–11.13**).

11.5 Synthesis of Amino Acids

a. The addition of nitrogen to the carbon chain is an important step in amino acid biosynthesis. Ammonia, nitrate, or N_2 can serve as the source of nitrogen.

b. Ammonia can be directly assimilated by the activity of transaminases and either glutamate dehydrogenase or the glutamine synthetase–glutamate synthase system (**figures 11.14–11.16**).

c. Nitrate is incorporated through assimilatory nitrate reduction catalyzed by the enzymes nitrate reductase and nitrite reductase (**figure 11.17**).

d. Nitrogen fixation is catalyzed by nitrogenase. Atmospheric molecular nitrogen is reduced to ammonia, which is then incorporated into amino acids (**figures 11.18** and **11.19**).

e. Microorganisms can use cysteine, methionine, and inorganic sulfate as sulfur sources. Sulfate must be reduced to sulfide before it is assimilated. This occurs during assimilatory sulfate reduction (**figure 11.21**).

f. Some amino acids are made directly by the addition of an amino group to a precursor metabolite, but most amino acids are made by pathways that are more complex. Many amino acid biosynthetic pathways are branched. Thus a single precursor metabolite can give rise to several amino acids (**figures 11.22** and **11.23**).

g. Anaplerotic reactions replace TCA cycle intermediates to keep the cycle in balance while it supplies precursor metabolites. The anaplerotic reactions include the glyoxylate cycle (**figure 11.24**).

11.6 Synthesis of Purines, Pyrimidines, and Nucleotides

a. Purines and pyrimidines are nitrogenous bases found in DNA, RNA, and other molecules. The nitrogen is supplied by certain amino acids that participate in purine and pyrimidine biosynthesis. Phosphorus is provided by either inorganic phosphate or organic phosphate.

b. Phosphorus can be assimilated directly by phosphorylation reactions that form ATP from ADP and P_i. Organic phosphorus sources are the substrates of phosphatases that release phosphate from the organic molecule.

c. The purine skeleton is synthesized beginning with ribose 5-phosphate and initially produces inosinic acid. Pyrimidine biosynthesis starts with carbamoyl phosphate and aspartate, and ribose is added after the skeleton has been constructed (**figures 11.25–11.28**).

11.7 Lipid Synthesis

a. Fatty acids are synthesized from acetyl-CoA, malonyl-CoA, and NADPH by fatty acid synthase. During synthesis, the intermediates are attached to the acyl carrier protein (**figure 11.29**). Double bonds can be added in two different ways.

b. Triacylglycerols are made from fatty acids and glycerol phosphate. Phosphatidic acid is an important intermediate in this pathway (**figure 11.30**).

c. Phospholipids such as phosphatidylethanolamine can be synthesized from phosphatidic acid by forming CDP-diacylglycerol, then adding an amino acid.

Critical Thinking Questions

1. Discuss the relationship between catabolism and anabolism. How does anabolism depend on catabolism?

2. In metabolism, important intermediates are covalently attached to carriers, as if to mark these as important so the cell does not lose track of them. Think about a hotel placing your room key on a very large ring. List a few examples of these carriers and indicate whether they are involved primarily in anabolism or catabolism.

3. Intermediary carriers are in a limited supply: when they cannot be recycled because of a metabolic block, serious consequences ensue. Think of some examples of these consequences.

Learn More

Learn more by visiting the Prescott website at www.mhhe.com/prescottprinciples, where you will find a complete list of references.

Genes: Structure, Replication, and Expression

12

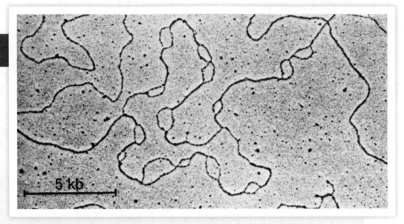

A micrograph of a replicating, eucaryotic chromosome.

anticodon   The base triplet on a tRNA that is complementary to the triplet codon on mRNA.

codon   A sequence of three nucleotides in mRNA that directs the incorporation of an amino acid during protein synthesis or signals the stop of translation.

DNA polymerase   An enzyme that synthesizes new DNA using a parental nucleic acid strand as a template.

gene   A DNA segment or sequence that codes for a polypeptide, rRNA, or tRNA.

genome   The full set of genes present in a cell or virus.

lagging strand   The strand of DNA that is discontinuously synthesized by the formation of short Okazaki fragments.

leader sequence   A sequence in a gene that corresponds to a nontranslated region at the 5′ end of mRNA.

leading strand   The strand of DNA that is synthesized continuously.

molecular chaperones   Proteins that assist in the folding and stabilization of other proteins or in directing newly synthesized proteins to secretion systems or other locations in the cell.

Okazaki fragments   Short stretches of polynucleotides produced during discontinuous DNA replication.

peptidyl transferase   The rRNA ribozyme (23S rRNA in *Bacteria*) that catalyzes the addition of an amino acid to the growing peptide chain during translation.

promoter   The region on DNA at the start of a gene to which RNA polymerase binds before beginning transcription.

reading frame   The way in which nucleotides in DNA and mRNA are grouped into codons for reading the message contained in the nucleotide sequence.

replication   The process by which an exact copy of a parental nucleic acid (DNA in cells; RNA in some viruses) is made with the parental molecule serving as a template.

replication fork   The Y-shaped structure where DNA is replicated; the arms of the Y contain template strand and a newly synthesized DNA copy.

replicon   A unit of the genome that contains an origin of replication and in which DNA is replicated.

replisome   A complex of proteins that replicates DNA.

RNA polymerase   The enzyme that catalyzes the synthesis of mRNA under the direction of a template.

Shine-Dalgarno sequence   A segment in the leader of bacterial and some archaeal mRNA that binds to a sequence on the 16S rRNA of the small ribosomal subunit; it helps properly orient the mRNA on the ribosome.

sigma factor   A protein that helps bacterial RNA polymerase core enzyme recognize the promoter.

stop codon   A codon that does not code for an amino acid but is a signal to stop protein synthesis; also called a nonsense codon.

template strand   A DNA strand that specifies the base sequence of a new complementary strand of DNA or RNA.

terminator   A sequence that marks the end of a gene and stops transcription.

transcription   The process by which single-stranded RNA with a base sequence complementary to the template strand of DNA (or RNA in some viruses) is synthesized.

translation   The process by which the genetic message carried by mRNA directs the synthesis of polypeptides with the aid of ribosomes and other cell constituents.

But the most important qualification of bacteria for genetic studies is their extremely rapid rate of growth. . . . a single *E. coli* cell will grow overnight into a visible colony containing millions of cells, even under relatively poor growth conditions. Thus, genetic experiments on *E. coli* usually last one day, whereas experiments on corn, for example, take months. It is no wonder that we know so much more about the genetics of *E. coli* than about the genetics of corn, even though we have been studying corn much longer.

—R. F. Weaver and P. W. Hedrick

In this chapter, we turn our attention to the synthesis of three major macromolecules—DNA, RNA, and proteins—from their constituent monomers. DNA serves as the storage molecule for the genetic instructions that enable organisms to carry out metabolism and reproduction. RNA functions in the expression of genetic information so that enzymes and other proteins can be made. These proteins are used to build cellular structures and to do other cellular work. The study of the synthesis of DNA, RNA, and protein falls into the realms of genetics and molecular biology.

In the mid-1800s the discipline of genetics was born from the work of Gregor Mendel, who studied the inheritance of various traits in pea plants. In the early twentieth century, Mendel's work was rediscovered and furthered by scientists working with fruit flies and plants such as corn. The use of microorganisms as models for genetic studies soon followed. Microorganisms, especially bacteria, have significant advantages as model organisms, in part because of their unique characteristics. One important feature is the nature of their genomes. The term **genome** refers to all DNA present in a cell or virus. Procaryotes normally have one set of genes; that is, they are haploid (1N). In addition, they often carry extrachromosomal genetic elements called plasmids. Eucaryotic organisms, including eucaryotic microorganisms, usually have two sets of genes, or are diploid (2N), and they rarely have plasmids. Viral genomes differ significantly from those of cellular organisms, and their genetics and molecular biology are discussed in chapters 5 and 24. **<<** *Procaryotic cytoplasm: Plasmids (section 3.3)*

We now review some of the most basic concepts of molecular genetics: how genetic information is stored and organized in the DNA molecule, the way in which DNA is replicated, gene structure, and how genes function (i.e., gene expression). Based on the foundation provided in this chapter, chapter 13 considers the regulation of gene expression. The regulation of gene expression is important because it links the **genotype** of an organism—the specific set of genes it possesses—to the **phenotype** of an organism—the collection of characteristics that are observable. Not all genes are expressed at the same time or in the same place, and the environment profoundly influences which genes are expressed at any given time. Finally, chapter 14 contains information on the nature of mutation, DNA repair, and genetic recombination. These three chapters provide the background needed for understanding microbial genomics (chapter 15) and recombinant DNA technology (chapter 16). Much of the information presented in chapters 12 through 14 will be familiar to those who have taken an introductory genetics course. Because of the importance of bacteria as model organisms, primary emphasis is placed on their genetics. The genetics of the *Archaea* is discussed in chapter 18.

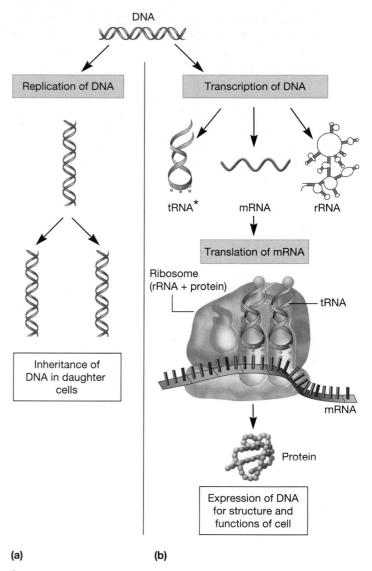

(a)   **(b)**

*The sizes of RNA are enlarged to show details.

Figure 12.1   Summary of the Flow of Genetic Information in Cells.   DNA serves as the storehouse for genetic information. (a) During cellular reproduction, DNA is replicated and passed to progeny cells. (b) To function properly, a cell must express the genetic information stored in DNA. This is accomplished when the genetic code is transcribed into mRNA molecules. The information in the mRNA is then translated into protein.

12.1   FLOW OF GENETIC INFORMATION

Biologists have long recognized a relationship among DNA, RNA, and protein, and this recognition has guided a vast amount of research over the past decades. The pathway from DNA to RNA and RNA to protein is conserved in all cellular forms of life and is often called the central dogma. **Figure 12.1** illustrates two essential concepts: the flow of genetic information from one generation to the next (replication); and the flow of information within a single cell, a process also called gene expression.

The transmission of genetic information from one generation to the next is shown in figure 12.1*a*. DNA functions as a storage molecule, holding genetic information for the lifetime of a cellular organism and allowing that information to be duplicated and passed on to its progeny. Synthesis of the duplicate DNA is directed by both strands of the parental molecule and is called **replication.** This process is catalyzed by DNA polymerase enzymes.

The genetic information stored in DNA is divided into units called **genes.** For an organism to function properly and reproduce, its genes must be expressed at the appropriate time and place. Gene expression begins with the synthesis of an RNA copy of the gene. This process of DNA-directed RNA synthesis is called **transcription** because the DNA base sequence is rewritten as an RNA base sequence. RNA polymerase enzymes catalyze transcription. Although DNA has two complementary strands, only one strand, the template strand, of a particular gene is transcribed. If both strands of a single gene were transcribed, two different RNA molecules would result and cause genetic confusion. However, different genes may be encoded on opposite strands, thus both strands of DNA can serve as templates for RNA synthesis, depending on the orientation of the gene on the DNA. Transcription yields three different types of RNA, depending on the gene transcribed. These are messenger RNA (mRNA), transfer RNA (tRNA), and ribosomal RNA (rRNA) (figure 12.1b).

During the last phase of gene expression, **translation,** genetic information in the form of an RNA base sequence in a messenger RNA (mRNA) is decoded and used to govern the synthesis of a polypeptide. Thus the amino acid sequence of a protein is a direct reflection of the base sequence in mRNA. In turn, the mRNA nucleotide sequence is complementary to a portion of the DNA genome. In addition to mRNA, translation also requires the activities of transfer RNA and ribosomal RNA. Thus all three types of RNA are involved in the production of protein, which is determined by the code present in the DNA.

1. Describe the general relationship between DNA, RNA, and protein.

2. What are the products of replication, transcription, and translation?

3. Until relatively recently, the "one gene–one protein" hypothesis was used to define the role of genes in organisms. Refer to figure 12.1 and explain why this description of a gene no longer applies.

12.2 NUCLEIC ACID STRUCTURE

The nucleic acids, DNA and RNA, are polymers of nucleotides (figure 12.2) linked together by phosphodiester bonds (figure 12.3a). However, DNA and RNA differ in terms of the nitrogenous bases they contain, the sugar component of their nucleotides, and whether they are double or single stranded. **Deoxyribonucleic acid (DNA)** contains the bases adenine, guanine, cytosine, and thymine. The sugar found in the nucleotides is deoxyribose (figure 12.2b), and DNA molecules usually are double stranded. **Ribonucleic acid (RNA),** on the other hand, contains the bases adenine, guanine, cytosine, and uracil (instead of thymine, although tRNA contains a modified form of thymine). Its sugar is ribose (figure 12.2b), and most RNA molecules are single stranded. However, an RNA strand can coil back on itself to form a hairpin-shaped structure with complementary base pairing and helical organization (p. 256, figure 12.19). The structure of DNA is described in more detail next.

DNA Structure

DNA molecules are very large and are usually composed of two polynucleotide chains coiled together to form a double helix 2.0 nm in diameter (figure 12.3). Each chain contains purine and pyrimidine deoxyribonucleosides joined by phosphodiester bonds (figure 12.3a). That is, a phosphate forms a bridge between a 3'-hydroxyl of one sugar and a 5'-hydroxyl of an adjacent sugar. Purine and pyrimidine bases are attached to the 1'-carbon of the deoxyribose sugars; they extend toward the middle of the cylinder formed by the two chains. (The numbers designating the carbons in the sugars are given a prime to distinguish them from the numbers designating the carbons and nitrogens in the nitrogenous bases.) The bases are stacked on top of each other in the center, one base pair every 0.34 nm. The purine adenine (A)

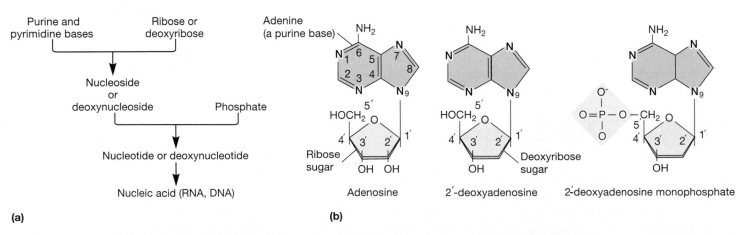

(a)   **(b)**

Figure 12.2 The Composition of Nucleic Acids. (a) A diagram showing the relationships of various nucleic acid components. Combination of a purine or pyrimidine base with ribose or deoxyribose gives a nucleoside (a ribonucleoside or deoxyribonucleoside). A nucleotide contains a nucleoside and one or more phosphates. Nucleic acids result when nucleotides are connected together in polynucleotide chains. (b) Examples of nucleosides—adenosine and 2'-deoxyadenosine–and the nucleotide 2'-deoxyadenosine monophosphate. The carbons of nucleoside and nucleotide sugars are indicated by numbers with primes to distinguish them from the carbons in the bases.

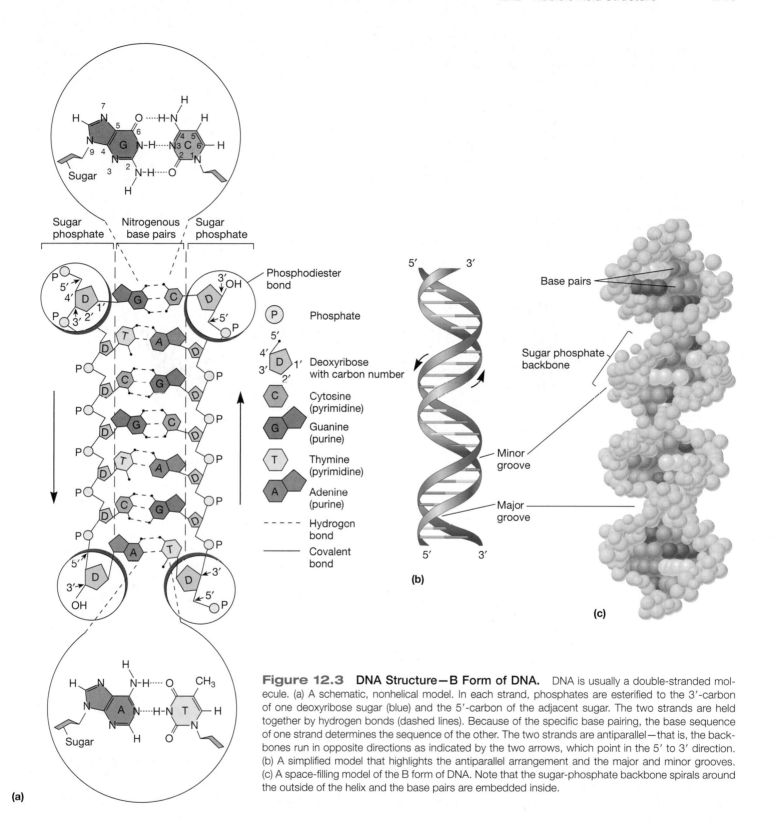

Figure 12.3 DNA Structure—B Form of DNA. DNA is usually a double-stranded molecule. (a) A schematic, nonhelical model. In each strand, phosphates are esterified to the 3'-carbon of one deoxyribose sugar (blue) and the 5'-carbon of the adjacent sugar. The two strands are held together by hydrogen bonds (dashed lines). Because of the specific base pairing, the base sequence of one strand determines the sequence of the other. The two strands are antiparallel—that is, the backbones run in opposite directions as indicated by the two arrows, which point in the 5' to 3' direction. (b) A simplified model that highlights the antiparallel arrangement and the major and minor grooves. (c) A space-filling model of the B form of DNA. Note that the sugar-phosphate backbone spirals around the outside of the helix and the base pairs are embedded inside.

of one strand is always paired with the pyrimidine thymine (T) of the opposite strand by two hydrogen bonds. The purine guanine (G) pairs with cytosine (C) by three hydrogen bonds. This AT and GC base pairing means that the two strands in a DNA double helix are **complementary.** In other words, the bases in one strand match up with those of the other according to specific base pairing rules. Because the sequences of bases in these strands encode genetic information, considerable effort has been devoted to determining the base sequences of DNA and RNA from many organisms, including a variety of microbes. >> *Microbial genomics (chapter 15)*

The two polynucleotide strands fit together much like the pieces in a jigsaw puzzle because of complementary base pairing. Inspection of figure 12.3b,c shows that the two strands are not positioned directly opposite one another in the helical cylinder. Therefore when the strands twist about one another, a wide major groove and narrower minor groove are formed by the backbone. Each base pair rotates 36° around the cylinder with respect to adjacent pairs so that there are 10.5 base pairs per turn of the helical spiral. Each turn of the helix has a vertical length of 3.4 nm. The helix is right-handed—that is, the chains turn counterclockwise as they approach a viewer looking down the longitudinal axis. The two backbones are antiparallel, which means they run in opposite directions with respect to the orientation of their sugars. One end of each strand has an exposed 5′-hydroxyl group, often with phosphates attached, whereas the other end has a free 3′-hydroxyl group (figure 12.3a). In a given direction, one strand is oriented 5′ to 3′ and the other, 3′ to 5′ (figure 12.3b).

The structure of DNA just described is that of the B form, the most common form in cells. Two other forms of DNA have been identified. The A form primarily differs from the B form in that it has 11 base pairs per helical turn rather than 10.5 and a vertical length of 2.6 Å rather than 3.4. Thus it is wider than the B form. The Z form is dramatically different, having a left-handed helical structure rather than right-handed as seen in the B and A forms. The Z form has 12 base pairs per helical turn and a vertical rise of 3.7 Å. Thus it is more slender than the B form. At this time, it is unclear whether the A form is found in cells. However, evidence exists that small portions of chromosomes can be in the Z form. The role, if any, for these stretches of Z DNA is unknown.

Organization of DNA in Cells

Although DNA exists as a double helix in all cells, its organization differs among cells in the three domains of life. DNA is organized in the form of a closed circle in all *Archaea* and most bacteria. This circular double helix is further twisted into supercoiled DNA (**figure 12.4**). In *Bacteria,* DNA is associated with basic proteins that appear to help organize it into a coiled, chromatinlike structure. << *Procaryotic cytoplasm: Nucleoid (section 3.3)*

DNA is much more highly organized in eucaryotes and is associated with a variety of proteins, the most prominent of which are **histones.** These are small, basic proteins rich in the amino acids lysine or arginine or both. There are five types of histones in most eucaryotic cells studied: H1, H2A, H2B, H3, and H4. Eight histone molecules (two each of H2A, H2B, H3, and H4) form an ellipsoid about 11 nm long and 6.5 to 7 nm in diameter (**figure 12.5**). DNA coils around the surface of the ellipsoid approximately 1 3/4 turns or 166 base pairs before proceeding on

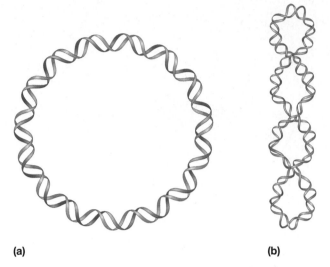

(a)　　　　　　　　　　　　　　**(b)**

Figure 12.4　DNA Forms. (a) The DNA double helix of most procaryotes is in the shape of a closed circle. (b) The circular DNA strands, already coiled in a double helix, are twisted a second time to produce supercoils.

to the next. This combination of histones plus DNA is called a **nucleosome.** DNA gently isolated from chromatin looks like a string of beads. The stretch of DNA between the beads (nucleosomes) is the linker region; it is associated with histone H1, which aids in folding DNA into more complex chromatin structures. When folding reaches a maximum, the chromatin takes the shape of the visible chromosomes seen in eucaryotic cells during mitosis and meiosis.

Although the *Archaea* share the procaryotic style of cellular organization with the *Bacteria,* some important differences exist. In many archaea, DNA is complexed with histone proteins. Like eucaryotic histones, archaeal histones form nucleoprotein complexes called archaeal nucleosomes. In these archaea, tetramers of histones (i.e., four histone proteins) interact with about 60 base pairs each, protecting the DNA from DNA-digesting nucleases. The structure of archaeal histones and their interaction with DNA strongly suggests that archaeal nucleosomes are evolutionarily similar to the eucaryotic nucleosomes formed by DNA and histones H3 and H4. >> *Microbial evolution (section 17.1)*

1. What are nucleic acids? How do DNA and RNA differ in structure?

2. What does it mean to say that the two strands of the DNA double helix are complementary and antiparallel? Examine figure 12.3b and explain the differences between the minor and major grooves.

3. What are histones and nucleosomes? Describe the way in which DNA is organized in the chromosomes of *Bacteria, Archaea,* and eucaryotes.

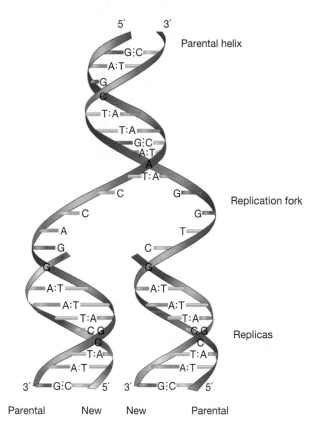

Figure 12.6 Semiconservative DNA Replication. The replication fork of DNA showing the synthesis of two progeny strands. Newly synthesized strands are purple. Each copy contains one new and one old strand. This process is called semiconservative replication.

Figure 12.5 Nucleosome Internal Organization. An illustration of how a string of nucleosomes, each associated with a histone H1, might be organized to form a highly supercoiled chromatin fiber. The nucleosomes are drawn as cylinders.

12.3 DNA REPLICATION

DNA replication is an extraordinarily important and complex process upon which all life depends. During DNA replication, the two strands of the double helix are separated; each then serves as a template for the synthesis of a complementary strand according to the base pairing rules. Each of the two progeny DNA molecules consists of one new strand and one old strand. Thus DNA replication is semiconservative (**figure 12.6**). DNA replication is also extremely accurate; *E. coli* makes errors with a frequency of only 10^{-9} or 10^{-10} per base pair replicated (or about one in a million [10^{-6}] per gene per generation). Despite its complexity and accuracy, replication is very rapid. In *Bacteria,* replication rates approach 750 to 1,000 base pairs per second. Eucaryotic replication is slower, about 50 to 100 base pairs per second.

In this section, we first discuss the various patterns of DNA replication observed in cells and viruses. We then consider the mechanism of DNA replication in *E. coli,* beginning with an examination of the replication machinery and then events at the replication fork.

Patterns of DNA Synthesis

Replication patterns are somewhat different in *Bacteria, Archaea,* and eucaryotes. For example, when the circular DNA chromosome of *E. coli* is copied, replication begins at a single point, the origin. Synthesis occurs at the **replication fork,** the place at which the DNA helix is unwound and individual strands are replicated. Two replication forks move outward from the origin until they have copied the whole **replicon**—the portion of the genome that contains an origin and is replicated as a unit. When the replication forks move around the circle, a structure shaped like the Greek letter theta (θ) is formed (**figure 12.7**). Because the bacterial chromosome is a single replicon, the forks meet on the other side and two separate chromosomes are released. Until recently, it was thought that all procaryotes have a single origin of replication. However, two members of the archaeal genus *Sulfolobus* have more than one origin; the reason for this is unknown.

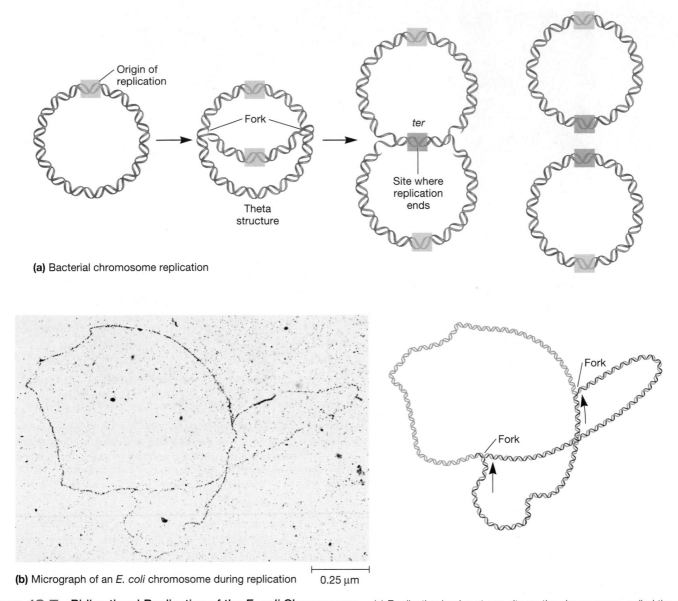

(a) Bacterial chromosome replication

(b) Micrograph of an *E. coli* chromosome during replication   0.25 µm

Figure 12.7   Bidirectional Replication of the *E. coli* Chromosome.   (a) Replication begins at one site on the chromosome, called the origin of replication. Two replication forks proceed in opposite directions from the origin until they meet at a site called the replication termination site (*ter*). The theta structure is a commonly observed intermediate of the process. (b) An autoradiograph of a replicating *E. coli* chromosome; about one-third of the chromosome has been replicated. To the right is a schematic representation of the chromosome. Parental DNA is blue; new DNA strands are purple, arrow represents direction of fork movement.

Eucaryotic DNA is much longer than procaryotic DNA, and many replication forks copy eucaryotic DNA simultaneously so that the molecule can be duplicated relatively quickly (**chapter opener**).

A different pattern of DNA replication occurs during *E. coli* conjugation, a type of genetic exchange mechanism observed in procaryotes. The pattern is called **rolling-circle replication,** and it also is observed during plasmid replication and the reproduction of some viruses (e.g., phage lambda). During rolling-circle replication, one strand is nicked and the free 3′-hydroxyl end is extended by replication enzymes (**figure 12.8**). The 3′ end is lengthened while the growing point rolls around the circular template and the

5′ end of the strand is displaced to form an ever-lengthening tail, much like the peel of an apple is displaced by a knife as an apple is pared. The single-stranded tail may be converted to the double-stranded form by complementary strand synthesis. This mechanism is particularly useful to viruses because it allows rapid, continuous production of many genome copies from a single initiation event.

Replication Machinery

Because DNA replication is essential to organisms, a great deal of effort has been devoted to understanding its mechanism. The replication of *E. coli* DNA, which requires at least 30 proteins,

Table 12.1	Components of the *E. coli* Replication Machinery
Protein	**Function**
DnaA protein	Initiation of replication; binds origin of replication (*oriC*)
DnaB protein	Helicase ($5'\rightarrow3'$); breaks hydrogen bonds holding two strands of double helix together; promotes DNA primase activity; involved in primosome assembly
DNA gyrase	Relieves supercoiling of DNA produced as DNA strands are separated by helicases; separates daughter molecules in final stages of replication
SSB proteins	Bind single-stranded DNA after strands are separated by helicases
DnaC protein	Helicase loader; helps direct DnaB protein (helicase) to DNA template
n' protein	Component of primosome; helicase ($3'\rightarrow5'$)
n protein	Primosome assembly; component of primosome
n" protein	Primosome assembly
I protein	Primosome assembly
DNA primase	Synthesis of RNA primer; component of primosome
DNA polymerase III holoenzyme	Complex of about 20 polypeptides; catalyzes most of the DNA synthesis that occurs during DNA replication; has $3'\rightarrow5'$ exonuclease (proofreading) activity
DNA polymerase I	Removes RNA primers; fills gaps in DNA formed by removal of RNA primer
Ribonuclease H	Removes RNA primers
DNA ligase	Seals nicked DNA, joining DNA fragments together
DNA replication terminus site-binding protein	Termination of replication
Topoisomerase IV	Segregation of chromosomes upon completion of DNA replication

(**table 12.1**) is probably best understood and is the focus of this discussion. The overall process in other bacteria, *Archaea,* and eucaryotes is thought to be similar.

Enzymes called **DNA polymerases** catalyze DNA synthesis. All known DNA polymerases catalyze the synthesis of DNA in the $5'$ to $3'$ direction. This is because the $3'$-hydroxyl group of the deoxyribose at the end of the growing DNA strand attacks the alpha phosphate (the phosphate closest to the $5'$ carbon) of the deoxynucleotide to be incorporated (**figure 12.9**). This results in the formation of a phosphodiester bond. The energy needed to form this covalent bond is provided by the release of the terminal diphosphate (the beta and gamma phosphates) from the nucleotide that is added to the growing chain. Thus deoxynucleotide (dNTPs: dATP, dTTP, dCTP, dGTP) serve as DNA polymerase substrates while deoxynucleoside monophosphates (dNMPs: dAMP, dTMP, dCMP, dGMP) are incorporated into the growing chain.

For DNA polymerase to catalyze the synthesis of DNA, it needs three things. The first is a template, which is read in the $3'$ to $5'$ direction and is used to direct the synthesis of a complementary DNA strand. The second is a primer (e.g., an RNA strand or a DNA strand) to provide a free $3'$-hydroxyl group to which nucleotides can be added (figure 12.9). The third is a set of

dNTPs. *E. coli* has five different DNA polymerases (DNA Polymerase I-V). DNA polymerase III plays the major role in replication, although it is assisted by DNA polymerase I.

DNA polymerase III holoenzyme is a complex of 10 proteins, including two core enzymes, each composed of three protein subunits (**figure 12.10**). The core enzymes are responsible for catalyzing DNA synthesis and proofreading the product to ensure fidelity of replication. A dimer of another subunit (tau) connects the two core enzymes. Associated with each core enzyme is a subunit called the β clamp. This protein tethers the core enzyme to one strand of the DNA molecule. A complex of proteins called the γ complex is responsible for loading the β clamp onto the DNA. Because there are two core enzymes, both strands of DNA are bound by a single DNA polymerase III holoenzyme.

In *E. coli,* replication begins when a collection of DnaA proteins binds to specific nucleotide sequences (DnaA boxes) within the **origin of replication.** The DnaA proteins use the energy from ATP hydrolysis to break or "melt" the hydrogen bonds between the DNA strands, thus making this localized region single stranded. Although this provides the initial template for replication, DNA polymerase III cannot by itself unwind and maintain the single-stranded DNA. These activities are provided by the

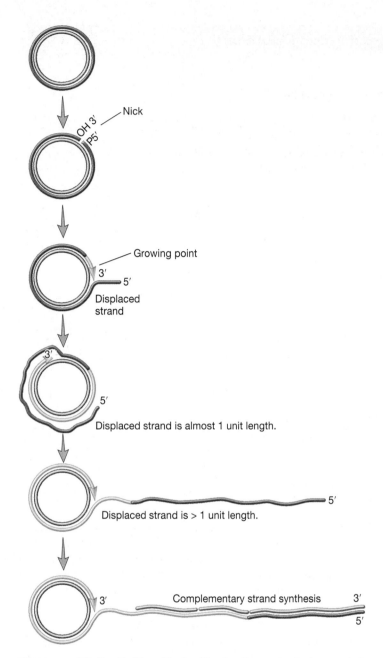

Figure 12.8 Rolling-Circle Replication. A single-stranded tail, often composed of more than one genome copy, is generated and can be converted to the double-stranded form by synthesis of a complementary strand. The "free end" of the rolling-circle strand is probably bound to the primosome. OH 3' is the 3'-hydroxyl and P 5' is the 5'-phosphate created when the DNA strand is nicked.

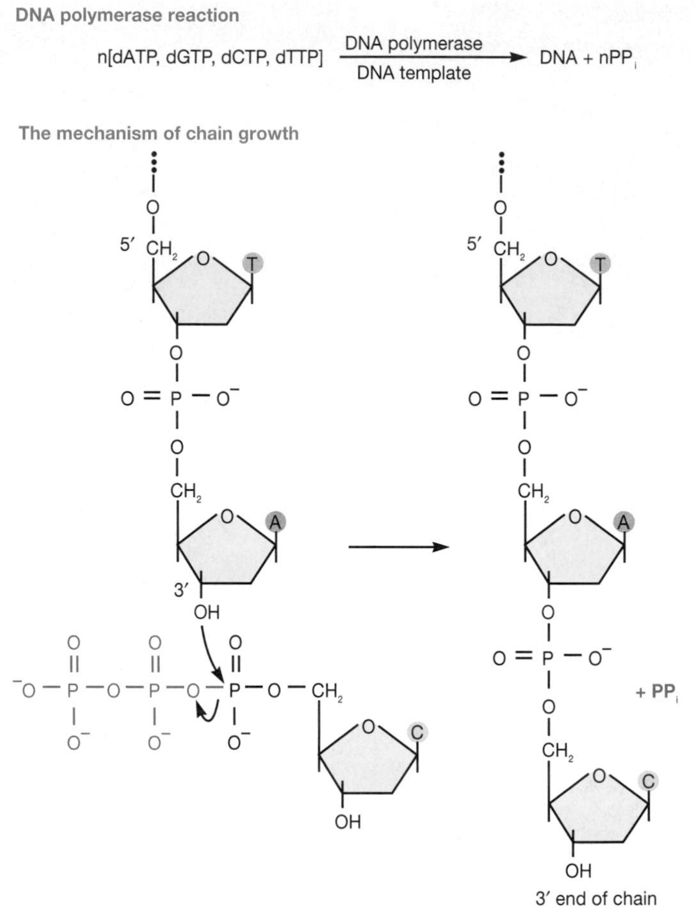

DNA polymerase reaction

$$n[dATP, dGTP, dCTP, dTTP] \xrightarrow[\text{DNA template}]{\text{DNA polymerase}} DNA + nPP_i$$

The mechanism of chain growth

Figure 12.9 The DNA Polymerase Reaction and Its Mechanism. The mechanism involves a nucleophilic attack by the hydroxyl of the 3' terminal deoxyribose on the alpha phosphate of the nucleotide substrate (in this example, adenosine attacks cytidine triphosphate).

action of other proteins found in the **replisome,** a huge complex of proteins that includes DNA polymerase III holoenzyme. These other proteins include helicases, single-stranded DNA binding proteins, and **topoisomerases (figure 12.11)**. **Helicases** are responsible for separating (unwinding) the DNA strands. These enzymes also use energy from ATP hydrolysis to unwind short stretches of helix just ahead of the replication fork. **Single-stranded DNA binding proteins (SSBs)** keep the strands apart once they have been separated, and topoisomerases relieve the tension generated by the rapid unwinding of the double helix (the replication fork may rotate as rapidly as 75 to 100 revolutions per second). This is important because rapid unwinding can lead to the formation of supercoils or supertwists in the helix, and these can impede replication if not removed. Topoisomerases change the structure of DNA by transiently breaking one or two strands without altering the nucleotide sequence of the DNA (e.g., a topoisomerase might tie or untie a knot in a DNA strand). **DNA gyrase** is an important topoisomerase in *E. coli.*

Once the template is prepared, the primer needed by DNA polymerase III can be synthesized. An enzyme called **primase** synthesizes a short RNA strand, usually around 10 nucleotides long and complementary to the DNA; this serves as the primer (figure 12.11). RNA is used as the primer because unlike DNA polymerase, RNA polymerases (such as primase) can initiate RNA synthesis without an existing 3'-OH. It appears that primase requires the assistance of several other proteins, and the complex of primase and its accessory proteins is called the **primosome** (table 12.1). The primosome is another important component of the replisome.

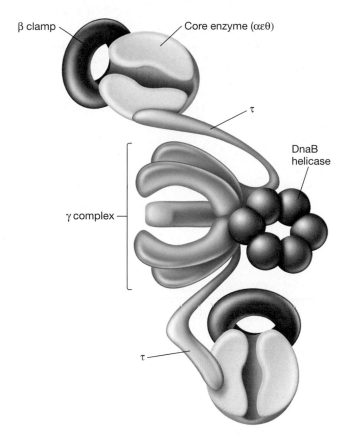

Figure 12.10 DNA Polymerase III Holoenzyme. The holoenzyme consists of two core enzymes (three subunits each; α, ε, θ, not shown) and several other subunits. The two tau (τ) subunits connect the two core enzymes to a large complex called the gamma (γ) complex. Each core enzyme is associated with a β clamp, which tethers a DNA template to each core enzyme.

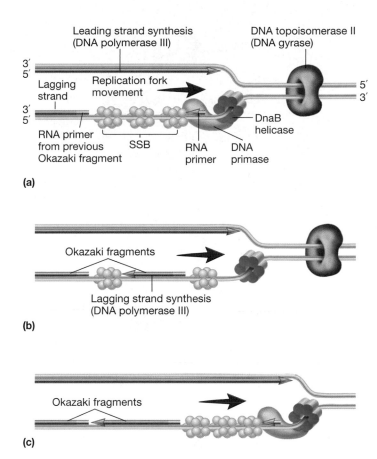

Figure 12.11 Other Replisome Proteins. A simplified diagram of DNA replication in *E. coli.* A single replication fork showing the activity of replisome proteins other than DNA polymerase III (not shown) is illustrated. DnaB helicase is responsible for separating the two strands of parental DNA. The strands are kept apart by single-stranded DNA binding proteins (SSB), which allows for synthesis of an RNA primer by DNA primase. DNA gyrase eases the strain introduced into the DNA double helix by helicase activity. Both leading strand and lagging strand synthesis are illustrated. The lagging strand is synthesized in short fragments called Okazaki fragments. A new RNA primer is required for the synthesis of each Okazaki fragment.

Because DNA polymerase enzymes must synthesize DNA in the 5′ to 3′ direction, only one of the strands, called the **leading strand,** can be synthesized continuously at its 3′ end as the DNA unwinds (figure 12.11). The other strand, called the **lagging strand,** cannot be extended in the same direction as the movement of the replication fork because there is no free 3′-OH to which a nucleotide can be added. As a result, the lagging strand is synthesized discontinuously in the 5′ to 3′ direction (i.e., in the direction opposite of the movement of the replication fork) and produces a series of fragments called **Okazaki fragments** after their discoverer, Reiji Okazaki. Discontinuous synthesis occurs as primase makes many RNA primers along the template strand. DNA polymerase III then extends these primers with DNA, and eventually the Okazaki fragments are joined to form a complete strand. Thus while the leading strand requires only one RNA primer (and only one primosome) to initiate synthesis, the lagging strand has many RNA primers (and primosomes) that must eventually be removed. Okazaki fragments are about 1,000 to 2,000 nucleotides long in *Bacteria* and approximately 100 nucleotides long in eucaryotic cells.

Events at the Replication Fork

The details of DNA replication are outlined in **figure 12.12.** In *E. coli,* DNA replication is initiated at specific nucleotides called the *oriC* locus (for *ori*gin of *c*hromosomal replication). Here we present replication as a series of discrete steps, but in the cell these events occur quickly and simultaneously on both the leading and lagging strands.

1. To initiate replication, as many as 40 DnaA proteins bind *oriC* while hydrolyzing ATP. Binding of DnaA proteins causes the DNA to bend around the protein complex, resulting in separation of the double-stranded DNA at regions within the origin having many A-T base pairs. Recall that adenines pair with thymines using only two hydrogen bonds, so A-T rich segments of DNA become single stranded more readily than do G-C–rich regions. Once the strands have separated, replication proceeds through the next four stages.

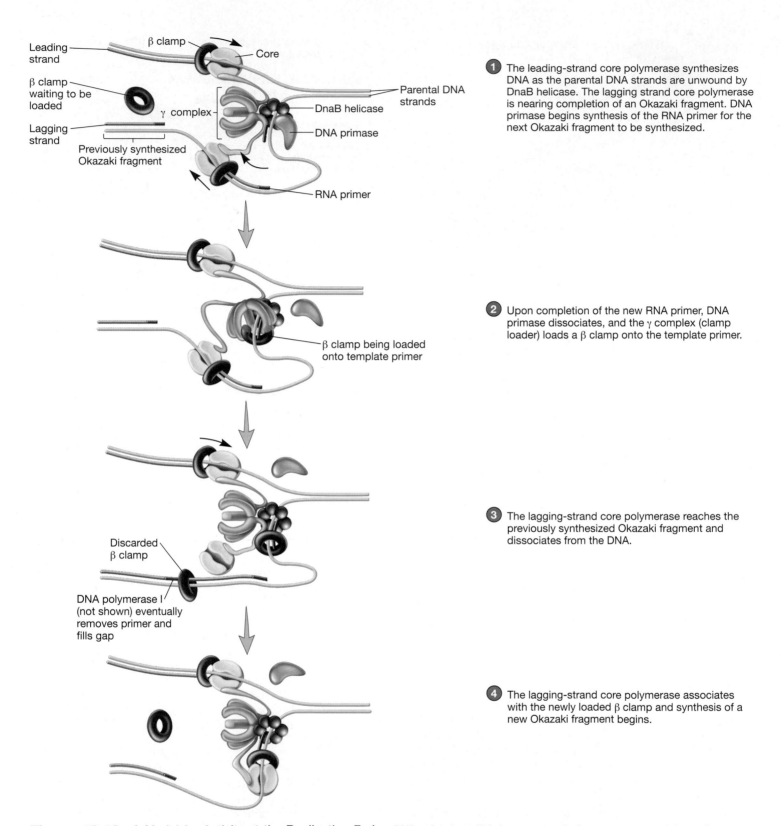

Figure 12.12   A Model for Activity at the Replication Fork.   DNA polymerase III holoenzyme and other components of the replisome are responsible for the synthesis of both leading and lagging strands. The arrows show the movement of each DNA core polymerase. After completion of each new Okazaki fragment, the old β sliding clamp is discarded and a new one loaded onto the template DNA (step 3). This is achieved by the activity of the γ complex (figure 12.10), which is also known as the clamp loader. Okazaki fragments are eventually joined together (figure 12.13) after removal of the RNA primer and synthesis of DNA to fill the gap, both catalyzed by DNA polymerase I; DNA ligase then seals the nick and joins the two fragments (figure 12.14).

2. Helicases unwind the helix with the aid of topoisomerases such as DNA gyrase (figure 12.12, step 1). It appears that DnaB protein is the helicase most actively involved in replication, but the n′ protein also may participate in unwinding. The single strands are kept separate by SSBs.

3. Primase synthesizes RNA primers as needed (figure 12.12, step 1). A single DNA polymerase III holoenzyme catalyzes both leading strand and lagging strand synthesis from the RNA primers. Lagging strand synthesis is particularly amazing because of the "gymnastic" feats performed by the replisome. It must discard old β clamps (figure 12.12, step 3), load new β clamps (figure 12.12, step 2), and tether the template to the core enzyme with each new round of Okazaki fragment synthesis. All of this occurs as DNA polymerase III is synthesizing DNA. Thus DNA polymerase III is a multifunctional enzyme.

4. After most of the lagging strand has been synthesized by the formation of Okazaki fragments, DNA polymerase I or (more rarely) RNaseH removes the RNA primer. DNA polymerase I does this because, unlike other DNA polymerases, it has 5′ to 3′ exonuclease activity—that is, it can snip off nucleotides one at a time starting at the 5′ end. Thus DNA polymerase I begins its exonuclease activity at the free end of the RNA primer. With the removal of each ribonucleotide, the adjacent 3′-OH from the deoxynucleotide is used by DNA polymerase I to fill the gap between Okazaki fragments (figure 12.13).

5. Finally, the Okazaki fragments are joined by the enzyme **DNA ligase,** which forms a phosphodiester bond between the 3′-OH of the growing strand and the 5′-phosphate of an Okazaki fragment (**figure 12.14**).

As we have seen, DNA polymerase III is a remarkable multi-protein complex, with several enzymatic activities. In *E. coli,* the polymerase component is encoded by the *dnaE* gene. *Bacillus subtilis,* a gram-positive bacterium that is another important experimental model, has a second polymerase gene called *dnaE$_{Bs}$.* Its protein product appears to be responsible for synthesizing the lagging strand. Thus while the overall mechanism by which DNA is replicated is highly conserved, there can be variations in replisome components.

Amazingly, DNA polymerase III, like all DNA polymerases, has an additional function that is critically important: **proofreading.** Proofreading is the removal of a mismatched base immediately after it has been added; its removal must occur before the next base is incorporated. The three subunits of the polymerase III core are α, ε, and θ. The α subunit has polymerase activity, and the ε subunit has 3′ to 5′ exonuclease activity. This

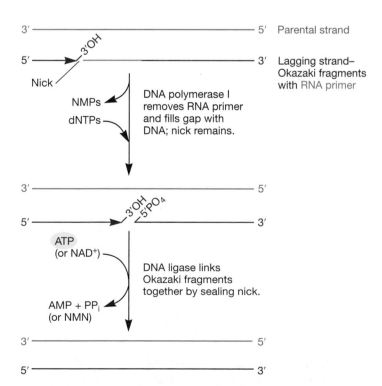

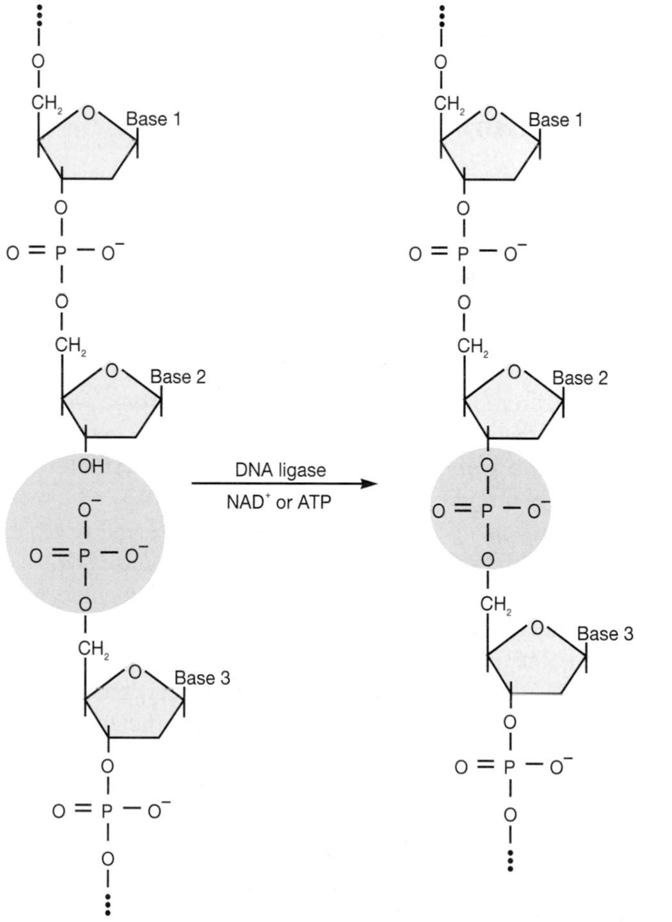

Figure 12.13 Completion of Lagging Strand Synthesis.

Figure 12.14 The DNA Ligase Reaction. The groups being altered are shaded in blue. Bacterial ligases use the pyrophosphate bond of NAD$^+$ as an energy source; many other ligases employ ATP.

activity enables the polymerase core to check each newly incorporated base to see that it forms stable hydrogen bonds. In this way, mismatched bases can be detected. If the wrong base has been mistakenly added, the ε subunit is able to remove it. The exonuclease activity can remove a mismatched base only as long as it is still at the 3' end of the growing strand. Once removed, holoenzyme backs up and adds the proper nucleotide in its place. DNA proofreading is not 100% efficient and, as discussed in chapter 14, the mismatch repair system is the cell's second line of defense.

Termination of Replication

In *E. coli,* DNA replication stops when the replisome reaches a termination site (*ter*) on the DNA. The Tus protein binds to the *ter* sites and halts progression of the forks. In many other bacteria, replication stops randomly when the forks meet. Regardless of how fork movement is stopped, there is often a problem to be solved by the replisome: separation of daughter molecules. When replication of a circular chromosome is complete, the two circular daughter chromosomes may remain intertwined. Such interlocked chromosomes are called **catenanes.** This is obviously a problem if each daughter cell is to inherit a single chromosome. Fortunately, topoisomerases solve the problem by temporarily breaking DNA molecules, so that the strands can be separated.

Replication of Linear Chromosomes

The fact that eucaryotic chromosomes are linear poses a problem during replication because of DNA polymerase's need for a primer that provides a free 3'-OH. At the ends (telomeres) of eucaryotic chromosomes, space is not available for synthesis of a primer on the lagging strand, and therefore it should be impossible to replicate the end of that strand. Over numerous rounds of DNA replication and cell division, this would lead to a progressively shortened chromosome. Ultimately the chromosome would lose critical genetic information, which would be lethal to the cell.

Eucaryotic cells have evolved the enzyme **telomerase** to solve this "end replication problem" (**figure 12.15**). Telomerase has two components: an internal RNA template and telomerase reverse transcriptase—an enzyme that can synthesize DNA using an RNA template. The internal RNA is complementary to the single strand of DNA jutting out from the end of the chromosome and acts as the template for DNA synthesis to elongate that strand (i.e., the 3'-OH of the telomere DNA strand serves as the primer for DNA synthesis). After being lengthened sufficiently, the single strand of telomere DNA can serve as the template for synthesis of the complementary strand by DNA polymerase III. Thus the length of the chromosome is maintained.

Telomerase has solved the problem of end replication for eucaryotes, but recall that some bacteria also have linear chromosomes. How do they replicate the ends of their chromosomes? Unfortunately little is known about the replication of linear bacterial chromosomes. However, a recent discovery in *Streptomyces,* an important group of soil bacteria, has led to

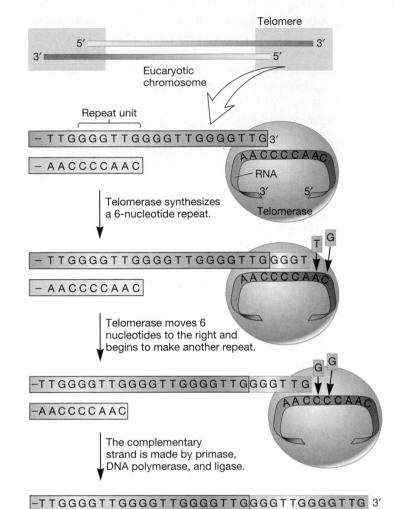

Figure 12.15   Replication of the Telomeres of Eucaryotic Chromosomes by Telomerase.   Telomerase contains an RNA molecule that can base pair with a small portion of the 3' overhang. The RNA serves as a template for DNA synthesis catalyzed by the reverse transcriptase activity of the enzyme. The 3'-OH of the telomere DNA serves as the primer and is lengthened. The process shown is repeated many times until the 3' overhang is long enough to serve as the template for the complementary telomere DNA strand.

speculation that a telomerase-like process may function in these bacteria. The ends of the linear chromosome of *Streptomyces coelicolor* are associated with a complex of proteins, including one with in vitro reverse transcriptase activity. No RNA has been found in the *Streptomyces* complex. Therefore it is unclear if the protein functions as a reverse transcriptase and what it might use as a template if it does.

1. Define the following: origin of replication, replicon, replication fork, primosome, and replisome.

2. Describe the nature and functions of the following replication components and intermediates: DNA

polymerases I and III, topoisomerase, DNA gyrase, helicase, single-stranded DNA binding proteins, Okazaki fragment, DNA ligase, leading strand, lagging strand, primase, and telomerase.

3. How does rolling-circle replication differ from the usual type of replication observed for cellular chromosomes?

4. Outline the steps involved in DNA synthesis at the replication fork. How do DNA polymerases correct their mistakes?

12.4 GENE STRUCTURE

DNA replication allows genetic information to be passed from one generation to the next. But how is the genetic information used? To answer that question, we must first look at how genetic information is organized. The basic unit of genetic information is the gene. The gene has been regarded in several ways. At first, it was thought that a gene contained information for the synthesis of one enzyme—the one gene–one enzyme hypothesis. This was modified to the one gene–one polypeptide hypothesis because of the existence of enzymes and other proteins composed of two or more different polypeptide chains coded for by separate genes. Historically, a segment of DNA that encodes a single polypeptide was termed a cistron; this term is still sometimes used. However, not all genes encode proteins; some code instead for rRNA and tRNA (figure 12.1). In addition, it is now known that some eucaryotic genes encode more than one protein. Thus a gene might be defined as a polynucleotide sequence that codes for a functional product (i.e., a polypeptide, tRNA, or rRNA).

When the nucleotide sequences of protein-coding genes are transcribed, the resulting mRNA can be "read" in discrete sets of three nucleotides, each set being a **codon.** Each codon codes for a single amino acid. The sequence of codons is "read" in only one way to produce a single product. That is, the code is not overlapping and there is a single starting point with one **reading frame**—that is, the way in which nucleotides are grouped into codons (**figure 12.16**). Each strand of DNA usually consists of gene sequences that do not overlap one another (**figure 12.17a**). However, there are exceptions to the rule. Some viruses such as the phage φX174 have overlapping genes (figure 12.17b), and parts of genes overlap in some bacterial genomes.

Procaryotic and viral gene structure differ greatly from that of eucaryotes. In procaryotic and viral systems, the coding information within a gene normally is continuous. However, in eucaryotic organisms, many genes contain coding information (exons) interrupted periodically by noncoding sequences (introns). The introns must be spliced out of the mRNA before the protein is made. This affords eucaryotes the ability to cut and paste mRNA molecules so that they can encode more than one polypeptide, a process known as **alternative splicing.** An interesting exception to this rule is eucaryotic histone genes, which lack introns. Because procaryotic and viral systems are the best characterized, the more detailed description of gene structure that follows focuses on E. coli genes.

Protein-Coding Genes

For genetic information in DNA to be used, it must first be transcribed to form an RNA molecule. The RNA product of a gene that codes for a protein is messenger RNA (mRNA). Although DNA is double stranded, only one strand of a gene directs RNA

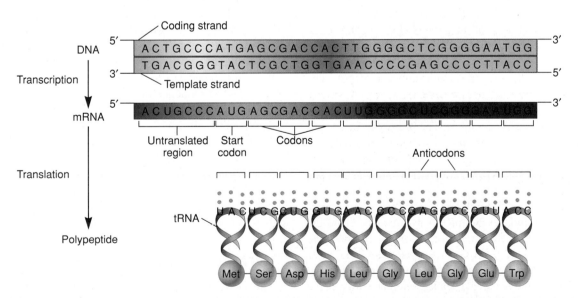

Figure 12.16 Reading Frame. During transcription, an mRNA complementary to the template strand of DNA is synthesized. The nucleotides in the mRNA are organized into groups of three, each group being a codon. The first codon translated into protein is the start codon. It establishes the reading frame and therefore the sequence of amino acids in the polypeptide chain that is made from the mRNA.

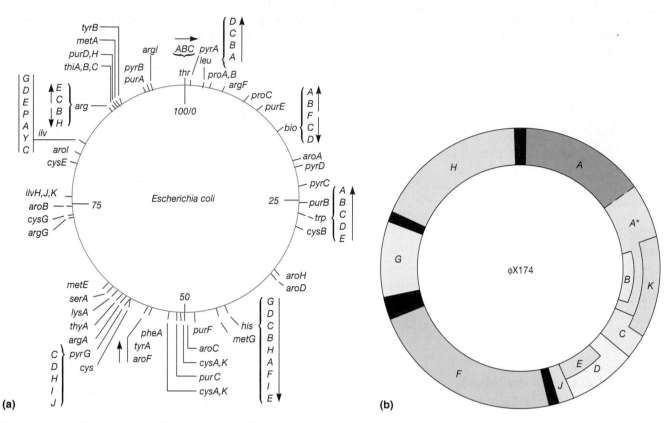

Figure 12.17   Chromosomal Organization in *Bacteria* and Viruses.   (a) Simplified genetic map of *E. coli*. The *E. coli* map is divided into 100 minutes. (b) The map of phage φX174 shows the overlap of gene *B* with *A*, *K* with *A* and *C*, and *E* with *D*. The solid regions are spaces lying between genes. Protein A* consists of the last part of protein A and arises from reinitiation of translation of gene *A* mRNA.

synthesis. This strand is called the **template strand,** and the complementary DNA strand is known as the coding strand because it is the same nucleotide sequence as the mRNA, except in DNA bases (**figure 12.18**). Because the mRNA is made from the 5′ to the 3′ end, the polarity of the DNA template strand is 3′ to 5′. Therefore the beginning of the gene is at the 3′ end of the template strand. An important site called the **promoter** is located at the start of the gene. The promoter is a recognition/binding site for RNA polymerase, the enzyme that synthesizes RNA. The promoter is neither transcribed nor translated; it functions strictly to orient RNA polymerase a specific distance from the first DNA nucleotide that will serve as a template. The promoter thus specifies which strand is to be transcribed and where transcription should begin. As we discuss in chapter 13, the sequences near the promoter are very important in regulating when and where a gene is transcribed.

The transcription start site (labeled +1 in figure 12.18) represents the first nucleotide in the mRNA synthesized from the gene. However, the initially transcribed portion of the gene does not necessarily code for amino acids. Instead it is a **leader sequence** that is transcribed into mRNA but is not translated into amino acids. In *Bacteria,* the leader sequence includes a region called the **Shine-Dalgarno sequence** that is important in the initiation of translation. The leader sometimes is also involved in regulation of transcription and translation. ▸▸ *Regulation of*

transcription elongation (section 13.3); Regulation at the level of translation (section 13.4)

Immediately next to (and downstream of) the leader is the most important part of the gene, the **coding region** (figure 12.18). The coding region typically begins with the template DNA sequence 3′-TAC-5′. This is transcribed into the codon 5′-AUG-3′, which in *Bacteria* codes for *N*-formylmethionine, a modified amino acid used to initiate protein synthesis. The remainder of the coding region is transcribed into a sequence of codons that specifies the sequence of amino acids for that particular protein. The coding region ends with a sequence that when transcribed is a **stop codon.** It signals the end of the protein and stops the ribosome during translation. The stop codon is immediately followed by the **trailer sequence,** which is transcribed but not translated. Just beyond the trailer (and sometimes slightly overlapping it) is a sequence that signals the RNA polymerase to stop transcription. The **terminator** is involved in dislodging the RNA polymerase from the template DNA.

tRNA and rRNA Genes

The DNA segments that code for tRNA and rRNA also are considered genes, although they give rise to RNA rather than protein. In *E. coli,* the genes for tRNA are fairly typical, consisting of a promoter and transcribed leader and trailer

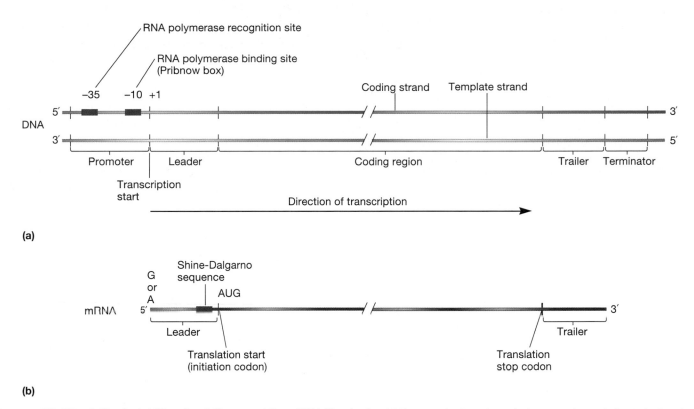

Figure 12.18 A Bacterial Structural Gene and Its mRNA Product. (a) The organization of a typical structural gene in bacteria. Leader and trailer sequences are included even though some genes lack one or both. Transcription begins at the +1 position in DNA and proceeds to the right, as shown. The template is read in the 3′ to 5′ direction. (b) Messenger RNA product of the gene shown in part a. The first nucleotide incorporated into mRNA is usually GMP or AMP. Translation of the mRNA begins with the AUG initiation codon. Regulatory sites are not shown.

sequences that are removed during the process of tRNA maturation (**figure 12.19a**). Genes coding for tRNA may code for more than a single tRNA molecule or type of tRNA. The segments coding for tRNAs are separated by short spacer sequences that are removed after transcription by ribonucleases—enzymes that degrade RNA; at least one of the ribonucleases is a ribozyme. In addition, many bacterial and archaeal tRNA genes contain introns that must be removed during tRNA maturation. << *Ribozymes (section 9.8)*

The genes for rRNA also are similar in organization to genes coding for proteins because they have promoters, trailers, and terminators (figure 12.19b). Interestingly all the rRNAs are transcribed as a single, large precursor molecule that is cut up by ribonucleases after transcription to yield the final rRNA products. *E. coli* pre-rRNA spacer and trailer regions even contain tRNA genes. Thus the synthesis of tRNA and rRNA involves posttranscriptional modification, a relatively rare process in procaryotes.

1. Define or describe the following: gene, template and coding strands, promoter, leader, coding region, reading frame, trailer, and terminator.

2. How do the genes of procaryotes and eucaryotes usually differ from each other?

3. Briefly discuss the general organization of tRNA and rRNA genes. How does their expression differ from that of structural genes with respect to posttranscriptional modification of the gene product?

12.5 TRANSCRIPTION

Synthesis of RNA under the direction of DNA is called transcription, and the RNA product has a sequence complementary to the DNA template directing its synthesis. Although adenine directs the incorporation of thymine during DNA replication, it usually codes for uracil during RNA synthesis. Transcription generates three kinds of RNA. **Transfer RNA (tRNA)** carries amino acids during protein synthesis, and **ribosomal RNA (rRNA)** molecules are components of ribosomes. **Messenger RNA (mRNA)** bears the message for protein synthesis. In *Bacteria* and *Archaea*, mRNA often bears coding information transcribed from adjacent genes. Therefore it is said to be polycistronic (**figure 12.20a**). Eucaryotic mRNAs, on the other hand, are usually monocistronic (figure 12.20b), containing

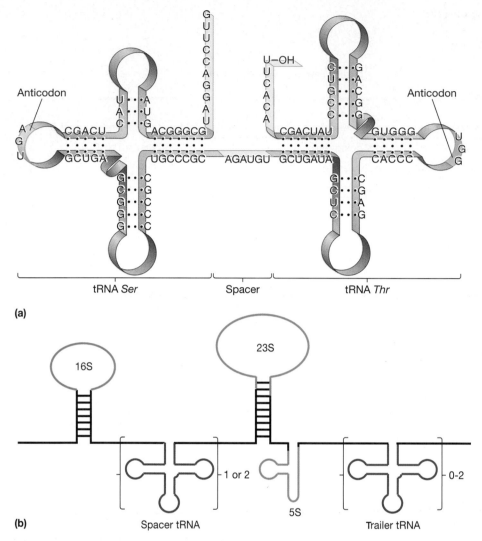

(a)

(b)   Spacer tRNA   5S   Trailer tRNA

Figure 12.19   tRNA and rRNA Genes.   (a) A tRNA precursor from *E. coli* that contains two tRNA molecules. The spacer and extra nucleotides at both ends are removed during processing. (b) The *E. coli* ribosomal RNA gene codes for a large transcription product that is cleaved into three rRNAs and one to three tRNAs. The 16S, 23S, and 5S rRNA segments are represented by blue lines, and tRNA sequences are placed in brackets. The seven copies of this gene vary in the number and kind of tRNA sequences.

Like DNA synthesis, RNA synthesis proceeds in a 5′ to 3′ direction with new nucleotides being added to the 3′ end of the growing chain at a rate of about 40 nucleotides per second at 37°C (**figure 12.21**). In both DNA and RNA polymerase reactions, pyrophosphate (PP$_i$) is produced. It is hydrolyzed to orthophosphate in a reaction catalyzed by the enzyme pyrophosphatase. Hydrolysis of pyrophosphate makes DNA and RNA synthesis irreversible. If the pyrophosphate level were too high, DNA and RNA would be degraded by a reversal of the polymerase reactions.

Most bacterial RNA polymerases contain five types of polypeptide chains: α, β, β′, ω, and σ (**figure 12.22**). The **RNA polymerase core enzyme** is composed of five chains (α$_2$, β, β′, and ω) and catalyzes RNA synthesis. The **sigma factor** (σ) has no catalytic activity but helps the core enzyme recognize the promoter. When sigma is bound to core enzyme, the six-subunit complex is termed **RNA polymerase holoenzyme.** Only holoenzyme can begin transcription, but core enzyme completes RNA synthesis once it has been initiated. The precise functions of the α, β, β′, and ω polypeptides are not yet clear. The α subunits seem to be involved in the assembly of the core enzyme, recognition of promoters, and interaction with some regulatory factors. The binding site for DNA is on β′, and the ω subunit seems to be involved in stabilizing the conformation of the β′ subunit. The β subunit binds ribonucleotide substrates.

information for a single polypeptide. The synthesis of bacterial mRNA is described first.

Transcription in *Bacteria*

RNA is synthesized under the direction of DNA by the enzyme **RNA polymerase.** The reaction is quite similar to that catalyzed by DNA polymerase (figure 12.9). ATP, GTP, CTP, and UTP are used to produce RNA complementary to the DNA template. Recall that these nucleotides contain ribose rather than deoxyribose (figure 12.2).

$$n[\text{ATP, GTP, CTP, UTP}] \xrightarrow[\text{DNA template}]{\text{RNA polymerase}} \text{RNA} + n\text{PP}_i$$

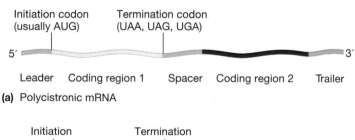

(a) Polycistronic mRNA

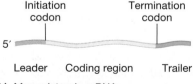

(b) Monocistronic mRNA

Figure 12.20   Polycistronic and Monocistronic mRNAs. Polycistronic mRNAs are commonly observed in *Bacteria* and *Archaea*. Eucaryotic genes give rise to monocistronic mRNAs.

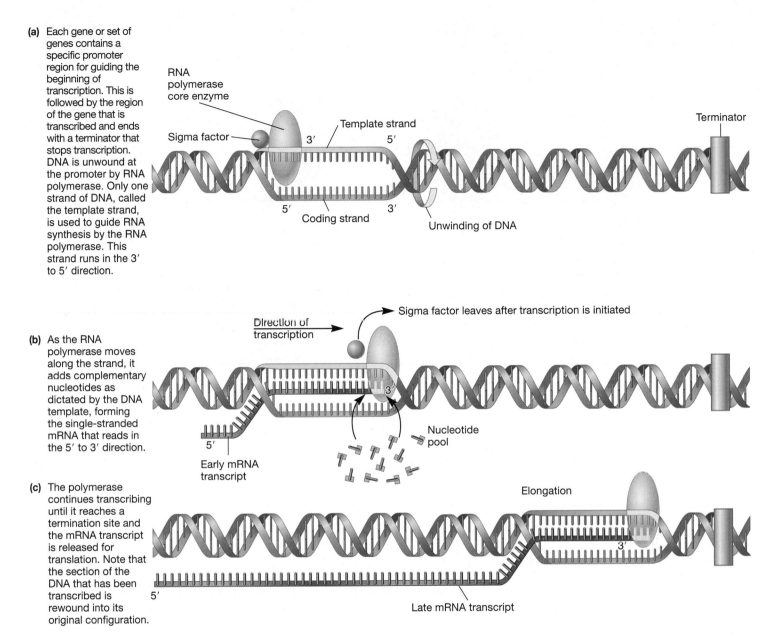

(a) Each gene or set of genes contains a specific promoter region for guiding the beginning of transcription. This is followed by the region of the gene that is transcribed and ends with a terminator that stops transcription. DNA is unwound at the promoter by RNA polymerase. Only one strand of DNA, called the template strand, is used to guide RNA synthesis by the RNA polymerase. This strand runs in the 3′ to 5′ direction.

RNA polymerase core enzyme

Sigma factor

Template strand

Coding strand

Unwinding of DNA

Terminator

3′ 5′

5′ 3′

(b) As the RNA polymerase moves along the strand, it adds complementary nucleotides as dictated by the DNA template, forming the single-stranded mRNA that reads in the 5′ to 3′ direction.

Direction of transcription

Sigma factor leaves after transcription is initiated

Nucleotide pool

5′

Early mRNA transcript

3′

(c) The polymerase continues transcribing until it reaches a termination site and the mRNA transcript is released for translation. Note that the section of the DNA that has been transcribed is rewound into its original configuration.

Elongation

3′

5′

Late mRNA transcript

Figure 12.21 Transcription in *Bacteria*.

Transcription involves three separate processes: initiation, elongation, and termination. Only a relatively short segment of DNA is transcribed (unlike replication in which the entire chromosome must be copied), and initiation begins when RNA polymerase holoenzyme binds to the promoter for the gene. Bacterial promoters have two characteristic features: a sequence of six bases (often TTGACA) about 35 base pairs before the transcription starting point and a TATAAT sequence called the **Pribnow box,** usually about 10 base pairs upstream of the transcriptional start site (**figure 12.23**, also figure 12.18). These regions are called the −35 and −10 sites, respectively, because these are their distances before or "upstream" of the first nucleotide to be transcribed (i.e., the +1 site). Because RNA polymerase holoenzyme recognizes the specific sequences at the −10 and −35 sites of all promoters, the nucleotide sequences must be similar. Thus they are called **consensus sequences.**

Once bound to the promoter, RNA polymerase is able to unwind the DNA without the aid of helicases (**figure 12.24**). The −10 site is rich in adenines and thymines, making it easier to break the hydrogen bonds that keep the DNA double stranded; when the DNA is unwound at this region, it is called an open complex. A region of unwound DNA equivalent to about 16–20 base pairs becomes the "transcription bubble," which moves with the RNA polymerase as it transcribes mRNA from the template DNA

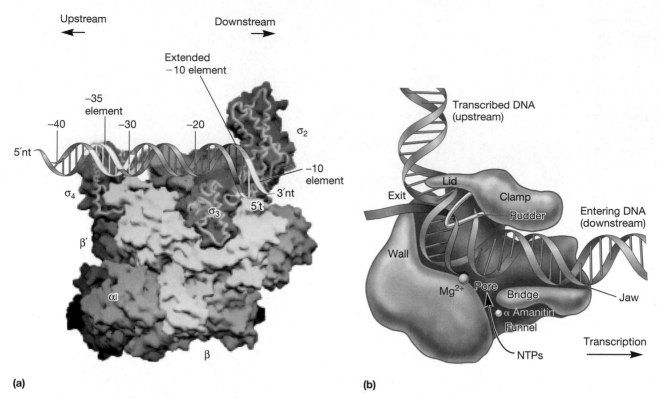

Figure 12.22 RNA Polymerase Structure. (a) The holoenzyme-DNA complex of the bacterium *Thermus aquaticus*. Protein surfaces that contact the DNA are in green and are located on the σ factor. The −10 and −35 elements in the promoter are in yellow. The internal active site is covered by the β subunit in this view. (b) A cutaway side view of yeast RNA polymerase II transcribing complex with the pathway of the nucleic acids and some of the more important parts shown. The enzyme is moving from left to right. A protein "wall" forces the DNA into a right-angle turn and aids in the attachment of nucleoside triphosphates to the growing 3′ end of the RNA. The newly synthesized RNA (red) is separated from the DNA template strand and exits beneath the rudder and lid of the RNA polymerase. The binding site of the inhibitor α-amanitin also is shown.

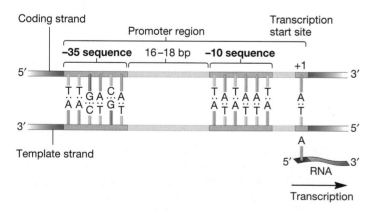

Figure 12.23 The Conventional Numbering System of Bacterial Promoters. The first nucleotide that acts as a template for transcription is designated +1. The numbering of nucleotides to the left of this spot is in a negative direction, while the numbering to the right is in a positive direction. For example, the nucleotide that is immediately to the left of the +1 nucleotide is numbered −1, and the nucleotide to the right of the +1 nucleotide is numbered +2. There is no zero nucleotide in this numbering system. In many bacterial promoters, sequence elements at the −35 and −10 regions play a key role in promoting transcription.

strand during elongation (**figure 12.25**). Within the transcription bubble, a temporary RNA:DNA hybrid is formed. As RNA polymerase progresses in the 3′ to 5′ direction along the DNA template, the sigma factor soon dissociates from core RNA polymerase and can aid another core enzyme initiate transcription. Core RNA polymerase continues to synthesize mRNA in the 5′ to 3′ direction, making it complementary and antiparallel to the template DNA. As elongation of the mRNA continues, single-stranded mRNA is released and the two strands of DNA behind the transcription bubble resume their double helical structure. As shown in figure 12.21, RNA polymerase is a remarkable enzyme capable of several activities, including unwinding the DNA, moving along the template, and synthesizing RNA.

Termination of transcription occurs when the core RNA polymerase dissociates from the template DNA. This is brought about by the terminator. Sequences just before procaryotic terminators often contain nucleotides that, when transcribed into RNA, form hydrogen bonds within the single-stranded RNA. This intrastrand base pairing creates a hairpin-shaped loop-and-stem structure. This structure appears to cause RNA polymerase to stop transcribing DNA.

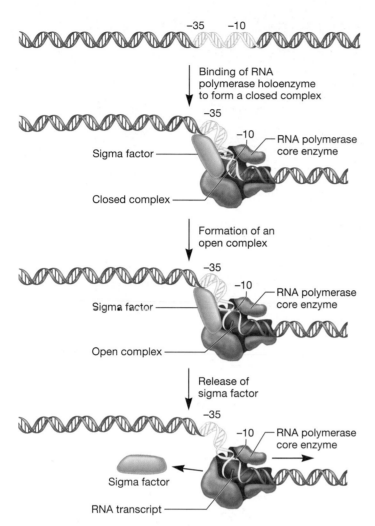

Figure 12.24 **The Initiation of Transcription in *Bacteria*.** The sigma factor of the RNA polymerase holoenzyme is responsible for positioning the core enzyme properly at the promoter. Sigma factor recognizes two regions in the promoter, one centered at –35 and the other centered at –10. Once positioned properly, the DNA at the –10 region unwinds to form an open complex. The sigma factor dissociates from the core enzyme after transcription is initiated.

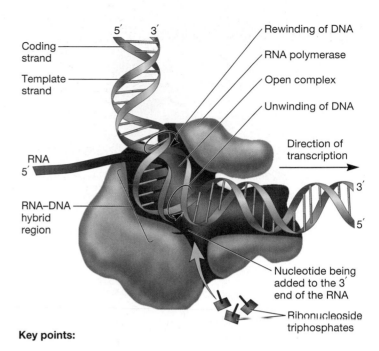

Key points:

- RNA polymerase slides along the DNA, creating an open complex as it moves.

- The template strand is used to make a complementary copy of RNA as an RNA–DNA hybrid.

- The RNA is synthesized in a 5′ to 3′ direction using ribonucleoside triphosphates as precursors. Pyrophosphate is released (not shown).

- The complementarity rule is the same as the AT/GC rule except that U is substituted for T in the RNA.

Figure 12.25 **The "Transcription Bubble."**

Transcription in Eucaryotes

Transcriptional processes in eucaryotic microorganisms (and in other eucaryotic cells) differ in several ways from bacterial transcription. There are three major RNA polymerases, not one as in *Bacteria*. RNA polymerase II is responsible for mRNA synthesis. Polymerases I and III synthesize rRNA and tRNA, respectively. RNA polymerase II is a large aggregate, at least 500,000 daltons in size, with about 10 or more subunits (figure 12.22*b*). The entering DNA is held in a clamp that closes down on it. A magnesium ion is located at the active site and the 9 base pair DNA-RNA hybrid in the transcription bubble is bound in a cleft formed by the two large polymerase subunits. The newly synthesized RNA exits the polymerase beneath the rudder and lid regions. The substrate nucleoside triphosphates probably reach the active site through a pore in the complex. Unlike bacterial polymerase, RNA polymerase II requires several transcription factors to recognize its promoters (**figure 12.28**). Eucaryotic promoters also differ from those in *Bacteria*. They have combinations of several elements. Three of the most common are the TATA box (located about 30 base pairs upstream of the start point), and the GC and CAAT boxes (located between 50 to 100 base pairs upstream of the start site). The general transcription factor TFIID plays an important role in transcription initiation in

There are two kinds of terminators. The first type causes intrinsic or rho-independent termination (**figure 12.26**). It features a stretch of about six uridine residues that follow the mRNA hairpin. Once the RNA polymerase pauses after formation of the hairpin loop in the mRNA, the A-U base pairs in the uracil-rich region of the terminator are too weak to hold the RNA:DNA duplex together and RNA polymerase falls off. The second kind of terminator lacks a poly-U region and often the hairpin. Instead it requires the aid of a protein called rho factor (ρ) and causes rho-dependent termination. It is thought that rho binds to mRNA and moves along the molecule until it reaches the RNA polymerase halted at the terminator (**figure 12.27**). The rho factor, which has hybrid RNA:DNA helicase activity, then causes RNA polymerase to dissociate from the mRNA, probably by unwinding the mRNA-DNA complex.

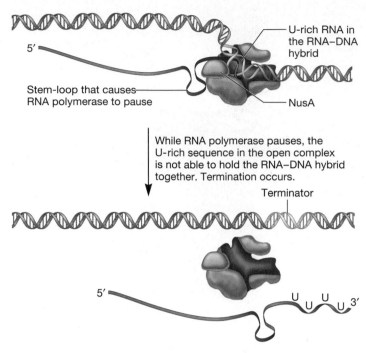

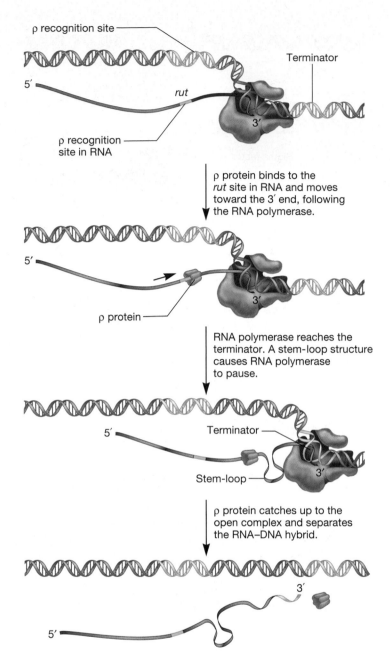

Figure 12.26   Intrinsic Termination of Transcription.   This type of terminator contains a U-rich sequence downstream from a stretch of nucleotides that can form a stem-loop structure. Formation of the stem loop in the newly synthesized RNA causes RNA polymerase to pause. This pausing is stabilized by the NusA protein. The U-A bonds in the uracil-rich region are not strong enough to hold the RNA and DNA together. Therefore, the RNA, DNA, and RNA polymerase dissociate and transcription stops.

eucaryotes (figure 12.28). This multiprotein complex contains the TATA-binding protein (TBP). TBP has been shown to bend the DNA sharply on attachment, making the DNA more accessible to other transcription factors. A variety of other general transcription factors, promoter specific factors, and promoter elements have been discovered in different eucaryotic cells. Each eucaryotic gene seems to be regulated differently, and more research will be required to understand fully the regulation of eucaryotic gene transcription.

Unlike bacterial mRNA, eucaryotic mRNA arises from **post-transcriptional modification** of large RNA precursors, about 5,000 to 50,000 nucleotides long, sometimes called pre-mRNA (**figure 12.29**). As pre-mRNA is synthesized, a 5′ cap is added. The 5′ cap is the unusual nucleotide 7-methylguanosine. It is attached to the 5′-hydroxyl of the pre-mRNA by a triphosphate linkage. After synthesis is completed, the pre-mRNA is modified by the addition of a 3′ poly-A tail, about 200 nucleotides long. The functions of the 5′ cap and poly-A tail are not completely clear. The 5′ cap on eucaryotic mRNA may promote the initial binding of ribosomes to the mRNA. It also may protect the mRNA from enzymatic attack. Poly-A protects mRNA from rapid enzymatic degradation. The poly-A tail must be shortened to about 10 nucleotides before mRNA can be degraded. Poly-A also seems to aid in mRNA translation.

As noted earlier, many eucaryotic genes contain **exons** (expressed sequences), which are found in mRNA molecules and are translated into protein, and **introns** (intervening sequences), which are

Figure 12.27   Rho-Factor (ρ)-Dependent Termination of Transcription.   The *rut* site stands for *rho utilization* site.

transcribed but not translated. Introns are removed from the pre-mRNA transcript by a process called RNA splicing (**figure 12.30**). Splicing of pre-mRNA occurs in a large complex called a **spliceo-some.** It is composed of the pre-mRNA, several small nuclear ribonucleoproteins (snRNPs—pronounced "snurps"), and several non-snRNP splicing factors. The snRNPs consist of small nuclear RNA (snRNA) molecules (about 60 to 300 nucleotides long) associated with proteins. Some snRNPs recognize and bind exon-intron junctions. Exon-intron junctions have a GU sequence at the intron's 5′ boundary and an AG sequence at its 3′ end. Thus the intron's borders are clearly marked for accurate removal.

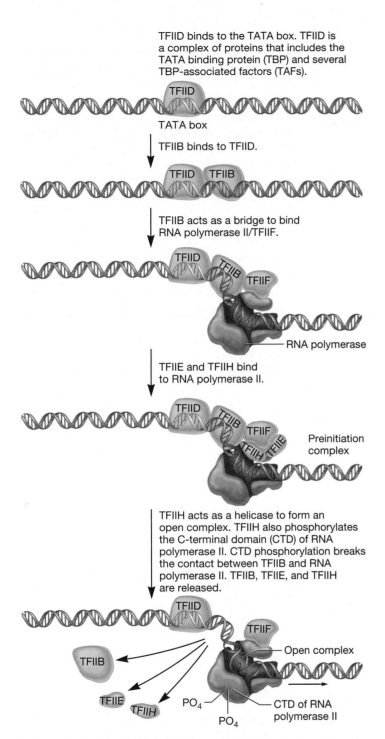

TFIID binds to the TATA box. TFIID is a complex of proteins that includes the TATA binding protein (TBP) and several TBP-associated factors (TAFs).

TFIID

TATA box

TFIIB binds to TFIID.

TFIID TFIIB

TFIIB acts as a bridge to bind RNA polymerase II/TFIIF.

TFIID TFIIB TFIIF

RNA polymerase

TFIIE and TFIIH bind to RNA polymerase II.

TFIID TFIIB TFIIF TFIIH TFIIE

Preinitiation complex

TFIIH acts as a helicase to form an open complex. TFIIH also phosphorylates the C-terminal domain (CTD) of RNA polymerase II. CTD phosphorylation breaks the contact between TFIIB and RNA polymerase II. TFIIB, TFIIE, and TFIIH are released.

TFIID TFIIF

TFIIB

TFIIE TFIIH PO$_4$ PO$_4$

Open complex

CTD of RNA polymerase II

Figure 12.28 Initiation of Transcription in Eucaryotes. The TATA box is a major component of eucaryotic promoters. *TATA binding protein (TBP)*, which is a component of a complex of proteins called TFIID (*transcription factor IID*), binds the TATA box. Note that numerous other transcription factors are required for initiation of transcription, unlike *Bacteria*, where only the sigma factor is needed. Initiation of transcription in *Archaea* is similar to that seen in eucaryotes.

Sometimes a pre-mRNA is spliced so that different patterns of exons remain. This alternative splicing allows a single gene to code for more than one protein. The splice pattern determines which protein is synthesized. Splice patterns can be cell-type specific or

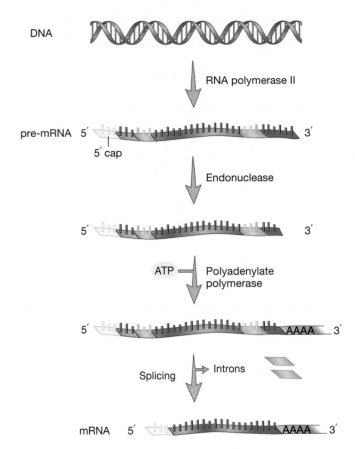

DNA

RNA polymerase II

pre-mRNA 5′ 3′

5′ cap

Endonuclease

5′ 3′

ATP Polyadenylate polymerase

5′ AAAA 3′

Splicing Introns

mRNA 5′ AAAA 3′

Figure 12.29 Eucaryotic mRNA Synthesis. The production of eucaryotic messenger RNA. The 5′ cap is added shortly after synthesis of the mRNA begins.

determined by the needs of the cell. The importance of alternative splicing in multicellular eucaryotes was clearly demonstrated when it was discovered that the human genome has only about 20,000 genes rather than the anticipated 100,000. It is thought that alternative splicing is one mechanism by which human cells produce their vast array of proteins using fewer genes.

Transcription in *Archaea*

Transcription in the *Archaea* is similar to and distinct from what is observed in *Bacteria* and eucaryotes. Each archaeon has a single RNA polymerase responsible for transcribing all genes in the cell (as in *Bacteria*). However, the RNA polymerase is larger and contains more subunits, many of which are similar to subunits in RNA polymerase II of eucaryotes. The promoters of archaeal genes are similar to those of eucaryotes in having a TATA box; binding of the archaeal RNA polymerase to its promoter requires a TATA-binding protein, just as in eucaryotes. Like the eucaryotic counterpart, the archaeal RNA polymerase also needs several additional transcription factors to function properly. Furthermore, some archaeal genes have introns, which must be removed by posttranscriptional processing. Finally, the mRNA molecules produced by transcription in *Archaea* are usually polycistronic, as in *Bacteria*. This intriguing mixture of bacterial and eucaryotic features in the *Archaea* fuels a

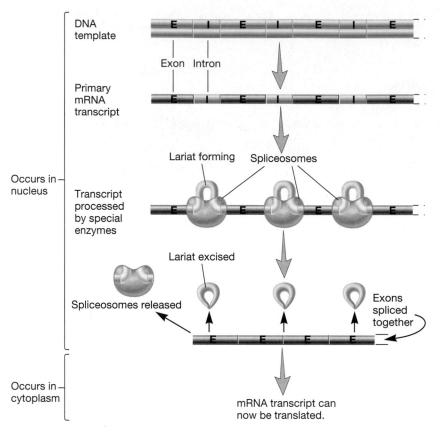

Figure 12.30   Splicing of Eucaryotic mRNA Molecules.

called translation because it is a decoding process. The information encoded in the language of nucleic acids must be rewritten in the language of proteins. Therefore, before we discuss protein synthesis, we examine the nature of the genetic code.

The genetic code, presented in RNA form, is summarized in **table 12.2**. Note that there is **code degeneracy.** That is, there are up to six different codons for a given amino acid. Only 61 codons, the sense codons, direct amino acid incorporation into protein. The remaining three codons (UGA, UAG, and UAA) are involved in the termination of translation and are called stop or **nonsense codons.** Despite the existence of 61 sense codons, there are not 61 different tRNAs, one for each codon. The 5′ base in the anticodon can vary, but generally, if the bases in the second and third anticodon positions complement the first two bases of the mRNA codon, an aminoacyl-tRNA with the proper amino acid will bind to the mRNA-ribosome complex. This pattern is evident on inspection of changes in the amino acid specified with variation in the third position (table 12.2). This somewhat loose base pairing is known as **wobble,** and it relieves cells of the need to synthesize so many tRNAs (**figure 12.31**). Wobble also decreases the effects of mutations. **>>** *Mutations and their chemical basis (section 14.1)*

great deal of speculation about the evolution of the three domains of life. **>>** *Microbial evolution (section 17.1); Introduction to the* Archaea: *Genetics and molecular biology (section 18.1)*

1. Define the following: polycistronic mRNA, RNA polymerase core enzyme, sigma factor, RNA polymerase holoenzyme, and rho factor.

2. Define or describe posttranscriptional modification, pre-mRNA, 3′ poly-A sequence, 5′ cap, split or interrupted genes, exon, intron, RNA splicing, snRNA, snRNP, and spliceosome.

3. How does bacterial RNA polymerase differ from bacterial DNA polymerase? How does RNA polymerase know when to begin and end transcription?

4. How do bacterial RNA polymerases and promoters differ from those of *Archaea* and eucaryotes?

12.6   THE GENETIC CODE

The final step in the expression of genes that encode proteins is translation. The mRNA nucleotide sequence is translated into the amino acid sequence of a polypeptide chain. Protein synthesis is

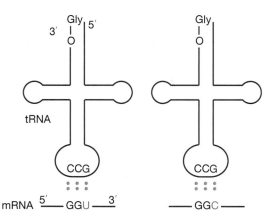

(a) Base pairing of one glycine tRNA with two codons due to wobble

Glycine mRNA codons: GGU, GGC, GGA, GGG (5′ ⟶ 3′)

Glycine tRNA anticodons: CCG, CCU, CCC (3′ ⟶ 5′)

(b) Glycine codons and anticodons

Figure 12.31   Wobble and Coding.   The use of wobble in coding for the amino acid glycine. (a) Because of wobble, G in the 5′ position of the anticodon can pair with either C or U in the 3′ position of the codon. Thus two codons can be recognized by the same tRNA. (b) Because of wobble, only three tRNA anticodons are needed to translate the four glycine (Gly) codons.

Table 12.2	The Genetic Code

<table>
<tr><th rowspan="2"></th><th colspan="8">Second Position</th><th rowspan="2"></th></tr>
<tr><th colspan="2">U</th><th colspan="2">C</th><th colspan="2">A</th><th colspan="2">G</th></tr>
<tr>
<td rowspan="4">U</td>
<td>UUU</td><td rowspan="2">Phe</td>
<td>UCU</td><td rowspan="4">Ser</td>
<td>UAU</td><td rowspan="2">Tyr</td>
<td>UGU</td><td rowspan="2">Cys</td>
<td rowspan="4"></td>
</tr>
<tr><td>U</td></tr>
</table>

First Position (5′ End)[a]	Second Position U	Second Position C	Second Position A	Second Position G	Third Position (3′ End)
U	UUU, UUC → Phe; UUA, UUG → Leu	UCU, UCC, UCA, UCG → Ser	UAU, UAC → Tyr; UAA, UAG → STOP	UGU, UGC → Cys; UGA → STOP; UGG → Trp	U, C, A, G
C	CUU, CUC, CUA, CUG → Leu	CCU, CCC, CCA, CCG → Pro	CAU, CAC → His; CAA, CAG → Gin	CGU, CGC, CGA, CGG → Arg	U, C, A, G
A	AUU, AUC, AUA → Ile; AUG → Met	ACU, ACC, ACA, ACG → Thr	AAU, AAC → Asn; AAA, AAG → Lys	AGU, AGC → Ser; AGA, AGG → Arg	U, C, A, G
G	GUU, GUC, GUA, GUG → Val	GCU, GCC, GCA, GCG → Ala	GAU, GAC → Asp; GAA, GAG → Glu	GGU, GGC, GGA, GGG → Gly	U, C, A, G

[a]The code is presented in the RNA form. Codons run in the 5′ to 3′ direction.

12.7 TRANSLATION

Translation involves decoding mRNA and covalently linking amino acids together to form a polypeptide. Just as DNA and RNA synthesis proceed in one direction (5′ to 3′), so too does protein synthesis. Polypeptide synthesis begins with the amino acid at the end of the chain with a free amino group (the N-terminal) and moves in the C-terminal direction. The ribosome is the site of protein synthesis. Protein synthesis is not only quite accurate but also very rapid. In *E. coli,* synthesis occurs at a rate of at least 900 residues per minute; eucaryotic translation is slower, about 100 residues per minute. >> *Proteins (appendix I)*

Cells that grow quickly must use each mRNA with great efficiency to synthesize proteins at a sufficiently rapid rate. The two subunits of the ribosome (the 50S subunit and the 30S subunit in *Bacteria* and *Archaea;* 60S and 40S in eucaryotes) are free in the cytoplasm if protein is not being synthesized. They come together to form the complete ribosome only when translation occurs. Frequently mRNAs are simultaneously complexed with several ribosomes, each ribosome reading the mRNA message and synthesizing a polypeptide. At maximal rates of mRNA use, there may be a ribosome every 80 nucleotides along the mRNA or as many as 20 ribosomes simultaneously reading an mRNA that codes for a 50,000 dalton polypeptide. A complex of mRNA with several ribosomes is called a **polyribosome** or polysome (**figure 12.32**). Polysomes are present in both procaryotes and eucaryotes. *Bacteria* and *Archaea* can further increase the efficiency of gene expression by coupling transcription and translation (figure 12.32b). While RNA polymerase is synthesizing an mRNA, ribosomes can already be attached to the mRNA so that transcription and translation occur simultaneously. Coupled transcription and translation is possible in procaryotes because a nuclear envelope does not separate the translation machinery from DNA as it does in eucaryotes.

Transfer RNA and Amino Acid Activation

The first stage of protein synthesis is **amino acid activation,** a process in which amino acids are attached to tRNA molecules. Transfer RNA molecules are about 70 to 95 nucleotides long and possess several characteristic structural features. These features

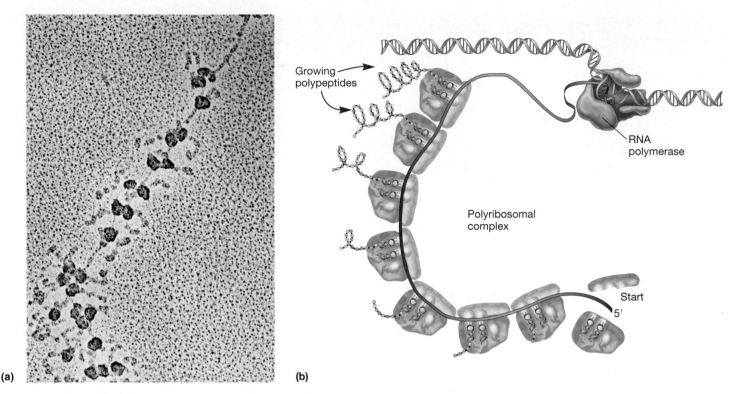

Figure 12.32   Coupled Transcription and Translation in Procaryotes.   (a) A transmission electron micrograph showing coupled transcription and translation. (b) A schematic representation of coupled transcription and translation. As the DNA is transcribed, ribosomes bind the free 5′ end of the mRNA. Thus translation is started before transcription is completed. Note that there are multiple ribosomes bound to the mRNA, forming a polyribosome.

become apparent when the tRNA is folded in such a way to maximize intramolecular base pairing. When represented two-dimensionally, intramolecular base pairing forms a cloverleaf conformation (**figure 12.33***a*). In cells, tRNAs are folded into an L-shaped, three-dimensional structure (figure 12.33*b*). One important feature of tRNAs is the acceptor stem, which holds the activated amino acid. The 3′ end of all tRNAs has the same —C—C—A sequence, and in all cases the amino acid is attached to the terminal adenylic acid. Another important feature of the tRNA is the **anticodon** triplet. The anticodon is complementary to the mRNA codon triplet and is located on the anticodon arm (figure 12.33*a*). Two other large arms are readily observed in the cloverleaf representation: the D or DHU arm has the unusual pyrimidine nucleoside dihydrouridine; and the T or TΨC arm has ribothymidine (T) and pseudouridine (Ψ), both of which are unique to tRNA. Finally, the cloverleaf has a variable arm whose length changes with the overall length of the tRNA; the other arms are fairly constant in size.

Amino acids are activated for protein synthesis through a reaction catalyzed by **aminoacyl-tRNA synthetases** (**figure 12.34**). Just as is true of DNA and RNA synthesis, the reaction is driven to completion when the pyrophosphate product is hydrolyzed to two orthophosphates. The amino acid is attached to the 3′-hydroxyl of the terminal adenylic acid on the tRNA by a high-energy bond and is readily attached to the growing peptide chain. The amino acid is said to be activated because it is ready to be donated to a growing polypeptide chain.

There are at least 20 aminoacyl-tRNA synthetases, each specific for a single amino acid and its tRNAs (cognate tRNAs). It is critical that each tRNA attach the corresponding amino acid because if an incorrect amino acid is attached to a tRNA, it will be incorporated into a polypeptide in place of the correct amino acid. The protein synthetic machinery recognizes only the anticodon of the aminoacyl-tRNA and cannot tell whether the correct amino acid is attached. Some aminoacyl-tRNA synthetases proofread just like DNA polymerases do. If the wrong amino acid is attached to tRNA, the enzyme hydrolyzes the amino acid from the tRNA rather than release the incorrect product.

Ribosome Structure

Protein synthesis takes place on ribosomes that serve as workbenches, with mRNA acting as the blueprint. Procaryotic ribosomes have a sedimentation value of 70S and a mass of 2.8 million daltons. A rapidly growing *E. coli* cell may have as many as 15,000 to 20,000 ribosomes, about 15% of the cell mass.

Each of the two subunits of the procaryotic ribosome is an extraordinarily complex structure constructed from one or two rRNA molecules and many polypeptides (**figure 12.35**). The region of the ribosome directly responsible for translation is

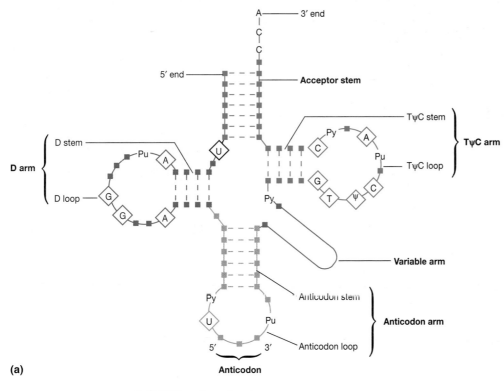

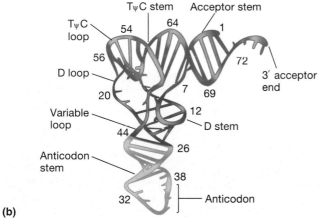

Figure 12.33   tRNA Structure.   (a) The two-dimensional cloverleaf structure for tRNA in procaryotes and eucaryotes. Bases found in all tRNAs are in diamonds; purine and pyrimidine positions in all tRNAs are labeled Pu and Py, respectively. (b) The three-dimensional structure of tRNA. The various regions are distinguished with different colors.

called the translational domain (figure 12.35*d*). Both subunits contribute to this domain. The growing peptide chain emerges from the large subunit at the exit domain. This is located on the side of the subunit opposite the central protuberance (figure 12.35*b*). X-ray diffraction studies have now confirmed this general picture of ribosome structure (figure 12.35*e–g*).

Ribosomal RNA is thought to have three roles. It obviously contributes to ribosome structure. The 16S rRNA of the 30S subunit is needed for the initiation of protein synthesis in *Bacteria*. The 3' end of the 16S rRNA binds to a site on the mRNA called the Shine-Dalgarno sequence, which is located in the **ribosome-**

binding site (RBS). This helps position the mRNA on the ribosome. The 16S rRNA also binds a protein needed to initiate translation (initiation factor 3) and the 3' CCA end of aminoacyl-tRNA. Finally, it appears that the 23S rRNA has a catalytic role in protein synthesis.

Initiation of Protein Synthesis

Like transcription, protein synthesis is divided into three stages: initiation, elongation, and termination. The initiation of protein synthesis is very elaborate. Apparently the complexity is necessary to ensure that the ribosome does not start synthesizing a polypeptide chain in the middle of a gene   a disastrous error.

In the initiation stage, *E. coli* and most other bacteria begin protein synthesis with a modified aminoacyl-tRNA,   *N*-formylmethionyl-tRNA^{fMet} **(figure 12.36).** Because the α-amino is

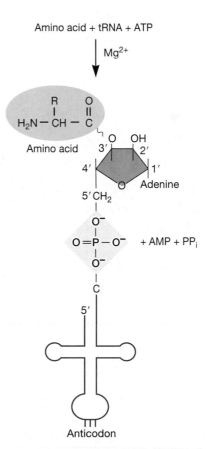

Figure 12.34   Aminoacyl-tRNA Synthetase Reaction.   The amino acid is attached by the appropriate aminoacyl-tRNA synthetase to the 3'-hydroxyl of adenylic acid by a high-energy bond (red).

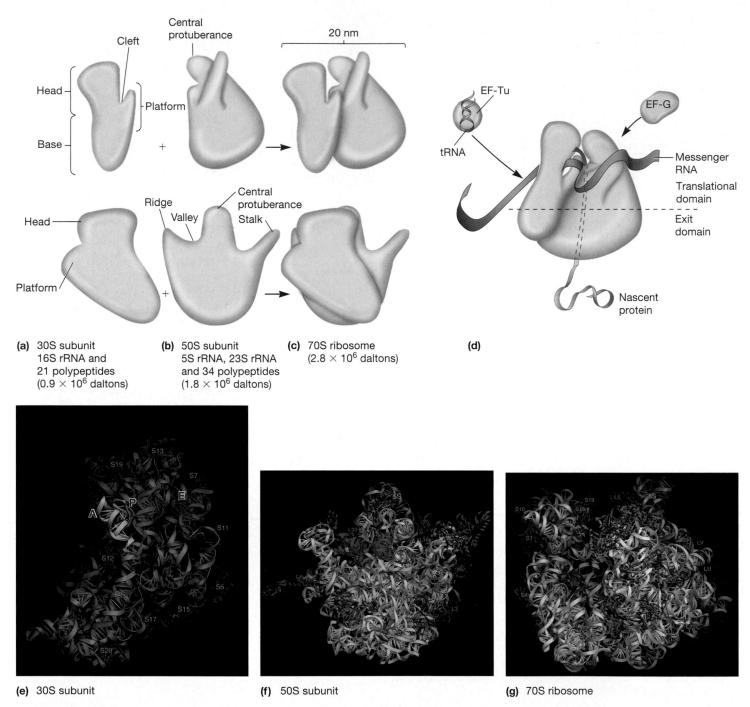

(a) 30S subunit
16S rRNA and
21 polypeptides
(0.9×10^6 daltons)

(b) 50S subunit
5S rRNA, 23S rRNA
and 34 polypeptides
(1.8×10^6 daltons)

(c) 70S ribosome
(2.8×10^6 daltons)

(d)

(e) 30S subunit

(f) 50S subunit

(g) 70S ribosome

Figure 12.35   Procaryotic Ribosome Structure.   Parts (a)–(d) illustrate *E. coli* ribosome organization; parts (e)–(g) show the molecular structure of the *Thermus thermophilus* ribosome. (a) The 30S subunit. (b) The 50S subunit. (c) The complete 70S ribosome. (d) A diagram of ribosomal structure showing the translational and exit domains. The locations of elongation factor and mRNA binding are indicated. The growing peptide chain probably remains unfolded and extended until it leaves the large subunit. (e) Interior interface view of the 30S subunit of the *T. thermophilus* 70S ribosome showing the positions of the A, P, and E site tRNAs. (f) Interior interface view of the *T. thermophilus* 50S subunit and portions of its three tRNAs. (g) The complete *T. thermophilus* 70S ribosome viewed from the right-hand side with the 30S subunit on the left and the 50S subunit on the right. The anticodon arm of the A site tRNA is visible in the interface cavity. The components in figures (e)–(g) are colored as follows: 16S rRNA, cyan; 23S rRNA, gray; 5S rRNA, light blue; 30S proteins, dark blue; 50S proteins, magenta; and A, P, and E site tRNAs (gold, orange, and red, respectively).

blocked by a formyl group, this aminoacyl-tRNA can be used only for initiation. When methionine is to be added to a growing polypeptide chain, a normal methionyl-tRNA^{Met} is employed. Eucaryotic protein synthesis (except in the mitochondrion and chloroplast) and archaeal protein synthesis begin with a special initiator methionyl-tRNAMet. Although most bacteria start protein synthesis with *N*-formylmethionine, the formyl group is not retained but is hydrolytically removed. In fact, one to three amino acids may be removed from the amino terminal end of the polypeptide after synthesis.

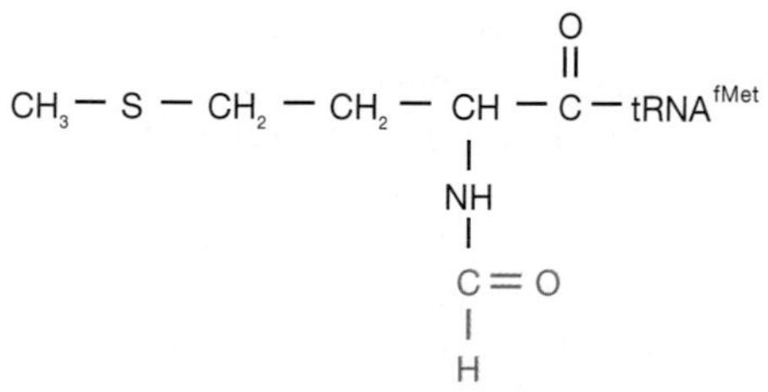

$$CH_3 - S - CH_2 - CH_2 - CH - \overset{\overset{\textstyle O}{\|}}{C} - tRNA^{fMet}$$

with NH below CH, and C=O, H below that.

Figure 12.36 Bacterial Initiator tRNA. The initiator aminoacyl-tRNA, *N*-formylmethionyl-tRNA^fMet, is used by *Bacteria*. The formyl group is in color. *Archaea* and eucaryotes use methionyl-tRNA for initiation.

The initiation stage is crucial for the translation of mRNA into the correct polypeptide (**figure 12.37**). In *Bacteria,* it begins when initiator *N*-formylmethionyl-tRNA^fMet (fMet-tRNA) binds to a free 30S ribosomal subunit. The 16S rRNA within the 30S subunit possesses nucleotide sequences at its 3′ end that are complementary to the Shine-Dalgarno sequence in the leader sequence of the mRNA. By aligning the Shine-Dalgarno sequence with 16S rRNA of the 30S ribosomal subunit, the **initiator codon** (AUG or sometimes GUG, though GUG is a not as good an initiator) specifically binds with the fMet-tRNA anticodon. This ensures that the codon for the

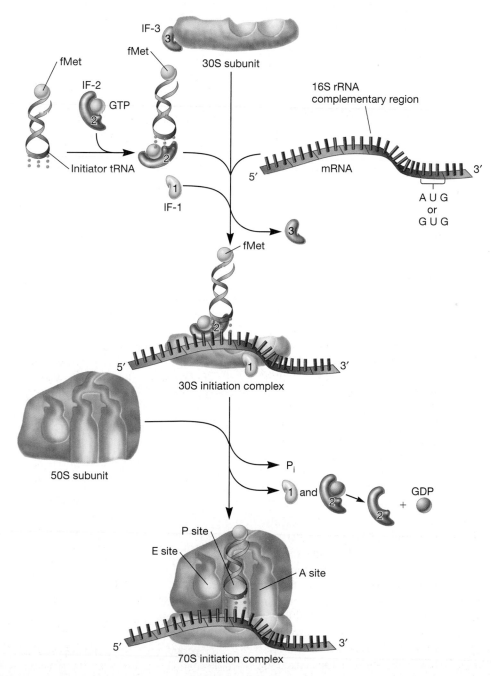

Figure 12.37 Initiation of Protein Synthesis. The initiation of protein synthesis in *Bacteria*. The following abbreviations are employed: IF-1, IF-2, and IF-3 stand for initiation factors 1, 2, and 3; initiator tRNA is *N*-formylmethionyl-tRNA^fMet. The ribosomal locations of initiation factors are depicted for illustration purposes only. They do not represent the actual initiation factor binding sites.

initiator fMet-tRNA will be translated first. When the 50S subunit of the ribosome binds to the 30S subunit-mRNA, an active ribosome-mRNA complex is formed, with the fMet-tRNA positioned at the peptidyl or P site (see description of the elongation cycle).

In *Bacteria,* three protein initiation factors are required (figure 12.37). Initiation factor 3 (IF-3) prevents 30S subunit binding to the 50S subunit and promotes the proper mRNA binding to the 30S subunit. IF-2 binds GTP and fMet-tRNA and directs the attachment of fMet-tRNA to the P site of the 30S subunit. GTP is hydrolyzed during association of the 50S and 30S subunits. IF-1, appears to be needed for release of IF-2 and GDP from the completed 70S ribosome. IF-1 may aid in the binding of the 50S subunit to the 30S subunit. It also blocks tRNA binding to the A site. Eucaryotes require more initiation factors; otherwise the process is quite similar to that of *Bacteria.*

Elongation of the Polypeptide Chain

Every amino acid addition to a growing polypeptide chain is the result of an elongation cycle composed of three phases: aminoacyl-tRNA binding, the transpeptidation reaction, and translocation. The process is aided by proteins called **elongation factors.** In each turn of the cycle, an amino acid corresponding to the proper mRNA codon is added to the C-terminal end of the polypeptide chain as the ribosome moves down the mRNA in the 5′ to 3′ direction. The bacterial elongation cycle is described next.

The ribosome has three sites for binding tRNAs: (1) the **peptidyl** or **donor site** (**P site**), (2) the **aminoacyl** or **acceptor site** (**A site**), and (3) the **exit site** (**E site**). At the beginning of an elongation cycle, the P site is filled with either *N*-formylmethionyl-tRNA^{fMet} or peptidyl-tRNA, and the A and E sites are empty (**figure 12.38**). Messenger RNA is bound to the ribosome in such a way that the proper codon interacts with the P site tRNA (e.g., an AUG codon for fMet-tRNA). The next codon is located within the A site and is ready to accept an aminoacyl-tRNA.

The first phase of the elongation cycle is the aminoacyl-tRNA binding phase. The aminoacyl-tRNA corresponding to the codon in the A site is inserted so its anticodon is aligned with the codon on the mRNA. When GTP is bound to EF-Tu, this elongation factor is in its active state, and it delivers aminoacyl-tRNA to the A site. This is followed by GTP hydrolysis, and the EF-Tu · GDP complex leaves the ribosome. EF-Tu · GDP is converted to EF-Tu · GTP with the aid of a second elongation factor, EF-Ts. Subsequently another aminoacyl-tRNA binds to EF-Tu · GTP (figure 12.38).

Aminoacyl-tRNA binding to the A site initiates the second phase of the elongation cycle, the transpeptidation reaction (figure 12.38 and **figure 12.39**). Transpeptidation is catalyzed by the **peptidyl transferase** activity of the 23S rRNA ribozyme, which is part of the 50S ribosomal subunit. The α-amino group of the A site amino acid nucleophilically attacks the α-carboxyl group of the C-terminal amino acid on the P site tRNA (figure 12.39). The peptide chain attached to the tRNA in the P site is transferred to the A site as a peptide bond is formed between the chain and the incoming amino acid. No extra energy source is required for

peptide bond formation because the bond linking an amino acid to tRNA is high in energy (figure 12.34).

The final phase in the elongation cycle is **translocation.** Three things happen simultaneously: (1) the peptidyl-tRNA moves about 20 Å from the A site to the P site; (2) the ribosome moves one codon along mRNA so that a new codon is positioned in the A site; and (3) the empty tRNA moves from the P site to the E site and subsequently leaves the ribosome. Ribosomal proteins are involved in these tRNA movements. The intricate process also requires the participation of the EF-G or translocase protein and GTP hydrolysis.

Termination of Protein Synthesis

Protein synthesis stops when the ribosome reaches one of three nonsense codons—UAA, UAG, and UGA (**figure 12.40**). The nonsense (stop) codon is found on the mRNA immediately before the trailer region. Three release factors (RF-1, RF-2, and RF-3) aid the ribosome in recognizing these codons. Because there is no cognate tRNA for a nonsense codon, the ribosome stops. Peptidyl transferase hydrolyzes the bond linking the polypeptide to the tRNA in the P site, and the polypeptide and the empty tRNA are released. GTP hydrolysis seems to be required during this sequence, although it may not be needed for termination in *Bacteria.* Next, the ribosome dissociates from its mRNA and separates into 30S and 50S subunits. IF-3 binds to the 30S subunit to prevent it from reassociating with the 50S subunit until the proper stage in initiation is reached. Thus ribosomal subunits associate during protein synthesis and separate afterward. The termination of eucaryotic protein synthesis is similar except that only one release factor appears to be active.

Protein synthesis is a very expensive process. Three GTP molecules probably are used during each elongation cycle, and two ATP high-energy bonds are required for amino acid activation (ATP is converted to AMP rather than to ADP). Therefore five high-energy bonds are required to add one amino acid to a growing polypeptide chain. GTP also is used in initiation and termination of protein synthesis (figures 12.37 and 12.40). Presumably this large energy expenditure is required to ensure the fidelity of protein synthesis. Very few mistakes can be tolerated.

1. In which direction are polypeptides synthesized? What is a polyribosome and why is it useful?

2. Briefly describe the structure of transfer RNA and relate this to its function. How are amino acids activated for protein synthesis, and why is the specificity of the aminoacyl-tRNA synthetase reaction so important?

3. What are the translational and exit domains of the ribosome? Contrast procaryotic and eucaryotic ribosomes in terms of structure. What roles does ribosomal RNA have?

4. Describe the nature and function of the following: fMet-tRNA, initiator codon, IF-3, IF-2, IF-1, elongation cycle, peptidyl and aminoacyl sites, EF-Tu, EF-Ts, transpeptidation reaction, peptidyl transferase, translocation, EF-G or translocase, nonsense codon, and release factors.

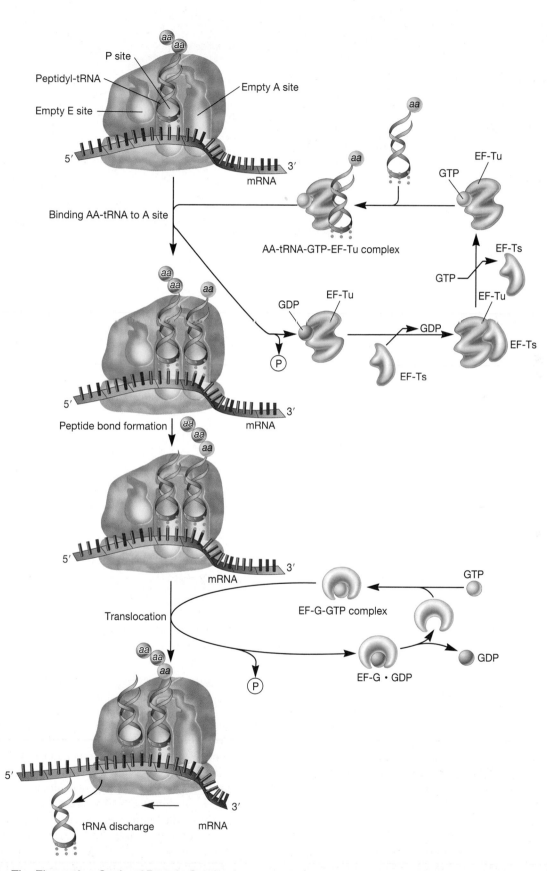

Figure 12.38 The Elongation Cycle of Protein Synthesis. The ribosome possesses three sites, a peptidyl or donor site (P site), an aminoacyl or acceptor site (A site), and an exit site (E site). The arrow below the ribosome in the translocation step shows the direction of mRNA movement.

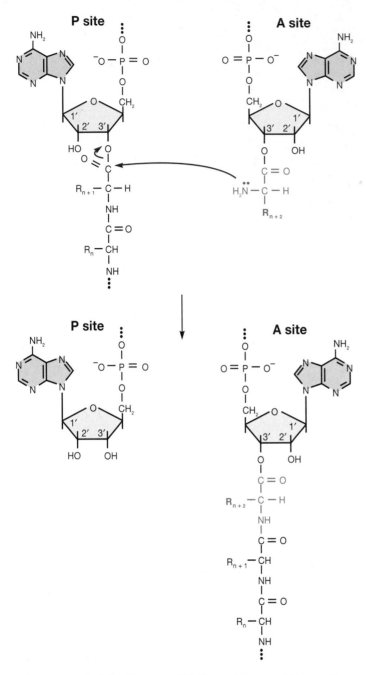

Figure 12.39 **Transpeptidation.** The peptidyl transferase reaction. The peptide grows by one amino acid and is transferred to the A site.

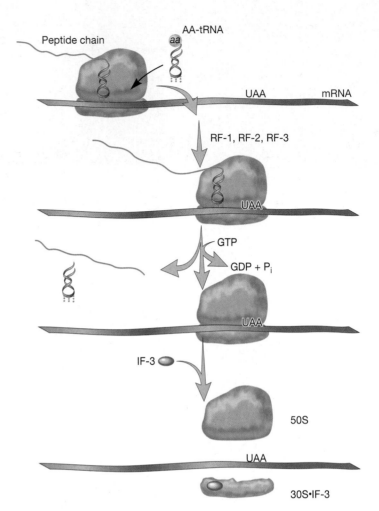

Figure 12.40 **Termination of Protein Synthesis in** *Bacteria*. Although three different nonsense codons can terminate chain elongation, UAA is most often used for this purpose. Three release factors (RF) assist the ribosome in recognizing nonsense codons and terminating translation. GTP hydrolysis is probably involved in termination. Transfer RNAs are in pink.

<table>
<tr><td>

12.8 PROTEIN MATURATION AND SECRETION

As a polypeptide emerges from a ribosome, it is not yet ready to assume its cellular functions. Protein function depends on its three-dimensional shape. Proteins must be properly folded and in some cases associate with other protein subunits to generate a functional enzyme (e.g., DNA and RNA polymerases are multimeric proteins). In addition, proteins must be delivered to the

</td><td>

proper subcellular or extracellular site. We now discuss these posttranslational events.

Protein Folding and Molecular Chaperones

Although the amino acid sequence of a polypeptide determines its final conformation, helper proteins aid the newly formed or nascent polypeptide in folding to its proper functional shape. These proteins, called **molecular chaperones** or chaperones, recognize only unfolded polypeptides or partly denatured proteins and do not bind to normal, functional proteins. Their role is essential because the cytoplasm is filled with nascent polypeptide chains and proteins. Under such conditions, it is quite likely that new polypeptide chains will fold improperly and aggregate to form nonfunctional complexes. Molecular chaperones suppress incorrect folding and may reverse any incorrect folding that has already taken place. They are so important that chaperones are present in all cells.

</td></tr>
</table>

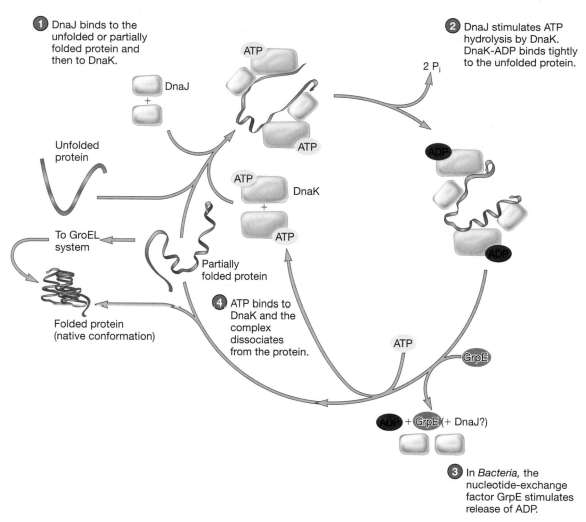

Figure 12.41   Chaperones and Polypeptide Folding.   The involvement of bacterial chaperones in the proper folding of a newly synthesized polypeptide chain is depicted in this diagram. Three possible outcomes of a chaperone reaction cycle are shown. A native protein may result, the partially folded polypeptide may bind again to DnaK and DnaJ, or the polypeptide may be transferred to GroEL and GroES.

Several chaperones and cooperating proteins aid proper protein folding in *Bacteria*. The process has been well studied in *E. coli* and involves at least four chaperones—DnaK, DnaJ, GroEL, and GroES—and the stress protein GrpE. After a sufficient length of nascent polypeptide extends from the ribosome, DnaJ binds to the unfolded chain (**figure 12.41**). DnaK, which is complexed with ATP, then attaches to the polypeptide. These two chaperones prevent the polypeptide from folding improperly as it is synthesized. When synthesis of the polypeptide is complete, the GrpE protein binds to the chaperone-polypeptide complex and causes DnaK to release ADP. DnaJ may also be released at this step. Then ATP binds to DnaK, and DnaK dissociates from the polypeptide. The polypeptide has been folding during this sequence of events and may have reached its final native conformation. If it is still only partially folded, it can bind DnaJ and DnaK again and repeat the process or be transferred to another set of chaperones, GroEL and GroES, where the final folding takes place. As with DnaK, ATP binding to GroEL and ATP hydrolysis change the chaperone's affinity for the folding polypeptide and regulate polypeptide binding and release

(polypeptide release is ATP-dependent). GroES binds to GroEL and assists in its binding and release of the refolding polypeptide.

Chaperones were first discovered because they dramatically increase in concentration when cells are exposed to high temperatures, metabolic poisons, and other stressful conditions. Thus many chaperones are called **heat-shock proteins.** When an *E. coli* culture is switched from 30 to 42°C, the concentrations of some 20 different heat-shock proteins increase greatly within about 5 minutes. If the cells are exposed to a lethal temperature, the heat-shock proteins are still synthesized but most other proteins are not. Thus chaperones protect the cell from thermal damage and other stresses as well as promote the proper folding of new polypeptides. For example, DnaK protects *E. coli* RNA polymerase from thermal inactivation in vitro. In addition, DnaK reactivates thermally inactivated RNA polymerase, especially if ATP, DnaJ, and GrpE are present. GroEL and GroES also protect intracellular proteins from aggregation. As one would expect, large quantities of chaperones are present in hyperthermophiles such as *Pyrodictium occultum,* an archaeon that grows at temperatures as high as 110°C. *P. occultum* has a chaperone similar to GroEL of *E. coli.* This archaeal chaperone hydrolyzes ATP most rapidly at 100°C and makes up almost three-quarters of the cell's soluble protein when *P. occultum* grows at 108°C.

Research indicates that procaryotes and eucaryotes may differ with respect to the timing of protein folding. In terms of conformation, proteins are composed of compact, self-folding, structurally independent regions. These regions, normally around 100 to 300 amino acids in length, are called domains. Larger proteins such as immunoglobulins (important proteins in the immune response) may have two or more domains that are linked by less structured portions of the polypeptide chain. In eucaryotes, domains fold independently right after being synthesized by the ribosome. It appears that procaryotic polypeptides, in contrast, do not fold until after the complete chain has been synthesized. Only then do the individual domains fold. This difference in timing may account for the observation that chaperones seem to be more important in the folding of procaryotic proteins. Folding a whole polypeptide is more complex than folding one domain at a time and would require the aid of chaperones. >> *Antibodies (section 29.7)*

Protein Splicing

A further level of complexity in the formation of proteins has been discovered. Some microbial proteins are spliced after translation. In protein splicing, a part of the polypeptide is removed before the polypeptide folds into its final shape. Self-splicing proteins begin as larger precursor proteins composed of one or more internal intervening sequences called **inteins** (about 130 and 600 amino acids in length) flanked by external sequences called **exteins** (figure 12.42*a*). Inteins are removed in an autocatalytic process involving a branched intermediate (figure 12.42*b*). Thus far, more than 130 inteins in 34 types of self-splicing proteins have been discovered. Some examples are an ATPase in the yeast *Saccharomyces cerevisiae,* the RecA protein of *Mycobacterium tuberculosis,* and a DNA polymerase in the archaeon *Pyrococcus.* Thus self-splicing proteins are present in all three domains of life.

Protein Secretion in Procaryotes

The membranes of the procaryotic cell envelope (i.e., the plasma membrane of *Archaea* and gram-positive bacteria, and the plasma and outer membranes of gram-negative bacteria) present a considerable barrier to the movement of large molecules into or out of the cell. Yet many important structures are located outside the wall. Thus materials inside the cell, including proteins, must be moved across the cell envelope to build these structures.

Protein secretion poses different difficulties depending on the structure of the microbial cell wall. For gram-positive bacteria to secrete proteins, the proteins must be transported across the plasma membrane. Once across the plasma membrane, the protein either passes through the relatively porous peptidoglycan into the external environment or becomes embedded in or attached to the peptidoglycan. Likewise, secreted archaeal proteins must be moved

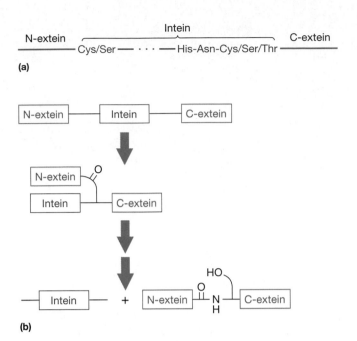

Figure 12.42   Protein Splicing.   (a) A generalized illustration of intein structure. The amino acids that are commonly present at each end of the inteins are shown. Note that many are thiol or hydroxyl-containing amino acids. (b) An overview of the proposed pattern or sequence of splicing. The precise mechanism is not yet known but presumably involves the hydroxyls or thiols located at each end of the intein.

across or into their unique cell walls. Gram-negative bacteria have more hurdles to jump when they secrete proteins. They, too, must transport the proteins across the plasma membrane, but to complete the secretion process, the proteins must be able to escape attack from protein-degrading enzymes in the periplasmic space and then must be transported across the outer membrane.

Common Secretion Pathways

In procaryotes, the major pathway for transporting proteins across the plasma membrane is the Sec-dependent (*secretion-dependent*) pathway (**figure 12.43**). In gram-negative bacteria, proteins can be transported across the outer membrane by several different mechanisms, some of which bypass the Sec-dependent pathway, moving proteins directly from the cytoplasm to the outside of the cell (**figure 12.44**). All protein secretion pathways described here require the expenditure of energy at some step in the process. The energy is usually supplied by the hydrolysis of high-energy molecules such as ATP and GTP. However, the proton motive force also sometimes plays a role. << *ATP (section 9.4); Electron transport and oxidative phosphorylation (section 10.5)*

The **Sec-dependent pathway,** sometimes called the general secretion pathway, is highly conserved, having been identified in all three domains of life (figure 12.43). It translocates unfolded proteins across the plasma membrane or integrates them into the membrane itself. Proteins to be transported across the plasma membrane by this pathway are synthesized as presecretory proteins called preproteins. The amino-terminus of the preprotein has a **signal peptide,** which is recognized by the Sec machinery. Soon

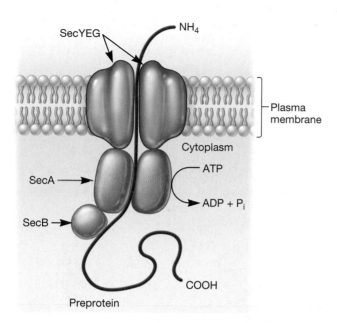

Figure 12.43 The Sec-Dependent Pathway of *E. coli.*

after the signal peptide is synthesized, chaperone proteins (e.g., SecB) bind it. This helps delay protein folding, thereby helping the preprotein reach the Sec transport machinery in the conformation needed for transport. Evidence exists that translocation of proteins can begin before the completion of their synthesis by ribosomes. Certain Sec proteins (SecY, SecE, and SecG) are thought to form a channel in the membrane through which the preprotein passes. Another protein (SecA) binds to the SecYEG proteins and the SecB-preprotein complex. SecA acts as a motor, using the energy released from ATP hydrolysis to translocate the preprotein through the plasma membrane. When the preprotein emerges from the plasma membrane, free from chaperones, an enzyme called signal peptidase removes the signal peptide. The protein then folds into the proper shape, and disulfide bonds are formed when necessary.

The **ABC protein secretion pathway,** which derives its name from *ATP binding cassette,* is ubiquitous in procaryotes—that is, it is present in gram-positive and gram-negative bacteria as well as *Archaea.* It is sometimes called the **type I protein secretion pathway** (figure 12.44). In pathogenic gram-negative bacteria, it is involved in the secretion of toxins (α-hemolysin), as well as proteases, lipases, and specific peptides. Secreted proteins usually contain C-terminal secretion signals that help direct the newly synthesized protein to the type I machinery, which spans the plasma membrane, the periplasmic space, and the outer membrane. These systems translocate proteins in one step across both membranes, bypassing the Sec-dependent pathway. Gram-positive bacteria use a modified version of the type I system to translocate proteins across the plasma membrane. Analysis of the *Bacillus subtilis* genome has identified 77 ABC transporters. This may reflect the fact that ABC transporters move a wide variety of solutes in addition to proteins, including sugars and amino acids, as well as exporting antibiotics from the cell interior.

In *Bacteria* and some archaea, another plasma membrane translocation system called the **Tat pathway** can move folded proteins across the plasma membrane. In gram-negative bacteria, these proteins are then delivered to the type II system, which is described next. Only proteins that feature two, or "twin," arginine residues in their signal sequence are transported by the Tat system—in fact, *Tat* stands for *twin arginine translocase.* The TatBC proteins are thought to recognize the signal peptide of the protein to be secreted, while the TatA protein appears to form the channel through which the protein is secreted.

Protein Secretion in Gram-Negative Bacteria

Currently five protein secretion pathways have been identified in gram-negative bacteria (figure 12.44). Gram-negative bacteria use the type II and type V pathways to transport proteins across the outer membrane after the protein has first been translocated across the plasma membrane by the Sec-dependent pathway. The type I and type III pathways do not interact with proteins that are first translocated by the Sec system, so they are said to be Sec independent. The type IV pathway sometimes is linked to the Sec-dependent pathway but usually functions on its own.

The **type II protein secretion pathway** is present in a number of plant and animal pathogens, including *Erwinia carotovora, Klebsiella pneumoniae, Pseudomonas aeruginosa,* and *Vibrio cholerae.* It is responsible for the secretion of proteins such as degradative enzymes (e.g., pullulanases, cellulases, pectinases, proteases, and lipases), as well as other proteins such as cholera toxin and pili proteins. Type II systems are quite complex and can contain as many as 12 to 14 proteins, most of which appear to be integral membrane proteins (figure 12.44). Even though some components of type II systems span the plasma membrane, they apparently translocate proteins only across the outer membrane. In most cases, the Sec-dependent pathway first translocates the protein across the plasma membrane and then the type II system completes the secretion process.

The **type V protein secretion pathways** are the most recently discovered protein secretion systems. They, too, rely on the Sec-dependent pathway to move proteins across the plasma membrane. However, once in the periplasmic space, many of these proteins are able to form a channel in the outer membrane through which they transport themselves; these proteins are referred to as autotransporters. Other proteins are secreted by the type V pathway with the aid of a separate helper protein.

Several gram-negative pathogens have the **type III protein secretion pathway,** another secretion system that bypasses the Sec-dependent pathway. These systems have been intensely studied because they inject virulence factors directly into the plant and animal host cells these pathogens attack. These virulence factors include toxins, phagocytosis inhibitors, stimulators of cytoskeleton reorganization in the host cell, and promoters of host cell suicide (apoptosis). Type III systems also transport other proteins, including (1) some of the proteins from which the system is built, (2) proteins that regulate the secretion process, and (3) proteins that aid in the insertion of secreted proteins into target cells. Type III systems are structurally complex and

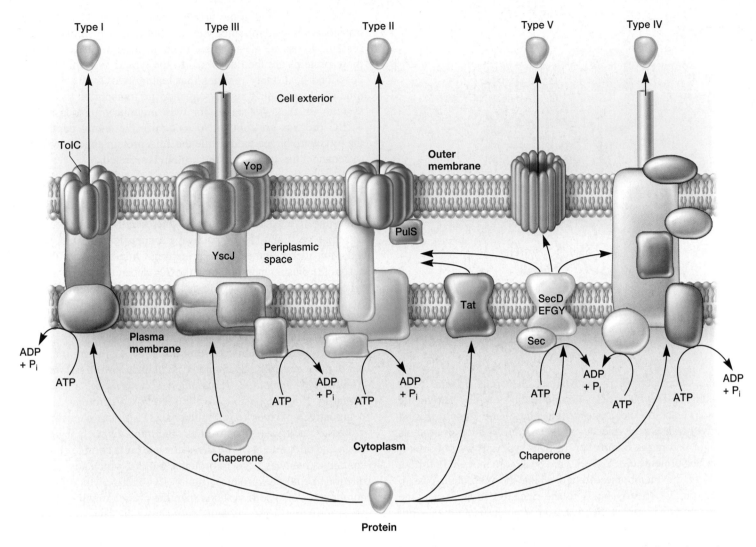

Figure 12.44 The Protein Secretion Systems of Gram-Negative Bacteria. The five secretion systems of gram-negative bacteria are shown. The Sec-dependent and Tat pathways deliver proteins from the cytoplasm to the periplasmic space and are found in archaea and gram-positive bacteria also. The type II, type V, and sometimes type IV systems complete the secretion process begun by the Sec-dependent pathway. The Tat system appears to deliver proteins only to the type II pathway. The type I and type III systems bypass the Sec-dependent and Tat pathways, moving proteins directly from the cytoplasm, through the outer membrane, to the extracellular space. The type IV system can work either with the Sec-dependent pathway or can work alone to transport proteins to the extracellular space. Proteins translocated by the Sec-dependent pathway and the type III pathway are delivered to those systems by chaperone proteins.

often shaped like a syringe (figure 12.44). The slender, needle-like portion extends from the cell surface; a cylindrical base is connected to both the outer membrane and the plasma membrane and looks somewhat like the flagellar basal body *(see figures 3.36 and 30.5)*. It is thought that proteins move through a translocation channel. Important examples of bacteria with type III systems are *Salmonella, Yersinia, Shigella, E. coli, Bordetella, Pseudomonas aeruginosa,* and *Erwinia.* The participation of type III systems in bacterial virulence is further discussed in chapter 30.

Type IV protein secretion pathways are unique in that they are used to secrete proteins as well as to transfer DNA from a donor bacterium to a recipient during a process called bacterial conjugation. Type IV systems are composed of many different

proteins, and like the type III systems, these proteins form a syringelike structure. Type IV systems and conjugation are described in more detail in chapter 14.

1. What are molecular chaperones and heat-shock proteins? Describe their functions.

2. Give the major characteristics and functions of the protein secretion pathways described in this section.

3. Which secretion pathway is most widespread?

4. What is a signal peptide? Why do you think a protein's signal peptide is not removed until after the protein is translocated across the plasma membrane?

Summary

12.1 Flow of Genetic Information

a. DNA serves as the storage molecule for genetic information. DNA replication is the process by which DNA is duplicated so that it can be passed on to the next generation **(figure 12.1).**

b. During transcription, genetic information in DNA is rewritten as an RNA molecule. The three products of transcription are messenger RNA, ribosomal RNA, and transfer RNA.

c. Translation converts genetic information in the form of a messenger RNA molecule into a polypeptide. Ribosomal RNA and transfer RNA participate in the decoding of genetic information during translation.

12.2 Nucleic Acid Structure

a. DNA differs in composition from RNA in having deoxyribose and thymine rather than ribose and uracil.

b. DNA is double stranded, with complementary AT and GC base pairing between the strands. The strands run antiparallel and are twisted into a right-handed double helix **(figure 12.3).**

c. RNA is normally single stranded, although it can coil upon itself and base pair to form hairpin structures.

d. In almost all procaryotes, DNA exists as a closed circle that is twisted into supercoils **(figure 12.4).** In *Bacteria,* DNA is associated with basic proteins but not with histones.

e. Eucaryotic DNA is associated with five types of histone proteins. Histones associate to form ellipsoidal octamers around which the DNA is coiled to produce the nucleosome **(figure 12.5).** The DNA of many archaea is complexed with archaeal histones.

12.3 DNA Replication

a. Most circular procaryotic DNAs are copied by two replication forks moving around the circle to form a theta-shaped (θ) figure **(figure 12.7).**

b. A rolling-circle mechanism is employed to replicate some plasmids and viral genomes **(figure 12.8).**

c. Eucaryotic DNA has many replicons and replication origins.

d. The replisome is a huge complex of proteins and is responsible for DNA replication.

e. DNA polymerase enzymes catalyze the synthesis of DNA in the 5′ to 3′ direction while reading the DNA template in the 3′ to 5′ direction **(figure 12.11).**

f. The double helix is unwound by helicases with the aid of topoisomerases such as DNA gyrase. DNA binding proteins keep the strands separate.

g. DNA polymerase III holoenzyme synthesizes a complementary DNA copy beginning with a short RNA primer made by the enzyme primase.

h. The leading strand is replicated continuously, whereas DNA synthesis on the lagging strand is discontinuous and forms Okazaki fragments **(figures 12.11 and 12.12).**

i. DNA polymerase I excises the RNA primer and fills in the resulting gap. DNA ligase then joins the fragments together **(figures 12.13 and 12.14).**

j. Telomerase is responsible for replicating the ends of eucaryotic chromosomes **(figure 12.15).**

12.4 Gene Structure

a. A gene may be defined as the nucleic acid sequence that codes for a polypeptide, tRNA, or rRNA.

b. The template strand of DNA carries genetic information and directs the synthesis of the RNA transcript.

c. The gene also contains a promoter, a coding region and a terminator; it may have a leader and a trailer **(figure 12.18).**

d. The genes for tRNA and rRNA often code for a precursor that is subsequently processed to yield several products **(figure 12.19).**

e. RNA polymerase binds to the promoter region, which contains RNA polymerase recognition and RNA polymerase binding sites **(figures 12.21, 12.23,** and **12.24).**

12.5 Transcription

a. RNA polymerase synthesizes RNA that is complementary to the DNA template strand **(figure 12.21).**

b. The sigma factor helps the bacterial RNA polymerase bind to the promoter region at the start of a gene **(figure 12.24).**

c. A terminator marks the end of a gene. The protein rho is needed for RNA polymerase release from some terminators **(figures 12.26 and 12.27).**

d. In eucaryotes, RNA polymerase II synthesizes pre-mRNA, which then undergoes posttranscriptional modification by RNA cleavage and addition of a 3′ poly-A sequence and a 5′ cap to generate mRNA **(figure 12.29).**

e. Many eucaryotic genes are split or interrupted genes that have exons and introns. Exons are joined by RNA splicing, which is carried out by spliceosomes **(figure 12.30).**

12.6 The Genetic Code

a. Genetic information is carried in the form of 64 nucleotide triplets called codons **(table 12.2);** 61 sense codons direct amino acid incorporation, and three stop or nonsense codons terminate translation.

b. The code is degenerate—that is, there is more than one codon for most amino acids.

12.7 Translation

a. In translation, ribosomes attach to mRNA and synthesize a polypeptide beginning at the N-terminal end. A polysome or polyribosome is a complex of mRNA with several ribosomes **(figure 12.32).**

b. Amino acids are activated for protein synthesis by attachment to the 3′ end of transfer RNAs. Activation requires ATP, and the reaction is catalyzed by aminoacyl-tRNA synthetases **(figure 12.34).**

c. Ribosomes are large, complex organelles composed of rRNAs and many polypeptides. Amino acids are added to a growing peptide chain at the translational domain of the ribosome **(figure 12.35).**

d. Protein synthesis begins with the binding of fMet-tRNA (*Bacteria*) **(figure 12.36)** or an initiator methionyl-tRNA^{Met} (eucaryotes and *Archaea*) to an initiator codon on mRNA and to the two ribosomal subunits. This involves the participation of protein initiation factors **(figure 12.37).**

e. In the elongation cycle, the proper aminoacyl-tRNA binds to the A site with the aid of EF-Tu and GTP **(figure 12.38).** Then the transpeptidation reaction is catalyzed by peptidyl transferase **(figure 12.39).** Finally, during translocation, the peptidyl-tRNA moves to the P site and the ribosome moves down the mRNA by one codon. Translocation requires GTP and EF-G or translocase. The empty tRNA leaves the ribosome by way of the exit site.

f. Protein synthesis stops when a nonsense codon is reached. Bacteria require three release factors for codon recognition and ribosome dissociation from the mRNA **(figure 12.40).**

12.8 Protein Maturation and Secretion

a. Procaryotic proteins may not fold until completely synthesized, whereas eucaryotic protein domains fold as they leave the ribosome. Molecular chaperones are involved in ensuring that proteins fold properly or are delivered to their destination site **(figure 12.41).**

b. Some proteins are self-splicing and excise portions of themselves before folding into their final shape **(figure 12.42).**

c. Many proteins must be transported across the plasma membrane of procaryotes. The most commonly used mechanism is the Sec-dependent pathway, which is found in all microbes **(figure 12.43).** ABC-transporters and the Tat pathway are also common to both bacteria and archaea **(figure 12.44).**

d. Gram-negative bacteria have evolved additional systems for translocating proteins across the outer membrane of the cell wall.

Critical Thinking Questions

1. Many scientists say that RNA was the first of the information molecules (i.e., RNA, DNA, protein) to arise during evolution. Given the information in this chapter, what evidence is there to support this hypothesis?

2. *Streptomyces coelicolor* has a linear chromosome. Interestingly there are no genes that encode essential proteins near the ends of the chromosome in this bacterium. Why do you think this is the case?

3. You have isolated several *E. coli* mutants:

    Mutant #1 has a mutation in the -10 region of the promoter of a structural gene encoding an enzyme needed for synthesis of the amino acid serine.

    Mutant #2 has a mutation in the -35 region in the promoter of the same gene.

    Mutant #3 is a double mutant with mutations in both the -10 and -35 region of the promoter of the same gene.

    Only Mutant #3 is unable to make serine. Why do you think this is so?

4. Suppose that you have isolated a microorganism from a soil sample. Describe how you would go about determining the nature of its genetic material.

Learn More

Learn more by visiting the Prescott website at www.mhhe.com/prescottprinciples, where you will find a complete list of references.

Regulation of Gene Expression

13

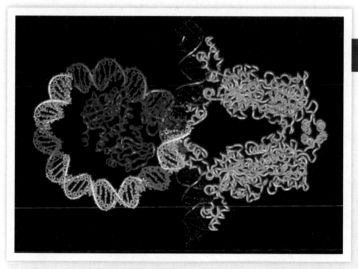

The *lac* repressor (pink) and catabolite activator protein (blue) are bound to the *lac* operon.

activator protein   A regulatory protein that can bind a specific site on DNA and promote transcription.

antisense RNA   A single-stranded RNA with a base sequence complementary to a segment of another RNA molecule that can specifically bind to the target RNA and alter its activity.

attenuation   A mechanism for the regulation of transcription elongation of some bacterial operons by aminoacyl-tRNAs.

catabolite repression   Inhibition of the synthesis of several catabolic enzymes in response to the presence of preferred carbon and energy sources such as glucose.

constitutive gene   A gene that is consistently expressed at the same level.

corepressor   A small molecule that binds and activates a repressor protein and thereby inhibits the transcription of a repressible gene.

global regulatory systems   Genetic regulatory mechanisms that control a suite of genes simultaneously.

housekeeping gene   A gene that is needed for normal, vegetative growth.

inducer   A small molecule that stimulates the transcription of an inducible gene either by binding and inactivating a repres-

sor protein or by binding and activating an activator protein.

inducible gene   A gene whose expression can be increased.

negative transcriptional control   A mechanism that uses a repressor protein to alter the level of expression of a gene or genes.

operator   The segment of bacterial DNA to which a repressor protein binds.

operon   The sequence of bases in bacterial and archaeal DNA that contains one or more structural genes, transcribed from a single promoter.

positive transcriptional control   A mechanism that uses an activator protein to regulate the level of expression of a gene or genes.

quorum sensing   The process in which microbes monitor their population density and regulate expression of density-dependent genes by sensing the levels of signal molecules that are released by the microorganisms.

regulon   A collection of genes or operons that is controlled by a common regulatory protein.

repressible gene   A gene whose level of expression can be decreased.

repressor protein   A regulatory protein that can bind a specific site on DNA and inhibit transcription.

riboswitch   A site in the leader of an mRNA molecule that interacts with a metabolite or other small molecule, causing the leader to change its folding pattern; this change alters either transcription levels or translation levels.

structural gene   A gene that codes for the synthesis of a polypeptide or polynucleotide (i.e., rRNA, tRNA) with a nonregulatory function.

two-component signal transduction system   A signal transduction system that uses the transfer of phosphoryl groups to control gene transcription and protein activity; it has two major components: a sensor kinase and a response regulator.

The particular field which excites my interest is the division between the living and the nonliving, as typified by, say, proteins, viruses, bacteria, and the structure of chromosomes. The eventual goal, which is somewhat remote, is the description of these activities in terms of their structure, i.e., the spatial distribution of their constituent atoms, in so far as this may prove possible. This might be called the chemical physics of biology.

—Francis Crick

The gram-positive soil bacterium *Bacillus subtilis* senses that the nutrient levels in its environment are decreasing, and it must determine if it should initiate sporulation. An *Escherichia coli* cell is in an environment rich in carbon and energy sources, and it must determine which to use and when to use them. A pathogen is transmitted from a stream to the intestinal tract of its animal host, and it must adjust to the warmer temperature, increased nutrient supply, and defenses of the host. These are just a few examples of situations to which microbes must respond. To make the most efficient use of the resources in the current environment and their own cellular machinery, microbes must respond to changes by altering physiological and behavioral processes. How is this accomplished?

The control of cellular processes by regulating the activity of enzymes and other proteins is a fine-tuning mechanism: it acts rapidly to adjust metabolic activity from moment to moment. Microorganisms also are able to control the expression of their genome, although over longer intervals. For example, the *E. coli* chromosome can code for about 4,500 polypeptides, yet not all are produced at the same time. Regulation of gene expression serves to conserve energy and raw materials, to maintain balance between the amounts of various cell proteins, and to acclimate to long-term environmental change. Thus control of gene expression complements the regulation of enzyme activity. **<<** *Posttranslational regulation of enzyme activity (section 9.10)*

In this chapter, we explore the various mechanisms organisms use to regulate gene expression. We begin with a brief discussion of the many levels at which regulation can occur. We then introduce some important examples of the regulation of transcription initiation, transcription elongation, and translation. Finally, we examine how cells use these various regulatory mechanisms to control suites of genes in response to changes in their environments.

13.1   LEVELS OF REGULATION OF GENE EXPRESSION

Figure 13.1 summarizes the expression of bacterial, archaeal, and eucaryotic genes and highlights points in the process where regulation often occurs. Although the overall processes

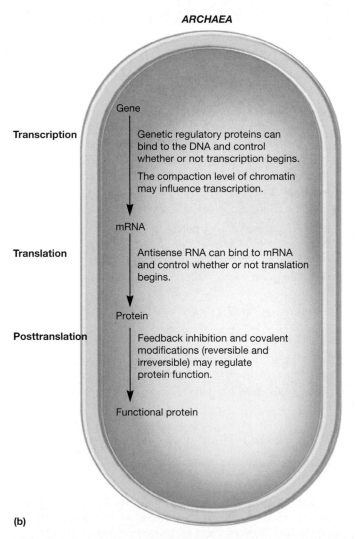

Figure 13.1   Gene Expression and Common Regulatory Mechanisms in the Three Domains of Life.

Figure 13.1 *(Continued)* *EUCARYA*

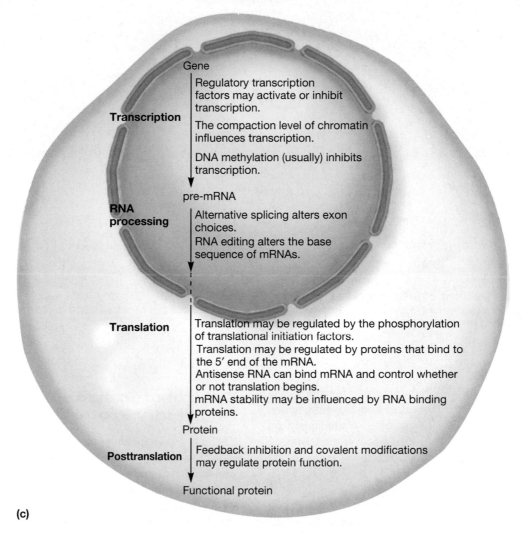

(c)

of transcription and translation in the three domains of life are similar, some differences affect gene expression. Thus regulation of gene expression is somewhat different in each domain. Our focus in this chapter is on well-understood bacterial regulatory processes. We begin our discussion by introducing two phenomena: induction of enzyme synthesis and repression of enzyme synthesis. Induction and repression provided the first models for gene regulation. These models involve the action of regulatory proteins, and the notion that gene expression is regulated solely by proteins persisted for many years. Eventually it was clearly demonstrated that RNA molecules also could have regulatory functions. Induction and repression also demonstrate the regulation of transcription initiation. Although many regulatory processes occur at this level, numerous regulatory mechanisms occur at other levels. We describe some of these mechanisms as well.

13.2 REGULATION OF TRANSCRIPTION INITIATION

Induction and repression are historically important, as they were the first regulatory processes to be understood in any detail. In this section, we first describe these phenomena and then examine the underlying regulatory events.

Induction and Repression of Enzyme Synthesis

Many enzymes are produced almost all of the time because they catalyze reactions in the cell that are needed routinely. These enzymes include those of the central metabolic pathways. Their

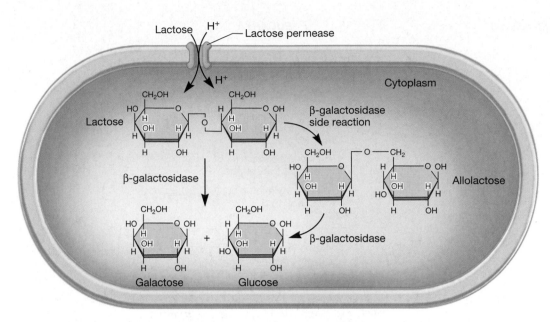

Figure 13.2 **The Reactions of β-Galactosidase.** The main reaction catalyzed by β-galactosidase is the hydrolysis of lactose, a disaccharide, into the monosaccharides galactose and glucose. The enzyme also catalyzes a minor reaction that converts lactose to allolactose. Allolactose acts as the inducer of β-galactosidase synthesis.

functions are often referred to as "housekeeping functions" and the genes that encode them are often called **housekeeping genes.** Those housekeeping genes that are expressed continuously are said to be **constitutive genes.** Many genes, however, are expressed only when needed. The β-galactosidase gene is an example of a regulated gene.

β-galactosidase catalyzes the hydrolysis of the disaccharide sugar lactose to glucose and galactose (**figure 13.2**). When *E. coli* grows with lactose as its only carbon source, each cell contains about 3,000 β-galactosidase molecules, but it has less than three molecules in the absence of lactose. The enzyme β-galactosidase is an inducible enzyme—that is, its level rises in the presence of a small effector molecule called an **inducer** (in this case, the lactose derivative allolactose). Likewise, the genes that encode inducible enzymes such as β-galactosidase are referred to as **inducible genes.**

β-galactosidase is an enzyme that functions in a catabolic pathway and many catabolic enzymes are inducible enzymes. On the other hand, the genes for enzymes involved in biosynthetic pathways are often called **repressible genes,** and their products are called repressible enzymes. For instance, an amino acid present in the surroundings may inhibit the formation of enzymes responsible for its biosynthesis. This makes sense because the microorganism does not need the biosynthetic enzymes for a particular substance if it is already available. Generally, repressible enzymes are necessary for synthesis and are present unless the end product of their pathway is available. Inducible enzymes, in contrast, are required only when their substrate is available; they are missing in the absence of the inducer.

Control of Transcription Initiation by Regulatory Proteins

The action of regulatory proteins is most often responsible for induction and repression. Regulatory proteins are DNA-binding proteins that can exert either negative or positive control. **Negative transcriptional control** occurs when the binding of the protein to DNA inhibits initiation of transcription. Regulatory proteins that act in this fashion are called **repressor proteins. Positive transcriptional control** occurs when the binding of the protein to DNA promotes transcription initiation. These proteins are called **activator proteins.**

Repressor and activator proteins usually act by binding DNA at specific sites. In *Bacteria,* repressor proteins bind a region called the **operator,** which usually overlaps or is downstream of the promoter (i.e., closer to the coding region) (**figure 13.3a,b**). When bound, the repressor protein either blocks binding of RNA polymerase to the promoter or prevents its movement. Activator proteins bind **activator-binding sites** (figure 13.3c,d). These are often upstream of the promoter (i.e., farther away from the coding region). Binding of an activator to its regulatory site generally promotes RNA polymerase binding.

Repressor and activator proteins must exist in both active and inactive forms if transcription initiation is to be controlled appropriately. The activity of regulatory proteins is modified by small effector molecules, most of which bind the regulatory protein noncovalently. Figure 13.3 shows the four basic ways in which the interactions of an effector and a regulatory protein can affect transcription. (1) For negatively controlled inducible genes (e.g., those encoding enzymes needed for catabolism

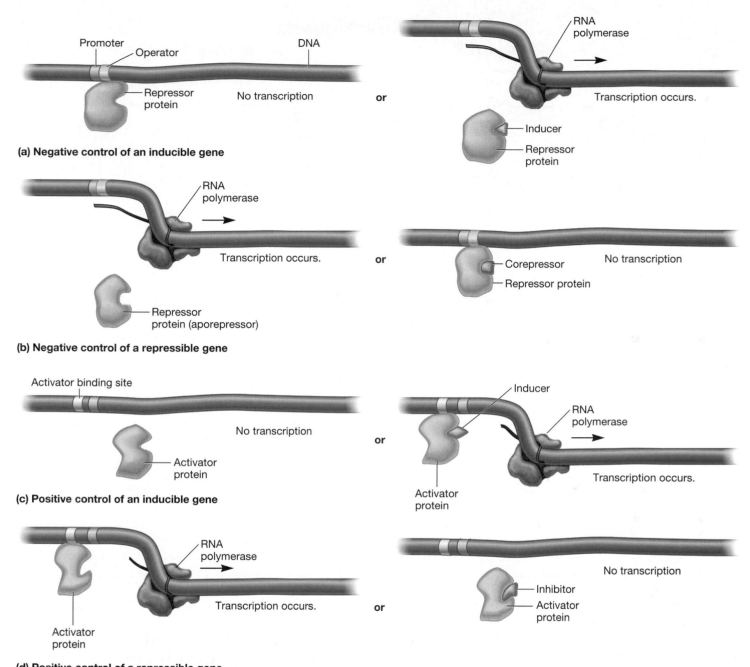

Figure 13.3 **Action of Bacterial Regulatory Proteins.** Bacterial regulatory proteins have two binding sites, one for a small effector molecule and one for DNA. The binding of the effector molecule changes the regulatory protein's ability to bind DNA. (a) In the absence of an inducer, the repressor protein blocks transcription. The presence of an inducer prevents the repressor from binding DNA and transcription occurs. (b) In the absence of a corepressor, the repressor is unable to bind DNA and transcription occurs. When the corepressor is bound to the repressor, the repressor is able to bind DNA and transcription is blocked. (c) The activator protein is only able to bind DNA and activate transcription when it is bound in the inducer. (d) The activator binds DNA and promotes transcription unless the inhibitor is present. When the inhibitor is present, the activator undergoes a conformational change that prevents it from binding DNA; this inhibits transcription.

of a sugar), the repressor protein is active and prevents transcription when the substrate of the pathway is not available (figure 13.3a). It is inactivated by binding of the inducer (e.g., the substrate of the pathway). (2) For negatively controlled repressible genes (e.g., those encoding enzymes needed for the synthesis of an amino acid), the repressor protein is initially

synthesized in an inactive form called the **aporepressor.** It is activated by binding of the **corepressor** (figure 13.3b). For repressible enzymes that function in a biosynthetic pathway, the corepressor is often the product of the pathway (e.g., an amino acid). (3) The activator of a positively regulated inducible gene is activated by the inducer (figure 13.3c), whereas (4) the activator

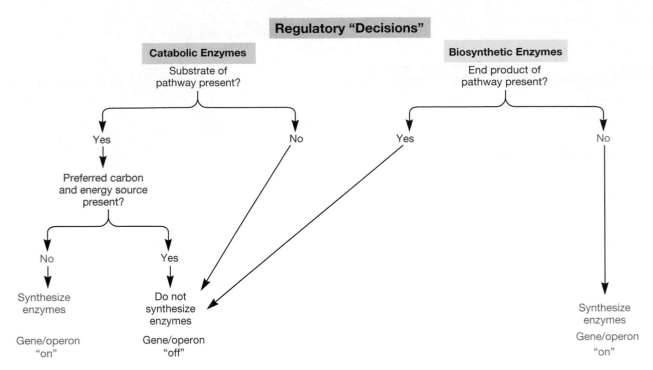

Figure 13.4   Examples of Regulatory "Decisions" Made by Cells.

of a positively regulated repressible gene is inactivated by an inhibitor (figure 13.3*d*). **<<** *Posttranslational regulation of enzyme activity: Allosteric regulation (section 9.10)*

Before we continue our discussion, we must consider two general aspects of regulation. The first is that gene expression is rarely an all-or-nothing phenomenon; it is a continuum. Inhibition of transcription usually does not mean that genes are "turned off" (though this terminology is frequently used). Rather it means the level of mRNA synthesis is decreased significantly and in most cases is occurring at very low levels. In other words, many promoters of regulated genes and operons are considered "leaky," in that there is always some low, basal level of transcription. The second aspect of regulation is the "decision-making" process used by microbial cells. Although cells do not have thought processes, it is convenient to think of regulation in this way. Consider the regulatory "decisions" made by an *E. coli* cell. It need only synthesize the enzymes of a specific catabolic pathway if the substrate of the pathway is present in the environment and a preferred carbon and energy source (e.g., glucose) is not (**figure 13.4**). Preferred carbon and energy sources usually are more easily catabolized or yield more energy. Thus it is to the cell's advantage to use the preferred

source before another source. Conversely, synthesis of the enzymes involved in biosynthetic pathways is inhibited when the end product of the pathway is present.

Recall that functionally related bacterial and archaeal genes are often transcribed from a single promoter. The **structural genes**—the genes coding for polypeptides—are simply lined up together on the DNA, and a single, polycistronic mRNA carries all the messages. The sequence of bases coding for one or more polypeptides, together with the promoter and operator or activator-binding sites, is called an **operon.** Many operons have been discovered and studied. Three well-studied operons are discussed next. They demonstrate different ways that regulatory proteins can be used to control gene expression at the level of transcription initiation.

Lactose Operon: Negative Transcriptional Control of Inducible Genes

The best-studied negative control system is the lactose (*lac*) operon of *E. coli*. The *lac* operon contains three structural genes controlled by the *lac* repressor, which is encoded by *lacI* (**figure 13.5**). One

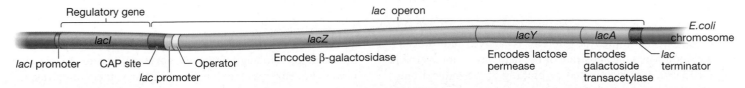

Figure 13.5   The *lac* Operon.   The *lac* operon consists of three genes: *lacZ, lacY,* and *lacA,* which are transcribed as a single unit from the *lac* promoter. The operon is regulated both negatively and positively. Negative control is brought about by the *lac* repressor, which is the product of the *lacI* gene. The operator is the site of *lac* repressor binding. Positive control results from the action of CAP. CAP binds the CAP site located just upstream from the *lac* promoter. CAP is partly responsible for a phenomenon called catabolite repression, an example of a global control network, in which numerous operons are controlled by a single protein.

gene codes for β-galactosidase; a second gene directs the synthesis of β-galactoside permease, the protein responsible for lactose uptake. The third gene codes for the enzyme β-galactoside trans-acetylase, whose function still is uncertain. The presence of the first two genes in the same operon ensures that the rates of lactose uptake and breakdown will vary together.

Lactose is one of many organic molecules *E. coli* can use as a carbon and energy source. It is wasteful to synthesize enzymes of the *lac* operon when lactose is not available. Therefore the cell only expresses this operon at high levels when lactose is the only carbon and energy source present in the environment; the *lac* repressor is responsible for inhibiting transcription when there is no lactose.

The *lac* repressor is a tetramer composed of four identical subunits. The tetramer is formed when two dimers interact. When lactose catabolism is not required, each dimer recognizes and tightly binds one of three different *lac* operator sites: O_1, O_2, and O_3 (**figure 13.6a**). O_1 is the main operator site and must be bound by the repressor if transcription is to be inhibited. When one dimer is at O_1 and another is at one of the two other operator sites, the dimers bring the two operator sites close together, with a loop of DNA forming between them. The binding of *lac* repressor is a two-step process. First, the repressor binds nonspecifically to DNA. Then it rapidly slides along the DNA until it reaches an operator site. A portion of the repressor fits into the major groove of operator-site DNA (figure 13.6b). Thus the shape of the repressor is ideally suited for specific binding to the DNA double helix.

How does the repressor inhibit transcription? The promoter to which RNA polymerase binds is located near the *lac* operator sites. When there is no lactose, the repressor binds O_1 and one of the other operator sites, bending the DNA in the promoter region. This prevents initiation of transcription either because RNA polymerase cannot access the promoter or because it is blocked from moving into the coding region (**figure 13.7a**). When lactose is available, it is taken up by the lactose permease. Once inside the cell, β-galactosidase converts lactose to allolactose, the inducer of the operon (figure 13.2). This occurs because there is always a low level of permease and β-galactosidase synthesis. Allolactose binds to the *lac* repressor and causes the repressor to change to an inactive shape that is unable to bind any operator sites. The inactivated repressor leaves the DNA and transcription occurs (figure 13.7b).

Close examination of figures 13.5 and 13.6 clearly shows that the regulation of the *lac* operon is not as simple as has just been described. That is because the *lac* operon is regulated by a second regulatory protein called *catabolite activator protein* (CAP). CAP functions in a global regulatory network that allows *E. coli* to use glucose preferentially over all other carbon and energy sources by a mechanism called catabolite repression. The use of two different regulatory proteins to control the synthesis of an operon illustrates another important point about regulatory processes—that there are often layers of regulation of any operon. As described in section 13.5, the use of two regulatory proteins generates a continuum of expression levels. The highest levels of transcription occur when lactose is available and glucose is not; the lowest levels occur when lactose is not available and glucose is.

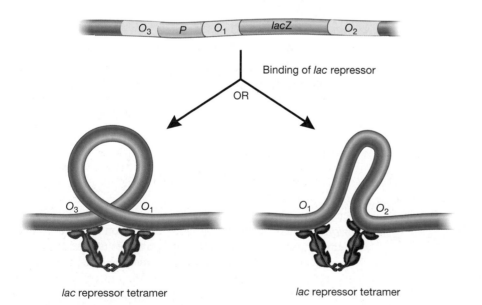

Binding of *lac* repressor

OR

O_3 O_1

lac repressor tetramer

O_1 O_2

lac repressor tetramer

(a) Possible DNA loops caused by the binding of the *lac* repressor

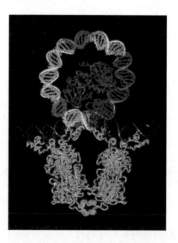

(b) Proposed model of the *lac* repressor binding to O_1 and O_3 (red) based on crystallography studies

Figure 13.6 The *lac* Operator Sites. The *lac* operon has three operator sites: O_1, O_2, and O_3 (a). O_1 is the same operator shown in figure 13.5. As shown in (a) and (b), the *lac* repressor (violet) binds O_1 and one of the other operator sites, forming a DNA loop. The DNA loop contains the −35 and −10 binding sites (green) recognized by RNA polymerase. Thus these sites are inaccessible and transcription is blocked. The DNA loop also contains the CAP binding site and CAP (blue) is shown bound to the DNA (b). When the *lac* repressor is bound to the operator, CAP is unable to activate transcription

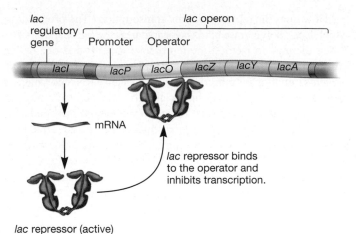

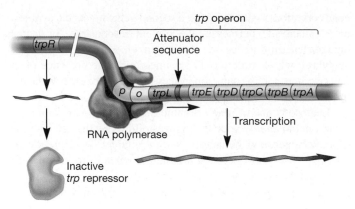

(a) Low trytophan levels, transcription of the entire *trp* operon occurs

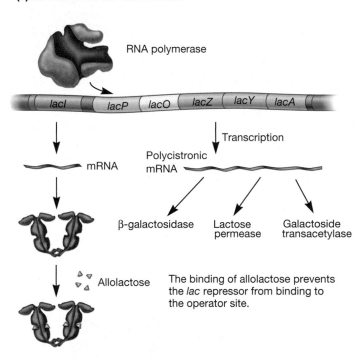

(b) Lactose present

Figure 13.7   Regulation of the *lac* Operon by the *lac* Repressor.   (a) The *lac* repressor is active and can bind the operator as long as the inducer of the operon, allolactose, is not present. Binding of the repressor to the operator inhibits transcription of the operon by RNA polymerase. (b) When lactose is available, some of it is converted to allolactose by β-galactosidase. When sufficient amounts of allolactose are present, it binds and inactivates the *lac* repressor. The repressor leaves the operator and RNA polymerase is free to initiate transcription.

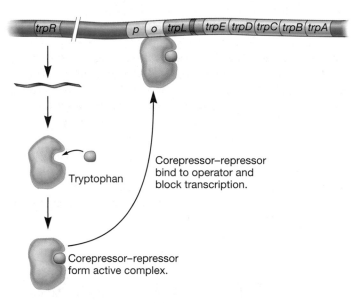

(b) High tryptophan levels, repression occurs

Figure 13.8   Regulation of the *trp* Operon by Tryptophan and the *trp* Repressor.   The *trp* repressor is inactive when first synthesized and therefore is unable to bind the operator. It is activated by the binding of tryptophan, which serves as the corepressor. (a) When tryptophan levels are low, the repressor is inactive and transcription occurs. The enzymes encoded by the operon catalyze the reactions needed for tryptophan biosynthesis. (b) When tryptophan levels are sufficiently high, it binds the repressor. The repressor-corepressor complex binds the operator and transcription of the operon is inhibited.

Tryptophan Operon: Negative Transcriptional Control of Repressible Genes

The tryptophan (*trp*) operon of *E. coli* consists of five structural genes that encode enzymes needed for synthesis of the amino acid tryptophan (**figure 13.8**). It is regulated by the *trp*

repressor, which is encoded by the *trpR* gene. Because the enzymes encoded by the *trp* operon function in a biosynthetic pathway, it is wasteful to make the enzymes needed for tryptophan synthesis when tryptophan is readily available. This is especially true because tryptophan is the most difficult amino acid to make, consuming many materials and considerable energy. Therefore, the operon functions only when tryptophan is not present and must be made de novo from precursor molecules (figure 13.4). To accomplish this regulatory goal, the *trp* repressor is synthesized in an inactive form that cannot bind the *trp* operator as long as tryptophan levels are low (figure 13.8*a*). When tryptophan levels increase, tryptophan acts as a corepressor, binding the repressor and activating it. The repressor-corepressor complex then binds the operator, blocking transcription initiation (figure 13.8b).

Like the *lac* operon, the *trp* operon is subject to another layer of regulation. In addition to being controlled at the level of transcription initiation by the *trp* repressor, expression of the *trp* operon is also controlled at the level of transcription elongation by a process called attenuation. This mode of regulation is discussed in section 13.3.

Arabinose Operon: Transcriptional Control by a Protein That Acts Both Positively and Negatively

Many regulatory proteins are versatile and can function as repressors for one operon and activators for others. The regulation of the *E. coli* arabinose (*ara*) operon illustrates how the same protein can function either positively or negatively, depending on the environmental conditions. The *ara* operon encodes enzymes needed for the catabolism of arabinose to xylulose 5-phosphate, an intermediate of the pentose phosphate pathway. The *ara* operon is regulated by AraC, which can bind three different regulatory sequences: *araO_2*, *araO_1*, and *araI* (**figure 13.9**). When arabinose is not present, one molecule of AraC binds *araI* and another binds *araO_2*. The two AraC proteins interact, causing the DNA to bend. This prevents RNA polymerase from binding to the promoter of the *ara* operon, thereby blocking transcription. In these conditions, AraC acts as a repressor (figure 13.9a). However, when arabinose is present, it binds AraC and prevents AraC molecules from interacting. This breaks the DNA loop. Furthermore, binding of two AraC-arabinose complexes to the *araI* site promotes transcription. Thus when arabinose is present, AraC acts as an activator (figure 13.9b). The *ara* operon, like the *lac* operon, is also subject to catabolite repression (section 13.5). << *Breakdown of glucose to pyruvate: Pentose phosphate pathway (section 10.3).*

Two-Component Regulatory Systems and Phosphorelay Systems

The activity levels of the *lac* repressor, *trp* repressor, and AraC protein are controlled by metabolites of those pathways. However, many environmental conditions do not produce a metabolite that can interact directly with a regulatory protein. These include temperature, osmolarity, and oxygen levels. How do organisms sense and respond to such stimuli? Many genes and operons are regulated in response to these types of signals by regulatory proteins that function in **two-component signal transduction systems.** These systems link events occurring outside the cell to gene expression inside the cell.

Two-component signal transduction systems are found in all three domains of life and are named after the two proteins that govern the regulatory pathway. The first is a **sensor kinase protein** that spans the cytoplasmic membrane so that part of it is exposed to the extracellular environment (periplasm, in

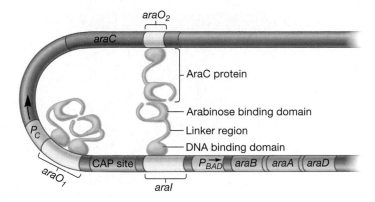

(a) Operon inhibited in the absence of arabinose

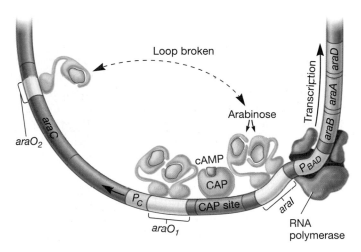

(b) Operon activated in the presence of arabinose

Figure 13.9 Regulation of the *ara* Operon by the AraC Protein. The AraC protein can act both as a repressor and as an activator, depending on the presence or absence of arabinose. (a) When arabinose is not available, the protein acts as a repressor. Two AraC proteins are involved. One binds the *araI* site and the other binds the *araO_2* site. The two proteins interact in such a way that the DNA between the two operator sites is bent, making it inaccessible to RNA polymerase. (b) When arabinose is present, it binds AraC, disrupting the interaction between the two AraC proteins. Subsequently, two AraC proteins, each bound to arabinose, form a dimer, which binds to the *araI* site. The AraC dimer functions as an activator and transcription occurs.

gram-negative bacteria) while another part is exposed to the cytoplasm (**figure 13.10**). In this way, it can sense specific changes in the environment and communicate information to the cell's interior. The second component is the **response-regulator protein.** The response regulator is a DNA-binding protein that when activated by the sensor kinase can act as either an activator or a repressor. As the former, it promotes transcription of genes or operons whose expression is needed for adaptation to the detected environmental stimulus. As the latter, it inhibits transcription of genes or operons whose products are no longer needed.

The regulation of the ratio of OmpF and OmpC porin proteins in *E. coli* is one of the best-understood two-component signal transduction systems. Recall that the outer membrane of gram-negative bacteria contains channels made of porin proteins

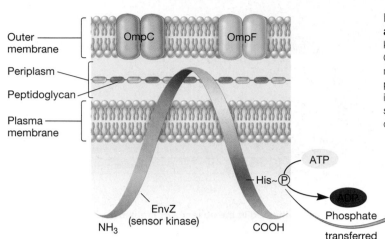

Figure 13.10 Two-Component Signal Transduction System and the Regulation of Porin Proteins. In this system, the sensor kinase protein EnvZ loops through the cytoplasmic membrane so that both its C- and N-termini are in the cytoplasm. When EnvZ senses an increase in osmolarity, it autophosphorylates a histidine residue at its C-terminus. EnvZ then passes the phosphoryl group to the response regulator OmpR, which accepts it on an aspartic acid residue located in its N-terminus. This activates OmpR so that it is able to bind DNA and repress *ompF* expression and enhance that of *ompC*.

(see figure 3.27). The two most important porins in *E. coli* are OmpF and OmpC (Omp for *outer membrane protein*). OmpC pores are slightly smaller and are made when the bacterium grows at high osmotic pressures. It is the dominant porin when *E. coli* is in the intestinal tract. The larger OmpF pores are favored when *E. coli* grows in a dilute environment; OmpF allows solutes to diffuse into the cell more readily. The cell must maintain a constant level of porin protein in the membrane, with the relative levels of the two porins corresponding to the osmolarity of the medium. Clearly, the cell must have a way of sensing changes in osmolarity so that *ompF* and *ompC* transcription can be adjusted as needed.

The sensor kinase in the OmpF:OmpC two-component regulatory system is the EnvZ protein (Env for cell *env*elope). It is an integral membrane protein anchored to the membrane by two membrane-spanning domains. EnvZ is looped through the membrane such that a central domain protrudes into the periplasm, while the amino and carboxyl termini are exposed to the cytoplasm (figure 13.10). The response-regulator protein, OmpR, is a soluble, cytoplasmic protein that regulates transcription of the *ompF* and *ompC* structural genes. The N-terminal end of OmpR is called the receiver domain because it possesses a specific aspartic acid residue that accepts the signal (a phosphoryl group) from the sensor kinase. Upon receipt of the signal, the C-terminal end of OmpR is able to regulate transcription by binding DNA. At low osmolarity, EnvZ is inactive, but when EnvZ senses that osmolarity has increased, EnvZ phosphorylates itself (autophosphorylation) on a specific histidine residue. This phosphoryl group is quickly transferred to the N-terminus of OmpR. Once OmpR is phosphorylated, it is able to regulate transcription of the porin genes so that *ompF* transcription is repressed and *ompC* transcription is activated.

Two-component signal transduction systems are simple in design: the signal recognized by the sensor kinase is directly transduced (sent) to the response regulator that mediates the required changes in gene expression; in many cases, numerous genes and operons may be regulated by the same response regulator. The effectiveness of two-component systems is illustrated by their abundance: most procaryotic cells use a variety of two-component

signal transduction systems to respond to an array of environmental stresses. For example, members of the soil bacterial genus *Streptomyces* have at least 80 such systems!

Two-component signal transduction systems involve a simple phosphorelay where the sensor kinase transfers a phosphoryl group directly to the response-regulator protein. However, there are instances when more proteins participate in the transfer of phosphoryl groups. These longer pathways are called **phosphorelay systems.** An important and well-studied phosphorelay system functions during sporulation in *Bacillus subtilis* and is described in section 13.5. It should be noted that some phosphorelay systems control protein activity rather than gene transcription. An example of this type of system is chemotaxis in *E. coli,* which is described in chapter 9.

1. Many genes and operons are regulated at the level of transcription initiation. What do you think are the advantages to regulating proteins before they are made rather than during or after translation?

2. What are induction and repression? How do bacteria use them to respond to changing nutrient supplies?

3. Define repressor protein, activator protein, operator, activator-binding site, inducer, corepressor, structural gene, and operon.

4. Using figure 13.4 as a guide, trace the "decision-making" pathway of an *E.coli* cell that is growing in a medium containing arabinose but lacking tryptophan.

5. *E. coli* has two phosphate uptake systems (one with high affinity for phosphate, the other with a low affinity). Describe how a two-component regulatory system might be used by this microbe to regulate phosphate transport.

13.3 REGULATION OF TRANSCRIPTION ELONGATION

Organisms can also regulate transcription by controlling the termination of transcription. In this type of regulation, transcription is initiated but prematurely stopped depending on the environmental conditions and the needs of the organism. Attenuation was the first demonstration of this level of regulation. It was discovered in the 1970s by studies of the *trp* operon. More recently, riboswitches have been discovered. These regulatory sequences in the leader of an mRNA both sense and respond to environmental conditions by either prematurely terminating transcription or blocking translation. Both attenuation and riboswitches are described in this section.

Attenuation

As noted in section 13.2, the tryptophan (*trp*) operon of *E. coli* is under the control of a repressor protein, and excess tryptophan inhibits transcription of operon genes by acting as a corepressor and activating the repressor protein. Although the operon is regulated mainly by repression, the continuation of transcription also is controlled. That is, there are two decision points involved in transcriptional control, the initiation of transcription and the continuation of transcription past the leader region.

This additional level of control serves to adjust levels of transcription in a more subtle fashion. When the repressor is not active, RNA polymerase begins transcription of the leader region but often does not progress to the first structural gene in the operon. Instead, transcription is terminated within the leader region; this is called **attenuation.** The ability to attenuate transcription is based on the nucleotide sequences in the leader region and on the fact that in procaryotes, transcription is coupled with translation *(see figure 12.32).* The leader of the *trp* operon mRNA is unusual in that it is translated. The product, which has never been isolated, is called the leader peptide. In addition to encoding the leader peptide, the leader contains **attenuator** sequences (**figure 13.11**). When transcribed, these sequences form stem-loop secondary structures in the newly formed mRNA. We define these sequences numerically (regions 1, 2, 3, and 4). When regions 1 and 2 pair with one another (1:2; figure 13.11*a*), they form a secondary structure called the pause loop, which causes RNA polymerase to slow down. The pause loop forms just prior to the formation of the terminator loop, which is made when regions 3 and 4 base pair (3:4; figure 13.11*a*). A poly(U) sequence follows the 3:4 terminator loop, just as it does in rho-independent transcriptional terminators *(see figure 12.26).* However, in this case, the terminator is in the leader rather than at the end of the gene. Another stem-loop structure can be formed in the leader region by the pairing of regions 2 and 3 (2:3, figure 13.11*b*). The formation of this antiterminator loop prevents the generation of both the 1:2 pause and 3:4 terminator loops.

How do these various loops control transcription termination? Three scenarios describe the process. In the first, translation is not coupled to transcription because protein synthesis is not occurring.

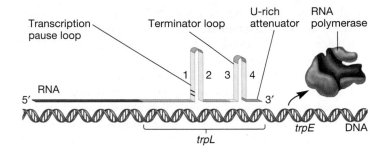

(a) No translation occurring

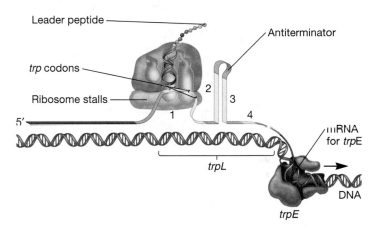

**(b) Translation occurring, low tryptophan levels,
2:3 forms ⟶ transcription continues**

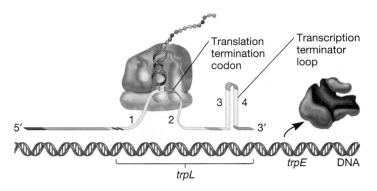

**(c) Translation occurring, high tryptophan levels,
3:4 forms ⟶ transcription is terminated**

Figure 13.11 Attenuation of the *trp* Operon. (a) When protein synthesis has slowed, transcription and translation are not tightly coupled. Under these conditions, the most stable form of the mRNA occurs when region 1 hydrogen bonds to region 2 (RNA polymerase pause loop) and region 3 hydrogen bonds to region 4 (transcription terminator or attenuator loop). The formation of the transcription terminator causes transcription to stop just beyond *trpL* (*trp* leader). (b) When protein synthesis is occurring, transcription and translation are coupled, and the behavior of the ribosome on *trpL* influences transcription. If tryptophan levels are low, the ribosome pauses at the *trp* codons in *trpL* because of insufficient amounts of charged tRNA^{trp}. This blocks region 1 of the mRNA, so that region 2 can hydrogen bond only with region 3. Because region 3 is already hydrogen bonded to region 2, the 3:4 terminator loop cannot form. Transcription proceeds and the *trp* biosynthetic enzymes are made. (c) If tryptophan levels are high, translation of *trpL* progresses to the stop codon, blocking region 2. Regions 3 and 4 can hydrogen bond and transcription terminates.

In other words, no ribosome is associated with the mRNA. In this scenario, the pause and terminator loops form, stopping transcription before RNA polymerase reaches the *trpE* gene (figure 13.11*a*).

In the next two scenarios, translation and transcription are coupled; that is, a ribosome associates with the leader mRNA as the rest of the mRNA is being synthesized. The interaction between RNA polymerase and the nearest ribosome determines which stem-loop structures are formed. As a ribosome translates the mRNA, it follows the RNA polymerase. Among the first several nucleotides of region 1 are two tryptophan (*trp*) codons; this is unusual because normally there is only one tryptophan residue per 100 amino acids in *E. coli* proteins. If tryptophan levels are low, there will not be enough charged tRNA^trp to fill the A site when the ribosome encounters the two *trp* codons, and it will stall (figure 13.11*b*). Meanwhile RNA polymerase continues to transcribe mRNA, moving away from the stalled ribosome. The presence of the ribosome on region 1 prevents region 1 from base pairing with region 2. As RNA polymerase continues, region 3 is transcribed, enabling the formation of the 2:3 antiterminator loop. This prevents the formation of the 3:4 terminator loop. Because the terminator loop is not formed, RNA polymerase is not ejected from the DNA and transcription continues into the *trp* biosynthetic genes. If, on the other hand, there is plenty of tryptophan in the cell, there will be an abundance of charged tRNA^trp and the ribosome will translate the two *trp* codons in the leader peptide sequence without hesitation. Thus the ribosome remains close to the RNA polymerase. As RNA polymerase and the ribosome continue through the leader, regions 1 and 2 are transcribed and readily form a pause loop. Then regions 3 and 4 are transcribed, the terminator loop forms, and RNA polymerase is ejected from the DNA template. Finally, the presence of a UGA stop codon between regions 1 and 2 causes early termination of translation (figure 13.11*c*). Although the leader peptide is synthesized, it appears to be rapidly degraded. ◂◂ *The genetic code (section 12.6)*

Attenuation's usefulness is apparent. If the bacterium is deficient in an amino acid other than tryptophan, protein synthesis will slow and tryptophanyl-tRNA will accumulate. Transcription of the tryptophan operon will be inhibited by attenuation. When the bacterium begins to synthesize protein rapidly, tryptophan may be scarce and the concentration of tryptophanyl-tRNA may be low. This would reduce attenuation activity and stimulate operon transcription, resulting in larger quantities of the tryptophan biosynthetic enzymes. Acting together, repression and attenuation can coordinate the rate of synthesis of amino acid biosynthetic enzymes with the availability of amino acid end products and with the overall rate of protein synthesis. When tryptophan is present at high concentrations, any RNA polymerases not blocked by the activated repressor protein probably will not get past the attenuator sequence. Repression decreases transcription about seventyfold and attenuation slows it another eight- to tenfold; when both mechanisms operate together, transcription can be slowed about 600-fold.

Attenuation is important in regulating at least five other operons that encode amino acid biosynthetic enzymes. In all cases, the leader peptide sequences resemble the tryptophan system in organization. For example, the leader peptide sequence of the histidine operon codes for seven histidines in a row and is followed by an attenuator that is a terminator sequence.

Riboswitches

Regulation by riboswitches (also called sensory RNAs) is a specialized form of transcription attenuation that involves mRNA folding but not ribosome behavior. In this case, if the leader of an mRNA is folded one way, transcription continues; if folded another, transcription is terminated. The leader region is called a **riboswitch** because it turns transcription on or off. What makes riboswitches unique and exciting is that the mRNA alters its folding pattern in direct response to the binding of an effector molecule—a capability previously thought to be associated only with proteins.

The riboswitch that regulates the riboflavin (*rib*) biosynthetic operon of *B. subtilis* will serve as our example (**figure 13.12**). The production of riboflavin biosynthetic enzymes is repressed by flavin mononucleotide (FMN), which is derived from riboflavin. When transcription of the *rib* operon begins, sequences in the leader region of the mRNA fold into a structure called the RFN-element. This

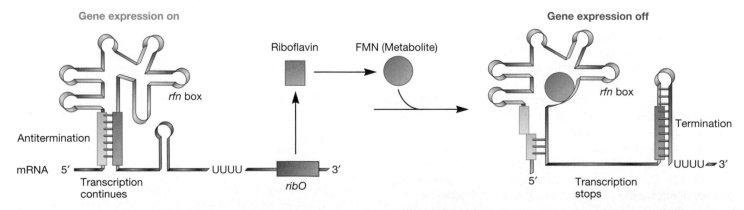

Figure 13.12 **Riboswitch Control of the Riboflavin (*rib*) Operon of *Bacillus subtilis*.** The *rib* operon produces enzymes needed for the synthesis of riboflavin, a component of flavin mononucleotide (FMN). Binding of FMN to the rfn (*rifampin*) box in the leader of *rib* mRNA causes a change in mRNA folding, which results in the formation of a transcription terminator and cessation of transcription.

Table 13.1	Regulation of Gene Expression by Riboswitches		
System	**Microbe(s)**	**Target genes encode:**	**Effector & Regulatory Response**
T box	Many gram-positive bacteria	Amino acid biosynthetic enzymes	Uncharged tRNA; anticodon base pairs to 5′ end of mRNA, preventing formation of transcriptional terminator
Vitamin B$_{12}$ element	*E. coli*	Cobalamine biosynthetic enzymes	Adenosylcobalamine (AdoCbl) binds to *btuB* mRNA and blocks translation
THI box	*Rhizobium etli* *E. coli* *B. subtilis*	Thiamin (Vitamin B$_1$) biosynthetic and transport proteins	Thiamin pyrophosphate (TPP) causes either premature transcriptional termination (*R. etli, B. subtilis*) or blocks ribosome binding (*E. coli*)
RFN-element	*B. subtilis*	Riboflavin biosynthetic enzymes	Flavin mononucleotide (FMN) cases premature transcriptional termination
S box	Low G + C gram-positive bacteria	Methionine biosynthetic enzymes	S-adenosylmethionine (SAM) causes premature transcriptional termination

element binds FMN and in doing so alters the folding of the leader region, creating a terminator that stops transcription.

Controlling transcription attenuation with sensory RNAs is an important method used by gram-positive bacteria to regulate amino acid-related genes. As in the case of the *rib* operon, the leader regions of these mRNAs contain a regulatory element. In this case, the region is called the T box. T box sequences give rise to competing terminator and antiterminator loops. The development of either a terminator or an antiterminator is determined by the binding of uncharged tRNA corresponding to the relevant amino acid. For instance, expression of a tyrosyl-tRNA synthetase gene (i.e., a gene that encodes the enzyme that links tyrosine to a tRNA molecule) is governed by the presence of tRNA^{Tyr}. When the level of charged tRNA^{Tyr} falls, the anticodon of an uncharged tRNA binds directly to the "specifier sequence" codon in the leader of the mRNA. At the same time, the antiterminator loop is stabilized by base pairing between sequences in the loop and the acceptor end of the tRNA, which normally binds the amino acid. This prevents formation of the terminator structure, and transcription of the tyrosyl-tRNA synthetase gene continues. Genomic analysis suggests that the T box mechanism may be involved in regulating over 300 genes and operons. Some other genes that bear sensory RNA in their leader regions are listed in **table 13.1.** Other riboswitches have been shown to function at the level of translation. They are described in the section 13.4.

13.4 REGULATION AT THE LEVEL OF TRANSLATION

It appears that in general, the riboswitches found in gram-positive bacteria function by transcriptional termination, whereas the riboswitches discovered in gram-negative bacteria regulate the translation of mRNA. Translation is usually regulated by block-

ing its initiation. In addition, some small RNA molecules can control translation initiation. Both are described in this section.

Regulation of Translation by Riboswitches

Similar to the riboswitches described in section 13.3, riboswitches that function at the translational level contain effector-binding elements at the 5′ end of the mRNA. Binding of the effector molecule alters the folding pattern of the mRNA leader, which often results in occlusion of the Shine-Dalgarno sequence and other elements of the ribosome-binding site. This inhibits ribosome binding and initiation of translation (**figure 13.13**). An example of this type of regulation is observed for the thiamine biosynthetic operons of numerous bacteria and some archaea. The leader regions of thiamine operons contain a structure called the THI-element, which can bind thiamin pyrophosphate. Binding of thiamin pyrophosphate to the THI-element causes a conformational change in the leader region that sequesters the Shine-Dalgarno sequence and blocks translation initiation.

Regulation of Translation by Small RNA Molecules

A large number of RNA molecules have been discovered that do not function as mRNAs, tRNAs, or rRNAs. Microbiologists often refer to them as **small RNAs (sRNAs)** or as noncoding RNAs (ncRNAs). In *E. coli,* there are more than 40 sRNAs, ranging in size from around 40 to 400 nucleotides. It is thought that eucaryotes may have hundreds to thousands of sRNAs with lengths from 21 to over 10,000 nucleotides. Although some sRNAs have been implicated in the regulation of DNA replication and transcription, many function at the level of translation.

In *E. coli,* most sRNAs regulate translation by base pairing to the leader region of a target mRNA. Thus they are complementary

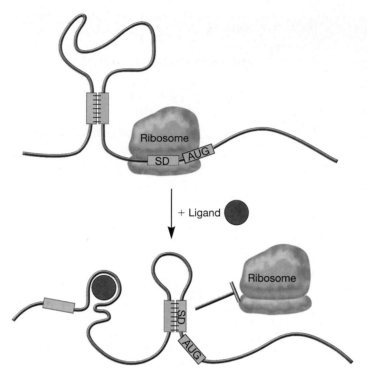

Figure 13.13 **Regulation of Translation by a Riboswitch.** In the absence of a relevant metabolite, an effector binding site is formed in the leader of the mRNA (red) when complementary sequences (orange box and green box) hydrogen bond. This folding pattern exposes important sequences in the ribosome-binding site (e.g., the Shine-Dalgarno sequence; blue box) and translation occurs. When the appropriate effector molecule is present, it binds the leader, disrupting the existing structure and creating a new structure with the ribosome-binding-site sequences. Thus the ribosome-binding site becomes inaccessible and translation is blocked.

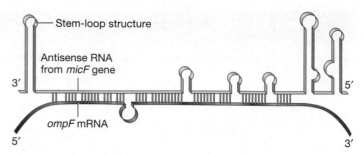

Figure 13.14 **Regulation of Translation by Antisense RNA.** The *ompF* mRNA encodes the porin OmpF. Translation of this mRNA is regulated by the antisense RNA MicF, the product of the *micF* gene. MicF is complementary to the *ompF* mRNA and, when bound to it, prevents translation.

high levels at the same time as OmpC protein. Some other antisense RNAs are listed in **table 13.2**.

1. Define attenuation. What are the functions of the leader region and ribosome in attenuation?

2. Describe how attenuation activity would vary when *E. coli* is placed in a tryptophan-rich medium, where it rapidly consumes the tryptophan within 8 hours of growth.

3. What are riboswitches? How are they similar to attenuation as described for the *trp* operon? How do they differ? How do riboswitches differ from sRNAs.

4. How do you think *micF* transcription is regulated?

to the mRNA and are called **antisense RNAs.** It seems intuitive that by binding to the leader, antisense RNAs would block ribosome binding and inhibit translation. Indeed, many antisense RNAs work in this manner. However, some antisense RNAs actually promote translation upon binding to the mRNA. Whether inhibitory or activating, most *E. coli* antisense RNAs work with a protein called Hfq to regulate their target RNAs. The Hfq protein is an RNA chaperone—that is, it interacts with RNA to promote changes in its structure. In addition, the Hfq protein may promote RNA-RNA interactions.

The regulation of synthesis of OmpF and OmpC porin proteins provides an example of translation control by an antisense RNA. In addition to regulation by the OmpR protein described previously (figure 13.10), expression of the *ompF* gene is regulated by an antisense RNA called MicF RNA, the product of the *micF* gene (*mic* for *m*RNA-*i*nterfering *c*omplementary RNA). The MicF RNA is complementary to *ompF* at the translation initiation site (**figure 13.14**). It base pairs with *ompF* mRNA and represses translation. MicF RNA is produced under conditions such as high osmotic pressure or the presence of some toxic material, both of which favor *ompC* expression. Production of MicF RNA helps ensure that OmpF protein is not produced at

13.5 GLOBAL REGULATORY SYSTEMS

Thus far, we have been considering the function of isolated operons. However, organisms must respond rapidly to a wide variety of changing environmental conditions and cope with such stressors as nutrient deprivation, desiccation, and major temperature fluctuations. They also have to compete successfully with other organisms for scarce nutrients and use these nutrients efficiently. These challenges require regulatory systems that can rapidly control many operons at the same time. Regulatory systems that affect many genes and pathways simultaneously are called **global regulatory systems.**

Although it is usually possible to regulate all the genes of a metabolic pathway in a single operon, there are good reasons for more complex global systems. Some processes involve too many genes to be accommodated in a single operon. For example, the machinery required for protein synthesis is composed of 150 or more gene products, and coordination requires a regulatory network that controls many separate operons. Sometimes two levels of regulation are required because individual operons must be controlled independently but also cooperate with other operons. Regulation of sugar catabolism in *E. coli* is a good example. *E. coli*

Table 13.2		Regulation of Gene Expression by Small Regulatory RNAs	
Small RNA	**Size**	**Bacterium**	**Function**
RhyB	90 nt[1]	*E. coli*	Represses translation of iron-containing proteins (e.g., *sodB*) when iron availability is low
Spot 42	109 nt	*E. coli*	Inhibits translation of *galK* mRNA (encodes galactokinase)
Rpr A	105 nt	*E. coli*	Promotes translation of *rpoS* mRNA; antisense repressor of global negative regulator H-NS (involved in stress responses)
MicF	109 nt	*E. coli*	Inhibits *ompF* mRNA translation
OxyS	109 nt	*E. coli*	Inhibits translation of transcriptional regulator *fhlA* mRNA and *rpoS* mRNA (encodes σ^s, a stationary phase sigma factor)
DsrA	85 nt	*E. coli*	Increases translation of *rpoS* mRNA
CsrB	366 nt	*E. coli*	Inhibits CsrA, a translational regulatory protein that positively regulates flagella synthesis, acetate metabolism, and glycolysis
RNAIII	512 nt	*Staphylococcus aureus*	Activates genes encoding secreted proteins (e.g., α hemolysin); Represses genes encoding surface proteins
RNA α	650 nt	*Vibrio anguillarum*	Decreased expression of Fat, an iron uptake protein
RsmB'	259 nt	*Erwinia carotovora* subsp. *carotovora*	Stabilizes mRNA of virulence proteins (e.g., cellulases, proteases, pectinolytic enzymes)

[1]nt: nucleotides

uses glucose when it is available; in such a case, operons for other catabolic pathways are repressed. If glucose is unavailable and another nutrient is present, only the appropriate operon is activated.

Global regulatory systems are so complex that a specialized nomenclature is used to describe the various kinds. Perhaps the most basic type is the **regulon.** A regulon is a collection of genes or operons that is controlled by a common regulatory protein. Usually the operons are associated with a single pathway or function (e.g., the production of heat-shock proteins or the catabolism of glycerol). A somewhat more complex situation is seen with a modulon. This is an operon network under the control of a common global regulatory protein but whose constituent operons also are controlled separately by their own regulators. A good example of a modulon is catabolite repression, which is discussed on p. 292. The most complex global systems are referred to as stimulons. A stimulon is a regulatory system in which all operons respond together in a coordinated way to an environmental stimulus. It may contain several regulons and modulons, and some of these may not share regulatory proteins. For instance, the genes involved in a response to phosphate limitation are scattered among several regulons and are part of one stimulon.

Mechanisms Used for Global Regulation

Global regulation is complex and often involves more than one regulatory mechanism. Most global regulatory networks are controlled by one or more regulatory proteins. Two-component regulatory systems and phosphorelay systems also play important roles in global control. In *Bacteria,* many global regulatory networks make use of **alternate sigma factors,** which can immediately change expression of many genes as they direct RNA polymerase to specific subsets of a bacterium's genome. This is possible because RNA polymerase core enzyme needs the assistance of a sigma factor to bind a promoter and initiate transcription. Each sigma factor recognizes promoters that differ in sequence, especially at the −10 and −35 positions. The specific sequences recognized by a given sigma factor are called its consensus sequences. When a complex process requires a radical change in transcription or a precisely timed sequence of transcription, it may be regulated by a series of sigma factors. << *Transcription (section 12.5)*

E. coli synthesizes several sigma factors (**table 13.3**). Under normal conditions, a sigma factor called σ^{70} directs RNA polymerase activity. (The superscript number or letter indicates the size or function of the sigma factor; 70 stands for 70,000 Da.) When flagella and chemotactic proteins are needed, *E. coli* produces σ^F (σ^{28}). σ^F then binds its consensus sequences in promoters of genes whose products are needed for flagella biosynthesis and chemotaxis. If the temperature rises too high, σ^H (σ^{32}) is produced and stimulates the formation of about 17 heat-shock proteins that protect the cell from thermal destruction. Importantly each sigma factor has its own set of promoters to which it binds.

In the discussions that follow, we describe three global regulatory networks. The first is the catabolite repression modulon, which involves regulation of transcription by both repressors and activators. The second is quorum sensing, which was introduced

Table 13.3	*E. coli* Sigma Factors
Sigma Factor	**Genes Transcribed**
σ^{70}	Genes needed during exponential growth
σ^{S}	Genes needed during the general stress response and during stationary phase
σ^{E}	Genes needed to restore membrane integrity and the proper folding of membrane proteins
σ^{H} (σ^{32})	Genes needed to protect against heat shock and other stresses, including genes encoding chaperones that help maintain or restore proper folding of cytoplasmic proteins and proteases that degrade damaged proteins
FecI σ	Genes that encode the iron citrate transport machinery in response to iron starvation and the availability of iron citrate
σ^{F} (σ^{28})	Genes involved in flagellum assembly
σ^{60}	Genes involved in nitrogen metabolism

in chapter 7. Finally, we examine sporulation in the gram-positive bacterium *B. subtilis*. Regulation of endospore formation involves numerous control mechanisms, including phosphorelay and sequential use of alternative sigma factors.

Catabolite Repression

If *E. coli* grows in a medium containing both glucose and lactose, it uses glucose preferentially until the sugar is exhausted. Then after a short lag, growth resumes with lactose as the carbon source (**figure 13.15**). This biphasic growth pattern is called **diauxic**

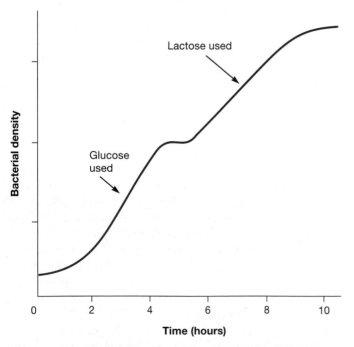

Figure 13.15 Diauxic Growth. The diauxic growth curve of *E. coli* grown with a mixture of glucose and lactose. Glucose is first used, then lactose. A short lag in growth is present while the bacteria synthesize the enzymes needed for lactose use.

growth. The cause of diauxic growth is complex and not completely understood, but **catabolite repression** plays a part. The enzymes for glucose catabolism are constitutive. However, operons that encode enzymes required for the catabolism of carbon sources that must first be modified before entering glycolysis (e.g., the *lac* operon) are regulated by catabolite repression. These include the *ara, mal* (maltose), and *gal* (galactose) operons, as well as the *lac* operon. Collectively, these can be called catabolite operons, and their expression is coordinately (or globally) repressed when glucose is plentiful.

The coordinated regulation of catabolite operons is brought about by **catabolite activator protein (CAP),** which is also called cyclic AMP receptor protein (CRP). CAP exists in two states: it is active when the small cyclic nucleotide **3′, 5′-cyclic adenosine monophosphate (cAMP; figure 13.16)** is bound, and it is inactive when it is free of cAMP. The levels of cAMP are controlled by the enzyme adenyl cyclase, which converts ATP to cAMP and PP$_i$. Adenyl cyclase is active only when little or no glucose is available. Thus the level of cAMP varies inversely with that of glucose: when glucose is unavailable and the catabolism of another sugar might be needed, the amount of cAMP in the cell increases, allowing cAMP to bind to and activate CAP.

All catabolite operons contain a CAP binding site, and CAP must be bound to this site before RNA polymerase can bind the promoter and begin transcription. Upon binding, CAP bends the DNA within two helical turns, which stimulates transcription (figure 13.6*b* and **figure 13.17**). Thus all catabolite operons are controlled by two regulatory proteins: the regulatory protein specific to each operon (e.g., *lac* repressor and AraC protein) and CAP. In the case of the *lac* operon, if glucose is absent and lactose is present, the inducer allolactose will bind to and inactivate the *lac* repressor protein, CAP will be in the active form (with cAMP bound), and transcription will proceed (**figure 13.18*a***). However, if glucose and lactose are both in short supply, even though CAP binds to the *lac* promoter, transcription will be

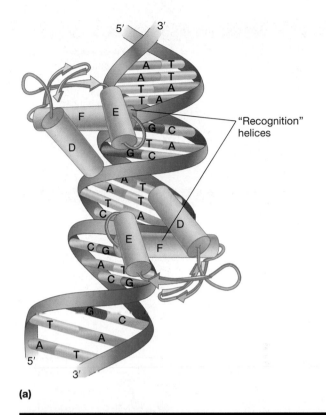

Figure 13.16 Cyclic Adenosine Monophosphate (cAMP). The phosphate extends between the 3′ and 5′ hydroxyls of the ribose sugar. The enzyme adenyl cyclase forms cAMP from ATP.

inhibited by the presence of the repressor protein, which remains bound to the operator in the absence of inducer (figure 13.18c). Dual control ensures that the *lac* operon is expressed only when lactose catabolic genes are needed.

We have seen how CAP controls catabolite operons; now let us turn our attention to the regulation of the levels of cAMP. The decrease in cAMP levels that occurs when glucose is present is due to the effect of the phosphoenolpyruvate: phosphotransferase system (PTS) on the activity of adenyl cyclase. Recall from chapter 6 that in the PTS, a phosphoryl group is transferred by a series of proteins from phosphoenolpyruvate (PEP) to glucose, which then enters the cell as glucose 6-phosphate *(see figure 6.7)*. When glucose is present, enzyme IIA transfers the phosphoryl group to enzyme IIB, which phosphorylates glucose. However, when glucose is absent, the phosphoryl groups from PEP are transferred to enzyme IIA but are not transferred to enzyme IIB. The phosphorylated form of enzyme IIA accumulates. This form of the enzyme activates adenyl cyclase, stimulating cAMP production. << *Uptake of nutrients: Group translocation (section 6.6)*

Catabolite repression is of considerable advantage to *E. coli.* It will use the most easily catabolized sugar (glucose) first rather than synthesize the enzymes necessary for catabolism of another

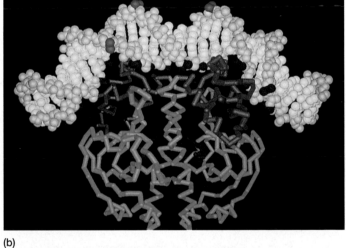

(b)

Figure 13.17 CAP Structure and DNA Binding. (a) The CAP dimer binding to DNA at the *lac* operon promoter. The recognition helices fit into two adjacent major grooves on the double helix. (b) A model of the *E. coli* CAP-DNA complex derived from crystal structure studies. The cAMP-binding domain is in blue and the DNA-binding domain, in purple. The cAMP molecules bound to CAP are in red. Note that the DNA is bent when complexed with CAP.

carbon and energy source. Catabolite repression is used by a variety of bacteria to regulate numerous metabolic pathways.

Quorum Sensing

Cell-to-cell communication among procaryotes occurs by the exchange of small molecules often termed signals or signaling molecules. The exchange of signaling molecules is essential in

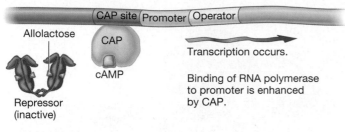

(a) Lactose but no glucose

Transcription occurs.

Binding of RNA polymerase to promoter is enhanced by CAP.

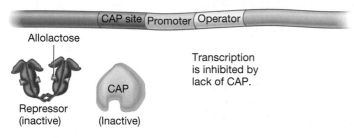

(b) Lactose and glucose

Transcription is inhibited by lack of CAP.

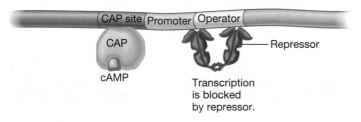

(c) Neither lactose nor glucose

Transcription is blocked by repressor.

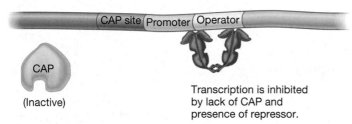

(d) Glucose but no lactose

Transcription is inhibited by lack of CAP and presence of repressor.

Figure 13.18   Regulation of the *lac* Operon by the *lac* Repressor and CAP.   A continuum of *lac* mRNA synthesis is brought about by the action of CAP, an activator protein, and the *lac* repressor. (a) When lactose is available and glucose is not, the repressor is inactivated and cAMP levels increase. Cyclic AMP binds CAP, activating it. CAP binds the CAP binding site near the *lac* promoter and facilitates binding of RNA polymerase. Under these conditions, transcription occurs at maximal levels. (b) When both lactose and glucose are available, both CAP and the *lac* repressor are inactive. Because RNA polymerase cannot bind the promoter efficiently without the aid of CAP, transcription levels are low. (c) When neither glucose nor lactose is available, both CAP and the *lac* repressor are active. In this situation, both proteins are bound to their regulatory sites. CAP binding enhances the binding of RNA polymerase to the promoter. However, the repressor blocks transcription. Transcription levels are low. (d) When glucose is available and lactose is not, CAP is inactive and the *lac* repressor is active. Thus RNA polymerase binds inefficiently, and those polymerase molecules that do bind are blocked by the repressor. This condition results in the lowest levels of transcription observed for the *lac* operon.

the coordination of gene expression in microbial populations. This was first recognized in the marine bioluminescent bacterium *Vibrio fischeri,* which produces light only if cells are at high density. It has since been discovered that intercellular communication plays an essential role in the regulation of genes whose products are needed for the establishment of virulence, symbiosis, biofilm production, plasmid transfer, and morphological differentiation in a wide range of microorganisms. Here we describe how signals that are secreted by a group of cells can regulate the genetic expression of that population. Our focus is on the regulation of a single operon. However, it should be kept in mind that **quorum sensing** can regulate multiple genes and operons. ◀◀ *Microbial growth in natural environments: Cell-cell communication within microbial populations (section 7.6)*

Quorum sensing in *V. fischeri* and many other gram-negative bacteria uses an **N-acylhomoserine lactone (AHL)** signal (**figure 13.19**). Synthesis of this small molecule is catalyzed by an enzyme called AHL synthase, the product of the *luxI* gene. The *luxI* gene is subject to positive autoregulation. That is to say, transcription of *luxI* increases as AHL accumulates in the cell. This is accomplished through the transcriptional activator LuxR, which is active only when it binds AHL. Thus a simple feedback loop is created. Without AHL-activated LuxR, the *luxI* gene is transcribed only at basal levels. AHL freely diffuses out of the cell and accumulates in the environment. As cell density increases, the concentration of AHL outside the cell eventually exceeds that inside the cell, and the concentration gradient is reversed. As AHL flows back into the cell, it binds and activates LuxR. LuxR can now activate high-level transcription of *luxI* and the genes whose products are needed for bioluminescence *(luxCDABEG)*. Quorum sensing is often called **autoinduction** and the AHL signal is termed **autoinducer (AI)** to reflect the autoregulatory nature of this system.

Another kind of quorum sensing depends on an elaborate, two-component signal transduction system. It is found in both gram-negative and gram-positive bacteria, including *Staphylococcus aureus, Ralstonia solanacearum, Salmonella enterica, Vibrio cholerae,* and *E. coli.* It has been best studied in the bioluminescent bacterium *Vibrio harveyi.* Unlike *V. fischeri, V. harveyi* responds to two autoinducer molecules: AI-1 and AI-2. AI-1 is a homoserine lactone and its synthesis depends on the *luxM* gene. AI-2 is furanosylborate, a small molecule that contains a boron atom—quite an unusual component in an organic molecule *(see figure 7.27).* Its synthesis relies on the product of the *luxS* gene. As shown in **figure 13.20,** AI-1 and AI-2 are secreted by the cell, which then uses separate proteins called LuxN and LuxPQ to detect their presence. At low cell density in the absence of either AI-1 or AI-2, LuxN and LuxPQ autophosphorylate and converge on a single phosphotransferase protein called LuxU. LuxU accepts phosphates from each sensor kinase and then phosphorylates the response regulator LuxO. Phosphorylated LuxO in turn activates the transcription of genes encoding several small RNAs that destabilize *luxR* mRNA. Because LuxR is a transcriptional activator of the operon *luxCDABE,* which encodes proteins needed for bioluminescence, cells do not make light at low cell density. An interesting

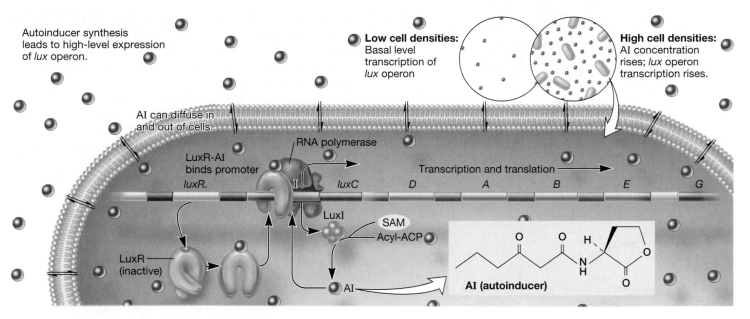

Figure 13.19 Quorum Sensing in V. fischeri. The AHL signaling molecule diffuses out of the cell; when cell density is high, AHL diffuses back into the cell where it binds to and activates the transcriptional regulator LuxR. Active LuxR then stimulates transcripton of the gene coding for AHL synthase (*luxI*) as well as the genes encoding proteins needed for light production.

thing happens as cell and autoinducer densities increase: Lux N binds AI-1 and LuxPQ binds AI-2, and the proteins switch from functioning as kinases to phosphatases, proteins that dephosphorylate rather than phosphorylate their substrates. The flow of phosphates is now reversed; LuxO is inactivated by dephosphorylation and *luxR* mRNA is translated. LuxR now activates transcription of *luxCDABE* and light is produced. Careful inspection of figure 13.20 reveals that another set of genes is controlled by the AI-1–AI-2 system of *V. harveyi*. In this microbe, genes for a type III protein secretion system (TTSS) are controlled in the opposite manner as those for bioluminescence.

LuxS-type autoinducers are produced by a number of bacteria, but the precise AI-2 structure is specific for each species. Nonetheless, bacteria of different species can "talk" to each other because the products of the LuxS enzymes spontaneously rearrange. Thus in a bacterial community consisting of several AI-2–producing bacterial species, AI-2 molecules in the environment interconvert. This means that individual cells may be responding to their own signal and that produced by other species. LuxS-producing bacteria can also interfere with each other's communication. This has been shown experimentally. When *V. harveyi* and *E. coli* are grown together, *E. coli* consumes AI-2 produced by *V. harveyi,* thereby inhibiting light production and promoting TTSS production at high cell density. Thus although the regulation of gene expression is generally considered in the context of a single cell or at least a single species, it appears that in nature, microbes are responding not only directly to the environment but to each other as well.

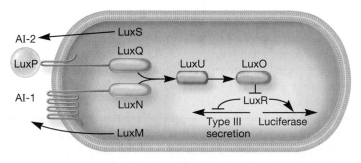

Figure 13.20 Quorum Sensing in V. harveyi. Two autoinducing signals, AI-1 and AI-2, are produced. At low cell density, the two-component signal transduction system consisting of the signal kinases LuxPQ and LuxN initiate a phosphorelay that results in inhibition of the transcriptional regulator LuxR. Without LuxR, bioluminescence genes are repressed but genes for a type III secretion system (TTSS) are transcribed. At high cell density, LuxPQ and LuxN function as phosphatases, reversing the flow of phosphates. This results in activation of LuxR, transcriptional activation of the bioluminescence operon and repression of the TTSS genes.

Sporulation in *Bacillus subtilis*

As discussed in chapter 3, endospore formation is a complex process that involves asymmetric division of the cytoplasm to yield a large mother cell and a smaller forespore, engulfment of the forespore by the mother cell, and construction of additional layers of spore coverings (**figures 13.21***a* and *3.46*). Sporulation takes approximately 8 hours. It is controlled by phosphorelay, posttranslational modification of proteins, numerous transcription initiation regulatory proteins, and alternate sigma factors. The latter are particularly important. When growing vegetatively, *B. subtilis* RNA polymerase uses sigma factors σ^A and σ^H to recognize genes for normal survival. However, when cells sense

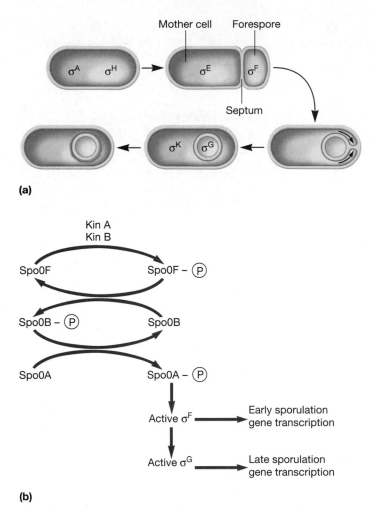

(a)

(b)

Figure 13.21 Genetic Regulation of Sporulation in *Bacillus subtilis.* (a) The initiation of sporulation is governed in part by the activities of two spatially separated sigma factors. σ^F is located in the forespore, while σ^E is confined to the mother cell. These sigma factors direct the initiation of transcription of genes whose products are needed for early events in sporulation. Later, σ^G and σ^K are localized to the developing endospore and mother cell, respectively. They control the expression of genes whose products are involved in the later steps of sporulation. (b) The activation of σ^F is accomplished through a phosphorelay system that is triggered by the activation of the sensor kinase protein KinA. When KinA senses starvation, it autophosphorylates a specific histidine residue. The phosphoryl group is then passed in relay fashion from SpoOF to SpoOB and finally to SpoOA.

a starvation signal, a cascade of events is initiated that results in the production of alternative sigma factors that are differentially expressed in the developing endospore and mother cell.

Initiation of sporulation is controlled by the protein SpoOA, a response-regulator protein that is part of a phosphorelay system (figure 13.21*b*). Sensor kinases associated with this system detect environmental stimuli that trigger sporulation. One of the most important sensor kinases is KinA, which senses nutrient starvation. When *B. subtilis* finds its nutrients are depleted, KinA autophosphorylates a specific histidine residue. The phosphoryl group is then transferred to an aspartic acid residue on SpoOF. However, SpoOF cannot directly regulate gene expression;

instead, SpoOF donates the phosphoryl group to a histidine on SpoOB. SpoOB in turn relays the phosphoryl group to SpoOA. Phosphorylated SpoOA positively controls genes needed for sporulation and negatively controls genes that are not needed. In response to SpoOA, the expression of over 500 genes is altered. Among the genes whose expression is stimulated by SpoOA is *sigF*, the gene encoding sigma factor σ^F, and *spoIIGB*, the gene encoding an inactive form of σ^E (pro-σ^E).

When sporulation starts, the chromosome has replicated, with one copy remaining in the mother cell and another to be partitioned in the forespore. Shortly after the formation of the spore septum, σ^F is found in the forespore, and pro-σ^E is localized to the mother cell. Pro-σ^E is cleaved by a protease to form active σ^E. The two sigma factors, σ^F and σ^E, bind to the promoters of genes needed in the forespore and mother cell, respectively. There they direct the expression of genes whose products are needed for the early steps of endospore formation. One of the many genes σ^F regulates is a gene that encodes another sigma factor, σ^G, which will replace σ^F in the developing endospore. Likewise, σ^E directs the transcription of a mother-cell-specific sigma factor, σ^K. Like σ^E, σ^K is first produced in an inactive form, pro-σ^K. Upon activation of pro-σ^K by proteolysis, σ^K ensures that genes encoding late-stage sporulation products are transcribed. Overall, temporal regulation is achieved because σ^F and σ^E direct transcription of genes needed early in the sporulation process, while σ^G and σ^K are needed for the transcription of genes whose products are needed later. In addition, spatial control of gene expression is accomplished because σ^F and σ^G are located in the forespore and σ^E and σ^K are found only in the mother cell.

13.6 REGULATION OF GENE EXPRESSION IN *EUCARYA* AND *ARCHAEA*

As is the case in *Bacteria,* the regulation of gene expression in *Eucarya* and *Archaea* can occur at transcriptional, translational, and posttranslational levels (figure 13.1). Much of the work on gene regulation in *Eucarya* has focused on transcription initiation. More recently regulation by small RNA molecules has attracted considerable attention. Unfortunately our understanding of the regulation of archaeal gene expression lags considerably behind what we know for *Eucarya* and *Bacteria.* However, some intriguing discoveries are briefly introduced here.

Transcription initiation in *Eucarya* involves numerous transcription factors (*see figure 12.28*). Many transcription factors, such as TFIID, are general transcription factors that are part of the machinery common to transcription initiation of all eucaryotic genes. On the other hand, **regulatory transcription factors** are specific to one or more genes and alter the rate of transcription. Those transcription factors that function as activators bind regulatory sites called **enhancers,** whereas those that function as repressors bind sites called **silencers (figure 13.22)**. After binding an enhancer or silencer, regulatory transcription factors act indirectly to increase or decrease the rate of transcription. Many regulatory transcription factors control transcription initiation by interacting with general transcription factors, in particular TFIID (figure 13.22*a*) and a multisubunit protein complex called mediator (figure 13.22*b*).

Another regulatory mechanism observed in eucaryotes (and *Bacteria*) is the use of sRNA molecules (p. 289) to control gene expression. Many sRNAs act as antisense RNAs and function at the level of translation, as described in section 13.4. Some eucaryotic antisense RNAs are much smaller than typical

bacterial antisense RNAs and are called microRNAs (miRNAs). Some sRNA molecules are important components of the spliceosome, where they contribute to the selection of splice sites used during mRNA processing. In doing so, different proteins can be made at certain times in the life cycle of the organism by combining different exons.

The regulation of gene expression in the *Archaea* has garnered a great deal of interest. This is because the archaeal transcription and translation machinery is most similar to that of the *Eucarya,* yet it functions in cells with typical, bacteria-like genome organization. The question being asked is whether archaeal regulation of gene expression is more like bacterial regulation or

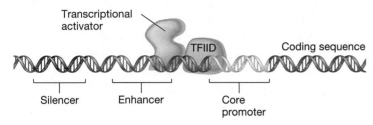

The transcriptional activator recruits TFIID to the core promoter and/or activates its function. Transcription will be activated.

The transcriptional repressor inhibits the binding of TFIID or inhibits its function. Transcription is repressed.

(a) Regulatory transcription factors and TFIID

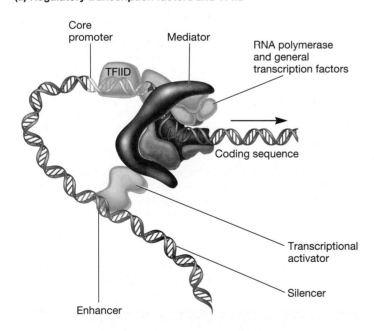

The transcriptional activator interacts with mediator. This enables RNA polymerase to form a preinitiation complex that can proceed to the elongation phase of transcription.

The transcriptional repressor interacts with mediator so that transcription is repressed.

(b) Regulatory transcription factors and mediator

Figure 13.22 The Activity of Eucaryotic Regulatory Transcription Factors.   Regulatory transcription factors do not exert their effects directly on RNA polymerase. Instead they act via other proteins, most commonly the general transcription factors TFIID and a protein called mediator. Mediator's role in transcription is to aid RNA polymerase in switching from the initiation stage of transcription to the elongation stage. (a) A regulatory protein acting through TFIID. Activators could influence transcription-enhancing TFIID recruitment of RNA polymerase to the promoter. Repressors would inhibit this ability. (b) A regulatory protein acting through the mediator protein. An activator would stimulate mediator activity; a repressor would decrease mediator activity.

more like eucaryotic regulation. Thus far, the answer is mixed. Most of the archaeal regulatory proteins function much like bacterial activators and repressors—that is, they bind DNA sites near the promoter and enhance or block binding of RNA polymerase, respectively. However, a few seem to function more like eucaryotic regulatory transcription factors in that they bring about their effects by interacting with DNA binding proteins, such as the TATA-binding protein (*see figure 12.28*). Small RNA molecules also have been identified in some archaea; their role in regulation is still being elucidated. >> *Introduction to the Archaea: Genetics and molecular biology (section 18.1)*

1. What are global regulatory systems and why are they necessary? Briefly differentiate between regulons, modulons, and stimulons.

2. What is diauxic growth? Explain how catabolite repression causes diauxic growth.

3. Describe the events that occur in each of the following growth conditions: *E. coli* in a medium containing glucose but not lactose; in a medium containing both

sugars; in a medium containing lactose but no glucose; and in a medium containing neither sugar.

4. What would be the phenotype of a *V. fischeri* mutant that could not regulate *luxI*, so that it was constantly producing autoinducer at high levels?

5. Why do you think bacteria use quorum sensing to regulate genes needed for virulence? How might this reason be related to the rationale behind using quorum sensing to establish a symbiotic relationship?

6. Briefly describe how a phosphorelay system and sigma factors are used to control sporulation in *B. subtilis*. Give one example of posttranslational modification as a means to regulate this process.

7. How are bacterial regulatory proteins and eucaryotic regulatory transcription factors similar? How do they differ?

8. What regulatory sequences in bacterial genomes are analogous to the enhancers and silencers observed in eucaryotic genomes?

Summary

13.1 Levels of Regulation of Gene Expression

a. Regulation of gene expression can be controlled at many levels, including transcription initiation, transcription elongation, translation, and posttranslation (**figure 13.1**).

b. The three domains of life differ in terms of their genome structure and the steps required to complete gene expression. These differences affect the regulatory mechanisms they use.

13.2 Regulation of Transcription Initiation

a. Induction and repression of enzyme levels are two important regulatory phenomena. They usually occur because of the activity of regulatory proteins.

b. Regulatory proteins are DNA-binding proteins. When bound to DNA, they can either inhibit transcription (negative control) or promote transcription (positive control). Their activity is modulated by small effector molecules called inducers, corepressors, and inhibitors (**figure 13.3**).

c. Repressors are responsible for negative control. They block transcription by binding an operator and interfering with the binding of RNA polymerase to its promoter or by blocking the movement of RNA polymerase after it binds DNA.

d. Activator proteins are responsible for positive control. They bind DNA sequences called activator-binding sites and, in doing so, promote binding of RNA polymerase to its promoter.

e. The *lac* operon of *E. coli* is an example of a negatively controlled inducible operon. When there is no lactose in the surroundings, the *lac* repressor is active and transcription is blocked. When lactose is available, it is converted to allolactose by the enzyme β-galactosidase. Allolactose acts as the inducer of the *lac* operon by binding the repressor and inactivating it. The inactive repressor cannot bind the operator and transcription occurs (**figure 13.7**).

f. The *trp* operon of *E. coli* is an example of a negatively controlled repressible operon. When tryptophan is not available, the *trp* repressor is inactive and transcription occurs. When tryptophan levels are high, tryptophan acts as a corepressor and binds the *trp* repressor, activating it. The *trp* repressor binds the operator and blocks transcription (**figure 13.8**).

g. The *ara* operon of *E. coli* is an example of an inducible operon that is regulated by the dual-function regulatory protein AraC. AraC functions as a repressor when arabinose is not available. It functions as an activator when arabinose, the inducer, is available (**figure 13.9**).

h. Some regulatory proteins are members of two-component signal transduction systems and phosphorelay systems. These systems have a sensor kinase that detects an environmental change. The sensor kinase transduces the environmental signal to the response-regulator protein either directly (two-component system) or indirectly (phosphorelay) by transferring a phosphoryl group to it. The response regulator then activates genes needed to adapt to the new environmental conditions and inhibits expression of those genes that are not needed (**figure 13.10**).

13.3 Regulation of Transcription Elongation

a. In the tryptophan operon, a leader region lies between the operator and the first structural gene (**figure 13.11**). It codes for the synthesis of a leader peptide and contains an attenuator, a rho-independent termination site. Synthesis of the leader peptide by a ribosome while RNA polymerase is transcribing the leader region regulates transcription: therefore the tryptophan operon is expressed only when there is insufficient tryptophan available. This mechanism of transcription control is called attenuation.

b. The leader regions of some mRNA molecules can bind metabolites that act as effector molecules. Binding of the metabolite to the mRNA causes a change in the leader structure, which can terminate transcription. This regulatory mechanism is called a riboswitch **(figure 13.12).**

13.4 Regulation at the Level of Translation

a. Some riboswitches regulate gene expression at the level of translation. For these riboswitches, the binding of a small molecule to specific sequences in the leader region of the mRNA alters leader structure and prevents ribosome binding **(figure 13.13).**

b. Translation can also be controlled by antisense RNAs. These small RNA molecules are noncoding. They base pair to the mRNA and usually inhibit translation **(figure 13.14).**

13.5 Global Regulatory Systems

a. Global regulatory systems control many operons simultaneously and help microbes respond rapidly to a wide variety of environmental challenges.

b. Global regulatory systems often involve many layers of regulation. Mechanisms such as regulatory proteins, alternate sigma factors, two-component signal transduction systems, and phosphorelay systems are often used.

c. Diauxic growth is observed when *E. coli* is cultured in the presence of glucose and another sugar such as lactose **(figure 13.15).** This growth pattern is the result of catabolite repression, where glucose is used preferentially over other sugars. Operons that are part of the catabolite repression system are regulated by the activator protein CAP. CAP activity is modulated by cAMP, which is produced only when glucose is not available. Thus when there is no glucose, CAP is active and promotes transcription of operons needed for the catabolism of other sugars **(figure 13.18).**

d. Quorum sensing is a type of cell-to-cell communication mediated by small signaling molecules such as *N*-acyl-homoserine lactone (AHL). Quorum sensing couples cell density to regulation of transcription. Well-studied quorum-sensing systems include the regulation of bioluminescence in *Vibrio* spp. **(figures 13.19** and **13.20).** Other systems regulate virulence genes and biofilm formation.

e. Endospore formation in *B. subtilis* is another example of a global regulatory system. Two important regulatory mechanisms used during sporulation are a phosphorelay system that is important in initiation of sporulation and the use of alternate sigma factors **(figure 13.21).**

13.6 Regulation of Gene Expression in *Eucarya* and *Archaea*

a. Regulatory transcription factors are used by *Eucarya* to control transcription initiation. They can exert either positive or negative control **(figure 13.22).**

b. Antisense RNAs are used by *Eucarya* to regulate translation.

c. Microbiologists know relatively little about archaeal regulation of gene expression. Some archaea use regulatory proteins that are similar to bacterial activators and repressors to control initiation of transcription.

Critical Thinking Questions

1. Attenuation affects anabolic pathways, whereas repression affects either anabolic or catabolic pathways. Provide an explanation for this.

2. Describe the phenotype of the following *E. coli* mutants when grown in two different media: glucose only and lactose only. Explain the reasoning behind your answer.

   a. A strain with a mutation in the gene encoding the *lac* repressor; the mutant repressor cannot bind allolactose.

   b. A strain with a mutation in the gene encoding CAP; the mutant form of CAP binds but cannot release cAMP.

   c. A strain in which the Shine-Dalgarno sequence has been deleted from the gene encoding adenyl cyclase.

3. What would be the phenotype of an *E. coli* strain in which the tandem *trp* codons in the leader region were mutated so that they coded for serine instead?

4. What would be the phenotype of a *B. subtilis* strain whose gene for σ^G has been deleted? Consider the ability of the mutant to survive in nutrient-rich versus nutrient-depleted conditions.

5. Propose a mechanism by which a cell might sense and respond to levels of Na^+ in its environment.

Learn More

Learn more by visiting the Prescott website at www.mhhe.com/prescottprinciples, where you will find a complete list of references.

14

Chapter Glossary

Mechanisms of Genetic Variation

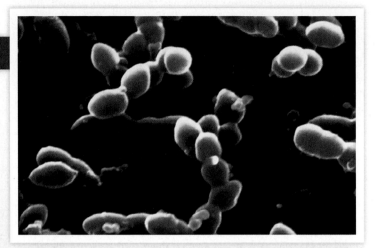

The scanning electron micrograph shows *Streptococcus pneumoniae*, the bacterium first used to study transformation and obtain evidence that DNA is the genetic material of organisms.

allele   An alternative form of a gene.

auxotroph   An organism with a mutation that causes it to lose the ability to synthesize an essential nutrient; the organism must obtain the nutrient or a precursor from its surroundings.

competent cell   A procaryotic cell that can take up free DNA and incorporate it into its genome during transformation.

conjugation   The form of gene transfer and recombination in procaryotes that requires direct cell-to-cell contact.

excision repair   A type of DNA repair mechanism in which damaged DNA is excised and replaced, using the complementary strand as a template.

F factor (F plasmid)   The fertility factor; a plasmid that carries genes for bacterial conjugation and makes its *E. coli* host the gene donor during conjugation.

frameshift mutation   Mutations arising from the loss or gain of a base or DNA segment, leading to a change in the codon reading frame.

generalized transduction   The transfer of any part of a procaryotic genome when the DNA fragment is packaged within a virus capsid by mistake.

Hfr strain   An *E. coli* strain that donates its genes with high frequency to a recipient cell during conjugation because the F factor is integrated into the donor's chromosome.

homologous recombination   Recombination involving two DNA molecules that are similar in nucleotide sequence.

horizontal (lateral) gene transfer (HGT)   The process by which genes are transferred from one mature, independent organism to another.

insertion sequence   A simple transposon that contains genes only for transposition.

mismatch repair system   A type of DNA repair in which a portion of a newly synthesized strand of DNA containing mismatched base pairs is removed and replaced, using the parental strand as a template.

missense mutation   A point mutation that changes a codon for one amino acid into a codon for another.

mutagen   A chemical or physical agent that causes mutations.

mutation   A heritable change in the genetic material.

nonsense mutation   A mutation that converts a sense codon to a nonsense or stop codon.

point mutation   A mutation that changes a single base pair.

prototroph   A microorganism that requires the same nutrients as most of the members of its species.

recombination   The process in which a new recombinant chromosome is formed by combining genetic material from two organisms.

site-specific recombination   Recombination of nonhomologous genetic material with a chromosome at a specific site.

SOS response   A complex, inducible process that allows bacterial cells with extensive DNA damage to survive, although often in a mutated form.

specialized transduction   A transduction process in which only a specific set of bacterial or archaeal genes is carried to a recipient cell by a temperate virus.

suppressor mutation   A mutation that overcomes the effect of another mutation and produces the normal phenotype.

transduction   The transfer of genes between bacterial or archaeal cells by viruses.

transformation   A mode of gene transfer in procaryotes in which a piece of free DNA is taken up by a cell and stably maintained.

transposition   The movement of a piece of DNA around a cell's genome.

transposon   A mobile genetic element that carries the genes required for transposition.

wild type   The genotype that is most commonly found in nature; it can refer to either a specific strain of a microbial species or a particular gene.

Deep in the cavern of the infant's breast
The father's nature lurks, and lives anew.

—*Horace, Odes*

Chapters 12 and 13 introduce the fundamentals of molecular genetics—the way genetic information is organized, stored, replicated, and expressed. For life to exist with stability, it is essential that the nucleotide sequence of genes is not disturbed to any great extent. However, sequence changes do occur and can result in altered phenotypes. These changes may be detrimental, but those that are not are important in generating new variability in populations and in contributing to the process of evolution.

In this chapter, we focus on processes that contribute to genetic variation in populations of microbes. We begin with an overview of the chemical nature of mutations and the effects of mutations at both the molecular and organismal levels. The chapter continues with a discussion of DNA repair mechanisms. Although these have evolved to prevent the

occurrence of mutations, some cellular attempts to correct DNA damage actually generate mutations. Finally, we examine microbial recombination and gene transfer in *Bacteria*. These processes are the basis for microbial evolution and have practical implications in terms of antibiotic resistance and the development of new infectious agents.

14.1 MUTATIONS AND THEIR CHEMICAL BASIS

Mutations (Latin *mutare,* to change) were initially characterized as altered phenotypes, but they are now understood at the molecular level. Several types of mutations exist. Some mutations arise from the alteration of single pairs of nucleotides and from the addition or deletion of one or two nucleotide pairs in the coding regions of a gene. Such small changes in DNA are sometimes called microlesions, and the smallest of these are called **point mutations** because they affect only one base pair in a given location. Larger mutations (macrolesions) are less common. These include large insertions, deletions, inversions, duplications, and translocations of nucleotide sequences.

Mutations occur in one of two ways: (1) **Spontaneous mutations** arise occasionally in all cells and in the absence of any added agent. (2) On the other hand, **induced mutations** are the result of exposure to a **mutagen,** which can be either a physical or a chemical agent. Mutations are characterized according to either the kind of genotypic change that has occurred or their phenotypic consequences. In this section, the molecular basis of mutations and mutagenesis is first considered. Then the phenotypic effects of mutations are discussed.

Spontaneous Mutations

Spontaneous mutations result from errors in DNA replication or from the action of mobile genetic elements such as transposons. A few of the more prevalent mechanisms are described here.

Replication errors can occur when the nitrogenous base of a template nucleotide takes on a rare tautomeric form. Tautomerism is the relationship between two structural isomers that are in chemical equilibrium and readily change into one another. Nitrogenous bases typically exist in the keto form. However, they can at times take on either an imino or enol form (**figure 14.1a**). These tautomeric shifts

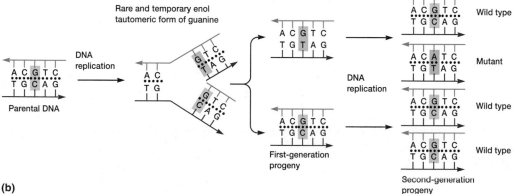

Figure 14.1 Tautomerization and Transition Mutations. Errors in replication due to base tautomerization. (a) Normally AT and GC pairs are formed when keto groups participate in hydrogen bonds. In contrast, enol tautomers produce AC and GT base pairs. The alteration in the base is shown in blue. (b) Mutation as a consequence of tautomerization during DNA replication. The temporary enolization of guanine leads to the formation of an AT base pair in the mutant, and a GC to AT transition mutation occurs. The process requires two replication cycles. Mutation only occurs if the abnormal first-generation GT base pair is missed by repair mechanisms. Wild type is the form of the gene before mutation occurred.

change the hydrogen-bonding characteristics of the bases, allowing purine for purine or pyrimidine for pyrimidine substitutions that can eventually lead to a stable alteration of the nucleotide sequence (figure 14.1*b*). Such substitutions are known as **transition mutations** and are relatively common. On the other hand, **transversion mutations,** mutations where a purine is substituted for a pyrimidine or a pyrimidine for a purine, are rarer due to the steric problems of pairing purines with purines and pyrimidines with pyrimidines.

Replication errors can also result in the insertion and deletion of nucleotides. These mutations generally occur where there is a short stretch of repeated nucleotides. In such a location, the pairing of template and new strand can be displaced by the distance of the repeated sequence, leading to insertions or deletions of bases in the new strand (**figure 14.2**).

Spontaneous mutations can originate from lesions in DNA as well as from replication errors. For example, it is possible for purine nucleotides to be depurinated—that is, to lose their base. This results in the formation of an apurinic site, which does not base pair normally and may cause a transition-type mutation after the next round of replication. Likewise, pyrimidines can be lost, forming an apyrimidinic site. Other lesions are caused by reactive forms of oxygen such as oxygen free radicals and peroxides produced during aerobic metabolism. For example, guanine can be converted to 8-oxo-7,8-dihydrodeoxyguanine, which often pairs with adenine rather than cytosine during replication.

Induced Mutations

Virtually any agent that damages DNA, alters its chemistry, or in some way interferes with its functioning will induce mutations. Mutagens can be conveniently classified according to their mode of action. Three common types of chemical mutagens are base analogs, DNA-modifying agents, and intercalating agents. A number of physical agents (e.g., radiation) are mutagens that damage DNA.

Base analogs are structurally similar to normal nitrogenous bases and can be incorporated into the growing polynucleotide chain during replication (**table 14.1**). Once in place, these compounds typically exhibit base pairing properties different from the bases they replace and can eventually cause a stable mutation. A widely used base analog is 5-bromouracil, an analog of thymine. It undergoes a tautomeric shift from the normal keto form to an enol much more frequently than does a normal base. The enol tautomer forms hydrogen bonds like cytosine, pairing with guanine rather than adenine. The mechanism of action of other base analogs is similar to that of 5-bromouracil.

There are many **DNA-modifying agents**—mutagens that change a base's structure and therefore alter its base pairing characteristics. Some of these mutagens are selective; they preferentially react with certain bases and produce a specific kind of DNA damage. For example, methyl-nitrosoguanidine is an alkylating agent that adds methyl groups to guanine, causing it to mispair with thymine (**figure 14.3**). A subsequent round of replication can then result in a GC-AT transition. Hydroxylamine is another example of a DNA-modifying agent. It hydroxylates the

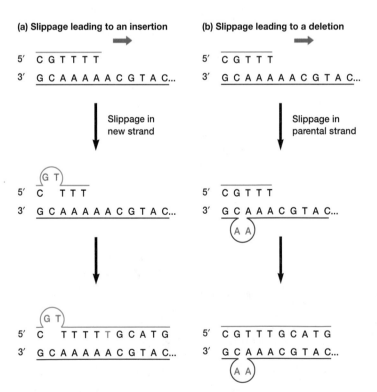

Figure 14.2   Insertions and Deletions. A mechanism for the generation of insertions and deletions during replication. The direction of replication is indicated by the blue arrow. In each case, there is strand slippage resulting in the formation of a small loop that is stabilized by hydrogen bonding in the repetitive sequence, the AT stretch in this example. (a) If the new strand slips, an addition of one T results. (b) Slippage of the parental strand yields a deletion (in this case, a loss of two Ts).

Table 14.1	Examples of Mutagens
Mutagen	**Effect(s) on DNA Structure**
Chemical	
5-Bromouracil	Base analog
2-Aminopurine	Base analog
Ethyl methanesulfonate	Alkylating agent
Hydroxylamine	Hydroxylates cytosine
Nitrogen mustard	Alkylating agent
Nitrous oxide	Deaminates bases
Proflavin	Intercalating agent
Acridine orange	Intercalating agent
Physical	
UV light	Promotes pyrimidine dimer formation
X rays	Causes base deletions, single-strand nicks, cross-linking, and chromosomal breaks

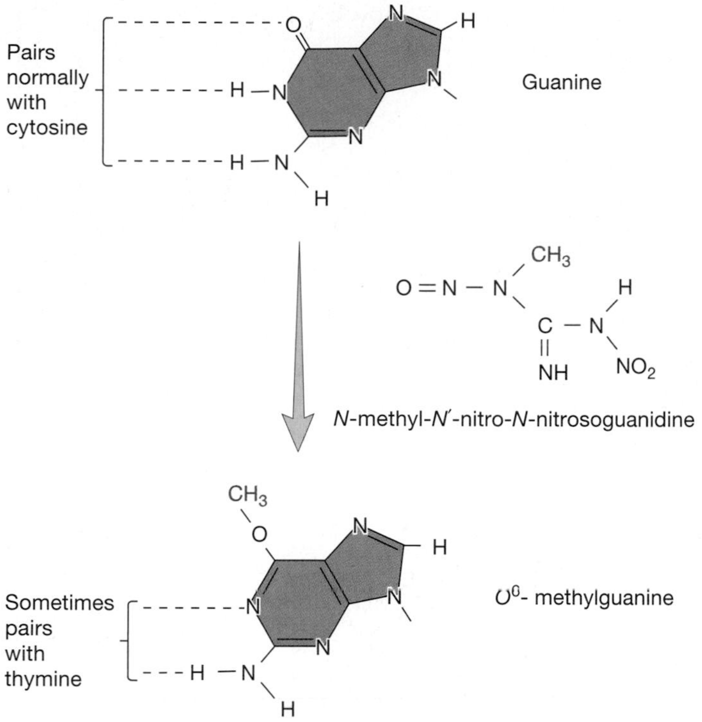

Figure 14.3 Methyl-Nitrosoguanidine Mutagenesis. Mutagenesis by methyl-nitrosoguanidine due to the methylation of guanine.

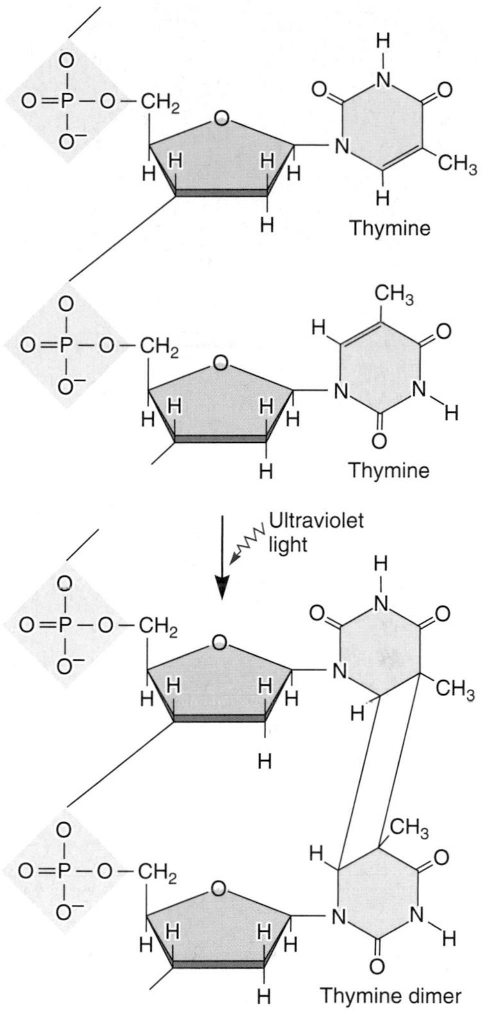

Figure 14.4 Thymine Dimer. Thymine dimers are formed by ultraviolet radiation.

C-4 nitrogen of cytosine *(see figure 12.3)*, causing it to base pair like thymine.

Intercalating agents distort DNA to induce single nucleotide pair insertions and deletions. These mutagens are planar and insert themselves (intercalate) between the stacked bases of the helix. This results in a mutation, possibly through the formation of a loop in DNA. Intercalating agents include acridines such as proflavin and acridine orange.

Many mutagens, and indeed many carcinogens, damage bases so severely that hydrogen bonding between base pairs is impaired or prevented and the damaged DNA can no longer act as a template for replication. For instance, UV radiation generates cyclobutane dimers, usually thymine dimers, between adjacent pyrimidines (**figure 14.4**). Other examples are ionizing radiation and carcinogens such as the fungal toxin aflatoxin B1 and other benzo(a)pyrene derivatives.

Effects of Mutations

The effects of a mutation can be described at the protein level and in terms of observed phenotypes. In all cases, the impact is readily noticed only if it produces a change in phenotype. In general, the more prevalent form of a gene and its associated phenotype is called the **wild type.** A mutation from wild type to a mutant form is called a **forward mutation (table 14.2).** A forward mutation can be reversed by a second mutation that restores the wild-type phenotype. When the second mutation is at the same site as the original mutation, it is called a **reversion mutation.** A true

reversion converts the mutant nucleotide sequence back to the wild-type sequence. If the second mutation is at a different site than the original mutation, it is called a **suppressor mutation.** Suppressor mutations may be within the same gene (intragenic suppressor mutation) or in a different gene (extragenic suppressor mutation). Because point mutations are the most common types of mutations, their effects are the focus here.

Mutations in Protein-Coding Genes

Point mutations in protein-coding genes can affect protein structure in a variety of ways. Point mutations are named according to if and how they change the encoded protein. The most common types of point mutations are silent mutations, missense mutations, nonsense mutations, and frameshift mutations. Examples of each are shown in table 14.2.

Silent mutations change the nucleotide sequence of a codon but do not change the amino acid encoded by that codon. This is possible because of the degeneracy of the genetic code. Therefore,

Table 14.2   Types of Point Mutations

Type of Mutation	Change in DNA	Example
Forward Mutations		
None	None	5'-A-T-G-A-C-C-T-C-C-C-G-A-A-A-G-G-3' Met - Thr - Ser - Pro - Lys - Gly
Silent	Base substitution	5'-A-T-G-A-C-A-T-C-C-C-G-A-A-A-G-G-G-3' Met - Thr - Ser - Pro - Lys - Gly
Missense	Base substitution	5'-A-T-G-A-C-C-T-G-C-C-G-A-A-A-G-G-G-3' Met - Thr - Cys - Pro - Lys - Gly
Nonsense	Base substitution	5'-A-T-G-A-C-C-T-C-C-C-G-T-A-A-G-G-G-3' Met - Thr - Ser - Pro - STOP!
Frameshift	Insertion/deletion	5'-A-T-G-A-C-C-T-C-C-G-C-G-A-A-A-G-G-G-3' Met - Thr - Ser - Ala - Glu - Arg
Reverse Mutations		
True reversion	Base substitution	5'-A-T-G-A-C-C-T-C-C $\xrightarrow{\text{forward}}$ A-T-G-C-C-C-T-C-C $\xrightarrow{\text{reverse}}$ A-T-G-A-C-C-T-C-C Met - Thr - Ser Met - Pro - Ser Met - Thr - Ser
	Base substitution	5'-A-T-G-A-C-C-T-C-C $\xrightarrow{\text{forward}}$ A-T-G-A-C-C-T-G-C $\xrightarrow{\text{reverse}}$ A-T-G-A-C-C-A-G-C Met - Thr - Ser Met - Thr - Cys Met - Thr - Ser
Equivalent Reversion	Base substitution	5'-A-T-G-A-C-C-T-C-C $\xrightarrow{\text{forward}}$ A-T-G-C-C-C-T-C-C $\xrightarrow{\text{reverse}}$ A-T-G-C-T-C-T-C-C Met - Thr - Ser Met - Pro - Ser Met - Leu - Ser (polar amino acid) (nonpolar amino acid) (polar amino acid) pseudo-wild type
Suppressor Mutations		
Frameshift of opposite sign (intragenic suppressor)	Insertion/deletion	5'-A-T-G-A-C-C-T-C-C-C-G-A-A-A-G-G-G-3' Met - Thr - Ser - Pro - Lys - Gly └─ Forward mutation 5'-A-T-G-A-C-C-T-C-C-G-C-G-A-A-A-G-G-G-3' Met - Thr - Ser - Ala - Glu - Arg └─ Suppressor mutation (deletion) 5'-A-T-G-A-C-C-C-G-A-A-A-G-G-G-3' Met - Thr - Pro - Pro - Lys - Gly
Extragenic suppressor Nonsense suppressor		Gene (e.g., for tyrosine tRNA) undergoes mutational event in its anticodon region that enables it to recognize and align with a mutant nonsense codon (e.g., UAG) to insert an amino acid (tyrosine) and permit completion of translation.
Physiological suppressor		A defect in one chemical pathway is circumvented by another mutation—for example, one that opens up another chemical pathway to the same product, or one that permits more efficient uptake of a compound produced in small quantities because of the original mutation.

when there is more than one codon for a given amino acid, a single base substitution may result in the formation of a new codon for the same amino acid. For example, if the codon CGU were changed to CGC, it would still code for arginine even though a mutation had occurred. When there is no change in the protein, there is no change in the phenotype of the organism. << *The genetic code (section 12.6)*

Missense mutations involve a single base substitution that changes a codon for one amino acid into a codon for another. For example, the codon GAG, which specifies glutamic acid, could be changed to GUG, which codes for valine. The effects of missense mutations vary. They alter protein structure, but the effect of this change may range from complete loss of activity to no change at all. This is because the effect of missense mutations on protein function depends on the type and location of the amino acid substitution. For instance, replacement of a nonpolar amino acid in the protein's interior with a polar amino acid can drastically alter the protein's three-dimensional structure and therefore its function. Similarly the replacement of a critical amino acid at the active site of an enzyme often destroys its activity. However, the replacement of one polar amino acid with another at the protein surface may have little or no effect. Missense mutations play a very important role in providing new variability to drive evolution because they often are not lethal and therefore remain in the gene pool. >> *Proteins (appendix I)*

Nonsense mutations convert a sense codon (i.e., one that codes for an amino acid) to a nonsense codon (i.e., a stop codon: one that does not code for an amino acid). This causes the early termination of translation and therefore results in a shortened polypeptide. Depending on the location of the mutation in the gene, the phenotype may be more or less severely affected. Most proteins retain some function if they are shortened by only one or two amino acids; complete loss of normal function usually results if the mutation occurs closer to the beginning or middle of the gene.

Frameshift mutations arise from the insertion or deletion of one or two base pairs within the coding region of the gene. Since the code consists of a precise sequence of triplet codons, the addition or deletion of fewer than three base pairs causes the reading frame to be shifted for all codons downstream. Frameshift mutations usually are very deleterious and yield mutant phenotypes resulting from the synthesis of nonfunctional proteins. In addition, frameshift mutations often produce a stop codon so that the peptide product is shorter as well as different in sequence. Of course if the frameshift occurs near the end of the gene or if there is a second frameshift shortly downstream from the first that restores the reading frame, the phenotypic effect might not be as drastic. A second nearby frameshift that restores the proper reading frame is an example of an intragenic suppressor mutation (table 14.2).

As noted previously, changes in protein structure can lead to changes in protein function, which in turn can alter the phenotype of an organism in several different ways. Morphological mutations change the microorganism's colonial or cellular morphology. Lethal mutations, when expressed, result in the death of the microorganism. Because a microbe must be able to grow to be isolated and studied, lethal mutations are recovered only if they are recessive in diploid organisms or are conditional mutations in haploid organisms. **Conditional mutations** are those that are expressed only under certain environmental conditions. For example, a conditional lethal mutation in *E. coli* might not be expressed under permissive conditions such as low temperature but would be expressed under restrictive conditions such as high temperature. Thus the mutant would grow normally at cooler temperatures but would die at high temperatures.

Biochemical mutations are those causing a change in the biochemistry of the cell. Since these mutations often inactivate a biosynthetic pathway, they frequently eliminate the capacity of the mutant to make an essential molecule such as an amino acid or nucleotide. A strain bearing such a mutation has a conditional phenotype: it is unable to grow on medium lacking that molecule but grows when the molecule is provided. Such mutants are called **auxotrophs,** and they are said to be auxotrophic for the molecule they cannot synthesize. If the wild-type strain from which the mutant arose is a chemoorganotroph able to grow on a minimal medium containing only salts (to supply needed elements such as nitrogen and phosphorus) and a carbon source, it is called a **prototroph.** Another type of biochemical mutant is the resistance mutant. These mutants have acquired resistance to some pathogen, chemical, or antibiotic. Auxotrophic and resistance mutants are quite important in microbial genetics due to the ease of their detection and their relative abundance.

Mutations in Regulatory Sequences

Some of the most interesting and informative mutations studied by microbial geneticists are those that occur in the regulatory sequences responsible for controlling gene expression. Constitutive lactose operon mutants in *E. coli* are excellent examples. Many of these mutations map in the operator site and produce altered operator sequences that are not recognized by the repressor protein. Therefore the operon is continuously transcribed, and β-galactosidase is always synthesized. Mutations in promoters also have been identified. If the mutation renders the promoter sequence nonfunctional, the mutant will be unable to synthesize the product even though the coding region of the structural gene is completely normal. Without a fully functional promoter, RNA polymerase rarely transcribes a gene as well as wild type. << *Regulation of transcription initiation (section 13.2)*

Mutations in tRNA and rRNA Genes

Mutations in tRNA and rRNA alter the phenotype of an organism through disruption of protein synthesis. In fact, these mutants often are initially identified because of their slow growth. On the other hand, a suppressor mutation involving tRNA restores normal (or near normal) growth rates. Here a base substitution in the anticodon region of a tRNA allows the insertion of the correct amino acid at a mutant codon (table 14.2).

1. List three ways in which spontaneous mutations might arise.

2. How do the mutagens 5-bromouracil, methyl-nitrosoguanidine, proflavin, and UV radiation induce mutations?

3. Give examples of intragenic and extragenic suppressor mutations.

4. Sometimes a point mutation does not change the phenotype. List all the reasons why this is so.

5. Why might a missense mutation at a protein's surface not affect the phenotype of an organism, while the substitution of an internal amino acid does?

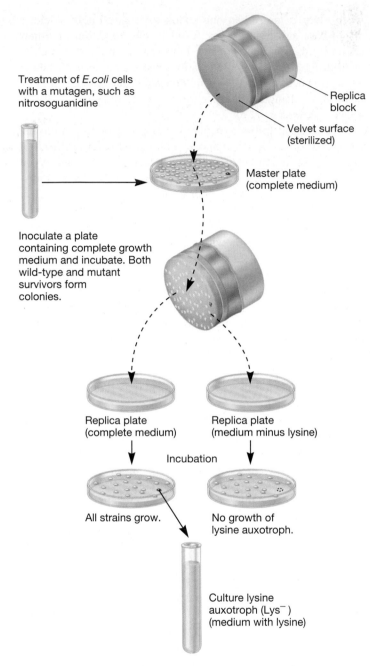

Treatment of *E.coli* cells with a mutagen, such as nitrosoguanidine

Replica block

Velvet surface (sterilized)

Master plate (complete medium)

Inoculate a plate containing complete growth medium and incubate. Both wild-type and mutant survivors form colonies.

Replica plate (complete medium)

Replica plate (medium minus lysine)

Incubation

All strains grow.

No growth of lysine auxotroph.

Culture lysine auxotroph (Lys⁻) (medium with lysine)

Figure 14.5 Replica Plating. The use of replica plating to isolate a lysine auxotroph. After growth of a mutagenized culture on a complete medium a piece of sterile velvet is pressed on the plate surface to pick up bacteria from each colony. Then the velvet is pressed to the surface of other plates and organisms are transferred to the same position as on the master plate. After the location of Lys⁻ colonies growing on the replica with complete medium are determined, the auxotrophs can be isolated and cultured.

14.2 DETECTION AND ISOLATION OF MUTANTS

To study microbial mutants, they must be readily detected, even when they are rare, and then efficiently isolated from wild-type organisms and other mutants that are not of interest. Microbial geneticists typically increase the likelihood of obtaining mutants by using mutagens to increase the rate of mutation. The rate can increase from the usual one mutant per 10^7 to 10^{11} cells to about one per 10^3 to 10^6 cells. Even at this rate, mutations are rare and carefully devised means for detecting or selecting a desired mutation must be used. This section describes some techniques used in mutant detection, selection, and isolation.

Mutant Detection

When collecting mutants of a particular organism, the wild-type characteristics must be known so that an altered phenotype can be recognized. A suitable detection system for the mutant phenotype also is needed. The use of such detection systems is called screening. Screening for mutant phenotypes in procaryotes and other haploid organisms is straightforward because any mutation usually can be seen immediately. Some screening procedures require only examination of colony morphology. For instance, if albino mutants of a normally pigmented bacterium are being studied, detection simply requires visual observation of colony color. Other screening methods are more complex. For example, the **replica plating** technique is used to screen for auxotrophic mutants. It distinguishes between mutants and the wild-type strain based on their ability to grow in the absence of a particular biosynthetic end product (**figure 14.5**). A lysine auxotroph, for instance, grows on lysine-supplemented media but not on a medium lacking an adequate supply of lysine because it cannot synthesize this amino acid.

Once a screening method is devised, mutants are collected. However, mutant collection can present practical problems. Consider a search for the albino mutants mentioned previously. If the mutation rate were around one in a million, on average a million or more organisms would have to be tested to find one

albino mutant. This probably would require several thousand plates. The task of isolating auxotrophic mutants in this way would be even more taxing with the added labor of replica plating. Thus, if possible, it is more efficient to use a selection system employing some environmental factor to separate mutants from wild-type microorganisms. Examples of selection systems are described next.

Mutant Selection

An effective selection technique uses incubation conditions under which the mutant grows, because of properties conferred by the mutation, whereas the wild type does not. Selection methods often involve reversion mutations or the development of resistance to an environmental stress. For example, if the intent is to isolate revertants from a lysine auxotroph (Lys⁻), the approach is quite easy. A large population of lysine auxotrophs is plated on minimal medium lacking lysine, incubated, and examined for colony formation. Only cells that have mutated to restore the ability to manufacture lysine will grow on minimal medium. Several million cells can be plated on a single petri dish, but only the rare revertant cells will grow. Thus many cells can be tested for mutations by scanning a few petri dishes for growth. This method has proven very useful in determining the relative mutagenicity of many substances.

Methods for selecting mutants resistant to a particular environmental stress follow a similar approach. Often wild-type cells are susceptible to virus attack or antibiotic treatment, so it is possible to grow the microbe in the presence of the agent and look for surviving organisms. Consider the example of a phage-sensitive wild-type bacterium. When it is cultured in medium lacking the virus and then plated on selective medium containing viruses, any colonies that form are resistant to virus attack and very likely are mutants in this regard. This type of selection can be used for virtually any environmental parameter; resistance to antibiotics and specific temperatures are commonly used.

Substrate utilization mutations also are employed in bacterial selection. Many bacteria use only a few primary carbon sources. With such bacteria, it is possible to select mutants by plating a culture on medium containing an alternate carbon source. Any colonies that appear can use the substrate and are probably mutants.

Mutant screening and selection methods are used for purposes other than understanding more about the nature of genes or the biochemistry of a particular microorganism. One very important role of mutant selection and screening techniques is in the study of carcinogens. The next section briefly describes one of the first and perhaps best known of the carcinogen testing systems.

Carcinogenicity Testing

An increased understanding of the mechanisms of mutation and their role in cancer has stimulated efforts to identify environmental carcinogens. The observation that many carcinogenic agents also are mutagenic was the basis for development of the Ames test by Bruce Ames in the 1970s. The Ames test is a mutational reversion assay employing several strains of *Salmonella enterica* serovar Typhimurium, each of which has a different mutation in the histidine biosynthesis operon; that is to say, they are histidine auxotrophs. The bacteria also have mutational alterations of their cell walls that make them more permeable to test substances. To further increase assay sensitivity, the strains are defective in the ability to repair DNA correctly.

In the Ames test, these tester strains of *Salmonella* are plated with the substance being tested and the appearance of visible colonies followed (**figure 14.6**). To ensure that DNA replication can take place in the presence of the potential mutagen, the bacteria and test substance are mixed in dilute molten top agar to which a trace of histidine has been added. This molten mix is then poured on top of minimal agar plates and incubated for 2 to 3 days at 37°C. All of the histidine auxotrophs grow for the first few hours in the presence of the test compound until the histidine is depleted. This is necessary because replication is required for the development of a mutation (figure 14.1). Once the histidine supply is exhausted, only revertants that have mutationally regained the ability to synthesize histidine continue to grow and produce visible colonies. These colonies need only be counted and compared to controls in order to estimate the relative mutagenicity of the compound: the more colonies, the greater the mutagenicity.

A mammalian liver extract is also often added to the molten top agar prior to plating. The extract converts potential carcinogens into electrophilic derivatives that readily react with DNA. This process occurs naturally when foreign substances are metabolized in the liver. Because bacteria do not have this activation system, addition of the liver extract promotes the same kind of enzymatic transformations that occur in mammals. Many potential carcinogens, such as aflatoxins, are not actually carcinogenic until they are modified in the liver. The addition of the extract shows which compounds have intrinsic mutagenicity and which need activation after uptake. Despite the use of liver extracts, only about half the potential animal carcinogens are detected by the Ames test.

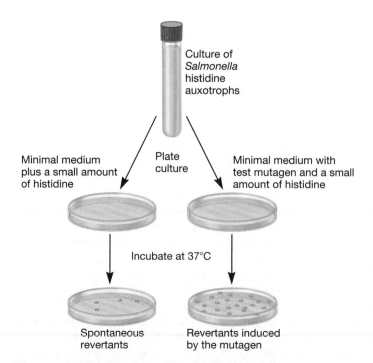

Culture of *Salmonella* histidine auxotrophs

Minimal medium plus a small amount of histidine

Plate culture

Minimal medium with test mutagen and a small amount of histidine

Incubate at 37°C

Spontaneous revertants

Revertants induced by the mutagen

Figure 14.6   The Ames Test for Mutagenicity.

1. Describe how replica plating is used to detect and isolate auxotrophic mutants.

2. Why are mutant selection techniques generally preferable to screening methods?

3. Briefly discuss how reversion mutations, resistance to an environmental factor, and the ability to use a particular nutrient can be employed in mutant selection.

4. Describe how you would isolate a mutant that required histidine for growth and was resistant to penicillin. The wild type is a prototroph.

5. What is the Ames test and how is it carried out? What assumption concerning mutagenicity and carcinogenicity is it based upon?

14.3 DNA REPAIR

Because replication errors and mutagens can alter nucleotide sequences, a microorganism must be able to repair any changes that might be lethal. In addition to **proofreading** by DNA polymerases during replication—the removal of an incorrect nucleotide immediately after its addition to the growing end of the chain—several other mechanisms can restore the sequence or repair damaged DNA. Repair in *E. coli* is best understood and is briefly described in this section. ◄◄ *DNA replication (section 12.3)*

Excision Repair

Excision repair corrects damage that causes distortions in the double helix. Two types of excision repair systems have been described: nucleotide excision repair and base excision repair. They both use the same approach to repair: remove the damaged portion of a DNA strand and use the intact complementary strand as the template for synthesis of new DNA. They are distinguished by the enzymes used to correct DNA damage.

In **nucleotide excision repair,** a repair enzyme called UvrABC endonuclease removes damaged nucleotides and some nucleotides on either side of the lesion. The resulting single-stranded gap, about 12 nucleotides long, is filled by DNA polymerase I, and DNA ligase joins the fragments (**figure 14.7**). This system can remove thymine dimers (figure 14.4) and repair almost any other injury that produces a detectable distortion in DNA.

Base excision repair employs DNA glycosylases to remove damaged or unnatural bases yielding apurinic or apyrimidinic (AP) sites. Enzymes called AP endonucleases recognize the damaged DNA and nick the backbone at the AP site (**figure 14.8**). DNA polymerase I removes the damaged region, using its 5′ to 3′ exonuclease activity. It then fills in the gap, and DNA ligase joins the DNA fragments.

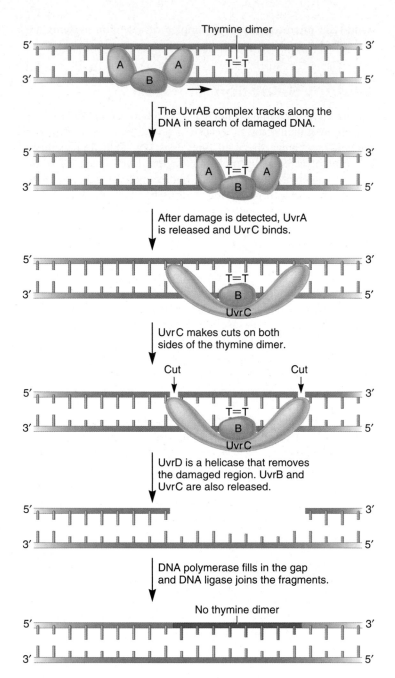

Figure 14.7 Nucleotide Excision Repair in *E. coli*.

Direct Repair

Thymine dimers and alkylated bases often are corrected by **direct repair. Photoreactivation** repairs thymine dimers (figure 14.4) by splitting them apart with the help of visible light. This photochemical reaction is catalyzed by the enzyme photolyase. Methyls and some other alkyl groups that have been added to the O^6 position of guanine can be removed with the help of an enzyme known as alkyltransferase or methylguanine methyltransferase. Thus damage to guanine from mutagens such as methyl-nitrosoguanidine (figure 14.3) can be repaired directly.

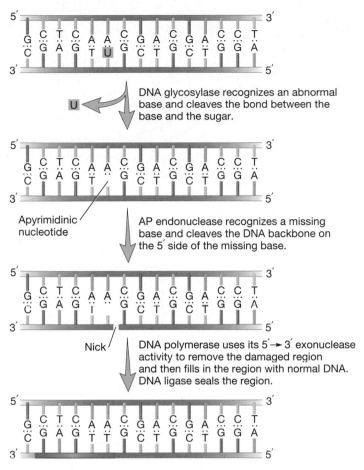

Figure 14.8   Base Excision Repair.

Mismatch Repair

Despite the accuracy of DNA polymerase and continual proof-reading, errors still are made during DNA replication. Remaining mismatched bases are usually detected and repaired by the **mismatch repair** system in *E. coli* (**figure 14.9**). The mismatch correction enzyme MutS scans the newly replicated DNA for mismatched pairs. Another enzyme, MutH, removes a stretch of newly synthesized DNA around the mismatch. A DNA polymerase then replaces the excised nucleotides, and the resulting nick is sealed by DNA ligase. In this regard, mismatch repair is similar to excision repair.

Successful mismatch repair depends on the ability of enzymes to distinguish between old and newly replicated DNA strands. This distinction is possible because newly replicated DNA strands lack methyl groups on their bases, whereas older DNA has methyl groups on the bases of both strands. **DNA methylation** is catalyzed by DNA methyltransferases and results in three different products: *N*6-methyladenine, 5-methylcytosine, and *N*4-methylcytosine. After strand synthesis, the *E. coli* DNA adenine *m*ethyltransferase (DAM) methylates adenine bases in GATC sequences to form *N*6-methyladenine. For a short time after the replication fork has passed, the new strand lacks methyl groups while the template strand is methylated. In other words,

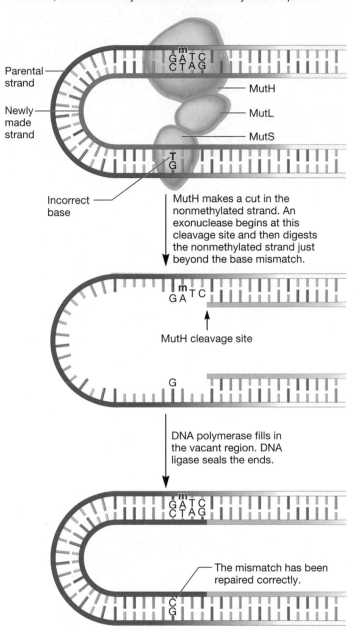

Figure 14.9   Methyl-Directed Mismatch Repair in *E. coli*.
MutS slides along the DNA and recognizes base mismatches in the double helix. MutL binds to MutS and acts as a linker between MutS and MutH. The DNA must loop for this interaction to occur. The role of MutH is to identify the methylated strand of DNA, which is the nonmutated parental strand. The methylated adenine is designated with an m.

the DNA is temporarily hemimethylated. The repair system cuts out the mismatch from the unmethylated strand.

Recombinational Repair

Recombinational repair corrects damaged DNA in which both bases of a pair are missing or damaged, or where there is a gap opposite a lesion. In this type of repair, the **RecA protein** cuts a piece of template DNA from a sister molecule and puts it into the gap or uses it to replace a damaged strand (**figure 14.10**). Although procaryotes are haploid, another copy of the damaged segment often is available because either it has recently been replicated or the cell is growing rapidly and has more than one copy of its chromosome. Once the template is in place, the remaining damage can be corrected by another repair system.

SOS Response

Despite having multiple repair systems, sometimes the damage to an organism's DNA is so great that the normal repair mechanisms just described cannot repair all the damage. As a result, DNA synthesis stops completely. In such situations, a global control network called the **SOS response** is activated. The SOS response, like recombinational repair, depends on the activity of the RecA protein. RecA binds to single- or double-stranded DNA breaks and gaps generated by cessation of DNA synthesis. RecA binding initiates recombinational repair. Simultaneously RecA takes on a proteolytic function that destroys a repressor protein called LexA. LexA negatively regulates the function of many genes involved in DNA repair and synthesis. Destruction of LexA increases transcription of genes for excision repair and recombinational repair, in particular. The first genes to be transcribed are those that encode the Uvr proteins needed for nucleotide excision repair (figure 14.7). Then expression of genes involved in recombinational repair is further increased. To give the cell time to repair its DNA, the protein SfiA is produced; SfiA blocks cell division. Finally, if the DNA has not been fully repaired after about 40 minutes, a process called **translesion DNA synthesis** is triggered. In this process, DNA polymerase IV (also known as DinB) and DNA polymerase V (UmuCD) synthesize DNA across gaps and other lesions (e.g., thymine dimers) that had stopped DNA polymerase III. However, because an intact template does not exist, these DNA polymerases often insert incorrect bases. Furthermore, they lack proofreading activity. Therefore even though DNA synthesis continues, it is highly error prone and results in the generation of numerous mutations. The SOS response is so named because it is a response made in a life-or-death situation. The response increases the likelihood that some cells will survive by allowing DNA synthesis to continue. For the cell, the risk of dying because of failure to replicate DNA is greater than the risk posed by the mutations generated by this error-prone process.

1. Compare and contrast each of the following repair processes: excision repair, recombinational repair, direct repair, and SOS response.
2. Explain how the following DNA alterations and replication errors would be corrected (there may be more than one way): base addition errors by DNA polymerase III during replication, thymine dimers, AP sites, methylated guanines, and gaps produced during replication.

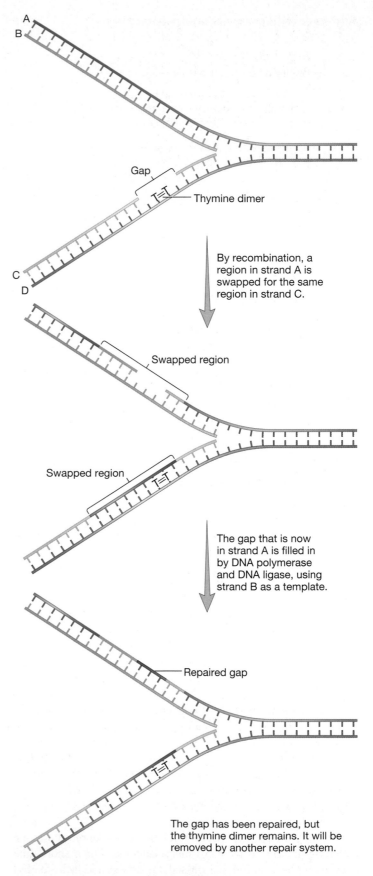

Gap

T=T — Thymine dimer

By recombination, a region in strand A is swapped for the same region in strand C.

Swapped region

Swapped region — T=T

The gap that is now in strand A is filled in by DNA polymerase and DNA ligase, using strand B as a template.

Repaired gap

The gap has been repaired, but the thymine dimer remains. It will be removed by another repair system.

Figure 14.10 Recombinational Repair.

14.4   CREATING GENETIC VARIABILITY

As discussed in section 14.1, the consequences of mutations can range from no effect to being lethal, depending not only on the nature of the mutation but also on the environment in which the organism lives. Thus all mutations are subject to selective pressure, and this determines if a mutation will persist in a population. Each mutant form is called an **allele,** an alternate form of the gene. Mutant alleles, as well as the wild-type allele, can be combined with other genes, leading to an increase in the genetic variability within a population. Each genotype in a population can be selected for or selected against. Organisms with genotypes, and therefore phenotypes, that are best suited to the environment survive and are most likely to pass on their genes. Shifts in environmental pressures can lead to changes in the population and ultimately result in the evolution of new species. The mechanisms by which new combinations of genes are generated are the topic of this section. All involve **recombination,** the process in which one or more nucleic acid molecules are rearranged or combined to produce a new nucleotide sequence. This is normally accompanied by a phenotypic change. Geneticists refer to organisms produced following a recombination event as recombinant organisms or simply **recombinants.**

Horizontal Gene Transfer in Procaryotes

The transfer of genes from parents to progeny is called vertical gene transfer. In eucaryotes, vertical gene transfer also introduces genetic variation into populations. With each new generation, recombinant progeny are formed when gametes from parents fuse during sexual reproduction. However, procaryotes do not reproduce sexually. This suggests that genetic variation in populations of procaryotes should be relatively limited, only occurring with the advent of a new mutation and its transfer to the next generation. However, this is not the case. Procaryotes have evolved three different mechanisms for creating recombinants. These mechanisms are referred to collectively as **horizontal (lateral) gene transfer (HGT).** HGT is distinctive from vertical gene transfer because genes from one independent, mature organism are transferred to another, often creating a stable recombinant having characteristics of both the donor and the recipient.

HGT has been important in the evolution of many species, and it is still commonplace in many environments. Furthermore, clear examples are known of DNA transfer from one species to distantly related species. The importance of HGT cannot be overstated. For instance, it has been demonstrated that procaryotes sharing an ecological niche (e.g., the human gut) can exchange genes and this alters the nature of the microbial community in a habitat. Another important example is the transfer of groups of genes encoding virulence factors from pathogenic bacteria to other bacteria. These groups of genes are called pathogenicity islands, and their discovery has provided valuable insights into the evolution of pathogenic bacteria. Finally, HGT plays a critical role in the evolution and spread of antibiotic-resistance genes among pathogenic bacteria. >> *Microbial evolution (section 17.1)*

During HGT, a piece of donor DNA, the **exogenote,** enters a recipient cell. The transfer can occur in three ways: direct transfer between two cells temporarily in physical contact (conjugation), transfer of a naked DNA fragment (transformation), and transport of DNA by viruses (transduction). If the exogenote contains genes already present in the recipient, the recipient will become temporarily diploid for those genes. This partially diploid cell is called a **merozygote** (figure 14.11). The exogenote has four possible fates in the recipient. First, when the exogenote has a sequence homologous to one in the recipient's chromosome **(endogenote),** integration may occur. That is, the donor's DNA may pair with the recipient's DNA and be incorporated to yield a recombinant genome. The recombinant then reproduces, yielding a population of stable recombinants. Second, if the exogenote is able to replicate itself (e.g., it is a plasmid), it may persist outside the endogenote. When the recipient reproduces, the exogenote replicates and a population of stable recombinants is formed.

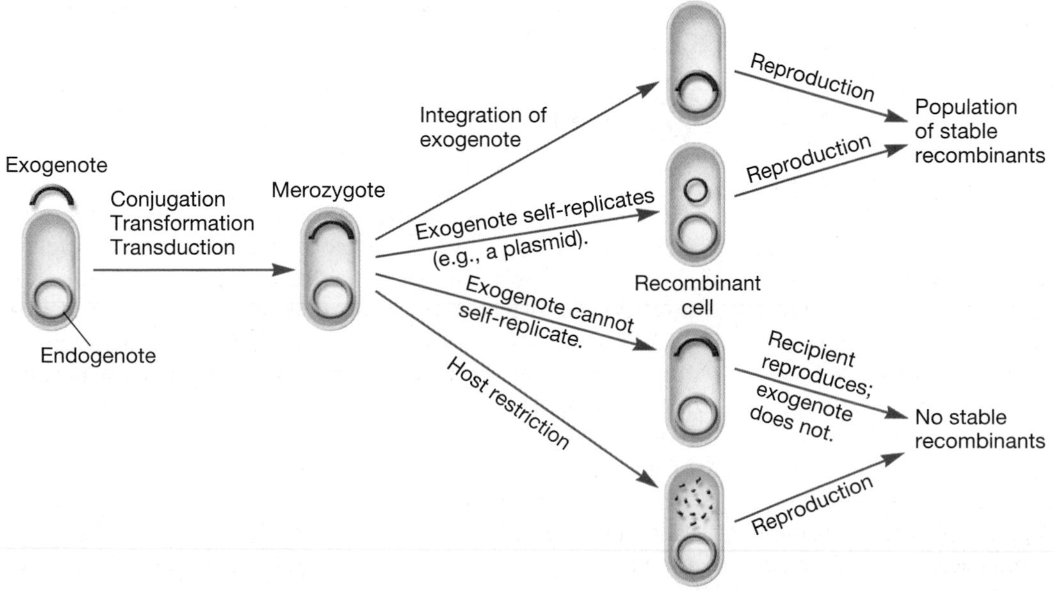

Figure 14.11   The Production and Fate of Merozygotes.

Table 14.3	*E. coli* Homologous Recombination Proteins
Protein	**Description**
Rec BCD	Recognizes double-stranded breaks and then generates single-stranded regions at the break site that are involved in strand invasion
Single-strand binding protein	Prevents excessive strand degradation by RecBCD
RecA	Promotes strand invasion and displacement of complementary strand to generate D loop
RecG	Helps form Holliday junctions and promotes branch migration
RuvABC	Endonuclease that binds Holliday junctions, promotes branch migration, and cuts strands in the Holliday junction in order to separate chromosomes

Third, the exogenote remains in the cytoplasm but is unable to replicate. When the recipient reproduces, the progeny cells lack the exogenote and it is eventually lost from the population. Finally, a process called host restriction may occur. In this process, host cell nucleases degrade the exogenote, thereby preventing the formation of a recombinant cell.

Recombination at the Molecular Level

Although different processes are used in eucaryotes and procaryotes to create recombinant organisms, the mechanisms of recombination at the molecular level are remarkably similar. Three types of recombination are observed: homologous recombination, site-specific recombination, and transposition.

Homologous recombination, the most common recombination event, usually involves a reciprocal exchange between a pair of DNA molecules with the same nucleotide sequence. It can occur anywhere on the chromosome, and it results from DNA strand breakage and reunion leading to crossing-over. Homologous recombination is carried out by the products of the *rec* genes, including the RecA protein, which is also important for DNA repair (table 14.3). The most widely accepted model of homologous recombination is the **double-stranded break model** (figure 14.12). It proposes that duplex DNA with a double-stranded break is processed to create DNA with single-stranded ends. RecA promotes the insertion of one single-stranded end into an intact, homologous piece of DNA. This is called strand invasion. As can be seen in figure 14.12, strand invasion results in the formation of two gaps in the two parent DNA molecules. The gaps are filled, yielding a structure with **heteroduplex DNA;** that is, it contains strands derived from both parent molecules. The two parental DNA molecules are now linked together by two structures called Holliday junctions. These structures move along the DNA molecule during branch migration until they are finally cut and the two DNA molecules are separated. Depending on how this occurs, the resulting DNA molecules

will be either recombinant or nonrecombinant. In some cases, a nonreciprocal form of homologous recombination occurs (**figure 14.13**). In nonreciprocal homologous recombination, a piece of genetic material is inserted into the chromosome through the incorporation of a single strand to form a stretch of heteroduplex DNA.

The other two types of recombination do not depend on long regions of sequence homology. **Site-specific recombination** is particularly important in the integration of viral genomes into host chromosomes. The enzymes responsible for this event are often specific for sequences within the particular virus and its host. **Transposition** can occur at many sites in the genome and is discussed in more detail in section 14.5.

1. Distinguish among the three forms of recombination mentioned in this section.
2. What four fates can DNA have after entering a bacterium?

14.5 TRANSPOSABLE ELEMENTS

The chromosomes of procaryotes, viruses, and eucaryotic cells contain pieces of DNA that can move and integrate into different sites in the chromosomes. Such movement is called transposition, and it plays important roles in the generation of new gene combinations. DNA segments that carry the genes required for transposition are **transposable elements** or **transposons,** sometimes called "jumping genes." Unlike other processes that reorganize DNA, transposition does not require extensive areas of homology between the transposon and its destination site. Transposons were first discovered in the 1940s by Barbara McClintock during her studies on maize genetics (a discovery for which she was awarded the Nobel Prize in 1983). They have been most intensely studied in *Bacteria*.

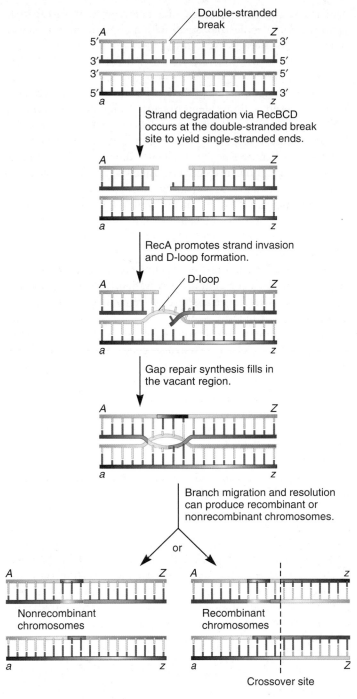

Figure 14.12 The Double-Stranded Break Model of Homologous Recombination.

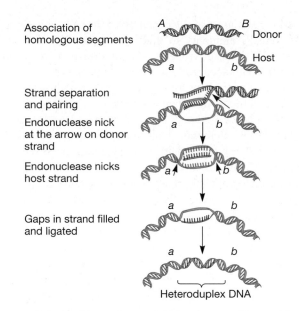

Figure 14.13 Nonreciprocal Homologous Recombination. The Fox model for nonreciprocal homologous recombination. This mechanism has been proposed for the recombination occurring during transformation in some bacteria.

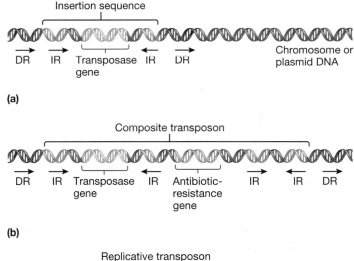

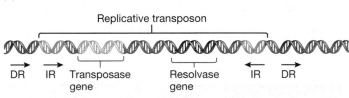

Figure 14.14 Transposable Elements. All transposable elements contain common elements. These include inverted repeats (IRs) at the ends of the element and a transposase gene. (a) Insertion sequences consist only of IRs on either side of the transposase gene. (b) Composite transposons and (c) replicative transposons contain additional genes. Insertion sequences and composite transposons move by simple (cut-and-paste) transposition. Replicative transposons move by replicative transposition. DRs, direct repeats in host DNA, flank a transposable element.

The simplest transposable elements are **insertion sequences,** or IS elements for short (**figure 14.14a**). An IS element is a short sequence of DNA (around 750 to 1,600 base pairs [bp] in length). It contains only the gene for the enzyme transposase, and it is bounded at both ends by inverted repeats—identical or very similar sequences of nucleotides in reversed orientation.

Inverted repeats are usually about 15 to 25 base pairs long and vary among IS elements so that each type of IS has its own characteristic inverted repeats. **Transposase** is required for transposition and accurately recognizes the ends of the IS. Each IS element is named by giving it the prefix IS followed by a number. IS elements have been observed in a variety of bacteria and some archaea.

Transposable elements also can contain genes in addition to those required for transposition (e.g., antibiotic resistance or toxin genes). These elements often are called **composite transposons.** Some composite transposons consist of a central region containing the extra genes, flanked on both sides by IS elements that are identical or very similar in sequence (figure 14.14*b*). Others are simpler, bounded only by short, inverted repeats, while the coding region contains both transposition genes and the extra genes. It is believed that composite transposons are formed when two IS elements associate with a central segment containing one or more genes. This association could arise if an IS element replicates and moves only a gene or two down the chromosome. Composite transposon names begin with the prefix Tn. Some properties of selected composite transposons are given in **table 14.4.**

The process of transposition in procaryotes occurs by two basic mechanisms. **Simple transposition,** also called **cut-and-paste transposition,** involves transposase-catalyzed excision of the transposon, followed by cleavage of a new target site and ligation of the transposon into this site (**figure 14.15**). Target sites are specific sequences about five to nine base pairs long. When a transposon inserts at a target site, the target sequence is duplicated so that short, direct-sequence repeats flank the transposon's terminal inverted repeats.

The second transposition mechanism is **replicative transposition.** In this mechanism, the original transposon remains at the parental site on the chromosome and a replicate is inserted at the target DNA site (**figure 14.16**). The transposition of Tn*3* is a well-studied example of replicative transposition. In the first stage, DNA containing Tn*3* fuses with the target DNA to form a cointegrate molecule (figure 14.16, step 1). This process requires the Tn*3* transposase enzyme coded for by the *tnpA* gene (**figure 14.17**). Note that the cointegrate has two copies of Tn*3*. In the second stage, the cointegrate is resolved to yield two DNA molecules, each with a copy of the transposon (figure 14.16, step 3). Resolution involves a crossover and is catalyzed by a resolvase enzyme coded for by the *tnpR* gene (figure 14.17).

Transposable elements produce a variety of important effects. They can insert within a gene to cause a mutation or stimulate DNA rearrangement, leading to deletions of genetic material. Because some transposons carry stop codons or termination sequences, when transposed into genes they may block translation or transcription, respectively. Likewise, other transposons carry promoters and can activate genes near the point of insertion. Thus some transposons can turn genes on or off. Transposons also are located in plasmids and participate in such processes as plasmid fusion, insertion of plasmids into chromosomes, and plasmid evolution.

The role of transposons in plasmid evolution is of particular note. Plasmids can contain several different transposon-target sites. Therefore transposons frequently move between plasmids. Of concern is the fact that many transposons contain antibiotic-resistance genes. Thus as they move from one plasmid to another, resistance genes are introduced into the target plasmid, creating a resistance (R) plasmid. Multiple drug-resistance plasmids can arise from the accumulation of transposons in a plasmid (figure 14.17). Many R plasmids are able to move from one cell to another during conjugation, which spreads the resistance genes throughout a population. Finally, because transposons also move between plasmids and chromosomes, drug resistance genes can exchange between these two molecules, resulting in the further spread of antibiotic resistance.

Some transposons bear transfer genes and can move between bacteria through the process of conjugation, as discussed in section 14.7. A well-studied example of a **conjugative transposon** is Tn*916* from *Enterococcus faecalis.* Although Tn*916* cannot replicate autonomously, it can transfer itself from *E. faecalis* to a variety of recipients and integrate into their chromosomes. Because it carries a gene for tetracycline resistance, this conjugative transposon also spreads drug resistance.

Table 14.4	The Properties of Selected Composite Transposons				
Transposon	Length (bp)	Terminal Repeat Length		Terminal Module	Genetic Markers
Tn*3*	4,957	38			Ampicillin resistance
Tn*501*	8,200	38			Mercury resistance
Tn*1681*	2,061			IS*1*	Heat-stable enterotoxin
Tn*2901*	11,000			IS*1*	Arginine biosynthesis

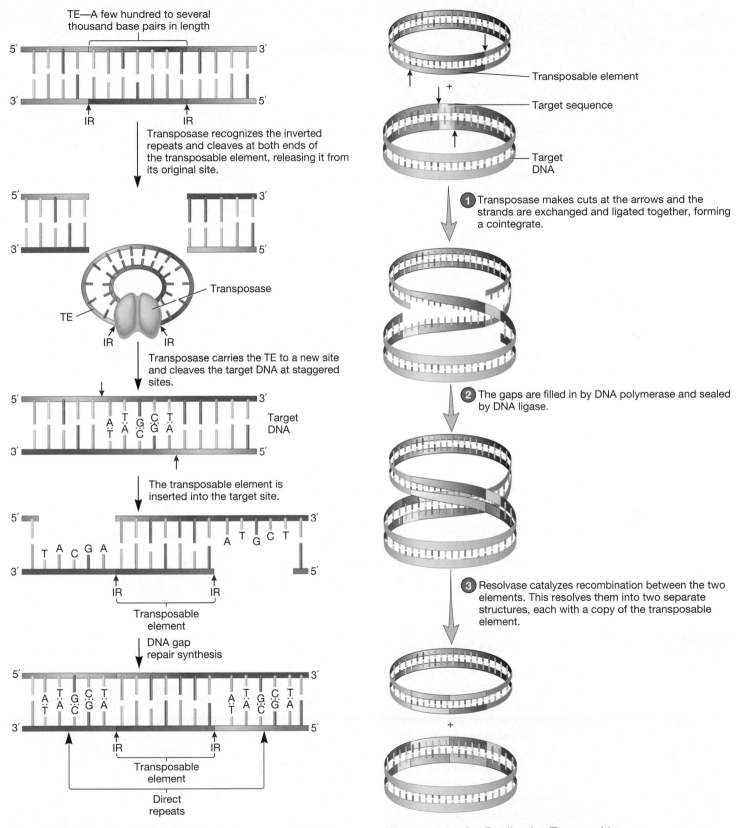

Figure 14.15   Simple Transposition.   TE, transposable element; IR, inverted repeat.

Figure 14.16   Replicative Transposition.

Figure 14.17 The Tn3 Composite Transposon within an R Plasmid. Tn3 is a replicative transposon that contains the gene for β-lactamase (*bla*), an enzyme that confers resistance to the antibiotic ampicillin (amp). The arrows below the Tn3 genes indicate the direction of transcription. Tn3 can be found in the resistance plasmid R1, where it is inserted into another transposable element, Tn4. Tn4 carries genes that provide resistance to streptomycin (Sm) and sulfonamide (Su). The plasmid also carries resistance genes for kanamycin (Km) and chloramphenicol (Cm). The RTF region of R1 codes for proteins needed for plasmid replication and transfer.

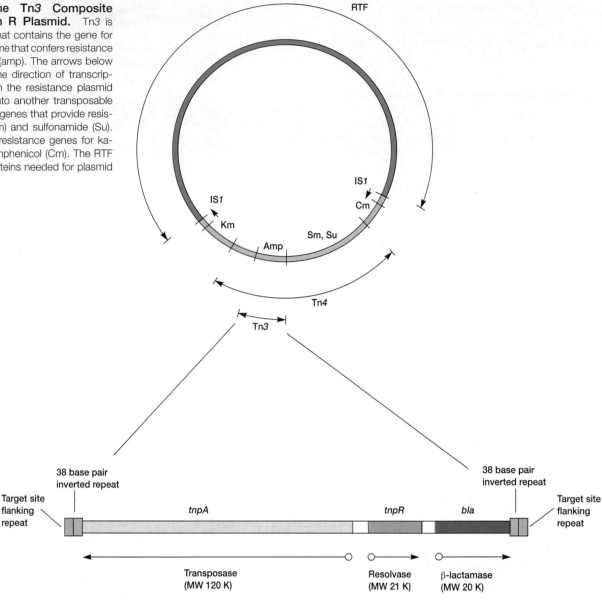

14.6 BACTERIAL PLASMIDS

Conjugation, the transfer of DNA between bacteria involving direct contact, depends on the presence of a plasmid. Recall from chapter 3 that **plasmids** are small, double-stranded DNA molecules that can exist independently of host chromosomes. They have their own replication origins and autonomously replicate and are stably inherited. Some plasmids are **episomes,** plasmids that can exist either with or without being integrated into host chromosomes. Although many plasmid types exist, our concern here is with **conjugative plasmids.** These plasmids can transfer copies of themselves to other bacteria during the process of conjugation, which is discussed in section 14.7.

Perhaps the best-studied conjugative plasmid is **F factor.** It plays a major role in conjugation in *E. coli,* and it was the first conjugative plasmid to be described (**figure 14.18**). The F factor is about 100,000 bases long and bears genes responsible for cell attachment and plasmid transfer between specific *E. coli.* Most of the information required for plasmid transfer is located in the *tra* operon, which contains at least 28 genes. Many of these direct the formation of sex pili that attach the F⁺ cell (the donor cell containing an F plasmid) to an F⁻ cell (**figure 14.19**). Other gene products aid DNA transfer. In addition, the F factor has several IS elements that assist plasmid integration into the host cell's chromosome. Thus the F factor is an episome that can exist outside the bacterial chromosome or can be integrated into it (**figure 14.20**).

1. Compare and contrast plasmids and transposable elements. Compare and contrast insertion sequences, composite transposons, and replicative transposons.

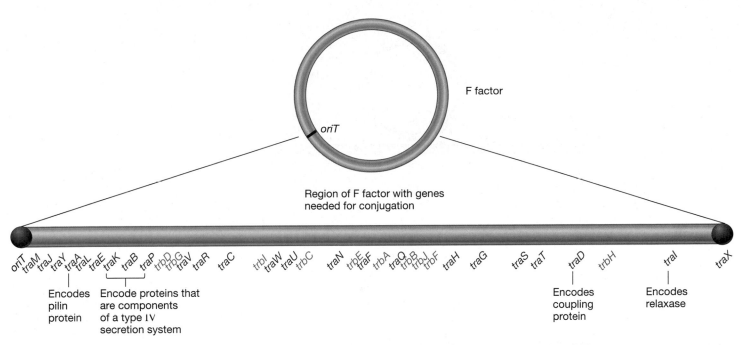

Figure 14.18   The F plasmid.   Genes that play a role in conjugation are shown, and some of their functions are indicated. The plasmid also contains three insertion sequences and a transposon. The site for initiation of rolling-circle replication and gene transfer during conjugation is *oriT*.

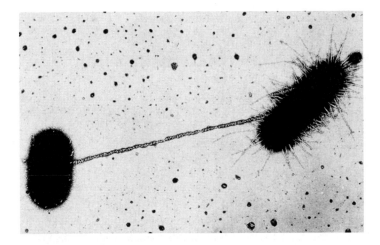

Figure 14.19   Bacterial Conjugation.   An electron micrograph of two *E. coli* cells in an early stage of conjugation. The F⁺ cell to the right is covered with fimbriae, and a sex pilus connects the two cells.

2. What is simple (cut-and-paste) transposition? What is replicative transposition? How do the two mechanisms of transposition differ? What happens to the target site during transposition?

3. What effect would you expect the existence of transposable elements and plasmids to have on the rate of microbial evolution? Give your reasoning.

4. How do multiple-drug-resistant plasmids often arise?

14.7 BACTERIAL CONJUGATION

The initial evidence for bacterial **conjugation,** the transfer of DNA by direct cell-to-cell contact, came from an elegant experiment performed by Joshua Lederberg and Edward Tatum in 1946. They mixed two auxotrophic strains, incubated the culture for several hours in nutrient medium, and then plated it on minimal medium. To reduce the chance that their results were due to simple reversion, they used double and triple auxotrophs on the assumption that two or three simultaneous reversions would be extremely rare. When recombinant prototrophic colonies appeared on the minimal medium after incubation, they concluded that the two auxotrophs were able to associate and undergo recombination.

Lederberg and Tatum did not directly prove that physical contact of the cells was necessary for gene transfer. This evidence was provided by Bernard Davis (1950), who constructed a U tube consisting of two pieces of curved glass tubing fused at the base to form a U shape with a fritted glass filter between the halves. The filter allowed the passage of media but not bacteria. The U tube was filled with nutrient medium and each side inoculated with a different auxotrophic strain of *E. coli.* During incubation, the medium was pumped back and forth through the filter to ensure medium exchange between the halves. When the bacteria were plated on minimal medium, Davis discovered that if the two auxotrophic strains were separated from each other by the filter, gene transfer did not take place. Therefore direct contact was required for the recombination that Lederberg and Tatum had observed. F factor-mediated conjugation is one of the best-studied conjugation systems. It is the focus of this section.

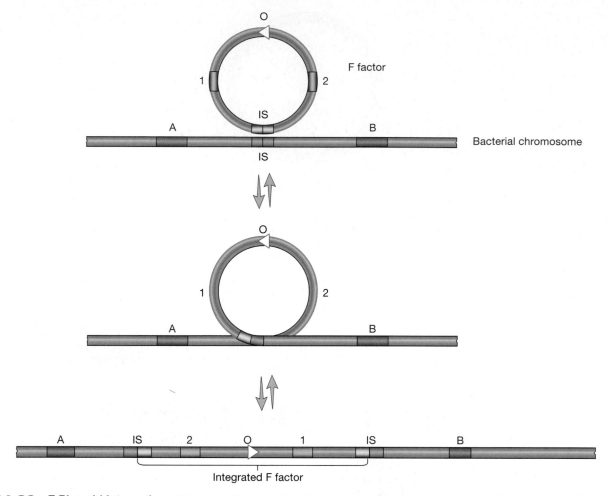

Figure 14.20 F Plasmid Integration.   The reversible integration of an F plasmid or factor into a host bacterial chromosome. The process begins with association between plasmid and bacterial insertion sequences. The O arrowhead (white) indicates the site at which oriented transfer of chromosome to the recipient cell begins. A, B, 1, and 2 represent genetic markers.

F⁺ × F⁻ Mating

In 1952 William Hayes demonstrated that the gene transfer observed by Lederberg and Tatum was polar. That is, there were definite donor (F⁺, or fertile) and recipient (F⁻, or nonfertile) strains, and gene transfer was nonreciprocal. He also found that in F⁺ × F⁻ mating, the progeny were only rarely changed with regard to auxotrophy (i.e., chromosomal genes usually were not transferred). However, F⁻ strains frequently became F⁺.

These results are readily explained in terms of the F factor described in section 14.6 (figure 14.18). The F⁺ strain contains an extrachromosomal F factor carrying the genes for sex pilus formation and plasmid transfer. The **sex pilus** is used to establish contact between the F⁺ and F⁻ cells (**figure 14.21a**). Once contact is made, the pilus retracts, bringing the cells into close physical contact. The F⁺ cell prepares for DNA transfer by assembling a type IV secretion apparatus, using many of the same genes used for sex pilus biogenesis; the sex pilus is embedded in the secretion structure (**figure 14.22**). The F factor then replicates by a rolling-circle mechanism (*see figure 12.8*). Replication is initiated by a complex

of proteins called the relaxosome, which nicks one strand of the F factor at a site called *oriT* (for origin of transfer). Relaxase, an enzyme associated with the relaxosome, remains attached to the 5′ end of the nicked strand. As F factor is replicated, the displaced strand and the attached relaxase enzyme move through the type IV secretion system to the recipient cell. During plasmid transfer, the entering strand is copied to produce double-stranded DNA. The recombination frequency is low because chromosomal genes are rarely transferred with the independent F factor.

Hfr Conjugation

Not long after the discovery of F⁺ × F⁻ mating, a second type of F factor–mediated conjugation was discovered. In this type of conjugation, the donor transfers chromosomal genes with great efficiency but does not change the recipient bacteria into F⁺ cells. Because of the *high f*requency of *r*ecombinants produced by this mating, it is referred to as **Hfr conjugation** and the donor is called an **Hfr strain.** Hfr strains contain the F factor integrated into their chromosome, rather than free in the cytoplasm (figure 14.21b). When

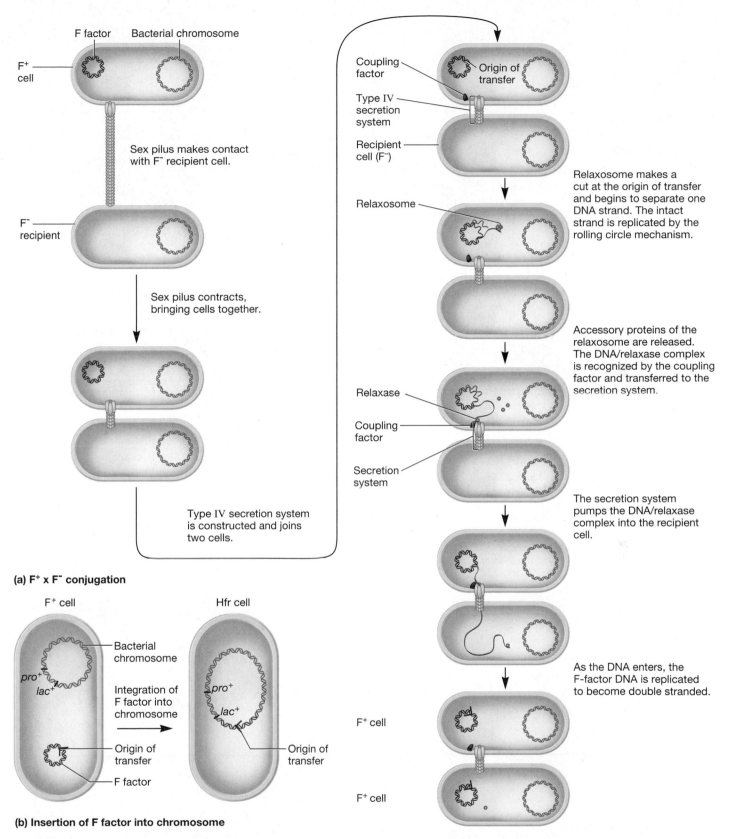

(a) F⁺ x F⁻ conjugation

F factor   Bacterial chromosome

F⁺ cell

Sex pilus makes contact with F⁻ recipient cell.

F⁻ recipient

Sex pilus contracts, bringing cells together.

Type IV secretion system is constructed and joins two cells.

Coupling factor   Origin of transfer

Type IV secretion system

Recipient cell (F⁻)

Relaxosome makes a cut at the origin of transfer and begins to separate one DNA strand. The intact strand is replicated by the rolling circle mechanism.

Relaxosome

Accessory proteins of the relaxosome are released. The DNA/relaxase complex is recognized by the coupling factor and transferred to the secretion system.

Relaxase

Coupling factor

Secretion system

The secretion system pumps the DNA/relaxase complex into the recipient cell.

As the DNA enters, the F-factor DNA is replicated to become double stranded.

F⁺ cell

F⁺ cell

(b) Insertion of F factor into chromosome

F⁺ cell   Hfr cell

Bacterial chromosome

pro⁺
lac⁺

Integration of F factor into chromosome

pro⁺
lac⁺

Origin of transfer

Origin of transfer

F factor

Figure 14.21 F Factor-Mediated Conjugation. The F factor encodes proteins for building the sex pilus and the type IV secretion system that transfers DNA from the donor to the F⁻ recipient. One protein, the coupling factor, is thought to guide the DNA to the secretion system. (a) During F⁺ × F⁻ conjugation, only the F factor is transferred because the plasmid is extrachromosomal. The recipient cell becomes F⁺. (b) Integration of the F factor into the chromosome creates an Hfr cell. (c) During Hfr × F⁻ conjugation, some plasmid genes and some chromosomal genes are transferred to the recipient. Note that only a portion of the F factor moves into the recipient. Because the entire plasmid is not transferred, the recipient remains F⁻. In addition, the incoming DNA must recombine into the recipient's chromosome if it is to be stably maintained.

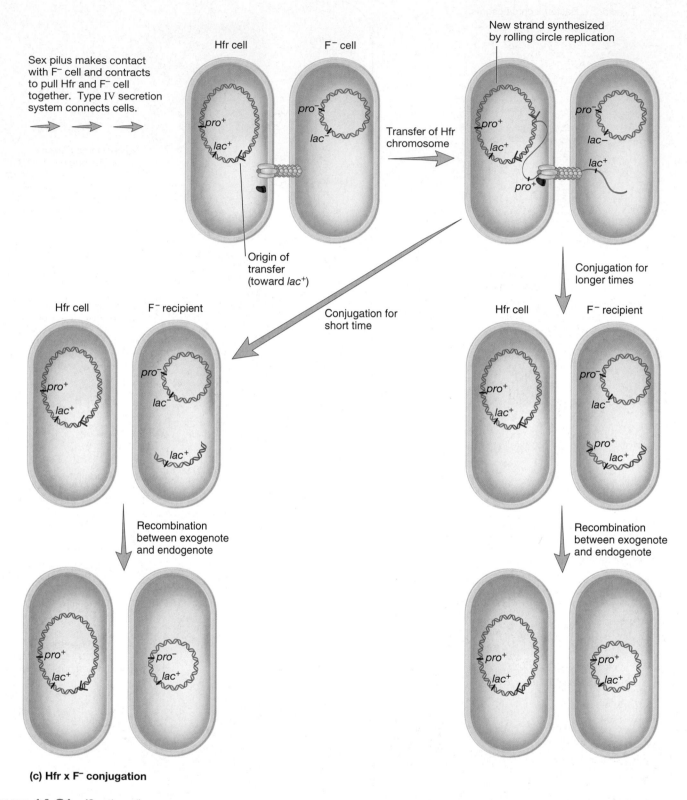

(c) **Hfr x F⁻ conjugation**

Figure 14.21 *(Continued)*

integrated, the F plasmid's *tra* operon is still functional; the plasmid can direct the synthesis of pili, carry out rolling-circle replication, and transfer genetic material to an F⁻ recipient cell. However, rather than transferring itself, the F factor directs the transfer of the host

chromosome. DNA transfer begins when the integrated F factor is nicked at its origin of transfer site. As it is replicated, the chromosome moves to the recipient (figure 14.21c). Only part of the F factor is initially transferred and the host chromosome follows.

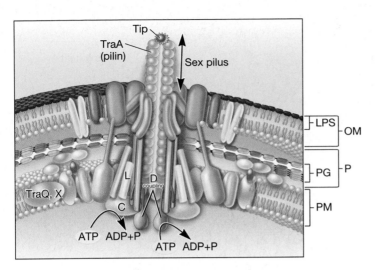

Figure 14.22 The Type IV Secretion System Encoded by F Factor. The F factor-encoded type IV secretion system is composed of numerous Tra proteins, including TraA proteins, which form the sex pilus, and TraD, which is the coupling factor. Some Tra proteins are located in the plasma membrane (PM); others extend into the periplasm (P) and pass through the peptidoglycan layer (PG) into the outer membrane (OM) and its lipopolysaccharide (LPS) layer.

Transfer of the entire chromosome with the integrated F factor requires about 100 minutes in *E. coli,* but the connection between the cells usually breaks before this process is finished. Thus a complete F factor is rarely transferred, and the recipient remains F^-.

As mentioned earlier, when an Hfr strain participates in conjugation, bacterial genes are frequently transferred to the recipient. Gene transfer can be in either a clockwise or a counterclockwise direction around the circular chromosome, depending on the orientation of the integrated F factor. After the replicated donor chromosome enters the recipient cell, it may be degraded or incorporated into the F^- genome by recombination.

F′ Conjugation

Because the F plasmid is an episome, it can leave the bacterial chromosome and resume status as an autonomous F factor. Sometimes during excision, the plasmid makes an error and picks up a portion of the chromosome. Because it is now genotypically distinct from the original F factor, it is called an **F′ plasmid** (figure 14.23*a*). A cell containing an F′ plasmid retains all of its genes, although some of them are on the plasmid. It mates only with an F^- recipient and $F′ \times F^-$ conjugation is similar to an $F^+ \times F^-$ mating. Once again, the plasmid is transferred as it is copied by rolling-circle replication. Bacterial genes on the chromosome are not transferred (figure 14.23*b*), but bacterial genes on the F′ plasmid are transferred. These genes need not be incorporated into the recipient chromosome to be expressed. The recipient becomes F′ and is a partially diploid merozygote because the same bacterial genes present on the F′ plasmid are also found on the recipient's chromosome. In this way, specific bacterial genes may spread rapidly throughout a bacterial population.

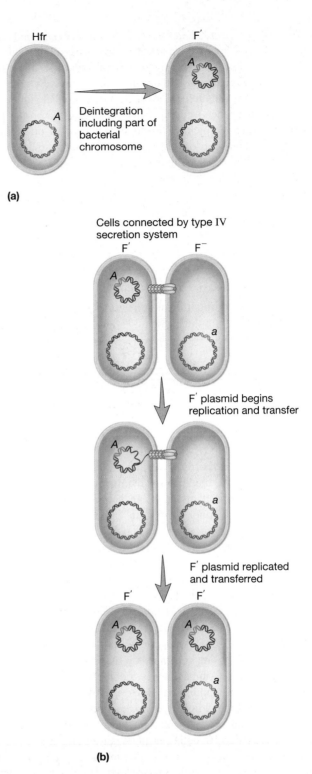

Figure 14.23 F′ Conjugation. (a) Due to an error in excision, the *A* gene of an Hfr cell is picked up by the F factor. (b) During conjugation the *A* gene is transferred to a recipient, which becomes diploid for that gene (i.e., Aa).

Other Examples of Bacterial Conjugation

Although most research on plasmids and conjugation has been done using *E. coli* and other gram-negative bacteria, conjugative plasmids are present in gram-positive bacterial genera such as

Bacillus, Streptococcus, Enterococcus, Staphylococcus, and *Streptomyces.* Much less is known about these systems. It appears that fewer transfer genes are involved, possibly because a sex pilus may not be required for plasmid transfer. For example, *Enterococcus faecalis* recipient cells release short peptide chemical signals that activate transfer genes in donor cells containing the proper plasmid. Donor and recipient cells directly adhere to one another through special plasmid-encoded proteins released by the activated donor cell. Plasmid transfer then occurs.

1. What is bacterial conjugation and how was it discovered?
2. Distinguish between F^+, Hfr, and F^- strains of *E. coli* with respect to their physical nature and role in conjugation.
3. Describe how $F^+ \times F^-$ and Hfr conjugation processes proceed, and distinguish between the two in terms of mechanism and the final results.
4. What is F′ conjugation and how does the F′ plasmid differ from a regular F plasmid?

14.8 BACTERIAL TRANSFORMATION

The second way DNA can move between bacteria is through transformation, discovered by Fred Griffith in 1928. **Transformation** is the uptake by a cell of DNA, either a plasmid or a fragment of linear DNA, from the medium and the maintenance of the DNA in the recipient in a heritable form. In natural transformation, the DNA comes from a donor bacterium. The process is random, and any portion of the donor's genome may be transferred.

When bacteria lyse, they release considerable amounts of DNA into the surrounding environment. These fragments may be relatively large and contain several genes. If a fragment contacts a **competent cell**—a cell that is able to take up DNA and be transformed—the DNA can be bound to the cell and taken inside (**figure 14.24**a). The transformation frequency of very competent cells is around 10^{-3} for most genera when an excess of DNA is used. That is, about one cell in every thousand will take up and integrate the gene. Competency is a complex phenomenon and depends on several conditions. Bacteria need to be in a certain stage of growth; for example, *Streptococcus pneumoniae* becomes competent during the exponential phase when the population reaches about 10^7 to 10^8 cells per ml. When a population becomes competent, bacteria such as *S. pneumoniae* secrete a small protein called the competence factor that stimulates the production of 8 to 10 new proteins required for transformation. Natural transformation has been discovered so far only in certain genera including *Streptococcus, Bacillus, Thermoactinomyces, Haemophilus, Neisseria, Moraxella, Acinetobacter, Azotobacter, Helicobacter,* and *Pseudomonas.* Gene transfer by this process occurs in soil and aquatic ecosystems and may be an important route of genetic exchange in biofilm and other microbial communities.

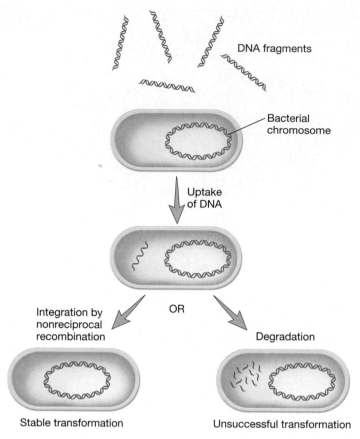

(a) Transformation with DNA fragments

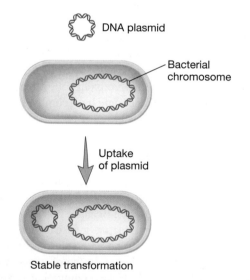

(b) Transformation with a plasmid

Figure 14.24 Bacterial Transformation. Transformation with (a) DNA fragments and (b) plasmids. Transformation with a plasmid often is induced artificially in the laboratory. The transforming DNA is in purple, and integration is at a homologous region of the genome.

The mechanism of transformation has been intensively studied in *S. pneumoniae* (**figure 14.25**). A competent cell binds a double-stranded DNA fragment if the fragment is moderately

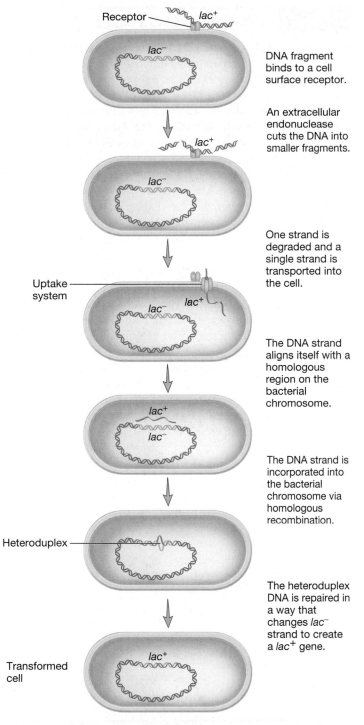

Receptor — — lac^+

DNA fragment binds to a cell surface receptor.

lac^-

An extracellular endonuclease cuts the DNA into smaller fragments.

lac^+

lac^-

One strand is degraded and a single strand is transported into the cell.

Uptake system

lac^- lac^+

The DNA strand aligns itself with a homologous region on the bacterial chromosome.

lac^+

lac^-

The DNA strand is incorporated into the bacterial chromosome via homologous recombination.

Heteroduplex

The heteroduplex DNA is repaired in a way that changes lac^- strand to create a lac^+ gene.

lac^+

Transformed cell

Figure 14.25   Bacterial Transformation as Seen in *S. pneumoniae*.

large; the process is random, and donor fragments compete with each other. The DNA then is cleaved by endonucleases to double-stranded fragments about 5 to 15 kilobases in size. DNA uptake requires energy expenditure. One strand is hydrolyzed by an envelope-associated exonuclease during uptake; the other strand associates with small proteins and moves through the plasma membrane. The single-stranded fragment

can then align with a homologous region of the genome and be integrated, probably by a mechanism similar to that depicted in figure 14.13.

Transformation in *Haemophilus influenzae,* a gram-negative bacterium, differs from that in *S. pneumoniae* in several respects. *H. influenzae* does not produce a protein factor to stimulate the development of competence, and it takes up DNA from only closely related species (*S. pneumoniae* is less particular about the source of its DNA). Double-stranded DNA, complexed with proteins, is taken in by membrane vesicles. The specificity of *H. influenzae* transformation is due to an 11 base pair sequence that is repeated over 1,400 times in *H. influenzae* DNA. DNA must have this sequence to be bound by a competent cell.

The protein complexes that take up free DNA must be able to move it through gram-negative and gram-positive cell walls. As expected, the machinery is quite large and complicated, and appears related to protein secretion systems. **Figure 14.26a** illustrates the complex used by the gram-negative bacterium *Neisseria gonorrhoeae.* PilQ aids in the movement across the outer membrane, and the pilin complex (PilE) moves the DNA through the periplasm and peptidoglycan. ComE is a DNA binding protein; N is the nuclease that degrades one strand before the DNA enters the cytoplasm

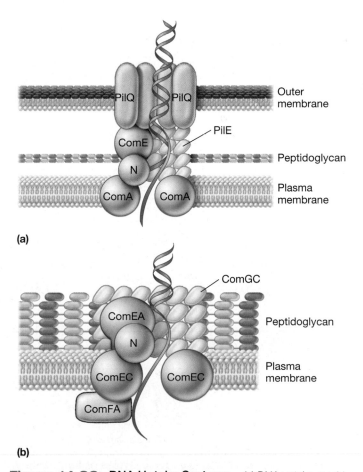

(a)

PilQ   PilQ   Outer membrane

PilE

ComE

N   Peptidoglycan

ComA   ComA   Plasma membrane

(b)

ComGC

ComEA

N   Peptidoglycan

ComEC   ComEC   Plasma membrane

ComFA

Figure 14.26   DNA Uptake Systems.   (a) DNA uptake machinery in *N. gonorrhoeae.* (b) Uptake machinery in *B. subtilis.*

through the transmembrane channel formed by ComA. The machinery in the gram-positive bacterium *Bacillus subtilis* is depicted in figure 14.26b. It is localized to the poles of the cell, and many of the components are similar to those of *N. gonorrhoeae:* the pilin complex (ComGC), DNA binding protein (ComEA), nuclease (N), and channel protein (ComEC). ComFA is a DNA translocase that moves the DNA into the cytoplasm. A gram-negative equivalent of ComFA has not been identified yet in *N. gonorrhoeae.*

Microbial geneticists exploit transformation to move DNA (usually recombinant DNA) into cells. Because many species, including *E. coli,* are not naturally transformation competent, these bacteria must be made artificially competent by certain treatments. Two common techniques are electrical shock and exposure to calcium chloride. Both approaches render the cell membrane more permeable to DNA, and both have been used to make artificially competent *E. coli* cells. To increase the transformation frequency with *E. coli,* strains that lack one or more nucleases are used. These strains are especially important when transforming the cells with linear DNA, which is vulnerable to attack by nucleases. It is easier to transform bacteria with plasmid DNA since plasmids can replicate within the host and are not as easily degraded as are linear fragments (figure 14.24b).

1. Define transformation and competence.

2. Describe how transformation occurs in *S. pneumoniae.* How does the process differ in *H. influenzae?*

3. Discuss two ways in which artificial transformation can be used to place functional genes within bacterial cells.

14.9 TRANSDUCTION

The third mode of bacterial gene transfer is **transduction.** It is a frequent mode of horizontal gene transfer in nature and is mediated by viruses. Recall from chapter 5 that viruses are structurally simple, often composed of just a nucleic acid genome protected by a protein coat called the capsid. They are unable to replicate autonomously. Instead, they infect and take control of a host cell, forcing the host to make many copies of the virus. Viruses that infect bacteria are called bacteriophages, or phages for short. **Virulent bacteriophages** replicate in their bacterial host immediately after entry. After the number of replicated phages reaches a certain number, they cause the host to lyse, so they can be released and infect new host cells (**figure 14.27**). Thus this process is called the **lytic cycle. Temperate bacteriophages,** on the other hand, do not immediately kill their host. Many temperate phages enter the host bacterium and insert their genomes into the bacterial chromosome. The inserted viral genome is called a **prophage.** The host bacterium is unharmed by this, and the phage genome is passively replicated as the host cell's genome is replicated. The relationship between these viruses and their host is called **lysogeny,** and bacteria that have been lysogenized are called **lysogens.** Temperate phages can remain inactive in their hosts for many generations. However, they can be induced to switch to a lytic cycle of growth under certain conditions, including UV irradiation. When this occurs, the prophage is excised from the bacterial genome and the lytic cycle proceeds.

Transduction is the transfer of procaryotic genes by viruses. The genes are packaged in the virus because of errors made during its life cycle. The virus containing these genes then

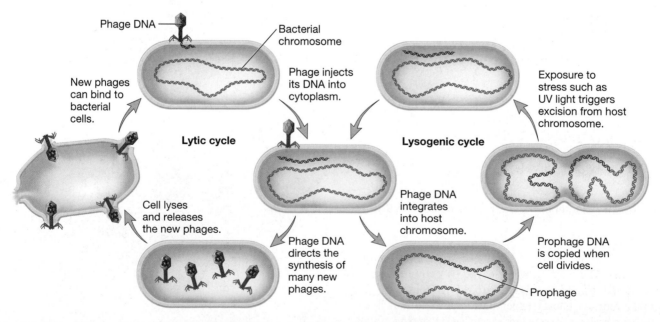

Figure 14.27 Lytic and Lysogenic Cycles of Temperate Phages.   Virulent phages undergo only the lytic cycle. Temperate phages have two phases to their life cycles. The lysogenic cycle allows the genome of the virus to be replicated passively as the host cell's genome is replicated. Certain environmental factors such as UV light can cause a switch from the lysogenic cycle to the lytic cycle. In the lytic cycle, new virions are made and released when the host cell lyses. Virulent phages are limited to just the lytic cycle.

transfers them to a recipient cell. Two kinds of bacterial transduction have been described: generalized and specialized.

Generalized Transduction

Generalized transduction occurs during the lytic cycle of virulent and some temperate phages. Any part of the bacterial genome can be transferred (**figure 14.28**). During the assembly stage, when the viral chromosomes are packaged into capsids, random fragments of the partially degraded bacterial chromosome may mistakenly be packaged. Because the capsid can contain only a limited quantity of DNA, the viral DNA is left behind. The quantity of bacterial DNA carried depends primarily on the size of the capsid. The P22 phage of *Salmonella enterica* serovar Typhimurium can carry about 1% of the bacterial genome; the P1 phage of *E. coli*

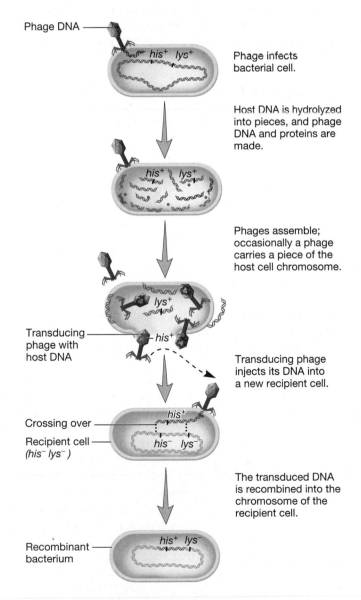

Phage DNA

his+ *lys*+ — Phage infects bacterial cell.

his+ *lys*+ — Host DNA is hydrolyzed into pieces, and phage DNA and proteins are made.

Phages assemble; occasionally a phage carries a piece of the host cell chromosome.

lys+

Transducing phage with host DNA — *his*+

Transducing phage injects its DNA into a new recipient cell.

his+

Crossing over

Recipient cell (*his*− *lys*−) — *his*− *lys*−

The transduced DNA is recombined into the chromosome of the recipient cell.

his+ *lys*−

Recombinant bacterium

The recombinant bacterium has a genotype (*his*+*lys*−) that is different from recipient bacterial cell (*his*− *lys*−).

Figure 14.28 Generalized Transduction in Bacteria.

and a variety of gram-negative bacteria can package about 2.0 to 2.5% of the genome. The resulting virus often injects the DNA into another bacterial cell but cannot initiate a lytic cycle. This phage is known as a generalized transducing particle and is simply a carrier of genetic information from the original bacterium to another cell. As in transformation, once the DNA fragment has been injected, it must be incorporated into the recipient cell's chromosome to preserve the transferred genes. The DNA remains double stranded during transfer, and both strands are integrated into the endogenote. About 70 to 90% of the transferred DNA is not integrated but often is able to remain intact temporarily and be expressed. Abortive transductants are bacteria that contain this nonintegrated, transduced DNA and are partial diploids.

Specialized Transduction

In **specialized transduction,** only specific portions of the bacterial genome are carried by transducing particles. Specialized transduction is made possible by an error in the lysogenic life cycle of phages that insert their genomes into a specific site in the host chromosome. When a prophage is induced to leave the host chromosome, excision is sometimes carried out improperly. The resulting phage genome contains portions of the bacterial chromosome (about 5 to 10% of the bacterial DNA) next to the integration site, much like the situation with F' plasmids (**figure 14.29**). A transducing particle genome usually is defective because it lacks some of its genome. However, it will inject bacterial genes into another bacterium, even though the defective phage cannot reproduce without assistance. The bacterial genes may become stably incorporated under the proper circumstances.

The best-studied example of specialized transduction is carried out by the *E. coli* phage lambda. The lambda genome inserts into the host chromosome at specific locations known as attachment or *att* sites (**figure 14.30**). The phage *att* sites and bacterial *att* sites are similar and can complex with each other. The *att* site for lambda is next to the *gal* and *bio* genes on the *E. coli* chromosome; consequently when lambda excises incorrectly to generate a specialized transducing particle, these bacterial genes are most often present. The lysate, or product of cell lysis, resulting from the induction of lysogenized *E. coli* contains normal phage and a few defective transducing particles. These particles are called lambda *dgal* if they carry the galactose utilization genes or lambda *dbio* if they carry the bio from the other side of the *att* site (figure 14.30). **>>** *Bacteriophage lambda: A temperate bacteriophage (section 24.2)*

1. Describe generalized transduction and how it occurs. What is an abortive transductant?

2. What is specialized transduction and how does it come about?

3. How might one tell whether horizontal gene transfer was mediated by generalized or specialized transduction?

4. Why doesn't a cell lyse after successful transduction with a temperate phage?

5. Describe how conjugation, transformation, and transduction are similar. How are they different?

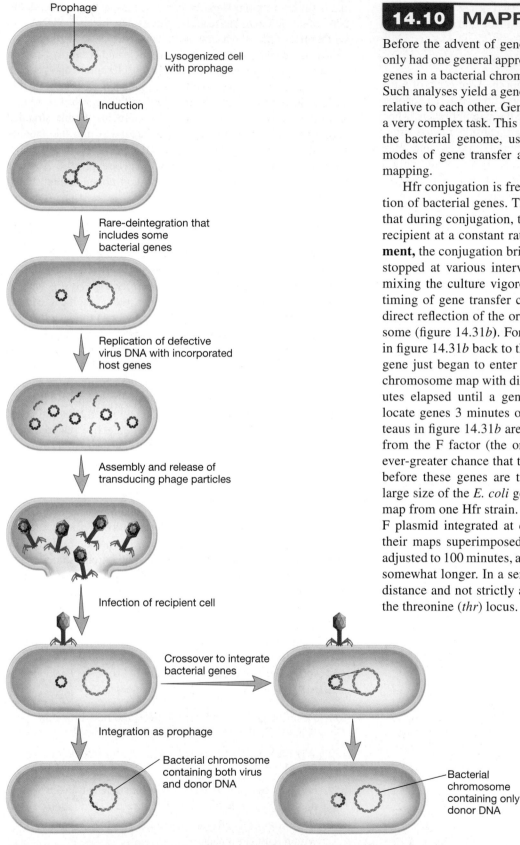

Figure 14.29   Specialized Transduction by a Temperate Bacteriophage.   Recombination can produce two types of transductants.

14.10 MAPPING THE GENOME

Before the advent of genome sequencing, microbial geneticists only had one general approach for elucidating the organization of genes in a bacterial chromosome—to carry out linkage analysis. Such analyses yield a genetic map showing the position of genes relative to each other. Genetic mapping using linkage analysis is a very complex task. This section surveys approaches to mapping the bacterial genome, using *E. coli* as an example. All three modes of gene transfer and recombination have been used in mapping.

Hfr conjugation is frequently used to map the relative location of bacterial genes. This technique rests on the observation that during conjugation, the chromosome moves from donor to recipient at a constant rate. In an **interrupted mating experiment,** the conjugation bridge is broken and Hfr × F⁻ mating is stopped at various intervals after the start of conjugation by mixing the culture vigorously (**figure 14.31a**). The order and timing of gene transfer can be determined because they are a direct reflection of the order of genes on the bacterial chromosome (figure 14.31b). For example, extrapolation of the curves in figure 14.31b back to the x-axis gives the time at which each gene just began to enter the recipient. The result is a circular chromosome map with distances expressed in terms of the minutes elapsed until a gene is transferred. This technique can locate genes 3 minutes or more apart. The heights of the plateaus in figure 14.31b are lower for genes that are more distant from the F factor (the origin of transfer) because there is an ever-greater chance that the sex pilus will spontaneously break before these genes are transferred. Because of the relatively large size of the *E. coli* genome, it is not possible to generate a map from one Hfr strain. Therefore several Hfr strains with the F plasmid integrated at different locations must be used and their maps superimposed on one another. The overall map is adjusted to 100 minutes, although complete transfer may require somewhat longer. In a sense, minutes are an indication of map distance and not strictly a measure of time. Zero time is set at the threonine (*thr*) locus.

Gene linkage, or the proximity of two genes on a chromosome, can be determined from transformation by measuring the frequency with which two or more genes simultaneously transform a recipient cell. Consider the case for cotransformation by two genes. In theory, a bacterium could simultaneously receive two genes, each carried on a separate DNA fragment. However, it is much more likely that genes residing on the same fragment will be simultaneously transferred. If two genes are closely linked on the

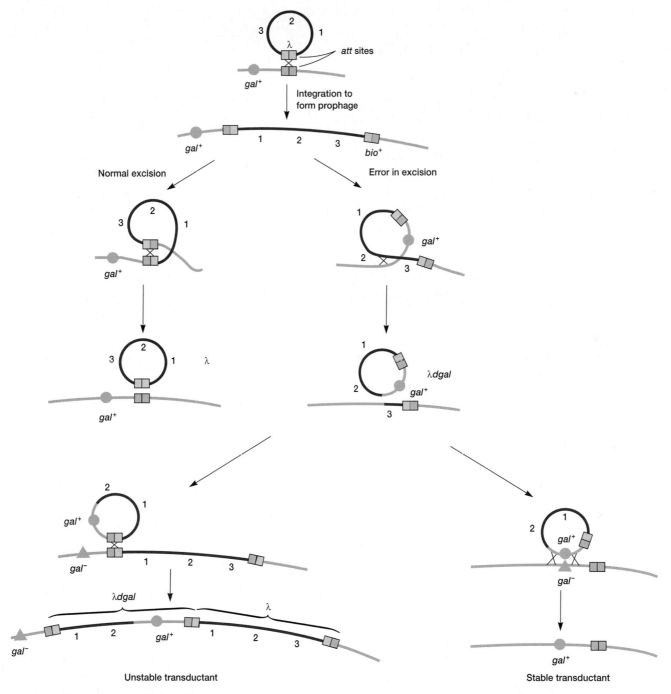

Figure 14.30   The Mechanism of Transduction for Phage Lambda and *E. coli.* Integrated lambda phage lies between the *gal* and *bio* genes. When it excises normally (top left), the new phage is complete and contains no bacterial genes. Rarely excision occurs asymmetrically (top right), and either the *gal* or *bio* genes are picked up and some phage genes are lost (only aberrant excision involving the *gal* genes is shown). The result is a defective lambda phage that carries bacterial genes and can transfer them to a new recipient.

chromosome, then they should be able to cotransform. The closer the genes are together, the more often they will be carried on the same fragment and the higher will be the frequency of cotransformation. If genes are spaced a great distance apart, they will be carried on separate DNA fragments and the frequency of double transformants will equal the product of the individual transformation frequencies.

Generalized transduction can be used to obtain linkage information in much the same way as transformation. Linkages usually are expressed as cotransduction frequencies, using the argument that the closer two genes are to each other, the more likely they both will reside on the DNA fragment incorporated into a single phage capsid. The *E. coli* phage P1 is often used in such mapping because it can randomly transduce up to 1 to 2% of the genome (**figure 14.32**).

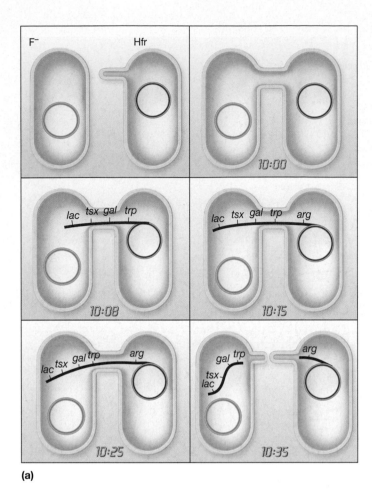

(a)

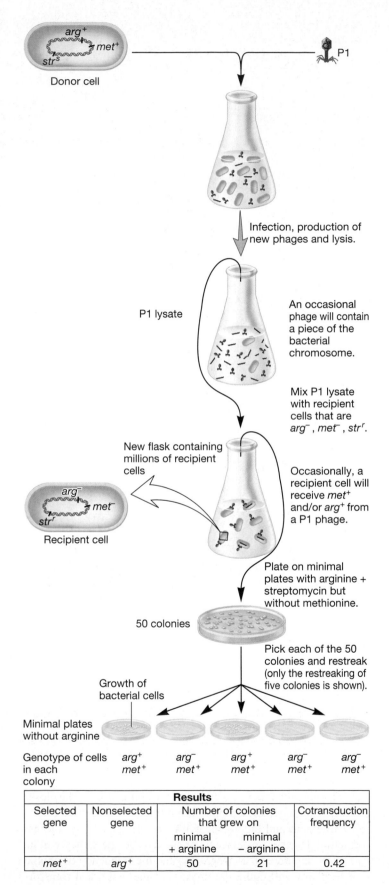

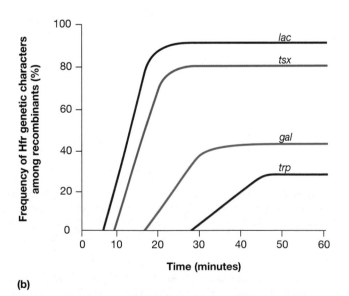

(b)

Figure 14.31 An Interrupted Mating Experiment. An interrupted mating experiment using Hfr × F⁻ conjugation. (a) The linear transfer of genes is stopped by breaking the conjugation bridge to study the sequence of gene entry into the recipient cell. (b) An example of the results obtained by an interrupted mating experiment. The gene order is *lac-tsx-gal-trp*.

Results				
Selected gene	Nonselected gene	Number of colonies that grew on		Cotransduction frequency
		minimal + arginine	minimal – arginine	
met⁺	*arg⁺*	50	21	0.42

Figure 14.32 A Cotransduction Experiment. In this experiment, the donor is able to synthesize the amino acids arginine and methionine (*arg*⁺ and *met*⁺) but is killed by the antibiotic streptomycin (*str*ˢ). The recipient is unable to synthesize arginine and methionine but is resistant to streptomycin (*str*ʳ). The phage lysate made by infecting the donor bacterium is mixed with the recipient bacterium. The mixture is then plated onto a medium containing streptomycin but lacking methionine. Therefore the only cells able to grow are those recipient cells that have received the functional methionine gene from the donor. The colonies that grow are then tested to see if they also received the gene for arginine biosynthesis from the donor. This is determined by plating the cells on a minimal medium lacking arginine. Only those cells that can synthesize arginine grow.

A simplified genetic map of *E. coli* K12 is given in **figure 14.33**. Because conjugation data are not high resolution and cannot be used to position genes that are very close together, the map was developed using interrupted mating data combined with those from cotransduction and cotransformation studies. Data from recombination studies also were used. Such analyses are no longer performed in *E. coli* and other microbes for which a genome sequence has been published.

How does the map generated by classical approaches compare to the actual nucleotide sequence of the genome (i.e., a physical map of the genome)? Using classical genetic approaches, researchers mapped about 2,200 genes of *E. coli* K12, whereas genome sequencing has revealed about 4,300 possible genes. Thus genetic analysis defined over half of the potential genes. The genetic map approximates the physical map, but they do not correspond perfectly. This is because the genetic map is derived from genetic linkage frequencies that do not correlate exactly with the number of nucleotides that separate two genes. Roughly speaking, one minute of the *E. coli* genetic map corresponds to 40,000 bases of DNA sequence.

1. Why is it necessary to use several different techniques in genome mapping? How is this done in practice?

2. Describe how you would precisely locate the *recA* gene and show that it was between 58 and 58.5 minutes on the *E. coli* chromosome.

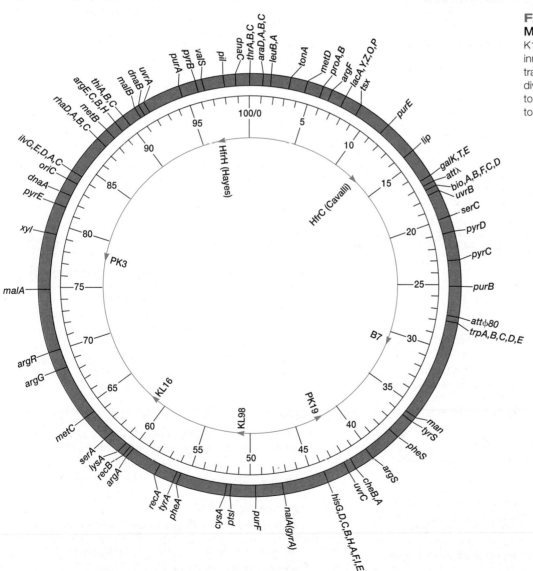

Figure 14.33 *E. coli* Genetic Map. A circular genetic map of *E. coli* K12 with the location of selected genes. The inner circle shows the origin and direction of transfer of several Hfr strains. The map is divided into 100 minutes, the time required to transfer the chromosome from an Hfr cell to F⁻ at 37°C.

Summary

14.1 Mutations and Their Chemical Basis

a. A mutation is a stable, heritable change in the nucleotide sequence of the genetic material.

b. Spontaneous mutations can arise from replication errors (transition, transversion, and insertion and deletion of nucleotides), and from DNA lesions (apurinic sites, apyrimidinic sites, oxidation of DNA) (**figures 14.1 and 14.2**).

c. Induced mutations are caused by mutagens. Mutations may result from the incorporation of base analogs, specific mispairing due to alterations of a base caused by DNA-modifying agents, the presence of intercalating agents, and severe damage to the DNA caused by exposure to radiation (**table 14.1**).

d. Mutations are usually recognized when they cause a change from the more prevalent wild-type phenotype. A mutant phenotype can be restored to wild type by either reversions or suppressor mutations (**table 14.2**).

e. There are four important types of point mutations: silent mutations, missense mutations, nonsense mutations, and frameshift mutations (**table 14.2**).

f. Mutations can affect phenotype in numerous ways. Some major types of mutations categorized based on their effects on phenotype are morphological, lethal, conditional, biochemical, and resistance mutations.

14.2 Detection and Isolation of Mutants

a. A sensitive and specific screening method is needed for detecting and isolating mutants. An example is replica plating for the detection of auxotrophs (**figure 14.5**).

b. One of the most effective techniques for isolating mutants is to select for mutants by adjusting environmental conditions so that the mutant will grow while the wild type does not.

c. Because many carcinogens are also mutagenic, one can test for mutagenicity with the Ames test and use the results as an indirect indication of carcinogenicity (**figure 14.6**).

14.3 DNA Repair

a. Cells have multiple mechanisms for correcting mispaired and damaged DNA.

b. Excision repair systems remove damaged portions from a single strand of DNA (e.g., thymine dimers) and use the other strand as a template for filling in the gap (**figures 14.7 and 14.8**).

c. Direct repair systems correct damaged DNA without removing damaged regions. For instance, during photoreactivation, thymine dimers are repaired by splitting the two thymines apart. This is catalyzed in the presence of light by the enzyme photolyase.

d. Mismatch repair is similar to excision repair, except that it replaces mismatched base pairs (**figure 14.9**).

e. Recombinational repair removes damaged DNA by recombination of the damaged DNA with a normal DNA strand elsewhere in the cell (**figure 14.10**).

f. When DNA damage is severe, DNA replication is halted. This triggers the SOS response. During the SOS response, genes of the repair systems are transcribed at a higher rate. In addition, special DNA polymerases are produced. These are able to replicate damaged DNA. However, they do so without a proper template and therefore create mutations.

14.4 Creating Genetic Variability

a. In recombination, genetic material from two different DNA molecules is combined to form a new hybrid molecule.

b. Horizontal gene transfer is an important mechanism for creating genetic diversity in procaryotes. It is a one-way process in which the exogenote is transferred from the donor to a recipient. In many transfers, the exogenote must be integrated into the endogenote to be stably maintained (**figure 14.11**).

c. There are three types of recombination: homologous recombination, site-specific recombination, and transposition (**figures 14.12 and 14.13**).

14.5 Transposable Elements

a. Transposons or transposable elements are DNA segments that move about the genome in a process known as transposition.

b. There are three types of transposable elements: insertion sequences, composite transposons, and replicative transposons (**figure 14.14**).

c. Simple (cut-and-paste) transposition and replicative transposition are two distinct mechanisms of transposition (**figures 14.15 and 14.16**).

d. Transposable elements can cause mutations, turn genes on and off, aid plasmid insertion, and carry antibiotic resistance genes.

14.6 Bacterial Plasmids

a. Plasmids are small, autonomously replicating DNA molecules that can exist independent of the host chromosome.

b. Episomes are plasmids that can be reversibly integrated with the host chromosome.

c. The F factor is one type of conjugative plasmid; that is, it is able to transfer itself from one bacterium to another (**figure 14.18**).

14.7 Bacterial Conjugation

a. Conjugation is the transfer of genes between bacteria that depends upon direct cell-to-cell contact. F factor conjugation is mediated by a sex pilus and a type IV secretion system.

b. In $F^+ \times F^-$, mating, the F factor remains independent of the chromosome and a copy is transferred to the F^- recipient; donor genes are not usually transferred (**figure 14.21a**).

c. Hfr strains transfer bacterial genes to recipients because the F factor is integrated into the host chromosome. A complete copy of the F factor is not often transferred (**figure 14.21b, c**).

d. When the F factor leaves an Hfr chromosome, it occasionally picks up some bacterial genes to become an F′ plasmid, which readily transfers these genes to other bacteria (**figure 14.23**).

14.8 Bacterial Transformation

a. Transformation is the uptake of naked DNA by a competent cell and its incorporation into the genome (**figures 14.24 and 14.25**).

b. Only certain bacterial species are naturally transformation competent. Other species can be made competent by artificial means.

14.9 Transduction

a. Bacterial viruses or bacteriophages can reproduce and destroy the host cell (lytic cycle) or become a latent prophage that remains within the host (lysogenic cycle) (**figure 14.27**).

b. Transduction is the transfer of genes by viruses.

c. In generalized transduction, any host DNA fragment can be packaged in a virus capsid and transferred to a recipient **(figure 14.28)**.

d. Certain temperate phages carry out specialized transduction by incorporating bacterial genes during prophage induction and then donating those genes to another bacterium **(figure 14.29)**.

14.10   Mapping the Genome

a. The bacterial genome can be mapped by following the order of gene transfer during Hfr conjugation **(figure 14.31)**; transformational and transductional mapping techniques also may be used **(figure 14.32)**.

b. The genomes of many microbes have been sequenced. For these microbes, genetic mapping is no longer perfomed.

Critical Thinking Questions

1. Mutations are often considered harmful. Give an example of a mutation that would be beneficial to a microorganism. What gene would bear the mutation? How would the mutation alter the gene's role in the cell, and what conditions would select for this mutant allele?

2. Mistakes made during transcription affect the cell but are not considered "mutations." Why not?

3. Given what you know about the differences between bacterial and eucaryotic cells, give two reasons why the Ames test detects only about half of potential carcinogens, even when liver extracts are used.

4. Suppose that transduction took place when a U-tube experiment was conducted. How would you confirm that a virus was passed through the filter and transduced the recipient?

5. Suppose that you carried out a U-tube experiment with two auxotrophs and discovered that recombination was not blocked by the filter but was stopped by treatment with deoxyribonuclease. What gene transfer process is responsible? Why would it be best to use double or triple auxotrophs in this experiment?

6. What would be the evolutionary advantage of having a period of natural "competence" in a bacterial life cycle? What would be possible disadvantages?

Learn More

Learn more by visiting the Prescott website at www.mhhe.com/prescottprinciples, where you will find a complete list of references.

Microbial Genomics

Chapter Glossary

bioinformatics The interdisciplinary field that manages and analyzes large biological data sets, including genome and protein sequences.

coding sequence (CDS) A nucleotide sequence that encodes or is thought to encode a protein.

comparative genomics The analysis of genomes from different organisms to look for significant differences and similarities.

complementary DNA (cDNA) A DNA copy of an RNA molecule (e.g., a DNA copy of an mRNA).

DNA microarrays Solid supports that have DNA attached in an organized grid pattern; used to evaluate gene expression.

environmental genomics Also called **metagenomics,** the study of genomes recovered from natural samples without first isolating members of the microbial community and growing them in pure cultures.

functional genomics The analysis of genome transcripts and the proteins they encode.

genome annotation The process of determining the location and potential function of specific genes and genetic elements in a genome sequence.

genomics The study of the molecular organization of genomes, their information content, and the gene products they encode.

in silico analysis The study of physiology and genetics through the examination of nucleic acid and amino acid sequence.

open reading frame (ORF) A sequence of DNA not interrupted by a stop codon and with an apparent promoter and ribosome binding site at the 5′ end and a terminator at the 3′ end.

orthologue A gene found in the genomes of two or more different organisms that share a common ancestry and are presumed to have similar functions.

paralogue Two or more genes in the genome of a single organism that arose through duplication of a common ancestral gene.

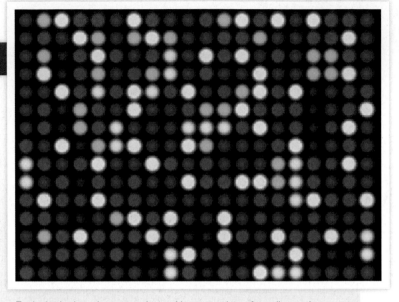

Each dot in the microarray pictured here consists of an oligonucleotide fragment of a single gene bound to a glass slide. Gene expression of two types of cells (e.g., one wild-type and one mutant) can be compared by labeling the cDNA from each cell with either a red or green fluorescent label and allowing the cDNAs to bind to homologous sequences attached to the microarray. The color of each spot reveals the relative level of expression of each gene.

phylotype A taxon that is characterized only by its nucleic acid sequence; generally discovered during metagenomic analysis.

proteome The complete collection of proteins that an organism produces.

transcriptome All the messenger RNA that is transcribed from the genome of an organism under a given set of circumstances.

whole-genome shotgun sequencing An approach to genome sequencing in which the complete genome is broken into random fragments, which are then individually sequenced. The fragments are then placed in the proper order based on overlapping nucleotide sequences.

A prerequisite to understanding the complete biology of an organism is the determination of its entire genome sequence.

—J. Craig Venter, et al.

Chapters 12–14 provide the foundations of microbial molecular genetics. In this chapter, we apply these concepts to the ongoing revolution in genomics. We begin with a general overview of the topic, followed by an introduction to DNA sequencing techniques. Next, the whole-genome shotgun sequencing method is briefly described. Genome function and the analysis of the transcripts and proteins produced by microbes are then explored. We focus on annotation, DNA microarrays, and proteomics. Finally, comparative and environmental genomics (metagenomics) are reviewed.

15.1   INTRODUCTION

Genomics is the study of the molecular organization of genomes, their information content, and the gene products they encode. It is a broad discipline, which may be divided into at least three general areas. **Structural genomics** involves sequencing the genome, identifying genes and potential genes, and analyzing the three-dimensional structure of the proteins encoded. **Functional genomics** is concerned with the way in which the genome functions. That is, it examines the transcripts produced by the genome and the array of proteins they encode. A third area of study is **comparative genomics,** in which genomes from different organisms are compared to look for significant differences and similarities. This helps identify important, conserved regions of the genome in an effort to discern patterns in function and regulation. These data also provide information about microbial evolution, particularly with respect to phenomena such as horizontal gene transfer.

Genomics is an exciting and growing field that has changed the ways in which key questions in microbial physiology, genetics, ecology, and evolution are pursued. Prior to the advent of genomics, analysis of gene expression was limited to the identification of a small subset of transcripts and proteins. As we will see, genomic analysis enables scientists to study the cell in a holistic way by capturing a snapshot of the entire pool of mRNA transcripts or proteins. The cell can now be viewed as a network of interconnected circuits, not as a series of individual pathways. Further, genomics provides a window into entire microbial communities: microbial ecologists no longer need to confine their studies to the tiny fraction of microorganisms that have been cultivated. Finally, our understanding of the evolution of all organisms can be illuminated by the insights we gain into procaryotic evolution from genomic approaches. In these ways and more, genomics has truly revolutionized biology.

15.2   DETERMINING DNA SEQUENCES

Techniques for sequencing DNA were developed in 1977 by Alan Maxam and Walter Gilbert (collaborating on one technique), and Frederick Sanger. Sanger's method is most commonly used and is discussed here. This method involves the synthesis of a new strand of DNA using the DNA to be sequenced as a template. The reaction begins when single strands of template DNA are mixed with primer (a short piece of DNA complementary to the region to be sequenced), DNA polymerase, the four deoxynucleoside triphosphates (dNTPs), and dideoxynucleoside triphosphates (ddNTPs). ddNTPs differ from dNTPs in that the 3′ carbon lacks a hydroxyl group (**figure 15.1**). In such a reac-

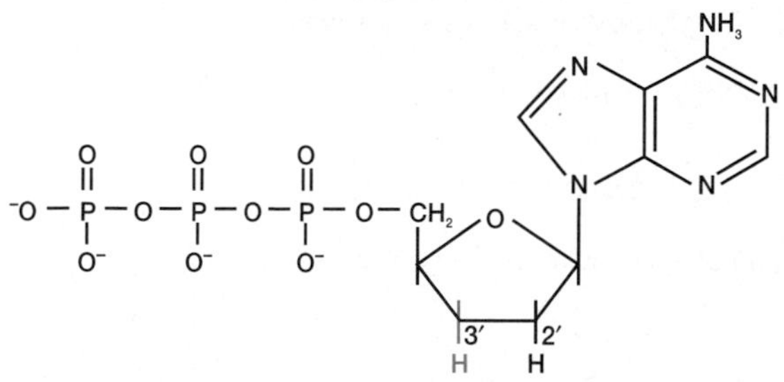

Figure 15.1   Dideoxyadenosine Triphosphate (ddATP).   Note the lack of a hydroxyl group on the 3′ carbon, which prevents further chain elongation by DNA polymerase.

tion mixture, DNA synthesis will continue until a ddNTP, rather than a dNTP, is added to the growing chain. Without a 3′-OH group to attack the 5′-PO$_4$ of the next dNTP to be incorporated, synthesis stops (*see figure 12.9*). Indeed, Sanger's technique is frequently referred to as the **chain-termination DNA sequencing method.** To obtain sequence information, four separate synthesis reactions must be prepared, one for each ddNTP (**figure 15.2**). When each DNA synthesis reaction is stopped, a collection of DNA fragments of varying lengths has been generated. The reaction prepared with ddATP produces fragments ending with an A, those with ddTTP produce fragments with T termini, and so forth. If the DNA is to be manually sequenced, radioactive dNTPs are used and each reaction is electrophoresed in a separate lane on a polyacrylamide gel. During gel electrophoresis, charged molecules are placed in an electrical field and allowed to migrate toward the pole with the opposite charge. Because DNA is negatively charged, it is loaded into wells at the negative pole of the gel and migrates toward the positive. The molecules separate because they move at different rates due to differences in charge and size. Each fragment's migration rate is inversely proportional to the log of its molecular weight. That is to say, the smaller a fragment is, the faster it moves through the gel. Because synthesis proceeds with the addition of a nucleotide to the 3′-OH of the primer, the dideoxynucleotide at the end of the shortest fragment is assigned as the 5′ end of the DNA sequence, while the largest fragment is the 3′ end. In this way, the DNA sequence is read directly from the gel from the smallest to the largest fragment.   ▸▸ *Gel electrophoresis (section 16.3)*

DNA is more often prepared for automated sequencing. Here the four reaction mixtures can be combined and loaded into a single lane of a gel because each ddNTP is labeled with a different colored fluorescent dye (figure 15.2b). These fragments are then electrophoresed and a laser beam determines the order in which they exit the gel. A chromatogram is generated in which the amplitude of each spike represents the fluorescent intensity of each particular fragment (figure 15.2c). The corresponding DNA sequence is listed above the chromatogram.

Fully automated capillary electrophoresis DNA analyzers are required for large projects. These sequencing machines are very fast and can run for 24 hours without operator attention. As many as 96

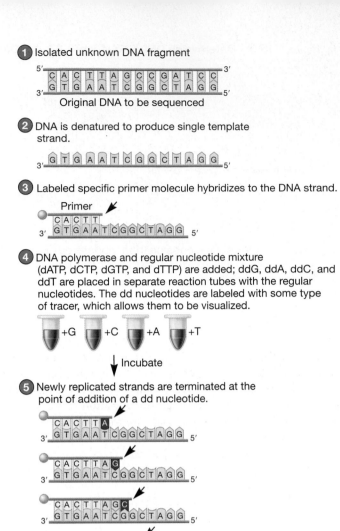

① Isolated unknown DNA fragment

5′ CACTTAGCCGATCC 3′
3′ GTGAATCGGGCTAGG 5′

Original DNA to be sequenced

② DNA is denatured to produce single template strand.

3′ GTGAATCGGCTAGG 5′

③ Labeled specific primer molecule hybridizes to the DNA strand.

Primer
CACTT
3′ GTGAATCGGCTAGG 5′

④ DNA polymerase and regular nucleotide mixture (dATP, dCTP, dGTP, and dTTP) are added; ddG, ddA, ddC, and ddT are placed in separate reaction tubes with the regular nucleotides. The dd nucleotides are labeled with some type of tracer, which allows them to be visualized.

+G   +C   +A   +T

Incubate

⑤ Newly replicated strands are terminated at the point of addition of a dd nucleotide.

CACTTA
3′ GTGAATCGGCTAGG 5′

CACTTAG
3′ GTGAATCGGCTAGG 5′

CACTTAGC
3′ GTGAATCGGCTAGG 5′

CACTTAGCC
3′ GTGAATCGGCTAGG 5′

(a)

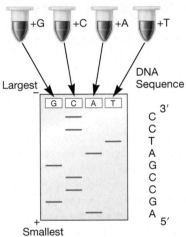

⑥ Schematic view of how all possible positions on the fragment are occupied by a labeled nucleotide

5′ AGCCGATCC 3′
AGCCGATC
AGCCGAT
AGCCGA
AGCCG
AGCC
AGC
AG
A

+G   +C   +A   +T

DNA Sequence

Largest −

| G | C | A | T |

3′
C
C
T
A
G
C
C
G
A
5′

+ Smallest

⑦ Running the reaction tubes in four separate gel lanes separates them by size and nucleotide type. Reading from bottom to top, one base at a time, provides the correct DNA sequence.

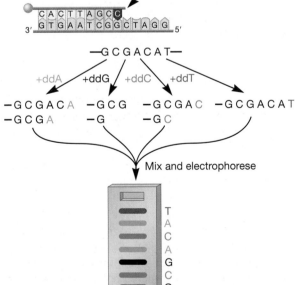

—GCGACAT—

+ddA   +ddG   +ddC   +ddT

−GCGACA   −GCG   −GCGAC   −GCGACAT
−GCGA   −G   −GC

Mix and electrophorese

T
A
C
A
G
C
G

(b)

Figure 15.2 The Sanger Method of DNA Sequencing. (a) Steps 1–6 are used for both manual and automated sequencing. Step (7) shows preparation of a gel for manual sequencing in which radiolabeled ddNTPs are used. (b) Part of an automated DNA sequencing run. Here the ddNTPs are labeled with fluorescent dyes. (c) Data generated during an automated DNA sequencing run. Bases 493 to 580 are shown.

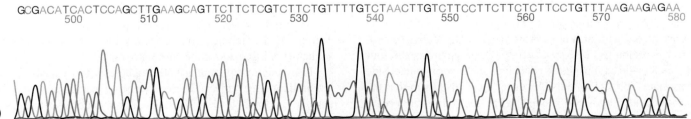

GCGACATCACTCCAGCTTGAAGCAGTTCTTCTCGTCTTCTGTTTTGTCTAACTTGTCTTCCTTCTTCTCTTCCTGTTTAAGAAGAGAA
    500   510   520   530   540   550   560   570   580

(c)

samples can be sequenced simultaneously, making it possible to sequence as many as 1 million bases per day, per sequencer. This level of automation, involving many sequencers running at the same time, is needed for the completion of whole-genome sequences.

15.3   WHOLE-GENOME SHOTGUN SEQUENCING

Efficient methods for sequencing whole genomes were not available until 1995, when J. Craig Venter, Hamilton Smith, and their collaborators developed **whole-genome shotgun sequencing** and the computer software needed to assemble sequence data into a complete genome. They used their new method to sequence the genomes of the bacteria *Haemophilus influenzae* and *Mycoplasma genitalium.* This was a significant accomplishment because prior to this, only a few viral genomes had been fully sequenced. These are much smaller than the genome of *H. influenzae,* which contains about 1,743 genes and 1,830,137 base pairs, or about 1.8 Mb (*million base pairs*). Venter and Smith's contribution to biology has ushered in what has been called the genomic era. Within 10 years, the number of complete genomes published grew from 2 to 250 with over 500 ongoing genome sequencing projects. **Table 15.1** lists just a few microbial genomes.

The process of whole-genome shotgun sequencing is fairly complex when considered in detail, but the following summary gives a general idea of the procedure. For simplicity, this approach may be broken into four stages: library construction, random sequencing, fragment alignment and gap closure, and editing.

1. **Library construction.** The DNA molecules are randomly broken into fragments using ultrasonic waves; the fragments are then purified (**figure 15.3**). These fragments are next attached to plasmid or cosmid vectors and isolated. Special *Escherichia coli* strains lacking restriction enzymes are transformed with the plasmids to produce a library of the plasmid clones.

   >> *Cloning vectors and creating recombinant DNA (section 16.4)*

2. **Random sequencing.** After the clones are prepared and the DNA purified, thousands of DNA fragments are sequenced with automated sequencers, employing dye-labeled primers. Normally the primers recognize the plasmid DNA sequences adjacent to the DNA insert. Almost all stretches of the genome are sequenced multiple times to increase the accuracy of the final results.

3. **Fragment alignment and gap closure.** Using computer analysis, the DNA sequence of each fragment is assembled into longer stretches of sequence. Two fragments are joined together to form a larger stretch of DNA if the sequences at their ends overlap and match. This overlap

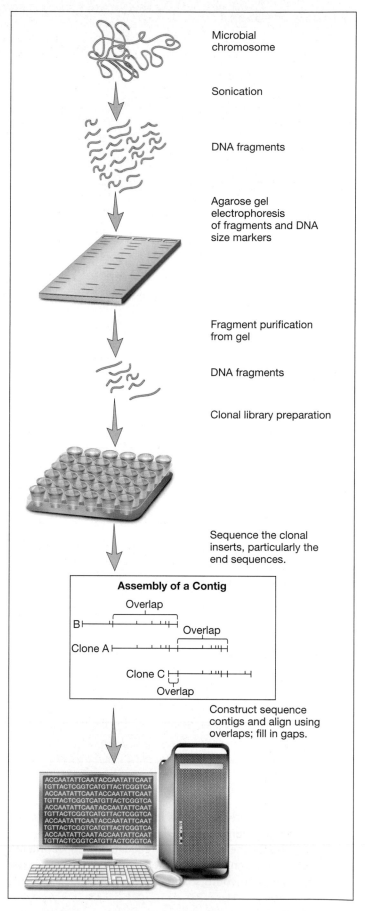

Figure 15.3   Whole-Genome Shotgun Sequencing.

Table 15.1	Examples of Complete, Published Microbial Genomes				
Genome	**Domain[a]**	**Number of Strains Sequenced**	**Size (Mb)**	**Approximate Number of Genes**	**% G + C**
Agrobacterium tumefaciens	B	2	4.92	4,550–4,660	60
Aquifex aeolicus	B	1	1.55	1,522	43
Bacillus anthracis	B	4	5.09–5.23	5,300	36
Borrelia burgdorferi	B	1	1.44	1,283	28
Campylobacter jejuni	B	1	1.64	1,838	31
Caulobacter crescentus	B	1	4.02	3,737	62–67
Chlamydophila pneumoniae	B	4	1.23	1,000–1,110	40
Chlorobium tepidum	B	1	2.15	2,255	57
Clostridium perfringens	B	1	3.03	2,660	29
Escherichia coli	B	6	4–5.45	4,300–5,370	50
Geobacter sulfurreducens PCA	B	1	3.81	3,447	61
Halobacterium sp. NRC-1	A	1	2.01	2,058	68
Methanobacterium thermoautotrophicum	A	1	1.75	1,869	49
Methanocaldococcus jannaschii	A	1	1.66	1,728	31
Mycobacterium tuberculosis	B	2	4.40	4,000–4,189	65
Mycoplasma pneumoniae	B	1	0.82	688	40
Nanobacterium equitans	A	1	0.49	536	32
Prochlorococcus marinus	B	5	1.66–2.41	1,716–2,273	31–51
Pseudomonas aeruginosa	B	1	6.26	5,566	67
Pyrococcus abyssi	A	1	1.77	1,784	44
Rickettsia prowazekii	B	1	1.11	834	29
Saccharomyces cerevisiae	E	1	12.14	6,600	38
Salmonella enterica serovar Typhimurium	B	1	4.86	4,452	50–53
Staphylococcus aureus	B	9	2.80–2.90	2,560–2,890	33
Streptococcus pneumoniae	B	2	2.16	2,043–2,125	40
Streptomyces coelicolor	B	1	8.67	7,825	72
Sulfolobus tokodaii	A	1	2.69	2,826	33
Synechocystis sp.	B	1	3.57	3,167	47
Thermoplasma acidophilum	A	1	1.56	1,478	46
Yersinia pestis	B	3	4.60–4.65	4,000–4,100	48

[a]The following abbreviations are used: A, *Archaea;* B, *Bacteria;* E, *Eucarya.*

comparison process results in a set of larger, contiguous nucleotide sequences called **contigs.**

Finally, the contigs are aligned in the proper order to form the complete genome sequence. If gaps exist between two contigs, sometimes fragments with ends in two adjacent contigs are available. These fragments are analyzed and the gaps filled in with their sequences. When this approach is not possible, a variety of other techniques are used to align contigs and fill gaps.

4. **Editing.** The sequence is then carefully proofread to resolve any ambiguities or frameshift mutations in the sequence. Proofreading is accomplished by ensuring that the sequences of the two DNA strands are complementary.

Using this approach, it took less than 4 months to sequence the *M. genitalium* genome (about 500,000 base pairs in size). The shotgun technique was also used to sequence the human genome. Researchers have also sequenced (and continue to sequence) the genomes of many important microbial pathogens and microbes of environmental relevance. Once an organism's genome has been sequenced, the level of inquiry and the pace of research are greatly enhanced. The following sections describe only some of the ways a genome sequence can be used to learn more about an organism.

1. What is the goal of each of the three general areas of genomics?
2. Why is the Sanger technique of DNA sequencing also called the chain-termination method?
3. How would one recognize a gap in the genome sequence following whole-genome shotgun sequencing?

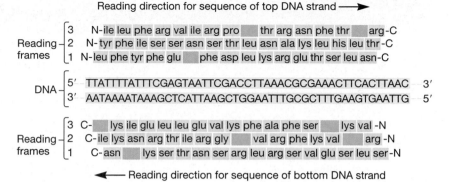

Figure 15.4   Finding Potential Protein Coding Genes. Annotation of genomic sequence requires that both strands of DNA be translated from the 5′ to 3′ direction in each of three possible reading frames. Stop codons are shown in green.

15.4   BIOINFORMATICS

The analysis of entire genomes generates not only a tremendous amount of nucleotide sequence data but also a rapidly growing volume of information regarding genome content, structure, and arrangement, as well as data detailing protein structure and function. This has led to the development of the field of **bioinformatics,** which combines biology, mathematics, computer science, and statistics. Determining the location and nature of genes or presumed genes on a newly sequenced genome is a complex process called annotation. Once genes have been identified, bioinformaticists can perform computer, or **in silico analysis,** to further examine the genome.

Obviously, obtaining nucleotide sequences without any understanding of the location and nature of individual genes would be a pointless exercise. The goal of **genome annotation** is to identify every potential (putative) protein-coding gene as well as each rRNA and tRNA coding gene. A protein-coding gene is usually first recognized as an **open reading frame (ORF);** to find all ORFs, both strands of DNA must be analyzed. A procaryotic ORF is generally defined as a sequence of at least 100 codons that it is not interrupted by a stop codon and has terminator sequences at the 3′ end. Only if these elements are present is an ORF considered a putative gene (**figure 15.4**). Differences in gene structure between the *Archaea* and

Bacteria genes make the annotation of archaeal genomes more challenging than those of bacteria. >> *Introduction to the* Archaea: *Genetics and molecular biology (section 18.1)*

ORFs presumed to encode proteins are called *coding* **sequences (CDS).** Bioinformaticists have developed algorithms to compare the sequence of predicted CDS with those in large databases containing nucleotide and amino acid sequences of known proteins. If an ORF matches one in the database, it is assumed to encode the same protein or type of protein. Although such comparisons are not without errors, they can provide tentative functional assignments for about 40 to 80% of the CDS in a given genome. The remaining fall into two classes: (1) **Conserved hypothetical proteins** are encoded by ORFs that have matches in the database but no function has yet been assigned. (2) **Proteins of unknown function** are the products of ORFs unique to that organism. However, as more genomes are published, future comparisons may reveal a match in another organism. Finally, it is helpful to organize identified genes according to product function or location in the cell, such as ribosomal and transfer RNAs, lipid metabolism, energy metabolism, cell wall–associated, etc. (**table 15.2**). Clearly bioinformatics is a dynamic field and its continued development is crucial for further progress in structural, functional, and comparative genomics.

15.5   FUNCTIONAL GENOMICS

Functional genomics seeks to explain how genes and genomes operate. The base-by-base comparison of two or more gene sequences is called **alignment.** Alignment of nucleotides on the same genome (i.e., from a single organism) may show that the nucleotide sequences of two or more genes are so alike that they most probably arose through gene duplication; such genes are called **paralogues.** Alignments of genes found in two or more different organisms may reveal that they are so strikingly similar that they are predicted to have the same function; these genes are

Table 15.2   Estimated Number of Genes Involved in Various Cell Functions[a]

Gene Function	Escherichia coli K12	Bacillus subtilis	Mycoplasma genitalium	Treponema pallidum	Rickettsia prowazekii	Chlamydia trachomatis	Methanocaldococcus jannaschii	Pyrococcus abyssi
Approximate total number of genes[b]	4,289	4,100	484	1,040	834	894	1,728	1,784
Cellular processes[c]	190	374	6	77	40	46	26	64
Cell envelope components	172	185	29	53	74	45	25	106
Transport and binding proteins	315	400	33	59	49	58	56	140
DNA metabolism	102	122	30	51	63	48	53	63
Transcription	41	114	13	25	26	18	21	37
Protein synthesis	122	161	90	99	104	133	118	108
Regulatory functions	176	293	5	22	26	12	19	66
Energy metabolism[d]	368	439	33	54	89	56	158	180
Central intermediary metabolism[e]	73	96	7	6	19	12	19	79
Amino acid biosynthesis	114	143	0	7	13	18	64	76
Fatty acid and phospholipid metabolism	67	84	8	11	22	27	9	18
Purines, pyrimidines, nucleosides, and nucleotides	77	81	19	21	19	14	37	51
Biosynthesis of cofactors and prosthetic groups	100	113	5	15	24	27	50	53

[a]Data adapted from TIGR (The Institute for Genomic Research) databases.
[b]The number of genes with known or hypothetical functions.
[c]Genes involved in cell division, chemotaxis and motility, detoxification, transformation, toxin production and resistance, pathogenesis, adaptations to atypical conditions, etc.
[d]Genes involved in amino acid and sugar catabolism, polysaccharide degradation and biosynthesis, electron transport and oxidative phosphorylation, fermentation, glycolysis/gluconeogenesis, pentose phosphate pathway, Entner-Doudoroff, pyruvate dehydrogenase, TCA cycle, photosynthesis, chemoautotrophy, etc.
[e]Amino sugars, phosphorus compounds, polyamine biosynthesis, sulfur metabolism, nitrogen fixation, nitrogen metabolism, etc.

called **orthologues.** Because the DNA code is degenerate, such alignments are generally performed after the genes coding sequence has been translated to amino acids.

The translated amino acid sequences of putative genes are also analyzed to gain an understanding of potential protein structure and function. Often a short pattern of amino acids, called a motif or domain, will represent a functional unit within a protein, such as the active site of an enzyme. For instance, **figure 15.5** shows the C-terminal domain of the cell division protein MinD from a number of microbes. Because these amino acids are found in such a wide range of organisms, they are considered phylogenetically well conserved. In this case, the conserved region is predicted to form a coil needed for proper localization of the protein to the membrane.

Functional genomics has been used to study many pathogens and microbes of environmental importance. *Haemophilus influenzae,* which causes both pneumonia and meningitis, has a relatively small genome—1.8 Mb—with about 1,750 genes. Microbial genomes are often drawn as a **physical map (figure 15.6).** It is helpful to color-code genes according to category. For instance, the *H. influenzae* genome reveals that about one-third of the genes are of unknown function (denoted by white regions on the map) and about 65 genes have regulatory functions (dark blue regions). One can gain an understanding about specific processes

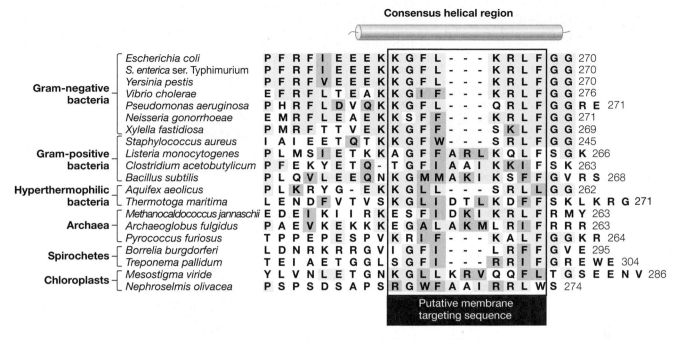

Figure 15.5 Analysis of Conserved Regions of Phylogenetically Well-Conserved Proteins. C-terminal amino acid residues of MinD from 20 organisms and chloroplasts are aligned to show strong similarities. Amino acid residues identical to *E. coli* are boxed in yellow, and conservative substitutions (e.g., one hydrophobic residue for another) are boxed orange. The number of the last residue shown relative to the entire amino acid sequence is shown at the extreme right of each line.

from such analysis. For example, *H. influenzae* lacks the genes to encode all the enzymes for a complete TCA cycle. On the other hand, given that this is a species capable of natural transformation, it is still rather surprising that it contains 1,465 copies of the recognition sequence used in DNA uptake during transformation. << *Bacterial transformation (section 14.8)*

Functional genomics can be used to assemble a relatively complete picture of a microbe's physiology (**figure 15.7**). This is of particular interest for the causative agent of syphilis, *Treponema pallidum.* Because it is not possible to grow *T. pallidum* outside the human body, we know little about its metabolism or the way it avoids host defenses. This has prevented the development of a vaccine for syphilis. The sequencing and annotation of the *T. pallidum* genome reveal that *T. pallidum* is metabolically crippled. It can use carbohydrates as an energy source but lacks TCA cycle and oxidative phosphorylation enzymes. *T. pallidum* also lacks many biosynthetic pathways (e.g., for enzyme cofactors, fatty acids, nucleotides, and some electron transport proteins) and must rely on molecules supplied by its host. In fact, about 5% of its genes code for transport proteins. Given the lack of several critical pathways, it is not surprising that the pathogen has not been cultured successfully. The genes for surface proteins are of particular interest. *T. pallidum* has a family of surface protein genes characterized by many repetitive sequences. Some have speculated that these genes might undergo recombination to generate new surface proteins, enabling the organism to avoid attack by the immune system. It may be possible to develop a vaccine for syphilis using some of these surface proteins. It may also be possible to identify strains of *T. pallidum* using these sur-face proteins, which would be of great importance in syphilis epidemiology. The annotated genome should ultimately help us understand how *T. pallidum* causes syphilis. >> *Phylum* Spirochaetes *(section 19.6)*

Evaluation of RNA-Level Gene Expression: Microarray Analysis

Once the identity and function of the genes that comprise a genome have been analyzed, the key question remains, "Which genes are expressed at any given time?" Prior to the genomic era, researchers could identify only a limited number of genes whose expression was altered under specific circumstances. However, the development of **DNA microarrays** now allows scientists to look at the expression level of a vast collection of genes at once. DNA microarrays are solid supports, usually of glass or silicon, upon which DNA is attached in an organized grid fashion. Each spot of DNA, called a **probe,** represents a single gene or ORF. The location of each gene on the grid is carefully recorded so that when analyzed, the genetic identity of each spot is known.

Of the several ways in which microarrays can be constructed, two techniques are most commonly employed. **Spotted arrays** are prepared by the robotic application of a probe to the chip. The probe may be a polymerase chain reaction (PCR) product, **complementary DNA (cDNA),** or a short DNA fragment within the gene or ORF, called an **oligonucleotide.** When eucaryotic genomes are to be analyzed, oligonucleotide probes are called **expressed *sequence tags* (ESTs)** because each is derived from cDNA, which lacks introns because it is DNA that

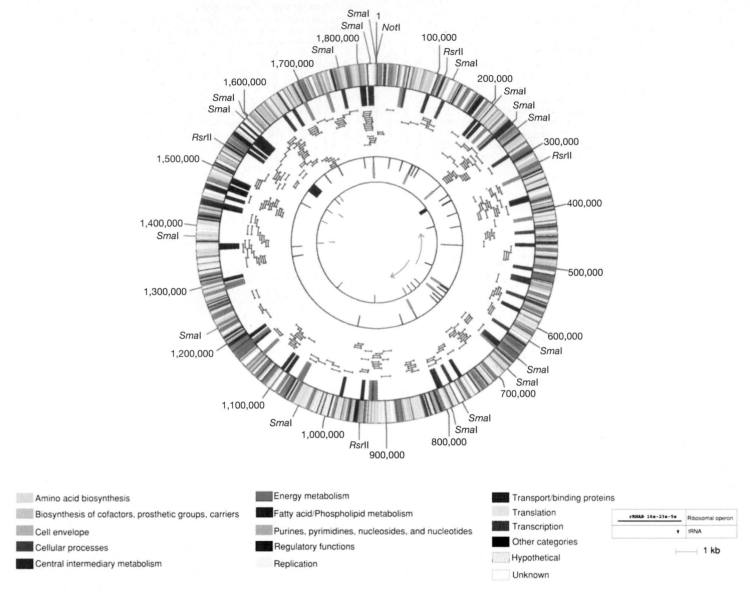

Figure 15.6   Map of the *Haemophilus influenzae* Genome.   The predicted coding regions in the outer concentric circle are indicated with colors representing their functional roles. The outer perimeter shows the *NotI, RsrII*, and *SmaI* restriction sites. The inner concentric circle shows regions of high G + C content (red and blue) and high A + T content (black and green). The third circle shows the coverage by λ clones (blue). The fourth circle shows the locations of rRNA operons (green), tRNAs (black), and the Mu-like prophage (blue). The fifth circle shows simple tandem repeats and the probable origin of replication (outward pointing green arrows). The red lines are potential termination sequences.   *Reprinted with permission from Fleischman, R. D., et al. 1995. Whole-genome random sequencing and assembly of* Haemophilus influenzae Rd. Science 269:496–512. Figure 1, page 507, and The Institute of Genomic Research.*

is copied from processed mRNA *(see figure 16.5)*. Spotted arrays are especially useful in the study of eucaryotic microbes, which have much larger genomes than procaryotes. Often these eucaryotic genomes are not fully sequenced because of cost constraints. The genes to be represented by DNA on spotted arrays can be carefully selected, enabling the development of custom-made microarrays.

Commercially prepared microarrays are usually prepared by a technique known as photolithography. Here a mask is laid over the chip, and light is used to control the synthesis of the oligonucleotide directly on the chip surface. Each hole in the

mask will eventually hold many copies of a different oligonucleotide, and each mask can control the synthesis of several hundred thousand squares. A single microarray can contain hundreds of thousands of different probes, and each probe is present in millions of copies. Commercial microarrays are available with probes for every expressed gene or ORF on the genomes of a number of organisms. Microbes include *E. coli* (about 4,200 ORFs), pathogens such as *Helicobacter pylori* (1,590 ORFs) and *Mycobacterium tuberculosis* (4,000 ORFs), as well as the yeast *Saccharomyces cerevisiae* (approximately 6,600 ORFs).

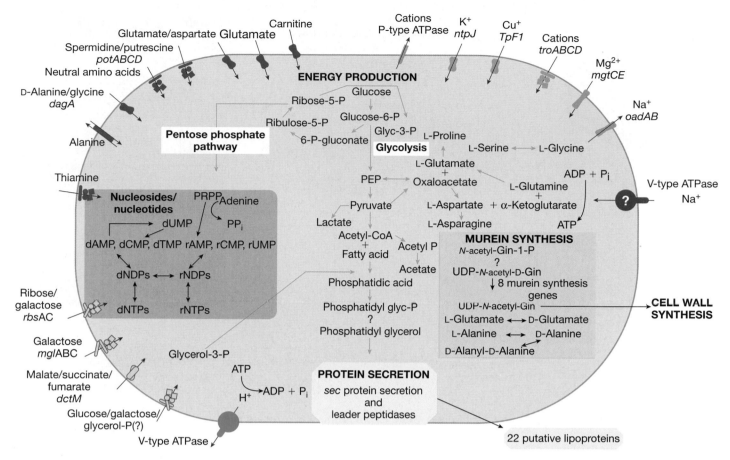

Figure 15.7 **Metabolic Pathways and Transport Systems of *Treponema pallidum*.** This depicts *T. pallidum* metabolism as deduced from genome annotation. Note the limited biosynthetic capabilities and extensive array of transporters. Although glycolysis is present, the TCA cycle and respiratory electron transport are lacking. Question marks indicate where uncertainties exist or expected activities have not been found.

The analysis of gene expression using microarray technology, like many other molecular genetic techniques, is based on hybridization between the single-stranded probe DNA and the nucleic acids to be analyzed, often called the targets, which may be either mRNA or single-stranded cDNA (**figure 15.8**). The target nucleotides are labeled with fluorescent dyes and incubated with the chip under conditions that ensure proper binding of target mRNA (or cDNA) to its complementary probe. Unbound target is washed off and the chip is then scanned with laser beams. Fluorescence at each spot or probe indicates that mRNA hybridized. Analysis of the color and intensity of each probe shows which genes were expressed.

DNA microarray analysis can be used to determine which genes are expressed during cellular differentiation or as the result of mutation. One common application of microarray technology in microbiology is to determine the genes whose expression is changed (either up- or downregulated) in response to environmental changes (figure 15.8). In this sort of analysis, two different-colored fluorochromes are used. For instance, the pathogen *Helicobacter pylori* dwells in the stomach where it causes ulcers. One might want to know which *H. pylori* genes are repressed or induced upon exposure to acidic conditions. To determine this, total cellular mRNA from bacteria grown at

a neutral pH is prepared and tagged with a green fluorochrome to serve as a control or reference, while mRNA from cells exposed to acidic medium is labeled red. The green (reference) and red (experimental) mRNA samples are then mixed and hybridized to the same microarray. After the unattached mRNA is washed off, the microarray is scanned and the image is computer analyzed. A yellow spot or probe indicates that roughly equal numbers of green and red mRNA molecules were bound, so there was no change in the level of gene expression for that gene. If a target is red, more mRNA from bacteria grown under the experimental acidic conditions was present when the two mRNAs were mixed, thus this gene was induced. Conversely, a green target indicates that the gene was repressed upon exposure to acid stress. Careful image analysis is used to determine the relative intensity of each spot, so that the magnitude of induction or repression of each gene whose expression is altered can be approximated.

Microarray experiments yield a vast amount of information that must be organized in some meaningful fashion. One common approach is to group genes with similar function or patterns of regulation together. As shown in **figure 15.9**, **hierarchical cluster analysis** groups induced genes (red spots) separately from repressed genes (green spots), while genes

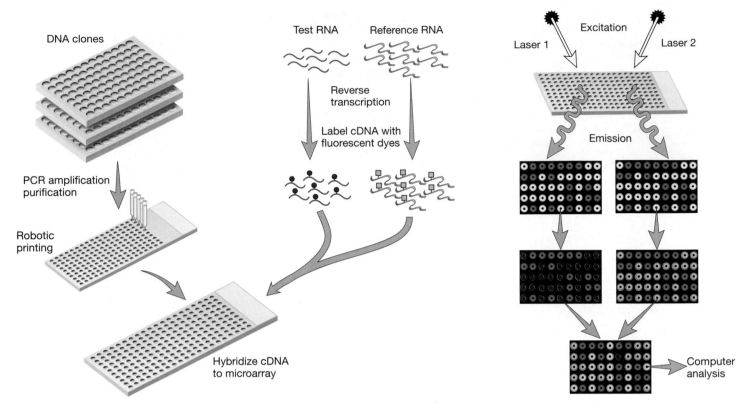

Figure 15.8 A Microarray System for Monitoring Gene Expression. Cloned genes from an organism are amplified by PCR, and after purification, samples are applied by a robotic printer to generate a spotted microarray. To monitor enzyme expression, mRNA from test and reference cultures are converted to cDNA by reverse transcriptase and labeled with two different fluorescent dyes. The labeled mixture is hybridized to the microarray and scanned using two lasers with different excitation wavelengths. The fluorescence responses are measured as normalized ratios that show whether the test gene response is higher or lower than that of the reference.

whose expression remains unaltered are shown in black. In the experiment shown here, the bacterium *Deinococcus radiodurans* was studied. This microbe has the remarkable ability to survive γ-radiation at doses many times above that needed to kill humans. Ionizing radiation causes double-stranded breaks in DNA—the most lethal form of DNA damage. *D. radiodurans* is able to reassemble its genome after it has been fragmented into thousands of pieces. Its genome consists of two circular chromosomes (2.6 Mb and 0.4 Mb), a megaplasmid (0.18 Mb), and a small plasmid (45,704 base pairs). Surely, it was thought, *D. radiodurans* must be superb in executing DNA repair. But, surprisingly, *D. radiodurans* has fewer DNA repair genes than does *E. coli.* To understand these findings, microbiologists used microarrays with about 94% of *D. radiodurans* genes represented to examine the **transcriptome** (all the mRNA present) following radiation treatment. These results were then scanned for genes with similar functions and then grouped by relatedness (degree of relatedness is statistically quantified by a correlation coefficient, or "r value"). Such analysis confirmed that the DNA repair gene *recA* and genes involved in DNA replication and recombination are dramatically upregulated following irradiation. In addition, genes whose products direct cell wall metabolism and cellular transport, and many genes whose protein products are unknown are

also induced. These results make it clear that *D. radiodurans'* survival at such high levels of radiation is more complex than originally thought, requiring the coordination of a complex network of processes involving both DNA repair and metabolic activity. **>>** Deinococcus-Thermus *(section 19.2)*

1. What is genome annotation? Why do you think it requires knowledge of mathematics and statistics as well as biology and computer science?

2. How can genomic sequencing be used to ask specific questions about the physiology of a given microbe? Explain a surprising result obtained from genomic analysis of a microbe.

3. Describe how microarrays are constructed and used to analyze gene expression. How might the following scientists use DNA microarrays in their research?

  (a) A microbial ecologist who is interested in how the soil microbe *Rhodopseudomonas palustris* degrades the toxic compound 3-chlorobenzene.

  (b) A medical microbiologist who wants to learn about how the pathogen *Salmonella* survives within a host cell.

Time (h)

0 0.5 1.5 3 5 9 12 16 24

Gene#, putative function[a]

Ratio (fold)[b]

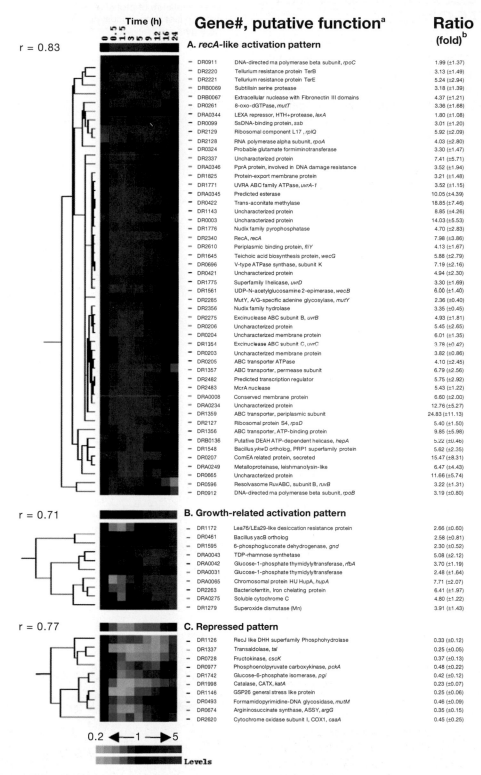

r = 0.83

A. recA-like activation pattern

Gene	Putative function	Ratio (fold)[b]
DR0911	DNA-directed rna polymerase beta subunit, rpoC	1.99 (±1.37)
DR2220	Tellurium resistance protein TerB	3.13 (±1.49)
DR2221	Tellurium resistance protein TerE	5.24 (±2.94)
DRB0069	Subtilisin serine protease	3.18 (±1.39)
DRB0067	Extracellular nuclease with Fibronectin III domains	4.37 (±1.21)
DR0261	8-oxo-dGTPase, mutT	3.36 (±1.68)
DRA0344	LEXA repressor, HTH+protease, lexA	1.80 (±1.08)
DR0099	SsDNA-binding protein, ssb	3.01 (±1.20)
DR2129	Ribosomal component L17, rplQ	5.92 (±2.09)
DR2128	RNA polymerase alpha subunit, rpoA	4.03 (±2.80)
DR0324	Probable glutamate formiminotransferase	3.30 (±1.47)
DR2337	Uncharacterized protein	7.41 (±5.71)
DRA0346	PprA protein, involved in DNA damage resistance	3.52 (±1.94)
DR1825	Protein-export membrane protein	3.21 (±1.48)
DR1771	UVRA ABC family ATPase, uvrA-1	3.52 (±1.15)
DRA0345	Predicted esterase	10.05 (±4.39)
DR0422	Trans-aconitate methylase	18.85 (±7.46)
DR1143	Uncharacterized protein	8.85 (±4.26)
DR0003	Uncharacterized protein	14.03 (±5.53)
DR1776	Nudix family pyrophosphatase	4.70 (±2.83)
DR2340	RecA, recA	7.98 (±3.86)
DR2610	Periplasmic binding protein, fliY	4.13 (±1.67)
DR1645	Teichoic acid biosynthesis protein, wecG	5.88 (±2.79)
DR0696	V-type ATPase synthase, subunit K	7.19 (±2.16)
DR0421	Uncharacterized protein	4.94 (±2.30)
DR1775	Superfamily I helicase, uvrD	3.30 (±1.69)
DR1561	UDP-N-acetylglucosamine 2-epimerase, wecB	6.00 (±1.40)
DR2285	MutY, A/G-specific adenine glycosylase, mutY	2.36 (±0.40)
DR2356	Nudix family hydrolase	3.35 (±0.45)
DR2275	Excinuclease ABC subunit B, uvrB	4.93 (±1.81)
DR0206	Uncharacterized protein	5.45 (±2.65)
DR0204	Uncharacterized membrane protein	6.01 (±1.35)
DR1354	Excinuclease ABC subunit C, uvrC	3.78 (±0.42)
DR0203	Uncharacterized membrane protein	3.82 (±0.86)
DR0205	ABC transporter ATPase	4.10 (±2.45)
DR1357	ABC transporter, permease subunit	6.79 (±2.56)
DR2482	Predicted transcription regulator	5.75 (±2.92)
DR2483	McrA nuclease	5.43 (±1.22)
DRA0008	Conserved membrane protein	6.60 (±2.00)
DRA0234	Uncharacterized protein	12.76 (±5.27)
DR1359	ABC transporter, periplasmic subunit	24.83 (±11.13)
DR2127	Ribosomal protein S4, rpsD	5.40 (±1.50)
DR1356	ABC transporter, ATP-binding protein	9.85 (±5.98)
DRB0136	Putative DEAH ATP-dependent helicase, hepA	5.22 (±0.46)
DR1548	Bacillus ykwD ortholog, PRP1 superfamily protein	5.62 (±2.35)
DR0207	ComEA related protein, secreted	15.47 (±8.31)
DRA0249	Metalloproteinase, leishmanolysin-like	6.47 (±4.43)
DR0665	Uncharacterized protein	11.66 (±5.74)
DR0596	Resolvasome RuvABC, subunit B, ruvB	3.22 (±1.31)
DR0912	DNA-directed rna polymerase beta subunit, rpoB	3.19 (±0.80)

r = 0.71

B. Growth-related activation pattern

Gene	Putative function	Ratio (fold)[b]
DR1172	Lea76/LEa29-like desiccation resistance protein	2.66 (±0.60)
DR0461	Bacillus yacB ortholog	2.58 (±0.81)
DR1595	6-phosphogluconate dehydrogenase, gnd	2.30 (±0.52)
DRA0043	TDP-rhamnose synthetase	5.08 (±2.12)
DRA0042	Glucose-1-phosphate thymidylyltransferase, rfbA	3.70 (±1.19)
DRA0031	Glucose-1-phosphate thymidylyltransferase	2.48 (±1.64)
DRA0065	Chromosomal protein HU HupA, hupA	7.71 (±2.07)
DR2263	Bacterioferritin, iron chelating protein	6.41 (±1.97)
DRA0275	Soluble cytochrome C	4.80 (±1.22)
DR1279	Superoxide dismutase (Mn)	3.91 (±1.43)

r = 0.77

C. Repressed pattern

Gene	Putative function	Ratio (fold)[b]
DR1126	RecJ like DHH superfamily Phosphohydrolase	0.33 (±0.12)
DR1337	Transaldolase, tal	0.25 (±0.05)
DR0728	Fructokinase, cscK	0.37 (±0.13)
DR0977	Phosphoenolpyruvate carboxykinase, pckA	0.48 (±0.22)
DR1742	Glucose-6-phosphate isomerase, pgi	0.42 (±0.12)
DR1998	Catalase, CATX, katA	0.23 (±0.07)
DR1146	GSP26 general stress like protein	0.25 (±0.06)
DR0493	Formamidopyrimidine-DNA glycosidase, mutM	0.46 (±0.09)
DR0674	Argininosuccinate synthase, ASSY, argG	0.35 (±0.15)
DR2620	Cytochrome oxidase subunit I, COX1, caaA	0.45 (±0.25)

0.2 ◄—1—► 5

Levels

Figure 15.9 Hierarchical Cluster Analysis of Gene Expression of D. radiodurans Following Exposure to γ-Radiation. Each row of colored strips represents a single gene and the color indicates the level of expression over nine time intervals. The far-left column is the control and thus is black (at control levels of expression). The level of induction or repression relative to the control value is indicated as the ratio (fold). Each group of genes has been scored for relatedness and a "tree" has been generated on the far left of the clusters, with the indicated correlation coefficient (r value). A large number of genes encoding DNA repair, synthesis, and recombination proteins are induced upon radiation. These are grouped together. Fewer genes that encode proteins involved in metabolism are induced. Finally, genes involved in other aspects of metabolism are repressed.

15.6 PROTEOMICS

Genome function can be studied at the translation level as well as the transcription level. The entire collection of proteins that an organism produces is called its **proteome.** Thus **proteomics** is the study of the proteome or the array of proteins an organism can produce. Proteomics provides information about genome function that mRNA studies cannot because a direct correlation between mRNA and the pool of cellular proteins does not always exist. This is because mRNA can be unstable (and therefore not detected) as well as because posttranslational regulation may determine whether or not a protein is active. Much of the research in this area is referred to as **functional proteomics.** It is focused on determining the function of different cellular proteins, how they interact with one another, and the ways in which they are regulated.

Although new techniques in proteomics are continuously being developed, we will focus briefly only on the most common approach, **two-dimensional gel electrophoresis.** In this procedure, a mixture of proteins is separated using two different electrophoretic procedures (dimensions). This permits the visualization of thousands of cellular proteins that would not otherwise be separated in a single electrophoretic dimension. As shown in **figure 15.10,** the first dimension makes use of **isoelectric focusing,** in which proteins move through a pH gradient (e.g., pH 3 to 10). First, the protein mixture is applied to an acrylamide gel in a tube with a stable pH gradient and electrophoresed. Each protein moves along the pH gradient until the protein's net charge is zero and the protein stops moving. The pH at this point is equal to the protein's **isoelectric point.** Thus the first dimension separates the proteins based on their content of ionizable amino acids. The second dimension is SDS *p*oly*a*crylamide *g*el *e*lectrophoresis (SDS-PAGE). SDS (sodium dodecyl sulfate) is an anionic detergent that denatures proteins and coats the polypeptides with a negative charge. After the isoelectric gel has been completed, the tube gel is soaked in SDS buffer and then placed at the edge of an SDS-PAGE gel. A voltage is then applied. Under these circumstances, polypeptides are separated according to their **molecular weight**—that is, the smallest polypeptide will travel fastest and farthest. Two-dimensional gel electrophoresis can resolve thousands of proteins; each protein is visualized as a spot of varying intensity depending on its cellular abundance. Radiolabeled proteins are often used, enabling greater sensitivity so that newly synthesized proteins can be distinguished. Computer analysis is used to compare two-dimensional gels from microbes grown under different conditions or to compare wild-type and mutant strains. Websites have been developed for the deposition of such images, allowing researchers access to valuable and ever-growing databases.

Two-dimensional gel electrophoresis is even more powerful when coupled with **mass spectrometry (MS).** The unknown protein spot is cut from the gel and cleaved into fragments by treatment with proteolytic enzymes. Then the fragments are analyzed by a mass spectrometer and the mass of the fragments is plotted. This mass fingerprint can be used to estimate the probable

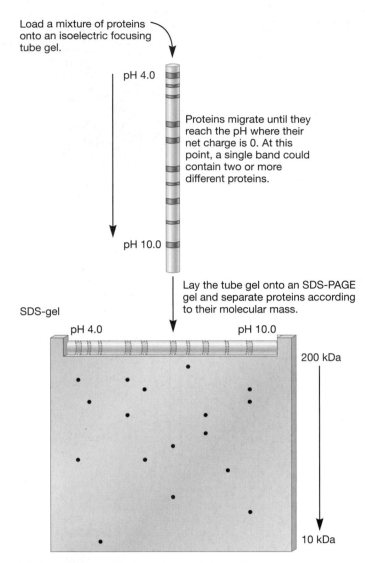

(a) The technique of 2-dimensional gel electrophoresis

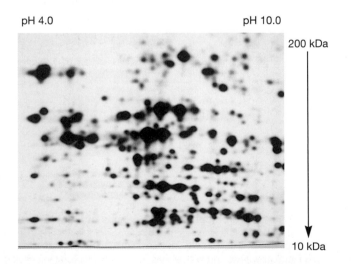

(b) An autoradiograph of a 2-dimensional gel. Each protein is a discrete spot.

Figure 15.10   **Two-Dimensional Gel Electrophoresis.**

amino acid composition of each fragment and tentatively identify the protein. Sometimes proteins or collections of fragments are run through two mass spectrometers in sequence, a process known as **tandem MS (figure 15.11)**. The first spectrometer separates proteins and fragments, which are further fragmented. The second spectrometer then determines the amino acid sequence of each smaller fragment. The sequence of a whole protein often can be determined by analysis of such fragment sequence data. Alternatively, if the genome of the organism has been sequenced, only a partial amino acid sequence is needed. Computer analysis is then used to compare this amino acid sequence with the predicted translated sequences of all the annotated ORFs on the organism's genome. In this way, both the protein and the gene that encodes it can be identified. Further investigation of the protein may rely on one of the large databases of protein sequences that enable comparative analysis. Comparing amino acid sequences can provide information regarding the protein structure, function, and evolution.

A second branch of proteomics is called **structural proteomics.** Here the focus is on determining the three-dimensional structures of many proteins and using these to predict the structures of other proteins and protein complexes. The assumption is that proteins fold into a limited number of shapes and can be grouped into families of similar structures. When a number of protein structures are determined for a given family, the patterns of protein structure organization or protein-folding rules will be known. Then computational biologists (i.e., bioinformaticists) use this information and the amino acid sequence of a newly discovered protein to predict its final shape, a process known as **protein modeling.**

In addition to proteomics, a variety of other "-omics" are studied. For instance, the goal of **metabolomics** is to identify all the small-molecule metabolites present in the cell at a given point in time. Just as the proteome provides a snapshot of all the proteins made by a cell at a specific moment in time, the metabolome enables a researcher to evaluate its physiological status. **Lipidomics** is used to determine a cell's lipid profile at a particular time. Both metabolomics and lipidomics rely heavily on the use of chromatography and as such are beyond the scope of this chapter. Nonetheless, the development of these fields illustrates the dynamic and ever-changing nature of microbiology and the need to take a holistic view if one seeks to understand the structure, function, and behavior of cells.

1. Why does two-dimensional gel electrophoresis allow the visualization of many more cellular proteins than seen in electrophoresis in a single dimension?

2. What is the role of mass spectrometry in proteomics?

3. What is the difference between functional and structural proteomics? How do you think structural proteomics might be used in vaccine development?

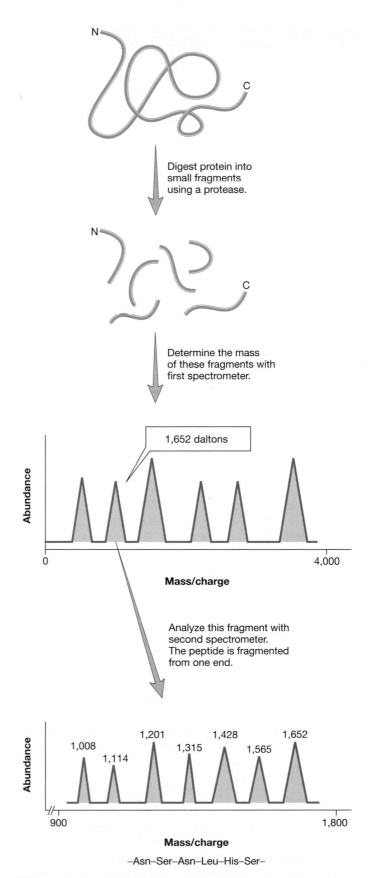

Figure 15.11   The Use of Tandem Mass Spectrometry to Determine the Amino Acid Sequence of a Peptide.

15.7 COMPARATIVE GENOMICS

Comparative genomics is simply comparing the genome of one organism with that of others. As already noted, the identification of orthologues is essential in assigning function to the proteins encoded by putative genes. Comparative genomics can also provide information about the physical structure of the genome, such as the presence of transposable elements, operons, and repeat elements. Indeed, one very striking insight is the discovery that microbial genomes are not as static as once thought. In fact, procaryotic genomes are amazingly fluid with a substantial portion of the genome transferred between cells by **lateral** or **horizontal gene transfer (HGT).** Broadly defined, HGT is the exchange of genetic material between organisms that need not be of similar evolutionary lineages. Genome analysis has revealed that HGT is frequently mediated by phages and that lysogeny may be the rule, rather than the exception. In fact, some procaryotes carry multiple prophages. It has become clear that HGT is a major evolutionary force in short-term procaryotic evolution and long-term speciation. For instance, it is thought that *E. coli* may have acquired the lactose *(lac)* operon from another microbe and thus became capable of colonizing the mammalian colon, where milk sugar is a common carbon source. On a larger scale, the methane-producing archaeon *Methanosarcina mazei* appears to have acquired about one-third of its genes from other procaryotes. ◄◄ *Mechanisms of genetic variation (chapter 14)*

Because HGT is so pervasive, comparative genomics can inform scientists as to the origin and prevalence of particular phenotypic traits. An example is provided by the genome sequence of *Picrophilis torridus.* This archaeon grows optimally at a pH of 0.7 and a temperature of 65°C. It thrives in hot, acidic, sulfataric fields, a habitat it shares with other extremophilic archaea, such as *Thermoplasma* and the more distantly related *Sulfolobus* (**figure 15.12a**). In most cases, microbes living at low pH are able to maintain a neutral internal pH, but this is not the case for *P. torridus,* whose intracellular pH is around 4.6. This suggests it has evolved a unique strategy to prevent irreversible macromolecular damage. By comparing the genes and their products found only in these acidophiles, researchers can focus on potential adaptive strategies (figure 15.12b). Understanding the mechanisms by which these microbes survive under these circumstances is not purely academic, as proteins capable of withstanding harsh treatment have industrial importance. ►► *The* Archaea *(chapter 18)*

Comparative genomics can also provide great insight into genes and gene products that are associated with virulence when pathogen genomes are analyzed. Such insights are particularly needed in understanding *Mycobacterium tuberculosis,* the causative agent of tuberculosis (TB). About one-third of the human population has TB. After establishing residence in immune system cells in the lung, *M. tuberculosis* often remains in a dormant state until the host's immune system is compromised. It then goes on to cause active disease, killing about 2 million people annu-

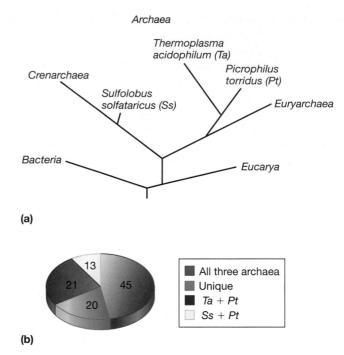

(a)

(b)

Figure 15.12 Comparative Genomics of *Picrophilus torridus.* (a) Simplified phylogenetic tree with the positions of the three acidiphilic, thermophilic archaea *P. torridus, T. acidophilum,* and *S. sulfataricus.* (b) ORFs found in *P. torridus, T. acidophilum,* and *S. sulfataricus,* ORFs shared between *P. torridus* and *T. acidophilum* (Ta + Pt), or *S. sulfataricus* (Ss + Pt), and those that are unique to *P. torridus* (Unique).

ally. Unfortunately, *M. tuberculosis* is becoming ever more drug resistant. The annotated *M. tuberculosis* genome was published in 1998; at that time it was predicted to have 3,974 genes. In 2002 its genome was reexamined and, based largely on inspection of small ORFs, an additional 82 genes were identified. The number of published genomes during the intervening four years is reflected by the change in the number of genes of unknown function: in 1998 there were 606 genes for which no function or orthologue could be assigned; by 2002 comparative genomic analysis found that there were only 272 such genes. ►► *Suborder* Corynebacterineae: *Family* Mycobacteriaceae *(section 22.4)*

The *M. tuberculosis* genome has been compared to the genomes of two relatives—*M. leprae,* which causes leprosy, and *M. bovis,* the causative agent of TB in a wide range of animals, including cows and humans. The genomes of *M. bovis* and *M. tuberculosis* are most similar—about 99.5% identical at the sequence level. However, the *M. bovis* genome is missing 11 separate regions, making its genome slightly smaller (4.3 Mb vs. 4.4 Mb). The sequence dissimilarities involve the inactivation of some genes, leading to major differences in the way the two bacteria respond to environmental conditions. This may account for the host range differences between these two closely related pathogens.

The divergence between *M. tuberculosis* and *M. leprae* is even more striking. The *M. leprae* genome is a third smaller than that of *M. tuberculosis.* About half the genome is devoid of functional genes. Instead, there are over 1,000 degraded, nonfunctional genes called **pseudogenes.** In total, *M. leprae* seems to have lost as many as 2,000 genes during its career as an intracellular parasite. This is a common finding in obligate intracellular pathogens because there is no selective pressure to maintain genes whose products are required for independent growth. *M. leprae* even lacks some of the enzymes required for energy production and DNA replication. This might explain why the bacterium has such a long doubling time, about two weeks in mice. One hope from genomic analysis is that critical surface proteins can be discovered and used to develop a sensitive test for early detection of leprosy. This would allow immediate treatment of the disease before nerve damage occurs.

Comparative genomics can be used to look at more recent evolutionary events. The rise of antibiotic-resistant bacteria is an area of vital investigation. The staphylococci are of particular concern in this regard. These gram-positive microbes cause an estimated 1 million serious infections each year. The two predominant opportunistic pathogens are *Staphylococcus epidermidis* and *S. aureus. S. epidermidis* is commonly found on the skin and has emerged as a serious pathogen only in recent years. It has been found to infect implanted medical devices such as artificial heart valves. In contrast, *S. aureus* is more aggressive and can cause conditions that range from minor skin infections to life-threatening abscesses, heart infections, and toxic shock syndrome. Beginning in the 1960s, methicillin and other semisynthetic penicillins were frequently prescribed to treat staphylococcal infections, giving rise to the development of methicillin-resistant *S. aureus* (MRSA) and *S. epidermidis* (MRSE). By 2005, 60% of *S. aureus* clinical isolates were resistant to methicillin; some strains are resistant to up to 20 different antibiotics! The only drug effective against such multiply resistant strains was vancomycin, but *S. aureus* strains resistant to vancomycin have now been isolated. ▷▷ *Antimicrobial chemotherapy (chapter 31)*

Strains of *S. aureus* and *S. epidermidis* that were initially isolated between 1960 and 1998 vary in their resistance to a number of antibiotics and their virulence levels. The genomes of a number of such strains have been used in comparative genomic analysis to track the evolution of antibiotic resistance and virulence. Most genes that are specific to an individual strain appear to have been introduced by horizontal transfer through prophages, transposons, insertion sequences, and plasmids. For instance, it appears that an *Enterococcus faecalis* transposon introduced vancomycin resistance into *S. aureus.* Both *S. aureus* and *S. epidermidis* have acquired the genes that encode a capsule made of glutamate polymers from the gram-positive pathogen *Bacillus anthracis,* the causative agent of anthrax. In all three species, the polyglutamate capsule is a major virulence factor.

Finally, comparative genomics can help elucidate phylogenetic relationships between microbes. An interesting example of this involves the genome of *Bacillus anthracis,* which underwent intense scrutiny following a series of letter-based bioterrorism attacks in the United States in the fall of 2001.

Although the genes that encode the components of the anthrax toxin are on a plasmid, the *B. anthracis* chromosome has a number of virulence-enhancing genes with orthologues in the pathogens *B. cereus, B. thuringiensis,* and *Listeria monocytogenes.* Although the *B. anthracis* genome is most similar to that of *B. cereus* (a cause of food poisoning), the orthologues found on the *B. thuringiensis* genome are particularly interesting. *B. thuringiensis* is an insect pathogen and genomic evidence suggests that *B. anthracis* may be derived from an insect-infecting ancestor. *L. monocytogenes* is an intracellular human pathogen, and the genes found in both organisms may allow these microbes to survive within host immune cells called macrophages. Unlike the genomes of many other bacterial species, there is little variation in the nucleotide sequences among different strains of *B. anthracis.* However, careful comparative genome sequencing of several key strains provides clues regarding the origin of the *B. anthracis* strain used in the 2001 bioterrorism attacks (**figure 15.13**). ▷▷ *Bioterrorism preparedness (section 33.9)*

1. Define horizontal gene transfer. How might HGT be partly responsible for the rapid rise in antibiotic-resistant microbes?

2. For what types of microorganisms is extensive gene loss common? What is the most likely explanation for this phenomenon?

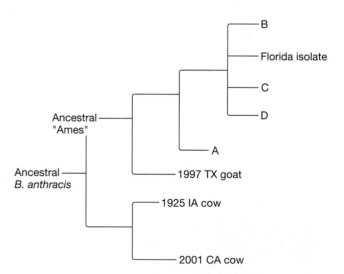

Figure 15.13   Proposed Phylogenetic Tree of the *B. anthracis* Ames Isolate Used in U.S. Bioterror Attacks in 2001. Isolates A–D are laboratory strains; other isolates are indicated by state and year of initial cultivation. Isolates B and C are identical to the isolate recovered from the first victim of the attacks (Florida isolate). Isolate D differs only by the insertion of a single adenine in a key region used for comparison. Laboratory strain A has two additional bases and has lost one of the toxin-encoding plasmids. This whole-genome analysis demonstrated the presence of four genetic elements on the genome that vary, despite previous analysis suggesting that these strains were identical or nearly identical.

15.8 ENVIRONMENTAL GENOMICS

It is clear that structural, functional, and comparative genomics have revolutionized the study of microbial physiology and genetics. Equally important is the growing field of **environmental genomics,** also called **metagenomics.** While the dominant role of microorganisms in driving the nutrient cycles that support life on Earth has long been recognized, efforts to comprehensively understand microbial communities have been stymied by the fact that only about 1% of all procaryotes have been cultured in laboratory conditions. New genomic techniques offer cultivation-independent approaches to studying microbial biodiversity. Environmental genomics can be used to take a census of microbial populations, as well as to discern the presence and abundance of certain classes of genes. That is to say, genomics can ask, "Who is there and what are they doing?" To do this, DNA fragments are extracted directly from the environment and cloned into plasmid vectors. In this way, a library of environmental DNA fragments can be maintained and amplified (**figure 15.14**). Alternatively, certain genes may be obtained by PCR amplification of DNA fragments derived from environmental samples. For this approach, knowledge of the gene's nucleotide sequence is required; this is common when amplifying genes that encode 16S rRNA for taxonomic purposes. In either case, one produces a stable source of nucleotide sequences reflecting the diversity of microbes growing in nature, not just those that can be grown in the laboratory. The nucleotides are then sequenced and analyzed, or expressed in a microbial host and screened for a specific function, such as the production of novel antimicrobial compounds. >> *Polymerase chain reaction (section 16.2); Techniques for determining microbial taxonomy and phylogeny (section 17.4); Microbial ecology and its methods: an overview (section 25.2)*

One field that metagenomics has revolutionized is marine microbiology. An average of 1 million microbial cells can be found per milliliter of seawater. While it has long been recognized that marine microbes account for the majority of the oceans' biomass, where they perform about half of the global photosynthesis, it has been difficult to study their taxonomic and metabolic diversity. In 2000 environmental genomics led to the discovery of a new procaryotic gene that encodes a protein in the rhodopsin family. Rhodopsins convert light energy directly into

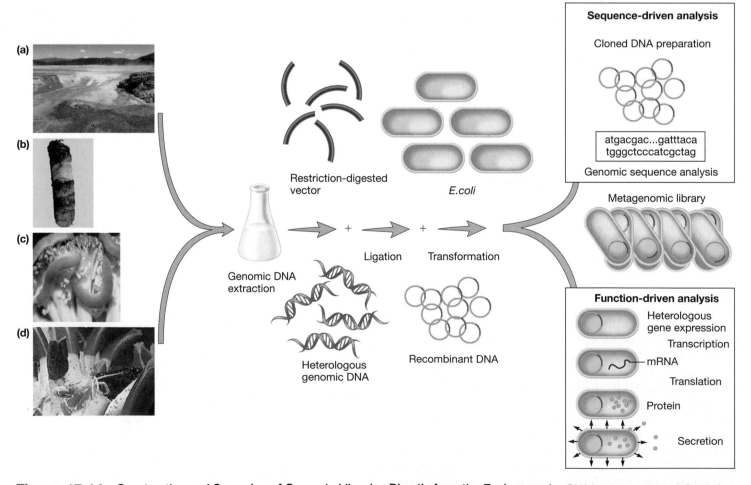

Figure 15.14 Construction and Screening of Genomic Libraries Directly from the Environment. DNA has been extracted directly from (a) bacterial mats at Yellowstone National Park, (b) soil samples from Alaska, (c) cabbage white butterfly larvae, and (d) tube worms from hydrothermal vents. The DNA is cloned into suitable vectors and transformed into a bacterial host. Sequences or gene products are then analyzed.

a transmembrane proton gradient, generating a proton motive force that fuels the production of ATP. It had long been held that the *Archaea* were the only procaryotes to produce a protein in the rhodopsin family. When rhodopsin-like genes were amplified from a variety of procaryotic taxa, they were called proteorhodopsins because they were found in γ-proteobacteria (genes for proteorhodopsin have since been found in other microbes as well). The nature of microbial metabolic diversity in the sea is now being reconsidered as it is estimated that at least 13% of marine microbes may have the genes to encode rhodopsin-based, light-driven proton pumps. Likewise, marine nitrogen budgets may need to be recalculated based on the discovery that the gene encoding nitrogenase, the enzyme that converts gaseous N_2 to ammonia, is present in far greater numbers among marine cyanobacteria than previously thought. << *Phototrophy (section 10.12);* >> *Phylum* Euryarchaeota *(section 18.3); Biogeochemical cycling: The nitrogen cycle (section 25.1)*

An ambitious metagenomics project was performed by J. Craig Venter, Hamilton Smith, and colleagues. They wanted to determine the procaryotic biodiversity of the Sargasso Sea, that portion of the Atlantic Ocean that surrounds Bermuda. They collected seawater and used filtration to exclude viruses and most eucaryotes. An environmental genomic library was prepared from DNA extracted from the remaining seawater. After sequencing over 1 billion base pairs, followed by manual and computer analysis to determine sequence relatedness, it was reported that at least 1,800 "genomic species," called **phylotypes,** were represented. Among these, about 145 phylotypes were previously unknown and may represent new species. Phylogenetic diversity was further evaluated by PCR amplification of genes, such as the recombinase gene *recA,* the 16S rRNA gene, and the gene that encodes RNA polymerase β subunit *(rpoB).* The sequences for these genes are highly conserved, making them good candidates for assessing species diversity **(figure 15.15).** Remarkably, Venter, Smith, and their collaborators report the discovery of 1.2 million previously unknown genes (however, this number is controversial), including over 700 new proteorhodopsin-like photoreceptors from taxa not previously known to possess light-harvesting capabilities. Certainly these results demonstrate the power of metagenomics and show that much more study is needed before we can fully appreciate microbial diversity.

1. What is a phylotype?

2. How might environmental genomics be used in expanding our knowledge of terrestrial microbial communities? How might environmental genomics be used by the biotechnology industry to develop new medically or industrially important natural products?

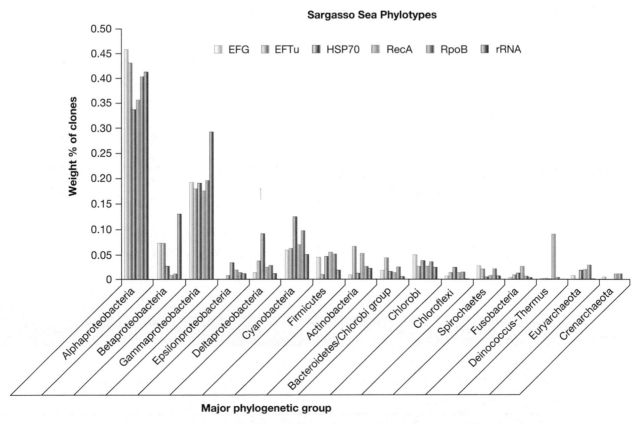

Figure 15.15 Phylogenetic Diversity of Sargasso Sea Microbes. The relative abundance (weight % of clones) of each group of microbes is shown according to the specific conserved gene that was used for analysis. The genes used were those encoding elongation factor G (EFG), elongation factor Tu (EfTu), heat shock protein 70 (HSP70), recombinase A (RecA), RNA polymerase B (RpoB), and the gene that encodes 16S rRNA.

Summary

15.1 Introduction

a. Genomics is the study of the molecular organization of genomes, their information content, and the gene products they encode. It may be divided into three broad areas: structural genomics, functional genomics, and comparative genomics.

15.2 Determining DNA Sequences

a. DNA fragments are normally sequenced using dideoxynucleotides and the Sanger chain termination technique (**figure 15.2**).

15.3 Whole-Genome Shotgun Sequencing

a. Most often microbial genomes are sequenced using the whole-genome shotgun technique of Venter, Smith, and collaborators. Four stages are involved: library construction, sequencing of randomly produced fragments, fragment alignment and gap closure, and editing the final sequence (**figure 15.3**).

15.4 Bioinformatics

a. Analysis of vast amounts of genome data requires sophisticated computer software; these analytical procedures are a part of the discipline of bioinformatics.

b. Bioinformatics enables the comparison of genes within genomes to identify paralogues and of genes between different organisms to identify orthologues.

c. Annotation of genomes can be used to identify many genes and their function, but the functional role of 20 to 60% of ORFs on a given genome usually cannot be discerned.

15.5 Functional Genomics

a. Functional genomics is used to reveal genome structure and function relationships (**figure 15.5**).

b. The physical map of microbe is often displayed so that genes of common function are easily identified (**figure 15.6**).

c. DNA microarrays can be used to assess gene expression as a measure of individual gene transcripts (mRNA). Gene expression can be determined for mutant versus wild-type strains or for a given organism under specific environmental conditions (**figure 15.8**).

15.6 Proteomics

a. The entire collection of proteins that an organism can produce is its proteome, and its study is called proteomics.

b. The proteome is often analyzed by two-dimensional gel electrophoresis, in which the total cellular protein pool can be visualized. In many cases, the amino acid sequence of individual proteins is determined by mass spectrometry; if this is coupled to genomics, both a protein of interest and the gene that encodes it can be identified (**figures 15.10** and **15.11**).

c. Structural proteomics seeks to model the three-dimensional structure of proteins based on computer analysis of amino acid sequence data.

15.7 Comparative Genomics

a. Comparing genome sequences reveals information about genome structure and evolution, including the importance of lateral gene transfer.

b. Comparative genomics is an important tool in discerning how microbes have adapted to particular ecological niches and in developing new therapeutic agents.

15.8 Environmental Genomics

a. Environmental genomics enables the study of the biodiversity and metabolic potential of microbial communities without culturing the individual microbes (**figures 15.14** and **15.15**).

Critical Thinking Questions

1. Propose an experiment that can be done easily with a DNA microarray that would have required years to do before this new technology.

2. What are the pitfalls of searches for homologous genes and proteins?

3. You are developing a new vaccine for a pathogen. You want your vaccine to recognize specific cell-surface proteins. Explain how you will use genome analysis to identify potential protein targets. What functional genomics approaches will you use to determine which of these proteins is produced when the pathogen is in its host?

Learn More

Learn more by visiting the Prescott website at www.mhhe.com/prescottprinciples, where you will find a complete list of references.

Biotechnology and Industrial Microbiology

Chapter Glossary

A scientist examines DNA following agarose gel electrophoresis. Each bright band is a fragment of DNA stained with ethidium bromide, so that upon illumination with ultraviolet light, the DNA fluoresces.

biotechnology The processes in which living organisms are manipulated, particularly at the molecular genetic level, to form useful products.

cloning The generation of a large number of genetically identical DNA molecules.

cloning vector A DNA molecule that can replicate independent of the host chromosome and transport a piece of inserted foreign DNA, such as a gene, into a recipient cell. It may be a plasmid, phage, cosmid, or artificial chromosome.

complementary DNA (cDNA) A DNA copy of an RNA molecule (e.g., a DNA copy of an mRNA).

cosmid A plasmid vector with lambda phage *cos* sites that can be packaged in a phage capsid; it is useful for cloning large DNA fragments.

expression vector A special cloning vector used to express a recombinant gene in host cells; the gene is transcribed and its protein synthesized.

fermenter A large vessel used to grow microbes in industrial settings. Physical and biological parameters are carefully controlled.

gel electrophoresis A technique that separates molecules according to charge and size.

genetic engineering The deliberate modification of an organism's genetic content by changing its genome.

genomic library A collection of clones that contains fragments that represent the complete genome of an organism.

heterologous gene expression The cloning, transcription, and translation of a gene that has been introduced (cloned) into an organism that normally does not possess the gene.

polymerase chain reaction (PCR) An in vitro technique used to synthesize large quantities of specific nucleotide sequences from small amounts of genetic material. It employs oligonucleotide primers complementary to specific sequences in the target gene and special heat-stable DNA polymerases.

primer A short piece of DNA or RNA that is hybridized to DNA so that DNA synthesis can be started, or "primed."

probe A short, labeled segment of RNA or DNA complementary in base sequence to part of another nucleic acid; used to identify or isolate a particular "target" nucleic acid from a mixture through its ability to bind specifically with the target.

recombinant DNA technology The techniques used in carrying out genetic engineering; they involve the identification and isolation of a specific gene, the insertion of the gene into a vector such as a plasmid to form a recombinant molecule, and the production of large quantities of the gene and its product.

restriction enzymes Enzymes that cleave viral DNA at specific points to protect the cell from virus infection; they are used in vitro to carry out genetic engineering.

The recombinant DNA breakthrough has provided us with a new and powerful approach to the questions that have intrigued and plagued man for centuries.

—Paul Berg

Although human beings have been altering the genetic makeup of organisms for centuries by selective breeding, only recently has the direct manipulation of DNA been possible. The deliberate modification of an organism's genetic information by directly changing the sequence of nucleic acids in its genome is called **genetic engineering** and is accomplished by a collection of methods known as **recombinant DNA technology.** The generation of a large number of genetically identical DNA molecules is called **cloning.** The most commonly used steps to clone a gene or other DNA element are outlined in **figure 16.1.** First, the DNA of interest is identified and isolated (figure 16.1, steps 1 and 2). Once purified, the gene or genes are fused with another piece of DNA called a cloning vector to form recombinant DNA molecules (step 3). These are propagated by insertion into an organism that may not even be in the same domain as the original gene donor but will nonetheless express the gene (step 4).

Although the term has several definitions, here **biotechnology** refers to those processes in which living organisms are manipulated, particularly at the molecular genetic level, to form useful products. In this chapter, we introduce the techniques used in biotechnology by briefly discussing each step in the gene cloning process. We then turn our attention to **industrial microbiology**—the use of microbes to manufacture important compounds or the use of microbes as products in their own right. These two topics are inextricably linked by the use of biotechnological approaches in industrial microbiology.

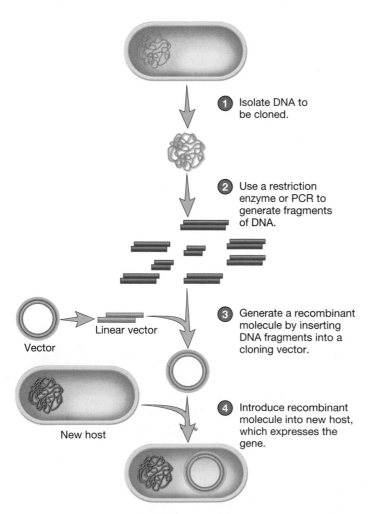

Figure 16.1 **Steps in Cloning a Gene.** Each step shown in this overview is discussed in more detail in this chapter.

① Isolate DNA to be cloned.

② Use a restriction enzyme or PCR to generate fragments of DNA.

Linear vector

Vector

③ Generate a recombinant molecule by inserting DNA fragments into a cloning vector.

New host

④ Introduce recombinant molecule into new host, which expresses the gene.

16.1 KEY DEVELOPMENTS IN RECOMBINANT DNA TECHNOLOGY

Recombinant DNA technology opened up new areas of research and applied biology. Indeed, the pace of research and discovery in the last half century has been remarkable (**table 16.1**). Many of these early discoveries are essential to biotechnology as it is practiced today; these are now discussed.

Restriction Enzymes

Recombinant DNA is DNA with a new sequence formed by joining fragments from two or more different sources. One of the first breakthroughs leading to recombinant DNA technology was the discovery of bacterial enzymes that make cuts in double-stranded DNA by Werner Arber and Hamilton Smith in the late 1960s. These enzymes, known as **restriction enzymes** or restriction endonucleases, recognize and cleave specific sequences about four to eight base pairs long (**figure 16.2**). Restriction enzymes recognize specific DNA sequences called recognition sites. Each restriction enzyme has its own recognition site. Hundreds of different restriction enzymes have been purified and are commercially available. Type I and type III endonucleases identify their unique recognition sites and then cleave DNA at a defined distance from it. The more common type II endonucleases cut DNA directly at their recognition sites. These enzymes can be used to prepare DNA fragments containing specific genes or portions of genes. For example, the restriction enzyme EcoRI, isolated by Herbert Boyer in 1969 from *Escherichia coli,* cleaves DNA between G and A in the base sequence 5′-GAATTC-3′ (**figure 16.3**). Because DNA is antiparallel, this sequence is reversed on the complementary strand of DNA. When EcoRI cleaves between the G and A residues, unpaired 5′-AATTC-3′ remains at the end of each strand. The complementary bases on two EcoRI-cut fragments can hydrogen bond, thus EcoRI and other endonucleases like it generate cohesive or **sticky ends.** In contrast, cleavage by restriction enzymes such as AluI and HaeIII leave blunt ends. A few restriction enzymes and their recognition sites are listed in **table 16.2.** Note that each enzyme is named after the bacterium from which it is purified.

Genetic Cloning and cDNA Synthesis

An important advance came in 1972, when David Jackson, Robert Symons, and Paul Berg reported that they had successfully generated recombinant DNA molecules. They allowed the sticky ends of fragments to anneal—that is, to base pair with one another—and then covalently joined the fragments with the enzyme DNA ligase. Within a year, plasmid vectors that carry foreign DNA fragments during gene cloning had been developed and combined with

Table 16.1	Some Milestones in Biotechnology and Recombinant DNA Technology
1958	DNA polymerase purified
1970	A complete gene synthesized in vitro Discovery of the first sequence-specific restriction endonuclease and the enzyme reverse transcriptase
1972	First recombinant DNA molecules generated
1973	Use of plasmid vectors for gene cloning
1975	Southern blotting technique for detecting specific DNA sequences
1976	First prenatal diagnosis using a gene-specific probe
1977	Methods for rapid DNA sequencing Discovery of "split genes" and somatostatin synthesized using recombinant DNA
1978	Human genomic library constructed
1979	Insulin synthesized using recombinant DNA First human viral antigen (hepatitis B) cloned
1982	Commercial production by *E. coli* of genetically engineered human insulin Isolation, cloning, and characterization of a human cancer gene Transfer of gene for rat growth hormone into fertilized mouse eggs
1983	Engineered Ti plasmids used to transform plants
1985	Tobacco plants made resistant to the herbicide glyphosate through insertion of a cloned gene from *Salmonella* Development of the polymerase chain reaction technique
1987	Insertion of a functional gene into a fertilized mouse egg cures the shiverer mutation disease of mice, a normally fatal genetic disease
1988	The first successful production of a genetically engineered staple crop (soybeans)
1989	First field test of a genetically engineered virus (a baculovirus that kills cabbage looper caterpillars)
1990	Production of the first fertile corn transformed with a foreign gene (a gene for resistance to the herbicide bialaphos)
1991	Development of transgenic pigs and goats capable of manufacturing proteins such as human hemoglobin First test of gene therapy on human cancer patients
1994	The Flavr Savr tomato introduced, the first genetically engineered whole food approved for sale Fully human monoclonal antibodies produced in genetically engineered mice
1995	*Haemophilus influenzae* genome sequenced
1997	Human clinical trials of antisense drugs and DNA vaccines begun
1998	First cloned mammal (the sheep Dolly)
2003	Completion of the draft of the human genome
2005	Reconstruction of 1918 influenza virus

foreign DNA (figure 16.4). Recombinant plasmids replicate within a bacterial host and maintain the cloned fragment of DNA.

Once genes could be recombined into cloning vectors, biologists sought to clone specific genes from various organisms. However, it was evident that cloning eucaryotic DNA into procaryotic hosts would be problematic. This is because eucaryotic pre-mRNA must be processed (e.g., introns spliced out), and procaryotes lack the molecular machinery to perform this task. In 1970, Howard Temin and David Baltimore independently discovered the enzyme that solved this dilemma. They isolated the enzyme **reverse transcriptase (RT)** from retroviruses. These viruses have an RNA genome

that is copied into DNA prior to replication. The mechanism by which reverse transcriptase accomplishes this is outlined in figure 16.5. By using processed mRNA as a template for **complementary DNA (cDNA)** synthesis, RNA processing is not required when cloned cDNA is expressed. **<<** *Viral multiplication (section 5.3)*

Southern Blotting

Another problem early biotechnologists faced was the inability to distinguish the fragment of DNA possessing the gene of interest from the numerous chromosomal fragments produced

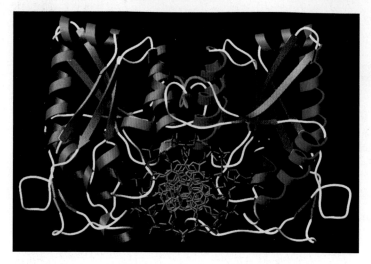

Figure 16.2 Restriction Endonuclease Binding to DNA. The structure of BamHI binding to DNA viewed down the DNA axis. The enzyme's two subunits lie on each side of the DNA double helix. The α-helices are in green, the β conformations in purple, and DNA is in orange.

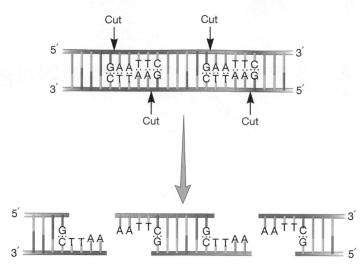

Figure 16.3 Restriction Endonuclease Action. The cleavage catalyzed by the restriction endonuclease *Eco*RI. The enzyme makes staggered cuts on the two DNA strands to form sticky ends.

Table 16.2	Some Type II Restriction Endonucleases and Their Recognition Sequences		
Enzyme	**Microbial Source**	**Recognition Sequence[a]**	**End Produced**
AluI	*Arthrobacter luteus*	5′AGCT 3′ 3′ TCGA 5′	5′ AG CT 3′ 3′ TC GA 5′
BamHI	*Bacillus amyloliquefaciens* H	5′ GGATCC 3′ 3′ CCTAGG5′	5′ G GATCC 3′ 3′ CCTAG G 5′
EcoRI	*Escherichia coli*	5′ GAATTC 3′ 3′ CTTAAG 5′	5′ G AATTC 3′ 3′ CTTAA G 5′
HaeIII	*Haemophilus aegyptius*	5′ GGCC 3′ 3′ CCGG 5′	5′ GG CC 3′ 3′ CC GG 5′
HindIII	*Haemophilus influenzae* d	5′ AAGCTT 3′ 3′ TTCGAA 5′	5′ A AGCTT 3′ 3′ TTCGA A 5′
NotI	*Nocardia otitidis-caviarum*	5′ GCGGCCGC 3′ 3′ CGCCGGCG 5′	5′ GC GGCCGC 3′ 3′ CGCCGG CG 5′
PstI	*Providencia stuartii*	5′ CTGCAG 3′ 3′ GACGTC 5′	5′ CTGCA G3′ 3′ G ACGTC 5′
SalI	*Streptomyces albus*	5′ GTCGAC 3′ 3′ CAGCTG 5′	5′ G TCGAC 3′ 3′ CAGCT G 5′

[a]The arrows indicate the sites of cleavage on each strand.

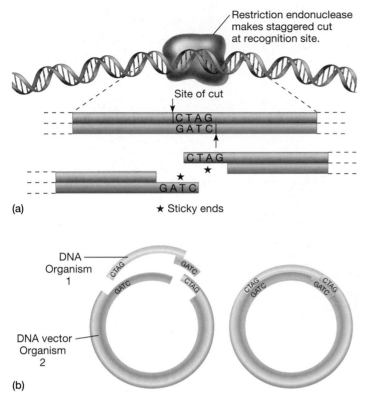

(a)

★ Sticky ends

(b)

Figure 16.4   Recombinant Plasmid Construction. (a) A restriction endonuclease recognizes and cleaves DNA at its specific recognition site. Cleavage produces sticky ends that accept complementary tails for gene splicing. (b) The sticky ends can be used to join DNA from different organisms by cutting it with the same restriction enzyme, ensuring that all fragments have complementary ends.

by restriction enzyme digestion. In 1975, Edwin Southern solved this problem with his **Southern blotting technique.** This procedure enables the detection of specific DNA fragments from a mixture of DNA molecules (**figure 16.6**). In this procedure, DNA fragments are first separated by size with agarose gel electrophoresis (see section 16.3). The fragments are then denatured (rendered single stranded) and transferred to a nylon membrane and treated so that each fragment is firmly bound to the filter at the same position as on the gel. The transfer occurs when buffer flows through the gel and the membrane, as shown in figure 16.6. Alternatively, the negatively charged DNA fragments can be electrophoresed from the gel onto the blotting membrane. The filter is bathed with a solution containing a radioactive **probe,** which is a fragment of labeled, single-stranded nucleic acid that is complementary to the DNA of interest. The DNA to which the probe hydrogen bonds is now radioactive and is readily detected by **autoradiography.** In this technique, a sheet of photographic film is placed over the filter. When developed, bands appear wherever the radioactive probe is bound because the energy released by the isotope causes the formation of dark-silver grains. Nonradioactive probes may also be used to detect

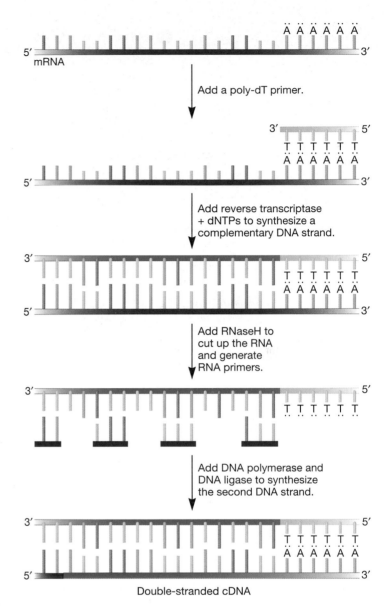

Double-stranded cDNA

Figure 16.5   Synthesis of cDNA. A poly-dT primer anneals to the 3′ end of mRNAs. Reverse transcriptase then catalyzes the synthesis of a complementary DNA strand (cDNA). RNaseH digests the mRNA into short pieces that are used as primers by DNA polymerase to synthesize the second DNA strand. The 5′ to 3′ exonuclease function removes all of the RNA primers except the one at the 5′ end (because there is no primer upstream from this site). This RNA primer can be removed by the subsequent addition of another RNase. After the double-stranded cDNA is made, it can be inserted into vectors, as described in figure 16.4.

specific DNAs. They often are more rapidly detected and are safer than using radioisotopes.

By the late 1970s the techniques for cloning DNA were harnessed to produce recombinant human insulin, and by 1982 commercial production of insulin from genetically engineered *E. coli* began. This was an important development for several reasons: first, diabetic individuals no longer had to depend on insulin from pigs or other animals; second, it demonstrated the

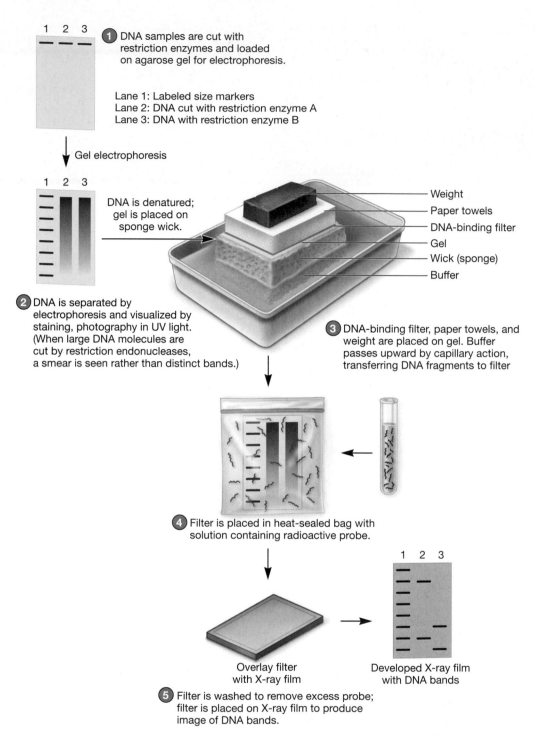

Figure 16.6   The Southern Blotting Technique.

commercial feasibility of using recombinant DNA to make a better product.

1. Describe restriction enzymes, sticky ends, and blunt ends. Can you think of a cloning situation where blunt-ended DNA might be more useful than DNA with sticky ends?

2. What is cDNA? How does it differ from the DNA isolated from a procaryote?

3. What is the purpose of Southern blotting? How is a probe selected? Why do you think the Southern blotting technique was an important breakthrough when it was introduced?

16.2 POLYMERASE CHAIN REACTION

So far, the manipulation of DNA purified from living cells has been reviewed. However, the ability to synthesize short pieces of DNA called oligonucleotides (Greek *oligo,* few or scant) was another important advance. Oligonucleotides are generally between 15 and 30 nucleotides long and can be either RNA or DNA. The generation of oligonucleotides using DNA synthesizer machines is a process that evolved over a number of years (the first report of chemically synthesized DNA was published in 1955, just two years after Watson and Crick resolved the structure of DNA). In contrast, the **polymerase chain reaction (PCR),** invented by Kary Mullis in the early 1980s, exploded onto the biotechnology landscape. Why is PCR so important? Quite simply, it enables the rapid synthesis of many, many copies of a specific DNA fragment from a complex mixture of DNA. Researchers can thus obtain large quantities of specific pieces of DNA for experimental and diagnostic purposes.

Figure 16.7 outlines how the PCR technique works. Suppose that one wishes to make large quantities of a particular DNA sequence, a process known as DNA or gene **amplification.** The first step is to synthesize DNA fragments with sequences identical to those flanking the targeted sequence. This is accomplished with a DNA synthesizer. These oligonucleotides are usually about 20 nucleotides long and serve as DNA **primers** for DNA synthesis. The primers are one component of the reaction mixture, which also contains the target DNA (often copies of an entire genome), a thermostable DNA polymerase, and each of the four deoxyribonucleoside triphosphates (dNTPs). PCR requires a series of repeated reactions, called cycles. Each cycle has three steps that are precisely executed in a machine called a thermocycler. In the first step, the DNA containing the sequence to be amplified is heat denatured to make it single stranded. Next, the temperature is lowered so that the primers can hydrogen bond or anneal to the DNA on both sides of the target sequence. Because the primers are very small and are present in excess, the targeted DNA strands anneal to the primers rather than to each other. Finally, DNA polymerase extends the primers and synthesizes copies of the target DNA sequence using dNTPs. Only polymerases able to function at the high temperatures employed in the PCR technique can be used. Two popular enzymes are the **Taq polymerase** from the thermophilic bacterium *Thermus aquaticus* and the Vent polymerase from *Thermococcus litoralis.* At the end of one cycle, the targeted sequences on both strands have been copied. When the three-step cycle is repeated (figure 16.7), the two strands from the first cycle are copied to produce four fragments. These are amplified in the third cycle to yield eight double-stranded products. Thus, each cycle increases the number of target DNA molecules exponentially. Depending on the initial concentration of the template DNA and other parameters such as the G + C composition of the DNA to be amplified, it is theoretically possible to produce about 1 million copies of targeted DNA sequence after 20 cycles and over 1 billion after

30 cycles. Pieces ranging in size from less than 100 base pairs to several thousand base pairs in length can be amplified, and the initial concentration of target DNA can be as low as 10^{-20} to 10^{-15} M. ◄◄ *DNA replication (section 12.3)*

PCR is most frequently used in one of two ways. If one wants to generate large quantities of a specific piece of DNA, the reaction products are collected and purified at the end of a designated number of cycles. The final number of DNA fragments amplified is not quantitative, meaning that the amount of final product does not always reflect the amount of template DNA present before amplification. In contrast, **real-time PCR** is quantitative. That is, it allows one to ask how much DNA or RNA template (which is converted to DNA with reverse transcriptase) is present in a given sample. This is accomplished by adding a fluorescently labeled probe to the reaction mixture and measuring its signal during the initial cycles. This is when the rate of DNA amplification is logarithmic. However, as the PCR cycles continue, substrates are consumed and polymerase efficiency declines. So although the amount of product increases, its rate of synthesis is no longer exponential (this is why end-point collection of PCR products is not quantitative). Thermocyclers specifically designed for real-time PCR record the amount of PCR product generated as it occurs, thus the term real-time PCR. Gene expression studies often rely on real-time PCR, because mRNA transcripts can be copied and amplified by reverse transcriptase (RT). Therefore the procedure monitors the level of transcription of the gene targeted by the primers.

PCR is an essential tool in many areas of molecular biology, medicine, and biotechnology. As shown in **figure 16.8,** when PCR is used to obtain DNA for cloning, a number of steps traditionally employed are no longer required. PCR is also used to generate DNA for nucleotide sequencing. Because the primers used in PCR target specific DNA, PCR can isolate specific fragments of DNA (e.g., genes) from solutions that contain many different genomes such as soil, water, and blood. For instance PCR is used in metagomics to amplify specific genes and has become an essential part of certain diagnostic tests, including those for AIDS, Lyme disease, chlamydia, tuberculosis, hepatitis, human papillomavirus, and other infectious agents and diseases. The tests are rapid, sensitive, and specific. PCR is particularly valuable in detecting genetic diseases such as sickle cell anemia, phenylketonuria, and muscular dystrophy. The technique is also employed in forensic science, where it is used in criminal cases as part of DNA fingerprinting technology. It is possible to exclude or incriminate suspects using extremely small samples of biological material discovered at the crime scene. ◄◄ *Environmental genomics (section 15.8)*

16.3 GEL ELECTROPHORESIS

Agarose or polyacrylamide **gel electrophoresis** is routinely used to separate DNA fragments. When DNA molecules are placed at the negative end of an electrical field, they migrate toward the

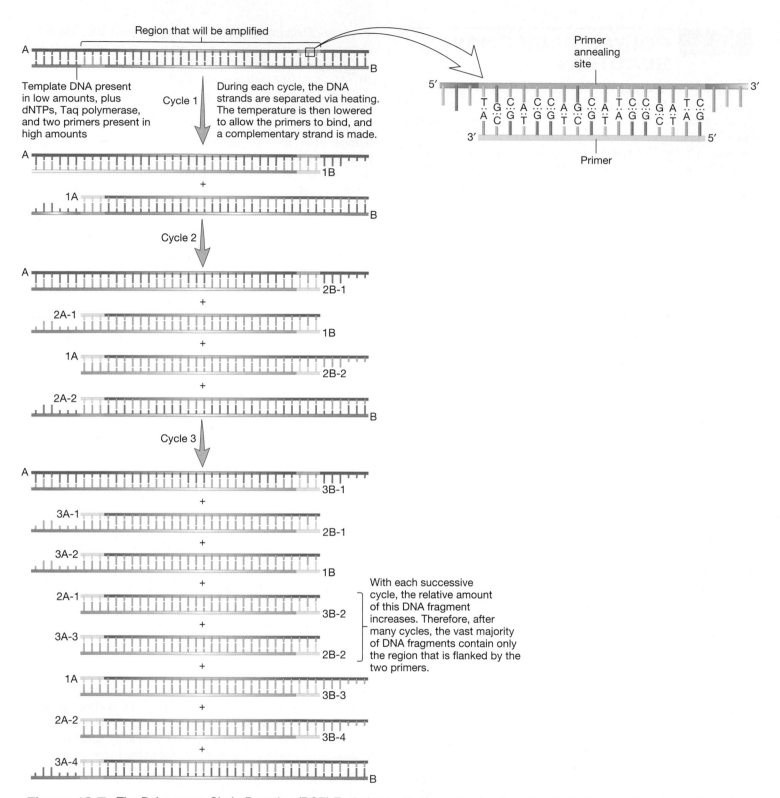

Figure 16.7 The Polymerase Chain Reaction (PCR) Technique. During each cycle, oligonucleotides that are complementary to the ends of the targeted DNA sequence bind to the DNA and act as primers for the synthesis of this DNA region. The primers used in actual PCR experiments are usually 15 to 20 nucleotides in length. The region between the two primers is typically hundreds of nucleotides in length, not just several nucleotides as shown here. The net result of PCR is the synthesis of many copies of DNA in the region that is flanked by the two primers.

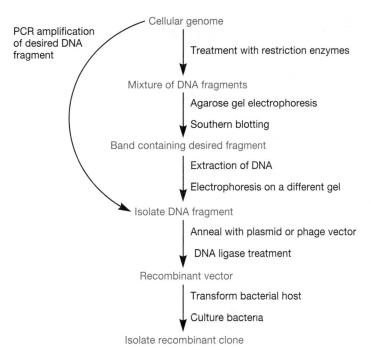

Figure 16.8 Cloning Cellular DNA Fragments. The preparation of a recombinant clone from isolated DNA fragments or DNA generated by PCR.

pole with a positive charge (**figure 16.9**). Each fragment's migration rate is determined by its molecular weight so that the smaller a fragment is, the faster it moves through the gel. Migration rate is also a function of gel density. In practice, this means that higher concentrations of gel material (agarose or acrylamide) provide better resolution of small fragments and vice versa. DNA that has not been digested with restriction enzymes is usually supercoiled. For this and other reasons, DNA is usually cut with restriction enzymes prior to electrophoresis. Small DNA molecules usually yield only a few bands because there are few restriction enzyme recognition sites. If the DNA fragment is large or an entire chromosome is digested, many such sites are present and the DNA is cut in numerous places. When such DNA is electrophoresed, it produces a smear representing many thousands of DNA fragments of similar sizes that cannot be individually resolved. The region of the gel containing the desired DNA fragment must then be located using the Southern blotting technique (figure 16.6).

16.4 CLONING VECTORS AND CREATING RECOMBINANT DNA

Recombinant DNA technology depends on the propagation of many copies of the nucleotide sequence of choice. To accomplish this, genes or other genetic elements are inserted into **cloning vectors**

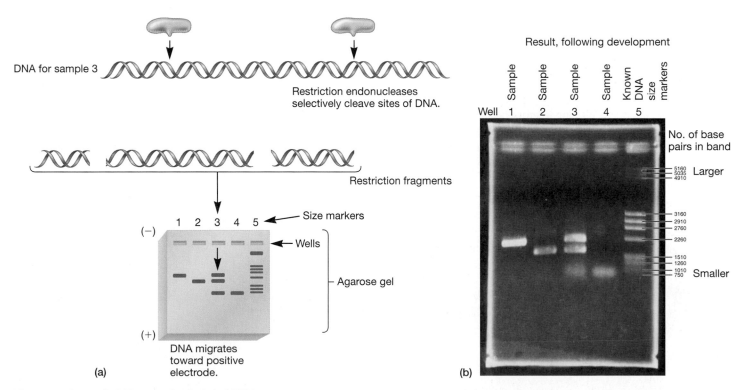

Figure 16.9 Gel Electrophoresis of DNA. (a) After cleavage into fragments, DNA is loaded into wells on one end of an agarose gel. When an electrical current is passed through the gel (from the negative pole to the positive pole), the DNA, being negatively charged, migrates toward the positive pole. The larger fragments, measured in numbers of base pairs, migrate more slowly and remain nearer the wells than the smaller (shorter) fragments. (b) An actual developed and stained gel reveals a separation pattern of the fragments of DNA. The size of a given DNA band can be determined by comparing it to a known set of molecular weight markers (lane 5) called a ladder.

Vector	Insert Size (kb, 1 kb = 1,000 bp)	Example	Features
Plasmid	<20 kb	pBR322, pUC19	Replicates independently of microbial chromosome so many copies may be maintained in a single cell
Bacteriophage	9–25 kb	λ 1059, λ gt11, M13mp18, EMBL3	Packaged into lambda phage particles; single-stranded DNA viruses such as M13 have been modified (e.g., M13mp18) to generate either double- or single-stranded DNA in the host
Cosmids	30–47 kb	pJC720, pSupercos	Can be packaged into lambda phage particles for efficient introduction into bacteria, then replicates as a plasmid
PACs (P1 artificial chromosomes)	75–100 kb	pPAC	Based on the bacteriophage P1 packaging mechanism
BACs (bacterial artificial chromosomes)	75–300 kb	pBAC108L	Modified F plasmid that can carry large DNA inserts; very stable within the cell
YACs (yeast artificial chromosomes)	100–1,000 kb	pYAC	Can carry largest DNA inserts, replicates in *Saccharomyces cerevisiae*

that replicate in a host organism. There are four major types of vectors: plasmids, bacteriophages and other viruses, cosmids, and artificial chromosomes (**table 16.3**). Each type has its own advantages, so the selection of the proper cloning vector is critical to the success of any cloning experiment. Most engineered vectors share three features: an origin of replication; a region of DNA that bears unique restriction sites, called a multicloning site or polylinker; and a selectable marker. These elements are described below in the discussion of plasmids, the most frequently used cloning vectors.

Plasmids

Plasmids make excellent cloning vectors because they replicate autonomously and are easy to purify. They can be introduced into microbes by conjugation or transformation. Many different plasmids are used in biotechnology, all derived from naturally occurring plasmids that have been genetically engineered (**figure 16.10**). << *Bacterial plasmids (section 14.6); Bacterial conjugation (section 14.7); Bacterial transformation (section 14.8)*

Origin of Replication

The **origin of replication** (*ori*) allows the plasmid to replicate in the microbial host independently of the chromosome. pUC19, an *E. coli* plasmid, is said to have a high copy number because it replicates about 100 times in the course of one generation. High copy number is often important because it facilitates plasmid purification and can dramatically increase the amount of cloned gene product produced by the cell. Some plasmids have two origins of replication, each recognized by

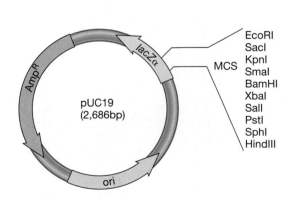

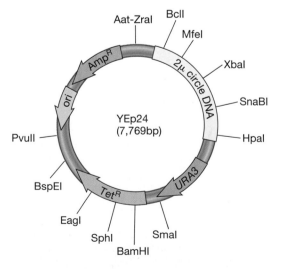

Figure 16.10 The Cloning Vectors pUC19 and YEp24. Restriction sites that are present only once in each vector are shown. pUC19 replicates only in *E. coli,* while YEp24 replicates in both *E. coli* and *S. cerevisiae.*

different host organisms. These plasmids are called **shuttle vectors** because they can move or "shuttle" from one host to another. YEp24 is a shuttle vector that can replicate in yeast (*Saccharomyces cerevisiae*) and in *E. coli* because it has the 2μ circle yeast replication element and *E. coli* origin of replication (figure 16.10).

Selectable Marker

Following the uptake of vector by host cells, one must be able to discriminate between cells that successfully obtained vector from those that did not. Furthermore, one must be able to continue to select for the presence of plasmid, otherwise the host cell may stop replicating it. This is achieved by the presence of a gene that encodes a protein that is needed for the cell to survive under certain, selective conditions. Such a gene is called a selectable marker. In the case of pUC19, the selectable marker encodes the ampicillin resistance enzyme (*amp^R*, sometimes called *bla,* for β-*lac*tamase). The shuttle vector YEp24 bears both the *amp^R* gene for selection in *E. coli* and URA3, which encodes a protein essential for uracil biosynthesis in yeast. Therefore when in *S. cerevisiae,* this plasmid must be maintained in uracil auxotrophs.

Multicloning Site (MCS) or Polylinker

A region of restriction enzyme cleavage sites found only once in the plasmid is essential for the insertion of foreign DNA. Cleavage at a unique restriction site generates a linear plasmid. Cleavage of the gene to be cloned with the same restriction enzyme results in compatible sticky ends, so that it may be inserted (ligated) into the **multicloning site (MCS).** Alternatively, two different, unique sites within the MCS may be cleaved and the DNA sequence between the two sites replaced with cloned DNA (**figure 16.11**). In either case, the plasmid and the DNA to be inserted are incubated in the presence of the enzyme DNA ligase so that when compatible sticky ends hydrogen bond, phosphodiester bonds can be generated between the cloned DNA fragment and the vector. This requires the input of energy, thus ATP is added to this in vitro ligation reaction (*see figure 12.14*).

pUC19 has a number of unique restriction sites in its MCS (figure 16.10); this provides a number of cleavage options, making it easier to obtain the same, or compatible, sticky ends in both vector and the DNA to be inserted. In pUC19, the MCS is located within the 5′ end of the *lacZ* gene, which encodes β-galactosidase (β-Gal). This enzyme cleaves the dissacharide lactose into galactose and glucose. When DNA has been cloned into the MCS, the *lacZ* gene is no longer intact, so a functional enzyme is not produced. This can be detected by the color of colonies: cells turn blue when β-Gal splits the alternative substrate, X-Gal (5-bromo-4-chloro-3-indolyl-β-D-galactopyranoside), which is included in the medium (figure 16.11). This is important because the ligation of foreign DNA into a vector is never 100% efficient. Thus when the ligation mixture is introduced into host cells, one must be able to distinguish cells that carry just plasmid from those that carry plasmid into which DNA was successfully inserted. In the case of pUC19, all *E. coli* cells that take up plasmid (with or without insert) are selected for their resistance to ampicillin (Amp^R). Among these, colonies with plasmid lacking DNA insert will be blue (due to the presence of functional *lacZ* gene), while those transformed with pUC19 into which DNA was successfully cloned will be white. There are a number of other clever ways in which cells with vector versus those with vector plus insert can be differentiated; the detection of blue versus white colonies is a common approach.

Phage Vectors

Phage vectors are engineered phage genomes that have been genetically modified to include useful restriction enzyme recognition sites for the insertion of foreign DNA. Once DNA has been inserted, the recombinant phage genome is packaged into viral capsids and used to infect host cells. The resulting phage lysate consists of thousands of phage particles that carry cloned DNA as well as the genes needed for host lysis. Two commonly used vectors are derived from the bacteriophages T7 and lambda (λ), both of which have double-stranded DNA genomes. Although these phages infect *E. coli*, phage vectors have been engineered for a number of different bacterial host species. << *Structure of viruses (section 5.2)*

Cosmids

Cosmids were developed when it became clear that cloning vectors were needed that could tolerate larger fragments of cloned DNA (table 16.3). Unlike phages and plasmids, cosmids do not exist in nature. Instead, these engineered vectors have been constructed to contain features from both. Cosmids have a selectable marker and MCS from plasmids, and a *cos* site from phage. In phage, the *cos* site is where multiple copies of phage genome are linked prior to packaging. Cleavage at the *cos* sites yields single genomes that are the right size for packaging. Cosmids take advantage of the fact that the only requirement for phage heads to package DNA is two *cos* sites on a linear DNA molecule or a single *cos* site on a circular one (*see figure 24.8*). As long as the cosmid with its cloned DNA is the appropriate size (about 37 to 52 kb), it will be packaged. The phage is then used to introduce the recombinant DNA into *E. coli,* where it replicates as a plasmid.

Artificial Chromosomes

Artificial chromosomes are special cloning vectors used when particularly large fragments of DNA must be cloned, as when constructing a genomic library or sequencing an organism's entire genome. In fact, **bacterial artificial chromosomes (BACs)** were crucial to the timely completion of the human genome project. Like natural chromosomes, artificial chromosomes replicate only once per cell cycle. **Yeast artificial chromosomes (YACs)** were developed first and consist of a yeast telomere at each end (TEL), a centromere sequence (CEN),

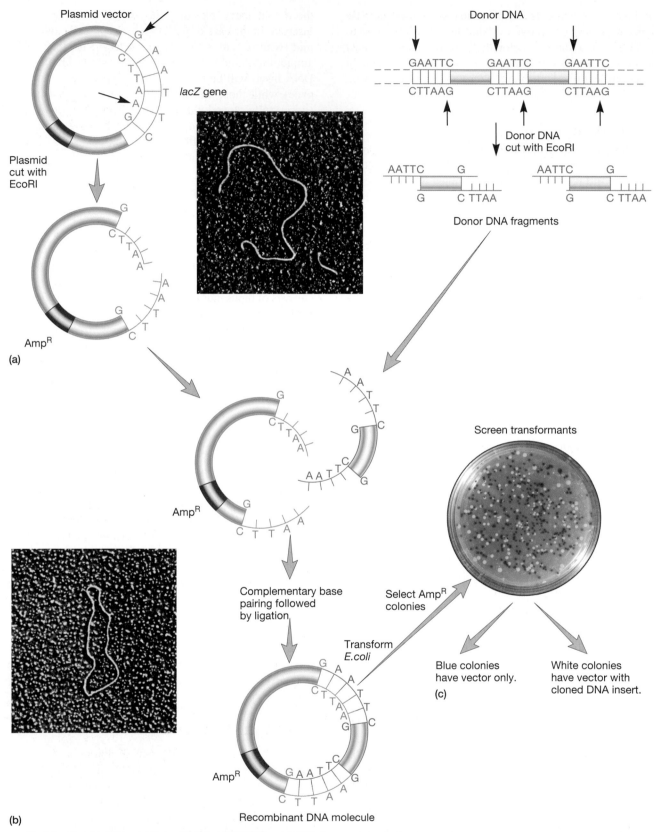

Figure 16.11 Recombinant Plasmid Construction and Cloning. The construction and cloning of a recombinant plasmid vector using an antibiotic resistance gene to select for the presence of the plasmid. The interruption of the *lacZ* gene by cloned DNA is used to detect vectors with insert. The scale of the sticky ends of the fragments and plasmid has been enlarged to illustrate complementary base pairing. (a) The electron micrograph shows a plasmid that has been cut by a restriction enzyme and a donor DNA fragment. (b) The micrograph shows a recombinant plasmid. (c) After transformation, *E. coli* cells are plated on medium containing ampicillin and X-Gal so that only ampicillin resistant transformants grow; X-Gal enables the visualization of colonies that were transformed with recombinant vector (vector + insert, white colonies).

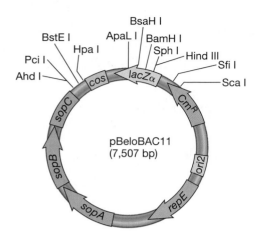

(a) Bacterial artificial chromosome (BAC)

| TEL | TRP1 | ARS | CEN | MCS | URA3 | TEL |

(b) Yeast artificial chromosome (YAC)

Figure 16.12 Artificial Chromosomes Can Be Used as Cloning Vectors. (a) The bacterial artificial chromosome pBeloBAC11 and (b) a yeast artificial chromosome.

a yeast origin of replication (ARS, *a*utonomously, *r*eplicating *s*equence), a selectable marker such as URA3, and an MCS to facilitate the insertion of foreign DNA (**figure 16.12**). YACs are used when extraordinarily large DNA pieces (up to 1,000 kb; table 16.3) are to be cloned. BACs were developed, in part, because YACs tend to be unstable and may recombine with host chromosomes, thereby rearranging the cloned DNA. Although BACs accept smaller DNA inserts than do YACs (up to 300 kb), they are generally more stable. BACs are based on the F fertility factor of *E. coli*. The example shown in figure 16.12 is typical in that it includes genes that ensure a replication complex will be formed (*repE*), as well as proper partitioning of one newly replicated BAC to each daughter cell (*sopA, sopB,* and *sopC*). It also includes features common to many plasmids such as an MCS within the *lacZ* gene for blue/white colony screening and a selectable marker, in this case for resistance to the antibiotic chloramphenicol (Cm^R).

1. Briefly describe the polymerase chain reaction. Explain the differences between reactions in which the products are collected after a defined number of cycles and real-time PCR. Suggest an application of each approach.

2. What is electrophoresis? How is it used in Southern blotting?

3. How are plasmids, cosmids, and artificial chromosomes structurally different? What are some of the different purposes served by each?

4. Explain selection for antibiotic resistance followed by blue versus white screening of colonies containing recombinant plasmids. Why must both antibiotic selection and color screening be used? What would you conclude if, after transforming a ligation mixture into *E. coli*, only blue colonies were obtained?

16.5 CONSTRUCTION OF GENOMIC LIBRARIES

The DNA to be cloned can be obtained in several ways. It can be synthesized by PCR, or it can be located on the chromosome by Southern blotting. However, PCR amplification of a gene requires foreknowledge of its nucleotide sequence (or at least sequences flanking the gene), and a suitable probe must be obtained for Southern blotting. In both cases, once the DNA fragment is purified, it is cloned using a procedure like that described for recombinant plasmids and shown in figure 16.11. However, what if researchers wanted to clone a gene but had no idea what its DNA sequence might be? A genomic library must then be constructed and screened.

The goal of **genomic library** construction is to have an organism's genome cut into separate fragments with each fragment cloned into a separate vector. Ideally the entire genome is represented; that is to say, the sum of the different fragments equals the whole genome. In this way, specific groups of genes can be analyzed and isolated. The construction of a genomic library begins with cleaving the genome into small pieces by a restriction endonuclease (**figure 16.13**). These genomic DNA fragments are then either cloned into vectors and introduced into a microbe or packaged into phage particles that are used to infect the host. In either case, many thousands of different clones— each with a different genomic DNA insert—are created.

To select the desired clone from the library, it is necessary to know something about the function of the target gene or genetic element. If the genomic library has been inserted into a microbe that expresses the foreign gene, it may be possible to assay each clone for a specific protein or phenotype. For example, if one is studying a newly isolated soil bacterium and wants to find genes that encode enzymes needed for the biosynthesis of the amino acid alanine, the library could be expressed in an *E. coli* or *Bacillus subtilis* alanine auxotroph (figure 16.13). Recall that alanine auxotrophs require the addition of this amino acid to the medium. Following introduction of the genomic library into host cells, those that now grow without alanine would be good candidates for the genomic library fragment that possesses the alanine biosynthetic genes. Success with this approach depends on the assumption that the function of the cloned gene product is similar in both organisms. If this is not the case, the host must be the same species from which the library was prepared. In this example, a soil bacterial mutant lacking the gene in question (e.g., an alanine auxotroph) is used as the genomic library host. The genetic complementation of a deficiency in the host cell is

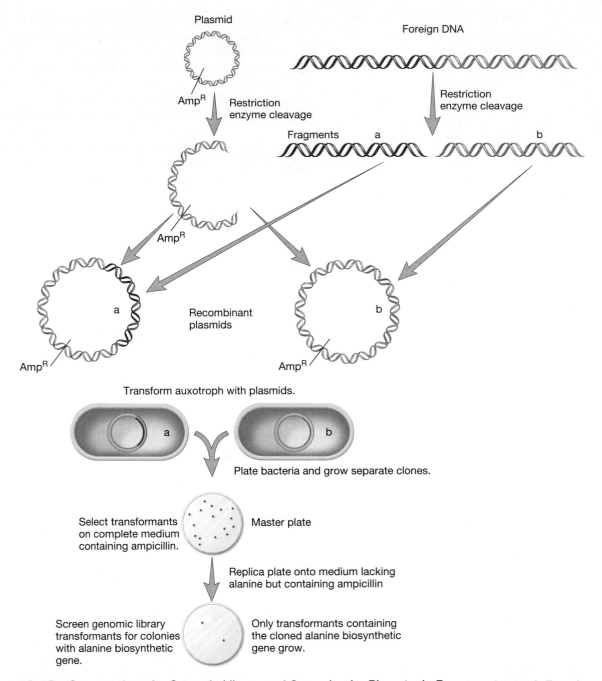

Figure 16.13 **Construction of a Genomic Library and Screening by Phenotypic Rescue.** A genomic library is made by cloning fragments of an organism's entire genome into a vector. For simplicity, only two possible recombinant vectors are shown. In reality, a large mixture of vectors with inserts is generated. This mixture is then introduced into a suitable host. Phenotypic rescue is one way to screen the colonies for the gene of interest. It involves using a host with a genetic defect that can be complemented or "rescued" by the expression of a specific gene that has been cloned.

sometimes called **phenotypic rescue.** << *Detection and isolation of mutants (section 14.2)*

If a genomic library is prepared from a eucaryote in an effort to isolate a structural gene, a cDNA library is usually constructed. In this way, introns are not present in the genomic library. Instead, only the protein-coding regions of the genome are cloned. cDNA is prepared (figure 16.5) and cloned into a

suitable vector. After the library is introduced into the host microbe, it may be screened by phenotypic rescue or by hybridization with an oligonucleotide, as described for Southern blotting. In some cases, neither phenotypic rescue nor hybridization with a probe is possible. Then the researcher must develop a novel approach that suits the particular set of circumstances to screen the genomic library.

16.6 INTRODUCING RECOMBINANT DNA INTO HOST CELLS

In cloning procedures, the selection of a host organism is as important as the choice of cloning vector. *E. coli* is the most frequent procaryotic host and *S. cerevisiae* is most common among eucaryotes. Host microbes that have been engineered to lack restriction enzymes and the recombination enzyme RecA make better hosts because it is less likely that the newly acquired DNA will be degraded or recombined with the host chromosome. There are several ways to introduce recombinant DNA into a host microbe. Transformation and electroporation are two commonly employed techniques. Often the host microbe does not have the capacity to be transformed naturally. This is the case with *E. coli* and most gram-negative bacteria as well as many gram-positive bacteria. In these cases, the host cells may be rendered competent by treatment with divalent cations and artificially transformed by heat shocking the cells. << *Bacterial transformation (section 14.8)*

Electroporation is a technique that is simple and has wide application to a number of host organisms, including plant and animal cells. In this procedure, cells are mixed with the recombinant DNA and exposed to a brief pulse of high-voltage electricity. The plasma membrane becomes temporarily permeable and DNA molecules are taken up by some of the cells. The cells are then grown on media that select for the presence of the cloning vector, as described in section 16.4.

16.7 EXPRESSING FOREIGN GENES IN HOST CELLS

When a gene from one organism is cloned into another, it is said to be a **heterologous gene.** Heterologous genes are not always expressed in the host cell without further modification of the recombinant vector. To be transcribed, the recombinant gene must have a promoter that is recognized by the host RNA polymerase. Translation of its mRNA depends on the presence of leader sequences and mRNA modifications that allow proper ribosome binding. These are quite different in eucaryotes and procaryotes. For instance, if the host is a procaryote and the gene has been cloned from a eucaryote, a procaryotic leader must be provided and introns removed.

The problems of expressing recombinant genes in host cells are largely overcome with the help of special cloning vectors called **expression vectors.** These vectors contain the necessary transcription and translation start signals in addition to convenient multicloning sites. Some expression vectors contain regulatory regions of the *lac* operon so that the expression of the cloned genes can be controlled in the same manner as the operon. << *Global regulatory systems: Catabolite repression (section 13.5)*

Somatostatin, the 14-residue hypothalamic peptide hormone that helps regulate human growth, provides an example of useful cloning and protein production. The gene for somatostatin was initially synthesized by chemical methods. Besides the 42 bases coding for somatostatin, the synthetic gene contains a codon for methionine at the 5′ end (which corresponds to the N-terminal end of the peptide) and two stop codons at the opposite end. To aid insertion into the plasmid vector, the 5′ ends of the gene were extended to form sticky ends complementary to those formed by the EcoRI and BamHI restriction enzymes. A plasmid cloning vector was cut with both EcoRI and BamHI to remove a part of the plasmid DNA. The synthetic gene was then inserted into the vector by taking advantage of its sticky ends (**figure 16.14**). Finally, a fragment containing the initial part of the *lac* operon (including the promoter, operator, ribosome binding site, and much of the β-galactosidase gene) was inserted upstream to, or at the 5′ end of, the somatostatin gene. The plasmid thus contained the somatostatin gene fused in the proper orientation to the remaining portion of the β-galactosidase gene.

When this recombinant plasmid is transformed into *E. coli*, the somatostatin gene is transcribed with the β-galactosidase gene

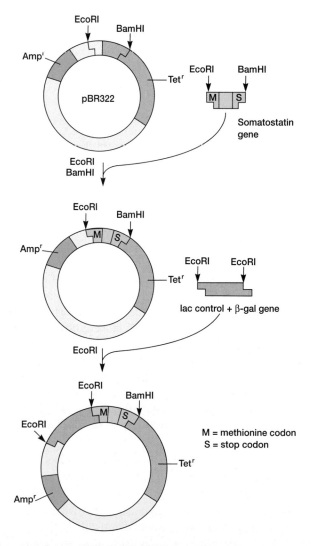

Figure 16.14   Cloning the Somatostatin Gene. An overview of the procedure used to synthesize a recombinant plasmid containing the somatostatin gene. Amp^r = ampicillin and Tet^r = tetracycline resistance genes.

Table 16.4	Some Human Peptides and Proteins Synthesized by Genetic Engineering
Peptide or Protein	**Potential Use**
α_1-antitrypsin	Treatment of emphysema
α-, β-, and γ-interferons	As antiviral, antitumor, and anti-inflammatory agents
Blood-clotting factor VIII	Treatment of hemophilia
Calcitonin	Treatment of osteomalacia
Epidermal growth factor	Treatment of wounds
Erythropoetin	Treatment of anemia
Growth hormone	Growth promotion
Insulin	Treatment of diabetes
Interleukins-1, 2, and 3	Treatment of immune disorders and tumors
Macrophage colony stimulating factor	Cancer treatment
Relaxin	Aid to childbirth
Serum albumin	Plasma supplement
Somatostatin	Treatment of acromegaly
Streptokinase	Anticoagulant
Tissue plasminogen activator	Anticoagulant
Tumor necrosis factor	Cancer treatment

fragment to generate a single mRNA. Translation generates a protein consisting of the hormone peptide fused to the β-galactosidase fragment by a methionine residue. Treatment of the fusion protein with cyanogen bromide, which cleaves peptide bonds at methionine residues, breaks the peptide chain at the methionine and releases the hormone. Once free, the peptide folds properly to become active. Because production of the fusion protein is under the control of the *lac* operon, it is easily regulated. Many proteins have been produced since the synthesis of somatostatin. Examples include human growth hormone, interferons, and proteins used in vaccine production (**table 16.4**). In addition, the fusion of one protein to another has become a useful research tool (**Techniques & Applications 16.1**).

1. What is a genomic library? Describe two ways in which a genomic library might be screened for the clone of interest.
2. How can one prevent recombinant DNA from undergoing recombination in a bacterial host cell?
3. List several reasons why a cloned gene might not be expressed in a host cell. What is an expression vector?

16.8 MICROORGANISMS USED IN INDUSTRIAL MICROBIOLOGY

Industrial microbiology harnesses the capacity of microbes to synthesize compounds that have important applications in medicine, agriculture, food preparation, and other industrial processes. These compounds are commonly called **natural products.** To obtain natural products, it is first necessary to identify or create a microorganism that carries out the desired process in the most efficient manner. This microorganism or its cloned genes are then used, either in a controlled environment such as a fermenter or in complex natural systems such as soils or waters, to achieve specific goals.

Thus the first task for an industrial microbiologist is to find a suitable microorganism, one that is genetically stable, easy to maintain and grow, and well suited for extraction or separation of desired products. A wide variety of approaches are available, ranging from isolating microorganisms from the environment to using sophisticated molecular techniques to modify existing microorganisms. Here we present some of the commonly used methods, demonstrating the usefulness of classical techniques and the advances that molecular biology have offered.

Finding Microorganisms in Nature

Until relatively recently, microbes used in industrial microbiology were cultured from natural materials such as soil samples, waters, and spoiled bread and fruit. Cultures from all areas of the world continue to be examined to identify new strains with desirable characteristics. Interest in hunting for new microorganisms, or **bioprospecting,** continues today.

Less than 1% of the microbial species estimated to exist in most environments has been isolated or cultured. With increased interest in microbial diversity, microbial ecology, and especially in microorganisms from extreme environments, microbiologists are exploring new ways to grow these previously uncultured microbes. In addition, the application of metagenomics to industrial microbiology is being pursued so that genes encoding potentially useful proteins might be obtained without having to purify and grow the microbe of origin. << *Environmental genomics (section 15.8)*

Genetic Manipulation of Microorganisms

Genetic manipulations are used to produce microorganisms with new and desirable characteristics. The classical methods of genetic exchange coupled with recombinant DNA technology play a vital role in the development of cultures for industrial microbiology.

Mutagenesis

Once a promising microorganism is found, a variety of techniques can be used for its improvement, including chemical, ultraviolet light, and transposon mutagenesis. For example, the

Techniques & Applications

16.1   Visualizing Proteins with Green Fluorescence

What does a jellyfish that lives in the cold waters of the northern Pacific have to do with biotechnology? It turns out, a lot. The jellyfish, *Aequorea victoria,* produces a protein called green fluorescent protein (GFP) that scientists have adopted to visualize gene expression and protein localization in living cells. GFP is encoded by a single gene that, when translated, undergoes self-catalyzed modification to generate a strong, green fluorescence. This means that it is easily cloned and expressed in any organism. And GFP isn't just green anymore; site-directed mutagenesis of the GFP gene has generated a variety of proteins that glow throughout the blue-green-yellow spectrum.

One widely used GFP application is to tag proteins so their cellular location can be seen with a light microscope. Formerly, electron microscopy (EM) was the only method by which the location of proteins could be visualized. Sample preparation for EM involves harsh treatment with solvents and cellular dehydration—procedures that can damage cell structures and lead to artifacts. In contrast, when a protein is tagged with GFP, the timing of protein production and its localization can be followed in living cells. To accomplish this, a structural gene is genetically fused to the GFP gene to create a chimeric protein—a protein that consists of two parts: the protein being studied and GFP. Of course, care must be taken to ensure that the protein fusion still functions like the original protein. One way to do this is to test for phenotypic rescue of a mutant lacking the structural gene of interest.

GFP protein fusion technology has been used to examine some of the most basic questions in biology. One example is cell division, or cytokinesis. Genetic evidence clearly shows that the tubulin-like protein FtsZ is key for new septum formation in most procaryotic cells. A special type of cell division that occurs during the formation of an endospore has been intensively studied in *Bacillus subtilis.* In this case, septum formation does not generate two equally sized daughter cells. Instead, asymmetric division gives rise to two different progeny: the endospore and the mother cell. Asymmetric cell division means that rather than FtsZ assembling in the center of the cell, it is polymerized at the pole that gives rise to the endospore. Originally it was hypothesized that the cell accomplishes this by preventing FtsZ ring assembly in the middle of the cell while simultaneously activating FtsZ polymerization sites near the poles (**box figure**).

The fusion of the *B. subtilis ftsZ* structural gene to the gene encoding GFP generated a surprising result: the deployment of FtsZ to the cell pole is a much more dynamic process. Time-lapse photomicroscopy shows that when the cell switches from vegetative growth and binary fission to endospore formation, FtsZ forms a spiral-like filament that moves from the cell center to the poles. Careful analysis reveals that many cells appear to have FtsZ spirals that are more

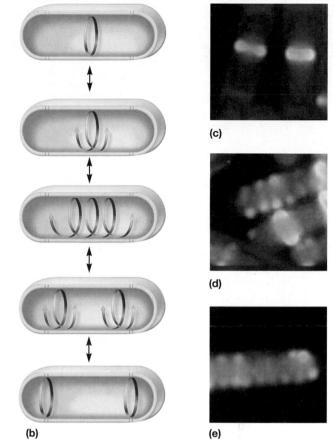
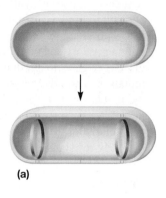

Asymmetric Septum Formation in *Bacillus subtilis*. (a) The originally hypothesized notion that FtsZ was redirected to cell poles. (b) FtsZ-GFP fusion analysis reveals that FtsZ forms spiral structures that migrate to the poles. (c) A photomicrograph of *B. subtilis* cells expressing the FtsZ-GFP fusion protein while growing vegetatively, therefore undergoing binary fission. (d) FtsZ-GFP is delocalized and forms a spiral toward the poles in the cell in upper half of the image, while the cell beneath it has formed a midcell septum. (e) The cell on the right has formed an asymmetric septum in preparation for sporulation, while the cell on the left has formed an FtsZ-GFP spiral. Each cell is about 5 μm in length.

abundant in one half of the cell. This suggests that the accumulation of a critical level of FtsZ at one pole before another may determine endospore placement. Surely GFP fusions will help to resolve this and other questions regarding protein localization. ◄◄ *Bacterial cell cycle (section 7.1); Global regulatory systems: Sporulation in* Bacillus subtilis *(section 13.5)*

From Ben-Yehuda, S., and Losick, R. 2002. Asymmetric cell division in B. subtilis *involves a spiral-like intermediate of the cytokinetic protein FtsZ. Cell. 109:257–66.*

first cultures of *Penicillium notatum,* which had to be grown in stationary vessels, produced low concentrations of penicillin. In 1943 strain NRRL 1951 of *Penicillium chrysogenum* was isolated and further improved through chemical and UV mutagenesis (**figure 16.15**). Today most penicillin is produced with *Penicillium chrysogenum* grown in aerobic stirred fermenters, yielding 55 times more penicillin than the original static cultures. ◄◄ *Mutations and their chemical basis (section 14.1)*

A more targeted approach to mutagenesis can be performed using **site-directed mutagenesis.** In this technique, the nucleotide sequence of a specific gene can be altered. An oligonucleotide of about 20 nucleotides that contains the desired nucleotide sequence change is synthesized. The altered oligonucleotide is allowed to bind to a single-stranded copy of the complete gene. DNA polymerase is added to the gene-primer complex. The polymerase extends the primer and replicates the remainder of the target gene to produce a new gene copy with the desired mutation. The DNA is then cloned to yield large quantities of the mutant protein for study. Often minor amino acid substitutions have been found to lead to unexpected changes in protein characteristics, resulting in new products such as more environmentally resistant enzymes.

Protoplast Fusion

Protoplast fusion is another technique that can be used to promote genetic variability in microorganisms and plant cells. In this method, protoplasts—cells lacking a cell wall—are prepared by growing the cells in an isotonic solution while treating them with enzymes that degrade the cell wall (e.g. lysozyme for bacteria, chitinase for yeasts). The protoplasts of cells of the same or even different species can be fused (joined) during coincubation. The cell wall is then regenerated using osmotic stabilizers such as sucrose. Protoplast fusion is inherently mutagenic and it is especially useful for prompting recombination. For example, when protoplasts of *Penicillium roquefortii* were fused with those of *P. chrysogenum,* new industrially useful *Penicillium* strains were created.

Transfer of Genetic Information Between Different Organisms

The transfer and expression of genes between different organisms can give rise to novel metabolic processes and products. This is part of the rapidly developing field of **combinatorial biology.** For instance, genes for antibiotic production are usually clustered. Using combinatorial biology, the genes that encode specific enzymes in the production of one antibiotic may be added to the gene cluster that encodes enzymes that synthesize a different antibiotic (in a different microbe) so that a novel molecule is produced.

When functional genes from one organism are transcribed and translated in another, it is called **heterologous gene expression.** Heterologous gene expression enables the production of specific proteins and peptides without contamination by other products that might be synthesized in the original organism. This approach can decrease the time and cost of recovering and purifying a product. For example, pediocin, a bacteriocin, is expressed in a yeast used in wine fermentation for the purpose of controlling bacterial contaminants. Another major advantage of engineered protein production is that only biologically active stereoisomers are produced. This specificity is required to avoid the possible harmful side effects of inactive stereoisomers. Examples of a few heterologous systems are shown in **table 16.5.**

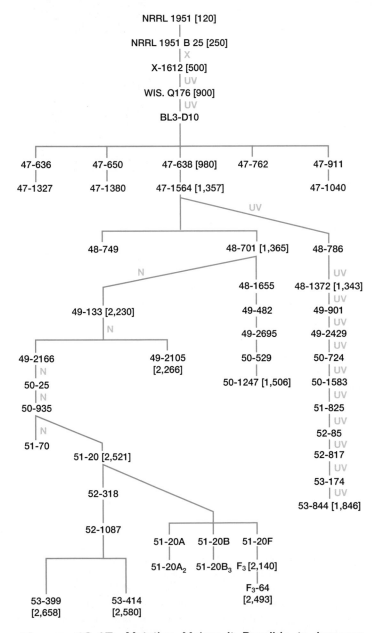

Figure 16.15 Mutation Makes It Possible to Increase Fermentation Yields. A "genealogy" of the mutation processes used to increase penicillin yields with *Penicillium chrysogenum* using X-ray treatment (X), UV treatment (UV), and mustard gas (N). By using these mutational processes, the yield was increased from 120 International Units (IU) to 2,580 IU, a 20-fold increase. Unmarked transfers were used for mutant growth and isolation. Yields in international units/ml in brackets.

Table 16.5	Heterologous Gene Expression to Improve Processes and Products	
Property or Product Transferred	**Microorganism Used**	**Combinatorial Process**
Ethanol production	*Escherichia coli*	Integration of pyruvate decarboxylase and alcohol dehydrogenase II from *Zymomonas mobilis*
1,3-Propanediol production	*E. coli*	Introduction of genes from the *Klebsiella pneumoniae dha* region into *E. coli* makes possible anaerobic 1,3-propanediol production
Cephalosporin precursor synthesis	*Penicillium chrysogenum*	Production of precursors by incorporation of the expandase gene of *Cephalosoporin acremonium* into *Penicillium* by transformation
Lactic acid production	*Saccharomyces cerevisiae*	A muscle bovine lactate dehydrogenase gene (LDH-A) expressed in *S. cerevisiae*

Adapted from Ostergaard, S., Olsson, L., and Nielson, J. 2000 . Metabolic engineering of *Saccharomyces cerevisiae*. *Microbiol. Mol. Biol. Rev.* 64(1):34–50.

Modification of Gene Expression

In addition to inserting new genes in organisms, it also is possible to increase product yield by modifying regulatory molecules or the DNA sites to which they bind. These approaches make it possible to overproduce a wide variety of products, such as antibiotics, amino acids, and enzymes of industrial importance.

The modification of gene expression also can be used to intentionally alter metabolic pathways by inactivation or deregulation of specific genes. Understanding metabolic pathways makes it possible to design or alter a pathway that will be most efficient by avoiding slower or energetically more costly routes. This approach has been used to improve penicillin production by **metabolic pathway engineering (MPE).** MPE can also be used to isolate compounds with altered structures. For instance, variants of the antibiotic erythromycin are produced by blocking specific biochemical steps in its biosynthetic pathway. These altered products can then be tested for their possible antimicrobial effects. In addition, this approach enables a better understanding of the structure-function relationships of natural products.

Protein Evolution

One of the newest approaches for creating novel metabolic capabilities in a given microorganism is protein evolution, which employs forced evolution, adaptive mutations, and in vitro evolution. **Forced evolution** and **adaptive mutation** involve the application of specific environmental stresses to "force" microorganisms to mutate and adapt, thus creating microorganisms with new biological capabilities. Adaptive mutation is thought to induce the SOS response, thus these genetic lesions tend to involve changes in genome structure such as deletions and inversions. << *DNA repair: The SOS response (section 14.3)*

In vitro evolution starts with purified nucleic acids rather than a whole organism. DNA templates (e.g., mutagenized versions of genes whose product is of interest) are transcribed in vitro by a phage RNA polymerase into RNA molecules that are selected based on their capacity to perform a specific function (e.g., demonstrate ribozyme activity or bind a specific fragment of DNA). The enzyme reverse transcriptase is then used to copy the selected RNA molecules into cDNA, which can then be amplified by PCR. After a number of such cycles, a gene that might be of industrial importance will "evolve."

In vitro protein evolution was introduced in the 1970s. In this technique, random or targeted mutations are made, and the DNA is transcribed and translated in vitro. The mutant proteins are then screened for the desired activity, such as stability under reducing conditions or increased ability to bind a specific molecule (i.e., enhanced affinity to a ligand). Before the development of **high-throughput screening (HTS),** biotechnologists were limited by the time it took to screen each new molecule. However, HTS now enables the rapid selection of a single desirable molecule from tens of thousands of compounds. HTS employs a combination of robotics and computer analysis to screen samples (proteins, natural compounds, and whole cells), usually in 96-well microtiter plates, for a specific trait. The combination of molecular biological approaches and HTS has propelled biotechology to a new level of efficiency not previously envisioned.

1. Why is the recovery of previously uncultured microorganisms from the environment an important goal?

2. Define protoplast fusion and list types of microorganisms used in this process.

3. What is combinatorial biology and what is the basic approach used in this technique? Besides antibiotics, what other types of biologically active molecules do you think might be created using combinatorial biology?

4. Why is it important to produce specific isomers of products for use in animal and human health?

5. What types of recombinant DNA techniques are being used to modify gene expression in microorganisms?

6. Define metabolic pathway engineering, forced evolution, and adaptive mutation.

7. What is high-throughput screening and why has it become so important?

(a)

16.9 MICROORGANISM GROWTH IN CONTROLLED ENVIRONMENTS

For many industrial processes, microorganisms must be grown using specifically designed media under carefully controlled conditions. The development of appropriate culture media and the growth of microorganisms under industrial conditions are the subjects of this section.

It is first necessary to clarify terminology. The term **fermentation,** used in a physiological sense in earlier sections of the book, is employed in a much more general way in relation to industrial microbiology and biotechnology. To industrial microbiologists, fermentation means the mass culture of microorganisms (or plant and animal cells). Industrial fermentations require the development of appropriate culture media and the transfer of small-scale technologies to a much larger scale. In fact, the success of an industrially important microbial product rests on the process of scale-up. This is when a procedure developed in a small flask is modified for use in a large fermenter. The microenvironment of the small culture must be maintained despite increases in the culture volume. If a successful transition is made from a process originally developed in a 250 ml Erlenmeyer flask to a 100,000 liter reactor, then the process of scale-up has been carried out successfully.

Microorganisms are often grown in stirred fermenters or other mass culture systems. Stirred **fermenters** can range in size from 3 liters to 100,000 liters or larger, depending on production requirements. A typical industrial stirred fermentation unit is illustrated in **figure 16.16.** Not only must the medium be sterilized but aeration, pH adjustment, sampling, and process monitoring must be carried out under rigorously controlled conditions. When required, foam control agents must be added, especially with high-protein media. Computers are used to monitor outputs from probes that determine microbial biomass, levels of critical metabolic products, pH, input and exhaust gas composition, and other parameters. Environmental conditions can be changed or held constant over time, depending on the requirements of the particular process.

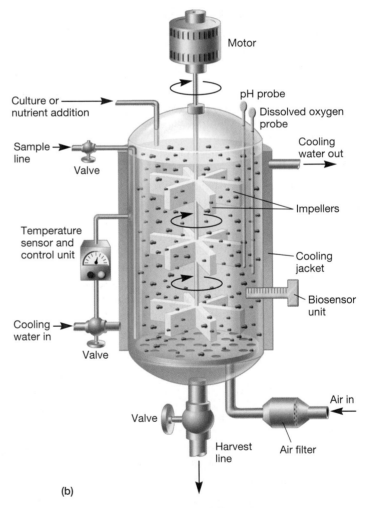

(b)

Figure 16.16 Industrial Stirred Fermenters. (a) Large fermenters used by a pharmaceutical company for the microbial production of antibiotics. (b) Details of a fermenter unit. This unit can be run under oxic or anoxic conditions, and nutrient additions, sampling, and fermentation monitoring can be carried out under aseptic conditions. Biosensors and infrared monitoring can provide real-time information on the course of the fermentation. Specific substrates, metabolic intermediates, and final products can be detected.

Frequently a critical component in the medium, often the carbon source, is added continuously—a process called **continuous feed**—so that the microorganism will not have excess substrate available at any given time. This is particularly important with glucose and other carbohydrates. If excess glucose is present at the beginning of a fermentation, it can be catabolized to yield ethanol, which is lost as a volatile product and reduces the final yield. This can occur even under oxic conditions.

Other culturing approaches exist, among them continuous culture techniques using chemostats. These can sometimes markedly improve cell outputs and rates of substrate use because microorganisms can be maintained in logarithmic phase. However, continuous active growth is undesirable in many industrial processes because the product of interest is not made during exponential growth. Microbial products often are classified as primary and secondary metabolites (**figure 16.17**). **Primary metabolites** consist of compounds related to the synthesis of microbial cells during balanced growth. They include amino acids, nucleotides, and fermentation end products such as ethanol and organic acids. In addition, industrially useful enzymes, either associated with the microbial cells or exoenzymes, often are synthesized by microorganisms during log phase growth. **Secondary metabolites** usually accumulate during the period of nutrient limitation or waste product accumulation that follows the active growth phase. These compounds have only a limited relationship to the synthesis of cell materials and normal growth. Most antibiotics and the mycotoxins fall into this category. << *Growth curve (section 7.2); Continuous culture of microorganisms (section 7.4)*

1. What is the objective of the scale-up process and why is it so critical?

2. What parameters can be monitored in a modern, large-scale industrial fermentation?

3. What is the difference between primary and secondary metabolites? What kind of system (fermenter vs. chemostat) would most likely be used for the production of the streptomycete antibiotic streptomycin? Explain your answer.

16.10 MAJOR PRODUCTS OF INDUSTRIAL MICROBIOLOGY

Products generated by industrial microbiology have profoundly changed our lives and life spans. They include industrial and agricultural products, food additives, products for human and animal health, and biofuels (**table 16.6**). Antibiotics and other natural products used in medicine and health have made major contributions to the improved well-being of animal and human populations. Only major products in each category are discussed here.

Antibiotics

Many antibiotics are produced by microorganisms, predominantly by actinomycetes in the genus *Streptomyces* and by filamentous fungi. The synthesis of penicillin and its derivatives illustrates the critical role of medium formulation and environmental control in the production of these important compounds. Although penicillin is less widely used because of antibiotic resistance, this drug produced by *Penicillium chrysogenum* is an excellent example of a fermentation for which careful adjustment of the medium composition is used to achieve maximum yields. Provision of the slowly hydrolyzed disaccharide lactose, in combination with limited nitrogen availability, stimulates a greater accumulation of penicillin after growth has stopped (**figure 16.18**). The same result can be achieved by using a slow, continuous feed of glucose. If a particular penicillin is needed, the specific precursor is added to the medium. For example, phenylacetic acid is added to maximize production of penicillin G, which has a benzyl side chain (*see figure 31.5*). The fermentation pH is maintained around neutrality by the addition of sterile alkali, which ensures maximum

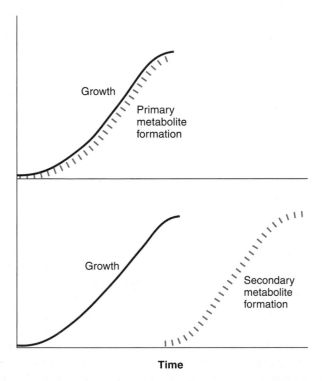

Figure 16.17 Primary and Secondary Metabolites. Depending on the particular organism, the desired product may be formed during or after growth. Primary metabolites are formed during the active growth phase, whereas secondary metabolites are formed after growth is completed.

Table 16.6	Major Microbial Products and Processes of Interest in Industrial Microbiology
Substances	**Microorganisms**
Industrial Products	
Ethanol (from glucose)	*Saccharomyces cerevisiae*
Ethanol (from lactose)	*Kluyveromyces fragilis*
Acetone and butanol	*Clostridium acetobutylicum*
2,3-butanediol	*Enterobacter, Serratia*
Enzymes	*Aspergillus, Bacillus, Mucor, Trichoderma*
Agricultural Products	
Gibberellins	*Gibberella fujikuroi*
Food Additives	
Amino acids (e.g., lysine)	*Corynebacterium glutamicum*
Organic acids (citric acid)	*Aspergillus niger*
Nucleotides	*Corynebacterium glutamicum*
Vitamins	*Ashbya, Eremothecium, Blakeslea*
Polysaccharides	*Xanthomonas*
Medical Products	
Antibiotics	*Penicillium, Streptomyces, Bacillus*
Alkaloids	*Claviceps purpurea*
Steroid transformations	*Rhizopus, Arthrobacter*
Insulin, human growth hormone, somatostatin, interferons	*Escherichia coli, Saccharomyces cerevisiae,* and others (recombinant DNA technology)
Biofuels	
Hydrogen	Photosynthetic microorganisms
Methane	*Methanobacterium*
Ethanol	*Zymomonas, Thermoanaerobacter*

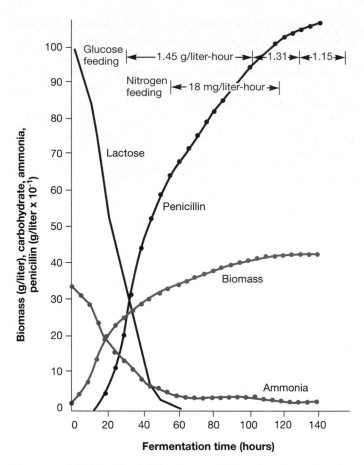

Figure 16.18 Penicillin Fermentation Involves Precise Control of Nutrients. The synthesis of penicillin begins when nitrogen from ammonia becomes limiting. After most of the lactose (a slowly catabolized disaccharide) has been degraded, glucose (a rapidly used monosaccharide) is added along with a low level of nitrogen. This stimulates maximum transformation of the carbon sources to penicillin. The scale factor is presented using the convention recommended by the American Society for Microbiology. That is, a number on the axis should be multiplied by 0.10 to obtain the true value.

Amino Acids

Amino acids such as lysine and glutamic acid are used in the food industry as nutritional supplements in bread products and as flavor-enhancing compounds such as monosodium glutamate (MSG). Amino acid production is typically carried out by regulatory mutants, which have a reduced ability to limit synthesis of a specific amino acid or a key intermediate. Wild-type microorganisms avoid overproduction of biochemical intermediates by the careful regulation of cellular metabolism. Production of glutamic acid and several other amino acids is carried out using mutants of *Corynebacterium glutamicum* that lack, or have only a limited ability to process, the TCA cycle intermediate α-ketoglutarate to succinyl-CoA (*see appendix II*). A controlled low biotin level and the addition of fatty acid derivatives result in increased membrane permeability and excretion of high concentrations of glutamic acid. The impaired

stability of the newly synthesized penicillin. Once the fermentation is completed, normally in six to seven days, the broth is separated from the fungal mycelium and processed by absorption, precipitation, and crystallization to yield the final product. This basic product can then be modified by chemical procedures to yield a variety of semisynthetic penicillins. >> *Antimicrobial chemotherapy (chapter 31)*

bacteria use the glyoxylate pathway to meet their needs for essential biochemical intermediates, especially during the growth phase. After growth becomes limited because of changed nutrient availability, an almost complete molar conversion (or 81.7% weight conversion) of isocitrate to glutamate occurs. ◀◀ *Synthesis of amino acids (section 11.5)*

Organic Acids

Citric, acetic, lactic, fumaric, and gluconic acids are major products. Until microbial processes were developed, the major source of citric acid was citrus fruit. Organic acid production by microorganisms is important in industrial microbiology and illustrates the effects of trace metal levels on organic acid synthesis and excretion.

The essence of citric acid fermentation involves limiting the amounts of trace metals such as manganese and iron to stop *Aspergillus niger* growth at a specific point in the fermentation. The medium often is treated with ion exchange resins to ensure low and controlled concentrations of available metals and is carried out in aerobic stirred fermenters. Generally, high sugar concentrations (15 to 18%) are used, and copper has been found to counteract the inhibition of citric acid production by iron above 0.2 ppm. The success of this fermentation depends on the regulation and functioning of the glycolytic pathway and the tricarboxylic acid cycle. After the active growth phase, when the substrate level is high, citrate synthase activity increases and the activities of aconitase and isocitrate dehydrogenase decrease. This results in citric acid accumulation and excretion by the stressed microorganism.

Biopolymers

Biopolymers are microbially produced polymers, primarily polysaccharides, used to modify the flow characteristics of liquids and to serve as gelling agents. These are employed in many areas of the pharmaceutical and food industries.

Biopolymers include (1) dextrans, which are used as blood expanders and absorbents; (2) *Erwinia* polysaccharides used in paints; (3) polyesters, derived from *Pseudomonas oleovorans,* which are a feedstock for specialty plastics; (4) cellulose microfibrils, produced by an *Acetobacter* strain, that are used as a food thickener; (5) polysaccharides such as scleroglucan that are used by the oil industry as drilling mud additives; and (6) xanthan polymers, which enhance oil recovery by improving water flooding and the displacement of oil. This use of xanthan gum, produced by *Xanthomonas campestris,* represents a large potential market for this microbial product.

Biosurfactants

Biosurfactants are amphiphilic molecules that possess both hydrophobic and hydrophilic regions; thus they partition at the interface between fluids that differ in polarity, such as oil and water. For this reason, they are used for emulsification, increasing detergency, wetting and phase dispersion, as well as solubilization. These properties are especially important in bioremediation, oil spill dispersion, and enhanced oil recovery. The most widely used microbially produced biosurfactants are glycolipids. These are carbohydrates that bear long-chain aliphatic acids or hydroxy aliphatic acids. They can be isolated as extracellular products from a variety of microorganisms, including pseudomonads and yeasts.

Many biosurfactants also have antibacterial and antifungal activity. The amphiphilic nature of these molecules promotes their ability to disrupt plasma membranes. Some biosurfactants even inactivate enveloped viruses. In addition, their capacity to prevent microbial adhesion to sites of invasion has generated interest in their use as potential protective agents. Although these and other medical applications of biosurfactants are still being investigated, these molecules hold promise for the future. They can be genetically engineered with relative ease, and the continued rise in resistance to antimicrobial agents has given rise to a tremendous need for new products.

Biofuels

The need to become independent of fossil fuels is driven by both political and environmental concerns. This has accelerated interest in and use of biofuels—fuel (chiefly ethanol) that is obtained by the fermentation of plant material. While corn is currently the substrate of choice, the use of crop residues could significantly boost biofuel yields. Crop residues are the plant material that is usually left in the field after harvest, and it consists of cellulose and hemicellulose. These polysaccharides are polymers of five different hexoses and pentoses: glucose, xylose, mannose, galactose, and arabinose. While no microorganism naturally ferments all five sugars, a *Saccharomyces cerevisiae* strain has been engineered to ferment xylose and an *E. coli* strain that expresses *Zymomonas mobilis* genes is able to ferment all these sugars. Another area of research focuses on degrading the cellulose and hemicellulose to release these monomers. This is commonly done by heating the plant material and treating it with acid, which is both expensive and corrosive. Work to harvest cellulase- and hemicellulase-producing fungi as well as bioprospecting for enzymes from thermoacidophiles are ongoing in an effort to replace the harsh thermochemical approach with a biological treatment.

Bioconversion Processes

Bioconversions, also known as **microbial transformations** or **biotransformations,** are minor changes in molecules, such as the insertion of a hydroxyl or keto function, or the saturation or desaturation of a complex cyclic structure, that are carried out by nongrowing microorganisms. The microorganisms thus act as biocatalysts. Bioconversions have many advantages over chemical procedures. A major advantage is stereochemical; the biologically active form of a product is made. Enzymes also carry out very specific reactions under mild conditions, and larger

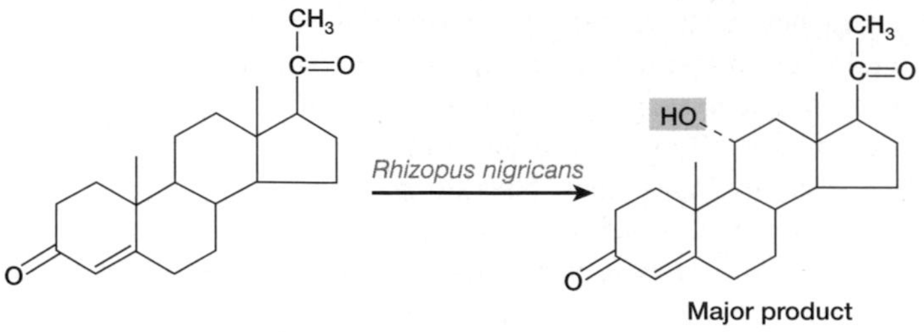

Figure 16.19   Biotransformation to Modify a Steroid.   Hydroxylation of progesterone in the 11α position by *Rhizopus nigricans*. The steroid is dissolved in acetone before addition to the pregrown fungal culture.

water-insoluble molecules can be transformed. Unicellular bacteria, actinomycetes, yeasts, and molds have been used in various bioconversions. The enzymes responsible for these conversions can be intracellular or extracellular. Cells can be produced in batch or continuous culture and then dried for direct use, or they can be prepared in more specific ways to carry out desired bioconversions.

A typical bioconversion is the hydroxylation of a steroid (**figure 16.19**). In this example, the water-insoluble steroid is dissolved in acetone and then added to the reaction system that contains the pregrown microbial cells. The course of the modification is monitored, and the final, soluble product is extracted from the medium and purified.

1. What critical limiting factors are used in the fermentation of penicillin? Why do you think the microbe must be stressed to produce high levels of drug?

2. What are regulatory mutants and how were they used to increase the production of glutamic acid by *Corynebacterium*?

3. What is the principal limitation created to stimulate citric acid accumulation by *Aspergillus niger*? In this case, is citric acid a primary or secondary metabolite?

4. Discuss the major uses for biopolymers and biosurfactants.

5. What are bioconversions or biotransformations? What kinds of molecular changes result from these processes?

16.11   RECOMBINANT DNA TECHNOLOGY IN AGRICULTURE

Cloned genes can be inserted into plant as well as animal cells. A plasmid from the plant pathogenic bacterium *Agrobacterium tumefaciens* is chiefly responsible for the successful genetic engineering of plants. In natural situations, infection of dicotyledonous (dicot) plant cells by the bacterium transforms them into tumor cells, and crown gall disease develops. The tumor forms because of the insertion of bacterial genes into the plant cell genome, and only strains of *A. tumefaciens* possessing a large conjugative plasmid called the Ti (*tumor inducing*) plasmid are pathogenic (*see figure 26.27*). This is because the Ti plasmid, which carries genes for virulence and the synthesis of substances involved in the regulation of plant growth, is transferred to the plant cell. The genes that induce tumor formation reside between two 23-base pair direct-repeat sequences. This region is known as T-DNA and is very similar to a transposon. The Ti plasmid bears genes for the synthesis of plant growth hormones (an auxin and a cytokinin) and an amino acid derivative called opine that serves as a nutrient source for the invading bacteria. In diseased plant cells, T-DNA is inserted into the chromosomes at various sites and is stably maintained in the cell nucleus. ◄◄ *Transposable elements (section 14.5);* ►► *Microorganism associations with vascular plants:* Agrobacterium (*section 26.2*)

When the molecular nature of crown gall disease was recognized, it became clear that the Ti plasmid and its T-DNA had great potential as a vector for the insertion of recombinant DNA into plant chromosomes. The Ti plasmid has been genetically engineered to improve its utility as a cloning vector. Nonessential regions, including the tumor-inducing genes, have been deleted, while elements found in other vectors (e.g., selectable marker and multicloning site; figure 16.10) have been added. Genes required for the actual infection of the plant cell by the plasmid are retained. The gene or genes of interest are inserted into the T-DNA region between the direct repeats needed for integration into the plant genome (**figure 16.20**). Then the plasmid is returned to *A. tumefaciens,* plant culture cells are infected with the bacterium, and transformants are selected by screening for antibiotic resistance (or another trait encoded by T-DNA). Finally, whole plants are regenerated from the transformed cells.

While the *A. tumefaciens* Ti plasmid can be used to modify dicots such as potato, tomato, celery, lettuce, and alfalfa, it cannot be transferred naturally to monocots such as corn, wheat, and other grains. However, genetic modification of the Ti plasmid so that it can be productively transferred to monocots is an active area of research. The creation of new procedures for inserting DNA into plant cells may well lead to the use of recombinant DNA techniques with many important crop plants.

Another application of recombinant DNA technology to agriculture involves the use of bacteria, fungi, and viruses as **bioinsecticides** and **biopesticides.** These are defined as biological agents, such as microbes or their components, which can be used to kill susceptible insects. Bacterial agents include a variety of *Bacillus* species; however, *B. thuringiensis* is most widely used. This bacterium is only weakly toxic to insects as a vegetative cell, but during sporulation, it produces an intracellular protein toxin crystal, the parasporal body, that acts as a microbial insecticide for specific insect groups. The parasporal crystal, after

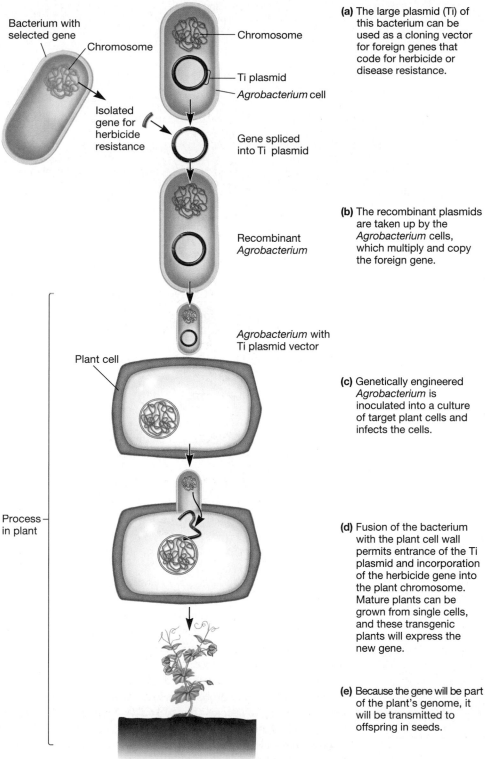

(a) The large plasmid (Ti) of this bacterium can be used as a cloning vector for foreign genes that code for herbicide or disease resistance.

(b) The recombinant plasmids are taken up by the *Agrobacterium* cells, which multiply and copy the foreign gene.

(c) Genetically engineered *Agrobacterium* is inoculated into a culture of target plant cells and infects the cells.

(d) Fusion of the bacterium with the plant cell wall permits entrance of the Ti plasmid and incorporation of the herbicide gene into the plant chromosome. Mature plants can be grown from single cells, and these transgenic plants will express the new gene.

(e) Because the gene will be part of the plant's genome, it will be transmitted to offspring in seeds.

Figure 16.20   Bioengineering of Plants.   Most techniques employ a genetically modified strain of the natural tumor-producing bacterium called *Agrobacterium tumefaciens*.

form a hexagonal-shaped pore through the midgut cell, as shown in figure 16.21*d*. This leads to the loss of osmotic balance and ATP, and finally to cell lysis. **>>** *Class* Bacilli: *Order* Bacillales *(section 21.4)*

B. thuringiensis has become an important industrial microbe. The toxin is a widely used insecticide known as **Bt,** and the toxin genes can be used to genetically modify plants. The manufacture of Bt begins by the growth of *B. thuringiensis* in fermenters. The spores and crystals are released into the medium when cells lyse. The medium is then centrifuged and converted to a dust or wettable powder for application to plants. This insecticide has been used on a worldwide basis for over 40 years. Unlike chemical insecticides, Bt does not accumulate in the soil or in nontarget animals. Rather it is readily lost from the environment by microbial and abiotic degradation.

In the United States, genetically modified (GM) crops are common. Genes that confer resistance to herbicides and insects have been introduced into a variety of crops, including soybeans, cotton, and corn. Soybeans have also been engineered to have a lower saturated fat content. Other examples of genetically engineered commercial crops are canola, potato, squash, and tomato. Unlike other genetically modified organisms (GMOs), transgenic plants expressing the *B. thuringiensis* toxin gene, *cry* (for *cry*stal), have generally been well accepted. In 1996 commercialized Bt-corn, Bt-potato, and Bt-cotton were introduced into the United States, and soon farmers in other countries such as Australia, Canada, China, France, Indonesia, Mexico, Spain, and Ukraine followed. The widespread acceptance of these plants reflects the history of safe application of Bt as an insecticide without adverse environmental or health impacts. In addition, it is well understood that the Cry protein can only be activated in the target insect. Long-term studies have shown that Bt is nontoxic to mammals and is not an allergen in humans. One potential problem, the horizontal gene flow of the *cry* gene to weeds and other plants, has not been reliably demonstrated.

exposure to alkaline conditions in the insect hindgut, fragments to release protoxin. After this reacts with a protease enzyme, the active toxin is produced (**figure 16.21**). Six of the active toxin units integrate into the plasma membrane (figure 16.21*b,c*) to

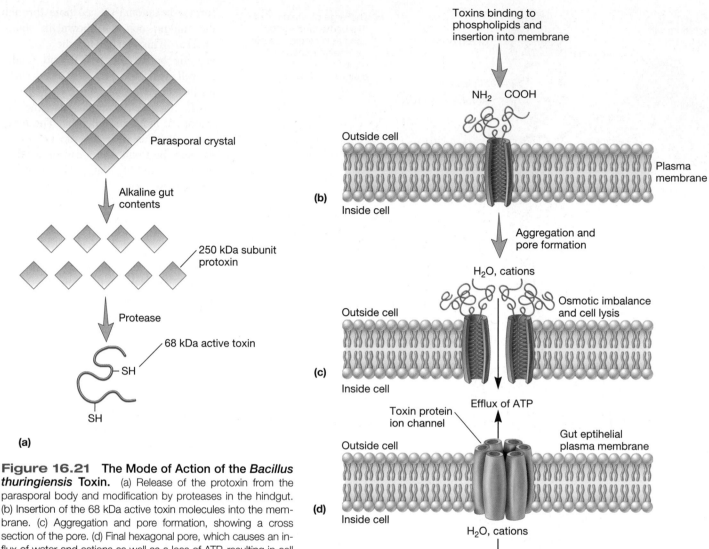

Figure 16.21 The Mode of Action of the *Bacillus thuringiensis* Toxin. (a) Release of the protoxin from the parasporal body and modification by proteases in the hindgut. (b) Insertion of the 68 kDa active toxin molecules into the membrane. (c) Aggregation and pore formation, showing a cross section of the pore. (d) Final hexagonal pore, which causes an influx of water and cations as well as a loss of ATP, resulting in cell imbalance and lysis.

1. What is the Ti plasmid and how is modified for the genetic modification of plants?

2. How is the Bt toxin produced and why is it so widely accepted?

3. Can you think of other traits that might be useful, either to the farmer or the consumer, that could be introduced into plants?

16.12 MICROBES AS PRODUCTS

So far, we have discussed the use of microbial products to meet defined goals. However, microbial cells can be marketed as valuable products. Perhaps the most common example is the inoculation

of legume seeds with rhizobia to ensure efficient nodulation and nitrogen fixation, as discussed in chapter 26. Here we introduce several other microbes and microbial structures that are of industrial or agricultural relevance.

Diatoms have aroused the interest of nanotechnologists. These photosynthetic protists produce intricate silica shells that differ according to species (**figure 16.22**). Nanotechnologists are interested in diatoms because they create precise structures at the micrometer scale. Three-dimensional structures in nanotechnology are currently built plane by plane, and meticulous care must be taken to etch each individual structure to its final, exact shape. Diatoms, on the other hand, build directly in three dimensions and do so while growing exponentially. There have been a number of ideas and approaches to harness these microbial "factories," but one technique is especially fascinating. Diatoms shells are incubated at 900°C in an atmosphere of magnesium for several hours. Amazingly, this results in an atom-for-atom substitution of

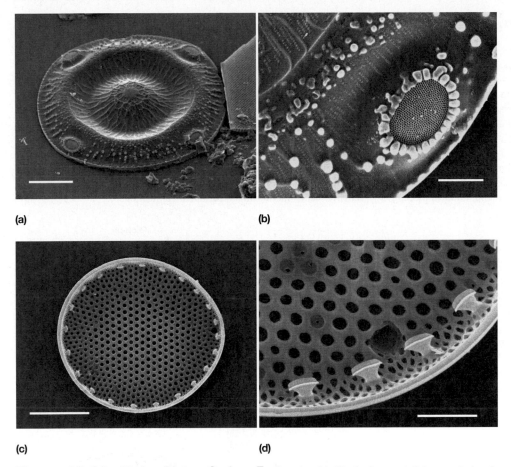

(a)

(b)

(c)

(d)

Figure 16.22   Marine Diatom Surface Features.   (a) *Glyphodiscus stellatus;* scale bar is 20 μm. (b) *G. stellatus* close-up; scale bar is 5 μm. (c) *Roperia tesselata;* scale bar is10 μm. (d) *R. tesselata* close-up; scale bar is 5 μm.

currents (**figure 16.23**). Biosensors are used to measure specific components in beer, to monitor pollutants, to detect flavor compounds in food, and to study environmental processes such as changes in biofilm concentration gradients. It is possible to measure the concentration of substances from many different environments. Recently biosensors have been developed using immunochemical-based detection systems. These biosensors detect pathogens, herbicides, toxins, proteins, and DNA. Since the bioterrorism attacks of 2001, the U.S. government has stepped up funding for research and development of biosensors capable of detecting minute levels of potential airborne pathogens. Many of these biosensors are based on the use of a streptavidin-biotin recognition system (**Techniques & Applications 16.2**).

1. Why are diatoms and magnetotactic bacteria of interest to nanotechnologists? What do you think might be special challenges when growing these microbes for industrial applications?

2. What are biosensors and how do they detect substances?

3. In what areas are biosensors being used to assist in chemical and biological monitoring efforts?

silicon with magnesium without loss of 3D structure. Thus silicon oxide, which is of little use in nanotechnology, is converted to highly useful magnesium oxide. **>>** *Protist classification: Stramenopiles (section 23.2)*

Magnetotactic bacteria are also of interest to nanotechnologists. Magnetosomes are formed by certain bacteria that convert iron to magnetite. The sizes and shapes of the magnetosomes differ among species, but like diatom shells, they are perfectly formed despite the fact that they are only tens of nanometers in diameter. Although tiny, magnetic beads can be chemically synthesized, they are not as precisely formed and lack the membrane that surrounds magnetosomes. This membrane enables the attachment of useful biological molecules such as enzymes and antibodies. Potential applications for magnetosomes currently under investigation include their use as a contrast medium to improve magnetic resonance tomography (MRI) and as biological probes to detect cancer at early stages. **<<** *Procaryotic cytoplasm: Inclusion bodies (section 3.3)*

Another rapidly developing area of biotechnology is that of **biosensor** production. In this field of bioelectronics, living microorganisms (or their enzymes or organelles) are linked with electrodes, and biological reactions are converted into electrical

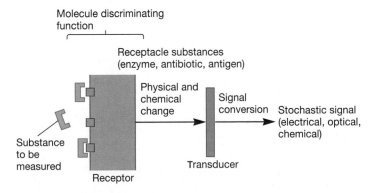

Figure 16.23   Biosensor Design.   Biosensors are finding increasing applications in medicine, industrial microbiology, and environmental monitoring. In a biosensor, a biomolecule or whole microorganism carries out a biological reaction, and the reaction products are used to produce an electrical signal.

16.2   Streptavidin-Biotin Binding and Biotechnology

Egg white contains many proteins and glycoproteins with unique properties. One of the most interesting, which binds tenaciously to biotin, was isolated in 1963. This glycoprotein, called avidin due to its "avid" binding of biotin, was suggested to play an important role: making egg white antimicrobial by "tying up" the biotin needed by many microorganisms. Avidin, which functions best under alkaline conditions, has the highest known binding affinity between a protein and a ligand. Several years later, scientists at Merck & Co., Inc., discovered a similar protein produced by the actinomycete *Streptomyces avidini,* which binds biotin at a neutral pH and does not contain carbohydrates. These characteristics make this protein, called streptavidin, an ideal binding

agent for biotin, and it has been used in an almost unlimited range of applications, as shown in the **box figure**. The streptavidin protein is joined to a probe. When a sample is incubated with the biotinylated binder, the binder attaches to any available target molecules. The presence and location of target molecules can be determined by treating the sample with a streptavidin probe because the streptavidin binds to the biotin on the biotinylated binder, and the probe is then visualized. This detection system is employed in a wide variety of biotechnological applications, including use as a nonradioactive probe in hybridization studies and as a critical component in biosensors for a wide range of environmental monitoring and clinical applications.

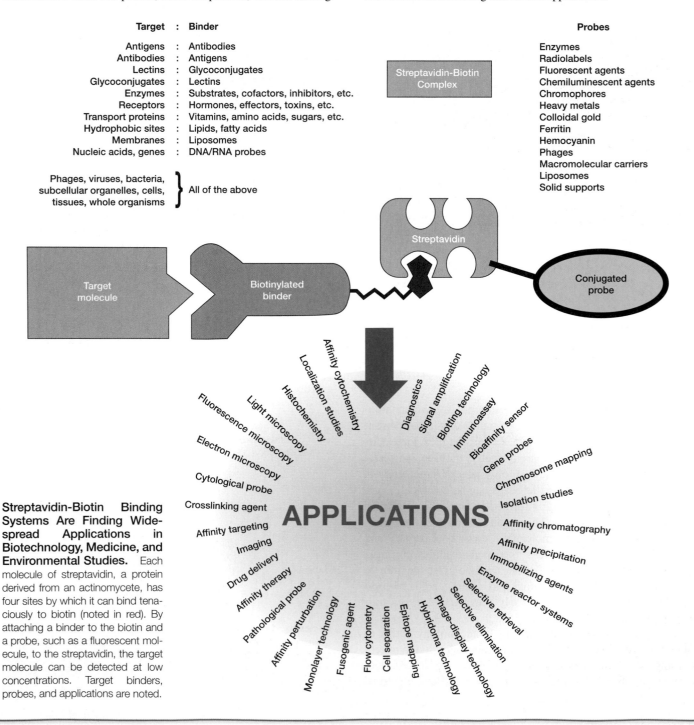

Streptavidin-Biotin Binding Systems Are Finding Widespread Applications in Biotechnology, Medicine, and Environmental Studies. Each molecule of streptavidin, a protein derived from an actinomycete, has four sites by which it can bind tenaciously to biotin (noted in red). By attaching a binder to the biotin and a probe, such as a fluorescent molecule, to the streptavidin, the target molecule can be detected at low concentrations. Target binders, probes, and applications are noted.

Summary

16.1 Key Developments in Recombinant DNA Technology

a. Genetic engineering became possible after the discovery of restriction enzymes and reverse transcriptase, and the development of essential methods in nucleic acid chemistry such as the Southern blotting technique.

b. Restriction enzymes are important because they cut DNA as specific sequences, thereby releasing fragments of DNA that can be cloned or otherwise manipulated (**figure 16.3** and **table 16.2**).

16.2 The Polymerase Chain Reaction

a. The polymerase chain reaction allows small amounts of specific DNA sequences to be increased in concentration thousands of times (**figure 16.7**).

16.3 Gel Electrophoresis

a. Gel electrophoresis is used to separate molecules according to charge and size.

b. DNA fragments are separated on agarose and acrylamide gels. Because DNA is acidic, it migrates from the negative to the positive end of a gel (**figure 16.9**).

16.4 Cloning Vectors and Creating Recombinant DNA

a. There are four types of cloning vectors: plasmids, phages and viruses, cosmids, and artificial chromosomes. Cloning vectors generally have at least three components: an origin of replication, a selectable marker, and a multicloning site or polylinker (**table 16.3; figures 16.10** and **16.12**).

b. The most common approach to cloning is to digest both vector and DNA to be inserted with the same restriction enzyme or enzymes so that compatible sticky ends are generated. The vector and DNA to be cloned are then incubated in vitro in the presence of DNA ligase, which catalyzes the formation of phosphodiester bonds once the DNA fragment inserts into the vector.

c. Once the recombinant plasmid has been introduced into host cells, cells carrying vector must be selected. This is often accomplished by allowing the growth of only antibiotic-resistant cells because the vector bears an antibiotic-resistance gene. Cells that took up vector with inserted DNA can be distinguished from those that contain only vector in several ways. Often a blue versus white colony phenotype is used; this is based on the presence or absence of a functional *lacZ* gene, respectively (**figure 16.11**).

16.5 Construction of Genomic Libraries

a. It is sometimes necessary to find a gene without the knowledge of the gene's DNA sequence. A genomic library is constructed by cleaving an organism's genome into many fragments, each of which is cloned into a vector to make a unique recombinant plasmid.

b. Genomic libraries can be screened for the gene of interest by either phenotypic rescue (genetic complementation) or DNA hybridization with an oligonucleotide probe. However, there are instances when a novel approach to screening the library for a specific gene must be devised (**figure 16.13**).

16.6 Inserting Recombinant DNA into Host Cells

a. The bacterium *E. coli* and the yeast *S. cerevisiae* are the most common host species.

b. DNA can be introduced into microbes by transformation or electroporation.

16.7 Expressing Foreign Genes in Host Cells

a. An expression vector has the necessary features to express any recombinant gene it carries.

b. If a eucaryotic gene is to be expressed, cDNA is used because it lack introns.

c. Many useful products have been synthesized using recombinant DNA technology (**table 16.4**).

16.8 Microorganisms Used in Industrial Microbiology

a. Industrial microbiology has been used to manufacture such products as antibiotics, amino acids, and organic acids, and has had many important positive effects on animal and human health.

b. Historically, work in this area has been carried out using microorganisms isolated from nature or modified by the use of classic mutation techniques. Biotechnology involves the use of molecular techniques to modify and improve microorganisms.

c. Finding new microorganisms in nature for use in biotechnology is a continuing challenge. Only about 1% of the observable microbial community has been grown, but major advances in growing uncultured microbes are being made.

d. Selection and mutation continue to be important approaches for identifying new microorganisms (**figure 16.15**).

e. Site-directed mutagenesis and protein engineering are used to modify genes and their expression. These approaches are leading to new and often different products with new properties.

f. Protein evolution is of increasing interest. This involves exploiting microbial responses to stress in adaptive mutation and forced evolution, with the hope of identifying microorganisms with new properties. Alternatively, an in vitro approach can be used.

16.9 Microorganism Growth in Controlled Environments

a. Microorganisms can be grown in controlled environments of various types using fermenters and other culture systems. If defined constituents are used, growth parameters can be chosen and varied in the course of growing a microorganism. This approach is used particularly for the production of amino acids, organic acids, and antibiotics.

16.10 Major Products of Industrial Microbiology

a. A wide variety of compounds are produced in industrial microbiology that impact our lives in many ways. These include antibiotics, amino acids, organic acids, biopolymers, and biosurfactants (**tables 16.4–16.6**).

b. Microorganisms also can be used as biocatalysts to carry out specific chemical reactions **(figure 16.19).**

16.11 Recombinant DNA Technology in Agriculture

a. Many plants are genetically modified through the use of the Ti plasmid of *Agrobacterium tumefaciens.* The plasmid can be engineered to introduce genes whose products confer new traits to plants **(figure 16.20).**

b. *Bacillus thuringiensis* is an important biopesticide, and the Bt gene has been incorporated into several important crop plants **(figure 16.21).**

16.12 Microbes as Products

a. Microorganisms are being used in a wide range of biotechnological applications such as nanotechnology and biosensors **(figures 16.22 and 16.23).**

Critical Thinking Questions

1. Could the Southern blotting technique be applied to RNA? If so, how might this be done?

2. Initial attempts to perform PCR were carried out using the DNA polymerase from *E. coli.* What was the major difficulty?

3. What advantage might there be in creating a genomic library first rather than directly isolating the desired DNA fragment?

4. You have cloned a structural gene required for riboflavin synthesis in *E. coli.* You find that an *E. coli* riboflavin auxotroph carrying the cloned gene on a vector makes less riboflavin than does the wild-type strain. Why might this be the case?

5. Suppose that you inserted a plasmid vector carrying a human interferon gene into *E. coli,* but none of the transformed bacteria produced interferon. Give as many plausible reasons as possible for this result.

6. What further technological approaches will be required to culture microorganisms that have not yet been grown? Consider the roles of nutrient fluxes, communication molecules, and competition.

Learn More

Learn more by visiting the Prescott website at www.mhhe.com/prescottprinciples, where you will find a complete list of references.

Microbial Evolution, Taxonomy, and Diversity

The stromatolites shown here are layered rocks formed by incorporation of minerals into microbial mats. Fossilized stromatolites indicate that microorganisms existed early in Earth's history.

endosymbiotic hypothesis The hypothesis that mitochondria and chloroplasts arose when bacteria established an endosymbiotic relationship with ancestral cells and then evolved into organelles.

G + C content The percent of an organism's genome that consists of guanine and cytosine (G + C); a parameter that is used in taxonomic analysis.

genotypic classification The use of genetic data to construct a classification scheme for the identification of an unknown species or the phylogeny of a group of microbes.

melting temperature (T_m) The temperature at which double-stranded DNA becomes single stranded; determined by the G + C content of the genome.

natural classification A classification system that arranges organisms into groups whose members share many characteristics.

Chapter Glossary

oligonucleotide signature sequences Short, conserved nucleotide sequences that are specific for a phylogenetically defined group of organisms.

phenetic system A classification system that groups organisms together based on the similarity of their observable characteristics.

phylogenetic or phyletic classification systems A classification system based on evolutionary relationships rather than the general similarity of characteristics.

phylogenetic tree A graph made of nodes and branches, much like a tree in shape, that shows phylogenetic and evolutionary relationships between groups of organisms.

polyphasic taxonomy An approach in which taxonomic schemes are developed using a wide range of phenotypic and genotypic information.

RNA world The hypothesis that the first replicating entity was an RNA molecule.

strain Microbes that are descendants of a single, pure microbial culture. A single species may have many strains.

systematics The scientific study of organisms with the ultimate objective of characterizing and arranging them in an orderly manner; often considered synonymous with taxonomy.

taxon A group into which related organisms are classified.

taxonomy The science of biological classification; it consists of three parts: classification, nomenclature, and identification.

universal phylogenetic tree A phylogenetic tree that considers the evolutionary relationships among organisms from all three domains of life: *Bacteria, Archaea,* and *Eucarya.*

What's in a name? That which we call a rose by any other name would smell as sweet. . . .

—W. Shakespeare

Perhaps the most fascinating aspect of the microbial world is its extraordinary diversity. It seems that almost every possible shape, size, physiology, and life-style can be found. In this section of the text, we focus on microbial diversity. Chapter 17 introduces the general principles of microbial evolution and taxonomy. This is followed by a five-chapter (18–22) survey of the most important procaryotic groups. Our survey of microbial diversity then moves to an introduction of eucaryotic microorganisms and finally that of viruses (chapters 23 and 24, respectively).

17.1   MICROBIAL EVOLUTION

Biological diversity is usually thought of in terms of plants and animals; yet the variety of microbial life forms is huge and largely unexplored. Consider the metabolic diversity of microorganisms—this alone suggests that the number of habitats occupied by microbes vastly exceeds that of all larger organisms. How has microbial life been able to radiate to such an astonishing level of diversity? To answer this question, one must consider microbial evolution. The field of microbial evolution, like any other scientific endeavor, is based on the formulation of hypotheses, the gathering and analysis of data, and the reformation of hypotheses based on newly acquired evidence. That is to say, the study of microbial evolution is based on the scientific method. To be sure, it is sometimes more difficult to amass evidence when considering events that occurred millions, and often billions, of years ago, but the advent of molecular biology has offered scientists a living record of life's ancient history. This chapter describes the outcome of this scientific research.

The Origin of Life

Dating meteorites through the use of radioisotopes places our planet at an estimated 4.5 to 4.6 billion years old. However, conditions on Earth for the first 100 million years or so were far too harsh to sustain any type of life. The first direct evidence of cellular life was discovered in 1977 in a geologic formation in South Africa known as the Swartkoppie chert, a granular type of silica. These microbial fossils as well as those from the Archaean Apex chert of Australia have been dated at about 3.5 billion years old (**figure 17.1**). Despite these findings, the microbial fossil record is understandably sparse. Thus to piece together the very early events that led to the origin of life, biologists must rely primarily on indirect evidence. Each piece of evidence must fit together like a jigsaw puzzle for a coherent picture to emerge.

The First Self-Replicating Entity: The RNA World

The origin of life rests on a single question: How did early cells arise? No one can say for certain; however, it seems likely that the first self-replicating entity was much simpler than even the most primitive modern, living cells. Before there was life, Earth was a very different place: hot and anoxic, with an atmosphere rich in gases such as hydrogen, methane, carbon dioxide, nitrogen, and ammonia. Earth's surface was a prebiotic soup in which chemicals reacted with one another, randomly "testing" the stability of the resulting molecules. To account for the evolution of life, one must consider which of three essential cellular molecules—DNA, RNA, or protein—developed first and thereby holds the key to understanding all that followed. For life to evolve, a molecule was needed that could both replicate and perform work. Proteins are capable of performing cellular work but cannot replicate, while just the opposite is true of DNA. A

possible solution to this problem was suggested in 1981 when Thomas Cech discovered self-splicing RNA in the protist *Tetrahymena*. Three years later, Sidney Altman found that RNaseP in *Escherichia coli* is an RNA molecule that cleaves phosphodiester bonds. RNA molecules that possess catalytic activity are called **ribozymes,** and to some, the ability of RNA to catalyze biochemical reactions suggests a precellular **"RNA world,"** a term coined by Walter Gilbert in 1986. This hypothesis posits that the first self-replicating molecule was RNA, which is capable of storing, copying, and expressing genetic information, and possesses enzymatic activity as well. In this version of early life, various forms of molecules were assembled and destroyed over roughly half a billion years, until ultimately an entity, or **probiont,** something like modern RNA enclosed in a lipid vesicle was generated (**figure 17.2**). ◀◀ *Ribozymes (section 9.8)*

An additional requirement for early cellular life was the development of a lipid membrane. It is well known that amphipathic lipids (i.e., polar at one end and nonpolar at the other) will spontaneously form liposomes—vesicles bounded by a lipid bilayer. A fascinating experiment performed by Marin Hanczyc, Shelly Fujikawa, and Jack Szostak in 2003 showed that clay triggers the formation of liposomes that actually grow and divide. Considering this together with the data on RNA synthesis and activity, it seems plausible that probionts may have been liposomes containing RNA molecules.

Apart from its ability to replicate and perform enzymatic activities, the function of RNA suggests its ancient origin. Consider that much of the cellular pool of RNA in modern cells exists in the ribosome, a structure that consists largely of rRNA and uses mRNA and tRNA to construct proteins. In fact, rRNA itself catalyzes peptide bond formation during protein synthesis. Thus RNA seems to be well poised for its importance in the development of proteins. Because RNA and DNA are structurally similar, RNA could have given rise to double-stranded DNA. It is suggested that once DNA evolved, it became the storage facility for cellular functions because it provided a more chemically stable structure. Two other pieces of evidence support the RNA world hypothesis: the fact that the energy currency of the cell, ATP, is a ribonucleotide, and the more recent discovery that RNA can regulate gene expression. So it would seem that proteins, genes, and cellular energy can be traced back to RNA. ◀◀ *Riboswitches (sections 13.3 and 13.4)*

However, some are skeptical of the RNA world hypothesis. They claim conditions on Earth 4 billion years ago would have prevented the stable formation of ribose, phosphate, purines, and pyrimidines—all needed to construct RNA. In fact, while purine bases have been generated abiotically in a heated mixture of hydrogen cyanide and ammonia, scientists have so far been unable to synthesize pyrimidines. Another problem with the RNA world hypothesis is the instability of RNA. In 1996 James Ferris and colleagues were able to overcome the problem of RNA degradation by adding the clay mineral montmorillonite to a solution of chemically charged nucleotides. They showed that the rate of RNA synthesis was faster than its degradation. Later it was shown

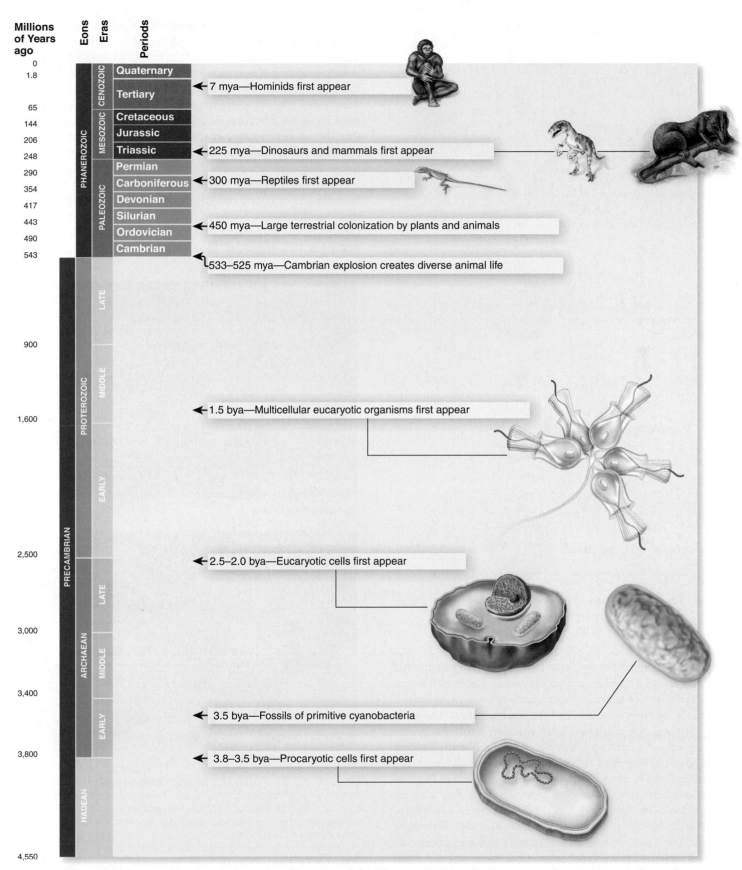

Figure 17.1 The Geological Timescale and an Overview of the History of Life on Earth.   mya = million years ago; bya = billion years ago.

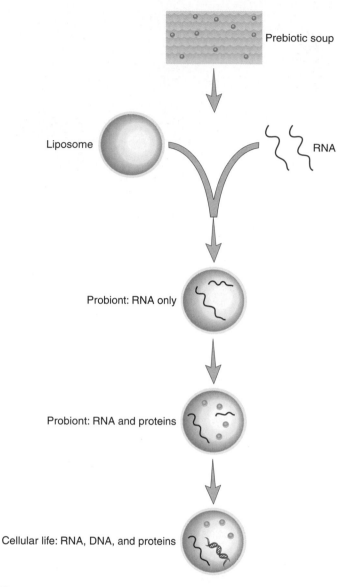

Figure 17.2 The RNA World Hypothesis for the Origin of Life.

Early cellular life, although primitive compared to modern life, was still relatively complex. Cells had to derive energy from a harsh, anoxic environment. When scientists attempt to reconstruct the nature of very ancient life, they look to extant (living) microbes for clues. For instance, it is thought that the FeS-based metabolism seen in some hyperthermophilic archaea may be a remnant of the first form of chemiosmosis. Here it is suggested that the energy-yielding reaction $FeS + H_2S \rightarrow FeS_2 + H_2$ provided the reducing power (H_2) to produce a proton motive force. Photosynthesis also appears to have evolved early in Earth's history. Fossil evidence places the evolution of cyanobacteria and oxygenic photosynthesis at about 3.5 billion years ago. **Stromatolites** are layered rocks, often domed, that are formed by the incorporation of mineral sediments into microbial mats dominated by cyanobacteria (**figure 17.3**). Recent

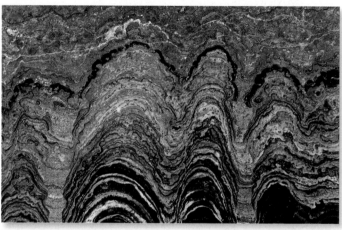

(a) Fossil stromatolite

(b) Modern stromatolites

Figure 17.3 Fossil and Modern Stromatolites: Evidence of Autotrophic Cyanobacteria. Each stromatolite is a rocklike structure, typically 1 m in diameter. (a) Section of a fossilized stromatolite. These layers are mats of mineralized cyanobacteria, one layer on top of the other. The existence of fossil stromatolites provides evidence of early autotrophic organisms, which produced organic molecules and oxygen near the beginning of the history of life on Earth. (b) Modern stromatolites in western Australia.

that amino acids in solution with the minerals hydroxyapatite and illite could also polymerize into polypeptides of about 50 amino acid residues. To some, these experiments provide experimental evidence for the biosynthesis of early organic polymers. However, one additional problem with the RNA world hypothesis concerns the ability of early RNA to self-replicate. Recall that in modern cells, RNA is synthesized by the enzyme RNA polymerase—a protein. Replication of early RNA without a protein was presumably accomplished by an ancient ribozyme. So far, no such ribozyme has been found, nor has it been generated experimentally. Thus although it is clear that microbial life ultimately emerged from a random mixture of chemicals, the actual mechanism by which the first cell-like entity arose is a controversy that may never be resolved.

evidence has shown that some fossilized stromatolites formed in a similar fashion. The presence of oxygen was critical because it enabled the evolution of a wider variety of energy-capturing strategies, including aerobic respiration. << *Electron transport and oxidative phosphorylation (section 10.5); Anaerobic respiration (section 10.6)*

Ironically the study of the most ancient organisms is one of the youngest disciplines in the biological sciences. The ability to culture and examine microorganisms was developed only about 150 years ago. Almost immediately, early microbiologists attempted to classify microbes and organize them according to possible relationships to one another. Two important elements not understood until the late twentieth century made this especially difficult. First, only about 1% of all microbes have been cultured in the laboratory. Second, the most accurate assessment of evolutionary relationships between organisms is obtained by comparing nucleotide and amino acid sequences. Prior to the advent of sequence-based techniques, it was impossible to discern evolutionary relationships among microorganisms.

The Three Domains of Life

The most important sequence-based investigation was initiated in 1977 by Carl Woese and his collaborator George Fox. They compared the **small subunit ribosomal RNAs (SSU rRNAs)** from a variety of organisms to determine that all living organisms belong to one of three domains: *Archaea, Bacteria,* and *Eucarya,* thereby separating the procaryotes or "Monera" into two groups. This initial observation has been further refined and substantiated by additional biochemical and genetic evidence. Recall from chapter 3 that most bacteria have cell walls with peptidoglycan and membrane lipids with ester-linked, straight-chained fatty acids that resemble eucaryotic membrane lipids (**table 17.1**). The *Archaea* differ from the *Bacteria* in many respects and resemble the *Eucarya* in some ways. Although the *Archaea* are described more fully in chapter 18, here we note that they differ from the *Bacteria* in lacking peptidoglycan in their cell walls, although some archaea have a peptidoglycan-like polymer called pseudomurein or pseudopeptidoglycan. They also possess (1) membrane lipids with ether-linked branched aliphatic chains, (2) transfer RNAs without thymidine in the T or TΨC arm, (3) distinctive RNA polymerase enzymes, and (4) ribosomes of different composition and shape. Thus although the *Archaea* and *Bacteria* have a similar cell architecture, they vary considerably at the molecular level. Both groups differ from eucaryotes in their cell ultrastructure and many other properties. However, the *Bacteria* and the *Archaea* share some biochemical properties with eucaryotic cells. For example, *Bacteria* and *Eucarya* have ester-linked membrane lipids; *Archaea* and *Eucarya* are similar with respect to some components of information flow (i.e., DNA, RNA, and protein synthesis).

Several views exist regarding the evolutionary history of microbes. We will first consider the **universal phylogenetic tree** as proposed by Norman Pace (**figure 17.4**). This analysis is based on SSU rRNA sequence analysis of organisms from all three domains of life. Importantly a similar tree can be constructed from the nucleotide sequences of any gene whose product is involved in information flow, as long as that gene is found in all three domains. Although the details of phylogenetic tree construction and the use of SSU rRNAs to measure relatedness are discussed in more detail later (p. 395), the general concept is not difficult to understand. In this case, 16S and 18S rRNA sequences from a diverse collection of procaryotes and eucaryotes, respectively, are aligned from the 5′ end to the 3′ end and homologous residues are compared in a pair-wise fashion. Each nucleotide sequence difference is counted and serves to represent some evolutionary distance between the organisms. When data from a large number of organisms are compared, the evolutionary relatedness between organisms can be determined but not the rate at which one organism diverged from another. Simply stated, rather than measuring time, the branches on such a tree measure the evolutionary distance between organisms. The concept is analogous to a map that accurately shows the distance between two cities but because of many factors (traffic, road conditions, etc.), one cannot determine the time needed to travel that distance. Evolutionary distance is measured along each line: the longer the line, the more evolutionarily diverged are the two organisms (or types of organisms) at each end.

What does the universal phylogenetic tree tell us about the origin of life? Close to the center is a line labeled "Root." This is where the data indicate the last common ancestor to all three domains should be placed (there are no branches here because there is no such extant organism). The root, or origin of modern life, is on the bacterial branch, so it appears that the *Archaea* and the *Eucarya* evolved independently, separate from the *Bacteria.* Following the lines of descent away from the root, toward the *Archaea* and the *Eucarya,* it is evident that they shared common ancestry but diverged and became separate domains. The common evolution of these two forms of life is still evident in the manner in which the *Archaea* and the *Eucarya* process genetic information. For instance, the RNA polymerases of the *Eucarya* and the *Archaea* resemble each other, to the exclusion of *Bacteria.* Thus the universal phylogenetic tree presents a picture in which all life, regardless of eventual domain, arose from a single, common ancestor. One can envision the universal tree of life as a real tree that grows from a single seed.

To be sure, alternative hypotheses exist. Some suggest that the *Bacteria* and the *Eucarya* arose independently and the *Archaea* are a mosaic, having combined traits of the other two. This view is based largely on the observation that while the *Archaea* share large numbers of genes with the *Eucarya,* they have an even larger number of genes in common with the *Bacteria.* Another interpretation suggests that the first eucaryotes arose from two ancient procaryotic ancestors: an archaeon and a bacterium. This notion is further explored here.

Table 17.1	Comparison of *Bacteria, Archaea, and Eucarya*		
Property	*Bacteria*	*Archaea*	*Eucarya*
Membrane-Enclosed Nucleus with Nucleolus	Absent	Absent	Present
Complex Internal Membranous Organelles	Absent	Absent	Present
Cell Wall	Almost always have peptidoglycan containing muramic acid	Variety of types, no muramic acid; some have pseudomurein	No muramic acid
Membrane Lipid	Have ester-linked, straight-chained fatty acids	Have ether-linked, branched aliphatic chains	Have ester-linked, straight-chained fatty acids
Gas Vesicles	Present	Present	Absent
Transfer RNA	Thymine present in most tRNAs *N*-formylmethionine carried by initiator tRNA	No thymine in T or TΨC arm of tRNA Methionine carried by initiator tRNA	Thymine present Methionine carried by initiator tRNA
Polycistronic mRNA	Present	Present	Absent
mRNA Introns	Absent	Absent	Present
mRNA Splicing, Capping, and Poly A Tailing	Absent	Absent	Present
Ribosomes			
Size	70S	70S	80S (cytoplasmic ribosomes)
Elongation factor 2 reaction with diphtheria toxin	Does not react	Reacts	Reacts
Sensitivity to chloramphenicol and kanamycin	Sensitive	Insensitive	Insensitive
Sensitivity to anisomycin	Insensitive	Sensitive	Sensitive
DNA-Dependent RNA Polymerase			
Number of enzymes	One	One	Three
Structure	Simple subunit pattern (6 subunits)	Complex subunit pattern similar to eucaryotic enzymes (8–12 subunits)	Complex subunit pattern (12–14 subunits)
Rifampicin sensitivity	Sensitive	Insensitive	Insensitive
RNA Polymerase II Type Promoters	Absent	Present	Present
Metabolism			
Similar ATPase	No	Yes	Yes
Methanogenesis	Absent	Present	Absent
Nitrogen fixation	Present	Present	Absent
Chlorophyll-based photosynthesis	Present	Absent	Present[a]
Chemolithotrophy	Present	Present	Absent

[a]Present in chloroplasts (of bacterial origin).

As we have seen, genes from both the *Archaea* and the *Bacteria* can be found in eucaryotic chromosomes, but how did they get there? While the universal phylogenetic tree indicates that common genes reflect a single common ancestor, the **genome fusion hypothesis** posits that an archaeon and a bacterium joined to form the first eucaryote (**figure 17.5a**). In this scenario, ancient archaeal cells and primitive gram-negative α-proteobacteria developed a close symbiotic relationship. Eventually the two

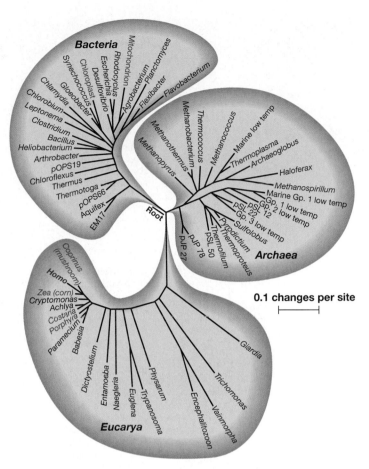

Figure 17.4 **Universal Phylogenetic Tree.** These evolutionary relationships are based on rRNA sequence comparisons. Length of branches indicates evolutionary relationships between organisms but not time. Microbes are printed in black.

cells fused and bacterial genes for metabolism and archaeal genes for replication, transcription, and translation were retained. Alternatively it has been suggested that the bacterium was an endosymbiont of the archaeon (figure 17.5*b*). >> *The* Proteobacteria *(chapter 20)*

The Endosymbiotic Origin of Mitochondria and Chloroplasts

In contrast to the unresolved origin of the nucleus, the **endosymbiotic hypothesis** is generally accepted as the origin of mitochondria and chloroplasts. That endosymbiosis was responsible for the development of these organelles (regardless of the exact mechanism) is supported by the fact that both organelles have bacterial-like ribosomes and most have a single, circular chromosome. Indeed, inspection of figure 17.4 shows that mitochondria and chloroplasts belong to the bacterial lineage. Important evidence for the origin of mitochondria comes from the genome sequence of the α-proteobacterium *Rickettsia*

prowazekii, an obligate intracellular parasite. Its genome is more closely related to that of modern mitochondrial genomes than to any other bacterium. Mitochondria are believed to have descended from an α-proteobacterial ancestor that became engulfed in a precursor cell and provided a function that was essential to the host (figure 17.5). It may be that oxygen toxicity was eliminated because the intracellular bacterium used aerobic respiration to generate ATP. In return, the host provided nutrients and a safe place to live. These bacterial endosymbionts evolved to become mitochondria. This hypothesis also accounts for the evolution of chloroplasts from an endosymbiotic cyanobacterium. Presently the cyanobacterium *Prochloron* has become a favorite candidate as the extant relative of the endosymbiotic cyanobacterium that gave rise to green algae and plant chloroplasts. This microbe lives within marine invertebrates and is the only procaryote to have both chlorophyll *a* and *b* but not phycobilins. This makes *Prochloron* most similar to chloroplasts. >> *Photosynthetic bacteria: Phylum* Cyanobacteria *(section 19.3)*

Recently an additional endosymbiosis theory has been advanced. The **hydrogen hypothesis** asserts that the endosymbiont was an anaerobic α-proteobacterium that produced H_2 and CO_2 as end products of fermentation. In the absence of an external H_2 source, the host became dependent on the bacterium, which made ATP by substrate-level phosphorylation. Ultimately the endosymbiont evolved into one of two organelles. If the endosymbiont developed the capacity to perform aerobic respiration, it evolved into a mitochondrion. However, in those cases where the endosymbiont did not acquire the ability to respire, it evolved into a **hydrogenosome**—an organelle found in some extant protists that produce ATP by fermentation (*see figure 4.20*). Some believe that hydrogenosomes might be derived from mitochondria, but three lines of evidence support the alternative idea that hydrogenosomes and mitochondria arose from the same ancestral organelle: (1) The heat-shock proteins of α-proteobacteria, mitochondria, and hydrogenosomes are closely related; (2) subunits of the mitochondrial enzyme NADH dehydrogenase are active in the hydrogenosomes of the protist *Trichomonas vaginalis;* and (3) the primitive genome found in the hydrogenosomes of the protist *Nyctotherus ovali*s encodes components of a mitochondrial electron transport chain. Taken together, these data suggest that mitochondria and hydrogenosomes are aerobic and anaerobic versions of the same ancestral organelle.

Finally, the endosymbiotic theory put forth by Lynn Margulis and her colleagues combines elements of the endosymbiotic origin of mitochondria with the genome fusion hypothesis. The **serial endosymbiotic theory (SET)** calls for the development of eucaryotes in a series of discrete endosymbiotic steps. This theory suggests that motility evolved first through endosymbiosis between anaerobic spirochetes and another anaerobe. Next, nuclei are thought to have formed by the development of internal membranes. These early nucleated forms would have been similar to modern protists with hydrogenosomes. The endosymbiotic events needed for the

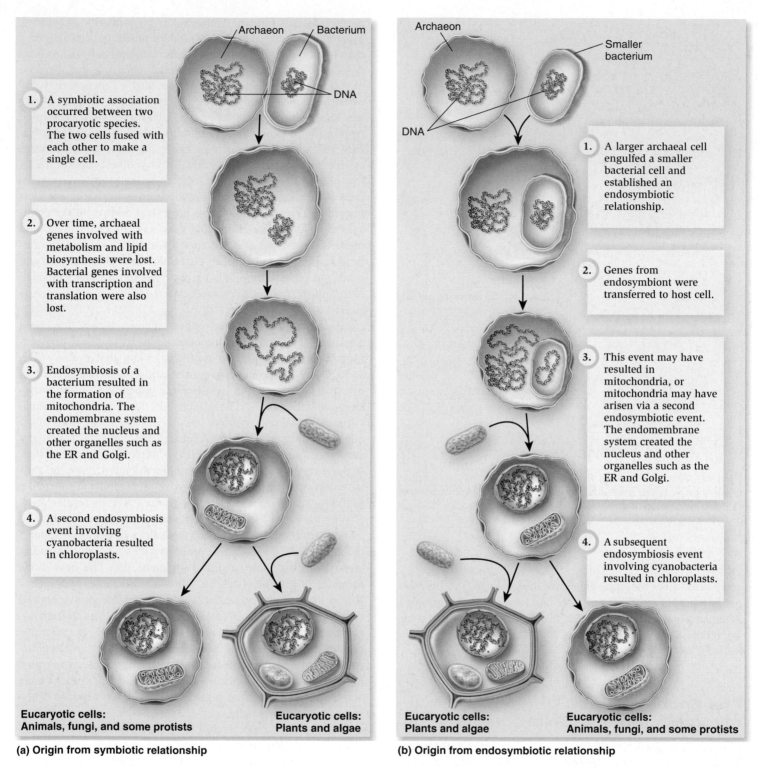

(a) Origin from symbiotic relationship

(b) Origin from endosymbiotic relationship

Figure 17.5   The Genome Fusion Hypothesis for the Origin of Eucaryotic Cells.

evolution of mitochondria are thought to have occurred later, giving rise to early fungi and animal cells, with subsequent endosymbiotic events leading to the development of chloroplasts and plants.

Evolutionary Processes

Clearly figure 17.4 demonstrates the astounding level of microbial diversity reflecting hundreds of millions of years of evolution. The application of Darwin's theory of natural selection to microbial evolution requires special consideration. **Anagenesis,** also known as **microevolution,** refers to small, random genetic changes that occur over generations to slowly drive either speciation or extinction, both of which are forms of **macroevolution.** Neither microevolution nor macroevolution occur at a constant rate. Instead, the fossil record shows that the slow and steady pace of evolution is periodically interrupted by rapid bursts of speciation driven by abrupt changes in the environment. This phenomenon is called **punctuated equilibria** and was introduced by Niles Eldredge and Steven Jay Gould. The theory of punctuated equilibria is one important reason why evolutionary distance, as measured by the similarity of genes in living organisms, provides little or no information regarding when evolutionary divergence occurred.

Procaryotic evolution results in the generation of microbial diversity upon which selective processes determine the development of new species. Recall that genetic diversity in the *Archaea* and the *Bacteria* must occur asexually. Thus heritable genetic changes in these organisms are introduced principally by two mechanisms: mutation and horizontal (lateral) gene transfer (HGT). Genome sequencing has revealed that HGT, particularly in the form of transposon- and phage-mediated gene transfer (transduction), appears to be more important than once thought. In addition, model studies designed to assess competition between microbial populations has led to some surprising observations. It had been thought that very small genetic differences between microbial populations of the same species were of little evolutionary significance. However, laboratory experiments demonstrate that when selection is applied, very small genetic differences can result in one population overtaking another. These recent analyses help illuminate the potential mechanisms by which the vast level of microbial diversity came about and will help guide future studies.
<< *Mechanisms of genetic variation (chapter 14); Comparative genomics (section 15.7)*

1. Why is RNA thought to be the first self-replicating biomolecule?

2. How is the evidence that the *Archaea* and the *Bacteria* share common genes interpreted in the universal tree of life? In the genome fusion hypotheses? Which theory do you favor? Why?

3. Explain the endosymbiotic hypothesis of the origin of mitochondria and chloroplasts. List two pieces of evidence that support this hypothesis.

4. What is a hydrogenosome? Why is it thought that mitochondria and hydrogenosomes arose from a single, common progenitor organelle? What is an alternative hypothesis?

5. What is the difference between macroevolution and microevolution? Define punctuated equilibria. Why does this theory help to explain why the universal phylogenetic tree cannot illustrate the rate at which the evolution of organisms occurred?

17.2 INTRODUCTION TO MICROBIAL CLASSIFICATION AND TAXONOMY

Microbiologists are faced with the daunting task of understanding the diversity of life forms that cannot be seen with the naked eye but can live seemingly anywhere on Earth. One of the first tools needed to survey this level of diversity is a reliable classification system. **Taxonomy** (Greek *taxis,* arrangement or order, and *nomos,* law, or *nemein,* to distribute or govern) is defined as the science of biological classification. In a broader sense, it consists of three separate but interrelated parts: classification, nomenclature, and identification. Once a classification scheme is selected, it is used to arrange organisms into groups called taxa (s., **taxon**) based on mutual similarity. **Nomenclature** is the branch of taxonomy concerned with the assignment of names to taxonomic groups in agreement with published rules. **Identification** is the practical side of taxonomy—the process of determining if a particular isolate belongs to a recognized taxon and, if so, which one. The term **systematics** is often used for taxonomy. However, many taxonomists define systematics in more general terms as the scientific study of organisms with the ultimate object of characterizing and arranging them in an orderly manner. Any study of the nature of organisms, when the knowledge gained is used in taxonomy, is a part of systematics. Thus systematics encompasses disciplines such as morphology, ecology, epidemiology, biochemistry, molecular biology, and physiology.

One of the oldest classification systems, called **natural classification,** arranges organisms into groups whose members share many characteristics. The Swedish botanist Carl von Linné, or Carolus Linnaeus as he often is called, developed the first natural classification in the middle of the eighteenth century. It was based largely on anatomical characteristics and was a great improvement over previously employed artificial systems because knowledge of an organism's position in the scheme provided information about many of its properties. For example, classification of humans as mammals denotes that they have hair, self-regulating body temperature, and milk-producing mammary glands in the female.

When natural classification is applied to higher organisms, evolutionary relationships become apparent simply because the morphology of a given structure (e.g., wings) in a variety of organisms (ducks, songbirds, hawks) suggests how that structure might have been modified to adapt to specific environments or behaviors. However, the traditional taxonomic assignment of microbes was not necessarily rooted in evolutionary relatedness. For instance, bacterial pathogens and microbes of industrial importance were historically given names that described the diseases they cause or the processes they perform (e.g., *Vibrio cholerae, Clostridium tetani,* and *Lactococcus lactis*). Although these labels are of practical use, they do little to guide the taxonomist concerned with the vast majority of microbes that are neither pathogenic nor of industrial consequence. Our recent understanding of the evolutionary relationships among microbes now serves as the theoretical underpinning for taxonomic classification.

In practice, determination of the genus and species of a newly discovered procaryote is based on **polyphasic taxonomy.** This approach includes phenotypic, phylogenetic, and genotypic features. To understand how all of these data are incorporated into a coherent profile of taxonomic criteria, we must first consider the individual components.

Phenetic Classification

For a very long time, microbial taxonomists relied exclusively on a **phenetic system,** which organizes organisms according to mutual similarity of their phenotypic characteristics. This classification system succeeded in bringing order to biological diversity and clarified the function of morphological structures. For example, because motility and flagella are always associated in particular microorganisms, it is reasonable to suppose that flagella are involved in at least some types of motility. Although phenetic studies can reveal possible evolutionary relationships, they do not depend on phylogenetic analysis. They compare many traits without assuming that any features are more phylogenetically important than others—that is, unweighted traits are employed in estimating general similarity. Obviously the best phenetic classification is one constructed by comparing as many attributes as possible. Organisms sharing many characteristics make up a single group or taxon.

Phylogenetic Classification

With the publication in 1859 of Charles Darwin's *On the Origin of Species,* biologists began developing **phylogenetic** or **phyletic classification systems** that sought to compare organisms on the basis of evolutionary relationships. The term **phylogeny** (Greek *phylon,* tribe or race, and *genesis,* generation or origin) refers to the evolutionary development of a species. Scientists realized that when they observed differences and similarities between organisms as a result of evolutionary processes, they also gained insight into the history of life on Earth. However, for much of the twentieth century, microbiologists could not effectively employ phylogenetic classification systems, primarily because of the lack of a good fossil record. When Woese and Fox proposed using rRNA nucleotide sequences to assess evolutionary relationships among microorganisms, the door opened to the resolution of long-standing inquiries regarding the origin and evolution of the majority of life forms on Earth—the microbes. The validity of this approach is now widely accepted, and there are currently over 200,000 different 16S and 18S rRNA sequences in the international databases GenBank and the Ribosomal Database Project (RDP-II). As discussed later (p. 395), the power of rRNA as a phylogenetic and taxonomic tool rests on the features of the rRNA molecule that make it a good indicator of evolutionary history and the ever-increasing size of the rRNA sequence database.

Genotypic Classification

Currently the genotype of a microbe can be evaluated in taxonomic terms in many ways. Some of these techniques are discussed in section 17.4. In general, **genotypic classification** seeks to compare the genetic similarity between organisms. Individual genes or whole genomes can be compared. Since the 1970s it has been widely accepted that procaryotes whose genomes are at least 70% homologous belong to the same species. Unfortunately, this 70% threshold value was established to avoid disrupting existing species assignments; it is not based on theoretical considerations of species identity. However, the genetic data obtained using newer molecular approaches often concur with these older assignments.

1. What is a natural classification?
2. What is polyphasic taxonomy and what three types of data does it consider?
3. Consider the finding that bacteria that can perform anoxygenic photosynthesis belong to several different phylogenetic lineages. How do you think these bacteria were originally classified and what types of data do you think were key in making the most recent taxonomic assignments?

17.3 | TAXONOMIC RANKS

The classification of microbes involves placing them within hierarchical taxonomic levels. Microbes in each level or rank share a common set of specific features. The ranks are arranged in a non-overlapping hierarchy so that each level includes not only the traits that define the rank above it but also a new set of more restrictive traits (**figure 17.6**). The highest rank is the domain, and all procaryotes belong to either the *Bacteria* or the *Archaea.* Within each domain, each microbe is assigned (in descending order) to a phylum, class, order, family, genus, and species. Some procaryotes are also given a subspecies designation. Microbial groups at each level have a specific suffix indicative of that rank or level. Microbiologists often use informal names in place of formal, hierarchical ones. Typical examples of such names are purple bacteria, spirochetes, methane-oxidizing bacteria, sulfate-reducing bacteria, and lactic acid bacteria. As we shall see, these informal names may not have taxonomic significance as they can include species from several phyla. A good example of this is the sulfur bacteria.

The basic taxonomic group in microbial taxonomy is the **species.** Taxonomists working with higher organisms define the term species differently than do microbiologists. Species of higher organisms are groups of interbreeding or potentially interbreeding natural populations that are reproductively isolated from other groups. This is a satisfactory definition for organisms capable of sexual reproduction but fails with procaryotes. As we have discussed, procaryotic species are characterized by phenotypic,

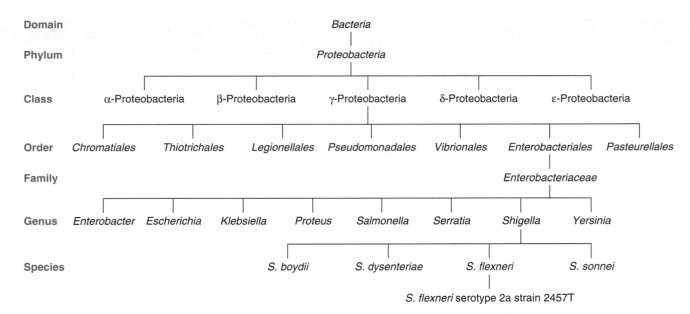

Figure 17.6 Hierarchical Arrangement in Taxonomy.   In this example, members of the genus *Shigella* are placed within higher taxonomic ranks. Not all classification possibilities are given for each rank to simplify the diagram. Note that *-ales* denotes order and *-ceae* indicates family.

genotypic, and phylogenetic criteria. A **procaryotic species** is a collection of strains that share many stable properties and differ significantly from other groups of strains. A **strain** consists of the descendants of a single, pure microbial culture. The definition of a procaryotic species is subjective and can be interpreted in many ways. With an increasing amount of genome sequence data, some have argued that the definition of a procaryotic species needs further revision. Perhaps a species should be the collection of organisms that share the same sequences in their core housekeeping genes (genes coding for products that are required by all cells and which are usually continually expressed). It will take much more work to resolve this complex issue. Whatever the definition, ideally a species also should be phenotypically distinguishable from other similar species.

Strains within a species may be described in a number of different ways. **Biovars** are variant strains characterized by biochemical or physiological differences, **morphovars** differ morphologically, and **serovars** have distinctive antigenic properties. For each species, one strain is designated as the **type strain.** It is usually one of the first strains studied and often is more fully characterized than other strains; however, it does not have to be the most representative member. The type strain for the species is called the type species and is the nomenclatural type or the holder of the species name. A nomenclatural type is a device to ensure permanence of names when taxonomic rearrangements take place. When nomenclature revisions occur, the type species must remain within the genus of which it is the nomenclatural type. Only those strains very similar to the type strain or type species are included in a species. Each species is assigned to a **genus,** the next rank in the taxonomic hierarchy. A genus is a well-defined group of one or more species that is clearly separate from other genera. In practice, considerable subjectivity occurs in assigning

species to a genus, and taxonomists may disagree about the composition of genera.

Microbiologists name microorganisms by using the **binomial system** of Linnaeus. The Latinized, italicized name consists of two parts. The first part, which is capitalized, is the generic name, and the second is the uncapitalized species name or specific epithet (e.g., *Yersinia pestis,* the causative agent of plague). The species name is stable; the oldest epithet for a particular organism takes precedence and must be used. In contrast, a generic name can change if the organism is assigned to another genus because of new information. For example, some members of the genus *Streptococcus* were placed into two new genera, *Enterococcus* and *Lactococcus,* based on rRNA analysis and other characteristics. Thus *Streptococcus faecalis* is now *Enterococcus faecalis.* Often the name will be shortened by abbreviating the genus name with a single capital letter, for example *Y. pestis.* A new procaryotic species cannot be recognized until it has been published in the *International Journal of Systematic and Evolutionary Microbiology;* until that time, the new species name will appear in quotation marks. *Bergey's Manual of Systematic Bacteriology* contains the currently accepted system of procaryotic taxonomy and is discussed in section 17.7.

1. What is the difference between a procaryotic species and a strain?

2. Define morphovar, serovar, and type strain.

3. Which is the correct way to write this microbe's name: *bacillus subtilis,* Bacillus subtilis, *Bacillus Subtilis,* or *Bacillus subtilis?* Identify the genus name and the species name.

17.4 TECHNIQUES FOR DETERMINING MICROBIAL TAXONOMY AND PHYLOGENY

Many different approaches are used in classifying and identifying microorganisms. For clarity, these have been divided into two groups: classical and molecular. Methods often employed in routine laboratory identification of bacteria are covered in chapter 32, Clinical Microbiology and Immunology.

Classical Characteristics

Classical approaches to taxonomy make use of morphological, physiological, biochemical, ecological, and genetic characteristics. These characteristics have been employed in microbial taxonomy for many years. They are quite useful in routine identification and may provide phylogenetic information as well.

Morphological Characteristics

Morphological features are important in microbial taxonomy for many reasons. Morphology is easy to study and analyze, particularly in eucaryotic microorganisms and the more complex procaryotes. In addition, morphological comparisons are valuable because structural features depend on the expression of many genes, are usually genetically stable, and normally (at least in eucaryotes) do not vary greatly with environmental changes. Thus morphological similarity often is a good indication of phylogenetic relatedness.

Many different morphological features are employed in the classification and identification of microorganisms (**table 17.2**). Although the light microscope has always been a very important tool, its resolution limit of about 0.2 μm reduces its usefulness in viewing smaller microorganisms and structures. The transmission and scanning electron microscopes, with their greater resolution, have immensely aided the study of all microbial groups. << *Microscopes and the study of microbial structure (chapter 2)*

Physiological and Metabolic Characteristics

Physiological and metabolic characteristics are very useful because they are directly related to the nature and activity of microbial enzymes and transport proteins. Because proteins are gene products, analysis of these characteristics provides an indirect comparison of microbial genomes. **Table 17.3** lists some of the most important of these properties.

Ecological Characteristics

The ability of a microorganism to colonize a specific environment is of taxonomic value. Some microbes may be very similar in many other respects but inhabit different ecological niches, suggesting they may not be as closely related as first

---

Table 17.2 Some Morphological Features Used in Classification and Identification

Feature	Microbial Groups
Cell shape	All major groups[a]
Cell size	All major groups
Colonial morphology	All major groups
Ultrastructural characteristics	All major groups
Staining behavior	Bacteria, some fungi
Cilia and flagella	All major groups
Mechanism of motility	Gliding bacteria, spirochetes, protists
Endospore shape and location	Endospore-forming bacteria
Spore morphology and location	Bacteria, protists, fungi
Cellular inclusions	All major groups
Colony color	All major groups

[a]Used in classifying and identifying at least some bacteria, archaea, fungi, and protists.

Table 17.3 Some Physiological and Metabolic Characteristics Used in Classification and Identification

Carbon and nitrogen sources

Cell wall constituents

Energy sources

Fermentation products

General nutritional type

Growth temperature optimum and range

Luminescence

Mechanisms of energy conversion

Motility

Osmotic tolerance

Oxygen relationships

pH optimum and growth range

Photosynthetic pigments

Salt requirements and tolerance

Secondary metabolites formed

Sensitivity to metabolic inhibitors and antibiotics

Storage inclusions

suspected. Some examples of taxonomically important ecological properties are life cycle patterns; the nature of symbiotic relationships; the ability to cause disease in a particular host; and habitat preferences such as requirements for temperature, pH, oxygen, and osmotic concentration. Many growth requirements are considered physiological characteristics as well (table 17.3). **>>** *Microbial interactions (section 27.1)*; **<<** *Influences of environmental factors on growth (section 7.5)*

Genetic Analysis

Because most eucaryotes are able to reproduce sexually, classical genetic analysis has been quite useful in the classification of these organisms. Although procaryotes do not reproduce sexually, the study of chromosomal gene exchange through transformation, conjugation, and transduction is sometimes useful in their classification.

Transformation can occur between different procaryotic species but only rarely between genera. The demonstration of transformation between two strains therefore provides evidence of a close relationship. Transformation studies have been carried out with several genera including *Bacillus, Micrococcus, Haemophilus,* and *Rhizobium.* Despite the usefulness of transformation, its results are sometimes hard to interpret because an absence of transformation may result from factors other than major differences in DNA sequence. **<<** *Bacterial transformation (section 14.8)*

Conjugation studies also yield taxonomically useful data, particularly with the enteric bacteria. For example, *Escherichia* can undergo conjugation with the genera *Salmonella* and *Shigella* but not with *Proteus* or *Enterobacter.* These observations fit with other data showing that the first three of these genera are more closely related to one another than to *Proteus* and *Enterobacter.* **<<** *Bacterial conjugation (section 14.7)*; **>>** *Class* Gammaproteobacteria: *Order* Enterobacteriales *(section 20.3)*

Plasmids are important taxonomically because they can confound the analysis of phenotypic traits. Most microbial genera carry plasmids and some plasmids are passed from one microbe to another with relative ease. When such plasmids encode a phenotypic trait (or traits) that is being used to develop a taxonomic scheme, the investigator may assume that the trait is encoded by chromosomal genes. Thus a microbe's phenetic characteristics are misunderstood and its relative degree of relatedness to another microbe may be overestimated. For example, hydrogen sulfide production and lactose fermentation are very important in the taxonomy of the enteric bacteria, yet genes for both traits can be borne on plasmids as well as bacterial chromosomes. One must take care to avoid errors as a result of plasmid-borne traits.

Molecular Characteristics

It is hard to overestimate how the study of DNA, RNA, and proteins has advanced our understanding of microbial evolution and taxonomy. Evolutionary biologists studying plants and animals draw from a rich fossil record to assemble a history of morphological changes; in these cases, molecular approaches supplement such data. In contrast, microorganisms have left almost no fossil record, so molecular analysis is the only feasible means of collecting a large and accurate data set from a number of microbes. When scientists are careful to make only valid comparisons, phylogenetic inferences based on molecular approaches provide the most robust analysis of microbial evolution.

Nucleic Acid Base Composition

Microbial genomes can be directly compared, and taxonomic similarity can be estimated in many ways. The first, and possibly the simplest, technique to be employed is the determination of DNA base composition. Recall that DNA contains four purine and pyrimidine bases: adenine (A), guanine (G), cytosine (C), and thymine (T). Base-pairing rules dictate that the (G + C)/(A + T) ratio or **G + C content**—the percent of G + C in DNA—reflects the base sequence and varies with sequence changes as follows:

$$\text{Mol\% G} + \text{C} = \frac{G + C}{G + C + A + T} \times 100$$

The base composition of DNA can be determined in several ways. Although the G + C content can be ascertained after hydrolysis of DNA and analysis of its bases with high-performance liquid chromatography (HPLC), physical methods are easier and more often used. The G + C content often is determined from the **melting temperature (T_m)** of DNA. In double-stranded DNA, three hydrogen bonds join GC base pairs and two bonds connect AT base pairs. As a result, DNA with a greater G + C content has more hydrogen bonds, and its strands separate at higher temperatures—that is, it has a higher melting point. DNA melting can be easily followed spectrophotometrically because the absorbance of DNA at 260 nm increases during strand separation. When a DNA sample is slowly heated, the absorbance increases as hydrogen bonds are broken and reaches a plateau when all the DNA has become single stranded (**figure 17.7**). The midpoint of the rising curve

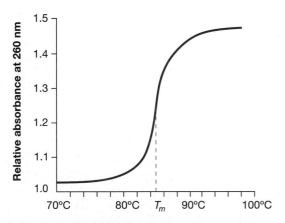

Figure 17.7   A DNA Melting Curve.   The T_m is indicated.

gives the melting temperature, a direct measure of the G + C content.

The G + C content of many microorganisms has been determined (table 17.4). The G + C content of DNA from animals and higher plants averages around 40% and ranges between 30 and 50%. In contrast, the DNA of both eucaryotic and procaryotic microorganisms varies greatly in G + C content; procaryotic G + C content is the most variable, ranging from around 25 to almost 80%. Despite such a wide range of variation, the G + C content of strains within a particular species is constant. If two organisms differ in their G + C content by more than about 10%, their genomes have quite different base sequences, indicating that they are not closely related. On the other hand, it is not safe to assume that organisms with very similar G + C contents also have similar DNA base sequences because two very different base sequences can be constructed from the same proportions of A + T and G + C base pairs. Only if two microorganisms also are alike phenotypically does their similar G + C content suggest close relatedness.

G + C content data are valuable in at least two ways. First, they can confirm a taxonomic scheme developed using other data. If organisms in the same taxon are too dissimilar in G + C content, the taxon probably should be divided. Second, G + C content appears to be useful in characterizing procaryotic genera because the variation within a genus is usually less than 10%, even though the content may vary greatly between genera. For example, *Staphylococcus* has a G + C content of 30 to 38%, whereas *Micrococcus* DNA has 64 to 75% G + C, yet these two genera of gram-positive cocci have many other features in common.

Nucleic Acid Hybridization

The similarity between genomes can be compared more directly by use of nucleic acid hybridization studies. If the genomes of two microbial isolates are heated to become single-stranded (ss) DNA and then cooled and held at a temperature about 25°C below the T_m, strands with complementary base sequences will reassociate to form stable dsDNA. However, noncomplementary strands will remain unpaired. Because strands with similar but not identical sequences associate to form less temperature-stable dsDNA hybrids, incubation of the mixture at 30 to 50°C below the T_m allows hybrids of more diverse ssDNAs to form. Incubation at 10 to 15°C below the T_m permits hybrid formation only with almost identical strands.

In one of the more widely used hybridization techniques, nylon filters with bound nonradioactive DNA strands are incubated at the appropriate temperature with radioactive ssDNA fragments. After the fragments hybridize with the membrane-bound ssDNA, the membrane is washed to remove any nonhybridized ssDNA and its radioactivity is measured. The quantity of radioactivity bound to the filter reflects the amount of hybridization and thus the similarity of the DNA sequences. The degree of similarity or homology is expressed as the percent of experimental DNA radioactivity retained on the filter compared with the percent of homologous DNA radioactivity bound under the same conditions (table 17.5 provides examples). Two strains whose DNAs show at least 70% relatedness under optimal hybridization conditions and less than a 5% difference in T_m often, but not always, are considered members of the same species.

Table 17.4 Representative G + C Content of Microorganisms

Organism	Percent G + C	Organism	Percent G + C	Organism	Percent G + C
Procaryotes		*Rickettsia*	29–33	*Plasmodium berghei*	41
Anabaena	39–44	*Spirochaeta*	51–65	*Spirogyra*	39
Caulobacter	62–65	*Staphylococcus*	30–38	*Stentor polymorphus*	45
Chlamydia	41–44	*Streptococcus*	33–44	*Trichomonas*	29–34
Chlorobium	49–58	*Streptomyces*	69–73	*Trypanosoma*	45–59
Clostridium	21–54	*Sulfolobus*	31–37	*Volvox carteri*	50
Deinococcus	62–70	*Thiobacillus*	52–68		
Escherichia	48–59			**Fungi**	
Halobacterium	66–68	**Protists**		*Agaricus bisporus*	44
Methanobacterium	32–50	*Acetabularia mediterranea*	37–53	*Amanita muscaria*	57
Micrococcus	64–75	*Amoeba proteus*	66	*Aspergillus niger*	52
Mycobacterium	62–70	*Chlamydomonas*	60–68	*Candida albicans*	33–35
Myxococcus	68–71	*Chlorella*	43–79	*Coprinus lagopus*	52–53
Neisseria	48–56	*Dictyostelium*	22–25	*Mucor rouxii*	38
Oscillatoria	40–50	*Euglena gracilis*	46–55	*Neurospora crassa*	52–54
Pseudomonas	58–69	*Nitzschia angularis*	47	*Rhizopus nigricans*	47
Rhodospirillum	62–66	*Paramecium* spp.	29–39	*Saccharomyces cerevisiae*	36–42

Table 17.5	Comparison of *Neisseria* Species by DNA Hybridization Experiments	
Membrane-Attached DNA[a]	**Percent Homology**[b]	
Neisseria meningitidis	100	
N. gonorrhoeae	78	
N. sicca	45	
N. flava	35	

Source: Data from Staley, T. E., and Colwell, R. R. 1973. Applications of molecular genetics and numerical taxonomy to the classification of bacteria. *Annu. Rev. Ecol. and Systematics,* 8: 282.

[a]The experimental membrane-attached nonradioactive DNA from each species was incubated with radioactive *N. meningitidis* DNA, and the amount of radioactivity bound to the membrane was measured. The more radioactivity bound, the greater the homology between DNA sequences.

[b]*N. meningitidis* DNA bound to experimental DNA/Amount bound to membrane attached *N. meningitidis* DNA × 100

If DNA molecules are very different in sequence, they will not form a stable, detectable hybrid. Therefore DNA-DNA hybridization is used to study only closely related microorganisms. More distantly related organisms can be compared by carrying out DNA-RNA hybridization experiments using radioactive ribosomal or transfer RNA. Distant relationships can be detected because rRNA and tRNA genes represent only a small portion of the total DNA genome and have not evolved as rapidly as most other microbial genes. The technique is similar to that employed for DNA-DNA hybridization: membrane-bound DNA is incubated with radioac-tive rRNA, washed, and counted. An even more accurate measurement of homology is obtained by finding the temperature required to dissociate and remove half the radioactive rRNA from the membrane; the higher this temperature, the stronger the rRNA-DNA complex and the more similar the sequences. ◄◄ *Gene structure: tRNA and rRNA genes (section 12.4)*

Nucleic Acid Sequencing

Despite the usefulness of G + C content determination and nucleic acid hybridization studies, rRNAs from small ribosomal subunits (16S and 18S rRNAs from procaryotes and eucaryotes, respectively) have become the molecules of choice for inferring microbial phylogenies and making taxonomic assignments at the genus level. The small subunit rRNAs (SSU rRNAs) are almost ideal for studies of microbial evolution and relatedness because they play the same role in all microorganisms. In addition, because the ribosome is absolutely necessary for survival and the SSU rRNAs are part of the complex ribosome structure, the genes encoding these rRNAs cannot tolerate large mutations. Thus these genes change very slowly with time and do not appear to be subject to horizontal gene transfer, an important factor in comparing sequences. The utility of SSU rRNAs is extended by the presence of certain sequences that are variable among organisms and other regions that are quite stable. The variable regions enable comparison between closely related microbes, whereas the stable sequences allow the comparison of distantly related microorganisms.

Comparative analysis of SSU rRNA sequences from thousands of organisms has demonstrated the presence of **oligonucleotide signature sequences (figure 17.8)**. These are short, conserved

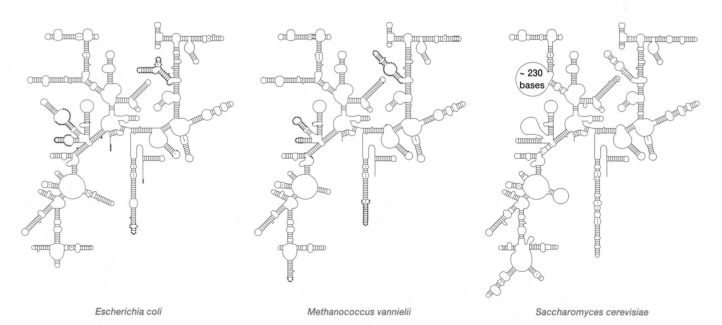

Escherichia coli   *Methanococcus vannielii*   *Saccharomyces cerevisiae*

Figure 17.8 Small Ribosomal Subunit RNA. Representative examples of rRNA secondary structures from the three primary domains: *Bacteria* (*Escherichia coli*), *Archaea* (*Methanococcus vannielii*), and *Eucarya* (*Saccharomyces cerevisiae*). The red dots mark positions where *Bacteria* and *Archaea* normally differ. *Source: Data from Woese, C. P. 1987.* Microbiological Rev. *51(2):221–27.*

nucleotide sequences that are specific for a phylogenetically defined group of organisms. Thus the signature sequences found in *Bacteria* are rarely or never found in *Archaea* and vice versa. Likewise, the 18S rRNA of eucaryotes also bears signature sequences that are specific to the domain *Eucarya*. Either complete rRNAs or, more often, specific rRNA fragments can be compared. The proper alignment of SSU rRNA nucleotide sequences and the application of computer algorithms enable sequence comparison between any number of organisms. When comparing rRNA sequences between two microorganisms, their relatedness can be represented by an association coefficient, or S_{ab} value. The higher the S_{ab} values, the more closely the organisms are related to each other. If the sequences of the 16S rRNAs of two organisms are identical, the S_{ab} value is 1.0. After S_{ab} values have been determined, a computer calculates the relatedness of the organisms and summarizes their relationships in a tree or dendrogram (e.g., figure 17.4).

The ability to amplify regions of rRNA genes (rDNA) by the polymerase chain reaction (PCR) and sequence the DNA using automated sequencing technology has greatly increased the efficiency by which SSU rRNA sequences can be obtained. PCR can be used to amplify rDNA from the genomes of organisms because conserved nucleotide sequences flank the regions of interest. In practice, this means that PCR primers are readily available or can be generated to amplify rDNA from both cultured and uncultured microbes. As noted, the Ribosomal Database Project has over 200,000 sequences. << *Determining DNA sequences (section 15.2); Polymerase chain reaction (section 16.2)*

Signature sequences are present in genes other than those encoding ribosomal RNA. Many genes have insertions or deletions of specific lengths and sequences at fixed positions, and a particular insertion or deletion may be found exclusively among all members of one or more phyla. R. S. Gupta refers to these as conserved indels. These signature sequences are particularly useful in phylogenetic studies when they are flanked by conserved regions. In such cases, observed changes in the signature sequence cannot be due to sequence misalignments. The signature sequences located in some highly conserved housekeeping genes do not appear greatly affected by lateral gene transfers and, like SSU rRNA, can be employed in phylogenetic analysis.

DNA sequences can also be used to determine species and strain, in addition to genus. However, such resolution requires the analysis of genes that evolve more quickly than those that encode rRNA. In fact, rather than using a single gene, five to seven conserved housekeeping genes can be sequenced and compared in a technique called **multilocus sequence typing (MLST).** Multiple genes are usually examined to avoid misleading results that can arise through lateral gene transfer. Because as many as 30 different versions, or alleles, of each gene can exist, the finding that two microbial isolates share the same allele for multiple genes is very strong evidence that the two strains are closely related, perhaps even the same strain. MLST was

originally developed to discriminate among pathogenic strains but has become more broadly applied to microbial taxonomy. Although MLST is helpful for differentiating isolates at the strain and species levels, the data become too difficult to interpret at higher taxonomic levels.

Genomic Fingerprinting

A group of techniques called **genomic fingerprinting** can also be used to classify microbes and help determine phylogenetic relationships. Unlike the molecular analyses so far discussed, genomic fingerprinting does not involve nucleotide sequencing. Instead, it employs the capacity of restriction endonucleases to recognize specific nucleotide sequences. Thus the pattern of DNA fragments generated by endonuclease cleavage (called restriction fragments) as examined by gel electrophoresis is a direct representation of nucleotide sequence (*see figure 16.9 and table 16.2*). The comparison of restriction fragments of specific genes or DNA fragments is the basis of **restriction fragment length polymorphism (RFLP)** analysis. Sometimes, rather than determining the nucleotide sequence of the SSU rRNA genes from a group of microbes, the rRNA genes (or other genes) undergo RFLP analysis. That is, the genes are digested by endonucleases and their restriction fragment patterns are compared. When rRNA genes are analyzed in this way, it is called **ribotyping.**

Highly conserved and repetitive DNA sequences present in many copies in the genomes of most gram-negative and some gram-positive bacteria can also be used to help identify microbes. There are three families of repetitive sequences: the 154 bp BOX elements, the 124–127 bp enterobacterial repetitive intergenic consensus (ERIC) sequence, and 35–40 bp repetitive extragenic palindromic (REP) sequences. These sequences are generally found at distinct sites between genes—that is, they are intergenic. Because they are conserved among genera, oligonucleotide primers can be used to specifically amplify the repetitive sequences by PCR. Different primers are used for each type of repetitive element, and the results are classified as arising from BOX-PCR, ERIC-PCR, or REP-PCR (**figure 17.9**). In each case, the amplified fragments from many microbial samples can be resolved and visualized on an agarose gel. Each lane of the gel corresponds to a single bacterial isolate, and the pattern created by many samples resembles a UPC bar code. The "bar code" is then computer analyzed using pattern recognition software as well as software that calculates phylogenetic relationships. Because DNA fingerprinting enables identification to the level of species, subspecies, and often strain, it is valuable not only in the study of microbial diversity but in the identification of human, animal, and plant pathogens as well.

Amino Acid Sequencing

The amino acid sequences of proteins directly reflect mRNA sequences and so represent the genes coding for their synthesis.

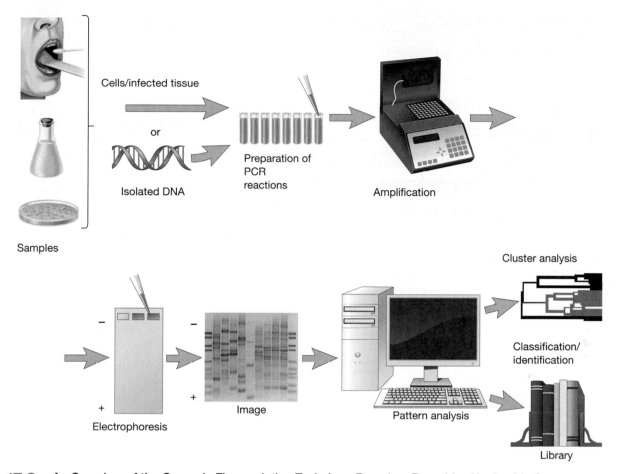

Cells/infected tissue

or

Isolated DNA

Preparation of PCR reactions

Amplification

Samples

Cluster analysis

Classification/ identification

Pattern analysis

Library

Electrophoresis

Image

Figure 17.9   An Overview of the Genomic Fingerprinting Technique Based on Repetitive Nucleotide Sequences.

However, the value of a given protein in taxonomic and phylogenetic studies varies. The sequences of proteins with dissimilar functions often change at different rates; some sequences change quite rapidly, whereas others are very stable. Therefore the most direct approach is to determine the amino acid sequence of proteins with the same function. If these sequences are similar, the organisms possessing them may be closely related. The sequences of cytochromes and other electron transport proteins, histones and heat-shock proteins, transcription and translation proteins, and a variety of metabolic enzymes have been used in taxonomic and phylogenetic studies. In contrast, rapidly evolving proteins, such as the outer surface proteins of the syphilis pathogen *Treponema pallidum,* are not appropriate for taxonomic or phylogenetic purposes. Suitable proteins may offer some advantages over rRNA comparisons. A sequence of 20 amino acids has more information per site than a sequence of four nucleotides. Protein sequences are less affected by organism-specific differences in G + C content than are DNA and RNA sequences.

There are several ways to compare proteins. Indirect methods of comparing proteins frequently have been traditionally employed. Specifically, the electrophoretic mobility of proteins has been used to study relationships at the species and subspecies levels. In addition, antibodies can discriminate between very similar proteins, and immunologic techniques can be used to compare proteins from different microorganisms. More recently, the use of mass spectrometry has been adopted for relatively rapid amino acid sequencing, which enables the direct comparison of amino acid sequences among proteins of interest. **Figure 17.10** shows the taxonomic utility of several kinds of molecular analyses including protein profiling; with the exception of genome sequencing, it is clear that a combination of approaches is best for identification at the species level or lower. ≪ *Proteomics (section 15.6)*

1. What are the advantages of using each major group of characteristics (morphological, physiological/metabolic, ecological, genetic, and molecular) in classification and identification? How is each group related to the nature and expression of the genome? Give examples of each type of characteristic.

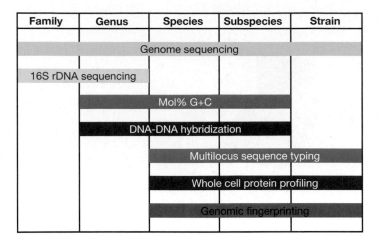

Figure 17.10   Relative Taxonomic Resolution of Various Molecular Techniques.

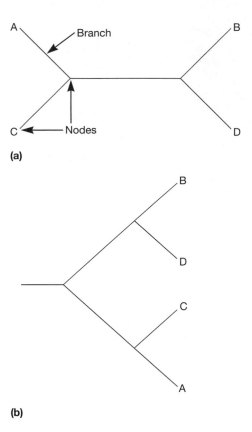

Figure 17.11   Examples of Phylogenetic Trees.   (a) Unrooted tree joining four taxonomic units. (b) Rooted tree.

2. What modes of genetic exchange in procaryotes have proved taxonomically useful? Why are plasmids important in bacterial taxonomy?

3. What is the G + C content of DNA, and how can it be determined through melting temperature studies?

4. Why is it not safe to assume that two microorganisms with the same G + C content belong to the same species? In what ways are G + C content data taxonomically valuable?

5. Describe how nucleic acid hybridization studies are carried out using membrane-bound DNA. Why might one wish to vary the incubation temperature during hybridization?

6. How are rRNA sequencing studies conducted, and why is rRNA so suitable for determining relatedness?

7. How is ribotyping similar to rRNA sequence analysis? How do the two techniques differ?

8. List some proteins used in phylogenetic and taxonomic studies. Why are they useful?

17.5   PHYLOGENETIC TREES

Microbial taxonomy is changing rapidly. This is caused by ever-increasing knowledge of the biology of microorganisms, remarkable advances in computer technology, and the use of molecular characteristics to determine phylogenetic relationships between microorganisms. It is common for phylogenetic relationships to be illustrated in the form of branched diagrams or trees. A **phylogenetic tree** is a graph made of branches that connect nodes (**figure 17.11**). The nodes represent taxonomic units such as species or genes; the external nodes at the end of the

branches represent living (extant) organisms. As in the universal phylogenetic tree (figure 17.4), the length of the branches represents the number of molecular changes that have taken place between the two nodes. Importantly, a tree may be unrooted or rooted. An unrooted tree (figure 17.11a) simply represents phylogenetic relationships but does not provide an evolutionary path. Figure 17.11a shows that A is more closely related to C than it is to either B or D, but it does not specify the common ancestor for the four species or the direction of change. In contrast, the rooted tree (figure 17.11b) gives a node that serves as the common ancestor and shows the development of the four species from this root. It is much more difficult to develop a rooted tree. For example, there are 15 possible rooted trees that connect four species but only three possible unrooted trees.

Phylogenetic trees are developed by comparing nucleotide or amino acid sequences. To compare two molecules, their sequences must first be aligned so that similar parts match up. The object is to align and compare homologous sequences, ones that are similar because they had a common origin in the past. This is not an easy task, and computers and fairly complex mathematics must be employed to minimize the number of gaps and mismatches in the sequences being compared. << *Bioinformatics (section 15.4)*

Once the molecules have been aligned, the number of positions that vary in the sequences are determined. These data are used to calculate a measure of the difference between the sequences. Often the difference is expressed as the **evolutionary distance.** This is simply a quantitative indication of the number of positions that differ between two aligned macromolecules. Statistical adjustments are made for back mutations and multiple substitutions that may have occurred. Organisms are then clustered together based on similarity in the sequences. The most similar organisms are clustered together, then compared with the remaining organisms to form a larger cluster associated together at a lower level of similarity or evolutionary distance. The process continues until all organisms are included in the tree.

Phylogenetic relationships also can be estimated by techniques such as **parsimony analysis.** In this approach, relationships are determined by estimating the minimum number of sequence changes required to give the final sequences being compared. It is presumed that evolutionary change occurs along the shortest pathway with the fewest changes or steps from an ancestor to the organism in question. The tree or pattern of relationships is favored that is simplest and requires the fewest assumptions.

1. Define phylogenetic tree and evolutionary distance.

2. What is the difference between an unrooted and a rooted tree? Examine figure 17.4: is it rooted or unrooted? Explain your answer.

17.6 THE MAJOR DIVISIONS OF LIFE

The division of all living organisms into three domains—*Archaea, Bacteria,* and *Eucarya*—has become widely accepted among microbiologists. Although in this text the universal phylogenetic tree as proposed by Norman Pace is emphasized (figure 17.4), some scientists contend that the *Bacteria* arose well before the *Archaea* and the *Eucarya,* as shown in **figure 17.12a**. Yet another interpretation is shown in the eocyte tree (figure 17.12b), which proposes that sulfur-dependent, extremely thermophilic procaryotes called eocytes (Greek *eo,* dawn, and *cyta,* hollow vessel) form a separate group more closely related to eucaryotes than the

Archaea. The existence of such organisms as a separate domain or group has been met with considerable skepticism. Finally, the idea that the *Eucarya* arose from a fusion of a bacterium and an archaeon is illustrated in figure 17.12c. There are many reasons biologists are unable to agree on a single model. For instance, the selection of genes to be compared can have an impact on the resulting tree. When genes that encode proteins used for the storage and processing of genetic information are compared, trees that are consistent with those obtained by SSU rRNA analysis are generated (e.g., figure 17.4). On the other hand, when proteins involved in metabolism are compared, trees that place the *Bacteria* and the *Archaea* as closest relatives are generated. Yet the comparison of other housekeeping activities leads to trees that place the *Archaea* and the *Eucarya* most closely related (figure 17.12a). Other factors that can give rise to incongruent trees include unrecognized gene duplications that occurred before the domains formed, leading to confusing patterns. Unequal rates of evolution can distort the trees. Phylogenetically important information may have been lost in some molecular sequences. Significant sequence variation may exist between the same molecules from different strains of the same species. Unless several strains are analyzed, false conclusions may be drawn. Thus inaccurate universal trees may result when only the sequences from a few molecules are employed.

One of the biggest challenges in constructing a satisfactory tree is widespread, frequent HGT. Genome sequence studies have shown that extensive HGT occurs within and between domains. Eucaryotes possess genes from both bacteria and archaea, and there has been frequent gene swapping between the two procaryotic domains. It appears that at least some bacteria even have acquired eucaryotic genes. Although a variety of mechanisms may be responsible, it has been suggested that much of this gene movement occurs by way of virus-mediated transfer. Clearly the pattern of microbial evolution is not as linear and treelike as previously thought. This has prompted the development of trees that attempt to display HGT (**figure 17.13**). Such trees resemble a web or network with many lateral branches linking various trunks, each branch representing the transfer of one or a few genes. Instead of having a single main trunk or common ancestor at its base, these trees have

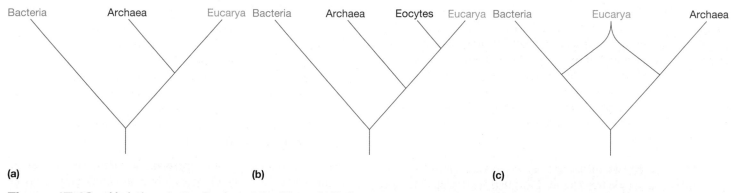

(a)   (b)   (c)

Figure 17.12   Variations in the Design of the "Tree of Life."   These three alternative phylogenetic trees are discussed in the text.

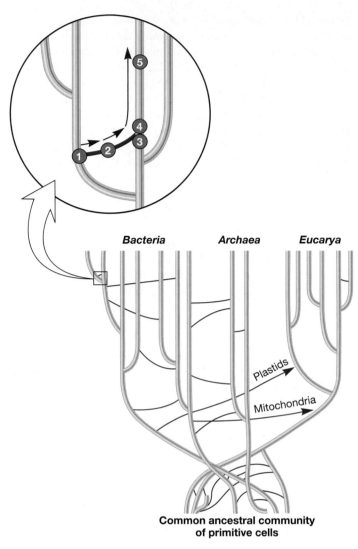

Figure 17.13 Universal Phylogenetic Tree with Lateral Gene Transfers. The effect of HGT on the evolution of life results in a tree with weblike interconnections that complicate the emergence of the three domains of life. The insert displays the series of events needed to give rise to the stable inheritance of a gene in a new organism.

several trunks or groups of primitive cells that contribute to the original gene pool. Although extensive gene transfer occurs between the *Archaea* and the *Bacteria*, the *Eucarya* seldomly participated in lateral gene transfer after the formation of fungi, plants, and animals. It is possible that eucaryotic cells originated in a complex process involving many gene transfers from both bacteria and archaea. This hypothesis is consistent with the formation of mitochondria and chloroplasts by endosymbiosis with α-proteobacteria and cyanobacteria, respectively. Presumably the three domains remain separate because many more gene transfers occur within each domain than between them.

1. Describe the two major alternatives to Pace's universal phylogenetic tree (figure 17.4) that are depicted in

figure 17.12*a–c*. Why have there been difficulties in developing an accurate tree?

2. Discuss the effect of frequent horizontal gene transfer on phylogenetic trees.

17.7 BERGEY'S MANUAL OF SYSTEMATIC BACTERIOLOGY

In 1923 David Bergey, professor of bacteriology at the University of Pennsylvania, and four colleagues published a classification of bacteria that could be used for identification of bacterial species, *Bergey's Manual of Determinative Bacteriology*. This manual is now in its ninth edition. In 1984 the first edition of *Bergey's Manual of Systematic Bacteriology* was published. It contained descriptions of all procaryotic species then identified (**Microbial Diversity & Ecology 17.1**). The second edition will consist of five volumes; the first volume was published in 2001 and the second in 2005; volumes 3 and 4 are to be published in 2008; and the last volume is also forthcoming.

Enormous progress has been made in procaryotic taxonomy since the first volume of *Bergey's Manual of Systematic Bacteriology* was published. In particular, the sequencing of rRNA, DNA, and proteins has made phylogenetic analysis of procaryotes feasible. Thus while microbial classification in the first edition was phenetic (based on phenotypic characterization), the second edition of *Bergey's Manual* is largely phylogenetic. Although gram-staining properties are generally considered phenetic characteristics, they also play a role in the phylogenetic classification of microbes. Some of the major differences between gram-negative and gram-positive bacteria are summarized in **table 17.6**.

The second edition has more ecological information about individual taxa. It does not group all the clinically important procaryotes together as the first edition did. Instead, pathogenic species are placed phylogenetically and thus scattered throughout the following five volumes.

Volume 1, *The Archaea and the Deeply Branching and Phototrophic Bacteria*
Volume 2, *The Proteobacteria*
Volume 3, *The Low G + C Gram-Positive Bacteria*
Volume 4, *The High G + C Gram-Positive Bacteria*
Volume 5, *The Planctomycetes, Spirochaetes, Fibrobacteres, Bacteroidetes, Fusobacteria, Chlamydiae, Acidobacteria, Verrumicrobia, and Dictyoglomi* (Volume 5 also will contain a section that updates descriptions and phylogenetic arrangements that have been revised since publication of volume 1.)

Table 17.7 summarizes the organization of the second edition and indicates where the discussion of a particular group may be found in this textbook.

Microbial Diversity & Ecology

17.1 "Official" Nomenclature Lists—A Letter from *Bergey's*

On a number of occasions lately, the impression has been given that the status of a bacterial taxon in *Bergey's Manual of Systematic Bacteriology* or *Bergey's Manual of Determinative Bacteriology* is in some sense official. Similar impressions are frequently given about the status of names in the Approved List of Bacterial Names and in the Validation Lists of newly proposed names that appear regularly in the *International Journal of Systematic Bacteriology*. It is therefore important to clarify these matters.

There is no such thing as an official classification. *Bergey's Manual* is not "official"—it is merely the best consensus at the time, and although great care has always been taken to obtain a sound and balanced view, there are also always regions in which data are lacking or confusing, resulting in differing opinions and taxonomic instability. When *Bergey's Manual* disavows that it is an official classification, many bacteriologists may feel that the solid earth is trembling. But many areas are in fact reasonably well established. Yet taxonomy is partly a matter of judgment and opinion, as is all science, and until new information is available, different bacteriologists may legitimately hold different views. They cannot be forced to agree to any "official classification." It must be remembered that, as yet, we know only a small percentage of the bacterial species in nature. Advances in technique also reveal new lights on bacterial relationships. Thus we must expect that existing boundaries of groups will have to be redrawn in the future, and it is expected that molecular biology, in particular, will imply a good deal of change over the next few decades.

The position with the Approved Lists and the Validation Lists is rather similar. When bacteriologists agreed to make a new start in bacteriological nomenclature, they were faced with tens of thousands of names in the literature of the past. The great majority were useless, because, except for about 2,500 names, it was impossible to tell exactly what bacteria they referred to. These 2,500 were therefore retained in the Approved Lists. The names are only approved in the sense that they were approved for retention in the new bacteriological nomenclature. The remainder lost standing in the nomenclature, which means they do not have to be considered when proposing new bacterial names (although names can be individually revived for good cause under special provisions).

The new International Code of Nomenclature of Bacteria requires all new names to be validly published to gain standing in the nomenclature, either by being published in papers in the *International Journal of Systematic Bacteriology* or, if published elsewhere, by being announced in the Validation Lists. The names in the Validation Lists are therefore valid only in the sense of being validly published (and therefore they must be taken account of in bacterial nomenclature). The names do not have to be adopted in all circumstances; if users believe the scientific case for the new taxa and validly published names is not strong enough, they need not adopt the names. For example, *Helicobacter pylori* was immediately accepted as a replacement for *Campylobacter pylori* by the scientific community, whereas *Tatlockia micdadei* had not generally been accepted as a replacement for *Legionella micdadei*. Taxonomy remains a matter of scientific judgment and general agreement.

From Sneath, P. H. A., and Brenner, D. J. 1992. Official nomenclature lists. *ASM News.* 58(4):175. Copyright © by the American Society for Microbiology. Reprinted by permission.

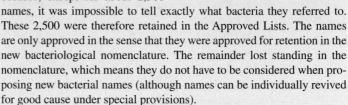

Table 17.6	Some Characteristic Differences between Gram-Negative and Gram-Positive Bacteria	
Property	**Gram-Negative Bacteria**	**Gram-Positive Bacteria**
Cell wall	Gram-negative type wall with inner 2–7 nm peptidoglycan layer and outer membrane (7–8 nm thick) of lipid, protein, and lipopolysaccharide.	Gram-positive type wall with a homogeneous, thick cell wall (20–80 nm) composed mainly of peptidoglycan. Other polysaccharides and teichoic acids may be present.
Cell shape	Spheres, ovals, straight or curved rods, helices or filaments; some have sheaths or capsules.	Spheres, rods, or filaments; may show true branching
Reproduction	Binary fission, sometimes budding	Binary fission, filamentous forms grow by tip extension
Metabolism	Phototrophic, chemolithoautotrophic, or chemoorganoheterotrophic	Usually chemoorganoheterotrophic, a few phototrophic
Motility	Motile or nonmotile. Flagella placement can be varied. Motility may also result from the use of axial filaments (spirochetes) or gliding motility.	Most often nonmotile; have peritrichous flagella when motile
Appendages	Can produce several types of appendages—pili and fimbriae, prosthecae, stalks	Usually lack appendages (may have spores on hyphae)
Endospores	Cannot form endospores	Some groups

Table 17.7	Organization of *Bergey's Manual of Systematic Bacteriology*	
Taxonomic Rank	**Representative Genera**	**Textbook Coverage**
Volume 1. *The Archaea and the Deeply Branching and Phototrophic Bacteria*		
Domain *Archaea*		
Phylum *Crenarchaeota*		
Class I. *Thermoprotei*	*Thermoproteus, Pyrodictium, Sulfolobus*	pp. 411–413
Phylum *Euryarchaeota*		
Class I. *Methanobacteria*	*Methanobacterium*	pp. 413–416
Class II. *Methanococci*	*Methanococcus*	
Class III. *Methanomicrobia*	*Methanomicrobium*	
Class IV. *Halobacteria*	*Halobacterium, Halococcus*	pp. 416–417
Class V. *Thermoplasmata*	*Thermoplasma, Picrophilus, Ferroplasma*	p. 417
Class VI. *Thermococci*	*Thermococcus, Pyrococcus*	p. 417
Class VII. *Archaeoglobi*	*Archaeoglobus*	p. 417
Class VIII. *Methanopyri*	*Methanopyrus*	
Domain *Bacteria*		
Phylum *Aquificae*	*Aquifex, Hydrogenobacter*	p. 421
Phylum *Thermotogae*	*Thermotoga, Geotoga*	p. 421
Phylum *Thermodesulfobacteria*	*Thermodesulfobacterium*	
Phylum *Deinococcus-Thermus*	*Deinococcus, Thermus*	pp. 421–422
Phylum *Chrysiogenetes*	*Chrysogenes*	
Phylum *Chloroflexi*	*Chloroflexus, Herpetosiphon*	p. 424
Phylum *Thermomicrobia*	*Thermomicrobium*	
Phylum *Nitrospira*	*Nitrospira*	
Phylum *Deferribacteres*	*Geovibrio*	
Phylum *Cyanobacteria*	*Prochloron, Synechococcus, Pleurocapsa, Oscillatoria, Anabaena, Nostoc, Stigonema*	pp. 425–428
Phylum *Chlorobi*	*Chlorobium, Pelodictyon*	p. 424
Volume 2. *The Proteobacteria*		
Phylum *Proteobacteria*		
Class I. *Alphaproteobacteria*	*Rhodospirillum, Rickettsia, Caulobacter, Rhizobium, Brucella, Nitrobacter, Methylobacterium, Beijerinckia, Hyphomicrobium*	pp. 440–448
Class II. *Betaproteobacteria*	*Neisseria, Burkholderia, Alcaligenes, Comamonas, Nitrosomonas, Methylophilus, Thiobacillus*	pp. 448–453
Class III. *Gammaproteobacteria*	*Chromatium, Leucothrix, Legionella, Pseudomonas, Azotobacter, Vibrio, Escherichia, Klebsiella, Proteus, Salmonella, Shigella, Yersinia, Haemophilus*	pp. 453–466
Class IV. *Deltaproteobacteria*	*Desulfovibrio, Bdellovibrio, Myxococcus, Polyangium*	pp. 467–471
Class V. *Epsilonproteobacteria*	*Campylobacter, Helicobacter*	pp. 471–472
Volume 3. *The Low G + C Gram-Positive Bacteria*		
Phylum *Firmicutes*		
Class I. *Clostridia*	*Clostridium, Peptostreptococcus, Eubacterium, Desulfotomaculum, Heliobacterium, Veillonella*	pp. 479–482
Class II. *Mollicutes*	*Mycoplasma, Ureaplasma, Spiroplasma, Acholeplasma*	pp. 475–477
Class III. *Bacilli*	*Bacillus, Caryophanon, Paenibacillus, Thermoactinomyces, Lactobacillus, Streptococcus, Enterococcus, Listeria, Leuconostoc, Staphylococcus*	pp. 483–497

(Continued)

Table 17.7	Organization of *Bergey's Manual of Systematic Bacteriology (Continued)*	
Volume 4. The High *G + C* Gram-Positive Bacteria		
Phylum *Actinobacteria*		
Class *Actinobacteria*	*Actinomyces, Micrococcus, Arthrobacter, Corynebacterium, Mycobacterium, Nocardia, Actinoplanes, Propionibacterium, Streptomyces, Thermomonospora, Frankia, Actinomadura, Bifidobacterium*	pp. 500–515
Volume 5. *The Planctomycetes, Spirochaetes, Fibrobacteres, Bacteriodetes, Fusobacteria, Chlamydiae, Acidobacteria, Verrucomicrobia,* and *Dictyoglomi*		
Phylum *Planctomycetes*	*Planctomyces, Gemmata*	p. 429
Phylum *Chlamydiae*	*Chlamydia*	pp. 429–432
Phylum *Spirochaetes*	*Spirochaeta, Borrelia, Treponema, Leptospira*	pp. 432–436
Phylum *Fibrobacteres*	*Fibrobacter*	
Phylum *Acidobacteria*	*Acidobacterium*	
Phylum *Bacteroidetes*	*Bacteroides, Porphyromonas, Prevotella, Flavobacterium, Sphingobacterium, Flexibacter, Cytophaga*	pp. 436–437
Phylum *Fusobacteria*	*Fusobacterium, Streptobacillus*	
Phylum *Verrucomicrobia*	*Verrucomicrobium*	
Phylum *Dictyoglomi*	*Dictyoglomus*	
Phylum *Gemmatimonadetes*	*Gemmatimonas*	

Summary

17.1 Microbial Evolution

a. Precellular life may have been an "RNA world" because RNA has the capacity to both replicate and catalyze chemical reactions (**figure 17.2**).

b. Living organisms can be divided into three domains: the *Eucarya,* the *Bacteria,* and the *Archaea* (**table 17.1**).

c. The origin of eucaryotic cells is an unsettled question. The root of the universal phylogenetic tree suggests *Bacteria, Archaea,* and *Eucarya* have a single common *Bacteria* ancestor but that the *Archaea* and the *Eucarya* evolved independently of the *Bacteria* (**figure 17.4**).

d. The endosymbiotic theory asserts that mitochondria and chloroplasts evolved from an endosymbiotic α-proteobacterium and cyanobacterium, respectively. Hydrogenosomes and mitochondria are probably derived from a single, common ancestor.

17.2 Introduction to Microbial Classification and Taxonomy

a. Taxonomy, the science of biological classification, is composed of three parts: classification, nomenclature, and identification.

b. A polyphasic approach is used to classify microbes. This incorporates information gleaned from genetic, phenotypic, and phylogenetic analysis.

17.3 Taxonomic Ranks

a. Taxonomic ranks are arranged in a nonoverlapping hierarchy (**figure 17.6**).

b. The definition of species is different for sexually and asexually reproducing organisms. A procaryotic species is a collection of strains that have many stable properties in common and differ significantly from other groups of strains.

c. Microorganisms are named according to the binomial system.

17.4 Techniques for Determining Microbial Taxonomy and Phylogeny

a. The classical approach to determining microbial taxonomy and phylogeny includes the use of morphological, physiological, metabolic, ecological, and genetic characteristics.

b. The study of transformation and conjugation in bacteria is sometimes taxonomically useful. Plasmid-borne traits can cause errors in bacterial taxonomy if care is not taken.

c. The G + C content of DNA is easily determined and taxonomically valuable because it is an indirect reflection of the base sequence.

d. Nucleic acid hybridization studies are used to compare DNA or RNA sequences and thus determine genetic relatedness.

e. Nucleic acid sequencing is the most powerful and direct method for comparing genomes. The sequences of 16S and 18S rRNA are used most often in phylogenetic studies of procaryotic and eucaryotic microbes, respectively **(figure 17.8).** Complete microbial genomes are now being sequenced and compared.

f. Amino acid sequence of some proteins can be taxonomically and phylogenetically relevant, although the value of each protein must be assessed individually.

17.5 Phylogenetic Trees

a. Phylogenetic relationships often are shown in the form of branched diagrams called phylogenetic trees **(figure 17.11).** Trees may be either rooted or unrooted and are created in several different ways.

b. The sequences of rRNA, DNA, and proteins are used to produce phylogenetic trees. Often members of a group will have a unique characteristic rRNA sequence that distinguishes them from members of other taxonomic groups.

17.6 The Major Divisions of Life

a. Most microbiologists favor the three-domain system, although discerning specific relationships between taxa is complicated by frequent horizontal genetic transfer, particularly between the *Bacteria* and the *Archaea* **(figure 17.13).**

17.7 *Bergey's Manual of Systematic Bacteriology*

a. *Bergey's Manual of Systematic Bacteriology* gives the accepted system of procaryotic taxonomy.

b. The second edition of *Bergey's Manual* provides phylogenetic classifications. Procaryotes are divided between two domains and 26 phyla **(table 17.7).** Comparisons of nucleic acid sequences, particularly 16S rRNA sequences, are the foundation of this classification.

Critical Thinking Questions

1. What experiments could be designed in a modern microbiology and/or chemistry lab to test the RNA world hypothesis?

2. Compare the findings of the universal phylogenetic tree and the genome fusion hypothesis. Debate the pros and cons of each.

3. Consider the fact that the use of 16S rRNA sequencing as a taxonomic and phylogenetic tool has resulted in tripling the number of procaryotic phyla. Why do you think the advent of this genetic technique has expanded the currently accepted number of microbial phyla?

4. Procaryotes were classified phenetically in the first edition of *Bergey's Manual of Systematic Bacteriology.* What do you think are the advantages and disadvantages of the phylogenetic classification used in the second edition?

5. Discuss the problems in developing an accurate phylogenetic tree. Is it possible to create a completely accurate universal phylogenetic tree?

6. Why is the current procaryotic classification system likely to change considerably? How would one select the best features to use in identification of unknown procaryotes and determination of relatedness?

Learn More

Learn more by visiting the Prescott website at www.mhhe.com/prescottprinciples, where you will find a complete list of references.

The *Archaea*

18

Archaea are often found in extreme environments such as this geyser in Yellowstone National Park. The orange color is due to the carotenoid pigments of thermophilic archaea.

Archaea The domain that contains procaryotes with isoprenoid glycerol diether or diglycerol tetraether lipids in their membranes and archaeal rRNA (among many other features).

Chapter Glossary

bacteriorhodopsin A transmembranous protein to which retinal is bound; it functions as a light-driven proton pump driving photophosphorylation without chlorophyll or bacteriochlorophyll. Found in the purple membrane of halophilic archaea.

halophile A microorganism that requires high levels of sodium chloride for growth.

hyperthermophile A microorganism that grows optimally at temperatures greater than 85°C.

methanogen Strictly anaerobic archaea that derive energy by converting CO_2, H_2, formate, acetate, and other compounds to either methane or methane and CO_2.

pseudomurein A peptidoglycan-like polymer found in some archaeal cell walls. It is distinguished from peptidoglycan by the presence of L-amino acids, *N*-acetyltalosaminuronic acid (NAT) and $\beta(1 \rightarrow 3)$ glycosidic bonds.

sensory rhodopsin A form of rhodopsin found in halobacteria and cyanobacteria that senses the spectral quality of light.

thermoacidophiles Microorganisms that grow best at high temperatures and low pH.

As is often the case, epoch-making ideas carry with them implicit, unanalyzed assumptions that ultimately impede scientific progress until they are recognized for what they are. So it is with the prokaryote-eukaryote distinction. Our failure to understand its true nature set the stage for the sudden shattering of the concept when a "third form of life" was discovered in the late 1970s, a discovery that actually left many biologists incredulous. Archaebacteria, as this third form has come to be known, have revolutionized our notion of the prokaryote, have altered and refined the way in which we think about the relationship between prokaryotes and eukaryotes . . . and will influence strongly the view we develop of the ancestor that gave rise to all extant life.

—C. R. Woese and R. S. Wolfe

Comparison of the sequences of rRNA from a great variety of organisms shows that organisms can be divided into three domains: *Bacteria, Archaea,* and *Eucarya* (*see figure 17.4*). Some of the most important features of these domains are summarized in table 17.1. Because the *Archaea* are different from both *Bacteria* and eucaryotes, we first review their taxonomy and most distinctive properties. We then survey the two phyla recognized by *Bergey's Manual,* the *Crenarchaeota* and *Euryarchaeota*.

18.1 INTRODUCTION TO THE *ARCHAEA*

Like the *Bacteria*, the **Archaea** are quite diverse, both in morphology and physiology. They may be spherical, rod-shaped, spiral, lobed, cuboidal, triangular, plate-shaped, irregularly shaped, or pleomorphic. Some are single cells, whereas others form filaments or aggregates. They range in diameter from 0.1 to over 15 μm, and some filaments can grow up to 200 μm in length. They can stain either gram positive or gram negative, but they have unique cell walls—different from that of the *Bacteria*. Multiplication may be by binary fission, budding, fragmentation, or other mechanisms. The *Archaea* are just as diverse physiologically. They can be aerobic, facultatively anaerobic, or strictly anaerobic. Nutritionally, they range from chemolithoautotrophs to organotrophs. They include psychrophiles, mesophiles, and hyperthermophiles that can grow above 100°C.

Archaea have typically been considered microbes of "extreme environments," or "extremophiles." Indeed, many archaea inhabit niches with very high or low temperatures or pH, concentrated salts, or that are completely anoxic. However, terms such as extreme and hypersaline reflect a human perspective, meaning that they are situations where humans could not survive. On the contrary, most of Earth (the oceans) is an "extreme environment" where it is very cold (about 4°C), dark, and under high pressure. Many archaea are well adapted to these environments, where they can grow to high numbers. For instance, crenarchaeotes constitute at least 34% of the procaryotic biomass in at least some Antarctic coastal waters. In some hypersaline environments, their populations become so dense that the brine is red with archaeal pigments. Nonetheless, the notion that archaea are exclusively "extremophiles" is no longer valid as archaea are known to inhabit a variety of soils and ocean waters. In addition, some are symbionts in the digestive tracts of animals, but to date, no pathogenic archaea have been described. **>>** *Microorganisms in natural environments (chapter 26)*

Archaeal Taxonomy

As shown in **table 18.1**, the *Archaea* can be divided into five major groups based on physiological and morphological differences. On the basis of phylogenetic evidence, *Bergey's Manual* divides the *Archaea* into the phyla *Euryarchaeota* (Greek *eurus,* wide, and *archaios,* ancient or primitive) and *Crenarchaeota* (Greek *crene,* spring or fount, and *archaios*) (**figure 18.1**). The euryarchaeotes are given this name because they occupy many different ecological niches and have a variety of metabolic patterns. The methanogens, extreme halophiles, sulfate reducers, and many extreme thermophiles with sulfur-dependent metabolism are *Euryarchaeota.* The crenarchaeotes are thought to resemble the ancestor of the *Archaea,* and almost all the well-characterized species are thermophiles or hyperthermophiles. However, metagenomic evidence indicates that mesophilic crenarchaeotes constitute a significant fraction of the oceanic plankton.

Table 18.1	**Characteristics of the Major Archaeal Physiological Groups**	
Group	**General Characteristics**	**Representative Genera**
Methanogenic archaea	Strict anaerobes. Methane is the major metabolic end product. S^0 may be reduced to H_2S without yielding energy production. Cells possess coenzyme M, factors 420 and 430, and methanopterin.	*Methanobacterium* *Methanococcus* *Methanomicrobium* *Methanosarcina*
Archaeal sulfate reducers	Irregular gram-negative staining coccoid cells. H_2S formed from thiosulfate and sulfate. Autotrophic growth with thiosulfate and H_2. Can grow heterotrophically. Traces of methane also formed. Extremely thermophilic and strictly anaerobic. Possess factor 420 and methanopterin but not coenzyme M or factor 430.	*Archaeoglobus*
Extremely halophilic archaea	Rods, cocci, or irregular shaped cells, that may include pyramids or cubes. Stain gram negative or gram positive but like all archaea lack peptidoglycan. Primarily chemoorganoheterotrophs. Most species require sodium chloride ≥ 1.5 M, but some survive in as little as 0.5 M. Most produce characteristic bright-red colonies; some are unpigmented. Neutrophilic to alkalophilic. Generally mesophilic; however, at least one species is known to grow at 55°C. Possess either bacteriorhodopsin or halorhodopsin and can use light energy to produce ATP.	*Halobacterium* *Halococcus* *Natronobacterium*
Cell wall-less archaea	Pleomorphic cells lacking a cell wall. Thermoacidophilic and chemoorganotrophic. Facultatively anaerobic. Plasma membrane contains a mannose-rich glycoprotein and a lipoglycan.	*Thermoplasma*
Extremely thermophilic S^0-metabolizers	Gram-negative staining rods, filaments, or cocci. Obligately thermophilic (optimum growth temperature between 70–110°C). Usually strict anaerobes but may be aerobic or facultative. Acidophilic or neutrophilic. Autotrophic or heterotrophic. Most are sulfur metabolizers. S^0 reduced to H_2S anaerobically; H_2S or S^0 oxidized to H_2SO_4 aerobically.	*Desulfurococcus* *Pyrodictium* *Pyrococcus* *Sulfolobus* *Thermococcus* *Thermoproteus*

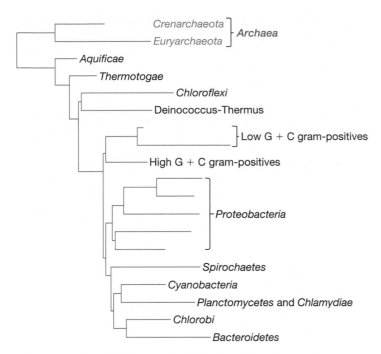

Figure 18.1 Phylogenetic Relationships Among Procaryotes. The *Archaea* are highlighted.

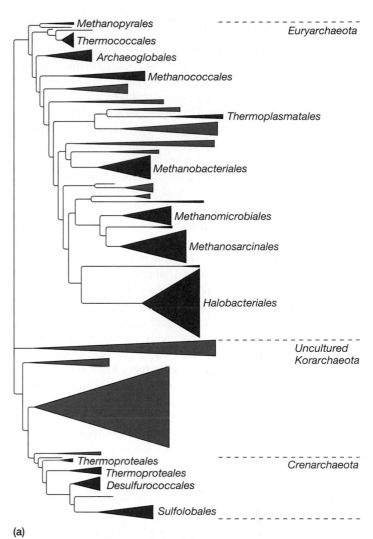

(a)

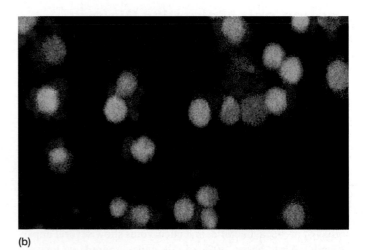

(b)

Figure 18.2 Phylogeny of the Archaeal Domain.   (a) Based on 16S rRNA nucleotide sequence analysis of cultured species (*Crenarchaeota, Euryarchaeota*) and uncultured samples (*Korarchaeaota*), the *Archaea* can be divided into at least three phyla. (b) Parasitic cells of *Nanoarchaeum equitans* attached to the surface of the crenarchaeote *Ignicoccus*. Confocal laser scanning micrograph in which *Ignicoccus* are stained green; *N. equitans* cells are red.

Whereas all known archaea currently in culture belong to either the *Crenarchaeota* or the *Euryarcheota,* an ever-growing collection of 16S rRNA nucleotide sequences cloned directly from the environment suggests that a third phylum of *Archaea* exists. Tentatively called the *Korarchaeota* (from the Greek word for "young man"), this phylum is gaining acceptance, although it is not clear if this designation will hold up to more complete analysis if and when any of these microbes are cultivated (**figure 18.2***a*). Yet a fourth phylum has been suggested by the discovery of the hyperthermophilic archaeon *Nanoarchaeum equitans* from submarine vents. This small microbe (0.4 μm) grows while attached to a crenarcheote belonging to the genus *Ignicoccus*, an autotrophic, sulfur-reducing microbe (figure 18.2*b*). *N. equitans* has not been cultured axenically, so little is known about its physiology. However, because *N. equitans* and *Ignicoccus* vary in size, it is possible to isolate cells for genome analysis. The small size of the *N. equitans* genome (0.49 Mb) suggests that this microbe has lived in association with other organisms for a long time—long enough for it to have lost genes for lipid, nucleotide, amino acid, and enzyme cofactor biosynthesis. The loss of essential biosynthetic genes and its limited catabolic capabilities indicates that *N. equitans* maintains a parasitic relationship with its host, making it the only known archaeal parasite.

Thus although there are currently two accepted archaeal phyla, it is uncertain how many phyla will eventually be recognized. This state of flux in archaeal phylogeny demonstrates how dynamic microbial taxonomy can be. The use of molecular probes to dissect microbial communities, combined with enhanced culture techniques, ensures that phylogenetic analysis will continue to evolve.

Archaeal Cell Walls and Membranes

As discussed in chapter 3, archaea can stain either gram positive or gram negative, even though they lack the muramic acid and D-amino acids that make up peptidoglycan. Without the constraints of the conserved molecule peptidoglycan, archaeal cell walls can be quite diverse. For instance, some methanogenic archaea have **pseudomurein** (a peptidoglycan-like polymer that is cross-linked with L-amino acids; *see figure 3.30*), while others contain a complex polysaccharide similar to the chondroitin sulfate of animal connective tissue. Interestingly, some hyperthermophilic archaea and methanogens have protein walls. << *Archaeal cell walls (section 3.5)*

One of the most distinctive archaeal features is their membrane lipids. As shown in table 17.1, the *Archaea* differ from both the *Bacteria* and the *Eucarya* in having branched chain hydrocarbons attached to glycerol by ether (rather than ester) linkages (*see figures 3.9–3.11*). Thermophilic archaea sometimes link two glycerol groups to form long tetraethers. Diether side chains are usually 20 carbons long, and tetraether chains contain 40 carbon atoms. However, cells can adjust chain lengths by cyclizing the chains to form cyclopentane rings. These rings are more densely packed within the membrane, making them more stable at high temperatures. In fact, thermophilic archaea increase the number of cyclopentane rings as growth temperature increases. Cyclopentane ring-containing lipids have also been discovered in nonthermophilic *Crenarchaeota*. These lipids, called **crenarchaeol,** are unique to these organisms so they are used as a biomarker for the presence of crenarchaeotes in natural environments such as marine plankton. Polar phospholipids, sulfolipids, and glycolipids are also found in archaeal membranes. << *Procaryotic cell membranes (section 3.2)*

Genetics and Molecular Biology

Some features of archaeal genetics are similar to those in the *Bacteria,* while others more closely resemble the *Eucarya.* The genomes of some archaea are significantly smaller than those of many bacteria. For instance, while the genome of the gram-positive bacterium *Bacillus subtilis* is 4.20 million base pairs (Mb), the crenarchaeote *Pyrobaculum aerophilum* genome is 2.22 Mb and that of *Methanothermobacter thermoautotrophicus* (formerly *Methanobacterium thermoautotrophicum*), a euryarchaeote, is 1.75 Mb. A sign of archaeal diversity is the variation in G + C content, from about 21% to 68%. To date, it appears that the *Archaea* have few plasmids.

Comparative genomics between the completely sequenced genomes of archaea, bacteria, and eucaryotes show several apparent trends. First, about 30% of all genes shared exclusively between archaea and eucaryotes encode proteins involved in transcription, translation, or DNA metabolism. In contrast, a large number of the genes shared only between *Bacteria* and *Archaea* are involved in metabolic pathways. In addition, evidence exists for horizontal gene transfer between these two domains, especially between thermophilic bacteria and archaea.

A small number of genes are found in all three domains, but these do not seem to fit any specific pattern.

Archaeal DNA replication appears to be a complex mixture of eucaryotic and procaryotic features. Like *Bacteria,* all characterized archaea have circular chromosomes and replication appears to be bidirectional. However, in archaeal genomes that have been sequenced, the replication origin is flanked by genes encoding the eucaryotic-like initiation protein Cdc6/Orc1 and at least a few archaea have multiple origins. While it was originally thought that archaeal replication proteins were uniformly eucaryotic-like, further genome analysis reveals that some replication proteins are similar to those in *Bacteria,* while still others are uniquely archaeal. Some archaeal chromosomes differ from *Bacteria* in having eucaryotic-like histone proteins that bind DNA to form nucleosome-like structures.

Transcription in the *Archaea* likewise blends bacterial and eucaryotic features. Archaeal RNA polymerases consist of at least 10 subunits that are highly homologous to eucaryotic subunits. Also, like eucaryotic nuclear RNA polymerase, archaeal RNA polymerases do not efficiently recognize promoter regions without the aid of additional proteins. Instead, promoter recognition depends on at least two eucaryotic-like proteins: the *TATA-box-binding protein* (TBP) and *transcription factor B* (TFB). It is therefore not surprising that many archaeal promoters are similar to certain eucaryotic promoters, possessing a TATA box (a 7-bp sequence found about 25 bp before the transcriptional start site) preceded by a purine-rich region called the *B responsive element* (BRE). In eucaryotes, the BRE is the site to which transcription factor IIB binds. It is thought that archaeal TFB and TBP bind the BRE region of DNA as a prerequisite for the assembly of RNA polymerase subunits prior to the initiation of transcription (**figure 18.3**).

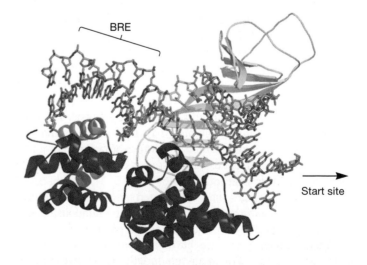

Figure 18.3 Archaeal Promoters Resemble Those of Eucaryotes. The crystal structure of the ternary complex between TBP, the carboxyl terminus of TFB, and a region of DNA containing a TATA box and BRE. DNA is shown in gray; TBP is the yellow ribbon structure; and TFB is magenta, with its recognition helix in turquoise. TBP = TATA-box binding protein; TFB = transcription factor B; BRE = B-responsive element.

However, archaeal mRNA appears to be similar to bacterial mRNA in that it is polycistronic and there is no evidence for mRNA splicing. ◄◄ *Transcription: Transcription in the* Archaea *(section 12.5)*

Finally, the translational machinery in the *Archaea* is unique. Unlike both *Bacteria* and eucaryotes, the TΨC arm of archaeal tRNA lacks thymine and contains pseudouridine or 1-methyl-pseudouridine. The archaeal initiator tRNA carries methionine as does the eucaryotic initiator tRNA. Although archaeal ribosomes are 70S, similar to bacterial ribosomes, electron microscopy studies show that their shape is quite variable and sometimes differs from that of both bacterial and eucaryotic ribosomes. They resemble eucaryotic ribosomes in their sensitivity to the antibiotic anisomycin and insensitivity to chloramphenicol and kanamycin. Furthermore, archaeal elongation factor 2 reacts with diphtheria toxin like the eucaryotic EF-2. ◄◄ *Translation (section 12.7)*

Archaeal protein secretion also has both bacterial and eucaryotic features. All three domains have signal recognition particles (SRPs) that target new proteins to translocation sites, but the archaeal SRP differs from those in the other two domains. As in the *Bacteria,* the archaeal SRP binds to the signal sequence of a preprotein and can direct it to the Sec-dependent protein secretion pathway for transport through the plasma membrane. The archaeal Sec-dependent pathway proteins, however, more closely resemble those of the eucaryotic pathway than the bacterial proteins. After the preprotein is moved across the membrane, its signal sequence is removed by a signal peptidase that resembles a subunit of the eucaryotic peptidase. ◄◄ *Protein maturation and secretion (section 12.8)*

Metabolism

Not surprisingly, in view of their variety of life-styles, archaeal metabolism varies greatly among the members of different groups. Some archaea are heterotrophs; others are autotrophic. A few carry out rhodopsin-based phototrophy.

Archaeal carbohydrate metabolism is best understood. The enzyme 6-phosphofructokinase has not been found in any archaea, and they do not seem to degrade glucose by way of the Embden-Meyerhof pathway. However, some hyperthermophiles appear to have a modified Embden-Meyerhof pathway that involves several novel enzymes, including an ADP-dependent phosphofructokinase. Extreme halophiles and thermophiles catabolize glucose using a modified form of the Entner-Doudoroff pathway in which the initial intermediates are not phosphorylated. All archaea that have been studied can oxidize pyruvate to acetyl-CoA. However, they lack the pyruvate dehydrogenase complex present in eucaryotes and respiratory bacteria and use the enzyme pyruvate oxidoreductase for this purpose. Halophiles and the extreme thermophile *Thermoplasma* seem to have a functional tricarboxylic acid cycle. Methanogens do not catabolize glucose to any significant extent, and so it is not surprising that they lack a complete tricarboxylic acid cycle. Evidence for functional respiratory chains has been obtained in halophiles and thermophiles. ◄◄ *Breakdown of glucose to pyruvate (section 10.3); Tricarboxylic acid cycle (section 10.4)*

Very little is known in detail about biosynthetic pathways in the *Archaea*. Evidence suggests that the synthetic pathways for amino acids, purines, and pyrimidines are similar to those in other organisms. Some methanogens can fix atmospheric dinitrogen. Many archaea, including halophiles and methanogens, use a reversal of the Embden-Meyerhof pathway to synthesize glucose, and at least some methanogens and extreme thermophiles employ glycogen as their major reserve material. ◄◄ *Synthesis of sugars and polysaccharides (section 11.4); Synthesis of amino acids (section 11.5)*

Autotrophy is widespread among the methanogens and extreme thermophiles, and CO_2 fixation occurs in more than one way. *Thermoproteus* and possibly *Sulfolobus* incorporate CO_2 by the reductive tricarboxylic acid cycle (**figure 18.4a**). This pathway is also present in the green sulfur bacteria. Methanogenic archaea and probably most extreme thermophiles incorporate CO_2 by the reductive acetyl-CoA pathway (figure 18.4b). A similar pathway also is present in acetogenic bacteria and autotrophic sulfate-reducing bacteria.

Thermostability

The ability of many archaea to grow at very high temperatures has stimulated much research. In particular, the thermostability of proteins has garnered a lot of interest because proteins that function at high temperature can be used in a variety of industrial applications. Recall that at high temperature proteins denature, that is they unfold. However thermophilic proteins remain folded at temperatures that would denature similar proteins from mesophiles. Structural strategies used by thermophiles include increased hydrophobicity of the protein core, an elevated number of ionic interactions on the protein surface, increased atomic packing density, additional hydrogen bonds, and shorter surface loop structures. Relatively minor changes in amino acid sequences can bring about these changes. For instance, thermostable proteins have relatively higher levels of valine, glutamate, and lysine, and fewer glutamine and valine residues. In addition to these changes, hyperthermophiles produce a specific class of chaperones at very high temperatures that function to refold partially denatured proteins. ◄◄ *Protein maturation and secretion (section 12.8)*

Thermophiles also use more than one mechanism to prevent DNA from denaturing (becoming single stranded). Interestingly, hyperthermophiles have a reverse DNA gyrase that induces positive, rather than negative, supercoils into DNA. Positive supercoiling dramatically enhances DNA thermostability; indeed, reverse DNA gyrase is found only in hyperthermophiles. Some hyperthermophiles also increase the solute concentration in their cytoplasm. This helps prevent the loss of purines and pyrimidines that can occur at high temperature. It has also been suggested that archaeal histones contribute to genome thermostability. ◄◄ *Nucleic acid structure (section 12.2); DNA replication (section 12.3)*

Figure 18.4 Mechanisms of Autotrophic CO$_2$ Fixation. (a) The reductive tricarboxylic acid cycle. The cycle is reversed with ATP and reducing equivalents [H] to form acetyl-CoA from CO$_2$. The acetyl-CoA may be carboxylated to yield pyruvate, which can then be converted to glucose and other compounds. This sequence appears to function in *Thermoproteus neutrophilus*. (b) The synthesis of acetyl-CoA and pyruvate from CO$_2$ in *Methanothermobacter thermoautotrophicus*. One carbon comes from the reduction of CO$_2$ to a methyl group, and the second is produced by reducing CO$_2$ to carbon monoxide through the action of the enzyme CO dehydrogenase (E$_1$). The two carbons are then combined to form an acetyl group. Corrin-E$_2$ represents the cobamide-containing enzyme involved in methyl transfers. Special methanogen coenzymes and enzymes are described in figures 18.11 and 18.12.

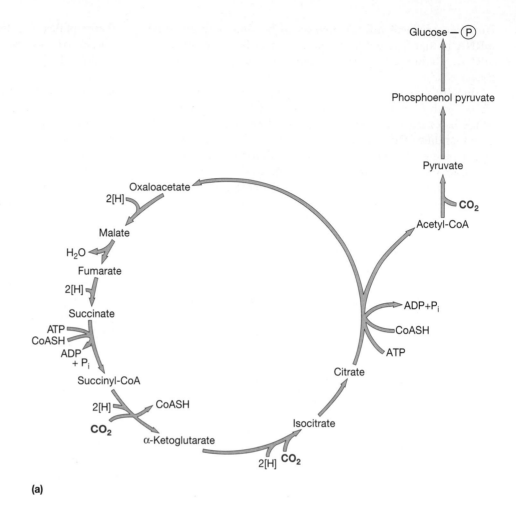

(a)

(b)

1. Briefly describe the major ways in which *Archaea* differ from *Bacteria* and eucaryotes.

2. How are the phyla *Euryarchaeota* and *Crenarchaeota* distinguished?

3. How do archaeal cell walls differ from those of the *Bacteria*? What is pseudomurein?

4. In what ways do archaeal membrane lipids differ from those of *Bacteria* and eucaryotes? How do these differences contribute to the survival of thermophilic and hyperthermophilic archaea?

5. List the differences between *Archaea* and the other domains with respect to DNA replication, transcription, and translation.

6. Briefly describe the ways in which archaea degrade and synthesize glucose. In what two unusual ways do they incorporate CO$_2$?

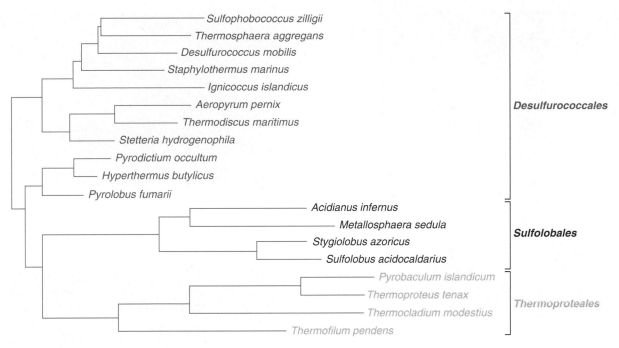

Pyrobaculum islandicum
Thermoproteus tenax
Thermocladium modestius
Thermofilum pendens

Thermoproteales

Pyrodictium occultum
Hyperthermus butylicus
Pyrolobus fumarii

Acidianus infernus
Metallosphaera sedula
Stygiolobus azoricus
Sulfolobus acidocaldarius

Sulfolobales

Sulfophobococcus zilligii
Thermosphaera aggregans
Desulfurococcus mobilis
Staphylothermus marinus
Ignicoccus islandicus
Aeropyrum pernix
Thermodiscus maritimus
Stetteria hydrogenophila

Desulfurococcales

Figure 18.5 **The Phylum *Crenarchaeota*.** A phylogenetic tree developed with 16S rRNA data for crenarchaeote type species. The order *Caldisphaerales* is not shown.

18.2 PHYLUM *CRENARCHAEOTA*

The phylum *Crenarchaeota* has only one class, *Thermoprotei*, which is divided into four orders and six families (**figure 18.5**). The order *Thermoproteales* contains gram-negative-staining anaerobic to facultative, hyperthermophilic rods. Members of the order *Sulfolobales* are coccus-shaped thermoacidophiles. The order *Desulfurococcales* contains gram-negative-staining coccoid or disk-shaped hyperthermophiles. They grow either chemolithotrophically by hydrogen oxidation or organotrophically by fermentation or respiration with sulfur as the electron acceptor. The order *Caldisphaerales* has only one genus, *Caldisphaera,* whose members are thermoacidophilic, aerobic, heterotrophic cocci.

Many of these thermophiles are sulfur dependent. The sulfur may be used either as an electron acceptor in anaerobic respiration or as an electron donor by lithotrophs. Many are strict anaerobes. They grow in geothermally heated water or soils that contain elemental sulfur. These environments are scattered all over the world. Examples are the sulfur-rich hot springs in Yellowstone National Park and the waters surrounding areas of submarine volcanic activity (**figure 18.6**). Such habitats are sometimes called solfatara. These archaea can be very thermophilic and often are classified as **hyperthermophiles.** The most extreme example was isolated from an active hydrothermal vent in the northeast Pacific Ocean. This is one of three novel isolates that constitute a new genus in the *Pyrodictiaceae* family. Its optimum growth temperature is about 105°C, but even autoclaving this microbe at 121°C for one hour fails to kill it. It is strictly anaerobic, using Fe(III) as a terminal electron acceptor and H_2 or formate as electron donors and energy sources (**figure 18.7**).

Figure 18.6 **Habitat for Thermophilic Archaea.** The Sulfur Cauldron in Yellowstone National Park. The water is at its boiling point and very rich in sulfur. *Sulfolobus* grows well in such habitats.

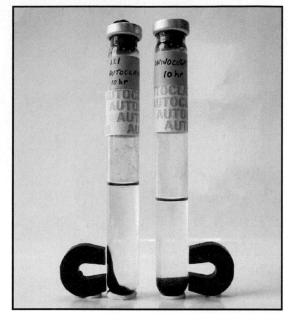

(a)

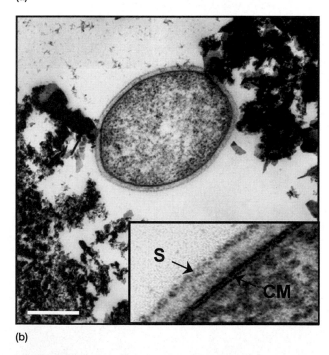

(b)

Figure 18.7 An Extremely Hyperthermophilic Crenarchaeote. (a) A member of the family *Pyrodictiaceae* (tube on left) grows following autoclaving at 121°C as shown by its ability to reduce Fe(III) to magnetite when incubated anaerobically (tube on right is sterile control). (b) A transmission electron micrograph shows the single layer cell envelope (S) and plasma membrane. Scale bar =1 µm.

At present, *Crenarchaeota* contains 25 genera; two of the better-studied genera are *Sulfolobus* and *Thermoproteus*. Members of the genus *Sulfolobus* stain gram negative and are aerobic, irregularly lobed, spherical archaea with a temperature optimum around 70 to 80°C and a pH optimum of 2 to 3 (**figure 18.8**). For this reason, they are **thermoacidophiles.** Their cell walls contain

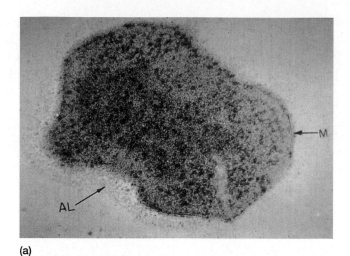

(a)

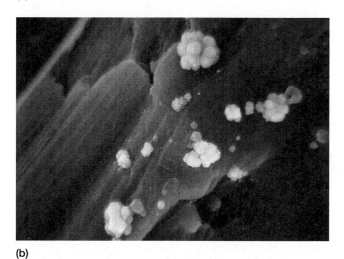

(b)

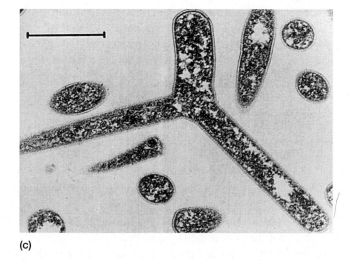

(c)

Figure 18.8 *Sulfolobus* and *Thermoproteus*. (a) A thin section of *Sulfolobus brierleyi*. The archaeon, about 1 µm in diameter, is surrounded by an amorphous layer (AL) instead of a well-defined cell wall; the plasma membrane (M) is also visible. (b) A scanning electron micrograph of a colony of *Sulfolobus* growing on the mineral molybdenite (MoS$_2$) at 60°C. At pH 1.5–3, the organism oxidizes the sulfide component of the mineral to sulfate and solubilizes molybdenum. (c) Electron micrograph of *Thermoproteus tenax*. Bar = 1 µm.

lipoprotein and carbohydrate. They grow chemolithotrophically on sulfur granules in hot acidic springs and soils while oxidizing the sulfur to sulfuric acid (figure 18.8*b*). Oxygen is the normal electron acceptor, but ferric iron may be used. Sugars and amino acids such as glutamate also serve as carbon and energy sources.

Thermoproteus is a long, thin rod that can be bent or branched (figure 18.8*c*). Its cell wall is composed of glycoprotein. *Thermoproteus* is a strict anaerobe and grows at temperatures from 70 to 97°C and pH values between 2.5 and 6.5. It is found in hot springs and other hot aquatic habitats rich in sulfur. It can grow organotrophically and oxidize glucose, amino acids, alcohols, and organic acids with elemental sulfur as the electron acceptor during anaerobic respiration. It will also grow chemolithotrophically using H_2 and S^0 as electron donors. Carbon monoxide or CO_2 can serve as the sole carbon source.

Although the *Crenarchaeota* are notorious for their life at high temperatures and acidic pH, the presence of the crenarchaeote-specific lipid crenarchaeol and sequence analysis of DNA fragments derived directly from environmental samples reveal that this phylum is more widespread in nature. Recall that only a small fraction of microbes have been grown in culture, so the ability to analyze microbial communities using molecular techniques is an important way to truly understand microbial diversity. Such studies reveal that the *Crenarchaeota* have significant populations in marine plankton from polar, temperate, and tropical waters. Crenarchaeotes also appear to inhabit rice paddies, soils, and freshwater lake sediments, and at least two symbiotic species have been isolated, one from a cold water sea cucumber and another from a marine sponge. ◄◄ *Environmental genomics (section 15.8)*

18.3 PHYLUM *EURYARCHAEOTA*

The *Euryarchaeota* is a very diverse phylum with many genera (figure 18.9). We discuss five major physiologic groups within the euryarchaeotes.

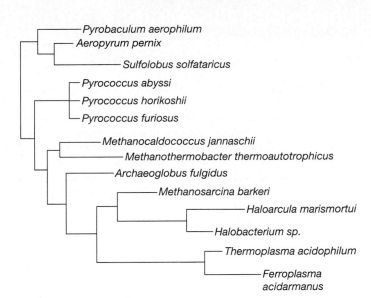

Figure 18.9 **The Phylum *Euryarchaeota*.** A phylogenetic tree developed from the nucleotide sequences of both small and large subunit rRNA sequences.

Methanogens

Methanogens are strict anaerobes that obtain energy by converting CO_2, H_2, formate, methanol, acetate, and other compounds to either methane or methane and CO_2. They are autotrophic when growing on H_2 and CO_2. Some methanogens can live autotrophically by forming acetyl-CoA from two molecules of CO_2 and then converting the acetyl-CoA to pyruvate and other products (figure 18.4*b*).

The methanogens are the largest group of cultured archaea. There are five orders (*Methanobacteriales, Methanococcales, Methanomicrobiales, Methanosarcinales,* and *Methanopyrales*) and 26 genera, which differ greatly in overall shape, 16S rRNA sequence, cell wall chemistry and structure, membrane lipids, and other features. For example, methanogens construct three different types of cell walls. Several genera have walls with pseudomurein; other walls contain either proteins or heteropolysaccharides. The morphology of two representative methanogens is shown in **figure 18.10**, and selected properties of representative genera are presented in **table 18.2**. It should be noted that although almost all archaea in these orders are methanogens, methanotrophs (i.e., organisms that use methane as a carbon and energy source, rather than generate methane) have recently been discovered in the *Methanosarcinales* (**Microbial Diversity & Ecology 18.1**).

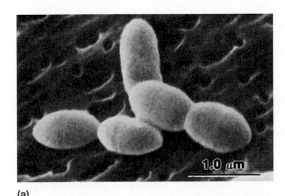

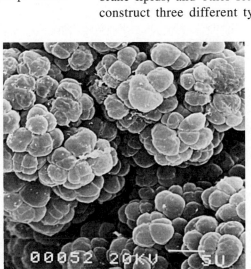

Figure 18.10 Selected Methanogens. (a) *Methanobrevibacter smithii.* (b) *Methanosarcina mazei*; SEM. Bar = 5 μm.

Microbial Diversity & Ecology

18.1 Methanotrophic *Archaea*

The marine environment may contain as much as 10,000 billion tons of methane hydrate buried in the ocean floor, around twice the amount of all known fossil fuel reserves. Although some methane rises toward the surface, it often is used before it escapes from the sediments in which it is buried. This is fortunate because methane is a much more powerful greenhouse gas than carbon dioxide. If the atmosphere were flooded with methane, Earth could become too hot to support life as we know it. The reason for this disappearance of methane in sediments has been unclear until a recent discovery.

By using fluorescent probes for specific DNA sequences, an assemblage of archaea and bacteria was discovered in anoxic, methane-rich sediments. These clusters of procaryotes contain a core of about 100 cells from the order *Methanosarcinales* surrounded by a layer of sulfate-reducing bacteria related to the *Desulfosarcina* (**box figure**). These two groups appear to cooperate metabolically in such a way that methane is anaerobically oxidized and sulfate reduced; perhaps the bacteria use waste products of methane oxidation to derive energy from sulfate reduction. Isotope studies show that the archaea feed on methane and the bacteria get much of their carbon from the archaea. These methanotrophs may be crucial contributors to Earth's carbon cycle because it is thought that they oxidize as much as 300 million tons of methane annually.

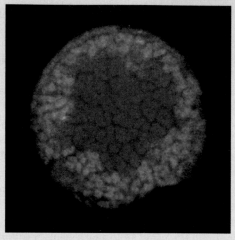

Methane-Consuming Archaea. A cluster of methanotrophic archaea, stained red by a specific fluorescent probe, surrounded by a layer of bacteria labeled by a green fluorescent probe.

| **Table 18.2** | **Selected Characteristics of Representative Genera of Methanogens** |

Genus	Morphology	% G + C	Wall Composition	Gram Reaction	Motility	Methanogenic Substrates Used
Order *Methanobacteriales*						
Methanobacterium	Long rods or filaments	32–61	Pseudomurein	+ to variable	−	$H_2 + CO_2$, formate
Methanothermus	Straight to slightly curved rods	33	Pseudomurein with an outer protein S-layer	+	+	$H_2 + CO_2$
Order *Methanococcales*						
Methanococcus	Irregular cocci	29–34	Protein	−	+	$H_2 + CO_2$, formate
Order *Methanomicrobiales*						
Methanomicrobium	Short curved rods	45–49	Protein	−	+	$H_2 + CO_2$, formate
Methanogenium	Irregular cocci	52–61	Protein or glycoprotein	−	−	$H_2 + CO_2$, formate
Methanospirillum	Curved rods or spirilla	47–52	Protein	−	+	$H_2 + CO_2$, formate
Order *Methanosarcinales*						
Methanosarcina	Irregular cocci, packets	36–43	Heteropolysaccharide or protein	+ to variable	−	$H_2 + CO_2$, methanol, methylamines, acetate

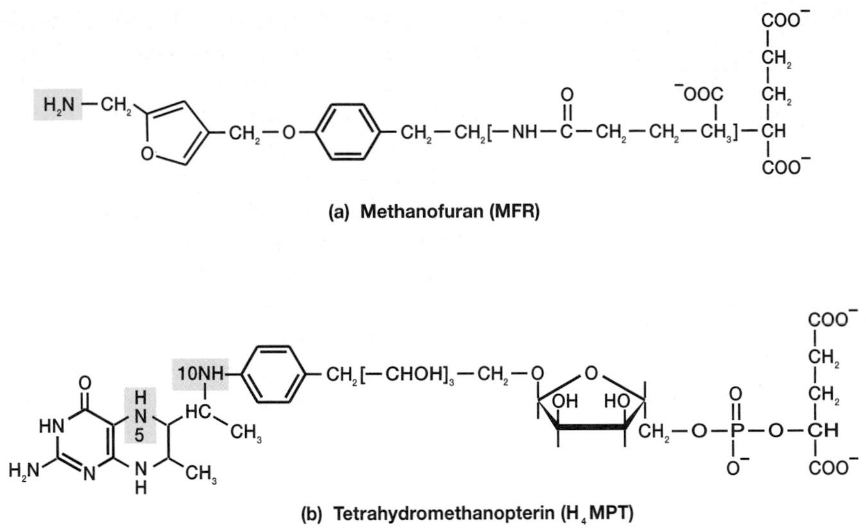

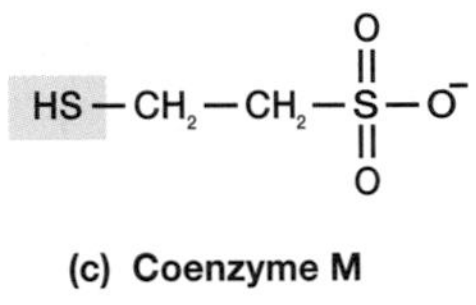

(a) Methanofuran (MFR)

(b) Tetrahydromethanopterin (H₄MPT)

(c) Coenzyme M

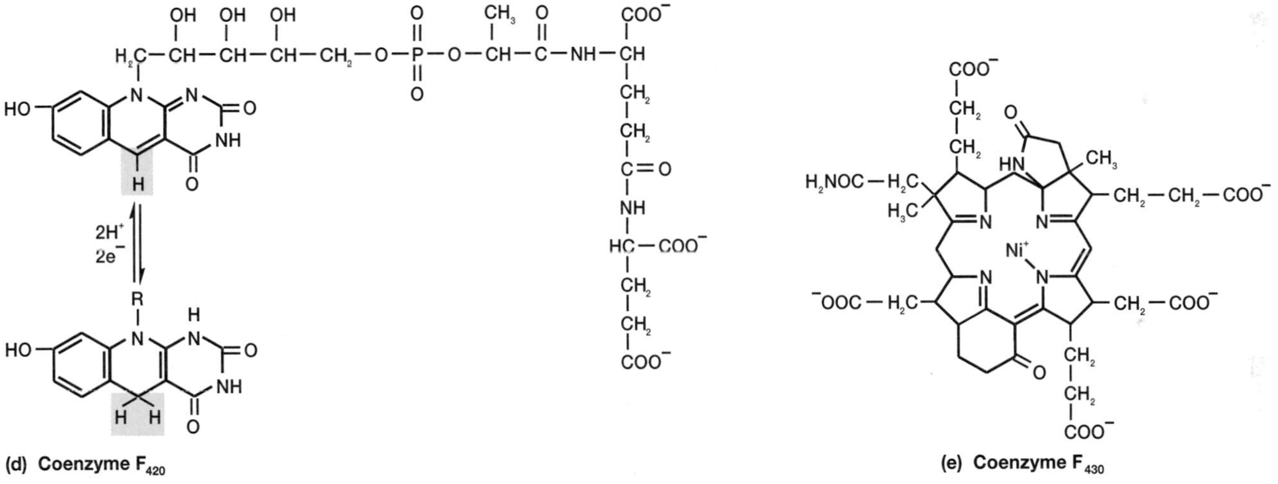

(d) Coenzyme F₄₂₀

(e) Coenzyme F₄₃₀

Figure 18.11   Methanogen Coenzymes. (a) Coenzyme MFR, (b) H₄MPT, and (c) coenzyme M are used to carry one-carbon units during methanogenesis. MFR and a simpler form of H₄MPT called methanopterin (MPT; not shown) also participate in the synthesis of acetyl-CoA. The portions of the coenzymes that carry the one-carbon units are shown in blue. H₄MPT carries carbon units on nitrogens 5 and 10, like the more common enzyme tetrahydrofolate. (d) Coenzyme F₄₂₀ participates in redox reactions. The part of the molecule that is reversibly oxidized and reduced is highlighted. (e) Coenzyme F₄₃₀ participates in reactions catalyzed by the enzyme methyl-CoM methylreductase.

One of the most unusual methanogenic groups is the class *Methanopyri*. It has one order, *Methanopyrales,* one family, and a single genus, *Methanopyrus*. This hyperthermophilic, rod-shaped methanogen has been isolated from a marine hydrothermal vent. *Methanopyrus kandleri* has a temperature minimum at 84°C and an optimum of 98°C; it will grow at temperatures up to 110°C. *Methanopyrus* occupies the deepest and most ancient branch of the euryarchaeotes. Perhaps methanogenic archaeal ancestors were among the earliest organisms. They certainly seem well adapted to living under conditions similar to those presumed to have existed on a young Earth.

As might be inferred from the methanogens' ability to produce methane anaerobically, their metabolism is unusual. These procaryotes contain several unique cofactors: tetrahydromethanopterin (H₄MPT), methanofuran (MFR), coenzyme M (2-mercaptoethanesulfonic acid), coenzyme F₄₂₀, and coenzyme F₄₃₀ (**figure 18.11**). The first three cofactors bear the single carbon unit when

CO_2 is reduced to CH_4. F_{420} carries electrons and protons, and F_{430} is a nickel tetrapyrrole serving as a cofactor for the enzyme methyl-CoM methylreductase. The pathway for methane synthesis is thought to function as shown in **figure 18.12**. It appears that ATP synthesis is linked with methanogenesis by electron transport, proton pumping, and a chemiosmotic mechanism. The means by which a proton motive force is generated in not known but may involve the oxidation of H_2 on the outer surface of the membrane. In addition, the transfer of the methyl groups from methyl-H₄MPT to HS-CoM releases sufficient energy for the uptake of sodium ions. It has been suggested that the resulting sodium motive force also drives ATP synthesis. Thus two types of ATP synthases appear to be present—one driven by proton movement and another coupled to Na^+ translocation. ◀◀ *Electron transport and oxidative phosphorylation (section 10.5)*

Methanogens thrive in anoxic environments rich in organic matter: the rumen and intestinal system of animals, freshwater

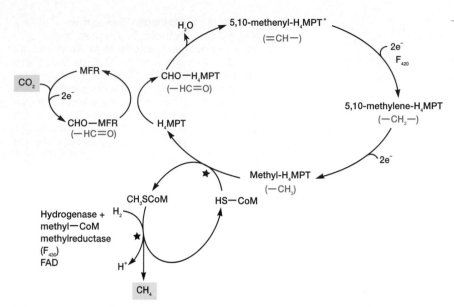

Figure 18.12 Methane Synthesis. Pathway for CH_4 synthesis from CO_2 in *M. thermoautotrophicus.* Cofactor abbreviations: methanopterin (MPT), methanofuran (MFR), and 2-mercaptoethanesulfonic acid or coenzyme M (CoM). The nature of the carbon-containing intermediates leading from CO_2 to CH_4 are indicated in parentheses. Stars indicate sites of energy conservation.

and marine sediments, swamps and marshes, hot springs, anoxic sludge digesters, and even within anaerobic protozoa. Methanogens often are of ecological significance. The rate of methane production can be so great that bubbles of methane sometimes rise to the surface of a lake or pond. Rumen methanogens are so active that a cow can belch 200 to 400 liters of methane a day. >> *Microbial interactions: The rumen ecosystem (section 27.1)*

Methanogenic archaea are potentially of great practical importance since methane is a clean-burning fuel and an excellent energy source. Anaerobic digester microbes degrade particulate wastes such as sewage sludge to H_2, CO_2, and acetate. CO_2-reducing

methanogens form CH_4 from CO_2 and H_2, while aceticlastic methanogens cleave acetate to CO_2 and CH_4 (about two-thirds of the methane produced by an anaerobic digester comes from acetate). A kilogram of organic matter can yield up to 600 liters of methane. It is quite likely that future research will greatly increase the efficiency of methane production and make methanogenesis an important source of pollution-free energy. >> *Wastewater treatment (section 35.2)*

Methanogenesis also can be an ecological problem. Methane absorbs infrared radiation and thus is a greenhouse gas, and atmospheric methane concentrations have been rising over the last 200 years. Methane production may significantly promote future global warming; this subject is discussed in chapter 26.

Halobacteria

The **extreme halophiles** or **halobacteria,** order *Halobacteriales*, are another major euryarchaeal group, currently with 17 genera in one family, the *Halobacteriaceae* (**figure 18.13**). Most are aerobic chemoorganotrophs with respiratory metabolism. Extreme halophiles demonstrate a wide variety of nutritional capabilities. The first **halophiles** were isolated from salted fish in the 1880s and required complex nutrients for growth. More recent isolates grow best in defined media, using carbohydrates or simple compounds such as glycerol, acetate, or pyruvate as their carbon source. Halophiles can be motile or nonmotile and are found in a variety of cell shapes. These include cubes and pyramids in addition to rods and cocci.

The most obvious distinguishing trait of this family is its absolute dependence on a high concentration of NaCl. These procaryotes

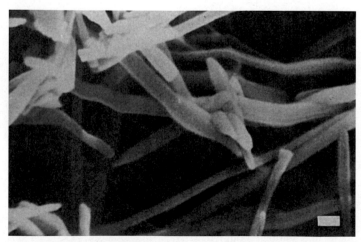

(a) *Halobacterium salinarium*

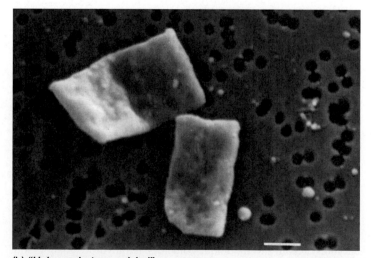

(b) *"Haloquadratum walsbyi"*

Figure 18.13 Examples of Halobacteria. (a) *Halobacterium.* A young culture that has formed long rods; SEM. Bar = 1 µm. (b) *"Haloquadratum walsbyi"*; SEM. Bar = 1 µm.

require at least 1.5 M NaCl (about 8%, wt/vol) and usually have a growth optimum at about 3 to 4 M NaCl (17 to 23%). They will grow at salt concentrations approaching saturation (about 36%). The cell walls of most halobacteria are so dependent on the presence of NaCl that they disintegrate when the NaCl concentration drops below 1.5 M. Thus halobacteria only grow in high-salinity habitats such as marine salterns and salt lakes such as the Dead Sea between Israel and Jordan, and the Great Salt Lake in Utah. Halobacteria often have red-to-yellow pigmentation from carotenoids that are probably used as protection against strong sunlight. They can reach such high population levels that salt lakes, salterns, and salted fish actually turn red. In fact, halophiles are used in the production of many salted food products.

Probably the best-studied member of the family is *Halobacterium salinarium*. This archaeon is unusual because it produces a protein called **bacteriorhodopsin** that can trap light energy without the presence of chlorophyll. Structurally similar to the rhodopsin found in the mammalian eye, bacteriorhodopsin functions as a light-driven proton pump. Like all members of the rhodopsin family, bacteriorhodopsin has two distinct features: (1) a chromophore that is a derivative of retinal (an aldehyde of vitamin A), which is covalently attached to the protein by a Schiff base with the amino group of lysine (**figure 18.14**); and (2) seven membrane-spanning domains connected by loops on either side of the membrane with the retinal resting within the membrane. Bacteriorhodopsin molecules form aggregates in a modified region of the plasma membrane called the **purple membrane.** When retinal absorbs light, the double bond between carbons 13 and 14 changes from a trans to a cis configuration and the Schiff base loses a proton. Protons move across the plasma membrane to the periplasmic space during these alterations, and the Schiff base changes are directly involved in this movement (figure 18.14). Bacteriorhodopsin undergoes several conformational changes during the photocycle. These conformational changes also are involved in proton transport. The light-driven proton pumping generates a pH gradient that can be used to power the synthesis of ATP by a chemiosmotic mechanism. << *Electron transport and oxidative phosphorylation (section 10.5); Rhodopsin-based phototrophy (section 10.12)*

Halobacterium has three additional rhodopsins, each with a different function. Halorhodopsin uses light energy to transport chloride ions into the cell and maintain a 4 to 5 M intracellular KCl concentration. The two additional rhodopsins are called sensory rhodopsin I (SRI) and SRII. **Sensory rhodopsins** act as photoreceptors; in this case, one for red light and one for blue. They control flagellar activity to position the organism optimally in the water column. *Halobacterium* moves to a location of high light intensity that lacks the ultraviolet light that would be lethal.

Surprisingly, it now appears rhodopsin is widely distributed among procaryotes. DNA sequence analysis of uncultivated marine bacterioplankton reveals the presence of rhodopsin genes among α- and β-proteobacteria and the *Bacteroidetes*. This newly discovered rhodopsin is called **proteorhodopsin.** Cyanobacteria also have sensory rhodopsin proteins, which like SRI and SRII of the halobacteria, sense the spectral quality of light. All these procaryotic rhodopsins conserve the seven transmembrane helices through the cell membrane and the lysine residue that forms the Schiff base linkage with retinal. >> *Marine and freshwater microbiology (section 26.1)*

Thermoplasms

Procaryotes in the class *Thermoplasmata* are thermoacidophiles that lack cell walls. At present, three genera, *Thermoplasma, Picrophilus*, and *Ferroplasma* are known. They are sufficiently different from one another to be placed in separate families, *Thermoplasmataceae, Picrophilaceae,* and *Ferroplasmataceae.*

Thermoplasma grows in refuse piles of coal mines. These piles contain large amounts of iron pyrite (FeS), which is oxidized to sulfuric acid by chemolithotrophic bacteria. As a result, the piles become very hot and acidic. This is an ideal habitat for *Thermoplasma* because it grows best at 55 to 59°C and pH 1 to 2. Although it lacks a cell wall, its plasma membrane is strengthened by large quantities of diglycerol tetraethers, lipid-containing polysaccharides, and glycoproteins. The organism's DNA is stabilized by association with archaeal histones that condense the DNA into structures resembling eucaryotic nucleosomes. At 59°C, *Thermoplasma* takes the form of an irregular filament, whereas at lower temperatures it is spherical. The cells may be flagellated and motile.

Picrophilus is even more unusual than *Thermoplasma*. It originally was isolated from moderately hot solfataric fields in Japan. *Picrophilus* has an S-layer outside its plasma membrane. The cells grow as irregularly shaped cocci, around 1 to 1.5 μm in diameter, and have large cytoplasmic cavities that are not membrane bounded. *Picrophilus* is aerobic and grows between 47 and 65°C with an optimum of 60°C. It is most remarkable in its pH requirements: it grows only below pH 3.5 and has a growth optimum at pH 0.7. Growth even occurs at about pH 0.

Extremely Thermophilic S°-Reducers

This physiological group contains the class *Thermococci*, with one order, *Thermococcales*. The *Thermococcales* are strictly anaerobic and can reduce sulfur to sulfide. They are motile by flagella and have optimum growth temperatures around 88 to 100°C. The order contains one family and three genera, *Thermococcus, Paleococcus*, and *Pyrococcus*.

Sulfate-Reducing *Euryarchaeota*

Euryarchaeal sulfate reducers are found in the class *Archaeoglobi* and the order *Archaeoglobales*. This order has only one family and three genera. *Archaeoglobus* contains gram-negative-staining, irregular coccoid cells with cell walls consisting of glycoprotein subunits. It can extract electrons from a variety of electron donors (e.g., H_2, lactate, glucose) and reduce sulfate, sulfite, or thiosulfate to sulfide. Elemental sulfur is not used as an acceptor. *Archaeoglobus* is extremely thermophilic (the optimum is about 83°C) and has been isolated from marine hydrothermal vents. It is not only unusual in being able to reduce sulfate, unlike

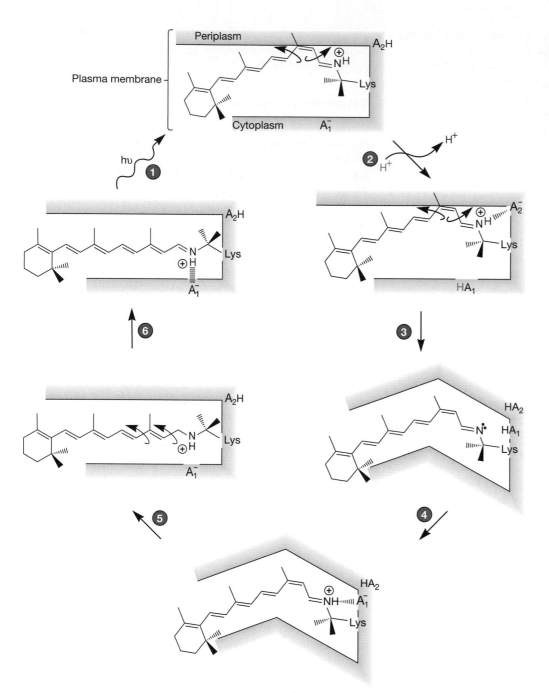

Figure 18.14 The Photocycle of Bacteriorhodopsin. In this hypothetical mechanism, the retinal component of bacteriorhodopsin is buried in the membrane and retinal interacts with two amino acids, A_1 and A_2 (aspartates 96 and 85), that can reversibly accept and donate protons. A_2 is connected to the cell exterior, and A_1 is closer to the cell interior. Light absorption by retinal in step 1 triggers an isomerization from 13-*trans*-retinal to 13-*cis*-retinal. The retinal then donates a proton to A_2 in steps 2 and 3, while A_1 is picking up another proton from the interior and A_2 is releasing a proton to the outside. In steps 4 and 5, retinal obtains a proton from A_1 and isomerizes back to the 13-*trans* form. The cycle is then ready to begin again after step 6.

other archaea, but it also possesses the methanogen coenzymes F_{420} and methanopterin.

1. What are thermoacidophiles and where do they grow? In what ways do they use sulfur in their metabolism? Briefly describe *Sulfolobus* and *Thermoproteus*.

2. Generally characterize methanogenic archaea and distinguish them from other groups.

3. Briefly describe how methanogens produce methane and the roles of their unique cofactors in this process.

4. Where does one find methanogens? Discuss their ecological and practical importance.

5. Where are the extreme halophiles found and what is unusual about their cell walls and growth requirements?

6. What is the purple membrane and what pigment does it contain?

7. How is *Thermoplasma* able to live in acidic, very hot coal refuse piles when it lacks a cell wall? How is its DNA stabilized? What is so remarkable about *Picrophilus*?

8. Characterize *Archaeoglobus*. In what way is it similar to the methanogens and how does it differ from other extreme thermophiles?

Summary

18.1 Introduction to the *Archaea*

a. The *Archaea* are highly diverse with respect to morphology, reproduction, physiology, and ecology. Although best known for their growth in anoxic, hypersaline, and high-temperature habitats, they also inhabit marine Arctic, temperate, and tropical waters.

b. *Archaea* may be divided into five groups: methanogenic archaea, sulfate reducers, extreme halophiles, cell wall-less archaea, and extremely thermophilic S^0-metabolizers (**table 18.1**).

c. The second edition of *Bergey's Manual* divides the *Archaea* into two phyla, the *Crenarchaeota* and *Euryarchaeota;* however, this may eventually be revised to include additional phyla (**figure 18.2**).

d. Archaeal cell walls do not contain peptidoglycan and differ from bacterial walls in structure. They may be composed of pseudomurein, polysaccharides, or glycoproteins and other proteins.

e. Archaeal membrane lipids differ from those of other organisms in having branched chain hydrocarbons connected to glycerol by ether links. Bacterial and eucaryotic lipids have glycerol connected to fatty acids by ester bonds.

f. Their tRNA, ribosomes, elongation factors, RNA polymerases, and other components distinguish *Archaea* from *Bacteria* and eucaryotes.

g. Although much of archaeal metabolism appears similar to that of other organisms, the *Archaea* differ with respect to glucose catabolism, pathways for CO_2 fixation, and the ability of some to synthesize methane (**figure 18.4**).

18.2 Phylum *Crenarchaeota*

a. The extremely thermophilic S^0-metabolizers in the phylum *Crenarchaeota* depend on sulfur for growth and are frequently acidophiles. The sulfur may be used as an electron acceptor in anaerobic respiration or as an electron donor by chemolithotrophs. Many are strict anaerobes and grow in geothermally heated soil and water that is rich in sulfur.

b. More recent metagenomic evidence suggests that *Crenarchaeota* are not confined to extreme environments.

18.3 Phylum *Euryarchaeota*

a. The phylum *Euryarchaeota* contains five major groups: methanogens, halobacteria, the thermoplasms, extremely thermophilic S^0-reducers, and sulfate-reducing archaea.

b. Methanogenic archaea are strict anaerobes that can obtain energy through the synthesis of methane. They have several unusual cofactors that are involved in methanogenesis (**figures 18.11 and 18.12**).

c. Extreme halophiles or halobacteria are aerobic chemoheterotrophs that require at least 1.5 M NaCl for growth. They are found in habitats such as salterns, salt lakes, and salted fish.

d. *Halobacterium salinarum* can carry out phototrophy without chlorophyll or bacteriochlorophyll by using bacteriorhodopsin, which employs retinal to pump protons across the plasma membrane (**figure 18.14**).

e. The thermophilic archaeon *Thermoplasma* grows in hot, acidic coal refuse piles and survives despite the lack of a cell wall. Another thermoplasm, *Picrophilus*, can grow at pH 0.

f. The class *Thermococci* contains extremely thermophilic organisms that can reduce sulfur to sulfide.

g. Sulfate-reducing archaea are placed in the class *Archaeoglobi*. The extreme thermophile *Archaeoglobus* differs from other archaea in using a variety of electron donors to reduce sulfate. It also contains the methanogen cofactors F_{420} and methanopterin.

Critical Thinking Questions

1. Some believe that the *Archaea* should be not be separate from the *Bacteria* because both groups are procaryotic. Do you agree or disagree? Give your reasoning.

2. Explain why the fixation of CO_2 by *Thermoproteus* and possibly by *Sulfolobus* using a reductive reversal of the TCA cycle is not photosynthesis. Recently members of the ε-proteobacteria that inhabit hydrothermal vents have been shown to fix CO_2 using the reductive TCA cycle. How could you determine if this was an instance of horizontal gene transfer or convergent evolution?

3. When the temperature increases, some procaryotes change their shapes from elongated rods into spheres. Suggest one reason for this change.

4. Why would ether linkages be more stable in membranes than ester lipids? How would the presence of tetraether linkages stabilize a thermophile's membrane?

5. Suppose you wished to isolate procaryotes from a hot spring in Yellowstone National Park. How would you go about it?

Learn More

Learn more by visiting the Prescott website at www.mhhe.com/prescottprinciples, where you will find a complete list of references.

19

Chapter Glossary

akinetes Specialized, nonmotile, dormant, thick-walled resting cells formed by some cyanobacteria.

anoxygenic photosynthesis Photosynthesis that does not oxidize water to produce oxygen; a form of photosynthesis characteristic of purple and green photosynthetic bacteria and heliobacteria.

axial fibrils or **periplasmic flagella** The flagella that lie under the outer sheath and extend from both ends of the spirochete cell to overlap in the middle and form the axial filament.

carboxysome Polyhedral inclusion bodies thought to contain the CO_2-fixation enzyme ribulose 1,5-bisphosphate carboxylase; found in cyanobacteria, nitrifying bacteria, and thiobacilli.

chancre The primary lesion of syphilis occurring at the site of entry of the infection.

chlamydiae Obligate intracellular bacteria that have a unique mode of reproduction.

chlorosomes Elongated, intramembranous vesicles found in green sulfur and nonsulfur bacteria; contain light-harvesting pigments. Sometimes called chlorobium vesicles.

cyanobacteria A large group of gram-negative bacteria that carry out oxygenic photosynthesis using a system like that present in photosynthetic eucaryotes.

elementary body (EB) A small, dormant body that serves as the agent of transmission between host cells in the chlamydial life cycle.

gliding motility A type of motility in which a microbial cell moves along a solid surface.

green nonsulfur bacteria Anoxygenic photosynthetic bacteria that contain bacteriochlorophylls *a* and *c*; usually photoheterotrophic and display gliding motility; include members of the phylum *Chloroflexi*.

green sulfur bacteria Anoxygenic photosynthetic bacteria that contain bacteriochlorophylls *a,* plus *c, d,* or *e;* photolitho-

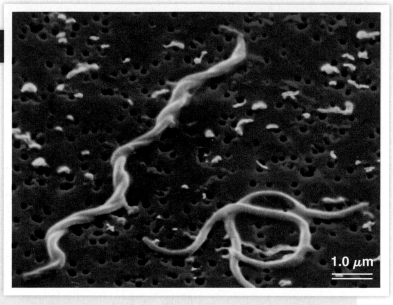

Scanning electron micrograph of *Borrelia burgdorferi,* which causes Lyme disease. These slender spirochetes can grow to 20 µm in length.

autotrophic; use H_2, H_2S, or S^0 as electron donor; include members of the phylum *Chlorobi.*

heterocysts Specialized cells of cyanobacteria that are the sites of nitrogen fixation.

oxygenic photosynthesis Photosynthesis that oxidizes water to form oxygen; the form of photosynthesis characteristic of plants, protists, and cyanobacteria.

phototaxis The ability of certain phototrophic bacteria to move, either by gliding or swimming motility, in response to a light source.

phycobilisomes Special particles on the membranes of cyanobacteria that contain photosynthetic pigments and electron transport chains.

reticulate body (RB) The cellular form in the chlamydial life cycle whose role is growth and reproduction within the host cell.

There are wide areas of the bacteriological landscape in which we have so far detected only some of the highest peaks, while the rest of the beautiful mountain range is still hidden in the clouds and the morning fogs of ignorance. The goal is still lying on the ground, but we have to bend down to grasp it.

—*M. Dworkin et al., Preface to* The Prokaryotes

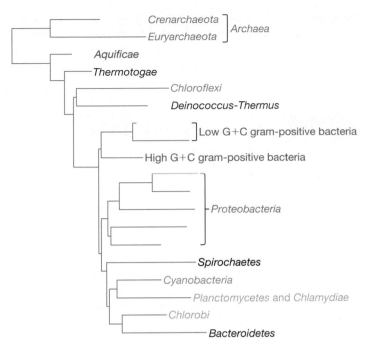

Figure 19.1 Phylogenetic Relationships Among Procaryotes. The *Deinococcus-Thermus* group and other nonproteobacterial gram-negatives are highlighted.

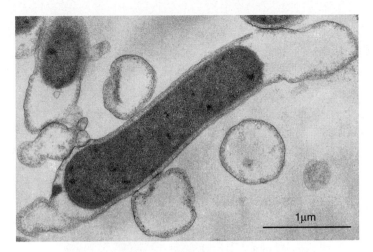

Figure 19.2   *Thermotoga maritima.* Note the loose sheath extending from each end of the cell.

Volumes 1 and 5 of *Bergey's Manual of Systematic Bacteriology* describe a wide variety of other procaryotic groups that are members of the second domain: *Bacteria*. This chapter is devoted to 10 of these bacterial phyla. Their phylogenetic locations are depicted in **figure 19.1**. We follow the general organization and perspective of the second edition of *Bergey's Manual* in most cases. In describing each bacterial group, we include aspects such as distinguishing characteristics, morphology, reproduction, physiology, metabolism, and ecology. The taxonomy of each major group is summarized, and representative species are discussed. In cases where species within a group are the causative agents of important infectious diseases, we also review the disease process.

19.1 AQUIFICAE AND THERMOTOGAE

Thermophilic microbes are found in both the bacterial and archaeal domains, but to date, all hyperthermophilic procaryotes (those with optimum growth temperatures above 85°C) belong to the *Archaea*. The phyla *Aquificiae* and *Thermotoga* are two examples of bacterial thermophiles.

The phylum *Aquificae* is thought to represent the deepest or oldest branch of *Bacteria* (see figure 17.4). It contains one class, one order, and eight genera. Two of the best-studied genera are *Aquifex* and *Hydrogenobacter*. *Aquifex pyrophilus* is a gram-negative, microaerophilic rod. It is thermophilic with a temperature optimum of 85°C and a maximum of 95°C. *Aquifex* is a chemo-lithoautotroph that captures energy by oxidizing hydrogen, thiosulfate, and sulfur with oxygen as the terminal electron

acceptor. Because *Aquifex* and *Hydrogenobacter* are thermophilic chemolithoautotrophs, it has been suggested that the original bacterial ancestor was probably thermophilic and chemolithoautotrophic. ≪ *Chemolithotrophy (section 10.11)*

The second deepest branch is the phylum *Thermotogae*, which has one class, one order, and six genera. The members of the genus *Thermotoga* (Greek *therme*, heat; Latin *toga*, outer garment), like *Aquifex*, are thermophiles with a growth optimum of 80°C and a maximum of 90°C. They are gram-negative rods with an outer sheathlike envelope (like a toga) that can extend or balloon out from the ends of the cell (**figure 19.2**). They grow in active geothermal areas found in marine hydrothermal systems and terrestrial solfataric springs. In contrast to *Aquifex*, *Thermotoga* is a chemoheterotroph with a functional glycolytic pathway that can grow anaerobically on carbohydrates and protein digests.

The genome of *Aquifex* is about a third the size of the *Escherichia coli* genome and, as expected, contains the genes required for chemolithoautotrophy. The *Thermotoga* genome is somewhat larger and has genes for sugar degradation. About 24% of its coding sequences are similar to archaeal genes; this proportion is greater than that of other bacteria, including *Aquifex* (16% similarity) and may be due to lateral (horizontal) gene transfer. ≪ *Comparative genomics (section 15.7)*

19.2 DEINOCOCCUS-THERMUS

The phylum *Deinococcus-Thermus* contains the class *Deinococci* and the orders *Deinococcales* and *Thermales*. There are only three genera in the phylum; the genus *Deinococcus* is best studied. Deinococci are spherical or rod-shaped and nonmotile. They often are associated in pairs or tetrads (**figure 19.3a**) and are aerobic, mesophilic, and catalase positive; usually they can produce acid from only a few sugars. Although they stain gram

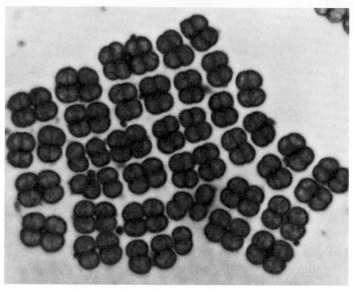

(a)

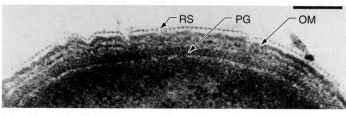

(b)

Figure 19.3 The Deinococci. (a) A *Deinococcus radiodurans* microcolony showing cocci arranged in tetrads (average cell diameter 2.5 µm). (b) The cell wall of *D. radiodurans* with a regular surface protein array (RS), a peptidoglycan layer (PG), and an outer membrane (OM). Bar = 100 nm.

positive, their cell wall is layered with an outer membrane like gram-negative bacteria (figure 19.3*b*). They also differ from gram-positive cocci in having L-ornithine in their peptidoglycan, lacking teichoic acid, and having a plasma membrane with large amounts of palmitoleic acid rather than phosphatidylglycerol phospholipids. Almost all strains are extraordinarily resistant to both desiccation and radiation; they can survive as much as 3 to 5 million rad of radiation (an exposure of 100 rad can be lethal to humans).

Much remains to be discovered about the biology of these bacteria. Deinococci can be isolated from ground meat, feces, air, freshwater, and other sources, but their natural habitat is not yet known. Their great resistance to radiation results from their ability to repair a severely damaged genome, which consists of a circular chromosome, a megaplasmid, and a small plasmid. When exposed to high levels of radiation, the genome is broken into many fragments. Within 12 to 24 hours, the genome is pieced back together, ensuring viability. It is unclear how this is accomplished. Genome studies have shown that *D. radiodurans* has a very efficient DNA repair system; however, novel DNA repair genes have not yet been reported (*see figure 15.9*). It appears that

the ability to accumulate high levels of Mn(II) may help protect the microbe from high levels of radiation-induced toxic oxygen species. DNA repair by this fascinating microbe is discussed more fully in chapter 15 (section 15.5).

19.3 PHOTOSYNTHETIC BACTERIA

There are three groups of gram-negative photosynthetic bacteria: purple bacteria, green bacteria, and cyanobacteria (**table 19.1**). The cyanobacteria differ fundamentally from the green and purple photosynthetic bacteria in being able to carry out **oxygenic photosynthesis.** They have photosystems I and II, use water as an electron donor, and generate oxygen during photosynthesis. In contrast, purple and green bacteria have only one photosystem and use **anoxygenic photosynthesis.** Because they are unable to use water as an electron source, they employ reduced molecules such as hydrogen sulfide, sulfur, hydrogen, and organic matter as their electron source for the reduction of NAD(P)$^+$ to NAD(P)H. Consequently many purple and green bacteria form sulfur granules. Purple sulfur bacteria accumulate granules within their cells, whereas green sulfur bacteria deposit the sulfur granules outside their cells. The purple nonsulfur bacteria normally use organic molecules as an electron source. Differences also occur in photosynthetic pigments, the organization of photosynthetic membranes, nutritional requirements, and oxygen relationships. ◀◀ *Phototrophy (section 10.12)*

Photosynthetic microbes contribute a significant amount of fixed carbon in a variety of habitats, and the differences in photosynthetic pigments and oxygen requirements among the photosynthetic bacteria have important ecological consequences. As shown in **figure 19.4,** the chlorophylls (Chl), bacteriochlorophylls (Bchl), and their associated accessory pigments have distinct absorption spectra. Oxygenic cyanobacteria and photosynthetic protists dominate the aerated upper layers of freshwater and marine microbial communities, where they absorb large amounts of red and blue light. Below these microbes, the anoxygenic purple and green photosynthetic bacteria inhabit the deeper anoxic zones that are rich in hydrogen sulfide and other reduced compounds that can be used as electron donors. Their bacteriochlorophyll and accessory pigments enable them to use light in the far-red spectrum that is not used by other photosynthetic organisms. In addition, the bacteriochlorophyll absorption peaks at about 350 to 550 nm, allowing them to grow at greater depths because shorter wavelength light can penetrate water farther. As a result, when the water is sufficiently clear, a layer of green and purple bacteria develops in the anoxic, hydrogen sulfide-rich zone.

Bergey's Manual places photosynthetic bacteria into seven major groups distributed between five bacterial phyla. The phylum *Chloroflexi* contains the green nonsulfur bacteria, and the phylum *Chlorobi,* the green sulfur bacteria. The cyanobacteria are placed in their own phylum, *Cyanobacteria.* Purple bacteria are divided between three groups. Purple sulfur bacteria are

| Table 19.1 | Characteristics of the Major Groups of Gram-Negative Photosynthetic Bacteria | | | | |

Characteristic	Anoxygenic Photosynthetic Bacteria				Oxygenic Photosynthetic Bacteria
	Green Sulfur[a]	Green Nonsulfur[b]	Purple Sulfur	Purple Nonsulfur	Cyanobacteria
Major photosynthetic pigments	Bacteriochlorophylls *a* plus *c, d,* or *e* (the major pigment)	Bacteriochlorophylls *a* and *c*	Bacteriochlorophyll *a* or *b*	Bacteriochlorophyll *a* or *b*	Chlorophyll *a* plus phycobiliproteins *Prochlorococcus* has divinyl derivatives of chlorophyll *a* and *b*
Morphology of photosynthetic membranes	Photosynthetic system partly in chlorosomes that are independent of the plasma membrane	Chlorosomes present when grown anaerobically	Photosynthetic system contained in spherical or lamellar membrane complexes that are continuous with the plasma membrane	Photosynthetic system contained in spherical or lamellar membrane complexes that are continuous with the plasma membrane	Thylakoid membranes lined with phycobilisomes
Photosynthetic electron donors	H_2, H_2S, S^0	Photoheterotrophic donors—a variety of sugars, amino acids, and organic acids; photoautotrophic donors—H_2S, H_2	H_2, H_2S, S^0	Usually organic molecules: sometimes reduced sulfur compounds or H_2	H_2O
Sulfur deposition	Outside of the cell	N/A[c]	Inside the cell[d]	Outside of the cell in a few cases	N/A
Nature of photosynthesis	Anoxygenic	Anoxygenic	Anoxygenic	Anoxygenic	Oxygenic (some are also facultatively anoxygenic)
General metabolic type	Obligate anaerobic photolithoautotrophs	Usually photoheterotrophic; sometimes photoautotrophic or chemoheterotrophic (when aerobic and in the dark)	Obligate anaerobic photolithoautotrophs	Usually anaerobic photoorgano-heterotrophs; some facultative photolithoautotrophs (in the dark, chemo-organoheterotrophs)	Aerobic photo-lithoautotrophs
Motility	Nonmotile; some have gas vesicles	Gliding	Motile with polar flagella; some are peritrichously flagellated	Motile with polar flagella or nonmotile; some have gas vesicles	Nonmotile, swimming motility without flagella or gliding motility; some have gas vesicles
Percent G + C	48–58	53–55	45–70	61–72	35–71
Phylum or class	*Chlorobi*	*Chloroflexi*	γ-proteobacteria	α-proteobacteria, β-proteobacteria (*Rhodocyclus*)	*Cyanobacteria*

[a]Characteristics of *Chlorobi.*
[b]Characteristics of *Chloroflexus.*
[c]N/A. Not applicable
[d]With the exception of *Ectothiorhodospira.*

placed in the γ-proteobacteria, families *Chromatiaceae* and *Ectothiorhodospiraceae.* The purple nonsulfur bacteria are distributed between the α-proteobacteria (five different families) and one family of the β-proteobacteria. Finally, the gram-positive heliobacteria in the phylum *Firmicutes* are also photosynthetic. There appears to have been considerable horizontal transfer of photosynthetic genes among the five phyla. At least 50 genes related to photosynthesis are common to all five. In this chapter, we describe the cyanobacteria and green bacteria; purple bacteria are discussed in chapter 20, while heliobacteria are featured in chapter 21.

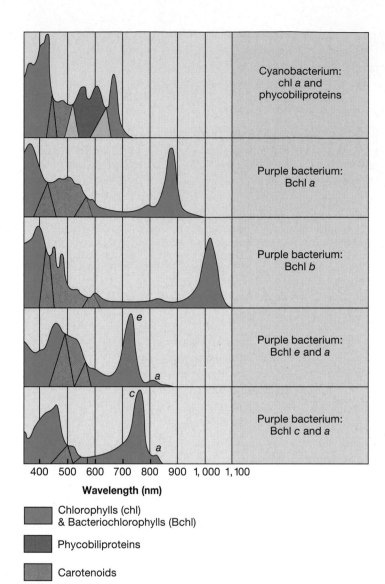

Chlorophylls (chl)
& Bacteriochlorophylls (Bchl)

Phycobiliproteins

Carotenoids

Figure 19.4 Photosynthetic Pigments. Absorption spectra of five photosynthetic bacteria showing the differences in absorption maxima and the contributions of various accessory pigments.

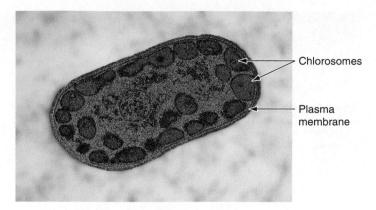

Figure 19.5 The Green Sulfur Bacterium *Chlorobium*. Light-harvesting chlorosomes surround the inside perimeter of the cytoplasmic membrane. Antenna bacteriochlorophylls (Bchl *c, d,* or *e*) transfer light energy to reaction center Bchl *a* in the plasma membrane.

Phylum *Chlorobi*

The phylum *Chlorobi* has only one class (*Chlorobia*), order (*Chlorobiales*), and family (*Chlorobiaceae*). The **green sulfur bacteria** are a small group of obligately anaerobic photolithoautotrophs that use hydrogen sulfide, elemental sulfur, and hydrogen as electron sources. The elemental sulfur produced by sulfide oxidation is deposited outside the cell. Their photosynthetic pigments are located in ellipsoidal vesicles called chlorosomes or chlorobium vesicles, which are attached to the plasma membrane but are not continuous with it (**figure 19.5**). **Chlorosomes** are the most efficient light-harvesting complexes found in nature. The chlorosome membrane is not a normal lipid bilayer. Instead, bacteriochlorophyll molecules are grouped into lateral arrays held together by carotenoids and lipids. Chlorosomes contain accessory (antenna) Bchl pigments *b, d,* or *e*; reaction centers with Bchl *a*

are located in the plasma membrane, where they obtain energy from chlorosome pigments. These bacteria flourish in the anoxic, sulfide-rich zones of lakes. Although they lack flagella and are nonmotile, some species have gas vesicles to adjust their depth for optimal light and hydrogen sulfide. Those without vesicles are found in sulfide-rich muds at the bottom of lakes and ponds.

The green sulfur bacteria are very diverse morphologically. They may be rods, cocci, or vibrios; some grow singly, and others form chains and clusters. They are either grass-green or chocolate-brown in color. Representative genera are *Chlorobium, Prosthecochloris,* and *Pelodictyon.*

Phylum *Chloroflexi*

The phylum *Chloroflexi* has both photosynthetic and nonphotosynthetic members. *Chloroflexus* is the major representative of the photosynthetic **green nonsulfur bacteria.** However, the term green nonsulfur is a misnomer because not all members of this group are green, and some can use sulfur. *Chloroflexus* is a filamentous, gliding, thermophilic bacterium that often is isolated from neutral to alkaline hot springs, where it grows in the form of orange-reddish mats, usually in association with cyanobacteria. It possesses small chlorosomes with accessory Bchl *c*. Its light-harvesting complexes contain Bchl *a* and are in the plasma membrane. Metabolically, it is similar to that of the purple nonsulfur bacteria. *Chloroflexus* can carry out anoxygenic photosynthesis with organic compounds as carbon sources or grow aerobically as a chemoheterotroph. It doesn't appear closely related to any other bacterial group based on 16S rRNA studies and is a deep and ancient branch of the bacterial tree (*see figure 17.4*). Genomic analysis of *Chloroflexus aurantiacus* should help elucidate the origin of photosynthesis.

The nonphotosynthetic, gliding, rod-shaped or filamentous bacterium *Herpetosiphon* also is included in this phylum. *Herpetosiphon* is an aerobic chemoorganotroph with respiratory metabolism that uses oxygen as the electron acceptor. It can be isolated from freshwater and soil habitats.

1. What are the major distinguishing characteristics of *Aquifex, Thermotoga,* and the deinococci? What is thought to contribute to the radiation resistance of the deinococci?

2. How do oxygenic and anoxygenic photosynthesis differ from each other? What is the ecological significance of these differences?

3. In general terms, give the major characteristics of the following groups: purple sulfur bacteria, purple nonsulfur bacteria, and green sulfur bacteria. How do purple and green sulfur bacteria differ?

4. What are chlorosomes or chlorobium vesicles?

5. Compare the green nonsulfur (*Chloroflexi*) and green sulfur bacteria (*Chlorobi*). Which are more metabolically versatile? What are the ecological implications of this metabolic flexibility?

Phylum *Cyanobacteria*

The **cyanobacteria** are the largest and most diverse group of photosynthetic bacteria. There is little agreement about the number of cyanobacterial species. Older classifications list as many as 2,000 or more species. *Bergey's Manual* describes 56 genera. Cyanobacterial diversity is reflected in the G + C content of the group, which ranges from 35 to 71%. Although cyanobacteria are gram-negative bacteria, their photosynthetic system closely resembles that of the eucaryotes, and endosymbiotic cyanobacteria are thought to have evolved into chloroplasts. It follows that all cyanobacteria and most photosynthetic eucaryotes have chlorophyll *a* and photosystems I and II, and thereby perform oxygenic photosynthesis. Like the red algae, cyanobacteria use phycobiliproteins as accessory pigments. Photosynthetic pigments and electron transport chain components are located in thylakoid membranes lined with particles called **phycobilisomes** (**figure 19.6**). These contain phycobilin pigments, particularly

phycocyanin and phycoerythrin, which transfer energy to photosystem II. Carbon dioxide is assimilated through the Calvin cycle; the enzymes needed for this process are thought to be found in internal structures called **carboxysomes.** The reserve carbohydrate is glycogen. Sometimes they store extra nitrogen as polymers of arginine and aspartic acid in **cyanophycin** granules. Phosphate is stored in polyphosphate granules (**figure 19.7**). Because cyanobacteria lack the enzyme α-ketoglutarate dehydrogenase, they do not have a fully functional citric acid cycle. The pentose phosphate pathway plays a central role in their carbohydrate metabolism. Although many cyanobacteria are obligate photolithoautotrophs, a few can grow slowly in the dark as chemoheterotrophs by oxidizing glucose and a few other sugars. Under anoxic conditions, *Oscillatoria limnetica* oxidizes hydrogen sulfide instead of water and carries out anoxygenic photosynthesis much like the green photosynthetic bacteria.

Cyanobacteria also vary greatly in shape and appearance. They range in diameter from about 1 to 10 μm and may be unicellular, exist as colonies of many shapes, or form filaments called trichomes (**figure 19.8***a*). A **trichome** is a row of bacterial cells that are in close contact with one another over a large area. Although many cyanobacteria appear blue-green because of phycocyanin, isolates from the open ocean are red or brown because of the pigment phycoerythrin. Cyanobacteria modulate the relative amounts of these pigments in a process known as chromatic adaptation. When orange light is sensed, phycocyanin production is stimulated, while blue and blue-green light promote the production of phycoerythrin. It is thought that sensory rhodopsins may play a role in signaling

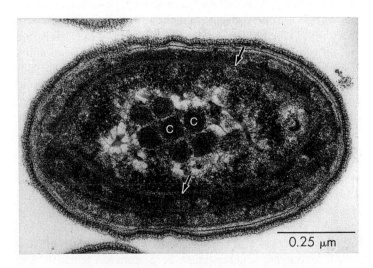

Figure 19.6 Cyanobacterial Thylakoids and Phycobilisomes. Marine *Synechococcus* with thylakoids (arrows). Carboxysomes are labeled c.

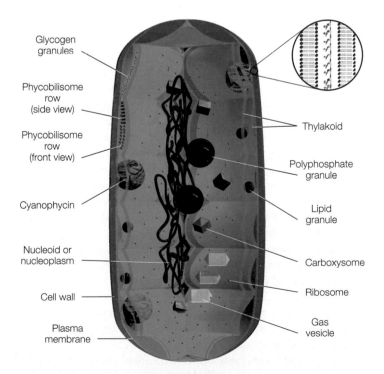

Glycogen granules

Phycobilisome row (side view)

Phycobilisome row (front view)

Cyanophycin

Nucleoid or nucleoplasm

Cell wall

Plasma membrane

Thylakoid

Polyphosphate granule

Lipid granule

Carboxysome

Ribosome

Gas vesicle

Figure 19.7 **Cyanobacterial Cell Structure.** Schematic diagram of a vegetative cell. The insert shows an enlarged view of the envelope with its outer membrane and peptidoglycan. *Copyright © Hartwell T. Crim, 1998.*

(a) *Oscillatoria*

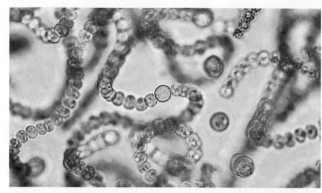

(c) *Nostoc*

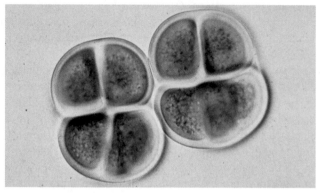

(b) *Chroococcus turgidus*

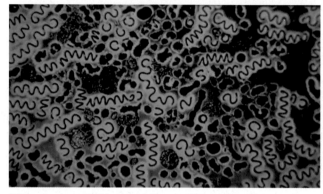

(d) *Anabaena spiroides* and *Microcystis aeruginosa*

Figure 19.8 **Oxygenic Photosynthetic Bacteria.** Representative cyanobacteria. (a) *Oscillatoria* trichomes seen with Nomarski interference-contrast optics (×250). (b) *Chroococcus turgidus,* two colonies of four cells each (×600). (c) *Nostoc* with heterocysts (×550). (d) The cyanobacteria *Anabaena spiroides* and *Microcystis aeruginosa.* The spiral *A. spiroides* is covered with a thick gelatinous sheath (×1,000).

the spectral quality of light (*see p. 417*). Many cyanobacterial species use gas vacuoles to position themselves in optimum illumination in the water column—a form of **phototaxis.** Gliding motility is used by other cyanobacteria. Although cyanobacteria lack flagella, about one-third of the marine *Synechococcus* strains swim at rates up to 25 µm/sec by an unknown mechanism. Swimming motility is not used for phototaxis; instead, it appears to be used in chemotaxis toward simple nitrogenous compounds such as urea. ◀◀ *Bacteria motility and chemotaxis (section 3.7)*

Cyanobacteria show great diversity with respect to reproduction and employ a variety of mechanisms: binary fission, budding, fragmentation, and multiple fission. In the last process, a cell enlarges and then divides several times to produce many smaller progeny, which are released upon the rupture of the parental cell. Fragmentation of filamentous cyanobacteria can generate small, motile filaments called **hormogonia.** Some species develop **akinetes,** specialized, dormant, thick-walled resting cells that are resistant to desiccation (**figure 19.9a**). These later germinate to form new filaments.

Many filamentous cyanobacteria fix atmospheric nitrogen by means of special cells called **heterocysts** (figure 19.9a,b). Around 5 to 10% of the cells develop into heterocysts when these cyanobacteria are deprived of both nitrate and ammonia,

their preferred nitrogen sources. Within these specialized cells, photosynthetic membranes are reorganized and the proteins that make up photosystem II and phycobilisomes are degraded. Photosystem I remains functional to produce ATP, but no oxygen is generated. This inability to generate O_2 is critical because the enzyme nitrogenase is extremely oxygen sensitive. Heterocysts develop thick cell walls, which slow or prevent O_2 diffusion into the cell, and any O_2 present is consumed during respiration. Heterocyst structure and physiology ensure that these specialized cells remains anaerobic; the heterocyst is dedicated to nitrogen fixation and does not replicate. Nutrients are obtained from adjacent vegetative cells, while the heterocysts contribute fixed nitrogen in the form of amino acids. Nitrogen fixation also is carried out by some cyanobacteria that lack heterocysts. Some fix nitrogen under dark, anoxic conditions in microbial mats. Planktonic forms such as *Trichodesmium* fix nitrogen and contribute significantly to the marine nitrogen budget. ◀◀ *Synthesis of amino acids: Nitrogen fixation (section 11.5);* ▶▶ *Biogeochemical cycling: Nitrogen cycle (section 25.1)*

The classification of cyanobacteria is still unsettled. At present all taxonomic schemes must be considered tentative, and while many genera have been assigned, species names have not yet been designated in most cases. *Bergey's Manual* divides the

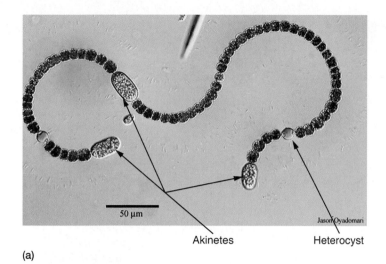

50 μm

Jason Oyadomari

Akinetes   Heterocyst

(a)

(b)

O₂ CO₂   N₂   CO₂ O₂

CH₂O   NH₃   CH₂O

Glutamine

Vegetative   Heterocyst   Vegetative
cells   cells

(c)

Figure 19.9 Cyanobacteria with Akinetes and Heterocysts.
(a) The filamentous cyanobacterium *Anabaenea* with terminal and subterminal akinetes and heterocysts. (b) *Anabaena* heterocysts (arrow) and
vegetative cells. (c) Heterocysts fix dinitrogen to form glutamine, which is
exchanged with adjacent vegetative cells for carbohydrate. Heterocysts
use only photosystem I, so photosynthesis is anoxygenic in these differentiated cells.

cyanobacteria into five subsections (**table 19.2**). These are distinguished using cell or filament morphology and reproductive
patterns. Some other properties important in cyanobacterial characterization are ultrastructure, genetic characteristics, physiology
and biochemistry, and habitat/ecology (preferred habitat and
growth habit). Subsection I contains unicellular rods or cocci.

Most are nonmotile and all reproduce by binary fission or budding. Organisms in subsection II are also unicellular, though
several individual cells may be held together in an aggregate by an
outer wall. Members of this group reproduce by multiple fission
to form spherical, very small, reproductive cells called **baeocytes,**
which escape when the outer wall ruptures. Some baeocytes disperse through gliding motility. The other three subsections contain
filamentous cyanobacteria. Filaments are often surrounded by a
sheath or slime layer. Cyanobacteria in subsection III form
unbranched trichomes composed only of vegetative cells, whereas
the other two subsections produce heterocysts in the absence of an
adequate nitrogen source and also may form akinetes.
Heterocystous cyanobacteria are subdivided into those that form
unbranched filaments (subsection IV) and those that divide in a
second plane to produce branches or aggregates (subsection V).

The cyanobacteria that are collectively referred to as
prochlorophytes (genera *Prochloron, Prochlorococcus,* and
Prochlorothrix) are distinguished by the presence of chlorophyll *a*
and *b* and the lack of phycobilins. This pigment arrangement imparts
a grass-green color to these microbes. As the only procaryotes to
possess chlorophyll *b,* the ancestors of unicellular prochlorophytes
are considered to be the best candidates as the endosymbionts that
gave rise to chloroplasts. Considerable controversy has surrounded
their phylogeny, but small subunit rRNA sequence analysis places
them with the cyanobacteria. << *Microbial evolution: The endosymbiotic origin of mitochondria and chloroplasts (section 17.1)*

The three recognized prochlorophyte genera are quite different from one another. *Prochloron* was first discovered as an
extracellular symbiont growing either on the surface or within
the cloacal cavity of marine colonial ascidian invertebrates. These
bacteria are single-celled, spherical, and from 8 to 30 μm in
diameter. Their mol% of G + C is 31 to 41. *Prochlorothrix* is free
living, has cylindrical cells that form filaments, and grows in
freshwater. Its DNA has a higher G + C content (53 mol%).

Prochlorococcus marinus, which is less than 1 μm in diameter, flourishes in marine plankton. It differs from other
prochlorophytes in having divinyl chlorophyll *a* and *b* and α-
carotene instead of chlorophyll *a* and β-carotene. During the
summer, it reaches concentrations of 5×10^5 cells per milliliter.
It is one of the most numerous of the marine plankton and may be
the most abundant oxygenic photosynthetic organism on Earth.

Another indication of the vast diversity among the cyanobacteria is the wide range of habitats they occupy. Thermophilic
species may grow at temperatures of up to 75°C in neutral to alkaline hot springs. Because these photoautotrophs are so hardy, they
are primary colonizers of soils and surfaces that are devoid of
plant growth. Some unicellular forms even grow in the fissures of
desert rocks. In nutrient-rich warm ponds and lakes, surface cyanobacteria such as *Anacystis* and *Anabaena* can reproduce rapidly
to form blooms (**figure 19.10**). The release of large amounts of
organic matter upon the death of the bloom microorganisms stimulates the growth of chemoheterotrophic bacteria. These microbes
subsequently deplete available oxygen. This kills fish and other
organisms. Some species can produce toxins that kill livestock
and other animals that drink the water. Other cyanobacteria (e.g.,

Table 19.2 Characteristics of the Cyanobacterial Subsections

Subsection	General Shape	Reproduction and Growth	Heterocysts	% G + C	Other Properties	Representative Genera
I	Unicellular rods or cocci; nonfilamentous aggregates	Binary fission, budding	−	31–71	Nonmotile or swim without flagella	*Chroococcus* *Gleocapsa* *Prochlorococcus* *Synechococcus*
II	Unicellular rods or cocci; may be held together in aggregates	Multiple fission to form baeocytes	−	40–46	Some baeocytes are motile	*Pleurocapsa* *Dermocarpella* *Chroococcidiopsis*
III	Filamentous, unbranched trichome with only vegetative cells	Binary fission in a single plane, fragmentation	−	34–67	Usually motile	*Lyngbya* *Oscillatoria* *Prochlorothrix* *Spirulina* *Pseudanabaena*
IV	Filamentous, unbranched trichome may contain specialized cells	Binary fission in a single plane, fragmentation to form hormogonia	+	38–47	Often motile, may produce akinetes	*Anabaena* *Cylindrospermum* *Nostoc* *Calothrix*
V	Filamentous trichomes either with branches or composed of more than one row of cells	Binary fission in more than one plane, hormogonia formed	+	42–44	May produce akinetes; greatest morphological complexity and differentiation in cyanobacteria	*Fischerella* *Stigonema* *Geitleria*

Oscillatoria) are so pollution resistant and characteristic of freshwaters with high organic matter content that they are used as water pollution indicators. >> *Marine and freshwater microbiology: Microorganisms in coastal marine systems (section 26.1).*

Cyanobacteria are particularly successful in establishing symbiotic relationships with other organisms. For example, they are the photosynthetic partner in some lichen associations. Cyanobacteria are symbionts with protozoa and fungi, and nitrogen-fixing species form associations with a variety of plants (liverworts, mosses, gymnosperms, and angiosperms). >> *Microbial interactions (section 27.1)*

1. What are the major characteristics of the cyanobacteria that distinguish them from other photosynthetic organisms.

2. What is a trichome and how does it differ from a simple chain of cells?

3. Briefly discuss the ways in which cyanobacteria reproduce.

4. Describe how a vegetative cell, a heterocyst, and an akinete are different. How are heterocysts modified to carry out nitrogen fixation?

5. Give the features of the five major cyanobacterial groups.

6. Compare the prochlorophytes with other cyanobacteria. Why do you think the phylogeny of the prochlorophytes has been so difficult to establish?

7. List some important positive and negative impacts cyanobacteria have on humans and the environment.

Figure 19.10 Bloom of Cyanobacteria and Algae in a Eutrophic Pond.

19.4 PHYLUM *PLANCTOMYCETES*

The phylum *Planctomycetes* contains one class, one order, and four genera. The planctomycetes are morphologically unique bacteria, having compartmentalized cells (**figure 19.11**). Although each species is unique, all follow a basic cellular organization that includes a plasma membrane closely surrounded by the cell wall, which lacks peptidoglycan. The largest internal compartment, the intracytoplasmic membrane, is separated from the plasma membrane by a peripheral, ribosome-free region called the paryphoplasm. The nucleoid of the planctomycete *Gemmata obscuriglobus* is located in the nuclear body, which is enclosed in a double membrane. The species *Candidatus* "Brocadia anammoxidans" does not localize its DNA within a nuclear body; however, it has another compartment, the anammoxosome. This is the site of anaerobic ammonia oxidation, a unique and recently discovered form of chemolithotrophy in which ammonium ion (NH_4^+) serves as the electron donor and nitrite (NO_2^-) as the terminal electron acceptor—it is reduced to nitrogen gas (N_2). Note that the nomenclature of this microbe is still in flux so in some cases, species names are enclosed in quotation marks. >> *Biogeochemical cycling: Nitrogen cycle (section 25.1)*

The genus *Planctomyces* attaches to surfaces through a stalk and holdfast; the other genera in the order lack stalks. Most of these bacteria have life cycles in which sessile cells bud to produce motile swarmer cells. The swarmer cells are flagellated and swim for a while before attaching to a substrate and beginning reproduction.

adverse affects. It is thought that these hosts represent a natural reservoir for the *Chlamydiae*.

The phylum *Chlamydiae* has one class, one order, four families, and only six genera. The genus *Chlamydia* is by far the most

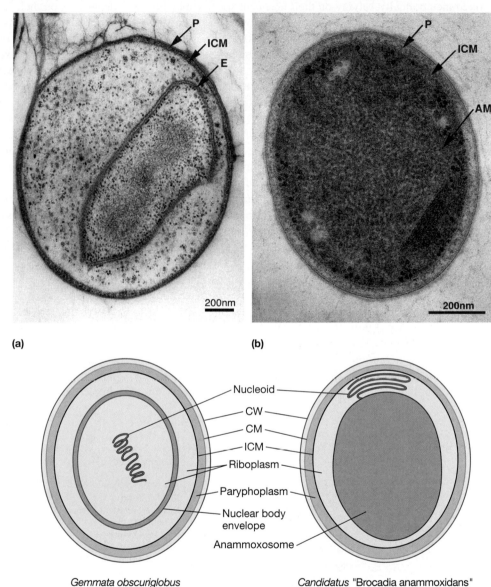

(a)

(b)

(c)

Figure 19.11 Planctomycete Cellular Compartmentalization. (a) An electron micrograph of *Gemmata obscuriglobus* showing the nuclear body envelope (E), the intracytoplasmic membrane (ICM), and the paryphoplasm (P). (b) An electron micrograph of the anaerobic ammonia-oxidizing planctomycete *Candidatus* "Brocadia anammoxidans". The anammoxosome is labeled AM. (c) Schematic drawings corresponding to (a) and (b): cell wall (CW), cytoplasmic membrane (CM).

19.5 PHYLUM *CHLAMYDIAE*

The gram-negative **chlamydiae** are obligate intracellular parasites. That is, they must grow and reproduce within host cells. Although their ability to cause disease is widely recognized, many species grow within protists and animal cells without

important and best studied; it is the focus of our attention. *Chlamydiae* are nonmotile, coccoid bacteria, ranging in size from 0.2 to 1.5 μm. They reproduce only within cytoplasmic vesicles of host cells by a unique developmental cycle involving the formation of two cell types: elementary bodies and reticulate bodies. Although their envelope resembles that of other gram-negative bacteria, the cell wall differs in lacking muramic acid and a peptidoglycan layer. Elementary bodies achieve osmotic stability by cross-linking their outer membrane proteins, and possibly periplasmic proteins, with disulfide bonds. *Chlamydiae* are extremely limited metabolically, relying on their host cells for key metabolites. This is reflected in the size of their genome. It is relatively small at 1.0 to 1.3 Mb; the G + C content is 41 to 44%.

Chlamydial reproduction is unique. Although they undergo binary fission, *Chlamydia* is one of only a few bacteria that lacks the cell division protein FtsZ. Reproduction begins with the attachment of an **elementary body (EB)** to the host cell surface (**figure 19.12**). Elementary bodies are 0.2 to 0.6 μm in diameter, contain electron-dense nuclear material and a rigid cell wall, and are infectious. The host cell phagocytoses the EB, which is held in inclusion bodies where the EB reorganizes to form a **reticulate body (RB)**. The RB is specialized for reproduction rather than infection. Reticulate bodies are 0.5 to 1.5 μm in diameter and have less dense nuclear material and more ribosomes than EBs; their walls are also more flexible. About 8 to 10 hours after infection, the reticulate body undergoes binary fission and RB reproduction continues until the host cell dies. Interestingly a chlamydia-filled inclusion can become large enough to be seen in a light microscope and even fill the host cytoplasm. After 20 to 25 hours, RBs begin to differentiate into infectious EBs and continue this process until the host cell lyses and releases the chlamydiae EBs 48 to 72 hours after infection.

Chlamydial metabolism is very different from that of other gram-negative bacteria. It had been thought that chlamydiae cannot catabolize carbohydrates or synthesize ATP. *Chlamydia psittaci,* one of the best-studied species, lacks both flavoprotein and cytochrome electron transport chain carriers but has a membrane translocase that acquires host ATP in exchange for ADP. Thus chlamydiae seem to be energy parasites that are completely dependent on their hosts for ATP. However, the *C. trachomatis* genome sequence indicates that the bacterium may be able to synthesize at least some ATP. Although there are two genes for ATP/ADP translocases, there also are genes for substrate-level phosphorylation, electron transport, and oxidative phosphorylation. When supplied with precursors from the host, RBs can synthesize DNA, RNA, glycogen, lipids, and proteins. Presumably the RBs have porins and active membrane transport proteins, but little is known about these. They also can synthesize at least some amino acids and coenzymes. The EBs have minimal metabolic activity and cannot take in ATP or synthesize proteins. They are designed exclusively for transmission and infection. Six important diseases are caused by the chlamydiae; these are now discussed.

Chlamydial Pneumonia

Chlamydial pneumonia is caused by *Chlamydophila* (formerly *Chlamydia*) *pneumoniae*. Clinically, infections are generally mild; pharyngitis, bronchitis, and sinusitis commonly accompany some lower respiratory tract involvement. Infections with *C. pneumoniae* are common but sporadic; about 50% of adults have antibody to the chlamydiae. Evidence suggests that *C. pneumoniae* is primarily a human pathogen directly transmitted from human to human by droplet (respiratory) secretions. *C. pneumoniae* infections have

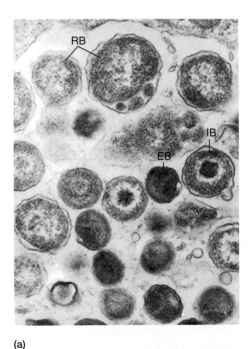

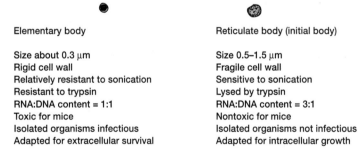

Elementary body

Size about 0.3 μm
Rigid cell wall
Relatively resistant to sonication
Resistant to trypsin
RNA:DNA content = 1:1
Toxic for mice
Isolated organisms infectious
Adapted for extracellular survival

Reticulate body (initial body)

Size 0.5–1.5 μm
Fragile cell wall
Sensitive to sonication
Lysed by trypsin
RNA:DNA content = 3:1
Nontoxic for mice
Isolated organisms not infectious
Adapted for intracellular growth

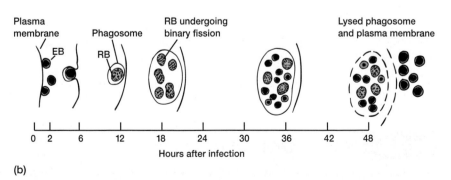

(a)   (b)

Figure 19.12   The Chlamydial Life Cycle.   (a) An electron micrograph of an inclusion body containing a mixture of small, black elementary bodies (EB), larger reticulate bodies (RB), and intermediate bodies (IB), a chlamydial cell intermediate in morphology between EB and RB. The RBs appear gray and granulated due to a high concentration of ribosomes. (b) A schematic representation of the infectious cycle of chlamydiae.

been linked with coronary artery disease as well as vascular disease at other sites. Following a demonstration of *C. pneumoniae*-like particles in atherosclerotic plaque tissue by electron microscopy, *C. pneumoniae* genes and antigens have been detected in artery plaque. But the microorganism has rarely been recovered in cultures of atheromatous tissue (i.e., artery plaque). As a result of these findings, the possible etiologic role of *C. pneumoniae* in coronary artery disease and systemic atherosclerosis is under intense scrutiny.

Inclusion Conjunctivitis

Inclusion conjunctivitis is an acute, infectious disease caused by *Chlamydia trachomatis* serotypes D–K, and it occurs throughout the world. It is characterized by a copious mucous discharge from the eye, an inflamed and swollen conjunctiva, and the presence of large inclusion bodies in the host cell cytoplasm. In inclusion conjunctivitis of the newborn, the chlamydiae are acquired during passage through an infected birth canal. The disease appears 7 to 12 days after birth. Erythromycin eyedrops given prophylactically to prevent ophthalmia neonatorum also prevent chlamydial inclusion conjunctivitis. If the chlamydiae colonize an infant's nasopharynx and tracheobronchial tree, pneumonia may result. Adult inclusion conjunctivitis is acquired by contact with infective genital tract discharges.

Lymphogranuloma Venereum

Lymphogranuloma venereum (LGV) is a sexually transmitted disease caused by *Chlamydia trachomatis* serotypes L1–L3. LGV proceeds through three phases. (1) In the primary phase, a small ulcer appears several days to several weeks after a person is exposed to the chlamydiae. The ulcer may appear on the penis in males or on the labia or vagina in females. The ulcer heals quickly and leaves no scar. (2) The secondary phase begins 2 to 6 weeks after exposure, when the chlamydiae infect lymphoid cells, causing the regional lymph nodes to become enlarged and tender; such nodes are called buboes (**figure 19.13**). Systemic symptoms such as fever, chills, and anorexia are common. (3) If the disease is not treated, a late phase ensues. This results from fibrotic changes and abnormal lymphatic drainage that produces fistulas (abnormal passages leading from an abscess or a hollow organ to the body surface or from one hollow organ to another) and urethral or rectal strictures (a decrease in size). An untreatable fluid accumulation in the penis, scrotum, or vaginal area may result.

Nongonococcal Urethritis

Nongonococcal urethritis (NGU) is any inflammation of the urethra not due to the bacterium *Neisseria gonorrhoeae*. This condition is caused both by nonmicrobial factors such as catheters and drugs and by infectious microorganisms. The most important causative agents are *C. trachomatis, Ureaplasma urealyticum, Mycoplasma hominis, Trichomonas vaginalis, Candida albicans,* and herpes simplex viruses. Most infections are acquired sexually, and of these, approximately 50% are *Chlamydia* infections. NGU

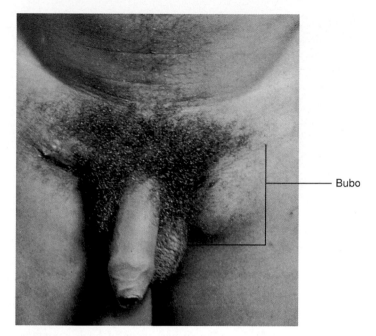

Figure 19.13   Lymphogranuloma Venereum. The bubo in the left inguinal area is draining.

is endemic throughout the world, with an estimated 10 million Americans infected.

Symptoms of NGU vary widely. Males may have few or no manifestations of disease; however, complications can exist. These include a urethral discharge, itching, and inflammation of the male reproductive structures. Females may be asymptomatic or have a severe infection leading to *pelvic inflammatory disease* (PID), which often leads to sterility. *Chlamydia* may account for as many as 200,000 to 400,000 cases of PID annually in the United States. In the pregnant female, a chlamydial infection is especially serious because it is directly related to miscarriage, stillbirth, inclusion conjunctivitis, and infant pneumonia.

Trachoma

Trachoma (Greek *trachoma,* roughness) is a contagious disease caused by *Chlamydia trachomatis* serotypes A–C. It is one of the oldest known infectious diseases of humans and is the greatest single cause of blindness throughout the world. Probably over 500 million people are infected and 20 million blinded each year by this chlamydia. In endemic areas, most children are chronically infected within a few years of birth. Active disease in adults over age twenty is three times as frequent in females as in males because of mother-child contact. Although trachoma is uncommon in the United States, except among Native Americans in the Southwest, it is widespread in Asia, Africa, and South America.

Trachoma is transmitted by contact with inanimate objects such as soap and towels, by hand-to-hand contact that carries *C. trachomatis* from an infected eye to an uninfected eye, or by flies. The disease begins abruptly with an inflamed conjunctiva. This leads to an inflammatory cell exudate and necrotic eyelash

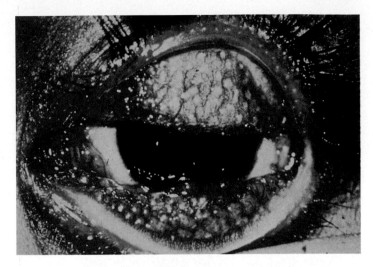

Figure 19.14   Trachoma.   An active infection showing marked follicular hypertrophy of the eyelid. The inflammatory nodules cover the thickened conjunctiva of the eye.

follicles (**figure 19.14**). The disease usually heals spontaneously. However, with reinfection, secondary infections, vascularization of the cornea, or scarring of the conjunctiva can occur. If scar tissue accumulates over the cornea, blindness results.

Psittacosis (Ornithosis)

Psittacosis (ornithosis) is a worldwide infectious disease of birds that is transmissible to humans. Diseases that are transmitted from animals to humans are called **zoonoses** (s., zoonosis). Psittacosis was first described in association with parrots and parakeets, both of which are psittacine birds. The disease is now recognized in many other birds—among them, pigeons, chickens, ducks, and turkeys—and the general term *ornithosis* (Latin *ornis*, bird) is used.

Ornithosis is caused by *Chlamydophilia psittaci*. Humans contract this disease either by handling infected birds or by inhaling dried bird excreta that contains viable *C. psittaci*. Ornithosis is recognized as an occupational hazard within the poultry industry, particularly to workers in turkey-processing plants. After entering the respiratory tract, the chlamydiae are transported to the cells of the liver and spleen. They multiply within these cells and then invade the lungs, where they cause inflammation, hemorrhaging, and pneumonia. With antibiotic therapy, the mortality rate is about 2%. Fewer than 100 cases of ornithosis are reported annually in the United States.

1. Describe the *Planctomycetes* and their distinctive morphological and metabolic properties.
2. Briefly describe the steps in the chlamydial life cycle. Why do you think *Chlamydiae* differentiate into specialized cell types for infection and reproduction?
3. How does chlamydial metabolism differ from that of other bacteria?
4. How does an infant acquire inclusion conjunctivitis?
5. List the different diseases caused by serotypes of *C. trachomatis*. What do you think the ability of different serotypes to cause specific disease indicates about the evolution of these strains?

19.6   PHYLUM SPIROCHAETES

The phylum *Spirochaetes* (Greek *spira*, a coil, and *chaete*, hair) contains gram-negative, chemoheterotrophic bacteria distinguished by their structure and mechanism of motility. *Bergey's Manual* divides the phylum *Spirochaetes* into one class, one order (*Spirochaetales*), and three families (*Spirochaetaceae, Serpulinaceae,* and *Leptospiraceae*). At present, there are 13 genera in the phylum. Spirochetes can be anaerobic, facultatively anaerobic, or aerobic. Carbohydrates, amino acids, long-chain fatty acids, and long-chain fatty alcohols may serve as carbon and energy sources. **Table 19.3** summarizes some of the more distinctive properties of selected genera.

Spirochetes are morphologically unique. They are slender, long bacteria (0.1 to 3.0 μm by 5 to 250 μm) with a flexible, helical shape (**figure 19.15** and chapter opener). The central protoplasmic cylinder, which contains cytoplasm and the nucleoid, is bounded by a plasma membrane and a gram-negative cell wall (**figure 19.16**). Two to more than a hundred flagella, called **axial fibrils** or **periplasmic flagella,** extend from both ends of the cylinder and often overlap one another in the center third of the cell. The whole complex of periplasmic flagella, called the **axial filament,** lies inside a flexible outer sheath. The outer sheath contains lipid, protein, and carbohydrate and varies in structure between different genera. Its precise function is unknown, but the sheath is essential because spirochetes die if it is damaged or removed. The outer sheath of *Treponema pallidum* has few proteins exposed on its surface. This allows the syphilis spirochete to avoid attack by host antibodies. *T. pallidum* is also discussed in chapter 15 (*see p. 339 and figure 15.7*).

Motility in the spirochetes is uniquely adapted to movement through viscous solutions. Like the external flagella of other bacteria, the axial fibrils rotate. This results in the rotation of the outer sheath in the opposite direction relative to the periplasmic cylinder, and the cell moves in a corkscrewlike movement through the medium (**figure 19.17**). Fibril rotation can also enable cell flexing and crawling on solid substrates. Thus, unlike other bacteria flagella, axial fibrils mediate movement through liquids and on solid surfaces.

Spirochetes are exceptionally diverse ecologically and grow in habitats ranging from mud to the human mouth. Members of the genus *Spirochaeta* are free-living and often grow in anoxic and

Table 19.3	Characteristics of Spirochete Genera				
Genus	Dimensions (μm) and Flagella	G + C Content (mol%)	Oxygen Relationship	Carbon + Energy Source	Habitats
Spirochaeta	0.2–0.75 × 5–250; 2–40 periplasmic flagella (almost always 2)	51–65	Facultatively anaerobic or anaerobic	Carbohydrates	Aquatic and free-living
Cristispira	0.5–3.0 × 30–180; ≥100 periplasmic flagella	N.A.[a]	Facultatively anaerobic?	N.A.	Mollusk digestive tract
Treponema	0.1–0.4 × 5–20; 2–16 periplasmic flagella	25–53	Anaerobic or microaerophilic	Carbohydrates or amino acids	Mouth, intestinal tract, and genital areas of animals; some are pathogenic (syphilis, yaws)
Borrelia	0.2–0.5 × 3–20; 14–60 periplasmic flagella	27–32	Anaerobic or microaerophilic	Carbohydrates	Mammals and arthropods; pathogens (relapsing fever, Lyme disease)
Leptospira	0.1 × 6–24; 2 periplasmic flagella	35–53	Aerobic	Fatty acids and alcohols	Free-living or pathogens of mammals, usually located in the kidney (leptospirosis)
Leptonema	0.1 × 6–20; 2 periplasmic flagella	51–53	Aerobic	Fatty acids	Mammals
Brachyspira	0.2 × 1.7–6.0; 8 periplasmic flagella	25–27	Anaerobic	N.A.	Mammalian intestinal tract
Serpulina	0.3–0.4 × 7–9; 16–18 periplasmic flagella	25–26	Anaerobic	Carbohydrates and amino acids	Mammalian intestinal tract

[a]N.A., information not available.

sulfide-rich freshwater and marine environments. Some species of the genus *Leptospira* grow in oxic water and moist soil. Some spirochetes form symbiotic associations with other organisms and are found in a variety of locations: the hindguts of termites and wood-eating roaches, the digestive tracts of mollusks (*Cristispira*) and mammals, and the oral cavities of animals (*Treponema denticola, T. oralis*). Spirochetes from termite hindguts and freshwater sediments can fix nitrogen. Evidence exists that they contribute significantly to the nitrogen nutrition of termites. Spirochetes coat the surfaces of many protozoa from termite and wood-eating roach hindguts. For example, the flagellate *Myxotricha paradoxa* is covered with slender spirochetes (0.15 by 10 μm in length) that are firmly attached and help move the protozoan. Several spirochetes are important human pathogens; we discuss two of these—*Treponema pallidum*, the causative agent of syphilis, and the spirochetes that cause Lyme disease.

Syphilis

Venereal syphilis (Greek *syn*, together, and *philein*, to love) is a contagious, sexually transmitted disease caused by *Treponema pallidum* subsp. *pallidum* (*T. pallidum*). **Congenital syphilis** is the disease acquired in utero from the mother.

T. pallidum enters the body through mucous membranes or minor breaks or abrasions of the skin. It migrates to the regional lymph nodes and rapidly spreads throughout the body. The disease is not highly contagious, and there is only about a 1 in 10 chance of acquiring it from a single exposure to an infected sex partner.

Three recognizable stages of syphilis occur in untreated adults (**figure 19.18**). In the primary stage, after an incubation period of

(a) *Cristispira*

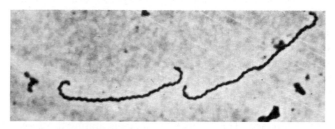

(b) *Leptospira interrogans*

Figure 19.15 The Spirochetes. Representative examples. (a) *Cristispira* sp. from a clam; phase contrast (×2,200). (b) *Leptospira interrogans* (×2,200).

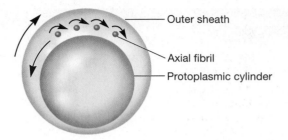

Figure 19.17 Spirochete Motility. A hypothetical mechanism for spirochete motility.

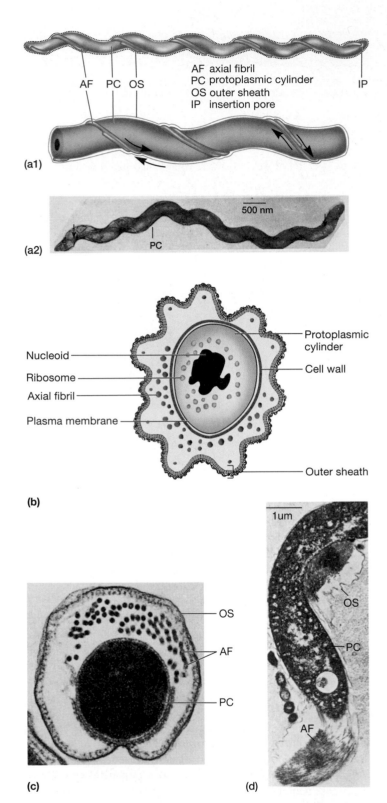

Figure 19.16 Spirochete Morphology. (a1) A surface view of spirochete structure as interpreted from electron micrographs. (a2) A longitudinal view of *Treponema zuelzerae* with axial fibrils extending most of the cell length. (b) A cross section of a typical spirochete showing morphological details. (c) Electron micrograph of a cross section of *Clevelandina* from the termite *Reticulitermes flavipes* showing the outer sheath, protoplasmic cylinder, and axial fibrils (×70,000). (d) Longitudinal section of *Cristispira* showing the outer sheath (OS), the protoplasmic cylinder (PC), and the axial fibrils (AF).

about 10 days to 3 weeks or more, the initial symptom is a small, painless, reddened ulcer, or **chancre** (French *canker,* a destructive sore) with a hard ridge that appears at the infection site and contains spirochetes (**figure 19.19***a*). Contact with the chancre during sexual contact may result in disease transmission. In about one-third of the cases, the disease does not progress further and the chancre disappears. However, the spirochetes typically enter the bloodstream and are distributed throughout the body.

Within 2 to 10 weeks after the primary lesion appears, the disease may enter the secondary stage, which is characterized by a highly variable skin rash (figure 19.19*b*). Other symptoms during this stage include the loss of patches of hair, malaise, and fever. Both the chancre and the rash lesions are infectious.

After several weeks, the disease becomes latent. During the latent period, the disease is not normally infectious, except for possible transmission from mother to fetus. After many years, a tertiary stage develops in about 40% of untreated individuals with secondary syphilis. During this stage, degenerative lesions called **gummas** (figure 19.19*c*) form in the skin, bone, and nervous system as the result of hypersensitivity reactions. This stage also is characterized by a great reduction in the number of spirochetes in the body. Involvement of the central nervous system may result in tissue loss that can lead to cognitive deficits, blindness, a "shuffle" walk (tabes), or insanity. Many of these symptoms have been associated with such well-known people as Al Capone, Francisco Goya, Henry VIII, Adolf Hitler, Scott Joplin, Friedrich Nietzsche, Franz Schubert, Oscar Wilde, and Kaiser Wilhelm.

Treatment in the early stages of the disease is easily accomplished with penicillin. Later stages of syphilis are more difficult to treat and require much larger doses over a longer period. For example, in neurosyphilis cases, treponemes occasionally survive such drug treatment. Immunity to syphilis is not complete, and subsequent infections can occur. An estimated 12 million new cases occur each year. In the United States, fewer than 10,000 cases of primary and secondary syphilis in the civilian population and about 500 cases of congenital syphilis are reported annually. The highest incidence is among those twenty to thirty-nine years of age. Importantly syphilis incidence has increased worldwide as a secondary infection in HIV-positive individuals; conversely untreated syphilis makes a person more susceptible to HIV infection.

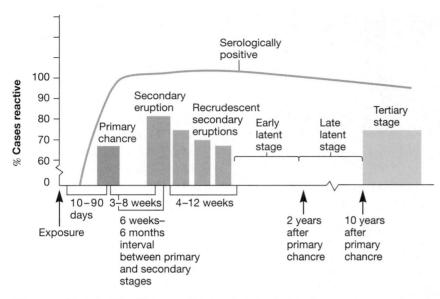

Figure 19.18 **The Course of Untreated Syphilis.**

Lyme Disease

Lyme disease (LD, Lyme borreliosis) was first observed and described in 1975 among people living in Old Lyme, Connecticut. It is now the most common tick-borne disease in the United States, with about 17,000 cases reported annually. The disease is also present in Europe and Asia.

The Lyme spirochetes responsible for this disease comprise at least three species, currently designated *Borrelia burgdorferi, B. garinii,* and *B. afzelii.* Deer and field mice are the natural hosts. In the northeastern United States, *B. burgdorferi* is transmitted to

humans by the bite of the infected deer tick (*Ixodes scapularis;* **figure 19.20***a*). On the Pacific Coast, especially in California, the reservoir is a dusky-footed woodrat and the tick, *I. pacificus.* Important prevention strategies are listed in **table 19.4**.

Lyme disease is a complex illness with three major stages. The initial, localized stage occurs a week to 10 days after an infectious tick bite. The illness often begins with an expanding, ring-shaped skin lesion with a red outer border and partial central clearing called erythema migrans (figure 19.20*b*). This often is accompanied by flulike symptoms (malaise and fatigue, headache, fever, and chills). Often the tick bite is unnoticed, and the skin lesion may be missed due to skin coloration or its obscure location, such as on the scalp. Thus treatment, which is usually effective at this stage, may not be given because the illness is assumed to be "just the flu."

The second, disseminated stage may appear weeks or months after the initial infection. It consists of several symptoms such as neurological abnormalities, heart inflammation, and bouts of arthritis (usually in the major joints such as the elbows or knees). Inflammation that produces organ damage is initiated and possibly perpetuated by the immune response to one or more spirochetal proteins. Finally, like the progression of syphilis, the late stage may appear years later. Infected individuals may develop neuron demyelination with symptoms resembling Alzheimer's disease and multiple sclerosis.

It is important to seek early diagnosis and treatment of Lyme disease. Treatment with amoxicillin or tetracycline early in the illness results in prompt recovery and prevents arthritis

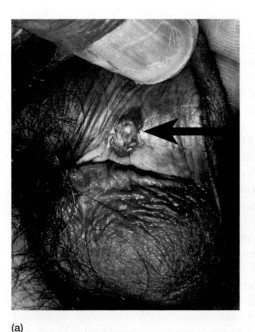

(a)

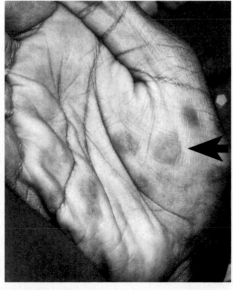

(b)

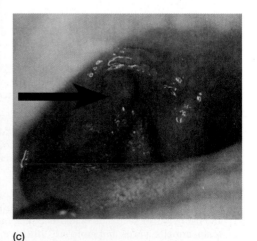

(c)

Figure 19.19 **Syphilis.** (a) Primary syphilitic chancre of the penis. (b) Palmar lesions of secondary syphilis. (c) Ruptured gumma and ulcer of upper hard palate of the mouth.

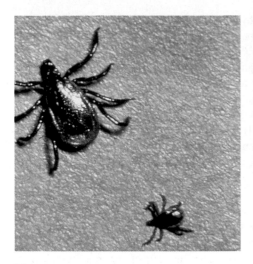

(a)

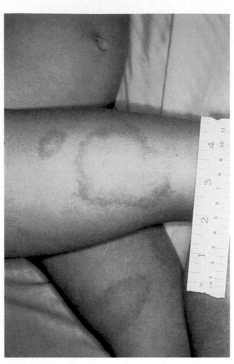

(b)

Figure 19.20   Lyme Disease.   (a) The vector for *Borrelia burgdorferi* in the northeastern U.S. is the tick *Ixodes scapularis.* An unengorged adult (bottom) is about the size of pinhead, and an engorged adult (top) can reach the size of a jelly bean. (b) The typical rash (erythema migrans) showing concentric rings around the initial site of the tick bite.

and other complications. If nervous system involvement is suspected, ceftriaxone is used because it can cross the blood-brain barrier.

1. Define the following: protoplasmic cylinder, axial fibrils or periplasmic flagella, axial filament, and outer sheath. Draw and label a diagram of spirochete morphology, locating these structures.

2. Compare periplasmic flagella with the flagella of other bacteria. Why do you think spirochete motility is especially well suited for movement through viscous fluids?

3. Compare the three stages of syphilis and Lyme disease. Why do you think both diseases are so hard to treat once they progress beyond the primary stage?

Table 19.4	Lyme Disease Prevention Strategies

Persons who are active in an area where Lyme disease or other tick-borne zoonoses occur should keep in mind the following points.

1. It takes a minimum of 24 hours of attachment and feeding for bacterial transmission to occur; thus prompt removal of attached ticks will greatly reduce the risk of infection. To remove an embedded tick, use tweezers to grasp the tick as close as possible to the skin and then pull with slow, steady pressure in a direction perpendicular to the skin.

2. Because each deer tick life cycle stage usually occurs at a certain time, there are periods when an individual should be most aware of the risk of infection. The most dangerous times are May through July, when the majority of nymphal deer ticks are present and the risk of transmission is greatest.

3. In the woods, wear light-colored pants and good shoes. Tuck the cuffs of your pants into long socks to deny ticks easy entry under your clothes. After coming out of the woods, check all clothes for ticks and exposed skin.

4. Repellents containing high concentrations of DEET (diethyltoluamide) or permanone are available over the counter and are very noxious to ticks. Premethrin kills ticks on contact but is approved only for use on clothing.

5. Immediately after being in a high-risk area, examine your body for ticks, bite marks, swellings, and redness. Taking a shower and using lots of soap aids in this examination. Areas such as the scalp, armpits, navel, and groin are difficult to examine effectively but are preferred sites for tick attachment. Pay special attention to these parts of the body.

19.7   PHYLUM *BACTEROIDETES*

The phylum *Bacteroidetes* is very diverse and seems most closely related to the phylum *Chlorobi.* The phylum has three classes (*Bacteroides, Flavobacteria,* and *Sphingobacteria*), 12 families, and 63 genera. The class *Bacteroides* contains anaerobic, gram-negative, nonsporing, motile or nonmotile rods of various shapes. These bacteria are chemoheterotrophic and usually produce a mixture of organic acids as fermentation end products. They do not reduce sulfate or other sulfur compounds. The genera are identified using properties such as general shape, motility and flagellation pattern, and fermentation end products. These bacteria grow in habitats such as the oral cavity and intestinal tract of vertebrates and the rumen of ruminants. **>>** *Microbial interactions: The rumen ecosystem (section 27.1); Normal microbiota of the human body: Large intestine (colon) (section 27.3)*

Although difficulty culturing these anaerobes has hindered our understanding of them, they are clearly widespread and important. Often they benefit their host. *Bacteroides ruminicola* is a major component of the rumen flora; it ferments starch, pectin, and other carbohydrates. About 30% of the bacteria isolated from human feces are members of the genus *Bacteroides,* and these organisms may provide extra nutrition by degrading cellulose, pectins, and other complex carbohydrates (*see figures 27.14 and 27.15*). The family also is involved in human disease. Members of the genus *Bacteroides* are associated with diseases of major organ systems, ranging from the central nervous system to the skeletal system. *B. fragilis* is a particularly common anaerobic pathogen found in abdominal, pelvic, pulmonary, and blood infections.

Another important group in the *Bacteroidetes* is the class *Sphingobacteria.* These bacteria often have sphingolipids in their cell walls. Within this class, the genera *Cytophaga, Sporocytophaga,* and *Flexibacter* differ from each other in morphology, life cycle, and physiology. Bacteria of the genus *Cytophaga* are slender rods, often with pointed ends (**figure 19.21***a*). *Sporocytophaga* is similar to *Cytophaga* but forms spherical resting cells called microcysts (figure 19.21*b*). *Flexibacter* produces long, flexible threadlike cells when young (figure 19.21*c*) and is unable to use complex polysaccharides. Often colonies of these bacteria are yellow to orange because of carotenoid or flexirubin pigments. Some of the flexirubins are chlorinated, which is unusual for biological molecules.

Members of the genera *Cytophaga* and *Sporocytophaga* are aerobes that actively degrade complex polysaccharides. Soil cytophagas digest cellulose; both soil and marine forms attack chitin, pectin, and keratin. Some marine species also degrade agar, a component of seaweed. Cytophagas play a major role in the mineralization of organic matter and can cause great damage to exposed fishing gear and wooden structures. They also are a major component of the bacterial population in sewage treatment plants and presumably contribute significantly to the waste treatment process. Although most cytophagas are free-living, some can be isolated from vertebrate hosts and are pathogenic. *Cytophaga columnaris* and others cause diseases such as columnaris disease, cold water disease, and fin rot in freshwater and marine fish. **>>** *Wastewater treatment (section 35.2)*

The gliding motility so characteristic of these organisms is quite different from flagellar motility. **Gliding motility** is present in a wide diversity of taxa: fruiting and nonfruiting aerobic chemoheterotrophs, cyanobacteria, green nonsulfur bacteria, and at least two gram-positive genera (*Heliobacterium* and *Desulfonema*). Gliding bacteria lack flagella and are stationary while suspended in liquid medium. When in contact with a surface, they glide along and some leave a slime trail. Movement can be very rapid; some cytophagas travel 150 μm in a minute, whereas filamentous gliding bacteria may reach speeds of more than 600 μm/minute. Young organisms are the most motile, and motility often is lost with age. Low nutrient levels usually stimulate gliding. The gliding mechanism is not well understood.

Gliding motility gives a bacterium many advantages. Many aerobic chemoheterotrophic gliding bacteria actively digest insoluble macromolecular substrates such as cellulose and chitin, and gliding motility is ideal for searching these out. Because many of the digestive enzymes are cell wall associated, the bacteria must be in contact with insoluble nutrient sources; gliding motility makes this possible. Gliding movement is well adapted to drier habitats and to movement within solid masses such as soil, sediments, and rotting wood that are permeated by small channels. Finally, gliding bacteria, like flagellated bacteria, can position themselves at optimal levels of oxygen, hydrogen sulfide, light intensity, temperature, and other factors that influence growth and survival. **<<** *Bacterial motility and chemotaxis (section 3.7)*

1. List the major properties of the class *Bacteroides.*
2. How do these bacteria benefit and harm their hosts?
3. List three advantages of gliding motility.
4. Briefly describe the following genera: *Cytophaga, Sporocytophaga,* and *Flexibacter.*
5. Why are the cytophagas ecologically important?

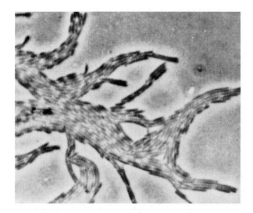

(a) *Cytophaga*

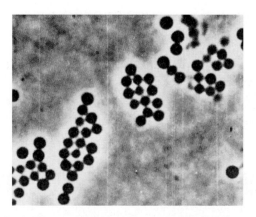

(b) *Sporocytophaga myxococcoides* microcysts

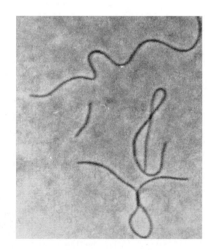

(c) *Flexibacter elegans*

Figure 19.21 Nonphotosynthetic, Nonfruiting, Gliding Bacteria. Representative members of the order *Cytophagales.* (a) *Cytophaga* sp. (×1,150). (b) *Sporocytophaga myxococcoides,* mature microcysts (×1,750). (c) Long thread cells of *Flexibacter elegans* (×1,100).

Summary

19.1 *Aquificae* and *Thermotogae*

a. *Aquifex* and *Thermotoga* are hyperthermophilic gram-negative rods that represent the two deepest or oldest phylogenetic branches of the *Bacteria*.

19.2 *Deinococcus-Thermus*

a. Members of the order *Deinococcales* are aerobic cocci and rods that are distinctive in their unusually great resistance to desiccation and radiation.

19.3 Photosynthetic Bacteria

a. Cyanobacteria carry out oxygenic photosynthesis; purple and green bacteria use anoxygenic photosynthesis.

b. The four most important groups of purple and green photosynthetic bacteria are the purple sulfur bacteria, the purple nonsulfur bacteria, the green sulfur bacteria, and the green nonsulfur bacteria (**table 19.1**).

c. The bacteriochlorophyll pigments of purple and green bacteria enable them to live in deeper, anoxic zones of aquatic habitats.

d. The phylum *Chlorobi* includes the green sulfur bacteria—obligately anaerobic photolithoautotrophs that use hydrogen sulfide, elemental sulfur, and hydrogen as electron sources.

e. Green nonsulfur bacteria such as *Chloroflexus* are placed in the phylum *Chloroflexi. Chloroflexus* is a filamentous, gliding thermophilic bacterium that is metabolically similar to the purple nonsulfur bacteria.

f. Cyanobacteria carry out oxygenic photosynthesis by means of a photosynthetic apparatus similar to that of the eucaryotes. Phycobilisomes contain the light-harvesting pigments phycocyanin and phycoerythrin (**figure 19.7**).

g. Cyanobacteria reproduce by binary fission, budding, multiple fission, and fragmentation by filaments to form hormogonia. Some produce dormant akinetes.

h. Many nitrogen-fixing cyanobacteria form heterocysts, specialized cells in which nitrogen fixation occurs (**figure 19.9**).

i. *Bergey's Manual* divides the cyanobacteria into five subsections and includes the prochlorophytes in the phylum *Cyanobacteria* (**table 19.2**).

19.4 Phylum *Planctomycetes*

a. The phyla *Planctomycetes* and *Chlamydiae* lack peptidoglycan in their walls. The *Planctomycetes* have unusual cellular compartmentalization and some perform the anammox reaction (**figure 19.11**).

19.5 Phylum *Chlamydiae*

a. Chlamydiae are nonmotile, coccoid, gram-negative bacteria that lack peptidoglycan and must reproduce within the cytoplasmic vacuoles of host cells by a life cycle involving elementary bodies (EBs) and reticulate bodies (RBs) (**figure 19.12**).

b. Chlamydial infections include chlamydial pneumonia, inclusion conjunctivitis, lymphogranuloma venereum, nongonococcal urethritis, and trachoma.

19.6 Phylum *Spirochaetes*

a. The spirochetes are slender, long, helical, gram-negative bacteria that are motile because of the axial filament underlying an outer sheath or outer membrane (**figures 19.15–19.17**).

b. Spirochetes include important human pathogens, including the causative agents of syphilis and Lyme disease.

19.7 Phylum *Bacteroidetes*

a. Members of the class *Bacteroides* are obligately anaerobic, chemoheterotrophic, nonsporing, motile or nonmotile rods of various shapes. Some are important rumen and intestinal symbionts; others can cause disease.

b. Gliding motility is present in a diverse range of bacteria, including the sphingobacteria and cytophagas.

c. Cytophagas degrade proteins and complex polysaccharides and are active in the mineralization of organic matter.

Critical Thinking Questions

1. The cyanobacterium *Anabaena* grows well in liquid medium that contains nitrate as the sole nitrogen source. Suppose you transfer some of these filaments to the same medium except it lacks nitrate and other nitrogen sources. Describe the morphological and physiologial changes you would observe.

2. Many types of movement are employed by bacteria in these phyla. Review them and propose mechanisms by which energy (ATP or proton gradients) might drive the locomotion.

3. Propose two experimental approaches you might use to examine the mechanism by which *Cytophaga* glides.

Learn More

Learn more by visiting the Prescott website at www.mhhe.com/prescottprinciples, where you will find a complete list of references.

The *Proteobacteria*

20

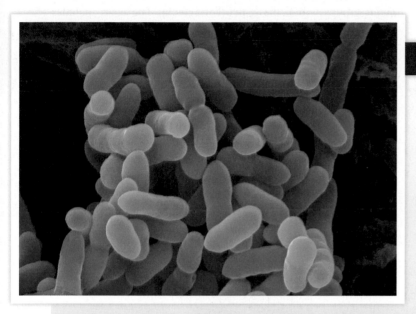

The β-proteobacterium *Burkholderia pseudomallei,* also called "the Great Mimicker," is the causative agent of melioidosis or glanders. Prevalent primarily in tropical climates, glanders can manifest as lung, blood, kidney, or heart infections, as well as skin abscesses.

α-proteobacteria The class of proteobacteria that contains most of the oligotrophic proteobacteria. Some have unusual metabolic modes such as methylotrophy, chemolithotrophy, and nitrogen fixing ability; many have distinctive morphological features.

β-proteobacteria The class of proteobacteria whose members are similar to the α-proteobacteria metabolically but tend to use substances that diffuse from organic matter decomposition in anoxic zones.

budding A means of asexual reproduction in which the daughter cell is smaller than the parent; found in yeast and some bacteria.

δ-proteobacteria The class of proteobacteria that includes chemoorganotrophic bacteria that usually are either predators of other bacteria or anaerobes that generate sulfide from sulfate and sulfite.

enteric bacteria or **enterobacteria** Members of the family *Enterobacteriaceae* (γ-proteobacteria, facultatively anaerobic, straight rods with simple nutritional requirements); the term is also used for bacteria that live in the intestinal tract.

Chapter Glossary

ε-proteobacteria The class of proteobacteria that includes slender rods, some of which are medically important (*Campylobacter* and *Helicobacter*).

fruiting body A specialized structure that holds sexually or asexually produced spores; found in fungi and some bacteria (e.g., the myxobacteria).

γ-proteobacteria The largest of the five classes of proteobacteria. This subgroup is very diverse physiologically; many important genera are facultatively anaerobic chemoorganotrophs.

holdfast A structure produced by some bacteria (e.g., *Caulobacter*) that attaches them to a solid object.

methylotroph A microbe that uses reduced one-carbon compounds such as methane and methanol as its sole source of carbon and energy.

nitrification The oxidation of ammonia to nitrate.

nitrifying bacteria Chemolithotrophic bacteria found in several families within the *Proteobacteria* that oxidize ammonia to nitrite and nitrite to nitrate.

prostheca An extension of a bacterial cell, including the plasma membrane and cell wall, that is narrower than the mature cell. Found in the genus *Caulobacter,* an important model bacterium.

Proteobacteria A large phylum of gram-negative bacteria that 16S rRNA sequence comparisons show to be phylogenetically related; *Proteobacteria* contain the purple photosynthetic bacteria and their relatives and are composed of the α, β, γ, δ, and ε classes.

purple nonsulfur bacteria Anoxygenic photosynthetic bacteria belonging to the α- and β-proteobacteria. Most are anaerobic photoorganoheterotrophs; some are facultative photolithoautotrophs (i.e., in the dark, they function as chemoorganoheterotrophs).

purple sulfur bacteria Obligate anaerobic photolithoautotrophic bacteria belonging to the γ-proteobacteria.

sheath A hollow, tubelike structure surrounding a chain of cells and present in several genera of bacteria.

Microbes is a vigitable, an'ivry man is like a conservatory full iv millyons iv these potted plants.

—*Finley Peter Dunne*

In this chapter, we introduce the bacteria that are covered in volume 2 of the second edition of *Bergey's Manual of Systematic Bacteriology.* This volume is devoted entirely to the **proteobacteria.** This is the largest and most diverse group of bacteria; there are over 500 known genera. Although 16S rRNA studies show that they are phylogenetically related, proteobacteria are remarkably diverse. The morphology of these gram-negative bacteria ranges from simple rods and cocci to genera with prosthecae, buds, and even fruiting bodies. Physiologically, they include photoautotrophs, chemolithotrophs, and chemoheterotrophs. No obvious overall pattern in metabolism, morphology, or reproductive strategy characterizes proteobacteria.

Comparison of 16S rRNA sequences has revealed five lineages of descent within the phylum *Proteobacteria: Alphaproteobacteria, Betaproteobacteria, Gammaproteobacteria, Deltaproteobacteria,* and *Epsilonproteobacteria* (**figure 20.1**). However, new sequence data suggest that the separation between the *Betaproteobacteria* and *Gammaproteobacteria* is less distinct than once thought. If further analysis supports this, the *Betaproteobacteria* may be considered a subgroup of the *Gammaproteobacteria.* Members of the purple photosynthetic bacteria are found among the α-, β-, and γ-proteobacteria. This has led to the proposal that the proteobacteria arose from a single photosynthetic ancestor, presumably similar to the purple bacteria. Subsequently photosynthesis would have been lost by various lines, and new metabolic capacities were acquired as these bacteria adapted to different ecological niches. Although the phylum is currently considered monophyletic, this has recently been questioned.

In discussing this extremely diverse group of microbes, we review the morphology, physiology, and where appropriate, their role as pathogens. However, it should be emphasized that although some proteobacteria are important agents of disease, the vast majority are not—indeed, many make life on Earth possible.

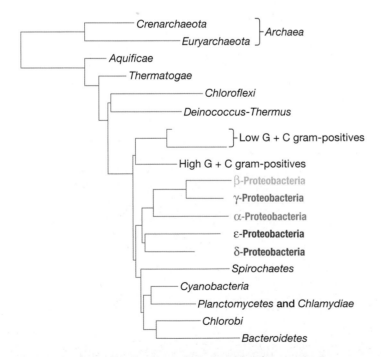

Figure 20.1 **Phylogenetic Relationships Among the Procaryotes.** The *Proteobacteria* are highlighted.

20.1 CLASS *ALPHA-PROTEOBACTERIA*

The class *Alphaproteobacteria* has seven orders and 20 families. **Figure 20.2** illustrates the phylogenetic relationships among major groups within the **α-proteobacteria.** The class includes most of the oligotrophic proteobacteria (those capable of growing at low nutrient levels). Some have unusual metabolic modes such as methylotrophy—the ability to grow using methane as a carbon source (*Methylobacterium*), chemolithotrophy (*Nitrobacter*), and the ability to fix nitrogen (*Rhizobium*). Members of genera such as *Rickettsia* and *Brucella* are important pathogens. Many genera are characterized by distinctive morphology such as prosthecae. **Table 20.1** summarizes the general characteristics of many of the bacteria discussed in the following sections.

Purple Nonsulfur Bacteria

All the purple bacteria use anoxygenic photosynthesis, possess bacteriochlorophylls *a* or *b,* and have their photosynthetic apparatus in membrane systems that are continuous with the plasma

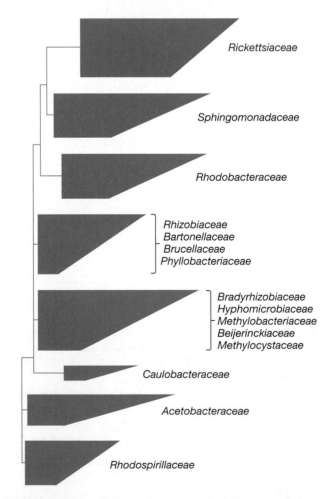

Figure 20.2 **Phylogenetic Relationships Among Major Families Within the α-Proteobacteria.** The relationships are based on 16S rRNA sequence data.

Table 20.1	Characteristics of Selected α-Proteobacteria

Genus	Dimensions (μm) and Morphology	G + C Content (mol%)	Genome size (Mb)	Oxygen Requirement	Other Distinctive Characteristics
Agrobacterium	0.6–1.0 × 1.5–3.0; motile, nonsporing rods with peritrichous flagella	57–63	2.5	Aerobic	Chemoorganotroph that can invade plants and cause tumors
Caulobacter	0.4–0.6 × 1–2; rod- or vibrioid-shaped with a flagellum and prostheca and holdfast	62–65	4.0	Aerobic	Heterotrophic and oligotrophic; asymmetric cell division
Hyphomicrobium	0.3–1.2 × 1–3; rod-shaped or oval with polar prosthecae	59–65	Nd[a]	Aerobic	Reproduces by budding; methylotrophic
Nitrobacter	0.5–0.9 × 1.0–2.0; rod- or pear-shaped, sometimes motile by flagella	59–62	3.4	Aerobic	Chemolithotroph, oxidizes nitrite to nitrate
Rhizobium	0.5–1.0 × 1.2–3.0; motile rods with flagella	57–66	5.1	Aerobic	Invades leguminous plants to produce nitrogen-fixing root nodules
Rhodospirillum	0.7–1.5 wide; spiral cells with polar flagella	62–64	4.4	Anaerobic, microaerobic, aerobic	Photoheterotroph under anoxic conditions
Rickettsia	0.3–0.5 × 0.8–2.0; short nonmotile rods	29–33	1.1–1.3	Aerobic	Obligate intracellular parasite

[a]Nd: Not determined; genome not yet sequenced

membrane. Most are motile by polar flagella. All purple nonsulfur bacteria are α-proteobacteria, with the exception of *Rhodocyclus* (β-proteobacteria). << *Photosynthetic bacteria (section 19.3)*

The **purple nonsulfur bacteria** are exceptionally flexible in their choice of an energy source. Normally they grow anaerobically as photoorganoheterotrophs; they trap light energy and employ organic molecules as both electron and carbon sources (table 20.1). Although they are called nonsulfur bacteria, some species can oxidize very low, nontoxic levels of sulfide to sulfate, but they do not oxidize elemental sulfur to sulfate. In the absence of light, most purple nonsulfur bacteria grow aerobically as chemoorganoheterotrophs, but some species carry out fermentations anaerobically. Oxygen inhibits bacteriochlorophyll and carotenoid synthesis so that cultures growing aerobically in the dark are colorless.

Purple nonsulfur bacteria vary considerably in morphology (**figure 20.3**). They may be spirals (*Rhodospirillum*), rods (*Rhodopseudomonas*), half circles or circles (*Rhodocyclus*), or they may even form prosthecae and buds (*Rhodomicrobium*). Because of their metabolism, they are most prevalent in the mud and water of lakes and ponds with abundant organic matter and low sulfide levels. Marine species also exist.

Rhodospirillum and *Azospirillum* (both in the family *Rhodospirillaceae*) are among several bacterial genera capable of forming **cysts** in response to nutrient limitation. These resting cells differ from the well-characterized endospores made by gram-positive bacteria such as *Bacillus* and *Clostridium*. Although cysts are also very resistant to desiccation, they are less tolerant of other environmental stresses such as heat and UV light. They have a thick outer coat and store an abundance of poly-β-hydroxybutyrate (PHB). Cyst-forming bacteria are not limited to α-proteobacteria; for instance, *Azotobacter*, a γ-proteobacterium, also forms cysts. << *Procaryotic cytoplasm: Inclusion bodies (section 3.3); Bacterial endospores (section 3.8)*

Rickettsia and *Coxiella*

The genus *Rickettsia* is placed in the order *Rickettsiales* and family *Rickettsiaceae* of the α-proteobacteria, whereas *Coxiella* is in the order *Legionellales* and family *Coxiellaceae* of the γ-proteobacteria. However, here we discuss *Rickettsia* and *Coxiella* together because of their similarity in life-style, despite their phylogenetic distance.

These bacteria are rod-shaped, coccoid, or pleomorphic with typical gram-negative walls and no flagella. Although their size varies, they tend to be very small. For example, *Rickettsia* is 0.3 to 0.5 μm in diameter and 0.8 to 2.0 μm long; *Coxiella* is 0.2 to 0.4 μm by 0.4 to 1.0 μm. All species are parasitic or mutualistic. The parasitic forms grow in vertebrate erythrocytes, macrophages, and vascular endothelial cells. Often they also live in blood-sucking arthropods such as fleas, ticks, mites, or lice, which serve as vectors or primary hosts. >> *Microbial interactions (section 27.1)*

(a) *Rhodospirillum rubrum*

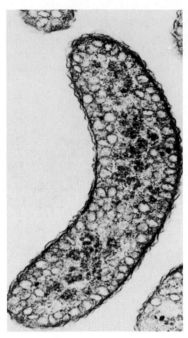

(b) *R. rubrum*

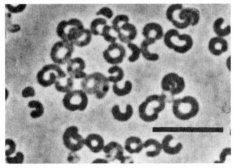

(c) *Rhodocyclus purpureus*

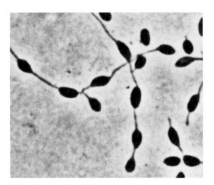

(d) *Rhodomicrobium vannielii*

Figure 20.3   Typical Purple Nonsulfur Bacteria.   (a) *Rhodospirillum rubrum;* phase contrast (×410). (b) *R. rubrum* grown anaerobically in the light. Note vesicular invaginations of the cytoplasmic membranes; transmission electron micrograph (×51,000). (c) *Rhodocyclus purpureus;* phase contrast. Bar = 10 μm. (d) *Rhodomicrobium vannielii* with vegetative cells and buds; phase contrast.

Rickettsias are very different from most other bacteria in physiology and metabolism. They lack glycolytic pathways and do not use glucose as an energy source but rather oxidize glutamate and tricarboxylic acid cycle intermediates such as succinate. The rickettsial plasma membrane has carrier-mediated transport systems, and host cell nutrients and coenzymes are absorbed and directly used. For example, rickettsias take up both NAD^+ and uridine diphosphate glucose. Their membrane also has a carrier that exchanges ADP for external ATP. Thus host ATP may provide much of the energy needed for growth. This metabolic dependence explains why many of these organisms must be cultivated in the yolk sacs of chick embryos or in tissue culture cells. Genome sequencing shows that *R. prowazekii* is similar in many ways to mitochondria, suggesting that these organelles arose from an endosymbiotic association with an ancestor of

Rickettsia. ◄◄ *Microbial evolution: Endosymbiotic origin of mitochondria and chloroplasts (section 17.1)*

Because these genera include important human pathogens, their reproduction and metabolism have been intensively studied. They enter the host cell by inducing phagocytosis. Members of the genus *Rickettsia* immediately escape the phagosome and reproduce by binary fission in the cytoplasm (**figure 20.4a,b**). In contrast, *Coxiella* remains within the phagosome after it has fused with a lysosome and actually reproduces within the phagolysosome (figure 20.4c). Eventually the host cell bursts, releasing new organisms. Besides incurring damage from cell lysis, the host is harmed by the toxic effects of bacterial cell walls (wall toxicity appears related to the mechanism of penetration into host cells). It is therefore not surprising that these orders contain many important pathogens, as we now discuss. ►► *Phagocytosis (section 28.3)*

Epidemic (Louse-Borne) Typhus

Epidemic (louse-borne) typhus is caused by *Rickettsia prowazekii,* which is transmitted from person to person by the body louse. In the United States, a reservoir of *R. prowazekii* also exists in the southern flying squirrel. When a louse feeds on an infected rickettsemic person, the rickettsias infect the insect's gut and multiply, and large numbers of organisms appear in the insect's feces in about a week. When a louse takes a blood meal, it defecates. The irritation causes the affected individual to scratch the site and contaminate the bite wound with rickettsias. The rickettsias then spread by way of the bloodstream and infect the endothelial cells of the blood vessels, causing a vasculitis (inflammation of the blood vessels). This produces an abrupt headache, fever, and muscle aches. A rash begins on the upper trunk and spreads. Without treatment, recovery can take about 2 weeks, though mortality rates are very high (around 50%), especially in the elderly. Recovery from the disease gives a vigorous immunity and also protects the person from murine typhus (see the section on endemic (murine) typhus).

The importance of louse control and good public hygiene is shown by the prevalence of typhus epidemics during times of war and famine, when there is crowding and little attention to the maintenance of proper sanitation. For example, around 30 million cases of typhus fever and 3 million deaths occurred in the Soviet Union and Eastern Europe between 1918 and 1922. Napoleon's retreat from Russia in 1812 may have been partially provoked by typhus and dysentery epidemics that ravaged the French army. Fewer than 25 cases of epidemic typhus are reported in the United States each year.

Endemic (Murine) Typhus

The etiologic agent of endemic (murine) typhus is *Rickettsia typhi.* It occurs in isolated areas around the world, including southeastern and Gulf Coast states, especially Texas. The disease

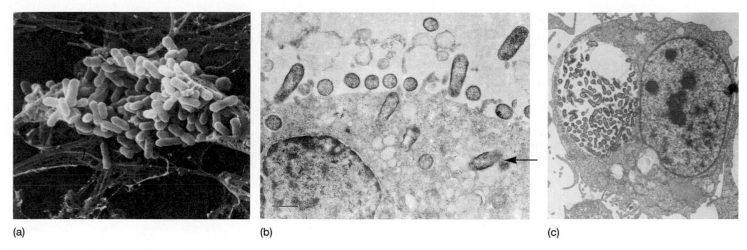

(a)   (b)   (c)

Figure 20.4 *Rickettsia* **and** *Coxiella.* Rickettsial morphology and reproduction. (a) A human fibroblast filled with *Rickettsia prowazekii.* (b) *Rickett-siae* attachment to endothelial cell and subsequent phagocytosis. *Rickettsiae* leave a disrupted phagosome (arrow) and enter the cytoplasmic matrix. (c) *Coxiella burnetti* growing within fibroblast vacuoles (×9,000).

occurs sporadically in individuals who come into contact with rats and their fleas (*Xenopsylla cheopi*). The disease is nonfatal in the rat and is transmitted from rat to rat by fleas. When an infected flea takes a human blood meal, it defecates. Its feces are heavily laden with rickettsias, which infect humans by contaminating the bite wound.

The clinical manifestations of murine (Latin *mus, muris,* mouse or rat) typhus are similar to those of epidemic typhus except that they are milder in degree and the mortality rate is much lower—less than 5%. Few cases of endemic typhus are reported in the United States each year, resulting in its removal from the reportable disease list. >> *The epidemiology of infectious disease (chapter 33)*

Rocky Mountain Spotted Fever

Rocky Mountain spotted fever is caused by *Rickettsia rickettsii.* Although this disease was originally detected in the Rocky Mountain area, most cases now occur east of the Mississippi River. The disease is transmitted by ticks and usually occurs in people who are or have been in tick-infested areas. There are two principal vectors: *Dermacentor andersoni,* the wood tick, is distributed in the Rocky Mountain states and is active during the spring and early summer. *D. variabilis,* the dog tick, has assumed greater importance and is almost exclusively confined to the eastern half of the United States. Unlike other rickettsias, *R. rickettsii* can pass from generation to generation of ticks through their eggs in a process known as transovarian passage. No humans or mammals are needed as reservoirs for the continued propagation of this rickettsia in the environment.

When humans contact infected ticks, the rickettsias are either deposited on the skin (if the tick defecates after feeding) and then subsequently rubbed or scratched into the skin, or the rickettsias are deposited into the skin as the tick feeds. Once inside the skin, the rickettsias enter the endothelial cells of small blood vessels, where they multiply and produce a characteristic vasculitis.

The disease is characterized by the sudden onset of a headache, high fever, chills, and a skin rash (**figure 20.5**) that initially appears on the ankles and wrists and then spreads to the trunk of the body. If the disease is not treated, the rickettsias can destroy the blood vessels in the heart, lungs, or kidneys and cause death. Usually, however, severe pathological changes are avoided by antibiotic therapy (chloramphenicol, chlortetracycline), supportive therapy, and the development of immune resistance. Roughly 400 to 800 cases of Rocky Mountain spotted fever are reported annually in the United States.

Ehrlichiosis

Ehrlichiosis is caused by the α-proteobacterium *Ehrlichia chaffeensis.* Members of the genus *Ehrlichia* are related to the genus *Rickettsia* and placed in the order *Rickettsiales. E. chaffeensis* is

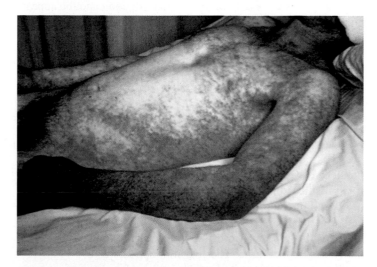

Figure 20.5 Rocky Mountain Spotted Fever. Typical rash occurring on the arms and chest consists of generally distributed, sharply defined macules.

transmitted from dogs and white-tailed deer, the primary reservoirs, to humans, primarily by the Lone Star tick (*Amblyomma americanum*). Once inside the human body, *E. chaffeensis* infects circulating white blood cells (monocytes), causing a nonspecific febrile illness (human monocytic ehrlichiosis, HME) that resembles Rocky Mountain spotted fever. More than 200 cases of ehrlichiosis are reported in the United States annually.

A form of ehrlichiosis caused by a species different from *E. chaffeensis* was discovered in 1994. Human granulocytic ehrlichiosis (HGE) is transmitted by deer ticks (*Ixodes scapularis*) and possibly dog ticks (*Dermacentor variabilis*), and has been found in 30 states, particularly in the southeastern, northern, and central United States. The disease is characterized by the rapid onset of fever, chills, headaches, and muscle aches. Both HME and HGE can be successfully treated with antibiotic therapy (doxycycline).

Q Fever

Q fever (Q for *query* because the cause of the fever was not known for some time) is caused by the γ-proteobacterium *Coxiella burnetii*. *C. burnetii* can survive outside host cells by forming a resistant, endospore-like body. It does not need an arthropod vector for transmission. This bacterium infects both wild animals and livestock. Cattle, sheep, and goats are the primary reservoirs. In animals, ticks (many species) transmit *C. burnetii,* whereas human transmission is primarily by inhalation of dust contaminated with bacteria from dried animal feces or urine, or consumption of unpasteurized milk. The disease can occur in epidemic form among slaughterhouse workers and sporadically among farmers and veterinarians. Each year, fewer than 30 cases of Q fever are reported in the United States.

In humans, Q fever is an acute illness characterized by the sudden onset of severe headache, malaise, confusion, sore throat, chills, sweats, nausea, chest pain, myalgia (muscle pain), and high fever. It is rarely fatal, but endocarditis—inflammation of the heart muscle—occurs in about 10% of the cases. Five to 10 years may elapse between the initial infection and the appearance of the endocarditis. During this interval, the bacteria reside in the liver and often cause hepatitis.

Caulobacteraceae and *Hyphomicrobiaceae*

A number of the proteobacteria are not simple rods or cocci but have some sort of appendage. These bacteria have interesting life cycles that feature a prostheca or reproduction by budding. A **prostheca** (pl., prosthecae), also called a **stalk**, is an extension of the cell, including the plasma membrane and cell wall that is narrower than the mature cell. **Budding** is distinctly different from the binary fission normally used by bacteria. The bud first appears as a small protrusion at a single point and enlarges to form a mature cell. Most or all of the bud's cell envelope is newly synthesized. In contrast, portions

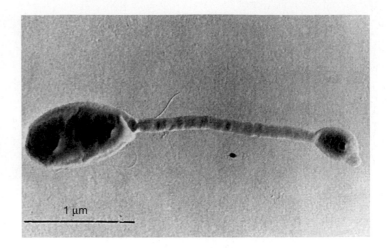

Figure 20.6 Prosthecate, Budding Bacteria. *Hyphomicrobium facilis* with hypha and young bud.

of the parental cell envelope are shared with the progeny cells during binary fission. Finally, the parental cell retains its identity during budding, and the new cell is often smaller than its parent. In binary fission, the parental cell disappears as it forms progeny of equal size. The families *Caulobacteraceae* and *Hyphomicrobiaceae* of the α-proteobacteria contain two of the best studied prosthecate genera: *Caulobacter* and *Hyphomicrobium.*
<< *Bacterial cell cycle (section 7.1)*

The genus *Hyphomicrobium* contains chemoheterotrophic, aerobic, budding bacteria that frequently attach to solid objects in freshwater, marine, and terrestrial environments. The vegetative cell measures about 0.5 to 1.0 by 1 to 3 μm (**figure 20.6**). At the beginning of the reproductive cycle, the mature cell produces a prostheca (also called a hypha) 0.2 to 0.3 μm in diameter that grows to several μm in length (**figure 20.7**). The nucleoid divides, and a copy moves into the hypha while a bud forms at its end. As

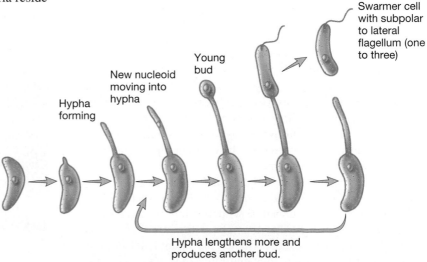

Hypha forming

New nucleoid moving into hypha

Young bud

Swarmer cell with subpolar to lateral flagellum (one to three)

Hypha lengthens more and produces another bud.

Figure 20.7 The Life Cycle of *Hyphomicrobium.*

the bud matures, it produces one to three flagella, and a septum divides the bud from the hypha. The bud is finally released as an oval- to pear-shaped swarmer cell, which swims off, then settles down and begins budding. The mother cell may bud several times at the tip of its hypha.

Hyphomicrobium also has distinctive physiology and nutrition. Sugars and most amino acids do not support abundant growth; instead, *Hyphomicrobium* grows on ethanol and acetate, and flourishes with one-carbon compounds such as methanol, formate, and formaldehyde. That is, it is a facultative **methylotroph** and can derive both energy and carbon from reduced one-carbon compounds. It is so efficient at acquiring one-carbon molecules that it can grow in a medium without an added carbon source (presumably the medium absorbs sufficient atmospheric carbon compounds). *Hyphomicrobium* may comprise up to 25% of the total bacterial population in oligotrophic or nutrient-poor freshwater habitats.

Bacteria in the genus *Caulobacter* alternate between polarly flagellated rods and cells that possess a prostheca and a **holdfast,** by which they attach to solid substrata (**figure 20.8**). Incredibly, the material secreted at the end of the *Caulobacter cresentus* hold-

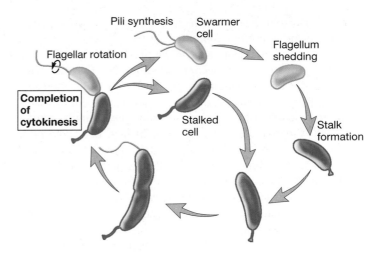

Figure 20.9 *Caulobacter* **Life Cycle.** Stalked cells attached to a substrate undergo asymmetric binary fission, producing a stalked and a flagellated cell, called a swarmer cell. The swarmer cell swims freely and makes pili until it settles, ejects its flagella, and forms a stalk. Only stalked cells can divide.

fast is the strongest biological adhesion molecule known—a sort of bacterial superglue. Caulobacters are usually isolated from freshwater and marine habitats with low nutrient levels, but they also are present in the soil. They often adhere to bacteria, photosynthetic protists, and other microorganisms, and may absorb nutrients released by their hosts. The prostheca differs from that of *Hyphomicrobium* in that it lacks cytoplasmic components and is composed almost totally of the plasma membrane and cell wall. It grows longer in nutrient-poor media and can reach more than 10 times the length of the cell body. The prostheca may improve the efficiency of nutrient uptake from dilute habitats by increasing surface area; it also gives the cell extra buoyancy.

The life cycle of *Caulobacter* is unusual (**figure 20.9**). When ready to reproduce, the cell elongates and a single polar flagellum forms at the end opposite the prostheca. The cell then undergoes asymmetric transverse binary fission to produce a flagellated swarmer cell that swims away. The swarmer, which cannot reproduce, comes to rest, ejects its flagellum, and forms a new prostheca on the formerly flagellated end. The new stalked cell then starts the cycle anew. This process takes about 2 hours to complete. The species *C. cresentus* has become an important model organism in the study of microbial development and the bacterial cell cycle.

Order *Rhizobiales*

The order *Rhizobiales* of the α-proteobacteria contains 11 families with a great variety of phenotypes. This includes the family *Hyphomicrobiaceae*, which has already been discussed. Another important family in this

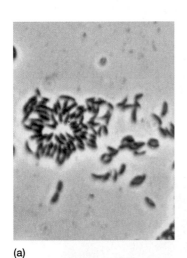

(a)

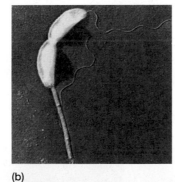

(b)

(c)

(d)

1 μm

Figure 20.8 *Caulobacter* **Morphology and Reproduction.** (a) Rosettes of cells adhering to each other by their prosthecae; phase contrast (×600). (b) A cell dividing to produce a swarmer (×6,030). Note prostheca and flagellum. (c) A stalked cell and a flagellated swarmer cell (×6,030). (d) A rosette of cells as seen in the electron microscope.

order is Rhizobiaceae, which includes the aerobic genera *Rhizobium* and *Agrobacterium*.

Members of the genus *Rhizobium* are 0.5 to 0.9 by 1.2 to 3.0 μm motile rods that become pleomorphic (having many shapes) under adverse conditions (**figure 20.10**). Cells often contain poly-β-hydroxybutyrate inclusions. They grow symbiotically within root nodule cells of legumes as nitrogen-fixing bacteroids (figure 20.10*b; see also figure 26.23*). In fact, *Leguminosae* is the most successful plant family on Earth, with over 18,000 species. Their proliferation reflects their capacity to establish symbiotic relationships with bacteria that form nodules on their roots. Within the nodules, the microbes reduce or fix atmospheric nitrogen into ammonia, making it directly available to the plant host. The process by which bacteria perform this fascinating and important symbiosis is discussed in chapter 26. << *Synthesis of amino acids: Nitrogen fixation (section 11.5);* >> *Microorganisms in terrestrial environments: Microorganism associations with vascular plants (section 26.2)*

The genus *Agrobacterium* is placed in the family *Rhizobiaceae* but differs from *Rhizobium* because it is a plant pathogen. Agrobacteria invade the crown, roots, and stems of many plants and transform plant cells into autonomously proliferating tumor cells (*see figures 26.26 and 26.27*). The best-studied species is *A. tumefaciens,* which enters many broad-leaved plants through wounds and causes crown gall disease (**figure 20.11**). The ability to produce tumors depends on the presence of a large Ti (for *t*umor *i*nducing) plasmid. Tumor production by *Agrobacterium* is discussed in greater detail in chapter 26, while the use of the Ti plasmid in the genetic engineering of plants is covered in chapter 16.

The order *Rhizobiales* also includes the family Brucellaceae. The genus *Brucella* is an important human and animal pathogen. **Brucellosis,** also called **undulant fever,** is a zoonotic disease—a disease transmitted from animals to humans. It is usually caused by tiny, faintly staining coccobacilli of the species *B. abortus, B. melitensis, B. suis,* or *B. canis. Brucella* is commonly transmitted through consumption of contaminated animal products or abrasions of the skin from handling infected mammals (cattle, sheep, goats, pigs, and rarely from dogs). The most common route is ingestion of contaminated milk products. Direct person-to-person spread of brucellosis is extremely rare. However, infants may be infected through their mother's breast milk; sexual transmission of brucellosis has also been reported. Although uncommon, transmission of brucellosis may also occur through transplantation of contaminated blood or tissue. A recent controversy has erupted in the northwestern United States over the transmission of *Brucella,* endemic in the wild bison and elk populations, to otherwise *Brucella*-free cattle.

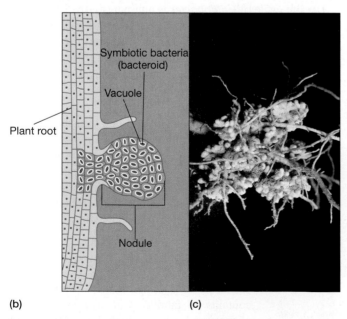

(a)

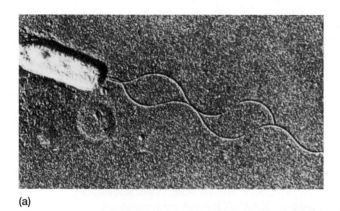

(b)   (c)

Figure 20.10 Rhizobium. (a) *Rhizobium leguminosarum* with two polar flagella (×14,000). (b) Diagram of differentiated bacteria, called bacteroids, in the nodule of a legume root. (c) Nodules on a legume root.

Figure 20.11 Agrobacterium. Crown gall tumor of a tomato plant caused by *Agrobacterium tumefaciens*.

Brucellosis, in the acute (8 weeks from onset) form, presents as nonspecific, flulike symptoms, including fever, sweats, malaise, anorexia, headache, myalgia, and back pain. In the undulant (rising and falling) form, symptoms of brucellosis include undulant fevers, arthritis, and testicular inflammation. Neurologic symptoms may occur acutely in up to 5% of the cases. In the chronic form (1 year from onset), brucellosis symptoms may include chronic fatigue, depression, and arthritis. Mortality is low, less than 2%. Sequellae of brucellosis are variable and include granulomatous hepatitis, peripheral arthritis, spondylitis, anemia, leukopenia, thrombocytopenia, meningitis, uveitis, optic neuritis, papilledema, and endocarditis. The insidious disease and significant morbidity associated with *Brucella* infection have prompted its restriction as a select agent; the public health impact, should it be used as a bioweapon, would be substantial.

Nitrifying Bacteria

The taxonomy of the aerobic chemolithotrophic bacteria, those bacteria that derive energy and electrons from reduced inorganic compounds, is quite complex. Normally these bacteria employ CO_2 as their carbon source and thus are chemolithoautotrophs, but some can function as chemolithoheterotrophs and use reduced organic carbon sources. In

Bergey's Manual, the chemolithotrophic proteobacteria are distributed between the α-, β-, and γ-proteobacteria. The nitrifying bacteria are found in all three classes. << *Chemolithotrophy (section 10.11)*

The **nitrifying bacteria** are a very diverse collection of bacteria. *Bergey's Manual* places nitrifying genera in several families: *Nitrobacter* in the *Bradyrhizobiaceae*, α-proteobacteria; *Nitrosomonas* and *Nitrosospira* in the *Nitrosomonadaceae*, γ-proteobacteria; *Nitrococcus* in the *Ectothiorhodospiraceae*, γ-proteobacteria; and *Nitrosococcus* in the *Chromatiaceae*, γ-proteobacteria. All are aerobic, gram-negative organisms with the ability to capture energy from the oxidation of either ammonia or nitrite. However, they differ considerably in other properties (**table 20.2**). Nitrifiers may be rod-shaped, ellipsoidal, spherical, spirillar, or lobate, and they may possess either polar or peritrichous flagella. Often they have extensive membrane complexes in their cytoplasm (**figure 20.12**). Identification is based on properties such as their preference for nitrite or ammonia, their general shape, and the nature of any cytomembranes present.

Nitrifying bacteria make important contributions to the nitrogen cycle. In soil, sewage disposal systems, and freshwater and marine habitats, the β-proteobacteria *Nitrosomonas, Nitrosospira* and *Nitrosolobus,* and the γ-proteobacterium *Nitrosococcus* oxidize ammonia to nitrite. In the same niches,

---

Table 20.2	Selected Characteristics of Representative Nitrifying Bacteria					
Species	Cell Morphology and Size (μm)	Reproduction	Motility	Cytomembranes	G + C Content (mol%)	Habitat
Ammonia-Oxidizing Bacteria						
Nitrosomonas europaea (β-proteobacteria)	Rod; 0.8–1.1 × 1.0–1.7	Binary fission	−	Peripheral, lamellar	50.6–51.4	Soil, sewage, freshwater, marine
Nitrosococcus oceani (γ-proteobacteria)	Coccoid; 1.8–2.2 in diameter	Binary fission	+; 1 or more subpolar flagella	Centrally located parallel bundle, lamellar	50.5	Obligately marine
Nitrosospira briensis (β-proteobacteria)	Spiral; 0.3–0.4 in diameter	Binary fission	+ or −; 1 to 6 peritrichous flagella	Rare	53.8–54.1	Soil
Nitrite-Oxidizing Bacteria						
Nitrobacter winogradskyi (α-proteobacteria)	Rod, often pear-shaped; 0.5–0.9 × 1.0–2.0	Budding	+ or −; 1 polar flagellum	Polar cap of flattened vesicles in peripheral region of the cell	61.7	Soil, freshwater, marine
Nitrococcus mobilis (γ-proteobacteria)	Coccoid; 1.5–1.8 in diameter	Binary fission	+; 1 or 2 subpolar flagella	Tubular cytomembranes randomly arranged in cytoplasm	61.3 (1 strain)	Marine

From Brenner, D. J.; Krieg, N. R.; and Staley, J. T. Eds. 2005. *Bergey's Manual to Systemic Bacteriology,* 2d ed. Vol. 2: *The Proteobacteria.* Garrity, G. M. Ed-in-Chief. New York: Springer.

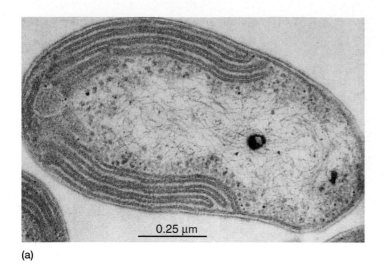

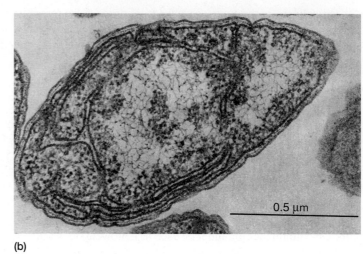

(a)

(b)

Figure 20.12   Representative Nitrifying Bacteria.   (a) *Nitrobacter winogradskyi.* Note the polar cap of cytomembranes (×213,000). (b) *N. europaea* with extensive cytoplasmic membranes (×81,700).

members of the α- and γ-proteobacteria *Nitrococcus* and *Nitrobacter,* respectively, as well as species of the nonproteobacterium *Nitrospira* then oxidize nitrite to nitrate. The whole process of converting ammonia to nitrite and then to nitrate is called **nitrification,** and it occurs rapidly in oxic soil treated with fertilizers containing ammonium salts. Nitrate is readily used by plants, but it is also rapidly lost through leaching of water-soluble nitrates and by denitrification to nitrogen gas, so the benefits gained from nitrification can be fleeting. >> *Biogeochemical cycling: Nitrogen cycle (section 25.1)*

1. Describe the general properties of the α-proteobacteria.

2. Discuss the characteristics and physiology of the purple nonsulfur bacteria. Where would one expect to find them growing?

3. Briefly describe the characteristics and life cycle of the genus *Rickettsia.*

4. How is epidemic typhus spread? Ehrlichiosis? Murine typhus? In what ways are they similar?

5. What is unusual about the physiology of *Hyphomicrobium?* How does this influence its ecological distribution?

6. What unique features of the *Caulobacter* life cycle do you think make this microbe such an attractive model system?

7. How do *Agrobacterium* and *Rhizobium* differ in lifestyle? What effect does *Agrobacterium* have on plant hosts?

8. Give the major characteristics of the nitrifying bacteria and discuss their ecological importance. How does the metabolism of *Nitrobacter* differ from that of *Nitrosomonas?*

20.2   CLASS BETAPROTEOBACTERIA

The **β-proteobacteria** overlap the α-proteobacteria metabolically but tend to use substances that diffuse from organic decomposition in anoxic habitats. Some of these bacteria use hydrogen, ammonia, methane, volatile fatty acids, and similar substances. As with the α-proteobacteria, there is considerable metabolic diversity; the β-proteobacteria may be chemoheterotrophs, photolithotrophs, methylotrophs, and chemolithotrophs.

The class *Betaproteobacteria* has seven orders and 12 families. **Figure 20.13** shows the phylogenetic relationships among

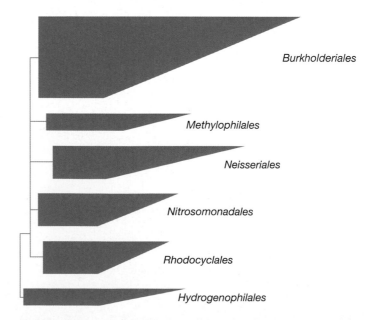

Figure 20.13   Phylogenetic Relationships Among Major Groups Within the β-Proteobacteria.   The relationships are based on 16S rRNA sequence data.

Table 20.3 Characteristics of Selected β-Proteobacteria

Genus	Dimensions (μm) and Morphology	G + C Content (mol%)	Genome size (Mb)	Oxygen Requirement	Other Distinctive Characteristics
Bordetella	0.2–0.5 × 0.5–2.0; nonmotile coccobacillus	66–70	3.7–5.3	Aerobic	Requires organic sulfur and nitrogen; mammalian parasite
Burkholderia	0.5–1.0 × 1.5–4; straight rods with single flagella or a tuft at the pole	59–69.5	4.1–7.2	Aerobic, some capable of anaerobic respiration with NO_3^-	Poly-β-hydroxybutyrate as reserve; can be pathogenic
Leptothrix	0.6–1.5 × 2.5–15; straight rods in chains with sheath, free cells flagellated	68–71	Nd[a]	Aerobic	Sheaths encrusted with iron and manganese oxides
Neisseria	0.6–1.9; cocci in pairs with flattened adjacent sides	48–56	2.2–2.3	Aerobic	Inhabitant of mucous membranes of mammals
Nitrosomonas	Size varies with strain; spherical to ellipsoidal cells with intracytoplasmic membranes	45–54	2.8	Aerobic	Chemolithotroph that oxidizes ammonia to nitrite
Sphaerotilus	1.2–2.5 × 2–10; single chains of cells with sheaths, may have holdfasts	70	Nd	Aerobic	Sheaths not encrusted with iron and manganese oxides
Thiobacillus	0.3–0.5 × 0.9–4; rods, often with polar flagella	52–68	Nd	Aerobic	All chemolithotrophic, oxidizes reduced sulfur compounds to sulfate, some also chemoorganotrophic

[a]Nd: Not determined; genome not yet sequenced

major groups within the β-proteobacteria, and **table 20.3** summarizes the general characteristics of many of the bacteria discussed in this section. Here, we discuss genera with important human pathogens: *Neisseria, Burkholderia,* and *Bordetella.*

Order *Neisseriales*

The order *Neisseriales* has one family, *Neisseriaceae,* with 15 genera. The best-known and most intensely studied genus is *Neisseria.* Members of this genus are nonmotile, aerobic, gramnegative cocci that most often occur in pairs with adjacent sides flattened. They may have capsules and fimbriae (often referrred to as pili). The genus is chemoorganotrophic and produces the enzymes oxidase and catalase. The genus is best known for the diseases it causes: meningitis and gonorrhea.

Meningococcal Meningitis

Meningitis (Greek *meninx,* membrane, and *-itis,* inflammation) is an inflammation of the brain or spinal cord meninges (membranes). Based on the specific cause, it can be divided into **bacterial (septic) meningitis** and **aseptic meningitis syndrome.** As shown in **table 20.4**, there are many causes of the aseptic meningitis syndrome, only some of which can be treated with antimicrobial agents. Thus accurate identification of the causative

Table 20.4 Causative Agents of Meningitis by Diagnostic Category

Type of Meningitis	Causative Agent
Bacterial (Septic) Meningitis	
	Streptococcus pneumoniae *Neisseria meningitidis* *Haemophilus influenzae* type b Group B streptococci *Listeria monocytogenes* *Mycobacterium tuberculosis* *Nocardia asteroides* *Staphylococcus aureus* *Staphylococcus epidermidis*
Aseptic Meningitis Syndrome	
Agents requiring antimicrobials	Fungi Amoebae *Treponema pallidum* Mycoplasmas Leptospires
Agents requiring other treatments	Viruses Cancers Parasitic cysts Chemicals

agent is essential for proper treatment of the disease. Bacterial meningitis can be diagnosed by a Gram stain and culture of bacteria from cerebral spinal fluid (CSF). Rapid tests are also used.

Bacterial meningitis can be caused by various gram-positive and gram-negative bacteria. However, three organisms tend to be associated with meningitis more frequently than others: *Streptococcus pneumoniae, Neisseria meningitidis,* and *Haemophilus influenzae* (serotype b). *H. influenzae* is discussed with the γ-proteobacteria (section 20.3), and *S. pneumoniae* is reviewed in chapter 21. *N. meningitidis,* often referred to as the **meningococcus,** is a normal inhabitant of the human nasopharynx (5 to 15% of humans carry the nonpathogenic serotypes). Most disease-causing *N. meningitidis* strains belong to serotypes A, B, C, Y, and W-135. In general, serotype A strains are the cause of epidemic disease in developing countries, while serotype C and W-135 strains are responsible for meningitis outbreaks in the United States. Infection results from airborne transmission of the bacteria in respiratory secretions, typically through close contact with a primary carrier. The disease process is initiated by pili-mediated colonization of the nasopharynx by pathogenic bacteria. The bacteria cross the nasopharyngeal epithelium (typically through endocytosis) and invade the bloodstream (meningococcemia), where they proliferate. They then cross the blood-brain barrier to enter the CSF, where they produce inflammation of the meninges.

Symptoms caused by *N. meningitidis* are variable, depending on the degree of bacterial dissemination. The usual symptoms of meningitis include an initial respiratory illness or sore throat interrupted by one of the meningeal syndromes: vomiting, headache, lethargy, confusion, and stiffness in the neck and back. Once bacterial meningitis is suspected, specific antibiotics are administered immediately. In fact, antibiotics are often administered prophylactically to patient contacts; if left untreated, meningitis is fatal.

Two vaccines are currently available: the meningococcal polysaccharide (MPSV4) and the meningococcal conjugate vaccine (MCV4). Both vaccines are effective against serotypes A, C, Y, and W-135. Vaccination is recommended for all college students living in residence halls. MCV4 is recommended for preteens, teens, and adults less than fifty-five years of age. MPSV4 should be used for children two to ten years of age and adults over fifty-five who are at risk.

Gonorrhea

Gonorrhea (Greek *gono,* seed, and *rhein,* to flow) is an acute, infectious, sexually transmitted disease of the mucous membranes of the genitourinary tract, eye, rectum, and throat caused by *N. gonorrhoeae.* These bacteria are also referred to as **gonococci** (s., gonococcus; Greek *gono* and *coccus,* berry) and have a worldwide distribution. Over 500,000 cases are reported annually in the United States; the actual incidence is significantly higher.

Once inside the body, the gonococci attach to the microvilli of mucosal cells by means of pili and protein II, which function

as adhesins. This attachment prevents the bacteria from being washed away by normal cervical and vaginal discharges or by the flow of urine. They are then phagocytosed by the mucosal cells and may even be transported through the cells to the intercellular spaces and subepithelial tissue. Phagocytes, such as neutrophils, also may contain gonococci inside vesicles (**figure 20.14**). Because the gonococci are intracellular at this time, the host's defenses have little effect on the bacteria. Following penetration of the bacteria, the host tissue responds locally by the infiltration of immune cells, including white blood cells, and antibody-secreting plasma cells. These cells are later replaced by fibrous tissue that may lead to urethral closing, or stricture, in males. **>>** *Cells, tissues, and organs of the immune system (section 28.2); Phagocytosis (section 28.3)*

In males, the incubation period is 2 to 8 days. The onset consists of a urethral discharge of yellow, creamy pus, and frequent, painful urination that is accompanied by a burning sensation. In females, the cervix is the principal site infected. The disease is more insidious in females and few are aware of any symptoms. However, some symptoms may begin 7 to 21 days after infection. These are generally mild; some vaginal discharge may occur. The gonococci also can infect the Fallopian tubes and surrounding tissues, leading to **pelvic inflammatory disease (PID).** This occurs in 10 to 20% of infected females. Gonococcal PID results in scar formation in the Fallopian tubes; it is a major cause of sterility and ectopic pregnancies. Gonococci disseminate most often during menstruation, when there is an increased concentration of free iron available to the bacteria. In both genders, disseminated gonococcal infection with bacteremia may occur. This can lead to involvement of the joints (gonorrheal arthritis), heart (gonorrheal endocarditis), or pharynx (gonorrheal pharyngitis). Gonorrheal eye infections can occur in newborns as they pass through an infected birth canal. The resulting disease is called **ophthalmia neonatorum,** or **conjunctivitis of the newborn.** This was once a leading cause of blindness in many parts of the world. To prevent this disease, antibiotic or silver nitrate in dilute

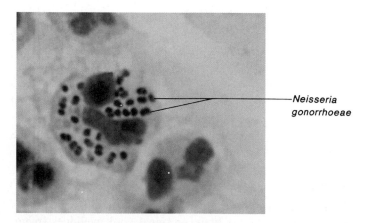

Figure 20.14 Gonorrhea. Gram stain of male urethral exudate showing *Neisseria gonorrhoeae* (diplococci) inside a PMN; light micrograph (×500). Although the presence of gram-negative diplococci in exudates is a probable indication of gonorrhea, the bacterium should be isolated and identified.

solution is placed in the eyes of newborns, as required by law in the United States and many other nations.

Penicillin-resistant strains of gonococci occur worldwide, and tetracycline-resistant *N. gonorrhoeae* also have developed. Recently the Centers for Disease Control and Prevention (CDC) has recommended discontinuation of quinolones to treat gonorrhea, as quinolone-resistant strains are increasing. Instead, it recommends ceftriaxone or cefixime. >> *Drug resistance (section 31.6)*

Order *Burkholderiales*

The order *Burkholderiales* contains four families, three of them with well-known genera. The genus *Burkholderia* is placed in the family *Burkholderiaceae*. This genus was established when *Pseudomonas* was divided into at least seven genera based on rRNA data: *Acidovorax, Aminobacter, Burkholderia, Comamonas, Deleya, Hydrogenophaga,* and *Methylobacterium*. Members of the genus *Burkholderia* are gram-negative, aerobic, nonfermentative, nonsporing, mesophilic straight rods (**chapter opener**). With the exception of one species, all are motile with a single polar flagellum or a tuft of polar flagella. Catalase is produced, and they often are oxidase positive. Most species use poly-β-hydroxybutyrate (PHB) as their carbon reserve. One of the most important species is *B. cepacia,* which can degrade over 100 different organic molecules and is very active in recycling organic materials in nature. Originally described as the plant pathogen that causes onion rot, it has emerged in the last 20 years as a major nosocomial pathogen. It is a particular problem for cystic fibrosis patients. Two other species, *B. mallai* and *B. pseudomallei,* are human pathogens that could be misused as bioterrorism agents. >> *Bioterrorism preparedness (section 33.9)*

Surprisingly, two genera within the *Burkholderiaceae* family are capable of forming nitrogen-fixing symbioses with legumes much like the rhizobia that belong to the α-proteobacteria. Genome analysis of nitrogen-fixing *Burkholderia* and *Ralstonia* isolates reveals the presence of nodulation (*nod*) genes that are very similar to those of the rhizobia. This suggests a common genetic origin. It is thought these β-proteobacteria gained the capacity to form symbiotic, nitrogen-fixing nodules with legumes through lateral gene transfer.

Some members of the order *Burkholderiales* have a **sheath**—a hollow, tubelike structure surrounding a chain of cells. Sheaths often are close fitting, but they are never in intimate contact with the cells they enclose. Some contain ferric or manganic oxides. They have at least two functions. Sheaths help bacteria attach to solid surfaces and acquire nutrients from slowly running water as it flows past, even if it is nutrient-poor. Sheaths also protect against predators such as protozoa.

Two well-studied sheathed genera are in the family *Comamonadaceae. Sphaerotilus* forms long, sheathed chains of rods, 0.7 to 2.4 μm by 3 to 10 μm, attached to

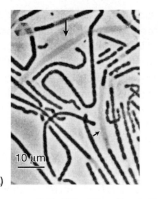
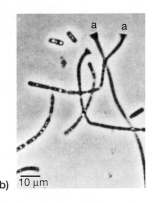

Figure 20.15 Sheathed Bacteria, *Sphaerotilus natans.* (a) Sheathed chains of cells and empty sheaths (arrows). (b) Chains with holdfasts (indicated by the letter a) and individual cells containing poly-β-hydroxybutyrate granules.

submerged plants, rocks, and other solid objects, often by a holdfast (**figure 20.15**). Single swarmer cells with a bundle of subpolar flagella escape the filament and form a new chain after attaching to a solid object at another site. *Sphaerotilus* grows best in slowly running freshwater polluted with sewage or industrial waste. It grows so well in activated sewage sludge that it sometimes forms tangled masses of filaments and interferes with the proper settling of sludge. *Leptothrix* characteristically deposits large amounts of iron and manganese oxides in its sheath (**figure 20.16**). This seems to protect it and allow *Leptothrix* to grow in the presence of high concentrations of soluble iron compounds.

The family *Alcaligenaceae* contains the genus *Bordetella*. This genus is composed of gram-negative, aerobic coccobacilli, about 0.2 to 0.5 μm by 0.5 to 2.0 μm in size. *Bordetella* is a

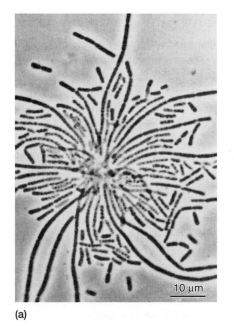
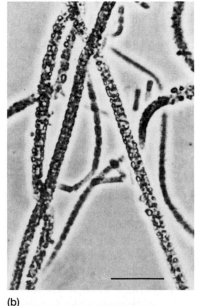

Figure 20.16 Sheathed Bacteria, *Leptothrix* Morphology. (a) *L. lopholea* trichomes radiating from a collection of holdfasts. (b) *L. cholodnii* sheaths encrusted with MnO₂.

chemoorganotroph with respiratory metabolism that requires organic sulfur and nitrogen (amino acids) for growth. It is a mammalian parasite that multiplies in respiratory epithelial cells. *B. bronchiseptica* is the causative agent of the canine condition kennel cough, and *B. pertussis* is a nonmotile, encapsulated species that causes pertussis, although *B. parapertussis* is a closely related species that causes a milder form of the disease.

Pertussis

Pertussis (Latin *per,* intensive, and *tussis,* cough), sometimes called **"whooping cough,"** is characterized by fever, malaise, uncontrollable cough, and cyanosis (bluish skin color resulting from inadequate tissue oxygenation). The disease gets its name from the characteristically prolonged and paroxysmal coughing that ends in an inspiratory gasp, or whoop. Pertussis is a highly contagious, vaccine-preventable disease that primarily affects children. It has been estimated that over 95% of the world's population has experienced either mild or severe symptoms of the disease. Around 300,000 people die from the disease each year. However, fewer than 7,000 cases and 10 deaths occur annually in the United States.

Transmission occurs by inhalation of the bacterium in droplets released from an infectious person. The incubation period is 7 to 14 days. Once inside the upper respiratory tract, the bacteria colonize the cilia of the mammalian respiratory epithelium through fimbrial-like structures called filamentous hemagglutinins that bind to phagocytes. *B. pertussis* produces several toxins. The most important is the **pertussis toxin (PTx).** PTx is a two-component, AB exotoxin (*see figure 30.6*). Many pathogens make AB toxins; in all cases, the B subunit binds to host cell surfaces and transports the A subunit into the cytoplasm of the cell. Once in the host cytoplasm, the A subunit is activated to have catalytic activity that is toxic to the host cell. The B subunit of *B. pertussis* is composed of five polypeptides (S2–S5; there are two S4 subunits) that bind to specific carbohydrates on cell surfaces. The A subunit (S1) of *B. pertussis* is an ADP ribosyl transferase—it transfers the ADP ribosyl moiety of NAD^+ to a membrane-bound, regulatory G protein, G_i. G_i normally inhibits eucaryotic adenylate cyclase, which catalyzes the conversion of ATP to cyclic AMP (cAMP). Thus the net effect of PTx on a cell is an increase in intracellular levels of cAMP. *B. pertussis* also produces three other toxins: extracytoplasmic invasive adenylate cyclase, tracheal cytotoxin, and dermonecrotic toxin, which destroys epithelial tissue. Working together, the tracheal cytotoxin and pertussis toxin also provoke the secretory cells in the respiratory tract to produce nitric oxide, which kills nearby ciliated cells, inhibiting removal of bacteria and mucus. The secretion of thick mucus also impedes ciliary action.

Pertussis is divided into three stages: (1) the catarrhal stage, so named because of the mucous membrane inflammation, which is insidious and resembles the common cold; (2) the paroxysmal stage, which is characterized by prolonged coughing sieges. During this stage, the infected person tries to cough up the

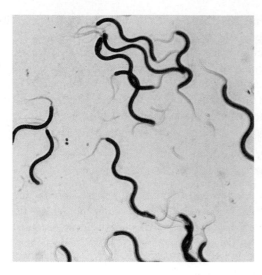

Figure 20.17 **The Genus *Spirillum*.** *Spirillum volutans* with bipolar flagella visible (×450).

mucous secretions by making 5 to 15 rapidly consecutive coughs followed by the characteristic whoop—a hurried, deep inspiration. The catarrhal and paroxysmal stages last about 6 weeks; and (3) the convalescent stage may take several months.

Order *Nitrosomonadales*

A number of chemolithotrophs are found in the order *Nitrosomonadales.* Two genera of nitrifying bacteria (*Nitrosomonas* and *Nitrosospira*) are members of the family *Nitrosomonadaceae* but were discussed earlier with other genera of nitrifying bacteria (pp. 447–448). The stalked chemolithotroph *Gallionella* is in this order. The family *Spirillaceae* has one genus, *Spirillum* (**figure 20.17**).

Order *Hydrogenophilales*

This small order contains *Thiobacillus,* one of the best-studied chemolithotrophs and most prominent of the colorless sulfur bacteria. Like the nitrifying bacteria, **colorless sulfur bacteria** are a highly diverse group. Many are unicellular rod-shaped or spiral, sulfur-oxidizing bacteria that can be either nonmotile or flagellated (**table 20.5**). *Bergey's Manual* divides the colorless sulfur bacteria between two classes; for example, *Thiobacillus* and *Macromonas* are β-proteobacteria, whereas *Thiomicrospira, Thiobacterium, Thiospira, Thiothrix, Beggiatoa,* and others are γ-proteobacteria. Only some of these bacteria have been isolated and studied in pure culture. Most is known about the genera *Thiobacillus* and *Thiomicrospira. Thiobacillus* is a gram-negative rod, and *Thiomicrospira* is a long spiral cell; both have polar flagella (**figure 20.18**). They differ from many of the nitrifying bacteria in that they lack extensive internal membrane systems.

The metabolism of *Thiobacillus* has been intensely studied. It grows aerobically by oxidizing a variety of inorganic sulfur compounds (elemental sulfur, hydrogen sulfide, thiosulfate) to sulfate. ATP is produced by a combination of oxidative phosphorylation

Table 20.5	Colorless Sulfur-Oxidizing Genera				
Genus	Cell Shape	Motility; Location of Flagella	G + C Content (mol%)	Sulfur Deposit[a]	Nutritional Type
Thiobacillus	Rods	+; polar	62–67	Extracellular	Obligate or facultative chemolithotroph
Thiomicrospira	Spirals, comma, or rod shaped	− or +; polar	39.6–49.9	Extracellular	Obligate chemolithotroph
Thiobacterium	Rods embedded in gelatinous masses	−	N.A.[b]	Intracellular[c]	Probably chemoorganoheterotroph
Thiospira	Spiral rods, usually with pointed ends	+; polar (single or in tufts)	N.A.	Intracellular	Unknown
Macromonas	Rods, cylindrical or bean shaped	+; polar tuft	67	Intracellular[c]	Probably chemoorganoheterotroph

[a]When hydrogen sulfide is oxidized to elemental sulfur.
[b]N.A., data not available.
[c]May use sulfur oxidation to detoxify H_2O_2.

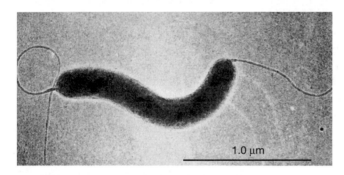

Figure 20.18 ***Thiomicrospira pelophila.*** A colorless sulfur bacterium, with polar flagella.

and substrate-level phosphorylation by means of adenosine 5'-phosphosulfate (*see figure 11.20*). Although most *Thiobacillus* spp. use CO_2 as the major carbon source, *T. novellus* and a few other species can grow heterotrophically. Some species are very flexible metabolically. For example, *T. ferrooxidans* also uses ferrous iron as an electron donor and produces ferric iron as well as sulfuric acid. *T. denitrificans* even grows anaerobically by reducing nitrate to nitrogen gas. Interestingly, some other sulfur-oxidizing bacteria such as *Thiobacterium* and *Macromonas* probably do not derive energy from sulfur oxidation. They may use the process to detoxify metabolically produced hydrogen peroxide.

Sulfur-oxidizing bacteria have a wide distribution and great practical importance. *Thiobacillus* grows in soil and aquatic habitats, both freshwater and marine. In marine habitats, *Thiomicrospira* is more important than *Thiobacillus.* Because of their great acid tolerance (*T. thiooxidans* grows at pH 0.5 and cannot grow above pH 6), these bacteria prosper in habitats they have acidified by sulfuric acid production, even though most other organisms cannot. The production of large quantities of sulfuric acid and ferric iron by *T. ferrooxidans* corrodes concrete and pipe structures. Thiobacilli often cause extensive acid and metal pollution when they release metals from mine wastes.

However, sulfur-oxidizing bacteria also are beneficial. They may increase soil fertility when they oxidize elemental sulfur to sulfate. Thiobacilli are used in processing low-grade metal ores because of their ability to leach metals from ore. >> *Biogeochemical cycling: Sulfur cycle (section 25.1)*

1. Describe the general properties of the β-proteobacteria.
2. Briefly describe the following genera and their practical importance: *Neisseria, Burkholderia,* and *Bordetella.*
3. What three species cause bacterial meningitis? Why is it so important to differentiate between aseptic and bacterial meningitis?
4. Infection with *N. gonorrhea* does not confer life-long protection from reinfection. Why do you think this is the case?
5. What is a sheath and of what advantage is it?
6. Describe the catalytic activity of the ADP-ribosyl transferase produced by *B. pertussis.*
7. How do colorless sulfur bacteria obtain energy by oxidizing sulfur compounds?
8. List several positive and negative impacts sulfur-oxidizing bacteria have on the environment and human activities.

20.3 CLASS *GAMMA-PROTEOBACTERIA*

The **γ-proteobacteria** constitute the largest subgroup of proteobacteria with an extraordinary variety of physiological types. Many important genera are chemoorganotrophic and facultatively anaerobic. Other genera contain aerobic chemoorganotrophs, photolithotrophs, chemolithotrophs, or methylotrophs. According to some DNA-rRNA hybridization studies, the γ-proteobacteria are composed of several deeply branching groups. One consists of the purple sulfur bacteria; a second includes the intracellular parasites *Legionella* and *Coxiella*. The two largest groups contain a wide variety of nonphotosynthetic genera. These microbes can be grouped into rRNA superfamilies based on the degree of sequence conservation among rRNA (*see figure 17.8*). Ribosomal RNA superfamily I includes the families *Vibrionaceae, Enterobacteriaceae,* and *Pasteurellaceae*. These bacteria use the Embden-Meyerhof and pentose phosphate pathways to catabolize carbohydrates. Most are facultative anaerobes. Ribosomal RNA superfamily II contains mostly aerobes that often use the Entner-Doudoroff and pentose phosphate pathways to catabolize many different kinds of organic molecules. The genera

Pseudomonas, Azotobacter, Moraxella, and *Acinetobacter* belong to this superfamily. **<<** *Techniques for determining microbial taxonomy and phylogeny (section 17.4)*

The exceptional diversity of these bacteria is evident from the fact that *Bergey's Manual* divides the class γ-proteobacteria into 14 orders and 28 families. **Figure 20.19** illustrates the phylogenetic relationships among major groups and selected γ-proteobacteria, and **table 20.6** outlines the general characteristics of some of the bacteria discussed in this section.

Purple Sulfur Bacteria

We have seen that the purple photosynthetic bacteria are distributed between three subgroups of the proteobacteria. Most of the purple nonsulfur bacteria are α-proteobacteria and were discussed in section 20.1. Because the purple sulfur bacteria are γ-proteobacteria, they are described here.

Bergey's Manual divides the **purple sulfur bacteria** into two families: the *Chromatiaceae* and *Ectothiorhodospiraceae* in the order *Chromatiales*. The family *Ectothiorhodospiraceae* contains eight genera. *Ectothiorhodospira* has red, spiral-shaped, polarly flagellated cells that deposit sulfur globules externally (**figure 20.20**). Internal photosynthetic membranes are organized as lamellar stacks. The majority of purple sulfur bacteria are in the family *Chromatiaceae,* which contains 26 genera.

The purple sulfur bacteria are strict anaerobes and usually photolithoautotrophs. *Chromatiaceae* oxidize hydrogen sulfide to sulfur and deposit it internally as sulfur granules (usually within invaginated pockets of the plasma membrane); often they eventually oxidize the sulfur to sulfate. Hydrogen also may serve as an electron donor. *Thiospirillum, Thiocapsa,* and *Chromatium* are typical purple sulfur bacteria (**figure 20.21**). They are found in anoxic, sulfide-rich zones of lakes, marshes, and lagoons where large blooms can occur under certain conditions (**figure 20.22**).

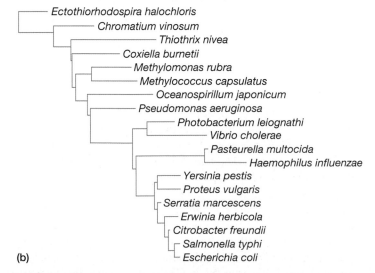

(a)

Aeromonadaceae (Aeromonas)
Alteromonadaceae (Shewanella)
Enterobacteriaceae (Escherichia, Salmonella, Shigella, Proteus)
Pasteurellaceae (Pasteurella, Haemophilus)
Succinivibrionaceae (Ruminobacter)
Vibrionaceae (Vibrio, Photobacterium)

Halomonadaceae (Halomonas)

Oceanospirillaceae (Oceanospirillum)

Pseudomonadaceae (Pseudomonas, Azotobacter)

Moraxellaceae (Moraxella, Acinetobacter)

Methylococcaceae (Methylococcus, Methylomonas)

Francisellaceae (Francisella)

Piscirickettsiaceae (Hydrogenovibrio, Thiomicrospira)

Legionellaceae (Legionella)

Betaproteobacteria

Xanthomonadales (Xanthomonas)

Chromatiales (Chromatium, Thiococcus, Thiospirillum)

(b)

Ectothiorhodospira halochloris
Chromatium vinosum
Thiothrix nivea
Coxiella burnetii
Methylomonas rubra
Methylococcus capsulatus
Oceanospirillum japonicum
Pseudomonas aeruginosa
Photobacterium leiognathi
Vibrio cholerae
Pasteurella multocida
Haemophilus influenzae
Yersinia pestis
Proteus vulgaris
Serratia marcescens
Erwinia herbicola
Citrobacter freundii
Salmonella typhi
Escherichia coli

Figure 20.19 Phylogenetic Relationships Among γ-Proteobacteria.   (a) The major phylogenetic groups based on 16S rRNA sequence comparisons. Representative genera are given in parentheses. Each tetrahedron in the tree represents a group of related organisms; its horizontal edges show the shortest and longest branches in the group. Multiple branching at the same level indicates that the relative branching order of the groups cannot be determined from the data. (b) The relationships of a few species based on 16S rRNA sequence data.   *Source: The Ribosomal Database Project.*

Table 20.6	Characteristics of Selected γ-Proteobacteria			
Genus	Dimensions (μm) and Morphology	G + C Content (mol%)	Oxygen Requirement	Other Distinctive Characteristics
Azotobacter	1.5–2.0; ovoid cells, pleomorphic, peritrichous flagella or nonmotile	63.2–67.5	Aerobic	Can form cysts; fix nitrogen nonsymbiotically
Beggiatoa	1–200 × 2–10; colorless cells form filaments, either single or in colonies	35–39	Aerobic or microaerophilic	Gliding motility; can form sulfur inclusions with hydrogen sulfide present
Chromatium	1–6 × 1.5–16; rod-shaped or ovoid, straight or slightly curved, polar flagella	48–50	Anaerobic	Photolithoautotroph that can use sulfide; sulfur stored within the cell
Ectothiorhodospira	0.7–1.5 in diameter; vibrioid- or rod-shaped, polar flagella	61.4–68.4	Anaerobic, some aerobic or microaerophilic	Internal lamellar stacks of membranes; deposits sulfur granules outside cells
Escherichia	1.1–1.5 × 2–6; straight rods, peritrichous flagella or nonmotile	48–59	Facultatively anaerobic	Mixed acid fermenter; formic acid converted to H_2 and CO_2, lactose fermented, citrate not used
Haemophilus	<1.0 in width, variable lengths; coccobacilli or rods, nonmotile	37–44	Aerobic or facultatively anaerobic	Fermentative; requires growth factors present in blood; parasites on mucous membranes
Leucothrix	Long filaments of short cylindrical cells, usually holdfast is present	46–51	Aerobic	Dispersal by gonidia, filaments don't glide; rosettes formed; heterotrophic
Methylococcus	0.8–1.5 × 1.0–1.5; cocci with capsules, nonmotile	59–65	Aerobic	Can form a cyst; uses methane, methanol, and formaldehyde as sole carbon and energy sources
Photobacterium	0.8–1.3 × 1.8–2.4; straight, plump rods with polar flagella	39–44	Facultatively anaerobic	Two species can emit blue-green light; Na^+ needed for growth
Pseudomonas	0.5–1.0 × 1.5–5.0; straight or slightly curved rods, polar flagella	58–69	Aerobic or facultatively anaerobic	Respiratory metabolism with oxygen or nitrate as acceptor; some use H_2 or CO as energy source
Vibrio	0.5–0.8 × 1.4–2.6; straight or curved rods with sheathed polar flagella	38–51	Facultatively anaerobic	Fermentative or respiratory metabolism; sodium ions stimulate or are needed for growth; oxidase positive

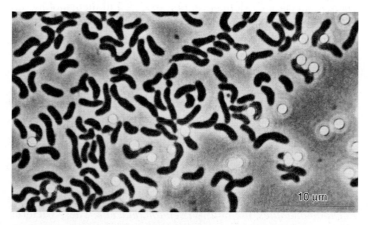

Figure 20.20 Purple Bacteria. *Ectothiorhodospira mobilis;* light micrograph. Note external sulfur globules.

Order *Thiotrichales*

The order *Thiotrichales* contains three families, the largest of which is the family *Thiotrichaceae.* This family has several genera that oxidize sulfur compounds (see the colorless sulfur bacteria [p. 453] and chapter 10 for sulfur oxidation and chemolithotrophy). Morphologically both rods and filamentous forms are present.

Two of the best-studied gliding genera in this family are *Beggiatoa* and *Leucothrix. Beggiatoa* is microaerophilic and grows in sulfide-rich habitats such as sulfur springs, freshwater with decaying plant material, rice paddies, salt marshes, and marine sediments. Its filaments contain short, disklike cells and lack a sheath (**figure 20.23**). *Beggiatoa* is very versatile metabolically. It oxidizes hydrogen sulfide to form large sulfur grains located in pockets formed by invaginations of the plasma membrane. *Beggiatoa* can subsequently oxidize the sulfur to

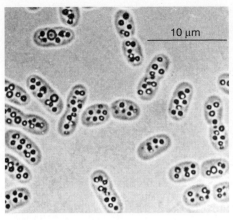

(a) *Chromatium vinosum*

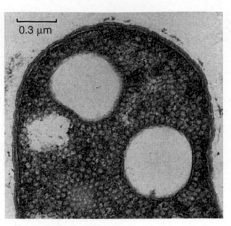

(b) *C. vinosum*

Figure 20.21 Typical Purple Sulfur Bacteria. (a) *Chromatium vinosum* with intracellular sulfur granules. (b) Electron micrograph of *C. vinosum*. Note the intracytoplasmic vesicular membrane system. The large white areas are the former sites of sulfur globules.

Figure 20.22   Purple Photosynthetic Sulfur Bacteria.   Purple photosynthetic sulfur bacteria growing in a marsh.

Figure 20.23   *Beggiatoa alba.*   A light micrograph showing part of a colony (×400). Note the dark sulfur granules within many of the filaments.

sulfate and donate electrons to the electron transport chain in energy production. Many strains also can grow heterotrophically with acetate as a carbon source, and some incorporate CO_2 autotrophically.

Leucothrix is an aerobic chemoorganotroph that forms filaments or trichomes up to 400 μm long (**figure 20.24**). It is usually marine and is attached to solid substrates by a holdfast. *Leucothrix* has a complex life cycle in which it is dispersed by the formation of gonidia. Rosette formation often is seen in culture (figure 20.24c). *Thiothrix* is a related genus that forms sheathed filaments and releases gonidia from the open end of the sheath (**figure 20.25**). In contrast with *Leucothrix*, *Thiothrix* is a chemolithotroph that oxidizes hydrogen sulfide and deposits sulfur granules internally. It also requires organic carbon for growth. *Thiothrix* grows in sulfide-rich flowing water and activated sludge sewage systems.

Order *Methylococcales*

The single family in this order is *Methylococcaceae.* It contains rods, vibrios, and cocci that use methane, methanol, and other reduced one-carbon compounds as their sole carbon and energy sources under aerobic or microaerobic (low oxygen) conditions. They are methylotrophs, different from bacteria that use methane exclusively as their carbon and energy source, which are called methanotrophs. The family contains seven genera, including *Methylococcus* (spherical, nonmotile cells) and *Methylomonas* (straight, curved, or branched rods with a single, polar flagellum). When oxidizing methane, the bacteria contain complex arrays of intracellular membranes. Almost all are capable of forming cysts. Methanogenesis from archaea using substrates such as H_2 and CO_2 is widespread in anoxic soil and water, and methylotrophic bacteria grow above these habitats all over the world.

Methane-oxidizing bacteria use methane as a source of both energy and carbon. Methane is first oxidized to methanol by the enzyme methane monooxygenase. The methanol is then oxidized to formaldehyde by methanol dehydrogenase, and the electrons from this oxidation are donated to an electron transport chain for ATP synthesis. Formaldehyde can be assimilated into cell material by the activity of either of two pathways, one involving the formation of the amino acid serine and the other proceeding through the synthesis of sugars such as fructose 6-phosphate and ribulose 5-phosphate.

Order *Pseudomonadales*

Pseudomonas is the most important genus in the order *Pseudomonadales,* the family *Pseudomonaceae.* These bacteria are straight or slightly curved rods, 0.5 to 1.0 μm by 1.5 to

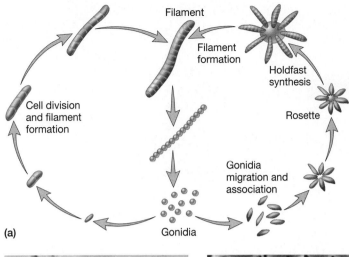

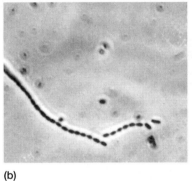

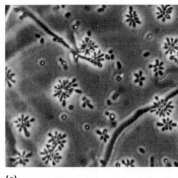

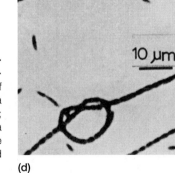

Figure 20.24 Morphology and Reproduction of *Leucothrix mucor.* (a) Life cycle of *L. mucor.* (b) Separation of gonidia from the tip of mature filament; phase contrast (×1,400). (c) Gonidia aggregating to form rosettes; phase contrast (×950). (d) A knot formed by a *Leucothrix* filament.

of a fluorescent pigment, pathogenicity, the presence of arginine dihydrolase, and glucose utilization. For example, the fluorescent subgroup does not accumulate PHB and produces a diffusible, water-soluble, yellow-green pigment that fluoresces under UV radiation. *P. aeruginosa, P. fluorescens, P. putida,* and *P. syringae* are members of this group.

The pseudomonads have a number of practical impacts, including:

1. Many can degrade an exceptionally wide variety of organic molecules. Thus they are very important in mineralization processes (the microbial breakdown of organic materials to inorganic substances) in nature and in sewage treatment. The fluorescent pseudomonads can use approximately 80 different substances as their carbon and energy sources. >> *Microorganisms in terrestrial environments (section 26.2)*

2. Several species (e.g., *P. aeruginosa*) are important experimental subjects. Many advances in microbial physiology and biochemistry have come from their study. For example, the study of *P. aeruginosa* has significantly advanced our understanding of how bacteria form biofilms and the role of extracellular signaling in bacterial communities and pathogenesis. The genome of *P. aeruginosa* has an unusually large number of genes for catabolism, nutrient transport, the efflux of organic molecules, and metabolic regulation. This may explain its ability to grow in many environments and resist antibiotics. << *Microbial growth in natural environments: Biofilms (section 7.6); Global regulatory systems: Quorum sensing (section 13.5)*

3. Some pseudomonads are major animal and plant pathogens. *P. aeruginosa* infects people with low resistance such as cystic fibrosis patients. It also invades burns and causes urinary tract infections. *P. syringae* is an important plant pathogen.

4. Pseudomonads such as *P. fluorescens* are involved in the spoilage of refrigerated milk, meat, eggs, and seafood because they grow at 4°C and degrade lipids and proteins.

5.0 μm in length, and are motile by one or several polar flagella (**figure 20.26**). These chemoheterotrophs usually carry out aerobic respiration. Sometimes nitrate is used as the terminal electron acceptor in anaerobic respiration. All pseudomonads have a functional tricarboxylic acid cycle and can oxidize substrates completely to CO_2. Most hexoses are degraded by the Entner-Doudoroff pathway rather than the Embden-Meyerhof pathway. << *Breakdown of glucose to pyruvate (section 10.3); Tricarboxylic acid cycle (section 10.4);* >> *Appendix II.*

The genus *Pseudomonas* is an exceptionally heterogeneous taxon currently composed of about 60 species. Many can be placed in one of seven rRNA homology groups. The three best characterized groups are subdivided according to properties such as the presence of poly-β-hydroxybutyrate (PHB), the production

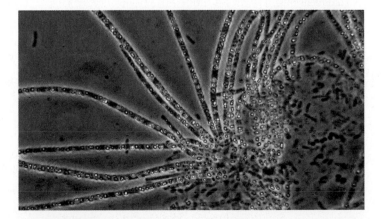

Figure 20.25 *Thiothrix.* A *Thiothrix* colony viewed with phase-contrast microscopy (×1,000).

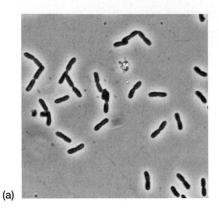
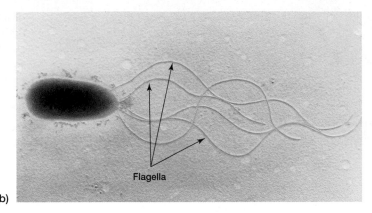

(a)

(b)

Flagella

Figure 20.26 The Genus *Pseudomonas.* (a) A phase-contrast micrograph of *Pseudomonas* cells. (b) A transmission electron micrograph of *Pseudomonas putida* with five polar flagella, each flagellum about 5–7 μm in length.

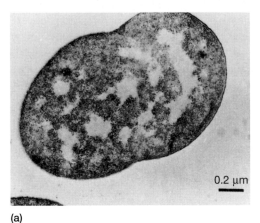
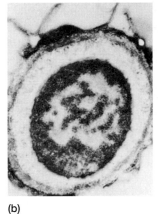

0.2 μm

(a)

(b)

Figure 20.27 *Azotobacter.* (a) Electron micrograph of vegetative *A. chroococcum.* (b) *Azotobacter* cyst.

The genus *Azotobacter* also is in the family *Pseudomonadaceae.* The genus contains, ovoid bacteria, 1.5 to 2.0 μm in diameter, that may be motile by peritrichous flagella. The cells are often pleomorphic, ranging from rods to coccoid shapes, and form cysts as the culture ages (**figure 20.27**). The genus is aerobic, catalase positive, and fixes nitrogen nonsymbiotically. *Azotobacter* is widespread in soil and water.

Order *Vibrionales*

Three closely related orders of the γ-proteobacteria contain a number of important bacterial genera. Each order has only one family of facultatively anaerobic gram-negative rods. **Table 20.7** summarizes the

Table 20.7	Characteristics of Families of Facultatively Anaerobic Gram-Negative Rods		
Characteristics	**Enterobacteriaceae**	**Vibrionaceae**	**Pasteurellaceae**
Cell dimensions	0.3–1.0 × 1.0–6.0 μm	0.3–1.3 × 1.0–3.5 μm	0.2–0.4 × 0.4–2.0 μm
Morphology	Straight rods; peritrichous flagella or nonmotile	Straight or curved rods; polar flagella; lateral flagella may be produced on solid media	Coccoid to rod-shaped cells, sometimes pleomorphic; nonmotile
Physiology	Oxidase negative	Oxidase positive; all can use D-glucose as sole or principal carbon source	Oxidase positive; heme and/or NAD often required for growth; organic nitrogen source required
G + C content	38–60%	38–51%	38–47%
Symbiotic relationships	Some parasitic on mammals and birds; some species are plant pathogens	Most not pathogens; several inhabit light organs of marine organisms	Parasites of mammals and birds
Representative genera	*Escherichia, Shigella, Salmonella, Citrobacter, Klebsiella, Enterobacter, Erwinia, Serratia, Proteus, Yersinia*	*Vibrio, Photobacterium*	*Pasteurella, Haemophilus*

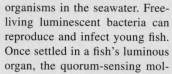
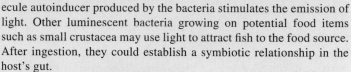

20.1 Bacterial Bioluminescence

Several species in the genera *Vibrio* and *Photobacterium* can emit light of a blue-green color. The enzyme luciferase catalyzes the reaction and uses reduced flavin mononucleotide, molecular oxygen, and a long-chain aldehyde as substrates.

$$FMNH_2 + O_2 + RCHO \xrightarrow{\text{luciferase}} FMN + H_2O + RCOOH + light$$

The evidence suggests that an enzyme-bound, excited flavin intermediate is the direct source of luminescence. Because the electrons used in light generation are probably diverted from the electron transport chain and ATP synthesis, the bacteria expend considerable energy on luminescence. It follows that luminescence is regulated and can be turned off or on under the proper conditions.

Much speculation arises about the role of bacterial luminescence and its value to bacteria, particularly because it is such an energetically expensive process. Luminescent bacteria occupying the luminous organs of fish do not emit light when they grow as free-living organisms in the seawater. Free-living luminescent bacteria can reproduce and infect young fish. Once settled in a fish's luminous organ, the quorum-sensing molecule autoinducer produced by the bacteria stimulates the emission of light. Other luminescent bacteria growing on potential food items such as small crustacea may use light to attract fish to the food source. After ingestion, they could establish a symbiotic relationship in the host's gut.

The mechanism by which autoinducer regulates light production in these marine bacteria is an important model for understanding quorum sensing in many gram-negative bacteria, including a number of pathogens. << *Microbial growth in natural environments: Cell-cell communication within microbial populations (section 7.6); Global regulatory systems: Quorum sensing (section 13.5)*

---

distinguishing properties of the families *Enterobacteriaceae*, *Vibrionaceae*, and *Pasteurellaceae* from the orders *Enterbacteriales*, *Vibrionales*, and *Pasteurellales*, respectively.

The order *Vibrionales* contains only one family, the *Vibrionaceae*. Members of the family *Vibrionaceae* are straight or curved, flagellated rods (**figure 20.28**). Most are oxidase positive, and all use D-glucose as their sole or primary carbon and energy source (table 20.7). The majority are aquatic microorganisms, widespread in freshwater and the sea. The family has eight genera: *Vibrio, Photobacterium, Salinivibrio, Listonella, Allomonas, Enterovibrio, Catencoccus,* and *Grimontia*.

Some members of the family are unusual in being bioluminescent. *Vibrio fischeri, V. harveyi,* and at least two species of *Photobacterium* are marine bacteria capable of bioluminescence. They emit a blue-green light because of the activity of the enzyme luciferase (**Microbial Diversity & Ecology 20.1**). The peak emission of light is usually between 472 and 505 nm, but one strain of *V. fischeri* emits yellow light with a major peak at 545 nm. Although many of these bacteria are free-living, *V. fischeri, V. harveyi, P. phosphoreum,* and *P. leiognathi* live symbiotically in the luminous organs of fish (**figure 20.29**) and squid (*see figure 7.28*).

Several vibrios are important pathogens. *Vibrio parahaemolyticus* can cause gastroenteritis in humans following consumption of contaminated seafood. *V. anguillarum* and others are responsible for fish diseases. We now discuss *V. cholerae* and the disease it causes, cholera.

Cholera

Throughout recorded history, cholera (Greek *chole,* bile) has caused seven pandemics in various areas of the world, especially in Asia, the Middle East, and Africa. The disease has been rare in the United States since the 1800s, but an endemic focus is believed to exist on the Gulf Coast of Louisiana and Texas.

Although there are many *V. cholerae* serogroups, only O1 and O139 have caused epidemics. *V. cholerae* O1 is divided into two serotypes, Inaba and Ogawa, and two biotypes, classic and El Tor. In 1961 the El Tor biotype emerged as an important cause of cholera pandemics, and in 1992 the newly identified strain *V. cholerae* O139 emerged in Asia. In Calcutta, India, serogroup O139 of *V. cholerae* has displaced El Tor *V. cholerae* serogroup O1, an event that has never before happened in the recorded history of cholera. Interestingly, the cholera toxin gene is carried by the lysogenic CTX filamentous bacteriophage. The receptor for the phage is the *t*oxin *c*oregulated *p*ilus (TCP)—the same structure used to colonize the host's gut. Thus *V. cholerae* is an excellent example of how horizontal transfer of genes can confer pathogenicity.

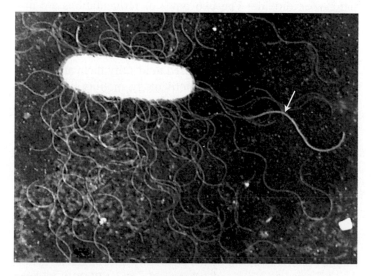

Figure 20.28 **The *Vibrionaceae*.** Electron micrograph of *Vibrio alginolyticus* grown on agar, showing a sheathed polar flagellum (arrow) and unsheathed lateral flagella (×18,000).

Figure 20.29 **Bioluminescence.** (a) A photograph of the Atlantic flashlight fish *Kryptophanaron alfredi.* The light area under the eye is the fish's luminous organ, which can be covered by a lid of tissue. (b) Ultrathin section of the luminous organ of a fish, *Equulities novaehollandiae,* with the bioluminescent bacterium *Photobacterium leiognathi,* PL.

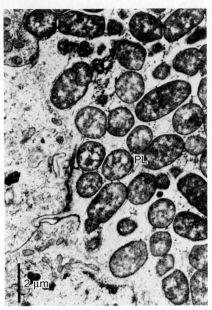

(a)

(b)

cholera are reported each year in the United States.

Order *Enterobacteriales*

The family *Enterobacteriaceae* is the largest of the families listed in table 20.7. It contains peritrichously flagellated or nonmotile, facultatively anaerobic, straight rods with simple nutritional requirements. The order *Enterobacteriales* has only one family, *Enterobacteriaceae,* with 44 genera. The relationship between *Enterobacteriales* and the orders *Vibrionales* and *Pasteurellales* can be seen by inspecting figure 20.19.

The metabolic properties of the *Enterobacteriaceae* are very useful in characterizing its constituent genera. Members of the family, often called **enterobacteria** or **enteric bacteria** (Greek *enterikos,* pertaining to the intestine), all degrade sugars by means of the Embden-Meyerhof pathway and cleave pyruvic acid to yield formic acid. Those enteric bacteria that produce large amounts of gas during sugar fermentation, such as *Escherichia* spp., have the formic hydrogenlyase complex that degrades formic acid to H_2 and CO_2. The family can be divided into two groups based on their fermentation products. The majority (e.g., the genera *Escherichia, Proteus, Salmonella,* and *Shigella*) carry out mixed acid fermentation and produce mainly lactate, acetate, succinate, formate (or H_2 and CO_2), and ethanol. In contrast, *Enterobacter, Serratia, Erwinia,* and *Klebsiella* are butanediol fermenters. The major products of butanediol fermentation are butanediol, ethanol, and carbon dioxide. The two types of fermentations are distinguished by the methyl red and Voges-Proskauer tests, respectively. << *Fermentation (section 10.7)*

Because the enteric bacteria are so similar in morphology, biochemical tests are normally used to identify them after a preliminary examination of their morphology, motility, and growth responses (**figure 20.30** provides a simple example). Some more commonly used tests are those for the type of fermentation, lactose and citrate utilization, indole production from tryptophan, urea hydrolysis, and hydrogen sulfide production. For example, lactose fermentation occurs in *Escherichia* and *Enterobacter* but not in *Shigella, Salmonella,* or *Proteus.* **Table 20.8** summarizes a few of the biochemical properties useful in distinguishing between genera of enteric bacteria. The mixed acid fermenters are located on the left in this table and the butanediol fermenters on the right. The usefulness of biochemical tests in identifying enteric bacteria is shown by the popularity of commercial identification systems, such as the Enterotube and API 20-E systems that are based on these tests. >> *Identification of microorganisms from specimens (section 32.2)*

Cholera is transmitted by ingesting food or water contaminated by fecal material from infected individuals. Once the bacteria enter the body, the incubation period is 12 to 72 hours. The bacteria adhere to the intestinal mucosa of the small intestine, where they are not invasive but secrete choleragen, a cholera toxin. Choleragen is an AB toxin (*see figure 30.6*). The A subunit enters the intestinal epithelial cells and activates the enzyme adenylate cyclase by the addition of an ADP-ribosyl group in a way similar to that employed by pertussis toxin. This results in hypersecretion of water and chloride ions while inhibiting absorption of sodium ions. The infected person loses massive quantities of fluid and electrolytes, causing abdominal muscle cramps, vomiting, fever, and watery diarrhea. The voided fluid is often referred to as "rice-water-stool" because of the flecks of mucus floating in it. The diarrhea can be so profuse that a person can lose 10 to 15 liters of fluid during the infection. Death may result from the elevated concentrations of blood proteins, caused by reduced fluid levels, which leads to circulatory shock and collapse.

Evidence indicates that passage through the human host enhances infectivity, although the exact mechanism is unclear. Before *V. cholerae* exits the body in watery stools, some unknown aspect of the intestinal environment stimulates the activity of certain bacterial genes. These genes, in turn, seem to prepare the bacteria for ever more effective colonization of their next victims, possibly fueling epidemics. *V. cholerae* can also be free-living in warm, alkaline, and saline environments.

Treatment is by oral rehydration therapy with NaCl plus glucose to stimulate water uptake by the intestine; antibiotics may also be given. The most reliable control methods are based on proper sanitation, especially of water supplies. The mortality rate without treatment is often over 50%, but with treatment and supportive care, it is less than 1%. Fewer than 20 cases of

Figure 20.30 **Identification of Enterobacterial Genera.** A dichotomous key to selected genera of enteric bacteria based on motility and biochemical characteristics. The following abbreviations are used: ONPG, *o*-nitrophenyl-β-D-galactopyranoside (a test for β-galactosidase); DNase, deoxyribonuclease; Gel. Liq., gelatin liquefaction; and VP, Voges-Proskauer (a test for butanediol fermentation).

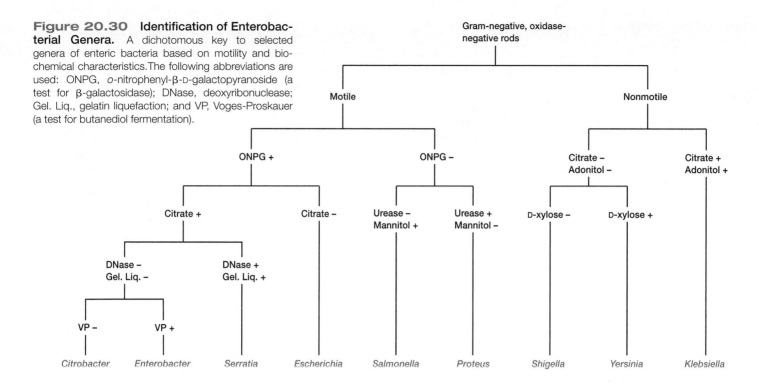

Members of the *Enterobacteriaceae* are so common, widespread, and important that they are probably more often seen in laboratories than any other bacteria. *Escherichia coli* is undoubtedly the best-studied bacterium and the experimental organism of choice for many microbiologists. It is an inhabitant of the colon of humans and other warm-blooded animals, and it is quite useful in the analysis of water for fecal contamination. Members of the genus *Erwinia* are major pathogens of crop plants and cause blights, wilts, and several other plant diseases. Some strains cause gastroenteritis or urinary tract infections. Several genera contain very important human pathogens, and we now discuss several of these infectious diseases.

Plague

In the southwestern part of the United States, plague (Latin *plaga*, pest) occurs primarily in wild ground squirrels, chipmunks, mice, and prairie dogs. However, massive human epidemics occurred in Europe during the Middle Ages, where the disease was known as the Black Death due to black-colored, subcutaneous hemorrhages. Infections now occur in humans only sporadically or in limited outbreaks. In the United States, 10 to 20 cases are reported annually; the mortality rate is about 14%.

The causative agent, *Yersinia pestis*, is transmitted from rodent to human by the bite of an infected flea, direct contact with infected animals or their products, or inhalation of contaminated airborne droplets (**figure 20.31***a*). *Y. pestis* is initially spread by contact with flea-infested animals but can then spread from person to person by airborne transmission. Once in the human body, the bacteria multiply in the blood and lymph.

An important factor in the virulence of *Y. pestis* is its ability to survive and proliferate inside phagocytic cells rather than being killed by them. One of the ways it accomplishes this is by its modification of host cell behavior. Like other gram-negative extracellular pathogens, *Y. pestis* uses a type III secretion system to deliver effector proteins. Type III bacterial secretion systems are specialized for the direct export of virulence factors into target host cells. They are comprised of 20 to 30 different proteins. Some of the proteins form a needlelike structure connected to a basal body. The needle penetrates the target cell membrane and delivers to the target cell other proteins, including those that function as virulence factors (*see figure 30.5*). Virulence factors typically subvert normal host cell functions to benefit the invading bacterium, such as the ability to acquire iron, adhere to host cells, or invade them. In the case of *Y. pestis*, plasmid-encoded *y*ersinal *o*uter membrane proteins (YOPs) are secreted into phagocytic cells to counteract natural defense mechanisms and help the bacteria multiply and disseminate in the host. **>>** *Overview of bacterial pathogenesis: Pathogenicity islands (section 30.3)*

Symptoms—besides subcutaneous hemorrhages—include fever, chills, headache, extreme exhaustion, and the appearance of enlarged lymph nodes called **buboes** (hence another old name, **bubonic plague**) (figure 20.31*c*). In 50 to 70% of the untreated cases, death follows in 3 to 5 days from toxic conditions caused by the large number of bacilli in the blood.

Pneumonic plague arises (1) from primary exposure to infectious respiratory droplets from a person or animal with respiratory plague, or (2) secondary to hematogenous spread in a patient with bubonic or septicemic plague. The mortality rate for this kind of plague is almost 100% if it is not recognized within 12 to 24 hours, at which time antibiotic therapy must begin. Obviously great care must be taken to prevent the spread of

Table 20.8	Some Characteristics of Selected Genera in the *Enterobacteriaceae*				
Characteristics	*Escherichia*	*Shigella*	*Salmonella*	*Citrobacter*	*Proteus*
Methyl red	+	+	+	+	+
Voges-Proskauer	−	−	−	−	d
Indole production	(+)	d	−	d	d
Citrate use	−	−	(+)	+	d
H₂S production	−	−	(+)	d	(+)
Urease	−	−	−	(+)	+
β-galactosidase	(+)	d	d	+	−
Gas from glucose	+	−	(+)	+	+
Acid from lactose	+	−	(−)	d	−
Phenylalanine deaminase	−	−	−	−	+
Lysine decarboxylase	(+)	−	(+)	−	−
Ornithine decarboxylase	(+)	d	(+)	(+)	d
Motility	d	−	(+)	+	+
Gelatin liquifaction (22°C)	−	−	−	−	+
% G + C	48–59	49–53	50–53	50–52	38–41
Genome size (Mb)	4.6–5.5	4.6	4.5–4.9	Nd^d	Nd
Other characteristics	1.1–1.5 × 2.0–6.0 µm; peritrichous flagella when motile	No gas from sugars	0.7–1.5 × 2–5 µm; peritrichous flagella	1.0 × 2.0–6.0 µm; peritrichous flagella	0.4–0.8 × 1.0–3.0 µm; peritrichous flagella

^{a}(+) usually present
^{b}(−) usually absent
^{c}d, strains or species vary in possession of characteristic
^{d}Nd: Not determined; genome not yet sequenced

airborne infections to personnel who care for pneumonic plague patients. A vaccine is available for persons at high risk.

Salmonellosis

Salmonellosis (Salmonella gastroenteritis) is caused by over 2,000 *Salmonella* serovars. Based on DNA homology studies, all known *Salmonella* are thought to belong to a single species, *Salmonella enterica,* although the taxonomy of this bacterium remains controversial. The most frequently isolated serovars from humans are Typhimurium and Enteritidis. (Serovar names are not italicized and the first letter is capitalized.)

The initial source of the bacterium is the intestinal tracts of birds and other animals. Humans acquire the bacteria from contaminated foods such as beef products, poultry, eggs, egg products, or water. Around 45,000 cases a year are reported in the United States, but there actually may be as many as 2 to 3 million cases annually.

Once the bacteria are in the body, the incubation time is only about 8 to 48 hours. The disease results when the bacteria multiply and invade the intestinal mucosa, where they produce an enterotoxin and a cytotoxin that destroy the epithelial cells. Abdominal pain, cramps, diarrhea, nausea, vomiting, and fever are the most prominent symptoms, which usually persist for 2 to 5 days but can last for several weeks. During the acute phase of the disease, as many as 1 billion *Salmonella* can be found per gram of feces. Most adult patients recover, but the loss of fluids can cause problems for children and elderly people.

Typhoid Fever

Typhoid (Greek *typhodes,* smoke) fever is caused by *Salmonella enterica* serovar Typhi and is acquired by ingestion of food or water contaminated by feces of infected humans or person-to-person contact. In earlier centuries, the disease occurred in great epidemics. A milder form of the disease, paratyphoid fever, is

Yersinia	Klebsiella	Enterobacter	Erwinia	Serratia
+	(+)[a]	(−)[b]	+	d[c]
− (37°C)	(+)	+	(+)	+
d	d	−	(−)	(−)
(−)	(+)	+	(+)	+
−	−	−	(+)	−
d	(+)	(−)	−	−
+	(+)	+	+	+
(−)	(+)	(+)	(−)	d
(−)	(+)	(+)	d	d
—	−	(−)	(−)	—
(−)	(+)	d	−	d
d	−	(+)	−	d
− (37°C)	−	+	+	+
(−)	−	d	d	(+)
46–50	53–58	52–60	50–54	52–60
4.6	Nd	Nd	5.1	5.1
0.5–0.8 × 1.0–3.0 μm; peritrichous flagella when motile	0.3–1.0 × 0.6–6.0 μm; capsulated	0.6–1.0 × 1.2–3.0 μm; peritrichous flagella	0.5–1.0 × 1.0–3.0 μm; peritrichous flagella; plant pathogens and saprophytes	0.5–0.8 × 0.9–2.0 μm; peritrichous flagella; colonies often pigmented

caused by serovars Paratyphi A, B, and C of *S. enterica* subspecies *enterica.*

In the small intestine, the incubation period is about 10 to 14 days. The bacteria colonize the small intestine, penetrate the epithelium, and spread to the lymphoid tissue, blood, liver, and gallbladder. Symptoms include fever, headache, abdominal pain, anorexia, and malaise, which last several weeks. Bacteria then reinfect the gastrointestinal tract, producing abdominal pain and diarrhea. After approximately 3 months, most individuals stop shedding bacteria in their feces. However, a few individuals continue to shed *Salmonella* for extended periods but show no symptoms. In these carriers, the bacteria continue to grow in the gallbladder and reach the intestine through the bile duct.

Shigellosis

Shigellosis, or bacillary dysentery, is a diarrheal illness resulting from an acute inflammatory reaction of the intestinal tract caused by the four species of the genus *Shigella*. About 20,000 to 25,000 cases a year are reported in the United States, and around 600,000 deaths a year worldwide are due to bacillary dysentery.

Shigella is restricted to human hosts. *S. sonnei* is the usual pathogen in the United States and Britain, but *S. flexneri* is also fairly common. The organism is transmitted by the fecal-oral route—primarily by food, fingers, feces, and flies (the four "F's")—and is most prevalent among children, especially one- to four-year-olds. The infectious dose is only around 10 to 100 bacteria. In the United States, shigellosis is a particular problem in day-care centers and crowded custodial institutions.

The shigellae are intracellular parasites that multiply within the villus cells of the colonic epithelium. The bacteria induce the Peyer's patch cells to phagocytose them. After being ingested, the bacteria disrupt the phagosome membrane and are released into the cytoplasm, where they reproduce. They then invade adjacent mucosal cells. *Shigella* initiate an inflammatory reaction in the mucosa. Both endotoxins and exotoxins may participate in disease progression, but the bacteria do not usually spread beyond the colonic epithelium. The watery stools often contain blood, mucus, and pus. In severe cases, the colon can become ulcerated. Virulent *Shigella* produce a heat-labile AB exotoxin known as the **shiga-toxin** (Sxt, formerly called verotoxin). The complete

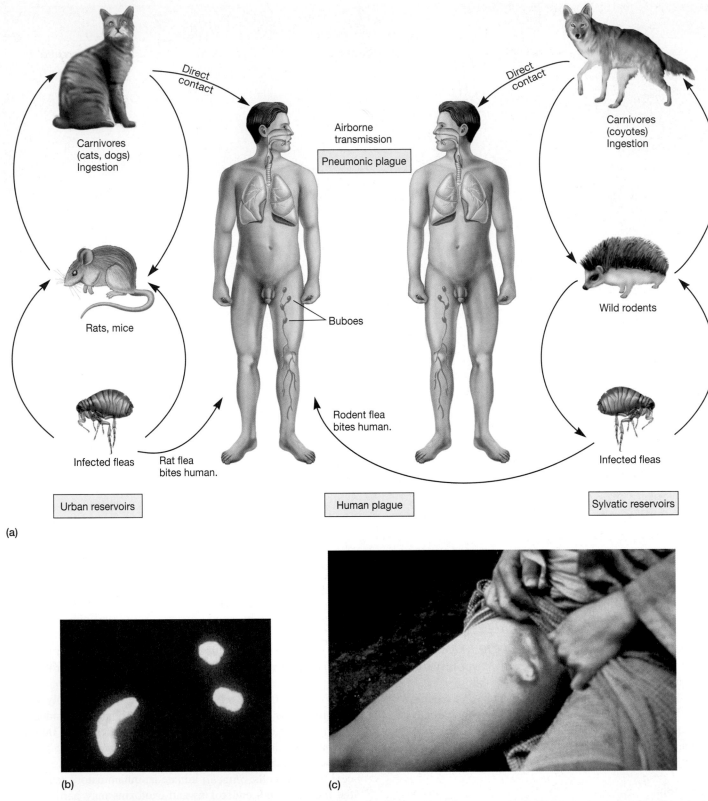

(a)

(b)

(c)

Figure 20.31 Plague.   (a) Plague is spread to humans through (1) the urban cycle and rat fleas, (2) the sylvatic cycle, centers on wild rodents and their fleas, or (3) by airborne transmission from an infected person leading to pneumonic plague. Dogs, cats, and coyotes can also acquire the bacterium by ingestion of infected animals. (b) *Yersinia pestis* can be stained with fluorescent antibodies for identification. (c) Enlarged lymph nodes called buboes are characteristic of *Yersinia* infection.

toxin molecule is composed of one A protein surrounded by five B proteins. The B proteins attach to host vascular cells, stimulating the internalization of the whole toxin. The A subunit protein is subsequently released from the B protein units and binds to host ribosomes, inhibiting protein synthesis. A specific target of the B protein seems to be the glomerular endothelium; toxin action on these cells leads to kidney failure. Like *Y. pestis,* shigellae also use a type III secretion system to deliver specific virulence factors to target epithelial cells as well. **>>** *Phagocytosis (section 28.3); Toxigenicity (section 30.4)*

The incubation period usually ranges from 1 to 3 days, and the organisms are shed over a period of 1 to 2 weeks. The disease normally is self-limiting in adults and lasts an average of 4 to 7 days; in infants and young children, it may be fatal. Fluid and electrolyte replacement are usually sufficient; antibiotics may not be required in mild cases, although they can shorten the duration of symptoms and limit transmission to family members. Sometimes, particularly in malnourished infants and children, neurological complications and kidney failure occur. Antibiotic-resistant strains are becoming a problem. Prevention is a matter of good personal hygiene and the maintenance of a clean water supply.

Traveler's Diarrhea and *Escherichia coli* Infections

Millions of people travel yearly from country to country. Unfortunately, a large percentage of these travelers acquire a rapidly acting, dehydrating condition called traveler's diarrhea. This diarrhea results from an encounter with certain viruses, bacteria, or protozoa usually absent from the traveler's normal environment. One of the major causative agents is *E. coli.* This bacterium circulates in the resident population, typically without causing symptoms due to the immunity afforded by previous exposure. Contaminated food and water are the major means by which the bacteria are spread, leading to the warning: "Boil it, peel it, cook it, or forget it."

Although the vast majority of *E. coli* strains are nonpathogenic members of the normal intestinal flora, some strains may cause diarrheal disease by several mechanisms. Six categories or strains of diarrheagenic *E. coli* are now recognized (**figure 20.32**): entero*t*oxigenic *E. coli* (ETEC), entero*i*nvasive *E. coli* (EIEC), entero*h*emorrhagic *E. coli* (EHEC), entero*p*athogenic *E. coli* (EPEC), entero*agg*regative *E. coli* (EAggEC), and *d*iffusely *a*dhering *E. coli* (DAEC).

The **enterotoxigenic *E. coli* (ETEC)** strains produce one or both of two distinct enterotoxins, which are responsible for the diarrhea and are distinguished by their heat stability: heat-stable enterotoxin (ST) and heat-labile enterotoxin (LT) (figure 20.32*a*). The genes for ST and LT production and colonization factors are usually plasmid-borne and acquired by horizontal gene transfer. ST binds to a glycoprotein receptor that is coupled to guanylate cyclase on the surface of intestinal epithelial cells. Activation of guanylate cyclase stimulates the production of cyclic guanosine monophosphate (cGMP), which leads to the secretion of electrolytes and water into the lumen of the small intestine, manifested

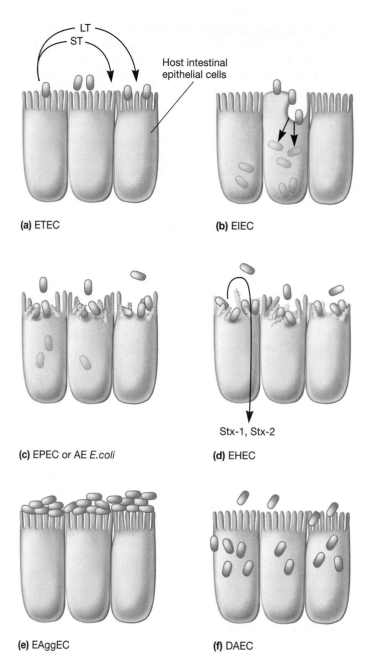

Figure 20.32   Six Classes of Diarrheagenic *E. coli.* Each class of diarrhea-causing *E. coli* can be classified by the nature of its interaction with host intestinal epithelial cells.

as watery diarrhea. LT binds to specific gangliosides on the epithelial cells and activates membrane-bound adenylate cyclase, which leads to increased production of cyclic adenosine monophosphate (cAMP) through the same mechanism employed by cholera toxin. Again, the result is hypersecretion of electrolytes and water into the intestinal lumen.

The **enteroinvasive *E. coli* (EIEC)** strains cause diarrhea by penetrating and multiplying within the intestinal epithelial cells (figure 20.32*b*). The ability to invade epithelial cells is associated with the presence of a large plasmid; EIEC may also produce a cytotoxin and an enterotoxin.

The **enteropathogenic *E. coli* (EPEC)** strains attach to the brush border of intestinal epithelial cells and cause a specific type of cell damage called effacing lesions (figure 20.32*c*). Effacing lesions or attaching-effacing (AE) lesions represent destruction of brush border microvilli adjacent to adhering bacteria. This cell destruction leads to the subsequent diarrhea. As a result of this pathology, the term AE *E. coli* is used to describe true EPEC strains. It is now known that AE *E. coli* is an important cause of diarrhea in children residing in developing countries.

The **enterohemorrhagic *E. coli* (EHEC)** strains carry the bacteriophage-encoded genetic determinants for shigalike toxin (Stx-1 and Stx-2 proteins; figure 20.32*d*). EHEC also produce AE lesions, causing hemorrhagic colitis with severe abdominal pain and cramps followed by bloody diarrhea. Stx-1 and Stx-2 (previously called verotoxins 1 and 2) have also been implicated in the extraintestinal disease hemolytic uremic syndrome, a severe hemolytic anemia that leads to kidney failure. It is believed these toxins kill vascular endothelial cells. A major strain of EHEC is *E. coli* **O157:H7,** which has caused many outbreaks of hemorrhagic colitis in the United States since it was first recognized in 1982. Currently an estimated 73,000 *E. coli* O157:H7 cases occur in the United States each year, resulting in about 60 deaths. Other serotypes of *E. coli* can cause similar disease, but they do not typically carry the genes for shigalike toxin. However, most laboratories do not test for non-O157 strains, so the actual incidence of EHEC is underreported.

The **enteroaggregative *E. coli* (EAggEC)** strains adhere to epithelial cells in localized regions, forming clumps of bacteria with a "stacked brick" appearance (figure 20.32*e*). Conventional extracellular toxins have not been detected in EAggEC, but unique lesions are seen in epithelial cells, suggesting the involvement of toxins.

The **diffusely adhering *E. coli* (DAEC)** strains adhere over the entire surface of epithelial cells and usually cause disease in immunologically naive or malnourished children (figure 20.32*f*). It has been suggested that DAEC may have an as yet undefined virulence factor.

Order *Pasteurellales*

The family *Pasteurellaceae* in the order *Pasteurellales* differs from the *Vibrionales* and the *Enterobacteriales* in several ways (table 20.7). Most notably, they are small (0.2 to 0.4 μm in diameter) and nonmotile, normally oxidase positive, have complex nutritional requirements, and are parasitic in vertebrates. The family contains seven genera: *Pasteurella, Haemophilus, Actinobacillus, Lonepinella, Mannheimia, Phocoenobacter,* and *Gallibacterium.*

As might be expected, members of this family are best known for the diseases they cause in humans and many animals. *Pasteurella multocida* and *P. haemolytica* are important animal pathogens. *P. multocida* is responsible for fowl cholera, which kills many chickens, turkeys, ducks, and geese each year. *P. haemolytica* is at least partly responsible for pneumonia in cattle, sheep, and goats (e.g., "shipping fever" in cattle). *H. influenzae* serotype b is a major human pathogen that causes a variety of diseases, including sinusitis, pneumonia, and bronchitis. It can disseminate to the bloodstream and cause a bacteremia. *H. influenzae* serotype b can cross into the CSF, resulting in inflammation of the meninges (meningitis). *H. influenzae* disease (including pneumonia and meningitis) is primarily observed in children younger than five. It is estimated that *H. influenzae* serotype b causes at least 3 million cases of serious disease and several hundreds of thousands of deaths each year globally.

Fortunately, a sharp reduction in the incidence of *H. influenzae* serotype b infections began in the mid-1980s due to administration of the *H. influenzae* type b conjugate vaccine, antibiotic (rifampin) prophylaxis of disease contacts, and the availability of more efficacious therapeutic agents. From 1987 through 1999, the incidence of invasive infection among U.S. children under five years of age declined by 95%. Three to 6 percent of all *H. influenzae* infections are fatal. Furthermore, up to 20% of surviving patients have permanent hearing loss or other long-term sequelae. Currently, all children should be vaccinated with the *H. influenzae* type b conjugate vaccine at the age of two months.

1. Describe the general properties of the γ-proteobacteria.
2. What are the major characteristics of the purple sulfur bacteria? Contrast the families *Chromatiaceae* and *Ectothiorhodospiraceae*.
3. Describe the genera *Beggiatoa, Leucothrix,* and *Thiothrix.*
4. In what habitats would one expect to see the *Methylococcaceae* growing and why?
5. What is a methylotroph? How do methane-oxidizing bacteria use methane as both an energy source and a carbon source?
6. Give the major distinctive properties of the genera *Pseudomonas* and *Azotobacter.*
7. Why are the pseudomonads such important bacteria? What is mineralization?
8. List the major distinguishing traits of the families *Vibrionaceae, Enterobacteriaceae,* and *Pasteurellaceae.*
9. Briefly describe bioluminescence and the way it is produced.
10. Many consider cholera as the most severe form of gastroenteritis. Why do you think this is so?
11. Into what two groups can the enteric bacteria be placed based on their fermentation patterns?
12. What is the usual source of the bacterium responsible for salmonellosis? Shigellosis? Where and how does *Shigella* infect people?
13. Describe a typhoid carrier. How does one become a carrier?
14. What are some specific causes of traveler's diarrhea? Briefly describe the six major types of pathogenic *E. coli*.

20.4   CLASS DELTA-PROTEOBACTERIA

Although the **δ-proteobacteria** are not a large assemblage of genera, they show considerable morphological and physiological diversity. These bacteria can be divided into two general groups, all of them chemoorganotrophs. Some genera are predators, such as the bdellovibrios and myxobacteria. Others are anaerobes that generate sulfide from sulfate and sulfur while oxidizing organic nutrients. The class has eight orders and 20 families. **Figure 20.33** illustrates the phylogenetic relationships among major groups within the δ-proteobacteria, and **table 20.9** summarizes the general properties of some representative genera.

Orders *Desulfovibrionales, Desulfobacterales,* and *Desulfuromonadales*

Desulfovibrionales, Desulfobacterales, and *Desulfuromonadales* are a diverse group of sulfate- or sulfur-reducing bacteria that are united by their anaerobic nature and the ability to reduce elemental

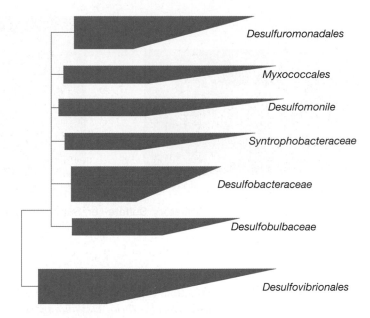

Figure 20.33   Phylogenetic Relationships Among Major Groups Within the δ-Proteobacteria.   The relationships are based on 16S rRNA sequence.

Table 20.9	Characteristics of Selected δ- and ε-Proteobacteria			
Class Genus	Dimensions (μm) and Morphology	G + C Content (mol%)	Oxygen Requirement	Other Distinctive Characteristics
δ-Proteobacteria				
Bdellovibrio	0.2–0.5 × 0.5–1.4; comma-shaped rods with a sheathed polar flagellum	49.5–51	Aerobic	Preys on other gram-negative bacteria where it grows in the periplasm, alternates between predatory and intracellular reproductive phases
Desulfovibrio	0.5–1.5 × 2.5–10; curved or sometimes straight rods, motile by polar flagella	46.1–61.2	Anaerobic	Oxidizes organic compounds to acetate and reduces sulfate or sulfur to H_2S
Desulfuromonas	0.4–0.9 × 1.0–4.0; straight or slightly curved or ovoid rods, lateral or subpolar flagella	54–62	Anaerobic	Reduces sulfur to H_2S, oxidizes acetate to CO_2, forms pink or peach-colored colonies
Myxococcus	0.4–0.7 × 2–8; slender rods with tapering ends, gliding motility	68–71	Aerobic	Forms fruiting bodies with microcysts not enclosed in a sporangium
Stigmatella	0.7–0.8 × 4–8; straight rods with tapered ends, gliding motility	67–68	Aerobic	Stalked fruiting bodies with sporangioles containing myxospores (0.9–1.2 × 2–4 μm)
ε-Proteobacteria				
Campylobacter	0.2–0.8 × 0.5–5; spirally curved cells with a single polar flagellum at one or both ends	29–47	Microaerophilic	Carbohydrates not fermented or oxidized; oxidase positive and urease negative; found in intestinal tract, reproductive organs, and oral cavity of animals
Helicobacter	0.2–1.2 × 1.5–10; helical, curved, or straight cells with rounded ends; multiple, sheathed flagella	24–48	Microaerophilic	Catalase and oxidase positive; urea rapidly hydrolyzed; found in the gastric mucosa of humans and other animals

(a) *Desulfovibrio saprovorans*

(b) *Desulfovibrio gigas*

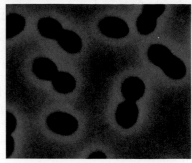

(c) *Desulfobacter postgatei*

Figure 20.34 **The Dissimilatory Sulfate- or Sulfur-Reducing Bacteria.** Representative examples. (a) Phase-contrast micrograph of *Desulfovibrio saprovorans* with PHB inclusions (×2,000). (b) *Desulfovibrio gigas;* phase contrast (×2,000). (c) *Desulfobacter postgatei;* phase contrast (×2,000).

sulfur or sulfate and other oxidized sulfur compounds to hydrogen sulfide during anaerobic respiration (**figure 20.34**). The best-studied sulfate-reducing genus is *Desulfovibrio; Desulfuromonas* uses only elemental sulfur as an acceptor. ◄◄ *Anaerobic respiration (section 10.6)*

These bacteria are very important in the cycling of sulfur within the ecosystem. Because significant amounts of sulfate are present in almost all aquatic and terrestrial habitats, sulfate-reducing bacteria are widespread and active in locations made anoxic by microbial digestion of organic materials. *Desulfovibrio* and other sulfate-reducing bacteria thrive in habitats such as muds and sediments of polluted lakes and streams, sewage lagoons and digesters, and waterlogged soils. *Desulfuromonas* is most prevalent in anoxic marine and estuarine sediments. It also can be isolated from methane digesters and anoxic hydrogen sulfide–rich muds of freshwater habitats. It uses elemental sulfur, but not sulfate, as its electron acceptor. Often sulfate and sulfur reduction are apparent from the smell of hydrogen sulfide and the blackening of water and sediment by iron sulfide. Hydrogen sulfide production in waterlogged soils can kill animals, plants, and microorganisms. Sulfate-reducing bacteria negatively impact industry because of their primary role in the anaerobic corrosion of iron in pipelines, heating systems, and other structures. ►► *Biogeochemical cycling: Sulfur cycle (section 25.1)*

Order *Bdellovibrionales*

The order *Bdellovibrionales* has only the family *Bdellovibrionaceae* and four genera. The genus *Bdellovibrio* (Greek *bdella,* leech) contains aerobic gram-negative, curved rods with polar flagella (**figure 20.35**). The flagellum is unusually thick due to the presence of a sheath that is continuous with the cell wall. *Bdellovibrio* has a distinctive life-style: it preys on other gram-negative bacteria and alternates between a nongrowing predatory phase and an intracellular reproductive phase.

The life cycle of *Bdellovibrio* is complex, although it requires only 1 to 3 hours for completion (**figure 20.36**). The free bacterium swims along very rapidly (about 100 cell lengths per second) until it collides violently with its prey. It attaches to the bacterial

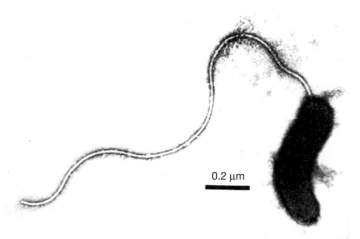

0.2 µm

Figure 20.35 *Bdellovibrio* **Morphology.** Negatively stained *Bdellovibrio bacteriovorus* with its sheathed polar flagellum.

surface, begins to rotate as fast as 100 revolutions per second, and bores a hole through the host cell wall in 5 to 20 minutes by releasing several hydrolytic enzymes. Its flagellum is lost during penetration of the cell.

After entry, *Bdellovibrio* takes control of the host cell and grows between the cell wall and plasma membrane while the host cell loses its shape and rounds up. The predator quickly inhibits host DNA, RNA, and protein synthesis, and disrupts the host's plasma membrane so that cytoplasmic constituents leak out of the cell. The growing bacterium uses host amino acids as its carbon, nitrogen, and energy source. It employs host fatty acids and nucleotides directly in biosynthesis, thus saving carbon and energy. The bacterium rapidly grows into a long filament under the cell wall and then divides into many smaller, flagellated progeny, which escape upon host cell lysis. Such multiple fission is rare in procaryotes.

The *Bdellovibrio* life cycle resembles that of bacteriophages in many ways. Not surprisingly, when *Bdellovibrio* is plated out on agar with host bacteria, plaques will form in the bacterial lawn. This technique is used to isolate pure strains and count the number of viable organisms just as with phages.

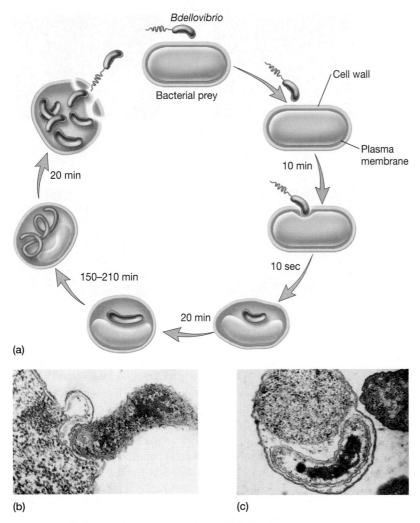

(a)

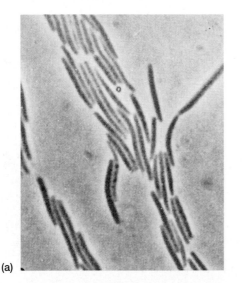

(b)

(c)

Figure 20.36   The Life Cycle of *Bdellovibrio.* (a) A general diagram showing the complete life cycle. (b) *Bdellovibrio bacteriovorus* penetrating the cell wall of *E. coli* (×55,000). (c) A *Bdellovibrio* encapsulated between the cell wall and plasma membrane of *E. coli* (×60,800).

Order *Myxococcales*

The myxobacteria are gram-negative, aerobic soil bacteria characterized by gliding motility, a complex life cycle with the production of fruiting bodies, and the formation of spores called myxospores. Their G + C content is around 67 to 71%, significantly higher than that of most gliding bacteria. Myxobacterial cells are rods, about 0.4 to 0.7 μm by 2 to 8 μm long, and may be either slender with tapered ends or stout with rounded, blunt ends (**figure 20.37**). The order *Myxococcales* is divided into six families based on the shape of vegetative cells, myxospores, and sporangia.

Most myxobacteria are micropredators or scavengers. They secrete an array of digestive enzymes that lyse bacteria and yeasts. Many myxobacteria also secrete antibiotics, which may kill their prey. The digestion products, primarily small peptides, are absorbed. Most myxobacteria use amino acids as their major source of carbon, nitrogen, and energy. All are chemoheterotrophs with respiratory metabolism.

The myxobacterial life cycle is quite distinctive and in many ways resembles that of the cellular slime molds (**figure 20.38**). In the presence of a food supply, myxobacteria glide along a solid surface, feeding and leaving slime trails. During this stage, the cells often form a swarm and move in a coordinated fashion. Some species congregate to produce a sheet of cells that moves rhythmically to generate waves or ripples. When their nutrient supply is exhausted, the myxobacteria aggregate and differentiate into a fruiting body. ◄◄ *Bacterial motility and chemotaxis (section 3.7)*

The life cycle of the species *Myxococcus xanthus* has been well studied. Development in this microbe is induced by nutrient limitation and involves the exchange of at least five different extracellular signaling molecules that

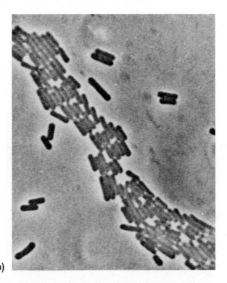

(a)

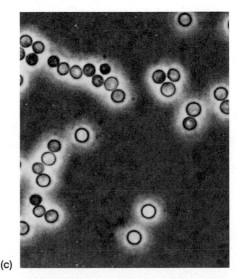

(c)

(b)

Figure 20.37   Gliding, Fruiting Bacteria (Myxobacteria). Myxobacterial cells and myxospores. (a) *Stigmatella aurantiaca* (×1,200). (b) *Chondromyces crocatus* (×950). (c) Myxospores of *Myxococcus xanthus* (×1,100). All photographs taken with a phase-contrast microscope.

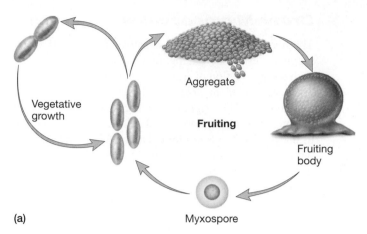

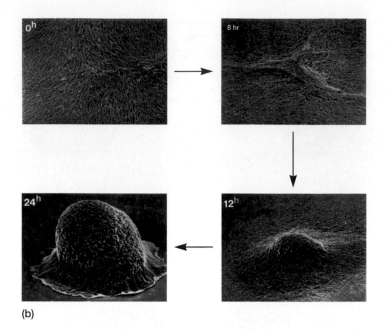

Figure 20.38 **Life cycle of *Myxococcus xanthus*.** (a) When nutrients are plentiful, *M. xanthus* grows vegetatively. However, when nutrients are depleted, a complex exchange of extracellular signaling molecules triggers the cells to aggregate and form fruiting bodies. Most of the cells within a fruiting body will become resting myxospores that will not germinate until nutrients are available. (b) Scanning electron micrographs taken during aggregate (0–12 hours) and fruiting body (24 hours) formation.

allow the cells to communicate with one another. Fruiting body development also requires gliding motility. Two types of gliding have been characterized in *M. xanthus*: adventurous (A) motility is propelled by the extrusion of a gel-like material from the rear pole, and social (S) motility is governed by the production of retractable pili from the front end of the cell. When the pili retract, the cell creeps forward. This type of motility was originally called social motility because it is only observed in cells that are close together. It is now known that cell-to-cell contact is required for S motility because cells share outer membrane lipoproteins involved in pili secretion.

A variety of new proteins are synthesized during **fruiting body** formation. Fruiting bodies range in height from 50 to 500 μm and often are colored red, yellow, or brown by carotenoid pigments.

Each species forms a characteristic fruiting body. They vary in complexity from simple globular objects made of about 100,000 cells (*Myxococcus*) to the elaborate, branching, treelike structures formed by *Stigmatella* and *Chondromyces* (**figure 20.39**). Some cells develop into dormant myxospores that frequently are enclosed in walled structures called sporangioles or sporangia.

Myxospores are dormant and desiccation-resistant; they may survive up to 10 years under adverse conditions. The use of fruiting bodies provides further protection for the myxospores and assists in their dispersal. (The myxospores often are suspended above the soil surface.) Because myxospores are kept together within the fruiting body, a colony of myxobacteria automatically develops when the myxospores are released and germinate. This communal organization may be advantageous because myxobacteria obtain nutrients by secreting hydrolytic enzymes and

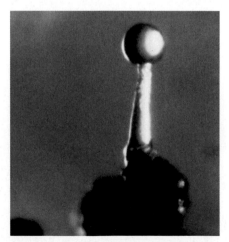

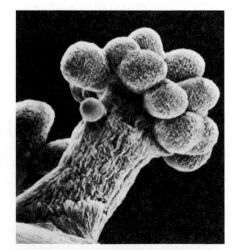

(a) *Myxococcus fulvus*   **(b)** *Myxococcus stipitatus*   **(c)** *Chondromyces crocatus*

Figure 20.39 **Myxobacterial Fruiting Bodies.** (a) *Myxococcus fulvus.* Fruiting bodies are about 150–400 μm high. (b) *Myxococcus stipitatus.* The stalk is as tall as 200 μm. (c) *Chondromyces crocatus* viewed with the SEM. The stalk may reach 700 μm or more in height.

absorbing soluble digestive products. A mass of myxobacteria can produce enzyme concentrations sufficient to digest their prey more easily than can an individual cell. Extracellular enzymes diffuse away from their source, and an individual cell has more difficulty overcoming diffusional losses than a swarm of cells.

Myxobacteria are found in soils worldwide. They are most commonly isolated from neutral soils or decaying plant material such as leaves and tree bark, and from animal dung. Although they grow in habitats as diverse as tropical rain forests and the Arctic tundra, they are most abundant in warm areas.

1. Briefly characterize the δ-proteobacteria.

2. Describe the metabolic specialization of the dissimilatory sulfate- or sulfur-reducing bacteria. Why are they important?

3. Characterize the genus *Bdellovibrio* and outline its life cycle in detail. Why do you think it preys only on gram-negative bacteria?

4. Briefly describe the myxobacterial life cycle. What are fruiting bodies and myxospores? Why do you think extracellular signaling molecules are required for fruiting body formation?

20.5 CLASS *EPSILON-PROTEOBACTERIA*

The **ε-proteobacteria** are the smallest of the five proteobacterial classes. They all are slender gram-negative rods, which can be straight, curved, or helical. The ε-proteobacteria have one order, *Campylobacterales,* and three families: *Campylobacteraceae, Helicobacteraceae,* and the recently added *Nautiliaceae.* Two pathogenic genera, *Campylobacter* and *Helicobacter,* are microaerophilic, motile, helical or vibrioid, gram-negative rods. Table 20.9 summarizes some of the characteristics of these two genera.

The genus *Campylobacter* contains both nonpathogens and species pathogenic for humans and other animals. *C. fetus* causes reproductive disease and abortions in cattle and sheep. It is associated with a variety of conditions in humans ranging from septicemia (pathogens or their toxins in the blood) to enteritis (inflammation of the intestinal tract). *C. jejuni* is a slender, gram-negative, motile, curved rod found in the intestinal tract of animals. It causes an estimated 2 million human cases of **Campylobacter gastroenteritis**—inflammation of the intestine—or **campylobacteriosis** and subsequent diarrhea in the United States each year. Studies with chickens, turkeys, and cattle have shown that as much as 50 to 100% of a flock or herd of these birds or animals excrete *C. jejuni.* These bacteria also can be isolated in high numbers from surface waters. They are transmitted to humans by contaminated food and water, contact with infected animals, or anal-oral sexual activity.

The incubation period is 2 to 10 days. *C. jejuni* invades the epithelium of the small intestine, causing inflammation, and also secretes an exotoxin that is antigenically similar to the cholera toxin.

Symptoms include diarrhea, high fever, severe inflammation of the intestine along with ulceration, and bloody stools. *C. jejuni* infection has also been linked to Guillain-Barré syndrome, a disorder in which the body's immune system attacks peripheral nerves, resulting in life-threatening paralysis. **>>** *Toxigenicity: Exotoxins (section 30.4)*

The disease is usually self-limited, and treatment is supportive; fluids, electrolyte replacement, and the antibiotic erythromycin may be used in severe cases. Recovery usually takes from 5 to 8 days. Prevention and control involve good personal hygiene and food-handling precautions, including pasteurization of milk and thorough cooking of poultry.

The other pathogenic ε-proteobacterial genus is *Helicobacter.* There are at least 23 species of *Helicobacter,* all isolated from the stomachs and upper intestines of humans, dogs, cats, and other mammals. In developing countries, 70 to 90% of the population is infected; the rate in developed countries ranges from 25 to 50%. Most infections are probably acquired during childhood, but the precise mode of transmission is unclear. The major human pathogen is *Helicobacter pylori,* which causes gastritis and peptic ulcer disease. In addition to *H. pylori* causing peptic ulcers, there are strong, positive correlations between gastric cancer rates and *H. pylori* infection rates in certain populations; it has been classified as a Class I carcinogen by the World Health Organization.

H. pylori colonizes only gastric mucus-secreting cells, beneath the gastric mucous layers, and surface fimbriae are believed to be one of the adhesins associated with this process (**figure 20.40**). *H. pylori* binds to Lewis B antigens (which are part of the blood group antigens that determine blood group O) and to the monosaccharide sialic acid, also found in the

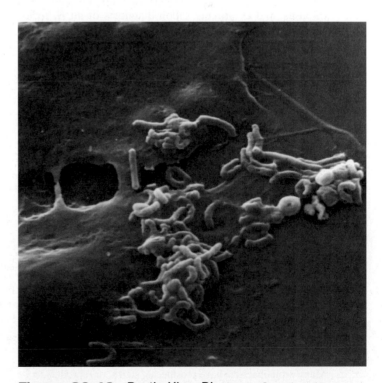

Figure 20.40 Peptic Ulcer Disease. Scanning electron micrograph (×3,441) of *Helicobacter pylori* adhering to gastric cells.

(a)

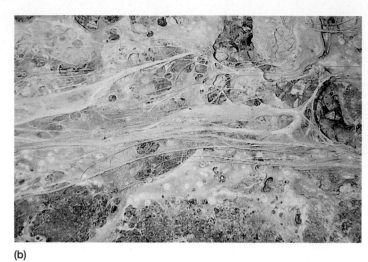

(b)

Figure 20.41 ε-Proteobacteria Dominate Filamentous Microbial Mats in a Wyoming Sulfidic Cave Spring. (a) A channel formed by the spring appears white due to the high density of filamentous ε-proteobacteria. The stake in the center of photo (arrow) is about 25 cm high. (b) Filamentous ε-proteobacteria within the springs.

glycoproteins on the surface of gastric epithelial cells. The bacterium moves into the mucous layer to attach to mucus-secreting cells. Movement into the mucous layer may be aided by the fact that *H. pylori* is a strong producer of urease. Urease activity is thought to create a localized alkaline environment when hydrolysis of urea produces ammonia. The increased pH may protect the bacterium from gastric acid until it is able to grow under the layer of mucus in the stomach. The potential virulence factors responsible for epithelial cell damage and inflammation probably include proteases, phospholipases, and cytotoxins. Upon diagnosis of an *H. pylori* infection, the goal of treatment is the complete elimination of the organism. Treatment is two-pronged: use of drugs to decrease stomach acid and antibiotics to kill the bacteria.

The ε-proteobacteria are now recognized to be more metabolically and ecologically diverse than previously thought. For instance, filamentous microbial mats in anoxic, sulfide-rich cave springs are dominated by members of the ε-proteobacteria

(**figure 20.41**). The inclusion of a new family, the *Nautiliaceae,* in *Bergey's Manual* is a result of the recent isolation of two genera of moderately thermophilic (optimum growth temperature about 55°C) chemolithoautotrophs from deep-sea hydrothermal vents. Members of the genera *Nautilia* and *Caminibacter* are strict anaerobes that oxidize H_2 and use sulfur as an electron acceptor. Species are found as either freely living or as symbionts of vent macrofauna. **>>** *Microbial interactions: Sulfide-based mutualisms (section 27.1)*

1. Briefly describe the properties of the ε-proteobacteria.
2. What is the public health significance of *Campylobacter* prevalence in poultry?
3. Describe how *H. pylori* can survive the acidic conditions of the stomach.

Summary

20.1 Class *Alphaproteobacteria*

a. The purple nonsulfur bacteria can grow anaerobically as photoorganoheterotrophs and often aerobically as chemoorganoheterotrophs. They are found in aquatic habitats with abundant organic matter and low sulfide levels.

b. Rickettsias are obligate intracellular parasites responsible for many diseases including typhus, Rocky Mountain spotted fever, and ehrlichiosis. They have numerous transport proteins in their plasma membranes and make extensive use of host cell nutrients, coenzymes, and ATP.

c. Many proteobacteria have prosthecae, stalks, or reproduction by budding. Most of these bacteria are placed among the α-proteobacteria.

d. Two examples of budding or appendaged bacteria are *Hyphomicrobium* (budding bacteria that produce swarmer cells) and *Caulobacter* (bacteria with prosthecae and holdfasts) (**figures 20.6–20.9**).

e. *Rhizobium* carries out nitrogen fixation, whereas *Agrobacterium* causes the development of plant tumors. Both are in the family *Rhizobiaceae* (**figures 20.10** and **20.11**).

f. Chemolithotrophic bacteria derive energy and electrons from reduced inorganic compounds. Nitrifying bacteria are aerobes that oxidize either ammonia or nitrite to nitrate and are responsible for nitrification (**table 20.2**).

20.2 Class *Betaproteobacteria*

a. The genus *Neisseria* contains nonmotile, aerobic, gram-negative cocci that usually occur in pairs. They colonize mucous membranes and cause several human diseases, including meningitis and gonorrhea.

b. *Sphaerotilus, Leptothrix,* and several other genera have sheaths, hollow tubelike structures that surround chains of cells without being in intimate contact with the cells

c. The colorless sulfur bacteria such as *Thiobacillus* oxidize elemental sulfur, hydrogen sulfide, and thiosulfate to sulfate while generating energy chemolithotrophically.

20.3 Class *Gammaproteobacteria*

a. The γ-proteobacteria are the largest subgroup of proteobacteria with great variety in physiological types (**table 20.6** and **figure 20.19**).

b. The purple sulfur bacteria are anaerobes and usually photolithoautotrophs. They oxidize hydrogen sulfide to sulfur and deposit the granules internally (**figure 20.21**).

c. Bacteria such as *Beggiatoa* and *Leucothrix* grow in long filaments or trichomes (**figures 20.23** and **20.24**). Both genera have gliding motility. *Beggiatoa* is primarily a chemolithotroph and *Leucothrix,* a chemoorganotroph.

d. The *Methylococcaceae* are methylotrophs; they use methane, methanol, and other reduced one-carbon compounds as their sole carbon and energy sources.

e. The genus *Pseudomonas* contains straight or slightly curved, gram-negative, aerobic rods that are motile by one or several polar flagella and do not have prosthecae or sheaths.

f. The pseudomonads participate in natural mineralization processes, are major experimental subjects, cause many diseases, and often spoil refrigerated food.

g. The most important facultatively anaerobic, gram-negative rods are found in three families: *Vibrionaceae, Enterobacteriaceae,* and *Pasteurellaceae* (**table 20.7**).

h. The *Enterobacteriaceae,* often called enterobacteria or enteric bacteria, are gram-negative, peritrichously flagellated or nonmotile, facultatively anaerobic, straight rods with simple nutritional requirements.

i. The enteric bacteria are usually identified by a variety of physiological tests and are very important experimental organisms and pathogens of plants and animals (**table 20.8**).

j. Enteric bacteria include important human pathogens such as *Yersinia pestis* (which causes plague), *Salmonella, Shigella,* and some strains of *E. coli* (**figures 20.31** and **20.32**).

20.4 Class *Deltaproteobacteria*

a. The δ-proteobacteria contain chemoorganotrophic, gram-negative bacteria that are anaerobic and can use elemental sulfur and oxidized sulfur compounds as electron acceptors in anaerobic respiration (**table 20.9**). They are very important in sulfur cycling in the ecosystem. Other δ-proteobacteria are predatory aerobes.

b. *Bdellovibrio* is an aerobic curved rod with sheathed polar flagellum that preys on other gram-negative bacteria and grows within their periplasmic space (**figures 20.35** and **20.36**).

c. *Myxobacteria* are gram-negative, aerobic soil bacteria with gliding motility and a complex life cycle that leads to the production of dormant myxospores held within fruiting bodies (**figures 20.37–20.39**).

20.5 Class *Epsilonproteobacteria*

a. The ε-proteobacteria are the smallest of the proteobacterial classes and contain two important pathogenic genera: *Campylobacter* and *Helicobacter.* These are microaerophilic, motile, helical or vibrioid, gram-negative rods (**table 20.9**).

b. Recently a new family, the *Nautiliaceae,* has been added. Many of these bacteria are chemolithoautotrophs from deep-sea hydrothermal vent ecosystems.

Critical Thinking Questions

1. *Helicobacter pylori* produces large quantities of urease. Urease catalyzes the reaction:

$$H_2N-\overset{\overset{\textstyle O}{\|}}{C}-NH_2 \rightarrow CO_2 + 2NH_3$$

Suggest why this allows *H. pylori* to inhabit the acidic habitat of the gastric mucosa.

2. Methylotrophs oxidize methane to methanol, then to formaldehyde, and finally into acetate. Suggest mechanisms by which the bacterium protects itself from the toxic effects of the intermediates, methanol and formaldehyde.

3. *Bdellovibrio* is an intracellular predator. Once it invades the periplasm, it manages to inhibit many aspects of host metabolism. Suggest a mechanism by which this inhibition could occur so rapidly.

4. Why might the ability to form dormant cysts be of great advantage to *Agrobacterium* but not as much to *Rhizobium?*

5. Why are gliding, budding, and appendaged bacteria distributed among so many different sections in *Bergey's Manual?*

Learn More

Learn more by visiting the Prescott website at www.mhhe.com/prescottprinciples, where you will find a complete list of references.

The Low G + C Gram-Positive Bacteria

21

Chapter Glossary

α-hemolysis A greenish zone of partial clearing around a bacterial colony growing on blood agar, resulting from incomplete hemolysis and hemoglobin breakdown.

anthrax A highly infectious animal disease that can be transmitted to humans through skin infection, inhalation, and ingestion. Caused by *Bacillus anthracis,* a member of the order *Bacillales.*

atypical pneumonia A lung infection that can range in severity from nearly asymptomatic to serious pneumonia. Often caused by *Mycobacterium pneumoniae,* a member of the class *Mollicutes.*

β-hemolysis A zone of complete clearing around a bacterial colony growing on blood agar, resulting from complete hemolysis and hemoglobin breakdown.

botulism A life-threatening intoxication resulting from a toxin produced by *Clostridium botulinum,* a member of the class *Clostridia.*

coagulase An enzyme that induces blood clotting; it is characteristically produced by pathogenic staphylococci.

gas gangrene A necrotizing infection of skeletal muscle caused by *Clostridium perfringens,* a member of the class *Clostridia.*

group A streptococci (GAS) An important group of human pathogens that includes members of the genus *Streptococcus* (order *Lactobacillales*) that form a capsule and produce extracellular degradative enzymes, streptokinases (enzymes that result in dissolution of blood clots), the white blood cell cytolysins streptolysin O and S, and M protein.

heterolactic fermentation The fermentation of sugars to form lactate and other products such as ethanol and CO_2.

homolactic fermentation The fermentation of sugars almost completely to lactic acid.

impetigo A common skin infection caused by *Staphylococcus aureus* and *Streptococcus pyogenes,* both members of the class *Bacilli.*

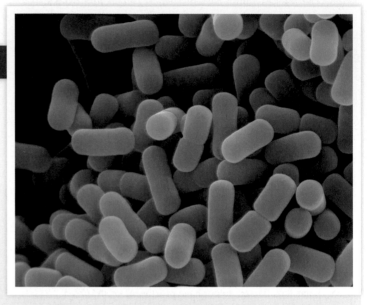

Lactobacillus plantarum is a gram-positive microbe that is used in fermentation of plant material, as in the preparation of sauerkraut and pickles.

lactic acid bacteria (LAB) Members of the order *Lactobacillales* that generate lactic acid as their major or sole fermentation product.

Lancefield grouping system A method by which streptococci can be placed in serologically distinguishable groups.

listeriosis A food-borne disease that causes invasive syndromes such as meningitis, sepsis, and stillbirth; caused by *Listeria monocytogenes,* class *Bacilli.*

mycoplasmas Bacteria that are members of the class *Mollicutes* that a lack cell walls and cannot synthesize peptidoglycan precursors; most require sterols for growth.

necrotizing fasciitis Inflammation and destruction of the sheath covering skeletal muscle caused by invasive group A streptococci.

streptococcal pharyngitis (strep throat) An infection of the pharynx by group A *Streptococcus pyogenes.*

tetanus An intoxication resulting from infection with *Clostridium tetani* (class *Clostridia*); causes spastic paralysis.

toxic shock syndrome (TSS) A staphylococcal disease caused by toxic shock syndrome toxin-1, staphylococcal enterotoxin B, or enterotoxin C1 (also known as superantigens).

We noted, after having grown the bacterium through a series of such cultures, each fresh culture being inoculated with a droplet from the previous culture, that the last culture of the series was able to multiply and act in the body of animals in such a way that the animals developed anthrax with all the symptoms typical of this affection. Such is the proof, which we consider flawless, that anthrax is caused by this bacterium.

—*Louis Pasteur*

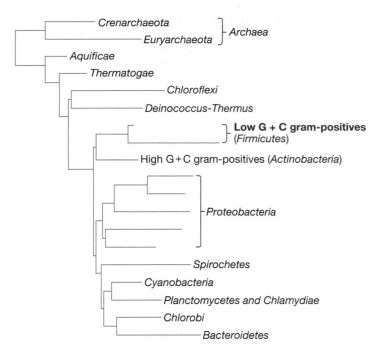

Figure 21.1 Phylogenetic Relationships Among the Procaryotes. The low G + C gram-positive bacteria are highlighted.

Chapters 21 and 22 introduce the gram-positive bacteria. These bacteria were historically grouped on the basis of their general shape (e.g., rods, cocci, or irregular) and their ability to form endospores. However, analysis of phylogenetic relationships within the gram-positive bacteria shows that they are divided into a low G + C group and a high G + C, or actinobacterial, group (**figure 21.1**). *Bergey's Manual of Systematic Bacteriology* places the low G + C gram-positive bacteria in volume 3. This volume describes over 1,300 species placed in 255 genera. ◄◄ *Bacterial cell walls: Gram-positive cell walls (section 3.4)*

The low G + C gram-positive bacteria are placed in the phylum *Firmicutes* and divided into three classes: *Clostridia, Mollicutes,* and *Bacilli.* The phylum *Firmicutes* is large and complex; it has 10 orders and 34 families. The mycoplasmas, class *Mollicutes,* are also considered low G + C gram positives despite their lack of a cell wall. Ribosomal RNA data indicate that the mycoplasmas are closely related to the lactobacilli. In this chapter, we focus on the mycoplasmas, *Clostridium* and its relatives, and the bacilli and lactobacilli. **Figure 21.2** shows the phylogenetic relationships among some of the bacteria reviewed in this chapter.

21.1 CLASS *MOLLICUTES* (THE MYCOPLASMAS)

The class *Mollicutes* has five orders and six families. The best-studied genera are found in the orders *Mycoplasmatales* (*Mycoplasma, Ureaplasma*), *Entomoplasmatales* (*Entomoplasma, Mesoplasma, Spiroplasma*), *Acholeplasmatales* (*Acholeplasma*), and *Anaeroplasmatales* (*Anaeroplasma, Asteroleplasma*). **Table 21.1** summarizes some of the major characteristics of these genera.

Members of the class *Mollicutes* are commonly called **mycoplasmas.** Although they evolved from ancestors with gram-positive cell walls, they now lack cell walls and cannot synthesize peptidoglycan precursors. Thus they are penicillin resistant but susceptible to lysis by osmotic shock and detergent treatment. Because they are bounded only by a plasma membrane, these procaryotes are pleomorphic and vary in shape from spherical or pear-shaped organisms, about 0.3 to 0.8 μm in diameter, to branched or helical filaments (**figure 21.3**). Some mycoplasmas (e.g., *M. genitalium*) have a specialized terminal structure that projects from the cell and gives them a flask or pear shape. This structure aids in attachment to eucaryotic cells. They are among the smallest bacteria capable of self-reproduction. The majority are nonmotile, but some can glide along liquid-covered surfaces. Most species require sterols for growth, which they often obtain from their host cell as cholesterol. Sterols are an essential component of the mycoplasma plasma membrane, where they may facilitate osmotic stability. Most species are facultative anaerobes, but a few are obligate anaerobes. When growing on agar, most form colonies with a "fried egg" appearance because they grow into the agar surface at the center while spreading outward on the surface at the colony edges (**figure 21.4**). Their genomes are among the smallest found in procaryotes, ranging from 0.7 to 1.7 Mb (table 21.1). The sequenced genomes of the human pathogens *Mycoplasma genitalium, M. pneumoniae,* and *Ureaplasma urealyticum* have fewer than 1,000 genes, suggesting a minimal genome size for a free-living existence.

Metabolically, the mycoplasmas are incapable of synthesizing a number of macromolecules. In addition to requiring sterols, they also need fatty acids, vitamins, amino acids, purines, and pyrimidines. Some produce ATP by the Embden-Meyerhof pathway and lactic acid fermentation. Others catabolize arginine or urea to generate ATP. The pentose phosphate pathway seems to be functional in at least some mycoplasmas; none appear to have the complete tricarboxylic acid cycle.

Mycoplasmas are remarkably widespread and can be isolated from animals, plants, the soil, and even compost piles. Although their complex growth requirements can make their growth in pure (axenic) cultures difficult, about 10% of the mammalian cell cultures in use are probably contaminated with mycoplasmas. This seriously interferes with tissue culture experiments. In animals, mycoplasmas colonize mucous membranes and joints, and often are associated with diseases of the respiratory and urogenital tracts. Mycoplasmas cause several major diseases in livestock, for example, contagious bovine pleuropneumonia in cattle (*M. mycoides*), chronic respiratory disease in chickens (*M. gallisepticum*), and pneumonia in swine (*M. hyopneumoniae*). Spiroplasmas have been isolated from insects, ticks, and a variety of plants. They cause disease in citrus plants, cabbage, broccoli, corn, honeybees, and other hosts. Arthropods may often act as vectors and carry the spiroplasmas between plants. In humans, *U. urealyticum* and *M. hominis* are common parasitic microorganisms of the genital tract, and their transmission is related to sexual activity. Both mycoplasmas can opportunistically cause inflammation of the reproductive organs of males and females. In addition, *Ureaplasma urealyticum* is associated with premature delivery of newborns, as well as neonatal meningitis and pneumonia. *M. pneumoniae* causes primary atypical pneumonia in humans, which is discussed next.

Mycoplasmal Pneumonia

Typical pneumonia has a bacterial origin (most frequently *Streptococcus pneumoniae*) with fairly consistent signs and symptoms. If the symptoms of pneumonia are different from what is typically observed, the disease is called **atypical pneumonia.** One cause of atypical pneumonia is *Mycoplasma pneumoniae,* a mycoplasma with worldwide distribution. Transmission involves close contact and airborne droplets. The disease is fairly common and mild in infants and small children; serious disease is seen principally in older children and young adults.

M. pneumoniae usually infects the upper respiratory tract and subsequently moves to the lower respiratory tract, where it attaches to respiratory mucosal cells. It then produces peroxide, which may be a virulence factor, but the exact mechanism of pathogenesis is unknown. A change in mucosal cell nucleic acid synthesis has been observed. This disease varies in severity from asymptomatic to a serious pneumonia. The latter is accompanied by death of the surface mucosal cells, lung infiltration, and congestion. Initial symptoms include headache, weakness, a low-grade fever, and a predominant, characteristic cough. The disease and its symptoms usually persist for weeks. Antibiotics (tetracyclines or erythromycin) are effective in treatment. There are no preventive measures. The mortality rate is less than 1%.

1. What morphological feature distinguishes the mycoplasmas? In what class are they found? Why have they been placed with the low G + C gram-positive bacteria?

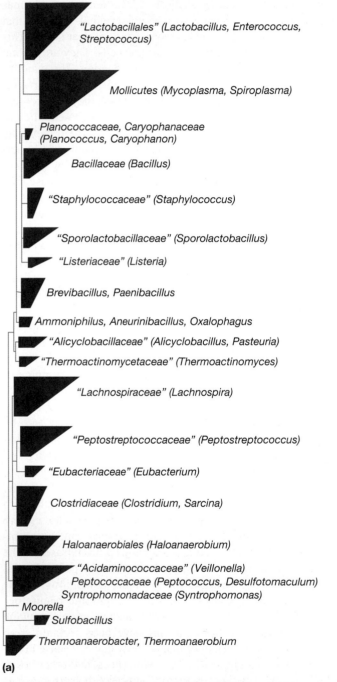

(a)

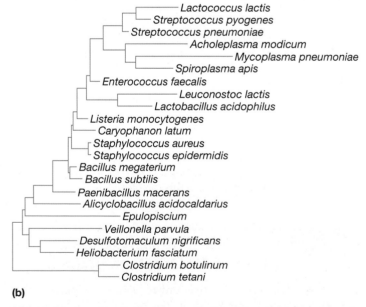

(b)

Figure 21.2 **Phylogenetic Relationships in the Phylum *Firmicutes* (Low G + C Gram Positives).** (a) The major phylogenetic groups with representative genera in parentheses. Each tetrahedron in the tree represents a group of related organisms; its horizontal edges show the shortest and longest branches in the group. Multiple branching at the same level indicates that the relative branching order of the groups cannot be determined from the data. The quotation marks around some names indicate that they are not formally approved taxonomic names. (b) The relationships of a few species based on 16S rRNA sequence date.   *Source: The Ribosomal Database Project.*

Table 21.1	Properties of Some Members of the Class *Mollicutes*					
Genus	No. of Recognized Species	G + C Content (mol%)	Genome Size (Mb)	Sterol Requirement	Habitat	Other Distinctive Features
Acholeplasma	13	26–36	1.50–1.65	No	Vertebrates, some plants and insects	Optimum growth 30–37°C
Anaeroplasma	4	29–34	1.50–1.60	Yes	Bovine or ovine rumen	Oxygen-sensitive anaerobes
Asteroleplasma	1	40	1.50	No	Bovine or ovine rumen	Oxygen-sensitive anaerobes
Entomoplasma	5	27–29	0.79–1.14	Yes	Insects, plants	Optimum growth 30°C
Mesoplasma	12	27–30	0.87–1.10	No	Insects, plants	Optimum growth 30°C; sustained growth in serum-free medium only with 0.04% Tween 80
Mycoplasma	104	23–40	0.60–1.35	Yes	Humans, animals	Optimum growth usually 37°C
Spiroplasma	22	25–30	0.94–2.20	Yes	Insects, plants	Helical filaments; optimum growth at 30–37°C
Ureaplasma	6	27–30	0.75–1.20	Yes	Humans, animals	Urea hydrolysis

Adapted from J. G. Tully, et al., "Revised Taxonomy of the Class *Mollicutes*" in *International Journal of Systematic Bacteriology*, 43(2):378–85. Copyright © 1993 American Society for Microbiology, Washington, D.C. Reprinted by permission.

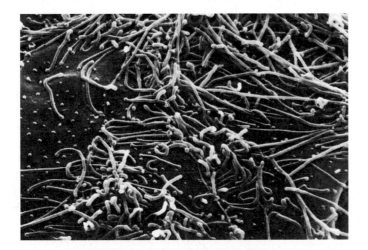

Figure 21.3 The *Mycoplasmas*. A scanning electron micrograph of *Mycoplasma pneumoniae* shows its pleomorphic nature. (×26,000).

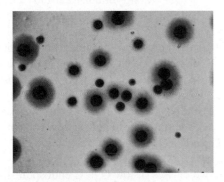

Figure 21.4 *Mycoplasma* Colonies. Note the "fried egg" appearance; colonies stained before photographing (×100).

2. What might mycoplasmas use sterols for?

3. What do you think the relationship between *Mycoplasma* genome size and growth requirements might be?

4. Where are mycoplasmas found in animals? List several animal and human diseases caused by them. What kinds of organisms do spiroplasmas usually infect?

21.2 PEPTIDOGLYCAN AND ENDOSPORE STRUCTURE

The gram-positive bacteria have traditionally been classified largely on the basis of observable characteristics such as cell shape, the clustering and arrangement of cells, the presence or absence of endospores, oxygen relationships, fermentation patterns, and peptidoglycan chemistry. Because of the importance of peptidoglycan and endospores in these bacteria, we briefly discuss these two important components.

Peptidoglycan structure varies considerably among different gram-positive groups. Many gram-positive bacteria (and most gram-negatives) have a peptidoglycan structure in which meso-diaminopimelic acid in position 3 is directly linked through its free amino group with the free carboxyl of the terminal D-alanine of an adjacent peptide chain (**figure 21.5a**). These include *Bacillus, Clostridium, Corynebacterium, Mycobacterium,* and *Nocardia*. In some other gram-positive bacteria, lysine is substituted for diaminopimelic acid in position 3, and the peptide subunits of the glycan

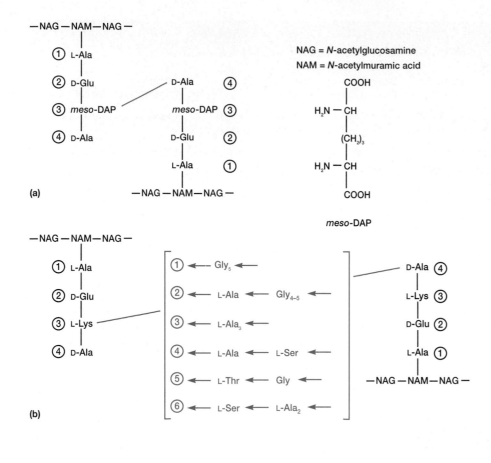

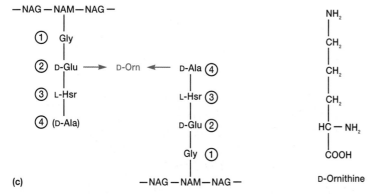

Figure 21.5 **Representative Examples of Peptidoglycan Structure.** (a) The peptidoglycan with a direct cross-linkage between positions 3 and 4 of the peptide subunits, which is present in most gram-negative and many gram-positive bacteria. (b) Peptidoglycan with lysine in position 3 and an interpeptide bridge. The bracket contains six typical bridges: (1) *Staphylococcus aureus*, (2) *S. epidermidis*, (3) *Micrococcus roseus* and *Streptococcus thermophilus*, (4) *Lactobacillus viridescens*, (5) *Streptococcus salvarius*, and (6) *Leuconostoc cremoris*. The arrows indicate the polarity of peptide bonds running in the C to N direction. (c) An example of the cross-bridge extending between positions 2 and 4 from *Corynebacterium poinsettiae*. The interbridge contains a D-diamino acid like ornithine, and L-homoserine (L-Hsr) is in position 3. The abbreviations and structures of amino acids in the figure are found in appendix I.

chains are cross-linked by interpeptide bridges containing monocarboxylic L-amino acids or glycine, or both (figure 21.5b). Many genera including *Staphylococcus, Streptococcus, Micrococcus, Lactobacillus,* and *Leuconostoc* have this type of peptidoglycan. The high G + C gram-positive genus *Streptomyces* and several other actinobacterial genera have replaced mesodiaminopimelic

acid with L,L-diaminopimelic acid in position 3 and have one glycine residue as the interpeptide bridge. The plant pathogenic corynebacteria (also a high G + C gram-positive) provide another example of peptidoglycan variation. In some of these bacteria, the interpeptide bridge connects positions 2 and 4 of the peptide subunits rather than 3 and 4 (figure 21.5c). Because the interpeptide bridge connects the carboxyl groups of glutamic acid and alanine, a diamino acid such as ornithine is used in the bridge. Many other variations in peptidoglycan structure are found, including other interbridge structures and large differences in the frequency of cross-linking between glycan chains. Bacilli and most gram-negative bacteria have fewer cross-links between chains than do gram-positive bacteria such as *Staphylococcus aureus* in which almost every muramic acid is cross-linked to another. These structural variants are often characteristic of particular groups and are therefore taxonomically useful. ◄◄ *Synthesis of sugars and polysaccharides: Synthesis of peptidoglycan (section 11.4)*

Bacterial endospores are assembled within the differentiating mother cell, which lyses to release the free spore (*see figures 3.43–3.47*). Mature spores have a complex structure with an outer coat, cortex, and inner spore membrane surrounding the protoplast (**figure 21.6**). They contain dipicolinic acid, are very heat resistant, and can remain dormant and viable for very long periods (**Microbial Tidbits 21.1**). Although endospore-forming bacteria are distributed widely, they are primarily soil inhabitants. Soil conditions are often extremely variable, and endospores are an obvious advantage in surviving periods of dryness or nutrient deprivation. In one well-documented experiment, spores remained viable for about 70 years. It has also been reported that viable *Bacillus sphaericus* spores have been recovered from Dominican bees that were encased in 25- to 40-million-year-old amber. Usually endospores are observed either in the light microscope after spore staining or by phase-contrast microscopy of unstained cells. They also can be detected by heating a culture at 70 to 80°C for 10 minutes followed by incubation in the proper growth medium. Because only endospores and some thermophiles survive such heating, bacterial growth tentatively confirms their presence. ◄◄ *Bacterial endospores (section 3.8); Preparation and staining of specimens (section 2.3)*

There are two classes of low G + C endospore-forming bacteria: *Clostridia* (the clostridia and relatives), and *Bacilli* (the bacilli and lactobacilli). Each class includes both rods and cocci, and are discussed next.

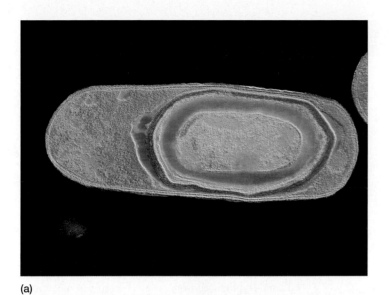

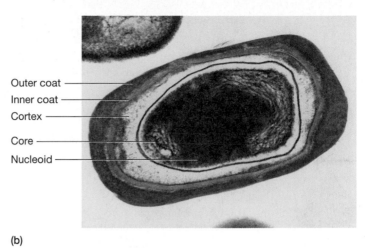

Outer coat
Inner coat
Cortex
Core
Nucleoid

(a)   (b)

Figure 21.6 Bacterial Endospores. (a) A colorized cross section of a *Bacillus subtilis* cell undergoing sporulation. The oval in the center is an endospore that is almost mature; when it reaches maturity, the mother cell will lyse to release it. (b) A cross section of a mature *B. subtilis* spore showing the cortex and spore coat layers that surround the core. The endospore in (a) is 1.3 μm; the spore in (b) is 1.2 μm.

Microbial Tidbits

21.1   Spores in Space

During the nineteenth-century argument over the question of the evolution of life, the panspermia hypothesis became popular. According to this hypothesis, life did not evolve from inorganic matter on Earth but arrived as viable "seeds" that escaped from another planet. The British astronomer Fred Hoyle later revived the hypothesis based on his study of the absorption of radiation by interstellar dust. Hoyle maintained that dust grains were initially viable bacterial cells that had been degraded and that the beginning of life on Earth was due to the arrival of bacterial spores that had survived their trip through space.

More recently Peter Weber and J. Mayo Greenberg from the University of Leiden in the Netherlands have studied the effect of very high vacuum, low temperature, and UV radiation on the survival of *Bacillus subtilis* spores. Their data suggest that spores within an interstellar molecular cloud might be able to survive between 4.5 to 45 million years. Molecular clouds move through space at speeds sufficient to transport spores between solar systems in this length of time. Although these results do not prove the panspermia hypothesis, they are consistent with the possibility that bacteria might be able to travel between planets capable of supporting life.

21.3   CLASS *CLOSTRIDIA*

The class *Clostridia* has a very wide variety of gram-positive bacteria distributed into three orders and 11 families. The characteristics of some of the more important genera are summarized in **table 21.2**. Phylogenetic relationships are shown in figure 21.2.

Desulfotomaculum is an endospore-forming genus that reduces sulfate and sulfite to hydrogen sulfide during anaerobic respiration (**figure 21.7**). Although it stains gram negative, electron microscopic studies have shown that *Desulfotomaculum* has a gram-positive–type cell wall. This concurs with phylogenetic studies that place it with the low G + C gram positives.

The heliobacteria are an excellent example of the diversity in this class. The genera *Heliobacterium* and *Heliophilum* are a group of unusual anaerobic, photosynthetic bacteria characterized by the presence of bacteriochlorophyll *g*. They have a photosystem I–type reaction center like the green sulfur bacteria but have no intracytoplasmic photosynthetic membranes; pigments are contained in the plasma membrane. Like *Desulfotomaculum*, they have a gram-positive–type cell wall with lower than normal peptidoglycan content, and they stain gram negative. Some heliobacteria form endospores. ◄◄ *Phototrophy (section 10.12)*

The phylogenetic placement of the genus *Veillonella* bears mentioning. Although these bacteria stain gram negative, *Bergey's Manual* places them in the family *Acidominococcaceae*, in the order *Clostridiales*. Members of the genus *Veillonella* are anaerobic, chemoheterotrophic cocci ranging in diameter from about 0.3 to 2.5 μm. Usually they are diplococci (often with their adjacent sides flattened), but they may exist as single cells, clusters,

Table 21.2   Characteristics of *Clostridia* and Relatives

Genus	Dimensions (µm), Morphology, and Motility	G + C Content (mol%)	Oxygen Relationship	Other Distinctive Characteristics
Clostridium	0.3–2.0 × 1.5–20; rod-shaped, often pleomorphic, nonmotile or peritrichous flagella	22–55	Anaerobic	Does not carry out dissimilatory sulfate reduction; usually chemoorganotrophic, fermentative, and catalase negative; forms oval or spherical endospores
Desulfotomaculum	0.3–1.5 × 3–9; straight or curved rods, peritrichous or polar flagella	37–50	Anaerobic	Reduces sulfate to H_2S, forms subterminal to terminal endospores; stains gram negative but has gram-positive wall, catalase negative
Heliobacterium	1.0 × 4–10; rods that are frequently bent, gliding motility	52–55	Anaerobic	Photoheterotrophic with bacteriochlorophyll *g;* stains gram negative but has gram-positive wall, some form endospores
Veillonella	0.3–0.5; cocci in pairs, short chains, and masses; nonmotile	36–43	Anaerobic	Stains gram negative; pyruvate and lactate fermented but not carbohydrates; acetate, propionate, CO_2, and H_2 produced from lactate; parasitic in mouths, intestines, and respiratory tracts of animals

or chains. All have complex nutritional requirements and ferment substances such as carbohydrates, lactate and other organic acids, and amino acids to produce gas (CO_2 and often H_2) plus a mixture of volatile fatty acids. They are parasites of homeothermic (warm-blooded) animals.

Like many groups of anaerobic bacteria, members of this genus have not been thoroughly studied. Some species are part of the normal biota of the mouth, the gastrointestinal tract, and the urogenital tract of humans, and other animals. For example, *Veillonella* is plentiful on the tongue surface and dental plaque of humans, and it can be isolated from the vagina. *Veillonella* is unusual in growing well on organic acids such as lactate, pyruvate, and malate while being unable to ferment glucose and other carbohydrates. It is well adapted to the oral environment because it can use the lactic acid produced from carbohydrates by the streptococci and other oral bacteria. *Veillonella* are found in infections of the head, lungs, and the female genital tract, but their precise role in such infections is unclear.

The largest genus in the class *Clostridia* is *Clostridium.* It includes obligately anaerobic, fermentative, gram-positive bacteria that form endospores (**figure 21.8**). The genus contains well over 100 species in several distinct phylogenetic clusters and may be subdivided into several genera in the future. Members of the genus *Clostridium* have great practical impact. Because they are anaerobic and form heat-resistant endospores, they are responsible for many cases of food spoilage, even in canned foods. Clostridia often can ferment amino acids to produce ATP by oxidizing one amino acid and using another as an electron acceptor in a process called the **Stickland reaction** (*see figure 10.18*). This reaction generates ammonia, hydrogen sulfide, fatty acids, and amines during the anaerobic decomposition of proteins. These products are responsible for many unpleasant odors arising during putrefaction.

Although clostridia are industrially valuable (for example, *C. acetobutylicum* is used to manufacture butanol), the pathogenic species that produce toxins are most well known. We

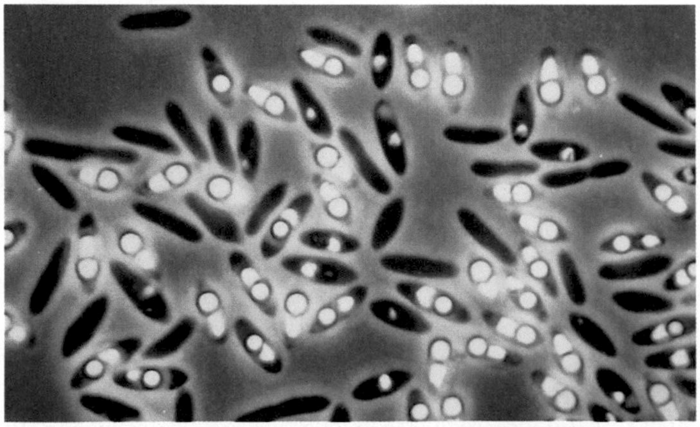

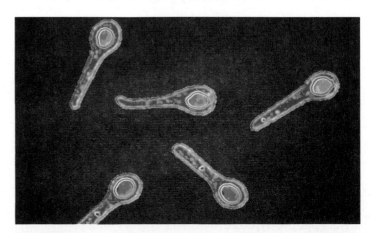

Figure 21.7 *Desulfotomaculum.* *Desulfotomaculum acetoxidans* with spores (bright spheres); phase contrast (×2,000).

Figure 21.8 *Clostridium tetani.* This pathogen makes spores that are round and terminal (×3,000).

discuss three of these species next: *C. botulinum,* which causes botulism; *C. tetani,* the causative agent of tetanus; and *C. perfringens,* which is responsible for gas gangrene and food poisoning.

Botulism

Food-borne **botulism** (Latin *botulus,* sausage) is a form of food poisoning. The most common source of infection is home-canned food that has not been heated sufficiently to kill contaminating *C. botulinum* spores. The spores then germinate, and a toxin is produced during anaerobic vegetative growth. If the food is later eaten without adequate cooking, the active toxin causes disease.

The botulinum toxin is a neurotoxin that binds to the synapses of motor neurons (**figure 21.9**). It selectively cleaves the synaptic vesicle membrane protein synaptobrevin, thus preventing exocytosis (release from the cell) and release of the neurotransmitter acetylcholine. As a consequence, muscles do not contract in response to motor neuron activity, and flaccid

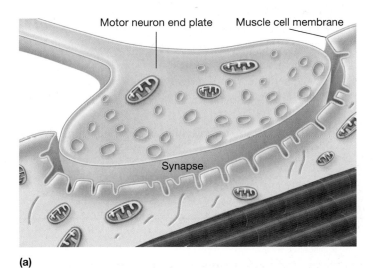

(a)

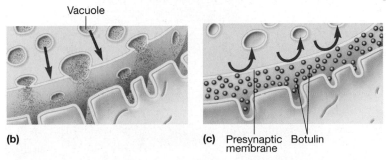

(b)   (c) Presynaptic   Botulin
          membrane

Figure 21.9   The Physiological Effects of Botulism Toxin.   (a) The relationship between the motor neuron and the muscle at the neuromuscular junction. (b) In the normal state, acetylcholine released at the synapse crosses to the muscle and creates an impulse that stimulates muscle contraction. (c) In botulism, the toxin enters the motor end plate and attaches to the presynaptic membrane, where it blocks release of the chemical. This prevents impulse transmission, and keeps the muscle from contracting.

paralysis results. Symptoms of botulism occur within 12 to 72 hours of toxin ingestion and include blurred vision, difficulty in swallowing and speaking, muscle weakness, nausea, and vomiting. Treatment relies on supportive care and polyvalent antitoxin. If untreated, one-third of the patients die of either respiratory or cardiac failure within a few days. Fewer than 100 cases of botulism occur in the United States annually.

Infant botulism is the most common form of botulism in the United States and is confined to children under a year of age. Approximately 100 cases are reported each year. It appears that ingested spores, which may be naturally present in honey or house dust, germinate in the infant's intestine. *C. botulinum* then multiplies and produces the toxin. The infant becomes constipated, listless, generally weak, and eats poorly. Death may result from respiratory failure. It is therefore recommended that infants should not be fed honey.

Tetanus

The spores of *C. tetani,* the causative agent of **tetanus** (Greek *tetanos,* to stretch) are commonly found in hospital environments, in soil and dust, and in the feces of many farm animals and humans (figure 21.8). Transmission to humans is associated with skin wounds. Any break in the skin can allow *C. tetani* spores to enter. If the oxygen tension is low enough, the spores germinate and release the neurotoxin tetanospasmin. Tetanospasmin is an endopeptidase that, like botulism's toxin, selectively cleaves synaptobrevin. However in this case, the release of inhibitory neurotransmitters (gamma-aminobutyric acid and glycine) at synapses within the spinal cord motor nerves is blocked. The result is uncontrolled stimulation of skeletal muscles (spastic paralysis). A second secreted toxin, tetanolysin, is a hemolysin that aids in tissue destruction.

Early in the course of the disease, tetanospasmin causes tension or cramping and twisting in skeletal muscles surrounding the wound and tightness of the jaw muscles. With more advanced disease, there is **trismus** (lockjaw), an inability to open the mouth due to spasm of the masseter muscles. Facial muscles may go into spasms, producing the characteristic expression known as risus sardonicus. Spasms or contractions of the trunk and extremity muscles may be so severe that there is boardlike rigidity, painful tonic convulsions, and opisthotonus—backward bowing of the back so that the heels and back approach each other (**figure 21.10**). Death usually results from spasms of the diaphragm and intercostal respiratory muscles.

The case fatality rate in generalized tetanus ranges from 30 to 90% because tetanus treatment is not very effective. Therefore prevention is all important and depends on (1) active immunization using tetanus toxoid, (2) proper care of wounds contaminated with soil, (3) prophylactic use of antitoxin, and (4) administration of penicillin. Around 50 cases of tetanus are reported annually in the United States, the majority of which are in intravenous drug users.

Figure 21.10   The Toll of Tetanus.   Sir Charles Bell's portrait (c. 1821) of a soldier wounded in the Peninsular War in Spain shows opisthotonus resulting from tetanus.

Gas Gangrene or Clostridial Myonecrosis

Clostridium perfringens, C. novyi, and *C. septicum* are termed the histotoxic clostridia. They produce a necrotizing infection of skeletal muscle called **gas gangrene** (Greek *gangraina,* an eating sore) or clostridial myonecrosis (Greek *myos,* muscle, and *nekrosis,* deadness); however, *C. perfringens* is the most common cause. Analysis of the *C. perfringens* genome sequence reveals that the microbe possesses the genes for fermentation with gas production but lacks genes encoding enzymes for the TCA cycle or a respiratory chain. Nonetheless, *C. perfringens* has an extraordinary doubling time of only 8 to 10 minutes when in the human host.

Histotoxic clostridia occur in the soil worldwide and also are part of the normal endogenous microflora of the human large intestine. Contamination of injured tissue with spores from soil containing histotoxic clostridia or bowel flora is the usual means of transmission. Infections are commonly associated with wounds resulting from abortions, automobile accidents, military combat, or frostbite. If the spores germinate in anoxic tissue, the bacteria grow and secrete α-toxin, which breaks down muscle tissue. Growth often results in the accumulation of gas (mainly hydrogen as a result of carbohydrate fermentation) and of the toxic breakdown products of skeletal muscle tissue (**figure 21.11**). >> *Toxigenicity: Exotoxins (section 30.4)*

Clinical manifestations include severe pain, edema, drainage, and muscle necrosis. The pathology arises from progressive skeletal muscle necrosis

due to the effects of α-toxin, a lecithinase. This disrupts cell membranes, leading to cell lysis. Other enzymes produced by the bacteria degrade collagen and tissue, facilitating spread of the disease.

Gas gangrene is a medical emergency that requires extensive surgical debridement (removal of all dead tissue), the administration of antitoxin, and antimicrobial therapy with penicillin and tetracycline. Hyperbaric oxygen therapy (the use of high concentrations of oxygen at elevated pressures) also is considered effective. The oxygen saturates the infected tissue and thereby prevents the growth of the obligately anaerobic clostridia. Amputation of limbs often is necessary to prevent further spread of the disease.

1. Describe, in a diagram, the chemical composition and structure of the peptidoglycan found in gram-negative bacteria and many gram-positive genera.

2. How do the cell walls of bacilli and most gram-negative bacteria differ from those of gram-positive bacteria such as *S. aureus* with respect to cross-linking frequency?

3. What is a bacterial endospore? Give its most important properties and two ways to demonstrate its presence. Why do you think reports of viable spores from ancient sources have been met with skepticism?

4. Give the general characteristics of *Desulfotomaculum* and the heliobacteria *Veillonella* and *Clostridium.* Briefly discuss why each is interesting or of practical importance.

5. What do you think is the evolutionary significance of the discovery of a photosynthetic genus within the low G + C gram-positive bacteria?

6. Compare and contrast the toxins made by *C. botulinum* and *C. tetani.*

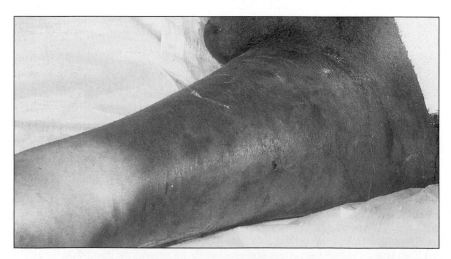

Figure 21.11   Gas Gangrene (Clostridial Myonecrosis).   Necrosis of muscle and other tissues results from the numerous toxins produced by *Clostridium perfringens.*

21.4 CLASS *BACILLI*

The second edition of *Bergey's Manual* gathers a large variety of gram-positive bacteria into one class, *Bacilli,* and two orders, *Bacillales* and *Lactobacillales.* These orders contain 17 families and over 70 gram-positive genera representing cocci, endospore-forming rods and cocci, and nonsporing rods. The biology of some members of the order *Bacillales* is described first, then important representatives of the order *Lactobacillales* are considered. The phylogenetic relationships between some of these organisms are shown in figure 21.2, and the characteristics of selected genera are summarized in **table 21.3**.

Order *Bacillales*

The genus *Bacillus,* family *Bacillaceae,* is the largest in the order *Bacillales.* The genus contains endospore-forming, chemoheterotrophic rods that are usually motile with peritrichous

Table 21.3	Characteristics of Members of the Class *Bacilli*				
Genus	Dimensions (µm), Morphology, and Motility	G + C Content (mol%)	Genome Size (Mb)	Oxygen Relationship	Other Distinctive Characteristics
Bacillus	0.5–2.5 × 1.2–10; straight rods, peritrichous flagella, spore-forming	32–69	4.2–5.4	Aerobic or facultative	Catalase positive; chemoorganotrophic
Caryophanon	1.5–3.0 × 10–20; multicellular rods with rounded ends, peritrichous flagella, nonsporing	41–46	Nd[a]	Aerobic	Acetate only major carbon source; catalase positive; trichome cells have greater width than length, trichomes can be in short chains
Enterococcus	0.6–2.0 × 0.6–2.5; spherical or ovoid cells in pairs or short chains, nonsporing, sometimes motile	34–42	3.2	Facultative	Ferments carbohydrates to lactate with no gas; complex nutritional requirements; catalase negative; occurs widely, particularly in fecal material
Lactobacillus	0.5–1.2 × 1.0–10; usually long, regular rods, nonsporing, rarely motile	32–53	1.9–3.3	Facultative or microaerophilic	Fermentative, at least half the end product is lactate; requires rich, complex media; catalase and cytochrome negative
Lactococcus	0.5–1.2 × 0.5–1.5; spherical or ovoid cells in pairs or short chains, nonsporing, nonmotile	38–40	2.4	Facultative	Chemoorganotrophic with fermentative metabolism; lactate without gas produced; catalase negative; complex nutritional requirements; in dairy and plant products
Leuconostoc	0.5–0.7 × 0.7–1.2; cells spherical or ovoid, in pairs or chains; nonmotile and nonsporing	38–44	Nd	Facultative	Requires fermentable carbohydrate and nutritionally rich medium for growth; fermentation produces lactate, ethanol, and gas; catalase and cytochrome negative
Staphylococcus	0.9–1.3; spherical cells occurring singly and in irregular clusters, nonmotile and nonsporing	30–39	2.5–2.8	Facultative	Chemoorganotrophic with both respiratory and fermentative metabolism; usually catalase positive; associated with skin and mucous membranes of vertebrates
Streptococcus	0.5–2.0; spherical or ovoid cells in pairs or chains, nonmotile and nonsporing	34–46	1.8–2.2	Facultative	Fermentative, producing mainly lactate and no gas; catalase negative; commonly attack red blood cells (α- or β-hemolysis); complex nutritional requirements; commensals or parasites on animals
Thermoactinomyces	0.4–1.0 in diameter; branched, septate mycelium resembles those of actinomycetes	52–54.8	Nd	Aerobic	Usually thermophilic; true endospores form singly on hyphae; numerous in decaying hay, vegetable matter, and compost

[a]Nd: Not determined; genome not yet sequenced.

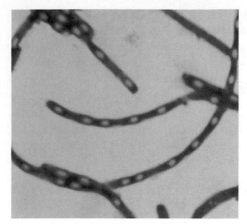

(a) *Bacillus anthracis*

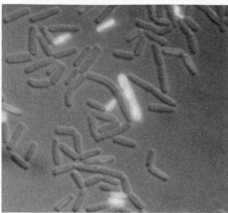

(b) *B. cereus*

Figure 21.12 Bacillus. (a) *B. anthracis,* spores elliptical and central (×1,600). (b) *B. cereus* stained with SYTOX Green nucleic acid stain and viewed by epifluorescence and differential interference contrast microscopy. The cells that glow green are dead.

flagella (**figure 21.12**). It is aerobic, or sometimes facultative, and catalase positive. Many species once included in this genus have been placed in other families and genera based on rRNA sequence data. Some examples of organisms that were formerly in the genus *Bacillus* are *Paenibacillus alvei, P. macerans,* and *P. polymyxa.*

Other changes to the order include the addition of some genera. For instance, the genus *Thermoactinomyces* has historically been classified as an actinomycete because it forms filaments that differentiate from soil-associated substrate hyphae into upwardly growing, aerial hyphae. However, phylogenetic analysis places it with the low G + C microbes in the order *Bacillales,* family *Thermoactinomycetaceae.* Its G + C content (52–55 mol%) is considerably lower than that of the *Actinobacteria.* This genus is thermophilic and grows between 45 and 60°C; it forms single spores on both its aerial and substrate hyphae (**figure 21.13**). *Thermoactinomyces* is commonly found in damp haystacks, compost piles, and other high-temperature habitats. Unlike actinomycete exospores, *Thermoactinomyces* forms true endospores that can survive at 90°C for 30 minutes (figure 21.13*b*). They are formed within hyphae and appear to have typical endospore structure, including the presence of calcium and dipicolinic acid. *Thermoactinomyces vulgaris* from haystacks, grain storage silos, and compost piles is a causative agent of farmer's lung, an allergic disease of the respiratory system. Spores from *Thermoactinomyces vulgaris* were recovered from the mud of a Minnesota lake and found to be viable after about 7,500 years of dormancy.

Bacillus subtilis, the type species for the genus, is the most well-studied gram-positive bacterium. It is nonpathogenic and a terrific model organism for the study of gene regulation, cell division, quorum sensing, and cellular differentiation. Its 4.2-Mb genome was one of the first genomes to be completely sequenced and reveals a number of interesting elements. For instance, several families of genes have been expanded by gene duplication; the largest such family encodes ABC transporters—the most frequent class of protein in *B. subtilis.* There are 18 genes that encode sigma factors. Recall that the use of alternative sigma subunits of RNA polymerase is one way in which bacteria regulate gene expression. In this case, many of the sigma factors govern spore formation and other responses to stressful conditions. The genome contains genes for the catabolism of many diverse carbon sources, antibiotic synthesis, and natural transformation competence. There are at least 10 integrated prophages or remnants of prophages. ◄◄ *Bacterial endospores (section 3.8); Protein maturation and secretion (section 12.8); Global regulations systems: Sporulation in* Bacillus subtilis *(section 13.5); Comparative genomics (section 15.7)*

Many species of *Bacillus* are of considerable importance. Some produce the antibiotics bacitracin, gramicidin, and polymyxin. *B. cereus* (figure 21.12*b*) causes some forms of food poisoning. Several species are used as insecticides. For example, *B. thuringiensis* and *B. sphaericus* form a solid protein crystal, the **parasporal body,** next to their endospores during spore formation (**figure 21.14**). As discussed in chapter 16, the

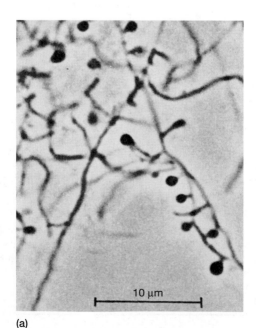

10 μm

(a)

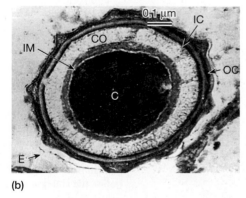

(b)

Figure 21.13 *Thermoactinomyces.* (a) *Thermoactinomyces vulgaris* aerial mycelium with developing endospores at tips of hyphae. (b) Thin section of a *T. sacchari* endospore. E, exosporium; OC, outer spore coat; IC, inner spore coat; CO, cortex; IM, inner forespore membrane; C, core.

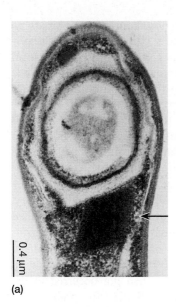

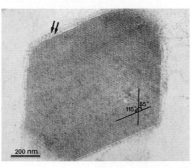

Figure 21.14 The Parasporal Body. (a) An electron micrograph of a *B. sphaericus* sporulating cell containing a parasporal body (arrow) just beneath the endospore. (b) The crystalline parasporal body at a higher magnification. The crystal is surrounded by a two-layered envelope (arrows).

B. thuringiensis parasporal body contains protein toxins that kill over 100 species of moths by dissolving in the alkaline gut of caterpillars and destroying the epithelium. The *B. sphaericus* parasporal body contains proteins toxic for mosquito larvae and may be useful in controlling the mosquitos that carry the malaria parasite *Plasmodium*. *B. anthracis* is the causative agent of the disease anthrax, which can affect both farm animals and humans. << *Microbes as products (section 16.12)*

Anthrax (Greek *anthrax,* coal) is a highly infectious animal disease that can be transmitted to humans by direct contact with infected animals (cattle, goats, sheep) or their products, especially hides. *B. anthracis* spores can remain viable in soil and animal products for decades (figure 21.12*a*). Although *B. anthracis* is one of the most molecularly monomorphic (of one shape) bacteria, it is now possible to separate all known strains into five categories (providing some clues to their geographic sites of origin) based on the number of tandem repeats in various genes (*see figure 15.13*).

Human infection is usually through a cut or abrasion of the skin, resulting in **cutaneous anthrax;** however, inhaling spores may result in **pulmonary anthrax,** also known as woolsorter's disease. If spores reach the gastrointestinal tract, **gastrointestinal anthrax** may result. *B. anthracis* bacteremia can develop from any form of anthrax. The principal virulence factors of *B. anthracis* are encoded on two plasmids—one involved in the synthesis of a polyglutamyl capsule that inhibits phagocytosis and the other bearing the genes for the synthesis of its exotoxins (a complex exotoxin system composed of three proteins: protective antigen [PA], edema factor [EF], and lethal factor [LF]).

For a successful infection, *B. anthracis* must evade the host's innate immune system by killing phagocytic macrophages. Macrophages have many receptors (capillary morphogenesis protein-2) on their plasma membranes to which the PA portion of the anthrax exotoxin system attaches. Attachment continues until seven PA-receptor complexes gather in a doughnut-shaped ring

(**figure 21.15***a*). The ring acts like a syringe, boring through the plasma membrane of the macrophage. It then binds EF and LF, and the entire complex is engulfed by the macrophages' plasma membrane and shuttled to an endosome inside the cell (*see figure 4.9*). Once there, the PA molecules form a pore that pierces the endosomal membrane, and EF and LF enter the cytoplasm. EF has adenylate cyclase activity so intracellular cAMP increases. Toxin activity results in fluid release, or the formation of edema. Additionally LF prevents the transcription factor nuclear factor B (NFκB) from regulating numerous cytokine and other immunity genes needed to promote macrophage survival. As thousands of macrophages die, they release their lysosomal contents, leading to fever, internal bleeding, septic shock, and rapid death.

In humans, more than 95% of naturally occurring anthrax is the cutaneous form. The incubation period for cutaneous anthrax is 1 to 15 days. Infection is initiated with the introduction of the spores through a break in the skin. After ingestion by macrophages at the site of entry, the spores germinate and give rise to vegetative cells, which multiply extracellularly, each forming capsule and exotoxins. Skin infections initially resemble insect bites, then develop into a papular vesicle (figure 21.15*b*), and finally into an ulcer with a necrotic center called an eschar (figure 21.15*c*). The eschar dries and falls off in 1 to 2 weeks with little scarring. Without antibiotic treatment, mortality can be as high as 20%. With proper antibiotic treatment, mortality for cutaneous anthrax is very rare.

In inhalation anthrax, the spores (1 to 2 μm in diameter) are inhaled and lodge in the alveolar spaces, where they are engulfed by alveolar macrophages. The spores survive phagocytosis and germinate within the endosome; the bacteria then spread to regional lymph nodes and eventually the bloodstream. Pulmonary anthrax results in massive pulmonary edema, hemorrhage, and respiratory arrest. Once the bacteria enter the bloodstream, they begin producing exotoxin. The medial lethal inhalation dose for humans is estimated to be about 8,000 spores.

The classic clinical description of inhalation anthrax is that of a two-phase illness. In the initial phase, which follows an incubation period of 1 to 6 days, the disease appears as a nonspecific illness characterized by mild fever, malaise, nonproductive cough, and some chest pain. The second phase begins abruptly and involves a higher fever, acute dyspnea (shortness of breath), and cyanosis (oxygen deficiency). This stage progresses rapidly, with septic shock and associated hypothermia; death occurs within 24 to 36 hours from respiratory failure. Antibiotic treatment is successful only if begun before a critical concentration of toxin has accumulated. One reason this form of anthrax is so difficult to treat is that the symptoms appear after *B. anthracis* has already multiplied and started to produce large amounts of exotoxin. Thus although antibiotics may kill the bacterium or suppress its growth, the exotoxin can eventually kill the patient. Sixteen of the 18 cases of inhalation anthrax reported in the United States between 1900 and 1978 were fatal. Five of the 22 cases in 2001 were fatal.

The symptoms of gastrointestinal anthrax appear 2 to 5 days after the ingestion of undercooked meat containing spores and include nausea, vomiting, fever, and abdominal pain. The

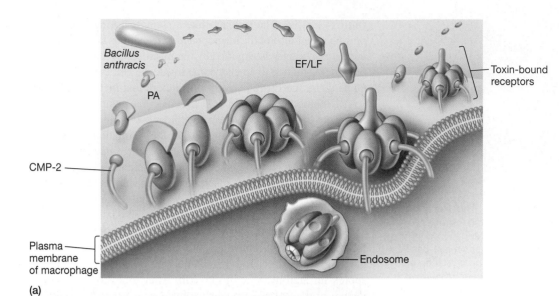

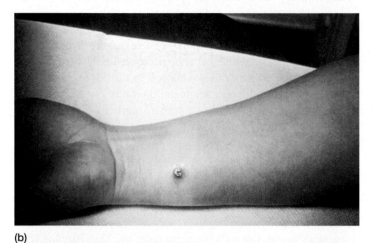

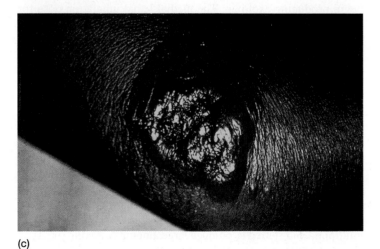

Figure 21.15 Anthrax. (a) A protein called protective antigen (PA) delivers two other proteins, edema factor (EF) and lethal factor (LF), to the capillary morphogenesis protein-2 (CMP-2) receptor on the cell membrane of a target macrophage where PA, EF, and LF are transported to an endosome. PA then delivers EF and LF from the endosome into the cytoplasm of the macrophage where they exert their toxic effects. (b) A cutaneous anthrax papule will ulcerate and necrose into (c) an eschar.

manifestations progress rapidly to severe, bloody diarrhea. The primary lesions are ulcerative, enabling *B. anthracis* to become blood-borne. Mortality is greater than 50 percent.

Between 20,000 and 100,000 cases of anthrax are estimated to occur worldwide annually; in the United States, the annual incidence was 127 cases in the early part of the twentieth century. However, it subsequently declined to less than 1 case per year—a rate maintained for 20 years. Until 2001 a case of inhalation anthrax had not occurred in the United States for more than 20 years. The 2001 occurrence of 22 cases of anthrax has spotlighted the real concern about anthrax as a weapon of bioterrorism.

Vaccination of animals, primarily cattle, is an important control measure. However, people with a high occupational risk, such as those who handle infected animals or their products, including hides and wool, should be immunized. U.S. military personnel also receive the vaccine.

Family *Staphylococcaceae*

The family *Staphylococcaceae* contains four genera, the most important of which is the genus *Staphylococcus*. Members of this genus are facultatively anaerobic, nonmotile, gram-positive cocci, 0.5 to 1.5 μm in diameter, occurring singly, in pairs, and in tetrads, and characteristically dividing in more than one plane to form irregular clusters (**figure 21.16**). They are usually catalase positive and oxidase negative, ferment glucose, and have teichoic acid in their cell walls. Staphylococci cause a variety of human diseases, as we review next.

Staphylococcal Diseases

Staphylococci are among the most important bacteria that cause disease in humans. They are normal inhabitants of the upper respiratory tract, skin, intestine, and vagina. Staphylococci, with pneumococci (*Streptococcus pneumoniae*) and other streptococci, are members of a group of invasive gram-positive bacteria known as the **pyogenic** (or pus-producing) cocci. These bacteria cause various suppurative, or pus-forming, diseases in humans (e.g., boils, carbuncles, folliculitis, impetigo contagiosa, scalded-skin syndrome).

Staphylococci can be divided into pathogenic and relatively nonpathogenic strains based on the synthesis of the enzyme **coagulase.** The coagulase-positive species *S. aureus* is the most important human pathogen in this genus. Coagulase-negative staphylococci (CoNS) such as *S. epidermidis* are nonpigmented and generally less invasive. However, they have increasingly been associated (as opportunistic pathogens) with serious nosocomial infections. ◄◄ *Comparative genomics (section 15.7)*

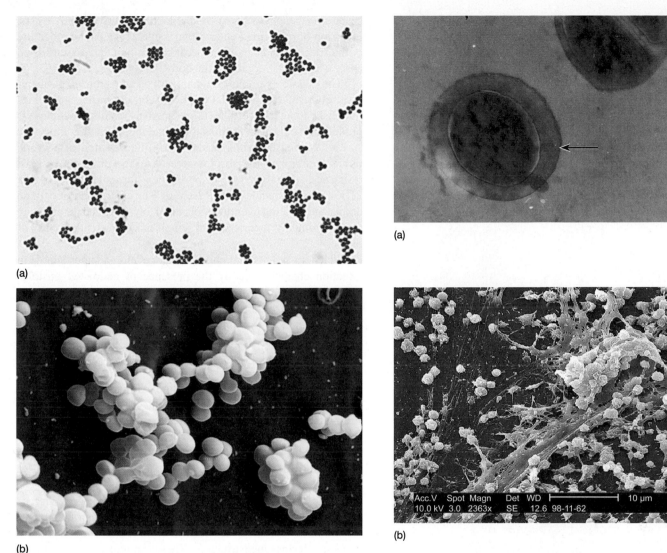

Figure 21.16 *Staphylococcus*. (a) *Staphylococcus aureus*, Gram-stained smear (×1,500). (b) *S. aureus* cocci arranged like clusters of grapes; color-enhanced scanning electron micrograph. Each cell is about 1 µm in diameter.

Figure 21.17 Slime and Biofilms. (a) Cells of *S. aureus,* one of which produces a slime layer (arrowhead; transmission electron microscopy, ×10,000). (b) A biofilm on the inner surface of an intravenous catheter. Extracellular polymeric substances, mostly polysaccharides, surround and encase the staphylococci (scanning electron micrograph, ×2,363).

Staphylococci are further classified into slime producers and non-slime producers. The ability to produce slime has been proposed as a marker for pathogenic strains of staphylococci (**figure 21.17***a*). Slime is a viscous, extracellular glycoconjugate that allows these bacteria to adhere to smooth surfaces such as prosthetic medical devices and catheters. Scanning electron microscopy has clearly demonstrated that biofilms (figure 21.17*b*) consisting of staphylococci encased in a slimy matrix are formed in association with biomaterial-related infections. Slime also appears to inhibit immune cell (neutrophil) chemotaxis, phagocytosis, and certain antimicrobial agents. ≪ *Microbial growth in natural environments: Biofilms (section 7.6)*

Staphylococci, harbored by either an asymptomatic carrier or a person with the disease (i.e., an active carrier), can be spread by the hands, expelled from the respiratory tract, or transported in or on animate and inanimate objects. Staphylococci can produce disease in almost every organ and tissue of the body (**figure 21.18**). For the most part, however, staphyloccal disease occurs in people whose defensive mechanisms have been compromised, such as hospital patients.

Staphylococci produce disease through their ability to multiply and spread widely in tissues and through their production of many virulence factors (**table 21.4**). Some of these factors are exotoxins, and others are enzymes thought to be involved in staphylococcal invasiveness. Many toxin genes are carried on plasmids; in some cases, genes responsible for pathogenicity reside on both a plasmid and the host chromosome. The pathogenic capacity of a particular *S. aureus* strain is due to the combined effect of extracellular factors and toxins, together with the invasive properties of the strain. At one end of the disease spectrum is staphylococcal food poisoning, caused solely by the ingestion of preformed enterotoxin (table 21.4). At the other end of the spectrum are staphylococcal bacteremia and disseminated abscesses in most organs of the body.

The classic example of a staphylococcal lesion is the localized abscess (**figure 21.19a,b**). When *S. aureus* becomes established in a hair follicle, tissue necrosis results. Coagulase is produced and forms a fibrin wall around the lesion that limits the spread. Within the center of the lesion, liquefaction of necrotic tissue occurs, and the abscess spreads in the direction of least resistance. The abscess may be either a furuncle (boil) (figure 21.19c) or a carbuncle (figure 21.19d). The central necrotic tissue drains, and healing eventually occurs. However, the bacteria may spread from any focus through lymph or blood to other parts of the body.

Newborn infants and children can develop a superficial skin infection characterized by the presence of encrusted pustules (figure 21.19e). This disease, called impetigo contagiosa, is caused by *S. aureus* or group A streptococci. It is contagious and can spread rapidly through a nursery or school. It usually occurs in areas where sanitation and personal hygiene are poor.

Staphylococcal scalded skin syndrome (SSSS) is another common staphylococcal disease (figure 21.19f). SSSS is caused by strains of *S. aureus* that produce the exfoliative toxin, or exfoliatin. This protein is usually plasmid encoded, although in some strains the toxin gene is on the bacterial chromosome. In this disease, the epidermis peels off to reveal a red area underneath—thus the name of the disease. SSSS is seen most commonly in infants and children, and neonatal nurseries occasionally suffer large outbreaks of the disease.

Toxic shock syndrome (TSS) is a staphylococcal disease with potentially serious consequences. Most cases of this syndrome have occurred in females who used superabsorbent tampons during menstruation. These tampons can trigger a change in the normal vaginal flora, thereby enabling the growth of toxin-producing *S. aureus*. The toxin associated with this syndrome is also produced in men and in nonmenstruating women by *S. aureus* present at sites other than the genital area (e.g., in surgical wound infections). Toxic shock syndrome is characterized by low blood pressure, fever, diarrhea, an extensive skin rash, and shedding of the skin. TSS results from the massive overproduction of cytokines by T cells induced by the TSST-1 protein (or staphylococcal enterotoxins B and C1). Cytokines are chemical signals produced by immune cells, including T cells that regulate a variety of cellular responses (*see figure 28.23*). TSST-1 stimulates T-cell responses in the absence of specific antigen by binding to both class II MHC receptors and T-cell receptors. TSST-1 and other proteins having this property are called superantigens. Superantigens activate 5 to 30% of the total T-cell population, whereas specific antigens activate only 0.01 to 0.1% of the T-cell population. The net effect of cytokine overproduction is circulatory collapse leading to shock and multiorgan failure. Tumor necrosis factor α and interleukins 1 and 6 are strongly associated with superantigen-induced shock. Mortality rates for TSS are 30 to 70%, and morbidity due to surgical

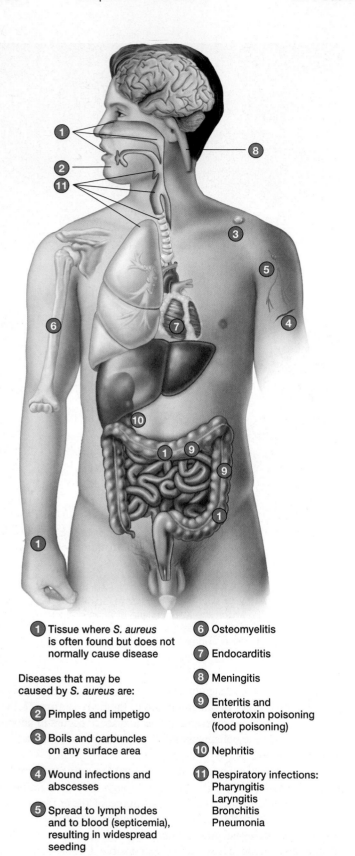

Figure 21.18 Staphylococcal Diseases. The anatomical sites of the major staphylococcal infections of humans are indicated by the corresponding numbers.

1 Tissue where *S. aureus* is often found but does not normally cause disease

Diseases that may be caused by *S. aureus* are:

2 Pimples and impetigo

3 Boils and carbuncles on any surface area

4 Wound infections and abscesses

5 Spread to lymph nodes and to blood (septicemia), resulting in widespread seeding

6 Osteomyelitis

7 Endocarditis

8 Meningitis

9 Enteritis and enterotoxin poisoning (food poisoning)

10 Nephritis

11 Respiratory infections: Pharyngitis Laryngitis Bronchitis Pneumonia

Table 21.4	Various Enzymes and Toxins Produced by Staphylococci
Product	**Physiological Action**
β-lactamase	Breaks down penicillin
Catalase	Converts hydrogen peroxide into water and oxygen, and reduces killing by phagocytosis
Coagulase	Reacts with prothrombin to form a complex that can cleave fibrinogen and cause the formation of a fibrin clot; fibrin may also be deposited on the surface of staphylococci, which may protect them from destruction by phagocytic cells; coagulase production is synonymous with invasive pathogenic potential.
DNase	Destroys DNA
Enterotoxins	Are divided into heat-stable toxins of six known types (A, B, C1, C2, D, E); responsible for the gastrointestinal upset typical of food poisoning
Exfoliative toxins A and B	Cause loss of the surface layers of the skin in scalded skin syndrome; superantigens
Hemolysins	Alpha hemolysin destroys erythrocytes and causes skin destruction. Beta hemolysin destroys erythrocytes and sphingomyelin around nerves.
Hyaluronidase	Also known as spreading factor; breaks down hyaluronic acid between cells, allowing for penetration and spread of bacteria
Panton-Valentine leukocidin	Inhibits phagocytosis by granulocytes and can destroy these cells by forming pores in their phagosomal membranes
Lipases	Break down lipids
Protein A	Is antiphagocytic by competing with neutrophils for the Fc portion of specific opsonins
Proteases	Break down proteins
Toxic shock syndrome toxin-1	Is associated with the fever, shock, and multisystem involvement of toxic shock syndrome; a superantigen

debridement and amputation is very high. Approximately 150 cases of toxic shock syndrome are reported annually in the United States. >> *Chemical mediators in nonspecific (innate) resistance: Cytokines (section 28.6); T-cell biology: Superantigens (section 29.5)*

Staphylococcal food poisoning is the major type of food intoxication in the United States. It is caused by ingestion of improperly stored or cooked food (particularly foods such as ham, processed meats, chicken salad, pastries, ice cream, and hollandaise sauce) in which *S. aureus* has been allowed to incubate. Because staphylococci produce heat-stable enterotoxins, the food can be extensively and properly cooked, killing the bacteria without destroying the toxin. Thirteen different enterotoxins have been identified; enterotoxins A, B, C1, C2, D, and E are the most common. Enterotoxins A and B are superantigens. Typical symptoms include severe abdominal pain, cramps, diarrhea, vomiting, and nausea. The onset of symptoms is rapid (usually 1 to 8 hours) and of short duration (usually less than 24 hours). The mortality rate of staphylococcal food poisoning is negligible among healthy individuals. >> *Toxigenicity (section 30.4)*

Family *Listeriaceae*

The family *Listeriaceae* includes two genera: *Brochothrix* and *Listeria*. *Brochothrix* is most commonly found in meat but is not pathogenic. However, as discussed next, *Listeria* is an important

cause of disease. The genus contains short rods that are aerobic or facultative, catalase positive, and motile by peritrichous flagella.

Listeriosis

Listeria monocytogenes can be isolated from soil, vegetation, and many animal reservoirs. Human disease due to *L. monocytogenes* generally occurs in pregnant or immunosuppressed people. A substantial number of cases of human **listeriosis** are attributable to the food-borne transmission of *L. monocytogenes*. *Listeria* outbreaks have been traced to sources such as contaminated milk, soft cheeses, vegetables, and meat. This microbe is a particular threat because it grows at temperatures used for refrigeration (4–6°C). Unlike the gastrointestinal illnesses of many of the food-borne pathogens, *L. monocytogenes* causes invasive syndromes such as meningitis, sepsis, and stillbirth.

L. monocytogenes is an intracellular pathogen, a characteristic consistent with its predilection for causing illness in persons with deficient cell-mediated immunity. This bacterium can be found as part of the normal gastrointestinal microbiota in healthy individuals. In immunosuppressed individuals, invasion, intracellular multiplication, and cell-to-cell spread of the bacterium appear to be mediated through proteins such as internalin, the hemolysin listeriolysin O, and phospholipase C. *L. monocytogenes* also uses host cell actin filaments to move within and between cells (*see*

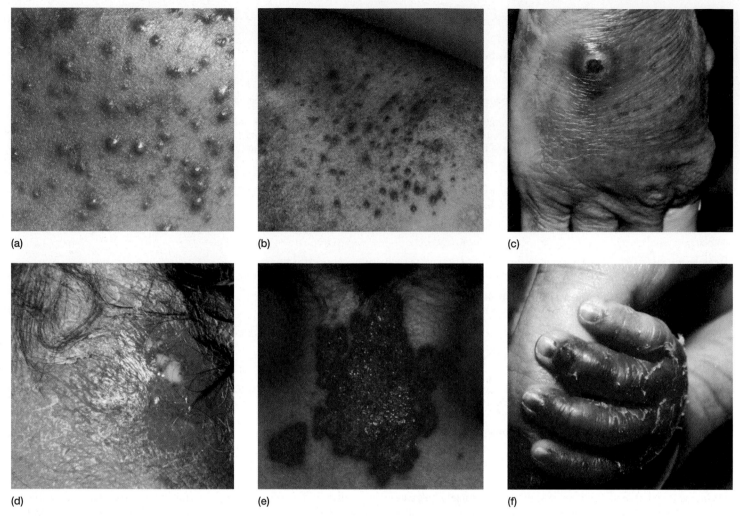

(a)

(b)

(c)

(d)

(e)

(f)

Figure 21.19   Staphylococcal Skin Infections.   (a) Superficial folliculitis in which raised, domed pustules form around hair follicles. (b) In deep folliculitis, the microorganism invades the deep portion of the follicle and dermis. (c) A furuncle arises when a large abscess forms around a hair follicle. (d) A carbuncle consists of a multilocular abscess around several hair follicles. (e) Impetigo on the neck of two-year-old male. (f) Scalded skin syndrome in a one-week-old premature male infant. Reddened areas of skin peel off, leaving scalded-looking moist areas.

figure 30.4). The increased risk of infection in pregnant women may be due to both systemic and local immunological changes associated with pregnancy. For example, local immunosuppression at the maternal-fetal interface of the placenta may facilitate intrauterine infection following transient maternal bacteremia.

Order *Lactobacillales*

Many members of the order *Lactobacillales* produce lactic acid as their major or sole fermentation product and are sometimes collectively called **lactic acid bacteria.** *Streptococcus, Enterococcus, Lactococcus, Lactobacillus,* and *Leuconostoc* are all members of this group. Lactic acid bacteria do not form spores and are usually nonmotile. They depend on sugar fermentation for energy. They lack cytochromes and obtain energy by substrate-level phosphorylation rather than by electron transport and oxidative phosphorylation. Nutritionally, they are fastidious, and many vitamins, amino acids, purines, and pyrimidines must be supplied because of their limited biosynthetic capabilities. Lactic acid bacteria usually are catego-

rized as facultative anaerobes, but some classify them as aerotolerant anaerobes. **<<** *Influences of environmental factors on growth: Oxygen concentration (section 7.5); Fermentation (section 10.7)*

The largest genus in this order is *Lactobacillus,* with about 100 species. *Lactobacillus* includes rods and some coccobacilli (**figure 21.20**). All lack catalase and cytochromes and produce lactic acid as their main or sole fermentation product (*see figure 10.17*). Lactobacilli carry out either **homolactic fermentation** using the Embden-Meyerhof pathway or **heterolactic fermentation** with the phosphoketolase pathway. They grow optimally under slightly acidic conditions, at a pH between 4.5 and 6.4. The genus is found on plant surfaces and in dairy products, meat, water, sewage, beer, fruits, and many other materials. Lactobacilli also are part of the normal flora of the human body in the mouth, intestinal tract, and vagina. They are rarely pathogenic.

Lactobacillus is indispensable to the food and dairy industry. Lactobacilli are used in the production of fermented vegetable foods (sauerkraut, pickles, silage), sourdough bread, Swiss and other hard cheeses, yogurt, and sausage. Yogurt is probably the most popular

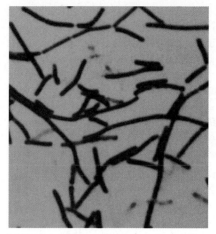

(a) *Lactobacillus acidophilus*

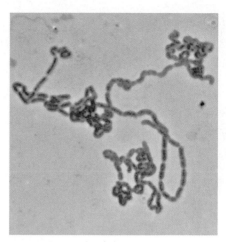

(b) *L. delbrueckii* subsp. *bulgaricus*

Figure 21.20 Lactobacillus. (a) *L. acidophilus* (×1,000). (b) *L. delbrueckii* subspecies *bulgaricus;* phase contrast (×600).

fermented milk product in the United States and, as described in chapter 34, *Streptococcus thermophilus* and *Lactobacillus delbrueckii* subspecies *bulgaricus* are used in its production.

At least one species, *L. acidophilus,* is sold commercially as a probiotic agent that may provide some health benefits for the consumer. On the other hand, some lactobacilli also create problems. They sometimes are responsible for spoilage of beer, wine, milk, and meat because their metabolic end products contribute undesirable flavors and odors.

Leuconostoc, family *Leuconostocaceae,* contains facultative gram-positive cocci, which may be elongated or elliptical and arranged in pairs or chains (**figure 21.21**). Leuconostocs lack catalase and cytochromes and carry out heterolactic fermentation by converting glucose to D-lactate and ethanol or acetic acid by means of the phosphoketolase pathway (**figure 21.22**). They can be isolated from plants, silage, and milk. The genus is used in wine production, in the fermentation of vegetables such as cabbage (sauerkraut) and cucumbers (pickles), and in the manufacture of buttermilk, butter, and cheese. *L. mesenteroides* synthesizes dextrans from sucrose and is important in industrial dextran production. *Leuconostoc* species are involved in food spoilage and tolerate high sugar concentra-

tions so well that they grow in syrup and are a major problem in sugar refineries.

Enterococcaceae and *Streptococcaceae* are important families of chemoheterotrophic, mesophilic, nonsporing cocci. They occur in pairs or chains when grown in liquid media (**figure 21.23**) and usually are nonmotile. They all ferment sugars to produce lactic acid but no gas—that is, they carry out homolactic fermentation. A few species are anaerobic rather than facultative. The enterococci such as *E. faecalis* are normal residents of the intestinal tracts of humans and most other animals. *E. faecalis* is an opportunistic pathogen that can cause urinary tract infections

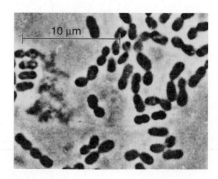

Figure 21.21 Leuconostoc. *Leuconostoc mesenteroides;* phase-contrast micrograph.

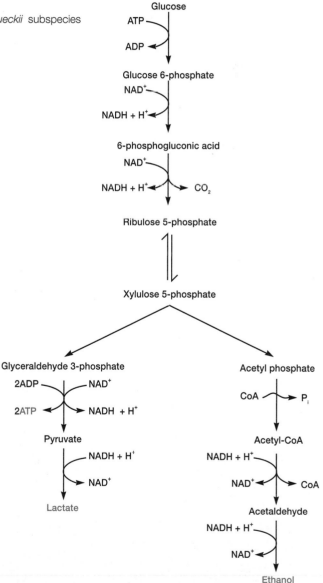

Figure 21.22 Heterolactic Fermentation and the Phosphoketolase Pathway. The phosphoketolase pathway converts glucose to lactate, ethanol, and CO_2.

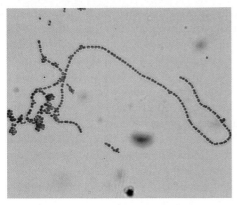

(a) *Streptococcus pyogenes*

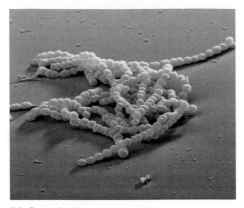

(b) *S. agalactiae*

Figure 21.23 Streptococcus.   (a) *Streptococcus pyogenes* (×900). (b) *Streptococcus agalactiae,* the cause of Group B streptococcal infections. Note the long chains of cells; color-enhanced scanning electron micrograph (×4,800).

and endocarditis. Enterococci are major agents in the horizontal transfer of antibiotic-resistance genes.

The family *Streptococcaceae* includes only two genera: *Lactococcus* and *Streptococcus. Lactococcus* spp. ferment sugars to lactic acid and can grow at 10°C but not at 45°C. *L. lactis* is widely used in the production of buttermilk and cheese because it can curdle milk and add flavor through the synthesis of diacetyl and other products. **>>** *Microbiology of fermented foods (section 34.6)*

The genus *Streptococcus* is large and complex. All species in the genus are faculatively anaerobic and catalase negative. They have been divided into three groups: pyogenic streptococci, oral streptococci, and other streptococci. Many bacteria originally placed within the genus have been moved to two other genera, *Enterococcus* and *Lactococcus.* Some major characteristics of these three genera are summarized in **table 21.5**. **Table 21.6** lists a few properties of selected genera.

Many characteristics are used to identify these cocci. One of their most important taxonomic characteristics is the ability to lyse erythrocytes when growing on blood agar, an agar medium

Table 21.5	Classification of the Streptococci, Entercocci, and Lactococci		
Characteristics	**Streptococcus**	**Enterococcus**	**Lactococcus**
Predominant arrangement (most common first)	Chains, pairs	Pairs, chains	Pairs, short chains
Capsule/slime layer	+	−	−
Habitat	Mouth, respiratory tract	Gastrointestinal tract	Dairy products
Growth at 45°C	Variable	+	−
Growth at 10°C	Variable	Usually +	+
Growth at 6.5% NaCl broth	Variable	+	+
Growth at pH 9.6	Variable	+	−
Hemolysis	Usually β (pyogenic) or α (oral)	α, β, −	Usually −
Serological group (Lancefield)	Variable (A–O)	Usually D	Usually N
Mol% G + C	34–46	34–42	38–40
Representative species	Pyogenic streptococci	*E. faecalis*	*L. lactis*
	S. agalactiae, S. pyogenes, S. equi, S. dysgalactiae	*E. faecium*	*L. raffinolactis*
	Oral streptococci	*E. avium*	*L. plantarum*
	S. gordonil, S. salvarius, S. sanguis, S. oralis	*E. durans*	
	S. pneumoniae, S. mitis, S. mutans	*E. gallinarum*	
	Other streptococci		
	S. bovis, S. thermophilus		

| Table 21.6 | Properties of Selected Streptococci and Relatives |

	Pyogenic Streptococci		Oral Streptococci		Enterococci	Lactococci
Characteristics	*S. pyogenes*	*S. pneumoniae*	*S. sanguis*	*S. mutans*	*E. faecalis*	*L. lactis*
Growth at 10°C	−[a]	−	−	−	+	+
Growth at 45°C	−	−	d	d	+	−
Growth at 6.5% NaCl	−	−	−	−	+	−
Growth at pH 9.6	−	−	−	−	+	−
Growth with 40% bile	−	−	d	d	+	+
α-hemolysis	−	+	+	−	−	d
β-hemolysis	+	−	−	−	+	−
Arginine hydrolysis	+	+	+	−	+	d
Hippurate hydrolysis	−	−	−	−	+	d
Mol% G + C of DNA	35–39	30–39	40–46	36–38	34–38	39
Genome size (Mb)	1.8	2.2	Nd[b]	2.0	3.2	2.3

Modified from *Bergey's manual of systematic bacteriology,* vol. 2, ed. P. H. A. Sneath, et al. Copyright © 1986 Williams and Wilkins, Baltimore, MD. Reprinted by permission.
[a]Symbols: +, 90% or more of strains positive; −, 10% or less of strains positive; d, 11–89% of strains are positive.
[b]Nd: Not determined; genome not yet sequenced.

containing 5% sheep or horse blood (**figure 21.24**). In **α-hemolysis,** a 1 to 3 mm greenish zone of incomplete hemolysis forms around the colony; **β-hemolysis** is characterized by a clear zone of complete lysis. In addition, other hemolytic patterns are sometimes seen. Serological studies are also very important in identification because these genera often have distinctive cell wall antigens. Polysaccharide and teichoic acid antigens found in the cell wall or between the wall and the plasma membrane are used to identify these cocci, particularly pathogenic β-hemolytic streptococci, by the **Lancefield grouping system.** Biochemical and physiological tests are essential in identification (e.g., growth temperature preferences, carbohydrate fermentation patterns, acetoin production, reduction of litmus milk, sodium chloride and bile salt tolerance, and the ability to hydrolyze arginine, esculin, hippurate, and starch). Sensitivity to bacitracin, sulfa drugs, and optochin (ethylhydrocuprein) also are used to identify particular species. Some of these techniques are being replaced by molecular genetic approaches such as multilocus sequence typing (MSLT). >> *Clinical microbiology and immunology (chapter 32);* << *Techniques for determining microbial taxonomy and phylogeny: Molecular characteristics (section 17.4)*

Pathogenic streptococci are commonly called strep; *Streptococcus pyogenes* (group A β-hemolytic streptococci) is one of the most important bacterial pathogens. The different serotypes of *group A streptococci* (**GAS**) produce (1) extracellular enzymes that break down host molecules; (2) streptokinases, enzymes that activate a host-blood factor that dissolves blood clots; (3) the cytolysins streptolysin O and streptolysin S, which kill host leukocytes; and (4) capsules and M protein, which help to retard phagocytosis (**figure 21.25**). M protein is the major virulence factor of GAS. It is a filamentous protein anchored in the streptococcal cell membrane, which facilitates attachment to host cells and prevents opsonization by complement protein C3b. M protein types 1, 3, 12, and 28 are commonly found in patients with streptococcal toxic shock and multiorgan failure. >> *Chemical mediators in nonspecific (innate) resistance: Complement (section 28.6)*

S. pyogenes is widely distributed among humans; some become asymptomatic carriers. Individuals with acute infections may spread the pathogen, and transmission can occur through respiratory droplets as direct or indirect contact. Vaccines are not available for streptococcal diseases other than streptococcal pneumonia because of the large number of serotypes. The best control measure is prevention of transmission. In the following sections, some of the more important human streptococcal diseases are discussed.

Impetigo and Erysipelas

The most frequently diagnosed skin infection caused by *S. pyogenes* is **impetigo** (impetigo also can be caused by *Staphylococcus aureus*). Impetigo is a superficial cutaneous infection, most commonly seen in children, usually located on the face, and characterized by crusty lesions and vesicles surrounded by a red

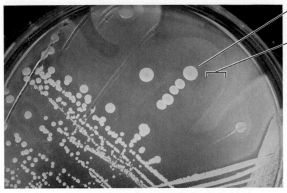

Inner zone of
hemolysis

Outer zone of
hemolysis

(a)

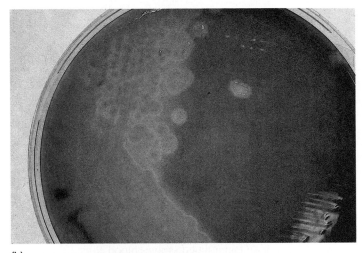

(b)

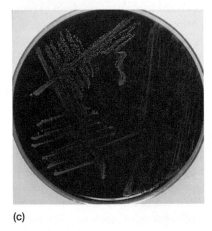

(c)

**Figure 21.24   Staphylococcal and Streptococcal Hemolytic
Patterns.**   (a) *Staphylococcus aureus* on blood agar, illustrating β-hemolysis.
(b) *Streptococcus pneumoniae* on blood agar, illustrating α-hemolysis.
(c) *Staphylococcus epidermidis* on blood agar with no hemolysis.

border (figure 21.19*e*). Impetigo is most common in late summer
and early fall. The drug of choice for impetigo is penicillin.

Erysipelas (Greek *erythros*, red, and *pella*, skin) is an acute
infection and inflammation of the dermal layer of the skin. It

occurs primarily in infants and people over thirty years of age
with a history of streptococcal sore throat. The skin often develops
painful reddish patches that enlarge and thicken with a sharply
defined edge (**figure 21.26**). Recovery usually takes a week or
longer if no treatment is given. The drugs of choice for the
treatment of erysipelas are erythromycin and penicillin. Erysipelas
may recur periodically at the same body site for years.

Invasive Streptococcus A Infections

In the nineteenth century, invasive *S. pyogenes* infections were a
major cause of morbidity and mortality. However, during the
twentieth century, the incidence of severe group A streptococcal
infections declined, especially with the arrival of antibiotic therapy. However, in the mid-1980s there was a worldwide increase
in group A streptococcal sepsis; clusters of rheumatic fever were
reported from locations within the United States, and a streptococcal toxic shocklike syndrome emerged. In fact, a virulent
"strep A" infection killed *Sesame Street* muppeteer Jim Henson
in 1990, and in 1994 the bacterium made headlines with articles
on "the flesh-eating invasive disease."

The development of invasive strep A disease appears to
depend on the presence of specific virulent strains (M-1 and M-3
serotypes, e.g.) and predisposing host factors (surgical or nonsur-
gical wounds, diabetes, and other underlying medical problems).
A life-threatening infection begins when invasive strep A strains
penetrate a mucous membrane or take up residence in a wound
such as a bruise. This infection can quickly lead either to **necrotizing fasciitis** (Greek *nekrosis*, deadness, Latin *fascis*, band or
bandage, and *–itis*, inflammation), which destroys the sheath covering skeletal muscles, or to **myositis** (Greek *myos*, muscle, and
–itis), the inflammation and destruction of skeletal muscle and fat
tissue (**figure 21.27**). Because necrotizing fasciitis and myositis
arise and spread so quickly, they have been colloquially called
"galloping gangrene."

Rapid treatment is necessary to reduce the risk of death, and
penicillin G remains the treatment of choice. In addition, surgical
removal of dead and dying tissue may be needed in more advanced
cases of necrotizing fasciitis. It is estimated that approximately
10,000 cases of invasive strep A infections occur annually in the
United States, and between 5 and 10% of them are associated
with necrotizing conditions.

One reason invasive strep A strains are so deadly is that about
85% of them carry the genes for the production of *s*treptococcal
*p*yrogenic *e*xotoxins A and B (Spe exotoxins). Exotoxin A acts as
a superantigen, stimulating T cells to produce abnormally large
quantities of cytokines. These cytokines damage the endothelial
cells that line blood vessels, causing fluid loss and rapid tissue
death from a lack of oxygen. Another pathogenic mechanism
involves the secretion of exotoxin B, a cysteine protease (a proteolytic enzyme that has a cysteine residue in the active site).
This protease rapidly destroys tissue by breaking down
proteins. ▶▶ *T-cell biology: Superantigens (section 29.5)*

Since 1986 it has been recognized that invasive strep A infections can also trigger a ***toxic shocklike syndrome* (TSLS),**

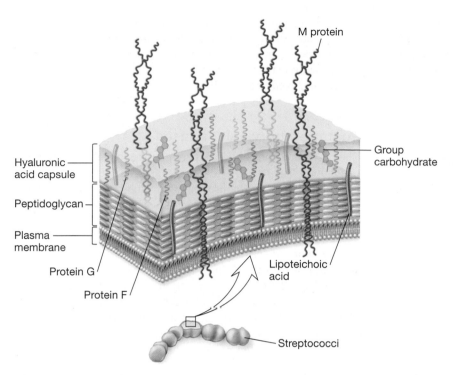

Figure 21.25 Streptococcal Cell Envelope. The M protein is a major virulence factor for streptococci. It facilitates bacterial attachment to host cells and has antiphagocytic activity. Protein G prevents attack by antibodies because it binds to the Fc portion, preventing the antigen-binding site from bacterial capture. Protein F is an epithelial cell attachment factor.

Because group A streptococci are less contagious than cold or flu viruses, infected individuals do not pose a major threat to people around them. The best preventive measures are simple ones such as covering food, washing hands, and cleansing and medicating wounds.

Streptococcal Pharyngitis

Streptococcal pharyngitis is one of the most common bacterial infections of humans and is commonly called strep throat. The β-hemolytic, group A streptococci are spread by droplets of saliva or nasal secretions. The incubation period in humans is 2 to 4 days.

The action of the strep bacteria in the throat (pharyngitis) or on the tonsils (tonsillitis) stimulates an inflammatory response and the lysis of white and red blood cells. An inflammatory exudate consisting of cells and fluid is released from the blood vessels and deposited in the surrounding tissue, although only about 50% of patients with strep pharyngitis present with an exudate. This is accompanied by a general feeling of discomfort or malaise, fever, and headache. Prominent physical manifestations include redness, edema, and lymph node enlargement in the throat. Signs and symptoms alone are not diagnostic because viral infections have a similar presentation. Several common rapid test kits are available for diagnosing strep throat. In the absence of complications, the disease can be self-limiting and may disappear within a week. However, antibiotic treatment (penicillin G benzathine or a macrolide antibiotic for penicillin-allergic people) can shorten the

characterized by a precipitous drop in blood pressure, failure of multiple organs, and a very high fever. TSLS is caused by an invasive strep A that produces one or more of the streptococcal pyrogenic exotoxins. TSLS has a mortality rate of over 30%.

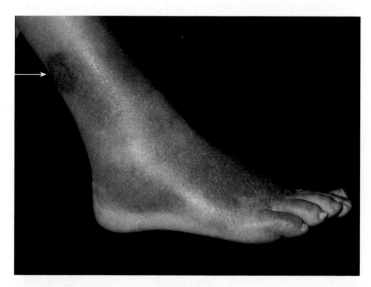

Figure 21.26 Erysipelas. Notice the bright, raised, rubbery lesion at the site of initial entry (white arrow) and the spread of the inflammation to the foot. The reddening is caused by toxins produced by the streptococci as they invade new tissue.

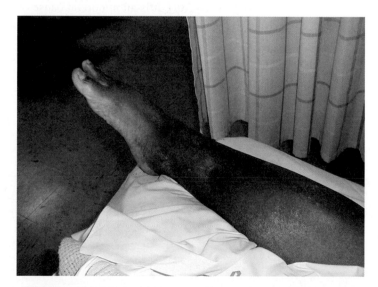

Figure 21.27 Necrotizing Fasciitis. Rapidly advancing streptococcal disease can lead to large, necrotic sites, sometimes with blisters that rupture and expose the dying tissue. This is often called flesh-eating disease or necrotizing fasciitis.

infection and clinical syndromes, and is especially important in children for the prevention of complications such as rheumatic fever and glomerulonephritis, as discussed shortly. Infections in older children and adults tend to be milder and less frequent due in part to the immunity they have developed against the many serotypes encountered in early childhood.

Streptococcal Pneumonia

Streptococcal pneumonia is considered an opportunistic infection—that is, it is contracted from one's own normal microbiota. It is caused by *S. pneumoniae,* normally found in the upper respiratory tract. However, disease usually occurs only in those individuals with predisposing factors such as viral infections of the respiratory tract, physical injury to the tract, alcoholism, or diabetes. About 60 to 80% of all respiratory diseases known as pneumonia are caused by *S. pneumoniae.* An estimated 150,000 to 300,000 people in the United States contract this form of pneumonia annually, and between 13,000 to 66,000 deaths result.

The primary virulence factor of *S. pneumoniae* is its capsular polysaccharide, which is composed of hyaluronic acid. The production of large amounts of hyaluronic capsular polysaccharide plays an important role in protecting the organism from ingestion and killing by phagocytes. Pathogenesis is due to the rapid multiplication of the bacteria in alveolar spaces (**figure 21.28***a*). The bacteria also produce the toxin pneumolysin, which destroys host cells. The alveoli fill with blood cells and fluid, and become inflamed. The sputum is often rust-colored because of blood coughed up from the lungs. The onset of symptoms is usually abrupt, with chills; hard, labored breathing; and chest pain. Antibiotic therapy (cefotaxime, ofloxacin, and ceftriaxone) has contributed to a greatly reduced mortality rate.

S. pneumoniae is also associated with sinusitis, conjunctivitis, and otitis media (figure 21.28*b*). It is an important cause of bacteremia (blood infections) and meningitis. Penicillin- and tetracycline-resistant strains of *S. pneumoniae* are now in the United States. Pneumococcal vaccines are available for people who are at greater risk for exposure (e.g., college students and people in chronic-care facilities).

Poststreptococcal Diseases

The poststreptococcal diseases are glomerulonephritis and rheumatic fever. They occur 1 to 4 weeks after an acute streptococcal infection. These two nonsuppurative (non-pus-producing) diseases are the most serious problems associated with streptococcal infections in the United States.

Glomerulonephritis, also called **Bright's disease,** is an inflammatory disease of the renal glomeruli—membranous structures within the kidney where blood is filtered. Damage probably results from the deposition of antigen-antibody complexes, possibly involving the streptococcal M protein, in the glomeruli. The complexes cause destruction of the glomerular membrane, allowing proteins and blood to leak into the urine. Clinically the affected person exhibits edema, fever, hypertension, and hematuria (blood in the urine). The disease occurs primarily among schoolage children. The incidence of glomerulonephritis in the United States is less than 0.5% of streptococcal infections. Penicillin G or erythromycin can be given for any residual streptococci. However, there is no specific therapy once kidney damage has occurred. About 80 to 90% of all cases undergo slow, spontaneous healing of the damaged glomeruli, whereas the others develop a chronic form of the disease. The latter may require a kidney transplant or lifelong renal dialysis. **>>** *Immune disorders: Type III hypersensitivity (section 29.11)*

Rheumatic fever is an autoimmune disease characterized by inflammatory lesions involving the heart valves, joints, subcuta-

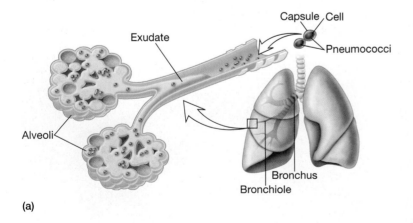

(a)

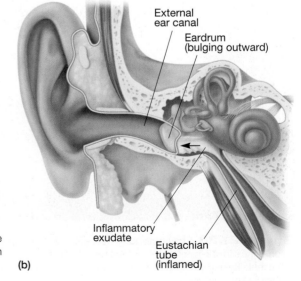

(b)

Figure 21.28 **Steptococcal Infections.** (a) The lower lung showing bronchiole and alveoli, where streptococci can cause pneumonia. (b) Streptococci can also travel through the Eustacian tube to enter the middle ear, leading to infection (otitis media).

neous tissues, and central nervous system. It usually results from a prior streptococcal pharyngitis. The exact mechanism of rheumatic fever development remains unknown. However, it has been associated with specific M strains. The disease occurs most frequently among children six to fifteen years of age and manifests itself through a variety of signs and symptoms, making diagnosis difficult. In the United States, rheumatic fever has become very rare (less than 0.05% of streptococcal infections). It occurs 100 times more frequently in tropical countries. Although rheumatic fever is rare, it is still the most common cause of permanent heart valve damage in children.

1. Briefly describe the genus *Thermoactinomyces*, with particular emphasis on its unique features. What disease does it cause?

2. List the major properties of the genus *Bacillus*. What practical impacts does it have on society? Define parasporal body.

3. Describe the genus *Staphylococcus*. On what basis are pathogenic and nonpathogenic species and strains distinguished?

4. List the major properties of the genus *Lactobacillus*. In what ways is it important in the food and dairy industries?

5. Describe the major distinguishing characteristics of the following taxa: *Streptococcus, Enterococcus, Lactococcus,* and *Leuconostoc*.

6. Of what practical importance is *Leuconostoc?* What are lactic acid bacteria?

7. What are α-hemolysis, β-hemolysis, and the Lancefield grouping system?

8. Describe the streptococcal exotoxins and how they are involved in virulence.

9. Which streptococcal disease is most prevalent? Why do you think this is the case?

Summary

21.1 Class *Mollicutes* (The Mycoplasmas)

a. Mycoplasmas stain gram-negative because they lack cell walls and cannot synthesize peptidoglycan precursors; however, they are phylogenetically related to the gram-positive bacteria (**figures 21.2** and **21.3**).

b. Many mycoplasmas require sterols for growth. They are some of the smallest bacteria capable of self-reproduction and usually grow on agar to give colonies a "fried egg" appearance (**figure 21.4**).

21.2 Peptidoglycan and Endospore Structure

a. Peptidoglycan structure often differs between groups in taxonomically useful ways. Most variations are in amino acid 3 of the peptide subunit or in the interpeptide bridge (**figure 21.5**).

b. Endospores resist desiccation and heat; they are used by bacteria to survive harsh conditions, especially in the soil (**figure 21.6**).

21.3 Class *Clostridia*

a. *Desulfotomaculum* is an anaerobic, endospore-forming genus that reduces sulfate to sulfide during anaerobic respiration (**table 21.2**).

b. The heliobacteria are anaerobic, photosynthetic bacteria with bacteriochlorophyll *g*. Some form endospores.

c. The family *Veillonellaceae* contains anaerobic cocci that stain gram-negative. Some are parasites of warm-blooded animals.

d. Members of the genus *Clostridium* are anaerobic gram-positive rods that form endospores and do not carry out dissimilatory sulfate reduction. They are responsible for food spoilage, botulism, tetanus, and gas gangrene.

21.4 Class *Bacilli*

a. The class *Bacilli* is divided into two orders: *Bacillales* and *Lactobacillales*.

b. *Thermoactinomyces* is a thermophile that forms a mycelium and true endospores (**figure 21.13**). It causes allergic reactions and leads to farmer's lung.

c. The genus *Bacillus* contains aerobic and facultative, catalase-positive, endospore-forming, chemoheterotrophic rods that are usually motile with peritrichous flagella. Species of *Bacillus* synthesize antibiotics and insecticides, and cause food poisoning and anthrax (**figure 21.15**).

d. Members of the genus *Staphylococcus* are facultatively anaerobic, nonmotile, gram-positive cocci that form irregular clusters (**figure 21.16**). They grow on the skin and mucous membranes of warm-blood animals, and some are important human pathogens that produce a variety of enzymes and toxins (**table 21.4**).

e. Several important genera such as *Lactobacillus* and *Listeria* contain regular, nonsporing, rods. *Lactobacillus* carries out lactic acid fermentation and is extensively used in the food and dairy industries. *Listeria* is an important agent in food poisoning. *Leuconostoc* carries out heterolactic fermentation using the phosphoketolase pathway (**figure 21.22**) and is involved in the production of fermented vegetable products, buttermilk, butter, and cheese.

f. The genera *Streptococcus, Enterococcus,* and *Lactococcus* contain cocci arranged in pairs and chains that are usually facultative and carry out homolactic fermentation (**tables 21.5** and **21.6**). Some important species are the pyogenic coccus *S. pyogenes, S. pneumoniae,* and the enterococcus *E. faecalis* and lactococcus *L. lactis*.

Critical Thinking Questions

1. Many low G + C bacteria are parasitic. The dependence on a host might be a consequence of the low G + C content. Elaborate on this concept.

2. How might one go about determining whether the genome of *M. genitalium* is the smallest compatible with a parasitic existence?

3. Account for the ease with which anaerobic clostridia can be isolated from soil and other generally aerobic niches.

4. Cells of *Bacillus subtilis* that are committed to sporulate secrete a signaling protein that prevents sister cells from entering sporulation and a killing factor that causes them to lyse. The nutrients released by these cells are consumed by the "killer" cells, providing the nutrients needed to complete spore formation. It has been proposed that this provides evidence that sporulation is a "last resort" for survival by this microbe. Do you agree that the capacity to cannibalize sister cells is evidence that the *B. subtilis* uses spore formation as last resort? Explain your answer, being careful not to project human terms onto the microbes (i.e., anthropomorphize).

Learn More

Learn more by visiting the Prescott website at www.mhhe.com/prescottprinciples, where you will find a complete list of references.

The High G + C Gram-Positive Bacteria

22

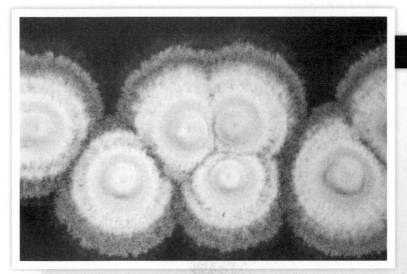

Colonies of the actinomycete *Streptomyces coelicolor* develop an aerial mycelium, imparting a white, fuzzy appearance. The streptomycetes produce most of the antibiotics currently used; *S. coelicolor* synthesizes at least four antibiotics, including actinorhodin, which turns the agar a deep purple color as shown here.

Chapter Glossary

actinobacteria A general term used to refer to all high G + C gram-positive bacteria that are placed in the phylum *Actinobacteria.*

actinomycete High G + C gram-positive, aerobic bacteria that produce filamentous hyphae and differentiate to produce asexual spores.

aerial mycelium A mat of hyphae formed by actinomycetes that grows above the substrate, imparting a fuzzy appearance to colonies.

diphtheria A vaccine-preventable disease caused by *Corynebacterium diphtheriae,* a member of the suborder *Corynebacterineae.*

hyphae The tubular, filamentous cellular structures formed by many actinomycetes and most fungi.

leprosy An insidious skin disease cause by *Mycobacterium leprae,* a member of the suborder *Corynebacterineae.*

madurose The sugar derivative 3-*O*-methyl-D-galactose, which is characteristic of several actinomycete genera that are collectively called maduromycetes.

***M. avium* complex (MAC)** A disease that chiefly affects immunocompromised individuals, caused by *Mycobacterium avium* and *M. intracellulare,* members of the suborder *Corynebacterineae.*

mycolic acids Complex 60 to 90 carbon fatty acids with a hydroxyl on the β-carbon and an aliphatic chain on the α-carbon; found in the cell walls of mycobacteria.

nocardioforms Bacteria that resemble members of the genus *Nocardia;* they develop a substrate mycelium that readily fragments into rods and coccoid elements.

snapping division A distinctive type of binary fission resulting in an angular or a palisade arrangement of cells, which is characteristic of the genera *Arthrobacter* and *Corynebacterium.*

streptomycetes Any high G + C gram-positive bacterium of the genera *Kitasatospora, Streptomyces,* and *Streptoverticillium.* The term is also often used to refer to other closely related families, including *Streptosporangiaceae* and *Nocardiopsaceae.*

substrate mycelium In the actinomycetes, hyphae that are on the surface and may penetrate into the solid medium on which the microbes are grown.

tuberculosis A disease that infects one-third of the world's population, caused principally by *Mycobacterium tuberculosis* but also by *M. bovis* and *M. africanum,* members of the suborder *Corynebacterineae.*

Actinomycetes . . . may be a nuisance, as when they decompose rubber products, grow in aviation fuel, produce odorous substances that pollute water supplies, or grow in sewage-treatment plants where they form thick clogging foams. . . . In contrast, actinomycetes are the producers of most of the antibiotics.

—H. A. Lechevalier and M. P. Lechevalier

Chapter 22, the last of the survey chapters on bacteria, describes the high G + C gram-positive bacteria (**figure 22.1**). They are found in volume 4 of *Bergey's Manual of Systematic Bacteriology.* Many of these bacteria are called actinomycetes. **Actinomycetes** are gram-positive, aerobic bacteria but are distinctive because they have filamentous hyphae that differentiate to produce asexual spores. Many closely resemble fungi in overall morphology. Presumably this resemblance results partly from adaptation to the

same habitats. In this chapter, we first summarize the general characteristics of the actinomycetes. Representatives are described next, with emphasis on morphology, taxonomy, reproduction, and general importance. The actinomycetes (s., actinomycete) are a diverse group, but they share many properties.

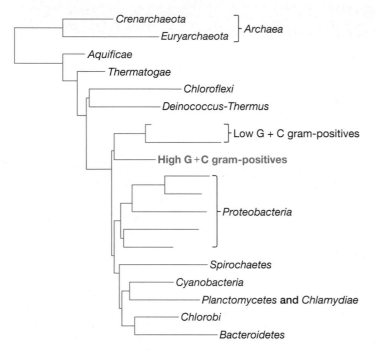

Figure 22.1 Phylogenetic Relationships Among Procaryotes. The high G + C gram-positive bacteria are highlighted.

22.1 GENERAL PROPERTIES OF THE ACTINOMYCETES

The actinomycetes are a fascinating group of microorganisms. They are the source of most of the antibiotics used in medicine today. They also produce metabolites that are used as anticancer drugs, antihelminthics (e.g., ivermectin given to dogs to prevent heartworm), and drugs that suppress the immune system in patients who have received organ transplants. This practical aspect of the actinomycetes is linked very closely to their mode of growth. Like the myxobacteria, the prosthecate bacteria, and several other microbes described in previous chapters, the actinomycetes have a complex life cycle. The life cycle of many actinomycetes includes the development of filamentous cells, called **hyphae,** and spores. When growing on a solid substratum such as soil or agar, the actinomycetes develop a branching network of hyphae (**figure 22.2**). The hyphae grow both on the surface of the substratum and into it to form a dense hyphal mat called a **substrate mycelium.** Septae usually divide the hyphae into long cells (20 μm and longer) containing several nucleoids. In many actinomycetes, substrate hyphae differentiate into upwardly growing hyphae to form an **aerial mycelium** that extends above the substratum. It is at this time that medically useful compounds are formed. Because the physiology of the

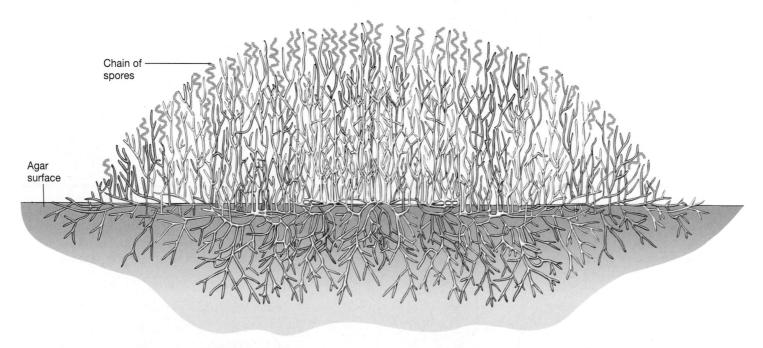

Figure 22.2 An Actinomycete Colony. The cross section of an actinomycete colony with living (blue) and dead (white) hyphae. The substrate mycelium and aerial mycelium with chains of spores are shown.

actinomycete has switched from actively growing vegetative cells into this special cell type, these compounds are often called **secondary metabolites.**

The aerial hyphae septate to form thin-walled spores. These spores are considered **exospores** because they do not develop within a mother cell like the endospores of *Bacillus* and *Clostridium.* If the spores are located in a sporangium, they may be called **sporangiospores.** The spores can vary greatly in shape (**figure 22.3**). Like spore formation in other bacteria, actinomycete sporulation is usually in response to nutrient deprivation. In general, actinomycete spores are not particularly heat resistant but withstand desiccation well, so they have considerable adaptive value. Most actinomycetes are not motile, and spores are dispersed by wind or adhering to animals; in this way, they may find a new habitat to provide needed nutrients. In the few motile genera, motility is confined to flagellated spores.

Actinomycetes also have great ecological significance. They are primarily soil inhabitants and are widely distributed. They can degrade an enormous variety of organic compounds, and are extremely important in the mineralization of organic matter. Although most actinomycetes are free-living microorganisms, a few are pathogens of humans, other animals, and some plants.

Actinomycete cell wall composition varies greatly among different groups and is of considerable taxonomic importance. Four major cell wall types can be distinguished according to three features of peptidoglycan composition and structure: the amino acid in the tetrapeptide side chain position 3, the presence of glycine in interpeptide bridges, and peptidoglycan sugar content (**table 22.1**; *see also figure 21.5*). Whole cell extracts of actinomycetes with wall types II, III, and IV also contain characteristic sugars that are useful in identification. Some other taxonomically valuable properties are the morphology and color of the mycelium and sporangia, the surface features and arrangement of spores, the percent G + C in DNA, the phospholipid composition of cell membranes, and spore heat resistance. Of course 16S rRNA sequences have

proven valuable. Several actinomycete genomes have been sequenced, including *Mycobacterium tuberculosis, M. leprae, Streptomyces coelicolor,* and *S. avermitilis.* ◄◄ *Bacterial cell walls (section 3.4); Peptidoglycan and endospore structure (section 21.2)*

It has been clear for some time that some of the phenotypic traits traditionally used to determine actinomycete taxonomy do not always fit with 16S rRNA sequence data. High G + C gram-positive bacteria are those gram-positive bacteria with a DNA base composition above approximately 50 mol% G + C. Based on 16S rRNA sequence data, all are placed in the phylum *Actinobacteria* and classified as shown in **figure 22.4.** The phylum is large

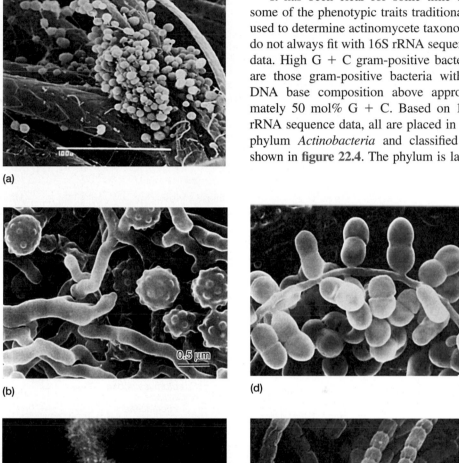

(a)

(b)

(d)

(c)

(e)

Figure 22.3 **Examples of Actinomycete Spores as Seen in the Scanning Electron Microscope.** (a) Spores of *Pilimelia columellifera* on mouse hair (×520). (b) *Micromonospora echinospora.* (c) A chain of hairy streptomycete spores. (d) *Microbispora rosea,* paired spores on hyphae. (e) Spore chain of *Kitasatosporia setae.*

Table 22.1	**Actinomycete Cell Wall Types and Whole Cell Sugar Patterns**			
Cell Wall Type	**Diaminopimelic Acid Isomer**	**Glycine in Interpeptide Bridge**	**Characteristic Sugars**	**Representative Genera**
I	L, L	+	NA[a]	*Nocardioides, Streptomyces*
II	*meso*	+	NA	*Micromonospora, Pilimelia, Actinoplanes*
III	*meso*	–	NA	*Actinomadura, Frankia*
IV	*meso*	–	Arabinose, galactose	*Saccharomonospora, Nocardia*
Sugar Pattern Types[b]		**Characteristic Sugars**		**Representative Genera**
A		Arabinose, galactose		*Nocardia, Rhodococcus, Saccharomonospora*
B		Madurose[c]		*Actinomadura, Streptosporangium, Dermatophilus*
C		None		*Thermomonospora, Actinosynnema, Geodermatophilus*
D		Arabinose, xylose		*Micromonospora, Actinoplanes*

[a]NA, either not applicable or no diagnostic sugars.
[b]Characteristic sugar patterns are present only in wall types II-IV, those actinomycetes with *meso*-diaminopimelic acid.
[c]Madurose is 3-*O*-methyl-D-galactose.

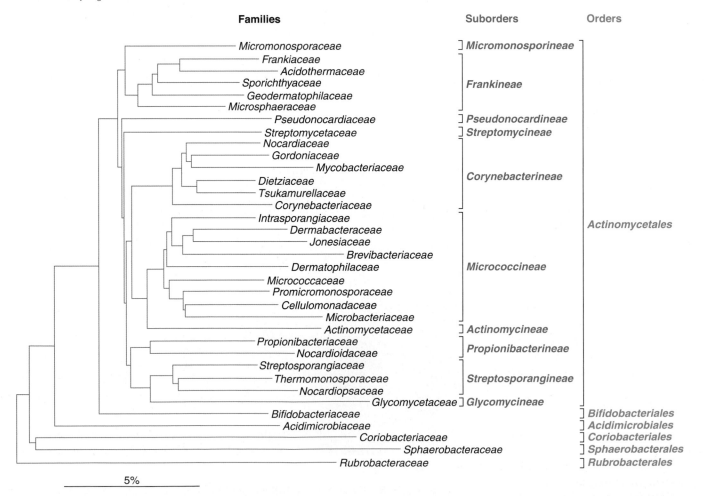

Figure 22.4   Classification of the Phylum *Actinobacteria*.   The phylogenetic relationships between orders, suborders, and families based on 16S rRNA data are shown. The bar represents 5 nucleotide substitutions per 100 nucleotides.   *Source: (a) Stackebrandt, E., Rainey F. A., and Ward-Rainey, N. L. 1997. Proposal for a new hierarchic classification system,* Actinobacteria, *classis nov. Fig. 3, p. 482.* Int. J. Syst. Bacteriol. *47(2):479–91.*

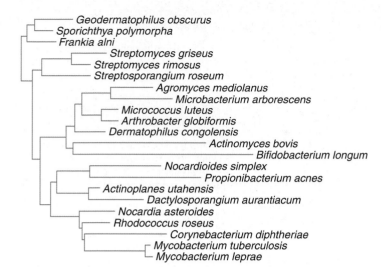

Figure 22.5 Phylogenetic Relationships Among Selected High G + C Gram-Positive Bacteria. The relationships among a few species based on 16S rRNA sequence data are shown. *Source: The Ribosomal Database Project.*

and very complex; it contains 1 class (*Actinobacteria*), 5 sub-classes, 6 orders, 14 suborders, and 44 families. In this system, the **actinobacteria** are composed of the actinomycetes and their high G + C relatives. **Figure 22.5** shows the phylogenetic relationships among selected actinobacterial representatives, and **table 22.2** summarizes the characteristics of some of the genera discussed in this chapter. << *Techniques for determining microbial taxonomy and phylogeny: Molecular characteristics (section 17.4)*

Most of the genera discussed in the following survey are in the subclass *Actinobacteridae* and order *Actinomycetales,* which is divided into 10 suborders. We focus on several of these suborders. The order *Bifidobacteriales* also is briefly described.

22.2 SUBORDER *ACTINOMYCINEAE*

There is one family with five genera in the suborder *Actinomycineae.* These include *Actinomyces, Actinobaculum, Arcanobacterium, Mobiluncus,* and *Varibaculum.* Most are irregularly shaped, nonsporing, gram-positive rods with aerobic or facultative metabolism. The rods may be straight or slightly curved and usually have swellings, club shapes, or other deviations from normal rod-shape morphology.

Members of the genus *Actinomyces* are either straight or slightly curved rods, or slender filaments with true branching (**figure 22.6**). The rods and filaments may have swollen or clubbed ends. They are either facultative or strict anaerobes that require CO_2 for optimal growth. The cell walls contain lysine but not diaminopimelic acid or glycine. *Actinomyces* species are normal inhabitants of mucosal surfaces of humans and other warm-blooded animals; the oral cavity is their preferred habitat. *A. bovis* causes lumpy jaw in cattle. *Actinomyces* is responsible for actinomycoses, ocular infections, and periodontal disease in humans.

22.3 SUBORDER *MICROCOCCINEAE*

The suborder *Micrococcineae* has 14 families and a wide variety of genera. Two of the best-known genera are *Micrococcus* and *Arthrobacter.* The genus *Micrococcus* contains aerobic, catalase-positive cocci that occur mainly in pairs, tetrads, or irregular clusters and are usually nonmotile (**figure 22.7**). Unlike many other actinomycetes, *Micrococcus* does not undergo morphological differentiation. Micrococci colonies often are yellow, orange, or red. They are widespread in soil, water, and on mammalian skin, which may be their normal habitat.

The genus *Arthrobacter* contains aerobic, catalase-positive rods with respiratory metabolism and lysine in its peptidoglycan. Its most distinctive feature is a rod-coccus growth cycle (**figure 22.8**). When *Arthrobacter* grows in exponential phase, the bacteria are irregular, branched rods. When these rods divide, they may undergo snapping movements or **snapping division.** As they enter stationary phase, the cells change to a coccoid form. Upon transfer to fresh medium, the coccoid cells differentiate to form actively growing rods. Although arthrobacters often are isolated from fish, sewage, and plant surfaces, their most important habitat is the soil, where they constitute a significant component of the microbial flora. They are well adapted to this niche because they are very resistant to desiccation and nutrient deprivation. This genus is unusually flexible nutritionally and can even degrade some herbicides and pesticides.

The mechanism of snapping division has been studied in *Arthrobacter.* These bacteria have a two-layered cell wall, and only the inner layer grows inward to generate a transverse wall dividing the new cells. The completed transverse wall or septum thickens and puts tension on the outer wall layer, which still holds the two cells together. Eventually increasing tension ruptures the outer layer at its weakest point, and a snapping movement tears the outer layer apart around most of its circumference. The new cells now rest at an angle to each other and are held together by the remaining portion of the outer layer, which acts as a hinge.

A third genus in this suborder is *Dermatophilus. Dermatophilus* (type IIIB) also forms packets of motile spores with tufts of flagella. It is a facultative anaerobe and a parasite of mammals responsible for the skin infection streptothricosis.

1. Define actinomycete, substrate mycelium, aerial mycelium, and exospore. Explain how these structures confer a survival advantage.

2. Describe how cell wall structure and sugar content are used to classify the actinomycetes. Include a brief description of the four major wall types.

3. Why are the actinomycetes of such practical interest?

4. Describe the phylum *Actinobacteria* and its relationship to the actinomycetes.

Table 22.2   Characteristics of Actinobacteria

Genus	Dimensions (µm), Morphology and Motility	G + C Content (mol%)	Oxygen Relationship	Other Distinctive Characteristics
Actinoplanes	Nonfragmenting, branching mycelium with little aerial growth; sporangia formed; motile spores with polar flagella	72–73	Aerobic	Hyphae often in palisade arrangement; highly colored; type II cell walls; found in soil and decaying plant material
Arthrobacter	0.8–1.2 × 1.0–8.0; young cells are irregular rods, older cells are small cocci; usually nonmotile	59–70	Aerobic	Has rod-coccus growth cycle; metabolism respiratory; catalase positive; mainly in soil
Bifidobacterium	0.5–1.3 × 1.5–8; rods of varied shape, usually curved; nonmotile	55–67	Anaerobic	Cells can be clubbed or branched, pairs often in V arrangement; ferment carbohydrates to acetate and lactate, but no CO_2; catalase negative
Corynebacterium	0.3–0.8 × 1.5–8.0; straight or slightly curved rods with tapered or clubbed ends; nonmotile	51–63	Facultatively anaerobic	Cells often arranged in a V formation or in palisades of parallel cells; catalase positive and fermentative; metachromatic granules
Frankia	0.5–2.0 in diameter; vegetative hyphae with limited-to-extensive branching and no aerial mycelium; multilocular sporangia formed	66–71	Aerobic to microaerophilic	Sporangiospores nonmotile; usually fixes nitrogen; type III cell walls; most strains are symbiotic with angiosperm plants and induce nodules
Micrococcus	0.5–2.0 diameter; cocci in pairs, tetrads, or irregular clusters; usually nonmotile	64–75	Aerobic	Colonies usually yellow or red; catalase positive with respiratory metabolism; primarily on mammalian skin and in soil
Mycobacterium	0.2–0.6 × 1.0–10; straight or slightly curved rods, sometimes branched; acid-fast; nonmotile and nonsporing	62–70	Aerobic	Catalase positive; can form filaments that are readily fragmented; walls have high lipid content; in soil and water; some parasitic
Nocardia	0.5–1.2 in diameter; rudimentary to extensive vegetative hyphae that can fragment into rod-shaped and coccoid forms	64–72	Aerobic	Aerial hyphae formed; catalase positive; type IV cell wall; widely distributed in soil
Propionibacterium	0.5–0.8 × 1–5; pleomorphic nonmotile rods, may be forked or branched; nonsporing	53–67	Anaerobic to aerotolerant	Fermentation produces propionate and acetate, and often gas; catalase positive
Streptomyces	0.5–2.0 in diameter; vegetative mycelium extensively branched; aerial mycelium forms chains of three to many spores	69–78	Aerobic	Forms discrete lichenoid or leathery colonies that often are pigmented; uses many different organic compounds as nutrients; soil organisms

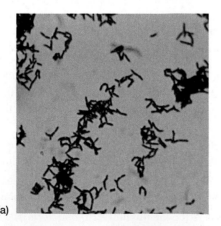

(a)

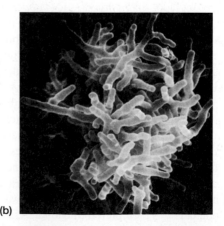

(b)

Figure 22.6   **Representatives of the Genus *Actinomyces.***   (a) *A. naeslundii;* Gram stain (×1,000). (b) *Actinomyces;* scanning electron micrograph (×18,000). Note the filamentous nature of the colony.

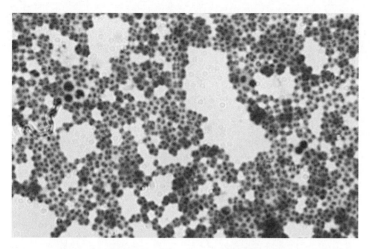

Figure 22.7 *Micrococcus. Micrococcus luteus,* methylene blue stain (×1,000).

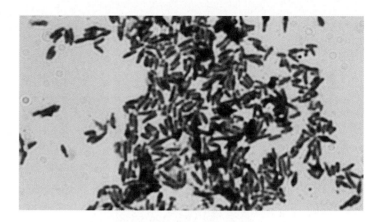

Figure 22.9 *Corynebacterium diphtheriae.* Note the irregular shapes of individual cells, the angular associations of pairs of cells, and palisade arrangements (×1,000). These gram-positive rods do not form endospores.

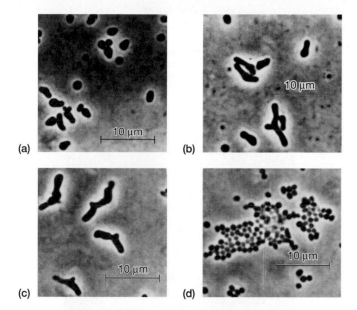

(a)   (b)   (c)   (d)

Figure 22.8 The Rod-Coccus Growth Cycle. The rod-coccus cycle of *Arthrobacter globiformis* when grown at 25°C. (a) Rods are outgrowing from cocci 6 hours after inoculation. (b) Rods after 12 hours of incubation. (c) Bacteria after 24 hours. (d) Cells after reaching stationary phase (3-day incubation). The cells used for inoculation resembled these stationary-phase cocci.

22.4 SUBORDER *CORYNEBACTERINEAE*

The suborder *Corynebacterineae* contains seven families with several well-known genera. Three of the most important genera are *Corynebacterium, Mycobacterium,* and *Nocardia.*

Family *Corynebacteriaceae*

The family *Corynebacteriaceae* has one genus, *Corynebacterium,* which includes aerobic and facultative, catalase-positive, straight to slightly curved rods, often with tapered ends. Club-shaped forms are also seen. The bacteria often remain partially attached after snapping division, resulting in angular arrangements of the cells or a **palisade arrangement** in which rows of cells are lined up side by side (**figure 22.9**). *Corynebacteria* form metachromatic granules, and their walls have meso-diaminopimelic acid. Although some species are harmless saprophytes, many corynebacteria are plant or animal pathogens. We next discuss *C. diphtheriae,* the causative agent of diphtheria in humans.

Diphtheria (Greek *diphthera,* membrane, and *–ia,* condition) is an acute, contagious disease. *C. diphtheriae* is well-adapted to airborne transmission by way of nasopharyngeal secretions because it is very resistant to drying. Once within the respiratory system, bacteria that carry the prophage β containing the *tox* gene produce diphtheria toxin; *tox*⁺ phage β infection of *C. diphtheriae* is required for toxin production. Diphtheria toxin is an exotoxin that causes an inflammatory response and the formation of a grayish pseudomembrane on the pharynx and respiratory mucosa (**figure 22.10**). The pseudomembrane consists of dead host cells and cells of *C. diphtheriae.* Diphtheria toxin is absorbed into the circulatory system and distributed throughout the body, where it may cause destruction of cardiac, kidney, and nervous tissues by inhibiting protein synthesis. The toxin is composed of two polypeptide subunits: A and B. The A subunit consists of the catalytic domain; the B subunit is composed of the receptor and transmembrane domains (*see figure 30.6*). The receptor domain attaches to the heparin-binding epidermal growth factor receptor on the surface of various eucaryotic cells. Once bound, the A subunit enters the cytoplasm by endocytosis. The transmembrane domain of the B subunit embeds itself into the target cell membrane, causing the catalytic domain to be cleaved and translocated into the cytoplasm. The cleaved catalytic domain becomes an active enzyme, catalyzing the attachment of ADP-ribose (from NAD⁺) to elongation factor-2 (EF-2). A single enzyme (i.e., catalytic domain) can exhaust the entire supply of cellular EF-2 within hours, resulting in protein synthesis inhibition and cell death. ▶▶ *Toxigenicity: AB toxins (section 30.4)*

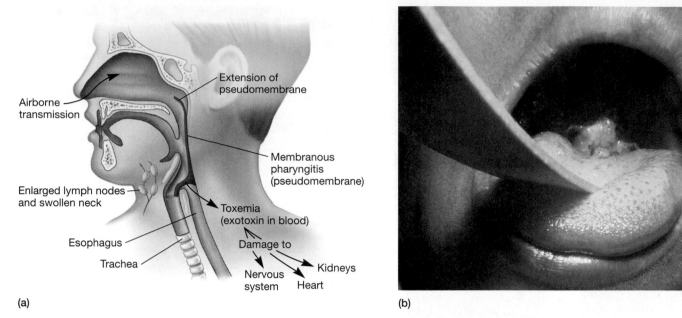

(a)   (b)

Figure 22.10   Diphtheria Pathogenesis.   (a) Diphtheria is a well-known, exotoxin-mediated infectious disease caused by *Corynebacterium diphtheriae*. The disease is an acute, contagious, febrile illness characterized by local oropharyngeal inflammation and pseudomembrane formation. If the exotoxin gets into the blood, it is disseminated and can damage the peripheral nerves, heart, and kidneys. (b) The clinical appearance includes gross inflammation of the pharynx and tonsils marked by grayish patches (a pseudomembrane) and swelling of the entire area.

Typical symptoms of diphtheria include a thick mucopurulent (containing both mucus and pus) nasal discharge, pharyngitis (sore throat), fever, cough, paralysis, and death. *C. diphtheriae* can also infect the skin, usually at a wound or skin lesion, causing a slow-healing ulceration termed cutaneous diphtheria.

Because the toxin as well as the pathogen must be inhibited, treatment involves the administration of diphtheria antitoxin to neutralize any unabsorbed exotoxin in the patient's tissues and antibiotics (penicillin and erythromycin) to treat the infection. Prevention is by active immunization; indeed, diphtheria mainly affects unvaccinated, people living in crowded conditions. Most cases occur in tropical areas and involve people over thirty years of age who have a weakened immunity to the diphtheria. Since 1980, fewer than six diphtheria cases have been reported annually in the United States, and most occur in nonimmunized individuals. >> *Control of epidemics: Vaccines and immunization (section 33.8)*

Family *Mycobacteriaceae*

The family *Mycobacteriaceae* contains the genus *Mycobacterium*, which is composed of slightly curved or straight rods that sometimes branch or form filaments (**figure 22.11**). Mycobacterial filaments differ from those of actinomycetes because they readily fragment into rods and coccoid bodies. They are aerobic and catalase positive. Mycobacteria grow very slowly and must be incubated for 2 to 40 days after inoculation on a solidified complex medium to form a visible colony. Their cell walls have a very high lipid content and contain waxes with 60 to 90 carbon **mycolic acids.** These are complex fatty acids with a hydroxyl

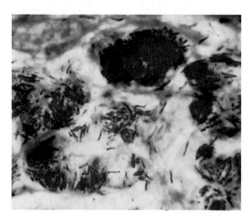

Figure 22.11   The Mycobacteria.   *Mycobacterium leprae.* Acid-fast stain (×400). Note the masses of red mycobacteria within blue-green host cells.

group on the β-carbon and an aliphatic chain attached to the α-carbon (**figure 22.12**). The presence of mycolic acids and other lipids outside the peptidoglycan layer makes mycobacteria **acid-fast** (basic fuchsin dye cannot be removed from the cell by acid alcohol treatment). Extraction of wall lipid with alkaline ethanol destroys acid-fastness. << *Preparation and staining of specimens: Differential staining (section 2.3)*

Although some mycobacteria are free-living saprophytes, they are best known as animal pathogens. *M. bovis* causes tuberculosis in cattle, other ruminants, and primates. Because this bacterium can produce tuberculosis in humans, dairy cattle are tested for the disease yearly; milk pasteurization kills the patho-

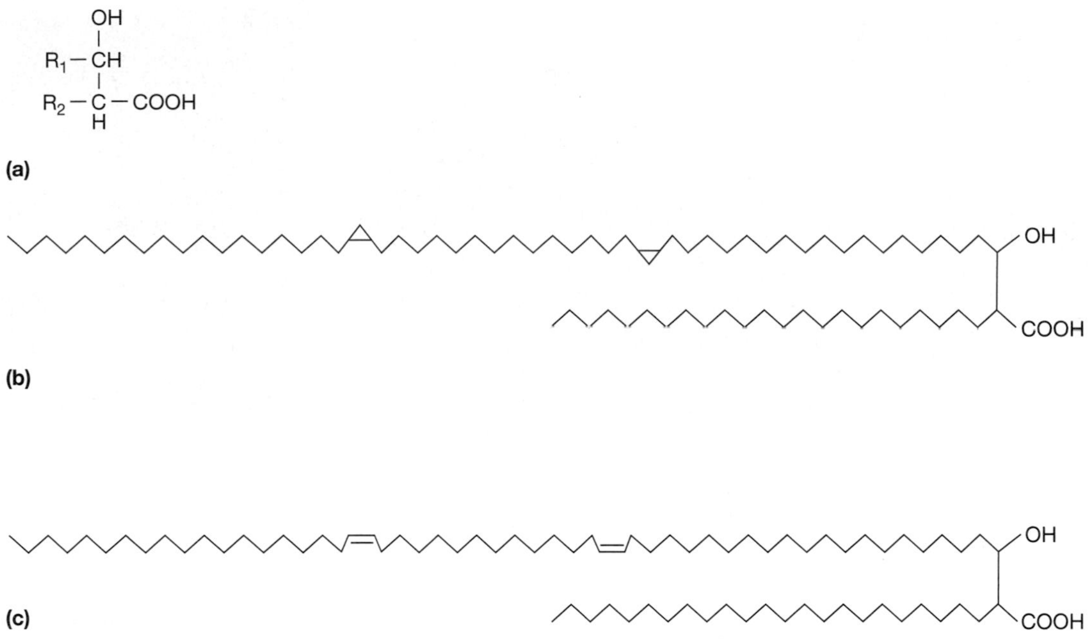

Figure 22.12 **Mycolic Acid Structure.** (a) The generic structure of mycolic acids, a family that includes over 500 different types. (b) A mycolic acid with two cyclopropane rings and (c) an unsaturated mycolic acid.

gen. Prior to widespread milk pasteurization, contaminated milk was a source of transmission. Other mycobacterial pathogens, including *M. tuberculosis*, the chief cause of tuberculosis in humans, are discussed next.

Mycobacterium tuberculosis Infections

Over a century ago, Robert Koch identified *M. tuberculosis* as the causative agent of **tuberculosis (TB).** At the time, TB was rampant, causing one-seventh of all deaths in Europe and one-third of deaths among young adults. Today TB remains a global health problem of enormous dimension. It is estimated that one-third of the world's human population is infected (i.e., at least 2 billion people), with 9 million new cases and 2 million deaths per year (**figure 22.13***a*).

The bacteria are phagocytosed by macrophages in the lungs, where they survive the normal antimicrobial processes (figure 22.13*b*). In fact, macrophages that have phagocytosed mycobacteria often die attempting to destroy them. Other immune effector cells are recruited to the site of infection by cytokines released from the responding macrophages. Together, and in response to several mycobacterial products, a hypersensitivity response results in the formation of small, hard nodules called **tubercles** composed of bacteria, macrophages, T cells, and various human proteins (figure 22.13*c*). Tubercles are characteristic of tuberculosis and give the disease its name. The disease process usually stops at this stage, but the bacteria often remain alive within macrophage phagosomes. In some cases, the disease may become active, even after many years of latency. The incubation period is about 4 to 12 weeks, and the disease develops slowly. The symptoms of tuberculosis are fever, fatigue, and weight loss.

A cough, which is characteristic of pulmonary involvement, may result in expectoration of bloody sputum.

M. tuberculosis (Mtb) does not produce classic virulence factors such as toxins, capsules, and fimbriae. Instead, Mtb has some unique products and properties that contribute to its virulence. Mycolic acids and other cell wall components (lipoarabinomannan, trehalose dimycolate, and phthiocerol dimycocerosate) are directly toxic to eucaryotic cells and create a hydrophobic barrier around the bacterium that facilitates impermeability and resistance to antimicrobial agents, resistance to killing by acidic and alkaline compounds, resistance to osmotic lysis, and resistance to lysozyme, an enzyme that normally degrades peptidoglycan. Cell wall glycolipids also associate with mannose, giving Mtb control over entry into macrophages and exploiting the macrophage mannose receptors. Once inside, Mtb can inhibit phagosome-lysosome fusion by altering the phagosome membrane. Resistance to oxidative killing, inhibition of phagosome-lysosome fusion, and inhibition of diffusion of lysosomal enzymes are just some of the mechanisms that help explain the survival of Mtb inside macrophages. >> *Phagocytosis (28.3)*

In time, the tubercle may change to a cheeselike consistency and is then called a caseous lesion. If such lesions calcify, they are called **Ghon complexes,** which show up prominently in a chest X ray. (Often the primary lesion is called the Ghon's tubercle or Ghon's focus.) Sometimes the tubercle lesions liquefy and form air-filled tuberculous cavities. From these cavities, the bacteria can spread to new foci of infection throughout the body (figure 22.13*d*). This spreading is often called **miliary tuberculosis** due to the many tubercles the size of millet seeds that are formed in the infected tissue. It also may be called **reactivation**

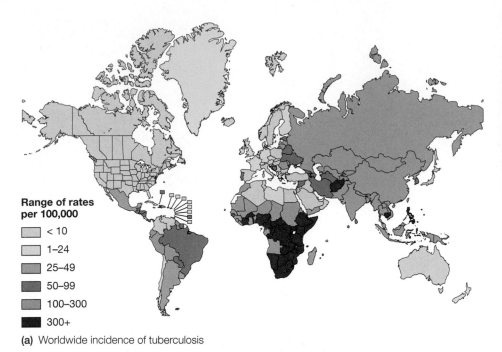

Range of rates per 100,000

	< 10
	1–24
	25–49
	50–99
	100–300
	300+

(a) Worldwide incidence of tuberculosis

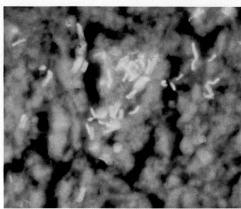

(b) *M. tuberculosis* in sputum

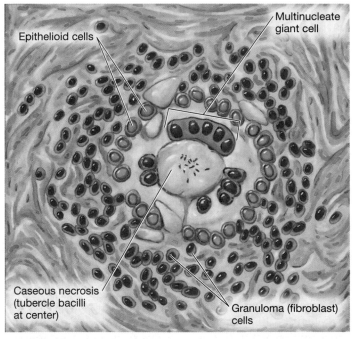

Multinucleate giant cell

Epithelioid cells

Caseous necrosis (tubercle bacilli at center)

Granuloma (fibroblast) cells

(c) A tubercle

Figure 22.13 Tuberculosis. (a) Tuberculosis is a significant global disease. (b) *Mycobacteria* are recovered in the sputum of tuberculosis patients and can be identified using a fluorescent acid-fast stain. (c) In the lungs, tuberculosis is identified by the tubercle, a granuloma of white blood cells, bacteria, fibroblasts, and epithelioid cells. The center of the tubercle contains caseous (cheesy) pus and bacteria.

tuberculosis because the bacteria have been reactivated in the initial site of infection.

TB must be treated with antimicrobial therapy. Several drugs are administered simultaneously (e.g., isoniazid [INH], plus rifampin, ethambutol, and pyrazinamide). These drugs are administered for 6 to 9 months as a way of decreasing the possibility that the bacterium develops drug resistance. However, **multidrug-resistant strains of tuberculosis (MDR-TB)** have developed and are spreading. A multidrug-resistant strain is defined as *M. tuberculosis* that is resistant to INH and rifampin, with or without resistance to other drugs. Between 2003 and 2004 in the United States, MDR-TB increased 13.3%, the largest yearly increase in over a decade. Inadequate therapy is the most common means by which resistant bacteria are acquired, and patients who have previously undergone therapy are presumed to harbor MDR-TB until proven otherwise. Recently strains of Mtb have emerged that are extremely drug-resistant (XDR); they are resistant to INH, rifampin, and the secondary antibiotics. MDR-TB and XDR-TB can be fatal.

Drug resistant TB arises because tubercle bacilli have spontaneous, predictable rates of chromosomal mutations that confer resistance to drugs. These mutations are unlinked; hence resistance to one drug is not associated with resistance to an unrelated drug. The observation that these mutations are not linked is the cardinal principle underlying TB chemotherapy. For example, mutations that cause resistance to isoniazid or rifampin occur roughly in 1 in 10^8 replications of *M. tuberculosis*. The likelihood of spontaneous mutations causing resistance to both isoniazid and rifampin is the product of these probabilities, or 1 in 10^{16}. However, the biological mechanisms of resistance break down when chemotherapy is inadequate. In the circumstances of monotherapy, erratic drug ingestion, omission of one or more drugs, suboptimal dosage, poor drug absorption, or an insufficient number of active drugs in a regimen, a susceptible strain of *M. tuberculosis* may become resistant to multiple drugs within a matter of months. This is why many public health organizations worldwide practice *d*irectly *o*bserved *t*reatment *s*hort course (DOTS), in which each dose is taken in the presence of a health care worker.

In the United States, TB occurs most commonly among the homeless, elderly, and malnourished, or among alcoholic males,

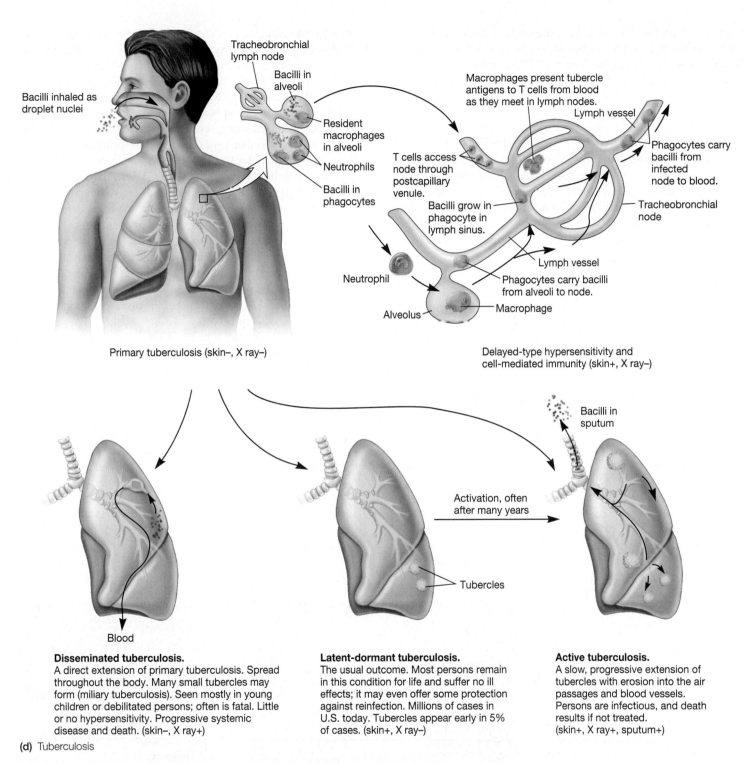

Primary tuberculosis (skin–, X ray–)

Delayed-type hypersensitivity and cell-mediated immunity (skin+, X ray–)

Activation, often after many years

Disseminated tuberculosis.
A direct extension of primary tuberculosis. Spread throughout the body. Many small tubercles may form (miliary tuberculosis). Seen mostly in young children or debilitated persons; often is fatal. Little or no hypersensitivity. Progressive systemic disease and death. (skin–, X ray+)

Latent-dormant tuberculosis.
The usual outcome. Most persons remain in this condition for life and suffer no ill effects; it may even offer some protection against reinfection. Millions of cases in U.S. today. Tubercles appear early in 5% of cases. (skin+, X ray–)

Active tuberculosis.
A slow, progressive extension of tubercles with erosion into the air passages and blood vessels. Persons are infectious, and death results if not treated. (skin+, X ray+, sputum+)

(d) Tuberculosis

Figure 22.13 **(*continued*).**   (d) Possible outcomes of tuberculosis infection.

minorities, immigrants, prison populations, and Native Americans. Between 1999 and 2005, the incidence of tuberculosis in the United States steadily declined to about 14,000 cases; about 1,000 deaths reported each year. During this time, slightly more than half of U.S. cases were in foreign-born persons. Most cases in the United States are acquired from other humans through droplet nuclei and the respiratory route. The majority of active cases result from the reactivation of old, dormant infections.

Worldwide, TB is caused by *M. bovis* and *M. africanum,* in addition to *M. tuberculosis.* This is likely due to closer interactions between people and infected livestock. Since AIDS has become pandemic, there has been a dramatic increase in the number of global TB cases. Because a close association exists between AIDS and TB, further spread of HIV infection among populations with a high prevalence of TB infection is resulting in dramatic increases in TB. Importantly, TB is the direct cause of death of over half of all AIDS patients worldwide.

M. avium–M. intracellulare Infections

An extremely large group of mycobacteria are normal inhabitants of soil, water, and house dust. Two of these are noteworthy pathogens in the United States—M. avium and M. intracellulare, which are so closely related that they are referred to as the **M. avium complex (MAC).** Globally, M. tuberculosis has remained more prevalent in developing countries, whereas MAC has become the most common cause of mycobacterial infections in the United States.

These mycobacteria are found worldwide and infect a variety of insects, birds, and animals. Both the respiratory and the gastrointestinal tracts have been proposed as entry portals for MAC; however, person-to-person transmission is not very efficient. MAC causes a pulmonary infection in humans similar to that caused by M. tuberculosis. Pulmonary MAC is more common in non-AIDS patients, particularly in elderly persons with preexisting pulmonary disease. The gastrointestinal tract is thought to be the most common site of colonization and dissemination in AIDS patients. In the United States, disseminated infection with MAC occurs in 15 to 40% of AIDS patients with CD4$^+$ cell counts of less than 100 per mm^3; it produces disabling symptoms, including fever, malaise, weight loss, night sweats, and diarrhea. Carefully controlled epidemiological studies have shown that MAC shortens survival by 5 to 7 months among persons with AIDS. With more effective antiviral therapy for AIDS and prolonged survival, the number of cases of disseminated MAC is likely to increase substantially, and its contribution to AIDS mortality will increase. **>>** *Viruses with single-stranded RNA genomes (Group VI-retroviruses): HIV (section 24.7) Cells, tissues, and organ of the immune system (section 28.2)*

Leprosy

Leprosy (Greek *lepros*, scaly, scabby, rough), or Hansen's disease, is a severely disfiguring skin disease caused by M. leprae. The only reservoirs of proven significance are humans, although armadillos seem to naturally harbor mycobacteria that are genetically identical to M. leprae. The disease most often occurs in tropical countries, where there are more than 11 million cases. An estimated 4,000 cases exist in the United States with approximately 100 new cases reported annually. Transmission of leprosy is not completely understood but most likely occurs when individuals are exposed for prolonged periods to infected individuals who shed large numbers of M. leprae. Mechanical abrasion of the skin may also facilitate transmission. Nasal secretions probably are the infectious material for family contacts.

The incubation period is about 3 to 7 years but may be much longer, and the disease progresses slowly due to exceedingly slow growth of the organisms. M. leprae grow primarily in macrophages found in cooler body regions (e.g., the digits). However, they also grow in nerve and skin cells, where they cause severe cellular damage. Growth in Schwann cells, which surround peripheral nerve axons, results in loss of sensation. The earliest symptom of leprosy is usually a slightly pigmented skin eruption several centimeters in diameter (**figure 22.14a**). Approximately 75% of all individuals with this early, solitary lesion heal spontaneously thanks to a cell-mediated immune response to M. leprae. However, in some individuals, this immune response may be too weak and one of two distinct forms of the disease occurs: tuberculoid or lepromatous leprosy.

Tuberculoid (neural) leprosy is a mild, nonprogressive form of leprosy associated with a delayed-type hypersensitivity reaction to antigens on the surface of M. leprae. It is characterized by damaged nerves resulting in regions of skin that have lost sensation and are surrounded by a border of nodules. Afflicted individuals who do not develop hypersensitivity have a relentlessly progressive and widespread form of the disease, called **lepromatous leprosy,** in which large numbers of M. leprae in macrophages and skin cells form granulomas (lepromas, figure 22.14b). The engorged skin cells, especially of the face, bulge outward and droop. This leads to a progressive loss of facial features, fingers, toes, and other structures as disfiguring nodules form all over the body. Nerves are also infected but are usually are less damaged than those in tuberculoid leprosy. Identification and treatment of patients with leprosy is the key to control. Children of presumably contagious parents should be

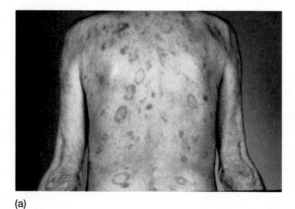

(a)

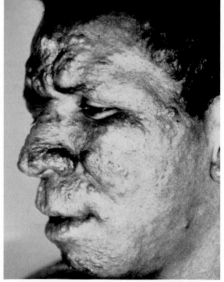

(b)

Figure 22.14 Leprosy. (a) In tuberculoid leprosy, the skin manifests shallow, painless lesions. (b) Mycobacterial infection of the nose, chin, and brows produces facial deformation, typical of lepromatous leprosy.

given chemoprophylactic drugs until treatment of their parents has made them noninfectious. >> *Immune disorders: Type IV hypersensitivity (section 29.11)*

Family *Nocardiaceae*

The family *Nocardiaceae* is composed of two genera, *Nocardia* and *Rhodococcus*. These bacteria develop a substrate mycelium that readily breaks into rods and coccoid elements (**figure 22.15**). They also form an aerial mycelium that rises above the substratum and may produce conidia. Almost all are strict aerobes. Most species have peptidoglycan with meso-diaminopimelic acid and no peptide interbridge. The wall usually contains a carbohydrate composed of arabinose and galactose, and mycolic acids are present. These and related genera that resemble members of the genus *Nocardia* (named after Edmond Nocard [1850–1903], French bacteriologist and veterinary pathologist) are collectively called **nocardioforms.**

Nocardia is distributed worldwide in soil and aquatic habitats. They are involved in the degradation of hydrocarbons and waxes, and can contribute to the biodeterioration of rubber joints in water and sewage pipes. Although most are free-living saprophytes, some species, particularly *N. asteroides,* are opportunistic pathogens that cause nocardiosis in humans and other animals. People with low resistance due to other health problems, such as individuals with HIV-AIDS, are most at risk. The lungs are typically infected; other organs and the central nervous system may be invaded as well.

Rhodococcus is widely distributed in soils and aquatic habitats. It is of considerable interest because members of the genus can degrade an enormous variety of molecules such as petroleum hydrocarbons, detergents, benzene, polychlorinated biphenyls (PCBs), and various pesticides. It may be possible to use rhodococci to remove sulfur from fuels, thus reducing air pollution by sulfur oxide emissions.

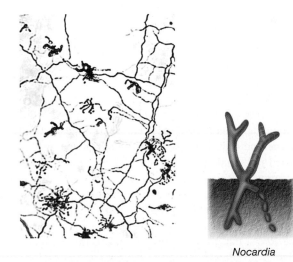

Nocardia

Figure 22.15 *Nocardia.* *Nocardia asteroides,* substrate mycelium and aerial mycelia with conidia illustration and light micrograph (×1,250).

1. Describe the major characteristics of *Corynebacterium* and *Mycobacterium.* Include comments on their normal habitat and importance.
2. What is snapping division?
3. Compare and contrast TB, MAC, and leprosy.
4. Describe the treatment of TB. How do multidrug-resistant strains of *M. tuberculosis* arise?
5. What is a nocardioform, and how can the group be distinguished from other actinomycetes?

22.5 SUBORDER *MICROMONOSPORINEAE*

The suborder *Micromonosporineae* contains only one family, *Micromonosporaceae.* Genera include *Micromonospora, Dactylosporangium, Pilimelia,* and *Actinoplanes.* Often the family is collectively referred to as the actinoplanetes (Greek *actinos,* a ray or beam, and *planes,* a wanderer). They have an extensive substrate mycelium and are type IID cells (tables 22.1 and 22.2). Often the hyphae are highly colored, and diffusible pigments may be produced. Normally an aerial mycelium is absent or rudimentary. Spores are usually formed within a sporangium raised above the surface of the substratum at the end of a special hypha called a sporangiophore. The spores can be either motile or nonmotile. These bacteria vary in the arrangement and development of their spores. Some genera (*Actinoplanes, Pilimelia*) have spherical, cylindrical, or irregular sporangia with a few to several thousand spores per sporangium (figure 22.3a and **figure 22.16**). The spores are arranged in coiled or parallel chains (figure 22.16b). *Dactylosporangium* forms club-shaped, finger-like, or pyriform sporangia with one to six spores (figure 22.16c,d). *Micromonospora* bears single spores, which often occur in branched clusters of sporophores (figure 22.3b).

Actinoplanetes grow in almost all soil habitats, ranging from forest litter to beach sand. They also flourish in freshwater, particularly in streams and rivers (probably because of abundant oxygen and plant debris). Some have been isolated from the ocean. The soil-dwelling species may have an important role in the decomposition of plant and animal material. *Pilimelia* grows in association with keratin. *Micromonospora* actively degrades chitin and cellulose, and produces antibiotics such as gentamicin.

22.6 SUBORDER *PROPIONIBACTERINEAE*

The suborder *Propionibacterineae* contains two families and 14 genera. The genus *Propionibacterium* contains pleomorphic, nonmotile, nonsporing rods that are often club-shaped with one end tapered and the other end rounded. Cells also may be coccoid or branched. They can be single, in short chains, or in clumps. The genus is facultatively anaerobic or aerotolerant; lactate and

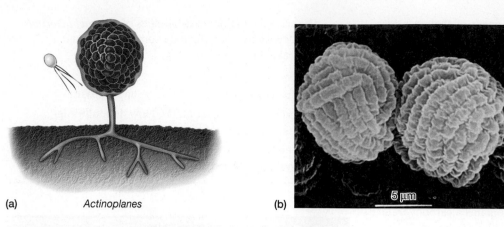

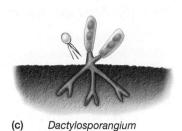

(a) *Actinoplanes*

(b)

(c) *Dactylosporangium*

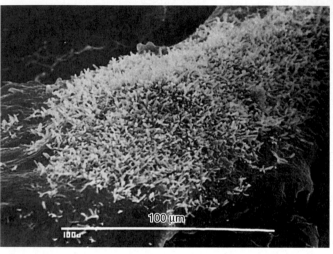

(d)

Figure 22.16 Family *Micromonosporaceae*. Actinoplanete morphology. (a) *Actinoplanes* structure. (b) A scanning electron micrograph of mature *Actinoplanes* sporangia. (c) *Dactylosporangium* structure. (d) A *Dactylosporangium* colony covered with sporangia.

sugars are fermented to produce large quantities of propionic and acetic acids, and often carbon dioxide. *Propionibacterium* is usually catalase positive. The genus is found growing on the skin and in the digestive tract of animals, and in dairy products such as cheese. *Propionibacterium* fermentation gives Swiss cheese its characteristic flavor. *P. acnes* is involved with the development of body odor and acne vulgaris.

22.7 SUBORDER *STREPTOMYCINEAE*

The suborder *Streptomycineae* has only one family, *Streptomycetaceae,* and three genera, the most important of which is *Streptomyces*. These bacteria have aerial hyphae that divide in a single plane to form chains of multiple nonmotile spores with surface texture ranging from smooth to spiny and warty (figure 22.3*e*; **figures 22.17** and **22.18**). All have a type I cell wall and a G + C content of 69 to 78%. Filaments grow by

tip extension rather than by fragmentation. Members of this family and similar bacteria are often called **streptomycetes** (Greek *streptos,* bent or twisted, and *myces,* fungus).

Streptomyces is a large genus; there are around 150 species. Members of the genus are strict aerobes and form chains of nonmotile spores (figures 22.17*c* and 22.18). *Streptomyces* species are determined by means of a mixture of morphological and physiological characteristics, including the following: the color of the aerial and substrate mycelia, spore arrangement, surface features of individual spores, carbohydrate use, antibiotic production, melanin synthesis, nitrate reduction, and the hydrolysis of urea and hippuric acid.

Streptomycetes are very important both ecologically and medically. The natural habitat of most streptomycetes is the soil, where they may constitute from 1 to 20% of the culturable population. In fact, the odor of moist earth is largely the result of streptomycete production of volatile substances such as geosmin. Streptomycetes play a major role in mineralization. They are flexible nutritionally and can degrade recalcitrant (resistant) substances such as pectin, lignin, chitin, keratin, latex, agar, and aromatic compounds. Streptomycetes are best known for their synthesis of a vast array of antibiotics.

Selman Waksman's discovery that *S. griseus* (**figure 22.19*a***) produces streptomycin was an enormously important contribution to science and public health. Streptomycin was the first drug to effectively combat tuberculosis, and in 1952 Waksman earned the Nobel Prize. In addition, this discovery set off a massive search resulting in the isolation of new *Streptomyces* species that produce other compounds of medicinal importance. In fact, since that time, the streptomycetes have been found to produce over 10,000 bioactive compounds. Hundreds of these natural products are now used in medicine and industry; about two-thirds of the antimicrobial agents used in human and veterinary medicine are derived from the streptomycetes. Examples include amphotericin B, chloramphenicol, erythromycin, neomycin, nystatin, and tetracycline. Some *Streptomyces* species produce more than one antibiotic. Antibiotic-producing bacteria have genes that encode proteins (or antisense RNA) that make them resistant to the antibiotics they synthesize. >> *Antimicrobial chemotherapy (chapter 31)*

The genome of *Streptomyces coelicolor,* which produces four antibiotics and serves as a model species for research, has been sequenced. At 8.67 Mbp, it is one of the largest procaryotic

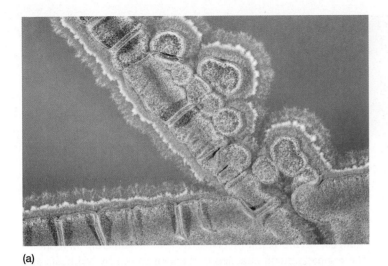

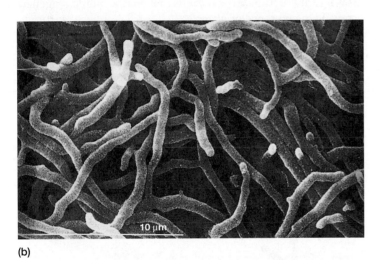

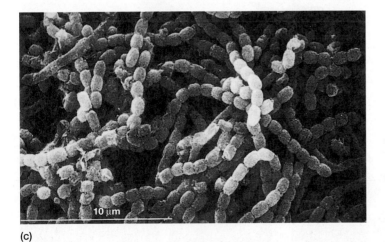

Figure 22.17 Streptomyces Development. (a) Growth of *Streptomyces coelicolor* on a solid substrate results in the formation of vegetative hyphae that differentiate into aerial hyphae to form an aerial mycelium, which imparts a white, fuzzy appearance to the colony surface. The blue background is caused by the diffusion of the pigmented polyketide antibiotic actinorhodin into the agar. (b) *S. coelicolor* vegetative hyphae are straight with branches. (c) Chains of *S. coelicolor* spores that will eventually pinch off and be released into the environment.

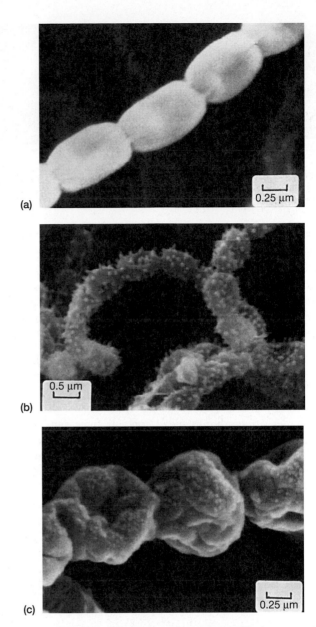

Figure 22.18 Streptomycete Spore Chains. (a) Smooth spores of *S. niveus* (b) Spiney spores of *S. viridochromogenes.* (c) Warty spores of *S. pulcher.* Scanning electron micrographs.

genomes. Its large number of genes (7,825) no doubt reflects the number of proteins required to undergo a complex life cycle. Many genes are devoted to regulation, with an astonishing 65 predicted RNA polymerase sigma subunits and about 80 two-component regulatory systems. The ability to exploit a variety of soil nutrients is also demonstrated by the presence of a large number of ABC transporters, the Sec and two Tat protein translocation systems, and secreted degradative enzymes. Finally, genes were discovered that are thought to encode an additional 18 secondary metabolites. ◄◄ *Protein maturation and secretion (section 12.8); Regulation of transcription initiation: Two-component regulatory systems (section 13.2); Global regulatory systems (section 13.5)*

Although most streptomycetes are nonpathogenic saprophytes, a few are associated with plant and animal diseases. *Streptomyces*

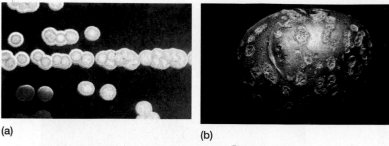

(a)   **(b)**

Figure 22.19 Streptomycetes of Practical Importance. (a) *Streptomyces griseus*. Colonies of the actinomycete that produces streptomycin. (b) *S. scabies* growing on a potato.

scabies causes scab disease in potatoes and beets (figure 22.19*b*). *S. somaliensis* is the only streptomycete known to be pathogenic to humans. It is associated with actinomycetoma, an infection of subcutaneous tissues that produces lesions and leads to swelling, abscesses, and even bone destruction if untreated. *S. albus* and other species have been isolated from patients with various ailments and may be pathogenic.

22.8 SUBORDER *STREPTOSPORANGINEAE*

The suborder *Streptosporangineae* contains three families and 16 genera. The family includes the maduromycetes, which have type III cell walls and the sugar derivative **madurose** (3-*O*-methyl-D-galactose) in whole cell homogenates. Their G + C content is 64 to 74 mol%. Aerial hyphae bear pairs or short chains of spores, and the substrate hyphae are branched (**figure 22.20**). Some genera form sporangia; spores are not heat resistant. Like *S. somaliensis*, *Actinomadura* is another actinomycete associated with the disease actinomycetoma. *Thermomonospora* produces single spores on the aerial mycelium or on both the aerial and substrate mycelia. It has been isolated from moderately high-temperature habitats such as compost piles and hay; it can grow at 40 to 48°C.

22.9 SUBORDER *FRANKINEAE*

The suborder *Frankineae* includes the genera *Frankia* and *Geodermatophilus*. Both form multilocular sporangia, characterized by clusters of spores when a hypha divides both transversely and longitudinally. (Multilocular means having many cells or compartments.) They have type III cell walls, although the cell extract sugar patterns differ. The G + C content varies from 57 to 75 mol%. *Geodermatophilus* (type IIIC) has motile spores and is an aerobic soil organism. *Frankia* (type IIID) forms nonmotile spores in a sporogenous body (**figure 22.21**). It grows in symbiotic association

with the roots of at least eight families of higher nonleguminous plants (e.g., alder trees) and is a microaerophile that can fix atmospheric nitrogen.

The roots of infected plants develop nodules containing *Frankia* that fix nitrogen so efficiently that a plant such as an alder can grow in the absence of combined nitrogen (e.g., NO_3^-) (figure 22.21*b*). Within the nodule cells, *Frankia* forms branching hyphae with globular vesicles at their ends, which may be the sites of nitrogen fixation. The nitrogen-fixation process resembles that of *Rhizobium* in that it is oxygen sensitive and requires molybdenum and cobalt. ➤➤ *Microorganisms in terrestrial environments: Nitrogen fixation (section 26.2)*

Another genus in this suborder, *Sporichthya*, is one of the strangest of the actinomycetes. It lacks a substrate mycelium. The hyphae remain attached to the substratum by holdfasts and grow upward to form aerial mycelia that release motile, flagellate spores in the presence of water.

22.10 ORDER *BIFIDOBACTERIALES*

The order *Bifidobacteriales* has one family, *Bifidobacteriaceae*, and 10 genera (five of which have unknown affiliation). *Falcivibrio* and *Gardnerella* are found in the human genital/

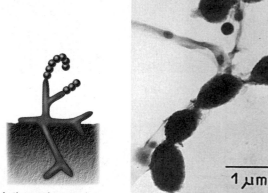

(a) *Actinomadura madurae*

(b)   *Streptosporangium*

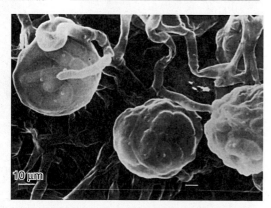

Figure 22.20 Maduromycetes. (a) *Actinomadura madurae* morphology; illustration and electron micrograph of a spore chain (×16,500). (b) *Streptosporangium* morphology; illustration and micrograph of *S. album* on oatmeal agar with sporangia and hyphae; SEM.

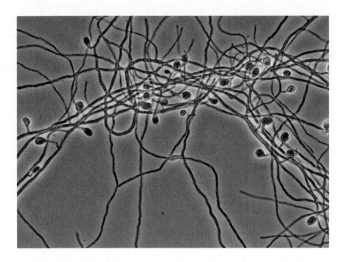

(a) **(b)**

Figure 22.21 *Frankia.* (a) A phase contrast micrograph showing hyphae, multilocular sporangia, and spores. (b) Nodules of the alder *Alnus rubra*.

urinary tract; *Gardnerella* is thought to be a major cause of bacterial vaginitis. *Bifidobacterium* probably is the best-studied genus. Bifidobacteria are nonmotile, nonsporing, gram-positive rods of varied shapes that are slightly curved and clubbed; often they are branched (**figure 22.22**). The rods can be single or in clusters and V-shaped pairs. *Bifidobacterium* is anaerobic and actively ferments carbohydrates to produce acetic and lactic acids but not carbon dioxide. It is found in the mouth and intestinal tract of warm-blooded vertebrates, in sewage, and in insects. *B. bifidum* is a pioneer colonizer of the human intestinal tract, particularly when babies are breast fed. A few *Bifidobacterium* infections have been reported in humans, but the genus does not appear to be a major cause of disease. In fact, several species are sold as probiotic agents. >> *Microbiology of fermented foods (section 34.6)*

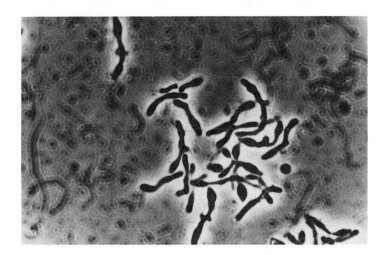

Figure 22.22 *Bifidobacterium.* *Bifidobacterium bifidum*; phase-contrast photomicrograph (×1,500).

1. Give the distinguishing properties of the actinoplanetes.
2. Describe the genus *Propionibacterium* and comment on its practical importance.
3. Describe the major properties of the genus *Streptomyces*.
4. Describe three ways in which *Streptomyces* is of ecological importance. Why do you think *Streptomyces* produce antibiotics?
5. Briefly describe the genera of the suborder *Streptosporangineae*. What is madurose? Why is *Actinomadura* important?
6. Describe *Frankia* and discuss its importance.
7. Characterize the genus *Bifidobacterium*. Where is it found and why is it significant?

Summary

22.1 General Properties of the Actinomycetes

a. Actinomycetes are aerobic, gram-positive bacteria that form branching, usually nonfragmenting, hyphae and asexual spores (**figure 22.2**).

b. The asexual spores borne on aerial mycelia are called spores if they are at the tip of hyphae or sporangiospores if they are within sporangia.

c. Actinomycetes have several distinctively different types of cell walls and often also vary in terms of the sugars present in cell extracts. Properties such as color and morphology are also taxonomically useful (**tables 22.1** and **22.2**).

d. *Bergey's Manual* classifies the high G + C bacteria phylogenetically using 16S rRNA data. The phylum *Actinobacteria* contains the actinomycetes and their high G + C relatives (**figure 22.4**).

22.2 Suborder *Actinomycineae*

a. The suborder *Actinomycineae* contains the genus *Actinomyces,* members of which are irregularly shaped, nonsporing rods that can cause disease in cattle and humans.

22.3 Suborder *Micrococcineae*

a. The suborder *Micrococcineae* includes the genera *Micrococcus, Arthrobacter,* and *Dermatophilus. Arthrobacter* has an unusual rod-coccus growth cycle and carries out snapping division **(figures 22.7** and **22.8).**

22.4 Suborder *Corynebacterineae*

a. The genera *Corynebacterium, Mycobacterium,* and *Nocardia* are placed in the suborder *Corynebacterineae.*

b. *Corynebacterium diphtheriae* causes the disease diphtheria **(figure 22.10).**

c. Mycobacteria form either rods or filaments that readily fragment. Their cell walls have a high lipid content and mycolic acids; the presence of these lipids makes them acid-fast **(figures 22.11** and **22.12).**

d. *Mycobacterium tuberculosis,* as well as *M. bovis* and *M. africanum,* cause tuberculosis. About one-third of the world's population is infected with either latent or active TB **(figure 22.13).** *M. leprae* is the causative agent of leprosy **(figure 22.14).**

e. Nocardioform actinomycetes have hyphae that readily fragment into rods and coccoid elements, and often form aerial mycelia with spores **(figure 22.15).**

22.5 Suborder *Micromonosporineae*

a. The suborder *Micromonosporineae* has genera that include *Micromonospora* and *Actinoplanes.* These actinomycetes have an extensive substrate mycelium and form special aerial sporangia **(figure 22.16).** They are present in soil, freshwater, and the ocean. The soil forms are probably important in decomposition.

22.6 Suborder *Propionibacterineae*

a. The genus *Propionibacterium* is in the suborder *Propionibacterineae.* Members of this genus are common skin and intestinal inhabitants, and are important in cheese manufacture and the development of acne vulgaris.

22.7 Suborder *Streptomycineae*

a. The suborder *Streptomycineae* includes the genus *Streptomyces.* Members of this genus have type I cell walls and aerial hyphae bearing chains of 3 to 50 or more nonmotile spores **(figures 22.17** and **22.18).**

b. Streptomycetes are important in the degradation of resistant organic material in the soil and produce many useful antibiotics. A few cause diseases in plants and animals. However, they are most notable for the number of antibiotics and other medically and industrially important compounds they produce.

22.8 Suborder *Streptosporangineae*

a. Many genera in suborder *Streptosporangineae* have the sugar derivative madurose and type III cell walls.

22.9 Suborder *Frankineae*

a. The genera *Frankia* and *Geodermatophilus* are placed in the suborder *Frankineae.* They produce clusters of spores at hyphal tips and have type III cell walls. *Frankia* grows in symbiotic association with nonleguminous plants and fixes nitrogen **(figure 22.21).**

22.10 Order *Bifidobacteriales*

a. The genus *Bifidobacterium* is placed in the order *Bifidobacteriales.* This irregular, anaerobic rod is one of the first colonizers of the intestinal tract in nursing babies.

Critical Thinking Questions

1. Even though these are high G + C organisms, there are regions of the genome that must be more AT-rich. Suggest a few such regions and explain why they must be more AT-rich.

2. Choose two different species in the phylum *Actinobacteria* and investigate their physiology and ecology. Compare and contrast these two organisms. Can you determine why having a high G + C genomic content might confer an evolutionary advantage?

3. *Streptomyces coelicolor* is studied as a model system for cellular differentiation. Some of the genes involved in sporulation contain a rare codon not used in vegetative genes. Suggest how *Streptomyces* might use the rare codon to regulate sporulation.

4. Suppose that you discovered a nodulated plant that could fix atmospheric nitrogen. How might you show that a bacterial symbiont was involved and that *Frankia* rather than *Rhizobium* was responsible?

Learn More

Learn more by visiting the Prescott website at www.mhhe.com/prescottprinciples, where you will find a complete list of references.

Eucaryotic Microbes

23

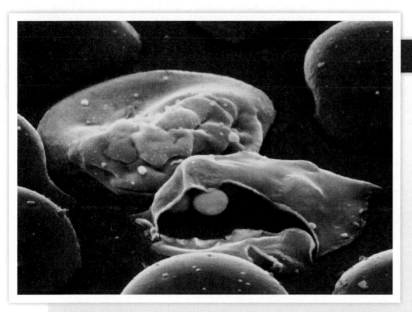

Malaria parasites (yellow) bursting out of red blood cells. Malaria has played an important part in the rise and fall of nations (see the chapter opening quote), and has killed untold millions the world over. Despite the combined efforts of 102 countries to eradicate malaria, it remains the most important disease in the world today in terms of lives lost and economic burden.

ascomycetes   A division of fungi that form ascospores.

apicomplexans   Protists that lack special locomotor organelles but have an apical complex and a spore-forming stage. All are intra- or extracellular parasites of animals.

basidiomycetes   A division of fungi in which the spores are borne on club-shaped organs called basidia.

chytrids   A term used to describe the *Chytridiomycota,* which are simple terrestrial and aquatic fungi that produce motile zoospores with single, posterior, whiplash flagella.

ciliates (*Ciliophora*)   Protists that move by means of rapidly beating cilia and belong to the *Alveolata* (super group *Chromalveolata*).

Chapter Glossary

cyst   The inactive or resting stage of a protozoan.

diatoms (*Bacillariophyta*)   Photosynthetic protists with siliceous cell walls called frustules. They constitute a substantial fraction of the phytoplankton.

dinoflagellates (*Dinoflagellata*)   Photosynthetic protists characterized by two flagella used in swimming in a spinning pattern.

filopodia   Long, narrow pseudopodia found in certain protists, for example, the *Rhizaria.*

holozoic nutrition   A nutritional strategy in which food items (such as bacteria) are endocytosed with the subsequent formation of a food vacuole or phagosome.

lobopodia   Rounded pseudopodia found in some amoeboid protists.

mycosis (pl., mycoses)   Any disease caused by a fungus.

osmotrophy   A form of nutrition in which soluble nutrients are absorbed through the cytoplasmic membrane; found in procaryotes, fungi, and some protists.

reticulopodia   Netlike pseudopodia found in certain protists, for example, *Foraminifera.*

saprophyte   An organism that takes up nonliving organic nutrients in dissolved form and usually grows on decomposing organic matter.

sporozoite   The motile, infective stage of apicomplexan protists, including *Plasmodium,* the causative agent of malaria.

trophozoite   The active, motile feeding stage of a protozoan.

yeast   A unicellular, uninuclear fungus that reproduces either asexually by budding or fission, or sexually through spore formation.

zygomycetes   A division of fungi that usually has a coenocytic mycelium with chitinous cell walls and lacks motile spores. Sexual reproduction normally involves the formation of zygospores.

Historians believe that malaria has probably had a greater impact on world history than any other infectious disease, influencing the outcome of wars, various population movements, and the development and decline of various civilizations.

—*Lynne S. Garcia*

So far, our discussion of microbial diversity has focused on the procaryotes. In this chapter, we turn our attention to protists and fungi. We follow the higher-level classification scheme of eucaryotes as proposed by the International Society of Protistologists, Our goal is to survey the vast diversity in the eucaryotic microbial world with an emphasis on the adaptations that make these microorganisms successful competitors. The general structure and reproduction of eucaryotic microbes are reviewed in chapter 4.

Eucaryotic microorganisms suffer from a confused taxonomic history. The kingdom Protista, as presented in many introductory

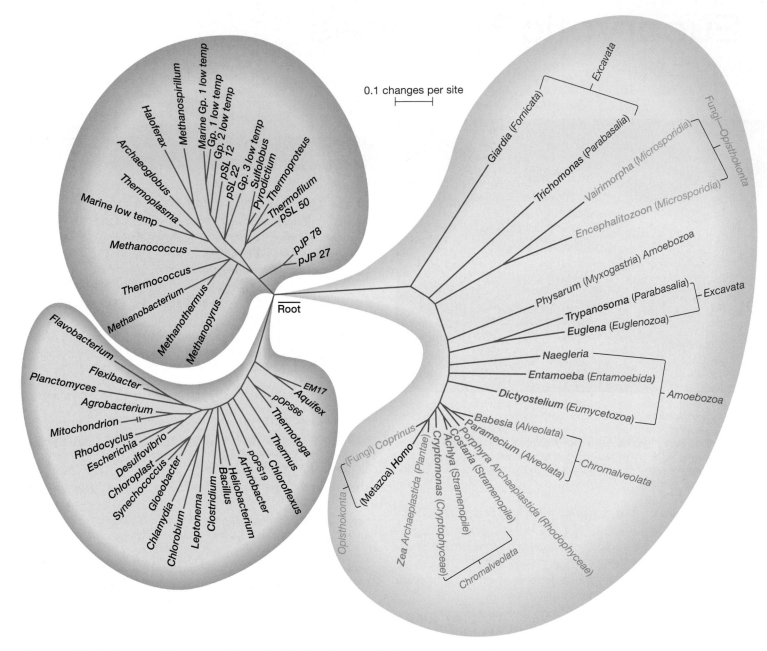

Figure 23.1 **Universal Phylogenetic Tree, Highlighting Eucaryotic Microorganisms.** Recent molecular phylogenetic evidence in combination with morphological and biochemical data has resulted in the establishment of super groups among the *Eucarya* (each super group is highlighted with a different color). This new, higher-order classification scheme is not entirely congruent with the placement of the protists on the universal tree of life; the latter is based entirely on analysis of SSU rRNA. Nonetheless, it demonstrates that the protists are a highly polyphyletic group. The fungi are monophyletic; *Caprinus* is shown as the representative fungal group. The *Microsporidia* have a confused taxonomic history but are now considered by most to be fungi.

biology texts, is an artificial grouping of over 64,000 different single-celled life forms that lack common evolutionary heritage; that is to say, they are polyphyletic (**figure 23.1**). In fact, the protists are unified only by what they lack: absent is the level of tissue organization found in the more evolved fungi, plants, and animals. The term **protozoa** (s., protozoan; Greek *protos,* first, and *zoon,* animal) traditionally refers to chemoorganotrophic protists; protozoology is the study of protozoa. The term **algae** can be used to describe photosynthetic protists. It was originally used to refer to all "simple aquatic plants," but this term has no phylogenetic utility. The study of photosynthetic protists (algae) is often referred to as **phycology** and is the realm of both botanists and protistologists.

The study of all protists, regardless of their metabolic type, is called **protistology.**

By contrast, the *Fungi* are monophyletic and are sometimes referred to as the true fungi or *Eumycota* (Greek *eu,* true, and *mykes,* fungus). Microbiologists use the term **fungus** (pl., fungi; Latin *fungus,* mushroom) to describe eucaryotic organisms that are spore-bearing, have absorptive nutrition, lack chlorophyll, and reproduce sexually and asexually. Scientists who study fungi are **mycologists** and the scientific discipline devoted to fungi is called **mycology.** The study of fungal toxins and their effects is called **mycotoxicology,** and the diseases caused by fungi in animals are known as **mycoses** (s., mycosis).

23.1 INTRODUCTION

Most eucaryotes are microbes, so it is no surprise that the vast diversity of protists is a function of their capacity to thrive in a wide variety of habitats. Their one common requirement is moisture because all are susceptible to desiccation. Most protists are free living and inhabit freshwater or marine environments. Many terrestrial chemoorganotrophic forms can be found in decaying organic matter and in soil, where they are important in recycling the essential elements nitrogen and phosphorus. Others are planktonic—floating free in lakes and oceans. Planktonic microbes (both procaryotic and eucaryotic) are responsible for a majority of the nutrient cycling that occurs in these ecosystems.

Every major group of protists includes species that live in association with other organisms. For instance, some photosynthetic protists associate with fungi to form lichens, while others live with corals, where they provide fixed carbon to the coral animal. Thousands of others are parasites and cause important diseases in humans and domesticated animals. Finally, protists are useful in biochemical and molecular biological research. Not only do they display an amazing array of unique adaptations, many biochemical pathways used by protists are found in other eucaryotes, making them useful model organisms.

Unlike protists, fungi are primarily terrestrial organisms, although a few are found in aquatic ecosystems. They have a global distribution from polar to tropical regions. With procaryotes and a few other groups of chemoorganotrophic organisms, fungi act as decomposers, a role of enormous significance. They degrade complex organic materials in the environment to simple organic compounds and inorganic molecules. In this way, carbon, nitrogen, phosphorus, and other critical constituents of dead organisms are released and made available for living organisms. Many fungi are pathogenic and infect plants and animals. Over 5,000 species attack economically valuable crops, garden plants, and many wild plants, and about 20 new human fungal pathogens are documented each year. Fungi also form beneficial relationships with other organisms. For example, the vast majority of vascular plant roots form associations with fungi, called mycorrhizae. >> *Microorganisms in terrestrial environments: Mycorrhizae (section 26.2); Microbial interactions (section 27.1)*

Fungi, especially the yeasts, are essential to many industrial processes involving fermentation. Examples include the making of bread, wine, beer, cheeses, and soy sauce. They are also important in the commercial production of many organic acids (citric, gallic) and certain drugs (ergometrine, cortisone), and in the manufacture of many antibiotics (penicillin, griseofulvin) and the immunosuppressive drug cyclosporin. In addition, fungi are important research tools in the study of fundamental biological processes. Cytologists, geneticists, biochemists, biophysicists, and microbiologists regularly use fungi in their research. The yeast *Saccharomyces cerevisiae* is the best understood eucaryotic cell. It has been a valuable model organism in the study of cell biology, genetics, and cancer. << *Biotechnology and industrial microbiology (chapter 16);* >> *Microbiology of food (chapter 34)*

Fungi and many chemoheterotrophic protists are **saprophytes,** securing nutrients from dead organic material by releasing degradative enzymes into the environment. They then absorb the soluble products—a process sometimes called **osmotrophy.** Other chemoheterotrophic protists employ **holozoic nutrition,** in which solid nutrients are acquired by phagocytosis. Photoautotrophic protists are strict aerobes and, like cyanobacteria, use photosystems I and II to perform oxygenic photosynthesis. Finally, it is difficult to classify the nutritional strategies of some protists. Certain protists simultaneously use both organic and inorganic carbon compounds. This strategy is sometimes called **mixotrophy.**

1. How can protists and fungi be defined?
2. What is the distribution of these microbes?
3. Why do you think the protists and fungi are such important model organisms?
4. Contrast and compare the nutritional strategies of the fungi and protists. Why do you think the mycelial morphology of the fungi makes them especially effective saprophytes?

23.2 PROTIST CLASSIFICATION

Ever since Antony van Leeuwenhoek described the first protozoan "animalcule" in 1674, the taxonomic classification of the protists has remained in flux. During the twentieth century, classification schemes were based on morphology rather than evolutionary relationships. Protists were often classified into four major groups based on their means of locomotion: flagellates (*Mastigophora*), ciliates (*Infusoria* or *Ciliophora*), amoebae (*Sarcodina*), and stationary forms (*Sporozoa*). Although nonprotistologists still use these terms, these divisions have no bearing on evolutionary relationships and should be avoided. It is now agreed that the old classification system is best abandoned, but little agreement exists on what should take its place. However, recent morphological, biochemical, and phylogenetic analyses have resulted in the development of a higher-level classification system for the eucaryotes. This scheme, as proposed by the International Society of Protistologists, is presented in **table 23.1**. This scheme does not use formal hierarchical rank designations such as class and order, reflecting the fact that protist taxonomy remains an area of active research.

Super Group *Excavata*

The *Excavata* includes some of the most primitive, or deeply branching, eucaryotes (figure 23.1). Most possess a cytostome characterized by a suspension-feeding groove with a posteriorly directed flagellum that is used to generate a feeding current. This enables the capture of small particles. Those that lack this morphological feature are presumed to have had it at one time during their evolution—that is to say, it is thought to have been secondarily lost.

Table 23.1 Classification of the Protists as Proposed by the International Society of Protistologists[a]

Super Group	Unifying Features	First Rank	General Description
Archaeplastida	Photosynthetic plastid with chlorophyll *a* from ancestral cyanobacterial endosymbiont, plastid later lost in some; uses starch as a storage product; usually have cell wall made of cellulose	*Glaucophyta*	Plastid is a cyanelle, which unlike chloroplasts has peptidoglycan between the two membranes; stacked thylakoids with chlorophyll *a* and phycobiliproteins and other pigments; includes flagellated and nonflagellated species or life cycle stages. Includes *Cyanophora*, *Glaucocystis*, and *Gloeochaeta*.
		Rhodophycae	Lacks flagellated stages, kinetosome, and centrioles, unstacked thylakoids; chloroplast without external endoplasmic reticulum; also called red algae, although traditional subgroups are no longer considered valid. Includes *Ceramium*, *Porphyra*, and *Sphaerococcus*.
		Chloroplastida	Pyrenoid often within plastid, which features chlorophylls *a* and *b*; cellulose-containing cell wall typical. Includes the *Charophya* (green algae), *Chara*, *Nitella*, and *Volvox*.
Amoebozoa	Amoeboid motility with lobopodia; naked or testate; mitochondria with tubular cristae; usually uninucleate, sometimes multinucleate; cysts common	*Tubulinea*	Naked or testate with tubular pseudopodia; lacks centrosomes; motility based on actinomyosin cytoskeleton; cytoplasmic microtubules rare; lacks flagellated stages. Includes *Amoeba*, *Hydramoeba*, *Flabellula*, and *Rhizamoeba*.
		Flabellinea	Flattened amoebae lacking tubular pseudopodia; motility based on actinomyosin cytoskeleton; lacks centrosome and flagellated stages. Includes *Dermamoeba*, *Podostoma*, *Sappinia*, and *Thecamoeba*.
		Stereomyxida	Branched or reticulate plasmodial organisms; trilaminate centrosome. Includes *Corallomyxe* and *Stereomyxa*.
		Acanthamoebidae	Thin glycocalyx; prominent subpseudopodia that narrow to a blunt or fine tip (acanthopodia); single flagellum; cysts usually have a double wall; motility based on actinomyosin cytoskeleton; centriole-like body observed. Includes *Acanthamoeba*, *Balamuthia*, and *Protacanthamoeba*.
		Entamoebida	No flagella or centrioles; lacks mitochondria, hydrogenosomes, and peroxisomes. Includes *Entamoeba*.
		Mastigamoebidae	Amoeboid with several pseudopodia; single flagellum projecting forward; single kinetosome and nucleus, although some multinucleate; lacks mitochondria; inhabits low-oxygen to anoxic, nutrient-rich environments. Includes *Mastigella* and *Mastigamoeba*.
		Pelomyxa	Multiple cilia; anaerobic; lacks mitochondria, hydrogenosomes, and peroxisomes; polymorphic life cycle with multinucleate stages; some are symbionts. Includes *Pelomyxa paulstris*.
		Eumycetozoa	Amoeboid organisms that produce fruiting body-like structures; both cellular and acellular slime molds. Includes *Dictyostelium*, *Physarum*, *Hemitichia*, and *Stemonitis*.

Super Group	Unifying Features	First Rank	General Description
Rhizaria	Possesses thin pseudopodia (filopodia)	Cercozoa	Biciliated and/or amoeboid; most have mitochondria with tubular cristae; many encyst; kinetosomes connected to nucleus with cytoskeleton. Includes *Cercomonas*, *Katabia*, *Medusetta*, and *Sagosphaera*.
		Haplosporidia	Plasmodial endoparasites of marine and freshwater animals; distinct mechanism of spore formation; mitochondria with tubular cristae. Includes *Haplosporidium*, *Minchinia*, and *Urosporidum*.
		Foraminifera	Filopodia with granular cytoplasm that forms a complex network of reticulopodia. Simplest forms are open tubes or hollow spheres; in others, shells, called tests, are divided into chambers added during growth. Tests made of organic compounds or inorganic particles cemented together, or crystalline calcite. Includes *Allogromia*, *Carpenteria*, *Globigerinella*, *Lana*, and *Textularia*.
		Gromia	Organic tests; branched filipodia; nongranular cytoplasm; flagellated dispersal cells or gametes. Includes *Gromia*.
		Radiolaria	Many species exhibit radial symmetry, from which the name is derived. All have a porous capsular cell wall through which axopodia project. Skeletons, when present, made of amorphous silica (opal) or strontium sulfate; morphology can be simple to ornate; mitochondria with tubular cristae; axopodia supported by internal microtubules. Includes *Acanthometra*, *Lophospyris*, and *Stauracon*.
Chromalveolata	Plastid from secondary endosymbiosis with an ancestral archaeplastid; plastid then lost in some; reacquired in others	Cryptophyceae	Auto-, mixo-, and heterotrophic forms with ejectosomes or trichocysts (dartlike structures used for defense); mitochondria with flat cristae; biflagellated; tubular channels and/or longitudinal groove lined with ejectosomes; chlorophyll a and c_2 if chloroplasts present. Includes *Campylomonas*, *Cryptomonas*, *Goniomonas*, and *Rhodomonas*.
		Haptophyta	Scales cover cell; solitary or colonial; motile cells biflagellate and usually have a haptonema (thin appendage between flagella used for prey capture and attachment to substrates); outer nuclear membrane continuous with chloroplast membrane; auto-, mixo-, and heterotrophic forms. Includes the coccoliths and *Diacronema*, *Isochrysis*, and *Phaeocystis*.
		Stramenopiles	Motile cells usually with two flagella, heterokont flagellation typical. Includes diatoms and labyrinthulids.
		Alveolata	Mixo- or heterotrophic. Includes dinoflagellates, *Apixcomplexa* (e.g., *Plasmodium*), and *Ciliphora* (e.g., *Paramecium* and *Stentor*).
Excavata	Suspension feeding groove (cytostome) present or presumed to have been lost; feed by a flagella-generated current	Fornicata	Lacks typical mitochondria; uninucleate; usually has a feeding groove. Includes *Giardia*.
		Malawimonas	Has mitochondria, two kinetosomes, and a single ventral flagellar vane. Includes *Malawinas*.
		Parabasalia	Has parabasal structure; striated parabasal fibers connect Golgi to flagellar apparatus; up to thousands of flagella; hydrogenosomes present. Includes *Calonympha*, *Holomastigotes*, *Spirotrichosoma*, and *Trichomonas*.
		Preaxostyla	Unicellular; flagellated; no mitochondria; heterotrophic. Includes *Dinenympha*, *Polymastix*, and *Streblomastix*.
		Jakobida	Two flagella placed at top of wide ventral feeding groove. Includes *Jakoba* and *Histiona*.
		Heterolobosea	Heterotrophic amoebae with eruptive pseudopodia; if flagellated, has 2 or 4 parallel flagella. Includes *Acrasis*, *Gruberella*, and *Rosculus*.
		Euglenozoa	One or two flagella inserted into apical or subapical pocket; usually with tubular feeding apparatus; two kinetosomes; mitochondria have discoid cristae. Includes *Dinema*, *Euglena*, *Leishmania*, *Trypanoplasma* and *Trypanosoma*.

[a]Adapted from: Adl, S. M.; Simpson, A. G. B.; Farmer, M. A.; Anderson, R. A.; Ancerson, O. R.; Barta J. R.; Browser, S. S.; et al. 2005. The new higher level classification of Eukaryotes with emphasis on the taxonomy of protists. *J Eukaryot. Microbiol.* 52:399–451.

Fornicata

Members of the *Fornicata* have flagella but lack mitochondria. Most are harmless symbionts; the few free-living forms are most often found in waters that are heavily polluted with organic nutrients. Only asexual reproduction by binary fission has been observed. The group includes the animal pathogens *Hexamida salmonis,* a troublesome fish parasite found in hatcheries and fish farms, and *H. meleagridis,* a turkey pathogen that is responsible for the annual loss of millions of dollars in poultry revenue.

The most important species in this group is *Giardia intestinalis,* discovered by van Leeuwenhoek in 1681 when he examined his own stools (**figure 23.2a**). This protist causes the very common intestinal disease **giardiasis,** which has a worldwide distribution. In the United States, this protist is the most common cause of epidemic waterborne diarrheal disease (about 30,000 cases yearly), affecting children more severely than adults.

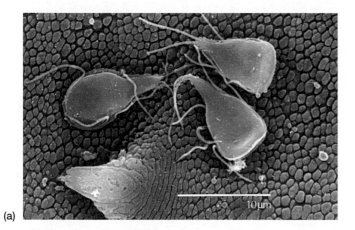

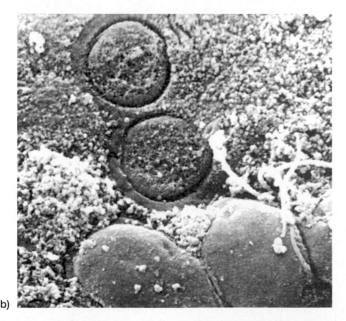

Figure 23.2 Scanning Electron Micrograph of *Giardia* sp.
(a) This pathogenic protist is shown attached to the parasitic flatworm *Echinostoma caproni.* (b) Upon detachment from the epithelium, the protozoa often leave clear impressions on the microvillus surface (upper circles); scanning electron micrographs.

Approximately 7% of the population are asymptomatic carriers who shed **cysts** in their feces. *G. intestinalis* is endemic in child day-care centers in the United States, with estimates of 5 to 15% of diapered children being infected. Transmission occurs most frequently by cyst-contaminated water supplies. Epidemic outbreaks have been recorded in wilderness areas, suggesting that humans may be infected from pristine stream water with *Giardia* harbored by rodents, deer, cattle, or household pets. This implies that human infections also can be a zoonosis. As many as 200 million humans may be infected worldwide.

Following ingestion, the cysts undergo excystment in the duodenum, forming trophozoites. The **trophozoites** inhabit the upper portions of the small intestine, where they attach to the intestinal mucosa by means of their sucking disks (figure 23.2b). The ability of the trophozoites to adhere to the intestinal epithelium accounts for the fact that they are rarely found in stools (van Leeuwenhoek was just lucky). It is thought that the trophozoites feed on mucous secretions and reproduce to form such a large population that they interfere with nutrient absorption by the intestinal epithelium. Giardiasis can be acute or chronic. Acute giardiasis is characterized by severe diarrhea, epigastric pain, cramps, voluminous flatulence ("passing gas"), and anorexia. Chronic giardiasis is characterized by intermittent diarrhea, with periodic appearance and remission of symptoms. ▶▶ *Water purification and sanitary analysis (section 35.1); Wastewater treatment (section 35.2)*

Parabasalia

Members of the *Parabasalia* are flagellated; most are endosymbionts of animals. Without a distinct cytostome, they use phagocytosis to engulf food items. Here, we consider two subgroups: the *Trichonymphida* and the *Trichomonadida. Trichonymphida* are obligate mutualists in the digestive tracts of wood-eating insects such as termites and wood roaches, where they secrete the enzyme cellulase needed for the digestion of wood particles, which they entrap with pseudopodia. One species, *Trichonympha campanula,* can account for up to one-third of the biomass of an individual termite. This species is particularly large for a protist (several hundred micrometers) and can bear several thousand flagella. Although asexual reproduction is the norm, a hormone called ecdysone produced by the host when molting triggers sexual reproduction.

Trichomonadida, or simply the trichomonads, do not require oxygen and possess hydrogenosomes rather than mitochondria (*see figure 4.20*). They undergo asexual reproduction only. They are symbionts of the digestive, reproductive, and respiratory tracts of many vertebrates, including humans. *Tritrichomonas foetus* is a cattle parasite and an important cause of spontaneous abortion in these animals. Four species infect humans: *Dientamoeba fragilis, Pentatrichomonas hominis, Trichomonas tenax,* and *T. vaginalis. D. fragilis* has recently been recognized as a cause of diarrhea, while *T. vaginalis* has long been known to be pathogenic. Found in the genitourinary tract of both men and women, most *T. vaginalis* strains are either not pathogenic or only mildly so. Nonetheless, the sexual transmission of pathogenic strains accounts for an

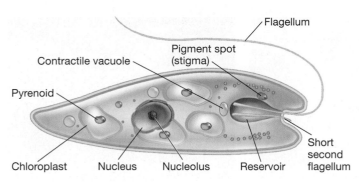

Figure 23.3 *Euglena:* **Principle Structures Found in this Euglenoid.** Notice that a short second flagellum does not emerge from the anterior invagination. In some euglenoids both flagella are emergent.

estimated 7 million cases of trichomoniasis annually in the United States and 180 million cases annually worldwide. **Trichomoniasis** triggers the accumulation of leukocytes at the site of the infection. In females, this usually results in a profuse, purulent vaginal discharge that is yellowish to light cream in color and characterized by a disagreeable odor. The discharge is accompanied by itching. Males are generally asymptomatic because of the trichomonacidal action of prostatic secretions; however, at times a burning sensation occurs during urination.

Euglenozoa

The *Euglenozoa* are commonly found in freshwater, although a few species are marine. About one-third of euglenids are photo-autotrophic; the remaining are free-living chemoorganotrophs. Most chemoorganotrophic forms are saprophytic, although a few parasitic species have been described. The representative genus is the photoautotroph *Euglena.* A typical *Euglena* cell (**figure 23.3**) is elongated and bounded by a plasmalemma. The pellicle consists of proteinaceous strips and microtubules; it is elastic enough to enable turning and flexing of the cell, yet rigid enough to prevent excessive alterations in shape. *Euglena* contains chlorophylls *a* and *b* together with carotenoids. The large nucleus contains a prominent nucleolus. The primary storage product is paramylon (a polysaccharide composed of β-1→3 linked glucose molecules), which is unique to euglenoids. A red eyespot called a **stigma** helps the organism orient to light and is located near an anterior reservoir. A contractile vacuole near the reservoir continuously collects water from the cell and empties it into the reservoir, thus regulating the osmotic pressure within the organism. Two flagella arise from the base of the reservoir, although only one emerges from the canal and actively beats to move the cell. Reproduction in euglenoids is by longitudinal mitotic cell division.

Several protists of medical relevance belong to the *Euglenozoa.* These include members of the genus *Leishmania,* which cause a group of conditions, collectively termed leishmaniasis. The trypanosomes are also of great importance. These microbes exist only as parasites of plants and animals, and have global significance. These microbes and the diseases they cause are discussed next.

Leishmaniasis Worldwide, 2 million new cases of leishmaniasis and about 60,000 deaths occur each year, with one-tenth of the world's population at risk of infection. The primary reservoirs of these parasites are canines and rodents. All species of *Leishmania* use female sand flies such as those of the genera *Lutzomyia* and *Phlebotomus* as intermediate hosts to transmit leishmanias from animals to humans or between humans. When an infected sand fly takes a human blood meal, it introduces flagellated promastigotes into the skin of the definitive (human) host. Athough the promastigotes are engulfed by macrophages, they multiply by binary fission and form small, nonmotile cells called amastigotes. These destroy the host cell, and they are then engulfed by other macrophages in which they continue to develop and multiply. *Leishmania braziliensis,* which has an extensive distribution in forest regions of tropical America, causes **mucocutaneous leishmaniasis** (espundia) (**figure 23.4***a*). The disease produces lesions involving the mouth, nose, throat, and skin, and results in extensive scarring and disfigurement. *Leishmania donovani* is endemic in large areas within northern China, eastern India, the Mediterranean countries, the Sudan, and Latin America. It produces **visceral leishmaniasis** (kala-azar), which involves the monocyte-macrophage system (*see figure 28.3*) and often results in intermittent fever and spleen and liver enlargement.

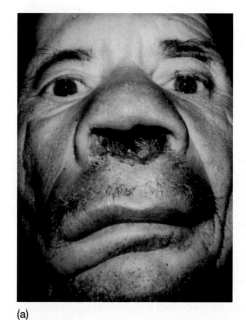
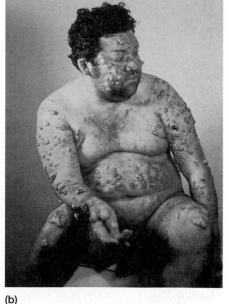

(a)   (b)

Figure 23.4 **Leishmaniasis.** (a) A person with mucocutaneous leishmaniasis, which has destroyed the nasal septum and deformed the nose and lips. (b) A person with diffuse cutaneous leishmaniasis.

Leishmania tropica and *L. mexicana* occur in the more arid regions of the Eastern Hemisphere and cause **cutaneous leishmaniasis.** *L. mexicana* is also found in the Yucatan Peninsula of Mexico and has been reported as far north as Texas. In this disease, a relatively small, red papule forms at the site of each insect bite—the inoculation site. These papules are frequently found on the face and ears. They eventually develop into crustated ulcers (figure 23.4*b*). Healing occurs with scarring and a permanent immunity.

Trypanosomiasis Another group of flagellated protists called trypanosomes cause the aggregate of diseases termed trypanosomiasis. *Trypanosoma brucei gambiense,* found in the rain forests of west and central Africa, is the agent of **West African sleeping sickness.** *T. brucei rhodesiense,* found in the upland savannas of east Africa, is the agent of **East African sleeping sickness.** Reservoirs for these trypanosomes are domestic

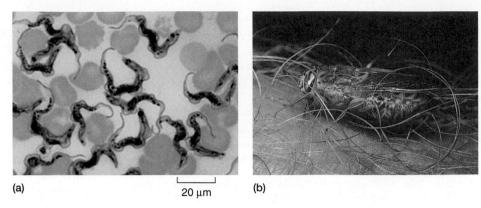

(a)   20 μm   (b)

Figure 23.5 **The Euglenozoan *Trypanosoma* and Its Insect Host.** (a) *Trypanosoma* among red blood cells. Note the dark-staining nuclei, anterior flagella, and undulating changeable shape (×500). (b) The tsetse fly, shown here sucking blood from a human arm, is an important vector of the *Trypanosoma* species that cause African sleeping sickness.

cattle and wild animals, within which the parasites cause severe malnutrition. Both species use tsetse flies (genus *Glossina*) as intermediate hosts (**figure 23.5**).

The trypanomastigote parasites are transmitted through the bite of the fly to humans (**figure 23.6**). The protists pass through

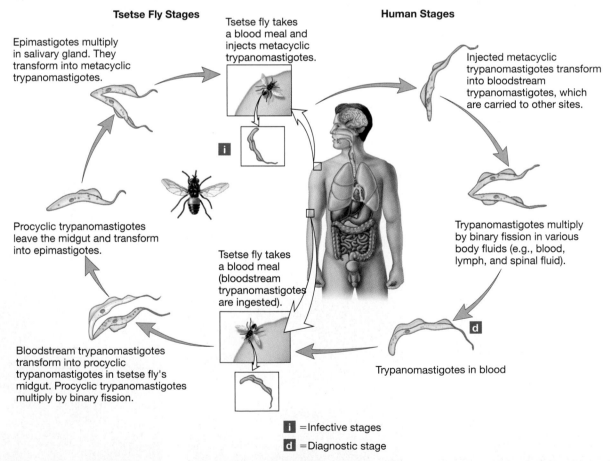

Tsetse Fly Stages

Epimastigotes multiply in salivary gland. They transform into metacyclic trypanomastigotes.

Tsetse fly takes a blood meal and injects metacyclic trypanomastigotes.

Human Stages

Injected metacyclic trypanomastigotes transform into bloodstream trypanomastigotes, which are carried to other sites.

Procyclic trypanomastigotes leave the midgut and transform into epimastigotes.

Trypanomastigotes multiply by binary fission in various body fluids (e.g., blood, lymph, and spinal fluid).

Tsetse fly takes a blood meal (bloodstream trypanomastigotes are ingested).

Bloodstream trypanomastigotes transform into procyclic trypanomastigotes in tsetse fly's midgut. Procyclic trypanomastigotes multiply by binary fission.

Trypanomastigotes in blood

i =Infective stages

d =Diagnostic stage

Figure 23.6 **Life Cycle of *Trypanosoma brucei.*** This trypanosome is transmitted to humans through the bite of the tsetse fly. The metacyclic trypanomastigotes enter the bloodstream where they disseminate to various tissue sites. They reproduce in the human and are ingested by tsetse flies as part of the blood meal. In the fly gut, they transform into procyclic trypanomastigotes and then into epimastigotes. Epimastogotes migrate to the salivary glands where they transform into metacyclic trypanomastigotes that can be passed on during feeding.

the lymphatic system and enter the bloodstream, and replicate by binary fission as they pass to other tissues. The tsetse fly bite is painful and can develop into a chancre. Victims may exhibit symptoms of fever, severe headaches, extreme fatigue, muscle and joint aches, irritability, and swollen lymph nodes. Some may develop a skin rash. The protists cause interstitial inflammation and necrosis within the lymph nodes and small blood vessels of the brain and heart. The disease gets its name from the fact that patients often exhibit lethargy—characteristically lying prostrate, drooling from the mouth, and showing insensitivity to pain; they also exhibit progressive confusion, slurred speech, seizures, and personality changes. In *T. brucei rhodesiense* infection, the disease develops so rapidly that infected individuals often die within a year. In *T. brucei gambiense* infection, invasion of the central nervous system causes necrotic damage, which gives rise to a variety of nervous disorders, including the characteristic sleeping sickness. Usually the victim dies in 2 to 3 years. Trypanosomiasis is such a problem in parts of Africa that millions of square miles are not fit for human habitation. Worldwide, over 40,000 new cases of both East and West African sleeping sicknesses occur each year. However, it is likely that the majority of cases are not reported due to the lack of public health infrastructure in endemic regions. As a result, more than 100,000 new cases per year are likely.

T. cruzi causes American trypanosomiasis, also called **Chagas' disease,** which occurs in the tropics and subtropics of continental America. The parasite uses the triatomine (kissing) bug as a vector (**figure 23.7**). As the triatomine bug takes a blood meal, the parasites are discharged in the insect's feces. Some trypanosomes enter the bloodstream through the wound and invade the liver, spleen, lymph nodes, and central nervous system. Cell invasion stimulates the trypanosome's transformation into amastigotes, resulting in clinical manifestations of infection. The bloodstream trypanomastigotes do not replicate, however, until they enter a cell (becoming amastigotes) or are ingested by the arthropod vector (becoming epimastigotes). In some parts of Latin America, a high percentage of heart disease is due to parasitized cardiac cells. There are 16 to 18 million new cases each year and over 50,000 deaths.

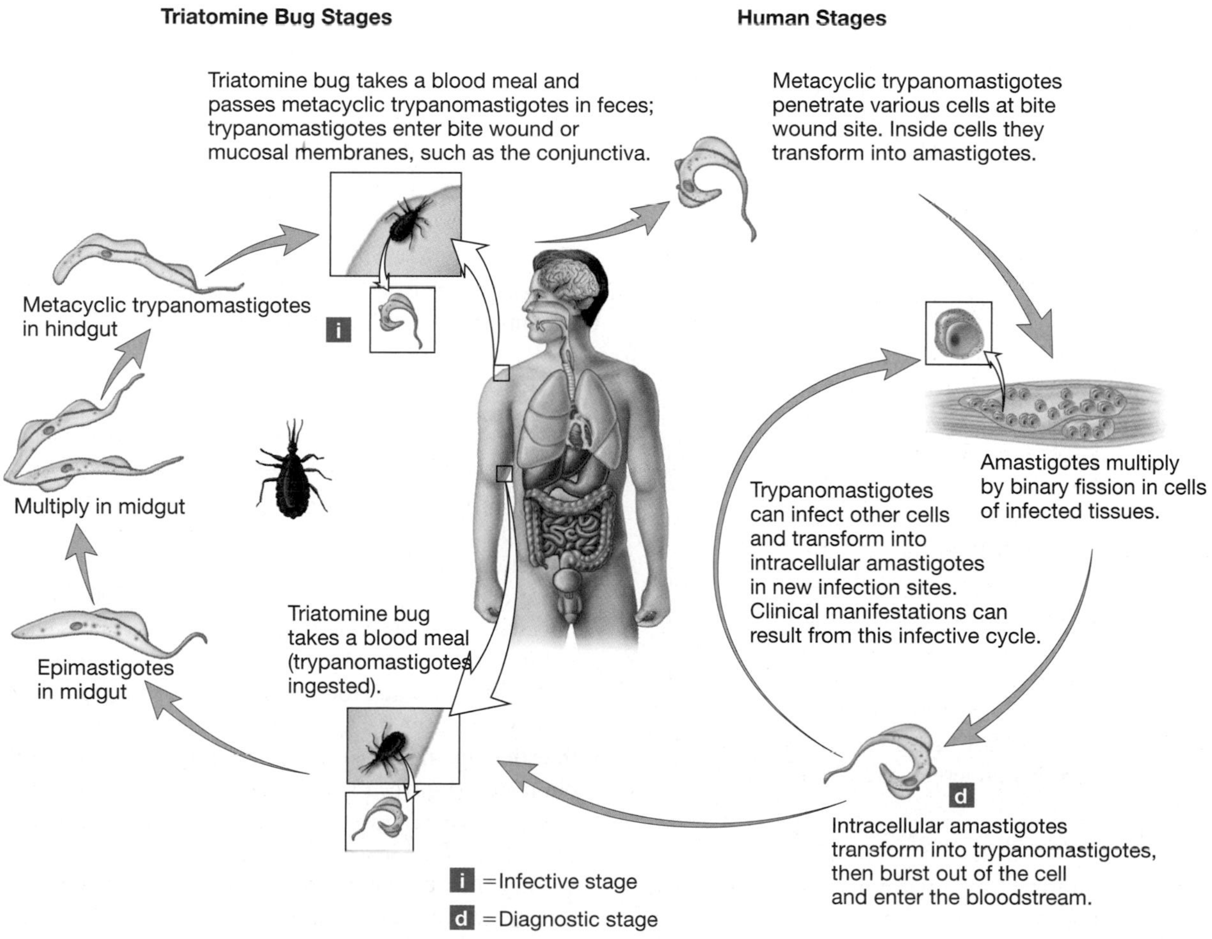

Triatomine Bug Stages

Triatomine bug takes a blood meal and passes metacyclic trypanomastigotes in feces; trypanomastigotes enter bite wound or mucosal membranes, such as the conjunctiva.

Metacyclic trypanomastigotes in hindgut

Multiply in midgut

Epimastigotes in midgut

Triatomine bug takes a blood meal (trypanomastigotes ingested).

Human Stages

Metacyclic trypanomastigotes penetrate various cells at bite wound site. Inside cells they transform into amastigotes.

Amastigotes multiply by binary fission in cells of infected tissues.

Trypanomastigotes can infect other cells and transform into intracellular amastigotes in new infection sites. Clinical manifestations can result from this infective cycle.

Intracellular amastigotes transform into trypanomastigotes, then burst out of the cell and enter the bloodstream.

i = Infective stage
d = Diagnostic stage

Figure 23.7 Life Cycle of *Trypanosoma cruzi.* This trypanosome is transmitted to humans from the triatomine bug. The metacyclic trypanomastigotes are shed in the bug feces, which enter the host through the bite wound or mucous membranes. The trypanomastigotes penetrate host cells and transform into amastigotes. After replication by binary fission, amastigotes transform into trypanomastigotes, which burst from their host cell to disseminate throughout the host. Blood-borne trypanomastigotes can be ingested by triatomine bugs as part of a blood meal, transform into epimastigotes in the bug's midgut, and transform into metacyclic trypanomastigotes in the hindgut.

All these trypanosomes have a thick glycoprotein layer coating the cell surface. The chemical composition of the glycoprotein layer is switched cyclically, expressing only one of 1,000 to 2,000 variable antigens at any given time. This process, known as **antigenic variation,** enables the parasite's escape from host immune surveillance. It is therefore not surprising that there are no vaccines for either Chagas' disease or African sleeping sickness and the few drugs available for treatment are not particularly effective. The annotated genome sequences of *T. cruzi* and *T. brucei* were reported in 2005. The release of these genomes allows scientists to compare the genetic elements and mechanisms by which these protists so successfully evade the host's immune system. This may help identify new drug and vaccine targets. << *Comparative genomics (section 15.7)*

1. What are some features that distinguish the *Parabasalia* from the *Euglenozoa?*

2. What is the function of the stigma in *Euglena?* How does this protist maintain osmotic balance?

3. What *Euglenozoa* genera cause disease? What adaptations make these protists successful pathogens?

4. Examine the life cycles of *T. brucei* and *T. cruzi* (figures 23.6 and 23.7). Compare the reproduction of these microbes in their insect hosts and in their human hosts. What are some of the similarities and differences?

Super Group *Amoebozoa*

It is clear that the amoeboid form arose independently numerous times from various flagellated ancestors. Thus some amoebae are placed in the super group *Amoebozoa,* while others are in the *Rhizaria.* One of the morphological hallmarks of amoeboid motility is the use of **pseudopodia** (meaning "false feet") for both locomotion and feeding (**figure 23.8**). Pseudopodia can be rounded (**lobopodia**), long and narrow (**filopodia**), or form a netlike mesh (**reticulopodia**). Amoebae that lack a cell wall or other supporting structures and are surrounded only by a plasma membrane are called **naked amoebae.** In contrast, the plasma membrane of a **testate amoeba** is covered by material that is either made by the protist itself or collected by the organism from the environment. Binary fission is the usual means of asexual division, although some amoebae form cysts that undergo multiple fission.

Tubulinea

The *Tubulinea* inhabit almost any environment where they will remain moist; this includes glacial meltwater, marine plankton, tidepools, lakes, and streams. Free-living forms are known to dwell in ventilation ducts and cooling towers, where they feed on microbial biofilms. Others are endosymbionts, commensals, or parasites of invertebrates, fishes, and mammals. Some harbor intracellular symbionts, including algae, bacteria, and viruses,

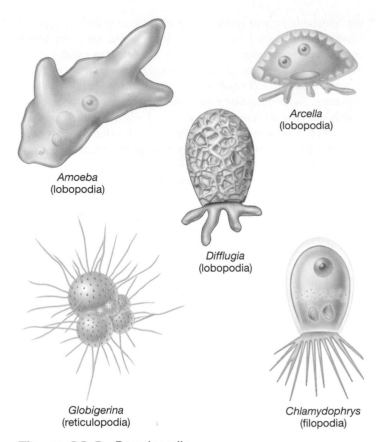

Amoeba (lobopodia)

Arcella (lobopodia)

Difflugia (lobopodia)

Globigerina (reticulopodia)

Chlamydophrys (filopodia)

Figure 23.8 Pseudopodia.

but the nature of these relationships is not well understood. *Amoeba proteus,* a favorite among introductory biology laboratory instructors, is included in this group.

Entamoebida

Two species in the *Entamoebida* infect humans: the nonpathogenic *Entamoeba dispar* and the pathogenic *E. histolytica. E. histolytica* is responsible for **amebiasis** (amebic dysentery). This very common parasite is endemic in warm climates where adequate sanitation and effective personal hygiene are lacking. It is a major cause of parasitic death worldwide; about 500 million people are infected and as many as 100,000 die of amebiasis each year.

Infection occurs by ingestion of mature cysts from fecally contaminated water, food, or hands, or from fecal exposure during sexual contact. After excystment in the lower region of the small intestine, the metacyst divides rapidly to produce eight small trophozoites (**figure 23.9**). These trophozoites move to the large intestine, where they can invade host tissue, live as commensals in the lumen of the intestine, or undergo encystment. In many hosts, trophozoites remain in the intestinal lumen, resulting in an asymptomatic carrier state with cysts shed with the feces.

If trophozoites invade the intestinal tissues, they multiply rapidly and spread laterally, while feeding on erythrocytes, bacteria, and yeasts. The invading trophozoites destroy the epithelial lining of the large intestine by producing a cysteine protease. This protease is a virulence factor of *E. histolytica* and may play a role

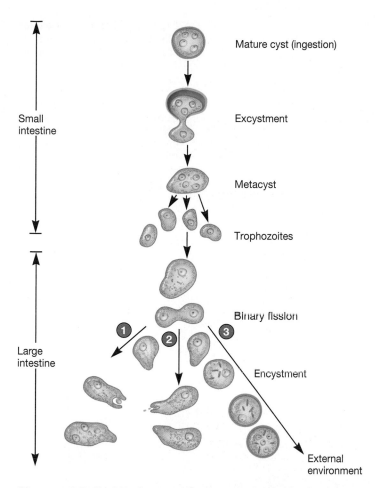

Figure 23.9 **Life Cycle of *Entamoeba histolytica.*** Infection occurs by the ingestion of a mature cyst. Excystment occurs in the lower region of the small intestine, and the metacyst rapidly divides to give rise to eight small trophozoites (only four are shown). These enter the large intestine, undergo binary fission, and may (1) invade the host tissues, (2) live in the lumen of the large intestine without invasion, or (3) undergo encystment and pass out of the host in the feces.

in intestinal invasion by degrading the extracellular matrix and circumventing the host immune response. Lesions (ulcers) are characterized by minute points of entry into the mucosa and extensive enlargement of the lesion after penetration into the submucosa. *E. histolytica* also may invade and produce lesions in other tissues, especially the liver, to cause hepatic amebiasis. However, all extraintestinal amebic lesions are secondary to those established in the large intestine. The symptoms of amebiasis are highly variable, ranging from an asymptomatic infection to fulminating dysentery (exhaustive diarrhea accompanied by blood and mucus), appendicitis, and abscesses in the liver, lungs, or brain.

Eumycetozoa

First described in the 1880s, the *Eumycetozoa* or "slime molds" have been classified as plants, animals, and fungi. As we examine their morphology and behavior, the source of this confusion should become apparent. Recent analysis of certain proteins (e.g., elongation factor EF-1, α-tubulin, and actin) as well as

physiological, behavioral, biochemical, and developmental data point to a monophyletic group (figure 23.1). The *Eumycetozoa* includes the *Myxogastria* and *Dictyostelia.* The **acellular slime mold,** or *Myxogastria,* life cycle includes a distinctive stage when the organisms exist as streaming masses of colorful protoplasm that creep along in amoeboid fashion over moist, rotting logs, leaves, and other organic matter, which they degrade (**figure 23.10**). Their name derives from the lack of individual cell membranes from which a large, multinucleate mass called a plasmodium is formed; there can be as many as 10,000 synchronously dividing nuclei within a single plasmodium (figure 23.10*b*). Feeding is by endocytosis. When starved or dried, the plasmodium develops ornate fruiting bodies. As these mature, they form stalks with cellulose walls that are resistant to environmental stressors (figure 23.10*c,d*). When conditions improve, spores germinate and release haploid amoeboflagellates. These fuse and as the resulting zygotes feed, nuclear division and synchronous mitotic divisions give rise to the multinucleate plasmodium.

The **cellular slime molds** (*Dictyostelia*) are strictly amoeboid and feed endocytically on bacteria and yeasts. Their complex life cycle involves true multicellularity, despite their primitive evolutionary status (**figure 23.11***a*). The species *Dictyostelium discoideum* is an attractive model organism. The vegetative cells move as a mass, sometimes called a pseudoplasmodium, because individual cells retain their cell membranes. Starved cells release cyclic AMP and a specific glycoprotein, which serve as molecular signals. Other cells sense these compounds and respond by forming an aggregate around the signal-producing cells (figure 23.11*b*). In this way large, motile, multicellular slugs develop and serve as precursors to fruiting body formation (figure 23.11*c*). Fruiting body morphogenesis commences when the slug stops and cells pile on top of each other. Cells at the bottom of this vertically oriented structure form a stalk by secreting cellulose, while cells at the tip differentiate into spores (figure 23.11*d,e*). Germinated spores become vegetative amoebae to start this asexual cycle anew.

Sexual reproduction in *D. discoideum* involves the formation of special spores call macrocysts. These arise by a form of conjugation that has some unusual features. First, a group of amoebae become enclosed within a wall of cellulose. Following conjugation a single, large amoeba forms and cannibalizes the remaining amoebae. The now giant amoeba matures into a macrocyst. Macrocysts can remain dormant within their cellulose walls for extended periods of time. Vegetative growth resumes after the diploid nucleus undergoes meiosis to generate haploid amoebae.

The 33.8-Mb genome of *D. discoideum* has been sequenced and annotated. Analysis of a number of proteins supports the notion that these soil-dwelling microbes are more primitive than the fungi. For instance, *D. discoideum* has 14 different histidine kinase receptor proteins; these proteins are generally thought to be distinctly procaryotic. Also of note is the presence of 40 genes that appear to be involved in cellulose biosynthesis or degradation. These genes could be involved in producing the cellulose the microbe needs during morphological differentiation or for degrading cellulose-containing microbes.

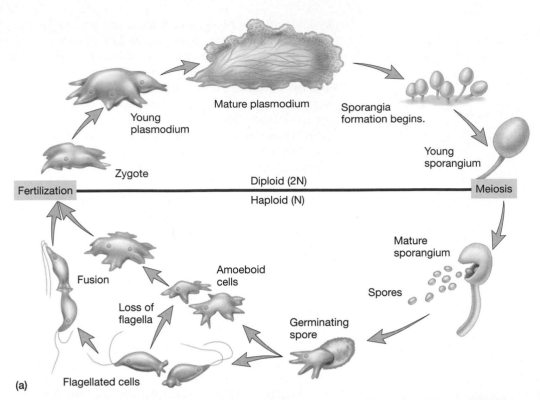

Young plasmodium

Mature plasmodium

Sporangia formation begins.

Young sporangium

Zygote

Fertilization

Diploid (2N)

Haploid (N)

Meiosis

Mature sporangium

Fusion

Amoeboid cells

Loss of flagella

Spores

Germinating spore

Flagellated cells

(a)

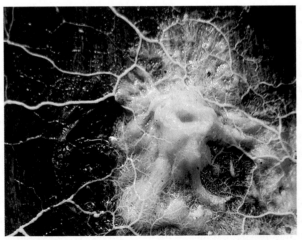

(b) *Physarum sp.*

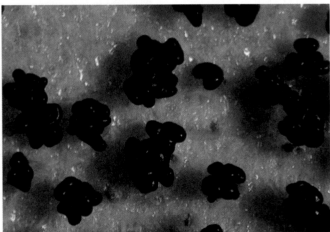

(c) *Physarum polycephalum*

(d) *Stemonitis*

Figure 23.10 Acellular Slime Molds. (a) The life cycle of a plasmodial slime mold includes sexual reproduction; when conditions are favorable for growth, the adult diploid forms sporangia. Following meiosis, the haploid spores germinate, releasing haploid amoeboid or flagellated cells that fuse. (b) Plasmodium of the slime mold *Physarum* sp. (×175). Sporangia of (c) *Physarum polycephalum,* and (d) *Stemonitis*.

Super Group *Rhizaria*

The *Rhizaria* are amoeboid in morphology and thus were historically grouped with the members of what we now call the *Amoebozoa*. However, molecular phylogenetic analysis makes it clear that the *Amoebozoa* and *Rhizaria* are not monophyletic. Morphologically, the *Rhizaria* can be distinguished by their filopodia, which can be simple, branched, or connected. Filopodia supported by microtubules are known as an **axopodia.** Axopodia protrude from a central region of the cell called the axoplast and are primarily used in feeding (**figure 23.12a**).

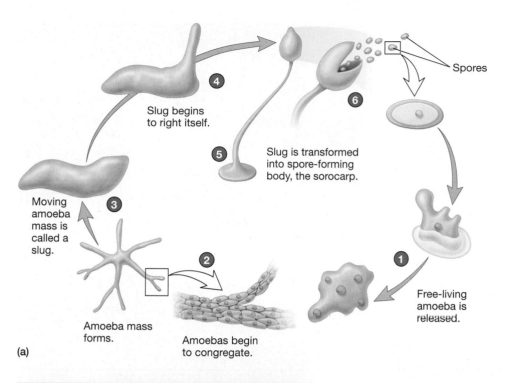

Slug begins to right itself.

Slug is transformed into spore-forming body, the sorocarp.

Spores

Moving amoeba mass is called a slug.

Amoeba mass forms.

Amoebas begin to congregate.

Free-living amoeba is released.

(a)

Radiolaria

Most *Radiolaria* have an internal skeleton made of siliceous material; however, members of the subgroup *Acantharia* have endoskeletons consisting of strontium sulfate. A few genera have an exoskeleton of siliceous spines or scales, while some lack a skeleton completely. Skeletal morphology is highly variable and often includes radiating spines that help the organisms float, as does the storage of oils and other low-density fluids (figure 23.12*b*).

The *Radiolaria* feed by endocytosis, using mucus-coated filopodia to entrap prey including bacteria, other protists, and even small invertebrates. Large prey items are partially digested extracellularly before becoming encased in a food vacuole. Many surface-dwelling radiolarians have algal symbionts thought to enhance their net carbon assimilation. The acantharians reproduce only sexually by consecutive mitotic and meiotic divisions that release

(b)

(c)

(d)

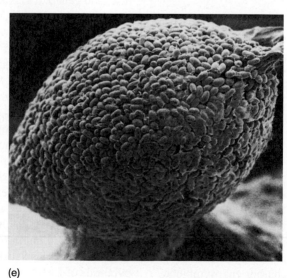

(e)

Figure 23.11 Development of *Dictyostelium discoideum,* a Cellular Slime Mold. (a) Life cycle. (b) Aggregating *D. discoideum* become polar and begin to move in an oriented direction in response to the molecular signal cAMP. (c) Slug begins to right itself and (d) forms a spore-forming body called a sorocarp. (e) Electron micrograph of a sorocarp showing individual spores (×1,800).

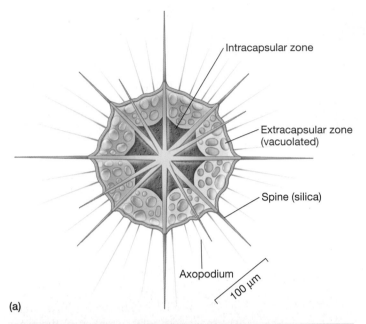

(a)

Intracapsular zone

Extracapsular zone (vacuolated)

Spine (silica)

Axopodium

100 µm

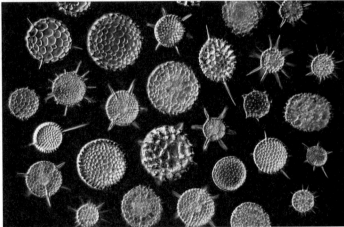

(b) Radiolarian shells

Figure 23.12   Radiolaria. (a) The radiolarian *Acanthometra elasticum* demonstrates the internal skeleton. (b) Radiolarian shells made of silica.

Figure 23.13   A Foraminiferan. The reticulopodia are seen projecting through pores in the calcareous test, or shell, of this protist.

hundreds of biciliated cells. Asexual reproduction (binary or multiple fission, or budding) is most common in other *Rhizaria,* but sexual reproduction can be triggered by nutrient limitation or a heavy feeding. In this case, two haploid nuclei fuse to form a diploid zygote encased in a cyst from which it is released when survival conditions improve.

Foraminifera

The *Foraminifera* (also called forams) range in size from roughly 20 µm to several centimeters. Their filopodia are arranged in a branching network called reticulopodia. They have characteristic tests arranged in multiple chambers that are sequentially added as the protist grows (**figure 23.13**).

Reticulopodia bear vesicles at their tips that secrete a sticky substance used to trap prey. Many species harbor endosymbiotic algae that migrate out of reticulopodia (without being eaten) to expose themselves to more sunlight. These algae contribute significantly to foram nutrition.

Foraminiferan life cycles can be complex. While some smaller species reproduce only asexually by budding or multiple fission, larger forms frequently alternate between sexual and asexual phases. During the sexual phase, flagellated gametes pair, fuse, and generate asexual individuals (agamonts). Meiotic division of the agamonts gives rise to haploid gamonts. There are several mechanisms by which gamonts return to the diploid condition. For instance, a variety of forams release flagellated gametes that become fertilized in the open water. In others, two or more gamonts attach to one another, enabling gametes to fuse within the chambers of the paired tests. When the shells separate, newly formed agamonts are released.

Foraminifera are found in marine and estuarine habitats. Some are planktonic, but most are benthic. Foraminiferin tests accumulate on the seafloor, where they create a fossil record dating back to the Early Cambrian (543 million years ago), which is helpful in oil exploration. Their remains, or ooze, can be up to hundreds of meters deep in some tropical regions. Foram tests make up most modern-day chalk, limestone, and marble, and are familiar to most as the white cliffs of Dover in England. They also formed the stones used to build the great pyramids of Egypt.

1. Describe filopodia, lobopodia, and reticulopodia form and function.

2. List at least one mechanism used by *Entamoeba histolytica* to successfully infect a human host.

3. Why do you think the slime molds have been so hard to classify?

4. What is a plasmodium? How does it differ between the acellular and cellular slime molds?

5. Describe the life cycle of *Dictyostelium discoideum*. Why is this organism a good model for the study of cellular differentiation and coordinated cell movement?

6. What adaptations do the planktonic radiolaria have to help them float?

7. Compare the means by which radiolaria use axopodia with the way foraminifera use reticulopodia.

8. Describe the forms of sexual reproduction in the forams.

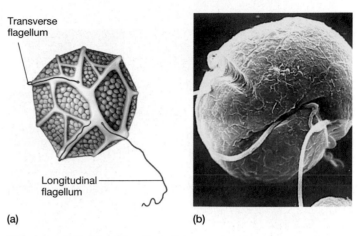

(a)   (b)

Figure 23.14 Dinoflagellates. (a) *Gonyaulax.* (b) Scanning electron micrograph of *Gymnodinium* (×4,000). Notice the plates of cellulose and two flagella: one in the transverse groove and the other projecting outward.

Super Group *Chromalveolata*

The *Chromalveolata* are diverse and include autotrophic, mixotrophic, and heterotrophic protists. They are united in plastid origin, which appears to have been acquired by endosymbiosis with an ancestral archaeplastid. Here, we introduce three subgroups: the *Alveolata, Stramenopiles,* and *Haptophyta,* some members of which have been previously considered to be orders or super groups.

Alveolata

The *Alveolata* is a large group that includes the *Dinoflagellata* (dinoflagellates), *Ciliophora,* and *Apicomplexa.* We begin our discussion with the **dinoflagellates**—a large group most commonly found in marine plankton, where some species are responsible for the phosphorescence sometimes seen in seawater. Their nutrition is complex; photoautotrophy, heterotrophy, and mixotrophy are all observed. Most are saprophytic (either entirely or as facultative chemoorganotrophs), but some also use endocytosis. Each cell bears two, distinctively placed flagella: one is wrapped around a transverse groove (the girdle), and the other is draped in a longitudinal groove (the sulcus; **figure 23.14**). The orientation and beating patterns of these flagella cause the cell to spin as it is propelled forward; the name *dinoflagellate* is derived from the Greek *dinein,* "to whirl." Many dinoflagellates are covered with cellulose plates that are secreted within alveolar sacs that lie just under the plasma membrane. These forms are said to be thecate or armored; those with empty alveoli are called athecate or naked and include the luminescent genus *Noctiluca.* Most dinoflagellates are free living, although some form important associations with other organisms. Endosymbiotic dinoflagellates that live as undifferentiated cells occasionally send out motile cells called **zooxanthellae.** The most well-known zooxanthella belongs to the genus *Symbiodinium.* These are photosynthetic endosymbionts of reef-building coral. They provide fixed carbon to the coral animal and help maintain the internal chemical environment needed for the coral to secrete its calcium carbonate exoskeleton. Dinoflagellates are also responsible for toxic "red tides" that harm other organisms, including humans. **>>** *Marine and freshwater microbiology: Microorganisms in coastal marine systems (section 26.1)*

The **ciliates (*Ciliophora*)** include about 12,000 species. All are chemoorganotrophic and range from about 10 μm to 4.5 mm long. They inhabit both benthic and planktonic communities in marine and freshwater systems, as well as moist soils. As their name implies, *Ciliophora* employ many cilia for locomotion and feeding. The cilia are generally arranged either in longitudinal rows (**figure 23.15**) or in spirals around the body of the organism. They beat with an oblique stroke, causing the protist to revolve as it swims. Ciliary beating is so precisely coordinated that they can go both forward and backward. There is great variation in shape, and most ciliates do not look like the slipper-shaped *Paramecium.* Some species, including *Vorticella,* attach to substrates by a long stalk. *Stentor* attaches to substrates and stretches out in a trumpet shape to feed. A few species have tentacles for the capture of prey. Some can discharge toxic, threadlike darts called toxicysts, which are used in capturing prey. A striking feature of the *Ciliophora* is their ability to entrap many particles in a short time by the action of the cilia around the buccal cavity. Food first enters the cytostome and passes into phagocytic vacuoles that fuse with lysosomes after detachment from the cytostome and the vacuole's contents are digested. After the digested material has been absorbed into the cytoplasm, the vacuole fuses with the cytoproct and waste material is expelled.

Most ciliates have two types of nuclei: a large **macronucleus** and a smaller **micronucleus.** The micronucleus is diploid and contains the normal somatic chromosomes. It divides by mitosis and transmits genetic information through meiosis and sexual reproduction. Macronuclei are derived from micronuclei by a complex series of steps. Within the macronucleus are many chromatin bodies, each containing many copies of only one or two

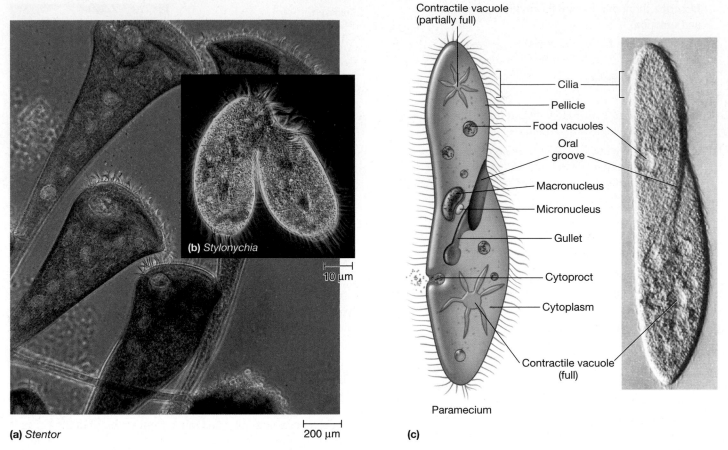

Figure 23.15   The *Ciliophora*.   (a) *Stentor,* a large, vase-shaped, freshwater protozoan. (b) Two *Stylonychia* conjugating. (c) Structure of *Paramecium*, adjacent to an electron micrograph.

genes. Macronuclei are thus polyploid and divide by elongating and then constricting. They produce mRNA to direct protein synthesis, maintain routine cellular functions, and control normal cell metabolism.

Some ciliates reproduce asexually by transverse binary fission, forming two equal daughter cells. The most common means of sexual reproduction among ciliates is conjugation. In this process, there is an exchange of gametes between paired cells of complementary mating types (conjugants). A well-studied example is *Paramecium caudatum* (**figure 23.16**). At the beginning of conjugation, two ciliates unite, fusing their pellicles at the contact point. The macronucleus in each is degraded. The individual micronuclei undergo meiosis to form four haploid pronuclei, three of which disintegrate. The remaining pronucleus divides again mitotically to form two gametic nuclei—a stationary one and a migratory one. The migratory nuclei pass into the respective conjugants. Then the ciliates separate, the gametic nuclei fuse, and the resulting diploid zygote nucleus undergoes three rounds of mitosis. The eight resulting nuclei have different fates: one nucleus is retained as a micronucleus; three others are destroyed; and the four remaining nuclei develop into macronuclei. Each separated conjugant now undergoes cell division. Eventually progeny with one macronucleus and one micronucleus are formed.

Although most ciliates are free living, symbiotic forms do exist. Some live as harmless commensals; for example,

Entodinium is found in the rumen of cattle and *Nyctotherus* occurs in the colon of frogs. Other ciliates are strict parasites; for example, *Balantidium coli* lives in the intestines of mammals, including humans, where it can cause dysentery. *Ichthyophthirius* lives in freshwater, where it can attack many species of fish, producing a disease known as "ick."

All **apicomplexans** are either intra- or intercellular parasites of animals and are distinguished by a unique arrangement of fibrils, microtubules, vacuoles, and other organelles, collectively called the apical complex, which is located at one end of the cell (**figure 23.17a**). This unique combination of organelles is designed to penetrate host cells. Motility (flagellated or amoeboid) is confined to the gametes and zygotes of a few species. Apicomplexans have complex life cycles in which certain stages sometimes occur in one host and other stages occur in a different host. The life cycle has both asexual (clonal) and sexual phases, and is characterized by an alternation of haploid and diploid generations. The clonal and sexual stages are haploid, except for the zygote. The motile, infective stage is called the **sporozoite.** When this haploid form infects a host, it differentiates into a **gamont;** male and female gamonts pair and undergo multiple fission, which produces many gametes. Released gametes pair, fuse, and form zygotes. Each zygote secretes a protective covering and is then considered a spore. Within the spore, the nucleus undergoes meiosis (restoring the haploid condition) followed by mitosis

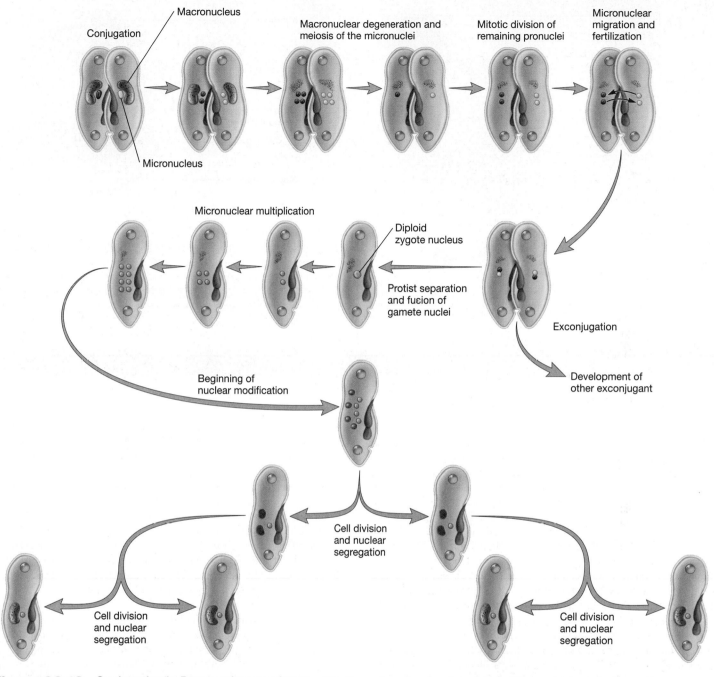

Figure 23.16 **Conjugation in *Paramecium caudatum.*** After the conjugants separate, only one of the exconjugants is shown; however, a total of eight new protists result from each conjugation.

to generate eight sporozoites ready to infect a new host (figure 23.17*b*).

A number of apicomplexans are important infectious agents. The most significant is *Plasmodium,* which causes malaria in some 500 million people annually, with a yearly death toll of 1 to 3 million. *Eimera* is the causative agent of cecal coccidiosis in chickens, a condition that costs hundreds of millions of dollars in lost animals each year in the United States. Toxoplasmosis, caused by members of the genus *Toxoplasma,* is transmitted either by consumption of undercooked meat or by fecal contamination from a cat's litterbox. Cryptosporidia are responsible for

cryptosporidiosis, an infection that begins in the intestines but can disseminate to other parts of the body. Cryptosporidiosis and another apicomplexan parasite, *Cyclospora,* have become problematic for AIDS patients and other immunocompromised individuals. We now consider these human pathogens and the diseases they cause in more detail.

Malaria The most important human pathogen among the protists is *Plasmodium,* the causative agent of malaria (**Disease 23.1**). Human malaria is caused by four species of *Plasmodium: P. falciparum, P. malariae, P. vivax,* and *P. ovale.* The life cycle of

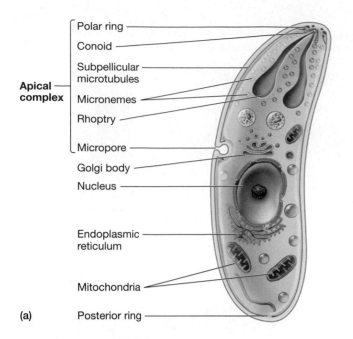

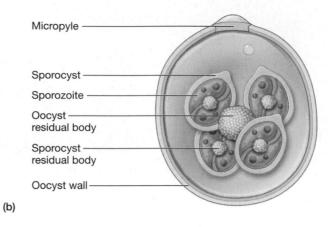

(a)

(b)

Figure 23.17   The Apicomplexan Cell.   (a) The vegetative cell, or merozoite, illustrating the apical complex, which consists of the polar ring, conoid, rhoptries, subpellicular microtubules, and micropore. (b) The infective oocyte of *Eimeria*. The oocyst is the resistant stage and has undergone multiple fission after zygote formation (sporogony).

Disease

23.1   A Brief History of Malaria

No other single infectious disease has had the impact on humans that malaria has had. The first references to its periodic fever and chills can be found in early Chaldean, Chinese, and Hindu writings. In the late fifth century BCE, Hippocrates described certain aspects of malaria. In the fourth century BCE, the Greeks noted an association between individuals exposed to swamp environments and the subsequent development of periodic fever and enlargement of the spleen (splenomegaly). In the seventeenth century, the Italians named the disease *mal' aria* (bad air) because of its association with the ill-smelling vapors from the swamps near Rome. At about the same time, the bark of the quinaquina (cinchona) tree of South America was used to treat the intermittent fevers, although it was not until the mid-nineteenth century that quinine was identified as the active alkaloid. The major epidemiological breakthrough came in 1880, when French army surgeon Charles Louis Alphonse Laveran observed gametocytes in fresh blood. Five years later, the Italian histologist Camillo Golgi observed the multiplication of the asexual blood forms. In the late 1890s, Patrick Manson postulated that malaria was transmitted by mosquitoes. Sir Ronald Ross, a British army surgeon in the Indian Medical Service, subsequently observed developing plasmodia in the intestine of mosquitoes, supporting Manson's theory. Using birds as experimental models, Ross definitively established the major features of the life cycle of *Plasmodium* and received the Nobel Prize in 1902.

Human malaria is known to have contributed to the fall of the ancient Greek and Roman empires. Troops in both the U.S. Civil War and the Spanish-American War were severely incapacitated by the disease. More than 25% of all hospital admissions during these wars were malaria patients. During World War II, malaria epidemics severely threatened both the Japanese and Allied forces in the Pacific. The same can be said for the military conflicts in Korea and Vietnam.

In the twentieth century, efforts were directed toward understanding the biochemistry and physiology of malaria, controlling the mosquito vector, and developing antimalarial drugs. In the 1960s, it was demonstrated that resistance to *P. falciparum* among West Africans was associated with the presence of hemoglobin-S (Hb-S) in their erythrocytes. Hb-S differs from normal hemoglobin-A by a single amino acid, valine, in each half of the Hb molecule. Consequently these erythrocytes—responsible for sickle cell disease—have a low binding capacity for oxygen. Because the malarial parasite has a very active aerobic metabolism, it cannot grow and reproduce within these erythrocytes.

In 1955 the World Health Organization began a worldwide malarial eradication program that finally collapsed by 1976. Among the major reasons for failure were the development of resistance to DDT by mosquitoes and the development of resistance to chloroquine by strains of *Plasmodium*. Scientists are exploring new approaches, such as the development of vaccines and more potent drugs. In 2002 the complete DNA sequences of *P. falciparum* and *Anopheles gambiae* (the mosquito that most efficiently transmits this parasite to humans in Africa) were determined. Together with the human genome sequence, researchers now have in hand the genetic blueprints for the parasite, its vector, and its victim. This has made possible a holistic approach to understanding how the parasite interacts with the human host, leading to new antimalarial strategies, including vaccine design. Overall, no greater achievement for microbiology could be imagined than the control of malaria—a disease that has caused untold misery throughout the world since antiquity and remains one of the world's most serious infectious diseases.

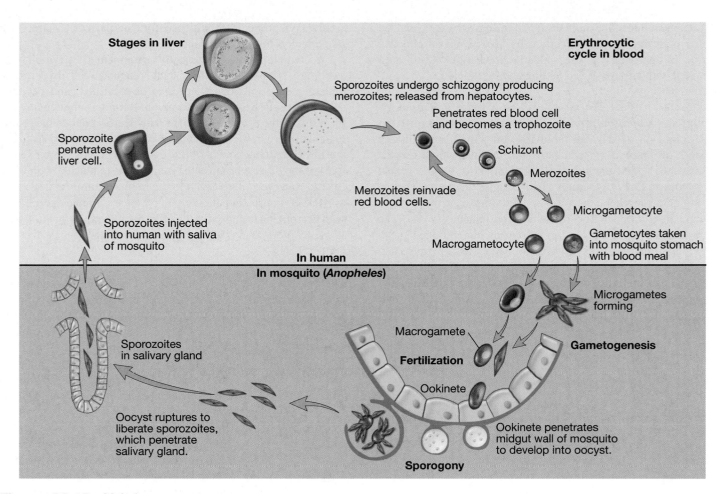

Figure 23.18 Malaria. Life cycle of *Plasmodium vivax*.

P. vivax is shown in **figure 23.18**. The parasite first enters the bloodstream through the bite of an infected female *Anopheles* mosquito. As she feeds, the mosquito injects a small amount of saliva containing an anticoagulant along with small haploid sporozoites. The sporozoites in the bloodstream immediately enter hepatic cells of the liver. In the liver, they undergo multiple asexual fission (schizogony) and produce merozoites. After being released from the liver cells, the merozoites attach to erythrocytes and penetrate these cells. Once inside the erythrocyte, *Plasmodium* begins to enlarge as a uninucleate cell termed a trophozoite. The trophozoite's nucleus then divides asexually to produce a schizont that has 6 to 24 nuclei. The schizont divides and produces mononucleated merozoites. Eventually the erythrocyte lyses, releasing the merozoites into the bloodstream to infect other erythrocytes. This erythrocytic stage is cyclic and repeats itself approximately every 48 to 72 hours or longer, depending on the species of *Plasmodium* involved. Occasionally merozoites differentiate into macrogametocytes and microgametocytes, which do not rupture the erythrocyte. When ingested by a mosquito, these develop into female and male gametes, respectively. The infected erythrocytes lyse in the mosquito's gut, and the gametes fuse to form a diploid zygote called the ookinete. The ookinete migrates to

the mosquito's gut wall, penetrates, and forms an oocyst. In a process called sporogony, the oocyst undergoes meiosis and forms sporozoites, which migrate to the salivary glands of the mosquito. The cycle is now complete, and when the mosquito bites another human host, the cycle begins again.

The pathological changes caused by malaria involve not only erythrocytes but also the spleen and other organs. Classic symptoms first develop with the synchronized release of merozoites and erythrocyte debris into the bloodstream, resulting in the malarial paroxysms—shaking chills, then burning fever followed by sweating. Several of these paroxysms constitute an attack. After one attack, there is a remission that lasts from a few weeks to several months, then there is a relapse. Between paroxysms, the patient feels normal. Anemia can result from the loss of erythrocytes, and the spleen and liver often enlarge. Children and nonimmune individuals can die of cerebral malaria.

Although malaria can be treated with antiparasitic agents, a vaccine would offer the best tool to fight this disease. For many years, a malarial vaccine has seemed out of reach; however, a credible vaccine against malaria was reported in 2005. Clinical trials showed that Mosquirix provided partial protection against malaria for 18 to 21 months in children one to four years of age.

The vaccine reduced the number of malaria cases by 29% and the number of severe malaria infections by 50%. Until an effective vaccine is approved for use, malaria prevention is still best attempted with the use of bed netting and insecticides.

Toxoplasmosis Toxoplasmosis is a disease caused by the protist *Toxoplasma gondii*. This apicomplexan protist has been found in nearly all animals and most birds. Large numbers of *Toxoplasma* can be found in cats, which are the definitive host, and are required for completion of the sexual cycle (**figure 23.19**). Animals shed oocysts in the feces; the oocysts enter another host by way of the nose or mouth; and the parasites colonize the intestine. Toxoplasmosis also can be transmitted by the ingestion of raw or undercooked meat, congenital transfer, blood transfusion, or a tissue transplant. Toxoplasmosis gained public notice when it was discovered that in pregnant women the protist can also infect the fetus, causing serious congenital defects or death. Most cases of toxoplasmo-

sis are asymptomatic. Adults usually complain of an infectious mononucleosis-like syndrome.

Acute toxoplasmosis is usually accompanied by lymph node swelling (lymphadenopathy) with reticular cell hyperplasia (enlargement). Pulmonary necrosis, myocarditis, and hepatitis caused by tissue necrosis are common. Retinitis (inflammation of the retina of the eye) is associated with necrosis due to the proliferation of the parasite within retinal cells. Toxoplasmosis is found worldwide, although most of those infected do not exhibit symptoms of the disease because the immune system usually prevents illness. However, *Toxoplasma* in the immunocompromised, such as AIDS or transplant patients, can produce a unique encephalitis with necrotizing lesions accompanied by inflammatory infiltrates. It causes more than 3,000 congenital infections per year in the United States.

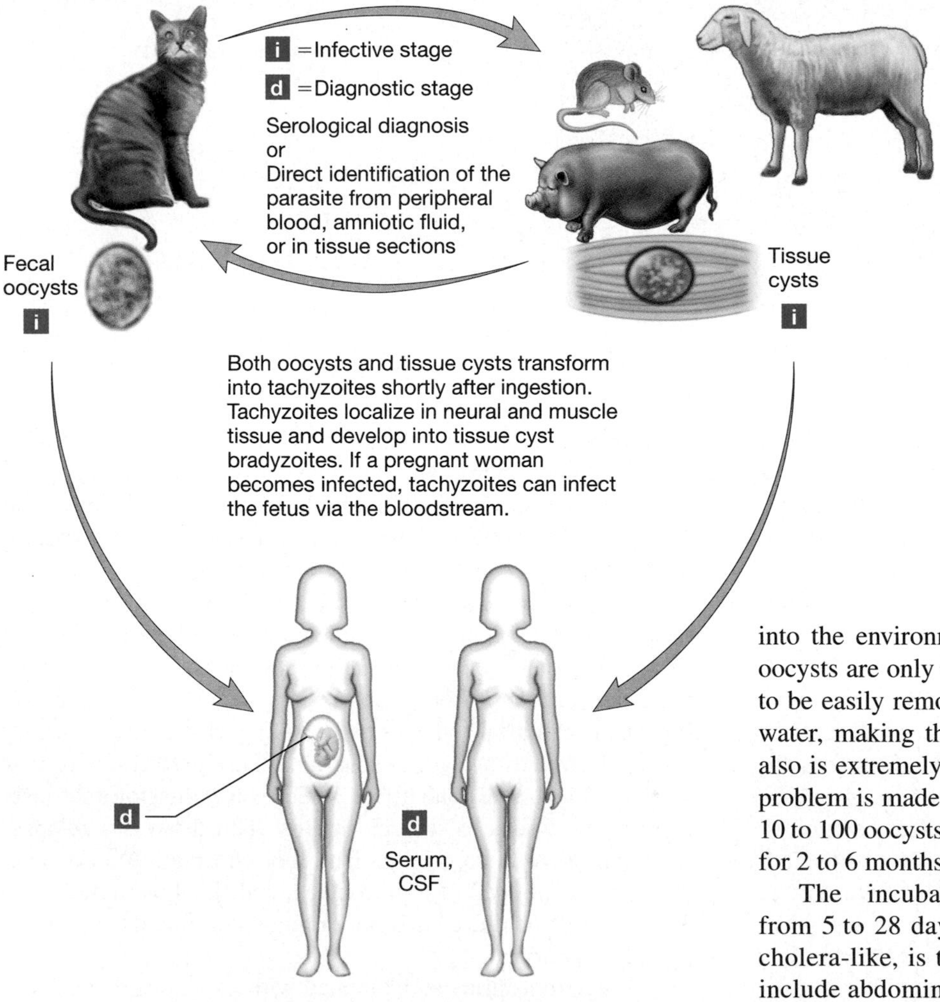

Cryptosporidiosis The first case of human cryptosporidiosis was reported in 1976. The protist responsible was identified as *Cryptosporidium parvum*. In 1993 *C. parvum* contaminated the Milwaukee, Wisconsin, water supply and caused severe diarrheal disease in about 400,000 individuals. *Cryptosporidium* ("hidden spore cysts") is found in about 90% of sewage samples, 75% of river waters, and 28% of drinking waters.

Figure 23.19 Life Cycle of *Toxoplasma gondii*. *Toxoplasma gondii* is a protozoan parasite of numerous mammals and birds. However, the cat, the definitive host, is required for completion of the sexual cycle. Oocysts are shed in the feces of infected animals where they may be ingested by another host. Ingested oocysts transform into tachyzoites, which migrate to various tissue sites via the bloodstream.

C. parvum is a common apicomplexan found in the intestine of many birds and mammals. When these animals defecate, oocysts are shed into the environment. If a human ingests food or water that is contaminated with the oocysts, excystment occurs within the small intestine, and sporozoites enter epithelial cells and develop into merozoites. Some of the merozoites subsequently undergo sexual reproduction to produce zygotes. The zygotes differentiate into thick-walled oocysts, whose release into the environment begins the life cycle again. Because the oocysts are only 4 to 6 μm in diameter, they are much too small to be easily removed by the sand filters used to purify drinking water, making them a public health concern. *Cryptosporidium* also is extremely resistant to disinfectants such as chlorine. The problem is made even worse by the low infectious dose, around 10 to 100 oocysts, and the fact that the oocysts may remain viable for 2 to 6 months in a moist environment.

The incubation period for cryptosporidiosis ranges from 5 to 28 days. Diarrhea, which characteristically may be cholera-like, is the most common symptom. Other symptoms include abdominal pain, nausea, fever, and fatigue. No chemotherapy is available and patients are simply rehydrated. Although the disease usually is self-limiting in healthy individuals, patients with late-stage AIDS or who are immunocompromised in other ways may develop prolonged, severe, and life-threatening diarrhea.

Cyclosporiasis Cyclosporiasis is caused by the unicellular axicomplexan *Cyclospora cayetanensis. Cyclospora* was first identified in 1979. The disease is more common in tropical and subtropical environments, although it has been reported in most countries. In Canada and the United States, cyclosporiasis has been responsible for affecting over 3,600 people in at least 11 food-borne outbreaks since 1990. A number of cyclosporiasis outbreaks have been linked to contaminated produce.

The protist infects the small intestine with a mean incubation period of approximately 1 week. Because cyclosporan oocysts in feces are not infective, direct oral-fecal transmission does not occur (**figure 23.20**). Instead, the oocysts must differentiate into sporozoites after days or weeks at temperatures between 22 and 32°C. Sporozoites can enter the food chain when oocyst-contaminated water is used to wash fruits and vegetables prior to their transport to market. Once ingested, the sporozoites are freed from the oocysts and invade intestinal epithelial cells, where they replicate asexually. Sexual development is completed when sporozoites mature into new oocysts and are released into the intestinal lumen to be shed with the feces.

Cyclosporiasis presents with frequent, sometimes explosive, diarrhea. Often the patient exhibits loss of appetite, cramps, and bloating due to substantial gas production, nausea, vomiting, fever, fatigue, and substantial weight loss. Patients may report symptoms for days to weeks with decreasing frequency and then relapses. The disease is treated with a combination of antimicrobial agents and rehydration therapy.

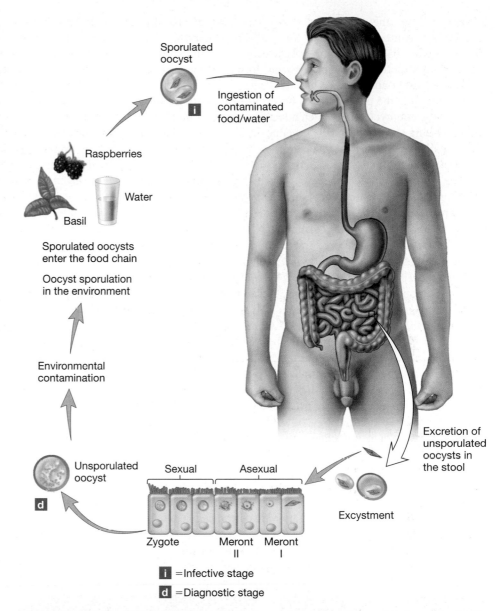

Figure 23.20 **Life Cycle of *Cyclospora cayetanensis.*** Unsporulated oocysts, shed in the feces of infected animals, sporulate and contaminate food and water. Once ingested, the sporulated oocysts excyst, penetrate host cells, and reproduce by binary fission (merogony). Sexual stages unite to form zygotes resulting in new oocysts (unsporulated) that shed in the feces. *Some of the elements in this figure were created based on an illustration by Ortega et al. 1998.* Cyclospora cayetanensis. *In* Advances in Parasitology: Opportunistic protozoa in humans, 399–418. *San Diego: Academic Press.*

1. What are zooxanthellae? Describe the relationship between *Symbiodinium* and its coral host.

2. What is the morphology of typical *Ciliophora?* Why do you think these are the fastest-moving protists?

3. Describe conjugation as it occurs in the *Ciliophora.* What is the fate of the micronucleus and the macronucleus during this process?

4. What is the apical complex seen in apicomplexans?

5. Why is the life cycle of apicomplexans so difficult to study? What do you think the implications of the complicated *Plasmodium* life cycle are for the development of a malaria cure or vaccine?

Stramenopiles

The large and diverse *Stramenopiles* group includes photosynthetic protists such as the diatoms, brown and golden algae (the *Chrysophyceae*), as well as chemoorganotrophic (saprophytic) genera such as the öomycetes (*Peronosporomycetes*), laby-rinthulids (slime nets), and the *Hyphochytriales*. The *Stramenopiles* also include brown seaweeds and kelp that form large, rigid structures and macroscopic forms that were once considered fungi and plants. One unifying feature of this very diverse taxon is the possession of **heterokont flagella** at some point in the life cycle. This is characterized by two flagella—one extending anteriorly and the other posteriorly. These flagella bear small hairs with a unique, three-part morphology; *stramenopila* means "straw hair."

The **diatoms** (***Bacillariophyta***) possess chlorophylls *a* and c_1/c_2, and the carotenoid fucoxanthin. When fucoxanthin is the dominant pigment, the cells have a golden-brown color. Their major carbohydrate reserve is chrysolaminarin (a polysaccharide storage product composed principally of $\beta(1{\rightarrow}3)$–linked glucose residues). Diatoms have a distinctive, two-piece cell wall of silica called a **frustule**. Diatom frustules are composed of two halves or thecae that overlap like a petri dish (**figure 23.21***a*). The larger half is the epitheca, and the smaller half is the hypotheca. Diatom frustules are composed of crystallized silica [$Si(OH)_4$] with very fine markings (figure 23.21*b*). They have distinctive, and often exceptionally beautiful, patterns that are unique for each species. Frustule morphology is very useful in diatom identification, and diatom frustules have a number of practical applications. The fine detail and precise morphology of these frustules have made them attractive for nanotechnology. << *Microbes as products (section 16.12)*

Although the majority of diatoms are strictly photoautotrophic, some are facultative chemoorganotrophs, absorbing carbon-containing molecules through the holes in their walls. The vegetative cells of diatoms are diploid and can be unicellular, colonial, or filamentous. They lack flagella and have a single, large nucleus and smaller plastids. Reproduction consists of the organism dividing asexually, with each half then constructing a new theca within the old one. Because the epitheca and hypotheca are of different sizes, each time the hypotheca is used as a template to construct a new hypotheca, the diatom gets smaller. However, when a cell has diminished to about 30% of its original size, sexual reproduction is usually triggered. The diploid vegetative cells undergo meiosis to form gametes, which then fuse to produce a zygote. The zygote develops into an auxospore, which increases in size again and forms a new wall. The mature auxospore eventually divides mitotically to produce vegetative cells with frustules of the original size.

Diatoms are found in freshwater lakes, ponds, streams, and throughout the world's oceans. Marine planktonic diatoms produce 40 to 50% of the organic carbon in the ocean; they are therefore very important in global carbon cycling. In fact, marine diatoms are thought to contribute as much fixed carbon as all rain forests combined. >> *Biogeochemical cycling: Carbon cycle (section 25.1)*

The complete genome of the marine planktonic diatom *Thalassiosira pseudonana* was recently sequenced. Its genome consists of 24 chromosomes (34 Mb), a plastid genome (0.139 Mb), and a mitochondrial genome (0.044 Mb). It contains novel genes for silicic acid transport and silica-based cell wall synthesis. Also discovered were genes for scavenging iron, multiple nitrate and ammonium transporters, and the enzymes for urea metabolism. The annotation of this genome will help scientists discover how this microbe has adapted so successfully to its nutrient-limited environment.

A group of protists once considered true fungi and traditionally called **öomycetes,** meaning "egg fungi," were recently assigned the name **peronosporomycetes.** They differ from true fungi in a number of ways, including their cell wall composition (cellulose and β-glucan instead of chitin) and the fact that they are diploid throughout their life cycle. When undergoing sexual reproduction, they form a relatively large egg cell (öogonium) that is fertilized by either a sperm cell or a smaller gametic cell (called an antheridium) to produce a zygote. When the zygote germinates, the asexual zoospores display heterokont flagellation.

Peronosporomycetes such as *Saprolegnia* and *Achlya* are saprophytes that grow as cottony masses on dead algae and animals, mainly in freshwater environments. Some öomycetes are parasitic on the gills of fish. *Peronospora hyoscyami* is responsible for "blue mold" on tobacco plants, and grape downy mildew is caused by *Plasmopara viticola*. Certainly the most famous öomycete is *Phytophthora infestans,* which attacked the European potato crop in the mid-1840s, spawning the Irish famine. The original classification of *P. infestans* as a fungus was misleading, and for decades farmers attempted to control its growth with fungicide, to which it is (of course) resistant. This protist continues to take its toll; potato blight costs some $5 billion annually worldwide.

(a) *Cyclotella meneghiniana*

(b)

Figure 23.21 Diatoms. (a) The silaceous epitheca and hypotheca of the diatom *Cyclotella meneghiniana* fit together like a petri dish. (b) A variety of diatoms show the intricate structure of the silica cell wall.

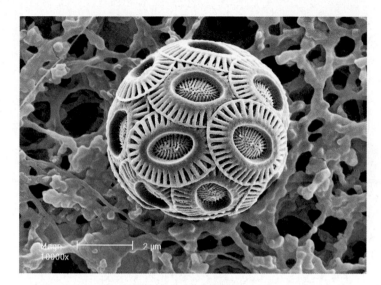

Figure 23.22 **The Haptophyte *Emiliana huxleyi*.** Note the ornamental scales made of calcite.

Labyrinthulids also have a complex taxonomic history: like the *Peronosporomycetes*, they were formerly considered fungi. However, molecular phylogenetic evidence combined with the observation that they form heterokont flagellated zoospores places them among the *Stramenopiles*. The more familiar, non-flagellated stage of the life cycle features spindle-shaped cells that form complex colonies that glide rapidly along an ectoplasmic net made by the organism. This net is actually an external network of calcium-dependent contractile fibers made up of actinlike proteins that facilitate the movement of cells. Their feeding mechanism is like that of fungi: osmotrophy aided by the production of extracellular degradative enzymes. In marine habitats, the genus *Labyrinthula* grows on plants and algae and is thought to play a role in the "wasting disease" of eelgrass, an important intertidal plant.

Haptophyta

An important subgroup of the *Haptophyta* is the *Coccolithales*. These photosynthetic protists bear ornate calcite scales called coccoliths (**figure 23.22**). Together with the *Foraminifera*, the **coccolithophores** precipitate calcium carbonate ($CaCO_3$) in the ocean, thereby influencing Earth's carbon budget (*see figure 26.2*). Cells are usually biflagellated and possess a unique organelle called a haptonema, which is somewhat similar to a flagellum but differs in microtubule arrangement. One species, *Emiliania huxleyi*, has been studied extensively. Like all

coccolithophores, it is planktonic. High concentrations or blooms of *E. huxleyi* can significantly alter nutrient flux by emitting sulfur (as dimethyl sulfide) to the atmosphere and sequestering calcium carbonate in the sediments. Other coccolithophore species are known to cause toxic blooms. >> *Marine and freshwater microbiology: Nutrient cycling in marine and freshwater environments (section 26.1)*

Super Group *Archaeplastida*

The *Archaeplastida* includes all organisms with a photosynthetic plastid that arose through an ancient endosymbiosis with a cyanobacterium. It thus includes all higher plants as well as many protist species. << *Microbial evolution: The endosymbiotic origin of mitochondria and chloroplasts (section 17.1)*

Chloroplastida

The *Chloroplastida* are often referred to as green algae (Greek *chloros,* green). These phototrophs grow in fresh and salt water, in soil, and on and within other organisms. They have chlorophylls *a* and *b* along with specific carotenoids, and they store carbohydrates such as starch. Many have cell walls made of cellulose. They exhibit a wide diversity of body forms, ranging from unicellular to colonial, filamentous, membranous or sheetlike, and tubular types (**figure 23.23**). Some species have a holdfast structure that anchors them to the substratum. Both asexual and sexual reproduction are observed.

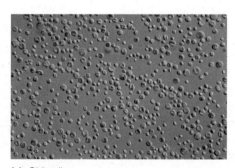
(a) *Chlorella*

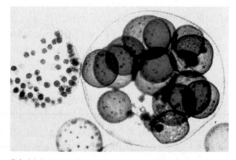
(b) *Volvox*

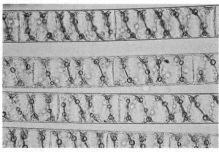
(c) *Spirogyra*

(d) *Acetabularia*

Figure 23.23 ***Chlorophyta* (Green Algae); Light Micrographs.** (a) *Chlorella,* a unicellular nonmotile Chlorophyte (×160). (b) *Volvox,* which demonstrates colonial growth (×450). (c) *Spirogyra* (×100). Four filaments are shown. Note the ribbonlike, spiral chloroplasts within each filament. (d) *Acetabularia,* the mermaid's wine goblet.

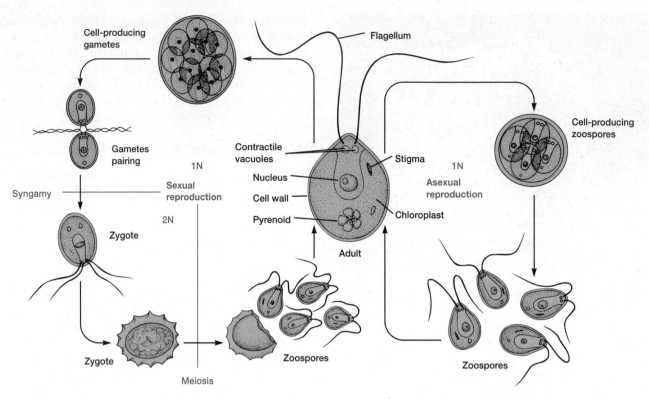

Figure 23.24 *Chlamydomonas:* **The Structure and Life Cycle of This Motile Green Alga.** During asexual reproduction, all structures are haploid; during reproduction, only the zygote is diploid.

Chlamydomonas is a member of the subgroup *Chlorophyta* (**figure 23.24**). Individuals have two flagella of equal length at the anterior end by which they move rapidly in water. Each cell has a single haploid nucleus, a large chloroplast, a conspicuous pyrenoid, and a stigma that aids the cell in phototactic responses. Two small contractile vacuoles at the base of the flagella function as osmoregulatory organelles. *Chlamydomonas* reproduces asexually by producing zoospores through cell division. Sexual reproduction occurs when some products of cell division act as gametes and fuse to form a four-flagellated, diploid zygote that ultimately loses its flagella and enters a resting phase. Meiosis occurs at the end of this resting phase and produces four haploid cells that give rise to adults.

The chlorophyte *Prototheca moriformis* causes the disease protothecosis in humans and animals. *Prototheca* cells are fairly common in the soil, and it is from this site that most infections occur. Severe systemic infections, such as massive invasion of the bloodstream, have been reported in animals. The subcutaneous type of infection is more common in humans. It starts as a small lesion and spreads slowly through the lymph glands, covering large areas of the body.

1. Explain the unique structural features of the diatoms. How does their morphology play a role in the alternation between asexual and sexual reproduction?

2. Why do you think the öomycetes and the labyrinthulids were formerly considered fungi?

3. What is the ecological importance of the coccolithophores?

4. Compare the morphology of *Chlamydomonas* with that of *Euglena* (figure 23.3).

23.3 CHARACTERISTICS OF THE FUNGAL DIVISIONS

The *Fungi* is an enormous group of organisms; about 90,000 fungal species have been described, and some estimates suggest that 1.5 million species may exist. It is thus not surprising that the taxonomy of these organisms has been revised numerous times. Most recently the application of molecular techniques, including sequence comparisons of small subunit rRNA and conserved proteins, has led a better understanding of the phylogenetic relationships between various fungal groups. For example, the division *Deuteromycetes* (also known as *fungi imperfecti*) is no longer recognized. Here, we present eight fungal groups: the *Chytridiomycetes, Zygomycota, Glomeromycota, Ascomycota, Basidiomycota,* urediniomycetes, ustilaginomycetes, and *Microsporidia*. The urediniomycetes and the ustilaginomycetes are commonly considered *Basidiomycota,* but evidence suggests that they may be taxonomically distinct. **Table 23.2** surveys all eight groups; we focus most of our discussion on the better-known groups.

Table 23.2	Abbreviated Classification of the Fungi as Proposed by International Society of Protistologists[a]	
Subclass	**Characteristics**	**Examples**
Chytridiomycota	Flagellated cells in at least one stage of life cycle; may have one or more flagella. Cell walls with chitin and β-1,3-1,6-glucan; glycogen is used as a storage carbohydrate. Sexual reproduction often results in a zygote that becomes a resting spore or sporangium; saprophytic or parasitic. Chytrid subdivisions include *Blastocladiales, Monoblepharidales, Neocallimastigaceae, Spizellomycetales,* and the *Chytridiales.*	*Allomyces* *Blastocladiella* *Coelomomyces* *Physoderma* *Synchytrium*
Zygomycota	Thalli usually filamentous and nonseptate, without cilia; sexual reproduction gives rise to thick walled zygospores that are often ornamented. Includes seven subdivisions: *Basidiobolus, Dimargaritales, Endogonales, Entomophthorales, Harpellales, Kickxellales, Mucorales,* and *Zoopagales.* Human pathogens found among the *Mucorales* and *Entomophthorales.*	*Amoebophilus* *Mucor* *Phycomyces* *Rhizopus* *Thamnidium*
Ascomycota	Sexual reproduction involves meiosis of a diploid nucleus in an ascus, giving rise to haploid ascospores; most also undergo asexual reproduction with the formation of conidiospores with specialized aerial hyphae called conidiophores. Many produce asci within complex fruiting bodies called ascocarps. Includes saprophytic, parasitic forms; many form mutualisms with phototrophic microbes to form lichens. Four monophyletic subdivisions *Saccharomycetes, Pezizomycotina, Taphrinomycotina,* and *Neolecta.*	*Ascobolus* *Aspergillis* *Candida* *Crinula* *Neurospora* *Penicillium* *Pneumocystis* *Saccharomyces*
Basidiomycota	Includes many common mushrooms and shelf fungi. Sexual reproduction involves formation of a basidium (small, club-shaped structure that typically forms spores at the ends of tiny projections) within which haploid basidiospores are formed. Usually 4 spores per basidium but can range from 1 to 8. Sexual reproduction involves fusion with opposite mating type resulting in a dikaryotic mycelium with parental nuclei paired but not initially fused. No subdivisions recognized.	*Agaricus* *Boletes* *Dacrymyces* *Lycoperdon* *Polyporus* *Russula* *Tremella*
Urediniomycota	Mycelial or yeast forms. Sexual reproduction involves fusion of parental nuclei in probasidium followed by meiosis in a separate compartment. Many are plant pathogens called rusts, animal pathogens, nonpathogenic endophytes, and rhizosphere species. No subdivisions recognized.	*Caeoma* *Melampsora* *Uromyces*
Ustilaginomycota	Plant parasites that cause rusts and smuts. Mycelial in parasitic phase; meiospores formed on septate or aseptate basidia; cell wall principally composed of glucose. No subdivisions recognized.	*Malassezia* *Tilletia* *Ustilago*
Glomeromycota	Filamentous, most are endomycorrhizal, arbuscular; lack cilium; form asexual spores outside of host plant; lack centrioles, conidia, and aerial spores. No subdivisions recognized.	*Acaulospora* *Entrophospora* *Glomus*
Microsporidia	Obligate intracellular parasites usually of animals. Lack mitochondria, peroxisomes, kinetosomes, cilia, and centrioles; spores have an inner chitin wall and outer wall of protein; produce a tube for host penetration. Subdivisions currently uncertain.	*Amblyospora* *Encephalitozoon* *Enterocytozoon* *Nosema*

[a]Adapted from: Adl, S. M.; Simpson, A. G. B.; Farmer, M. A.; Anderson, R. A.; Anderson, O. R.; Barta, J. R.; Bowser, S. S.; et al. 2005. The new higher level classification of Eukaryotes with emphasis on the taxonomy of protists. *J. Eukaryot. Microbiol.* 52:399–451.

Chytridiomycota

The simplest of the fungi belong to the *Chytridiomycota,* or **chytrids.** They are unique among fungi in the production of a zoospore with a single, posterior, whiplash flagellum (**figure 23.25a**). This is considered a primitive feature that was lost in more evolved fungi. Free-living members of this taxon are saprophytic, living on plant or animal matter in freshwater, mud, or soil. Parasitic forms infect aquatic plants and animals, includ-

ing insects (figure 23.25b). A few are found in the anoxic rumen of herbivores.

Chytridiomycota display a variety of life cycles involving both asexual and sexual reproduction. Members of this group are microscopic in size and may consist of a single cell, a small multinucleate mass, or a true mycelium with hyphae capable of penetrating porous substrates. Many are capable of degrading cellulose and even keratin, which enables the degradation of crustacean exoskeletons. The genus *Allomyces* is used to study morphogenesis.

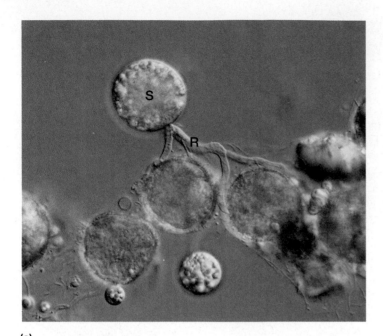

(a)

(b)

Figure 23.25 The *Chytridiomycota.* (a) A chytrid with a single round sporangium (S) that is about 50 μm in diameter. In this image, the fungus is growing between brown pollen grains, which it will eventually penetrate with its branched rhizoids (R). (b) Parasitic chytrids attached to the surface of a photosynthetic protist (a green alga).

Zygomycota

The *Zygomycota* contains fungi called **zygomycetes.** Most live on decaying plant and animal matter in the soil; a few are parasites of plants, insects, other animals, and humans. The hyphae of zygomycetes are coenocytic, with many haploid nuclei. Asexual spores develop in sporangia at the tips of aerial hyphae and are usually wind dispersed. Sexual reproduction produces tough, thick-walled zygotes called zygospores that can remain dormant when the environment is too harsh for growth of the fungus.

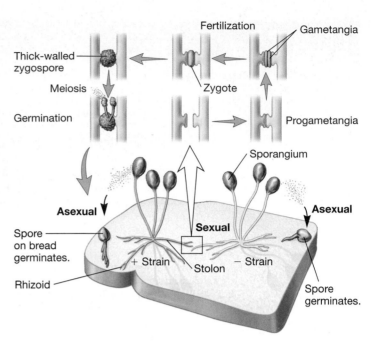

Figure 23.26 The *Zygomycota.* Diagrammatic representation of the life cycle of *Rhizopus stolonifer.* Both the sexual and asexual phases are illustrated.

The mold *Rhizopus stolonifer* is a common member of this division. This fungus grows on the surface of moist, carbohydrate-rich foods, such as breads, fruits, and vegetables. On breads, for example, *Rhizopus* hyphae can rapidly cover the surface. Hyphae called rhizoids extend into the bread and absorb nutrients (**figure 23.26**). Other hyphae (stolons) become erect, then arch back into the substratum, forming new rhizoids. Still others remain erect and produce at their tips asexual sporangia filled with black spores, giving the mold its characteristic color. Each spore, when liberated, can germinate to start a new mycelium.

Rhizopus usually reproduces asexually, but if food becomes scarce or environmental conditions unfavorable, sexual reproduction occurs. Sexual reproduction requires compatible strains of opposite mating types. When the two mating strains are close, hormones are produced that cause their hyphae to form projections called progametangia that mature into gametangia. After fusion of the gametangia, the nuclei of the two gametes fuse, forming a zygote. The zygote develops a thick, rough, black coat and becomes a dormant zygospore. Meiosis often occurs at the time of germination; the zygospore then splits open and produces a hypha that bears an asexual sporangium to begin the cycle again.

Glomeromycota

Considered zygomycetes by some, **glomeromycetes** are of critical ecological importance because most are mycorrhizal symbionts of vascular plants. **Mycorrhizal fungi** form important associations with the roots of almost all herbaceous plants and tropical trees. As described in section 27.1, this is considered a mutualistic relationship because both the host plant and the fungus benefit: the fungus helps protect its host from stress and delivers soil nutrients to the plant, which in turn provides carbohydrate to the fungus. >> *Microorganisms in terrestrial environments: Mycorrhizae (section 26.2)*

Only asexual reproduction is known to occur in glomeromycetes. Spores are produced and germinate when in contact with the roots of a suitable host plant. Specialized flat hyphae called **appressoria** (s., appressorium) are formed. These enable penetration and subsequent reproduction within the host. Propagation can also occur by fragmentation and colonization of hyphae from the soil or a nearby plant.

1. What are the *Chytridiomycetes?* How do they differ from other fungi?
2. Describe how a typical zygomycete reproduces? What are some beneficial uses for zygomycetes?
3. Compare the function of rhizoids and stolens in *Rhizopus stolonifer.*

Ascomycota

The *Ascomycota,* or **ascomycetes,** are commonly known as **sac fungi.** Ascomycetes are ecologically important in freshwater, marine, and terrestrial habitats because they degrade many chemically stable organic compounds, including lignin, cellulose, and collagen. Many species are quite familiar and economically important (**figure 23.27**). For example, most of the red, brown, and blue-green molds that cause food spoilage are ascomycetes. The powdery mildews that attack plant leaves and the fungi that cause chestnut blight and Dutch elm disease are ascomycetes. Many yeasts as well as edible morels and truffles are also ascomycetes. The pink bread mold *Neurospora crassa* is an extremely important research tool in genetics and biochemistry.

The ascomycetes are named for their characteristic reproductive structure, the saclike **ascus** (pl., asci; Greek *askos,* sac). Many ascomycetes are yeasts. The term **yeast** is used to refer to unicellular fungi that reproduce asexually by either budding or binary fission (**figure 23.28a**); the life cycle of the yeast *Saccharomyces cerevisiae* is well understood. This ascomycete alternates between haploid and diploid states (figure 23.28b). As

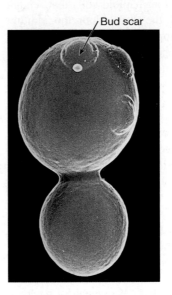

(a) *Saccharomyces cerevisiae:* budding division

Figure 23.28 The Life Cycle of the Yeast *Saccharomyces cerevisiae.* (a) Budding division results in asymmetric septation and the formation of a smaller daughter cell. (b) When nutrients are abundant, haploid and diploid cells undergo mitosis and grow vegetatively. When nutrients are limited, diploid *S. cerevisiae* cells undergo meiosis to produce four haploid cells that remain bound within a common cell wall, the ascus. Upon the addition of nutrients, two haploid cells of opposite mating types (a and α) fuse to create a diploid cell.

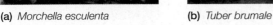

(a) *Morchella esculenta* (b) *Tuber brumale*

Figure 23.27 The *Ascomycota.* (a) The common morel, *Morchella esculenta,* is one of the choicest edible fungi. It fruits in the spring. (b) The black truffle, *Tuber brumale,* is highly prized for its flavor by gourmet cooks. Truffles are mycorrhizal associations on oak trees.

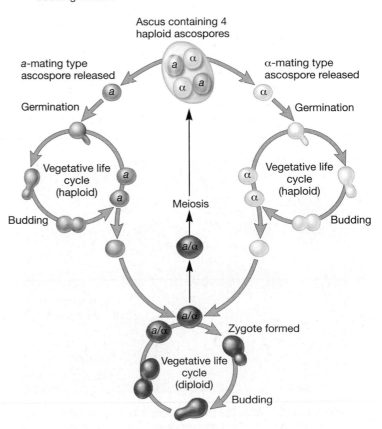

(b) *S. cerevisiae* life cycle

long as nutrients remain plentiful, haploid and diploid cells undergo mitosis to produce haploid and diploid daughter cells, respectively. Each daughter cell leaves a scar on the mother cell as it separates, and daughter cells bud only from unscarred regions of the cell wall. When a mother cell has no more un-scarred cell wall remaining, it can no longer reproduce and will senesce (die). When nutrients are limited, diploid *S. cerevisiae* cells undergo meiosis to produce four haploid cells that remain bound within a common cell wall, the acsus. Upon the addition of nutrients, two haploid cells of opposite mating types (a and α) come into contact and fuse to create a diploid. Typically only cells of opposite mating types can fuse; this process is tightly regulated by the action of pheromones—chemical signals that are exchanged by the haploid cells.

Filamentous ascomycetes form septate hyphae. Asexual repro-duction is common in these ascomycetes and is associated with the production of **conidiospores** (**figure 23.29**). Sexual reproduction also involves ascus formation, with each ascus usually bearing eight haploid ascospores, although some species can produce over 1,000 (**figure 23.30***a*). In the more complex ascomycetes, ascus formation is preceded by the development of special ascogenous hyphae into which pairs of nuclei migrate (figure 23.30*b*). One nucleus of each pair originates from a "male" mycelium (anther-idium) or cell and the other from a "female" organ or cell (ascogonium) that has fused with it. As the ascogenous hyphae grow, the paired nuclei divide so that there is one pair of nuclei in each cell. After the ascogenous hyphae have matured, nuclear fusion occurs at the hyphal tips in the ascus mother cells. The dip-loid zygote nucleus then undergoes meiosis, and the resulting four haploid nuclei divide mitotically again to produce a row of eight nuclei in each developing ascus. These nuclei are walled off from one another. Thousands of asci may be packed together in a cup- or flask-shaped fruiting body called an ascocarp. When the ascospores mature, they often are released from the asci with great force. If the mature ascocarp is jarred, it may appear to belch puffs of "smoke"

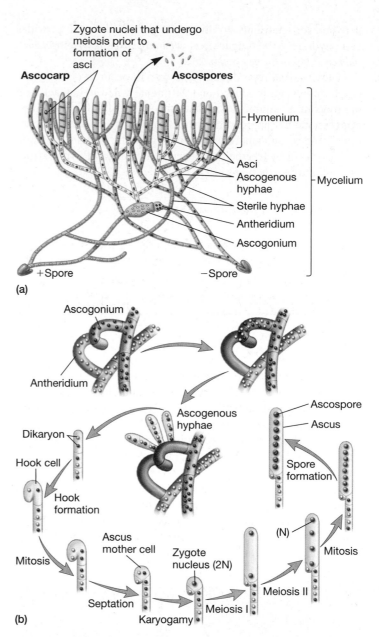

Figure 23.30 The Typical Life Cycle of a Filamentous Ascomycete. Sexual reproduction involves the formation of asci and ascospores. Within the ascus, karyogamy is followed by meiosis to produce the ascospores. (a) Sexual reproduction and ascocarp morphology of a cup fungus. (b) The details of sexual reproduction in ascogenous hyphae. The nuclei of the two mating types are represented by unfilled and filled circles.

consisting of thousands of ascospores. Upon reaching a suitable environment, the ascospores germinate and start the cycle anew.

Most human fungal pathogens are ascomycetes. Many are opportunistic pathogens; systemic fungal infections can be life threatening. Next we discuss some of these fungi and the myco-ses they cause.

Candidiasis

Candidiasis is a mycosis caused by the dimorphic fungus *Candida albicans* (**figure 23.31***a*) or *C. glabrata*. In contrast to the other

Figure 23.29 Asexual Reproduction in *Ascomyota*. Char-acteristic conidiospores of *Aspergillus* as viewed with the scanning electron microscope (×1,200).

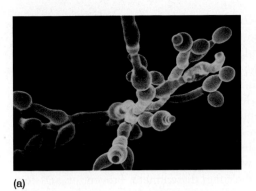
(a)

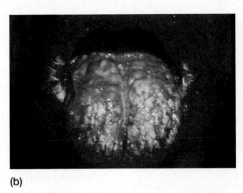
(b)

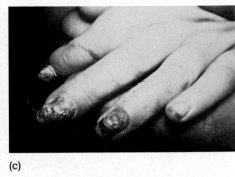
(c)

Figure 23.31 Opportunistic Mycoses Caused by *Candida albicans*. (a) Scanning electron micrograph of the yeast form (×10,000). Notice that some of the cells are reproducing by budding. (b) Thrush, or oral candidiasis, is characterized by the formation of white patches on the mucous membranes of the tongue and elsewhere in the oropharyngeal area. These patches form a pseudomembrane composed of spherical yeast cells, leukocytes, and cellular debris. (c) Paronychia and onychomycosis of the hands.

pathogenic fungi, *C. albicans* and *C. glabrata* are members of the normal microbiota within the gastrointestinal tract, respiratory tract, vaginal area, and mouth. In healthy individuals, they do not produce disease because growth is suppressed by other microbiota and other host resistance mechanisms. However, if anything upsets the normal microbiota and immune competency, *Candida* may multiply rapidly and produce candidiasis. Most *Candida* infections involve the skin or mucous membranes. This occurs because *Candida* is a strict aerobe and finds such surfaces suitable for growth. *Candida* species are important nosocomial pathogens and can represent almost 10% of nosocomial bloodstream infections in some hospitals.

Like many fungi, including some that cause diseases, *Candida* is dimorphic—that is, it has two forms. **Dimorphic fungi** change from the yeast (Y) form in the animal to the mold or mycelial (M) form in the external environment in response to changes in various environmental factors (nutrients, CO_2 tension, temperature). This is called the **YM shift.** The YM shift is so important in pathogenesis that the mechanisms that control this process may be good targets for new antifungal agents. And since no other mycotic pathogen produces as diverse a spectrum of human diseases as does *Candida,* this microbe is the subject of much research.

Oral candidiasis, or **thrush** (figure 23.31*b*), is a fairly common disease in newborns. It appears as many small, white flecks that cover the tongue and mouth. If the mother's vagina is heavily colonized with *Candida,* the upper respiratory tract of the newborn becomes colonized during passage through the birth canal. Because newborns lack a normal microbiota, thrush occurs. **Paronychia** and **onychomycosis** are associated with *Candida* infections of the subcutaneous tissues of the digits and nails, respectively (figure 23.31*c*). These infections usually result from continued immersion in water.

Intertriginous candidiasis usually involves areas of the body with opposed skin surfaces that are warm and moist: axillae, groin, skin folds. **Candidal vaginitis** can result when the yeast is transmitted from infected males and as a complication of diabetes, antibiotic therapy, oral contraceptives, pregnancy, or any other factor that compromises the female host. Normally the omnipres-

ent vaginal lactobacilli control *Candida* by the low pH they create. However, if their numbers decrease, *Candida* may proliferate, causing a curdlike, yellow-white discharge from the vaginal area associated with intense itching and discomfort. *Candida* can be transmitted to males during intercourse and lead to **balanitis,** an infection of the glans penis. In this case, candidiasis can also be considered a sexually transmitted disease.

Like any fungal infection, if *Candida* invades the blood or disseminates to visceral organs, as occasionally seen in immunocompromised patients, the condition is difficult to treat. Because fungi are eucaryotes, they have fewer unique molecular targets for antimicrobial therapy. Although significant advances have been made in the last decade in the development of antifungal agents, systemic candidiasis has a mortality rate of almost 50%.
>> *Antifungal drugs (section 31.7)*

Blastomycosis

Blastomycosis is the systemic mycosis caused by *Blastomyces dermatitidis,* another dimorphic fungi that undergoes a YM shift. It is found predominately in moist soil enriched with decomposing organic debris, as in the Mississippi and Ohio River basins. *B. dermatitidis* is endemic in parts of the south-central, southeastern, and midwestern United States. Additionally, microfoci have been reported in Central and South America, and in parts of Africa. The disease occurs in three clinical forms: cutaneous, pulmonary, and disseminated. The initial infection begins when blastospores are inhaled into the lungs. The fungus can then spread rapidly, especially to the skin, where cutaneous ulcers and abscess formation occur (**figure 23.32**). Mortality is about 5%. There are no preventive or control measures.

Histoplasmosis

Histoplasmosis is caused by *Histoplasma capsulatum* var. *capsulatum,* a facultative, parasitic fungus that grows intracellularly. It appears as a small, budding yeast in humans, where it grows within phagocytic cells. In culture at 25°C, it grows as a mold, producing

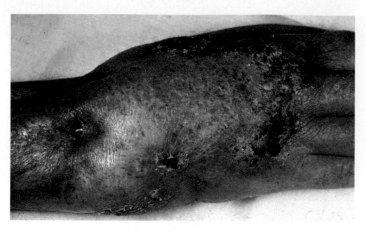

Figure 23.32 Blastomycosis. A mycosis of the forearm caused by *Blastomyces dermatitidis.*

small microconidia (1 to 5 µm in diameter) that are borne singly at the tips of short conidiophores (*see figure 4.23d*). Large macroconidia (8 to 16 µm in diameter) are also formed on conidiophores (**figure 23.33**). The mycelial form is found in soils throughout the world; its growth is supported by nutrients it obtains from bird or bat excrement. When contaminated soil is disturbed, the microconidia can become airborne and cause infection if inhaled. Histoplasmosis is not, however, transmitted from person to person. Within the United States, histoplasmosis is endemic within the Mississippi, Kentucky, Tennessee, Ohio, and Rio Grande river basins. More than 80% of the people who reside in parts of these areas have antibodies against the fungus. It has been estimated that in endemic areas of the United States, about 500,000 individuals are infected annually: 50,000 to 200,000 become ill; 3,000 require hospitalization; and about 50 die. The total number of infected individuals may be over 40 million in the United States alone.

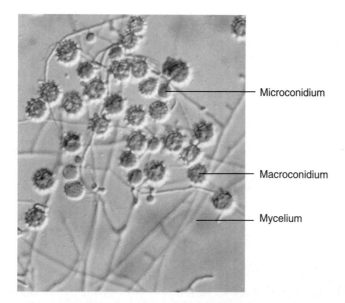

Microconidium

Macroconidium

Mycelium

Figure 23.33 Morphology of *Histoplasma capsulatum* var. *capsulatum.* Mycelia, microconidia, and chlamydospores as found in the soil. These are the infectious particles; light micrograph (×125).

Histoplasmosis is a common disease among poultry farmers, spelunkers (people who explore caves), and bat guano miners (bat guano is used as fertilizer).

Because histoplasmosis is a disease of the monocyte-macrophage system, many organs of the body can be infected (*see figure 28.3*). However, more than 95% of "histo" cases have either no symptoms or mild symptoms such as coughing, fever, and joint pain. Lesions may appear in the lungs and show calcification; the disease may resemble tuberculosis. Most infections resolve without treatment with antifungal agents, and only rarely do the fungi disseminate. Prevention and control involve wearing protective clothing and masks before entering or working in infested habitats. >> *Antifungal drugs (section 31.7)*

Aspergillosis

Of all the fungi that cause disease in human hosts, none is as widely distributed in nature as the *Aspergillus* species. It is found wherever organic debris occurs, especially in soil, decomposing plant matter, household dust, building materials, some foods, and water. *Aspergillus fumigatus* is the usual cause of aspergillosis. *A. flavus* is the second most important species, particularly in invasive disease of immunosuppressed patients. Invasive disease typically results in pulmonary infection (with fever, chest pain, and cough) that disseminates to the brain, kidney, liver, bone, or skin. In immunocompetent patients, *Aspergillus* typically causes allergic sinusitis, allergic bronchitis, or a milder, localized bronchopulmonary infection.

Inhalation of *Aspergillus* conidiospores can lead to several types of pulmonary aspergillosis. One type is allergic aspergillosis. Infected individuals may develop an immediate allergic response and suffer asthma attacks when exposed to fungal antigens on the conidiospores. In bronchopulmonary aspergillosis, *Aspergillus* often can be cultured from the sputum, although the fungus rarely invades the tissue. Pulmonary involvement may lead to colonizing aspergillosis, in which *Aspergillus* forms colonies within the lungs that develop into "fungus balls" called aspergillomas. These consist of a tangled mass of hyphae growing in a circumscribed area. This can be dangerous as the fungus may spread, producing disseminated aspergillosis in a variety of tissues and organs. In patients whose resistance is severely compromised, invasive aspergillosis may occur and fill the lung with fungal hyphae. >> *Immune disorders: Hypersensitivities (section 29.11)*

Aflatoxins

Aflatoxins are fungal toxins (mycotoxins) produced as secondary metabolites by the fungi *Aspergillus flavus* and *A. parasiticus.* The two fungi grow as saprophytes. They have a worldwide distribution but are more common in tropical climates that have wide variations in rainfall, humidity, and temperature. They commonly contaminate corn, grains, peanuts, and other food crops during their growth, harvest, storage, or processing. An estimated 4.5 billion people in developing countries may be exposed chronically to aflatoxins through their diet. Exposure to aflatoxin B_1 is known to cause both chronic and acute hepatocellular injury and typically leads to liver cancer. Indeed, aflatoxins are extremely carcinogenic, mutagenic, and immunosuppressive. Approximately 18 different types of aflatoxins exist; several notable aflatoxins are aflatoxins B_1, B_2, G_1, and G_2. *A. flavus* typically produces aflatoxins B_1 and B_2. *A. parasiticus* produces aflatoxins G_1 and G_2, in addition to B_1 and B_2. *A. flavus* and *A. parasiticus* also produce minute amounts of aflatoxins M_1 and M_2, metabolites of B_1 and B_2, respectively. Both species also produce parasiticol and aflatoxicol, also aflatoxins. Aflatoxins are difuranocoumarins and classified in two broad groups according to their chemical structure: the difurocoumarocyclopentenone aflatoxins B_1, B_2, M_1, M_2, and aflatoxicol; and the difurocoumarolactone aflatoxins G_1, G_2, GM_1, GM_2, and B_3 (*see figure 34.5*). The aflatoxins fluoresce strongly at 365 nm (ultraviolet light); B_1 and B_2 fluoresce blue, and G_1 and G_2 fluoresce green. Lethality studies suggest increasing toxicities as $G_2 < B_2 < G_1 < B_1$.

>> *Microbial growth and food spoilage (section 34.2)*

Stachybotrys

Stachybotrys is an asexually reproducing, cellulolytic (hydrolyzes cellulose) fungus with a worldwide distribution. DNA and mitochondrial gene sequencing suggests that the ordinal and familial placement of *Stachybotrys* is within the *Pezizomycotina*, a subdivision of the *Ascomycota*. It is commonly isolated from soil and frequently grows in water-damaged buildings. It is implicated as a cause of "sick building syndrome," a serious and sometimes fatal disease of humans. *Stachybotrys* produces satratoxins, also known as trichothecene mycotoxins. Trichothecene mycotoxins are potent inhibitors of DNA, RNA, and protein synthesis. They induce inflammation and disrupt alveolar surfactant phospholipids. These toxins may lead to pathological changes in tissues resulting in physical symptoms of cough, chest tightness, fever, myalgia, and allergic responses.

Superficial Mycoses

The superficial mycoses are extremely rare in the United States, and most occur in the tropics. The fungi responsible are limited to the outer surface of hair and skin, and hence are called superficial. Infections of the hair shaft are collectively called **piedras** (Spanish for stone because they are associated with the hard nodules formed by mycelia on the hair shaft). For example, black piedra is caused by the ascomycete *Piedraia hortae* and forms hard, black nodules on the hairs of the scalp (**figure 23.34**). White piedra is caused by the basidiomycete *Trichosporon beigelii* and forms light-colored nodules on the beard and mustache.

Some superficial mycoses are called **tineas** (Latin for grub, larva, worm)—the specific type is designated by a modifying term. Tineas are superficial fungal infections involving the outer layers of the skin, nails, and hair. Tinea versicolor is also caused by a basidiomycete, *Malassezia furfur*, which forms brownish-red scales on the skin of the trunk, neck, face, and arms. Treatment involves removal of the skin scales with a cleansing agent and removal of the infected hairs. Good personal hygiene prevents these infections.

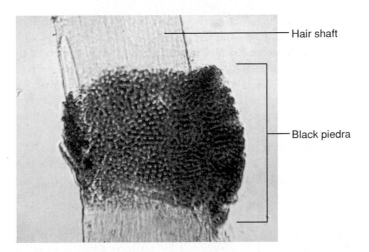

Figure 23.34 Superficial Mycosis: Black Piedra. Hair shaft infected with *Piedraia hortae;* light micrograph (×200).

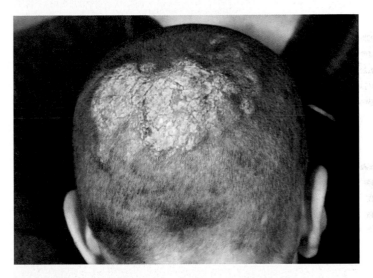

Figure 23.35 Cutaneous Mycosis: Tinea Capitis. Ringworm of the head caused by *Microsporum audouinii.*

Cutaneous Mycoses

Cutaneous mycoses—also called **dermatomycoses, ringworms, or tineas**—occur worldwide and represent the most common fungal diseases in humans. Three genera of cutaneous fungi, or dermatophytes, are involved in these mycoses: *Epidermophyton, Microsporum,* and *Trichophyton.* Tinea barbae (Latin *barba,* the beard) is an infection of beard hair caused by *Trichophyton mentagrophytes* or *T. verrucosum.* It is predominantly a disease of men who live in rural areas and acquire the fungus from infected animals. Tinea capitis (Latin *capita,* the head) is an infection of scalp hair (**figure 23.35**). It is characterized by loss of hair, inflammation, and scaling. Tinea capitis is primarily a childhood disease caused by *Trichophyton* or *Microsporum* species. Person-to-person transmission of the fungus occurs frequently when poor hygiene and overcrowded conditions exist. The fungus also occurs in domestic animals, which can transmit it to humans.

Tinea corporis (Latin *corpus,* the body) is a dermatophytic infection that can occur on any part of the skin (**figure 23.36**). The disease is characterized by circular, red, well-demarcated, scaly, vesiculopustular lesions accompanied by itching. Tinea corporis is caused by *Trichophyton rubrum, T. mentagrophytes,* or *Microsporum canis.* Transmission of the fungus is by direct contact with infected animals and humans or by indirect contact through fomites (inanimate objects). Tinea cruris (Latin *crura,* the leg) is a dermatophytic infection of the groin (**figure 23.37**). The pathogenesis and clinical manifestations are similar to those of tinea corporis. The responsible fungi are *Epidermophyton floccosum, T. mentagrophytes,* or *T. rubrum.* Factors predisposing one to recurrent disease are moisture, occlusion, and skin trauma. Wet bathing suits, athletic supporters (jock itch), tight-fitting slacks, panty hose, and obesity are frequently contributing factors. Tinea pedis (Latin *pes,* the foot), also known as athlete's foot, and tinea manuum (Latin *manus,* the hand) are dermatophytic infections of the feet and hands, respectively. Clinical symptoms vary from a fine scale to a vesiculopustular eruption. Itching is frequently present. Warmth, humidity, trauma, and occlusion increase susceptibility to infection. Most infections are caused by *T. rubrum, T. mentagrophytes,* or *E. floccosum.* Finally, tinea unguium (Latin *unguis,* nail) is a dermatophytic infection of the nail bed. In this disease, the nail becomes discolored and then thickens. The nail plate rises and separates from the nail bed. *Trichophyton rubrum* or *T. mentagrophytes* are the causative fungi.

Subcutaneous Mycoses

The dermatophytes that cause subcutaneous mycoses are normal saprophytic inhabitants of soil and decaying vegetation. Because they are unable to penetrate the skin, they must be introduced into the subcutaneous tissue by a puncture wound. Most infections involve barefooted agricultural workers. Once in the subcutaneous tissue, the disease develops slowly—often over a period of years. During this time, the fungi produce a nodule that eventually ulcerates, and the organisms spread along lymphatic channels, producing more subcutaneous nodules. At times, such nodules drain to the skin surface. The administration of systemic antifungal agents and surgical excision are the usual treatments.

One type of subcutaneous mycosis is **chromoblastomycosis.** The nodules are dark brown. This disease is caused by the black molds *Phialophora verrucosa* and *Fonsecaea pedrosoi.* These fungi exist worldwide, especially in tropical and subtropical regions. Most infections involve the legs and feet (**figure 23.38a**). Another subcutaneous mycosis is **maduromycosis,** caused by *Madurella mycetomatis,* which is distributed worldwide and is also prevalent in the tropics. Because the fungus destroys subcutaneous tissue and produces serious deformities, the resulting infection is often called a **eumycotic mycetoma,** or fungal tumor (figure 23.38b). One form of mycetoma, known as Madura foot, occurs through skin abrasions acquired while walking barefoot on contaminated soil. Finally, a third type of subcutaneous mycosis, **sporotrichosis,** is caused by the dimorphic fungus *Sporothrix schenckii.* The disease occurs throughout the world and is the most common subcutaneous mycotic disease in the United States. The fungus can be found in soil and on plants. Infection occurs by a puncture wound from a thorn or splinter contaminated with the fungus. After an incubation period of 1 to 12 weeks, a small, red papule arises and begins to ulcerate (figure 23.38c). New lesions appear along lymph channels and can remain localized or spread throughout the body, producing extracutaneous sporotrichosis.

Pneumocystis Pneumonia

Pneumocystis is a fungus found in the lungs of a wide variety of mammals. Like many eucaryotic microbes, its taxonomic identity has been keenly debated. Once considered a protist and then

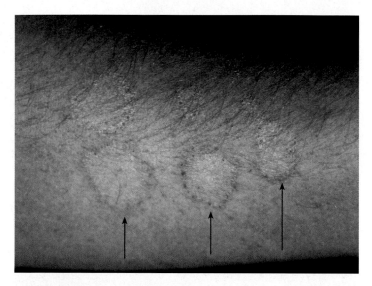

Figure 23.36   Cutaneous Mycosis: Tinea Corporis.   Ringworm of the body—in this case, the forearm—caused by *Trichophyton mentagrophytes.* Notice the circular patches (arrows).

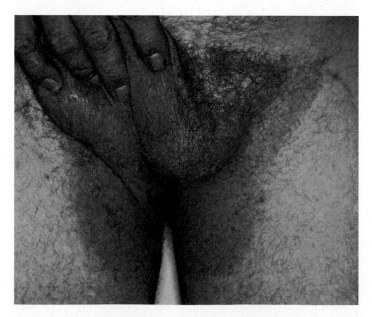

Figure 23.37   Cutaneous Mycosis: Tinea Cruris.   Ringworm of the groin caused by *Epidermophyton floccosum.*

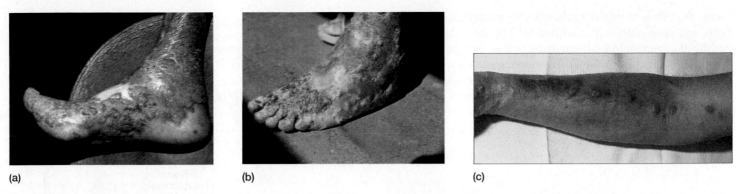

(a)   **(b)**   **(c)**

Figure 23.38 Subcutaneous Mycosis. (a) Chromoblastomycosis of the foot caused by *Fonsecaea pedrosoi.* (b) Eumycotic mycetoma of the foot caused by *Madurella mycetomatis.* (c) Sporotrichosis of the arm caused by *Sporothrix schenckii.*

a fungus occupying an intermediate position between the asco-mycetes and basidiomyces, most recent molecular analysis shows that *Pneumocystis* is an ascomycete. The life cycle of *Pneumocystis* is presented in **figure 23.39**, although some aspects of its development are not well known. The disease that this fungus causes has been called ***Pneumocystis* pneumonia** or ***Pneumocystis carinii* pneumonia (PCP).** Its name was changed

to *Pneumocystis jiroveci* in honor of the Czech parasitologist, Otto Jirovec, who first described this pathogen in humans. In medical literature, the acronym PCP for the disease has been retained despite the loss of the old species name. PCP now stands for *Pneumocystis* pneumonia.

Serological data indicate that most humans are exposed to *Pneumocystis* by age three or four. As with other opportunistic

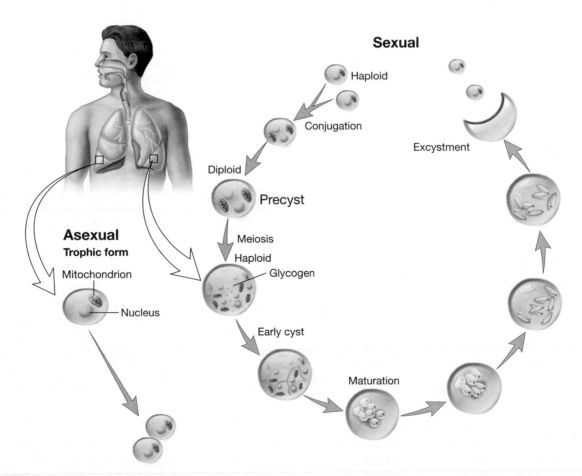

Figure 23.39 Life Cycle of *Pneumocystis jiroveci.* The *Pneumocystis* fungus exhibits both sexual and asexual reproductive stages. Sexual reproduction occurs when haploid cells undergo conjugation followed by meiosis. Asexual reproduction occurs by binary fission. Both forms of reproduction can occur within the human host.

fungi, PCP occurs almost exclusively in immunocompromised hosts and premature, malnourished infants. Both the organism and the disease remain localized in the lungs—even in fatal cases. Within the lungs, *Pneumocystis* causes the alveoli to fill with a frothy exudate. Patients are treated with a combination of oxygen therapy and antifungal agents. Prevention and control are through prophylaxis with drugs in susceptible persons.

1. Describe the ascomycete life cycle. How are the ascomycetes useful to humans?

2. How do yeasts reproduce sexually? Asexually? Why do you think *Saccharomyces cerevisiae* has become such an important model organism?

3. What parts of the human body can be affected by *Candida* infections?

4. Why are systemic fungal infections potentially life-threatening?

5. Describe two piedras that affect humans.

6. In what ways are bronchopulmonary aspergillosis and PCP similar and different?

Basidiomycota

The *Basidiomycota* includes the **basidiomycetes,** commonly known as **club fungi.** Examples include jelly fungi, rusts, shelf fungi, stinkhorns, puffballs, toadstools, mushrooms, and bird's nest fungi. Basidiomycetes are named for their characteristic structure or cell, the **basidium,** which is involved in sexual reproduction (**figure 23.40**). A basidium (Greek *basidion,* small base) is produced at the tip of hyphae and normally is club shaped. Two or more basidiospores are produced by the basidium, and basidia may be held within fruiting bodies called **basidiocarps.**

The basidiomycetes affect humans in many ways. Most are saprophytes that decompose plant debris, especially cellulose and lignin. For example, the common fungus *Polyporus squamosus* forms large, shelflike structures that project from the lower portion of dead trees, which they help decompose. The fruiting body can reach 60 cm in diameter and has many pores, each lined with basidia that produce basidiospores. Thus a single fruiting body can produce millions of spores. Many mushrooms are used as food throughout the world. The cultivation of the mushroom *Agaricus campestris* is a multimillion-dollar business (*see figure 34.16*).

Of course not all mushrooms are edible. Many mushrooms produce alkaloids that act as either poisons or hallucinogens. One such example is the "death angel" mushroom, *Amanita phalloides.* Two toxins isolated from this species are phalloidin and α-amanitin. Phalloidin primarily attacks liver cells, where it binds to plasma membranes, causing them to rupture and leak their contents. Alpha-amanitin attacks the cells lining the stomach and small intestine, and is responsible for the severe gastrointestinal symptoms associated with mushroom poisoning.

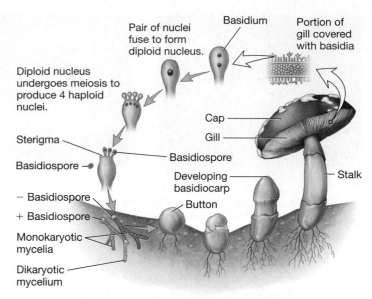

Figure 23.40 **The *Basidiomycota.*** The life cycle of a typical soil basidiomycete starts with a basidiospore germinating to produce a monokaryotic mycelium (a single nucleus in each septate cell). The mycelium grows and spreads throughout the soil. When this primary mycelium meets another monokaryotic mycelium of a different mating type, the two fuse to initiate a dikaryotic secondary mycelium. The secondary mycelium is divided by septa into cells, each of which contains two nuclei, one of each mating type. This dikaryotic mycelium is eventually stimulated to produce basidiocarps. A solid mass of hyphae forms a button that pushes through the soil, elongates, and develops a cap. The cap contains many platelike gills, each of which is coated with basidia. The two nuclei in the tip of each basidium fuse to form a diploid zygote nucleus, which immediately undergoes meiosis to form four haploid nuclei. These nuclei push their way into the developing basidiospores, which are then released at maturity.

The basidiomycete *Cryptococcus neoformans* is an important human and animal pathogen. It produces the disease called **cryptococcosis,** a systemic infection primarily involving the lungs and central nervous system. This fungus always grows as a large, budding yeast, and the production of an elaborate capsule is an important virulence factor for the microbe. In the environment, *C. neoformans* is a saprophyte with a worldwide distribution. Aged, dried pigeon droppings are an apparent source of infection. Cryptococcosis is found in approximately 15% of AIDS patients. The fungus enters the body by the respiratory tract, causing a minor pulmonary infection that is usually transitory. However, some pulmonary infections spread to the skin, bones, viscera, and the central nervous system. Once the nervous system is involved, cryptococcal meningitis usually results. Cryptococcosis must be treated with systemic antifungal agents.

Urediniomycota and Ustilaginomycota

Often considered *Basidiomycota,* both the urediniomycetes and the ustilaginomycetes include important plant pathogens causing "rusts" and "smuts." In addition, some urediniomycetes include human pathogens, and some are virulent plant pathogens that cause extensive damage to cereal crops, destroying millions of dollars worth of

Figure 23.41 _Ustilago maydis._ This pathogen causes tumor formation in corn. Note the enlarged tumors releasing dark teliospores in place of normal corn kernels.

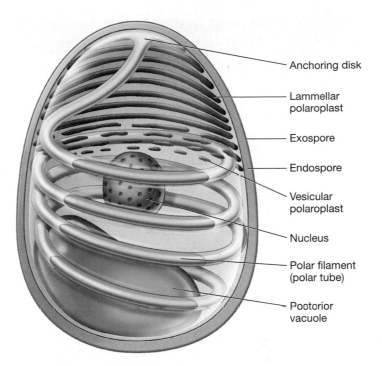

- Anchoring disk
- Lammellar polaroplast
- Exospore
- Endospore
- Vesicular polaroplast
- Nucleus
- Polar filament (polar tube)
- Posterior vacuole

Figure 23.42 The Unique Structure of a Microsporidian Spore. Upon germination in a host cell, the polar tube is ejected with enough force to pierce the host cell membrane, allowing the fungus to gain entry.

crops annually. These fungi do not form large basidiocarps. Instead, small basidia arise from hyphae at the surface of the host plant. The hyphae grow either intra- or extracellularly in plant tissue.

The ustilaginomycete _Ustilago maydis_ is a common corn pathogen that has become a model organism for plant smuts (**figure 23.41**). It is dimorphic; in plant-associated fungi, the YM shift is reversed so that the mycelial form occurs in the plant and the yeastlike form in the external environment. Yeastlike saprophytic _U. maydis_ cells can be easily grown in the laboratory. In nature, the yeast form must mate to produce infectious, filamentous dikaryons that depend on the host plant for continued development. Once a plant is infected, _U. maydis_ forms appressoria within the host. This triggers the plant to form tumors, in which the fungus proliferates and eventually produces diploid spores called **teliospores.** Upon germination, cells undergo meiosis and haploid sporidia are released, causing the infection to spread from plant to plant.

Microsporidia

Of all the fungi, perhaps the _Microsporidia_ have had the most confused taxonomic history. They have been considered protists and are sometimes still cited as such. However, molecular analysis of ribosomal RNA and specific proteins such as α- and β-tubulin shows that they are related to fungi; they are perhaps "curious fungi." Unlike other fungi, they lack mitochondria, peroxisomes, and centrioles. Importantly, they are obligate intracellular parasites that infect insects, fish, and humans. In particular, they infect immunocompromised individuals, especially those with HIV/AIDS. Common human pathogens include _Enterocystozoon bieneusi,_ which causes diarrhea and pneumonia (depending on whether it was acquired through ingestion or

inhalation, respectively), and _Encephalitozoon cuniculi,_ which causes encephalitis and nephritis (kidney disease).

Microsporidia morphology is also unique among eucaryotes. Small spores of 1 to 40 μm are viable outside the host. Depending on the species, spores may be spherical, rod-like, or egg- or crescent-shaped. Spore germination is triggered by a signal from the host and results in the expulsion of a tightly packed organelle called the polar tube or filament (**figure 23.42**). The polar tube is ejected with enough force to pierce the host cell membrane, which permits the parasite's entry. Once inside the host cell, the microsporidian undergoes a developmental cycle that differs among the various microsporidian species. However, in all cases more spores are produced and eventually take over the host cell. Although this can have catastrophic consequences for humans, the use of microsporidia that infect destructive and pathogen-bearing insects is under development as a biocontrol measure.

1. Describe the life cycle of a typical basidiomycete. Discuss their importance.

2. How does _Ustilago maydis_ trigger tumor formation? How is tumor formation advantageous to the fungi?

3. Why do you think _Microsporidia_ are still sometimes considered protists? Describe the germination of a microsporidian spore.

Summary

23.1 Introduction

a. Protists and fungi are widespread in the environment, found wherever water, suitable organic nutrients, and an appropriate temperature occur.

b. Protists are important components of many terrestrial, aquatic, and marine ecosystems, where they contribute to nutrient cycling. Many are parasitic in humans and animals, and some have become very useful in the study of molecular biology.

c. Fungi are important decomposers that break down organic matter; live as parasites on animals, humans, and plants; play a role in many industrial processes; and are used as research tools in the study of fundamental biological processes. Fungi secrete enzymes outside their body structure and absorb the digested food.

d. Most fungi are saprophytes and grow best in moist, dark habitats. These chemoorganoheterotrophs are usually aerobic; some yeasts are fermentative.

23.2 Protist Classification

a. Protist phylogeny is the subject of active research and debate. The classification scheme presented here is that of the International Society of Protistologists (**table 23.1**).

b. The super group *Excavata* includes the *Fornicata*, the *Parabasalia*, and the *Euglenozoa*. Most have a cytoproct and use a flagellum for suspension feeding.

c. The human pathogen *Giardia* is a member of the *Fornicata*. It is considered one of the most primitive eucaryotes (**figure 23.2**).

d. Most *Parabasalia* are flagellated endosymbionts of animals. They include the obligate mutualists of wood-eating insects such as *Trichonympha* and the human pathogen *Trichomonas*.

e. Many of the members of the *Euglenozoa* are photoautotrophic (**figure 23.3**). The remainder are chemoorganotrophs, of which most are saprophytic. Important human pathogens include members of the genus *Trypanosoma*. Trypanosomes cause a number of important human diseases including leishmaniasis, Chagas' disease, and African sleeping sickness (**figures 23.4–23.7**).

f. Amoeboid forms use pseudopodia, which can be lobopodia, filopodia, or reticulopodia (**figure 23.8**). Amoebae that bear external plates are called testate; those without plates are called naked or atestate amoebae.

g. Members of the *Amoebozoa* subclasses *Tubulinea* and *Entamoebida* include a number of endosymbiotic and parasitic protists, including *Entamoeba histolytica*, a major cause of parasitic death worldwide (**figure 23.9**).

h. The *Amoebozoa* subclass *Eumycetozoa* includes the acellular and cellular slime molds. The acellular slime molds form a large mass of protoplasm, called a plasmodium, in which individual cells lack a cell membrane (**figure 23.10**). The cellular slime molds produce a pseudoplasmodium, and each cell within has a cell wall. *Dictyostelium discoideum* is a cellular slime mold that is used as a model organism in the study of chemotaxis, cellular development, and cellular behavior (**figure 23.11**).

i. The *Rhizaria* are amoeboid forms that include the *Radiolaria*, which have filopodia, and the *Foraminifera*, which bear netlike reticulopodia and tests that can be ornate. Most foraminifera are benthic and their tests accumulate on the ocean floor, where they are useful in oil exploration (**figures 23.12** and **23.13**).

j. The super group *Chromalveolata* is diverse. It includes the *Alveolata*, which consists of the dinoflagellates, stramenopiles, ciliates, and the apicomplexans.

k. The dinoflagellates are a large group of nutritionally complex protists. Most are marine and planktonic. They are known for their phosphorescence and for causing toxic blooms.

l. The *Ciliophora* are chemorganotrophic protists that use cilia for locomotion and feeding. In addition to asexual reproduction, conjugation is used in sexual reproduction (**figure 23.16**).

m. Apicomplexans are parasitic with complex life cycles. The motile, infective stage is called the sporozoite (**figure 23.17**).

n. The most important apicomplexan is *Plasmodium*, which causes malaria (**figure 23.18**).

o. The *Stramenopiles* are extremely diverse and includes diatoms, golden and brown algae, the öomycetes, and labyrinthulids. Diatoms are found in fresh- and saltwater and are important components of marine plankton (**figure 23.21**). The öomycetes and labyrinthulids were once thought to be fungi.

p. The haptophytes include the coccolithophores, planktonic photosynthetic protists that contribute to the global carbon budget by precipitating calcium carbonate for their ornate scales (**figure 23.22**).

q. The *Archaeplastida* include the *Chlorophyta*, also known as green algae. All are photosynthetic with chlorophylls *a* and *b* along with specific carotenoids. They exhibit a wide range of morphologies (**figure 23.23**).

23.3 Characteristics of the Fungal Divisions

a. *Chytridiomycota* produce motile spores. Most are saprophytic; some reside in the rumen of herbivores (**figure 23.25**).

b. Zygomycetes are coenocytic. Most are saprophytic. One example is the common bread mold, *Rhizopus stolonifer*. Sexual reproduction occurs through conjugation between strains of opposite mating type (**figure 23.26**).

c. *Glomeromycota* include the mycorrhizal fungi that grow in association with plant roots. They serve to increase plant nutrient uptake.

d. *Ascomycota* are known as the sac fungi because they form a sac-shaped reproductive structure called an ascus. Ascomycetes can have either a yeast morphology (**figure 23.28**) or a mold morphology (**figure 23.29**). Many ascomycetes—particularly pathogens—are dimorphic, switching between the two morphologies.

e. A variety of mycoses are caused by ascomycetes, including candidiasis, aspergillosis, the tineas, and pneumocystis pneumonia. Alfatoxins also contribute to human disease.

f. *Basidiomycota* are the club fungi. They are named after their basidium that produces basidiospores (**figure 23.40**).

g. Urediniomycetes and ustilaginomycetes are often considered *Basidiomycota*. Genera from both groups include important plant pathogens (**figure 23.41**).

h. *Microsporidia* have a unique morphology and are still sometimes considered protists. They include virulent human pathogens; some infect other vertebrates and insects (**figure 23.42**).

Critical Thinking Questions

1. Why do you think our knowledge of the biology of protists has lagged so far behind that of procaryotes, fungi, and higher eucaryotes?

2. Which of the protists discussed in this chapter do you think are the most evolutionarily advanced or derived? Explain your reasoning.

3. Vaccine development for diseases caused by protists (e.g., malaria, Chagas' disease) has been much less successful than that for bacterial diseases. Discuss one biological reason and one geopolitical reason for this fact.

4. What are some logical targets to exploit in treating animals or plants suffering from fungal infections? Compare these with the targets you would use when treating infections caused by bacteria.

5. Fungi and some protists tend to reproduce asexually when nutrients are plentiful and conditions are favorable for growth but reproduce sexually when environmental or nutrient conditions are not favorable. Why is this an evolutionarily important and successful strategy?

6. Some fungi can be viewed as coenocytic organisms that exhibit little differentiation. When differentiation does occur, such as in the formation of reproductive structures, it is preceded by septum formation. Why does this occur?

7. Both bacteria and fungi are major environmental decomposers. Obviously competition exists in a given environment, but fungi usually have an advantage. What characteristics specific to the fungi provide this advantage?

Learn More

Learn more by visiting the Prescott website at www.mhhe.com/prescottprinciples, where you will find a complete list of references.

Viral Diversity

Chapter Glossary

baculoviruses   Rod-shaped, enveloped double-stranded (ds) DNA viruses of helical symmetry that infect insects.

Epstein-Barr virus   A dsDNA virus in the family *Herpesviridae;* causes infectious mononucleosis (mono).

Hepatitis A virus   A naked, icosahedral, plus-strand RNA virus of the genus *Hepatovirus* in the family *Picornaviridae;* causes a type of hepatitis transmitted by fecal-oral contamination.

Hepatitis B virus   A dsDNA virus with a gapped genome in the family *Hepadnaviridae;* infections may be asymptomatic or associated with cirrhosis and liver cancer.

Hepatitis C virus   An enveloped, plus-strand RNA virus in the family *Flaviviridae* that causes serious liver disease.

herpes simplex virus type 1 (HSV-1)   A dsDNA virus that is a member of the family *Herpesviridae;* causes cold sores.

herpes simplex virus type 2 (HSV-2)   A dsDNA virus that is a member of the family *Herpesviridae;* causes genital herpes.

human immunodeficiency virus (HIV)   The single-stranded (ss) RNA virus of the family *Retroviridae* that causes AIDS.

Human parvovirus B19   A small, ssDNA virus that causes a suite of diseases including fifth disease in young children.

influenza viruses   A group of minus-strand RNA viruses with segmented genomes that belong to the family *Orthomyxoviridae;* cause influenza (the flu).

Measles virus   An enveloped minus-strand RNA virus in the genus *Morbillivirus* and the family *Paramyxoviridae;* causes measles.

Poliovirus   A plus-strand RNA virus in the family *Picornaviridae;* causes poliomyelitis.

rotaviruses   dsRNA viruses belonging to the family *Reoviridae* that are one cause of acute viral gastroenteritis.

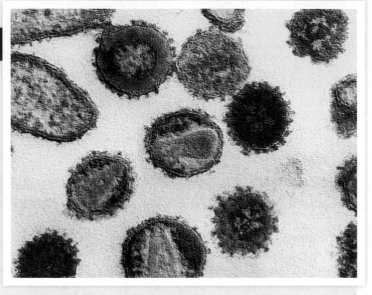

The human immunodeficiency virus (HIV) is an enveloped retrovirus. Protein spikes are embedded in the membrane and the ssRNA genome is enclosed within a core.

Rabies virus   A minus-strand RNA virus belonging to the genus *Lyssavirus* in the family *Rhabdoviridae;* causes rabies.

Rubella virus   A plus-strand RNA virus in the family *Togaviridae;* causes rubella (German measles).

SARS coronoavirus   A plus-strand RNA virus that causes severe acute respiratory syndrome (SARS).

Tobacco mosaic virus **(TMV)**   A filamentous, plus-strand RNA virus of plants.

varicella-zoster virus (VZV)   A dsDNA virus that is a member of the family *Herpesviridae;* causes chickenpox and shingles.

Variola virus   A large dsDNA virus in the family *Poxviridae;* causes smallpox.

West Nile virus   A plus-strand RNA flavivirus that causes West Nile fever (encephalitis).

You might wonder how such naive outsiders get to know about the existence of bacterial viruses. Quite by accident, I assure you. Let me illustrate by reference to an imaginary theoretical physicist, who knew little about biology in general, and nothing about bacterial viruses in particular. . . . Suppose now that our imaginary physicist, the student of Niels Bohr, is shown an experiment in which a virus particle enters a bacterial cell and 20 minutes later the bacterial cell is lysed and 100 virus particles are liberated. He will say: "How come, one particle has become 100 particles of the same kind in 20 minutes? That is very interesting. Let us find out how it happens! . . . Is this multiplying a trick of organic chemistry which the organic chemists have not yet discovered? Let us find out."

—*Max Delbrück*

Our discussion of microbial diversity could not be complete without a review of viruses. Indeed, recent enumeration of viruses in natural environments reveals that they are the most abundant microbes on Earth. In this chapter, we first review viral taxonomy; our discussion of viral diversity is then organized according to the Baltimore system of viral classification—that is, by the mechanisms used to replicate the genome and synthesize mRNA. For each Baltimore group, we present a brief overview of the strategies used, as well as the complete replication cycle of at least one virus. We follow this with examples of other important viruses. Whenever possible, we discuss viruses of procaryotes, plants, and insects. However, the emphasis is on viruses that infect humans because these have been well studied due to their medical importance.

24.1   PRINCIPLES OF VIRUS TAXONOMY

The classification of viruses is in a much less satisfactory state than that of cellular microorganisms. In part, this is due to a lack of knowledge of their origin and evolutionary history. In 1971 the International Committee on Taxonomy of Viruses (ICTV) developed a uniform classification system. Since then the number of viruses and taxonomic categories has continued to expand. In its eighth report, the ICTV described almost 2,000 virus species and placed them in 3 orders, 73 families, 9 subfamilies, and 287 genera (**table 24.1**). The committee places greatest weight on specific properties to define families: nucleic acid type, nucleic acid strandedness, the sense (positive or negative) of ssRNA genomes (*see figure 5.10*), presence or absence of an envelope, symmetry of the capsid, and dimensions of the virion and capsid. Virus order names end in *-virales;* virus family names in *-viridae;* subfamily names, in *-virinae;* and genus names, in *-virus.* An example of this nomenclature scheme is shown in **figure 24.1**, and the diversity of viral morphology and genome structure among viruses that infect procaryotes, animals, and plants is illustrated in **figure 24.2**. $\ll$ *Structure of viruses (section 5.2)*

Although the ICTV reports are the official authority on viral taxonomy, many virologists find it useful to use an alternative classification scheme devised by Nobel laureate David Baltimore. The Baltimore system complements the ICTV system but focuses on the viral genome and the process used to synthesize viral mRNA. Baltimore's system recognizes seven groups (**table 24.2**). This system helps virologists (and microbiology students) simplify the vast array of viral life cycles into a relatively small number of basic types; thus we use it here.

As is the case with the taxonomy of cellular life forms, the taxonomy of viruses is rapidly changing as more and more viruses are isolated and viral genomes are sequenced. Genome analyses have been useful in establishing evolutionary relationships among viruses and have led to the creation of new virus families and genera.

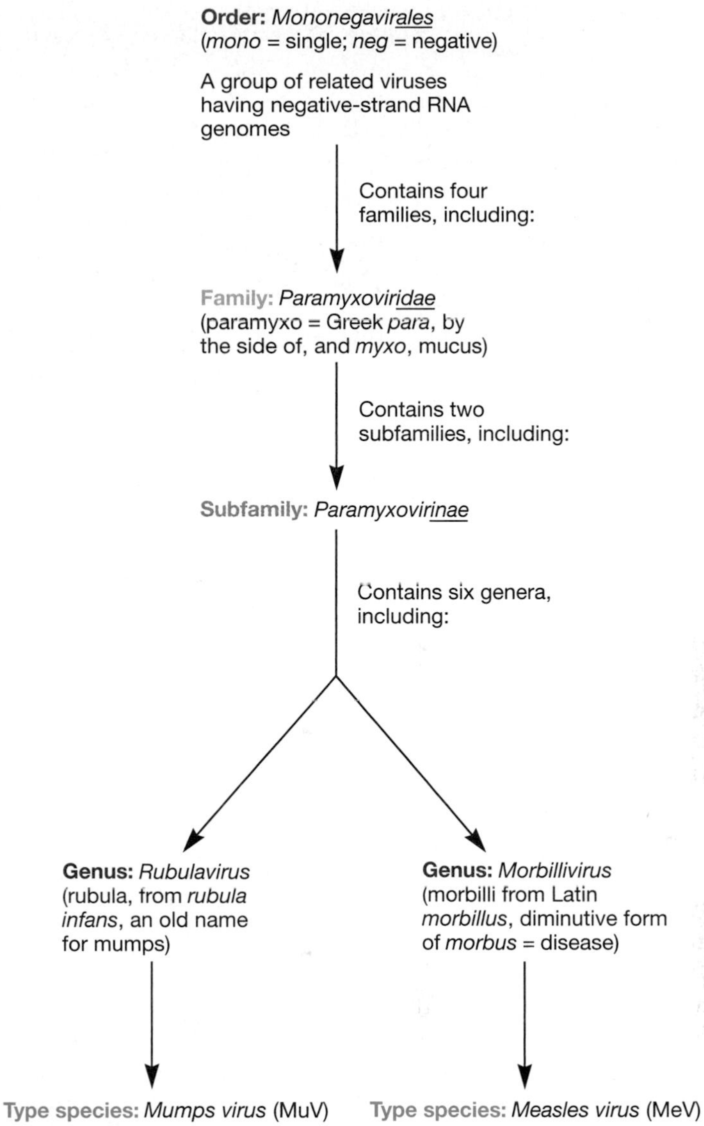

Order: *Mononegavirales*
(*mono* = single; *neg* = negative)

A group of related viruses having negative-strand RNA genomes

Contains four families, including:

Family: *Paramyxoviridae*
(paramyxo = Greek *para*, by the side of, and *myxo*, mucus)

Contains two subfamilies, including:

Subfamily: *Paramyxovirinae*

Contains six genera, including:

Genus: *Rubulavirus*
(rubula, from *rubula infans*, an old name for mumps)

Genus: *Morbillivirus*
(morbilli from Latin *morbillus*, diminutive form of *morbus* = disease)

Type species: *Mumps virus* (MuV)   **Type species:** *Measles virus* (MeV)

Figure 24.1   The Naming of Viruses.   Because of the difficulty in establishing evolutionary relationships, most virus families have not been placed into an order. Virus names are derived from various aspects of their biology and history, including the features of their structure, diseases they cause, and locations where they were first identified or recognized.

1. List some characteristics used in classifying viruses. Which seem to be the most important?
2. What are the endings for the names of virus families, subfamilies, and genera?

24.2   VIRUSES WITH DOUBLE-STRANDED DNA GENOMES (GROUP I)

Group I is perhaps the largest group of known viruses; most bacteriophages have double-stranded (ds) DNA genomes as do a number of important vertebrate viruses, including the

Table 24.1 Some Common Virus Groups and Their Characteristics

ICTV Taxon (Baltimore System Group)[a]	Genome Size (kbp or kb)	Nucleic Acid	Strandedness	Capsid Symmetry[b]	Presence of Envelope	Size of Capsid (nm)[c]	Host Range[d]
Picornaviridae (IV)	7–8	RNA	Single	I	−	22–30	A
Togaviridae (IV)	10–12	RNA	Single	I	+	40–70(e)	A
Retroviridae (VI)	7–12	RNA	Single	C	+	100(e)	A
Orthomyxoviridae (V)	10–15	RNA	Single	H	+	9(h), 80–120(e)	A
Paramyxoviridae (V)	15	RNA	Single	H	+	18(h), 125–250(e)	A
Coronaviridae (IV)	27–31	RNA	Single	H	+	14–16(h), 80–160(e)	A
Rhabdoviridae (V)	11–15	RNA	Single	H	+	18(h), 70–80 × 130–240 (bullet shaped)	A
Bromoviridae (IV)	8–9	RNA	Single	I,B	−	26–35; 18–26 × 30–85	P
Tobamovirus (IV)	7	RNA	Single	H	−	18 × 300	P
Leviviridae [Qβ] (IV)	3–4	RNA	Single	I	−	26–27	B
Reoviridae (III)	19–32	RNA	Double	I	−	70–80	A,P
Cystoviridae (III)	13	RNA	Double	I	+	100(e)	B
Parvoviridae (II)	4–6	DNA	Single	I	−	20–25	A
Geminiviridae (II)	3–6	DNA	Single	I	−	18 × 30 (paired particles)	P
Microviridae (II)	4–6	DNA	Single	I	−	25–35	B
Inoviridae (II)	7–9	DNA	Single	H	−	6 × 900–1,900	B
Polyomaviridae (I)	5	DNA	Double	I	−	40	A
Papillomaviridae (I)	7–8	DNA	Double	I	−	55	A
Adenoviridae (I)	28–45	DNA	Double	I	−	60–90	A
Iridoviridae (I)	140–383	DNA	Double	I	−	130–180	A
Herpesviridae (I)	125–240	DNA	Double	I	+	100, 180–200(e)	A
Poxviridae (I)	130–375	DNA	Double	C	+	200–260 × 250–290(e)	A
Baculoviridae (I)	80–180	DNA	Double	H	+	40 × 300(e)	A
Hepadnaviridae (VII)	3	DNA	Double	C	+	28 (core), 42(e)	A
Caulimoviridae (I)	8	DNA	Double	I,B	−	50; 30 × 60–900	P
Corticoviridae (I)	9	DNA	Double	I	−	60	B
Myoviridae (I)	39–169	DNA	Double	Bi	−	80 × 110, 110[e]	B, Arch
Lipothrixviridae (I)	16	DNA	Double	H	+	38 × 410	Arch

[a]ICTV = International Committee on Virus Taxonomy. The ICTV and Baltimore clarification systems are discussed in section 24.1.

[b]Types of symmetry: I, icosahedral; H, helical; C, complex; Bi, binal; B, bacilliform.

[c]Diameter of helical capsid (h); diameter of enveloped virion (e).

[d]Host range: A, animal; P, plant; B, bacterium; Arch, archaeon.

[e]The first number is the head diameter; the second number, the tail length.

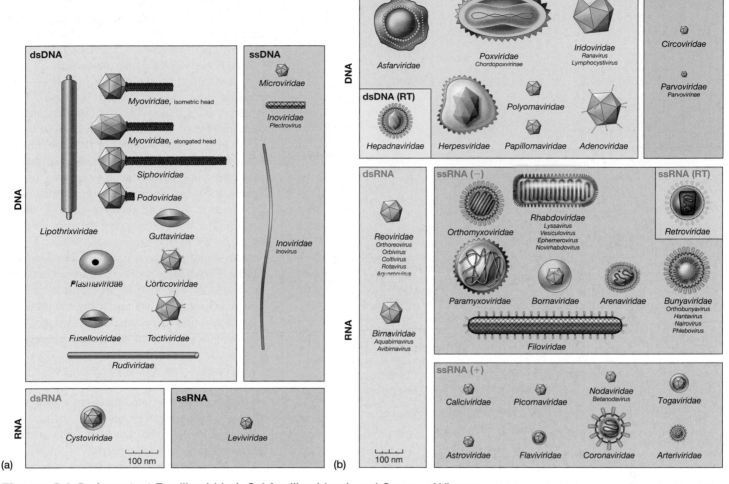

Figure 24.2 Important Families (-*idae*), Subfamilies (-*inae*), and Genera of Viruses. (a) Viruses that infect procaryotes. The *Myoviridae* have contractile tails. *Plasmaviridae* are pleomorphic. *Tectiviridae* have distinctive double capsids. *Corticoviridae* have lipid-containing capids. (b) Viruses that infect vertebrates. Many of these viruses are enveloped viruses.

herpesviruses, poxviruses, and several insect viruses. The pattern of multiplication for dsDNA viruses is shown in **figure 24.3**. It is similar to what occurs in cells; therefore some dsDNA viruses can rely entirely on their host's polymerases to synthesize viral nucleic acids.

Bacteriophage T4: A Virulent Bacteriophage

The life cycle of T4 bacteriophage (family *Myoviridae*) serves as our example of a virulent dsDNA phage. Virulent bacteriophages are capable only of the lytic cycle. That is, their infection of a host always ends with cell lysis. As with any virus, the first step of viral infection is attachment (adsorption) to the host cell surface. T4 phage attachment begins when a tail fiber contacts the appropriate receptor (**figure 24.4a**). As more tail fibers make contact, the baseplate settles down on the surface (figure 24.4b). Binding is probably due to electrostatic interactions and is influenced by pH and the presence of ions such as Mg^{2+} and Ca^{2+}.

After the baseplate is seated firmly on the cell surface, the baseplate and sheath change shape, and the tail sheath reorganizes so that it shortens from a cylinder 24 rings long to one of 12 rings (figure 24.4c,d). As the sheath becomes shorter and wider, the central tube is pushed through the bacterial cell wall. The baseplate contains the protein gp5, which has lysozyme activity. This aids in the penetration of the tube through the peptidoglycan layer. Finally, the linear DNA is extruded from the head, probably through the tail tube, and into the host cell (figure 24.4e,f). The tube may interact with the plasma membrane to form a pore through which DNA passes.

Within 2 minutes after injection of T4 DNA into a host *E. coli* cell, the *E. coli* RNA polymerase starts synthesizing T4 mRNA (**figure 24.5**). This mRNA is called early mRNA because it is made before viral DNA is made. Within 5 minutes, viral DNA synthesis commences, catalyzed by a virus-encoded **DNA-dependent DNA polymerase.** DNA replication is initiated from several origins of replication and proceeds bidirectionally from each. Viral DNA replication is followed by the synthesis

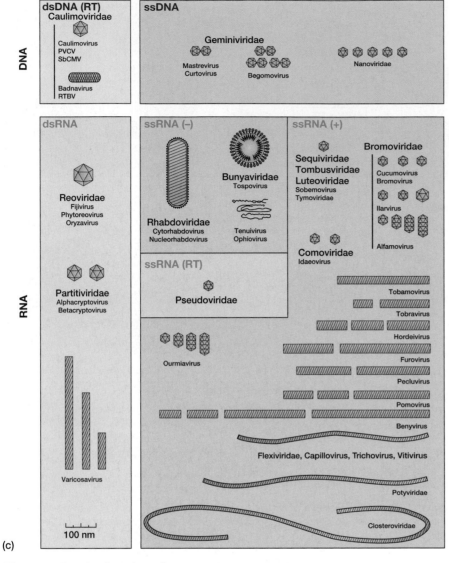

Figure 24.2 (continued). (c) Viruses that infect plants. RT stands for reverse transcriptase, another name for an RNA-dependent DNA polymerase.

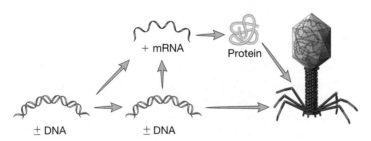

Figure 24.3 Reproductive Strategy of Double-Stranded DNA Viruses. Because the genome of double-stranded DNA viruses is similar to the host, genome replication and mRNA synthesis closely resemble that of the host cell and can involve the use of host polymerases, viral polymerases, or both. The DNA serves as the template for DNA replication and mRNA synthesis. Translation of the mRNA by the host cell's translation machinery yields viral proteins, which are assembled with the viral DNA to make mature virions. These are eventually released from the host.

of late mRNAs, which are important in later stages of the infection.

As with all viruses, the expression of viral genes is temporally ordered. T4 controls the expression of its genes by regulating the activity of the *E. coli* RNA polymerase. Initially T4 genes are transcribed by the regular host RNA polymerase and the sigma factor σ^{70} (*see table 13.3*). After a short interval, a viral enzyme catalyzes the transfer of the chemical group ADP-ribose from NAD to an α-subunit of RNA polymerase (*see figure 12.22*). This modification of the host enzyme helps inhibit the transcription of host genes and promotes viral gene expression. Later the second α-subunit receives an ADP-ribosyl group. This turns off some of the early T4 genes but not before the product of one early gene, *motA,* stimulates transcription of somewhat later genes. One of these later genes encodes the sigma factor gp55. This viral sigma factor helps RNA polymerase bind to late promoters and transcribe late genes, which become active around 10 to 12 minutes after infection. << *Global regulatory systems (section 13.5)*

The tight regulation of expression of T4 genes is aided by the organization of the T4 genome, in which genes with related functions—such as the genes for phage head or tail fiber construction—are usually clustered together. Early and late genes also are clustered separately on the genome; they are even transcribed in different directions—early genes in the counterclockwise direction and late genes, clockwise. A considerable portion of the T4 genome encodes products needed for its replication, including all the protein subunits of its replisome and enzymes needed to prepare for synthesis of DNA (**figure 24.6**). Some of these enzymes synthesize an important component of T4 DNA, hydroxymethylcytosine (HMC) (**figure 24.7**). HMC is a modified nucleotide that replaces cytosine in T4 DNA. Once HMC is synthesized, replication ensues by a mechanism similar to that seen in *Bacteria*. After T4 DNA has been synthesized, it is glucosylated by the addition of glucose to the HMC residues. Glucosylated HMC residues protect T4 DNA from attack by *E. coli* endonucleases called **restriction enzymes,** which would otherwise cleave the viral DNA at specific points and destroy it. This bacterial defense mechanism is called **restriction.** Other phages also chemically modify their DNA to protect against host restriction. << *Key developments in recombinant DNA technology: Restriction enzymes (section 16.1)*

The linear dsDNA genome of T4 shows what is called terminal redundancy—that is, a base sequence is repeated at each end of the molecule. These two characteristics contribute to the formation of long DNA molecules called **concatemers,** which are composed of several genome units linked together in the

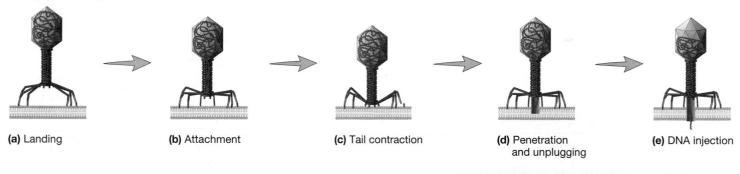

Table 24.2	The Baltimore System of Virus Classification
Group	**Description**
I	Double-stranded DNA genome *genome replication: dsDNA → dsDNA* *mRNA synthesis: dsDNA → mRNA*
II	Single-stranded DNA genome *genome replication: ssDNA → dsDNA → ssDNA* *mRNA synthesis: ssDNA → dsDNA → mRNA*
III	Double-stranded RNA genome *replication: dsRNA → ssRNA → dsRNA* *mRNA synthesis: dsRNA → mRNA*
IV	Plus-strand RNA genome *replication: +RNA → −RNA → +RNA* *mRNA synthesis: +RNA = mRNA*
V	Negative-strand RNA genome *replication: −RNA → +RNA → −RNA* *mRNA synthesis: −RNA → mRNA*
VI	Single-stranded RNA genome *replication: ssRNA → dsDNA → ssRNA* *mRNA synthesis: ssRNA → dsDNA → mRNA*
VII	Double-stranded gapped DNA genome *replication: gapped dsDNA → dsDNA → +RNA →* *−DNA → gapped dsDNA* *mRNA synthesis: gapped dsDNA → dsDNA → mRNA*

The assembly of T4 phage is an exceptionally complex self-assembly process that involves viral proteins and some host cell factors. Late mRNA transcription begins about 9 minutes after injection of T4 DNA into *E. coli* (figure 24.5). Late mRNA products include phage structural proteins, proteins that help with phage assembly without becoming part of the virion, and proteins involved in cell lysis and phage release. These proteins are used in four fairly independent subassembly lines that ultimately converge to generate a mature T4 virion.

DNA packaging within T4 is accomplished by a complex of proteins sometimes called the "packasome." The packasome includes a set of proteins called the terminase complex, which generates double-stranded ends at the ends of the concatemers created during viral genome replication. These double-stranded ends are needed for packaging the T4 genome. The packasome then moves the viral genome into the head, and the concatemer is cut when the phage head is filled with DNA—a DNA molecule roughly 3% longer than the length of one set of T4 genes.

Finally, the virions are released so that they can infect new cells and begin the cycle anew. T4 lyses *E. coli* when about 150 virus particles have accumulated in the host cell. T4 encodes two proteins to accomplish this. The first, called holin, creates holes in the *E. coli* plasma membrane. The second enzyme, termed T4 lysozyme, degrades peptidoglycan in the host's cell wall. Thus the activity of holin enables T4 lysozyme to move from the cytoplasm to the peptidoglycan so that both the plasma membrane and the cell wall are destroyed.

(a) Landing **(b)** Attachment **(c)** Tail contraction **(d)** Penetration and unplugging **(e)** DNA injection

same orientation (**figure 24.8**). Why does this occur? As discussed in chapter 12, the ends of most linear DNA molecules cannot be replicated without the help of the enzyme telomerase. T4 and its *E. coli* host do not have telomerase activity. Therefore, each progeny DNA molecule has single-stranded 3′ ends. These ends participate in homologous recombination with double-stranded regions of other progeny DNA molecules, generating the concatemers. During assembly, concatemers are cleaved such that the genome packaged in the capsid is slightly longer than the T4 gene set. Thus each progeny virus has a genome unit that begins with a different gene. However, if each genome of the progeny viruses were circularized, the sequence of genes in each virion would be the same. Therefore the T4 genome is said to be circularly permuted, and the genetic map of T4 is drawn as a circular molecule.

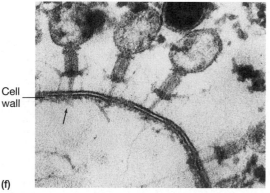

Cell wall

(f)

Figure 24.4 T4 Phage Adsorption and DNA Injection. (a–e) Adsorption and DNA injection is mediated by the phage's tail fibers and base plate as shown here. (f) An electron micrograph of an *E. coli* cell being infected by T-even phages. These phages have injected their nucleic acid through the cell wall and now have empty capsids.

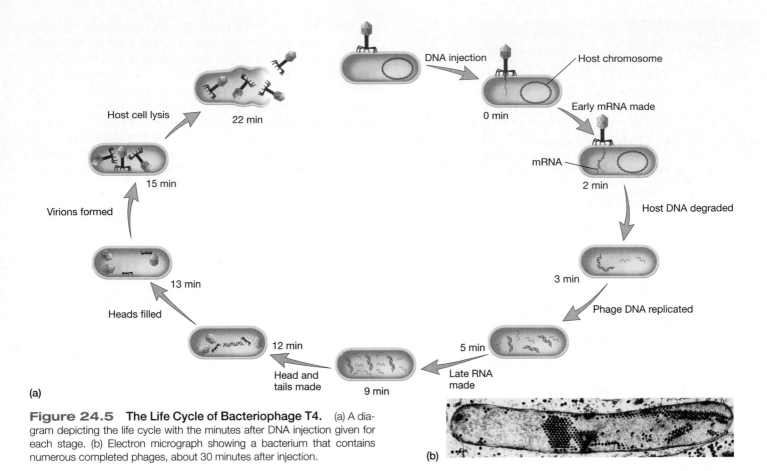

DNA injection — Host chromosome

0 min

Early mRNA made

mRNA

2 min

Host DNA degraded

3 min

Phage DNA replicated

5 min

Late RNA made

9 min

Head and tails made

12 min

Heads filled

13 min

Virions formed

15 min

Host cell lysis

22 min

(a)

(b)

Figure 24.5 **The Life Cycle of Bacteriophage T4.** (a) A diagram depicting the life cycle with the minutes after DNA injection given for each stage. (b) Electron micrograph showing a bacterium that contains numerous completed phages, about 30 minutes after injection.

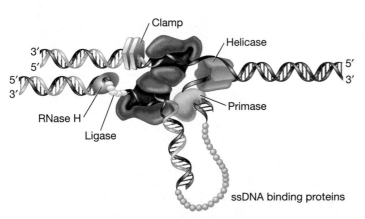

Clamp

Helicase

3′
5′

5′
3′

5′
3′

RNase H

Ligase

Primase

ssDNA binding proteins

Figure 24.6 **A Model of the T4 Replisome.** T4 encodes most of the proteins needed to replicate its dsDNA genome, including components of the T4 replisome. This figure illustrates the viral replisome at the replication fork. Compare this model with the bacterial replisome shown in figure 12.12.

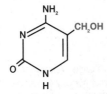

Figure 24.7 **5-Hydroxymethylcytosine (HMC).** In T4 DNA, the HMC often has glucose attached to its hydroxyl.

Bacteriophage Lambda: A Temperate Bacteriophage

Unlike bacteriophage T4 and other "T-even" phages, temperate bacteriophages, such as phage lambda (family *Siphoviridae*), can enter either the lytic or lysogenic cycle upon infecting a host cell. If a temperate virus enters the lysogenic cycle, its dsDNA genome often is integrated into the host's chromosome, where it resides as a prophage until conditions for induction occur (*see figure 5.17*). Upon induction, the viral genome is excised from the host genome and directs the initiation of the lytic cycle. << *Types of viral infections: Infections of procaryotic cells (section 5.4); Transduction (section 14.9)*

How does a temperate phage "decide" which reproductive cycle to follow? The process by which phages make this decision is best illustrated by bacteriophage lambda (λ), which infects the K12 strain of *E. coli*. As shown in **figure 24.9**, λ has an icosahedral head 55 nm in diameter and a noncontractile tail with a thin tail fiber at its end. Its DNA genome is a linear molecule with cohesive ends—single-stranded stretches, 12 nucleotides long, that are complementary to each other and can base pair.

Like most bacteriophages, λ attaches to its host and then injects its genome into the cytoplasm, leaving the capsid outside. Once inside the cell, the linear genome is circularized when the two cohesive ends base pair with each other; the breaks in the strands are sealed by the host cell's DNA ligase (**figure 24.10**).

Figure 24.8   The Terminally Redundant, Circularly Permuted Genome of T4.   The formation of concatemers during replication of the T4 genome is an important step in phage reproduction. During assembly of the virions, the phage head is filled with DNA cleaved from the concatemer. Because slightly more than one set of T4 genes is packaged in each head, each virion contains a different DNA fragment (note that the ends of the fragments are different). However, if each genome were circularized, the sequence of genes would be the same.

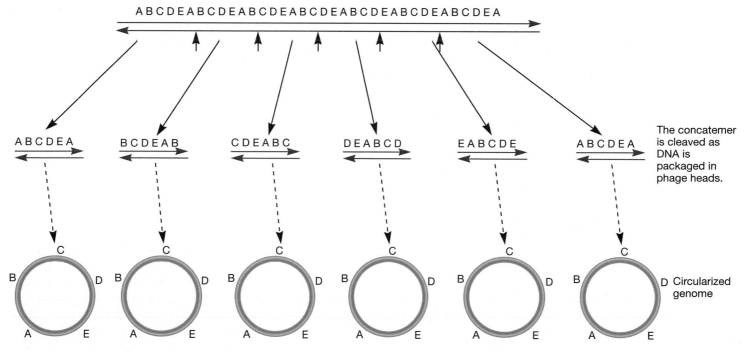

A B C D E A   Parent DNA

Replication yields progeny molecules with single-stranded 3′ ends.

A B C D E A

Homologous recombination between 6 to 10 progeny molecules creates a concatemer.

A B C D E A B C D E A B C D E A B C D E A B C D E A B C D E A

The concatemer is cleaved as DNA is packaged in phage heads.

A B C D E A   B C D E A B   C D E A B C   D E A B C D   E A B C D E   A B C D E A

Circularized genome

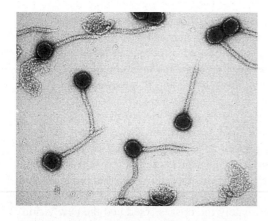

Figure 24.9   Bacteriophage Lambda.

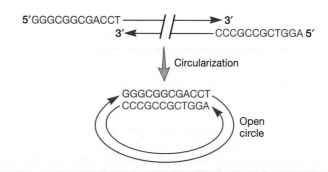

5′ GGGCGGCGACCT ——— 3′
           3′ ——— CCCGCCGCTGGA 5′

↓ Circularization

GGGCGGCGACCT
CCCGCCGCTGGA

Open circle

Figure 24.10   Lambda Phage DNA.   A diagram of lambda phage DNA showing its 12 base, single-stranded cohesive ends and the circularization their complementary base sequences make possible.

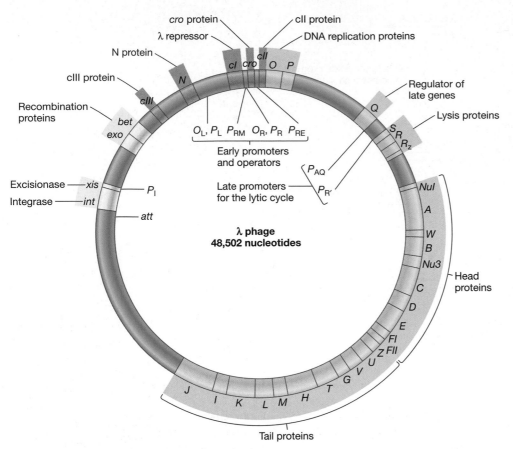

Figure 24.11 **The Genome of Phage Lambda (λ).** The genes are color-coded according to function. Those genes shaded in orange encode regulatory proteins that determine if the lytic or lysogenic cycle will be followed. Genes involved in establishing lysogeny are shaded in yellow. Those required for the lytic cycle are shaded in tan. Important promoters and operators are also noted.

The λ genome has been carefully mapped, and over 40 genes have been located (**figure 24.11**). Most genes are clustered according to their function, with separate groups involved in head synthesis, tail synthesis, lysogeny, DNA replication, and cell lysis. This organization is important because once the genome is circularized, a cascade of regulatory events occurs that determine if the phage pursues a lytic cycle or establishes lysogeny. Regulation of appropriate genes is facilitated by clustering and coordinated transcription from the same promoters. Below, we first introduce the key regulatory molecules used by phage λ in its decision-making process. This is followed by a more detailed examination of the events in which they participate.

The cascade of events leading to either lysogeny or the lytic cycle involves a number of regulatory proteins that function as repressors or activators or both. Two regulatory proteins are of particular importance: the **lambda repressor** (product of the *cI* gene) and the **Cro protein** (product of the *cro* gene). The λ repressor promotes lysogeny, and the Cro protein promotes the lytic cycle. In essence, the decision to pursue lysogeny versus a lytic cycle is the result of a race between the production of these two proteins. If λ repressor prevails, the production of Cro protein is inhibited and lysogeny occurs; if the Cro protein prevails, the production of λ repressor is inhibited and the lytic cycle occurs. This is because the λ repressor prevents transcription of viral genes, while Cro does just the opposite: it ensures viral gene expression.

The λ repressor is 236 amino acids long and folds into a dumbbell shape with globular domains at each end. One domain binds DNA while the other binds another λ repressor molecule to form a dimer. The dimer is the most active form of the repressor. Lambda repressor binds two operator sites, O_L and O_R, thereby blocking transcription of most viral genes (**figure 24.12** and **table 24.3**). When bound at O_L, it represses transcription from the promoter P_L (promoter leftward). Likewise, when bound at O_R, it represses transcription in the rightward direction from P_R (promoter rightward). However, it also activates transcription in the leftward direction from the *cI* promoter P_{RM} (RM stands for *repressor maintenance*). Recall that *cI* encodes the λ repressor. Thus λ repressor controls its own synthesis.

As noted earlier, if λ repressor wins the race with Cro protein, lysogeny is established and the λ genome is integrated into the host chromosome. Integration is catalyzed by the enzyme **integrase**, the product of the *int* gene, and takes place at a site in the host chromosome called the attachment site (*att; see figure 14.33*). A homologous site is present on the phage genome, so the phage and bacterial *att* sites can base pair with each other. The bacterial site is located between the galactose (*gal*) and biotin (*bio*) operons, and as a result of integration, the circular λ genome becomes a linear stretch of DNA located between these two host operons (*see figure 14.30*). The prophage can remain integrated indefinitely, being replicated as the bacterial genome is replicated.

The Cro protein is composed of 66 amino acids and, like λ repressor, forms a dimer that binds the operator sites O_R and O_L, blocking transcription from the P_R and P_L promoters. If Cro protein wins the race with λ repressor, it blocks synthesis of λ repressor and prevents integration of the λ genome into the host chromosome. By the time synthesis of λ repressor is blocked, another regulatory protein called Q protein has accumulated. Q promotes transcription from a promoter called P_R', and in the presence of Q protein, the genes encoding viral structural proteins, as well other proteins needed for virus assembly and host lysis, are transcribed. Ultimately the host is lysed and the new virions are released.

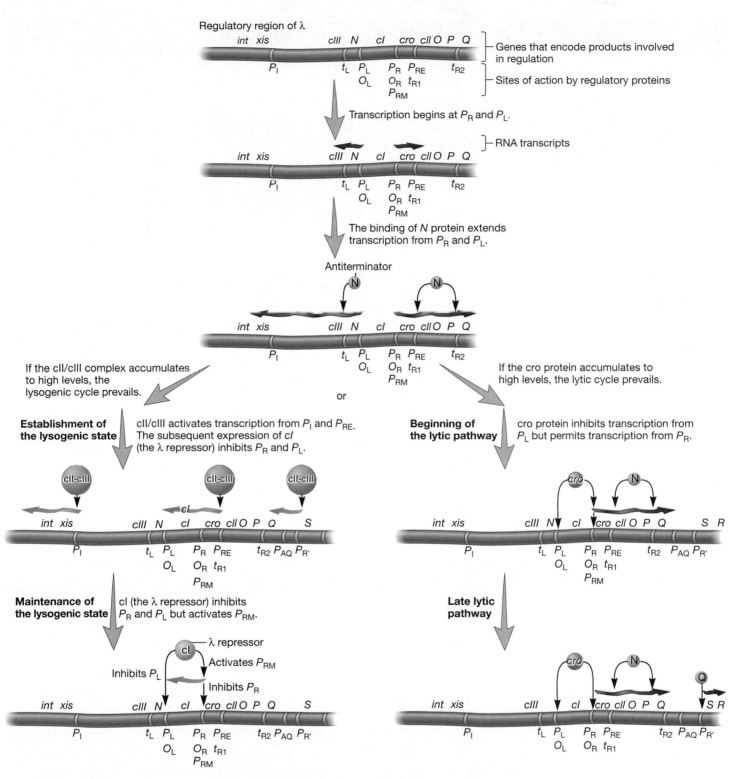

Figure 24.12 **The Decision-Making Process for Establishing Lysogeny or the Lytic Pathway.** The initial transcripts are synthesized by the host RNA polymerase. These encode the N protein and the Cro protein. The N protein is an antiterminator that allows transcription to proceed past the terminator sequences t_L, t_{R1}, and t_{R2}. This allows transcription of other regulatory genes as well as the *xis* and *int* genes. The latter genes encode the enzymes excisionase and integrase, respectively. The left side of the figure illustrates what occurs if lysogeny is established. The right side of the figure shows the lytic pathway.

Table 24.3	Functions of Lambda Promoters and Operators	
Promoter or Operator	**Name Derivation**	**Function**
P_L	**P**romoter **L**eftward	Promoter for transcription of *N, cIII, xis,* and *int* genes; important in establishing lysogeny
O_L	**O**perator **L**eftward	Binding site for lambda repressor and Cro protein; binding by lambda repressor maintains lysogenic state; binding by Cro protein prevents establishment of lysogeny
P_R	**P**romoter **R**ightward	Promoter for transcription of *cro, cII, O, P,* and *Q* genes; Cro, O, P, and Q proteins are needed for lytic cycle; CII protein helps establish lysogeny
O_R	**O**perator **R**ightward	Binding site for lambda repressor and Cro protein; binding by lambda repressor maintains lysogenic state; binding by Cro allows transcription to occur
P_{RE}	**P**romoter for Lambda **R**epressor **E**stablishment	Promoter for *cI* gene (lambda repressor gene); recognized by CII protein, a transcriptional activator; important in establishing lysogeny
P_I	**P**romoter for **I**ntegrase Gene	Transcription from P_I generates mRNA for integrase protein but not excisionase; recognized by the transcriptional activator CII; important for establishing lysogeny
P_{AQ}	**P**romoter for **A**nti-**Q** mRNA	Transcription from P_{AQ} generates an antisense RNA that binds Q mRNA, preventing its translation; recognized by the transcriptional activator CII; important for establishing lysogeny
P_{RM}	**P**romoter for **R**epressor **M**aintenance	Promoter for transcription of lambda repressor gene (cI); activated by lambda repressor; important in maintaining lysogeny
$P_{R'}$	**P**romoter **R**ightward'	Promoter for transcription of viral structural genes; activated by Q protein; important in lytic cycle

With this brief introduction, we can now examine the decision-making process more closely. The sequence of events unfolds as follows. Once the λ genome is circularized within the host cytoplasm, transcription is initiated by host **DNA-dependent RNA polymerase** at promoters P_R and P_L (figure 24.12). Very early in the infection, only the *N* and *cro* genes are expressed, and transcription is terminated at the end of these two genes. However, once the *N* protein is synthesized, it functions as an antiterminator so that RNA polymerase continues transcription beyond the *N* and *cro* genes. Thus from P_R, the *cII* and DNA replication genes *O, P,* and *Q* are transcribed (the *Q* gene encodes the Q protein); from P_L, *cIII* and all the genes through the excision (*xis*) and integration (*int*) genes are transcribed. The synthesis of the CII and CIII proteins is critical to the next step in the infection—it is at this point where the choice to enter lysogeny or the lytic cycle is made.

CII is a transcriptional activator protein that recognizes the promoter P_{RE} (promoter for λ repressor *e*stablishment) and initiates transcription of the *cI* gene, which encodes the λ repressor. It also recognizes the promoters P_I and P_{AQ}. Transcription leftward from the P_I promoter synthesizes mRNA encoding the enzyme integrase, which catalyzes the insertion of lambda DNA into the *E. coli* chromosome. Transcription leftward from P_{AQ} synthesizes an antisense RNA that is complementary to the mRNA encoding the Q protein. Recall that Q protein is needed for expression of viral structural and lysis genes. The activity of the CII protein is influenced by environmental factors. For instance, CII protein is particularly susceptible to proteases.

When *E. coli* is in nutrient-rich conditions, many proteases are formed and CII is more likely to be degraded. The CIII protein's role is to protect CII from degradation. However, its protection is not complete, and in certain conditions, CII is inactivated whether CIII is present or not.

At this point in the infection, several genes are being transcribed and their messages as well as the proteins they encode are accumulating within the host cell. These include Cro protein, λ repressor, CII protein, CIII protein, integrase, N protein, and Q protein. It is now that the race between the λ repressor and Cro protein begins in earnest. Recall that both Cro protein and λ repressor bind the regulatory site O_R. If Cro binds O_R, it blocks transcription of λ repressor. If λ repressor binds O_R, it promotes its own synthesis but blocks synthesis of the Cro protein. Because synthesis of Cro protein begins before synthesis of λ repressor, initially the amount of Cro protein exceeds the amount of λ repressor. However, Cro protein binds O_R less tightly than does λ repressor. Thus it takes a higher concentration of Cro protein in the cell to bind O_R and block the binding of λ repressor. If the CII protein is plentiful (i.e., it is not being degraded by host proteases), the amount of λ repressor in the cell will be sufficient to win the race for the O_R regulatory site. Thus integrase will catalyze integration of the λ genome into the host chromosome. Furthermore, antiQ RNA will be plentiful. Hybridization of this antisense molecule to the mRNA encoding Q protein leads to the destruction of Q protein message. Therefore the action of CII, λ repressor, and antiQ RNA represses the synthesis of all viral proteins except λ repressor. If CII protein is not plentiful (i.e., the cell is in a nutrient-rich environment and many proteases are present within the cell, leading to degradation of CII), then Cro protein will accumulate to a sufficient level to outcompete λ repressor for the O_L and O_R sites. Transcription of the λ repressor genes and other genes that function to establish lysogeny will be blocked and the lytic cycle will proceed.

We have now considered the regulatory processes that dictate whether lysogeny is established or the lytic cycle is pursued. However, how does induction reverse this process? Recall that induction usually occurs in response to environmental factors such as UV light or chemical mutagens that damage DNA. This damage alters the activity of the host cell's **RecA** protein. As we describe in chapter 14, RecA plays important roles in recombination and DNA repair processes. When activated by DNA damage, RecA interacts with λ repressor, causing the repressor to cleave itself. As more and more repressor proteins destroy themselves, transcription of the *cI* gene is decreased, further lowering the amount of λ repressor in the cell. Eventually the level becomes so low that transcription of the *xis, int,* and *cro* genes begins. The *xis* gene encodes the protein excisionase. It binds integrase, causing it to reverse the integration process, and the prophage is freed from the host chromosome. As λ repressor levels decline, the Cro protein levels increase. Eventually, synthesis of λ repressor is completely blocked and the lytic cycle proceeds to completion.

Our attention has been on λ phage, but there are many other temperate phages. Most, like λ, exist as integrated prophages in the lysogen. However, not all temperate phages integrate into the host chromosome at specific sites. Bacteriophage Mu uses a transposition mechanism to integrate randomly into the genome. It then expresses a repressor protein that inhibits lytic growth. Furthermore, integration is not an absolute requirement for lysogeny. The *E. coli* phage P1 is similar to λ in that it circularizes after infection and begins to manufacture repressor. However, it remains as an independent circular DNA molecule in the lysogen and is replicated at the same time as the host chromosome. When *E. coli* divides, P1 DNA is apportioned between the daughter cells so that all lysogens contain one or two copies of the phage genome.

1. Explain why the T4 phage genome is circularly permuted.

2. Precisely how, in molecular terms, is a bacterial cell made lysogenic by a temperate phage such as phage λ?

3. How is a prophage induced to become active again?

4. Describe the roles of the λ repressor, Cro protein, RecA protein, integrase, and excisionase in lysogeny and induction.

5. How do the temperate phages Mu and P1 differ from lambda phage?

So far, our discussion of Group I viruses has focused on bacteriophages. We now turn our attention to some important dsDNA viruses of eucaryotes, including humans, insects, and protists. Our discussion of viruses that infect human cells also reviews many of the hallmarks of the diseases they cause.

Herpesviruses

Herpesviruses, members of the family *Herpesviridae,* are subdivided into three subfamilies (alpha-, beta-, and gammaherpesviruses) and a group of unclassified herpesviruses. Herpesvirus virions are 120–200 nm diameter, somewhat pleomorphic, and enveloped. The envelope contains distinct viral projections (called spikes) that are regularly dispersed over the surface. Their genomes are linear, about 160,000 base pairs long, and contain 50 to 100 genes. The nucleocapsid is icosahedral with DNA wrapped around a spool-like fibrillar core. When herpesviruses target cells of vertebrate hosts, some bind to epithelial cells, others to neurons. Host cell binding is mediated by specific interaction with the envelope spikes.

Herpesviruses begin their replication cycles with receptor-mediated attachment and entry into the host cell (**figure 24.13**). Immediately upon release of the viral DNA into the host nucleus, the DNA circularizes and is transcribed by host DNA-dependent RNA polymerase to form mRNAs, which direct the synthesis of several early proteins. These are mostly regulatory proteins and the enzymes required for replication of viral DNA (figure 24.13, steps 1 and 2). Replication of the genome with a virus-specific DNA-dependent DNA polymerase begins in the cell nucleus within 4 hours after infection (figure 24.13, step 3). Host DNA synthesis gradually slows during a lethal infection.

The term herpes is derived from the Greek word meaning "to creep," reflecting clinical observations of latent recurring infections that progress slowly. Of note is the fact that herpesvirus infection appears to be widespread within most socioeconomic populations with 80 to 90% testing positive for antibody to herpesviruses. All herpesviruses become latent in the host, as discussed in the following examples.

Two of the most commonly known alphaherpesviruses are **herpes simplex virus type 1 (HSV-1)** and **herpes simplex virus type 2 (HSV-2) (figure 24.14)**. Clinical descriptions of herpes date back to the time of Hippocrates (circa 400 BC). Transmission is through direct contact of epithelial tissue surfaces with the virus. HSV-1 was initially thought to cause **cold sores** or **fever blisters** and HSV-2 to cause **genital herpes.** However, both viruses can infect either tissue site. The herpesvirus genome must first enter an epithelial cell for the initiation of infection. The initial association is between proteoglycans of the epithelial cell surface and viral glycoproteins. This is followed by a specific interaction with one of several cellular receptors collectively termed HVEMs for *h*erpes*v*irus *e*ntry *m*ediators. The viral capsid, along with some associated proteins, then migrates along the cellular microtubule transport machinery to nuclear envelope pores. This "docking" is thought to result in the viral DNA being injected through the nuclear envelope pores while the capsid remains in the cytoplasm.

Active and latent phases have been identified within an infected host. After an incubation period of about a week, the active phase begins. During the active phase, the virus multiplies explosively; between 50,000 and 200,000 new virions are produced from each infected cell. During this replication cycle, herpesvirus inhibits its host cell's metabolism and degrades host

Figure 24.13 Generalized Life Cycle for Herpes Simplex Virus Type 1. Only one possible pathway for release of the virions is shown.

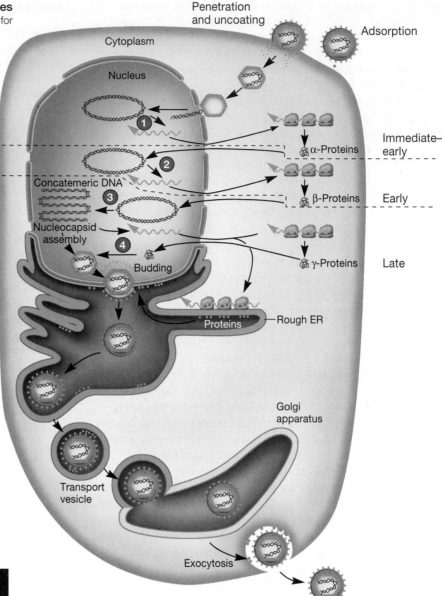

① Circularization of genome and transcription of immediate-early genes

② α-proteins (products of immediate-early genes) stimulate transcription of early genes.

③ β-proteins (products of early genes) function in DNA replication, yielding concatemeric DNA. Late genes are transcribed.

④ γ-proteins (products of late genes) participate in virion assembly.

DNA. As a result, the cell dies. Such an active infection may be symptom-free, or painful blisters in the infected tissue may occur. Indeed, apparently healthy people can transmit HSV to other hosts or their newborns. This is especially true of transmission with HSV-2; it can be passed to sexual contacts even when there are no clinical signs of infection. Blisters involving the epidermis and surface mucous membranes of the lips, mouth, and gums (gingivostomatitis) are referred to as **herpes labialis (figure 24.15a)**. Blisters involving the epidermis and surface mucous membranes of the genitals and perianal region

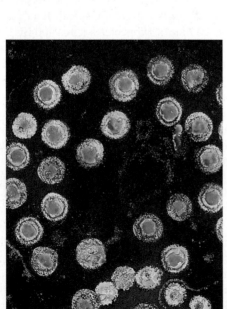

Figure 24.14 HSV-2. Herpes simplex virus type 2 inside an infected cell.

are referred to as genital herpes (figure 24.15b). The blisters are the result of cell lysis and the development of a local inflammatory response; they contain fluid and infectious virions. Fever, headache, muscle aches and pains, a burning sensation, and genital soreness are frequently present during the active phase.

Although blisters usually heal within a week, after a primary infection in cells around the mouth, the virus travels to the trigeminal nerve ganglion. The virus travels to the sacral ganglion after infection of cells in the genital region. In both cases, the retreat to the nerve ganglion results in a latent state for the lifetime of the infected host, apparently to prevent detection by the host immune system. Stressful stimuli such as excessive sunlight, fever, trauma, chilling, emotional stress, and hormonal changes can reactivate the virus. Once reactivated, the virus moves from the nerve ganglion down a peripheral nerve and back to epithelial cells to produce the active phase.

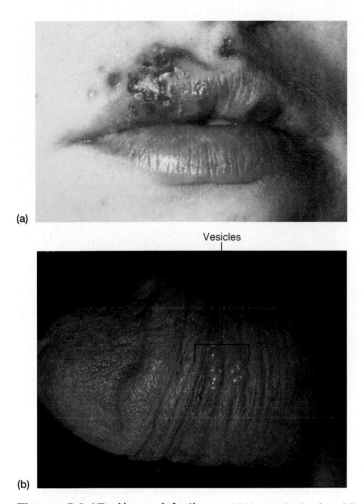

(a)

Vesicles

(b)

Figure 24.15   Herpes Infections. (a) Herpes simplex fever blisters on the lip, caused by herpes simplex type 1 virus. (b) Herpes vesicles on the penis. The vesicles contain fluid that is infectious.

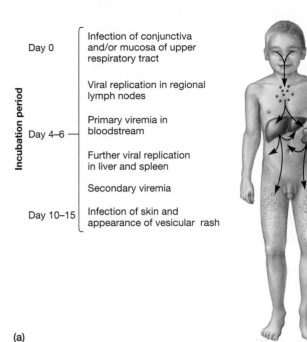

Incubation period		
Day 0	Infection of conjunctiva and/or mucosa of upper respiratory tract	
	Viral replication in regional lymph nodes	
Day 4–6	Primary viremia in bloodstream	
	Further viral replication in liver and spleen	
	Secondary viremia	
Day 10–15	Infection of skin and appearance of vesicular rash	

(a)

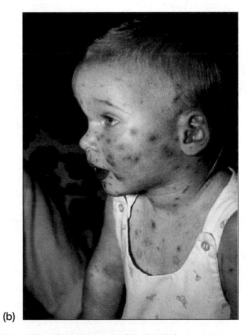

(b)

Figure 24.16   Chickenpox (Varicella). (a) Course of infection. (b) Typical vesicular skin rash. This rash occurs all over the body but is heaviest on the trunk and diminishes in intensity toward the periphery.

Primary and recurring infections with HSV also may occur in the eyes, causing herpetic keratitis (inflammation of the cornea)— currently a major cause of blindness in the United States, and in infants during vaginal delivery, leading to **congenital (neonatal) herpes.** Congenital herpes is one of the most life-threatening of all infections in newborns, affecting approximately 1,500 to 2,200 babies per year in the United States. It can result in neurological involvement as well as blindness. As a result, any pregnant female who has active or new genital herpes should have a cesarian section instead of delivering vaginally. The HSV-2 virus is also associated with a higher-than-normal rate of cervical cancer and miscarriages.

Another alphaherpesvirus is the **varicella-zoster virus (VZV),** causative agent of **chickenpox** (varicella) and **shingles** (herpes zoster). Following an incubation period of 10 to 23 days, small vesicles erupt on the face or upper trunk, fill with pus, rupture, and become covered by scabs (**figure 24.16**). Healing of the vesicles occurs in about 10 days. During this time, intense itching often occurs.

Chickenpox is highly contagious, primarily infecting children two to seven years of age. The virus is acquired by droplet inhalation

into the respiratory system. In the prevaccine era, about 4 million cases of chickenpox occurred annually in the United States, resulting in approximately 11,000 hospitalizations and 100 deaths.

Individuals who recover from chickenpox are subsequently immune to this disease; however, they are not free of the virus, as viral DNA resides in a dormant (latent) state within the nuclei of cranial nerves and sensory neurons in the dorsal root ganglia. This viral DNA is maintained in infected cells but virions cannot

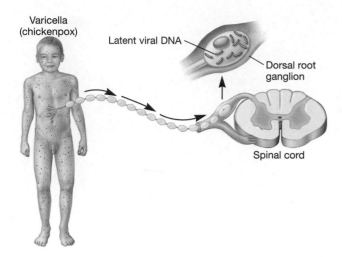

(a) Primary infection—Chickenpox

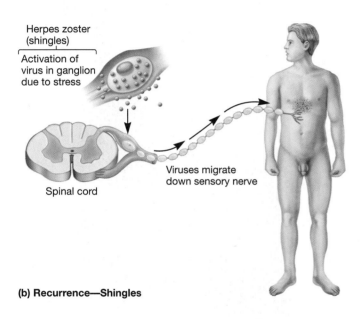

(b) Recurrence—Shingles

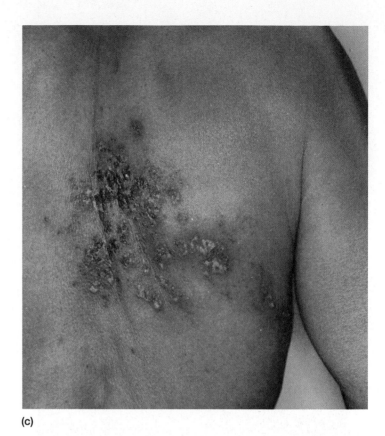

(c)

Figure 24.17 Pathogenesis of the Varicella-Zoster Virus. (a) After an initial infection with varicella (chickenpox), the viruses migrate up sensory peripheral nerves to their dorsal root ganglia, producing a latent infection. (b) When a person becomes immunocompromised or is under stress, the viruses may be activated. (c) They migrate down sensory nerve axons, initiate viral replication, and produce painful vesicles. Since these vesicles usually appear around the trunk of the body, the name *zoster* (Greek for girdle) is used.

be detected (**figure 24.17a**). When the infected person becomes immunocompromised by such factors as age, neoplastic diseases, organ transplants, AIDS, or stress, the viruses may become activated (figure 24.17b). They migrate down sensory nerves, initiate viral multiplication, and produce painful vesicles because of sensory nerve damage (figure 24.17c). This syndrome is called **postherpetic neuralgia.** The reactivated form of chickenpox is called shingles, or **herpes zoster.** Most cases occur in people over fifty years of age. More than 500,000 cases of herpes zoster occur annually in the United States.

The **Epstein-Barr virus (EBV)** is the etiologic agent of **infectious mononucleosis (mono).** EBV is a gammaherpesvirus. After a brief bout of multiplication in epithelial cells, new viruses are shed and infect immune cells called memory B cells. Infected B cells rapidly proliferate and take on an atypical appearance (Downey cells) that is useful in diagnosis (**figure 24.18**). The disease is manifested by enlargement of the lymph nodes and spleen, sore throat, headache, nausea, general weakness and tiredness,

and a mild fever that usually peaks in the early evening. The disease lasts for 1 to 6 weeks and is self-limited. Because EBV occurs in oropharyngeal secretions, it can be spread by mouth-to-mouth contact (thereby acquiring the name "kissing disease") or shared drinking bottles and glasses. A person gets infected when the virus from someone else's saliva makes its way into epithelial cells lining the throat. The peak incidence of mononucleosis occurs in people fifteen to twenty-five years of age. Collegiate populations have a high incidence of the disease. About 50% of college students have no immunity, and approximately 15% of these can be expected to contract mononucleosis.

Papillomaviruses

Human papillomavirus (HPV) is the name given to a group of DNA viruses that includes more than 100 different strains, some of which are oncogenic (cancer-associated) (**figure 24.19a**). They differ in terms of the types of epithelium they infect; some

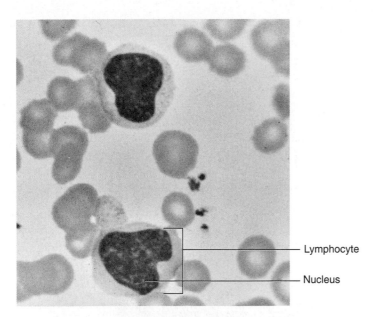

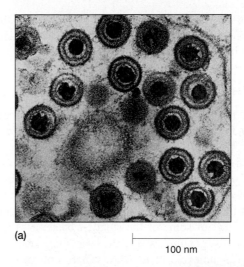

(a)

100 nm

Figure 24.18 Evidence of Epstein-Barr Infection in the Blood Smear of a Patient with Infectious Mononucleosis. Note the abnormally large lymphocytes containing indented nuclei with light discolorations.

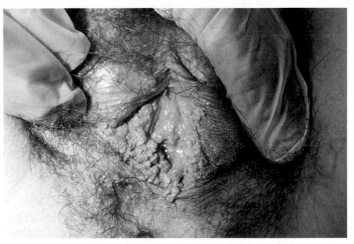

(b)

Figure 24.19 Human Papilloma Virus. (a) Transmission electron micrograph of HPV. (b) Genital warts of the vaginal labia.

infect cutaneous sites, whereas others infect mucous membranes. More than 30 of these viruses are sexually transmitted; they can cause genital warts—soft, pink cauliflowerlike growths that occur on the genital area of women (figure 24.19*b*) and men. They are the most common sexually transmitted disease in the United States today.

HPV strains are designated by numbers and can be divided into "high-risk" (oncogenic) and "low-risk" (nononcogenic) strains. Four high-risk strains (6, 11, 16, and 18) have been reported to be responsible for 70% of cervical cancers and 90% of genital warts. Specifically they have been found in association with invasive cancers of the cervix, vulva, penis, or anus (and other sites). Other high-risk strains have been found to be associated with cancers no more than 1% of the time. These other strains cause benign or low-grade cervical cell changes and genital warts but are rarely, if ever, associated with invasive cancer.

More than 50% of sexually active men and women are infected with HPV at some point in their lives. In the United States, approximately 20 million people age fifteen to forty-nine years (mostly women, representing approximately 15% of the population) are currently infected with HPV. Approximately 6.2 million people in the United States become infected each year. Frequent sexual contact is the most consistent predictor of HPV infection; the number of sexual partners is proportionately linked to the risk of HPV infection. About half of the infected individuals are sexually active teens and young adults (fifteen to twenty-four years of age). HPV is typically transmitted through direct contact, especially during vaginal or anal sex. Other types of genital contact in the absence of penetration (oral-genital, manual-genital, and genital-genital contact) can also lead to infection with HPV, but infection by these

routes is less common. In most cases, infections with HPV are not serious. Most HPV infections are asymptomatic, transient, and clear on their own without treatment. However, in some individuals, HPV infections result in genital warts weeks or months after infection.

A quadrivalent vaccine against four strains of HPV (6, 11, 16, and 18) was licensed in the United States in June 2006. The vaccine is made from noninfectious HPV-like particles and was recommended by the U.S. Advisory Committee on Immunization Practices for prophylactic use in females nine to twenty-six years of age. The vaccine has been found to be safe and to cause no serious side effects in studies of over 11,000 females of the target age. Clinical trials in HPV-naive women sixteen to twenty-six years of age have demonstrated 100% efficacy in preventing cervical precancers caused by the four HPV strains and nearly 100% efficacy in preventing vulvar and vaginal precancers and genital warts caused by the same four strains. While it is possible that vaccination of males with the vaccine may offer direct health benefits, currently no data support use of the HPV vaccine in males. The duration of vaccine protection is unknown, although the vaccine remains effective for at least 5 years. No evidence exists of waning immunity during that time period.

Variola virus

Smallpox (variola) is a highly contagious illness of humans caused by the orthopoxvirus, variola major. The *Variola virus* belongs to the family *Poxviridae,* which includes *Vaccinia virus* also known (as cowpox or the smallpox vaccine virus), *Monkeypox virus,* and *Molluscum contagiosum virus.* The virion is large, brick-shaped, and contains a dumbell-shaped core. Interestingly the size of the smallpox virus (300 by 250 to 200 nm) is slightly larger than that of some of the smallest bacteria—for example, *Chlamydia.* The genome, which has over 200 genes, consists of a single, linear molecule of dsDNA, which replicates in the host cell's cytoplasm. These viruses enter through receptor-mediated endocytosis in coated vesicles; the central core escapes from the endosome and enters the cytoplasm. The core contains a DNA-dependent RNA polymerase. The polymerase synthesizes early mRNAs, one of which directs the production of an enzyme that completes virus uncoating. DNA polymerase and other enzymes needed for DNA replication are also synthesized early in the reproductive cycle, and genome replication begins about 1.5 hours after infection. About the time DNA replication starts, transcription of late genes is initiated. Many late proteins are structural proteins used in capsid construction. The complete reproductive cycle in poxviruses takes about 24 hours.

Smallpox was once one of the most prevalent of all diseases. It was a universally dreaded scourge for more than 3 millennia, with case fatality rates of 20 to 50%. Since the advent of immunization with the *Vaccinia virus* and because of concerted efforts by the World Health Organization, smallpox has been eradicated throughout the world—the greatest public health achievement ever. Eradication was possible because smallpox has obvious clinical features, virtually no asymptomatic carriers, humans are its only natural host, and it has a short period of infectivity (3 to 4 weeks). However, the vaccine has a fairly high rate of complications, making the decision to vaccinate large populations (e.g., military personnel) as protection against potential biowarfare or terrorism very difficult.

There are two clinical forms of smallpox. In both forms, characteristic symptoms of infection include acute onset of fever at 101°F (38.3°C) followed by a rash that features firm, deep-seated vesicles or pustules in the same stage of development without other apparent cause (**figure 24.20**). Disease caused by **variola major** is more severe and the most common form of smallpox, with a more extensive rash and higher fever. Historically this form of smallpox had an overall fatality rate of about 33%, with significant illness in those who did not die. **Variola minor** is a less common form of smallpox, with much less severe disease and death rates of 1% or less. Variola is generally transmitted by direct and fairly prolonged face-to-face contact.

Great concern exists that the smallpox virus could be used as a weapon by terrorists. An accidental or deliberate release of

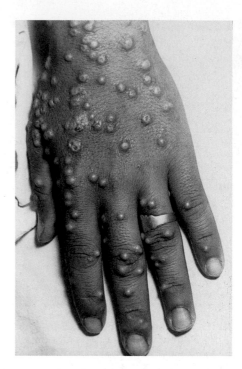

Figure 24.20 Smallpox. A hand showing a single crop of smallpox vesicles.

smallpox virus would be catastrophic in an unimmunized population and could cause a major pandemic. Because smallpox vaccination has not been performed routinely since about 1972, the population of susceptible persons is now large. Currently less than half the world's population has been exposed to either smallpox or to the vaccine. Thus if an outbreak occurred, prompt recognition and institution of control measures would be paramount. **>>** *Bioterrorism preparedness (section 33.9)*

Insect Viruses

Several important Group I viruses infect insects. The *Iridoviridae* are icosahedral viruses with lipid in their capsids and a linear dsDNA genome. They are responsible for diseases in the crane fly and some beetles. The group's name comes from the observation that larvae of infected insects can have an iridescent coloration due to the presence of crystallized virions in their fat bodies.

Many viral infections of insects are accompanied by the formation of inclusion bodies within the infected cells. The *Baculoviridae,* or simply **baculoviruses,** are rod-shaped, enveloped viruses of helical symmetry that include granulosis viruses and nuclear polyhedrosis viruses. Granulosis viruses form granular inclusions in the cytoplasm of infected cells, while nuclear polyhedrosis infections produce polyhedral inclusion bodies in the nucleus or cytoplasm. Both types of inclusion bodies are protein in nature and enclose one or more virions (**figure 24.21**). Insect larvae are infected when they feed on leaves contaminated with inclusion bodies. Polyhedral bodies protect the virions against heat, low pH, and many chemicals; the viruses can remain viable in the soil for years. However, when exposed to the alkaline contents of insect guts, the inclusion bodies dissolve to

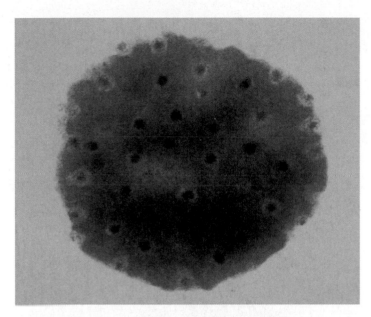

Figure 24.21 Polyhedral Inclusion Bodies. A section of a cytoplasmic polyhedron from a gypsy moth *(Lymantria dispar)*. The occluded virus particles with dense cores are clearly visible (×50,000).

liberate the virions, which then infect midgut cells. Some viruses remain in the midgut, while others spread throughout the insect. Just as with bacterial and vertebrate viruses, insect viruses can persist in a latent state within the host for generations while producing no disease symptoms. The disease may reappear following exposure to chemicals, thermal shock, or even a change in the insect's diet.

Much of the current interest in insect viruses arises from their promise as biological control agents for insect pests. Baculoviruses have received the most attention for at least three reasons. First, they attack only invertebrates and have considerable host specificity; this means that they should be safe for nontarget organisms. Second, because they are encased in protective inclusion bodies, these viruses have a good shelf life and better viability when dispersed in the environment. Finally, they are well suited for commercial production because they often reach extremely high concentrations in larval tissue (as high as 10^{10} viruses per larva). The use of nuclear polyhedrosis viruses for the control of the cotton bollworm, Douglas fir tussock moth, gypsy moth, alfalfa looper, and European pine sawfly has either been approved by the U.S. Environmental Protection Agency or is being considered.

Protist Viruses

Viruses that infect photosynthetic protists are Group I viruses that belong to the family *Phycodnaviridae*. Four genera are recognized by the ICTV. The viruses of photosynthetic protists that have been studied have polyhedral capsids. One virus of *Uronema gigas* resembles many bacteriophages in having a tail. Another interesting group of viruses infects *Chlorella* strains that are endosymbionts of the ciliated protist *Paramecium busaria*. These viruses have very large genomes and encode proteins not usually encoded by other viral genomes. They, along with *Mimivirus* described next, have fostered considerable discussion about what features distinguish cellular organisms from acellular entities.

Very few viruses that infect protozoa (chemoorganoheterotrophic protists) have been studied. One of the few is a giant dsDNA virus, named *Mimivirus*, discovered in the amoeba *Acanthamoeba polyphaga*. The virus is 400 nm in diameter and has a genome about 800 kilobase pairs in size; thus its genome is larger than that of some bacteria. *Mimivirus* is distantly related to the *Poxviridae* and *Phycodnaviridae*.

1. Why do cold sores reoccur throughout the lifetime of an HSV-1–infected individual?
2. In what part of the host cell does the HSV-2 genome replicate? Where does the viral genome reside during the latent phase of infection?
3. How do the latent infections of EBV and CMV differ from those of the other herpesviruses discussed?
4. Many DNA viruses rely on host enzymes for replication and transcription. Why do you think *Variola virus* uses its own DNA-dependent RNA polymerase and DNA polymerase?
5. What is the function of viral inclusion bodies?

24.3 VIRUSES WITH SINGLE-STRANDED DNA GENOMES (GROUP II)

Most DNA viruses are double-stranded, but several important viruses with single-stranded (ss) DNA genomes have been described. The life cycles of ssDNA viruses are similar to those of dsDNA viruses with one major exception. An additional step must occur in the synthesis stage because the ssDNA genome needs to be converted to a dsDNA molecule. A few Group II viruses are discussed next.

Bacteriophages φX174 and fd

The life cycle of φX174 (family *Microviridae*) begins with its attachment to the cell wall of its *E. coli* host. The circular ssDNA genome is injected into the cell, while the protein capsid remains outside the cell. The φX174 ssDNA genome has the same base sequence as viral mRNA and is therefore plus-strand DNA. For either transcription or genome replication to occur, the phage DNA must be converted to a double-stranded **replicative form (RF)** (**figure 24.22**). This is catalyzed by the host's DNA polymerase. The RF directs the synthesis of more RF copies and plus-strand DNA, both by rolling-circle replication (*see figure 12.8*). After assembly of virions, the host is lysed by a viral

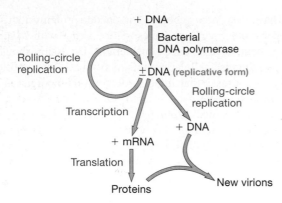

Figure 24.22   **The Reproduction of φX174, a Plus-Strand DNA Phage.**

enzyme (enzyme E) that blocks peptidoglycan synthesis. Enzyme E inhibits the activity of the bacterial protein MraY, which catalyzes the transfer of peptidoglycan precursors to lipid carriers (*see figure 11.11*). Blocking cell wall synthesis weakens the host cell wall, causing the cell to lyse and release the progeny virions.

Although the **fd phage** (family *Inoviridae*) also has a circular, positive-strand DNA genome, it behaves quite differently from φX174 in many respects. It is shaped like a long fiber about 6 nm in diameter by 900 to 1,900 nm in length. Its ssDNA lies in the center of the filament and is surrounded by a tube made of a coat protein organized in a helical arrangement. The virus infects F$^+$, Hfr, and F′ *E. coli* cells by attaching to the tip of the pilus; the DNA enters the host along or possibly through the F factor–encoded sex pilus with the aid of an adsorption protein. As with φX174, a replicative form is first synthesized and then transcribed. A phage-coded protein then aids in replication of the phage DNA by rolling-circle replication. A distinguishing feature of fd and other filamentous phages (e.g., Pf1 phage of *Pseudomonas aeruginosa*) is that they do not kill their host cell. Instead, they establish a relationship in which new virions are continually released by a secretory process. Filamentous phage coat proteins are first inserted into the membrane. The coat then assembles around the viral DNA as it is secreted through the host plasma membrane (**figure 24.23**). Although the host cell is not lysed, it grows and divides at a slightly reduced rate.

Human Parvovirus B19

Since its discovery in 1974, *Human parvovirus B19* (family *Parvoviridae,* genus *Erythrovirus*) has emerged as a significant human pathogen. B19 virions are uniform, icosahedral, naked particles approximately 23 nm in diameter. Parvoviruses have a genome composed of one ssDNA molecule of about 5,000 bases. Parvoviruses are among the simplest of the DNA viruses. The genome is so small that it directs the synthesis of only two capsid and two nonstructural proteins that play key roles in viral replication. Even so, the virus must resort to the use of overlapping genes

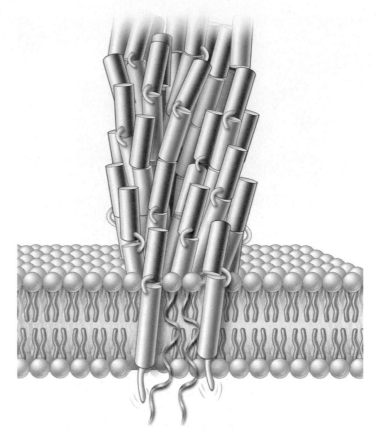

Figure 24.23   **Release of Pf1 Phage.**   The Pf1 phage is a filamentous bacteriophage that is released from *Pseudomonas aeruginosa* without lysis. In this illustration, the blue cylinders are hydrophobic α-helices that span the plasma membrane, and the red cylinders are amphipathic helices that lie on the membrane surface before virus assembly. In each protomer, the two helices are connected by a short, flexible peptide loop (yellow). It is thought that the blue helix binds with circular, single-stranded viral DNA (green) as it is extruded through the membrane. The red helix simultaneously attaches to the growing virus coat that projects from the membrane surface. Eventually the blue helix leaves the membrane and also becomes part of the capsid.

to fit these genes into such a small molecule. Since the genome does not code for any enzymes, the virus must use host cell enzymes for all biosynthetic processes. Thus viral DNA can only be replicated in the nucleus during the S phase of the cell cycle, when the host cell replicates its own DNA. Because the viral genome is single stranded and linear, the host DNA polymerase must be tricked into copying it. By using a self-complementary sequence at the ends of the viral DNA, the parvovirus genome folds back on itself to form a primer for replication (**figure 24.24**). This is recognized by the host DNA polymerase and DNA replication ensues. ◀◀ *DNA replication (section 12.3)*

A spectrum of disease is caused by parvovirus B19 infection, ranging from mild symptoms (fever, headache, chills, malaise) in normal persons, and **erythema infectiosum** in children (**fifth disease**) to a joint disease syndrome in adults. More serious diseases include aplastic crisis in persons with sickle cell disease and autoimmune hemolytic anemia, and pure red cell aplasia due

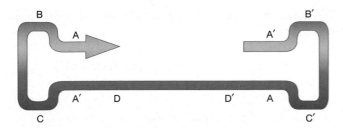

Figure 24.24 The Secondary Structure of the Parvovirus ssDNA Genome. The linear ssDNA genome of parvoviruses exhibits intrastrand base pairing that results in the formation of double-stranded regions at each end of the molecule. This provides a primer for DNA synthesis by the host DNA polymerase.

to persistent B19 virus infection in immunocompromised individuals. The B19 parvovirus can also infect the fetus, resulting in anemia, fetal hydrops (the accumulation of fluid in tissues), and spontaneous abortion. It is assumed that the natural mode of infection is by the respiratory route. The average incubation period is 4 to 14 days. Approximately 20% of infected individuals are asymptomatic, and a smaller percentage of infected individuals have symptoms for up to three weeks. Infection typically results in a life-long immunity to B19.

1. Why is it necessary for some ssDNA viruses to manufacture a replicative form?
2. From the point of view of the virus, compare the advantages and disadvantages of host cell lysis, as seen in ϕX174, with the continuous release of phage without lysis, as with fd phage.
3. How does parvovirus B19 "trick" the host DNA polymerase into replicating its genome? Why must this occur in the nucleus?

24.4 VIRUSES WITH DOUBLE-STRANDED RNA GENOMES (GROUP III)

Because their host cells have dsDNA genomes, viruses with RNA genomes cannot rely on host cell enzymes for genome replication or mRNA synthesis. Instead, Group III, IV, and V viruses use a viral enzyme called **RNA-dependent RNA polymerase** to complete their life cycles. When an RNA-dependent RNA polymerase is used to replicate the RNA genome, it is often referred to as a replicase. When it is used to synthesize mRNA, it is often said to have transcriptase activity. Most RNA viruses use the identical enzyme to carry out both functions.

Among viruses with RNA genomes, those with dsRNA appear to be least abundant. These viruses share a common reproductive strategy (**figure 24.25**). Here, we discuss two representative dsRNA viruses: a bacteriophage and a vertebrate virus.

Bacteriophage ϕ6

Several dsRNA phages have been discovered and form the family *Cystoviridae*. The bacteriophage **ϕ6** of *Pseudomonas syringae* pathovar Phaseolicola (previously called *Pseudomonas phaseolicola*), a plant pathogen, is the best studied. It is an unusual bacteriophage for several reasons. One is that it is an enveloped phage. Within the envelope is a nucleocapsid containing a segmented genome consisting of three distinct dsRNA molecules and an RNA-dependent RNA polymerase. The life cycle of the virus also has unusual features. Like some other phages, ϕ6 attaches to the side of a pilus. However, ϕ6 uses an envelope protein to facilitate adsorption. Retraction of the pilus brings the phage into contact with the outer membrane of its gram-negative host. The viral envelope then fuses with the cell's outer membrane, a process mediated by another envelope protein. Fusion of the two membranes delivers the nucleocapsid into the periplasmic space. Here, a protein associated with the nucleocapsid digests the peptidoglycan, allowing the nucleocapsid to cross this layer of the cell wall. Finally, the intact nucleocapsid enters the host cell by a process that resembles endocytosis. Because bacterial cells do not have the proteins and other factors needed for endocytosis, it is thought that viral proteins mediate this mechanism of entry.

Once inside the host, the viral RNA polymerase acts as a **transcriptase,** catalyzing synthesis of viral mRNA from each dsRNA segment. The enzyme also acts as a replicase, synthesizing plus-strand RNA from each segment. These are enclosed within newly formed capsid proteins, where they serve as templates for the synthesis of the complementary negative strand, regenerating the dsRNA genome. Once the nucleocapsid is completed, a nonstructural viral protein called P12 functions in surrounding the nucleocapsid with a plasma membrane-derived envelope. Interestingly the enveloped nucleocapsid is located within the cytoplasm. Finally, additional viral proteins are added to the envelope and the host cell is lysed, releasing the mature virions.

Rotaviruses

Viewed by electron microscopy, rotaviruses (family *Reoviridae*) have a characteristic wheel-like appearance (the name *rotavirus* is derived from the Latin *rota,* meaning wheel). They are naked

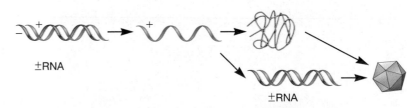

±RNA

±RNA

Figure 24.25 Reproductive Strategy of Double-Stranded RNA Viruses. Both strands can serve as templates for synthesis of mRNA by transcriptase. The negative strand of the viral genome also serves as the template for synthesis of positive strands by replicase. The positive strands are used by replicase to synthesize negative strands, thus replicating the dsRNA genome.

viruses with a double shell, and their genomes are composed of 11 segments of dsRNA (**figure 24.26**). The RNA segments code for six structural and five nonstructural proteins. These virions are very stable in the environment, often remaining infective for several days. The primary mode of transmission is fecal-oral; most infections result from the ingestion of fecal material contaminating water or food, or from contaminated surfaces.

Infection with rotaviruses, as with other viruses, is most common during the cooler months. The clinical manifestations typically range from asymptomatic to a relatively mild diarrhea with headache and fever, to a severe, watery, nonbloody diarrhea with abdominal cramps; symptoms are most severe in young patients. Vomiting is almost always present, especially in children. Viral gastroenteritis is usually self-limited. Treatment is designed to provide relief through the use of oral fluid replacement with isotonic liquids, analgesics, and antiperistaltic agents. Symptoms usually last for 1 to 5 days and recovery often leads to incomplete immunity, although repeated infections may be less severe than the first.

Diarrheal diseases are the leading cause of childhood deaths (5 to 10 million deaths per year) in developing countries where malnutrition is common. Current estimates are that viral gastroenteritis produces 30 to 40% of the cases of infectious diarrhea in the United States, far outnumbering documented cases of bacterial and protozoan diarrhea (the cause of approximately 40% of self-reported cases of diarrhea remains unknown). Rotavirus is responsible for the hospitalization of approximately 55,000 children in the United States and for the death of over 600,000 children worldwide annually. Viral gastroenteritis is seen most frequently in infants one to eleven months of age, where the virus attacks the epithelial cells of the upper intestinal

villi, causing malabsorption, impairment of sodium transport, and diarrhea.

1. Describe the life cycle of φ6 phage. What makes this phage unusual when compared with other bacteriophage?
2. How does the rotavirus structure give rise to its name?

24.5 VIRUSES WITH PLUS-STRAND RNA GENOMES (GROUP IV)

Group IV viruses have nonsegmented plus-strand RNA genomes that can act as mRNA and be translated upon entry into the host cell. All of these viruses replicate in the host cytoplasm and synthesize a viral RNA replicase. For some viruses, the replicase is used to synthesize negative-strand RNAs, which are then used to make more plus-strand RNAs (**figure 24.27**). In some cases, this occurs by way of a double-stranded intermediate called the replicative form (RF). There are a number of important positive-strand animal viruses (e.g., the viruses that cause polio, SARS, and hepatitis A), and most plant viruses have plus-stranded RNA genomes.

Plus-strand RNA viruses of vertebrates exhibit three levels of gene expression complexity. Picornaviruses and caliciviruses are the simplest in that their genomes serve as a single mRNA, they have no virion projections, and the virion is naked. The genome is a single open reading frame that is translated into a large polypeptide that is cleaved into final products. A few of these products have dual roles, displaying one activity as a partially cleaved protein and a second activity once fully cleaved. The mRNA genome of picornaviruses has an extremely long 5′ end that directly interacts with the small ribosomal subunit of the host, eliminating the need for a 5′ cap. RNA synthesis is primed by a protein. The 5′ ends of calicivirus genomes are usually capped by genome-linked proteins or a methylated nucleotide (in the case of hepatitis E virus). The 3′ ends have a poly (A) tract. Notable examples of the picornaviruses are *Poliovirus*

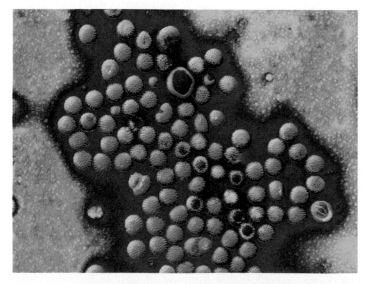

Figure 24.26   Viral Gastroenteritis. Electron micrograph of rotaviruses in a human gastroenteritis stool filtrate (×90,000). Note the wheel-like appearance of the icosahedral capsids that surround double-stranded RNA within each virion.

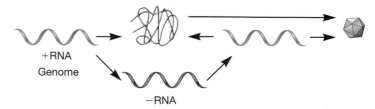

Figure 24.27 **Reproductive Strategy of Plus-Strand RNA Viruses.** The plus-strand RNA genome can serve directly as mRNA. One of the first viral proteins synthesized is RNA replicase, which replicates the genome, sometimes via a double-stranded replicative form. Transcriptase is responsible for synthesizing additional mRNA molecules.

and *Hepatitis A virus*. The newsworthy noroviruses (often the cause of severe gastrointestinal illness on cruise ships) are examples of caliciviruses.

The second level of plus-strand RNA virus complexity is demonstrated by the flaviviruses and the togaviruses. The morphology of these viruses is spherical to pleomorphic, with a size range of 40 to 70 nm. The genome is monopartite of 10 to 12 kilobases long, exhibiting a 5′ cap and 3′ poly (A) tract (some of the flaviviruses do not use a 3′ poly [A] tract). Additionally these viruses are enveloped. Their gene expression differs from that of the previous group in that they express a subgenomic mRNA (mRNA transcribed from only a portion of the genome), which also is translated into a polyprotein from which three to four structural genes are expressed. Both virus families contain viruses transmitted through insects. Examples of flaviviruses include the *Yellow fever virus*, West Nile fever virus, and the *Hepatitis C virus*. The *Rubella virus* is an example of a togavirus.

The coronaviruses represent the final level of complexity of gene expression in the plus-strand RNA viruses. Their virions are roughly spherical or kidney-shaped (some can also assume a rod shape) ranging from 120 to 150 nm in diameter. Coronaviruses are enveloped and have distinct, club-shaped spikes that are evenly dispersed around the envelope. The spikes extend from the virion to give the illusion of a halo, or corona, around the virus. The nucleocapsid is rod-shaped with helical symmetry. The 5′ end of the genome is capped and the 3′ end has a poly (A) tract. The genome encodes five or six structural proteins expressed separately through separate mRNA. Splicing is not involved in gene expression, but rather an initiation-release-reinitiation method of RNA synthesis is used. mRNA and thus gene product concentration are controlled by the efficiency of reinitiation. The SARS coronavirus is one of the more well-known coronaviruses.

Bacteriophages MS2 and Qβ

Bacteriophages MS2 and Qβ, family *Leviviridae,* are small, tailless, icosahedral viruses. MS2 and Qβ have only three or four genes and are genetically the simplest phages known. In MS2, one protein is involved in phage adsorption to the host cell (and possibly also in virion construction or maturation). The other three genes code for a coat protein, RNA replicase, and a protein needed for host cell lysis.

MS2 and Qβ attach to the side of the F pilus of their *E. coli* host. Retraction of the pilus brings the virions close to the outer membrane of the cell, from which they gain entry. As with many bacteriophages, the capsids of these viruses remain outside the cell and only the RNA genome enters. Once the viral RNA replicase is synthesized, it uses the plus-strand genome to synthesize an RF. The replicase then uses the RF to synthesize thousands of copies of plus-strand RNA. Some are used to make more RF in order to accelerate replication of the genome. Other plus-strand RNAs act as mRNA. Eventually plus-strand RNAs are incorporated into the capsid and mature virions are released by lysis.

Tobacco Mosaic virus

Most plant viruses are RNA viruses and, of these, plus-strand RNA viruses are most common. *Tobacco mosaic virus* (**TMV**) is the best studied plus-strand RNA plant virus and is the focus of this discussion. TMV virions are filamentous with coat proteins arranged in a helical pattern. Plant viruses usually enter the host through an abrasion or wound on the plant; biting insects are often involved in transmission of the virus. Following entry into its host, the TMV RNA genome is not immediately transcribed and translated, even though their plus-strand genomes can serve as mRNA. Rather mRNA is generated by a mechanism that is not well understood. The resulting mRNAs encode several proteins, including the coat protein and an RNA-dependent RNA polymerase. Thus TMV, like phage and animal plus-strand RNA viruses, can replicate its own genome. However, this is not the case with all positive-strand RNA plant viruses. Most plants contain an enzyme with RNA-dependent RNA polymerase activity. Thus it is possible that some plus-strand RNA plant viruses make use of this host enzyme.

After the coat protein and RNA genome of TMV have been synthesized, they spontaneously assemble into complete TMV virions in a highly organized process (**figure 24.28**). The protomers come together to form disks composed of two layers of protomers arranged in a helical spiral. Association of coat protein with TMV RNA begins at a specific assembly initiation site close to the 3′ end of the genome. The helical capsid grows by the addition of protomers, probably as disks, to the end of the rod. As the rod lengthens, the RNA passes through a channel in its center and forms a loop at the growing end. In this way, the RNA easily fits as a spiral into the interior of the helical capsid.

Multiplication of plant viruses within their host depends on the virus's ability to spread throughout the plant. Viruses can move long distances through the plant vasculature; usually they travel in the phloem. The spread of plant viruses in nonvascular tissue is hindered by the presence of tough cell walls. TMV thus spreads slowly, about 1 mm per day or less, moving from cell to cell through the plasmodesmata. These slender cytoplasmic strands extend through holes in adjacent cell walls and join plant cells by narrow bridges. Viral "movement proteins" are required for transfer from cell to cell. The TMV movement protein accumulates in the plasmodesmata, but the way in which it promotes movement is not well understood.

Several cytological changes can take place in TMV-infected cells. Viral infections of plants often produce microscopically visible intracellular inclusions, usually composed of virion aggregates. Hexagonal crystals of almost pure TMV virions sometimes develop in TMV-infected cells. In addition, the host cell chloroplasts become abnormal and often degenerate, while new chloroplast synthesis is inhibited.

Hepatitis A virus

Hepatitis A (infectious hepatitis) usually is transmitted by fecal contamination of food, drink, or shellfish that live in contaminated water and contain the virus in their digestive system. The

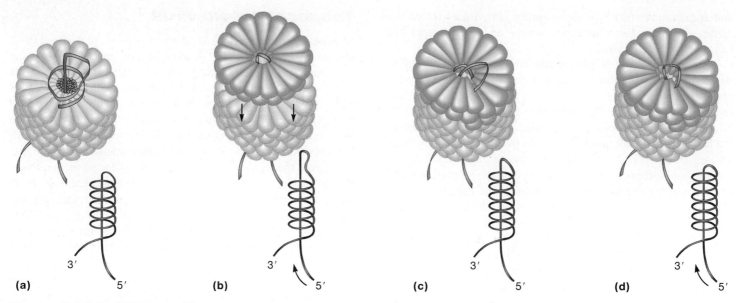

(a) 3′ 5′   **(b)** 3′ 5′   **(c)** 3′ 5′   **(d)** 3′ 5′

Figure 24.28 **TMV Assembly.** The elongation phase of *Tobacco mosaic virus* nucleocapsid construction. The lengthening of the helical capsid through the addition of a protein disk to its end is shown in a sequence of four illustrations; line drawings depicting RNA behavior are included. The RNA genome inserts itself through the hole of an approaching disk and then binds to the groove in the disk as it locks into place at the end of the cylinder.

disease is caused by the **Hepatitis A virus** (**HAV**) of the genus *Hepatovirus* in the family *Picornaviridae*. We should note that while all hepatitis viruses can cause liver disease, not all are taxonomically related (**table 24.4**). Once in the digestive system, the viruses multiply within the intestinal epithelium. Usually only mild intestinal symptoms result and infections in children are usually asymptomatic. Occasionally viremia (the presence of viruses in the blood) occurs and the viruses may spread to the liver. The viruses reproduce in the liver, enter the bile duct, and are released into the small intestine. This explains why feces are so infectious. After about a 4 week incubation period, symptoms develop and include anorexia, general malaise, nausea, diarrhea,

fever, and chills. If the liver becomes infected, jaundice ensues. Most cases resolve in 4 to 6 weeks and yield a strong immunity, although some patients relapse, exhibiting symptoms for 6 months or more. Fortunately the mortality rate is low (less than 1%). Approximately 40 to 80% of the U.S. population has antibodies, though few are aware of having had the disease. Control of infection is by simple hygienic measures, the sanitary disposal of excreta, and the HAV vaccine. The number of new cases has been dramatically reduced since the introduction of the hepatitis A vaccine in the 1990s. However, the increasing number of sexually transmitted HAV outbreaks among men who have sex with other men is of continuing concern.

Table 24.4 **Characteristics of Hepatitides Caused by Hepatotropic Viruses**

Viral Disease	Genome	Classification	Transmission	Outcome	Prevention
Hepatitis A	RNA	*Picornaviridae, Hepatovirus*	Fecal-oral	Subclinical, acute infection	Killed HAV vaccine
Hepatitis B	DNA	*Hepadnaviridae, Orthohepadnavirus*	Blood, needles, body secretions, placenta, sexually	Subclinical, acute chronic infection; cirrhosis; primary hepatocarcinoma	Recombinant HBV vaccines
Hepatitis C	RNA	*Flaviviridae, Hepacivirus*	Blood, sexually	Subclinical, acute chronic infection; primary hepatocarcinoma	Routine screening of blood
Hepatitis D	RNA	Virusoid	Blood, sexually	Superinfection or coinfection with HBV	HBV vaccine
Hepatitis E	RNA	Unassigned	Fecal-oral	Subclinical, acute infection (but high mortality in pregnant women)	Improve sanitary conditions
Hepatitis G	RNA	*Flaviviridae*	Sexually, parenterally	Chronic liver inflammation	HBV vaccine

Poliovirus

Poliomyelitis (Greek *polios,* gray, and *myelos,* marrow or spinal cord), polio, or infantile paralysis is caused by the ***Poliovirus*** and was first described in England in 1789 as a "leg-wasting" disease of children, although a pictograph from ancient Egypt clearly depicts a man with what appears to be polio. The virus is an enterovirus—a transient inhabitant of the gastrointestinal tract—and a member of the family *Picornaviridae. Poliovirus* is composed of three different subtypes. Importantly immunity to one subtype does not confer immunity to the others; the vaccine is therefore tripartite. Like other viruses transmitted by the fecal-oral route, *Poliovirus* is very stable, especially at acidic pH, and can remain infectious for relatively long periods in food and water—its main routes of transmission. The incubation period ranges from 6 to 20 days.

Once ingested, the virus multiplies in the mucosa of the throat or small intestine. From these sites, the virus invades the tonsils and lymph nodes of the neck and terminal portion of the small intestine. Generally there are either no symptoms or a brief illness characterized by fever, headache, sore throat, vomiting, and loss of appetite. The virus sometimes enters the bloodstream causing viremia. In most cases (more than 99%), the viremia is transient and clinical disease does not result. However, in a minority of cases, the viremia persists and the virus enters the central nervous system and causes paralytic polio. The virus has a high affinity for anterior horn motor nerve cells of the spinal cord. Once inside these cells, it multiplies and destroys the cells; this results in motor and muscle paralysis. Since the licensing of the tripartite Salk vaccine (1955) and the tripartite Sabin vaccine (1962), the incidence of polio has decreased markedly. No wild polioviruses exist in the United States. An ongoing global polio eradication effort has been very successful. However, sporadic cases are reported, mostly in areas where religious views and misinformation diminish vaccination efforts and civil wars interrupt public health efforts. Nonetheless, it is likely that polio will be the next human disease to be completely eradicated.

Noroviruses

Noroviruses are a group of related caliciviruses that cause acute gastroenteritis in humans. *Norovirus* is the genus name now used for this group of viruses previously described as "Norwalk-like viruses" and named after the original *Norwalk virus* strain, which was responsible for an outbreak of gastroenteritis in Norwalk, Ohio, in 1968. There are at least five norovirus genogroups, which in turn are divided into at least 31 genetic clusters. Noroviruses are transmitted primarily through the fecal-oral route. Environmental and fomite contamination may also be a source of virus. Transmission due to aerosolization of vomitus (presumably resulting in droplet contamination) also occurs. There are no data to suggest that infection occurs through the respiratory tract. The Centers for Disease Control and Prevention (CDC) estimate that at least 50% of all food-borne outbreaks of gastroenteritis can be attributed to noroviruses. >> *The infectious disease cycle: Story of a disease (section 33.5)*

Noroviruses are highly contagious, with as few as 10 viral particles being sufficient to cause disease. The incubation period for norovirus-associated gastroenteritis in humans is typically 24 to 48 hours. However, there have been reports of cases that have occurred within 12 hours of exposure. The signs and symptoms of norovirus infection are nausea, acute-onset vomiting, and watery, nonbloody diarrhea with abdominal cramps. Low-grade fever has been reported. Vomiting is more common in children. Symptoms typically last 24 to 60 hours and may result in significant dehydration, especially among the young and elderly. Recovery is usually complete with rare occurrences of serious, long-term sequelae. Meticulous personal hygiene by infected individuals and their caregivers is necessary to control additional infections.

Hepatitis C virus

Hepatitis C is caused by the enveloped ***Hepatitis C virus*** **(HCV),** which has an 80 nm diameter and a lipid coat, and contains a linear ssRNA genome. It is a member of the family *Flaviviridae.* HCV is classified into multiple genotypes. It is transmitted by contact with virus-contaminated blood, by the fecal-oral route, by in utero transmission from mother to fetus, sexually, or through organ transplantation. HCV is found worldwide. Prior to routine screening, HCV accounted for more than 90% of hepatitis cases developed after a blood transfusion. Worldwide, hepatitis C has reached epidemic proportions, with more than 1 million new cases reported annually. In the United States, nearly 4 million persons are infected and 25,000 new cases occur annually. Currently, HCV is responsible for about 8,000 deaths annually in the United States, where it is also the leading reason for liver transplantation. Antiviral therapy can eliminate the virus in 50% of those infected with genotype 1 and in 80% of those infected with genotype 2 or 3.

West Nile Fever Virus

West Nile fever (encephalitis) is caused by a flavivirus that occurred primarily in the Middle East, Africa, and Southwest Asia. The disease was first discovered in 1937 in the West Nile district of Uganda. In 1999 the virus appeared unexpectedly in the New York, causing seven deaths among 62 confirmed human encephalitis cases and extensive mortality in a variety of domestic and exotic birds. It probably crossed the Atlantic in an infected bird, mosquito, or human traveler.

By 2003 46 U.S. states reported ***West Nile virus*** **(WNV)** infections in over 9,800 people, resulting in 264 deaths. By the start of 2006, every state in the continental United States reported *West Nile virus* in either animals or humans. It is also now reported in Mexico, South America, and the Caribbean. WNV is transmitted predominately to humans by *Culex* spp. mosquitoes that feed on infected birds. Mosquitoes harbor the greatest concentration of virus in the early fall; there is a peak of disease in late August to early September.

The risk of disease then decreases as the mosquitoes die when the weather becomes colder. Although many people are bitten by WNV-infected mosquitoes, most do not know they have been exposed. Most infected individuals remain asymptomatic or exhibit only mild, flulike symptoms. Data from the outbreak in Queens, New York, suggest that 2.6% of the population was infected, 20% of the infected people developed mild illness, and only 0.7% of the infected people developed meningitis or encephalitis.

Human-to-human transmission has been reported through blood and organ donation. However, the risk of acquiring WNV infection from donated blood or organs has greatly diminished since the introduction of a PCR-based detection assay in 2003. No data suggest that WNV transmission to humans occurs from handling infected birds (live or dead), but barrier protection is suggested in handling potentially infected animals. Currently no human vaccine exists to prevent WNV infection. Mosquito abatement and the use of repellents such as DEET appear to be the only control measures. ◁◁ *Polymerase chain reaction (section 16.2)*

Rubella virus

Rubella (Latin *rubellus,* reddish) was first described in Germany in the 1800s and as such was subsequently called "German measles." It is a moderately contagious disease that occurs primarily in children five to nine years of age. However, because rubella is usually such a mild disease in children, no treatment is indicated. All children and women of childbearing age who have not been previously exposed to rubella should be vaccinated. Because routine vaccination began in the United States in 1969, fewer than 1,000 cases of rubella and 10 cases of congenital rubella currently occur annually.

Rubella is caused by the ***Rubella virus,*** the only member of the genus *Rubivirus* in the family *Togaviridae;* unlike other members of the *Togaviridae,* this genus is not transmitted by arthropods. Rubella is worldwide in distribution and occurs more frequently during the winter and spring months. This virus is spread in droplets shed from the respiratory secretions of infected individuals. Once the virus is inside the body, the incubation period ranges from 12 to 23 days. A rash of small red spots (**figure 24.29**), usually lasting no more than 3 days, and a low-grade fever are the normal symptoms. Rubella can be a disastrous disease in the first trimester of pregnancy (congenital rubella syndrome) and can lead to fetal death, premature delivery, or a wide array of congenital defects that affect the heart, eyes, and ears.

SARS Coronavirus

Severe acute respiratory syndrome (SARS) is a highly contagious viral disease caused by a novel coronavirus known as the **SARS coronavirus (SARS-CoV).** The virus causes a febrile (>100.4°F or 38°C) lower respiratory tract illness. Sudden, severe illness in otherwise healthy individuals is a hallmark of the disease. Other symptoms may include headache; mild, flulike discomfort; and body aches. SARS patients may develop a dry cough after a few days, and most will develop pneumonia. About 10 to 20% of patients have diarrhea. If not detected early, this

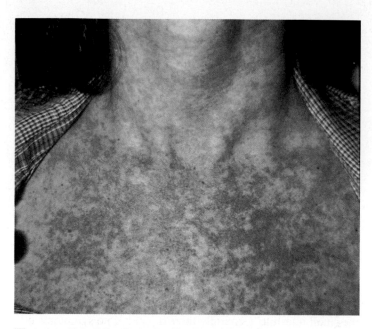

Figure 24.29   Rubella (German Measles).   This disease is characterized by a rash of red spots. Notice that the spots are not raised above the surrounding skin as in measles (rubeola; see figure 24.33).

disease can be fatal even with supportive care. SARS is transmitted by close contact with respiratory secretions (droplet spread).

The initial outbreak of SARS appears to have originated in China in late 2002, and it spread rapidly to at least 29 other countries by summer 2003. The outbreak resulted in 8,098 persons with possible SARS, including 744 deaths reported by the World Health Organization. There were 373 possible SARS cases in the United States; however, SARS-CoV identification has been confirmed in only 8 of them. Seven of the eight cases were likely due to exposure during international travel, and the eighth case was probably due to exposure to one of the other seven. The genome sequence of SARS-CoV strains provides an explanation for the apparent disappearance of the disease (**Microbial Diversity & Ecology 24.1**). The 2003 SARS epidemic demonstrated to the world the ease with which a viral infectious agent can spread. Rapid detection and prevention measures were pursued during and after the outbreak. Diligent screening for signs of fever or respiratory disease at airports and the initiation of SARS-CoV vaccine trials are two examples of protective measures. No specific treatment is currently approved.

1. Why can many plant viruses with RNA genomes be more dependent on their host for reproduction than procaryotic and animal viruses?

2. How does TMV spread from one host cell to another?

3. Define viremia; what is the importance of this condition?

4. Prepare a table that compares viral taxonomy, route of infection, and control measures for hepatitis A and C.

5. What do you think were two important public health lessons learned from the 2003 SARS outbreak?

Microbial Diversity & Ecology

24.1 SARS: Evolution of a Virus

In November 2002 a mysterious pneumonia was seen in the Guangdong Province of China, but the first case of this new type of pneumonia was not reported until February 2003. Thanks to the ease of global travel, it took only a couple of months for the pneumonia to spread to more than 25 countries in Asia, Europe, and North and South America. This newly emergent pneumonia was labeled severe acute respiratory syndrome (SARS) and its causative agent was identified as a previously unrecognized member of the coronavirus family, the SARS-CoV. Almost 10% of the roughly 8,000 people with SARS died. However, once the epidemic was contained, the virus appeared to "die out," and with the exception of a few mild, sporadic cases in 2004, no additional cases have been identified. How does a newly emergent virus arise? What does it mean when a virus "dies out"?

We can answer these questions thanks to the availability of the complete SARS-CoV genome sequence and the power of molecular modeling. Coronaviruses are large, enveloped viruses with positive-strand RNA genomes. They are known to infect a variety of mammals and birds. Researchers suspected that SARS-CoV had "jumped" from its animal host to humans, so samples of animals at open markets in Guangdong were taken for nucleotide sequencing. These studies revealed that catlike animals called masked palm civets (*Paguma larvata*) harbored variants of the SARS-CoV. Although thousands of civets were then slaughtered, further studies failed to find widespread infection of domestic or wild civets. In addition, experimental infection of civets with human SARS-CoV strains made these animals ill, making the civet an unlikely candidate for the reservoir species. Such a species would be expected to harbor SARS-CoV without symptoms so that it could efficiently spread the virus.

Bats are hosts of several viruses spread from animals to people (zoonotic viruses), including the emerging Hendra and Nipah viruses that have been found in Australia and East Asia, respectively. Thus it was perhaps not too surprising when in 2005 two groups of international scientists independently demonstrated that Chinese horseshoe bats (genus *Rhinolophus*) are the natural reservoir of a SARS-like coronavirus. When the genomes of the human and bat viruses are aligned, 92% of the nucleotides are identical. More revealing is alignment of the translated amino acid sequences of the proteins encoded by each virus. The amino acid sequences are 96 to 100% identical for all proteins except the receptor-binding spike proteins, which are only 64% identical. The SARS-CoV spike protein mediates both host cell surface attachment and membrane fusion. Thus a mutation of the spike protein allowed the virus to "jump" from bat host cells to those of another species. It is not clear if the SARS-CoV was transmitted directly to humans (bats are eaten as a delicacy and bat feces are a traditional Asian cure for asthma) or if transmission to humans occurred through infected civets.

The relationship between the viruses found in civets and humans has also been studied in detail and offers insight into why there have been no additional cases of SARS since 2004 (at least as this book went to press). The region of the SARS-CoV spike protein that binds to the host receptor, *a*ngiotensin-*c*onverting *e*nzyme-2 (ACE2), forms a shallow pocket into which ACE2 rests. The region of the spike protein that makes this pocket is called the receptor-binding domain (RBD). Of the approximately 220 amino acids within the RBD, only four differ between civet and human. Two of these amino acids appear to be critical. As shown in the **box figure**, compared to the spike RBD in the SARS-CoV that caused the 2002–2003 epidemic, the civet spike has a serine (S) substituted for a threonine (T) at position 487 (T487S) and a lysine (K) at position 479 instead of asparagine (N), N479K. This causes a 1,000-fold decrease in the capacity of the virus to bind to human ACE2. Furthermore, the spike found in SARS-CoV isolated from patients in 2003 and 2004 also has a serine at position 487 as well as a proline (P) for leucine (L) substitution at position 472 (L472P). These amino acid substitutions could be responsible for the reduced virulence of the virus found in these more recent infections. In other words, these mutations could be the reason the SARS virus appears to have "died out."

Meanwhile a SARS vaccine based on the virulent 2002–2003 strain is being tested. This raises additional questions. Does the original virulent SARS-CoV strain still exist? Will the most recently identified SARS-CoV continue to evolve into less virulent forms? If not, will this vaccine be effective in preventing another highly infective SARS outbreak? Unfortunately, these questions cannot be easily answered.

Receptor Activity

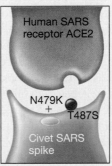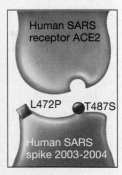

(a) Good   **(b)** Poor   **(c)** Poor

Host Range of SARS-CoV Is Determined by Several Amino Acid Residues in the Spike Protein. **(a)** The spike protein of the SARS-CoV that caused the SARS epidemic in 2002–2003 fits tightly to the human host cell receptor ACE2. **(b)** The civet SARS-CoV has two different amino acids at positions 479 and 487. This spike protein binds very poorly to human ACE2, thus the receptor is only weakly activated. **(c)** The spike protein on the human SARS-CoV that was isolated from patients in 2003 and 2004 also differs from that seen in the epidemic-causing SARS-CoV by two amino acids. This SARS-CoV variant caused only mild, sporadic cases.

Sources: Li, F.; Li, W.; Farzan, M.; and Harrison, S. C. 2005. Structure of SARS coronavirus spike receptor-binding domain complexed with receptor. Science 309:1864–68.

Li, W.; Shi, Z.; Yu, M.; Ren, W.; Smith, C.; Epstein, J. H.; and Wang, H., et al. 2005. Bats are natural reservoirs of SARS-like coronavirus. Science 310:676–79.

24.6 VIRUSES WITH MINUS-STRAND RNA GENOMES (GROUP V)

Negative-strand RNA viruses seem to be a more recent evolutionary development, sharing a basic genome arrangement and significant sequence similarities. In general, the virions are spherical to pleomorphic, measure 80 to 120 nm in diameter (influenza viruses can also assume a filamentous shape, 200–300 nm long and 20 nm in diameter), and are enveloped. The grouping of negative-strand RNA viruses includes those that have segmented and unsegmented genomes. Segmented genomes are composed of multiple linear pieces of RNA. All minus-strand RNA genomes have self-complementary 3′ and 5′ ends, and the transcripts have a 5′ cap and a 3′ poly (A) tail. Segmented genomes may have evolved from unsegmented ones; they may have been created by the reduction of redundant genetic regions. The unsegmented, negative-strand RNA genome is arranged with genes in a highly conserved order. The genes are tandemly linked and typically separated by nontranscribed intergenic sequences. The negative-strand viruses include the families *Rhabdoviridae* (e.g., *Rabies virus*), the *Filoviridae* (e.g., Marburg and Ebola viruses), and the *Paramyxoviridae* (e.g., *Measles virus*). Of the segmented negative-strand RNA genomes, the *Bunyaviridae* (e.g., hantaviruses) have three segments transcribing six proteins, and the *Orthomyxoviridae* (e.g., influenza viruses) have seven or eight segments transcribing 10 proteins. The *Bunyaviridae* replicate in the host cell cytoplasm, while the *Orthomyxoviridae* are the only RNA viruses that replicate in the host nucleus.

The genomes of negative-strand RNA viruses cannot function as mRNA. Therefore these viruses must bring at least one RNA-dependent RNA polymerase into the host cell during entry. Initially the viral genome serves as the template for mRNA synthesis (**figure 24.30**). Later, the virus switches from mRNA synthesis to genome replication, as the RNA-dependent RNA polymerase synthesizes a distinct plus-strand RNA for replication. During this phase of the replication cycle, the plus-strand RNA molecules synthesized from the minus-strand genome serve as templates for the manufacture of new negative-strand RNA genomes. In this section, we focus on viruses that are pathogenic to humans as these are best studied.

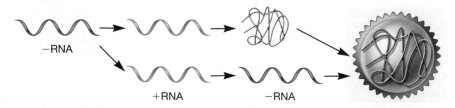

−RNA

+RNA −RNA

Figure 24.30 **Reproductive Strategy of Negative-Strand RNA Viruses.** An RNA-dependent RNA polymerase enters the host cell at the same time the negative-strand RNA genome enters. The genome serves as a template for synthesis of mRNA. Later in the infection, the negative-strand genome is used for plus-strand synthesis. These plus strands then act as templates for replication of the negative-strand genomes.

Rabies virus

Rabies (Latin *rabere*, rage or madness) is caused by a number of different strains of highly neurotropic viruses belonging to the genus *Lyssavirus* (Greek *lyssa*, rage or rabies), family *Rhabdoviridae* (**figure 24.31a**). *Rabies virus* multiplies in the salivary glands of an infected host. It is transmitted to humans or other animals by the bite of an infected animal whose saliva contains the virus; by aerosols of the virus that can be spread in caves where bats dwell; or by contamination of scratches, abrasions, open wounds, and mucous membranes with saliva from an infected animal. After inoculation, a region of the virion's glycoprotein envelope spike attaches to the plasma membrane of nearby skeletal muscle cells, the virus enters the cells, and multiplies. When the concentration of the virus in the muscle is sufficient, the virus enters the nervous system through unmyelinated sensory and motor terminals; the binding site is the nicotinic acetylcholine receptor.

The virus spreads by retrograde axonal flow at 8 to 20 mm per day until it reaches the spinal cord, when the first specific symptoms of the disease—pain or paresthesia at the wound site—may occur. Within brain neurons, the virus produces characteristic **Negri bodies**—masses of viruses or unassembled viral subunits that are visible in the light microscope. Rapidly progressive encephalitis develops as the virus quickly disseminates through the central nervous system. The virus then spreads throughout the body along the peripheral nerves, including those in the salivary glands, where it is shed in the saliva.

Improvements in prevention during the past 50 years have led to almost complete elimination of indigenously acquired rabies in the United States, where rabies is primarily a disease of feral animals, especially bats. Most wild animals can become infected with rabies, but susceptibility varies according to species (figure 24.31b). Worldwide, almost all cases of human rabies are attributed to dog bites. In developing countries where canine rabies is still endemic, rabies accounts for up to 40,000 deaths per year.

Symptoms of rabies usually begin 2 to 16 weeks after viral exposure and include anxiety, irritability, depression, fatigue, loss of appetite, fever, and sensitivity to light and sound. The disease quickly progresses to a stage of paralysis. In about 50% of all cases, intense and painful spasms of the throat and chest muscles occur when the victim swallows liquids. The mere sight, thought, or smell of water can set off spasms. Consequently, rabies has been called hydrophobia (fear of water). Death results from destruction of the regions of the brain that regulate breathing. Safe and effective vaccines against rabies are available; however, to be effective they must be given soon after the person has been infected. Veterinarians and laboratory personnel, who have a high risk of exposure to rabies, usually are immunized every 2 years. In the United States, fewer than 10 cases of rabies occur yearly in humans, although about 8,000 cases of animal rabies are reported each year from various sources (figure 24.31b). Prevention and control involves preexposure vaccination of

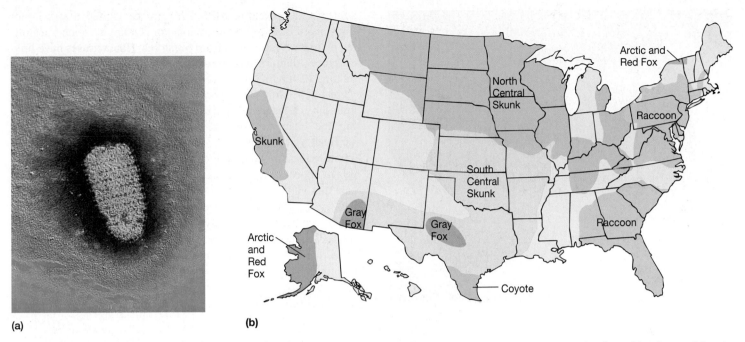

(a)   **(b)**

Figure 24.31   Rabies.   (a) Electron micrograph of *Rabies virus* (yellow) (×36,700). Note the bullet shape. The external surface of the virus contains glycoprotein spikes that bind specifically to cellular receptors. (b) In the United States, rabies is found in terrestrial animals in 10 distinct geographic areas. In each area a particular species is the reservoir and one of five antigenic variants of the virus predominates, as illustrated by the five different colors. Although not shown, another eight viral variants are found in insectivorous bats and cause sporadic cases of rabies in terrestrial animals throughout the country. Absence of a strain does not imply absence of rabies.

dogs and cats, postexposure vaccination of humans, and preexposure vaccination of humans at special risk.

Ebola Virus

Ebola hemorrhagic fever is caused by the Ebola virus, first recognized near the Ebola River in the Democratic Republic of the Congo in Africa. As a member of the *Filoviridae,* it is one of the four known subtypes; Ebola-Zaire, Ebola-Sudan, and Ebola-Ivory Coast cause disease in humans, and Ebola-Reston appears to cause disease only in nonhuman primates (**figure 24.32**).

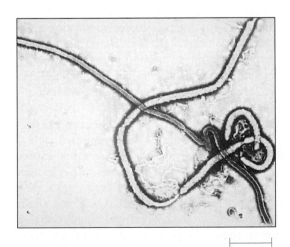

100 nm

Figure 24.32   Ebola Virus.

Infection with Ebola is severe and approximately 80% fatal. The incubation period for hemorrhagic fever caused by Ebola ranges from 2 to 21 days and is characterized by abrupt fever, headache, joint and muscle aches, sore throat, and weakness, followed by diarrhea, vomiting, and stomach pains. Signs of the infection include rash, red eyes, bleeding, and hiccups; these symptoms alert of internal hemorrhage. The reservoir of Ebola appears to be three species of fruit bats that are native only to Africa. Contact with an infected animal (bats, apes, or other primates) and subsequent transmission to other humans likely initiates an outbreak. Transmission can be from direct contact with the blood and secretions from an infected person or clinical samples. Exposure can also occur through contact with bodies of Ebola victims. No standard treatment for Ebola infection exists. Patients receive supportive therapy. This consists of balancing patients' fluids and electrolytes, maintaining their oxygen status and blood pressure, and treating them for any complicating infections. Experimental vaccines are currently being evaluated and show promise in nonhuman primate models.

Measles virus

Measles (rubeola: Latin *rubeus,* red) is a highly contagious skin disease that is endemic throughout most of the world. Although measles is no longer endemic in the United States, we discuss measles because it remains of global importance. *Measles virus* is an enveloped virus, in the genus *Morbillivirus* and the family *Paramyxoviridae.* The virus enters the body through the respiratory

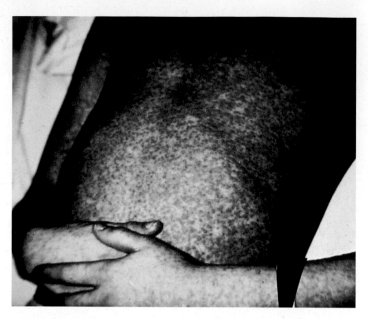

Figure 24.33 **Measles (Rubeola).** The rash of small, raised spots is typical of measles. The rash usually begins on the face and moves downward to the trunk.

tract or the conjunctiva of the eyes. It's receptor is a membrane protein called CD46, which functions in the immune response.

The incubation period is usually 10 to 21 days, and the first symptoms begin about the tenth day with nasal discharge, cough, fever, headache, and conjunctivitis. Within 3 to 5 days, skin eruptions occur as faintly pink, blotchy lesions with pustules that are at first discrete but gradually become confluent (**figure 24.33**). The rash normally lasts about 5 to 10 days. Lesions of the oral cavity include the diagnostically useful bright-red **Koplik's spots** with a bluish-white speck in the center of each. Koplik's spots represent a viral exanthem (a skin eruption) occurring in the form of macules or papules as a result of the viral infection. Very infrequently a progressive degeneration of the central nervous system called **subacute sclerosing panencephalitis** occurs. No specific treatment is available for measles. Immunization is recommended for all children (*see table 33.3*). Despite the availability of a measles, mumps, rubella (MMR) vaccine, measles infects about 50 million people and kills about 4 million a year worldwide. Serious outbreaks of measles are still reported in North America and Europe, reflecting significant numbers of people who have declined the MMR vaccine due to an unsubstantiated fear of vaccine-associated autism.

Hantaviruses

Hantaviruses (family *Bunyaviridae*) cause several diseases in humans, including Hantavirus pulmonary syndrome (HPS). HPS is a rare but potentially deadly disease that is typically transmitted to humans by inhalation of virus particles shed in urine, feces, or saliva of infected rodents. HPS was first recognized in 1993 and has since been identified throughout the United States. Rodent control in and around the home remains the primary

strategy for preventing HPS. HPS in the United States is not transmitted from person to person, nor is it known to be transmitted by rodents purchased from pet stores. Hantaviruses have lipid envelopes that are susceptible to most disinfectants. The length of time hantaviruses can remain infectious in the environment varies depending on environmental conditions. Temperature, humidity, exposure to sunlight, and even the rodent's diet (affecting the chemistry of rodent urine) strongly influence viral survival. Viability of dried virus has been reported at room temperature for 2 to 3 days. Hantaviruses are shed in body fluids but do not appear to cause disease in their reservoir hosts. Data indicate that viral transfer may occur through biting as field studies suggest that viral transmission in rodent populations occurs horizontally and more frequently between males. A specific treatment for HPS is unknown. Supportive therapy is used to treat symptoms, including balancing the patient's fluids and electrolytes, maintaining oxygen status and blood pressure, replacing lost blood and clotting factors, and treating for other complicating infections.

Influenza (Flu) Viruses

Influenza (Italian, to be influenced by the stars—*un influenza di freddo*), or the **flu,** is a respiratory system disease caused by viruses that belong to the family *Orthomyxoviridae*. There are four groups: **influenza viruses A, B,** and **C,** and **Thogoto viruses** (which will not be discussed further). They contain 7 to 8 segments of linear RNA, with a genome length from 12,000 to 15,000 nucleotides (*see figure 5.4*). The virus is acquired by inhalation or ingestion of virus-infected respiratory secretions during droplet tranmission. During an incubation period of 1 to 2 days, the virus adheres to the epithelium of the respiratory system and replicates. The **neuraminidase (NA)** present in envelope spikes may hydrolyze the mucus that covers the epithelium. The virus attaches to the epithelial cell by its **hemagglutinin (HA)** spike protein, causing part of the cell's plasma membrane to bulge inward, seal off, and form a vesicle (receptor-mediated endocytosis). This encloses the virus in an endosome (**figure 24.34**). The hemagglutinin molecule in the virus's envelope undergoes a dramatic conformational change when the endosomal pH decreases. The hydrophobic ends of the hemagglutinin spring outward and extend toward the endosomal membrane. After they contact the membrane, fusion occurs and the nucleocapsids are released into the cytoplasm.

Once the genome and its associated RNA-dependent RNA polymerase enter the host cell, the genome serves as the template for mRNA synthesis (figure 24.34, step 1). Later, the virus switches from mRNA synthesis to genome replication. During this phase of the life cycle, the plus-strand RNA molecules synthesized from the minus-strand genome segments serve as templates for new negative-strand RNA genomes (figure 24.34, step 4). The viruses exit the host cell by budding and thus acquire their envelope (figure 24.34, step 6; *see also figure 5.15*).

Influenza A infections are responsible for the majority of clinical influenza cases, with influenza B accounting for approximately 3% of flu in the United States. Influenza A infections usually peak

① The endonuclease activity of the PB1 protein cleaves the cap and about 10 nucleotides from the 5′ end of host mRNA (cap snatching). The fragment is used to prime viral mRNA synthesis by the RNA-dependent RNA polymerase activity of the PB1 protein.

② Viral mRNA is translated. Early products include more NP and PB1 proteins.

③ RNA polymerase activity of the PB1 protein synthesizes +ssRNA from genomic −ssRNA molecules.

④ RNA polymerase activity of the PB1 protein synthesizes new copies of the genome using +ssRNA made in step 3 as templates. Some of these new genome segments serve as templates for the synthesis of more viral mRNA. Later in the infection, they will become progeny genomes.

⑤ Viral mRNA molecules transcribed from other genome segments encode structural proteins such as hemagglutinin (HA) and neuraminidase (NA). These messages are translated by ER-associated ribosomes and delivered to the cell membrane.

⑥ Viral genome segments are packaged as progeny virions bud from the host cell.

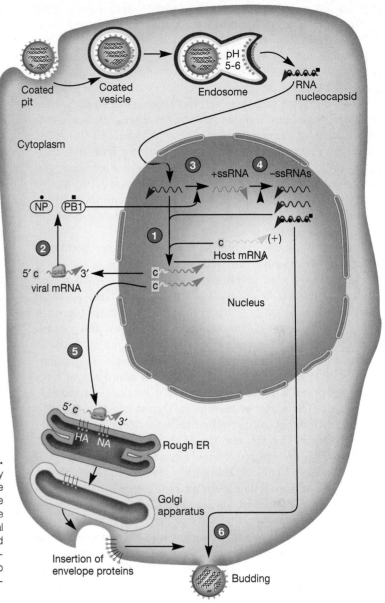

Figure 24.34 Simplified Life Cycle of an Influenza Virus. Several steps have been eliminated for simplicity and clarity. After entry by receptor-mediated endocytosis, the virus envelope fuses with the endosome membrane, releasing the nucleocapsids into the cytoplasm (each genome segment is associated with nucleocapsid proteins [NP] and PB1 to form the nucleocapsid). The nucleocapsids enter the nucleus, where synthesis of viral mRNA and genomes occurs. Critical to the production of viral mRNA and genomes is the enzyme RNA-dependent RNA polymerase. This is one activity of the PB1 enzyme. The other is endonuclease activity, which is used to cleave the 5′ ends from host mRNA. Steps 1 through 6 illustrate the remaining steps of the virus's life cycle.

in the winter and involve 10% or more of the population, with rates as high as 50 to 75% in school-age children. Influenza A viruses are widely distributed, infect a variety of mammal and bird hosts, and are further classified into subtypes (or strains) based on their membrane surface glycoproteins, HA and NA. There are 16 HA and 9 NA antigenic forms known; reassortment produces various HA/NA subtypes of influenza. All subtype combinations infect birds. Influenza A viruses having H1, H2, and H3 HA antigens, along with N1 and N2 NA antigens, are predominant in nature. H1N1 viruses appeared in 1918 and were replaced in 1957 by H2N2 subtypes as the predominant subtype. The H2N2 viruses were subsequently replaced by H3N2 as the principal subtypes in 1968. The H1N1 subtype reappeared in 1977 and cocirculates today with H3N2 in humans and with H2N1, H3N2, H5N2, H7N2, H7N3, H7N7, H9N2, H10N7, and H5N1 viruses in birds.

The standard nomenclature system for influenza virus subtypes includes the following information: group (A, B, or C), host of origin, geographic location, strain number, and year of original isolation. Antigenic descriptions of the HA and NA are given in parentheses for type A. The host of origin is not indicated for human isolates—for example, A/Hong Kong/03/68 (H3N2). However, the host origin is given for other isolates—for example, A/swine/Iowa/15/30 (H1N1).

One of the most important features of the influenza viruses is the frequency with which changes in their spike proteins occur. The spike proteins are antigenic; that is, they elicit an immune response. Because of these changes, the virus escapes detection by the host's immune system. If the variation in the spike proteins is small, it is called **antigenic drift** (*see figure 33.6*). Antigenic drift results from the accumulation of mutations of

HA and NA in a single strain of flu virus within a geographic region. This is a constant and gradual process causing local increases in the number of flu infections. **Antigenic shift** is a large antigenic change resulting from the reassortment of genome segments. This occurs when two different strains of flu viruses—one from an animal (i.e., bird) and the other from a human—infect the same host cell and segments from each strain are incorporated into a single new capsid. This generates a new combination of segments, thus a new genome and flu strain. Because a greater change occurs with antigenic shift than antigenic drift, antigenic shifts can yield major epidemics and pandemics (global epidemics). Antigenic variation occurs almost yearly with *Influenza A virus,* less frequently with the B virus, and has not been demonstrated in the C virus. >> *Recognition of an epidemic (section 33.4)*

Animal hosts are critical to the epidemiology of human influenza. For example, rural China is one region of the world where chickens, pigs, and humans live in close, crowded conditions. Influenza is widespread in chickens, and although chickens cannot usually transmit the virus to humans, they can transfer it to pigs. Pigs can transfer it to humans, and humans back to pigs. Thus reassortment of genome segments between human and avian strains occurs in pigs, leading to major antigenic shifts. This explains why influenza continues to be a major epidemic disease and frequently produces worldwide pandemics. The worst pandemic on record occurred in 1918 and is now thought to have killed between 50 million and 100 million people. This disaster, traced to the "Spanish influenza" virus (*see figure 33.10*), was followed by pandemics of "Asian flu" (1957), "Hong Kong flu" (1968), and "Russian flu" (1977). The names reflect popular impressions of where the episodes began, although all are now thought to have originated in China.

The **H5N1 subtype** (also know as bird flu and avian influenza) appears to be the most likely candidate to initiate another influenza pandemic (global epidemic). The H5N1 subtype was responsible for six deaths in 1997, although it had been sporadically detected prior to that. Throughout 2003 to 2007, however, H5N1 was responsible for a substantial number of bird infections, resulting in the culling of 200 million birds. The birds were destroyed to help prevent transmission of the virus to susceptible humans. Yet, by late October 2007, H5N1 was responsible for 333 confirmed cases of influenza and 204 deaths in 12 countries. Because of the potentially severe consequences of pandemic H5N1 infection of humans, the World Health Organization coordinates an international effort to monitor virus epidemiology, biology, and control efforts. The National Institutes of Health began clinical trials of an avian flu vaccine in 2005, but it is unknown if this vaccine will provide sufficient protection against the final version of the flu strain capable of human-to-human transmission, if and when such a strain emerges.

Influenza is characterized by chills, fever (usually 102°F, 39°C), headache, malaise, cough, sore throat, and general muscular aches and pains. These symptoms arise from the death of respiratory epithelial cells, probably due to attacks by immune system cells called activated T cells. These symptoms are more debilitating than are symptoms of the common cold. Recovery usually occurs in 3 to 7 days. While influenza alone usually is not fatal, death may result from pneumonia caused by secondary bacterial invaders. As with many other viral diseases, only the symptoms of influenza usually are treated. Certain antiviral agents have been shown to reduce the duration and symptoms of type A influenza if administered during the first two days of illness (*see figure 31.20*). >> *Antiviral drugs (section 31.8)*

The mainstay for prevention of influenza since the late 1940s has been vaccines that contain inactivated viruses (i.e., viruses that are unable to infect a host and reproduce). These vaccines are recommended for most people and strongly recommended for the chronically ill, individuals over age sixty-five, residents of nursing homes, and health-care workers in close contact with people at risk, because clinical disease in these patients is most likely to be severe. Because of influenza's high genetic variability, efforts are made each year to incorporate new virus subtypes into the vaccine. Even when no new subtypes are identified in a given year, annual immunization is still recommended because immunity generated by vaccination typically lasts only 1 to 2 years. >> *Control of epidemics (section 33.8)*

1. What are Koplik's spot and how do they arise?
2. What function do the influenza virus neuraminidase and hemagglutinin glycoproteins serve?
3. Trace the reproduction of an influenza virus starting with host cell attachment and ending with the exit of virions.
4. Why is antigenic shift most prevalent in viruses with segmented genomes?

24.7 VIRUSES WITH SINGLE-STRANDED RNA GENOMES (GROUP VI—RETROVIRUSES)

Although retroviruses have positive-strand RNA genomes, their genomes do not function as mRNA. Instead, retroviruses first convert their ssRNA genomes into dsDNA using an **RNA-dependent DNA polymerase** also called **reverse transcriptase** (**figure 24.35**). The dsDNA then integrates into the host's DNA, where it can serve as a template for mRNA synthesis. The dsDNA also serves as the template for synthesis of the plus-strand RNA genome. The synthesis of both is catalyzed by the host cell's DNA-dependent RNA polymerase.

Numerous retroviruses have been identified and studied. However, the **human immunodeficiency virus (HIV),** the cause

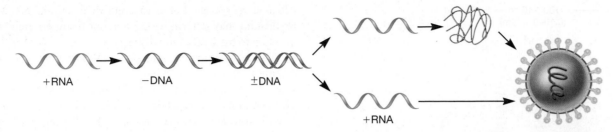

Figure 24.35   Reproductive Strategy of Retroviruses.   Retroviruses are Group VI viruses. These viruses have a plus-strand RNA genome that is first converted into dsDNA by the enzyme reverse transcriptase. The viral dsDNA integrates into the host chromosome, where it serves as the template for synthesis of viral mRNA and viral genomes. Both are synthesized using the host cell's DNA-dependent RNA polymerase.

of **AIDS (acquired immune deficiency syndrome),** is of particular interest. AIDS is now recognized as the great pandemic of the second half of the twentieth century. Because of its global importance, we focus exclusively on HIV in this section, although there are a number of other retroviruses whose molecular biology is far less complicated than that of HIV.

First described in 1981, AIDS is the result of an infection by HIV, a member of the genus *Lentivirus* within the family *Retroviridae*. In the United States, AIDS is caused primarily by HIV-1 (some cases result from an HIV-2 infection). HIV-1 is an enveloped virus. The envelope surrounds an outer shell, which encloses a cylindrical core (capsid) (**figure 24.36a**). The core contains two copies of the HIV RNA genome and several enzymes. Thus far, 10 virus-specific proteins have been discovered.

Once inside the body, the gp120 viral envelope protein binds to host cells that have a surface glycoprotein called CD4 (figure 24.36b). These cells include $CD4^+$ T cells (also called T-helper cells), macrophages, dendritic cells, and monocytes—all cells with key roles in host defenses. Dendritic cells are present throughout the body's mucosal surfaces, and it is possible that these are the first cells infected by HIV in sexual transmission. The virus requires a coreceptor in addition to CD4. Macrophage-tropic strains, which seem to predominate early in the disease and infect both macrophages and $CD4^+$ T cells, also require the CCR5 protein, which normally functions as a receptor for signaling molecules of the immune system called chemokines. A second chemokine receptor, called CXCR-4 or fusin, is used by $CD4^+$ T-cell-tropic strains that are active at later stages of infection. These strains induce the formation of syncytia—multiple host cells that are fused together. Individuals with two defective copies of the CCR5 gene do not seem to develop AIDS; apparently the virus cannot infect their cells. People with one good copy of the CCR5 gene do get AIDS but survive several years longer than those with no mutation. >> *Cells, tissues, and organs of the immune system (section 28.2); T-cell biology (section 29.5)*

Entry into the host cell begins when the envelope fuses with the plasma membrane, and the virus releases its core into the cytoplasm (**figure 24.37a**). Inside the infected cell, the core protein dissociates from the RNA, and it is copied into a single strand of DNA by the RNA-dependent DNA polymerase activity of the

reverse transcriptase enzyme. The RNA is next degraded by another reverse transcriptase component, ribonuclease H, and the DNA strand is duplicated to form a double-stranded DNA copy of the original RNA genome. A complex of the double-stranded DNA (the provirus) and the integrase enzyme moves into the nucleus. Then the proviral DNA is integrated into the cell's DNA through a complex sequence of reactions catalyzed by the enzyme integrase. The provirus can force the cell to synthesize viral mRNA (figure 24.37b). Some of the mRNA is translated to produce viral proteins by the cell's ribosomes. Some of the proteins have been shown to affect host cell function; for example, the protein NEF decreases synthesis of host proteins called major histocompatibility (MHC) class I proteins. These proteins are vital for immune cell recognition of host cells invaded by pathogens. Thus decreased synthesis interferes with host immune responses. Viral proteins and the complete HIV-1 RNA genome are then assembled into new nucleocapsids that bud from the infected host cell (figure 24.37c). Eventually the host cell dies from repeated budding, immune destruction, or induction of apoptosis. >> *Recognition of foreignness: Major histocompatibility complex (section 29.4)*

Once a person becomes infected with HIV, the course of disease varies greatly. Some "rapid progressors" develop clinical AIDS and die within 2 to 3 years. A small percentage of "long-term nonprogressors" remain relatively healthy for at least 10 years after infection. For the majority of HIV-infected individuals, HIV infection progresses to AIDS in 8 to 10 years. The CDC has developed a classification system for the stages of HIV-related conditions: acute, asymptomatic, chronic symptomatic, and AIDS.

The acute infection stage occurs 2 to 8 weeks after HIV infection. About 70% of individuals in this stage experience a brief illness referred to as acute retroviral syndrome, with symptoms that may include fever, malaise, headache, macular rash, weight loss, lymph node enlargement (lymphadenopathy), and oral candidiasis (**figure 24.38a**). During this stage, the virus multiplies rapidly and disseminates to lymphoid tissues throughout the body, until an acquired immune response can be generated to bring virus replication under control. During the acute infection stage, levels of HIV may reach 10^5 to 10^6 copies of viral RNA per milliliter of plasma. It is believed that the extent to which the

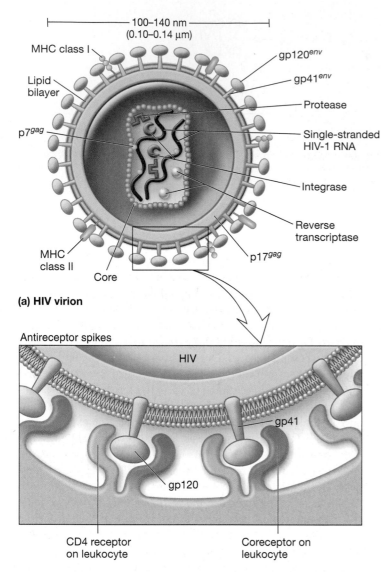

(a) HIV virion

- MHC class I
- Lipid bilayer
- p7^{gag}
- MHC class II
- Core
- gp120^{env}
- gp41^{env}
- Protease
- Single-stranded HIV-1 RNA
- Integrase
- Reverse transcriptase
- p17^{gag}
- 100–140 nm (0.10–0.14 µm)

Antireceptor spikes

HIV

- gp41
- gp120
- CD4 receptor on leukocyte
- Coreceptor on leukocyte

(b) HIV attachment to host cell

Figure 24.36   The HIV-1 Virion. (a) The HIV-1 virion is an enveloped structure containing 72 external spikes. These spikes are formed by the two major viral-envelope proteins, gp120 and gp41. (gp stands for glycoprotein—proteins linked to sugars—and the number refers to the mass of the protein, in thousands of daltons.) The HIV-1 lipid bilayer is also studded with various host proteins, including class I and class II major histocompatibility complex molecules, acquired during virion budding. The cone-shaped core of HIV-1 contains four nucleocapsid proteins (p24, p17, p9, p7), each of which is proteolytically cleaved from a 53 kDa *gag* precursor by the HIV-1 protease. The phosphorylated p24 polypeptide forms the chief component of the inner shell of the nucleocapsid, whereas the p17 protein is associated with the inner surface of the lipid bilayer and stabilizes the exterior and interior components of the virion. The p7 protein binds directly to genomic RNA through a zinc finger structural motif and together with p9 forms the nucleoid core. The retroviral core contains two copies of genomic RNA associated with the viral enzymes, reverse transcriptase, integrase, ribonuclease, and protease. (b) The snug attachment of HIV glycoprotein molecules (gp41 and gp120) to their specific receptors on a human cell membrane. These receptors are CD4 and a co-receptor called CXCR-4 (fusin). They permit docking with the host cell and fusion with the cell membrane.

immune response is able to control this initial burst of virus replication may determine the amount of time required for progression to the next clinical stages.

The asymptomatic stage of HIV infection may last from 6 months to 10 years or more in some individuals. During this stage, the levels of detectable HIV in the blood decrease, but the virus continues to reproduce, particularly in lymphoid tissues. Even before any changes in CD4$^+$ T cells can be detected, the virus may affect certain immune functions, such as memory cell responses to common antigens including tetanus toxoid or *Candida albicans.*

During the chronic symptomatic stage, which can last for months to years, viral multiplication continues and the number of CD4$^+$ T cells in the blood begins to significantly decrease. These cells are critically important in the generation of acquired immunity, including the production of host proteins called antibodies, which circulate in the blood and lymph, and help bring about the destruction of pathogens. Thus individuals at this stage develop a variety of symptoms, including fever, weight loss, malaise, fatigue, anorexia, abdominal pain, diarrhea, headaches, and lymphadenopathy. Paradoxically some patients develop increased antibody production during this stage, perhaps due to generalized immune system dysfunction. These antibodies, however, do little to protect the host from infection. As CD4$^+$ T cell numbers continue to decline, some patients develop opportunistic infections such as oral candidiasis (figure 24.38*a*) or Kaposi's sarcoma (figure 24.38*b*). Indeed, a variety of disease processes are associated with the progression of AIDS (**table 24.5**).

Experimental evidence supports several potential mechanisms of CD4$^+$ T cell depletion by HIV, although the most important one remains enigmatic. The mechanisms include: (1) direct cytopathic effects of HIV on T-helper cells, (2) formation of syncytia, (3) immune-mediated destruction of HIV-infected cells, and (4) effects of viral products (such as gp120) on uninfected cells. The cytopathic effect may be due to the disruption of plasma membrane integrity and function by excessive budding of virus. Insertion of HIV proviral DNA into the host cell's DNA can disrupt cell function, destroying the host T cells. Expression of gp120 on virus-infected host cells may interact with the CD4 receptors on T-helper cells, causing the cells to fuse and form multinucleated syncytia that eventually die. Moreover, free gp120 proteins released from infected cells may bind to CD4 on uninfected cells and induce those cells to undergo apoptosis. Finally, multiple components of the immune system (cytotoxic T cells, NK cells, complement, and antibody-dependent cellular cytotoxicity) may contribute to the continuing destruction of virus-infected CD4$^+$ T cells, resulting in acquired immune deficiency. Other factors besides CD4$^+$ cell destruction may also contribute to AIDS pathogenesis. For example, HIV may also inhibit or destroy dendritic cells. HIV mutates exceptionally rapidly and thus could evade and eventually overwhelm the host immune system. HIV may disrupt the balance between the various types of T cells, also altering immune system integrity. Eventually viral reproduction outpaces the host's attempts to control it, resulting in clinical AIDS.

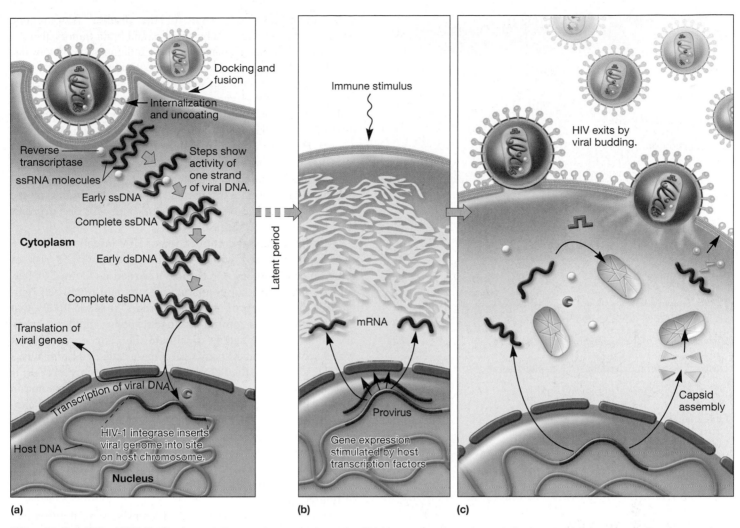

Figure 24.37   HIV Life Cycle.   (a) The virus is adsorbed onto the CD4$^+$ host cell and membrane fusion occurs releasing the core into the cytoplasm. Reverse transcriptase catalyzes the synthesis of a single complementary strand of DNA. This DNA serves as a template for synthesis of double-stranded DNA. The dsDNA can be inserted into the host chromosome as a provirus (latency). (b) The provirus genes are transcribed. (c) Viral mRNA is translated into virus components (capsid, reverse transcriptase, spike proteins), and the virus is assembled. Mature virus particles bud from the host cell, taking host membrane as their envelope.

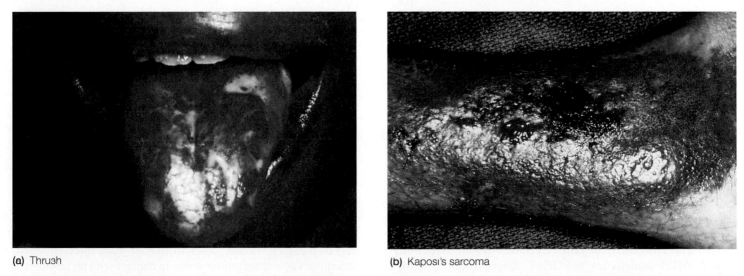

(a) Thrush

(b) Kaposi's sarcoma

Figure 24.38   Some Diseases Associated with AIDS.   (a) Candidiasis of the oral cavity and tongue (thrush) caused by *Candida albicans*. (b) Kaposi's sarcoma on the arm of an AIDS patient. The flat purple tumors can occur in almost any tissue and are frequently multiple.

Table 24.5 Disease Processes Associated with AIDS

Candidiasis of bronchi, trachea, or lungs

Candidiasis, esophageal

Cervical cancer, invasive

Coccidioidomycosis, disseminated or extrapulmonary

Cryptosporidiosis, chronic intestinal (>1 month's duration)

Cyclospora, diarrheal disease

Cytomegalovirus disease (other than liver, spleen, or lymph nodes)

Cytomegalovirus retinitis (with loss of vision)

Encephalopathy, HIV-related

Herpes simplex: chronic ulcer(s) (>1 month's duration); or bronchitis, pneumonitis, or esophagitis

Histoplasmosis, disseminated or extrapulmonary

Isosporiasis, chronic intestinal (>1 month's duration)

Kaposi's sarcoma

Lymphoma, Burkitt's (or equivalent term)

Lymphoma, immunoblastic (or equivalent term)

Lymphoma, primary, of brain

Mycobacterium avium complex or *M. kansasii*

Mycobacterium tuberculosis, any site

Mycobacterium, other species or unidentified species

Pneumocystis pneumonia

Pneumonia, recurrent

Progressive multifocal leukoencephalopathy

Salmonella septicemia, recurrent

Toxoplasmosis of brain

Wasting syndrome

Source: Data from *MMWR* 41 (No. RR17). 1993 Revised Classification System for HIV Infection and Expanded Surveillance Case Definition for AIDS Among Adolescents and Adults.

In addition to its devastating effects on the immune system, HIV infection can lead to disease of the central nervous system because virus-infected macrophages can cross the blood-brain barrier. The classical symptoms of central nervous system disease in AIDS patients are headaches, fevers, subtle cognitive changes, abnormal reflexes, and ataxia (irregularity of muscular action). Dementia and severe sensory and motor changes charac-

terize more advanced stages of the disease. Autoimmune neuropathies, cerebrovascular disease, and brain tumors also are common. Histological changes include inflammation of neurons, nodule formation, and demyelination. Evidence indicates that these neurological changes are correlated with higher levels of HIV-1 antigen, HIV-1 genome, or both in central nervous system cells. In AIDS dementia, macrophages and glial cells (supporting cells of the nervous system) are primarily infected and bud new viruses. However, it is unlikely that direct infection of these cells by HIV-1 is responsible for the symptoms. More likely the symptoms arise through either the secretion of viral proteins or viral induction of cytokines, important signaling molecules of the immune system. These proteins bind to glial cells and neurons, causing neurological disease.

Another potential complication of HIV infection is cancer. Individuals infected with HIV-1 have an increased risk of three types of tumors: (1) **Kaposi's sarcoma** (figure 24.38*b*), (2) carcinomas of the mouth and rectum, and (3) B-cell lymphomas or lymphoproliferative disorders. It seems likely that the depression of the initial immune response enables secondary tumor-causing agents to initiate the cancers. In 1994 it was discovered that Kaposi's sarcoma-associated herpesvirus (KSHV, since classified as *Human herpesvirus 8* or HHV-8) is a virus that is consistently present in Kaposi's sarcoma and in primary effusion (body cavity-based) lymphomas. These cancers occur most frequently in AIDS patients. KSHV is a gammaherpesvirus with homology to herpesvirus saimiri and Epstein-Barr virus, both of which are transmissible through contact with the saliva of an infected individual. Both KSHV and, to a less extent, EBV can cause cancerous transformation of lymphocytes.

Although HIV has been studied intensively for the past two and a half decades, its origin only now seems clear. Molecular epidemiology data indicate that HIV-1 arose from the SIVcpz retrovirus harbored by the chimpanzee *Pan troglodytes troglodytes* (Ptt). The deduced evolutionary sequence suggests that SIVcpz ancestors recombined when they crossed between several species of nonhuman primates, then into chimpanzees, and finally into humans. Stable viral infections in humans occurred on at least three occasions, with several groups of mutated SIVcpz viruses adapting to humans residing within chimp habitats. Full-length genome analyses indicate that HIV-1 groups M, N, and O are most similiar to SIVcpz viruses of Ptt, evolving from separate SIVcpz lineages. However, only the SIVcpz strain now referred to as group M HIV-1 gave rise to the virus causing the global AIDS pandemic. Notably, group M HIV-1 has diverged into several subtypes (clades) indicated as A-K. Subtype B was the first to appear in the United States and it remains the predominant type (80%) through the Americas.

Epidemiologically, AIDS occurs worldwide, although the HIV-2 strain predominates in West Africa (**figure 24.39**). HIV may be passed from one person to another when infected blood, semen, or vaginal secretions come in contact with an uninfected person's broken skin or mucous membranes. In the developing world, AIDS affects men and women alike, with many women

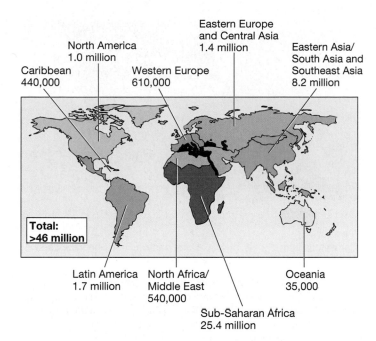

North America
1.0 million

Caribbean
440,000

Eastern Europe
and Central Asia
1.4 million

Western Europe
610,000

Eastern Asia/
South Asia and
Southeast Asia
8.2 million

Total:
>46 million

Latin America
1.7 million

North Africa/
Middle East
540,000

Oceania
35,000

Sub-Saharan Africa
25.4 million

Figure 24.39 Distribution of HIV/AIDS in Adults by Continent or Region. The figure shows data from a 2003 United Nations report. According to recent estimates by the UN, the number of HIV/AIDS cases may be over 46 million *Source of data: UNAIDS.*

getting AIDS from their husbands who have multiple sex partners. In the United States, the groups most at risk for acquiring AIDS are (in descending order) men who have sex with other men, intravenous drug users, heterosexuals who have sex with drug users and prostitutes, children born of infected mothers, and infants breast-fed by HIV-infected women.

At present no cure exists for AIDS. Primary treatment is directed at reducing the viral load and disease symptoms, and at treating opportunistic infections and malignancies. The antiviral drugs currently approved for use in HIV disease are of four types: (1) nucleoside reverse transcriptase inhibitors (NRTIs) are analogues that inhibit the enzyme reverse transcriptase as it synthesizes viral DNA; (2) nonnucleoside reverse transcriptase inhibitors (NNRTIs); (3) protease inhibitors (PIs), which work by blocking the activity of the HIV protease and thus interfere with virion assembly; and (4) fusion inhibitors (FIs)—a newer category of drugs that prevent HIV entry into cells. The most successful treatment approach in combating HIV/AIDS is to use drug combinations. In many patients the virus disappears from the patient's blood with proper treatment and drug-resistant strains do not seem to arise. HIV can remain dormant in memory T cells, survive drug cocktails, and reactivate. Thus patients are not completely cured with drug treatment. It should be noted that side effects can be very severe, and treatment is prohibitively expensive for those without medical insurance. Globally, the vast majority of HIV-positive individuals do not have access to effective combination therapy.

The development of a vaccine has been a long-sought research goal. Such a vaccine would ideally (1) stimulate the production of neutralizing antibodies, which can bind to the virus envelope and prevent it from entering host cells; and (2) promote the formation of cytotoxic T cells (CTLs), which can destroy cells infected with virus. Among the many problems encountered in developing an HIV vaccine is the fact that the envelope proteins of the virus continually change their antigenic properties.

Many HIV researchers continue to take great interest in HIV-infected persons who are long-term nonprogressors. These individuals maintain $CD4^+$ T cell counts of at least 600 per microliter of blood, have less than 5,000 copies of HIV RNA per milliliter of blood, and have remained this way for more than 10 years after documented infection (even in the absence of antiviral agents). At least three explanations of this phenomenon have been proposed: long-term nonprogressors (1) may react with a more effective immune response (cytotoxic T cells and neutralizing antibody) to relatively conserved proteins; (2) may have been initially infected with an attenuated strain; or (3) may have predisposing genetic differences that prevent or inhibit infection, as in the example of the CCR5 mutation already discussed.

1. What is the function of each of the following HIV products: gp120, reverse transcriptase, ribonuclease H, and integrase?

2. Describe the four stages or classifications of HIV infection. Correlate each with virus activity in the host.

3. How is HIV thought to form syncytia? Why is this harmful to the host?

4. List some other conditions associated with HIV infection.

5. Find out your HIV (and hepatitis C) status.

24.8 VIRUSES WITH GAPPED DNA GENOMES (GROUP VII)

The **hepadnaviruses** such as *Hepatitis B virus* (**HBV**) are quite different from other DNA viruses with respect to genome replication. They have circular dsDNA genomes that consist of one complete but nicked strand and a complementary strand that has a large gap—that is, it is incomplete (**figure 24.40**). After infecting the cell, the virus's gapped DNA is released into the nucleus. There, host repair enzymes fill the gap and seal the nick, yielding a covalently closed, circular DNA. Transcription of viral genes occurs in the nucleus using host RNA polymerase and yields several mRNAs, including a large 3.4 kilobase RNA known as the pregenome (plus-strand RNAs). The RNAs move to the cytoplasm, and the mRNAs are translated to produce viral proteins, including core

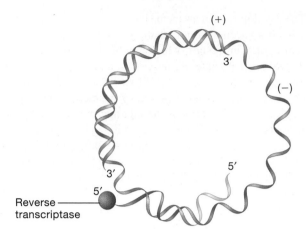

(+)

3′

(−)

3′

5′

5′

Reverse
transcriptase

Figure 24.40   The Gapped Genome of Hepadnaviruses.
The genomes of hepadnaviruses are unusual in several respects. The negative strand of the dsDNA molecule is complete but nicked. The enzyme
reverse transcriptase is attached to its 5′ end. The positive strand is gapped;
that is, it is incomplete (shown in purple). A short stretch of RNA is attached
to its 5′ end (shown in green). Upon entry into the host nucleus, the nick in
the negative strand is sealed and the gap in the positive strand is filled, yielding a covalently closed, circular dsDNA molecule.

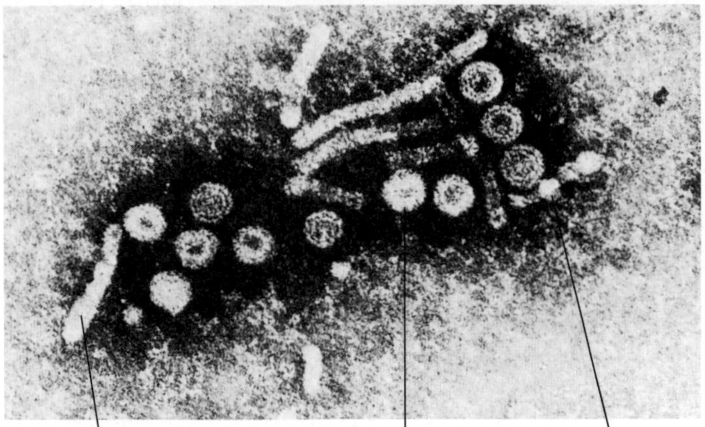

Filamentous form
(22 nm diameter)

Dane particle
(42 nm diameter)

Spherical particle
(22 ±2 nm diameter)

Figure 24.41   Hepatitis B Virus in Serum.   Electron micrograph (×210,000) showing the three distinct types of hepatitis B antigenic
particles.The spherical particles and filamentous forms are small spheres or
long filaments without an internal structure, and only two of the three characteristic viral envelope proteins appear on their surface. Dane particles are
the complete, infectious virion.

proteins and a polymerase having three activities (DNA polymerase, reverse transcriptase, RNase H). Then the RNA pregenome
associates with the polymerase and core protein to form an immature core particle. Reverse transcriptase subsequently reverse
transcribes the RNA using a protein primer to form a minus-strand
DNA from the pregenome RNA. After almost all the pregenome
RNA has been degraded by RNase H, the remaining RNA fragment serves as a primer for synthesis of the gapped dsDNA
genome, using the minus-strand DNA as template. Finally, the
nucleocapsid is completed and the progeny virions are released.

HBV is classified as an *Orthohepadnavirus* within the family
Hepadnaviridae. The infecting virus is a 42 nm spherical particle
(containing DNA and DNA polymerase) called the **Dane particle.** The viral genome is 3.2 kb in length, consisting of four
partially overlapping, open reading frames that encode viral proteins. Viral replication takes place predominantly in hepatocytes.
The infecting virus encases its double-shelled Dane particles
within membrane envelopes coated with hepatitis B surface antigen. The inner nucleocapsid core antigen encloses a single
molecule of double-stranded HBV DNA and an active DNA
polymerase. Interestingly serum from individuals infected with
hepatitis B contains three distinct antigenic particles: the Dane
particle, a spherical 22 nm particle, and tubular or filamentous
particles that vary in length (**figure 24.41**). The role of these other
particles is debated.

The clinical signs of hepatitis B vary widely. Most cases are
asymptomatic. However, sometimes fever, loss of appetite,
abdominal discomfort, nausea, fatigue, and other symptoms
gradually appear following an incubation period of 1 to 3 months.

The virus infects liver hepatic cells and causes liver tissue degeneration and the release of liver-associated enzymes (transaminases)
into the bloodstream. This is followed by jaundice, the accumulation of bilirubin (a breakdown product of hemoglobin) in the skin
and other tissues with a resulting yellow appearance. Chronic
hepatitis B infection also causes the development of primary liver
cancer, known as hepatocellular carcinoma.

Hepatitis B virus is normally transmitted through blood or
other body fluids (saliva, sweat, semen, breast milk, urine, feces)
and body fluid–contaminated equipment (including shared intravenous needles). The virus can also pass through the placenta to the
fetus of an infected mother. The number of new HBV cases in
the United States has declined over the last decade in part due to
the availability of a vaccine. However, there are currently about
1.25 million chronically infected Americans. In the United States,
about 5,000 persons die yearly from hepatitis-related cirrhosis
and about 1,000 die from HBV-related liver cancer. (HBV is second only to tobacco as a known cause of human cancer.)
Worldwide, HBV infects over 200 million people.

1. Describe the HBV genome. How is it converted to
   covalently, closed, circular DNA in the host?

2. Trace the HBV reproductive cycle, paying particular
   attention to localization within the host cell during the
   biosynthetic stage.

3. How is HBV transmitted? What are the complications
   of hepatitis B infection?

Summary

24.1 Principles of Virus Taxonomy

a. Currently viruses are classified with a taxonomic system placing primary emphasis on the type and strandedness of viral nucleic acids, and on the presence or absence of an envelope **(table 24.1)**.

b. The Baltimore system of virus classification is used by many virologists to organize viruses based on their genome type and the mechanisms used to synthesize mRNA and replicate their genomes **(table 24.2)**.

24.2 Viruses with Double-Stranded DNA Genomes (Group I)

a. The T4 phage is a virulent bacteriophage that causes lytic infections of *E. coli*. After attachment to a specific receptor site on the bacterial surface, T4 injects its dsDNA into the cell **(figures 24.4 and 24.5)**. T4 DNA contains hydroxymethylcytosine (HMC) in place of cytosine, and glucose is often added to the HMC to protect the phage DNA from attack by host restriction enzymes **(figure 24.7)**. T4 DNA replication produces concatemers, long strands of several genome copies linked together **(figure 24.8)**.

b. Lambda (λ) phage is a temperate bacteriophage. It can establish lysogeny rather than pursuing a lytic infection. During lysogeny, the viral DNA, called a prophage or provirus, is replicated as the cell's genome is replicated. Lysogeny is reversible, and the prophage can be induced to become active again and lyse its host. This highly regulated process is an important model system for regulatory processes. The two most important proteins involved in regulating the choice between lysogeny and a lytic cyle are the λ repressor and the Cro protein. A race between synthesis of λ repressor and that of the Cro protein determines which path is followed. If the Cro protein level rises high enough in time, λ repressor synthesis is blocked and the lytic cycle initiated; otherwise, all genes other than the λ repressor gene are repressed and the cell becomes a lysogen **(figure 24.12)**.

c. Herpesviruses are a large group of dsDNA viruses. They cause acute infections such as cold sores, genital herpes, chickenpox, and mononucleosis. This is followed by a lifelong latent infection **(figure 24.13)**.

d. Some human papillomaviruses cause genital warts. Within this group are viruses associated with cervical cancer. A vaccine is available against the four high risk strains.

e. *Variola virus* is a large virus responsible for smallpox. The eradication of smallpox is the greatest public health achievement ever.

f. Many insect viruses form inclusion bodies, making them attractive candidates as insecticides.

24.3 Viruses with Single-Stranded DNA Genomes (Group II)

a. φX174 is an example of a ssDNA bacteriophage. Its replication involves the formation of a dsDNA replicative form **(figure 24.22)**.

b. fd phage is filamentous phage that upon infection is continuously released by the host without causing lysis.

c. The parvovirus B19 causes a spectrum of diseases. Its genome is replicated by host DNA polymerase in the host cell's nucleus.

24.4 Viruses with Double-Stranded RNA Genomes (Group III)

a. Group III viruses use a viral enzyme with RNA-dependent RNA polymerase activity to synthesize mRNA (transcriptase activity) and replicate their genomes (replicase activity) **(figure 24.25)**.

b. The bacteriophage φ6 is an unusual phage in that it is enveloped and enters the host bacterium through a process that resembles endocytosis.

c. Rotaviruses are one cause of viral gastroenteritis. They are a major cause of diarrhea in children and are responsible for a large number of deaths in developing countries.

24.5 Viruses with Plus-Strand RNA Genomes (Group IV)

a. The genomes of Group IV viruses serve directly as mRNA molecules. Among the first viral proteins synthesized is an RNA-dependent RNA polymerase that replicates the plus-strand RNA genome, sometimes by forming a dsRNA replicative intermediate. The negative RNA strands produced by the RNA-dependent RNA polymerase can either be used to make more genomes or more mRNA **(figure 25.27)**.

b. Bacteriophages MS2 and Qβ are small phages with only a few genes. They enter the host cell by attaching to the F pilus of their *E. coli* host.

c. TMV is like most known plant viruses in that it has a plus-strand RNA genome. The TMV nucleocapsid forms spontaneously when disks of coat protein protomers complex with the RNA **(figure 24.28)**.

d. Hepatitis A is caused by HAV, a naked, icosahedral virus. The disease is transmitted by the fecal-oral route.

e. *Poliovirus* first enters the gastrointestinal tract; in a small number of cases, it can enter the central nervous system and cause paralysis. Worldwide polio vaccination is ongoing in an effort to eliminate this disease.

f. Noroviruses are a group of related, highly contagious viruses that cause gastroenteritis.

g. Hepatitis C is caused by HCV and is a life-threatening condition.

h. West Nile fever and SARS are newly emerging diseases; however, whereas West Nile virus continues to cause encephalitis, the SARS-CoV appears to have become less infectious.

i. *Rubella virus* is the causative agent of rubella, also called German measles.

j. Severe acute respiratory syndrome (SARS) is caused by the SARS coronavirus.

24.6 Viruses with Minus-Strand RNA Genomes (Group V)

a. For Group V viruses to synthesize mRNA and replicate their genomes, their virions must carry an RNA-dependent RNA polymerase, which enters the host cells as the genome does. The polymerase first synthesizes mRNA from the negative-strand genome. Later, it is used to replicate the genome by way of a plus-strand intermediate **(figure 24.30)**.

b. *Rabies virus* is an enveloped virus that uses glycoprotein spikes to attach to the skeletal muscle cells from which it spreads to the central nervous system.

c. Ebola hemorrhagic fever is caused by Ebola virus and is nearly 80% fatal.

d. *Measles virus* is an enveloped virus. Despite the availability of a vaccine, measles continues to be of global concern.

e. Hantavirus pulmonary syndrome is caused by a hantavirus, typically transmitted from rodents or their droppings to humans.

f. Influenza viruses have segmented genomes. Host cell entry involves both the NA and HA glycoproteins, which are also important in classifying virus subtypes. The viruses bring their own RNA-dependent RNA polymerase into the host cell for transcription and genome replication; they exit by budding **(figure 24.34).**

24.7   Viruses with Single-Stranded RNA Genomes (Group VI—Retroviruses)

a. Retroviruses replicate their genomes and synthesize mRNA via a dsDNA intermediate. The dsDNA is formed by an RNA-dependent DNA polymerase called reverse transcriptase **(figure 24.35).**

b. HIV is an enveloped retrovirus with a cylindrical core that contains two copies of its genome and several enzymes, including reverse transcriptase, an RNA-dependent DNA polymerase. Upon infection, its ssRNA genome is converted to dsDNA, which is then integrated into the host genome **(figure 24.37).** Integrated HIV provirus can remain latent, in which case the patient is asymptomatic, or be actively transcribed, which will result in disease.

c. HIV infection can be classified as one of four stages: acute, asymptomatic, chronic symptomatic, and AIDS.

d. There are several mechanisms by which HIV depletes CD4$^+$ T cells: cytopathic effects, syncytia formation, immune-mediated destruction of HIV-infected cells, and the effects of viral products on uninfected cells.

e. In addition to destroying the immune system, HIV is associated with a variety of other conditions, including Kaposi's sarcoma, carcinomas of the mouth and rectum, B-cell cancers, and central nervous system disease.

f. There is no cure for AIDS; however, effective combination antiviral drug therapy is available to individuals in developed countries.

24.8   Viruses with Gapped DNA Genomes (Group VII)

a. Hepadnaviruses, including *Hepatitis B virus,* have dsDNA genomes that consist of one complete but nicked strand and an incomplete (i.e., gapped) complementary strand. Upon infection, the host cell repairs the gap and seals the nick to generate a covalently, closed, circular viral genome **(figure 24.40).**

b. HBV in the blood of infected individuals consists of three separate particles **(figure 24.41).**

c. Chronic hepatitis B causes liver cancer.

Critical Thinking Questions

1. No temperate RNA phages have yet been discovered. How might this absence be explained?

2. The choice between lysogeny and lysis is influenced by many factors. How would external conditions such as starvation or crowding be "sensed" and communicated to the transcriptional machinery and influence this choice?

3. The most straightforward explanation as to why the endolysin of T4 is expressed so late in infection is that its promoter is recognized by the gp55 alternative sigma factor. Propose a different explanation.

4. Several characteristics of AIDS render it particularly difficult to detect, prevent, and treat effectively. Discuss two of them. Contrast the disease with polio and smallpox.

5. In terms of molecular genetics, why is influenza such a prevalent viral infection in humans?

6. Will it be possible to eradicate many viral diseases in the same way as smallpox? Why or why not?

Learn More

Learn more by visiting the Prescott website at www.mhhe.com/prescottprinciples, where you will find a complete list of references.

Biogeochemical Cycling and the Study of Microbial Ecology

25

Some microorganisms live in environments where most known organisms cannot survive. These strands of iron-oxidizing *Ferroplasma* were discovered growing at pH 0 in an abandoned mine near Redding, California. This hardy microorganism has only a plasma membrane to protect itself from the rigors of this harsh environment.

assimilatory reduction   The reduction of an inorganic molecule to incorporate it into organic material. No energy is made available during this process.

biogeochemical cycling   The oxidation and reduction of substances carried out by living organisms and abiotic processes that results in the cycling of elements within and between different parts of ecosystems (the soil, aquatic environments, and atmosphere).

Chapter Glossary

community   An assemblage of different types of organisms or a mixture of different populations.

denitrification   The reduction of nitrate to gaseous products, primarily nitrogen gas, during anaerobic respiration.

dissimilatory reduction   The use of a substance as an electron acceptor in energy generation. The acceptor (e.g., sulfate or nitrate) is reduced but not incorporated into organic matter during biosynthetic processes.

enrichment culture   The growth of a specific microbe or group of physiologically similar microbes from natural samples by including selective features (light, temperature, nutrients) to promote the growth of the desired microorganisms while counter-selecting for other microbes that may be found in the same environment.

fluorescent in situ hybridization (FISH)   A technique for identifying certain genes or organisms in which specific DNA fragments are labeled with fluorescent dye and hybridized to the chromosomes of interest.

microelectrodes   Thin electrodes that can measure a variety of parameters (e.g., pH, O_2, H_2) at very high spatial resolution.

mineralization   The conversion of organic nutrients into inorganic material during microbial growth and metabolism.

nitrification   The oxidation of ammonia to nitrate.

nitrogen fixation   The metabolic process in which atmospheric molecular nitrogen (N_2) is reduced to ammonia; carried out by cyanobacteria, *Rhizobium*, and other nitrogen-fixing procaryotes.

population   An assemblage of organisms of the same type.

Everything is everywhere, the environment selects.

—M. W. Beijerinck

In previous chapters, microorganisms are usually considered as isolated entities. However, microorganisms exist in **communities.** Recent estimates suggest that most microbial communities have between 10^{10} to 10^{17} individuals representing at least 10^7 different taxa. How can such huge populations exist and, moreover, survive together in a productive fashion? To answer this question, one must know which microbes are present and how they interact—that is, one must study **microbial ecology.** In this chapter, we begin our consideration of this multidisciplinary field by discussing the process of elemental cycling—the microbe-mediated exchange of nutrients that makes life on Earth possible. This is sometimes called **environmental microbiology (Microbial**

Microbial Diversity & Ecology

25.1 Microbial Ecology versus Environmental Microbiology

The term *microbial ecology* is now used in a general way to describe the presence and contributions of microorganisms, through their activities, to the places where they are found. Students of microbiology should be aware that much of the information on microbial communities and their contributions to soils, waters, and associations with plants, now described by this term, would have been considered "environmental microbiology" in the past. Thomas Brock has given a useful definition of microbial ecology: "Microbial ecology is the study of the behavior and activities of microorganisms in their natural environments." He continues by pointing out that "microbes are small; their environments also are small." In these small environments or "microenvironments," other kinds of microorganisms (and macroorganisms) often also are present, a critical point that was emphasized by Sergei Winogradsky in 1947.

Environmental microbiology, in comparison, relates primarily to microbial processes that occur in soil, water, or food, as examples. It is not concerned with the particular "microenvironment" where the microorganisms actually are functioning but with the broader-scale effects of microbial presence and activities. One can study these microbially mediated processes and their possible global impacts at the scale of "environmental microbiology" without knowing about the specific microenvironment (and the organisms functioning there) where these processes actually take place. However, it is critical to be aware that microbes function in their localized environments and affect ecosystems at greater scales, including causing global-level effects. In the last decades, the term *microbial ecology* largely has lost its original meaning, and recently the statement has been made that *microbial ecology* has become a catchall term. As you read various textbooks and scientific papers, possible differences of meaning between microbial ecology and environmental microbiology should be kept in mind.

Diversity & Ecology 25.1). We then present an overview of some of the more important tools and techniques used to study microbial ecology. This chapter thus provides the foundation for a more detailed review of microbial communities in nature (chapter 26).

25.1 BIOGEOCHEMICAL CYCLING

Microorganisms, in the course of their growth and metabolism, interact with each other as they transfer nutrients from one to another, to larger organisms, and to the atmosphere. This **nutrient cycling** includes the exchange of carbon, nitrogen, phosphorus, sulfur, iron, and manganese. Nutrient cycling is called **biogeochemical cycling** when applied to the environment, where it involves both biological and chemical processes and is of global importance. Nutrients are transformed and exchanged, often by oxidation-reduction reactions that can alter the chemical and physical characteristics of the nutrients. All of the biogeochemical cycles are linked, and the metabolism-related transformations of these nutrients make life on Earth possible. << *Oxidation-reduction reactions (section 9.5)*

The major reduced and oxidized forms of the most important elements are noted in **table 25.1**. Significant gaseous components occur in the carbon and nitrogen cycles and, to a lesser extent, in the sulfur cycles. Some microorganisms often can fix gaseous forms of carbon and nitrogen compounds. In other cycles, such as those for phosphorus and iron, there is no gaseous component.

Carbon Cycle

Carbon is present in reduced forms, such as methane (CH_4) and organic matter, and in more oxidized forms, such as carbon monoxide (CO) and carbon dioxide (CO_2). The major pools in an integrated, simplified carbon cycle are shown in **figure 25.1**. Electron donors (e.g., hydrogen, which is a strong reductant) and electron acceptors (e.g., O_2) influence the course of biological and chemical reactions involving carbon. Hydrogen can be produced when organic matter is degraded, especially when fermentation occurs under anoxic conditions. Although carbon cycles continuously from one form to another, for the sake of clarity, we shall say that the cycle "begins" with carbon fixation—the conversion of CO_2 into organic matter. Plants such as trees and crops are often thought of as the principal CO_2-fixing organisms, but at least half the carbon on Earth is fixed by microbes, particularly marine photosynthetic procaryotes and protists (e.g., the cyanobacteria *Prochlorococcus* and *Synechococcus,* and diatoms, respectively). Carbon is also fixed by chemolithoautotrophic microbes. All fixed carbon enters a common pool of organic matter that can then be oxidized back to CO_2 through aerobic or anaerobic respiration and fermentation.

Alternatively, inorganic (CO_2) and organic carbon can be reduced anaerobically to methane (CH_4). Methane is produced by archaea in anoxic habitats. In a water column, the anoxic zone where methane is produced is often below an oxic zone. Therefore, as methane moves up the water column, it is oxidized before reaching the atmosphere. However, in many situations, such as in rice paddies without an overlying oxic water zone, the methane is released directly to the atmosphere, thus contributing to atmospheric methane increases. Rice paddies, ruminant animals, coal mines, sewage treatment plants, landfills, and marshes are important sources of methane. Even the archaea such as *Methanobrevibacter* in the guts of termites contribute to global methane production. << *Phylum* Euryarchaeota: *Methanogens (section 18.3)*

In the carbon cycle depicted in figure 25.1, no distinction is made between the different types of organic matter formed and

| Table 25.1 | The Major Forms of Carbon, Nitrogen, Sulfur, and Iron Important in Biogeochemical Cycling | | | | |

Cycle	Significant Gaseous Component Present?	Major Forms and Valences			
		Reduced Forms	Intermediate Oxidation State Forms		Oxidized Forms
C	Yes	Methane: CH_4 (-4)	Carbon monoxide CO $(+2)$		CO_2 $(+4)$
N	Yes	Ammonium: NH_4^+, organic N (-3)	Nitrogen gas: N_2 (0)	Nitrous oxide N_2O $(+1)$ Nitrite: NO_2^- $(+3)$	Nitrate: NO_3^- $(+5)$
S	Yes	Hydrogen sulfide: H_2S, SH groups in organic matter (-2)	Elemental sulfur: S^0 (0)	Thiosulfate: $S_2O_3^{2-}$ $(+2)$ Sulfite: SO_3^{2-} $(+4)$	Sulfate: SO_4^{2-} $(+6)$
Fe	No	Ferrous iron: Fe^{2+} $(+2)$			Ferric Iron: Fe^{3+} $(+3)$

Note: The carbon, nitrogen, and sulfur cycles have significant gaseous components, and these are described as gaseous nutrient cycles. The iron cycle does not have a gaseous component, and this is described as a sedimentary nutrient cycle. Major reduced, intermediate oxidation state, and oxidized forms are noted, together with valences.

degraded. This is a marked oversimplification because organic matter varies widely in physical characteristics and in the biochemistry of its synthesis and degradation. Organic matter varies in terms of elemental composition, structure of basic repeating units, linkages between repeating units, and physical and chemical characteristics. Its degradation is influenced by a series of factors. These include (1) nutrient availability; (2) abiotic conditions (e.g., pH, oxidation-reduction potential, O_2, osmotic conditions), and (3) the microbial community present.

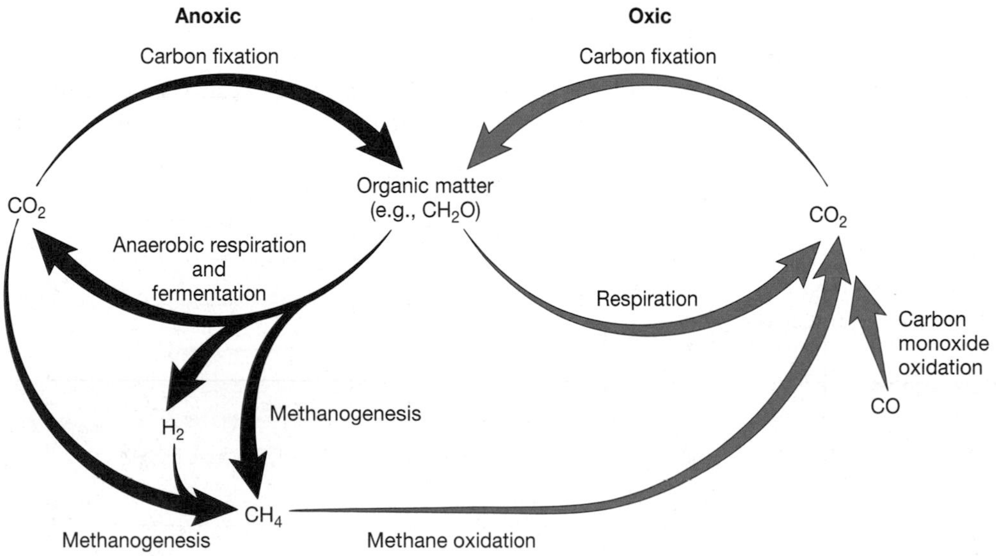

The major complex organic substrates used by microorganisms are summarized in table 25.2. Because microorganisms require each macronutrient, if an environment is enriched in one nutrient but relatively deficient in another, the nutrients may not be completely recycled into living biomass. For instance, chitin, protein, and nucleic acids contain nitrogen in large amounts. If these substrates are used for growth, the excess nitrogen and other minerals that are not used in the formation of new microbial biomass are released to the environment in the process of **mineralization.** This is the process by which organic matter is decomposed to release simpler, inorganic compounds (e.g., CO_2, NH_3, CH_4, H_2).

The other complex substrates listed in table 25.2 contain only carbon, hydrogen, and oxygen. If microorganisms are to grow by using these substrates, they must acquire the remaining nutrients they need for biomass synthesis from the environment. This is often very difficult, as the concentration of nitrogen, phosphorus, and iron may be very low. The inability to assimilate sufficient levels of one or more macronutrients may then limit the growth of a given population. For instance, in open-ocean microbial communities, growth of many autotrophic microbes is often nitrogen limited. In other words, if higher concentrations of usable nitrogen (NO_3^-, NH_4^+) were available, the rate of growth of individual microbes

Figure 25.1 The Basic Carbon Cycle in the Environment. Carbon fixation can occur through the activities of photoautotrophic and chemoautotrophic microorganisms. Methane can be produced from inorganic substrates ($CO_2 + H_2$) or from organic matter. Carbon monoxide (CO)—produced by sources such as automobiles and industry—is returned to the carbon cycle by CO-oxidizing bacteria. Aerobic processes are noted with blue arrows, and anaerobic processes are shown with red arrows.

Table 25.2	Complex Organic Substrate Characteristics that Influence Decomposition and Degradability									
			Elements Present in Large Quantity					**Degradation**	
Substrate	**Basic Subunit**	**Linkages (if Critical)**	**C**	**H**	**O**	**N**	**P**	**With O$_2$**	**Without O$_2$**
Starch	Glucose	α(1→4) α(1→6)	+	+	+	−	−	+	+
Cellulose	Glucose	β(1→4)	+	+	+	−	−	+	+
Hemicellulose	C6 and C5 monosaccharides	β(1→4), β(1→3), β(1→6)	+	+	+	−	−	+	+
Lignin	Phenylpropene	C–C, C–O bonds	+	+	+	−	−	+	−
Chitin	*N*-acetylglucosamine	β(1→4)	+	+	+	+	−	+	+
Protein	Amino acids	Peptide bonds	+	+	+	+	−	+	+
Hydrocarbon	Aliphatic, cyclic, aromatic		+	+	−	−	−	+	+/−
Lipids	Glycerol, fatty acids; some contain phosphate and nitrogen	Esters, ethers	+	+	+	+	+	+	+
Microbial biomass		Varied	+	+	+	+	+	+	+
Nucleic acids	Purine and pyrimidine bases, sugars, phosphate	Phosphodiester and *N*-glycosidic bonds	+	+	+	+	+	+	+

would increase, as would their overall population size. Those nutrients that are converted into biomass become temporarily "tied up" from nutrient cycling; this is sometimes called nutrient **immobilization.**

Most carbon substrates can be degraded easily with or without oxygen present, but this is not always the case. The exceptions are hydrocarbons and lignin. Hydrocarbons are unique in that microbial degradation most often involves the initial addition of molecular O$_2$. However, anaerobic degradation of hydrocarbons with sulfate or nitrate as electron acceptors can occur. This proceeds more slowly and only in microbial communities that have been exposed to these compounds for extended periods. Nonetheless, under oxic or anoxic conditions, the relative insolubility of hydrocarbons reduces their availability for microbial degradation. >> *Biodegradation and bioremediation by natural communities (section 35.3)*

Lignin, an important structural component in mature plant materials, is a complex amorphous polymer linked by carbon-carbon and carbon-ether bonds (*see figure 26.17*). Lignin is structurally very heterogeneous (i.e., there is no single lignin molecule), so no single degradative enzyme can attack lignin using a uniform strategy. Fungi and the streptomycetes degrade lignin by oxidative depolymerization, a process that occurs under oxic conditions. Lignin's diminished biodegradability under anoxic conditions results in accumulation of lignified materials, including the formation of peat bogs and muck soils.

Oxygen availability also affects the final products that accumulate when organic substrates have been

processed and mineralized by microorganisms. Under oxic conditions, oxidized products such as nitrate, sulfate, and carbon dioxide will result from microbial degradation of complex organic matter (**figure 25.2**). In comparison, under anoxic conditions, reduced end products tend to accumulate, including ammonium ion, sulfide, and methane.

The carbon cycle has come under intense scrutiny in the last decade or so. This is because CO$_2$ levels in the atmosphere have risen from their preindustrial concentration of about 278 parts per million (ppm) to about 380 ppm in 2006. This represents an increase of about one-third, and CO$_2$ levels continue to rise at a rate of about 1.9 ppm per year. Like CO$_2$, methane is a greenhouse gas and its atmospheric concentration is

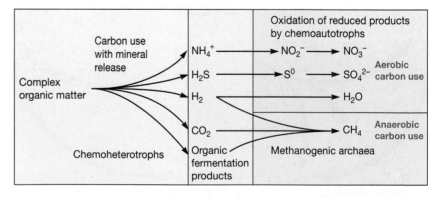

Figure 25.2 The Influence of Oxygen on Organic Matter Decomposition. Microorganisms form different products when breaking down complex organic matter aerobically than they do under anoxic conditions. Under oxic conditions, oxidized products accumulate, while reduced products accumulate anaerobically. These reactions also illustrate that the waste products of one group of microorganisms may be used by a second type of microorganism.

increasing about 1% per year, from 0.7 to 1.7 ppm since the early 1700s. These changes are clearly the result of the combustion of fossil fuels and altered land use. The term **greenhouse gas** describes the ability of these gases to trap heat within Earth's atmosphere, leading to a documented increase in the planet's mean temperature. Indeed, over the past 100 years, Earth's average temperature has increased by 0.6°C and continues to rise at a rapid rate. The increase in CO_2 levels would be even more dramatic if it were not for the removal of large quantities of CO_2 from the atmosphere by carbon fixation in both marine and terrestrial ecosystems. >> *Marine and freshwater microbiology: Microorganisms in the photic zone of the open ocean (section 26.1); Microorganisms in terrestrial environments: Soil microorganisms and global warming (section 26.2)*

1. What is biogeochemical cycling?
2. Define mineralization and immobilization, and give examples.
3. What is unique about lignin and its degradation?
4. What C, N, and S forms will accumulate after anaerobic degradation of organic matter? Compare these with the forms that accumulate after aerobic degradation.

Nitrogen Cycle

Like the carbon cycle, cycling of nitrogenous materials makes life on Earth possible. A simplified nitrogen cycle is presented in **figure 25.3**. Again, we begin our discussion of this cycle with the fixation of the inorganic gaseous element (N_2) to its organic form (e.g., amino acids). **Nitrogen fixation** is a uniquely procaryotic

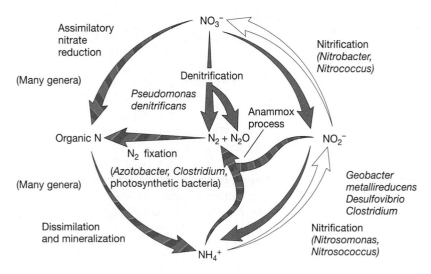

Figure 25.3   A Simplified Nitrogen Cycle.   Reactions that occur predominantly under oxic conditions are noted with open arrows. Anaerobic processes are noted with solid purple arrows. Processes occurring under both oxic and anoxic conditions are marked with striped arrows. Important genera contributing to the nitrogen cycle are noted.

process; apart from a limited amount of nitrogen fixation that occurs during lightning strikes, all naturally produced organic nitrogen is of procaryotic origin (the manufacture of fertilizer is another source of nitrogen fixation). Although the nitrogenase enzyme is sensitive to oxygen, nitrogen fixation can be carried out under oxic and anoxic conditions. Microbes such as *Azotobacter* and the cyanobacterium *Trichodesmium* fix nitrogen aerobically, while free-living anaerobes such as members of the genus *Clostridium* fix nitrogen anaerobically. Perhaps the best-studied nitrogen-fixing microbes are the bacterial symbionts of leguminous plants, including *Rhizobium,* its α-proteobacterial relatives, and some recently discovered β-proteobacteria (e.g., *Burkholderia* and *Ralstonia*). Other bacterial symbionts fix nitrogen as well. For instance, the actinomycete *Frankia* fixes nitrogen while colonizing many types of woody shrubs, and the heterocystous cyanobacterium *Anabaenea* fixes nitrogen when in association with the water fern *Azolla.* << *Synthesis of amino acids: Nitrogen fixation (section 11.5); Photosynthetic bacteria: Phylum* Cyanobacteria *(section 19.3);* >> *Microorganisms in terrestrial environments: Nitrogen fixation (section 26.2)*

The product of N_2 fixation is ammonia (NH_3), which is immediately incorporated into organic matter as an amine. These amine N-atoms are eventually introduced into proteins, nucleic acids, and other biomolecules. The N cycle continues with the degradation of these molecules into ammonium (NH_4^+) within mixed assemblages of microbes. One important fate of this ammonium is its conversion to nitrate (NO_3^-), a process called **nitrification.** This is a two-step chemolithotrophic process whereby ammonium is first oxidized to nitrite (NO_2^-), which is then oxidized to nitrate. Different genera of bacteria mediate the two steps of nitrification. For example, *Nitrosomonas* and *Nitrosococcus* play important roles in the first step, and *Nitrobacter* and related bacteria carry out the second step. In addition, *Nitrosomonas eutropha* has been found to oxidize ammonium ion anaerobically to nitrite and nitric oxide (NO) using nitrogen dioxide (NO_2) as an acceptor in a denitrification-related reaction.

The production of nitrate is important because it can be reduced and incorporated into organic compounds; this process is known as assimilatory nitrate reduction. The use of nitrate as a source of organic nitrogen is an example of **assimilatory reduction.** Alternatively, for some microbes nitrate serves as a terminal electron acceptor during anaerobic respiration; this is a form of **dissimilatory reduction.** In this case, nitrate is removed from the ecosystem and returned to the atmosphere as dinitrogen gas (N_2) through a series of reactions that are collectively known as **denitrification.** This dissimilatory process usually involves heterotrophs such as *Pseudomonas denitrificans.* The major products of denitrification include nitrogen gas (N_2) and nitrous oxide (N_2O), although nitrite (NO_2^-) also can accumulate. Alternatively, nitrate can be transformed to ammonia in dissimilatory reduction by a variety of bacteria, including *Geobacter metallireducens, Desulfovibrio* spp., and *Clostridium* spp. << *Anaerobic respiration (section 10.6); Synthesis of amino acids: Nitrogen assimilation (section 11.5)*

A recently identified form of nitrogen conversion is called the **anammox reaction** (anoxic *amm*onium *ox*idation). In this anaerobic reaction, chemolithotrophs use ammonium ion (NH_4^+) as the electron donor and nitrite (NO_2^-) as the terminal electron acceptor; it is reduced to nitrogen gas (N_2). In effect, the anammox reaction is a shortcut to N_2, proceeding directly from ammonium and nitrite without having to cycle first through nitrate (figure 25.3). Although this reaction was known to be energetically possible, microbes capable of performing the anammox reaction were only recently documented. The discovery that marine bacteria perform the anammox reaction in the anoxic waters just below oxygenated regions in the open ocean solved a long-standing mystery. For many years, microbiologists wondered where the "missing" NH_4^+ could be—mass calculations did not agree with experimentally derived nitrogen measurements. The discovery that planctomycete bacteria oxidize measurable amounts of NH_4^+ to N_2, thereby removing it from the marine ecosystem, has necessitated a reevaluation of nitrogen cycling in the open ocean. << *Phylum* Planctomycetes *(section 19.4)*

---

1. Why is nitrogen fixation important and under what circumstances does it occur?

---

2. Describe the two-step process that makes up nitrification. Why do you think nitrification requires two different types of bacteria?

3. What is the difference between assimilatory nitrate reduction and denitrification? Which reaction is performed by most microbes, and which is a more specialized metabolic capability?

4. Describe the anammox reaction. Why do you think it was so difficult for microbiologists to discover the microbes that perform this reaction?

Phosphorus Cycle

Unlike the carbon and nitrogen cycles, the phosphorus cycle has no gaseous component (**figure 25.4**). Biogeochemical cycling of phosphorus is important for a number of reasons. All living cells require phosphorus for nucleic acids, lipids, and some polysaccharides. However, most environmental phosphorus is present in low concentrations, locked within Earth's crust. Thus phosphorus frequently limits growth.

Unlike carbon and nitrogen, which can be obtained from the atmosphere, phosphorus is derived solely from the weathering of phosphate-containing rocks. Therefore in soil, phosphorus exists in both inorganic and organic forms. Organic phosphorus includes not only that found in biomass but also that in materials such as humus and other organic compounds. The phosphorus in these organic materials is recycled by microbial activity. Inorganic phosphorus is negatively charged, so it complexes readily with cations in the environment, such as iron, aluminum, and calcium. These compounds are relatively insoluble, and their dissolution is pH dependent such that it is most available to plants and microbes between pH 6 and 7. Under such conditions, these organisms rapidly convert phosphate to its organic form so that it becomes available to animals. The microbial transformation of phosphorus features the transformation of simple orthophosphate (PO_4^{3-}), which bears phosphorus in the +5 valence state, to more complex forms. These include the polyphosphates seen in metachromatic granules as well as more familiar macromolecules. Note that the phosphorus in all of these organic forms remains in the +5 valence state. << *Procaryotic cytoplasm: Inclusion bodies (section 3.3) Synthesis of purines, pyrimidines, and nucleotides: Phosphorus assimilation (section 11.6)*

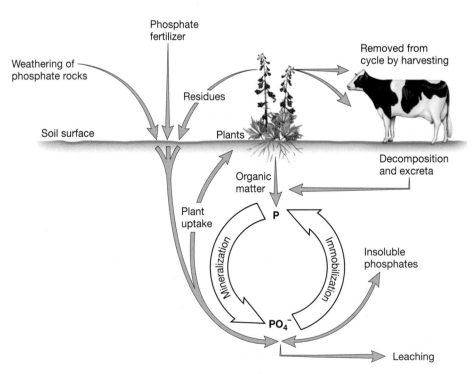

Figure 25.4 A Simplified Phosphorus Cycle. Phosphorus enters soil and water through the weathering of rocks, phosphate fertilizer, and surface residue of plant degradation. Plants and microbes rapidly convert inorganic phosphorus to its organic form, causing immobilization. However, much of the soil phosphorus can leach great distances or complex with cations to form relatively insoluble compounds.

Sulfur Cycle

Microorganisms contribute greatly to the sulfur cycle; a simplified version is shown

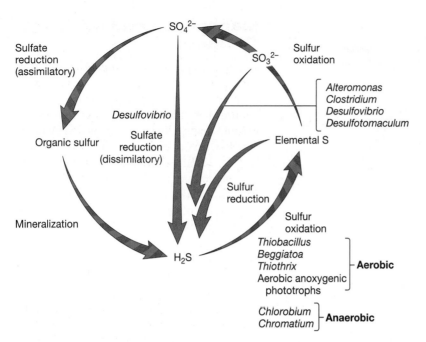

Figure 25.5   A Simplified Sulfur Cycle.   Photosynthetic and chemosynthetic microorganisms contribute to the environmental sulfur cycle. Sulfate and sulfite reductions carried out by *Desulfovibrio* and related microorganisms, noted with purple arrows, are dissimilatory processes. Elemental sulfur reduction to sulfide is carried out by *Desulfuromonas,* thermophilic archaea, or cyanobacteria in hypersaline sediments. Sulfate reduction also can occur in assimilatory reactions, resulting in organic sulfur forms. Sulfur oxidation can be carried out by a wide range of aerobic chemotrophs and by aerobic and anaerobic phototrophs.

out **aerobic anoxygenic photosynthesis (AAP).** Unlike other anoxygenic phototrophs, these aerobic bacteria are photoheterotrophs that metabolize carbon and other organic materials using O_2-dependent respiration. They also contain bacteriochlorophyll *a* and carotenoid pigments and are able to fix CO_2 during photosynthesis, thereby enhancing their growth. They are found in marine and freshwater environments and are often components of microbial mat communities. Important genera include the α-proteobacteria *Erythromonas, Roseococcus, Porphyrobacter,* and *Roseobacter,* although recently it has been discovered that a wider diversity of bacteria are capable of AAP.

Minor compounds in the sulfur cycle play major roles in biology. An excellent example is dimethylsulfonopropionate, which is used by bacterioplankton (floating bacteria) as a sulfur source for protein synthesis and is transformed to dimethylsulfide, a volatile sulfur form that can affect atmospheric processes.

When pH and oxidation-reduction conditions are favorable, several key transformations in the sulfur cycle also occur as the result of chemical reactions in the absence of microorganisms. An important example of such an abiotic process is the oxidation of sulfide to elemental sulfur. This takes place rapidly at a neutral pH, with a half-life of approximately 10 minutes for sulfide at room temperature.

Iron Cycle

The iron cycle principally features the interchange of ferrous iron (Fe^{2+}) to ferric iron (Fe^{3+}) (**figure 25.6**). In environments with a neutral pH that are fully aerated, Fe^{2+} quickly oxidizes to Fe^{3+}, which then forms insoluble precipitates. Thus microbial iron oxidation under these conditions is not favored, although earlier literature suggested that the genera *Sphaerotilus* and *Leptothrix* could oxidize iron at neutral pH; these microbes are now known to be chemoorganotrophs. By contrast, the chemolithotrophic genus *Gallionella* is capable of iron oxidation at neutral pH but only at reduced oxygen levels. Under acidic conditions, Fe^{2+} is more readily available, so in these environments, *Thiobacillus ferrooxidans* and the thermophile *Sulfolobus* oxidize iron. All these microbes use O_2 as the terminal electron acceptor.

Ferrous iron can also be oxidized under anoxic conditions with nitrate as the electron acceptor. One interesting anaerobic microbe is *Dechlorosoma suillum,* which oxidizes Fe^{2+} using chlorate and perchlorate as electron acceptors. Because perchlorate is a major component of explosives and rocket propellents, it is a frequent contaminant at retired munitions facilities and military bases. Thus *D. suillum* may be used in the bioremediation (biological cleanup) of such sites. This process also occurs in aquatic sediments with depressed levels of oxygen and may be another route by which large zones of oxidized iron have accumulated in environments with lower oxygen levels. Banded iron

in **figure 25.5**. Recall that sulfide can serve as an electron source for both photosynthetic microorganisms and chemolithoautotrophs such as *Thiobacillus;* it is converted to elemental sulfur and sulfate. When sulfate diffuses into reduced habitats, it provides an opportunity for different groups of microorganisms to carry out **sulfate reduction.** For example, when a usable organic electron donor is present, *Desulfovibrio* uses sulfate as its terminal electron acceptor during anaerobic respiration. Dissimilatory sulfate reduction—the use of sulfate as an external electron acceptor—results in sulfide accumulation in the environment. In comparison, the reduction of sulfate for use in amino acid and protein biosynthesis is described as assimilatory sulfate reduction. Other microorganisms have been found to carry out dissimilatory elemental sulfur (S^0) reduction. These include *Desulfuromonas,* thermophilic archaea, and cyanobacteria in hypersaline sediments. Sulfite (SO_3^{2-}) is another critical intermediate that can be reduced to sulfide by a wide variety of microorganisms, including *Alteromonas* and *Clostridium,* as well as *Desulfovibrio* and *Desulfotomaculum. Desulfovibrio* is usually considered an obligate anaerobe. Research, however, has shown that this interesting organism also respires using oxygen under hypoxic conditions (dissolved oxygen level of 0.04%).

In addition to the important photolithotrophic sulfur oxidizers such as *Chromatium* and *Chlorobium,* which function under strict anoxic conditions in deep water columns, some bacteria carry

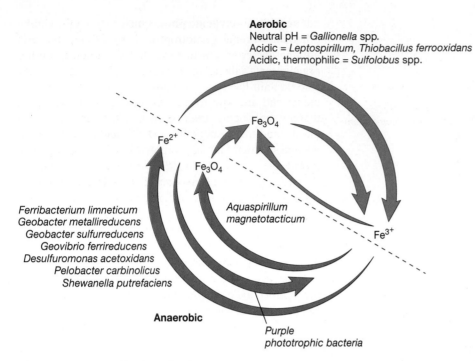

Aerobic
Neutral pH = *Gallionella* spp.
Acidic = *Leptospirillum, Thiobacillus ferrooxidans*
Acidic, thermophilic = *Sulfolobus* spp.

Fe$_3$O$_4$

Fe^{2+}

Fe$_3$O$_4$

Aquaspirillum magnetotacticum

Fe^{3+}

Ferribacterium limneticum
Geobacter metallireducens
Geobacter sulfurreducens
Geovibrio ferrireducens
Desulfuromonas acetoxidans
Pelobacter carbinolicus
Shewanella putrefaciens

Anaerobic

Purple phototrophic bacteria

Figure 25.6 The Iron Cycle. A simplified iron cycle with examples of microorganisms contributing to these oxidation and reduction processes. In addition to ferrous ion (Fe^{2+}) oxidation and ferric ion (Fe^{3+}) reduction, magnetite (Fe$_3$O$_4$), a mixed valence iron compound formed by magnetotactic bacteria, is important in the iron cycle. Different microbial groups carry out the oxidation of ferrous ion depending on environmental conditions.

formation that occurred when atmospheric oxygen levels were beginning to increase at the end of the Precambrian era may be evidence of increased iron bacteria activity. >> *Biodegradation and bioremediation by natural communities (section 35.3); << Microbial evolution (section 17.1)*

Iron reduction occurs under anoxic conditions resulting in the accumulation of ferrous ion. Although many microorganisms can reduce small amounts of iron during their metabolism, most iron reduction is carried out by specialized iron-respiring microorganisms such as *Geobacter metallireducens, Geobacter sulfurreducens, Ferribacterium limneticum,* and *Shewanella putrefaciens,* which can obtain energy for growth on organic matter using ferric iron as the electron acceptor. << *Anaerobic respiration (section 10.6)*

In addition to these relatively simple reductions to ferrous ion, some magnetotactic bacteria such as *Aquaspirillum magnetotacticum* transform extracellular iron to the mixed valence iron oxide mineral magnetite (Fe$_3$O$_4$) and construct intracellular magnetic compasses. Magnetotactic bacteria may be described as magneto-aerotactic bacteria because they are thought to use magnetic fields to migrate to the position in a bog or swamp where the oxygen level is best suited for their functioning. Furthermore, dissimilatory iron-reducing bacteria accumulate magnetite as an extracellular product. << *Procaryotic cytoplasm: Inclusion bodies (section 3.3)*

Manganese and Mercury Cycles

The importance of microorganisms in manganese cycling is becoming much better appreciated. The manganese cycle involves the transformation of manganous ion (Mn^{2+}) to MnO$_2$ (equivalent to manganic ion [Mn^{4+}]), which occurs in hydrothermal vents and bogs (**figure 25.7**). *Leptothrix, Arthrobacter,* and *Pedomicrobium* are important in Mn^{2+} oxidation. *Shewanella, Geobacter,* and other chemoorganotrophs can carry out the complementary manganese reduction process.

The mercury cycle is of particular interest and illustrates many characteristics of those metals that can be methylated. Mercury compounds were widely used in industrial processes over the centuries. A devastating situation developed in southwestern Japan when large-scale mercury poisoning occurred in the Minamata Bay region because of industrial mercury released into the marine environment. Inorganic mercury that accumulated in bottom muds of the bay was methylated by anaerobic bacteria of the genus *Desulfovibrio* (**figure 25.8**). Such methylated mercury forms are volatile and lipid soluble, and the mercury concentrations increased in the food chain by the process of **biomagnification** (the progressive accumulation of refractile compounds by successive trophic levels). Fish containing high levels of mercury were ultimately ingested by humans—the "top consumers"—leading to severe neurological disorders, particularly among children.

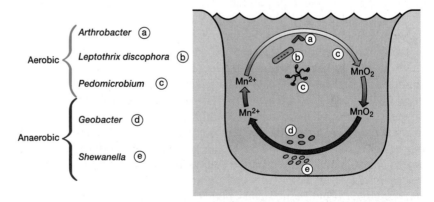

Aerobic
{
Arthrobacter (a)
Leptothrix discophora (b)
Pedomicrobium (c)
}

Anaerobic
{
Geobacter (d)
Shewanella (e)
}

Mn^{2+} MnO$_2$
Mn^{2+} MnO$_2$

Figure 25.7 The Manganese Cycle, Illustrated in a Stratified Lake. Microorganisms make many important contributions to the manganese cycle. After diffusing from anoxic (pink) to oxic (blue) zones, manganous ion (Mn^{2+}) is oxidized chemically and by many morphologically distinct microorganisms in the oxic water column to manganic oxide—MnO$_2^{(IV)}$, valence equivalent to 4+. When the MnO$_2^{(IV)}$ diffuses into the anoxic zone, bacteria such as *Geobacter* and *Shewanella* carry out the complementary reduction process. Similar processes occur across oxic/anoxic transitions in soils, muds, and other environments.

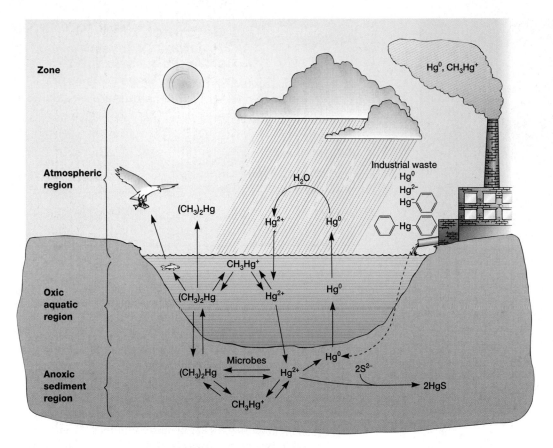

Figure 25.8 The Mercury Cycle. Interactions between the atmosphere, oxic water, and anoxic sediment are critical. Microorganisms in anoxic sediments, primarily *Desulfovibrio,* can transform mercury to methylated forms that can be transported to water and the atmosphere. These methylated forms also undergo biomagnification. The production of volatile elemental mercury (Hg^0) releases this metal to waters and the atmosphere. Sulfide, if present in the anoxic sediment, can react with ionic mercury to produce less soluble HgS.

1. Trace the fate of a single phosphorus atom from a rock to a stream and back to the earth.

2. Describe the difference between assimilatory and dissimilatory sulfate reduction.

3. Why is microbial oxidization of Fe^{2+} favored in acidic conditions?

4. What are some important microbial genera that contribute to manganese cycling?

5. How can microbial activity render some metals more or less toxic to warm-blooded animals?

25.2 MICROBIAL ECOLOGY AND ITS METHODS: AN OVERVIEW

Microbial ecologists study natural microbial communities that exist in soils or waters, or in association with other organisms, including humans. Regardless of the habitat they study, these scientists seek to answer several fundamental questions. First, they are interested in the microbial **population:** How many microbes are present? What genera and species are represented in the ecosystem? Once these questions have been addressed, community dynamics can be explored: How do these microorganisms interact with one another, with multicellular eucaryotes, and with abiotic factors in the environment? The multidisciplinary nature of microbial ecology has generated an enormous assortment of approaches, including microscopic, cultural, physical, chemical, and molecular techniques. Some of these methods are now discussed.

Examination of Microbial Populations

Traditionally, microbial populations have been studied by obtaining isolates in pure (axenic) culture. This has generated a great deal of insight into the microbial morphology, physiology, and genetics of a variety of organisms. Unfortunately, until it was realized that only a small faction (less than 1%) of microbes have been cultured in the laboratory, it led to some misunderstandings in terms of the degree of community diversity and the relative importance of specific genera. While it is still widely recognized that an axenic culture is the "gold standard" for microbial analysis, most microbiologists agree that to understand natural microbial populations, one must also consider the diversity of life-forms found within each specific niche. Thus as we discuss the examination of microbial populations, we review both culture- and nonculture-based approaches.

Staining Techniques

The most direct way to assess microbial populations and community structure is to observe microbes in nature. This can be carried out in situ using immersed slides or electron microscope grids placed in a location of interest, which are then recovered later for observation. Samples taken from the environment often are examined in the laboratory using classical cellular stains, fluorescent stains, or fluorescent molecular probes. The fluorescent stain DAPI (4′, 6-*di*amido-2-*p*henyl*i*ndole) is commonly used to visualize microbes in environmental, food, and clinical samples. It specifically labels nucleic acids (i.e., both DNA and RNA), and little sample preparation is required prior

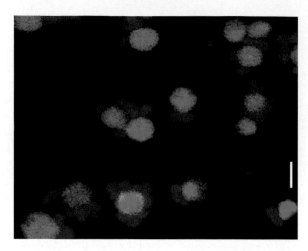

Figure 25.9 **The Use of FISH to Study Microbial Populations.** In this study, fluorescently tagged 16S rRNA probes were used to study the physical relationships between two archaea: a larger host, a member of the genus *Ignicoccus* (*green*), and a nanosized (~400 nm) hyperthermophilic symbiont named *Nanoarchaeum equitans* (*red*). Bar = 1 μm.

to its application. Thus DAPI staining is a convenient way to enumerate all the microbes in a specimen.

However, often a microbiologist will want to enumerate a specific genus or type of microbe. In these cases, a technique called *f*luorescent *i*n *s*itu *h*ybridization **(FISH)** can be used to identify specific microbes. Here, fluorescently labeled oligonucleotides (small pieces of DNA) that are known to be specific to the microbe of interest are used to label natural samples. The oligonucleotides are also called probes. Regions within the gene encoding **small subunit (SSU) ribosomal RNA** (16S for procaryotes, 18S for eucaryotes) are commonly used probes. As discussed in chapter 17, certain sequences within rRNA are unique to different microbial genera (*see figure 17.8*). This specificity has made FISH a popular tool in clinical diagnostics and food microbiology, as well as microbial ecology. For example, FISH was used to discover the association between the archaea *Nanoarchaeum equitans* and *Ignicoccus* **(figure 25.9).** Here, two fluorescent tags—one green, the other red—were attached to 16S rRNA probes, each having a different genus-specific nucleotide sequence. In this way, it was found that *N. equitans,* which is only 0.4 μm in diameter, grows on the larger archaeon as an obligate symbiont. *Nanoarchaeum* has one of the smallest procaryotic genomes found to date—only 0.5 megabases—and may represent a new phylum within the *Archaea* domain (*see figure 18.2*). Thus the combination of direct observation and molecular probes made it possible to document the life-style of this unusual archaeal symbiont and extend our understanding of archaeal diversity. << *Introduction to the* Archaea: *Archaeal taxonomy (section 18.1);* >> *Identification of microorganisms from specimens (section 32.2); Detection of food-borne pathogens (section 34.5)*

Culturing Techniques

It is now well understood that the vast majority of microbes have not been grown under laboratory conditions **(table 25.3).** The

Table 25.3	Estimates of the Percent Cultured Microorganisms from Various Environments
Environment	**Estimated Percent Cultured**
Seawater	0.001–0.100
Freshwater	0.25
Mesotrophic lake	0.1–1.0
Unpolluted estuarine waters	0.1–3.0
Activated sludge	1–15
Sediments	0.25
Soil	0.3

Source: Cowan, D. A. 2000. Microbial genomes—the untapped resource. *Tibtech* 18:14–16, table 2, p. 15.

discrepancy between the number of microbial cells observed by microscopic examination and the number of colonies that can be cultivated from the same natural sample has been called the **great plate count anomaly (GPCA).** The two most obvious reasons for the GPCA are that some of the organisms observed upon microscopic examination of natural samples may be truly nonviable. Alternatively, the right conditions for their growth in the lab have not yet been created. This has led to the description of such potentially viable microbes as being "nonculturable." A microbe is deemed **viable but nonculturable (VBNC)** if it shows motility or the presence of dividing cells when directly observed, is known to grow in a natural environment, or is stained with dyes that discriminate between live and dead cells **(figure 25.10).** Despite the

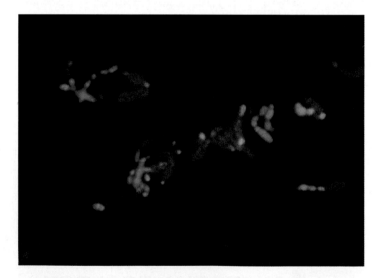

Figure 25.10 **Assessment of Microbial Viability by Use of Direct Staining.** By using differential staining methods, it is possible to estimate the portion of cells in a given population that are viable. In the LIVE/DEAD *Bac*Light Bacterial Viability procedure, two stains are used: a membrane-permeable green fluorescent, nucleic acid stain, and propidium iodide (red) that penetrates only cells with damaged membranes. In this *Bifidobacterium* culture, living cells stain green, while dead and dying cells stain red.

difficulties, the isolation and growth of microorganisms in axenic (pure) culture remains of fundamental importance.

A variety of standardized growth media can be used in viable count procedures, which are based on colony formation and enumeration. These media are satisfactory if one is isolating microbes that are commonly studied; however, it is often impossible to cultivate novel species or microbes rarely studied on these media. If it is necessary to isolate specific groups of microbes or to attempt to search for organisms with new capabilities, **enrichment culture** techniques can be used. These techniques are based on the expansion of the microenvironment to allow abundant growth of a microorganism formerly restricted to a small ecological niche while inhibiting growth of others. This approach plays a central role in finding new microbes. The success of an enrichment culture depends on a good understanding of the specific niche the microbe of interest inhabits and the physiological features that set that microbe apart from others. A simple example would be the isolation of anoxygenic photoautotrophs such as green sulfur bacteria. One might incubate a natural sample thought to harbor these bacteria in anoxic conditions with a source of reduced sulfur, carbonate (for CO_2), and light of a long wavelength. The absence of organic carbon selects against heterotrophs, while the anoxic conditions limit the growth of cyanobacteria and other, eucaryotic oxygenic photosynthetic microorganisms. ◄◄ *Photosynthetic bacteria (section 19.3)*

Enrichment cultures rarely yield axenic populations of microbes and are generally regarded as the first step in developing a pure culture. However, many microorganisms exist in nature as microbial assemblages, particularly protists and cyanobacteria. When such microorganisms are finally studied as axenic cultures, their morphological and physiological characteristics may change due to the lack of growth factors and vitamins formerly provided by other organisms.

A variety of new culture techniques have been applied that address specific needs of microorganisms growing in diverse environments. For example, in the **extinction culture technique,** natural microbial communities are diluted to a known number between one and 10 cells in sterile water from the same environment. Microbial growth is monitored by flow cytometry, a method by which individual cells are fluorescently labeled and counted by laser beams. Because this method is labor-intensive, a high-throughput version of this method can be used to isolate cultures in small volumes (e.g., 50 μl) of media (**figure 25.11**). In this way, a large number of culture attempts (and conditions) can be established in microtiter dishes and examined over a relatively short period of time. The development of new culture techniques requires an understanding of microbial physiology and ecology, as well as perseverance and a bit of luck.

It is often not possible to obtain an axenic culture, but single-cell isolations enable the examination of uncontaminated specimens. This is accomplished by "optical tweezers"—a laser beam is used to drag a microbe away from its neighbors—coupled with micromanipulation. With a micromanipulator, a desired cell or cellular organelle is drawn up into a micropipette after direct observation (**figure 25.12**). Once the microbe is isolated, PCR

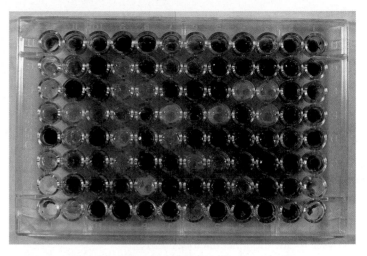

Figure 25.11 High-Throughput Screening Enrichment Cultures. A 96-well microtiter plate has been inoculated with natural samples at various dilutions. In many cases, the successful enrichment of pigment-producing actinomycetes can be seen by the development of a dark purple color.

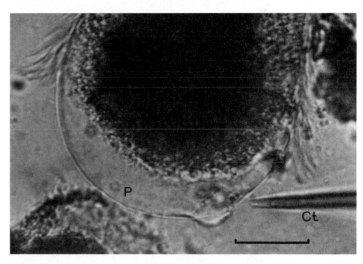

Figure 25.12 Micromanipulation for Isolation of Single Cells. Recovery of an endosymbiotic mycoplasma from single cell of the protist *Koruga bonita* by micromanipulation (bar = 10 μm). Protist (P), capillary tube (Ct).

amplification of the DNA from the individual cell or cell organelle provides sequence data for phylogenetic analysis. For example, it has been possible to establish the phylogenetic relationship of a mycoplasma recovered from the flagellate *Koruga bonita* by micromanipulation.

Consideration of microbial ecology on the scale of the individual cell has led to important ecological insights. It is now evident that there is surprising heterogeneity in what have been assumed to be homogenous microbial populations. Cells of a genetically uniform population do not have similar phenotypic attributes—the phenomenon of phenotypic or population heterogeneity. This is important in understanding responses of microorganisms in complex environments and particularly in disease processes.

Molecular Techniques

Because so few microbes have been grown in culture, the importance of molecular techniques to identify and quantify microorganisms is paramount. Nucleic acids are routinely recovered from soils and waters by direct extraction techniques. However, it has been found that the DNA obtained can vary depending on the method employed. A second problem, especially with muds and soils, is that one has no knowledge of the source of the bulk-extracted DNA. Some of the DNA recovered by this approach may not be derived from living organisms. Third, the DNA may have been recovered from nonfunctional propagules such as fungal or bacterial spores, or other resting structures.

With these caveats in mind, microbial ecologists regularly use SSU rRNA analysis to identify the microbes that populate a community. SSU rRNA can be amplified by the polymerase chain reaction (PCR) directly from samples of soil, water, or other natural material (e.g., sputum or blood in a clinical setting). The use of specific primers that target either archaeal, bacterial, or eucaryotic SSU rRNA genes (or other conserved genes) enables researchers to use PCR to obtain a sufficient number of nucleic acid fragments for DNA sequencing. ◀◀ *Environmental genomics (section 15.8); Polymerase chain reaction (section 16.2); Techniques for determining microbial taxonomy and phylogeny: Molecular characteristics (section 17.4)*

Once nucleotide sequences are obtained, they can be compared with sequences from the SSU rRNA genes isolated from other microbes by using several different databases. In this way, microbial ecologists can get a reasonable idea of the identity of the microbes that occupy a specific niche. Because whole organisms are not isolated and studied, it is said that specific **phylotypes** or unique SSU rRNA genes have been identified. Sometimes other genes besides SSU rRNA analysis are used. For example, a structural gene that confers a specific metabolic capability might be used as a means of determining not only the diversity of phylotypes but also potential community function (*see figure 15.15*).

As discussed in chapter 17, it is sometimes desirable to avoid DNA sequencing and obtain a DNA fingerprint instead. DNA fingerprints generate a barcodelike image of the population by separating the amplified rRNA gene

fragments on an agarose gel. One microbial population can then be compared with another (differing in either time or space). However, if one uses a single pair of primers to amplify the same region of a SSU rRNA gene from a population of genomes, the PCR products will have very similar molecular weights and thus may appear as a single band on an agarose gel. How can one separate such a collection of PCR products to obtain specific patterns of bands on a gel? **Denaturing gradient gel electrophoresis (DGGE)** can be used (**figure 25.13**). This approach relies on the fact that although the fragments are about the same size, they differ in nucleotide sequence. Recall that the temperature at which

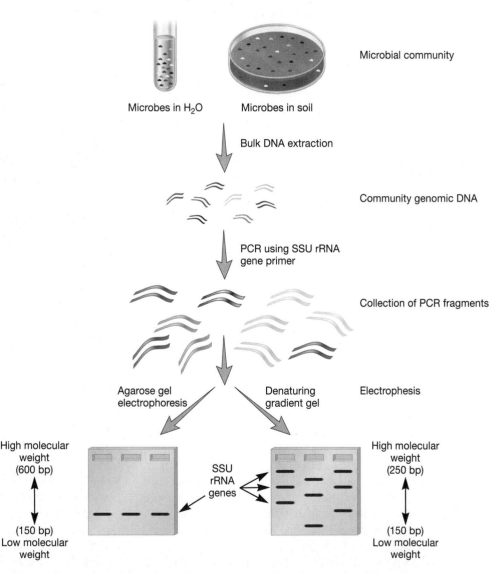

Figure 25.13 Denaturing Gradient Gel Electrophoresis (DGGE). The identification of phylotypes starts with the extraction of DNA from a microbial community and the PCR amplification of the gene of choice, usually that which encodes SSU rRNA. Because the majority of amplified DNA fragments have about the same molecular weight, when visualized by agarose gel electrophoresis they appear identical as shown in the gel on the left. However, DGGE uses a gradient of DNA denaturing agents to separate the fragments based on the condition under which they become single stranded. When a fragment is denatured, it stops migrating through the gel matrix (gel on right). Individual DNA fragments can then be cut out of the gel and cloned, and the nucleotide sequence determined.

double-stranded DNA becomes single stranded varies with the G + C content; that is, it varies with DNA sequence. DGGE takes advantage of the observation that DNA of different nucleotide sequences will denature at varying rates, although a gradient of chemicals that denature the DNA (usually urea and formamide), rather than temperature, is used. In this technique, a mixture of DNA fragments is placed in a single well in a gradient gel. As electrophoresis proceeds, fragments will migrate until they become denatured. What appeared to be a single DNA fragment (band) on a nongradient gel will resolve into separate fragments (multiple bands) by DGGE. Other techniques for generating DNA fingerprints of microbial populations include temperature gradient gel electrophoresis (TGGE; much like DGGE except a temperature gradient is used), single strand conformation polymorphism (SSCP), and terminal restriction fragment length polymorphism (T-RFLP). SSCP separates fragments based on the altered melting temperature that arises from differing secondary structure (i.e., internal base pairing) in single-stranded DNA. RFLP analysis is discussed in chapter 17 (*see p. 396*). << *Gel electrophoresis (section 16.3); Determining DNA sequences (section 15.2)*

The importance of microbial communities has led to vigorous debate regarding the magnitude of procaryotic diversity in recent years. SSU rRNA provides some insight regarding species diversity, but it is generally agreed that this technique is not well suited to determine the true number of species present in a given ecosystem. Nonetheless, it is clear that a culture-independent technique is the only valid strategy. One approach is to "count" the number of genomes. This is accomplished through **DNA reassociation.** When DNA is rendered single stranded by heating, it will spontaneously reanneal (become double stranded) when allowed to cool. The rate at which DNA reanneals depends on its size: the larger the fragment (or chromosome), the longer it takes. Determining procaryotic diversity based on rates of DNA reassociation rests on the notion that DNA extracted from the environment can be viewed as a giant genome; the length of time it takes to reanneal can be divided by the length of time it takes for the average procaryotic genome to reanneal. This then gives researchers a general idea of how many genomes are present. The use of this technique has revealed that on the whole, microbial communities in soil are more diverse than those in aquatic ecosystems and that unspoiled environments have more microbial species than those that are polluted (**table 25.4**). Most recently the application of mathematical models to analyze new and previously reported data suggests that microbial diversity may be even greater than previously imagined, with over a million different species in a single pristine soil community.

Examination of Microbial Community Activity

Microbial growth rates in complex systems can be measured directly. Changes in microbial numbers are followed over time, and the frequency of dividing cells is also used to estimate

Table 25.4	Procaryotic Diversity	
DNA Source	**Abundance (cells/cm³)[a]**	**Genome Equivalents[b]**
Forest soil	4.8×10^9	6,000
Pasture soil	1.8×10^{10}	3,500–8,800
Arable soil	2.1×10^{10}	140–350
Marine fish farm	7.7×10^9	50
Hypersaline pond (22% salinity)	6.0×10^9	7

[a]Direct cell counts using fluorescence microscopy
[b]DNA reassociation rate of DNA; genome equivalents are based on the assumption that the average procaryotic genome is about the size of the *E. coli* genome (4.1 Mb).
From Torsvik, V.; Ovraes, L.; and Thingstad, T. F. 2002. *Science* 296:1064–66.

production. Alternatively, the incorporation of radiolabeled components such as thymidine (a DNA constituent, usually labeled with ^{3}H) into microbial biomass provides information about growth rates and microbial turnover. This approach is most easily applied to studies of aquatic microorganisms; it is more difficult to directly observe and extract terrestrial microbial communities.

Microbial communities also can be described in terms of their structure and the nutrients contained in the community. Aquatic microorganisms can be recovered directly by filtration; the volume, dry weight, or chemical content of the microorganisms can then be measured. If it is not possible to directly measure the cell density, carbon, nitrogen, phosphorus, or an organic constituent of the cells (such as lipids or ergosterol) can be determined. This will give a single-value estimate of the microbial community. Such chemical measurements may be used either directly or expressed as microbial biomass.

To determine community activity, it is necessary to have a good understanding of the microenvironment inhabited by the microbial community in question. In many ecosystems, these microhabitats are studied by **microelectrodes**—electrodes capable of measuring pH, oxygen tension, H_2, H_2S, or N_2O. The tips of these electrodes range from 2 to 100 μm, enabling measurements in increments of a millimeter or less. Because these parameters include electron donor and acceptors, microbial activities can be inferred. These electrodes were first developed in the study of microbial mats (**figure 25.14**). These highly striated communities of photoautotrophs and chemoheterotrophs develop in saline lakes, marine intertidal zones, and hot springs.

In determining community function, it is often instructive to explore the cycling of a particular elemental species. In this case, **stable isotope analysis** may be used. Stable isotopes are those forms of elements that differ in atomic weight because they bear different numbers of neutrons but are not radioactive (i.e., they do not readily decay and are therefore stable). For example, ^{14}N and ^{15}N both occur in nature, but ^{14}N is found in

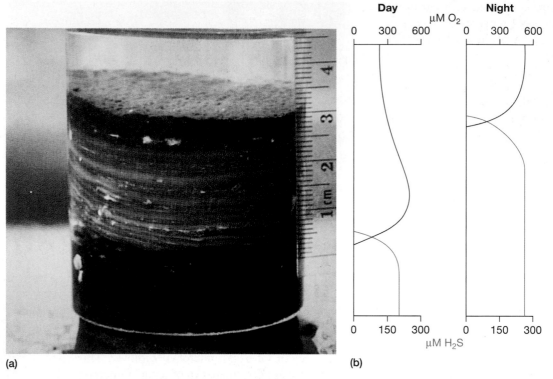

(a)

(b)

Figure 25.14 **Microelectrode Analysis of Microbial Mats.** (a) Microorganisms, through their metabolic activities, can create environmental gradients resulting in layered ecosystems. A vertical section of a hot spring (55°C) microbial mat, showing the various layers of microorganisms. (b) The measurement of oxygen and H_2S with microelectrodes demonstrates the diurnal shift in gas production: during the day, oxygen produced by cyanobacteria in the upper layers inhibits anaerobic respiration, so sulfate reduction is confined to the lower (dark) strata. At night, atmospheric oxygen has a limited range of diffusion into the mat and H_2S diffuses upward, where it can be used as a reductant by purple photosynthetic bacteria.

$^{13}CO_2$, was then used to identify the microbes. Recall that methanogenic microbes are archaea; the microbes found in this study have not yet been cultured but were identified by phylotype as members of "Rice Cluster-I" (RC-I). These methanogens rely on other microbes to ferment photosynthetically derived compounds excreted from plant roots. The H_2 produced by fermentation is then used by the RC-I archaea to reduce CO_2 to CH_4. This study demonstrates the value and importance of culture-independent approaches to understanding microbial community ecology and physiology. **<<** *Phylum Euryarchaeota: Methanogens (section 18.3)*

Microbial ecologists also use **reporter microbes** to characterize the physical microenvironment on the scale of an individual bacterium (around 1 to 3 μm). This is done by constructing cells with reporter genes, often based on green fluorescent protein (GFP), that change their fluorescence in response to environmental and physiological alterations. Such microbes are used to measure oxygen availability, UV radiation dose, pollutant or toxic chemical effects, and stress. For example, when microbes that contain a moisture stress reporter gene have less available water, an increase occurs in GFP-based fluorescence. **<<** *Techniques & Applications 16.1: Visualizing proteins with green fluorescence*

great abundance, whereas ^{15}N is rare. Organisms discriminate between stable isotopes, and for any given element, the lighter (in this case, ^{14}N) is preferentially incorporated into biomass, a phenomenon known as **isotopic fractionation.** For instance, one might want to explore the fate of NO_3^- in a soil ecosystem. Samples of the soil would be incubated in the presence of $^{15}NO_3^-$. One would then sample the soil microorganisms and measure the ratio of ^{15}N to ^{14}N relative to the ratio generally found in nature (a standard ratio). An organism that assimilates the $^{15}NO_3^-$ would be enriched for ^{15}N when compared to the standard.

Another technique called **stable isotope probing** can be used to examine nutrient cycling in microbial communities as well as identify the microbes taking up the element of interest. As an example, this technique was employed to explore methanogenesis in the rice paddy soil ecosystem. Researchers built **microcosms** (small incubation chambers with conditions that mimic the natural setting) to simulate natural rice paddies and introduced $^{13}CO_2$. Gaseous $^{13}CH_4$ was collected and RNA was isolated from the soil. ^{12}C-containing RNA was separated from the ^{13}C-RNA on the basis of differing buoyant densities. ^{13}C-containing rRNA, which could only be synthesized by those bacteria that assimilated the

1. What is FISH? Describe an example of the application of FISH in each of the following disciplines: microbial ecology, clinical diagnostics, and food inspection.

2. Explain the rationale for using conditions that select for the desired microbe and against all others in an enrichment culture. Describe an enrichment culture designed to isolate nitrogen-fixing soil bacteria.

3. What are stable isotopes? Design an experiment in which you plan on determining whether or not denitrification is taking place in a soil ecosystem.

Summary

25.1 Biogeochemical Cycling

a. Microorganisms—functioning with plants, animals, and the environment—play important roles in nutrient cycling, which is also termed biogeochemical cycling. Assimilatory processes involve incorporation of nutrients into the organism's biomass during metabolism; dissimilatory processes, in comparison, involve the release of nutrients to the environment after metabolism.

b. Biogeochemical cycling involves oxidation and reduction processes, and changes in the concentrations of gaseous cycle components, such as carbon, nitrogen, phosphorus, and sulfur, can result from microbial activity **(figures 25.1–25.8).**

c. Major organic compounds used by microorganisms differ in structure, linkage, elemental composition, and susceptibility to degradation under oxic and anoxic conditions. Lignin is degraded only under oxic conditions, a fact that has important implications in terms of carbon retention in the biosphere.

25.2 Microbial Ecology and Its Methods: An Overview

a. A variety of staining techniques are used to observe microbes in natural environments. Fluorescent in situ hybridization (FISH) labels specific microbes that possess a specific nucleotide sequence, usually a region of the SSU rRNA gene **(figure 25.9).**

b. Although most microbes cannot be grown in pure culture, enrichment techniques are invaluable in isolating microorganisms.

c. The analysis of rRNA from natural samples can be accomplished after amplification by PCR (to determine nucleotide sequence) or by DNA fingerprinting **(figure 25.13).**

d. DNA reassociation is also used to evaluate the size and complexity of microbial communities. This technique considers the community as a collection of genomes.

e. Microelectrodes can be used to determine physical and biological parameters such as pH, O_2, and H_2S concentrations in microniches **(figure 25.14).**

f. Stable isotope analysis is based on the fact that organisms discriminate between heavy and light naturally occurring isotopes. It is often used to monitor nutrient flux through ecosystems.

Critical Thinking Questions

1. Examine the carbon cycle shown in figure 25.1. What do you think are some major biogenic sources of CO_2 emission into the atmosphere? What are the major biogenic sinks of atmospheric CO_2?

2. How might you show that a microorganism found in a particular extreme environment is actually growing there?

3. How might you attempt to grow in the laboratory a chemolithoautotroph that uses ferrous iron as an electron donor and oxygen as an electron acceptor?

4. Both acetate and CO_2 can be used by methanogenic archaea to generate methane (CH_4). How would you determine the source of carbon for methanogenesis in a waterlogged peat?

Learn More

Learn more by visiting the Prescott website at www.mhhe.com/prescottprinciples, where you will find a complete list of references.

Microorganisms in Natural Environments

26

Chapter Glossary

bacteroid A modified bacterial cell that carries out nitrogen fixation within the root nodule cells of legumes.

carbonate equilibrium system The balance of CO_2, bicarbonate, and carbonate that keeps the ocean buffered between pH 7.6 and 8.2.

dissolved organic matter (DOM) Nutrients that are available in the soluble, or dissolved, state.

epiphyte An organism that grows on the surface of a plant.

eutrophic Pertaining to a nutrient-enriched environment.

marine snow Flocculent suspended particles in the ocean's water column; consist primarily of fecal pellets, diatom frustules, and other materials that are not easily degraded.

methanotroph An organism with the ability to grow on methane as the sole carbon source.

microbial loop The cycling of organic matter synthesized by photosynthetic microorganisms through the activity of other microbes, such as bacteria and protozoa. This process "loops" nutrients back for reuse by the primary producers and makes some organic matter unavailable to higher consumers.

mycorrhizal fungi Fungi that form stable, mutualistic relationships on (ectomycorrhizal) or in (endomycorrhizal) the root cells of vascular plants. The plants provide carbohydrate for the fungi, while the fungal hyphae extend into the soil and bring nutrients to the plants.

oligotrophic environment An environment containing low levels of nutrients, particularly those that support microbial growth.

particulate organic matter (POM) Nutrients that are not dissolved or soluble, including microorganisms or their cellular debris after senescence or viral lysis.

photosynthate Nutrient material that leaks from phototrophic organisms. It contributes to the pool of dissolved organic matter that is available to microorganisms.

phytoplankton Phototrophic prokaryotes and protists that drift in the upper, sunlit regions of marine and freshwater ecosystems.

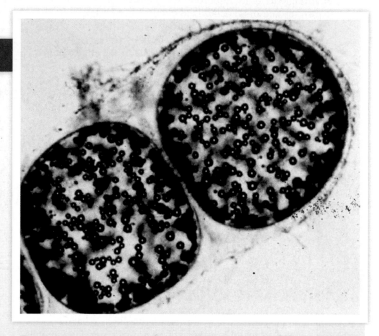

The giant marine bacterium, *Thiomargarita namibiensis*, about 100 to 300 μm in diameter, accumulates sulfur from sediments in its refractive sulfur granules and nitrate from overlying waters to support its growth.

primary production The incorporation of carbon dioxide into organic matter by photosynthetic organisms and chemoautotrophic bacteria (i.e., primary producers).

rhizobia Any one of a number of α- and β-proteobacteria that form symbiotic nitrogen-fixing nodules on the roots of leguminous plants.

rhizoplane The surface of a plant root.

rhizosphere A region around the plant root where materials released from the root promote microbial growth.

root nodule A knoblike structure on roots that contains endosymbiotic nitrogen-fixing bacteria (e.g., *Rhizobium* or *Bradyrhizobium*).

SAR11 The most abundant microbe on Earth, this lineage of α-proteobacteria has been found in almost all marine ecosystems studied; includes the SAR11 isolate, *Pelagibacter ubique*.

virioplankton Viruses that drift in marine and freshwater environments; such viruses are very abundant.

They [the leaves] that waved so loftily, how contentedly, they return to dust again and are laid low, resigned to lie and decay at the foot of the tree and afford nourishment to new generations of their kind, as well as to flutter on high!

—*Henry D. Thoreau*

All microbes are aquatic—even those that live on land. Procaryotes, protists, and most fungi require at least a thin film of liquid for replication. In this chapter, we turn our attention to the microbial ecology of aquatic and terrestrial environments. The oceans cover over two-thirds of the planet and contain all but 3% of its water. Freshwater environments are the source of our drinking water and are thus required for terrestrial life. Terrestrial microbes help drive the global biogeochemical cycles and play essential roles in agriculture and in maintaining environmental quality. Microbiologists studying these systems are examining some of the most important and life-sustaining microbial communities on Earth. We begin with a description of the marine environment and its biomes. We then introduce several freshwater microbial ecosystems. The second half of the chapter focuses on soil and plant-associated microbial communities.

26.1 MARINE AND FRESHWATER MICROBIOLOGY

It is an exciting time for aquatic microbiology—the development of molecular microbial ecology, novel culturing techniques, remote sensing, and deep-sea exploration has propelled this discipline to a new age of discovery. Recent reports have revealed a level of microbial diversity not previously imagined as well as the impact of microbes in maintaining a balanced ecosystem. We now realize the importance of microbes in addressing problems such as global warming and pollution. In addition, these new technologies have advanced our understanding of the food webs that govern the world's fisheries. Thus the microbiology of lakes, streams, and oceans is of enormous interest and significance.

Water as a Microbial Habitat

The nature of water as a microbial habitat depends on a number of physical factors such as temperature, pH, and light penetration. One of the most important of these is dissolved oxygen content. The flux rate of oxygen through water is about 1/10,000 times less than its rate through air. However, in some marine habitats, the limits to oxygen diffusion can be offset by the increased solubility of oxygen at colder temperatures and increasing atmospheric pressures. Thus, for the very deep ocean, the dissolved oxygen concentration actually increases with depth, even though the air-water interface can be literally miles away (figure 26.1). On the other hand, tropical lakes and summertime-temperate lakes may become oxygen limited only meters below the surface. In this case, aerobic microbes consume the surface-associated oxygen faster than it can be replenished. This frequently leads to the formation of hypoxic or anoxic zones in these aquatic environments. This enables anaerobic microbes, both chemotrophic and phototrophic, to grow in the lower regions of lakes where light can penetrate.

The second major gas in water, CO_2, plays many important roles in chemical and biological processes. The pH of unbuffered distilled water is determined by dissolved CO_2 in equilibrium with the air and is approximately 5.0 to 5.5. The pH of freshwater systems such as lakes and streams, which are usually only weakly buffered, is therefore controlled by terrestrial input (e.g., minerals that may be either acidic or alkaline) and the rate at which CO_2 is removed by photosynthesis. When autotrophic organisms such as diatoms use CO_2, the pH of the water is often increased.

In contrast, seawater is strongly buffered by the balance of CO_2, bicarbonate (HCO_3^-), and carbonate (CO_3^{2-}). Atmospheric CO_2 enters the oceans and either is converted to organic carbon by photosynthesis or reacts with seawater to form carbonic acid

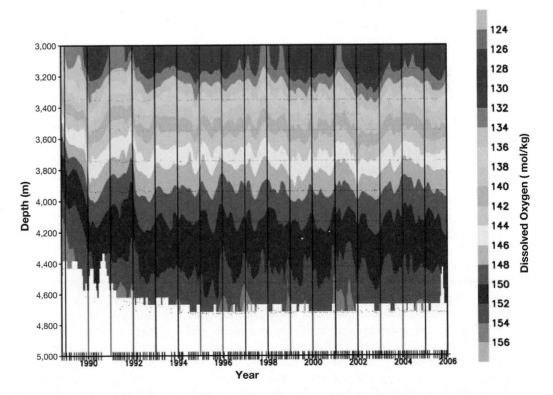

Figure 26.1 Oxygen versus Depth in the Deep Ocean. Dissolved oxygen measured in water samples between 3,000 and 5,000 m at Station Aloha, in the Hawaii Ocean Time series program. Oxygen concentration increases with depth due to increased oxygen solubility in cold waters and at high pressure.

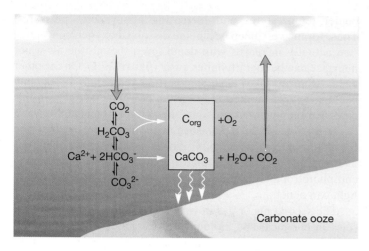

Figure 26.2 **The Carbonate Equilibrium System.** Atmospheric CO_2 enters seawater and is converted to organic carbon (C_{org}) or is converted to carbonic acid (H_2CO_3) that rapidly dissociates into the weak acids bicarbonate (HCO_3^-) and carbonate (CO_3^{2-}). Calcium carbonate ($CaCO_3$), a solid, precipitates to the seafloor, where it helps form a carbonate ooze. This system keeps seawater buffered at about pH 8.0.

(H_2CO_3), which quickly dissociates to form bicarbonate and carbonate (**figure 26.2**):

$$CO_2 + H_2O \rightleftharpoons H_2CO_3 \rightleftharpoons H^+ + HCO_3^- \rightleftharpoons 2H^+ + CO_3^{2-}$$

The oceans are effectively buffered between pH 7.6 and 8.2 by this **carbonate equilibrium system.** Much like the buffer one might use in a chemistry experiment, the pH of seawater is determined by the relative concentrations of the weak acids bicarbonate and carbonate. The equilibrium of these reactions has taken on new importance. Many oceanographers predict that the pH of the ocean will drop by 0.35 units by 2100 unless effective means of limiting greenhouse gas emissions (in particular, CO_2) are implemented. Recall that the pH scale is logarithmic; it is unclear what the implications of this change in carbonate equilibrium will mean for the marine environment and indeed for all of life on Earth.

Other gases also are important in aquatic environments. These include N_2, used as a nitrogen source by nitrogen fixers; H_2, which is both a waste product and a vital substrate; and methane (CH_4). These gases vary in their water solubility; methane is the least soluble of the three. Methane, produced under anoxic conditions either accumulates in the benthic (bottom) environment (e.g., natural gas deposits), or diffuses up through the water column, where it is oxidized by methanotrophs or released to the atmosphere. This eliminates the problem of waste accumulation that occurs with many microbial metabolic products, such as organic acids and ammonium ion.

Light is also critical for the health of marine and freshwater ecosystems. Like all life on Earth, microbes in these environments depend on **primary producers**—autotrophic organisms—to provide organic carbon. In streams, lakes, and coastal marine systems, macroscopic algae and plants are the chief primary producers, and organic carbon enters these systems in terrestrial runoff. The

situation is very different in the open ocean, where all organic carbon is the product of microbial autotrophy. In fact, about half of all the organic carbon on Earth is the result of this microbial carbon fixation. Water from the surface to the depth to which light penetrates with sufficient intensity to support these important autotrophs is called the **photic zone.** We see a marked difference in the depth of the photic zone when we compare nearshore waters with the open ocean. In lakes and estuaries where the water is turbid, the photic zone may be only a meter or two deep. This is in sharp contrast to nutrient-depleted areas such as the open ocean and many tropical areas where the water seems "crystal clear." In these regions, the photic zone ranges from 150 to 200 m.

Solar radiation warms the water and this can lead to thermal stratification. Warm water is less dense than cool water, so as the sun heats the surface in tropical and temperate waters, a **thermocline** develops. A thermocline can be thought of as a mass of warmer water "floating" on top of cooler water. These two water masses remain separate until there is either a substantial mixing event, such as a severe storm, or in temperate climates, the onset of autumn. As the weather cools, the upper layer of warm water becomes cooled and the two water masses mix. This is often associated with a pulse of nutrients from the lower, darker waters to the surface. This can trigger a sudden and rapid increase in the population of certain microbes and a bloom may develop. This is considered more fully in the discussion of microorganisms in coastal marine systems (p. 612).

1. What factors influence oxygen solubility? How is this important in considering aquatic environments?

2. Describe the buffering system that regulates the pH of seawater. What might be the implications of this stable buffering on microbial evolution?

3. What is the photic zone? How and why does it differ in lakes and coastal ecosystems versus the open ocean?

4. What is a thermocline?

Nutrient Cycling in Marine and Freshwater Environments

Obviously many differences exist between nearshore and open-ocean environments. From a microbial point of view, lakes, estuaries, and other coastal regions have relatively high rates of primary production. They therefore must have a higher influx and turnover (or reuse) of essential nutrients. In these regions, nearby agricultural and urban activity frequently generates runoff that provides substantial nutrients (especially nitrogen and phosphate) to these environments. In contrast, nutrient levels are very low in the open ocean, which is unaffected by rivers, streams, and terrestrial runoff. Here, nitrogen, phosphorus, iron, and even silica, which diatoms need to construct their frustules, can be limiting. ◄◄ *Protist classification: Stramenopiles (section 23.2)*

The major source of organic matter in illuminated surface waters is photosynthetic activity, primarily from **phytoplankton** (Greek *phyto,* plant, and *planktos,* wandering)—autotrophic organisms that float in the photic zone. Common planktonic cyanobacterial genera are *Prochlorococcus* and *Synechococcus,* which can reach densities of 10^4 to 10^5 cells per milliliter in the photic zone. These **picoplankton** (planktonic microbes between 0.2 and 2.0 μm in size) can represent 20 to 80% of the total phytoplankton biomass. Larger eucaryotic autotrophs, especially diatoms, also contribute a significant fraction of fixed carbon to these ecosystems.

As they grow and fix CO_2 to form organic matter, phytoplankton acquire needed nitrogen and phosphorus from the surrounding water. In marine systems, the nutrient composition of the water affects the final carbon-nitrogen-phosphorus (C:N:P) ratio of the phytoplankton, which is termed the **Redfield ratio,** named for the oceanographer Alfred Redfield. A commonly used value for this ratio is 106 parts C, 16 parts N, and 1 part P. This ratio is important when following nutrient dynamics, especially mineralization and immobilization processes, and for studying the sensitivity of oceanic photosynthesis to atmospheric additions of CO_2, nitrogen, sulfur, and iron.

As primary producers, microbes play an essential role in cycling other nutrients. Traditionally the interaction of organisms at different trophic levels has been depicted as a food chain in which primary producers are most numerous, followed by herbivores, which are then consumed by carnivores. Such diagrams generally show microorganisms functioning strictly as decomposers, mineralizing most of the waste products produced in the ecosystem. However, this simplified version of trophic interactions does not adequately describe the important and varied roles of microbes. Consider that microbial ecologists estimate at least 6×10^{30} microbial cells reside on the planet at any given time. This unseen biomass far exceeds that of all other organismal groups combined. If microbes functioned only to degrade and mineralize organic material to their inorganic forms, these essential elements would be in danger of being irreversibly removed from the ecosystem. Instead, microbes interact with several trophic levels, serving to recycle nutrients many times within the community before any given element is either mineralized or sinks to the sediments below.

The microbial loop describes the many roles that microbes serve (**figure 26.3**). For example, eucaryotic and procaryotic microbes can consume **dissolved organic matter (DOM)** that larger organisms cannot degrade. Larger plants and animals, the liquid waste of zooplankton, and material that leaks from the phytoplankton, sometimes called **photosynthate,** all contribute to a common pool of DOM. Viruses are also a source of DOM. Marine viruses can be present at concentrations up to 10^8 per

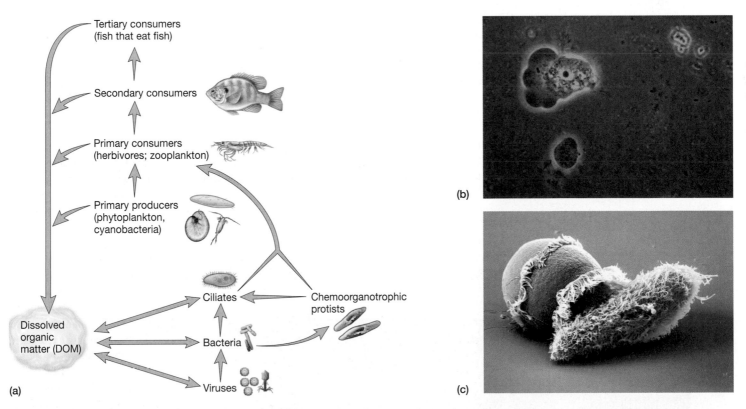

Figure 26.3 The Microbial Loop. (a) Microorganisms play vital roles in ecosystems as primary producers, decomposers, and primary consumers. All organisms contribute to a common pool of dissolved organic matter (DOM) that is consumed by microbes. Viruses contribute DOM by lysing their hosts, and procaryotes are consumed by protists, which also consume other protists. These microbes are then consumed by herbivores that often select their food items by size, thereby ingesting both heterotrophic and autotrophic microbes. Thus nutrient cycling is a complex system driven in large part by microbes. (b) Protists consume bacteria; in this case a naked amoeba is consuming the cyanobacterium *Synechococcus,* which fluoresces red. (c) Protists consume protists; here the ciliate *Didinium* (gold) is preying upon another ciliate, *Paramecium.*

milliliter; the lysis of their host cells contributes significantly to the return of nutrients into the microbial loop (p. 616). Protists, including flagellates and ciliates, consume smaller microbes (figure 26.3*b,c*), which can be thought of as **particulate organic matter (POM).** Because protists are then consumed by zooplankton, both DOM and POM are recycled for use at a number of trophic levels.

1. Describe the differences in nutrient input in coastal ecosystems as compared to the open ocean.
2. What is picoplankton and why is it important?
3. How does the microbial loop differ from a food chain?

Microorganisms in Coastal Marine Systems: Estuaries and Salt Marshes

An estuary is a semienclosed coastal region where a river meets the sea. Estuaries are defined by tidal mixing between freshwater and saltwater. They feature a characteristic salinity profile called a **salt wedge** (figure 26.4). Salt wedges are formed because saltwater is denser than freshwater, so seawater sinks below overlying freshwater. As the contribution from the incoming river increases and that of the ocean decreases, the relative amount of seawater declines with the estuary's increased distance from the sea. The distance the salt wedge intrudes up the estuary is not static. Most estuaries undergo large-scale tidal flushing; this forces organisms to adapt to changing salt concentrations on a daily basis. Microbes that live under such conditions combat the resulting osmotic stress by adjusting their intracellular osmolarity to limit the difference with that of the surrounding water. Most protists and fungi produce osmotically active carbohydrates for this purpose, whereas procaryotic microbes regulate internal concentrations of potassium or special amino acids such as ecoine and betaine. Thus most microbes that inhabit estuaries are halotolerant, which is distinct from halophilic. **Halotolerant** microbes can withstand significant changes in salinity; halophilic microorganisms have an absolute requirement for high salt concentrations.

Estuaries are unique in many respects. Their calm, nutrient-rich waters serve as nurseries for juvenile forms of many commercially important fish and invertebrates. However, despite their importance to the commercial fishing industry, estuaries are among the most polluted marine environments. They are the receptacles of waste dumped in rivers and pollutants discharged during industrial processes. Recall that from the seventeenth century through most of the twentieth century, industries dumped wastes without fear of punitive consequences. The cleanup of rivers and estuaries contaminated with industrial wastes such as polychlorinated biphenyls (PCBs) continues to this day. Often industrial pollution includes organic materials that can be used as nutrients. This further magnifies the problem because chemoorganotrophic microbes consume available oxygen, forming anoxic dead zones. Such anoxic regions, which are devoid of almost all macroscopic life, are now present in the Gulf of Mexico, Chesapeake Bay, and fjords entering the North Sea. **>>** *Biodegradation and bioremediation by natural communities (section 35.3)*

Organic pollution can also create the opposite problem: too much growth. However, in this case, a single microbial species, either algal or cyanobacterial, grows at the expense of all other organisms in the community. This phenomenon, called a **bloom,** often results from the introduction of nutrients combined with mixing sediments. If the microbes produce a toxic product or are in themselves toxic to other organisms such as shellfish or fish, the term **harmful algal bloom (HAB)** is used. Some HABs are called red tides because the microbial density is so great that the water becomes red or pink (the color of the algae). The number of HABs has dramatically increased in the last decade or so, although the reasons for this are unclear. HABs are sometimes responsible for killing large numbers of fish or marine mammals. For instance, off the coast of California, an HAB was responsible for sudden, large-scale deaths of sea lions. In this case, the bloom species was a diatom of the genus *Pseudonitzschia.* Anchovies consumed the diatoms and a potent neurotoxin, domoic acid, accumulated in the fish. The mammals were poisoned after they ingested large quantities of the fish, an important component of the sea lion diet.

HABs are often caused by dinoflagellates. Some bloom-causing dinoflagellates produce potent neurotoxins. Dinoflagellates of the genus *Alexandrium* produce a toxin responsible for most of the paralytic shellfish poisoning (PSP), which affects humans as well as

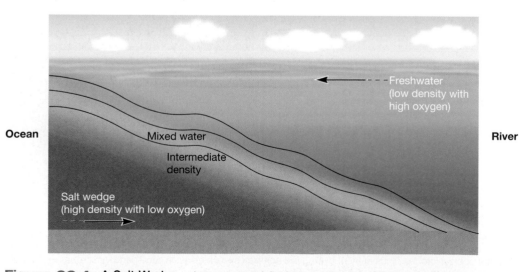

Figure 26.4   A Salt Wedge. An estuary contains both freshwater and saltwater. Because seawater is denser than freshwater, the water masses do not mix. Rather, the seawater remains below the freshwater with the relative amount of seawater decreasing in the upper reaches of the estuary.

other animals in coastal, temperate North America. The toxin produced by the dinoflagellate *Karenia brevis* killed a number of endangered manatees and bottlenose dolphins in 2002 in a bloom in Florida. Another dinoflagellate, *Pfiesteria piscicida,* has become a problem in the Chesapeake Bay and regions south. This protist produces lethal lesions in fish (**figure 26.5**) and has

had a devastating effect on the local fishery industry. Exposure to this microbe may also cause neurological damage to humans. Our understanding of this microbe has been limited by a number of factors, including problems in isolating its toxin. << *Protist classification: Alveolata and Stramenopiles (section 23.2)*

Salt marshes generally differ from estuaries in that they lack freshwater input from a single source. Smaller streams enter salt marshes, which are flatter and have a wider expanse of sediment and plant life exposed at low tide. The microbial communities within salt marsh sediments are very dynamic. These ecosystems can be modeled in **Winogradsky columns** (**figure 26.6**), named after the pioneering microbial ecologist Sergei Winogradsky (1856–1953). A Winogradsky column is easily constructed using a glass cylinder into which sediment is placed and then overlaid with saltwater. When the top of the column is sealed, much of the cylinder becomes anoxic. The addition of shredded newspaper introduces a source of cellulose that is degraded to fermentation products by the genus *Clostridium.* With these fermentation products available as electron donors, *Desulfovibrio* and other microbes use sulfate as a terminal electron acceptor to produce hydrogen sulfide (H_2S). The H_2S diffuses upward toward the oxygenated zone, creating a stable hydrogen sulfide gradient. In this gradient, the photoautotrophs *Chlorobium* and *Chromatium* develop as visible olive green and purple zones, respectively. These lithoautotrophs use H_2S as an electron source and CO_2, from sodium carbonate, as a carbon source. Above this region, the purple nonsulfur bacteria of the genera *Rhodospirillum* and *Rhodopseudomonas* can grow. These photoorganotrophs use organic matter as an electron donor under anoxic conditions and function in a zone where the sulfide level is lower. Both O_2 and H_2S may be present higher in the column, allowing specially adapted microorganisms to function. These include the chemolithotrophs *Beggiatoa* and *Thiothrix,* which use reduced sulfur compounds as electron donors and O_2 as an acceptor. In the upper portion of the column, diatoms and cyanobacteria may be visible. << *Anaerobic respiration (section 10.6); Phototrophy (section 10.12); Photosynthetic bacteria (section 19.3); Class* Alphaproteobacteria: *Purple nonsulfur bacteria (section 20.1); Class* Gammaproteobacteria *(section 20.3); Class* Clostridia *(section 21.3)*

Figure 26.5 *Pfiesteria* **Lesions.** Lesions on menhaden resulting from parasitism by the dinoflagellate *Pfiesteria piscicida.*

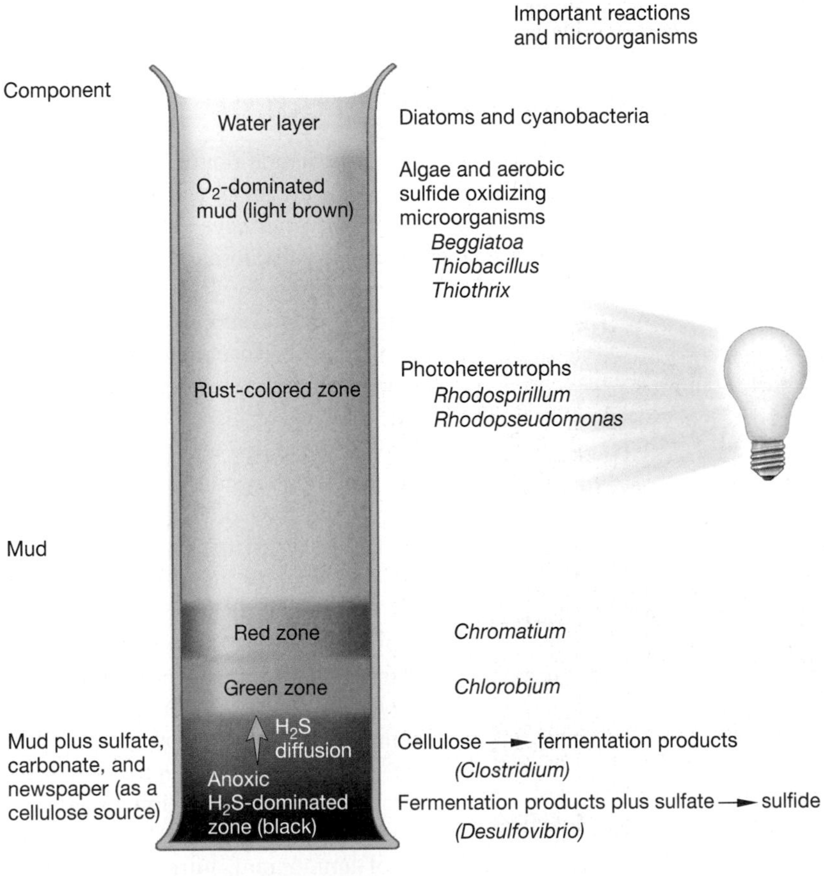

Figure 26.6 The Winogradsky Column. A microcosm in which microorganisms and nutrients interact over a vertical gradient. Fermentation products and sulfide migrate up from the reduced lower zone, and oxygen penetrates from the surface. This creates conditions similar to those in a lake or salt marsh with nutrient-rich sediments. Light is provided to simulate the penetration of sunlight into the anoxic lower region, which allows photosynthetic microorganisms to develop.

1. Define salt wedge and explain its influence on estuarine microbial communities.

2. Consider the fact that during droughts, rivers that normally flow into an estuary have a significantly diminished flow. Examine figure 26.4 and explain how this could result in salt intrusion into a public water supply. What do you think the consequences would be on estuarine microbial communities?

3. Name two groups of protists known to cause HABs in marine ecosystems. What hazards do HABs pose for other marine organisms and humans?

4. Describe the ecosystem that develops within a Winogradsky column. Trace the flow of electron donors and acceptors.

Microorganisms in the Photic Zone of the Open Ocean

Sometimes called "the invisible rain forest," the upper 200 to 300 m of the open ocean are home to a diverse collection of photosynthetic microbes. Metagenomic projects have demonstrated that this region hosts a vastly more diverse population of microbes than previously imagined (*see figure 15.15*). The use of satellite imagery to measure chlorophyll (**figure 26.7**) shows that although chlorophyll levels are lower in the sea than on land, the sheer volume of the oceans accounts for the fact that about half the world's photosynthesis is performed by marine microbes. The open ocean is an **oligotrophic environment**—that is to say, nutrient levels are very low. Because the influx of new nutrients is limited, primary productivity (and thus the whole ecosystem) depends on rapid recycling of nutrients. Unlike terrestrial ecosystems, at least 90% of the nutrients required by the microbial community are recycled within the microbial loop (figure 26.3).

Global warming has focused intense scrutiny on determining how much CO_2 phytoplankton can "draw down" out of the atmosphere and sequester in the benthos. As shown in **figure 26.8**, there is a constant exchange of CO_2 at the ocean surface, as well as limited export of carbon to the seafloor. Organic matter

escapes the photic zone and falls through the depths as **marine snow (figure 26.9)**. This material gets its name from its appearance, sometimes seen in video images of mid- and deep-ocean water, where drifting flocculent particles look much like snow. Marine snow consists primarily of fecal pellets, diatom frustules, and other materials that are not easily degraded. During its fall to the bottom, marine snow is colonized by microbes and mineralization continues. By the time it reaches the seafloor, less than 1% of photosynthetically derived materials remain unaltered. Once there, only a tiny fraction will be buried in the sediments; most will return to the surface in upwelling regions (figure 26.8). An urgent question is whether or not the oceans can draw down (remove and bury) more CO_2 as atmospheric concentrations continue to increase. One suggested approach to slowing global warming is to fertilize specific regions of the oceans. These regions, known as **high-nutrient, low-chlorophyll (HNLC) areas,** are limited by iron. A number of experiments have been performed wherein large transects of the southern Pacific were fertilized with iron to trigger diatom blooms. However, it is not apparent that this will be effective: while all studies show a temporary increase in primary production, only some show a measurable loss of carbon from the photic zone; others report no increase in CO_2 drawdown. Furthermore, many scientists are skeptical about the long-term effects of altering such a large and important ecosystem. And finally, many doubt that the amount of CO_2 that could be removed will be significant enough to impact the problem of global warming.

The cycling of other elements, in addition to carbon, occurs within the photic zone. In general, the open ocean is limited by nitrogen, not iron. Two recent discoveries have led marine microbiologists and oceanographers to reexamine the traditionally accepted nitrogen cycle. First, the filamentous cyanobacterium *Trichodesmium* fixes far more N_2 than previously thought. Second, unicellular cyanobacteria collected at sea are also fixing N_2. These microbes were known to fix nitrogen under laboratory conditions, but researchers assumed they did not do so in the open ocean. Thus there appears to be at least two recently identified sources of "new" organic nitrogen. The other recent discovery that has revolutionized our understanding of nitrogen cycling is the presence of bacteria that perform the anammox reaction below the photic zone, where oxygen concentrations reach a minimum. In nitrogen-limited areas, it is generally understood that there is a net loss of ammonium, nitrate, and nitrite. A large fraction of this loss has been attributed to denitrification (the anaerobic reduction of nitrate to N_2) occurring at the oxygen-minimum zones. It now

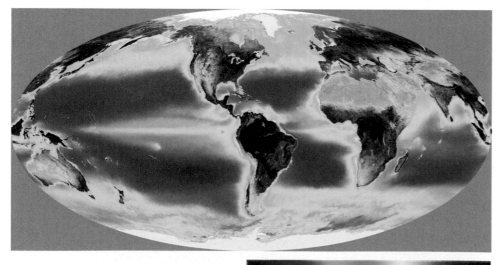

Figure 26.7 **Mean Annual Surface Chlorophyll Levels.** Global chlorophyll as measured by ocean satellites. The dark-blue regions correspond to subtropical gyres (massive spirals of water) where nutrient levels are especially low; note scale below image. *Image courtesy of NASA-Goddard Space Flight Center and the Orbital Sciences Corporation.*

Chlorophyll *a* Concentration (mg / m³)

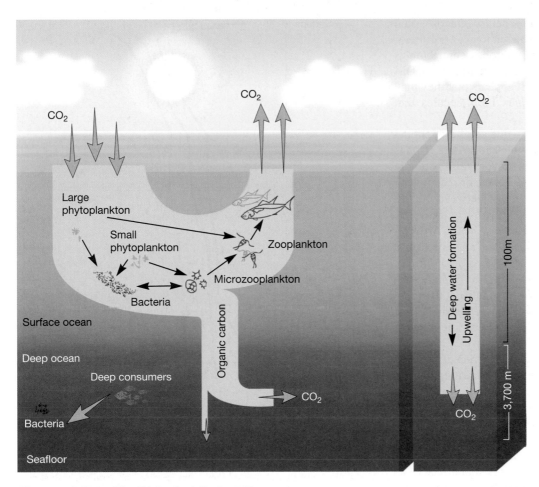

Figure 26.8 **The Biological Carbon Pump.** The vast majority of the carbon fixed by microbial autotrophs in the open ocean remains in the photic zone. However, a small fraction is "pumped" to the seafloor, which in turn returns much of it to the surface in regions where deep and surface waters mix (upwelling regions). The flux of CO_2 in and out of the world's oceans is in equilibrium, but this equilibrium may be upset by the increase in atmospheric CO_2, the hallmark of global warming. Note depth scale on right.

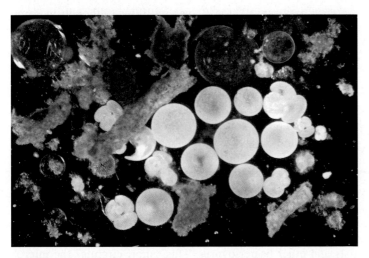

Figure 26.9 **Marine Snow.** The export of organic matter out of the photic zone to deeper water occurs through the sinking of marine snow. Shown here is material collected in a sediment trap at 5,367 m on the Sohm Abyssal Plain in the Sargasso Sea. It includes cylindrical fecal pellets, planktonic tests (round white objects), transparent snail-like pteropod shells, radiolarians, and diatoms.

appears that bacteria capable of the anammox reaction—the anaerobic oxidation of NH_4^+ to N_2—are responsible for much of the loss of nitrogen that could otherwise support life. These bacteria are members of the interesting phylum *Planctomycetes*. Once again, the urgency to understand the global oceanic nitrogen budget reflects concern over increasing levels of atmospheric CO_2, global warming, and the capacity of the world's oceans to remove more CO_2 while maintaining overall ecosystem equilibrium. << *Photosynthetic bacteria: Phylum* Cyanobacteria *(section 19.3); Phylum* Planctomycetes *(section 19.4); Biogeochemical cycling: Nitrogen cycle (section 25.1)*

It is clear that when researchers consider microbial activities in the open ocean, they are assessing global processes. So perhaps it is not surprising that the most abundant group of monophyletic organisms on Earth is marine. Members of the α-proteobacterial clade called **SAR11** (named after the *Sar*gasso Sea, where they were first detected) have been detected by rRNA gene cloning from almost all open-ocean samples taken worldwide. In addition, they have been found at depths of 3,000 m, in coastal waters, and even in some freshwater lakes. Using fluorescence in situ hybridization (FISH; *see figure 25.9*), SAR11 bacteria have been found to constitute 25 to 50% of the total procaryotic community in the surface waters in both nearshore and open-ocean samples. Indeed, SAR11 bacteria are estimated to constitute about 25% of all microbial life on the planet.

But what exactly are SAR11 microbes and what are they doing? The answer to these questions took about a decade to emerge because a member of the SAR11 clade could not be coaxed into growing in the laboratory. In 2002, several SAR11 bacteria were isolated in pure culture, and one isolate was named *Pelagibacter ubique*. *P. ubique* is vibrioid (comma-shaped) and only 0.4 to 0.9 μm in length, making it one of the smallest known, free-living microbes. Unlike most microbes in culture, which grow to densities exceeding 10^8 cells per milliliter, SAR11 isolates stop replicating after reaching about 10^6 cells per milliliter. Curiously, this is the density at which they are generally found in nature, suggesting that natural factors in seawater somehow control population growth. By measuring how much and how fast SAR11 microbial isolates assimilate radiolabeled amino acids, glucose, and complex biomolecules, scientists have begun to get a picture of their role in the microbial loop. These microbes contribute as much as 50% to the bacterial biomass production and DOM flux in some marine environments, and it appears that they may selectively degrade the kinds of complex biomolecules that comprise marine snow.

The genome of *P. ubique* is 1.31 Mb—the smallest genome of any independently replicating cell sequenced to date. This is partly because the SAR11 genome is devoid of "genomic waste." There are no pseudogenes, phage genes, or recent gene duplications, and the number of nucleotides between coding regions is very limited. Nonetheless, *P. ubique* has the genes necessary for the Entner-Doudoroff pathway, the TCA cycle, and a complete electron transport chain. It has adapted to life in the oligotrophic ocean by encoding a number of high-affinity nutrient transport systems and maintaining its small size, thereby optimizing its surface area-to-volume ratio. It uses respiration to capture energy, but it also has a proteorhodopsin proton pump that may also contribute to the generation of a proton motive force.

One of the most interesting discoveries in recent years is the widespread presence of large numbers of marine archaea, once thought to inhabit only extreme environments. In fact, the *Archaea* are ubiquitous in the marine environment—they have been found in polar and tropical regions, and in estuarine, planktonic, and deep-sea communities. Archaea are also found in freshwater environments. 16S rRNA analysis, the detection of archaeal specific lipids (e.g., crenarchaeol), and direct visualization of archaea using epifluorescence have revealed that at least 20% of oceanic picoplankton are *Crenarchaeota*. Many of these populations are closely related, suggesting that a shared collection of adaptations has been fine-tuned to meet the needs of the different niches within specific picoplankton communities. The distribution of archaea relative to that of bacteria differs widely with the particular ecosystem. **Figure 26.10** shows typical results for the open ocean. Bacteria are most numerous in the upper 150 to 200 m (i.e., the photic zone), but archaea increase in relative abundance with depth until they approximate, or sometimes exceed, bacteria.

As mentioned in our discussion of the microbial loop, viruses are important members of marine and freshwater microbial communities. In fact, **virioplankton** are the most numerous members of marine ecosystems. However, quantifying viruses is tricky: the traditional method of plaque formation requires knowledge of both virus and its host, and the ability to grow the host in the laboratory. Because so few microbes have been cultured, this prevents the measurement of virus diversity by examining actual virus infection. Instead, virus particles may be visualized directly. This requires the concentration of many tens of liters of water for direct examination by transmission electron microscopy. Because this does not prove that any given virus can actually infect a host cell, viruses enumerated in this way are called *virus/like particles* (**VLPs**). Using this approach, the average VLP density in sea-

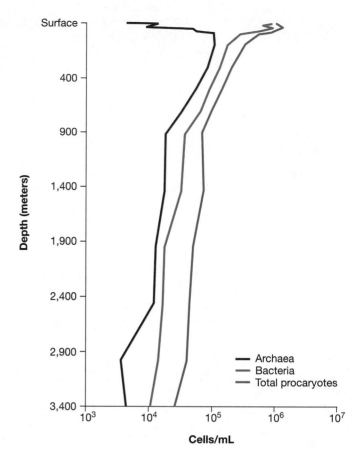

Figure 26.10 **Archaea Are Plentiful in Ocean Depths.** The distribution of archaea and bacteria, together with an estimate of total procaryotes, over a depth of 3,400 m is shown at a Pacific Ocean location. These results indicate that archaea make up a significant part of the observable picoplankton below the surface zone.

water is between 10^6 to 10^7 per milliliter (although in some cases it may be closer to 10^8 per milliliter); their numbers decline to roughly 10^6 below about 250 m. Marine viruses are so abundant that VLPs are now recognized as the most abundant microbes on Earth. As might be expected by their numbers, these viruses are very diverse, including single- and double-stranded RNA and DNA viruses that infect archaea, bacteria, and protists. Viral lysis of bacteria and archaea in the field is difficult to measure but appears to account for between 10 and 50% of the total mortality of the procaryotic members of the marine microbial community. Computer modeling and model experiments indicate that viruses contribute to nutrient cycling by accelerating the rate at which their microbial hosts are converted to POM and DOM, thereby "feeding" other microorganisms without first making them available for protists and other bacteriovores. This "short-circuits" the microbial loop (**figure 26.11**).

Genome sequencing of DNA cloned directly from marine ecosystems suggests that phages are important vectors for horizontal gene transfer. In fact, it has been calculated that in the oceans, phage-mediated gene transfer occurs at an astounding

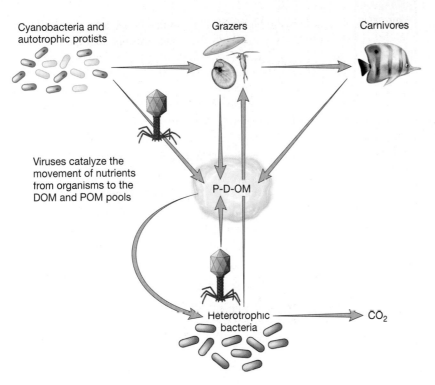

Cyanobacteria and
autotrophic protists

Grazers

Carnivores

Viruses catalyze the
movement of nutrients
from organisms to the
DOM and POM pools

P-D-OM

Heterotrophic
bacteria

CO_2

Figure 26.11   The Role of Viruses in the Microbial Loop.   Viral lysis of autotrophic and heterotrophic microbes accelerates the rate at which these microbes are converted to particulate and dissolved organic matter (P-D-OM). This is thought to increase net community respiration and decrease the efficiency of nutrient transfer to higher trophic levels.

rate of 20 billion times per second. Recently the importance of phage-mediated lateral gene transfer was demonstrated when it was discovered that cyanophages that infect the cyanobacteria *Synechococcus* and *Prochlorococcus* carry the structural gene for a photosynthetic reaction center protein. These phages are presumably shuttling this essential gene between the two genera of cyanobacteria and may thereby play a critical role in the evolution of these important primary producers.

Microorganisms in Benthic Marine Environments

The majority of the Earth's crust is under the sea, which means that of all the world's microbial ecosystems, we know the least about the largest. However, the combination of deep-ocean drilling projects and the exploration of geologically active sites, such as submarine volcanoes and hydrothermal vents, has revealed that the study of ocean sediments, or benthos, can be rewarding and surprising. Marine sediments range from the very shallow to the deepest trenches, from dimly illuminated to completely dark, and from the newest sediment on Earth to material that is millions of years old. The temperature of such sediments depends on their proximity of geologically active areas. Hydrothermal vent communities with large and diverse invertebrates, some of which depend on endosymbiotic chemolithotrophic bacteria, have been intensely investigated since their exciting discovery in the late 1970s. These microbes are

discussed in chapter 27. Because the vast majority of Earth's crust lies at great depth far from geothermally active regions, most benthic marine microbes live under high pressure, without light, and at temperatures between 1°C to 4°C.

Deep-ocean sediments were once thought to be devoid of all life and were therefore not thought worth the considerable effort it takes to study them. In fact, it took an international consortium of scientists to organize the Ocean Drilling Project in 1985, which has now been expanded to the Integrated Ocean Drilling Program through 2013. Researchers aboard the research vessel *JOIDES Resolution* drill cores of sediments from water depths up to 8,200 m (at its deepest, the ocean is about 11,000 m). Microbiologists now know that far from being sterile, it appears the total subsurface (intraterrestrial) microbial biomass equals that of all terrestrial and marine plants. This is possible because benthic marine microbes inhabit not just the surface of the seafloor but within sediments to a depth of least 0.6 km. To survive at these depths, microorganisms must be able to tolerate atmospheric pressures up to 1,100 atm (pressure increases about 1 atm/10 m depth). Such microbes are said to be **barophilic** (Greek, *baro*, weight, and *philein*, to love). Some are obligate barophiles and must be cultured in special hyperbaric incubation chambers. In fact, scientists have yet to find the subterranean depth limit of microbial growth. It appears that this value will not be governed by pressure; rather, it will be determined by temperature. It seems unlikely that the maximum temperature at which life can be sustained has been identified.

One outcome of deep-ocean sediment exploration has been the discovery of **methane hydrates.** These pools of trapped methane are produced by archaea that convert acetate to methane. This accumulates in latticelike cages of crystalline water 500 m or more below the sediment surface in many regions of the world's oceans. The formation of methane hydrates requires both cold temperatures and high pressure. This discovery is very significant because there may be up to 10^{13} metric tons of methane hydrate worldwide—80,000 times the world's current known natural gas reserve.   ◀◀ *Phylum Euryarchaeota (section 18.3).*

It is also exciting that recent deep-sea sediment drilling has turned our understanding of bacterial energetics literally upside down. As discussed in chapter 10, it is generally understood that anaerobic respiration occurs such that there is preferential use of available terminal electron acceptors. Acceptors that yield the most energy (more negative ΔG) from the oxidation of NADH or an inorganic reduced compound (e.g., H_2, H_2S) will be used before those that produce a less negative ΔG (*see table 9.1 and figure 9.5*). Thus following oxygen depletion, nitrate will be reduced; then manganese, iron, sulfate, and finally carbon dioxide. When sediment cores up to 420 m deep were collected off the coast of Peru, this predictable profile of electron acceptors and their microbial-derived

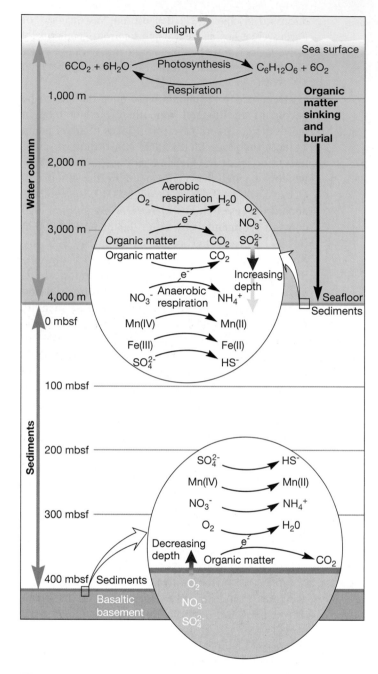

Figure 26.12 Microbial Activity in Deep-Ocean Sediments.
At the surface of the seafloor, reduction of oxidized substrates that serve as electron acceptors in anaerobic respiration follows a predictable stratification based on thermodynamic considerations. This sequence is just the opposite in very deep subsurface sediments, suggesting a source of electron acceptors from deep within Earth's crust. Meters below seafloor, mbsf.

reduced products was observed within the upper strata of the sediments (**figure 26.12**). However, when these signature chemical compounds were measured at great depth, the profile was reversed. This suggests the presence of unknown sources of these electron acceptors at subsurface depths of more than 420 m. In addition, contrary to our long-held notion of thermodynamic limits, methane formation (methanogenesis) and iron and manganese reduction

co-occur in sediments with high sulfate concentrations. Although the identity of the microbes that make up this community awaits further study, it is clear that with densities of 10^8 cells per gram of sediment at the seafloor surface and 10^4 cells per gram in deep subsurface sediments, these communities are important. << *Free energy and reactions (section 9.3); Oxidation-reduction reactions (section 9.5); Anaerobic respiration (section 10.6)*

1. What is marine snow? Why is it important in CO_2 drawdown?
2. Name two sources of organic nitrogen in the open ocean that have only recently been recognized.
3. Why do you think that despite its great abundance, SAR11 was not discovered until the late twentieth century?
4. Describe the role of marine viruses in the microbial loop.
5. What are methane hydrates? Why are they important?
6. Explain what is meant by "upside-down microbial energetics" as described for some deep subsurface ocean sediments.

Microorganisms in Glaciers and Permanently Frozen Lakes

We begin our discussion of freshwater microbes with those that reside in ice that has remained frozen for thousands of years. Although this may seem like an extreme environment, it is important to note that a majority of the Earth's surface never exceeds a temperature of 5°C. This includes polar regions, the deep ocean, as well as high-altitude terrestrial locations throughout the world. Surprisingly, microbes within glaciers are not dormant. Rather, evidence that active microbial communities exist in these environments has been growing over the last decade. In fact, this is an exciting time in glacial microbiology; determining the diversity in these systems and assessing the role of these microorganisms in biogeochemical cycling can involve novel and creative techniques. The results may be of great consequence because glaciers have traditionally been regarded as areas that do not contribute to the global carbon budget. In addition, since the discovery of ice on Mars and on Jupiter's moon Europa, astrobiologists have become very interested in ice-dwelling psychrophilic microbes. << *Influences of environmental factors on growth: Temperature (section 7.5)*

Among the frozen landscapes of interest to microbiologists are permanently frozen lakes, such as Antarctica's McMurdo Dry Valley Lakes, where the ice is 3 to 6 m deep. Life in these ecosystems depends on the photosynthetic activity of microbial psychrophiles. In contrast, lakes that lie below glaciers are blocked from solar radiation. These communities are driven by chemosynthesis. One of the most intriguing and well-studied Antarctic habitats is Lake Vostok, one of 68 lakes located 3 to 4 km below the East Antarctic Ice Sheet (**figure 26.13**). Geothermal heating, pressure, and the insulation of the overlying

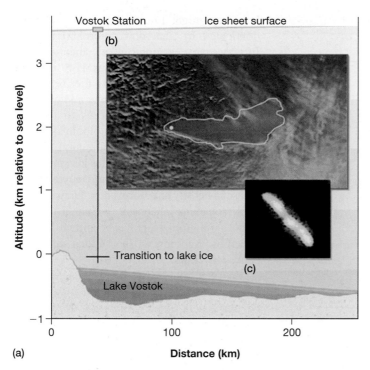

(a)

Figure 26.13 The East Antarctic Ice Sheet and Lake Vostok. (a) Lake Vostok lies beneath several kilometers of the ice sheet. (b) The deep drilling station from which ice and lake water samples were obtained is indicated by the yellow dot. The yellow outline identifies a smooth plateau of snow and ice that floats on top of Lake Vostok. The surrounding rough ice is formed when the ice sheet moves over bedrock rather than liquid water. (c) Microbes have been found in both the overlying ice and the lake water, as seen in this epifluorescent image.

ice keep these lakes in a liquid state. It is thought that Lake Vostok was formed approximately 420,000 years ago and that its water is about a million years old. This stable environment supports a number of microbes, including proteobacteria and actinomycetes. The overlying ice also harbors similar microbes, although at lower population densities. To find out if these microbial populations are active, radiolabeled substrates, including ^{14}C-acetate and ^{14}C-glucose, were added to samples incubated at an Antarctic laboratory. Indeed, the respiration of these compounds, measured as ^{14}C-CO_2, demonstrates that these interesting communities are dynamic.

Microorganisms in Streams and Rivers

As freshwater glaciers melt, their waters enter streams and rivers. This marks a departure from an environment that is stable on a geologic time scale to one that is extremely changeable. The continuous flow of water in streams and all but the largest rivers prevents the development of significant planktonic communities. Instead, most of the microbial biomass is attached to surfaces. Depending on the size of the stream or river, the source of nutrients may vary. The source may be in-stream, called **autochthonous** production based on photosynthetic microorganisms.

Nutrients also may come from outside the stream, including runoff sediment from riparian areas (the edge of a river) or leaves and other organic matter falling directly into the water. Such nutrients are called **allochthonous.** Chemoorganotrophic microorganisms metabolize the available organic material, recycling nutrients within the ecosystem. Autotrophic microorganisms grow using the minerals released from the organic matter. This leads to the production of O_2 during the daylight hours; respiration occurs at night, resulting in diurnal oxygen shifts. Thus when the amount of organic matter added to streams and rivers does not exceed the system's oxidative capacity, productive streams and rivers are maintained.

The capacity of streams and rivers to process added organic matter is limited. If too much organic matter is added, the system is said to be **eutrophic,** and the rate of respiration exceeds that of photosynthesis. Thus the water may become anoxic. This can be the case in streams and rivers adjacent to urban and agricultural areas. The release of inadequately treated municipal wastes and other materials from a specific location along a river or stream represents **point source pollution.** These additions of organic matter can produce distinct and predictable changes in the microbial community and available oxygen, creating an oxygen sag curve in which oxygen is depleted just downstream of the pollution source (**figure 26.14**). Runoff from agriculturally active fields and feedlots is an example of **nonpoint source pollution.** This can also cause disequilibrium in the microbial community leading to algal or cyanobacterial blooms.

1. Describe the Lake Vostok ecosystem. Why is this a chemosynthesis-based, rather than a photosynthetically based, microbial community?
2. What is an oxygen sag curve? What changes in a river cause this effect?
3. What are point and nonpoint source pollution? Can you think of examples in your community?

Microorganisms in Lakes

Lakes offer a completely different set of physical and biological features for microbial communities. Lakes vary in nutrient status. Some are oligotrophic (**figure 26.15a**), others are eutrophic (figure 26.15b). Nutrient-poor lakes remain oxic throughout the year, and seasonal temperature shifts usually do not result in distinct oxygen stratification. In contrast, **eutrophic** lakes usually have bottom sediments that are rich in organic matter. In thermally stratified lakes, the **epilimnion** (warmer, upper layer) is oxic, while the **hypolimnion** (colder, bottom layer) often is anoxic (particularly if the lake is nutrient-rich). The epilimnion and hypolimnion are separated by a thermocline.

Because little mixing occurs between the epilimnion and the hypolimnion, the bottom waters may become deprived of oxygen. This is a permanent situation in tropical eutrophic lakes and occurs in the summer in temperate eutrophic lakes. If nutrient

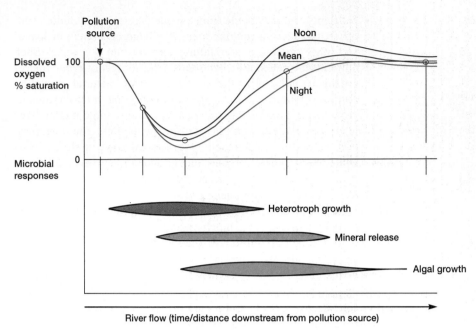

Figure 26.14 The Dissolved Oxygen Sag Curve. Microorganisms and their activities can create gradients over distance and time when nutrients are added to rivers. An excellent example is the dissolved oxygen sag curve, caused when organic wastes are added to a clean river system. During the later stages of self-purification, the phototrophic community will again become dominant, resulting in diurnal changes in river oxygen levels.

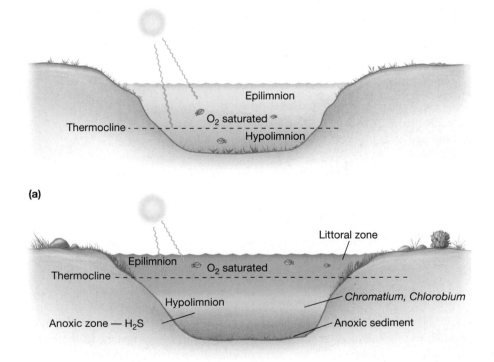

Figure 26.15 Oligotrophic and Eutrophic Lakes. Lakes can have different levels of nutrients, ranging from low nutrient to extremely high nutrient systems. The comparison of (a) an oligotrophic (nutrient-poor) lake, which is oxygen saturated and has a low microbial population, with (b) a eutrophic (nutrient-rich) lake. The eutrophic lake has a bottom sediment layer and can have an anoxic hypolimnion. As microbial biomass increases with nutrient levels, light penetration is diminished. Thus the bottom of eutrophic lakes may become dark, anoxic, and even poisonous from H$_2$S production.

levels are high, this benthic zone becomes dominated by anaerobic microbial activity. In very warm eutrophic lakes, anaerobic microbes release gases such as H$_2$S into the water. In addition, human activities (e.g., septic and agricultural runoff) may add high levels of nitrogen and phosphorus to the lake waters. This can result in a bloom of algae, plants, or bacteria in the epilimnion. During autumn cooling, temperate lakes lose their thermocline because cooling surface waters increase in density and storms mix the two layers. This sometimes happens within a 24-hour period; if bottom water has become filled with anaerobic byproducts, the sudden upwelling causes fish kills.

In oligotrophic lakes, cyanobacteria that are capable of nitrogen fixation may bloom. Several genera, notably *Anabaena*, *Nostoc*, and *Cylindrospermum*, can fix nitrogen under oxic conditions. The genus *Oscillatoria*, using hydrogen sulfide as an electron donor for photosynthesis, can fix nitrogen under anoxic conditions. If both nitrogen and phosphorus are present, cyanobacteria compete with algae. Cyanobacteria function more efficiently in alkaline waters (pH 8.5 to 9.5) and higher temperatures (30 to 35°C). By using CO$_2$ at rapid rates, cyanobacteria also increase the pH, making the environment less suitable for protists. << *Photosynthetic bacteria: Phylum Cyanobacteria (section 19.3)*

Cyanobacteria have additional competitive advantages. Many produce hydroxamates, which bind iron, making this important trace nutrient less available for protists. Some cyanobacteria also resist predation because they produce toxins. In addition, some synthesize odor-producing compounds that affect the quality of drinking water. However, both cyanobacteria and protists (algae) can contribute to massive blooms in strongly eutrophied lakes. Lake management can improve the situation by removing or sealing bottom sediments or adding coagulating agents to speed sedimentation.

1. What terms can be used to describe the different parts of a lake?

2. What are some important effects of eutrophication on lakes?

3. Why do you think cyanobacteria are so important in waters that have been polluted by phosphorus additions?

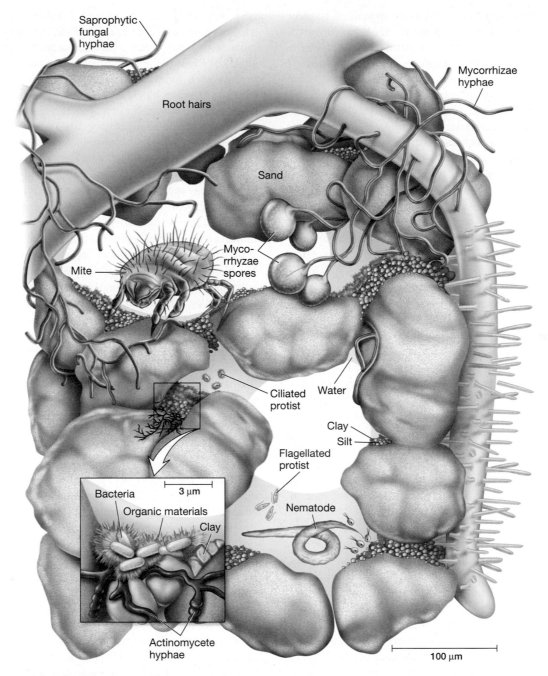

Figure 26.16 **The Soil Habitat.** A typical soil habitat contains a mixture of clay, silt, and sand along with soil organic matter. Roots, animals (e.g., nematodes and mites), as well as chemoorganotrophic bacteria consume oxygen, which is rapidly replaced by diffusion within the soil pores where the microbes live. Note that two types of fungi are present: mycorrhizal fungi, which derive their organic carbon from their symbiotic partners—plant roots; and saprophytic fungi, which contribute to the degradation of organic material.

26.2 MICROORGANISMS IN TERRESTRIAL ENVIRONMENTS

A soil scientist would describe soil as weathered rock combined with organic matter and nutrients. An agronomist would point out that soil supports plant life. However, a microbial ecologist knows that the formation of organic matter and the growth of plants depend on the microbial community within the soil. As we shall see,

terrestrial microbiology—the study of those microbes found in soil and associated with plants—is an important and dynamic field.

Soils as a Microbial Habitat

Most soils are dominated by inorganic geological materials, which are modified by the biotic community, including microorganisms and plants, to form soils. The spaces between soil particles are critical for the movement of water and gases (**figure 26.16**). Total pore space, and thus gas diffusion, is determined by the texture of the soil. For instance, sandy soils have larger pore spaces than do clay soils, so sandy soils tend to drain more quickly. Pores are also critical because they provide the optimum environment for microbial growth. Here, microbes are within thin water films on particle surfaces where oxygen is present at high levels and can be easily replenished by diffusion. The oxygen concentrations and flux rates in pores and channels are high, whereas within water-filled zones, the oxygen flux rate is much lower. For example, particles as small as about 2.0 μm can be oxic on the outside and anoxic on the inside.

Depending on the physical characteristics of the soil, rainfall or irrigation may rapidly change a soil from being well aerated to an environment with isolated pockets of water, which are "miniaquatic" habitats. If flooding continues, a waterlogged soil can be created that is more like lake sediment. If oxygen consumption exceeds that of oxygen diffusion, waterlogged soils can become anoxic. Shifts in water content and gas fluxes also affect the concentrations of CO_2, CO, and other gases. These changes are accentuated in the smaller pores where many bacteria are found. The roots of plants growing in aerated soils also consume oxygen and release CO_2, influencing the concentrations of these gases in the root environment.

Soils can be divided into two general categories: A **mineral soil** contains less than 20% organic carbon, whereas an **organic soil** possesses at least this amount. By this definition, the vast majority of Earth's soils are mineral. The importance of organic matter within soils cannot be underestimated. **Soil organic matter (SOM)** helps to retain nutrients, maintain soil structure, and hold water for plant use. SOM is subject to gains and losses, depending on changes in environmental conditions and agricultural management practices. Plowing and other disturbances expose SOM to more oxygen, leading to extensive microbiological degradation of organic matter. Irrigation causes periodic wetting and drying, which can also lead to increased degradation of SOM, especially at higher temperatures.

Microbial degradation of plant material results in the evolution of CO_2 and the incorporation of plant carbon into additional microbial biomass. However, a small fraction of the decomposed plant material remains in the soil as SOM. When considering this material, it is convenient to divide the SOM into humic and nonhumic fractions (**table 26.1**). Nonhumic SOM has not undergone significant biochemical degradation. It can represent up to about 20% of the soil organic matter. Humic SOM, or humus, results when the products of microbial metabolism have undergone chemical transformation within the soil. Although it has no precise chemical composition, humus can be described as a complex blend of phenolic compounds, polysaccharides, and proteins. The recalcitrant nature of this material to degradation is evident by ^{14}C dating: the average age of most SOM ranges from 150 to 1,500 years.

The degradation of plant material and the development of SOM can be thought of as a three-step process. First, easily degraded compounds such as soluble carbohydrates and proteins are broken down. About half the carbon is respired as CO_2 and the remainder is rapidly incorporated into new biomass. During the second stage, complex carbohydrates, such as the plant structural polysaccharide cellulose, are degraded. Fungi and bacterial genera such as *Streptomyces, Pseudomonas,* and *Bacillus* produce extracellular cellulase enzymes that break down cellulose into two to three glu-

cose units called cellobiose and cellotriose, respectively. These smaller compounds are readily degraded and assimilated as glucose monomers. Finally, very resistant material, in particular lignin, is attacked. **Lignin** is an important structural component of woody plants. While its exact structure differs among plant species, the common building block is the phenylpropene unit. This consists of a hydroxylated six-carbon aromatic benzene ring and a three-carbon linear side chain (**figure 26.17**). A single lignin molecule can consist of up to 600 cross-linked phenylpropene units. It is therefore not surprising that lignin degradation is much slower than that of cellulose. Basidiomycete fungi and actinomycetes (e.g., *Streptomyces* spp.) are capable of extracellular lignin degradation. These microbes produce extracellular phenoloxidase enzymes needed for aerobic lignin degradation. Lignin decomposition is also limited by the physical nature of the material. For example, healthy woody plants are saturated with sap, which limits oxygen diffusion. In addition, high ethylene and CO_2 levels and the presence of phenolic and terpenoid compounds retard the growth of lignin-degrading microbes. It follows that no more than 10% of the carbon found in lignin is recycled into new microbial biomass. Lignin can be degraded anaerobically, but this process is very slow, so lignin tends to accumulate in wet, poorly oxygenated soils, such as peat bogs.

Nitrogen is another important element in soil ecosystems. Nitrogen in soil is often considered in relation to the soil carbon content as the organic **carbon to nitrogen ratio (C/N ratio).** A C/N ratio of 20 or less results in loss of soluble nitrogen from the system. Conversely ratios above 20 enable microbes to convert ammonium and nitrate to biomass (e.g., proteins and nucleic acids). Many soils are nitrogen limited; this is why each year tons of nitrogen fertilizer are added to agricultural soils. The major types of nitrogen fertilizers used in agriculture are liquid ammonia and ammonium nitrate. Ammonium ion usually is added because it will be attracted to the negatively charged clays in a soil and be retained on the clay surfaces until used as a nutrient by the plants. However, the nitrifier populations in a soil can oxidize the ammonium ion to nitrite and nitrate, and these anions can be leached from the plant environment and enter surface waters and groundwaters. ◀◀ *Biogeochemical cycling: Nitrogen cycle (section 25.1)*

Phosphorus in fertilizers also is critical. The binding of this anionic fertilizer component to soils depends on the cation exchange capacity and soil pH. As the soil's phosphorus sorption capacity is reached, the excess, together with phosphorus that moves to lakes, streams, and estuaries with soil erosion, can stimulate the growth of freshwater organisms, particularly nitrogen-fixing cyanobacteria, in the process of eutrophication.

1. What is the importance of soil pores?
2. List three reasons why soil organic matter (SOM) is important.
3. What is the difference between humic and nonhumic SOM? Which is more abundant in most soils? Why?
4. Describe the three phases of plant degradation and SOM formation.

Table 26.1	Fractions of Soil Organic Matter	
SOM Fraction	**Definition**	**Physical Appearance**
Humic substances	High molecular-weight organic material produced by secondary synthesis reactions	Dark brown to black
Nonhumic substances	Unaltered remains of plants, animals, and microbes, from which macromolecules have not yet been extracted	Light brown
Humic acid	Organic matter extracted from soils by various reagents (often dilute alkali treatment) that is then precipitated by acidification	Dark brown to black
Fulvic acid	Soluble organic matter that remains after humic acid extraction	Yellow
Humin	SOM that cannot be extracted from soil with dilute alkali	

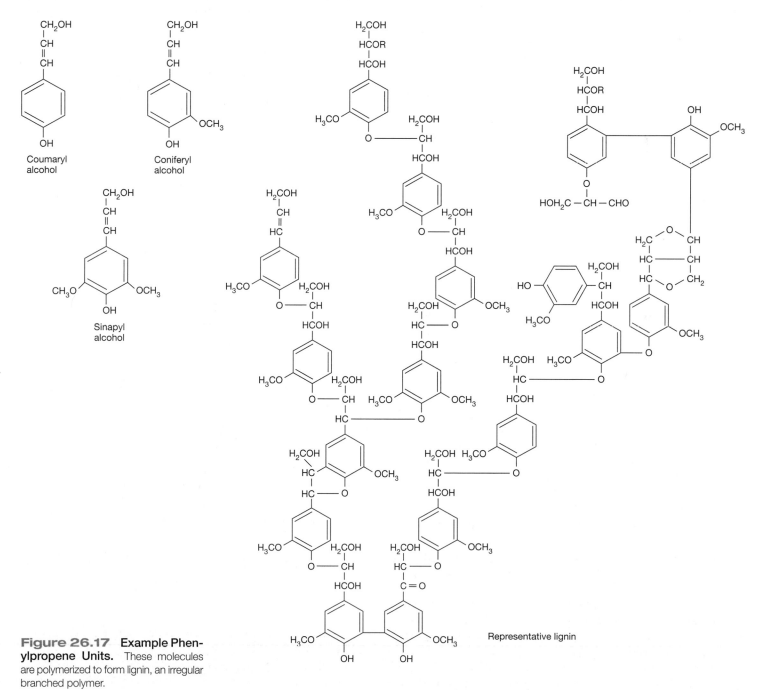

Figure 26.17 Example Phenylpropene Units. These molecules are polymerized to form lignin, an irregular branched polymer.

5. What are possible effects of nitrogen-containing fertilizers on microbial communities?

6. Most nitrogen fertilizer is added as ammonium ion. Why is this preferred over nitrate?

7. Why might enrichment of freshwater with phosphorus be even more critical than nitrogen enrichment?

Microorganisms in the Soil Environment

If we look at a soil in greater detail (figure 26.16), we find that bacteria, archaea, fungi, and protists use different functional strategies to take advantage of this complex physical matrix. Most soil procaryotes are located on the surfaces of soil particles, where water and nutrients are in their immediate vicinity. Procaryotes are found most frequently on surfaces within smaller soil pores (2 to 6 μm in diameter). Here, they are probably less liable to be eaten by protozoa, unlike those located on the exposed outer surface of a sand grain or organic matter particle.

Filamentous fungi, in comparison, bridge open areas between soil particles or aggregates and are exposed to high levels of oxygen. These fungi tend to darken and form oxygen-impermeable structures called sclerotia and hyphal cords. This is particularly important for basidiomycetes, which form such structures as an oxygen-sealing mechanism. Within these structures, the filamentous fungi move nutrients and water over great distances, including across air spaces, a unique part of their functional strategy. << *Characteristics of the fungal divisions:* Basidiomycota *(section 23.3)*

The microbial populations in soils can be very high. In a surface soil, the bacterial population can approach 10^9 to 10^{10} cells per gram dry weight of soil as measured microscopically. Fungi can be present at up to several hundred meters of hyphae per gram of soil. We tend to think of soil fungi as small structures such as the mushrooms sprouting on our lawns. However, the vast majority of fungal biomass is below ground. For instance, an individual clone of the fungus *Armillaria bulbosa,* which lives associated with tree roots in hardwood forests, covers about 30 acres in the Upper Peninsula of Michigan. It is estimated to weigh a minimum of 100 tons (an adult blue whale weighs about 150 tons) and be at least 1,500 years old. Thus some fungal mycelia are among the largest and most ancient living organisms on Earth.

Soil microbial communities appear to be more diverse than those found in most freshwater and marine environments. The *Crenarchaeota* have been discovered in soil by analysis of 16S rRNA sequences extracted from natural habitats. Similar analysis has revealed that microorganisms are present and prolific in subsurface environments, including oil reservoirs. Hyperthermophilic archaea have been found in such harsh subsurface environments and are probably indigenous to these poorly understood regions. << *Phylum* Crenarchaeota *(section 18.2)*

The coryneforms, nocardioforms, and streptomycetes (**table 26.2**) are an important part of the soil microbial community. These

Table 26.2	Easily Cultured Gram-Positive Irregular Branching and Filamentous Bacteria Common in Soils	
Bacterial Group	**Representative Genera**	**Comments and Characteristics**
Coryneforms	*Arthrobacter*	Rod-coccus cycle
	Cellulomonas	Important in degradation of cellulose
	Corynebacterium	Club-shaped cells
Mycobacteria	*Mycobacterium*	Acid-fast
Nocardioforms	*Nocardia*	Rudimentary branching
Streptomycetes	*Streptomyces*	Aerobic filamentous bacteria
Bacilli	*Thermoactinomyces*	Higher temperature growth

gram-positive bacteria play a major role in the degradation of hydrocarbons, older plant materials, and soil humus. In addition, some members of these groups actively degrade pesticides. The filamentous actinomycetes, primarily of the genus *Streptomyces,* produce an odor-causing compound called geosmin, which gives soils their characteristic earthy odor. Polyprosthecate bacteria such as the genera *Verrucomicrobium, Pedomicrobium,* and *Prosthecobacter* are present in soils at high levels. With their small size, and the difficulties involved in culturing them, they have been largely overlooked in assessments of soil microbial diversity. << *High G + C gram-positive bacteria (chapter 22)*

Nutrients are regenerated in soils through a microbial loop that differs from that which operates in aquatic ecosystems. A major distinction is that plants rather than microbes account for most primary production in nearly all terrestrial systems. But much like the microbial loop in marine waters, microbes in soils rapidly recycle the organic material derived from plants and animals, including the many nematodes and insects, which function to reduce the size of solid organic matter. In turn, microbes themselves are preyed upon by soil protists, whose numbers can reach 100,000 per gram of soil. This makes microbial organic matter available to other trophic levels. Another difference between marine and terrestrial microbial loops reflects the physical and biological properties of soil. Degradative enzymes released by plants, insects, and other animals do not rapidly diffuse away. Instead, they represent a significant contribution to the biological activity in soil ecosystems by contributing to the many hydrolytic degradation reactions needed for the assimilation and recycling of nutrients.

1. What are the differences in preferred soil habitats between bacteria and filamentous fungi?

2. What types of archaea have been detected in soils?

3. How does the microbial loop function in soils? Compare this with its function in aquatic systems.

Microorganisms and the Formation of Different Soils

Soil formation is the result of the combined action of weathering and colonization of geologic material by microbes. The microbial community that initially colonizes a newly disturbed environment will bring changes to the local environment by degrading and recycling organic material. The major types of plant-soil systems are shown in **figure 26.18** and are discussed here.

Tropical and Temperate Region Soils

In warm, moist tropical soils, organic matter is decomposed very quickly, and mobile inorganic nutrients can be leached out of the surface soil environment, causing a rapid loss of fertility. To limit nutrient loss, many tropical plant root systems penetrate the rapidly decomposing litter layer. As soon as organic material and minerals are released during decomposition, the roots take them up to avoid losses in leaching. Thus it is possible to recycle nutrients before they are lost due to water movement through the soil (figure 26.18*a*). With deforestation, there is no leaf litter so nutrients are not recycled, leading to their loss from the soil and decreased soil fertility.

Tropical plant-soil communities are often used in **slash-and-burn agriculture.** The vegetation on a site is chopped down and burned to release the trapped nutrients. For a few years, until the minerals are washed from the soils, crops can be grown. When the minerals are lost from these low organic matter soils, the farmer must move to a new area and start over by again cutting and burning the native plant community. Slash-and-burn agriculture is stable if there is sufficient time for the plant community to regenerate

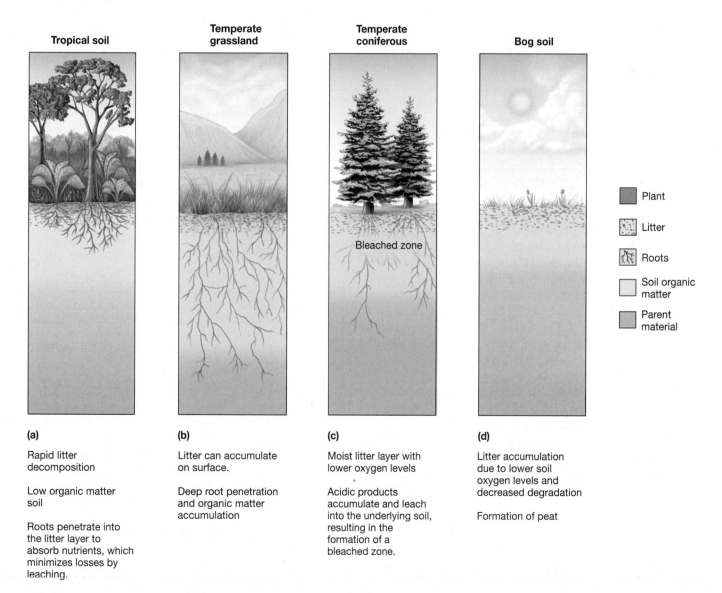

Tropical soil

Temperate grassland

Temperate coniferous

Bog soil

Bleached zone

Plant

Litter

Roots

Soil organic matter

Parent material

(a)

Rapid litter decomposition

Low organic matter soil

Roots penetrate into the litter layer to absorb nutrients, which minimizes losses by leaching.

(b)

Litter can accumulate on surface.

Deep root penetration and organic matter accumulation

(c)

Moist litter layer with lower oxygen levels

Acidic products accumulate and leach into the underlying soil, resulting in the formation of a bleached zone.

(d)

Litter accumulation due to lower soil oxygen levels and decreased degradation

Formation of peat

Figure 26.18 **Examples of Tropical and Temperate Region Plant-Soil Biomes.** In these figures, the characteristics of (a) tropical, (b) temperate grassland, (c) temperate coniferous forest, and (d) bog soils are illustrated.

before it is again cut and burned. If the cycle is too short, rapid and almost irreversible degradation of the soil occurs.

In many temperate region soils, in contrast, the decomposition rates are less than that of primary production, leading to litter accumulation. Deep root penetration in temperate grasslands results in the formation of fertile soils, which provide a valuable resource for the growth of crops in intensive agriculture (figure 26.18*b*).

The soils in many cooler coniferous forest environments suffer from an excessive accumulation of organic matter as plant litter (figure 26.18*c*). In winter, when moisture is available, the soils are cool, and this limits decomposition. In summer, when the soils are warm, water is not as available for decomposition. Organic acids are produced in the cool, moist litter layer, and they leach into the underlying soil. These acids solubilize soil components such as aluminum and iron, and a bleached zone may form. Litter continues to accumulate, and fire becomes the major means by which nutrient cycling is maintained. Controlled burns have become part of the environmental management in this type of plant-soil system.

Bog soils provide a unique set of conditions for microbial communities (figure 26.18*d*). In these soils, decomposition is slowed by the waterlogged, predominantly anoxic conditions, which lead to peat accumulation. When such areas are drained, they become more oxic and SOM is degraded, resulting in soil subsidence. Under oxic conditions, the lignin-cellulose complexes of the accumulated organic matter are more susceptible to decomposition by filamentous fungi.

Cold Moist Soils

Soils in cold environments, whether in Arctic, Antarctic, or alpine regions, are of extreme interest because of their wide distribution and impacts on global-level processes. The colder soil temperatures at these sites decrease the rates of decomposition and plant growth. In these cases, SOM accumulates, and plant growth can become limited due to the immobilization of nutrients. Often, below the plant growth zone, these soils are permanently frozen. These permafrost soils hold about 11% of the Earth's soil carbon and 95% of its organically bound nutrients. These soils are very sensitive to physical disturbance and pollution, and the widespread exploration of such areas for oil and minerals can have long-term effects on their structure and function.

In water-saturated bog areas, oxygen limitation means that bacteria are more important than fungi in decomposition processes, and there is decreased degradation of lignified materials. As in other soils, the nutrient cycling processes of nitrification, denitrification, nitrogen fixation, and methane synthesis and utilization, although occurring at slower rates, can have major impacts on global gaseous cycles, as discussed later (pp. 635–637).

Desert Soils

Soils of hot and cold, arid and semiarid deserts depend on periodic and infrequent rainfall. When it rains, water can puddle in low areas and be retained on the soil surface by microbial communities called **desert crusts.** These consist of cyanobacteria (e.g., *Anabaena, Microcoleus, Nostoc,* and *Scytonema*) and associated microbes. The depth of the photosynthetic layer is perhaps

1 mm, and the cyanobacterial filaments and slime link the sand particles, which change the surface soil albedo (the amount of sunlight reflected), water infiltration rate, and susceptibility to erosion. These crusts are quite fragile, and vehicle damage can be evident for decades. After a rainfall, nitrogen fixation begins within approximately 25 to 30 hours, and when the rain evaporates or drains, the crust dries up and nitrogen is released for use by other microorganisms and the plant community. << *Photosynthetic bacteria: Phylum* Cyanobacteria *(section 19.3)*

Geologically Heated Hyperthermal Soils

Geologically heated soils are found in such areas as Iceland, the Kamchatka peninsula in eastern Russia, Yellowstone National Park, and many mining waste sites. These soils are populated by bacteria and archaea, many of which are chemolithoautotrophs. A wide variety of chemoorganotrophic genera also are found in these environments; these include the aerobes *Thermomicrobium* and *Thermoleophilum,* and the anaerobes *Thermosipho* and *Thermotoga.* An important microorganism found in heated mining wastes is *Thermoplasma.* Such geothermal soils have been of great interest as a source of new microbes to use in biotechnology, and the search for new, unique microorganisms in these areas is intensifying all over the world. << *Phylum* Euryarchaeota: *Thermoplasms (section 18.3)*

1. Characterize each major soil type discussed in this section in terms of the balance between primary production and organic matter decomposition.
2. What is slash-and-burn agriculture? Describe the roles of microorganisms in this process.
3. What is unique about bogs in terms of organic matter degradation?
4. Describe desert crusts. What types of microorganisms function in these unique environments?
5. What unique microbial genera are found in geothermally heated soils?

Microorganism Associations with Vascular Plants

The vast majority of soil microbes are heterotrophic, so it should come as no surprise that many have evolved close relationships with plants, the major source of terrestrial primary production. Many microbe-plant interactions are commensalistic—they do no harm to the plant, whereas the microbe gains some advantage. Many other important interactions are beneficial to both the microorganism and the plant (i.e., they are mutualistic). Finally, other microbes are plant pathogens and parasitize their plant hosts. In all cases, the microbe and the plant have established the capacity to communicate. The microbe detects and responds to plant-produced chemical signaling molecules. This generally triggers the release of microbial compounds that are in turn recognized by the plant, thereby beginning a two-way "conversation" that employs a molecular lexicon. Once a microbe-plant relationship is initiated, microbes and plants continue to monitor the physiology of their partner and adjust their own activities accordingly. The nature of the signaling molecules and the mechanisms by which both plants and microbes respond involve exciting multidisciplinary research in soil microbiology, ecology, molecular biology, genetics, and biochemistry. ◄◄ *Microbial growth in natural environments: Cell-cell communication within microbial populations (section 7.6);* ►► *Microbial interactions (section 27.1)*

Microbe-plant interactions can be broadly divided into two classes: microbes that live on the surface of plants, called **epiphytes** and those that colonize internal plant tissues, called **endophytes.** Further, we can consider microbes that live in the above ground, or aerial, surfaces of plants separately from those that inhabit below-ground plant tissues. We begin our discussion of microbe-plant interactions by first introducing the microbial communities associated with aerial regions of plants. We then turn our attention to two important microbe-root symbioses—the mycorrhizal fungi and the nitrogen-fixing rhizobia. Finally, we consider several microbial plant pathogens.

Phyllosphere Microorganisms

The environment of the aerial portion of a plant, called the **phyllosphere,** was once thought to be too hostile to support a stable microbial community. Leaves and stems undergo frequent and rapid changes in humidity, UV exposure, and temperature. This in turn results in fluctuations of the leaching of organic material (primarily simple sugars) that could support a microbial population. It is now known that the phyllosphere is home to a diverse assortment of microbes, including bacteria, filamentous fungi, yeasts, and photosynthetic and heterotrophic protists. Numerically, it appears that the γ-proteobacteria *Pseudomonas syringae* and *Erwinia,* and *Pantoea* are most important. Another abundant bacterial genus, *Sphingomonas,* produces pigments that function like suncreen so it can survive the high levels of UV irradiation occurring on these plant surfaces. This bacterium, also common in soils and waters, can occur at 10^8 cells per gram of plant tissue. It often represents a majority of the culturable species.

Rhizosphere and Rhizoplane Microorganisms

Plant roots receive between 30 to 60% of the net photosynthesized carbon. Of this, an estimated 40 to 90% enters the soil as a wide variety of materials including alcohols, ethylene, sugars, amino and organic acids, vitamins, nucleotides, polysaccharides, and enzymes. These materials create a unique environment for soil microorganisms called the **rhizosphere.** The plant root surface, termed the **rhizoplane,** also provides an exceptional environment for microorganisms, as these gaseous, soluble, and particulate materials move from the plant to the soil. Rhizosphere and rhizoplane microbial community composition and function change when these newly available substrates become available. In addition, rhizosphere and rhizoplane microorganisms serve as labile sources of nutrients for other organisms, creating a soil microbial loop and thereby playing critical roles in organic matter synthesis and degradation.

A wide range of microbes in the rhizosphere can promote plant growth, orchestrated by their ability to communicate with plants using complex chemical signals. Some of these signaling com-

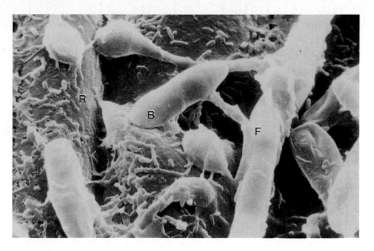

Figure 26.19   Root Surface Microorganisms. Plant roots release nutrients that allow intensive development of bacteria and fungi on and near the plant root surface, the rhizoplane. A scanning electron micrograph shows bacteria and fungi growing on a root surface. R = root surface; B = bacterium; F = fungal hypha.

pounds include auxins, gibberellins, glycolipids, and cytokinins, and are beginning to be fully appreciated in terms of their biotechnological potential. Plant growth-promoting rhizobacteria include the genera *Pseudomonas* and *Achromobacter.* These can be added to the plant, even in the seed stage, if the bacteria produce the required surface attachment proteins. The genes that control the expression of these attachment proteins are of great interest to agricultural biotechnologists.

A critical process that occurs on the surface of the plant, and particularly in the root zone, is associative nitrogen fixation, in which nitrogen-fixing microorganisms are on the surface of the plant root, the rhizoplane (**figure 26.19**), as well as in the rhizosphere. This process is carried out by members of the genera *Azotobacter, Azospirillum,* and *Acetobacter.* These bacteria contribute to nitrogen accumulation by tropical grasses. However, evidence suggests that their major contribution may not be nitrogen fixation but the production of growth-promoting hormones that increase root hair development, thereby enhancing plant nutrient uptake. This is an area of research that is particularly important in tropical agricultural areas.

1. Define rhizosphere, rhizoplane, and associative nitrogen fixation.
2. What unique stresses does a microorganism on a leaf face but not a microbe in the soil?
3. What is the importance of plant growth-promoting bacteria?
4. What important genera are involved in associative nitrogen fixation?

Mycorrhizae

Mycorrhizae (derived from the Greek meaning "fungus root") are mutualistic relationships that develop between most plants and a limited number of fungal species. Both partners in mutualistic

Microbial Diversity & Ecology

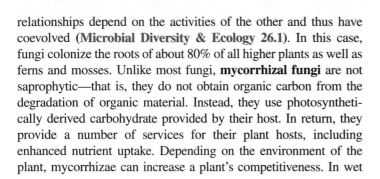

26.1 Mycorrhizae and the Evolution of Vascular Plants

Fossil evidence shows that endomycorrhizal symbioses were as frequent in vascular plants during the Devonian period, about 400 million years ago, as they are today. As a result, some botanists have suggested that the evolution of this type of association may have been a critical step in allowing colonization of land by plants. During this period, soils were poorly developed, and as a result, mycorrhizal fungi were probably significant in aiding the uptake of phosphorus and other nutrients. Even now, those plants that start to colonize extremely nutrient-poor soils survive much better if they have endomycorrhizae. Thus it may have been a symbiotic association of plants and fungi that initially colonized the land and led to our modern vascular plants.

relationships depend on the activities of the other and thus have coevolved (**Microbial Diversity & Ecology 26.1**). In this case, fungi colonize the roots of about 80% of all higher plants as well as ferns and mosses. Unlike most fungi, **mycorrhizal fungi** are not saprophytic—that is, they do not obtain organic carbon from the degradation of organic material. Instead, they use photosynthetically derived carbohydrate provided by their host. In return, they provide a number of services for their plant hosts, including enhanced nutrient uptake. Depending on the environment of the plant, mycorrhizae can increase a plant's competitiveness. In wet environments, they increase the availability of nutrients, especially phosphorus. In arid environments, where nutrients do not limit plant functioning to the same degree, the mycorrhizae aid in water uptake, allowing increased transpiration rates in comparison with nonmycorrhizal plants. >> *Microbial interactions: Mutualism (section 27.1)*

Mycorrhizae can be broadly classified as **endomycorrhizae**—those with fungi that enter the root cells, or as **ectomycorrhizae**—those that remain extracellular, forming a sheath of interconnecting filaments (hyphae) around the roots. Although all six types of mycorrhizae are detailed in **table 26.3**, we confine our discussion

Table 26.3 Mycorrhizal Associations

Mycorrhizal Classification	Fungi Involved	Plants Colonized	Fungal Structural Features	Fungal Function
Ectomycorrhizae	Basidiomycetes, including those with large fruiting bodies (e.g., toadstools); some ascomycetes	~90% of trees and woody plants in temperate regions; fungal-plant colonization is often species specific	Hartig net, mantle or sheath; rhizomorphs; root hair development is usually limited.	Nutrient (N and P) uptake and transfer
Arbuscular	Glomeromycetes, in particular six genera of the order *Glomales*	Wild and crop plants, tropical trees; fungal-plant colonization is not highly specific	Arbuscules: hyphae-filled invaginations of cortical root cell	Nutrient (N and P) uptake and transfer; facilitate soil aggregation; promote seed production; reduce pest and nematode infection; increase drought and disease resistance
Ericaceous	Ascomycetes Basidiomycetes	Low evergreen shrubs, heathers	Some intracellular, some extracellular	Mineralization of organic matter
Orchidaceous	Basidiomycetes	Orchids	Hyphal coils called pelotons within host tissue	Some orchids are non-photosynthetic and others produce chlorophyll when mature; these organisms are almost completely dependent on mycorrhizae for organic carbon and nutrients.
Ectendomycorrhizae	Ascomycetes	Conifers	Hartig net with some intracellular hyphae	Nutrient uptake and mineralization of organic matter
Monotropoid mycorrhizae	Ascomycetes Basidiomycetes	Flowering plants that lack chlorophyll (*Monotropaceae;* e.g., Indian pipe)	Hartig net one cell deep in the root cortex	Nutrient uptake and transfer

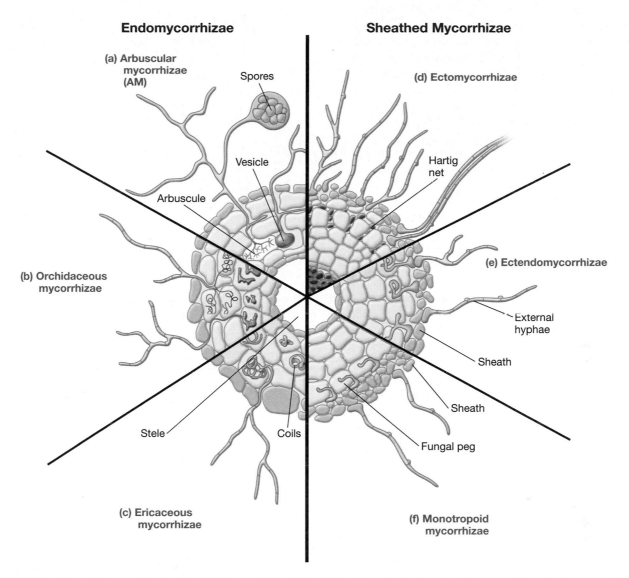

Endomycorrhizae   **Sheathed Mycorrhizae**

(a) Arbuscular mycorrhizae (AM)

Spores

Vesicle

Arbuscule

(b) Orchidaceous mycorrhizae

Stele   Coils

(c) Ericaceous mycorrhizae

(d) Ectomycorrhizae

Hartig net

(e) Ectendomycorrhizae

External hyphae

Sheath

Sheath

Fungal peg

(f) Monotropoid mycorrhizae

Figure 26.20   Mycorrhizae.   Fungi can establish mutually beneficial relationships with plant roots, called mycorrhizae. Root cross sections illustrate different mycorrhizal relationships.

to the two most important types: ectomycorrhizae and the endomycorrhizae called arbuscular mycorrhizae. **<<** *Characteristics of the fungal divisions (section 23.3)*

The ectomycorrhizae (ECM) are formed by both ascomycete and basidomycete fungi. ECM colonize almost all trees in cooler climates. Their importance arises from their ability to transfer essential nutrients, especially phosphorus and nitrogen, to the root. The development of ECM starts with the growth of a fungal mycelium around the root. As the mycelium thickens, it forms a sheath or mantle so that the entire root may be covered by the fungal mycelium (**figure 26.20d**). Most ECM produce signaling molecules that limit the growth of root hairs, thus ECM-colonized roots often appear blunt and covered in fungi (**figure 26.21**). From the root surface, the fungi extend hyphae into the soil; these filaments may aggregate to form rhizomorphs, which are often visible to the naked eye.

Figure 26.21   Ectomycorrhizae as Found on Roots of a Pine Tree.   Typical irregular branching of the white, smooth mycorrhizae is evident.

Hyphae on the inner side of the sheath penetrate between (but not within) the cortical root cells, forming a characteristic meshwork of hyphae called the **Hartig net** (figure 26.20*d*). Soil nutrients taken up by rhizomorphs must first pass through the hyphal sheath and then into the Hartig net filaments, which form numerous contacts with root cells. This results in efficient, two-way transfer of soil nutrients to the plant and carbohydrates to the fungus. This relationship has evolved to the point that some plants synthesize sugars such as mannitol and trehalose that cannot be used by the plants and can only be assimilated by their fungal symbionts.

Arbuscular mycorrhizae (AM) are the most common type of mycorrhizae. They can be found in association with many tropical plants and, importantly, with most crop plants. AM fungi belong to the division *Glomeromycota* and have not yet been grown in pure culture without their plant hosts. These microbes enter root cells between the plant cell wall and invaginations in the plasma membrane (figure 26.20*a*). So, although AM are endomycorrhizae, they do not breach the root cell membrane. Instead, treelike hyphal networks called **arbuscules** develop within the folds of the plasma membrane. Individual arbuscules are transient; they last at most two weeks. AM can be vigorous colonizers: a 5 cm segment of root can support the growth of as many as eight species, and hyphae from a single germinated spore can simultaneously colonize multiple roots from unrelated plant species.

AM are believed to provide a number of services to their plant hosts, including protection from disease, drought, nematodes, and other pests. Their capacity to transfer phosphorus to roots has been well documented, and recently the nature of their transfer of nitrogen has been explored in detail. Stable isotope experiments were performed in which AM-colonized plants (and the appropriate negative controls lacking mycorrhizae) were treated with $^{15}NO_3^-$ and $^{15}NH_4^+$ (**figure 26.22**). These forms of nitrogen are incorporated into fungal tissue through the glutamine synthetase–glutamate synthase pathway (GS-GOGAT; *see figure 11.15*). Thus ^{15}N-containing glutamine (which is converted to arginine) is recovered. However, prior to transferring nitrogen to host cells, intracellular fungal hyphae degrade the amino acids, transferring only the $^{15}NH_4^+$. Thus while the fungus provides its host with much needed ammonium, it retains the carbon skeleton it needs for nitrogen uptake. $\ll$ *Synthesis of amino acids: Nitrogen assimilation (section 11.5)*

Bacteria are also associated with the mycorrhizal fungi. As the external hyphal network radiates out into the soil, a mycorrhizosphere is formed due to the flow of carbon from the plant into the mycorrhizal hyphal network and then into the surrounding soil. In addition, "mycorrhization helper bacteria" can play a role in the development of mycorrhizal relationships with ectomycorrhizal fungi. Bacterial symbionts also are found in the cytoplasm of AM fungi. These organisms appear to be related to *Burkholderia cepacia*. It has been suggested that these "trapped" bacteria contribute to the nitrogen metabolism of the plant-fungal complex by assisting with the synthesis of essential amino acids.

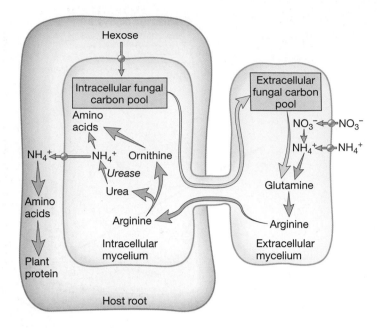

Figure 26.22 Nitrogen Exchange Between Arbuscular Mycorrhizal Fungi and Host Plant. Nitrate and ammonium are taken up by the fungal mycelium that is outside the host plant cell (extracellular mycelium) and converted to arginine. This amino acid is transferred to the mycelium within the host plant cell and broken down so that only the ammonium enters the plant.

1. Describe the two-way relationship between mycorrhizal fungi and the plant host.

2. List three major differences between arbuscular mycorrhizae and ectomycorrhizae.

3. What is the function of the rhizomorph and the Hartig net?

4. Describe the uptake and transfer of ammonium by arbuscular mycorrhizae to the plant host. Why do you think only the ammonium is transferred?

5. Propose two potential functions for mycorrhization helper bacteria.

Nitrogen Fixation

The enzymatic conversion of gaseous nitrogen (N_2) to ammonia (NH_3) often occurs as part of a symbiotic relationship between bacteria and plants. These symbioses produce more than 100 million metric tons of fixed nitrogen annually and are a vital part of the global nitrogen cycle. In addition, symbiotic nitrogen fixation accounts for more than half of the nitrogen used in agriculture. The provision of fixed nitrogen enables the growth of host plants in soils that would otherwise be nitrogen limiting and simultaneously reduces loss of nitrogen by denitrification and leaching. It follows that nitrogen fixation, particularly that by members of the genus *Rhizobium* and related α-proteobacteria in association with their leguminous host plants, has been the subject of intense investigation.

The Rhizobia Several bacterial genera are able to form nitrogen-fixing nodules with legumes. These include the α-proteobacteria *Allorhizobium, Azorhizobium, Bradyrhizobium, Mesorhizobium, Sinorhizobium,* and *Rhizobium.* Collectively these genera are often called the **rhizobia.** Recently the phylogenetic diversity of the rhizobia has been extended by the discovery that the β-proteobacteria *Burkholderia caribensis* and *Ralstonia taiwanensis* also form nitrogen-fixing nodules on legumes. Here, we discuss the general process of nodulation followed by molecular details that have been revealed largely though studies of the genus *Rhizobium.* << *Class* Alphaproteobacteria *(section 20.1); Class* Betaproteobacteria *(section 20.2)*

Rhizobia live freely in the soil. In nitrogen-sufficient soils, plants secrete ethylene-mediated compounds that inhibit nodulation. But even in nitrogen-starved soils, when rhizobia approach the plant root, they are assumed to be an alien invader. The plant responds with an oxidative burst, producing a mixture of compounds that can contain superoxide radicals, hydrogen peroxide, and N_2O. This redox-based oxidative burst, involving glutathione and homoglutathione, is critical in determining the fate of the infection process. The rhizobia, if they are to be effective colonizers, must use antioxidant defenses to survive and become symbionts. Only rhizobia and related genera with sufficient antioxidant abilities are able to proceed to the next step in the infection process.

The plant roots also release flavonoid inducer molecules that stimulate rhizobial colonization of the root surfaces (**figure 26.23a**). In response to this molecular message, rhizobia produce their own signaling compounds called **Nod factors.** The precise structure of individual Nod factors depends on the bacterial species, but all consist of four to five units of β-1,4 linked N-acetyl-D-glucosamine bearing an acyl chain at the nonreducing terminal residue and a sulfate attached to the reducing end (figure 26.23c; note highlighted moieties). Upon receipt of the Nod factor message, gene expression in the outer (epidermal) cells of the roots is altered so that the root hairs become deformed. In some cases, the root hairs will curl to resemble a shepherd's crook, entrapping bacteria (figure 26.23c,d). In these regions, the plant cell wall is locally modified, the plant plasma membrane invaginates, and new plant material is laid down. These modifications lead to the development of a bacteria-filled, tubelike structure called the infection thread (figure 26.23e,f). The infection thread grows toward the base of the root hair cell to a region called the primordium. Division of these cells ultimately gives rise to the **root nodule.** When bacteria are released from the infection thread into the primodium, they remain surrounded by a plant cell membrane called the peribacteroid membrane (figure 26.23h). It is here that each bacterial cell differentiates into the nitrogen-fixing form called a **bacteroid.** Bacteroids are terminally differentiated—they can neither divide nor revert back to the nondifferentiated state. Further growth and differentiation lead to the development of a structure called a **symbiosome.** Recall that the nitrogenase enzyme is very sensitive to oxygen. To help protect the nitrogenase, a protein called **leghemoglobin** is produced. This binds to oxygen and helps maintain microaerobic conditions within the mature nodule (figure 26.23i, j). Leghemoglobin is similar in structure to the hemoglobin found in animals; however, it has a higher affinity for oxygen. Interestingly, the protein moiety is encoded by plant genes, whereas the heme group is the product of bacterial genes. << *Synthesis of amino acids: Nitrogen fixation (section 11.5)*

The symbiosomes within mature root nodules are the site of nitrogen fixation. Within these nodules, the differentiated bacteriods reduce atmospheric N_2 to ammonia, and in return they receive carbon and energy in the form of dicarboxylic acids from their host legume. It had long been thought the symbionts transferred ammonia to the host plant, but more recent evidence shows that a more complex cycling of amino acids occurs. Apparently the plant provides certain amino acids to the bacteroids so that they do not need to assimilate ammonia. In return, the bacteroids shuttle amino acids (which bear the newly fixed nitrogen) back to the plant. This creates an interdependent relationship, providing selective pressure for the evolution of mutualism.

The molecular mechanisms by which both the legume host and the rhizobial symbionts establish productive nitrogen-fixing bacteriods within nodules continues to be an intense area of research. A major goal of biotechnology is to introduce nitrogen fixation genes into plants that do not normally form such associations. It has been possible to produce modified lateral roots on nonlegumes such as rice, wheat, and oilseed rape, which are invaded by nitrogen-fixing bacteria. Although these modified root structures have not yet been found to fix useful amounts of nitrogen, they enhance rice production, and intense work is expected to continue in this area.

Stem-Nodulating Rhizobia Other associations of nitrogen-fixing microorganisms with plants also occur. A particularly interesting association is caused by stem-nodulating rhizobia, found primarily in tropical legumes (**figure 26.24**). These nodules form at the base of adventitious roots branching out of the stem just above the soil surface, and because they contain oxygen-producing photosynthetic tissues, they have unique mechanisms to protect the oxygen-sensitive nitrogen fixation enzymes. One microorganism that forms such root and stem nodules is *Azorhizobium caulinodans,* which forms nodules on the tropical legume *Sesbania rostrata.* It has been shown that some of these stem-nodulating rhizobia are photosynthetic. Thus they can obtain their energy not only from the plant's organic compounds but also from light.

Actinorhizae Another example of symbiotic nitrogen fixation occurs between the actinomycete *Frankia* and eight nonleguminous host plant families. These bacterial associations with plant roots are called actinorhizae or actinorhizal relationships (**figure 26.25**). *Frankia* fixes nitrogen and is important particularly in trees and shrubs. For example, these associations occur in areas where Douglas fir forests have been clear-cut and in bog and heath environments where bayberries and alders are dominant. The nodules of some plants (*Alnus, Ceanothus*) are as large as baseballs. The nodules of *Casuarina* (Australian pine) approach soccer ball size.

Members of the genus *Frankia* are slow growing and were impossible to culture apart from the plant until 1978. Since

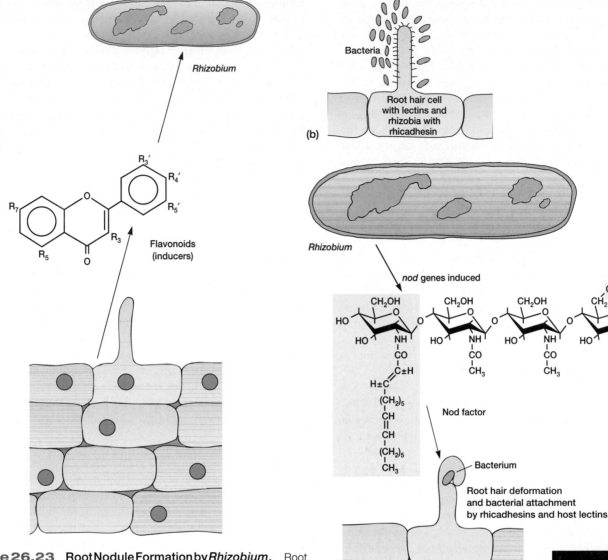

Figure 26.23 Root Nodule Formation by *Rhizobium*. Root nodule formation on legumes by *Rhizobium* is a complex process that produces the nitrogen-fixing symbiosis. (a) The plant root releases flavonoids that stimulate the production of various Nod metabolites by *Rhizobium*. Many different Nod factors control infection specificity. (b) Attachment of *Rhizobium* to root hairs involves specific bacterial proteins called rhicadhesins and host plant lectins that affect the pattern of attachment and *nod* gene expression. (c) Structure of a typical Nod factor that promotes root hair curling and plant cortical cell division. The bioactive portion (nonreducing *N*-fatty acyl glucosamine) is highlighted. These Nod factors enter root hairs and migrate to their nuclei. (d) A plant root hair covered with *Rhizobium* and undergoing curling.

then, this actinomycete has been grown on specialized media supplemented with metabolic intermediates such as pyruvate. Major advances in understanding the physiology, genetics, and molecular biology of these microorganisms are now taking place.

1. List several bacteria that are considered rhizobia.

2. Describe the communication system between a rhizobial bacterium and its legume host.

3. What is the function of the infection thread? Why do you think it is important that the bacteria do not enter plant cell cytoplasm until they reach the primordium?

4. What does the term terminally differentiated mean? Can you think of other cells that are also terminally differentiated?

5. How is leghemoglobin made and what is its function?

6. What is the major contribution of *Frankia* to its plant hosts? Which types of plants are infected?

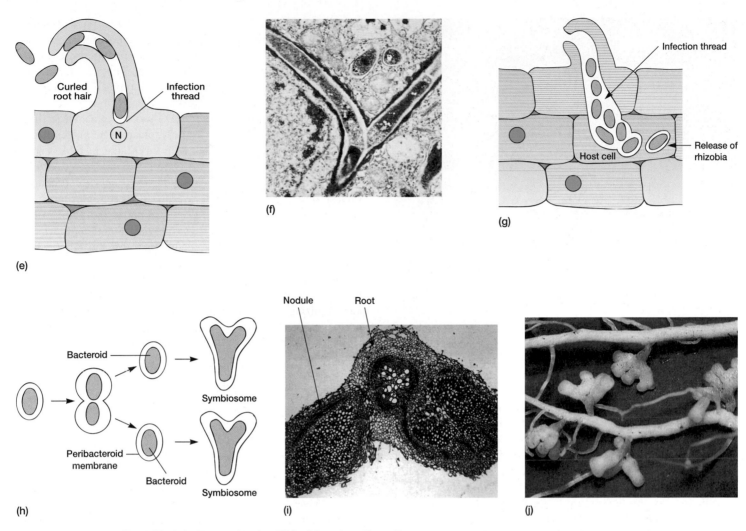

(e) (f) (g)

(h) (i) (j)

Figure 26.23 **Root Nodule Formation by *Rhizobium* (*continued*).** (e) Initiation of bacterial penetration into the root hair cell and infection thread growth coordinated by the plant nucleus "N." (f) A branched infection thread shown in an electron micrograph. (g) Cell-to-cell spread of *Rhizobium* through transcellular infection threads followed by release of rhizobia and infection of host cells. (h) Formation of bacteroids surrounded by plant-derived peribacteroid membranes and differentiation of bacteroids into nitrogen-fixing symbiosomes. The bacteria change morphologically and enlarge around seven to 10 times in volume. The symbiosome contains the nitrogen-fixing bacteroid, a peribacteroid space, and the peribacteroid membrane. (i) Light micrograph of two nodules that develop by cell division (×5). This section is oriented to show the nodules in longitudinal axis and the root in cross section. (j) *Sinorhizobium meliloti* nitrogen-fixing nodules on roots of white sweet clover *(Melilotus alba)*.

Figure 26.24 **Stem-Nodulating Rhizobia.** Nitrogen-fixing rhizobia also can form nodules on stems of some tropical legumes. Nodules formed on the stem of a tropical legume by a stem-nodulating *Rhizobium*.

Figure 26.25 **Actinorhizae.** *Frankia*-induced actinorhizal nodules in *Ceanothus* (buckbrush).

Figure 26.26 *Agrobacterium.* *Agrobacterium*-caused tumor on a *Kalanchoe* sp. plant.

Agrobacterium

Clearly some microbe-plant interactions are beneficial for both partners. However, others involve microbial pathogens that harm or even kill their host. ***Agrobacterium tumefaciens*** is an α-proteobacterium that has been studied intensely for several decades. Initially research focused on how this microbe causes crown gall disease, which results in the formation of tumorlike growths in a wide variety of plants (**figure 26.26**). Within the last 20 years or so, however, *A. tumefaciens* has become one of biotechnology's most important tools. The molecular genetics by which this pathogen infects its host is the basis for plant genetic engineering. ◀◀ *Recombinant DNA technology in agriculture (section 16.11); Class* Alphaproteobacteria *(section 20.1)*

The genes for plant infection and virulence are encoded on an *A. tumefaciens* plasmid, called the **Ti (tumor-inducing) plasmid.** These genes include 21 *vir* genes (*vir* stands for virulence), which are found in six separate operons. Two of these genes, *virD1* and *virD2,* encode proteins that excise a separate region of the Ti plasmid, called **T DNA.** After excision, this T DNA fragment is integrated into the host plant's genome. Once incorporated into a plant cell's genome, T DNA directs the overproduction of phytohormones that cause unregulated growth and reproduction of plant cells, thereby generating a tumor or gall in the plant.

The *vir* genes are not expressed when *A. tumefaciens* is living saprophytically in the soil. Instead, they are induced by the presence of plant phenolics and monosaccharides present in an acidic (pH 5.2–5.7) and cool (below 30°C) environment (**figure 26.27a**). The microbe usually infects its host through a wound. Upon reception of the plant signal, a two-component signal transduction system is activated: VirA is a sensor kinase that, in the presence of a phenolic signal, phosphorylates the response regulator VirG.

Activated VirG then induces transcription of the other *vir* genes. This enables the bacterial cell to become adequately positioned relative to the plant cell, at which point the *virB* operon expresses the apparatus that will transfer the T DNA. This transfer is similar to bacterial conjugation and involves a type IV secretion system. After VirD1 and VirD2 excise the T DNA from the Ti plasmid, the T DNA, with the VirD2 protein attached to the 5' end, is delivered to the plant cell cytoplasm. The protein VirE2 is also transferred, and together with VirD2, the T DNA is shepherded to the plant cell nucleus, where it is integrated into the host's genome. Here, the T DNA has two specific functions. First, it directs the host cell to overproduce phytohormones that cause tumor formation. Second, it stimulates the plant to produce special amino acid and sugar derivatives called opines (figure 26.27b). Opines are not metabolized by the plant but *A. tumefaciens* is attracted to opines; chemotaxis of bacteria from the surrounding soil population will further advance the infection because the bacterium can use opines as sources of carbon, energy, nitrogen, and, in some cases, phosphorus. ◀◀ *Regulation of transcription initiation: Two-component signal transduction systems (section 13.2); Bacterial conjugation (section 14.7)*

Other Plant Pathogens

In addition to *Agrobacterium,* many other bacteria cause an array of spots, blights, wilts, rots, cankers, and galls, as shown in **table 26.4**. The soft rots caused by the enterobacteria *Erwinia chrysanthemi* and *E. carotovora* have significant economic impact. These bacteria digest plant tissue by producing extracellular enzymes that degrade pectin, cellulose, and proteins. These exoenzymes are secreted by type II and type I secretion systems; mutants lacking these secretion systems are no longer pathogenic. Similarly, proteobacteria belonging to the genera *Ralstonia, Pseudomonas, Pantoea,* and *Xanthomonas* rely on type III secretion systems to deliver virulence proteins. Although these microbes cause a diverse collection of plant diseases, all colonize the spaces between plant cells, rather than invading the cells themselves, to kill their hosts. Another group of important plant pathogens includes the wall-less phytoplasms that infect vegetable and fruit crops such as sweet potatoes, corn, and citrus. ◀◀ *Protein maturation and secretion (section 12.8)*

As discussed in chapters 5, 23, and 24, protists, fungi, viruses, and virusoids can be devastating plant pathogens. Examples are the fungus *Puccinia graminis,* which causes wheat rust, and the öomycete *Phytophthora infestans,* which was responsible for the Irish potato famine, and *Tobacco mosaic virus* (TMV), the first virus to be characterized. A virus of particular interest in terms of plant-pathogen interactions is a hypovirus (family *Hypoviridae*) that infects the fungus *Cryphonectria parasitica,* the cause of chestnut blight. Based on pioneering studies carried out in Italy and France, workers in Connecticut and West Virginia noted that if they infected the fungus with this hypovirus, the rate and occurrence of blight

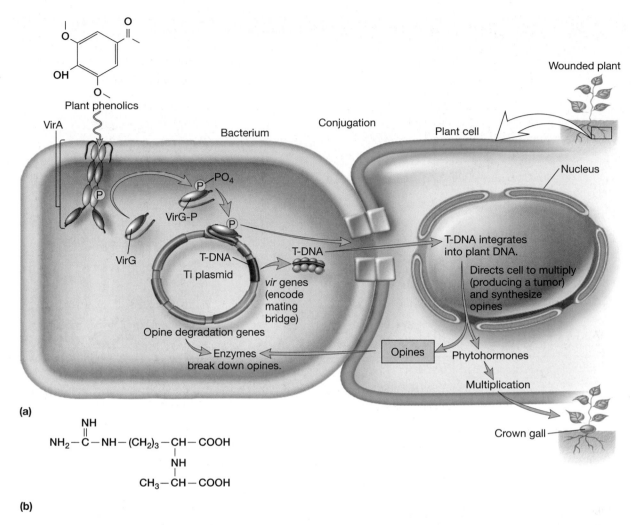

Figure 26.27   Functions of Genes Carried on the *Agrobacterium* Ti Plasmid.   (a) Genes carried on the Ti plasmid of *Agrobacterium* control tumor formation by a two-component regulatory system that stimulates formation of the mating bridge and excision of the T-DNA. The T-DNA is moved into the plant cell where it integrates into the plant DNA. T-DNA encodes plant hormones that cause the plant cells to divide, producing the tumor. (b) The tumor cells produce opines that can serve as a carbon source for the infecting *Agrobacterium*. Ultimately a crown gall is formed on the stem of the wounded plant above the soil surface.

were decreased. They are hoping to treat trees with the less lethal virus strains and eventually transform the indigenous lethal strains of *Cryphonectria* into more benign fungi.

1. Discuss the nature and importance of the Ti plasmid.

2. What functions do the members of the two-component system play in infection of a plant by *Agrobacterium?* What are the roles of phenolics and opines in this infection process?

3. What kinds of exoenzymes are produced by some plant pathogens?

4. How are plant pathologists attempting to control chestnut blight?

Soil Microorganisms and Global Climate Change

Soil microorganisms, like marine microbes, can have major effects on global fluxes of a variety of gases. These gases include those that are "relatively stable" and those that are "reactive." Relatively stable gases that are influenced by microbial activities include carbon dioxide, nitrous oxide, nitric oxide, and methane. Microorganisms also contribute to the flow of reactive gases such as ammonia, hydrogen sulfide, and dimethylsulfide. These reactive gases tend to be produced in more waterlogged environments. **≪** *Biogeochemical cycling (section 25.1)*

Atmospheric gases such as carbon dioxide, nitrous oxide, nitric oxide, and methane are **greenhouse gases.** The production and consumption of greenhouse gases can be influenced by a range of human activities. These include plant

Table 26.4	Major Plant Diseases Caused by Bacteria	
Symptoms	**Examples**	**Pathogen**
Spots and blights	Wildfire (tobacco)	*Pseudomonas syringae* pv.[a] *tabaci*
	Haloblight (bean)	*P. syringae* pv. *phaseolica*
	Citrus blast	*P. syringae* pv. *syringae*
	Leaf spot (bean)	*P. syringae* pv. *syringae*
	Blight (rice)	*Xanthomonas campestris* pv. *oryzae*
	Blight (cereals)	*X. campestris* pv. *translucens*
	Spot (tomato, pepper)	*X. campestris* pv. *vesicatoria*
	Ring rot (potato)	*Clavibacter michiganensis* pv. *sepedonicum*
Vascular wilts	Wilt (tomato)	*C. michiganensis* pv. *michiganensis*
	Stewart's wilt (corn)	*Erwina stewartii*
	Fire blight (apples)	*E. amylovora*
	Moko disease (banana)	*P. solanacearum*
Soft rots	Black rot (crucifers)	*X. campestris* pv. *campestris*
	Soft rots (numerous)	*E. carotovora* pv. *carotovora*
	Black leg (potato)	*E. carotovora* pv. *atroseptica*
	Pink eye (potato)	*P. marginalis*
	Sour skin (onion)	*P. cepacia*
Canker	Canker (stone fruit)	*P. syringae* pv. *syringae*
	Canker (citrus)	*X. campestris* pv. *citri*
Galls	Crown galls (numerous)	*Agrobacterium tumefaciens*
	Hairy root	*A. rhizogenes*
	Olive knot	*P. syringae* pv. *savastonoi*

[a]pv., pathovar, a variety of microorganisms with phytopathogenic properties.
Source: From Lengler, J. W.; Drews, G.; and Schlegel, H. G. 1999. *Biology of the prokaryotes,* Blackwell Science, Malden, Mass., table 34.4.

fertilization and automobiles, conversion of soils to agricultural use, and landfills.

Just as marine primary production helps prevent an even more rapid increase in atmospheric CO_2, terrestrial forests are a tremendous CO_2 "sink." During the 1990s, terrestial ecosystems sequestered about 3 gigatons of carbon per year (a gigaton is 1 billion metric tons). And just as some think that autotrophic phytoplankton might be able to consume much of the excess atmospheric CO_2, there has been much speculation regarding the capacity of forests to take up additional CO_2. Some believe that plants will assimilate the extra CO_2, resulting in an accelerated average rate of growth. This optimistic view holds that increases in CO_2 will be buffered by forest ecosystems (however, the rate of deforestation must also be considered). An international team of scientists recently addressed this question by growing a variety of trees for several years under higher concentrations of CO_2. Although they measured plant growth stimulation, it was coupled to an increase in respiration by soil microbes. Recall that as respiration rates increase, so does the quantity of CO_2 released. We don't know if these findings can be extrapolated to a world where CO_2 levels continue to increase, but if so, forest ecosystems may sequester less carbon than predicted.

Methane is a greenhouse gas of increasing concern that can be derived from a variety of sources. These include ruminants, rice paddies, landfills, and even the methanogenic archaea that inhabit termite guts. In well-drained, oxic soils, aerobic bacteria called **methanotrophs** oxidize methane, whereas in water-saturated soils, methane may be produced faster than it can be used by methanotrophs. Based on analyses of gas bubbles in glacier ice cores, the levels of methane in the atmosphere remained essentially constant until about 400 years ago. Since then the methane level has increased two and a half times, to the present level of 1.7 parts per million (ppm) by volume. Considering these trends, there is a worldwide interest in understanding the factors that control methane synthesis and use by microorganisms.

Terrestrial environments, the oceans, and biomass burning are all significant sources of another important greenhouse gas,

Table 26.5	Global Emissions of Chloromethane
Source	**Inputs to Atmosphere (105 tons/year)[a]**
Natural Sources	
Terrestrial processes	0–20
Ocean fluxes	3–20
Biomass burning	4–14
Anthropogenic Sources	0–3

[a]Estimated atmospheric inputs from natural and anthropogenic sources.
Source: From Watling, R., and Harper, D. B. 1998. *Mycol. Res.* 102(7):769–87.

chloromethane (CH_3Cl). Basidiomycete decomposition of woody plants results in its release; the global input of CH_3Cl to the atmosphere from plant decomposition is thought to be 160,000 tons, 75% of which is derived from tropical and subtropical soils (**table 26.5**). It is estimated that 15 to 20% of the chlorine-catalyzed ozone destruction is due to naturally produced chlorohydrocarbons. << *Characteristics of the fungal divisions: Basidiomycota (section 23.3)*

Addition of nitrogen-containing fertilizers also affects atmospheric gas exchange processes in a soil. Nitrogen additions stimulate the production of the nitrification intermediates NO and N_2O, which are also critical greenhouse gases. NO also appears to be required for *Nitrosomonas eutropha* to carry out nitrification. The oxidation of NH_4^+ involves the formation of hydroxylamine (NH_2OH) and NO as intermediates. NO reacts with oxygen to give NO_2, which then can repeat the process in this reaction:

$$NH_4^+ \; NO_2^- \rightleftharpoons NH_2OH + NO + H^+$$

$$2NO + O_2 \rightleftharpoons 2NO_2 \text{ (nitrogen dioxide)}$$

In this sequence, molecular oxygen does not react with NH_4^+ but with NO. If NO is absent, the reaction will not proceed.

In addition to generating and consuming greenhouse gases, microbes also respond to them. In fact, scientists question how global warming may change patterns of infectious disease outbreaks in humans and other animals. Recently a significant clue was provided by ecologists studying the extinction of 67% of the 110 species of harlequin frogs (*Atelopus*) native to tropical America, which has occurred over the last 20 years (**figure 26.28**). Evidence supports the hypothesis that these frogs have succumbed to a pathogenic chytrid fungus (*Batrachochytrium dendrobatidis*) whose range has expanded in response to warmer temperatures. This is not the first account of a pathogen responding to climate change. Pine blister rust, caused by the fungus *Cronartium ribicola*, is on the rise in mountainous regions of North America because warmer temperatures enable its vector, the mountain pine beetle (*Dendroctonus poderosae*) to complete its life cycle in one, rather than two, years. Epidemiologists are currently monitoring the geography of infectious disease outbreaks in an effort to model potential patterns on a warmer planet.

1. List some greenhouse gases. Discuss their origins.
2. Discuss the possible role of forests in the control of CO_2.
3. What microbial processes occur in soils to both produce and degrade methane?
4. Describe the role of fungi in the extinction of harlequin frogs.

The Subsurface Biosphere

For many years, it was thought that life could exist only in the thin veneer of Earth's surface and that any microbes recovered from sediments hundreds of meters deep were contaminants obtained during sampling. This view was drastically altered in the 1980s when the U.S. Department of Energy started looking for novel ways to clean up toxic waste. The agency began funding studies that applied modern technologies to sample the deep subsurface biosphere. Subsequent reports of microbes at great depth gained credibility, and international teams of geologists and microbiologists have since recovered microbes from thousands of meters below Earth's surface. The application of culture-independent techniques to quantify the numbers and diversity of microbes has revealed that subsurface microbes constitute about one-third of Earth's living biomass. This realization has made deep subsurface microbiology an exciting and active field.

Microbial processes take place in different subsurface regions, including (1) the shallow subsurface where water flowing from the surface moves below the plant root zone; (2) regions where organic matter, originating from the Earth's surface in times past, has been transformed by chemical and biological processes to yield coal (from land plants), kerogens (from marine and

Figure 26.28 Chytrid Fungi Appear to Be Responsible for the Extinction of Many Species of Harlequin Frogs. This species of Panamanian golden frog can still be seen, but many tree frog species have been eliminated as a result of fungal infection. The fungus (*Batrachochytrium dendrobatidis*) has been able to expand its range in response to warmer temperatures.

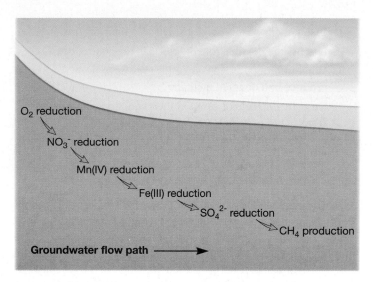

Figure 26.29 The Shallow Subsurface Biosphere. The shallow subsurface, as in a stable sediment, showing the distribution with depth of electron acceptors that can occur in an oxic pristine aquifer. In oxic sediments, the electron acceptors will be distributed with the most energetically favorable (oxygen) near the surface and the least energetically favorable at the lower zones of the geological structure. *Source: Lovey, D. K., 1991. Dissimilatory Fe (III) and Mn (IV) reduction. Microbiol. Rev. 55:259–87.*

freshwater microorganisms), and oil and gas; and (3) zones where methane is being synthesized as a result of microbial activity.

In the shallow subsurface, surface waters often move through aquifers—porous geological structures below the plant root zone. In a pristine system with an oxic surface zone, the electron acceptors used in catabolism are distributed from the most oxidized and energetically favorable (oxygen) near the surface to the least favorable (in which CO_2 is used in methanogenesis) in lower zones (**figure 26.29**).

In subsurface regions where organic matter from the Earth's surface has been buried and processed by thermal and possibly biological processes, kerogen and coals break down to yield gas and oil. After their generation, these mobile products, predominantly hydrocarbons, move upward into the more porous geological structures where microorganisms can be active. Chemical signature molecules from plant and microbial biomass are present in these petroleum hydrocarbons.

Below these zones lie vast regions where methane is present in geological structures; this methane is continuously being released to the overlying strata. Based on stable carbon isotopes analysis, methane made by archaea using H_2 as an energy source and CO_2 as both electron acceptor and carbon source will have less ^{13}C isotope than methane produced abiotically. As shown in **figure 26.30**, methane is not depleted of the heavier carbon isotope in the underlying hotter abiogenic zone, indicating that this is of chemical and thermal origin. Interestingly, the origin of H_2 for deep subsurface methanogenesis is thought to be geological. However, the actual origin of this almost limitless supply of hydrogen has not been determined with certainty.

Microorganisms also appear to be growing in intermediate-depth, oil-bearing structures. Recent studies indicate that active procaryotic assemblages are present in high-temperature (60 to 90°C) oil reservoirs, including such genera as *Thermotoga*, *Thermoanaerobacter,* and *Thermococcus*. The archaeal genera are dominated by methanogens. Thus microbial activities may be occurring above or in the "deep hot biosphere," a term suggested by Thomas Gold to describe this poorly understood region.

The discovery of deep subsurface microbes has a variety of implications. Evidence suggests that these microbes have been trapped in this environment for at least 80 million years, perhaps as long as 160 million years. They have evolved to exist in a stable environment that is anoxic and without sunlight; some chemolithoautotrophs survive only on H_2, CO_2, and water. The presence of primary production in the absence

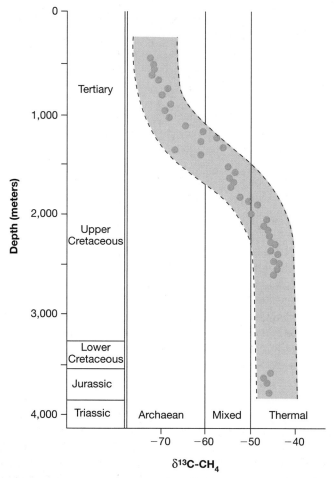

Figure 26.30 Methane Synthesis by Archaea in the Subsurface. Stable isotope techniques show that microbially mediated production of methane occurs in the subsurface. The decrease in the occurrence of the ^{13}C isotope of carbon indicates that the methane is produced by microorganisms down to a depth of 1,500 m under the floor of the North Sea. Below 2,000 m, the methane does not have a lower frequency of the ^{13}C isotope of carbon, indicating that it is formed by abiotic processes. The $\delta^{13}C$ value gives an indication of the relative proportion of ^{13}C to ^{12}C in the sample. The more negative scale values signify a decreased presence of the heavier isotope in the upper zone at this location.

of sunlight suggests that autotrophy might also be possible in the subsurface sediments of Mars. Exploring these possibilities, along with learning how metabolically active Earth's deep subsurface microbes are, remains the next challenge.

1. What types of microbial activities have been observed in the deep subsurface?

2. What happens in terms of microbiological processes when organic matter leaches from the surface into the subsurface?
3. Why are stable isotope analyses so important in studies of microbe-geological interactions?
4. What microbial genera have been observed in oil field materials?

Summary

26.1 Marine and Freshwater Microbiology

a. Most water on the Earth is marine (97%). The majority of this is cold (2 to 5°C) and at high pressure. Freshwater is a minor but important part of Earth's biosphere.

b. Oxygen solubility and diffusion rates in surface waters are limited; waters are low oxygen diffusion rate environments, in comparison with soils. Carbon dioxide, nitrogen, hydrogen, and methane are also important gases for microbial activity in waters.

c. The carbonate equilibrium system keeps the oceans buffered at pH 7.6 to 8.2 (**figure 26.2**).

d. The penetration of light into surface waters determines the depth of the photic zone. Warming the surface waters can lead to the development of a thermocline.

e. The microbial loop describes the transfer of nutrients between trophic levels while taking into account the multiple contributions of microbes to recycling nutrients. Nutrients are recycled so efficiently, the majority remain in the photic zone (**figure 26.3**).

f. Tidal mixing in estuaries, as characterized by a salt wedge, is osmotically stressful to microbes in this habitat. Thus they have evolved mechanisms to cope with rapid changes in salinity (**figure 26.4**).

g. Coastal regions such as estuaries and salt marshes can be the sites of harmful algal blooms.

h. Autotrophic microbes in the photic zone within the open ocean account for about one-half of all the carbon fixation on Earth.

i. The carbon and nitrogen budgets of the open-ocean photic zone are intensely studied because of their implications for controlling global warming (**figure 26.8**).

j. The α-proteobacteria clade SAR11 are the most abundant organisms on Earth and demonstrate unique adaptations to life in the oligotrophic open ocean.

k. Archaea are important components of the microbial community. Viruses are present at high concentrations in many waters and occur at ten fold higher levels than the bacteria. In marine systems, they may play a major role in nutrient turnover (**figure 26.11**).

l. Sediments deep beneath the ocean's surface are home to about one-half of the world's procaryotic biomass.

m. Methane hydrates, the result of psychrophilic archaeal methanogenesis under extreme atmospheric pressure, may contain more natural gas than is currently found in known reserves.

n. Glaciers and permanently frozen lakes are sites of active microbial communities. The East Antarctic Ice Sheet and Lake Vostok, which lies beneath it, are productive study sites (**figure 26.13**).

o. Nutrient sources for streams and rivers may be autochthonous or allochthonous. Often allochthonous inputs include urban, industrial, and agricultural runoff.

p. Lakes can be oligotrophic or eutrophic. Eutrophication can cause increased growth of chemoorganotrophic microbes, and the system may become anoxic (**figure 26.15**).

26.2 Microorganisms in Terrestrial Environments

a. In an ideal soil, microorganisms function in thin water films that have close contact with air. Miniaquatic environments can form within soils (**figure 26.16**).

b. Soil organic matter (SOM) helps retain nutrients and water, and maintain soil structure. It can be divided into humic and nonhumic material (**table 26.1**).

c. Bacteria and fungi in soils have different functional strategies. Fungi tend to develop on the surfaces of aggregates, whereas microcolonies of bacteria are commonly associated with smaller pores.

d. Soils form under many conditions. In all cases, organic matter accumulation occurs through the direct activities of primary producers or by the import of preformed organic materials. Soils can be formed in regions such as the Antarctic where there are no vascular plants (**figure 26.18**).

e. Mycorrhizal relationships (plant-fungal associations) are varied and complex. Six basic types can be observed, including endomycorrhizal and sheathed/ectomycorrhizal types. The hyphal network of the mycobiont can lead to the formation of a mycorrhizosphere (**figure 26.20; table 26.3**).

f. The *Rhizobium*-legume symbiosis is one of the best-studied examples of plant-microorganism interactions. This interaction is mediated by complex chemicals that serve as communication signals (**figure 26.23**).

g. The actinomycete *Frankia* forms nitrogen-fixing symbioses with some trees and shrubs.

h. *Agrobacterium* establishes a complex communication system with its plant host into which it transfers a fragment of DNA. Genes on this DNA encode proteins that result in the formation of plant tumors or galls (**figures 26.26** and **26.27**).

i. Microorganisms can play major roles in the dynamics of greenhouse gases such as carbon dioxide, nitrous oxide, nitric oxide, and methane. Microorganisms can contribute to both the production and consumption of these gases.

j. The subsurface includes at least three zones: the shallow subsurface; the zone where gas, oil, and coal have accumulated; and the deep subsurface, where methane synthesis occurs (**figures 26.29** and **26.30**).

Critical Thinking Questions

1. How might it be possible to cleanse an aging eutrophic lake? Consider chemical, biological, and physical approaches as you formulate your plan.

2. *Clostridium botulinum,* the causative agent of botulism, sometimes causes fish kills in lakes where people swim. Currently there are few, if any, monitoring procedures for this potential source of disease transmission to humans. Do you think a monitoring program is needed, and, if so, how would you implement such a program?

3. Do you think fertilization of the ocean with iron to increase CO_2 drawdown is a good approach to controlling global warming? Why or why not? Why do you think similar experiments yield different results?

4. Tropical soils throughout the world are under intense pressure in terms of agricultural development. What land use and microbial approaches might be employed to better maintain this valuable resource?

5. Why might vascular plants have developed relationships with so many types of microorganisms? These molecular-level interactions show many similarities when microbe-plant and microbe-human interactions are considered. What does this suggest concerning possible common evolutionary relationships?

6. How might you maintain organisms from the deep hot subsurface under their in situ conditions when trying to culture them? Compare this problem with that of working with microorganisms from deep marine environments.

7. Soil bacteria such as *Streptomyces* produce the bulk of known antibiotics. Look up the competitors for *Streptomyces,* the types of antibiotics these bacteria produce, and how the compounds are effective against competitors (what are the physiological targets?). Would you expect aquatic/marine bacteria to be major producers of antibiotics? Why or why not?

Learn More

Learn more by visiting the Prescott website at www.mhhe.com/prescottprinciples, where you will find a complete list of references.

Microbial Interactions

27

A "garden" of tube worms (*Riftia pachyptila*) at the Galápagos Rift hydrothermal vent site (depth 2,550 m). Each worm grows to more than a meter in length thanks to endosymbiotic chemolithoautotrophic bacteria, which provide carbohydrate to these gutless worms. The bacteria use the Calvin cycle to fix CO_2 and H_2S as an electron donor.

amensalism   A relationship in which the product of one organism has a negative effect on another organism.

commensalism   Living on or within another organism without injuring or benefiting the other organism. Usually the smaller of the two organisms is the **commensal.**

competition   An interaction between two organisms attempting to use the same resource (nutrients, space, etc.).

competitive exclusion principle   When two competing organisms overlap in resource use, one of the organisms will be more successful than the other.

consortium   A physical association of two different organisms, usually beneficial to both organisms.

cooperation   A positive but not obligatory interaction between two different organisms.

ectosymbiont   An organism that lives on or outside the body of another organism in a symbiotic association.

endosymbiont   An organism that lives within the body of another organism in a symbiotic association.

gnotobiotic   Animals that are germfree (microorganism free) or live in association with one or more known microorganisms.

microbial flora or **microbiota**   The typical or normal microorganisms found on and in a eucaryotic host.

mutualism   A type of symbiosis in which both partners gain from the association and are unable to survive without it. The **mutualist** and the host are metabolically dependent on each other.

normal microbial flora or **microbiota**   The microorganisms normally associated with a particular tissue or structure.

opportunistic microorganism or **pathogen**   A microorganism that is usually free-living or a part of the host's normal microbiota but that may become pathogenic under certain circumstances, such as when the immune system is compromised.

parasitism   A type of symbiosis in which one organism benefits from the other and the host is usually harmed.

pathogen   Any virus, bacterium, or other infectious agent that causes disease.

symbiosis   The living together or close association of two dissimilar organisms; each of these organisms is known as a **symbiont.**

syntrophism   The association in which the growth of one organism either depends on or is improved by the provision of one or more growth factors or nutrients by a neighboring organism. Sometimes both organisms benefit.

. . . every organic being is related, in the most essential yet often hidden manner, to that of all other organic beings, with which it comes into competition for food or residence, or from which it has to escape, or on which it preys.

—*Charles Darwin*

Our discussion of microbial ecology has so far considered microbial communities in complex ecosystems. However, the ecology of microorganisms also involves the physiology and behavior of microbes as they interact with one another and with higher organisms. In this chapter, we begin by defining types of microbial interactions and present a number of illustrative examples. We conclude this chapter with a consideration of perhaps the most intimate, yet still relatively unexplored, microbial habitat: the human body.

We must begin our discussion by defining the term **symbiosis.** Although symbiosis is often used in a nonscientific sense to mean a mutually beneficial relationship, here we use the term in its original broadest sense, as an association of two or more different species of organisms, as suggested by H. A. deBary in 1879.

27.1 MICROBIAL INTERACTIONS

Microorganisms can associate physically with other organisms in a variety of ways. One organism can be located on the surface of another, as an **ectosymbiont** or, one organism can be located within another organism as an **endosymbiont.** While the simplest microbial interactions involve two members, a **symbiont** and its host, a number of interesting organisms host more than one symbiont. The term **consortium** can be used to describe this physical relationship. These physical associations can be intermittent and cyclic or permanent. Examples of intermittent and cyclic associations of microorganisms with plants and marine animals are shown in **table 27.1.** Important human diseases, including listeriosis, malaria, leptospirosis, legionellosis, and vaginosis, also involve such intermittent and cyclic symbioses. Interesting permanent relationships also occur between bacteria and animals, as shown in **table 27.2.** In these cases, an important characteristic of the host animal is conferred by the permanent bacterial symbiont.

Although it is possible to observe microorganisms in these varied physical associations with other organisms, the fact that there is some type of physical contact provides no information on the nature of the interactions that might be occurring. These interactions include mutualism, cooperation, commensalism, predation, parasitism, amensalism, and competition (**figure 27.1**). These interactions are now discussed.

Table 27.1	Intermittent and Cyclical Symbioses of Microorganisms with Plants and Marine Animals	
Symbiosis	**Host**	**Cyclical Symbiont**
Plant-bacterial	*Gunnera* (tropical angiosperm)	*Nostoc* (cyanobacterium)
	Azolla (rice paddy fern)	*Anabaena* (cyanobacterium)
	Phaseolus (bean)	*Rhizobium* (N_2 fixer)
	Ardisia (angiosperm)	*Protobacterium*
Marine animals	Coral coelenterates	*Symbiodinium* (dinoflagellate)
	Luminous fish	*Vibrio, Photobacterium*

Adapted from Margulis, L., and Chapman, M. J. 1998. Endosymbioses: Cyclical and permanent in evolution. *Trends in Microbiology* 6(9):342–46, tables 1, 2, and 3.

Mutualism

Mutualism (Latin *mutuus,* borrowed or reciprocal) defines the relationship in which some reciprocal benefit accrues to both partners. This is an obligatory relationship in which the **mutualist** and the host are dependent on each other. In many cases, the individual organisms will not survive when separated. Several examples of mutualism are presented next.

Microorganism-Insect Mutualisms

Mutualistic associations are common between insects and microbes. This is related to the foods they use, which often include plant sap or animal fluids lacking in essential vitamins and amino acids. The required vitamins and amino acids are provided by bacterial symbionts in exchange for a secure habitat and ample nutrients (**Microbial Diversity & Ecology 27.1**). The aphid is an excellent example of this mutualistic relationship. Cells of this insect harbor the γ-proteobacterium *Buchnera aphidicola,* and a mature insect contains literally millions of these bacteria in its body. *B. aphidicola* provides its host with 10 essential amino acids, and if the

Table 27.2	Examples of Permanent Bacterial-Animal Symbioses and the Characteristics Contributed by the Bacterium to the Symbiosis	
Animal Host	**Symbiont**	**Symbiont Contribution**
Sepiolid squid (*Euprymna scolopes*)	Luminous bacterium (*Vibrio fischeri*)	Luminescence
Medicinal leech (*Hirudo medicinalis*)	Enteric bacterium (*Aeromonas veronii*)	Blood digestion
Aphid (*Schizaphis graminum*)	Bacterium (*Buchnera aphidicola*)	Amino acid synthesis
Nematode worm (*Heterorhabditis* spp.)	Luminous bacterium (*Photorhabdus luminescens*)	Predation and antibiotic synthesis
Shipworm mollusk (*Lyrodus pedicellatus*)	Gill cell bacterium	Cellulose digestion and nitrogen fixation

Source: From Ruby, E. G. 1999. Ecology of a benign "infection": Colonization of the squid luminous organ by *Vibrio fischeri*. In *Microbial ecology and infectious disease,* E. Rosenberg, editor, American Society for Microbiology, Washington, D.C., 217–31, table 1.

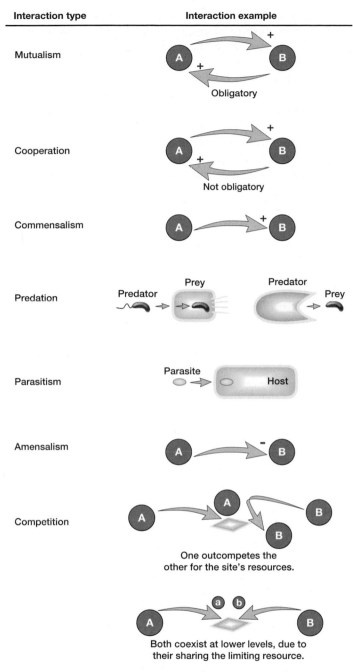

Interaction type	Interaction example

Mutualism — Obligatory

Cooperation — Not obligatory

Commensalism

Predation — Prey / Predator / Predator / Prey

Parasitism — Parasite / Host

Amensalism

Competition — One outcompetes the other for the site's resources.

Both coexist at lower levels, due to their sharing the limiting resource.

Figure 27.1 Microbial Interactions. Basic characteristics of symbiotic interactions that can occur between different organisms.

insect is treated with antibiotics, it dies. Likewise, *B. aphidicola* is an obligate mutualistic symbiont. The inability of either partner to grow without the other indicates that the two organisms underwent coevolution, or have evolved together. It is estimated that the *B. aphidicola*–aphid endosymbiosis was established about 150 million years ago. The genomic sequences of two different endosymbiotic *B. aphidicola* strains reveal extreme genomic stability. These strains diverged 50 to 70 million years ago, and since that time, there have been no gene duplications, translocations, inversions, or genes acquired by horizontal transfer. The genomes are small, only 0.64 Mb each, with 93%

of genes common to both strains. Furthermore, only two genes have orthologues in the taxonomically related *E. coli*. This tremendous degree of stability implies that although the initial acquisition of the endosymbiont by ancestral aphids enabled their use of an otherwise deficient food source (sap), the bacteria have not continued to expand the ecological niche of their insect host through the acquisition of new traits that might be advantageous elsewhere. ≪ *Bioinformatics: Genome annotation (section 15.4); Microbial evolution (section 17.1)*

The protozoan-termite relationship is another classic example of mutualism in which flagellated protists live in the gut of wood roaches and termites (**figure 27.2***a*). The protists engulf wood

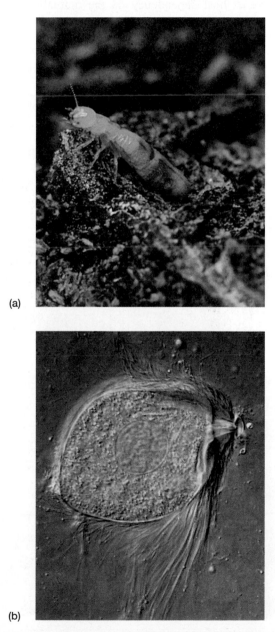

(a)

(b)

Figure 27.2 Mutualism. Light micrographs of (a) a worker termite of the genus *Reticulitermes* eating wood (×10) and (b) *Trichonympha,* a multiflagellated protozoan from the termite's gut (×135). The ability of *Trichonympha* to break down cellulose allows termites to use wood as a food source.

particles ingested by their host, digest the cellulose, and metabolize it to acetate and other products (figure 27.2*b*). Termites then use the acetate released by the protists as their true carbon source. Because the host is almost always incapable of synthesizing cellulases (enzymes that catalyse the hydrolysis of cellulose), it depends on the mutualistic protists for its existence.

Zooxanthellae

Many marine invertebrates (sponges, jellyfish, sea anemones, corals, ciliated protists) harbor endosymbiotic dinoflagellates called zooxanthellae within their tissue (**figure 27.3*a***). Because the degree of host dependency on the mutualistic protist is somewhat variable, only one well-known example is presented. ≪ *Protist classification:* Alveolata *(section 23.2)*

The hermatypic (reef-building) corals (figure 27.3*b*) satisfy most of their energy requirements using their zooxanthellae,

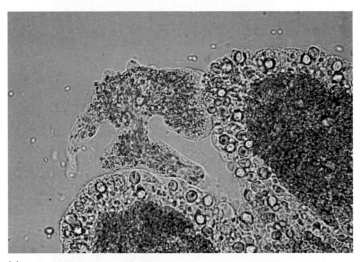

(a)

(b)

Figure 27.3 Zooxanthellae. (a) Zooxanthellae (green) within the tip of a hydra tentacle (×150). (b) The green color of this rose coral (*Manilina*) is due to the abundant zooxanthellae within its tissues.

which are found at densities between 5×10^5 and 5×10^6 cells per square centimeter of coral animal. In exchange for up to 95% of their photosynthate (fixed carbon), zooxanthellae receive nitrogenous compounds, phosphates, CO_2, and protection from UV light from their hosts. This efficient form of nutrient cycling and tight coupling of trophic levels accounts for the stunning success of reef-building corals in developing vibrant ecosystems. However, during the past several decades, the number of coral bleaching events has increased dramatically. **Coral bleaching** is defined as a loss of either the photosynthetic pigments from the zooxanthellae or the expulsion of the protists by the coral. It has been determined that damage to photosystem II of the zooxanthellae generates reactive oxygen species (ROS); it is these ROS that appear to be the direct cause of damage (recall that photosystem II uses water as the electron source resulting in the evolution of oxygen). Coral bleaching appears to be caused by a variety of stressors, but one important factor is temperature. Temperature increases as small as 2°C above the average summer maxima can trigger coral bleaching. Sadly, evidence suggests that many corals and their zooxanthellae will be unable to evolve quickly enough to keep pace with the predicted increases in ocean temperatures if global warming continues unchecked. ≪ *Phototrophy: Light reaction in oxygenic photosynthesis (section 10.12)*

Sulfide-Based Mutualisms

Tube worm–bacterial relationships exist several thousand meters below the surface of the ocean, where the Earth's crustal plates are spreading apart (**figure 27.4**). Vent fluids are anoxic, contain high concentrations of hydrogen sulfide, and can reach a temperature of 350°C. However, because of increased atmospheric pressure, the water does not boil. The seawater surrounding these vents has sulfide concentrations around 250 µM and temperatures 10 to 20°C above the ambient seawater temperature of about 2°C.

Giant (>1 m in length), red, gutless tube worms (*Riftia* spp. **figure 27.5*a***) near these hydrothermal vents provide an example of a remarkably successful form of mutualism in which chemolithotrophic bacterial endosymbionts are maintained within specialized cells of the tube worm host (figure 27.5*b,c,d*). *Riftia* tube worms live at the interface between hot, anoxic, sulfide-containing fluids of the vents and cold, oxygenated seawater. To provide both reduced sulfur and oxygen to their bacterial endosymbionts, the worm's blood contains a unique kind of hemoglobin, which accounts for the bright-red plume extending out of their tubes. Hydrogen sulfide (H_2S) and O_2 are removed from the seawater by the worm's hemoglobin and delivered to a special organ called the trophosome. The trophosome is packed with chemolithotrophic bacterial endosymbionts that fix CO_2 using the Calvin cycle (*see figure 11.5*) with electrons provided by H_2S. The CO_2 is carried to the endosymbionts in three ways: (1) freely in the bloodstream, (2) bound to hemoglobin, and (3) as organic acids such as malate and succinate. When these acids are decarboxylated, they release CO_2. This process is similar to carbon fixation by

27.1 *Wolbachia pipientis*: The World's Most Infectious Microbe?

Most people have never heard of the bacterium *Wolbachia pipientis,* but this rickettsia infects more organisms than does any other microbe. It is known to infect a variety of crustaceans, spiders, mites, millipedes, and parasitic worms and may infect more than a million insect species worldwide. *Wolbachia* inhabits the cytoplasm of these animals where it apparently does no harm. To what does *Wolbachia* owe its extraordinary success? Quite simply, this endosymbiont is a master at manipulating its hosts' reproductive biology. In some cases, it can even change the sex of the infected organism.

Wolbachia is transferred from one generation of host to the next through the eggs of infected female hosts. So to survive, this microbe must ensure the fertilization and viability of infected eggs while decreasing the likelihood that uninfected eggs survive. In the 1970s, scientists noticed that if a male wasp infected with *Wolbachia* mated with an uninfected female, few if any of the uninfected offspring survived. However, if infected females of the same wasp species mated with either infected or uninfected males, all of the eggs were viable—and infected with *Wolbachia*. Although scientists had no clear understanding of the cellular or molecular mechanisms involved, they suggested that "cytoplasmic incompatibility" might be responsible. They proposed that the cytoplasm of infected sperm was toxic to uninfected eggs but that eggs carrying *Wolbachia* produced an antidote to the hypothetical poison.

It was not until the 1990s that the mechanism of cytoplasmic incompatibility became clear. In the wasp *Nasonia vitripennis,* cytoplasmic incompatibility involves timing, not toxins. Specifically, when infected sperm fertilize uninfected eggs, the sperm chromosomes try to align with those of the egg while the egg's chromosomes are still confined to the pronucleus. These eggs ultimately divide as if never fertilized and develop into males. However, chromosomes behave normally when a male and infected female mate. This yields a normal sex distribution and all progeny are infected with *Wolbachia*.

In other infected species, scientists discovered that in addition to cytoplasmic incompatibility, *Wolbachia* has developed other means to ensure its endurance (**Box figure**). In some insect species, the endosymbiont simply kills all the male offspring and induces parthenogenesis of infected females—that is, the mothers simply clone themselves. This limits genetic diversity but allows 100% transmission of *Wolbachia* to the next generation. In still other insects, the microbe allows the birth of males but then modifies their hormones so that the males become feminized and produce eggs.

Wolbachia's ability to manipulate the reproduction of its hosts has led some to question the bacterium's potential role in speciation. It was noticed that two North American wasp species (*N. giraulti* and *N. longicornis*), each carrying a different strain of *Wolbachia,* appeared to be morphologically, behaviorally, and, most importantly, genetically similar. Predictably, when the two species mate with each other, there are no viable offspring. But when wasps are treated with an antibiotic to cure them of their *Wolbachia* infections, the two wasp species mate and produce viable, fertile offspring. Could it be that while there are no genetic barriers to reproduction, the presence of two different *Wolbachia* endosymbionts forms a reproductive wall that could ultimately drive the evolution of new species?

Finally, *Wolbachia* may be driving more than speciation. They are required for the embryogenesis of filarial nematodes, which cause diseases such as elephantitis and river blindness. *Onchocerca volvulus* is the filarial nematode that causes river blindness. It is transmitted by blackflies in Africa, Latin America, and Yemen, with a worldwide incidence of about 18 million people. When a fly bites a person, the nematode establishes its

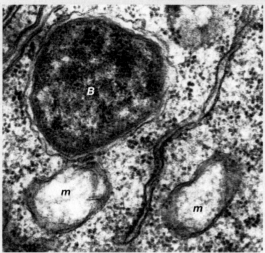

***Wolbachia pipientis* Within the Egg Cytoplasm of the Ant *Gnamptogenys menadensis*.** In this insect, *Wolbachia* is maternally transmitted, so the bacterium has evolved mechanisms to manipulate the sex distribution of the offspring so that the host produces mostly females. The *Wolbachia* cell is indicated by B and host mitochondria are labeled m.

River blindness. This is the second-leading cause of blindness worldwide. Evidence suggests that it is not the nematode but its endosymbiont, *Wolbachia pipientis*, that causes the severe inflammatory response that leaves many, like the man shown here, blind.

home in small nodules beneath the skin, where it can survive for as long as 14 years. During that time, it releases millions of larvae, many of which migrate to the eye. Eventually the host mounts an inflammatory response that results in progressive vision loss (**Box figure**). It is now recognized that this inflammatory response is principally directed at the *Wolbachia* infecting the nematodes, not the worms themselves. German researchers discovered that in patients treated with a single course of antibiotic to kill the endosymbiont, nematode reproduction stopped. While inflammatory

damage cannot be reversed, disease progression is halted. This may prove a more effective and cost-efficient means of treatment than the current anti-parasitic method that must be repeated every six months.

Our understanding of the distribution of *Wolbachia* and the mechanistically clever ways it has evolved to ensure its survival will continue to grow. The microbe may ultimately become a valuable tool in investigating the complexities of speciation as well as the key to curing a devastating disease.

Sources: St. Andre, A; Blackwell, N. M.; Hall, L. R.; Hoerauf, A.; Brattig, N. W.; Volkman, L.; Taylor, M. J.; Ford, L.; Hise, A. G.; Lass, J. H.; Diaconu, E.; and Pearlman, E. 2002. The role of endosymbiotic Wolbachia *bacteria in the pathogenesis of river blindness.* Science 295:1892–95.

Zimmer, C. 2001. Wolbachia: A tale of sex and survival. Science 292:1093–95.

plants and cyanobacteria, but it occurs in the deepest, darkest reaches of the ocean. This mutualism enables *Riftia* to grow to an astounding size in densely packed communities and the bacteria (which have not yet been cultured in the laboratory) reach densities of up to 10^{11} cells per gram of worm tissue.

Methane-Based Mutualisms

Other unique food chains involve methane-fixing microorganisms. By converting methane to carbohydrate, these bacteria perform the first step in providing organic matter for consumers. **Methanotrophic bacteria** are capable of using methane as a sole carbon source. They occur as intracellular symbionts of methane-vent mussels. In these mussels, the thick, fleshy gills are filled with bacteria. In the Barbados Trench, methanotrophic carnivorous sponges have been discovered in a mud volcano at a depth of 4,943 m. Abundant methanotrophic symbionts were confirmed by the presence of enzymes related to methane oxidation in sponge tissues. These sponges are not satisfied with just bacterial symbionts; they also trap swimming prey.

Methanotrophic microorganisms are also important in other ecosystems. For example, methanotrophic endosymbionts help reduce the flux of methane from peat bogs; wetlands are the largest natural source of this greenhouse gas. Sphagnum moss, the principal plant in peat bogs (and a favorite among florists), can grow when submerged in water. Methanotrophic α-proteobacteria living within the outer cortex cells of sphagnum stems oxidize methane as it diffuses through the water column:

$$CH_4 + 2O_2 \rightleftharpoons CO_2 + 2H_2O$$

The resulting CO_2 is then readily fixed by the plant, which uses the Calvin cycle:

$$2CO_2 + 2H_2O \rightleftharpoons 2CH_2O + 2O_2$$

This enables extremely efficient carbon recycling within this ecosystem:

$$CH_4 + CO_2 \rightleftharpoons 2CH_2O$$

The Rumen Ecosystem

Ruminants are the most successful and diverse group of mammals on Earth today. Examples include cattle, deer, elk, bison, water buffalo, camels, sheep, goats, giraffes, and caribou. These animals spend vast amounts of time chewing their cud—a small ball of partially digested grasses that the animal has consumed but not yet completely digested. It is thought that the ruminants evolved an "eat now, digest later" strategy because their grazing can often be interrupted by predator attacks.

These herbivorous animals have stomachs that are divided into four chambers (**figure 27.6**). The upper part of the ruminant stomach is expanded to form a large pouch called the rumen and a smaller, honeycomb-like region, the reticulum. The lower portion is divided into an antechamber, the omasum, followed by the "true" stomach, the abomasum. The rumen is a highly muscular, anaerobic fermentation chamber where grasses eaten by the animal are digested by a diverse microbial community that includes bacteria, archaea, fungi, and protists. This microbial community is large—about 10^{12} organisms per milliliter of digestive fluid. When the animal eats plant material, it is mixed with saliva and swallowed without chewing to enter the rumen. Here microbial attack and further mixing coats the grass with microbes, reducing it to a pulpy, partially digested, mass. At this point the mass moves into the reticulum, where it is regurgitated as cud, chewed, and reswallowed by the animal. As this process proceeds, the grass becomes progressively more liquefied and flows out of the rumen into the omasum and then the abomasum. Here the nutrient-enriched grass material meets the animal's digestive enzymes, and soluble organic and fatty acids are absorbed into the animal's bloodstream.

The microbial community in the rumen is extremely dynamic. The rumen is slightly warmer than the rest of the animal, and with a redox potential of about −30 mV, all resident microorganisms must carry out anaerobic metabolism. Very specific interactions occur within the microbial community. One group of bacteria produces extracellular cellulases that cleave the $\beta(1{\rightarrow}4)$ linkages between the successive D-glucose molecules that form plant cellulose. The D-glucose is then

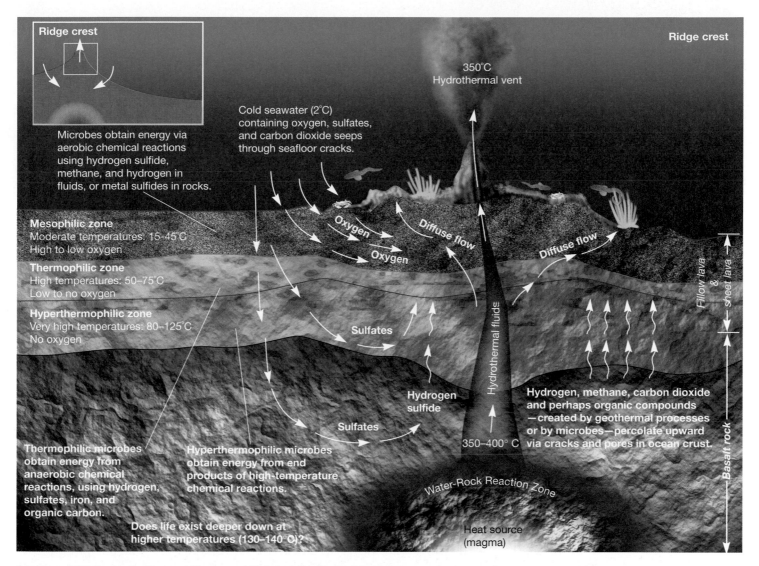

Figure 27.4 Hydrothermal Vents and Related Geological Activity. The chemical reactions between seawater and rocks that occur over a range of temperatures on the seafloor supply the carbon and energy that support a diverse collection of microbial communities in specific niches within the vent system.

fermented to organic acids such as acetate, butyrate, and propionate. These organic acids, as well as fatty acids, are the true energy source for the animal. In some ruminants, the processing of organic matter stops at this stage. In others, such as cows, acetate, CO_2, and H_2 are used by methanogenic archaea to generate methane (CH_4), a greenhouse gas. In fact, a single cow can produce as much as 200 to 400 liters of CH_4 per day. The animal releases this CH_4 by a process called eructation (Latin *eructare,* to belch)—and although it may seem preposterous, ruminants contribute a significant fraction of global CH_4. Although methanogens consume acetate that could be used by their animal hosts, they provide most of the vitamins needed by the ruminant. In fact, rumen microbes are so effective in fortifying the grass consumed by the animal, most ruminants, unlike humans, have no required dietary amino acids. << *Fermentation (section 10.7)*

1. What is the critical characteristic of a mutualistic relationship?

2. How might one test to see if an insect-microbe relationship is mutualistic?

3. What is the role of *Riftia* hemoglobin to the success of the tube worm–endosymbiont mutualistic relationship?

4. How is the *Riftia* endosymbiont similar to cyanobacteria? How is it different?

5. Describe how the sphagnum-methanotroph mutualism results in efficient carbon cycling.

6. What structural features of the rumen make it suitable for an herbivorous diet?

7. Why is it important that the rumen is a reducing environment?

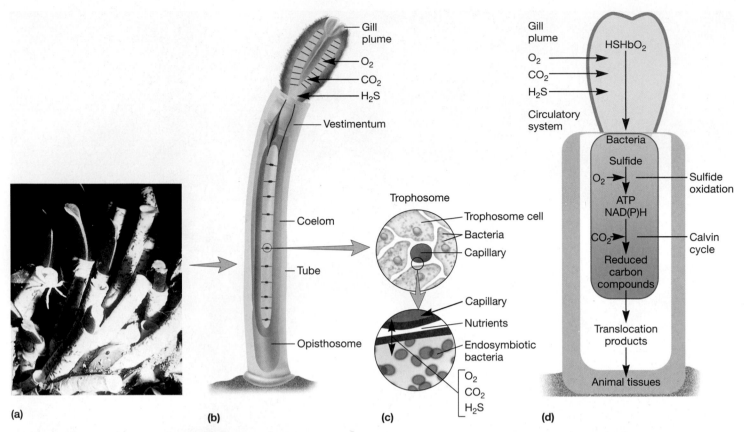

Figure 27.5 **The Tube Worm–Bacterial Relationship.** (a) A community of tube worms (*Riftia pachyptila*) at the Galápagos Rift hydrothermal vent site (depth 2,550 m). Each worm is more than a meter in length and has a 20 cm gill plume. (b, c) Schematic illustration of the anatomical and physiological organization of the tube worm. The animal is anchored inside its protective tube by the vestimentum. At its anterior end is a respiratory gill plume. Inside the trunk of the worm is a trophosome consisting primarily of endosymbiotic bacteria, associated cells, and blood vessels. At the posterior end of the animal is the opisthosome, which anchors the worm in its tube. (d) Oxygen, carbon dioxide, and hydrogen sulfide are absorbed through the gill plume and transported to the blood cells of the trophosome. Hydrogen sulfide is bound to the worm's hemoglobin ($HSHbO_2$) and carried to the endosymbiont bacteria. The bacteria oxidize the hydrogen sulfide and use some of the released energy to fix CO_2 in the Calvin cycle. Some of the reduced carbon compounds synthesized by the endosymbiont are translocated to the animal's tissues.

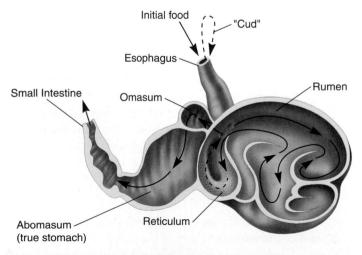

Figure 27.6 **Ruminant Stomach.** The stomach compartments of a cow. The microorganisms are active mainly in the rumen. Arrows indicate direction of food movement.

Cooperation

Cooperation and commensalism are two positive but not obligatory types of symbioses found widely in the microbial world (figure 27.1). They involve syntrophic relationships. **Syntrophism** (Greek *syn*, together, and *trophe*, nourishment) is an association in which the growth of one organism either depends on or is improved by growth factors, nutrients, or substrates provided by another organism growing nearby. Sometimes both organisms benefit.

A cooperative relationship is not obligatory, and for most microbial ecologists, this nonobligatory aspect differentiates cooperation from mutualism. Unfortunately, it is often difficult to distinguish obligatory from nonobligatory because that which is obligatory in one habitat may not be in another (e.g., the laboratory). Nonetheless, the most useful distinction between cooperation and mutualism is the observation that cooperating organisms can be separated from one another and remain viable, although they may not function as well.

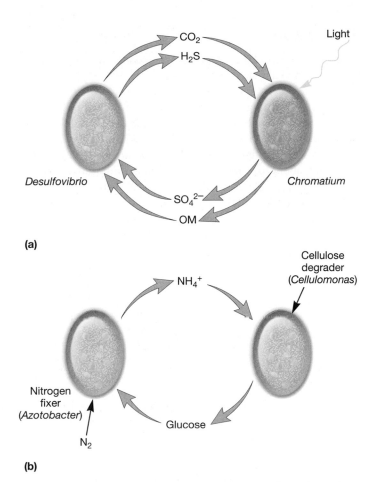

(a)

(b)

Figure 27.7 Examples of Cooperative Symbiotic Processes. (a) The organic matter (OM) and sulfate required by *Desulfovibrio* are produced by the *Chromatium* in its photosynthesis-driven reduction of CO_2 to organic matter and oxidation of sulfide to sulfate. (b) *Azotobacter* uses glucose provided by a cellulose-degrading microorganism such as *Cellulomonas,* which uses the nitrogen fixed by *Azotobacter.*

Figure 27.8 A Marine Worm–Bacterial Cooperative Relationship. *Alvinella pompejana,* a 10 cm long worm, forms a cooperative relationship with bacteria that grow as long threads on the worm's surface. The bacteria and *Alvinella* are found near the black smoker-heated water fonts.

1. How does cooperation differ from mutualism? What might be some of the evolutionary implications of both types of symbioses?

2. What is syntrophism? Is physical contact required for this relationship?

3. Why is *Alvinella* a good example of cooperative microorganism-animal interactions?

Two examples of a cooperative relationship include the association between *Desulfovibrio* and *Chromatium,* in which the carbon and sulfur cycles are linked (**figure 27.7a**), and the interaction of a nitrogen-fixing microorganism with a cellulolytic organism such as *Cellulomonas* (figure 27.7b). In the second example, the cellulose-degrading microorganism liberates glucose from the cellulose, which can be used by nitrogen-fixing microbes.

Cooperative interactions are also found between microorganisms and eucaryotic hosts. In some cases, the carbon fixed by sulfide-dependent autotrophic filamentous microbes serves as the carbon and energy source for its host. Some of the most interesting examples include the polychaete worms *Alvinella pompejana* (**figure 27.8**), the Pompeii worm, and *Paralvinella palmiformis,* the Palm worm. Both have filamentous bacteria on their dorsal surfaces. These filamentous bacteria tolerate high levels of metals such as arsenic, cadmium, and copper. When growing on the surface of the animal, they may provide protection from these toxic metals, as well as thermal protection; in addition, they appear to be used as a food source.

Commensalism

Commensalism (Latin *com,* together, and *mensa,* table) is a relationship in which one symbiont, the **commensal,** benefits while the other (sometimes called the host) is neither harmed nor helped, as shown in figure 27.1. This is a unidirectional process. Often both the host and the commensal "eat at the same table." The spatial proximity of the two partners permits the commensal to feed on substances captured or ingested by the host, and the commensal often obtains shelter by living either on or in the host. The commensal is not directly dependent on the host metabolically, so when it is separated from the host experimentally, it can survive without the addition of factors of host origin.

Commensalistic relationships between microorganisms include situations in which the waste product of one microorganism is a substrate for another species. One good example is nitrification—the oxidation of ammonium ion to nitrate. Nitrification occurs in two steps: first, microorganisms such as *Nitrosomonas* oxidize ammonium to nitrite, and second, nitrite is oxidized to nitrate by *Nitrobacter* and similar bacteria. *Nitrobacter* benefits from its association with *Nitrosomonas* because it uses nitrite to obtain energy for growth. A second example of this type of relationship is found in anoxic methanogenic ecosystems such as sludge digesters, anoxic freshwater aquatic sediments, flooded soils, and the rumen ecosystem. Obligate proton-reducing acetogens live in these environments. These microbes oxidize ethanol, butyrate, propionate, and other fermentation end products to H_2, CO_2, and acetate (hence the name "acetogen"). Although the enzymology is not well understood, it is clear that these reactions are thermodynamically unfavorable under standard conditions, that is, they have a positive ΔG. However, in the presence of methanogens, which consume the H_2 as it is being generated, the reaction becomes thermodynamically possible. This is called **interspecies hydrogen transfer.** Various fermentative bacteria produce low molecular-weight fatty acids that can be degraded by anaerobic bacteria such as *Syntrophobacter* to produce H_2 as follows:

$$\text{Propionic acid} \rightarrow \text{acetate} + CO_2 + H_2$$

Syntrophobacter uses protons ($H^+ + H^+ \rightarrow H_2$) as terminal electron acceptors in ATP synthesis. The products H_2 and CO_2 are then used by methanogenic archaea such as *Methanospirillum*:

$$4H_2 + CO_2 \rightarrow CH_4 + 2H_2O$$

By synthesizing methane, *Methanospirillum* maintains a low H_2 concentration in the immediate environment of both microbes. Continuous removal of H_2 promotes further fatty acid fermentation and H_2 production. Because increased H_2 production and consumption stimulate the growth rates of *Syntrophobacter* and *Methanospirillum*, both participants in the relationship benefit. Interestingly, it was long held that these reactions were performed by a single microbe.

Commensalistic associations also occur when one microbial group modifies the environment to make it better suited for another organism. The synthesis of acidic waste products during fermentation stimulates the proliferation of more acid-tolerant microorganisms, which may be only a minor part of the microbial community at neutral pH. A good example is the succession of microorganisms during milk spoilage. Biofilm formation provides another example. The colonization of a newly exposed surface by one type of microorganism (an initial colonizer) makes it possible for other microorganisms to attach to the microbially modified surface. << *Microbial growth in natural environments: Biofilms (section 7.6)*

Commensalism also is important in the colonization of the human body and the surfaces of other animals and plants. The microorganisms associated with an animal's skin and body orifices can use volatile, soluble, and particulate organic compounds from the host as nutrients. Under most conditions, these microbes do not cause harm. However, if the host organism is stressed or the skin is punctured, these normally commensal microorganisms may become pathogenic by entering a different environment. These interactions are discussed in more detail in section 30.2.

1. How does commensalism differ from cooperation?
2. Why is nitrification a good example of a commensalistic process?
3. What is interspecies hydrogen transfer, and why is this beneficial to both producers and consumers of hydrogen?

Predation

As is the case with larger organisms, predation among microbes involves a predator species that attacks and usually kills its prey. Over the last several decades, microbiologists have discovered a number of fascinating bacteria that survive by their ability to prey upon other microbes. Several of the best examples are *Bdellovibrio*, *Vampirococcus*, and *Daptobacter* (**figure 27.9**).

Bdellovibrio is an active hunter that is vigorously motile, swimming about looking for susceptible gram-negative bacterial prey. Upon sensing such a cell, *Bdellovibrio* swims faster until it collides with the prey cell. It then bores a hole through the outer membrane of its prey and enters the periplasmic space. As it grows, it forms a long filament that eventually septates to produce progeny bacteria. Lysis of the prey cell releases new *Bdellovibrio*

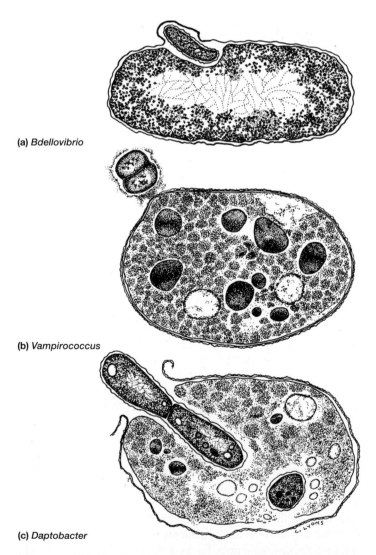

(a) *Bdellovibrio*

(b) *Vampirococcus*

(c) *Daptobacter*

Figure 27.9 Examples of Predatory Bacteria Found in Nature. (a) *Bdellovibrio,* a periplasmic predator that penetrates the cell wall and grows outside the plasma membrane, (b) *Vampirococcus* has a unique epibiotic mode of attacking a prey bacterium, and (c) *Daptobacter* showing its cytoplasmic location as it attacks a susceptible bacterium.

cells. *Bdellovibrio* will not attack mammalian cells, and gram-negative prey bacteria have never been observed to acquire resistance to *Bdellovibrio* attack. This has raised interest in the use of *Bdellovibrio* as a "probiotic" to treat infected wounds. Although this has not yet been tried, one can imagine that with the rise in antibiotic-resistant pathogens, such forms of treatments may become viable alternatives. << *Class* Deltaproteobacteria *(section 20.4)*

Although *Vampirococcus* and *Daptobacter* also kill their prey, they gain entry in a less-dramatic fashion. *Vampirococcus* attaches itself as an epibiont to the outer membrane of its prey (figure 27.9*b*). It then secretes degradative enzymes that result in the release of the prey's cytoplasmic contents. In contrast, *Daptobacter* penetrates the prey cell and consumes the cytoplasmic contents directly (figure 27.9*c*).

A surprising finding is that predation has many beneficial effects, especially when one considers interactive populations of predators and prey. Simple ingestion and assimilation of a prey bacterium can lead to increased rates of nutrient cycling, critical for the functioning of the microbial loop (*see figure 26.3*). Ingestion and short-term retention of bacteria also are critical for ciliated protists functioning in the rumen, where methanogenic bacteria contribute to the health of the ciliates by decreasing toxic hydrogen levels by using H_2 to produce methane, which then is passed from the rumen.

Predation also can provide a protective, high-nutrient environment for particular prey. Protists ingest the gram-positive bacterium *Legionella* and protect this important pathogen from chlorine, which often is used in an attempt to control *Legionella* in cooling towers and air-conditioning units. *L. pneumophila* has been found to have a greater potential to invade macrophages and epithelial cells after predation, indicating that ingestion not only provides protection but also may enhance pathogenicity. A similar phenomenon of survival in protozoa has been observed for *Mycobacterium avium*, a pathogen of worldwide concern. These protective aspects of predation have major implications for survival and control of disease-causing microorganisms in biofilms present in water supplies and air-conditioning systems.

Clearly predation in the microbial world is not straightforward. It often has a fatal and final outcome for an individual prey organism, but it can have a wide range of beneficial effects on prey populations. Regardless of the outcome, predation is critical in the functioning of natural environments.

Parasitism

Parasitism is one of the most complex microbial interactions; the line between parasitism and predation is difficult to define (figure 27.1). This is a relationship between two organisms in which one benefits from the other and the host is usually harmed. It can involve nutrient acquisition, physical maintenance in or on the host, or both. In parasitism, there is always some coexistence between host and parasite. This is because a host that dies immediately after parasite invasion may prevent the microbe from reproducing to sufficient numbers to ensure colonization of a

new host. But what happens if the host-parasite equilibrium is upset? If the balance favors the host (perhaps by a strong immune defense or antimicrobial therapy), the parasite loses its habitat and may be unable to survive. On the other hand, if the equilibrium is shifted to favor the parasite, the host becomes ill and, depending on the specific host-parasite relationship, may die. One good example is the disease typhus. This disease is caused by *Rickettsia typhi*, which is harbored in fleas that live on rats. It is transmitted to humans who are bitten by such fleas. Humans often live in association with rats, and in such communities, there is always a small number of people with typhus—that is to say, typhus is endemic. However, during times of war or when people are forced to become refugees, lack of sanitation and overcrowding result in an increased number of rat-human interactions. Typhus can then reach epidemic proportions. During the Crimean War (1853–1856), about 213,000 men were killed or wounded in combat, while over 850,000 were sickened or killed by typhus.

On the other hand, a controlled parasite-host relationship can be maintained for long periods of time. For example, lichens (**figure 27.10**) are the association between specific ascomycetes (a fungus) and certain genera of either green algae or cyanobacteria. The fungal partner is termed the **mycobiont** and the algal or cyanobacterial partner, the **phycobiont.** The fungus obtains nutrients from its partner by projections of fungal hyphae called haustoria, which penetrate the phycobiont cell wall. It also uses the O_2 produced by the phycobiont in carrying out respiration.

Figure 27.10 Lichens. Crustose (encrusting) lichens growing on a granite post.

In turn, the fungus protects the phycobiont from high light intensities, provides water and minerals, and creates a firm substratum within which the phycobiont can grow protected from environmental stress. In the past, the lichen symbiosis was considered to be a mutualistic interaction. However, this appears to be a controlled parasitism between the fungi and phycobiont, as some phycobiotic cyanobacteria and algae grow more quickly when cultured alone. << *Characteristics of fungal divisions: Ascomycota (section 23.3); Photosynthetic bacteria: Phylum Cyanobacteria (section 19.3)*

An important aspect of many symbiotic relationships, including parasitism, is that over time, the symbiont, once it has established a relationship with the host, will tend to discard excess, unused genomic information, a process called **genomic reduction.** This is clearly the case with the aphid endosymbiont *Buchnera aphidicola* (p. 642), and it has also occurred with the parasitic bacterium *Mycobacterium leprae* and the microsporidium *Encephalitozoon cuniculi.* The latter organism, which parasitizes a wide range of animals, including humans, now can only survive inside the host cell. << *Suborder Corynebactineae: Leprosy (section 22.4); Characteristics of the fungal divisions: Microsporidia (section 23.3)*

1. Define predation and parasitism. How are these similar and different?

2. How can a predator confer positive benefits on its prey? Think of the responses of individual organisms versus populations as you consider this question.

3. What are examples of parasites that are important in microbiology?

4. What is a lichen? Discuss the benefits the phycobiont and mycobiont provide each other.

Amensalism

Amensalism describes the adverse effect that one organism has on another organism (figure 27.1). This is a unidirectional process based on the release of a specific compound by one organism that has a negative effect on another organism. A classic example of amensalism is the microbial production of antibiotics that can inhibit or kill another, susceptible microorganism **(figure 27.11a).** Community complexity is demonstrated by the capacity of attine ants (ants belonging to a New World tribe) to take advantage of an amensalistic relationship between an actinomycete and the parasitic fungi *Escovopsis.* This amensalistic relationship enables the ant to maintain a mutualism with another fungal species, *Leucocoprini.* Amazingly, these ants cultivate a garden of *Leucocoprini* for their own nourishment (figure 27.11b). To prevent the parasitic fungus *Escovopsis* from decimating their fungal garden, the ants also promote the growth of an actinomycete of the genus *Pseudonocardia,* which produces an antimicrobial compound that inhibits the growth of

Escovopsis. This unique amensalistic process appears to have evolved 50 to 65 million years ago in South America. Thus this relationship has been subject to millions of years of coevolution, such that particular groups of ants cultivate specific strains of fungi that are then subject to different groups of *Escovopsis* parasites. In addition, the ants have developed intricate crypts within their exoskeletons for the growth of the antibiotic-producing *Pseudonocardia.* As shown in figure 27.11c, these crypts have been modified throughout the ants' evolutionary history. The most primitive "paleo-attine" ants carry the bacterium on their forelegs, "lower" and "higher" attines have evolved special plates on their ventral surfaces, while the entire surface of the most recent attines, leaf-cutter ants of the genus *Acromyrmex,* are covered with the bacterium. Related ants that do not cultivate fungal gardens (e.g., *Atta* sp.) do not host *Pseudonocardia.* This unique multipartner relationship has enabled scientists to explore the behavioral, physiological, and structural aspects of the organisms involved.

Competition

Competition arises when different organisms within a population or community try to acquire the same resource, whether this is a physical location or a particular limiting nutrient (figure 27.1). If one of the two competing organisms can dominate the environment, whether by occupying the physical habitat or by consuming a limiting nutrient, it will overtake the other organism. This phenomenon was studied by E. F. Gause, who in 1934 described it as the **competitive exclusion principle.** He found that if two competing ciliates overlapped too much in terms of their resource use, one of the two protist populations was excluded. In chemostats, competition for a limiting nutrient may occur among microorganisms with transport systems of differing affinity. This can lead to the exclusion of the slower-growing population under a particular set of conditions. If the dilution rate is changed, the previously slower-growing population may become dominant. Often two microbial populations that appear to be similar nevertheless coexist. In this case, they share the limiting resource (space, a limiting nutrient) and coexist while surviving at lower population levels. << *Continuous culture of microorganisms: Chemostats (section 7.4)*

1. What is the origin of the term amensalism?

2. The production of antimicrobial agents and their effects on target species has been called a "microbial arms race." Explain what you think this phrase means and its implications for the attine ant system.

3. What is the competitive exclusion principle? Can you think of another example where this principle is demonstrated in the natural world?

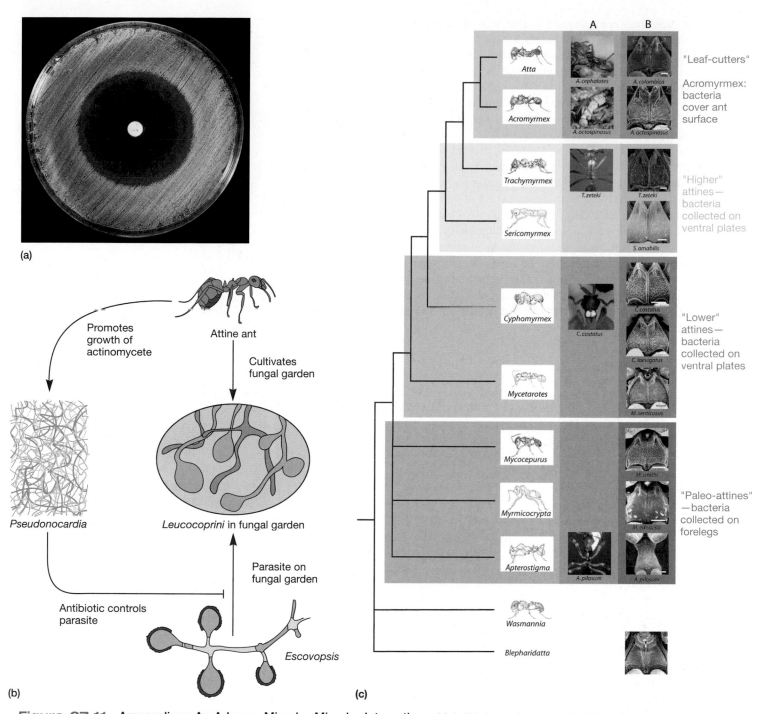

Figure 27.11 Amensalism: An Adverse Microbe-Microbe Interaction. (a) Antibiotic production and inhibition of growth of a susceptible bacterium on an agar medium. (b) A diagram describing the use of antibiotic-producing streptomycetes by ants to control fungal parasites in their fungal garden. (c) Coevolution of attine ants and the antibiotic-producing *Pseudonocardia* has resulted in specialized localization of the bacterium on the ant. This rooted tree illustrates the phylogeny of fungus-growing ants; column A shows the placement of the bacteria on the ants' body, column B presents scanning electron micrographs of the areas colonized by the microbe.

27.2 HUMAN-MICROBE INTERACTIONS

As we have seen, many microorganisms live much of their lives in a special ecological relationship: an important part of their environment is a member of another species. We now turn our attention to microorganisms normally associated with the human body, the normal **microbial flora** or **microbiota.** The human body is a diverse environment in and on which specific niches are formed, thus the normal flora may be discussed as the microbial ecology of a human. In fact, the average adult carries 10 times more microbial cells (10^{14}) than human cells (average about

10^{13}). Interactions between hosts and microbes are dynamic, permitting niche fulfillment that maximizes benefit to the microbe. Tolerating a normal flora likewise suggests that the host derives benefit. Acquisition of a normal microbial flora represents a selective process, where a niche may be defined by cellular receptors, surface properties, or secreted products. In humans, microbial niche variations are also related to age, gender, diet, nutrition, and developmental stage.

The survival of a host, such as a human, depends upon an elaborate network of defenses that keeps harmful microorganisms and other foreign material from entering the body. Should they gain access, additional host defenses are summoned to prevent them from establishing another type of relationship, one of parasitism or pathogenicity. **Pathogenicity** (Greek *pathos,* emotion or suffering, and *gennan,* to produce) is the ability to produce pathologic changes or disease. A **pathogen** is any disease-producing microorganism. Here, we introduce the normal human microbiota, which function not as pathogens but as symbionts that are part of the host's first line of defense against harmful infectious agents.

Gnotobiotic Animals

To determine the role of the normal microorganisms associated with a host and evaluate the consequences of colonization, it is possible to deliver an animal by cesarian section and raise that animal in the absence of microorganisms—that is, germfree. These microorganism-free animals provide suitable experimental models for investigating the interactions of animals and their microbial flora. Comparing animals possessing normal microbiota (conventional animals) with germfree animals permits the elucidation of many complex relationships between microorganisms, hosts, and specific environments. Germfree experiments also extend and challenge the microbiologist's "pure culture concept" to in vivo research.

The term **gnotobiotic** (Greek *gnotos,* known, and *biota,* the flora and fauna of a region) has been defined in two ways. Some think of a gnotobiotic environment or animal as one in which all the microbiota are known; they distinguish it from one that is truly germfree. We shall use the term in a more inclusive sense. Gnotobiotic refers to a microbiologically monitored environment or animal that is germfree (axenic [*a,* without, and Greek *xenos,* a stranger]) or in which the identities of all microbiota are known.

Gnotobiotic animals and systems have become commonplace in research laboratories (**figure 27.12**). Germfree animals are usually more susceptible to pathogens because without normal commensal microbiota, pathogenic microorganisms establish themselves very easily. The number of microorganisms necessary to infect a germfree animal and produce a diseased state is much smaller. Conversely, germfree animals are almost completely resistant to the intestinal protozoan *Entamoeba histolytica,* the cause of amebic dysentery, because it is deprived of its bacterial food source. Germfree animals also do not show any dental caries or plaque formation, unless inoculated with cariogenic (caries or cavity-

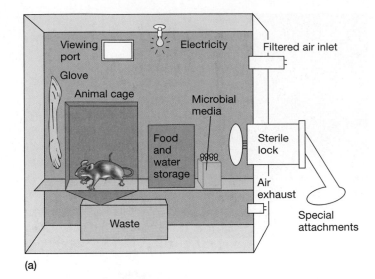

(a)

(b)

Figure 27.12 **Raising Gnotobiotic Animals.** (a) Schematic of a gnotobiotic isolator. The microbiological culture media monitor the sterile environment. (b) Gnotobiotic isolators for rearing colonies of small mammals.

causing) streptococci and fed a high-sucrose diet. ≪ *Protist classification: Supergroup* Amoebozoa (*section 23.2*)

1. Define gnotobiotic.
2. Compare a germfree mouse to a normal one with regard to overall susceptibility to pathogens. What benefits does an animal gain from its microbiota?

27.3 NORMAL MICROBIOTA OF THE HUMAN BODY

Development of a lifelong symbiotic relationship with microbes begins during birth. The human fetus in utero is usually free of microorganisms. The infant's exposure to vaginal mucosa, skin, hair, food, and other nonsterile objects quickly results in the acquisition of a predominantly commensal normal flora. The microbial population stabilizes during the first week or two of life. Colonization of the newborn varies with respect to its environment. The newborn acquires external flora from those who provide its care, while internal flora are acquired through its diet. Bifidobacteria represent more than 90% of the culturable intestinal bacteria in breast-fed infants, with *Enterobacteriaceae* and enterococci in smaller proportions. This suggests that human milk may act as a selective medium for nonpathogenic bacteria, as bottle-fed babies appear to have a much smaller proportion of intestinal bifidobacteria. Switching to cow's milk or solid food (mostly polysaccharide) appears to result in the loss of bifidobacteria predominance, as *Enterobacteriaceae,* enterococci, bacteroides, lactobaccili, and clostridia increase in number.

In a healthy human, regardless of age, the internal tissues (e.g., brain, blood, cerebrospinal fluid, muscles) are normally free of microorganisms. Conversely, the surface tissues (e.g., skin and mucous membranes) are constantly in contact with environmental microorganisms and become readily colonized by various microbial species. The mixture of microorganisms regularly found at any anatomical site is referred to as the **normal microbiota,** the indigenous microbial population, the **microflora,** or the normal flora. For consistency, the term normal microbiota is used in this chapter. An overview of the microbiota native to different regions of the body is presented next (**figure 27.13**). Because procaryotes make up most of the normal microbiota, they are emphasized over the fungi (mainly yeasts) and protists.

There are many reasons to understand the normal human microbiota. Three specific examples include:

- An understanding of the different microorganisms at particular locations provides greater insight into the possible infections that might result from injury to these body sites.

- Knowledge of the normal microbiota helps in understanding the causes and consequences of colonization and growth by microorganisms normally absent at a specific body site.

- An increased awareness of the role that these normal microbiota play in stimulating the host immune response can be gained. This is important because the immune system provides protection against potential pathogens.

Skin

The adult human is covered with approximately 2 m^2 of skin. It has been estimated that this surface area supports about 10^{12} bacteria. Commensal microorganisms living on or in the skin can be either resident (normal) or transient microbiota. Resident organisms normally grow on or in the skin. Their presence becomes fixed in well-defined distribution patterns. Those that are temporarily pres-

ent are transient microorganisms. Transients usually do not become firmly entrenched and are unable to multiply.

It should be emphasized that the skin is a mechanically strong barrier to microbial invasion. Few microorganisms can penetrate the skin because its outer layer consists of thick, closely packed cells called keratinocytes. In addition to direct resistance to penetration, continuous shedding of the outer epithelial cells removes many of those microorganisms adhering to the skin surface.

The anatomy and physiology of the skin vary from one part of the body to another, and the normal resident microbiota reflect these variations. The skin surface or epidermis is not a favorable environment for microbial colonization. In addition to a slightly acidic pH, a high concentration of sodium chloride, a lack of moisture in many areas, and certain inhibitory substances (bactericidal or bacteriostatic) on the skin help control microbial colonization. For example, the sweat glands release **lysozyme** (muramidase), an enzyme that lyses *Staphylococcus epidermidis* and other gram-positive bacteria by hydrolyzing the $\beta(1\rightarrow4)$ glycosidic bond connecting *N*-acetylmuramic acid and *N*-acetylglucosamine in the bacterial cell wall peptidoglycan (*see figure 28.17*). Sweat glands also produce antimicrobial peptides called cathelicidins (Latin *catharticus*, to purge, and *cida*, to kill) that help protect against infectious agents by forming pores in bacterial plasma membranes. << *Bacterial cell walls (section 3.4);* >> *Chemical mediators in nonspecific (innate) resistance (section 28.6)*

The oil glands secrete complex lipids that may be partially degraded by the enzymes from certain gram-positive bacteria (e.g., *Propionibacterium acnes*). These bacteria can change the secreted lipids to unsaturated fatty acids such as oleic acid that have strong antimicrobial activity against gram-negative bacteria and some fungi. Some of these fatty acids are volatile and may be associated with a strong odor. Therefore many deodorants contain antibacterial substances that act selectively against gram-positive bacteria to reduce the production of volatile unsaturated fatty acids and body odor.

Most skin bacteria are found on superficial cells, colonizing dead cells, or closely associated with the oil and sweat glands. Secretions from these glands provide the water, amino acids, urea, electrolytes, and specific fatty acids that serve as nutrients primarily for *S. epidermidis* and aerobic corynebacteria. Gram-negative bacteria generally are found in the moister regions. The yeasts *Pityrosporum ovale* and *P. orbiculare* normally occur on the scalp.

The most prevalent bacterium in the oil glands is the gram-positive, anaerobic, lipophilic rod *Propionibacterium acnes.* This bacterium usually is harmless; however, it is associated with the skin disease acne vulgaris. Acne commonly occurs during adolescence when the endocrine system is very active. Hormonal activity stimulates an overproduction of sebum, a fluid secreted by the oil glands. A large volume of sebum accumulates within the glands and provides an ideal microenvironment for *P. acnes.* In some individuals, this accumulation triggers an inflammatory response that causes redness and swelling of

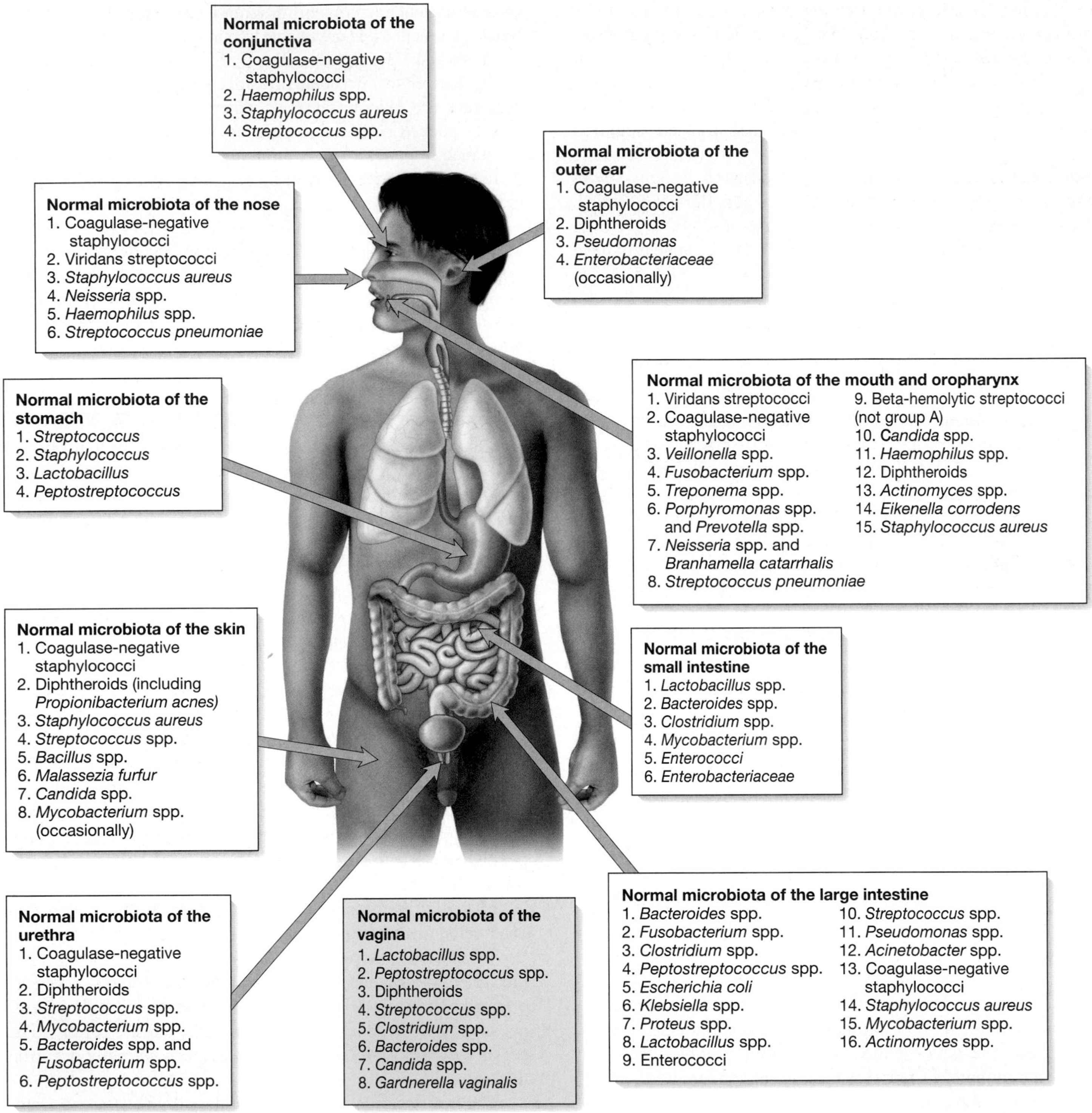

Normal microbiota of the conjunctiva
1. Coagulase-negative staphylococci
2. *Haemophilus* spp.
3. *Staphylococcus aureus*
4. *Streptococcus* spp.

Normal microbiota of the outer ear
1. Coagulase-negative staphylococci
2. Diphtheroids
3. *Pseudomonas*
4. *Enterobacteriaceae* (occasionally)

Normal microbiota of the nose
1. Coagulase-negative staphylococci
2. Viridans streptococci
3. *Staphylococcus aureus*
4. *Neisseria* spp.
5. *Haemophilus* spp.
6. *Streptococcus pneumoniae*

Normal microbiota of the stomach
1. *Streptococcus*
2. *Staphylococcus*
3. *Lactobacillus*
4. *Peptostreptococcus*

Normal microbiota of the mouth and oropharynx
1. Viridans streptococci
2. Coagulase-negative staphylococci
3. *Veillonella* spp.
4. *Fusobacterium* spp.
5. *Treponema* spp.
6. *Porphyromonas* spp. and *Prevotella* spp.
7. *Neisseria* spp. and *Branhamella catarrhalis*
8. *Streptococcus pneumoniae*
9. Beta-hemolytic streptococci (not group A)
10. *Candida* spp.
11. *Haemophilus* spp.
12. Diphtheroids
13. *Actinomyces* spp.
14. *Eikenella corrodens*
15. *Staphylococcus aureus*

Normal microbiota of the skin
1. Coagulase-negative staphylococci
2. Diphtheroids (including *Propionibacterium acnes*)
3. *Staphylococcus aureus*
4. *Streptococcus* spp.
5. *Bacillus* spp.
6. *Malassezia furfur*
7. *Candida* spp.
8. *Mycobacterium* spp. (occasionally)

Normal microbiota of the small intestine
1. *Lactobacillus* spp.
2. *Bacteroides* spp.
3. *Clostridium* spp.
4. *Mycobacterium* spp.
5. Enterococci
6. *Enterobacteriaceae*

Normal microbiota of the urethra
1. Coagulase-negative staphylococci
2. Diphtheroids
3. *Streptococcus* spp.
4. *Mycobacterium* spp.
5. *Bacteroides* spp. and *Fusobacterium* spp.
6. *Peptostreptococcus* spp.

Normal microbiota of the vagina
1. *Lactobacillus* spp.
2. *Peptostreptococcus* spp.
3. Diphtheroids
4. *Streptococcus* spp.
5. *Clostridium* spp.
6. *Bacteroides* spp.
7. *Candida* spp.
8. *Gardnerella vaginalis*

Normal microbiota of the large intestine
1. *Bacteroides* spp.
2. *Fusobacterium* spp.
3. *Clostridium* spp.
4. *Peptostreptococcus* spp.
5. *Escherichia coli*
6. *Klebsiella* spp.
7. *Proteus* spp.
8. *Lactobacillus* spp.
9. Enterococci
10. *Streptococcus* spp.
11. *Pseudomonas* spp.
12. *Acinetobacter* spp.
13. Coagulase-negative staphylococci
14. *Staphylococcus aureus*
15. *Mycobacterium* spp.
16. *Actinomyces* spp.

Figure 27.13   Normal Microbiota of a Human. A compilation of microorganisms that constitute normal microbiota encountered in various body sites.

the gland's duct and produces a comedo (pl., comedones), a plug of sebum and keratin in the duct. Inflammatory lesions (papules, pustules, nodules) commonly called "blackheads" or "pimples" can result when pores or ducts clog with sebum or bacteria. *P. acnes* produces lipases that hydrolyse the sebum triglycerides into free fatty acids. Free fatty acids are especially irritating because they can enter the dermis and promote inflammation. Because *P. acnes* is extremely sensitive to tetracycline, this antibiotic may aid acne sufferers. Retin A and accutane, synthetic forms of vitamin A, are also used.

1. Why is it important to understand the normal human microbiota?
2. Why is the skin not usually a favorable microenvironment for colonization by bacteria?
3. How do microorganisms contribute to body odor?
4. What physiological role does *Propionibacterium acnes* play in the establishment of acne vulgaris?

Nose and Nasopharynx

The normal microbiota of the nose is found just inside the nostrils. *Staphylococcus aureus* and *S. epidermidis* are the predominant bacteria present and are found in approximately the same numbers as on the skin of the face.

The nasopharynx, that part of the pharynx lying above the level of the soft palate, may contain small numbers of potentially pathogenic bacteria such as *Streptococcus pneumoniae, Neisseria meningitidis,* and *Haemophilus influenzae.* Diphtheroids, a large group of nonpathogenic gram-positive bacteria that resemble *Corynebacterium,* are commonly found in both the nose and nasopharynx.

Oropharynx

The oropharynx is that division of the pharynx lying between the soft palate and the upper edge of the epiglottis. The most important bacteria found in the oropharynx are the various alpha-hemolytic streptococci (*S. oralis, S. milleri, S. gordonii, S. salivarius*); large numbers of diphtheroids; *Branhamella catarrhalis*; and small gram-negative cocci related to *N. meningitidis.* The palatine and pharyngeal tonsils harbor a similar microbiota, except within the tonsillar crypts, where there is an increase in *Micrococcus* and the anaerobes *Porphyromonas, Prevotella,* and *Fusobacterium.*

Respiratory Tract

The upper and lower respiratory tracts (trachea, bronchi, bronchioles, alveoli) do not have a normal microbiota. This is because microorganisms are removed in at least three ways. First, a continuous stream of mucus is generated by the goblet cells. This entraps microorganisms, and the ciliated epithelial cells continually move the entrapped microorganisms out of the respiratory tract. Second, alveolar macrophages phagocytize and destroy microorganisms. Finally, a bactericidal effect is exerted by the enzyme lysozyme, present in the nasal mucus.

Eye and External Ear

At birth and throughout human life, a small number of bacterial commensals are found on the conjunctiva of the eye. The predominant bacterium is *S. epidermidis,* followed by *S. aureus, Haemophilus* spp., and *S. pneumoniae.*

The normal microbiota of the external ear resembles that of the skin, with coagulase-negative staphylococci and *Corynebacte-*

rium predominating. Mycological studies show the following fungi to be normal microbiota: *Aspergillus, Alternaria, Penicillium, Candida,* and *Saccharomyces.*

Mouth

The normal microbiota of the mouth or oral cavity contains organisms that resist mechanical removal by adhering to surfaces such as the gums and teeth. Those that cannot attach are removed by the mechanical flushing of the oral cavity contents to the stomach, where they are destroyed by hydrochloric acid. The continuous desquamation (shedding) of epithelial cells also removes microorganisms. Those microorganisms able to colonize the mouth find a very comfortable environment due to the availability of water and nutrients, the suitability of pH and temperature, and the presence of many other growth factors. << *Microbial growth in natural habitats: Biofilms (section 7.6)*

Soon after an infant is born, the mouth is colonized by microorganisms from the surrounding environment. Initially the microbiota consists mostly of the genera *Streptococcus, Neisseria, Actinomyces, Veillonella, Lactobacillus,* and some yeasts. Most microorganisms that invade the oral cavity initially are aerobes and obligate anaerobes. When the first teeth erupt, anaerobes (*Porphyromonas, Prevotella,* and *Fusobacterium*) become dominant due to the anoxic nature of the space between the teeth and gums. As the teeth grow, *Streptococcus parasanguis* and *S. mutans* attach to their enamel surfaces; *S. salivarius* attaches to the buccal and gingival epithelial surfaces and colonizes the saliva. These streptococci produce a glycocalyx and various other adherence factors that enable them to attach to oral surfaces. The presence of these bacteria contributes to the eventual formation of dental plaque, caries, gingivitis, and periodontal disease. >> *Polymicrobial disease (section 30.5)*

Stomach

As noted earlier, many microorganisms are washed from the mouth into the stomach. The very acidic pH (2 to 3) of the gastric contents kills most microorganisms. As a result the stomach usually contains less than 10 viable bacteria per milliliter of gastric fluid. These are mainly *Streptococcus, Staphylococcus, Lactobacillus, Peptostreptococcus,* and yeasts such as *Candida* spp. Microorganisms may survive if they pass rapidly through the stomach or if the organisms ingested with food are particularly resistant to gastric pH (e.g., mycobacteria).

Small Intestine

The small intestine is divided into three anatomical areas: duodenum, jejunum, and ileum. The duodenum (the first 25 cm of the small intestine) contains few microorganisms because of the combined influence of the stomach's acidic juices and the inhibitory action of bile and pancreatic secretions that are added here. Of the bacteria present, gram-positive cocci and rods comprise most of the microbiota. *Enterococcus faecalis,* lactobacilli, diphtheroids, and the yeast *Candida albicans* are occasionally found in the jejunum. In the distal portion of the small intestine (ileum), the microbiota begins to take on the characteristics of the colon microbiota. It is within the ileum that the pH becomes more alkaline. As a result anaerobic gram-negative bacteria and members of the family *Enterobacteriaceae* become established.

Large Intestine (Colon)

The large intestine or colon has the largest microbial community in the body. Microscopic counts of feces approach 10^{12} organisms per gram wet weight. Over 400 different species have been isolated from human feces. The microbiota consists primarily of anaerobic, gram-negative bacteria and gram-positive rods. Not only are the vast majority of microorganisms anaerobic, but many different species are present in large numbers. Several studies have shown that the ratio of anaerobic to facultative anaerobic bacteria is approximately 300 to 1. Besides the many bacteria in the large intestine, the yeast *Candida albicans* and certain protozoa may occur as harmless commensals. The protists *Trichomonas hominis, Entamoeba hartmanni, Endolimax nana,* and *Iodamoeba butschlii* are common inhabitants.

   The importance of the microbes living within the human colon, which can be likened to an anaerobic bioreactor, has prompted a number of investigations using culture-independent molecular approaches. Recent 16S rRNA analysis of microbes shed in feces, as well as microbes collected from gut epithelium, reveals that the majority of colonic procaryotes are currently uncultivated. However, one bacterium, *Bacteroides thetaiotaomicron,* has been the focus of recent interest. This microbe is well suited for survival in the gut, where it is able to degrade complex dietary polysaccharides. Genome analysis reveals that *B. thetaiotaomicron* has a large collection of genes that encode proteins needed for the acquisition and metabolism of carbohydrates. It resides in a specific microenvironment: rather than adhering to the intestinal epithelium, it produces substrate-specific binding proteins that allow it to colonize exfoliated host cells, food particles, and even sloughed mucus (**figure 27.14**). It is thought that such attachment helps retain the microbes in the gut, and once they are bound, the induced expression of extracellular hydrolases enables efficient digestion. Of course, the diversity and density of microbes within the colon suggest that such "nutrient rafts" are colonized by a community of bacteria. For example, methanogenic bacteria are thought to remove the products of fermentation by converting H_2 and CO_2 to methane, just as they do in the rumen microbial community.

   Various physiological processes move the microbiota through the colon so an adult eliminates about 3×10^{13} microorganisms daily. These processes include peristalsis and desquamation of the surface epithelial cells to which microorganisms are attached, and continuous flow of mucus that carries adhering microorganisms with it. To maintain homeostasis of the microbiota, the body must continually replace lost microorganisms. The bacterial population in the human colon usually doubles once or twice a day.

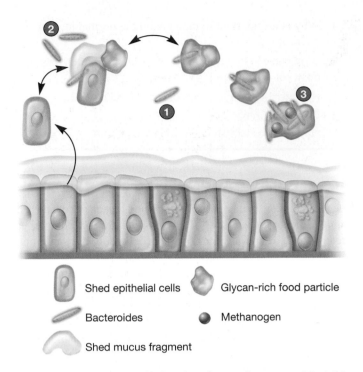

Figure 27.14 *Bacteriodes thetaiotaomicron* as a Model for Colon Microbial Physiology and Community Dynamics. (1) *B. thetaiotaomicron* is rapidly eliminated if it remains planktonic; however, (2) it efficiently adheres to substrates within the lumen of the gut rather than the gut itself (3) where, together with other microbes including methanogenic archaea, it degrades complex carbohydrates.

Legend: Shed epithelial cells · Bacteroides · Shed mucus fragment · Glycan-rich food particle · Methanogen

Under normal conditions, the resident microbial community is self-regulating. Competition and mutualism between different microorganisms and between the microorganisms and their host serve to maintain a status quo. However, if the intestinal environment is disturbed, the normal microbiota may change greatly. Disruptive factors include stress, altitude changes, starvation, parasitic organisms, diarrhea, and use of antibiotics. Finally, it should be emphasized that the actual proportions of the individual bacterial populations within the indigenous microbiota depend largely on a person's diet.

Genitourinary Tract

The upper genitourinary tract (kidneys, ureters, and urinary bladder) is usually free of microorganisms. In both the male and female, a few bacteria (*S. epidermidis, E. faecalis,* and *Corynebacterium* spp.) usually are present in the distal portion of the urethra.

   In contrast, the adult female genital tract, because of its large surface area and mucous secretions, has a complex microbiota that constantly changes with the female's menstrual cycle. The major microorganisms are the acid-tolerant lactobacilli, primarily *Lactobacillus acidophilus*, often called Döderlein's bacillus. They ferment the glycogen produced by the vaginal epithelium, forming lactic acid. As a result the pH of the vagina and cervix is maintained between 4.4 and 4.6, inhibiting other microorganisms.

1. What are the most common microorganisms found in the nose? The oropharynx? The nasopharynx? The lower respiratory tract? The mouth? The eye? The external ear? The stomach? The small intestine? The colon? The genitourinary tract?

2. Why is the colon considered a large fermentation vessel?

3. What physiological processes move the microbiota through the gastrointestinal tract?

4. Describe the microbiota of the upper and lower female genitourinary tract.

The Relationship Between Normal Microbiota and the Host

The interaction between a host and a microorganism is a dynamic process in which each partner acts to maximize its survival. In some instances, after a microorganism enters or contacts a host, a positive mutually beneficial relationship occurs that becomes integral to the health of the host. These microorganisms become the normal microbiota. In other instances, the microorganism causes deleterious effects on the host; the end result may be disease or even death of the host. >> *Pathogenicity of microorganisms (chapter 30)*

Our environment is teeming with microorganisms and we come in contact with many of them every day. Some of these microorganisms are pathogenic—that is, they cause disease. Yet these pathogens are at times prevented from causing disease by competition provided by the normal microbiota. In general, the normal microbiota uses space, resources, and nutrients needed by pathogens. In addition, its members may produce chemicals that repel invading pathogens. For instance, the lactobacilli in the female genital tract maintain a low pH and inhibit colonization by pathogenic bacteria and yeast, and the corynebacteria on the skin produce fatty acids that inhibit colonization by pathogenic bacteria. These are excellent examples of amensalism.

Products made by colonic bacteria (such as vitamins B and K) also benefit the host. Interestingly, studies using germfree animals suggest a strong correlation between the establishment of a stable microbial flora and the induction of immune competency. For example, the introduction of normal fecal flora to germfree rodents stimulates the production and secretion of angiogenin-4, an antimicrobial peptide of intestinal Paneth cells. Furthermore, the reconstitution of germfree rodents with flora from conventionally raised siblings causes the abnormal gut-associated lymphoid (GALT) tissue and intestinal lamina propria to resemble that of the conventional animals (i.e., their

lymphoid tissues and immunity are normalized). Even cell wall fragments from gram-positive bacteria can induce these changes. This normalization also includes an increase in the local lymphocyte populations and increased mucosal antibody production. >> *Cells, tissues, and organs of the immune system (section 28.2)*

Although normal microbiota offer some protection from invading pathogens, its members may themselves become pathogenic and produce disease under certain circumstances and then are termed **opportunistic microorganisms** or **pathogens.** These opportunistic microorganisms are adapted to the noninvasive mode of life defined by the limitations of the environment in which they are living. If removed from these environmental restrictions and introduced into the bloodstream or tissues, disease can result. For example, streptococci of the viridans group are the most common resident bacteria of the mouth and oropharynx. If they are introduced into the bloodstream in large numbers (e.g., following tooth extraction or a tonsillectomy), they may settle on deformed or prosthetic heart valves and cause endocarditis.

Opportunistic microorganisms often cause disease in compromised hosts. A **compromised host** is seriously debilitated and has a lowered resistance to infection. There are many causes of this condition, including malnutrition, alcoholism, cancer, diabetes, leukemia, another infectious disease, trauma from surgery or an injury, an altered normal microbiota from the prolonged use of antibiotics, and immunosuppression by various factors (e.g., drugs, viruses [HIV], hormones, and genetic deficiencies). For example, *Bacteroides* species are one of the most common residents in the large intestine (figure 27.14) and are quite harmless in that location. If introduced into the peritoneal cavity or into the pelvic tissues as a result of trauma, they cause suppuration (the formation of pus) and bacteremia (the presence of bacteria in the blood). Importantly, the normal microbiota are harmless and are often beneficial in their normal location in the host and in the absence of coincident abnormalities. However, they can produce disease if introduced into foreign locations or compromised hosts.

1. Describe two examples of the normal microbiota benefiting a host.

2. Explain how the principle of competitive exclusion is used by normal host microbiota in preventing the establishment of pathogens.

3. How would you define an opportunistic microorganism or pathogen? A compromised host?

Summary

27.1 Microbial Interactions

a. Symbiotic interactions include mutualism (mutually beneficial and obligatory), cooperation (mutually beneficial, not obligatory), and commensalism (product of one organism can be used beneficially by another organism). Predation involves one

organism (the predator) ingesting/killing a larger or smaller prey, parasitism (a longer-term internal maintenance of another organism or acellular infectious agent), and amensalism (a microbial product inhibiting another organism). Competition involves organisms competing for space or a limiting nutrient. This can lead to dominance of one organism or coexistence of both at lower populations **(figure 27.1).**

b. A consortium is a physical association of organisms that have a mutually beneficial relationship based on positive interactions.

c. Mutual advantage is central to many organism-organism interactions. These interactions can be based on material transfers related to energetics or physical protection. With several important mutualistic interactions, chemolithotrophic microorganisms play a critical role in making organic matter available for use by an associated organism (e.g., endosymbionts in *Riftia*) **(figure 27.5)**.

d. The rumen is an excellent example of a mutualistic interaction between an animal and a complex microbial community. In this microbial community, complex plant materials are broken down to simple organic compounds that can be absorbed by the ruminant, as well as forming waste gases such as methane that are released to the environment.

e. Commensalism or syntrophism simply means growth together. It does not require physical contact and involves a mutually positive transfer of materials, such as interspecies hydrogen transfer.

f. Cooperative interactions are beneficial for both organisms but are not obligatory. Important examples are marine animals, including *Alvinella* and *Paralvinella,* that involve interactions with hydrogen sulfide-oxidizing chemotrophs.

g. Predation and parasitism are closely related. Predation has many beneficial effects on populations of predators and prey. These include the microbial loop (returning minerals immobilized in organic matter to mineral forms for reuse by chemotrophic and photosynthetic primary producers), protection of prey from heat and damaging chemicals, and possibly aiding pathogenicity.

27.2 Human-Microbe Interactions

a. Animals and environments that are germfree or have one or more known microorganisms are termed gnotobiotic. Gnotobi- otic animals and techniques provide good experimental systems with which to investigate the interactions of animals and specific species or microorganisms **(figure 27.12)**.

27.3 Normal Microbiota of the Human Body

a. Various microbes have adapted to specific niches found on the human host. These niches are uniquely able to support microbe growth by maintaining a relatively constant environment **(figure 27.13)**.

b. Commensal microorganisms living on or in the skin can be characterized as either transients or residents.

c. The normal microbiota of the oral cavity is composed of those microorganisms able to resist mechanical removal.

d. The stomach contains very few microorganisms due to its acidic pH.

e. The distal portion of the small intestine and the entire large intestine have the largest microbial community in the body. Over 400 species have been identified, the vast majority of them anaerobic.

f. The upper genitourinary tract is usually free of microorganisms. In contrast, the adult female genital tract has a complex microbiota.

g. In some instances, after a microorganism contacts or enters a host, a positive mutually beneficial relationship occurs and becomes integral to the health of the host. In other instances, the microorganism may produce disease or even death of the host.

h. Many of the normal host microbiota compete with pathogenic microorganisms.

i. An opportunistic microorganism is generally harmless in its normal environment but may become pathogenic when moved to a different body location or in a compromised host.

Critical Thinking Questions

1. Describe an experimental approach to determine if a plant-associated microbe is a commensal or a mutualist.

2. Some patients who take antibiotics for acne develop yeast infections of the mouth or genital-urinary tract. Explain.

3. How does knowing the anatomical location of commensal flora help clinicians diagnose infection?

4. Compare and contrast the microbial communities that reside in a ruminant with those in the human gut.

Learn More

Learn more by visiting the Prescott website at www.mhhe.com/prescottprinciples, where you will find a complete list of references.

Nonspecific (Innate) Host Resistance

28

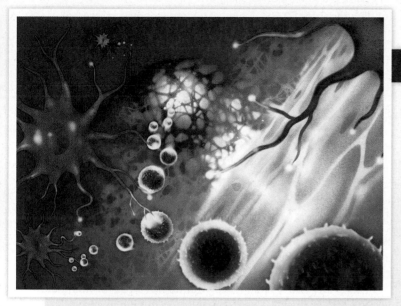

The immune system is a myriad of cells, tissues, and soluble factors that cooperate to defend against foreign invaders.

antibody (immunoglobulin)   A glycoprotein made by plasma cells (mature B cells) in response to the introduction of an antigen. The antibody-binding region of an antibody molecule assumes a final configuration that is the three-dimensional mirror image of the antigen that stimulated its synthesis. Thus the antibody can bind to the antigen with exact specificity.

antigen   A foreign substance (such as a protein, nucleoprotein, polysaccharide, or sometimes a glycolipid) that induces an immune response.

basophil   A white blood cell in the granulocyte lineage that is weakly phagocytic. Importantly, it synthesizes and stores vasoactive molecules (e.g., histamine) that are released in response to external triggers.

B lymphocyte or **B cell**   A type of lymphocyte derived from bone marrow stem cells that matures into an immunologically competent cell. Following interaction with an antigen, it becomes a **plasma cell,** which synthesizes and secretes antibodies.

chemokine   A glycoprotein having potent leukocyte activation and chemotactic activity.

complement system   Plasma proteins that bind to foreign materials to remove them from the host.

cytokine   A general term for proteins released by cells of the immune system in response to specific stimuli. Cytokines influence the activities of other cells.

dendritic cell   An antigen-presenting cell that has long membrane extensions resembling the dendrites of neurons. These cells are found in the lymph nodes, spleen, and thymus (inter-

digitating dendritic cells); skin (Langerhans cells); and other tissues (interstitial dendritic cells).

interferon (IFN)   A glycoprotein that has nonspecific antiviral activity by stimulating cells to produce antiviral proteins, which inhibit the synthesis of viral RNA and proteins. Interferons also regulate the growth, differentiation, and function of a variety of immune system cells.

interleukin   A cytokine produced by macrophages and T cells that regulates cellular growth and differentiation, particularly of lymphocytes.

lymph node   A small secondary lymphoid organ that contains lymphocytes, macrophages, and dendritic cells. It serves as a site for filtration and removal of foreign antigens and for the activation and proliferation of lymphocytes.

lymphokine   A biologically active glycoprotein secreted by activated lymphocytes, especially sensitized T cells. It acts as an intercellular mediator of the immune response and transmits signals affecting cell growth, differentiation, and behavior.

macrophage   A large, mononuclear phagocytic cell, present in various tissues. Macrophages are derived from **monocytes.** They phagocytose and destroy pathogens; macrophages also activate B cells and T cells.

membrane attack complex (MAC)   Complement components (C5b–C9) that assemble to create a pore in the plasma membrane of a target cell, leading to target cell lysis.

monocyte-macrophage system   The collection of fixed phagocytic cells (including macrophages, monocytes, and specialized endothelial cells) located in the liver, spleen, lymph nodes, and bone marrow. This system is an important component of the host's nonspecific defense against pathogens.

monokine   A protein (cytokine) produced by mononuclear phagocytes (monocytes or macrophages) that mediates immune responses.

natural killer (NK) cell   A type of white blood cell that has a lineage independent of the granulocyte, B-cell, and T-cell lineages. NK cells are capable of killing virus-infected and malignant cells.

neutrophil   A mature white blood cell in the granulocyte lineage formed in bone marrow. It has a nucleus with three to five lobes and is phagocytic.

nonspecific immune response   General defense mechanisms that are inherited as part of the innate structure and function of each animal; also known as nonspecific, innate, or natural immunity.

opsonization The coating of foreign substances by antibody proteins, complement proteins, or fibronectin to make the substances more readily recognized by phagocytic cells.

phagocytosis The endocytotic process in which a cell encloses large particles in a membrane-delimited phagocytic vacuole or **phagosome.**

specific immune response (acquired, adaptive, or **specific immunity)** Refers to the type of specific (adaptive) immunity that develops after exposure to an antigen.

T lymphocyte or **T cell** A type of lymphocyte derived from bone marrow stem cells that matures into an immunologically competent cell under the influence of the thymus. T cells are involved in a variety of cell-mediated immune reactions and support B cell growth and development.

toll-like receptor (TLR) A type of pattern recognition receptor on phagocytes such as macrophages that triggers the proper response to different classes of pathogens. It signals the production of transcription factor NFκB, which stimulates formation of cytokines, chemokines, and other defense molecules.

Half the secret of resistance is cleanliness, the other half is dirtiness.

—*Anonymous*

The integrity of any eucaryotic organism depends not only on the proper expression of its genes but also on its freedom from invading microorganisms. Commensal relationships aside, when microorganisms inhabit a multicellular host, competition for resources at the cellular level occurs. In this chapter, we explore the mechanisms by which humans (and other mammalian hosts) defend themselves against such microbial invasion.

28.1 OVERVIEW OF HOST RESISTANCE

To establish an infection, an invading microorganism must first overcome many surface barriers, such as skin, degradative enzymes, and mucus that have either direct antimicrobial activity or inhibit attachment of the microorganism to the host. Because neither the surface of the skin nor the mucus-lined body cavities are ideal environments for the vast majority of microorganisms, most **pathogens** must breach these barriers to reach underlying tissues. Any microorganism that penetrates these barriers encounters two levels of resistance: other nonspecific resistance mechanisms and the specific immune response.

Vertebrates (including humans) are continuously exposed to microorganisms that can cause disease. Fortunately these animals are equipped with an immune system that protects against adverse consequences of this exposure. The **immune system** is composed of widely distributed cells, tissues, and organs that recognize foreign substances, including microorganisms. Together they act to neutralize or destroy them.

Immunity

The term *immunity* (Latin *immunis,* free of burden) refers to the general ability of a host to resist infection or disease. **Immunology** is the science that is concerned with immune responses to foreign chal-

lenge and how these responses are used to resist infection. It includes the distinction between "self" and "nonself" and all the biological, chemical, and physical aspects of the immune response.

There are two fundamentally different types of immune responses to an invading microorganism or foreign material. The **nonspecific immune response** is also known as **nonspecific resistance** and **innate** or **natural immunity;** it offers resistance to any microorganism or foreign material encountered by the vertebrate host. It includes general mechanisms inherited as part of the innate structure and function of each animal (such as skin, mucus, and constituitively produced antimicrobial mediators such as lysozyme), and acts as a first line of defense. The nonspecific immune response defends against foreign particles equally and lacks immunological memory—that is, nonspecific responses occur to the same extent each time a microorganism or foreign body is encountered.

In contrast, the **specific immune responses,** also known as **acquired, adaptive,** or **specific immunity,** resist a particular foreign agent. The effectiveness of specific immune responses increases on repeated exposure to foreign agents such as viruses, bacteria, or toxins; that is to say, specific responses have "memory." Substances that are recognized as foreign and provoke immune responses are called **antigens.** The presence of foreign antigens causes specific cells to replicate and manufacture a variety of proteins that function to protect the host. One such cell, the B cell, produces and secretes glycoproteins called antibodies. **Antibodies** bind to specific antigens and inactivate them or contribute to their elimination. Other immune cells become activated to destroy cells harboring intracellular pathogens. The nonspecific and specific responses work together to eliminate pathogenic microorganisms and other foreign agents.

The distinction between the innate and adaptive systems is, to a degree, artificial. Although innate systems predominate immediately upon initial exposure to foreign substances, multiple bridges occur between innate and adaptive immune system components (**figure 28.1**). Importantly, there is a variety of cells that function in

Host Defenses

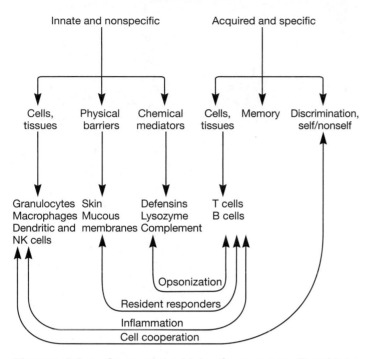

Figure 28.1   Some of the Major Components That Make Up the Mammalian Immune System. Double-headed arrows indicate potential bridging events that unite innate and acquired forms of immunity.

both innate and adaptive immunity. These cells are known as the **white blood cells (WBCs),** or leukocytes. Blood cell development occurs in the bone marrow of mammals during the process of **hematopoesis.** Some WBCs function in the innate system, whereas others are part of an adaptive immune response. The WBCs form the basis for immune responses to invading microbes and foreign substances. Many of these cells reside in specific tissues and organs. Some tissues and organs provide supportive functions in nurturing the cells so that they can mature and respond correctly to antigens.

The bridges that interconnect the innate and specific immune responses are numerous and present challenges for any author trying to describe immunity. We begin our discussion with an introduction to the various cells, tissues, and organs of the innate defenses. In this discussion, reference will be made to processes and chemical mediators that are described in detail either later in this chapter or in chapter 29 (specific defenses). For your convenience, cross-references to the detailed discussion are provided. Thus the discussion of innate immunity proceeds from the general to the ever-more specific and detailed.

1. Define each of the following: immune system, immunity, immunology, antigen, antibody.

2. Compare and contrast the specific and nonspecific immune responses.

28.2 CELLS, TISSUES, AND ORGANS OF THE IMMUNE SYSTEM

The immune system is an organization of molecules, cells, tissues, and organs, each with a specialized role in defending against viruses and other microorganisms, cancer cells, and other foreign entities. Immune system cells and tissue are now considered.

Cells of the Immune System

The cells responsible for both nonspecific and specific immunity are the **leukocytes** (Greek *leukos,* white, and *kytos,* cell). All leukocytes originate from pluripotent stem cells in the fetal liver and in the bone marrow of the animal host (**figure 28.2**). Pluripotent stem cells have not yet committed to differentiating into one specific cell type. When they migrate to other body sites, some differentiate into hematopoietic stem cells that are destined to become blood cells. When stimulated to undergo further development, some leukocytes become residents within tissues, where they respond to local trauma. These cells may sound the alarm that signals invasion by foreign organisms. Other leukocytes circulate in body fluids and are recruited to the sites of infection after the alarm has been raised. The average adult has approximately 7,400 leukocytes per cubic millimeter of blood (**table 28.1**). This average value shifts substantially during an immune response. In defending the host against pathogenic microorganisms, leukocytes cooperate with each other first to recognize the pathogen as an invader and then to destroy it. These different leukocytes are now briefly examined.

Granulocytes

Granulocytes have irregularly shaped nuclei with two to five lobes. Their cytoplasm has granules that contain reactive substances that kill microorganisms and enhance inflammation (figure 28.2). Three types of granulocytes exist: basophils, eosinophils, and neutrophils. Because of the irregularly shaped nuclei, neutrophils are also called **polymorphonuclear neutrophils,** or **PMNs.**

Table 28.1	Normal Adult Blood Count	
Cell Type	**Cells/mm³**	**Percent WBC**
Red blood cells	5,000,000	
Platelets	250,000	
White blood cells	7,400	100
Neutrophils	4,320	60
Lymphocytes	2,160	30
Monocytes	430	6
Eosinophils	215	3
Basophils	70	1

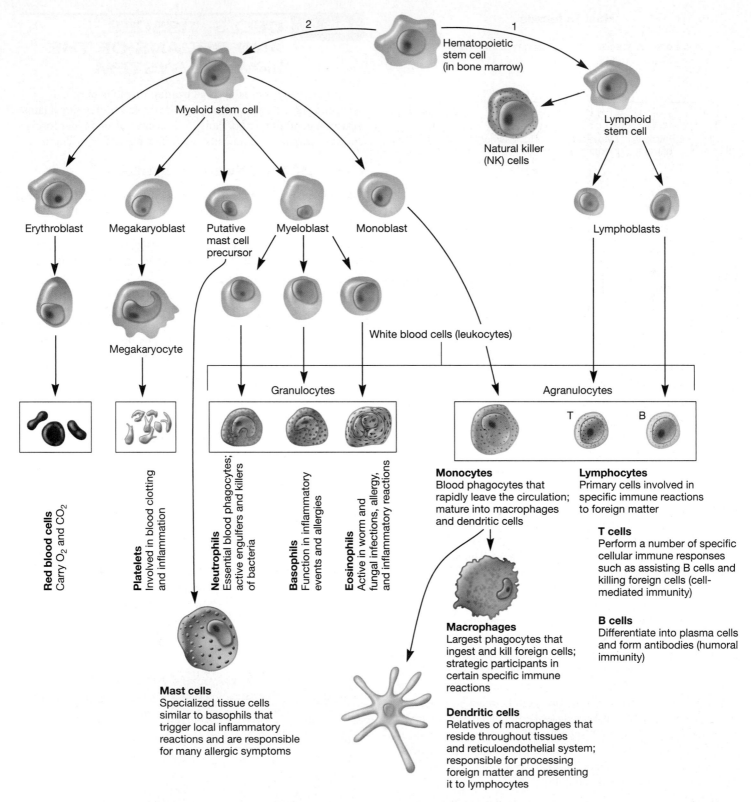

Red blood cells
Carry O_2 and CO_2

Platelets
Involved in blood clotting and inflammation

Neutrophils
Essential blood phagocytes; active engulfers and killers of bacteria

Basophils
Function in inflammatory events and allergies

Eosinophils
Active in worm and fungal infections, allergy, and inflammatory reactions

Monocytes
Blood phagocytes that rapidly leave the circulation; mature into macrophages and dendritic cells

Lymphocytes
Primary cells involved in specific immune reactions to foreign matter

T cells
Perform a number of specific cellular immune responses such as assisting B cells and killing foreign cells (cell-mediated immunity)

B cells
Differentiate into plasma cells and form antibodies (humoral immunity)

Macrophages
Largest phagocytes that ingest and kill foreign cells; strategic participants in certain specific immune reactions

Dendritic cells
Relatives of macrophages that reside throughout tissues and reticuloendothelial system; responsible for processing foreign matter and presenting it to lymphocytes

Mast cells
Specialized tissue cells similar to basophils that trigger local inflammatory reactions and are responsible for many allergic symptoms

Figure 28.2 The Different Types of Human Blood Cells. Pluripotent stem cells in the bone marrow divide to form two blood cell lineages: (1) the lymphoid stem cell gives rise to B cells that become antibody-secreting plasma cells, T cells that become activated T cells, and natural killer cells. (2) The common myeloid progenitor cell gives rise to the granulocytes (neutrophils, eosinophils, basophils), monocytes that give rise to macrophages and dendritic cells, an unknown precursor that gives rise to mast cells, megakaryocytes that produce platelets, and the erythroblast that produces erythrocytes (red blood cells).

Basophils (Greek *basis,* base, and *philein,* to love) have irregularly shaped nuclei with two lobes, and granules that stain bluish-black with basic dyes (figure 28.2). Basophils are nonphagocytic cells that release specific compounds from their cytoplasmic granules in response to certain types of stimulation. These molecules include histamine, prostaglandins, serotonin, and leukotrienes. Because these compounds influence the tone and diameter of blood vessels, they are termed **vasoactive mediators.** Basophils (and mast cells) possess high-affinity receptors for the type of antibody associated with allergic responses. When these cells become coated with this type of antibody, binding of antigen to the antibody triggers the secretion of these vasoactive mediators. As discussed in chapter 29, these molecules play a major role in certain allergic responses such as eczema, hay fever, and asthma. >> *Antibodies (section 29.7); Immune disorders: Hypersensitivities (section 29.11)*

Eosinophils (Greek *eos,* dawn, and *philien*) have a two-lobed nucleus connected by a slender thread of chromatin and granules that stain red with acidic dyes (figure 28.2). Unlike basophils, eosinophils migrate from the bloodstream into tissue spaces, especially mucous membranes. They are important in the defense against protozoan and helminth parasites, mainly by releasing cationic peptides (p. 679) and reactive oxygen intermediates (p. 672) into the extracellular fluid. These molecules damage the parasite plasma membrane, destroying it. Eosinophils also play a role in allergic reactions, as they have granules containing histaminase and aryl sulphatase, downregulators of the inflammatory mediators histamine and leukotrienes, respectively. Thus their numbers often increase during allergic reactions, especially type 1 hypersensitivities (*see p. 716*).

Neutrophils (Latin *neuter,* neither, and *philien*) are phagocytic cells with a nucleus that has three to five lobes connected by slender threads of chromatin. Neutrophils have inconspicuous organelles known as the primary and secondary granules, which contain lytic enzymes and bactericidal substances. Primary granules contain peroxidase, lysozyme, defensins, and various hydrolytic enzymes, whereas the smaller secondary granules have collagenase, lactoferrin, cathelicidins, and lysozyme. These granules help digest foreign material after it is phagocytosed. Neutrophils also use oxygen-dependent and oxygen-independent pathways that generate additional antimicrobial substances to kill ingested microorganisms. Like macrophages, neutrophils have receptors for antibodies and complement proteins (p. 680) and are highly phagocytic. However, unlike macrophages, neutrophils do not reside in healthy tissue but circulate in blood so they can rapidly migrate to the site of tissue damage and infection, where they become the principal phagocytic and microbicidal cells. Neutrophils and their antimicrobial compounds are described in more detail in the context of phagocytosis (section 28.3) and the inflammatory response (section 28.4).

Mast Cells

Mast cells are bone marrow–derived cells that differentiate in the blood and connective tissue. Although they contain granules with histamine and other pharmacologically active substances similar to those in basophils, they arise from a different cellular lineage (figure 28.2). Mast cells, along with basophils, are important in the development of allergies and hypersensitivities.

Monocytes and Macrophages

Monocytes and macrophages are highly phagocytic and make up the **monocyte-macrophage system** (figure 28.3). Although the specifics of phagocytosis will be discussed shortly, recall that it involves the engulfment of large particles and microorganisms that are then enclosed in a phagocytic vacuole. **Monocytes** (Greek *monos,* single, and *cyte,* cell) are mononuclear leukocytes with an ovoid- or kidney-shaped nucleus and granules in the cytoplasm (figure 28.2). They are produced in the bone marrow and enter the blood, circulate for about eight hours, enlarge, migrate to the tissues, and mature into macrophages or dendritic cells.

Because **macrophages** (Greek *macros,* large, and *phagein,* to eat) are derived from monocytes, they are also classified as mononuclear phagocytic leukocytes. However, they are larger than monocytes, contain more organelles that are critical for phagocytosis, and have a plasma membrane covered with

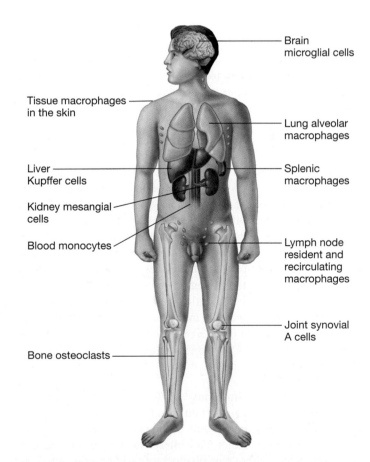

Figure 28.3 The Monocyte-Macrophage System. This system consists of tissue (such as found within the liver, spleen, and lymph nodes) containing "fixed" or immobile phagocytes that have specific names depending on their location.

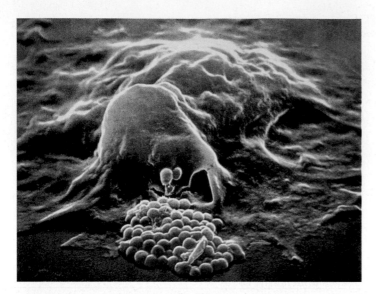

Figure 28.4   Phagocytosis by a Macrophage.   One type of nonspecific host resistance involves white blood cells called macrophages and the process of phagocytosis. This scanning electron micrograph (×3,000) shows a macrophage devouring a colony of bacteria. Phagocytosis is one of many nonspecific defenses humans and other animals use to combat microbial pathogens.

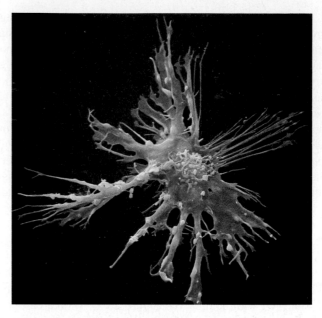

Figure 28.5   The Dendritic Cell.   The dendritic cell was named for its cellular extensions, which resemble the dendrites of nerve cells. Dendritic cells reside in most tissue sites where they survey their local environments for pathogens and altered host cells.

microvilli (**figure 28.4**). Macrophages have surface molecules that function as receptors to recognize common components of pathogens. These receptors include mannose and fucose receptors, and a special class of molecules called toll-like receptors (p. 670), which bind lipopolysaccharide (LPS), peptidoglycan, fungal cell wall components called zymosan, viral nucleic acids, and foreign DNA (i.e., pathogen-associated molecular patterns). Macrophages also have receptors for antibodies and serum glycoproteins known as complement (p. 680). Both antibody and complement proteins can coat microorganisms or foreign material and enhance their phagocytosis. This enhancement is termed opsonization and is discussed in further detail in section 28.6. Macrophages spread throughout the animal body and take up residence in specific tissues; they are given special names depending on their location (figure 28.3). Because macrophages are highly phagocytic, their function in nonspecific resistance is discussed in more detail in the context of phagocytosis. >> *Antigens (section 29.2); Antibodies (section 29.7)*

Dendritic Cells

Dendritic cells constitute only 0.2% of white blood cells in the blood but play an important role in nonspecific resistance (**figure 28.5**). They are present in the skin and mucous membranes of the nose, lungs, and intestines, where they readily contact invading pathogens, phagocytose and process antigens, and display foreign antigens on their surface. This process is known as **antigen presentation** and is discussed in more detail in section 28.2.

Like macrophages, dendritic cells recognize specific *p*athogen-*a*ssociated *m*olecular *p*atterns (PAMPs) on microorganisms (p. 670). These molecular patterns enable dendritic cells to distinguish between potentially harmful microbes and other host molecules. After the pathogen is recognized, the dendritic cell's pattern recognition receptors (PRRs) bind the pathogen and phagocytose it. The dendritic cells then migrate to lymphoid tissues where, as activated cells, they present antigen to T cells. Antigen presentation triggers the activation of T cells, which is critical for the initiation and regulation of an effective specific immune response. Thus not only do dendritic cells destroy invading pathogens as part of the innate response, but they also help trigger specific immune responses. >> *T cell biology (section 29.5)*

1. Describe the structure and function of each of the following blood cells: basophil, eosinophil, neutrophil, monocyte, macrophage, and dendritic cell. Which cells are phagocytic?

2. What is the significance of the respective blood cell percentages in blood?

Lymphocytes

Lymphocytes (Latin *lympha,* water, and *cyte,* cell) are the major cells of the specific immune system. Lymphocytes can be divided into three populations: T cells, B cells, and null cells (which include special cells called natural killer, or NK, cells). Lymphocytes leave

Figure 28.6 The Development and Function of B and T Lymphocytes. B cells and T cells arise from the same cell lineage but diverge into two different functional types. Immature B cells and T cells are indistinguishable by histological staining. However, they express different proteins on their surfaces that can be detected by immunohistochemistry. Additionally, the final secreted products of mature B and T cells can be used to identify the cell type.

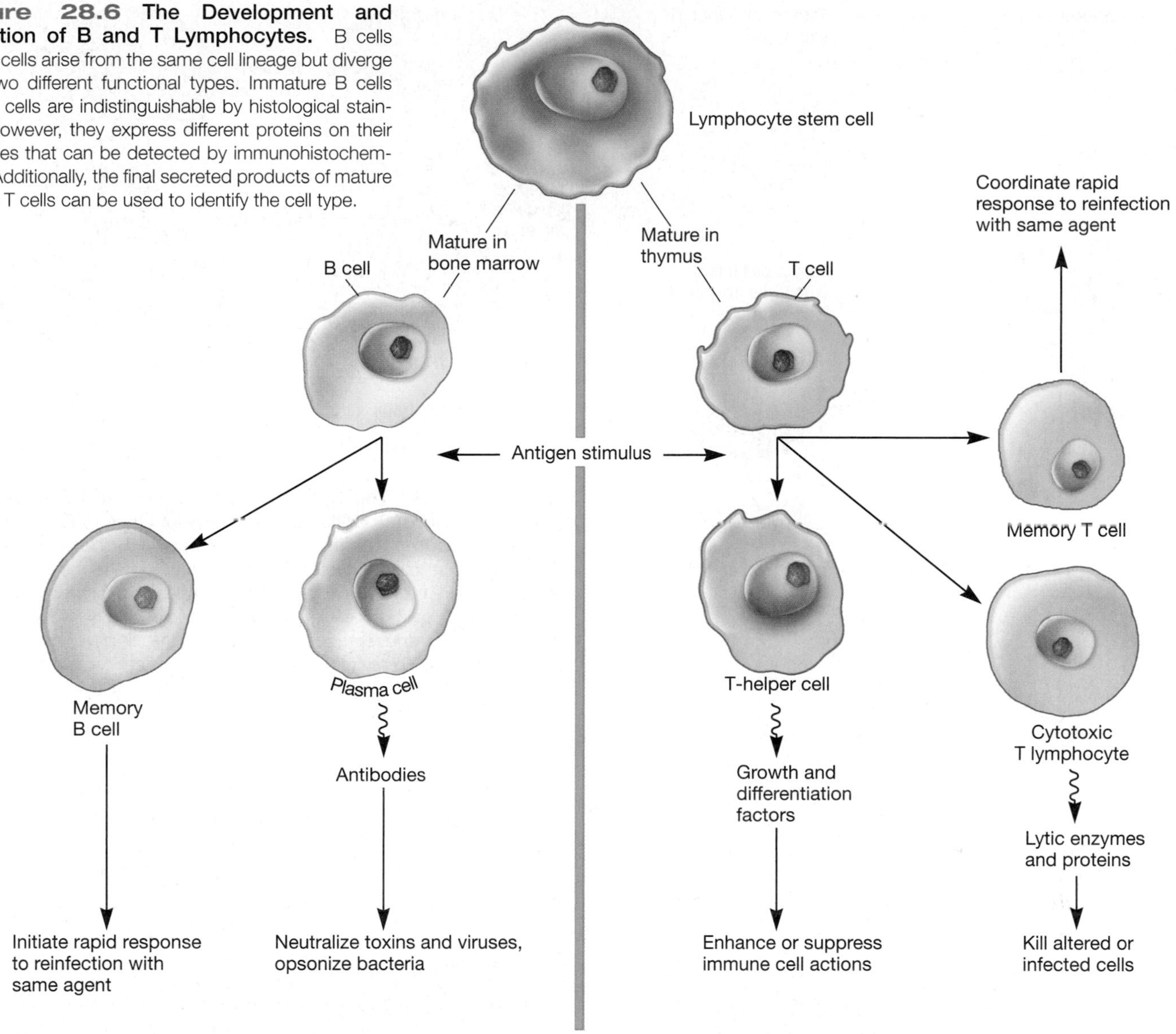

Lymphocyte stem cell

Coordinate rapid response to reinfection with same agent

B cell

Mature in bone marrow

Mature in thymus

T cell

Antigen stimulus

Memory T cell

Memory B cell

Plasma cell

T-helper cell

Cytotoxic T lymphocyte

Antibodies

Growth and differentiation factors

Lytic enzymes and proteins

Initiate rapid response to reinfection with same agent

Neutralize toxins and viruses, opsonize bacteria

Enhance or suppress immune cell actions

Kill altered or infected cells

the bone marrow in a kind of cellular stasis—not actively replicating like other somatic cells. B and T lymphocytes differentiate from their respective lymphoid stem cell precursor cells and are then blocked from continuing mitosis. This is important because it ensures that their gene products are made only when they are needed. In general, lymphocytes require a specific antigen to bind to a surface receptor (the B-cell receptor on B cells or the T-cell receptor on T cells) so that cellular activation can occur. Activation stimulates the cell to enter mitosis. Once activated, the lymphocytes continue to replicate as they circulate throughout the host, leaving several clones of activated lymphocytes to populate various lymphoid tissues. In addition to activated cells, some of the replicated lymphocytes are inhibited from further replication, waiting to be activated by the same antigen sometime later in the life of the host. These cells are called **memory cells.**

After **B lymphocytes** or **B cells** reach maturity within the bone marrow, they circulate in the blood and disperse into vari-

ous lymphoid organs, where they become activated. The activated B cell becomes more ovoid. Its nuclear chromatin condenses and numerous folds of endoplasmic reticulum become more visible. A mature, activated B cell is called a **plasma cell,** which secretes large quantities of antibodies. Some of these antibodies can directly neutralize toxins and viruses, and are important in stimulating an efficient phagocytic response (**figure 28.6**). ▶▶ *Actions of antibodies (section 29.8)*

Lymphocytes destined to become **T lymphocytes** or **T cells** leave the bone marrow and mature in the thymus gland. They can remain in the thymus, circulate in the blood, or reside in lymphoid organs such as the lymph nodes and spleen, like B cells do. Also like B cells, T cells require a specific antigen to bind to their receptor to signal the continuation of replication. Unlike B cells, however, T cells do not secrete antibodies. Activated T cells produce and secrete proteins called **cytokines** (figure 28.6 and p. 684). Cytokines can have various effects on other cells, including

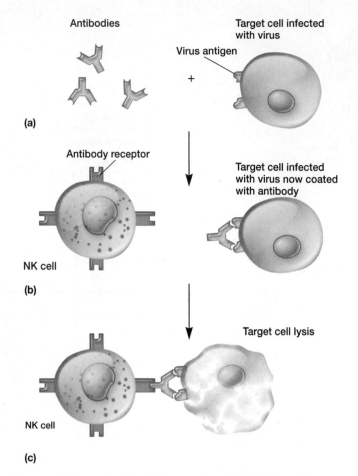

(a)

Antibodies

Target cell infected with virus

Virus antigen

+

(b)

Antibody receptor

Target cell infected with virus now coated with antibody

NK cell

(c)

Target cell lysis

NK cell

Figure 28.7 **Antibody-Dependent Cell-Mediated Cytotoxicity.** (a) In this mechanism, antibodies bind to a target cell infected with a virus. (b) NK cells have specific antibody receptors on their surface. (c) When the NK cells encounter infected virus infected cells coated with antibody, they kill the target cell.

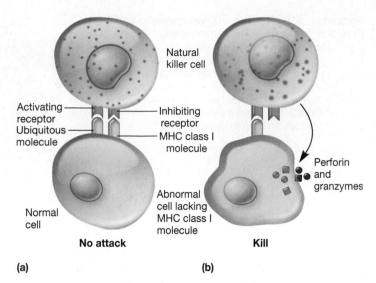

Natural killer cell

Activating receptor

Inhibiting receptor

Ubiquitous molecule

MHC class I molecule

Perforin and granzymes

Normal cell

Abnormal cell lacking MHC class I molecule

No attack

Kill

(a)

(b)

Figure 28.8 **The System Used by Natural Killer Cells to Recognize Normal Cells and Abnormal Cells That Lack the Major Histocompatability Complex Class I Surface Molecule.** (a) The killer-activating receptor recognizes a ubiquitous molecule on the plasma membrane of a normal cell. Since the killer-inhibitory receptor recognizes the MHC class I molecule, there is no attack. (b) In the absence of the inhibitory signal, the receptor issues an order to the NK cell to attack and kill the abnormal cell. The cytotoxic granules of the NK cell contain perforin and granzymes. With no inhibitory signal, the granules release their contents, killing the abnormal cell.

as when some viruses or cancers overtake the cell, the NK cell kills it by releasing pore-forming proteins and cytotoxic enzymes called **granzymes.** Together the pore-forming proteins and the granzymes cause the target cell to lyse (**figure 28.8**). >> *Recognition of foreignness (section 29.4)*

1. What is a plasma cell?
2. What is the purpose of T-cell secretion of cytokines?
3. Discuss the role of NK cells in protecting the host. What are the two mechanisms NK cells use to kill other cells?

Organs and Tissues of the Immune System

Based on function, the organs and tissues of the immune system can be divided into primary or secondary lymphoid organs and tissues (**figure 28.9**). The primary organs and tissues are where immature lymphocytes mature and differentiate into antigen-sensitive B and T cells. The thymus is the primary lymphoid organ for T cells, and the bone marrow is the primary lymphoid tissue for B cells. The secondary organs and tissues serve as areas where lymphocytes may encounter and bind antigen, whereupon they proliferate and differentiate into fully active, antigen-specific effector cells. The spleen is a secondary lymphoid organ, and the lymph nodes and mucosal-associated tissues (GALT—gut-associated lymphoid tissue, and SALT—skin-associated lymphoid tissues) are the secondary lymphoid tissues. The thymus, bone marrow, lymph

other T cells, B cells, granulocytes, and other somatic cells. In some cases, cytokines stimulate cells to mature and differentiate, secrete new effector products, and even die. Because cytokines are so powerful, it is important that their release is tightly regulated. B and T cells must be activated by specific antigens; therefore they are part of the adaptive or specific immune system. B cells and T cells are discussed further in chapter 29.

Natural killer (NK) cells are a small population of large, nonphagocytic granular lymphocytes that play an important role in innate immunity (figure 28.2). The major NK cell function is to destroy malignant cells and cells infected with microorganisms. They recognize their targets in one of two ways. They can bind to antibodies that coat infected or malignant cells; thus the antibody bridges the two cell types. This process is called **antibody-dependent cell-mediated cytotoxicity (ADCC)** (**figure 28.7**) and can result in the death of the target cell. The second way that NK cells recognize infected cells and cancer cells relies on the presence of specialized proteins on the surface of all nucleated host cells, known as the class I major histocompatibility (MHC) antigen. If a host cell loses this MHC protein,

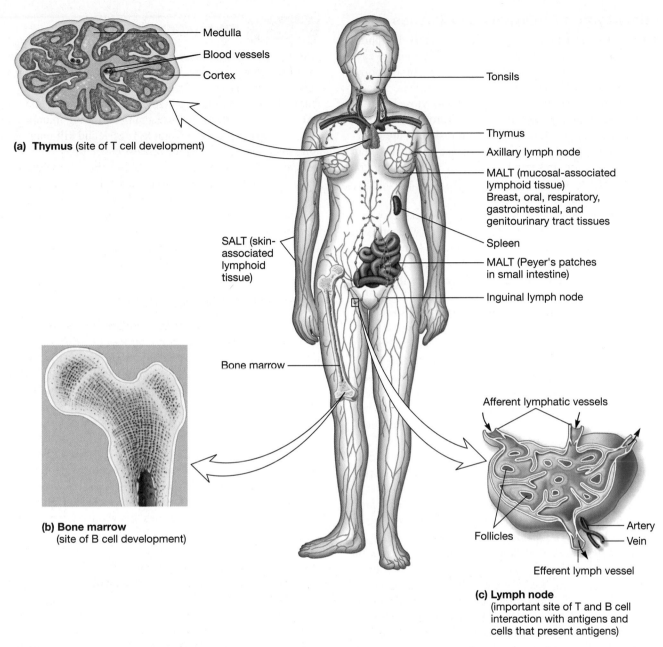

Figure 28.9 Anatomy of the Lymphoid System. Lymph is distributed through a system of lymphatic vessels, passing through many lymph nodes and lymphoid tissues. For example, (a) the thymus is involved in T-cell development and atrophies with age; (b) the bone marrow is the site of B-cell development; and (c) lymph enters a lymph node through the afferent lymph vessels, percolates through and around the follicles in the node, and leaves through the efferent lymphatic vessels. The lymphoid follicles are the site of cellular interactions and extensive immunologic activity.

nodes, and spleen are now discussed in more detail. GALT and SALT are described in section 28.5 as part of the host's physical and mechanical barriers.

Primary Lymphoid Organs and Tissues

The **thymus** is a highly organized lymphoid organ located above the heart. Precursor cells from the bone marrow migrate into the outer cortex of the thymus, where they proliferate. As they mature and acquire T-cell surface markers, about 90% die. This is due to a process known as **thymic selection** in which T cells that recog-

nize host (self) antigens are destroyed. The remaining 10% move into the medulla of the thymus (figure 28.9a), become mature T cells, and subsequently enter the bloodstream. These T cells recognize non host (non-self) antigens.

In mammals, the **bone marrow** (figure 28.9b) is the site of B-cell maturation. Like thymic selection during T-cell maturation, a selection process within the bone marrow eliminates B cells bearing self-reactive antigen receptors. In birds, undifferentiated lymphocytes move from the bone marrow to the bursa of Fabricius, where B cells mature; this is where B cells were first identified and how they came to be known as "B" (for bursa) cells.

Secondary Lymphoid Organs and Tissues

The **spleen** is the most highly organized secondary lymphoid organ. The spleen is a large organ located in the abdominal cavity (figure 28.9). It filters the blood and traps blood-borne microorganisms and antigens. Once trapped by splenic macrophages or dendritic cells, the pathogen is phagocytosed, killed, and digested. The resulting peptide antigens (protein fragments consisting of less than about 50 amino acid residues) are delivered to the macrophage or dendritic cell surface within a protein receptor, where they are presented to B and T cells. This is the most common way B and T lymphocytes become activated to carry out their immune functions.

Lymph nodes lie at the junctions of lymphatic vessels, where phagocytic macrophages and dendritic cells trap pathogens and antigens (figure 28.9c). They then phagocytose the foreign material and present antigen to lymphocytes. Antigen presented to the class of T lymphocytes called T-helper cells activates these cells to release cytokines needed for B-cell activation. Thus it is within lymph nodes that B cells differentiate into memory cells and antibody-secreting plasma cells. >> *B-cell biology (section 29.6)*

Lymphoid tissues are found throughout the body and act as regional centers of antigen sampling and processing (figure 28.9). Lymphoid tissues are found as highly organized or loosely associated cellular complexes. Some lymphoid cells are closely associated with specific tissues such as skin (skin-associated lymphoid tissue, or SALT) and mucous membranes (mucosal-associated lymphoid tissue, or MALT). SALT and MALT are good examples of highly organized lymphoid tissues, typically seen histologically as macrophages surrounded by specific areas of B and T lymphocytes and sometimes dendritic cells. Loosely associated lymphoid tissue is best represented by the bronchial-associated lymphoid tissue, or BALT, characterized by the lack of cellular partitioning. The primary role of these lymphoid tissues is to efficiently organize leukocytes to increase interaction between the innate and the acquired arms of the immune response. In other words, the lymphoid tissues serve as the interface between the nonspecific (innate) and specific (acquired) immunity of a host. Thus as we shall see, a microbe attempting to invade a potential host is greeted by nonspecific, physical, chemical, and granulocyte barriers that are designed to kill the invader, digest the carcass into small antigens, and assist the lymphocytes in formulating long-term protection against the next invasion. We now examine the phagocytic processes in more detail and then consider how the host integrates many of the innate immune activities into a substantial barrier to microbial invasion, known as the inflammatory response.

1. Briefly describe each of the primary lymphoid organs and tissues.

2. What is the function of the spleen? A lymph node? The thymus? What is the importance of thymic selection?

3. Injury to the spleen can lead to its removal. What impact would this have on host defenses?

28.3 PHAGOCYTOSIS

During their lifetimes, humans and other vertebrates encounter many microbial species, but only a few of these can grow and cause serious disease in otherwise healthy hosts. Phagocytic cells (monocytes, tissue macrophages, dendritic cells, and neutrophils) are an important early defense against invading microorganisms. These phagocytic cells recognize, ingest, and kill many extracellular microbial species by the process called **phagocytosis** (Greek *phagein,* to eat, *cyte,* cell, and *osis,* a process) (**figure 28.10**). Phagocytic cells use two basic molecular mechanisms for the recognition of microorganisms: (1) opsonin-independent (nonopsonic) recognition and (2) opsonin-dependent (opsonic) recognition. The phagocytic process can be greatly enhanced by opsonization. We discuss nonopsonic recognition here and reserve our discussion of opsonic recognition for section 28.6.

Pathogen Recognition

The **opsonin-independent mechanism** is a receptor-based system wherein components common to many different pathogens are recognized to activate phagocytes (figure 28.10a). Phagocytic cells recognize pathogens by several means but appear to exploit a common signaling system to respond. One recognition mode, termed lectin phagocytosis, is based on the binding of a microbial lectin (Latin *legere,* to select or choose), a protein that specifically binds or cross-links carbohydrates to a carbohydrate moiety of a cell receptor (**figure 28.11**). A second mode results from protein-protein interactions between the peptide sequence arginine-glycine-aspartic acid (RGD) on the cell surface of microorganisms and RGD receptors found on all phagocytes. Third, hydrophobic interactions between bacteria and phagocytic cells also promote phagocytosis. A particular microbial species can express multiple binding sites, each recognized by a distinct receptor present on phagocytic cells.

A fourth type of interaction also involves the recognition of microbial antigen and plays a crucial role in nonspecific host resistance. This recognition strategy is based on the detection of conserved molecular structures called ***p*athogen-*a*ssociated *m*olecular *p*atterns (PAMPs).** PAMPs are unique to microorganisms and are invariant among microorganisms of a given class. The most well-known examples of PAMPs are the lipopolysaccharide (LPS) of gram-negative bacteria and the peptidoglycan of gram-positive bacteria. These and other PAMPs are recognized by receptors on phagocytic cells called ***p*attern *r*ecognition *r*eceptors (PRRs).** Because PAMPs are produced only by microorganisms, they are perceived by the phagocytic cells of the innate immune system as molecular signatures of infection.

Toll-like Receptors

Several structurally and functionally distinct classes of PRRs evolved in phagocytic cells to induce various host defensive pathways. For example, secreted PRRs bind to microbial cells and mark them for destruction. Another class of PRRs function

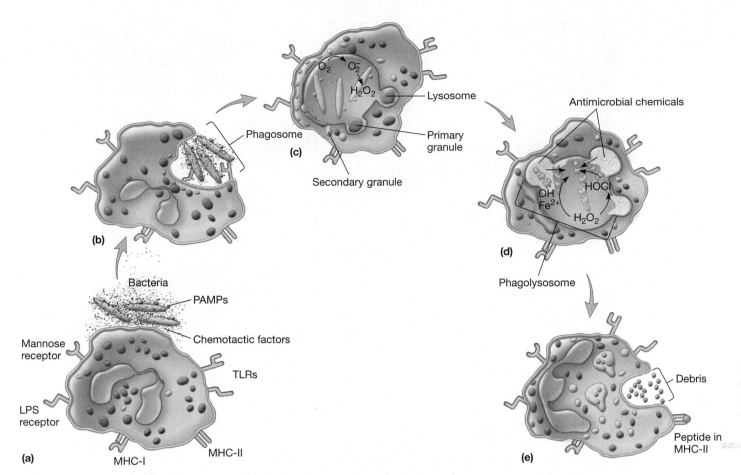

Figure 28.10   Phagocytosis.   (a) Receptors on a phagocytic cell, such as a macrophage, and the corresponding PAMPs participating in phagocytosis. The process of phagocytosis includes (b) ingestion, (c) participation of primary and secondary granules, and O_2-dependent killing events, (d) intracellular digestion, and (e) exocytosis. LPS receptor: lipopolysaccharide receptor; TLRs: toll-like receptors; MHC-I: class I major histocompatibility protein; MHC-II: class II major histocompatibility protein; PAMPs: pathogen-associated molecular patterns.

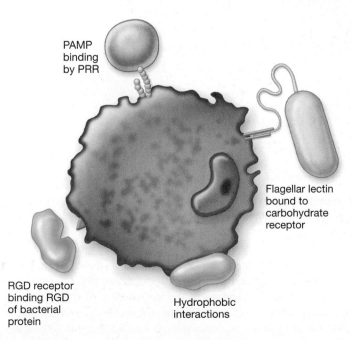

Figure 28.11   Some of the Possible Mechanisms by Which a Macrophage Can Recognize and Capture Microbes.

exclusively as signaling receptors. These receptors are known as ***toll-like receptors*** **(TLRs)** (**figure 28.12**). TLRs recognize and bind unique PAMPs of different classes of pathogens (viruses, bacteria, or fungi) and subsequently communicate that binding to the host cell nucleus to initiate appropriate gene expression and host response. There are at least 10 distinct proteins in this family of mammalian receptors. For example, TLR-4 signals the presence of bacterial lipopolysaccharide and heat-shock proteins. TLR-9 signals the dinucleotide CpG motif present on DNA released by dying bacteria. TLR-2 signals the presence of bacterial lipoproteins and peptidoglycans. Binding of TLRs triggers an evolutionarily ancient signaling pathway that activates transcription factor NFκB through the degradation of its inhibitor, IκB. NFκB then induces expression of a variety of genes, including genes for cytokines, chemokines, and costimulatory molecules that play essential roles in calling forth and directing the adaptive immune response later in an infection. Thus binding of specific microbial components to phagocyte receptors is an important first step in phagocytosis. Once bound, the microbe or its components can be internalized as part of a phagosome that is then united with a lysosome to facilitate microbial killing and digestion.

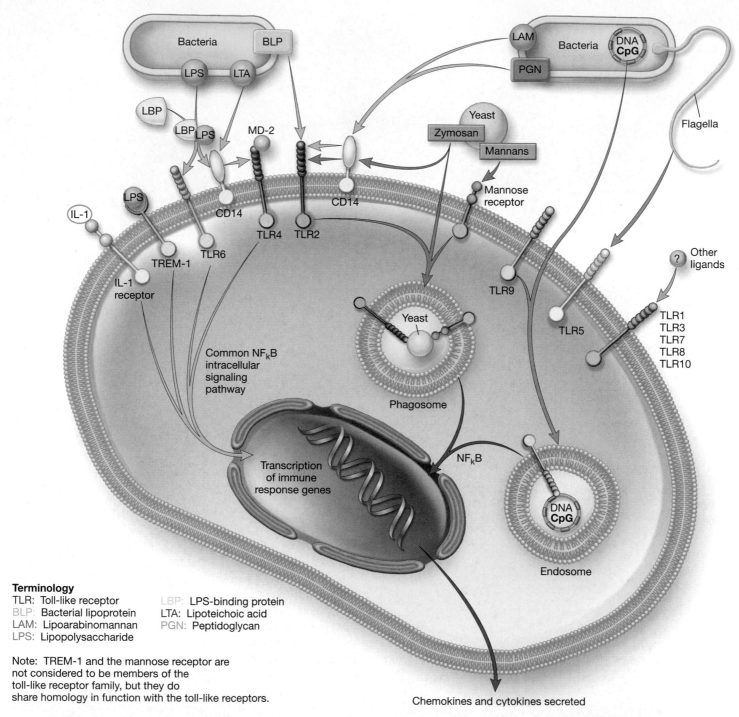

Figure 28.12 Recognition of Pathogen-Associated Molecular Patterns (PAMPs) by Toll-like Receptors (TLRs). PAMP binding of TLR results in a signaling process that upregulates gene expression. A common NFκB signal transduction pathway is used.

Intracellular Digestion

Once ingested by phagocytosis, microorganisms in membrane-enclosed vesicles are delivered to a **lysosome** by fusion of the phagocytic vesicle, called a **phagosome,** with the lysosome membrane, forming a new vacuole called a **phagolysosome** (figure 28.10c, d). Lysosomes deliver a variety of hydrolases such as lysozyme, phospholipase A$_2$, ribonuclease, deoxyribo-

nuclease, and proteases. The activity of these degradative enzymes is enhanced by the acidic vacuolar pH. Collectively, these enzymes participate in the destruction of the entrapped microorganisms. In addition to these oxygen-independent lysosomal hydrolases, macrophage and neutrophil lysosomes contain oxygen-dependent enzymes that produce toxic **reactive oxygen intermediates (ROIs)** such as the superoxide

Table 28.2	Formation of Reactive Oxygen Intermediates
Oxygen Intermediate	**Reaction**
Superoxide $(O_2^-\cdot)$	$$NADPH + 2 O_2 \xrightarrow{\text{NADPH oxidase}} 2 O_2^-\cdot + H^+ + NADP^+$$
Hydrogen peroxide (H_2O_2)	$$2 O_2^-\cdot + 2 H^+ \xrightarrow{\text{Superoxide dismutase}} H_2O_2 + O_2$$
Hypochlorous acid $(HOCl)$	$$H_2O_2 + Cl^- \xrightarrow{\text{Myeloperoxidase}} HOCl + OH^+$$
Singlet oxygen $(^1O_2)$	$$ClO^- + H_2O_2 \xrightarrow{\text{Peroxidase}} {}^1O_2 + Cl^- + H_2O$$
Hydroxyl radical (OH^-)	$$O_2^-\cdot + H_2O_2 \xrightarrow{\text{Peroxidase}} 2OH^- + O_2$$

radical $(O_2^-\cdot)$, hydrogen peroxide (H_2O_2), singlet oxygen $(^1O_2)$, and hydroxyl radical (OH^-). Neutrophils also contain the heme-protein myeloperoxidase, which catalyzes the production of hypochlorous acid. Some reactions that form ROIs are shown in **table 28.2**. These reactions result from the **respiratory burst** that accompanies the increased oxygen consumption and ATP generation needed for phagocytosis. These reactions occur within the lysosome as soon as the phagosome is formed; lysosome fusion is not necessary for the respiratory burst. ROIs are effective in killing invading microorganisms. << *Influences of environmental factors on growth: Oxygen concentration (section 7.5)*

Macrophages, neutrophils, and mast cells have also been shown to form **reactive nitrogen intermediates (RNIs).** These molecules include nitric oxide (NO) and its oxidized forms, nitrite (NO_2^-) and nitrate (NO_3^-). The RNIs are very potent cytotoxic agents and may be either released from cells or generated within cell vacuoles. Nitric oxide is probably the most effective RNI. Macrophages produce it from the amino acid arginine. Nitric oxide can block cellular respiration by complexing with the iron in electron transport proteins. Macrophages use RNIs in the destruction of a variety of infectious agents as well as to kill tumor cells.

Neutrophil granules contain a variety of other microbicidal substances such as several cationic peptides, the bactericidal permeability-increasing protein (BPI), and the family of broad-spectrum antimicrobial peptides including defensins. These substances are compartmentalized strategically so as to locate them for extracellular secretion or delivery to phagocytic vacuoles. Susceptible microbial targets include a variety of gram-positive and gram-negative bacteria, yeasts and molds, and some viruses. Defensins act against bacteria and fungi by permeabilizing cell membranes. They form voltage-dependent membrane channels that allow ionic efflux. Antiviral activity involves direct neutralization of enveloped viruses, so they can no longer bind host cell receptors; nonenveloped viruses are not affected by defensins. Defensins are described in more detail in section 28.6.

Exocytosis

Once the microbial invaders have been killed and digested into small antigenic fragments, the phagocyte may do one of two things. Neutrophils tend to expel the microbial fragments by the process of **exocytosis** (figure 28.10*e*). This is essentially a reverse of the phagocytic process whereby the phagolysosome unites with the cell membrane, resulting in the extracellular release of the microbial fragments. By contrast, macrophages and dendritic cells become **antigen-presenting cells.** This is accomplished by passing some of the microbial fragments from the phagolysosome to the endoplasmic reticulum. Here, the peptide components of the fragments are united with glycoproteins destined for the cell membrane. The glycoproteins bind the peptides so that they are presented outward from the cell once the glycoprotein is secured in the cell membrane. Antigen presentation is critical because it enables wandering lymphocytes to evaluate killed microbes (as antigens) and be activated. Thus antigen presentation links a nonspecific immune response (phagocytosis) to a specific immune response (lymphocyte activation). >> *Recognition of foreignness (section 29.4)*

1. Once a phagolysosome forms, how is the entrapped microorganism destroyed?
2. What is the purpose of the respiratory burst that occurs within macrophages and other phagocytic cells? Describe the nature and function of reactive oxygen and nitrogen intermediates.
3. How do macrophages and dendritic cells become antigen-presenting cells?

28.4 INFLAMMATION

We next turn our attention to how innate immune cells perceive an impending invasion by pathogens so they can be recruited for host defense. In this regard, the process of inflammation is key. Inflammation (Latin, *inflammatio,* to set on fire) is an important nonspecific defense reaction to tissue injury, such as that caused by a pathogen or wound. Acute inflammation is the immediate response of the body to injury or cell death. The gross features were described over 2,000 years ago and are still known as the cardinal signs of inflammation: redness (*rubor*), warmth (*calor*), pain (*dolor*), swelling (*tumor*), and altered function (*functio laesa*).

The **acute inflammatory response** begins when injured tissue cells release chemical signals (chemokines) that activate the inner lining (endothelium) of nearby capillaries (**figure 28.13**). Within the capillaries, **selectins** (a family of cell adhesion molecules) are displayed on the activated endothelial cells. These

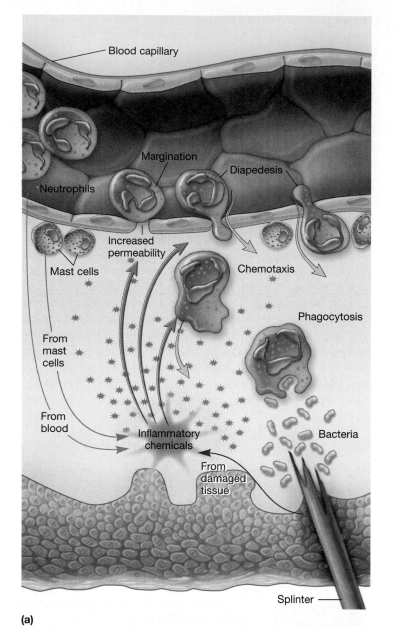

(a)

(b)

adhesion molecules attract and attach wandering neutrophils to the endothelial cells. This slows the neutrophils and causes them to roll along the endothelium, where they encounter the inflammatory chemicals that act as activating signals (figure 28.13b). These signals activate **integrins** (adhesion receptors) on the neutrophils. The integrins then attach tightly to the selectins, causing the neutrophils to stick to the endothelium and stop rolling (margination). The neutrophils now undergo dramatic shape changes, squeeze through the endothelial wall (diapedesis) into the interstitial tissue fluid, migrate to the site of injury (extravasation), and attack the pathogen or other cause of the tissue damage. Neutrophils and other leukocytes are attracted to the infection site by chemotactic factors, which are also called **chemotaxins.** They include substances released by bacteria, endothelial cells, mast cells, and tissue breakdown products. Depending on the severity and nature of tissue damage, other types of leukocytes (e.g., lymphocytes, monocytes, and macrophages) may follow the neutrophils.

The release of inflammatory mediators from injured tissue cells sets into motion a cascade of events that results in the development of the signs of inflammation. These mediators increase the acidity in the surrounding extracellular fluid, which activates the extracellular enzyme **kallikrein** (**figure 28.14**). Kallikrein cleavage releases the peptide bradykinin from its long precursor chain. Bradykinin then binds to receptors on the capillary wall, opening the junctions between cells and allowing fluid, red blood cells, and infection-fighting leukocytes to leave the capillary and enter the infected tissue. Simultaneously, bradykinin binds to mast cells in the connective tissue associated with most small blood vessels. This activates mast cells by causing an influx of calcium ions, which leads to degranulation and release of preformed mediators such as histamine. If nerves in the infected area are damaged, they release substance P, which also binds to mast cells, boosting preformed-mediator release. Histamine in turn makes the intercellular junctions in the capillary wall wider so that more fluid, leukocytes, kallikrein, and bradykinin move out, causing swelling or edema. Bradykinin then binds to nearby capillary cells and stimulates the production of prostaglandins (PGE_2 and PGF_2) to promote tissue swelling in the infected area. Prostaglandins also bind to free nerve endings, making them fire and start a pain impulse.

Activated mast cells also release a small molecule called arachidonic acid, the product of a reaction catalyzed by phospholipase A_2. Arachidonic acid is metabolized by mast cell to form potent mediators, including prostaglandins E_2 and F_2, thromboxane, slow-reacting substance (SRS), and leukotrienes

Figure 28.13 Physiological Events of the Acute Inflammatory Response. (a) At the site of injury (splinter), chemical messengers are released from the damaged tissue, mast cells, and the blood plasma. These inflammatory chemicals stimulate neutrophil migration, diapedesis, chemotaxis, and phagocytosis. (b) Neutrophil integrins interact with endothelial selectins (1) to facilitate margination (2) and diapedesis (3).

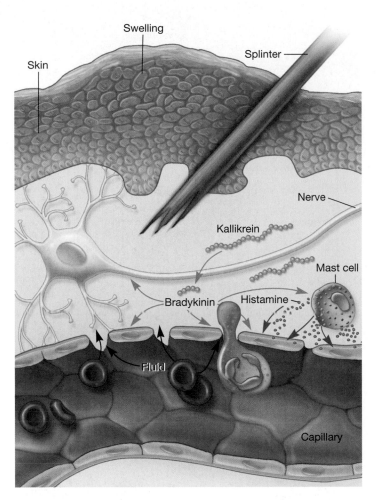

Figure 28.14 Tissue Injury Results in the Recruitment of Kallikrein, from Which Bradykinin Is Released. Bradykinin acts on endothelial and nerve cells resulting in edema and pain, respectively. It also stimulates mast cells to release histamine. Histamine acts on endothelial cells, further increasing blood leakage into injured tissue sites.

(LTC_4 and LTD_4). These mediators play specific roles in the inflammatory response. During acute inflammation, the offending pathogen is neutralized and eliminated by a series of important events:

1. The increase in blood flow and capillary dilation bring into the area more antimicrobial factors and leukocytes that destroy the pathogen. Dead host cells also release antimicrobial factors.

2. Blood leakage into tissue spaces increases the temperature and further stimulates the inflammatory response and may inhibit microbial growth.

3. A fibrin clot often forms and may limit the spread of the invaders.

4. Phagocytes collect in the inflamed area and phagocytose the pathogen. In addition, chemicals stimulate the bone marrow to release neutrophils and increase the rate of granulocyte production.

Chronic Inflammation

In contrast to acute inflammation, which is a rapid and transient process, chronic inflammation, which lasts at least two weeks, is a slow process characterized by the formation of new connective tissue, and it usually causes permanent tissue damage. Chronic inflammation can occur as a distinct process without much acute inflammation. The persistence of bacteria can stimulate chronic inflammation. For example, mycobacteria, which include species that cause tuberculosis and leprosy, have cell walls with a very high lipid and wax content, making them relatively resistant to phagocytosis and intracellular killing. These bacteria and a number of other microbial pathogens, including some protozoa such as *Leishmania*, can survive within macrophages. In addition, some bacteria produce toxins that stimulate tissue-damaging reactions even after bacterial death. << *Suborder* Corynebacterineae *(section 22.4); Protist classification: Excavata (section 23.2)*

Chronic inflammation is characterized by a dense infiltration of lymphocytes and macrophages. If the macrophages are unable to protect the host from tissue damage, the body attempts to wall off and isolate the site by forming a **granuloma** (Latin, *granulum,* a small particle; Greek, *oma,* to form). Granulomas are formed when neutrophils and macrophages are unable to destroy the microorganism during inflammation. Infections caused by some microbial pathogens as well as helminth parasites and large antibody-antigen complexes (as in rheumatoid arthritis) result in granuloma formation and chronic inflammation.

1. What major events occur during an inflammatory reaction, and how do they contribute to pathogen destruction?

2. How does chronic inflammation differ from acute inflammation?

28.5 PHYSICAL BARRIERS IN NONSPECIFIC (INNATE) RESISTANCE

With few exceptions, a potential microbial pathogen invading a human host immediately confronts a vast array of nonspecific defense mechanisms (**figure 28.15**). Although the effectiveness of some individual mechanisms is not great, collectively their defense is formidable. Many direct factors (nutrition, physiology, fever, age, genetics) and equally as many indirect factors (personal hygiene, socioeconomic status, living conditions) influence all host-microbe relationships. In addition to these direct and indirect factors, a vertebrate host has some specific physical and mechanical barriers.

Physical and Mechanical Barriers

Physical and mechanical barriers, along with the host's secretions (flushing mechanisms), are the first line of defense against microorganisms. Protection of the most important body surfaces by these mechanisms is discussed next.

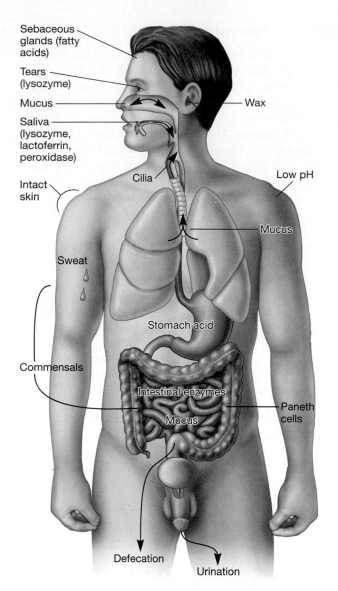

Figure 28.15 Host Defenses. Some nonspecific host defense mechanisms that help prevent entry of microorganisms into the host's tissues.

Skin

Intact skin contributes greatly to nonspecific host resistance. It forms a very effective mechanical barrier to microbial invasion. Its outer layer consists of thick, closely packed cells called **keratinocytes,** which produce keratins. Keratins are scleroproteins (i.e., insoluble proteins) that make up the main components of hair, nails, and the outer skin cells. These outer skin cells shed continuously, removing microorganisms that manage to adhere to their surface. The skin is slightly acidic (around pH 5 to 6) due to skin oil, secretions from sweat glands, and organic acids produced by commensal staphylococci. It also contains a high concentration of sodium chloride and is subject to periodic drying.

Despite the skin's defenses, at times pathogenic microorganisms gain access to the tissue under the skin surface. Here, they encounter a specialized set of cells called the *skin-associated lymphoid tissue* (**SALT**) (**figure 28.16**). The major function of

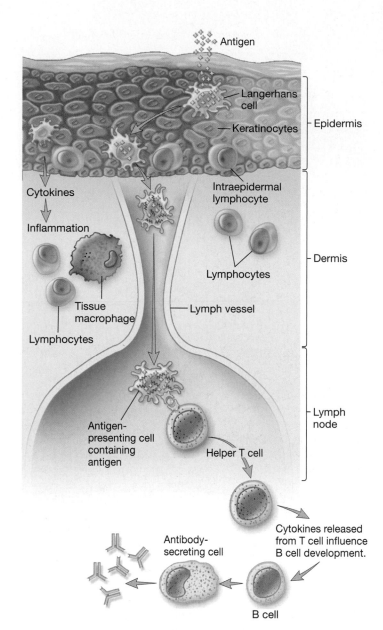

Figure 28.16 Skin-Associated Lymphoid Tissue (SALT). Keratinocytes make up 90% of the epidermis. They are capable of secreting cytokines that cause an inflammatory response to invading pathogens. Langerhans cells internalize antigen and move to a lymph node where they differentiate into dendritic cells that present antigen to helper T cells. The intraepidermal lymphocytes may function as T cells that can activate B cells to induce an antibody response.

SALT is to confine microbial invaders to the area immediately underlying the epidermis and to prevent them from gaining access to the bloodstream. One type of SALT cell is the **Langerhans cell,** a specialized myeloid cell that can phagocytose antigens. Once the Langerhans cell has internalized the antigen, it migrates from the epidermis to nearby lymph nodes, where it differentiates into a mature dendritic cell. Recall that dendritic cells can present antigens and activate nearby lymphocytes to induce the acquired immune system. This dendritic cell–lymphocyte interaction illustrates another bridge between the innate and acquired immune systems.

The epidermis also contains another type of SALT cell called the **intraepidermal lymphocyte** (figure 28.16). These cells are strategically located in the skin so that they can intercept any antigens that breach the first line of defense. Most of these specialized SALT cells are T cells. Unlike other T cells, they have limited receptor diversity and have likely evolved to recognize common skin pathogen patterns. A large number of tissue macrophages (figure 28.3) are also located in the dermal layer of the skin and phagocytose most microorganisms they encounter.

Mucous Membranes

The mucous membranes of the eye (conjunctiva) and the respiratory, digestive, and urogenital systems withstand microbial invasion because the intact stratified squamous epithelium and mucus secretions form a protective covering that resists penetration and traps many microorganisms. This mechanism contributes to nonspecific immunity. Furthermore, many mucosal surfaces are bathed in specific antimicrobial secretions. For example, cervical mucus, prostatic fluid, and tears are toxic to many bacteria. One antibacterial substance in these secretions is **lysozyme** (muramidase), an enzyme that lyses bacteria by hydrolyzing the $\beta(1 \rightarrow 4)$ bond connecting N-acetylmuramic acid and N-acetylglucosamine of the bacterial cell wall peptidoglycan—especially in gram-positive bacteria (**figure 28.17**). These mucous secretions also contain specific immune proteins that help prevent the attachment of microorganisms. In addition, they possess significant amounts of the iron-binding protein, lactoferrin. **Lactoferrin** is released by activated macrophages and polymorphonuclear leukocytes (PMNs). It sequesters iron from the plasma, reducing the amount of iron available to invading microbial pathogens and limiting their ability to multiply. Finally, mucous membranes produce **lactoperoxidase,** an enzyme that catalyzes the production of superoxide radicals, a reactive oxygen intermediate that is toxic to many microorganisms (table 28.2). << *Bacterial cell walls: Peptidoglycan structure (section 3.4)*

Like the skin, mucous membranes also have a specialized immune barrier called *mucosal-associated lymphoid tissue* **(MALT).** There are several types of MALT. The system most stud-

ied is the *gut-associated lymphoid tissue* **(GALT).** GALT includes the tonsils, adenoids, diffuse lymphoid areas along the gut, and specialized regions in the intestine called Peyer's patches. Less well-organized MALT also occurs in the respiratory system and is called *bronchial- associated lymphoid tissue* **(BALT);** the diffuse MALT in the urogenital system does not have a specific name. MALT operates by two basic mechanisms. First, when an antigen arrives at the mucosal surface, it contacts an **M cell** (**figure 28.18a**).

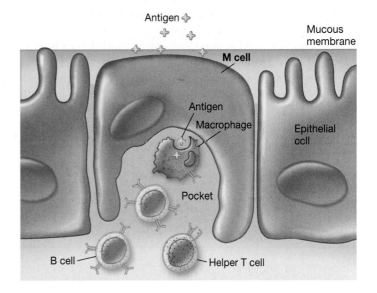

(a)

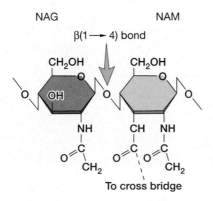

Figure 28.17   Action of Lysozyme on the Cell Wall of Gram-Positive Bacteria.   In the structure of the cell wall peptidoglycan backbone, the $\beta(1 \rightarrow 4)$ bonds connect alternating N-acetylglucosamine (NAG) and N-acetylmuramic acid (NAM) residues. The chains are linked by cross-bridges. Lysozyme splits the molecule where indicated by the arrow.

(b)

Figure 28.18   Function of M Cells in Mucosal-Associated Immunity.   (a) Structure of an M cell located between two epithelial cells in a mucous membrane. The M cell endocytoses the pathogen and releases it into the pocket containing helper T cells, B cells, and macrophages. It is within the pocket that the pathogen often is destroyed. (b) The antigen is transported by the M cell to the organized lymphoid follicle containing B cells. The activated B cells mature into plasma cells, which produce antibodies that are released into the lumen, where they react with antigen that caused their production.

The M cell lacks the brush border of microvilli found on adjacent columnar epithelial cells. It resides above a large epithelial pocket containing B cells, T cells, and macrophages. When an antigen contacts the M cell, it is endocytosed and released into the pocket. Macrophages engulf the antigen or pathogen and try to destroy it. An M cell also can endocytose an antigen and transport it to a cluster of cells called an organized lymphoid follicle (figure 28.18b). The B cells within this follicle recognize the antigen and mature into antibody-producing plasma cells. The plasma cells leave the follicle and secrete mucous membrane–associated antibody. The antibody is then transported into the lumen of the gut, where it interacts with the antigen that caused its production. Similar to SALT, GALT intra- and interepithelial lymphocytes are strategically distributed so the likelihood of antigen detection is increased should the intestinal membrane be breached. >> *Antibodies (section 29.7)*

1. Why is the skin such a good first line of defense against pathogenic microorganisms?
2. How do intact mucous membranes resist microbial invasion of the host?
3. Describe SALT function in the immune response.
4. How do M cells function in MALT?

Respiratory System

The mammalian respiratory system has formidable defense mechanisms. The average person inhales at least eight microorganisms a minute, or 10,000 each day. Once inhaled, a microorganism must first survive and penetrate the air-filtration system of the upper and lower respiratory tracts. Because the airflow in these tracts is very turbulent, microorganisms are deposited on the moist, sticky mucosal surfaces. Microbes larger than 10 μm usually are trapped by hairs and cilia lining the nasal cavity. The cilia in the nasal cavity beat toward the pharynx, so that mucus with its trapped microorganisms is moved toward the mouth and expelled (**figure 28.19**). Humidification of the air within the nasal cavity causes many hygroscopic (attracting moisture from the air) microorganisms to swell, and this aids phagocytosis. Microbes smaller than 10 μm often pass through the nasal cavity and are trapped by the **mucociliary blanket** that coats the mucosal surfaces of lower portions of the respiratory system. The trapped microbes are transported by ciliary action—the mucociliary escalator—that moves them away from the lungs. Coughing and sneezing reflexes clear the respiratory system of microorganisms by expelling air forcefully from the lungs through the mouth and nose, respectively. Salivation also washes microorganisms from the mouth and nasopharyngeal areas into the stomach. Microorganisms that succeed in reaching the alveoli of the lungs encounter a population of fixed phagocytic cells called **alveolar macrophages** (figure 28.3). These cells can ingest and kill most bacteria by phagocytosis.

Gastrointestinal Tract

Most microorganisms that reach the stomach are killed by gastric juice (a mixture of hydrochloric acid, proteolytic enzymes,

Nasal cavity

Nostril

Oral cavity

Pharynx

Epiglottis

Larynx

Trachea

Bronchus

Bronchioles

Right lung   Left lung

Cilia   Microvilli

Figure 28.19   The Bronchial-Associated Lymphoid Tissue (BALT).   The respiratory tract is lined with a mucous membrane of ciliated epithelial cells. Cilia (×5,000) sweep particles upward toward the throat to be expectorated.

and mucus). The very acidic gastric juice (pH 2 to 3) is sufficient to destroy most organisms and their toxins, although exceptions exist (protozoan cysts, *Helicobacter pylori, Clostridium,* and *Staphylococcus* toxins). However, organisms embedded in food particles are protected from gastric juice and reach the small intestine. There microorganisms often are damaged by various pancreatic enzymes, bile, enzymes in intestinal secretions, and the GALT system. **Peristalsis** (Greek *peri,* around, and *stalsis,* contraction) and the normal loss of columnar epithelial cells act in concert to purge intestinal microorganisms. In addition, the normal microbiota of the large intestine (*see figure 27.13*) is extremely important in preventing the establishment of pathogenic organisms. For example, the metabolic products (e.g, fatty acids) of many normal commensals in the intestinal tract prevent unwanted microorganisms from becoming established. Other normal microbiota outcompete potential pathogens for attachment sites and nutrients. The mucous membranes of the intestinal tract contain cells called **Paneth cells**. These cells produce lysozyme (figure 28.17) and peptides called **cryptins.** Cryptins are toxic for some bacteria, although their mode of action is not known.

Genitourinary Tract

Under normal circumstances, the kidneys, ureters, and urinary bladder of mammals are sterile. Urine within the urinary bladder also is sterile. However, in both the male and female, a few bacteria are usually present in the distal portion of the urethra (*see figure 27.13*). The factors responsible for this sterility are complex. In addition to removing microbes by flushing action, urine kills some bacteria due to its low pH and the presence of urea and other metabolic end products (uric acid, hippuric acid, indican, fatty acids, mucin, enzymes). The kidney medulla is so hypertonic that few organisms can survive there. In males, the anatomical length of the urethra (20 cm) provides a distance barrier that excludes microorganisms from the urinary bladder. Conversely, the short urethra (5 cm) in females is more readily traversed by microorganisms; this explains why urinary tract infections are 14 times more common in females than in males.

The vagina has another unique defense. Under the influence of estrogens, the vaginal epithelium produces increased amounts of glycogen that acid-tolerant *Lactobacillus acidophilus* bacteria degrade to form lactic acid. Normal vaginal secretions contain up to 10^8 of these bacteria per milliliter. Thus an acidic environment (pH 3 to 5) unfavorable to most organisms is established. Cervical mucus also has some antibacterial activity.

The Eye

The conjunctiva is a specialized, mucus-secreting epithelial membrane that lines the interior surface of each eyelid and the exposed surface of the eyeball. It is kept moist by the continuous flushing action of tears (lacrimal fluid) from the lacrimal glands. Tears contain large amounts of lysozyme, lac-

toferrin, and antibody, and thus provide chemical as well as physical protection.

1. Describe the different antimicrobial defense mechanisms that operate within the respiratory system of mammals.
2. What factors operate within the gastrointestinal system that help prevent the establishment of pathogenic microorganisms?
3. Except for the anterior portion of the urethra, why is the genitourinary tract a sterile environment?

28.6   CHEMICAL MEDIATORS IN NONSPECIFIC (INNATE) RESISTANCE

Mammalian hosts have a chemical arsenal with which to combat the continuous onslaught of microorganisms. Some of these chemicals (gastric juices, salivary glycoproteins, lysozyme, oleic acid on the skin, urea) have already been discussed with respect to the specific body sites they protect. In addition, blood, lymph, and other body fluids contain a potpourri of defensive chemicals such as defensins and other polypeptides.

Antimicrobial Peptides

Cationic Peptides

Antimicrobial cationic peptides appear to be highly conserved through evolution. We will only discuss those peptides found in humans. There are three generic classes of cationic peptides whose biological activity is related to their ability to damage bacterial plasma membranes. This is accomplished by electrostatic interactions with membranes—the formation of ionic pores or transient gaps, thereby altering membrane permeability.

The first group of cationic peptides includes those that are linear, alpha-helical peptides that lack cysteine amino acid residues. An important example is **cathelicidin,** a peptide that arises from a precursor protein having a C-terminus bearing the mature peptide of some 12 to 80 amino acids. Cathelicidins are produced by a variety of cells (e.g., neutrophils, respiratory epithelial cells, and alveolar macrophages), and substantial heterogeneity exists between cathelicidins made by various cells. >> *Protein structure (appendix I)*

A second group, the **defensins,** are peptides that are rich in arginine and cysteine, and disulfide linked. The group is composed of various structural motifs with an approximate average molecular weight of 4,000 Daltons. In mammals, defensins have antiparallel beta sheet structures with beta hairpin loops containing cationic amino acids. Two types of defensins have been reported in humans—alpha and beta. Alpha defensins tend to be peptides of 29 to 35 amino acid residues, whereas beta defensins are usually 36 to 42 amino acids in length and are found in the primary granules of neutrophils, intestinal Paneth cells, and intestinal and respiratory epithelial cells.

A third group contains larger peptides that are enriched for specific amino acids and exhibit regular structural repeats. **Histatin,** one such peptide isolated from human saliva, has antifungal activity. Histatin is a 24 to 38 amino acid peptide, heavily enriched with histidine, that translocates to the fungal cytoplasm, where it targets mitochondria.

Other natural antimicrobial products include fragments from (1) histone proteins, (2) lactoferrin, and (3) chemokines. A number of antibacterial peptides are produced by bacteria as well. The most notable of these are the bacteriocins.

Bacteriocins

As noted previously, the first line of defense against microorganisms is the host's anatomical barrier, consisting of the skin and mucous membranes. These surfaces are colonized by normal microbiota, which by themselves provide a biological barrier against uncontrolled proliferation of foreign microorganisms. Many of the bacteria that are part of the normal microflora synthesize and release toxic proteins called bacteriocins that are lethal to other strains of the same species as well as other closely related bacterial species. Bacteriocin peptides range from about 900 to 5,800 Daltons and can be cationic, neutral, or anionic. Bacteriocins may give their producers, which are naturally immune to the antibacterial products they make, an adaptive advantage against other bacteria. Ironically, they sometimes increase bacterial virulence by damaging host cells such as mononuclear phagocytes. Bacteriocins are produced by gram-negative and gram-positive bacteria. For example, *Escherichia coli* synthesizes bacteriocins called colicins, which are encoded by genes on several different plasmids (ColB, ColE1, ColE2, ColI, and ColV). Some colicins bind to specific receptors on the cell envelope of sensitive target bacteria and cause cell lysis, attack specific intracellular sites such as ribosomes, or disrupt energy production. Other examples include the lantibiotics produced by genera such as *Streptococcus, Bacillus, Lactococcus,* and *Staphylococcus.* These antimicrobial peptides act as defensive effector molecules protecting the bacterial flora and its human host. << *Normal microbiota of the human body (section 27.3)*

1. How do cationic peptides function against grampositive bacteria?
2. How do bacteriocins function?

Complement

Complement was discovered many years ago as a heat-labile component of human blood plasma that augments phagocytosis. This activity was said to "complement" the antibacterial activity of antibody; hence the name complement. It is now known that the **complement system** is composed of over 30 serum proteins that have a complex (and somewhat confusing) nomenclature.

This system has three major physiological activities: (1) defending against bacterial infections by facilitating and enhancing phagocytosis through opsonization, chemotaxis, activation of leukocytes, and lysis of bacterial cell walls; (2) bridging innate and adaptive immunity by the augmentation of antibody responses and enhancement of immunologic memory; and (3) disposing of wastes such as immune complexes, the products of inflammatory injury, and dead host cells.

To achieve these activities, we need to revisit the idea of opsonization, introduced in section 28.3. **Opsonization** (Greek *opson,* to prepare victims for) is a process in which microorganisms or other particles are coated by serum components, thereby preparing them for recognition and ingestion by phagocytic cells. Molecules that function in this capacity are collectively known as opsonins. In the opsonin-dependent recognition mechanism, the host serum components function as a bridge between the microorganism and the phagocyte. They act by binding to the surface of the microorganism at one end and to specific receptors on the phagocyte surface at the other **(figure 28.20).** Some of the complement proteins are opsonins because they bind microbial cells, coating them for recognition by phagocytes (figure 28.20*b*). Additionally, other complement proteins are strong chemotactic signals that recruit phagocytes to the site of their activation. Still other complement proteins puncture cell membranes to cause lysis. Interestingly, one of the several triggers that can activate

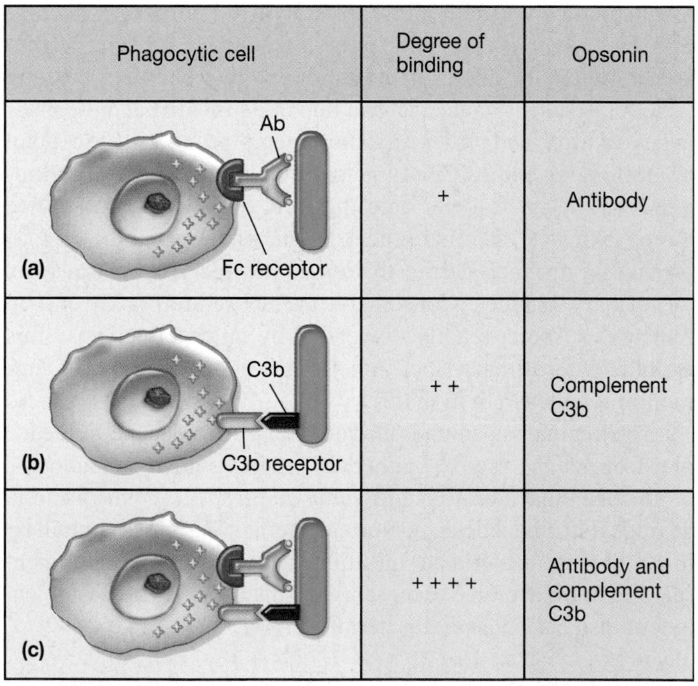

Phagocytic cell	Degree of binding	Opsonin
(a) Ab — Fc receptor	+	Antibody
(b) C3b — C3b receptor	+ +	Complement C3b
(c)	+ + + +	Antibody and complement C3b

Figure 28.20 Opsonization. (a) The intrinsic ability of a phagocyte to bind to a microorganism is enhanced if the microorganism elicits the formation of antibodies (Ab) that act as a bridge to attach the microorganism to the Fc antibody receptor on the phagocytic cell. (b) If the activated complement (C3b) is bound to the microbe, the degree of binding is further increased by the C3b receptor. (c) If both antibody and C3b opsonize, binding is maximal.

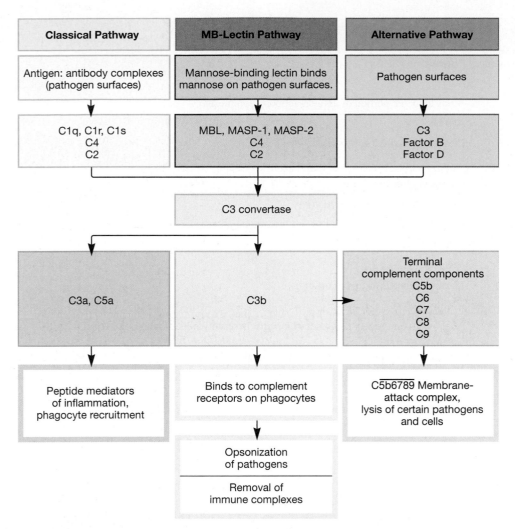

Figure 28.21 **The Main Components and Actions of Complement.** Complement activation involves a series of enzymatic reactions that culminate in the formation of C3 convertase, which cleaves complement component C3 into C3b and C3a. The production of the C3 convertase is where the three pathways converge. C3a is a peptide mediator of local inflammation. C3b binds covalently to the bacterial cell and opsonizes the bacteria, enabling phagocytes to internalize them. C5a and C5b are generated by the cleavage of C5 by a C5 convertase. C5a is also a powerful peptide mediator of inflammation. C5b promotes the terminal components of complement to assemble into a membrane-attack complex.

The **alternative complement pathway** (figure 28.21) plays an important role in the nonspecific immune defense against intravascular invasion by bacteria and some fungi. The alternative pathway is initiated in response to bacterial molecules with repetitive structures such as lipopolysaccharide (LPS). It begins with cleavage of C3 into fragments C3a and C3b by a blood enzyme. These fragments are initially produced at a slow rate, and free C3b is rapidly cleaved into inactive fragments by another protein called Factor I. However, C3b becomes stable when it binds to the LPS of gram-negative bacterial cell walls or to aggregates of antibodies. A protein in blood termed Factor B adsorbs to bound C3b and is cleaved into two fragments by Factor D (the bar indicates an activated enzyme complex), leading to the formation of active enzyme C3bBb. This complex is called the C3 convertase of the alternative pathway because it cleaves more C3 to C3a and C3b, thereby increasing the rate at which C3 is converted. C3bBb is further stabilized by a second blood protein, properdin, which allows another addition of C3b, forming C5 convertase (C3bBb3b). This convertase then cleaves C5 to C5a and C5b. The two proteins C6 and C7 rapidly bind to C5b, forming a C5b67 complex that is stabilized by binding to a membrane. C8 and C9 then bind, forming the **membrane attack complex (MAC)** (C5b6789), which creates a pore in the plasma membrane or outer membrane of the target cell (**figure 28.22**). If the cell is eucaryotic, Na⁺ and H₂O enter through the pore and the cell lyses. Lysozyme can pass through pores in the outer membrane of gram-negative cell walls and digest the peptidoglycan, thus weakening the cell wall and aiding lysis. In contrast, gram-positive bacteria resist the cytolytic action of the membrane attack complex because they lack an exposed outer membrane and have a thick peptidoglycan protecting the plasma membrane. Unfortunately, eucaryotic host cell membranes are also susceptible to attack by complement proteins, and bystander lysis is a potential consequence of complement activation.

the complement process is the recognition of specific antibody on a target cell. All together, the complement activities unite the nonspecific and specific arms of the immune system to assist in the killing and removal of invading pathogens.

Complement proteins are produced in an inactive form; they become active following enzymatic cleavage. There are three pathways of complement activation: the alternative, lectin, and classical pathways (**figure 28.21**). Although they employ similar mechanisms, specific proteins are unique to the first part of each pathway (**table 28.3**). Each complement pathway is activated in a cascade fashion: the activation of one component results in the activation of the next. Thus complement proteins are poised for immediate activity when the host is challenged by an infectious agent.

The generation of complement fragments C3a and C5a leads to several important inflammatory effects. For example, binding of C3a and C5a to their cellular receptors induces some cells to release other biological mediators. These amplify

Table 28.3		Some Important Proteins of the Complement Cascade
Protein	**Fragment**	**Function**
Recognition Unit		
C1	q	Binds to the Fc portion of antigen-antibody complexes
	r	Activates C1s
	s	Cleaves C4 and C2 due to its enzymatic activity
Activation Unit		
C2		Causes viral neutralization
C3	a	Anaphylatoxin, immunoregulatory
	b	Key component of the alternative pathway and major opsonin in serum
	e	Induces leukocytosis
C4	a	Anaphylatoxin
	b	Causes viral neutralization; opsonin
Membrane Attack Unit		
C5	a	Anaphylatoxin; principal chemotactic factor in serum; induces neutrophil attachment to blood vessel walls
	b	Initiates membrane attack
C6 C7 C8 C9		Participate with C5b in formation of the membrane attack complex that lyses targeted cells
Alternative Pathway		
Factor B		Causes macrophage spreading on surfaces; precursor of C3 convertase
Factor $\overline{D}$		Cleaves Factor B to form active $\overline{C3bBb}$ in alternative pathway
Properdin		Stabilizes alternative pathway C3 convertase
Regulatory Proteins		
Factor H		Promotes C3b breakdown and regulates alternative pathway
Factor I		Degrades C3b and regulates alternative pathway
C4b binding protein		Inhibits assembly and accelerates decay of $\overline{C4bC2a}$
C1 INH complex		Binds to and dissociates C1r and C1s from C1
S protein		Binds fluid-phase $\overline{C5b67}$; prevents membrane attachment

the inflammatory signals of C3a and C5a by dilating vessels, increasing permeability, stimulating nerves, and recruiting phagocytic cells. C5a induces a directed, chemotactic migration of neutrophils to the site of complement activation; neutrophils follow the diffusion gradient. Macrophages in the area can synthesize even more complement components to interact with the invading microbe. All of these defensive events promote the ingestion and ultimate destruction of the microbe by neutrophils and macrophages.

The **lectin complement pathway** (also called the mannan-binding lectin pathway) also begins with the activation of C3

convertase. However, in this case a lectin, a protein that binds to specific carbohydrates, initiates the proteolytic cascade. When macrophages ingest viruses, bacteria, or other foreign material, they release chemicals that stimulate liver cells to secrete acute phase proteins such as **mannose-binding protein (MBP).** Because mannose, in certain three-dimensional configurations, is a major component of bacterial cell walls and of some virus envelopes and antigen-antibody complexes, MBP binds to these components. MBP enhances phagocytosis and is therefore an opsonin. When MBP is bound to the *MBP-a*ssociated *s*erine esterase (MASP), it activates the same C3 convertase found in

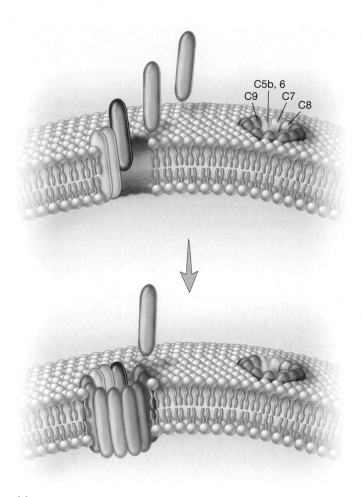

(a)

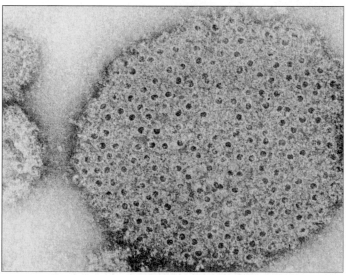

(b)

Figure 28.22   The Membrane Attack Complex. The membrane attack complex (MAC) is a tubular structure that forms a transmembrane pore in the target cell's membrane. (a) This representation shows the subunit architecture of the membrane attack complex. The transmembrane channel is formed by a C5b678 complex and 10 to 16 polymerized molecules of C9. (b) MAC pores appear as craters or donuts by electron microscopy.

the alternative complement pathway. Thus the lectin pathway activates the same complement cascade that the classical and alternative pathways do. However, it uses a mechanism that is independent of antibody-antigen interactions (the classical pathway), and it does not require interaction of complement with pathogen surfaces (alternative pathway).

This overview of the alternative and lectin complement pathways (figure 28.21) provides a basis for consideration of the function of complement as an integrated system during an animal's defensive effort. Bacteria arriving at a local tissue site interact with components of the alternative pathway, resulting in the generation of biologically active fragments, opsonization of the bacteria, and initiation of the lytic sequence. If the bacteria persist or if they invade the animal a second time, antibody responses also will activate the classical complement pathway.

Activation of the **classical complement pathway** can occur in response to some microbial products (lipid A, staphylococcal protein A, etc.). However, it is usually initiated by the interaction of antibodies with an antigen (figure 28.21). Antibody secretion is part of an acquired immune response (and is discussed in chapter 29). It is important to emphasize that antibodies are glycoproteins that bind to specific antigens. This binding triggers the C1 complement component, composed of three proteins (q, r, and s), to attach to the antibody through its C1q subcomponent. In the presence of calcium ions, a trimolecular complex (C1qrs + antigen + antibody) with esterase activity is rapidly formed. The activated C1s subcomponent attacks and cleaves its natural substrates in serum (C2 and C4). This leads to binding of a portion of each molecule (C2a and C4b) to the antigen-antibody-complement complex and the release of C4a and C2b fragments. With the binding of C2a to C4b, an enzyme with trypsinlike proteolytic activity is generated. The natural substrate for this enzyme is C3; thus C2a4b is a C3 convertase. Just as we saw with the lectin and alternative pathways, the C3 convertase cleaves C3 into a bound subcomponent C3b and a C3a soluble component. This sets in motion the activation of the complement cascade, which leads to the formation of the membrane attack complex, opsonins, and the release of mediators that influence inflammation. Thus the three complement pathways have three different initiating processes. However, their common outcomes of opsonization, stimulation of inflammatory mediators, and lysis of membrane-bound microorganisms achieve the goal of innate host defense against foreign invaders.

1. What effect does the formation of the membrane attack complex have on eucaryotic cells? Procaryotic cells? Host cells?

2. How is the alternative pathway activated? The lectin pathway?

3. What role do complement fragments C3a and C5a play in an animal's defense against gram-negative bacteria?

4. How is the classical complement pathway activated?

Table 28.4	The Four Cytokine Families	
Family	**Examples**	**Functions**
Chemokines	IL-8, RANTES[a], MIP (macrophage inflammatory protein)	Cytokines that are chemotactic and chemokinetic for leukocytes. They stimulate cell migration and attract phagocytic cells and lymphocytes. Chemokines play a central role in the inflammatory response.
Hematopoietins	Epo (erythropoietin), various colony-stimulating factors	Cytokines that stimulate and regulate the growth and differentiation processes involved in blood cell formation (hematopoiesis)
Interleukins	IL-1 to IL-18	Cytokines produced by lymphocytes and monocytes that regulate the growth and differentiation of other cells, primarily lymphocytes and hematopoietic stem cells. They often also have other biological effects.
Tumor necrosis factor (TNF) family	TNF-α, TNF-β, Fas ligand	Cytokines that are cytotoxic for tumor cells and have many other effects such as promoting inflammation, fever, and shock; some can induce apoptosis

[a]RANTES: *Regulated on activation, normal T expressed and secreted*; also called CCL5; member of the IL-8 cytokine superfamily.

Cytokines

Defense against viruses, microorganisms and their products, parasites, and cancer cells is mediated by both nonspecific immunity and specific immunity. Cytokines are required for regulation of both of these immune responses. *Cytokine* (Greek *cyto,* cell, and *kinesis,* movement) is a generic term for any soluble protein or glycoprotein released by one cell population that acts as an intercellular (between cells) mediator or signaling molecule. When released from mononuclear phagocytes, these proteins are called **monokines;** when released from T lymphocytes, they are called **lymphokines;** when produced by a leukocyte and the action is on another leukocyte, they are **interleukins;** and if their effect is to stimulate the growth and differentiation of immature leukocytes in the bone marrow, they are called **colony-stimulating factors (CSFs).** Cytokines have been grouped into the following categories or families: chemokines, hematopoietins, interleukins, and members of the tumor necrosis factor (TNF) family. Some examples of these cytokine families are listed in **table 28.4.** Cytokines can affect the same cell responsible for their production (an autocrine function) or nearby cells (a paracrine function), or they can be distributed by the circulatory system to distant target cells (an endocrine function). Their production is induced by nonspecific stimuli such as a viral, bacterial, or parasitic infection; cancer; inflammation; or the interaction between a T cell and antigen. Some cytokines also can induce the production of other cytokines.

Cytokines produce biological actions only when they bind to specific receptors on the surface of target cells. Most cells have hundreds to a few thousand cytokine receptors, but a maximal cellular response results even when only a small number of these are occupied by a cytokine. This is because the affinity of cytokine receptors for their specific cytokines is very high, and consequently cytokines are effective at very low concentrations. Cytokine binding activates specific intracellular signaling path-

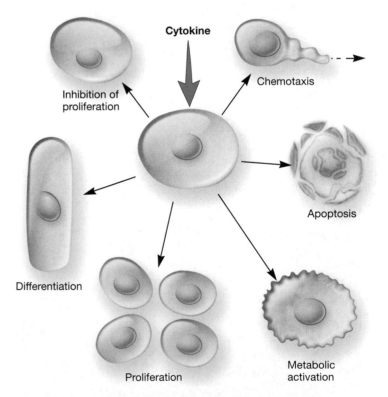

Figure 28.23   Range of Biological Actions That Cytokines Have on Eucaryotic Cells.   Chemokines are one family of cytokines that induce leukocyte chemotaxis and migration. Other cytokines activate cell metabolism and synthesis. This can lead to the synthesis of a wide range of proteins including cyclooxygenase II, proteolytic enzymes, NO synthase, and various adhesion receptors. In addition, other cytokines can cause proliferation, inhibition of cell proliferation, or apoptosis.

ways that switch on genes encoding proteins essential for appropriate cellular functions. For example, cytokine binding may result in the cell's production of other cytokines, cell-to-cell

Table 28.5 — Some of the Cytokines That Mediate Immune Responses

Cytokine	Cell Source	Functions
IL-1 (interleukin-1)	Monocytes/macrophages, endothelial cells, fibroblasts, neuronal cells, glial cells, keratinocytes, epithelial cells	Produces a wide variety of effects on the differentiation and function of cells involved in inflammatory and immune effector responses; also affects central nervous and endocrine systems. It is an endogenous pyrogen.
IL-2 (interleukin-2, T-cell growth factor)	T cells (T_H1)	Stimulates T-cell proliferation and differentiation; enhances cytolytic activity of NK cells; promotes proliferation and immunoglobin secretion of activated B cells
IL-3 (interleukin-3)	T cells, keratinocytes, neuronal cells, mast cells	Stimulates the production and differentiation of macrophages, neutrophils, eosinophils, basophils, and mast cells
IL-4 (interleukin-4, B-cell growth factor-1 [BCGF-1], B-cell stimulatory factor-1 [BCSF-1])	T cells (T_H2), macrophages, mast cells, basophils, B cells	Induces the differentiation of naive $CD4^+$ T cells into T-helper cells; induces the proliferation and differentiation of B cells; exhibits diverse effects on T cells, monocytes, granulocytes, fibroblasts, and endothelial cells
IL-5 (interleukin-5)	T cells (T_H2)	Growth and activation of B cells and eosinophils; chemotactic for eosinophils
IL-6 (interleukin-6, cytotoxic T-cell differentiation factor, B-cell differentiation factor)	T_H2 cells, monocytes/macrophages, fibroblasts, hepatocytes, endothelial cells, neuronal cells	Activates hematopoietic cells; induces growth of T cells, B cells, hepatocytes, keratinocytes, and nerve cells; stimulates the production of acute-phase proteins
IL-8 (interleukin-8)	Monocytes, endothelial cells, fibroblasts, alveolar epithelium, T cells, keratinocytes, neutrophils, hepatocytes	Chemoattractant for PMNs and T cells; causes PMN degranulation and expression of receptors; inhibits adhesion of PMNs to cytokine-activated endothelium; promotes migration of PMNs through nonactivated endothelium
IL-10 (interleukin-10)	T cells (T_H2), B cells, macrophages, keratinocytes	Reduces the production of IFN-γ, IL-1, TNF-α, and IL-6 by macrophages; in combination with IL-3 and IL-4, causes mast cell growth; in combination with IL-2, causes growth of cytotoxic T cells and differentiation of $CD8^+$ cells
IFNs α/β (interferons α/β)	T cells, B cells, monocytes/macrophages, fibroblasts	Antiviral activity, antiproliferative; stimulates macrophage activity; increases MHC class I protein expression on cells; regulates the development of the specific immune response
IFN-γ (interferon-γ)	T cells (T_H1, CTLs), NK cells	Activation of T cells, macrophages, neutrophils, and NK cells; antiviral and antiproliferative activities; increases class I and II MHC molecule expression on various cells
TNF-α (tumor necrosis factor-α [cachectin])	T cells, macrophages and NK cells	A wide variety of effects due to its ability to mediate expression of genes for growth factors and cytokines, transcription factors, receptors, inflammatory mediators, and acute-phase proteins; plays a role in host resistance to infection by serving as an immunostimulant and mediator of the inflammatory response; cytotoxic for tumor cells
TNF-β (tumor necrosis factor-β [lymphotoxin])	T cells, B cells	Same as TNF-α
G-CSF (granulocyte colony-stimulating factor)	T cells, macrophages, neutrophils	Enhances the differentiation and activation of neutrophils
M-CSF (macrophage colony-stimulating factor)	T cells, neutrophils, macrophages, fibroblasts, endothelial cells	Stimulates various functions of monocytes and macrophages, promotes the growth and development of macrophage colonies from undifferentiated precursors

adhesion receptors, proteases, lipid-synthesizing enzymes, and nitric oxide synthase (the production of nitric oxide has potent antimicrobial activity). In addition, cytokines can activate cell proliferation or cell differentiation (**figure 28.23**). They also can inhibit cell division and cause apoptosis (programmed cell death).

Chemokines, one type of cytokine, stimulate chemotaxis and chemokinesis (i.e., they direct cell movement) and thus play an important role in the acute inflammatory response (figure 28.13). Some examples of important cytokines and their functions are given in **table 28.5**.

Figure 28.24 **The Antiviral Action of Interferon.** Interferon (IFN) synthesis and release is often induced by a virus infection or double-stranded RNA (dsRNA). Interferon binds to a ganglioside receptor on the plasma membrane of a second cell and triggers the production of enzymes that render the cell resistant to virus infection. The two most important such enzymes are oligo(A) synthetase and a protein kinase. When an interferon-stimulated cell is infected, viral protein synthesis is inhibited by an active endoribonuclease that degrades viral RNA. An active protein kinase phosphorylates and inactivates the initiation factor eIF-2 required for viral protein synthesis.

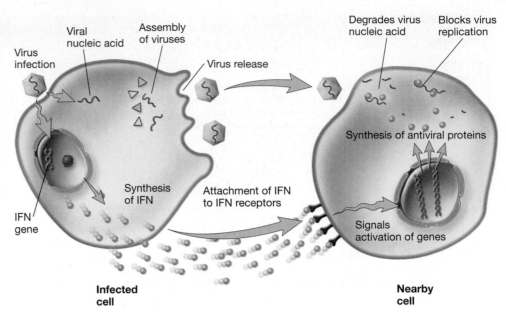

Interferons (IFNs) are a group of related low molecular-weight, regulatory cytokines produced by certain eucaryotic cells in response to a viral infection. Several classes of interferons are recognized: IFN-γ is a family of 20 different molecules that can be synthesized by virus-infected leukocytes, antigen-stimulated T cells, and natural killer cells (table 28.5). IFN-α/β is derived from virus-infected fibroblasts. Although interferons do not prevent virus entry into host cells, they prevent viral replication and assembly, thereby limiting viral infection (**figure 28.24**).

Another group of noteworthy cytokines are **endogenous pyrogens,** which elicit fever in the host. From a physiological point of view, fever results from disturbances in hypothalamic thermoregulatory activity, leading to an increase of the thermal "set point." In adult humans, fever is defined as an oral temperature above 98.6°F (37°C). The most common cause of a fever is a viral or bacterial infection (or bacterial toxins). Examples of these pyrogens include interleukin-1 (IL-1), IL-6, and tissue necrosis factor (TNF); all are produced by host macrophages in response to pathogenic microorganisms. After their release, these pyrogens circulate to the hypothalamus and induce neurons to secrete prostaglandins. Prostaglandins reset the hypothalamic thermostat to a higher temperature, and temperature-regulating reflex mechanisms then act to bring the core body temperature up to this new setting. IL-1 also causes proliferation, maturation, and activation of T and B cells, which in turn augment the immune response of the host to the pathogen.

The fever induced by a microorganism augments the host's defenses in three ways: (1) it stimulates leukocytes so that they can destroy the microorganism; (2) it enhances the specific activity of the immune system; and (3) it enhances microbiostasis (growth inhibition) by decreasing available iron to the microorganism. Evidence suggests that some hosts are able to redistribute the iron during a fever in an attempt to withhold it from the microorganism (hypoferremia). Conversely, the virulence of many microorganisms is enhanced with increased iron availability (hyperferremia). For example, gonococci, the causative agent of gonorrhea, spread most often during menstruation, a time in which an increased concentration of free iron is available to these bacteria.

Acute-Phase Proteins

Macrophages release cytokines (IL-1, IL-6, IL-8, TNF-α, etc.) upon activation by bacteria, which stimulate the liver to rapidly produce acute-phase proteins. These include C-reactive protein (CRP), mannose-binding lectin (MBL), and surfactant proteins A (SP-A) and D (SP-D), all of which can bind bacterial surfaces and act as opsonins. CRP can interact with C1q to activate the classical complement pathway. MBL activates the alternative complement pathway. SP-A, SP-D, and C1q are collectins—proteins composed of a collagen-like motif connected by α-helices to globular binding sites (**figure 28.25**). Thus these proteins (along with others) police host tissues by binding to and assisting in the removal of bacteria.

1. Describe the role of cytokines and interferons in innate immunity.
2. How do interferons render cells resistant to viruses?
3. How might acute phase reactants assist in pathogen removal?
4. How can a fever be beneficial to a host?
5. What is the role of collectins in innate immunity?

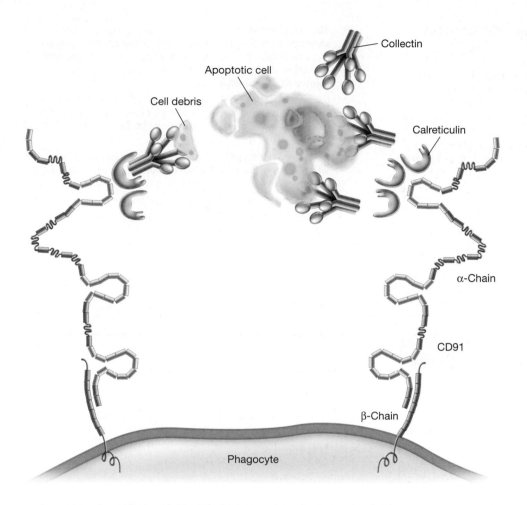

Collectin

Apoptotic cell

Cell debris

Calreticulin

α-Chain

CD91

β-Chain

Phagocyte

Figure 28.25 Collectins Are Molecular Scavengers. Collectins (also known as defense collagens) are a family of similar proteins that bind cellular debris and dying cells through their globular head groups. Their collagenous tails are then recognized by calreticulin associated with α-2 macroglobulin (CD91) on the surface of phagocytes.

Summary

28.1 Overview of Host Resistance

a. There are two fundamentally different types of immune responses to invading microorganisms and foreign material. The nonspecific (innate) response offers resistance to any microorganism or foreign material. It includes general mechanisms that are a part of the animal's innate structure and function. The nonspecific system has no immunological memory—that is, nonspecific responses occur to the same extent each time. In contrast, the specific (adaptive) response resists a particular foreign agent; moreover, specific immune responses improve on repeated exposure to the agent.

28.2 Cells, Tissues, and Organs of the Immune System

a. The cells responsible for both nonspecific and specific immunity are the white blood cells called leukocytes (**figure 28.2**). Examples include monocytes and macrophages, dendritic cells, granulocytes, and mast cells.

b. Immature undifferentiated lymphocytes are generated in the bone marrow, mature, and become committed to a particular antigenic specificity within the primary lymphoid organs and tissues. In mammals, T cells mature in the thymus and B cells in the bone marrow. The thymus is the primary lymphoid organ; the bone marrow is the primary lymphoid tissue (**figure 28.6**).

c. Natural killer cells are a small population of large, nonphagocytic lymphocytes that destroy cancer cells and cells infected with microorganisms (**figures 28.7** and **28.8**).

d. The secondary lymphoid organs and tissues serve as areas where lymphocytes may encounter and bind antigens, then they proliferate and differentiate into fully mature, antigen-specific effector cells. The spleen is a secondary lymphoid organ, and the lymph nodes and mucosal-associated tissues (GALT and SALT) are the secondary lymphoid tissue (**figure 28.9**).

28.3 Phagocytosis

a. Phagocytosis involves the recognition, ingestion, and destruction of pathogens by lysosomal enzymes, superoxide radicals, hydrogen peroxide, defensins, RNIs, and metallic ions. Phagocytic cells use two basic mechanisms for the recognition of microorganisms: opsonin-dependent and opsonin-independent (**figure 28.10**).

28.4 Inflammation

a. Inflammation is one of the host's nonspecific defense mechanisms to tissue injury that may be caused by a pathogen. Inflammation can either be acute or chronic (**figures 28.13** and **28.14**).

28.5 Physical Barriers in Nonspecific (Innate) Resistance

a. Many direct factors (age, nutrition) or general barriers contribute in some degree to all host-microbe relationships. At times they favor the establishment of the microorganism; at other times they provide some measure of general defense to the host.

b. Physical and mechanical barriers along with host secretions are the host's first line of defense against pathogens. Examples include the skin and mucous membranes; and the epithelia of the respiratory, gastrointestinal, and genitourinary systems (**figures 28.15, 28.16, 28.18,** and **28.19**).

28.6 Chemical Mediators in Nonspecific (Innate) Resistance

a. Mammalian hosts have specific chemical barriers that help combat the continuous onslaught of pathogens. Examples include cationic peptides, bacteriocins, cytokines, interferons, endogenous pyrogens, acute-phase proteins, and complement.

b. The complement system is composed of a large number of serum proteins that play a major role in the animal's defensive immune response. There are three pathways of complement activation: the classical, alternative, and lectin pathways (**figure 28.21** and **table 28.3**).

c. Cytokines are required for regulation of both the nonspecific and specific immune responses. Cytokines have a broad range of actions on eucaryotic cells (**figure 28.23** and **tables 28.4** and **28.5**).

d. Interferons are a group of cytokines that respond in a defensive way to viral infections, double-stranded RNA and many pathogens capable of intracellular growth (**figure 28.24**).

e. Fever induced by a microorganism augments the host's defenses in three ways: it stimulates leukocytes so they can destroy the invading microorganism; it enhances microbiostasis by decreasing iron available to the microorganism; and it enhances the specific activity of the immune system.

Critical Thinking Questions

1. Some pathogens invade cells, others invade tissue spaces. Explain how the nonspecific immune response differs for both types of pathogens.

2. How might the various antimicrobial chemical factors be developed into new methods to control infectious disease?

3. How might a scientist use selective gene "knock-outs" to test the role of the toll-like receptor proteins?

4. Some infectious microbes have evolved to become complement resistant. What modifications to the microbial cell would be needed to be confer complement resistance?

Learn More

Learn more by visiting the Prescott website at www.mhhe.com/prescott principles, where you will find a complete list of references.

Specific (Adaptive) Immunity

29

Nude (athymic) mice have a genetic defect (*nu* mutation) that affects thymus gland development. Thus T cells do not form. However, they do have a B-cell component. This unique deficiency provides animals in which to study B/T-cell dichotomy and environmental influences on the maturation and differentiation of T cells, as well as many different immune disorders.

acquired immune tolerance   The ability to produce antibodies against nonself antigens while "tolerating" (not producing antibodies against) self antigens.

allergen   An antigen that induces an allergic response.

antibody or **immunoglobulin (Ig)**   A glycoprotein made by plasma cells (mature B cells) in response to the introduction of an antigen. The antigen-binding region of an antibody molecule is the three-dimensional mirror image of the antigen that stimulated its synthesis. Thus, the antibody can bind to the antigen with exact specificity. There are five classes of antibody: IgG, IgM, IgA, IgD, and IgE.

antigen   A substance (such as a protein, nucleoprotein, polysaccharide, or glycolipid) to which lymphocytes respond; also known as an immunogen because it induces the immune response.

antigen-presenting cell (APC)   Cells that take in protein antigens, process them, and present antigen fragments to B cells and T cells in conjunction with class II MHC molecules so that the cells are activated. Macrophages, B cells, dendritic cells, and Langerhans cells can act as APCs.

antigen processing   The hydrolytic digestion of antigens to produce antigen fragments. Antigen fragments can be collected by class I or class II MHC molecules and presented on the surface of a cell. Antigen processing can occur by proteasome action on antigens that entered a cell by means other than phagocytosis (e.g., viral infection). This is known as **endogenous antigen processing,** and the processed antigen is presented in a class I MHC molecule. Antigen processing can also occur during phagocytosis as antigens are degraded within the phagolysosome. This is known as **exogenous antigen processing,** and the antigen is presented in a class II MHC molecule.

antitoxin   An antibody to a microbial toxin that combines specifically with the toxin, thereby neutralizing it.

autoimmune disease   A disease in which self-reactive T and B cells are activated, causing the immune system to attack self antigens and leading to tissue or organ damage.

B-cell receptor (BCR)   A transmembrane Ig complex on the surface of a B cell that binds antigen stimulating the B cell. It is composed of a membrane-bound Ig, usually monomeric IgM, complexed with another membrane protein (the Ig-α/Ig-β heterodimer).

cellular (cell-mediated) immunity   The type of immunity mediated by T cells, including T-helper (T_H) cells and cytotoxic T lymphocytes (CTLs).

class switching   The change in Ig isotype (or class) secretion that results during B-cell and then plasma cell differentiation.

clonal selection   The process by which an antigen, when bound to the best-fitting B-cell receptor, activates that B cell, resulting in the synthesis of antibody against that antigen and clonal expansion of the B cells.

cluster of differentiation molecules (CDs) or antigens   Functional cell surface proteins or receptors that can be used to identify leukocyte subpopulations.

cytotoxic T lymphocyte (CTL)   A type of T cell that recognizes antigen in class I MHC molecules, destroying the cell on which the antigen is displayed. Also called **CD8+ T cell.**

epitope   An area of an antigen that stimulates the production of and combines with specific antibodies; also known as the antigenic determinant site.

humoral (antibody-mediated) immunity   The type of immunity that results from the presence of antibodies in blood and lymph.

major histocompatibility complex (MHC)   A chromosome locus encoding the histocompatibility antigens. **Class I MHC** molecules are cell surface glycoproteins present on all nucleated cells and present endogenous antigens to CD8+ T cells; **class II MHC** glycoproteins are on antigen-presenting cells and present exogenous antigens to CD4+ T cells.

memory cell   An inactive lymphocyte clone derived from a sensitized B or T cell capable of an accentuated response to a subsequent antigen exposure.

negative selection   The process by which lymphocytes that recognize host (self) antigens undergo apoptosis or become anergic (inactive).

perforin pathway   The use by CTLs and NK cells of perforin protein, which polymerizes to form membrane pores, to help destroy cells during cell-mediated cytotoxicity.

plasma cell   A mature, differentiated B lymphocyte that synthesizes and secretes antibody.

superantigen   A toxic microbial protein that stimulates T cells to proliferate and release cytokine much more extensively than do normal antigens (e.g., streptococcal scarlet fever toxins and staphylococcal toxic shock syndrome toxin-1).

T-cell receptor (TCR)   The receptor on the T-cell surface consisting of two antigen-binding peptide chains associated with a large number of other glycoproteins.

T-dependent antigen   An antigen that effectively stimulates a B-cell response only with the aid of T-helper cells that produce interleukin-2 and B-cell growth factor.

T-helper (T_H) cell   A cell that is needed for T cell–dependent antigens to be effectively presented to B cells. It also promotes cell-mediated immune responses. There are three classes of functional T_H cells: T_H1 cells interact with cytotoxic T lympho-cytes, and T_H2 cells typically interact with B cells and T_H17 cells recruit neutrophils.

T-independent antigen   An antigen that triggers a B cell into antibody production without T-cell cooperation.

toxin neutralization   The inactivation of toxins by specific antibodies, called antitoxins, that react with them.

type I hypersensitivity   A form of immediate hypersensitivity arising from the binding of antigen to IgE attached to mast cells, which then release anaphylaxis mediators such as histamine. Examples: hay fever, asthma, and food allergies.

type II hypersensitivity   A form of immediate hypersensitivity involving the binding of antibodies to antigens on cell surfaces, followed by destruction of the target cells (e.g., through complement attack, phagocytosis, or agglutination).

type III hypersensitivity   A form of immediate hypersensitivity resulting from the exposure to excessive amounts of antigens to which antibodies bind. These antibody-antigen complexes activate complement and trigger an acute inflammatory response with subsequent tissue damage.

type IV hypersensitivity   A delayed hypersensitivity response (appears 24 to 48 hours after antigen exposure) that results from the binding of antigen to activated T lymphocytes, which then release cytokines and trigger inflammation and macrophage attacks that damage tissue (e.g., contact dermatitis from poison ivy, leprosy, and tertiary syphilis).

viral neutralization   An antibody-mediated process in which IgG, IgM, and IgA antibodies bind to some viruses during their extracellular phase and inactivate or neutralize them.

The remarkable capacity of the immune system to respond to many thousands of different substances with exquisite specificity saves us all from certain death by infection.

—Martin C. Raff

Chapter 28 discusses nonspecific host resistance and the innate mechanisms by which the host is protected from invading microorganisms. Recall that the innate resistance system responds to a foreign substance in the same manner and to the same magnitude each time, and that its activation can assist in the formation of specific immune responses. In chapter 29, we continue our discussion of the immune response by describing the specific (adaptive) responses used to protect the host. Although all infants are born with the capacity to acquire specific immunity, it requires sufficient time to fully develop. Unlike innate immunity, upon subsequent exposure to the same substance (antigen), activation of a specific immune response is significantly faster and stronger than that of the initial response. A fully mature immune response will therefore involve cooperation between the host's ever-present innate resistance mechanisms and inducible specific responses.

After a brief overview of how the specific immune responses work, we launch into chapter 29 by first defining the nature of molecules (antigens) that elicit specific immune reactions, including the methods by which specific immunity can be induced, and the role recognition of "self" plays in a host's detection of invaders. We then elaborate on the structure and function of T and B cells, along with their effector products. Finally, we end the chapter with a brief overview of how the host tolerates itself and the disorders that occur when the host does not.

29.1   OVERVIEW OF SPECIFIC (ADAPTIVE) IMMUNITY

The specific (adaptive) immune system of vertebrates has three major functions: (1) to recognize anything that is foreign to the body ("nonself"); (2) to respond to this foreign material; and (3) to remember the foreign invader. The recognition response is highly specific. The immune system is able to distinguish one pathogen from another, to identify cancer cells, and to discriminate the body's own "self" proteins and cells as different from "nonself" proteins, cells, tissues, and organs. After recognition of an invader has occurred, the specific immune system responds by activating and amplifying specific lymphocytes to attack it. This is called an **effector response.** A successful effector response either eliminates the foreign material or renders it harmless to the host, thus preventing disease. If the same invader is encountered at a later time, the immune system is prepared to mount a more intense and rapid memory or **anamnestic response** that eliminates the invader once again and protects the host from disease. Four characteristics distinguish specific immunity from nonspecific (innate) resistance:

1. **Discrimination between self and nonself.** The specific immune system almost always responds selectively to nonself and produces specific responses against the stimulus. This is possible because host cells express a unique protein on their surface, marking them as residents of that host, or as "self." Thus the introduction of materials lacking that unique self marker results in their destruction by the host.

2. **Diversity.** The system is able to generate an enormous diversity of molecules such as antibodies that recognize trillions of different foreign substances.

3. **Specificity.** Immunity is also selective in that it can be directed against one particular pathogen or foreign substance (among trillions); the immunity to this one pathogen or substance usually does not confer immunity to others.

4. **Memory.** When reexposed to the same pathogen or substance, the body reacts so quickly that there is usually no noticeable pathogenesis. By contrast, the reaction time for nonspecific defenses is just as long for later exposures to a given antigen as it was for the initial one.

Two branches or arms of specific immunity are recognized (**figure 29.1**): humoral (antibody-mediated) immunity and cellular (cell-mediated) immunity. **Humoral (antibody-mediated) immunity,** named for the fluids or "humors" of the body, is based on the action of soluble glycoproteins called antibodies that occur in body fluids and on the plasma membranes of B lymphocytes. Circulating antibodies bind to microorganisms, toxins, and extracellular viruses, neutralizing them or "tagging" them for destruction by mechanisms as described in section 29.8. **Cellular (cell-mediated) immunity** is based on the action of specific kinds of T lymphocytes that directly attack cells infected with viruses or parasites, transplanted cells or organs, and cancer cells. T cells can lyse these cells or release chemicals (cytokines) that enhance specific immunity and nonspecific defenses such as phagocytosis and inflammation. Because the activity of the

acquired immune response is so potent, it is imperative that T and B cells consistently discriminate between self and nonself with great accuracy. How they accomplish this is discussed next.

29.2 ANTIGENS

The immune system distinguishes between self and nonself through an elaborate recognition process. During their development, B and T cells that would recognize components of their host (self-determinants) are induced to undergo apoptosis (programmed cell death). This ensures that the host will have only lymphocytes that produce specific immunologic reactions against foreign materials and organisms. Self and nonself substances that elicit an immune response and react with the products of that response are called **antigens.** Antigens include molecules such as proteins, nucleoproteins, polysaccharides, and some glycolipids. While the term immunogen (*immun*ity *gen*erator) is a more precise descriptor for a substance that elicits a specific immune response, antigen is used more frequently. Most antigens are large, complex molecules with a molecular weight generally greater than 10,000 Daltons (Da). The ability of a molecule to function as an antigen depends on its size, structural complexity, chemical nature, and degree of foreignness to the host. << *Cells, tissues, and organs of the immune system: Lymphocytes (section 28.2)*

Each antigen can have several **antigenic determinant sites,** or **epitopes** (**figure 29.2**). Epitopes are the regions or sites of the antigen that bind to a specific antibody or T-cell receptor. Chemically, epitopes include sugars, organic acids and bases, amino acid side chains, hydrocarbons, and aromatic groups. The number of epitopes on the surface of an antigen is its **valence.** The valence determines the number of antibody molecules that can combine with the antigen at one time. If one determinant site is present, the antigen is monovalent. Most antigens, however, have more than one copy of the same epitope and are termed multivalent. Multivalent antigens generally elicit a stronger immune response than do monovalent antigens. As we will see, each antibody molecule has at least two antigen-binding sites, so multivalent antigens can be "cross-linked" by antibodies, a phenomenon that can result in precipitation or agglutination of antigen. **Antibody affinity** relates to the strength with which an antibody binds to its antigen at a given antigen-binding site. Affinity tends to increase during the course of an immune response and is discussed in section 29.7. The **avidity** of an antibody relates to its overall ability to bind antigen at all antigen-binding sites.

Haptens

Many small organic molecules are not antigenic by themselves but become antigenic if they bond to a larger carrier molecule such as a protein. These small antigens are called **haptens**

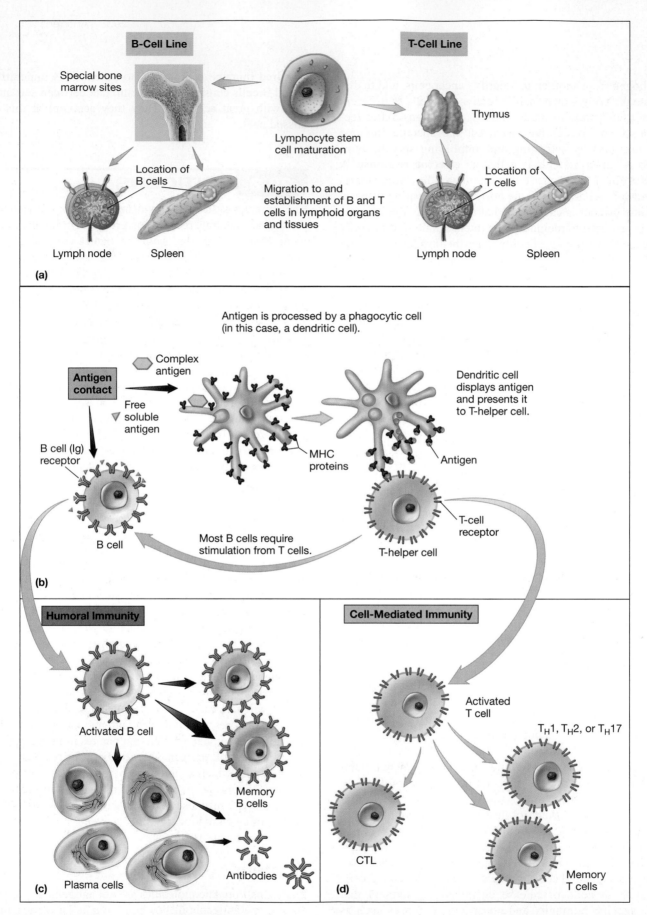

Figure 29.1 Acquired Immune System Development. (a) Lymphocyte stem cells develop into B- and T-cell precursors that migrate to the bone marrow or thymus, respectively. Mature B and T cells seed secondary lymphoid tissues. (b) Lymphocyte receptor binding of antigen activates B and T cells to become effector cells. (c) B lymphocytes develop into memory cells and antibody-secreting plasma cells. (d) T cells develop into memory cells, helper T cells, and cytotoxic T cells.

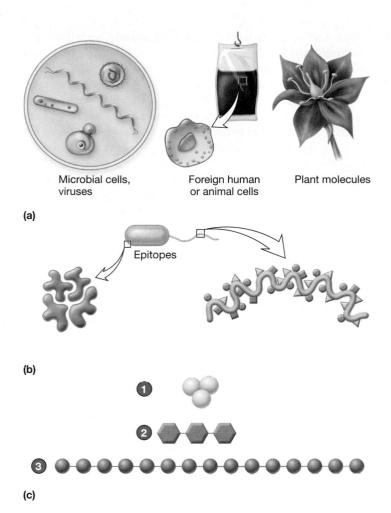

(a)

Microbial cells,
viruses | Foreign human
or animal cells | Plant molecules

Epitopes

(b)

(c)

Figure 29.2 Antigen Characteristics Are Numerous and Diverse. (a) Whole cells and viruses make good immunogens. **(b)** Complex molecules with several epitopes make good immunogens. **(c)** Poor immunogens include small molecules not attached to a carrier molecule (*1*), simple molecules (*2*), and large but repetitive molecules (*3*).

(Latin *haptein,* to grasp). When lymphocytes are stimulated by combined hapten-carrier molecules, they can react to either the hapten or the larger carrier molecule. This occurs because the hapten functions as one epitope of the carrier. When the carrier is processed and presented to T cells, responses to both the hapten and the carrier protein can be elicited. As a result, both hapten-specific and carrier-specific antibodies can be made. One example of a hapten is penicillin. By itself penicillin is a small molecule that is not antigenic. However, when it is combined with certain serum proteins of sensitive individuals, the resulting molecule becomes immunogenic, activates lymphocytes, and initiates a severe and sometimes fatal allergic immune reaction. In these instances, the hapten is acting as an antigenic determinant on the carrier molecule.

Cluster of Differentiation Molecules

Lymphocytes and other immune cells bear cell surface proteins that have specific roles in intercellular communication and are called *cluster of differentiation* (**CDs**) **molecules** or **antigens.** CDs are cell surface proteins and many are receptors. CDs have both biological and diagnostic significance. They can be measured in situ and from peripheral blood, biopsy samples, or other body fluids. They often are used in a classification system to differentiate between leukocyte subpopulations. To date, over 300 CDs have been characterized. **Table 29.1** summarizes some of the functions of several CDs.

The presence of various CDs on the cell's surface can be used to determine the cell's identity. For example, it has been established that the CD4 molecule is a cell surface receptor for human immunodeficiency virus (HIV; the virus that causes AIDS), and CD34 is the cell surface indicator of stem cells. As we will see, using the CD antigen system to name cells is more efficient than describing all of a cell's functions. We also use this approach in naming specific cell types as we discuss their relative functions in immunity.

Table 29.1	Functions of Some Cluster of Differentiation (CD) Molecules
Molecule	**Function**
CD1 a,b,c	MHC class I-like receptor used for lipid antigen presentation
CD3 $\delta,\varepsilon,\gamma$	T-cell antigen receptor
CD4	MHC class II coreceptor on T cells, monocytes, and macrophages; HIV-1 and HIV-2 (gp120) receptor
CD8	MHC class I coreceptor on cytotoxic T cells
CD11 a, b, c, d	α-subunits of integrin found on various myeloid and lymphoid cells; used for binding to cell adhesion molecules
CD19	B-cell antigen coreceptor
CD34	Stem cell protein that binds to sialic acid residues
CD45	Tyrosine phosphatase common to all hematopoeitic cells
CD56	NK cell and neural cell adhesion molecule

1. Distinguish between self and nonself substances.
2. Define and give several examples of an antigen. What is an antigenic determinant site or epitope?
3. Define hapten and CD antigens.
4. Give some examples of the biological significance of cluster of differentiation molecules (CDs).

29.3 TYPES OF SPECIFIC IMMUNITY

Acquired immunity refers to the type of specific immunity a host develops after exposure to foreign substances or after transfer of antibodies or lymphocytes from an immune donor. Acquired immunity can be obtained actively or passively by natural or artificial means (**figure 29.3**).

Naturally Acquired Immunity

Naturally acquired active immunity occurs when an individual's immune system contacts a foreign stimulus (antigen) such as a pathogen that causes an infection. The immune system responds by producing antibodies and activated lymphocytes that inactivate or destroy the pathogen. The immunity produced can be either lifelong, as with measles or chickenpox, or last for only a few years, as with influenza. **Naturally acquired passive immunity** involves the transfer of antibodies from one host to another. For example, some of a pregnant woman's antibodies pass across the placenta to her fetus. If the female is immune to diseases such as polio or diphtheria, this placental transfer also gives her fetus and newborn temporary immunity to these diseases. Certain other antibodies can pass from a mother to her offspring in the first secretions (called colostrum) from the mammary glands. These maternal antibodies are essential for providing immunity to the newborn for the first few weeks or months of life, until the child's own immune system matures. Naturally acquired passive immunity generally lasts only a short time (weeks to months, at most).

Artificially Acquired Immunity

Artificially acquired active immunity results when an animal is vaccinated, that is, intentionally exposed to a foreign material and induced to form antibodies and activated lymphocytes. A vaccine may consist of a preparation of killed microorganisms; living, weakened (attenuated) microorganisms; genetically engineered organisms or their products; or inactivated bacterial toxins (toxoids) that are administered to induce immunity artificially. Vaccines and immunizations are discussed in detail in section 33.8.

Artificially acquired passive immunity results when antibodies or lymphocytes that have been produced by one host are introduced into another. Although this type of immunity is

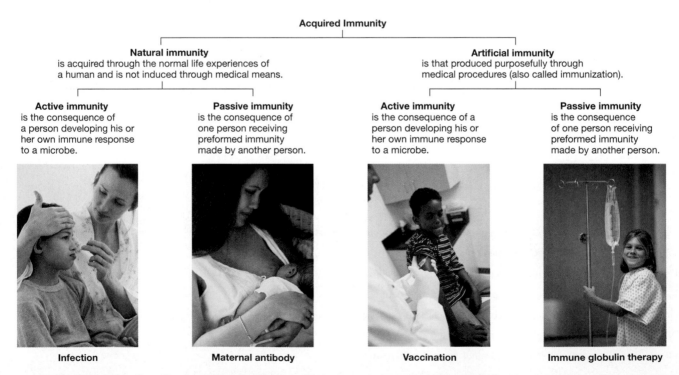

Figure 29.3 Immunity Can Be Acquired by Various Means. Naturally acquired immunity as well as artificially acquired immunity can be either active or passive.

immediate, it is short-lived, lasting only a few weeks to a few months. An example would be botulinum antitoxin produced in a horse and given to a human suffering from botulism food poisoning or a bone marrow transplant given to a patient with genetic immunodeficiency.

1. What are the three related activities mediated by the specific immune systems?
2. What distinguishes specific immunity from nonspecific resistance?
3. What are the two arms of specific immunity?
4. Of the four types of acquired immunity, which do you think your immune system has undergone?

29.4 RECOGNITION OF FOREIGNNESS

Distinguishing between self and nonself is essential in maintaining host integrity. This distinction must be highly specific and selective so invading pathogens are eliminated but host tissue is not destroyed. An important extension of this exquisite survival mechanism is seen in the modern use of tissue transplantation, where organs, tissues, and cells from unrelated persons are carefully matched to the self-recognition markers on the recipient's cells. While entire textbooks are devoted to this subject, we only discuss the aspects of foreignness recognition that assist us in understanding why and how lymphocytes respond. Without the ability to recognize foreign materials, lymphocytes have no reason to differentiate into effector cells.

Recall that each cell of a particular host needs to be identified as a member of that host so it can be distinguished from foreign invaders. To accomplish this, each cell must express proteins that mark it as a resident of that host. Furthermore, it is not enough to simply identify resident (self) cells; in addition, effective cooperation between cells must occur so that efficient information sharing and selective effector activities occur. Such a system has evolved in mammals and is encoded in the major histocompatibility gene complex.

The **major histocompatibility complex (MHC)** is a collection of genes on chromosome 6 in humans. This term is derived from the Greek word for tissue (*histo*) and the ability to get along (*compatibility*). The human MHC is called the **human leukocyte antigen (HLA)** complex. HLA molecules can be divided into three classes: class I molecules are found on all types of nucleated body cells; class II molecules appear only on cells that can process nonself materials and present antigens to other cells (i.e., macrophages, dendritic cells, and B cells); and class III molecules include various secreted proteins that have immune functions. However, unlike class I and II MHC molecules, class III molecules are mostly secreted

products that are not required for the discrimination between self and nonself. Furthermore, the class III MHC molecules are not membrane proteins, are not related to class I or II molecules, and have no role in antigen presentation. We will not discuss them further.

Each individual has two sets of MHC genes—one from each parent, and both are expressed (i.e., they are codominant). Thus a person expresses many different HLA products. The HLA proteins differ among individuals; the closer two people are related, the more similar are their HLA molecules. In addition, many forms of each HLA gene exist. This is because multiple alleles of each gene have arisen by high gene mutation rates, gene recombination, and other mechanisms (i.e., each gene locus is polymorphic).

Class I MHC molecules comprise HLA types A, B, and C, and serve to identify almost all cells of the body as "self." They consist of a complex of two protein chains, one with a mass of 45,000 Daltons (Da), known as the alpha chain, and the other with a mass of 12,000 Da called β_2-microglobulin (**figure 29.4a**). Only the alpha chain spans the plasma membrane, but both chains interact to form an antigen-binding site. Because MHC class I proteins are found on all nucleated cells (i.e., only red blood cells lack them), they stimulate an immune response when cells from one host are introduced into another host with different class I molecules. This is the basis for MHC typing when a patient is being prepared for an organ transplant.

Class II MHC molecules are produced only by certain white blood cells, such as activated macrophages, dendritic cells, mature B cells, some T cells, and certain cells of other tissues. Importantly, class II molecules are required for T-cell communication with macrophages, dendritic cells, and B cells. Class II MHC molecules are also transmembrane proteins consisting of α and β chains of mass 34,000 Da and 28,000 Da, respectively (**figure 29.4b**). Both chains combine to form a three-dimensional protein pocket, the antigen-binding pocket, into which a nonself peptide fragment can be captured for presentation to other cells of the immune system (immunocytes). Although MHC class I and class II molecules are structurally distinct, both fold into similar shapes. Each MHC molecule has a deep groove into which a short peptide derived from a foreign substance can bind (**figure 29.4c,d**). As discussed in section 29.5, foreign peptides (antigen fragments) in the MHC groove must be present to activate T cells, which in turn activate other immunocytes.

By binding and presenting foreign peptides, class I and class II molecules inform the immune system of the presence of nonself. These peptides arise in different places within cells as the result of **antigen processing.** Class I molecules bind to peptides that originate in the cytoplasm. Foreign peptides within the cytoplasm of mammalian cells come from replicating viruses or other intracellular pathogens, or are the result of cancerous transformation. These intracellular antigenic proteins are digested by a cytoplasmic structure called the proteasome (*see figure 4.8*) as part of the natural process by which a

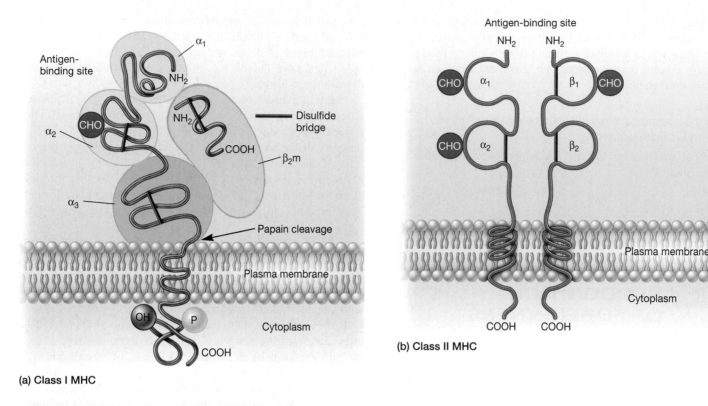

(a) Class I MHC

(b) Class II MHC

(c)

(d)

Figure 29.4   The Membrane-Bound Class I and Class II Major Histocompatibility Complex Molecules.   (a) The class I molecule is a heterodimer composed of the alpha protein, which is divided into three domains: α_1, α_2, and α_3, and the protein β_2 microglobulin (β_2m). (b) The class II molecule is a heterodimer composed of two distinct proteins called alpha and beta. Each is divided into two domains α_1, α_2 and β_1, β_2 respectively. (c) This space-filling model of a class I MHC protein illustrates that it holds shorter peptide antigens (blue) than does a class II MHC. (d) The difference is because the peptide binding site of the class I molecule is closed off, whereas the binding site of the class II molecules is open on both ends.

cell continually renews its protein contents. Specific transport proteins are used to pump the resulting short peptide fragments from the cytoplasm into the endoplasmic reticulum (ER). Within the ER, the class I MHC alpha chain is synthesized and associates with β_2-microglobulin. The class I MHC molecule and antigenic peptide are then carried to and anchored in the plasma membrane. This process, known as **endogenous antigen processing,** enables the host cell to present the antigen to a subset of T cells called CD8[+], or cytotoxic T lymphocytes. CD8[+] T cells bear a receptor that is specific for class I MHC molecules that are presenting antigen; as will be discussed in section 29.5, these T cells bind and ultimately kill infected host cells.

Class II MHC molecules bind to fragments that initially come from antigens outside the cell, thus they undergo **exogenous antigen processing.** This pathway functions with bacteria, viruses, and toxins that have been taken up by endocytosis.

An **antigen-presenting cell (APC),** such as a macrophage, dendritic cell, or B cell, takes in the antigen or pathogen by receptor-mediated endocytosis or phagocytosis and produces antigen fragments by digestion in the phagolysosome. Fragments then combine with preformed class II MHC molecules and are delivered to the cell surface. The peptide can now be recognized by CD4[+] T-helper cells. Unlike CD8[+] T cells, CD4[+] T cells do not directly kill target cells. Instead, they respond in two distinct ways. One is to proliferate, thereby increasing the number of CD4[+] cells that can react to the antigen. Some of these cells will become memory T cells, which can respond to subsequent exposures to the same antigen. The second response is to secrete cytokines that either directly inhibit the pathogen that produced the antigen or recruit and stimulate other cells to join in the immune response. << *Phagocytosis (section 28.3); Chemical mediators in nonspecific (innate) resistance: Cytokines (section 28.6)*

29.5 T-CELL BIOLOGY

For acquired immunity to develop, T cells and B cells must be activated. T cells are major players in the cell-mediated immune response (figure 29.1) and have a major role in B-cell activation. They are immunologically specific, can carry a vast repertoire of immunologic memory, and can function in a variety of regulatory and effector ways. Because of their paramount importance, we discuss them first.

T-Cell Receptors

T cells respond to antigen fragments presented in the MHC molecules. Therefore they have specific **T-cell receptors (TCRs)** for antigens on their plasma membrane surface. The receptor site is composed of two parts: an α polypeptide chain and a β polypeptide chain (**figure 29.5a**). Each chain is stabilized by disulfide bonds. The receptor is anchored in the plasma membrane, and parts of the α and β chains extend into the cytoplasm. The recognition sites of the T-cell receptor extend away from the membrane and have a terminal variable section complementary to antigen fragments.

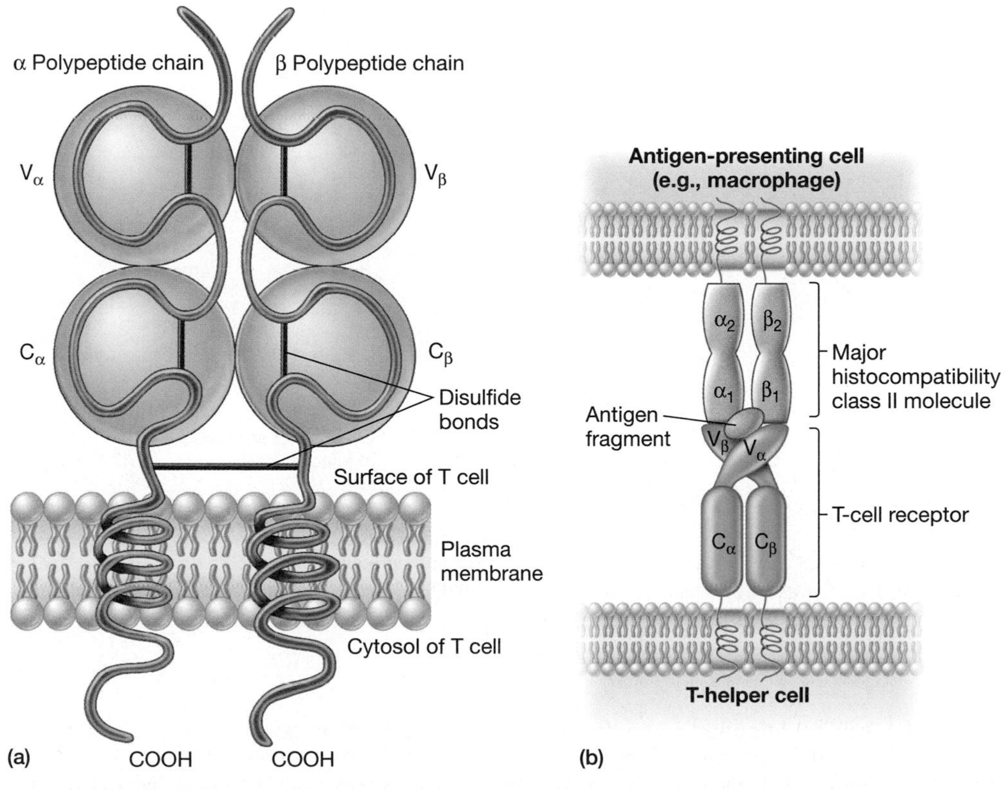

Types of T Cells

T cells originate from stem cells in the bone marrow, but T-cell precursors migrate to the thymus for further differentiation. This includes destruction of T cells that recognize self antigens (so-called self-reactive T cells). Like all lymphocytes, T cells that have survived the process of development are called mature cells, but they are also considered to be "naïve" cells because they have not yet been activated by a specific MHC-antigen peptide combination. This activation of T cells involves specific molecular signaling events inside the cell, which is discussed in the section on T-cell activation. Once activation occurs, T cells proliferate to form **memory cells,** as well as activated or **effector cells,** which carry out specific functions to protect the host against the invading antigen. The two major types of T cells, the T-helper (T_H) cells and the cytotoxic T lymphocytes (CTLs), are discussed first, then some of the details of T-cell activation are examined (**table 29.2**).

T-Helper Cells

T-helper (T_H) cells, also known as CD4$^+$ T cells, are activated by antigen presented by class II MHC molecules on APCs (e.g., macrophages and dendritic cells). They can be further subdivided into T_H0 cells, T_H1 cells, T_H2 cells, and T_H17 cells. **T_H0 cells** are simply undifferentiated precursors of T_H1, T_H2, and T_H17 cells, whereas T_H1, T_H2, and T_H17 cells are distinguished by the types of transcription factors and cytokines they produce, and the cells with which they interact (**figure 29.6**). Activated **T_H1 cells** promote cytotoxic T lymphocyte (CTL) activity, activate macrophages, and mediate inflammation by producing interleukin (IL)-2, interferon (IFN)-γ, and tumor necrosis factor (TNF)-α when the transcription factor T-bet is active. These cytokines are also responsible for delayed-type (type IV) hypersensitivity reactions, in which host cells and tissues are damaged nonspecifically by activated T cells (p. 721). **T_H2 cells** tend to stimulate antibody responses in general and defend against helminth parasites when the GATA-3 transcription factor upregulates cytokines IL-4, IL-5, IL-6, IL-10, and IL-13 expression. An overabundance of T_H2 type responses may also be involved in promoting allergic reactions. T_H17 cells are found predominantly in the skin and intestinal epithelia where they respond to bacterial invasion by producing IL-17 (and IL-22) under the regulation of the transcription factor

Figure 29.5 **The Role of the T-Cell Receptor Protein in T-Helper Cell Activation.** (a) The proposed overall structure of the antigen receptor site on a T-cell plasma membrane. (b) An antigen-presenting cell begins the activation process by displaying an antigen fragment (e.g., peptide) within histocompatibility molecules. A T-helper cell is activated after the variable region of its receptor (designated V_α and V_β) reacts with the antigen fragment in a class II MHC molecule on the presenting cell surface.

Table 29.2 Some Functional Examples of T-Cell Subpopulations

Type	Function
CD4$^+$ T-helper (T$_H$0) cell	Precursor cell activated by specific antigen presented on class II MHC to differentiate into T$_H$1 or T$_H$2 cells
T$_H$1 cell	Produces IL-2, IFN-γ, and TNF-α, which activate macrophages and CTLs to promote cellular immune responses
T$_H$2 cell	Produces IL-4, IL-5, IL-6, IL-10, and IL-13 to promote B-cell maturation and humoral immune responses
CD8$^+$ T cell	Precursor cell activated by specific antigen presented on MHC-I to differentiate into cytotoxic lymphocyte
Cytotoxic T lymphocyte (CTL), also called cytotoxic T cell	Kills cells expressing foreign specific antigen on class I MHC by perforin and granzyme release or induction of apoptosis by CD95L (Fas ligand)

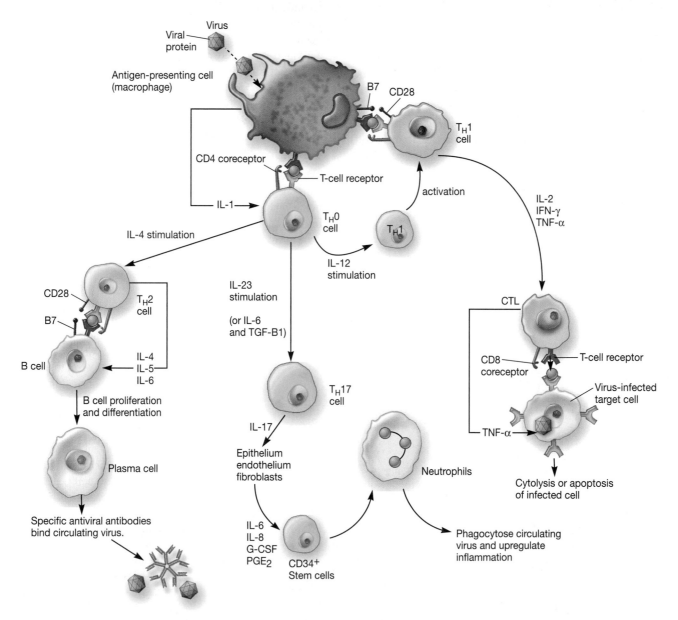

Figure 29.6 Three Potential T-Cell Responses to Viral Infection. Phagocytosis of a virus by a macrophage results in the presentation of viral protein fragments to naïve T$_H$ cells, in association with class II MHC molecules. Once activated, the T$_H$0 cell may differentiate into T$_H$1, T$_H$2 or T$_H$17 cells, depending on the growth and differentiation factors in its environment (IL-12, IL-4 or IL-23, respectively). T$_H$1 cells secrete IL-2, IFN-γ, and TNF-α. IL-2 regulates the proliferation of cytotoxic T cells (CTLs). Once the CTL differentiates into an activated effector cell, it attacks and causes programmed cell death of a virus-infected cell by either the perforin or Fas pathways. T$_H$2 cells secrete IL-4, IL-5, and IL-6, followed by IL-10 and IL-13 (not shown), which causes B cell proliferation and differentiation into antibody secreting plasma cells, secreting specific antiviral antibodies. T$_H$17 cells secrete IL-17 which stimulates epithelial cells, endothelial cells and fibroblasts to produce various growth and differentiation factors that stimulate CD34$^+$ stem cells to differentiate into neutrophils. Neutrophils phagocytose circulating virus and upregulate inflammation to assist in virus removal.

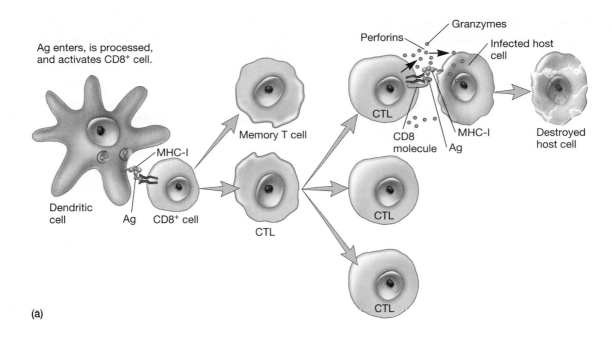

Ag enters, is processed, and activates CD8⁺ cell.

Dendritic cell

Ag   MHC-I   CD8⁺ cell

Memory T cell

CTL

CTL

CTL

CTL

Perforins   Granzymes

Infected host cell

CD8 molecule   MHC-I   Ag

Destroyed host cell

(a)

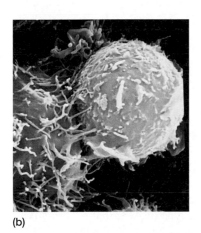

(b)

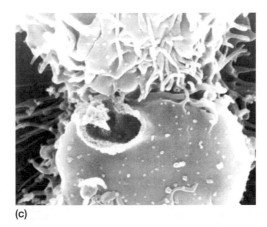

(c)

Figure 29.7 How an Effector Cytotoxic T Cell Destroys a Virus-Infected Target Cell. (a) Naïve CD8⁺ T cells are activated when they are exposed to antigen within a class I MHC molecule on an antigen-presenting cell (e.g., a dendritic cell). Antigen activation leads to development of effector CTL and memory cells, which subsequently react with antigen expressed in class I MHC molecules of any host cell to destroy it. T-cell cytotoxicity often involves the perforin pathway and leads to apoptosis and cytolysis. (b) A cytotoxic T cell (left) contacts a target cell (right) (×5,700). (c) The T cell secretes perforin that forms pores in the target cell's plasma membrane. These pores allow the contents of the target cell to leak out and granzymes to enter and induce apoptosis. Ag, antigen.

RORγt. T$_H$17 cells respond to invading bacteria by recruiting neutrophils and inducing a strong inflammatory response. In addition, T$_H$17 cells have been implicated in several autoimmune diseases. Allergic and hypersensitivity reactions are discussed in section 29.11. ◄◄ *Chemical mediators in nonspecific (innate) resistance: Cytokines (section 28.6)*

Cytotoxic T Lymphocytes

Cytotoxic T lymphocytes (CTLs) are CD8⁺ T cells that function to destroy host cells that have been infected by an intracellular pathogen, such as a virus. Activation of CTLs can be thought of as a two-step process. First, naïve CD8⁺ cells must interact with an APC that has processed the antigen and presents it to the immature CTL on its class I MHC molecule (**figure 29.7**). Note that unlike CD4⁺ cells, CD8⁺ cells interact with APCs through their class I MHCs. This is important because CD8⁺ cells then mature into CTLs that can respond to the same antigen as presented in the class I MHC of any host cells infected by the same intracellular pathogen. All host cells that present the same anti-

gen are thus targeted for destruction. Once activated, these CTLs kill target cells in at least two ways: the perforin pathway and the CD95 pathway.

In the **perforin pathway,** binding of the CTL to the target cell triggers movement of cytoplasmic granules toward the part of the plasma membrane that is in contact with the target cell. These CTL granules fuse with the plasma membrane, releasing molecules called perforin and granzymes into the intercellular space. Perforin, which has considerable homology to the C9 component of complement that forms the membrane attack complex, polymerizes in the target cell's membrane to form pores. These allow the granzymes to enter the target cell, where they induce programmed cell death (apoptosis). ◄◄ *Chemical mediators in nonspecific (innate) resistance: Complement (section 28.6)*

In the **Fas-FasL,** or **CD95, pathway,** the activated CTL increases expression of a protein called Fas ligand (FasL; also known as CD95L) on its surface. FasL can interact with the transmembrane Fas protein receptor found on the target cell surface.

This induces the target cell to undergo apoptosis. By inducing target cell apoptosis rather than cell lysis, both the perforin and the CD95 pathways stimulate membrane changes that are thought to trigger phagocytosis and destruction of the apoptotic cell by macrophages. Thus any infectious agent (e.g., a virus) that caused the cell to be initially attacked by the CTL is also destroyed. If the target cell were simply to be lysed, any infectious agent it harbored could potentially be released unharmed and infect surrounding cells.

T-Cell Activation

To respond to a foreign substance, lymphocytes must be activated by binding a specific antigen. Antigen binding within the lymphocyte receptor initiates a signaling cascade involving other membrane-bound proteins and intracellular messengers. However, lymphocyte proliferation, differentiation, and expression of specific cytokine genes occur only when a second signal is communicated along with the antigen. A general discussion of this process in T cells now follows.

All naïve T cells, whether CD4$^+$ or CD8$^+$ cells, require two signals to be activated by antigen. **Signal 1** occurs when an antigen fragment, presented in a MHC molecule of an antigen-presenting cell (APC), fills the appropriate T-cell receptor. In the case of T$_H$ cells, antigen is presented by class II MHC molecules, which triggers CD4$^+$ coreceptors on the T$_H$ cell to interact with the antigen-bound MHC molecule (**figure 29.8**). For CTLs to be activated, an endogenous (cytoplasmic) antigen is presented on class I MHC molecules and the CD8$^+$ coreceptor on the CTL interacts with the antigen-bound MHC molecule on the APC. In both cases, the T cell coreceptor assists with signal 1 recognition. << *Organelles of the biosynthetic-secretory and endocytic pathways (section 4.4)*

In addition to signal 1, both naïve cell types require a second, costimulatory **signal 2** to become activated. A T cell receiving signal 1 only will often become **anergic,** or unresponsive to that antigen. More than one factor may contribute to signal 2, but the most important seems to be the **B7 (CD80) protein** on the surface of an APC, which binds to the CD28 receptor on the T cell (figure 29.8). One type of APC that is particularly good at stimulating naïve T cells is the dendritic cell. This type of phagocytic cell expresses high levels of B7 constitutively (at all times). Thus the combination of signals 1 and 2 provided by a mature dendritic cell presenting the antigen fragment stimulates molecular events inside the T cell, which causes it to proliferate and differentiate.

In T$_H$ cells, signal 1 stimulates a signal transduction pathway that results in the production of cytokines, in particular interleukin-2 (IL-2). First, signal 1 activates a tyrosine kinase located in the cytoplasm (figure 29.8). In eucaryotic cells, membrane-bound tyrosine kinases (which add phosphate groups to the amino acid tyrosine in proteins) are

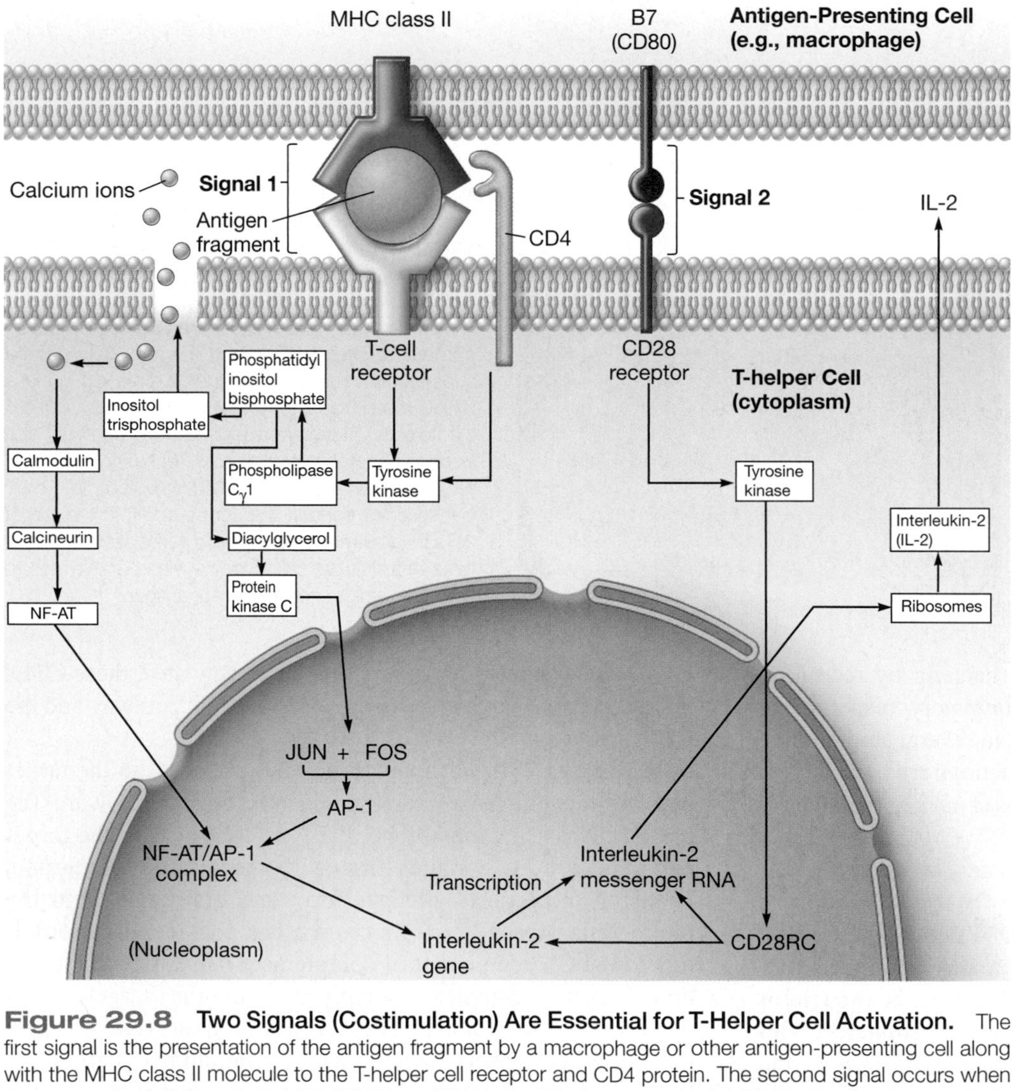

Figure 29.8 Two Signals (Costimulation) Are Essential for T-Helper Cell Activation. The first signal is the presentation of the antigen fragment by a macrophage or other antigen-presenting cell along with the MHC class II molecule to the T-helper cell receptor and CD4 protein. The second signal occurs when the macrophage presents the B7 (CD80) protein to the T-helper cell with its CD28 protein receptor. Both signals send information into the cytoplasm of the T-helper cell. The first signal causes interleukin-2 mRNA to be produced. The second signal boosts the production of interleukin-2 to effective concentrations. The gene for interleukin-2 cannot be transcribed unless the NF-AT/AP-1 complex, and other transcription factors (e.g., CD28RC) are present. All these factors must be produced anew or activated when the T-helper cell is stimulated through its antigen-specific receptor.

often used to initiate a series of intracellular events that culminates with a protein entering the nucleus and directing a change in gene expression. In this case, two proteins are stimulated to enter the nucleus. When protein kinase C moves into the nucleus, it catalyzes the formation of a protein complex called AP-1. The protein NF-AT also migrates into the nucleoplasm, where it binds to the newly formed AP-1, forming NF-AT/AP-1 complex. This functions as a transcription factor, causing IL-2 mRNA to be expressed. IL-2 is a T-cell growth factor. Its production stimulates T-cell differentiation in an autocrine-like manner (i.e., the IL-2 stimulates differentiation of the T cell that produced it). Thus signal 1 triggers the production of this important cytokine required for further T-cell development.

Signal 2, mediated by the receptor CD28 and its ligand B7, activates a different tyrosine kinase, causing the formation of the transcription factor CD28RC (figure 29.8). It stabilizes the mRNA transcribed from the IL-2 gene, thereby further increasing production of IL-2. T_H1 cells that are activated by these two signals secrete large amounts of IL-2, which can also activate nearby cytotoxic T cells in a paracrine-like fashion (figure 29.6).

$CD8^+$ cell activation occurs in a manner similar to that of T_H cells. Importantly, once signals 1 and 2 trigger $CD8^+$ T cells to differentiate into CTLs, they can respond to antigen presented by the class I MHC of other target cells (typically virus-infected or cancerous cells) without the stringent B7 costimulation required for their initial activation. The B7 of a mature dendritic cell is sufficient to convey signal 2 to the $CD8^+$ cell. However, as noted, other APCs may not produce sufficient B7 to communicate signal 2. In these cases, activated T_H cells are required to stimulate the APC to generate a different signal 2. In this way T_H cells help $CD8^+$ cells become activated. For example, APCs can be stimulated by T_H cells to produce 4-IBBL (CD137 ligand). 4-IBBL binds to the 4-IBB receptor of signal 1–stimulated $CD8^+$ cell to complete their activation. The activated $CD8^+$ cell then synthesizes and secretes IL-2 to drive its own proliferation and differentiation. Overall, once $CD8^+$ cells have been activated by two signals, they differentiate into CTLs, which can rapidly respond to host cells infected with an intracellular pathogen (i.e., a target cell) by simply recognizing foreign antigen fragments within the target cell's class I MHC protein and docking on the MHC-antigen complex.

Superantigens

Several bacterial and viral proteins can provoke a drastic and harmful response when they are exposed to T cells. These proteins are known as **superantigens** because they "trick" a huge number of T cells into activation when no specific antigen has triggered them. Superantigens accomplish this by bridging class II MHC molecules on APCs to T-cell receptors (TCRs) in the absence of a specific antigen in the MHC binding site. This interaction allows many different T cells with different antigen specificities to become activated. The consequence of this nonspecific activation is the release of massive quantities of cytokines from $CD4^+$ T cells, leading to organ failure and suppression of specific immune responses. Thus numerous T cells (nearly 30%) can be activated by superantigen to overproduce cytokines such as TNF-α and interleukins 1 and 6, resulting in endothelial damage, circulatory shock, and multiorgan failure. Superantigens can be considered virulence factors whose effects contribute to microbial pathogenicity. Examples of superantigens include the staphylococcal enterotoxins (which can cause food poisoning) and the toxin that causes toxic shock syndrome. Because of these activities, staphylococcal enterotoxin B has been added to the U.S. government's select agent list as a potential agent of terrorism. The devastating effects superantigens have on the host serve to emphasize the importance of tightly regulating a normal immune response.

1. What is the function of an antigen-presenting cell? What is a T-cell receptor and how is it involved in T-cell activation?
2. What are MHCs and HLAs? Describe the roles of the three MHC classes.
3. Describe antigen processing. How does this process differ for endogenous and exogenous antigens?
4. Briefly describe the cytotoxic T cell, its general role, how it is activated, and the two ways in which it destroys target cells.
5. Outline the functions of a T-helper cell. How do T_H1, T_H2, and T_H17 cells differ in function? Briefly describe how T_H cells are activated by costimulation versus superantigen.

29.6 B-CELL BIOLOGY

Stem cells in the bone marrow produce B-cell precursors (figure 29.1). As with T cells, a specific antigen must activate B cells to proliferate into mature, antibody-secreting plasma cells. Prior to antigen stimulation, B cells produce a special form of antibody that is attached to their cell membrane and oriented so that the part of the antibody that binds to antigen is facing outward, away from the cell (**figure 29.9**). These cell-surface, transmembrane antibodies (also known as immunoglobulins) act as receptors for the one specific antigen that will activate that particular B cell. When an antigen is captured by the immunoglobulin receptor, the receptor communicates this capture to the nucleus through a signal transduction pathway similar to that described for T cells. On a molecular level, the immunoglobulin receptor molecules on the B-cell surface associate with other proteins known as the Ig-α/Ig-β heterodimer proteins (similar to how the T-cell receptor interacts with

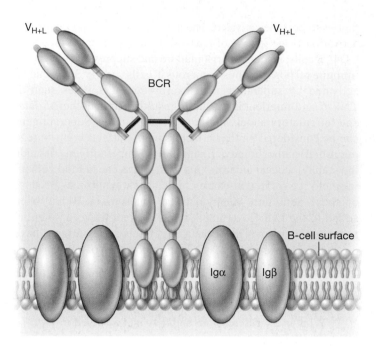

Figure 29.9 The Membrane-Bound B-Cell Receptor (BCR). The BCR is composed of monomeric IgM antibody and the co-receptors Igα and Igβ. A B cell is activated after the variable region of one receptor (designated V_{H+L}) binds an antigen fragment attached to the class II MHC molecule of an activated T_H2 cell (T-dependent activation) or after the variable regions of (two or more) receptors are bridged by an antigen (T-independent activation).

CD4 or CD8). Together the transmembrane immunoglobulin and the heterodimer protein complexes are called **B-cell receptors (BCRs).**

Each B cell may have as many as 50,000 BCRs on its surface. While each individual human carries BCRs specific for as many as 10^{13} different antigens, each individual B cell possesses BCRs specific for only one particular epitope on an antigen. Therefore a host produces at least 10^{13} different, undifferentiated B cells. These naïve B cells circulate in the blood awaiting activation by specific antigenic epitopes. Upon activation, B cells differentiate into antibody-producing **plasma cells** and memory cells (figure 29.1).

So far we have introduced the mechanism that activates B cells and the fact that activated B cells secrete antibody. However, it is important to note that B cells also internalize the antigen-bound receptor to present its three-dimensional configuration to other cells—that is, BCRs that have captured their antigenic epitope are able to trigger endocytosis of that antigen, leading to antigen processing inside the B cell. As is the case with macrophages and dendritic cells, a small antigen fragment is then presented on the surface of the B cell but in association with class II MHC molecules. Thus B cells have two immunological roles: (1) they proliferate and differentiate into memory cells and plasma cells, which respond to antigens by making antibodies, and at the same time (2) they can act as antigen-presenting cells.

B-Cell Activation

In general, an activated B cell requires growth and differentiation factors supplied by other cells. Recall that T-helper cells produce cytokines, some of which act on B cells to assist in their growth and differentiation. Although activation of the B cell is typically antigen-specific, it can additionally be T-cell dependent or T-cell independent. This distinction reflects additional supportive activity provided by T-helper cells beyond the release of initial cytokine growth factors.

T-Dependent Antigen Triggering

In most cases, B cells that are specific for a given epitope on an antigen (e.g., epitope X) cannot develop into plasma cells that secrete antibody (anti-X) without the collaboration of T-helper cells, a process called **T-dependent antigen triggering.** In other words, binding of epitope X to the B cell may be necessary, but it is not usually sufficient for B-cell activation. Antigens that elicit a response with the aid of T-helper cells are called **T-dependent antigens.** Examples include bacteria, foreign red blood cells, certain proteins, and hapten-carrier combinations.

The basic mechanism for T-dependent antigen triggering of a B cell is illustrated in **figure 29.10** and involves three cells: (1) a macrophage or other APC that processes and presents the antigen; (2) a T-helper cell that recognizes and responds to the antigen; and (3) a B cell specific for the antigen. When these cells and the antigen are present, the following events take place: (1) The APC (e.g., macrophage) presents the antigen in its class II MHC to the T-helper cell (signal #1 to the T cell). (2) Costimulation is provided by the B7-CD28 interaction (signal #2) between the APC and the T cell, resulting in T-cell differentiation into T_H2 cells, proliferation, and cytokine production. (3) These T_H2 cells then directly associate with B cells that display the same antigen-MHC complex that was presented on the APC. This promotes the secretion of additional cytokines by the T_H2 cell. (4) This interaction causes the B cells to proliferate and differentiate into plasma cells, which start producing antibodies.

As noted, the B cell also requires the antigen, recognized through its BCR (signal #1 for the B cell), to help trigger B-cell proliferation and differentiation into a plasma cell. Thus B cells, like T cells, require two signals: antigen-BCR interaction (signal 1) and T-cell cytokines (signal 2). This is a very effective process: one plasma cell can synthesize more than 10 million antibody molecules per hour! In addition, once the antigen binds to surface BCRs, B cells become more effective antigen presenters than macrophages especially at low antigen concentrations. Here, B cells bind antigen, take it up by receptor-mediated endocytosis, and present it to T-helper cells to activate them. In this situation, B cells and T-helper cells activate each other.

T-Independent Antigen Triggering

A few specific antigens can trigger B cells into antibody production without T-cell cooperation. These are called **T-independent antigens,** and their stimulation of B cells is known as **T-independent antigen triggering.** Examples include bacterial lipopolysaccharides, certain tumor-promoting agents, antibodies against other antibodies, and antibodies to certain B-cell differentiation antigens. The T-independent antigens tend to be polymeric—that is, they are composed of repeating polysaccharide or protein subunits. The resulting antibody generally has a relatively low affinity for antigen.

The mechanism for activation by T-independent antigens probably depends on their polymeric structure. Large molecules present a large array of identical epitopes to a B cell specific for that determinant. The repeating epitopes cross-link membrane-bound BCRs such that cell activation occurs and antibody is secreted. However, without T cell help, the B cell cannot alter its antibody production, and no memory B cells are formed. Thus T-independent B-cell activation is less effective than T-dependent B-cell activation: the antibodies produced have a low affinity for antigen and no immunologic memory is formed.

1. What are B-cell receptors? How are they involved in B-cell activation?

2. Briefly compare and contrast B cells and T cells with respect to their formation, structure, and roles in the immune response.

3. How does antigen-antibody binding occur? What is the basis for antibody specificity?

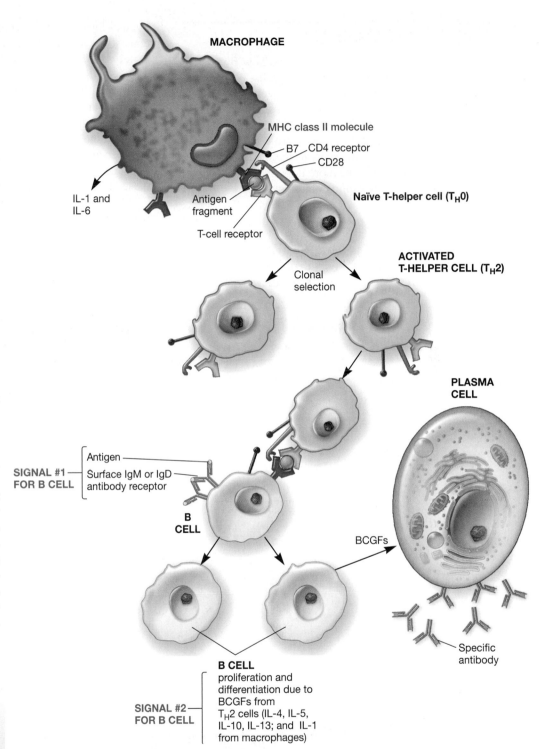

Figure 29.10 T-Dependent Antigen Triggering of a B Cell. Schematic diagram of the events occurring in the interactions between macrophages, T-helper cells, and B cells that produce humoral immunity. Many cytokines (e.g., IL-1, IL-4, IL-5, IL-6, IL-10, IL-13) stimulate B-cell proliferation. Cytokines such as IL-2, IL-4, IL-6, and IL-13 stimulate B-cell differentiation into plasma cells.

4. How does T-independent antigen triggering of B cells differ from T-dependent triggering?

29.7   ANTIBODIES

So far we have discussed antibody function and how B cells are activated to become antibody-secreting plasma cells. We now turn our attention to antibody structure and how a seemingly infinite diversity of antibodies can be synthesized by a single individual. An **antibody** or **immunoglobulin (Ig)** is a glycoprotein. There are five classes of human antibodies; all share a basic structure that varies in order to accomplish specific biological functions. **Table 29.3** summarizes some of the more important physiochemical properties of the human immunoglobulin classes.

Immunoglobulin Structure

All immunoglobulin molecules have a basic structure composed of four polypeptide chains: two identical heavy and two identical light chains connected to each other by disulfide bonds (**figure 29.11**). Each light chain polypeptide usually consists of about 220 amino acids and has a mass of approximately 25,000 Da. Each heavy chain consists of about 440 amino acids and has a mass of about 50,000 to 70,000 Da. The heavy chains are structurally distinct for each immunoglobulin class or subclass. Both light (L) and heavy (H) chains contain two different regions. The **constant (C) regions** (C_L and C_H) have amino acid sequences that do not vary significantly between antibodies of the same class. The **variable (V) regions** (V_L and V_H) have different amino acid sequences, because these regions fold together to form the antigen-binding sites.

The four chains are arranged in the form of a flexible "Y" with a hinge region. This hinge allows the antibody molecule to be more flexible, adjusting to the different spatial arrangements of epitopes on antigens. The stalk of the Y is termed the **crystallizable fragment (Fc)** and can bind to a host cell by interacting with the cell surface Fc receptor. The top of the Y

Table 29.3   Physicochemical Properties of Human Immunoglobulin Classes

Property	IgG[a]	IgM	IgA[b]	IgD	IgE
Heavy chain	γ_1	μ	α_1	δ	ε
Mean serum concentration (mg/ml)	9	1.5	3.0	0.03	0.00005
Percent of total serum antibody	80–85	5–10	5–15	<1	0.002–0.05
Valency	2	5(10)	2(4)	2	2
Mass of heavy chain (kDa)	51	65	56	70	72
Mass of entire molecule (kDa)	146	970	160[c]	184	188
Placental transfer	+	–	–	–	–
Half-life in serum (days)[d]	23	5	6	3	2
Complement activation					
Classical pathway	++	+++	–	–	–
Alternative pathway	–	–	+	–	–
Induces mast cell degranulation	–	–	–	–	+
% carbohydrate	3	7–10	7	12	11
Major characteristics	Most abundant Ig in body fluids; neutralizes toxins; opsonizes bacteria; activates complement; transplacental antibody	First to appear after antigen stimulation; very effective agglutinator; expressed as membrane-bound antibody on B cells	Secretory antibody; protects external surfaces	Present on B-cell surface; B-cell recognition of antigen	Anaphylactic-mediating antibody; resistance to helminths

[a]Properties of IgG subclass 1.
[b]Properties of IgA subclass 1.
[c]sIgA = 360 – 400 kDa
[d]Time required for half of the antibodies to disappear.

consists of two **antigen-binding fragments (Fab)** that bind with compatible epitopes. The Fc fragments are composed only of constant regions, whereas the Fab fragments have both constant and variable regions. Both the heavy and light chains contain several homologous units of about 100 to 110 amino acids. Within each unit, called a domain, disulfide bonds form a loop of approximately 60 amino acids (figure

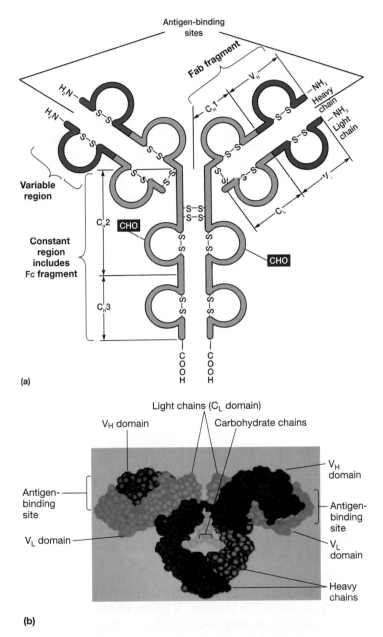

(a)

(b)

Figure 29.11   Immunoglobulin (Antibody) Structure.   (a) An immunoglobulin molecule consists of two identical light chains and two identical heavy chains held together by disulfide bonds, which form domains. All light chains contain a single variable domain (V_L) and a single constant domain (C_L). Heavy chains contain a variable domain (V_H) and either three or four constant domains (C_H1, C_H2, C_H3, and C_H4). The variable regions (V_H, V_L), when folded together in three-dimensions, form the antigen-binding sites. (b) A computer-generated model of antibody structure showing the arrangement of the four polypeptide chains.

29.11). Interchain disulfide bonds also link heavy and light chains together.

The light chain may be either of two distinct forms called kappa (κ) and lambda (λ). These can be distinguished by the amino acid sequence of the constant (C) portion of the chain. In humans, the constant regions of all chains are identical. However, four similar protein sequences are possible, reflecting four slightly different subtypes. Regardless of immunoglobulin class, each antibody molecule produced by a sole B cell will contain either κ or λ light chains but never both. The light-chain variable (V) domain contains hypervariable regions, or complementarity-determining regions (CDRs), that differ in amino acid sequence more frequently than the rest of the variable domain. These figure prominently in determining antigen specificity.

The amino-terminal domain of the heavy chain has a pattern of variability similar to that of the variable (V) region of the light chain domains and is termed the V_H domain. The other domains of the heavy chains are termed constant (C) domains. The constant domains of the heavy chain form the constant (C_H) region. The amino acid sequence of this region determines the classes of heavy chains. In humans, there are five classes of heavy chains designated by lowercase Greek letters: gamma (γ), alpha (α), mu (μ), delta (δ), and epsilon (ε), and usually written as G, A, M, D, or E. The properties of these heavy chains determine, respectively, the five immunoglobulin (Ig) classes—IgG, IgA, IgM, IgD, and IgE (table 29.3). Each immunoglobulin class differs in its general properties, half-life, distribution in the body, and interaction with other components of the host's defensive systems. Furthermore, there are variants of immunoglobulins that can be classified as follows: (1) **Isotypes** are the variations in the heavy-chain constant regions associated with the different classes that are normally present in all individuals. Therefore there are five isotypes corresponding to the five antibody classes. (2) **Allotypes** are genetically controlled, allelic forms of immunoglobulin molecules that are not present in all individuals. They arise by genetic recombination (pp. 708–710). (3) **Idiotypes** are individual, specific immunoglobulin molecules that differ in the hypervariable region of the Fab portion due to mutations that occur during B-cell development. These variations of immunoglobulin structure reflect the diversity of antibodies generated by the immune response.

1. What is the variable region of an antibody? The hypervariable or complementarity-determining region? The constant region?

2. What is the function of the Fc region of an antibody? The Fab region?

3. Name the two types of antibody light chains.

4. What determines the class of heavy chain of an antibody? Name the five immunoglobulin classes.

5. Distinguish among isotype, allotype, and idiotype.

Immunoglobulin Function

Each end of the immunoglobulin molecule has a unique role. The Fab region is concerned with binding to antigen, whereas the Fc region mediates binding to Fc receptors found on various cells of the immune system, or the first component of the classical complement system. The binding of an antibody to an antigen usually does not destroy the antigen or the microorganism, cell, or agent to which it is attached. Rather the antibody serves to mark and identify the nonself agent as a target for immunological attack and to activate nonspecific immune responses that can destroy the target.

An antigen binds to an antibody at the antigen-binding site within the Fab region of the antibody. More specifically, a pocket is formed by the folding of the V_H and V_L regions (figure 29.11b). At this site, specific amino acids contact the antigen's epitope and form multiple noncovalent bonds between the antigen and amino acids of the binding site. Because binding is due to weak, noncovalent bonds such as hydrogen bonds and electrostatic attractions, the antigen's shape must exactly match that of the antigen-binding site. If the shape of the epitope and binding site are not truly complementary, the antibody will not effectively bind the antigen. Thus a lock-and-key mechanism operates; however, in at least one case, the antigen-binding site changes shape when it complexes with the antigen (an induced-fit mechanism). In either case, antibody specificity results from the nature of antibody-antigen binding.

Phagocytes have Fc receptors for immunoglobulin on their surface, so bacteria that are covered with antibodies are better targets for phagocytosis by neutrophils and macrophages. This is termed **opsonization.** Other cells, such as natural killer cells, destroy antibody-coated cells through a process called antibody-dependent cell-mediated cytotoxicity (*see figure 28.7*). Immune destruction also is promoted by antibody-induced activation of the classical complement system. << *Chemical mediators in nonspecific (innate) immunity: Complement (section 28.6)*

Immunoglobulin Classes

Immunoglobulin γ, or **IgG,** is the major immunoglobulin in human serum, accounting for 80% of the immunoglobulin pool (**figure 29.12a**). IgG is present in blood plasma and tissue fluids. The IgG class acts against bacteria and viruses by opsonizing the invaders and neutralizing toxins and viruses (p. 712). It is also one of the two immunoglobulin classes that activate complement by the classical pathway. IgG is the only immunoglobulin molecule able to cross the placenta and provide natural immunity in utero and to the neonate at birth.

There are four human IgG isotypes (IgG1, IgG2, IgG3, and IgG4) that vary chemically in their heavy-chain composition and the number and arrangement of interchain disulfide bonds (figure 29.12b). About 65% of the total serum IgG is IgG1 and 23% is IgG2. Differences in biological function have been noted in these isotypes. For example, IgG2 antibodies are opsonic and develop in response to toxins. IgG1 and IgG3, upon recognition of their specific antigens, bind to Fc receptors expressed on

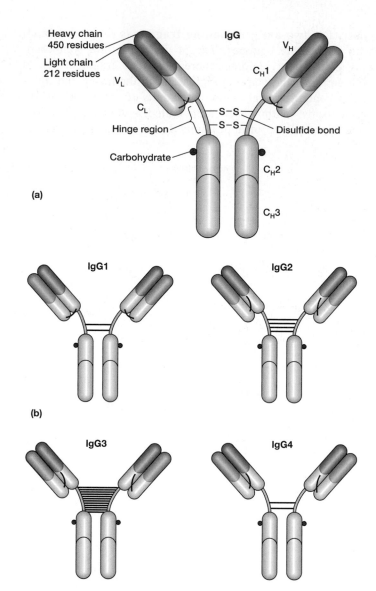

Figure 29.12 Immunoglobulin G. (a) The basic structure of human IgG. (b) The structure of the four human IgG subclasses. Note the arrangement and numbers of disulfide bonds (shown as thin black lines). Carbohydrate side chains are shown in red.

neutrophils and macrophages. This increases phagocytosis by these cells. The IgG4 antibodies function as skin-sensitizing immunoglobulins.

Immunoglobulin μ, or **IgM,** accounts for about 10% of the immunoglobulin pool. It is usually a polymer of five monomeric units (pentamer), each composed of two heavy chains and two light chains (**figure 29.13**). The monomers are arranged in a pinwheel array with the Fc ends in the center, held together by disulfide bonds and a special J (joining) chain. IgM is the first immunoglobulin made during B-cell maturation, and individual IgM monomers are expressed on B cells, serving as the antibody component of the BCR. Pentameric IgM is secreted into serum during a primary antibody response (as is discussed shortly). IgM tends to remain in the bloodstream, where it agglutinates (or clumps) bacteria, activates

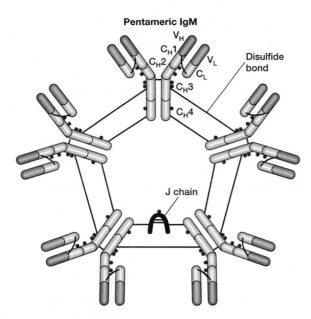

Figure 29.13 Immunoglobulin M. The pentameric structure of human IgM. The disulfide bonds linking peptide chains are shown in black; carbohydrate side chains are in red. Note that 10 antigen-binding sites are present.

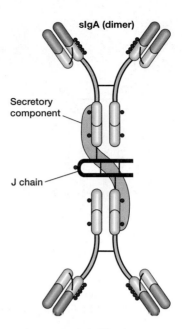

Figure 29.14 Immunoglobulin A. The dimeric structure of human secretory IgA. Notice the secretory component (tan) wound around the IgA dimer and attached to the constant domain of each IgA monomer. Carbohydrate side chains are shown in red.

complement by the classical pathway, and enhances the ingestion of pathogens by phagocytic cells.

Although most IgM appears to be pentameric, around 5% or less of human serum IgM exists in a hexameric form. This molecule contains six monomeric units but seems to lack a J chain. Hexameric IgM activates complement up to 20 times more effectively than does the pentameric form. It has been suggested that bacterial cell wall antigens such as gram-negative lipopolysaccharides may directly stimulate B cells to form hexameric IgM without a J chain. If this is the case, the immunoglobulins formed during primary immune responses are less homogeneous than previously thought.

Immunoglobulin α, or **IgA,** accounts for about 15% of the immunoglobulin pool. Some IgA is present in the serum as a monomer. However, IgA is most abundant in mucus secretions, where it is a dimer held together by a J chain (**figure 29.14**). IgA has special features that are associated with secretory mucosal surfaces. IgA, when transported from the mucosal-associated lymphoid tissue (MALT) to mucosal surfaces, acquires a protein called the secretory component. **Secretory IgA (sIgA),** as the modified molecule is called, is the primary immunoglobulin of MALT. Secretory IgA is also found in saliva, tears, and breast milk. In these fluids and related body areas, sIgA plays a major role in protecting surface tissues against infectious microorganisms by the formation of an immune barrier. For example, in the intestine, sIgA attaches to viruses, bacteria, and protozoan parasites. This prevents pathogen adherence to mucosal surfaces and invasion of host tissues, a phenomenon known as **immune exclusion.** In addition, sIgA binds to antigens within the mucosal layer of the small intestine; subsequently the antigen-sIgA complexes are excreted

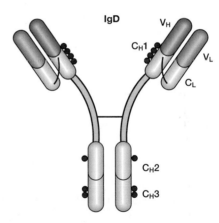

Figure 29.15 Immunoglobulin D. The structure of human IgD. The disulfide bonds linking protein chains are shown in black; carbohydrate side chains are in red.

through the adjacent epithelium into the gut lumen. This rids the body of locally formed immune complexes and decreases their access to the circulatory system. Secretory IgA also plays a role in the alternative complement pathway. ≪ *Chemical mediators in nonspecific (innate) immunity: Complement (section 29.6)*

Immunoglobulin δ, or **IgD,** is an immunoglobulin found in trace amounts in blood serum. It has a monomeric structure (**figure 29.15**) similar to that of IgG. IgD antibodies are abundant in combination with IgM on the surface of B cells and thus are part of the B-cell receptor complex. Therefore their function is to signal the B cell to start antibody production upon initial antigen binding.

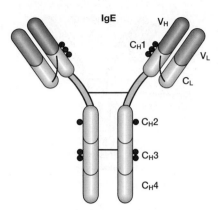

Figure 29.16 Immunoglobulin E. The structure of human IgE.

Immunoglobulin ε, or **IgE** (figure 29.16), makes up only a small percent of the total immunoglobulin pool. The skin-sensitizing and anaphylactic antibodies belong to this class. The Fc portion of IgE can bind to special FcE receptors on mast cells, eosinophils, and basophils. When two IgE molecules on the surface of these cells are cross-linked by binding to the same antigen, the cells degranulate. This degranulation releases histamine and other mediators of inflammation. It also stimulates eosinophilia (an excessive number of eosinophils in the blood) and gut hypermotility (increased rate of movement of the intestinal contents), which aid in the elimination of helminthic parasites. Thus although IgE is present in small amounts, this class of antibodies has potent biological capabilities, as is discussed in section 29.11.

1. Explain the different functions of IgM when it is bound to B cells versus when it is soluble in serum.

2. Describe the major functions of each immunoglobulin class.

3. Why is the structure of IgG considered the model for all five immunoglobulin classes?

4. Which immunoglobulin can cross the placenta?

5. Which immunoglobulin is most prevalent in the immunoglobulin pool? The least prevalent?

Antibody Kinetics

The synthesis and secretion of antibody can also be evaluated with respect to time. Monomeric IgM serves as the B-cell receptor for antigen, and pentameric IgM is secreted after B-cell activation. Furthermore, under the influence of T-helper cells, IgM-secreting plasma cells may stop producing and secreting IgM in favor of another antibody class (e.g., IgG, IgA, or IgE). This is known as **class switching.** These events take time to unfold.

The Primary Antibody Response

When an individual is exposed to an antigen (e.g., an infection or vaccine), there is an initial lag phase, or latent period, of several days to weeks before an antibody response is mounted. During this latent period, no Ag-specific antibody can be detected in the blood (figure 29.17). Once B cells have differentiated into plasma cells, antibody is secreted and can be detected. This explains why antibody-based HIV tests, for example, are not accurate until weeks after exposure. The **antibody titer,** which is a measurement of serum antibody concentration (the reciprocal of the highest dilution of an antiserum that gives a positive reaction in the test being used), then rises logarithmically to a plateau during the second, or log, phase. In the plateau phase, the antibody titer stabilizes. This is followed by a decline phase, during which antibodies are naturally metabolized or bound to the antigen and cleared from the circulation. During the primary antibody response, IgM appears first, then switches to another antibody class, usually IgG. The affinity of the antibodies for the antigen's determinants is low to moderate during the primary antibody response.

The Secondary Antibody Response

The primary antibody response primes the immune system so that it possesses specific immunological memory through its clones of memory B cells. Upon secondary antigen challenge, as occurs when an individual is reexposed to a pathogen or receives a vaccine booster, B cells mount a heightened secondary, or anamnestic (Greek *anamnesis,* remembrance), response to the same antigen (figure 29.17). Compared to the primary antibody response, the secondary antibody response has a much shorter lag phase and a more rapid log phase, persists for a longer plateau period, attains a higher IgG titer, and produces antibodies with a higher affinity for the antigen.

Diversity of Antibodies

One unique property of antibodies is their remarkable diversity. According to current estimates, each human can synthesize antibodies that can bind to more than 10^{13} (10 trillion) different epitopes. How is this diversity generated? The answer is threefold: (1) rearrangement of antibody gene segments, called combinatorial joining; (2) generation of different codons during antibody gene splicing; and (3) somatic mutations.

Combinatorial joining of immunoglobulin loci occurs because these genes are split or interrupted into many gene segments. The genes that encode antibody proteins in precursor B cells contain a small number of exons, close together on the same chromosome, that determine the constant (C) region of the light chains. Separated from them but still on the same chromosome is a larger cluster of segments that determines the variable (V) region of the light chains. During B-cell differentiation, exons for the constant region are joined to one segment of the variable region. This occurs by recombination and is

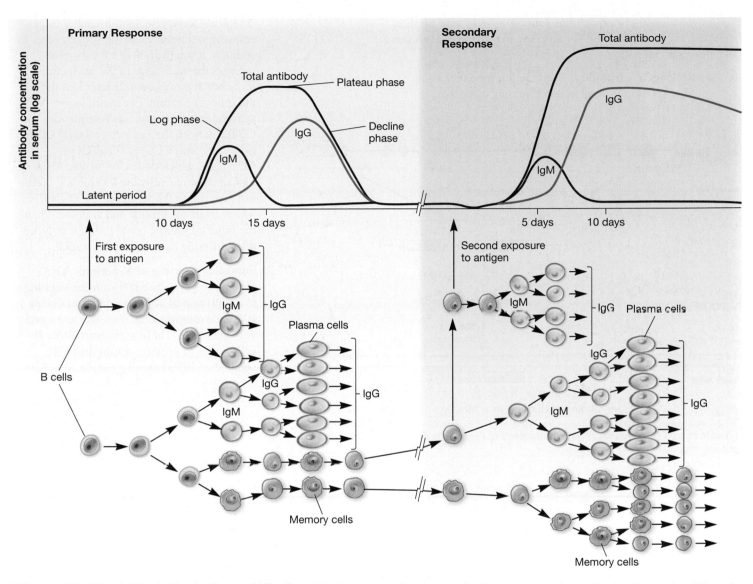

Figure 29.17 **Antibody Production and Kinetics.** The four phases of a primary antibody response correlate to the clonal expansion of the activated B cell, differentiation into plasma cells, and secretion of the antibody protein. The secondary response is much more rapid, and total antibody production is nearly 1,000 times greater than that of the primary response.

mediated by specific enzymes called RAG-1 and RAG-2. This splicing process joins the coding regions for constant and variable regions of light chains to produce a complete light chain of an antibody. A similar splicing produces a complete heavy-chain antibody gene.

Because the light-chain genes actually consist of three parts and the heavy-chain genes consist of four, the formation of a finished antibody molecule is slightly more complicated than previously outlined. The germ line DNA for the light-chain gene contains multiple coding sequences called V and J (joining) regions (**figure 29.18**). During the development of a B cell in the bone marrow, the RAG enzymes join one V gene segment with one J segment. This DNA joining process is termed combinatorial joining because it can create many combinations of the V and J regions. In addition, an enzyme called

*t*erminal *d*eoxynucleotidyl *t*ransferase (tdt) inserts nucleotides at the V-J junction, creating additional diversity. When the light-chain gene is transcribed, transcription continues through the DNA region that encodes the constant portion of the gene. RNA splicing subsequently joins the V, J, and C regions, creating mRNA.

Combinatorial joining in the formation of a heavy-chain gene occurs by means of DNA splicing of the heavy-chain counterparts of V and J along with a third set of D (diversity) sequences (**figure 29.19a**). Initially, all heavy chains have the μ type of constant region. This corresponds to antibody class IgM (figure 29.19b). If a particular B cell is reexposed to its antigen, another DNA splice joins the VDJ region with a different constant region that can subsequently change the class of antibody produced by the B cell (figure 29.19c)—the phenomenon of class switching.

As mentioned, combinatorial joining is only one of three mechanisms by which antibody diversity is generated. The other two processes include:

1. **Splice-site variability:** The junction for either VJ or VDJ splicing in combinatorial joining can occur

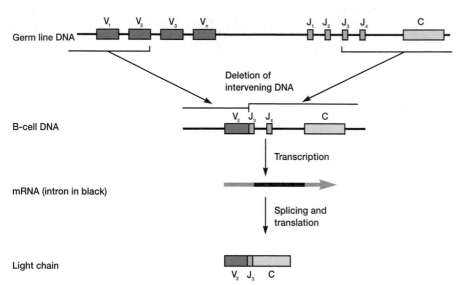

Figure 29.18 Light-Chain Production in a Mouse. One V segment is randomly joined with one J-C region by deletion of the intervening DNA. The remaining J segments are eliminated from the RNA transcript during RNA processing. An intron is a segment of DNA occurring between expressed regions of genes.

between different nucleotides and thus generate different codons in the spliced gene. In addition, the activation of tdt can greatly increase the variability of the nucleotide sequence at the VJ or VDJ junctions during the splicing process. For example, one VJ splicing event can join the V sequence CCTCCC with the J sequence TGGTGG in two ways: CCTCCC + TGGTGG = CCGTGG, which codes for proline and tryptophan. Alternatively, the VJ splicing event can give rise to the sequence CCTCGG, which codes for proline and arginine. Thus the same VJ joining could produce polypeptides differing in a single amino acid.

2. **Somatic mutation of V regions:** The V regions of germ-line DNA are susceptible to a high rate of somatic mutation during B-cell development in response to an antigen challenge. These mutations allow B-cell clones to produce antibodies with somewhat different polypeptide sequences.

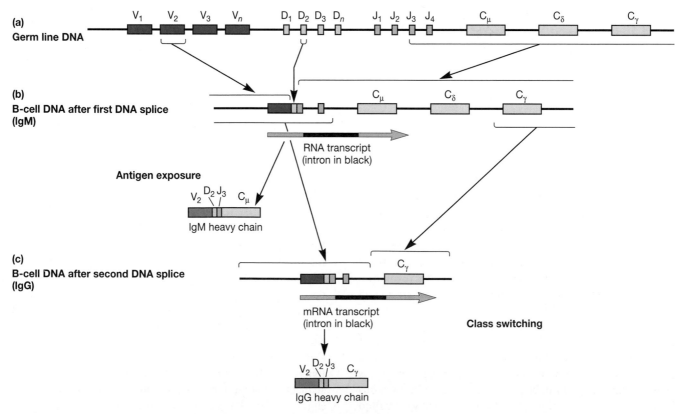

Figure 29.19 The Formation of a Gene for the Heavy Chain of an Antibody Molecule.

1. How many chromosomes encode for antibody production in humans?
2. What is the name of each part of the gene that encodes for the different regions of antibody chains?
3. Describe what is meant by combinatorial joining of V, D, and J gene segments.
4. In addition to combinatorial joining, what other two processes play a role in antibody diversity?

Clonal Selection

As noted previously, combinatorial joinings, somatic mutations, and variations in the splicing process generate the great variety of antibodies produced by mature B cells. From a large, diverse B-cell pool, specific cells are stimulated by antigens to reproduce and form B-cell clones containing the same genetic information. This is known as **clonal selection.** Clonal selection is important because it accounts for immunological specificity and memory.

The clonal selection theory has four components or tenets. The first tenet is that there exists a pool of lymphocytes that can bind to a tremendous range of epitopes (**figure 29.20**). The process of how antibody diversity is generated for B cells is well understood. Because some of the B cells generated by this process will produce antibodies that can react with self-epitopes, the second tenet of the theory is that these self-reactive cells are eliminated at an early stage of development. Indeed, this has been shown to be true for developing B cells (in the bone marrow) and self-reacting T cells (in the thymus). The third tenet is that once a lymphocyte has been released into the body and is exposed to its specific antigen, it proliferates to form a **clone** (a population of identical cells derived from a single parent cell). Note that this clone has been "selected" by exposure to specific antigen, hence the name of the theory. The final tenet states that all clonal cells react with the same antigenic epitope that stimulated its formation. However, the cells may differentiate to have somewhat different functions. Figure 29.21 shows this process for a B cell, which, after proliferating in response to antigen exposure, forms two different cell populations, antibody-producing plasma cells and memory cells.

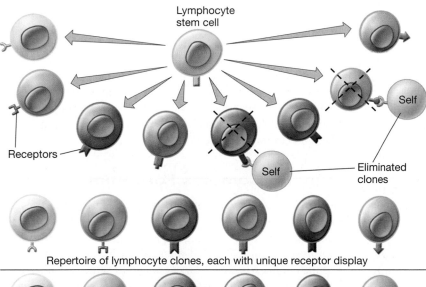

Lymphocyte stem cell

Receptors

Self

Self

Eliminated clones

Repertoire of lymphocyte clones, each with unique receptor display

Clonal selection

Lymphocytes in lymphatic tissues

Entry of antigen

Immune response against antigen

(a) Antigen-Independent Period

1 During development of early lymphocytes from stem cells, a given stem cell undergoes rapid cell division to form numerous progeny.

During this period of cell differentiation, random rearrangements of the genes that code for cell surface protein receptors occur. The result is a large array of genetically distinct cells, called clones, each clone bearing a different receptor that is specific to react with only a single type of foreign molecule or antigen.

2 At the same time, any lymphocyte clones that have a specificity for self molecules and could be harmful are eliminated from the pool of diversity. This is called immune tolerance.

3 The specificity for a single antigen molecule is programmed into the lymphocyte and is set for the life of a given clone. The end result is an enormous pool of mature but naïve lymphocytes that are ready to further differentiate under the influence of certain organs and immune stimuli.

(b) Antigen-Dependent Period

4 Lymphocytes come to populate the lymphatic organs, where they will finally encounter antigens. These antigens will become the stimulus for the lymphocytes' final activation and immune function. Entry of a specific antigen selects only the lymphocyte clone or clones that carry matching surface receptors. This will trigger an immune response, which varies according to the type of lymphocyte involved.

Figure 29.20 Lymphocyte Clonal Expansion. (a) Cell populations expand and are restricted based on their MHC expression. (b) They are further expanded when activated by a specific antigen.

Plasma cells are literally protein factories that produce about 2,000 antibodies per second in their brief 5- to 7-day life span. Memory B cells can initiate the antibody-mediated immune response upon detecting the particular antigen specific for their B cell receptors. These memory cells circulate more actively from blood to lymph and live much longer (years or even decades) than do plasma cells. Memory cells are responsible for the immune system's rapid secondary antibody response (figure 29.17) to the same antigen. Finally, memory B cells and plasma cells are usually not produced unless the B cell has interacted with and received cytokine signals from activated T-helper cells (figure 29.10). In addition to providing a theoretical basis for understanding how the adaptive immune system can amplify its responses to specific antigens, clonal selection is now widely accepted as the explanation for the differences between primary and secondary antibody responses. It has also led to the development of **monoclonal antibody (mAb)** technology (**Techniques & Applications 29.1**).

29.8 ACTION OF ANTIBODIES

The antigen-antibody interaction is a bimolecular association that exhibits exquisite specificity. The in vivo interactions that occur in vertebrate animals are absolutely essential in protecting the animal against the continuous onslaught of microorganisms and their products, and cancer cells. This occurs partly because the antibody coats the invading foreign material, marking it for enhanced recognition by other components of the innate and adaptive immune systems. The mechanisms by which antibodies achieve this are now discussed.

Neutralization

Some bacteria produce extracellular toxins that contribute to their pathogenic effects. Immunity to such a disease (e.g., diphtheria or anthrax) depends on the production of specific antibodies that inactivate the toxins produced by the bacteria. This process is called **toxin neutralization** (**figure 29.21**). Once neutralized, the toxin-antibody complex either is unable to attach to receptor sites on host target cells and is unable to enter the cell, or it is ingested by macrophages. For example, diphtheria toxin, a heterodimer, inhibits protein synthesis after binding to the cell surface by its B subunit and subsequent passage of its active A subunit into the cytoplasm of the target cell. The antibody blocks the toxic effect by inhibiting the entry of the A subunit by binding to the B fragment. An antibody capable of neutralizing a toxin or antiserum containing neutralizing antibody against a toxin is called **antitoxin.** >> *Toxigenicity: AB toxins (section 30.4)*

IgG, IgM, and IgA antibodies can bind to some viruses during their extracellular phase and inactivate them. This antibody-mediated viral inactivation is called **viral neutralization.** Fixation of the classical pathway complement component C3b to a virus aids the neutralization process. Viral neutralization prevents a viral infection due to the inability of the virus to bind to and enter its target cell.

The capacity of bacteria to colonize the mucosal surfaces of mammalian hosts depends in part on their ability to adhere to mucosal epithelial cells. Secretory IgA (sIgA) antibodies inhibit certain bacterial adherence-promoting factors. Thus sIgA can protect the host against infection by some pathogenic bacteria and perhaps by other microorganisms on mucosal surfaces by neutralizing their adherence to host cells.

Immune reactions against protozoan and helminthic parasites are only partially understood. Parasites that have a tissue-invasive phase in their life cycle often are associated with both eosinophilia and elevated IgE levels. Evidence suggests that, in the presence of elevated IgE, eosinophils can bind to the parasites and discharge their granules. Degranulation releases lytic and inflammatory mediators that neutralize and even kill parasites.

Opsonization

Phagocytes have an intrinsic ability to bind directly to microorganisms by pattern recognition and nonspecific cell surface receptors, engulf the microorganisms, form phagosomes, and digest the microorganisms. This phagocytic process can be greatly enhanced by opsonization. As noted in section 28.6, opsonization is the process by which microorganisms or other foreign particles are coated with antibody or complement and thus prepared for "recognition" and ingestion by phagocytic cells. Opsonizing antibodies, especially IgG1 and IgG3, bind to Fc receptors on the surface of macrophages and neutrophils. This binding provides the phagocyte with a method for the specific capture of antigens. In other words, the antibody forms a bridge between the phagocyte and the antigen, thereby increasing the likelihood of its phagocytosis (*see figure 28.20*).

Immune Complex Formation

Because antibodies have at least two antigen-binding sites and most antigens have at least two antigenic determinants, cross-linking can occur, producing large aggregates termed **immune complexes** (figure 29.21). If the antigens are soluble molecules and the complex becomes large enough to settle out of solution, a **precipitation** (Latin *praecipitare,* to cast down) or **precipitin reaction** occurs and is caused by a **precipitin** antibody. When the immune complex involves the cross-linking of cells or particles, an **agglutination reaction** occurs and the responsible antibody is called an **agglutinin.** These immune complexes are more rapidly phagocytosed in vivo than free antigens.

The extent of immune complex formation depends on the relative concentrations of the precipitin antibody and antigen. If there is a large excess of antibody, separate antibody molecules usually bind to each antigenic determinant and a less insoluble network or lattice forms. When antigen is present in excess, two separate antigen molecules tend to bind to each antibody and network development or cross-linking is inhibited. The ratio of

Techniques & Applications

29.1   Monoclonal Antibody Technology

The value of antibodies as tools for locating or identifying antigens is well established. For many years, antiserum extracted from human or animal blood was the main source of antibodies for tests and therapy, but most antiserum is problematic. It contains polyclonal antibodies, meaning it is a mixture of different antibodies because it reflects dozens of immune reactions from a wide variety of B-cell clones. This characteristic is to be expected, because several immune reactions may be occurring simultaneously, and even a single species of microbe can stimulate several different types of antibodies. Certain applications in immunology require a pure preparation of monoclonal antibodies (mAbs) that originate from a single clone and have a single specificity for antigen.

The technology for producing monoclonal antibodies is possible by hybridizing cancer cells and activated B cells in vitro. This technique began with the discovery that tumors isolated from multiple myelomas (a blood cancer) in mice consist of identical plasma cells. These monoclonal plasma cells secrete a strikingly pure form of antibody with a single specificity and continue to divide indefinitely. Immunologists recognized the potential in these plasma cells and devised a hybridoma approach to creating mAb. The basic idea behind this approach is to hybridize or fuse a myeloma cell with a normal plasma cell from a mouse spleen to create an immortal cell that secretes a supply of functional antibodies with a single specificity.

The introduction of this technology has the potential for numerous biomedical applications. Monoclonal antibodies have provided immunologists with excellent standardized tools for studying the immune system and for expanding disease diagnosis and treatment. Most of the successful applications thus far use mAbs in in vitro diagnostic testing and research. Although injecting monoclonal antibodies to treat human disease is an exciting prospect, this therapy has been stymied because most mAbs are of mouse origin, and many humans will develop hypersensitivity to them. However, using genetic engineering, human antibody constant regions are cloned to mouse antibody-binding regions to create a hybrid antibody that is highly specific but less likely to cause hypersensitivity reactions.

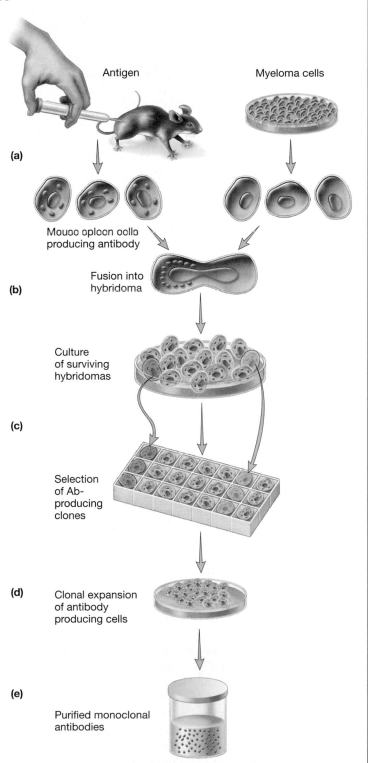

Monoclonal Antibody Formation.   **(a)** A mouse is inoculated with an antigen having the desired specificity, and activated cells are isolated from its spleen. A special strain of mouse provides the myeloma cells. **(b)** The two cell populations are mixed with polyethylene glycol, which causes some cells in the mixture to fuse and form hybridomas. **(c)** Surviving cells are cultured and separated into individual wells. **(d)** Tests are performed on each hybridoma to determine specificity of the antibody (Ab) it secretes. **(e)** A hybridoma with the desired specificity is grown in tissue culture; antibody is then isolated and purified.

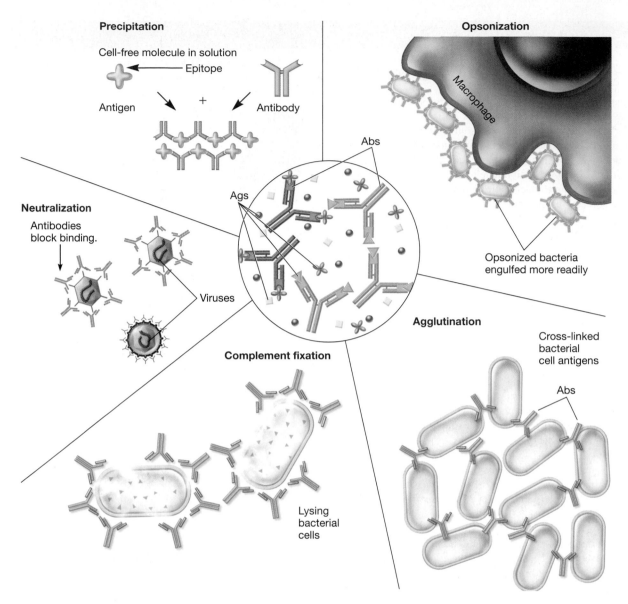

Precipitation

Cell-free molecule in solution

Epitope

Antigen + Antibody

Neutralization

Antibodies block binding.

Viruses

Complement fixation

Lysing bacterial cells

Ags

Abs

Opsonization

Macrophage

Opsonized bacteria engulfed more readily

Agglutination

Cross-linked bacterial cell antigens

Abs

Figure 29.21 Consequences of Antigen-Antibody Binding. Immune complexes form when soluble antigens (Ag) bind soluble antibody (Ab), resulting in precipitation. Opsonization occurs when antibody binds to antigens on larger molecules or cells to be recognized by phagocytic cells. Agglutination results when insoluble antigens (such as viral or bacterial cells) are cross-linked by antibody. The classical complement cascade can be activated by immune complexes; this is called complement fixation. Neutralization results when antibody binds to antigens, preventing the antigen from binding to host cells.

antibody and antigen is said to be in the **equivalence zone** when their concentration is optimal for the formation of a large network of interconnected antibody and antigen molecules. When in the equivalence zone, all antibody and antigen molecules precipitate or agglutinate as an insoluble complex.

Immune complexes can be used experimentally or diagnostically outside the animal body (in vitro). This has led to the development of a variety of immunological assays that can detect the presence of either antibody or antigen. These assays are important in the diagnosis of diseases; in the identification of specific viruses, bacteria, and parasites; in monitoring the level of the humoral response and immunologic problems; and in identifying molecules of medical and biological interest.

Immunological assays differ in their speed and sensitivity; some are qualitative, whereas others are quantitative. >> *Clinical immunology: Immunoprecipitation (section 32.3)*

1. How does toxin neutralization occur? Viral neutralization?

2. How does opsonization inhibit microbial adherence to eucaryotic cells?

3. Describe an immune complex. What is the equivalence zone?

29.9 SUMMARY: THE ROLE OF ANTIBODIES AND LYMPHOCYTES IN IMMUNE DEFENSE

Although both the humoral and cellular arms of the specific immune response have been considered separately, it is important to understand that the host response to any particular pathogen may involve complex interactions between the host and the pathogen, as well as the components of both nonspecific and specific immunity. The next section summarizes the defense mechanisms vertebrate hosts use against viral and bacterial pathogens. At times, however, these defenses are not enough to protect the host because pathogens have evolved mechanisms to circumvent many of the host's defenses. This is the subject of chapter 30.

Immune Responses to Viral Infections

Resistance to viral infections involves humoral immunity, interferon sensitization of host cells, and cell-mediated immunity.

1. **Interferons** are important in resistance in the early stage of viral infection. Interferon-stimulated cells shut down viral protein synthesis and destroy viral mRNA (*see figure 28.24*). Some interferons also stimulate the activity of T cells (figure 29.6) and natural killer cells (*see figure 28.7*), thus accelerating the immune response to a viral infection. << *Cells, tissues, and organs of the immune system (section 28.2)*

2. **Cell-mediated immunity** to viruses is a major resistance mechanism when enveloped viruses modify host cell membranes and bud from the surface (e.g., herpesvirus, poxvirus, and influenza virus). Activated lymphocytes can recognize and destroy virus-infected cells by detecting changes in surface molecules. Cytotoxic T lymphocytes (CTLs) destroy virus-infected cells by inducing apoptosis of the infected cell through the release of the CD95L peptide (FasL) and the production of granzymes and perforin, which form channels through the plasma membrane of infected cells, resulting in cytolysis (*see figure 28.8*). The class I MHC proteins are involved in T-cell recognition of infected cells (figure 29.6). Cells displaying both viral antigens and the proper class I MHC will be destroyed. CTLs are also involved in the destruction of cancer cells, a process known as **immune surveillance.**

3. **Humoral responses** are activated when viruses are released from host cells and thus can be detected by macrophages and other immune system cells. Binding of antibodies to virus particles has two important outcomes. First, it can neutralize viruses, thereby interfering with their adsorption and entrance into host cells (figure 29.21). This limits spread of the infection.

Antibodies also act as opsonins (*see figure 28.20*) and enhance phagocytosis of the viruses. << *Phagocytosis (section 28.3); Chemical mediators in nonspecific (innate) resistance: Complement (section 28.6)*

Immune Responses to Bacterial Infections

If a bacterium successfully breaches the physical barriers (skin and mucous membranes) that serve as the host's first line of defense, then other innate defenses, as well as specific immune responses, are elicited. Inflammation, complement activation, and humoral immunity are more important than cell-mediated immunity for those bacteria that remain outside host cells. However, for intracellular bacterial pathogens, cell-mediated responses are also important.

1. The **inflammatory response** helps destroy bacterial pathogens. In addition, it recruits macrophages to the site of bacterial invasion. These APCs not only ingest the bacteria but also signal its presence to T-helper lymphocytes.

2. **T_H2 cells are formed, which help activate B cells, triggering humoral responses.** When antibodies bind the bacteria, several outcomes are possible: (a) opsonization, (b) agglutination, (c) neutralization, and (d) complement activation. IgG is an opsonin that aids in the phagocytosis of bacteria by macrophages and granulocytes. IgM and IgG agglutinate bacterial pathogens, thus limiting their spread and enhancing the efficiency of phagocytosis (figure 29.21). Some antibodies act as antitoxins and neutralize bacterial exotoxins—toxins that are secreted by bacteria. Activation of complement by the classical pathway can result in opsonization of the bacteria by the C3b and C4b components of the system, formation of the membrane attack complex by the C5b-9 components (*see figure 28.22*), and enhancement of the inflammatory response by C3a, C5a, and C5b67. These complement proteins attract neutrophils and macrophages to the site of the infection.

3. **Cell-mediated responses** by activated macrophages and T cells (figure 29.6) are also important, particularly in resisting intracellular bacterial pathogens. Activated T cells secrete several cytokines that have a variety of effects. Among these, (a) interferon-gamma (IFN-γ) is a major factor that stimulates macrophages to become "angry" and more effectively phagocytose and destroy pathogens; (b) the macrophage chemotactic factor and migration inhibition factor attract more macrophages and keep them in the area of infection after arrival; and (c) interleukin-2 (IL-2) stimulates the proliferation of activated T cells to increase the population of cells involved in the cell-mediated immune response. It also increases the effectiveness of cytotoxic T cells and NK cells by promoting the synthesis of other cytokines by T cells.

29.10   ACQUIRED IMMUNE TOLERANCE

Acquired immune tolerance is the body's ability to produce T cells and antibodies against nonself antigens such as microbial antigens, while "tolerating" (not responding to) self antigens. Some of this tolerance arises early in embryonic life, when immunologic competence is being established. Three general tolerance mechanisms have been proposed: (1) negative selection by clonal deletion, (2) the induction of anergy, and (3) inhibition of the immune response by T cells with suppressor/regulatory function.

Negative selection by clonal deletion removes lymphocytes that recognize any self antigens that are present. These cells are eliminated by apoptosis. T-cell tolerance induced in the thymus and B-cell tolerance in the bone marrow is called **central tolerance.** However, another mechanism is needed to prevent immune reactions against self antigens, termed autoimmunity, because many antigens are tissue-specific and are not present in the thymus or bone marrow.

Mechanisms occurring elsewhere in the body are collectively referred to as **peripheral tolerance.** These supplement central tolerance. Peripheral tolerance is thought to be based largely on incomplete activation signals given to the lymphocyte when it encounters self antigens in the periphery of the body. This mechanism leads to a state of unresponsiveness called **anergy,** which is associated with impaired intracellular signaling and apoptosis. Many autoreactive B cells undergo clonal deletion or become anergic as they mature in the bone marrow. The deletion of self-reactive B cells also takes place in secondary lymphoid tissue such as the spleen and lymph nodes. Since B cells recognize native antigen, there is no need for the participation of MHC molecules in these processes. For those self antigens present at relatively low concentrations, immunologic tolerance is often maintained only within the T-cell population. This is sufficient to sustain tolerance because it denies the help essential for antibody production by self-reactive B cells. T cells with suppressor activity have been defined as cells that can specifically inhibit responses of other T cells in an antigen-specific manner, although their existence has not been conclusively proven.

1. Describe the three ways acquired immune tolerance develops in the vertebrate host.
2. How would you define anergy?

29.11   IMMUNE DISORDERS

As in any system in a vertebrate animal, disorders also occur in the immune system. Immune disorders can be categorized as hypersensitivities, autoimmune diseases, transplantation (tissue) rejection, and immunodeficiencies. Each of these is now discussed.

Hypersensitivities

Hypersensitivity is an exaggerated specific immune response that results in tissue damage and is manifested in the individual on a second or subsequent contact with an antigen. Hypersensitivity reactions can be classified as either immediate or delayed. Obviously, immediate reactions appear faster than delayed ones, but the main difference between them is the nature of the immune response to the antigen. Realizing this fact in 1963, Peter Gell and Robert Coombs developed a classification system for reactions responsible for hypersensitivities. Their system correlates clinical symptoms with information about immunologic events that occur during hypersensitivity reactions. The **Gell-Coombs classification** system divides hypersensitivity into four types: I, II, III, and IV.

Type I Hypersensitivity

An **allergy** (Greek *allos,* other, and *ergon,* work) is one kind of **type I hypersensitivity** reaction. Allergic reactions occur when an individual who has produced IgE antibody in response to the initial exposure to an antigen (**allergen**) subsequently encounters the same allergen. Upon initial exposure to a soluble allergen, B cells are stimulated to differentiate into plasma cells and produce specific IgE antibodies with the help of T_H cells (**figure 29.22**). This IgE is sometimes called a **reagin,** and the individual has a hereditary predisposition for its production. Once synthesized, IgE binds to the Fc receptors of mast cells (basophils and eosinophils can also be bound) and sensitizes these cells, making the individual sensitized to the allergen. When a subsequent exposure to the allergen occurs, the allergen attaches to the surface-bound IgE on the sensitized mast cells, causing mast cell degranulation.

Degranulation releases physiological mediators such as histamine, leukotrienes, heparin, prostaglandins, PAF (*p*latelet-*a*ctivating *f*actor), ECF-A (*e*osinophil *c*hemotactic *f*actor of *a*naphylaxis), and proteolytic enzymes. These mediators trigger smooth muscle contractions, vasodilation, increased vascular permeability, and mucus secretion (figure 29.22). The inclusive term for these responses is **anaphylaxis** (Greek *ana,* up, back again, and *phylaxis,* protection). Anaphylaxis can be divided into systemic and localized reactions.

Systemic anaphylaxis is a generalized response that is immediate due to a sudden burst of mast cell mediators. Usually there is respiratory impairment caused by smooth muscle constriction in the bronchioles. The arterioles dilate, which greatly reduces arterial blood pressure and increases capillary permeability with rapid loss of fluid into the tissue spaces. These physiological changes can be rapid and severe enough to be fatal within a few minutes from reduced venous return, asphyxiation, reduced blood pressure, and circulatory shock. Common examples of allergens that can produce systemic anaphylaxis include drugs (penicillin), passively administered antisera, peanuts, and insect venom from the stings or bites of wasps, hornets, or bees.

Localized anaphylaxis is called an **atopic** ("out of place") **reaction.** The symptoms that develop depend primarily on the route by which the allergen enters the body. Hay fever (allergic rhinitis) is a good example of atopy involving the upper respiratory tract. Initial exposure involves airborne allergens—such as plant pollen, fungal spores, animal hair and dander, and house dust mites—that sensitize mast cells located within the

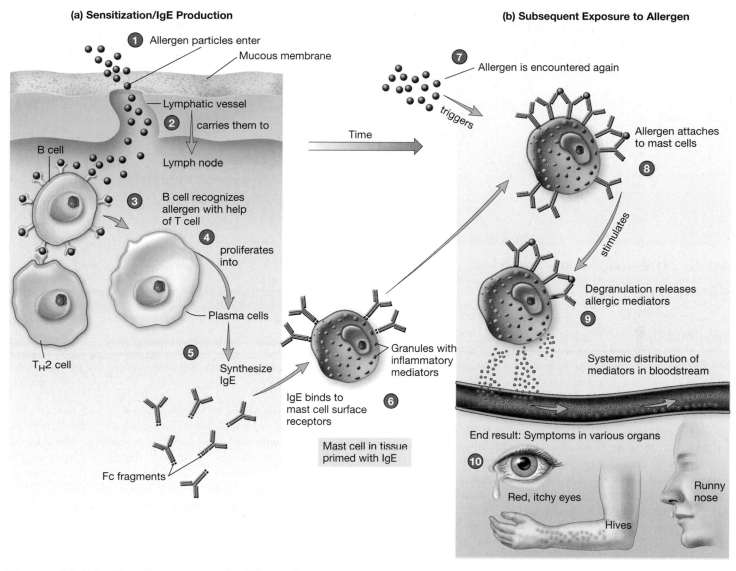

(a) Sensitization/IgE Production

1. Allergen particles enter
- Mucous membrane
2. Lymphatic vessel carries them to
- Lymph node
- B cell
3. B cell recognizes allergen with help of T cell
4. proliferates into
- Plasma cells
- T$_H$2 cell
5. Synthesize IgE
- Fc fragments
6. IgE binds to mast cell surface receptors
- Granules with inflammatory mediators
- Mast cell in tissue primed with IgE

Time

(b) Subsequent Exposure to Allergen

7. Allergen is encountered again
- triggers
8. Allergen attaches to mast cells
- stimulates
9. Degranulation releases allergic mediators
- Systemic distribution of mediators in bloodstream
10. End result: Symptoms in various organs
- Red, itchy eyes
- Hives
- Runny nose

Figure 29.22 **Type I Hypersensitivity (Allergic Response).** (a) The initial contact (sensitization) of lymphocytes by small-protein allergens at mucous membranes results in T$_H$2 cell–assisted antibody class switching; plasma cells secrete IgE antibody. The IgE binds to its receptor on tissue mast cells (1–6). (b) Subsequent exposure to the same allergens results in their capture by the cell-bound IgE (7), triggering mast cell degranulation (8, 9). Characteristic signs and symptoms (hives, swelling, itching, etc.) of allergy ensue (10).

mucous membranes of the respiratory tract. Reexposure to the allergen causes a localized anaphylactic response: itchy and tearing eyes, congested nasal passages, coughing, and sneezing. Antihistamine drugs are used to help alleviate these symptoms.

Skin testing can be used to identify the antigen responsible for allergies. These tests involve inoculating small amount of the suspect allergens into the skin. Sensitivity to an antigen is shown by a rapid inflammatory reaction characterized by redness, swelling, and itching at the site of inoculation (**figure 29.23**). The affected area in which the allergen–mast cell reaction takes place is called a wheal and flare reaction site. Once the responsible allergen has been identified, the individual should avoid contact with it. If this is not possible, **desensitization** is warranted. This procedure consists of a series of allergen doses injected beneath the skin to stimulate the production of IgG antibodies rather than IgE antibodies.

The circulating IgG antibodies can then act to intercept and neutralize allergens before they have time to react with mast cell–bound IgE. Desensitizations are about 65 to 75% effective in individuals whose allergies are caused by inhaled allergens.

Type II Hypersensitivity

Type II hypersensitivity is generally called a **cytolytic** or **cytotoxic reaction** because it results in the destruction of host cells, either by lysis or toxic mediators. In type II hypersensitivity, IgG or IgM antibodies are inappropriately directed against cell surface or tissue-associated antigens. They usually stimulate the classical complement pathway and a variety of effector cells (**figure 29.24**). The antibodies interact with complement (C1q) and the effector cells through their Fc regions. The damage mechanisms are a

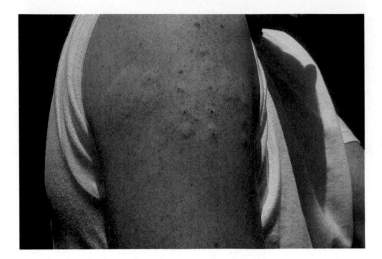

Figure 29.23 In Vivo Skin Testing. Skin prick tests with grass pollen in a person with summer hay fever. Notice the various reactions with increasing dosages (from top to bottom).

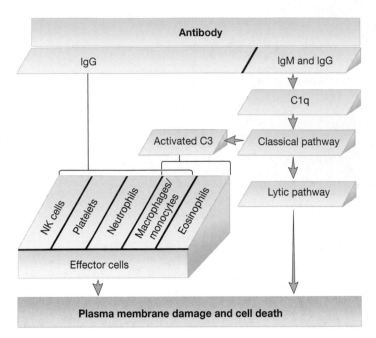

Figure 29.24 Type II Hypersensitivity. The action of antibody occurs through effector cells or the membrane attack complex, which damages target cell plasma membranes, causing cell destruction.

reflection of the normal physiological processes involved in interaction of the immune system with pathogens. Classical examples of type II hypersensitivity reactions are the response exhibited by a person who receives a transfusion with blood from a donor with a different blood group and erythroblastosis fetalis.

Blood transfusion was often fatal prior to the discovery of distinct blood types by Karl Landsteiner in 1904. Landsteiner's observation that sera from one person could agglutinate the blood cells of another person led to his identification of four distinct types of human blood. The red blood cell types were subsequently determined to result from cell surface glycoproteins, now called the **ABO blood groups** (**figure 29.25a**). Blood types are genetically inherited as the A, B, or O alleles respectively encoding the A- or B-type glycoprotein, or no glycoprotein at all. Thus homozygous expression of A or B alleles results in type A or type B blood, respectively. Heterozygous expression of the codominant A and B alleles results in type AB blood. Heterozygous expression of the dominant A or B alleles with the O allele results in type A or type B blood, respectively. Homozygous expression of the O allele results in type O blood. AB glycoproteins are self antigens. Thus AB reactive lymphocytes of the developing host are destroyed during the negative selection process. However, the lymphocytes specific for AB glycoproteins not expressed by the host remain to be activated upon exposure to those specific antigens (which are ubiquitously distributed throughout nature). Consequently, type A hosts produce anti-B antibodies, type B hosts produce anti-A antibodies, and type O hosts produce both anti-A and anti-B antibodies. Type O individuals are considered "universal donors" because their red blood cells lack A and B surface antigens. Conversely, type AB hosts produce neither anti-A nor anti-B antibodies, so such individuals are called "universal recipients." The type II hypersensitivity reaction seen in blood transfusion occurs as complement is activated by cross-linking antibodies (figure 29.25b).

Blood typing can be accomplished by a slightly more sophisticated method of Landsteiner's process whereby blood from one host is mixed with antibodies specific for type A or type B blood. Agglutination of red cells by antibody (figure 29.25c) is used as a diagnostic tool to determine blood type. Another red blood cell antigen often reported with the ABO type was discovered during experiments with rhesus monkeys. The so-called Rh factor (or D antigen) is determined by the expression of two alleles, one dominant (coding for the factor) and one recessive (not coding for the factor). Thus homozygous or heterozygous expression of the Rh allele confers the antigen (indicated as Rh$^+$). Expression of two recessive alleles results in the designation of Rh$^-$ (no Rh factor). Incompatibility between Rh$^-$ mothers and their Rh$^+$ fetus can result in maternal anti-Rh antibodies destroying fetal blood cells (**figure 29.26**). This type II hypersensitivity is called **erythroblastosis fetalis.** Control of this potentially fatal hemolytic disease of the newborn can be mitigated if the mother is passively immunized with anti-Rh factor antibodies, or RhoGAM.

Type III Hypersensitivity

Type III hypersensitivity involves the formation of immune complexes (**figure 29.27**). Normally these complexes are phagocytosed effectively by the fixed monocytes and macrophages of the monocyte-macrophage system. In the presence of excess amounts of some soluble antigens, the antigen-antibody complexes may not be efficiently removed. Their accumulation can lead to a hypersensitivity reaction from complement that triggers a variety of inflammatory processes. The antibodies of type III reactions are primarily IgG. The inflammation caused by immune complexes and cells responding to such inflammation can result in significant damage,

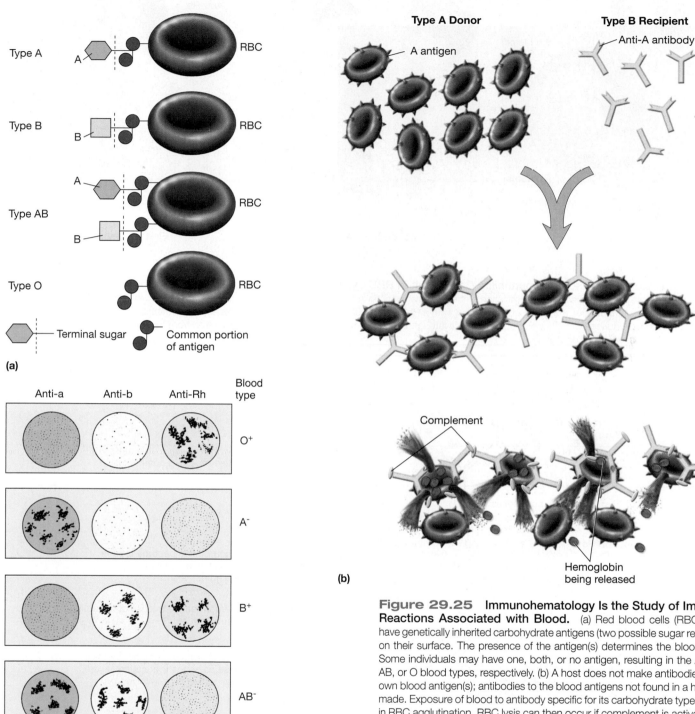

Figure 29.25   Immunohematology Is the Study of Immune Reactions Associated with Blood.   (a) Red blood cells (RBCs) can have genetically inherited carbohydrate antigens (two possible sugar residues) on their surface. The presence of the antigen(s) determines the blood type. Some individuals may have one, both, or no antigen, resulting in the A or B, AB, or O blood types, respectively. (b) A host does not make antibodies to its own blood antigen(s); antibodies to the blood antigens not found in a host are made. Exposure of blood to antibody specific for its carbohydrate type results in RBC agglutination. RBC lysis can then occur if complement is activated by the antibody-agglutinated cells. (c) RBC agglutination with specific antibody is the basis for blood typing. Another molecule, the rhesus (Rh) factor, is another major RBC antigen that is typed to determine blood compatibility.

especially of blood vessels (vasculitis), kidney glomerular basement membranes (glomerulonephritis), joints (arthritis), and skin (systemic lupus erythematosus).

Type IV Hypersensitivity

Type IV hypersensitivity involves delayed, cell-mediated immune reactions. A major factor in the type IV reaction is the time required for T cells to migrate to and accumulate near the antigens. Both T_H and CTL cells can elicit type IV hypersensitivity reactions, depending on the pathway in which the antigen is processed and presented. These events usually take a day or more to plateau.

In general, type IV reactions occur when antigens, especially those binding to serum proteins or tissue cells, are processed and presented to T cells. If the antigen is phagocytosed, it will be

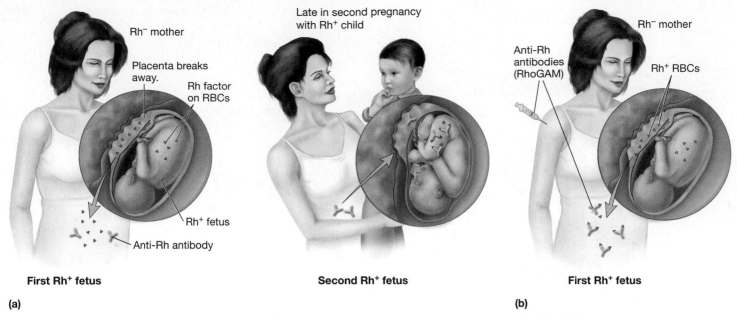

Figure 29.26 Rh Factor Incompatibility Can Result in RBC Lysis. (a) A naturally occurring blood cell incompatibility results when a Rh⁺ fetus develops within a Rh⁻ mother. Initial sensitization of the maternal immune system occurs when fetal blood passes the placental barrier. In most cases, the fetus develops normally. However, a subsequent pregnancy with a Rh⁺ fetus results in a severe, fetal hemolysis. (b) Anti-Rh antibody (RhoGAM) can be administered to Rh⁻ mothers during pregnancy to help bind, inactivate, and remove any Rh factor that may be transferred from the fetus. In some cases, RhoGAM is administered before sensitization occurs.

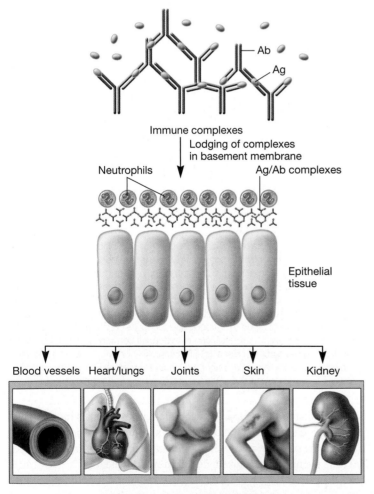

Steps:

1 Antibody combines with excess soluble antigen, forming large quantities of Ab/Ag complexes.

2 Circulating immune complexes become lodged in the basement membrane of epithelia in sites such as kidney, lungs, joints, skin.

3 Fragments of complement cause release of histamine and other mediator substances.

4 Neutrophils migrate to the site of immune complex deposition and release enzymes that cause severe damage to the tissues and organs involved.

Major organs that can be targets of immune complex deposition

Figure 29.27 Type III Hypersensitivity. Circulating immune complexes may lodge at various tissue sites where they activate complement and subsequently cause tissue cell lysis. Complement activation also results in granulocyte recruitment and release of their mediators, causing further injury to the tissue. Increased vascular permeability, resulting from granulocyte mediators, allows immune complexes to be deposited deeper into tissue sites where platelet recruitment leads to microthrombi (blood clots), which impair blood flow, resulting in additional tissue damage.

Figure 29.28 Type IV (or Delayed-Type) Hypersensitivity. The mechanism of type IV hypersensitivity is illustrated by the tuberculin skin test, used to determine exposure to *M. tuberculosis*. Injection of the tuberculin antigens into the skin of individuals previously sensitized by *M. tuberculosis* results in the localized recruitment of macrophages and T_H1 cells over 12 to 48 hours. The T_H1 cells are activated by the antigens presented by the class II MHC molecules on the macrophages. T_H1 cells then secrete inflammatory cytokines, which increase vascular permeability and recruit other immune cells, resulting in a visible swelling at the injection site.

presented to T_H cells by the class II MHC molecules on the APC. This activates the T_H1 cell, causing it to proliferate and secrete cytokines such as IFN-γ and TNF-α. If the antigen is lipid soluble, it can cross the cell membrane to be processed within the cytosol and be presented to CTL cells by class I MHC molecules. CTLs secrete cytokines and kill the cell that is presenting the antigen. Regardless of the cytokine source, they stimulate the expression of adhesion molecules on the local endothelium and increase vascular permeability, allowing fluid and cells to enter the tissue space. Cytokines also stimulate keratinocytes of the skin to release their own cytokines. Together, the cytokines attract lymphocytes, macrophages, and basophils to the affected tissue, exacerbating the inflammation. Extensive tissue damage may result. Examples of type IV hypersensitivities include tuberculin hypersensitivity (the TB skin test; **figure 29.28**), allergic contact dermatitis, some autoimmune diseases, transplantation rejection, and killing of cancer cells.

In **tuberculin hypersensitivity,** a partially purified protein called tuberculin, which is obtained from the bacterium that causes tuberculosis, is injected into the skin of the forearm (figure 29.28). The response in a tuberculin-positive individual begins in about 8 hours, and a reddened area surrounding the injection site becomes indurated (firm and hard) within 12 to 24 hours. The T_H1 cells that migrate to the injection site are responsible for the induration. The reaction reaches its peak in 48 hours and then subsides. The size of the induration is directly related to the amount of antigen that was introduced and to the degree of hypersensitivity of the tested individual. Several important chronic diseases involve cell and tissue destruction by type IV hypersensitivity reactions. These diseases are caused by

viruses, mycobacteria, protozoa, and fungi that produce chronic infections in which the macrophages and T cells are continually stimulated. Examples are leprosy, tuberculosis, leishmaniasis, candidiasis, and herpes simplex lesions. These infectious diseases are discussed in chapters 22–24.

Allergic contact dermatitis is a type IV reaction caused by haptens that combine with proteins in the skin to form the allergen that elicits the immune response (**figure 29.29**). The haptens are the antigenic determinants, and the skin proteins are the carrier molecules for the haptens. Examples of these haptens include cosmetics, plant materials (catechol molecules from poison ivy and poison oak), topical chemotherapeutic agents, metals, and jewelry (especially jewelry containing nickel).

1. Discuss the mechanism of type I hypersensitivity reactions and how these can lead to systemic and localized anaphylaxis.
2. What causes a wheal and flare reaction site?
3. Why are type II hypersensitivity reactions called cytolytic or cytotoxic?
4. What characterizes a type III hypersensitivity reaction? Give an example.
5. Characterize a type IV hypersensitivity reaction.

Autoimmune Diseases

As discussed in section 29.1, the body is normally able to distinguish its own self antigens from foreign nonself antigens and usually does not mount an immunologic attack against itself. This phenomenon is called immune tolerance. At times the body loses tolerance and mounts an abnormal immune attack, either with antibodies or T cells, against its own self antigens.

It is important to distinguish between autoimmunity and autoimmune disease. Autoimmunity often is benign, whereas autoimmune disease often is fatal. **Autoimmunity** is characterized only by the presence of serum antibodies that react with self antigens. These antibodies are called autoantibodies. The formation of autoantibodies is a normal consequence of aging; is readily inducible by infectious agents, organisms, or drugs; and is potentially reversible (it disappears when the offending "agent" is removed or eradicated). **Autoimmune disease** results from the activation of self-reactive T and B cells that, following stimulation by genetic or environmental triggers, cause actual tissue damage (**table 29.4**). Examples include rheumatoid arthritis and type I diabetes mellitus.

Four factors influence the development of autoimmune disease. Two major factors are genetic and viral. The third factor is endocrine—the effect of hormones. The fourth factor is psycho-neuro-immunological—the influence of stress and neurochemicals on the immune response. Overall, these factors can affect gene expression, which directly or indirectly interferes with important immunoregulatory actions. Although their causal mechanism is not well known, autoimmune diseases may involve viral or bacterial infections. Some investigators believe that the release of abnormally large quantities of antigens may occur when the infectious agent causes tissue damage. The same agents also may cause body proteins to change into forms that stimulate antibody production or T-cell activation. Simultaneously, the capacity of T cells to limit this type of reaction seems to be repressed. Many autoimmune diseases have a genetic component. For example, there is a well-documented association between an individual's susceptibility to Graves' disease (which causes hyperthyroidism) and the neurodegenerative disease multiple sclerosis and specific determinants on the major histocompatibility complex.

Transplantation (Tissue) Rejection

Transplants between genetically different individuals within a species are termed **allografts** (Greek *allos,* other). Some transplanted tissues do not stimulate an immune response. For example, a transplanted cornea is rarely rejected because lymphocytes do not circulate into the anterior chamber of the eye. This site is considered an immunologically privileged site. Another example of a privileged tissue is the heart valve, which in fact, can be transplanted from a pig to a human without stimulating an immune response. Such a graft between different species is termed a **xenograft** (Greek *xenos,* strayed).

Transplanting tissue that is not immunologically privileged generates the possibility that the recipient's cells will recognize the donor's tissues as foreign. This triggers the recipient's immune mechanisms, which may destroy the donor tissue. Such a response is called a **tissue rejection reaction.** A tissue rejection reaction can occur by two different mechanisms. First, foreign class II MHC molecules on transplanted tissue, or the "graft," are recognized by host T-helper cells, which aid cytotoxic T cells in graft destruction (**figure 29.30**). Cytotoxic T cells then recognize the graft because it bears foreign class I MHC molecules. This response is much like the activation of CTLs by virally infected host cells. A second mechanism involves the T-helper cells reacting to the graft (transplanted tissue) and releasing cytokines. The cytokines stimulate macrophages to enter, accumulate within the graft, and destroy it. The MHC molecules play a dominant role in tissue rejection reactions because of their unique association with the recognition system of T cells. Unlike antibodies, T cells cannot recognize or react directly with non-MHC molecules (viruses, allergens). They recognize these molecules only in association with, or complexed to, an MHC molecule.

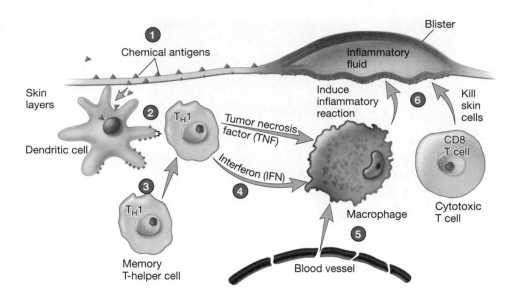

1. Lipid-soluble catechols are absorbed by the skin.

2. Dendritic cells close to the epithelium pick up the allergen, process it,

3. Previously sensitized T_H1 ($CD4^+$) cells recognize the presented allergen.

4. Sensitized T_H1 cells are activated and secrete cytokines (IFN, TNF).

5. These cytokines attract macrophages and cytotoxic T cells to the site.

6. Macrophages release mediators that stimulate a strong, local inflammatory reaction. Cytotoxic T cells directly kill cells and damage the skin. Fluid-filled blisters result.

Figure 29.29 Contact Dermatitis. In contact dermatitis to poison ivy, a person initially becomes exposed to the allergen, predominantly 3-n-pentadecyl-catechol found in the resinous sap material uroshiol, which is produced by the leaves, fruit, stems, and bark of the poison ivy plant. The catechol molecules, acting as haptens, combine with high molecular weight skin proteins. After 7 to 10 days, sensitized T cells are produced and give rise to memory T cells. Upon second contact, the catechols bind to the same skin proteins, and the memory T cells become activated in only 1 to 2 days, leading to an inflammatory reaction (contact dermatitis).

Because class I MHC molecules are present on every nucleated cell in the body, they are important targets of the rejection reaction. The greater the antigenic difference between class I molecules of the recipient and donor tissues, the more rapid and severe the rejection reaction is likely to be. However, the reaction can sometimes be minimized if recipient and donor tissues are matched as closely as possible. Most recipients are not 100% matched to their donors, so immunosuppressing drugs are used to prevent host-mediated rejection of the graft.

Organ transplant recipients also can develop **graft-versus-host disease.** This occurs when the transplanted tissue contains immune cells that recognize host antigens and attack the host. The immunosuppressed recipient cannot control the response of the grafted tissue. Graft-versus-host disease is a common problem in bone marrow transplants. The transplanted bone marrow contains many mature, postthymic T cells. These cells recognize

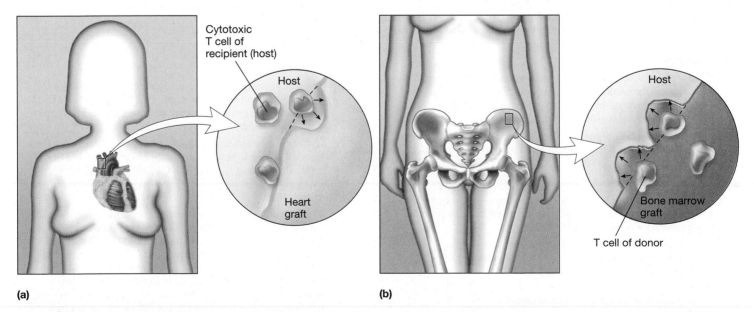

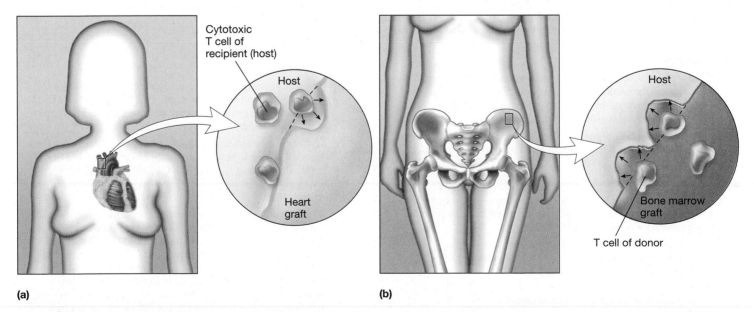

Table 29.4 Some Autoimmune Diseases in Humans

Disease	Autoantigen	Pathophysiology
Acute rheumatic fever	Streptococcal cell wall antigens; antibodies cross-react with cardiomyocytes	Arthritis, scarring of heart valves, myocarditis
Autoimmune hemolytic anemia	Rh blood group, I antigen	Red blood cells are destroyed by complement and phagocytosis, anemia
Autoimmune thrombocytopenia purpura	Platelet integrin	Perfuse bleeding
Goodpasture's syndrome	Basement membrane collagen	Glomerulonephritis, pulmonary hemorrhage
Graves' disease	Thyroid-stimulating hormone receptor	Hyperthyroidism
Multiple sclerosis	Myelin basic protein	Demyelination of axons
Myasthenia gravis	Acetycholine receptor	Progressive muscular weakness
Pemphigus vulgaris	Cadherin in epidermis	Skin blisters
Rheumatoid arthritis	Unknown synovial joint antigen	Joint inflammation and destruction
Systemic lupus erythematosus	DNA, histones, ribosomes	Arthritis, glomerulonephritis, vasculitis, rash
Type 1 diabetes mellitus	Pancreatic beta cell antigen	Beta cell destruction

the host MHC antigens and attack the immunosuppressed recipient's normal tissue cells. Currently one way to prevent graft-versus-host disease is to deplete the bone marrow of mature T cells by using immunosuppressive techniques. Examples include drugs that attack T cells (azathioprine, methotrexate, and cyclophosphamide), immunosuppressive drugs (cyclosporin, tacrolimus, and rapamycin), anti-inflammatory drugs (corticosteroids), irradiation of the lymphoid tissue, and antibodies directed against T-cell antigens.

Immunodeficiencies

Defects in one or more components of the immune system can result in its failing to recognize and respond properly to antigens. Such **immunodeficiencies** can make a person more prone to infection than those people capable of a complete and active immune response. Despite the increase in knowledge of functional derangements and cellular abnormalities in the various immunodeficiency disorders, the fundamental biological errors responsible for them remain largely unknown. To date, most

(a) **(b)**

Figure 29.30 Potential Transplantation Reactions. (a) Donated tissues, not from an identical twin, contain cellular MHC proteins that are recognized as foreign by the recipient host (host-versus-graft disease). The tissue is then attacked by host CTLs, resulting in its damage and rejection. (b) Donated tissue may also contain immune cells that react against host antigens. Recognition of a foreign host by donor CTLs results in graft-versus-host disease.

Table 29.5	Some Congenital Immune Deficiencies in Humans	
Condition	Symptoms	Cause
Chronic granulomatous disease	Defective monocytes and neutrophils leading to recurrent bacterial and fungal infections	Failure to produce reactive oxygen intermediates due to defective NADPH oxidase
X-linked agammaglobulinemia	Plasma cell or B-cell deficiency and inability to produce adequate specific antibodies	Defective B-cell differentiation due to loss of tyrosine kinase
DiGeorge syndrome	T-cell deficiency and very poor cell-mediated immunity	Lack of thymus or a poorly developed thymus
Severe combined immunodeficiency disease (SCID)	Both antibody production and cell-mediated immunity impaired due to a great reduction of B- and T-cell levels	Various mechanisms (e.g., defective B- and T-cell maturation because of X-linked gene mutation; absence of adenosine deaminase in lymphocytes)

genetic errors associated with these immunodeficiencies are located on the X chromosome and produce primary or congenital immunodeficiencies (**table 29.5**). Other immunodeficiencies can be acquired because of infections by immunosuppressive microorganisms, such as HIV.

1. What is an autoimmune disease and how might it develop?

2. What is an immunologically privileged site and how is it related to transplantation success?

3. How does a tissue rejection reaction occur? How might a patient about to receive a bone marrow transplant be prepared? Explain why this is necessary.

4. Describe an immunodeficiency. How might immunodeficiencies arise?

Summary

29.1 Overview of Specific (Adaptive) Immunity

a. The specific (adaptive) immune response system consists of lymphocytes that can recognize foreign molecules (antigens) and respond to them. Two branches or arms of immunity are recognized: humoral (antibody-mediated) immunity and cellular (cell-mediated) immunity (**figure 29.1**).

b. Acquired immunity refers to the type of specific (adaptive) immunity that a host develops after exposure to a suitable antigen.

29.2 Antigens

a. An antigen is a substance that stimulates an immune response and reacts with the products of that response. Each antigen can have several antigenic determinant sites called epitopes that stimulate production of and combine with specific antibodies (**figure 29.2**).

b. Haptens are small organic molecules that are not antigenic by themselves but can become antigenic if bound to a larger carrier molecule.

29.3 Types of Specific (Adaptive) Immunity

a. Immunity can be acquired by natural means—actively through infection or passively through receipt of preformed antibodies, as through colostrums (**figure 29.3**).

b. Immunity can be acquired by artificial means—actively through immunization or passively through receipt of preformed antibodies, as with antisera.

29.4 Recognition of Foreignness

a. MHC molecules are cell surface proteins coded by a group of genes termed the major histocompatibility complex. Class I MHC proteins are found on all nucleated cells of mammals.

Class II MHC proteins are only expressed on B cells and cells that can phagocytose foreign materials and organisms. The human MHC gene products are called the human leukocyte antigens (HLA) (**figure 29.4**).

b. Class I MHC proteins collect foreign peptides processed by the proteasome and present them to cytotoxic T cells.

c. Class II MHC proteins collect foreign peptides processed by the phagosome and present them to T-helper cells.

29.5 T-Cell Biology

a. T cells are pivotal elements of the immune response. T cells have antigen-specific receptor proteins (**figure 29.5**).

b. Antigen-presenting cells—macrophages, dendritic cells, and B cells—take in foreign antigens or pathogens, process them, and present antigenic fragments complexed with MHC molecules to T-helper cells (**figure 29.6**).

c. Cytotoxic T lymphocytes recognize target cells such as virus-infected cells that have foreign antigens and class I MHC molecules on their surface. The CTLs then attack and destroy the target cells using the CD95 pathway, the perforin pathway, or both (**figure 29.7**).

d. T cells control the development of other cells, including effector B and T cells. T-helper cells ($CD4^+$) regulate cell behavior; cytotoxic T cells ($CD8^+$) also regulate cell behavior, but in addition, they can kill altered host cells directly. There are four subsets of T-helper cells: T_H1, T_H2, T_H17 and T_H0. T_H1 cells produce various cytokines and are involved in cellular immunity. The T_H2 cells also produce various cytokines but are involved in humoral immunity. T_H17 cells primarily produce IL-17 and induce inflammation by recruiting neutrophils. T_H0 cells are simply undifferentiated precursors of T_H1, T_H2 and T_H17 cells.

29.6 B-Cell Biology

a. B cells defend against antigens by differentiating into plasma cells that secrete antibodies into the blood and lymph, providing humoral or antibody-mediated immunity.

b. B cells can be stimulated to divide or differentiate to secrete antibody when triggered by the appropriate signals.

c. B cells have receptor immunoglobulins on their plasma membrane surface that are specific for given antigenic determinants. Contact with the antigenic determinant is required for the B cell to divide and differentiate into plasma cells and memory cells (**figures 29.9** and **29.10**).

29.7 Antibodies

a. Antibodies (immunoglobulins) are a group of glycoproteins present in the blood, tissue fluids, and mucous membranes of vertebrates. All immunoglobulins have a basic structure composed of four polypeptide chains (two light and two heavy) connected to each other by disulfide bonds (**figure 29.11**). In humans, five immunoglobulin classes exist: IgG, IgA, IgM, IgD, and IgE (**figures 29.12–29.16** and **table 29.3**).

b. The primary antibody response in a host occurs following initial exposure to the antigen. This response has lag, log, plateau, and decline phases. Upon secondary antigen challenge, the B cells mount a heightened and accelerated anamnestic response (**figure 29.17**).

c. Antibody diversity results from the rearrangement and splicing of the individual gene segments on the antibody-coding chromosomes, somatic mutations, the generation of different codons during splicing, and the independent assortment of light- and heavy-chain genes (**figures 29.18** and **29.19**).

d. Immunologic specificity and memory are partly explained by the clonal selection theory (**figure 29.20**).

e. Hybridomas result from the fusion of lymphocytes with myeloma cells. These cells produce a single monoclonal antibody. Monoclonal antibodies have many uses (**Techniques & Applications 29.1**).

29.8 Action of Antibodies

a. Various types of antigen-antibody reactions occur in vertebrates and initiate the participation of other body processes that determine the ultimate fate of the antigen. For example,

the complement system can be activated, leading to cell lysis, phagocytosis, chemotaxis, or stimulation of the inflammatory response. Other defensive antigen-antibody interactions include toxin neutralization, viral neutralization, adherence inhibition, opsonization, and immune complex formation (**figure 29.21**).

29.9 Summary: The Role of Antibodies and Lymphocytes in Immune Defense

a. The response of a vertebrate host to any particular pathogen may involve a complex set of responses. Both humoral and cellular immune responses can join nonspecific (innate) defenses to ensure a maximal survival advantage against viral and bacterial pathogens.

29.10 Acquired Immune Tolerance

a. Acquired immune tolerance is the ability of a host to react against nonself antigens while tolerating self antigens. It can be induced in several ways.

29.11 Immune Disorders

a. When the immune response occurs in an exaggerated form and results in tissue damage to the individual, the term hypersensitivity is applied. There are four types of hypersensitivity reactions, designated as types I through IV (**figures 29.22–29.28**).

b. Autoimmune diseases result when self-reactive T and B cells attack the body and cause tissue damage. A variety of factors can influence the development of autoimmune disease.

c. The immune system can act detrimentally and reject tissue transplants. There are different types of transplants. Xenografts involve transplants of privileged tissue between different species, and allografts are transplants between genetically different individuals of the same species.

d. Immunodeficiency diseases are a diverse group of conditions in which an individual's susceptibility to various infections is increased; several severe diseases can arise because of one or more defects in the specific (adaptive) or nonspecific (innate) immune response.

Critical Thinking Questions

1. Why do you think antibodies are proteins rather than polysaccharides or lipids? List all properties of proteins that make them suitable molecules from which to make antibodies.

2. How did the clonal selection theory inspire the development of monoclonal antibody techniques?

3. Why do MHC, TCR, and BCR molecules require accessory proteins or coreceptors for a signal to be sent within the cell?

4. Why do you think two signals are required for B- and T-cell activation but only one signal is required for activation of an APC?

5. Most immunizations require multiple exposures to the vaccine (i.e., boosters). Why is this the case?

Learn More

Learn more by visiting the Prescott website at www.mhhe.com/prescottprinciples, where you will find a complete list of references.

Pathogenicity of Microorganisms

30

Chapter Glossary

AB toxins The structure and activity of many exotoxins; the B subunit binds the toxin to a cell, and the A subunit disrupts cell function when it enters the cell.

apoptosis, also called **programmed cell death** The fragmentation of a cell into membrane-bound particles that are eliminated by phagocytosis; a normal mechanism that preserves homeostasis.

bacteremia The presence of viable bacteria in the blood.

colonization In the context of microbial pathogenicity, colonization is the establishment of a site of microbial reproduction on or in a host.

cytotoxin A toxin that acts upon specific cells; cytotoxins are named according to the cell for which they are specific (e.g., neurotoxin).

endotoxin The lipid A component of gram-negative bacterial cell wall lipopolysaccharide. Systemic effects of endotoxin are sometimes referred to as septic shock.

enterotoxin A toxin specifically affecting the cells of the intestinal mucosa, causing vomiting and diarrhea.

exotoxin A heat-labile, toxic protein produced by a bacterium and usually released into the bacterium's surroundings.

infectious dose 50 (ID_{50}) and lethal dose 50 (LD_{50}) Refers to the number of organisms or dose that will infect or kill, respectively, 50% of an experimental group of hosts within a specified time period.

intoxication A disease **(toxemia)** that results from the presence of a specific toxin in the host.

leukocidin A microbial toxin that can damage or kill leukocytes.

noncytopathic virus A virus that does not kill its host cell by viral release–induced lysis.

opportunistic pathogen An organism that is part of the host's normal microbiota but is able to cause disease when the host is immunocompromised or when the organism has gained access to other tissue sites.

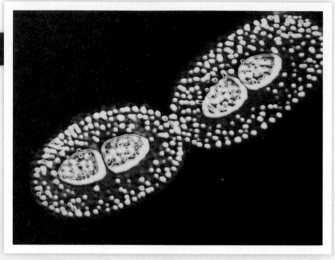

Three *Streptococcus pneumoniae,* each surrounded by a slippery mucoid capsule (shown as a layer of white spheres around the diplococcus bacteria). The polysaccharide capsule is vital to the pathogenicity of this bacterium because it prevents phagocytic cells from accomplishing phagocytosis.

pathogenicity island A 10 to 200 Kb segment of DNA in some pathogens that contains the genes responsible for virulence.

phospholipase An enzyme that hydrolyzes a specific ester bond in phospholipids.

polymicrobial disease Infectious disease caused by more than one type of microbe that interact to cause a specific pathology.

primary pathogen An organism that causes disease in a healthy host by direct infection.

septicemia A disease associated with the presence of pathogens or bacterial toxins in the blood.

toxin A microbial product (e.g., a protein) or component (e.g., lipid A) that injures another cell or organism.

toxoid A bacterial exotoxin that has been modified so that it is no longer toxic but is still immunogenic, stimulating antitoxin formation in a host.

viremia The presence of viruses in the blood stream.

virulence factor A bacterial product or structure that contributes to virulence or pathogenicity.

Pathogenicity is not the rule. Indeed, it occurs so infrequently and involves such a relatively small number of species, considering the huge population of bacteria on earth, that it has a freakish aspect. Disease usually results from inconclusive negotiations for symbiosis, an overstepping of the line by one side or the other, a biological misinterpretation of borders.

—*Lewis Thomas*

Chapter 27 introduces the concept of symbiosis and deals with several of its subordinate categories, including commensalism and mutualism. In this chapter, the process of parasitism is presented along with one of its possible consequences—pathogenicity. The parasitic way of life is so successful, it has evolved independently in nearly all groups of organisms. Understanding host-parasite relationships requires an interdisciplinary approach, drawing on knowledge of cell biology, microbiology, entomology, immunology, ecology, and zoology. This chapter examines the parasitic way of life in terms of health and disease with an emphasis on viral and bacterial disease mechanisms.

30.1 HOST-PARASITE RELATIONSHIPS

Relationships between two organisms can be very complex. A larger organism that supports the survival and growth of a smaller organism is called the host. Technically, **parasites** are those organisms that live on or within a host organism and are metabolically dependent on the host. Unfortunately, the term parasite has other meanings. It is often used to mean a protozoan or helminth living within a host. However, any organism that causes disease is a parasite. Even commensals, such as those associated with the gut, can become parasites when they are present in a location within the host other than the site they normally colonize. These organisms are often referred to as opportunists.

Microbiologists can define infectious disease by the **host-parasite relationship,** which is complex and dynamic. When a parasite is growing and multiplying within or on a host, the host is said to have an **infection.** The nature of an infection can vary widely with respect to severity, location, and number of organisms involved (**table 30.1**). An infection may or may not result in overt disease. An infectious disease is any change from a state of health in which part or all of the host body is not capable of carrying on its normal functions due to the presence of a parasite or its products. Any organism or agent that produces such a disease is also known as a **pathogen** (Greek *patho,* disease, and *gennan,* to produce). Its ability to cause disease is called **pathogenicity.** A **primary pathogen** is any organism that causes disease in a healthy host by direct interaction. Conversely, an **opportunistic pathogen** refers to an organism that is part of the host's normal microbiota but is able to cause disease when the host is immunocompromised or when the organism has gained access to other tissue sites.

At times an infectious organism can enter a **latent state** in which no shedding of the organism occurs (i.e., the organism is not infectious at that time) and no symptoms are present within the host. This latency can be either intermittent or quiescent. Intermittent latency is exemplified by the herpesvirus that causes cold sores (fever blisters). After an initial infection, the symptoms subside. However, the virus remains in nerve tissue

Table 30.1	Various Types of Infections
Type	**Definition**
Abscess	A localized infection with a collection of pus surrounded by an inflamed area
Acute	Short but severe
Bacteremia	Presence of viable bacteria in the blood
Chronic	Persisting over a long time
Focal	Existing in circumscribed areas
Fulminating	Infectious agent multiplying with great intensity
Iatrogenic	Caused as a result of health care
Latent	Persisting in tissues for long periods, during most of which there are no symptoms
Localized	Restricted to a limited region or to one or more anatomical areas
Nosocomial	Developed during a stay at a hospital or other clinical care facility
Opportunistic	Resulting from endogenous microbiota, especially when host resistance is very low
Polymicrobial	More than one type of organism present simultaneously
Pyogenic	Resulting in pus formation
Sepsis	(1) The condition resulting from the presence of bacteria or their toxins in blood or tissues; the presence of pathogens or their toxins in the blood or other tissues (2) Systemic response to infection; this systemic response is manifested by two or more of the following conditions as a result of infection: temperature, >38 or <36°C; heart rate, >90 beats per min; respiratory rate, >20 breaths per min, or pCO_2, <32 mm Hg; leukocyte count, >12,000 cells per ml^3, or >10% immature (band) forms
Septicemia	Blood poisoning associated with persistence of pathogenic organisms or their toxins in the blood
Septic shock	Sepsis with hypotension despite adequate fluid resuscitation, along with the presence of perfusion abnormalities that may include, but are not limited to, organ dysfunction, hypoperfusion, hypotension, lactic acidosis, oliguria, or an acute alteration in mental status
Subclinical	No detectable symptoms or manifestations occurring (covert)
Systemic	Spread throughout the body
Toxemia	Condition arising from toxins in the blood
Zoonotic	Caused by a microorganism normally found in animals other than humans

$$\text{Infection (infectious disease)} = \frac{\text{No. of organisms} \times \text{Virulence}}{\text{Host resistance}}$$

Figure 30.1 **Mathematical Expression of Infection.** Infection or infectious disease can be evaluated by determining the relative contributions of the number of organisms, their virulence, and the host resistance. Organism number reflects the infectious dose and the rate at which the organism can reproduce. Virulence reflects the total number of virulence factors expressed in the host. Host resistance is a function of immune status (immunizations, nutrition, previous exposure, etc.) that may be combined with the effects of chemotherapeutic intervention.

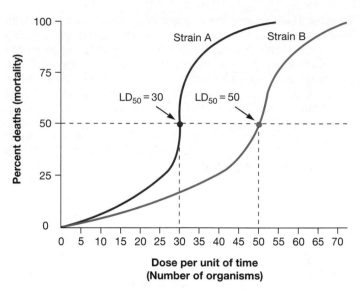

Figure 30.2 **Determination of the LD_{50} of a Pathogenic Microorganism.** Various doses of a specific pathogen are introduced into experimental host animals. Deaths are recorded and a graph constructed. In this example, the graph represents the susceptibility of host animals to two different strains of a pathogen—strain A and strain B. For strain A, the LD_{50} is 30, and for strain B, it is 50. Hence strain A is more virulent than strain B.

and can be cyclically activated weeks or years later by factors such as stress or sunlight. In quiescent latency, the organism persists but remains inactive for long periods of time, usually for years. For example, the varicella-zoster virus causes chickenpox in children and remains after the disease has subsided. In adulthood, under certain conditions, the same virus may erupt into a disease called shingles. ◄◄ *Viruses with double-stranded DNA genomes (section 24.2)*

The outcome of most host-parasite relationships depends on three main factors: (1) the number of microorganisms infecting the host, (2) the degree of pathogenicity (or virulence) of the organism, and (3) the host's defenses or degree of resistance (**figure 30.1**). Usually, the more pathogenic organisms within a given host, the greater the likelihood that the microbes will overcome or evade the host immune defenses and cause disease. However, a few organisms can cause disease if they are extremely virulent or if the host's resistance is low. Such infections can be a serious problem among hospitalized patients with very low resistance.

The term **virulence** (Latin *virulentia,* from virus, poison) refers to the degree or intensity of pathogenicity. As mentioned previously, pathogenicity is a general term that refers to an organism's potential to cause disease. Various physical and chemical characteristics (such as structures that facilitate attachment and molecules that bypass host defenses) contribute to pathogenicity and thus are called **virulence factors.** Virulence is determined by the degree that the pathogen causes damage, including invasiveness and infectivity. **Invasiveness** is the ability of the organism to spread to adjacent or other tissues. **Infectivity** is the ability of the organism to establish a discrete, focal point of infection. Another major aspect of pathogenic potential is toxigenicity. **Toxigenicity** is the pathogen's ability to produce toxins—chemical substances that damage the host and produce disease. Virulence is often measured experimentally by determining the **lethal dose 50 (LD_{50})** or the **infectious dose 50 (ID_{50}).** These values refer to the dose or number of pathogens that either kill or infect, respectively, 50% of an experimental group of hosts within a specified period (**figure 30.2**).

Disease can also result from causes other than toxin production. Sometimes a host triggers an exaggerated immunological response (**immunopathology**) upon a second or chronic exposure to a microbial antigen. These hypersensitivity reactions damage the host even though the pathogen does not produce a toxin. Tuberculosis is a good example of the involvement of hypersensitivity reactions in disease. Some diseases result from autoimmune responses. For instance, a viral or bacterial pathogen may stimulate the immune system to attack host tissues because it carries antigens that resemble those of the host, a phenomenon known as molecular mimicry. Streptococcal infections may cause rheumatic fever in this way. ◄◄ *Immune disorders: Hypersensitivities (section 29.11)*

1. Define parasitic organism, infection, infectious disease, pathogenicity, virulence, invasiveness, infectivity, pathogenic potential, and toxigenicity.

2. What factors determine the outcome of most host-parasite relationships?

30.2 **PATHOGENESIS OF VIRAL DISEASES**

The fundamental process of viral infection is the expression of the viral replicative cycle in a host cell (*see section 5.3*). The steps for the infectious process involving viruses usually include the following:

1. Maintain a reservoir. A reservoir is a place to live and multiply before and after causing an infection.

2. Enter a host.

3. Contact and enter susceptible cells.

4. Replicate within the cells.

5. Release from host cells (immediately or delayed).

6. Evade the host's immune response.

7. Spread to adjacent cells.

8. Be either cleared from the body of the host, establish a persistent infection, or kill the host.

9. Be shed back into the environment.

The success by which a virus accomplishes each of these steps contributes to its overall pathogenicity, as we now discuss.

Maintaining a Reservoir

Like other infectious agents, viruses must reside somewhere before they are transmitted to a specific host or tissue site. Because most viruses are limited to the type of host they can infect (animal viruses infect animals, plant viruses infect plants, and so on), they must gain access to a susceptible host so that they can replicate. Thus the most common **reservoirs** of human viruses are humans and other animals. Most often, viruses are transmitted from reservoir (a human host) to host (another human) to cause noticeable infection in a relatively short time frame. Because viruses require viable host cells in which to replicate, the source or reservoir may harbor large numbers of viral particles that can infect equally large numbers of new hosts upon their release. However, some viruses may not be able to leave their reservoirs or may leave at a very slow rate. Because the source or reservoir of a pathogen is part of the infectious disease cycle, this aspect of pathogenicity is discussed in detail in chapter 33, which covers the epidemiology of infectious diseases. << *Microbial Diversity & Ecology 24.1: SARS: Evolution of a virus*

Contact, Entry, and Primary Replication

The first step in the infectious process is the attachment and entrance of the virus into a susceptible host and the host's cells. Entrance may be accomplished through one of the body surfaces: skin, respiratory system, gastrointestinal system, urogenital system, or the conjunctiva of the eye. Other viruses enter the host by sexual contact, needle sticks, blood transfusions and organ transplants, or insect **vectors**—organisms that transmit the pathogen from one host to another.

Regardless of the method of entry into the host organism, viral infection begins when the viral particle penetrates a host cell to gain access to the cell's replicative machinery. This process is called **adsorption,** or the attachment to the cell surface. Recall that adsorption occurs because viruses produce specific protein ligands that bind to host cell receptors embedded within their plasma membranes. Host specificity for the virus is a function of viral gene expression; the virus must express the ligand so as to dock with a specific host cell. Protein ligands are usually positioned on the virus to maximize contact with the cell. Enveloped viruses use spikes—viral proteins that protrude from their membrane. The ligands of naked viruses are part of their capsid proteins.

Each viral ligand binds only to a complementary receptor on the host cell surface. Binding of a virus to its receptor typically results in penetration of the cell or the delivery of viral nucleic acid to the cytoplasm of the cell. In the case of human viruses, the viral genome enters the host cell by (1) endocytosis and the release of the genome from the capsid (uncoating), as with *Poliovirus* and the poxviruses; or (2) fusion of the viral envelope with the cell membrane and subsequent uncoating, as with influenza virus (*see figure 5.11*).

Some viruses remain localized, replicating only in cells that constitute one tissue or organ (e.g., respiratory or gastrointestinal infections). Others spread to sites distant from the point of entry and replicate at these sites. For example, the *Poliovirus* enters through the gastrointestinal tract but produces disease in the central nervous system. << *Viruses with plus-strand RNA genomes (section 24.5)*

Release from Host Cells

As reviewed in chapter 5, viruses use two distinct release mechanisms to exit their host cells. One mechanism, host cell lysis, is very dramatic and results in relatively large numbers of virions leaving the host cell at the same time and host cell death (*see figure 5.15*). The other release mechanism, called budding or "blebbing," occurs when a newly formed nucleocapsid pushes against the host cell membrane until the membrane evaginates and pinches off behind the virus. The released virus is coated with host cell membrane, which now composes the viral envelope (*see figure 5.16*). Release of viral particles by budding is a slower process than lysis; exiting viral particles take relatively small amounts of host cell membrane. The host cell can replenish its membrane, permitting continued virus release over the life of the infected cell.

Evasion of Host Defenses by Viruses

The evasion of viral host defenses begins when the virus first infects the host. However, for the virus to cause a successful infection (from the viral point of view), it must be able to avoid host immunity so it can spread to a sufficient number of host cells to amplify the number of virions; this is when disease ensues. We thus discuss the evasion of host immunity as a prerequisite to viral spread.

Viruses have evolved a variety of ways to suppress or evade the host's immune response. For instance, some viruses mutate and change antigenic sites (antigenic drift) on the virion capsids (e.g., the influenza virus) or may downregulate the level of expression of viral cell surface proteins (e.g., the herpesvirus). Other viruses (HIV) infect cells (T cells) of the immune system and diminish their function. HIV as well as the *Measles virus* and cytomegalovirus cause fusion of host cells. This allows these viruses to move from an infected cell to an uninfected cell without exposure to the antibody-containing fluids of the host. The herpesvirus infects neurons that express little or no major histocompatibility complex molecules. The adenovirus produces proteins that inhibit major histocompatibility complex function. Finally, *Hepatitis B* virus-infected cells produce large amounts of antigens not associated with the complete virus. These antigens serve as decoys, binding available neutralizing antibody so that there is insufficient free antibody to bind with the complete virion. These are just a few examples of the ways in which viruses evade host defenses; our understanding of these and other mechanisms is advancing through genomics and the analysis of specific gene products. ◄◄ *Viral diversity (chapter 24); Recognition of foreignness (section 29.4)*

Viral Spread and Cell Tropism

Mechanisms of viral spread vary, but the most common routes are the bloodstream and the lymphatic system. The presence of viruses in the blood is called **viremia.** In some instances, spread is by way of nerves (e.g., rabies, herpes simplex, and varicella-zoster viruses).

Viruses exhibit cell, tissue, and organ specificities. These specificities are called **tropisms** (Greek *trope,* turning). A tropism by a specific virus usually reflects the presence of specific cell surface receptors on the eucaryotic host cell for that virus.

Virus-Host Interactions

The interaction between a virus and its host cell can result in a variety of effects. Viruses can be either cytopathic or noncytopathic. **Cytopathic viruses** are those that ultimately kill the host cell; the result is often local necrosis. Alternatively, cytopathic viruses can trigger **apoptosis,** or programmed cell death, which culminates in the death of the host cell, often before viral replication can occur (*see figure 5.21*). Although both involve death of host cells, necrosis and apoptosis are very different. Apoptosis is a normal process in multicellular organisms. It is used during development to remove cells or tissues that are no longer needed. It involves nuclear degeneration, the partial digestion of many cell proteins by proteolytic enzymes called caspases, and the formation of apoptotic bodies (membrane-enclosed cell components), which are subsequently phagocytosed by macrophages. Unlike necrosis, the apoptotic cell does not lyse and release its contents. Rather, apoptosis is a controlled dismantling of the cell that results in cell death. In the case of viral infection, apoptosis also prevents virus replication and spread. Some viruses stimulate apoptosis but use special viral proteins to prolong the process long enough to complete viral replication. This ensures that the virus can continue to infect new host cells.

Noncytopathic viruses do not immediately produce cell death and result in latent or persistent infections. As a result, noncytopathic viruses can be considered either productive or nonproductive. **Productive noncytopathic viruses** produce persistent infection with the release of only a few new viral particles at a time. **Nonproductive viruses** do not actively make virus at detectable levels for a period of time, so are said to cause latent infection. However, nonproductive viruses may be triggered to a reactivated (productive) state by environmental stressors or other factors.

As anyone who has ever had the flu or a cold knows, clinical illness may result from virus-host cell interactions. Some tissues, such as intestinal epithelium, can quickly regenerate when damaged by viruses. Thus they are easily repaired following cellular damage. In contrast, tissues of the nervous system are limited in their ability to regenerate and are thus difficult to repair following damage by viruses. Moreover, infection of cells with some viruses can result in the integration of viral DNA. In a few cases, this can cause them to transform into cancerous cells. This is the result of viral DNA interference with host DNA growth-cycle regulation. ◄◄ *Types of viral infections (section 5.4)*

Virus Shedding

The last step in the infectious process is shedding of the virus back into the environment. This is necessary to maintain a source of viruses in a population of hosts. Shedding often occurs from the same body surface used for entry. During this period, an infected host is infectious (contagious) and can spread the virus. In some viral infections, such as a rabies infection, the infected human is the final host because virus shedding does not occur.

1. For a virus to cause disease, certain steps are usually accomplished. Briefly describe each of these steps.
2. If you were to design an antiviral drug, which step or steps in the viral life cycle would you target? Explain your answer.
3. What are the four most common patterns of viral infections? Describe apoptosis and its role in viral infections.

30.3 OVERVIEW OF BACTERIAL PATHOGENESIS

The steps for infections by pathogenic bacteria usually include the following:

1. Maintain a reservoir.
2. Initial transport to and entry into the host.
3. Adhere to, colonize, or invade host cells or tissues.
4. Evade host defense mechanisms.
5. Multiply (grow) or complete its life cycle on or in the host or the host's cells.
6. Damage the host.
7. Leave the host and return to the reservoir or enter a new host.

Table 30.2	Bacterial Adherence Factors That Play a Role in Infectious Diseases
Adherence Factor	**Description**
Fimbriae	Filamentous structures that help attach bacteria to other bacteria or to solid surfaces
Glycocalyx or capsule	A layer of exopolysaccharide fibers with a distinct outer margin that surrounds many cells; inhibits phagocytosis and aids in adherence; when the layer is well organized and not easily washed off, it is called a capsule
Pili	Filamentous structures that bind procaryotes together for the transfer of genetic material
S layer	The outermost regularly structured layer of cell envelopes of some bacteria that may promote adherence to surfaces
Slime layer	A bacterial film that is less compact than a capsule and is removed easily
Teichoic and lipoteichoic acids	Cell wall components in gram-positive bacteria that aid in adhesion

The first five factors influence the degree of infectivity and invasiveness. Toxigenicity plays a major role in the sixth. These determinants are now discussed in more detail.

Maintaining a Reservoir of the Bacterial Pathogen

All bacterial pathogens must have at least one reservoir. The most common reservoirs for human pathogens are other humans, animals, and the environment. The increasing impingement of humans on the environment and increased selective pressure on microorganisms have played a significant role in the substantial change in reservoirs over the last 100 years. Since the source or reservoir of the pathogen is part of the infectious disease cycle, this aspect of pathogenicity is discussed in detail in chapter 33, which covers the epidemiology of infectious diseases. >> *The infectious disease cycle (section 33.5)*

Transport of the Bacterial Pathogen to the Host

An essential feature in the development of an infectious disease is the initial transport of the bacterial pathogen to the host. The most obvious means is direct contact—from host to host (coughing, sneezing, body contact). Bacteria also are transmitted indirectly in a variety of ways. Infected hosts shed bacteria into their surroundings. Once in the environment, bacteria can be deposited on various

surfaces from which they can be either resuspended into the air or indirectly transmitted to a host. Soil, water, and food are indirect vehicles that harbor and transmit bacteria to hosts. Arthropod vectors and **fomites** (inanimate objects that harbor and transmit pathogens) also are involved in the spread of many bacteria.

Attachment and Colonization by the Bacterial Pathogen

After being transmitted to an appropriate host, the bacterial pathogen must be able to adhere to and colonize host cells or tissues. In this context, **colonization** means the establishment of a site of microbial reproduction on or within a host. It does not necessarily result in tissue invasion or damage. Colonization depends on the ability of the bacteria to survive in the new (host) environment and to compete successfully with host cells and the host's normal microbiota for essential nutrients. Specialized structures that allow bacteria to compete for surface attachment sites also are necessary for colonization.

Bacterial pathogens adhere with a high degree of specificity to particular tissues. Adherence structures such as pili and fimbriae (**table 30.2**) and specialized adhesion molecules on the bacterial cell surface that bind to complementary receptor sites on the host cell surface (**figure 30.3**) facilitate bacterial attachment to host cells. These bacterial products and structural components contribute to virulence, thus they are one type of virulence factor. << *Components external to the cell wall (section 3.6)*

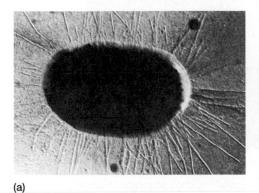
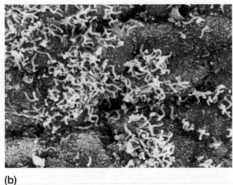
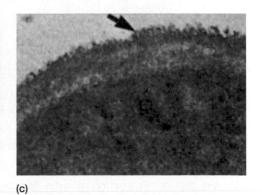

(a)   (b)   (c)

Figure 30.3   Microbial Adherence.   (a) Transmission electron micrograph of fimbriated *Escherichia coli* (×16,625). (b) Scanning electron micrograph of epithelial cells with adhering vibrios (×1,200). (c) *Candida albicans* fimbriae (arrow) are used to attach the fungus to vaginal epithelial cells.

Evasion of Host Defenses by Bacteria

Most bacteria do not cause disease, in part because they are eliminated before colonization by the host's normal flora or its immune system. However, pathogens have evolved many mechanisms to evade host defenses. Although these mechanisms come into play upon initial host exposure, they become paramount in preventing further proliferation of bacteria. Thus successful pathogens can elude both the initial host responses that are part of innate immunity as well as those governed by the adaptive immune system.

Evading the Complement System

To evade the activity of complement, some bacteria have capsules (**chapter opening figure**) that prevent complement activation. Some gram-negative bacteria can lengthen the O chains in their lipopolysaccharide to prevent complement activation. Others such as *Neisseria gonorrhoeae* generate **serum resistance.** These bacteria have modified lipooligosaccharides on their surface that interfere with proper formation of the membrane attack complex (*see figure 28.22*) during the complement cascade. The virulent forms of *N. gonorrhoeae* that exhibit serum resistance are able to spread throughout the body of the host and cause systemic disease, whereas those *N. gonorrhoeae* that lack serum resistance remain localized in the genital tract. $\ll$ *Chemical mediators in nonspecific (innate) resistance: Complement (section 28.6)*

Resisting Phagocytosis

Before a phagocytic cell can engulf a bacterium, it must first directly contact the bacterial cell surface. Some bacteria such as *Streptococcus pneumoniae, Neisseria meningitidis,* and *Haemophilus influenzae* can produce a slippery mucoid capsule that prevents the phagocyte from effectively contacting the bacterium. Other bacteria evade phagocytosis by producing specialized surface proteins such as the M protein on *S. pyogenes*. Like capsules, these proteins interfere with adherence between a phagocytic cell and the bacterium. $\ll$ *Phagocytosis (section 28.3)*

Bacterial pathogens use other mechanisms to resist phagocytosis. For example, *Staphylococcus* produces leukocidins that destroy phagocytes before phagocytosis can occur. *Streptococcus pyogenes* releases a protease that cleaves the C5a complement factor and thus inhibits complement's ability to attract phagocytes to the infected area.

Survival Inside Phagocytic Cells

Some bacteria have evolved the ability to survive inside neutrophils, and macrophages. These microorganisms are very pathogenic because they are impervious to an important innate resistance mechanism of the host. One method of evasion is to escape from the phagosome before it merges with the lysosome, as seen with *Listeria monocytogenes, Shigella,* and *Rickettsia*. These bacteria use actin-based motility to move within mammalian host cells and spread between them. Upon lysing the phagosome, they gain access to the host cell cytoplasm, where they activate the assembly of an actin tail using host cell actin and other cytoskeletal proteins (**figure 30.4a**). The actin tails propel the bacteria through the cytoplasm of the infected cell to its surface, where they push out against the plasma membrane and form protrusions (figure 30.4b). The protrusions are engulfed by adjacent cells, and the bacteria once again enter phagosomes and escape into the cytoplasm. In this way, the infection spreads to adjacent cells. The lysosomes never have a chance to merge with the phagosomes. Another approach is to resist the toxic products released into the phagolysosome after fusion occurs. A good example of a bacterium that is resistant to the lysosomal enzymes is *Mycobacterium tuberculosis,*

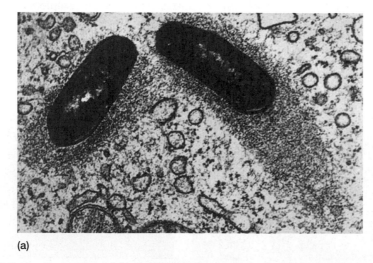
(a)

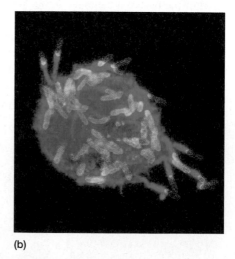
(b)

Figure 30.4 **Formation of Actin Tails by Intracellular Bacterial Pathogens.** (a) Transmission electron micrograph of *Lysteria monocytogenes* in a host macrophage. The bacterium has polymerized host actin into a long tail that it uses for intracellular propulsion and to move from one host cell to another. (b) *Burkholderia pseudomallei* (stained red) also forms actin tails as shown in this confocal micrograph. Note that the actin tails enable the bacterial cells to be propelled out of the host cell.

probably at least partly because of its waxy external layer. Still other bacteria prevent fusion of phagosomes with lysosomes (e.g., *Chlamydia*).

Evading the Specific Immune Response

To evade the specific immune response, some bacteria (e.g., *S. pyogenes*) produce capsules that are not antigenic because they resemble host tissue components. *N. gonorrhoeae* can evade the specific immune response by two mechanisms: (1) it makes genetic variations in its pili through a process called phase variation, resulting in altered pilin protein sequence and expression. This renders specific antibodies useless against the new pili, and adherence to host tissue occurs; and (2) it produces IgA proteases that destroy secretory IgA and allow adherence. Finally, some bacteria produce proteins (such as staphylococcal protein A and protein G of *S. pyogenes*) that interfere with antibody-mediated opsonization by binding to the Fc portion of immunoglobulins.

Invasion of Host Tissues

Entry into tissues is a specialized strategy used by many bacterial pathogens for survival and multiplication. Pathogens often actively penetrate the host's mucous membranes and epithelium after attachment to the epithelial surface. This may be accomplished through production of lytic substances that alter the host tissue by (1) attacking the extracellular matrix and basement membranes of integuments and intestinal linings, (2) degrading carbohydrate-protein complexes between cells or on the cell surface (the glycocalyx), or (3) disrupting the cell surface.

At times a bacterial pathogen can penetrate the epithelial surface by passive mechanisms not related to the pathogen itself. Examples include (1) small breaks, lesions, or ulcers in a mucous membrane that permit initial entry; (2) wounds, abrasions, or burns on the skin's surface; (3) arthropod vectors that create small wounds while feeding and (4) tissue damage caused by other organisms (e.g., a dog bite).

Once under the mucous membrane, the bacterial pathogen may penetrate to deeper tissues and continue disseminating throughout the body of the host. One way the pathogen accomplishes this is by producing specific structures or enzymes that promote spreading (**table 30.3**). These products represent other types of virulence factors. Bacteria may also enter the small terminal lymphatic capillaries that surround epithelial cells. These capillaries merge into large lymphatic vessels that eventually drain into the circulatory system. Once the circulatory system is reached, the bacteria have access to all organs and systems of the host.

Bacterial invasiveness varies greatly among pathogens. For example, *Clostridium tetani* (cause of tetanus) produces a variety of virulence factors (e.g., toxins and proteolytic enzymes) but is considered noninvasive because it does not spread from one tissue to another. *Bacillus anthracis* (cause of anthrax) and *Yersinia pestis* (cause of plague) also produce substantial virulence factors (e.g., capsules and toxins) but are highly invasive. Members of the genus *Streptococcus* span the spectrum of virulence factors and invasiveness.

Growth and Multiplication of the Bacterial Pathogen

For a bacterial pathogen to be successful in growth and reproduction (colonization), it must find an appropriate environment (e.g., nutrients, pH, temperature, redox potential) within the host. Those areas of the host's body that provide the most favorable conditions will harbor the pathogen and allow it to grow and multiply to produce an infection. Infectious bacteria are considered **extracellular pathogens** if, during the course of disease, they remain in tissues and fluids but never enter host cells. For instance, some bacteria can actively grow and multiply in the blood. The presence of viable bacteria in the bloodstream is called **bacteremia.** The infectious disease process caused by bacteria or their toxins in the blood is termed **septicemia** (Greek *septikos,* produced by putrefaction, and *haima,* blood). *Yersinia pestis* is a good example of an extremely virulent extracellular pathogen.

Some bacteria are able to grow and multiply within various cells of a host and are called **intracellular pathogens:** they can be further subdivided into two groups. **Facultative intracellular pathogens** are those organisms that can reside within the cells of the host or in the environment. An example is *Brucella abortus,* which is capable of growth and replication within macrophages, neutrophils, and trophoblast cells (cells that surround the developing embryo). However, facultative intracellular pathogens can also be grown in pure culture without host cell support. In contrast, **obligate intracellular pathogens** are incapable of growth and multiplication outside a host cell. Examples of obligate intracellular bacterial pathogens include *Chlamydia* and the rickettsia. These microbes cannot be grown in the laboratory outside of their host cells.

Leaving the Host

The last determinant of a successful bacterial pathogen is its ability to leave the host and enter either a new host or a reservoir. Unless a successful escape occurs, the disease cycle will be interrupted and the microorganism will not be perpetuated. Most bacteria employ passive escape mechanisms. Passive escape occurs when a pathogen or its progeny leave the host in feces, urine, droplets, saliva, or desquamated cells.

Regulation of Bacterial Virulence Factor Expression

Some pathogenic bacteria have adapted to both the free-living state and an environment within a human host. In the adaptive process, these pathogens have evolved complex signal transduction pathways to regulate the genetic expression of virulence factors only when in the host. Common factors that control virulence genes include temperature, osmolality, available iron,

Table 30.3	Microbial Products (Virulence Factors) Involved in Bacterial Pathogen Dissemination	
Product	**Organism Involved**	**Mechanism of Action**
Coagulase	*Staphylococcus aureus*	Coagulates (clots) fibrinogen. The clot protects the pathogen from phagocytosis and isolates it from other host defenses.
Collagenase	*Clostridium* spp.	Breaks down collagen that forms the framework of connective tissues; allows the pathogen to spread
Deoxyribonuclease (along with calcium and magnesium)	Group A streptococci, staphylococci, *Clostridium perfringens*	Lowers viscosity of exudates, giving the pathogen more mobility
Elastase and alkaline protease	*Pseudomonas aeruginosa*	Cleaves laminin associated with basement membranes
Hemolysins	Staphylococci, streptococci, *Escherichia coli*, *Clostridium perfringens*	Lyse erythrocytes; make iron available for microbial growth
Hyaluronidase	Groups A, B, C, and G streptococci, staphylococci, clostridia	Hydrolyzes hyaluronic acid, a constituent of the extracellular matrix that cements cells together and renders the intercellular spaces amenable to passage by the pathogen
Hydrogen peroxide (H_2O_2) and ammonia (NH_3)	*Mycoplasma* spp., *Ureaplasma* spp.	Are produced as metabolic wastes. These are toxic and damage epithelia in respiratory and urogenital systems
Immunoglobulin A protease	*Streptococcus pneumoniae*	Cleaves immunoglobulin A into Fab and Fc fragments
Lecithinase or phospholipase	*Clostridium* spp.	Destroys the lecithin (phosphatidylcholine) component of plasma membranes, allowing pathogen to spread
Leukocidins	Staphylococci, pneumococci, streptococci	Pore-forming exotoxins that kill leukocytes; cause degranulation of lysosomes within leukocytes, which decreases host resistance
Porins	*Salmonella enterica* serovar Typhimurium	Inhibit leukocyte phagocytosis by activating the adenylate cyclase system
Protein A Protein G	*Staphylococcus aureus* *Streptococcus pyogenes*	Located on cell wall. Immunoglobulin G (IgG) binds to either protein A or G by its Fc end, thereby preventing complement from interacting with bound IgG.
Pyrogenic exotoxin B (cysteine protease)	Group A streptococci, (*Streptococcus pyogenes*)	Degrades proteins
Streptokinase (fibrinolysin, staphylokinase)	Group A, C, and G streptococci, staphylococci	A protein that binds to plasminogen and activates the production of plasmin, thus digesting fibrin clots; this allows the pathogen to move from the clotted area

pH, specific ions, and nutrients. Interestingly, not all virulence factors are encoded by genes on the bacterial chromosome. Indeed some reside on lysogenic phage genomes (prophages) or plasmids. Several examples are now presented.

Corynebacterium diphtheriae causes diphtheria, and the gene for diphtheria toxin is carried on the temperate B phage. Thus the toxin is produced only by strains lysogenized by the phage. Toxin expression is regulated by iron. Expression of the virulence genes of *Bordetella pertussis,* which causes whooping cough, is enhanced when the bacteria grow at body temperature (37°C) and suppressed when grown at a lower temperature. Finally, the genes for cholera toxin, produced by *Vibrio cholerae,* are also carried on a temperate phage (CTX phage). Cholera toxin synthesis is regulated by many environmental factors; for instance, expression is higher at pH 6 than at pH 8 and higher at 30°C than at 37°C.

Pathogenicity Islands

The horizontal transfer of virulence genes to bacteria appears to have a rich evolutionary history, as made clear by the presence of genes that encode major virulence factors on large segments of bacterial chromosomal and plasmid DNA, called **pathogenicity islands.** Many bacteria (e.g., *Yersinia* spp., *Pseudomonas aeruginosa, Shigella flexneri, Salmonella,* enteropathogenic *Escherichia coli*) carry at least one pathogenicity island. Pathogenicity islands generally increase microbial virulence and are absent in nonpathogenic members of the same genus or species. Pathogenicity islands can be recognized by several common sequence characteristics: (1) The 3′ and 5′ ends of the islands contain insertion-like elements, suggesting their promiscuity as mobile genetic elements. (2) The G + C nucleotide content of pathogenicity islands

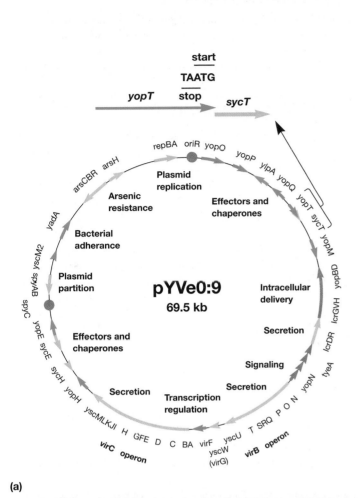

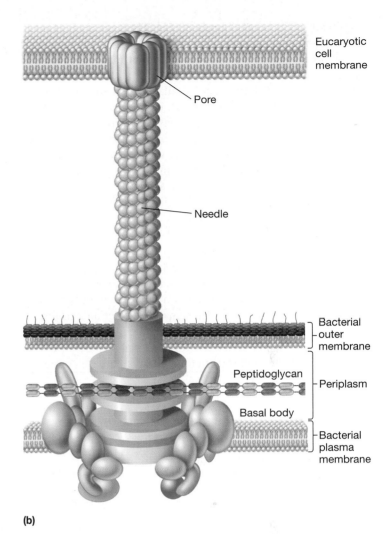

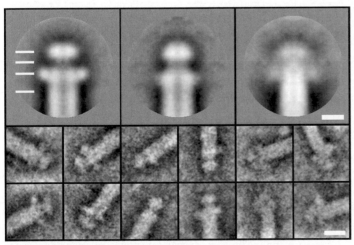

(c)

Figure 30.5   Type III Secretion System.   (a) The type III secretion system (TTSS) and other virulence genes of *Yersinia* are encoded on the pYV plasmid. The TTSS genes encoding the *Yersinia* outer proteins (Yop) are homologous to many of the genes encoding flagellar proteins. (b) X-ray fiber diffraction resolves the injectisome as a helical structure. (c) Scanning tunneling electron microscopy reveals the injectisome tip, indicating how it may lock into the translocator pore on the target cell.

differs significantly from that of the remaining bacterial genome. (3) The pathogenicity island DNA also exhibits several open reading frames, suggesting other putative genes. Interestingly, pathogenicity islands are typically associated with genes that encode tRNA. An excellent example of virulence genes carried on a pathogenicity island are those involved in protein secretion. So far, five pathways of protein secretion (types I to V) have been described in gram-negative bacteria. A set of approximately 25 genes encodes a **type III secretion system (TTSS)** that enables gram-negative bacteria to secrete and inject virulence proteins into the cytoplasm of eucaryotic host cells. ◄◄ *Protein maturation and secretion (section 12.8)*

Many gram-negative bacteria that live in close relationships with host organisms are able to modulate host activities by secreting proteins directly into the interior of the host cell using the TTSS. Perhaps the best studied TTSS is that of *Yersinia pestis* and *Y. enterocolitica*, which cause bubonic plague and gastroenteritis, respectively. Both bacteria use the same plasmid-encoded TTSS consisting of the Yop (*Yersinia* outer protein) secretion (Ysc) injectisome and secreted Yop products (**figure 30.5a**). The TTSS injectisome is composed of a basal body and a needle (figure 30.5b). The basal body is made from a number of proteins that are homologous (having similar amino acid sequences) to proteins that make up the basal body of bacterial flagella. This suggests that the injectisome is held in the bacterial envelope, similar to the ring system that holds a flagellum (see figure 3.36). The injectisome employs an ATPase that "energizes" the transport of other TTSS proteins, called "effectors," through "translocator pores" (formed by YopB and YopD proteins), into the host cell. X-ray fiber diffraction of TTSS of *Shigella* demonstrates that the needle component of the TTSS has a helical arrangement. Scanning tunneling electron microscopy of the injectisome needle reveals a characteristic tip (head, neck, and base) through which a central channel is seen (figure 30.5c).

The TTSS is triggered specifically by contact with host cells, which helps avoid inappropriate activation of host defenses. Secretion of these virulence proteins into a host cell allows the pathogen to subvert the host cell's normal signal transduction system. Redirection of cellular signal transduction can disarm host immune responses or reorganize the cytoskeleton, thus establishing subcellular niches for bacterial colonization and facilitating "stealth and interdiction" of host defense communication lines.

1. What are some ways in which bacterial pathogens are transmitted to their hosts? Define vector and fomite.
2. Describe several specific adhesins by which bacterial pathogens attach to host cells.
3. What are virulence factors?
4. If you were annotating the genomic sequence of a bacterium, how would you recognize a pathogenicity island?

30.4 TOXIGENICITY

Two distinct categories of disease can be recognized based on the role of the bacteria in the disease-causing process: infections and intoxications. Host damage in an **infection** results primarily from the pathogen's growth and reproduction (or invasiveness). **Intoxications** are diseases that result from a specific toxin (e.g., botulinum toxin) produced by bacteria. A **toxin** (Latin *toxicum*, poison) is a substance, such as a metabolic product of the organism, that alters the normal metabolism of host cells with deleterious effects on the host. The term **toxemia** refers to the condition caused by toxins that have entered the blood of the host. Some toxins are so potent that even if the bacteria that produced them are eliminated (e.g., by antibiotic chemotherapy), the disease conditions persist. Toxins produced by bacteria can be divided into two main categories: exotoxins and endotoxins. The primary characteristics of the two groups are compared in **table 30.4**.

Exotoxins

Exotoxins are soluble, heat-labile proteins that usually are released into the surroundings as the bacterial pathogen grows. Most exotoxins are produced by gram-positive bacteria, although some gram-negative bacteria also make exotoxins. Often exotoxins may

Table 30.4	Characteristics of Exotoxins and Endotoxins	
Characteristic	**Exotoxins**	**Endotoxins**
Chemical composition	Protein, often with two components (A and B)	Lipopolysaccharide complex on outer membrane; lipid A portion is toxic
Disease examples	Botulism, diphtheria, tetanus	Gram-negative infections, meningococcemia
Effect on host	Highly variable between different toxins	Similar for all endotoxins
Fever	Usually do not produce fever	Produce fever by induction of interleukin-1 and TNF
Genetics	Frequently carried by plasmids	Synthesized directly by chromosomal genes
Heat stability	Most are heat sensitive and inactivated at 60–80°C	Heat stable to 250°C
Immune response	Antitoxins provide host immunity; highly antigenic	Weakly immunogenic; immunogenicity associated with polysaccharide
Location	Usually excreted outside the living cell	Part of outer membrane of gram-negative bacteria
Production	Produced by gram-positive and negative bacteria	Found only in gram-negative bacteria; released on bacterial death and some liberated during growth
Toxicity	Highly toxic and fatal in nanogram quantities	Less potent and less specific than exotoxin; cause septic shock
Toxoid production	Converted to antigenic, nontoxic toxoids used to immunize (e.g., tetanus toxoid)	Toxoids cannot be made.

travel from the site of infection to other body tissues or target cells, where they exert their effects.

Exotoxins are usually synthesized by specific bacteria that often have plasmids or prophages bearing the toxin genes. They are associated with specific diseases and often are named for the disease they produce (e.g., the diphtheria toxin). Exotoxins are among the most lethal substances known; they are toxic in nanogram-per-kilogram concentrations (e.g., botulinum toxin) but are typically heat-labile (inactivated at 60 to 80°C). Exotoxins exert their biological activity by specific mechanisms. As proteins, the toxins are highly immunogenic and can stimulate the production of neutralizing antibodies called antitoxins. The toxins can also be inactivated by formaldehyde, iodine, and other chemicals to form immunogenic **toxoids** (e.g., tetanus toxoid). In fact, the tetanus vaccine is a solution of tetanus toxoid.

Exotoxins can be grouped into four types based on their structure and physiological activities. (1) One type is the AB toxin, which gets its name from the fact that the B portion of the toxin binds to a host cell receptor and is separate from the A portion, which enters the cell and has enzyme activity that causes the toxicity (**figure 30.6a**). (2) A second type, which also may be an AB toxin, consists of those toxins that affect a specific host site (nervous tissue [neurotoxins], the intestines [enterotoxins], general tissues [cytotoxins]). The general properties of some AB exotoxins are presented in (**table 30.5**). (3) A third type does not have separable A and B portions and acts by disorganizing host cell membranes. (4) Finally, exotoxins called superantigens act by stimulating T cells directly to release cytokines. Examples of these types are now discussed.

AB Toxins

AB toxins are composed of an enzymatic subunit (A) that is responsible for the toxic effect once inside the host cell and a binding subunit (B) (figure 30.6). Isolated A subunits are enzymatically active but lack binding and cell entry capability, whereas isolated B subunits

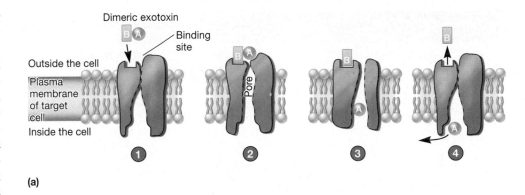

(a)

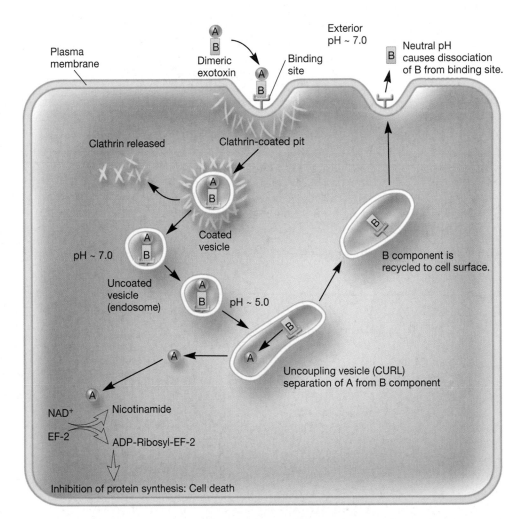

(b)

Figure 30.6   Two AB Exotoxin Transport Mechanisms.   (a) Subunit B of the dimeric exotoxin (AB) binds to a specific membrane receptor of a target cell [1]. A conformational change [2] generates a pore [3] through which the A subunit crosses the membrane and enters the cytosol, followed by recreation [4] of the binding site. (b) Receptor-mediated endocytosis of the diphtheria toxin involves the dimeric exotoxin binding to a receptor-ligand complex that is internalized in a clathrin-coated pit that pinches off to become a coated vesicle. The clathrin coat depolymerizes, resulting in an uncoated endosome vesicle. The pH in the endosome decreases due to the H^+-ATPase activity. The low pH causes A and B components to separate. An endosome in which this separation occurs is sometimes called a CURL (*c*ompartment of *u*ncoupling of *r*eceptor and *l*igand). The B subunit is then recycled to the cell surface. The A subunit moves through the cytosol, catalyzes the ADP-ribosylation of EF-2 (elongation factor 2) and inhibits protein synthesis, leading to cell death.

| Table 30.5 | Properties of Some AB Model Bacterial Exotoxins | | | | | |

Toxin	Organism	Gene Location	Subunit Structure	Target Cell Receptor	Enzymatic Activity	Biologic Effects
Anthrax toxins	*Bacillus anthracis*	Plasmid	Three separate proteins (EF, LF, PA)[a]	Capillary morphogenesis protein 2 (CMP-2) and tumor endothelium marker 8 (TEM8)	EF is a calmodulin-dependent adenylate cyclase; LF is a zinc-dependent protease that cleaves a host signal transduction molecule (MAPKK).	EF + PA: increase target cell cAMP level, localized edema; LF + PA: alter cell signaling; death of target cells
Bordetella adenylate cyclase toxin	*Bordetella* spp.	Chromosomal	A-B[b]	CR3 integrin (CD11–CD18)	Calmodulin-activated adenylate cyclase	Increase target cell cAMP level; decrease ATP production; modify cell function or cell death
Botulinum toxin	*Clostridium botulinum*	Phage	A-B[c]	Synaptic vesicle 2 (SV2)	Zinc-dependent endoprotease cleavage of presynaptic protein (SNARE)	Decrease in peripheral, presynaptic acetylcholine release; flaccid paralysis
Cholera toxin	*Vibro cholera*	Phage	A-5B[d]	Ganglioside (GM$_1$)	ADP ribosylation of adenylate cyclase regulatory protein, G$_S$	Activation of adenylate cyclase, increase cAMP level; secretory diarrhea
Diphtheria toxin	*Corynebacterium diphtheriae*	Phage	A-B[e]	Heparin-binding, EGF-like growth factor precursor	ADP ribosylation of elongation factor 2	Inhibit protein synthesis; cell death
Heat-labile enterotoxins[f]	*E. coli*	Plasmid	———————— Similar or Identical to Cholera Toxin ————————			
Pertussis toxin	*Bordetella pertussis*	Chromosomal	A-5B[g]	Asparagine-linked oligosaccharide and lactosylceramide sequences	ADP ribosylation of signal-transducing G proteins	Block signal transduction mediated by target G proteins
Pseudomonas exotoxin A	*P. aeruginosa*	Chromosomal	A-B	α$_2$-Macroglobulin/LDL receptor	—— Similar or Identical to Diphtheria Toxin ——	
Shiga toxin	*Shigella dysenteriae*	Chromosomal	A-5B[h]	Globotriaosylceramide (Gb$_3$)	RNA *N*-glycosidase	Inhibit protein synthesis, cell death
Shiga-like toxin 1	*Shigella* spp., *E. coli*	Phage	———————— Similar or Identical to Shiga Toxin ————————			
Tetanus toxin	*C. tetani*	Plasmid	A-B[c]	Ganglioside (GT$_1$ and/or GD$_{1b}$)	Zinc-dependent endopeptidase cleavage of synaptobrevin	Decrease neurotransmitter release from inhibitory neurons; spastic paralysis

Adapted from Mandell, G. L., et al., *Principles and Practice of Infectious Diseases,* 3d ed. Copyright © 1990 Churchill-Livingstone, Inc., Medical Publishers, New York, NY. Reprinted by permission.

[a] The binding component (known as protective antigen [PA]) catalyzes/facilitates the entry of either edema factor (EF) or lethal factor (LF).

[b] Apparently synthesized as a single polypeptide with binding and catalytic (adenylate cyclase) domains

[c] Holotoxin is apparently synthesized as a single polypeptide and cleaved proteolytically as diphtheria toxin; subunits are referred to as L: light chain, A equivalent; H: heavy chain, B equivalent.

[d] The A subunit is proteolytically cleaved into A1 and A2, with A1 possessing the ADP-ribosyl transferase activity; the binding component is made up of five identical B units.

[e] Holotoxin is synthesized as a single polypeptide and cleaved proteolytically into A and B components held together by disulfide bonds.

[f] The heat-labile enterotoxins of *E. coli* are now recognized to be a family of related molecules with identical mechanisms of action.

[g] The binding portion is made up of two dissimilar heterodimers labeled S2-S3 and S2-S4 that are held together by a bridging peptide, SS.

[h] Subunit composition and structure are similar to cholera toxin.

bind to target cells but are nontoxic and biologically inactive. The B subunit recognizes and binds to specific receptors on the target cell or tissue (table 30.5). The B subunit therefore determines what cell type the toxin will affect.

Several mechanisms for the entry of A subunits or fragments into target cells have been proposed. In one mechanism, the B subunit inserts into the plasma membrane and creates a pore through which the A subunit enters (figure 30.6a). In another mechanism, entry is by receptor-mediated endocytosis (figure 30.6b).

The mechanism of action of an AB toxin can be quite complex, as shown by the example of diphtheria toxin (figure 30.6b). This toxin is a protein of about 62,000 Daltons. Upon B subunit binding, it is taken into the cell through the formation of a clathrin-coated vesicle. The toxin then enters the vesicle membrane and the two subunits are separated. The A subunit escapes into the cytosol, where it catalyzes the addition of an ADP-ribose group to the eucaryotic elongation factor EF2 that aids in translocation during protein synthesis. The substrate for this reaction is the coenzyme NAD^+.

$$NAD^+ + EF2 \rightarrow ADP\text{-ribosyl-EF2} + nicotinamide$$

The modified EF2 protein can no longer participate in the elongation cycle of protein synthesis, so the cell dies. This A subunit activity is called **ADP-ribosylation,** and it is seen in the A subunit of a number of toxins; however, the specific host molecule to which the ADP-ribose group is attached differs.

A variation of this AB toxin is the **cytolethal distending toxin (CDT)** produced by *Campylobacter* spp. CDT is a tripartite complex encoded by three tandem genes; *cdtA, cdtB,* and *cdtC.* CDT binding and internalization appear to be encoded by the *cdtA* and *cdtC* genes, whereas the active component is encoded by the *cdtB* gene. The predicted amino acid sequence of CdtB is homologous to proteins having deoxyribonuclease (DNase) I activity. However, the mechanism of CDT in disease is unclear. *C. jejuni* has all three genes. In culture with epithelial cells, the CDT of *C. jejuni* induces a progressive epithelial cell distension resulting from an irreversible blockage of the cell cycle at the G2/M phase. This leads to oversized cells without cell division (distension) and cell death.

Specific Host Site Exotoxins

The second type of exotoxin is categorized on the basis of the site affected: neurotoxins (nerve tissue), enterotoxins (intestinal mucosa), and **cytotoxins** (general tissues). Some of the bacterial pathogens that produce these exotoxins are presented in table 30.5: neurotoxins (botulinum toxin and tetanus toxin), enterotoxins (cholera toxin, *E. coli* heat-labile toxins), and cytotoxins (diphtheria toxin, Shiga toxin). Note that many AB toxins are also host site specific, thus these categories are not mutually exclusive.

Neurotoxins usually are ingested as preformed toxins that affect the nervous system and indirectly cause enteric (pertaining to the small intestine) symptoms. Examples include staphylococcal enterotoxin B, *Bacillus cereus* emetic toxin (Greek *emetos,* vomiting), and botulinum toxin (*see figure 21.9*).

True **enterotoxins** (Greek *enter,* intestine) have a direct effect on the intestinal mucosa and elicit profuse diarrhea. Cholera tox-

in (choleragen), an AB toxin, has been studied extensively. The B subunit is made of five parts arranged as a donut-shaped ring. The B subunit ring anchors itself to the epithelial cell's plasma membrane and then inserts the smaller A subunit into the cell. The A subunit ADP-ribosylates and thereby activates tissue adenylate cyclase to increase intestinal cyclic AMP (cAMP) concentrations. High concentrations of cAMP provoke the movement of massive quantities of water and electrolytes from intestinal cells into the lumen of the gut. To maintain osmotic homeostasis, the cell then releases this water, resulting in severe diarrhea (cholera victims can lose 20% of their water per day). << *Class* Gammaproteobacteria: *Order* Vibrionales *(section 20.3)*

Membrane-Disrupting Exotoxins

The third type of exotoxin lyses host cells by disrupting the integrity of the plasma membrane. There are two subtypes of membrane-disrupting exotoxins. The first is a protein that binds to the cholesterol portion of the host cell plasma membrane, inserts itself into the membrane, and forms a channel (pore) (**figure 30.7a**). This causes the cytoplasmic contents to leak

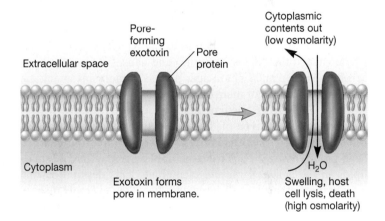

(a)

(b)

Figure 30.7 Two Subtypes of Membrane-Disrupting Exotoxins. (a) A channel-forming (pore-forming) type of exotoxin inserts itself into the normal host cell membrane and makes an open channel (pore). Formation of multiple pores causes cytoplasmic contents to leave the cell and water to move in, leading to cellular lysis and death of the host cell. (b) A phospholipid-hydrolyzing phospholipase exotoxin destroys membrane integrity. The exotoxin removes the charged polar head groups from the phospholipid part of the host cell membrane. This destabilizes the membrane and causes the host cell to lyse.

out. Also, because the osmolality of the cytoplasm is higher than that of the extracellular fluid, there is a sudden influx of water into the cell, causing it to swell and rupture. Two specific examples of this type of membrane-disrupting exotoxin are now presented.

Some pathogens produce membrane-disrupting toxins that kill phagocytic leukocytes. These are termed **leukocidins** (*leuko*cyte and Latin *caedere,* to kill). Most leukocidins are produced by pneumococci, streptococci, and staphylococci. Because these exotoxins destroy leukocytes, host resistance is decreased. Other toxins, called **hemolysins** (Greek *haima,* blood, and *lysis,* dissolution), also can be secreted by pathogenic bacteria. Many hemolysins probably form pores in the plasma membrane of erythrocytes through which hemoglobin or ions are released. Hemolysins may provide iron to the hemolysing pathogens; hemolysins are substantial virulence factors. Streptolysin-O (SLO) is a hemolysin produced by *Streptococcus pyogenes.* It is inactivated by O_2 (hence the "O" in its name). When blood agar plates are incubated anaerobically, SLO causes a complete zone of clearing around the bacterial colony. This is called β hemolysis, whereas a partial clearing of the blood (leaving a greenish halo of hemoglobin) is called α hemolysis. Streptolysin-S (SLS) is also produced by *S. pyogenes* but is insoluble and bound to the bacterial cell. It is O_2 stable (hence the "S" in its name) and causes β hemolysis on aerobically incubated blood-agar plates. Hemolysins attack the plasma membranes of many cells, not just erythrocytes. In fact, SLO and SLS are also leukocidins.

The second subtype of membrane-disrupting toxins are the **phospholipase** enzymes. Phospholipases remove the charged head group (figure 30.7*b*) from the lipid portion of the phospholipids in the host-cell plasma membrane. This destabilizes the membrane so that the cell lyses and dies. One example disease to which phospholipases contribute is gas gangrene caused by *Clostridium perfringens.* In this disease, the phospholipase activity of α-toxin almost completely destroys the local population of white blood cells that are drawn in by inflammation to fight the infection.

Superantigens

The mechanism by which superantigens provoke as many as 30% of a person's T cells to release cytokines is discussed in chapter 29 (*see p. 701*). The best-studied superantigen is the staphylococcal enterotoxin B (SEB), which also exhibits superantigen activity at nanogram concentrations. It is so potent, it is classified as a select agent because it has the potential to be misused as a bioterror agent. Like all superantigens, SEB exerts its activity by bridging the unfilled class II MHC molecules of antigen-presenting cells to T-cell receptors. Because no processed antigen is involved, many T cells are activated at once. This activation of T cells results in normal cytokine release; however, because so many T cells release cytokines, cells and tissues are overwhelmed. Cytokines stimulate endothelial dam-

age, circulatory shock, and multiorgan failure. ◀◀ *T-cell biology: Superantigens (section 29.5)*

Roles of Exotoxins in Disease

Humans are exposed to bacterial exotoxins in three main ways: (1) ingestion of preformed exotoxin, (2) bacterial colonization of a mucosal surface followed by exotoxin production, and (3) colonization of a wound or abscess followed by local exotoxin production. Each of these is now briefly discussed.

In the first way, the exotoxin is produced by bacteria growing in food. When food is consumed, the preformed exotoxin is also consumed. An example is staphylococcal food poisoning caused solely by the ingestion of preformed enterotoxin. Because the bacterium (*Staphylococcus aureus*) cannot colonize the gut, it passes through the body without producing any more exotoxin; thus, this type of bacterial disease is self-limiting.

In the second way, bacteria colonize a mucosal surface but do not invade underlying tissue or enter the bloodstream. The toxin either causes disease locally or enters the bloodstream and is distributed systemically where it can cause disease at distant sites. For example cholera caused by *Vibrio cholerae.* Once the bacteria enter the body, they adhere to the intestinal mucosa. They are not invasive but secrete the cholera toxin, causing prolific diarrhea.

The third example of exotoxins in disease pathogenesis occurs when bacteria grow in a wound or abscess. The exotoxin causes local tissue damage or kills phagocytes that enter the infected area. A disease of this type is gas gangrene in which the α-toxin of *Clostridium perfringens* lyses red and white blood cells, induces edema, and causes tissue destruction in the wound.

1. What is the difference between an infectious disease and an intoxication? Define toxemia.

2. Describe some general characteristics of exotoxins.

3. Describe the biological mechanisms and effects of several bacterial exotoxins.

4. What is the mode of action of a leukocidin? Of a hemolysin? Name the two types of hemolysins.

5. What are the three main roles exotoxins have in human disease pathogenesis?

Endotoxins

Gram-negative bacteria have **lipopolysaccharide (LPS)** in the outer membrane of their cell wall that, under certain circumstances, is toxic to specific hosts. This LPS is called an **endotoxin** because it is bound to the bacterium and is released when the microorganism lyses (**Techniques & Applications 30.1**). Some is also released during bacterial multiplication. The toxic component of the LPS is the lipid portion, called lipid A. Lipid A is not a single macromolecular structure but appears to be a complex array of lipid residues (*see figure 3.26*). Lipid A is heat

Techniques & Applications

30.1   Detection and Removal of Endotoxins

Bacterial endotoxins plagued the pharmaceutical industry and medical device producers for years. For example, administration of drugs contaminated with endotoxins resulted in complications—even death—to patients. In addition, endotoxins can be problematic for individuals and firms working with cell cultures and genetic engineering. The result has been the development of sensitive tests and methods to identify and remove these endotoxins. The procedures must be very sensitive to trace amounts of endotoxins. Most firms have set a limit of 0.25 endotoxin units (E.U.), 0.025 ng/ml, or less as a release standard for their drugs, media, or products.

One of the most accurate tests for endotoxins is the in vitro *Limulus* amoebocyte lysate (LAL) assay. The assay is based on the observation that when an endotoxin contacts the clot protein from circulating amoebocytes of the horseshoe crab (*Limulus*), a gel-clot forms. The assay kits contain calcium, proclotting enzyme, and procoagulogen. The proclotting enzyme is activated by bacterial endotoxin (lipopolysaccharide) and calcium to form

active clotting enzyme. Active clotting enzyme then catalyzes the cleavage of procoagulogen into polypeptide subunits (coagulogen). The subunits join by disulfide bonds to form a gel-clot. Spectrophotometry is then used to measure the protein precipitated by the lysate. The LAL test is sensitive at the nanogram level but must be standardized against U.S. Food and Drug Administration Bureau of Biologics endotoxin reference standards. Results are reported in endotoxin units per milliliter and reference made to the particular reference standards used.

Removal of endotoxins presents more of a problem than their detection. Those present on glassware or medical devices can be inactivated if the equipment is heated at 250°C for 30 minutes. Soluble endotoxins range in size from 20 kDa to large aggregates with diameters up to 0.1 μm. Thus they cannot be removed by conventional filtration systems. Manufacturers have developed special filtration systems and filtration cartridges that retain endotoxins to help alleviate contamination problems.

stable and toxic in nanogram amounts but only weakly immunogenic. << *Bacterial cell walls: Gram-negative cell walls (section 3.4)*

Unlike the structural and functional diversity of exotoxins, the lipid A of various gram-negatives produces similar systemic effects regardless of the microbe from which it is derived. These include fever (i.e., endotoxin is pyrogenic), shock, blood coagulation, weakness, diarrhea, inflammation, intestinal hemorrhage, and fibrinolysis (enzymatic breakdown of fibrin, the major protein component of blood clots).

The characteristics of endotoxins and exotoxins are contrasted in table 30.4. The main biological effect of lipid A is an indirect one, mediated by host molecules and systems rather than by lipid A itself. For example, endotoxins can initially activate Hageman Factor (blood clotting factor XII), which in turn activates up to four humoral systems: coagulation, complement, fibrinolytic, and kininogen systems. Endotoxins also indirectly induce a fever in the host by causing macrophages to release **endogenous pyrogens** that reset the hypothalamic thermostat. One important endogenous pyrogen is the cytokine interleukin-1 (IL-1). Other cytokines released by macrophages, such as the tumor necrosis factor, also produce fever. The net effect is often called **septic shock** and can also be induced by certain pathogenic fungi and gram-positive bacteria (**figure 30.8**). << *Chemical mediators in nonspecific (innate) resistance: Cytokines (section 28.6)*

Mycotoxins are protein toxins produced as secondary metabolites by fungi. For example, *Aspergillus flavus* and *A. parasiticus* produce aflatoxins, and *Stachybotrys* produces satratoxins, also known as trichothecene mycotoxins. These fungi commonly contaminate food crops and water-damaged buildings, respectively. An estimated 4.5 billion people in developing countries may be exposed chronically to aflatoxins through their diet. Exposure to

aflatoxins is known to cause both chronic and acute liver disease and liver cancer. Aflatoxins are extremely carcinogenic, mutagenic, and immunosuppressive. Approximately 18 different types of aflatoxins exist. Aflatoxins are difuranocoumarins and classified in two broad groups according to their chemical structure. Aflatoxins fluoresce strongly at 365 nm (ultraviolet light) appearing blue or green, depending on the aflatoxin chemistry. The *Stachybotrys* trichothecene mycotoxins are potent inhibitors of DNA, RNA, and protein synthesis. They induce inflammation, disrupt surfactant phospholipids in the lungs, and may lead to pathological changes in tissues.

The fungus *Claviceps purpurea* also produces toxic substances. The products are generically referred to as ergots, reflecting the name of the tuberlike structure of the fungi. The ergot is a fungal resting stage and is composed of a compact mass of hyphae. The ergots from various *Claviceps* spp. produce alkaloids that have varying physiological effects on humans. One such alkaloid is lysergic acid, a psychotropic hallucinogen. The ergot alkaloids have long been suspected as the cause of St. Anthony's fire in the eighth to sixteenth centuries in Europe and the hallucinations associated with the Salem witch trials of early American infamy.

1. Describe the chemical structure of the LPS endotoxin.

2. List some general characteristics of endotoxins.

3. How do gram-negative endotoxins induce fever in a mammalian host?

4. Describe some of the physiological effects associated with mycotoxin poisoning.

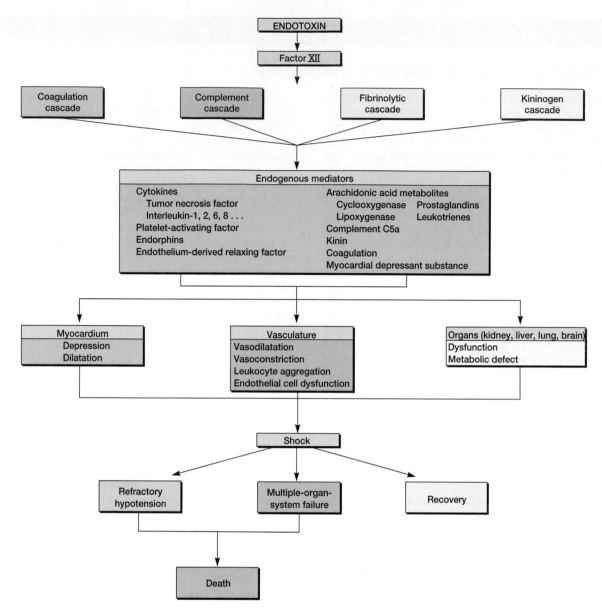

Figure 30.8 The Septic Shock Cascade. Gram-negative bacterial endotoxin triggers the biochemical events that lead to such serious complications as shock, adult respiratory distress syndrome, and disseminated intravascular coagulation.

30.5 POLYMICROBIAL DISEASES

It is now recognized that many infectious diseases involve the interactions of more than one infectious agent. These conditions are called **polymicrobial diseases.** These interactions work together to cause a specific pathology as well as induce a variety of host immune responses. Polymicrobial diseases can be polyviral (involving multiple virus species), polybacterial, combined viral and bacterial infections, polymycotic or protozoan (caused by more than one pathogenic fungi or protist, respectively), and diseases that arise from microbe-induced immunosuppression or inflammatory responses. Examples of polyviral conditions include coinfections with more than one type of hepatitis or retrovirus.

Polybacterial diseases include abscesses, vaginosis (vaginal infections), trauma induced infections, and necrotizing soft tissue infections. As we now discuss, dental plaque and periodontal disease are very common polymicrobial infections.

Dental Plaque

The human tooth has a natural defense mechanism against bacterial colonization that complements the protective role of saliva. The hard enamel surface selectively absorbs acidic glycoproteins (mucins) from saliva, forming a membranous layer called the acquired enamel pellicle. This pellicle is an organic covering with a net negative charge. Because most bacteria also have a net negative charge, a natural repulsion occurs between the tooth surface and bacteria in the oral cavity. Unfortunately, this natural defense mechanism breaks down when dental plaque forms.

Dental plaque formation begins with the initial colonization of the pellicle by *Streptococcus gordonii, S. oralis,* and *S. mitis* (**figure 30.9**). These bacteria selectively adhere to the pellicle by specific ionic, hydrophobic, and lectinlike interactions. Once the tooth surface is colonized, subsequent attachment of other bacteria results from a variety of specific coaggregation reactions (**figure 30.10**). Coaggregation is the result of cell-to-cell recognition between genetically distinct bacteria. Many of these interactions are mediated by a lectin on one bacterium that interacts with a complementary carbohydrate on another bacterium. The most important species at this stage are *Actinomyces viscosus, A. naeslundii,* and *Streptococcus gordonii.* After these species colonize the pellicle, a microenvironment is created that facilitates the attachment of *S. mutans* and *S. sobrinus* to these initial colonizers (**figure 30.11**). The polymicrobial environment within dental plaque is an excellent example of a biofilm. The succession of various organisms and their interdependence are

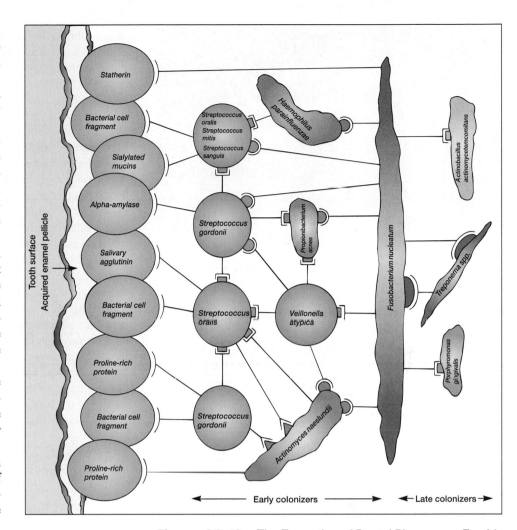

Figure 30.10   The Formation of Dental Plaque on a Freshly Cleaned Tooth Surface.   The proposed temporal relationship of bacterial accumulation and multigeneric coaggregation during the formation of dental plaque on the acquired enamel pellicle. Early tooth surface colonizers coaggregate with each other, and late colonizers coaggregate with each other. With a few exceptions, early colonizers do not recognize late colonizers. After the tooth surface is covered with the earliest colonizers, each newly added bacterium becomes a new surface for recognition by unattached bacteria.

pivotal features of a biofilm. ◄◄ *Microbial growth in natural environments: Biofilms (section 7.6)*

S. mutans and *S. sobrinus* produce extracellular enzymes that hydrolyze sucrose (table sugar) into fructose and glucose. The fructose is used in fermentation. Other exoenzymes called glucosyltransferases polymerize the glucose moiety into a heterogeneous group of extracellular, water-soluble glucan polymers and other polysaccharides. These act like cement to bind bacterial cells together, forming a plaque ecosystem. In fact, dental plaque is one of the densest collections of bacteria in the body. Once plaque becomes established, the surface of the tooth becomes anoxic. This leads to the growth of strictly anaerobic bacteria such as *Bacteroides melaninogenicus, B. oralis,* and *Veillonella alcalescens,* especially between opposing teeth and in the dental-gingival crevices (figure 30.11).

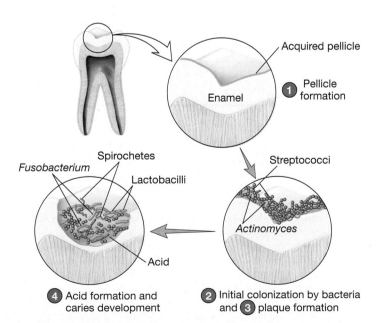

Figure 30.9   Stages in Plaque Development and Cariogenesis.   A microscopic view of pellicle and plaque formation, acidification, and destruction of tooth enamel.

Figure 30.11 The Microscopic Appearance of Plaque. Scanning electron micrograph of plaque with long filamentous forms and "corn cobs" that are mixed bacterial aggregates.

After the microbial plaque ecosystem develops, bacteria produce lactic and possibly acetic and formic acids from sucrose and other sugars. Because plaque is not permeable to saliva, the acids are not diluted or neutralized, and they demineralize the enamel to produce a lesion on the tooth. It is this chemical lesion that initiates dental decay.

Periodontal Disease

Periodontal disease refers to a diverse group of inflammatory diseases that affect the periodontium and is the most common chronic infection in adults. The periodontium is the supporting structure of a tooth and includes the cementum, the periodontal membrane, the bones of the jaw, and the gingivae (gums). The gingiva is dense fibrous tissue and its overlying mucous membrane that surrounds the necks of the teeth. The gingiva helps to hold the teeth in place. Disease is initiated by the formation of subgingival plaque, the plaque that forms at the dentogingival margin and extends down into the gingival tissue. Colonization of the subgingival region is aided by the ability of *Porphyromonas gingivalis* to adhere to substrates such as adsorbed salivary molecules, matrix proteins, epithelial cells, and bacteria in biofilms on teeth and epithelial surfaces. Binding to these substrates is mediated by *P. gingivalis* fimbrillin, the structural subunit of the major fimbriae. *P. gingivalis* does not use sugars as an energy source but requires hemin as a source of iron and peptides for energy and growth. The bacterium produces at least three hemagglutinins and five proteases to satisfy these requirements. It is the proteases that are responsible for the breakdown of the gingival tissue. A number of other bacterial species contribute to tissue damage. The result is an initial inflammatory reaction known as periodontitis, which is caused by the host's immune response to both the plaque bacteria and the tissue destruction. This leads to swelling of the tissue and the formation of periodontal pockets. Bacteria colonize these pockets and cause more inflammation, which leads to the formation of a periodontal abscess; bone destruction, or periodontosis; inflammation of the gingiva, or gingivitis; and general tissue necrosis. If the condition is not treated, the tooth may fall out of its socket. ◄◄ *Phylum* Bacteroidetes *(section 19.7)*

1. What is a polymicrobial disease? Can you think of a polymicrobial disease not listed or discussed here?

2. How do odontopathogens interact to cause dental plaque? Could dental plaque be caused be any one of these microbes alone?

3. How does the host contribute to periodontal disease?

Summary

30.1 Host-Parasite Relationships

a. Parasitism is a type of symbiosis between two species in which the smaller organism is physiologically dependent on the larger one, termed the host. The parasitic organism usually harms its host in some way.

b. An infection is the colonization of the host by a parasitic organism. An infectious disease is the result of the interaction between the parasitic organism and its host, causing the host to change from a state of health to one of disease. Any organism that produces such a disease is a pathogen.

c. Pathogenicity refers to the quality or ability of an organism to produce pathological changes or disease. Virulence refers to the degree or intensity of pathogenicity of an organism and is measured experimentally by the LD_{50} or ID_{50} (**figure 31.2**).

30.2 Pathogenesis of Viral Diseases

a. The fundamental process of viral infection is the expression of the viral replicative cycle in a host cell. To produce disease, a virus enters a host, comes into contact with susceptible cells, and reproduces.

b. Viruses spread to adjacent cells when they are released by either host cell lysis or budding.

c. Host cell damage caused by viruses stimulates a host immune response involving neutralizing antibodies for free virions and activated killer cells for intracellular viruses. Many viruses have evolved mechanisms to evade these host immune responses.

d. Recovery from infection results when the virus has either been cleared from the body of the host, establishes a persistent infection, or kills the host.

e. The viral infection cycle is complete when the virus is shed back into the environment to be acquired by another host.

30.3 Overview of Bacterial Pathogenesis

a. Pathogens or their products can be transmitted to a host by either direct or indirect means. Transmissibility is the initial requisite in the establishment of an infectious disease.

b. Special adherence factors allow pathogens to bind to specific receptor sites on host cells and colonize the host (**table 30.2 and figure 30.3**).

c. Pathogens can enter host cells by both active and passive mechanisms. Once inside, they can produce specific products or enzymes that promote dissemination throughout the body of the host. These are termed virulence factors (**table 30.3**).

d. The pathogen generally is found in the area of the host's body that provides the most favorable conditions for its growth and multiplication.

e. During coevolution with human hosts, some pathogenic bacteria have evolved complex signal transduction pathways to regulate the genes necessary for virulence.

f. Many bacteria are pathogenic because they have large segments of DNA called pathogenicity islands that carry genes responsible for virulence.

30.4 Toxigenicity

a. Intoxications are diseases that result from the entrance of a specific toxin into a host. The toxin can induce the disease in the absence of the toxin-producing organism. Toxins produced by pathogens can be divided into two main categories: exotoxins and endotoxins (**table 30.4**).

b. Exotoxins are soluble, heat-labile, potent, toxic proteins produced by the pathogen. They can be divided into four types: (1) the AB toxins, (2) specific host site toxins (neurotoxins, enterotoxins, cytotoxins), (3) toxins that disrupt plasma membranes of host cells (leukocidins, hemolysins, and phospholipases), and (4) superantigens.

c. Most exotoxins conform to the AB model in which the A subunit is enzymatic and the B subunit is the binding portion (**table 30.5**). Several mechanisms exist by which the A component enters target cells (**figure 30.6**).

d. Bacterial exotoxins cause disease in a human host in three main ways: (1) ingestion of preformed exotoxin, (2) colonization of a mucosal surface followed by exotoxin production, and (3) colonization of a wound followed by local exotoxin production.

e. Endotoxins are heat-stable, toxic substances that are part of the cell wall lipopolysaccharide of gram-negative bacteria. Most endotoxins function by initially activating Hageman Factor, which in turn activates one to four humoral systems.

30.5 Polymicrobial Diseases

a. Some infectious diseases involve multiple pathogenic microbes that interact to elicit specific symptoms.

b. Dental infections are well-studied and common polymicrobial diseases.

c. Dental plaque formation begins on a tooth with the initial colonization of the acquired enamel pellicle by three streptococcal species. Other bacteria then become attached and form a plaque ecosystem (**figure 30.9**). The bacteria produce acids that cause a chemical lesion on the tooth and initiate dental decay or caries.

d. Periodontal disease is a group of diverse clinical entities that affect the periodontium. Disease is initiated by the formation of subgingival plaque, which leads to tissue inflammation known as periodontitis and to periodontal pockets. Bacteria that colonize these pockets can cause an abscess, periodontosis, gingivitis, and general tissue necrosis (**figures 30.10 and 30.11**).

Critical Thinking Questions

1. Why does a parasitic organism not have to be a parasite?

2. In general, infectious diseases that are commonly fatal are newly evolved relationships between the parasitic organism and the host. Why is this so?

3. Explain the observation that different pathogens infect different parts of the host.

4. Intracellular bacterial infections present a particular difficulty for the host. Why is it harder to defend against these infections than against viral infections and extracellular bacterial infections?

Learn More

Learn more by visiting the Prescott website at www.mhhe.com/prescottprinciples, where you will find a complete list of references.

Antimicrobial Chemotherapy

31

Chapter Glossary

aminoglycoside antibiotic   An antibiotic that contains a cyclohexane ring and amino sugars; this antibiotic binds to the small ribosomal subunit and inhibits protein synthesis.

antimetabolite   A compound that blocks metabolic pathway function by competitively inhibiting a key enzyme's use of a metabolite because it closely resembles the normal enzyme substrate.

β-lactam   Refers to antibiotics containing a β-lactam ring that inhibit bacterial cell wall synthesis. Includes the penicillins and cephalosporins.

β-lactam ring   The cyclic chemical structure composed of three carbon atoms and one nitrogen atom.

β-lactamase or **penicillinase**   An enzyme that hydrolyzes the β-lactam ring, rendering the antibiotic inactive.

broad-spectrum drugs   Chemotherapeutic agents that are effective against many different kinds of pathogens.

chemotherapeutic agent   Compounds used in the treatment of disease that destroy pathogens or inhibit their growth at concentrations low enough to avoid doing undesirable damage to the host. Includes **antibiotics,** which are natural microbial products, and **antimicrobials,** which may be naturally or chemically synthesized.

cidal   Having the capacity to cause cell death.

dilution susceptibility test   A method by which antibiotics are evaluated for their ability to inhibit bacterial growth in vitro. Bacteria are added to serially diluted antibiotics and incubated. Tubes in which bacteria fail to grow suggest antibiotic concentrations that are bacteriocidal or bacteriostatic.

integron   A genetic element with an attachment site for site-specific recombination and an integrase gene. It can capture genes and gene cassettes.

Kirby-Bauer method   A disk diffusion test to determine the susceptibility of a microorganism to chemotherapeutic agents.

macrolide antibiotic   An antibiotic containing a macrolide ring, a large lactone ring with multiple keto and hydroxyl groups, linked to one or more sugars; inhibits bacterial growth by binding to the 50S ribsomal subunit, thereby blocking protein synthesis.

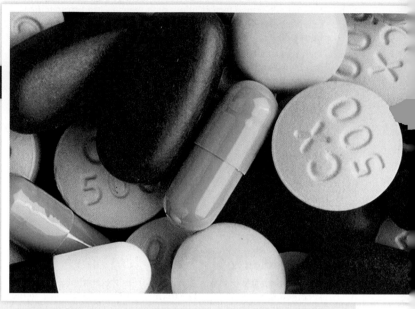

Many antimicrobial medications are available to combat infections. Nonetheless, they fall into a limited number of classes based on their modes of action.

minimal inhibitory concentration (MIC)   The lowest concentration of a drug that prevents the growth of a particular microorganism.

minimal lethal concentration (MLC)   The lowest concentration of a drug that kills a particular microorganism.

narrow-spectrum drugs   Chemotherapeutic agents that are effective only against a limited variety of microorganisms.

quinolones   A class of broad-spectrum antibiotics, derived from nalidixic acid, that bind to bacterial DNA gyrase, inhibiting DNA replication and transcription. This group of antibiotics is bacteriocidal.

R plasmids   Plasmids bearing one or more drug-resistant genes.

static   Having the capacity to limit microbial growth but not cause microbial death.

tetracyclines   A family of antibiotics with a common four-ring structure that are isolated from the genus *Streptomyces* or produced semisynthetically; all are related to chlortetracycline or oxytetracycline.

therapeutic index   The ratio between the toxic dose and the therapeutic dose of a drug, used as a measure of the drug's relative safety.

It was the knowledge of the great abundance and wide distribution of actinomycetes, which dated back nearly three decades, and the recognition of the marked activity of this group of organisms against other organisms that led me in 1939 to undertake a systematic study of their ability to produce antibiotics.

—Selman A. Waksman

Modern medicine depends on chemotherapeutic agents—chemical agents that are used to treat disease. Ideally, chemotherapeutic agents used to treat infectious disease destroy pathogenic microorganisms or inhibit their growth at concentrations low enough to avoid undesirable damage to the host. Most of these agents are **antibiotics** (Greek *anti,* against, and *bios,* life), microbial products or their derivatives that can kill susceptible microorganisms or inhibit their growth. Drugs such as the sulfonamides are sometimes called antibiotics although they are synthetic chemotherapeutic agents, not microbially synthesized. This chapter introduces the principles of **antimicrobial** chemotherapy and briefly reviews the characteristics of selected antibacterial, antifungal, antiprotozoan, and antiviral drugs.

31.1 THE DEVELOPMENT OF CHEMOTHERAPY

The modern era of chemotherapy began with the work of the German physician **Paul Ehrlich** (1854–1915). Ehrlich was fascinated with dyes that specifically bind to and stain microbial cells. He reasoned that one of the dyes could be a chemical that would selectively destroy pathogens without harming human cells—a "magic bullet." By 1904 Ehrlich found that the dye trypan red was active against the trypanosome that causes African sleeping sickness (*see figure 23.5*) and could be used therapeutically. Subsequently Ehrlich and a young Japanese scientist named Sahachiro Hata tested a variety of arsenicals on syphilis-infected rabbits and found that arsphenamine was active against the syphilis spirochete. Arsphenamine was made available in 1910 under the trade name Salvarsan and paved the way to the testing of hundreds of compounds for their selective toxicity and therapeutic potential.

In 1927 the German chemical industry giant I. G. Farbenindustrie began a long-term search for chemotherapeutic agents under the direction of **Gerhard Domagk.** Domagk had screened a vast number of chemicals for other "magic bullets" and discovered that Prontosil Red, a new dye for staining leather, protected mice completely against pathogenic streptococci and staphylococci without apparent toxicity. Jacques and Therese Trefouel later showed that the body metabolized the dye to sulfanilamide. Domagk received the 1939 Nobel Prize in Physiology or Medicine for his discovery of sulfonamides, or sulfa drugs.

Penicillin, the first true antibiotic, was initially discovered in 1896 by a twenty-one-year-old French medical student named Ernest Duchesne. His work was forgotten until **Alexander Fleming** accidentally rediscovered penicillin in September 1928. After returning from a weekend vacation, Fleming noticed that a petri plate of *Staphylococcus* also had mold growing on it and there were no bacterial colonies surrounding it (**figure 31.1**). Although the precise events are still unclear, it has been suggested that a *Penicillium notatum* spore had contaminated the petri dish before it had been inoculated with the staphylococci. The mold apparently grew before the bacteria and produced penicillin. The bacteria nearest the fungus were lysed. Fleming correctly deduced that the mold produced a diffusible substance, which he called penicillin. Unfortunately, Fleming could not demonstrate

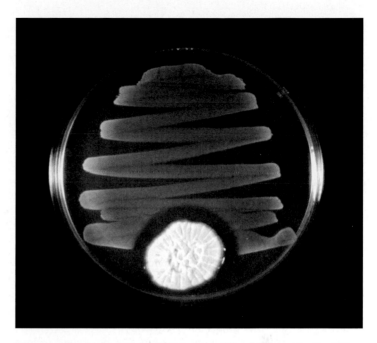

Figure 31.1 Bacteriocidal Action of Penicillin. The *Penicillium* mold colony secretes penicillin that kills *Staphylococcus aureus* that was streaked nearby.

that penicillin remained active in vivo long enough to destroy pathogens and thus dropped the research.

In 1939 **Howard Florey,** a professor of pathology at Oxford University, was in the midst of testing the bactericidal activity of many substances. After reading Fleming's paper on penicillin, one of Florey's coworkers, **Ernst Chain,** obtained the *Penicillium* culture from Fleming and set about purifying the antibiotic. **Norman Heatley,** a biochemist, was enlisted to help. He devised the original assay, culture, and purification techniques needed to produce crude penicillin for further experimentation. When purified penicillin was injected into mice infected with streptococci or staphylococci, almost all the mice survived. Florey and Chain's success was reported in 1940, and subsequent human trials were equally successful. Fleming, Florey, and Chain received the Nobel Prize in 1945 for the discovery and production of penicillin.

The discovery of penicillin stimulated the search for other antibiotics. **Selman Waksman,** while at Rutgers University, announced in 1944 that he and his associates had found a new antibiotic, streptomycin, produced by the actinomycete *Streptomyces griseus.* This discovery arose from the careful screening of about 10,000 strains of soil bacteria and fungi. The importance of streptomycin cannot be understated, as it was the first drug that could successfully treat tuberculosis. Waksman received the Nobel Prize in 1952, and his success led to a worldwide search for other antibiotic-producing soil microorganisms. Microorganisms producing chloramphenicol, neomycin, terramycin, and tetracycline were isolated by 1953.

The discovery of chemotherapeutic agents and the development of newer, more powerful drugs has transformed modern medicine and greatly alleviated human suffering. Furthermore, antibiotics have proven exceptionally useful in microbiological research.

1. What are chemotherapeutic agents? Antibiotics?
2. Pasteur is often credited as saying, "Chance favors the prepared mind." How does this apply to Fleming?

31.2 GENERAL CHARACTERISTICS OF ANTIMICROBIAL DRUGS

As Ehrlich so clearly saw, to be successful a **chemotherapeutic agent** must have **selective toxicity:** it must kill or inhibit the microbial pathogen while damaging the host as little as possible. The degree of selective toxicity may be expressed in terms of (1) the therapeutic dose—the drug level required for clinical treatment of a particular infection, and (2) the toxic dose—the drug level at which the agent becomes too toxic for the host. The **therapeutic index** is the ratio of the toxic dose to the therapeutic dose. The larger the therapeutic index, the better the chemotherapeutic agent (all other things being equal).

A drug that disrupts a microbial function not found in host animal cells often has a greater selective toxicity and a higher therapeutic index. For example, penicillin inhibits bacterial cell wall peptidoglycan synthesis but has little effect on host cells because they lack cell walls; therefore penicillin's therapeutic index is high. A drug may have a low therapeutic index because it inhibits the same process in host cells or damages the host in other ways. The undesirable effects on the host, or side effects, are of many kinds and may involve almost any organ system (**table 31.1**). Because side effects can be severe, chemotherapeutic agents should be administered with great care.

Some bacteria and fungi are able to naturally produce many of the commonly employed antibiotics. In contrast, several important chemotherapeutic agents, such as sulfonamides, trimethoprim, ciprofloxacin, isoniazid, and dapsone, are synthetic—manufactured by chemical procedures independent of microbial activity. Some antibiotics are semisynthetic—natural antibiotics that have been structurally modified by the addition of chemical groups to make them less susceptible to inactivation by pathogens (e.g., ampicillin and methicillin). In addition, many semisynthetic drugs have a broader spectrum of antibiotic activity than does their parent molecule. This is particularly true of the semisynthetic penicillins (e.g., ampicillin, amoxycillin) versus the naturally produced penicillin G and penicillin V.

Drugs vary considerably in their range of effectiveness. Many are **narrow-spectrum drugs**—that is, they are effective only against a limited variety of pathogens (table 30.1). Others are **broad-spectrum drugs** that attack many different kinds of pathogens. Drugs may also be classified based on the general microbial group they act against: antibacterial, antifungal, antiprotozoan, and antiviral. A few agents can be used against more than one group; for example, sulfonamides are active against bacteria and some protozoa. Finally, chemotherapeutic agents can be either **cidal** or **static.** Static agents reversibly inhibit growth; if the agent is removed, the microorganisms will recover and grow again. ◄◄ *Growth curve: Senescence and death (section 7.2)*

Although a cidal agent kills the target pathogen, it may be static at low levels. The effect of an agent also varies with the target species: an agent may be cidal for one species and static for another. Because static agents do not directly destroy the pathogen, elimination of the infection depends on the host's own immunity mechanisms. A static agent may not be effective if the host is immunosuppressed.

Some idea of the effectiveness of a chemotherapeutic agent against a pathogen can be obtained from the **minimal inhibitory concentration (MIC).** The MIC is the lowest concentration of a drug that prevents growth of a particular pathogen. On the other hand, the **minimal lethal concentration (MLC)** is the lowest drug concentration that kills the pathogen. A cidal drug generally kills pathogens at levels only two to four times the MIC, whereas a static agent kills at much higher concentrations, if at all.

1. Define the following: selective toxicity, therapeutic index, side effect, narrow-spectrum drug, broad-spectrum drug, synthetic and semisynthetic antibiotics, cidal and static agents, minimal inhibitory concentration, and minimal lethal concentration.
2. How do semisynthetic antibiotics commonly differ from their parent molecules?
3. Use the MIC and MLC concepts to distinguish between cidal and static agents.

31.3 DETERMINING THE LEVEL OF ANTIMICROBIAL ACTIVITY

Determination of antimicrobial effectiveness against specific pathogens is essential for proper therapy. Testing can show which agents are most effective against a pathogen and give an estimate of the proper therapeutic dose.

Dilution Susceptibility Tests

Dilution susceptibility tests can be used to determine MIC and MLC values. Antibiotic dilution tests can be done in both agar and broth. In the broth dilution test, a series of broth tubes (usually Mueller-Hinton broth) containing antibiotic concentrations in the range of 0.1 to 128 μg/ml (two-fold dilutions) is prepared and inoculated with a standard density of the test organism. The lowest concentration of the antibiotic resulting in no growth after 16 to 20 hours of incubation is the MIC. The MLC can be ascertained if the tubes showing no growth are then cultured into fresh medium lacking antibiotic. The lowest antibiotic concentration from which the microorganisms do not grow when transferred to fresh medium is the MLC. The agar dilution test is very similar to the broth dilution test. Plates containing Mueller-Hinton agar and various amounts of antibiotic are inoculated and examined for growth. Several automated systems for susceptibility testing and MIC determination with broth or agar cultures have been developed.

Table 31.1		Properties of Some Common Antibacterial Drugs			
Antibiotic Group	**Primary Effect**	**Mechanism of Action**	**Members**	**Spectrum**	**Common Side Effects**
Cell Wall Synthesis Inhibition					
Penicillins	Cidal	Inhibit transpeptidation enzymes involved in cross-linking the polysaccharide chains of the bacterial cell wall peptidoglycan Activate cell wall lytic enzymes	Penicillin G, penicillin V, methicillin Ampicillin, carbenicillin	Narrow (gram-positive) Broad (gram-positive, some gram-negative)	Allergic responses (diarrhea, anemia, hives, nausea, renal toxicity)
Cephalosporins	Cidal	Same as above	Cephalothin, cefoxitin, cefaperazone, ceftriaxone	Broad (gram-positive, some gram-negative)	Allergic responses, thrombophlebitis, renal injury
Vancomycin	Cidal	Prevent transpeptidation of peptidoglycan subunits by binding to D-Ala-D-Ala amino acids at the end of peptide cross-bridges. Thus it has a different binding site than that of the penicillins.	Vancomycin	Narrow (gram-positive)	Ototoxic (tinnitus and deafness), nephrotoxic, allergic reactions
Protein Synthesis Inhibition					
Aminoglycosides	Cidal	Bind to small ribosomal subunit (30S) and interfere with protein synthesis by directly inhibiting synthesis and causing misreading of mRNA	Neomycin, kanamycin, gentamicin Streptomycin	Broad (gram-negative, mycobacteria) Narrow (aerobic gram-negative)	Ototoxic, renal damage, loss of balance, nausea, allergic responses
Tetracyclines	Static	Same as above	Oxytetracycline, chlortetracycline	Broad (including rickettsia and chlamydia)	Gastrointestinal upset, teeth discoloration, renal and hepatic injury
Macrolides	Static	Bind to 23S rRNA of large ribosomal subunit (50S) to inhibit peptide chain elongation during protein synthesis	Erythromycin, clindamycin	Broad (aerobic and anaerobic gram-positive, some gram-negative)	Gastrointestinal upset, hepatic injury, anemia, allergic responses
Chloramphenicol	Static	Same as above	Chloramphenicol	Broad (gram-positive and -negative, rickettsia and chlamydia	Depressed bone marrow function, allergic reactions
Nucleic Acid Synthesis Inhibition					
Quinolones and Fluoroquinolones	Cidal	Inhibit DNA gyrase and topoi-somerase II, thereby blocking DNA replication	Norfloxacin, ciprofloxacin, Levofloxacin	Narrow (gram-negatives better than gram-positives) Broad spectrum	Tendonitis, headache, lightheadedness, convulsions, allergic reactions
Rifampin	Cidal	Inhibits bacterial DNA-dependent RNA polymerase	R-Cin, rifacilin, rifamycin, rimactane, rimpin, siticox	*Mycobacterium* infections and some gram-negative such as *Neisseria meningitidis* and *Haemophilus influenzae* b	Nausea, vomiting, di-arrhea, fatigue, anemia, drowsiness, headache, mouth ulceration, liver damage
Cell Membrane Disruption					
Polymyxin B	Cidal	Bind to plasma membrane and disrupts its structure and perme-ability properties	Polymyxin B, polymyxin topical ointment	Narrow—mycobacterial infections, principally leprosy	Can cause severe kidney damage, drowsiness, dizziness

(continued)

Table 31.1		Properties of Some Common Antibacterial Drugs (continued)			
Antibiotic Group	Primary Effect	Mechanism of Action	Members	Spectrum	Common Side Effects
Antimetabolites					
Sulfonamides	Static	Inhibit folic acid synthesis by competing with *p*-aminobenzoic acid (PABA)	Silver sulfadiazine, sodium sulfacetamide, sulfamethoxazole, sulfanilamide, sulfasalazine, sulfisoxazole	Broad spectrum	Nausea, vomiting, and diarrhea; hypersensitivity reactions such as rashes, photosensitivity
Trimethoprim	Static	Block folic acid synthesis by inhibiting the enzyme tetrahydrofolate reductase	Trimethoprim (in combination with a sulfamethoxazole [1:5])	Broad spectrum	Same as sulfonamides, but less frequent
Dapsone	Static	Thought to interfere with folic acid synthesis	Dapsone	Narrow—mycobacterial infections, principally leprosy	Back, leg, or stomach pains; discolored fingernails, lips, or skin; breathing difficulties; fever; loss of appetite; skin rash; fatigue
Isoniazid	Cidal if bacteria are actively growing, static if bacteria are dormant	Exact mechanism is unclear, but it is thought to inhibit lipid synthesis (especially mycolic acid); putative enoyl-reductase inhibitor	Isoniazid	Narrow—mycobacterial infections, principally tuberculosis	Nausea, vomiting, liver damage, seizures, "pins and needles" in extremities (peripheral neuropathy)

Disk Diffusion Tests

If a rapidly growing microbe such as *Staphylococcus* or *Pseudomonas* is being tested, a disk diffusion technique may be used to save time and media. The principle behind the assay technique is fairly simple. When an antibiotic-impregnated disk is placed on agar previously inoculated with the test bacterium, the antibiotic diffuses radially outward through the agar, producing an antibiotic concentration gradient. The antibiotic is present at high concentrations near the disk and affects even minimally susceptible microorganisms (resistant organisms will grow up to the disk). As the distance from the disk increases, the antibiotic concentration decreases and only more susceptible pathogens are harmed. A clear zone or ring is present around an antibiotic disk after incubation if the agent inhibits bacterial growth. The wider the zone surrounding a disk, the more susceptible the pathogen is. Zone width also is a function of the antibiotic's initial concentration, its solubility, and its diffusion rate through agar. Thus zone width cannot be used to compare directly the effectiveness of different antibiotics.

Currently the disk diffusion test most often used is the **Kirby-Bauer method,** which was developed in the early 1960s at the University of Washington Medical School by William Kirby, A. W. Bauer, and their colleagues. Freshly grown bacteria are used to inoculate the entire surface of a Mueller-Hinton agar plate. After the agar surface has dried for about 5 minutes, the appropriate antibiotic test disks are placed on it, either with sterilized forceps or with a multiple applicator device (**figure 31.2**). The plate is immediately placed at 35°C. After 16 to 18 hours of incubation, the diameters of the zones of inhibition are measured to the nearest millimeter.

Kirby-Bauer test results are interpreted using a table that relates zone diameter to the degree of microbial resistance (**table 31.2**). The values in table 31.2 were derived by finding the MIC values and zone diameters for many different microbial strains. A plot of MIC (on a logarithmic scale) versus zone inhibition diameter (arithmetic scale) is prepared for each antibiotic (**figure 31.3**). These plots are then used to find the zone diameters corresponding to the drug concentrations actually reached in the body. If the zone diameter for the lowest level reached in the body is smaller than that seen with the test pathogen, the pathogen should have an MIC value low enough to be destroyed by the drug. A pathogen with too high an MIC value (too small a zone diameter) is resistant to the agent at normal body concentrations.

The Etest®

The Etest® from AB BIODISK may be used in sensitivity testing under some conditions. It is particularly convenient for use with anaerobic pathogens. Agar medium is streaked in three different directions with the test organism and plastic Etest® strips

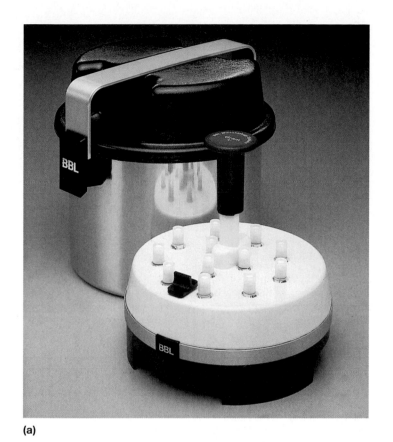

(a)

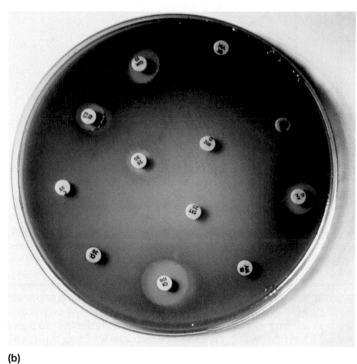

(b)

Figure 31.2   The Kirby-Bauer Method.   (a) A multiple antibiotic disk dispenser and (b) disk diffusion test results.

are placed on the surface so that they extend out radially from the center (**figure 31.4**). Each strip contains a gradient of an antibiotic and is labeled with a scale of MIC values. The lowest concentration in the strip lies at the center of the plate. After 24 to 48 hours of incubation, an elliptical zone of inhibition appears. As shown in figure 31.4, MICs are determined from the point of intersection between the inhibition zone and the strip's scale of MIC values.

1. How can dilution susceptibility tests and disk diffusion tests be used to determine microbial drug sensitivity?
2. Briefly describe the Kirby-Bauer test and its purpose.
3. What would you surmise if you examined a Kirby-Bauer assay and found individual colonies of the plated microbe growing within the zone of inhibition?
4. How is the Etest® carried out?

Table 31.2	Inhibition Zone Diameter of Selected Chemotherapeutic Drugs			
		Zone Diameter (Nearest mm)		
Chemotherapeutic Drug	Disk Content	Resistant	Intermediate	Susceptible
Carbenicillin (with *Proteus* spp. and *E. coli*)	100 μg	≤17	18–22	≥23
Carbenicillin (with *Pseudomonas aeruginosa*)	100 μg	≤13	14–16	≥17
Erythromycin	15 μg	≤13	14–17	≥18
Penicillin G (with staphylococci)	10 U[a]	≤20	21–28	≥29
Penicillin G (with other microorganisms)	10 U	≤11	12–21	≥22
Streptomycin	10 μg	≤11	12–14	≥15
Sulfonamides	250 or 300 μg	≤12	13–16	≥17

[a]One milligram of penicillin G sodium = 1,600 units (U).

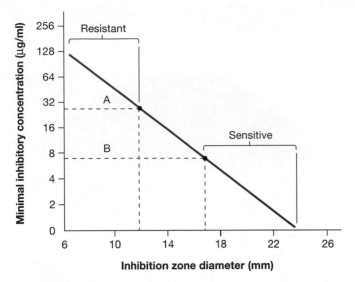

Figure 31.3 Interpretation of Kirby-Bauer Test Results.
The relationship between the minimal inhibitory concentrations of a hypothetical drug and the size of the zone around a disk in which microbial growth is inhibited. As the sensitivity of microorganisms to the drug increases, the MIC value decreases and the inhibition zone grows larger. Suppose that this drug varies from 7–28 μg/ml in the body during treatment. Dashed line A shows that any pathogen with a zone of inhibition less than 12 mm in diameter will have an MIC value greater than 28 μg/ml and will be resistant to drug treatment. A pathogen with a zone diameter greater than 17 mm will have an MIC less than 7 μg/ml and will be sensitive to the drug (see line B). Zone diameters between 12 and 17 mm indicate intermediate sensitivity and usually signify resistance.

31.4 ANTIBACTERIAL DRUGS

Since Fleming's discovery of penicillin, many antibiotics have been found that can damage pathogens in several ways. A few antibacterial drugs are described here and summarized in table 31.1, with emphasis on their mechanisms of action.

Inhibitors of Cell Wall Synthesis

The most selective antibiotics are those that interfere with bacterial cell wall synthesis. Drugs such as penicillins, cephalosporins, vancomycin, and bacitracin have a high therapeutic index because they target structures not found in eucaryotic cells. << *Bacterial cell walls (section 3.4)*

Penicillins

Most **penicillins** (e.g., penicillin G or benzylpenicillin) are derivatives of 6-aminopenicillanic acid and differ from one another with respect to the side chain attached to the amino group (**figure 31.5**). The most crucial feature of the molecule is the **β-lactam ring,** which is essential for bioactivity. Many penicillin-resistant bacteria produce **penicillinase** (also called **β-lactamase**), an enzyme that inactivates the antibiotic by hydrolyzing a bond in the **β-lactam** ring.

The structure of the penicillins resembles the terminal D-alanyl-D-alanine found on the peptide side chain of the pep-

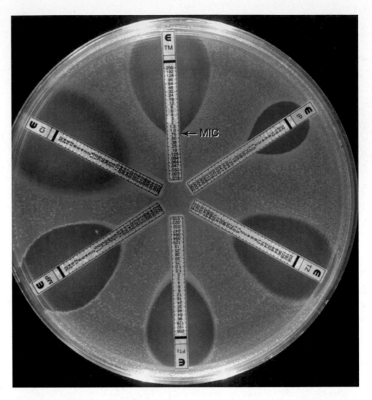

Figure 31.4 The Etest®. An example of a bacterial culture plate with Etest® strips arranged radially on it. The strips are arranged so that the lowest antibiotic concentration in each is at the center. The MIC concentration is read from the scale at the point it intersects the zone of inhibition as shown by the arrow in this example. Etest® is a registered trademark of AB BIODISK and patented in all major markets.

tidoglycan subunit. It is thought that this structural similarity blocks the enzyme catalyzing the transpeptidation reaction that forms the peptidoglycan cross-links (*see figure 11.11*). Thus formation of a complete cell wall is blocked, leading to osmotic lysis. This mechanism is consistent with the observation that penicillins act only on growing bacteria that are synthesizing new peptidoglycan. However, the mechanism of penicillin action is actually more complex. It has been discovered that penicillins bind to several periplasmic proteins (penicillin-binding proteins, or PBPs) and may also destroy bacteria by activating their own autolytic enzymes. However, penicillin may kill bacteria even in the absence of autolysins or murein hydrolases. Lysis could occur after bacterial viability has already been lost. Penicillin may stimulate proteins called bacterial holins to form holes or lesions in the plasma membrane, leading directly to membrane leakage and death. Murein hydrolases also could move through the holes, disrupt the peptidoglycan, and lyse the cell.

Penicillins differ from each other in several ways. The two naturally occurring penicillins, penicillin G and penicillin V, are narrow-spectrum drugs (figure 31.5). Penicillin G is effective against gonococci, meningococci, and several gram-positive pathogens such as streptococci and staphylococci. However, it must be administered by injection (parenterally) because it is destroyed by stomach acid. Penicillin V is similar to penicillin G

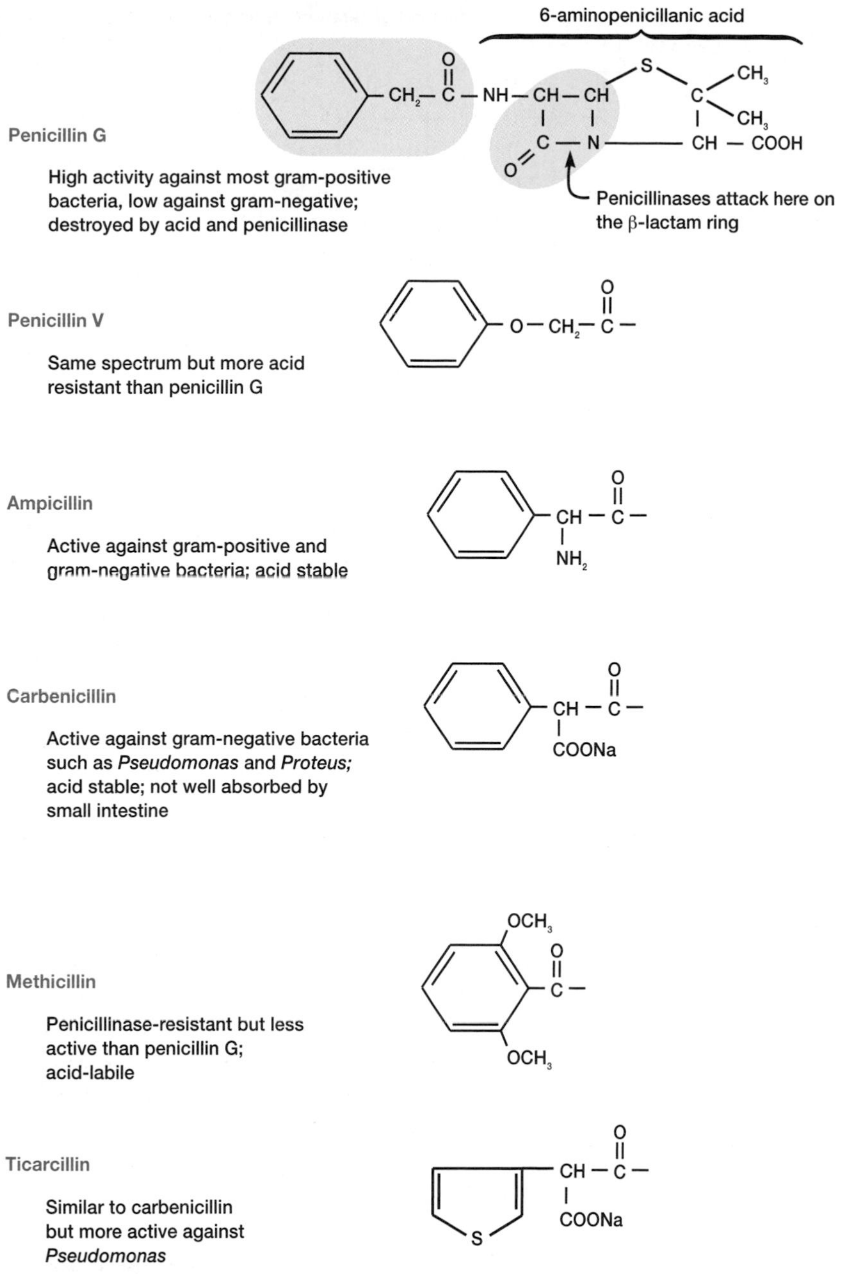

Penicillin G

High activity against most gram-positive bacteria, low against gram-negative; destroyed by acid and penicillinase

Penicillin V

Same spectrum but more acid resistant than penicillin G

Ampicillin

Active against gram-positive and gram-negative bacteria; acid stable

Carbenicillin

Active against gram-negative bacteria such as *Pseudomonas* and *Proteus*; acid stable; not well absorbed by small intestine

Methicillin

Penicillinase-resistant but less active than penicillin G; acid-labile

Ticarcillin

Similar to carbenicillin but more active against *Pseudomonas*

Figure 31.5 **Penicillins.** The structures and characteristics of representative penicillins. All are derivatives of 6-aminopenicillanic acid; in each case, the shaded portion of penicillin G is replaced by the side chain indicated. The β-lactam ring is also shaded (blue).

in spectrum of activity but can be given orally because it is more resistant to acid. The semisynthetic penicillins, on the other hand, have a broader spectrum of activity. Ampicillin can be administered orally and is effective against gram-negative bacteria such as *Haemophilus, Salmonella,* and *Shigella.* Carbenicillin and ticarcillin are potent against *Pseudomonas* and *Proteus.*

An increasing number of bacteria have become resistant to natural penicillins and many of the semisynthetic analogs. Physicians frequently employ specific semisynthetic penicillins that are not destroyed by β-lactamases to combat antibiotic-

resistant pathogens. These include methicillin (figure 31.5), nafcillin, and oxacillin. However, this practice has been confounded by the emergence of methicillin-resistant bacteria.

Although penicillins are the least toxic of the antibiotics, about 1 to 5% of the adults in the United States develop allergies to them. Occasionally, a person will die of a violent allergic response; therefore patients should be questioned about penicillin allergies before treatment is begun. << *Immune disorders: Hypersensitivities (section 29.11)*

Cephalosporins

Cephalosporins are a family of antibiotics originally isolated in 1948 from the fungus *Cephalosporium.* They contain a β-lactam structure that is very similar to that of the penicillins (**figure 31.6**). As might be expected from their structural similarities to penicillins, cephalosporins also inhibit the transpeptidation reaction during peptidoglycan synthesis. They are broad-spectrum drugs frequently given to patients with penicillin allergies (although about 10% of patients allergic to penicillin are also allergic to cephalosporins).

Cephalosporins are broadly categorized into four generations (groups of drugs that were sequentially developed) based on their spectrum of activity. First-generation cephalosporins are more effective against gram-positive pathogens than gram-negatives. Second-generation drugs, developed after the first generation, have improved effects on gram-negative bacteria with some anaerobe coverage. Third-generation drugs are particularly effective against gram-negative pathogens, and some reach the central nervous system. This is of particular note because many antimicrobial agents do not cross the blood-brain barrier. Finally, fourth-generation cephalosporins are broad spectrum with excellent gram-positive and gram-negative coverage and, like their third-generation predecessors, inhibit the growth of the difficult opportunistic pathogen *Pseudomonas aeruginosa.*

Vancomycin and Teicoplanin

Vancomycin is a glycopeptide antibiotic produced by *Streptomyces orientalis.* It is a cup-shaped molecule composed of a peptide linked to a disaccharide (**figure 31.7**). The peptide portion blocks the transpeptidation reaction by binding specifically to the D-alanine-D-alanine terminal sequence on the pentapeptide portion of peptidoglycan. The antibiotic is bactericidal for *Staphylococcus* and some members of the genera *Clostridium, Bacillus, Streptococcus,* and *Enterococcus.* It is given both orally and intravenously, and has been particularly important in the treatment of antibiotic-resistant staphylococcal and enterococcal infections. However, vancomycin-resistant strains of *Enterococcus* have become widespread and cases of resistant

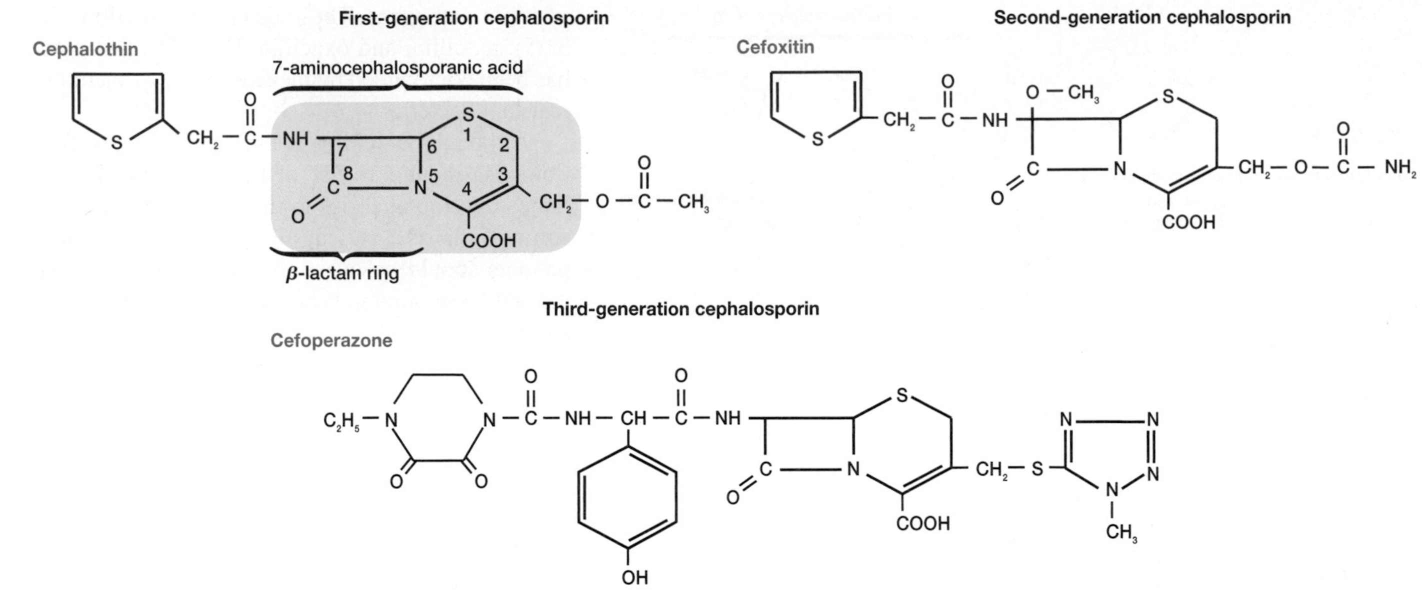

First-generation cephalosporin

Cephalothin

7-aminocephalosporanic acid

β-lactam ring

Second-generation cephalosporin

Cefoxitin

Third-generation cephalosporin

Cefoperazone

Figure 31.6   Cephalosporin Antibiotics.   These drugs are derivatives of 7-aminocephalosporanic acid and contain a β-lactam ring.

Staphylococcus aureus have appeared. In this case, resistance is conferred when bacteria change the terminal D-alanine to either D-lactate or a D-serine residue. Vancomycin resistance poses a serious public health threat: vancomycin has been considered the "drug of last resort" in cases of antibiotic-resistant *S. aureus*. Clearly newer drugs must be developed.

Teicoplanin, another glycopeptide antibiotic, is produced by the actinomycete *Actinoplanes teichomyceticus*. It is similar in structure and mechanism of action to vancomycin but has fewer side effects. It is active against staphylococci, enterococci, streptococci, clostridia, *Listeria,* and many gram-positive pathogens.

Protein Synthesis Inhibitors

Many antibiotics inhibit protein synthesis by binding with the procaryotic ribosome and other components of protein synthesis. Because these drugs discriminate between procaryotic and eucaryotic ribosomes, their therapeutic index is fairly high but not as high as that of cell wall inhibitors. Several different steps in protein synthesis can be affected: aminoacyl-tRNA binding, peptide bond formation, mRNA reading, and translocation.
<< *Translation (section 12.7)*

Aminoglycosides

Although considerable variation in structure occurs among several important **aminoglycoside antibiotics,** all contain a cyclohexane ring and amino sugars (**figure 31.8**). **Streptomycin,** kanamycin, neomycin, and tobramycin are synthesized by different species of the genus *Streptomyces,* whereas gentamicin comes from another actinomycete, *Micromonospora purpurea.* Aminoglycosides bind to the 30S (small) ribosomal subunit and interfere with protein synthesis by directly inhibiting the synthesis process and causing misreading of the mRNA.

These antibiotics are bactericidal and tend to be most effective against gram-negative pathogens. Streptomycin's usefulness has decreased greatly due to widespread drug resistance but may still be effective when other aminoglycosides are not tolerated

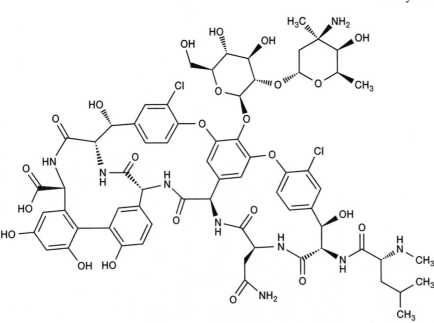

Figure 31.7   Vancomycin.

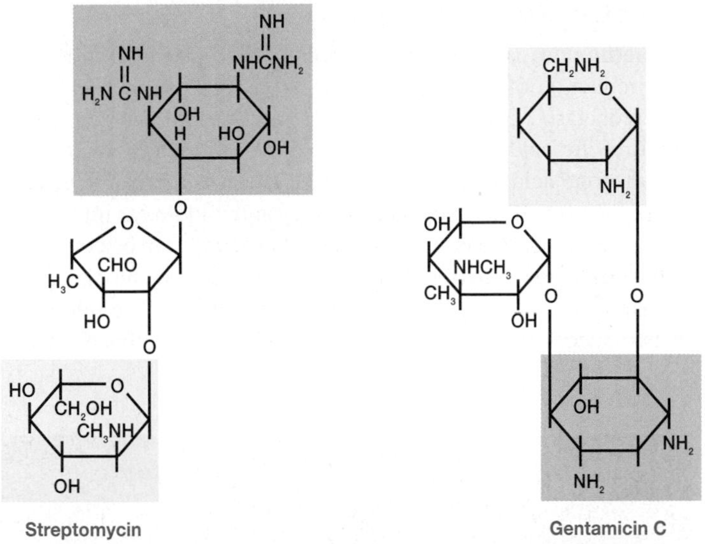

Figure 31.8 Representative Aminoglycoside Antibiotics. Identifying cyclohexane rings (blue) and amino sugars (yellow).

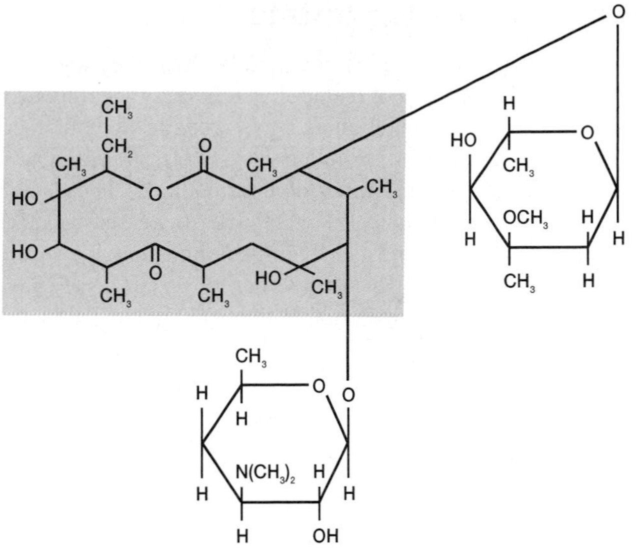

Figure 31.10 Erythromycin, a Macrolide Antibiotic. The 14-member lactone ring (tan) is connected to two sugars.

or are contraindicated due to interactions with other drugs (e.g., HIV protease inhibitors). Gentamicin is used to treat *Proteus, Escherichia, Klebsiella,* and *Serratia* infections. Aminoglycosides can be quite toxic, however, and can cause deafness, renal damage, loss of balance, nausea, and allergic responses.

Tetracyclines

The **tetracyclines** are a family of antibiotics with a common four-ring structure to which a variety of side chains are attached (**figure 31.9**). Oxytetracycline and chlortetracycline are produced naturally by *Streptomyces* species, whereas others are semisynthetic drugs. These antibiotics are similar to the aminoglycosides and combine with the 30S subunit of the ribosome. This inhibits the binding of aminoacyl-tRNA molecules to the A site of the ribosome. Their action is only bacteriostatic.

Tetracyclines are broad-spectrum antibiotics that are active against most bacteria, including rickettsias, chlamydiae, and mycoplasmas. Although their use has declined in recent years, they are still sometimes used to treat acne.

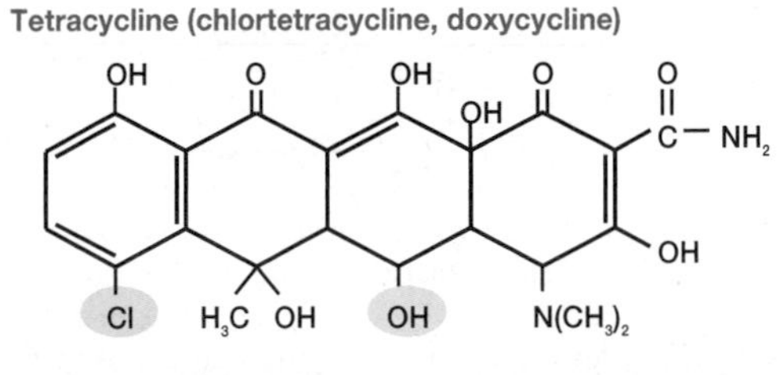

Figure 31.9 Tetracyclines. Three members of the tetracycline family. Tetracycline lacks both of the groups that are shaded. Chlortetracycline (aureomycin) differs from tetracycline in having a chlorine atom (blue): doxycycline consists of tetracycline with an extra hydroxyl (light blue).

Macrolides

The **macrolide antibiotics** contain 12- to 22-carbon lactone rings linked to one or more sugars (**figure 31.10**). Erythromycin binds to the 23S rRNA of the 50S ribosomal subunit to inhibit peptide chain elongation during protein synthesis. Erythromycin is a relatively broad-spectrum antibiotic effective against gram-positive bacteria, mycoplasmas, and a few gram-negative bacteria but is usually only bacteriostatic. It is used with patients who are allergic to penicillins and in the treatment of whooping cough, diphtheria, diarrhea caused by *Campylobacter,* and pneumonia from *Legionella* or *Mycoplasma* infections. Clindamycin is effective against a variety of bacteria, including staphylococci, and anaerobes such as *Bacteroides.* Azithromycin, which has surpassed erythromycin in use, is particularly effective against many bacteria including *Chlamydia trachomatis.*

Chloramphenicol

Chloramphenicol (**figure 31.11**) was first produced from cultures of *Streptomyces venezuelae* but it is now synthesized chemically. Like erythromycin, this antibiotic binds to 23S rRNA on the 50S ribosomal subunit to inhibit the peptidyl transferase reaction. It has a very broad spectrum of activity but, unfortunately, is quite toxic. The most common side effect is depression of bone marrow function, leading to aplastic anemia and a decreased number of white blood cells. Consequently, this antibiotic is used only in life-threatening situations when no other drug is adequate.

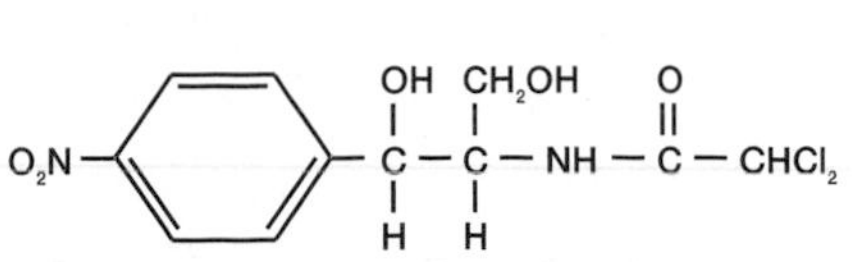

Figure 31.11 Chloramphenicol.

Metabolic Antagonists

Several valuable drugs act as **antimetabolites:** they antagonize, or block, the functioning of metabolic pathways by competitively inhibiting the use of metabolites by key enzymes. These drugs can act as structural analogs, molecules that are structurally similar to naturally occurring metabolic intermediates. These analogs compete with intermediates in metabolic processes because of their similarity but are just different enough that they prevent normal cellular metabolism. By preventing metabolism, they are bacteriostatic but broad spectrum; their removal reestablishes the metabolic activity.

Sulfonamides or Sulfa Drugs

Sulfonamides, or sulfa drugs, are structurally related to sulfanilamide, an analog of *p*-aminobenzoic acid, or PABA (**figure 31.12**). PABA is used in the synthesis of the cofactor folic acid (folate). When sulfanilamide or another sulfonamide enters a bacterial cell, it competes with PABA for the active site of an enzyme involved in folic acid synthesis, causing a decline in folate concentration. This decline is detrimental to the bacterium because folic acid is a precursor of purines and pyrimidines, the bases used in the construction of DNA, RNA, and other important cell constituents. The resulting inhibition of purine and pyrimidine synthesis leads to cessation of protein synthesis and DNA replication. Sulfonamides are selectively toxic for many bacterial and protozoan pathogens because these microbes manufacture their own folate and cannot effectively take up this cofactor, whereas humans do not synthesize folate (we must obtain it in our diet). Sulfonamides thus have a high therapeutic index.

The increasing resistance of many bacteria to sulfa drugs limits their effectiveness. Furthermore, as many as 5% of the patients receiving sulfa drugs experience adverse side effects, chiefly allergic responses such as fever, hives, and rashes.

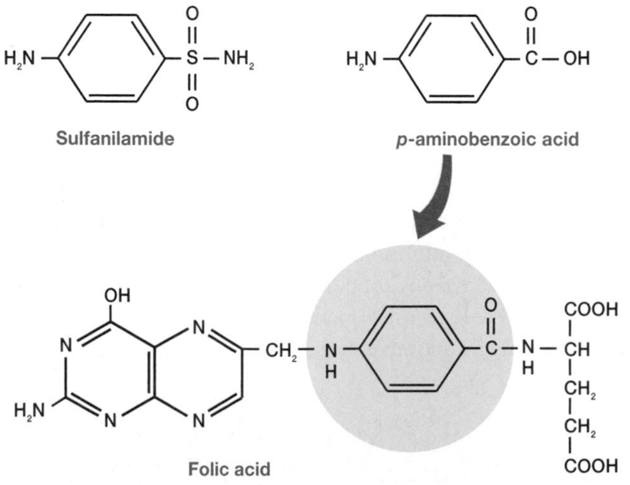

Trimethoprim

Trimethoprim is a synthetic antibiotic that also interferes with the production of folic acid. It does so by binding to dihydrofolate reductase (DHFR), the enzyme responsible for converting dihydrofolic acid to tetrahydrofolic acid, competing against the dihydrofolic acid substrate (**figure 31.13**). It is a broad-spectrum antibiotic often used to treat respiratory and middle ear infections, urinary tract infections, and traveler's diarrhea. It can be combined with sulfa drugs to increase efficacy of treatment by blocking two key steps in the folic acid pathway (**figure 31.14**). The inhibition of two successive steps in a single biochemical pathway means

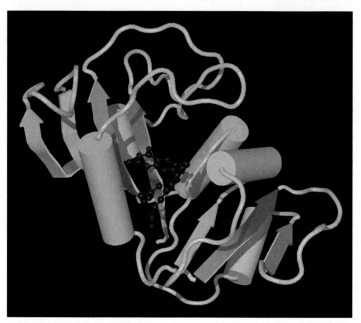

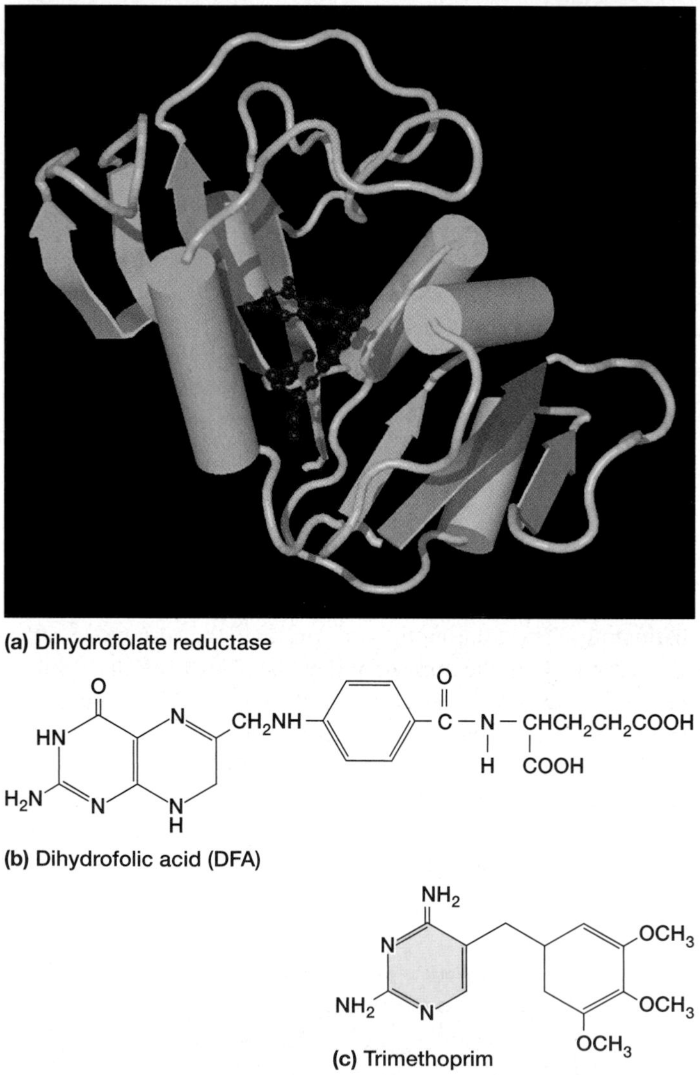

(a) Dihydrofolate reductase

(b) Dihydrofolic acid (DFA)

(c) Trimethoprim

Figure 31.13 Competitive Inhibition of Dihydrofolate Reductase (DHFR) by Trimethoprim. (a) DHFR structure and its interaction with dihydrofolic acid (DFA; red), its natural substrate. Note the chemical structure and how it fits into the active site of the enzyme. (b) DFA is the natural substrate for the DHFR enzyme of the folic acid pathway. (c) Trimethoprim mimics the structural orientation of the DFA and thus competes for the active site of the enzyme. This causes delayed or absent folic acid synthesis because the DFA cannot be converted to tetrahydrofolic acid when trimethoprim occupies the DHFR active site.

Sulfanilamide

p-aminobenzoic acid

Folic acid

Figure 31.12 Sulfanilamide. Sulfanilamide and its relationship to the structure of folic acid.

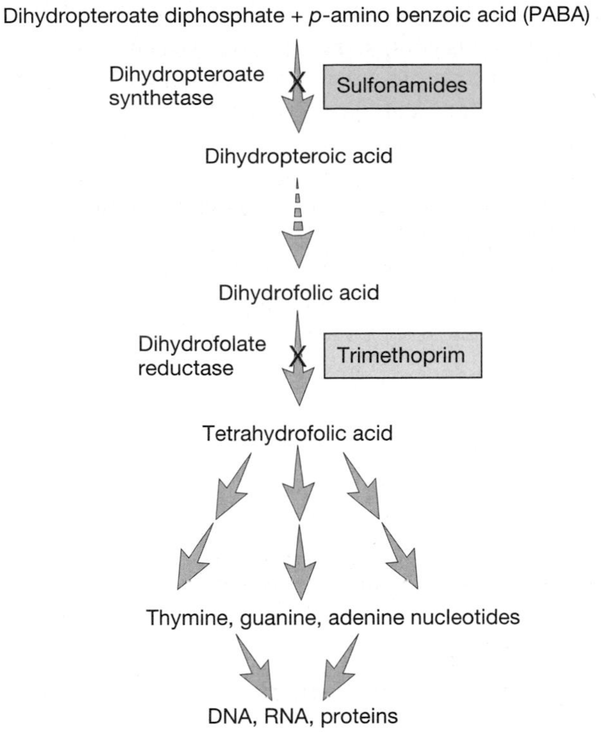

Dihydropteroate diphosphate + *p*-amino benzoic acid (PABA)

Figure 31.14 Synergistic Drug Interaction Between the Sulfonamides and Trimethoprim. Two successive steps in the biochemical pathway for folic acid synthesis are blocked by these drugs. Thus the efficacy of the drug combination is greater than that of either drug used alone.

that less of each drug is needed in combination than when used alone. This is termed a **synergistic drug interaction.**

Nucleic Acid Synthesis Inhibition

The antibacterial drugs that inhibit nucleic acid synthesis function by inhibiting (1) DNA polymerase and DNA helicase or (2) RNA polymerase to block replication or transcription, respectively. These drugs are not as selectively toxic as other antibiotics because procaryotes and eucaryotes do not differ greatly with respect to nucleic acid synthesis.

Quinolones

The **quinolones** are synthetic drugs that contain the 4-quinolone ring. The quinolones are important antimicrobial agents that inhibit nucleic acid synthesis. They are increasingly used to treat a wide variety of infections. The first quinolone, nalidixic acid (**figure 31.15**), was synthesized in 1962. Since that time, generations of fluoroquinolones have been produced. Three of these—ciprofloxacin, norfloxacin, and ofloxacin—are currently used in the United States, and more fluoroquinolones are being synthesized and tested.

Quinolones act by inhibiting the bacterial DNA gyrase and topoisomerase II. DNA gyrase introduces negative twist in DNA and helps separate its strands. Inhibition of DNA gyrase disrupts DNA replication and repair, bacterial chromosome separation during division, and other processes involving DNA.

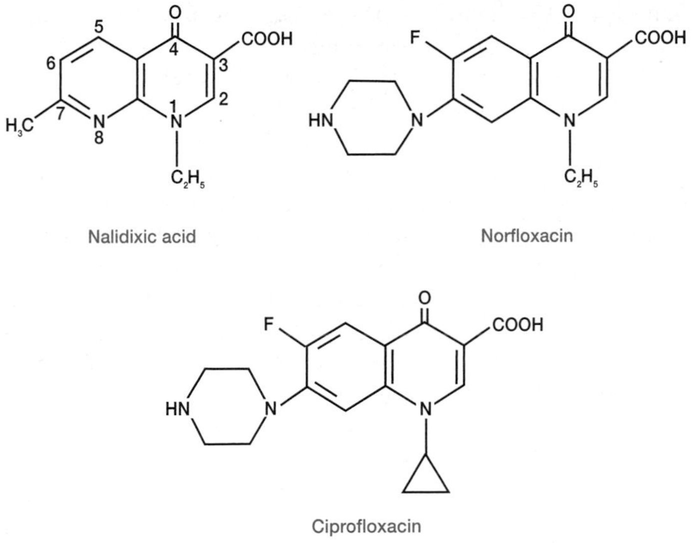

Figure 31.15 Quinolone Antimicrobial Agents. Ciprofloxacin and norfloxacin are newer generation fluoroquinolones. The 4-quinolone ring in nalidixic acid has been numbered.

Fluoroquinolones also inhibit topoisomerase II, another enzyme that untangles DNA during replication. It is not surprising that quinolones are bactericidal. ◄◄ *DNA replication (section 12.3)*

The quinolones are broad-spectrum antibiotics. They are highly effective against enteric bacteria such as *Escherichia coli* and *Klebsiella pneumoniae.* They can be used with *Haemophilus, Neisseria, P. aeruginosa,* and other gram-negative pathogens. The quinolones also are active against gram-positive bacteria such as *S. aureus, Streptococcus pyogenes,* and *Mycobacterium tuberculosis.* Thus they are used in treating a wide range of infections.

1. Explain five ways in which chemotherapeutic agents kill or damage bacterial pathogens.
2. Why do penicillins and cephalosporins have a higher therapeutic index than most other antibiotics?
3. Would there be any advantage to administering a bacteriostatic agent along with penicillins? Any disadvantage?
4. What are antimetabolites? Why do you think these are effective against protozoan pathogens while the other drugs presented here generally are not?

31.5 FACTORS INFLUENCING ANTIMICROBIAL DRUG EFFECTIVENESS

It is crucial to recognize that effective drug therapy is not a simple matter. Drugs may be administered in several different ways, and they do not always spread rapidly throughout the body or immediately kill all invading pathogens. A complex array of factors influences the effectiveness of drugs.

First, the drug must actually be able to reach the site of infection. Understanding the factors that control drug activity, stability, and metabolism in vivo are essential in drug formulation. For example, the mode of administration plays an important role. A drug such as penicillin G is not suitable for oral administration because it is relatively unstable in stomach acid. Some antibiotics—for example, gentamicin and other aminoglycosides—are not well absorbed from the intestinal tract and must be injected intramuscularly or given intravenously. Other antibiotics (neomycin, bacitracin) are so toxic that they can only be applied topically to skin lesions. Nonoral routes of administration are called **parenteral routes.** Even when an agent is administered properly, it may be excluded from the site of infection. For example, blood clots, necrotic tissue, or biofilms can protect bacteria from a drug, either because body fluids containing the agent may not easily reach the pathogens or because the agent is absorbed by materials surrounding them.

Second, the pathogen must be susceptible to the drug. Bacteria in biofilms or abscesses may be replicating very slowly and are therefore resistant to chemotherapy, because many agents affect pathogens only if they are actively growing and dividing. A pathogen, even though growing, may simply not be susceptible to a particular agent. To control resistance, drug cocktails can be used to treat some infections. A notable example of this is the use of clavulonic acid (to inactivate penicillinase) combined with ampicillin to treat penicillin-resistant bacteria.

Third, the chemotherapeutic agent must exceed the pathogen's MIC value if it is going to be effective. The concentration reached will depend on the amount of drug administered, the route of administration and speed of uptake, and the rate at which the drug is cleared or eliminated from the body. It makes sense that a drug will remain at high concentrations longer if it is absorbed over an extended period and excreted slowly.

Finally, chemotherapy has been rendered less effective and much more complex by the spread of drug-resistance genes and prevention of drug access by biofilm components.

1. What factors do you think must be considered when treating an infection present in a biofilm on a medical implant (e.g., an artificial hip) versus a skin infection caused by the same microbe?

2. What is parenteral administration of a drug? Why is it used?

31.6   Drug Resistance

The spread of drug-resistant pathogens is one of the most serious threats to public health in the twenty-first century (**Disease 31.1**). This section describes the ways in which bacteria acquire drug resistance and how resistance spreads within a bacterial population.

Disease

31.1   Antibiotic Misuse and Drug Resistance

The sale of antimicrobial drugs is big business. In the United States, millions of pounds of antibiotics valued at billions of dollars are produced annually. As much as 70% of these antibiotics are added to livestock feed.

Because of the massive quantities of antibiotics being prepared and used, an increasing number of diseases are resisting treatment due to the spread of drug resistance. A good example is *Neisseria gonorrhoeae,* the causative agent of gonorrhea. Gonorrhea was first treated successfully with sulfonamides in 1936, but by 1942 most strains were resistant and physicians turned to penicillin. Within 16 years, a penicillin-resistant strain emerged in Asia. A penicillinase-producing gonococcus reached the United States in 1976 and is still spreading in this country. Thus penicillin is no longer used to treat gonorrhea.

In late 1968 an epidemic of dysentery caused by *Shigella* broke out in Guatemala and affected at least 112,000 persons; 12,500 deaths resulted. The strains responsible for this devastation carried an R plasmid conferring resistance to chloramphenicol, tetracycline, streptomycin, and sulfonamide. In 1972 a typhoid epidemic swept through Mexico producing 100,000 infections and 14,000 deaths. It was due to a *Salmonella* strain with the same multiple-drug-resistance pattern seen in the previous *Shigella* outbreak.

Haemophilus influenzae type b is responsible for many cases of childhood pneumonia and middle ear infections, as well as respiratory infections and meningitis. It is now becoming increasingly resistant to tetracyclines, ampicillin, and chloramphenicol. Similarly, the worldwide rate of penicillin-nonsusceptible (i.e., resistant) *Streptococcus pneumoniae* (PNSP) continues to increase. There is a direct correlation between the daily use of antibiotics (expressed as defined daily dose [DDD] per day) and the percent of PNSP isolates cultured (**box figure**). This dramatic correlation is alarming. More alarming is the continued indiscriminant use of antibiotics in light of these data.

In 1946 almost all strains of *Staphylococcus* were penicillin sensitive. Today most hospital strains are resistant to penicillin G, and some are now also resistant to methicillin and gentamicin and only can be treated with vancomycin. Strains of *Enterococcus* have become resistant to most antibiotics, including vancomycin, and a few cases of vancomycin-resistant *S. aureus* have been reported in the United States and Japan.

It is clear from these and other examples (e.g., multi-drug resistant *Mycobacterium tuberculosis*) that drug resistance is an extremely serious public health problem. Much of the difficulty
(continued)

arises from drug misuse. Drugs frequently have been overused. It has been estimated that over 50% of the antibiotic prescriptions in hospitals are given without clear evidence of infection or adequate medical indication. Many physicians have administered antibacterial drugs to patients with colds, influenza, viral pneumonia, and other viral diseases. A recent study showed that over 50% of the patients diagnosed with colds and upper respiratory infections and 66% of those with chest colds (bronchitis) are given antibiotics, even though over 90% of these cases are caused by viruses. Frequently antibiotics are prescribed without culturing and identifying the pathogen or without determining bacterial sensitivity to the drug. Toxic, broad-spectrum antibiotics are sometimes given in place of narrow-spectrum drugs as a substitute for culture and sensitivity testing, with the consequent risk of dangerous side effects, opportunistic infections, and the selection of drug-resistant mutants. The situation is made worse by patients not completing their course of medication. When antibiotic treatment is ended too early, drug-resistant mutants may survive. Drugs are available without prescription to the public in many countries; people may practice self-administration of antibiotics and further increase the prevalence of drug-resistant strains.

The use of antibiotics in animal feeds is undoubtedly another contributing factor to increasing drug resistance. The addition of low levels of antibiotics to livestock feeds raises the efficiency and rate of weight gain in cattle, pigs, and chickens (partially because of infection control in overcrowded animal populations). However, this also increases the number of drug-resistant bacteria in animal intestinal tracts. Evidence exists for the spread of bacteria such as *Salmonella* from animals to human populations. In 1983, 18 people in four midwestern states were infected with a multiple-drug-resistant strain of *Salmonella newport*. Eleven were hospitalized for salmonellosis and one died. All 18 patients had been infected by eating hamburger from beef cattle fed subtherapeutic doses of chlortetracycline for growth promotion. Resistance to some antibiotics has been traced to the use of specific farmyard antibiotics. Avoparcin resembles vancomycin in structure, and virginiamycin resembles Synercid. There is good circumstantial evidence that extensive use of these two antibiotics in animal feed has led to an increase in vancomycin and Synercid resistance among enterococci. The use of the quinolone antibiotic enrofloxacin in swine herds appears to have promoted ciprofloxacin resistance in pathogenic strains of *Salmonella*. In 2005 the use of

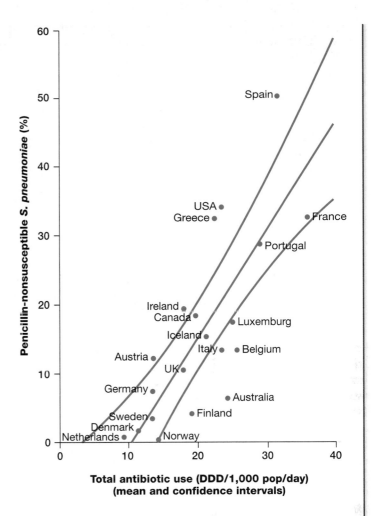

fluoroquinolones in U.S. poultry farming was banned in recognition of this public health threat.

The spread of antibiotic resistance can be due to quite subtle factors. For example, products such as soap and deodorants often now contain triclosan and other germicides. Evidence is increasing that the widespread use of triclosan actually favors an increase in antibiotic resistance (*see section 8.5*).

Mechanisms of Drug Resistance

The long-awaited "superbug" arrived in the summer of 2002. *S. aureus,* a common but sometimes deadly bacterium, had acquired a new antibiotic-resistance gene. The new strain was isolated from foot ulcers on a diabetic patient in Detroit, Michigan. **Methicillin-resistant *S. aureus* (MRSA)** had been well known as the bane of hospitals. This newer strain had developed resistance to vancomycin, one of the few antibiotics thought to still control *S. aureus.* This new vancomycin-resistant *S. aureus* (VRSA) strain also resisted most other antibiotics, including ciprofloxacin, methicillin, and penicillin. Isolated from the same patient was another dread of hospitals—**vancomycin-resistant enterococci (VRE)**. Genetic analyses revealed that the

patient's own vancomycin-sensitive *S. aureus* had acquired the vancomycin-resistance gene, *vanA,* from VRE through conjugation. So was born a new threat to the health of the human race.

<< *Bacterial conjugation (section 14.7); Bacterial plasmids (14.6)*

Bacteria have become resistant to antibiotics for a number of reasons (**figure 31.16**). Unfortunately, a particular resistance mechanism is not confined to a single class of drugs (**figure 31.17**). Two bacteria may use different resistance mechanisms to withstand the same chemotherapeutic agent. Furthermore, resistant mutants may arise spontaneously and are then selected for in the presence of the drug.

Pathogens often become resistant simply by preventing entrance of the drug. Many gram-negative bacteria are unaffected

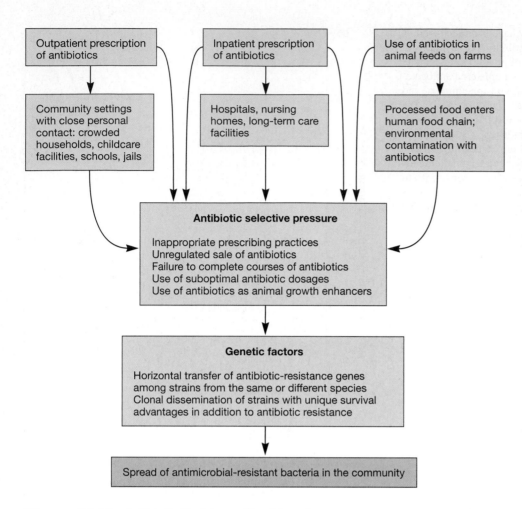

Figure 31.16 Antibiotic Resistance Has Many Sources. Incomplete and indiscriminant use of antibiotics in people and animals leads to increased selective pressure on bacteria. Bacteria capable of resisting antibiotics survive and spread these traits by horizontal gene transfer.

by penicillin G because it cannot penetrate the bacterial outer membrane. A decrease in permeability can lead to sulfonamide resistance. Mycobacteria resist many drugs because of the high content of mycolic acids (*see figure 22.12*) in a complex lipid layer outside their peptidoglycan. This layer is impermeable to most water-soluble drugs. << *Suborder* Corynebacterineae: *Genus* Mycobacterium *(section 22.4)*

A second resistance strategy is to pump the drug out of the cell after it has entered. Some pathogens have plasma membrane translocases, often called **efflux pumps,** that expel drugs. Because they are relatively nonspecific and pump many different drugs, these transport proteins often are multidrug-resistance pumps. Many are drug/proton antiporters—that is, protons enter the cell as the drug leaves. Such systems are present in *E. coli, P. aeruginosa,* and *S. aureus,* to name a few.

Many bacterial pathogens resist attack by **drug inactivation** through chemical modification. The best-known example is the hydrolysis of the β-lactam ring of penicillins by the enzyme penicillinase. Drugs also are inactivated by the addition of

chemical groups. For example, chloramphenicol contains two hydroxyl groups (figure 31.11) that can be modified by the addition of acetyl-CoA, a reaction catalyzed by the enzyme chloramphenicol acetyltransferase. Aminoglycosides (figure 31.8) can be modified and inactivated in several ways. For instance, acetyltransferases catalyze the acetylation of amino groups. Some aminoglycoside-modifying enzymes catalyze the addition to hydroxyl groups of either phosphates (phosphotransferases) or adenyl groups (adenyltransferases).

Because each chemotherapeutic agent acts on a specific target enzyme or cellular structure, resistance arises through **target modification.** We have already discussed this in the case of vancomycin resistance. The affinity of ribosomes for erythromycin and chloramphenicol also can be decreased by a change in the 23S rRNA to which they bind. This drastically reduces antibiotic binding. Antimetabolite action may be resisted through alteration of susceptible enzymes. In sulfonamide-resistant bacteria, the enzyme that uses *p*-aminobenzoic acid during folic acid synthesis (the dihydropteroic acid synthetase; figure 31.14) often has a much lower affinity for sulfonamides.

Finally, resistant bacteria may either use an **alternate pathway** to bypass the sequence inhibited by the agent or increase the production of the target metabolite. For example, some bacteria are resistant to sulfonamides simply because they use preformed folic acid from their surroundings rather than synthesize it themselves. Other strains increase their rate of folic acid production and thus counteract sulfonamide inhibition.

The Origin and Transmission of Drug Resistance

Genes for drug resistance may be present on bacterial chromosomes, plasmids, transposons, and integrons. Because they are often found on mobile genetic elements, they can freely exchange between bacteria. Spontaneous mutations in the bacterial chromosome, although they do not occur very often (with the exception of *M. tuberculosis*), can make bacteria drug resistant. Usually such mutations result in a change in the drug target; therefore the antibiotic cannot bind and inhibit growth. If a patient fails to take prescribed antibiotics as directed (e.g., does not complete the course of treatment), resistant mutants survive

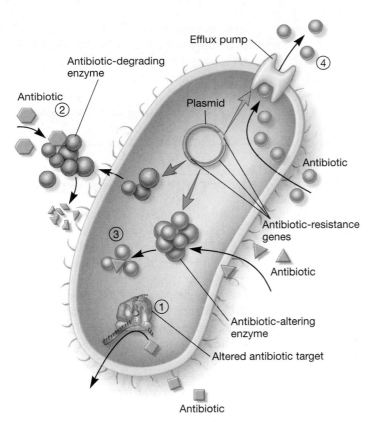

Figure 31.17 **Several Antibiotic Resistance Mechanisms.** Bacteria can resist the action of antibiotics by (1) preventing access to (or altering) the target of the antibiotic, (2) degrading the antibiotic, (3) altering the antibiotic, or (4) rapidly extruding the antibiotic.

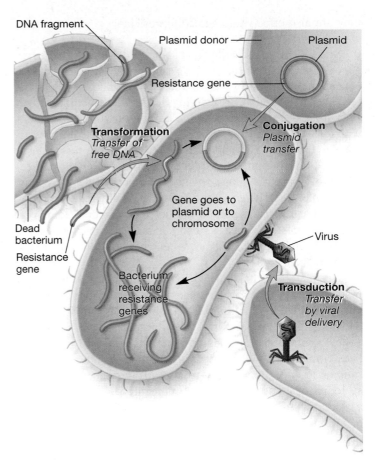

Figure 31.18 **Horizontal Gene Exchange.** Bacteria exchange genetic information, such as antibiotic resistance genes, through conjugation, transformation, or transduction.

and flourish because of their competitive advantage over nonresistant strains. Preventing the growth of such mutants is the rationale behind the direct observation of TB patients when taking each antibiotic dose. ≪ *Suborder* Corynebacterineae: *Family* Mycobacteriaceae *(section 22.4)*

Frequently a bacterial pathogen is drug resistant because it has a plasmid bearing one or more resistance genes; such plasmids are called **R plasmids** (resistance plasmids). Plasmid resistance genes often code for enzymes that destroy or modify drugs; for example, the hydrolysis of penicillin or the acetylation of chloramphenicol and aminoglycoside drugs. Plasmid-associated genes have been implicated in resistance to the aminoglycosides, choramphenicol, penicillins and cephalosporins, erythromycin, tetracyclines, sulfonamides, and others. Once a bacterial cell possesses an R plasmid, the plasmid (or its genes) may be transferred to other cells quite rapidly through normal gene exchange processes such as conjugation, transduction, and transformation (**figure 31.18**). Because a single plasmid may carry genes for resistance to several drugs, a pathogen population can become resistant to several antibiotics simultaneously, even though the infected patient is being treated with only one drug. ≪ *Bacterial conjugation (section 14.7); Transduction (section 14.9); Bacterial transformation (section 14.8); Bacterial plasmids (section 14.6)*

Antibiotic resistance genes can be located on genetic elements other than plasmids. Many composite transposons contain genes for antibiotic resistance and can move rapidly between plasmids and through a bacterial population. They are found in both gram-negative and gram-positive bacteria. Some examples and their resistance markers are Tn5 (kanamycin, bleomycin, streptomycin), Tn21 (streptomycin, spectinomycin, sulfonamide), Tn551 (erythromycin), and Tn4001 (gentamicin, tobramycin, kanamycin). Often several resistance genes are carried together as gene cassettes in association with a genetic element known as an integron. **Gene cassettes** are genetic elements that may exist as circular, nonreplicating DNA when moving from one site to another but that normally are a linear part of a transposon, plasmid, or bacterial chromosome. An **integron** is composed of an integrase gene and sequences for site-specific recombination. Thus integrons can capture genes and gene cassettes. Several cassettes can be integrated sequentially in an integron. Thus integrons also are important in spreading resistance genes. Finally, conjugative transposons, like composite transposons, can carry resistance genes. Because they are capable of moving between bacteria by conjugation, they are also effective in spreading resistance. ≪ *Transposable elements (section 14.5)*

Several strategies can be employed to discourage the emergence of drug resistance. The drug can be given in a high enough concentration to destroy susceptible bacteria and most spontaneous mutants that might arise during treatment. Sometimes two or even three different drugs can be administered simultaneously with the hope that each drug will prevent the emergence of resistance to the other. This approach is used in treating tuberculosis, HIV and malaria. Importantly, chemotherapeutic drugs, particularly broad-spectrum drugs, should be used only when necessary. If possible, the pathogen should be identified, drug sensitivity tests performed, and the proper narrow-spectrum drug employed. Patient compliance is just as important because completing a full course of antimicrobial therapy often prevents full mutation to resistant phenotypes.

Despite efforts to control the emergence and spread of drug resistance, the situation continues to worsen. Thus an urgent need exists for new antibiotics that microorganisms have never encountered. Pharmaceutical and biotechnology companies collect and analyze samples from around the world in a search for completely new antimicrobial agents. Structure-based or rational drug design is another option. If the three-dimensional structure of a susceptible target molecule such as an enzyme essential to microbial function is known, computer programs can be used to design drugs that precisely fit the target molecule. These drugs might be able to bind to the target and disrupt its function sufficiently to destroy the pathogen. Pharmaceutical companies are using these approaches to develop drugs for the treatment of AIDS, cancer, septicemia caused by lipopolysaccharide (LPS), and the common cold.

Information derived from the sequencing and analysis of pathogen genomes is also useful in identifying new targets for antimicrobial drugs. For example, genomics studies are providing data for research on inhibitors of both aminoacyl-tRNA synthetases and the enzyme that removes the formyl group from the N-terminal methionine during bacterial protein synthesis. The drug susceptibility of enzymes required for fatty acid synthesis is also being analyzed. ≪ *Microbial genomics (chapter 15)*

A most interesting response to the current crisis is the renewed interest in an idea first proposed early in the twentieth century by Felix d'Herelle, one of the discoverers of bacterial viruses or bacteriophages. d'Herelle proposed that bacteriophages could be used to treat bacterial diseases. Although most microbiologists did not pursue his proposal actively due to technical difficulties and the advent of antibiotics, Russian scientists developed the medical use of bacteriophages. Currently Russian physicians use bacteriophages to treat many bacterial infections. Bandages are saturated with phage solutions, phage mixtures are administered orally, and phage preparations are given intravenously to treat *Staphylococcus* infections. Several American companies are actively conducting research on phage therapy and preparing to carry out clinical trials. ≪ *Viruses and other acellular agents (chapter 5)*

1. Briefly describe the five major ways in which bacteria become resistant to drugs and give an example of each.
2. Define plasmid, R plasmid, integron, and gene cassette. How are R plasmids involved in the spread of drug resistance?
3. List several ways in which the development of antibiotic-resistant pathogens can be slowed or prevented.

31.7 ANTIFUNGAL DRUGS

Treatment of fungal infections generally has been less successful than that of bacterial infections largely because eucaryotic fungal cells are much more similar to human cells than are bacteria. Many drugs that inhibit or kill fungi are therefore quite toxic for humans and thus have a low therapeutic index. In addition, most fungi have a detoxification system that modifies many antifungal agents. Therefore antibiotics are fungistatic only as long as repeated application maintains high levels of unmodified antibiotic. Nonetheless, a few drugs are useful in treating many major fungal diseases. Effective antifungal agents frequently either extract membrane sterols or prevent their synthesis. Similarly, because fungi have cell walls made of chitin, the enzyme chitin synthase is the target for antifungals such as polyoxin D and nikkomycin.

Fungal infections are often subdivided into infections called superficial mycoses, subcutaneous mycoses, and systemic mycoses. Treatment for these types of disease is very different. Several drugs are used to treat superficial mycoses. Three drugs containing imidazole—miconazole, ketoconazole (**figure 31.19**), and clotrimazole—are broad-spectrum agents available as creams and solutions for the treatment of dermatophyte infections such as athlete's foot, and oral and vaginal candidiasis. They are thought to disrupt fungal membrane permeability and inhibit sterol synthesis. Nystatin (figure 31.19), a polyene antibiotic from *Streptomyces,* is used to control *Candida* infections of the skin, vagina, or alimentary tract. It binds to sterols and damages the membrane, leading to fungal membrane leakage. Griseofulvin (figure 31.19), an antibiotic formed by *Penicillium,* is given orally to treat chronic dermatophyte infections. It is thought to disrupt the mitotic spindle and inhibit cell division; it also may inhibit protein and nucleic acid synthesis. Side effects of griseofulvin include headaches, gastrointestinal upset, and allergic reactions. ≪ *Characteristics of the fungal divisions: Ascomycota (section 23.3)*

Systemic infections are very difficult to control and can be fatal. Three drugs commonly used against systemic mycoses are amphotericin B, 5-flucytosine, and fluconazole (figure 31.19). Amphotericin B from *Streptomyces* spp. binds to the sterols in fungal membranes, disrupting membrane permeability and causing leakage of cell constituents. It is quite toxic to humans and used only for serious, life-threatening infections. The synthetic oral antimycotic agent 5-flucytosine (5-fluorocytosine) is effective against most systemic fungi, although drug

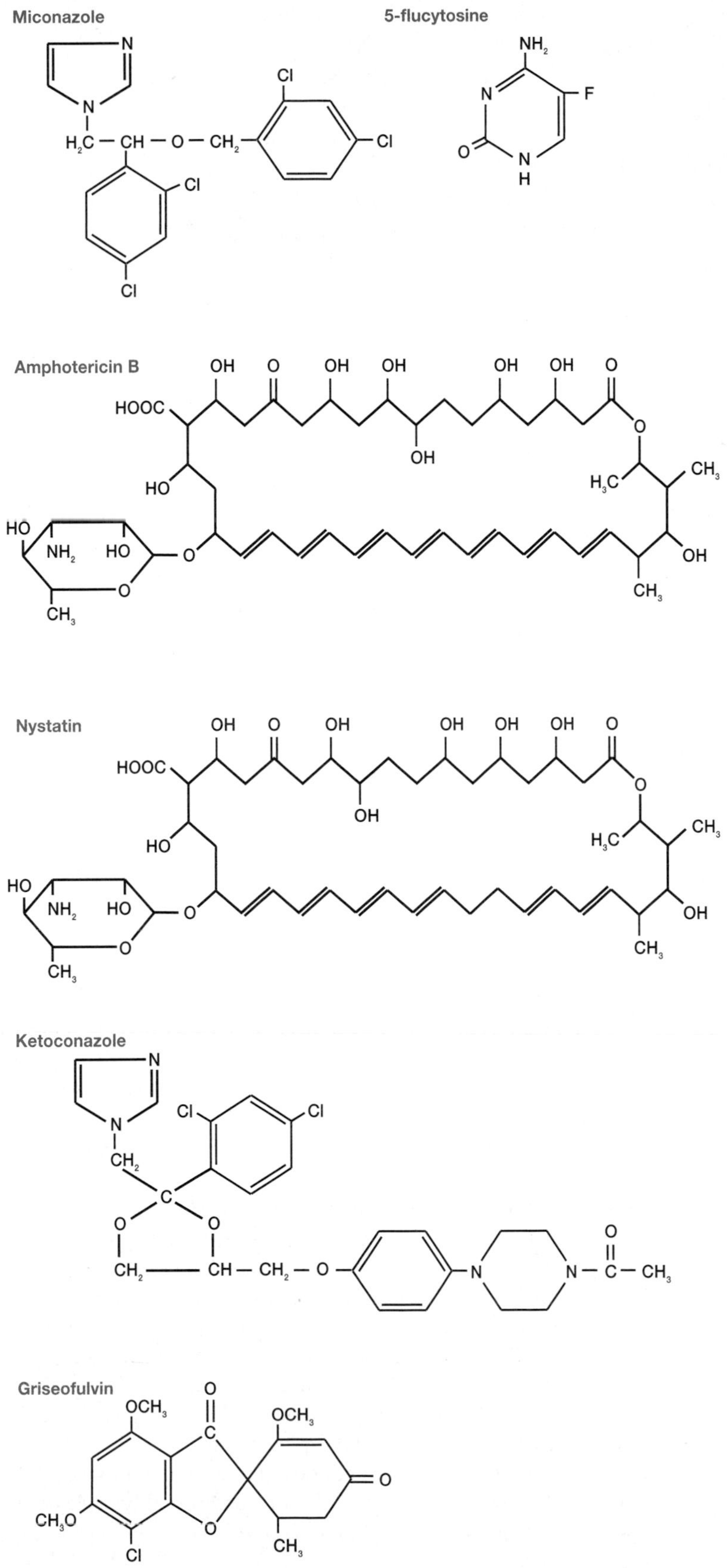

Figure 31.19 Antifungal Drugs. Six commonly used drugs are shown.

resistance often develops rapidly. The drug is converted to 5-fluorouracil by the fungi, incorporated into RNA in place of uracil, and disrupts RNA function. Its side effects include skin rashes, diarrhea, nausea, aplastic anemia, and liver damage. Atovaquone and pentamidine are used to treat *Pneumocystis jiroveci* (formerly called *P. carinii*). Some reports indicate that pentamidine interferes with *P. jiroveci* metabolism, although the drug only moderately inhibits glucose metabolism, protein synthesis, RNA synthesis, and intracellular amino acid transport in vitro. Fluconazole is used in the treatment of candidiasis, cryptococcal meningitis, and coccidioidal meningitis. Because adverse effects of fluconazole are relatively uncommon, it is used prophylactically to prevent life-threatening fungal infections in AIDS patients and other individuals who are severely immunosuppressed.

Subcutaneous mycoses, such as mycetoma and sporotrichosis, are typically treated with combinations of therapies that would be used for superficial and systemic mycoses. As with systemic mycoses, strict attention to potential toxic side effects is warranted.

1. Summarize the mechanism of action and the therapeutic use of the following antifungal drugs: miconazole, nystatin, griseofulvin, amphotericin B, and 5-flucytosine.

31.8 ANTIVIRAL DRUGS

Because viruses enter host cells and make use of host cell enzymes and constituents, it was long thought that a drug that blocked virus reproduction would be toxic for the host. However, inhibitors of virus-specific enzymes and life cycle processes have been discovered, and several drugs are used therapeutically. Some important examples are shown in **figure 31.20.** << *Viral multiplication (section 5.3)*

Most antiviral drugs disrupt either critical stages in the virus life cycle or the synthesis of virus-specific nucleic acids. **Amantadine** and **rimantadine** can be used to prevent influenza A infections. When given early in the infection (in the first 48 hours), they reduce the incidence of influenza by 50 to 70% in an exposed population. Amantadine blocks the penetration and uncoating of influenza virus particles. Adenine arabinoside or vidarabine disrupts the activity of DNA polymerase and several other enzymes involved in DNA and RNA synthesis and function. It is given intravenously or applied as an ointment to treat herpes infections. A third drug, acyclovir, is also used in the treatment of herpes infections. Upon phosphorylation, acyclovir resembles deoxyGTP and inhibits the viral DNA polymerase. Unfortunately, acyclovir-resistant strains of herpes have developed. Effective acyclovir derivatives and relatives are now available. Valacyclovir is

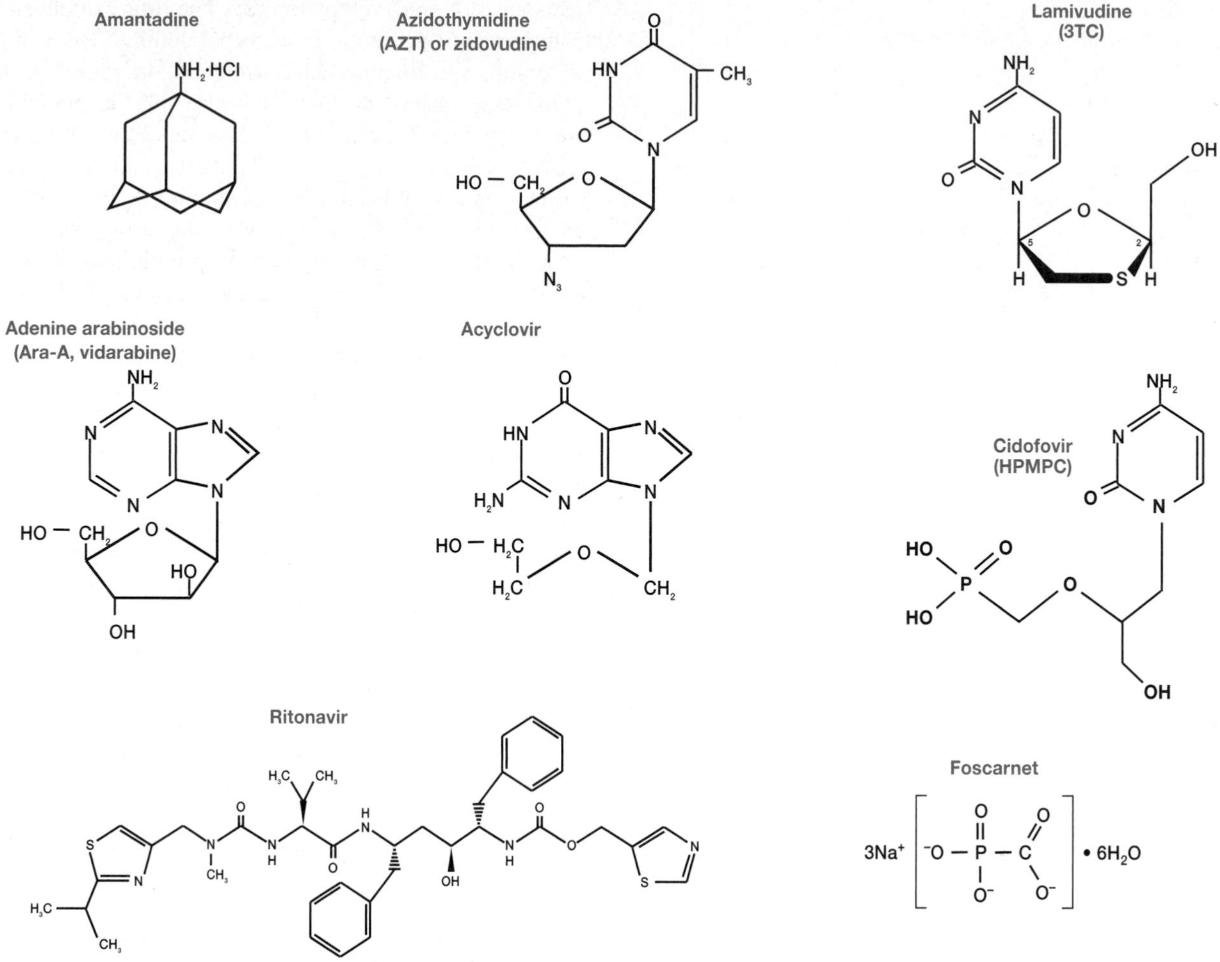

Figure 31.20 Representative Antiviral Drugs.

an orally administered prodrug form of acyclovir. Prodrugs are inactive until metabolized. Ganciclovir, penciclovir, and penciclovir's oral form, famciclovir, are effective in treatment of herpesviruses. Another kind of drug, foscarnet, inhibits the virus DNA polymerase in a different way. Foscarnet is an organic analog of pyrophosphate (figure 31.20) that binds to the polymerase active site and blocks the cleavage of pyrophosphate from nucleoside triphosphate substrates. It is used in treating herpes and cytomegalovirus infections. << *Viral diversity (chapter 24)*

Several broad-spectrum anti-DNA virus drugs have been developed. A good example is the drug HPMPC, or cidofovir (figure 31.20). It is effective against papovaviruses, adenoviruses, herpesviruses, iridoviruses, and poxviruses. The drug acts on the viral DNA polymerase as a competitive inhibitor and alternative substrate of dCTP. It has been used primarily against cytomegalovirus but also against herpes simplex and human papillomavirus infections.

Research on anti-HIV drugs has been particularly active. Many of the first drugs to be developed were reverse transcrip-

tase inhibitors such as **azidothymidine (AZT)** or zidovudine, lamivudine (3TC), didanosine (ddI), zalcitabine (ddC), and stavudine (d4T) (figure 31.20). These interfere with reverse transcriptase activity and therefore block HIV reproduction. More recently HIV protease inhibitors have also been developed. Three of the most used are saquinvir, indinavir, and ritonavir. **Protease inhibitors** are effective because HIV, like many viruses, translates multiple proteins as a single polypeptide. This polypeptide must then be cleaved into individual proteins required for virus replication. Protease inhibitors mimic the peptide bond that is normally attacked by the protease. The most successful HIV treatment regimen has been a cocktail of agents given at high dosages to prevent the development of drug resistance. For example, the combination of AZT, 3TC, and ritonavir is very effective in reducing HIV plasma concentrations almost to zero. However, the treatment does not eliminate proviral HIV DNA that still resides in memory T cells and possibly elsewhere. << *Viruses with single-stranded RNA genomes: Human immunodeficiency virus (section 24.7)*

Probably the most publicized antiviral agent has been **Tamiflu** (generically, oseltamivir phosphate). Tamiflu is a neuraminidase inhibitor that has received much attention in light of predictions of a twenty-first-century influenza pandemic, including avian influenza ("bird flu"). While Tamiflu is not a cure for neurominidase-expressing viruses, two clinical trials showed that patients who took Tamiflu were relieved of flu symptoms 1.3 days faster than patients who did not take Tamiflu. However, prophylactic use has resulted in viral resistance to Tamiflu. Tamiflu is not a substitute for yearly flu vaccination and frequent hand-washing.

31.9 ANTIPROTOZOAN DRUGS

The mechanism of action for most antiprotozoan drugs is not completely understood. Drugs such as chloroquine, atovaquone, mefloquine, iodoquinol, metronidazole, and nitazoxanide, for example, have potent antiprotozoan action, but a clear mechanism by which protozoan growth is inhibited is unknown. However, most antiprotozoan drugs appear to act on protozoan nucleic acid or some metabolic event.

Chloroquine is used to treat malaria. Several mechanisms of action have been reported. It can raise the internal pH, clump the plasmodial pigment, and intercalate into plasmodial DNA. Chloroquine also inhibits heme polymerase, an enzyme that converts toxic heme into nontoxic hemazoin. Inhibition of this enzyme leads to a buildup of toxic heme. Mefloquine is also used to treat malaria and has been found to swell the *Plasmodium falciparum* food vacuoles, where it may act by forming toxic complexes that damage membranes and other plasmodial components.

Metronidazole is used to treat *Entamoeba* infections. Anaerobic organisms readily reduce it to the active metabolite within the cytoplasm. Aerobic organisms appear to reduce it using ferrodoxin (a protein of the electron transport system). Reduced metronidazole interacts with DNA, altering its helical structure and causing DNA fragmentation; it prevents normal nucleic acid synthesis, resulting in cell death.

A number of antibiotics that inhibit bacterial protein synthesis are also used to treat protozoan infection. These include the aminoglycosides clindamycin and paromomycin. Aminoglycosides can be considered polycationic molecules that have a high affinity for nucleic acids. Specifically, aminoglycosides possess high affinities for RNAs. Different aminoglycoside antibiotics bind to different sites on RNAs. RNA binding interferes with the normal expression and function of the RNA, resulting in cell death.

Interference of eucaryotic electron transport is one common activity of some antiprotozoan drugs. Atovaquone is used to treat *Toxoplasma gondii*. It is an analog of ubiquinone, an integral component of the eucaryotic electron transport system. As an analog of ubiquinone, atovaquone can act as a competitive inhibitor and thus suppress electron transport. The ultimate metabolic effects of electron transport blockade include inhibited or delayed synthesis of nucleic acids and ATP. Another drug that interferes with electron transport is nitazoxanide, which is used to treat cryptosporidiosis. It appears to exert its effect through interference with the pyruvate:ferredoxin oxidoreductase. It has also been reported to form toxic free radicals once the nitro group is reduced intracellularly. Pyrimethamine and dapsone, used to treat *Toxoplasma* infections, appear to act in the same way as trimethoprim—interfering with folic acid synthesis by inhibition of dihydrofolate reductase.

As with other antimicrobial therapies, traditional drug development starts by identifying a unique target to which a drug can bind and thus prevent some vital function. A second consideration is often related to drug spectrum: how many different species have that target so that the proposed drug can be used broadly as a chemotherapeutic agent? This is also true for use of agents needed to remove protozoan parasites from their hosts. However, because protozoa are eucaryotes, the potential for drug action on host cells and tissues is greater than it is when targeting procaryotes. Most of the drugs used to treat protozoan infection have significant side effects; nonetheless, the side effects are usually acceptable when weighed against the parasitic burden.

1. Why do you think drugs that inhibit bacterial protein synthesis are also effective against some protists?
2. Why do you think malaria, like tuberculosis, is now treated with several drugs simultaneously?
3. What special considerations must be taken into account when treating infections caused by protozoan parasites?

Summary

31.1 The Development of Chemotherapy

a. The modern era of chemotherapy began with Paul Ehrlich's work on drugs against African sleeping sickness and syphilis. Other early pioneers were Gerhard Domagk, Alexander Fleming, Howard Florey, Ernst Chain, Norman Heatley, and Selman Waksman.

31.2 General Characteristics of Antimicrobial Drugs

a. An effective chemotherapeutic agent must have selective toxicity. A drug with great selective toxicity has a high therapeutic index and usually disrupts a structure or process unique to the pathogen. It has fewer side effects.

b. Antibiotics can be classified in terms of the range of target microorganisms (narrow spectrum versus broad spectrum); their source (natural, semisynthetic, or synthetic); and their general effect (static versus cidal) (**table 31.1**).

31.3 Determining the Level of Antimicrobial Activity

a. Antibiotic effectiveness can be estimated through the determination of the minimal inhibitory concentration and the minimal lethal concentration with dilution susceptibility tests. Tests such as the Kirby-Bauer test (a disk diffusion test) and the Etest® are often used to estimate a pathogen's susceptibility to drugs quickly (**figures 31.2–31.4**).

31.4 Antibacterial Drugs

a. Members of the penicillin family contain a β-lactam ring and disrupt bacterial cell wall synthesis, resulting in cell lysis (**figure 31.5**). Some, such as penicillin G, are usually administered by injection and are most effective against gram-positive bacteria. Others can be given orally (penicillin V), are broad spectrum (ampicillin, carbenicillin), or are usually penicillinase resistant (methicillin).

b. Cephalosporins are similar to penicillins and are given to patients with penicillin allergies (**figure 31.6**).

c. Vancomycin is a glycopeptide antibiotic that inhibits the transpeptidation reaction during peptidoglycan synthesis (**figure 31.7**). It is used against drug-resistant staphylococci, enterococci, and clostridia.

d. Aminoglycoside antibiotics such as streptomycin and gentamicin bind to the small ribosomal subunit, inhibit protein synthesis, and are bactericidal (**figure 31.8**).

e. Tetracyclines are broad-spectrum antibiotics having a four-ring nucleus with attached groups (**figure 31.9**). They bind to the small ribosomal subunit and inhibit protein synthesis.

f. Erythromycin is a bacteriostatic macrolide antibiotic that binds to the large ribosomal subunit and inhibits protein synthesis (**figure 31.10**).

g. Chloramphenicol is a broad-spectrum, bacteriostatic antibiotic that inhibits protein synthesis (**figure 31.11**). It is quite toxic and used only for very serious infections.

h. Sulfonamides or sulfa drugs resemble *p*-aminobenzoic acid and competitively inhibit folic acid synthesis (**figure 31.12**).

i. Trimethoprim is a synthetic antibiotic that inhibits the dihydrofolate reductase, which is required by organisms in the manufacture of folic acid (**figure 31.13**).

j. Quinolones are a family of bactericidal synthetic drugs that inhibit DNA gyrase and thus inhibit such processes as DNA replication (**figure 31.15**).

31.5 Factors Influencing Antimicrobial Drug Effectiveness

a. A variety of factors can greatly influence the effectiveness of antimicrobial drugs during use. These include route, effective concentration, and pathogen sensitivity.

31.6 Drug Resistance

a. Bacteria can become resistant to a drug by excluding it from the cell, pumping the drug out of the cell, enzymatically altering it, or modifying the target enzyme or structure so it is no longer affected by the drug. The genes for drug resistance may be found on the bacterial chromosome, a plasmid called an R plasmid, or other genetic elements such as transposons (**figures 31.17** and **31.18**).

b. Chemotherapeutic agent misuse fosters the increase and spread of drug resistance.

31.7 Antifungal Drugs

a. Because fungi are more similar to human cells than bacteria, antifungal drugs generally have lower therapeutic indexes than antibacterial agents and produce more side effects.

b. Superficial mycoses can be treated with miconazole, ketoconazole, clotrimazole, tolnaftate, nystatin, and griseofulvin (**figure 31.19**). Amphotericin B, 5-flucytosine, and fluconazole are used for systemic mycoses.

31.8 Antiviral Drugs

a. Antiviral drugs interfere with critical stages in the virus life cycle (amantadine, rimantadine, ritonavir) or inhibit the synthesis of virus-specific nucleic acids (zidovudine, adenine arabinoside, acyclovir) (**figure 31.20**).

31.9 Antiprotozoan Drugs

a. The mechanisms by which most drugs used to treat protozoan infection are unknown.

b. Some antiprotozoan drugs interfere with critical steps in nucleic acid synthesis, protein synthesis, electron transport, or folic acid synthesis.

Critical Thinking Questions

1. What advantage might soil bacteria and fungi gain from the synthesis of antibiotics?

2. You are the CEO of a biotechnology company that seeks to develop new antibiotics. Given that only 1% of all microbes have so far been cultured, which of the following approaches is your company going to use to find new drugs: (1) the development of new culture techniques; (2) metagenomics (*see section 15.8 and chapter 16*); or (3) the chemical synthesis of novel compounds based on specific microbial metabolic or structural targets? Be able to defend your choice.

3. Some advocate stockpiling the drug Tamiflu in the event of an influenza pandemic. Others point out that wealthy, Western nations would have an unfair advantage because developing nations (where the pandemic is most likely to start) would not have access to this expensive antiviral. Furthermore, some fear that indiscriminate use of the drug would promote the evolution of resistant flu strains. Given these caveats, do you think developed nations should stockpile Tamiflu for the protection and treatment of their citizens? Explain your answer.

4. How might the use of antibiotics as growth promoters in livestock contribute to antibiotic resistance among human pathogens?

5. A recent study found that 480 bacterial strains freshly isolated from the soil are resistant to at least six different antibiotics. In fact, some isolates are resistant to 20 different antibiotic drugs. Why do you think these bacteria, which are neither pathogenic nor exposed to human use of antibiotics, are resistant to so many drugs? What might be the implications for human bacterial pathogens?

6. You are a pediatrician treating a child with an upper respiratory infection that is clearly caused by a virus. The child's mother insists that you prescribe antibiotics—she's not leaving without them! How do you convince the child's mother that antibiotics will do more harm than good?

Learn More

Learn more by visiting the Prescott website at www.mhhe.com/prescottprinciples, where you will find a complete list of references.

Clinical Microbiology and Immunology

agglutination reaction The formation of an insoluble immune complex by the cross-linking of cells or particles.

bacteriophage (phage) typing Identification of bacterial strains based on their susceptibility to specific phages.

direct immunofluorescence Visualization technique using fluorescently-labeled antibodies to detect antigens.

enzyme-linked immunosorbent assay (ELISA) A serological assay in which bound antigen or antibody is detected by another antibody that is conjugated to an enzyme. The enzyme converts a colorless substrate to a colored product reporting the antibody capture of the antigen.

flow cytometry A tool for defining and enumerating cells. Fluorescently labeled cells pass single file through a capillary tube. Laser light detectors connected to a computer analyze the light patterns to identify cells.

hemadsorption The adherence of red blood cells to the surface of something, such as another cell or a virus.

immunoblotting The electrophoretic transfer of proteins from polyacrylamide gels to filters to demonstrate the presence of specific proteins through reaction with labeled antibodies.

immunodiffusion A technique involving the diffusion of antigen and antibody within a semisolid gel to produce a precipitin reaction in which they meet in proper proportions.

immunoelectrophoresis The electrophoretic separation of protein antigens followed by diffusion and precipitation in gels using antibodies against the separated proteins.

immunofluorescence A technique used to identify particular antigens microscopically in cells or tissues by the binding of a fluorescent antibody conjugate.

immunoprecipitation A reaction involving soluble antigens reacting with antibodies to form a large aggregate that precipitates out of solution.

indirect immunofluorescence A visualization technique using fluorescently-labeled antibodies to detect the presence of other antibodies made in response to a specific antigen.

monoclonal antibody (mAb) An antibody of a single type that is produced by a population of genetically identical plasma cells.

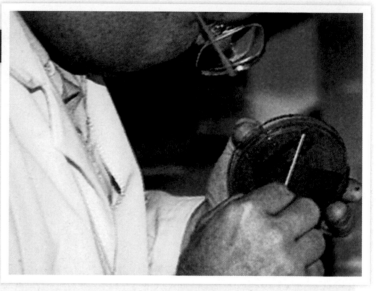

The major objective of the clinical microbiologist is to isolate and identify pathogens from clinical specimens rapidly. In this illustration, a clinical microbiologist is picking up suspect bacterial colonies for biochemical, immunologic, or molecular testing.

plasmid fingerprinting A technique used to identify microbial isolates as belonging to the same strain because they contain the same plasmids as demonstrated by endonuclease digestion, which generates fragments of the same molecular weight.

Quellung reaction The increase in visibility or the swelling of the capsule of a microorganism in the presence of antibodies against capsular antigens.

radioimmunoassay (RIA) A very sensitive assay that uses a purified radioisotope-labeled antigen or antibody to compete for antibody or antigen with unlabeled standard antigen or test antigen in experimental samples to determine the concentration of a substance in the samples.

ribotyping The identification of bacterial strains based on the nucleotide sequence of the rRNA.

serotyping A technique or serological procedure used to differentiate between strains (serovars or serotypes) of microorganisms that have differences in antigenic composition.

viral hemagglutination The clumping or agglutination of red blood cells caused by some viruses.

The pure culture is the foundation of all research on infectious disease.

—*Robert Koch*

Pathogens, particularly bacteria and yeasts, coexist with harmless microorganisms on or in the host. The challenge for clinical microbiologists is to identify these pathogens as the cause of infectious diseases. A variety of approaches may be used based on morphological, biochemical, immunologic, and molecular procedures. Time is a significant factor, especially in life-threatening situations, and advances in technology continue to increase the speed at which definitive identifications can be made. Even in the absence of a culture, polymerase chain reaction (PCR) and immunologic tests can be used to detect pathogens in specimens such as blood and sputum. In the final analysis, the patient's well-being and health can benefit significantly from information provided by the clinical laboratory—the subject of this chapter.

32.1 OVERVIEW OF THE CLINICAL MICROBIOLOGY LABORATORY

The major goal of the **clinical microbiologist** is to isolate and identify pathogenic microorganisms from clinical specimens rapidly. The purpose of the clinical laboratory is to provide the physician with information concerning the presence or absence of microorganisms that may be involved in the infectious disease process. Clinical microbiology is interdisciplinary, and the clinical microbiologist must have a working knowledge of microbial biochemistry and physiology, immunology, molecular biology, genomics, and the dynamics of host-parasite relationships. Importantly, tests developed to exploit the antigen-antibody binding capabilities, the focus of clinical immunology, can often detect microorganisms in specimens by identifying microbial antigens and quantifying the type and amount of responding antibody.

In clinical microbiology, a clinical specimen represents a portion or quantity of human material that is tested, examined, or studied to determine the presence or absence of particular microorganisms (**figure 32.1**). Safety for the patients, hospital, and laboratory staff is of utmost importance. The guidelines presented in **Techniques & Applications 32.1** were established by the Centers for Disease Control and Prevention (CDC) to address areas of personal protection and specimen handling. Other important concerns regarding specimens need emphasis:

1. The specimen selected should adequately represent the diseased area and also may include additional sites (e.g., urine and blood specimens) to isolate and identify potential agents of the particular disease process.

2. A quantity of specimen adequate to allow a variety of diagnostic testing should be obtained.

3. Attention must be given to specimen collection to avoid contamination from the many varieties of microorganisms indigenous to the skin, mucous membranes, and environment (*see figure 27.13*).

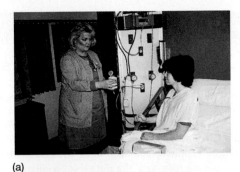

(a)

(b)

Figure 32.1 Collection of Patient Specimen for Microorganism Testing in a Clinical Laboratory. (a) The identification of the microorganism begins at the patient's beside. The nurse is giving instructions to the patient on how to obtain a sputum specimen. (b) The specimen is sent to the laboratory to be processed. Notice that the specimen and worksheet are in different bags.

4. The specimen should be collected in appropriate containers and forwarded promptly to the clinical laboratory.

5. If possible, the specimen should be obtained before antimicrobial agents have been administered to the patient.

32.2 IDENTIFICATION OF MICROORGANISMS FROM SPECIMENS

The clinical microbiology laboratory can provide preliminary or definitive identification of microorganisms based on (1) microscopic examination of specimens, (2) study of the growth and biochemical characteristics of isolated microorganisms (pure cultures), (3) immunologic tests that detect antibodies or microbial antigens, (4) bacteriophage typing (restricted to research settings and the CDC), and (5) molecular methods.

Microscopy

Wet-mount, heat-fixed, or chemically fixed specimens can be examined with an ordinary bright-field microscope. Examination can be enhanced with either phase-contrast or dark-field microscopy. The latter is the procedure of choice for the detection of

Techniques & Applications

32.1 Standard Microbiological Practices

The identification of potentially fatal, blood-borne infectious agents (HIV, hepatitis B virus, and others) spurred the codification of standard microbiological practices to limit exposure to such agents. These standard microbiological practices are minimum guidelines that should be supplemented with other precautions based on the potential exposure risks and biosafety level regulations for the lab. Briefly:

1. Eating, drinking, manipulation of contact lenses, and the use of cosmetics, gum, and tobacco products are strictly prohibited in the lab.

2. Hair longer than shoulder length should be tied back. Hands should be kept away from face at all times. Items (e.g., pencils) should not be placed in the mouth while in the lab. Protective clothing (lab coat, smock, etc.) is recommended while in the lab. Exposed wounds should be covered and protected.

3. Lab personnel should know how to use the emergency eyewash and shower stations.

4. Work space should be disinfected at the beginning and completion of lab time. Hands should be washed thoroughly after any exposure and before leaving the lab.

5. Precautions should be taken to prevent injuries caused by sharp objects (needles, scalpels, etc.). Sharp instruments should be discarded for disposal in specially marked containers.

Recommended guidelines for additional precautions should reflect the laboratory's biosafety level (BSL). The following table defines the BSL for the four categories of biological agents and suggested practices.

BSL	Agents	Practices
1	Not known to consistently cause disease in healthy adults (e.g., *Lactobacillus casei, Vibrio fischeri*)	Standard Microbiological Practices
2	Associated with human disease, potential hazard if percutaneous injury, ingestion, mucous membrane exposure occurs (e.g., *Salmonella typhi, E. coli* O157:H7, *Staphylococcus aureus*)	BSL-1 practice plus: • Limited access • Biohazard warning signs • "Sharps" precautions • Biosafety manual defining any needed waste decontamination or medical surveillance policies
3	Indigenous or exotic agents with potential for aerosol transmission; disease may have serious or lethal consequences (e.g., *Coxiella burnetti, Yersinia pestis,* herpesviruses)	BSL-2 practice plus: • Controlled access • Decontamination of all waste • Decontamination of lab clothing before laundering • Baseline serum values determined in workers using BSL-3 agents
4	Dangerous/exotic agents that pose high risk of life-threatening disease, aerosol-transmitted lab infections; or related agents with unknown risk of transmission (e.g., variola major (smallpox virus), Ebola virus, hemorrhagic fever viruses)	BSL-3 practices plus: • Clothing change before entering • Shower on exit • All material decontaminated on exit from facility

spirochetes in skin lesions associated with early syphilis or Lyme disease. The fluorescence microscope can be used to identify certain acid-fast microorganisms (*Mycobacterium tuberculosis*) after they are stained with fluorochromes such as auramine-rhodamine. Some morphological and genetic features used in classification and identification of microorganisms are presented in section 17.4 and table 17.3. Direct microscopic examination of most specimens suspected of containing fungi can be made as well. Identification of hyphae in clinical specimens is a presumptive positive result for fungal infection. Definitive identification of most fungi is based on the morphology of reproductive structures (spores). ◀◀ *Light microscopes (section 2.2)*

Many stains that can be used to examine specimens for specific microorganisms have been described. Two of the more widely used bacterial stains are the Gram stain and the acid-fast stain. Because these stains are based on the chemical composition of cell walls, they are not useful in identifying bacteria without cell walls (e.g., mycoplasmas). Lactophenol aniline (cotton) blue is typically used to stain fungi from cultures. Fungal infections (i.e., mold and yeast infections) often are diagnosed by direct microscopic examination of specimens using fluorescence. For example, the identification of molds often can be made if a portion of the specimen is mixed with a drop of 10% Calcofluor White stain on a glass slide. Concentrated wet mounts of blood, stool, or urine specimens can be examined microscopically for the presence of eggs, cysts, larvae, or vegetative cells of parasites. D'Antoni's iodine (1%) is often used to stain internal structures of parasites. Blood smears for apicomplexan (malaria) and flagellate (trypanosome) parasites are stained with Giemsa. Refer to standard references, such as the *Manual of Clinical Microbiology* published by the American Society for Microbiology, for details about other reagents and staining procedures.

The field of clinical microbiology changed dramatically in the 1980s when immunologists created hybrid cells (hybridomas) that secrete antibodies and live a very long time (*see Techniques & Applications 29.1*). Recall that each **hybridoma** cell and its progeny normally produce a **monoclonal antibody (mAb)** of a single specificity. These antibodies recognize a single epitope and are therefore used for diagnostics. One such method, known as **immunofluorescence** or **immunohistochemistry**, results from the chemical attachment of fluorescent dyes called fluorochromes to mAbs; the mAb then binds to a single epitope and the fluorescent molecule "reports" that binding. This technique can be used to tag specific microorganisms in a clinical specimen. Examples of commonly used fluorochromes include rhodamine B and fluorescein isothiocyanate (FITC), which can be coupled to antibody molecules without changing the antibody's capacity to bind to a specific antigen. In the clinical microbiology laboratory, fluorescently labeled mAbs to viral or bacterial antigens have replaced polyclonal antisera for use in culture confirmation when accurate, rapid identification is required. With the use of sensitive techniques such as fluorescence microscopy, it is possible to perform antibody-based microbial identifications with improved accuracy, speed, and fewer organisms. ≪ *Antibodies (section 29.7)*

Two main kinds of fluorescent antibody assays are used: direct and indirect. **Direct immunofluorescence** involves fixing the specimen (cell or microorganism) containing the antigen of interest onto a slide (**figure 32.2a**). Fluorescein-labeled antibodies are then added to the slide and incubated. The slide is washed to remove any unbound antibody and examined with the fluorescence microscope (*see figure 2.11*) for a yellow-green fluorescence. The pattern of fluorescence reveals the antigen's location. Direct immunofluorescence is used to identify antigens such as those found on the surface of group A streptococci and to diagnose enteropathogenic *Escherichia coli, Neisseria meningitidis, Salmonella* spp., *Shigella sonnei, Listeria monocytogenes, Haemophilus influenzae* type b, and *Rabies virus*.

Indirect immunofluorescence (figure 32.2b) is used to detect the presence of antibodies in serum following an individual's exposure to microorganisms. In this technique, a known antigen is fixed onto a slide. The test antiserum is then added, and if the specific antibody is present, it reacts with antigen to form a complex. When fluorescein-labeled antibodies are added, they react with the fixed antibody. After incubation and washing, the slide is examined with the fluorescence microscope. The occurrence of fluorescence shows that antibody specific to the test antigen is present in the serum. Indirect immunofluorescence is used to identify the presence of *Treponema pallidum* antibodies in the diagnosis of syphilis as well as antibodies produced in response to other microorganisms.

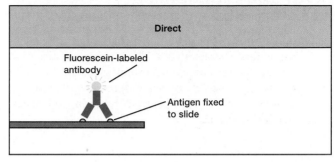

(a)

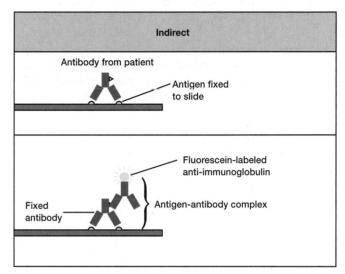

(b)

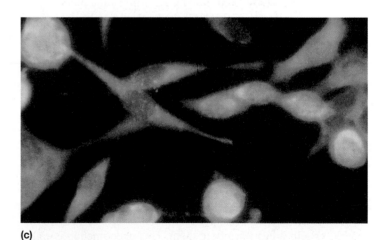

(c)

Figure 32.2 Direct and Indirect Immunofluorescence. (a) In the direct fluorescent-antibody (DFA) technique, the specimen containing antigen is fixed to a slide. Fluorescently labeled antibodies that recognize the antigen are then added, and the specimen is examined with a fluorescence microscope for yellow-green fluorescence. (b) The indirect fluorescent-antibody technique (IFA) detects antigen on a slide as it reacts with an antibody directed against it. The antigen-antibody complex is located with a fluorescent antibody that recognizes antibodies. (c) Several infected cells in a herpes simplex virus positive tissue culture.

Growth and Biochemical Characteristics

Many microorganisms can be identified by particular growth patterns and biochemical characteristics. These characteristics vary, depending on whether the clinical microbiologist is dealing with viruses, fungi (yeasts, molds), parasites (protozoa, helminths), common gram-positive or gram-negative bacteria, rickettsias, chlamydiae, or mycoplasmas.

Viruses

Viruses are identified by isolation in conventional cell (tissue) culture, by immunodiagnosis (fluorescent antibody, enzyme immunoassay, radioimmunoassay, latex agglutination, and immunoperoxidase), and by molecular detection methods such as nucleic acid probes and PCR amplification assays. Several types of systems are available for virus cultivation: cell cultures, embryonated hen's eggs, and experimental animals; these are discussed shortly.

Cell cultures are divided into three general classes:

1. **Primary cultures** consist of cells derived directly from tissues such as monkey kidney and mink lung cells that have undergone one or two passages (subcultures) since harvesting.

2. **Semicontinuous cell cultures** or low-passage cell lines are obtained from subcultures of a primary culture and usually consist of diploid fibroblasts that undergo a finite number of divisions.

3. **Continuous or immortalized cell cultures,** such as HEp-2 cells, are derived from transformed cells that are generally epithelial in origin. These cultures grow rapidly, are heteroploid (having a chromosome number that is not a simple multiple of the haploid number), and can be subcultured indefinitely.

Each type of cell culture favors the growth of a different array of viruses, just as bacterial culture media have differing selective and restrictive properties for growth of bacteria. Viral replication in cell cultures is detected in two ways: (1) by observing the presence or absence of cytopathic effects (CPEs) and (2) by hemadsorption. A cytopathic effect is an observable morphological change that occurs in cells because of viral replication. Examples include ballooning, binding together, clustering, or even death of the culture cells (*see figure 5.21*). During the incubation period of a cell culture, red blood cells can be added. Several viruses alter the plasma membrane of infected culture cells so that red blood cells adhere firmly to them. This phenomenon is called **hemadsorption**. ≪ *Viral multiplication (section 5.3)*

Embryonated chicken eggs can be used for virus isolation. There are three main routes of egg inoculation for virus isolation as different viruses grow best on different cell types: (1) the allantoic cavity, (2) the amniotic cavity, and (3) the chorioallantoic membrane. Egg tissues are inoculated with clinical specimens to determine the presence of virus; virus is revealed by the development of pocks on the chorioallantoic membrane, by the development of hemagglutinins in the allantoic and amniotic fluid, and by death of the embryo.

Laboratory animals, especially suckling mice, also may be used for virus isolation. Inoculated animals are observed for specific signs of disease or death. Several serological tests for viral identification make use of mAb-based immunofluorescence. These tests (figure 32.2c) detect viruses such as herpes simplex virus in tissue cultures.

1. Name two specimens for which microscopy would be used in the initial diagnosis of an infectious disease.
2. Name three general classes of cell cultures.
3. Explain two ways by which the presence of viral replication is detected in cell culture.
4. What are the advantages of using monoclonal antibody (mAb) immunofluorescence in the identification of viruses?

Fungi

Fungal cultures remain the standard for the recovery of fungi from patient specimens; however, the time needed to culture fungi varies anywhere from a few days to several weeks, depending on the organism. For this reason, fungal cultures demonstrating no growth should be maintained for a minimum of 30 days before they are discarded as a negative result. Cultures should be evaluated for rate and appearance of growth on at least one selective and one nonselective agar medium, with careful examination of colonial morphology, color, and dimorphism. Typically, the isolation of fungi is accomplished by concurrent culture of the specimen on media that is respectively supplemented and unsupplemented with antibiotics and cycloheximide. Antibiotics inhibit bacteria that may be in the specimen and cycloheximide inhibits saprophytic (living on decaying matter) molds. However, a number of media formulations are routinely used to culture specific fungi (**table 32.1**). Fungal

Table 32.1	Examples of Media Used to Isolate Fungi
Culture Medium	**Target Fungi**
Antibiotic agar	Fungi in polymicrobial specimens
Caffeine agar	*Aspergillus, Rhizopus,* and others
Chlamydospore agar	Dimorphic fungi
Cornmeal agar	Most fungi including pathogens
Malt agar	Ascomycetes
Malt extract agar	Basidiomycetes
Potato dextrose-yeast extract agar	Mushrooms
Sabouraud dextrose agar (SAB)	Most fungi
SAB + chloramphenicol + cyclohexamide	Dimorphic fungi and dermatophytes

serology (e.g., complement fixation and immunodiffusion) is designed to detect serum antibody but is limited to a few fungi. The cryptococcal latex antigen test is routinely used for the direct detection of *Cryptococcus neoformans* in serum and cerebrospinal fluid. In the clinical laboratory, nonautomated and automated methods for rapid identification (minutes to hours) are used to detect most yeasts. Any biochemical methods used to detect fungi should always be accompanied by morphological studies examining for pseudohyphae, yeast cell structure, chlamydospores, and so on.

Parasites

Culture of parasites from clinical specimens is not routine. Identification and characterization of ova, trophozoites, and cysts in the specimen result in the definitive diagnosis of parasitic infection. This is typically accomplished by direct microscopic evaluation of the clinical specimen. Typical histological staining of blood, negative staining of other body fluids, and immunofluorescence staining are routinely used in the identification of parasites. Some serological tests also are available.

Bacteria

Isolation and growth of bacteria are required before many diagnostic tests can be used to confirm the identification of the pathogen. The presence of bacterial growth usually can be recognized by the development of colonies on solid media or turbidity in liquid media. The time for visible growth to occur is an important variable in the clinical laboratory. For example, most pathogenic bacteria require only a few hours to produce visible growth, whereas it may take weeks for colonies of mycobacteria or mycoplasmas to become evident. The clinical microbiologist as well as the clinician should be aware of reasonable reporting times for various cultures.

The initial identity of a bacterial organism may be suggested by (1) the source of the culture specimen; (2) its microscopic appearance and gram reaction; (3) its pattern of growth on selective, differential, or metabolism-determining media; and (4) its hemolytic, metabolic, and fermentative properties on the various media (**table 32.2**; *see also table 17.4*). For example, methylene blue is often used to inhibit the growth of gram-positive bacteria, whereas phenylethyl alcohol is often used to inhibit gram-negative bacteria. Sheep blood–supplemented agars can be used to determine hemolytic capabilities.

After the microscopic and growth characteristics of a pure culture of bacteria are examined, specific biochemical tests can be performed. Some of the most common biochemical tests used to identify bacterial isolates are listed in **table 32.3**. Classic dichotomous keys are coupled with the biochemical tests for the identification of bacteria from specimens. Generally, fewer than 20 tests are required to identify clinical bacterial isolates to the species level (**figure 32.3**). ◄◄ *Microbial nutrition (chapter 6)*

Certain bacteria require special considerations. For instance, the rickettsias, chlamydiae, and mycoplasmas differ from other bacterial pathogens in a variety of ways. Rickettsias can be diagnosed by immunoassays or by isolation of the microorganism. Because isolation is both hazardous and expensive, immunological methods are preferred. Isolation of rickettsias and diagnosis of rickettsial diseases are generally confined to reference and specialized research laboratories.

Chlamydiae can be demonstrated in tissues and cell scrapings with Giemsa staining, which detects the characteristic intracellular inclusion bodies (*see figure 19.12*). Immunofluorescent staining of tissues and cells with monoclonal antibody reagents is a more sensitive and specific means of diagnosis. The most sensitive methods for demonstrating chlamydiae in clinical specimens involve nucleic acid sequencing and PCR-based methods. ◄◄ *Techniques for determining microbial taxonomy and phylogeny: Molecular characteristics (section 17.4)*

The most routinely used techniques for identification of the mycoplasmas are immunological (hemagglutinin), complement-fixing antigen-antibody reactions using the patient's sera and PCR. These microorganisms are slow growing; therefore positive results from isolation procedures are rarely available before 30 days—a long delay with an approach that offers little advantage over standard techniques. DNA probes are also used for the detection of *Mycoplasma pneumoniae* in clinical specimens.

---

1. How can fungi and parasites be detected in a clinical specimen? Rickettsias? Chlamydiae? Mycoplasmas?

2. How can a clinical microbiologist determine the initial identity of a bacterium?

3. Describe a dichotomous key that could be used to identify a bacterium.

Rapid Methods of Identification

Clinical microbiology has benefited greatly from technological advances in equipment, computer software and databases, molecular biology, and immunochemistry. With new technology, it has been possible to shift from the multistep methods previously discussed to unitary procedures and systems that incorporate standardization, speed, reproducibility, miniaturization, mechanization, and automation. These rapid identification methods can be divided into three categories: (1) manual biochemical "kit" systems, (2) mechanized/automated systems, and (3) immunologic systems.

One example of a "kit approach" biochemical system for the identification of members of the family *Enterobacteriaceae* and other gram-negative bacteria is the API 20E system. It consists of a plastic strip with 20 microtubes containing dehydrated biochemical substrates that can detect certain biochemical

Table 32.2	Isolation of Pure Bacterial Cultures from Specimens

Selective Media

A selective medium is prepared by the addition of specific substances to a culture medium that will permit growth of one group of bacteria while inhibiting growth of some other groups. These are examples:

Salmonella-Shigella agar (SS) is used to isolate *Salmonella* and *Shigella* species. Its bile salt mixture inhibits many groups of coliforms. Both *Salmonella* and *Shigella* species produce colorless colonies because they are unable to ferment lactose. Lactose-fermenting bacteria will produce pink colonies.

Mannitol salt agar (MS) is used for the isolation of staphylococci. The selectivity is obtained by the high (7.5%) salt concentration that inhibits growth of many groups of bacteria. The mannitol in this medium helps in differentiating the pathogenic from the nonpathogenic staphylococci, as the former ferment mannitol to form acid while the latter do not. Thus this medium is also differential.

Bismuth sulfite agar (BS) is used for the isolation of *Salmonella enterica* serovar Typhi, especially from stool and food specimens. *S. enterica* serovar Typhi reduces the sulfite to sulfide, resulting in black colonies with a metallic sheen.

The addition of blood, serum, or extracts to tryptic soy agar or broth will support the growth of many fastidious bacteria. These media are used primarily to isolate bacteria from cerebrospinal fluid, pleural fluid, sputum, and wound abscesses.

Differential Media

The incorporation of certain chemicals into a medium may result in diagnostically useful growth or visible change in the medium after incubation. These are examples:

Eosin methylene blue agar (EMB) differentiates between lactose fermenters and nonlactose fermenters. EMB contains lactose, salts, and two dyes—eosin and methylene blue. *E. coli*, which is a lactose fermenter, will produce a dark colony or one that has a metallic sheen. *S. enterica* serovar Typhi, a nonlactose fermenter, will appear colorless.

MacConkey agar is used for the selection and recovery of *Enterobacteriaceae* and related gram-negative rods. The bile salts and crystal violet in this medium inhibit the growth of gram-positive bacteria and some fastidious gram-negative bacteria. Because lactose is the sole carbohydrate, lactose-fermenting bacteria produce colonies that are various shades of red, whereas nonlactose fermenters produce colorless colonies.

Hektoen enteric agar is used to increase the yield of *Salmonella* and *Shigella* species relative to other microbiota. The high bile salt concentration inhibits the growth of gram-positive bacteria and retards the growth of many coliform strains.

Blood agar: addition of citrated blood to tryptic soy agar makes possible variable hemolysis, which permits differentiation of some species of bacteria. Three hemolytic patterns can be observed on blood agar.
1. α-hemolysis—greenish to brownish halo around the colony (e.g., *Streptococcus gordonii*, *Streptococcus pneumoniae*)
2. β-hemolysis—complete lysis of blood cells resulting in a clearing effect around growth of the colony (e.g., *Staphylococcus aureus* and *Streptococcus pyogenes*)
3. Nonhemolytic—no change in medium (e.g., *Staphylococcus epidermidis* and *Staphylococcus saprophyticus*)

Media to Determine Biochemical Reactions

Some media are used to test bacteria for particular metabolic activities, products, or requirements. These are examples:

Urea broth is used to detect the enzyme urease. Some enteric bacteria are able to break down urea, using urease, into ammonia and CO_2.

Triple sugar iron (TSI) agar contains lactose, sucrose, and glucose plus ferrous ammonium sulfate and sodium thiosulfate. TSI is used for the identification of enteric organisms based on their ability to attack glucose, lactose, or sucrose, and to liberate sulfides from ammonium sulfate or sodium thiosulfate.

Citrate agar contains sodium citrate, which serves as the sole source of carbon, and ammonium phosphate, the sole source of nitrogen. Citrate agar is used to differentiate enteric bacteria on the basis of citrate utilization.

Lysine iron agar (LIA) is used to differentiate bacteria that can either deaminate or decarboxylate the amino acid lysine. LIA contains lysine, which permits enzyme detection, and ferric ammonium citrate for the detection of H_2S production.

Sulfide, indole, motility (SIM) medium is used for three different tests. One can observe the production of sulfides, formation of indole (a metabolic product from tryptophan utilization), and motility. This medium is generally used for the differentiation of enteric organisms.

characteristics (**figure 32.4**). The biochemical substrates in the 20 microtubes are inoculated with a pure culture of bacteria evenly suspended in sterile physiological saline. After 5 to 12 hours of incubation, the 20 test results are converted to a seven- or nine-digit profile. This profile number can be used with a computer or a book called the *API Profile Index* to identify the bacterium.

Clinical laboratory scientists (medical technologists) are the trained and certified workforce that is the front line in laboratory-based disease detection. They staff the sentinel laboratories that receive patient specimens. The production of faster and more specific detection technologies has enabled the rapid and accurate identification of disease agents. However, the bioterror incidents of 2001 spawned a renewed

Table 32.3	Some Common Biochemical Tests Used by Clinical Microbiologists in the Diagnosis of Bacteria from a Patient's Specimen	
Biochemical Test	**Description**	**Laboratory Application**
Carbohydrate fermentation	Acid and/or gas are produced during fermentative growth with sugars or sugar alcohols.	Fermentation of specific sugars used to differentiate enteric bacteria as well as other genera or species
Casein hydrolysis	Detects the presence of caseinase, an enzyme able to hydrolyze milk protein casein. Bacteria that use casein appear as colonies surrounded by a clear zone	Used to cultivate and differentiate aerobic actinomycetes based on casein utilization. For example, *Streptomyces* uses casein and *Nocardia* does not.
Catalase	Detects the presence of catalase, which converts hydrogen peroxide to water and O_2	Used to differentiate *Streptococcus* ($-$) from *Staphylococcus* ($+$) and *Bacillus* ($+$) from *Clostridium* ($-$)
Citrate utilization	When citrate is used as the sole carbon source, this results in alkalinization of the medium.	Used in the identification of enteric bacteria. *Klebsiella* ($+$), *Enterobacter* ($+$), *Salmonella* (often $+$); *Escherichia* ($-$), *Edwardsiella* ($-$)
Coagulase	Detects the presence of coagulase. Coagulase causes plasma to clot.	This is an important test to differentiate *Staphylococcus aureus* ($+$) from *S. epidermidis* ($-$)
Decarboxylases (arginine, lysine, ornithine)	The decarboxylation of amino acids releases CO_2 and amine.	Used in the identification of enteric bacteria
Esculin hydrolysis	Tests for the cleavage of a glycoside	Used in the differentiation of *Staphylococcus aureus, Streptococcus mitis,* and others ($-$) from *S. bovis, S. mutans,* and enterococci ($+$)
β-galactosidase (ONPG) test	Demonstrates the presence of an enzyme that cleaves lactose to glucose and galactose	Used to separate enterics (*Citrobacter* $+$, *Salmonella* $-$) and to identify pseudomonads
Gelatin liquefaction	Detects whether or not a bacterium can produce proteases that hydrolyze gelatin and liquify solid gelatin medium	Used in the identification of *Clostridium, Serratia, Pseudomonas,* and *Flavobacterium*
Hydrogen sulfide (H_2S)	Detects the formation of hydrogen sulfide from the amino acid cysteine due to cysteine desulfurase	Important in the identification of *Edwardsiella, Proteus,* and *Salmonella*
IMViC (indole; methyl red; Voges-Proskauer; citrate)	The indole test detects the production of indole from the amino acid tryptophan. Methyl red is a pH indicator to determine whether the bacterium carries out mixed acid fermentation. VP (Voges-Proskauer) detects the production of acetoin. The citrate test determines whether or not the bacterium can use sodium citrate as a sole source of carbon.	Used to separate *Escherichia* (MR$+$, VP$-$, indole $+$) from *Enterobacter* (MR$-$, VP$+$, indole$-$) and *Klebsiella pneumoniae* (MR$-$, VP$+$, indole$-$); also used to characterize members of the genus *Bacillus*
Lipid hydrolysis	Detects the presence of lipase, which breaks down lipids into simple fatty acids and glycerol	Used in the separation of clostridia
Nitrate reduction	Detects whether a bacterium can use nitrate as an electron acceptor	Used in the identification of enteric bacteria, which are usually $+$
Oxidase	Detects the presence of cytochrome c oxidase that is able to reduce O_2 and artificial electron acceptors	Important in distinguishing *Neisseria* and *Moraxella* spp. ($+$) from *Acinetobacter* ($-$), and enterics (all $-$) from pseudomonads ($+$)
Phenylalanine deaminase	Deamination of phenylalanine produces phenylpyruvic acid, which can be detected colorimetrically.	Used in the characterization of the genera *Proteus* and *Providencia*
Starch hydrolysis	Detects the presence of the enzyme amylase, which hydrolyzes starch	Used to identify typical starch hydrolyzers such as *Bacillus* spp.
Urease	Detects the enzyme that splits urea to NH_3 and CO_2	Used to distinguish *Proteus, Providencia rettgeri,* and *Klebsiella pneumoniae* ($+$) from *Salmonella, Shigella* and *Escherichia* ($-$)

demand for "better, faster, and smarter" microbial detection and identification technologies. While nucleic acid-based detection systems, such as PCR, have garnered much attention as the basis of newer detection systems, antibody-based identification technologies are still considered more flexible and easier to modify. Traditional antibody-based detection technologies are being linked to sophisticated reporting systems that provide "med techs" with an ever-increasing array of cutting-edge technology. Examples of more recent microbial identification technologies include **biosensors** based on: (1) microfluidic

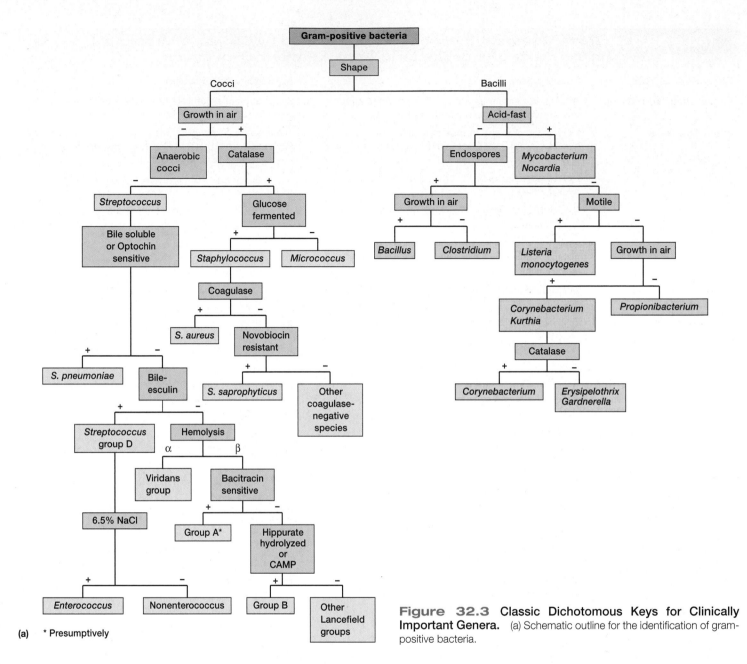

Figure 32.3 Classic Dichotomous Keys for Clinically Important Genera. (a) Schematic outline for the identification of gram-positive bacteria.

antigen sensors, (2) real time (20-minute) PCR, (3) highly sensitive spectroscopy systems, and (4) liquid crystal amplification of microbial immune complexes. Some of these technologies are being used as part of military sentinel detection programs; others are awaiting approval by various licensing agencies before being deployed in clinical laboratories. Additional technologies are expected as the demand for immediate, highly sensitive microbial detection increases globally. Thus the rapidly growing discipline of immunology has greatly aided the clinical microbiologist. Numerous technologies now exist that exploit the specificity and sensitivity of monoclonal antibodies to detect and identify microorganisms. These are briefly discussed here and expanded upon in section 32.3.

Monoclonal antibodies (mAbs) have many applications. For example, they are routinely used in the typing of tissue; in the identification and epidemiological study of infectious microorganisms, tumors, and other surface antigens; in the classification of leukemias; in the identification of functional populations of different types of T cells; and in the identification and mapping of antigenic determinants (epitopes) on proteins. Importantly, mAbs can be conjugated with molecules that provide colorimetric, fluorometric, or enzymatic activity to report the binding of the mAb to specific microbial antigens. Numerous microbial detection kits are available to screen clinical specimens for the presence of specific microorganisms **(table 32.4)**. mAbs have been produced against a wide variety of bacteria, viruses, fungi, and protozoans, and

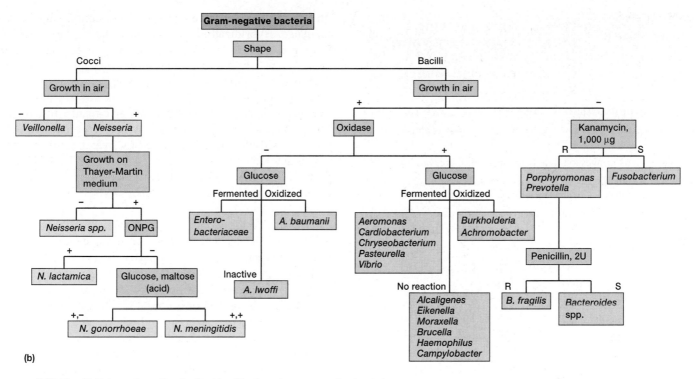

Figure 32.3 (b) Schematic outline for the identification of gram-negative bacteria.

have been made as cross-species or cross-genus reactivity to be used as an adjunct method in the taxonomic identification of microorganisms. Those mAbs that define species-specific antigens are extremely valuable in diagnostic reagents. Monoclonal antibodies that exhibit more restrictive specificity can be used to identify strains of biotypes within a species and in epidemiological studies involving the matching of microbial strains. Coupling sensitive visualization technologies such as fluorescence or scanning tunneling microscopy to mAb detection systems makes it possible to perform microbial identifications with improved accuracy, speed, and fewer organisms.

1. Describe in general how biochemical tests are used in the API 20E system to identify bacteria.

2. Why might cultures for some microorganisms be unavailable?

3. What are some of the benefits of monoclonal antibodies as diagnostic reagents?

Bacteriophage Typing

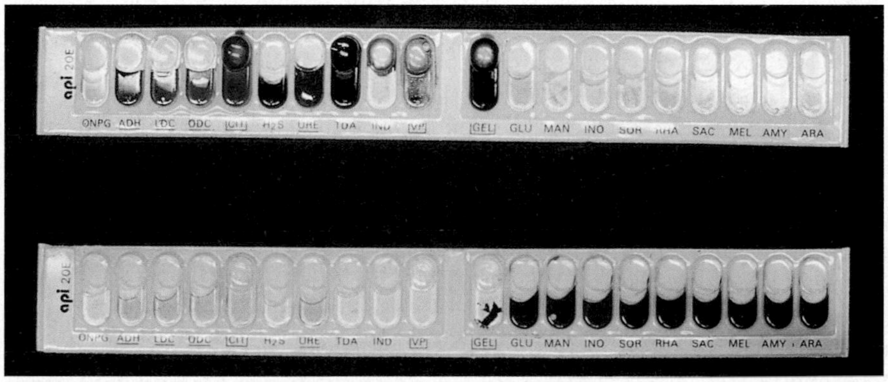

Bacteriophages are viruses that attack members of a particular bacterial species or strains within a species. Bacteriophage (phage) typing is based on the specificity of phage surface proteins for cell surface receptors. Only those bacteriophages that can attach to these surface receptors can infect bacteria and cause lysis. On a petri dish culture, lytic bacteriophages cause plaques on lawns of sensitive bacteria. These plaques represent infection by the virus (*see figure 5.19*). << *Viruses and other acellular agents (chapter 5)*

In **bacteriophage typing,** the clinical microbiologist inoculates the bacterium to be tested onto a petri plate. The plate is heavily and uniformly inoculated so that the bacteria will grow to form a solid lawn of cells. The plate is then marked off into squares (15 to 20 mm per side), and each square is inoculated with a drop of suspension from the different phages available for typing. After the plate is incubated for 24 hours, it is observed for plaques. The phage type is reported as a specific genus and species

Figure 32.4 A "Kit Approach" to Bacterial Identification. The API 20E manual biochemical system for microbial identification. (a) Positive and (b) negative results.

Table 32.4	**Some Common Rapid Immunologic Test Kits for the Detection of Bacteria and Viruses in Clinical Specimens**

Culturette Group A Strep ID Kit (Marion Scientific, Kansas City, Mo.)

The Culturette kit is used for the detection of group A streptococci from throat swabs.

Directigen (Hynson, Wescott, and Dunning, Baltimore, Md.)

The Directigen Meningitis Test kit is used to detect *H. influenzae* type b, *S. pneumoniae,* and *N. meningitidis* groups A and C.

The Directigen Group A Strep Test kit is used for the direct detection of group A streptococci from throat swabs.

Gono Gen (Micro-Media Systems, San Jose, Calif.)

The Gono Gen kit detects *Neisseria gonorrhoeae.*

OraQuick (OraSure Technologies, Bethleham, Pa.)
Detects HIV antibodies in saliva in 10 minutes.

Staphaurex (Wellcome Diagnostics, Research Triangle Park, N.C.)

Staphaurex screens and confirms *Staphylococcus aureus* in 30 seconds.

SureCell Herpes (HSV) Test (Kodak, Rochester, N.Y.)

Detects the herpes (HSV) 1 and 2 viruses in minutes.

followed by the types that can infect the bacterium. For example, the series 10/16/24 indicates that this bacterium is sensitive to phages 10, 16, and 24, and belongs to a collection of strains, called a phagovar, that have this particular phage sensitivity. Bacteriophage typing remains a tool of research and reference laboratories.

Molecular Genetic Methods

Some of the most accurate approaches to microbial identification are through the analysis of proteins and nucleic acids. Examples include comparison of proteins; physical, kinetic, and regulatory properties of microbial enzymes; nucleic acid–base composition; nucleic acid hybridization; and nucleic acid sequencing (*see figure 15.2*). Other molecular methods being widely used are nucleic acid probes, gas-liquid chromatography, and DNA fingerprinting. << *Techniques for determining microbial taxonomy and phylogeny (section 17.4)*

Nucleic acid–based diagnostic methods for the detection and identification of microorganisms have become routine in clinical microbiology laboratories. For example, DNA hybridization technology can identify a microorganism by probing its genetic composition. The use of cloned DNA as a probe is based on the capacity of single-stranded DNA to bind (hybridize) with a complementary nucleic acid sequence present in test specimens to form a double-stranded DNA hybrid (*see figure 16.6*). Thus DNA derived from one microorganism

(the probe) is used to search for others containing the same sequence. Hybridization reactions may be applied to purified DNA preparations, to bacterial colonies, or to clinical specimens such as tissue, serum, sputum, and pus. DNA probes have been developed that bind to complementary strands of ribosomal RNA. These DNA:rRNA hybrids are more sensitive than conventional DNA probes, give results in 2 hours or less, and require the presence of fewer microorganisms. DNA probe sensitivity can be increased by over 1 million times if the target DNA is first amplified using PCR. DNA:rRNA probes are available or are currently being developed for most clinically important microorganisms. << *Polymerase chain reaction (section 16.2)*

The nucleotide sequence of small subunit ribosomal RNA (rRNA) can be used to identify bacterial genera (*see figure 17.8*). Usually the rRNA encoding gene or gene fragment is amplified by PCR. After nucleotide sequencing, the rRNA gene nucleotide sequence is compared with those on international databases. This method of bacterial identification, called **ribotyping,** is based on the high level of 16s rRNA conservation among bacteria. As discussed in chapter 17, multiple "housekeeping" genes, instead of 16S rRNA genes, may be sequenced and analyzed in a technique called **multilocus sequence typing (MLST)** (*see p. 396*).

Genomic fingerprinting is also used in identifying pathogens. This does not involve nucleotide sequencing; rather, it compares the similarity of specific DNA fragments generated by restriction endonuclease digestion. BOX-, ERIC-, and REP-PCR are also described in section 17.4 (*see figure 17.9*). Plasmids—autonomously replicating extrachromosomal molecules of DNA—can also be used in DNA fingerprinting. **Plasmid fingerprinting** identifies microbial isolates of the same or similar strains; related strains often contain the same plasmids. In contrast, microbial isolates that are phenotypically distinct have different plasmid fingerprints. Plasmid fingerprinting of many *E. coli, Salmonella, Campylobacter,* and *Pseudomonas* strains and species has demonstrated that this method often is more accurate than phenotyping methods such as biotyping, antibiotic resistance patterns, phage typing, and serotyping. Just as in genomic fingerprinting, isolated plasmid DNA is cut with specific restriction endonucleases (*see figure 16.3*). The DNA fragments are then separated by gel electrophoresis to yield a pattern of fragments, which appear as bands in the gel. The molecular weight of each plasmid species can then be determined and patterns of restriction fragments compared (**figure 32.5**). << *Gel electrophoresis (section 16.3)*

---

1. What is the basis for bacteriophage typing?

2. How can nucleic acid–based detection methods be used by the clinical microbiologist?

3. How can a suspect bacterium be plasmid fingerprinted?

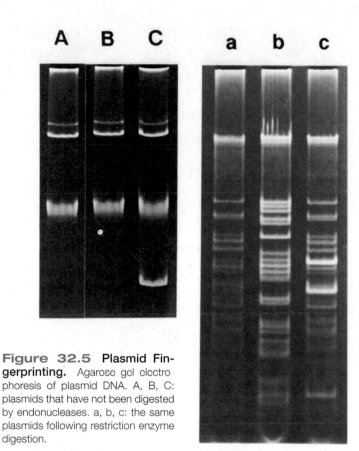

A   B   C   a   b   c

Figure 32.5 Plasmid Fingerprinting. Agarose gel electrophoresis of plasmid DNA. A, B, C: plasmids that have not been digested by endonucleases. a, b, c: the same plasmids following restriction enzyme digestion.

32.3 CLINICAL IMMUNOLOGY

The culturing of certain viruses, bacteria, fungi, and protozoa from clinical specimens may not be possible because the methodology remains undeveloped (e.g., *Treponema pallidum;* hepatitis A, B, C; and Epstein-Barr virus), is unsafe (rickettsias and HIV), or is impractical for all but a few microbiology laboratories (e.g., mycobacteria, strict anaerobes, *Borrelia*). Cultures also may be negative because of prior antimicrobial therapy. Under these circumstances, detection of antibodies or antigens may be quite valuable diagnostically.

Immunologic systems for the detection and identification of pathogens from clinical specimens are easy to use, give relatively rapid reaction end points, and are sensitive and specific with a low percentage of false positives and negatives. Some of the more popular immunologic rapid test kits for viruses and bacteria are presented in table 32.4.

Dramatic advances in clinical immunology have given rise to a marked increase in the number, sensitivity, and specificity of serological tests. This increase reflects a better understanding of (1) immune cell surface antigens (CD antigens), (2) lymphocyte biology, (3) the production of monoclonal antibodies, and (4) the development of sensitive antibody-binding reporter systems. For a number of reasons, the utility of these tests depends on proper test selection and timing of specimen collection. For instance, each individual's immunologic response to a microorganism is quite variable, making the interpretation of immunologic tests potentially difficult. For example, a single, elevated IgM titer does not distinguish between active and past infections. Rather an elevated IgG titer typically indicates an active infection, especially when subsidence of symptoms correlates with a fourfold (or greater) decrease in antibody titer. Furthermore, a lack of a measurable antibody titer may reflect an organism's lack of immunogenicity or insufficient time for an antibody response to develop following the onset of the infectious disease. Some patients are also immunosuppressed due to other disease processes or treatment procedures (e.g., cancer and AIDS patients) and therefore do not respond. In this section, some of the more common antibody-based techniques employed in the diagnosis of microbial and immunological diseases are discussed.

Serotyping

Serum is the liquid portion of blood (devoid of clotting factors) that contains many different components, especially the immunoglobulins or antibodies. **Serotyping** refers to the use of serum (antibodies) to specifically detect and identify other molecules. Serotyping can be used to identify specific white blood cells or the proteins on cell surfaces. Serotyping can also be used to differentiate strains (serovars or serotypes) of microorganisms that differ in the antigenic composition of a structure or product. The serological identification of a pathogenic strain has diagnostic value. Often the symptoms of infections depend on the nature of the cell products released by the pathogen. Therefore it is sometimes possible to identify a pathogen serologically by testing for cell wall antigens. For example, there are 90 different strains of *Streptococcus pneumoniae*, each unique in the nature of its capsular material. These differences can be detected by antibody-induced capsular swelling, termed the **Quellung reaction,** when the appropriate antiserum for a specific capsular type is used (**figure 32.6**).

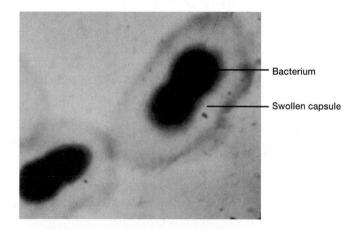

— Bacterium

— Swollen capsule

Figure 32.6 Serotyping. *Streptococcus pneumoniae* has reacted with a specific pneumococcal antiserum leading to capsular swelling (the Quellung reaction). The capsules seen around the bacteria indicated potential virulence.

Agglutination

An **agglutination reaction** occurs when an immune complex is formed by cross-linking cells or particles with specific antibodies (*see figure 29.22*). Agglutination reactions usually form visible aggregates or clumps, called **agglutinates,** that can be seen with the unaided eye. Direct agglutination reactions are very useful in the diagnosis of certain diseases. For example, the **Widal test** is a reaction involving the agglutination of typhoid bacilli when they are mixed with serum containing typhoid antibodies from an individual who has typhoid fever. << *Action of antibodies: Immune complex formation (section 29.8)*

Techniques have also been developed that employ microscopic synthetic latex spheres coated with antigens. These coated microspheres are extremely useful in diagnostic agglutination reactions. For example, microspheres are used in common pregnancy tests that detect the elevated level of human chorionic gonadotropin (hCG) hormone, which occurs in a woman's urine and blood early in pregnancy. Latex agglutination tests are also used to detect antibodies that develop during certain mycotic, helminthic, and bacterial infections, as well as in drug testing.

Hemagglutination usually results from antibodies cross-linking red blood cells through attachment to surface antigens and is routinely used in blood typing. In addition, some viruses can accomplish **viral hemagglutination.** For example, if a person has a certain viral disease, such as measles, antibodies will be present in the serum to react with the measles viruses and neutralize them. Normally hemagglutination occurs when measles viruses and red blood cells are mixed. However, red blood cells may be mixed first with a person's serum followed by the addition of virions. If no hemagglutination occurs, the serum antibodies have neutralized the measles viruses. This is considered a positive test result for the presence of virus-specific antibodies (**figure 32.7**). Hemagglutination inhibition tests are widely used to diagnose influenza, measles, mumps, mononucleosis, and other viral infections.

Agglutination tests are also used to measure antibody titer. In tube or well agglutination tests, a specific amount of antigen is added to a series of tubes or shallow wells in a microtiter plate (**figure 32.8**). Serial dilutions of serum (1/20, 1/40, 1/80, 1/160, etc.) containing the antibody are then added to each tube or well. The greatest dilution of serum showing an agglutination reaction is determined, and the reciprocal of this dilution is the serum antibody titer.

1. What is serology?
2. When would you use the Widal test?
3. Why does hemagglutination occur and how can it be used in the clinical laboratory?

Complement Fixation

When complement binds to an antigen-antibody complex, it becomes "fixed" and "used up." Complement fixation tests are very sensitive and can be used to detect extremely small

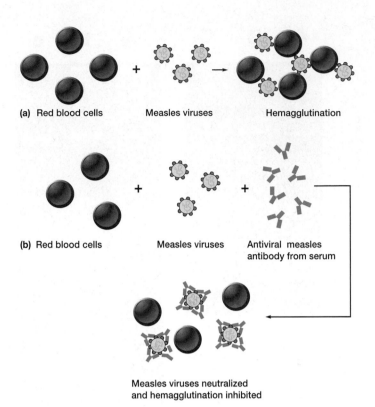

(a) Red blood cells   Measles viruses   Hemagglutination

(b) Red blood cells   Measles viruses   Antiviral measles antibody from serum

Measles viruses neutralized and hemagglutination inhibited

Figure 32.7 Viral Hemagglutination. (a) Certain viruses can bind to red blood cells causing hemagglutination. (b) If serum containing specific antibodies to the virus is mixed with the red blood cells, the antibodies will neutralize the virus and inhibit hemagglutination (a positive test).

amounts of an antibody for a suspect microorganism in an individual's serum. A known antigen is mixed with test serum lacking complement (**figure 32.9***a*). When immune complexes have had time to form, complement is added (figure 32.9*b*) to the mixture. If immune complexes are present, they will fix and consume complement. Afterward, sensitized indicator cells, usually sheep red blood cells previously coated with complement-fixing antibodies, are added to the mixture. If specific antibodies are present in the test serum and complement is consumed by the immune complexes, insufficient amounts of complement will be available to lyse the indicator cells. On the other hand, in the absence of antibodies, complement remains and lyses the indicator cells (figure 32.9*c*). Thus absence of lysis shows that specific antibodies are present in the test serum. Complement fixation was once used in the diagnosis of syphilis (the Wassermann test) and is still used as a rapid, inexpensive screening method in the diagnosis of certain viral, fungal, rickettsial, chlamydial, and protozoan diseases.

Enzyme-Linked Immunosorbent Assay

The *enzyme-linked immunosorbent assay* (**ELISA**) has become one of the most widely used serological tests for antibody or antigen detection. This test involves the linking of various "label" enzymes to either antigens or antibodies. Two basic methods are

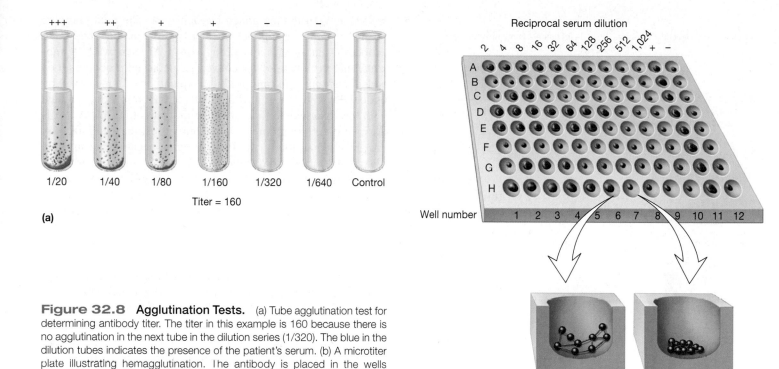

Figure 32.8   Agglutination Tests.   (a) Tube agglutination test for determining antibody titer. The titer in this example is 160 because there is no agglutination in the next tube in the dilution series (1/320). The blue in the dilution tubes indicates the presence of the patient's serum. (b) A microtiter plate illustrating hemagglutination. The antibody is placed in the wells (rows 1–10). Positive controls (row 11) and negative controls (row 12) are included. Red blood cells are added to each well. If sufficient antibody is present to agglutinate the cells, they sink as a mat to the bottom of the well. If insufficient antibody is present, they form a pellet at the bottom. Can you read the different titers in rows A–H?

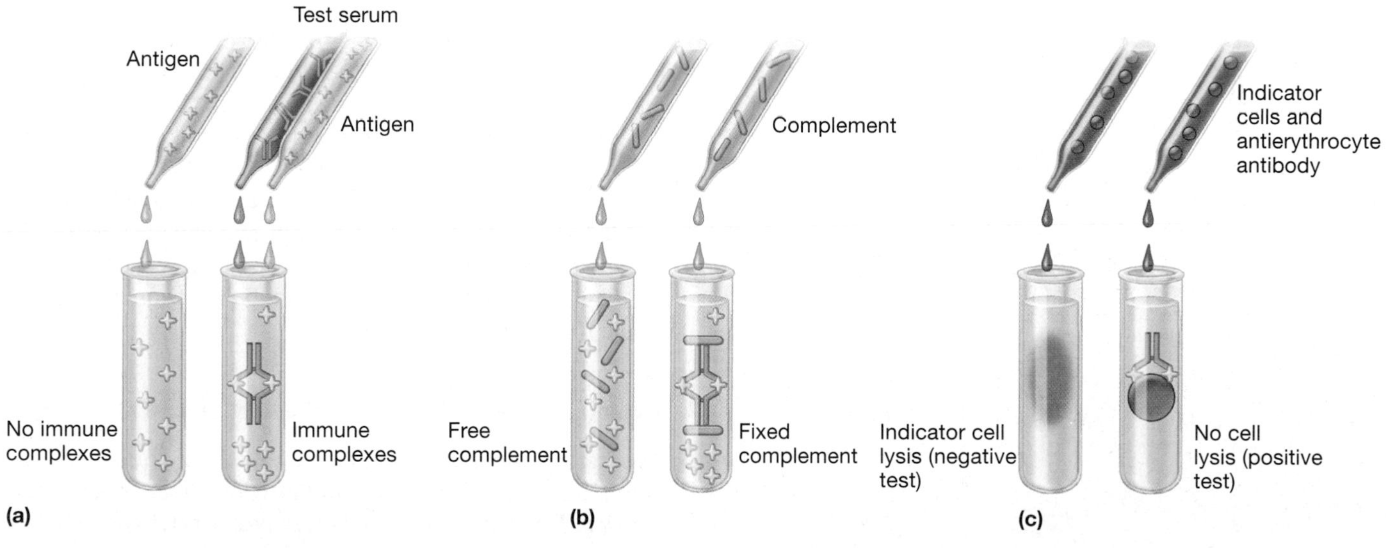

Figure 32.9   Complement Fixation.   (a) Test serum is added to one test tube. A fixed amount of antigen is then added to both tubes. If antibody is present in the test serum, immune complexes form. (b) When complement is added, if complexes are present, they fix complement and consume it. (c) Indicator cells and a small amount of anti-erythrocyte antibody are added to the two tubes. If there is complement present, the indicator cells will lyse (a negative test); if the complement has been consumed by the immune complex, no lysis will occur (a positive test).

used: the direct (also called the double antibody sandwich assay) and the indirect immunosorbent assay.

The double antibody sandwich assay is used for the detection of antigens (**figure 32.10a**). In this assay, specific antibody is placed in wells of a microtiter plate (or it may be attached to a membrane). The antibody is absorbed onto the walls, coating the plate. A test antigen (in serum, urine, etc.) is then added to each well. If the antigen reacts with the antibody, the antigen is retained when the well is washed to remove unbound antigen. A commercially prepared antibody-enzyme conjugate specific

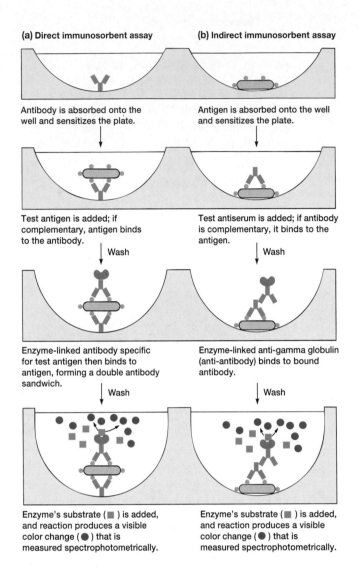

(a) Direct immunosorbent assay

Antibody is absorbed onto the well and sensitizes the plate.

Test antigen is added; if complementary, antigen binds to the antibody.

Wash

Enzyme-linked antibody specific for test antigen then binds to antigen, forming a double antibody sandwich.

Wash

Enzyme's substrate (■) is added, and reaction produces a visible color change (●) that is measured spectrophotometrically.

(b) Indirect immunosorbent assay

Antigen is absorbed onto the well and sensitizes the plate.

Test antiserum is added; if antibody is complementary, it binds to the antigen.

Wash

Enzyme-linked anti-gamma globulin (anti-antibody) binds to bound antibody.

Wash

Enzyme's substrate (■) is added, and reaction produces a visible color change (●) that is measured spectrophotometrically.

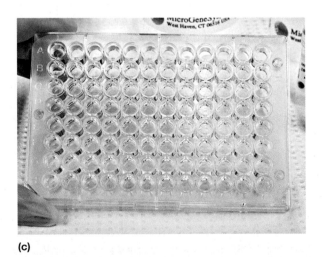

(c)

Figure 32.10 **The ELISA or EIA Test.** (a) The direct or double antibody sandwich method for the detection of antigens. (b) The indirect immunosorbent assay for detecting antibodies. (c) HIV antibody ELISA.

for the antigen is then added to each well. The final complex formed is an outer antibody-enzyme, middle antigen, and inner antibody—that is, it is a layered (Ab-Ag-Ab) sandwich. A substrate that the enzyme will convert to a colored product is then added, and any resulting product is quantitatively measured by optical density scanning of the plate. If the antigen has reacted with the absorbed antibodies in the first step, the ELISA test is positive (i.e., it is colored). If the antigen is not recognized by the absorbed antibody, the ELISA test is negative because the unattached antigen has been washed away, and no antibody-enzyme is bound (it is colorless). This assay is currently being used for the detection of *Helicobacter pylori* infections, brucellosis, salmonellosis, and cholera. Many other antigens also can be detected by the sandwich method. For example, some ELISA kits on the market can test for many different food allergens.

The indirect immunosorbent assay detects antibodies rather than antigens. In this assay, antigen in appropriate buffer is incubated in the wells of a microtiter plate (figure 32.10*b*) and is adsorbed onto the walls of the wells. Free antigen is washed away. Test antiserum is added, and if specific antibody is present, it binds to the antigen. Unbound antibody is washed away. Alternatively the test sample can be incubated with a suspension of latex beads that have the desired antigen attached to their surface. After allowing time for antibody-antigen complex formation, the beads are trapped on a filter and unbound antibody is washed away. An anti-antibody that has been covalently coupled to an enzyme, such as horseradish peroxides, is added next. The antibody-enzyme complex (the conjugate) binds to the test antibody, and after unbound conjugate is washed away, the attached ligand is visualized by the addition of a **chromogen.** A chromogen is a colorless substrate acted on by the enzyme portion of the ligand to produce a colored product. The amount of test antibody is quantitated in the same way as an antigen is in the double antibody sandwich method. The indirect immunosorbent assay currently is being used to test for antibodies to HIV and *Rubella virus* (German measles), and to detect certain drugs in serum. For example, antigen-coated latex beads are used in the OraQuick HIV-1 test to detect HIV serum antibodies in about 10 minutes.

Immunoblotting (Western Blotting)

Another immunologic technique used in the clinical microbiology laboratory is immunoblotting, also known as Western blotting. **Immunoblotting** involves polyacrylamide gel electrophoresis of a protein specimen followed by transfer of the separated proteins to sheets of nitrocellulose or polyvinyl difluoride. Protein bands are then visualized by treating the nitrocellulose sheets with solutions of enzyme-tagged antibodies. This procedure demonstrates the presence of common and specific proteins among different strains of microorganisms. Immunoblotting also can be used to show strain-specific immune responses to microorganisms, to serve as an important diagnostic indicator of a recent infection with a particular strain

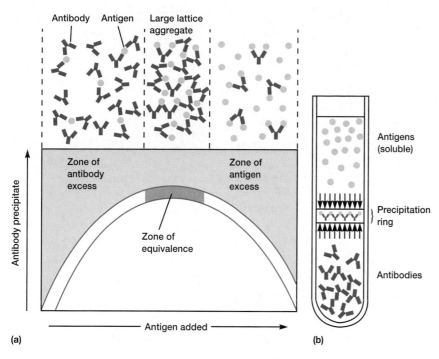

Figure 32.11 Immunoprecipitation. (a) Graph showing that a precipitation curve is based on the ratio of antigen to antibody. The zone of equivalence represents the optimal ratio for precipitation. (b) A precipitation ring test. Antibodies and antigens diffuse toward each other in a test tube. A precipitation ring is formed at the zone of equivalence.

of microorganism, and to allow for prognostic implications with severe infectious diseases.

Immunoprecipitation

The **immunoprecipitation** technique detects soluble antigens that react with antibodies called **precipitins.** The precipitin reaction occurs when bivalent or multivalent antibodies and antigens are mixed in the proper proportions. The antibodies link the antigen to form a large antibody-antigen network or lattice that settles out of solution when it becomes sufficiently large (**figure 32.11a**). Immunoprecipitation reactions occur only at the equivalence zone when there is an optimal ratio of antigen to antibody so that an insoluble lattice forms. If the precipitin reaction takes place in a test tube (figure 32.11b), a precipitation ring forms in the area in which the optimal ratio or equivalence zone develops. ◀◀ *Action of antibodies: Immune complex formation (section 29.8)*

Immunodiffusion

Immunodiffusion refers to a precipitation reaction that occurs between an antibody and antigen in an agar gel medium. Two techniques are routinely used: single radial immunodiffusion and double diffusion in agar.

The **single radial immunodiffusion (RID) assay** or Mancini technique quantitates antigens. Antibody is added to agar, then

the mixture is poured onto slides and allowed to set. Wells are cut in the agar and known amounts of standard antigen added. The unknown test antigen is added to a separate well (**figure 32.12a**). The slide is left for 24 hours or until equilibrium has been reached, during which time the antigen diffuses out of the wells to form insoluble complexes with antibodies. The size of the resulting precipitation ring surrounding various dilutions of antigen is proportional to the amount of antigen in the well (the wider the ring, the greater the antigen concentration). This is because the antigen's concentration drops as it diffuses farther out into the agar. The antigen forms a precipitin ring in the agar when its level has decreased sufficiently to reach equivalence and combine with the antibody to produce a large, insoluble network. This method is commonly used to quantitate serum immunoglobulins, complement proteins, and other substances.

The **double diffusion agar assay (Öuchterlony technique)** is based on the principle that diffusion of both antibody and antigen (hence double diffusion) through agar can form stable and easily observable immune complexes. Test solutions of antigen and antibody are added to the separate wells punched in agar. The solutions diffuse outward, and when antigen and the appropriate antibody meet, they combine and precipitate at the equivalence zone, producing an indicator line (or lines) (figure 32.12b). The visible line of precipitation permits a comparison of antigens for identity (same antigenic determinants), partial identity (cross-reactivity), or nonidentity against a given selected antibody. For example, if a V-shaped line of precipitation forms, this demonstrates that the antibodies bind to the same antigenic determinants in each antigen sample and are identical. If one well is filled with a different antigen that shares some but not all determinants with the first antigen, a Y-shaped line of precipitation forms, demonstrating partial identity. In this reaction, the stem of the Y, called a spur, is formed if those antigen or antigenic determinants absent in the first well but present in the second one (antigen A in figure 32.12b) react with the diffusing antibodies. If two completely unrelated antigens are added to the wells, either a single straight line of precipitation forms between the two wells or two separate lines of precipitation form, creating an X-shaped pattern, a reaction of nonidentity.

Immunoelectrophoresis

Some antigen mixtures are too complex to be resolved by simple diffusion and precipitation. Greater resolution is obtained by the technique of **immunoelectrophoresis** in which antigens are first separated based on their electrical charge, then visualized by the precipitation reaction. In this procedure, antigens are separated by electrophoresis in an agar gel. Positively charged proteins move to

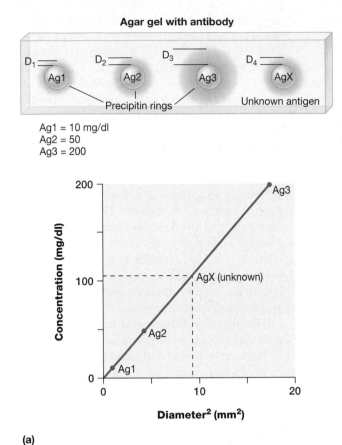

Ag1 = 10 mg/dl
Ag2 = 50
Ag3 = 200

(a)

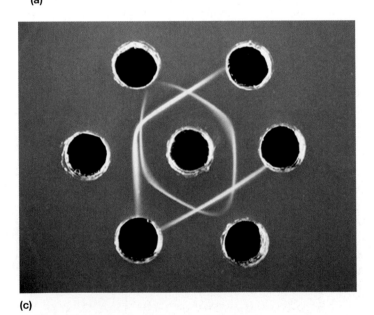

(c)

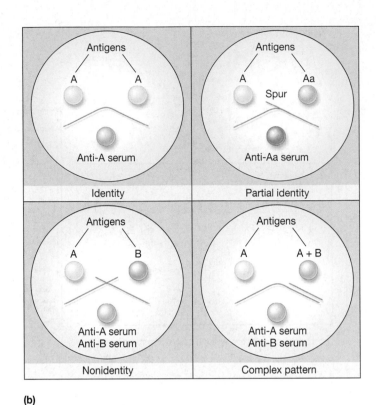

(b)

Figure 32.12 Immunodiffusion. (a) Single radial immunodiffusion assay. Three standard solutions of different antigen concentrations (Ag1, Ag2, Ag3) and an unknown (AgX) are placed on agar. After equilibration, the ring diameters are measured. Usually the square of the diameter of the standard rings is plotted on the x-axis and the antigen concentration on the y-axis. From this standard curve, the concentration of an unknown can be determined. (b) Double diffusion agar assay showing characteristics of identity (top left), reaction of nonidentity (bottom left), partial identity (top right), and a complex pattern (bottom right). (c) Experimental results of identity, partial identity, and nonidentity.

1. What does a negative complement fixation test show? A positive test?

2. What are the two types of ELISA methods and how do they work? What is a chromogen?

3. Specifically, when do immunoprecipitation reactions occur?

4. Name two types of immunodiffusion tests and describe how they operate.

5. Describe the immunoelectrophoresis technique.

the negative electrode, and negatively charged proteins move to the positive electrode (**figure 32.13a**). A trough is then cut next to the wells (figure 32.13b) and filled with antibody. When the plate is incubated, the antibodies and antigens diffuse and form precipitation bands or arcs (figure 32.13c) that can be better visualized by staining (figure 32.13d). This assay is used to separate the major blood proteins in serum for certain diagnostic tests. << *Gel electrophoresis (section 16.3)*

Flow Cytometry

Flow cytometry allows single- or multiple-microorganism detection in clinical samples in an easy, reliable, fast way. In flow cytometry, microorganisms are identified on the basis of their unique cytometric parameters or by means of certain dyes called fluorochromes that can be used either independently or bound to specific antibodies or oligonucleotides. The flow

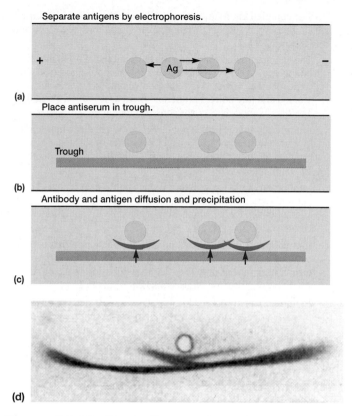

(a) Separate antigens by electrophoresis.

+   Ag   −

(b) Place antiserum in trough.

Trough

(c) Antibody and antigen diffusion and precipitation

(d)

Figure 32.13   Classical Immunoelectrophoresis. (a) Antigens are separated in an agar gel by an electrical charge. (b) Antibody (antiserum) is then placed in a trough cut parallel to the direction of the antigen migration. (c) The antigens and antibodies diffuse through the agar and form precipitin arcs. (d) After staining, better visualization is possible.

cytometer forces a suspension of cells through a laser beam and measures the light they scatter or the florescence the cells emit as they pass through the beam. For example, cells can be tagged with a fluorescent antibody directed against a specific surface antigen. As the stream of cells flows past the laser beam, each fluorescent cell can be detected, counted, and even separated from the other cells in the suspension. The cytometer also can measure a cell's shape, size, and content of DNA and RNA. This technique has enabled the development of quantitative methods to assess antimicrobial susceptibility and drug cytotoxicity in a rapid, accurate, and highly reproducible way. One outstanding contribution of this technique is the ability to detect the presence of heterogenous microbial populations with different responses to antimicrobial treatments.

Radioimmunoassay

The *radioimmunoassay* (**RIA**) technique has become an extremely important tool in biomedical research and clinical practice (e.g., in cardiology, blood banking, diagnosis of allergies, and endocrinology). Indeed, Rosalyn Yalow won the 1977 Nobel Prize in Physiology or Medicine for its development. RIA uses a purified antigen that is radioisotope-labeled and competes for antibody with unlabeled standard antigen or test antigen in experimental samples. The radioactivity associated with the antibody is then detected by means of radioisotope analyzers and autoradiography (photographic emulsions that show areas of radioactivity). If there is a lot of antigen in an experimental sample, it will compete with the radioisotope-labeled antigen for antigen-binding sites on the antibody, and little radioactivity will be bound. A large amount of bound radioactivity indicates that little antigen is present in the experimental sample.

1. Explain how flow cytometry is both qualitative and quantitative (i.e., how it can determine both the identity and the number of a specific cell type).
2. Describe the RIA technique.

Summary

32.1 Overview of the Clinical Microbiology Laboratory

a. The major focus of the clinical microbiologist is identifying microorganisms from clinical specimens accurately and rapidly. This may involve isolation of the microbe or the application of molecular or immunologic tests that do not require microbial cultures (**figure 32.1**).

32.2 Identification of Microorganisms from Specimens

a. The clinical microbiology laboratory can provide preliminary or definitive identification of microorganisms based on (1) microscopic examination of specimens; (2) growth and biochemical characteristics of microorganisms isolated from cultures; (3) immunologic techniques that detect antibodies or microbial antigens; (4) bacteriophage typing; and (5) molecular amplification and hybridization techniques.

b. Viruses are identified by isolation in living cells or immunologic tests. Several types of living cells are available: cell culture, embryonated hen's eggs, and experimental animals.

c. The culturing of many types of bacteria is routine; however, a few require special consideration. Rickettsial disease can be diagnosed immunologically. Chlamydiae can be demonstrated in tissue and cell scrapings with Giemsa stain, which detects the characteristic intracellular inclusion bodies. The most routinely used techniques for identification of the mycoplasmas are immunologic.

d. Identification of fungi often can be made if a portion of the specimen is mixed with a drop of 10% Calcofluor White stain. Wet mounts of stool specimens or urine can be examined microscopically for the presence of parasites.

e. Immunofluorescence is a process in which fluorochromes are irradiated with UV, violet, or blue light to make them fluoresce. These dyes can be coupled to an antibody. There are two main kinds of fluorescent antibody assays: direct and indirect (**figure 32.2**).

f. The initial identity of a bacterial organism may be suggested by (1) the source of the culture specimen; (2) its microscopic appearance; (3) its pattern of growth on selective, differential, enrichment, or characteristic media; and (4) its hemolytic, metabolic, and fermentative properties.

g. Rapid methods for microbial identification can be divided into three categories: (1) manual biochemical systems, (2) mechanized/automated systems, and (3) immunologic systems.

h. Bacteriophage typing for bacterial identification is based on the fact that phage surface receptors bind to specific cell surface receptors. On a petri plate culture, bacteriophages cause plaques on lawns of bacteria possessing the proper receptors.

i. Various molecular methods also can be used to identify microorganisms. Examples include nucleic acid-based detection, and genome and plasmid fingerprinting.

32.3 Clinical Immunology

a. Serotyping refers to serological procedures used to differentiate strains (serovars or serotypes) of microorganisms that have differences in the antigenic composition of a structure or product (**figure 32.6**).

b. Agglutination reactions in vitro usually form aggregates or clumps (agglutinates) that are visible with the naked eye. Tests have been developed, such as the Widal test, latex microsphere agglutination reaction, hemagglutination, and viral hemagglutination, to detect antigen as well as to determine antibody titer (**figures 32.7** and **32.8**).

c. The complement fixation test can be used to detect a specific antibody for a suspect microorganism in an individual's serum (**figure 32.9**).

d. The enzyme-linked immunosorbent assay (ELISA) involves linking various enzymes to either antigens or antibodies. Two basic methods are involved: the double antibody sandwich method and the indirect immunosorbent assay (**figure 32.10**). The first method detects antigens and the latter, antibodies.

e. Immunoblotting involves polyacrylamide gel electrophoresis of a protein specimen followed by transfer of the separated proteins to nitrocellulose sheets and identification of specific bands by labeled antibodies.

f. Immunoprecipitation reactions occur only when there is an optimal ratio of antigen and antibody to produce a lattice at the zone of equivalence, which is evidenced by a visible precipitate (**figure 32.11**).

g. Immunodiffusion refers to a precipitation reaction that occurs between antibody and antigen in an agar gel medium. Two techniques are routinely used: double diffusion in agar and single radial diffusion (**figure 32.12**).

h. In immunoelectrophoresis, antigens are separated based on their electrical charge, then visualized by precipitation and staining (**figure 32.13**).

i. Flow cytometry and fluorescence allow single- or multiple-microorganism detection based on their cytometric parameters or by means of certain dyes called fluorochromes.

Critical Thinking Questions

1. As more ways of identifying the characteristics of microorganisms emerge, the number of distinguishable microbial strains also seems to increase. Why do you think this is the case?

2. Why are miniaturized identification systems used in clinical microbiology? Describe one such system and its advantage over classic dichotomous keys.

3. It has been speculated that as good as nucleic acid tests are in identifying viruses, antibody-based identification tests will be just as specific but easier and faster to develop and get to the marketplace. Do you agree or disagree? Defend your answer.

4. ELISA tests usually use a primary and secondary antibody. Why? What are the necessary controls one would need to perform to ensure that the antibody specificities are valid (i.e., no false-positive or false-negative reactions)?

Learn More

Learn more by visiting the Prescott website at www.mhhe.com/prescottprinciples, where you will find a complete list of references.

The Epidemiology of Infectious Disease

33

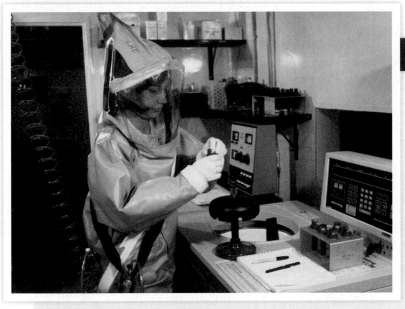

This laboratory worker at the Centers for Disease Control and Prevention is in the highest level of isolation (Level 4) to avoid contact with microorganisms and to prevent their escape into the environment.

airborne transmission The spread of an infectious organism by its suspension in the air, traveling over a meter or more from the source to the host.

antigenic drift A small change in the immunogenic character of an organism that allows it to evade immune system attack.

antigenic shift A major change in the immunogenic character of an organism; it is unrecognized by immune mechanisms.

carrier An infected individual who is a potential source of infection for others.

common-source epidemic An epidemic characterized by a sharp rise to a peak and then a less rapid decline in the number of individuals infected; it usually involves a single infectious source.

communicable disease A disease associated with a pathogen that can be transmitted from one host to another.

endemic disease A disease that is constantly present in a population, usually at a steady low frequency.

epidemic A disease that suddenly increases in occurrence above the normal level in a given population.

epidemiology The study of the factors determining and influencing the frequency and distribution of disease, injury, and other health-related events and their causes.

Chapter Glossary

fomite An inanimate object that is not in itself harmful but is able to harbor and transmit pathogens.

herd immunity The resistance of a population to infection and spread of an infectious agent due to the immunity of a high percentage of the population.

incubation period The period after pathogen entry into a host and before signs and symptoms appear.

morbidity rate The number of individuals who become ill as a result of a particular disease within a susceptible population during a specific time period.

mortality rate The ratio of the number of deaths from a given disease to the total number of cases of the disease.

nosocomial infection An infection that is acquired while a patient is in a hospital or other clinical care facility.

pandemic An increase in the occurrence of a disease within a large and geographically widespread population (often refers to a worldwide epidemic).

prodromal stage The period during the course of a disease in which signs and symptoms appear, but are not yet distinctive enough to make an accurate diagnosis.

reservoir A site, alternate host, or carrier that normally harbors pathogenic organisms and serves as a source from which other individuals can be infected.

toxoid A bacterial exotoxin that has been modified so that it is no longer toxic but will still stimulate antitoxin formation when injected into a person or animal.

vaccine A preparation of either killed microorganisms; living, weakened (attenuated) microorganisms; or inactivated bacterial toxins (toxoids); given to induce an immune response and protect the individual against a pathogen or a toxin.

vector A living organism, usually an arthropod or other animal, that transfers an infective agent between hosts.

vehicle An inanimate substance or medium that transmits a pathogen.

zoonoses A disease of animals that can be transmitted to humans.

Epidemics of infectious disease are often compared with forest fires. Once fire has spread through an area, it does not return until new trees have grown up. Epidemics in humans develop when a large population of susceptible individuals is present. If most individuals are immune, then an epidemic will not occur.

—*Andrew Cliff and Peter Haggett*

In this chapter, we describe the epidemiological parameters that are studied in the infectious disease cycle. The practical goal of epidemiology is to establish effective recognition, control, prevention, and eradication measures within a given population. Because emerging and reemerging diseases and pathogens, as well as the threat of bioterrorism, are worldwide concerns, these topics are covered here. Nosocomial (hospital-acquired) infections have increased in recent years, and a brief synopsis of their epidemiology also is presented.

By definition, **epidemiology** (Greek *epi,* upon; *demos,* people or population; and *logy,* study) is the science that evaluates the occurrence, determinants, distribution, and control of health and disease in a defined human population (**figure 33.1**). Any individual who practices epidemiology is an **epidemiologist.** Epidemiologists are, in effect, disease detectives. Their major concerns are the discovery of the factors essential to disease occurrence and the development of methods for disease prevention. In the United States, the **Centers for Disease Control and Prevention** (CDC), headquartered in Atlanta, GA, serves as the national agency for developing and carrying out disease prevention and control, environmental health, and health promotion and education activities. Its worldwide counterpart is the **World Health Organization** (WHO), located in Geneva, Switzerland.

The science of epidemiology originated and evolved in response to the great epidemic diseases such as cholera, typhoid fever, smallpox, influenza, and yellow fever (**Historical Highlights 33.1**). More recent epidemics of ebola, HIV/AIDS, cryptosporidiosis, enteropathogenic *Escherichia coli,* SARS, and avian influenza have underscored the importance of epidemiology in preventing global catastrophies caused by infectious diseases. Today its scope encompasses all diseases: infectious diseases, genetic abnormalities, metabolic dysfunction, malnutrition, neoplasms, psychiatric disorders, and aging. This chapter emphasizes only infectious disease epidemiology.

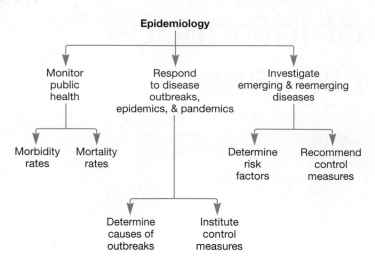

Figure 33.1 Epidemiology. Epidemiology is a multifaceted science that investigates diseases to discover their origin, evaluates diseases to assess their risk, and controls diseases to prevent future outbreaks.

33.1 EPIDEMIOLOGICAL TERMINOLOGY

When a disease occurs occasionally and at irregular intervals in a human population, it is a **sporadic disease** (e.g., bacterial meningitis). When it maintains a steady, low-level frequency at a moderately regular interval, it is an **endemic** (Greek *endemos,* dwelling in the same people) disease (e.g., the common cold).

Historical Highlights

33.1   John Snow—The First Epidemiologist

Much of what we know today about the epidemiology of cholera is based on the classic studies conducted between 1849 and 1854 by the British physician John Snow. During this period, a series of cholera outbreaks occurred in London, England, and Snow set out to find the source of the disease. Some years earlier when he was still a medical apprentice, Snow had been sent to help during an outbreak among coal miners of cholera. His observations convinced him that the disease was usually spread by unwashed hands and shared food, not by "bad" air or casual direct contact.

Thus when the outbreak of 1849 occurred, Snow believed that cholera was spread among the poor in the same way as among the coal miners. He suspected that water, and not unwashed hands and shared food, was the source of the cholera infection among the wealthier residents. Snow examined official death records and discovered that most of the victims in the Broad Street area had lived close to the Broad Street water pump or had been in the habit of drinking from it. He concluded that cholera was spread by drinking water from the Broad Street pump, which was contaminated with raw sewage containing the disease agent. When the pump handle was removed, the number of cholera cases dropped dramatically.

In 1854 another cholera outbreak struck London. Part of the city's water supply came from two different suppliers: the Southwark and Vauxhall Company, and the Lambeth Company. Snow interviewed cholera patients and found that most of them purchased their drinking water from the Southwark and Vauxhall Company. He also discovered that this company obtained its water from the Thames River below locations where Londoners had discharged their sewage. In contrast, the Lambeth Company took its water from the Thames before the river reached the city. The death rate from cholera was over eightfold lower in households supplied with Lambeth Company water. Water contaminated by sewage was transmitting the disease. Finally, Snow concluded that the cause of the disease must be able to multiply in water. Thus he nearly recognized that cholera was caused by a microorganism, though Robert Koch did not discover the causative bacterium (*Vibrio cholerae*) until 1883.

To commemorate these achievements, the John Snow Pub now stands at the site of the old Broad Street pump. Those who complete the Epidemiologic Intelligence Program at the Centers for Disease Control and Prevention receive an emblem bearing a replica of a barrel of Whatney's Ale—the brew dispensed at the John Snow Pub.

Hyperendemic diseases gradually increase in occurrence frequency beyond the endemic level but not to the epidemic level (e.g., the common cold during winter months). An **outbreak** is the sudden, unexpected occurrence of a disease, usually focally or in a limited segment of a population (e.g., Legionnaires' disease). An **epidemic** (Greek *epidemios,* upon the people), on the other hand, is an outbreak affecting many people at once (i.e., there is a sudden increase in the occurrence of a disease above the expected level). Influenza is an example of a disease that may occur suddenly and unexpectedly in a family and often achieves epidemic status in a community. The first case in an epidemic is called the **index case.** Finally, a **pandemic** (Greek *pan,* all) is an increase in disease occurrence within a large population over a very wide region (the world). Usually pandemic diseases spread among continents. The global H1N1 influenza outbreak of 1918 is a good example.

33.2 MEASURING FREQUENCY

To determine if an outbreak, epidemic, or pandemic is occurring, epidemiologists measure disease frequency at single time points and over time. They then use statistics to analyze the data and determine risk factors and other factors associated with disease. **Statistics** is the branch of mathematics dealing with the collection, organization, and interpretation of numerical data. As a science particularly concerned with rates and the comparison of rates, epidemiology was the first medical field in which statistical methods were extensively used.

Measures of frequency usually are expressed as fractions. The numerator is the number of individuals experiencing the event—infection or other problem—and the denominator is the number of individuals in whom the event could have occurred, that is, the population at risk. The fraction is a proportion or ratio but is commonly called a rate because a time period is specified. (A rate also can be expressed as a percentage.) In population statistics, rates usually are stated per 1,000 individuals, although other powers of 10 may be used for particular diseases (e.g., per 100 for very common diseases and per 10,000 or 100,000 for uncommon diseases).

A **morbidity rate** measures the number of individuals who become ill due to a specific disease within a susceptible population during a specific time interval. The rate is commonly determined when the number of new cases of illness in the general population is known from clinical reports. It is calculated as follows:

$$\text{Morbidity rate} = \frac{\text{number of new cases of a disease}}{\text{Number of individuals in the population}}$$

For example, if in one month 700 new cases of influenza per 100,000 individuals occur, the morbidity rate would be expressed as 700 per 100,000 or 0.7%.

The **prevalence rate** refers to the total number of individuals infected in a population at any one time no matter when the disease began. The prevalence rate depends on both the incidence rate and the duration of the illness.

The **mortality rate** is the relationship between the number of deaths from a given disease and the total number of cases of the disease. The mortality rate is a simple statement of the proportion of all deaths that are assigned to a single cause. It is calculated as follows:

$$\text{Mortality rate} = \frac{\text{number of deaths due to a given disease}}{\substack{\text{size of the total population} \\ \text{with the same disease}}}$$

For example, if 15,000 deaths occurred due to AIDS in a year and the total number of people infected was 30,000, the mortality rate would be 15,000 per 30,000 or 1 per 2 or 50%.

The determination of morbidity, prevalence, and mortality rates helps public health personnel in directing health care efforts to control the spread of infectious diseases. For example, a sudden increase in the morbidity rate of a particular disease may indicate a need for the implementation of preventive measures designed to reduce mortality.

1. What is epidemiology?
2. What terms are used to describe the occurrence of a disease in a human population?
3. Define morbidity rate, prevalence rate, and mortality rate.

33.3 RECOGNITION OF AN INFECTIOUS DISEASE IN A POPULATION

An **infectious disease** is a disease resulting from an infection by microbial agents such as viruses, bacteria, fungi, protozoa, and helminths. A **communicable disease** is an infectious disease that can be transmitted from person to person (not all infectious diseases are communicable; e.g., rabies is an infectious disease acquired through contact with a rabid animal). The manifestations of an infectious or communicable disease can range from mild to deadly, depending on the agent and host. An epidemiologist studying an infectious disease is concerned with the causative agent, the source or reservoir of the disease agent, how it is transmitted, what host and environmental factors can aid development of the disease within a defined population, and how best to control or eliminate the disease. These factors describe the natural history or cycle of an infectious disease (section 33.5).

Epidemiologists recognize an infectious disease in a population by using various surveillance methods. Surveillance is a dynamic activity that includes gathering information on the development and occurrence of a disease, collating and analyzing the data, summarizing the findings, and using the information to select control methods (**figure 33.2**). Some combination of these surveillance methods is used most often for the:

1. Generation of morbidity data from case reports

2. Collection of mortality data from death certificates

3. Investigation of actual cases

4. Collection of data from reported epidemics

5. Field investigation of epidemics

6. Review of laboratory results: for example, surveys of a population for antibodies against the agent and specific microbial serotypes, skin tests, cultures, and stool analyses

7. Population surveys using statistical sampling to determine who has the disease

8. Use of animal and vector disease data

9. Collection of information on the use of specific biologics—antibiotics, antitoxins, vaccines, and other prophylactic measures

10. Use of demographic data on population characteristics such as human movements during a specific time of the year

11. Use of remote sensing and geographic information systems

As noted, surveillance may not always require the direct examination of cases. However, to accurately interpret surveillance data and study the course of a disease in individuals, epidemiologists and other medical professionals must be aware of the pattern of infectious diseases. Often infectious diseases have characteristic signs and symptoms. **Signs** are objective changes in the body, such as a fever or rash, that can be directly observed. **Symptoms** are subjective changes, such as pain and loss of appetite, that are personally experienced by the patient. The term symptom is often used in a broader scope to include the clinical signs. A **disease syndrome** is a set of signs and symptoms that are characteristic of the disease. Frequently additional laboratory tests are required for an accurate diagnosis because symptoms and readily observable signs may not be sufficient for diagnosis.

The course of an infectious disease usually has a characteristic pattern and can be divided into several phases (**figure 33.3**). Knowledge of the pattern is essential in accurately diagnosing the disease. The **incubation period** is the time between pathogen entry and the development of signs and symptoms. The pathogen is spreading but has not reached a sufficient level to cause clinical manifestations. This period's length varies with disease. Next, the **prodromal stage** occurs with an onset of signs and symptoms that are not yet specific enough to make a diagnosis. However, the patient often is contagious. This is followed by the **illness period,** when the disease is most severe and displays characteristic signs and symptoms. The immune response is typically triggered; B and T cells become active. Finally, during the **period of decline,** the signs

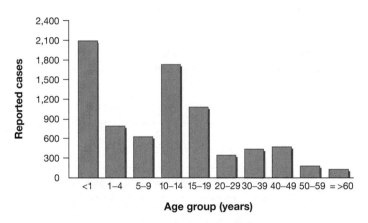

(a) Pertussis—Reported cases by age group, United States, 2000

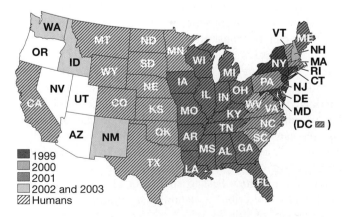

(b) West Nile virus cases were first reported in 1999 among a variety of animals and humans. Colors track its rapid progress across the United States in just 4 years.

Figure 33.2 Graphic Representation of Epidemiological Data. The Centers for Disease Control and Prevention collects and evaluates a number of parameters related to human health and disease. The data are then presented in various formats including (a) age and (b) geographic region.

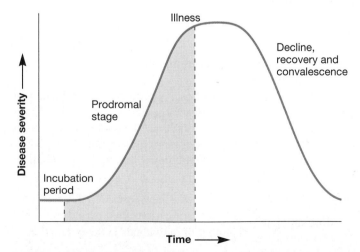

Figure 33.3 The Course of an Infectious Disease. Most infectious diseases occur in four stages. The duration of each stage is a characteristic feature of each disease. The shaded area represents when a disease is typically communicable.

Historical Highlights

33.2   "Typhoid Mary"

In the early 1900s there were thousands of typhoid fever cases, and many people died of the disease. Most of these cases arose when people drank water contaminated with sewage or ate food handled by or prepared by individuals who were shedding the typhoid fever bacterium (*Salmonella enterica* serovar Typhi). The most famous carrier of the typhoid bacterium was Mary Mallon.

Between 1896 and 1906, Mary Mallon worked as a cook in seven homes in New York City. Twenty-eight cases of typhoid fever occurred in these homes while she worked in them. As a result, the New York City Health Department had Mallon arrested and admitted to an isolation hospital on North Brother Island in New York's East River. Examination of her stools showed that she

was shedding large numbers of typhoid bacteria, though she exhibited no external symptoms of the disease. An article published in 1908 in the *Journal of the American Medical Association* referred to Mallon as "Typhoid Mary," an epithet by which she is still known today. She was released when she pledged not to cook for others or serve food to them, Mallon changed her name and began to work as a cook again. For five years she managed to avoid capture while continuing to spread typhoid fever. Eventually the authorities tracked her down. She was held in custody for 23 years until she died in 1938. As a lifetime carrier, Mary Mallon was positively linked with 10 outbreaks of typhoid fever, 53 cases, and 3 deaths.

---

and symptoms begin to disappear. The recovery stage often is referred to as convalescence.

Remote Sensing and Geographic Information Systems: Charting Infectious Diseases

Remote sensing and geographic information systems are map-based tools that can be used to study the distribution, dynamics, and environmental correlates of microbial diseases. **Remote sensing (RS)** is the gathering of data—digital images of Earth's surface from satellites and output from biological sensors, for example—transforming the data into maps. A **geographic information system (GIS)** is a data management system that organizes and displays digital map data from RS and facilitates the analysis of relationships between mapped features. Statistical relationships often exist between mapped features and diseases in natural host or human populations. Examples include the location of the habitats of the malaria parasite and mosquito vectors in Mexico and Asia, Lyme disease in the United States, and African trypanosomiasis in both humans and livestock in the southeastern United States. RS and GIS may also permit the assessment of human risk from pathogens such as *Sin Nombre virus* (the virus that causes hantavirus pulmonary syndrome in North America). RS and GIS are most useful if disease dynamics and distributions are clearly related to mapped environmental variables. For example, if a microbial disease is associated with certain vegetation types or physical characteristics (e.g., elevation, precipitation), RS and GIS can identify regions where risk is relatively high.

After an infectious disease has been recognized in a population, epidemiologists correlate the disease outbreak with a specific organism—its exact cause must be discovered (**Historical Highlights 33.2**). At this point, the clinical or diagnostic microbiology laboratory enters the investigation. Its purpose is to isolate and identify the organism responsible for the disease.

33.4   RECOGNITION OF AN EPIDEMIC

Two major types of epidemic are recognized: common source and propagated. A **common-source epidemic** is characterized as having reached a peak level within a short period of time (1 to 2 weeks) followed by a moderately rapid decline in the number of infected patients (**figure 33.4a**). This type of epidemic usually results from a single common contaminated source such as food (food poisoning) or water (Legionnaires' disease). By contrast, a **propagated epidemic** is characterized by a relatively slow and prolonged rise and then a gradual decline in the number of

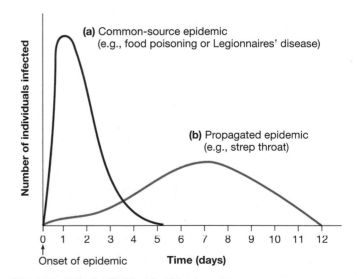

Figure 33.4 Epidemic Curves. (a) In a common-source epidemic, there is a rapid increase up to a peak in the number of individuals infected and then a rapid but more gradual decline. Cases usually are reported for a period that equals approximately one incubation period of the disease. (b) In a propagated epidemic, the curve has a gradual rise and then a gradual decline. Cases usually are reported over a time internal equivalent to several incubation periods of the disease.

individuals infected (figure 33.4*b*). This type of epidemic usually results from the introduction of a single infected individual into a susceptible population. The initial infection is then propagated to others in a gradual fashion until many individuals within the population are infected. An example is the increase in strep throat cases that coincides with new populations of sensitive children who arrive in classrooms. One infected child is sufficient to initiate the epidemic.

To understand how epidemics are propagated, consider **figure 33.5**. At time 0, all individuals in this population are susceptible to a hypothetical pathogen. The introduction of an infected individual initiates the epidemic outbreak (lower curve), which spreads to reach a peak by day 15. As individuals recover from the disease, they become immune and no longer transmit the pathogen (upper curve). The number of susceptible individuals therefore decreases. The decline in the number of susceptibles to the threshold density (the minimum number of individuals necessary to continue propagating the disease) coincides with the peak of the epidemic wave, and the incidence of new cases declines because the pathogen cannot propagate itself.

Herd immunity is the resistance of a population to infection and pathogen spread because of the immunity of a large percentage of the population. The larger the proportion of those immune, the smaller the probability of effective contact between infective and susceptible individuals—that is, many contacts will be immune, and thus the population will exhibit a group resistance. A susceptible member of such an immune population enjoys an immunity that is not of his or her own making but instead arises because of membership in the group.

At times public health officials immunize large portions of the susceptible population in an attempt to maintain a high level of herd immunity. Any increase in the number of susceptible

individuals may result in an endemic disease becoming epidemic. The proportion of immune to susceptible individuals must be constantly monitored because new susceptible individuals continually enter a population through migration and birth.

Pathogens cause endemic diseases because infected humans continually transfer them to others (e.g., sexually transmitted diseases) or because they continually reenter the human population from animal reservoirs (e.g., rabies). Other pathogens continue to evolve and may produce epidemics (e.g., AIDS, influenza virus [A strain], and *Legionella* bacteria). One way in which a pathogen changes is by **antigenic shift,** a major genetically determined change in the antigenic character of a pathogen (**figure 33.6**). An antigenic shift can be so extensive that the pathogen is no longer recognized by the host's immune system. For example, influenza viruses frequently change by recombination from one antigenic type to another. Antigenic shift also occurs through the hybridization of different influenza virus serovars; two serovars of a virus intermingle to form a new antigenic type. Hybridization may occur between an animal strain and a human strain of the virus. Even though resistance in the human population becomes so high that the virus can no longer spread (herd immunity), it can be transmitted to animals, where the hybridization takes place. Smaller antigenic changes also can take place by mutations in pathogenic strains that help the pathogen avoid host immune responses. These smaller changes are called **antigenic drift.** ◄◄ *Viruses with minus-strand RNA genomes (section 24.6)*

Whenever antigenic shift or drift occurs, the population of susceptible individuals increases because the immune system does not recognize the new mutant strain. If the percentage of susceptible people is above the threshold density (figure 33.5), the level of protection provided by herd immunity will decrease and the morbidity rate will increase. For example, the morbidity rates of influenza among schoolchildren may reach epidemic levels if the number of susceptible people rises above 30% for the whole population. As a result, the goal of public health agencies is to make sure that at least 70% of the population is immunized against infectious diseases to provide the herd immunity necessary for the protection of those who are not immunized.

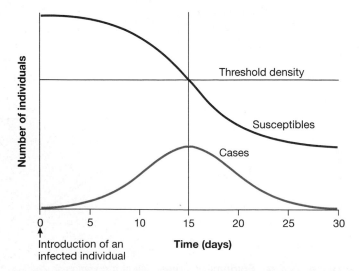

Figure 33.5 The Spread of an Imaginary Propagated Epidemic. The lower curve represents the number of cases and the upper curve the number of susceptible individuals. Notice the coincidence of the peak of the epidemic wave with the threshold density of susceptible people.

1. How can epidemiologists recognize an infectious disease in a population? Define sign, symptom, and disease syndrome. What are the four phases seen during the course of an infection?

2. How can remote sensing and geographic information systems chart infectious diseases?

3. Differentiate between common-source and propagated epidemics.

4. Explain herd immunity. How does this protect the community?

5. What is the significance of antigenic shift and drift in epidemiology?

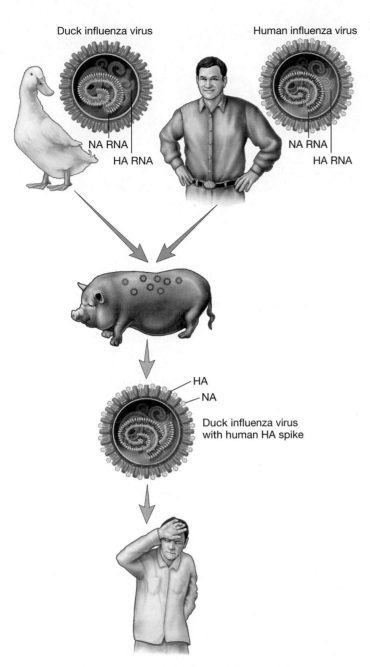

Figure 33.6 Antigenic Shift in Influenza Viruses. Close habitation of humans, pigs, and ducks permits the multiple infection of the pigs with both human and duck influenza viruses. Exchange of genetic information then results in a new strain of influenza that is new to humans; humans have no immunity to the new strain.

33.5 THE INFECTIOUS DISEASE CYCLE: STORY OF A DISEASE

To continue to exist, a pathogen must reproduce and be disseminated among its hosts. Thus an important aspect of infectious disease epidemiology is a consideration of how reproduction and dissemination occur. The **infectious disease cycle** or **chain of infection** represents these events in the form of an intriguing epidemiological mystery story (**figure 33.7**).

What Pathogen Caused the Disease?

The first link in the infectious disease cycle is the pathogen. After an infectious disease has been recognized in a population, epidemiologists correlate the disease outbreak with a specific pathogen. This is where Koch's postulates and modifications of them are used to determine the etiology or cause of an infectious disease. At this point, the clinical or diagnostic microbiology laboratory enters the investigation. Its purpose is to isolate and identify the pathogen that caused the disease and to determine the pathogen's susceptibility to antimicrobial agents or methods that may assist in its eradication. << *Golden age of microbiology: Koch's postulates (section 1.5); Clinical microbiology and immunology (chapter 32)*

What Was the Source or Reservoir of the Pathogen?

The source or reservoir of a pathogen is the second link in the infectious disease cycle. If the source or reservoir of the infection can be eliminated or controlled, the infectious disease cycle itself will be interrupted and transmission of the pathogen will be prevented.

A **source** is the location from which the pathogen is immediately transmitted to the host, either directly through the environment or indirectly through an intermediate agent. The source can be either animate (e.g., humans or animals) or inanimate (e.g., water, soil, or food). The period of infectivity is the time during which the source is infectious or is disseminating the pathogen.

A **reservoir** is the site or natural environmental location in which the pathogen normally resides. It is also the site from which a source acquires the pathogen or where direct infection of the host can occur. Thus a reservoir sometimes functions as a source. Reservoirs also can be animate or inanimate. Much of the time, human hosts are the most important animate sources of the pathogen and are called carriers. A **carrier** is an infected individual who is a potential source of infection for others. Carriers play an important role in the epidemiology of disease (Historical Highlights 33.1 and 33.2).

Convalescent, healthy, and incubatory carriers may harbor the pathogen for only a brief period (hours, days, or weeks) and then are called casual, acute, or transient carriers. If they harbor the pathogen for long periods (months, years, or life), they are called chronic carriers.

Infectious diseases that can be transmitted from animals to humans are termed **zoonoses** (Greek *zoon*, animal, and *nosos*, disease); thus animals also can serve as reservoirs. Humans contract the pathogen by several mechanisms: coming into direct contact with diseased animal flesh (e.g., tularemia); drinking contaminated cow's milk (e.g., tuberculosis and brucellosis); inhaling dust particles contaminated by animal excreta or products (e.g., Q fever, hantavirus pulmonary infection, anthrax); or eating insufficiently cooked infected flesh (e.g., anthrax, trichinosis). In addition, being bitten by arthropod vectors (organisms that spread

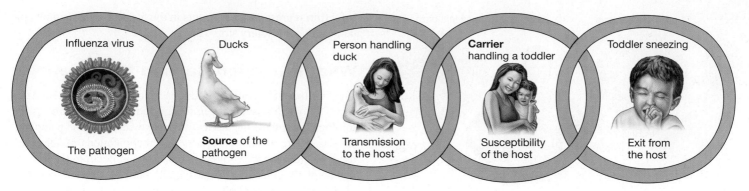

Figure 33.7   Proposed Chain of Avian Influenza Infection.

disease from one host to another) such as mosquitoes, ticks, fleas, mites, or biting flies (e.g., equine encephalomyelitis, malaria, Lyme disease, Rocky Mountain spotted fever, and plague); or being bitten by a diseased animal (e.g., rabies) can lead to infection.

Table 33.1 lists some common zoonoses found in the Western Hemisphere. This table is noninclusive in scope; it merely abbreviates the enormous spectrum of zoonotic diseases that are relevant to human epidemiology. Domestic animals are the most common source of zoonoses because they live in greater proximity to humans than do wild animals. Diseases of wild animals that are transmitted to humans tend to occur sporadically because close contact is infrequent. Other major reservoirs of pathogens are water, soil, and food. These reservoirs are discussed in detail in chapters 34 and 35.

How Was the Pathogen Transmitted?

To maintain an active infectious disease in a human population, the pathogen must be transmitted from one host or source to another. Transmission is the third link in the infectious disease cycle and occurs by four main routes: airborne, contact, vehicle, and vector-borne (**figure 33.8**).

Airborne Transmission

Because air is not a suitable medium for the growth of pathogens, any pathogen that is airborne must have originated from a source such as humans, other animals, plants, soil, food, or water. In **airborne transmission,** the pathogen is suspended in the air and travels over a meter or more from the source to the host. The pathogen can be contained within droplet nuclei or dust. **Droplet nuclei** can be small particles, 1 to 4 μm in diameter, that result from the evaporation of larger particles (5 μm or more in diameter) called droplets. Droplet nuclei can remain airborne for hours or days and travel long distances. Chicken pox and measles are examples of droplet-spread diseases. Lung and many systemic mycoses are examples of transmission in dust.

When animals or humans are the source of the airborne pathogen, it usually is propelled from the respiratory tract into the air by coughing, sneezing, or vocalization. For example,

enormous numbers of moisture droplets are aerosolized during a typical sneeze (**figure 33.9**). Each droplet is about 10 μm in diameter and initially moves about 100 m/sec or more than 200 mi/hr!

Dust also is an important route of airborne transmission. At times a pathogen adheres to dust particles and contributes to the number of airborne pathogens when the dust is resuspended by some disturbance. A pathogen that can survive for relatively long periods in or on dust creates an epidemiological problem, particularly in hospitals, where dust can be the source of hospital-acquired (nosocomial) infections. **Table 33.2** summarizes some human airborne pathogens and the diseases they cause.

Contact Transmission

Contact transmission implies the coming together or touching of the source or reservoir of the pathogen and the host. Contact can be direct or indirect. Direct contact implies an actual physical interaction with the infectious source (figure 33.8). This route is frequently called person-to-person contact. Person-to-person transmission occurs primarily by touching, kissing, or sexual contact; by contact with oral secretions or body lesions (e.g., herpes and boils); by nursing mothers (e.g., staphylococcal infections); and through the placenta (e.g., AIDS, syphilis). Some infectious pathogens also can be transmitted by direct contact with animals or animal products (e.g., *Salmonella* and *Campylobacter*).

Indirect contact refers to the transmission of the pathogen from the source to the host through an intermediary—most often an inanimate object. The intermediary is usually contaminated by an animate source. Common examples of intermediary inanimate objects include thermometers, eating utensils, drinking vessels, stethoscopes, and neckties. *Pseudomonas* bacteria are easily transmitted by the indirect route. This mode of transmission is often also considered a form of vehicle transmission (p. 796).

In droplet spread, the pathogen is carried on particles smaller than 5 μm. The route is through the air for a very short distance—usually less than a meter. As a result, droplet transmission of a pathogen depends on the proximity of the source and the host. Contact with oral secretions may also result when droplet nuclei contaminate body surfaces that touch mucous membranes (e.g., respiratory secretions on hands that contact eyes).

Table 33.1	Infectious Organisms in Nonhuman Reservoirs That May Be Transmitted to Humans		
Disease	**Etiologic Agent**	**Usual or Suspected Nonhuman Host**	**Usual Method of Human Infection**
Anthrax	*Bacillus anthracis*	Cattle, horses, sheep, swine, goats, dogs, cats, wild animals, birds	Inhalation or ingestion of spores; direct contact
Babesiosis	*Babesia bovis, B. divergens, B. microti, B. equi*	*Ixodes* ticks of various species	Bite of infected tick
Brucellosis (undulant fever)	*Brucella melitensis, B. abortus, B. suis, B. canis*	Cattle, goats, swine, sheep, horses, mules, dogs, cats, fowl, deer, rabbits	Milk; direct or indirect contact
Campylobacteriosis	*Campylobacter fetus, C. jejuni*	Cattle, sheep, poultry, swine, pets	Contaminated water and food
Cat-scratch disease	*Bartonella henselae*	Cats, dogs	Cat or dog scratch
Cryptosporidiosis	*Cryptosporidium* spp.	Farm animals, pets	Contaminated water
Encephalitis (St. Louis)	Arbovirus	Birds	Mosquito
Giardiasis	*Giardia intestinalis*	Rodents, deer, cattle, dogs, cats	Contaminated water
Glanders	*Burkholderia mallei*	Horses	Skin contact; inhalation
Hantavirus pulmonary syndrome	Pulmonary syndrome hantavirus	Deer mice	Contact with the saliva, urine, or feces of deer mice; aerosolized viruses
Influenza	Influenza virus	Water fowl, pigs	Direct contact or inhalation
Leptospirosis	*Leptospira interrogans*	Dogs, rodents, wild animals	Direct contact with urine, infected tissue, and contaminated water
Listeriosis	*Listeria monocytogenes*	Sheep, cattle, goats, guinea pigs, chickens, horses, rodents, birds, crustaceans	Food-borne
Lyme disease	*Borrelia burgdorferi*	Ticks (*Ixodes scapularis* or related ticks)	Bite of infected tick
Melioidosis	*Burkholderia pseudomallei*	Rats, mice, rabbits, dogs, cats	Arthropod vectors, water, food
Plague (bubonic)	*Yersinia pestis*	Domestic rats, many wild rodents	Flea bite
Psittacosis	*Chlamydia psittaci*	Birds	Direct contact, respiratory aerosols
Q fever	*Coxiella burnetii*	Cattle, sheep, goats	Inhalation of contaminated soil and dust
Rabies	*Rabies virus*	Dogs, bats, opposums, skunks, raccoons, foxes, cats, cattle	Bite of rabid animal
Rocky Mountain spotted fever	*Rickettsia rickettsii*	Rabbits, squirrels, rats, mice, groundhogs	Tick bite
Salmonellosis	*Salmonella* spp. (except *S. typhosa*)	Fowl, swine, sheep, cattle, horses, dogs, cats, rodents, reptiles, birds, turtles	Direct contact; food
SARS	SARS coronavirus	Bats, civets	Contact with infected animal or person
Tuberculosis	*Mycobacterium bovis, M. tuberculosis, M. avium* complex, *M. africanum*	Cattle, horses, cats, dogs	Milk; direct contact
Tularemia	*Francisella tularensis*	Wild rabbits, most other wild and domestic animals	Direct contact with infected carcass, usually rabbit; tick bite, biting flies
Weil's disease (leptospirosis)	*Leptospira interrogans*	Rats, mice, skunks, opposums, wildcats, foxes, raccoons, shrews, bandicoots, dogs, cattle, swine	Through skin, drinking water, eating food
Yellow fever (jungle)	*Yellow fever virus*	Monkeys, marmosets, lemurs, mosquitoes	Mosquito

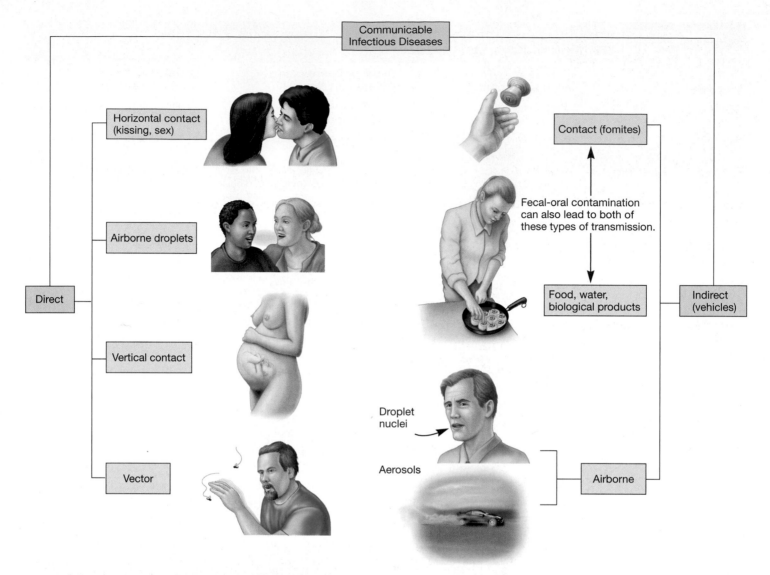

Figure 33.8   Transmission of Communicable Disease.   Infectious diseases can be transmitted from person to person by various direct and indirect methods.

Figure 33.9   A Sneeze.   High-speed photograph of an aerosol generated by an unstifled sneeze. The expelled particles are comprised of saliva and mucus laden with microorganisms. These airborne particles may be infectious when inhaled by a susceptible host. Even a surgical mask will not prevent the spread of all particles.

Vehicle Transmission

Inanimate materials or objects involved in pathogen transmission are called **vehicles.** In **common vehicle transmission,** a single inanimate vehicle or source serves to spread the pathogen to multiple hosts but does not support its reproduction. Examples include surgical instruments, bedding, and eating utensils. In epidemiology, these common vehicles are called **fomites** (s., fomes or fomite). A single source containing pathogens (e.g., blood, drugs, IV fluids) can contaminate a common vehicle that causes multiple infections. Food and water are important common vehicles for many human diseases.

Vector-Borne Transmission

Living transmitters of a pathogen are called **vectors.** Most vectors are arthropods (e.g., insects, ticks, mites, fleas) or vertebrates (e.g., dogs, cats, skunks, bats). **Vector-borne transmission** can be either external or internal. In external (mechanical) transmission,

Table 33.2	Some Airborne Pathogens and the Diseases They Cause in Humans
Microorganisms	**Disease**
Common virus name	
Varicella	Chickenpox
Influenza	Flu (Influenza)
Rubeola	Measles
Rubella	German measles
Mumps	Mumps
Polio	Poliomyelitis
Acute respiratory viruses	Viral pneumonia
Pulmonary syndrome hantavirus	Hantavirus pulmonary syndrome
Variola	Smallpox
Bacteria	
Actinomyces spp.	Lung infections
Bordetella pertussis	Whooping cough
Chlamydia psittaci	Psittacosis
Corynebacterium diphtheriae	Diphtheria
Mycoplasma pneumoniae	Atypical pneumonia
Mycobacterium tuberculosis	Tuberculosis
Neisseria meningitidis	Meningitis
Streptococcus spp.	Pneumonia, sore throat
Fungi	
Blastomyces spp.	Lung infections
Coccidioides spp.	Coccidioidomycosis
Histoplasma capsulatum	Histoplasmosis

the pathogen is carried on the body surface of a vector. Carriage is passive, with no growth of the pathogen during transmission. An example would be flies carrying *Shigella* organisms on their feet from a fecal source to a plate of food that a person is eating.

In internal transmission, the pathogen is carried within the vector by either a harborage or biologic transmission. In **harborage transmission,** the pathogen does not undergo morphological or physiological changes within the vector. An example would be the transmission of *Yersinia pestis* (the etiologic agent of plague) by the rat flea from rat to human. **Biologic transmission** implies that the pathogen does go through a morphological or physiological change within the vector. An example would be the developmental sequence of the malarial parasite inside its mosquito vector.

Why Was the Host Susceptible to the Pathogen?

The fourth link in the infectious disease cycle is the host. The susceptibility of the host to a pathogen depends on both the pathogenicity of the organism and the nonspecific and specific defense

mechanisms of the host. These susceptibility factors are the basis for chapters 28 and 29. In addition to host defense mechanisms, nutrition, genetic predisposition, and stress also influence host susceptibility to infection.

How Did the Pathogen Leave the Host?

The fifth and last link in the infectious disease cycle is release or exit of the pathogen from the host. From the point of view of the pathogen, successful escape from the host is just as important as its initial entry. Unless a successful escape occurs, the disease cycle will be interrupted and the pathogen will not be perpetuated. Escape can be active or passive, although sometimes a combination of the two occurs. Active escape takes place when a pathogen actively moves to a portal of exit and leaves the host. Examples include parasitic helminths that migrate through the body of their host, eventually reaching the surface and exiting. Passive escape occurs when a pathogen or its progeny leaves the host in feces, urine, droplets, saliva, or desquamated cells. Microorganisms usually employ passive escape mechanisms.

1. What are some epidemiologically important characteristics of a pathogen? What is a communicable disease?
2. Define source, reservoir, period of infectivity, and carrier.
3. What types of infectious disease carriers does epidemiology recognize?
4. Describe the four main types of infectious disease transmission and give examples of each.
5. Define droplet nuclei, vehicle, fomite, and vector.

33.6 VIRULENCE AND THE MODE OF TRANSMISSION

Evidence exists that a pathogen's virulence may be strongly influenced by its mode of transmission and ability to live outside its host. When the pathogen uses a mode of transmission such as direct contact, it cannot afford to make the host so ill that it will not be transmitted effectively. This is the case with the common cold, which is caused by rhinoviruses and several other respiratory viruses. If the virus reproduced too rapidly and damaged its host extensively, the person would be bedridden and not contact others. The efficiency of transmission would drop because rhinoviruses shed from the cold sufferer could not contact new hosts and would be inactivated by exposure. Cold sufferers must be able to move about and directly contact others. Thus virulence is low and people are not incapacitated by the common cold.

On the other hand, if a pathogen uses a mode of transmission not dependent on host health and mobility, then the person's health will not be a critical matter. The pathogen might be quite successful—that is, transmitted to many new hosts even though it kills its host relatively quickly. Host death will mean the end of any resident pathogens, but the species as a whole can spread and flourish as long as the increased transmission rate outbalances the loss due to host death. This situation may arise in several ways.

When a pathogen is transmitted by a vector, it will benefit by extensive reproduction and spread within the host. If pathogen levels are very high in the host, a vector such as a biting insect has a better chance of picking up the pathogen and transferring it to a new host. Indeed, pathogens transmitted by biting arthropods often are very virulent (e.g., malaria, typhus, sleeping sickness). It is important that such pathogens do not harm their vectors, and the vector generally remains healthy, at least long enough for pathogen transmission.

Virulence also is often directly correlated with a pathogen's ability to survive in the external environment. If a pathogen cannot survive well outside its host and does not use a vector, it depends on host survival and will tend to be less virulent. When a pathogen can survive for long periods outside its host, it can afford to leave the host and simply wait for a new one to come along. This seems to promote increased virulence. Host health is not critical, but extensive multiplication within the host will increase the efficiency of transmission. Good examples are tuberculosis and diphtheria. *Mycobacterium tuberculosis* and *Corynebacterium diphtheriae* survive for a long time, at least weeks to months, outside human hosts.

Human cultural patterns and behavior affect pathogen virulence. Waterborne pathogens such as *Vibrio cholerae* are transmitted through drinking water systems. They can be virulent because immobile hosts still shed pathogens, which frequently reach the water. This is why the establishment of an uncontaminated drinking water supply is critical in limiting a cholera outbreak. The same appears to be true of *Shigella* and dysentery. Often one of the best ways to reduce virulence may be to reduce the frequency of transmission.

33.7 EMERGING AND REEMERGING INFECTIOUS DISEASES AND PATHOGENS

Only a few decades ago, a grateful public trusted that science had triumphed over infectious diseases by building a fortress of health protection. Antibiotics, vaccines, and aggressive public health campaigns had yielded a string of victories over old enemies such as whooping cough, pneumonia, polio, and smallpox. In developed countries, people were lulled into believing that microbial threats were a thing of the past. Trends in the number of deaths caused by infectious diseases in the United States from 1900 through 1982 supported this conclusion (**figure 33.10**). However, this downward trend ended in 1982, and the death rate has since risen. The world has seen the global spread of AIDS, the resurgence of tuberculosis, and the appearance of new enemies such as hantavirus pulmonary syndrome, hepatitis C and E, Ebola virus, Lyme disease, cryptosporidiosis, and *E. coli* O157:H7. In addition, during this same time period:

- A "bird flu" virus that had never before attacked humans began to kill people in southeast Asia.

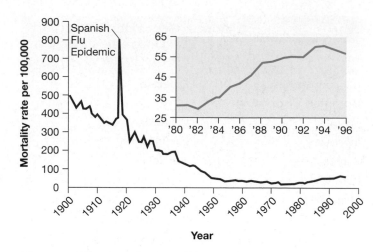

Figure 33.10 Infectious Disease Mortality in the United States Decreased Greatly During Most of the Twentieth Century. The insert is an enlargement of the right-hand portion of the graph and shows that the death rate from infectious diseases increased between 1980 and 1994.

- A new variant of a fatal prion disease of the brain, Creutzfeldt-Jakob disease, was identified in the United Kingdom, transmitted by beef from animals with "mad cow disease."

- *Staphylococcus* bacteria with resistance to methicillin and vancomycin, long the antibiotics of first choice and last resort, respectively, were seen for the first time.

- Several major multistate foodborne outbreaks occurred in the United States, including those caused by protists on raspberries, viruses on strawberries, and bacteria in produce, ground beef, cold cuts, and breakfast cereal.

- A new strain of tuberculosis that is resistant to many drugs and occurs most often in people infected with HIV arose in New York and other large cities.

By the 1990s the idea that infectious diseases no longer posed a serious threat to human health was obsolete. In the twenty-first century, it is clear that humans will continually be faced with both new infectious diseases and the reemergence of older diseases once thought to be conquered (e.g., tuberculosis, dengue hemorrhagic fever, yellow fever). The Centers for Disease Control and Prevention has defined these diseases as "new, reemerging, or drug-resistant infections whose incidence in humans has increased within the past three decades or whose incidence threatens to increase in the near future." Some of the most recent examples of these diseases are shown in (**figure 33.11**). The increased importance of emerging and reemerging infectious diseases has stimulated the establishment of a field called **systematic epidemiology,** which focuses on the ecological and social factors that influence the development of these diseases.

Why are viruses, bacteria, fungi, and parasites posing such a problem, despite dramatic advances in medical research, drug discovery, technology development, and sanitation? Many factors

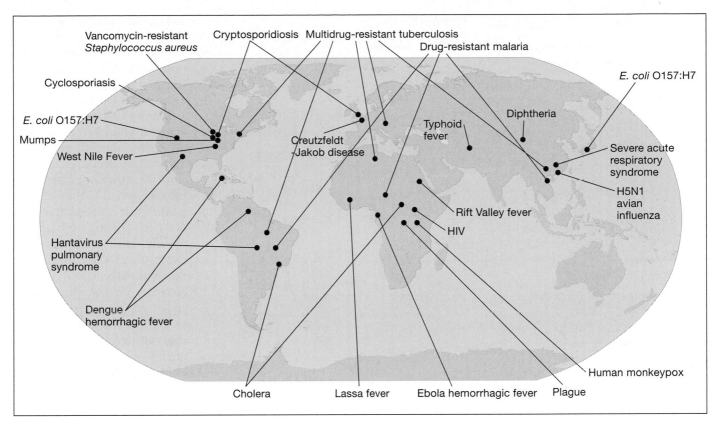

Figure 33.11   Some Examples of Emerging and Reemerging Infectious Diseases.   Although diseases such as HIV are indicated in only one or two significant locations, they are very widespread and a threat in many regions.

characteristic of the modern world undoubtedly favor the development and spread of these microorganisms and their diseases. Examples include:

1. Unprecedented worldwide population growth, population shifts, and urbanization

2. Increased international travel, transport (commerce), migration, and relocation of animals and food products

3. Changes in food processing, handling, and agricultural practices

4. Changes in human behavior, technology, and industry

5. Human encroachment on wilderness habitats that are reservoirs for insects and animals that harbor infectious agents

6. Microbial evolution (e.g., selective pressure) and the development of resistance to antibiotics and other antimicrobial drugs

7. Changes in ecology and climate

8. Modern medicine (e.g., immunosuppression)

9. Inadequacy of public infrastructure and vaccination programs

10. Social unrest, civil wars, and bioterrorism (section 33.9)

As population density increases in cities, the dynamics of microbial exposure and evolution increase in humans themselves.

Urbanization often crowds humans and increases exposure to microorganisms. Crowding leads to unsanitary conditions and hinders the effective implementation of adequate medical care, enabling more widespread transmission and propagation of pathogens. In modern societies, crowded workplaces, community-living settings, day-care centers, large hospitals, and public transportation all facilitate microbial transmission. Furthermore, land development and the exploration and destruction of natural habitats have increased the likelihood of human exposure to new pathogens and may put selective pressures on pathogens to adapt to new hosts and changing environments. The introduction of pathogens to a new environment or host can alter transmission and exposure patterns, leading to sudden proliferation of disease. For example, the spread of Lyme disease in New England probably is due partly to ecological disruption that eliminates predators of deer. An increase in deer and the deer tick populations provides a favorable situation for pathogen spread to humans. Whenever the environment is altered and new environments are created, this may not only confer a survival advantage but may also increase a pathogen's virulence and alter its drug susceptibility profile. When changes in climate or ecology occur, it should not be surprising to find changes in both beneficial and detrimental microorganisms. Global warming also affects microorganism selection and survival. Finally, mass migrations of refugees, workers, and displaced persons have led to a steady growth of urban centers at the expense of rural areas.

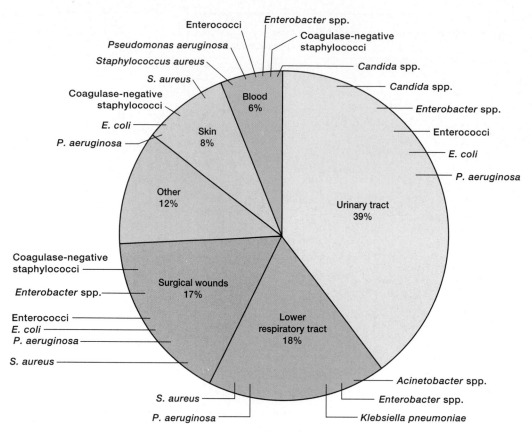

Figure 33.12 Nosocomial Infections. Relative frequency by body site. These data are from the National Nosocomial Infections Surveillance, which is conducted by the Centers for Disease Control and Prevention.

ceftriaxone (Rocephin), ceftazidime, and aztreonam (Azactam).

Without a doubt, the key factors responsible for the rise in drug-resistant pathogens have been the excessive or inappropriate use of antimicrobial therapy and the indiscriminant use of broad-spectrum antibiotics. Although the CDC acknowledges that it is too late to solve the resistance problem simply by using antimicrobial agents more prudently, it is no less true that the problem of drug resistance will continue to worsen if this does not occur. Also needed (especially in underdeveloped countries) is a renewed emphasis on alternate prevention and control strategies that prevailed in the years before antimicrobial chemotherapy. These include improved sanitation and hygiene, isolation of infected persons, antisepsis, and vaccination. << *Drug resistance (section 31.6)*

In this new millennium, the speed and volume of international travel are major factors contributing to the global emergence of infectious diseases. The spread of a new disease often used to be limited by the travel time needed to reach a new host population. If the travel time was sufficiently long, as when a ship crossed the ocean, the infected travelers would either recover or die before reaching a new population. Because travel by air has obliterated time between exposure and disease outbreak, a traveler can spread virtually any human disease in a matter of hours. Vehicles of human transport, such as aircraft and ships, also transport the infectious agents and their vectors.

Emerging and reemerging pathogens and their diseases are therefore the outcome of many different factors. Because the world is now so interconnected, we cannot isolate ourselves from other countries and continents. Changes in the disease status of one part of the world may well affect the health of the remainder. As Nobel laureate Joshua Lederberg has so eloquently stated, "The microbe that felled one child in a distant continent yesterday can reach your child today and seed a global pandemic tomorrow."

Microbiologists are all too familiar with the development of resistance to antibiotics used in human medicine. The distribution of nosocomial pathogens (**figure 33.12**) has changed throughout the antibiotic era. Hospital-acquired infections were dominated early on by staphylococci, which initially responded to penicillin. During subsequent years, the emergence of methicillin-resistant *Staphylococcus aureus* (MRSA) increased dramatically; similar patterns are emerging for penicillin-resistant *Streptococcus pneumoniae*. Vancomycin-resistant *S. aureus* infections are now a major nosocomial threat. Glycopeptide-resistant *Enterococcus* was first reported in the late 1980s; vancomycin-resistant enterococci (VRE) are now common in U.S. hospitals. Recently recognized nosocomial, gram-positive species include *Corynebacterium jeikeium* and *Rhodococcus equi*. The incidences of infections by the gram-negative pathogens *Pseudomonas aeruginosa* and *Acinetobacter* ssp. and the recently renamed gram-negative bacterial pathogens *Burkholderia cepacia* and *Stenotrophomonas maltophilia* have increased. With respect to the extended-spectrum β-lactam-resistant gram-negative bacilli, bacteria such as *Klebsiella pneumoniae*, *E. coli*, other *Klebsiella* spp., *Proteus* spp., *Morganella* spp., *Citrobacter* spp., *Salmonella* spp., and *Serratia marcescens* are resistant to penicillins; first-generation cephalosporins; and some third-generation cephalosporins such as cefotaxime (Claforan),

1. Describe how virulence and the mode of transmission may be related. What might cause the development of new human diseases?

2. How would you define emerging or reemerging infectious diseases?

3. What are some of the factors responsible for the emergence or reemergence of pathogens?

4. What are some key factors responsible for the rise in drug-resistant bacteria?

33.8 CONTROL OF EPIDEMICS

The development of an infectious disease is a complex process involving many factors, as is the design of specific epidemiological control measures. Epidemiologists must consider available resources and time constraints, adverse effects of potential control measures, and human activities that might influence the spread of the infection. Many times control activities reflect compromises among alternatives. To proceed intelligently, one must identify components of the infectious disease cycle that are primarily responsible for a particular epidemic. Control measures should be directed toward that part of the cycle that is most susceptible to control—the weakest link in the chain (figure 33.7).

There are three types of control measures. The first type is directed toward reducing or eliminating the source or reservoir of infection:

1. Quarantine and isolation of carriers

2. Destruction of an animal reservoir of infection

3. Treatment of water sewage to reduce contamination (see chapter 35)

4. Therapy that reduces or eliminates infectivity of the individual

The second type of control measure is designed to break the connection between the source of the infection and susceptible individuals. Examples include general sanitation measures:

1. Chlorination of water supplies

2. Pasteurization of milk and other beverages

3. Supervision and inspection of food and food handlers

4. Destruction of vectors

The third type of control measure reduces the number of susceptible individuals and raises the general level of herd immunity by immunization. Examples include:

1. Passive immunization to give a temporary immunity following exposure to a pathogen or when a disease threatens to take an epidemic form

2. Active immunization to protect the individual from the pathogen and the host population from the epidemic

Vaccines and Immunization

A **vaccine** (Latin *vacca,* cow) is a preparation of one or more microbial antigens used to induce protective immunity. **Immunization** is the result achieved by the successful delivery of vaccines; it stimulates immunity. Vaccination attempts to induce antibodies and activated T cells to protect a host from future infection. Many epidemics have been stayed by mass prophylactic immunization. Vaccines have eradicated smallpox, pushed polio to the brink of extinction, and spared countless individuals from hepatitis A and B, influenza, measles, rotavirus disease, tetanus, typhus, and other dangerous diseases. **Vaccinomics**, the application of genomics and bioinformatics to vaccine development, is bringing a fresh approach to the Herculean problem of making vaccines against various microorganisms and parasites. << *Microbial genomics (chapter 15)*

To promote a more efficient immune response, antigens in vaccines can be mixed with an **adjuvant** (Latin *adjuvans,* aiding), which enhances the rate and degree of immunization. Adjuvants can be any nontoxic material that prolongs antigen interaction with immune cells, assists in the antigen-presenting cell (APC) processing of antigens, or otherwise nonspecifically stimulates the immune response to the antigen. Several types of adjuvants can be used. Oil in water emulsions (Freund's incomplete adjuvant), aluminum hydroxide salts (alum), beeswax, and various combinations of bacteria (live or killed) are used in vaccine adjuvants. In most cases, the adjuvant materials trap the antigen, thereby promoting a sustained release as APCs digest and degrade the preparation. In other cases, the adjuvant activates APCs so that antigen recognition, processing, and presentation are more efficient.

The modern era of vaccines and immunization began in 1798 with Edward Jenner's use of cowpox as a vaccine against smallpox (**Historical Highlights 33.3**) and in 1881 with Louis Pasteur's rabies vaccine. Vaccines for other diseases did not emerge until later in the nineteenth century, when largely through a process of trial and error, methods for inactivating and attenuating microorganisms were improved and vaccines were produced. Vaccines were eventually developed against most of the epidemic diseases that plagued Western Europe and North America (e.g., diphtheria, measles, mumps, pertussis, German measles, and polio). Indeed, toward the end of the twentieth century, it seemed that the combination of vaccines and antibiotics would temper the problem of microbial infections. Such optimism was cut short by the emergence of new and previously unrecognized pathogens and antibiotic resistance among existing pathogens. Nevertheless, vaccination is still one of the most cost-effective weapons for preventing microbial disease.

Vaccination of most children should begin at about age two months (**table 33.3**). Before that age, they are protected by passive natural immunity from maternal antibodies. Further vaccination of teens and adults depends on their relative risk for exposure to infectious disease. Individuals living in close quarters (e.g., college students in residence halls, military personnel), the elderly, and individuals with reduced immunity (e.g., those with chronic and metabolic diseases) should receive vaccines for influenza, meningitis, and pneumonia. International travelers may be immunized against cholera, hepatitis A, plague, polio, typhoid, typhus, and yellow fever, depending on the country visited. Veterinarians, forest rangers, and others whose jobs involve contact with animals may be vaccinated against rabies, plague, and anthrax. Health-care workers are typically immunized against hepatitis B virus. The role of immunization as a protective therapy cannot be overstated; immunizations save lives.

Historical Highlights

33.3   The First Immunizations

Since the time of the ancient Greeks, it has been recognized that people who have recovered from plague, smallpox, yellow fever, and various other infectious diseases rarely contract the diseases again. The first scientific attempts at artificial immunizations were made in the late eighteenth century by Edward Jenner (1749–1823), who was a country doctor from Berkley, Gloucestershire, England. Jenner investigated the basis for the widespread belief of the English peasants that anyone who had vaccinia (cowpox) never contracted smallpox. Smallpox was often fatal—10 to 40% of the victims died—and those who recovered had disfiguring pockmarks. Yet most English milkmaids, who were readily infected with cowpox, had clear skin because cowpox was a relatively mild infection that left no scars.

It was on May 14, 1796, that Jenner extracted the contents of a pustule from the arm of a cowpox-infected milkmaid, Sarah Nelmes, and injected it into the arm of eight-year-old James Phipps. As Jenner expected, immunization with the cowpox virus caused only mild symptoms in the boy. When he subsequently inoculated the boy with smallpox virus (an act now considered completely unethical), the boy showed no symptoms of the disease. Jenner then inoculated large numbers of his patients with cowpox pus, as did other physicians in England and on the European continent (**box figure**). By 1800 the practice known as variolation had begun in America, and by 1805 Napoleon Bonaparte had ordered all French soldiers to be vaccinated.

Further work on immunization was carried out by Louis Pasteur (1822–1895). Pasteur discovered that if cultures of chicken cholera

Nineteenth-Century Physicians Performing Vaccinations on Children.

bacteria were allowed to age for 2 or 3 months, the bacteria produced only a mild attack of cholera when inoculated into chickens. Somehow the old cultures had become less pathogenic (attenuated) for the chickens. He then found that fresh cultures of the bacteria failed to produce cholera in chickens that had been previously inoculated with old, attenuated cultures. To honor Jenner's work with cowpox, Pasteur gave the name *vaccine* to any preparation of a weakened pathogen that was used (as was Jenner's "vaccine virus") to immunize against infectious disease.

Whole-Cell Vaccines

Many of the current vaccines used for humans that are effective against viral and bacterial diseases consist of whole microorganisms, termed **whole-cell vaccines.** These are either inactivated (killed) or attenuated (live but avirulent) (table 33.3). The major characteristics of these vaccines are compared in **table 33.4.** **Inactivated vaccines** are effective, but they are less immunogenic, so they often require several boosters and normally do not adequately stimulate cell-mediated immunity or secretory IgA production. In contrast, **attenuated vaccines** usually are given in a single dose and stimulate both humoral and cell-mediated immunity.

Even though whole-cell vaccines are considered the "gold standard" of vaccines, they can be problematic. For example, whole-organism vaccines fail to shield against some diseases. Attenuated vaccines can also cause full-blown illness in an individual whose immune system is compromised (e.g., AIDS patients, cancer patients undergoing chemotherapy, the elderly). These same individuals may also contract the disease from healthy people who have been vaccinated recently. Moreover, attenuated viruses can at times mutate in ways that restore virulence, as has happened in some monkeys given an attenuated simian form of the AIDS virus.

Acellular or Subunit Vaccines

A few of the common risks associated with whole-cell vaccines can be avoided by using only specific, purified macromolecules derived from pathogenic microorganisms. There are three general forms of **subunit vaccines:** (1) capsular polysaccharides, (2) recombinant surface antigens, and (3) inactivated exotoxins called **toxoids.** The purified microbial subunits or their secreted products can be prepared as nontoxic antigens to be used in the formulation of vaccines (**table 33.5**).

Recombinant-Vector and DNA Vaccines

Genes isolated from a pathogen that encode major antigens can be inserted into nonvirulent viruses or bacteria. Such recombinant microorganisms serve as vectors, replicating within the host and expressing the gene product of the pathogen-encoded antigenic proteins. The antigens elicit humoral immunity (i.e., antibody production) when they escape from the vector, and they also elicit cellular immunity when they are broken down and properly displayed on the cell surface (just as occurs when host cells harbor an active pathogen). Several microorganisms, such as adenovirus and attenuated *Salmonella,* have been used in the production of these **recombinant-vector vaccines.**

Table 33.3 Examples of Vaccines to Prevent Viral and Bacterial Diseases in Humans

Disease	Vaccine	Booster	Recommendation
Viral Diseases			
Chickenpox	Attenuated Oka strain (Varivax)	None	Children 12–18 months: older children who have not had chickenpox
Hepatitis A	Inactivated virus (Havrix)	6–12 months	International travelers
Hepatitis B	HB viral antigen (Engerix-B, Recombivax HB)	1–4 months 6–18 months	High-risk medical personnel: children, birth to 18 months and 11–12 years of age
Influenza A	Inactivated virus or live attenuated	Yearly	All persons
Measles, Mumps, Rubella	Attenuated viruses (combination MMR vaccine)	None	First dose 12–15 months, 2nd dose 4–6 years
Poliomyelitis	Attenuated (oral poliomyelitis vaccine, OPV) or inactivated virus	Adults as needed	First dose at 2 months, 2nd at 4 months, 3rd at 16–18 months, 4th at 4–6 years
Rabies	Inactivated virus	None	For individual in contact with wildlife, animal control personnel, veterinarians
Respiratory disease	Live attenuated adenovirus	None	Military personnel
Smallpox	Live attenuated *Vaccinia virus*	None	Laboratory, health-care, and military personnel
Yellow fever	Attenuated virus	10 years	Military personnel and individuals traveling to endemic areas
Bacterial Diseases			
Anthrax	Extracellular components of unencapsulated *B. anthracis*	None	Agricultural workers, veterinary, and military personnel
Cholera	Fraction of *Vibrio cholerae*	6 months	Individuals in endemic areas, travelers
Diphtheria, Pertussis, Tetanus	Diphtheria toxoid, killed *Bordetella pertussis,* tetanus toxoid (DPT vaccine) or with acellular *pertussis* (DtaP); or tetanus toxoid, reduced diphtheria toxoid, and acellular pertussis vaccine, adsorbed (Tdap)	10 years	Children from 2–3 months old to 12 years, and adults; children 10–18 years, at least 5 years after DPT series, should receive Tdap
Haemophilus influenzae type b	Polysaccharide-protein conjugate (HbCV) or bacterial polysaccharide (HbPV)	None	First dose at 2 months, 2nd at 4 months, 3rd at 6 months, 4th at 12–15 months
Meningococcal infections	*Neisseria meningitidis* polysaccharides of serotypes A/C/Y/W-135	None	Military; high-risk individuals; college students living in dormatories; elderly in nursing homes
Plague	Fraction of *Yersinia pestis*	Yearly	Individuals in contact with rodents in endemic areas
Pneumococcal pneumonia	Purified *S. pneumoniae* polysaccharide of 23 pneumococcal types	None	Adults over 50 with chronic disease
Q fever	Inactivated *Coxiella burnetii*	None	Workers in slaughter houses and meat-processing plants
Tuberculosis	Attenuated *Mycobacterium bovis* (BCG vaccine)	3–4 years	Individuals exposed to TB for prolonged periods of time; used in some countries, not licensed in the U.S.
Typhoid fever	*Salmonella enterica* Typhi Ty21a (live attenuated or polysaccharide)	None	Resident of and travelers to areas of endemic disease
Typhus fever	Killed *Rickettsia prowazekii*	Yearly	Scientists and medical personnel in areas where typhus is endemic

Table 33.4 — A Comparison of Inactivated (Killed) and Attenuated (Live) Vaccines

Major Characteristic	Inactivated Vaccine	Attenuated Vaccine
Booster shots	Multiple boosters required	Only a single booster, if any, required
Production	Virulent microorganism inactivated by chemicals or irradiation	Virulent microorganism grown under adverse conditions or passed through different hosts until avirulent
Reversion tendency	None	May revert to a virulent form
Stability	Very stable, even where refrigeration is unavailable	Less stable
Type of immunity induced	Humoral	Humoral and cell-mediated

Source: Adapted from Goldsby, R. A. Kindt, T. J. and Osborne, B. A. 2003. *Kuby Immunology,* New York: W. H. Freeman

Table 33.5 — Subunit Vaccines Currently Available for Human Use

Microorganism or Toxin	Vaccine Subunit
Capsular polysaccharide	
Haemophilus influenzae type b	Polysaccharide-protein conjugate (HbCV) or bacterial polysaccharide (HbPV)
Neisseria meningitidis	Polysaccharides of serotypes A/C/Y/W-135
Streptococcus pneumoniae	23 distinct capsular polysaccharides
Surface antigen	
Hepatitis B virus	Recombinant surface antigen (HbsAg)
Toxoids	
Corynebacterium diphtheriae toxin	Inactivated exotoxin
Clostridium tetani toxin	Inactivated exotoxin

On the other hand, **DNA vaccines** introduce DNA directly into the host cell. When injected into muscle cells, the DNA is taken into the nucleus and the pathogen's DNA fragment is expressed, generating foreign proteins to which the host's immune system responds. DNA vaccines are very stable; refrigeration is often unnecessary. At present, human trials are underway with several different DNA vaccines against malaria, AIDS, influenza, hepatitis B, and herpesvirus. DNA vaccines against a number of cancers (such as lymphomas, prostate, colon) are also being tested.

The Role of the Public Health System: Epidemiological Guardian

The control of an infectious disease relies heavily on a well-defined network of clinical microbiologists, nurses, physicians, epidemiologists, and infection control personnel who supply epidemiological information to a network of local, state, national, and international organizations. These individuals and organizations comprise the public health system. For example, each state has a public health laboratory that is involved in infection surveillance and control. The communicable disease section of a state laboratory includes specialized laboratory services for the examination of specimens or cultures submitted by physicians, local health departments, hospital laboratories, sanitarians, epidemiologists, and others. These groups share their findings with other health agencies in the state, the Centers for Disease Control and Prevention, and the World Health Organization.

33.9 BIOTERRORISM PREPAREDNESS

Bioterrorism is defined as "the intentional or threatened use of viruses, bacteria, fungi, or toxins from living organisms to produce death or disease in humans, animals, and plants." The use of biological agents to effect personal or political outcome is not new, and the modern use of biological agents is a reality. The most notable intentional uses of biological agents for criminal or terror intent are (1) the use of *Salmonella enterica* serovar Typhimurium in 10 restaurant salad bars (by the Rajneeshee religious cult in The Dalles, OR, 1984); (2) the intentional release of *Shigella dysentariae* in a hospital laboratory break room (perpetrator[s] still unknown, Texas, 1996); and (3) the use of *Bacillus anthracis* spores delivered through the U.S. postal system (perpetrator[s] still unknown, five eastern U.S. states, 2001). The *Salmonella*-contaminated salads resulted in 751 documented cases and 45 hospitalizations due to salmonellosis. The *Shigella* release resulted in eight confirmed cases and four hospitalizations for shigellosis. The *Bacillus* spores infected 22 people (11 cases of inhalation anthrax and 11 cases of cutaneous anthrax) and caused five deaths. The list of biological agents that could pose the greatest public health risk in the event of a bioterrorist attack is relatively short and includes viruses, bacteria, parasites, and toxins that, if acquired and properly disseminated, could become a difficult public health challenge in terms of limiting the numbers of casualties and controlling panic (**table 33.6**).

Among weapons of mass destruction, biological weapons can be more destructive than chemical weapons, including nerve gas. In certain circumstances, biological weapons can be as devastating as a nuclear explosion—a few kilograms of anthrax could kill as many people as a Hiroshima-size nuclear bomb. Biological agents are likely to be chosen as a means of localized attack (biocrime) or mass casualty (bioterrorism) for several reasons. They are mostly invisible, odorless, tasteless, and difficult to detect. Use of biological agents for terrorism also means that perpetrators may escape undetected as it may take hours to days before signs and symptoms of their use become evident. Additionally, the general public is not likely to be protected immunologically against agents that might be used in bioterrorism. Ultimately, the use of biological agents in terrorism results in fear, panic, and chaos.

In 1998 the U.S. government launched the first national effort to create a biological weapons defense. The initiatives included (1) the first-ever procurement of specialized vaccines and medicines for a national civilian protection stockpile; (2) invigoration of research and development in the science of biodefense; (3) investment of more time and money in genome sequencing, new vaccine research, and new therapeutic research; (4) development of improved detection and diagnostic systems; and (5) preparation of clinical microbiologists and clinical microbiology laboratories as members of first responder teams to respond in a timely manner to acts of bioterrorism. In 2002 the U.S. Congress enacted the Public Health Security and Bioterrorism Preparedness and Response Act, which identified "select" agents whose use is now tightly regulated. The agents are categorized (A, B, or C) based on (1) ease of dissemination, (2) communicability, and (3) morbidity and mortality. A final rule implementing the provisions of the act that govern the possession, use, and transport of biological agents that are considered likely to be used for biocrimes or bioterrorism was issued in 2005.

In 2003, Congress established the Department of Homeland Security to coordinate the defense of the United States against terrorist attacks. As one of many duties, the secretary of Homeland Security is responsible for maintaining a National Incident Management System to monitor large-scale hazardous events. Bioterrorism and other public health incidents are managed within this system. The Department of Health and Human Services has the initial responsibility for the national public health and will deploy assets as needed within the areas of its statutory responsibility (e.g., the Public Health Service Act and the Federal Food, Drug, and Cosmetic Act) while keeping the secretary of Homeland Security apprised during an incident of the nature of the response. The secretary of Health and Human Services directs the CDC to effect the necessary integration of public health activities.

Table 33.6	Pathogens and Toxins Defined by the CDC as Select Agents	
Category	Definition	Disease (Agent)
A	Easily disseminated or transmitted from person to person; high mortality rates; potential for major public health impact; cause public panic and social disruption; require special action for public health preparedness	**Anthrax** (*Bacillus anthracis*) **Botulism** (*Clostridium botulinum* toxin) **Plague** (*Yersinia pestis*) **Smallpox** (Variola major) **Tularemia** (*Francisella tularensis*) **Viral hemorrhagic fever** (filoviruses and arenaviruses)
B	Moderately easy to disseminate, moderate morbidity and mortality rates; require specific enhancements of CDC's diagnostic capacity and enhanced disease surveillance	**Brucellosis** (*Brucella* species) **Glanders** (*Burkholderia mallei*) **Melioidosis** (*Burkholderia pseudomallei*) **Psittacosis** (*Chlamydia psittaci*) **Q fever** (*Coxiella burnetii*) **Typhus fever** (*Rickettsia prowazekii*) **Viral encephalitis** (alphaviruses) **Toxemia** 　Ricin from castor beans 　Staphylococcal enterotoxin B 　Epsilon toxin *Clostridium perfringens* **Other** 　Water safety threats (e.g., *Vibrio cholerae, Cryptosporidium parvum*) 　Food safety threats (e.g., *Salmonella* spp., *E. coli* O157:H7, *Shigella* spp.)
C	Emerging pathogens that could be engineered for mass dissemination; potential for high morbidity and mortality rates; major health impact potential	*Nipah virus* Hantaviruses Tickborne hemorrhagic fever viruses Tickborne encephalitis viruses *Yellow fever virus* Multidrug-resistant and extremely drug resistant *Mycobacterium tuberculosis*

The events of September and October 2001 in the United States have changed the world. Global efforts to prevent terrorism, especially using biological agents, are evolving from cautious planning to proactive preparedness. In the United States, the CDC has partnered with academic institutions across the country to educate, train, and drill public health employees, first responders, and numerous environmental and health-care providers. Centers for Public Health Preparedness were established to bolster the overall response capability to bioterrorism. Another CDC-managed program that began in 1999, the Laboratory Response Network (LRN), serves to ensure an effective laboratory response to bioterrorism by helping to improve the nation's public health laboratory infrastructure through its partnership with the FBI and the Association of Public Health Laboratories (APHL). The LRN maintains an integrated network that links state and local public health, federal, military, and international laboratories so that a rapid and coordinated response to bioterrorism or other public health emergencies (including veterinary, agriculture, military, and water- and food-related) can occur.

In the absence of overt terrorist threats and without the ability to rapidly detect bioterrorism agents, it is likely that a bioterrorism act will be defined by a sudden spike in an unusual (nonendemic) disease reported to the public health system. Also, sudden increased numbers of zoonoses, diseased animals, or vehicle-borne illnesses may indicate bioterrorism. Important guidelines and standardized protocols prepared for all sentinel (local hospital, contract, clinic, etc.) laboratories to assist in the management of clinical specimens containing select agents have been established by the American Society for Microbiology in coordination with the CDC and the APHL.

1. In what three general ways can epidemics be controlled? Give one or two specific examples of each type of control measure.
2. Name some of the microorganisms that can be used to commit biocrimes. From this list, which pose the greatest risk for causing large numbers of casualties?
3. Why are biological weapons more destructive than chemical weapons?
4. What is the Public Health Security and Bioterrorism Preparedness and Response Act designed to do?

33.10 NOSOCOMIAL INFECTIONS

Nosocomial infections (Greek *nosos,* disease, and *komeion,* to take care of) result from pathogens acquired by patients while in a hospital or other clinical care facility (figure 33.12). Besides harming patients, nosocomial infections can affect nurses, physicians, aides, visitors, salespeople, delivery personnel, custodians, and anyone else who has contact with the hospital. Most nosocomial infections become clinically apparent while patients are still hospitalized; however, disease onset can occur after patients have been discharged. Infections that are incubating when patients are admitted to a hospital are not nosocomial; they are community acquired. However, because such infections can serve as a ready source or reservoir of pathogens for other patients or personnel, they are also considered in the total epidemiology of nosocomial infections. The CDC estimates that about 10% of all hospital patients (2 to 4 million people) acquire some type of nosocomial infection. Thus nosocomial infections represent a significant proportion of all infectious diseases acquired by humans.

Nosocomial diseases are usually caused by bacteria, most of which are noninvasive and part of the normal microbiota; viruses, protozoa, and fungi are rarely involved. Figure 33.12 summarizes the most common types of nosocomial infections and the most common nosocomial pathogens.

Source

The nosocomial pathogens that cause diseases come from either endogenous or exogenous sources. Endogenous sources are the patient's own microbiota; exogenous sources are microbiota other than the patient's. Endogenous pathogens are either brought into the hospital by the patient or are acquired when the patient becomes colonized after admission. In either case, the pathogen colonizing the patient may subsequently cause a nosocomial disease (e.g., when the pathogen is transported to another part of the body or when the host's resistance drops). If it cannot be determined that the specific pathogen responsible for a nosocomial disease is exogenous or endogenous, then the term autogenous is used. An autogenous infection is one that is caused by an agent derived from the microbiota of the patient, despite whether it became part of the patient's microbiota following his or her admission to the hospital.

Many potential exogenous sources exist in a hospital. Animate sources are the hospital staff, other patients, and visitors. Some examples of inanimate exogenous sources are food, computer keyboards, urinary catheters, intravenous and respiratory therapy equipment, and water systems (e.g., softeners, dialysis units, and hydrotherapy equipment).

Control, Prevention, and Surveillance

In the United States, nosocomial infections prolong hospital stays by 4 to 13 days, result in over $4.5 billion a year in direct hospital charges, and lead to over 20,000 direct and 60,000 indirect deaths annually. The enormity of this problem has led most hospitals to allocate substantial resources to developing methods and programs for the surveillance, prevention, and control of nosocomial infections.

All personnel involved in the care of patients should be familiar with basic infection control measures such as isolation policies of the hospital; aseptic techniques; proper handling of equipment, supplies, food, and excreta; and surgical wound care and dressings. To adequately protect their patients, hospital personnel must practice proper aseptic technique and handwashing procedures, and must wear gloves when contacting blood, mucous membranes, and secretions. Patients should be monitored with respect to the frequency, distribution, symptomatology, and other characteristics common to nosocomial infections. A dynamic control and surveillance program can be invaluable in preventing many nosocomial infections, patient discomfort, extended stays, and further expense. All hospitals desiring accreditation by the Joint Commission on Accreditation of Healthcare Organizations must have a designated individual directly responsible for developing and implementing policies governing control of infections and communicable diseases. This infection control practitioner periodically evaluates laboratory reports, patients' charts, and surveys to determine whether any increase has occurred in the frequency of particular infectious diseases or potential pathogens.

1. Describe a nosocomial infection and explain why such infections are important.

2. What two general sources are responsible for nosocomial infections? Give some specific examples of each general source.

3. What does an infection control practitioner do to control nosocomial infections?

Summary

33.1 Epidemiological Terminology

a. Epidemiology is the science that evaluates the determinants, occurrence, distribution, and control of health and disease in a defined population.

b. Specific epidemiological terminology is used to communicate disease incidence in a given population. Frequently used terms include sporadic disease, endemic disease, hyperendemic disease, epidemic, index case, outbreak, and pandemic.

33.2 Measuring Frequency: The Epidemiologist's Tools

a. Statistics is an important tool used in the study of modern epidemiology.

b. Epidemiological data can be obtained from such factors as morbidity, prevalence, and mortality rates.

33.3 Recognition of an Infectious Disease in a Population

a. An infectious disease is caused by microbial agents such as viruses, bacteria, fungi, protozoa, and helminths. A communicable disease can be transmitted from person to person.

b. The manifestations of an infectious disease can range from mild to severe to deadly, depending on the agent and host.

c. Surveillance is necessary for recognizing a specific infectious disease within a given population. This consists of gathering data on the occurrence of the disease, collating and analyzing the data, summarizing the findings, and applying the information to control measures.

d. It is important to recognize the signs and symptoms that compose a disease syndrome. This includes the characteristic course of the disease, including such aspects as the incubation period and the prodromal stage (**figure 33.3**).

e. Remote sensing and geographic information systems can be used to gather epidemiological data on the environment.

33.4 Recognition of an Epidemic

a. A common-source epidemic is characterized by a sharp rise to a peak and then a rapid but not as pronounced decline in the number of individuals infected (**figure 33.4**). A propagated epidemic is characterized by a relatively slow and prolonged rise and then a gradual decline in the number of individuals infected (**figure 33.5**).

b. Herd immunity is the resistance of a population to infection and pathogen spread because of the immunity of a large percentage of the individuals within the population.

33.5 The Infectious Disease Cycle: Story of a Disease

a. The infectious disease cycle or chain involves the characteristics of the pathogen, the source or reservoir of the pathogen, the transmission of the pathogen, the susceptibility of the host, the exit mechanism of the pathogen from the body of the host, and the pathogen's spread to a new reservoir or host (**figure 33.7**).

b. There are four major modes of transmission: airborne, contact, vehicle, and vector-borne (**figure 33.8**).

33.6 Virulence and the Mode of Transmission

a. The degree of virulence may be influenced by the pathogen's preferred mode of transmission. New human diseases may arise and spread because of ecosystem disruption, rapid transportation, human behavior, and other factors.

33.7 Emerging and Reemerging Infectious Diseases and Pathogens

a. Humans will continually be faced with both new infectious diseases and the reemergence of older diseases once thought to be conquered.

b. The CDC has defined these diseases as "new, reemerging, or drug-resistant infections whose incidence in humans has increased within the past two decades or whose incidence threatens to increase in the near future" (**figure 33.11**).

c. Many factors characteristic of the modern world undoubtedly favor the development and spread of these microorganisms and their diseases.

33.8 Control of Epidemics

a. The public health system consists of individuals and organizations that function in the control of infectious diseases and epidemics.

b. Vaccination is one of the most cost-effective weapons for microbial disease prevention, and vaccines constitute one of the greatest achievements of modern medicine.

c. Many of the current vaccines in use for humans **(table 33.3)** consist of whole organisms that are either inactivated (killed) or attenuated (live but avirulent).

d. Some of the risks associated with whole-cell vaccines can be avoided by using only specific purified macromolecules derived from pathogenic microorganisms. Currently there are three general forms of subunit or acellular vaccines: capsular polysaccharides, recombinant surface antigens, and inactivated exotoxins (toxoids) **(table 33.5)**.

e. A number of microorganisms have been used for recombinant-vector vaccines. The attenuated microorganism serves as a vector, replicating within the host and expressing the gene product of the pathogen-encoded antigenic proteins. The proteins can elicit humoral immunity when the proteins escape from the cells and cellular immunity when they are broken down and properly displayed on the cell surface.

f. DNA vaccines elicit protective immunity against a pathogen by activating both branches of the immune system: humoral and cellular.

g. Epidemiological control measures can be directed toward reducing or eliminating infection sources, breaking the connection between sources and susceptible individuals, or isolating the susceptible individuals and raising the general level of herd immunity by immunization.

33.9 Bioterrorism Preparedness

a. Among weapons of mass destruction, biological weapons are more destructive than chemical weapons. The list of biological agents that could pose the greatest public health risk in the event of a bioterrorist attack is short and includes viruses, bacteria, parasites, and toxins **(table 33.6)**.

33.10 Nosocomial Infections

a. Nosocomial infections are infections acquired within a hospital or other clinical care facility and are produced by a pathogen acquired during a patient's stay. These infections come from either endogenous or exogenous sources **(figure 33.12)**.

b. Hospitals must designate an individual to be responsible for identifying and controlling nosocomial infections. This person is known as the infection control practitioner.

Critical Thinking Questions

1. Why is international cooperation a necessity in the field of epidemiology? What specific problem can you envision if there were no such cooperation?

2. What common sources of infectious disease are found in your community? How can the etiologic agents of such infectious diseases spread from their source or reservoir to members of your community?

3. How could you prove that an epidemic of a given infectious disease was occurring?

4. How can changes in herd immunity contribute to an outbreak of a disease on an island?

5. College dormitories are notorious for outbreaks of flu and other infectious diseases. These are particularly prevalent during final exam weeks. Using your knowledge of the immune response and epidemiology, suggest practices that could be adopted to minimize the risks at such a critical time.

6. Why does an inactivated vaccine induce only a humoral response, whereas an attenuated vaccine induces both humoral and cell-mediated responses?

7. Why is a DNA vaccine delivered intramuscularly and not by intravenous or oral routes?

Learn More

Learn more by visiting the Prescott website at www.mhhe.com/prescottprinciples, where you will find a complete list of references.

Microbiology of Food

34

Large tanks used for wine production. Fermentations can be carried out in such open-air units in temperate regions. After completion of the fermentation, the fresh wine will be transferred to barrels for storage and aging.

aflatoxin   A polyketide secondary fungal metabolite that can cause cancer.

bacteriocin   A peptide produced by a bacterium that kills other closely related bacteria. Nisin is a bacteriocin used to prevent spoilage of certain foods.

enology   The science of wine making.

food-borne infection   Gastrointestinal illness caused by ingestion of microorganisms, followed by their growth within

the host. Symptoms arise from tissue invasion and toxin production.

food intoxication   Food poisoning caused by microbial toxins produced in a food prior to consumption. The presence of living bacteria is not required.

fumonisin   A family of toxins produced by mold belonging to the genus *Fusarium;* primarily affects corn and is known to be hepato- and nephrotoxic in animals.

GRAS   The designation "*generally regarded as safe*" is given to food additives and some preservatives.

lactic acid bacteria (LAB)   Strictly fermentative, gram-positive bacteria that produce lactic acid as the primary fermentation end product.

pasteurization   The process of heating milk and other liquids to destroy microorganisms that can cause spoilage or disease.

probiotic   Living microorganisms that confer a health benefit to the host when consumed or administered in sufficient quantites.

putrefaction   The microbial decomposition of organic matter, especially the anaerobic breakdown of proteins, with the production of foul-smelling compounds such as hydrogen sulfide and amines.

starter culture   An inoculum, consisting of a mixture of carefully selected microorganisms, used to start a commercial fermentation.

Tell me what you eat, and I will tell you what you are.

—*Jean Anthelme Brillat-Savarin*

Foods, microorganisms, and humans have had an interesting association that developed long before recorded history. On the one hand, microorganisms can be used to transform raw foods into gastronomic delights including chocolate, cheeses, pickles, sausages, soy sauce, wines, and beers. On the other hand, microorganisms can degrade food quality and lead to spoilage.

Sometimes a fine line exists between the microbial enhancement of foods and degradation and potential disease transmission. The detection and control of pathogens and food spoilage microorganisms are important parts of food microbiology. In this chapter, we consider the two opposing roles of microorganisms in food production and preservation.

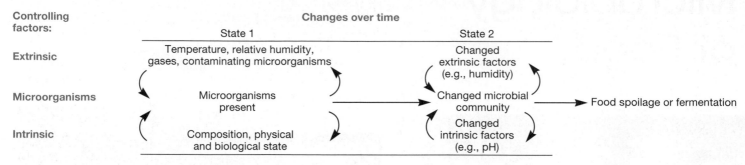

Figure 34.1   Intrinsic and Extrinsic Factors.   A variety of intrinsic and extrinsic factors can influence microbial growth in foods. Time-related successional changes occur in the microbial community and the food.

34.1   MICROORGANISM GROWTH IN FOODS

Foods, because they are nutrient-rich, are excellent environments for the growth of microorganisms. Microbial growth is controlled by factors related to the food itself, called intrinsic factors, and to the environment where the food is stored, described as extrinsic factors, as shown in **figure 34.1**.

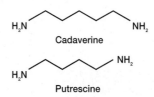

Figure 34.2   Cadaverine and Putrescine.   These foul-smelling amines are the products of protein degradation.

Intrinsic Factors

Food composition is a critical intrinsic factor that influences microbial growth. If a food consists primarily of carbohydrates, fungal growth predominates and spoilage does not result in major odors. Thus foods such as breads, jams, and some fruits first show spoilage by fungal growth. In contrast, when foods contain large amounts of proteins or fats (e.g., meat and butter), bacterial growth can produce a variety of foul odors—think of rotting eggs. The anaerobic breakdown of proteins is called **putrefaction**. It yields foul-smelling amine compounds such as cadaverine (imagine the origin of that name) and putrescine (**figure 34.2**). Degradation of fats ruins food as well. The production of short-chained fatty acids from fats renders butter rancid and foul smelling. Thus the major substrate present in a food helps determine the type of spoilage that may occur (**table 34.1**). ◀◀ *Influences of environmental factors on growth (section 7.5)*

The presence and availability of water also affect the ability of microorganisms to colonize foods. Simply by drying a food,

one can control or eliminate spoilage processes. Water can be made less available by adding solutes such as sugar and salt. Water availability is measured in terms of **water activity** (a_w). This represents the ratio of relative humidity of the air over a test solution compared with that of distilled water, which has an a_w of one. When large quantities of salt or sugar are added to food, most microorganisms are dehydrated by the hypertonic conditions and cannot grow (**table 34.2**). Even under these adverse conditions, osmophilic and xerophilic microorganisms may spoil food. Osmophilic (Greek *osmus,* impulse, and *philein,* to love) microbes grow best in or on media with a high osmotic concentration (e.g., jams and jellies), whereas xerophilic (Greek *xerosis,* dry, and *philein,* to love) microorganisms prefer a low a_w environment (e.g., dried fruits, cereals) and may not grow under high a_w conditions.

The pH and oxidation-reduction (redox) potential of a food also are critical. A low pH favors the growth of yeasts and molds. In neutral or alkaline pH foods, such as meats, bacteria are more dominant in spoilage and putrefaction. Furthermore,

Table 34.1	Differences in Spoilage Processes in Relation to Food Characteristics		
Substrate	**Food Example**	**Chemical Reactions or Processes**[a]	**Typical Products (and Effects)**
Pectin	Fruits	Pectinolysis	Methanol, uronic acids (loss of fruit structure, soft rots)
Proteins	Meat	Proteolysis, deamination	Amino acids, peptides, amines, H_2S, ammonia, indole (bitterness, souring, bad odor, sliminess)
Carbohydrates	Starchy foods	Hydrolysis, fermentations	Organic acids, CO_2, mixed alcohols (souring, acidification)
Lipids	Butter	Hydrolysis, fatty acid degradation	Glycerol and mixed fatty acids (rancidity, bitterness)

[a]Other reactions also occur during the spoilage of these substrates.

Table 34.2	Approximate Minimum Water Activity Relationships of Microbial Groups Important in Food Spoilage	
Organisms	a_w	
Groups		
Most spoilage bacteria	0.9	
Most spoilage yeasts	0.88	
Most spoilage molds	0.80	
Halophilic bacteria	0.75	
Xerophilic molds	0.61	
Osmophilic yeasts	0.61	

Adapted from James M. Jay. 2000. *Modern Food Microbiology*, 6th edition. Reprinted by permission of Aspen Publishers, Inc. Gaithersburg, MD. Tables 3–5, p. 42.

when meat products, especially broths, are cooked, they often have lower oxidation-reduction potentials—that is, they present a reducing environment for microbial growth. These products, with their readily available amino acids, peptides, and growth factors, are ideal media for the growth of anaerobes, including *Clostridium*.

The physical structure of a food also can affect the course and extent of spoilage. The grinding and mixing of foods such as sausage and hamburger increase the food surface area and distribute contaminating microorganisms throughout the food. This can result in rapid spoilage if such foods are stored improperly. In addition, some spoilage microorganisms have specialized enzymes that help them penetrate protective peels and rinds, especially after the fruits and vegetables have been bruised.

Many foods contain natural antimicrobial substances, including complex chemical inhibitors and enzymes. Coumarins found in fruits and vegetables exhibit antimicrobial activity. Cow's milk and eggs also contain antimicrobial substances. Eggs are rich in the enzyme lysozyme that can lyse the cell walls of contaminating gram-positive bacteria (*see figure 28.17*).

Herbs and spices often possess significant antimicrobial substances; generally fungi are more sensitive than most bacteria. Sage and rosemary are two of the most antimicrobial spices. Aldehydic and phenolic compounds that inhibit microbial growth are found in cinnamon, mustard, and oregano. Other important inhibitors are garlic, which contains allicin; cloves, which have eugenol; and basil, which contains rosmarinic acid. Nonetheless, spices can sometimes contain pathogenic and spoilage organisms. Enteric bacteria, *Bacillus cereus*, *Clostridium perfringens*, and *Salmonella* species have been detected in spices. Microorganisms can be eliminated or reduced by ethylene oxide sterilization. This treatment can result in *Salmonella*-free spices and herbs and a 90% reduction in the levels of general spoilage organisms. ≪ *The use of chemical agents in control (section 8.5)*

Extrinsic Factors

Temperature and relative humidity are important extrinsic factors in determining whether a food will spoil. At higher relative humidities, microbial growth is initiated more rapidly, even at lower temperatures (especially when refrigerators are not maintained in a defrosted state). When drier foods are placed in moist environments, moisture absorption can occur on the food surface, promoting microbial growth.

The atmosphere in which food is stored also is important. This is especially true with shrink-wrapped foods because many plastic films allow oxygen diffusion, which results in increased growth of surface-associated microorganisms. Excess CO_2 can decrease the solution pH, inhibiting microbial growth. Storing meat in a high CO_2 atmosphere inhibits gram-negative bacteria, resulting in a population dominated by the lactobacilli.

The observation that food storage atmosphere is important has led to the development of **modified atmosphere packaging (MAP)**. Modern shrink-wrap materials and vacuum technology make it possible to package foods with controlled atmospheres. These materials are largely impermeable to oxygen. This prolongs shelf life by a factor of two to five times compared to the same product packaged in air. With a CO_2 content of 60% or greater in the atmosphere surrounding a food, spoilage fungi will not grow, even if low levels of oxygen are present. Recently it has been found that high-oxygen MAP also may be effective. This is due to the formation of the superoxide (O_2^-) anion inside cells under these conditions. The superoxide anion is then transformed to highly toxic peroxide and hydroxyl radical, resulting in antimicrobial effects. Some products currently packaged using MAP technology include delicatessen meats and cheeses, pizza, grated cheese, some bakery items, and dried products such as coffee.

1. What are some intrinsic factors that influence food spoilage and how do they exert their effects?
2. What are the effects of food composition on spoilage processes?
3. Why might sausage and other ground meat products provide a better environment for the growth of food spoilage organisms than solid cuts of meats?
4. What extrinsic factors can determine whether food spoilage will occur?
5. What are the major gases involved in MAP? How are their concentrations varied to inhibit microbial growth?

34.2 MICROBIAL GROWTH AND FOOD SPOILAGE

Meat and dairy products, with their high nutritional value and easily metabolized carbohydrates, fats, and proteins, provide ideal environments for microbial spoilage. Proteolysis and putrefaction are typical results of microbial spoilage of such high-protein materials. Unpasteurized milk undergoes a predictable four-step microbial succession during spoilage: acid production by *Lactococcus lactis* subsp. *lactis* is followed by additional acid production associated with the growth of more acid-tolerant organisms such as *Lactobacillus*. At this point yeasts and molds become dominant and degrade the accumulated lactic acid, and the acidity gradually decreases. Eventually protein-digesting bacteria become active, resulting in a putrid odor and bitter flavor. The milk, originally opaque, eventually becomes clear (**figure 34.3**).

In comparison with meat and dairy products, most fruits and vegetables have a much lower protein and fat content and undergo a different kind of spoilage, which often is initiated by molds. These organisms have enzymes that contribute to the weakening and penetration of the protective outer skin. The readily degradable carbohydrates within favor spoilage by bacteria, especially bacteria that cause soft rots, such as *Erwinia carotovora,* which produces hydrolytic enzymes. The lack of reduced conditions enables aerobes and facultative anaerobes to contribute to the decomposition processes.

Food spoilage problems occur with minimally processed, concentrated frozen citrus products. These are prepared with little or no heat treatment, and major spoilage can be caused by *Lactobacillus* and *Leuconostoc* spp., which convert citrate to diacetyl. *Saccharomyces* and *Candida* can also spoil juices. Concentrated juice has a decreased water activity (a_w 0.8 to 0.83) and, when kept frozen, can be stored for long periods. However, when concentrated juices are diluted with water that contains spoilage organisms, or if the juice is stored in improperly washed containers, problems can occur. Also, microorganisms in the frozen concentrated juices can begin the spoilage process after the addition of water. Ready-to-serve (RTS) juices present other problems as the a_w values are sufficiently high to allow microbial growth. Although pasteurization results in some flavor loss, most juices are now routinely pasteurized (section 34.3).

Molds can rapidly grow on grains and corn when these products are stored in moist conditions. The moldy bread pictured in **figure 34.4***a* shows extensive fungal hyphal development and sporulation. The green growth most likely is *Penicillium;* the black growth is characteristic of *Rhizopus stolonifer* (*see figure 23.26*). Contamination of grains by the ascomycete *Claviceps purpura* causes **ergotism,** a toxic condition. Hallucinogenic alkaloids produced by this fungus can lead to altered behavior, abortion, and death if infected grains are eaten. Molds are also a special problem for tomatoes. Even the slightest bruising of the tomato skin, exposing the interior, will result in rapid fungal growth. This affects the quality of tomato products, including tomato juices and ketchups.

(a)

Figure 34.3 Spoilage of a Dairy Product. Fresh (left) and curdled (right) milk are shown. The spoilage process has produced acidic conditions that have denatured and precipitated the milk casein to yield separated curds and whey.

(b)

Figure 34.4 Food Spoilage. When foods are not stored properly, microorganisms can cause spoilage. Typical examples are fungal spoilage of (a) bread and (b) corn. Such spoilage of corn is called ear rot and can result in major economic losses.

Fungus-derived carcinogens include the aflatoxins and fumonisins. **Aflatoxins** are produced most commonly in moist grains and nut products. Aflatoxins were discovered in 1960, when 100,000 turkey poults died from eating fungus-infested peanut meal. *Aspergillus flavus* was found in the infected peanut meal, together with alcohol-extractable toxins termed aflatoxins. These flat-ringed planar compounds intercalate with nucleic acids and act as potent frameshift mutagens and carcinogens. This occurs primarily in the liver, where they are converted to unstable derivatives. At the present time, a total of 18 aflatoxins are known. The most important are shown in **figure 34.5**. Of these, aflatoxin B1 is the most common and the most potent carcinogen. Aflatoxins B1 and B2, after ingestion by lactating animals, are modified in the animal body to yield the aflatoxins M1 and M2. If cattle consume aflatoxin-contaminated feeds, aflatoxins also can appear in milk and dairy products. Besides their importance in grains, they have also been found in beer, cocoa, raisins, and soybean meal.

Diet appears to be related to aflatoxin exposure: the average aflatoxin intake in the typical European-style diet is 19 ng/day, whereas for some Asian diets it is estimated to be 103 ng/day. Aflatoxin sensitivity also can be influenced by prior disease exposure. Individuals who have had hepatitis B have a 30-times

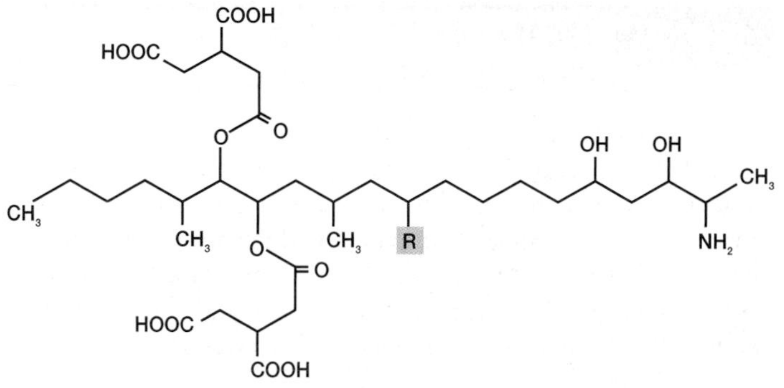

Figure 34.6 Fumonisin Structure. The basic structure of fumonisins FB1 and FB2 produced by *Fusarium moniliforme,* a fungal contaminant that can grow in improperly stored corn. A total of at least 10 different fumonisins have been isolated. These are strongly polar compounds that cause diseases in domestic animals and humans. FB1, R = OH; FB2, R = H.

higher risk of liver cancer upon exposure to aflatoxins than individuals who have not had this disease. This association illustrates an emerging link between inflammation and cancer. It has been observed that prevention of hepatitis B infections by vaccination and reduction of carrier populations helps to control the potential effects of aflatoxins in foodstuffs.

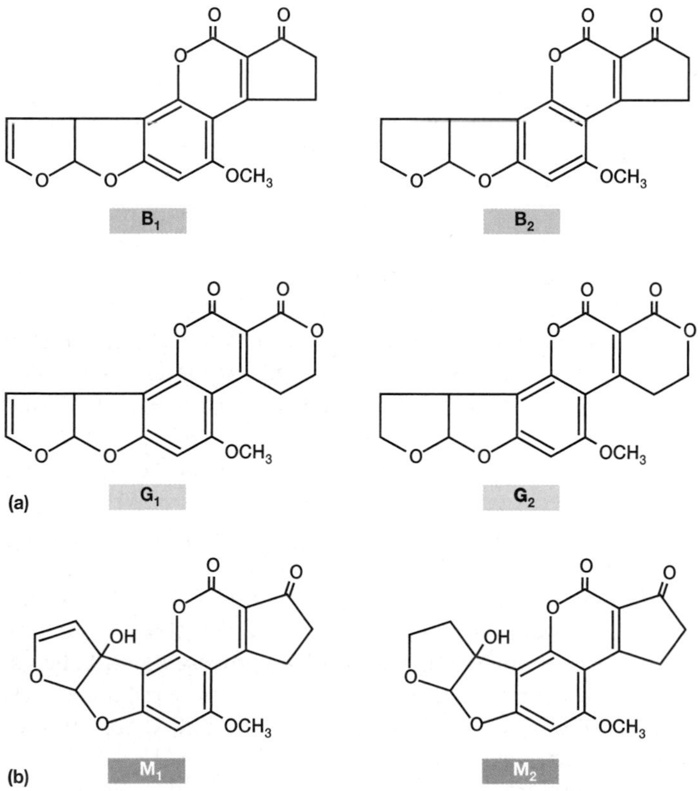

The **fumonisins** are fungal contaminants of corn that were first isolated in 1988. These are produced by *Fusarium moniliforme* and cause leukoencephalomalacia in horses (also called "blind staggers"—it is fatal within 2 to 3 days), pulmonary edema in pigs, and esophageal cancer in humans. The fumonisins inhibit ceramide synthase, a key enzyme for the proper use of fatty substances in the cell. This disrupts the synthesis and metabolism of sphingolipids, important compounds that influence a wide variety of cell functions. At least 10 different fumonisins exist; the basic structures of fumonisins FB1 and FB2 are shown in **figure 34.6**. Corn and corn-based feeds and foods, including cornmeal and corn grits, can be contaminated. Thus it is extremely important to store corn and corn products under dry conditions where these fungi cannot develop.

Eucaryotic microorganisms can synthesize potent toxins other than aflatoxins and fumonisins. Algal toxins contaminate fish and thus affect the health of marine animals higher in the food chain; they also contaminate shellfish and finfish, which are later consumed by humans. These toxins are produced during harmful algal blooms, which are discussed in section 26.1.

Figure 34.5 Aflatoxins. When *Aspergillus flavus* and related fungi grow on foods, carcinogenic aflatoxins can be formed. These have four basic structures. (a) The letter designations refer to the color of the compounds under ultraviolet light after extraction from the grain and separation by chromatography. The B₁ and B₂ compounds fluoresce with a blue color, and the G₁ and G₂ appear green. (b) The two type M aflatoxins are found in the milk of lactating animals that have ingested type B aflatoxins.

1. What fungal genus produces ergot alkaloids? What conditions are required for the synthesis of these substances?

2. Aflatoxins are produced by which fungal genus? How do they damage animals that eat the contaminated food?

3. What microbial genus produces fumonisins and why are these compounds of concern? If improperly stored, what are the major foods and feeds in which these chemicals might be found?

34.3 CONTROLLING FOOD SPOILAGE

With the beginning of agriculture and a decreasing dependence on hunting and gathering, the need to preserve surplus foods became essential to human survival. The use of salt as a meat preservative, the production of wines, and the preservation of fish and meat by smoking were introduced in Near Eastern civilization as early as 3000 BCE. But it was not until the nineteenth century that the microbial spoilage of food was studied systematically. Louis Pasteur established the modern era of food microbiology in 1857, when he showed that microorganisms cause milk spoilage. Pasteur's work in the 1860s proved that heat could be used to control spoilage organisms in wines and beers. ◄◄ *Golden age of microbiology (section 1.5)*

Foods can be preserved by a variety of methods. The goal of each method is to eliminate or reduce the populations of spoilage and disease-causing microorganisms while maintaining food quality. We now briefly discuss some of these techniques.

Removal of Microorganisms

Microorganisms can be removed from water, wine, beer, juices, soft drinks, and other liquids by filtration. This can keep bacterial populations low or eliminate them entirely. Removal of large particulates by prefiltration and centrifugation maximizes filter life and effectiveness. Several major brands of beer are filtered rather than pasteurized to better preserve the flavor and aroma of the original product.

Low Temperature

Refrigeration at 5°C retards microbial growth, although with extended storage, microorganisms eventually grow and produce spoilage. Slow microbial growth at temperatures below 10°C has been described, particularly with fruit juice concentrates, ice cream, and some fruits. Of particular concern is the growth of *Listeria,* which can grow at temperatures used for refrigeration. It should be kept in mind that refrigeration slows the metabolic activity of most microbes, but it does not lead to significant decreases in overall microbial populations.

High Temperature

Controlling microbial populations in foods by means of high temperatures can significantly limit disease transmission and spoilage. Heating processes, first used by Nicholas Appert in 1809, provide a safe means of preserving foods, particularly when carried out in commercial canning operations (**figure 34.7**). Canned food is heated in special containers called retorts at about 115°C for intervals ranging from 25 to over 100 minutes. The precise time and temperature depend on the nature of the food. Sometimes canning does not kill all microorganisms but only those that will spoil the food (e.g., remaining bacteria are unable to grow due to acidity of the food). After heat treatment, the cans are cooled as rapidly as possible, usually with cold

Figure 34.7 Food Preparation for Canning. Microbial control is important in processing and preserving many foods. Here a worker is pouring peas into a large, clean vat during the preparation, of vegetable soup. After preparation, the soup is transferred to cans. Each can is heated for a short period, sealed, processed at temperatures from 110 to 121°C in a canning retort to destroy spoilage microorganisms, and finally cooled.

water. Quality control and processing effectiveness are sometimes compromised, however, in home processing of foods, especially with less acidic (pH values greater than 4.6) products such as green beans or meats. ◄◄ *The use of physical methods in control (section 8.4)*

Despite efforts to eliminate spoilage microorganisms during canning, canned foods may become spoiled. This may be due to spoilage before canning, underprocessing during canning, and leakage of contaminated water through seams during cooling. Spoiled food can be altered in such characteristics as color, texture, odor, and taste. Organic acids, sulfides, and gases (particularly CO_2 and H_2S) may be produced. If spoilage microorganisms produce gas, both ends of the can will bulge outward. Sometimes the swollen ends can be moved by thumb pressure (soft swells); in other cases, the gas pressure is so great that the ends cannot be dented by hand (hard swells). However, swelling is not always due to microbial spoilage; acid in low pH foods may react with iron in the can to release hydrogen and generate a hydrogen swell.

Pasteurization involves heating food to a temperature that kills disease-causing microorganisms and substantially reduces the levels of spoilage organisms. When processing milk, beers, and fruit juices by conventional low-temperature holding (LTH) pasteurization, the liquid is maintained at 62.8°C for 30 minutes. Products can also be held at 72°C for 15 seconds, a high-temperature, short-time (HTST) process; milk can be treated at 138°C for 2 seconds for ultra-high-temperature (UHT) processing. Shorter-term processing results in improved flavor and extended product shelf life. The duration of pasteurization is based on the statistical probability that the number of remaining viable microorganisms will be below a certain level after a particular heating time at a specific temperature. These calculations are discussed in detail in section 8.4.

Water Availability

Dehydration, such as lyophilization to produce freeze-dried foods, is a common means of eliminating microbial growth. The modern process is simply an update of older procedures in which grains, meats, fish, and fruits were dried. The combination of free-water loss with an increase in solute concentration in the remaining water makes this type of preservation possible.

Chemical-Based Preservation

Various chemical agents can be used to preserve foods, and these substances are closely regulated by the U.S. Food and Drug Administration and are listed as being "*generally recognized as safe*" or **GRAS** (table 34.3). They include simple organic acids, sulfite, ethylene oxide as a gas sterilant, sodium nitrite, and ethyl formate. These chemical agents may damage the microbial plasma membrane or denature various cell proteins. Other compounds interfere with the functioning of nucleic acids, thus inhibiting cell reproduction.

Sodium nitrite is an important chemical used to help preserve ham, sausage, bacon, and other cured meats by inhibiting the growth of *Clostridium botulinum* and the germination of its spores. This protects against botulism and reduces the rate of spoilage. Besides increasing meat safety, nitrite decomposes to nitric acid, which reacts with heme pigments to keep the meat red in color. Concern about nitrite arises from the observation that it can react with amines to form carcinogenic nitrosamines.

Low pH can also be used to hinder microbial spoilage. For example, acetic and lactic acids inhibit listerial growth. Organic acids (1–3%) can be used to treat meat carcasses, and poultry can be cleansed with 10% lactic acid/sodium lactate buffer (pH 3) prior to packaging. In addition, low pH can increase the activity of other chemical preservatives. Sodium propionate is most effective at lower pH values, where it is primarily undissociated. Breads, with their low pH values, often contain sodium propionate as a preservative.

Radiation

The major method used for radiation sterilization of food is gamma irradiation from a cobalt-60 source; however, cesium-137 is used in some facilities. Gamma radiation has excellent penetrating power but must be used with moist foods because radiation is effective only if it can generate peroxides from water in the microbial cells, resulting in oxidation of sensitive cellular constituents. This process of **radappertization,** named after Nicholas Appert, can extend the shelf life of seafoods, fruits, and vegetables. To sterilize meat products, commonly 4.5 to 5.6 megarads are used.

Electron beams can also be used to irradiate foods. The electrons are generated electrically, so unlike gamma radiation, they can be turned on only when needed. Also, this approach does not generate radioactive waste. On the other hand, electron beams do not penetrate food items as deeply as does gamma radiation. It is important to note that regardless of the radiation source (gamma rays or electron beams), the food itself does not become radioactive.

Microbial Product–Based Inhibition

Interest is increasing in the use of bacteriocins for the preservation of foods. **Bacteriocins** are bactericidal proteins active against closely related bacteria, which bind to specific sites on the cell and often affect cell membrane integrity and function. The only currently approved product is nisin, a small amphiphilic peptide produced by some strains of *Lactococcus lactis*. It is nontoxic to

---

Table 34.3 Major Groups of Chemicals Used in Food Preservation

Preservatives	Approximate Maximum Use	Organisms Affected	Foods
Propionic acid/propionates	0.32%	Molds	Bread, cakes, some cheeses; inhibitor of ropy bread dough
Sorbic acid/sorbates	0.2%	Molds	Hard cheeses, figs, syrups, salad dressings, jellies, cakes
Benzoic acid/benzoates	0.1%	Yeasts and molds	Margarine, pickle relishes, apple cider, soft drinks, tomato ketchup, salad dressings
Parabens[a]	0.1%	Yeasts and molds	Bakery products, soft drinks, pickles, salad dressings
SO_2/sulfites	200–300 ppm	Insects and microorganisms	Molasses, dried fruits, wine, lemon juice (not used in meats or other foods recognized as sources of thiamine)
Ethylene/propylene oxides	700 ppm	Yeasts, molds, vermin	Fumigant for spices, nuts
Sodium diacetate	0.32%	Molds	Bread
Dehydroacetic acid	65 ppm	Insects	Pesticide on strawberries, squash
Sodium nitrite	120 ppm	Clostridia	Meat-curing preparations
Caprylic acid	—	Molds	Cheese wraps
Ethyl formate	15–200 ppm	Yeasts and molds	Dried fruits, nuts

From James M. Jay. 2000. *Modern Food Microbiology,* 6th edition. Reprinted by permission of Aspen Publishing, Frederick, MD.
[a]Methyl-, propyl-, and heptyl-esters of *p*-hydroxybenzoic acid.

humans and affects other gram-positive bacteria by binding to the lipid II portion of the growing peptidoglycan and forming pores in the plasma membrane (*see figure 11.11*). Nisin can be used particularly in low-acid foods to improve inactivation of *Clostridium botulinum* during the canning process or to inhibit germination of any surviving spores. << *Chemical mediators in nonspecific (innate) resistance: Bacteriocins (section 28.6)*

In 2006 the U.S. Food and Drug Adminstration approved the use of a preparation of six strains of **bacteriophages** that specifically infect and kill *Listeria monocytogenes*. The phages are present in a spray designed for use on ready-to-eat meats such as hot dogs and lunch meats. The spray is applied to the surface of the meats prior to packaging. The phages are present in equal concentration; using multiple phage types significantly reduces the chances of spontaneous phage resistance.

1. Describe the major approaches used in food preservation.

2. What types of chemicals can be used to preserve foods?

3. Under what conditions can gamma radiation be used to control microbial populations in foods and food preparation? What is radappertization?

4. How does nisin function? What bacterial genus produces this important peptide?

5. Consider the fact that the *Listeria* phages must be propagated in the pathogen prior to producing in the anti-*Listeria* spray. What safeguards do you think must be instituted to ensure the safety of the spray?

34.4 FOOD-BORNE DISEASES

Food-borne illnesses impact the entire world. In the United States, based on recent information from the Centers for Disease Control and Prevention, annual incidences of food-related diseases involve 76 million cases, of which only 14 million can be attributed to known pathogens. Food-borne diseases result in 325,000 hospitalizations and at least 5,000 deaths per year. Since 1942 the number of recognized food-borne pathogens has increased over fivefold. In most cases, these "new" pathogens are simply agents that now can be described, based on an improved understanding of microbial diversity. Recent estimates indicate that noroviruses, *Campylobacter jejuni,* and *Salmonella* are the major causes of food-borne diseases. In addition, *Escherichia coli* O157:H7 and *Listeria* are important food-related pathogens.

There are two primary types of food-related diseases: food-borne infections and food intoxications. All of these food-borne diseases are associated with poor hygienic practices. Whether by water or food transmission, the fecal-oral route is key, with the food providing the vital link between hosts. Fomites, such as sink faucets, drinking cups, and cutting boards, also play a role in the maintenance of the fecal-oral route of contamination.

Food-Borne Infection

A **food-borne infection** involves the ingestion of the pathogen, followed by growth in the host, including tissue invasion or the release of toxins. The major diseases of this type are summarized in **table 34.4**.

Salmonellosis results from ingestion of a variety of *Salmonella* serovars, particularly Typhimurium and Enteritidis. Gastroenteritis is the disease of most concern in relation to foods such as meats, poultry, and eggs, and the onset of symptoms occurs after an incubation time as short as 8 hours. *Salmonella* infection can arise from contamination by workers in food-processing plants and restaurants, as well in canning processes. << *Historical Highlights 33.2: Typhoid Mary.*

Campylobacter jejuni is considered a leading cause of acute bacterial gastroenteritis in humans. This important pathogen is often consumed in under- or uncooked poultry products. For example, transmission often occurs when kitchen utensils and containers are used for chicken preparation and then for salads. Contamination with as few as 10 viable *C. jejuni* cells can lead to the onset of diarrhea. *C. jejuni* also is transmitted by raw milk, and the organism has been found on various red meats. Thorough cooking of food prevents its transmission.

Listeriosis, caused by *Listeria monocytogenes,* was responsible for the largest meat recall in U.S. history. In 2002 a seven-state listeriosis outbreak was linked to deli meats and hot dogs produced at a single meat-processing plant in Pennsylvania. Pregnant women, the young and the old, and immunocompromised individuals are especially vulnerable to *L. monocytogenes* infections. In this outbreak, 7 deaths, 3 stillbirths, and 46 illnesses were caused by consumption of contaminated meats. Microbiologists matched the strain of *L. monocytogenes* found in the contaminated food products with samples obtained from floor drains in a Wampler, Inc. packaging plant. This prompted the recall of 27.4 million pounds of meats that had been distributed over a 5-month period to stores, restaurants, and school lunch programs. Following the outbreak, the plant closed for a month and the Wampler brand name was phased out. This episode prompted the U.S. Department of Agriculture (USDA) to step up its environmental testing program for *L. monocytogenes* so that it now tests plants that do not regularly submit data to the USDA. It also performs surprise inspections of those that do. The USDA advises people at risk of contracting listeriosis to avoid eating soft cheeses (e.g., feta, Brie, Camembert), refrigerated smoked meats such as lox, as well as deli meats and undercooked hot dogs. As a final note, the Wampler plant in Pennsylvania was closed in early 2006, having never recovered from the $100 million cost and the damage to its reputation caused by the 2002 outbreak.

Escherichia coli is an important food-borne disease organism. Enteropathogenic, enteroinvasive, and enterotoxigenic types can cause diarrhea (*see figure 20.32*). *E. coli* O157:H7, with its specific LPS O-antigen (O) and flagellar (H) antigen, is thought to have acquired enterohemorrhagic genes from *Shigella,* including the genes for shigalike toxins. This produced a pathogenic strain, first discovered in 1982, that is transmitted by the fecal-oral route. Its infectious dose appears to be only 500 bacteria. Enterohemorrhagic *E. coli* has been found in meat products such as hamburger and salami, in unpasteurized fruit drinks, on fruits and vegetables, and in untreated well water. Prevention of food contamination by *E. coli* O157:H7 is essential from the time of production until consumption. Hygiene must be monitored carefully in larger-volume slaughterhouses, where contact of meat with fecal material can occur, and fruits and vegetables should be handled with care. Caution also is essential at the point of use. For example, utensils used with raw foods should not contact cooked food; proper cleaning of cutting boards and utensils minimizes contamination.

Virus contamination is always a potential problem. This is based on transmission by water or by lack of hygiene in food preparation and direct contamination by food processors and handlers. Similar situations occur with protozoan pathogens. Virus contamination has become a severe problem on many cruise ships, where noroviruses have been involved in outbreaks, with person-to-person contact and possibly foods implicated in these avoidable occurrences.

An infectious agent of increasing worldwide concern with respect to food safety is the prion that causes **new variant Creutzfeldt-Jakob disease (vCJD).** This is one of a group of progressively degenerative neuronal diseases termed transmissible spongiform encephalopathies (TSEs) and is associated with beef cattle. It is often called "mad cow disease." A major problem in controlling new vCJD is the lack of reliable detection methods. The major means of vCJD transmission between animals is the use of mammalian tissue in ruminant animal feeds; at the present time, significant problems exist in detecting such prohibited animal products in ruminant feeds. ◄◄ *Prions (section 5.7)*

Foods that are transported and consumed in an uncooked state are an increasingly important source of food-borne infection. The problem becomes more serious with rapid movement of people and products around the world. International trade in uncooked foods, aided by rapid air transport, provides many opportunities for disease transmission. As examples we discuss sprouts, seafood, and raspberries. Sprouts, a popular and attractive garnish, are sometimes germinated and grown in waters containing pathogens. Contaminated alfalfa, bean, watercress, mungbean, mustard, and soybean sprouts can be major sources of typhoid and cholera.

Shellfish and finfish also present major concerns. Clams, oysters, and mussels are filter feeders that process several liters of water per day, leading to the potential concentration of toxins that may be produced during harmful algal blooms (*see p. 612*). Viruses introduced in raw or undertreated sewage can also accumulate in shellfish. Heavy rainfall in shellfish areas can cause runoff of pathogens from adjacent septic systems and contaminate coastal waters. Often it is necessary to ban shellfish harvesting until the animals void pathogens or toxins from their digestive systems. Alternatively shellfish from contaminated areas can be moved to clean waters to allow them to clean their digestive systems.

Raspberries provide an important example of another major problem: the rapid air transport of raw agricultural products around the world. Major outbreaks of *Cyclospora cayetanensis* poisoning have been traced to raspberries imported from Central America into the United States and Canada. In the growth and harvesting process, raspberries become contaminated, resulting in serious diarrhea in affected individuals. *Cyclospora* has a complex life cycle that is not fully understood at the present time. In comparison with *Giardia* and *Cryptosporidium,* which are infective immediately after being shed in feces, *Cyclospora* is not immediately infectious; sporulation or maturation requires 12 hours after release from the body. The mature infective cyst has two sporocysts (*see figure 23.20*), an important criterion for confirming the presence of this protist on foods or in the environment.

Food-Borne Intoxications

Microbial growth in food products also can result in **food intoxication;** the involved bacteria are summarized in table 34.4. Intoxication produces symptoms shortly after the food is consumed because growth of the disease-causing microorganism is not required. Toxins produced in the food can be associated with microbial cells or can be released from the cells.

Most *Staphylococcus aureus* strains cause a staphylococcal enteritis related to the synthesis of extracellular toxins. These are heat-resistant proteins, so heating does not usually render the food safe. The effects of the toxins are quickly felt, with disease symptoms occurring within 2 to 6 hours. The main reservoir of *S. aureus* is the human nasal cavity. Frequently *S. aureus* is transmitted to a person's hands and then is introduced into food during preparation. Growth and enterotoxin production usually occur when contaminated foods are held at room temperature for several hours.

Three gram-positive rods are known to cause food intoxications: *Clostridium botulinum, C. perfringens,* and *Bacillus cereus. C. botulinum* poisoning is discussed in chapter 21. However, here we note that baked potatoes served in aluminum foil can, even after washing, be contaminated with *C. botulinum,* which naturally occurs in the soil. If the foil-covered potatoes are not heated sufficiently in the baking process, surviving clostridia can proliferate after removal of the potatoes from the oven and rapidly produce toxins.

Clostridium perfringens food poisoning is one of the more widespread food intoxications. These microorganisms, which produce exotoxins, must grow to levels of approximately 10^6 bacteria per gram or higher in a food to cause disease. At least 10^8 bacteria must be ingested. They are common inhabitants of soil, water, food, spices, and the intestinal tract. Upon ingestion, the cells sporulate in the intestine. The enterotoxin is a spore-specific protein and is produced during the sporulation process. Enterotoxin can be detected in the feces of affected individuals. *C. perfringens* food poisoning can occur after meat products are heated, which results in O_2 depletion. If the foods are cooled slowly, growth of the microorganism can occur. At

Table 34.4 Some Bacteria That Cause Acute Bacterial Diarrhea and Food Poisoning

Organism	Incubation Period (Hours)	Vomiting	Diarrhea	Fever	Epidemiology
Staphylococcus aureus	1–8 (rarely, up to 18)	+++[a]	+	–[b]	Staphylococci grow in meats, dairy, and bakery products and produce enterotoxins.
Bacillus cereus	2–16	+++	++	–	Reheated fried rice causes vomiting or diarrhea.
Clostridium perfringens	8–16	±[c]	+++	–	Clostridia grow in rewarmed meat dishes.
Clostridium botulinum	18–24	±	Rare	–	Clostridia grow in anoxic foods and produce toxin.
Escherichia coli (enterohemorrhagic)	3–5 days	±	++	±	Generally associated with ingestion of undercooked ground beef, unpasteurized fruit juices and cider, and raw vegetables
Escherichia coli (enterotoxigenic strain)	24–72	±	++	–	Organisms grow in gut and are a major cause of traveler's diarrhea.
Vibrio parahaemolyticus	6–96	+	++	±	Organisms grow in seafood and in gut and produce toxin or invade.
Vibrio cholerae	24–72	+	+++	–	Organisms grow in gut and produce toxin.
Shigella spp.	24–72	±	++	+	Organisms grow in superficial gut epithelium. *S. dysenteriae* produces toxin.
Salmonella spp. (gastroenteritis)	8–48	±	++	+	Organisms grow in gut.
Salmonella enterica serovar Typhi (typhoid fever)	10–14 days	±	±	++	Bacteria invade the gut epithelium and reach the lymph nodes, liver, spleen, and gallbladder.
Clostridium difficile	Days to weeks after antibiotic therapy	–	+++	+	Antibiotic-associated colitis
Campylobacter jejuni	2–10 days	–	+++	++	Infection by oral route from foods, pets. Organism grows in small intestine.
Yersinia enterocolitica	4–7 days	±	++	+	Fecal-oral transmission, food-borne, animals infected

Adapted from Geo. F. Brooks, et al., *Medical Microbiology,* 21st edition. Copyright 1998 Appleton & Lange, Norwalk, CT. Reprinted by permission.
[a]+ indicates condition is present, number of symbols indicates severity
[b]– indicates condition is absent
[c]± indicates condition sometimes occurs

Pathogenesis	Clinical Features
Enterotoxins act on gut receptors that transmit impulses to medullary centers; may also act as superantigens.	Abrupt onset, intense vomiting for up to 24 hours, recovery in 24–48 hours. Occurs in persons eating the same food. No treatment usually necessary except to restore fluids and electrolytes. With incubation period of 2–8 hours, mainly vomiting
Enterotoxins formed in food or in gut from growth of *B. cereus*	Mainly diarrhea
Enterotoxins produced during sporulation in gut; causes hypersecretion.	Abrupt onset of profuse diarrhea; vomiting occasionally. Recovery usual without treatment in 1–4 days. Many clostridia in cultures of food and feces of patients
Toxin absorbed from gut and blocks acetylcholine release at neuromuscular junction.	Diplopia, dysphagia, dysphonia, difficulty breathing. Treatment requires clearing the airway, ventilation, and intravenous polyvalent antitoxin. Exotoxin present in food and serum. Mortality rate high
Toxins cause epithelial necrosis in colon; mild to severe complications.	Symptoms vary from mild to severe bloody diarrhea. The toxin can be absorbed, becoming systemic and producing hemolytic uremic syndrome, most frequently in children
Heat-labile (LT) and heat-stable (ST) enterotoxins cause hypersecretion in small intestine.	Usually abrupt onset of diarrhea; vomiting rare. A serious infection in newborns. In adults, "traveler's diarrhea" is usually self-limited in 1–3 days.
Toxin causes hypersecretion; vibrios invade epithelium; stools may be bloody.	Abrupt onset of diarrhea in groups consuming the same food, especially crabs and other seafood. Recovery is usually complete in 1–3 days. Food and stool cultures are positive
Toxin causes hypersecretion in small intestine. Infective dose $>10^5$ vibrios.	Abrupt onset of liquid diarrhea in endemic area. Needs prompt replacement of fluids and electrolytes. Tetracyclines shorten course of disease. Stool cultures positive
Organisms invade epithelial cells; blood, mucus, and neutrophils in stools. Infective dose $< 10^3$ organisms.	Abrupt onset of diarrhea, often with blood and pus in stools, cramps, tenesmus, and lethargy. Stool cultures are positive. Trimethoprim sulfamethoxazole, ampicillin, or chloramphenicol given in severe cases. Do not give opiates. Often mild and self-limited. Restore fluids
Superficial infection of gut, little invasion. Infective dose $>10^5$ organisms.	Gradual or abrupt onset of diarrhea and low-grade fever. Nausea, headache, and muscle aches common. Antimicrobials not administered unless systemic dissemination is suspected. Stool cultures are positive. Prolonged carriage is frequent.
Symptoms probably due to endotoxins and tissue inflammation; infective dose $\geq 10^7$ organisms.	Initially fever, headache, malaise, anorexia, and muscle pains. Fever may reach 40°C by the end of the first week of illness and lasts for 2 or more weeks. Diarrhea often occurs, and abdominal pain, cough, and sore throat may be prominent. Antibiotic therapy shortens duration of the illness.
Toxins causes epithelial necrosis in colon; pseudomembranous colitis.	Especially after abdominal surgery, abrupt bloody diarrhea and fever. Toxins in stool. Oral vancomycin useful in therapy; probiotic therapy may be used.
Invasion of mucous membrane; toxin production uncertain	Fever, diarrhea; PMNs and fresh blood in stool, especially in children. Usually self-limited. Special media needed for culture at 43°C. Erythromycin given in severe cases with invasion. Usual recovery in 5–8 days
Gastroenteritis or mesenteric adenitis; occasional bacteremia; toxin produced occasionally	Severe abdominal pain, diarrhea, fever; PMNs and blood in stool; polyarthritis, erythema nodosum, especially in children. Gentamicin used in severe cases.

45°C, enterotoxin can be detected 3 hours after growth is initiated. Onset of the symptoms—watery diarrhea, nausea, and abdominal cramps—usually occurs in about 8 to 16 hours.

Bacillus cereus is of concern in starchy foods. It can cause two distinct types of illnesses, depending on the type of toxin produced: an emetic illness characterized by nausea and vomiting with an incubation time of 1 to 6 hours, and a diarrheal type, with an incubation of 4 to 16 hours. The emetic type is often associated with boiled or fried rice, while the diarrheal type is associated with a wider range of foods.

1. Discuss the major characteristic of a food-borne infection in terms of the time required between ingestion of the pathogen and the onset of the disease. Why does this occur?

2. What common food is often related to *Campylobacter*-caused gastroenteritis? What means can be used to control the occurrence of this disease from this source?

3. What sources of *E.coli* O157:H7 have been of concern?

4. What are some of the major genera involved in food-borne intoxications?

5. How does a food-borne intoxication differ from a food-borne infection?

34.5 DETECTION OF FOOD-BORNE PATHOGENS

A major goal in maintaining food safety is the rapid detection of pathogenic microorganisms in order to curb outbreaks that can affect large populations. This is especially important because of widescale distribution of perishable foods. Standard culture techniques may require days to weeks for positive identification of pathogens. Identification is often complicated by the low numbers of pathogens compared with the background microflora. Furthermore the varied chemical and physical composition of foods can make isolation difficult. Fluorescent antibody, *enzyme-linked immunosorbant assays* (ELISAs), and radioimmunoassay techniques have proven of value. These can be used to detect small amounts of pathogen-specific antigens. ◄◄ *Clinical microbiology and immunology (chapter 32)*

Molecular techniques also are often used in identification. These methods are valuable for three purposes: (1) to detect the presence of a specific pathogen; (2) to detect viruses that cannot be grown conveniently; and (3) to identify slow-growing or non-culturable pathogens. ◄◄ *Biotechnology and industrial microbiology (chapter 16)*

Pathogens are frequently identified by detecting specific DNA or RNA base sequences with oligonucleotide probes that are specific for the pathogen of interest. The probes are labeled by linking them to a variety of enzymatic, isotopic, chromogenic, or luminescent/fluorescent markers. A major advantage

of their use is the speed with which specific microorganisms can be detected in a set of cultures, as shown in **figure 34.8**. In this example, a hydrophobic grid-membrane system has been used. *L. monocytogenes* cultures are detected having bound the radioactive probe, while other *Listeria* species do not show probe binding.

The important pathogen *E. coli* O157:H7 can be isolated and identified using selective culture media, rapid identification kits, rapid probe-based identification procedures, serotype-specific probes, and polymerase chain reaction (PCR) techniques. These molecular techniques enable the detection of a few target cells in large populations of background microorganisms. For example, by using PCR, as few as 10 toxin-producing *E. coli* cells can be detected in a population of 100,000 cells isolated from soft-cheese samples.

Because PCR is so sensitive, it can be used to detect as few as two colony-forming units of *Salmonella* (**figure 34.9**) or other target pathogens. It is possible to confirm bacterial contamination within 24 hours, whereas at least 3 to 4 days is needed for presumptive identification with culture procedures. Multiple pathogens can be detected in the same food sample through the recovery of PCR products of differing sizes, which can be separated electrophoretically. For instance, it is possible to detect *Campylobacter jejuni* and *Arcobacter butzleri* in the same sample within 8 hours. ◄◄ *Polymerase chain reaction (section 16.2); Gel electrophoresis (section 16.3)*

A major advance in the detection of food-borne pathogens is the use of standardized pathogen DNA patterns, or "food-borne pathogen fingerprinting." The Centers for Disease Control and Prevention has established a program, called PulseNet, in which pulsed-field gel electrophoresis (PFGE) is used under carefully controlled and duplicated conditions to determine the distinctive DNA pattern of each bacterial pathogen. With this

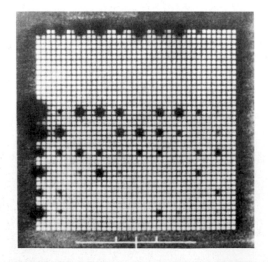

Figure 34.8 Molecular Probes and Food Microbiology. Autoradiogram of a radioactively-labeled *Listeria monocytogenes* probe against 100 *Listeria* cultures. Only the *Listeria monocytogenes* cultures show sequence homology and binding with the DNA probe, darkening the autoradiogram film. The other *Listeria* spp. do not react with the probe.

uniform procedure, it is possible to link pathogens associated with disease outbreaks in different parts of the world to a specific food source. Data from around the world are used in FoodNet, an active surveillance network, to follow nine major food-borne diseases. Using the FoodNet approach, it is possible to trace the course and cause of infection in days and not weeks. As an example, a *Shigella* outbreak in three different areas of North America was traced to Mexican parsley that had been tainted with polluted irrigation water. This program has resulted in more rapid establishment of epidemiological linkages and a decreased occurrence of many of these important food-borne diseases.

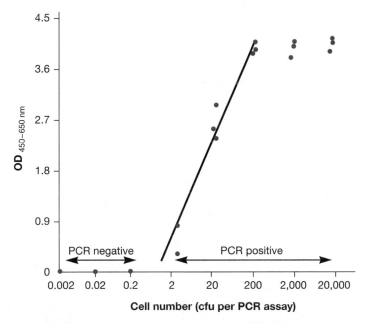

Figure 34.9 Polymerase Chain Reaction (PCR)–Based Pathogen Detection. Comparison of PCR sensitivity and growth for *Salmonella enterica* subsp. *enterica* serovar Agona detection. The Probalia PCR system can detect as few as two colony forming units (CFU) of the pathogen. OD = optical density.

1. How is the polymerase chain reaction used in pathogen detection?
2. How are PulseNet and FoodNet used in the surveillance of food-borne diseases?

34.6 MICROBIOLOGY OF FERMENTED FOODS

Fermentation has been a major way of preserving food for thousands of years. Microbial growth, either of natural or inoculated populations, causes chemical or textural changes to form a product that can be stored for extended periods. The fermentation process also is used to create new, pleasing food flavors and odors—such as chocolate (**Techniques & Application 34.1**).

The major fermentations used in food microbiology are the lactic, propionic, and alcoholic fermentations. These fermentations are carried out with a wide range of microbes, many of which have not been characterized. ≪ *Fermentation (section 10.7)*

Fermented Milks

Throughout the world, at least 400 different fermented milks are produced. The majority of fermented milk products rely on **lactic acid bacteria (LAB),** which include species belonging to the genera *Lactobacillus, Lactococcus, Leuconostoc,* and *Streptococcus* (**figure 34.10**). These are low G + C gram-positive bacteria that tolerate acidic conditions, are nonsporing, and are aerotolerant with a strictly fermentative metabolism. The art of fermentation developed long before the science, and fermented milks were produced for thousands of years before Louis Pasteur discovered lactic acid fermentation. Pasteur's work enabled the development of pure LAB starter cultures and the industrialization of milk fermentation. In this section, we discuss some major examples of the fermentation types listed in **table 34.5.** ≪ *Class Bacilli: Order* Lactobacillales *(section 21.4)*

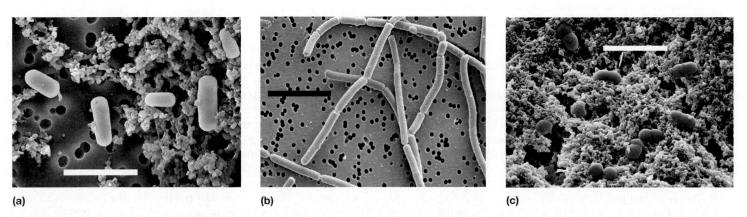

(a)   (b)   (c)

Figure 34.10 Lactic Acid Bacteria (LAB). Colorized scanning electron micrographs of LAB used as starter cultures. (a) *Lactobacillus helveticus.* (b) *Lactobacillus delbrueckii* subspecies *bulgaricus.* (c) *Lactococcus lactis.* The bacteria are supported by filters, seen as holes in the background. Scale bar = 5 μm.

Techniques & Applications

34.1   Chocolate: The Sweet Side of Fermentation

Chocolate could be characterized as the "world's favorite food," but few people realize that fermentation is an essential part of chocolate production. The Aztecs were the first to develop chocolate fermentation, serving a chocolate drink made from the seeds of the chocolate tree, *Theobroma cacao* (Greek *theos,* god, and *broma,* food, or "food of the gods"). Chocolate trees now grow in West Africa as well as South America.

The process of chocolate fermentation has changed very little over the past 500 years. Each tree produces large pods that each hold 30 to 40 seeds in a sticky pulp (**box figure**). Ripe pods are harvested and slashed open to release the pulp and seeds. The sooner the fermentation begins, the better the product, so fermentation occurs on the farm where the trees are grown. The seeds and pulp are placed in "sweat boxes" or in heaps in the ground and covered, usually with banana leaves.

Like most fermentations, this process involves a succession of microbes. First, a community of yeasts, including *Candida rugosa* and *Kluyveromyces marxianus,* hydrolyzes the pectin that covers the seeds and ferments the sugars to release ethyl alcohol and CO_2. As the temperature and the alcohol concentration increase, the yeasts are inhibited and lactic acid bacteria increase in number. The mixture is stirred to aerate the microbes and ensure an even temperature distribution. Lactic acid production drives the pH down; this encourages the growth of bacteria that produce acetic acid as a fermentation end product. Acetic acid is critical to the production of fine chocolate because it kills the sprout inside the seed and releases enzymes that cause further degradation of proteins and carbohydrates, contributing to the overall taste of the chocolate. In addition, acetate esters, derived from acetic acid, are important for the development of good flavor. Fermentation takes five to seven days. An experienced chocolate grower knows when the fermentation is complete: if it is stopped too soon, the chocolate will be bitter and astringent; on the other hand, if fermentation lasts too long, microbes start growing on the seeds instead of in the pulp. "Off-tastes" arise when the gram-positive bacterium *Bacillus* and the filamentous fungi *Aspergillis, Penicillium,* and *Mucor* hydrolyze lipids in the seeds to release short-chain fatty acids. As the pH begins to rise, bacteria of the genera *Pseudomonas, Enterobacter,* and *Escherichia* also contribute to bad tastes and odor.

After fermentation, the seeds, now called beans, are spread out to dry. Ideally this is done in the sun, although drying ovens are also used. The dried beans are brown and lack pulp. They are bagged and sold to chocolate manufacturers, who first roast the beans to further reduce the bitter taste and kill most of the microbes (some *Bacillus* spores may remain). The beans are then ground and the nibs—the inner part of each bean—are removed. The nibs are crushed into a thick paste called a chocolate liquor, which contains cocoa solids and cocoa butter but no alcohol. Cocoa solids are brown and have a rich flavor, and cocoa butter has a high fat content and is off-white in color. The two components are separated, and the cocoa solids can be sold as cocoa for baking and hot chocolate, while the cocoa butter is used to make white chocolate or sold to cosmetics companies for use in lipsticks and lotions. However, the bulk of these two components will be used to make chocolate. The cocoa solids and butter are reunited in controlled ratios, and sugar, vanilla, and other flavors are

(a)

(b)

Cocoa Fermentation. (a) Cocoa pods growing on the cocoa tree. Each pod is 13 to 15 cm in length and contains 30 to 40 seeds in a sticky white pulp. **(b)** Seeds and pulp are fermented in boxes covered with banana leaves for 5 to 7 days and then dried in the sun, as shown here. Chocolate cannot be produced without fermentation.

added. The better the fermentation, the less sugar needs to be added (and the more expensive the chocolate will be).

The final product, delicious chocolate, is a combination of over 300 different chemical compounds. This mixture is so complex that no one has yet been able to make synthetic chocolate that can compete with the natural fermented plant (note that artificial vanilla is readily available). Microbiologists and food scientists are studying the fermentation process to determine the role of each microbe. But like the chemists, they have had little luck in replicating the complex, imprecise fermentation that occurs on cocoa farms. In fact, the finest, most expensive chocolate starts as cocoa on farms where the details of fermentation have been handed down through generations. Chocolate production is truly an art as well as a science, while eating it is simply divine.

Table 34.5	Major Categories and Examples of Fermented Milk Products

Category	Typical Examples
I. Lactic fermentations	
Mesophilic	Buttermilk, Cultured buttermilk, *Långofil, Tëtmjolk, Ymer*
Thermophilic	Yogurt, *laban, zabadi, labneh, skyr* Bulgarian buttermilk
Probiotic	Biogarde, Bifighurt Acidophilus milk, *yakult* Cultura-AB
II. Yeast-lactic fermentations	Kefir, *koumiss,* acidophilus-yeast milk
III. Mold-lactic fermentations	*Viili*

Source: Table 3.1, p. 58. In B. A. Law, editor, 1997. *Microbiology and Biochemistry of Cheese and Fermented Milk,* 2nd ed. New York: Chapman and Hall.

Mesophilic Fermentations

Mesophilic milk fermentations result from similar manufacturing techniques, in which acid produced through microbial activity causes protein denaturation. To carry out the process, milk is typically inoculated with the desired **starter culture**—a carefully selected group of microbes used to initiate the fermentation. It is then incubated at optimum temperature (approximately 20 to 30°C). Microbial growth is stopped by cooling, and *Lactobacillus* spp. and *Lactococcus lactis* cultures are used for aroma and acid production. The organism *Lactococcus lactis* subspecies *diacetilactis* converts milk citrate to diacetyl, which gives a richer flavor to the finished product. The use of these microorganisms with skim milk produces cultured buttermilk, and when cream is used, sour cream is the result.

Thermophilic Fermentations

Thermophilic fermentations are carried out at temperatures around 45°C. An important example is yogurt production. Yogurt is one of the most popular fermented milk products in the United States and Europe. In commercial production, nonfat or low-fat milk is pasteurized, cooled to 43°C or lower, and inoculated with a 1:1 ratio of *Streptococcus salivarius* subspecies *thermophilus* (*S. thermophilus*) and *Lactobacillus delbrueckii* subspecies *bulgaricus* (*L. bulgaricus*). *S. thermophilus* grows more rapidly at first and renders the milk anaerobic and weakly acidic. *L. bulgaricus* then acidifies the milk even more. Acting together, the two species ferment almost all of the lactose to lactic acid and flavor the yogurt with diacetyl (*S. thermophilus*) and acetaldehyde (*L. bulgaricus*). Fruits or fruit flavors are pasteurized separately and then combined with the yogurt. Freshly prepared yogurt contains about 10^9 bacteria per gram.

Probiotics

The health benefits of fermented foods such as yogurt have been touted for a great number of years. However, only recently have rigorous studies explored the effects of certain bacteria that are either commensals or mutualists in the human intestine.

Microorganisms such as *Lactobacillus* and *Bifidobacterium* are being used in the rapidly developing area of **probiotics,** the addition of microorganisms to the diet to provide health benefits beyond basic nutritive value. The possible health benefits of the use of such microbial dietary adjuvants include immunomodulation, control of diarrhea, anticancer effects, and possible improvement of Crohn's disease (inflammatory bowel disease). These bacteria may also influence antigen presentation, uptake, and possible degradation. Probiotics have become a more attractive treatment option because the rate of antibiotic resistance among pathogens continues to climb. In addition, disease ecologists have come to recognize that intestinal microflora can be a contributing factor for certain conditions (e.g., Crohn's disease).

Acidophilus milk is produced by using *Lactobacillus acidophilus*. *L. acidophilus* may modify the microbial flora in the lower intestine, thus improving general health, and it often is used as a dietary adjunct, especially for lactose-intolerant persons. Many microorganisms in fermented dairy products stabilize the bowel microflora, and some appear to have antimicrobial properties. The exact nature and extent of health benefits of consuming fermented milks may involve minimizing lactose intolerance, lowering serum cholesterol, and possibly exhibiting anticancer activity. Several lactobacilli have antitumor compounds in their cell walls. Such findings suggest that diets including lactic acid bacteria, especially *L. acidophilus,* may contribute to the prevention of colon cancer. << *Normal microbiota of the human body: Large intestine (section 27.3)*

Another interesting group used in milk fermentations are the bifidobacteria. The genus *Bifidobacterium* contains irregular, nonsporing, gram-positive rods that may be club-shaped or forked at the end (**figure 34.11**). Bifidobacteria are nonmotile, anaerobic, and ferment lactose and other sugars to acetic and lactic acids. They are typical residents of the human intestinal tract, and many beneficial properties are attributed to them. Bifidobacteria are thought to help maintain the normal intestinal balance, while improving lactose tolerance; to possess antitumorigenic activity; and to reduce serum cholesterol levels. In addition, some believe that they promote calcium absorption and the synthesis of B-complex vitamins. It has also been suggested that bifidobacteria reduce or prevent the excretion of rotaviruses, a cause of diarrhea among children. *Bifidobacterium*-amended fermented milk products, including yogurt, are commercially available.

Yeast-Lactic Fermentation

Yeast-lactic fermentations include kefir, a product with an ethanol concentration of up to 2%. This unique fermented milk originated in the Caucasus Mountains and is produced east into

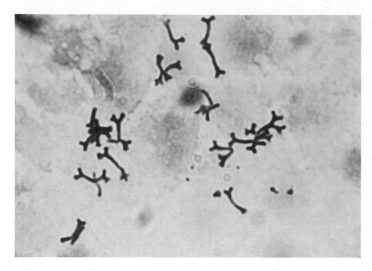

Figure 34.11 Bifidobacteria. Cultured milks are increasing in popularity. A light micrograph of *Bifidobacterium*, a microorganism thought to provide many health benefits.

Mongolia. Kefir products tend to be foamy and frothy, due to active carbon dioxide production. This process is based on the use of kefir "grains" as an inoculum. These are coagulated lumps of casein that contain yeasts, lactic acid bacteria, and acetic acid bacteria. In this fermentation, the grains are used to inoculate the fresh milk and then recovered at the end of the fermentation.

Originally, kefir was produced in leather sacks hung by the front door during the day, and passersby were expected to push and knead the sack to mix and stimulate the fermentation. Fresh milk could be added occasionally to maintain activity.

Mold-Lactic Fermentation

Mold-lactic fermentation results in a unique Finnish fermented milk called *viili*. The milk is placed in a cup and inoculated with a mixture of the fungus *Geotrichium candidum* and lactic acid bacteria. The cream rises to the surface, and after incubation at 18 to 20°C for 24 hours, lactic acid reaches a concentration of 0.9%. The fungus forms a velvety layer across the top of the final product, which also can be made with a bottom fruit layer.

Cheese Production

Cheese is one of the oldest foods, probably developed roughly 8,000 years ago. About 2,000 distinct varieties of cheese are produced throughout the world, representing approximately 20 general types (**table 34.6**). Often cheeses are classified based on texture or hardness as soft cheeses (cottage, cream, Brie), semi-soft cheeses (Muenster, Limburger, blue), hard cheeses (cheddar, Colby, Swiss), or very hard cheeses (Parmesan). All cheese results from a lactic acid fermentation of milk, which results in

Table 34.6	Major Types of Cheese and Microorganisms Used in Their Production	
	Contributing Microorganisms[a]	
Cheese (Country of Origin)	**Earlier Stages of Production**	**Later Stages of Production**
Soft, unripened		
Cottage	*Lactococcus lactis*	*Leuconostoc cremoris*
Cream	*L. cremoris, L. diacetylactis, Streptococcus salivarius* subspecies *thermophilus, L. delbrueckii* subspecies *bulgaricus*	
Mozzarella (Italy)	*S. thermophilus, L. bulgaricus*	
Soft, ripened		
Brie (France)	*Lactococcus lactis, L. cremoris*	*Penicillium camemberti, P. candidum, Brevibacterium linens*
Camembert (France)	*L. lactis, L. cremoris*	*Penicillium camemberti, B. linens*
Semisoft		
Blue, Roquefort (France)	*Lactococcus lactis, L. cremoris*	*P. roqueforti*
Brick, Muenster (United States)	*L. lactis, L. cremoris*	*B. linens*
Limburger (Belgium)	*L. lactis, L. cremoris*	*B. linens*
Hard, ripened		
Cheddar, Colby (Britain)	*Lactococcus lactis, L. cremoris*	*Lactobacillus casei, L. plantarum*
Swiss (Switzerland)	*L. lactis, L. helveticus, S. salivarius* subspecies *thermophilus*	*Propionibacterium shermanii, P. freudenreichii*
Very hard, ripened		
Parmesan (Italy)	*Lactococcus lactis, L. cremoris, S. salivarius* subspecies *thermophilus*	*L. delbrueckii* subspecies *bulgaricus*

[a]*Lactococcus lactis* stands for *L. lactis* subspecies *lactis*. *Lactococcus cremoris* is *L. lactis* subspecies *cremoris*, and *Lactococcus diacetilactis* is *L. lactis* subspecies *diacetilactis*.

coagulation of milk proteins and formation of a curd. Rennin, an enzyme from calf stomachs but now produced by genetically engineered microorganisms, can also be used to promote curd formation. After the curd is formed, it is heated and pressed to remove the watery part of the milk (called the whey), salted, and then usually ripened (**figure 34.12**). The cheese curd can be packaged for ripening with or without additional microorganisms.

Lactococcus lactis is used as a starter culture for a number of cheeses. Starter culture density is often over 10^9 colony-forming units (CFUs) per gram of cheese before ripening. However, the high salt, low pH, and the temperatures that characterize the cheese microenvironment reduce these numbers rather quickly. This enables other bacteria, sometimes called nonstarter lactic acid bacteria (NSLAB), to grow; their numbers can reach 10^7 to 10^9 CFUs/g after several months of aging. Thus both starter and nonstarter LAB contribute to the final taste, texture, aroma, and appearance of the cheese.

In some cases, molds are used to further enhance the cheese. Obvious examples are Roquefort and blue cheese. For these cheeses, *Penicillium roqueforti* spores are added to the curds just before the final cheese processing. Sometimes the surface of an already formed cheese is inoculated at the start of ripening; for example, Camembert cheese is inoculated with spores of *Penicillium camemberti*. The final hardness of the cheese is partially a function of the length of ripening. Soft cheeses are ripened for only about 1 to 5 months, whereas hard cheeses need 3 to 12 months, and very hard cheeses such as Parmesan require 12 to 16 months of ripening. The ripening process also

Figure 34.12 Cheddar Cheese Production. Cheddar, a village in England, has given its name to a cheese made in many parts of the world. "Cheddaring" is the process of turning and piling the curd to expel whey and develop desired cheese texture.

is critical for Swiss cheese; gas production by *Propionibacterium* contributes to final flavor development and hole or eye formation in this cheese. Some cheeses, such as Limburger, are soaked in brine to stimulate the development of specific fungi and bacteria.

Meat and Fish

A variety of meat products can be fermented: sausage, country-cured hams, salami, cervelat, Lebanon bologna, fish sauces (processed by halophilic *Bacillus* species), *izushi,* and *katsuobushi. Pediococcus acidilactici* and *Lactobacillus plantarum* are most often involved in sausage fermentations. *Izushi* is based on the fermentation of fresh fish, rice, and vegetables by *Lactobacillus* spp.; *katsuobushi* results from the fermentation of tuna by *Aspergillus glaucus*. These fermentations originated in Japan.

1. What are the major types of milk fermentations?
2. Briefly describe how buttermilk, sour cream, and yogurt are made.
3. What is unique about the morphology of *Bifidobacterium*? Why is it used in milk?
4. What major steps are used to produce cheese? How is the cheese curd formed in this process? What is whey? How does Swiss cheese get its holes?
5. Which fungal genus is often used in cheese making?

Wines and Champagnes

Wine production, or the focus of **enology** (Greek *oinos,* wine, and *ology,* the science of), starts with the collection of grapes, continues with their crushing and the separation of the liquid, called **must,** before fermentation, and concludes with a variety of storage and aging steps (**figure 34.13**). All grapes have white juices. To make a red wine from a red grape, the grape skins are allowed to remain in contact with the must before fermentation to release their skin-coloring components. Wines can be produced by using the natural grape skin microorganisms, but this natural mixture of bacteria and yeasts gives unpredictable fermentation results. To avoid this, fresh must is treated with a sulfur dioxide fumigant and a desired strain of the yeast *Saccharomyces cerevisiae* or *S. ellipsoideus* is added. After inoculation, the juice is fermented for 3 to 5 days at temperatures between 20 and 28°C. Depending on the alcohol tolerance of the yeast strain (the alcohol eventually kills the yeast that produced it), the final product may contain 10 to 14% alcohol. Clearing and development of flavor occur during the aging process. The malolactic fermentation is an important part of wine production. Grape juice contains high levels of organic acids, including malic and tartaric acids. If the levels of these acids are not decreased during the fermentation process, the wine will be too acidic and have poor stability and "mouth feel."

Processing step	Biological change

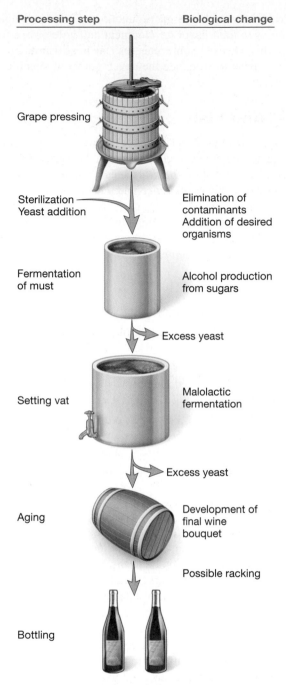

Grape pressing

Sterilization —
Yeast addition

Elimination of
contaminants
Addition of desired
organisms

Fermentation
of must

Alcohol production
from sugars

Excess yeast

Setting vat

Malolactic
fermentation

Excess yeast

Aging

Development of
final wine
bouquet

Possible racking

Bottling

Figure 34.13   Wine Making. Once grapes are pressed, the sugars in the juice (the must) can be immediately fermented to produce wine. Must preparation, fermentation, and aging are critical steps.

This essential fermentation is carried out by the bacteria *Leuconostoc oenos, L. plantarum, L. hilgardii, L. brevis,* and *L. casei.* The activities of these microbes transform malic acid (a four-carbon tricarboxylic acid) to lactic acid (a three-carbon monocarboxylic acid) and carbon dioxide. This results in deacidification, improvement of flavor stability, and, in some cases, the possible accumulation of bacteriocins in the wines. ◄◄ *Characteristics of the fungal division:* Ascomycota *(section 23.3); Fermentation (section 10.7)*

A critical part of wine making involves the choice of whether to produce a dry (no remaining free sugar) or a sweeter (varying amounts of free sugar) wine. This can be controlled by regulating the initial must sugar concentration. With higher levels of sugar, alcohol will accumulate and inhibit the fermentation before the sugar can be completely used, thus producing a sweeter wine. During final fermentation in the aging process, flavoring compounds accumulate and influence the bouquet of the wine.

Microbial growth during the fermentation process produces sediments, which are removed during **racking.** Racking can be carried out at the time the fermented wine is transferred to bottles or casks for aging or after the wine is placed in bottles.

Natural champagnes are produced when the fermentation is continued in bottles to produce a naturally sparkling wine. Sediments that remain are collected in the necks of inverted champagne bottles after the bottles have been carefully turned. The necks of the bottles are then frozen and the corks removed to disgorge the accumulated sediments. The bottles are refilled with clear champagne from another disgorged bottle, and the product is ready for final packaging and labeling.

Beers and Ales

Beer and ale production uses cereal grains such as barley, wheat, and rice. The complex starches and proteins in these grains must be hydrolyzed to a more readily usable mixture of simpler carbohydrates and amino acids. This process, known as **mashing,** involves germination of the barley grains and activation of their enzymes to produce a **malt** (**figure 34.14**). The malt is then mixed with water and the desired grains, and the mixture is transferred to the mash tun or cask in order to hydrolyze the starch to usable carbohydrates. Once this process is completed, the **mash** is heated with **hops** (dried flowers of the female vine *Humulus lupulis*), which were originally added to the mash to inhibit spoilage microorganisms (**figure 34.15**). The hops also provide flavor and assist in clarification of the wort. In this heating step, the hydrolytic enzymes are inactivated and the wort can be **pitched**—inoculated—with the desired yeast.

Most beers are fermented with bottom yeasts, related to *Saccharomyces pastorianus,* which settle at the bottom of the fermentation vat. The beer flavor also is influenced by the production of small amounts of glycerol and acetic acid. Bottom yeasts require 7 to 12 days of fermentation to produce beer with a pH of 4.1 to 4.2. With a top yeast, such as *Saccharomyces cerevisiae,* the pH is lowered to 3.8 to produce ales. Freshly fermented (green) beers are aged or lagered, and when they are bottled, CO_2 is usually added. Beer can be pasteurized at 140°F or higher or sterilized by passage through membrane filters to minimize flavor changes.

Distilled Spirits

Distilled spirits are produced by an extension of beer production processes. The fermented liquid is boiled, and the volatile components are condensed to yield a product with a higher alcohol content than beer. Rye and bourbon are examples of whiskeys. Rye whiskey must contain at least 51% rye grain, and bourbon

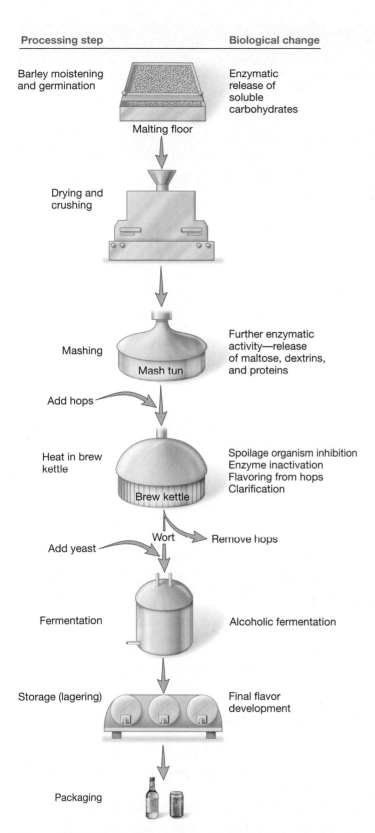

Processing step | **Biological change**

Barley moistening and germination — Malting floor — Enzymatic release of soluble carbohydrates

Drying and crushing

Mashing — Mash tun — Further enzymatic activity—release of maltose, dextrins, and proteins

Add hops

Heat in brew kettle — Brew kettle — Spoilage organism inhibition / Enzyme inactivation / Flavoring from hops / Clarification

Wort — Remove hops

Add yeast

Fermentation — Alcoholic fermentation

Storage (lagering) — Final flavor development

Packaging

Figure 34.14 Producing Beer. To make beer, the complex carbohydrates in the grain must first be transformed into a fermentable substrate. Beer production thus requires the important steps of malting and the use of hops and boiling for clarification, flavor development, and inactivation of malting enzymes to produce the wort.

Figure 34.15 Brew Kettles Used for Preparation of Wort. In large-scale processes, copper brew kettles can be used for wort preparation, as shown here at the Carlsberg Brewery, Copenhagen, Denmark.

must contain at least 51% corn. Scotch whiskey is made primarily of barley. Usually a **sour mash** is used; the mash is inoculated with a homolactic (lactic acid is the major fermentation product) bacterium such as *Lactobacillus delbrueckii* subspecies *bulgaricus* (figure 34.10*b*), which can lower the mash pH to around 3.8 in 6 to 10 hours. This limits the development of undesirable organisms. Vodka and grain alcohols are also produced by distillation. Gin is vodka to which resinous flavoring agents—often juniper berries—have been added to provide a unique aroma and flavor.

Production of Breads

Bread is one of the most ancient of human foods. The use of yeasts to leaven bread is carefully depicted in paintings from ancient Egypt, and a bakery at the Giza Pyramid area, from the year 2575 BCE, has been excavated. In breadmaking, yeast growth is carried out under aerobic conditions. This results in increased CO_2 production and minimum alcohol accumulation. The fermentation of bread involves several steps: α- and β-amylases present in the moistened dough release maltose and sucrose from starch. Then a baker's strain of the yeast *Saccharomyces cerevisiae*, which produces maltase, invertase, and zymase enzymes, is added. The CO_2 produced by the yeast results in the light texture of many breads, and traces of fermentation products contribute to the final flavor. Usually bakers add sufficient yeast to allow the bread to rise within 2 hours—the longer the rising time, the more additional growth by contaminating bacteria and fungi can occur, making the product less desirable. For instance, bread products can be spoiled by *Bacillus* species that produce ropiness. If the dough is baked after these organisms have grown, stringy and ropy bread will result, leading to decreased consumer acceptance.

Other Fermented Foods

Many other plant products can be fermented, as summarized in **table 34.7.** These include *sufu,* which is produced by the fermentation of tofu, a chemically coagulated soybean milk product. To

Table 34.7	Fermented Foods Produced from Fruits, Vegetables, Beans, and Related Substrates		
Foods	**Raw Ingredients**	**Fermenting Microorganisms**	**Area**
Coffee	Coffee beans	*Erwinia dissolvens, Saccharomyces* spp.	Brazil, Congo, Hawaii, India
Gari	Cassava	*Corynebacterium manihot, Geotrichum* spp.	West Africa
Kenkey	Corn	*Aspergillus* spp., *Penicillium* spp., lactobacilli, yeasts	Ghana, Nigeria
Kimchi	Cabbage and other vegetables	Lactic acid bacteria	Korea
Miso	Soybeans	*Aspergillus oryzae, Zygosaccharomyces rouxii*	Japan
Ogi	Corn	*Lactobacillus plantarum, Lactococcus lactis, Zygosaccharomyces rouxii*	Nigeria
Olives	Green olives	*Leuconostoc mesenteroides, Lactobacillus plantarum*	Worldwide
Ontjom	Peanut presscake	*Neurospora sitophila*	Indonesia
Peujeum	Cassava	Molds	Indonesia
Pickles	Cucumbers	*Pediococcus cerevisiae, L. plantarum, L. brevis*	Worldwide
Poi	Taro roots	Lactic acid bacteria	Hawaii
Sauerkraut	Cabbage	*L. mesenteroides, L. plantarum, L. brevis*	Worldwide
Soy sauce	Soybeans	*Aspergillus oryzae* or *A. soyae, Z. rouxii, Lactobacillus delbrueckii*	Japan
Sufu	Soybeans	*Actinimucor elegans, Mucor* spp.	China
Tao-si	Soybeans	*A. oryzae*	Philippines
Tempeh	Soybeans	*Rhizopus oligosporus, R. oryzae*	Indonesia, New Guinea, Surinam

Adapted from Jay, James M. 2000. *Modern Food Microbiology*, 6th ed. Reprinted by permission of Aspen Publishing, Frederick, MD.

carry out the fermentation, the tofu curd is cut into small chunks and dipped into a solution of salt and citric acid. After the cubes are heated to pasteurize their surfaces, the fungi *Actinimucor elegans* and some *Mucor* species are added. When a white mycelium develops, the cubes, now called *pehtze*, are aged in salted rice wine. This product has achieved the status of a delicacy in many parts of the Western world. Another popular product is tempeh, a soybean mash fermented by *Rhizopus*.

Sauerkraut or sour cabbage is produced from shredded cabbage. Usually the mixed microbial community of the cabbage is used. A concentration of 2.2 to 2.8% sodium chloride restricts the growth of gram-negative bacteria while favoring the development of the lactic acid bacteria. The primary microorganisms contributing to this product are *Leuconostoc mesenteroides* and *Lactobacillus plantarum*. A predictable microbial succession occurs in sauerkraut's development. The activities of the lactic acid-producing cocci usually cease when the acid content reaches 0.7 to 1.0%. At this point, *Lactobacillus plantarum* and *Lactobacillus brevis* continue to function. The final acidity is generally about pH 1.7, with lactic acid comprising 1.0 to 1.3% of the total acid in a satisfactory product.

Pickles are produced by placing cucumbers and components such as dill seeds in casks filled with a brine. The sodium chloride concentration begins at 5% and rises to about 16% in 6 to 9 weeks. The salt not only inhibits the growth of undesirable bacteria but also extracts water and water-soluble constituents from the cucumbers. These soluble carbohydrates are converted to lactic acid.

The fermentation, which can require 10 to 12 days, involves the growth of the gram-positive bacteria *L. mesenteroides, Enterococcus faecalis, Pediococcus acidilactici, L. brevis,* and *L. plantarum. L. plantarum* plays the dominant role in this fermentation process. Sometimes, to achieve more uniform pickle quality, natural microorganisms are first destroyed and the cucumbers are fermented using pure cultures of *P. acidilactici* and *L. plantarum*.

Grass, chopped corn, and other fresh animal feeds, if stored under moist anoxic conditions, will undergo a mixed acid fermentation that produces pleasant-smelling **silage.** Trenches or more traditional vertical steel or concrete silos are used to store the silage. The accumulation of organic acids in silage can cause rapid deterioration of these silos. Older wooden stave silos, if not properly maintained, allow the outer portions of the silage to become oxic, resulting in spoilage of a large portion of the plant material.

1. Describe and contrast the processes of wine and beer production.
2. How do champagnes differ from wines?
3. Describe how distilled spirits such as whiskey are produced.
4. How are bread, sauerkraut, and pickles produced? What microorganisms are most important in these fermentations?

34.7 MICROORGANISMS AS FOODS AND FOOD AMENDMENTS

A variety of bacteria, yeasts, and other fungi have been used as animal and human food sources. Mushrooms (e.g., *Agaricus bisporus*) are one of the most important fungi used directly as a food source. Large caves provide optimal conditions for their production (**figure 34.16**). Another popular microbial food supplement is the cyanobacterium *Spirulina* (*see front cover*). It is used as a food source in Africa and is sold in North American health food stores as a dried cake or powdered product.

Probiotic microbes can also be used as food amendments. Such microbes, primarily *Lactobacillus acidophilus,* are used in beef cattle feed. When the bacteria are sprayed on feed, the cattle that eat it appear to have markedly lower (60% in some experiments) carriage of the toxic *E. coli* strain O157:H7. This can make it easier to produce beef that will meet current standards for microbiological quality at the time of slaughter.

Probiotics are also used successfully with poultry. For instance, the USDA has designated a probiotic *Bacillus* strain for use with chickens as GRAS (p. 815). Feeding chickens a strain of *Bacillus subtilis* leads to increased body weight and feed conversion. There is also a reduction in coliforms and *Campylobacter* in the processed carcasses. It has been suggested that this probiotic decreases the need for antibiotics in poultry production and pathogen levels on farms. *Salmonella* can be controlled by spraying a patented blend of 29 bacteria, isolated from the chicken cecum, on day-old chickens. As they preen themselves, the chicks ingest the bacterial mixture, establishing a functional microbial

Figure 34.16 Mushroom Farming. Growing mushrooms requires careful preparation of the growth medium and control of environmental conditions. The mushroom bed is a carefully developed compost, which can be steam sterilized to improve mushroom growth.

community in the cecum and limiting *Salmonella* colonization of the gut in a process called competitive exclusion. In 1998 this product, called PREEMPT, was approved for use in the United States by the Food and Drug Administration.

1. What conditions are needed to have most efficient production of edible mushrooms?
2. How are probiotics used in agriculture?

Summary

34.1 Microorganism Growth in Foods

a. Most foods, especially when raw, provide an excellent environment for microbial growth. This growth can lead to spoilage or preservation, depending on the microorganisms present and environmental conditions.

b. The course of microbial development in a food is influenced by the intrinsic characteristics of the food itself—pH, salt content, substrates present, water presence and availability—and extrinsic factors, including temperature, relative humidity, and atmospheric composition.

c. Microorganisms can spoil meat, dairy products, fruits, vegetables, and canned goods in several ways. Spices, with their antimicrobial compounds, sometimes protect foods.

d. Modified atmosphere packaging (MAP) is used to control microbial growth in foods and to extend product shelf life. This process involves decreased oxygen and increased carbon dioxide levels in the space between the food surface and the wrapping material.

34.2 Microbial Growth and Food Spoilage

a. Food spoilage is a major concern globally. This can occur at any point in the food production process: growth, harvesting, transport, storage, or final preparation.

b. Fungi that grow in foods, especially cereals and grains, can produce important disease-causing chemicals, including the carcinogens aflatoxins (**figure 34.5**) and fumonisins (**figure 34.6**), and ergot alkaloids.

c. Algal toxins can be transmitted to humans by marine products. These can have severe amnesic, diarrhetic, and neurotoxic effects.

34.3 Controlling Food Spoilage

a. Foods can be preserved in a variety of physical and chemical ways, including filtration, alteration of temperature (cooling, pasteurization, sterilization), drying, the addition of chemicals, radiation, and fermentation.

b. Interest is increasing in using bacteriocins for food preservation. Nisin, a product of *Lactococcus lactis,* is the major substance approved for use in foods.

34.4 Food-Borne Diseases

a. Foods can be contaminated by pathogens at any point in the food production, storage, or preparation processes. Pathogens such as *Salmonella, Campylobacter, Listeria,* and *E. coli* can be transmitted by the food to the susceptible consumer, where they grow and cause disease or a food-borne infection (**table 34.4**). If the pathogen grows in the food before consumption and forms toxins that affect the food consumer without further microbial growth, the disease is a food-borne intoxication. Examples are intoxications caused by *Staphylococcus, Clostridium,* and *Bacillus.*

b. Noroviruses, *Campylobacter,* and *Salmonella* are thought to be the most important causes of food-borne illness. The cause of most food-borne illnesses is not known.

c. *E. coli* O157:H7 is an enterohemorrhagic bacterium that produces shigalike toxins, which especially affect the young. Proper food handling and thorough cooking are critical in control.

d. New variant Creutzfeldt-Jakob disease (vCJD) is of worldwide concern as a food-borne infectious agent, as it is related to the occurrence of "mad cow disease." The major means of vCJD transmission between animals is the use of mammalian tissue in ruminant animal feeds. Detecting such prohibited animal products in ruminant feeds is difficult.

e. Raw foods such as sprouts, seafood, and raspberries provide routes for disease transmission. Increases in the international shipment of fresh foods contribute to this problem.

34.5 Detection of Food-Borne Pathogens

a. Detection of food-borne pathogens is a major part of food microbiology. The use of immunological and molecular techniques such as DNA and RNA hybridization, PCR, and pulsed field gel electrophoresis often makes it possible to link disease occurrences to a common infection source. PulseNet and FoodNet programs are used to coordinate these control efforts.

34.6 Microbiology of Fermented Foods

a. Dairy products can be fermented to yield a wide variety of cultured milk products. These include mesophilic, therapeutic probiotic, thermophilic, lactoethanolic, and mold-lactic products.

b. Growth of lactic acid–forming bacteria, often with the additional use of rennin, can coagulate milk solids. These solids can be processed to yield a wide variety of cheeses, including soft unripened, soft ripened, semisoft, hard, and very hard types (**table 34.6**). Both bacteria and fungi are used in these cheese production processes.

c. Wines are produced from pressed grapes and can be dry or sweet, depending on the level of free sugar that remains at the end of the alcoholic fermentation (**figure 34.13**). Champagne is produced when the fermentation, resulting in CO_2 formation, is continued in the bottle.

d. Beer and ale are produced from cereals and grains. The starches in these substrates are hydrolyzed, in the processes of malting and mashing, to produce a fermentable wort. *Saccharomyces cerevisiae* is a major yeast used in the production of beer and ale (**figure 34.14**).

e. Many plant products can be fermented with bacteria, yeasts, and molds. Important products are breads, soy sauce, *sufu,* and tempeh (**table 34.7**). Sauerkraut and pickles are produced in a fermentation process in which natural populations of lactobacilli play a major role.

34.7 Microorganisms as Foods and Food Amendments

a. Microorganisms themselves can serve as an important food source. Mushrooms are one of the most important fungi used as a food source. *Spirulina,* a cyanobacterium, also is a popular food source sold in specialty stores.

b. Many microorganisms, including some of those used to ferment milks, can be used as food amendments or microbial dietary adjuvants. Several types of probiotic microorganisms are used successfully in poultry production.

Critical Thinking Questions

1. Fresh lemon slices are often served with raw or steamed seafood (oysters, crab, shrimp). From a food microbiology perspective, provide an explanation for their being served. Are there other examples in either the cooking or the serving of foods that not only enhance flavor but might have an antimicrobial strategy? Consider the example of marinades.

2. You are going through a salad line in a cafeteria at the end of the day. Which types of foods would you tend to avoid and why?

3. Why were aflatoxins not discovered before the 1960s? Do you think this was the first time they had grown in a food product to cause disease?

4. What advantage might the shigalike toxin give *E. coli* O157:H7? Can we expect to see other "new" pathogens appearing, and what should we do, if anything to monitor their development?

5. Keep a record of what you eat for a day or two. Determine if the food, beverages, and snacks you ate were produced (at any level) with the aid of microorganisms. Indicate at what levels microorganisms were deliberately used. Be sure to consider ingredients such as citric acid, which is produced at the industrial level by several species of fungi.

6. Colonization of a susceptible human is critical to food-borne disease microorganisms. How might it be possible to modify foods to decrease these attachment processes?

Learn More

Learn more by visiting the Prescott website at www.mhhe.com/prescottprinciples, where you will find a complete list of references.

Applied Environmental Microbiology

Biodegradation often can be facilitated by changing environmental conditions. Polychlorinated biphenyls (PCBs) are widespread industrial contaminants that accumulate in anoxic river muds. Although reductive dechlorination occurs under these conditions, oxygen is required to complete the degradation process. In this experiment, muds are being aerated to allow the final biodegradation steps to occur.

activated sludge Solid matter or sediment composed of actively growing microbes that participate in the aerobic portion of a biological sewage treatment process.

bioaugmentation Addition of pregrown microbial cultures to an environment to perform a specific task, for example, nitrogen fixation.

biochemical oxygen demand (BOD) The amount of oxygen used by organisms in water under certain standard conditions; it provides an index of the amount of microbially oxidizable organic matter present.

biodegradation The breakdown of a complex chemical through biological processes.

bioremediation The use of biologically mediated processes to remove pollutants from specific environments.

bulking sludge Sludges produced in sewage treatment that do not settle properly, usually due to the development of filamentous microorganisms.

chemical oxygen demand (COD) The amount of chemical oxidation required to convert organic matter in water and wastewater to CO_2.

cometabolism The modification of a compound not used for growth by a microorganism, which occurs in the presence of another organic material that serves as a carbon and energy source.

disinfection by-products (DBPs) Chlorinated organic compounds such as trihalomethanes formed during chlorine use for water disinfection. Many are carcinogens.

indicator organism An organism whose presence suggests the quality of a substance or environment, for example, the potential presence of pathogens.

membrane filter technique The use of a thin, porous filter to collect microorganisms from water, air, and food.

most probable number (MPN) The statistical estimation of the probable population in a liquid by diluting and determining end points for microbial growth.

primary treatment The first step of sewage treatment, in which physical settling and screening are used to remove particulate materials.

reductive dehalogenation The cleavage of carbon-halogen bonds by anaerobic bacteria.

secondary treatment The biological degradation of dissolved organic matter in the process of sewage treatment.

septic tank A tank used to process home sewage. Solid material settles out and is partially degraded by anaerobic bacteria as sewage slowly flows through the tank. The outflow is further treated or dispersed in aerobic soil.

settling basin A basin used during water purification to chemically precipitate fine particles, microorganisms, and organic material by coagulation or flocculation.

slow sand filter A bed of sand through which water slowly flows; the gelatinous microbial layer on the sand grain surfaces removes waterborne microorganisms.

sludge A general term for the precipitated solid matter produced during water and sewage treatment.

tertiary treatment The removal from sewage of inorganic nutrients, heavy metals, viruses, etc., by chemical and biological means after microbes have degraded dissolved organic material during secondary sewage treatment.

wastewater treatment The use of physical and biological processes to remove particulate and dissolved material from sewage and to control pathogens.

The microbe will have the last word.

—Louis Pasteur

In this final chapter, we use the term applied environmental microbiology to refer to the use of microbes in their natural environment to perform processes useful to humankind. Such processes include wastewater treatment and bioremediation. In developed countries, the processes and products of applied microbiology are taken for granted, but this is not true globally. For instance, clean drinking water and sanitary treatment of contaminated water are beyond reach for an alarmingly high number of people. According to the World Health Organization, over 1 billion people worldwide do not have access to safe, drinkable water and about 40% of the world's population lacks basic sanitation. We begin this chapter by presenting ways in which water can be purified so that it can be consumed without fear of disease transmission. We then describe several approaches to treat wastewater to keep our rivers, streams, lakes, and groundwater—often the source of drinking water—clean. The remainder of the chapter focuses on the use of microbes as tools to clean up toxic chemicals. Many concepts presented throughout the text are integrated in these important subjects.

35.1 WATER PURIFICATION AND SANITARY ANALYSIS

Many important human pathogens are maintained in association with living organisms other than humans, including many wild animals and birds. Some of these bacterial and protozoan pathogens can survive in water and infect humans (**table 35.1**). When waters are used for recreation or are a source of food, the possibility for disease transmission exists. In many cases, such waters are the source of drinking water.

Water purification is a critical link in controlling disease transmission in waters. As shown in **figure 35.1**, water purifica-

tion can involve a variety of steps, depending on the type of impurities in the raw water source. Usually municipal water supplies are purified by a process that consists of at least three or four steps. If the raw water contains a great deal of suspended material, it often is first routed to a **sedimentation basin** and held so that sand and other very large particles can settle out. The partially clarified water is then moved to a **settling basin** and mixed with chemicals such as alum (aluminium sulfate) and lime to facilitate further precipitation. This procedure is called **coagulation** or flocculation and removes microorganisms, organic matter, toxic contaminants, and suspended fine particles. The water is further purified by passing it through **rapid sand filters** to physically trap fine particles and flocs. This removes up to 99% of the bacteria. After filtration the water is disinfected. This step usually involves chlorination, but ozonation is becoming increasingly popular. When chlorination is employed, the chlorine dose must be large enough to leave residual free chlorine at a concentration of 0.2 to 2.0 mg/l. A concern is the creation of **disinfection by-products (DBPs)** such as trihalomethanes (THMs), formed when chlorine reacts with organic matter. Some DBPs are carcinogens.

This purification process removes or inactivates disease-causing bacteria and indicator organisms (coliforms). Unfortunately, the use of coagulants, rapid filtration, and chemical disinfection often does not remove *Giardia intestinalis cysts, Cryptosporidium* oocysts, *Cyclospora*, and viruses. *Giardia,* a cause of human diarrhea, is now recognized as the most commonly identified waterborne pathogen in the United States (*see figure 23.2*). More consistent removal of *Giardia* cysts, which are about 7 to 10 by 8 to 12 µm in size, can be achieved with **slow sand filters.** This treatment involves the slow passage of water through a bed of sand in which a microbial layer covers the surface

Table 35.1	**Water-Borne Bacterial Pathogens**	
Organism	Reservoir	Comments
Aeromonas hydrophila	Free living	Sometimes associated with gastroenteritis, cellulitis, and other diseases
Campylobacter	Bird and animal reservoirs	Major cause of diarrhea; common in processed poultry
Helicobacter pylori	Free living	Can cause chronic gastritis, peptic ulcers, gastric adenocarcinomas
Legionella pneumophila	Free living and associated with protozoa	Found in cooling towers, evaporators, condensers, showers, and other water sources
Leptospira	Infected animals	Hemorrhagic effects, jaundice
Mycobacterium	Infected animals and free living	Complex recovery procedure required
Pseudomonas aeruginosa	Free living	Swimmer's ear and related infections
Salmonella enteriditis	Animal intestinal tracts	Common in many waters
Vibrio cholerae	Free living	Found in many waters including estuaries
Vibrio parahaemolyticus	Free living in coastal waters	Causes diarrhea in shellfish consumers
Yersinia enterocolitica	Frequent in animals and in the environment	Waterborne gastroenteritis

Water purification steps

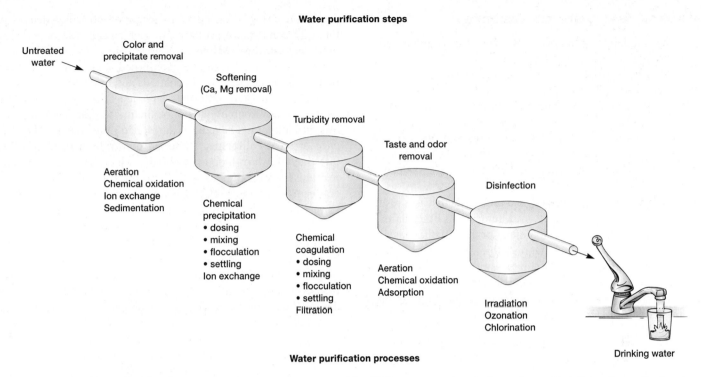

Water purification processes

Figure 35.1 Water Purification. Several alternatives can be used for drinking water treatment, depending on the initial water quality. A major concern is disinfection: chlorination can lead to the formation of *d*isinfection-*by*products (DBPs), including potentially carcinogenic *trihalo*methanes (THMs).

of each sand grain. Waterborne microorganisms are removed by adhesion to the gelatinous surface microbial layer (**Techniques & Applications 35.1**).

In the last few years, *Cryptosporidium* has become a significant problem. This protist is smaller than *Giardia* and is even more difficult to remove from water. A major source of *Giardia* and *Cryptosporidium* contamination is the Canada goose, a migratory bird with an expanding population. Each goose produces about 0.68 kg of manure per day, and geese are documented carriers of these important pathogens. The geese gather on golf courses, pastures, croplands, and vulnerable watersheds used for public water supplies, which of course adds to the problem.

Viruses in drinking water also must be inactivated or removed. Coagulation and filtration reduce virus levels about 90 to 99%.

Further inactivation of viruses by chemical oxidants, high pH, and photooxidation may yield a reduction as great as 99.9%. None of these processes, however, is considered sufficient protection. New standards for virus inactivation are being developed. Bacteriophages, which are easily grown and assayed, are used to monitor disinfection. If sufficient reductions in bacteriophage infectivity occur with a given disinfection process, it is assumed that viruses capable of infecting humans will also be reduced to satisfactory levels.

1. What steps are usually taken to purify drinking water?
2. Why is chlorination, although beneficial in terms of bacterial pathogen control, of environmental concern?
3. How can *Giardia* be removed from a water supply?

Techniques & Applications

35.1 Waterborne Diseases, Water Supplies, and Slow Sand Filtration

Slow sand filtration, in which drinking water is passed through a sand filter that develops a layer of microorganisms on its surface, has had a long and interesting history. After London's severe cholera epidemic of 1849, Parliament required that the entire water supply of London be passed through slow sand filters before use.

The value of this process was shown in 1892, when a major cholera epidemic occurred in Hamburg, Germany, and 10,000 lives were lost. The neighboring town, Altona, which used slow sand

filtration, did not have a cholera epidemic. Slow sand filters were installed in many cities in the early 1900s, but the process fell into disfavor with the advent of rapid sand filters, chlorination, and the use of coagulants such as alum. Slow sand filtration, a time-tested process, is regaining favor because of its filtration effectiveness and lower maintenance costs. Slow sand filtration is particularly effective for the removal of *Giardia* cysts. For this reason, slow sand filtration is used in many mountain communities where *Giardia* is a problem.

Sanitary Analysis of Waters

Monitoring and detecting indicator and disease-causing microorganisms are major parts of sanitary microbiology. Bacteria from the intestinal tract generally do not survive in aquatic environments, or are under physiological stress and gradually lose their ability to form colonies on differential and selective media. Their die-out rate depends on the water temperature, the effects of sunlight, the populations of other bacteria present, and the chemical composition of the water. Procedures have been developed to attempt to "resuscitate" these stressed coliforms using selective and differential media.

A wide range of viral, bacterial, and protozoan diseases result from the contamination of water with human and other animal fecal wastes (table 35.1). Although many of these pathogens can be detected directly, environmental microbiologists generally use **indicator organisms** as an index of possible water contamination by human pathogens. Researchers are still searching for the "ideal" indicator organism to use in sanitary microbiology. Among the suggested criteria for such an indicator are:

1. The indicator bacterium should be suitable for the analysis of all types of water: tap, river, ground, impounded, recreational, estuary, sea, and waste.

2. It should be present whenever enteric pathogens are present.

3. It should survive longer than the hardiest enteric pathogen.

4. It should not reproduce in the contaminated water, thereby producing an inflated value.

5. It should be harmless to humans.

6. Its level in contaminated water should have some direct relationship to the degree of fecal pollution.

7. The assay procedure for the indicator should have great specificity; in other words, other bacteria should not give positive results. In addition, the procedure should have high sensitivity and detect low levels of the indicator.

8. The testing method should be easy to perform.

Coliforms, including *Escherichia coli,* are members of the family *Enterobacteriaceae.* These bacteria make up about 10% of the intestinal microorganisms of humans and other animals and have found widespread use as indicator organisms. They lose viability in freshwater at slower rates than most of the major intestinal bacterial pathogens. When such "foreign" enteric indicator bacteria are not detectable in a specific volume (100 ml) of water, the water is considered potable (Latin *potabilis,* fit to drink).
<< *Class* Gammaproteobacteria: *Order* Enterobacteriales *(section 20.3)*

The coliform group includes *E. coli, Enterobacter aerogenes,* and *Klebsiella pneumoniae.* Coliforms are defined as facultatively anaerobic, gram-negative, nonsporing, rod-shaped bacteria that ferment lactose with gas formation within 48 hours at 35°C. The original test for coliforms involved the presumptive, confirmed, and completed tests, as shown in **figure 35.2.** The presumptive step is carried out by means of tubes inoculated with three different sample volumes to give an estimate of the ***most probable number*** **(MPN)** of coliforms in the water. The complete process, including the confirmed and completed tests, requires at least 4 days of incubations and transfers.

Unfortunately the coliforms include a wide range of bacteria whose primary source may not be the intestinal tract. To address this difficulty, tests have been developed that assay for the presence of **fecal coliforms.** These are coliforms derived from the intestine of warm-blooded animals, which grow at the more restrictive temperature of 44.5°C. To test for coliforms and fecal coliforms, and more effectively recover stressed coliforms, a variety of simpler and more specific tests has been developed.

The **membrane filter technique** has become a common and often preferred method of evaluating the microbiological characteristics of water. The water sample is passed through a membrane filter. The filter with its trapped bacteria is transferred to the surface of a solid medium or to an absorptive pad containing the desired liquid medium. Use of the proper medium enables the rapid detection of total coliforms, fecal coliforms, or fecal enterococci by the presence of their characteristic colonies. Samples can be placed on a less selective resuscitation medium or incubated at a less stressful temperature prior to growth under the final set of selective conditions. An example of a resuscitation step is the use of a 2-hour incubation on a pad soaked with lauryl sulfate broth, as is carried out in the LES Endo procedure. A resuscitation step often is needed with chlorinated samples, where the microorganisms are especially stressed. The advantages and disadvantages of the membrane filter technique are summarized in **table 35.2.**

Table 35.2	The Membrane Filter Technique for Evaluation of the Microbial Quality of Water
Advantages	
Reproducible	
Single-step results often possible	
Filters can be transferred between different media.	
Large volumes can be processed to increase assay sensitivity.	
Time savings are considerable.	
Ability to complete filtrations on site	
Lower total cost in comparison with MPN procedure	
Disadvantages	
High-turbidity waters limit volumes sampled.	
High populations of background bacteria cause overgrowth.	
Metals and phenols can adsorb to filters and inhibit growth.	

Source: Data from Clesceri, L. S., et al., 1998. *Standard Methods for the Examination of Water and Wastewater,* 20th ed. pp. 9–56. American Public Health Association, Washington, D.C.

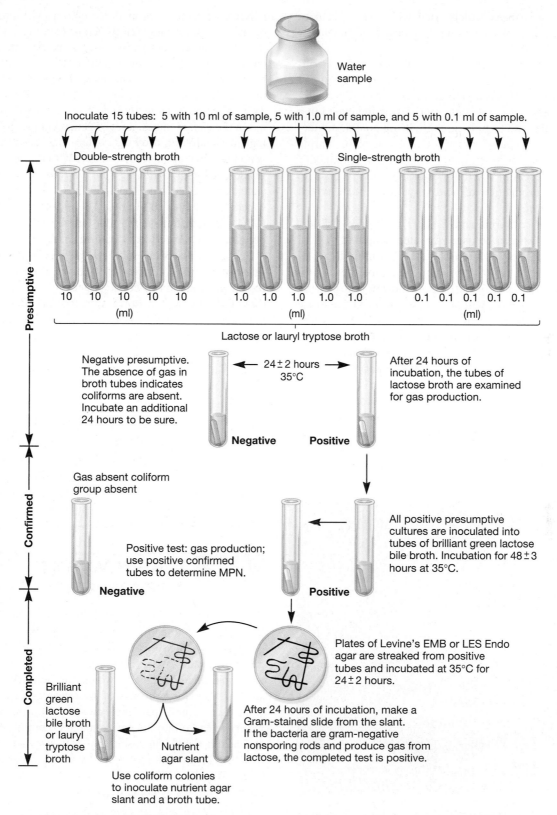

Figure 35.2 The Multiple-Tube Fermentation Test.   The multiple-tube fermentation technique has been used for many years for the sanitary analysis of water. Lactose broth tubes are inoculated with different water volumes in the presumptive test. Tubes that are positive for gas production are inoculated into brilliant green lactose bile broth in the confirmed test, and positive tubes are used to calculate the most probable number (MPN) value. The completed test is used to establish that coliform bacteria are present.

Membrane filters have been widely used with water that does not contain high levels of background organisms, sediment, or heavy metals.

More simplified tests for detecting coliforms and fecal coliforms are now available. The ***presence-absence (P-A) test*** can be used for coliforms. This is a modification of the MPN procedure in which a larger water sample (100 ml) is incubated in a single culture bottle with a triple-strength broth containing lactose broth, lauryl tryptose broth, and bromcresol purple indicator. The P-A test is based on the assumption that no coliforms should be present in 100 ml of drinking water. A positive test results in the production of acid (a yellow color), which requires confirmation.

To test for both coliforms and *E. coli,* the related **Colilert defined substrate test** can be used. A water sample of 100 ml is added to a specialized medium containing *o*-nitrophenyl-β-D-galactopyranoside (ONPG) and 4-methylumbelliferyl-β-D-glucuronide (MUG) as the only nutrients. If coliforms are present, the medium will turn yellow within 24 hours at 35°C due to the hydrolysis of ONPG, which releases *o*-nitrophenol, as shown in **figure 35.3**. To check for *E. coli,* the medium is observed under long-wavelength UV light for fluorescence. When *E. coli* is present, MUG is modified to yield a fluorescent product. If the test is negative for the presence of coliforms, the water is considered acceptable for human consumption. Current standards dictate that water be free of coliforms and fecal coliforms. If coliforms are present, fecal coliforms or *E. coli* must be tested for.

Molecular techniques are now used routinely to detect coliforms in waters and other environments, including foods. 16S

Figure 35.3 The Defined Substrate Test. This much simpler test is used to detect coliforms and fecal coliforms in single 100 ml water samples. The medium uses ONPG and MUG as defined substrates. (a) Uninoculated control. (b) Yellow color due to the presence of coliforms. (c) Fluorescent reaction due to the presence of fecal coliforms.

rRNA gene-targeted primers for coliforms enable the detection of one colony-forming unit (CFU) of *E. coli* per 100 ml of water, if a short enrichment step precedes the use of the polymerase chain reaction (PCR) amplification. This technique differentiates between nonpathogenic and enterotoxigenic strains, including the shigalike-toxin producing *E. coli* O157:H7. ◄◄ *Polymerase chain reaction (section 16.2)*

In the United States, a set of general guidelines for microbiological quality of drinking waters has been developed. The *maximum contaminant level goal (MCLG)* is zero for coliforms, viruses, *Cryptosporidium,* and *Giardia*. If unfiltered surface waters are used, one coliform test must be run each day when the waters have higher turbidities. The maximum contaminant level (MCL) for coliforms is set at no more than 5% positive samples per month, while no fecal coliforms or *E. coli* are permitted.

Other indicator microorganisms include **fecal enterococci.** These microbes are increasingly being used as an indicator of fecal contamination in brackish and marine water. In saltwater, these bacteria die at a slower rate than the fecal coliforms, providing a more reliable indicator of possible recent pollution. ◄◄ *Class* Bacilli: *Order* Lactobacillales *(section 21.4)*

1. What is an indicator organism, and what properties should it have?

2. How is a coliform defined? How does this definition relate to presumptive, confirmed, and completed tests?

3. How does one differentiate between coliforms and fecal coliforms in the laboratory?

4. Why do you think workers who perform these tests must practice excellent personal hygiene and be specifically trained in the implementation of these techniques?

35.2 WASTEWATER TREATMENT

Waters often contain high levels of organic matter from industrial, agricultural, and human wastes, which can be removed by the process of **wastewater treatment.** Depending on the effort given to this task, it may still produce waters containing nutrients and some microorganisms, which can be released to rivers and streams. Thus, the process of wastewater treatment, when performed at a municipal level, must be monitored to ensure that waters released into the environment do not pose environmental and health risks. Our discussion of wastewater treatment must therefore begin with the means by which water quality is monitored. We then discuss large-scale wastewater treatment processes, followed by home treatment systems.

Measuring Water Quality

Carbon removal during wastewater treatment can be measured several ways, including (1) as **total organic carbon (TOC),** (2) as chemically oxidizable carbon by the **chemical oxygen** demand (**COD**) test, or (3) as biologically usable carbon by the **biochemical oxygen demand (BOD)** test. The TOC includes all carbon, whether or not it can be used by microorganisms. It is measured by oxidizing the organic matter in a sample to CO_2 at high temperature in an oxygen stream. The resultant CO_2 is measured by infrared or potentiometric techniques. The COD gives a similar measurement, except that lignin often will not react with the oxidizing chemical, such as permanganate, that is used in this procedure.

The BOD test, in comparison, measures only the portion of the total carbon that can be oxidized by microorganisms in a 5-day period at 20°C. It is the amount of dissolved O_2 needed for microbial oxidation of biodegradable organic matter. When O_2 consumption is measured, the O_2 itself must be present in excess and not limit oxidation of nutrients such as nitrogen and phosphorus. To achieve this, the waste sample is diluted to ensure that at least 2 mg/l of O_2 are used while at least 1 mg/l of O_2 remains in the test bottle. Ammonia released during organic matter oxidation can also exert an O_2 demand in the BOD test, so nitrification or the **nitrogen oxygen demand (NOD)** is often inhibited by 2-chloro-6-(trichloromethyl) pyridine (nitrapyrin). In most BOD tests, nitrification is not a major concern. However, when treated effluents are analyzed, NOD can be a problem.

In terms of speed, the TOC is fastest, but it is less informative in terms of biological processes. The COD is slower and involves the use of wet chemicals with higher waste chemical disposal costs. The TOC, COD, and BOD provide different but complementary information on the carbon in a water sample. It is critical to note that these measurements are concerned only with carbon removal. They do not directly address the removal of minerals such as nitrate, phosphate, and sulfate. These minerals impact cyanobacterial and algal growth in lakes, rivers, and oceans by contributing to the process of eutrophication. ◀◀ *Marine and freshwater microbiology: Nutrient cycling (section 26.1)*

1. Compare TOC, COD, and BOD: how are these similar and different?

2. What factors can lead to a nitrogen oxygen demand (NOD) in water?

3. What minerals can contribute to eutrophication?

Wastewater Treatment Processes

An aerial photograph of a modern sewage treatment plant is shown in **figure 35.4***a*. Wastewater treatment involves a number of steps that are spatially segregated (figure 35.4*b*). The first three steps are called primary, secondary, and tertiary treatment (**table 35.3**). At the end of the process, the water is usually chlorinated (itself an emerging environmental and human health problem) before it is released.

(a)

(b)

Figure 35.4 An Aerial View of a Conventional Sewage Treatment Plant. Sewage treatment plants allow natural processes of self-purification that occur in rivers and lakes to be carried out under more intense, managed conditions in large concrete vessels. (a) A plant in New Jersey. (b) A diagram of flows in the plant.

1 = Primary clarifiers
2 = Activated sludge vessels
3 = Final clarifiers
4 = Chlorination vessels

Primary treatment physically removes 20 to 30% of the BOD present in particulate form. In this treatment, particulate material is removed by screening, precipitation of small particulates, and settling in basins or tanks. The resulting solid material is usually called **sludge.**

Secondary treatment promotes the biological transformation of dissolved organic matter into microbial biomass and carbon dioxide. About 90 to 95% of the BOD and many bacterial pathogens are removed by this process. Several approaches can be used in secondary treatment to remove dissolved organic matter. All of these techniques involve similar microbial activities. Under oxic conditions, dissolved organic matter will be transformed into additional microbial biomass plus carbon dioxide. When microbial growth is completed, under ideal conditions the microorganisms will aggregate and form stable flocs that settle. Minerals in the water also may be tied up in microbial biomass.

Table 35.3	Major Steps in Primary, Secondary, and Tertiary Treatment of Wastes
Treatment Step	**Processes**
Primary	Removal of insoluble particulate materials by settling, screening, addition of alum and other coagulation agents, and other physical procedures
Secondary	Biological removal of dissolved organic matter
	Trickling filters
	Activated sludge
	Lagoons
	Extended aeration systems
	Anaerobic digesters
Tertiary	Biological removal of inorganic nutrients
	Chemical removal of inorganic nutrients
	Virus removal/inactivation
	Trace chemical removal

As shown in **figure 35.5a**, a healthy settleable floc is compact. In contrast, poorly formed flocs have a network of filamentous microbes that retard settling (figure 35.5*b*).

When these processes occur with lower O$_2$ levels or with a microbial community that is too young or too old, unsatisfactory floc formation and settling can occur. The result is a **bulking sludge,** caused by the massive development of filamentous bacteria such as *Sphaerotilus* and *Thiothrix,* together with many poorly characterized filamentous organisms. These important filamentous bacteria form flocs that do not settle well and thus produce effluent quality problems. **<<** *Class* Betaproteobacteria: *Order* Burkholderiales *(section 20.2); Class* Gammaproteobacteria: *Order* Thiotrichales *(section 20.3)*

An aerobic **activated sludge** system (**figure 35.6a**) involves the horizontal flow of materials with recycling of sludge—the active biomass that is formed when organic matter is oxidized and degraded by microorganisms. Activated sludge systems can be designed with variations in mixing. In addition, the ratio of organic matter added to the active microbial biomass can be varied. A low-rate system (low nutrient input per unit of microbial biomass), with slower growing microorganisms, will produce an effluent with low residual levels of dissolved organic matter. A high-rate system (high nutrient input per unit of microbial biomass), with faster growing microorganisms, will remove more dissolved organic carbon per unit time but produce a poorer quality effluent.

Aerobic secondary treatment also can be carried out with a **trickling filter** (figure 35.6*b*). The waste effluent is passed over rocks or other solid materials upon which microbial biofilms have developed, and the microbial community degrades the organic waste. A sewage treatment plant can be operated to produce less sludge by employing the **extended aeration** process (figure 35.6*c*). Microorganisms grow on the dissolved organic matter, and the newly formed microbial biomass is eventually consumed to meet maintenance energy requirements. This requires extremely large aeration basins and long aeration times. In addition, with the biological self-utilization of the biomass, minerals originally present in the microorganisms are again released to the water.

All aerobic processes produce excess microbial biomass, or sewage sludge, which contains many recalcitrant organics. Often the sludges from aerobic sewage treatment, together with the materials settled out in primary treatment, are further treated by anaerobic digestion. **Anaerobic digesters** are large tanks designed to operate

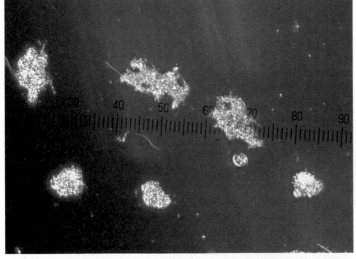

(a)

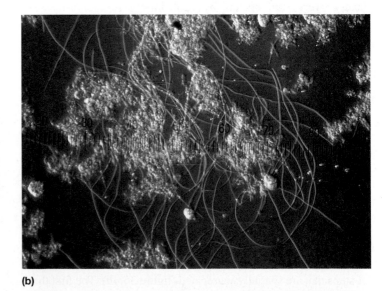

(b)

Figure 35.5   Proper Floc Formation in Activated Sludge.   Microorganisms play a critical role in the functioning of activated sludge systems. (a) The operation is dependent on the formation of settleable flocs. (b) If the plant does not run properly, poorly settling flocs can form due to such causes as low aeration, sulfide, and acidic organic substrates. These flocs do not settle properly because of their open or porous structure. As a consequence, the organic material is released with the treated water and lowers the quality of the final effluent.

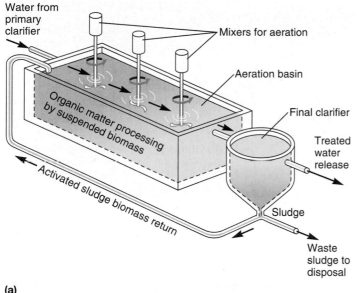

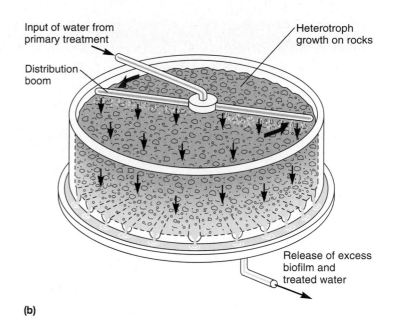

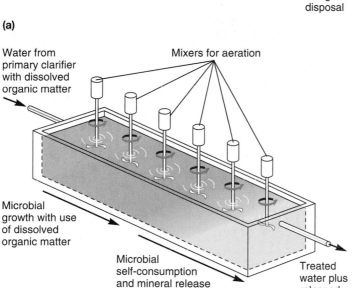

Figure 35.6 Aerobic Secondary Sewage Treatment. (a) Activated sludge with microbial biomass recycling. The biomass is maintained in a suspended state to maximize oxygen, nutrient, and waste transfer processes. (b) Trickling filter, where waste water flows over biofilms attached to rocks or other solid supports, resulting in transformation of dissolved organic matter to new biofilm biomass and carbon dioxide. Excess biomass and treated water flow to a final clarifier. (c) An extended aeration process, where aeration is continued beyond the point of microbial growth, allows the microbial biomass to self-consume due to microbial maintenance energy requirements. (The extended length of the reactor allows this process of biomass self-consumption to occur.) Minerals originally incorporated in the microbial biomass are released to the water as the process occurs.

with continuous input of untreated sludge and removal of the final, stabilized sludge product. Methane is vented and often burned for heat and electricity production. This digestion process involves three steps: (1) the fermentation of the sludge components to form organic acids, including acetate; (2) production of the methanogenic substrates: acetate, CO_2, and hydrogen; and finally, (3) methanogenesis by the methane producers. These methanogenic processes, summarized in **table 35.4**, involve critical balances between electron acceptors and donors. To function most efficiently, the hydrogen concentration must be maintained at a low level. If hydrogen and organic acids accumulate, methane production can be inhibited, resulting in a stuck digester. ◄◄ *Phylum* Euryarchaeota: *Methanogens (section 18.3)*

Anaerobic digestion has many advantages. Most of the microbial biomass produced in aerobic growth is used for methane production in the anaerobic digester. Also, because the process of methanogenesis is energetically very inefficient, the microbes must

consume about twice the nutrients to produce an equivalent biomass as that of aerobic systems. Consequently, less sludge is produced and it can be easily dried. Dried sludge removed from well-operated anaerobic systems can even be sold as organic garden fertilizer. However, sludge can be dangerous if the system is not properly managed because heavy metals and other environmental contaminants may be concentrated in it.

Tertiary treatment further purifies wastewaters. It is particularly important to remove nitrogen and phosphorus compounds that can promote eutrophication. Organic pollutants can be removed with activated carbon filters. Phosphate usually is precipitated as calcium or iron phosphate (e.g., by the addition of lime). To remove phosphorus, oxic and anoxic conditions are used alternately in a series of treatments, and phosphorus accumulates in microbial biomass as polyphosphate. Excess nitrogen may be removed by "stripping," volatilization of NH_3 at high pHs. Ammonia itself can be chlorinated to form dichloramine, which is then converted to

Table 35.4   Sequential Reactions in the Anaerobic Digester

Process Step	Substrates	Products	Major Microorganisms	
Fermentation	Organic polymers	Butyrate, propionate, lactate, succinate, ethanol, acetate,[a] H_2,[a] CO_2[a]	*Clostridium* *Bacteroides* *Peptostreptococcus*	*Peptococcus* *Eubacterium* *Lactobacillus*
Acetogenic reactions	Butyrate, propionate, lactate, succinate, ethanol	Acetate, H_2, CO_2	*Syntrophomonas* *Syntrophobacter*	
Methanogenic reactions	Acetate H_2 and HCO_3^-	$CH_4 + CO_2$ CH_4	*Methanosarcina* *Methanothrix* *Methanobrevibacter* *Methanomicrobium*	*Methanogenium* *Methanobacterium* *Methanococcus* *Methanospirillum*

[a]Methanogenic substrates produced in the initial fermentation step.

molecular nitrogen. In some cases, microbial processes can be used to remove nitrogen and phosphorus. A widely used process for nitrogen removal is denitrification. Here, nitrate, produced by microbes under aerobic conditions, is used as an electron acceptor under conditions of low oxygen with organic matter added as an energy source. Nitrate reduction yields nitrogen gas (N_2) and nitrous oxide (N_2O) as the major products. In addition to denitrification, the anammox process is also important. In this reaction, ammonium ion (used as the electron donor) is reacted with nitrite (the electron acceptor) produced by partial nitrification (i.e., the oxidation of ammonium to nitrite). The anammox process can convert up to 80% of the beginning ammonium ion to N_2 gas. Tertiary treatment is expensive and is usually not employed except where necessary to prevent obvious ecological disruption. << *Biogeochemical cycling: Nitrogen cycle (section 25.1)*

Wetlands are a vital natural resource and a critical part of our environment, and increasingly efforts are being made to protect these fragile aquatic communities from pollution. A major means of wastewater treatment is the use of constructed wetlands, where the basic components of natural wetlands (soils, aquatic plants, waters) are used as a functional waste treatment system. Constructed wetlands now are increasingly employed in the treatment of liquid wastes and for bioremediation, which is discussed in section 35.3. This system uses floating, emergent, or submerged plants, as shown in **figure 35.7**. The aquatic plants provide nutrients in the root zone, which support microbial growth. Especially with emergent plants, the root zone can be maintained in an anoxic state in which sulfide, produced by *Desulfovibrio* using root zone organic matter as an energy source, can trap metals. Constructed wetlands also are being

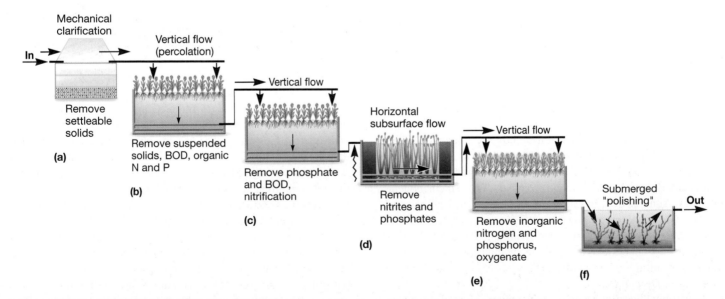

Figure 35.7 Constructed Wetland for Wastewater Treatment.   Multistage constructed wetland systems can be used for organic matter and phosphate removal. (a) Free-floating macrophytes (b,c,e), such as duckweed and water hyacinth, can be used for a variety of purposes. Emergent macrophytes (d), such as bulrush, allow surface flow as well as vertical and horizontal subsurface flow. Submerged vegetation (f), such as waterweed, allows final "polishing" of the water. These wetlands also can be designed for nitrification and metal removal from waters.

used to treat acid mine drainage and industrial wastes in many parts of the world.

1. Explain how primary, secondary, and tertiary treatments are accomplished.

2. What is bulking sludge? Name several important microbial groups that contribute to this problem.

3. What are the steps of organic matter processing that occur in anaerobic digestion? Why is acetogenesis such an important step?

4. After anaerobic digestion is completed, why is sludge disposal of concern?

5. Why might different aquatic plant types be used in constructed wetlands?

Home Treatment Systems

Groundwater—the water in gravel beds and fractured rocks below the surface soil—is a widely used but often unappreciated resource. In the United States, groundwater supplies at least 100 million people with drinking water, and in rural and suburban areas beyond municipal water distribution systems, 90 to 95% of all drinking water comes from this source.

Unfortunately our dependence on groundwater has not resulted in a corresponding understanding of microorganisms and microbiological processes that occur in this environment. Pathogenic microorganisms and dissolved organic matter are removed from water during subsurface passage through adsorption and trapping by fine sandy materials, clays, and organic matter. Microorganisms associated with these materials—including predators such as protozoa—can use the trapped pathogens as food. This results in purified water with a lower microbial population.

This combination of adsorption and biological predation is used in home treatment systems (**figure 35.8**). Conventional **septic tank** systems include an anaerobic liquefaction and digestion step that occurs in the septic tank, which functions as a simple anaerobic digester. This is followed by organic matter adsorption and entrapment of microorganisms in an aerobic leach field where biological oxidation occurs.

A septic tank may not operate correctly for several reasons. If the retention time of the waste in the septic tank is too short, undigested solids move into the leach field, plugging the system. If the leach field floods and becomes anoxic, biological oxidation does not occur, and effective treatment ceases. Other problems can occur, especially when a suitable soil is not present and the septic tank outflow from a conventional system drains too rapidly to the deeper subsurface. Fractured rocks and coarse gravel materials provide little effective adsorption or filtration. This may result in the contamination of well water with pathogens and the transmission of disease. In addition, nitrogen and phosphorus from the waste can pollute the groundwater. This leads to nutrient enrichment of ponds, lakes, rivers, and estuaries as the subsurface water enters these environmentally sensitive ecosystems.

Domestic and commercial on-site septic systems are now being designed with nitrogen and phosphorus removal steps. Nitrogen is usually removed by nitrification and denitrification, with organic matter provided by sawdust or a similar material. For phosphorus removal, a reductive iron dissolution process can be used. With the need to control nitrogen and phosphorus releases from septic systems, there is an increased emphasis on use of these and similar technologies.

Subsurface zones also can become contaminated with pollutants from other sources. Land disposal of sewage sludges, illegal dumping of septic tank pumpage, improper toxic waste disposal, and runoff from agricultural operations all contribute to groundwater contamination with chemicals and microorganisms. Many pollutants that reach the subsurface will persist and may affect the quality of groundwater for extended periods. Much research is being conducted to find ways to treat groundwater in place—in situ treatment. As will be discussed in section 35.3, microorganisms and microbial processes are critical in many of these remediation efforts.

1. What factors can limit microbial activity in subsurface environments? Consider the energetic and nutritional requirements of microorganisms in your answer.

2. How, in principle, are a conventional septic tank and leach field system supposed to work? What factors can reduce the effectiveness of this system?

3. What type of wastewater treatment system does your hometown use?

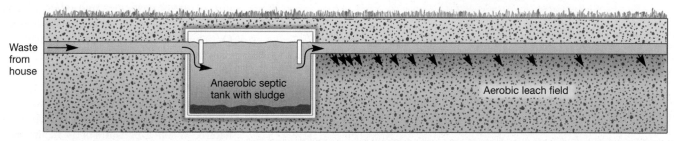

Figure 35.8 The Conventional Septic Tank Home Treatment System. This system combines an anaerobic waste liquefaction unit (the septic tank) with an aerobic leach field. Biological oxidation of the liquefied waste takes place in the leach field, unless the soil becomes flooded.

BIODEGRADATION AND BIOREMEDIATION BY NATURAL COMMUNITIES

The metabolic activities of microbes can also be exploited in complex natural environments such as waters, soils, or high organic matter–containing composts where the physical and nutritional conditions for microbial growth cannot be completely controlled and a largely unknown microbial community is present. Examples are (1) the use of microbial communities to carry out biodegradation, bioremediation, and environmental maintenance processes; and (2) the addition of microorganisms to soils or plants for the improvement of crop production. We discuss both of these applications in this section.

Biodegradation and Bioremediation Processes

Before discussing biodegradation processes carried out by natural microbial communities, it is important to consider definitions. **Biodegradation** has at least three outcomes (**figure 35.9**): (1) a minor change in an organic molecule leaving the main structure still intact, (2) fragmentation of a complex organic molecule in such a way that the fragments could be reassembled to yield the original structure, and (3) complete mineralization, which is

the transformation of organic molecules to inorganic forms. ≪ *Biogeochemical cycling (section 25.1)*

The removal of toxic industrial products in soils and aquatic environments has become a daunting and necessary task. Compounds such as perchloroethylene (PCE), trichloroethylene (TCE), and polychlorinated biphenyls (PCBs) are common contaminants. These compounds adsorb onto organic matter in the environment, making decontamination using traditional approaches difficult or ineffective. The use of microbes to transform these contaminants to nontoxic degradation products is called **bioremediation.** To understand how bioremediation takes place at the level of an ecosystem, we first must consider the biochemistry of biodegradation.

Degradation of complex compounds requires several discrete stages, usually performed by different microbes. Initially contaminants are converted to less-toxic compounds that are more readily degraded. The first step for many contaminants, including organochloride pesticides, alkyl solvents, and aryl halides, is **reductive dehalogenation.** This is the removal of a halogen substituent (e.g., chlorine, bromine, fluorine) while at the same time adding electrons to the molecule. This can occur in two ways. In hydrogenolysis, the halogen substituent is replaced by a hydrogen atom. Alternatively, dihaloelimination removes two halogen substituents from adjacent carbons while inserting an additional bond between the carbons. Both processes require an electron donor. The dehalogenation of PCBs uses electrons derived from water; alternatively hydrogen can be the electron donor for the dehalogenation of different chlorinated compounds. Major genera that carry out this process include *Desulfitobacterium, Dehalospirillum,* and *Desulfomonile.*

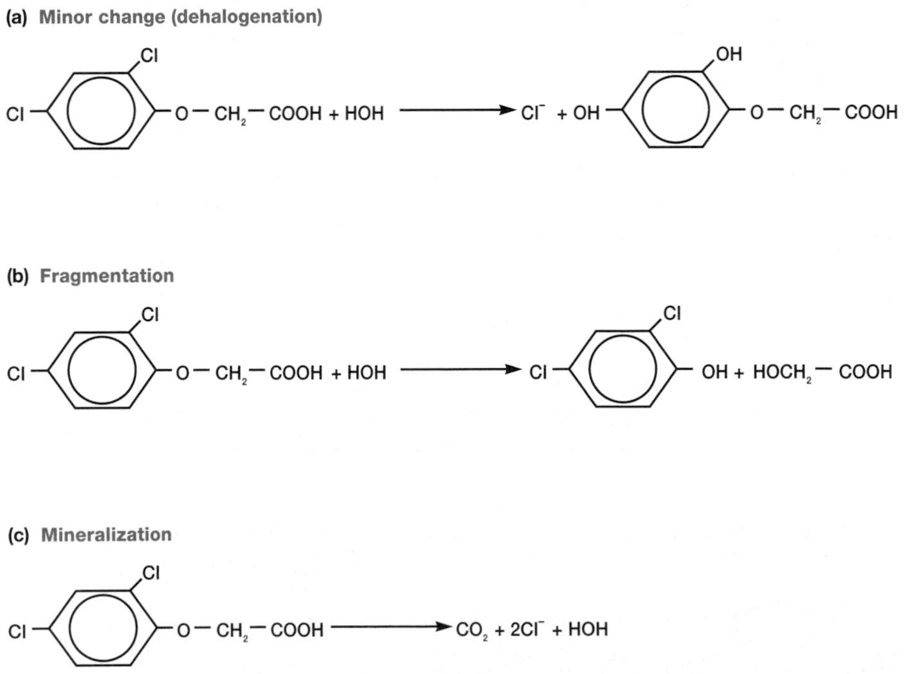

Reductive dehalogenation usually occurs under anoxic conditions. In fact, humic acids (polymeric residues of lignin decomposition that accumulate in soils and waters) have been found to play a role in anaerobic biodegradation processes. They can serve as terminal electron acceptors under what are called "humic acid-reducing conditions." The use of humic acids as electron acceptors has been observed with the anaerobic dechlorination of vinyl chloride and dichloroethylene.

Once the anaerobic dehalogenation steps are completed, degradation of the main structure of many pesticides and other xenobiotics often proceeds more rapidly in the presence of O_2. Thus the degradation of halogenated toxic compounds generally requires the action of several microbial genera, sometimes referred to as a consortium.

Structure and stereochemistry are critical in predicting the fate of a specific chemical in nature. When a constituent is in the *meta* as opposed to the *ortho* position, the compound will be degraded at a much slower rate. The

(a) Minor change (dehalogenation)

(b) Fragmentation

(c) Mineralization

Figure 35.9   Biodegradation Has Several Meanings. Biodegradation is a term that can be used to describe three major types of changes in a molecule. (a) A minor change in the functional groups attached to an organic compound, such as the substitution of a hydroxyl group for a chlorine group. (b) An actual breaking of the organic compound into organic fragments in such a way that the original molecule could be reconstructed. (c) The complete degradation of an organic compound to minerals.

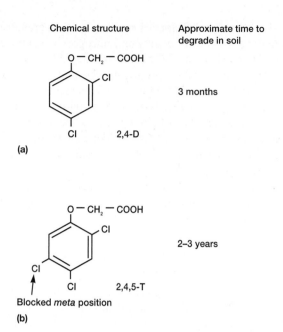

Chemical structure | Approximate time to degrade in soil

(a) 2,4-D — 3 months

(b) 2,4,5-T — 2–3 years

Blocked *meta* position

Figure 35.10 The *Meta* Effect and Biodegradation. Minor structural differences can have major effects on the biodegradability of chemicals. The *meta* effect is an important example. (a) Readily degradable 2,4-dichlorophenoxyacetic acid (2,4-D) with an exposed *meta* position on the ring degrades in several months; (b) recalcitrant 2,4,5-trichlorophenoxyacetic acid (2,4,5-T) with the blocked *meta* position can persist for years.

meta **effect** is shown in **figure 35.10**. This stereochemical difference is the reason that the common lawn herbicide 2,4-dichlorophenoxyacetic acid (2,4-D), with a chlorine in the *ortho* position, will be largely degraded in a single summer. In contrast, 2,4,5-trichlorophenoxyacetic acid, with a constituent in the *meta* position, will persist in the soils for several years and thus is used for long-term brush control.

An important aspect of managing biodegradation is the recognition that many of the compounds that are added to environments are **chiral,** or possess asymmetry and handedness. Microorganisms often can degrade only one isomer of a substance; the other isomer will remain in the environment. At least 25% of herbicides are chiral. Thus it is critical to add the herbicide isomer that is effective and also degradable. Studies have shown that microbial communities in different environments will degrade different isomers. Changes in environmental conditions and nutrient supplies can alter the patterns of chiral form degradation.

Microbial communities change their characteristics in response to the addition of inorganic or organic substrates. If a particular compound, such as an herbicide, is added repeatedly to a microbial community, the community adapts and faster rates of degradation can occur—a process of acclimation (**figure 35.11**). The adaptive process often is so effective that enrichment culture–based approaches, established on the principles elucidated by Beijerinck, can be used to isolate organisms with a desired set of capabilities. For example, a microbial community can become so efficient at herbicide degradation that herbicide effectiveness

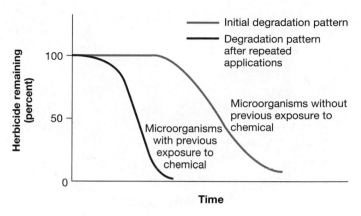

Figure 35.11 Repeated Exposure and Degradation Rate. Addition of an herbicide to a soil can result in changes in the degradative ability of the microbial community. Relative degradation rates for an herbicide after initial addition to a soil and after repeated exposure to the same chemical.

is diminished. To counteract this process, herbicides can be changed to alter the microbial community, thus preserving the effectiveness of the chemicals.

Degradation processes that occur in soils also can be used in large-scale degradation of hydrocarbons or wastes from agricultural operations. The waste material is incorporated into the soil or allowed to flow across the soil surface, where degradation occurs. Unfortunately such degradation processes do not always reduce environmental problems. In fact, the partial degradation or modification of an organic compound may not lead to decreased toxicity. An example of this problem is the microbial metabolism of 1,1,1-trichloro-2,2-bis-(p-chlorophenyl) ethane (DDT), a pesticide once commonly used in the United States. Degradation removes a chlorine function to give 1,1-dichloro-2,2-bis-(p-chlorophenyl) ethylene (DDE), which is still of environmental concern (**figure 35.12a**). Another important example is the degradation of trichloroethylene (TCE), a widely used solvent. If this

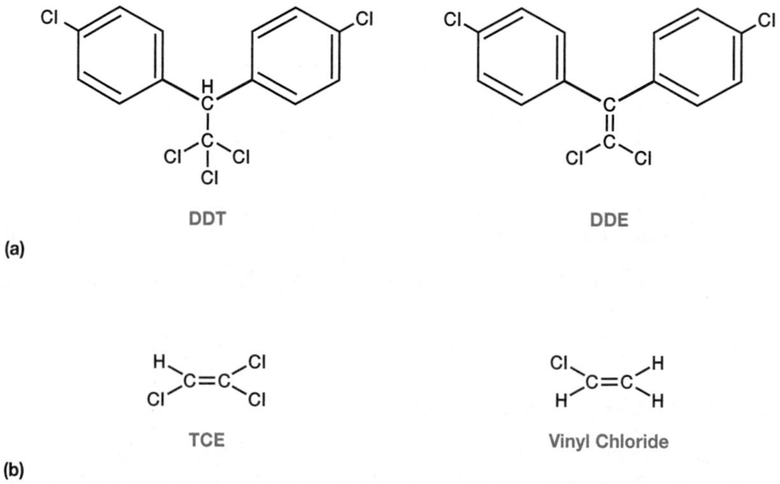

(a) DDT   DDE

(b) TCE   Vinyl Chloride

Figure 35.12 Toxic Degradation Products. The parent compounds DDT and TCE are metabolized to the toxins DDE and vinyl chloride, respectively.

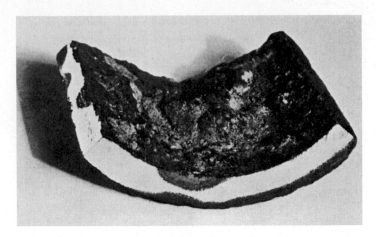

Figure 35.13 Microbial-Mediated Metal Corrosion. The microbiological corrosion of iron is a major problem. The graphitization of iron under a rust bleb on the pipe surface allows microorganisms, including *Desulfovibrio*, to corrode the inner surface.

is degraded under anoxic conditions, the dangerous carcinogen vinyl chloride can be formed (figure 35.12b)

Biodegradation also can lead to widespread damages and financial losses. Metal corrosion is a particularly important example. The microbially mediated corrosion of metals is particularly critical where iron pipes are used in waterlogged anoxic environments or in secondary petroleum recovery processes carried out at older oil fields. In these older fields, water is pumped down a series of wells to force residual petroleum to a central collection point. If the water contains low levels of organic matter and sulfate, anaerobic microbial communities can develop in rust blebs or tubercles (**figure 35.13**), resulting in punctured iron pipe and loss of critical pumping pressure. Microorganisms that use elemental iron as an electron donor during the reduction of CO_2 in methanogenesis contribute to the corrosion of soft iron. Because of the wide range of interactions that occur between microorganisms and metals, there is a critical need to develop strategies to deal with microbial corrosion problems.

1. Give alternative definitions for the term biodegradation.
2. What is reductive dehalogenation? Describe humic acids and the role they can play in anaerobic degradation processes.
3. Why is the *meta* effect important for understanding biodegradation?
4. Discuss chirality and its importance for understanding biodegradation.

Stimulating Biodegradation

Bioremediation usually involves stimulating the degradative activities of microorganisms already present in contaminated waters or soils because natural microbial communities often cannot carry out biodegradation processes at a desired rate due to limiting

physical or nutritional factors. For example, biodegradation may be limited by low levels of oxygen, nitrogen, phosphorus, and other needed nutrients. In these cases, it is necessary to determine the limiting factors and supply the needed materials or modify the environment. Often the addition of easily metabolized organic matter such as glucose increases biodegradation of recalcitrant compounds that are usually not used as carbon and energy sources by microorganisms. This process, termed **cometabolism,** can be carried out by simply adding easily catabolized organic matter and the compound to be degraded to a complex microbial community. Plants also may be used to provide the organic matter. Cometabolism is important in many different biodegradation systems.

Stimulating Hydrocarbon Degradation in Waters and Soils

Experience with oil spills in marine environments illustrates how effective stimulating natural microbial communities can be. In order to degrade dispersed hydrocarbons in the ocean, contact between microorganisms, the hydrocarbon substrate, and other essential nutrients must be maintained. To achieve this, pellets containing nutrients and an oleophilic (hydrocarbon soluble) preparation are used. This technique accelerates the degradation of different crude oil slicks by 30 to 40%, in comparison with control oil slicks where the additional nutrients are not available.

These bioremediation approaches are also used in soils and sediments. For example, a two-stage process can be used to degrade PCBs in river sediments. First, partial dehalogenation of the PCBs occurs naturally under anoxic conditions. Then the muds are aerated to promote the complete degradation of the less chlorinated residues (**chapter opening figure**).

Phytoremediation

Phytoremediation, or the use of plants to stimulate the extraction, degradation, adsorption, stabilization, or volatilization of contaminants, is becoming an important part of biodegradation technology. A plant provides nutrients that allow cometabolism to occur in the plant root zone or rhizosphere. Phytoremediation also includes plant contributions to degradation, immobilization, and volatilization processes. Transgenic plants may be employed in phytoremediation. Using cloning techniques with *Agrobacterium,* the *merA* and *merB* genes have been integrated into the mustard plant *Arabidopsis thaliana,* making it possible to transform extremely toxic organic mercury forms to elemental mercury, which is less of an environmental hazard. Transgenic tobacco plants have been constructed that express tetranitrate reductase, an enzyme from an explosive-degrading bacterium, thereby enabling the transgenic plants to degrade nitrate ester and nitro aromatic explosives. The genetically modified plants grow in solutions of explosives that unmodified plants cannot tolerate. << *Recombinant DNA technology in agriculture (section 16.11)*

Microbial genetics was recently shown to improve phytoremediation of toluene. Researchers noticed that microbes on the surface of the roots of yellow lupines only slowly degraded this toxin. Scientists reasoned that if endophytic bacteria (i.e., bacteria growing within the plant roots) were genetically modified to efficiently degrade toluene, their introduction into the plant would detoxify the surrounding soil. To accomplish this, conjugation was used to introduce a plasmid bearing the genes for toluene degradation into an endophytic strain of *Burkholderia cepacia.* Indeed, when the new toluene-degrading *B. cepacia* bacteria colonize the yellow lupine, the host plant is protected from the toxic effects of toluene and reduces the amount that escapes into the atmosphere. Experiments such as these suggest that the field of phytoremediation will continue to provide important new approaches to decontaminating soils.

Metal Bioleaching

Bioleaching is the use of microorganisms that produce acids from reduced sulfur compounds to create acidic environments that solubilize desired metals for recovery. This approach is used to recover metals from ores and mining tailings with metal levels too low for smelting. Bioleaching carried out by natural populations of *Leptospirillum*-like species and thiobacilli, for example, allow recovery of up to 70% of the copper in low-grade ores. As shown in **figure 35.14**, this involves the biological oxidation of copper present in these ores to produce soluble copper sulfate. In fact, this process is now used for copper mining.

It is apparent that nature will assist in bioremediation if given a chance. The role of microorganisms in biodegradation is now

better appreciated. An excellent example is the xenobiotic metabolism of the versatile fungus *Phanerochaete chrysosporium* (**Microbial Diversity & Ecology 35.2**).

1. What is cometabolism and why is this important for degradation processes?
2. What factors must one consider when attempting to stimulate the microbial degradation of an oil spill in a marine environment?
3. What role can microorganisms play in phytoremediation?
4. How is bioleaching carried out and what microbial genera are involved?

35.4 BIOAUGMENTATION

The acceleration of microbiological processes by the addition of known active microorganisms to soils, waters, or other complex systems is called **bioaugmentation.** For example, commercial culture preparations are available to facilitate silage formation and to improve septic tank performance.

With the development of the "superbug" by A. M. Chakrabarty in 1974, there was initial excitement as it was hoped that such an improved microorganism might be able to degrade hydrocarbon pollutants very effectively. Chakrabarty's "superbug" was a laboratory pseudomonad that had been transformed with plasmids that encoded enzymes needed for efficient degradation of several hydrocarbon compounds. However, a critical point not considered was the actual location, or microhabitat, where the microbe had to survive and function. Engineered microorganisms were added to soils and waters with the expectation that rates of degradation would be stimulated as these microorganisms established themselves. While such additions usually led to short-term increases in rates of the desired activity, after a few days the microbial community responses were similar in treated and control systems. The lack of effectiveness of such added cultures was due to at least three factors: (1) the attractiveness of laboratory-grown microorganisms as a food source for predators such as soil protozoa, (2) the inability of these added microorganisms to contact the compounds to be degraded, and (3) the failure of the added microorganisms to survive and compete with indigenous microorganisms. Such a modified microorganism may be less fit to compete and survive because of the additional energetic burden required to maintain the extra DNA.

Microorganism additions to natural environments can be more successful if the microorganism is added together with a microhabitat that gives the

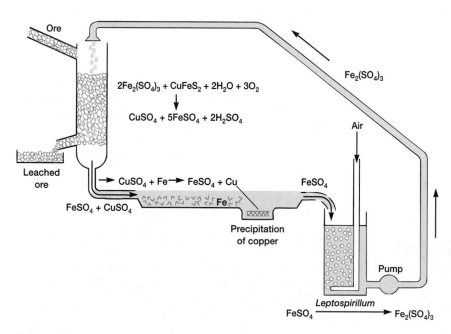

Figure 35.14 Copper Leaching from Low-Grade Ores. The chemistry and microbiology of copper ore leaching involve interesting complementary reactions. The microbial contribution is the oxidation of ferrous iron (Fe^{2+}) to ferric iron (Fe^{3+}). *Leptospirillum ferrooxidans* and related microorganisms are very active in this oxidation. The ferric iron then reacts chemically to solubilize the copper. The soluble copper is recovered by a chemical reaction with elemental iron, which results in an elemental copper precipitate.

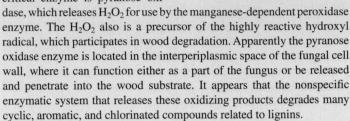

35.2 A Fungus with a Voracious Appetite

The basidiomycete *Phanerochaete chrysosporium* (the scientific name means "visible hair, golden spore") is a fungus with unusual degradative capabilities. This organism is termed a "white rot fungus" because of its ability to degrade lignin, a randomly linked, phenylpropene-based polymeric component of wood. The cellulosic portion of wood is attacked to a lesser extent, resulting in the characteristic white color of the degraded wood. This organism also degrades a truly amazing range of xenobiotic compounds (nonbiological foreign chemicals) using both intracellular and extracellular enzymes.

As examples, the fungus degrades benzene, toluene, ethylbenzene, and xylenes (the so-called BTEX compounds), chlorinated compounds such as 2,4,5-trichloroethylene (TCE), and trichlorophenols (figure 35.12). The latter are present as contaminants in wood preservatives and are used as pesticides. In addition, other chlorinated benzenes can be degraded with or without toluenes being present.

How does this microorganism carry out such feats? Apparently most xenobiotic degradation occurs after active growth, during secondary metabolic lignin degradation. Degradation of some compounds involves important extracellular enzymes including lignin peroxidase, manganese-dependent peroxidase, and glyoxal oxidase. A critical enzyme is pyranose oxidase, which releases H_2O_2 for use by the manganese-dependent peroxidase enzyme. The H_2O_2 also is a precursor of the highly reactive hydroxyl radical, which participates in wood degradation. Apparently the pyranose oxidase enzyme is located in the interperiplasmic space of the fungal cell wall, where it can function either as a part of the fungus or be released and penetrate into the wood substrate. It appears that the nonspecific enzymatic system that releases these oxidizing products degrades many cyclic, aromatic, and chlorinated compounds related to lignins.

We can expect to continue hearing of many new advances regarding this organism. Potentially valuable applications being studied include growth in bioreactors, where intracellular and extracellular enzymes can be maintained in the bioreactor while liquid wastes flow past the immobilized fungi.

organism physical protection, as well as possibly supplying nutrients. This makes it possible for the microorganism to survive in spite of the intense competitive pressures that exist in the natural environment, including pressure from protozoan predators. Microhabitats may be either living or inert. Specialized living microhabitats include the surface of a seed, a root, or a leaf. Here, higher nutrient fluxes and rates of initial colonization by the added microorganisms protect against the fierce competitive conditions in the natural environment. For example, to ensure that the nitrogen-fixing microbe *Rhizobium* is in close association with the legume, seeds are coated with the microbe using an oil-organism mixture or the bacteria are placed in a band under the seed where the newly developing primary root will penetrate. << *Microorganisms in terrestrial environments: The Rhizobia (section 26.2)*

Microorganisms can be added to natural communities together with protective inert microhabitats. As an example, if microbes are added to a soil with microporous glass, the survival of added microorganisms can be markedly enhanced. Other microbes have been observed to create their own microhabitats. Microorganisms in the water column overlying PCB-contaminated sand-clay soils have been observed to create their own "clay hutches" by binding clays to their outer surfaces with exopolysaccharides. Thus the application of principles of microbial ecology can facilitate the successful management of microbial communities in nature.

1. What factors might limit the ability of microorganisms, after addition to a soil or water, to persist and carry out desired functions?

2. What types of microhabitats can be used with microorganisms when they are added to a complex natural environment?

Summary

35.1 Water Purification and Sanitary Analysis

a. Water purification can involve the use of sedimentation, coagulation, chlorination, and rapid and slow sand filtration. Chlorination may lead to the formation of organic disinfection by-products, including trihalomethane compounds, which are potential carcinogens (**figure 35.1**).

b. *Cryptosporidium, Cyclospora,* viruses, and *Giardia* are of concern, as conventional water purification and chlorination will not always ensure their removal and inactivation to acceptable limits.

c. Indicator organisms are used to assess the presence of pathogenic microorganisms. Most probable number (MPN) and membrane filtration procedures are employed to estimate the number of indicator organisms present. The presence-absence (P-A) test is used for coliforms and defined substrate tests for coliforms and *E. coli* (**table 35.2; figures 35.2 and 35.3**).

d. Molecular techniques based on the polymerase chain reaction (PCR) can be used to detect waterborne pathogens such as shigalike-toxin producing *E. coli* O157:H7 when a preenrichment step is used.

35.2 Wastewater Treatment

a. The biochemical oxygen demand (BOD) test is an indirect measure of the organic matter that can be oxidized by the aerobic microbial community. The chemical oxygen demand (COD) and total organic carbon (TOC) tests provide information on carbon that is not biodegraded in the 5-day BOD test.

b. Conventional sewage treatment is a controlled intensification of natural self-purification processes, and it can involve primary, secondary, and tertiary treatment steps **(figure 35.4).**

c. Constructed wetlands involve the use of aquatic plants (floating, emergent, submerged) and their associated microorganisms for the treatment of liquid wastes **(figure 35.7).**

d. Home treatment systems operate on general self-purification principles. The conventional septic tank **(figure 35.8)** provides anaerobic liquefaction and digestion, whereas the aerobic leach field allows oxidation of the soluble effluent. These systems are now designed to provide nitrogen and phosphorus removal to lessen impacts of on-site sewage treatment systems on vulnerable marine and freshwaters.

35.3 Biodegradation and Bioremediation by Natural Communities

a. Microorganism growth in complex natural environments such as soils and waters is used to carry out environmental management processes, including bioremediation, plant inoculation, and other related activities.

b. Biodegradation is a critical part of natural systems mediated largely by microorganisms. This can involve minor changes in a molecule, fragmentation, or mineralization **(figure 35.9).**

c. Biodegradation can be influenced by many factors, including oxygen levels, humic acids, and the presence of readily usable organic matter. Reductive dehalogenation occurs best under anoxic conditions, and the presence of organic matter can facilitate modification of recalcitrant compounds in the process of cometabolism.

d. The structure of organic compounds influences degradation. If constituents are in specific locations on a molecule, as in the *meta* position **(figure 35.10),** or if varied structural isomers are present, degradation can be affected.

e. Degradation management can be carried out in place, whether this be large marine oil spills, soils, or the subsurface. Such large-scale efforts usually involve the use of natural microbial communities.

f. Plants can be used to stimulate biodegradation processes during phytoremediation. This can involve extraction, filtering, stabilization, and volatilization of pollutants.

35.4 Bioaugmentation

a. Microorganisms can be added to environments that contain complex microbial communities with greater success if living or inert microhabitats are used. *Rhizobium* is an important example of a microorganism added to a complex environment using a living microhabitat (the plant root).

Critical Thinking Questions

1. You wish to develop a constructed wetland for removing metals from a stream at one site, and at another site, you wish to treat acid mine drainage. How might you approach each of these problems?

2. What alternatives, if any, can one use for protection against microbiological infection when swimming in polluted recreational water? Assume that you are part of a water rescue team.

3. What possible alternatives could be used to eliminate N and P releases from sewage treatment systems? What suggestions could you make that might lead to new technologies?

4. Why, when a microorganism is removed from a natural environment and grown in the laboratory, will it usually not be able to effectively colonize its original environment if it is added back? Consider the nature of growth media used in the laboratory in comparison to growth conditions in a soil or water when attempting to understand this fundamental problem in microbial ecology.

Learn More

Learn more by visiting the Prescott website at www.mhhe.com/prescottprinciples, where you will find a complete list of references.

Appendix I: A Review of the Chemistry of Biological Molecules

Appendix I provides a brief summary of the chemistry of organic molecules with particular emphasis on the molecules present in microbial cells. Only basic concepts and terminology are presented; introductory textbooks in biology and chemistry should be consulted for a more extensive treatment of these topics.

ATOMS AND MOLECULES

Matter is made of elements that are composed of atoms. An element contains only one kind of atom and cannot be broken down to simpler components by chemical reactions. An atom is the smallest unit characteristic of an element and can exist alone or in combination with other atoms. When atoms combine, they form molecules. Molecules are the smallest particles of a substance. They have all the properties of the substance and are composed of two or more atoms.

Although atoms contain many subatomic particles, three directly influence their chemical behavior—protons, neutrons, and electrons. The atom's nucleus is located at its center and contains varying numbers of protons and neutrons (**figure AI.1**). Protons have a positive charge, and neutrons are uncharged. The mass of these particles and the atoms that they compose is given in terms of the atomic mass unit (AMU), which is 1/12 the mass of the most abundant carbon isotope. Often the term *dalton* (Da) is used to express the mass of molecules. It also is 1/12 the mass of an atom of ^{12}C or 1.661×10^{-24} grams. Both protons and neutrons have a mass of about 1 dalton. The atomic weight is the actual measured weight of an element and is almost identical to the mass number for the element, the total number of protons and neutrons in its nucleus. The mass number is indicated by a superscripted number preceding the element's symbol (e.g., ^{12}C, ^{16}O, and ^{14}N).

Negatively charged particles called electrons circle the atomic nucleus (**figure AI.1**). The number of electrons in a neutral atom equals the number of its protons and is given by the atomic number, the number of protons in an atomic nucleus. The atomic number is characteristic of a particular type of atom. For example, carbon has an atomic number of six, hydrogen's number is one, and oxygen's is eight (**table AI.1**).

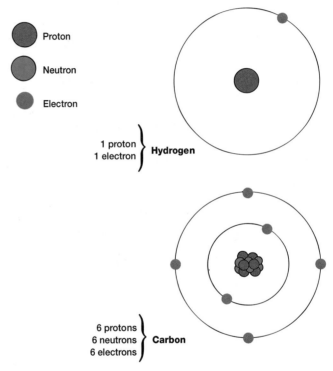

Figure AI.1 **Diagrams of Hydrogen and Carbon Atoms.** The electron orbitals are represented as concentric circles.

Table AI.1	Atoms Commonly Present in Organic Molecules			
Atom	Symbol	Atomic Number	Atomic Weight	Number of Chemical Bonds
Hydrogen	H	1	1.01	1
Carbon	C	6	12.01	4
Nitrogen	N	7	14.01	3
Oxygen	O	8	16.00	2
Phosphorus	P	15	30.97	5
Sulfur	S	16	32.06	2

The electrons move constantly within a volume of space surrounding the nucleus, even though their precise location in this volume cannot be determined accurately. This volume of space in which an electron is located is called its orbital. Each orbital can contain two electrons. Orbitals are grouped into shells of different energy that surround the nucleus. The first shell is closest to the nucleus and has the lowest energy; it contains only one orbital. The second shell contains four orbitals, one circular and three shaped like dumbbells (**figure AI.2a**). It can contain up to eight electrons. The third shell has even higher energy and holds more than eight electrons. Shells are filled beginning with the innermost and moving outward. For example, carbon has six electrons, two in its first shell and four in the second (figures AI.1 and AI.2b). The electrons in the outermost shell are the ones that participate in chemical reactions.

The most stable condition is achieved when the outer shell is filled with electrons. Thus the number of bonds an element can form depends on the number of electrons required to fill the outer shell. Since carbon has four electrons in its outer shell and the shell is filled when it contains eight electrons, it can form four covalent bonds (table AI.1).

CHEMICAL BONDS

Molecules are formed when two or more atoms associate through chemical bonding. Chemical bonds are attractive forces that hold together atoms, ions, or groups of atoms in a molecule or other substance. Many types of chemical bonds are present in organic molecules; three of the most important are covalent bonds, ionic bonds, and hydrogen bonds.

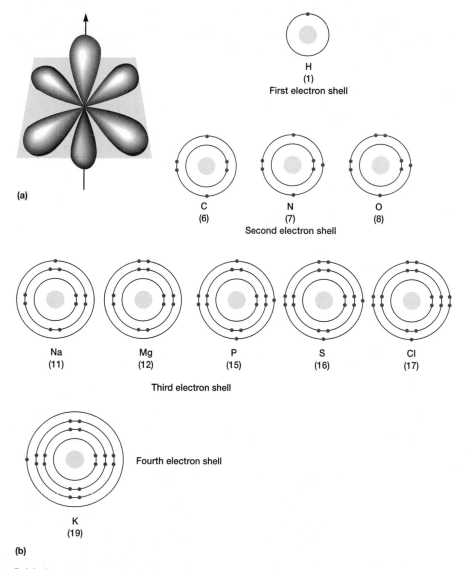

Figure AI.2   Electron Orbitals.   (a) The three dumbbell-shaped orbitals of the second shell. The orbitals lie at right angles to each other. (b) The distribution of electrons in some common elements. Atomic numbers are given in parentheses.

Figure AI.3 The Covalent Bond. A hydrogen molecule is formed when two hydrogen atoms share electrons.

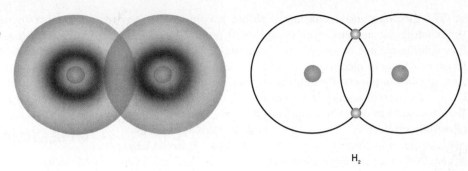

H_2

In covalent bonds, atoms are joined together by sharing pairs of electrons (**figure AI.3**). If the electrons are equally shared between identical atoms (e.g., in a carbon-carbon bond), the covalent bond is strong and nonpolar. When two different atoms such as carbon and oxygen share electrons, the covalent bond formed is polar because the electrons are pulled toward the more electronegative atom, the atom that more strongly attracts electrons (the oxygen atom). A single pair of electrons is shared in a single bond; a double bond is formed when two pairs of electrons are shared.

Atoms often contain either more or fewer electrons than the number of protons in their nuclei. When this is the case, they carry a net negative or positive charge and are called ions. Cations carry positive charges and anions have a net negative charge. When a cation and an anion approach each other, they are attracted by their opposite charges. This ionic attraction that holds two groups together is called an ionic bond. Ionic bonds are much weaker than covalent bonds and are easily disrupted by a polar solvent such as water. For example, the Na^+ cation is strongly attracted to the Cl^- anion in a sodium chloride crystal, but sodium chloride dissociates into separate ions (ionizes) when dissolved in water. Ionic bonds are important in the structure and function of proteins and other biological molecules.

When a hydrogen atom is covalently bonded to a more electronegative atom such as oxygen or nitrogen, the electrons are unequally shared and the hydrogen atom carries a partial positive charge. It will be attracted to an electronegative atom such as oxygen or nitrogen, which carries an unshared pair of electrons; this attraction is called a hydrogen bond (**figure AI.4**). Although an individual hydrogen bond is weak, there are so many hydrogen bonds in proteins and nucleic acids that they play a major role in determining protein and nucleic acid structure.

ORGANIC MOLECULES

Most molecules in cells are organic molecules, molecules that contain carbon. Since carbon has four electrons in its outer shell, it tends to form four covalent bonds in order to fill its outer shell with eight electrons. This property makes it possible to form chains and rings of carbon atoms that also can bond with hydrogen and other atoms (**figure AI.5**). Although adjacent carbons usually

Figure AI.4 Hydrogen Bonds. Representative examples of hydrogen bonds present in biological molecules.

(a) C_6H_{14} (Hexane)

(b) C_6H_{12} (Cyclohexane)

(c) C_6H_6 (Benzene)

Figure AI.5 Hydrocarbons. Examples of hydrocarbons that are (a) linear, (b) cyclic, and (c) aromatic.

are connected by single bonds, they may be joined by double or triple bonds. Rings that have alternating single and double bonds, such as the benzene ring, are called aromatic rings. The hydrocarbon chain or ring provides a chemically inactive skeleton to which more reactive groups of atoms may be attached. These reactive groups with specific properties are known as functional groups. They usually contain atoms of oxygen, nitrogen, phosphorus, or sulfur (**figure AI.6**) and are largely responsible for most characteristic chemical properties of organic molecules.

Organic molecules are often divided into classes based on the nature of their functional groups. Ketones have a carbonyl group within the carbon chain, whereas alcohols have a hydroxyl on the chain. Organic acids have a carboxyl group, and amines have an amino group (**figure AI.7**).

Organic molecules may have the same chemical composition and yet differ in their molecular structure and properties. Such molecules are called isomers. One important class of isomers is the stereoisomers. Stereoisomers have the same atoms arranged

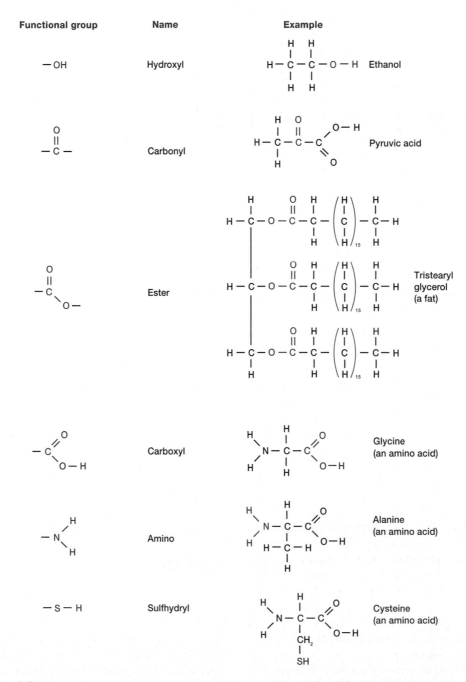

Functional group	Name	Example

Figure AI.6   Functional Groups.   Some common functional groups in organic molecules. The groups are shown in red.

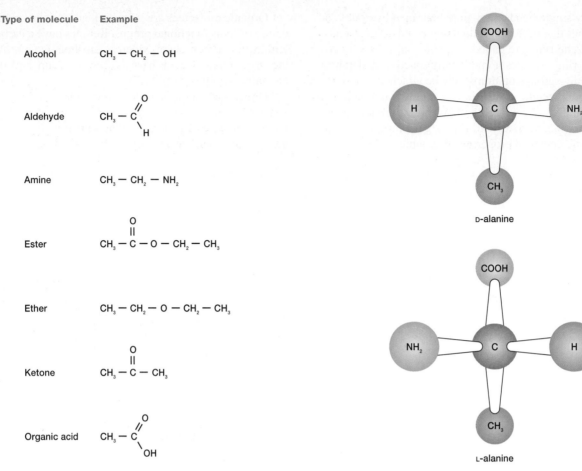

Type of molecule	Example
Alcohol	CH₃ — CH₂ — OH
Aldehyde	CH₃ — C (=O) H
Amine	CH₃ — CH₂ — NH₂
Ester	CH₃ — C(=O) — O — CH₂ — CH₃
Ether	CH₃ — CH₂ — O — CH₂ — CH₃
Ketone	CH₃ — C(=O) — CH₃
Organic acid	CH₃ — C(=O) OH

Figure AI.7 Types of Organic Molecules. These are classified on the basis of their functional groups.

D-alanine

L-alanine

Figure AI.8 The Stereoisomers of Alanine. The α-carbon is in gray. L-alanine is the form usually present in proteins.

in the same nucleus-to-nucleus sequence but differ in the spatial arrangement of their atoms. For example, an amino acid such as alanine can form stereoisomers (**figure AI.8**). L-Alanine and other L-amino acids are the stereoisomer forms normally present in proteins.

CARBOHYDRATES

Carbohydrates are aldehyde or ketone derivatives of polyhydroxy alcohols. The smallest and least complex carbohydrates are the simple sugars or monosaccharides. The most common sugars have five or six carbons (**figure AI.9**). A sugar in its ring form has two isomeric structures, the α and β forms, that differ in the orientation of the hydroxyl on the aldehyde or ketone carbon, which is called the anomeric or glycosidic carbon (**figure AI.10**). Microorganisms have many sugar derivatives in which a hydroxyl is replaced by an amino group or some other functional group (e.g., glucosamine).

Two monosaccharides can be joined by a bond between the anomeric carbon of one sugar and a hydroxyl or the anomeric carbon of the second (**figure AI.11**). The bond joining sugars is a glycosidic bond and may be either α or β depending on the orientation of the anomeric carbon. Two sugars linked in this way constitute a disaccharide. Some common disaccharides are maltose (two glucose molecules), lactose (glucose and galactose), and sucrose (glucose and fructose). If 10 or more sugars are linked together by glycosidic bonds, a polysaccharide is formed. For example, starch and glycogen are common polymers of glucose that are used as sources of carbon and energy (**figure AI.12**).

LIPIDS

All cells contain a heterogeneous mixture of organic molecules that are relatively insoluble in water but very soluble in nonpolar solvents such as chloroform, ether, and benzene. These molecules are called lipids. Lipids vary greatly in structure and include triacylglycerols, phospholipids, steroids, carotenoids, and many other types. Among other functions, they serve as membrane components, storage forms for carbon and energy,

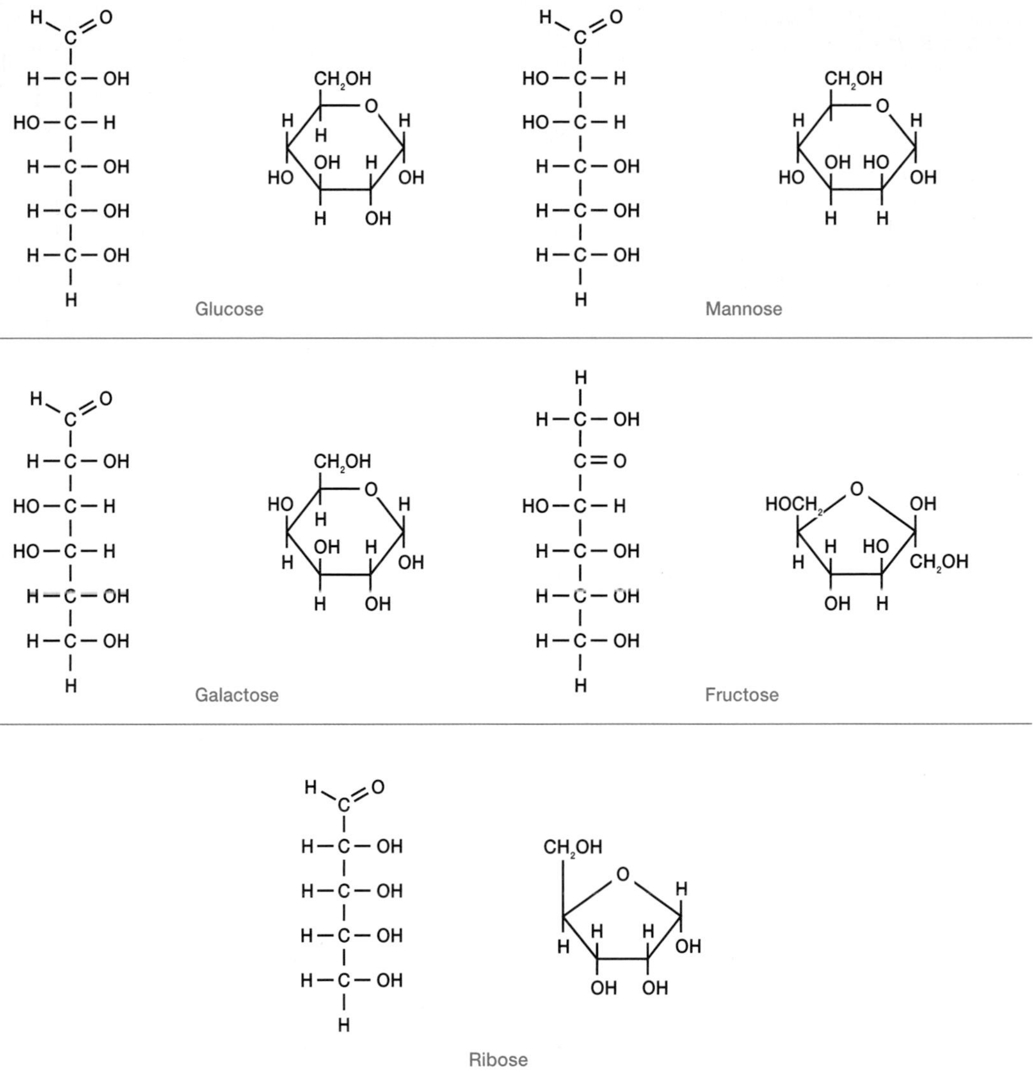

Figure AI.9 Common Monosaccharides. Structural formulas for both the open chains and the ring forms are provided.

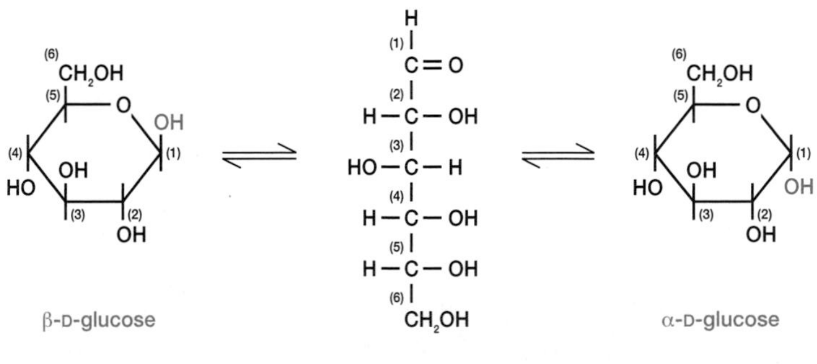

Glucose

Mannose

Galactose

Fructose

Ribose

Figure AI.10 The Interconversion of Monosaccharide Structures. The open chain form of glucose and other sugars is in equilibrium with closed ring structures (depicted here with Haworth projections). Aldehyde sugars form cyclic hemiacetals, and keto sugars produce cyclic hemiketals. When the hydroxyl on carbon one of cyclic hemiacetals projects above the ring, the form is known as a β form. The α form has a hydroxyl that lies below the plane of the ring. The same convention is used in showing the α and β forms of hemiketals such as those formed by fructose.

β-D-glucose

α-D-glucose

precursors of other cell constituents, and protective barriers against water loss.

Most lipids contain fatty acids, which are monocarboxylic acids that often are straight chained but may be branched. Saturated fatty acids lack double bonds in their carbon chains, whereas unsaturated fatty acids have double bonds. The most common fatty acids are 16 or 18 carbons long.

Two good examples of common lipids are triacylglycerols and phospholipids. Triacylglycerols are composed of glycerol esterified to three fatty acids (**figure AI.13a**). They are used to store carbon and energy. Phospholipids are lipids that contain at least one phosphate group and often have a nitrogenous constituent as well. Phosphatidylethanolamine is an important phospholipid frequently present in bacterial membranes

Figure AI.11 Common Disaccharides. (a) The formation of maltose from two molecules of an α-glucose. The bond connecting the glucose extends between carbons one and four, and involves the α form of the anomeric carbon. Therefore, it is called an α (1→4) glycosidic bond. (b) Sucrose is composed of a glucose and a fructose joined to each other through their anomeric carbons, and αβ (1→2) bond. (c) The milk sugar lactose contains galactose and glucose joined by a β (1→4) glycosidic bond.

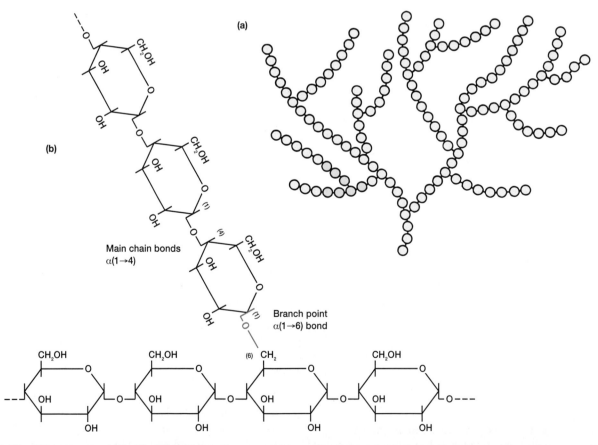

Figure AI.12   Glycogen and Starch Structure.   (a) An overall view of the highly branched structure characteristic of glycogen and most starch. The circles represent glucose residues. (b) A close-up of a small part of the chain (shown in blue in part *a*) revealing a branch point with its α (1→6) glycosidic bond, which is colored blue.

$$
\begin{array}{l}
\text{CH}_2-\text{O}-\overset{\displaystyle \overset{\text{O}}{\|}}{\text{C}}-\text{R} \\
\mid \qquad\quad\; \overset{\text{O}}{\|} \\
\text{CH}\;\; -\text{O}-\overset{\displaystyle \|}{\text{C}}-\text{R} \\
\mid \qquad\quad\; \overset{\text{O}}{\|} \\
\text{CH}_2-\text{O}-\overset{\displaystyle \|}{\text{C}}-\text{R}
\end{array}
$$

(a)

$$
\begin{array}{l}
\text{CH}_2-\text{O}-\overset{\displaystyle \overset{\text{O}}{\|}}{\text{C}}-\text{R} \\
\mid \qquad\quad\; \overset{\text{O}}{\|} \\
\text{CH}\;\; -\text{O}-\overset{\displaystyle \|}{\text{C}}-\text{R} \\
\mid \qquad\quad\; \overset{\text{O}}{\|} \\
\text{CH}_2-\text{O}-\overset{\displaystyle \|}{\text{P}}-\text{O}-\text{CH}_2-\text{CH}_2-\text{NH}_3 \\
\mid \qquad\qquad\qquad\qquad\qquad\quad + \\
\text{O}^-
\end{array}
$$

(b)

Figure AI.13 Examples of Common Lipids. (a) A triacylglycerol or neutral fat. (b) The phospholipid phosphatidylethanolamine. The R groups represent fatty acid side chains.

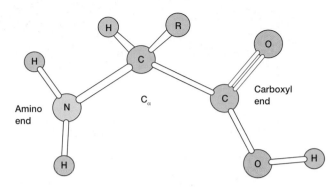

Figure AI.14 L-Amino Acid Structure. The uncharged form is shown.

PROTEINS

The basic building blocks of proteins are amino acids. An amino acid contains a carboxyl group and an amino group on its alpha carbon (**figure AI.14**). About 20 amino acids are normally found in proteins; they differ from each other with respect to their side chains (**figure AI.15**). In proteins, amino acids are linked together by peptide bonds between their carboxyls and α-amino groups to form linear polymers called peptides (**figure AI.16**). If a peptide contains more than 50 amino acids, it usually is called a polypeptide. Each protein is composed of one or more polypeptide chains and has a molecular weight greater than about 6,000 to 7,000.

Proteins have three or four levels of structural organization and complexity. The primary structure of a protein is the sequence of the amino acids in its polypeptide chain or chains. The structure of the polypeptide chain backbone is also considered part of the primary structure. Each different polypeptide has its

(figure AI.13*b*). It is composed of two fatty acids esterified to glycerol. The third glycerol hydroxyl is joined with a phosphate group, and ethanolamine is attached to the phosphate. The resulting lipid is very asymmetric with a hydrophobic nonpolar end contributed by the fatty acids and a polar, hydrophilic end. In cell membranes, the hydrophobic end is buried in the interior of the membrane, while the polar-charged end is at the membrane surface and exposed to water.

own amino acid sequence that is a reflection of the nucleotide sequence in the gene that codes for its synthesis. The polypeptide chain can coil along one axis in space into various shapes like the α-helix (**figure AI.17**). This arrangement of the polypeptide in space around a single axis is called the secondary structure. Secondary structure is formed and stabilized by the interactions of amino acids that are fairly close to one another on the polypeptide chain. The polypeptide with its primary and secondary structure can be coiled or organized in space along three axes to form a more complex, three-dimensional shape (**figure AI.18**). This level of organization is the tertiary structure (**figure AI.19**). Amino acids more distant from one another on the polypeptide chain contribute to tertiary structure. Secondary and tertiary structures are examples of conformation, molecular shape that can be changed by bond rotation and without breaking covalent bonds. When a protein contains more than one polypeptide chain, each chain with its own primary, secondary, and tertiary structure associates with the other chains to form the final molecule. The way in which polypeptides associate with each other in space to form the final protein is called the protein's quaternary structure (**figure AI.20**). The final conformation of a protein is ultimately determined by the amino acid sequence of its polypeptide chains.

Protein secondary, tertiary, and quaternary structure is largely determined and stabilized by many weak noncovalent forces such as hydrogen bonds and ionic bonds. Because of this, protein shape often is very flexible and easily changed. This flexibility is very important in protein function and in the regulation of enzyme activity. Because of their flexibility, however, proteins readily lose their proper shape and activity when exposed to harsh conditions. The only covalent bond commonly involved in the secondary and tertiary structure of proteins is the disulfide bond. The disulfide bond is formed when two cysteines are linked through their sulfhydryl groups. Disulfide bonds generally strengthen or stabilize protein structure.

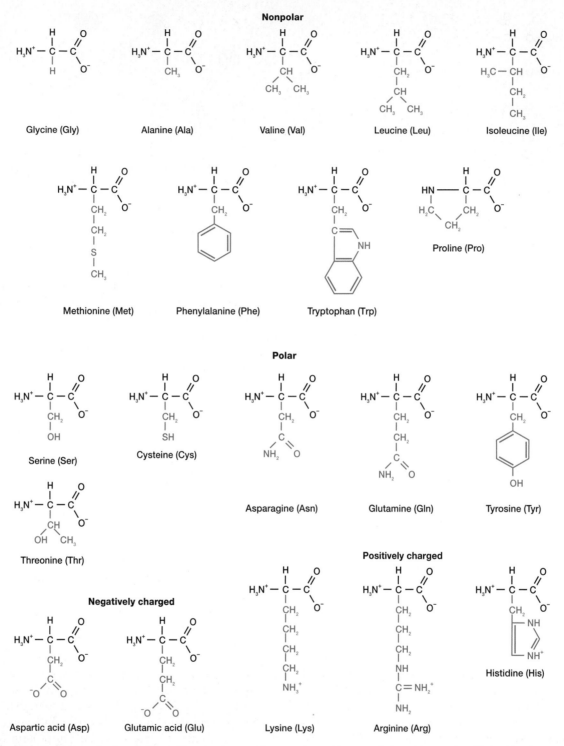

Figure AI.15 The Common Amino Acids. The structures of the α-amino acids normally found in proteins. Their side chains are shown in blue, and they are grouped together based on the nature of their side chains—nonpolar, polar, negatively charged (acid), or positively charged (basic). Proline is actually an imino acid rather than an amino acid.

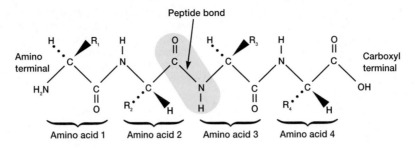

Figure AI.16   A Tetrapeptide Chain.   The end of the chain with a free α-amino group is the amino or N terminal. The end with the free α-carboxyl is the carboxyl or C terminal. One peptide bond is shaded in blue.

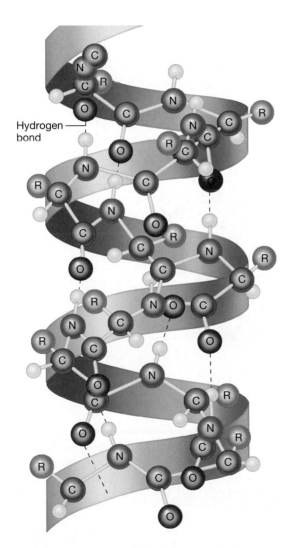

Figure AI.17   The α-Helix.   A polypeptide twisted into one type of secondary structure, the α-helix. The helix is stabilized by hydrogen bonds joining peptide bonds that are separated by three amino acids.

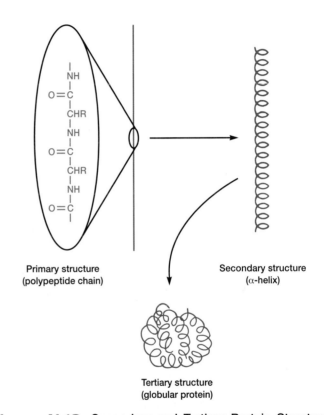

Figure AI.18   Secondary and Tertiary Protein Structures. The formation of secondary and tertiary protein structures by folding a polypeptide chain with its primary structure.

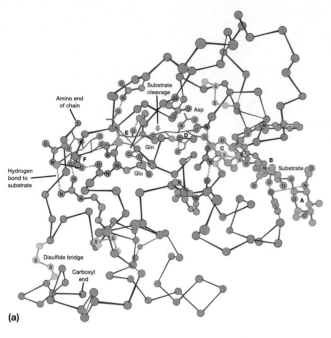

(a)

Figure AI.19 Lysozyme. The tertiary structure of the enzyme lysozyme. (a) A diagram of the protein's polypeptide backbone with the substrate hexasaccharide shown in blue. The point of substrate cleavage is indicated. (b) A space-filling model of lysozyme. The figure on the right shows the empty active site with some of its more important amino acids indicated. On the left, the enzyme has bound its substrate (in pink).

B²

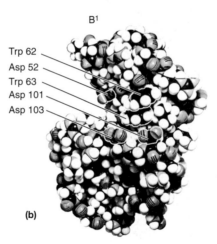

B¹

Trp 62
Asp 52
Trp 63
Asp 101
Asp 103

(b)

NUCLEIC ACIDS

The nucleic acids, deoxyribonucleic acid (DNA) and ribonucleic acid (RNA), are polymers of deoxyribonucleosides and ribonucleosides joined by phosphate groups. The nucleosides in DNA contain the purines adenine and guanine, and the pyrimidine bases thymine and cytosine. In RNA the pyrimidine uracil is substituted for thymine. Because of their importance for genetics and molecular biology, the chemistry of nucleic acids is introduced earlier in the text. The structure and synthesis of purines and pyrimidines are discussed in chapter 11 (pp. 235–236). The structures of DNA and RNA are described in chapter 12 (pp. 242–244).

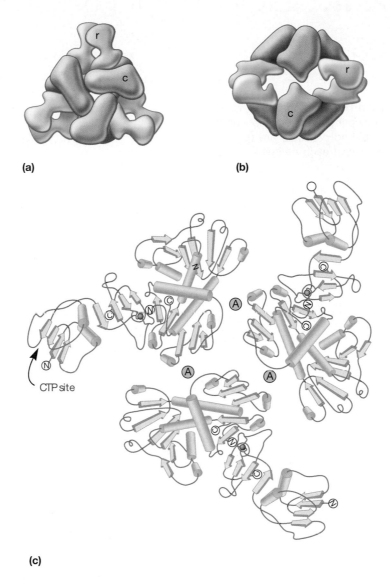

(a)

(b)

CTP site

(c)

Figure AI.20   An Example of Quaternary Structure.   The enzyme aspartate carbamoyltransferase from *Escherichia coli* has two types of sub-units, catalytic and regulatory. The association between the two types of subunits is shown: (a) a top view, and (b) a side view of the enzyme. The catalytic (C) and regulator (r) subunits are shown in different colors. (c) The peptide chains shown when viewed from the top as in (a). The active sites of the enzyme are located at the positions indicated by A. *(a and b) Adapted from Krause, et al., in Proceedings of the National Academy of Sciences, V. 82, 1985, as appeared in Biochemistry, 3d edition by Lubert Stryer. Copyright © 1975, 1981, 1988. Reprinted with permission of W. H. Freeman and Company. (c) Adapted from Kantrowitz, et al., in Trends in Biochemical Science, V. 5, 1980, as appeared in Biochemistry, 3d edition by Lubert Stryer. Copyright © 1975, 1981, 1988. Reprinted with permission of W. H. Freeman and Company.*

Appendix II: Common Metabolic Pathways

This appendix contains a few of the more important pathways discussed in the text, particularly those involved in carbohydrate catabolism. Enzyme names and final end products are given in color. Consult the text for a description of each pathway and its physiologic role.

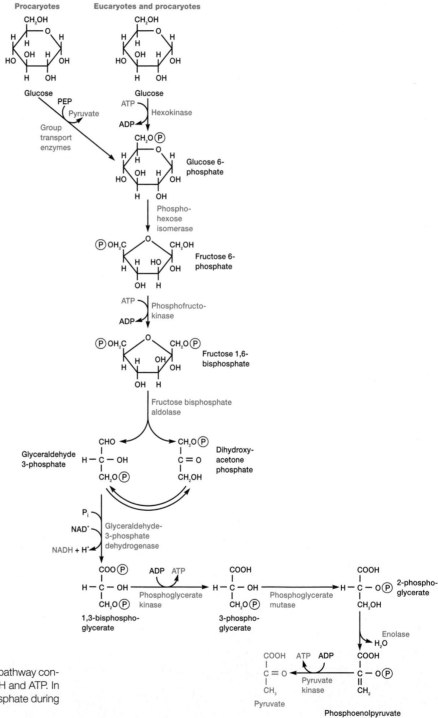

Figure AII.1 The Embden-Meyerhof Pathway. This pathway converts glucose and other sugars to pyruvate and generates NADH and ATP. In some procaryotes, glucose is phosphorylated to glucose 6-phosphate during group translocation transport across the plasma membrane.

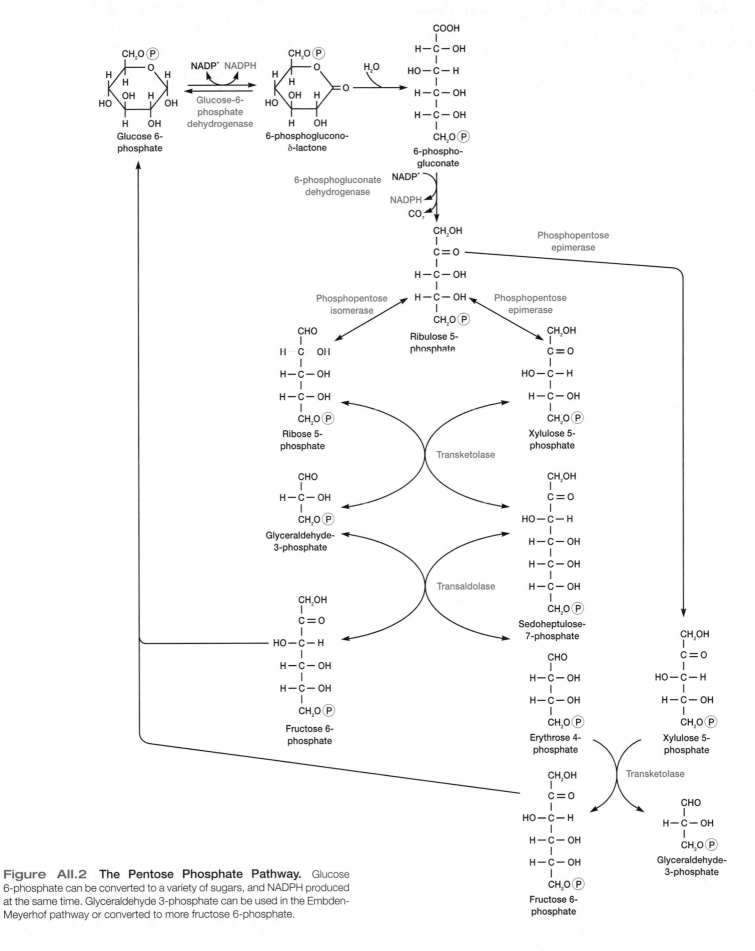

Figure AII.2 The Pentose Phosphate Pathway. Glucose 6-phosphate can be converted to a variety of sugars, and NADPH produced at the same time. Glyceraldehyde 3-phosphate can be used in the Embden-Meyerhof pathway or converted to more fructose 6-phosphate.

Figure AII.3   The Entner-Doudoroff Pathway.

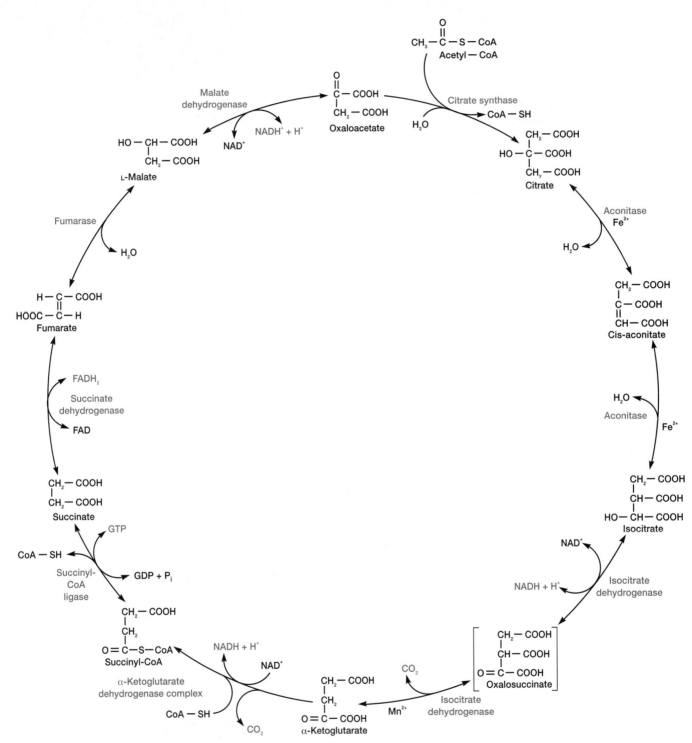

Figure AII.4   The Tricarboxylic Acid Cycle. Cis-aconitate and oxalosuccinate remain bound to aconitase and isocitrate dehydrogenase. Oxalosuccinate has been placed in brackets because it is so unstable.

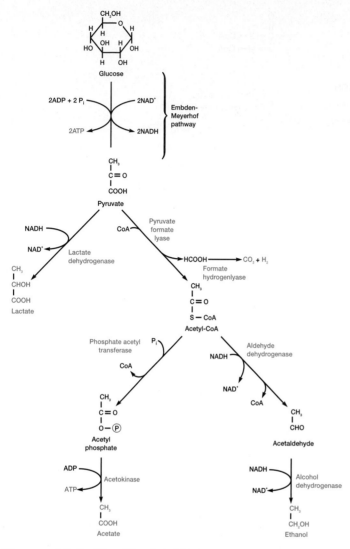

Figure AII.5   The Mixed Acid Fermentation Pathway.   This pathway is characteristic of many members of the *Enterobacteriaceae* such as *Escherichia coli*.

Glucose

2ADP + 2 P$_i$ 2NAD$^+$

Embden-Meyerhof pathway

2ATP   2NADH

CH$_3$
|
C=O
|
COOH
Pyruvate

NADH   CoA   Pyruvate formate lyase

NAD$^+$

Lactate dehydrogenase   HCOOH ——— CO$_2$ + H$_2$

CH$_2$ Formate hydrogenlyase
|
CHOH
| CH$_3$
COOH |
Lactate   C=O
                               |
                               S — CoA
                               Acetyl-CoA

Phosphate acetyl transferase   P$_i$ Aldehyde dehydrogenase

CoA   NADH

                      NAD$^+$

CH$_3$ CoA   CH$_3$
| |
C=O   CHO
|
O — (P)   Acetaldehyde
Acetyl phosphate

ADP   NADH

Acetokinase   Alcohol dehydrogenase

ATP   NAD$^+$

CH$_3$ CH$_3$
| |
COOH   CH$_2$OH
Acetate   Ethanol

Figure AII.6   The Butanediol Fermentation Pathway.   This pathway is characteristic of members of the *Enterobacteriaceae* such as *Enterobacter*. Other products may also be formed during butanediol fermentation.

Glucose

Embden-Meyerhof pathway   2NAD$^+$

2ADP + 2P$_i$ 2NADH + 2H$^+$

2ATP

CH$_3$
|
C=O
|
COOH
2 pyruvate

                              α-Acetolactate synthase
CO$_2$

CH$_3$
|
C=O
|
HO — C — COOH
|
CH$_3$
Acetolactate

                              Acetolactate decarboxylase
CO$_2$

CH$_3$
|
C=O
|
H — C — OH
|
CH$_3$
Acetoin

NADH + H$^+$ 2,3-butanediol dehydrogenase

NAD$^+$

CH$_3$
|
CHOH
|
CHOH
|
CH$_3$
2,3-butanediol

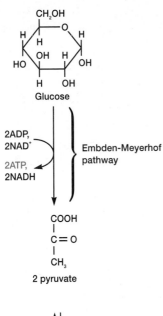

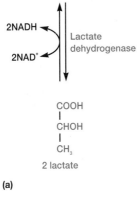

Glucose

2ADP,
2NAD⁺
$\longrightarrow$
2ATP,
2NADH

Embden-Meyerhof
pathway

2 pyruvate

2NADH
2NAD⁺

Lactate
dehydrogenase

2 lactate

(a)

Figure AII.7 Lactic Acid Fermentations.
(a) Homolactic fermentation pathway. (b) Heterolactic
fermentation pathways.

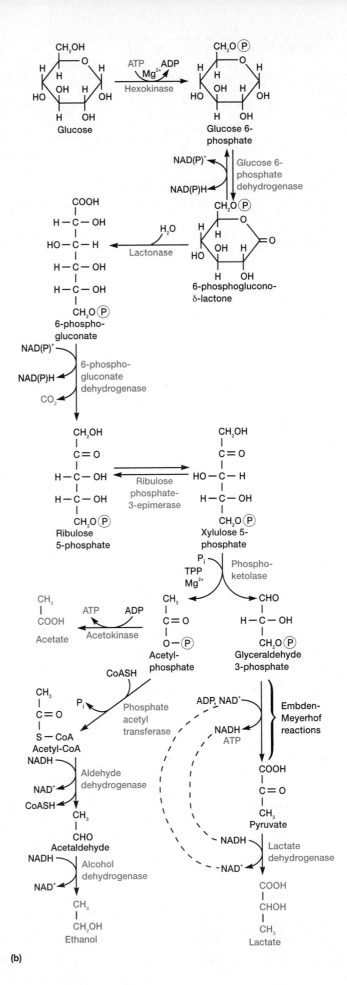

(b)

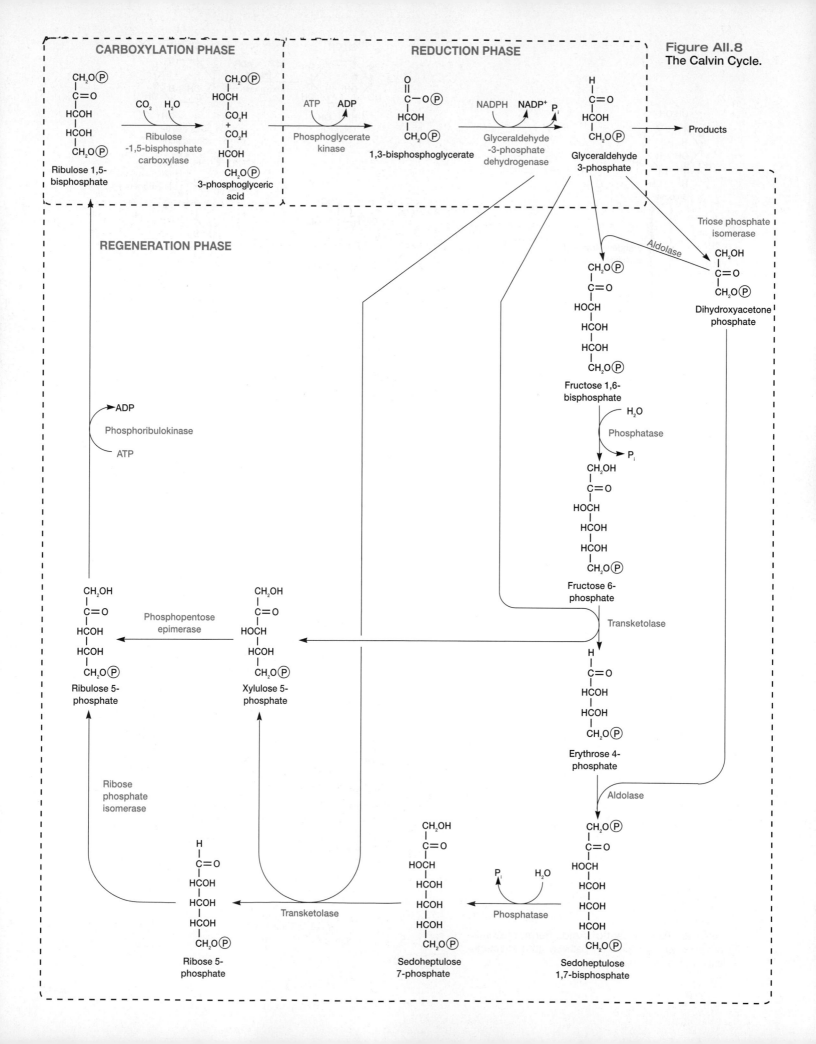

Figure AII.8
The Calvin Cycle.

Figure AII.9 The Pathway for Purine Biosynthesis. Inosinic acid is the first purine end product. The purine skeleton is constructed while attached to a ribose phosphate.

Glossary

Pronunciation Guide

Many of the boldface terms in this glossary are followed by a phonetic spelling in parentheses. These pronunciation aids usually come from *Dorland's Illustrated Medical Dictionary*. The following rules are taken from this dictionary and will help in using its phonetic spelling system.

1. An unmarked vowel ending a syllable (an open syllable) is long; thus, *ma* represents the pronunciation of *may; ne,* that of *knee; ri,* of *wry; so,* of *sew; too,* of *two;* and *vu,* of *view.*

2. An unmarked vowel in a syllable ending with a consonant (a closed syllable) is short; thus, *kat* represents *cat; bed, bed; hit, hit; not, knot; foot, foot;* and *kusp, cusp.*

3. A long vowel in a closed syllable is indicated by a macron; thus, *māt* stands for *mate; sēd,* for *seed; bīl,* for *bile; mōl,* for *mole; fūm,* for *fume;* and *fōōl,* for *fool.*

4. A short vowel that ends or itself constitutes a syllable is indicated by a breve; thus, *ĕ-fekt'* for *effect,* *ĭ-mūn'* for *immune,* and *ŭ-klōōd'* for *occlude.*

Primary (′) and secondary (″) accents are shown in polysyllabic words. Unstressed syllables are followed by hyphens. Some common vowels are pronounced as indicated here.

ə	sof<u>a</u>	ē	m<u>e</u>t	ŏ	g<u>o</u>t
ā	m<u>a</u>te	ī	b<u>i</u>te	ū	f<u>u</u>el
ă	b<u>a</u>t	ĭ	b<u>i</u>t	ŭ	b<u>u</u>t
ē	b<u>ea</u>m	ō	h<u>o</u>me		

From Dorland's Illustrated Medical Dictionary. *Copyright © 1988 W. B. Saunders, Philadelphia, Pa. Reprinted by permission.*

Microorganism Pronunciation Guide

Bacteria

Acetobacter (ah-se″to-bak′ter)
Acinetobacter (as″ĭ-net″o-bak′ter)
Actinomyces (ak″tĭ-no-mi′sēz)
Agrobacterium (ag″ro-bak-te′re-um)
Alcaligenes (al″kah-lij′ĕ-nēz)
Anabaena (ah-nab′ē-nah)
Arthrobacter (ar″thro-bak′ter)
Bacillus (bah-sil′lus)
Bacteroides (bak″tĕ-roi′dēz)
Bdellovibrio (del″o-vib′re-o)
Beggiatoa (bej″je-ah-to′ah)
Beijerinckia (bi″jer-ink′e-ah)
Bifidobacterium (bi″fid-o-bak-te′re-um)
Bordetella (bor″dē-tel′lah)
Borrelia (bŏ-rel′e-ah)
Brucella (broo-sel′lah)
Campylobacter (kam″pī-lo-bak′ter)
Caulobacter (kaw″lo-bak′ter)
Chlamydia (klah-mid′e-ah)
Chlorobium (klo-ro′be-um)
Chromatium (kro-ma′te-um)
Citrobacter (sit″ro-bak′ter)
Clostridium (klo-strid′e-um)
Corynebacterium (ko-ri″ne-bak-te′re-um)
Coxiella (kok″se-el′lah)
Cytophaga (si-tof′ah-gah)
Desulfovibrio (de-sul″fo-vib′re-o)
Enterobacter (en″ter-o-bak′ter)
Erwinia (er-win′e-ah)

Escherichia (esh″er-i′ke-ah)
Flexibacter (flek″sĭ-bak′ter)
Francisella (fran-sĭ-sel′ah)
Frankia (frank′e-ah)
Gallionella (gal″le-o-nel′ah)
Haemophilus (he-mof′ĭ-lus)
Halobacterium (hal″o-bak-te′re-um)
Hydrogenomonas (hi-dro″jĕ-no-mo′nas)
Hyphomicrobium (hi″fo-mi-kro′be-um)
Klebsiella (kleb″se-el′lah)
Lactobacillus (lak″to-bah-sil′lus)
Legionella (le″jun-el′ah)
Leptospira (lep″to-spi′rah)
Leptothrix (lep′to-thriks)
Leuconostoc (loo″ko-nos′tok)
Listeria (lis-te′re-ah)
Methanobacterium (meth″ah-no-bak-te′ re-um)
Methylococcus (meth″il-o-kok′-us)
Methylomonas (meth″il-o-mo′nas)
Micrococcus (mi″kro-kok′us)
Mycobacterium (mi″ko-bak-te′re-um)
Mycoplasma (mi″ko-plaz′mah)
Neisseria (nīs-se′re-ah)
Nitrobacter (ni″tro-bak′ter)
Nitrosomonas (ni-tro″so-mo′nas)
Nocardia (no-kar′de-ah)
Pasteurella (pas″tē-rel′ah)
Photobacterium (fo″to-bak-te′re-um)
Propionibacterium (pro″pe-on″e-bak-te′re-um)

Proteus (pro′te-us)
Pseudomonas (soo″do-mŏ′nas)
Rhizobium (ri-zo′be-um)
Rhodopseudomonas (ro″do-soo″do-mo′nas)
Rhodospirillum (ro″do-spi-ril′um)
Rickettsia (rĭ-ket′se-ah)
Salmonella (sal″mo-nel′ah)
Sarcina (sar′sĭ-nah)
Serratia (sĕ-ra′she-ah)
Shigella (shĭ-gel′ah)
Sphaerotilus (sfe-ro″tĭ-lus)
Spirillum (spi-ril′um)
Spirochaeta (spi″ro-ke′tah)
Spiroplasma (spi″ro-plaz′mah)
Staphylococcus (staf″ĭ-lo-kok′us)
Streptococcus (strep″to-kok′us)
Streptomyces (strep″to-mi′sēz)
Sulfolobus (sul″fo-lo′bus)
Thermoactinomyces (ther″mo-ak″tĭ-no-mi′sēz)
Thermoplasma (ther″mo-plaz′mah)
Thiobacillus (thi″o-bah-sil′us)
Thiothrix (thi′o-thriks)
Treponema (trep″o-ne′mah)
Ureaplasma (u-re′ah-plaz″ma)
Veillonella (va″yon-el′ah)
Vibrio (vib′re-o)
Xanthomonas (zan″tho-mo′nas)
Yersinia (yer-sin′e-ah)
Zoogloea (zo″o-gle′ah)

Microorganism Pronunciation Guide (*continued*)

Viruses

Vernacular, nonscientific virus names are not written in italics.

adenovirus (ad″ĕ-no-vi′rus)
arbovirus (ar″bo-vi′rus)
baculovirus (bak″u-lo-vi′rus)
coronavirus (kor″o-nah-vi′rus)
cytomegalovirus (si″to-meg″ah-lo-vi′rus)
Epstein-Barr virus (ep′stīn-bar′)
hepadnavirus (hep-ad″nə-vi′rus)
hepatitis virus (hep″ah-ti′tis)
herpesvirus (her″pēz-vi′rus)
influenza virus (in″flu-en′zah)
Measles virus (me′zelz)
Mumps virus (mumps)
orthomyxovirus (or″tho-mik″so-vi′rus)
papillomavirus (pap″i-lo″mah-vi′rus)
paramyxovirus (par″ah-mik″so-vi′rus)
parvovirus (par″vo-vi′rus)
picornavirus (pi-kor″nah-vi′rus)
Poliovirus (po″le-o-vi′rus)
polyomavirus (pol″e-o-mah-vi′rus)
poxvirus (poks-vi′rus)
Rabies virus (ra′bēz)
reovirus (re″o-vi′rus)
retrovirus (re″tro-vi′rus)
rhabdovirus (rab″do-vi′rus)
rhinovirus (ri″no-vi′rus)
rotavirus (ro′tah-vi″rus)
Rubella virus (roo-bel′ah)
togavirus (to″gah-vi′rus)
varicella-zoster virus (var″i-sel′ah zos′ter)
Variola virus (vah-ri′o-lah)

Fungi

Agaricus (ah-gar′i-kus)
Amanita (am″ah-ni′tah)
Arthrobotrys (ar″thro-bo′tris)
Aspergillus (as″per-jil′us)
Blastomyces (blas″to-mi′sēz)
Candida (kan′dĭ-dah)
Cephalosporium (sef″ah-lo-spo′re-um)
Claviceps (klav′ĭ-seps)
Coccidioides (kok-sid″e-oi′dēz)
Cryptococcus (krip″to-kok′us)
Epidermophyton (ep″ĭ-der-mof′ĭ-ton)
Fusarium (fu-sa′re-um)
Histoplasma (his″to-plaz′mah)
Microsporum (mi-kros′po-rum)
Mucor (mu′kor)
Neurospora (nu-ros′po-rah)
Penicillium (pen″ĭ-sil′e-um)
Phytophthora (fi-tof′tho-rah)
Pneumocystis (noo″mo-sis′tis)
Rhizopus (ri-zo′pus)
Saccharomyces (sak″ah-ro-mi′sēz)
Saprolegnia (sap″ro-leg′ne-ah)
Sporothrix (spo′ro-thriks)
Trichoderma (trik-o-der′mah)
Trichophyton (tri-kof′ĭ-ton)

Protists

Acanthamoeba (ah-kan″thah-me′bah)
Acetabularia (as″ĕ-tab″u-la′re-ah)
Amoeba (ah-me′bah)
Balantidium (bal″an-tid′e-um)
Chlamydomonas (klah-mid″do-mo′nas)
Chlorella (klo-rel′ah)
Cryptosporidium (krip″to-spo-rid′e-um)
Entamoeba (en″tah-me′bah)
Euglena (u-gle′nah)
Giardia (je-ar′de-ah)
Gonyaulax (gon″e-aw′laks)
Leishmania (lēsh-ma′ne-ah)
Naegleria (na-gle′re-ah)
Paramecium (par″ah-me′she-um)
Plasmodium (plaz-mo′de-um)
Prototheca (pro″to-the′kah)
Spirogyra (spi″ro-ji′rah)
Tetrahymena (tet″rah-hi′mĕ-nah)
Toxoplasma (toks″o-plaz′mah)
Trichomonas (trik″o-mo′nas)
Trypanosoma (tri″pan-o-so′mah)
Volvox (vol′voks)

AB toxins Exotoxins composed of two parts (A and B): The B portion is responsible for toxin binding to a cell but does not directly harm it; the A portion enters the cell and disrupts its function.

ABC protein secretion pathway *See* ATP-binding cassette transporters.

accessory pigments Photosynthetic pigments such as carotenoids and phycobiliproteins that aid chlorophyll in trapping light energy.

acellular slime mold Chemoorganotrophic protists with a distinctive life cycle that includes the formation of a multinucleate mass (plasmodium). Also called *Myxogastria.*

acetyl-CoA pathway A biochemical pathway used by methanogens to fix CO_2, and by acetogens to generate acetic acid.

acetyl-coenzyme A (acetyl-CoA) A combination of acetic acid and coenzyme A that is energy rich; it is produced by many catabolic pathways and is the substrate for the tricarboxylic acid cycle, fatty acid biosynthesis, and other pathways.

acid-fast Refers to bacteria (e.g., mycobacteria) that cannot be easily decolorized with acid alcohol after being stained with dyes such as basic fuchsin.

acid-fast staining A staining procedure that differentiates between bacteria based on their ability to retain a dye when washed with an acid alcohol solution.

acidic dyes Dyes that are anionic or have negatively charged groups such as carboxyls.

acidophile (as′id-o-fīl″) A microorganism that has its growth optimum between about pH 0 and 5.5.

acquired immune deficiency syndrome (AIDS) An infectious disease syndrome caused by the human immunodeficiency virus (HIV) and characterized by the loss of a normal immune response, followed by increased susceptibility to opportunistic infections and an increased risk of some cancers.

acquired immune tolerance The ability to produce antibodies against nonself-antigens while "tolerating" (not producing antibodies against) self-antigens.

acquired immunity Refers to the type of specific (adaptive) immunity that develops after exposure to a suitable antigen or is produced after antibodies are transferred from one individual to another.

actinobacteria (ak″tĭ-no-bak-tēr-e-ah) A group of gram-positive bacteria containing the actinomycetes and their high G+C relatives.

actinomycete (ak″tĭ-no-mi′sēt) An aerobic, gram-positive bacterium that forms branching filaments (hyphae) and asexual spores.

actinorhizae Associations between actinomycetes and plant roots.

activated sludge Solid matter or sediment composed of actively growing microorganisms that participate in the aerobic portion of a biological sewage treatment process.

activation energy The energy required to bring reacting molecules together to reach the transition state in a chemical reaction.

activator-binding site *See* activator protein.

activator protein A transcriptional regulatory protein that binds to a specific site on DNA (activator-binding site) and enhances transcription initiation.

active carrier An individual who has an overt clinical case of a disease and who can transmit the infection to others.

active immunization The induction of active immunity by natural exposure to a pathogen or by vaccination.

active site The part of an enzyme that binds the substrate to form an enzyme-substrate complex and catalyze the reaction. Also called the catalytic site.

active transport The transport of solute molecules across a membrane against a gradient; it requires a carrier protein and the input of energy.

acute carrier   *See* casual carrier.

acute infections   Infections with a rapid onset that last for a relatively short time.

acute viral gastroenteritis   An inflammation of the stomach and intestines caused by viruses.

acyclovir (a-si′klo-vir)   A synthetic purine nucleoside derivative with antiviral activity against herpes simplex virus.

N-**acylhomoserine lactone**   *See* quorum sensing.

adenine (ad′e-nēn)   A purine derivative, 6-aminopurine, found in nucleosides, nucleotides, coenzymes, and nucleic acids.

adenine arabinoside (vidarabine)   An antiviral agent used especially to treat keratitis and encephalitis caused by the herpes simplex virus.

adenosine diphosphate (ADP) (ah-den′o-se⁻n) The nucleoside diphosphate usually formed upon the breakdown of ATP when it provides energy for work.

adenosine 5′-triphosphate (ATP)   The triphosphate of the nucleoside adenosine, which is a high energy molecule and serves as the cell's major form of energy currency.

adhesin (ad-he′zin)   A molecular component on the surface of a microorganism that is involved in adhesion to a substratum (e.g., host tissue) or cell.

adjuvant   Material added to an antigen in a vaccine preparation that increases its immunogenicity.

aerial mycelium   The mat of hyphae formed by actinomycetes or fungi that grows above the substrate, imparting a fuzzy appearance to colonies.

aerobe   An organism that grows in the presence of atmospheric oxygen.

aerobic anoxygenic photosynthesis   Photosynthetic process in which electron donors such as organic matter or sulfide are used under aerobic conditions.

aerobic respiration   A metabolic process in which molecules, often organic, are oxidized with oxygen as the final electron acceptor.

aerotolerant anaerobes   Microbes that grow equally well whether or not oxygen is present.

aflatoxin (af″lah-tok′sin)   A polyketide secondary fungal metabolite that can cause cancer.

agar (ahg′ar)   A complex sulfated polysaccharide, usually from red algae, that is used as a solidifying agent in the preparation of culture media.

agglutinates   The visible aggregates or clumps formed by an agglutination reaction.

agglutination reaction   The formation of an insoluble immune complex by the cross-linking of cells or particles.

agglutinin   The antibody responsible for an agglutination reaction.

AIDS   *See* acquired immune deficiency syndrome.

airborne transmission   The type of transmission in which the pathogen is suspended in the air and travels over a meter or more from the source to the host.

akinetes   Specialized, nonmotile, dormant, thick-walled resting cells formed by some cyanobacteria.

alcoholic fermentation   A fermentation process that produces ethanol and CO_2 from sugars.

alga (al′gah)   A common term for several unrelated groups of photosynthetic eucaryotic microorganisms lacking multicellular sex organs (except for the charophytes) and conducting vessels. Most are now considered protists.

algicide (al′ji-sīd)   An agent that kills algae.

alignment   In the comparison of gene, RNA, and protein sequences, the arrangement of sequences such that a base-by-base (or amino acid by amino acid) comparison can be made and used to calculate the degree of similarity between sequences.

alkalophile (alkaliphile)   A microorganism that grows best at pHs from about 8.5 to 11.5.

allele   An alternative form of a gene.

allergen   An antigen that induces an allergic response.

allergic contact dermatitis   An allergic reaction caused by haptens that combine with proteins in the skin to form the allergen that stimulates the immune response.

allergy   *See* Type I hypersensitivity.

allochthonous (ăl′ək-thə-nəs)   Substances not native to a given environment (e.g., nutrient influx into freshwater ecosystems).

allograft   A transplant between genetically different individuals of the same species.

allosteric effector   *See* allosteric enzyme.

allosteric enzyme   An enzyme whose activity is altered by the noncovalent binding of a small effector (allosteric effector) at a regulatory site separate from the catalytic site; effector binding causes a conformational change in the enzyme's, catalytic site, which leads to enzyme activation or inhibition.

allotype   Allelic variants of antigenic determinant(s) found on antibody chains of some, but not all, members of a species, which are inherited as simple Mendelian traits.

alpha hemolysis   *See* hemolysis.

Alphaproteobacteria   One of the five classes of proteobacteria.

alternative complement pathway   An antibody-independent pathway of complement activation.

alternative splicing   The use of different exons during RNA splicing to generate different polypeptides from the same gene.

alveolar macrophage   A vigorously phagocytic macrophage located on the epithelial surface of the lung alveoli, where it ingests inhaled particulate matter and microorganisms.

amantadine (ah-man′tah-den)   An antiviral agent used to prevent type A influenza infections.

amebiasis (amebic dysentery) (am″e-bi′ah-sis) An infection with amoebae, often resulting in dysentery; usually it refers to an infection by *Entamoeba histolytica*.

amensalism   A relationship in which the product of one organism has a negative effect on another organism.

Ames test   A test that uses a special *Salmonella* strain to test chemicals for mutagenicity and potential carcinogenicity.

amino acid activation   The initial stage of protein synthesis in which amino acids are attached to transfer RNA molecules. The reaction is catalyzed by aminoacyl-tRNA synthetases.

aminoacyl or **acceptor site (A site)**   The ribosomal site that contains an aminoacyl-tRNA at the beginning of the elongation cycle during protein synthesis.

aminoacyl-tRNA synthetases   *See* amino acid activation.

aminoglycoside antibiotics   A group of antibiotics synthesized by *Streptomyces* and *Micromonospora,* which contain a cyclohexane ring and amino sugars; all aminoglycoside antibiotics bind to the small ribosomal subunit and inhibit protein synthesis.

amoeboid movement   Moving by means of cytoplasmic flow and the formation of pseudopodia.

amphibolic pathways   Metabolic pathways that function both catabolically and anabolically.

amphipathic   Term describing a molecule that has both hydrophilic and hydrophobic regions (e.g., phospholipids).

amphitrichous (am-fit′rĕ-kus)   A cell with a single flagellum at each end.

amphotericin B (am″fo-ter′i-sin)   An antifungal agent produced by *Streptomyces nodosus*.

amplification   In reference to the polymerase chain reaction, the process of making large quantities of a particular DNA sequence.

anabolism   The synthesis of complex molecules from simpler molecules with the input of energy and reducing power.

anaerobe (an-a′er-ōb)   An organism that grows in the absence of free oxygen.

anaerobic digesters (an″a-er-o′bik)   The microbiological treatment of sewage wastes under anaerobic conditions to produce methane.

anaerobic respiration   An energy-conserving process in which the electron transport chain acceptor is a molecule other than oxygen.

anagenesis   Changes in gene frequencies and distribution among species; the accumulation of small genetic changes within a population that introduces genetic variability but is not enough to result in either speciation or extinction.

anammox reaction   The coupled use of nitrite as an electron acceptor and ammonium ion as an electron donor under anoxic conditions to yield nitrogen gas.

anamnestic response (an″am-nes′tik)   The recall, or the remembering, by the immune system of a prior response to a given antigen.

anaphylaxis (an″ah-fĭ-lak′sis)   An immediate (type I) hypersensitivity reaction following exposure of a sensitized individual to the appropriate antigen.

anaplasia   The reversion of an animal cell to a more primitive, undifferentiated state.

anaplerotic reactions (an′ah-plĕ-rot′ik) Reactions that replenish depleted tricarboxylic acid cycle intermediates.

anergy (an′ər-je) A state of unresponsiveness to antigens. Absence of the ability to generate a sensitivity reaction to substances that are expected to be antigenic.

anoxic Without oxygen present.

anoxygenic photosynthesis Photosynthesis that does not oxidize water to produce oxygen.

antheridium (an″ther-id′e-um; pl., **antheridia**) A male gamete-producing organ, which may be unicellular or multicellular.

anthrax An infectious disease of warm-blooded animals, especially of cattle and sheep, caused by *Bacillus anthracis* that can be transmitted to humans.

antibiotic A microbial product or its derivative that kills susceptible microorganisms or inhibits their growth.

antibody (immunoglobulin) A glycoprotein made by plasma cells in response to the introduction of an antigen.

antibody affinity The strength of binding between an antigen and an antibody.

antibody-dependent cell-mediated cytotoxicity (ADCC) The killing of antibody-coated target cells by cells with Fc receptors that recognize the Fc region of the bound antibody.

antibody-mediated immunity *See* humoral immunity.

antibody titer An approximation of the antibody concentration required to react with an antigen.

anticodon The three bases on a tRNA that are complementary to the codon on mRNA.

antigen A substance (such as a protein, nucleoprotein, polysaccharide, or sometimes a glycolipid) to which lymphocytes respond.

antigen-binding fragment (Fab) "Fragment antigen binding." A monovalent antigen-binding fragment of an immunoglobulin molecule that consists of one light chain and part of one heavy chain, linked by interchain disulfide bonds.

antigen processing The hydrolytic digestion of antigens to produce antigen fragments, which are collected by class I or class II MHC molecules and presented on the surface of a cell.

antigenic determinant site *See* epitope.

antigenic drift A small change due to mutation in the antigenic character of an organism that allows it to avoid attack by the immune system.

antigenic shift A major change in the antigenic character of an organism that makes it unrecognized by host immune mechanisms.

antigen-presenting cells (APCs) Cells that take in protein antigens, process them, and present antigen fragments to T cells in conjunction with MHC molecules so that the cells are activated; includes macrophages, B cells, and dendritic cells.

antimetabolite A compound that blocks metabolic pathway function by competitively inhibiting a key enzyme because it closely resembles the normal enzyme substrate.

antimicrobial agent An agent that kills microorganisms or inhibits their growth.

antiport Coupled transport of two molecules in which one molecule enters the cell as the other leaves the cell.

antisense RNA A single-stranded RNA with a base sequence complementary to a segment of a target RNA molecule; when bound to the target RNA the target's activity is altered.

antisepsis The prevention of infection or sepsis.

antiseptic Chemical agent applied to tissue to prevent infection by killing or inhibiting pathogens.

antiserum Serum containing antibodies.

antitoxin An antibody to a microbial toxin, usually a bacterial exotoxin, that combines specifically with the toxin, in vivo and in vitro, neutralizing the toxin.

apical complex A set of organelles characteristic of members of the protist subdivision *Apicomplexa*: includes polar rings, subpellicular microtubules, conoid, rhoptries, and micronemes.

apicomplexan (a′pĭ-kom-plek′san) A protist that lacks special locomotor organelles but has an apical complex and a spore-forming stage. It is either an intra- or extracellular parasite of animals; a member of the taxon *Apicomplexa*.

apoenzyme (ap″o-en′zīm) The protein part of an enzyme that also has a nonprotein component.

apoptosis (ap″o-to′sis) Programmed cell death. A physiological suicide mechanism that results in fragmentation of a cell into membrane-bound particles that are eliminated by phagocytosis.

aporepressor The inactive form of a repressor protein; the repressor becomes active when the co-repressor binds to it.

appressorium A flattened region of hypha found in some plant-infecting fungi that aids in penetrating the host plant cell wall. Occurs in both pathogenic fungi and nonpathogenic mycorrhizal fungi.

arbuscular mycorrhizae (AM) fungi The mycorrhizal fungi in a fungus-root association that penetrate the outer layer of the root, grow intracellularly, and form characteristic much-branched hyphal structures called arbuscules.

arbuscules *See* arbuscular mycorrhizal fungi.

Archaea The domain that contains organisms with procaryotic cell structure, isoprenoid glycerol diether or diglycerol tetraether lipids in their membranes, and archaeal rRNA (among many differences).

arthroconidia (arthrospores) Fungal spores formed by fragmentation.

artificially acquired active immunity The type of immunity that results from immunizing an animal with a vaccine. The immunized animal produces its own antibodies and activated lymphocytes.

artificially acquired passive immunity The temporary immunity that results from introducing into an animal antibodies that have been produced either in another animal or by in vitro methods.

ascocarp A multicellular structure in ascomycetes lined with specialized cells called asci.

ascogenous hypha A specialized hypha that gives rise to one or more asci.

ascogonium (as″ko-go′ne-um; pl., **ascogonia**) The receiving (female) organ in ascomycetous fungi that, after fertilization, gives rise to ascogenous hyphae and later to asci and ascospores.

ascomycetes (as″ko-mi-se′tēz) A taxon of fungi that form ascospores.

ascospore (as′ko-spor) A spore contained or produced in an ascus.

ascus (pl. asci) A specialized cell, characteristic of the ascomycetes, in which two haploid nuclei fuse to produce a zygote, which often immediately divides by meiosis; at maturity, an ascus contains ascospores.

aseptic meningitis syndrome *See* meningitis.

aspergillosis (as″per-jil-o′sis) A fungal disease caused by species of *Aspergillus*.

assimilatory reduction (e.g., **assimilatory nitrate reduction** and **assimilatory sulfate reduction**) The reduction of an inorganic molecule to incorporate it into organic material. No energy is conserved during this process.

associative nitrogen fixation Nitrogen fixation by bacteria in the plant root zone (rhizosphere).

atomic force microscope *See* scanning probe microscope.

atopic reaction A type I hypersensitivity response caused by environmental allergens.

ATP-binding cassette transporters (ABC transporters) Transport systems that use ATP hydrolysis to drive translocation across the plasma membrane, can be used for nutrient uptake (ABC transporter) or for protein secretion (ABC protein secretion pathway).

ATP synthase An enzyme that catalyzes synthesis of ATP from ADP and P_i, using energy derived from the proton motive force.

attenuated vaccine Live, nonpathogenic organisms used to activate adaptive immunity. The nonpathogenic organisms grow in the vaccinated individual without producing serious clinical disease while stimulating lymphocytes to produce antibody and activated T cells.

attenuation 1. A mechanism for the regulation of transcription termination of some bacterial operons by aminoacyl-tRNAs. 2. A procedure that reduces or abolishes the virulence of a pathogen without altering its immunogenicity.

attenuator A factor-independent transcription termination site in the leader sequence that is involved in attenuation.

atypical pneumonia An acute respiratory disease characterized by high fever and coughing. It is atypical in that little fluid is found in the lungs; often caused by *Mycoplasma pneumoniae*.

autochthonous (ô-tŏk-thə-nəs) Substances (nutrients) that originate in a given environment. *See also* allochthonous.

autoclave An apparatus for sterilizing objects by the use of steam under pressure.

autogamy In the reproduction of certain protists, this form of self-fertilization involves the fusion of haploid nuclei or gametes derived from a single cell.

autogenous infection (aw-toj′e-nus) An infection that results from a patient's own microbiota, regardless of whether the infecting organism became part of the patient's microbiota subsequent to admission to a clinical care facility.

autoimmune disease A disease produced by the immune system attacking self-antigens.

autoimmunity A condition characterized by the presence of serum autoantibodies and self-reactive lymphocytes that may be benign or pathogenic.

autolysins (aw-tol′ĭ-sins or aw″to-lye′sins) Enzymes that partially digest peptidoglycan in growing bacteria so that the cell wall can be enlarged.

autophagosome A membrane-bound structure formed during autophagy that delivers materials to be digested to a late endosome.

autophagy Digestion of cell components carried out by eucaryotic cells.

autoradiography A procedure that detects radioactively labeled materials using a photographic process.

autotroph (aw′to-trōf) An organism that uses CO_2 as its sole or principal source of carbon.

auxotroph (awk′so-trōf) An organism with a mutation that causes it to lose the ability to synthesize an essential nutrient; because of the mutation, the organism must obtain the nutrient or a precursor from its surroundings.

avidity The combined strength of binding between an antigen and all the antibody-binding sites.

axenic (a-zen′ik) Not contaminated by any foreign organisms; the term is used in reference to pure microbial cultures or to germfree animals.

axial fibrils *See* periplasmic flagella.

axial filament *See* periplasmic flagella.

axopodium A thin, needlelike type of pseudopodium with a central core of microtubules.

bacille Calmette-Guerin (BCG) An attenuated form of *Mycobacterium tuberculosis* used in some countries as a vaccine for tuberculosis.

bacillus (bah-sil′lus) A rod-shaped bacterium.

bacteremia The presence of viable bacteria in the blood.

Bacteria The domain that contains procaryotic cells with primarily diacyl glycerol diesters in their membranes and with bacterial rRNA.

bacterial artificial chromosome (BAC) A cloning vector constructed from the *E. coli* F-factor plasmid.

bacterial (septic) meningitis *See* meningitis.

bactericide An agent that kills bacteria.

bacteriochlorophyll (bak-te″re-o-klo′ro-fil) A modified chlorophyll that serves as the primary light-trapping pigment in purple and green photosynthetic bacteria and heliobacteria.

bacteriocin (bak-te′re-o-sin) A protein produced by a bacterial strain that kills other closely related bacteria.

bacteriophage (bak-te′re-o-fāj″) A virus that uses bacteria as its host; often called a phage.

bacteriophage (phage) typing A technique in which strains of bacteria are identified based on their susceptibility to specific bacteriophages.

bacteriorhodopsin A transmembranous protein to which retinal is bound; it functions as a light-driven proton pump driving photophosphorylation without chlorophyll or bacteriochlorophyll. Found in the purple membrane of halophilic archaea.

bacteriostatic Inhibiting the growth and reproduction of bacteria.

bacteroid A modified, often pleomorphic, bacterial cell within the root nodule cells of legumes; after transformation into a symbiosome, it carries out nitrogen fixation.

baculoviruses Common name for viruses belonging to the family *Baculoviridae*.

baeocytes Small, spherical, reproductive cells produced by pleurocapsalean cyanobacteria through multiple fission.

balanced growth Microbial growth in which all cellular constituents are synthesized at constant rates relative to each other.

balanitis (bal″ah-ni′tis) Inflammation of the glans penis usually associated with *Candida* fungi; a sexually transmitted disease.

barophilic (bar″o-fil′ik) or **barophile** Organisms that prefer or require high pressures for growth and reproduction.

barotolerant Organisms that can grow and reproduce at high pressures but do not require them.

basal body The structure at the base of flagella that attaches them to the cell.

base analogs Molecules that resemble normal DNA nucleotides and can substitute for them during DNA replication, leading to mutations.

base excision repair *See* excision repair.

basic dyes Dyes that are cationic, or have positively charged groups, and bind to negatively charged cell structures.

basidiocarp (bah-sid′e-o-karp″) The fruiting body of a basidiomycete that contains the basidia.

basidiomycetes (bah-sid″e-o-mi-se′tēz) A division of fungi in which the spores are borne on club-shaped organs called basidia.

basidiospore (bah-sid′e-ō-spōr) A spore borne on the outside of a basidium following karyogamy and meiosis.

basidium (bah-sid′e-um; pl., **basidia**) A structure formed by basidiomycetes; it bears on its surface basidiospores (typically four) that are formed following karyogamy and meiosis.

basophil A weakly phagocytic white blood cell in the granulocyte lineage. It synthesizes and stores vasoactive molecules (e.g., histamine) that are released in response to external triggers (*see* mast cell).

batch culture A culture of microorganisms produced by inoculating a closed culture vessel containing a single batch of medium.

B cell (B lymphocyte) A type of lymphocyte that following interaction with antigen, becomes a plasma cell, which synthesizes and secretes antibody molecules involved in humoral immunity.

B-cell receptor (BCR) A transmembrane immunoglobulin complex on the surface of a B cell that binds an antigen and stimulates the B cell. It is composed of a membrane-bound immunoglobulin, usually IgD or a modified IgM, complexed with another membrane protein (the Ig-α /Ig-β heterodimer).

benthic Pertaining to the bottom of the sea or another body of water.

beta hemolysis *See* hemolysis.

β-lactam Referring to antibiotics containing a β-lactam ring (e.g., penicillins and cephalosporins).

β-lactam ring The cyclic chemical structure composed of one nitrogen and three carbon atoms. It has antibacterial activity, interfering with bacterial cell wall synthesis.

β lactamase An enzyme that hydrolyzes the β-lactam ring rendering the antibiotic inactive. Sometimes called penicillinase.

β-oxidation pathway The major pathway of fatty acid oxidation to produce NADH, $FADH_2$, and acetyl coenzyme A.

Betaproteobacteria One of the five classes of proteobacteria.

binal symmetry The symmetry of some virus capsids (e.g., those of complex phages) that is a combination of icosahedral and helical symmetry.

binary fission Asexual reproduction in which a cell or an organism separates into two identical daughter cells.

binomial system The nomenclature system in which an organism is given two names; the first is the capitalized generic name, and the second is the uncapitalized specific epithet.

bioaugmentation Addition of pregrown microbial cultures to an environment to perform a specific task.

biocatalysis The use of enzymes or whole microbes to perform chemical transformations on natural products.

biochemical oxygen demand (BOD) The amount of oxygen used by organisms in water under certain standard conditions; it provides an index of the amount of microbially oxidizable organic matter present.

biocrime The use of biological materials (organisms or their toxins) to subvert societal goals or laws.

biodegradation The breakdown of a complex chemical through biological processes that can result in minor loss of functional groups, fragmentation into smaller constitutents, or complete breakdown to carbon dioxide and minerals.

biofilms Organized microbial communities consisting of layers of microbial cells associated with surfaces, often with complex structural and functional characteristics.

biogeochemical cycling The oxidation and reduction of substances carried out by living organisms and abiotic processes that results in the cycling of elements within and between different parts of the ecosystem.

bioinformatics The interdisciplinary field that manages and analyzes large biological data sets, including genome and protein sequences.

bioinsecticide A pathogen that is used to kill or disable unwanted insect pests.

biologic transmission A type of vector-borne transmission in which a pathogen goes through some morphological or physiological change within the vector.

bioluminescence The production of light by living cells, often through the oxidation of molecules by the enzyme luciferase.

biomagnification The increase in concentration of a substance in higher-level consumer organisms.

biopesticide A microorganism or another biological agent used to control a specific pest.

bioprospecting The collection, cataloging, and analysis of organisms including microorganisms and plants with the intent of finding a useful application and/or to document biodiversity.

bioremediation The use of biologically mediated processes to remove or degrade pollutants from specific environments.

biosensor A device for the detection of a particular substance (an analyte) that combines a biological receptor with a physicochemical detector.

biosynthesis *See* anabolism.

biosynthetic reactions *See* anabolism.

biosynthetic-secretory pathway The process used by eucaryotic cells to synthesize proteins and lipids, followed by secretion or delivery to organelles or the plasma membrane.

bioterrorism The intentional or threatened use of microorganisms or toxins from living organisms to produce death or disease in humans, animals, and plants.

biotransformation (microbial transformation) The use of living organisms to modify substances that are not normally used for growth.

biovar Variant strains of microbes characterized by biological or physiological differences.

blastomycosis (blas″to-mi-ko′sis) A systemic fungal infection caused by *Blastomyces dermatitidis* and marked by suppurating tumors in the skin or by lesions in the lungs.

blastospores Fungal spores produced from a vegetative mother cell by budding.

B lymphocyte *See* B cell.

botulism (boch′oo-lizm) A form of food poisoning caused by a neurotoxin (botulin) produced by *Clostridium botulinum*; sometimes found in improperly canned or preserved food.

bright-field microscope A microscope that illuminates the specimen directly with bright light and forms a dark image on a brighter background.

Bright's disease *See* glomerulonephritis.

broad-spectrum drugs Chemotherapeutic agents that are effective against many different kinds of pathogens.

bronchial-associated lymphoid tissue (BALT) The type of defensive tissue found in the lungs. Part of the nonspecific (innate) immune system.

bronchial asthma An atopic allergy involving the lower respiratory tract.

brucellosis (broo′sə-lo′sĭs) A zoonotic disease caused by the α-proteobacterium *Brucella*. It is also known as Bang's disease, Malta fever, Mediterranean fever, Rock fever, and undulant fever.

bubo (bu′bo) A tender, inflamed, enlarged lymph node that results from a variety of infections.

bubonic plague *See* plague.

budding A vegetative outgrowth of yeast and some bacteria as a means of asexual reproduction; the daughter cell is smaller than the parent.

bulking sludge Sludge produced in sewage treatment that does not settle properly, usually due to the development of filamentous microorganisms.

butanediol fermentation A type of fermentation most often found in the family *Enterobacteriaceae* in which 2,3-butanediol is a major product; acetoin is an intermediate in the pathway and may be detected by the Voges-Proskauer test.

calorie The amount of heat needed to raise one gram of water from 14.5 to 15.5°C.

Calvin cycle The main pathway for the fixation (reduction and incorporation) of CO_2 into organic material by photoautotrophs and chemolithoautotrophs.

campylobacteriosis A disease caused by species of the bacterium *Campylobacter*, primarily *Campylobacter jejuni*, characterized by an inflammatory, sometimes bloody, diarrhea and may present as a dysentery syndrome.

cancer A malignant tumor that expands locally by invasion of surrounding tissues and systemically by metastasis.

candidal vaginitis Infection of the vagina caused by the fungus *Candida* sp.

candidiasis (kan″dĭ-di′ah-sis) An infection caused by *Candida* species.

capsid The protein coat or shell that surrounds a virion's nucleic acid.

capsomer The ring-shaped morphological unit of which icosahedral capsids are constructed.

capsule A layer of well-organized material, not easily washed off, lying outside the cell wall.

carbonate equilibrium system The interchange among CO_2, HCO_3^- and CO_3^{2-} that keeps oceans buffered between pH 7.6 to 8.2.

carboxysomes Polyhedral inclusion bodies that contain the CO_2 fixation enzyme ribulose 1,5-bisphosphate carboxylase.

cardinal temperatures The minimal, maximal, and optimum temperatures for growth.

carotenoids (kah-rot′e-noids) Pigment molecules, usually yellowish in color, that are often used to aid chlorophyll in trapping light energy during photosynthesis.

carrier An infected individual who is a potential source of infection for others and plays an important role in the epidemiology of a disease.

caseous lesion (ka′se-us) A lesion resembling cheese or curd; cheesy. Most caseous lesions are caused by *Mycobacterium tuberculosis*.

casual carrier An individual who harbors an infectious organism for only a short period.

catabolism That part of metabolism in which larger, more complex molecules are broken down into smaller, simpler molecules with the release of energy.

catabolite activator protein (CAP) A protein that regulates the expression of genes encoding catabolite repressible enzymes; also called cyclic AMP receptor protein (CRP).

catabolite repression Inhibition of the synthesis of several catabolic enzymes by a preferred carbon and energy source (e.g., glucose).

catalase An enzyme that catalyzes the destruction of hydrogen peroxide.

catalyst A substance that accelerates a reaction without being permanently changed itself.

catalytic site *See* active site.

catenanes (kăt′ə-nāns′) Circular, covalently closed nucleic acid molecules that are locked together like the links of a chain.

cathelicidins Antimicrobial cationic peptides that are produced by a variety of cells (e.g., neutrophils, respiratory epithelial cells, and alveolar macrophages).

catheter A tubular instrument for withdrawing fluids from a cavity of the body.

caveola (ka-ve-o′lə) A small flask-shaped invagination of the plasma membrane formed during caveolae-dependent endocytosis.

CD4+ cell *See* T-helper cell.

CD8+ cell *See* cytotoxic T lymphyocyte.

CD95 pathway *See* Fas-FasL pathway.

cell cycle The sequence of events in a cell's growth-division cycle between the end of one division and the end of the next.

cell envelope The plasma membrane plus all other structures outside it.

cell-mediated immunity The type of immunity that results from T cells coming into close contact with foreign cells or infected cells to destroy them.

cellular slime molds Protists with a vegetative phase consisting of amoeboid cells that aggregate to form a multicellular pseudoplasmodium. They belong to the taxon *Dictyostelia;* formerly considered fungi.

cellulitis (sel″u-li′tis) A diffuse spreading infection of subcutaneous skin tissue caused by streptococci, staphylococci, or other organisms. The tissue is inflamed with edema, redness, pain, and interference with function.

cellulose The major structural carbohydrate of plant cell walls; a linear $\beta(1 \rightarrow 4)$ glucan.

cell wall The strong structure that lies outside the plasma membrane; it supports and protects the membrane and gives the cell shape.

central metabolic pathways Those pathways central to the metabolism of an organism because they function catabolically and anabolically (e.g., glycolytic pathways and tricarboxylic acid cycle).

central tolerance The process by which immune cells are rendered inactive.

cephalosporin (sef″ah-lo-spōr′in) A group of β-lactam antibiotics derived from the fungus *Cephalosporium,* which share the 7-aminocephalosporanic acid nucleus.

Chagas' disease *See* trypanosomiasis.

chain termination DNA sequencing method A DNA-sequencing method that uses dideoxynucleotides to cause termination of DNA replication at random sites.

chancre (shang′ker) The primary lesion of syphilis occurring at the site of entry of the pathogen.

chaperone proteins Proteins that assist in the folding and stabilization of other proteins. Some are also involved in directing newly synthesized proteins to protein secretion systems or to other locations in the cell.

chemical fixation *See* fixation.

chemical oxygen demand (COD) The amount of chemical oxidation required to convert organic matter in water and wastewater to CO_2.

chemiosmotic hypothesis (kem″e-o-os-mot′ik) The hypothesis that proton and electrochemical gradients are generated by electron transport and then used to drive ATP synthesis by oxidative phosphorylation or photophosphorylation.

chemoheterotroph (ke″mo-het′er-o-trōf″) *See* chemoorganoheterotroph.

chemokine A type of cytokine that stimulates chemotaxis and chemokinesis.

chemolithoautotroph A microorganism that oxidizes reduced inorganic compounds to derive both energy and electrons; CO_2 is the carbon source. Also called chemolithotrophic autotroph.

chemolithoheterotroph A microorganism that uses reduced inorganic compounds to derive both energy and electrons; organic molecules are used as the carbon source.

chemolithotroph A microorganism that uses reduced inorganic compounds as a source of energy and electrons.

chemoorganoheterotroph An organism that uses organic compounds as sources of energy, electrons, and carbon for biosynthesis. Also called chemoheterotroph and chemoorganotrophic heterotroph.

chemoreceptors Proteins in the plasma membrane or periplasmic space that bind chemicals and trigger the appropriate chemotaxic response.

chemostat A continuous culture apparatus that feeds medium into the culture vessel at the same rate as medium containing microorganisms is removed; the medium in a chemostat contains one essential nutrient in a limiting quantity.

chemotaxis The pattern of microbial behavior in which the microorganism moves toward chemical attractants and away from repellents.

chemotherapeutic agents Compounds used in the treatment of disease that destroy pathogens or inhibit their growth.

chemotherapy The use of chemical agents to treat disease.

chemotrophs Organisms that obtain energy from the oxidation of chemical compounds.

chickenpox (varicella) A highly contagious skin disease, usually affecting 2- to 7-year-old children, it is caused by the varicella-zoster virus.

chiral (ki′rəl) Having handedness: consisting of one or another stereochemical form.

chitin (ki′tin) A tough, resistant, nitrogen-containing polysaccharide forming the walls of certain fungi, the exoskeleton of arthropods, and the epidermal cuticle of other surface structures of certain protists and animals.

chlamydiae (klə-mid′e-e) Members of the genera *Chlamydia* and *Chlamydiophila:* gram-negative, coccoid cells that reproduce only within the cytoplasmic vesicles of host cells.

chlamydial pneumonia A pneumonia caused by *Chlamydiophila pneumoniae.*

chloramphenicol (klo″ram-fen′ĭ-kol) A broad-spectrum antibiotic that is produced by *Streptomyces venzuelae* or synthetically; it binds to the large ribosomal subunit and inhibits the peptidyl transferase reaction.

chlorophyll The green photosynthetic pigment that consists of a large tetrapyrrole ring with a magnesium atom in the center.

chloroplast A eucaryotic plastid that contains chlorophyll and is the site of photosynthesis.

chlorosomes Elongated, intramembranous vesicles found in the green sulfur and nonsulfur bacteria; they contain light-harvesting pigments. Sometimes called chlorobium vesicles.

cholera An acute infectious enteritis caused by *Vibrio cholerae.*

choleragen (kol′er-ah-gen) The Cholera toxin.

chromatin The DNA-containing portion of the eucaryotic nucleus; the DNA is almost always complexed with histones. It can be very condensed (heterochromatin) or more loosely organized and genetically active (euchromatin).

chromoblastomycosis (kro″mo-blas″to-mi-ko′sis) A chronic fungal skin infection, producing wartlike nodules that may ulcerate; caused by *Phialophora verrucosa* or *Fonsecaea pedrosoi.*

chromogen (kro′me-jen) A colorless substrate that is acted on by an enzyme to produce a colored end product.

chromophore group A chemical group with double bonds that absorbs visible light and gives a dye its color.

chromosomes The bodies that have most or all of the cell's DNA and contain most of its genetic information (mitochondria and chloroplasts also contain DNA).

chronic carrier An individual who harbors a pathogen for a long time.

chytrids A term used to describe the *Chytridiomycota,* which are simple terrestrial and aquatic fungi that produce motile zoospores with single, posterior, whiplash flagella.

cilia Threadlike appendages extending from the surface of some protists that beat rhythmically to propel them; cilia are membrane-bound cylinders with a complex internal array of microtubules, usually in a 9 + 2 pattern.

citric acid cycle *See* tricarboxylic acid (TCA) cycle.

class I MHC molecule *See* major histocompatibility complex.

class II MHC molecule *See* major histocompatibility complex.

class switching The change in immunoglobulin isotype (class) secretion that results during B-cell and then plasma cell differentiation.

classical complement pathway The antibody-dependent pathway of complement activation; it leads to the lysis of pathogens and stimulates phagocytosis and other host defenses.

classification The arrangement of organisms into groups based on mutual similarity or evolutionary relatedness.

clathrin A fibrous protein located on the cytoplasmic side of eucaryotic plasma membranes; functions during clathrin-dependent endocytosis. *See also* endocytosis.

clonal selection The process by which an antigen binds to the best-fitting B-cell receptor, activating that B cell, resulting in the synthesis of antibody and clonal expansion.

clone (1) A group of genetically identical cells or organisms derived by asexual reproduction from a single parent. (2) A DNA sequence that has been isolated and replicated using a cloning vector.

clostridial myonecrosis (klo-strid′e-al mi″o-ne-kro′sis) Death of individual muscle cells caused by clostridia. Also called gas gangrene.

club fungi Common name for the *Basidiomycota.*

clue cells Vaginal epithelial cells covered with bacteria.

cluster of differentiation molecules (CDs) Functional cell surface proteins that can be used to identify leukocyte subpopulations (e.g., interleukin-2 receptor, CD4, CD8, CD25, and intercellular adhesion molecule-1).

coagulase (ko-ag′u-las) An enzyme that induces blood clotting; it is characteristically produced by pathogenic staphylococci.

coated vesicles The clathrin-coated vesicles formed by clathrin-dependent endocytosis.

coccidioidomycosis (kok-sid″e-oi″do-mi-ko′sis) A fungal disease caused by *Coccidioides immitis* that exists in dry, highly alkaline soils. Also known as valley fever, San Joaquin fever, or desert rheumatism.

coccolithophore Photosynthetic protists belonging to the phylum *Stramenopile.* They are characterized by coccoliths—intricate cell walls made of calcite.

coccus (kok′us, pl. **cocci,** kok′si) A roughly spherical procaryotic cell.

code degeneracy The presence of more than one codon for each amino acid.

coding sequences (CDS) In genomic analysis, open reading frames presumed to encode proteins but not tRNA or rRNA.

codon (ko′don) A sequence of three nucleotides in mRNA that directs the incorporation of an amino acid during protein synthesis or signals the stop of translation.

coenocytic (se″no-sit′ik) Refers to a multinucleate cell or hypha formed by repeated nuclear divisions not accompanied by cell divisions.

coenzyme A loosley bound cofactor that often dissociates from the enzyme active site after product has been formed.

coenzyme Q (CoQ, ubiquinone) *See* electron transport chain.

cofactor The nonprotein component of an enzyme; it is required for catalytic activity.

cold sore A lesion caused by the herpes simplex virus; usually occurs on the lips or nares. Also known as a fever blister or herpes labialis.

colicin (kol′ĭ-sin) A plasmid-encoded protein that is produced by enteric bacteria and binds to receptors on the cell envelope of sensitive target bacteria, where it may cause lysis or attack specific intracellular sites such as ribosomes.

coliform A gram-negative, nonsporing, facultative rod that ferments lactose with gas formation within 48 hours at 35°C.

colonization The establishment of a site of microbial reproduction on an inanimate surface or organism without necessarily resulting in tissue invasion or damage.

colony An assemblage of microorganisms growing on a solid surface; the assemblage often is directly visible.

colony forming units (CFU) The number of microorganisms that form colonies when cultured using spread plates or pour plates, an indication of the number of viable microorganisms in a sample.

colony stimulating factor (CSF) A protein that stimulates the growth and development of specific cell populations (e.g., granulocyte-CSF stimulates granulocytes to be made from their precursor stem cells).

colorless sulfur bacteria A diverse group of nonphotosynthetic proteobacteria that can oxidize reduced sulfur compounds such as hydrogen sulfide. Many are chemolithotrophs and derive energy from sulfur oxidation.

col plasmid A plasmid that encodes colicin.

combinatorial biology The creation of novel metabolic processes by transferring and expressing genes between organisms.

cometabolism The modification of a compound not used for growth by a microorganism, which occurs in the presence of another organic material that serves as a carbon and energy source.

commensal Living on or within another organism without injuring or benefiting the other organism.

commensalism A type of symbiosis in which one individual gains from the association (the commensal) and the other is neither harmed nor benefited.

common cold An acute, self-limiting, and highly contagious viral infection of the upper respiratory tract that produces inflammation, profuse discharge, and other symptoms.

common-source epidemic An epidemic characterized by a sharp rise to a peak and then a rapid but not as pronounced decline in the number of individuals infected; it usually involves a single contaminated source from which individuals are infected.

common vehicle transmission The transmission of a pathogen to a host by means of an inanimate medium or vehicle.

communicable disease A disease associated with a pathogen that can be transmitted from one host to another.

community An assemblage of different types of organisms or a mixture of different microbial populations.

comparative genomics Comparison of genomes from different organisms to discern differences and similarities.

compartmentation The differential distribution of enzymes and metabolites to separate cell structures or organelles.

compatible solute A low-molecular-weight molecule used to protect cells against changes in solute concentrations (osmolarity) in their habitat; it can exist at high concentrations within the cell and still be compatible with metabolism and growth.

competent cell A cell that can take up free DNA fragments and incorporate them into its genome during transformation.

competition An interaction between two organisms attempting to use the same resource (nutrients, space, etc.).

competitive exclusion principle Two competing organisms overlap in resource use, which leads to the exclusion of one of the organisms.

competitive inhibitor A molecule that inhibits enzyme activity by binding the enzyme's active site.

complementarity determining regions (CDRs) Hypervariable regions in an immunoglobulin protein that form the three-dimensional epitope binding sites.

complementary DNA (cDNA) A DNA copy of an RNA molecule (e.g., a DNA copy of an mRNA).

complement system A group of plasma proteins that plays a major role in an animal's immune response.

complex medium Culture medium that contains some ingredients of unknown chemical composition.

complex viruses Viruses with capsids having a complex morphology that is neither icosahedral nor helical.

composite transposon A mobile genetic element that contains genes encoding proteins other than those needed for transposition (e.g., antibiotic resistance genes).

compromised host A host with lowered resistance to infection and disease.

concatemer A long DNA molecule consisting of several genomes linked together in a row.

conditional mutations Mutations with phenotypes that are expressed only under certain environmental conditions.

confocal scanning laser microscope (CSLM) A light microscope in which monochromatic laser derived light scans across the specimen at a specific level. Stray light from above and below the plane of focus is blocked out to give an image with excellent contrast and resolution.

congenital (neonatal) herpes An infection of a newborn caused by transmission of a herpesvirus during vaginal delivery.

congenital rubella syndrome A wide array of congenital defects affecting the heart, eyes, and ears of a fetus during the first trimester of pregnancy and caused by *Rubella virus*.

congenital syphilis Syphilis acquired in utero from the mother.

conidiospore (ko-nid′e-o-spōr) An asexual, thin-walled spore borne on hyphae and not contained within a sporangium; it may be produced singly or in chains.

conidium (ko-nid′e-um; pl., **conidia**) *See* conidiospore.

conjugants Complementary mating types among protists that participate in a form of sexual reproduction called conjugation. (620)

conjugation (1) The form of gene transfer and recombination in procaryotes that requires direct cell-to-cell contact. (2) A complex form of sexual reproduction commonly employed by protists.

conjugative plasmid A plasmid that carries the genes that enable its transfer to other bacteria during conjugation (e.g., F plasmid).

conjugative transposon A mobile genetic element that is able to transfer itself from one bacterium to another by conjugation.

conjunctivitis of the newborn *See* ophthalmia neonatorum.

consensus sequence A commonly occurring sequence of nucleotides within a genetic element.

conserved hypothetical proteins In genomic analysis, proteins of unknown function for which at least one match exists in a database.

consortium A physical association of two different organisms, usually beneficial to both organisms.

constant region (C_L and C_H) The part of an antibody molecule that does not vary greatly in amino acid sequence among molecules of the same class, subclass, or type.

constitutive gene A gene that is expressed at nearly the same level at all times.

constitutive mutant A strain that produces an inducible enzyme continually, regardless of need, because of a mutation in either the operator or regulator gene.

constructed wetlands Intentional creation of marshland plant communities and their associated microorganisms for environmental restoration or to purify water by the removal of bacteria, organic matter, and chemicals as the water passes through the aquatic plant communities.

consumer An organism that feeds directly on living or dead animals, by ingestion or by phagocytosis.

contact transmission Transmission of the pathogen by contact of the source or reservoir of the pathogen with the host.

continuous culture system A culture system with constant environmental conditions maintained through continual provision of nutrients and removal of wastes. *See* chemostat and turbidostat.

contractile vacuole (vak′u-ōl) In protists and some animals, a clear fluid-filled vacuole that takes up water from within the cell and then contracts, releasing it to the outside through a pore in a cyclical manner. It functions primarily in osmoregulation and excretion.

convalescent carrier An individual who has recovered from an infectious disease but continues to harbor large numbers of the pathogen.

cooperation A positive but not obligatory interaction between two different organisms.

coral bleaching The loss of photosynthetic pigments by either physiological inhibition or expulsion of the coral photosynthetic endosymbiont, zooxanthellae, a dinoflagellate.

core polysaccharide *See* lipopolysaccharide.

corepressor A small molecule that binds a transcription repressor protein, thereby activating the repressor and inhibiting the synthesis of a repressible enzyme.

cosmid A plasmid vector with lambda phage *cos* sites that can be packaged in a phage capsid; it is useful for cloning large DNA fragments.

crenarchaeol Cyclopentane ring-containing lipids unique to nonthermophilic *Crenarchaeota.*

cristae (kris′te) Infoldings of the inner mitochondrial membrane.

crown gall disease A plant tumor (gall) caused by the α-proteobacterium *Agrobacterium,* most commonly *A. tumefaciens.*

cryptins Antimicrobial peptides produced by Paneth cells in the intestines.

cryptococcosis (krip″to-kok-o′sis) An infection caused by the basidiomycete, *Cryptococcus neoformans,* which may involve the skin, lungs, brain, or meninges.

cryptosporidiosis (krip″to-spo-rid″e-o′sis) Infection with protozoa of the genus *Cryptosporidium.* The most common symptoms are prolonged diarrhea, weight loss, fever, and abdominal pain.

crystallizable fragment (Fc) The stem of the Y portion of an antibody molecule. Cells such as macrophages bind to the Fc region; it also is involved in complement activation.

curing The loss of a plasmid from a cell.

cutaneous anthrax A form of anthrax involving the skin.

cutaneous leishmaniasis *See* leishmaniasis.

cyanobacteria A large group of gram-negative bacteria that carry out oxygenic photosynthesis using a system like that present in photosynthetic eucaryotes.

cyanophycin granules Inclusion bodies found in cyanobacteria that store nitrogen.

cyclic photophosphorylation (fo″to-fos″for-ĭ-la′shun) The formation of ATP when light energy is used to move electrons cyclically through an electron transport chain during photosynthesis.

cyclosporiasis A disease caused by infection with the protozoan parasite *Cyclospora cayetanensis.*

cyst A general term used for a specialized microbial cell enclosed in a wall. Cysts are formed by protists and a few bacteria. They may be dormant, resistant structures formed in response to adverse conditions or reproductive cysts that are a normal stage in the life cycle.

cytochromes (si′to-krōms) *See* electron transport chain.

cytokine (si′to-kīn) A general term for proteins released by a cell in response to inducing stimuli, which are mediators that influence other cells. Produced by lymphocytes, monocytes, macrophages, and other cells.

cytokinesis Processes that apportion the cytoplasm and organelles, synthesize a septum, and divide a cell into two daughter cells during cell division.

cytomegalovirus inclusion disease An infection caused by a cytomegalovirus and marked by nuclear inclusion bodies in enlarged infected cells.

cytopathic effect The observable change that occurs in cells as a result of viral replication.

cytoplasm All material in the cell enclosed by the plasma membrane, with the exception of the nucleus in eucaryotic cells.

cytoproct A specific site in certain protists (e.g., ciliates) where digested material is expelled.

cytosine (si′to-sēn) A pyrimidine 2-oxy-4-aminopyrimidine found in nucleosides, nucleotides, and nucleic acids.

cytoskeleton A network of microfilaments, microtubules, intermediate filaments, and other components in the cytoplasm of eucaryotic cells that helps give them shape.

cytostome A permanent site in a ciliate protist in which food is ingested.

cytotoxic T lymphocyte (CTL) A type of T cell that recognizes antigen in class I MHC molecules. Also called CD8$^+$ cell.

cytotoxin A toxin that has a specific toxic action upon cells; cytotoxins are named according to the cell for which they are specific (e.g., nephrotoxin).

Dane particle A 42 nm spherical particle that is one of three seen in *Hepatitis B virus* infections. The Dane particle is the complete virion.

dark-field microscopy Microscopy in which the specimen is brightly illuminated while the background is dark.

dark reaction *See* photosynthesis.

dark reactivation The excision and replacement of thymine dimers in DNA that occurs in the absence of light.

deamination The removal of amino groups from amino acids.

decimal reduction time (*D* or *D* value) The time required to kill 90% of the microorganisms or spores in a sample at a specified temperature.

decomposer An organism that breaks down complex materials into simpler ones, including the release of simple inorganic products.

defensin Specific peptides produced by neutrophils that permeabilize the outer and plasma membranes of certain microorganisms, thus killing them.

defined (synthetic) medium Culture medium made with components of known composition.

Deltaproteobacteria One of the five classes of proteobacteria.

denaturation A change in either protein shape or nucleic-acid structure.

denaturing gradient gel electrophoresis (DGGE) A technique by which DNA is rendered single-stranded (denatured) while undergoing electrophoresis so that DNA fragments of the same size can be separated according to nucleotide sequence rather than molecular weight.

dendritic cell An antigen-presenting cell that has long membrane extensions resembling the dendrites of neurons.

dendrogram A treelike diagram that is used to graphically summarize mutual similarities and relationships between organisms.

denitrification The reduction of nitrate to gaseous products, primarily nitrogen gas, during anaerobic respiration.

deoxyribonucleic acid (DNA) (de-ok″se-ri″bo-nu-kle′ik) A polynucleotide that constitutes the genetic material of all cellular organisms. It is composed of deoxyribonucleotides connected by phosphodiester bonds.

depth filters Filters composed of fibrous or granular materials that are used to decrease microbial load and sometimes to sterilize solutions.

dermatomycosis (der′ma-to-mi-ko′sis) A fungal infection of the skin; includes the various forms of tinea, sometimes used to specifically refer to athlete's foot (tinea pedis).

dermatophyte (der′mah-to-fit″) A fungus parasitic on the skin.

desensitization To make a sensitized or hypersensitive individual insensitive or nonreactive to a sensitizing agent (e.g., an allergen).

desert crust A crust formed by microbial binding of sand grains in the surface zone of desert soil; primarily involves cyanobacteria.

detergent An organic molecule, other than a soap, that serves as a wetting agent and emulsifier; it is normally used as cleanser, but some may be used as antimicrobial agents.

diatoms Photosynthetic protists with siliceous cell walls called frustules.

diauxic growth (di-awk′sik) A biphasic growth response in which a microorganism, when exposed to two nutrients, initially uses one of them for growth and then alters its metabolism to make use of the second.

differential interference contrast (DIC) microscope A light microscope that employs two beams of plane polarized light. The beams are combined after passing through the specimen and their interference is used to create the image.

differential media Culture media that distinguish between groups of microorganisms based on differences in their growth and metabolic products.

differential staining Staining procedures that divide bacteria into separate groups based on staining properties.

diffusely adhering *E. coli* (DAEC) Strains of *E. coli* that adhere over the entire surface of epithelial cells and usually cause diarrheal disease in immunologically naive and malnourished children.

dikaryotic stage (di-kar-e-ot′ik) In fungi, having pairs of nuclei within cells or compartments. Each cell contains two separate haploid nuclei, one from each parent.

dilution susceptibility tests A method by which antibiotics are evaluated for their ability to inhibit bacterial growth in vitro. A standardized concentration of bacteria is added to serially diluted antibiotics and incubated. Tubes lacking additional bacterial growth suggest antibiotic concentrations that are bacteriocidal or bacteriostatic.

dimorphic fungi Fungi able to switch back and forth between a yeast form and a filamentous form.

dinoflagellate (di″no-flaj′e-lāt) A photosynthetic protist characterized by two flagella used in swimming in a spinning pattern.

diphtheria (dif-the′re-ah) An acute, highly contagious childhood disease that generally affects the membranes of the throat and less frequently the nose; caused by *Corynebacterium diphtheriae*.

diplococcus (dip″lo-kok′us) A pair of cocci.

direct repair A type of DNA repair mechanism in which a damaged nitrogenous base is returned to its normal form (e.g., conversion of a thymine dimer back to two normal thymine bases).

disease A deviation or interruption of the normal structure or function of any part of the body that is manifested by a characteristic set of symptoms and signs.

disease syndrome A set of signs and symptoms that are characteristic of a disease.

disinfectant An agent, usually chemical, that disinfects; normally, it is employed only with inanimate objects.

disinfection The killing, inhibition, or removal of microorganisms that may cause disease. It usually refers to the treatment of inanimate objects with chemicals.

disinfection by-products (DBPs) Chlorinated organic compounds such as trihalomethanes formed during chlorine use for water disinfection. Many are carcinogens.

dissimilatory reduction (e.g., **dissimilatory nitrate reduction** and **dissimilatory sulfate reduction**) The use of a substance as an electron acceptor for an electron transport chain. The acceptor (e.g., sulfate or nitrate) is reduced but not incorporated into organic matter.

dissolved organic matter (DOM) Nutrients that are available in the soluble, or dissolved, state.

DNA gyrase A topoisomerase enzyme that relieves tension generated by the rapid unwinding of DNA during DNA replication.

DNA ligase An enzyme that joins two DNA fragments together through the formation of a new phosphodiester bond.

DNA microarrays Solid supports that have DNA attached in organized arrays and are used to evaluate gene expression.

DNA polymerase (pol-im′er-ās) An enzyme that synthesizes new DNA using a parental nucleic acid strand (usually DNA) as a template.

DNA vaccine A vaccine that contains DNA which encodes antigenic proteins.

double diffusion agar assay (Öuchterlony technique) An immunodiffusion reaction in which both antibody and antigen diffuse through agar to form stable immune complexes, which can be observed visually.

double-stranded break model A model for reciprocal homologous recombination that involves the formation of breaks in both strands of DNA and the activity of the RecA protein.

doubling time *See* generation time.

DPT (diphtheria-pertussis-tetanus) vaccine A vaccine containing three antigens that is used to immunize people against diphtheria, pertussis (whooping cough), and tetanus.

droplet nuclei Small particles (0 to 4 μm in diameter) that represent what is left from the evaporation of larger particles (10 μm or more in diameter) called droplets.

D value *See* decimal reduction time.

early mRNA Messenger RNA produced early in a viral infection that codes for proteins needed to take over the host cell and manufacture viral nucleic acids.

East African sleeping sickness *See* trypanosomiasis.

Ebola hemorrhagic fever (a′bo-lə) An often fatal acute infection caused by Ebola virus and characterized by fever, bleeding, and shock.

ecosystem A self-regulating biological community and its associated physical and chemical environment.

ectomycorrhizae A mutualistic association between fungi and plant roots in which the fungus surrounds the root tip with a sheath.

ectoplasm In some protists, the cytoplasm directly under the cell membrane (plasmalemma) is divided into an outer gelatinous region, the ectoplasm, and an inner fluid region, the endoplasm.

ectosymbiont An organism that lives on or outside the body of another organism in a symbiotic association.

ectosymbiosis A type of symbiosis in which one organism remains outside of the other organism.

effacing lesion The type of lesion caused by enteropathogenic strains of *E. coli* (EPEC) when the bacteria destroy the brush border of intestinal epithelial cells.

ehrlichiosis (ar-lik″e-o′sis) A tick-borne rickettsial disease caused by *Ehrlichia chaffeensis* that resembles Rocky Mountain spotted fever.

electron acceptor A compound that accepts electrons in an oxidation-reduction reaction. Often called an oxidizing agent or oxidant.

electron donor An electron donor in an oxidation-reduction reaction. Often called a reducing agent or reductant.

electron transport chain (ETC) A series of electron carriers that operate together to transfer electrons from donors to acceptors such as oxygen. Molecules involved in electron transport include nicotinamide adenine dinucleotide (NAD^+), NAD phosphate ($NADP^+$), cytochromes, heme proteins, non-heme proteins (e.g., iron-sulfur proteins and ferredoxin), coenzyme Q, flavin adenine dinucleotide (FAD), and flavin mononucleotide (FMN). Also called an electron transport system.

electrophoresis (e-lek″tro-fo-re′sis) A technique that separates substances through differences in their migration rate in an electrical field.

electroporation The application of an electric field to create temporary pores in the plasma membrane in order to render a cell temporarily transformation competent.

elementary body (EB) A small, dormant body that serves as the agent of transmission between host cells in the chlamydial life cycle.

elongation cycle The cycle in protein synthesis that results in the addition of an amino acid to the growing end of a peptide chain.

elongation factors Proteins that function in the elongation cycle of protein synthesis.

Embden-Meyerhof pathway (em′den mi′er-hof) A pathway that degrades glucose to pyruvate (i.e., a glycolytic pathway).

encystment The formation of a cyst.

endemic disease A disease that is commonly or constantly present in a population, usually at a relatively steady low frequency.

endemic (murine) typhus A form of typhus fever caused by *Rickettsia typhi* that occurs sporadically in individuals who come into contact with rats and their fleas.

endergonic reaction (end″er-gon′ik) A reaction that does not spontaneously go to completion as written; the standard free energy change is positive, and the equilibrium constant is less than one.

endocytic pathways *See* endocytosis.

endocytosis The process in which a cell takes up solutes or particles by enclosing them in vesicles pinched off from its plasma membrane. It often occurs at regions of the plasma membrane coated by proteins such as clathrin and caveolin. Endocytosis involving these proteins is called clathrin-dependent endocytosis and caveolae-dependent endocytosis, respectively. When the substance endocytosed first binds to a receptor, the process is called receptor-mediated endocytosis.

endogenote (en″do-je′nōt) The genome of a procaryotic cell that acts as a recipient during horizontal gene transfer; transferred DNA may integrate into the endogenote.

endogenous infection An infection by a member of an individual's own normal body microbiota.

endogenous pyrogen A host-derived chemical mediator (e.g., interleukin-1) that acts on the hypothalamus, stimulating a rise in core body temperature (i.e., it stimulates the fever response).

endomycorrhizae Referring to a mutualistic association of fungi and plant roots in which the fungus penetrates into the root cells.

endophyte A microorganism living within a plant, but not necessarily parasitic on it.

endoplasm *See* ectoplasm.

endoplasmic reticulum (ER) A system of membranous tubules and flattened sacs (cisternae) in the cytoplasm of eucaryotic cells. Rough endoplasmic reticulum (RER) bears ribosomes on its surface; smooth endoplasmic reticulum (SER) lacks them.

endosome A membranous vesicle formed by endocytosis. It undergoes a maturational process that starts with early endosomes and proceeds to late endosomes and finally lysosomes.

endospore An extremely heat- and chemical-resistant, dormant, thick-walled spore that develops within some gram-positive bacteria. It has a complex structure that includes (from outermost to innermost) exosporium, spore coat, cortex, spore cell wall, and spore core.

endosymbiont An organism that lives within the body of another organism.

endosymbiosis A type of symbiosis in which one organism is found within another organism.

endosymbiotic hypothesis The theory that mitochondria, hydrogenosomes, and chloroplasts arose when bacteria established an endosymbiotic relationship with ancestral procaryotic cells that then evolved into organelles.

endotoxin The lipid A component of lipopolysaccharide (LPS) that is released from gram-negative bacterial cell walls when the bacteria die. Nanogram quantities can induce fever, activate complement and coagulation cascades, act as a mitogen to B cells, and stimulate cytokine release from a variety of cells. Systemic effects of endotoxin are referred to as endotoxic shock.

end product inhibition *See* feedback inhibition.

energy The capacity to do work or cause particular changes.

enhancer A site in the DNA to which a eucaryotic activator protein binds.

enology The science of wine making.

enrichment culture The growth of specific microbes from natural samples by including selective features to promote gowth of the desired microbes while selecting against other microbes that might be in the same sample.

enteric bacteria (enterobacteria) Members of the family *Enterobacteriaceae*; also can refer to bacteria that live in the intestinal tract.

enteroaggregative *E. coli* (EAggEC) A toxin-producing strain of *E. coli* associated with persistent watery, bloody diarrhea and cramping in young children. EAggEC adhere to the intestinal mucosa and cause nonbloody diarrhea without invading or causing inflammation.

enterohemorrhagic *E. coli* (EHEC) (en′tər-o-hem″ə-raj′ik) EHEC strains of *E. coli* (O157:H7) produce several cytotoxins that provoke fluid secretion and diarrhea.

enteroinvasive *E. coli* (EIEC) EIEC strains of *E. coli* cause diarrhea by penetrating and binding to the intestinal epithelial cells. EIEC may also produce a cytotoxin and enterotoxin.

enteropathogenic *E. coli* (EPEC) EPEC strains of *E. coli* attach to the brush border of intestinal epithelial cells and cause effacing lesions that lead to diarrhea.

enterotoxigenic *E. coli* (ETEC) ETEC strains of *E. coli* that produce two plasmid-encoded enterotoxins: heat-stable enterotoxin (ST) and heat-labile enterotoxin (LT).

enterotoxin A toxin specifically affecting the cells of the intestinal mucosa, causing vomiting and diarrhea.

enthalpy Heat content of a system.

Entner-Doudoroff pathway A pathway that converts glucose to pyruvate and glyceraldehyde 3-phosphate by producing 6-phosphogluconate and then dehydrating it (i.e., a glycolytic pathway).

entropy A measure of the randomness or disorder of a system; a measure of that part of the total energy in a system that is unavailable for useful work.

envelope In virology, an outer membranous layer that surrounds the nucleocapsid in some viruses.

enveloped virus A virus that consists of a nucleocapsid enclosed within an envelope.

environmental genomics *See* metagenomics.

enzyme A protein catalyst with specificity for the reaction catalyzed and its substrates.

enzyme-linked immunosorbent assay (ELISA) A serological assay in which bound antigen or antibody is detected by another antibody that is conjugated to an enzyme. The enzyme converts a colorless substrate to a colored product reporting the antibody capture of the antigen.

eosinophil (e″o-sin′o-fil) A phagocytic, polymorphonuclear leukocyte that has a two-lobed nucleus and cytoplasmic granules that stain yellow-red.

epidemic A disease that suddenly increases in occurrence above the normal level in a given population.

epidemic (louse-borne) typhus A disease caused by *Rickettsia prowazekii* that is transmitted from person to person by the body louse.

epidemiologist A person who specializes in epidemiology.

epidemiology The study of the factors determining and influencing the frequency and distribution of disease, injury, and other health-related events and their causes in defined human populations.

epilimnion The upper, warmer layer of water in a stratified lake.

epiphyte An organism that grows on the surface of plants.

episome A plasmid that can exist either independently of the host cell's chromosome or be integrated into it.

epitheca (ep″ĭ-the′kah) The larger of two halves of a diatom frustule (shell).

epitope (ep′i-tōp) An area of an antigen that stimulates the production of, and combines with, specific antibodies; also known as the antigenic determinant site.

Epsilonproteobacteria One of the five classes of proteobacteria.

equilibrium The state of a system in which no net change is occurring and free energy is at a minimum; in a chemical reaction at equilibrium, the rates in the forward and reverse directions exactly balance each other out.

equilibrium constant (Keq) A value that relates the concentration of reactants and products to each other when a reaction is at equilibrium.

ergot The dried sclerotium of *Claviceps purpurea*. Also, an ascomycete that parasitizes rye and other higher plants, causing an animal disease called ergotism.

ergotism The disease or toxic condition caused by eating grain infected with ergot.

erysipelas (er″ĭ-sip′ĕ-las) An acute inflammation of the dermal layer of the skin, occurring primarily in infants and persons over 30 years of age with a history of streptococcal sore throat.

erythema infectiosum (er″ə-the′-mə) A disease in children caused by *Human parvovirus B19*, sometimes called fifth disease.

erythromycin (ĕ-rith″ro-mi′sin) An intermediate spectrum macrolide antibiotic produced by *Saccharopolyspora erythraea*.

Eucarya The domain that contains organisms composed of eucaryotic cells with primarily glycerol fatty acyl diesters in their membranes and eucaryotic rRNA.

eucaryotic cells Cells that have a membrane-delimited nucleus and differ in many other ways from procaryotic cells; protists, fungi, plants, and animals are all eucaryotic.

euglenids (u-gle′nids) A group of protists (super group *Excavata*) that includes chemoorganotrophs and photoautotrophs with chloroplasts containing chlorophyll *a* and *b*. They usually have a stigma and one or two flagella emerging from an anterior reservoir.

Eumycetozoa Protists called cellular and extracellular slime molds. *See also* acellular slime mold, cellular slime mold.

eumycotic mycetoma (mi″se-to′mah) *See* maduromycosis.

eutrophic A nutrient-enriched environment.

eutrophication The enrichment of an aquatic environment with nutrients.

evolutionary distance A quantitative indication of the number of positions that differ between two aligned macromolecules, and presumably a measure of evolutionary similarity between molecules and organisms.

excision repair A type of DNA repair mechanism in which a section of a strand of damaged DNA is excised and replaced, using the complementary strand as a template. Two types are recognized: base excision repair and nucleotide excision repair.

excystment The escape of one or more cells or organisms from a cyst.

exergonic reaction (ek″ser-gon′ik) A reaction that spontaneously goes to completion as written; the standard free energy change is negative, and the equilibrium constant is greater than one.

exfoliative toxin (exfoliatin) An exotoxin produced by *Staphylococcus aureus* that causes the separation of epidermal layers and the loss of skin surface layers. It produces the symptoms of the scalded skin syndrome.

exit site (E site) The location on a ribosome to which an empty (uncharged) tRNA moves from the P site before it finally leaves during protein synthesis.

exoenzymes Enzymes that are secreted by cells.

exogenote (eks″o-je′nōt) The piece of donor DNA that enters a procaryotic cell during horizontal gene transfer.

exon The region in a split (interrupted) gene that codes for pre-mRNA and is retained in mature mRNA.

exospore A spore formed outside the mother cell; often observed in actinobacteria.

exotoxin A heat-labile, toxic protein produced by a bacterium and usually released into the bacterium's surroundings.

exponential (log) phase The phase of the growth curve during which the microbial population is growing at a constant and maximum rate, dividing and doubling at regular intervals.

expressed sequence tag (EST) A partial gene sequence unique to a gene that can be used in microarray analysis to determine when a gene is expressed.

expression vector A cloning vector used to express a recombinant gene in host cells; the gene is transcribed and its protein synthesized.

exteins Polypeptide sequences of precursor self-splicing proteins that are joined together during formation of the final, functional protein. They are separated from one another by intein sequences.

extinction culture technique An approach used to study community structure in aquatic habitats.

extreme environment An environment in which physical factors such as temperature, pH, salinity, and pressure are outside of the normal range for growth of most microorganisms; these conditions allow unique organisms to survive and function.

extreme halophiles *See* halobacteria.

extremophiles Microorganisms that grow in extreme environments.

extrinsic factor An environmental factor such as temperature that influences microbial growth in food.

facilitated diffusion Diffusion across the plasma membrane that is aided by a carrier protein.

facultative anaerobes Microorganisms that do not require oxygen for growth, but grow better in its presence.

facultative psychrophile (fak′ul-ta″tiv si′kro fi″l) *See* psychrotroph.

Fas-FasL pathway One of two pathways used by cytotoxic T cells to kill target cells (e.g., those infected with a virus). Fas is another name for a protein called CD95 receptor, which is located on the target cell. FasL is the ligand to the CD95 receptor and is found on the cytotoxic T cell.

fatty acid synthase The multienzyme complex that makes fatty acids.

fecal coliform Coliforms whose normal habitat is the intestinal tract and that can grow at 44.5°C. They are used as indicators of fecal pollution of water.

fecal enterococci (en″ter-o-kok′si) Enterococci found in the intestine of humans and other warm-blooded animals.

feedback inhibition A negative feedback mechanism in which an end product inhibits the activity of an enzyme in the pathway leading to its formation.

fermentation An energy-yielding process in which an organic molecule is oxidized without an exogenous electron acceptor. Usually pyruvate or a pyruvate derivative serves as the electron acceptor.

fermenter In industrial microbiology, the large vessel used to culture microorganisms.

ferredoxin *See* electron transport chain.

fever A complex physiological response to disease mediated by pyrogenic cytokines and characterized by a rise in core body temperature and activation of the immune system.

fever blister *See* cold sore.

F factor The fertility factor, a plasmid that carries genes for bacterial conjugation and makes its *E. coli* host the gene donor during conjugation.

fifth disease *See* erythema infectiosum.

filopodia Long, narrow pseudopodia found in certain amoeboid protists.

fimbria (fim′bre-ah; pl., **fimbriae**) A fine, hair-like protein appendage on many gram-negative bacteria, some gram-positive bacteria, and some fungi. They attach cells to surfaces, and some are involved in twitching motility.

final host The host on or in which a parasite either attains sexual maturity or reproduces.

first law of thermodynamics Energy can be neither created nor destroyed (although it can be changed in form or redistributed).

fixation The process in which the internal and external structures of cells and organisms are preserved and fixed in position. Two methods are commonly used: heat fixation and chemical fixation.

flagellin (flaj′ĕ-lin) The protein used to construct the filament of a bacterial flagellum.

flagellum (flah-jel′um; pl., **flagella**) A thread-like appendage on many cells that is responsible for their motility. Bacterial flagella are composed of a flagellar filament, flagellar hook, and basal body.

flavin adenine dinucleotide (FAD) (fla′vin ad′ĕ-nēn) An electron carrying cofactor often involved in energy production (for example, in the tricarboxylic acid cycle and the β-oxidation pathway). *See also* electron transport chain.

flavin mononucleotide *See* electron transport chain.

flow cytometry A tool for defining and enumerating cells using a capillary tube to control cell movement and a laser to detect cell size and morphology.

fluid mosaic model The model of cell membranes in which the membrane is a lipid bilayer with integral proteins buried in the lipid and peripheral proteins more loosely attached to the membrane surface.

fluorescent in situ hybridization (FISH) A technique for identifying certain genes or organisms in which specific DNA fragments are labeled with fluorescent dye and hybridized to the chromosomes of interest.

fluorescence microscope A microscope that exposes a specimen to light of a specific wavelength and then forms an image from the fluorescent light produced.

fluorescent light The light emitted by a substance when it is irradiated with light of a shorter wavelength.

fluorochrome A fluorescent dye.

fomite (pl., **fomites**) An object that is not in itself harmful but is able to harbor and transmit pathogenic organisms. Also called fomes.

food-borne infection Gastrointestinal illness caused by ingestion of microorganisms, followed by their growth within the host.

food intoxication Food poisoning caused by microbial toxins produced in a food prior to consumption. The presence of living bacteria is not required.

food poisoning A general term usually referring to a gastrointestinal disease caused by the ingestion of food contaminated by pathogens or their toxins.

F₁ particle Component of ATP synthase.

F′ plasmid An F plasmid that carries some bacterial genes and transmits them to recipient cells when the F′ cell carries out conjugation.

forward mutation A mutation from the wild type to a mutant form.

fragmentation A type of asexual reproduction among filamentous microbes in which hyphae break into two or more parts, each of which forms new hyphae.

frameshift mutations Mutations arising from the loss or gain of a base or DNA segment, leading to a change in the codon reading frame and thus a change in the amino acids incorporated into protein.

free energy change The total energy change in a system that is available to do useful work as the system goes from its initial state to its final state at constant temperature and pressure.

fruiting body A specialized structure that holds sexually or asexually produced spores; found in fungi and in some bacteria (e.g., the myxobacteria).

frustule (frus′tūl) A silicified cell wall in the diatoms.

fueling reactions Chemical reactions that supply the ATP, precursor metabolites, and reducing power needed for biosynthesis.

fumonisins A family of toxins produced by molds belonging to the genus *Fusarium*. It primarily affects corn and it is known to be hepato- and nephrotoxic in animals.

functional genomics Genomic analysis concerned with determining the way a genome functions.

functional proteomics Analysis concerned with determining the function of proteins produced by a cell.

fungicide An agent that kills fungi.

fungistatic Inhibiting the growth and reproduction of fungi.

fungus (pl., **fungi**) Achlorophyllous, heterotrophic, spore-bearing eucaryotes with absorptive nutrition and a walled thallus.

gametangium (gam-e˘-tan′je-um; pl., **gametangia**) A structure that contains gametes or in which gametes are formed.

Gammaproteobacteria One of the five classes of proteobacteria.

gamonts Gametic cells formed by protists when they undergo sexual reproduction.

gas gangrene A type of gangrene that arises from lacerated wounds infected by anaerobic bacteria, especially species of *Clostridium*. As the bacteria grow, they release toxins and ferment carbohydrates to produce carbon dioxide and hydrogen gas.

gastritis Inflammation of the stomach.

gastroenteritis An acute inflammation of the lining of the stomach and intestines, characterized by anorexia, nausea, diarrhea, abdominal pain, and weakness. Also called enterogastritis.

gastrointestinal anthrax The intestinal form of anthrax characterized by nausea, loss of appetite, vomiting, fever, and followed by abdominal pain, vomiting of blood, and severe diarrhea.

gas vacuole A gas-filled vacuole found in cyanobacteria and some other aquatic bacteria and archaea that provides flotation. It is composed of gas vesicles, which are made of protein.

gas vesicle *See* gas vacuole.

gel electrophoresis The separation of molecules according to charge and size through a gel matrix.

gene A DNA segment or sequence that codes for a polypeptide, rRNA, or tRNA.

generalized transduction The transfer of any part of a procaryotic genome when the DNA fragment is packaged within a virus capsid by mistake.

general secretion pathway (GSP) *See* Sec-dependent pathway.

generation (doubling) time The time required for a microbial population to double in number.

genetic engineering The deliberate modification of an organism's genetic information by directly changing its nucleic acid genome.

genital herpes A sexually transmitted disease caused by herpes simplex virus type 2.

genome The full set of genes present in a cell or virus; all the genetic material in an organism.

genome annotation The process of determining the location and potential function of specific genes and genetic elements in a genome sequence.

genome fusion hypothesis A hypothesis that seeks to explain the origin of the nucleus. It posits that certain archaeal and bacterial genes were combined to form a single eucaryotic genome.

genomic fingerprinting A series of techniques based on restriction enzyme digestion patterns that enable the comparison of microbial species and strains and is thus useful in taxonomic identification.

genomic library The collection of clones that contains fragments which represent the complete genome of an organism.

genomic reduction The decrease in genomic information that occurs over evolutionary time as an organism or organelle becomes increasingly dependent on another cell or a host organism.

genomics The study of the molecular organization of genomes, their information content, and the gene products they encode.

genotype The specific set of alleles carried in the genome of an organism.

genotypic classification The use of genetic data to construct a classification scheme for the identification of an unknown species or the phylogeny of a group of microbes.

genus A well-defined group of one or more species that is clearly separate from other organisms.

German measles *See* rubella.

germicide An agent that kills pathogens and many nonpathogens but not necessarily bacterial endospores.

germination The stage following spore activation in which the spore breaks its dormant state. Germination is followed by outgrowth.

Ghon complex (gon) The initial focus of infection in primary pulmonary tuberculosis.

giardiasis (je″ar-di′ah-sis) A common intestinal disease caused by the parasitic protozoan *Giardia intestinalis*.

gliding motility A type of motility in which a microbial cell glides along a solid surface.

global regulatory systems Regulatory systems that simultaneously affect many genes.

glomeromycetes Common name for members of the fungal taxon *Glomeromycota*.

glomerulonephritis (glo-mer″u-lo-ne˘-fri′tis) An inflammatory disease of the renal glomeruli.

gluconeogenesis (gloo″ko-ne″o-jen′e-sis) The synthesis of glucose from noncarbohydrate precursors such as lactate and amino acids.

glutamine synthetase–glutamate synthase (GS-GOGAT) system A mechanism used by many microbes to incorporate ammonia.

glycocalyx (gli″ko-kal′iks) A network of polysaccharides extending from the surface of bacteria and other cells.

glycogen (gli′ko-jen) A highly branched polysaccharide containing glucose, which is used to store carbon and energy.

glycolysis (gli-kol′ĭ-sis) The conversion of glucose to pyruvic acid by use of the Embden-Meyerhof pathway, pentose phosphate pathway, or Entner-Doudoroff pathway.

glycolytic pathway A pathway that converts glucose to pyruvic acid (e.g., Embden-Meyerhof pathway).

glyoxylate cycle A modified tricarboxylic acid cycle in which the decarboxylation reactions are bypassed by the enzymes isocitrate lyase and malate synthase; it is used to convert acetyl-CoA to succinate and other metabolites.

gnotobiotic (no″to-bi-ot′ik) Animals that are germ-free (microorganism free) or live in association with one or more known microorganisms.

Golgi apparatus (gol′je) A membranous eucaryotic organelle composed of stacks (dictyosomes) of flattened sacs (cisternae) that is involved in packaging and modifying materials for secretion and many other processes.

gonococci (gon′o-kok′si) Bacteria of the species *Neisseria gonorrhoeae*—the organism causing gonorrhea.

gonorrhea An acute infectious sexually transmitted disease of the mucous membranes of the genitourinary tract, eye, rectum, and throat. It is caused by *Neisseria gonorrhoeae*.

Gram stain A differential staining procedure that divides bacteria into gram-positive and gram-negative groups based on their ability to retain crystal violet when decolorized with an organic solvent such as ethanol.

granulocyte A type of white blood cell that stores preformed molecules (enzymes and antimicrobial proteins) in vacuoles near the cell membrane.

granuloma Term applied to nodular inflammatory lesions containing phagocytic cells.

greenhouse gases Gases (e.g., CO_2, CH_4) released from Earth's surface through chemical and biological processes that interact with the chemicals in the stratosphere to decrease radiational cooling of Earth. This leads to global warming.

green nonsulfur bacteria Anoxygenic photosynthetic bacteria that contain bacteriochlorophylls *a* and *c;* usually photoheterotrophic and display gliding motility. Include members of the phylum *Chloroflexi*.

green sulfur bacteria Anoxygenic photosynthetic bacteria that contain bacteriochlorophylls *a*, plus *c, d,* or *e;* photolithoautotrophic; use H_2, H_2S, or S as electron donor. Include members of the phylum *Chlorobi*.

griseofulvin (gris″e-o-ful′vin) An antibiotic from *Penicillium griseofulvum* given orally to treat chronic dermatophytic infections of skin and nails.

group A streptococcus (GAS) A gram-positive, coccus-shaped bacterium often found in the throat and on the skin of humans, having the A group of surface carbohydrate. *See* streptococcal pharyngitis.

group B streptococcus (GBS) A gram-positive, coccus-shaped bacterium found occasionally on mucous membranes of humans, having the B group of surface carbohydrate.

group translocation A transport process in which a molecule is moved across a membrane by carrier proteins while being chemically altered at the same time (e.g., phosphoenolpyruvate: sugar phosphotransferase system).

growth An increase in cellular constituents or an increase in population size.

growth factors Organic compounds that must be supplied in the diet for growth because they are essential cell components or precursors of such components and cannot be synthesized by the organism.

guanine (gwan′in) A purine derivative, 2-amino-6-oxypurine found in nucleosides, nucleotides, and nucleic acids.

gumma (gum′ah) A soft, gummy tumor occurring in tertiary syphilis.

gut-associated lymphoid tissue (GALT) The defensive lymphoid tissue present in the intestines.

halobacteria (extreme halophiles) A group of archaea that depend on high NaCl concentrations for growth and do not survive at a concentration below about 1.5 M NaCl.

halophile A microorganism that requires high levels of sodium chloride for growth.

halotolerant The ability to withstand large changes in salt concentration.

hantavirus pulmonary syndrome (HPS) An infectious lung disease caused by at least four different hantaviruses.

hapten A molecule not immunogenic by itself that, when coupled to a macromolecular carrier, can elicit antibodies directed against itself.

harborage transmission The mode of transmission in which an infectious organism does not undergo morphological or physiological changes within the vector.

harmful algal bloom (HAB) In aquatic ecosystems, the growth of a single population of phototroph, either a protist (e.g., diatom, dinoflagellate) or a cyanobacterium, that produces a toxin that is poisonous to other organisms, sometimes including humans.

Hartig net The area of nutrient exchange between ectomycorrhizal fungal hyphae and plant host cells.

healthy carrier An individual who harbors a pathogen but is not ill.

heat fixation *See* fixation.

heat-shock proteins Proteins produced when cells are exposed to high temperatures or other stressful conditions. They protect the cells from damage and often aid in the proper folding of proteins.

helical capsid A viral capsid in the form of a helix.

helicases Enzymes that use ATP energy to unwind DNA ahead of the replication fork.

hemadsorption The adherence of red blood cells to the surface of something, such as another cell or a virus.

hemagglutination (hem″ah-gloo″tĭ-na′shun) The agglutination of red blood cells by hemagglutinins (e.g., antibodies or components of virus capsids).

hemagglutination assay A testing procedure based on a hemagglutination reaction.

hematopoesis The process by which blood cells develop into specific lineages from stem cells. Red and white blood cells and platelets develop from this process.

hemolysin (he-mol′ĭ-sin) A substance that causes hemolysis.

hemolysis The disruption of red blood cells and release of their hemoglobin. There are several types of hemolytic reactions when bacteria grow on blood agar. In α-hemolysis, a greenish zone of incomplete hemolysis forms around the colony. A clear zone of complete hemolysis without any obvious color change is formed during β-hemolysis.

hepadnaviruses Common name for viruses belonging to the family *Hepadnaviridae*.

hepatitis Any infection that results in inflammation of the liver. Also refers to liver inflammation.

hepatitis A (formerly infectious hepatitis) A type of hepatitis that is transmitted by fecal-oral contamination; it primarily affects children and young adults, especially in environments where there is poor sanitation and overcrowding. It is caused by the *Hepatitis A virus*.

hepatitis B (formerly serum hepatitis) A type of hepatitis caused by *Hepatitis B virus* (HBV) formerly called the "Dane particle." The virus is transmitted by body fluids.

hepatitis C A liver disease caused by a virus that is spread through infected blood.

hepatitis D (formerly delta hepatitis) The liver diseases caused by the hepatitis D virusoid in those individuals already infected with the *Hepatitis B virus*.

herd immunity The resistance of a population to infection and spread of an infectious agent due to the immunity of a high percentage of the population.

herpes labialis *See* cold sore.

herpes zoster A reactivated form of chickenpox that is also called shingles.

herpetic keratitis (her-pet′ik ker″ah-ti′tis) An inflammation of the cornea and conjunctiva of the eye resulting from a herpes simplex virus infection.

heterocysts Specialized cells of cyanobacteria that are the sites of nitrogen fixation.

heteroduplex DNA A double-stranded stretch of DNA formed by two slightly different strands that are not completely complementary.

heterokont flagella A pattern of flagellation found in the protist subdivision *Stramenopile*, featuring two flagella, one extending anteriorly and the other posteriorly.

heterolactic fermenters Microorganisms that ferment sugars to form lactate and other products such as ethanol and CO_2.

heterologous gene expression The cloning, transcription, and translation of a gene that has been introduced (cloned) into an organism that normally does not possess the gene.

heterotroph An organism that uses reduced, preformed organic molecules as its principal carbon source.

heterotrophic nitrification Nitrification carried out by chemoheterotrophic microorganisms.

hexon or **hexamer** A virus capsomer composed of six protomers.

hexose monophosphate pathway *See* pentose phosphate pathway.

Hfr conjugation Conjugation involving an Hfr strain and an F⁻ strain.

Hfr strain A strain of *Escherichia coli* that donates its genes with high frequency to a recipient cell during conjugation because the F factor is integrated into the donor's chromosome.

hierarchical cluster analysis The organization of microarray data such that induced and repressed genes are grouped separately.

high-efficiency particulate (HEPA) filter A depth filter constructed to remove 99.97% of particles that are ≤0.3 μm.

high-throughput screening (HTS) A system that combines liquid handling devices, robotics, computers, data processing, and a sensitive detection system to screen thousands of compounds for a single capability. It is often used by pharmaceutical companies to identify natural products that have potentially useful applications.

histatin An antimicrobial peptide composed of 24 to 38 amino acids, heavily enriched with histidine, that targets fungal mitochondria.

histone A small basic protein with large amounts of lysine and arginine that is associated with eucaryotic DNA in chromatin. Related proteins are observed in many archaeal species, where they form archaeal nucleosomes.

histoplasmosis A systemic fungal infection caused by *Histoplasma capsulatum* var *capsulatum*.

HIV protease inhibitor A drug that prevents HIV protease enzymes from cleaving HIV polyproteins into mature proteins required for HIV virion assembly.

holdfast A structure produced by some bacteria (e.g., *Caulobacter*) that attaches them to a solid object.

holoenzyme A complete enzyme consisting of the apoenzyme plus a cofactor.

holozoic nutrition Acquisition of nutrients (e.g., bacteria) by endocytosis and the subsequent formation of a food vacuole or phagosome.

homolactic fermenters Organisms that ferment sugars almost completely to lactic acid.

homologous recombination Recombination involving two DNA molecules that are very similar in nucleotide sequence; it can be reciprocal or nonreciprocal.

hopanoids Lipids found in bacterial membranes that are similar in structure and function to the sterols found in eucaryotic membranes.

horizontal (lateral) gene transfer (HGT/LGT) The process in which genes are transferred from one mature, independent organism to another. In procaryotes, transformation, conjugation, and transduction are the mechanisms by which HGT can occur.

hormogonia Small motile fragments produced by fragmentation of filamentous cyanobacteria; used for asexual reproduction and dispersal.

host An organism that harbors another organism.

host-parasite relationship The symbiosis between a pathogen and its host. The term parasite is used to imply pathogenicity in this context.

host restriction The degradation of foreign genetic material by nucleases after the genetic material enters a host cell.

human immunodeficiency virus (HIV) A lentivirus of the family *Retroviridae* that is the cause of AIDS.

human leukocyte antigen complex (HLA) A collection of genes on chromosome 6 that encodes proteins involved in host defenses; also called major histocompatibility complex.

humoral (antibody-mediated) immunity The type of immunity that results from the presence of soluble antibodies in blood and lymph.

hybridoma (hi″brĭ-do′mah) A fast-growing cell line produced by fusing a cancer cell (myeloma) to another cell, such as an antibody-producing cell.

hydrogen hypothesis An hypothesis that considers the origin of the eucaryotes through the development of the hydrogenosome. It suggests the organelle arose as the result of an endosymbiotic anaerobic bacterium that produced CO_2 and H_2 as the products of fermentation.

hydrogenosome An organelle found in some anaerobic protists that produce ATP by fermentation.

hydrophilic A polar substance that has a strong affinity for water (or is readily soluble in water).

hydrophobic A nonpolar substance lacking affinity for water (or which is not readily soluble in water).

3-hydroxypropionate cycle A pathway used by green nonsulfur bacteria and some archaea to fix CO_2.

hyperendemic disease A disease that has a gradual increase in occurrence beyond the endemic level but not at the epidemic level in a given population; also may refer to a disease that is equally endemic in all age groups.

hyperferremia Excessive iron in the blood.

hypersensitivity A condition in which the body reacts to an antigen with an exaggerated immune response that usually harms the individual. Also termed an allergy.

hyperthermophile (hi″per-ther′mo-fĭl) A microbe that has its growth optimum between 85°C and about 120°C. Hyperthermophiles usually do not grow well below 55°C.

hypha (hi′fah; pl., **hyphae**) The unit of structure of most fungi and some bacteria; a tubular filament.

hypoferremia Deficiency of iron in the blood.

hypolimnion The colder, bottom layer of water in a stratified lake.

hypotheca The smaller half of a diatom frustule.

hypoxic (hi pok′sik) Having a low oxygen level.

icosahedral capsid A viral capsid that has the shape of a regular polyhedron having 20 equilateral triangular faces and 12 corners.

identification The process of determining that a particular isolate or organism belongs to a recognized taxon.

idiotype A specific immunoglobulin that differs from others in the hypervariable region due to mutations that occur during B cell development.

IgA Immunoglobulin A; the class of immunoglobulins that is present in dimeric form in many body secretions (e.g., saliva, tears, and bronchial and intestinal secretions) and protects body surfaces. IgA also is present in serum.

IgD Immunoglobulin D; the class of immunoglobulins found on the surface of many B lymphocytes; thought to serve as an antigen receptor in the stimulation of antibody synthesis.

IgE Immunoglobulin E; the immunoglobulin class that binds to mast cells and basophils, and is responsible for type I or anaphylactic hypersensitivity reactions such as hay fever and asthma. IgE is also involved in resistance to helminth parasites.

IgG Immunoglobulin G; the predominant immunoglobulin class in serum. Has functions such as neutralizing toxins, opsonizing bacteria, activating complement, and crossing the placenta to protect the fetus and neonate.

IgM Immunoglobulin M; the class of serum antibody first produced during an infection. It is a large, pentameric molecule that is active in agglutinating pathogens and activating complement. The monomeric form is present on the surface of some B lymphocytes.

immobilization The incorporation of a simple, soluble substance into the body of an organism, making it unavailable for use by other organisms.

immune complex The product of an antigen-antibody reaction, which may also contain components of the complement system.

immune surveillance The process by which cells of the immune system police the host for non-self antigens.

immune system The defensive system in an animal consisting of the nonspecific (innate) and specific (adaptive) immune responses. It is composed of widely distributed cells, tissues, and organs that recognize foreign substances and microorganisms and acts to neutralize or destroy them.

immunity Refers to the overall general ability of a host to resist a particular disease; the condition of being immune.

immunization The deliberate introduction of foreign materials into a host to stimulate an adaptive immune response. *See* vaccine.

immunoblotting The electrophoretic transfer of proteins from polyacrylamide gels to filters to demonstrate the presence of specific proteins through reaction with labeled antibodies.

immunodeficiency The inability to produce a normal complement of antibodies or immunologically sensitized T cells in response to specific antigens.

immunodiffusion A technique involving the diffusion of antigen or antibody within a semisolid gel to produce a precipitin reaction. Often both the antibody and antigen diffuse through the gel; sometimes an antigen diffuses through a gel containing antibody.

immunoelectrophoresis (ĭ-mu″no-e-lek″tro-fo-re′sis; pl., **immunoelectrophoreses**) The electrophoretic separation of protein antigens followed by diffusion and precipitation in gels using antibodies against the separated proteins.

immunofluorescence A technique used to identify particular antigens microscopically in cells or tissues by the binding of a fluorescent antibody conjugate.

immunoglobulin (Ig) (im″u-no-glob′u-lin) *See* antibody.

immunology The study of host defenses against invading foreign materials, including pathogenic microorganisms, transformed or cancerous cells, and tissue transplants from other sources.

immunopathology The study of diseases or conditions resulting from immune reactions.

immunoprecipitation A reaction involving soluble antigens reacting with antibodies to form a large aggregate that precipitates out of solution.

immunotoxin A monoclonal antibody that has been attached to a specific toxin or toxic agent (antibody + toxin = immunotoxin) and can kill specific target cells.

impetigo (im″pə-ti′go) A superficial cutaneous disease, most commonly seen in children, characterized by crusty lesions, usually located on the face; the lesions typically have vesicles surrounded by a red border; caused by *Streptococcus pyogenes* or *Staphylococcus aureus*.

inclusion bodies (1) Granules of organic or inorganic material in the cytoplasm of bacteria. (2) Clusters of viral proteins or virions within the nucleus or cytoplasm of virus-infected cells.

inclusion conjunctivitis (kon-junk″tĭ-vi′tis) An infectious disease of the eye caused by *Chlamydia trachomatis*; causes inflammation and the occurrence of large inclusion bodies.

incubation period The period after pathogen entry into a host and before signs and symptoms appear.

incubatory carrier An individual who is incubating a pathogen but is not yet ill.

index case The first disease case in an outbreak or epidemic.

indicator organism An organism whose presence indicates the condition of a substance or environment, for example, the potential presence of pathogens. Coliforms are used as indicators of fecal pollution.

induced mutations Mutations caused by exposure to a mutagen.

inducer A small molecule that stimulates the synthesis of an inducible enzyme.

inducible enzyme   An enzyme whose level rises in the presence of a small molecule that stimulates its synthesis or activity.

inducible gene   A gene that encodes an inducible product.

induction   (1) In virology, the events that trigger a virus to switch from a lysogenic mode to a lytic pathway. (2) In genetics, an increase in gene expression.

infantile paralysis   *See* poliomyelitis.

infection   The invasion of a host by a microorganism with subsequent establishment and multiplication of the agent. An infection may or may not lead to overt disease.

infection thread   A tubular structure formed during the infection of a root by nitrogen-fixing bacteria. The bacteria enter the root by way of the infection thread and stimulate the formation of the root nodule.

infectious disease   Any change from a state of health in which part or all of the host's body cannot carry on its normal functions because of the presence of an infectious agent or its products.

infectious disease cycle (chain of infection)   The chain or cycle of events that describes how an infectious organism grows, reproduces, and is disseminated.

infectious dose 50 (ID_{50})   Refers to the dose or number of organisms that will infect 50% of an experimental group of hosts within a specified time period.

infectious mononucleosis (mono) (mon″o-nu″kle-o′sis)   An acute, self-limited infectious disease of the lymphatic system caused by Epstein-Barr virus and characterized by fever, sore throat, lymph node and spleen swelling, and the proliferation of monocytes and abnormal lymphocytes.

infectivity   Infectiousness; the state or quality of being infectious or communicable.

inflammation   A localized protective response to tissue injury or destruction. Acute inflammation is characterized by pain, heat, swelling, and redness in the injured area.

influenza (flu)   An acute viral infection of the respiratory tract, caused by three strains of influenza virus, labeled types A, B, and C, based on envelope antigens.

initiator codon   The first codon, usually AUG, in the coding region of mRNA.

innate or **natural immunity**   *See* nonspecific resistance.

insertion sequence   A simple transposon that contains genes only for those enzymes, such as transposase, that are required for transposition.

in silico **analysis**   The study of physiology and/or genetics through the examination of nucleic acid and amino acid sequence. *See* bioinformatics.

integral proteins   *See* plasma membrane.

integrase   An enzyme observed in some viruses; it catalyzes the integration of provirus DNA into the host chromosome.

integration   The incorporation of one DNA segment into a second DNA molecule to form a hy-

brid DNA; occurs during such processes as genetic recombination, episome incorporation into host DNA, and provirus insertion into the host chromosome.

integrins (in′tə-grinz)   Cellular adhesion receptors that mediate cell-cell and cell-substratum interactions, usually by recognizing linear amino acid sequences on protein ligands.

integron   A genetic element with an attachment site for site-specific recombination and an integrase gene. It can capture genes and gene cassettes.

inteins   Internal intervening sequences of precursor self-splicing proteins that are removed during formation of the final protein.

intercalating agents   Molecules that can be inserted between the stacked bases of a DNA double helix, thereby distorting the DNA and inducing insertion and deletion (i.e., frameshift) mutations.

interdigitating dendritic cell   Special dendritic cells in the lymph nodes that function as potent antigen-presenting cells and develop from Langerhans cells.

interferon (IFN) (in″tər-fēr′on)   A set of cytokines that stimulate cells to produce antiviral proteins. Others regulate growth, differentiation, or function of a variety of immune system cells.

interleukin (in″tər-loo′kin)   A cytokine produced by macrophages and T cells that regulates growth and differentiation, particularly of lymphocytes. Interleukins promote cellular and humoral immune responses.

intermediate filaments   Small protein filaments, about 8 to 10 nm in diameter, in the cytoplasm of eucaryotic cells that are important in cell structure.

intermediate host   The host that serves as a temporary but essential environment for development of a parasite and completion of its life cycle.

interspecies hydrogen transfer   The linkage of hydrogen production from organic matter by anaerobic heterotrophic microorganisms to the use of hydrogen by other anaerobes in the reduction of carbon dioxide to methane.

intertriginous candidiasis   A skin infection caused by *Candida* species. Involves those areas of the body, usually opposed skin surfaces, that are warm and moist (axillae, groin, skin folds).

intoxication   A disease that results from the entrance of a specific toxin into the body of a host. The toxin can induce the disease in the absence of the toxin-producing organism.

intraepidermal lymphocytes   T cells found in the epidermis of the skin that express the γδ T-cell receptor.

intranuclear inclusion body   A structure found within cells infected with cytomegalovirus.

intrinsic factors   Food-related factors such as moisture, pH, and available nutrients that influence microbial growth.

intron   A noncoding intervening sequence in a split (interrupted) gene that codes for pre-mRNA and is missing from the final RNA product.

invasiveness   The ability of a microorganism to enter a host, grow and reproduce within the host, and spread throughout its body.

iodophor   An antimicrobial agent consisting of an organic compound complexed with iodine.

ionizing radiation   Radiation of very short wavelength and high energy that causes atoms to lose electrons (i.e., ionize).

iron-sulfur (Fe-S) protein   *See* electron transport chain.

isoelectric focusing   An electrophoretic technique in which proteins are separated based on their isoelectric point.

isoelectric point   The pH value at which a protein or other molecule no longer carries a net charge.

isoenzyme   An enzyme that carries out the same catalytic function but differs in terms of its amino acid sequence, regulatory properties, or other characteristics.

isotopic fractionation   The preferential use of one stable isotope over another by microbes carrying out metabolic processes.

isotype   A variant form of an immunoglobulin that differs in the constant region of the heavy and light chains and occurs in every normal individual of a particular species.

J chain   A polypeptide present in polymeric IgM and IgA that links the subunits together.

kallikrein   An enzyme that acts on kininogen, releasing the active bradykinin protein.

Kaposi's sarcoma   A type of cancer associated with HIV infection.

keratitis   Inflammation of the cornea of the eye.

kinetoplast (ki-ne′to-plast)   A special structure in the mitochondrion of certain protists. It contains the mitochondrial DNA.

kinetosome   Intracellular microtubular structure that serves as the base of cilia in ciliated protists. Similar in structure to a centriole. Also referred to as a basal body.

Kirby-Bauer method   A disk diffusion test to determine the susceptibility of a microorganism to chemotherapeutic agents.

Koch's postulates   A set of rules for proving that a microorganism causes a particular disease.

Koplik's spots   Lesions of the oral cavity caused by the *Measles virus* that are characterized by a bluish white speck in the center of each.

Korarchaeota   A proposed phylum in the *Archaea* domain. To date, it is based entirely on uncultured microbes that have been identified through 16S rRNA nucleotide sequences cloned directly from the environment.

Krebs cycle   *See* tricarboxylic acid (TCA) cycle.

labyrinthulids   A subgroup of the stramenopiles characterized by having heterokont flagellated zoospores.

lactic acid fermentation   A fermentation that produces lactic acid as the sole or primary product.

lactoferrin   An iron-sequestering protein released from macrophages and neutrophils into plasma.

lager   Pertaining to the process of aging beers to allow flavor development.

lag phase   A period following the introduction of microorganisms into fresh culture medium when there is no increase in cell numbers or mass during batch culture.

lagging strand   The strand of DNA synthesized discontinuously during DNA replication.

laminar flow biological safety cabinets   Cabinets that use HEPA filters to project a curtain of sterile air across its opening, preventing microbes from entering or exiting.

Lancefield system (group)   One of the serologically distinguishable groups (e.g., group A and group B) into which streptococci can be divided.

Langerhans cell   Cell found in the skin that internalizes antigen and moves in the lymph to lymph nodes, where it differentiates into a dendritic cell.

late mRNA   Messenger RNA produced later in a viral infection, which codes for proteins needed in capsid construction and virus release.

latent viral infections   Viral infections in which the virus stops reproducing and remains dormant for a period before becoming active again.

lateral gene transfer   *See* horizontal gene transfer.

leader sequence   A sequence in a gene that lies between the promoter and the initiation codon. It is transcribed, becoming a nontranslated sequence at the 5′ end of mRNA. It aids in initiation and regulation of transcription.

leading strand   The strand of DNA that is synthesized continuously during DNA replication.

lectin complement pathway   An antibody independent pathway of complement activation that is initiated by microbial lectins (proteins that bind carbohydrates) and includes the C3–C9 components of the classical pathway.

leghemoglobin   A heme-containing pigment produced in leguminous plants. It is similar in structure to vertebrate hemoglobin and functions to protect nodule-forming, nitrogen-fixing bacteria from oxygen, which would poison their nitrogenase.

leishmanias (lēsh″ma′ne-ăs)   Trypanosomal protists of the genus *Leishmania.*

leishmaniasis (lēsh″mah-ni′ah-sis)   A group of human diseases caused by protists called leishmanias, which cause the disease; includes cutaneous (skin surface), mucocutaneous (mouth, nose, throat, skin), and visceral (monocyte-macrophage system) leishmaniasis.

lepromatous (progressive) leprosy   A relentless, progressive form of leprosy in which large numbers of *Mycobacterium leprae* develop in skin cells, killing the skin cells and resulting in the loss of features. Disfiguring nodules form all over the body.

leprosy (Hansen's disease)   A severe disfiguring skin disease caused by *Mycobacterium leprae.*

lethal dose 50 (LD$_{50}$)   Refers to the dose or number of organisms that will kill 50% of an experimental group of hosts within a specified time period.

leukocidin (loo″ko-si′din)   A microbial toxin that can damage or kill leukocytes.

leukocyte (loo′ko-sĭt)   Any white blood cell.

lichen (li′ken)   A symbiotic association of a fungus and either photosynthetic protists or cyanobacteria.

light reactions   *See* photosynthesis.

lignin   A complex organic molecule that is an important structural component of woody plants.

lipidomics   Determination of an organism's lipid content.

lipid raft   A microdomain in the plasma membrane that is enriched for particular lipids and proteins.

lipopolysaccharide (LPS) (lip″o-pol″e-sak′ah-rĭd)   A molecule containing both lipid and polysaccharide, which is important in the outer membrane of the gram-negative cell wall.

listeriosis (lis-ter″e-o′sis)   A sporadic disease of animals and humans, particularly those who are immunocompromised or pregnant, caused by the bacterium *Listeria monocytogenes.*

lithotroph   An organism that uses reduced inorganic compounds as its electron source.

lobopodia   Rounded pseudopodia found in some amoeboid protists.

log phase   *See* exponential phase.

lophotrichous (lo-fot′rĭ-kus)   A cell with a cluster of flagella at one or both ends.

Lyme disease (LD, Lyme borreliosis)   A tick-borne disease caused by the spirochete *Borrelia burgdorferi.*

lymph node   A small secondary lymphoid organ that contains lymphocytes, macrophages, and dendritic cells. It serves as a site for (1) filtration and removal of foreign antigens and (2) the activation and proliferation of lymphocytes.

lymphocyte   A nonphagocytic, mononuclear leukocyte that is an immunologically competent cell, or its precursor. Lymphocytes are present in the blood, lymph, and lymphoid tissues. *See* B cell and T cell.

lymphogranuloma venereum (LGV) (lim″fo-gran″u-lo′mah)   A sexually transmitted disease caused by *Chlamydia trachomatis* serotypes L$_1$–L$_3$, which affect the lymph organs in the genital area.

lymphokine   A glycoprotein cytokine (e.g., IL-1) secreted by activated lymphocytes, especially sensitized T cells.

lyophilization   Freezing and dehydrating samples as a means of preservation. Commonly referred to as freeze-drying.

lysis (li′sis)   The rupture or physical disintegration of a cell.

lysogenic (li-so-jen′ik)   *See* lysogens.

lysogenic conversion   A change in the phenotype of a bacterium due to the presence of a prophage.

lysogens (li′so-jens)   Bacterial and archaeal cells that carry a provirus and can produce viruses under the proper conditions.

lysogeny (li-soj′e-ne)   The state in which a viral genome remains within a bacterial or archaeal cell after infection and reproduces along with it rather than taking control of the host cell and destroying it.

lysosome   A spherical membranous eucaryotic organelle that contains hydrolytic enzymes and is responsible for the intracellular digestion of substances.

lysozyme   An enzyme that degrades peptidoglycan by hydrolyzing the β(1 → 4) bond that joins *N*-acetylmuramic acid and *N*-acetylglucosamine.

lytic cycle (lit′ik)   A viral life cycle that results in the lysis of the host cell.

macroelement   A nutrient that is required in relatively large amounts (e.g., carbon and nitrogen).

macroevolution   Major evolutionary change leading to either speciation or extinction.

macrolide antibiotic   An antibiotic containing a macrolide ring, a large lactone ring with multiple keto and hydroxyl groups, linked to one or more sugars.

macromolecule   A large molecule that is a polymer of smaller units joined together.

macromolecule vaccine   A vaccine made of specific, purified macromolecules derived from pathogenic microorganisms.

macronucleus   The larger of the two nuclei in ciliate protists. It is normally polyploid and directs the routine activities of the cell.

macrophage   The name for a large, mononuclear phagocytic antigen-presenting cell, present in blood, lymph, and other tissues.

maduromycosis (mah-du′ro-mi-ko′sis)   A subcutaneous fungal infection caused by *Madurella mycetomatis;* also termed an eumycotic mycetoma.

madurose   The sugar derivative 3-*O*-methyl-D-galactose, which is characteristic of several actinomycete genera that are collectively called maduromycetes.

magnetosomes   Magnetite particles in magnetotactic bacteria that are tiny magnets and allow the bacteria to orient themselves in magnetic fields.

maintenance energy   The energy a cell requires simply to maintain itself or remain alive and functioning properly. It does not include the energy needed for either growth or reproduction.

major histocompatibility complex (MHC)   A chromosome locus encoding the histocompatibility antigens and other components of the immune system. Class I MHC molecules are cell surface glycoproteins present on all nucleated cells; class II MHC glycoproteins are on antigen-presenting cells.

malaria   A serious infectious illness caused by the parasitic protozoan *Plasmodium,* and characterized by bouts of high chills and fever that occur at regular intervals.

malt   Grain soaked in water to soften it, induce germination, and activate its enzymes. The malt is then used in brewing and distilling.

mash   The soluble materials released from germinated grains and prepared as a microbial growth medium.

mashing   The process in which cereals are mixed with water and incubated in order to degrade their complex carbohydrates (e.g., starch) to more readily usable forms such as simple sugars.

mass spectrometry   A type of spectrometry that determines mass-to-charge ratio of ions formed from the molecule being analyzed. The ratio can be used to identify structures and determine sequences of proteins.

mast cell   A white blood cell that produces vasoactive molecules (e.g., histamine) and stores them in vacuoles near the cell membrane where they are released upon cell stimulation by external triggers.

mating type   A strain of a eucaryotic organism that can mate sexually with another strain of the same species. Commonly refers to strains of fungal mating types (MAT) (e.g., MAT α and MAT **a** of *Saccharomyces cerevisiae.*

M cell   Specialized cell of the intestinal mucosa and other sites (e.g., urogenital tract) that delivers antigen from its apical face to lymphocytes clustered within the pocket in its basolateral face.

mean generation (doubling) time   The time it takes a population to double in size.

mean growth rate constant (k)   The rate of microbial population growth expressed in terms of the number of generations per unit time.

measles (rubeola)   A highly contagious skin disease that is endemic throughout the world. It is caused by a virus in the family *Paramyxoviridae,* which enters the body through the respiratory tract or through the conjunctiva.

medical mycology   The discipline that deals with the fungi that cause human disease.

meiosis (mi-o′sis)   The type of cell division by which a diploid cell divides and forms four haploid cells.

melting temperature (T_m)   The temperature at which double-stranded DNA separates into individual strands; it is dependent on the G + C content of the DNA and is used to compare genetic material in microbial taxonomy.

membrane attack complex (MAC)   The complex of complement proteins (C5b–C9) that creates a pore in the plasma membrane of a target cell and leads to cell lysis.

membrane-disrupting exotoxin   A type of exotoxin that lyses host cells by disrupting the integrity of the plasma membrane.

membrane filter technique   The use of a thin porous filter made from cellulose acetate or some other polymer to collect microorganisms from water, air, and food.

memory cell   An inactive lymphocyte derived from a sensitized B or T cell capable of an accentuated response to a subsequent antigen exposure.

meningitis   Inflammation of the brain or spinal cord meninges (membranes). The disease can be divided into bacterial (septic) meningitis and aseptic meningitis syndrome (caused by nonbacterial sources).

meningococcus   Common name for *Neisseria meningitidis,* one of the causative agents of bacterial meningitis.

merozygote   A partially diploid procaryotic cell produced by horizontal gene transfer.

mesophile   A microorganism with a growth optimum around 20 to 45°C, a minimum of 15 to 20°C, and a maximum about 45°C or lower.

messenger RNA (mRNA)   Single-stranded RNA synthesized from a nucleic acid template (DNA in cellular organisms, RNA in some viruses) during transcription; mRNA binds to ribosomes and directs the synthesis of protein.

metabolic channeling   The localization of metabolites and enzymes in different parts of a cell.

metabolic control engineering   Modification of the controls for biosynthetic pathways without altering the pathways themselves in order to improve process efficiency.

metabolic pathway engineering (MPE)   The use of molecular techniques to improve the efficiency of pathways that synthesize industrially important products.

metabolism   The total of all chemical reactions in the cell; almost all are enzyme catalyzed.

metabolomics   Determination of the small metabolites present in a cell.

metachromatic granules   Granules of polyphosphate in the cytoplasm of some bacteria that appear a different color when stained with a blue basic dye. They are storage reservoirs for phosphate. Sometimes called volutin granules.

metagenomics   Also called environmental or community genomics, metagenomics is the study of genomes recovered from environmental samples without first isolating members of the microbial community and growing them in pure cultures.

metastasis (mě-tas′tah-sis)   The transfer of a disease such as cancer from one organ to another not directly connected with it.

methanogens (meth′ə-no-jens″)   Strictly anaerobic archaea that derive energy by converting CO_2, H_2, formate, acetate, and other compounds to either methane or methane and CO_2.

methanotroph   An organism that carries out methanotrophy.

methanotrophic bacteria   The ability to grow on methane as the sole carbon source.

methylotroph   A bacterium that uses reduced one-carbon compounds such as methane and methanol as its sole source of carbon and energy.

Michaelis constant (K_m) (mĭ-ka′lis)   A kinetic constant for an enzyme reaction that equals the substrate concentration required for the enzyme to operate at half maximal velocity.

microaerophile (mi″kro-a′er-o-fīl)   A microorganism that requires low levels of oxygen for growth, around 2 to 10%, but is damaged by normal atmospheric oxygen levels.

microbial ecology   The study of microorganisms in their natural environments, with a major emphasis on physical conditions, processes, and interactions that occur on the scale of individual microbial cells.

microbial flora (microbiota)   The microbes found in a habitat.

microbial loop   The cycling of organic matter synthesized by photosynthetic microorganisms among other microbes, such as bacteria and protozoa. This process "loops" organic nutrients, minerals, and carbon dioxide back for reuse by the primary producers and makes the organic matter unavailable to higher consumers.

microbial mat   A firm structure of layered microorganisms with complementary physiological activities that can develop on surfaces in aquatic environments.

microbial transformation   *See* biotransformation.

microbiology   The study of organisms that are usually too small to be seen with the naked eye. Special techniques are required to isolate and grow them.

microbivory   The use of microorganisms as a food source by organisms (e.g., protists) that can ingest or phagocytose them.

microcosms   Small incubation chambers with conditions that mimic those in a natural setting.

microenvironment   The immediate environment surrounding a microbial cell or other structure, such as a root.

microevolution   *See* anagenesis.

microfilaments   Protein filaments, about 4 to 7 nm in diameter, that are present in the cytoplasm of eucaryotic cells and play a role in cell structure and motion.

micronucleus   The smaller of the two nuclei in ciliate protists. Micronuclei are diploid and involved only in genetic recombination and the regeneration of macronuclei.

micronutrients   Nutrients such as zinc, manganese, and copper that are required in very small quantities for growth and reproduction. Also called trace elements.

microorganism   An organism that is too small to be seen clearly with the naked eye and is often unicellular, or if multicellular, does not exhibit a high degree of differentiation.

microsporidia   Primitive, fungal, obligate intracellular parasites of animals, primarily vertebrates. The infectious spore (0.5 to 2.0 μm) contains a coiled polar tubule used for injecting the spore contents into host cells where the sporoplasm undergoes mitotic division, producing more spores.

microtubules (mi″kro-tu′buls)   Small cylinders, about 25 nm in diameter, made of tubulin proteins and present in the cytoplasm and flagella of eucaryotic cells; they are involved in cell structure and movement.

miliary tuberculosis (mil′e-a-re)   An acute form of tuberculosis in which small tubercles are formed in a number of organs of the body because

M. tuberculosis is disseminated throughout the body by the bloodstream. Also known as reactivation tuberculosis.

mineralization The conversion of organic nutrients into inorganic material during microbial growth and metabolism.

mineral soil Soil that contains less than 20% organic carbon.

minimal inhibitory concentration (MIC) The lowest concentration of a drug that will prevent the growth of a particular microorganism.

minimal lethal concentration (MLC) The lowest concentration of a drug that will kill a particular microorganism.

minus (negative) strand A viral single-stranded nucleic acid that is complementary to the viral mRNA.

mismatch repair A type of DNA repair in which a portion of a newly synthesized strand of DNA containing mismatched base pairs is removed and replaced, using the parental strand as a template.

missense mutation A single base substitution in DNA that changes a codon for one amino acid into a codon for another.

mitochondrion (mi″to-kon′dre-on) The eucaryotic organelle that is the site of electron transport, oxidative phosphorylation, and pathways such as the Krebs cycle; it provides most of a nonphotosynthetic cell's energy under aerobic conditions.

mitosis A process that takes place in the nucleus of a eucaryotic cell and results in the formation of two new nuclei, each with the same number of chromosomes as the parent.

mixed acid fermentation A type of fermentation carried out by members of the family *Enterobacteriaceae* in which ethanol and a complex mixture of organic acids are produced.

mixotrophy A nutritional type that is a mixture of different types of metabolism (e.g., phototrophy and chemoorganotrophy).

modified atmosphere packaging (MAP) Addition of gases such as nitrogen and carbon dioxide to packaged foods in order to inhibit the growth of spoilage organisms.

mold Any of a large group of fungi that exist as multicellular filamentous colonies; also the deposit or growth caused by such fungi. Molds typically do not produce macroscopic fruiting bodies.

molecular chaperones See chaperone proteins.

monoclonal antibody (mAb) An antibody of a single type that is produced by a population of genetically identical plasma cells (a clone); a monoclonal antibody is typically produced from a cell culture derived from the fusion of a cancer cell and an antibody-producing cell (i.e., hybridoma).

monocyte A mononuclear phagocytic leukocyte; precursor of macrophages and dendritic cells.

monocyte-macrophage system The collection of fixed phagocytic cells (including macrophages, monocytes, and specialized endothelial cells) located in the liver, spleen, lymph nodes, and bone marrow; an important component of the host's general nonspecific (innate) defense against pathogens.

monokine A generic term for a cytokine produced by macrophages or monocytes.

monotrichous (mon-ot′rĭ-kus) Having a single flagellum.

morbidity rate Measures the number of individuals who become ill as a result of a particular disease within a susceptible population during a specific time period.

mordant A substance that helps fix dye on or in a cell.

morphovar A variant strain of a microbe characterized by morphological differences.

mortality rate The ratio of the number of deaths from a given disease to the total number of cases of the disease.

most probable number (MPN) The statistical estimation of the probable population in a liquid by diluting and determining end points for microbial growth.

mucociliary blanket The layer of cilia and mucus that lines certain portions of the respiratory system; it traps microorganisms and then transports them by ciliary action away from the lungs.

mucosal-associated lymphoid tissue (MALT) Organized and diffuse immune tissues found as part of the mucosal epithelium. It can be specialized to the gut (GALT) or the bronchial system or (BALT).

multicloning site (MCS) A region of DNA on a cloning vector that has a number of restriction enzyme recognition sequences to facilitate the introduction, or cloning, of a gene.

multidrug-resistant strains of tuberculosis (MDR-TB) *Mycobacterium tuberculosis* strains resistant to isoniazid and rifampin, with or without resistance to other drugs.

multilocus sequence typing (MLST) A method for genotypic classification of procaryotes within a single genus using nucleotide differences among five to seven housekeeping genes.

murein See peptidoglycan.

must The juices of fruits, including grapes, that can be fermented for the production of alcohol.

mutagen (mu′tah-jen) A chemical or physical agent that causes mutations.

mutation A permanent, heritable change in the genetic material.

mutualism A type of symbiosis in which both partners gain from the association and are metabolically dependent on each other.

mutualist An organism associated with another in an obligatory relationship that is beneficial to both.

mycelium (mi-se′le-um) A mass of branching hyphae found in fungi and some bacteria.

mycobiont The fungal partner in a lichen.

mycolic acids Complex 60 to 90 carbon fatty acids with a hydroxyl on the β-carbon and an aliphatic chain on the α-carbon; found in the cell walls of mycobacteria.

mycologist A person specializing in mycology; a student of mycology.

mycology The science and study of fungi.

mycoplasma Bacteria that are members of the phylum *Firmicutes,* class *Mollicutes,* and order *Mycoplasmatales;* they lack cell walls and cannot synthesize peptidoglycan precursors; most require sterols for growth.

mycorrhizal fungi Fungi that form stable, mutualistic relationships on (ectomycorrhizal) or in (endomycorrhizal) the roots of vascular plants.

mycosis (mi-ko′sis; pl., **mycoses**) Any disease caused by a fungus.

mycotoxicology (mi-ko′tok″si-kol′o-je) The study of fungal toxins and their effects on various organisms.

myeloma cell (mi″e-lo′mah) A tumor cell that is similar to the cell type found in bone marrow. Also, a malignant, neoplastic plasma cell that produces large quantities of antibodies and can be readily cultivated.

myositis (mi″o-si′tis) Inflammation of a striated or voluntary muscle.

myxobacteria A group of gram-negative, aerobic soil bacteria characterized by gliding motility, a complex life cycle with the production of fruiting bodies, and the formation of myxospores.

Myxogastria See acellular slime molds.

myxospores (mik′so-spōrs) Special dormant spores formed by the myxobacteria.

N-acylhomoserine lactone See quorum sensing.

naked virus A virus composed only of a nucleocapsid (i.e., lacking an envelope).

narrow-spectrum drugs Chemotherapeutic agents that are effective only against a limited variety of microorganisms.

natural attenuation The decrease in the level of an environmental contaminant that results from natural chemical, physical, and biological processes.

natural classification A classification system that arranges organisms into groups whose members share many characteristics and reflect as much as possible the biological nature of organisms.

natural killer (NK) cell A type of white blood cell that has a lineage independent of the granulocyte, B-cell, and T-cell lineages; part of the innate immune system.

naturally acquired active immunity The type of active immunity that develops when an individual's immunologic system comes into contact with an appropriate antigenic stimulus during the course of normal activities; it often arises as the result of recovering from an infection and lasts a long time.

naturally acquired passive immunity The type of temporary immunity that involves the transfer of antibodies from one individual to another.

necrotizing fasciitis (nek′ro-tiz″ing fas″e-i′tis) A disease that usually results from a severe invasive group A streptococcus infection. Necrotizing fasciitis is an infection of the subcutaneous soft tissues, particularly of fibrous tissue, and is most common on the extremities.

negative selection In immunology, the process by which lymphocytes that recognize host (self) antigens undergo apoptosis or become anergic (inactive).

negative staining A staining procedure in which a dye is used to make the background dark while the specimen is unstained.

negative transcriptional control Regulation of transcription by a repressor protein. When bound to the repressor-binding site, transcription is inhibited.

Negri bodies (na′gre) Masses of viruses or unassembled viral subunits found within the neurons of rabies-infected animals.

neoplasia Abnormal cell growth and reproduction due to a loss of regulation of the cell cycle; produces a tumor in solid tissues.

neuraminidase An enzyme that cleaves the chemical bond linking neuraminic acids to the sugars present on the surface of animal cells; in virology, one type of envelope spike on influenza viruses has neuraminidase activity and is used to identify different strains.

neurotoxin (nu″ro-tok′sin) A toxin that is poisonous to or destroys nerve tissue.

neutrophil A mature white blood cell in the granulocyte lineage. It has a nucleus with three to five lobes and is very phagocytic.

neutrophile A microorganism that grows best at a neutral pH range between pH 5.5 and 8.0.

niche The function of an organism in a complex system, including place of the organism, the resources used in a given location, and the time of use.

nicotinamide adenine dinucleotide (NAD⁺) (nik″o-tin′ah-mīd) An electron-carrying coenzyme; it is particularly important in catabolic processes and usually transfers electrons from an electron source to an electron transport chain.

nicotinamide adenine dinucleotide phosphate (NADP⁺) An electron-carrying coenzyme that most often participates as an electron carrier in biosynthetic metabolism.

nitrification The oxidation of ammonia to nitrate.

nitrifying bacteria Chemolithotrophic, gram-negative bacteria that are members of several families within the phylum *Proteobacteria* that oxidize ammonia to nitrite and nitrite to nitrate.

nitrogenase (ni′tro-jen-ās) The enzyme that catalyzes biological nitrogen fixation.

nitrogen fixation The metabolic process in which atmospheric molecular nitrogen (N_2) is reduced to ammonia; carried out by cyanobacteria, *Rhizobium,* and other nitrogen-fixing procaryotes.

nitrogen oxygen demand (NOD) The demand for oxygen in sewage treatment, caused by nitrifying microorganisms.

nocardioforms Bacteria that resemble members of the genus *Nocardia;* they develop a substrate mycelium that readily breaks up into rods and coccoid elements (a quality sometimes called fugacity).

nod factors Signaling compounds that alter gene expression in the plant host of a rhizobium.

nomenclature The branch of taxonomy concerned with the assignment of names to taxonomic groups in agreement with published rules.

noncompetitive inhibitor A chemical that inhibits enzyme activity by a mechanism that does not involve binding the active site of the enzyme.

noncyclic photophosphorylation (fo″to-fos″for-i-la′shun) The process in which light energy is used to make ATP when electrons are moved from water to NADP⁺ during oxygenic photosynthesis; both photosystem I and photosystem II are involved.

noncytopathic virus A virus that does not kill its host cell by viral release-induced lysis.

nongonococcal urethritis (NGU) (u″rə-thri′tis) Any inflammation of the urethra not caused by *Neisseria gonorrhoeae.*

nonheme iron protein *See* electron transport chain.

nonsense (stop) codon A codon that does not code for an amino acid but is a signal to terminate protein synthesis.

nonsense mutation A mutation that converts a sense codon to a nonsense (stop) codon.

nonspecific immune response (innate or natural immunity) *See* nonspecific resistance.

nonspecific resistance Refers to those general defense mechanisms that are inherited as part of the innate structure and function of each animal; also known as nonspecific, innate or natural immunity.

normal microbiota (indigenous microbial population, microflora, microbial flora) The microorganisms normally associated with a particular tissue or structure.

nosocomial infection (nos″o-ko′me-al) An infection that is acquired during the stay of a patient in a hospital or other type of clinical care facility.

nuclear envelope The complex double-membrane structure forming the outer boundary of the eucaryotic nucleus. It is covered by nuclear pores through which substances enter and leave the nucleus.

nuclear pore *See* nuclear envelope.

nucleic acid hybridization The process of forming a hybrid double-stranded DNA molecule using a heated mixture of single-stranded DNAs from two different sources; if the sequences are fairly complementary, stable hybrids will form.

nucleocapsid (nu″kle-o-kap′sid) The viral nucleic acid and its surrounding capsid; the basic unit of virion structure.

nucleoid An irregularly shaped region in the procaryotic cell that contains its genetic material.

nucleolus (nu-kle′o-lus) The organelle, located within the nucleus and not bounded by a membrane, that is the location of ribosomal RNA synthesis and the assembly of ribosomal subunits.

nucleoside (nu′kle-o-sīd″) A combination of ribose or deoxyribose with a purine or pyrimidine base.

nucleosome (nu′kle-o-sōm″) A complex of histones and DNA found in eucaryotic chromatin and some archaea; the DNA is wrapped around the surface of the beadlike histone complex.

nucleotide (nu′kle-o-tīd) A combination of ribose or deoxyribose with phosphate and a purine or pyrimidine base; a nucleoside plus one or more phosphates.

nucleotide excision repair *See* excision repair.

nucleus The eucaryotic organelle enclosed by a double-membrane envelope that contains the cell's chromosomes.

numerical aperture The property of a microscope lens that determines how much light can enter and how great a resolution the lens can provide.

nutrient A substance that supports growth and reproduction.

nystatin (nis′tah-tin) A polyene antibiotic from *Streptomyces noursei* that is used in the treatment of *Candida* infections of the skin, vagina, and alimentary tract.

O antigen A polysaccharide antigen extending from the outer membrane of some gram-negative bacterial cell walls; it is part of the lipopolysaccharide.

obligate aerobes Organisms that grow only in the presence of oxygen.

obligate anaerobes Microorganisms that cannot tolerate the presence of oxygen and die when exposed to it.

Okazaki fragments Short stretches of polynucleotides produced during discontinuous DNA replication.

oligonucleotide A short fragment of DNA or RNA, usually artificially synthesized, used in a number of molecular genetic techniques such as DNA sequencing, polymerase chain reaction, and Southern blotting.

oligonucleotide signature sequence Short, conserved nucleotide sequences that are specific for a phylogenetically defined group of organisms. The signature sequences found in small subunit rRNA molecules are most commonly used.

oligotrophic environment (ol″ĭ-go-trof′ik) An environment containing low levels of nutrients, particularly nutrients that support microbial growth.

oncogene (ong″ko-jēn) A gene whose activity is associated with the conversion of normal cells to cancer cells.

oncovirus A virus known to be associated with the development of cancer.

onychomycosis (on″i-ko-mi-ko′sis) A fungal infection of the nail plate producing nails that are opaque, white, thickened, friable, and brittle. Also called ringworm of the nails and tinea unguium.

oocyst (o′o-sist) Cyst formed around a zygote of malaria and related protozoa.

öomycetes (o″o-mi-se′tēz) A collective name for protists also known as the water molds. Formerly thought to be fungi.

open reading frame (ORF) A sequence of DNA not interrupted by a stop codon and with an apparent promoter and ribosome binding site at the 5′ end and a terminator at the 3′ end. It is usually determined by nucleic acid sequencing studies.

operator   The segment of DNA in a bacterial operon to which the repressor protein binds; it controls the expression of the genes adjacent to it.

operon   The sequence of bases in DNA that contains one or more structural genes together with the operator or activator-binding site controlling their expression.

ophthalmia neonatorum (of-thal′me-ah ne″o-nat-or-um)   A gonorrheal eye infection in a newborn, which may lead to blindness. Also called conjunctivitis of the newborn.

opportunistic microorganism or **pathogen**   A microorganism that is usually free-living or a part of the host's normal microbiota but which may become pathogenic under certain circumstances, such as when the immune system is compromised.

opsonization (op″so-ni-za′shun)   The coating of foreign substances by antibody, complement proteins, or fibronectin to make the substances more readily recognized by phagocytic cells.

optical tweezer   The use of a focused laser beam to drag and isolate a specific microorganism from a complex microbial mixture.

oral candidiasis   *See* thrush.

organelle   A structure within or on a cell that performs specific functions and is related to the cell in a way similar to that of an organ to the body of a multicellular organism.

organotrophs   Organisms that use reduced organic compounds as their electron source.

origin of replication   A site on a chromosome or plasmid where DNA replication is initiated.

orthologue   A gene found in the genomes of two or more different organisms that share a common ancestry. The products of orthologous genes are presumed to have similar functions.

O side chain   *See* O antigen.

osmophilic microorganisms   Microorganisms that grow best in or on media of high solute concentration.

osmotolerant   Organisms that grow over a fairly wide range of water activity or solute concentration.

osmotrophy   A form of nutrition in which soluble nutrients are absorbed through the cytoplasmic membrane; found in procaryotes, fungi, and some protists.

Öuchterlony technique   *See* double diffusion agar assay.

outbreak   The sudden, unexpected occurrence of a disease in a given population.

outer membrane   A membrane located outside the peptidoglycan layer in the cell walls of gram-negative bacteria.

oxidation-reduction (redox) reactions   Reactions involving electron transfers; the electron donor (reductant) gives electrons to an electron acceptor (oxidant).

oxidative burst   The generation of reactive oxygen species, primarily superoxide anion (O_2^-) and hydrogen peroxide (H_2O_2) by a plant or an animal, in response to challenge by a potential bacterial, fungal, or viral pathogen. *See also* respiratory burst.

oxidative phosphorylation (fos″for-ĭ-la′-shun)   The synthesis of ATP from ADP using energy made available during electron transport initiated by the oxidation of a chemical energy source.

oxygenic photosynthesis   Photosynthesis that oxidizes water to form oxygen; the form of photosynthesis characteristic of plants, protists and cyanobacteria.

palisade arrangement   Angular arrangements of bacteria whereby cells remain partially attached after snapping division.

pandemic   An increase in the occurrence of a disease within a large and geographically widespread population (often refers to a worldwide epidemic).

Paneth cell (pah′net)   A specialized epithelial cell of the intestine that secretes hydrolytic enzymes and antimicrobial proteins and peptides.

paralogue   Two or more genes in the genome of a single organism that arose through duplication of a common ancestral gene. The products of paralogous genes may have slightly different functions.

parasite   An organism that lives on or within another organism (the host) and benefits from the association while harming its host. Often the parasite obtains nutrients from the host.

parasitism   A type of symbiosis in which one organism benefits from the other and the host is usually harmed.

parasporal body   An intracellular, solid protein crystal made by the bacterium *Bacillus thuringiensis*. It is the basis of the bacterial insecticide Bt.

parenteral route (pah-ren′ter-al)   A route of drug administration that is nonoral (e.g., by injection).

parfocal   A microscope that retains proper focus when the objectives are changed.

paronychia (par″o-nik′e-ah)   Inflammation involving the folds of tissue surrounding the nail; usually caused by *Candida albicans*.

parsimony analysis   A method for developing phylogenetic trees based on the estimation of the minimum number of nucleotide or amino acid sequence changes needed to give the sequences being compared.

particulate organic matter (POM)   Nutrients that are not dissolved or soluble, generally referring to aquatic ecosystems. This includes microorganisms or their cellular debris after senscence or viral lysis.

passive diffusion   The process in which molecules move from a region of higher concentration to one of lower concentration as a result of random thermal agitation.

passive immunization   Temporary immunity brought about by the transfer of immune products, such as antibodies or sensitized T cells, from an immune vertebrate to a nonimmune one.

pasteurization   The process of heating liquids to destroy microorganisms that can cause spoilage or disease.

pathogen (path′o-jən)   Any virus, bacterium, or other agent that causes disease.

pathogen-associated molecular pattern (PAMP)   Conserved molecular structures that occur in patterns on microbial surfaces. The structures and their patterns are unique to particular types of microorganisms and invariant among members of a given microbial group.

pathogenicity (path″o-je-nis′ĭ-te)   The condition or quality of being pathogenic, or the ability to cause disease.

pathogenicity island   A 10 to 200 Kb segment of DNA in some pathogens that contains the genes responsible for virulence; often it codes for a type III secretion system that allows the pathogen to secrete virulence proteins and damage host cells.

pathogenic potential   The degree that a pathogen causes morbid signs and symptoms.

pattern recognition receptor (PRR)   A receptor found on macrophages and other phagocytic cells that binds to pathogen-associated molecular patterns on microbial surfaces.

pellicle (pel′ĭ-k′l)   A relatively rigid layer of proteinaceous elements just beneath the plasma membrane in many protists. The plasma membrane is sometimes considered part of the pellicle.

pelvic inflammatory disease (PID)   A severe infection of female reproductive organs caused by gonococci or chlamydiae.

penicillins   A group of antibiotics containing a β-lactam ring.

pentose phosphate pathway   A glycolytic pathway that oxidizes glucose 6-phosphate to ribulose 5-phosphate and then converts it to a variety of three to seven carbon sugars.

peplomer (spike) (pep′lo-mer)   A protein or protein complex that extends from the viral envelope and often is important in virion attachment to the host cell surface.

peptic ulcer disease   A gastritis caused by *Helicobacter pylori*.

peptide interbridge   A short peptide chain that connects the tetrapeptide chains in the peptidoglycan of some bacteria.

peptidoglycan (pep″tĭ-do-gli′kan)   A large polymer composed of long chains of alternating *N*-acetylglucosamine and *N*-acetylmuramic acid residues. The polysaccharide chains are linked to each other through connections between tetrapeptide chains attached to the *N*-acetylmuramic acids. It provides much of the strength and rigidity possessed by bacterial cell walls.

peptidyl or **donor site (P site)**   The site on the ribosome that contains the peptidyl-tRNA at the beginning of the elongation cycle during protein synthesis.

peptidyl transferase   The ribozyme that catalyzes the transpeptidation reaction in protein synthesis; in this reaction, an amino acid is added to the growing peptide chain.

peptones Water-soluble digests or hydrolysates of proteins that are used in the preparation of culture media.

perforin pathway The cytotoxic pathway that uses perforin protein, which polymerizes to form membrane pores that help destroy cells during cell-mediated cytotoxicity.

period of infectivity Refers to the time during which the source of an infectious disease is infectious or is disseminating the pathogen.

peripheral protein *See* plasma membrane.

peripheral tolerance The inhibition of effector lymphocyte activity resulting in the lack of a specific response.

periplasm (per′ĭ-plaz-əm) The substance that fills the periplasmic space.

periplasmic flagella The flagella that lie under the outer sheath and extend from both ends of the spirochete cell to overlap in the middle and form the axial filament. Also called axial fibrils, axial filaments, and endoflagella.

periplasmic space The space between the plasma membrane and the outer membrane in gram-negative bacteria, and between the plasma membrane and the cell wall in gram-positive bacteria.

peristalsis The muscular contractions of the gut that propel digested foods and waste through the intestinal tract.

peritrichous (pĕ-rit′rĭ-kus) A cell with flagella distributed over its surface.

permease (per′me-ās) A membrane-bound carrier protein or a system of two or more proteins that transports a substance across the membrane.

peronosporomycetes New name for the öomycetes, a subgroup of the *Stramenopila*, distinguished by the formaton of large egg cells when they reproduce.

pertussis (pər-tus′is) An acute, highly contagious infection of the respiratory tract, most frequently affecting young children, usually caused by *Bordetella pertussis* or *B. parapertussis*. Consists of peculiar paroxysms of coughing, ending in a prolonged crowing or whooping respiration; hence, the name whooping cough.

petri dish (pe′tre) A shallow dish consisting of two round, overlapping halves that is used to grow microorganisms on solid culture medium.

phage (fāj) *See* bacteriophage.

phagocytic vacuole (fag″o-sit′ik vak′u-ol) A membrane-delimited vacuole produced by cells carrying out phagocytosis. It is formed by the invagination of the plasma membrane and contains solid material.

phagocytosis (fag″o-si-to′sis) The endocytotic process in which a cell encloses large particles in a phagocytic vacuole and engulfs them.

phagolysosome (fag″o-li′so-sōm) The vacuole that results from the fusion of a phagosome with a lysosome.

phagosome A membrane-enclosed vacuole formed by the invagination of the cell membrane during endocytosis.

phagovar (fag′o-var) A specific phage type.

pharyngitis (far″in-ji′tis) Inflammation of the pharynx.

phase-contrast microscope A microscope that converts slight differences in refractive index and cell density into easily observed differences in light intensity.

phenetic system A classification system that groups organisms together based on the similarity of their observable characteristics.

phenol coefficient test A test to measure the effectiveness of disinfectants by comparing their activity against test bacteria with that of phenol.

phenotype Observable characteristics of an organism.

phenotypic rescue The identification of a cloned gene based on its ability to complement a genetic deficiency in the host cell.

phosphatase (fos′fah-tās″) An enzyme that catalyzes the hydrolytic removal of phosphate from molecules.

phospholipase An enzyme that hydrolyzes a specific ester bond in phospholipids.

phosphorelay system A mechanism for regulating either transcription or enzyme activity that involves the transfer of a phosphate group from one molecule to another.

photic zone Area of an aquatic habitat from the surface to the depth to which sufficient light penetrates, to support photosynthesis.

photolithoautotroph An organism that uses light energy, an inorganic electron source (e.g., H_2O, H_2, H_2S), and CO_2 as its carbon source. Also called photolithotrophic autotroph.

photoorganoheterotroph A microorganism that uses light energy, organic electron sources, and organic molecules as a carbon source. Also called photoorganotrophic heterotroph.

photophosphorylation The synthesis of ATP from ADP using energy made available by the absorption of light.

photoreactivation The process in which blue light is used by a photoreactivating enzyme to repair thymine dimers in DNA.

photosynthate Nutrient material that leaks from phototrophic organisms. Photosynthate contributes to the pool of dissolved organic matter that is available to microorganisms.

photosynthesis The trapping of light energy and its conversion to chemical energy (light reactions), which is then used to reduce CO_2 and incorporate it into organic forms (dark reactions).

photosystem I The photosystem in eucaryotic cells and cyanobacteria that absorbs longer wavelength light, usually greater than about 680 nm, and transfers the energy to chlorophyll P700 during photosynthesis; it is involved in both cyclic and noncyclic photophosphorylation.

photosystem II The photosystem in eucaryotic cells and cyanobacteria that absorbs shorter wavelength light, usually less than 680 nm, and transfers the energy to chlorophyll P680 during photosynthesis; it participates in noncyclic photophosphorylation.

phototaxis The ability of certain phototrophic organisms to move in response to a light source. In positive phototaxis, the microbe moves toward the light; in negative, it moves away.

phototrophs Organisms that use light as their energy source.

phycobiliproteins Photosynthetic pigments found in cyanobacteria that are composed of proteins with attached tetrapyrroles.

phycobilisomes Special particles on the membranes of cyanobacteria that contain photosynthetic pigments.

phycobiont (fi″ko-bi′ont) The photosynthetic protist or cyanobacterial partner in a lichen.

phycocyanin (fi″ko-si′an-in) A blue phycobiliprotein pigment used to trap light energy during photosynthesis.

phycoerythrin (fi″ko-er′i-thrin) A red photosynthetic phycobiliprotein pigment used to trap light energy.

phycology (fi-kol′o-je) The study of algae.

phyllosphere The surface of plant leaves.

phylogenetic or **phyletic classification system** (fi″lo-jĕ-net′ik, fi-let′ik) A classification system based on evolutionary relationships rather than general similarity of characteristics.

phylogenetic tree A graph made of nodes and branches, much like a tree in shape, that shows phylogenetic relationships between groups of organisms and sometimes also indicates the evolutionary development of groups.

phylogeny The evolutionary development of a species.

phylotype A taxon that is characterized only by its nucleic acid sequence; generally discovered during metagenomic analysis.

phytoplankton A community of floating photosynthetic organisms, composed of photosynthetic protists and cyanobacteria.

phytoremediation The use of plants and their associated microorganisms to remove, contain, or degrade environmental contaminants.

picoplankton Planktonic microbes between 0.2 and 2.0 μm in size, including the cyanobacterial genera *Prochlorococcus* and *Synechococcus*.

piedra (pe-a′drah) A fungal disease of the hair in which white or black nodules of fungi form on the shafts.

pitching Pertaining to inoculation of a nutrient medium with yeast, for example, in beer brewing.

plague An acute febrile, infectious disease caused by the γ-proteobacterium *Yersinia pestis*, which has a high mortality rate; the two major types are bubonic plague and pneumonic plague.

plankton Free-floating, mostly microscopic microorganisms that can be found in almost all waters.

plaque A clear area in a lawn of host cells that results from their lysis by viruses.

plaque assay A method used to determine the number of infectious virions in a population of viruses.

plaque-forming unit The unit of measure of a plaque assay. It usually represents a single infectious virion.

plasma cell A mature, differentiated B lymphocyte that synthesizes and secretes antibody.

plasmalemma The plasma membrane in protists.

plasma membrane The selectively permeable membrane surrounding the cell's cytoplasm; also called the cell membrane, plasmalemma, or cytoplasmic membrane. For most cells it is a lipid bilayer (some archaea have a lipid monolayer) with proteins embedded in it (integral proteins) and associated with the surface (peripheral proteins).

plasmid A double-stranded DNA molecule that can exist and replicate independently of the chromosome. A plasmid is stably inherited but is not required for the host cell's growth and reproduction.

plasmid fingerprinting A technique used to identify microbial isolates as belonging to the same strain because they contain the same number of plasmids with the identical molecular weights and similar phenotypes.

plasmodial (acellular) slime mold A member of the protist division *Amoebozoa (Myxogastria)* that exists as a thin, streaming, multinucleate mass of protoplasm, which creeps along in an amoeboid fashion.

plasmodium (pl., plasmodia) A stage in the life cycle of myxogastria protists; a multinucleate mass of protoplasm surrounded by a membrane. Also, a parasite of the genus *Plasmodium*.

plasmolysis (plaz-mol′ĭ-sis) The process in which water osmotically leaves a cell, which causes the cytoplasm to shrivel up and pull the plasma membrane away from the cell wall.

plastid A cytoplasmic organelle of algae and higher plants that contains pigments such as chlorophyll, stores food reserves, and often carries out processes such as photosynthesis.

pleomorphic (ple″o-mor′fik) Refers to cells or viruses that are variable in shape and lack a single, characteristic form.

plus strand (positive strand) A viral nucleic acid strand that is equivalent in base sequence to the viral mRNA.

***Pneumocystis* pneumonia** or ***Pneumocystis carinii* pneumonia (PCP)** (noo″mo-sis-tis) A type of pneumonia caused by the protist *Pneumocystis jiroveci*.

pneumonic plague *See* plague.

point mutation A mutation that affects only a single base pair.

polar flagellum A flagellum located at one end of an elongated cell.

poliomyelitis (polio or **infantile paralysis)** (po″le-o-mi″e-li′tis) An acute, contagious viral disease that attacks the central nervous system, injuring or destroying the nerve cells that control the muscles and sometimes causing paralysis.

poly-β-hydroxybutyrate (PHB) (hi-drok″se-bu′tĭ-rāt) A linear polymer of β-hydroxybutyrate used as a reserve of carbon and energy by many bacteria.

polymerase chain reaction (PCR) An in vitro technique used to synthesize large quantities of specific nucleotide sequences from small amounts of DNA. It employs oligonucleotide primers complementary to specific sequences in the target gene and special heat-stable DNA polymerases (e.g., Taq polymerase).

polymorphonuclear leukocyte (pol″e-mor″fo-noo′kle-ər) A leukocyte that has a variety of nuclear forms.

polyphasic taxonomy An approach in which taxonomic schemes are developed using a wide range of phenotypic and genotypic information.

polyribosome (pol″e-ri′bo-sōm) A complex of several ribosomes with a messenger RNA; each ribosome is translating the same message.

Pontiac fever A bacterial disease caused by *Legionella pneumophila* that resembles an allergic disease more than an infection. *See* Legionnaires' disease.

population An assemblage of organisms of the same type.

porin proteins Proteins that form channels across the outer membrane of gram-negative bacterial cell walls through which small molecules enter the periplasm.

positive transcriptional control Control of transcription by an activator protein. When the activator is bound to the activator-binding site, the level of transcription increases.

postherpetic neuralgia The severe pain after a herpes infection.

posttranscriptional modification The processing of the initial RNA transcript (pre-mRNA) to form mRNA.

potable Refers to water suitable for drinking.

pour plate A petri dish of solid culture medium with isolated microbial colonies growing both on its surface and within the medium, which has been prepared by mixing microorganisms with cooled, still liquid medium and then allowing the medium to harden.

precipitation (or precipitin) reaction The reaction of an antibody with a soluble antigen to form an insoluble precipitate.

precipitin The antibody responsible for a precipitation reaction.

precursor metabolites Intermediates of glycolytic pathways, TCA cycle, and other pathways that serve as starting molecules for biosynthetic pathways that generate monomers and other building blocks needed for synthesis of macromolecules.

pre-mRNA In eucaryotes, the RNA transcript of DNA made by RNA polymerase II; it is processed to form mRNA.

prevalence rate Refers to the total number of individuals infected at any one time in a given population regardless of when the disease began.

Pribnow box A base sequence in bacterial promoters that is recognized by RNA polymerase and is the site of initial polymerase binding.

primary amebic meningoencephalitis An infection of the meninges of the brain by the free-living amoebae *Naegleria* or *Acanthamoeba*.

primary (frank) pathogen Any organism that causes a disease in the host by direct interaction with or infection of the host.

primary metabolites Microbial metabolites produced during active growth of an organism.

primary producer Photoautotrophic and chemoautotrophic organisms that incorporate carbon dioxide into organic carbon and thus provide new biomass for the ecosystem.

primary production The incorporation of carbon dioxide into organic matter by photosynthetic organisms and chemoautotrophic organisms.

primary treatment The first step of sewage treatment, in which physical settling and screening are used to remove particulate materials.

primase *See* primosome.

primers *See* polymerase chain reaction and primosome.

primosome A complex of proteins that includes the enzyme primase, which is responsible for synthesizing the RNA primers needed for DNA replication.

prion (pre′on) An infectious agent consisting only of protein; prions cause a variety of spongiform encephalopathies such as scrapie in sheep and goats.

probe A short, labeled nucleic acid segment complementary in base sequence to part of another nucleic acid that is used to identify or isolate the particular nucleic acid from a mixture through its ability to bind specifically with the target nucleic acid.

probiont The hypothesized early RNA-based biological entity that gave rise to DNA-based cellular life-forms.

probiotic A living organism that may provide health benefits beyond its nutritional value when ingested.

procaryotic cells (pro″kar-e-ot′ik) Cells that lack a true, membrane-enclosed nucleus; *Bacteria* and *Archaea* are procaryotic and have their genetic material located in a nucleoid.

procaryotic species A collection of bacterial or archaeal strains that share many stable properties and differ significantly from other groups of strains.

prochlorophytes A group of cyanobacteria having both chlorophyll *a* and *b*, but lacking phycobilins.

prodromal stage (pro-dro′məl) The period during the course of a disease in which there is the appearance of signs and symptoms, but they are not yet distinctive and characteristic enough to make an accurate diagnosis.

programmed cell death (1) In eucaryotes, apoptosis. (2) In some bacteria, a mechanism proposed to account for the decline in cell numbers during the death phase of a growth curve.

promoter The region on DNA at the start of a gene that RNA polymerase binds to before beginning transcription.

proofreading The ability of DNA polymerase to check newly synthesized DNA to ensure that the correct nucleotides have been added.

propagated epidemic An epidemic that is characterized by a relatively slow and prolonged rise and then a gradual decline in the number of individuals infected. It usually results from the introduction of an infected individual into a susceptible population, and the pathogen is transmitted from person to person.

prophage (pro′fāj) The latent form of a temperate phage that remains within the lysogen, usually integrated into the host chromosome.

prostheca (pros-the′kah) An extension of a bacterial cell, including the plasma membrane and cell wall, that is narrower than the mature cell.

prosthetic group A tightly bound cofactor that remains at the active site of an enzyme during its catalytic activity.

protease (pro′te-ās) An enzyme that hydrolyzes proteins to their constituent amino acids. Also called a proteinase.

proteasome A large, cylindrical protein complex that degrades ubiquitin-labeled proteins to peptides in an ATP-dependent process. Also called 26S proteasome.

protein engineering The rational design of proteins by constructing specific amino acid sequences through molecular techniques, with the objective of modifying protein characteristics.

protein modeling The process by which the amino acid sequence of a protein is analyzed using software designed to predict the protein's three-dimensional structure.

protein splicing The posttranslational process in which part of a precursor polypeptide is removed before the mature polypeptide folds into its final shape; it is carried out by self-splicing proteins that remove inteins and join the remaining exteins.

Proteobacteria (pro″te-o-bak-tēr′e-ah) A large phylum of gram-negative bacteria that 16S rRNA sequence comparisons show to be phylogenetically related.

proteome The complete collection of proteins that an organism produces.

proteomics The study of the structure and function of cellular proteins.

proteorhodopsin A rhodopsin molecule produced by certain bacteria. It allows these bacteria to carry out rhodopsin-based prototrophy.

protist Unicellular and sometimes colonial eucaryotic organisms that lack cellular differentiation into tissues. Vegetative cell differentiation is limited to cells involved in sexual reproduction, alternate vegetative morphology, or resting states such as cysts. Protists vary in morphology and metabolism, including phototrophy, chemoorganotrophy, and mixotrophy. Many phototrophic and mixotrophic forms (e.g., diatoms and dinoflagellates) are frequently referred to as algae.

protistology The study of protists.

protomer An individual subunit of a viral capsid; a capsomer is made of protomers.

proton motive force (PMF) The force arising from a gradient of protons and a membrane potential that powers ATP synthesis and other processes.

proto-oncogene The normal cellular form of a gene that when mutated or overexpressed results in or contributes to malignant transformation of a cell.

protoplast A bacterial or fungal cell with its cell wall completely removed. It is spherical in shape and osmotically sensitive.

protoplast fusion The joining of cells that have had their walls weakened or completely removed.

prototroph A microorganism that requires the same nutrients as the majority of naturally occurring members of its species.

protozoan or **protozoon** (pro″to-zo′an, pl. **protozoa**) A unicellular eucaryotic chemoorganotrophic protist.

protozoology The study of protozoa.

proviral DNA Viral DNA that has been integrated into host cell DNA. In retroviruses, it is the double-stranded DNA copy of the RNA genome.

pseudogene A degraded, nonfunctional gene.

pseudomurein A complex polysaccharide observed in the cell walls of some archaea; it resembles peptidoglycan (murein) in structure and chemical make-up; also called pseudopeptidoglycan.

pseudopodium or **pseudopod** (soo″do-po′deum) A nonpermanent cytoplasmic extension of the cell by which amoeboid protists move and feed.

psychrophile (si′kro-fīl) A microorganism that grows well at 0°C and has an optimum growth temperature of 15°C or lower, and a temperature maximum around 20°C.

psychrotroph A microorganism that grows at 0°C but has a growth optimum between 20 and 30°C and a maximum of about 35°C.

pulmonary anthrax A form of anthrax involving the lungs. Also known as woolsorter's disease.

pulmonary syndrome hantavirus *See* hantavirus pulmonary syndrome.

puncutated equilibria The observation based on the fossil record that evolution does not proceed at a slow and linear pace but rather is periodically interrupted by rapid bursts of speciation and extinction driven by abrupt changes in environmental conditions.

pure culture A population of cells that are identical because they arise from a single cell.

purine (pu′rin) A heterocyclic, nitrogen-containing molecule with two joined rings that occurs in nucleic acids and other cell constituents (e.g., adenine and guanine).

purple membrane An area of the plasma membrane of *Halobacterium* that contains bacteriorhodopsin and is active in trapping light energy.

purple nonsulfur bacteria Anoxygenic photosynthetic bacteria belonging to the α- and β-proteobacteria.

purple sulfur bacteria Anoxygenic photosynthetic bacteria belonging to the γ-proteobacteria.

putrefaction The microbial decomposition of organic matter, especially the anaerobic breakdown of proteins, with the production of foul-smelling compounds such as hydrogen sulfide and amines.

pyogenic Pus-producing.

pyrenoid (pi′rĕ-noid) The differentiated region of the chloroplast that is a center of starch formation in some photosynthetic protists.

pyrimidine (pi-rim′i-dēn) A nitrogen-containing molecule with one ring that occurs in nucleic acids and other cell constituents; the most important pyrimidines are cytosine, thymine, and uracil.

Q fever An acute zoonotic disease caused by the bacterium *Coxiella burnetii*.

Quellung reaction The increase in visibility or the swelling of the capsule of a microorganism in the presence of antibodies against capsular antigens.

quinolones A class of broad-spectrum antibiotics, derived from nalidixic acid, that bind to bacterial DNA gyrase, inhibiting DNA replication. This group of antibiotics is bactericidal.

quorum sensing The process in which bacteria monitor their own population density by sensing the levels of signal molecules (e.g., *N*-acylhomoserine lactone) that are released by the microorganisms. When these signal molecules reach a threshold concentration, quorum-dependent genes are expressed.

rabies An acute infectious disease of the central nervous system that affects all warm-blooded animals (including humans). It is caused by *Rabies virus*.

racking The removal of sediments from wine bottles.

radappertization The use of gamma rays from a cobalt source for control of microorganisms in foods.

radioimmunoassay (RIA) (ra″de-o-im″u-no-as′a) A very sensitive assay technique that uses a purified radioisotope-labeled antigen or antibody to compete for antibody or antigen with unlabeled standard antigen or test antigen in experimental samples to determine the concentration of a substance in the samples.

rational drug design An approach to new drug development based on a specific cellular macromolecule or target.

reaction center chlorophyll pair The two chlorophyll molecules in a photosystem that are energized by the absorption of light and release electrons to an associated electron transport chain, thus initiating energy conservation by photophosphorylation.

reactivation tuberculosis *See* miliary tuberculosis.

reactive nitrogen intermediate (RNI) Charged nitrogen radicals.

reactive oxygen intermediate (ROI) Charged oxygen radicals; also called reactive oxygen species (ROS).

reading frame The way in which nucleotides in DNA and mRNA are grouped into codons for reading the message contained in the nucleotide sequence.

reagin (re'ah-jin) Antibody that mediates immediate hypersensitivity reactions. IgE is the major reagin in humans.

real-time PCR A type of polymerase chain reaction (PCR) that quantitatively measures the amount of template in a sample as the amount of fluorescently labeled amplified product.

RecA A protein found in all domains of life; it functions in DNA repair and recombination processes.

receptor-mediated endocytosis A type of endocytosis that involves the specific binding of molecules to membrane receptors followed by the formation of coated vesicles. The substances being taken in are concentrated during the process.

receptors Proteins that bind signaling molecules (ligands) thereby initiating cellular responses. Many viruses use host receptors to gain access to the cell.

recombinant DNA technology The techniques used in carrying out genetic engineering; they involve the identification and isolation of a specific gene, the insertion of the gene into a vector such as a plasmid to form a recombinant molecule, and the production of large quantities of the gene and its product.

recombinants Organisms produced following a recombination event.

recombinant-vector vaccine A vaccine produced by the introduction of one or more of a pathogen's genes into attenuated viruses or bacteria. The attenuated virus or bacterium serves as a vector, replicating within the vertebrate host and expressing the gene(s) of the pathogen. The pathogen's antigens induce an immune response.

recombination The process in which a new recombinant chromosome is formed by combining genetic material from two organisms.

recombinational repair A DNA repair process that repairs damaged DNA when there is no remaining template; a piece of DNA from a sister molecule is used.

Redfield ratio The carbon-nitrogen-phosphorus ratio of marine microorganisms. This ratio is important for predicting limiting factors for microbial growth.

red tides Usually population blooms of dinoflagellates that release red pigments and toxins, which can lead to paralytic shellfish poisoning. Also called harmful algal blooms (HABs).

reducing power Molecules such as NADH and NADPH that temporarily store electrons. The stored electrons are used in anabolic reactions such as CO_2 fixation and the synthesis of monomers (e.g., amino acids).

reductive dehalogenation The cleavage of carbon-halogen bonds by anaerobic bacteria that creates a strong electron-donating environment.

reductive TCA cycle A pathway used by some chemolithotrophs and some anoxygenic phototrophs to fix CO_2. It is essentially the TCA cycle running in the reverse direction; also called reverse citric acid cycle.

refractive index The ratio of the velocity of light in the first of two media to that in the second as it passes from the first to the second.

regulator T cell T cells that control the development of effector T cells and B cells. Two types exist: T-helper cells ($CD4^+$ cells) and T-suppressor cells.

regulatory mutants Mutant organisms that have lost the ability to limit synthesis of a product, which normally occurs by regulation of activity of an earlier step in the biosynthetic pathway.

regulatory transcription factors Eucaryotic proteins that specifically regulate transcription of one or more genes.

regulon A collection of genes or operons that is controlled by a common regulatory protein.

replica plating A technique for isolating mutants from a population by plating cells from each colony growing on a nonselective agar medium onto plates with selective media or environmental conditions, such as the lack of a nutrient or the presence of an antibiotic or a phage.

replicase An RNA-dependent RNA polymerase used to replicate the genome of an RNA virus.

replication The process in which an exact copy of parental DNA (or viral RNA) is made with the parental molecule serving as a template.

replication fork The Y-shaped structure where DNA is replicated. The arms of the Y contain template strand and a newly synthesized DNA copy.

replicative form (RF) A double-stranded nucleic acid that is formed from a single-stranded viral genome and used to synthesize new copies of the genome.

replicative transposition A transposition event in which a replicate of the transposon remains in its original site.

replicon A unit of the genome that contains an origin of replication and in which DNA is replicated.

replisome A protein complex or replication factory that copies the DNA double helix to form two daughter chromosomes.

reporter microbes Microbes that have been engineered to have certain genetic characteristics; they can be used to characterize the physical microenvironment.

repressible gene A gene that encodes an enzyme whose level drops in the presence of a small molecule, often an end product of its metabolic pathway.

repressor protein A protein coded for by a regulator gene that can bind to a repressor-binding site and inhibit transcription.

reservoir A site, alternate host, or carrier that normally harbors pathogenic organisms and serves as a source from which other individuals can be infected.

reservoir host An organism other than a human that is infected with a pathogen that can also infect humans.

residual body A lysosome after digestion of its contents has occurred. It contains undigested material.

resistance factor *See* R plasmid.

resolution The ability of a microscope to separate or distinguish between small objects that are close together.

respiration An energy-yielding process in which the energy substrate is oxidized using an exogenous or externally derived electron acceptor.

respiratory burst An increase in respiratory activity that occurs when an activated phagocytic cell increases its oxygen consumption to support the increased metabolic activity of phagocytosis. The burst generates highly toxic oxygen products such as singlet oxygen, superoxide radical, hydrogen peroxide, hydroxyl radical, and hypochlorite.

response-regulator protein *See* two-component signal transduction system.

restriction A process used by bacteria to defend the cell against viral infection. It is accomplished by enzymes that cleave the viral DNA.

restriction enzymes Enzymes produced by bacterial cells that cleave DNA at specific points. They evolved to protect bacteria from viral infection and are used in carrying out genetic engineering.

reticulate body (RB) The cellular form in the chlamydial life cycle whose role is growth and reproduction within the host cell.

reticulopodia Netlike pseudopodia found in certain amoeboid protists.

retroviruses A group of viruses with RNA genomes that carry the enzyme reverse transcriptase and form a DNA copy of their genome during their reproductive cycle.

reverse electron flow An energy-consuming process used by some chemolithotrophs and phototrophs to generate reducing power.

reverse transcriptase (RT) An RNA-dependent DNA polymerase that uses a viral RNA genome as a template to synthesize a DNA copy; this is a reverse of the flow of genetic information in cells, which proceeds from DNA to RNA.

reversible covalent modification A mechanism of enzyme regulation in which the enzyme's activity is either increased or decreased by the reversible covalent addition of a group such as phosphate or AMP to the protein.

reversion mutation A mutation in a mutant organism that changes its phenotype back to the wild type.

rheumatic fever (roo-mat'ik) An autoimmune disease characterized by inflammatory lesions involving the heart valves, joints, subcutaneous tissues, and central nervous system. The disease is associated with hemolytic streptococci in the body.

rhizobia Any one of a number of α- and β-proteobacteria that form symbiotic nitrogen-fixing nodules on the roots of leguminous plants.

rhizomorph A macroscopic, densely packed thread consisting of individual cells formed by some fungi. A rhizomorph can remain dormant and/or serve as a means of fungal dissemination.

rhizoplane The surface of a plant root.

rhizosphere A region around the plant root where materials released from the root increase the microbial population and its activities.

ribonucleic acid (RNA) (ri″bo-nu-kle′ik) A polynucleotide composed of ribonucleotides joined by phosphodiester bonds.

ribosomal RNA (rRNA) The RNA present in ribosomes. They contribute to ribosome structure and are also directly involved in the mechanism of protein synthesis.

ribosome (ri′bo-sōm) The organelle where protein synthesis occurs; the message encoded in mRNA is translated here.

ribosome-binding site (RBS) Nucleotide sequences on mRNA recognized by a ribosome, enabling the ribosome to orient properly on the mRNA and begin translation at the appropriate codon.

riboswitch A site in the leader of an mRNA molecule that interacts with a metabolite or other small molecule that causes the leader to change its folding pattern. In some riboswitches, this change can alter transcription; in others, it affects translation.

ribotyping The use of conserved rRNA sequences to probe chromosomal DNA for typing bacterial strains.

ribozyme An RNA molecule with catalytic activity.

ribulose 1,5-bisphosphate carboxylase (Rubisco) The enzyme that catalyzes incorporation of CO_2 in the Calvin cycle.

ringworm The common name for a fungal infection of the skin, even though it is not caused by a worm and is not always ring-shaped in appearance.

RNA-dependent DNA polymerase An enzyme that synthesizes DNA using an RNA template; also called reverse transcriptase.

RNA-dependent RNA polymerase An enzyme that synthesizes RNA using an RNA template.

RNA polymerase An enzyme that catalyzes the synthesis of RNA.

RNA polymerase core enzyme The portion of bacterial RNA polymerase that synthesizes RNA. It consists of all subunits except sigma factor.

RNA polymerase holoenzyme Bacterial RNA polymerase core enzyme plus sigma factor.

RNA silencing A response made by eucaryotic cells that protects them from infections caused by double-stranded RNA viruses.

RNA world The theory that posits that the first self-replicating molecule was RNA and this led to the evolution of the first primitive cell.

rolling-circle replication A mode of DNA replication in which the replication fork moves around a circular DNA molecule, displacing a strand to give a 5′ tail that is also copied to produce a new double-stranded DNA.

root nodule Gall-like structures on roots that contain endosymbiotic nitrogen-fixing bacteria (e.g., *Rhizobium* or *Bradyrhizobium*).

rough endoplasmic reticulum (RER) See endoplasmic reticulum.

R plasmids (R factors) Plasmids bearing one or more drug-resistance genes.

rubella (German measles) A moderately contagious skin disease that occurs primarily in children 5 to 9 years of age that is caused by *Rubella virus*, which is acquired by droplet inhalation.

rubeola See measles.

rumen The expanded upper portion or first compartment of the stomach of ruminants.

ruminant An herbivorous animal that has a stomach divided into four compartments and chews a cud consisting of regurgitated, partially digested food.

run The straight line movement of a bacterium.

sac fungi Common name for the *Ascomycota.*

salmonellosis (sal″mo-nel-o′sis) An infection with certain species of the genus *Salmonella*, usually caused by ingestion of food containing salmonellae or their products. Also known as *Salmonella* gastroenteritis or *Salmonella* food poisoning.

salt wedge A characteristic salinity profile observed in coastal regions where freshwater and salt water mix.

sanitization Reduction of the microbial population on an inanimate object to levels judged safe by public health standards.

saprophyte (sap′ro-fit) An organism that takes up nonliving organic nutrients in dissolved form and usually grows on decomposing organic matter.

saprozoic nutrition (sap″ro-zo′ik) Having the type of nutrition in which organic nutrients are taken up in dissolved form; normally refers to animals or protists.

SAR11 The most abundant group of microbes on Earth, this lineage of α-proteobacteria has been found in almost all marine ecosystems studied.

scaffolding proteins Proteins that are used to aid capsid construction that are not present in the mature virion.

scanning electron microscope (SEM) An electron microscope that scans a beam of electrons over the surface of a specimen and forms an image of the surface from the electrons that are emitted by it.

scanning probe microscope A microscope used to study surface features by moving a sharp probe over the object's surface (e.g., atomic force and scanning tunneling microscopes).

scanning tunneling microscope See scanning probe microscope.

Sec-dependent pathway A system found in all domains of life that can transport proteins through the plasma membrane or insert them into the membrane.

secondary metabolites Products of metabolism that are synthesized after growth has been completed. Antibiotics are considered secondary metabolites.

secondary treatment The biological degradation of dissolved organic matter in the process of sewage treatment; the organic material is either mineralized or changed to settleable solids.

second law of thermodynamics Physical and chemical processes proceed in such a way that the entropy of the universe (the system and its surroundings) increases to the maximum possible.

secretory IgA (sIgA) The primary immunoglobulin of the secretory immune system. *See* IgA.

secretory vacuole In protists and some animals, these organelles usually contain specific enzymes that perform various functions such as excystment. Their contents are released to the cell exterior during exocytosis.

segmented genome A viral genome that is divided into several parts or fragments, each usually coding for a single polypeptide.

selectable marker A gene whose wild-type or mutant phenotype can be determined by growth on specific media.

selectins A family of cell adhesion molecules that are displayed on activated endothelial cells; they mediate leukocyte binding to the vascular endothelium.

selective media Culture media that favor the growth of specific microorganisms; this may be accomplished by inhibiting the growth of undesired microorganisms.

selective toxicity The ability of a chemotherapeutic agent to kill or inhibit a microbial pathogen while damaging the host as little as possible.

self-assembly The spontaneous formation of a complex structure from its component molecules without the aid of special enzymes or factors.

sensor kinase protein See two-component signal transduction system.

sensory rhodopsin Microbial rhodopsins that sense the spectral quality of light. Found in halophilic archaea (SRI and SRII) and cyanobacteria. *See* bacteriorhodopsin.

sepsis Systemic response to infection manifested by two or more of the following conditions: temperature >38 or <36°C; heart rate >90 beats per min; respiratory rate >20 breaths per min, or pCO_2 <32 mm Hg; leukocyte count >12,000 cells per ml^3 or >10% immature (band) forms. Sepsis also has been defined as the presence of pathogens or their toxins in blood and other tissues.

septate Divided by a septum or cross wall; also with more or less regular occurring cross walls.

septation The process of forming a cross wall between two daughter cells during cell division.

septicemia (sep″tĭ-se′me-ah) A disease associated with the presence in the blood of pathogens or bacterial toxins.

septic shock Sepsis associated with severe hypotension despite adequate fluid resuscitation, along with the presence of perfusion abnormalities that may include but are not limited to lactic acidosis, oliguria, or an acute alteration in mental status.

septic tank A tank used to process domestic sewage. Solid material settles out and is partially degraded by anaerobic bacteria as sewage slowly flows through the tank. The outflow is further treated or dispersed in oxic soil.

septum (pl., septa) A partition or crosswall that occurs between two cells in a procaryotic or fungal filament, or that partitions structures such as spores. A septum also divides a parent cell into two daughter cells during binary fission.

serial endosymbiotic theory (SET) A theory of eucaryotic origin that suggests eucaryotes arose by a series of discrete endosymbiotic steps, each endosymbiont giving rise to a different organelle.

serology The branch of immunology that is concerned with in vitro reactions involving one or more serum constituents (e.g., antibodies and complement).

serotyping A technique or serological procedure used to differentiate between strains (serovars or serotypes) of microorganisms that have differences in the antigenic composition of a structure or product.

serovar A variant strain of a microbe that has distinctive antigenic properties.

serum (pl., **serums** or **sera**) The clear, fluid portion of blood lacking both blood cells and fibrinogen. It is the fluid remaining after coagulation of plasma, the noncellular liquid fraction of blood.

serum resistance The type of resistance that occurs with bacteria such as *Neisseria gonorrhoeae* because the pathogen interferes with membrane attack complex formation during the complement cascade.

settling basin A basin used during water purification to chemically precipitate out fine particles, microorganisms, and organic material by coagulation or flocculation.

severe acute respiratory syndrome (SARS) A disease caused by a coronavirus, characterized by fever, lower respiratory symptoms, and radiographic evidence of pneumonia.

sex pilus (pi′lus) A thin protein appendage required for bacterial mating or conjugation. The cell with sex pili donates DNA to recipient cells.

sheath A hollow tubelike structure surrounding a chain of cells and present in several genera of bacteria.

shigellosis (shĭ′gəl-o′sis) The diarrheal disease that arises from an infection with *Shigella* spp. Often called bacillary dysentery.

Shine-Dalgarno sequence A segment in the leader of bacterial mRNA and some archaeal mRNA that binds to a sequence on the 16S rRNA of the small ribosomal subunit. This helps properly orient the mRNA on the ribosome.

shingles (herpes zoster) A reactivated form of chickenpox caused by latent varicella-zoster virus.

shuttle vector A DNA vector (e.g., plasmid, cosmid) that has two origins of replication, each recognized by a different microorganism. Thus, the vector can replicate in both microbes.

siderophore (sid′er-o-for″) A small molecule that complexes with ferric iron and supplies it to a cell by aiding in its transport across the plasma membrane.

sigma factor A protein that helps bacterial RNA polymerase core enzyme recognize the promoter at the start of a gene.

sign An objective change in a diseased body that can be directly observed (e.g., a fever or rash).

signal peptide The amino-terminal sequence on a peptide destined for transport that delays protein folding and is recognized in cells by the Sec-dependent pathway machinery.

silage Fermented plant material with increased palatability and nutritional value for animals, which can be stored for extended periods.

silencer A site in the DNA to which a eucaryotic-repressor protein binds.

silent mutation A mutation that does not result in a change in the organism's proteins or phenotype even though the DNA base sequence has been changed.

simple (cut-and-paste) transposition A transposition event in which the transposon is cut out of one site and ligated into a new site.

single radial immunodiffusion (RID) assay An immunodiffusion technique that quantitates antigens by following their diffusion through a gel containing antibodies directed against the test antigens.

site-specific recombination Recombination of nonhomologous genetic material with a chromosome at a specific site.

skin-associated lymphoid tissue (SALT) The lymphoid tissue in the skin that forms a first-line defense as a part of nonspecific (innate) immunity.

S-layer A regularly structured layer composed of protein or glycoprotein that lies on the surface of many bacteria and archaea.

slime The viscous extracellular glycoproteins or glycolipids produced by some bacteria that is an easily removed, diffuse, unorganized layer of extracellular material that surrounds the cell.

slime layer A layer of diffuse, unorganized, easily removed material lying outside an archaeal or bacterial cell wall.

slime mold A common term for members of the protist divisions *Myxogastria* and *Dictyostelia*.

slow sand filter A bed of sand through which water slowly flows; the gelatinous microbial layer on the sand grain surface removes waterborne microorganisms, particularly *Giardia,* by adhesion to the gel. This type of filter is used in some water purification plants.

sludge A general term for the precipitated solid matter produced during water and sewage treatment; solid particles composed of organic matter and microorganisms that are involved in aerobic sewage treatment (activated sludge).

smallpox (variola) A highly contagious, often fatal disease caused by a poxvirus. Its most noticeable symptom was the appearance of blisters and pustules on the skin.

small RNAs (sRNAs) Small regulatory RNA molecules that do not function as messenger, ribosomal, or transfer RNAs.

small subunit rRNA (SSU rRNA) The rRNA associated with the small ribosomal subunit: 16S rRNA in procaryotes and 18S rRNA in eucaryotes.

smooth endoplasmic reticulum *See* endoplasmic reticulum.

snapping division A distinctive type of binary fission resulting in an angular or a palisade arrangement of cells, which is characteristic of the genera *Arthrobacter* and *Corynebacterium*.

SOS response A complex, inducible process that allows bacterial cells with extensive DNA damage to survive, although often in a mutated form; it involves cessation of cell division, upregulation of several DNA repair systems, and induction of translesion DNA synthesis.

source The location or object from which a pathogen is immediately transmitted to the host, either directly or through an intermediate agent.

Southern blotting technique The procedure used to isolate and identify DNA fragments from a complex mixture. The isolated, denatured fragments are transferred from an agarose gel to a nylon filter and identified by hybridization with probes.

specialized transduction A transduction process in which only a specific set of bacterial or archaeal genes is carried to a recipient cell by a temperate virus; the cell's genes are acquired because of a mistake in the excision of a provirus during the lysogenic life cycle.

species Species of higher organisms are groups of interbreeding or potentially interbreeding natural populations that are reproductively isolated. Procaryotic species are collections of strains that have many stable properties in common and differ significantly from other strains.

specific immune response (acquired, adaptive, or specific immunity) *See* acquired immunity.

spheroplast (sfēr′o-plast) A spherical cell formed by the weakening or partial removal of the rigid cell wall component (e.g., by penicillin treatment of gram-negative bacteria). Spheroplasts are usually osmotically sensitive.

spike *See* peplomer.

spirillum (spi-ril′um) A rigid, spiral-shaped bacterium.

spirochete (spi′ro-kēt) A flexible, spiral-shaped bacterium with periplasmic flagella.

spleen A secondary lymphoid organ where old erythrocytes are destroyed and blood-borne antigens are trapped and presented to lymphocytes.

spliceosome A complex of proteins that carries out RNA splicing.

split (interrupted) gene A structural gene with DNA sequences that code for the final RNA product (exons) separated by regions coding for RNA absent from the mature RNA (intervening sequences or introns).

spongiform encephalopathies Degenerative central nervous system diseases in which the brain has a spongy appearance; they are due to prions.

spontaneous generation An early belief that living organisms could develop from nonliving matter.

spontaneous mutations Mutations that occur due to errors in replication or the occurence of lesions in the DNA (e.g., apurinic sites).

sporadic disease A disease that occurs occasionally and at random intervals in a population.

sporangiospore (spo-ran′je-o-spōr) A spore borne within a sporangium.

sporangium (spo-ran′je-um; pl., **sporangia**) A saclike structure or cell, the contents of which are converted into one or more spores.

spore A differentiated, specialized cell that can be used for dissemination, for survival of adverse conditions because of its heat and dessication resistance, and/or for reproduction. Spores are usually unicellular and may develop into vegetative organisms or gametes. They may be produced asexually or sexually.

sporogenesis (spor′o-jen′ĕ-sis) *See* sporulation.

sporozoite The motile, infective stage of apicomplexan protists, including *Plasmodium,* the causative agent of malaria.

sporulation The process of spore formation.

spotted arrays *See* DNA microarrays.

spread plate A petri dish of solid culture medium with isolated microbial colonies growing on its surface, which has been prepared by spreading a dilute microbial suspension evenly over the agar surface.

sputum The mucus secretion from the lungs, bronchi, and trachea that is ejected (expectorated) through the mouth.

stable isotope probing A technique that uses stable isotopes to examine nutrient cycling and identify the microbes involved.

stalk A nonliving bacterial appendage produced by the cell and extending from it.

standard free energy change The free energy change of a reaction at 1 atmosphere pressure when all reactants and products are present in their standard states; usually the temperature is 25°C.

standard reduction potential A measure of the tendency of an electron donor to lose electrons in an oxidation-reduction (redox) reaction. The more negative the reduction potential of a compound, the better electron donor it is.

staphylococcal food poisoning (staf″i-lo-kok′al) A type of food poisoning caused by ingestion of improperly stored or cooked food in which *Staphylococcus aureus* has grown. The bacteria produce exotoxins that accumulate in the food.

staphylococcal scalded skin syndrome (SSSS) A disease caused by staphylococci that produce an exfoliative toxin. The skin becomes red (erythema) and sheets of epidermis may separate from the underlying tissue.

starter culture An inoculum, consisting of a mixture of microorganisms, used to start a commercial fermentation.

static Inhibiting or retarding the bacterial growth.

stationary phase The phase of microbial growth in a batch culture when population growth ceases and the growth curve levels off.

stem-nodulating rhizobia Rhizobia that produce nitrogen-fixing structures above the soil surface on plant stems.

sterilization The process by which all living cells, spores, viruses, and viroids are either destroyed or removed from an object or habitat.

Stickland reaction A type of fermentation in which an amino acid serves as the energy source and another amino acid serves as the electron acceptor.

stigma A photosensitive region on the surface of certain protists that is used in phototaxis.

strain A population of organisms that descends from a single organism or pure culture isolate.

streak plate A petri dish of solid culture medium with isolated microbial colonies growing on its surface, which has been prepared by spreading a microbial mixture over the agar surface, using an inoculating loop.

streptococcal pharyngitis (strep throat) Infection of the pharynx (throat) by the gram-positive bacterium *Streptococcus pyogenes. See* group A streptococcus.

streptococcal pneumonia An endogenous infection of the lungs caused by *Streptococcus pneumoniae.*

streptolysin-O (SLO) (strep-tol′ĭ-sin) A specific hemolysin produced by *Streptococcus pyogenes* that is inactivated by oxygen (hence, the "O" in its name). SLO causes beta-hemolysis of blood cells on agar plates incubated anaerobically.

streptolysin-S (SLS) A product of *Streptococcus pyogenes* that is bound to the bacterial cell but may sometimes be released. SLS causes beta hemolysis on aerobically incubated blood-agar plates and can act as a leukocidin by killing white blood cells that phagocytose the bacterial cell to which it is bound.

streptomycete Any high G+C gram-positive bacterium of the genera *Kitasatospora, Streptomyces,* and *Streptoverticillium.* The term is also often used to refer to other closely related families including *Streptosporangiaceae* and *Nocardiopsaceae.*

streptomycin (strep′to-mi″sin) A bactericidal aminoglycoside antibiotic produced by *Streptomyces griseus.*

strict anaerobes *See* obligate anaerobes.

stroma (stro′mah) The chloroplast matrix that is the location of the photosynthetic carbon dioxide fixation reactions.

stromatolite (stro″mah-to′līt) Domelike microbial mat communities consisting of filamentous photosynthetic bacteria and occluded sediments (often calcareous or siliceous). They usually have a laminar structure.

structural gene A gene that codes for the synthesis of a polypeptide or polynucleotide (i.e., rRNA, tRNA) with a nonregulatory function.

structural genomics Genomic analysis that involves sequencing the genome, identifying genes and potential genes, and predicting three-dimensional protein structure.

structural proteomics Proteomic analysis that focuses on determining the three-dimensional structure of many proteins and using these to predict the structure of other proteins and protein complexes.

subacute sclerosing panencephalitis Diffuse inflammation of the brain resulting from a persistent *Measles virus* infection.

substrate-level phosphorylation The synthesis of ATP from ADP by phosphorylation coupled with the exergonic breakdown of a high-energy organic substrate molecule.

substrate mycelium In the actinomycetes and fungi, hyphae that are on the surface and may penetrate into the solid medium on which the microbes are growing.

subsurface biosphere The region below the plant root zone where microbial populations can grow and function.

sulfate reduction The use of sulfate as an electron acceptor, which results in the accumulation of reduced forms of sulfur such as sulfide or incorporation of sulfur into organic molecules, usually as sulfhydryl groups.

sulfonamide (sul-fon′ah-mīd) A chemotherapeutic agent that has the SO_2-NH_2 group and is a derivative of sulfanilamide.

superantigen Toxic bacterial proteins that stimulate the immune system much more extensively than do normal antigens.

superinfection A new bacterial or fungal infection of a patient that is resistant to the drug(s) being used for treatment.

superoxide dismutase (SOD) (dis-mu′tas) An enzyme that protects many microorganisms by catalyzing the destruction of the toxic superoxide radical.

supportive media Culture media that are able to sustain the growth of many different kinds of microorganisms.

suppressor mutation A mutation that overcomes the effect of another mutation and produces the normal phenotype.

Svedberg unit (sfed′berg) The unit used in expressing the sedimentation coefficient; the greater a particle's Svedberg value, the faster it travels in a centrifuge.

swab A wad of absorbent material usually wound around one end of a small stick and used for applying medication or for removing material from an area; also, a dacron-tipped polystyrene applicator.

symbiosis The living together or close association of two dissimilar organisms, each of these organisms being known as a symbiont.

symbiosome The final nitrogen-fixing form of rhizobia within root nodule cells.

symport Linked transport of two substances in the same direction.

symptom A change during a disease that a person subjectively experiences (e.g., pain, fatigue, or loss of appetite). Sometimes the term symptom is used more broadly to include any observed signs.

syncytium A large, multinucleate cell formed by the fusion of numerous cells.

syndrome *See* disease syndrome.

syngamy The fusion of haploid gametes.

synthetic medium *See* defined medium.

syntrophism (sin′trōf-izəm) The association in which the growth of one organism either depends on, or is improved by, the provision of one or more growth factors or nutrients by a neighboring organism. Sometimes both organisms benefit.

syphilis (sif′ĭ-lis) *See* venereal syphilis.

systematic epidemiology   The field of epidemiology that focuses on the ecological and social factors that influence the development of emerging and reemerging infectious diseases.

systematics   The scientific study of organisms with the ultimate objective of characterizing and arranging them in an orderly manner; often considered synonymous with taxonomy.

systemic lupus erythematosus (loo′pus er″ĭ-them-ah-to′sus)   An autoimmune, inflammatory disease that may affect every tissue of the body.

tandem MS   The analysis of molelcules use two mass spectrometers sequentially.

Taq polymerase   A thermostable DNA polymerase used in the polymerase chain reaction.

Tat pathway   A system used to transport folded proteins across the plasma membrane; in gram-negative bacteria, the proteins are then delivered to a Type II secretion system for translocation across the outer membrane.

taxon   A group into which related organisms are classified.

taxonomy   The science of biological classification; it consists of three parts: classification, nomenclature, and identification.

TB skin test   Tuberculin hypersensitivity test for a previous or current infection with *Mycobacterium tuberculosis.*

T cell (T lymphocyte)   A type of lymphocyte that matures into an immunologically competent cell under the influence of the thymus.

T-cell antigen receptor (TCR)   The receptor on the T cell surface consisting of two antigen-binding peptide chains; it is associated with a large number of other glycoproteins. Binding of antigen to the TCR, usually in association with MHC, activates the T cell.

T-dependent antigen   An antigen that effectively stimulates B-cell response only with the aid of T-helper cells.

T DNA   A portion of the *Agrobacterium tumefaciens* Ti plasmid that is integrated into the host plant's chromosome.

teichoic acids (ti-ko′ik)   Polymers of glycerol or ribitol joined by phosphates; they are found in the cell walls of gram-positive bacteria.

teliospores   Diploid spores formed by fungi of the *Ustilaginomycota.*

telomerase   An enzyme in eucaryotes that is responsible for replicating the ends of chromosomes.

temperate phages   Bacteriophages that can establish a lysogenic relationship rather than immediately lysing their hosts.

template strand   A DNA or RNA strand that specifies the base sequence of a new complementary strand of DNA or RNA.

terminator   A sequence that marks the end of a gene and stops transcription.

terminus   (1) The end of a molecule (e.g., amino terminus of a protein). (2) In reference to DNA replication, the site on a procaryotic chromosome where DNA replication stops.

tertiary treatment   The removal from sewage of inorganic nutrients, heavy metals, viruses, etc., by chemical and biological means after microbes have degraded dissolved organic material during secondary sewage treatment.

test   A loose-fitting shell of an amoeba.

testate amoeba   An amoeba having a covering (test) made either by the protist or from materials collected from the environment.

tetanolysin (tet″ah-nol′ĭ-sin)   A hemolysin that aids in tissue destruction and is produced by *Clostridium tetani.*

tetanospasmin (tet″ah-no-spaz′min)   The neurotoxic component of the tetanus toxin, which causes the muscle spasms of tetanus.

tetanus   An often fatal disease caused by the anaerobic, spore-forming bacillus *Clostridium tetani* and characterized by muscle spasms and convulsions.

tetracyclines (tet″rah-si′klēns)   A family of antibiotics with a common four-ring structure that are isolated from the genus *Streptomyces* or produced semisynthetically; all are related to chlortetracycline or oxytetracycline.

T_H0 cell   Precursors to T_H1, T_H17, and T_H2 cells.

T_H 1 cell   A CD4$^+$ T-helper cell that secretes interferon-gamma, interleukin-2, and tumor necrosis factor, influencing phagocytic cells and cytotoxic T cells.

T_H 2 cell   A CD4$^+$ T-helper cell that secretes interleukin (IL)-4, IL-5, IL-6, IL-9, IL-10, and IL-13, influencing growth and differentiation of B cells.

thallus (thal′us)   A type of body that is devoid of root, stem, or leaf; characteristic of fungi.

T-helper (T_H) cell   A cell that is needed for T-cell-dependent antigens to be effectively presented to B cells. It also promotes cell-mediated immune responses.

therapeutic index   The ratio between the toxic dose and the therapeutic dose of a drug, used as a measure of the drug's relative safety.

thermal death time (TDT)   The shortest period of time needed to kill all the organisms in a microbial population at a specified temperature and under defined conditions.

thermoacidophiles   Microbes that grow best at acidic pHs and high temperatures.

thermocline   A layer of water characterized by a steep temperature gradient.

thermocycler   The instrument in which the polymerase chain reaction is performed.

thermodynamics   A science that analyzes energy changes in a system.

thermophile (ther′mo-fīl)   A microorganism that can grow at temperatures of 55°C or higher; the minimum is usually around 45°C.

thrush   Infection of the oral mucous membrane by the fungus *Candida albicans;* also known as oral candidiasis.

thylakoid (thi′lah-koid)   A flattened sac in the chloroplast stroma that contains photosynthetic pigments and the photosynthetic electron transport chain.

thymine (thi′min)   The pyrimidine 5-methyluracil that is found in nucleosides, nucleotides, and DNA.

thymus (thi′məs)   A primary lymphoid organ in the chest that is necessary in early life for the development of immunological functions, including T-cell maturation.

T-independent antigen   An antigen that triggers a B cell into immunoglobulin production without T-cell cooperation.

tinea (tin′e-ah)   A name applied to many different kinds of superficial fungal infections of the skin, nails, and hair, the specific type (depending on characteristic appearance, etiologic agent, and site) usually designated by a modifying term.

Ti plasmid   Tumor-inducing plasmid found in plant pathogenic species of the bacterium *Agrobacterium.*

titer (ti′ter)   Reciprocal of the highest dilution of an antiserum that gives a positive reaction in the test being used.

T lymphocyte   *See* T cell.

toll-like receptor (TLR)   A type of pattern recognition receptor on phagocytes such as macrophages that triggers the proper response to different classes of pathogens. It signals the production of transcription factor NFκB, which stimulates formation of cytokines, chemokines, and other defense molecules.

toxemia (tok-se′me-ah)   The condition caused by toxins in the blood of the host.

toxic shocklike syndrome (TSLS)   A disease caused by an invasive group A streptococcus infection that is characterized by a rapid drop in blood pressure, failure of many organs, and a very high fever.

toxic shock syndrome (TSS)   A staphylococcal disease that most commonly affects females who use ultra-absorbant tampons during menstruation. It is associated with the toxic shock syndrome toxin produced by strains of *Staphylococcus aureus.*

toxigenicity (tok″sĭ-jĕ-nis′i-tē)   The capacity of an organism to produce a toxin.

toxin   A microbial product or component that injures another cell or organism. Often the term refers to a protein, but toxins may be lipids and other substances.

toxin neutralization   The inactivation of toxins by specific antibodies, called antitoxins, that react with them.

toxoid   A bacterial exotoxin that has been modified so that it is no longer toxic but will still stimulate antitoxin formation when injected into a person or animal.

toxoplasmosis (tok″so-plaz-mo′sis)   A disease of animals and humans caused by the parasitic protozoan, *Toxoplasma gondii.*

trachoma (trah-ko′mah)   A chronic infectious disease of the conjunctiva and cornea, producing pain, inflammation and sometimes blindness. It is caused by *Chlamydia trachomatis* serotypes A–C.

trailer sequence   That portion of a gene downstream of the coding region. It is transcribed but not translated.

transaminase An enzyme that catalyzes a transamination reaction.

transamination The removal of an amino acid's amino group by transferring it to an α-keto acid acceptor.

transcriptase (trans-krip′tās) An enzyme that catalyzes transcription; in viruses with RNA genomes, this enzyme is an RNA-dependent RNA polymerase.

transcription The process in which single-stranded RNA with a base sequence complementary to the template strand of DNA or RNA is synthesized.

transcriptome All the messenger RNA that is transcribed from the genome of an organism under a given set of circumstances.

transduction The transfer of genes between bacterial or archaeal cells by viruses.

transfer host A host that is not necessary for the completion of a parasite's life cycle but is used as a vehicle for reaching a final host.

transfer RNA (tRNA) A small RNA that binds an amino acid and delivers it to the ribosome for incorporation into a polypeptide chain during protein synthesis.

transformation (1) A mode of gene transfer in procaryotes in which a piece of free DNA is taken up by a cell and integrated into its genome. (2) The conversion of a cell into a malignant, cancerous cell.

transgenic animal or **plant** An animal or plant that has gained new genetic information by the insertion of foreign DNA.

transient carrier *See* casual carrier.

transition mutations Mutations that involve the substitution of a different purine base for the purine present at the site of the mutation or the substitution of a different pyrimidine for the normal pyrimidine.

transition-state complex An intermediate in a chemical reaction that is a complex of the reactants and resembles both the reactants and the products.

translation Protein synthesis; the process by which the genetic message carried by mRNA directs the synthesis of polypeptides with the aid of ribosomes and other cell constituents.

translesion DNA synthesis A type of DNA synthesis that occurs during the SOS response. DNA synthesis occurs without an intact template and generates mutations as a result.

translocation During the elongation cycle of protein synthesis, the movement of the ribosome relative to the mRNA that repositions the A site of the ribosome over the next codon to be translated.

transmissible spongiform encephalopathies (TSE) Fatal, incurable, degenerative diseases of the brain caused by prions and characterized by deteriorating mental and physical abilities.

transmission electron microscope (TEM) A microscope in which an image is formed by passing an electron beam through a specimen and focusing the scattered electrons with magnetic lenses.

transovarian passage (trans″o-va′re-an) The passage of a microorganism such as a rickettsia from generation to generation of hosts through tick eggs. No humans or other mammals are needed as reservoirs for continued propagation.

transpeptidation (1) The reaction that forms the peptide cross-links during peptidoglycan synthesis. (2) The reaction that forms a peptide bond during the elongation cycle of protein synthesis.

transposable elements *See* transposon.

transposase An enzyme that catalyzes the movement of a mobile genetic element.

transposition The movement of a piece of DNA around the chromosome.

transposon (tranz-po′zon) A DNA element that carries the genes required for transposition and moves about the genome.

transversion mutations Mutations that result from the substitution of a purine base for the normal pyrimidine or a pyrimidine for the normal purine.

traveler's diarrhea A type of diarrhea resulting from ingestion of viruses, bacteria, or protozoa normally absent from the traveler's environment. A major pathogen is enterotoxigenic *Escherichia coli.*

triacylglycerol A lipid consisting of glycerol to which three fatty acids are attached.

tricarboxylic acid (TCA) cycle The cycle that oxidizes acetyl coenzyme A to CO_2 and generates NADH and $FADH_2$ for oxidation in an electron transport chain; the cycle also supplies precursor metabolites for biosynthesis.

trichome (tri′kōm) A row or filament of microbial cells that are in close contact with one another over a large area.

trichomoniasis (trik″o-mo-ni′ah-sis) A sexually transmitted disease caused by the parasitic protozoan *Trichomonas vaginalis.*

trickling filter A bed of rocks covered with a microbial film that aerobically degrades organic waste during secondary sewage treatment.

trimethoprim A synthetic antibiotic that inhibits production of folic acid by binding to dihydrofolate reductase. Trimethoprim has a wide spectrum of activity and is bacteriostatic.

trophozoite (trof″o-zo′īt) The active, motile feeding stage of a protozoan organism; in the malarial parasite, the stage of schizogony between the ring stage and the schizont.

tropism (1) The movement of living organisms toward or away from a focus of heat, light, or other stimulus. (2) The selective infection of certain organisms or host tissues by a virus; results from the distribution of the specific receptor for a virus in different organisms or certain tissues of the host.

trypanosome (tri-pan′o-sōm) A protozoan of the genus *Trypanosoma.* They are parasitic flagellate protozoa that often live in the blood of humans and other vertebrates and are transmitted by insect bites.

trypanosomiasis (tri-pan″o-so-mi′ah-sis) An infection with trypanosomes that live in the blood and lymph of the infected host; includes East African sleeping sickness, West African sleeping sickness, and Chaga's disease.

tubercle (too′ber-k′l) A small, rounded nodular lesion produced by *Mycobacterium tuberculosis.*

tuberculoid (neural) leprosy (too-ber′ku-loid) A mild, nonprogressive form of leprosy that is associated with delayed-type hypersensitivity to antigens on the surface of *Mycobacterium leprae.* It is characterized by early nerve damage and regions of the skin that have lost sensation and are surrounded by a border of nodules.

tuberculosis (TB) (too-ber″ku-lo′sis) An infectious disease of humans and other animals resulting from an infection by a species of *Mycobacterium* and characterized by the formation of tubercles and tissue necrosis, primarily as a result of host hypersensitivity and inflammation.

tuberculous cavity An air-filled cavity that results from a tubercle lesion caused by *M. tuberculosis.*

tularemia (too″lah-re′me-ah) A plaguelike disease of animals caused by the bacterium *Francisella tularensis* subsp. *tularensis* (Jellison type A), which may be transmitted to humans.

tumble Random turning or tumbling movements made by bacteria when they stop moving in a straight line.

tumor A growth of tissue resulting from abnormal new cell growth and reproduction (neoplasia).

tumor necrosis factor (TNF) A cytokine produced by activated macrophages, originally named for its cytotoxic effect on tumor cells.

tumor-suppressor proteins Proteins that regulate cell cycling or repair DNA. Their inactivation can contribute to the development of cancer.

turbidostat A continuous culture system equipped with a photocell that adjusts the flow of medium through the culture vessel to maintain a constant cell density or turbidity.

two-component signal transduction system A regulatory system that uses the transfer of phosphoryl groups to control gene transcription and protein activity. It has two major components: a sensor kinase and a response regulator protein.

two-dimensional gel electrophoresis An electrophoretic procedure that separates proteins using two procedures: isoelectric focusing and SDS-PAGE.

tyndallization The process of repeated heating and incubation to destroy bacterial spores.

type I hypersensitivity A form of immediate hypersensitivity arising from the binding of antigen to IgE attached to mast cells, which then release anaphylaxis mediators such as histamine. Examples: hayfever, asthma, and food allergies.

type II hypersensitivity A form of immediate hypersensitivity involving the binding of antibodies to antigens on cell surfaces followed by destruction of the target cells (e.g., through complement attack, phagocytosis, or agglutination).

type III hypersensitivity A form of immediate hypersensitivity resulting from the exposure to excessive amounts of antigens to which antibodies bind. These antibody-antigen complexes activate complement and trigger an acute inflammatory response with subsequent tissue damage.

type IV hypersensitivity   A delayed hypersensitivity response (it appears 24 to 48 hours after antigen exposure). It results from the binding of antigen to activated T lymphocytes, which then release cytokines and trigger inflammation and macrophage attacks that damage tissue. Type IV hypersensitivity is seen in contact dermititis from poison ivy, leprosy, and tertiary syphilis.

type I protein secretion pathway   *See* ABC protein secretion pathway.

type II protein secretion pathway   A system that transports proteins from the periplasm across the outer membrane of gram-negative bacteria.

type III protein secretion pathway   A system in gram-negative bacteria that secretes virulence factors and injects them into host cells.

type IV protein secretion pathway   A system in gram-negative bacteria that secretes proteins; some systems transfer DNA during bacterial conjugation.

type V protein secretion pathway   A pathway in gram-negative bacteria that translocates proteins across the outer membrane after they have been translocated across the plasma membrane by the Sec system.

type strain   The microbial strain that is the nomenclatural type or holder of the species name. A type strain will remain within that species should nomenclature changes occur.

typhoid fever   A bacterial infection transmitted by contaminated food, water, milk, or shellfish. The causative organism is *Salmonella enterica* serovar Typhi, which is present in human feces.

ultramicrobacteria   Bacteria that can exist normally in a miniaturized form or that are capable of miniaturization under low-nutrient conditions. They may be 0.2 μm or smaller in diameter.

ultraviolet (UV) radiation   Radiation of fairly short wavelength, about 10 to 400 nm, and high energy.

unbalanced growth   A type of growth observed in response to changes in a microbe's environment. The rates of synthesis of cell components vary relative to each other. *See* balanced growth.

undulant fever   A zoonotic disease caused by *Brucella* spp.

universal phylogenetic tree   A phylogenetic tree that considers the evolutionary relationships among organism from all three domains of life: *Bacteria*, *Archaea*, and *Eucarya*.

uracil   (u′rah-sil)   The pyrimidine 2,4-dioxypyrimidine, which is found in nucleosides, nucleotides, and RNA.

vaccine   A preparation of either killed microorganisms; living, weakened (attenuated) microorganisms; or inactivated bacterial toxins (toxoids) administered to induce development of the immune response and protect the individual against a pathogen or a toxin.

vaccinomics   The application of genomics and bioinformatics to vaccine development.

valence   The number of antigenic determinant sites on the surface of an antigen or the number of antigen-binding sites possessed by an antibody molecule.

vancomycin   A glycopeptide antibiotic obtained from *Nocardia orientalis* that is effective only against gram-positive bacteria and inhibits peptidoglycan synthesis.

variable region (V_L and V_H)   The region at the N-terminal end of immunoglobulin heavy and light chains whose amino acid sequence varies between antibodies of different specificity. Variable regions form the antigen-binding site.

vasculitis   (vas″ku-li′tis)   Inflammation of a blood vessel.

vector   (1) In genetic engineering, another name for a cloning vector. A DNA molecule that can replicate (a replicon) and transport a piece of inserted foreign DNA, such as a gene, into a recipient cell. It may be a plasmid, phage, cosmid or artificial chromosome. (2) In epidemiology, it is a living organism, usually an arthropod or other animal, that transfers an infective agent between hosts.

vector-borne transmission   The transmission of an infectious pathogen between hosts by means of a vector.

vehicle   An inanimate substance or medium that transmits a pathogen.

venereal syphilis   (ve-ne′re-al sif′ĭ-lis)   A contagious, sexually transmitted disease caused by the spirochete *Treponema pallidum*.

vesicular nucleus   The most common nuclear morphology seen in protists, characterized by a nucleus 1 to 10 μm in diameter, spherical, with a distinct nucleolus and uncondensed chromosomes.

viable but nonculturable (VBNC) microorganisms   Microbes that have been determined to be living in a specific environment but are not actively growing and cannot be cultured under standard laboratory conditions.

vibrio   (vib′re-o)   A rod-shaped bacterial cell that is curved to form a comma or an incomplete spiral.

viral hemagglutination   The clumping or agglutination of red blood cells caused by some viruses.

viral hemorrhagic fevers (VHF)   A group of illnesses caused by several distinct viruses, all of which cause symptoms of fever and bleeding in infected humans.

viral neutralization   An antibody-mediated process in which IgG, IgM, and IgA antibodies bind to some viruses during their extracellular phase and inactivate or neutralize them.

viremia   (vi-re′me-ə)   The presence of viruses in the blood stream.

viricide   (vir′i-sīd)   An agent that inactivates viruses so that they cannot reproduce within host cells.

virion   A complete virus particle that represents the extracellular phase of the virus life cycle; at the simplest, it consists of a protein capsid surrounding a single nucleic acid molecule.

virioplankton   Viruses that occur in waters; high levels are found in marine and freshwater environments.

viroid   An infectious agent that is a single-stranded RNA not associated with any protein; the RNA does not code for any proteins and is not translated.

virology   The branch of microbiology that is concerned with viruses and viral diseases.

virulence   The degree or intensity of pathogenicity of an organism as indicated by case fatality rates and/or ability to invade host tissues and cause disease.

virulence factor   A product, usually a protein or carbohydrate, that contributes to virulence or pathogenicity.

virulent (bacteriophages) viruses   Viruses that lyse their host cells during the reproductive cycle.

virus   An infectious agent having a simple acellular organization with a protein coat and a nucleic acid genome, lacking independent metabolism, and reproducing only within living host cells.

virusoid   An infectious agent that is composed only of single-stranded RNA that encodes some but not all proteins required for its replication; can be replicated and transmitted to a new host only if it infects a cell also infected by a virus that serves as a helper virus.

visceral leishmaniasis   See leishmaniasis.

vitamin   An organic compound required by organisms in minute quantities for growth and reproduction because it cannot be synthesized by the organism; vitamins often serve as enzyme cofactors or parts of cofactors.

volutin granules   (vo-lu′tin)   *See* metachromatic granules.

wart   An epidermal tumor of viral origin.

wastewater treatment   The use of physical and biological processes to remove particulate and dissolved material from sewage and to control pathogens.

water activity (a_w)   A quantitative measure of water availability in the habitat; the water activity of a solution is one-hundredth its relative humidity.

water mold   A common term for an öomycete.

West African sleeping sickness   See trypanosomiasis.

West Nile fever (encephalitis)   A neurological viral disease that is spread from birds to humans by mosquitoes.

white blood cell (WBC)   Blood cells having innate or acquired immune function. They are named for the white or buffy layer in which they are found when blood is centrifuged.

whole-cell vaccine   A vaccine made from complete pathogens, which can be of four types: inactivated viruses; attenuated viruses; killed microorganisms; and live, attenuated microbes.

whole-genome shotgun sequencing   An approach to genome sequencing in which the complete genome is broken into random fragments, which are then individually sequenced. Finally the fragments are placed in the proper order using computer programs.

whooping cough *See* pertussis.

Widal test (ve-dahl′) A test involving agglutination of typhoid bacilli when they are mixed with serum containing typhoid antibodies from an individual having typhoid fever; used to detect the presence of *Salmonella typhi* and *S. paratyphi*.

wild type Prevalent form of a gene or phenotype.

Winogradsky column A glass column with an anoxic lower zone and an oxic upper zone, which allows growth of microorganisms under conditions similar to those found in a nutrient-rich lake.

wobble The loose base pairing between an anticodon and a codon at the third position of the codon.

wort The filtrate of malted grains used as the substrate for the production of beer and ale by fermentation.

xenograft (zen″o-graft) A tissue graft between animals of different species.

xerophilic microorganisms (ze″ro-fil′ik) Microorganisms that grow best under low a_w conditions, and may not be able to grow at high a_w values.

yeast A unicellular, uninuclear fungus that reproduces either asexually by budding or fission, or sexually through spore formation.

yeast artificial chromosome (YAC) Engineered DNA that contains all the elements required to propagate a chromosome in yeast and which is used to clone foreign DNA fragments in yeast cells.

yellow fever An acute infectious disease caused by a flavivirus, which is transmitted to humans by mosquitoes. The liver is affected and the skin turns yellow.

YM shift The change in shape by dimorphic fungi when they shift from the yeast (Y) form in the animal body to the mold or mycelial form (M) in the environment.

zidovudine A drug that inhibits nucleoside reverse transcriptase (also known as AZT, ZDV, or retrovir) used as an anti-HIV treatment.

zoonosis (zo″o-no′sis; pl. **zoonoses**) A disease of animals that can be transmitted to humans.

zooxanthella (zo″o-zan-thel′ah) A dinoflagellate found living symbiotically within cnidarians (corals) and other invertebrates.

z value The increase in temperature required to reduce the decimal reduction time to one-tenth of its initial value.

zygomycetes (zi″go-mi-se′tez) A division of fungi that usually has a coenocytic mycelium with chitinous cell walls. Sexual reproduction normally involves the formation of zygospores.

zygospore A thick-walled, sexual, resting spore characteristic of the zygomycetous fungi.

zygote The diploid (2n) cell resulting from the fusion of male and female gametes.

Credits

Design Elements

Historical Highlights: © Stockbyte/Punchstock; Microbial Tibits: © Vol. 72 PhotoDisc/Getty; Disease Feature Box: © Vol. 83 PhotoDisc/Getty; Techniques & Applications Box: © Stockbyte/Getty; Microbial Diversity Box: © The McGraw-Hill Companies, Inc./John A. Karachewski, photographer.

Chapter 1

Opener: © John D. Cunningham/Visuals Unlimited; 1.3a: © Bettmann/Corbis; 1.3b(both): © Kathy Park Talaro/Visuals Unlimited; 1.3c: © Science VU/Visuals Unlimited; 1.4: © John D. Cunningham/Visuals Unlimited; 1.6: © Bettmann/Corbis; 1.7: North Wind Picture Archives.

Chapter 2

Opener: © Alfred Pasieka/Peter Arnold, Inc.; 2.3: Courtesy of Leica, Inc.; 2.7a: © Charles Stratton/Visuals Unlimited; 2.7b: © Robert Calentine/Visuals Unlimited; 2.7c: ASM Microbelibrary.org/Photo by William Ghiorse; 2.7d: © F. Widdel/Visuals Unlimited; 2.7e & 2.10: © M. Abbey/Visuals Unlimited; 2.12a: Courtesy of Molecular Probes, Eugene, OR; 2.12b: © Richard L. Moore/Biological Photo Service; 2.12c: © Evans Roberts; 2.13a: © Kathy Park Talaro; 2.13b: Harold J. Benson; 2.13c,d: © Jack Bostrack/Visuals Unlimited; 2.13e: © Manfred Kage/Peter Arnold, Inc.; 2.13f: © A.M. Siegelman/Visuals Unlimited; 2.13g: © David B. Fankhauser, University of Cincinnati Clermont College http://biology.clc.uc.edu/Fankhauser/; 2.14b: © Leon J. Le Beau/Biological Photo Service; 2.16a: © George J. Wilder/Visuals Unlimited; 2.16b: © Biology Media/Photo Researchers, Inc.; 2.17: © William Ormerod/Visuals Unlimited; 2.19a: © Harold Fisher; 2.19b: © Fred Hossler/Visuals Unlimited; 2.20: Courtesy of E.J. Laishley, University of Calgary; 2.22a: © David M. Philips/Photo Researchers, Inc.; 2.22b: © Paul W. Johnson/Biological Photo Service; 2.23: B. Pitts/P. Dirckx, MSU Center for Biofilm Engineering; 2.25: © Driscoll, Yougquist & Baldeschwieler, Caltech/SPL/Photo Researchers, Inc.; 2.27a,b: from Simon Scheuring (Scheuring S, Ringler P, Borgnia M, Stahlberg H, Müller DJ, Agre P, Engel A: "High resolution AFM topographs of the Escherichia coli water channel aquaporin," Z. *EMBO J.* 1999, 18:4981-4987).

Chapter 3

Opener: © M. Dworkin-H. Reichenbach/Phototake; 3.1a: © Bruce Iverson; 3.1b: Eye of Science/Photo Researchers, Inc.; 3.1c: ©

Arthur M. Siegelman/Visuals Unlimited; 3.2a: Centers for Disease Control; 3.2b: © Thomas Tottleben/Tottleben Scientific Company; 3.2c: Reprinted from *The Shorter Bergey's Manual of Determinative Bacteriology*, 8/e, John G. Holt, Editor, 1977 © Bergey's Manual Trust. Published by Williams & Wilkins Baltimore, MD; 3.2d: © David M. Phillips/Visuals Unlimited; 3.2e: Reprinted from *The Shorter Bergey's Manual of Determinative Bacteriology*, 8/e, John B. Holt, Editor, 1977 © Bergey's Manual Trust. Published by Williams & Wilkins Baltimore, MD; 3.2f: From Walther Stoeckenius: *Walsby's Square Bacterium: Fine structures of an Orthogonal Procaryote*; p. 36: © Dr. Leon J. Le Beau; 3.8a: American Society for Microbiology; 3.8b: Reprinted from *The Shorter Bergey's Manual of Determinative Bacteriology*, 8/e, John B. Holt, Editor, 1977 © Bergey's Manual Trust. Published by Williams & Wilkins Baltimore, MD; 3.12a,b: Image courtesy of Rut Carballido-López and Jeff Errington; 3.13a: © Ralph A. Slepecky/Visuals Unlimited; 3.13b: Courtesy of Daniel Branton, Harvard University; 3.13c: Reprinted from *The Shorter Bergey's Manual of Determinative Bacteriology*, 8/e, John B. Holt, Editor, 1977 © Bergey's Manual Trust. Published by Williams & Wilkins Baltimore, MD; 3.14: Y. Gorby; 3.15: Harry Noller, University of California, Santa Cruz; 3.16a: © CNRI/SPL/Photo Researchers, Inc.; 3.16b: © Dr. Gopal Murti/SPL/Photo Researchers, Inc.; 3.17(both): © T.J. Beveridge/Biological Photo Service; 3.22: Courtesy of M.R.J. Salton, NYU Medical Center; 3.26b: From M. Kastowsky, T. Gutberlet, and H. Bradaczek, Journal of Bacteriology, 774:4798-4806, 1992; 3.27a,b: Hiroshi Nikaido, *MMBR* 67(4):593-656 ASM/2003, Fig 2, p. 598; 3.29(both): From J.T. Staley, M.P. Bryant, N. Pfenning, and J.G. Holt (Eds.), *Bergey's Manual of Systematic Bacteriology*, Vol. 3. © 1989 Williams and Wilkins Co., Baltimore. Micrograph courtesy of D. Janekovic and W. Zillig; 3.31a,b: © John D. Cunningham/Visuals Unlimited; 3.32: © George Musil/Visuals Unlimited; 3.33: Dr. Robert G.E. Murray; 3.34: © Fred Hossler/Visuals Unlimited; 3.35a,b: © E.C.S. Chan/Visuals Unlimited; 3.35c: © George J. Wilder/Visuals Unlimited; 3.36a, 3.41a,b: Courtesy of Dr. Julius Adler; 3.44: American Society of Microbiology; 3.46(all): From A.N. Barker et al. (Eds.), *Spore Research*, 1974, pages 161-174, 1971 Academic Press; 3.47: American Society for Microbiology.

Chapter 4

Opener: © Natural History Museum, London; 4.1a: © Eric Grave/Photo Researchers, Inc.;

4.1b: © Carolina Biological Supply/Phototake; 4.1c: © Arthur M. Siegelman/Visuals Unlimited; 4.1d: © John D. Cunningham/Visuals Unlimited; 4.1e: © Tom E. Adams/Visuals Unlimited; 4.2: © Richard Rodewald/Biological Photo Service; 4.6: © Manfred Schliwa/Visuals Unlimited; 4.7a: © Henry C. Aldrich/Visuals Unlimited; 4.8b: U.S. Department of Energy Genomics: GTL Program http://www.ornl.gov/hgmis; 4.10: Courtesy of Dr. Garry T. Cole, Univ. of Texas Austin; 4.11: © Don Facett/Visuals Unlimited; 4.12b: *International Reivew of Cytology*, Vol. 68, p. 111. Academic Press, 1980; 4.13a: © Michael J. Dykstra/Visuals Unlimited; 4.13b: © Manfred Schliwa/Visuals Unlimited; 4.14a: Prentice Hall, Upper Saddle River, New Jersey; 4.16: National Research Council of Canada; 4.17: © Karl Aufderheide/Visuals Unlimited; 4.18a: © K.G. Murti/Visuals Unlimited; 4.19a: © Ralph A. Slepecky/Visuals Unlimited; 4.19b: © W.L. Dentler/Biological Photo Service; 4.20a,b: Hardy et al, "Trichomonas hydrogenosomes contain the NADH dehydrogenase module of mitochondrial complex I" *Nature* 2004 Vol. 432 p. 618 figures 1a & 1c; 4.22c: Courtesy of Dr. Garry T. Cole, Univ. of Texas at Austin.

Chapter 5

Opener: © Oliver Meckes/Photo Researchers, Inc.; Box 5.1(a-d): Haring, et al. "Independent virus development outside a host." *Nature* Vol. 436 Nature Publishing 2005 Figure 1 p. 1101; 5.3a: © Dennis Kunkel/Phototake; 5.3c: Courtesy of Gerald Stubbs and Keiichi Namba, Vanderbilt University; and Donald Caspar, Brandeis University; 5.4b: © Dennis Kunkel/CNRI/Phototake; 5.5a: © R. Feldman-Dan McCoy/Rainbow; 5.5b: Courtesy of Harold Fisher, University of Rhode Island and Robley Williams, University of California at Berkeley; 5.5c: © Science VU-NIH, R. Feldman/Visuals Unlimited; 5.8b: © Harold Fisher; 5.9a: © Bjornberg/Photo Researchers, Inc.; 5.9b: © Dr. Linda Stannard, UCT/Photo Researchers, Inc.; 5.9c: © R. Feldman-Dan McCoy/Rainbow; 5.14: © K.G. Murti/Visuals Unlimited; 5.15: © Lee D. Simon/Photo Researchers, Inc.; 5.20a: © Dr. Jack Griffith; 5.20b: © Terry Hazen/Visuals Unlimited; 5.21a-c: M. Cooney; 5.22: © Charles Marden Fitch.

Chapter 6

Opener: Dr. Eshel Ben-Jacob; 6.1a: © John D. Cunningham/Visuals Unlimited; 6.1b: From *ASM News* 53(2): cover, 187, American Society for Microbiology. Photo by H. Kaltwasser; 6.2a: © Woods Hole Oceanographic Institution; 6.2b: Image courtesy Mark Schneegurt; 6.9a,b,

6.10b, 6.11b: © Kathy Park Talaro; 6.13b: Image courtesy Mark Schneegurt; 6.13c: Dr. Eshel Ben-Jacob.

Chapter 7

Opener: © Science VU-D Foster, WHOI/Visuals Unlimited; 7.13a: Courtesy of Nalge Company; 7.13b: © B. Otero/Visuals Unlimited; 7.13c: Courtesy of Nalge Company; 7.22: Photo provided by ThermoForma of Marietta, Ohio; 7.25a: Dr. Joachim Reitner; 7.25b: Courtesy Nicholas G. Sotereanos, M.D.; 7.28a: Chris Frazee; 7.28b: Courtesy Dr. Margaret Jean McFall-Ngai.

Chapter 8

Opener: © Lee D. Simon/Photo Researchers, Inc.; 8.3a: Courtesy of AMSCO Scientific, Apex, NC; 8.4: © Raymond B. Otero/Visuals Unlimited; 8.6: © Fred Hossler/Visuals Unlimited; 8.7a: Photo provided by ThermoForma of Marietta, Ohio; 8.11a: Anderson Products, www.anpro.com.

Chapter 9

Opener: Courtesy of David Eisenberg, UCLA; 9.14a,b: John Wiley & Sons. Courtesy of Donald Voet; 9.21b: Courtesy Prof. Janine Maddock.

Chapter 10

Opener: © T.E. Adams/Visuals Unlimited.

Chapter 11

Opener: © John D. Cunningham/Visuals Unlimited.

Chapter 12

Opener: Kriegstein, H.J. & Hogness, D.S. "Mechanism of DNA replication in Drosophila chromosomes: structure of replication forks and evidence for bidirectionality," *PNAS* 1974 Vol 71. f.2 p. 137; 12.7b: From Cold Spring Harbor *Symposia of Quantitative Biology*, 28, p. 43 (1963); 12.22a: Reprinted with permission from Murakami, K.S.; Masuda, S.; and Darst, S.A. *Science 296*, May 17, 2002. Copyright 2002 by the AAAS. Image courtesy Seth Darst; 12.32a: Steven McKnight and Oscar L. Miller, Department of Biology, University of Virginia; 12.35e-g: Reprinted figures 2b,f,g from Yusupov, M. M.; Yusupov, G; Baucom, A; Lieberman, K; Earnest, T; Cate, J; and Noller, H. *Science* 292:883-896. May 4, 2001. Copyright 2001 AAAS.

Chapter 13

Opener & 13.6b: Lewis et al. "Crystal Structure of the Lactose Operon Repressor and Its Complexes with DNA and Inducer," *Science*, 1996 Mar 1, Vol. 271 cover; 13.17b: From S.C. Scnultz, G.C. Shields and T.A. Steitz, "Crystal structure of a CAP-DNA Complex. The DNA is Bent by 90 degrees," *Science* 253: 1001-1007, Aug 30, 1991. Copyright 1991 AAAS.

Chapter 14

Opener: © Oliver Meckes/Photo Researchers, Inc.; 14.19: Courtesy of Charles C. Brinton, Jr. and Judith Carnahan.

Chapter 15

Opener: © Alfred Pasieka/Photo Researchers, Inc.; 15.6: Reprinted with permission from Fleishchman, R.E. et al. 1995. "Whole-Genome Random Sequencing and Assembly of Haemophilus Influenzae Rd.," *Science* 269:496-512. Fig 1, page 507. © 1995 by the AAAS. Photo by the Institute for Genomic Research; 15.9: Liu, Zhou, Omelchenko, et al "Transcritomedyamics," *PNAS*, April 2003 vol. 100: 4191-96 Fig. 4193; 15.10b: © Tyne/Simon Fraser/Photo Researchers, Inc.; 15.14a-d: J. Hundelsman, "Metagenomics" *MMBR*, ASM Press Dec 2004 Vol. 68, fig 3 page 674.

Chapter 16

Opener: © Argus Fotoarchiv/Peter Arnold, Inc.; 16.2: Provided by Dr. A.K. Aggarwal (Mount Sinai School of Medicine) from Newman et al. *Science* 269, 656-663 (1995). © 1995 by the AAAS; 16.9b: © Kathy Park Talaro; 16.11a,b: © Huntington Potter and David Dressler/Time Life Pictures/Getty Images; 16.11c: Reprinted with permission of Edvotek,Inc. www.edvotek.com; Box 16.1c-e: From Cell 109: 257-266 Cell Press 2002. Courtesy Sigal Ben-Yehuda; 6.16a: Society for Industrial Microbiology; 16.22a: Mary Ann Tiffany; 16.22b-d: From Drum & Gordon, *Trends Biotechnology*, Vol. 21, No 8, 2003 pp. 235-328 Figure 2 p. 326, Elsevier.

Chapter 17

Opener: © Michael & Patricia Fogden/Minden Pictures; 17.3a: © Dirk Wiersma/SPL/Photo Researchers, Inc; 17.3b: © Roger Garwood & Trish Ainslie/Corbis.

Chapter 18

Opener: Lansing Prescott; 18.2b: Karl O. Stetter, M. Hohn; 18.3: Prepared by Stephen D. Bell, MRC Cancer Cell Unit, UK; 18.6: Lansing Prescott; 18.7a: Kazem Kashefi; 18.8a,b: © Corale L. Brierley/Visuals Unlimited; 18.8c: From J.T. Staley, M.P. Bryant, N. Pfenninging and J.G. Holt (Eds) *Bergey's Manual of Systematic Bacteriology*, Vol. 3. © 1989 Williams and Wilkins Co., Baltimore. Robinson, Dept. of Micro, U of Cal, LA; 18.7b: Kazem Kashefi; 18.10a: © Friederich Widdell/Visuals Unlimited; 18.10b: From J.T. Staley, M.P. Bryant, N. Pfenninging and J.G. Holt (Eds), *Bergey's Manual of Systematic Bacteriology*, Vol. © 1989 Williams and Wilkins Co., Baltimore; p. 414: From PNAS, Vol. 99, pages 7663-7668, fig 1, Orphan et al, May 2002. Image courtesy of Christopher H. House; 18.13b: Francisco Rodriguez-Valera; 18.13a: From J.T. Staley, M.P. Bryant, N. Pfenninging and J.G. Holt, *Bergey's Manual of Systematic Bacteriology*, Vol 3 © 1989 Williams and Wilkins Co., Baltimore,

Prepared by G. Bentzen photographed by the Laboratory of Clinical Electron Microscopy, U. of Bergen.

Chapter 19

Opener: Wadsworth Center, New York State Department of Health; 19.2: R. Huber, H. Konig, K.O. Stetter; 19.3a: From J.T. Staley, M.P. Bryant, N. Pfenning and J.G. Holt, *Bergey's Manual of Systematic Bacteriology*, Vol. 3 © 1989 Williams and Wilkins Co., Baltimore; 19.3b: From *Bergey's Manual of Systematic Bacteriology*, 2/e, vol. 1, Figure B4.4, part A page 400, Family I. Deinococcaceae, Battista, J.R., and Rainey, F.A. © Springer 2001. Image courtesy John R. Battista; 19.5: © Dr. T.J. Beveridge/Visuals Unlimited; 19.6: Joanne M. Willey, Ph.D.; 19.8a: © M.I. Walker/Photo Researchers, Inc.; 19.8b: © T.E. Adams/Visuals Unlimited; 19.8c: © Ron Dengler/Visuals Unlimited; 19.8d: © T.E. Adams/Visuals Unlimited; 19.9a: Jason K. Oyadomari, www.keweenawalgae.mtu.edu; 19.9b: Micrographia; 19.10: © John D. Cunningham/Visuals Unlimited; 19.11a,b: Image courtesy John A. Fuerst and Richard I. Webb, from "Novel Compartmentalisation in Planctomycete Bacteria," *Microsc & Microanal*, 2004 10 (Suppl 2); 19.12a: © David M. Phillips/Visuals Unlimited; 19.13 & 19.14: Armed Forces Institute of Pathology; 19.15a,b: Reprinted from *The Shorter Bergey's Manual of Determinative Bacteriology*, 8e, John G. Holt, Editor, 1977 © Bergey's Manual Trust. Published by Williams & Wilkins Baltimore, MD; 19.15b-d: From S.C. Holt, *Microbiological Reviews* 42(1):122, 1978 American Society for Microbiology; 19.19a-c: © Carroll H. Weiss/Camera M.D. Studios; 19.20a,b: © CDC/Peter Arnold, Inc.; 19.21a-c: From M.P. Starr, et al. (Eds), *The Prokaryotes*, Springer-Verlag.

Chapter 20

Opener: © Dennis Kunkel Microscopy, Inc.; 20.3a: © George J. Wilder/Visuals Unlimited; 20.3b: Reprinted from *The Shorter Bergey's Manual of Determinative Bacteriology*, 8e, John G. Holt, Editor, 1977 © Bergey's Manual Trust. Published by Williams & Wilkins Baltimore, MD; 20.3c: From M.P. Starr, et al. (Eds), *The Prokaryotes*, Springer Verlag; 20.3d: Reprinted from *The Shorter Bergey's Manual of Determinative Bacteriology*, 8e, John G. Holt, Editor, 1977 © Bergey's Manual Trust. Published by Williams & Wilkins Baltimore, MD; 20.5: Department of Health & Human Services, courtesy of Dr. W. Burgdorfer; 20.4a: University of Maryland School of Medicine; 20.4b: ASM MicrobeLibrary.org © Walker; 20.4c: Courtesy of Dr. K.E. Hechemy; 20.6: From J.T. Staley, M.P. Bryant, N. Pfenninging and J.G. Holt (Eds), *Bergey's Manual of Systematic Bacteriology*, Vol. 3. © 1989 Williams and Wilkins Co., Baltimore; 20.8a: © George J. Wilder/Visuals Unlimited; 20.8b,c: Courtesy of Jeanne S. Poindexter, Long Island

© John D. Cunningham/Visuals Unlimited; 23.22: © Natural History Museum, London; 23.25a,b: Martha Powell; 23.27.a.b: © John D. Cunningham/Visuals Unlimited; 23.28b: © J. Forsdyke, Gene Cox/SPL/Photo Researchers, Inc.; 23.29, 23.31a: © David M. Phillips/Visuals Unlimited; 23.31b,c: © Evertt S. Beneke/Visuals Unlimited; 23.32: Reprinted by permission of Upjohn Co. from E.S. Beneke et al., 1984 *Human Mycosis in Microbiology*; 23.33: © Arthur M. Siegelman/Visuals Unlimited; 23.34, 23.35: © Everett S. Beneke/Visuals Unlimited; 23.36, 23.37: © Carroll H. Weiss/Camera M.D. Studios; 23.38a: © Everett S. Beneke/Visuals Unlimited; 23.38b: Reprinted by permission of Upjohn Co. from E.S. Beneke et al., 1984 *Human Mycosis in Microbiology*; 23.38c: © Leonard Morse, M.D./Medical Images, Inc.; 23.41: JK Pataky, University of Illinois.

Chapter 24

Opener: © Science Photo Library/Photo Researchers, Inc.; 24.4f: © Lee D. Simon/Photo Researchers, Inc; 24.5b: George Chapman, Georgetown University; 24.9: © M. Wurtz/Photo Researchers, Inc.; 24.14: © CDC/Science Source/Photo Researchers; 24.15a,b: © Carroll H. Weiss/Camera M.D. Studios; 24.16b: © John D. Cunningham/Visuals Unlimited; 24.17c: © Carroll H. Weiss/Camera M.D. Studios; 24.18: Barbara O'Connor; 24.19a: © CDC/Science Source/Photo Researchers, Inc.; 24.19b: © Kenneth Greer/Visuals Unlimited; 24.20: Armed Forces Institute of Pathology; 24.21: © J.R. Adams/Visuals Unlimited; 24.26: © Eye of Science/Photo Researchers; 24.29: ©Carroll H. Weiss/Camera M.D. Studios; 24.31a: © Science Photo/Custom Medical Stock; 24.32: Courtesy of Dr. Frederick A. Murphy, Centers for Disease Control and Prevention; 24.33: Armed Forces Institute of Pathology; 24.38a: © Carroll H. Weiss/Camera M.D. Studios; 24.38b: © Science VU/Visuals Unlimited; 24.41: Courtesy of The National Institute of Health.

Chapter 25

Opener: Reprinted with permission from Edwards, K.J., Bond, P.L., Gihring, T.M., and Banfield, J.F. "An Archael Iron-oxidizing Extreme Acidophile Important in: Acid Mine Drainage," *Science* 287: 1796-2799. (10 March, 2000) Figure 3A, page 1798. © 2000 American Association for the Advancement of Science; Image courtesy of K.J. Edwards; 25.9: Reprinted with permission from *Nature* 417:63 H. Huber et al. © 2002 Nature; Prof. Dr. K.O. Stetter, Dr. R. Rachel, Dr. H. Huber, University of Regensburg, Germany; 25.10: Courtesy of Molecular Probes, Inc.; 25.12: Frohlich, J., and H. Koening, 1999. "Rapid Isolation of Single Microbial Cells from Mixed Natural and Laboratory Populations with Aid of a Micromanipulator System," *Applied Microbi-ology* 2:249-257. Figure 4 page 253. Urban and Fisher Verlag. Photo courtesy Dr. Helmut Koenig.; 25.11: Joanne M. Willey, Ph.D.; 25.14a: Y. Cohen and E. Rosenberg, *Microbial Mats*, Fig 1a p. 4 1986. American Society for Microbiology.

Chapter 26

Opener: Reprinted with permission from Schulz; H.N., Brinkhoff, T., Ferdelman, T.G., Hernandez Marine, M., Teske, A., and Jorgensen, B.B. 1999. "Dense Populations of a Giant Sulfur Bacterium in Namibian Shelf Sediments," *Science* 284, 493-495, Fig 1. © 1999 American Association for the Advancement of Science. Image courtesy of Heide Schulz; 26.3b: Jackie Parry; 26.3c: © Eye of Science/Photo Researchers, Inc.; 26.5: Burkholder Laboratory, North Carolina State University & Sea Grant National Media Relations; 26.7: NASA; 26.9: Photo by Susumu Honjo, Woods Hole Oceanographic Institution; 26.13b: Reprinted with permission from Vincent et al., *Science* 286: 2094 (1999). Copyright 2006 AAAS; 26.13c: Reprinted with permission from Karl et al., *Science* 286: 2144-47 (10 Dec 1999). Copyright 2006 AAAS; 26.19: © Sherman Thompson/Visuals Unlimited; 26.21: N.C. Schenck, *Methods & Principles of Micorrhizal Research*, © 1992 American Phytopathological Society. Photo courtesy of Dr. Hugh Wilcox; 26.23d: Courtesy of Ray Tully, U.S. Department of Agriculture; 26.23f: Courtesy of Dr. Ralph W.F. Hardy and the National Research Council of Canada; 26.23i, j: © John D. Cunningham/Visuals Unlimited; 26.24: Dr. Bernard Dreyfus; 26.25: © John D. Cunningham/Visuals Unlimited; 26.26: Courtesy of Dr. Sandor Sule, Plant Protection Institute, Hungary Academy of Sciences; 26.28: © Michael & Patricia Fogden/Minden Pictures.

Chapter 27

Opener: © Science VU/WHOI/Visuals Unlimited; 27.2a: © William J. Weber/Visuals Unlimited; 27.2b: © M. Abbey/Visuals Unlimited; 27.3a: © Stan Elems/Visuals Unlimited; 27.3b: © Bob DeGoursey/Visuals Unlimited; Box 27.1(top): © T. Wenseleers; Box 27.1(bottom): World Health Organization; 27.5a: © WHOI/Visuals Unlimited; 27.8: Craig Cary, University of Delaware; 27.10: © John D. Cunningham/Visuals Unlimited; 27.11a: © John Durham/SPL/Photo Researchers; 27.11c(all): From Currie et al, *Science* Vol. 311: 81-83, Jan 6 2006. Image courtesy Cameron Currie; 27.12b: © Science Source/Photo Researchers.

Chapter 28

Opener: ©Jim Dowdalls/Photo Researchers, Inc.; 28.4: © Lennart Nilsson/Albert Bonniers Forlag AB; 28.5: © David Scharf/Peter Arnold; 28.19b: © Ellen R. Dirksen/Visuals Unlimited; 28.22b: © Reprinted from Wendell F. Rosse et al., "Immune Lysis of Normal Human and Parozysmal Nicturnal Hemogloginuria (PNH) Red Blood Cells," *Journal of Experimental Medicine*, 123:969, 1966. Rockefeller University Press.

Chapter 29

Opener: © Science Source/Photo Researchers, Inc.; 29.3 (infection, antibody, vaccination): © PhotoDisc RF/Getty; (immune): © Creatas/PictureQuest; 29.4c,d: Courtesy of Dr. Paul Travers; 29.7b,c: Courtesy of Dr. Gilla Kaplan, The Rockefeller University; 29.11b: © R. Feldman-Dan McCoy/Rainbow; 29.23: © Stan Elms/Visuals Unlimited; 29.28: © Kathy Park Talaro.

Chapter 30

Opener: © Alfred Pasieka/Peter Arnold; 30.3a: © Stan Elms/Visuals Unlimited; 30.3b: © Veronika Burmeister/Visuals Unlimited; 30.3c: From M. Persi, J.C. Burham and J.L. Duhring, "Effects of Carbon Dioxide and pH on Adhesion of Candida albicans to Vaginal Epithelial Cells," *Infections and Immunity* 30 (s): 82-90, Oct. 1985. American Society for Microbiology; 30.4a: Daniel A. Portnoy, from 1989 *Journal of Cell Biology*, Rockefeller Press; 30.4b: Stevens, M.P., J.M. Stevens, R.L. Jeng, L.A. Taylor, M.W. Wood, P. Hawes, P. Monaghan, M.D. Welch & E.E. Galyov. 2005. "Identification of a bacterial factor required for actin based motility of Burkholderian pseudomallei," *Mol. Microbiol.* 56: 40-43. (Blackwell Publishing); 30.5c: Prof. Guy R. Cornelis; 30.11: © Stanley Flegler/Visuals Unlimited.

Chapter 31

Opener: © SPL/Photo Researchers; 31.1: Christine L. Case/Skyline College; 31.2a: Courtesy Becton-Dickinson Microbiology Systems; 31.2b: © Lauritz Jensen/Visuals Unlimited; 31.4: Etest® is a registered trademark of AB BIODISK and patented in all major markets; 31.13a: Alex MacKerell, Ph.D.

Chapter 32

Opener, 32.1a,b: © Raymond B. Otero/Visuals Unlimited; 32.2c: Courtesy of Genetic System Corporation; 32.4a,b: Analytab Products, A Division of BioMerieux, Inc.; 32.5: Centers for Disease Control and Prevention; 32.6: © Raymond B. Otero/Visuals Unlimited; 32.10c: © Hank Morgan/Science Source/Photo Researchers, Inc.; 32.12c: Department of Pathology, University of Cambridge; 32.13d: From N.R. Rose et al., *Manual of Clinical Laboratory Immunology*, 1992. American Society of Microbiology.

Chapter 33

Opener: Jim Gathany, Centers for Disease Control and Prevention; 33.9: © Runk/Schoenberger/Grant Heilman Photography; Box 33.3: © Historic VU-NIH/Visuals Unlimited.

Line Art / Tables / Illustrations

Index